Fundamentals of Tissue Engineering and Regenerative Medicine

Ulrich Meyer · Thomas Meyer
Jörg Handschel · Hans Peter Wiesmann
(Eds.)

Fundamentals of Tissue Engineering and Regenerative Medicine

Ulrich Meyer, Prof. Dr. med. dent. Dr. med.
Clinic for Maxillofacial and Plastic
Facial Surgery
Heinrich Heine University Düsseldorf
Moorenstraße 5
40225 Düsseldorf
Germany
E-mail: praxis@mkg-muenster.de

Jörg Handschel, Priv.-Doz., Dr. med. dent. Dr. med.
Clinic for Maxillofacial and Plastic
Facial Surgery
Heinrich Heine University Düsseldorf
Moorenstraße 5
40225 Düsseldorf
Germany
E-mail: handschel@med.uni-duesseldorf.de

Thomas Meyer, Prof. Dr. med. Dr. phil. Dr. rer. nat.
Department of Internal Medicine – Cardiology
University Hospital Marburg
Baldingerstraße 1
35033 Marburg
Germany
E-mail: meyert@med.uni-marburg.de

Hans Peter Wiesmann, Priv.-Doz., Dr. rer. medic.
Biomineralisation and Tissue Engineering Group
Department of Experimental Maxillofacial Surgery
University of Münster
Waldeyerstraße 30
48149 Münster
Germany
E-mail: wiesmann@life-rds.eu

ISBN: 978-3-540-77754-0   e-ISBN: 978-3-540-77755-7

DOI: 10.1007/ 978-3-540-77755-7

Library of Congress Control Number: 2008931995

© 2009 Springer-Verlag Berlin Heidelberg

This work is subject to copyright. All rights are reserved, whether the whole or part of the material is concerned, specifically the rights of translation, reprinting, reuse of illustrations, recitation, broadcasting, reproduction on microfilm or any other way, and storage in data banks. Duplication of this publication or parts thereof is permitted only under the provisions of the German Copyright Law of September 9, 1965, in it current version, and permission for use must always be obtained from Springer. Violations are liable to prosecution under the German Copyright Law.

The use of general descriptive names, registed names, trademarks etc. in this publication does not imply, even in the absence of a specific statement, that such names are exempt from the relevant protective laws and regulations and therefore free for general use.

Product liability: the publishers cannot guarantee the accuracy of any information about dosage and application contained in this book. In every individual case the user must check such information by consulting the relevant literature.

Cover design: Frido Steinen-Broo, eStudio Calamar, Spain
Reproduction, typesetting and production: le-tex publishing services oHG, Leipzig, Germany

Printed on acid-free paper

9 8 7 6 5 4 3 2 1

springer.com

Preface

The man-made creation of tissues, organs, or even larger organisms was for a long time a matter of myth and dream throughout the history of medicine. It now comes into clinical reality. Tissue engineering and regenerative medicine are the terms that are nowadays used to describe the approach to generate complex tissues and organs from simpler pieces. Both are multidisciplinary, young and emerging fields in biotechnology and medicine, which are expected to change patient treatment profoundly, generating and regenerating tissues and organs instead of just repairing them. There is much promise and expectation connected to this biomedical discipline regarding improved treatment possibilities, enhanced quality of the patient's life, and the ability to overcome in a future perspective the need for major grafting procedures. It is anticipated that this biotechnology has also a high economical impact on clinical medicine. To fulfil these expectations several challenges concerning scientific, technological, clinical, ethical, and also social issues need to be met. Basic research still requires the evaluation and elaboration of fundamental processes and procedures in multiple research fields. However, first bioengineered products have already been introduced in the markets, and much more are in the preclinical stage, and many companies are involved in this area.

In addition to having a therapeutic application, where the tissue is either grown in a patient or outside the patient and transplanted, tissue engineering can have diagnostic applications where the tissue is made in vitro and used for testing drug metabolism and uptake, toxicity, and pathogenicity. The foundation of tissue engineering/regenerative medicine for either therapeutic or diagnostic applications is the ability to exploit living cells in a variety of ways. Whereas tissue engineering is a more technical concept of tissue and organ reconstruction by the use of cells, scaffolds, and biomolecules, the term regenerative medicine is more focused on the support of self healing capabilities and the use of stem cells. Medicine-oriented stem cell research includes research that involves stem cells, whether from human, non-human, embryonic, fetal, or adult sources. It includes all aspects in which stem cells are isolated, derived, or cultured for purposes such as developing cell or tissue therapies, studying cellular differentiation, research to understand the factors necessary to direct cell specialization to specific pathways, and other developmental studies. In this sense it does not include transgenic studies, gene knock-out studies, nor the generation of chimeric animals.

Both concepts (tissue engineering and regenerative medicine) of cell, tissue, or organ regeneration and reconstruction are based on an multidisciplinary approach bringing together various scientific fields such as biochemistry, pharmacology, material science, cell biology, and engineering and clinical disciplines. The promising biotechnology, now introduced as a new clinical tool in the restoration of lost tissues or the healing of diseases, is assumed to change treatment regimes and to contribute significantly to clinical medicine in future decades. A lot of current limitations seem most likely be overcome in the near future, suggesting that tissue engineering as well as regenerative medicine strategies will replace other therapies in routine clinical practice.

The fast growth of the tissue engineering and regenerative medicine discipline is mirrored by the high number of excellent research papers covering all aspects of these fields. Additionally, numerous high quality books are available describing in detail different aspects of tissue engineering or regenerative medicine. Despite the fact that such literature is already available, we decided to edit a book on tissue engineering and regenerative medicine. There were three reasons for this decision: during our experimental and clinical work on tissue regeneration and reconstruction, with our main focus on bone and cartilage engineering, which we have done for more than a decade in our clinics as well as in our interdisciplinary biomineralization and tissue engineering research group, we observed that many specialists of the different fields, involved in approaching this area, had difficulties in overviewing the complexity of the field. We therefore intended to edit a comprehensive book covering all major aspects of this field. Secondly, during the last decade a shift and, at the same time, interdentation was seen between the tissue engineering field and the field of regenerative medicine (with a main focus on stem cell research). In recent years stem cell research and use was applied with tissue engineering techniques and the border between both areas therefore blurred. This fusion is mirrored also by the emergence of new societies (for regenerative medicine) or the renaming of the most influential society (Tissue Engineering and Regenerative Medicine Society, formerly the Tissue Engineering Society International). Therefore, there was a need to integrate both aspects in one book. Thirdly, as tissue engineering brings together basic researchers, mainly having a biological, biophysical, or material science-oriented background, with clinically oriented physicians, we found that they differed in the used "language." In this text book the contributors tried to use an uniform terminology as a common platform for discussions across the borders of medical subspecialities.

Fundamentals of Tissue Engineering and Regenerative Medicine is intended not only as a text for biomedical engineering students and students in all fields of tissue engineering and cell biology, and medical courses at basic and advanced levels, but also as a reference for research and clinical laboratories. In addition, a special aim of this book was to define the current state of tissue engineering and regenerative medicine approaches which are applied in the various clinical particualar specialities. We have therefore conceptualised the book according to a methodological approach (social, economical, and ethical considerations; basic biological aspects of regenerative medicine; classical methods of tissue engineering (cell, tissue, organ culture, scaffolds, bioreactors); and a medical discipline-oriented approach (application of these techniques in the various medical disciplines). Since during the last years these therapeutic options have been introduced in clinical treatment decisions, this book gives profound basic tissue engineering information (as how to generate and regenerate tissues and organs) and at the same time the medical specialist will find detailed information on the state of regenerative medicine in his/her discipline. The text of this book is supported by numerous tables, schematic illustrations, and photos in order to provide a better understanding of the information offered in this book. As the recent detailed knowledge in tissue engineering and regenerative medicine far exceeds the content of a book, we have tried to find a compromise between a comprehensive depiction of this new biomedical field and one that is manageable for the reader.

The expertise required to generate this book far exceeded that of it editors. No single expert, to date, is able to have detailed insight into all aspects of this fast growing and complex biomedical field. The content of the book represents the combined intellect and experience of more than one hundred researchers and clinicians, all of them outstanding specialists in their field. Their fundamental work has not only set the basis for the tremendous advances in this biotechnology field but has also given patients new and fascinating treatment options in clinical medicine.

Finally, we believe that, especially today, it is important to understand and reflect the current limitations of the field. The expectations must be aligned with scientific and, perhaps more importantly, ethical considerations and reflections. Given that stem cell use is a mainstay in regenerative medicine, a special focus is given to ethical as well as theological considerations. In addition to the impressive speed with which the advances in tissue engineering and regenerative medicine during the last decade have made a clinical impact on the treatment of many diseases, a fascinating aspect of this area of biotechnology is that it is a model of how basic biology is closely connected with and directly transferred to clinical medicine.

We hope this book will add further stimulus for all basic researchers and clinicians who are involved in investigating and applying tissue engineering and regenerative medicine techniques and will contribute to make this an attractive and reliable alternative treatment option in medicine.

Ulrich Meyer
Thomas Meyer
Jörg Handschel
Hans Peter Wiesmann

Contents

Part A General Aspects 1

I General and Ethical Aspects 3

    1 The History of Tissue Engineering and Regenerative Medicine in Perspective 5
        U. Meyer

    2 Economic Modeling and Decision Making in the Development and Clinical Application of Tissue Engineering 13
        P. Vavken, R. Dorotka, M. Gruber

    3 Ethical Issues in Tissue Engineering 23
        P. Gelhaus

    4 Sourcing Human Embryonic Tissue: The Ethical Issues 37
        C. Rehmann-Sutter, R. Porz, J. L. Scully

    5 Tissue Engineering and Regenerative Medicine. Their Goals, Their Methods and Their Consequences from an Ethical Viewpoint 47
        E. Schockenhoff

Part B Biological Considerations 57

II Tissue and Organ Differentiation 59

    6 Control of Organogenesis: Towards Effective Tissue Engineering 61
        M. Unbekandt, J. Davies

    7 Cytokine Signaling in Tissue Engineering 71
        T. Meyer, V. Ruppert, B. Maisch

    8 Influence of Mechanical Effects on Cells. Biophysical Basis 83
        D. Jones

Part C Engineering Strategies ... 89

III Engineering at the Genetic and Molecular Level ... 91

9 Towards Genetically Designed Tissues for Regenerative Medicine ... 93
W. Weber, M. Fussenegger

10 Posttranscriptional Gene Silencing ... 109
V. Ruppert, S. Pankuweit, B. Maisch, T. Meyer

11 Biomolecule Use in Tissue Engineering ... 121
R. A. Depprich

IV Engineering at the Cellular Level ... 137

12 Fetal Tissue Engineering: Regenerative Capacity of Fetal Stem Cells ... 139
P. Wu, D. Moschidou, N. M. Fisk

13 Embryonic Stem Cell Use ... 159
J. Handschel, U. Meyer, H. P. Wiesmann

14 The Unrestricted Somatic Stem Cell (USSC) From Cord Blood For Regenerative Medicine ... 167
G. Kögler

15 Mesenchymal Stem Cells: New Insights Into Tissue Engineering and Regenerative Medicine ... 177
F. Djouad, R. S. Tuan

16 Stem Cell Plasticity: Validation Versus Valedictory ... 197
N. D. Theise

V Engineering at the Tissue Level ... 209

17 Bone Tissue Engineering ... 211
U. Meyer, H. P. Wiesmann, J. Handschel, N. R. Kübler

18 Cartilage Engineering ... 233
J. Libera, K. Ruhnau, P. Baum, U. Lüthi, T. Schreyer, U. Meyer, H. P. Wiesmann, A. Herrmann, T. Korte, O. Pullig, V. Siodla

19 Muscle Tissue Engineering ... 243
M. P. Lewis, V. Mudera, U. Cheema, R. Shah

20	**Tendon and Ligament Tissue Engineering: Restoring Tendon/ Ligament and Its Interfaces**	255
	J. J. Lim, J. S. Temenoff	
21	**Neural Tissue Engineering and Regenerative Medicine**	271
	N. Zhang, X. Wen	
22	**Adipose Tissue Engineering**	289
	T. O. Acartürk	
23	**Intervertebral Disc Regeneration**	307
	J. Libera, Th. Hoell, H.-J. Holzhausen, T. Ganey, B. E. Gerber, E. M. Tetzlaff, R. Bertagnoli, H.-J. Meisel, V. Siodla	
24	**Tissue Engineering of Ligaments and Tendons**	317
	P. Vavken	
25	**Tissue Engineering of Cultured Skin Substitutes**	329
	R. E. Horch	
26	**Dental Hard Tissue Engineering**	345
	J. M. Mason, P. C. Edwards	
27	**Mucosa Tissue Engineering**	369
	G. Lauer	
28	**Tissue Engineering of Heart Valves**	381
	C. Lüders, C. Stamm, R. Hetzer	

VI	**Engineering at the Organ Level**		387
	29	**Breast Tissue Engineering**	389
		E. Geddes, X. Wu, C. W. Patrick Jr.	
	30	**Bioartificial Liver**	397
		J.-K. Park, S.-K. Lee, D.-H. Lee, Y.-J. Kim	
	31	**Pancreas Engineering**	411
		R. Cortesini, R. Calafiore	
	32	**Tissue-Engineered Urinary Bladder**	429
		A. M. Turner, J. Southgate	
	33	**Cell-Based Regenerative Medicine for Heart Disease**	441
		C. Stamm, C. Lüders, B. Nasseri, R. Hetzer	

Part D Technical Aspects 453

VII Biomaterial Related Aspects 455

34 Biomaterials 457
H. P. Wiesmann, U. Meyer

35 Biomaterial-Related Approaches: Surface Structuring 469
G. Jell, C. Minelli, M. M. Stevens

36 Mineralised Collagen as Biomaterial and Matrix for Bone Tissue Engineering 485
M. Gelinsky

37 Hydrogels for Tissue Engineering 495
J. Teßmar, F. Brandl, A. Göpferich

VIII Scaffold Related Aspects 519

38 Defining Design Targets for Tissue Engineering Scaffolds 521
S. J. Hollister, E. E. Liao, E. N. Moffitt, C. G. Jeong, J. M. Kemppainen

39 Scaffold Structure and Fabrication 539
H. P. Wiesmann, L. Lammers

40 Prospects of Micromass Culture Technology in Tissue Engineering 551
J. Handschel, H. P. Wiesmann, U. Meyer

IX Laboratory Aspects and Bioreactor Use 557

41 Laboratory Procedures – Culture of Cells and Tissues 559
C. Naujoks, K. Berr, U. Meyer

42 Bioreactors in Tissue Engineering: From Basic Research to Automated Product Manufacturing 595
D. Wendt, S. A. Riboldi

43 The Evolution of Cell Printing 613
B. R. Ringeisen, C. M. Othon, J. A. Barron, P. K. Wu, B. J. Spargo

44 Biophysical Stimulation of Cells and Tissues in Bioreactors 633
H. P. Wiesmann, J. Neunzehn, B. Kruse-Lösler, U. Meyer

45 Microenvironmental Determinants of Stem Cell Fate 647
R. L. Mauck, W-J. Li, R. S. Tuan

Part E Transplantation Issues 665

X Functional Aspects in Biological Engineering 667

46 Perfusion Effects and Hydrodynamics 669
R. A. Peattie, R. J. Fisher

47 Ex Vivo Formation of Blood Vessels 685
R. Y. Kannan, A. M. Seifalian

48 Biomechanical Function in Regenerative Medicine 693
B. David, J. Pierre, C. Oddou

49 Influence of Biomechanical Loads 705
U. Meyer, J. Handschel

XI Immune System Issues 719

50 Innate and Adaptive Immune Responses in Tissue Engineering 721
L. W. Norton, J. E. Babensee

51 Toll-Like Receptors: Potential Targets for Therapeutic Interventions 749
S. Pankuweit, V. Ruppert, B. Maisch, T. Meyer

XII Study Design Principles 757

52 Tissue Engineered Models for In Vitro Studies 759
C. R. McLaughlin, R. Osborne, A. Hyatt, M. A. Watsky, E. V. Dare,
B. B. Jarrold, L. A. Mullins, M. Griffith

53 In Vivo Animal Models in Tissue Engineering 773
J. Haier, F. Schmidt

54 Assessment of Tissue Responses to Tissue-Engineered Devices .. 781
K. Burugapalli, J. C. Y. Chan, A. Pandit

Part F Clinical Use 797

XIII Clinical Application 799

55 Evidence-based Application in Tissue Engineering and Regenerative Medicine 801
U. Meyer, J. Handschel

56 Tissue Engineering Applications in Neurology 815
E. L. K. Goh, H. Song, G.-Li Ming

57 Tissue Engineering in Maxillofacial Surgery 827
H. Schliephake

58 Tissue Engineering Strategies in Dental Implantology 839
U. Joos

59 Tissue Engineering Application in General Surgery 855
Y. Nahmias, M. L. Yarmush

60 Regeneration of Renal Tissues 869
T. Aboushwareb, J. J. Yoo, A. Atala

61 Tissue Engineering Applications in Plastic Surgery 877
M. D. Kwan, B. J. Slater, E. I. Chang, M. T. Longaker,
G. C. Gurtner

62 Tissue Engineering Applications for Cardiovascular Substitutes 887
M. Cimini, G. Tang, S. Fazel, R. Weisel, R.-K. Li

63 Tissue Engineering Applications in Orthopedic Surgery 913
A. C. Bean, J. Huard

64 Tissue Engineering and Its Applications in Dentistry 921
M. A. Ommerborn, K. Schneider, W. H.-M. Raab

65 Tissue Engineering Applications in Endocrinology 939
M. R. Hammerman

66 Regenerative Medicine Applications in Hematology 951
A. Wiesmann

67 The Reconstructed Human Epidermis Models in Fundamental Research 967
A. Coquette, Y. Poumay

XIV Clinical Handling and Regulatory Issues 977

68 Regulatory Issues 979
B. Lüttenberg

Subject Index 983

Contributors

Tamer Aboushwareb Wake Forest Institute for Regenerative Medicine, Wake Forest University Health Sciences, Medical Center Boulevard, Winston-Salem, NC 27157, USA

T. Oğuz Acartürk Department of Plastic, Reconstructive and Aesthetic Surgery, Çukurova University School of Medicine, Adana 01330, Turkey,
E-mail: toacarturk@yahoo.com

Anthony Atala Wake Forest Institute for Regenerative Medicine, Wake Forest University Health Sciences, Medical Center Boulevard, Winston-Salem, NC 27157, USA, E-mail: aatala@wfubmc.edu

Julia E. Babensee Wallace H. Coulter Department of Biomedical Engineering, Georgia Institute of Technology and Emory University, 313 Ferst Drive, Atlanta, GA 30332, USA, E-mail: julia.babensee@bme.gatech.edu

Jason A. Barron Stern, Kessler, Goldstein and Fox, LLC, Washington, DC 20005, USA, E-mail: jbarron@skgf.com

Peter Baum Gelenkklinik Gundelfingen, Alte Bundesstraße 29, 79194 Gundelfingen, Germany

Allison C. Bean Stem Cell Research Center, Children's Hospital of Pittsburgh, 3460 Fifth Avenue, 4100 Rangos Research Center, Pittsburgh, PA 15213, USA

Karin Berr Clinic for Maxillofacial and Plastic Facial Surgery, Heinrich Heine University, Moorenstraße 5, 40225 Düsseldorf, Germany,
E-mail: karinberr@yahoo.com

Rudolf Bertagnoli Pro Spine Center, St. Elisabeth Hospital, St. Elisabeth-Straße 23, 94315, Straubing, Germany

Ferdinand Brandl Department of Pharmaceutical Technology, University of Regensburg, Universitaetsstraße 31, 93040 Regensburg, Germany

Krishna Burugapalli Heinz Wolff Building, Kingston Lane, Brunel Institute of Bioengineering, Brunel University, Uxbridge, Middlesex, UB8 3PH, UK

Riccardo Calafiore Department of Internal Medicine, Perugia University of Perugia, Via E. Dal Pozzo, 06126 Perugia, Italy, E-mail: islet@unipg.it

Jeffrey C. Y. Chan National Centre for Biomedical Engineering Science, National University of Ireland, Galway, Galway, Ireland,
E-mail: jeffrey.chan@nuigalway.ie

Edward I. Chang Division of Plastic and Reconstructive Surgery, Department of Surgery, Stanford University School of Medicine, 257 Campus Drive, GK201, Stanford, CA 94305-5148, USA

Umber Cheema Tissue Repair and Engineering Center, UCL Institute of Orthopaedics and Musculoskeletal Science, Royal National Orthopaedic Hospital, Brockley Hill, Stanmore, Middlesex HA7 4LP, UK, E-mail: u.cheema@ucl.ac.uk

Massimo Cimini Division of Cardiovascular Surgery, Department of Surgery, Toronto General Research Institute, University of Toronto, Toronto, ON M5G 1L7, Canada

Alain Coquette Division of Applied Biology, SGS Life Science Services, Vieux Chemin du Poète 10, 1301 Wavre, Belgium,
E-mail: alain.coquette@sgs.com

Raffaello Cortesini Department of Pathology, Columbia University, 630 West 168 Street, P&S 14-401, New York, NY 10032, USA,
E-mail: rc238@columbia.edu

Emma V. Dare University of Ottawa, Department of Cellular and Molecular Medicine, The Ottawa Hospital, General Campus, 501 Smyth Road, Ottawa, ON K1H 8L6, Canada

Bertrand David Laboratoire Mécanique des Sols, Structures et Matériaux, UMR CNRS 8579, École centrale, Paris, France,
E-mail: bertrand.david@paris7.jussieu.fr

Jamie Davies Centre for Integrative Physiology, University of Edinburgh, George Square, Edinburgh EH8 9XB, UK, E-mail: jamie.davies@ed.ac.uk

Rita A. Depprich Clinic for Maxillofacial and Plastic Facial Surgery, University of Düsseldorf, Moorenstraße 5, 40225 Düsseldorf, Germany,
E-mail: depprich@med.uni-duesseldorf.de

Farida Djouad Cartilage Biology and Orthopaedics Branch, National Institute of Arthritis and Musculoskeletal and Skin Diseases, 50 South Drive, Room 1523, BMSC 8022, Bethesda, MD 20892-8022, USA

Ronald Dorotka Department of Orthopedic Surgery, Medical University of Vienna, Waehringer Guertel 18–20, 1090 Vienna, Austria,
E-mail: ronald.dorotka@meduniwien.ac.at

Paul C. Edwards Division of Oral Pathology, Medicine and Radiology, Department of Periodontics and Oral Medicine, University of Michigan School of Dentistry, 1011 N. University Ave., Office 2029E, Ann Arbor, MI 48109-1078, USA, E-mail: paulce@umich.edu

Shafie Fazel Division of Cardiovascular Surgery, Department of Surgery, Toronto General Research Institute, University of Toronto, Toronto, ON M5G 1L7, Canada

Nicholas M. Fisk Experimental Fetal Medicine Group, Institute of Reproductive and Developmental Biology, Imperial College, Hammersmith Campus, Du Cane Road, London W12 0NN, UK
and
University of Queensland Centre for Clinical Research, Brisbane, QLD 4029, Australia, E-mail: n.fisk@uq.edu.au

Robert J. Fisher Department of Chemical Engineering, Building 66, Room 446, Massachusetts Institute of Technology, Cambridge, MA 02139, USA, E-mail: rjfisher@mit.edu

Martin Fussenegger Institute for Chemical and Bioengineering, HCI F115, ETH Zurich, Wolfgang-Pauli-Straße 10, 8093 Zurich, Switzerland, E-mail: fussenegger@chem.ethz.ch

Tim Ganey Atlanta Medical Center, 303 Parkway Drive NE, Box 227, Atlanta, GA 30329, USA

Elizabeth Geddes The University of Texas, Houston, 1515 Holcombe Boulevard, TX 77030

Petra Gelhaus Institut für Ethik, Geschichte und Theorie der Medizin, Universitätsklinik Münster, Von-Esmarch-Straße 62, 48149 Münster, Germany, E-mail: gelhaus@uni-muenster.de

Michael Gelinsky Research Group Tissue Engineering and Biomineralisation, The Max Bergmann Center of Biomaterials Dresden, Technische Universität Dresden, Institute of Materials Science, Budapester Str. 27, 01069 Dresden, Germany, E-mail: gelinsky@tmfs.mpgfk.tu-dresden.de

Bruno E. Gerber Biological Repair, University Hospital Lewisham, Lewisham High Street, London, SE 13 GLH, UK

Eyleen L. K. Goh Institute for Cell Engineering, Department of Neurology, Johns Hopkins University School of Medicine, 733 N. Broadway, Broadway Research Building 706, Baltimore, MD 21205, USA, E-mail: egoh2@jhmi.edu
and
Duke-NUS Graduate Medical School Singapore, 2 Jalan Bukit Merah, Singapore 169547, Singapore

Achim Göpferich Department of Pharmaceutical Technology, University of Regensburg, Universitaetsstraße 31, 93040 Regensburg, Germany,
E-mail: Achim.Goepferich@chemie.uni-regensburg.de

May Griffith University of Ottawa Eye Institute, The Ottawa Hospital, General Campus, 501 Smyth Road, Ottawa, ON K1H 8L6, Canada,
E-mail: mgriffith@ohri.ca

Martin Gruber Department of Orthopedic Surgery, Medical University of Vienna, Waehringer Guertel 18–20, 1090 Vienna, Austria,
E-mail: martin.gruber@meduniwien.ac.at

Geoffrey C. Gurtner Children's Surgical Research Program, Division of Plastic and Reconstructive Surgery, Department of Surgery, Stanford University School of Medicine, 257 Campus Drive, GK201, Stanford, CA 94305-5148, USA,
E-mail: ggurtner@stanford.edu

Jörg Haier Clinic for Surgery, University of Münster, Waldeyerstraße 16, 48149 Münster, Germany, E-mail: joerg.haier@ukmuenster.de

Marc R. Hammerman Renal Division, Department of Medicine, Washington University School of Medicine, 660 S. Euclid Avenue, St. Louis, MO 63110, USA,
E-mail: mhammerm@wustl.edu

Jörg Handschel Clinic for Maxillofacial and Plastic Facial Surgery, University of Düsseldorf, Moorenstraße 5, 40225 Düsseldorf, Germany,
E-mail: handschel@med.uni-duesseldorf.de

Andreas Herrmann Department of Molecular Biophysics, Humboldt University of Berlin, Invalidenstraße 43, 10115 Berlin, Germany

Roland Hetzer Laboratory for Tissue Engineering, Department of Cardiothoracic and Vascular Surgery, Deutsches Herzzentrum Berlin, Campus Benjamin Franklin, Augustenburger Platz 1, 13353 Berlin, Germany,
E-mail: hetzer@dhzb.de

Thomas Hoell Spine Center Baden, Mittelbaden Hospital, Robert-Koch-Str. 70, 77815 Bühl, Germany

Scott J. Hollister Departments of Biomedical Engineering, Surgery and Mechanical Engineering, University of Michigan, 2200 Bonisteel Boulevard, Ann Arbor, MI 41809, USA, E-mail: scottho@umich.edu

Hans-Jürgen Holzhausen Institute for Pathology, University of Halle, Magdeburger Str. 14, 06112, Halle, Germany

Raymund E. Horch Department of Plastic and Hand Surgery, University of Erlangen-Nürnberg, Krankenhausstraße 12, 91054 Erlangen, Germany,
E-mail: horchrd@chir.imed.uni-erlangen.de

Johnny Huard Stem Cell Research Center, Children's Hospital of Pittsburgh, 3460 Fifth Avenue, 4100 Rangos Research Center, Pittsburgh, PA 15213, USA, E-mail: jhuard@pitt.edu

A. Hyatt University of Ottawa Eye Institute, The Ottawa Hospital, General Campus, 501 Smyth Road, Ottawa, ON K1H 8L6, Canada

Bradley B. Jarrold The Procter & Gamble Company, Beauty Technology Division, Miami Valley Innovation Center, Cincinnati, OH 45253, USA

Gavin Jell Department of Materials and Institute of Biomedical Engineering, Imperial College London, London SW7 2AZ, UK, E-mail: g.jell@imperial.ac.uk

Claire G. Jeong Scaffold Tissue Engineering Group and Department of Biomedical Engineering, The University of Michigan, 2208 Lurie Biomedical Engineering Building, 1101 Beal Avenue, Ann Arbor, MI 48109-2099, USA

David Jones Institute of Experimental Orthopaedics and Biomechanics, Philipps University of Marburg, Baldingerstraße, 35033 Marburg, Germany, E-mail: jones@med.uni-marburg.de

Ulrich Joos Clinic for Cranio Maxillofacial Surgery, University of Münster, Münster, Germany, E-mail: joos@eacmfs.org

Ruben Y. Kannan Biomaterials and Tissue Engineering Centre, Academic Division of Surgical and Interventional Sciences, University College London, Rowland Hill Street, London NW3 2PF, UK, E-mail: ykruben@yahoo.com

Jessica M. Kemppainen Scaffold Tissue Engineering Group and Department of Biomedical Engineering, The University of Michigan, 2208 Lurie Biomedical Engineering Building, 1101 Beal Avenue, Ann Arbor, MI 48109-2099, USA

Young-Jin Kim Biomedical Research Institute, Lifecord Inc, Yeoksam-dong, 708-33 Kangnam-gu, Seoul 139-919, South Korea, E-mail: jin@lifecord.co.kr

Gesine Kögler José Carreras Cord Blood Bank, Institute for Transplantation Diagnostics and Cell Therapeutics, Heinrich Heine University Medical Center, Moorenstraße 5, Bldg. 14.88, 40225 Düsseldorf, Germany, E-mail: koegler@itz.uni-duesseldorf.de

Thomas Korte Department of Molecular Biophysics, Humboldt University of Berlin, Invalidenstraße 43, 10115 Berlin, Germany

Birgit Kruse-Lösler Clinic for Cranio-, Maxillofacial Surgery, University of Münster, Waldeyerstr. 30, 48149 Münster

Norbert R. Kübler Clinic for Maxillofacial and Plastic Facial Surgery, University of Düsseldorf, Moorenstraße 5, 40225 Düsseldorf, Germany

Matthew D. Kwan Division of Plastic and Reconstructive Surgery, Department of Surgery, Stanford University School of Medicine, 257 Campus Drive, GK201, Stanford, CA 94305-5148, USA, E-mail: mkwan001@stanford.edu

Lydia Lammers Clinic for Cranio-, Maxillofacial Surgery, University of Münster, Waldeyerstr. 30, 48149 Münster, Germany

Günter Lauer Department of Oral and Maxillofacial Surgery, University Hospital Carl Gustav Carus Dresden, Fetscherstraße 74, 01307 Dresden, Germany, E-mail: guenter.lauer@uniklinikum-dresden.de

Doo-Hoon Lee Biomedical Research Institute, Lifecord Inc, Yeoksam-dong, 708-33 Kangnam-gu, Seoul 139-919, South Korea, E-mail: dhl@lifecord.co.kr

Suk-Koo Lee Department of Surgery, Samsung Medical Center, Sungkyunkwan University, Seoul 135-710, South Korea, E-mail: sklee@smc.samsung.co.kr

Mark P. Lewis Division of Biomaterials and Tissue Engineering, UCL Eastman Dental Institute, 256 Gray's Inn Road, London WC1X 8LD, UK, E-mail: m.lewis@eastman.ucl.ac.uk

Ren-Ke Li Division of Cardiovascular Surgery, Department of Surgery, Toronto General Research Institute, University of Toronto, Toronto, ON M5G 1L7, Canada, E-mail: renkeli@uhnres.utoronto.ca

Wan-Ju Li Department of Orthopedics and Rehabilitation, Department of Biomedical Engineering, University of Wisconsin, Madison, 600 Highland Avenue, K4/769 Clinical Science Center, Madison, WI 53792-7375, USA, E-mail: li@orthorehab.wisc.edu

Elly E. Liao Scaffold Tissue Engineering Group and Department of Biomedical Engineering, The University of Michigan, 2208 Lurie Biomedical Engineering Building, 1101 Beal Avenue, Ann Arbor, MI 48109-2099, USA

Jeanette Libera co.don AG, Warthestraße 21, 14513 Teltow, Germany, E-mail: jeanlibera@hotmail.com

Jeremy J. Lim Department of Biomedical Engineering, Georgia Institute of Technology and Emory University, 313 Ferst Drive, Atlanta, GA 30332, USA, E-mail: jeremy.lim@bme.gatech.edu

Michael T. Longaker Division of Plastic and Reconstructive Surgery, Department of Surgery, Stanford University School of Medicine, 257 Campus Drive, GK201, Stanford, CA 94305-5148, USA

Cora Lüders Cardiothoracic Surgery, Deutsches Herzzentrum Berlin, Augustenburger Platz 1, 13353 Berlin, Germany, E-mail: lueders@dhzb.de

Ursus Lüthi Sports Clinic Zurich, Tödistraße 49, 8002 Zürich, Switzerland

Beate Lüttenberg Tissue Engineering Laboratory, Clinic for Maxillofacial and Plastic Facial Surgery, Waldeyerstr. 30, 48149 Münster, Germany,
E-Mail: Bea.Luettenberg@ukmuenster.de
and
Centrum for Bioethics, University Münster, Von-Esmarch-Str. 62, 48149 Münster, Germany

Bernhard Maisch Department of Internal Medicine – Cardiology, University Hospital Marburg, Baldingerstraße 1, 35033 Marburg, Germany,
E-mail: Bernhard.Maisch@.med.uni-marburg.de

James M. Mason Molecular & Cellular Therapeutics and Gene Therapy Vector Laboratories, NS-LIJ Feinstein Institute for Medical Research, 350 Community Drive, Manhasset, NY 11030, USA, E-mail: jmason@nshs.edu

Robert L. Mauck McKay Orthopaedic Research Laboratory, Department of Orthopaedic Surgery, University of Pennsylvania, 424 Stemmler Hall, MC6081, 36th Street and Hamilton Walk, Philadelphia, PA 19104, USA,
E-mail: lemauck@mail.med.upenn.edu

Cristopher R. McLaughlin University of Ottawa Eye Institute, The Ottawa Hospital, General Campus, 501 Smyth Road, Ottawa, ON K1H 8L6, Canada

Hans-Jörg Meisel Neurosurgery, Bergmannstrost Hospital, Merseburger Straße 165, 06112 Halle, Germany

Thomas Meyer Department of Internal Medicine – Cardiology, University Hospital Marburg, Baldingerstraße 1, 35033 Marburg, Germany,
E-mail: meyert@med.uni-marburg.de

Ulrich Meyer Clinic for Maxillofacial and Plastic Facial Surgery, Heinrich Heine University, Moorenstraße 5, 40225 Düsseldorf, Germany,
E-mail: praxis@mkg-muenster.de

Caterina Minelli Department of Materials and Institute of Biomedical Engineering, Imperial College London, London SW7 2AZ, UK,
E-mail: c.minelli@imperial.ac.uk

Guo-Li Ming Institute for Cell Engineering, Department of Neurology, Johns Hopkins University School of Medicine, 733 N. Broadway, Broadway Research Building 706, Baltimore, MD 21205, USA

Erin N. Moffitt Scaffold Tissue Engineering Group and Department of Biomedical Engineering The University of Michigan, 2208 Lurie Biomedical Engineering Building, 1101 Beal Avenue, Ann Arbor, MI 48109-2099, USA

Dafni Moschidou Experimental Fetal Medicine Group, Institute of Reproductive and Developmental Biology, Imperial College, Hammersmith Campus, Du Cane Road, London W12 0NN, UK, E-mail: dafni.moschidou04@imperial.ac.uk

Vivek Mudera Tissue Repair and Engineering Center, UCL Institute of Orthopaedics and Musculoskeletal Science, Royal National Orthopaedic Hospital, Brockley Hill, Stanmore, Middlesex HA7 4LP, UK, E-mail: rmhkvim@ucl.ac.uk

Lisa A. Mullins The Procter & Gamble Company, Beauty Technology Division, Miami Valley Innovation Center, Cincinnati, OH 45253, USA

Yaakov Nahmias Center for Engineering in Medicine, Department of Surgery, Massachusetts General Hospital, Shriners Burns Hospital, Harvard Medical School, 51 Blossom Street, Boston, MA 02114, USA,
E-mail: nahmias.yaakov@mgh.harvard.ed

Boris Nasseri Cardiothoracic Surgery, Deutsches Herzzentrum Berlin, Augustenburger Platz 1, 13353 Berlin, Germany, E-mail: nasseri@dhzb.de

Christian Naujoks Department of Maxillofacial and Plastic Facial Surgery, University Hospital Düsseldorf, Moorenstraße 5, 40225 Düsseldorf, Germany,
E-mail: christian.naujoks@med.uni-duesseldorf.de

Jörg Neunzehn Klinik und Poliklinik für spezielle Mund-Kiefer-Gesichts-Chirurgie mit Institut für Experimentelle Zahnheilkunde, Waldeyerstrasse 30, 48149 Münster, Germany, E-mail: joerg.neunzehn@ukmuenster.de

Lori W. Norton Wallace H. Coulter Department of Biomedical Engineering, Georgia Institute of Technology and Emory University, 313 Ferst Drive, Atlanta, GA 30332, USA, E-mail: lori.norton@bme.gatech.edu

Christian Oddou Faculté des Sciences et Technologie, B2OA, CNRS 7052 et Universités Paris 7, 12 & 13, 61 Avenue du General de Gaulle, 94010 Creteil Cedex, France, E-mail: oddou@univ-paris12.fr

Michelle Alicia Ommerborn Department of Operative and Preventive Dentistry and Endodontics, Heinrich-Heine-University, Moorenstraße 5, 40225 Düsseldorf, Germany, E-mail: Ommerborn@med.uni-duesseldorf.de

Rosemarie Osborne The Procter & Gamble Company, Beauty Technology Division, Miami Valley Innovation Center, Cincinnati, OH 45253, USA

Christina M. Othon Chemistry Division, US Naval Research Laboratory, Washington, DC 20375, USA, E-mail: othon@nrl.navy.mil

Abhay Pandit National Centre for Biomedical Engineering Science, National University of Ireland, Galway, Ireland, E-mail: abhay.pandit@nuigalway.ie

Sabine Pankuweit Department of Internal Medicine – Cardiology, University Hospital Marburg, Baldingerstraße, 35043 Marburg, Germany,
E-mail: pankuwei@staff.uni-marburg.de

Jung-Keug Park Department of Chemical and Biochemical Engineering, Dongguk University, Center for Advanced Colloidal Materials (CACOM), E208, Wonheungkwan, 3-26, Pil-dong, Choong-gu, Seoul 100-715, South Korea,
E-mail: jkpark@dongguk.edu

Charles W. Patrick Jr. Office of Institutional Advancement, Southwestern Baptist Theological Seminary, Fort Worth, Tx 76122, USA,
E-mail: cpatrick@swbts.edu

Robert A. Peattie Department of Biomedical Engineering, Tufts University, 4 Colby Street, Medford, MA 02155, USA, E-mail: peattie@engr.orst.edu

Julien Pierre Faculté des Sciences et Technologie, B2OA, CNRS 7052 et Universités Paris 7-12-13, 61 Avenue du Général de Gaulle, 94010 Creteil Cedex, France, E-mail: j.pierre@univ-paris12.fr

Rouven Porz Ethikstelle, Inselspital, University Bern, 3010 Bern, Switzerland

Yves Poumay Cell and Tissue Laboratory, URPHYM, University of Namur (FUNDP), Rue de Bruxelles 61, 5000 Namur, Belgium

Oliver Pullig Clinic for Orthopaedics and Rheumatology, University Hospital Erlangen, Rathsbergerstraße 57, 91054 Erlangen, Germany

Wolfgang Hans-Michael Raab Department of Operative and Preventive Dentistry and Endodontics, Heinrich Heine University, Moorenstraße 5, 40225 Düsseldorf, Germany, E-mail: Raab@med.uni-duesseldorf.de

Christoph Rehmann-Sutter Ethics in Bioscience, University of Basel, Schönbeinstraße 20, 4056 Basel, Switzerland,
E-mail: christoph.rehmann-sutter@unibas.ch

Stefania Adele Riboldi Institute for Surgical Research and Hospital Management, University Hospital Basel, Hebelstraße 20, 4031 Basel, Switzerland,
E-mail: riboldis@uhbs.ch

Bradley R. Ringeisen Alternative Energy Section, Code 6113, US Naval Research Laboratory, 4555 Overlook Ave. SW, Washington, DC 20375, USA,
E-mail: bradley.ringeisen@nrl.navy.mil

Klaus Ruhnau St. Marien-Hospital Buer, Mühlenstraße 5–9, 45894 Gelsenkirchen, Germany

Volker Ruppert Department of Internal Medicine – Cardiology, University Hospital Marburg, Baldingerstraße 1, 35033 Marburg, Germany,
E-mail: ruppert@med.uni-marburg.de

Henning Schliephake Department of Oral and Maxillofacial Surgery, George Augusta University, Robert-Koch-Straße 40, 37075 Göttingen, Germany,
E-mail: schliephake.henning@med.uni-goettingen.de

Fabian Schmidt Clinic for Surgery, University of Münster, Waldeyerstraße 16, 48149 Münster, Germany

Kurt Schneider Department of Operative and Preventive Dentistry and Endodontics, Heinrich Heine University, Moorenstraße 5, 40225 Düsseldorf, Germany, E-mail: Kurt.Schneider@uni-duesseldorf.de

Eberhard Schockenhoff AB Moraltheologie, Universität Freiburg, 79085 Freiburg, Germany, E-mail: eberhard.schockenhoff@theol.uni-freiburg.de

Thomas Schreyer Evangelisches Krankenhaus Elisabethenstift GmbH, Landgraf-Georg-Straße 100, 64287 Darmstadt, Germany

Jackie Leach Scully School of Geography, Politics and Sociology, Newcastle University, 5th floor, Claremont Bridge, Claremont Road, Newcastle upon Type NE1 7RU, UK

Alexander M. Seifalian Biomaterials and Tissue Engineering Centre, Academic Division of Surgical and Interventional Sciences, University College London, Rowland Hill Street, London NW3 2PF, UK,
E-mail: a.seifalian@medsch.ucl.ac.uk

Rishma Shah Division of Biomaterials and Tissue Engineering, UCL Eastman Dental Institute, 256 Gray's Inn Road, London WC1X 8LD, UK,
E-mail: r.shah@eastman.ucl.ac.uk

Vilma Siodla Department of Neurosurgery, BG Clinic Bergmannstrost, Merseburgerstraße 165, 06112 Halle, Germany

Bethany J. Slater Division of Plastic and Reconstructive Surgery, Department of Surgery, Stanford University School of Medicine, 257 Campus Drive, GK201, Stanford, CA 94305-5148, USA

Hongjun Song Departments of Neurology and Neuroscience, Institute for Cell Engineering, Johns Hopkins University, 733 N. Broadway, Baltimore, MD 21205, USA, E-mail: shongju1@jhmi.edu

Jennifer Southgate The Jack Birch Unit of Molecular Carcinogenesis, Department of Biology, University of York, York YO10 5YW, UK,
E-mail: js35@york.ac.uk

Barry J. Spargo Chemistry Division, US Naval Research Laboratory, Washington, DC 20375, USA, E-mail: spargo@nrl.navy.mil

Christof Stamm Cardiothoracic Surgery, Deutsches Herzzentrum Berlin, Augustenburger Platz 1, 13353 Berlin, Germany, E-mail: stamm@dhzb.de

Molly M. Stevens Department of Materials and Institute of Biomedical Engineering, Imperial College of Science, Prince Consort Road, London SW7 2BP, UK,
E-mail: m.stevens@imperial.ac.uk

Gilbert Tang Division of Cardiovascular Surgery, Department of Surgery, Toronto General Research Institute, University of Toronto, Toronto, ON M5G 1L7, Canada, E-mail: gilbert.tang@utoronto.ca

Jörg Teßmar Department of Pharmaceutical Technology, University of Regensburg, Universitaetsstraße 31, 93040 Regensburg, Germany

Johnna S. Temenoff Department of Biomedical Engineering, Georgia Institute of Technology and Emory University, 313 Ferst Drive, Atlanta, GA 30332, USA E-mail: johnna.temenoff@bme.gatech.edu

Ernst M. Tetzlaff Orthopedic Clinic, Praxis für Orthopädie, Am Alten Markt, 22926 Ahrensburg, Germany

Neil David Theise Division of Digestive Diseases, Beth Israel Medical Center, First Avenue at 16th Street, New York, NY 10003, USA, E-mail: ntheise@chpnet.org

Rocky S. Tuan Cartilage Biology and Orthopaedics Branch, National Institute of Arthritis and Musculoskeletal and Skin Disorders, National Institutes of Health, Department of Health and Human Services, 50 South Drive, MSC 8022, Building 50, Room 1503, Bethesda, MD 20892-8022, USA, E-mail: tuanr@mail.nih.gov

Alexander M. Turner The Jack Birch Unit of Molecular Carcinogenesis, Department of Biology, University of York, York YO10 5YW, UK, E-mail: alexturner64@yahoo.co.uk

Mathieu Unbekandt Centre for Integrative Physiology, University of Edinburgh, George Square, Edinburgh EH8 9XB, UK, E-mail: munbekan@staffmail.ed.ac.uk

Patrick Vavken Department of Orthopedic Surgery, Children's Hospital Boston, Harvard Medical School, 300 Longwood Ave, Enders 1016, Boston, MA 02115, USA, E-mail: patrick.vavken@childrens.harvard.edu

Mitchell A. Watsky Department of Physiology, University of Tennessee Health Science Center, 894 Union Avenue, Memphis, TN 38163, USA

Wilfried Weber Institute for Chemical and Bioengineering, HCI F115, ETH Zurich, Wolfgang-Pauli-Straße 10, 8093 Zurich, Switzerland, E-mail: wilfried.weber@chem.ethz.ch

Richard Weisel Division of Cardiovascular Surgery, Department of Surgery, Toronto General Research Institute, University of Toronto, Toronto, ON M5G 1L7, Canada

Xuejun Wen Department of Cell Biology and Anatomy and Department of Orthopaedic Surgery, Medical University of South Carolina, Charleston, SC 29425, USA, E-mail: xuejun@musc.edu

David Wendt Institute for Surgical Research and Hospital Management, University Hospital Basel, Hebelstraße 20, 4031 Basel, Switzerland,
E-mail: dwendt@uhbs.ch

Anne Wiesmann Hämato-Onkologisches Zentrum Hamburg-Ost, Hamburger Str. 41, 21465 Reinbek, Germany, E-mail: anne.wiesmann@alice-dsl.net

Hans Peter Wiesmann Biomineralisation and Tissue Engineering Group, Department of Experimental Maxillofacial Surgery, University of Münster, Waldeyerstr. 30, 48149 Münster, Germany, E-mail: wiesmann@life-rds.eu

Pensée Wu Experimental Fetal Medicine Group, Institute of Reproductive and Developmental Biology, Imperial College, Hammersmith Campus, Du Cane Road, London W12 0NN, UK, E-mail: p.wu@imperial.ac.uk

Peter K. Wu Southern Oregon University, Ashland, OR 97520, USA,
E-mail: wu@sou.edu

Xuemei Wu Department of Biomedical Engineering, The University of Texas M. D. Anderson Cancer Center, 1515 Holcombe Blvd., Unit 193, Houston, TX 77030, USA, E-mail: xuewu@mdanderson.org

Martin L. Yarmush Center for Engineering in Medicine, Department of Surgery, Massachusetts General Hospital, Shriners Burns Hospital, Harvard Medical School, 51 Blossom Street, Boston, MA 02114, USA,
E-mail: nahmias.yaakov@mgh.harvard.ed

James J. Yoo Wake Forest Institute for Regenerative Medicine, Wake Forest University Health Sciences, Medical Center Boulevard, Winston-Salem, NC 27157, USA, E-mail: jyoo@wfubmc.edu

Ning Zhang Clemson-MUSC Bioengineering Program, Department of Bioengineering, Clemson University, Charleston, SC 29425, USA
and
Department of Cell Biology and Anatomy, Medical University of South Carolina, Charleston, SC 29425, USA

Part A

General Aspects

I General and Ethical Aspects

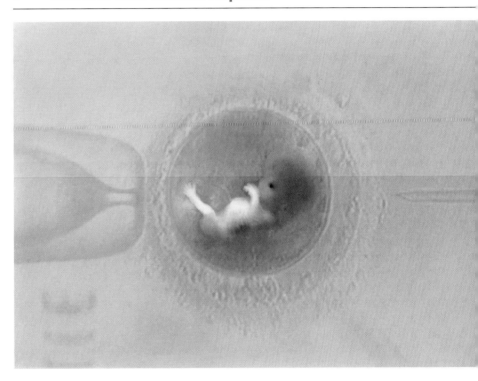

The History of Tissue Engineering and Regenerative Medicine in Perspective

U. Meyer

Contents

1.1 History 5
1.2 Future Prospects 11
    References 12

1.1 History

The artificial generation of tissues, organs, or even more complex living organisms was throughout the history of mankind a matter of myth and dream. During the last decades this vision became feasible and has been recently introduced in clinical medicine. Tissue engineering and regenerative medicine are terms for the field in biomedicine that deal with the transformation of these fundamental ideas to practical approaches. Several aspects of generating new tissues and organs out of small pieces of living specimens are now scientifically solved, but at this point it is unknown how much impact these new approaches will have on clinical medicine in the future. In this respect it seems important to recapitulate from where the visions and the work came, in order to speculate or predict where tissue engineering and regenerative medicine will head.

The concept of tissue engineering and regenerative medicine as measures to create more complex organisms from simpler pieces is deeply embedded in the people's imaginary world. A change in the vision, hope, and believe of how to create or regenerate complex organs or organisms can be observed during history as a mirror of the cultural history of mankind. Even the early history of men is related to the idea that independent life can be created without sexual reproduction. Stories from Greek mythology [the creation of persons without sexual reproduction, e.g., the generation of Prometheus (Fig. 1.1)] may be considered as early reports representing the idea of creating living creatures from living or nonliving specimens. The Biblical tale of Eve created from Adam's rib is a further and perhaps the most well-known

Fig. 1.1 The generation of Prometheus

Fig. 1.2 Healing of Justinian

Fig. 1.3 Theophrastus von Hohenheinm

example of this concept [1] (in a modern view a kind of hybrid cloning). A multitude of examples in literature and the arts mirrors the desire of humans to be able to create by themselves living individuals or at least parts of individuals. The envisioned measures to create life are influenced by the social, cultural, and scientific background of individual persons at that time.

The famous painting "Healing of Justinian" (Fig. 1.2) a visualization of the legend of St. Cosmas and St. Damien (278 AD) depicting the transplantation of a homograft limb onto an injured soldier, is one early instance of the vision of regenerative medicine. As humans progressed in the understanding of nature and as they developed more advanced culture techniques they envisioned the generation of living creatures by applying physicochemical or biological techniques. During the transformation from the Middle Ages to the Renaissance in Europe, there was the hope and belief by a number of scientists that through alchemy living organisms could be generated. Theophrastus von Hohenheim, better known as Paracelsus (Fig. 1.3), tried (and failed) to find a recipe to create human life by a mixture of chemical substances in a defined environment.

Johann Wolfgang von Goethe (1749–1832) deals in his fundamental work of literature *Faust* [2] with the relation of an individual (Faust) to knowledge, power, morality, and theology. One central theme in the struggle of Faust to be powerful is the deeply embedded wish to create life. The creation of the artificial being Homunculus in Goethe's *Faust* is a central part of the drama, by which Goethe reveals various transformational processes working in the human soul. In the famous laboratory scene of *Faust (Part II)* he describes the vision of men being able to create life by alchemy (Fig. 1.4), representing the irrepressible human dream of "engineering" life:

Look there's a gleam! – Now hope may be fulfilled,
That hundreds of ingredients, mixed, distilled –
And mixing is the secret – give us power
The stuff of human nature to compound
If in a limbeck we now seal it round
And cohobate with final care profound,
The finished work may crown this silent hour

Chapter 1 The History of Tissue Engineering and Regenerative Medicine in Perspective

Fig. 1.4 Depiction of Dr. Faustus and his Homunculus

many contemporary "Faustian" technologies, such as cloning, genetic, or stem cell techniques in modern tissue engineering and regenerative medicine. With respect to an historical view of tissue engineering, Faust is a representative of Northern European humanity striving for evolution from the scientific and ethical limitations and strictures of the 16th century Reformations to the new aspirations of humanity that Goethe saw developing during the 18th century Enlightenment era. He was attracted to the idea of creating life by adding substances to nonliving specimens, similar to visions of how God created Adam, visualized by the famous painting of Michelangelo (Fig. 1.5). Goethe struggles to weave the personal inner journey of Faust towards some enlightenment (described in the prologue):

I've studied now Philosophy,
And Jurisprudence, Medicine,
And even alas! Theology
All through and through with ardour keen!
Here now I stand, poor fool, and see
I'm just as wise as formerly.
Am called a Master, even Doctor too,
And now I've nearly ten years through
Pulled my students by their noses to and fro
And up and down, across, about,
And see there's nothing we can know!

It works! The substance stirs, is turning clearer!
The truth of my conviction passes nearer
The thing in Nature as high mystery prized,
This has our science probed beyond a doubt
What Nature by slow process organized,
That have we grasped, and crystallized it out.

The description of the creation of Homunculus is also of special concern today, since it is suggestive of

thereby being in the context of the collective social forces that are undergoing transformation through the historical processes of that time. As Faust deals with nearly all aspects and questions that arise in tissue engineering and regenerative medicine (and that are discussed in the first chapter of this book), it can

Fig. 1.5 Michelangelo's painting The Creation of Adam

be considered to be a timeless and always relevant consideration on the field of biomedicine.

Later on, as science and medicine progressed, a multitude of stories, reports, paintings, and films dealt with the idea that humans could create life by modern "scientific" measures. A prominent newer example in literature and film is the story of Frankenstein, written by Mary Shelley in 1818 (Fig. 1.6), describing the vitalization of a creature, reassembled from different body parts.

Parallel to the mythological, biblical, and fictional reports, various persons performed pioneering practical work to generate, heal, or regenerate body parts. The emergence of tissue engineering is, through their work, closely connected with the development of clinical medicine (prosthetics, reconstructive surgery, transplantation medicine, microsurgery) and biology (cell biology, biochemistry, molecular biology, genetics).

The mechanical substitution of body parts by nonvital prosthetic devices (metallic and ivory dentures, wooden legs) can be considered as early efforts to use biomaterials in reconstructive medicine. The first attempts to replace teeth in the sense of modern dental implantology seems to go back as early as in the Galileo-Roman period. The anthroposophic finding of a human skull, containing a metallic implant in the jaw [3], is indicative of early attempts of humans to regain lost function by tissue substitution. Leading areas of reconstructive medicine in clinical use were evident in the age before modern dentistry and orthopedics. Ambroise Pare` (1510–1590) described in his work *Dix livres de la chirurgie* [4] measures to reconstruct teeth, noses, and other parts of the body. A common method in the 18th century to replace teeth was the homologous transplantation of teeth in humans. John Hunter (1728–1793) investigated in his pioneering work the effect of transplantation not only at a clinical level (he claimed, that homologous transplanted teeth lasted for years in the host) but also performed animal experimental work on the fate of transplants, thereby setting the basis for a scientific approach on transplantation medicine [5].

A milestone in the modern view of tissue engineering was the use of skin grafts. The use of skin grafts is closely related to the work of the famous surgeon Johann Friedrich Dieffenbach (1792–1847). As he performed animal experimental and clinical work on skin transplantation (described in *Nonnulla de Regeneratione et Transplantatione* [6]), and as he also established ways to use pedicled skin flaps (since most of the clinical skin transplantation treatments failed), Dieffenbach is one of the modern founders of plastic and reconstructive surgery and can also be considered to be an early practitioner in transplantation medicine. Breakthroughs in the clinical use of skin grafts were made by Heinrich Christian Bünger, first successful autologous skin transplantation [7]; Jaques Reverdin (1842–1929), use of small graft islets; and Karl Thiersch (1827–1895), split thickness grafts [8, 9]. The high number of failures were overcome by the observation of Esser (1877–1964) that immobilization of transplants through the use of dental impression materials improves the fate of transplants in facial wound reconstruction. The clinical efforts reached through the combined use of surgical and dental techniques in reconstructive surgery and transplantation medicine led to the evolution of the dental- and medical-based Maxillofacial and Plastic Facial Surgery discipline. The foundation and establishment of this new specialty at the Westdeutsche Kieferklinik in Düsseldorf and the extensive experience in this center with injured soldiers during the

Fig. 1.6 Book cover of Frankenstein (Edition 1831)

First and Second World War led to significant improvements in tissue regeneration and reconstruction in the plastic and reconstructive surgery field. The underlying biological reason for the success of clinical skin transplantation by refining the transplantation approach (the shift from enlarged grafts to small cell-containing particles, the invention of fixation protocols) was in the beginning to a great extent unknown. Enlightenment into the biological mechanisms that accounted for the fate of transplants was provided by the fundamental biological work of Rudolf Virchow (1821–1902). He described in his *Cellularpathologie* [10] that tissue regeneration is dependent on cell proliferation. His work led not only to the investigation of tissue healing through cellular effects, but also to the cultivation of cells outside the body (in vitro, first suggested by Leo Loeb [11]). C.A. Ljunggren and J. Jolly were the first researchers to attempt to cultivate cells outside the body [12]. The milestone breakthrough in in vitro cell cultivation was reached by R.G. Harrison (1870–1959), demonstrating active growth of cells in culture [13]. Since that time, cell biology and especially in vitro cell culture became the mainstay of what can be considered classical tissue engineering [14].

Underlying in vitro cell culture with subsequent cell transplantation, modern tissue engineering and regenerative medicine is directly connected to microsurgery. Alexis Carrel (1873–1944) can be considered the founder of modern organ transplantation due to his work elaborating the methods of vascular anastomosis [15, 16]. The use of microvascular surgery was primarily performed in organ transplantation and plastic surgery. Whole organ transplantation or transplantation of body parts were made possible by this technique. E. Ullman (first kidney transplantation in animals [17]) and J.P. Merrill (first successful clinical kidney transplantation in identical twins) are directly related to the advances in transplantation medicine [18]. One of the most well-known milestones in organ transplantation medicine was the first heart transplantation 1967 by the South African surgeon Christiaan Barnard. His life—saving transplantation not only had extensive coverage in the newspapers at that time, but also raised an intensive and controversial debate on ethical issues in transplantation medicine [19]. Whereas the indications for microvascular tissue transplantation were extended towards plastic and reconstructive surgery (use of free myocutaneous, osteomyocutaneus, and other vessel containing flaps), and measures to perform microsurgery were refined and standardized, failures in clinical transplantation medicine were mainly based on immune incompatibilities (except for the use of autografts). The success of transplantation medicine, whether cells, tissues, or organs are in use was, and still is, to a great extent dependant on the immune state of graft and host. The science of immunomodulation and immunosupression is therefore still a critical aspect in all tissue engineering and regenerative medicine applications.

The term "tissue engineering" was up to the mid 1980s loosely applied in the literature in cases of surgical manipulation of tissues and organs or in a broader sense when prosthetic devices or biomaterials were used [20]. The term "tissue engineering" as it is nowadays used was introduced in medicine in 1987. The definition that was agreed on was: "Tissue Engineering is the application of the principles and methods of engineering and life sciences toward the fundamental understanding of structure-function relationships in normal and pathologic mammalian tissue and the development of biological substitutes to restore, maintain, or improve function." The early years of tissue engineering were based on cell and tissue culture approaches. W.T. Green undertook a number of experiments in the early 1970s to generate cartilage using a chondrocyte culture technique in combination with a "bone scaffold." Despite of his inability to generate new cartilage, he set the theoretical and practical concept to connect and coax cells with scaffolds. Innovations in this approach were made by Burke and Yannas through a laboratory and clinical collaboration between Massachusetts General Hospital and M.I.T. in Boston, aimed at generating skin by a culture of dermal fibroblasts or keratinocytes on protein scaffolds, and using it for the regeneration of burn wounds. A key point in tissue engineering was given by the close cooperation between Dr. Joseph Vacanti from Boston Children's Hospital and Dr. Robert Langer from M.I.T. Their article in Science [21], describing the new technology, may be referenced as the beginning of this new biomedical discipline. Later on, a high number of centers all over the world focused their research efforts towards this field. Tissue engineering was catapulted to the forefront of the public awareness with a BBC broadcast exploring the potential of tissue-engineered cartilage which included images of the now infamous "mouse with the human ear" on its back from the laboratory of Dr. Charles Vacanti at the University of Massachusetts Medical Center. The visual power of the photograph of the "auriculosaurus" helped to transfer the idea and vision of generating new tissues or organs from the imaginary world of human beings to the real world. Since that time, tissue engineering has been considered one of the most promising biomedical technologies of the century [22].

Regenerative medicine seems to be more difficult to define, as this term was used earlier but was less defined in the literature then the term "tissue engineering." It is now viewed by most biologists and physicists as a field where stem cells drive embryonic formation, or where inductive organizers induce a blastema to regenerate a tissue, aimed at reforming damaged tissues and organs in humans. It seems that a rigid definition of regenerative medicine is not constructive while the principle approaches that define the field are still being delineated. Stem cells, being at the center of expectations, hold great promise for the future of regenerative medicine. Stem cell plasticity and cloning, with nuclear transfer, transdifferentiation, and cell fusion as measures to modulate the stem cell differentiation pathway, is now a central issue in regenerative medicine [23]. During cloning, an adult nucleus is transplanted into an egg, which must erase the adult genome's epigenetic marks, so it can re-express every gene necessary to build a new animal. Robert Briggs and Thomas King were the first to demonstrate (based on a similar experiment proposed by Hans Spemann at the University of Freiburg as early as in 1938) how to clone frogs by replacing the nuclei of eggs with cells from tadpoles and adult intestinal epithelium. The further major step in cloning research was the cloning of two lambs (Megan and Morag) from embryonic cells in 1996. Cloning of the sheep Dolly [24] was, from a public awareness point of view, the key event in stem cell research. Ian Wilmut at the Roslin Institute near Edinburgh and his colleagues at PPL Therapeutics in East Lohian reported on February 27th, 1997 in *Nature* that they had produced a lamb named Dolly (Fig. 1.7), born the previous July, that was the first mammalian clone created using the genetic material from an adult cell. As soon as the story hit the front pages (the news was broke by the British Sunday newspaper *The Observer* four days ahead of *Nature's* publication) a public and media maelstrom ensued. Editors and writers of other newspaper went so far as to speculate

Fig. 1.7 Sheep Dolly and her first-born lamb Bonnie

that "It is the prospect of cloning people, creating armies of dictators, that will attract most attention." The creation of Dolly, cloned from an adult udder cell, overturned the idea that in mammals, developed cells could not reverse their fate. The news had a tremendous impact on society, mirrored by the fact that within days the President of the United States, the head of the European Commission, the Vatican, and many others were calling for a review of regulations on cloning research, if not an outright ban. It became obvious that the world was simply not prepared for the debate. The use of modern techniques like cloning or stem cell modulation was considered by a large part of the public to be a modern version of alchemy [25]. Japanese researchers reported in 1998 on the successful birth of eight cloned calves using adult cells from slaughterhouse entrails, raising the possibility that animals could be cloned for the quality of their meat (a situation close to the vision of the creation of Eve). During the last decade, scientists have shown that they are able to clone a variety of mammals including mice, rats, calves, cows, pigs, cats, and dogs. Some envisaged an industry of cloning applications, from the production of medicines in live bioreactors—cloned, genetically modified livestock—to the creation of herds of cloned animals that might one day be used as organ donors. But the low efficiency of the cloning process, reflecting the problems related to reprogramming a cell's DNA by the content of eggs, has stymied industrial development. New directions focus on the addition of chemicals or proteins to adult cell nuclei, in order to bypass the need for eggs altogether. Egg-free approaches may also enable what many see as the most promising potential application of cloning: the creation of human embryonic stem cells, or cells made from them, that could be used to treat human disease [26].

Ten years later, the ethical debate launched by Dolly and encouraged by science fiction stories has changed. After a decade, mammalian cloning is moving forward with central societal issues remaining unresolved. The recent situation has now been supplanted by a bioethical discussion that is more complex and more focused on the real situation in stem cell biology. One outcome of this discussion is the risk assessment, currently open for public consultation, by the US Food and Drug Administration (see: www.fda.gov/cvm/Clone RiskAssessment.htm). Researchers and physicists speculate that unless there is some unknown fundamental biological obstacle, and given wholly positive motivations, human reproductive cloning is an eventual certainty [27].

1.2
Future Prospects

The future of tissue engineering and regenerative medicine holds the promise of custom-made medical solutions for injured or diseased patients, with genetic (re-)engineering in the zygote or even earlier as one of the most promising and also most debatable aspects of this field. As the field of tissue engineering and regenerative medicine is established as a central discipline in biomedicine nowadays, it seems interesting to speculate where the field will head. As both areas have matured to the point that its research and clinical aspects can be conceptually categorized, various commercial or noncommercial organizations have tried to assess the future of the field. These assessments seem to be important for researchers, policy makers, regulators, funders, technology developers, and the biomedical industry. Among the different assessments, published in open-access or limited access documents (nearly all of them thinking along similar lines), a study performed by the *Tissue Engineering* Journal can be considered to be the most scientific approach [28].

The Journal undertook such an assessment through the evaluation of a questionnaire (using a modified Hoshin process) given to the editorial board of *Tissue Engineering*. The aim of the study was to identify a list of strategically important concepts to achieve clinically relevant solutions for medical problems (up to the year of 2021). The evaluation of the questionnaire revealed some important aspects for the future of tissue engineering and regenerative medicine. One important finding was that highly strategic issues are not at the forefront of the daily work. The most striking example was angiogenic control. The dominance of this issue over all other issues and its low level of present progress propelled it to the top of strategic concepts. The second most important area was assumed to be stem cell science. As the editors assumed that understanding and control of stem cell research may give way for a short conduit in a number of tissue engineering approaches, molecular biology and system biology research seemed to be similarly important in the strategic development of the field. In addition to angiogenic control and stem cell biology, cell sourcing, cell/tissue interaction, immunologic understanding and control, manufacturing and scale up, as well some other issues were considered to be important (in a decreasing manner). With respect to the near and immediate future, the dominant concept that was supported by the largest number of specialists was clinical understanding and interaction. It seemed important that, as the field is oriented towards clinical application, close collaboration between researchers and clinically working physicians is of utmost importance. A close cooperation, based ideally on a deep understanding of the "other's" field, is not only valuable for the establishment of engineered tissue design criteria but also to enhance the potential for the final introduction of such therapies into clinical practice at large. The present book, intended to give researchers and clinicians a common platform, is hopefully one tool to reach this important aim. The future will show when and how which of the multiple approaches in tissue engineering and regenerative medicine will withstand the proof of clinical usage of such therapies over time.

References

1. Fox E (1983) The Schoken Bible, vol 1. Schoken, New York
2. Goethe JW (1831/1959) Faust (part two) Act I: laboratory. Penguin, Harmondsworth
3. Crubézy E, Murail P, Girard L, Bernadou JP (1998) False teeth of the roman world. Nature 391(6662):29
4. Pare A (1634) The works of that famous Chirurgion Ambrose Parey. Cotes and Young, London
5. Hunter J (1771) The natural history of the human teeth. Johnson, London
6. Dieffenbach JF (1822) Nonnulla de regeneratione et transplantatione. Richter, Würzburg
7. Bünger HC (1823) Gelungener Versuch einer Nasenbildung aus einem völligen getrennten Hautstück aus dem Beine. J Chir Augenhkd 4:569
8. Thiersch K (1874) Ueber die feineren anatomischen Veränderungen beim Aufheilen von Haut auf Granulationen. Arch Klin Chir 37:318–324
9. Mangoldt F (1895) Die Ueberhäutung von Wundflächen und Wundhöhlen durch Epithelaussaat, eine neue Methode der Transplantation. Dtsch Med Wochenschr 21:798–799
10. Virchow R (1858) Die Cellularpathologie in ihrer Begründung auf physiologische und pathologische Gewebelehre. Hirschwald, Berlin
11. Loeb L (1897) Ueber die Entstehung von Bindegewebe, Leukocyten und roten Blutkörperchen aus Epithel und über eine Methode, isolierte Gewebsteile zu züchten. Stern, Chicago
12. Ljunggren CA (1898) Von der Fähigkeit des Hautepithels, ausserhalb des Organismus sein Leben zu behalten, mit Berücksichtigung der Transplantation. Dtsch Z Chir 47:608–615
13. Harrison RG (1910) The outgrowth of the nerve fiber as a mode of protoplasmic extension. J Exp Zool 9:787–846
14. Fell HB (1972) Tissue culture and it's contribution to biology and medicine. J Exp Biol 57:1–13
15. Carrel A, Burrows MT (1911) Cultivation of tissues in vitro and its technique. J Exp Med 13:387–398
16. Witkowski JA (1979) Alexis Carrel and the mysticism of tissue culture. J Med Hist 23:279–296
17. Ullmann E (1902) Experimentelle Nierentransplantation. Wien Klin Wochenschr 15:281–282
18. Merrill JP, Murray JE, Harrison JH (1956) Successful homotransplantation of human kidney between identical twins. J Am Med Assoc 160:277–282
19. Barnard C, Pepper CB (1969) Christiaan Barnard. One life. Timmins, Cape Town
20. Skalak R, Fox CF (1988) Tissue engineering. Liss, New York
21. Langer R, Vacanti JP (1993) Tissue Engineering. Science 260:920–926
22. Nerem RM (1991) Cellular engineering. Ann Biomed Eng 19:529–545
23. Wadman M (2007) Dolly: a decade on. Nature 445:800–801
24. Wilmut I, Schnieke AE, McWhir J, Kind AJ, Campbell KHS (1997) Viable offspring derived from fetal and adult mammalian cells. Nature 385:810–813
25. Rutenberg MS, Hamazaki T, Singh AM, Terada N (2004) Stem cell plasticity: beyond alchemy. Int J Hematol 79:15–21
26. Theise D, Wilmut I (2003) Cell plasticity: flexible arrangements. Nature 425:21
27. Nature editorial (2007) Dolly's legacy. Nature 445:795
28. Johnson PC, Mikos AG, Fisher JP, Jansen JA (2007) Strategic directions in tissue engineering. Tissue Eng 13:2827–2837

Economic Modeling and Decision Making in the Development and Clinical Application of Tissue Engineering

P. Vavken, R. Dorotka, M. Gruber

Contents

2.1 Evaluating Tissue Engineering Between Public Health and Microeconomics 13
2.2 Evaluation of a Treatment 13
2.3 Allocation of Tissue Engineering as a Health Care Resource 14
2.4 Economic Modeling of Cost-effectiveness . 15
2.5 Decision Analysis Models 16
2.6 Cartilage Tissue Engineering 16
2.7 Genitourinary Tissue Engineering 17
2.8 Conclusion 20
References 21

2.1 Evaluating Tissue Engineering Between Public Health and Microeconomics

Tissue engineering in regenerative medicine has become one of the most promising subjects in current medical research. A number of procedures based on or involving tissue engineering have already been approved for human application and are clinically available, and numerous others are on the edge of becoming so. During the development of a new procedure and the process of its approval, repeated evaluations, focusing on different aspects, are necessary. Initial studies focus on the effect of a treatment; subsequently its efficacy and effectiveness as well as safety profiles are established. Before a procedure will be directed towards clinical application, economic evaluations are done assessing whether further investment of scarce resources is justified. In public health such justification is consistent with showing a highly effective treatment and ruling out that the same investment could result in more productive research or a more effective treatment elsewhere, while from a microeconomic perspective potential revenue is addressed. Such argumentation is a source of much debate, but can hardly be avoided given a true scarcity of resources and increasing costs of health care. Due to the increasing importance of such issues, especially in a high end, multidisciplinary field such as tissue engineering, knowledge on the principles of economic evaluation has become a valuable addition to a researcher's expertise. It is the objective of this chapter to give a concise introduction to economic modeling and resulting decision-making strategies in health care in the evaluation of tissue engineering.

2.2 Evaluation of a Treatment

The evaluation of a new intervention comprises the assessment of four distinct parameters: Efficacy, Effectiveness, Efficiency, and Availability and Allocation. Efficacy describes the maximal effect under optimized conditions, for example, in an in vitro study or a meticulously monitored clinical trial using a per-protocol analysis. Per-protocol means that patients are analyzed depending on the treatment they eventually received and not per initial allocation. Effectiveness describes the effect of an intervention in a "real life" situation including noncompliance, losses to follow-up, wrong dosages, et cetera. Effectiveness can be assessed in clinical trials using intention-to-treat, i.e., data analysis retains all patients in the group of their initial allocation, regardless of crossing-over between study groups, losses to follow-up, and the like. While efficacy is of meaning for treatment development, effectiveness has the higher impact on clinical life and is considered more relevant for the purpose of subsequent economic evaluation.

In parallel, effectiveness is an important parameter in decision-making processes. Figure 2.1 shows the framework of evaluation and decision making for a treatment.

2.3
Allocation of Tissue Engineering as a Health Care Resource

As a resource in health care, tissue engineering should be evaluated in the light of four parameters: efficiency, equity, need, and appropriateness to guarantee balanced service provisions. Efficiency describes the effect of a procedure in relation to its costs. Efficiency will be described in detail later in this chapter. Equity stands for a fair distribution of allocations and can be divided into horizontal equity, i.e., treating the same those who have the same need, and vertical equity, i.e., treating differently those who have different levels of need, or in other words to prioritize urgent cases over nonurgent cases. Defining equity, in turn, begs the question of need. Need summarizes three subgroups, corporate need, the need for health, and the need for health care. While corporate need is determined in adjustment to pressure from interest groups affected, the need for health is a more universal parameter. It is established on population data and mirrors the need for health as a function of the prevalence of disease. The need for health care, however, exists when a patient is suffering from a disease that can be successfully addressed by an existing treatment. As such, tissue engineering applications may increase the need for health care by introducing new treatment options to existing diseases. Need assessment and market analysis for supply and demand are not necessarily the same. While their descriptions sometimes sound rather similar, need assessment is a public health instrument aiming towards equitable and effective use of (financial) resources, while a market analysis is the microeconomic way of estimating the potential of value generation, and does not always produce the same result. Appropriateness finally means that a patient receives the treatment he or she needs. As for any new therapy with not yet fully understood outcomes the appropriateness

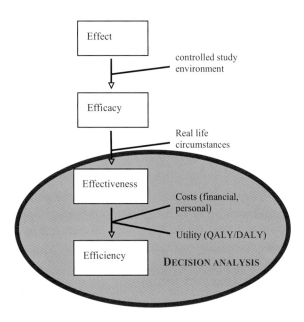

Fig. 2.1 Framework of the evaluation for a treatment. The general effect of a procedure is known and put into action in a specific stetting to establish the efficacy of this procedure for a given question. Taking "real life" circumstances, such as patient compliance, wrong dosage, losses to follow-up and so forth, into account, the effectiveness can be estimated. This value, in turn, is translated into utilities and/or put into relation with costs to establish the efficiency. All the latter parameters are finally used in decision analysis

of various tissue engineering treatments has been a matter of debate.

2.4
Economic Modeling of Cost-effectiveness

Economic models are needed to evaluate the relation between costs and effectiveness, usually in so-called cost-effectiveness analyses (CEA), which is both a generic term and the name of one specific method. There is a considerable variety of validated models for different purposes such as cost estimation and the characterization of cost-outcome relationships, which are clearly beyond the scope and intention of this book [8]. We want to focus on cost-effectiveness and cost-utility analyses for the purpose of this chapter. However, before any analysis can be done, the relevant parameters need to be identified.

The determination of the costs of a tissue engineering application is fairly difficult. Generally, total cost can be split into direct healthcare costs (e.g., procedure, hospital stay, medication, personnel, supplies), indirect healthcare costs (transportation from and to the hospital/office, parking, childcare while admitted to the hospital), and indirect costs such as productivity cost due to lost economic productivity because of impaired ability to work. Obviously cost estimates can differ considerably, depending on perspective and treatment at hand. Translating the cost of an intervention into current currency will produce different results for different markets, depending on the value and price of goods and work force. Finally, time has to be included as a factor. It is a commonly accepted fact that one unit of currency, e.g., one Euro, will lose in value over time. Hence a discount of costs over time, usually 3–5%, is used. This discount, however, does not account for inflation! Given the considerable paucity of data on clinical tissue engineering applications as far as financial burdens are concerned, most of these parameters have to be estimated. An important issue in such predictions is, again, need, since an urgently needed, and thus probably frequently used procedure or treatment will benefit from the economy of scale.

Effectiveness has already been mentioned above. Briefly, effectiveness describes the effect of a treatment in a clinical setting. To identify this parameter, sufficiently sized trials of flawless design and—equally important—studying clearly defined questions are required. It is important to scrutinize published estimations of effectiveness for their validity. Does the evidence at hand really focus on the relevant questions and are appropriate measures used? The brunt of tissue engineering applications proposes beneficial long-term effects, suggesting superiority over current treatment options, but have patients been sufficiently followed to prove this? In summary, effectiveness, although often considered a straight-forward and very objective measure compared to costs or utilities, offers ample opportunities for misconceptions and has to be assessed with much caution.

Utility tries to describe the value of an intervention to a person. While for example a 34% reduction of mortality is an excellent result for an intervention, this has little meaning to the individual patient. One of the most important aspects of utilities as a healthcare outcome is quality of life. Quality of life can be measured using validated scores, and utilities account for this factor by describing quality-adjusted life years (QALY) or disability-adjusted life years (DALY). Another approach to utilities that is used in decision analysis is directly assessing the value of an outcome by interrogating patients. Briefly, a patient is offered two options, e.g., suffering from disease or the status post a procedure, which are usually represented as doors. Each door has a distinct likelihood of achieving an outcome and the patient is asked to choose accordingly. Likelihoods are altered until the patient changes his mind, and this value is then used as the utility of the outcome. This is referred to as a reference lottery and, needless to say, it is very subjective, depending on personal, as well as societal preferences [8, 22]. It is a common option to do a reference lottery with medical specialists, especially for interventions with long-term effects or arcane complication potential, to account for bias arising from patient ignorance. Finally it is important to consider that utilities just like costs are a time-dependent variable and decline over time, not only because patients prefer achieving a specific state of health rather now than in one year, but also because nonrelated health events might interfere with or even stop this process. The worst-case scenario would be a patient dying from another disease or in an accident before eventually receiving a highly effective treatment. Hence a discount of utilities has to be included in all estimates.

Once the parameters' cost, effectiveness, and utility have been established, analyses are done by calculating the amount of money spent per unit of effectiveness or utility, respectively. Two important parameters can be deducted from these analyses. One is the incremental cost-effectiveness ratio (ICER) defined as the ratio of change in costs of a treatment to the change in effect of the treatment, or in simpler terms the additional effect of a treatment bought for an extra Euro/Dollar/Pound. Previous studies have estimated the upper border of ICER a society is willing to pay at €45,000 ($60,000 or £30,000) [8]. The other parameter is headroom, or the amount of extra cost for which a treatment is still cost-effective given its effectiveness. Considering these analyses it is important to note that these processes are not only dependent on the market. Regulatory issues differ significantly on both international and national levels and can affect costs heavily.

2.5
Decision Analysis Models

Information acquired in these studies can and should be used together with decision analysis. The classic model of decision analysis is the decision tree, representing a unidirectional decision support structure leading from a question at the left hand side to branches of possible outcomes on the right hand side. On the very end, the utilities of the individual outcomes are stated, so that the utility of every event and decision on the tree can be calculated by going backwards and multiplying utilities with probabilities. To account for the variability of the included parameters sensitivity analyses must be done. During such an analysis, the values of one or more variables are altered, and change in the final outcome is monitored. The major limitation of the decision tree, however, is its unidirectional nature. In various healthcare situations patients will rather move back and forth between states than travel down a line of distinct events. Such behavior is modeled in Markov Models representing cycles of events during which a patient may move to another health state, regress into an earlier situation, or remain in his current position. Sensitivity analyses for Markov models are done using Monte Carlo simulation, a method that repeatedly changes all variables at a time using random values form a predefined range to construct 95% confidence intervals for the outcome.

2.6
Cartilage Tissue Engineering

Osteoarthritis is a major burden in the population of developed countries, causing pain and suffering on the individual level and loss of productivity and extensive healthcare costs on the societal level. Total joint replacement is an outstandingly successful treatment for osteoarthritis of most joints, but is associated with complications such as component wear and loosening, especially in long use. Hence its use is considerably limited in younger and active patients. One of the earliest tissue engineering applications in clinical use was autologous chondrocyte implantation (ACI) in the management of cartilage defects, and a considerable amount of data is available to evaluate this method [18]. Brittberg published his groundbreaking paper in the New England Journal of Medicine in 1994 [5]. Originally, the method is described as harvesting chondrocytes from non-weight bearing areas of the cartilage, culturing these cells in vitro, and implanting them into a bioactive chamber created by covering the debrided cartilage defect with a periosteum flap. This original method has been further developed and currently biomaterials are in broad use in Europe, and awaiting FDA approval for use in the USA. Tissue engineering in cartilage repair is thus a good example of evaluating a clinically employed treatment.

Evaluating ACI as a healthcare resource, need assessment is fairly straightforward. Considering the high incidences of cartilage damage in our highly active present-day population, the need for health is obvious. Considering the painful and chronic nature of osteoarthritis and the resulting impairment and disability, the need for health care on the individual as well as societal level is imminent. Appropriateness of ACI should be determined in the face of its alternatives. Briefly, ACI should have a convincing immediate effect on pain and joint function, but this effect should also be long lasting.

A number of trials and reviews studied the effect of ACI compared to alternative treatment options. Data on efficacy and effectiveness are available from seven randomized controlled studies with a maximum follow-up of 5 years presenting clinical as well as histological findings [3, 4, 7, 12–14, 20, 21]. Long-term results are available from cohort studies for 9–11 years [18, 19]. Other studies have described costs, quality of life, and cost-utility of ACI. Lindahl, Wildner, and Minas provide costs for ACI in SEK, DM, and USD, respectively [15, 17, 24]. Cost-effectiveness and cost-utility analyses were presented by Clar, Minas, and Vavken [6, 17, 23]. The health technology assessment by Clar, et al. is the most extensive study and used short and long-term models comparing ACI with microfracture and mosaicplasty using QALY and time to total knee replacement as outcome [6]. For their short-term model they found that ACI would have to be 70–100% more effective than microfracture to be cost-effective. Over 10 years, however, only 10–20% would be required. Clar performed extensive sensitivity analyses to test the robustness of their result and found them to be dependent on quality of life assessments only.

From a decision analysis perspective these data are challenging. The considerable heterogeneity between individual studies and the lack of long-term data make predictive modeling difficult. A representative decision model was generated from the existing data. Briefly, data produced by Clar, Knutsen, Wood, and Saris were used to estimate outcome probabilities [6, 14, 20, 25]. Utilities were estimated by surgeons focusing on cartilage repair and based on feed-back from patients. Surgeons were primarily targeted in this reference lottery, because from our experience, there is insufficient knowledge on ACI among patients leading to an inability to grasp the meaning of the potential complications, and utility estimates are deviated by prejudice and spurious assumptions. The decision tree for this model is presented in Fig. 2.2. To test the robustness of these estimates, one-way sensitivity analyses of the effect of complication rates, the severity of complications, and the loss of utility due to complications can be done (Fig. 2.3a–c).

This model shows a higher utility for ACI compared to microfracture. Sensitivity analyses support this finding over a wide area of variance for included parameters. It should be noted that this model is an example of decision making in tissue engineering, and while it uses real evidence it is still incomplete and estimates might be subject to considerable change after addition of further parameters. For example, the model could be expanded taking revisions and second-line operations due to development of OA into account.

2.7
Genitourinary Tissue Engineering

Tissue engineering of urogenital tissue has been proposed by Atala as early as 1993, but has not yet drawn much attention [1, 2]. A number of conditions requiring urethroplasty or cystoplasty may profit from tissue engineering approaches. The need for cystoplasty as determined by a literature review is based prevailingly on cases of bladder carcinoma with approximately 10,000 to 50,000 case per year in the USA and UK [9, 10]. The current gold standard is radical cystectomy, followed by ileocystoplasty. Other indications of cystoplasty such as detrusor instability or postinflammatory scarring are unlikely to be promising niches for tissue engineering applications. Urethroplasty currently is performed in cases of strictures using buccal mucosa. In summary, urogenital tissue engineering is a good example for evaluating a tissue engineering application on the edge of clinical implementation.

The recent study by McAteer reviewed costs and utility of urogenital tissue engineering using headroom analyses [16]. The main problem for this study was to define cost and utilities since there is almost no pre-existing literature. McAteer showed a minimal financial headroom for engineering urethral tissue of £186 (€275, $375) based on a difference in QALY of 0.0062. For an engineered bladder, in turn, this paper found a headroom of £15,000 (€22,000, $30,000), mostly because of a rather high increase of quality of life with a difference in QALY of 0.5.

Decision tree for autologous chondrocyte implantation (ACI) verus microfracture (MFX) in cartilage repair

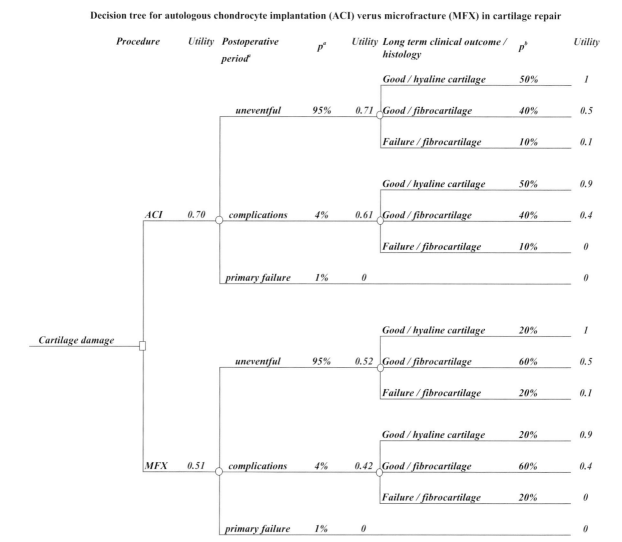

a Rates of complications from Wood et al. JBJS 2006; 88:503-507
b Probabiliites obtained from Clar et al. Health Technol Assess 2005;9(47)

Fig. 2.2 Possible decision tree for using autologous chondrocyte implantation (ACI) instead of microfracture (MFX). At the beginning there is the problem of "cartilage damage" leading to a decision node (represented as a *square*). This decision leads to chance nodes (*circles*) of possible outcomes, given with the probabilities p. The *blue figures* show the utilities of an outcome. The initial utilities on the *right side* were obtained from a reference lottery among orthopedic surgeons specializing in cartilage repair; the other utilities were calculated by multiplication with the corresponding probabilities

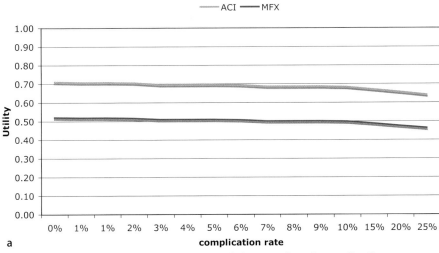

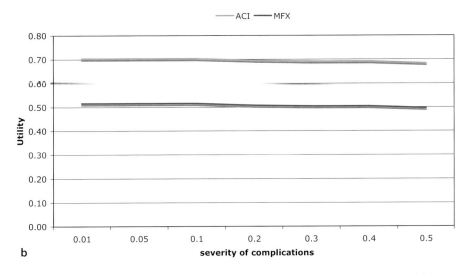

Fig. 2.3a–c The results from one-way sensitivity analyses testing the robustness of the findings from the decision tree. **a** and **b** show that autologous chondrocyte implantation (ACI) retains higher utilities regardless of complication rate and severity. A two-way sensitivity analysis including rates and severity into one model supports this finding (not shown). **c** *see next page*

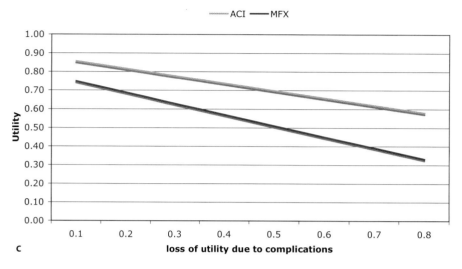

Fig. 2.3a–c *(continued)* **c** tests the assumption of lost utility due to fibrocartilage production instead of hyaline cartilage. This is probably the most important point in the sensitivity analyses since fibrocartilage formation has been associated with subsequent OA development. Given the lesser rates of such scarring in ACI, there is an obvious benefit for ACI that is multiplied by the loss of utility due to fibrocartilage, acting as a surrogate for OA in this case

This high difference derives from a rather long-lasting utility of 10 years and the fairly high number of comorbidities associated with ileocystoplasty.

In this case, decision analysis is based on financial headroom. Tissue engineering in urethroplasty seems to be a poor choice, but one must consider that the estimates for headroom are based on a number of assumptions, one among them is a maximum amount of money that will be made available for an additional effect. From this point of view, a benefit-cost analysis based on a person's willingness-to-pay (WTP) could show quite different results [11].

2.8
Conclusion

Economic evaluation has become an essential part of modern medicine. Given the limited resources and increasing demand of treatments at increasing costs, it is justified to strive for efficiency and balanced resource allocation. Public health uses models of economic and decision analysis to do so. However, it has to be remembered that these mechanisms are based on populations and the good of the many, whereas the interest of the physician during an actual decision focuses on the good of his or her patient at hand. This gradient of interests has to be compensated. Fortunately, health care is an economy of large averages and wide variance, and in considerably wide context high costs in one individual may be compensated by affordable management in others.

We wanted to stress the fact that economic modeling, given its necessity, should not be more than one part of a decision. Microeconomic approaches focusing on revenue should not outweigh equity and need, and quality, unlike costs and utility, must not be discounted. Such tendency might not develop from harmful intent, but health care is not an industry plant, and methods developed on well-understood and tightly controlled processes are bound to fail. Just as tissue engineering is an outstanding example of a multidisciplinary approach to a common problem, economic modeling in health care is not. This chapter thus aimed at succinctly presenting important principles of common models and processes to researchers and clinicians to stimulate an environment of equally informed discussion on a mutual level.

All models are simplified representations of reality and their predictions and estimates are bound on approximations. The quality of these approximations is crucial, thus extensive considerations of the effect of errors in these assumptions, i.e., sensitivity analyses, are pivotal. On the other hand, assumptions with proven sufficient robustness are readily available for model building, whereas the accumulation of real data might be prohibitively expensive financially or logistically.

Using autologous chondrocyte implantation in cartilage repair as an example, we could successfully establish a model that shows a high efficiency of this procedure compared to alternative options. Including rates and severity of complication, a decision analysis showed that this procedure is superior to microfracture based on a beneficial long-term result because of a higher production of hyaline cartilage, and its use in a clinical setting is recommend by these findings. This model, however, is based on a number of assumptions, such as proportion of patients with hyaline repair tissue, and its external validity might very well be affected. However, this is a general problem in all model-based analyses.

Tissue engineering in the genitourinary system is a beautiful example for an application on the verge of in vivo testing. Headroom analyses suggest a strong potential for a tissue-engineered bladder, especially because of ample complications in current techniques of ileocystoplasty. Urethroplasty, in turn, failed to achieve such a result. Notably, these results are based on predicted estimates of effectiveness and utility that might change due to future developments.

Tissue engineering holds a position at the intersection of high costs for R&D and clinical application on the one hand and highly effective treatments that might satisfy urgent health needs on the other. What stands out from the past is that negligence of sound economic models has led to the bankruptcy of two leading tissue engineering companies, Organogenesis and Advanced Tissue Sciences, both focusing on skin grafts, in 2002. During both the development and clinical use of tissue engineering, meticulous economic modeling considering the specifics of the tissue engineering market—slow product and reimbursement approval, high costs in research and development, and a surprisingly slow market—will be a key issue to successful implementation of a treatment, parallel to its proven effectiveness.

References

1. Atala A, Freeman MR, Vacanti JP, et al. (1993) Implantation in vivo and retrieval of artificial structures consisting of rabbit and human urothelium and human bladder muscle. J Urol 150:608–612
2. Atala A, Bauer SB, Soker S, et al. (2006) Tissue-engineered autologous bladders for patients needing cystoplasty. Lancet 367:1241–1246
3. Basad E, Stürz H, Steinmeyer J (2004) Die Behandlung chondraler Defekte mit MACI oder microfracture—erste Ergebnisse einer vergleichenden klinischen Studie. Orthop Prax 40:6–10
4. Bentley G, Biant LC, Carrington RWJ, et al. (2003) A prospective, randomized comparison of autologous chondrocyte implantation versus mosaicplasty for osteochondral defects in the knee. J Bone Joint Surg Br 85-B:223–230
5. Brittberg M (1994) Treatment of deep cartilage defects in the knee with autologous chondrocyte transplantation. N Eng J Med 331:889–895
6. Clar C, Cummins E, McIntyre L, et al. (2005) Clinical and cost-effectiveness of autologous chondrocyte implantation for cartilage defects in knee joints: a systematic review and economic evaluation. Health Technol Assess 9:1–82
7. Dozin B, Malpeli M, Cancedda R, et al. (2005) Comparative evaluation of autologous chondrocyte implantation and mosaicplasty: a multicentered randomized clinical trial. Clin J Sport Med 15:220–226
8. Edejer Tt, Baltussen R, Adam T, et al. (2003) WHO guide to cost-effectiveness analysis. WHO, Geneva
9. Frimberger D, Lin HK, Kropp BP (2006) The use of tissue engineering and stem cells in bladder regeneration. Regen Med 1:425–435
10. Guan Y, Ou L, Hu G, et al. (2008) Tissue engineering of urethra using human vascular endothelial growth factor gene-modified bladder urothelial cells. Artif Organs 32:91–99
11. Hammit JK (2002) QALYs versus WTP. Risk Anal 22:985–1001
12. Horas U, Pelinkovic D, Herr G, et al. (2003) Autologous chondrocyte implantation and osteochondral cylinder transplantation in cartilage repair of the knee joint: a prospective, comparative trial. J Bone Joint Surg Am 85:185–192
13. Knutsen G, Engebretsen L, Ludvigsen TC, et al. (2004) Autologous chondrocyte implantation compared with microfracture in the knee. A randomized trial. J Bone Joint Surg Am 86:455–464

14. Knutsen G, Drogset JO, Engebretsen L, et al. (2007) A randomized trial comparing autologous chondrocyte implantation with microfracture. Findings at five years. J Bone Joint Surg Am 89:2105–2112
15. Lindahl A, Brittberg M, Peterson L (2001) Health economics benefits following autologous chondrocyte transplantation for patients with focal chondral lesions of the knee. Knee Surg Sports Traumatol Arthrosc 9:358–363
16. McAteer H, Cosh E, Freeman G, et al. (2007) Cost-effectiveness analysis at the development phase of a potential health technology: examples based on tissue engineering of bladder and urethra. J Tissue Eng Regen Med 1:343–349
17. Minas T (1999) Chondral lesions of the knee: comparisons of treatments and treatment costs. Am J Orthop 28:374
18. Peterson L, Minas T, Brittberg M, et al. (2000) Two- to 9-year outcome after autologous chondrocyte transplantation of the knee. Clin Orthop Relat Res 212–234
19. Peterson L, Minas T, Brittberg M, et al. (2003) Treatment of osteochondritis dissecans of the knee with autologous chondrocyte transplantation: results at two to ten years. J Bone Joint Surg Am 85:17–24
20. Saris D (2007) Prospective multi-center randomized controlled trial of ChondroCelect (in an autologous chondrocyte transplantation procedure) versus microfracture (as procedure) in the repair of symptomatic cartilaginous defects of the femoral condyles of the knee. J Am Acad Orthop Surg, February 14–18, San Diego
21. Saris DB, Vanlauwe J, Victor J, et al. (2008) Characterized chondrocyte implantation results in better structural repair when treating symptomatic cartilage defects of the knee in a randomized controlled trial versus microfracture. Am J Sports Med 36:235–246
22. Thornton JG, Lilford RJ, Johnson N (1992) Decision analysis in medicine. BMJ 304:1099–1103
23. Vavken P, Gruber M, Dorotka R (2008) Tissue Engineering in orthopedic surgery—clinical effectiveness and cost effectiveness of autologous chondrocyte transplantation. Z Orthop Unfall 146:26–30
24. Wildner M, Shangha O, Behrend C (2000) Wirtschaftlichkeitsuntersuchung zur autologen Chondrozytentransplantation. Arthroskopie 13:123–131
25. Wood JJ, Malek MA, Frassica FJ, et al. (2006) Autologous cultured chondrocytes: adverse events reported to the United States Food and Drug Administration. J Bone Joint Surg Am 88:503–507

Ethical Issues in Tissue Engineering

P. Gelhaus

Contents

3.1 Introduction 23
3.2 Risk Assessment 24
3.2.1 Invasiveness 24
3.2.2 Importance of the Substituted Function 25
3.2.3 Methodological Risks 25
3.2.4 Alternatives 25
3.2.5 Unexpected Risks 26
3.3 Informed Consent 26
3.4 Status of the Embryo 27
3.5 Gene Therapy 28
3.6 Animal Research 29
3.7 Tissue and Organ Transplantation 30
3.7.1 Organ Supply 30
3.7.2 Further Social Implications 31
3.7.3 Histocompatibility 31
3.7.4 Remaining Problems 32
3.8 Enhancement 32
3.9 Conclusion 32
References 34

3.1 Introduction

Tissue Engineering (or regenerative medicine) is a rapidly developing field of research pursued by very different branches of medicine. The most results are published by orthopedists [28] and cardiologists, but meanwhile, there is nearly no special subject that does not investigate in this area. Furthermore, there is a multitude of methods and techniques that are involved in tissue engineering. Therefore, it is difficult to define or to describe the scope of "tissue engineering" in one sentence.

According to Langer and Vacanti, tissue engineering is "an interdisciplinary field that applies the principles of engineering and life sciences toward the development of biological substitutes that restore, maintain or improve tissue function or a whole organ [27]." It includes up to three elements: (1) isolated cells or cell substitutes, (2) tissue-inducing substances and (3) cells placed on or within matrices (scaffolds). This includes a multitude of scientific and therapeutic purposes in regenerative medicine, most spectacular perhaps the construction of whole autologous organs that would not only be a solution for the lack of donor organs, but would also answer the problem of adverse host response and minimize the risk of infection. In summary, there is nearly no method or technique that could not be part of tissue engineering.

Analogically, the ethical assessment of tissue engineering is complex, depending on the applied techniques, the kind (or in the case of embryos also the origin) of subjects and the intended goals as well as the social context. Tissue engineering involves a couple of highly controversially discussed issues like embryonic or stem cell research, gene therapy and enhancement, but it also includes unsuspicious, low-risk research like basic cell culture experiments. This article presents the most important topics to be considered in a careful overall evaluation of tissue engineering. Nevertheless, it is also important to remember that ethics is not just a corset to restrict and retard the scientific progress, but a method to identify the good and the worthwhile. The function of ethics is to canalize the progress in directions that are accepted by the society and to emphasize the most acceptable means on this way. One of the essential tasks of an ethicist apart from the evaluation of the chances and promises as well as the risks of a new technique is to check the alternatives. If ethics is merely understood as a killjoy and hindrance for real research, this is an indicator for a very one-sided perspective (either of the ethicist in question or of the investigator).

3.2
Risk Assessment

Any kind of clinical research has to consider the relation of risk and chance for the subjects. In the older versions of the Declaration of Helsinki, there was a strict segregation between research on patients and research on healthy subjects [52]. For patients, the unavoidable risk of the new treatment had to be weighed out by an acceptable chance of individual benefit. This made the invention of placebo-controlled branches of clinical trials difficult; only if no effective therapy was available, a placebo control was judged as morally acceptable. Healthy subjects, on the other hand, nearly never profit (concerning their health) from clinical trials. Though any reasonable precaution is also necessary in this case, including laboratory, toxicological and animal research as well as careful evaluation of the related literature, healthy subjects are considered as autonomous persons who can decide for themselves. If they are willing to take the risk of this clinical trial, the only essential premise is their informed consent. The older Declaration has protected patients more carefully, considering that they are more vulnerable and prone to persuasion because of their dependency of their physician. The new version of the Helsinki Declaration, on the contrary, has skipped the differentiation between therapeutic and nontherapeutic research and explicitly allows placebo-controlled research, though an adequate relation between risk and benefit still is required [52]. According to the new regulation, methodological requirements are better considered; the value of a higher validity of scientific evidence is appreciated. The special protection and regard of the patient, on the other hand, has accordingly to be diminished [19]. Actually, the equipoise of risk-benefit-relation in the different branches of the trial is no longer restricted to the individual, but a large scientific or therapeutic benefit may equate an advanced individual risk—informed consent presupposed. Ethically, these changes are controversial: the concessions to medical progress (and therefore to public welfare) are purchased by a minor level of care for the individual patient. On the other hand, the new charter on clinical ethics emphasizes more than previous regulations the responsibility of the physician to every individual patient. So the traditional conflict between duties of the physician and the investigator—usually in one person—are actualized and pointed [1].

On behalf of tissue engineering, a simple, general risk assessment is not possible. The purposes, techniques, and applications are too variable. Several aspects influence the evaluation.

3.2.1
Invasiveness

The risk of tissue engineering depends on the degree of invasiveness of the applied method. A mere in vitro experiment in order to gain knowledge of, e.g., signal transduction of a cell cluster is nearly riskless for the patient—general laboratory caution assumed. The ex vivo cultivation of skin cells is also safe, but the reimplantation implies risks of contamination, errors, incompatibilities, etc. Roughly speaking, the more steps of modification of the autologous cells are involved, the higher is the risk of adverse events. If more surgery is needed for the substitution, the risk of invasive surgery adds on to the experimental risk. The interactivity of the applied substances also influences the degree of invasiveness. Very bioactive ingredients also mean a deeper risk of invasiveness. Very small particles may also imply a special risk as in the case of nanotechnology, which is a promising branch especially in the construction of scaffolds [3].

In addition, the purposes of the new technique are important. Often, it is useful if a material is inert and minimizes interaction between the prosthesis and the surrounding body. This also minimizes the risk of adverse events [14, 37]. The rationale of tissue engineering however, is to get nearer to physiological processes and thus to optimize the interactivity. However, the more interactivity between materials and the body, the higher the risk of side effects. Then again, the nearer the interaction meets physiological mechanisms, the better are the chances of successful replacement of the defective function. An example for the search of an optimized risk-benefit relation is the development of degradable but otherwise not too bioactive scaffolds that are impregnated with growth factors in order to induce autologous cell growth [46]. Though very interactive at first—being prone to degradation and activating regeneration—the scaffolds reach the ideal of inactivity in the long run, leaving only physiological structures. But on the way to deliberately using physiological mechanisms, the risks are the higher the deeper the methods address the body's own physiological processes.

3.2.2
Importance of the Substituted Function

Another factor for the assessment of a new technique is the intended function of the engineered tissue. If the intended function is essential for the body, the supposed benefit of the therapy will be very impressive. On the other hand, a failure could be disastrous. The autologous replacement of defective heart valves, for example, might just as easily risk as save the life of a patient [45]. The replacement of a lacking tooth contrariwise, can be an aesthetic and functional advantage, but never justifies a similar amount of risk [4, 33]. But a life-threatening disease does not necessarily justify any kind of dangerous therapy. In fact, here a mere lack of success suffices for a very severe adverse event, while in a more modest case, a

failure may only mean that there is no improvement, but—apart from the intervention itself—also no aggravation. In addition, the precariousness of life can diminish the capacity of the patient to give a valid, autonomous informed consent, because he would accept every kind of experiment even if it only offers him a shade of hope. For a careful risk assessment, the impression "anything is better than this state" is not sufficient. Even in a hopeless situation, a reasonable expectation of amelioration is required.

3.2.3
Methodological Risks

Each method is associated with specific risks and dangers. Often it is possible to minimize the risk and use precautionary measures, but it is seldom possible to rule out completely the specific risks of the method. With regard to the different types of tissue engineering, there are some important sources of risk to be considered. One area of concern is the infection risk, either by contamination of the source material or by disease transmission from allogeneic cells, especially if a few, very successful cell lines are broadly distributed [7]. Every step of processing in the production can also be associated with microbial contamination, modification of the cells, mistaken identity of the cells, and general delivery of unwanted cells. With regard to scaffolds, there is the possibility of latent reaction to degradation products, toxicity of process additives or residues and, of course, a product performance that does not work in the expected way [45]. Stem cell therapy as well as gene therapy generally have the risk of inducing cancer because of uncontrolled cell growth due to activating oncogenes or by deactivating protective genes.

3.2.4
Alternatives

An often neglected field of risk-benefit evaluation is the open and careful analysis of possible alternative therapies. If there is a satisfactory "gold standard" for therapy, the situation is not too difficult.

The promises of the new therapy have to compensate for the higher risk of a not-so-well-tested intervention. But if there are many (or no) options of indicated therapies, a more individual evaluation of medical and personal advantages and disadvantages of each option is necessary, and this complex situation has to be measured with the new therapy approach. The ethical importance of this individual assessment contrasts the objective needs of statistical valid research. A gradual course of action from single patient use to elected patients and small controlled trials up to statistically valid trials is not only legally but also ethically claimed.

3.2.5
Unexpected Risks

Unexpected risks do not refer to the naivety of the investigator but to the remaining risks after all careful and reasonable evaluation of current knowledge, e.g., cancer and deadly infections due to viral gene transfer which were not unexpected risks of gene therapy before they occurred, but theoretically well-known dangers of the methodological approach. The life-shortening effect of some antiarrhythmics, on the contrary, was a real conceptual surprise and could count as an unexpected effect. The problem with this kind of risk is that they exist in addition to all calculable dangers and, per definition, are not expected and therefore not assessable. The more important are precautions not to ignore unforeseen effects and correlations, even if theoretical relations to the experiment seem to be improbable [51, 39]. That is the reason why usual protocols of clinical trials require that every adverse event has to be reported, even if common sense indicates that there is no causal connection to the tested intervention. In fact, unexpected risks are the main reason why an experimental therapy is always riskier than an established therapy; and though it is difficult to disclose about unforeseen risks, this part is one of the most important aspects of a valid informed consent. These general considerations concerning risk assessment of new therapy approaches fully apply to nearly all types of tissue engineering, though they are, of course, not specific to it.

3.3
Informed Consent

Every medical intervention requires the informed consent of the concerned patient. Exceptions can only be made in emergency cases when there is no opportunity to get an informed consent. In these cases, the assumed will of the patient is that his life is saved no matter what medical means are needed. Legislation strictly insists on a formal informed consent before elective surgery and more so before experimental interventions on human beings. Not only the cruelties of Nazi experiments but also ongoing unethical experiments, e.g., on prostitutes, prisoners, children, or uninformed patients also before and after World War II, have led to a strong emphasis on informed consent in order to prevent the worst excesses. The underlying moral principle is the respect for the autonomy of the patient. The mere legal, formal consent refers to certain ethical requirements. A valid informed consent usually contains the following elements:
1. Disclosure
2. Understanding
3. Voluntariness
4. Capacity
5. Consent

According to Faden and Beauchamp any kind of manipulation or compulsion by the physician is not acceptable. But a certain kind of pressure in a therapeutic context may be adequate in order to oppose inner compulsions due to the disease [15]. This does not apply in a research context. The voluntariness rests on the independence from the wishes, values, and interests of the investigator [30]. The problem is that an objective, uninterested informing of a subject by the investigator is nearly impossible [50]. If the investigator is not convinced by the promises of the new approach, it would be unethical to start the trial at all. But how can an investigator give unbiased information and a selection of all relevant facts if he has to be necessarily partial? This contradiction underlines the importance of an ethical awareness of the investigator. If the informed consent shall mirror the respect for the autonomy of the subject, the disclosing investigator has to be highly aware of the ideal of objectivity on the one hand and his own personal tendency to support the trial on the other hand.

The respect for autonomy alone cannot prevent all unethical research; irresponsible experiments may use extraordinarily courageous or careless subjects. By that, the investigators do not violate the autonomy of the subjects, but nevertheless other valid moral principles like beneficence, nonmaleficence, and justice are neglected [5]. Apart from that, the adequate regard for an informed consent is not always the only or a sufficient manner to respect the autonomy of a patient. It is, in any case, the condition sine qua non an ethical research on human subjects usually is impossible. I don't want to discuss the difficulties of assumed consent or consent by proxy in this chapter, but if the patient himself is capable of consent, this condition is not replaceable by any other means [8].

3.4
Status of the Embryo

Some of the most promising tissue engineering projects require embryonic stem cells. Embryonic stem cell research is ethically controversial because it presupposes the use (and destruction) of human embryos for basic research or for the benefit of other persons [40]. The ethical evaluation fundamentally depends on the status of the embryo and the concept of the beginning of human life. Genetically speaking, it is uncontestable that human life usually begins with the fusion of an oocyte and a spermatozoon. The moral status of the fertilized egg cell, on the other hand, is highly controversial. The extremes go from a moral status comparable to the status of adult, autonomous persons (or even better because of the innocence and vulnerability of the embryo) to a mere cell cluster without claim on any kind of moral care [44]. No scientific evidence can solve this problem, because it is not a matter of evidence but of normative attribution. The known facts of embryology are not the source of dissent. Neither side holds that human life does not start with the fertilization of an egg cell, but on the other hand, an egg cell is not equivalent to a full-grown adult and needs a lot of luck and development to become that. The fundamental question in this context is a moral one: what amount of consideration must be attributed to which kind of development of a possible future person? This formulation seems to imply a gradual answer, but that is not necessarily true. It is not inconsistent to attribute the full moral concern to the fertilized egg cell. Four types of arguments are often used to support this comprehensive position: the species membership, the continuity of the developing person, the commencement of individuality, and the potentiality of the zygote (the so-called SCIP-arguments). For each of these arguments there are, of course, counter-arguments [38]. The species argument, for example, is prone to the naturalistic fallacy by directly deriving normative judgments from biological facts. The continuity can only be seen a posteriori, because the majority of fertilized egg cells never succeed in becoming a born human, let alone an adult person. The definition of individuality is discussable, e.g., because of the possibility of twinning or fusion in the earliest stages of the embryo, but also because of the exclusive focus on the genetic aspect of individuality. The potentiality-argument tends to be vague, because it is difficult to specify the kind and statistical probability to take into account, because, e.g., "in an era of cloning, some degree of potentiality attaches to all bodily cells" [22]. If already the early embryo has got claims on full moral acceptance then it has not only a strong right to live (not merely in a research setting, but also in therapeutic reproductive circumstances and concerning abortion), but it also must not be only a means for the welfare of others without even consenting to the sacrifice. The prognosis for a societal agreement on these matters in a secular, pluralistic society is therefore poor [53].

Of course, there are other important arguments against embryonic stem cell research, e.g., the slippery slope from therapeutic cloning to reproductive cloning and designer babies according to the taste of the parents. Also the destabilization of family structures as a basis for our society is a concern. Further arguments are doubts about justice in gaining egg cells from poor women or dependant female researchers and in disregarding women in general (and therefore persons with full moral claims). Finally, social side effects in many colors and shades are a widespread source of concern. However, all these arguments are relative to certain circumstances and can, in principle, be ruled out by safety standards, laws, and supervision or be outweighed by higher values. This does not apply to the categorical moral status of the embryo. If embryos are fully moral persons, any kind of exploitative use is undoubtedly unethical. A compromise is hardly imaginable. On the other hand, there does not seem to be a stable majority in western society that is convinced of this high moral status of the early embryo regarding the blastocyst in question. Therefore, it may be useful to consider some differentiations and alternatives.

Incidentally, the term embryo does not fit very well, because no fertilized egg cell in stem cell research grows inside a maternal womb (ancient Greek "embryo": to grow inside). This terminological remark does not imply any certain moral judgment [12, 21]. In the worst case scenario, they would be aborted only for reasons of research (a rather theoretical setting). In the second worst case scenario, they are produced only for reasons of research. The ranking might not be undisputed: possibly the creation of a human being only for exploitation and murder is worse than the simple murder of it. On the other hand, the intentional creation may diminish the moral status of the cell; there has never been a realistic potential to grow to a full person. In the third worst case scenario, the embryo would be destroyed anyway, because it is an excessive embryo from reproductive therapy. The use for stem cell research would only add some possible good to the general situation. On the other hand, it could encourage immoral methods like reproductive medicine, because supernumerary embryos would not any longer only count as waste. A minor case of wrong-doing would be research on already existing cell-lines. True, it still presupposes the killing of human embryos, but it does not commit new crimes and, though it may be generally false to profit from past wrong [22], it offers the possibility to compromise with other positions without neglecting the fundamental status of the embryo. If even a deadline in the past is included, there should be no encouragement of further wrongdoing (such is the current but not undisputed legal situation in Germany) [31].

Still, the alternatives should not be neglected. If the hoped-for progress could be achieved by adult stem cell research, it would be the much better way, because no human would have to be killed for the results and informed consent could be gotten in advance [25]. But if there still is a need of embryonic stem cells, what about cloning using an unfertilized egg cell? If it would be possible, on the first view no embryo had to die. Looking closer, still a potential human being would be created, this time not by the fusion of egg cell and sperm, but by the hybridization of an egg and a body cell. The moral evaluation even from a conservative point of view is difficult. Another step in the direction of a difficult moral status is the current attempt to fuse animal egg cells with human DNA. The resulting chimeric stem cells take the categorical concerns seriously but nevertheless open doors that are possibly even more forbidden from a strictly conservative perspective.

On the other hand, a delay of scientific progress by alternative and less controversial methods is not available without a price. Many lives might be saved during the time between the possible success of embryonic and adult stem cell therapy. The problem is that all these promises necessarily lie in the future, so a reliable evaluation is not really available. Nevertheless, it is an open question how to weigh categorical concerns of only a part of the society against relative but also consistent and important opposite moral concerns.

3.5
Gene Therapy

Human gene therapy is still in the experimental phase nearly twenty years after the first officially accepted trials in 1989 and 1990. There have been some minor successes, e.g., in treating Severe Combined Immune Deficiency (SCID) and in thousands of treated patients (mainly cancer patients), but despite the high expectations of a causal and therefore low-risk and finally healing approach to eliminate genetic diseases, results have been disappointing. Currently, gene therapy usually is rather used pharmacologically and takes into account that the corrected genes seldom persist for a long time in the organism. Unfortunately, the theoretical risks of gene transfer such as implementation of cancer or serious adverse effects of viral vectors really have occurred. Until now, only somatic gene therapy ex vivo or in vivo has been tried in humans. One problem in addition to the deactivation of therapeutic genes is that for a therapeutic effect all or at least very many of the target cells have to be manipulated. Very effective vector systems are needed that therefore also carry a relatively high risk of adverse events. A germ-line therapy, on the other hand, could possibly be more effective as the inserted gene would be transferred to all developing somatic cells. According to the current state of the art, this approach as a therapeutic intention would be obsolete, because the amount of needed germ cells would usually make a selection of in vitro fertilized egg cells more effective. In addition, the risks of creating a genetically modified human being are at the moment unacceptably high considering that these experiments would produce humans on probation.

An interesting step in between is the gene transfer in cell cultures, in progenitor cells or in adult stem cells. Of course, also gene transfer in embryonic stem cells is imaginable, but one of the advantages of embryonic stem cells is their potential to be easily grown in culture and differentiated to a variety of cells, so the usefulness of genetic manipulation without endangering the specific advantages of embryonic stem cells must be judged critically. Even therapeutic cloning could count as a comprehensive form of gene therapy. All these experiments would also meet the definition of tissue engineering. With regard to risk assessment, the mentioned techniques tend to involve a lower risk than in vivo gene therapy, though of course the usual end of an ex vivo trial is the retransfer of the manipulated cell culture to the patient and thus, though some risks can be eliminated by purification and selection methods, the intrinsic gene transfer risk remains.

However, risk assessment alone is not a sufficient ethical analysis. As we have seen before, the origin of the manipulated cells also is important, and in this regard somatic cells like adult stem cells are much less problematic than embryonic stem cells. Also the amount of manipulation can be a factor; a neat addition or edition of one single disease-causing gene may be easier to tolerate than a large-scale transfer of DNA. On the other side, the somatic cell nuclear transfer (SCNT) or therapeutic cloning is hard to classify in this respect. On the one hand, the whole nuclear genome is transferred, on the other hand, just cellular parts of the initial cell remain; so is the transfer nearly complete or only rudimental from a genetic point of view? [11] Apart from the moral status of potentially consumed embryos, namely the status of the human genome itself is in question with regard to gene therapy. Again, the question of whether the human DNA is just one biochemical macromolecule among others or the deepest essence of human identity is highly controversial. In this respect, the scientific campaign to geneticism has been more successful, than current investigators can enjoy. The beauty and simplicity of the deciphered genetic code even temporarily has tempted researchers to (very unscientifically) claiming a "biogenetic dogma." Human omnipotence seemed to be near. Unfortunately, this exaggerated outlook has at the same time awoken many fears about using forbidden knowledge and playing god without the necessary carefulness and foresight [16]. Still, the public and even biology students learn genetic principles in a very simplified (and in tendency, geneticistic) way. No wonder, that a nearly religious or mythic awe results, with all consequences including moral concerns of an overly pragmatic use.

Nevertheless, even in a pluralistic society like ours, a general agreement with regard to the high moral value of helping the ill still holds. The arguments for limiting hopeful therapeutic research must be very convincing in order to overrule the prima facie ethical claim on medical improvement [16]. It would be helpful if the enthusiasm of the investigators for the potential of their research would not lead them to exaggerated prognoses that might not only gain more public attention and funding, but also backfire in waking justified fears in those who do not share the same utopias (e.g., intentional replacement of the evolution, or the total abolition of mortality and disease).

3.6
Animal Research

Animal research is another field of bioethical debate that is touched by tissue engineering. In this respect (like in certain aspects of gene therapy), tissue engineering tends to rather diminish moral concerns regarding existing methods and conventions. Animal research is a necessity in clinical research to lessen the risks of new therapies for human beings. On the other hand, animals are sentient living creatures and their use in painful and deadly experiments is ethically disputed. After all, moral status does not fall from the sky but is subject to reasonable grounds. According to some ethical accounts, ethical regard depends on reciprocity and well-understood self interest. In this case, animals have no moral rights and animal research is not problematic, though still there is no need to be crueler than necessary. But in interest- or need-oriented approaches the moral claim is equivalent to the possible good or evil an organism can experience and accordingly, animal research especially in higher developed animals is not tolerable [41, 42].

Tissue engineering offers the opportunity of designing specific human cell-lines for testing the effects of new pharmaceuticals and therefore could replace (at least in parts) the urge of animal research with their sometimes questionable evidence concerning human beings. In fact, animal research finds itself in a conflict between epistemology and ethics. On the one hand, the guinea-pigs should be as close to humans as possible in order to get valid findings, on the other hand, their resemblance to humans makes their moral negligence objectionable. Tissues in culture could be an answer to this problem, at least in some cases, as they are not sentient and only can have interests and needs in a very metaphorical way, but can produce findings that are often even more transferable to whole human beings than animal experiment results. Therefore, firms that offer designed cell cultures advertise with the moral advantage [10] and animal protectors appeal to use these alternatives whenever possible [32]. On the other hand, as for every experimental therapy approach, new tissue engineering projects require accessory animal experiments at every step of development from the testing of used materials, matrices, cells, and signaling substances over the proof of the therapy principle itself up to clinically relevant animal models [26]. Also the ethical considerations of the European Group on Ethics (EGE) recommend the legal ensuring of sufficient animal experiments [13]. So all things considered, the saving of animal lives by tissue engineering methods might be a marginal effect.

3.7
Tissue and Organ Transplantation

One of the greatest promises of tissue engineering is the replacement of the function of whole defective organs [23]. Up until now we have usually been dependent on donor organs of living or dead donors [2]. Although organ transplantation often is a life-saving, effective, and inevitable form of therapy, it still implies many practical and ethical problems such as lack of organs and therefore questions of justice and allocation, of pressure on possible donors, of altruism and duty, organ sale, xenotransplantation, and so on. Furthermore, there are problems of compatibility, graft failure, graft rejection, side effects of the immunosuppressive therapy, the mental and economic costs, and related difficulties. Also the near connection to the controversial brain death criterion, the implied mechanical understanding of the human body, the possible mixture of identities, and social implications often provoke uneasiness.

Different types of tissue engineering could give answers to or evade many of these difficulties. Therefore, the mere availability of alternatives to organs from cadaver donors could rule out many important problems.

3.7.1
Organ Supply

The gap between organ supply and organ demand and the search for alternatives is one of the engines of tissue engineering research. If enough organs could be bred without the need of donor organs, the considerations about ethically acceptable ways of enhancing the willingness to donation would be superfluous. The conflict about consent or dissent regulations, club models, living donors, and organ sale could completely be skipped. Currently, the discussion continues as heated as ever. Accepting the principle of organ transplantation as a good in general, the many possible methods to increase their number are still controversial. The advance consent of a dying person makes the transplantation relatively unproblematic, but what about the many cases where the wish of the potential donor is not known? Can relatives or friends make a valid surrogate decision? Is it morally acceptable, if a state presupposes the consent and requires a formal denial? Is the risk for living donors acceptable? Isn't it irresponsible for a physician if he mutilates and risks the life of a healthy person without possible benefit for the donor himself, only for the welfare of another patient? Even if the donor consents, what about his motives? Is the consent really voluntary, or has there been any kind of pressure by the family, the patient, the social surrounding? Is there an exaggerated hope for gratefulness? The setting where a woman donates a kidney for her husband who later divorces her for another woman is not illusionary. What is so unethical about organ sale? Why not let the donors profit from their sacrifice? And does not the common use of organ transplantation imply complicity even with criminal organ robbery? Or what about diminishing the demand and only allow transplantations to those who have been willing to donate their own organs in advance? Or prefer those who are useful and of merit for the society? Many questions of justice are posed by the general lack of organs, and the seemingly simple answer to solve the problem in a way that does result in the best possible distribution in the society with a special look on the worst-off (the Rawls' model) does not really offer a practical solution. All these and more sophisticated considerations—so is the hope regarding artificial organs—could perhaps be placed back in the theoretical sphere.

3.7.2
Further Social Implications

The common use of cadaver organs also threatens in a certain sense the dignity of the dead. On the one hand, the organs are no longer useful for the donor, on the other hand, the dying human body could easily be seen as a spare parts store. Piety to the dead could be diminished if from the moment of the brain death diagnosis all attention is no longer on the dying person but on the successful transfer of as many organs as possible. At least, a thorough explantation is not everybody's idea of a peaceful death.

The brain death criterion as such is a very controversial conception with two important distinct functions. It defines the human death as the death of the whole brain, which makes the death of humans fundamentally different to the death of all other living organisms. It specifies some important and typically human traits and locates humanity in the bodily equivalent of these traits. This action would be obscure if not for substantial pragmatic reasons. Both reasons are connected with high tech medicine. Firstly, by the advanced possibilities of technical replacement of many organ functions, death often is a measure of decision. Heart or respiration failure, e.g., are no longer practically equivalent with death, so reasonable grounds have to be found for the decision to stop all further therapy attempts and let the patient die. The irreversible destruction of the whole brain is not the only, but a very plausible reason to stop every form of therapy, so the death of the rest of the body will follow soon. The second reason for the need of the brain death criterion is the organ transplantation. If the death of the human organism and its single organs would not be dissociated, the concept of cadaver organ transplantation could not be verified. Especially the second practical connection of brain death and organ transplantation again and again provokes distrust in both [6]. For those who are not convinced by the brain death criterion (e.g., because of the segregation from human and other animals' death), the explantation of a brain-dead donor actually kills the donor. If tissue engineering would succeed in creating enough artificial organs to render the classical organ transplantation superfluous, the skepticism concerning the brain death criterion also would be strongly diminished.

3.7.3
Histocompatibility

A very promising branch of tissue engineering works with autologous cells of the patient. Whether the initial materials are somatic cells or adult stem cells of the patient, most practical complications of heterologous transplantations, especially the problem of immunosuppressive treatment, are avoided. The main resting risk is a possible contamination of the tissue und a resulting infection or rejection. Therapeutic cloning with DNA of the patient also shares these advantages. If in the farer future whole autologous organs could be bred, the concerns about a possible mixture of identities by implantation of organs of another individual would also be dissipated.

3.7.4
Remaining Problems

All these exciting perspectives of tissue engineering will not occur at one moment. But as long as only parts of these goals are realized, many of the advantages cannot supervene. It would not be sound to reckon all these hopes as if they were already or with absolute certainty in the near future true. Until the realization, the risk assessment should content itself with the directly expected consequences of the trial.

But even if it is possible to replace all classical transplantations by better tissue or organ engineering methods, the concern about a mechanical body model behind the whole regenerative medicine could not be refuted. But I doubt that this disadvantage alone would lead to a refusal of the technique in general. Rather, the complementary focus on the patient as a whole person should be encouraged. If we doubtlessly respect our patients as persons, we evidently are not prone to a mechanical body model although we have technical means to replace certain body functions.

3.8
Enhancement

While the visions of ameliorating transplantation medicine as part of therapeutic progress are nearly everywhere accepted, tissue engineering also can promise developments and progress that are not so uniformly approved. In principle, the improvement of cell or organ function does not necessarily stop with normal or average function [18]. The same techniques could also apply to better or optimal functioning. Sometimes, there is a short step from therapy to cosmetics, life style or doping. If medical knowledge and techniques are no longer used for the fight against diseases but for diverse wishes of the client, we usually speak of enhancement. The difference does not lie in the means but in the ends. This does not imply that enhancement generally is immoral, but it does not any longer possess the moral backing like medicine in the true sense. Also the risk assessment has to be even more careful, because the expected benefit is not as clear as in a medical context. What counts as improvement? Which size for female breasts is optimal? How deep may wrinkles be? Aesthetic and life style questions cannot hope for the same degree of agreement than the goals of medicine. The European group on Ethics (EGE) therefore recommends with regard to tissue engineering to carefully distinguish between medical and cosmetic use [13].

But also regarding anti-aging research, the border between medicine and life-style is not easily drawn. While the single degenerative symptoms of aging, e.g., arthrosis, are usually seen as diseases and therefore as candidates for therapy, the aging as such is part of the normal human lifespan. Prolonging life is a central end in medicine, but abolishing death at all (if that would be possible) is no medical goal. It would change the human condition in a basic way, and that is no part of medicine in a narrower sense.

Also, in sports, the classification of replacement or enhancement of function is ambiguous, e.g., according to news reports, in 2008 a successful Paralympics athlete was denied the approval for the Olympics in China, because his leg prosthesis would be too big an advantage compared to healthy sportsmen. And substitution therapy or doping with many drugs is a matter of dose and patient rather than of application of principle. In sports, doping is not only disapproved because of the side effects for the athletes, but above all because it is unfair to the honest competitors. If a similar conception of fairness also applies to enhancement in other social sectors is controversial. The liberal position holds that there is nothing wrong with improving human functions if the intervention is voluntary, while a conservative perspective is very skeptical about the motives behind and the worthiness of the aim. If the difficult bordering between disease and enhancement [17] leads to an uncritical medicalization of the social environment, it is wise to proceed with caution [20]. Otherwise, the consensus about the primacy of medical goals in many contexts is also endangered.

3.9
Conclusion

As a result from viewing the various different fields that are touched by tissue engineering research, we can give no simple sweeping statement. With regard to the variability of methods, goals, perspectives, and scientific branches, there can be no general yes or no to all types of regenerative medicine. It is even difficult to give a list of points to consider, because the range of involved techniques is so wide. For many basic cell culture experiments, the problems of justice in organ transplantation are totally irrelevant, whereas the construction of artificial organs may still be in such an early phase of animal experiments that a sophisticated analysis of risk and potential benefit in a clinical setting may seem absolutely premature. But every single experiment has also to be seen in a broader scientific and social context [9]. In a trial to give an order of the main lines of ethical analysis, it seems to be useful to distinguish between the origin of the material, the applied methods, the intended aims, and the social implications.

Classical philosophy of science often neglects the subject of research because of its development from mathematical logic and physics. In medical research, the origin of many chemical or biochemical substances is also of minor interest. In experiments on humans however, the origin of the research materials (or even subjects!) is much more important. Even the use of surgical waste requires at least the consent of the patient; the use of research animals presupposes a sound reasoning of its necessity; the use of embryos for research is highly controversial and the creation of human-animal-chimeric cell lines requires a careful awareness of implied ethical concerns.

The applied methods determine the kind and measure of risks and accordingly the carefulness that is necessary. One of the most important points to consider in the assessment is the availability of alternatives. There is a certain danger in the observation that for a hammer everything looks like a nail, i.e., the often applied techniques in a research unit seem to be more accessible than perhaps more appropriate ones that are not yet in use there. For a valid informed consent as well as for a good risk assessment, a thorough overview of relevant research approaches and of possible methods is inevitable.

The goals, on the other hand, determine the possible benefit of the research. It is important to distinguish between direct goals of the experiment at hand and fairer perspectives that envision the research and open the discussion for supporters of this progress and skeptics on its possible or probable implications. These larger goals are not totally irrelevant for the specific risk assessment, but because of their putative status they should be assessed with caution.

The social implications often refer to these larger goals and the evaluation of their worthiness. As long as the larger purpose is a therapeutic one (e.g., gene therapy, organ transplantation, etc.) it is generally nearly uncontested. Social visions (independence of sexual reproduction, e.g.) and the perspective on individual nontherapeutic improvement on the contrary, often don't find general approval [35]. The slippery slope argument is another kind of social reflection that does not refer to the intended aims but to necessary or possible pragmatic implications of the invention of the new technique. A very narrowly connected social implication of tissue engineering is the commercial use of human body parts, for example, whereas questions of ownership and patenting are unclear [7, 13, 29, 48]. These implications are at hand, but we can easily imagine further social developments that may not necessarily follow but still have a certain probability. The fear of a general moral decay and of the impossibility to stop at the further steps of development (e.g., concerning embryo research and reproductive cloning up to a generally manipulated and designed reproduction) can lead to a rejection also of techniques that are not controversial per se (e.g., somatic gene therapy). A third type of social reflection that has to be considered is the categorical form of objections. If these objections are valid, an ethical refusal is irrefutable. The social aspect therein is that usually—as in the case of a full moral status of early human embryos—not the whole society agrees in the categorical refusal. Unfortunately, categorical arguments are not debatable. If the ethical background thereof is important for society, a renunciation of the research in this point can be necessary. In this case, the fantasy of the politicians is challenged in order to find a reasonable compromise between the justified positions. An example of such an attempt is the German legislation on stem cell research. It forbids the creation of embryonic stem cells but allows stem cell research on existing cell lines that were created before the deadline of May 1, 2007. This takes the moral status of the embryo seriously but does not totally exclude embryonic stem cell research. Unfortunately, the compromise has the taste of double moral that transfers unethical research in foreign countries and does not really pay the price for a good conscience [31, 47].

Not each of these points applies to any kind of tissue engineering, of course, but working in therapeutic and therefore generally approved circumstances alone does not justify any kind of research either, so a careful individual ethical assessment is recommended. The reliance on research freedom can be misleading: though it is also a societal approved value [53], it is important to note that it is neither the only nor the highest good. In a general ethical assessment, freedom of research is one aspect among others.

Tissue engineering and its multiple perspectives is also an example for the power of research to resolve many medical, but also some social problems. It is always prudent to be cautious if social problems are approached by scientific means (e.g., concerning eugenics and the problem of how to deal appropriately with handicapped people), but since some social and conceptual problems are caused by scientific progress [36], it can be only advantageous if they are overcome by further progress, or if bad compromises are rendered superfluous.

References

1. ABIM, ACP-ASIM, European Federation of Internal Medicine (2002) Medical professionalism in the new millennium. A physician charter. Ann Internal Med 136(3):243–246
2. Abouna GM (2003) Ethical issues in organ and tissue transplantation. Med Princ Pract 12(1):54–69
3. Ach JS, Jömann N (2006) Ethical implications of nanobiotechnology. State-of-the-art survey of ethical issues related to nanobiotechnology. In: Siep L (eds) Nano-bioethics: ethical dimensions of nanobiotechnology. Münster, Germany: Lit pp 13–62
4. Baum BJ, Mooney DJ (2000) The impact of tissue engineering on dentistry. J Am Dental Assoc 131:309–318
5. Beauchamp TL, Childress JF (1994) Principles of biomedical ethics, 4th ed. Oxford University Press, New York
6. Birnbacher D (2007) Der Hirntod—eine pragmatische Verteidigung. Jahrbuch für Recht und Ethik 15:459–477
7. Black J (1997) Thinking twice about "tissue engineering". IEEE Eng Med Biol Mag 16(4):102–104
8. Buchanan AE, Brock DW (1990) Deciding for others: the ethics of surrogate decision making. University Press, Cambridge
9. Daar AS, Bhatt A, Court E, Singer A (2004) Stem cell research and transplantation: science leading ethics. Transplant Proc 36:2504–2506
10. de Brugerolle A (2007) SkinEthic Laboratories, a company devoted to develop and produce In vitro alternative methods to anima use. ALTEX 24(3):167–171
11. Denker HW (1999) Embryonic stem cells: an exciting field for basic research and tissue engineering, but also an ethical dilemma? Cells Tissues Organs 165:246–249
12. Dickens BM, Cook RJ (2007) Acquiring human embryos for stem cell research. Int J Gynaecol Obstet 96:67–71
13. European Group on Ethics (2004) Report of the European Group on Ethics on the Ethical Aspects of Human Tissue Engineered Products
14. Eyrich D, Brandl F, Appel B, Wiese H, Maier G, Wenzel M, Staudenmaier R, Goepferich A, Blunk T (2007) Long-term stable fibrin gels for cartilage engineering. Biomaterials 28:55–65
15. Faden RR, Beauchamp TL (1986) A history and theory of informed consent. Oxford, New York
16. Gelhaus P (2006) Gentherapie und Weltanschauung. Ein Überblick über die genethische Discussion. Wissenschaftlicher Buchverlag, Darmstadt
17. Gelhaus P (2007) Was bedeutet eigentlich "gesund" im medizinischen Umfeld? In: Jansen G, Schwarzfischer K (eds) Gesundheit-wozu? Jansen-Verlag, Lüneburg, pp 83–102
18. Gelhaus P (2008) Wie groß ist zu groß? Zur Funktionalität des Normalen. In: Groß D, Müller S, Steinmetzer J (eds) Normal—anders—krank? Akzeptanz, Stigmatisierung und Pathologisierung im Kontext der Medizin. Medizinisch-Wissenschaftliche Verlagsgesellschaft Berlin, pp 33–49
19. Gelhaus P, Middel CD (2002) "Individuelle Gesundheit versus public health?!" ein Streitgespräch zur Forschungsentwicklung am Beispiel der revidierten Deklaration von Helsinki (Edinburgh 2000). In: Brand A, Engelhardt DV, Simon A, Wehkamp KH, (eds) Individuelle Gesundheit versus public health? Münster, London, pp 96–112
20. Gordijn B (2004) Medizinische Utopien. Eine ethische Betrachtung. Vandenhoeck & Ruprecht Göttingen, pp 65–110
21. Green RM (2002) Benefiting from "evil": an incipient moral problem in human stem cell research. Bioethics 16(6):544–556
22. Green RM (2004) Ethical considerations. In: Lanza R, et al. (eds) Handbook of stem cells. Elsevier, Amsterdam, pp 759–764
23. Hammerman MR, Cortesini R (2004) Organogenesis and tissue engineering.Transpl Immunol 12:191–192
24. Heinonen M, Oila O, Nordström K (2005) Current issues in the regulation of human tissue-engineering products in the European Union. Tissue Eng 11(11/12):1905–1911
25. Henon PR:(2003) Human embryonic or adult stem cells: an overview on ethics and perspectives for tissue engineering. Adv Exp Med Biol 534:27–45
26. Hunziker EB (2003) Tissue engineering of bone and cartilage. From the preclinical model to the patient. Novartis Found Symp 249:70–78
27. Langer R, Vacanti JP (1993) Tissue engineering. Science 260:920–926
28. Lohmander LS (2003) Tissue engineering of cartilage: do we need it, can we do it, is it good and can we prove it? Novartis Found Symp 249:2–10
29. Longley D, Lawford P (2001) Engineering human tissue and regulation: confronting biology and law to bridge the gaps. Med Law Int 5:101–115

30. Meyer T (2005) Ethical considerations in bone and cartilage tissue engineering. Meyer U, Wiesmann HP (eds) Bone and cartilage engineering. Springer, Berlin, pp 203–206
31. Nationaler Ethikrat (Germany) (2007) Zur Frage einer Änderung des Stammzellgesetzes. Stellungnahme, www.ethikrat.org/stellungnahmen/pdf/stn-stammzellgesetz.pdf (access: 03–06–2008)
32. Nordgren A (2004) Moral imagination in tissue engineering research on animal models. Biomaterials 25:1723–1734
33. Reed JA, Patarca R (2006) Regenerative dental medicine: stem cells and tissue engineering in dentistry. J Environ Pathol Toxicol Oncol 25(3):537–569
34. San Martin A, Borlongan CV (2006) Cell transplantation: toward cell therapy. Cell Transplant 15:665–673
35. Satava RM (2002) Disruptive visions. Moral and ethical challenges from advanced technologies and issues for the new generation of surgeons. Surg Endosc 16:1403–1408
36. Satava RM (2003) Biomedical, ethical, and moral issues being forced by advanced medical technologies. Proc Am Philos Soc 147(3):246–258
37. Schilling AF, Linhart W, Filke S, Gebauer M, Schinke T, Rueger JM, Amling M (2004) Resorbability of bone substitute biomaterials by human osteoclasts. Biomaterials 25:3963–3972
38. Schöne-Seifert B, Ach JS, Siep L (2006) Totipotenz und Potentialität. Zum moralischen Status von Embryonen bei unterschiedlichen Varianten der Gewinnung humaner embryonaler Stammzellen. Gutachten für das Kompetenznetzwerk Stammzellforschung NRW. Jahrbuch für Wissenschaft und Ethik 11:261–321
39. Schwab IR, Johnson NT, Harkin DG (2006) Inherent risks associated with manufacture of bioengineered ocular surface tissue. Arch Ophtalmol 124:1734–1740
40. Siep L (2006) Ethical problems of stem cell research and stem cell transplantation, In: Deltas C, Kalokairinou EM, Rogge S (eds) Progress in science and the danger of hybris. Waxmann, Münster, pp 91–99
41. Singer P (1975) Animal liberation: a new ethics for our treatment of animals. Random House, New York
42. Singer P (1993) Practical Ethics 2nd ed. Cambridge, New York
43. Stanworth SJ, Newland AC (2001) Stem cells: progress in research and edging towards the clinical setting. Clin Med 1(5):378–382
44. Steinbock B (2007) Moral status, moral value and human embryos: implications for stem cell research. In: Steinbock B (ed) The Oxford handbook of bioethics. Oxford, New York, pp 416–440
45. Sutherland FWH, Mayer JE Jr (2003) Ethical and regulatory issues concerning engineered tissues for congenital heart repair. Semin Thorac Cardiovasc Surg Pediatr Card Surg Annu 6:152–163
46. Sylvester KG, Longaker MT (2004) Stem cells. Arch Surg 139:93–99
47. Takala T, Häyry M (2007) Benefiting from past wrongdoing, human embryonic stem cells and the fragility of the German legal position. Bioethics 21:150–159
48. Trommelmans L, Selling J, Dierickx K (2007) A critical assessment of the directive on tissue engineering of the European Union. Tissue Eng 13(4):667–672
49. Vats A, Tolley NS, Polak JM, Buttery LDK (2002) Stem cells: sources and applications. Clin Otolaryngol Allied Sci 27:227–232
50. Weyr S (1993) Informed consent: patient autonomy and physician beneficence within clinical medicine. Kluwer, Dordrecht
51. Williams D (2006) A registry for tissue engineering clinical trials, In: Med Device Technol 5:8–10
52. World Medical Association (1964, 1975, 1983, 1989, 1996, 2000, 2004) Declaration of Helsinki
53. Zoloth L (2004) Immortal cells, moral selves. In: Lanza R, et al. (eds.) Handbook of stem cells, vol 1. Elsevier, Amsterdam, pp 747–757

Sourcing Human Embryonic Tissue: The Ethical Issues

C. Rehmann-Sutter, R. Porz, J. L. Scully

Contents

4.1	Introduction	37
4.2	Actor Perspectives	38
4.3	The Political Community	38
4.4	The (Future) Patients	39
4.5	Embryo Donors	40
4.6	Egg/Sperm Donors	42
4.7	Embryo	43
4.8	Scientists and Medical Professionals	44
	References	45

4.1 Introduction

For some tissue engineering projects human embryonic cells are used. The feature of pluripotency makes them particularly interesting to study and desirable to use. The human embryo, however, is an entity that is in many ways special and precious. It is a being that is in a process of ongoing development—if transferred to the uterus and the conditions are favorable, it will develop into a fetus and be born as a child. Experiments with human embryos have been necessary and have stimulated ethical controversies in the context of reproductive technologies and genetic diagnostics. But stem cell research demands extensive and systematic sourcing of embryonic cells as a resource for new biotechnological developments. From this point of view it is not surprising that this research is surrounded by ethical questions.

Since the late 1990s, stem cells have become one of the key symbols in the public discourse of bioethics and biopolitics. More or less detailed regulations have been developed by political bodies, defining more or less restricted conditions under which stem cells may be retrieved from human embryos. The embryo is a prenatal—under in vitro conditions, a pregestational—stage in the life of the human organism. It is a marginal existence: life on the way to becoming. Underlying the debate about regulations is the problem of finding appropriate moral categories to describe life at its margins: is an embryo a thing, i.e., something without intrinsic moral rights, or an entity with a certain moral dignity, to which we have obligations, or even a person, not much different from a newborn, as some participants in the debate claim? The controversial ontology and thus the ambivalent moral status of this liminal entity called the embryo makes it difficult to craft justifications for when we are allowed to destroy an embryo for use as a source of the raw materials for regenerative medicine and tissue engineering.

Tissue engineering generates a complicated network of relationships and responsibilities between concrete people who adopt certain roles, like the "stem cell researcher" or the "embryo donor." In shaping these roles and these sometimes conflicting responsibilities, it also generates interfaces between different domains of action, such as the therapy–research interface in IVF clinics that have established close connections with stem cell laboratories. Let us take a closer look at those different actor perspectives.

4.2
Actor Perspectives

We can start with the following scenario: A woman donates an oocyte that is used to generate an embryo for an infertile couple. For some reason, the embryo is not used for in vitro fertilization (IVF) but becomes spare, and is donated by the couple to stem cell research. The embryo is disintegrated into cells and gives rise to a stem cell culture that is characterized in the laboratory, expanded, and offered to other laboratories through an international stem cell research network. In one of the laboratories that purchases this stem cell line, a treatment for a cell degenerative disease may be developed and offered to a patient.

If we look at this scenario, however simplified it may be compared to real cases, a series of actor perspectives can be identified which build up a complex network of relationships: (i) a germ cell donor, (ii) the team in the donation clinic, (iii) the couple undergoing IVF who are asked to donate the spare embryo, (iv) the staff in the reproductive medicine center, (v) the team of stem cell researchers, (vi) the stem cell research network, (vii) the laboratory that purchases the stem cell line, (viii) a patient and his or her medical care team, and (ix) state and private funding bodies, research and policy advising ethics committees, and the wider publics involved in an ethical discourse and biopolitics.

The cells that are transplanted into the patient's body are therefore not simply biological entities with certain properties, but also social entities that carry different meanings and provoke ethical concerns embedded in the concrete relationships. Cells are more than biology because they relate all these people within a network of mutually responsible practices. In the practices of regenerative medicine, new social roles find representatives, and the question arises of what it would mean to perform these roles excellently: not merely from a technical, but also from an ethical point of view.

The picture becomes even more complex when we acknowledge that many of the people involved are not acting as isolated individuals but in the context of institutions, some of them public (clinics, universities), others private (companies). And many of the concepts which they use to understand and solve the practical questions that arise, originate not in isolation but in different discourses, where language is developed, metaphors are used, and meaning is generated, sometimes in close relation with financial and other interests [1, 3, 20, 25]. It is not simply a question of rules that scientists and medical professionals must follow.

We organize our brief discussion of the ethical issues according to six different actor perspectives: the political community, the (future) patient, the embryo donors, the gamete donor, the embryo, and the scientists or medics in their different functions. These are obviously not the only relevant perspectives but they cover at least those of the most directly involved parties.

4.3
The Political Community

The public is involved in tissue donation not through direct acts, but because, ideally, the whole community is in agreement with whatever regulations/guidelines are produced. These regulations and procedures reflect the prevailing ethos and political culture of the community. Furthermore, the "private" attitudes of other, directly involved individuals such donors, patients, or researchers are inevitably formed by the "public" culture they inhabit.

From the legal perspective the first question to be answered is: should the state be involved at all? Should a legal regulation of any sort be put into force, or can the state leave this practice unregulated? In most countries where stem cell research developed the answer was preordained, because older regulations were already in place that either allowed or banned the utilization of embryos, including use for stem cell research. But as a question of principle, we think that one reason that speaks strongly in favor of regulation is that the "entry conditions" into the legal community for human life during development need to be clear. A system of norms cannot leave the question of membership unconsidered without loss of consistency.

The next decision is between three basic options: (1) forbidding the use of embryonic cells for research and development, (2) allowing use under certain conditions, and (3) allowing unrestricted use. All countries in which stem cell research developed chose some variant of option (2). The reasons may differ. One argument against a complete ban is that it is difficult to argue that an embryonic *cell* brought into tissue culture needs to be protected legally from being used for research or therapeutic purposes.

Within option (2) allowing the use of embryonic cells under certain restrictions, there are at least three further basic options. Roughly, they cover the variety of legal solutions that may be found in different countries. The lawgiver can specify conditions in terms of (a) the *time* during embryonic development, (b) the *conditions of access* to the embryo, or (c) a calendar *deadline* of generation of the stem cell line, or any combinations of these.

The USA (under President George W Bush) and Germany both chose the calendar deadline solution, setting the deadline before putting the rule into force and arguing that no embryo should be killed anywhere in the world because of research funded or allowed under the authority of this law. The UK (following the recommendations of the Warnock committee) and China have both chosen to set a maximal delay after fertilization. Research is allowed only up to 14 days post fertilization, and it is subject to strict oversight. The production of embryos for research is permitted in the UK, Sweden, Denmark, Finland, or Spain. France, Switzerland, and the US National Bioethics Advisory Committee under President Clinton opted for a variant of the middle option, and defined the conditions under which an embryo from IVF may become available for stem cell research [7]: these conditions include that it must be a spare embryo that was developed for the purpose of pregnancy but cannot be used for this purpose any more, and that the couple must have consented to donate the embryo to research.

4.4
The (Future) Patients

Some future patients might find the potential source of their treatment (embryonic tissue) ethically problematic. They may face a moral dilemma, if there is only one cure for a severe or life-threatening condition. Even those patients who can live with the fact that the cure involves embryonic tissue might still wish that their lives should not depend on cells that are produced under immoral or dubious circumstances. There is, therefore, a moral obligation towards the future patients to care about the ethics of stem cell science and to find open and transparent regulations that define conditions that can be accepted ethically by as many potential patients as possible [18].

This can be seen in parallel to transplantation medicine. An obligation not to procure organs under objectionable conditions derives from the therapeutic responsibility *for the patient*. In stem cell medicine, in contrast to transplantation medicine, there is no prevailing international consensus in sight, about exactly what are the acceptable circumstances. But even if solutions are influenced by cultural and religious factors, this therapeutic relationship to the future patients leads to the recognition that the community of stem cell scientists has a moral responsibility towards the public, that is those who are also potential patients. The responsibility includes discussing the conditions of embryo use openly and transparently, in order to enable public participation in the debate and clear public agreement or disagreement with the solution found.

There is also an interesting question here about the limits to knowledge and informed consent for medical treatments. Should we argue that all patients must always be (made) aware of the ethical dilemmas "hidden" in the treatment they are being given? Collective responsibility might imply this. Or, given that some of these patients may be *in extremis*, should we argue that beneficence demands that they be spared such remote issues?

Apart from the moral issues of embryo use there is also an obvious need to adhere to accepted ethical standards in clinical trials to ensure they are safe and fair for the study participants. This means that the source of the stem cells has to be ethically acceptable to the future patient(s). If a patient asks for information about the sources of the cells with which he or she is treated the question may arise of whether the patient should be given the names of the couple who produced the embryo. In transplantation medicine, the established practice is to keep organ donation anonymous. Knowing the identity of the person who once lived with the organ might be problematic for the patient learning to identify with her or his "new" body [8, 14]. A similar argument can be made in stem cell medicine.

From the patient's perspective justice issues may also arise. There are ethical concerns about who benefits from this technology, and what are the likely costs of developing and implementing it. The patient may question the fairness of distribution of risks and benefits: why should she or he accept higher risks than others while participating in a study with unclear benefit? Or why was he or she lucky enough to receive therapy where others do not? Or, if the patient is not among those who can profit from the new treatment, he or she may feel discriminated. Even in the developed world it is likely that at least at first, regenerative technologies will be available only to the more privileged sections of society. As always with highly sophisticated and expensive innovative healthcare, there is a debate to be had about global disparities in health standards and healthcare provision.

4.5 Embryo Donors

When a couple is undergoing an IVF treatment it may happen that one (or more) of the embryos they produce become "spare." This "supernumerary embryo" may arise when the team in the IVF clinic considers an embryo to be unsuitable for use on morphological grounds, or from some other circumstance that unexpectedly prevents the transfer of the embryo to the woman's uterus. In many countries (including the UK and Switzerland) there is a legal requirement to keep the number of "spare" embryos as low as possible, and corresponding clinical practices are in place. However, doctors and patients generally tend to want to create enough embryos to give a good chance that at least one of them will be suitable for transfer. If some embryos become "spare" the question arises of what to do with them. In some countries, for example the UK and the USA, it is legally permissible to freeze them. However, this only postpones the question of what to do with them to a later point in time: a couple in this situation will inevitably have to make some kind of decision about them (even if the decision is to refuse to make a decision, and leave it up to the clinic, which may mean the embryos are simply disposed of). Sooner or later the couple can become "embryo donors," even by default.

How people act in this new social role of potential embryo donors [21] depends among other things on how they view their embryos. Couples who perceive their embryo as purely "biological tissue" may have fewer problems in deciding what to do. If they consider it to be a "living entity" or a "potential child" to some degree, they may be confronted with additional hard questions. If they see the supernumerary embryo as having value as a "symbolic relic" of their IVF treatment, their line of thinking is likely to be quite different, and to be determined by whether or not the IVF treatment was ultimately successful or whether the experiences they had in course of their treatment were good ones. They may want to treat their spare embryo in a symbolic way itself, perhaps to give the story of their IVF treatment a "proper" ending, and to come to terms with the often strenuous process of the physically and emotionally demanding period of IVF cycles. Interview studies with couples undergoing IVF have shown that attitudes to the embryo are not pregiven and fixed, but rather are defined and acted upon in relation to their social context at different times and places [9, 17, 24]. This includes both *personal* time (the stage interviewees are at in their treatment process and in their parenting project) and space (whether the embryos are embodied, are posttransfer, or are in the clinic as fresh or frozen embryos), and *socio-cultural* time and space (in relation to legislation and to developments in science and medicine).

Hence, their vision of the spare embryo and the context of their previous experiences with reproductive medicine shape the rationale of how to deal with their new role as potential embryo donors. But acting in accordance with one's own vision might be inadequate, especially when this vision has to be adjusted to the different options of what to do with the spare embryos. Donation of embryos is only one possibility, and even "donation" might have different implications, depending on the purpose of the donation. Currently, at least five different options can be identified, illustrating the complexity of the situation a couple might be faced with after having undergone IVF:

1. Faced with a spare embryo, couples can decide to attempt another pregnancy. This is only an option if the spare embryo has been or can be frozen and is still viable. This decision is also dependent on

life circumstances, the goals of the couple, and the number of children they feel emotionally, psychologically, or financially capable of raising.

2. They can maintain the surplus ones frozen. But in some countries like the USA, freezing costs an annual fee, which might be out of some parents' reach, while in other countries couples are only legally allowed to freeze for a restricted number of years. At least in this scenario, couples in principle have the time to think about their frozen embryos and what they want to have done with them. Alternatively some couples may prefer not to think about them, until a decision is forced on them for legal or clinical reasons. Freezing policies differ. Under current Swiss law, for example, it is generally not allowed to freeze a zygote more than 24 h after insemination. Within the first 24 h however it is considered in law as a "impregnated egg:" a Swiss couple can therefore freeze "impregnated eggs," but not an embryo produced in an ongoing IVF treatment. These parameters may change when the law does. From the perspective of the couple, such arbitrariness can be puzzling.

3. If the decision was taken against having further children, and against further freezing, the couple may decide to thaw the surplus embryos and then dispose of them. This again sounds more straightforward in outline than the reality. For example, one couple might find the disposal to be loaded with an indefinite feeling of guilt, while for other infertile couples it might look like squandering a potential child. The couple may also be aware, or be made aware, that to researchers who are interested in stem cells, disposal would be a waste of a valuable good.

4. In countries such as the USA or Sweden, the couple can donate their surplus embryo to another couple. This option provides another couple with a chance to fulfill their wish for a child. Donation of an embryo to another couple creates a new kind of relationship between the two couples that needs to be carefully considered because conflicts can arise. If donation is anonymous the child, after having grown up, may still have a legal right to find out who her or his genetic parents were. Some couples opt for openness from the beginning and maintain a relationship with the biosocial parents and with the child. Others might not wish to do that. Or they may wish to but find that the recipient couple are against such openness.

5. In some countries, the couple can donate embryos to stem cell research. Couples have to undergo a process of informed consent, the concrete steps of which depend on cultural factors, medical history in the country, the current legal situation, the communication skills of the doctor or the research team involved, and the experience this particular team of doctors/researchers has acquired in this area of research. Questions have been raised about the appropriate level of information that is possible to provide in this context.

The International Society for Stem Cell Research [13] emphasizes that both giving information and obtaining consent should be performed by the research team, not by the doctors involved in IVF. A clear separation of the clinical IVF world from the world of research is important to make sure that donation of surplus embryos is in no way connected with the couple's IVF treatment (ISSCR Guidelines, para 11.4). This seems important in order to minimize the risk that patients donate out of personal gratitude to the doctor or are motivated by a sense of obligation to the IVF clinic.

The ISSCR further states that "the informed consent process should take into account language barriers and the educational level of the subject themselves" (ISSCR Guidelines, para 11.3). At first sight this makes perfect sense, but one has to remember the number of additional features that couples need to be informed about: the derivation process, the way the surplus embryo is "destroyed" in the stem cell lab, the fact that derived stem cell lines will be kept for years, that new lines can branch out of the original ones, that some cells might be transformed into engineered tissues, that the lines can be internationally exchanged, what kind of research the research team is working on, whether there is any potential conflict of interest or financial interests that must be disclosed, that no immediate benefit might arise out of these lines—to mention only some of the essential aspects. This is a lot of highly technical information for a couple who wanted to get a child by IVF. Suddenly they find themselves forced to become, at least to some extent, specialists in the background, goals, and conflicts of stem cell research and tissue engineering. The embryo donor has not become a donor for the sake of it, but out of a completely different life project: the desire for a child (by IVF) and a family. From the embryo donors' perspective, therefore, their original goal has no common ground with stem cell research. Nevertheless, in the request to consider donating, embryo donors now find themselves having to bridge the gap between their previous goal (getting a child) and research in regenerative medicine.

Arriving at one's own vision of the embryo, making one's way through different decisive situations, keeping clear the separation between IVF treatment and stem cell research, trying to understand the full implications of a dense and interwoven procedure of information and informed consent: becoming an ethically responsible embryo donor is not easy. Every step along the clinical procedures (which the patients may perceive as an inexorable conveyor belt) can produce vulnerabilities for the couple, particularly when they are going through this for the first time and are likely to have been completely unaware of this potential chain of events in their initial plans to have a child.

4.6
Egg/Sperm Donors

Research embryos can be generated in vitro with oocytes and sperm cells that have been donated for this purpose. The procedure may however not be legal everywhere. For example, it is forbidden in Germany and Switzerland, but permitted in the UK as long as the embryo is not grown beyond 14 days after fertilization. Some possible scenarios for the production of embryonic stem cells work even without sperm. The oocyte can be stimulated to divide by itself long enough to allow the derivation of embryonic stem cell lines (parthenogenesis). There may be technical limitations with this scenario, but they are beyond the focus of this chapter.

Many of the same issues apply to donors of eggs or sperm as apply to embryo donors, e.g., the transparency of communication and information, and informed consent. However, from the perspective of prospective donors, there are some significant differences [10]. Chief among these is the context in which "spare" eggs and sperm, and embryos, become available. In most countries where IVF is regulated, spare embryos are an unintended byproduct of IVF—they constitute a failure in some sense of IVF. Donation means giving away an embryo that already exists and might otherwise be destroyed. Egg and sperm donation is rather different: they are explicitly to be used for the production of an embryo for research purposes. The decision includes the question "Should an embryo be produced with my germ cells that will be used for the production of stem cells?" The eggs might be donated in or beyond the context of the donor's own IVF (depending on the regulatory framework). This makes a significant difference especially for the woman who has to undergo hormone stimulation and an operation in order to retrieve the oocytes. If however spare oocytes from a completed IVF treatment are donated, the woman has no extra physical burden. Sperm can be retrieved much easier (by masturbation), and sperm has been donated for many years for use in donor insemination procedures. Hence, egg/sperm donation points up the highly gendered nature of the different processes—ease of sperm donation, vs. egg donation involving potential risk to woman through hormonal (over)stimulation or egg extraction. This gender difference is also relevant to the question of whether sperm or eggs may be sold. A financial incentive may influence the woman to accept a health risk that she would otherwise reject. Alternatively, payment might be seen as no more than reasonable recompense for the burdens undertaken.

Other ethical considerations concern trade and traffic. In some countries there are agencies that do business in eggs and sperm. This is problematic for several reasons. One is the increased distance between donor and user and the lack of transparency that may result. Another is a fundamental question of whether trade in human body parts should be permitted at all. Trading transforms the traded object into a commodity and can be seen as inconsistent with the dignity of the bodies of women and men. There are unresolved ethical debates about the acceptability of payment for donation. Traditionally it was acceptable to pay sperm donors (generations of British medical students are widely believed to have supported themselves through medical school this way), but payment to egg donors has been more contentious.

The social meaning of the two functions also differs along gender lines. In part this is because of a longer history of sperm donation. There is something of a social role for a sperm donor, but not yet for an egg donor. In part this is because of social expectations about maternal and paternal roles: it may be seen as more acceptable for a man to donate in a way that may produce many offspring entirely unknown to him, but seen as "unmaternal" for women to do the same. And/or it can be seen as "natural" for women to donate "altruistically."

Further issues must be included here if the perspective is widened to a global one. Feminist bioethicists [11] have been among those pointing out the potential for exploitation of women as donors of oocytes and embryos. This exploitation may be emotional/psychological (e.g., when an expectation is placed on a woman that she will be altruistic in providing an egg). Equally significantly, women—especially economically disadvantaged women—may be tempted to become involved in an unregulated, even global trade in oocytes [5, 6]. Among the unresolved ethical issues here are (1) whether from this perspective there is an in principle objection to payment for provision of eggs or sperm (or embryos), and (2) whether women would best be prevented from exploitation by prohibiting the sale of gametes, or by permitting and properly regulating it.

4.7 Embryo

The embryo is not an agent like the future parents, or the scientific and medical professionals. Nor is it a subject with a moral perspective, as the child would be. But it is very difficult to define exactly what the embryo is. It is the focus of moral debates whose intransigence is rooted in differences in the description of the embryo. Some will say it is just an organized conglomerate of biologically very active cells and nothing more. They endorse a model of the embryo as essentially a "thing." Others will say that the embryo has a full set of genes, making it an early stage of human life, fully capable (other conditions given) to develop into a fetus and a child (and an adult). They therefore want the embryo to be treated as a subject morally and legally, and they want society to recognize the embryo's right to life. They endorse a model of the embryo as essentially a "person." A third group of people take an intermediary position and say, the embryo is unique: a very early stage in human development, highly dependent on the female organism in a variety of ways, perhaps building an organic unity within it over long period of time; it is an entity that will, in a later stage, after many transformations and only under favorable conditions, grow into a fetus and be born as a child. This, they argue, gives reason to treat the embryo with at least some dignity, perhaps a changing level of dignity, not as a person but not as just an assembly of materials either. They endorse a model of the embryo as an "object of respect."

This difficulty of descriptions, together with the different scripts that are implied with the descriptions, and not the presupposed "moral perspective" of the embryo as an agent, is the most important issue to be discussed in this section. The descriptions we chose (we have tried to group them into three basic "models" [16]) are always extrapolations and projections by other agents such as scientists, philosophers, politicians, and of course "ordinary people," who all have ideas about the beginning of human life, based somewhere in their world views. The embryo is an entity whose identity depends heavily on the categories we use to describe it.

What, then, is an embryo in moral terms? We cannot suggest a general and objective answer [2, 12, 15]. But without having complete agreement among ourselves, we tend to the third of the models described above [19] and we could provide reasons for the plausibility of our variants of this position and implausibility of the others, which might convince some but certainly not all of those endorsing the other models. What is more objective however is the differentiation of the practical situations that are generated in the context of IVF and stem cell technologies that may have an impact on our thinking about the moral position of the embryo.

Reproductive technologies introduce the complexity of different kinds of embryo (frozen, spare, in vitro, in utero, cloned, parthenogenetic, chimeric, etc.), made from natural or artificial (stem cell-derived) germ cells, which on being placed in a different practical context pose different moral questions. Some ethicists would even argue that they must be granted alternative moral status or can legitimately be treated in different ways, because they are different *kinds* of *entities*. Other ethicists would argue moral status must remain the same. Drawing on empirical data from interviews [9], we can hypothesize that parents/potential donors often seem to consider their embryos differently depending on the context—whether in a dish in front of them, frozen, before or after nidation, developing into a child or not.

An important ethical issue that must be discussed in all moral "models" of the embryo is this: Who has the obligation (and the right) to make decisions about its fate? How far can couples decide autonomously about the embryo that grows from germ cells taken from their bodies? What needs to be legally regulated by the state and what should better be left to soft-law policy regulations such as ethical guidelines or recommendations—or personal judgment? This question raises new issues in the debate about public and private arenas in modern life—a fundamental question in liberal societies about what is considered open to regulation by the state and what should be left to the individual. Traditionally, the law started from the assumption that its regulations are needed as soon as there are relationships between independent human beings—persons—under concern. Therefore, legal personhood—the capacity to have legal rights—in most legal systems begins at birth, or immediately before birth in order to protect the unborn infant from maltreatment during birth and in prenatal pediatric care. Reproductive technologies produce entities—embryos outside the body of a woman—and procedures—an arsenal of interventional laboratory technologies—that straddle the traditional boundary between family life (private) and healthcare provision (public).

Other issues are related more exclusively to only one of the "models." Scholars endorsing the strongest protective model who see the embryo as a person in its own right ask that the life of the embryo be protected, and oppose their use for stem cell research. But even in this position it is not entirely clear whether an embryo that has no chance of survival (e.g., embryos that have been positively diagnosed as carrying a genetic disease in preimplantation genetic diagnosis) and which would be left to die anyway may be used in stem cell research. The issue here is whether instrumentalization of the embryo in a situation where life cannot be protected is morally objectionable or not. Those who believe that embryonic life must be protected because the embryo is an entity with intrinsic dignity, could still endorse a view that instrumentalization *alone* (without sacrificing a life that could be saved) is ethically tolerable.

4.8
Scientists and Medical Professionals

Scientists and medical professionals are large and diverse groups of people. The goals, priorities and ethics of science are very different from the goals, priorities, and ethics of therapy. There needs to be clarity over this in order for medical professionals and scientists to understand the responsibilities, values, and the ethical questions in the constitution of their roles and also for patients to be treated appropriately.

Both medical professionals and scientists must adhere to conventional standards of ethical behavior in the field. There are a growing number of regulations and guidelines in the different countries that must be observed. In addition, the experience of research with human subjects has led to international standards of good clinical practice [4, 22, 26] that also apply to the interface between stem cell research and reproductive medicine.

For stem cell researchers, there are major forces that can operate against them remaining within appropriate ethical boundaries (more so than with medical professionals, where in principle the interests of the patient provide a strong counterbalance to many potentially unethical acts). For example, researchers face the drives of personal, institutional, even national ambition, scientific curiosity, a demand for rapid progress to fulfill public promises that have been made in this field, and also the increasing need to satisfy funding bodies (public or private) and secure further funding for research in expanding centers.

There are therefore ethical issues about commercial or political interests involved in funding research. Commercial funding is not automatically unethical but demands transparency in terms of involvement, and adequate governance.

Further ethical issues arise in the interaction between researchers and other groups, especially public and funding bodies. For example, there is a requirement for accuracy and integrity in the presentation of work [23] and the avoidance of hype. The latter is an obligation to the general public and is justified both through the legitimate interest of the general public in being adequately informed and through the particular responsibility of scientists to seek knowledge (not illusion) and pursue the path of truth.

With regard to economy and fairness we would like to conclude by referring to the issues raised for potential patients and by adding the following question: are researchers morally (co)responsible for the just allocation of the economic and medical results of what they do? Should it be of concern to researchers that patients locally and globally have unequal access to therapies, and/or will be unequally burdened by producing tissue for research? These wider perspective issues have traditionally been seen as beyond the remit of scientific or medical ethics, but there are increasing claims from within bioethics that it is legitimate to expect those involved to address such questions—especially those who in some way benefit from the process, like researchers.

In this chapter we have traced the ethical issues and the moral responsibilities in this extended field of practice. We have specifically not defended one of the substantial points of view on the status of the embryo, that is either the liberal or the conservative standpoint, but have tried to make the ethical discussion more accessible to those working in the fields of stem cell research and in tissue engineering. Ethical analysis can help to untangle some of the moral knots; seeing the problems from the perspectives of different involved actors, not just from an abstract theoretical point of view, will help as well.

Acknowledgment

We are grateful for support from the Swiss National Science Foundation and the Käthe Zingg-Schwichtenberg Fonds that made our research possible. For English revision of the chapter we thank Monica Buckland.

References

1. Andrews L, Nelkin D (2001) Body bazaar: the market for human tissue in the biotechnology age. Crown, New York
2. Anselm R, Körtner UHJ (eds) (2003) Streitfall Biomedizin. Urteilsfindung in christlicher Verantwortung. Vandenhoeck & Ruprecht, Göttingen
3. Bock V, Wülfingen B (2007) Genetisierung der Zeugung. Eine Diskurs- und Metaphernanalsyse reproduktionsgenetischer Zukünfte. Bielefeld: Transkript
4. CIOMS (2002) Council of International Organizations of Medical Sciences, International Ethical Guidelines for Biomedical Research Involving Human Subjects. CIOMS, Geneva. Available via www.cioms.ch/frame_guidelines_nov_2002.htm
5. Dickenson D (2002) Commodification of human tissue: implications for feminist and development ethics. Dev World Bioeth 2:55–63
6. Dickenson D (2007) Property in the body: feminist perspectives. Cambridge University Press, Cambridge
7. European Group on Ethics in Sciences and New Technologies to the European Commission (2007) Recommendations on the ethical review of hESC FP7 research projects. Office of Official Publications of the EC, Luxembourg
8. Franklin P (2003) The recipient's perspective. In: Morris P (ed) Ethical eye: transplants. Council of Europe Publishing, Strasbourg, pp 51–62
9. Haimes E, Porz R, Scully JL, Rehmann-Sutter C (2008) 'So, what is an embryo?' A comparative study of the views of those asked to donate embryos for hESC research in the UK and Switzerland. New Genet Soc 27(2):113–126
10. HFEA (2006) SEED Report. A report on the Human Fertilization and Embryology Authority's review of sperm, egg and embryo donation in the United Kingdom. Available via: www.hfea.gov.uk/en/492.html
11. Holland S (2001) Beyond the embryo: a feminist appraisal of the embryonic stem cell debate. In: Holland S, et al (eds) The human embryonic stem cell debate: science, ethics, and public policy. MIT, Cambridge, pp 73–86
12. Holland S, Lebacqz K, Zoloth L (eds) (2001) The human embryonic stem cell debate. science, ethics, and public policy. MIT Press, Cambridge, Mass
13. ISSCR (2006) International Society for Stem Cell Research, Guidelines for the Conduct of Human Embryonic Stem Cell Research. Available via: www.isscr.org/guidelines/ISSCRhESCguidelines2006.pdf
14. Le Breton D (2003) Identity problems and transplantations. In: Morris P (ed.) Ethical eye: transplants. Council of Europe Publishing, Strasbourg, pp 41–49
15. Lenzen W (ed) (2004) Wie bestimmt man den "moralischen Status" von Embryonen? Mentis, Paderborn
16. Maio G (2002) Welchen Respekt schulden wir dem Embryo? Die embryonale Stammzellforschung in medizinethischer Perspektive. Deutsche Medizinische Wochenschrift 127:160–163
17. Parry S (2006) (Re)constructing embryos in stem cell research. Soc Sci Med 62:2349–2359
18. Rehmann-Sutter C (2002) Why care about the ethics of therapeutic cloning? Differentiation 69:179–181

19. Rehmann-Sutter C (2006) Altered nuclear transfer, Genom-Metaphysik und das Argument der Potentialität. Die ethische Schutzwürdigkeit menschlicher Embryonen in vitro. Jahrbuch für Wissenschaft und Ethik 11:351–374
20. Rose N (2007) The politics of life itself. Biomedicine, and subjectivity in the twenty-first century. Princeton University Press, Princeton
21. Scully JL, Rehmann-Sutter C (2006) Creating donors: the 2005 Swiss law on donation of 'spare' embryos to hESC research. J Bioeth Inq 3:81–93
22. Smith T (1999) Ethics in medical research. A handbook of good practice. Cambridge University Press, Cambridge
23. Shamoo AE, Resnik DB (2003) Responsible conduct of research. Oxford University Press, Oxford
24. Svendsen MN, Koch L (2008) Unpacking the 'spare embryo': facilitating stem cell research in a moral landscape. Soc Stud Sci 38(1):93–110
25. Waldby, C, Mitchell R (2006) Tissue Economics. Blood, Orgns, and Cell Lines in Late Capitalism. Duke University Press, Durham
26. WMA (2004) The World Medical Association, The Declaration of Helsinki. Available via: www.wma.net/e/ethicsunit/helsinki.htm

Tissue Engineering and Regenerative Medicine. Their Goals, Their Methods and Their Consequences from an Ethical Viewpoint

E. Schockenhoff

Contents

5.1 Introduction 47
5.2 Criteria for Ethical Value Judgments 48
5.3 Application to the Procedures of Tissue Engineering and Regenerative Medicine ... 49
5.3.1 Justification of the Goals 50
5.3.2 Scrutinizing the Means and Methods 51
5.4 Responsibility for the Consequences 54

5.1 Introduction

An ethical analysis of scientific research projects is not a retrospective corollary that for reasons of completeness must not be omitted from a manual on the basics of tissue engineering and regenerative medicine. It is, rather necessitated by man's self-image as a moral protagonist, obligated to justify and take responsibility for his/her actions. The issue here is not solely the personal responsibility of the individual scientist or individual moral intuitions. Insofar as they promulgate the moral standards of the society concerned and its culturally accepted ethos, moral intuitions have *prima facie* an important function for moral judgments. Considerations of social acceptability, which particular research procedures may or may not find in a given historical situation, cannot, however, replace an ethical analysis of research practice.

This has to be subject to rational validation of arguments and practical conclusions; it does not ask about the chance of social acceptability or the actual practicability of research interests, but about normative validity, about what should be the case because it can be justified in a rational discourse with inter-subjectively verifiable, generalizable arguments. An ethical analysis accordingly targets not only the offering of advice to academic organizations or political decision-making bodies, but also the individual's capacity for moral judgment. It comprehends itself as a guide to better moral judgment, by encouraging critical reflection on one's own moral intuitions, checking their validity in a reflective process of argumentation, and ultimately revising them where appropriate (or finding them confirmed on a reflective meta-level).

Science and ethics were for a long time regarded as two consecutive processes: the thesis involved was that ethical reflection always relates retrospectively to what has first been researched and discovered, and already established itself in broad areas of scientific practice. The model of a successive sequence of research and ethics, however, proves inadequate for several different reasons. As in the fairytale of the hare and the tortoise, ethical reflection always comes too late here; it finds itself in the dilemma of either issuing ineffective prohibitions on action or justifying retrospectively what is a barely tolerable borderline case long since being practiced in numerous research laboratories. Moreover, in the paradigm of temporally successive research and ethical assessment, it is overlooked that in the execution of scientific practice itself both of them have to intermesh as soon as the researchers involved reflect on what they are doing and their clients consider their expectations. Splitting the research process into two successive subfunctions, which due to professionalization, furthermore, fall under different career categories (scientists and doctors on the one hand, philosophers, theologians, and jurists on the other), remains unsatisfactory for all the parties involved. This is why more interdisciplinary intermeshing of science and ethics is desirable. A model of this kind of research-concurrent ethical expertise aims at ensuring the integration of scientific research and ethical reflection at as early a stage as possible, so that this can already become prospectively effective in the choice of research objectives and research methods.

5.2 Criteria for Ethical Value Judgments

Human action differs from biological processes or physical phenomena in nature by reason of its intentional structure. It does not originate from causes, but is determined by reasons that guide the actor; it is teleological. Humans have to account to themselves and to others for the goals pursued in their actions. This constitutes a crucial differentiating feature between actions and natural processes. While a human being can and must be responsible for his/her actions, there is in the case of natural processes no court of appeal in a position to do this. This is why moral judgments that draw conclusions on the permissibility or inadmissibility of corresponding human actions directly from biological facts or the circumstance that something does not occur in nature are based on a category error. This is known as the naturalistic, or in the converse case as the normativistic fallacy.

From the circumstance that nature deals rather carelessly with the early embryonic phases of human life, it is accordingly erroneous to conclude that humans can follow nature's example in this regard; nor does the biological fact that nature among the mammals knows clone formation through embryo splitting entail a moral right of humans to imitate it. A description of biological processes can always lead only to descriptive insights, whereas moral judgments make prescriptive statements on what should or should not happen. Biological facts are, admittedly, morally relevant in that a knowledge of them is presupposed in moral judgments. Only thus can an ethical analysis verify the facts it is adducing. But biology does not provide any information about the criteria and evaluative standards on which the ethical judgment should be based. For this, a consensus is needed on man's self-image and the significance of his anthropological implications (physicality, imperfection, finitude, sociality) and his normative ideas of justice, which require that the perspectives of all parties involved be incorporated in one's own value judgment.

Every action is characterized by two aspects, each of which is significant for its moral assessment: an action can be teleologically regarded in terms of its externality, in *what* the actor does, or its internality, in *why* he/she is doing it. The intention of an action must not be confused with a mere wish or an emotional predilection for the contents of our wishes. When a person intends a particular action, he/she performs an act of inner self-determination, in which he/she makes a selection from among his/her wishes, attitudes and emotions, and decides the goals on which his/her actions are to be focused. In contrast to a mere wish, the conscious intent of an action presupposes a rational scrutiny of goals and their rational justification.

Although the intent of an action and the justification of the objectives being pursued in it play a paramount role among the criteria for moral judgments, it is not sufficient simply to ask for what purpose something is being done. An intended action has not yet been completely described if we merely state the goals we aim to achieve thereby. Rather, the moral assessment of actions must also include a judgment of the means by which we wish to attain our purposes. Within the double-poled structure of human actions, good motivation and the intent to achieve high-ranking goals relate only to one aspect, which needs to be supplemented by another. The moral assessment of an action's externality, of *what* we do in pursuing our goals, is indispensable, because others may be affected by our actions, whose rights are not nullified by the high-ranking desirability of the goals we are pursuing. Besides justifying our goals, then we are also called upon to legitimate the means selected or a particular method of research.

When the choice of means is examined, the permissible category will exclude those methods that imply a violation of human dignity or fundamental human rights. Due to the multifaceted meanings of the term "human dignity," an ethical judgment requires that the normative core content of human dignity be tightly defined as a minimalistic term, enabling a practical consensus to be reached. It accordingly violates the dignity of a human being to utilize him/her as a mere object of someone else's will ("prohibition on objectal relationships") and to use him/her solely for achieving extraneous purposes irrelative to his/her existence ("instrumentalization prohibition"). In relation to the research procedures of regenerative medicine, the question must accordingly be raised of whether the various forms of obtaining human stem cells make an embryo into a mere object at any phase of the production process, and reduce its existence to the achievement of extraneous purposes. A judgment on the moral permissibility of research with embryonic human stem cells accordingly presupposes an answer to the question of whether we regard the human embryo outside the mother's body as a human being, to whom we must, irrespective of the early stage of its development and its as yet unformed human shape, grant the fundamental rights to which every person is inherently entitled, i.e. without any further preconditions and the presence of additional characteristics.

Besides justification of the goals and a scrutiny of the means, the third criterion of ethical judgment requires responsibility to be taken for the foreseeable consequences of an action. Usually, human actions bring about not only the consequence intended as the goal, but also other consequences, perhaps also harmful ones, whose weighting has to be incorporated in the overall judgment on the action concerned. In medical ethics, responsibility for the consequences of your own actions is entailed by the maxims of consideration for the patient's well being and the avoidance of harm. In these classical constellations, the task is to weigh up therapeutically desirable consequences against unwanted side effects, both of which affect one and the same person, namely the patient, to whom the doctor is primarily obligated. Questions of responsibility for the consequences of science can, however, above and beyond the acceptance of therapeutic risks, also arise where one's own actions lead to consequential problems of morality, whose solution is not foreseeable at the juncture of decision-making, as is, for example, the case with knowledge of the high number of superfluous embryos in certain forms of in vitro fertilization or the foreseeable expansion of the indication spectrum following an initially restrictive approval policy for preimplantation diagnostics. But it is also possible that current ethical reservations will become groundless, because research is developing in a different direction or the research results can be clinically utilized in an ethically unobjectionable manner. Thus the ethical objections to the use of EHS cells could prove nugatory, if these are no longer required and the adult stem cells were to suffice for treating diseases.

5.3 Application to the Procedures of Tissue Engineering and Regenerative Medicine

The three-stage model of comprehensive ethical judgment applies for both individual actions in the interpersonal category and for the collective actions of social institutions and for the research activities of the scientific community. These are subject to the same conditions for justification as are applied to every category of action: neither should the scientific community have more stringent stipulations imposed upon it because it is researching into cellular biology, which opens up hitherto undreamed-of options for manipulating human life, nor should it receive dispensation from compliance with ethical principles and fundamental norms of human coexistence (e.g. the prohibition on killing), because its successes promise to make some of humankind's ancient dreams come true. Like all other people in their private and professional lives, scientists are tasked with justifying the goals of their actions, with scrutinizing the appropriate means for achieving them, and deliberating on the consequences of their actions. Application of this tripartite approach to research projects and therapeutic models being discussed in the field of tissue engineering and regenerative medicine leads to a differentiated result in terms of ethical analysis.

5.3.1
Justification of the Goals

On the teleological level, no plausible ethical objections can be discerned from an attempt to use stem cell research in order to create replacement tissue that performs the function of the diseased organism at the defective point. Even if these options for therapeutic utilization are still a long way off, they do denote high-ranking research objectives that in many areas of medicine may lead to new therapeutic approaches. Should it one day actually be possible to selectively control the reprogramming of human tissue cells, so that they develop into skin cells, liver cells, myocardial tissue, or cells of the brain and the central nervous system, advances of this kind in regenerative medicine would be unequivocally welcomed in the interests of patients. It would be tantamount to a paradigm shift in medical thinking if functional organ failures due to disease or accidents were no longer to be remedied by artificial prosthetics or transplantation of donors' organs but by stimulating the organism concerned to form the relevant types of tissue anew "on the spot." In some areas (e.g. in the case of coronary diseases of the heart or of liver, kidney and pulmonary diseases), this elegant therapeutic approach could replace the present-day standard therapy, while in others it could open the way to substantially better or the first actually significant treatment results.

The greatest hopes at present are that researchers will succeed in reprogramming the body's own tissue stem cells and using them for healing strokes, paraplegia or degenerative nerve diseases like Parkinson's, Alzheimer's, or multiple sclerosis. It is also conceivable that comparative studies will provide insights into the growth factors for tumor cells, enabling more efficient and patient-friendlier therapeutic approaches to be adopted in all areas of oncology. Finally, the prospects for regenerative medicine in the field of plastic surgery merit a mention: if, as is already the case in part ("skin from the tube"), tissue can be successfully grown in order to stimulate the body into forming new skin cells, patients could be liberated from mutilations or other bodily disfigurement after illness, accident, or burns. The gains in quality of life thus achieved for these patients can be properly appreciated only within the framework of a holistic concept of health, also incorporating psychological factors of health like emotional equilibrium, an intact ego, or self-esteem in terms of their importance for self-perceived human health. Precisely because humans, as physical–mental creatures, find their personal identities only in the interaction of body and soul, so that the body is simultaneously the self-expression of the person involved and the medium of his/her self-depiction in the social environment, health must not be reduced to a symptomless functioning of the organism in its somatic context.

One possible objection to the goals of generative medicine points to the utopian character of an expanded definition of the term "health." If they were one day to become reality, could not the options for growing new organs lead to a redefinition of the medical profession's self-image, focusing no longer on healing illnesses, but on the optimization of human nature in the sense of enhancement and anti-ageing medicine? Does not regenerative medicine also nourish the hope that with the aid of grown organs we shall ultimately be able to conquer death or at least repeatedly postpone it? The horror scenarios in which everyone stocks up with organic spares for his/her own body, and parents are held responsible for genetic optimization of their children even in the act of procreation, underline the relevant fears. But these are utopian hopes for the future, without any foundations in the anthropological condition of the human as a finite, flawed creature. Conquering the ageing process and death forever, and thus transcending the barriers of finite lifetimes in a new drive for potential immortality—this is a dream that people of every era have dreamed. But, as the shift in the disease spectrum and the emergence of new affluence-related diseases, in conjunction with the path-breaking successes of modern medicine since the 19th century, go to show, it is a vain dream, aiming to eradicate the *conditio humana*. The advances in medicine have at all times also aroused unrealistic expectations of healing, to be followed by subsequent disillusionment. When the prospects of regenerative medicine inspire people's imagination, and—for better or for worse—encourage utopian dream worlds or horror scenarios, philosophers and theologians are called upon to elucidate the meaning of life's finitude, limitations and imperfections for the comprehension of human existence. But this does not produce a normative argument against particular medical research concepts. The anthropological problems entailed by an ambitious medicine geared to improving human nature and eradicating its constitutive limitations could be adduced at any stage of scientific progress, and accordingly do not constitute a specific objection to the goals of regenerative medicine, but rather formulate a disquiet that places the ongoing innovations in medical thinking under a general suspicion.

5.3.2
Scrutinizing the Means and Methods

The insight that not everything that is desirable or technically feasible is also ethically acceptable denotes a potentially consensual starting point for ethical debates on biomedical issues. This consensus, however, gives way to controversial assessments as soon as the need arises to specifically define where the borderline runs between the morally permissible, the still-acceptable and the ethically counter-indicated. In most cases of bio-ethical conflict, the issue is not whether the goals of biomedical research are justified, but rather the ways in which they may be permissibly achieved. In the field of regenerative medicine, the paramount question is whether research into cellular biology and tissue-growing is to be performed only with the body's own adult stem cells, or whether it may also have recourse to embryonic stem cells. It exceeds the genuine competence of philosophical or theological ethics to make a value judgment on how far the development potential and differentiability of adult and embryonic stem cells are actually comparable; nor can the normative action sciences use their criteria to assess whether the findings obtained in an animal experiment on the reprogramming of a mouse's stem cells can be transferred to human beings. Ethics can neither itself decide controversial scientific issues nor may it simply assume as realistic the variant more favorable for it.

Ethical research, however, can from the available results of stem cell research draw conclusions on how far the original intrascientific estimations have been confirmed during the further course of the research process. One important argument for the alleged lack of alternatives to research with EHS cells was that the mechanism of reprogramming for differentiated body cells can be comprehended *only* in this way; another disadvantage of adult stem cells was perceived in the fact that their potential for reproducing and their ability to form cells of different tissue types were said to be very limited. Both arguments can be regarded as refuted since mouse cells have been successfully reprogrammed so as to be able to form tissue of all three blastodermic layers in a comparable way to EHS cells. Proof of principle has thus been achieved for such reprogramming of adult stem cells, usually regarded as a breakthrough. The successes announced in 2007 in obtaining induced pluripotent stem cells (= IPS cells), procured from connective-tissue cells, demonstrate that this is also a realistic research scenario applicable to humans. IPS cells created artificially possess the potential to develop into specialized types of tissue like nerve or liver cells, but are not totipotent, i.e. they cannot develop into a new individual as an adult human being. The advantage of such IPS cells over embryonic stem cells is that there are no ethical reservations about the method of obtaining them and the nature of their use.

What weight is attached to the ethical reservations against research with embryonic human stem cells? Since obtaining them with present-day production processes necessitates the destruction of human embryos, the answer to this question will depend on how the moral and legal status of human embryos is assessed. The decision on the status issue has to be taken independently of whether the result meets the interests of the research and scientific communities or not. But it is inadmissible to opt for a "balancing" process in which we assign to the embryo, in dependence on extraneous utilization aspirations, a moral and legal status that ignores the embryo its own perspective. Rather, the irreversible asymmetry of the judgmental level—we, who have already been born, are deciding upon the viewpoint under which we shall regard an embryo's early phases of life—obligates us to use particular caution and to provide advocatory representation of the embryo's concerns within the context of our own judgment. Only when this is done from an impartial standpoint, incorporating the interests of the research community in having a human life's entitlement to protection beginning at the latest possible juncture, can it be regarded as morally sound.

Advocatory representation of the embryo's position against the interests of the scientific community or patient groupings is a prescript of impartiality and thus of justice; it cannot be relativised by the statement that particularly high-ranking goods are at stake on the part of the scientific community, or that patients suffering from potentially fatal diseases are seeing their livelihood at risk. The moral right to kill in self-defense presupposes that the attacker poses a danger to the life of the person under threat, one that can be averted only by killing the aggressor. However, the embryo does not threaten anyone; on the contrary, it itself is an innocent creature requiring protection; one, moreover, that has been brought by specifically human action into the precarious situation of its current existential context. If, in specifying the temporal beginning of its entitlement to protection, there should be some latitude in terms of the human-biological basics (e.g. between the completion of the fertilization cascade and the beginning of nidation), this must not be tacitly utilized to the embryo's disadvantage. Ethical rationality rather dictates a search for the least arbitrary juncture from an impartial point of view. The findings of modern-day genetics, particularly the discovery of DNA and the process for its recombination in the process of fertilization, favor the conclusion that this least arbitrary juncture coincides with the fusion of the ovum and sperm cells. Creation of a new, unique genome constitutes a qualitative leap in which, compared with the separate existences of the ovum and sperm cells interacting in the process of procreation, something radically new and underivable comes into being. It accordingly seems reasonable to give preference to the completion of fertilization over later junctures that denote other maturation processes or the overcoming of critical danger zones. Compared to later definitions, it accordingly appears appropriate to regard the completion of fertilization as the juncture that marks the beginning of an individual human's life.

On the basis of current findings in developmental biology, it can be assumed that the embryo exists from the very beginning both as a specific species (as a *human*) and as a specific individual (as *this* human). From the moment of fertilization, it possesses the unique hereditary dispositions that it can unfold in a continuous process without any relevant disjunctions, provided it is offered the requisite environmental conditions. This also applies as a basic principle for the extracorporeal embryo. Its alleged lack of a chance to develop is not to be construed as an innate defect, as a missing quality, or as an intrinsic ontological deficit. Rather, a superfluous embryo is cut off from its chance to develop only because this chance has been withheld by its progenitors. The argumentation that a superfluous embryo is no longer destined for life and accordingly loses its moral and legal entitlement to protection, stands the logic of an admissible moral approach on its head: if the options provided by reproductive medicine enable human life to be created outside the mother's body, this must not be allowed to lead to our giving the in vitro embryo we have created a lesser degree of respect than is merited by an in utero embryo created in a natural act of procreation. It follows that parents and doctors must not treat the extracorporeal embryo as the end result of a production process, that they can dispose of as they please to suit their own ideas and interests. Since they have brought the embryo into its current position by their own actions and by using the new options of biotechnology, they are, on the contrary, obligated to ensure that it receives a chance of development comparable to the natural process of procreation.

The relationship of the progenitor to the embryo created should in moral terms not be one of unilateral exploitation, but one of recognition, anticipating the embryo's future opportunities in life. Given an approach of this nature, the artificiality of the embryo's creation does not change the fact that the biological process initiated by the progenitors can conceal the beginning of a story narrating the personal freedom of a subject whom they may subsequently encounter at a later stage of his/her development as a fully equal partner for interaction. When the empirical findings concerning the beginning of human embryos' development are interpreted in the light of normative premises such as human dignity, the principle of equality or the prohibition on killing, it emerges that human life, right from the start, i.e. from completion of fertilization, falls under the protection of human dignity, something to which every human being is entitled from the origination of his/her existence and which demands that his/her existence be respected for its own sake. Insofar as life is an ineluctable precondition for moral self-determination, and has to be regarded as the existential foundation for a person's origination and development, the dignity guaranteed by legal codes in democracies demands the assurance of effective protection for human life. Protection of human dignity and protection of human life are accordingly inseparable.

For the life of human embryos, it follows that even in the early phase of their existence and in their extracorporeal location they must remain outside the applicability of a balancing of goods. Since as far as the embryo is concerned, the issue is not one involving a greater or lesser degree of acceptable restrictions, but the totality of its existence, the concept of a graduated protection of life gives it no protection in cases of doubt. A graduated, slowly evolving entitlement to protection, by its very nature, cannot take effect in a possible balancing of goods against high-ranking goals, as basic scientific research or the possible saving of seriously ill patients indubitably are; it offers the embryo no protection, in view of the extreme danger to its existence. Permitting a balancing of possible goods in favor of high-ranking research goals would be tantamount to arbitrary unequal treatment of the embryo compared to born humans, violating the former's fundamental rights.

The phrase occasionally used by scientists as well, describing the embryo as a mere conglomeration of cells, which cannot as yet be a human being "like you and me," overlooks the circumstance that human dignity and the right to life are not tied to phenotypical characteristics (like the size of the bodily integument) or a particular phase of a human's life. If each and every individual is entitled to the rights owed to human beings by nature without their effective observance requiring further characteristics, abilities, or evidence of capability, the conclusion is irrefutable: neither the age (whether at an earlier or later juncture of ontogenesis) nor the location of an embryo (whether in vitro or in vivo) supplies a cogent criterion for differentiation that could justify its exploitational use for purposes of research. For recognition of its dignity and its right to life, it is immaterial whether a new human being exists as a zygote, as an embryo, as an infant, as an adolescent, as an adult at the zenith of his/her life, or as an elderly person. Entitlement to some civil freedoms like the right to vote or the right to testify comes only from a certain age, while others can be legally withdrawn again by reason of illness and accident (like the right to conduct your own affairs). But the graduation of civil rights status does not affect his/her inherent identity as a human being, which forms the basis for recognition of those fundamental rights that protect each human individual irrespective of all further differentiating factors.

These considerations lead to the result that for ethical reasons regenerative medicine should refrain from utilizing embryonic human stem cells. The trialling of promising research options may become a moral impossibility if it entails violating elementary rights and entitlements of others. We sympathize with the bandit Robin Hood, who steals not out of self-interest, but for the benefit of the poor: he is not a common thief, since thanks to his virtuous motivation his deeds take on a degree of dignity, nobility, and magnanimity. Nonetheless, we know that robbery is not a morally and legally permissible method for helping the poor. The same principle also applies in the moral conflicts that may arise in the field of regenerative medicine from the use of embryonic human stem cells. It is true that doctors and researchers are here pursuing high-ranking goals; they are conducting basic or therapeutic research in order to comprehend the origination of diseases and to develop new forms of treatment. An ethical assessment of particular research approaches, however, cannot be based solely on the researchers' intentions. A moral point of view rather is arrived at only when the concerns of all parties involved are impartially factored in. In this context, the rule of preference applies, that the preservation of elementary rights must take priority over possible assistance for others in the event of a conflict. The protection of fundamental rights—particularly the right to life, unequivocally owed to every innocent human being—outweighs the putative beneficial effects for others.

In the event of a collision between positive duties of assistance (virtuous duties) and negative duties of omission (legal duties) that cannot be resolved by adherence to a temporal sequence, in line with the motto "first what's urgent, then what's postponable," the no-harm principle must accordingly take priority over providing assistance. That one's own life may not be preserved at any price, i.e. in the event of a conflict also by the violation of another's rights, also applies for a seriously ill human being. The right to healing, which subsumes research and experimental utilization of new therapeutic procedures, comes up against its limits where its implementation would necessitate the destruction of another's life.

A moral assessment of research approaches demands an impartial stance, regarding all parties affected as equal participants in a shared life experience, and none of them as a mere object to be exploited by others. For those who may be the beneficiaries of biomedical research, the capacity for ethical judgments demands the ability to assess one's own expectations from the perspective of others. Specifically, this means: an ill person may wish to be healed and utilize every conceivable chance for healing. It is, however, a moral impossibility to wish, i.e. to intentionally, above and beyond merely wishing for the outcome, assent as an actual means to another person's suffering harm, and the harm he/she suffers being the price of one's own advantage, the obverse of one's own benefit, the necessary loss in one's own gain. Acquiring the capacity to make your own moral judgments begins with the perception that there are other people who stand in the way of asserting your interests and achieving your wishes, and who nonetheless possess a justified perspective of their own, which you as a morally acting person must not ignore even if you find yourself in an existential emergency.

5.4 Responsibility for the Consequences

For an ethical assessment of the question as to whether a course of action or a research project is permitted or prohibited, the presumptive consequences are also relevant. It is inadmissible to refuse their observance with a generalized reference to the speculative character of slippery-slope arguments. Rather, wherever the protection of high-ranking goods like human dignity and life is at stake, preventive responsibility ethics shall be given preference when assessing the probability of the feared consequences actually occurring. The rule that risk prognoses shall proceed from the less favorable prognosis (*Hans Jonas*) wherever high-ranking goods are irreversibly at risk does not, however, mean that this may simply be assumed without any empirical evidence. Rather, the predictions must be continually reviewed against actual developments, and appropriately corrected if they prove to be groundless.

In contrast to the case of cloning for biomedical research purposes, consequence-driven arguments meanwhile play only a minor role in evaluating research with adult or embryonic stem cells. While many people reject research cloning for the reason that they see it as entailing the risk of far-reaching instrumentalization of women as egg donors, or because they fear that the further development of cloning technology may encourage attempts to grow humans, developments in the field of stem cell research are currently being assessed less controversially. Now that IPS cells have been successfully obtained using a process that manages without the destruction of human embryos and does not create any totipotent development potential, leading researchers have announced that in future they will be more frequently opting for a cell programming approach (instead of cell nucleus transfer within the framework of cloning research). Those research groups, too, who despite all ethical reservations regard research with classical EHS cells as (still) indispensable are announcing that this is to be the case only for a transitional period until the requisite comparative experiments between humans and animals have been completed. Since in the case of a possible therapeutic use on humans, the utilization of adult stem cells anyway merits preference, due to their lack of an immunological rejection reaction and their lower tendency to tumor formation, the risk of an uncontrollably expanding production of embryonic human stem cells can at present be classified as improbable. In view of the advances being made by research in cellular biology and its successes in reprogramming mature tissue cells, the specter of industrialized mass production of embryonic stem cells in embryo factories lacks plausibility.

The most recent advances in reprogramming technology for adult body cells, however, also render obsolete the thinking with regard to the creation of what are known as research embryos. Consumptive research on embryos created specifically for this purpose is unequivocally a violation of human dignity, in that not only are human beings in the early stage of their development being used for extraneous purposes but are also being created solely for purposes of their instrumentalization. When even the decision to create a person is governed not by the wish for a pregnancy but from a preconceived intention to destroy the person concerned, his/her entire existence is reduced to the attainment of extraneous purposes. The reification of human life, which the prohibitions on objectal relationships and instrumentalization are designed to preclude, would here be exemplified in the highest conceivable degree of flagrancy, since a human being would be created solely for a purpose foreign to its own existence and reduced completely to its suitability for serving this purpose. Uncompromising rejection of the creation of research embryos merits special mention, because in the past there have repeatedly been voices raised that regarded even consumptive research on such embryos, either now or at a later juncture, as scientifically advisable and ethically acceptable. For the ethical reasons already outlined, propositions of this kind must be unequivocally rejected. The creation of human embryos for research purposes is expressly prohibited in Article 18.2 of the European Convention on Human Rights and Biomedicine.

An ethical assessment of alternative research concepts cannot make any statements of its own on their scientific suitability for achieving particular findings, but must in this regard rely on the professional competence of the scientists concerned and their (frequently divergent) evaluations. The most recent development in the field of reprogramming research, however, demonstrates that the prospect of being able to obtain the hoped-for findings in an ethically acceptable manner is greater than many people have hitherto assumed. This should strengthen the readiness among everyone involved to grant genuinely ethical considerations a greater weight in the development of research concepts. Ethically harmless research alternatives must from a moral perspective be given fundamental priority over those subject to severe ethical reservations. Scientific research groups are facing the question of what weighting they wish to give to considerations of morality when planning their research strategies. Their alternatives here are to opt right from the start for ethically harmless research concepts or to adopt a pragmatic approach and trust that the breaching of moral barriers will ultimately pay off, because the hoped-for success will retrospectively justify the means. The successes of regenerative medicine, however, show that researchers do not have to choose between morality and success, but that successful research is also possible in a morally acceptable manner, and has already produced impressive results. The more ethically acceptable research approaches compete with each other, the more probable appears the prospect that the hoped-for gains from advances in regenerative medicine will actually materialize.

Part B
Biological Considerations

II Tissue and Organ Differentiation

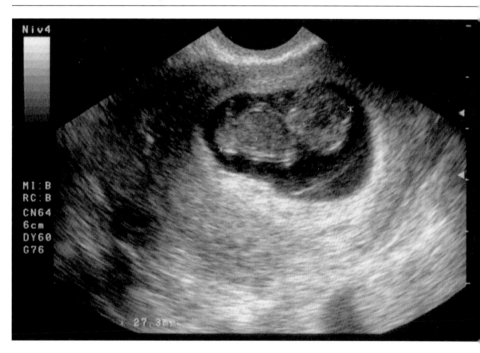

Control of Organogenesis: Towards Effective Tissue Engineering

M. Unbekandt, J. Davies

Contents

6.1 Introduction 61
6.2 Features of Natural Organogenesis 61
6.3 Engineered Organogenesis 63
6.3.1 Extracorporeal Tissue Engineering 63
6.3.2 Intracorporeal Tissue Engineering 64
6.4 A Longer-Term Vision for Tissue Engineering: Synthetic Morphology 66
References 67

6.1 Introduction

The word "Organogenesis" is defined as "the production and development of the organs of an animal or plant" [1]. In the context of medical research, it has traditionally been applied to the natural processes of fetal development but it is now beginning to be applied also to the creation of living organs, or organ substitutes, by artificial means. It is this latter meaning that is most relevant to this book and most of this chapter will therefore focus on artificial organogenesis. It will be helpful, though, to review the basic features of natural organogenesis first, because the most successful methods of artificial organogenesis tend to build on them.

6.2 Features of Natural Organogenesis

Each organ of the human body forms in its own way but decades of research, mainly in mice, have revealed some constant themes. One is that most organs develop relatively autonomously; rudiments of lungs, salivary glands, kidneys, prostates, etc., can be removed from embryos and placed, in isolation, in organ culture where they will grow organotypically in relatively simple media with no influence from other embryonic tissues [2]. In the context of a real embryo there is, of course, some communication between different organs of the body, which is responsible for keeping their development in step, amongst other things, but the ability of the organs to develop to a large extent in culture demonstrates the extent to which the information required for controlling organogenesis resides within an organ itself. This point is critical to the whole enterprise of tissue engineering.

Fully-formed organs contain many cell types and have complicated anatomies. Visceral organs, with which this review is mainly concerned, typically contain epithelial tubes, which may or may not be branched, smooth muscles, vascular endothelia cells, neurons, and stroma. Each of these broad categories may include several different cell types. For example, epithelial tubes may contain simple cells involved simply with building "plumbing" and also specialized cells that undertake various types of excretion or solute exchange. Their rudiments, though, are usually simple and consist of very few tissue types. The metanephric ("permanent") kidney, for example, consists initially of two tissues, an epithelial tubule called the ureteric bud and a surrounding mesenchyme. The epithelial tubule apparently consists of two cell types ("tip" and "stalk") and the mesenchyme may consist of just a single cell type. Over the course of development, these give rise to the approximately 100 distinct renal tissues listed in the recently-published ontology of the kidney [3].

The progressive differentiation of cell types during organogenesis, and their action in creating tissue shapes, is coordinated by a number of paracrine and autocrine signaling molecules. Many of these mediate communication between mesenchymal and epithelial components, mesenchyme-derived signals being required for epithelial development and vice versa. In many organs, epithelial development depends on mesenchymal secretion of members of the fibroblast growth factor (FGF) family, particularly FGFs 7 and 10 each of which signals mainly via FGFR2IIIb. FGF7 is used mainly in the seminal vesicle, salivary gland, and ventral prostate [4–7]. FGF10 is important in lung, thyroid, pituitary, prostate, salivary gland, pancreas, and lachrymal gland [6, 8–10]. FGFs are also of some importance to renal development [11] but here the most important pathway for mesenchymal-epithelial signaling uses the ligand GDNF and its receptor Ret [12–14]. Wnt molecules also control epithelial development in many organs, such as lung, kidney, and lachrymal gland [15, 16]. Members of the TGFβ superfamily, including TGFs themselves and BMPs, are also involved, sometimes acting positively and sometimes negatively on epithelial growth and branching [17–19]. Signals also pass from the epithelial components to the mesenchymal components. These signals include sonic hedgehog (shh) [20–22], Wnts [23] and FGFs [24].

The extracellular matrix is also very important to survival, differentiation, and morphogenesis of cells within developing organs [25]. Fibronectin is required for branching morphogenesis of epithelia in kidneys, lungs, and salivary glands [26–28] and interstitial collagens can also play an important role in defining the clefts that separate new branches [29]. Other widespread matrix components such as laminins are also required [30], as are more organ-specific molecules such as nephronectin [31]. Specific matrix receptors, such as integrins, are also important and organ development fails in their absence [30, 32]. The correct balance of matrix deposition and matrix destruction is also critical, and is controlled by secretion of proteases and protease inhibitors (Fig. 6.1) [33].

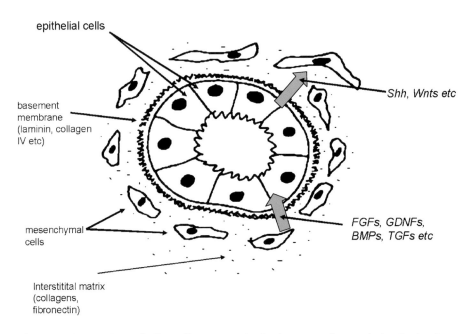

Fig. 6.1 Molecules that typically mediate communication between cell types during the development of typical organs

On the face of it, then, controlling organogenesis artificially would require an ability to manipulate signaling molecules, matrices, and matrix receptors to a degree far beyond the limitations of our current technology. It turns out, though, that developing organs make extensive use of adaptive self-organization; the genome does not specify their precise architecture, but rather it specifies the machinery that allows cells to organize themselves into organotypic (but not exactly identical) arrangements [34]. It is therefore possible to provide cells with some of their requirements, for example critical survival factors and a minimal matrix, and allow them to organize a full matrix and a full conversation of signaling molecules and differentiated states themselves. This is particularly true when one starts with uncommitted stem cells, but it to a large extent true even with primary cultures of cells from adult organs.

Epithelial cells from tubules in organs, for example, tend to associate with one another as an epithelial sheet even in two-dimensional cell culture and will, when provided with an appropriate three-dimensional matrix and a few signaling molecules, spontaneously organize themselves into cysts or tubules (Fig. 6.2a) [35–37]. Mixtures of epithelial and mesenchymal cells will, in the right culture conditions, tend to sort spontaneously to form epithelial cysts or tubes surrounded by mesenchymal tissue (Fig. 6.2b) [38]. Furthermore, they secrete the signaling molecules that they would secrete in normal development and this is potentially important if the cells are transplanted into a host animal. Production of growth factors such as VEGF, for example, can recruit and organize endothelial cells of a host to serve the needs of the transplant [39–42]. Each of the techniques for engineering organogenesis for clinical purposes that will be described below makes extensive use of cells' ability to organize themselves and/or cells of their hosts.

6.3
Engineered Organogenesis

The ultimate purpose of engineering organogenesis is to provide a new functional organ or tissue to substitute for a missing or damaged organ in a human patient. Although the science is young there is already a great variety of approaches being taken. They can be divided, for convenience, several different ways; one of the clearest is that between production of extracorporeal organ substitutes and intracorporeal ones.

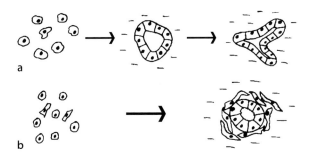

Fig 6.2a,b Spontaneous self-organization of cells from embryonic organs when placed in organ culture. **a** Epithelial cells from kidney, salivary gland, etc., will organize themselves to produce polarized cysts when placed in an appropriate three-dimensional matrices. These cysts will produce branching tubules in the presence of ramogens such as hepatocyte growth factor (HGF). **b** Mixtures of fetal lung cells placed in three-dimensional matrices organize themselves spontaneously into epithelial and mesenchymal compartments, with mesenchyme organized organotypically around the basal side of epithelial cysts

6.3.1
Extracorporeal Tissue Engineering

Extracorporeal devices support the life of a body from outside that body. The most primitive examples predate the advent of tissue engineering by many decades and are based on established techniques of mechanics, hydraulics, and electronics. Examples are the negative pressure ventilator ("iron lung," invented in 1928) and the renal dialysis machine (invented in 1943). Abiotic renal dialysis machines can perform the filtration function of the kidneys with respectable efficiency but they cannot easily perform the feedback-controlled biosynthetic functions of an intact kidney that are responsible for the carefully-controlled production of renin, erythropoietin, and the deiodination of thyroid hormone [43, 44].

Addition of a bioreactor containing renal cells to the dialysis circuit helps to substitute for these functions and improves the clinical performance of dialysis machines [45–47]. Renal cells require a proper substrate and, for this purpose, standard hemofiltration capillary tubes can be used, the renal cells being grown on the luminal surface.

This problem is even more acute for liver, the metabolic functions of which cannot be substituted by simple filtration. Some purely mechanochemical techniques such as toxin adsorption on to activated charcoal [48] can be helpful in acute conditions but regulative and biosynthetic functions of the liver can be provided only by living cells. The extracorporeal approach to this requires flow of substances between a patient's blood or plasma and hepatic cells maintained in a bioreactor. Liver cells will generally not function in suspension so need to be provided with a suitable substrate, and this needs to allow very high densities of cells to be maintained since a high mass of hepatocytes is needed to substitute for liver function in an animal as large as a human. In most current approaches, these needs are met by the use of microporous hollow fibers, the outsides of which are covered with human liver cells in primary culture, these cells generally coming from livers that are unsuitable for use in transplantation [49]. When mixtures of human liver cells are used, they generally associate spontaneously into aggregates that include parenchymal and nonparenchymal cells and that form sinusoid-like structures [50]. Bioreactors like these (which are used in combination with dialysis) can support patients for at least 144 h [51]. Similar bioreactors have also been produced with porcine hepatocytes and have been used to support human life [52]. An alternative to the use of capillary fibers is the use of micro-encapsulated aggregates of hepatocytes [53], although this is less common.

6.3.2
Intracorporeal Tissue Engineering

There are three main approaches to intracorporeal tissue engineering: (i) the use of individual cells, (ii) the use of cells already attached to a scaffold, and (iii) the use of developing organs (Fig. 6.3).

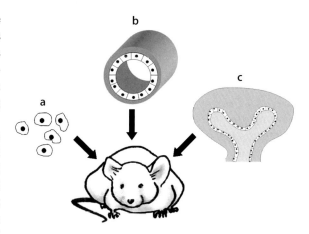

Fig. 6.3a–c The three main approaches currently under study for intracorporeal tissue engineering devices. **a** Cells can be injected intracorporeally to try to regenerate or repair an organ. **b** Cells can be seeded on an "organ-shaped" scaffold in order to give them a shape before intracorporeal transplantation. **c** Another approach consists in transplanting developing embryonic organs into adults

The first method generally works by injection of cell suspensions into the body, either topically or systemically. It relies on the assumption that, if the cells are placed in the right chemical and physical context, they will be able to undergo organogenesis and/or to integrate in organs during the natural healing process. For this purpose, stem cells are particularly promising candidates by virtue of their pluripotency and potential for proliferation. When, for example, a single mammary stem cell is injected into a mammary fat pad, it is able to reconstruct a functional mammary gland [54]. In culture, embryonic stem cells can form embryoid bodies that include cells that normally derive from each of the three embryonic germ layers [55], and they can also give rise to a rich variety of cell types in tissue engineering. When they are injected into embryonic kidneys growing in culture, they contribute to mesenchyme and integrate to epithelial tubules [56]. Observations like this have given rise to the idea of "preculturing" ES cells in environments that cause them to differentiate into the cell types characteristic of the organ to be regenerated or repaired, improving the efficiency of the regeneration process and possibly restricting the potential of the cells to make trouble in the body by differentiating into unwanted structures.

Preculture of ES cells need not be done in the complex environment of a cultured fetal organ, and at least for some cell types a cocktail of growth factors can be used (generally those produced in the normal developing organ). Again working in kidney, experimenters have used combinations of retinoic acid, BMP4, BMP7, or activin A to induce ES cells to differentiate into cell types that form tubules when injected into embryonic metanephros [57, 58]. Direct transfection of cells with transgenes can also be used to program them. This has been demonstrated by transfection of ES cells with *Wnt4*, a gene encoding for an autocrine signaling molecule involved in normal kidney development. This induces them to form a cell type that expresses the renal channel *aquaporin* 2. After injection under the cortex of adult mouse kidneys, these cells form teratomas containing tubular structures [59]. Clearly, the teratoma-forming tendency of the cells would not be clinically desirable, and if programming by transfection is to be used, genetic systems must be designed carefully so that they do not remain active after they are needed. The history of gene therapy is plagued with problems such as this. For example, the treatment of patients with severe X-linked immunodeficiency with bone marrow stem cells transfected with a retroviral vector led to a long-term significant improvement of their conditions for nine out of ten patients but also to leukemia for two of them [60].

The bone marrow is a source of circulating stem cells in the body that can repopulate the whole hematopoietic lineages of an irradiated host—indeed bone marrow transplantation is by far the most effective method of stem cell therapy in current clinical practice. Bone marrow cells can also contribute to the regeneration of injured liver and heart [61, 62] The potential of bone marrow-derived stem cells (BMSC) to contribute to repair of other organs has been extensively studied but the experiments have led to conflicting results. Whole bone marrow transplantation or by cell injection has suggested on one hand that these cells can contribute to kidney repair after ischemia/perfusion injury and to basal turnover [63, 64] and on the other hand that they do not contribute significantly to kidney repair [65–67]. All of the groups involved in these conflicting results are known and respected for careful experimental work, and the divergent results suggest that the behavior of bone marrow cells may depend critically on precise experimental conditions.

It is now generally believed that even adult organs contain niches of adult pluripotent stem cells that can be used for healing after an injury. In uninjured organs, stem cells cycle very slowly and they can therefore be identified, at least tentatively, by labeling all nuclei with a base analogue such as BrdU and then searching for cells that are still labeled after many weeks. In the kidney, a population of such slow-cycling, label-retaining cells exists in the renal papilla [68]. These cells proliferate after ischemia/reperfusion injury and are thought to contribute to the repair of the organ. Isolation of these cells, and their production in large numbers, could have an important impact for future kidney cell therapy and much work is currently being done to develop efficient means to do this.

The second broad approach to intracorporeal tissue engineering consists of the seeding and culture of cells on a scaffold (sometimes biodegradable) before the cell-and-scaffold assembly is introduced to the body. The scaffold is used to give cells the right shape and/or to isolate them from the other body parts. The engineered "tissues" thus produced are unlikely to have the full range of actions of their normal counterparts, but they can perform specific functions well. An example of their use is to deal with a problem suffered by patients on hemodialysis. Extracorporeal dialysis, as has been mentioned, is efficient at removing small molecules from the circulation but it cannot remove larger molecules, such as β2-microglobulin, that are normally removed by living kidney cells. In an attempt to deal with this problem, [69] a collagen sponge, impregnated with basic Fibroblast growth factor to promote vascularization, was implanted subcutaneously in nude mice and was injected with cells that expressed megalin, which promotes cellular uptake of β2-microglobulin. The implant led to the significantly better control of β2-microglobulin in anephric rats.

This "cell on a scaffold" technique can be used to reconstruct complete organs and has been successful for the reconstruction of bladders, at least in animals. In order to obtain cells for bladder reconstruction, a biopsy of an adult dog bladder was performed and the collected cells were cultured to increase their numbers. They were then seeded on a bladder-shaped scaffold and were transplanted into the body of a host dog, with effective reconstruction of a functional bladder in at least some cases [70]. For the kidney, a similar process has been used in animals, in conjunction with therapeutic cloning. Dermal cells were collected from a notch cut in the ear of a cow, and their nuclei from bovine dermal cells were transfected into ovocytes. The ovocytes were allowed to develop after insemination and the resulting fetuses were harvested after 56 days. The developing metanephric kidneys were removed from these fetuses and their cells isolated and seeded on cylindrical polycarbonate membranes. These were then transplanted back into the cow that was the original donor of the dermal cells (and would therefore not produce immunological rejection). On the subcutaneous, transplanted scaffold, tubular kidney structures and glomeruli formed and they secreted a fluid with properties similar to urine, in terms of as concentrations of urea and creatinine [71]. This technique although very promising, could not be of course be applied directly to humans because of the obvious ethical issues surrounding the fetuses that were brought into existence only to have their organs harvested.

The third broad approach to intracorporeal organ reconstruction consists in the transplantation of developing embryonic organs, or parts of them, into adult hosts. This differs from the techniques already described in that it retains the tissue relationships of the donor cells, rather than reducing them to a cell suspension. The technique has been extensively studied for the kidney (for an review, see [72]). Due to the absence of a vasculature during the first stages of kidney development, antigen presenting cells are almost absent of the developing metanephric kidney and transplanted organs are remarkably free from rejection. Parts of embryonic kidneys have been successfully implanted in the renal parenchyma [40] and integrated with the host kidney. Whole organs have also been transplanted under the renal capsule or intraperitoneally [73], and the survival time of anephric rats has been increased by these grafts [74]. The survival time is proportional to the number of transplanted embryonic metanephros, illustrating their efficient renal function [75].

From a clinical point of view, the supply of embryonic organ rudiments presents a problem similar to that of adult donor organs, even more so if they have to be at a specific developmental stage. There are also serious ethical, as well as practical, problems in using material from aborted human fetuses. There is therefore a great interest in producing "embryonic organs" in culture, either directly from stem cells or from subcultured tissues from an original fetal donor. An example of this subculture has been demonstrated for the progenitor of the urine collecting duct system of the kidney, the ureteric bud. Isolated rodent ureteric buds can be grown intact in three-dimensional gels, where their tubules branch as they would in life. Parts of these branches can be cut off and placed in new gels, where they generate new branching systems of their own [76]. This technique could be an efficient way to generate in vitro multiple renal structures from one metanephros by recombination with mesenchymal cells, which could be cultured independently from the epithelium [77], before transplantation. One fetal donor could therefore yield many "fetal organs."

At the moment, intracorporeal tissue engineering remains an immature field. Much good work has been done, and several efficient techniques have been demonstrated but these are each like well-formed pieces of a jig-saw puzzle that have yet to be assembled into the full clinical picture. Some pieces have yet to be made. One of the most serious problems faced by the field is that the cell types that seem to be easiest to work with and that yield the most impressive results are those that involve harvesting from fetuses or involve cloning, neither of which is desirable in a human context. Those that involve simpler cell lines offer less flexibility, although remarkable things have still been achieved and these may well be the first systems in regular clinical use.

6.4
A Longer-Term Vision for Tissue Engineering: Synthetic Morphology

All cells described so far have been essentially normal, or have been altered only by the introduction of a single gene (as in the *Wnt4* example). Tissue engineering using them is therefore limited to using properties already present in body cells. This may change. Recent years have brought very rapid advances in our ability to engineer complex genetic control systems, a field now called "synthetic biology" [78], and they have also brought a rapid increase in our understanding of basic morphogenetic mechanisms [34]. Placing cellular morphogenetic modules under the control of engineered control systems would allow the programming of cells to produce designed morphological responses to signals and environments that are outside the repertoire of normal cells. Very primitive steps in this endeavor, called "synthetic morphology" [79] have already been made and progress is likely to be rapid over the next decade or so.

Synthetic morphology offers the possibility of creating interfaces between the body and "lumps" of conventional tissue engineering (imagine cells programmed to make tubules that connect fluid drains of tissue engineered kidneys with the natural bladder). It also offers the possibility of interfacing the body to electromechanical devices (imagine neurons programmed to connect synthetic limbs to the motor and sensory roots of the CNS). Synthetic morphology makes a clearer break between tissue engineering and conventional surgery, in that it is re-engineering the developmental program itself rather than using it, as surgeons use the wound-healing program, as it is.

It takes only a little imagination to conceive applications of advanced tissue engineering that could create entirely novel tissues and organs. Many humans already alter their body for decorative purposes, for example by piercing or tattooing. Could tissue engineering be used for decorative or vain purposes, either in a mild way, producing an endogenous tattoo for example, or enlarging certain organs. Should it? Could it also be applied in a much more radical way, giving decorative "antlers" to a human who wants them, for example? Should it? Mice have already been engineered with additional photoreceptors from other animals, so that they gain the "full" color vision that (most of) we humans take for granted. Could mice, and then humans, be made sensitive to other wavelengths that no mammal can see, perhaps using ultraviolet sensors from insects, if not in eyes than at least as a diffuse sensation in organs in the skin? Should they? Experience suggests that when a technology developed for serious and noble purposes becomes available, the artistic side of the human spirit tends to apply it in quite unexpected ways (the relationship between medical pharmacology and recreational drug use, or between the World Wide Web developed for scientific purposes in CERN and cybersex are examples). There is no reason to suppose that a mature and effective technology of tissue engineering will be any different. It may therefore be sensible for pioneers of tissue engineering to maintain a dialogue with ethicists so that, for once, they and law-makers will not be reacting to advances in biology too late, and with suspicion.

References

1. The New Oxford Dictionary of English (1998) Oxford University Press, Oxford
2. Grobstein C (1953) Inductive epitheliomesenchymal interaction in cultured organ rudiments of the mouse. Science 118(3054):52–55
3. Little MH, Brennan J, Georgas K, Davies JA, Davidson DR, Baldock RA, et al. (2007) A high-resolution anatomical ontology of the developing murine genitourinary tract. Gene Expr Patterns 7(6):680–699
4. Alarid ET, Rubin JS, Young P, Chedid M, Ron D, Aaronson SA, et al. (1994) Keratinocyte growth factor functions in epithelial induction during seminal vesicle development. Proc Natl Acad Sci U S A 91(3):1074–1078
5. Morita K, Nogawa H (1999) EGF-dependent lobule formation and FGF7-dependent stalk elongation in branching morphogenesis of mouse salivary epithelium in vitro. Dev Dyn 215(2):148–154
6. Thomson AA (2001) Role of androgens and fibroblast growth factors in prostatic development. Reproduction 121(2):187–195
7. Sugimura Y, Foster BA, Hom YK, Lipschutz JH, Rubin JS, Finch PW, et al. (1996) Keratinocyte growth factor (KGF) can replace testosterone in the ductal branching morphogenesis of the rat ventral prostate. Int J Dev Biol 40(5):941–951
8. Bellusci S, Grindley J, Emoto H, Itoh N, Hogan BL (1997) Fibroblast growth factor 10 (FGF10) and branching morphogenesis in the embryonic mouse lung. Development 124(23):4867–4878
9. Ohuchi H, Hori Y, Yamasaki M, Harada H, Sekine K, Kato S, et al. (2000) FGF10 acts as a major ligand for FGF receptor 2 IIIb in mouse multi-organ development. Biochem Biophys Res Commun 277(3):643–649
10. Makarenkova HP, Ito M, Govindarajan V, Faber SC, Sun L, McMahon G, et al. (2000) FGF10 is an inducer and Pax6 a competence factor for lacrimal gland development. Development 127(12):2563–2572
11. Bates CM (2007) Role of fibroblast growth factor receptor signaling in kidney development. Pediatr Nephrol 22(3):343–349
12. Sainio K, Suvanto P, Davies J, Wartiovaara J, Wartiovaara K, Saarma M, et al. (1997) Glial-cell-line-derived neurotrophic factor is required for bud initiation from ureteric epithelium. Development 124(20):4077–4087
13. Schuchardt A, D'Agati V, Larsson-Blomberg L, Costantini F, Pachnis V (1994) Defects in the kidney and enteric

nervous system of mice lacking the tyrosine kinase receptor Ret. Nature 367(6461):380–383
14. Schuchardt A, D'Agati V, Pachnis V, Costantini F (1996) Renal agenesis and hypodysplasia in ret-k- mutant mice result from defects in ureteric bud development. Development 122(6):1919–1929
15. Dean CH, Miller LA, Smith AN, Dufort D, Lang RA, Niswander LA (2005) Canonical Wnt signaling negatively regulates branching morphogenesis of the lung and lacrimal gland. Dev Biol 286(1):270–286
16. Merkel CE, Karner CM, Carroll TJ (2007) Molecular regulation of kidney development: is the answer blowing in the Wnt? Pediatr Nephrol 22(11):1825–1838
17. Michos O, Goncalves A, Lopez-Rios J, Tiecke E, Naillat F, Beier K, et al. (2007) Reduction of BMP4 activity by gremlin 1 enables ureteric bud outgrowth and GDNF/WNT11 feedback signalling during kidney branching morphogenesis. Development 134(13):2397–2405
18. Ritvos O, Tuuri T, Eramaa M, Sainio K, Hilden K, Saxen L, et al. (1995) Activin disrupts epithelial branching morphogenesis in developing glandular organs of the mouse. Mech Dev 50(2–3):229–245
19. Dean C, Ito M, Makarenkova HP, Faber SC, Lang RA (2004) Bmp7 regulates branching morphogenesis of the lacrimal gland by promoting mesenchymal proliferation and condensation. Development 131(17):4155–4165
20. Bellusci S, Furuta Y, Rush MG, Henderson R, Winnier G, Hogan BL (1997) Involvement of sonic hedgehog (shh) in mouse embryonic lung growth and morphogenesis. Development 124(1):53–63
21. Yu J, Carroll TJ, McMahon AP (2002) Sonic hedgehog regulates proliferation and differentiation of mesenchymal cells in the mouse metanephric kidney. Development 129(22):5301–5312
22. Pepicelli CV, Lewis PM, McMahon AP (1998) Sonic hedgehog regulates branching morphogenesis in the mammalian lung. Curr Biol 8(19):1083–1086
23. Carroll TJ, Park JS, Hayashi S, Majumdar A, McMahon AP (2005) Wnt9b plays a central role in the regulation of mesenchymal to epithelial transitions underlying organogenesis of the mammalian urogenital system. Dev Cell 9(2):283–292
24. Karavanova ID, Dove LF, Resau JH, Perantoni AO (1996) Conditioned medium from a rat ureteric bud cell line in combination with bFGF induces complete differentiation of isolated metanephric mesenchyme. Development 122(12):4159–4167
25. Stabellini G, Calvitti M, Becchetti E, Carinci P, Calastrini C, Lilli C, et al. (2007) Lung regions differently modulate bronchial branching development and extracellular matrix plays a role in regulating the development of chick embryo whole lung. Eur J Histochem 51(1):33–42
26. Sakai T, Larsen M, Yamada KM (2003) Fibronectin requirement in branching morphogenesis. Nature 423(6942):876–881
27. Ye P, Habib SL, Ricono JM, Kim NH, Choudhury GG, Barnes JL, et al. (2004) Fibronectin induces ureteric bud cells branching and cellular cord and tubule formation. Kidney Int 66(4):1356–1364
28. Larsen M, Wei C, Yamada KM (2006) Cell and fibronectin dynamics during branching morphogenesis. J Cell Sci 119(Pt 16):3376–3384
29. Nakanishi Y, Ishii T (1989) Epithelial shape change in mouse embryonic submandibular gland: modulation by extracellular matrix components. Bioessays 11(6):163–167
30. Zent R, Bush KT, Pohl ML, Quaranta V, Koshikawa N, Wang Z, et al. (2001) Involvement of laminin binding integrins and laminin-5 in branching morphogenesis of the ureteric bud during kidney development. Dev Biol 238(2):289–302
31. Brandenberger R, Schmidt A, Linton J, Wang D, Backus C, Denda S, et al. (2001) Identification and characterization of a novel extracellular matrix protein nephronectin that is associated with integrin alpha8beta1 in the embryonic kidney. J Cell Biol 154(2):447–458
32. Wang R, Li J, Lyte K, Yashpal NK, Fellows F, Goodyer CG (2005) Role for beta1 integrin and its associated alpha3, alpha5, and alpha6 subunits in development of the human fetal pancreas. Diabetes 54(7):2080–2089
33. Gill SE, Pape MC, Leco KJ (2006) Tissue inhibitor of metalloproteinases 3 regulates extracellular matrix—cell signaling during bronchiole branching morphogenesis. Dev Biol 298(2):540–554
34. Davies J (2005) Mechanisms of morphogenesis. Academic Press, Oxford
35. Sakurai H, Barros EJ, Tsukamoto T, Barasch J, Nigam SK (1997) An in vitro tubulogenesis system using cell lines derived from the embryonic kidney shows dependence on multiple soluble growth factors. Proc Natl Acad Sci U S A 94(12):6279–6284
36. Wei C, Larsen M, Hoffman MP, Yamada KM (2007) Self-organization and branching morphogenesis of primary salivary epithelial cells. Tissue Eng 13(4):721–735
37. Soriano JV, Pepper MS, Nakamura T, Orci L, Montesano R (1995) Hepatocyte growth factor stimulates extensive development of branching duct-like structures by cloned mammary gland epithelial cells. J Cell Sci 108 (Pt 2):413–430
38. Schuger L, O'Shea KS, Nelson BB, Varani J (1990) Organotypic arrangement of mouse embryonic lung cells on a basement membrane extract: involvement of laminin. Development 110(4):1091–1099
39. Preminger GM, Koch WE, Fried FA, Mandell J (1980) Utilization of the chick chorioallantoic membrane for in vitro growth of the embryonic murine kidney. Am J Anat 159(1):17–24
40. Woolf AS, Palmer SJ, Snow ML, Fine LG (1990) Creation of a functioning chimeric mammalian kidney. Kidney Int 38(5):991–997
41. Tufro A (2000) VEGF spatially directs angiogenesis during metanephric development in vitro. Dev Biol 227(2):558–566
42. Sariola H, Ekblom P, Lehtonen E, Saxen L (1983) Differentiation and vascularization of the metanephric kidney grafted on the chorioallantoic membrane. Dev Biol 96(2):427–435

43. Eckardt KU (1996) Erythropoietin production in liver and kidneys. Curr Opin Nephrol Hypertens 5(1):28–34
44. Della BR, Kurtz A, Schricker K (1996) Regulation of renin synthesis in the juxtaglomerular cells. Curr Opin Nephrol Hypertens 5(1):16–19
45. Humes HD, Fissell WH, Weitzel WF, Buffington DA, Westover AJ, MacKay SM, et al. (2002) Metabolic replacement of kidney function in uremic animals with a bioartificial kidney containing human cells. Am J Kidney Dis 39(5):1078–1087
46. Tiranathanagul K, Brodie J, Humes HD (2006) Bioartificial kidney in the treatment of acute renal failure associated with sepsis. Nephrology (Carlton) 11(4):285–291
47. Humes HD, Buffington DA, Lou L, Abrishami S, Wang M, Xia J, et al. (2003) Cell therapy with a tissue-engineered kidney reduces the multiple-organ consequences of septic shock. Crit Care Med 31(10):2421–2428
48. Kramer L, Gendo A, Madl C, Ferrara I, Funk G, Schenk P, et al. (2000) Biocompatibility of a cuprophane charcoal-based detoxification device in cirrhotic patients with hepatic encephalopathy. Am J Kidney Dis 36(6):1193–1200
49. Sauer IM, Neuhaus P, Gerlach JC (2002) Concept for modular extracorporeal liver support for the treatment of acute hepatic failure. Metab Brain Dis 17(4):477–484
50. Zeilinger K, Sauer IM, Pless G, Strobel C, Rudzitis J, Wang A, et al. (2002) Three-dimensional co-culture of primary human liver cells in bioreactors for in vitro drug studies: effects of the initial cell quality on the long-term maintenance of hepatocyte-specific functions. Altern Lab Anim 30(5):525–538
51. Sauer IM, Zeilinger K, Obermayer N, Pless G, Grunwald A, Pascher A, et al. (2002) Primary human liver cells as source for modular extracorporeal liver support—a preliminary report. Int J Artif Organs 25(10):1001–1005
52. Irgang M, Sauer IM, Karlas A, Zeilinger K, Gerlach JC, Kurth R, et al. (2003) Porcine endogenous retroviruses: no infection in patients treated with a bioreactor based on porcine liver cells. J Clin Virol 28(2):141–154
53. Dixit V, Gitnick G (1998) The bioartificial liver: state-of-the-art. Eur J Surg Suppl 164(582):71–76
54. Shackleton M, Vaillant F, Simpson KJ, Stingl J, Smyth GK, Asselin-Labat ML, et al. (2006) Generation of a functional mammary gland from a single stem cell. Nature 439(7072):84–88
55. Desbaillets I, Ziegler U, Groscurth P, Gassmann M (2000) Embryoid bodies: an in vitro model of mouse embryogenesis. Exp Physiol 85(6):645–651
56. Steenhard BM, Isom KS, Cazcarro P, Dunmore JH, Godwin AR, St John PL, et al. (2005) Integration of embryonic stem cells in metanephric kidney organ culture. J Am Soc Nephrol 16(6):1623–1631
57. Bruce SJ, Rea RW, Steptoe AL, Busslinger M, Bertram JF, Perkins AC (2007) In vitro differentiation of murine embryonic stem cells toward a renal lineage. Differentiation 75(5):337–349
58. Kim D, Dressler GR (2005) Nephrogenic factors promote differentiation of mouse embryonic stem cells into renal epithelia. J Am Soc Nephrol 16(12):3527–3534
59. Kobayashi T, Tanaka H, Kuwana H, Inoshita S, Teraoka H, Sasaki S, Terada Y (2005) Wnt4-transformed mouse embryonic stem cells differentiate into renal tubular cells. Biochem Biophys Res Commun. 336(2):585-595.
60. Hacein-Bey-Abina S, Von Kalle C, Schmidt M, McCormack MP, Wulffraat N, Leboulch P, et al. (2003) LMO2-associated clonal T cell proliferation in two patients after gene therapy for SCID-X1. Science 302(5644):415–419
61. Lagasse E, Connors H, Al Dhalimy M, Reitsma M, Dohse M, Osborne L, et al. (2000) Purified hematopoietic stem cells can differentiate into hepatocytes in vivo. Nat Med 6(11):1229–1234
62. Orlic D, Kajstura J, Chimenti S, Bodine DM, Leri A, Anversa P (2003) Bone marrow stem cells regenerate infarcted myocardium. Pediatr Transplant 7(3):86–88
63. Kale S, Karihaloo A, Clark PR, Kashgarian M, Krause DS, Cantley LG (2003) Bone marrow stem cells contribute to repair of the ischemically injured renal tubule. J Clin Invest 112(1):42–49
64. Lin F, Cordes K, Li L, Hood L, Couser WG, Shankland SJ, et al. (2003) Hematopoietic stem cells contribute to the regeneration of renal tubules after renal ischemia-reperfusion injury in mice. J Am Soc Nephrol 14(5):1188–1199
65. Duffield JS, Park KM, Hsiao LL, Kelley VR, Scadden DT, Ichimura T, et al. (2005) Restoration of tubular epithelial cells during repair of the postischemic kidney occurs independently of bone marrow-derived stem cells. J Clin Invest 115(7):1743–1755
66. Krause DS, Theise ND, Collector MI, Henegariu O, Hwang S, Gardner R, et al. (2001) Multi-organ, multi-lineage engraftment by a single bone marrow-derived stem cell. Cell 105(3):369–377
67. Lin F, Moran A, Igarashi P (2005) Intrarenal cells, not bone marrow-derived cells, are the major source for regeneration in postischemic kidney. J Clin Invest 115(7):1756–1764
68. Oliver JA, Maarouf O, Cheema FH, Martens TP, al Awqati Q (2004) The renal papilla is a niche for adult kidney stem cells. J Clin Invest 114(6):795–804
69. Saito A, Kazama JJ, Iino N, Cho K, Sato N, Yamazaki H, et al. (2003) Bioengineered implantation of megalin-expressing cells: a potential intracorporeal therapeutic model for uremic toxin protein clearance in renal failure. J Am Soc Nephrol 14(8):2025–2032
70. Oberpenning F, Meng J, Yoo JJ, Atala A (1999) De novo reconstitution of a functional mammalian urinary bladder by tissue engineering. Nat Biotechnol 17(2):149–155
71. Lanza RP, Chung HY, Yoo JJ, Wettstein PJ, Blackwell C, Borson N, et al. (2002) Generation of histocompatible tissues using nuclear transplantation. Nat Biotechnol 20(7):689–696
72. Hammerman MR (2003) Tissue engineering the kidney. Kidney Int 63(4):1195–1204
73. Rogers S, Hammerman M (2004) Prolongation of life in anephric rats following de novo renal organogenesis. Organogenesis 1(1):22–25

74. Rogers SA, Lowell JA, Hammerman NA, Hammerman MR (1998) Transplantation of developing metanephroi into adult rats. Kidney Int 54(1):27–37
75. Marshall D, Dilworth MR, Clancy M, Bravery CA, Ashton N (2007) Increasing renal mass improves survival in anephric rats following metanephros transplantation. Exp Physiol 92(1):263–271
76. Steer DL, Bush KT, Meyer TN, Schwesinger C, Nigam SK (2002) A strategy for in vitro propagation of rat nephrons. Kidney Int 62(6):1958–1965
77. Barasch J, Yang J, Ware CB, Taga T, Yoshida K, Erdjument-Bromage H, et al. (1999) Mesenchymal to epithelial conversion in rat metanephros is induced by LIF. Cell 99(4):377–386
78. Heinemann M, Panke S (2006) Synthetic biology—putting engineering into biology. Bioinformatics 22(22):2790–2799
79. Davies JA (2008) Synthetic morphology: prospects for engineered, self-constructing anatomies. J Anat 212(6):707-719.

Cytokine Signaling in Tissue Engineering

T. Meyer, V. Ruppert, B. Maisch

Contents

7.1 Summary 71
7.2 Pleiotropic Effects of Cytokines 71
7.3 Components in Cytokine-Induced Signal Transduction 72
7.4 Molecular Events in JAK-STAT Signaling .. 74
7.5 Dynamic Regulation of STAT Transcriptional Activity 75
7.6 Transcriptional Activity of STAT Proteins .. 77
7.7 Blocking of the JAK-STAT Pathway Within the Context of Tissue Engineering .. 78
References 78

7.1 Summary

Signal transduction is based on the need of different cells to communicate with each other in order to coordinate their growth and differentiation. The mechanisms for such complex regulation include secretion of soluble signaling molecules, as well as direct contacts between cells. Signal transduction pathways usually converge in the nucleus where transcription factors execute distinct gene expression programs. Many, though not all transcription factors bind to a specific base sequence in the promoter region of genes. Signal transducer and activator of transcription (STAT) proteins constitute a family of cytokine-inducible transcription factors that modulate broadly diverse biological processes, including cell growth, differentiation, apoptosis, immune regulation, fetal development, and transformation. They were originally discovered as DNA-binding proteins mediating interferon signal transduction. STAT proteins transmit cytokine signals directly from the plasma membrane to the nucleus without the interplay of a second messenger. In response to extracellular ligands receptor-associated Janus kinases (JAKs) activate STATs by phosphorylation on a single tyrosine at the carboxy terminus of the molecule. Activated STAT molecules form dimers through reciprocal interactions between the SH2 domain of one monomer and the phosphorylated tyrosine residue of the other. In the nucleus they bind to sequence-specific DNA elements and modulate the expression of a broad range of target genes. Recently, it was shown that STAT proteins shuttle between the cytoplasm and nucleus both in the presence and absence of cytokine stimulation. In this review we summarize the principles of the JAK-STAT signaling circuit and briefly discuss putative pharmacological interventions thereof used in tissue engineering.

7.2 Pleiotropic Effects of Cytokines

Cytokines comprise a large number of secreted factors regulating cell growth, development, and immune responses. Stimulation with different cytokines results in overlapping effects in many cell types, and this redundancy is probably due to the fact that they share common receptor subunits and signal components.

Cytokines can be divided into two types: Type I cytokines include interleukins, colony-stimulating factors, neutrophic factors, and hormones, while type II cytokines include interferons and interleukin-10 [1]. Interferons mediate antiviral responses, inhibit proliferation and participate in immune surveillance and tumor suppression [2]. Due to their substantial pleiotropy, cytokines elicit a broad range of diverse effects that may either be injurious or protective, depending on the particular cytokine, its concentration in the local tissue microenvironment, and the duration of its action.

7.3
Components in Cytokine-Induced Signal Transduction

The intracellular effects of cytokines are mediated by the **J**anus **k**inase (JAK)-**s**ignal **t**ransducer and **a**ctivator of **t**ranscription (STAT) pathway. The JAK-STAT pathway is involved in a wide range of distinct cellular processes, including inflammation, apoptosis, cell-cycle regulation, and development, suggesting that it has a major impact on the control of cell fate in normal and pathophysiological states. This pathway is regarded as a paradigmatic model for direct signal transduction, because it transmits information received from extracellular polypeptide signals directly to target genes in the nucleus without the interplay of second messengers [3–5]. JAK proteins, named after the Roman two-headed mythical god Janus, function as cytosolic tyrosine kinases and induce a cascade of phosphorylation steps that finally result in the activation of STAT proteins. In human cells four different JAK proteins (termed JAK1, JAK2, JAK3, and TYK2) are expressed, each consisting of approximately 1,200 amino acid residues [6]. The JAK kinases exhibit different receptor affinities, but all transmit their signals through the recruitment of STAT transcription factors. Cell lines lacking either JAK1 or JAK2 expression are unable to mediate a response to interferon-γ, whereas those deficient in TYK2 fail to respond to interferon-α/β [7–9]. JAK1-deficient mice exhibited perinatal lethality and defective lymphoid development [10]. JAK2 knockout mice exhibited an embryonic lethal phenotype caused by defective erythropoiesis but intact lymphoid development [11, 12]. Targeted disruption of the *jak3* gene resulted in a severe combined immunodeficiency with reduced numbers of functional T and B lymphocytes, and dysregulated myelopoiesis [13–15]. JAK proteins share a characteristic feature of seven conserved domains, also referred to as JAK homology (JH) domains [6]. The carboxy-terminal JH1 domain appears to convey full catalytic activity, while the neighboring JH2 domain appears to be a pseudokinase domain with no enzymatic activity.

As their name implies, the STAT proteins have the dual function of transducing signals from the plasma membrane to the nucleus, where they modulate transcription of target genes. The STAT proteins, originally discovered as DNA-binding proteins engaged in interferon signaling, constitute a family of evolutionarily conserved transcription factors that in humans consists of seven members: STAT1, STAT2, STAT3, STAT4, STAT5a, STAT5b, and STAT6 with numerous splice variants The different members are activated by specific sets of cytokines, growth factors, or hormones.

STAT1 has been implicated as key molecule in interferon-induced antiviral defense and apoptotic cell death [5]. Stimulation with interferon-γ results in the formation of STAT1 homodimers that bind to GAS (gamma-activated site) elements in the promoter region of interferon-γ-activated genes. Knockout-mice lacking STAT1 expression are viable, but cannot respond to viral or bacterial infections, presumably due to a defect in interferon signaling. In these animals the expression of major histocompatibility complex (MHC) class II protein, interferon-regulatory factor-1 (IRF-1), guanylate-binding protein-1 (GBP-1), and the MHC class II transactivating protein (CIITA) is critically diminished.

In response to interferon-α, STAT1 primarily forms the complex transcription factor interferon-stimulated gene factor 3 (ISGF3) which also includes STAT2 and interferon-regulatory factor-9 (IRF-9, also termed p48). ISGF3 binds to interferon-stimulated response elements and modulates the expression of genes engaged in antiviral response. Knockout mice lacking STAT2 expression develop normally, but are susceptible to viral infections due to an unresponsive interferon-α/β signaling [16].

STAT3 is involved in promoting cell-cycle progression, cellular transformation, and proliferation [1]. STAT3 becomes tyrosine phosphorylated by a variety of cytokines, such as IL-6, IL-11, leukemia inhibitory factor (LIF), oncostatin M (OSM), ciliary neurotrophic factor, and cardiotrophin-1. Propagation of these cytokine signals requires a common receptor subunit, gp130, and various ligand-binding subunits. STAT3 is considered as an oncogene, because it is persistently activated in numerous tumor entities. In squamous cell carcinomas, constitutive STAT3 activation is due to aberrant epidermal growth factor signaling, and in multiple myeloma IL-6 signaling is abnormally regulated [17]. It induces the expression of cyclin D1, c-Myc, and Bcl-xl, thus accounting for its transforming potential. Embryos lacking STAT3 expression die early in embryogenesis prior to gastrulation [18]. Tissue-specific knockout models with STAT3-deficient T cells showed an impaired proliferation of T cells in response to IL-6, thus confirming the central role of IL-6 in STAT3 activation [19].

STAT4 shows remarkable specificity for IL-12 receptor family signaling (IL-12, IL-23, IL-27). It was shown that STAT4 knockout mice were deficient in T_H1 cells, but otherwise normal. The phenotype of STAT4 knockout mice closely resembles that of mice deficient in IL-12 expression [20, 21].

STAT5 was initially identified as a prolactin-induced transcription factor in mammary gland tissue and termed mammary gland factor (MGF) [22]. Later, it was shown that STAT5 consists of two highly related genes, which share 96% homology at the amino acid level [23]. In addition to activation by prolactin, STAT5 is also activated by IL-2, IL-3, IL-5, IL-7, IL-9, IL-15, GM-CSF, thrombopoietin, erythropoietin, and growth hormone. In knockout experiments, STAT5a was demonstrated to be required for mammary gland development and lactogenesis [24, 25]. Targeted disruption of the *stat5b* gene resulted in mice that lack sexual dimorphism of body growth rates and liver gene expression, probably resulting from altered response to growth hormone [25, 26]. The STAT5a/STAT5b double knockout mice exhibited defects in mammopoiesis and sexual dimorphism and had a profound deficit in peripheral T cells, which probably resulted from unresponsiveness to circulating IL-2 [27, 28]. Additionally these animals showed defective fetal liver erythropoiesis.

Tyrosine phosphorylation of STAT6 is induced in response to stimulation with IL-4 as well as IL-13 [29, 30]. As expected, mice lacking STAT6 expression show defects in IL-4-mediated functions including induction of CD23 and MHC class II, immunoglobulin switching to IgE, B-cell and T-cell proliferation, and T_H2 cell development [30–32].

Despite their functional diversity, all members of the STAT family share a conserved domain structure with an amino-terminal domain separated by a protease-sensitive linker peptide from the core domain and a carboxy-terminal transactivating domain [3, 4, 33] (Fig. 7.1). The amino-terminal domain of about 130 residues folds into a unique hook-shaped domain that promotes tetramer formation and resultant cooperative binding to DNA [34–36]. The large core domain encompasses several structurally distinct domains beginning at the amino-terminal end with a four-helix bundle, which is engaged in protein–protein interactions [37]. The DNA-binding domain displays an immunoglobulin fold and is required for sequence-specific DNA binding of phosphorylated STAT dimers [38]. The consecutive linker region consists of a unique all-alpha helical architecture that appears to participate in DNA binding [39]. The Src homology 2 (SH2) domain mediates binding to phosphotyrosine residues in cytokine receptors and is required for the formation of phosphorylated STAT dimers. The latter is achieved via reciprocal interaction between the phosphotyrosine of one monomer and the SH2 domain of the corresponding monomer [40]. The carboxy terminus is most divergent in size and sequence between the different STAT family members. In numerous splice variants this transactivating domain is deleted [41].

Fig. 7.1 Domain structure of STATs. Members of the family of STAT transcription factors share a common modular organization into functional domains. The N-domain (ND) is engaged in tetramerization on DNA, formation of dimers between unphosphorylated STAT monomers, tyrosine dephosphorylation, and nuclear import. The coiled–coil domain facilitates protein–protein interactions. The DNA-binding domain (DBD, missing in STAT2) contains a dimer-specific nuclear import signal and together with the adjacent linker domain (LD) interacts with genomic DNA. The SH2 domain (SH2) is involved in receptor recruitment and STAT dimerization through binding to the phosphotyrosine residue (p-Y) of the other monomer. The transactivation domain (TAD) contains a conserved serine residue (p-S) which is phosphorylated upon cytokine stimulation and is important for maximal transcriptional activity and rapid nuclear export of STAT1

7.4 Molecular Events in JAK-STAT Signaling

Ligand binding to the extracellular domain of cytokine receptors induces dimerization of the receptor molecules followed by auto-phosphorylation of the noncovalently attached JAK kinases (for an overview see Fig. 7.2). The activated JAKs then trans-phosphorylate tyrosine residues within the cytoplasmic domain of the cytokine receptor, thereby creating docking sites for the SH2 domain of STAT proteins [2, 4, 6]. Once STAT proteins are recruited to the intracellular receptor chains, receptor-associated JAKs catalyze the phosphorylation of a single tyrosine residue within their carboxy-terminal domain [3, 40, 41]. STAT proteins are then released from the receptor complex as phosphorylated homodimers or heterodimers (STAT1:STAT2, STAT1:STAT3) and subsequently migrate into the nucleus [41, 42]. Within the nucleus, they bind to specific DNA elements of the consensus sequence 5´-TTCN$_3$GAA-´3, termed **g**amma **a**ctivated **s**ites (GAS), in the promoters of target genes and modulate gene transcription [38]. Specific tyrosine phosphatases then dephosphorylate STAT proteins, for example, the phosphatase Tc45 inactivates STAT1 [43, 44].

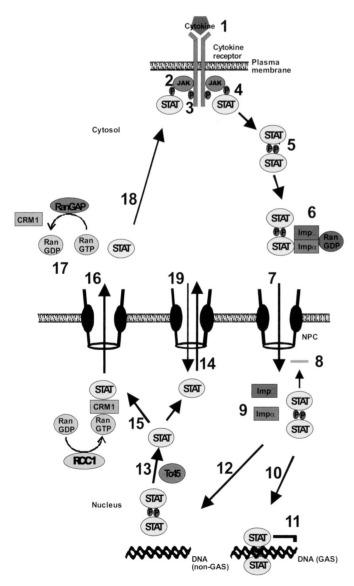

Fig. 7.2 Diagram depicting the signal transduction of the transcription factor STAT1. Ligand binding to cytokine receptors (*1*) leads to the auto-phosphorylation of JAK kinases (*2*) and the JAK-catalysed tyrosine phosphorylation of intracellular receptor chains (*3*). STAT1 is recruited to the receptor complex and subsequently phosphorylated on a single tyrosine residue (*4*). STAT1 dimers (*5*) form a complex with importin α, importin β, and RanGDP (*6*), which facilitates nuclear import through the nuclear pore complex (NPC) (*7*). Tyrosine-phosphorylated STAT1 is barred from nuclear exit (*8*). In the nucleus the import complex disassembles and liberates transcriptionally active STAT1 (*9*). STAT1 dimers can either recognize GAS sites in the promoter region of target genes (*10*) and modulate their expression (*11*) or alternatively bind to non-GAS sites (*12*), where they are rapidly inactivated by Tc45 phosphatase activity (*13*). The dephosphorylated proteins exit the nucleus either via energy-independent transport through the nuclear pore (*14*) or in association with the export receptor CRM1 (*15*) in a RanGTP-dependent manner (*16*). Nuclear import of phospho-STAT1 and export via CRM1 are driven by the asymmetric RanGDP/RanGTP distribution across the nuclear envelope that results from the differential localization of RanGTPase-activating protein (RanGAP) and the guanine nucleotide exchange factor RCC1. In the cytosol, the CRM1-containing export complex disassembles and liberates unphosphorylated STAT1 (*17*). The STAT1 molecules can then either be rephosphorylated at the receptor (*18*) or participate in constitutive, energy-independent nucleocytoplasmic shuttling through the NPC (*19*). Not included in the diagram is the extensive spatial reorganization within the STAT1 dimer from a DNA-bound "parallel" to an "anti-parallel" form

However, STAT activation is not mediated exclusively by cytokine receptors that lack intrinsic tyrosine kinase domains, but is also observed in cells stimulated with epidermal growth factor (EGF) or platelet-derived growth factor (PDGF) [17]. Both EGF and PDGF receptors are capable of directly phosphorylating STAT proteins in the absence of JAK activation [45]. STAT proteins may also be activated by nonreceptor tyrosine kinases such as Src and Abl. Mammalian cells transformed by oncogenic Src show constitutively tyrosine-phosphorylated STAT3 and negative-dominant forms of STAT3 block the transforming ability of Src, demonstrating the close correlation between STAT3 activation and oncogenic transformation [46].

7.5 Dynamic Regulation of STAT Transcriptional Activity

In resting cells, STAT proteins are localized predominantly in the cytosol [47]. Upon cytokine stimulation, STAT proteins accumulate in the nucleus for a few hours depending on the cytokine concentration and the cell type used, until the resting distribution is restored (Fig. 7.3). It has been revealed, however, that despite persisting nuclear accumulation the STATs are constantly shuttling between the cytosol and nucleus with high translocation rates [48–51]. Import into the nucleus occurs along two different translocation pathways: a direct association of unphosphorylated STAT molecules with components of the nuclear pore complex (nucleoporins) as well as binding of phospho-STAT1 to import factors, called importins (Fig. 7.2). Binding to importin requires the "parallel" orientation of both monomers within the STAT dimer [33, 52]. Nuclear import is independent of cytokine stimulation, since both phosphorylated and unphosphorylated STAT molecules enter the nucleus, albeit via different pathways. Translocation of the unphosphorylated protein requires neither metabolic energy nor transport factors, but follows a concentration gradient, indicating that it functions as facilitated diffusion. This carrier-free nucleocytoplasmic translocation is mediated through direct contacts between STAT proteins and nucleoporins located in the nuclear core complex [50].

In contrast, nuclear import of phosphorylated STAT1 is energy-dependent and based on the asymmetric distribution of the small GTPase Ran at the two sites of the nuclear membrane [53]. The directionality of active transport is given by the differential nucleotide binding of the small GTPase Ran at both sites of the nuclear membrane, which reflects the asymmetric localization of Ran nucleotide exchange factors across the nuclear envelope [54]. The intracellular RanGDP/RanGTP gradient across the nuclear envelope constitutes the driven force behind this active transport and enables the entry of STATs into the nucleus against a concentration gradient. Due to nucleotide hydrolysis by cytoplasmically localized RanGTPase-activating protein (RanGAP), the concentration of the guanosine triphosphate (GTP) form of Ran in the cytosol is low. In the nucleus, however, high levels of RanGTP are maintained by the guanine nucleotide exchange factor RCC1 which catalyses the conversion of RanGDP to RanGTP. Both the nuclear import of activated STAT dimers and export of the unphosphorylated STAT via the export receptor chromosomal region maintenance 1 (CRM1) rely on this asymmetric distribution of RanGTP.

STAT1 homodimers and STAT1-STAT2 heterodimers interact with importin α5 (NPI-1/hSrp1) [55, 56]. Tyrosine-phosphorylated STAT1 is imported into the nucleus in a complex with importin α5 and importin β, also known as karyopherin β [56, 57]. However, the stoichiometry of this import complex is not known. Binding to importin α5 occurs via an unusual dimer-specific nuclear localization signal (dsNLS) within the DNA-binding domain of STAT1 [48, 56–59]. The high RanGTP level in the nucleus promotes the disassembly of the import complex and facilitates the binding of the export receptor CRM1 to unphosphorylated STAT1 [60]. Also for IL-4 signaling, a continuous cycling of STAT6 is required [61].

Tyrosine-phosphorylated STAT1 is unable to exit the nucleus before it has been dephosphorylated [62]. Recently, it has been revealed that dephosphorylation of phosphotyrosine on STAT1 dimers requires extensive spatial reorientation of the monomers facilitated by the amino-terminal domain [52]. Additionally, we found that enzymatic dephosphorylation by Tc45 phosphatase is proportional to the sequence-specific dissociation rate of STAT1 from DNA [62]. STAT1 dimers bound to GAS sites on DNA are effectively protected from dephosphorylation and thus kept

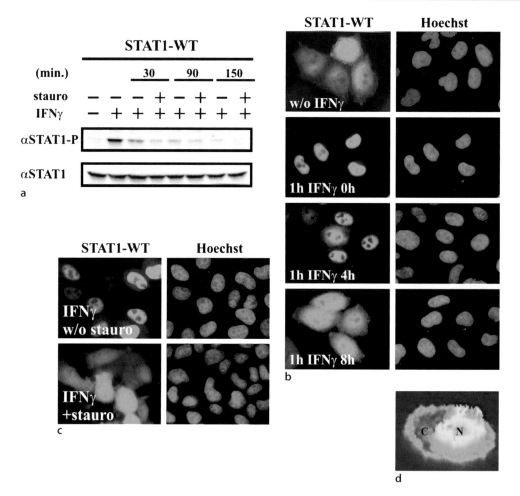

Fig. 7.3a–d Tyrosine phosphorylation and nuclear accumulation of STAT1 upon stimulation with interferon γ. **a** Time course of STAT1 tyrosine dephosphorylation. A Western blot from U3A cells expressing recombinant STAT1 carboxy-terminally fused to green fluorescent protein (STAT1-GFP) is shown. Cells were either unstimulated (-IFNγ) or stimulated for 45 min. with 5 ng/ml human interferon γ (+IFNγ). The cytokine-containing medium was then replaced for the indicated times with fresh medium containing no staurosporine (-Stauro) or 500 nM staurosporine (+Stauro). Whole cell extracts were analyzed by Western blotting with an anti-phospho-STAT1 antibody and reprobed with anti-STAT1 antibody. Note that the kinase inhibitor staurosporine significantly reduced the phosphorylation level of STAT1, demonstrating that persistent kinase activity is required for maintaining activated STAT1. **b** Time course of interferon-γ-induced nuclear accumulation of STAT1. HeLa cells expressing recombinant STAT1-GFP were either unstimulated (w/o IFNγ) or stimulated for 1 h with 5 ng/ml interferon γ (1h IFNγ). The cells were incubated in the absence of interferon γ for the indicated times. Fluorescence micrographs for the localization of STAT1-GFP and the corresponding Hoechst-stained nuclei are shown. **c** The blocking of kinase activity results in the breakdown of STAT1 nuclear accumulation. HeLa cells expressing STAT1-GFP were stimulated for 1 h with interferon γ and subsequently incubated in the absence (w/o stauro) or presence (+stauro) of 500 nM staurosporine for additional 2 h. The cells were then fixed and stained with Hoechst dye for the detection of nuclei. **d** Intracellular localization of STAT1 in an interferon-γ-stimulated cell. The spatial distribution of STAT1 tagged with green fluorescent protein (GFP) in a HeLa cell stimulated for 45 min. with interferon γ is shown. The fluorescence intensity plotted over the cell as measured by laser scanning microscopy is depicted. Note that STAT1-GFP accumulates adjacent to the plasma membrane and in the nucleus. *Light blue* indicates low concentration, *green* moderate concentration, and *yellow* and *red* high concentrations of STAT1-GFP

longer in a transcriptionally active state (Fig. 7.2). Thus, nucleocytoplasmic shuttling enables the coupling between receptor processes at the cell membrane and transcriptional regulation in the nucleus. Taken together, STAT signaling requires a constant cycle of cytosolic activation and subsequent deactivation in the nucleus that is linked to extensive spatial reorientation within the dimer [48, 49–52, 61].

7.6 Transcriptional Activity of STAT Proteins

For transcriptional activity, STAT proteins cooperate with numerous other transcription factors, such as Sp1, c-Jun, Fos, nuclear factor-κB (NF-κB), glucocorticoid receptor (GR), and interferon-regulatory factors (IRFs). For example, STAT1 and Sp1 interact synergistically on the ICAM promoter, STAT3, c-Jun and GR bind to the α2-macroglobulin promoter, and STAT5 and GR interact on the β-casein promoter [63–65]. STATs also interact with coactivators which function in chromatin remodeling, such as p300/CREB-binding protein (CBP), N-Myc interactor (Nmi), MCM-5, BRG1, and histone deacetylases (HDACs) [66–71]. In mammalian cells, CBP has been shown to play a key role in mediating transcriptional control via a number of other DNA-binding transcription factors, including CREB (cAMP-responsive element-binding protein), p53, NF-κB, steroid/thyroid hormone receptors, and MyoD. CBP has histone acetyltransferase activity, indicating that it can stimulate gene expression by acetylating histones and, therefore, opening up the chromatin structure. The coactivator CBP also directly interacts with various components of the basal transcription complex, such as TBP (TATA-binding protein), TFIIB, and the RNA polymerase holoenzyme, suggesting that it can bridge the gap between the STATs and the basal complex. For optimal transcriptional activity, STAT proteins require phosphorylation of a single serine residue in the carboxy-terminal region of the molecule. Figure 7.4 illustrates the transcriptional activity of STAT1 as demonstrated by a reporter gene assay and a viral plaque assay.

Among the target genes of tyrosine-phosphorylated STAT transcription factors are the suppressor of cytokine signaling (SOCS) proteins that form a negative feedback loop by inhibiting the activity of

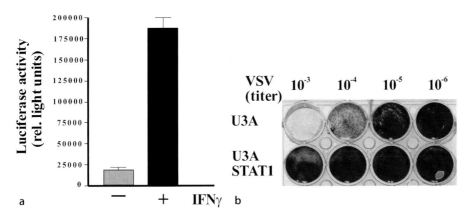

Fig. 7.4a,b Transcriptional responses of STAT1. **a** Interferon-γ-induced expression of a STAT1-responsive reporter as revealed by a luciferase reporter gene assay. U3A cells were transfected with three different plasmids coding for wild-type STAT1, a luciferase gene with three tandem STAT-binding sites in its promoter, and a constitutively expressed GAPDH gene used for normalization. The relative light units of luciferase activity in extracts from cells incubated for 6 h in the absence (–IFNγ) or presence of interferon γ (+IFNγ) are shown. **b** Expression of STAT1 protects against the lytic activity of vesiculo stomatitis virus (VSV). STAT1-deficient U3A cells and U3A cells reconstituted with STAT1 were infected with increasing titers of VSV ranging from 10^{-6} to 10^{-3}. Twelve hours later the cells were fixed and the living cells were stained with crystal violet

JAKs [72–74]. Other important negative regulators of the JAK-STAT pathway are cytosolic tyrosine phosphatases, which inactivate JAK kinases, such as SH2-containing phosphatase-1 (SHP-1), SHP2, protein-tyrosine-phosphatase-1B (PTP1B), CD45, and Tc45 [75–77]. Protein inhibitors of activated STAT (PIASs) localized in the nucleus have been described as negative regulators that block the DNA-binding activity of STAT dimers [78, 79].

7.7 Blocking of the JAK-STAT Pathway Within the Context of Tissue Engineering

The crucial role of the distinct STAT family members in balancing proliferative and apoptotic processes suggests that pharmacological interventions in the JAK-STAT pathway may be exploitable for tissue engineering. Recently, specific compounds have been identified that function as potent blockers of this important signaling pathway. Particularly, immunosuppressive JAK3 inhibitors have been introduced as a new class of anti-inflammatory drugs, which promise to prolong the survival rate of heterologous tissue-engineered transplants [80–85]. A potent JAK3 inhibitor, termed CP-690550, was shown to effectively prevent organ allograft rejection in mouse and cynomolgus models [82, 86]. Another promising approach to directly target STAT proteins is the development of specific phosphopeptides that mimic the tyrosine-phosphorylated SH2 domain-binding sequence of STATs. Peptidomimetics have been successfully tested for their efficacy as selective inhibitors of STAT activation in experimental animal studies. These compounds are derivates of the tyrosine-phosphorylated STAT3 SH2 domain binding sequence that contain the STAT3 sequence PYLKTK, where Y represents the phosphotyrosine [87–90]. Phosphorylation of the tyrosine residue is essentially required for the efficacy of this approach, as nonphosphorylated derivates consistently failed to inhibit STAT signaling. It was proposed that these synthetic peptide inhibitors exert their effects by competitive binding to the STAT SH2 domain, thereby blocking either dimerization or receptor recruitment. According to this assumption, peptidomimetics disrupt transcriptionally active STAT dimers, probably by causing their dissociation into functionally incompetent monomers, which do not bind to the promoters of target genes [91].

Furthermore, decoy oligonucleotides carrying a consensus STAT binding site have been successfully used for the neutralizing of STAT proteins [92–96]. This strategy allows direct reduction in STAT transcriptional activity on target promoters through competitive binding to high-affinity DNA elements. For example, administration of synthetic STAT1 decoy oligonucleotides results in a delayed acute rejection and a prolonged survival of cardiac allografts in transplanted rats [94]. Other small synthetic or naturally occurring compounds have been demonstrated to specifically reduce the transcriptional activity of STAT proteins [97–102].

In summary, blocking components of the JAK-STAT pathway may be exploitable to improve the outcome of tissue engineering. The success of such a pharmacological approach probably depends on the STAT family member that is neutralized and the role the STAT protein plays in the tissue-engineered organ. Future work will show whether and to what extent the JAK-STAT pathway and, in particular, the nucleocytoplasmic shuttling thereof, might be a prime target for therapeutic intervention. Thus, studies on blocking cytokine-driven transcription factors in the context of tissue engineering will surely be exciting and rewarding.

References

1. Calò V, Migliavacca M, Bazan V, Macaluso M, Buscemi M, Gebbia N, Russo A. STAT proteins: from normal control of cellular events to tumorigenesis. J Cell Physiol 2003;197:157-68
2. Stark GR, Kerr IM, Williams BR, Silverman RH, Schreiber RD. How cells respond to interferons. Annu Rev Biochem 1998;67:227-64
3. Ihle JN. The Stat family in cytokine signaling. Curr Opin Cell Biol 2001;13:211-17
4. Levy DE, Darnell JE Jr. Stats: transcriptional control and biological impact. Nat Rev Mol Cell Biol 2002;3:651-62
5. Shuai K, Liu B. Regulation of JAK-STAT signalling in the immune system. Nat Rev Immunol 2003;3:900-11
6. Rane SG, Reddy EP. Janus kinases: components of multiple signaling pathways. Oncogene 2000;19:5662-79

7. Velazquez L, Fellous M, Stark GR, Pellegrini S. A protein tyrosine kinase in the interferon α/β signalling pathway. Cell 1992;70:313-22
8. Müller M, Briscoe J, Laxton C, Guschin D, Ziemiecki A, Silvennoinen O, Harpur AG, Barbieri G, Witthuhn BA, Schindler C, Pellegrini S, Wilks AF, Ihle JN, Stark GR, Kerr IM. The protein tyrosine kinase JAK1 complements defects in interferon-α/β and -γ signal transduction. Nature 1993;366:129-35
9. Watling D, Gushin D, Müller M, Silvennoinen O, Witthuhn BA, Quelle FW, Rogers NC, Schindler C, Stark GR, Ihle JN, Kerr IM. Complementation by the protein tyrosine kinase JAK2 of a mutant cell line defective in interferon-gamma signal transduction. Nature 1993;366:166-70
10. Rodig SJ, Meraz MA, White JM, Lampe PA, Riley JK, Arthur CD, King KL, Sheehan KC, Yin L, Pennica D, Johnson EM Jr, Schreiber RD. Disruption of the *jak1* gene demonstrates obligatory and nonredundant roles of the Jaks in cytokine-induced biologic responses. Cell 1998;93:373-83
11. Neubauer H, Cumano A, Müller M, Wu H, Huffstadt U, Pfeffer K. Jak2 deficiency defines an essential developmental checkpoint in definitive hematopoiesis. Cell 1998;93:397-409
12. Parganas E, Wang D, Stravopodis D, Topham DJ, Marine JC, Teglund S, Vanin EF, Bodner S, Colamonici OR, van Deursen JM, Grosveld G, Ihle JN. Jak2 is essential for signaling through a variety of cytokine receptors. Cell 1998;93:385-95
13. Nosaka T, van Deursen JM, Tripp RA, Thierfelder WE, Witthuhn BA, McMickle AP, Doherty PC, Grosveld GC, Ihle JN. Defective lymphoid development in mice lacking Jak3. Science 1995;270:800-2
14. Park SY, Saijo K, Takahashi T, Osawa M, Arase H, Hirayama N, Miyake K, Nakauchi H, Shirasawa T, Saito T. Developmental defects of lymphoid cells in Jak3 kinase-deficient mice. Immunity 1995;3:771-82
15. Grossman WJ, Verbsky JW, Yang L, Berg LJ, Fields LE, Chaplin DD, Ratner L. Dysregulated myelopoiesis in mice lacking Jak3. Blood 1999;94:932-9
16. Park C, Li S, Cha E, Schindler C. Immune response in Stat2 knockout mice. Immunity 2000;13:795-804
17. Bromberg JF. Activation of STAT proteins and growth control. BioEssays 2001;23:161-9
18. Takeda K, Noguchi K, Shi W, Tanaka T, Matsumoto M, Yoshida N, Kishimoto T, Akira S. Targeted disruption of the mouse *Stat3* gene leads to early embryonic lethality. Proc Natl Acad Sci USA 1997;94:3801-4
19. Takeda K, Kaisho T, Yoshida N, Takeda J, Kishimoto T, Akira S. Stat3 activation is responsible for IL-6-dependent T cell proliferation through preventing apoptosis: generation and characterization of T cell-specific Stat3-deficient mice. J Immunol 1998;161:4652-60
20. Kaplan MH, Sun YL, Hoey T, Grusby MJ. Impaired IL-12 responses and enhanced development of Th2 cells in Stat4-deficient mice. Nature 1996;382:174-7
21. Thierfelder WE, van Deursen JM, Yamamoto K, Tripp RA, Sarawar SR, Carson RT, Sangster MY, Vignali DA, Doherty PC, Grosveld GC, Ihle JN. Requirement for Stat4 in interleukin-12-mediated responses of natural killer and T cells. Nature 1996;382:171-4
22. Wakao H, Gouilleux F, Groner B. Mammary gland factor (MGF) is a novel member of the cytokine regulated transcription factor gene family and confers the prolactin response. EMBO J 1994;13:2182-91
23. Liu X, Robinson GW, Gouilleux F, Groner B, Hennighausen L. Cloning and expression of Stat5 and an additional homologue (Stat5b) involved in prolactin signal transduction in mouse mammary tissue. Proc Natl Acad Sci USA 1995;92:8831-5
24. Liu X, Robinson GW, Wagner KU, Garrett L, Wynshaw-Boris A, Henninghausen L. Stat5a is mandatory for adult mammary gland development and lactogenesis. Genes Dev 1997;11:179-86
25. Teglund S, McKay C, Schuetz E, van Deursen JM, Stravopodis D, Wang D, Brown M, Bodner S, Grosveld G, Ihle JN. Stat5a and Stat5b proteins have essential and nonessential, or redundant, roles in cytokine responses. Cell 1998;93:841-50,
26. Udy GB, Towers RP, Snell RG, Wilkins RJ, Park SH, Ram PA, Waxman DJ, Davey HW. Requirement of STAT5b for sexual dimorphism of body rates and liver gene expression. Proc Natl Acad Sci USA 1997;94:7239-44
27. Moriggl R, Sexl V, Piekorz R, Topham D, Ihle JN. Stat5 activation is uniquely associated with cytokine signaling in peripheral T cells. Immunity 1999;11:225-30
28. Moriggl R, Topham DJ, Teglund S, Sexl V, McKay C, Wang D, Hoffmeyer A, van Deursen J, Sangster MY, Bunting KD, Grosveld GC, Ihle JN. Stat3 is required for IL-2-induced cell cycle progression of peripheral T cells. Immunity 1999;10:249-59
29. Hou J, Schindler U, Henzel WJ, Ho TC, Brasseur M, McKnight SL. An interleukin-4-induced transcription factor: IL-4 Stat. Science 1994;265:1701-6
30. Takeda K, Kamanaka M, Tanaka T, Kishimoto T, Akira S. Impaired IL-13-mediated functions of macrophages in STAT6-deficient mice. J Immunol 1996;157:3320-2
31. Kaplan MH, Schindler U, Smiley ST, Grusby MJ. Stat6 is required for mediating responses to IL-4 and for development of Th2 cells. Immunity 1996;4:313-9
32. Shimoda K, van Deursen J, Sangster MY, Sarawar SR, Carson RT, Tripp RA, Chu C, Quelle FW, Nosaka T, Vignali DA, Doherty PC, Grosveld G, Paul WE, Ihle JN. Lack of IL-4-induced Th2 response and IgE class switching in mice with disrupted *Stat6* gene. Nature 1996;380:630-3
33. Mao X, Ren Z, Parker GN, Sondermann H, Pastorello MA, Wang W, McMurray JS, Demeler B, Darnell JE Jr, Chen X. Structural bases of unphosphorylated STAT1 association and receptor binding. Mol Cell 2005;17:761-71
34. Shuai K, Liao J, Song MM. Enhancement of antiproliferative activity of gamma interferon by the specific inhibition of tyrosine dephosphorylation of Stat1. Mol Cell Biol 1996;16:4932-41
35. Vinkemeier U, Cohen SL, Moarefi I, Chait BT, Kuriyan J, Darnell JE Jr. DNA binding of in vitro activated Stat1-alpha, Stat1-beta and truncated Stat1: interaction between

NH2-terminal domains stabilizes binding of two dimers to tandem DNA sites. EMBO J 1996;15:5616-26
36. Meyer T, Hendry L, Begitt A, John S, Vinkemeier U. A single residue modulates tyrosine dephosphorylation, oligomerization, and nuclear accumulation of Stat transcription factors. J Biol Chem 2004;279:18998-9007
37. Horvath CM, Stark GR, Kerr IM, Darnell JE Jr. Interactions between STAT and non-STAT proteins in the interferon-stimulated gene factor 3 transcription complex. Mol Cell Biol 1996;16:6957-64
38. Horvath CM, Wen Z, Darnell JE Jr. A STAT protein domain that determines DNA sequence recognition suggests a novel DNA-binding domain. Genes Dev 1995;15:984-94
39. Yang E, Henriksen MA, Schaefer O, Zakharova N, Darnell JE Jr. Dissociation time from DNA determines transcriptional function in a STAT1 linker mutant. J Biol Chem 2002;277:13455-62
40. Shuai K, Stark GR, Kerr IM, Darnell JE Jr. A single phosphotyrosine residue of Stat91 required for gene activation by interferon-γ. Science 1993;261:1744-6
41. Schindler C, Shuai K, Prezioso VR, Darnell JE Jr. Interferon-dependent tyrosine phosphorylation of a latent cytoplasmic transcription factor. Science 1992;257:809-13
42. Shuai K, Horvath CM, Huang LH, Qureshi SA, Cowburn D, Darnell JE Jr. Interferon activation of the transcription factor Stat91 involves dimerization through SH2-phosphotyrosyl peptide interactions. Cell 1994;76:821-8
43. Haspel RL, Salditt-Georgieff M, Darnell JE Jr. The rapid inactivation of nuclear tyrosine phosphorylated Stat1 depends upon a protein tyrosine phosphatase. EMBO J 1996;15:6262-8
44. ten Hoeve J, de Jesus Ibarra-Sanchez M, Fu Y, Zhu W, Tremblay M, David M, Shuai K. Identification of a nuclear Stat1 protein tyrosine phosphatase. Mol Cell Biol 2002;22:5662-8
45. Akira S. Functional roles of STAT family proteins: lessons from knockout mice. Stem Cells 1999;17:138-46
46. Turkson J, Bowman T, Garcia R, Caldenhoven E, De Groot RP, Jove R. Stat3 activation by Src induces specific gene regulation and is required for cell transformation. Mol Cell Biol 1998;18:2545-52
47. Meyer T, Gavenis K, Vinkemeier U. Cell-type specific and tyrosine phosphorylation-independent nuclear presence of STAT1 and STAT3. Exp Cell Res 2002;272:45-55
48. Meyer T, Begitt A, Lödige I, van Rossum M, Vinkemeier U. Constitutive and IFNγ-induced nuclear import of STAT1 proceed through independent pathways. EMBO J 2002;21:344-54
49. Zeng R, Aoki Y, Yoshida M, Arai K, Watanabe S. Stat5B shuttles between cytoplasm and nucleus in a cytokine-dependent and -independent manner. J Immunol 2002;168:4567-75
50. Marg A, Shan Y, Meyer T, Meissner T, Brandenburg M, Vinkemeier U. Nucleocytoplasmic shuttling by nucleoporins Nup153 and Nup214 and CRM1-dependent nuclear export control the subcellular distribution of latent Stat1. J Cell Biol 2004;165:823-33
51. Pranada AL, Metz S, Herrmann A, Heinrich PC, Müller-Newen G. Real time analysis of STAT3 nucleocytoplasmic shuttling. J Biol Chem 2004;279:15114-23
52. Mertens C, Zhong M, Krishnaraj R, Zou W, Chen X, Darnell JE Jr. Dephosphorylation of phosphotyrosine on STAT1 dimers requires extensive spatial reorientation of the monomers facilitated by the N-terminal domain. Genes Dev 2006;20:3372-81
53. Sekimoto T, Nakajima K, Tachibana T, Hirano T, Yoneda Y. Interferon-γ-dependent nuclear import of Stat1 is mediated by the GTPase activity of Ran/TC4. J Biol Chem 1996;271:31017-20
54. Bischoff FR, Scheffzek K, Ponstingl H. How Ran is regulated. Results Probl Cell Differ 2002;35:49-66
55. Sekimoto T, Imamoto N, Nakajima K, Hirano T, Yoneda Y. Extracellular signal-dependent nuclear import of Stat1 is mediated by nuclear pore-targeting complex formation with NPI-1, but not Rch1. EMBO J 1997;16:7067-77
56. Fagerlund R, Melén K, Kinnunen L, Julkunen I. Arginine/lysine-rich nuclear localization signals mediate interactions between dimeric STATs and importin α5. J Biol Chem 2002;277:30072-8
57. McBride KM, Banninger G, McDonald C, Reich NC. Regulated nuclear import of the STAT1 transcription factor by direct binding of importin-α. EMBO J 2002;21:1754-63
58. Melén K, Kinnunen L, Julkunen I. Arginine/lysine-rich structural element is involved in interferon-induced nuclear import of STATs. J Biol Chem 2001;276:16447-55
59. Melén K, Fagerlund R, Franke J, Köhler M, Kinnunen L, Julkunen I. Importin α nuclear localization signal binding sites for STAT1, STAT2, and influenza A virus nucleoprotein. J Biol Chem 2003;278:28193-200
60. McBride KM, McDonald C, Reich NC. Nuclear export signal located within the DNA-binding domain of the STAT1 transcription factor. EMBO J 2000;19:6196-206
61. Andrews RP, Ericksen MB, Cunningham CM, Daines MO, Hershey GK. Analysis of the life cycle of stat6. Continuous cycling of STAT6 is required for IL-4 signaling. J Biol Chem 2002;277:36563-9
62. Meyer T, Marg A, Lemke P, Wiesner B, Vinkemeier U. DNA binding controls inactivation and nuclear accumulation of the transcription factor Stat1. Genes Dev 2003;17:1992-2005
63. Look DC, Pelletier MR, Tidwell RM, Roswit WT, Holtzman MJ. Stat1 depends on transcriptional synergy with Sp1. J Biol Chem. 1995;270:30264-7
64. Stöcklin E, Wissler M, Gouilleux F, Groner B. Functional interactions between Stat5 and the glucocorticoid receptor. Nature 1996;383:726-8
65. Zhang X, Wrzeszczynska MH, Horvath CM, Darnell JE Jr. Interacting regions in Stat3 and c-Jun that participate in cooperative transcriptional activation. Mol Cell Biol 1999;19:7138-46
66. Zhang JJ, Vinkemeier U, Gu W, Chakravarti D, Horvath CM, Darnell JE Jr. Two contact regions between Stat1 and CBP/p300 in interferon γ signaling. Proc Natl Acad Sci USA 1996;93:15092-6
67. Zhang JJ, Zhao Y, Chait BT, Lathem WW, Ritzi M, Knippers R, Darnell JE Jr. Ser727-dependent recruitment of MCM5 by Stat1α in IFN-γ-induced transcriptional activation. EMBO J 1998;17:6963-71

68. Zhu M, John S, Berg M, Leonard WJ. Functional association of Nmi with Stat5 and Stat1 in IL-2- and IFNγ-mediated signaling. Cell 1999;96:121-30
69. Huang M, Qian F, Hu Y, Ang C, Li Z, Wen Z. Chromatin-remodelling factor BRG1 selectively activates a subset of interferon-α-inducible genes. Nat Cell Biol 2002;4:774-81
70. Nusinzon I, Horvath CM. Interferon-stimulated transcription and innate antiviral immunity require deacetylase activity and histone deacetylase 1. Proc Natl Acad Sci USA 2003;100:14742-7
71. Ni Z, Karaskov E, Yu T, Callaghan SM, Der S, Park DS, Xu Z, Pattenden SG, Bremner R. Apical role for BRG1 in cytokine-induced promoter assembly. Proc Natl Acad Sci USA 2005;102:14611-6
72. Endo TA, Masuhara M, Yokouchi M, Suzuki R, Sakamoto H, Mitsui K, Matsumoto A, Tanimura S, Ohtsubo M, Misawa H, Miyazaki T, Leonor N, Taniguchi T, Fujita T, Kanakura Y, Komiya S, Yoshimura A. A new protein containing an SH2 domain that inhibits JAK kinases. Nature 1997;387:921-4
73. Starr R, Willson TA, Viney EM, Murray LJ, Rayner JR, Jenkins BJ, Gonda TJ, Alexander WS, Metcalf D, Nicola NA, Hilton DJ. A family of cytokine-inducible inhibitors of signalling. Nature 1997;387:917-21
74. Naka T, Fujimoto M, Kishimoto T. Negative regulation of cytokine signaling: STAT-induced STAT inhibitor. Trends Biochem Sci 1999;24:394-8
75. You M, Yu DH, Feng GS. Shp-2 tyrosine phosphatase functions as a negative regulator of the interferon-stimulated JAK-STAT pathway. Mol Cell Biol 1999;19:2416-24
76. Aoki N, Matsuda T. A nuclear protein tyrosine phosphatase TC-PTP is a potential negative regulator of the PRL-mediated signaling pathway: dephosphorylation and deactivation of signal transducer and activator of transcription 5a and 5b by TC-PTP in nucleus. Mol Endocrinol 2002;16:58-69
77. Simoncic PD, Lee-Loy A, Barber DL, Tremblay ML, McGlade CJ. The T cell protein tyrosine phosphatase is a negative regulator of Janus family kinases 1 and 3. Curr Biol 2002;12:446-53
78. Chung CD, Liao J, Liu B, Rao X, Jay P, Berta P, Shuai K. Specific inhibition of Stat3 signal transduction by PIAS3. Science 1997;278:1803-5
79. Shuai K. Modulation of STAT signaling by STAT-interacting proteins. Oncogene 2000;19:2638-44
80. O'Shea JJ, Pesu M, Borie DC, Changelian PS. A new modality for immunosuppression: targeting the JAK/STAT pathway. Nat Rev Drug Discov 2004;3:555-564
81. Rakesh K, Agrawal DK. Controlling cytokine signaling by constitutive inhibitors. Biochem Pharmacol 2005;70:649-57
82. Changelian PS, Flanagan ME, Ball DJ, Kent CR, Magnuson KS, Martin WH et al. Prevention of organ allograft rejection by a specific Janus kinase 3 inhibitor. Science 2003;302:875-8
83. Conklyn M, Andresen C, Changelian P, Kudlacz E. The JAK3 inhibitor CP-690550 selectively reduces NK and CD8+ cell numbers in cynomolgus monkey blood following chronic oral dosing. J Leukoc Biol 2004;76:1248-55
84. Desrivières S, Kunz C, Barash I, Vafaizadeh V, Borghouts C, Groner B. The biological functions of the versatile transcription factors STAT3 and STAT5 and new strategies for their targeted inhibition. J Mammary Gland Biol Neoplasia 2006;11:75-87
85. Ferrajoli A, Faderl S, Ravandi F, Estrov Z. The JAK-STAT pathway: a therapeutic target in hematological malignancies. Curr Cancer Drug Targets 2006;6:671-9
86. Kudlacz E, Perry B, Sawyer P, Conklyn M, McCurdy S, Brissette W, Flanagan M, Changelian P. The novel JAK-3 inhibitor CP-690550 is a potent immunosuppressive agent in various murine models. Am J Transplant 2004;4:51-7
87. Turkson J, Ryan D, Kim JS, Zhang Y, Chen Z, Haura E, Laudano A, Sebti S, Hamilton AD, Jove R. Phosphotyrosyl peptides block Stat3-mediated DNA binding activity, gene regulation, and cell transformation. J Biol Chem 2001;276:45443-55
88. Ren Z, Cabell LA, Schaefer TS, McMurray JS. Identification of a high-affinity phosphopeptide inhibitor of Stat3. Bioorg Med Chem Lett 2003;13:633-6
89. Lui VW, Boehm AL, Koppikar P, Leeman RJ, Johnson D, Ogagan M, Childs E, Freilino M, Grandis JR. Antiproliferative mechanisms of a transcription factor decoy targeting signal transducer and activator of transcription (STAT) 3: the role of STAT1. Mol Pharmacol 2007;71:1435-43
90. Catalano RD, Johnson MH, Campbell EA, Charnock-Jones DS, Smith SK, Sharkey AM. Inhibition of Stat3 activation in the endometrium prevents implantation: a nonsteroidal approach to contraception. Proc Natl Acad Sci USA 2005;102:8585-90
91. Turkson J, Kim JS, Zhang S, Yuan J, Huang M, Glenn M, Haura E, Sebti S, Hamilton AD, Jove R. Novel peptidomimetic inhibitors of signal transducer and activator of transcription 3 dimerization and biological activity. Mol Cancer Ther 2004;3:261-9
92. Quarcoo D, Weixler S, Groneberg D, Joachim R, Ahrens B, Wagner AH, Hecker M, Hamelmann E. Inhibition of signal transducer and activator of transcription 1 attenuates allergen-induced airway inflammation and hyperreactivity. J Allergy Clin Immunol 2004;114:288-95
93. Xi S, Gooding WE, Grandis JR. In vivo antitumor efficacy of STAT3 blockade using a transcription factor decoy approch: implication for cancer therapy. Oncogene 2005;24:970-9
94. Hölschermann H, Stadlbauer TH, Wagner AH, Fingerhuth H, Muth H, Rong S, Güler F, Tillmanns H, Hecker M. STAT-1 and AP-1 decoy oligonucleotide therapy delays acute rejection and prolongs cardiac allograft survival. Cardiovasc Res 2006;71:527-36
95. Hückel M, Schurigt U, Wagner AH, Stöckigt R, Petrow PK, Thoss K, Gajda M, Henzgen S, Hecker M, Bräuer R. Attenuation of murine antigen-induced arthritis by treatment with a decoy oligonucleotide inhibiting signal transducer and activator of transcription-1 (STAT-1). Arthritis Res Ther 2006;8:R17
96. Stojanovic T, Scheele L, Wagner AH, Middel P, Bedke J, Lautenschläger I, Leister I, Panzner S, Hecker M. STAT-1 decoy oligonucleotide improves microcirculation and reduces acute rejection in allogeneic rat small bowel transplants. Gene Ther 2007;14:883-90

97. Song H, Wang R, Wang S, Lin J. A low-molecular-weight compound discovered through virtual database screening inhibits Stat3 function in breast cancer cells. Proc Natl Acad Sci USA 2005;102:4700-5
98. Siddiquee K, Zhang S, Guida WC, Blaskovich MA, Greedy B, Lawrence HR, Yip ML, Jove R, McLaughlin MM, Lawrence NJ, Sebti SM, Turkson J. Selective chemical probe inhibitor of Stat3, identified through structure-based virtual screening, induces antitumor activity. Proc Natl Acad Sci USA 2007;104:7391-9
99. Natarajan C, Bright JJ. Curcumin inhibits experimental allergic encephalomyelitis by blocking IL-12 signaling through Janus kinase-STAT pathway in T lymphocytes. J Immunol 2002;169:6506-13
100. Lee YK, Isham CR, Kaufman SH, Bible KC. Flavopiridol disrupts STAT3/DNA interactions, attenuates STAT3-directed transcription, and combines with the Jak kinase inhibitor AG490 to achieve cytotoxic synergy. Mol Cancer Ther 2006;5:138-48
101. Aggarwal BB, Sethi G, Ahn KS, Sandur SK, Pandey MK, Kunnumakkara AB, Sung B, Ichikawa H. Targeting signal-transducer-and-activator-of-transcription-3 for prevention and therapy of cancer: modern target but ancient solution. Ann NY Acad Sci 2006;1091:151-69
102. Lynch RA, Etchin J, Battle TE, Frank DA. A small-molecule enhancer of signal transducer and activator of transcription 1 transcriptional activity accentuates the antiproliferative effects of IFN-γ in human cancer cells. Cancer Res 2007;67:1254-61

Influence of Mechanical Effects on Cells. Biophysical Basis

D. Jones

Contents

8.1	Introduction	83
8.2	Differences in Response	83
8.2.1	What Are the Downstream Transduction Mechanisms?	84
8.2.2	Why Is There a Lag in Response?	84
8.3	Stretch Sensors and Fluid Shear Flow Sensors	85
8.3.1	Other Mechanisms	85
8.4	Which Cells?	85
8.4.1	What Properties Must a Bone Mechanosensor Have to Fit the Known Data?	86
	References	87

8.1 Introduction

Mechanosensing is a phenomenon where a deformation is transduced by a biophysical mechanism into a biochemical response that can elicit movement and/or changes in gene expression. Many systems are well understood in various organisms and organs, for example hearing (vibration) and touch, which involve sensing nerve cells. However for tissues in many multicellular organisms especially those involved in feedback coupling for fitness, such as muscle, lungs, and blood circulation the phenomenon is not well understood at the biochemical level and especially not for the biophysical level. The possible biophysical mechanisms are the subject of this chapter specifically.

There are a number of outstanding issues that still need resolving:
1. How many cells and cell types contribute to mechanosensing in the tissue?
2. What is the biophysical nature of the mechanosensor or mechanosensors?
3. How many biochemical transduction mechanisms are there?
4. How does mechanosensing result in the overall tissue response?
5. Is there a differentiation between physiological load and hyper-physiological load?

Here we briefly review mechanosensing and then specifically consider the present state of the art in bone cells.

8.2 Differences in Response

The first thing to point out in bone mechanosensing is that there are significant differences between individuals in both response to loading and response to unloading and that these differences are genetic. Most data comes from animal studies, but studies in humans show the same complex pattern. For instance, different strains of mice have different breaking strengths of bone [1]. Some of the strains in this study were also studied by Judex et al. [2]

finding that different strains of mice react differently to use and disuse, BALB/cByJ and C57BL/6J responding to low level stimulation and C3H/HeJ not at all while BALB/cByJ lost significantly more than the other two when in disuse, C3H/HeJ being a weak bone and BALB/cByJ being amongst the strongest in the Wergedal study. This was shown subsequently to be due to inheritance of several different loci on several chromosomes, indicating multiple factors causing these effects some of which have been identified [3]. The underlying biological mechanisms of use are constant within a genetic group but vary strongly in the population. Either this means that there are two mechanosensors (one for use the other for disuse) or one sensor that is regulated downstream differently by a complex group of factors, hormones, and cytokines. The complex pattern of inheritance of multiple gene products above indicate that each function (bone formation or resorption) has a separate downstream regulation.

Partially reflecting this, human studies by Rubin et al. [3] have shown differences in response related to the body mass (the more adipose tissue the less the increase, but thinner women lose more bone mass in general). What is not known is are these differences due to variation in the mechanosensor per se or the downstream transduction?

8.2.1
What Are the Downstream Transduction Mechanisms?

In most vertebrate cells studies the biochemical signaling pathways seems to be more or less the same in all cells from most tissues so far studied. However although we can see often an immediate (within a second) cytoskeletal movement response to applied stretches, the first biochemical transduction mechanism (release of intracellular free calcium) is first detected 90–120 sec after the stimulus and often only after repeated stimulation. (The other second messengers are dependent on this but happen at the same time.)

A phospholipase C (β2) is activated after a pause of 90–120 sec of a 1 or 10 Hz stretch [4]. Immediately many second messenger systems are activated. The IP3 released by the PLC releases intracellular free calcium activating cGMP and subsequently iNOS; PKC activated by PLC stimulates ERK I and II, PLA2, COX thus prostaglandin synthesis. These in turn lead to a plethora of further downstream signaling and gene activation. What is not known and for which there is no consensus, is the nature and location of the mechanosensor(s). Some very puzzling properties that are known do not fit any of the present hypotheses. One consideration to be taken into account by the ion channel theorists is that downstream biochemical pathway signaling from ion channel activation involving any the above second messengers are unknown, after 40 years of investigation. In fact the only cells in which ion channel activity is known to release biochemical signaling molecules (and those involved in bone signaling) are nerve cells. This will be discussed further below.

8.2.2
Why Is There a Lag in Response?

The first and perhaps most basic mechanosensing system possessed by all cells to the same degree is that for osmotic pressure. Osmotic control of cells, including bacteria, is sensed mechanically directly in the membrane which transfers the energy (Joules) into ion channels. In this specific case one can talk perhaps correctly about membrane strain (at least at the membrane protein interface) and interactions with specific ion channels but nowhere else. These mechanosensors are well investigated and described in the literature at the biophysical and biochemical level [5]. Confusion often exists, however, in the minds of people investigating mechanosensing as to the difference between the osmotic sensing system and other forms of mechanosensing. A distinction exists between them and the biophysics of the loading mechanosensor, the biological transduction is also quite distinct. Direct stimulation of the membrane for instance, or interference with ion channels may directly affect this osmo-regulating mechanism and lead to confusion in the results. It is in fact very difficult to make a direct link between an ion channel regulation and downstream biochemical signal such as those mentioned above. However as we shall seem it is difficult in any case to make a direct link to mechanical stimulation and downstream biochemical signaling. Ion channels are well known however to be involved in release of neurotransmitters in nerve cells. Nerve cells are also mechanosensors in many organs, so might be a possible candidate for the main target of mechanotransduction as opposed to other cell types.

Cells use mechanosensing as part of a mechanism to apply force to the substrate. Most fibroblast-like cells will sense over very tiny areas of their membrane the substrate stiffness and react to this by moving, generally onto the stiffer surface [10]. In this case the sensing is through the actin cytoskeleton (primary cilia above are tubulin structures). That this mechanosensor could also be involved is an attractive possibility.

8.3
Stretch Sensors and Fluid Shear Flow Sensors

In vertebrates kidneys possess fluid shear flow sensors derived from the primary cilium. Other tissue types such as the oviduct and lungs possess cilia, which are active organs of movement. Cilia also possess a sensing function and feedback control [6]. The so-called primary cilia also exists in all cell types, being, during cell division, one half of the apparatus for chromosome separation and is located at the cell surface. It has been suggested that, as in the cilia of the kidney, primary cilia act as shear flow sensors in bone cells [7, 8]. Their studies preclude any involvement of ion channels and calcium influx, which also contradicts many other studies. However, any fluid force acting on a cell must also deform it. It is therefore difficult to distinguish between the two biophysical mechanisms, detection at the fluid shear surface or detection or at the cytoplasm shear surface (internal).

8.3.1
Other Mechanisms

Other mechanisms that have been suggested are through caveolae, for example [9]. However these structures might be involved as part of any other mechanosensing structures, including the primary cilium and ion channel modification.

8.4
Which Cells?

All cells are mechanosensitive. It is a question as to what type of deformation, amplitude, and frequency. It is not true to say that all deformation parameters are equal!

The question can be reformulated: what cells are most important in the physiological regulation of bone? The fact that all cells are sensitive makes distinguishing one from the other difficult. It could also be that all cell types play a role! It has become fashionable to believe that osteocytes are the mechanosensing system in bone.

Skerry first noticed that osteocytes seemed to respond extremely quickly to mechanical loads in bone [11] when he noticed a rapid increase in glucose 6-phosphate dehydrogenase. Whether this particular enzyme was the one actually measured is not so important as to the speed in which the osteocytes responded. Osteocytes produce many signaling molecules some of which are specific to the osteocyte including sclerostin, which inhibits Wnt signaling and bone formation. Animals that lack this protein have much increased bone formation. We could thus speculate that single point mutation polymorphisms can be at least partially responsible for the genetic difference found in animals mentioned above [12]. Noticing that osteocytes are in an extensive network and believing that gap junctions [13] formed a syncytial-like pathway through the bone, many investigators also believe that communication is through the gap junctions between osteocytes. However gap junctions do not mediate communication. Shirmmacher et al. [14] investigated properties of bone gap junctions and like all other gap junctions they are not strongly coupled. Firstly, only relatively small molecules pass through (circa MW 300); they are only electrochemically 50% coupled; there is no amplification mechanism so any signal passes only 3 cells before dying away and many signals such as NO and Ca^{2+} are very short lived. As we see below very small but high frequency signals also stimulate bone so the sensor is most likely not flow at all! In any case flows over the osteocyte are not measured at this time.

One type of cell present extensively in bone with known long transmission and amplification systems and known to have specialized forms for mechanosensing, which also have an extensive nerve-like network in bone and also produce bone stimulating factors as a result of ion channel activity is—nerve cells themselves!

There is no obvious logical reason to exclude any cell from the process of tissue regulation by mechanical loading. Most cells as pointed out above are mechanosensitive. The role of gap junctions is to smooth signaling; thus, more cells that sense the more regulated the response will be.

However, many aspects of bone mechanosensing are contradictory. If we assume that fluid shear flow is the main sensing mechanism, i.e., the same as in endothelial cells, then how is it that bone responds also to very small amplitudes? A very curious finding is that the smaller the amplitude the more repetitions are needed and that these repetitions can have a high frequency such that $30 \times 3,000\,\mu str$ cycles at 1 Hz can be replaced by $3,000 \times 30\,\mu str$ cycles at 100 Hz.

Using the Zetos system [15], which applies defined compression cycles to bone, we applied a typical walking pattern of deformation. Using a fast Fourier transform (FFT) the waveform was analyzed as to the frequency and energy spectrum. Applying a software filter to split the waveform into high and low frequencies (in this case above and below 30 Hz) we found that each half of the waveform was equally potent in stimulating stiffness increases.

With very tiny amplitudes of $30\,\mu\varepsilon$ and less there can be no flow inside the bone. The fact that many repetitions are needed—even at high amplitudes (around 36 at $3,000\,\mu\varepsilon$) means that the cells have a memory of the previous deformation. Separating the repetitive stimulus period (in stretching or bending experiments) with short pauses increases the response. This also emphasizes that there is a memory effect [16]. Over a wide range of amplitudes and frequencies this relationship is linear, which indicates that it is the stored energy (Joules J) that activates; above a certain threshold formation is promoted over breakdown. This is true over a very large frequency range—from 1 Hz into ultrasound. (The biochemical transduction mechanism of ultrasound is also through PLC-PKC [17].) The cell thus integrates the amount of energy applied over a short time (several minutes per day). The cells have a memory which is short. These findings imply a relaxation period whereby the cells become insensitive and then sensitive again after a short period, however, storing the memory of previous stimulations. By what mechanism can the energy of tiny vibrations over an enormously large frequency range be stored? What is the mechanism of energy accumulation?

As far as can be said from the limited amount of work done on the amplitude/cycle number/frequency relationship it appears to be linear. In other words, 300 cycles (at 1 Hz) (i.e., 5 min) and $3,000\,\mu str$ can be equated to 30,000 cycles at $30\,\mu str$ over 5 min) and 300,000 cycles at $3\,\mu str$ (also over 5 min) and 3,000,000 cycles—i.e., 1, 10, and 100 and 1 kHz and beyond, at least to 45 kHz. At 120 Hz Rubin and colleagues are stimulating bone on vibrating platforms. Thus small vibrations in the tissue (thus the cells) of $30\,\mu str$ or 0.003% across a cell is roughly 0.5 nm. This is within the thermal motion of the molecules (1–4 nm with a force of 1–4 pN). So far there are good reasons to believe that the critical energy accumulation to stimulate must be within a period of 15–30 min, not spread out over a longer period. Thus the frequency is also a critical factor and this implies that it is power (energy in Joules per unit of time) that is critical and not the energy per se.

8.4.1
What Properties Must a Bone Mechanosensor Have to Fit the Known Data?

1. Sensitive to a very wide frequency range from less than 1 Hz to 45 kHz.
2. Amplitudes from 3,000 µstr at 1 Hz to 0.3 µstr at 10 kHz.

3. The number of cycles range from 300 at 1 Hz and 3,000 μstr to several hundred thousand at the higher frequency/lower amplitude (this appears to be a linear relationship).
4. The stimulation can take place in just 5 min per day.
5. One stimulation is not enough.

What are the possibilities for a mechanical accumulator? Well, a wind-up watch or the winding mechanism of a crossbow are mechanisms to store energy. If the spring has a relaxation time (made of a slowly recycling material for instance) the energy must be stored in a short period of time within this relaxation time—a ratchet is a possibility. Feynman proposed the idea of a molecular ratchet some years ago [18]. In this case his idea was to show that the idea of converting Brownian motion into useful work would not be feasible.

If we consider the power and not just energy, then small vibrations which are very fast contain much more power than the local Brownian motion and thus is above the thermal background. In other words, we are adding energy into a system at a rate faster than the relaxation time. One place where such a mechanism could be placed is in the actin myosin contraction network at the base of the cell.

An alternative mechanism which also fits the data is based on biochemical storage. Many second messengers have a limited life span, they are broken down or in the case of calcium ions pushed out of the cell or returned to internal stores. One could suppose that the mechanosensitive cell is affected by a small disturbance and a small number of signaling molecules are produced; these will have a half-life of around 40 min. Repeated stimulation results in an increasing number of signaling molecules until a threshold is reached and the mechanosignaling process is triggered. This occurs against a background of the signaling molecules being reduced in the cell by the normal processes. Thus larger disturbances don't have to occur so often and smaller ones do.

Presently many groups seem to be busy in gathering data in order to prove one hypothesis or the other rather than actually testing the various hypotheses, so it might be a long time before there is any significant progress in this area.

References

1. Wergedal JE, Sheng MH, Ackert-Bicknell CL, Beamer WG, Baylink DJ. Genetic variation in femur extrinsic strength in 29 different inbred strains of mice is dependent on variations in femur cross-sectional geometry and bone density. Bone. 36(1):111-22, 2005
2. Judex S, SquireRG-M, DonahueL-R, Rubin C. Genetically Based Influences on the Site-Specific Regulation of Trabecular and Cortical Bone Morphology J. of Bone and Mineral Research,:19:600-606, 2004
3. Rubin C, Recker, R, Diane Cullen D, John Ryaby J, McCabe J, McLeod K Prevention of Postmenopausal Bone Loss by a Low-Magnitude, High-Frequency Mechanical Stimuli: A Clinical Trial Assessing Compliance, Efficacy, and Safety. Journal of Bone and Mineral Research 19:343-350, 2004
4. Hoberg M, Gratz HH, Noll M, Jones DB Mechanosensitivity of human osteosarcoma cells and phospholipase C β2 expression BBRC 333 142-149, 2005
5. Sukharev, S I, Blount, P, Martinac, B, Blattner FR, Kung, C. A large-conductance mechanosensitive channel in E. coli encoded by mscL alone. Nature 368, 265-268, 1994
6. Satir P. The Physiology Of Cilia and Mucociliary Interactions Annu. Rev Physiol 52:137-155, 1990
7. Jones D Leivseth G and Tenbosch J. Mechano-Reception in Osteoblast-like Cells. Biochemistry and Cell Biology 73:525-534, 1995
8. Malone A, Anderson T, Tummala P, Kwon RY. Johnston TR, Stearns T R. Jacobs CR Primary cilia mediate mechanosensing in bone cells by a calcium-independent mechanism PNAS 104:13325-13330, 2007
9. Spisni E. Mechanosensing role of caveolae and caveolar constituents in human endothelial cells. J Cell Physiol 197:198-204, 2003
10. Lo, C-M Wang H-B Dembo M and Yu-li Wang. Cell Movement Is Guided by the Rigidity of the Substrate Biophys J,79:144-152, 2000
11. Skerry TM, Bitensky L, Chayen J, Lanyon LEJ. Early strain-related changes in enzyme activity in osteocytes following bone loading in vivo. Bone Miner Res (5):783-788 1989
12. ten Dijke P, Krause C, de Gorter DJ, Löwik CW, van Bezooijen RL. Osteocyte-derived sclerostin inhibits bone formation:its role in bone morphogenetic protein and Wnt signaling. J Bone Joint Surg Am 1:31-5, 2008
13. Doty SB. Morphological evidence of gap junctions between bone cells. Calcif Tissue Int. 33(5):509-12, 1981
14. Schirrmacher K, Schmitz I, Winterhager E, Traub O Brummer F, Jones D & Bingmann D Characterization of gap junctions between osteoblast-like cells in culture. Calcif.Tissue Int. 51:285-290, 1992
15. David V et al. Ex Vivo Bone Formation in Bovine Trabecular Bone Cultured in a Dynamic 3D Bioreactor Is Enhanced by Compressive Mechanical Strain Tissue Engineering 14:117-126, 2008
16. Skerry TM One mechanostat or many. modifications of the site-specific response of bone to mechqnical loading in nature and nurture. J Musculoskeletal Neuronal Interactions 6:122-127, 2006

17. Jones DB, Fischler H Ultrasound stimulates mitosis and the PI-PLC-PKC pathway in a dose dependant manner in osteoblast-like cells. Calcif Tiss Int 44 (supp):S-97 P5, 1989

18. Feynman RP, Leighton RB, Sands M. The Feynman Lectures on Physics Addison-Wesley, Reading, Massachusetts, 1966, Vol. I, Chapter 46

Part C
Engineering Strategies

III Engineering at the Genetic and Molecular Level

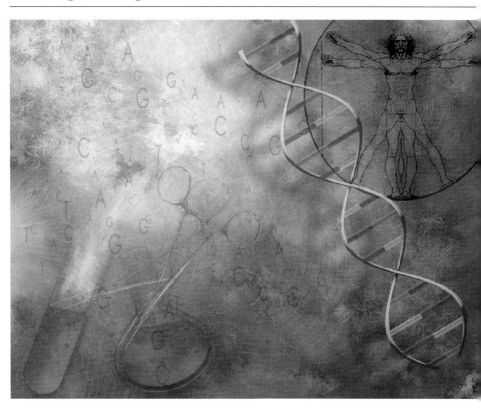

Towards Genetically Designed Tissues for Regenerative Medicine

W. Weber, M. Fussenegger

Contents

9.1　Introduction 93
9.2　Inducible Expression Control in Mammalian Cells 93
9.2.1　Regulation of Gene Expression at the Transcriptional Level 94
9.2.2　Regulation of Gene Expression at the Translational Level 98
9.2.3　Multiregulated Multigene Engineering 99
9.2.4　Synthetic Gene Networks—Emulating Nature's Design Principles 102
9.2.5　Cell-to-Cell Communication Systems 103
9.3　Delivery of Genetic Switches 103
9.4　Gene-based Tissue Engineering 103
9.5　Outlook 103
References 106

9.1 Introduction

Integration of fundamental research, elucidating cell differentiation of synthetically designed trigger-inducible promoters and of viral transduction technology, is the basis of tissue engineering for regenerative medicine [29]. Fundamental research has revealed numerous factors that play a role in targeted cell differentiation, such as bone morphogenetic proteins, BMPs [37], angiogenesis-related factors (VEGF, angiopoietin [71]) or cell-cycle regulators (cyclin-dependent kinase inhibitors [10]). These proteins can be introduced efficiently and safely into various cell types through the development of viral vectors, which are most commonly derived from adeno- and retroviruses [74]. However, as in small-molecule-based medicine, the dose of the therapeutic agent (i.e., the transgene) differentiates a poison from a remedy (Paracelsus, 1493–1541), thereby necessitating genetic tools for adjustment of the transgene expression levels into the therapeutic window. To achieve this, a variety of adjustable expression systems have been designed and have proven essential in almost all fields of mammalian cell technology including functional genomics [23], gene therapy [74], tissue engineering [73], drug discovery [2, 88], synthetic biology [32, 36] as well as the manufacturing of protein therapeutics [6, 13]. This chapter provides an overview of current inducible expression systems with a special focus on systems suitable for tissue engineering applications. Together with an overview of currently used viral vectors this chapter explains how the discovered proteins can be applied in gene-based tissue engineering for regenerative medicine.

9.2 Inducible Expression Control in Mammalian Cells

Trigger-inducible switches to control the expression of (sets of) target genes in tissue engineering applications must meet the following criteria:

- The inducer molecule, used to trigger gene expression, must be licensed for use in humans, must be devoid of pleiotropic side effects, and must have pharmacokinetic parameters to enable rapid penetration into the target tissue as well as rapid clearance from the tissue.
- The system should enable adjustable gene expression ranging from no or only negligible levels of expression in the OFF state to maximal expression, comparable to that of strong constitutive promoters in the ON state. Furthermore, the system should enable all intermediate expression levels by adjusting the concentration of the inducer molecule accordingly.
- Expression control should be reversible so that transgene expression can be timed for optimum control of cell growth, differentiation, and death.
- The molecular setup of the system should be compact (optimally one or two vectors) and should be compatible with different types of viral vectors for efficient transfer of the target gene.
- The regulating components of the system should not interfere with host cell physiology and should exhibit low immunogenicity.

Regulated target protein production can be achieved by inducible interference at any level of the protein biosynthesis chain, at the genome level through inducible excision of expression units [8, 28], at the transcriptional level by controlling the activity of synthetic or artificial promoters [9, 74], and at the translational level by interfering with translation or mRNA stability [9, 74], by inducibly blocking secretion [57] or finally by modulating the target protein through small molecules [58]. While the control of protein expression at the genome, secretion, or protein levels has been implemented successfully for specific applications, the highest compliance of the inducible expression systems with the above-mentioned selection criteria is found in regulation at the transcriptional and translational levels and in genetic networks based on these designs, which are discussed below.

9.2.1
Regulation of Gene Expression at the Transcriptional Level

Gene expression at the transcriptional level can be modulated either by reversibly blocking constitutive promoters or by conditionally activating minimal promoters devoid of enhancer elements according to the approaches described below (Fig. 9.1, Table 9.1).

- Small molecule-responsive DNA-binding proteins (Fig. 9.1a,b). Small molecule-regulated binding between allosteric bacterial repressor proteins and cognate DNA operator sequences can be used to regulate the activity of constitutive or minimal promoters. Constitutive promoters can be repressed by placing the operator sequence within or downstream of the promoter sequence so that binding of the repressor protein interferes with RNA polymerase binding or progression. Small-molecule-triggered release of the repressor protein from the operator sequence clears the DNA for subsequent transcription. The transcription-repressing effect of DNA-binding proteins can further be enhanced by fusion to trans-silencing domains like the Krüppel-associated box protein (KRAB, [3]) or a deacetylase [5] (Fig. 9.1a). On the other hand, minimal promoters (comprising a TATA box but devoid of enhancer sequences) can be activated by small-molecule-inducible binding of bacterial repressor proteins fused to transcription-activating domains (e.g., VP16 of *Herpes simplex*, [64]) to cognate operator sequences located 5´ of the promoter sequence (Fig. 9.1b). Inducible release of the DNA-binding protein-transactivator fusion deactivates promoter function, thereby resulting in transcriptional silence.
- Small-molecule-regulated protein dimerization (Fig. 9.1c). Protein pairs that can be dimerized by small drugs (e.g., FKBP and FRB dimerized by rapamycin derivatives or streptavidin and Avi-Tag dimerized by biotin [56, 80]) can be used to inducibly reconstitute functional transcription factors by fusing the first dimerization partner to a DNA-binding protein (e.g., yeast Gal4 or artificial zinc finger proteins) and the second to a transcriptional activation domain (e.g., VP16).

The target promoter consists of a minimal promoter fused in 3′ of an operator sequence specific to the DNA-binding protein (Fig. 9.1c).

- Small-molecule-triggered protein sequestration (Fig. 9.1c). Steroid hormone receptors are sequestrated by the cytoplasmatic heat shock protein 90 (Hsp90) and are released there from upon contact with their cognate steroid hormone, migrate into the nucleus and trigger activation of their cognate target promoters. In order to avoid interference with the endogenous hormone network, the hormone receptors were (i) mutated to be exclusively responsive to synthetic hormone analogs like tamoxifen or mifepristone, (ii) fused to DNA-binding domains recognizing specific operator sequences fused upstream of minimal promoters and (iii) fused to strong transactivation domains like VP16 or human p65 of NF-κB (Fig. 9.1d).

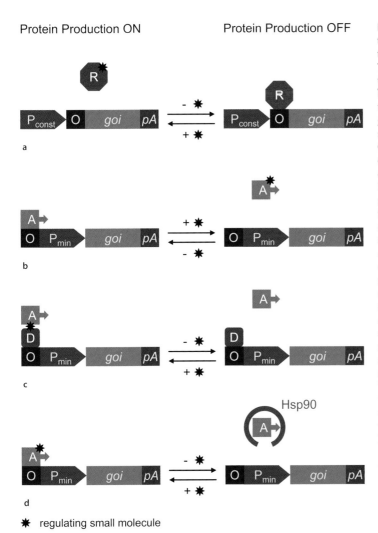

Fig. 9.1a–d Expression control strategies at the transcriptional level. Implementation of the different regulation strategies is given in Table 9.1. **a** Repressor-based expression control. A constitutive promoter (P_{const}) is blocked by binding a repressor protein (R) to an operator sequence (O) between the promoter and the gene of interest (*goi*). Addition of a regulating small molecule triggers dissociation of R and O and de-represses P_{const}. **b** Activator-based expression control. A minimal promoter (P_{min}, devoid of enhancer sequences) is activated by a transactivator (A) bound to its operator sequence (O) upstream of P_{min} resulting in transcription of the gene of interest (*goi*). Addition of a regulating small molecule dissociates A from O and deactivates P_{min}. The reverse configuration where A binds O in the presence of the inducing molecule has also been realized (e.g., rtTA, rCymR, AlcR, Table 9.1). **c** Dimerization-based expression control. Addition of a regulating small molecule dimerizes a DNA-binding protein (D) with an activation domain (A) resulting in activation of the minimal promoter (P_{min}) fused to an operator site (O) specific to the DNA-binding protein. In the absence of the dimerizing molecule, A cannot bind D and P_{min} remains silent. **d** Sequestration-based expression control. In the absence of a regulating small molecule the transactivator (A) is sequestrated in the cytoplasm by the heat shock protein 90 (Hsp90). Addition of the inducer triggers dissociation of A from Hsp90, translocation to the nucleus, binding to the specific operator sequence O, activation of the minimal promoter P_{min} and expression of the gene of interest (*goi*)

Table 9.1 Concepts for inducible expression control in mammalian cells. The regulation principle of the different systems (Figs. 9.1–9.3) is indicated in the first column

Principle	System	Mechanism	Ref.
9.1a	E-ON	The repressor E(-KRAB) binds the operator ETR8 in the absence of macrolide antibiotics and blocks transcription from the constitutive simian virus 40 promoter. In the presence of macrolide antibiotics the repressor is relieved and transcription resumes.	[81]
9.1a	PIP-ON	The repressor PIP(-KRAB) binds the operator PIR3 in the absence of streptogramin antibiotics and blocks transcription from the constitutive simian virus 40 promoter. In the presence of streptogramin antibiotics the repressor is relieved and transcription resumes.	[14]
9.1a	Q-mate repressor-based	The repressor CymR binds the operator CuO in the absence of cumate and blocks transcription from the constitutive cytomegalovirus promoter. In the presence of cumate the repressor is relieved and transcription resumes.	[47]
9.1a	Q-ON	The repressor ScbR-KRAB binds the operator OscbR8 in the absence of the butyrolactone SCB1 and blocks transcription from the constitutive simian virus 40 promoter. In the presence of SCB1 the repressor is relieved and transcription resumes.	[82]
9.1a	TET-ON, repressor-based	The repressor tTS-H4 (TetR-HDAC4) binds the operator tetO in the absence of tetracycline and blocks transcription from the constitutive HPRT promoter. In the presence of tetracycline the repressor is relieved and transcription resumes.	[5]
9.1a	T-REX	The repressor TetR binds the operator tetO2 in the absence of tetracycline and blocks transcription from the constitutive cytomegalovirus promoter. In the presence of tetracycline the repressor is relieved and transcription resumes.	[84]
9.1b	AIR	In the presence of acetaldehyde the transactivator AlcR binds its operator OAlcA and activates the minimal human cytomegalovirus promoter.	[79]
9.1b	E-OFF	In the absence of macrolide antibiotics the transactivator E-VP16 binds its operator ETR and activates the minimal human cytomegalovirus promoter.	[81]
9.1b	GyrB	The presence of coumermycin dimerizes the bacterial gyrase B subunits fused to the p65 transactivation domain and a phage λ-derived DNA-binding domain. The dimerized proteins bind their operator and activate the minimal cytomegalovirus promoter.	[87]
9.1b	PIP-OFF	In the absence of streptogramin antibiotics the transactivator PIP-VP16 binds its operator PIR and activates the minimal Drosophila heat shock protein 90 promoter.	[14]
9.1b	Q-mate activation-based	In the presence of cumate the transactivator rCymR binds its operator CuO6 and activates the minimal human cytomegalovirus promoter.	[47]
9.1b	QuoRex	In the absence of the butyrolactone SCB1 the transactivator ScbR-VP16 binds its operator OscbR and activates the minimal human cytomegalovirus promoter.	[83]
9.1b	TET-OFF	In the absence of tetracycline antibiotics the transactivator TetR-VP16 (tTA) binds its operator tetO7 and activates the minimal human cytomegalovirus promoter.	[22]
9.1b	TET-ON	In the presence of doxycycline the transactivator rTetR-VP16 (rtTA) binds its operator tetO7 and activates the minimal human cytomegalovirus promoter. The rtTA mutant rtTA2S-M2 shows superior regulation profiles.	[24,66]
9.1c	Biotin	In the presence of biotin, the avitagged VP16 transactivation domain is biotinylated by E.coli biotin ligase BirA. Biotinylated VP16 binds to streptavidin (SA) fused to the tetracycline repressor TetR. The dimerized protein complex (TetR-SA-Biotin-avitag-VP16) binds to the tetracycline operator tetO7 and activates the minimal cytomegalovirus promoter.	[77,80]
9.1c	Rapamycin	In the presence of rapamycin analogs FKBP fused to a DNA-binding zinc finger domain dimerizes with FRB fused to the human p65 transactivator. The dimerized protein complex (zinc finger-FKBP-Rapamycin-FRB-p65) is bound to its cognate operator element and activates the minimal cytomegalovirus promoter.	[56]

Table 9.1 *(continued)* Concepts for inducible expression control in mammalian cells. The regulation principle of the different systems (Figs. 9.1–9.3) is indicated in the first column

Principle	System	Mechanism	Ref.
9.1d	Estrogen	The estrogen receptor ligand-binding domain (ER-LBD), fused to the Gal4 DNA-binding domain and the VP16 transactivation domain, is sequestrated by the heat shock protein 90 (Hsp90) in the cytoplasm. In the presence of estrogen (or its analog tamoxifene), the tripartite fusion protein is dissociated from Hsp90, migrates to the nucleus, binds its Gal4 operator and activates a minimal adenovirus-derived promoter.	[7]
9.1d	Mifepristone (RU486)	The mutated progesterone receptor ligand-binding domain (PR-LBD), fused to the Gal4 DNA-binding domain and the VP16 transactivation domain is sequestrated by the heat shock protein 90 (Hsp90) in the cytoplasm. In the presence of mifepristone, the tripartite fusion protein is dissociated from Hsp90, migrates to the nucleus, binds its Gal4 operator and activates a minimal thymidine kinase promoter.	[70]
9.2a	E.REX, TET, QuoRex	The minimal cytomegalovirus promoters and polyadenylation sites in the E.REX, TET, and QuoRex systems were engineered for expression of short hairpin RNAs (shRNA), thereby enabling inducible shRNA under the control of the macrolide-, tetracycline- and butyrolactone-responsive promoters.	[38,61]
9.2b	Ribozyme	A self-cleaving hammerhead ribozyme (rz) in the 5´ untranslated region of the messenger RNA triggers mRNA destruction. In the presence of toyocamycin the ribozyme in inhibited, thereby preserving mRNA and enabling translation of the protein of interest.	[86]
9.2c	Neomycin	A stop codon downstream from the start codon triggers premature termination of protein translation. In the presence of neomycin the stop signal is overridden and the full-length protein of interest is translated.	[48]
9.3a	E.REX	One vector-based inducible shRNA expression. An intron-encoded shRNA was included in the macrolide-responsive transactivator coding sequence. In the absence of macrolide antibiotics, E-VP16 is expressed, binds its operator ETR and activates the minimal cytomegalovirus promoter in a positive feedback loop. At the same time the intron is eliminated by splicing, resulting in a functional shRNA to knock down the target gene mRNA. Addition of macrolide antibiotics interrupts the positive feedback loop by dissociating E-VP16 from its operator ETR.	

AlcR, *Aspergillus nidulans*-derived acetaldehyde-responsive transactivator; Avitag, synthetic biotinylation signal; BirA, *E.coli* biotin ligase; CuO, CymR-specific operator element; CymR, *Pseudomonas putida*-derived cumate-responsive repressor; E, *E.coli*-derived macrolide-responsive repressor; ER-LBD, estrogen receptor ligand-binding domain; ETR, E-specific operator (optionally multimerized, e.g., 8x); FKBP, FK-binding protein; FRB, FKBP-rapamycin-binding domain; GyrB, bacterial gyrase B subunit; HDAC4, human deacetylase 4; HPRT, hypoxanthine phosphoribosyl transferase; Hsp90, heat shock protein 90; KRAB, Krüppel-associated box protein-derived transcriptional repressor; O_{AlcA}, AlcR-specific operator element; O_{ScbR}, ScbR-specific operator element (optionally multimerized, e.g., 8x); p65, transactivation domain of human NF-κB; PIP, *Streptomyces pristinaespiralis*-derived streptogramin-responsive repressor; PIR, PIP-specific promoter element; PR-LBD, progesterone receptor ligand-binding domain; rCymR, *Pseudomonas putida*-derived cumate-responsive repressor engineered to bind its operator CuO in the presence of cumate; rTetR, *Tn10*-derived tetracycline-responsive repressor engineered to bind its operator tetO in the presence of doxycycline; rtTA, rTetR-VP16 fusion; rz, self-cleaving hammerhead ribozyme; SA, streptavidin; SCB1, butyrolactone inducer specific to ScbR; ScbR, *Streptomyces coelicolor*-derived butyrolactone-responsive repressor; shRNA, short hairpin RNA; tetO, TetR-specific operator element (optionally multimerized, e.g., 7x); TetR, *Tn10*-derived tetracycline-responsive repressor; tTA, TetR-VP16 fusion protein; VP16, *Herpes simplex* viral protein 16-derived transactivation domain

9.2.2
Regulation of Gene Expression at the Translational Level

Inducible expression approaches targeting mRNA are suitable for specific applications but commonly fail to show regulation comparable to that achieved by the transcription control systems (Fig. 9.2, Table 9.1).

- Inducible knockdown of the target gene mRNA by inducible expression of short hairpin RNAs (shRNA) (Fig. 9.2a). The conventional transcription control systems can be modified for the expression of shRNA, which specifically targets and primes complementary mRNA sequences for destruction [38, 61]. This approach is especially useful when endogenous genes are to be knocked down, e.g., in differentiation studies.
- Small-molecule-responsive ribozymes (Fig. 9.2b). Regulation at the mRNA level can be achieved by the integration of self-cleaving hammerhead ribozymes into the untranslated regions, resulting in autocatalytic mRNA destruction, which can be prevented by the addition of small molecule ribozyme inhibitors, thereby preserving the integrity of translation-competent mRNA [85, 86].
- Modulation of translation termination (Fig. 9.2c). Placing a stop codon within the open reading frame results in premature termination of translation and synthesis of nonfunctional protein products. Addition of aminoglycoside antibiotics (e.g., neomycin) overrides the stop codon and restores translation of the full-length protein [48].

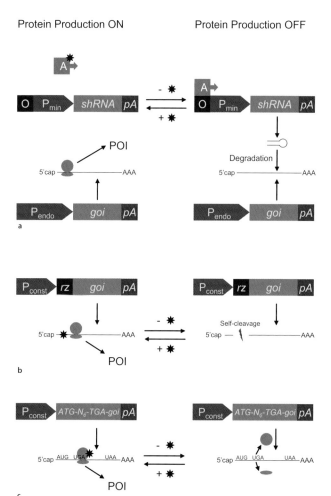

Fig. 9.2a–c Expression control strategies at the mRNA level. Implementation of the different regulation strategies is given in Table 9.1. **a** Coupled transcription and translation control. In the absence of the regulating small molecule, the transactivator (A) binds to its operator (O) and activates transcription from the minimal promoter (P_{min}) resulting in transcription of short hairpin RNA (*shRNA*). shRNA triggers degradation of complementary mRNA sequences of endogenous genes of interest (goi) transcribed under the control of an endogenous promoter (P_{endo}). In the presence of the regulating small molecule A is dissociated from O and shRNA expression is silent thereby preserving the integrity of goi mRNA and subsequent production of the protein of interest (POI). **b** Ribozyme-controlled gene expression. A hammerhead ribozyme (rz) placed in the 5′ untranslated region triggers self-cleavage and destruction of mRNA, thus shutting down translation of the gene of interest (*goi*). Inhibition of the ribozyme by adding regulating small molecules preserves mRNA integrity and translation of the goi to the protein of interest (POI). **c** Regulation by modulation of translational termination. A UGA stop codon, six nucleotides (N_6) downstream of the start codon (AUG), triggers premature termination of translation. Addition of aminoglycosides, such as neomycin, induces the overriding of the stop codon by the ribosomes, resulting in translation of full-length protein of interest (POI)

9.2.3 Multiregulated Multigene Engineering

Cellular differentiation and growth are rarely controlled by a single gene product, therefore necessitating the availability of genetic tools for the expression of multiple genes, either at the same time or for dual regulation, using mutually compatible expression control systems (Fig. 9.3).

- Coordinated expression of several genes (Fig. 9.3a). For the expression of several genes under the control of one inducible promoter, multicistronic expression vectors (pTRIDENT vectors [11, 78]) were developed, similar to those

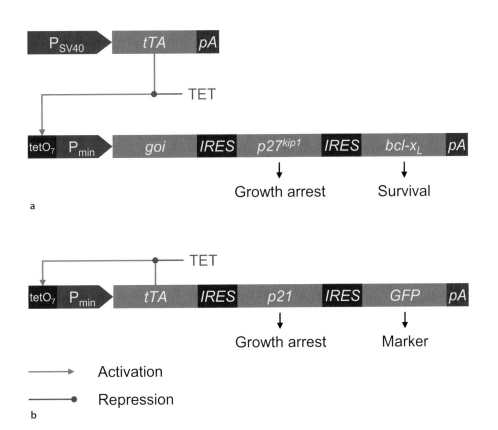

Fig. 9.3a–f Multigene-multiregulated expression control and synthetic gene networks. **a** Coordinated expression of several genes under the control of one inducible promoter. The simian virus 40 promoter (P_{SV40}) drives expression of the tetracycline-responsive transactivator (*tTA*, TetR-VP16), which, in the absence of tetracycline, binds to its operator $tetO_7$ and triggers activation of the minimal promoter P_{min} driving transcription of a multicistronic mRNA. The gene in the first cistron is translated in a classical cap-dependent manner, whereas initiation of translation of the genes in the second and third cistrons is mediated by internal ribosome entry sites (*IRES*), which trigger binding of the ribosome, independent of a 5′ cap structure. Such multicistronic vectors can be applied for multigene engineering strategies like the concomitant expression of a gene of interest (e.g., differentiation factor), a growth-arrest gene ($p27^{Kip1}$) to limit cell expansion and a survival factor ($bcl-x_L$) to reduce apoptosis [13]. **b** One vector-based autoregulated inducible expression. Multicistronic vectors can be applied for autoregulated designs, where the inducible promoter triggers expression of its own transactivator. Leaky mRNA transcripts, driven by P_{min}, are translated into the transactivator *tTA*, which binds its operator $tetO_7$ and further activates P_{min} in a positive feedback cycle. The positive feedback can be interrupted by addition of tetracycline (TET). Inducible expression of tTA is linked to the genes of interest by IRES elements (see Fig. 9.3a) on the same mRNA and can, for example, be applied for inducible growth arrest (mediated by p21) and for expression of a transfection marker (green fluorescent protein, *GFP*) [12]

in bacteria, where several proteins are translated from one mRNA. In these multicistronic expression vectors, the gene in the first cistron is translated in a classical cap-dependent manner, while translation of the downstream genes relies on internal ribosome entry sites (*IRES*), which promote ribosome assembly and translation (Fig. 9.3a).

- Autoregulated expression concepts (Fig. 9.3b). The above-described inducible expression systems commonly rely on two expression units, one containing the inducible promoter and one the transcriptional modulator (transrepressor or transactivator). In order to reduce genetic complexity and to facilitate implementation of inducible expression concepts, autoregulated designs have been developed, where the inducible promoter expresses its own transactivator in a positive feedback loop [12, 40] (Fig. 9.3b). Leaky transcripts, arising from the residual activity of the minimal promoter, trigger formation of a few transactivator molecules, which bind to and activate the inducible promoter leading to the production of more transactivator molecules. Addition of the

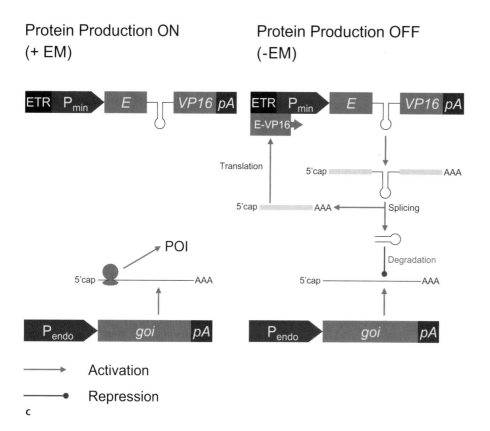

Fig. 9.3a–f (*continued*) Multigene-multiregulated expression control and synthetic gene networks **c** One vector-based expression of short hairpin mRNAs (shRNA) for inducible knockdown of target genes. Leaky transcripts from the minimal promoter (P_{min}) trigger translation of E-VP16, which binds the operator ETR and further activates P_{min} in a positive feedback loop. E-VP16 contains an intron-encoded shRNA, which is eliminated by splicing yielding translation-competent mRNA as well as free shRNA, triggering the degradation of complementary mRNA molecules transcribed by endogenous promoters (P_{endo}). Addition of erythromycin (+EM) disrupts the positive feedback loop and neither E-VP16 nor shRNA is produced thereby preserving endogenous mRNA and restoring translation of the protein of interest (POI) (see also Table 9.1)

regulating component, however, disrupts the positive feedback loop and results in transcriptional silence (Fig. 9.3b). This approach was recently used to construct an inducible one-promoter-based short hairpin RNA expression unit (Fig. 9.3c), a significant simplification, comparable to hitherto established two-vector-based inducible knockdown strategies [38].

- Differentially regulated gene expression (Fig. 9.3d). Independent control of several genes, such as growth or differentiation factors, requires mutually compatible inducible expression systems. The inducible expression systems responsive to macrolide, streptogramin, and tetracycline antibiotics are compatible and enable independently regulated expression of up to three transgenes within one cell [81]. Accordingly, a palette of expression vectors has been constructed for the easy installation of dual-regulated gene expression in mammalian cells (pDuoRex vectors, [15, 18, 46, 78]), which has been successfully validated in the controlled differentiation of C2C12 myoblasts into osteoblasts or adipocytes [16, 17] (Fig. 9.3d).

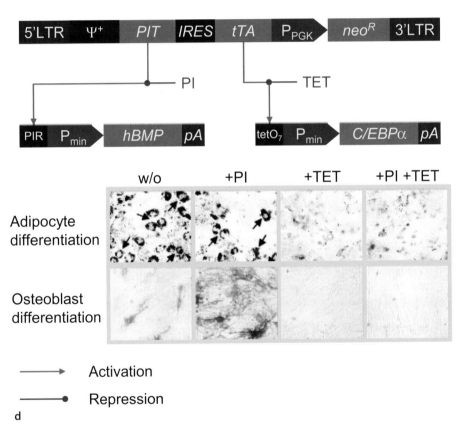

Fig. 9.3a–f (*continued*) Multigene-multiregulated expression control and synthetic gene networks **d** Dual-regulated expression for independent control of two transgenes. Mutually compatible inducible expression systems (e.g., TET and PIP systems, Table 9.1) can be used to independently regulate expression of two genes within one cell. A dicistronic (Fig. 9.3a) retroviral vector is used for expression of the transactivators PIT and tTA, which bind to their respective operator sequences PIR and tetO$_7$ and trigger activation of the minimal promoters (P_{min}) driving expression of human bone morphogenetic protein 2 (BMP-2) or of the CAAT enhancer binding protein α (C/EBPα). Addition of the streptogramin antibiotic pristinamycin I (PI) or of tetracycline (TET) disrupts PIT and tTA binding to their operators and deactivates transcription. Differential expression control of BMP-2 and C/EBPα enabled lineage control of C2C12 myoblast into adipocytes (triglyceride droplet staining) or osteoblasts (alkaline phosphatase staining) [17]

9.2.4
Synthetic Gene Networks—Emulating Nature's Design Principles

Expression of growth- and differentiation-controlling factors in mammalian cells is regulated by endogenous gene networks governing cell fate in space and time. Such regulatory networks display fundamental regulation motifs such as positive and negative feedback, signal transduction and amplification, hysteretic switching as well as epigenetic bistability [72]. The advent of synthetic biology has enabled the reconstruction of such genetic circuitries [25, 27, 30] by using well-defined modular building blocks such as transcriptional modulators and responsive promoters.

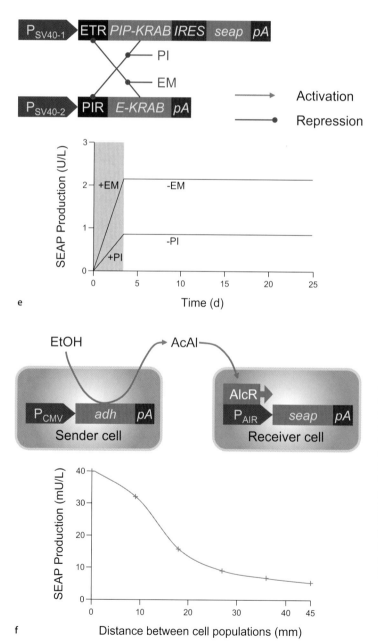

Fig. 9.3a–f (*continued*) Multigene-multiregulated expression control and synthetic gene networks **e** Bi-stable toggle switch for long-term sustained gene expression even in the absence of a regulating molecule. The first simian virus 40 promoter (P_{SV40-1}) triggers expression of the repressor PIP-KRAB and the reporter gene (SEAP, human placental secreted alkaline phosphatase) in a dicistronic configuration (Fig. 9.3a). PIP-KRAB binds to its operator PIR and represses P_{SV40-2} on the second plasmid, thereby preventing expression of E-KRAB, which in turn would bind to ETR and repress P_{SV40-1}. Therefore, P_{SV40-1} remains derepressed while P_{SV40-2} does not turn on unless a regulating compound is added. Addition of pristinamycin I (PI) for a short period of time inactivates PIP-KRAB and therefore derepresses P_{SV40-2}, which now triggers production of E-KRAB and represses P_{SV40-1}, leading to the sustained activation of P_{SV40-2}, even after the removal of regulating PI. The activity of the promoters can be switched back by adding erythromycin (EM) for transient inactivation of E-KRAB. **f** Cell-to-cell communication system. A sender cell, metabolically engineered for alcohol dehydrogenase (adh) production under the control of a cytomegalovirus immediate early promoter, converts ethanol in the medium into acetaldehyde, which diffuses via gas or the liquid phase to receiver cells, where it triggers activation of the promoter P_{AIR} by its transactivator AlcR (Table 9.1), resulting in the production of a reporter gene (SEAP, secreted alkaline phosphatase). It was shown [75] that this cell-to-cell communication can be used to translate the distance between the sender and receiver cell population into a graded expression readout, a mechanism which emulates the gradient-controlled differentiation of the developing embryo

In pioneering studies, antibiotic-responsive promoters were interconnected to perform Boolean logic operations [34] or to function as signal transduction and amplification cascades [35, 33]. Mathematical modeling of transcription control elements further enabled the rational design of higher-order networks, resulting in hysteretic switches [32] as well as gene circuitries showing programmable time-delay expression kinetics [76, 80]. A recent study describes an epigenetic bi-stable gene switch, which enables sustained gene expression even after removal of the inducing molecule, which initially triggered its expression state [36]. This scenario resembles a light switch, where short impulses reverse the output signal and might prove especially useful in applications where the transgene is expressed and silenced for prolonged periods. This design was realized by the interconnection of two repressible promoters in such a way that one promoter triggers the expression of a repressor specific to the second promoter [36] (Fig. 9.3e). Short-term inactivation of one or the other repressor by administering of the cognate antibiotic relieves promoter repression and reverses the gene expression status [36].

9.2.5
Cell-to-Cell Communication Systems

Cell-to-cell communication by means of hormones, cytokines, or gradient-forming signal molecules governs synchronization of the different organs in the body as well as growth and differentiation during embryogenesis (Fig. 9.3f). A recent study described the first synthetic communication system between mammalian cells based on sender cells metabolically engineered to produce acetaldehyde and receiver/effector cells harboring acetaldehyde-responsive promoters for perceiving the acetaldehyde signal and responding by expression of the appropriate transgene [75]. It was demonstrated that sender cells produce an acetaldehyde gradient around the sender cell colony, which is subsequently converted into a gene-expression gradient, a mechanism which mimics signal molecule production, diffusion, and translation into gene expression in the developing embryo (Fig. 9.1f) [59, 75].

9.3
Delivery of Genetic Switches

For the efficient and safe transfer of genetic information into mammalian cells, tissues, and whole organisms, viral transduction systems have prevailed, mainly lentiviral and adenoviral systems displaying highly efficient gene transfer, long-term stability, and acceptable safety profiles. Table 9.2 gives the criteria for selecting the viral vector, which might best fit a given application.

9.4
Gene-based Tissue Engineering

The functional characterization of differentiation- and growth-regulating factors in fundamental research together with the design of synthetic promoters for trigger-inducible gene activation control in space and time were driving forces in the development of gene-based tissue engineering studies in recent years. Attractive targets in those studies were the vascular system, where angiogenesis can be efficiently modulated by well-defined growth factors (VEGF, endostatin, [1, 52, 62]) or bone regeneration of critically sized calvarial effects, where de novo bone formation can be triggered by inducible expression of bone morphogenetic protein 2 [19]. In both cases, transgene expression must be controlled precisely in order to prevent severe side effects like uncontrolled vascularization or excessive ossification of the surrounding tissues. Table 9.3 presents an overview of gene-based tissue engineering studies for regenerative medicine.

9.5
Outlook

Fundamental research on cellular differentiation and growth, the design of synthetic inducible promoters and gene networks as well as the advent of clinically validated viral gene transfer vectors have progressed significantly within the last few years.

The three disciplines can now be combined to design new approaches in tissue engineering for regenerative medicine in vivo or ex vivo. The first approaches to be implemented in the clinics will most probably be based on the regulated expression of a single differentiation factor, for example in the fields of angiogenesis or bone formation, where research and development are most advanced. Depending on the results of these first studies, next-generation molecular interventions will enable the regeneration of tissues by means of multiple differentiation factors under control the of (semi-) synthetic gene networks for an optimum therapeutic outcome.

Acknowledgements

We thank Marcia Schoenberg for critical comments on the manuscript.

Table 9.2 Viral gene delivery tools for therapeutic applications. Frequency in clinical trials was retrieved from http://www.abedia.com/wiley/vectors.php

Property	Adeno-associated virus	Adenovirus	Lentivirus	Onco-retrovirus
Compatibility with inducible expression system	E.REX; Rapamycin, [69]; TET, [26]; Ribozyme, [86]	Mifepristone, [4]; PIP, [21]; Q-mate, [47]; Rapamycin, [68]; Tamoxifen, [62]; TET, [21]	AIR, [79]; Biotin, (Weber et al., submitted); E.REX, [42]; Mifepristone, [60]; Neomycin, [48]; PIP, [43]; Rapamycin, [31]; TET, [39]	E.REX, [78]; Mifepristone, [51]; PIP, [46]; Rapamycin, [53]; TET, [65]
Capacity	4.5 kb	5–8 kb (30–35, gutless, high capacity)	8 kb	8.5 kb
Long-term stability	Episomal, occasional stable integration	Episomal, transient in dividing cells	Stable integration	Stable integration
Safety concerns	Insertional mutagenesis	Inflammation	Insertional mutagenesis	Insertional mutagenesis
Side effects	Low, pre-primed patients possible	Pre-primed patients, possible inflammatory/immune response	Low	Low
Tropism	Broad, not suitable for hematopoetic cells	Broad	Broad	Dividing cells only
Frequency used in gene therapy trials	3.7%	26%	23%	

Table 9.3 Inducible gene expression technology for tissue engineering and regenerative medicine

System	Vector	Transgene	Application	Ref.
E-OFF	Lentiviral	VEGF	Inducible expression of vascular endothelial growth factor on the chicken chorioallantoic membrane conditionally triggered angiogenesis.	[42]
Hypoxia	Naked DNA	Heme oxygenase-1	Transfection of a hypoxia-inducible expression vector for heme oxygenase-1 into the left anterior ventricular wall and subsequent surgically induced myocardial infarction resulted in infarct-specific expression of heme oxygenase-1 and significantly reduced apoptosis in the infarct region.	[63]
PIP-OFF	Lentiviral	VEGF	Inducible expression of vascular endothelial growth factor on the chicken chorioallantoic membrane conditionally triggered angiogenesis.	[43]

Table 9.3 *(continued)* Inducible gene expression technology for tissue engineering and regenerative medicine

System	Vector	Transgene	Application	Ref.
PIP-OFF/TET-OFF	Retroviral/Transfection	myoD and msx-1	Differential expression of myoD and msx-1 by means of mutually compatible PIP-OFF and TET-OFF expression technologies enabled the inducible differentiation of C2C12 myoblasts into the myogenic, osteogenic, or adipogenic lineage.	[16]
PIP-OFF/TET-OFF	Retroviral/Transfection	C/EBP-α and BMP-2	Dual-regulated expression of BMP-2 and C/EBP-α using the PIP-OFF and TET-OFF systems enabled differential differentiation of C2C12 myoblasts into adipocytes or osteoblasts.	[17]
Rapamycin	Retroviral	Engineered thrombopoietin receptor	Isolated canine CD34+ cells were transduced with a thrombopoietin receptor engineered for multimerization and activation in the presence of the rapalog dimerizer. Following reimplantation, expansion of the transduced cells could be triggered by administration of the rapalog.	[49]
Tamoxifen	Adenoviral	Endostatin	Inhibition of VEGF-mediated retinal neovascularization and detachment of the retina in mice.	[62]
TET-OFF	Adenoviral	BMP-2	C9 mesenchymal stem cells were engineered for doxycycline-regulated expression of BMP-2. Implantation of these cells either into the tibialis anterior muscles or into the joints of CB17-severe combined immunedeficient bg mice induced cartilage and bone filled with bone marrow after 10 days.	[50]
TET-OFF	Electroporation	pdx-1	Production of insulin-secreting cells from embryonic stem cells. Embryonic stem cells were transfected with tetracycline-regulated expression vectors for pdx-1. Induction of pdx-1 resulted in the expression of insulin 2, glucokinase, neurogenin 3, and somatostatin in the resulting differentiated cells.	[44]
TET-OFF	Lentiviral	Human CNTF	Ciliary neurotrophic factor triggers neuroprotection in the quinolinic acid model of rats with Huntington's disease.	[54]
TET-OFF	Lentiviral	Mutated Huntingtin	Reversible expression of mutated huntingtin in the rat striatum reversibly induced Huntington's disease phenotypes.	[55]
TET-OFF	Retroviral	Runx2	Primary skeletal myoblasts were engineered for tetracycline-regulated Runx2 expression. In vitro induction of Runx2 triggered osteoblastic differentiation as monitored by alkaline phosphatase activity and matrix mineralization. In vivo inducible mineralization was achieved by intramuscular implantation of engineered myoblasts and application of tetracycline in the drinking water.	[20]
TET-ON	AAV	BMP-2	Bone regeneration by inducible expression of bone morphogenetic protein 2 in mice with critically sized calvarial defects.	[19]
TET-ON	Cells from transgenic mice	SV40 T antigen	Correction of hyperglycemia in diabetic mice transplanted with beta cells reversibly immortalized by inducible expression of the simian virus 40 T antigen.	[41]
TET-ON	Lentiviral	Tyrosine hydroxylase	Prototype treatment of Parkinson's disease by inducible expression of tyrosine hydroxylase in the rat striatum.	[67]
TET-ON	Retroviral	HPV-16-E6/E7	Primary human corneal epithelial (HCE) cells were transduced with retroviral vectors for doxycycline-inducible expression of human papilloma virus 16-E6-E7 genes for tight regulation of proliferation and normal differentiation as a model for scarcely available HCEs.	[45]
T-REX	Cell transfer	EGF	Regulated expression of the human epidermal growth factor in a porcine wound model.	[84]

AAV, adeno-associated virus; *BMP-2*, bone morphogenetic protein 2; *CNTF*, ciliary neurotropic factor; *EGF*, epidermal growth factor; *HCE*, human corneal epithelial cells; *HPV*, human papilloma virus; *SV40*, simian virus 40; Rapalog, molecule analog to rapamycin; *VEGF*, vascular endothelial growth factor

References

1. Abruzzese RV, Godin D, Mehta V et al. (2000) Ligand-dependent regulation of vascular endothelial growth factor and erythropoietin expression by a plasmid-based autoinducible GeneSwitch system. Mol Ther 2:276–287
2. Aubel D, Morris R, Lennon B et al. (2001) Design of a novel mammalian screening system for the detection of bioavailable, non-cytotoxic streptogramin antibiotics. J Antibiot (Tokyo) 54:44–55
3. Bellefroid EJ, Poncelet DA, Lecocq PJ et al. (1991) The evolutionarily conserved Kruppel-associated box domain defines a subfamily of eukaryotic multifingered proteins. Proc Natl Acad Sci U S A 88:3608–3612
4. Bhat RA, Stauffer B, Komm BS et al. (2004) Regulated expression of sFRP-1 protein by the GeneSwitch system. Protein Expr Purif 37:327–335
5. Bockamp E, Christel C, Hameyer D et al. (2007) Generation and characterization of tTS-H4:a novel transcriptional repressor that is compatible with the reverse tetracycline-controlled TET-ON system. J Gene Med 9:308–318
6. Boorsma M, Nieba L, Koller D et al. (2000) A temperature-regulated replicon-based DNA expression system. Nat Biotechnol 18:429–432
7. Braselmann S, Graninger P, Busslinger M (1993) A selective transcriptional induction system for mammalian cells based on Gal4-estrogen receptor fusion proteins. Proc Natl Acad Sci U S A 90:1657–1661
8. Brocard J, Feil R, Chambon P et al. (1998) A chimeric Cre recombinase inducible by synthetic,but not by natural ligands of the glucocorticoid receptor. Nucleic Acids Res 26:4086–4090
9. Fussenegger M (2001) The impact of mammalian gene regulation concepts on functional genomic research, metabolic engineering, and advanced gene therapies. Biotechnol Prog 17:1–51
10. Fussenegger M, Bailey JE (1998) Molecular regulation of cell-cycle progression and apoptosis in mammalian cells:implications for biotechnology. Biotechnol Prog 14:807–833
11. Fussenegger M, Mazur X, Bailey JE (1998) pTRIDENT, a novel vector family for tricistronic gene expression in mammalian cells. Biotechnol Bioeng 57:1–10
12. Fussenegger M, Moser S, Mazur X et al. (1997) Autoregulated multicistronic expression vectors provide one-step cloning of regulated product gene expression in mammalian cells. Biotechnol Prog 13:733–740
13. Fussenegger M, Schlatter S, Datwyler D et al. (1998) Controlled proliferation by multigene metabolic engineering enhances the productivity of Chinese hamster ovary cells. Nat Biotechnol 16:468–472
14. Fussenegger M, Morris RP, Fux C et al. (2000) Streptogramin-based gene regulation systems for mammalian cells. Nat Biotechnol 18:1203–1208
15. Fux C, Fussenegger M (2003) Toward higher order control modalities in mammalian cells-independent adjustment of two different gene activities. Biotechnol Prog 19:109–120
16. Fux C, Langer D, Fussenegger M (2004) Dual-regulated myoD- and msx1-based interventions in C2C12-derived cells enable precise myogenic/osteogenic/adipogenic lineage control. J Gene Med 6:1159–1169
17. Fux C, Mitta B, Kramer BP et al. (2004) Dual-regulated expression of C/EBP-alpha and BMP-2 enables differential differentiation of C2C12 cells into adipocytes and osteoblasts. Nucleic Acids Res 32:e1
18. Fux C, Moser S, Schlatter S et al. (2001) Streptogramin- and tetracycline-responsive dual regulated expression of p27(Kip1) sense and antisense enables positive and negative growth control of Chinese hamster ovary cells. Nucleic Acids Res 29:E19
19. Gafni Y, Pelled G, Zilberman Y et al. (2004) Gene therapy platform for bone regeneration using an exogenously regulated, AAV-2-based gene expression system. Mol Ther 9:587–595
20. Gersbach CA, Le Doux JM, Guldberg RE et al. (2006) Inducible regulation of Runx2-stimulated osteogenesis. Gene Ther 13:873–882
21. Gonzalez-Nicolini V, Fussenegger M (2005) A novel binary adenovirus-based dual-regulated expression system for independent transcription control of two different transgenes. J Gene Med 7:1573–1585
22. Gossen M, Bujard H (1992) Tight control of gene expression in mammalian cells by tetracycline-responsive promoters. Proc Natl Acad Sci U S A 89:5547–5551
23. Gossen M, Bujard H (2002) Studying gene function in eukaryotes by conditional gene inactivation. Annu Rev Genet 36:153–173
24. Gossen M, Freundlieb S, Bender G et al. (1995) Transcriptional activation by tetracyclines in mammalian cells. Science 268:1766–1769
25. Guido NJ, Wang X, Adalsteinsson D et al. (2006) A bottom-up approach to gene regulation. Nature 439:856–860
26. Haberman RP, McCown TJ (2002) Regulation of gene expression in adeno-associated virus vectors in the brain. Methods 28:219–226
27. Hasty J, McMillen D, Collins JJ (2002) Engineered gene circuits. Nature 420:224–230
28. Hunter NL, Awatramani RB, Farley FW et al. (2005) Ligand-activated Flpe for temporally regulated gene modifications. Genesis 41:99–109
29. Kelm JM, Kramer BP, Gonzalez-Nicolini V et al. (2004) Synergies of microtissue design, viral transduction and adjustable transgene expression for regenerative medicine. Biotechnol Appl Biochem 39:3–16
30. Kobayashi H, Kaern M, Araki M et al. (2004) Programmable cells:interfacing natural and engineered gene networks. Proc Natl Acad Sci U S A 101:8414–8419
31. Kobinger GP, Deng S, Louboutin JP et al. (2005) Pharmacologically regulated regeneration of functional human pancreatic islets. Mol Ther 11:105–111
32. Kramer BP, Fussenegger M (2005) Hysteresis in a synthetic mammalian gene network. Proc Natl Acad Sci U S A 102:9517–9522
33. Kramer BP, Weber W, Fussenegger M (2003) Artificial regulatory networks and cascades for discrete multilevel transgene control in mammalian cells. Biotechnol Bioeng 83:810–820
34. Kramer BP, Fischer C, Fussenegger M (2004) BioLogic gates enable logical transcription control in mammalian cells. Biotechnol Bioeng 87:478–484

35. Kramer BP, Fischer M, Fussenegger M (2005) Semi-synthetic mammalian gene regulatory networks. Metab Eng 7:241–250
36. Kramer BP, Viretta AU, Daoud-El-Baba M et al. (2004) An engineered epigenetic transgene switch in mammalian cells. Nat Biotechnol 22:867–870
37. Leboy PS (2006) Regulating bone growth and development with bone morphogenetic proteins. Ann N Y Acad Sci 1068:14–18
38. Malphettes L, Fussenegger M (2004) Macrolide- and tetracycline-adjustable siRNA-mediated gene silencing in mammalian cells using polymerase II-dependent promoter derivatives. Biotechnol Bioeng 88:417–425
39. Markusic D, Oude-Elferink R, Das AT et al. (2005) Comparison of single regulated lentiviral vectors with rtTA expression driven by an autoregulatory loop or a constitutive promoter. Nucleic Acids Res 33:e63
40. Mazur X, Eppenberger HM, Bailey JE et al. (1999) A novel autoregulated proliferation-controlled production process using recombinant CHO cells. Biotechnol Bioeng 65:144–150
41. Milo-Landesman D, Surana M, Berkovich I et al. (2001) Correction of hyperglycemia in diabetic mice transplanted with reversibly immortalized pancreatic beta cells controlled by the tet-on regulatory system. Cell Transplant 10:645–650
42. Mitta B, Weber CC, Fussenegger M (2005) In vivo transduction of HIV-1-derived lentiviral particles engineered for macrolide adjustable transgene expression. J Gene Med 7:1400–1408
43. Mitta B, Weber CC, Rimann M et al. (2004) Design and in vivo characterization of self-inactivating human and non-human lentiviral expression vectors engineered for streptogramin-adjustable transgene expression. Nucleic Acids Res 32:e106
44. Miyazaki S, Yamato E, Miyazaki J (2004) Regulated expression of pdx-1 promotes in vitro differentiation of insulin-producing cells from embryonic stem cells. Diabetes 53:1030–1037
45. Mohan RR, Possin DE, Mohan RR et al. (2003) Development of genetically engineered tet HPV16-E6/E7 transduced human corneal epithelial clones having tight regulation of proliferation and normal differentiation. Exp Eye Res 77:395–407
46. Moser S, Rimann M, Fux C et al. (2001) Dual-regulated expression technology: a new era in the adjustment of heterologous gene expression in mammalian cells. J Gene Med 3:529–549
47. Mullick A, Xu Y, Warren R et al. (2006) The cumate gene-switch: a system for regulated expression in mammalian cells. BMC Biotechnol 6:43
48. Murphy GJ, Mostoslavsky G, Kotton DN et al. (2006) Exogenous control of mammalian gene expression via modulation of translational termination. Nat Med 12:1093–1099
49. Neff T, Horn PA, Valli VE et al. (2002) Pharmacologically regulated in vivo selection in a large animal. Blood 100:2026–2031
50. Noel D, Gazit D, Bouquet C et al. (2004) Short-term BMP-2 expression is sufficient for in vivo osteochondral differentiation of mesenchymal stem cells. Stem Cells 22:74–85
51. Pollett JB, Zhu YX, Gandhi S et al. (2003) RU486-inducible retrovirus-mediated caspase-3 overexpression is cytotoxic to bcl-xL-expressing myeloma cells in vitro and in vivo. Mol Ther 8:230–237
52. Pollock R, Giel M, Linher K et al. (2002) Regulation of endogenous gene expression with a small-molecule dimerizer. Nat Biotechnol 20:729–733
53. Pollock R, Issner R, Zoller K et al. (2000) Delivery of a stringent dimerizer-regulated gene expression system in a single retroviral vector. Proc Natl Acad Sci U S A 97:13221–13226
54. Regulier E, Pereira de Almeida L, Sommer B et al. (2002) Dose-dependent neuroprotective effect of ciliary neurotrophic factor delivered via tetracycline-regulated lentiviral vectors in the quinolinic acid rat model of Huntington's disease. Hum Gene Ther 13:1981–1990
55. Regulier E, Trottier Y, Perrin V et al. (2003) Early and reversible neuropathology induced by tetracycline-regulated lentiviral overexpression of mutant huntingtin in rat striatum. Hum Mol Genet 12:2827–2836
56. Rivera VM, Clackson T, Natesan S et al. (1996) A humanized system for pharmacologic control of gene expression. Nat Med 2:1028–1032
57. Rivera VM, Wang X, Wardwell S et al. (2000) Regulation of protein secretion through controlled aggregation in the endoplasmic reticulum. Science 287:826–830
58. Rossi F, Charlton CA, Blau HM (1997) Monitoring protein-protein interactions in intact eukaryotic cells by beta-galactosidase complementation. Proc Natl Acad Sci U S A 94:8405–8410
59. Roth S, Stein D, Nusslein-Volhard C (1989) A gradient of nuclear localization of the dorsal protein determines dorsoventral pattern in the Drosophila embryo. Cell 59:1189–1202
60. Sirin O, Park F (2003) Regulating gene expression using self-inactivating lentiviral vectors containing the mifepristone-inducible system. Gene 323:67–77
61. Szulc J, Wiznerowicz M, Sauvain MO et al. (2006) A versatile tool for conditional gene expression and knockdown. Nat Methods 3:109–116
62. Takahashi K, Saishin Y, Saishin Y et al. (2003) Intraocular expression of endostatin reduces VEGF-induced retinal vascular permeability, neovascularization, and retinal detachment. Faseb J 17:896–898
63. Tang YL, Tang Y, Zhang YC et al. (2005) A hypoxia-inducible vigilant vector system for activating therapeutic genes in ischemia. Gene Ther 12:1163–1170
64. Triezenberg SJ, Kingsbury RC, McKnight SL (1988) Functional dissection of VP16, the trans-activator of herpes simplex virus immediate early gene expression. Genes Dev 2:718–729
65. Unsinger J, Kroger A, Hauser H et al. (2001) Retroviral vectors for the transduction of autoregulated, bidirectional expression cassettes. Mol Ther 4:484–489
66. Urlinger S, Baron U, Thellmann M et al. (2000) Exploring the sequence space for tetracycline-dependent transcriptional activators: novel mutations yield expanded range and sensitivity. Proc Natl Acad Sci U S A 97:7963–7968
67. Vogel R, Amar L, Thi AD et al. (2004) A single lentivirus vector mediates doxycycline-regulated expression of transgenes in the brain. Hum Gene Ther 15:157–165

68. Wang J, Voutetakis A, Zheng C et al. (2004) Rapamycin control of exocrine protein levels in saliva after adenoviral vector-mediated gene transfer. Gene Ther 11:729–733
69. Wang J, Voutetakis A, Papa M et al. (2006) Rapamycin control of transgene expression from a single AAV vector in mouse salivary glands. Gene Ther 13:187–190
70. Wang Y, O'Malley BW, Jr., Tsai SY et al. (1994) A regulatory system for use in gene transfer. Proc Natl Acad Sci U S A 91:8180–8184
71. Weber CC, Cai H, Ehrbar M et al. (2005) Effects of protein and gene transfer of the angiopoietin-1 fibrinogen-like receptor-binding domain on endothelial and vessel organization. J Biol Chem 280:22445–22453
72. Weber W, Fussenegger M (2002) Artificial mammalian gene regulation networks-novel approaches for gene therapy and bioengineering. J Biotechnol 98:161–187
73. Weber W, Fussenegger M (2004) Approaches for trigger-inducible viral transgene regulation in gene-based tissue engineering. Curr Opin Biotechnol 15:383–391
74. Weber W, Fussenegger M (2006) Pharmacologic transgene control systems for gene therapy. J Gene Med 8:535–556
75. Weber W, Daoud El-Baba M, Fussenegger M (2007) Synthetic ecosystems based on airborne inter- and intrakingdom communication. Proc Natl Acad Sci U S A:In press
76. Weber W, Kramer BP, Fussenegger M (2007) A genetic time delay circuitry in mammalian cells. Biotechnol Bioeng: in press.
77. Weber W, Bacchus W, Gruber F et al. (2007) A novel vector platform for vitamin H-inducible transgene expression in mammalian cells. Submitted for publication
78. Weber W, Marty RR, Keller B et al. (2002) Versatile macrolide-responsive mammalian expression vectors for multiregulated multigene metabolic engineering. Biotechnol Bioeng 80:691–705
79. Weber W, Rimann M, Spielmann M et al. (2004) Gas-inducible transgene expression in mammalian cells and mice. Nat Biotechnol 22:1440–1444
80. Weber W, Stelling J, Rimann M et al. (2007) A synthetic time-delay circuit in mammalian cells and mice. Proc Natl Acad Sci U S A 104:2643–2648
81. Weber W, Fux C, Daoud-el Baba M et al. (2002) Macrolide-based transgene control in mammalian cells and mice. Nat Biotechnol 20:901–907
82. Weber W, Malphettes L, de Jesus M et al. (2005) Engineered Streptomyces quorum-sensing components enable inducible siRNA-mediated translation control in mammalian cells and adjustable transcription control in mice. J Gene Med 7:518–525
83. Weber W, Schoenmakers R, Spielmann M et al. (2003) Streptomyces-derived quorum-sensing systems engineered for adjustable transgene expression in mammalian cells and mice. Nucleic Acids Res 31:e71
84. Yao F, Svensjo T, Winkler T et al. (1998) Tetracycline repressor, tetR, rather than the tetR-mammalian cell transcription factor fusion derivatives, regulates inducible gene expression in mammalian cells. Hum Gene Ther 9:1939–1950
85. Yen L, Magnier M, Weissleder R et al. (2006) Identification of inhibitors of ribozyme self-cleavage in mammalian cells via high-throughput screening of chemical libraries. Rna 12:797–806
86. Yen L, Svendsen J, Lee JS et al. (2004) Exogenous control of mammalian gene expression through modulation of RNA self-cleavage. Nature 431:471–476
87. Zhao HF, Boyd J, Jolicoeur N et al. (2003) A coumermycin/novobiocin-regulated gene expression system. Hum Gene Ther 14:1619–1629
88. Zhao HF, Kiyota T, Chowdhury S et al. (2004) A mammalian genetic system to screen for small molecules capable of disrupting protein-protein interactions. Anal Chem 76:2922–2927

Posttranscriptional Gene Silencing

V. Ruppert, S. Pankuweit, B. Maisch, T. Meyer

Contents

10.1	Introduction	109
10.2	The RNA Interference Process	109
10.3	Cellular Functions of CAR	110
10.4	Small Interfering RNAs for Silencing the CAR Gene	112
10.5	Gene Silencing of CAR Protects from CVB3-induced Cell Death	116
10.6	Potential Role of siRNA Technology in Regenerative Medicine	117
	References	118

10.1 Introduction

Regenerative medicine is an emerging field that has been extensively studied and includes modern approaches for gene targeting. Recent findings have highlighted the effectiveness of gene silencing in therapeutic settings [1–7]. The key therapeutic advantage of using RNA interference (RNAi) lies in an ability to downregulate the expression of an undesired gene product of known sequence specifically and potently [8, 9]. RNA interference was first discovered in the nematode *Caenorhabditis elegans* a decade ago, and was later shown to occur also in mammalian cells in response to double-stranded small interfering RNAs (siRNA) of approximately 21 nucleotides in length that serve as effector molecules for sequence-specific gene silencing [10, 11]. In recent years, rapid progress has been made towards the use of RNAi as a therapeutic modality against a broad spectrum of human diseases. RNAi-based techniques promise to refine the clinical outcome of tissue-engineered products designed for clinical application. However, real hurdles need to be overcome to translate the therapeutic potential of the RNAi approach into clinical practice. In the following we will first give a very short overview of the mechanistic understanding of RNAi and then present our own data on the RNAi-guided knockdown of the CAR gene, which encodes the cellular receptor for coxsackie- and adenoviruses. Not included in this review are the actions of ribozymes and antisense oligonucleotides, which are also used to pharmacologically suppress the expression of a targeted gene product.

10.2 The RNA Interference Process

RNA interference is defined as a cellular process by which double-stranded RNA directs mRNA for destruction in a sequence-dependent manner. Exogenous siRNAs target complementary mRNA for transcript cleavage and degradation in a process known as posttranscriptional gene silencing (PTGS). The first event in RNAi involves the enzymatic fragmentation of cytoplasmic double-stranded RNA into 21- to 23-bp nucleotide duplexes that hybridize with the target mRNA [12]. Cleavage is carried out by an RNase-III endonuclease called Dicer, which is complexed with TAR-RNA binding protein (TRBP) and protein activator of protein kinase PKR (PACT). The Dicer-TRBP-PACT complex directs siRNAs to the RNA-induced silencing complex, termed RISC [13]. The core components of RISC are the Argonaute (AGO) family members, and in humans only Ago2 possesses an active catalytic domain for cleavage activity [14]. The guide strand is bound within the catalytic, RNase H-like PIWI domain of Ago2 at the 5′ end and a PIWI-Argonaute-Zwille (PAZ) domain that recognizes the 3′ end of the siRNA [15, 16]. Argonaute2 cleaves the target mRNA molecules between bases 10 and 11 relative to the 5′ end of the antisense siRNA strand. Effective posttranscriptional gene silencing requires perfect or near-perfect Watson-Crick base pairing between the mRNA transcript and the antisense or guide strand of the siRNA [13]. Partial complementarity between siRNA and target mRNA may repress translation or destabilize the transcript if the binding mimics microRNA (miRNA) interactions with target sites.

MicroRNAs are the endogenous substrates for the RNAi machinery, and RNA interference functions as a regulatory mechanism in eukaryotic cells for homology-dependent control of gene activity. Endogenously encoded primary microRNAs (pri-miRNAs) are transcribed by RNA polymerase II and initially processed in the nucleus by Drosha-DGCR8 (DiGeorge syndrome critical region gene 8) to generate precursor miRNA (pre-miRNA) [17, 18]. These pre-miRNA molecules consist of approximately 70 nucleotide-long stem-loop structures and are exported to the cytoplasm by exportin-5 in a Ran-dependent manner [19, 20]. In the cytosol, the pre-miRNAs bind to a complex containing Dicer, TRBP, and PACT. After cleavage, the approximately 22-nucleotide-long, mature miRNAs are then loaded into the AGO2 and RISC and recognize target sequences in the 3′ untranslated region of mRNA to direct translational repression. In the case of complete sequence complementarity with the target mRNA, the miRNA appears to direct cleavage of the mRNA transcript through RISC activity.

Much progress has been made in the last few years in the employment of RNAi-based techniques for interfering with gene expression in vivo and, indeed, the potential to specifically knock-out target genes offers new alternatives for clinical applications. Cellular proteins that are involved in viral attachment or replication appear to be interesting targets for an in vivo RNAi approach, since this promises to directly reduce the burden of a viral challenge. Considerable attention has been paid to the cellular receptor for adenoviruses, since these viruses have gained widespread interest as vectors for therapeutic gene delivery [21]. Thus, in the following we focus on the silencing of the CAR gene, that encodes the cellular receptor for coxsackie B- and adenoviruses in human cells.

10.3
Cellular Functions of CAR

The cellular receptor for coxsackie B viruses and adenoviruses is a 346-amino acid protein with a single membrane-spanning domain which belongs to the immunoglobulin superfamily [22–25]. The extracellular domain is composed of two immunoglobulin-like domains (D1 and D2), and the amino-terminal D1 domain contains the primary attachment site for these viruses. The membrane-spanning domain separates the extracellular domain of 216 residues from a 107-residue intracellular domain [23]. The cytoplasmic domain contains a site for palmitylation, potential phosphorylation sites, and a carboxy-terminal hydrophobic peptide motif that interacts with PDZ-domain proteins. The presence of the cytoplasmic domain is required for CAR-dependent inhibition of cellular growth [26]. In several human cancers including bladder and prostate carcinoma, and glioblastoma multiforme, CAR expression is downregulated, and reconstitution of CAR has resulted in decreased cell proliferation and tumorigenicity in animals [27–29]. In malignant glioma cells, CAR acts as a tumor suppressor [30].

The coxsackie virus and adenovirus receptor (CAR) was first identified as a cellular surface protein involved in attachment and uptake of a group B coxsackie virus, and was later recognized as a cell adhesion molecule that mediates the formation of homotypic cell–cell contacts [22, 31, 32]. Cohen and colleagues have shown that CAR plays a role as a transmembrane component of the epithelial cell tight junction [32]. Adenoviruses interact with CAR through an elongated fiber protein which projects from the virus capsid. Binding to adenoviruses occurs with high affinity (1 nM) via a globular fiber knob at a site that has been involved in CAR dimerization, suggesting that fiber interactions disrupt CAR-mediated intercellular adhesion [23, 33].

The expression of CAR appears to be developmentally regulated. Kashimura and coauthors found an age-dependent expression of CAR in rats [34]. During the late embryonic and early postnatal period there is a high expression particularly in the brain, heart, and skeletal muscle, while in the later postnatal period the expression of CAR gradually decreases [34, 35]. Despite this gradual decrease during postnatal development, CAR is still detectable in adults [32]. Immunohistochemistry showed that CAR is localized on the cell surface of cardiomyocytes in immature rat hearts. In contrast, in adult hearts CAR is detected predominantly in intercalated discs (Fig. 10.1). The tissue distribution of human CAR with higher expression levels in the heart, brain, and pancreas is consistent with the tropism of coxsackie virus B3, which causes myocarditis, meningoencephalitis, and pancreatitis.

From knock-out mice we have learned that CAR plays an essential role in normal heart development. Mice lacking CAR expression die in utero between day E11.5 and E13.5 possibly due to insufficient heart function [36, 37]. Chen and colleagues found that cardiomyocyte-specific CAR deletion results in hyperplasia of the embryonic left ventricle and abnormalities of sinuatrial valves [38]. It was reported that the knock-out mice had a reduced density and disturbed intracellular arrangement of myofibrils as well as mitochondrial abnormalities in their hearts, as determined by ultrastructural analysis [37, 38].

Expression of CAR in the heart can be reinduced in rats by immunization of purified pig cardiac myosin [39]. In experimental autoimmune myocarditis, the expression of CAR is increased on damaged myocardium during the recovery phase, suggesting a possible role for CAR in tissue remodeling [35, 40]. Myocardial infarction in rats induced by permanent ligation of the left anterior descending coronary artery resulted in a locally confined CAR upregulation in the infarct zone [35]. While CAR expression is low in normal adult human heart, strong upregulation of CAR was observed in the majority of heart biopsies obtained from patients with dilated cardiomyopathy, where it is mainly restricted to intercalated discs and the sarcolemma of cardiomyocytes [41]. Expression of CAR in diseased hearts suggests a functional role for CAR in mediating cell–cell contacts between cardiomyocytes and tissue remodeling [41, 42]. However, redistribution of viral attachment factor CAR in these diverse pathological circumstances may create a predisposition for recurrent cardiac infections with coxsackieviruses and adenoviruses, and thus be a major determinant of susceptibility to viral myocarditis.

Coxsackie viruses are positive-sense single-stranded members of the enteroviruses group of the family *Picornaviridae* which cause a broad spectrum of clinically relevant diseases including acute and chronic myocarditis [43–45]. It is well known that acute myocarditis can be induced in mice following inoculation with coxsackie virus B3 (CVB3) and that CVB infection leads to cardiac myocyte cell death in vitro [46, 47]. However, not all mouse strains are susceptible to CVB3-induced myocarditis [48]. For example, DBA/2 and A/J (H-2^a) mice usually develop a severe course of acute viral myocarditis, whereas C57Bl/6 mice are resistant to CVB3-induced heart damage. It has been reported that infants are more susceptible to CVB3 infection than young children or adults, and that the mortality in human infants due to CVB infection is comparably high [49].

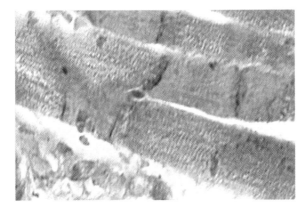

Fig. 10.1 Immunohistochemical detection of coxsackie B- and adenovirus receptor (CAR) in murine heart tissue. The section was counterstained with haematoxylin-eosin. Note the predominant localization of CAR in intercalated discs

10.4
Small Interfering RNAs for Silencing the CAR Gene

In this article we examined the effect of CAR gene silencing using RNA interference (RNAi) on coxsackie virus infection. HeLa cells were transfected with double-stranded siRNA against the human CAR gene followed by infection with CVB3. Coxsackie virus B3 stocks were prepared by one passage in HeLa cells. After threefold freezing and thawing, the virus was clarified by centrifugation. At the beginning of each experiment the virus titer was determined by plaque assays. HeLa cells cultured in Quantum 101 medium were transfected with the corresponding siRNAs. The siRNA sequences were selected from a BLAST search of the National Center for Biotechnology Information's expressed sequence tag library in order to ensure that they targeted only the desired genes (Fig. 10.2). A CAR-specific double-stranded siRNA and two controls (21-mer) were synthesized by QIAGEN. The final concentration of all siRNA was 20 µM in provided buffer. For complex formation, RNAiFect transfection reagent (QIAGEN) was added to the diluted siRNAs and mixed by pipetting up and down. The samples were incubated for 15 min at room temperature to allow the formation of transfection complexes. After complex formation, the transfection solution was added drop-wise onto the cells and the cells were incubated at 37°C under normal conditions. On the next day the cells were infected with CVB3 at a titer of approximately 1 PFU/cell. Twenty-four hours and two days (48 h) after infection, cell lysates from the 6 well-plates were collected and stored in a –80°C freezer. Cell viability assays were carried out in 96 well-plates.

For real-time RT-PCR analysis total RNA was extracted from HeLa cells using the QIAGEN RNeasy Kit according to the manufacture's instructions, including a DNase digestion (Serva). The mRNA of the CAR gene from HeLa cells was amplified with specific primer pairs by one-step real-time PCR using an iCycler (BioRad). Polymerase chain reaction was carried out according to the manufacturer's recommendations in a total volume of 25 µl, containing 2.5 µl total RNA, 2.5 µl SYBR green and either 7.5 µM CAR or GAPDH (house keeping gene) specific forward and reverse primers (Table 10.1). The protocol used included reverse transcription at 50°C for 30 min, denaturation at 95°C for 15 min, and amplification steps repeated 40 times (95°C for 45 sec, annealing at 60°C for 45 sec, and extension at 72°C for 45 sec). A melting curve analysis was run after final amplification via a temperature gradient from 55 to 94°C in 0.5°C increment steps measuring fluorescence at each temperature for a period of 10 sec. All reactions were carried out in at least duplicate for each sample. The relative expression of CAR and lamin A/C transcripts were normalized to the level of GAPDH. Using the BioRad iQ iCycler system software, the threshold (Ct) at which the cycle numbers were measured was adjusted to areas of exponential

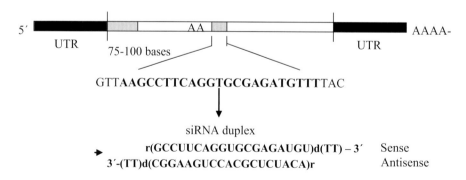

Fig. 10.2 Sequence of CAR-specific siRNA

Table 10.1 Primers used for the real-time PCR

Name	Forward primer	Reverse primer	T_{Ann}*
CAR	5′-gCCCACTTCATggTTAgCAg-3′	5′-TACggCTCTTTggAggTggC-3′	60°C
Lamin A/C	5′-AgCACTgCTCTCAgTgAgAA-3′	5′-CTCTCAAACTCACgCTgCTT-3′	59°C
GAPDH	5′-gAAggTgAAggTCggAgT C-3′	5′-gAAgATggTgATgggATTTC-3′	60°C

T_{Ann}, annealing temperature

amplification of the traces. The ΔΔ-method was used to determine comparative expression level by applying the formula $2^{-(\Delta Ct\ target\ -\ \Delta Ct\ reference\ sample)}$, as described [50].

HeLa cells treated with CAR siRNA or nonsilencing siRNA were lysed in 9M UREA buffer containing 20 mM deoxycholate, 10 mM EDTA, 3% NP40, 1% Trasylol and 1 mM phenylmethylsulfonyl fluoride (PMSF), pH 7.4. After measuring the concentrations of each sample using a DC protein assay kit from Bio-Rad, 2 μg of protein were separated by SDS-polyacrylamide gel electrophoresis (PAGE) and analyzed by immunoblotting with a primary anti-CAR antibody (Santa Cruz) and an appropriate secondary antibody. GAPDH expression was detected using an anti-GAPDH antibody. Bound immunoreactivity was detected by chemiluminescence (Pierce) and densitometry was performed using computer software (Quantity One, PDI).

Cell viability was measured by using 3-(4,5-dimethylthiazol-2-yl)-2,5-diphenyl-tetrazolium bromide (MTT). The cells were incubated with 10 μl MTT solution for 2 h. The plates were then centrifuged for 10 min at 500 g, the supernatant was discarded and the formazan salts formed were extracted with a solution containing 12.5% w/v SDS and 45% di-methylformamide, pH 4.7. After extraction, the optical density at 570 nm was measured using an enzyme-linked immunosorbent assay (ELISA) reader. The absorbance of sham-infected cells was defined as the value of 100% survival, and measured absorbance for either siRNA-treated, nontreated, or control cells was normalized to the ratio of the sham-infected sample. Data were analyzed using the SigmaStat Advisory Statistics for scientists (SYSTAT). For comparison of CAR levels in siRNA-treated and untreated samples, an unpaired, two-tailed Student's t-test was used. A p-value of less than or equal to 0.05 was considered significant.

To demonstrate that CAR expression plays a crucial role in the susceptibility to CVB3 infection, CAR-specific siRNA was transfected in HeLa cells that are highly permissive for this viral strain. For assessing the efficacy of gene silencing, the technique of one-step SYBR green real-time PCR was applied. A typical result from real-time PCR and the corresponding melting curves are shown in Fig. 10.3. The parameter C_t (threshold cycle) was defined as the cycle number at which the fluorescence emission exceeds the fixed threshold. SYBR green is a fluorogenic minor groove-binding dye that exhibits little fluorescence when in solution but emits a strong fluorescent signal upon binding to double-stranded DNA [51]. Thus, incorporated SYBR green functions as a nonsequence-specific fluorescent intercalating agent which directly measures amplicon production (including nonspecific amplification and formation of primer dimers). The melting point of the CAR amplicon was set at 83.5°C and GAPDH at 83°C.

Figures 10.4 and 10.5 demonstrate the gene silencing effects of different siRNAs in HeLa cells as determined by real-time PCR. Clearly, the results showed a considerable downregulation of either lamin A/C or CAR mRNA, depending on the siRNA used. The mRNA expression of the lamin A/C gene was downregulated to less than 20% (17.6±6.4%) in cells transfected with lamin A/C-specific siRNA. However, there was no such effect in either untransfected cells or cells transfected with control siRNA and CAR-specific siRNA, respectively (Fig. 10.4). Transfection of HeLa cells with CAR-specific siRNA reduced the expression level of CAR mRNA to less than one third (32.5±4.3%) as compared to controls (Fig. 10.5). Forty-eight hours after transfection, CAR mRNA expression dropped to 7% and then increased slightly to 10% one day later (Fig. 10.6). No CAR downregulation was found in cells transfected with nonspecific or lamin A/C-specific siRNA (Figs. 10.5, 10.6).

Western blot analysis confirmed the real-time RT-PCR results. One day after CAR siRNA transfection, protein expression of CAR decreased to 48% as compared to samples transfected with nonsilencing siRNA. Forty-eight hours after transfection the expression levels of mRNA and protein were similarly reduced. CAR protein expression then dropped to 8% and was below the detection threshold 72 h after transfection. The time course of gene silencing as determined by immunoblotting is shown in Fig. 10.6.

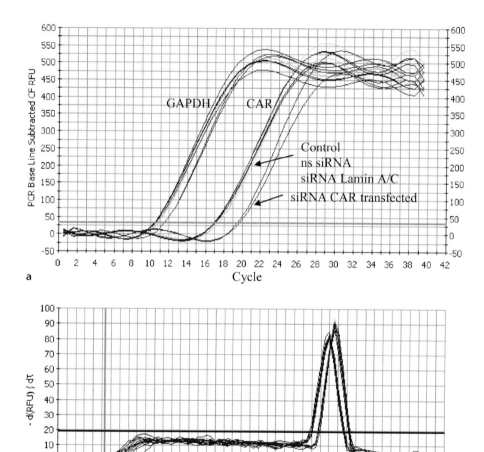

Fig. 10.3 Results of a SYBR green real-time RT-PCR experiment with the corresponding melting curves. **a** The calculated threshold using the maximum correlation coefficient approach was set at 32.9. Baseline cycles were determined automatically. Weighted mean digital filtering was applied, and global filtering was off. **b** Melting curves of the PCR fragments of the house keeping gene GAPDH (melting temperature 83°C) and CAR (83.5°C) are shown

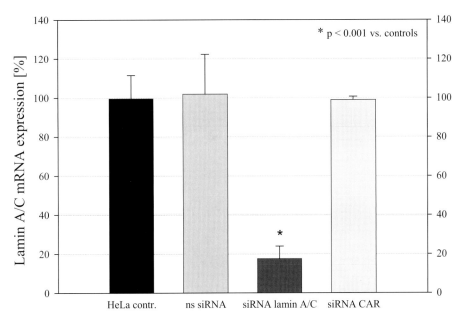

Fig. 10.4 Lamin A/C mRNA expression in HeLa cells 24 h after lamin A/C siRNA transfection (ns siRNA = nonsilencing siRNA). Lamin A/C expression decreased significantly in cells transfected with lamin A/C-specific siRNA. In HeLa cells treated with either nonsilencing siRNA or CAR siRNA there was no such reduction in lamin A/C mRNA expression

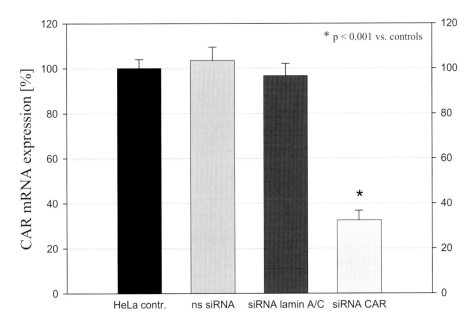

Fig. 10.5 CAR mRNA expression in HeLa cells 24 h after siRNA transfection (ns siRNA = nonsilencing siRNA). Cells treated with CAR-specific siRNA showed a significantly reduced expression of CAR. In contrast, CAR expression was not silenced in cells treated with either nonsilencing siRNA or laminA/C siRNA

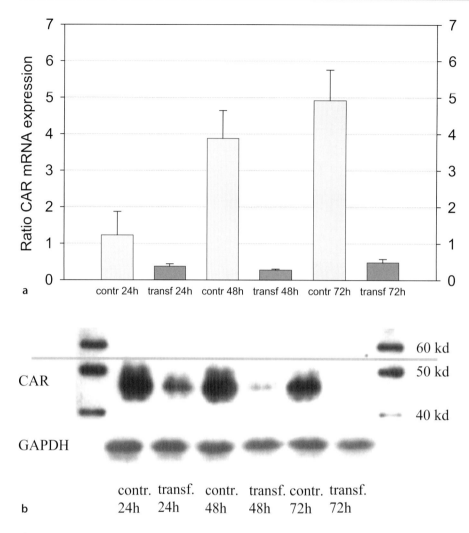

Fig. 10.6 Time course of CAR gene silencing following transfection of HeLa cells with CAR-specific siRNA as determined by real-time RT-PCR. **a** Note that CAR mRNA expression was reduced over at least 72 h following transfection. Western blot analysis confirmed the reduced CAR expression. **b** Seventy-two hours posttransfection, CAR expression was below the detection threshold

10.5
Gene Silencing of CAR Protects from CVB3-induced Cell Death

The influence of the transfection solution RNAiFect on the viability of HeLa cells is demonstrated in Fig. 10.7. Two concentrations of the transfection solution were tested: total and half of the concentration recommended by the manufacturer. The viability of HeLa cells without any treatment was set at 100%. After incubation with the transfection solution there was a slight dose-dependent decrease in the percentage of viable cells. The presence of siRNAs in the transfection solution did not further impair cell viability as compared to cells incubated with the transfection solution alone (data not shown).

Finally we tested the effect of CAR downregulation on the susceptibility to viral infection. Therefore, we first silenced CAR expression in HeLa cells using siRNA technology, and then infected the cells with CVB3 at a dose of 1 PFU/cell. Coxsackie virus B3 is known to induce a cytopathic effect in cultured human cells, as demonstrated by a decrease in cell

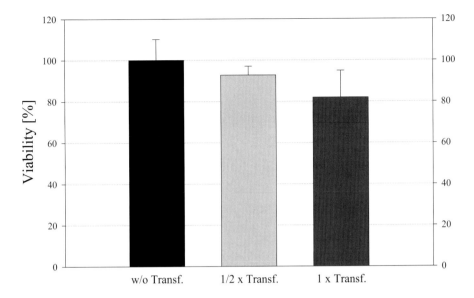

Fig. 10.7 Viability of HeLa cells after incubation with different concentrations of the transfection solution. Cells were either untransfected or transfected with half and the total concentration of the transfection solution, respectively, as recommended by the manufacture (QIAGEN)

viability [45]. Twenty-four hours after CVB3 infection we did not observe a significant reduction in the viability of CAR siRNA-treated cells as compared to control cells (data not shown). However, 48 h after CVB3 infection we detected significant differences in the viability of the differentially treated cells (Fig. 10.8). Upon infection with CVB3 we observed significantly impaired cell viability in cells with unaltered CAR expression. Interestingly, silencing of the CAR gene resulted in a partial resistance towards the cytopathic effects of CVB3 (Fig. 10.8). Neither untransfected cells nor cells transfected with nonspecific or lamin A/C-specific siRNA resisted CVB3 infection, suggesting that knocking down CAR gene expression efficiently protected cells from virus-induced cell lysis.

10.6
Potential Role of siRNA Technology in Regenerative Medicine

In the presented study, we employed the method of RNA interference to specifically knock down the receptor for coxsackieviruses and adenoviruses in a human cell line. Double-stranded small interfering RNAs were successfully exploited to inhibit the expression of CAR in a sequence-specific manner [52, 53]. The potential of RNA interference to inhibit virus propagation has been well established [53, 54]. Transfection of HeLa cells with CVB3-specific siRNA prior to infection blocked viral replication and resulted in a decreased cytopathic effect. We targeted the human CAR receptor, which is the key molecule regulating coxsackieviral attachment and cellular uptake. Our data support the hypothesis that knocking down the expression of a target gene may be a successful strategy to refine the properties of a cell line that is used for implantation in clinical settings. Although CAR may not be a good candidate for such an RNAi-based gene approach, our results suggest that this in vitro technology offers some potential for future clinical application in regenerative medicine. Inhibiting gene expression in a target-specific manner appears to be a promising approach to ameliorate human cells in vitro before they are incorporated into cell-based devices for therapeutic purposes.

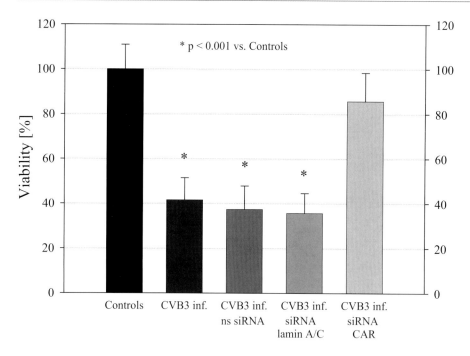

Fig. 10.8 Downregulation of CAR using siRNA transfection protects HeLa cells from infection with coxsackie virus. Only cells transfected with CAR-specific siRNA were rescued from infection with coxsackie virus B3. The mean value of untransfected HeLa cells was set at 100% (ns siRNA = nonsilencing siRNA)

Acknowledgement

The authors thank Verena Koch and Marlies Crombach for expert technical assistance. The research on this subject was in part funded by the BMBF German competence net on heart failure. V. Ruppert was supported by the Verein zur Förderung der Kardiologie Marburg and T. Meyer by a grant from the Deutsche Forschungsgemeinschaft and Deutsche Krebshilfe.

References

1. Leung RK, Whittaker PA. RNA interference: from gene silencing to gene-specific therapeutics. Pharmacol Therapeutics 2005;107:222–239
2. Aagaard L, Rossi JJ. RNAi therapeutics: Principles, prospects and challenges. Adv Drug Deliv Rev 2007;59:75–86
3. de Fougerolles A, Vornlocher HP, Maraganore J, Lieberman J. Interfering with disease: a progress report on siRNA-based therapeutics. Nat Rev Drug Discov 2007;6,443–453
4. Gewirtz AM. On future's doorstep: RNA interference and the pharmacopeia of tomorrow. J Clin Invest 2007;117:3612–3614
5. Kim DH, Rossi JJ. Strategies for silencing human disease using RNA interference. Nat Rev Genet 2007;8:173–184
6. Kumar LD, Clarke AR. Gene manipulation through the use of small interfering RNA (siRNA): from in vitro to in vivo applications. Adv Drug Deliv Rev 2007;59:87–100
7. Mao CP, Lin YY, Hung CF, Wu TC. Immunological research using RNA interference technology. Immunol 2007;121:295–307
8. Elbashir SM, Harborth J, Lendeckel W, Yalcin A, Weber K, Tuschl T. Duplexes of 21-nucleotide RNAs mediate RNA interference in cultured mammalian cells. Nature 2001;411:494–498
9. Tuschl T. RNA interference and small interfering RNAs. Chembiochem 2001;2:239–245
10. Fire A, Xu S, Montgomery MK, Kostas SA, Driver SE, Melo CC. Potent and specific genetic interference by double-stranded RNA in *Caenorhabditis elegans*. Nature 1998;391:806–811
11. Sen GL, Blau HM. A brief history of RNAi: the silence of the genes. FASEB J 2006;20:1293–1299
12. Zamore PD, Tuschl T, Sharp PA, Bartel DP. RNAi: double-stranded RNA directs the ATP-dependent cleavage of mRNA at 21 to 23 nucleotide intervals. Cell 2000;101:25–33

13. Martinez J, Patkaniowska A, Urlaub H, Lührmann R, Tuschl T. Single-stranded antisense siRNA guide target RNA cleavage in RNAi. Cell 2002;110:563–574
14. Liu J, Carmell MA, Rivas FV, Marsden CG, Thomson JM, Song JJ, Hammond SM, Joshua-Tor L, Hannon GJ. Argonaute2 is the catalytic engine of mammalian RNAi. Science 2004; 305:1437–1441
15. Matranga C, Tomari Y, Shin C, Bartel DP, Zamore PD. Passenger-strand cleavage facilitates assembly of siRNA into Ago2-containing RNAi enzyme complexes. Cell 2005;123:607–620
16. Rand TA, Petersen S, Du F, Wang X. Argonaute2 cleaves the anti-guide strand of siRNA during RISC activation. Cell 2005;123:621–629
17. Lee Y, Ahn C, Han J, Choi H, Kim J, Yim J, Lee J, Provost P, Rådmark O, Kim S, Kim VN. The nuclear RNase III Drosha initiates microRNA processing. Nature 2003;425:415–419
18. Han J, Lee Y, Yeom KH, Kim YK, Jin H, Kim VN. The Drosha-DGCR8 complex in primary microRNA processing. Genes Dev 2004;18:3016–3027
19. Yi R, Qin Y, Macara IG, Cullen BR. Exportin-5 mediates the nuclear export of pre-microRNAs and short hairpin RNAs. Genes Dev 2003;17:3011–3016
20. Lund E, Güttinger S, Calado A, Dahlberg JE, Kutay U. Nuclear export of microRNA precursors. Science 2004;303:95–98
21. Hutchin ME, Pickles RJ, Yarbrough WG. Efficiency of adenovirus-mediated gene transfer to oropharyngeal epithelial cells correlates with cellular differentiation and human coxsackie and adenovirus receptor expression. Hum Gene Ther 2000;11:2365–2375
22. Bergelson JM, Cunningham JA, Droguett G, Kurt-Jones EA, Krithivas A, Hong JS, Horwitz MS, Crowell RL, Finberg RW. Isolation of a common receptor for Coxsackie B viruses and adenoviruses 2 and 5. Science 1997;275:1320–1323
23. Coyne CB, Bergelson JM. CAR: A virus receptor within the tight junction. Adv Drug Deliv Rev 2005;57:869–882
24. Hauwel M, Furon E, Gasque P. Molecular and cellular insights into the coxsackie-adenovirus receptor: role in cellular interactions in the stem cell niche. Brain Res Rev 2005;48:265–272
25. Raschperger E, Thyberg J, Pettersson S, Philipson L, Fuxe J, Pettersson RF. The coxsackie- and adenovirus receptor (CAR) is an in vivo marker for epithelial tight junctions, with a potential role in regulating permeability and tissue homeostasis. Exp Cell Res 2006;312:1566–1580
26. Okegawa T, Pong RC, Li Y, Bergelson JM, Sagalowsky AI, Hsieh JT. The mechanism of the growth-inhibitory effect of coxsackievirus and adenovirus receptor (CAR) on human bladder cancer: a functional analysis of CAR protein structure. Cancer Res 2001;61:6592–6600
27. Okegawa T, Li Y, Pong RC, Bergelson JM, Zhou J, Hsieh JT. The dual impact of coxsackie and adenovirus receptor expression on human prostate cancer gene therapy. Cancer Res 2000;60:5031–5036
28. Sachs MD, Rauen KA, Ramamurthy M, Dodson JL, De Marzo AM, Putzi MJ, Schoenberg MP, Rodriguez R. Integrin α_v and coxsackie adenovirus receptor expression in clinical bladder cancer. Urol 2002;60:531–536
29. Fuxe J, Liu L, Malin S, Philipson L, Collins VP, Pettersson RF. Expression of the coxsackie and adenovirus receptor in human astrocytic tumors and xenografts. Int J Cancer 2003;103:723–729
30. Kim M, Sumerel LA, Belousova N, Lyons GR, Carey DE, Krasnykh V, Douglas JT. The coxsackievirus and adenovirus receptor acts as a tumour suppressor in malignant glioma cells. Br J Cancer 2003;88,1411–1416
31. Honda T, Saitoh H, Masuko M, Katagiri-Abe T, Tominaga K, Kozakai I, Kobayashi K, Kumanishi T, Watanabe YG, Odani S, Kuwano R. The coxsackie virus-adenovirus receptor protein as a cell adhesion molecule in the developing mouse brain. Brain Res Mol Brain Res 2000;77:19–28
32. Cohen CJ, Shieh JT, Pickles RJ, Okegawa T, Hsieh JT, Bergelson JM. The coxsackievirus and adenovirus receptor is a transmembrane component of the tight junction. Proc Natl Acad Sci USA 2001;98:15191–15196
33. Walters RW, Freimuth P, Moninger TO, Ganske I, Zabner J, Welsh MJ. Adenovirus fiber disrupts CAR-mediated intercellular adhesion allowing virus escape. Cell 2002;110:789–799
34. Kashimura T, Kodama M, Hotta Y, Hosoya J, Yoshida K, Ozawa T, Watanabe R, Okura Y, Kato K, Hanawa H, Kuwano R, Aizawa Y. Spatiotemporal changes of coxsackievirus and adenovirus receptor in rat hearts during postnatal development and in cultured cardiomyocytes of neonatal rat. Virchows Arch 2004;444:283–292
35. Fechner H, Noutsias M, Tschoepe C, Hinze K, Wang X, Escher F, Pauschinger M, Dekkers D, Vetter R, Paul M, Lamers J, Schultheiss HP, Poller W. Induction of coxsackievirus-adenovirus-receptor expression during myocardial tissue formation and remodeling. Identification of a cell-to-cell contact-dependent regulatory mechanism. Circulation 2003;107:876–882
36. Asher DR, Cerny AM, Weiler SR, Horner JW, Keeler ML, Neptune MA, Jones SN, Bronson RT, DePinho RA, Finberg RW. Coxsackievirus and adenovirus receptor is essential for cardiomyocyte development. Genesis 2005;42:77–85
37. Dorner AA, Wegmann F, Butz S, Wolburg-Buchholz K, Wolburg H, Mack A, Nasdala I, August B, Westermann J, Rathjen FG, Vestweber D. Coxsackievirus-adenovirus receptor (CAR) is essential for early embryonic cardiac development. J Cell Sci 2005;118:3509–3521
38. Chen JW, Zhou B, Yu QC, Shin SJ, Jiao K, Schneider MD, Baldwin HS, Bergelson JM. Cardiomyocyte-specific deletion of the coxsackievirus and adenovirus receptor results in hyperplasia of the embryonic left ventricle and abnormalities of sinuatrial valves. Circ Res 2006;98:923–930
39. Ito M, Kodama M, Masuko M, Yamaura M, Fuse K, Uesugi Y, Hirono S, Okura Y, Kato K, Hotta Y, Honda T, Kuwano R, Aizawa Y. Expression of coxsackievirus and adenovirus receptor in hearts of rats with experimental autoimmune myocarditis. Circ Res 2000;86:275–280
40. Bowles NE, Javier Fuentes-Garcia F, Makar KA, Li H, Gibson J, Soto F, Schwimmbeck PL, Schultheiss HP, Pauschinger M. Analysis of the coxsackievirus B-adeno-

virus receptor gene in patients with myocarditis or dilated cardiomyopathy. Mol Genet Metab 2002;77:257–259

41. Noutsias M, Fechner H, de Jonge H, Wang X, Dekkers D, Houtsmuller AB, Pauschinger M, Bergelson J, Warraich R, Yacoub M, Hetzer R, Lamers J, Schultheiss HP, Poller W. Human coxsackie-adenovirus receptor is colocalized with integrins $\alpha_v\beta_3$ and $\alpha_v\beta_5$ on the cardiomyocyte sarcolemma and upregulated in dilated cardiomyopathy. Implications for cardiotropic viral infections. Circulation 2001;104:275–280

42. Zen K, Liu Y, McCall IC, Wu T, Lee W, Babbin BA, Nusrat A, Parkos CA. Neutrophil migration across tight junctions is mediated by adhesive interactions between epithelial coxsackie and adenovirus receptor and a junctional adhesion molecule-like protein on neutrophils. Mol Biol Cell 2005;16:2694–2703

43. Bowles NE, Olsen EG, Richardson PJ, Archard LC. Detection of Coxsackie-B-virus-specific RNA sequences in myocardial biopsy samples from patients with myocarditis and dilated cardiomyopathy. Lancet 1986;327:1120–1123

44. Andréoletti L, Hober D, Becquart P, Belaich S, Copin MC, Lambert V, Wattré P. Experimental CVB3-induced chronic myocarditis in two murine strains: evidence of interrelationships between virus replication and myocarditis damage in persistent cardiac infection. J Med Virol 1997;52:206–214

45. Ahn J, Joo CH, Seo I, Kim D, Kim YK, Lee H. All CVB serotypes and clinical isolates induce irreversible cytopathic effects in primary cardiomyocytes. J Med Virol 2005a;75:290–294

46. Herzum M, Ruppert V, Küytz B, Jomaa H, Nakamura I, Maisch B. Coxsackievirus B3 infection leads to cell death of cardiac myocytes. J Mol Cell Cardiol 1994;26:907–913

47. Tracy S, Höfling K, Pirruccello S, Lane PH, Reyna SM, Gauntt CJ. Group B coxsackie virus myocarditis and pancreatitis: connection between viral virulence phenotypes in mice. J Med Virol 2000;62:70–81

48. Opavsky MA, Martino T, Rabinovitch M, Penninger J, Richardson C, Petric M, Trinidad C, Butcher L, Chan J, Liu PP. Enhanced ERK-1/2 activation in mice susceptible to coxsackie virus-induced myocarditis. J Clin Invest 2002;109:1561–1569

49. Kaplan MH. Coxsackie virus infection in children under 3 months of age. In: M. Bendinelli H. Friedman (Eds) Coxsackie virus: A General Update. New York, NY, Plenum Press, 1988; 241–252

50. Pfaffl MW. A new mathematical model for relative quantification in real-time RT-PCR. Nucleic Acids Res 2001;29:e45

51. Morrison TB, Weis JJ, Wittwer CT. Quantification of low-copy transcripts by continuous SYBR Green I monitoring during amplification. Biotechniques 1998;24:954–958

52. Ahn J, Jun ES, Lee HS, Yoon SY, Kim D, Joo CH, Kim YK, Lee H. A small interfering RNA targeting coxsackievirus B3 protects permissive HeLa cells from viral challenge. J Virol 2005b;79:8620–8624

53. Werk D, Schubert S, Lindig V, Grunert HP, Zeichhardt H, Erdmann VA, Kurreck J. Developing an effective RNA interference strategy against a plus-strand RNA virus: silencing of coxsackievirus B3 and its cognate coxsackievirus-adenovirus receptor. Biol Chem 2005;386:857–863

54. Merl S, Michaelis C, Jaschke B, Vorpahl M, Seidl S, Wessely R. Targeting 2A protease by RNA interference attenuates coxsackieviral cytopathogenicity and promotes survival in highly susceptible mice. Circulation 2005;111:1583–1592

Biomolecule Use in Tissue Engineering

R. A. Depprich

Contents

11.1 Introduction 121
11.2 Epidermal Growth Factors (EGFs) 123
11.3 Fibroblast Growth Factors (FGFs) 124
11.4 Platelet-derived Growth Factors (PDGFs) . 125
11.5 Insulin-like Growth Factors (IGFs) 126
11.6 Transforming Growth Factor Beta (TGF-ß) 126
11.7 Bone Morphogenetic Proteins (BMPs) 127
References 130

11.1 Introduction

Tissue engineering is an interdisciplinary field in biomedical engineering that aims to regenerate new biological material to replace diseased or damaged tissues or organs. Tissue engineering is based on principles of cellular and molecular developmental biology and morphogenesis guided by bioengineering and biomechanics. The three key ingredients for both morphogenesis and tissue engineering are inductive signals (regulatory biomolecules, morphogens), responding cells and extracellular matrix (scaffolds) [108]. Morphogens regulate the proliferation and differentiation of cells, whereas scaffolds serve as a carrier and delivery system for the morphogens and simultaneously produce and influence the microenvironment [113]. Regulatory biomolecules, such as differentiation or growth factors and cytokines, are released by many different sorts of cells in a diverse manner (endocrine, autocrine, paracrine, juxtacrine or intracrine), targeted at a particular cell or cells to carry out a specific reaction (Tables 11.1, 11.2) [75]. The term cytokine is generally reserved to describe factors associated with cells involved in the immune system, but in many instances the term growth factor is used as a synonym for cytokines [101]. When a growth factor binds to a target cell receptor an intracellular signal transduction system is activated that finally reaches the nucleus and produces a biological response [79]. These ligand-receptor interactions are very specific and vary from the simple binding of one ligand to one particular cellular receptor to complex interactions with one or more ligands binding to one or more receptors. Additionally, the ligand-receptor interactions are even more complicated as different variants of the same growth factor may bind to a single receptor, or different growth factor receptors may be activated by a single ligand [53].

Fibroblast growth factors (FGFs) comprise a family of related mitogens, which, in vertebrates, includes 22 members [102]. Their biological functions are mediated through four high-affinity transmembrane receptor tyrosine kinases, known as FGF receptors (FGFRs). The FGF receptors consist of four known members (FGFR1–FGFR4) with overlapping affinities for the various members of the FGF family [61]. Recently a fifth receptor has been identified [124]. The various FGFs can bind to these receptors with different affinities. Binding of FGF-2 to the high affinity receptors FGF receptor-1 (the flg gene product) and FGF receptor-2 (the bek gene product) results in autophosphorylation of the receptor and signaling to the cell [30]. Alternative splicing in the extracellular domain of FGFR1, FGFR2 and FGFR3

Table 11.1 Biomolecules in tissue engineering

Growth factor	Source	Receptor	Function
Epidermal growth factors (EGFs)	Saliva, plasma, urine and most other body fluids	Tyrosine kinase	Mitogen for ectodermal, mesodermal and endodermal cells, promotes proliferation and differentiation of epidermal and epithelial cells
Fibroblast growth factors (FGFs)	Macrophages, mesenchymal cells, chondrocytes, osteoblasts	Tyrosine kinase	Proliferation of mesenchymal cells, chondrocytes and osteoblasts
Platelet-derived growth factors (PDGFs)	Platelets, macrophages, endothelial cells, fibroblasts, glial cells, astrocytes, myoblasts, smooth muscle cells	Tyrosine kinase	Proliferation of mesenchymal cells, osteoblasts and fibroblasts, macrophage chemotaxis
Insulin-like growth factors (IGFs)	Liver, bone matrix, osteoblasts, chondrocytes, myocytes	Tyrosine kinase	Proliferation and differentiation of osteoprogenitor cells
Transforming growth factor beta (TGF-ß)	Platelets, bone, extracellular matrix	Serine threonine sulfate	Stimulates proliferation of undifferentiated mesenchymal cells
Bone morphogenetic proteins (BMPs)	Bone extracellular matrix, osteoblasts, osteoprogenitor cells	Serine threonine sulfate	Differentiation of -mesenchymal cells into chondrocytes and osteoblasts -osteoprogenitor cells into osteoblasts influences embryonic development

Table 11.2 Effects of growth factors on different cells

Growth factor	Fibroblast proliferation	Osteoblast proliferation	Mesenchymal cell differentiation	Vascularization	Extracellular matrix synthesis
EGFs	++	–	++	+	–
FGFs	++	++	–	++	–
PDGFs	++	++	–	+ (indirect effect)	–
IGFs	+	++	–	–	++
TGF-ß	+ or –	+ or –	–	+ (indirect effect)	++
BMPs	–	±	++	++ (indirect effect)	±

++ greatly increased, + increased, – no or negative effect

further enhances the complexity of the FGFR system, since this splicing action generates receptor variants, which differ in their ligand-binding specificities [61, 64]. The FGF receptor-1 and FGF receptor-2 can exist in forms containing either two or three immunoglobulin-like domains in the extracellular portion of the molecule [111].

Members of the transforming growth factor beta (TGF-ß) superfamily, which include bone morphogenetic proteins (BMPs), bind to a heteromeric receptor complex with intrinsic serine/threonine kinase activity. This receptor complex consists of two type I and type II receptors, which are structurally similar, with glycosylated cysteine-rich extracellular regions, short transmembrane parts and intracellular serine/threonine kinase domains [87, 107]. At present, 12 type I and seven type II receptors with different affinity for the members of the TGF-ß family have been identified [116]. Type I receptors, but not type II receptors, have a region rich in glycine and serine residues (GS domain) in the juxtamembrane domain [126]. The type II receptor is capable of binding to the ligand, and has a constitutively active kinase, however, it cannot propagate a signal without the type I receptor. Each member of the TGF-ß superfamily binds to a characteristic combination of type I and type II receptors, both of which are needed for signaling as the type I receptor can only form a complex with a ligand that is already bound to the type II receptor [141]. The intracellular signaling pathways that are induced by the activated receptor ligand complexes involve a family of signaling molecules called Smad proteins which are divided into three subclasses: R-Smads (receptor-activated Smads), Co-Smads (common mediator Smads), and I-Smads (inhibitory Smads). Whereas binding of BMPs to the receptor complex initiates interaction with the R-Smads Smad1, Smad5 and Smad8, Smad2 and Smad3 are phosphorylated after stimulation by TGF-ß or activin [53, 87, 99]. Activated R-Smads form heteromeric (preferentially trimeric) complexes with Co-Smads (Smad4) and translocate into the nucleus where they bind to the specific sequences in the promoters of target genes and activate transcription of those genes. Smad6 and Smad7 function as general inhibitors of TGF-ß, activin and BMP signaling, whereas Smad6 specifically inhibits BMP signaling [53, 94, 95, 107, 142].

11.2
Epidermal Growth Factors (EGFs)

The aim of tissue engineering is the regeneration of defective or lost tissues or organs in which the main focus was initially concentrated on material based approaches and now focuses on cell-based devices. These constructs are likely to encompass additional families of growth factors, evolving biological scaffolds and incorporation of mesenchymal stem cells [105]. The EGF family contains several peptide growth factors, among these are the epidermal growth factor (EGF), transforming growth factor-α (TGFα), heparin-binding EGF-like growth factor (HB-EGF), amphiregulin (AR), betacellulin (BTC), epiregulin (EPR), epigen and the four neuregulins. NRG-1 is also known as Neu differentiation factor (NDF), heregulin (HRG), acetylcholine receptor-inducing activity (ARIA) and glial growth factor (GGF) [31]. The biological activities of the different members of the EGF family are similar. All the members have been shown to induce cell migration, differentiation and gene expression, and are involved in processes such as angiogenesis, wound healing, bone reabsorption, atherosclerosis, blastocyst implantation and tumor growth [22, 73]. Cras-Meneur et al. [26] investigated the effect of epidermal growth factor (EGF) on proliferation and differentiation of undifferentiated pancreatic embryonic cells in vitro. It was demonstrated that EGF expanded the pool of embryonic pancreatic epithelial precursor cells but at the same time repressed the differentiation into endocrine tissue. Once expanded and after removal of EGF, these precursor cells differentiated and formed endocrine cells by a default pathway. Similar effects were found when EGF was added to cultures of embryonic or adult precursor cells derived from the nervous system. It was shown that EGF acts as a mitogen for precursor cells inducing their proliferation while repressing their differentiation. When EGF was removed, cell differentiation into mature cells (neurons, glial cells) occurred [47, 109, 127]. The presence of EGF and hepatocyte growth factor (HGF) led to the differentiation of putative hepatic stem cells derived from adult rats into mature hepatocytes [52]. A combination of EGF and HGF showed synergistic effects on the proliferation of monolayers of hepatocytes [11]. Mixed cultures of hepatocytes and nonparenchymal cells growing as clusters of cells. Collagen coated plastic beads in roller bottles, subsequently placed in Matrigel formed three-dimensional ducts and sheets of mature hepatocytes surrounded by nonparenchymal cells. Under the influence of EGF and HGF neither growth factor added separately was sufficient to maintain prolonged viability of the epithelial cells [91]. For the purpose of tissue engineering of oropharyngeal mucosa Blaimauer et al. [10] investigated the effects of EGF and keratinocyte growth factor (KGF) on the growth of oropharyngeal keratinocytes in a coculture with autologous fibroblasts in a three-dimensional matrix (Matrigel). The investigators revealed that a physiologic EGF concentration was best for promoting cell growth and that EGF did not change any of the growth characteristics of oropharyngeal keratinocytes, with the exception of the speed of growth. To develop a decellularized bone-anterior cruciate ligament (ACL)-bone allograft for treatment of ACL disruption the impact of EGF on cellular ingrowth of primary ACL fibroblasts in decellularized ligaments was examined by Harrison and Gratzer [50]. No increase of cellular ingrowth by the addition of EGF to the culture medium of seeded cells could be revealed. In contrast EGF photo-immobilized onto polystyrene culture plates using UV irradiation increased ACL cell proliferation in proportion to the amount of added EGF. The investigators concluded that photo-immobilized EGF induced rapid proliferation of ACL fibroblast cells by artificial juxtacrine stimulation and speculated that similar EGF immobilization onto bioabsorbable materials (e.g., polyglycolic acid or polylactic acid) might contribute to a new therapy for the treatment of ACL injuries [138].

11.3
Fibroblast Growth Factors (FGFs)

Acidic FGF (aFGF or FGF-1) and basic FGF (bFGF or FGF-2) the prototype members of the FGF family are ubiquitous cytokines found in many tissues [137]. FGFs show a very high affinity to heparin and are therefore also sometimes referred to as heparin binding growth factors [68]. FGFs have effects on multiple cell types derived from mesoderm and neuroectoderm, including endothelial cells [45, 123]. Because FGF-1 binds to the same receptor as FGF-2 FGF-1 displays more or less the same spectrum of activities. It is, however, generally approximately 50-100-fold less active than FGF-2 in similar assays [33]. FGFs are key regulators of the various stages of embryonic development, as revealed by their expression pattern and the phenotypes observed in FGF and FGFR knockout animals [102]. As FGF-1 and FGF-2 stimulate angiogenesis and wound healing, numerous investigations were made to induce vascularization and fibrous tissue formation [6, 39, 48, 96, 144]. Ribatti and coworkers investigated the role of endogenous and exogenous fibroblast growth factor-2 (FGF-2) in the wound healing reparative processes, utilizing the chick embryo chorioallantoic membrane (CAM) as an in-vivo model of wound healing. The results showed that the application of exogenous recombinant FGF-2 greatly accelerated the wound repair occurring approximately 24 h earlier than in untreated CAMs, stimulating angiogenesis, fibroblast proliferation, and macrophage infiltration [110]. To improve wound healing, full-thickness skin defects were enhanced via the use of FGF-1 using a collagen scaffold in an animal experiment. FGF-1 delivered from a collagen scaffold showed promising results [103]. With the purpose of developing functional tissue-engineered cardiovascular structures eventually designed for implantation, Fu and coworkers studied the effects of FGF-2 and TGF-ß on maturation of human pediatric aortic cell culture. The findings demonstrated the best results with the addition of FGF-2 enhancing cell proliferation and collagen synthesis on the biodegradable polymer, leading to the formation of more mature, well organized tissue engineered structures [38]. To replace diseased or damaged cartilage by in vitro engineered, viable cells or graft tissues is a novel therapeutic approach. Cartilage tissue engineering by sequential exposure of chondrocytes to FGF-2 during 2D expansion and BMP-2 during three-dimensional (3D) cultivation demonstrated the positive effects of FGF-2 enhancing the responsiveness to chondrogenic factors (e.g., BMP-2), thus improving the size and composition of engineered cartilage generated by subsequent 3D cultivation on PGA scaffolds [85]. An experimental study using chondrocytes either unstimulated or treated with several growth factors to regenerate full thickness defects in rabbit joint cartilage showed that treatment of chondrocyte cultures with FGF-2 had a stabilizing effect on the differentiated state of the cells in implanted grafts compared with BMPs and TGF-ß. Only FGF had a clear beneficial effect to the graft tissues after 1 month [134]. Igai and coworkers investigated whether implantation of a gelatin sponge, releasing FGF-2 slowly into a tracheal cartilage defect, would induce regeneration of autologous tracheal cartilage in an animal study. The authors observed in the FGF-2 group regenerated cartilage in all dogs [56, 57]. FGFs also enhance the intrinsic osteogenic potential but the nature of FGF action is complex and the biological effect of FGFs may depend on the differentiation stage of osteoblasts, interaction with other cytokines, or the length and mode of exposure to factors [34]. In an animal experimental study no clear benefit of using knitted polylactide scaffolds combined with FGF-1 on the healing of calvarial critical size defects in rats could be revealed [44]. A recent field of research is the combination of FGFs with stem cells to regenerate new tissues. It was demonstrated that FGF-2 stimulates adipogenic differentiation of human adipose-derived stem cells [66], chondrogenic phenotypic differentiation on bone marrow-derived mesenchymal cells [24] and that mesenchymal stem cells cultured with FGF-2 and rhBMP-2 could act as a substitute for autograft in lumbar arthrodesis [92].

11.4
Platelet-derived Growth Factors (PDGFs)

Platelet-derived growth factor (PDGF), a dimeric molecule that was originally identified in platelets, exists in several subtypes consisting of homodimers or heterodimers of the PDGF-A and PDGF -B gene products [51]. PDGF-B chains are intrinsically more active than PDGF-A subunits [19]. Up to the present, many cell types, including macrophages, endothelial cells, fibroblasts, glial cells, astrocytes, myoblasts, and smooth muscle cells, have subsequently been determined to synthesize PDGF. Many different cell types, particularly those of mesenchymal origin including periodontal ligament cells and osteoblasts, respond to PDGF [90]. The primary effect of PDGF is that of a mitogen, initiating cell division. PDGF plays an important role during mammalian organogenesis and its overexpression has been linked to different types of fibrotic disorders and malignancies [1, 49, 54, 135]. In contrast to many other cytokines PDGF is not released into the circulation and the biological half-life is less than 2 min after intravenous administration [13].

With the aim of optimizing the in-vitro culture conditions for ligament tissue engineering, Fawzi-Grancher and coworkers investigated the effects of PDGF-AB on fibroblast proliferation in different substrates. Application of PDGF-AB resulted in significant increase in cell proliferation on all the substrates, the highest increase was found on gelatin coated silicon sheet culture [35]. Regeneration of periodontal structures lost during periodontal diseases constitutes a complex biological process regulated among others by interactions between cells and growth factors. The observation that PDGF is chemotactic for periodontal ligament cells promoting collagen and total protein synthesis [18] and reduces the inhibitory effects of lipopolysaccharide on gingival fibroblast proliferation [5] led to intense research for periodontal regeneration involving the application of PDGF [14, 41, 84, 104]. An in vitro evaluation of the mitogenic effect of PDGF-BB on human periodontal ligament cells cultured with various bone allografts was performed by Papadopoulos et al. [104]. The combination of human demineralized freeze-dried allografts of cortical (DFDBA) or cancellous (DFBA) bone with PDGF-BB provoked a statistically significant increase in periodontal ligament cell proliferation compared with cell/control cultures with PDGF-BB, while non-demineralized freeze-dried allografts (FBA) from cancellous bone had no similar effect. [104]. The investigation of experimental periodontal grafting materials consisting of beta-tricalcium phosphate (ß-TCP) or $CaSO_4$ scaffolds enriched with PDGF-BB revealed that PDGF-BB enhances the bone graft materials osteogenic properties by promoting cellular ingrowth into the osseous defect and bone matrix [7, 76, 100]. Recently a growth-factor enhanced matrix (GEM) has become available for clinical use. A combination of β-TCP with rhPDGF-BB (*GEM21S*) was approved by the FDA in November 2005 for the treatment of intrabony and furcation periodontal defects and gum tissue recession associated with periodontal defects [83]. As one of the major problems of the topical administration of PDGF is the maintaining of therapeutic protein levels at the defect site, research was focused on PDGF gene transfer to a greater extend [3, 16, 63, 149]. Anusaksathien and coworkers evaluated the effect of PDGF-A and PDGF-B gene transfer to human gingival fibroblasts on ex-vivo repair in three-dimensional collagen lattices. PDGF-A and PDGF-B gene expression was maintained for at least 10 days. PDGF-B gene transfer stimulated potent increases in cell repopulation and defect fill above that of PDGF-A and corresponding controls [2]. A recent experimental study demonstrated the preparation of porous chitosan/coral composites combined with plasmid PDGF-B gene for periodontal tissue engineering. These gene-activated scaffolds were colonized by human periodontal ligament cells and implanted subcutaneously into athymic mice. The results demonstrated an increased proliferation of human periodontal ligament cells and increased expression of PDGF-B compared with cells on the pure coral scaffolds [148]. A novel approach to fabricate tissue engineering scaffolds capable of controlled PDGF delivery was demonstrated by Wei et al. [133]. Microspheres containing PDGF-BB were incorporated into 3D nano-fibrous scaffolds. The incorporation of the microspheres into the scaffolds significantly reduced the initial burst release. The biological activity of released PDGF-BB was evidenced by stimulation of human gingival fibroblast DNA synthesis in vitro.

11.5
Insulin-like Growth Factors (IGFs)

IGF-I and IGF-II sharing approximately 50% structural homology to proinsulin were isolated initially as serum factors with insulin-like activities that could not be neutralized by antibodies directed against insulin [36, 115]. The majority of circulating IGF-I and IGF-II is produced by the liver, although various tissues have the capability to synthesize these peptides locally, including kidney, heart, lung, fat tissues, and various glandular tissues. IGF-I is produced also by chondroblasts, fibroblasts, and osteoblasts. The hepatic synthesis of IGF-I is largely growth hormone dependent, whereas the synthesis of IGF-II is relatively independent of growth hormone. Both forms of IGF are mitogenic in vitro for a number of mesodermal cell types. IGFs are essential for normal embryonic and postnatal growth, and play an important role in the function of a healthy immune system, lymphopoiesis, myogenesis and bone growth among other physiological functions [27, 37, 121, 125]. IGF-I is a potent mitogen for fibroblasts and chondrocytes, its insulin-like molecular structure causes it to mimic the effects of insulin on adipocytes and muscle cells [9, 128]. IGF-I is also important for bone remodeling and maintenance of skeletal mass and plays a significant role in age-related osteoporosis. IGF-I is also reported to be expressed during fracture healing and to stimulate it, suggesting a role as an autocrine/paracrine factor potentiating bone regeneration. [150]. IGF-II is functionally similar to IGF-I, but is thought to be relatively more important in fetal growth [46]. Several investigations evaluated the effects of IGF-I alone or in combination with other growth factors on cartilage tissue engineering [25, 59, 88]. Three-dimensional scaffolds and IGF-I in combination with the synergistic growth factor TGF-ß have been proven to be effective stimulants for chondrogenesis in articular cartilage tissue regeneration [29]. Polymeric microspheres were investigated as a delivery system that releases IGF-I in a controlled manner by Elisseeff and coworkers for cartilage tissue engineering. Statistically significant changes in glycosaminoglycan production and increased cell content compared with the control groups were observed after a 14 day incubation period with IGF-I and IGF-I/TGF-ß microspheres [32]. In contrast, Veilleux and Spector did not find any beneficial effects of IGF-I compared with FGF-2 or the combination of both growth factors on chondrocyte differentiation in type II collagen-glycosaminoglycan scaffolds in vitro [130]. In a recent publication, the potential of a novel 3D hydrogel scaffold made of a thermoreversible gelation polymer was investigated as a delivery system for chondrocytes, IGF-1 and/or TGF-ß. TGP was demonstrated a potential scaffold material in the generation of tissue-engineered cartilage in vitro [146]. Westreich and coworkers tested the subcutaneous site as a recipient bed for the engineering of cartilage in vivo. After implantation of fibrin glue and autologous chondrocytes cartilage formation occurred. The addition of FGF and IGF-I had no beneficial influence on cartilage yield [136]. Recently, Capito and Spector used a collagen glycosaminoglycan scaffold as a nonviral gene delivery vehicle for facilitating gene transfer to seeded adult articular chondrocytes to produce a local overexpression of IGF-I for enhancing cartilage regeneration. The results showed noticeably elevated IGF-I expression levels and enhanced cartilage formation compared with the controls [20]. Several investigators evaluated the effects of IGF-I in the chondrogenesis of different sorts of stem cells. Not only IGF-I but also TGF-ß and FGF-2 have been proven to be potent chondroinductive growth factors on chondrogenesis of stem cells [40, 58, 69, 81].

11.6
Transforming Growth Factor Beta (TGF-ß)

The transforming growth factor-beta (TGF-ß) family includes a large number of proteins of related homology in addition to the TGF-ßs themselves. The TGF-ßs exist in five isoforms, three of which are expressed in mammals and referred to as TGF-ß1, TGF-ß2, and TGF-ß3. The TGF-ßs and their receptors are ubiquitously expressed in normal tissues and most cell lines and play critical roles in growth regulation and development [67, 86]. The three major activities of TGF-ßs include inhibition of cell proliferation, enhancement of extracellular matrix deposition and the exhibition of complex immunoregulatory properties. TGF-ßs act primarily on epithelial and lymphoid cells, and directly counteract the effects of many different mitogenic growth factors. Each cell type that responds to TGF-ß has specific TGF-ß receptors. The cellular responses to the TGF-ßs vary with the cell type, growth factor dose, presence of other growth factors, and various other conditions [72]. TGF-ß1 is known to be produced locally during bone development and regeneration. It is one of the most important factors in the bone environment, helping to retain the balance between the dynamic processes of bone resorption and bone formation [60]. It is secreted by both bone marrow stromal cells and osteoblasts and is stored in the bone matrix, from which it is released and activated on bone resorption [77]. TGF-ß1 has been studied as a potential induction factor for bone tissue engineering by several investigators [12, 43, 78, 80, 82]. However, the in-vitro action of TGF-ß1 seems to depend strongly on culture conditions, dosages, the type of osteoblastic cells employed, and their stage of maturation, as several in-vitro studies on the effects of TGF-ß1 provided widely divergent results. Recently, Lee and coworkers published the use of collagen/chitosan composite microgranules loaded with TGF-ß1 for in-vivo bone regeneration in rabbit calvarial defects. After 4 weeks, the TGF-ß1-loaded microgranules showed a higher bone-regenerative capacity compared with the TGF-ß1-unloaded microgranules [74]. In a similar study, chitosan scaffolds containing microspheres were investigated as carriers for controlled TGF-ß1 release for cartilage tissue engineering. After 3 weeks, a significant increase was observed when the chondrocytes were cultured on scaffolds containing microspheres loaded with TGF-ß1 in contrast to the chondrocytes grown on scaffolds without microspheres [17]. As TGF-ß1 appears to have positive effects on cartilage differentiation and repair this growth factor was employed for cartilage tissue engineering at which the tissue responses depend on the dose of TGF-ß and length of exposure [55, 93]. Mainly, two types of scaffolds, solid types (e.g., porous body, mesh, sponge, and unwoven fabric) and hydrogels in combination with TGF-ß, were investigated for cartilage tissue engineering. Hydrogel materials were demonstrated to be more effective in retaining chondrocyte functions or promoting matrix synthesis, when compared with monolayer cultures. The hydrogel types of scaffolds have further been shown to reconstruct the 3D environment for the chondrocytes in cartilage tissue engineering [145]. Park et al. [106] evaluated injectable hydrogel composites in combination with TGF-ß1 for cartilage tissue engineering. Bovine chondrocytes embedded in hydrogels co-encapsulating gelatin microparticles loaded with TGF-β1 showed a statistically significant increase in cellular proliferation compared with controls. With the aim to generate cardiovascular replacement tissues, Sales and coworkers investigated the ability of noninvasively isolated blood-derived endothelial progenitor cells to secrete extracellular matrix on scaffolds in response to TGF-ß1. The authors demonstrated significant extracellular matrix production and phenotypic change of endothelial progenitor cells to a cell type with characteristics of valvular interstitial cells [117]. The effect of TGF-ß1 alone or in combination with other growth factors for tissue engineering of ligament was also investigated by several authors [35, 89, 97]. Jenner and coworkers showed that rhTGF-ß1 and to a lesser extent GDF-5, can stimulate human bone marrow stromal cells cultured onto woven, bioabsorbable, 3D scaffolds to form collagenous soft tissues [62]. A first clinical case was published by Becerra et al. [8] employing a recombinant human TGF-ß1 containing an auxiliary collagen binding domain (rhTGF-ß1-F2) dexamethasone (DEX) and beta-glycerophosphate (beta-GP). Autologous bone marrow-derived cells were exposed to rhTGF-ß1-F2 and expanded in vitro. After loading into porous ceramic scaffolds and transplantation into the bone defect of a 69-year-old man the culture-expanded cells differentiated into bone tissue.

11.7
Bone Morphogenetic Proteins (BMPs)

Current research on bone tissue regeneration has emerged from a pioneering study on bone growth more than 4 decades ago. In 1964, Marshall Urist discovered that implantation of dried demineralized bone powder into the muscle of a rabbit could stimulate the growth of new bone. Urist and collaborators named the proteinaceous substance bone morphogenetic protein (BMP) [129]. In 1981, A.H. Reddi and his workgroup were able to extract a soluble protein component of the inductive matrix. Neither the soluble component nor the residual collagen matrix could induce new bone growth on their own, but bone inductivity recurred after recombination of the extract and the matrix. This finding proved that a soluble osteogenic fraction and a solid carrier were necessary factors for bone regeneration [118]. In 1987, Reddi et al. [119] published the successful clearance and characterization of an inductive factor they named osteogenin (this factor was later referred to as BMP-3). At about the same time, Wozney and coworkers successfully isolated and characterized more inductive factors (BMP-2, BMP-4, BMP-5, BMP-6, BMP-7) and succeeded in expression of the recombinant human proteins [23, 139]. To date, mainly through their sequence homology with other BMPs, approximately 20 BMP family members have been identified. Most of these play important roles in embryogenesis and morphogenesis of various tissues and organs. After birth, the BMPs play roles in tissue repair and regeneration. With the exception of BMP-1, which is a cysteine-rich zinc-peptidase, the BMP family members are the largest subgroup of the TGF-ß superfamily of growth and differentiation factors (Table 11.3). Although BMPs were originally isolated and identified from bone, they are expressed

Table 11.3 The BMP family

Name	Synonym	Potential function	Amino acid homology in %
BMP-2	BMP-2A	Bone and cartilage morphogenesis	60
BMP-3	Osteogenin	Bone induction	n.a.
BMP-3B	GDF-10	Bone induction	n.a.
BMP-4	BMP-2B	Bone and cartilage morphogenesis	58
BMP-5	-	Bone morphogenesis	88
BMP-6	Vgr-1	Cartilage hypertrophy	87
BMP-7	OP-1	Bone differentiation	100
BMP-8	BMP-8A, OP-2	Bone induction	74
BMP-8B	OP-3	Bone and cartilage development	67
BMP-9	GDF-2	Bone induction	51
BMP-10	-	Tabeculation of the heart	47
BMP-11	GDF-11	Neurogenesis	36
BMP-12	GDF-7, CDMP-3	Ligament and tendon development	n.a.
BMP-13	GDF-6, CDMP-2	Cartilage development and hypertrophy	53
BMP-14	GDF-5, CDMP-1, CDMP-2	Mesenchyme aggregation and chondrogenesis	51
BMP-15	CDMP-1, GDF 9B	Oocyte and follicular development	n.a.
BMP-16	-	Bone and cartilage development	n.a.

DMP cartilage derived morphogenetic protein

OP osteogenic protein

TGF-ß transforming growth factor ß

Vgr vegetal (protein) related

n.a. not applicable

in most other tissues. As the original characteristic of the BMP family is its unique ability to induce new bone formation, most studies evaluated the efficacy of BMPs (most commonly utilized OP-1/BMP-7 or BMP-2) for bone tissue regeneration [70, 116]. The BMPs induce the differentiation of resident mesenchymal cells into chondroblasts and osteoblasts. The responding cells can be provided by the surrounding soft tissue (e.g., muscle and vessels), the periosteum, bone marrow and the associated stroma, for tissue engineering applications in bone [140]. A variety of matrix systems are utilized in conjunction with BMPs for bone tissue engineering. The four major categories of investigated BMP matrix materials include natural polymers (e.g., collagen, alginate, hyaluronic acid), inorganic materials (e.g., calcium phosphate cements, hydroxyapatite, tricalcium phosphate), synthetic polymers {PDLLA [poly(D,L-lactide)], PEG [poly(ethylene glycol)], PLLA [poly(L-lactic acid)]} and composites of these materials. Autograft or allograft carriers have also been used. Carrier configurations range from simple depot delivery systems to more complex systems mimicking the extracellular matrix structure and function. Alternative BMP delivery systems are composed of viral vectors, genetically altered cells, conjugated factors and small molecules [114, 122].

BMP-2 is the most widely used growth factor to stimulate osteoblastic differentiation [147]. In 2004, the genetically engineered bone graft product, recombinant human bone morphogenetic protein-2/absorbable collagen sponge (rhBMP-2/ACS), was approved by the U.S. Food and Drug Administration (FDA) for the treatment of acute, open tibia shaft fractures in adults [65]. In the same year, OP-1 (BMP-7), in combination with bovine collagen, was also approved by the FDA. It is intended for use in making a new posterolateral spinal fusion in patients who have had a failed spinal fusion surgery and are not able to provide their own bone or bone marrow for grafting because of a condition such as osteoporosis, diabetes, or smoking [21].

Collagen is a well-established and most commonly used natural material for controlled delivery of BMPs. Several investigations demonstrated bone regeneration using rhBMP-2 and collagenous scaffolds in critical size defects. Würzler and coworkers showed mandibular reconstruction with an osteoinductive implant following an extensive continuity resection of the lower jaw in Göttinger mini-pigs [143]. Boyne and coworkers used rhBMP-2 in a collagenous carrier to regenerate hemimandibulectomy defects in elderly sub-human primates. The authors emphasized that aged non-human primates, chronologically comparable with 80-year-old humans, respond as favorably to rhBMP-2 as do the middle-aged animals. Extrapolating the results to the clinical level, the authors would expect that rhBMP-2 would produce a comparable result in the regeneration of large hemimandibulectomy-type defects in clinical human patients [15]. With the aim to repair an extended mandibular discontinuity defect, Warnke and coworkers grew a custom vascularized bone graft in a man. An ideal shaped titanium mesh cage filled with bone mineral blocks, the patient's bone marrow and rhBMP-7 was initially implanted in the latissimus dorsi muscle of an adult male patient and 7 weeks later transplanted as a free bone-muscle flap to repair the mandibular defect. The scintigraphic and radiological results showed new bone formation and bone remodeling and mineralization inside the mandibular transplant [132]. Recently, Swedish researchers published the reconstruction of a frontal bone defect by creating a bone graft in the latissimus dorsi muscle using BMP-2 and a polyamide template. A combination of heparin with bovine type I collagen, hyaluronic acid, and fibrin was used as carrier for BMP-2 [4].

As a result of refinements in genetic engineering, genetically modified BMPs were developed. Kübler and coworkers succeeded in production of a non-natural BMP-variant (EHBMP-2) with osteoinductive properties by expression in E. coli through specific mutation of the amino acid sequence [71]. Depprich and coworkers demonstrated that modification of the heparin-binding site by alteration of the amino acid sequence of BMP-2 increased osteoinductive activity. The authors postulated a longer retention period of BMP-2 in the tissue and thus a better bioavailability of the osteoinductive protein [28]. To investigate the effects of BMP on the development of stem cells is the main topic in tissue engineering research. As BMP signals play key roles throughout embryology, stem and progenitor cells show reactions upon exposure to BMP ligands, but the final outcome of a BMP signal is basically unpredictable and differs enormously between previously studied cell types [131]. In addition BMP, Wnt, and Notch pathways have been implicated to play key roles in self-renewal and differentiation of hematopoietic, intestinal, and epidermal stem cells [120]. BMP-2 has been shown to rapidly initiate the commitment of pluripotential mesenchymal stem cells to osteoprogenitor cells (preosteoblasts) and promote their maturation into osteoblasts [112]. Zur Nieden and coworkers demonstrated that BMP-2 can induce chondro-, osteo- and adipogenesis in embryonic stem cells, but this is depending on supplementary co-factors and on the concentration of BMP [151]. Recently, osteogenic differentiation of rabbit mesenchymal stem cells in a 3D hybrid scaffold containing hydroxyapatite and BMP-2 was published. The homogeneous bone formation observed throughout the hybrid scaffolds was greatly influenced by the addition of BMP-2 [98]. A new approach in tissue engineering is the use of distinct gene-based strategies. Gersbach and coworkers used primary skeletal myoblasts overexpressing either BMP-2 growth factor or Runx2 transcription factor. The authors' findings revealed that retroviral delivery of BMP-2 or Runx2 stimulated differentiation into an osteoblastic phenotype in vitro and in vivo, but BMP-2 stimulated osteoblastic markers faster and to a greater extent than Runx2. The authors emphasize the complexity of gene therapy-based bone substitutes as an integrated relationship of differentiation state, construct maturation, and paracrine signaling of osteogenic cells [42].

References

1. Alvarez RH, Kantarjian HM, Cortes JE (2006) Biology of platelet-derived growth factor and its involvement in disease. Mayo Clin Proc 81:1241-1257
2. Anusaksathien O, Webb SA, Jin QM, Giannobile WV (2003) Platelet-derived growth factor gene delivery stimulates ex vivo gingival repair. Tissue Eng 9:745-756
3. Anusaksathien O, Jin Q, Zhao M, Somerman MJ, Giannobile WV (2004) Effect of sustained gene delivery of platelet-derived growth factor or its antagonist (PDGF-1308) on tissue-engineered cementum. J Periodontol 75:429-440
4. Arnander C, Westermark A, Veltheim R, Docherty-Skogh AC, Hilborn J, Engstrand T (2006) Three-dimensional technology and bone morphogenetic protein in frontal bone reconstruction. J Craniofac Surg 17:275-279
5. Bartold PM, Narayanan AS, Page RC (1992) Platelet-derived growth factor reduces the inhibitory effects of lipopolysaccharide on gingival fibroblast proliferation. J Periodontal Res 27:499-505
6. Bastaki M, Nelli EE, Dell'Era P, Rusnati M, Molinari-Tosatti MP, Parolini S, Auerbach R, Ruco LP, Possati L, Presta M (1997) Basic fibroblast growth factor-induced angiogenic phenotype in mouse endothelium. A study of aortic and microvascular endothelial cell lines. Arterioscler Thromb Vasc Biol 17:454-464
7. Bateman J, Intini G, Margarone J, Goodloe S, Bush P, Lynch SE, Dziak R (2005) Platelet-derived growth factor enhancement of two alloplastic bone matrices. J Periodontol 76:1833-1841
8. Becerra J, Guerado E, Claros S, Alonso M, Bertrand ML, Gonzalez C, Andrades JA (2006) Autologous human-derived bone marrow cells exposed to a novel TGF-beta1 fusion protein for the treatment of critically sized tibial defect. Regen Med 1:267-278
9. Binz K, Schmid C, Bouillon R, Froesch ER, Jurgensen K, Hunziker EB (1994) Interactions of insulin-like growth factor I with dexamethasone on trabecular bone density and mineral metabolism in rats. Eur J Endocrinol 130:387-393
10. Blaimauer K, Watzinger E, Erovic BM, Martinek H, Jagersberger T, Thurnher D (2006) Effects of epidermal growth factor and keratinocyte growth factor on the growth of oropharyngeal keratinocytes in coculture with autologous fibroblasts in a three-dimensional matrix. Cells Tissues Organs 182:98-105
11. Block GD, Locker J, Bowen WC, Petersen BE, Katyal S, Strom SC, Riley T, Howard TA, Michalopoulos GK (1996) Population expansion, clonal growth, and specific differentiation patterns in primary cultures of hepatocytes induced by HGF/SF, EGF and TGF alpha in a chemically defined (HGM) medium. J Cell Biol 132:1133-1149
12. Bonewald L (2002) Transforming growth factor beta. In: Bilezikian JP, Raisz LG, Rodan GA (eds) 2nd edn. Academic Press, San Diego, pp 903-918
13. Bowen-Pope DF, Malpass TW, Foster DM, Ross R (1984) Platelet-derived growth factor in vivo: levels, activity, and rate of clearance. Blood 64:458-469
14. Boyan LA, Bhargava G, Nishimura F, Orman R, Price R, Terranova VP (1994) Mitogenic and chemotactic responses of human periodontal ligament cells to the different isoforms of platelet-derived growth factor. J Dent Res 73:1593-1600
15. Boyne PJ, Salina S, Nakamura A, Audia F, Shabahang S (2006) Bone regeneration using rhBMP-2 induction in hemimandibulectomy type defects of elderly sub-human primates. Cell Tissue Bank 7:1-10
16. Breitbart AS, Mason JM, Urmacher C, Barcia M, Grant RT, Pergolizzi RG, Grande DA (1999) Gene-enhanced tissue engineering: applications for wound healing using cultured dermal fibroblasts transduced retrovirally with the PDGF-B gene. Ann Plast Surg 43:632-639
17. Cai DZ, Zeng C, Quan DP, Bu LS, Wang K, Lu HD, Li XF (2007) Biodegradable chitosan scaffolds containing microspheres as carriers for controlled transforming growth factor-beta1 delivery for cartilage tissue engineering. Chin Med J (Engl) 120:197-203
18. Canalis E, McCarthy TL, Centrella M (1989) Effects of platelet-derived growth factor on bone formation in vitro. J Cell Physiol 140:530-537
19. Canalis E, Varghese S, McCarthy TL, Centrella M (1992) Role of platelet derived growth factor in bone cell function. Growth Regul 2:151-155

20. Capito RM, Spector M (2007) Collagen scaffolds for nonviral IGF-1 gene delivery in articular cartilage tissue engineering. Gene Ther 14:721-732
21. Carlisle E, Fischgrund JS (2005) Bone morphogenetic proteins for spinal fusion. Spine J 5:240S-249S
22. Carpenter G, Cohen S (1979) Epidermal growth factor. Annu Rev Biochem 48:193-216
23. Celeste AJ, Iannazzi JA, Taylor RC, Hewick RM, Rosen V, Wang EA, Wozney JM (1990) Identification of transforming growth factor beta family members present in bone-inductive protein purified from bovine bone. Proc Natl Acad Sci U S A 87:9843-9847
24. Chiou M, Xu Y, Longaker MT (2006) Mitogenic and chondrogenic effects of fibroblast growth factor-2 in adipose-derived mesenchymal cells. Biochem Biophys Res Commun 343:644-652
25. Chua KH, Aminuddin BS, Fuzina NH, Ruszymah BH (2004) Interaction between insulin-like growth factor-1 with other growth factors in serum depleted culture medium for human cartilage engineering. Med J Malaysia 59 Suppl B:7-8
26. Cras-Meneur C, Elghazi L, Czernichow P, Scharfmann R (2001) Epidermal growth factor increases undifferentiated pancreatic embryonic cells in vitro: a balance between proliferation and differentiation. Diabetes 50:1571-1579
27. Denley A, Cosgrove LJ, Booker GW, Wallace JC, Forbes BE (2005) Molecular interactions of the IGF system. Cytokine Growth Factor Rev 16:421-439
28. Depprich R, Handschel J, Sebald W, Kubler NR, Wurzler KK (2005) Comparison of the osteogenic activity of bone morphogenetic protein (BMP) mutants. Mund Kiefer Gesichtschir 9:363-368
29. Detamore MS, Athanasiou KA (2005) Evaluation of three growth factors for TMJ disc tissue engineering. Ann Biomed Eng 33:383-390
30. Dionne CA, Crumley G, Bellot F, Kaplow JM, Searfoss G, Ruta M, Burgess WH, Jaye M, Schlessinger J (1990) Cloning and expression of two distinct high-affinity receptors cross-reacting with acidic and basic fibroblast growth factors. Embo J 9:2685-2692
31. Dreux AC, Lamb DJ, Modjtahedi H, Ferns GA (2006) The epidermal growth factor receptors and their family of ligands: their putative role in atherogenesis. Atherosclerosis 186:38-53
32. Elisseeff J, McIntosh W, Fu K, Blunk BT, Langer R (2001) Controlled-release of IGF-I and TGF-beta1 in a photopolymerizing hydrogel for cartilage tissue engineering. J Orthop Res 19:1098-1104
33. Engele J, Bohn MC (1992) Effects of acidic and basic fibroblast growth factors (aFGF, bFGF) on glial precursor cell proliferation: age dependency and brain region specificity. Dev Biol 152:363-372
34. Fakhry A, Ratisoontorn C, Vedhachalam C, Salhab I, Koyama E, Leboy P, Pacifici M, Kirschner RE, Nah HD (2005) Effects of FGF-2/-9 in calvarial bone cell cultures: differentiation stage-dependent mitogenic effect, inverse regulation of BMP-2 and noggin, and enhancement of osteogenic potential. Bone 36:254-266
35. Fawzi-Grancher S, De Isla N, Faure G, Stoltz JF, Muller S (2006) Optimisation of biochemical condition and substrates in vitro for tissue engineering of ligament. Ann Biomed Eng 34:1767-1777
36. Froesch ER, Buergi H, Ramseier EB, Bally P, Labhart A (1963) Antibody-suppressible and nonsuppressible insulin-like activities in human serum and their physiologic significance. an insulin assay with adipose tissue of increased precision and specificity. J Clin Invest 42:1816-1834
37. Froesch ER (1993) IGFs: function and clinical importance. J Intern Med 234:533-534
38. Fu P, Sodian R, Luders C, Lemke T, Kraemer L, Hubler M, Weng Y, Hoerstrup SP, Meyer R, Hetzer R (2004) Effects of basic fibroblast growth factor and transforming growth factor-beta on maturation of human pediatric aortic cell culture for tissue engineering of cardiovascular structures. Asaio J 50:9-14
39. Fujita M, Ishihara M, Shimizu M, Obara K, Nakamura S, Kanatani Y, Morimoto Y, Takase B, Matsui T, Kikuchi M, Maehara T (2007) Therapeutic angiogenesis induced by controlled release of fibroblast growth factor-2 from injectable chitosan/non-anticoagulant heparin hydrogel in a rat hindlimb ischemia model. Wound Repair Regen 15:58-65
40. Fukumoto T, Sperling JW, Sanyal A, Fitzsimmons JS, Reinholz GG, Conover CA, O'Driscoll SW (2003) Combined effects of insulin-like growth factor-1 and transforming growth factor-beta1 on periosteal mesenchymal cells during chondrogenesis in vitro. Osteoarthritis Cartilage 11:55-64
41. Gamal AY, Mailhot JM (2000) The effect of local delivery of PDGF-BB on attachment of human periodontal ligament fibroblasts to periodontitis-affected root surfaces--in vitro. J Clin Periodontol 27:347-353
42. Gersbach CA, Guldberg RE, Garcia AJ (2007) In vitro and in vivo osteoblastic differentiation of BMP-2-, Runx2-engineered skeletal myoblasts. J Cell Biochem 100:1324-1336
43. Gombotz WR, Pankey SC, Bouchard LS, Ranchalis J, Puolakkainen P (1993) Controlled release of TGF-beta 1 from a biodegradable matrix for bone regeneration. J Biomater Sci Polym Ed 5:49-63
44. Gomez G, Korkiakoski S, Gonzalez MM, Lansman S, Ella V, Salo T, Kellomaki M, Ashammakhi N, Arnaud E (2006) Effect of FGF and polylactide scaffolds on calvarial bone healing with growth factor on biodegradable polymer scaffolds. J Craniofac Surg 17:935-942
45. Gospodarowicz D, Mescher AL, Birdwell CR (1978) Control of cellular proliferation by the fibroblast and epidermal growth factors. Natl Cancer Inst Monogr 109-130
46. Graves DT, Cochran DL (1990) Mesenchymal cell growth factors. Crit Rev Oral Biol Med 1:17-36
47. Gritti A, Frolichsthal-Schoeller P, Galli R, Parati EA, Cova L, Pagano SF, Bjornson CR, Vescovi AL (1999) Epidermal and fibroblast growth factors behave as mitogenic regulators for a single multipotent stem cell-like population from the subventricular region of the adult mouse forebrain. J Neurosci 19:3287-3297
48. Gualandris A, Rusnati M, Belleri M, Nelli EE, Bastaki M, Molinari-Tosatti MP, Bonardi F, Parolini S, Albini A, Morbidelli L, Ziche M, Corallini A, Possati L, Vacca A,

Ribatti D, Presta M (1996) Basic fibroblast growth factor overexpression in endothelial cells: an autocrine mechanism for angiogenesis and angioproliferative diseases. Cell Growth Differ 7:147-160
49. Hammacher A, Hellman U, Johnsson A, Ostman A, Gunnarsson K, Westermark B, Wasteson A, Heldin CH (1988) A major part of platelet-derived growth factor purified from human platelets is a heterodimer of one A and one B chain. J Biol Chem 263:16493-16498
50. Harrison RD, Gratzer PF (2005) Effect of extraction protocols and epidermal growth factor on the cellular repopulation of decellularized anterior cruciate ligament allografts. J Biomed Mater Res A 75:841-854
51. Hart CE, Bowen-Pope DF (1990) Platelet-derived growth factor receptor: current views of the two-subunit model. J Invest Dermatol 94:53S-57S
52. He ZP, Tan WQ, Tang YF, Feng MF (2003) Differentiation of putative hepatic stem cells derived from adult rats into mature hepatocytes in the presence of epidermal growth factor and hepatocyte growth factor. Differentiation 71:281-290
53. Heldin CH, Miyazono K, ten Dijke P (1997) TGF-beta signalling from cell membrane to nucleus through SMAD proteins. Nature 390:465-471
54. Hirst SJ, Barnes PJ, Twort CH (1996) PDGF isoform-induced proliferation and receptor expression in human cultured airway smooth muscle cells. Am J Physiol 270:L415-428
55. Hunziker EB, Driesang IM, Morris EA (2001) Chondrogenesis in cartilage repair is induced by members of the transforming growth factor-beta superfamily. Clin Orthop Relat Res S171-181
56. Igai H, Chang SS, Gotoh M, Yamamoto Y, Misaki N, Okamoto T, Yamamoto M, Tabata Y, Yokomise H (2006) Regeneration of canine tracheal cartilage by slow release of basic fibroblast growth factor from gelatin sponge. Asaio J 52:86-91
57. Igai H, Yamamoto Y, Chang SS, Yamamoto M, Tabata Y, Yokomise H (2007) Tracheal cartilage regeneration by slow release of basic fibroblast growth factor from a gelatin sponge. J Thorac Cardiovasc Surg 134:170-175
58. Im GI, Jung NH, Tae SK (2006) Chondrogenic differentiation of mesenchymal stem cells isolated from patients in late adulthood: the optimal conditions of growth factors. Tissue Eng 12:527-536
59. Indrawattana N, Chen G, Tadokoro M, Shann LH, Ohgushi H, Tateishi T, Tanaka J, Bunyaratvej A (2004) Growth factor combination for chondrogenic induction from human mesenchymal stem cell. Biochem Biophys Res Commun 320:914-919
60. Janssens K, ten Dijke P, Janssens S, Van Hul W (2005) Transforming growth factor-beta1 to the bone. Endocr Rev 26:743-774
61. Jaye M, Schlessinger J, Dionne CA (1992) Fibroblast growth factor receptor tyrosine kinases: molecular analysis and signal transduction. Biochim Biophys Acta 1135:185-199
62. Jenner JM, van Eijk F, Saris DB, Willems WJ, Dhert WJ, Creemers LB (2007) Effect of transforming growth factor-beta and growth differentiation factor-5 on proliferation and matrix production by human bone marrow stromal cells cultured on braided poly lactic-co-glycolic Acid scaffolds for ligament tissue engineering. Tissue Eng 13:1573-1582
63. Jin Q, Anusaksathien O, Webb SA, Printz MA, Giannobile WV (2004) Engineering of tooth-supporting structures by delivery of PDGF gene therapy vectors. Mol Ther 9:519-526
64. Johnson DE, Williams LT (1993) Structural and functional diversity in the FGF receptor multigene family. Adv Cancer Res 60:1-41
65. Jones AL, Bucholz RW, Bosse MJ, Mirza SK, Lyon TR, Webb LX, Pollak AN, Golden JD, Valentin-Opran A (2006) Recombinant human BMP-2 and allograft compared with autogenous bone graft for reconstruction of diaphyseal tibial fractures with cortical defects. A randomized, controlled trial. J Bone Joint Surg Am 88:1431-1441
66. Kakudo N, Shimotsuma A, Kusumoto K (2007) Fibroblast growth factor-2 stimulates adipogenic differentiation of human adipose-derived stem cells. Biochem Biophys Res Commun 359:239-244
67. Khalil N (1999) TGF-beta: from latent to active. Microbes Infect 1:1255-1263
68. Klagsbrun M (1990) The affinity of fibroblast growth factors (FGFs) for heparin; FGF-heparan sulfate interactions in cells and extracellular matrix. Curr Opin Cell Biol 2:857-863
69. Koay EJ, Hoben GM, Athanasiou KA (2007) Tissue Engineering with Chondrogenically-differentiated Human Embryonic Stem Cells. Stem Cells
70. Korchynsky O, ten Dijke P (2002) Bone morphogenetic protein receptors and their nuclear effectors in bone formation. In: Vukicevic S, Sampath KT (eds) Bone morphogenetic proteins: from laboratory to clinical practice. Birkhäuser, Basel, pp 31-60
71. Kubler NR, Wurzler K, Reuther JF, Faller G, Sieber E, Kirchner T, Sebald W (1999) EHBMP-2. Initial BMP analog with osteoinductive properties. Mund Kiefer Gesichtschir 3 Suppl 1:S134-S139
72. Lawrence DA (2001) Latent-TGF-beta: an overview. Mol Cell Biochem 219:163-170
73. Lee DC, Fenton SE, Berkowitz EA, Hissong MA (1995) Transforming growth factor alpha: expression, regulation, and biological activities. Pharmacol Rev 47:51-85
74. Lee JY, Kim KH, Shin SY, Rhyu IC, Lee YM, Park YJ, Chung CP, Lee SJ (2006) Enhanced bone formation by transforming growth factor-beta1-releasing collagen/chitosan microgranules. J Biomed Mater Res A 76:530-539
75. Lee SJ (2000) Cytokine delivery and tissue engineering. Yonsei Med J 41:704-719
76. Lee YM, Park YJ, Lee SJ, Ku Y, Han SB, Klokkevold PR, Chung CP (2000) The bone regenerative effect of platelet-derived growth factor-BB delivered with a chitosan/tricalcium phosphate sponge carrier. J Periodontol 71:418-424
77. Lieb E, Milz S, Vogel T, Hacker M, Dauner M, Schulz MB (2004) Effects of transforming growth factor beta1 on bonelike tissue formation in three-dimensional cell culture. I. Culture conditions and tissue formation. Tissue Eng 10:1399-1413

78. Lieb E, Vogel T, Milz S, Dauner M, Schulz MB (2004) Effects of transforming growth factor beta1 on bonelike tissue formation in three-dimensional cell culture. II: Osteoblastic differentiation. Tissue Eng 10:1414-1425
79. Lieberman JR, Daluiski A, Einhorn TA (2002) The role of growth factors in the repair of bone. Biology and clinical applications. J Bone Joint Surg Am 84-A:1032-1044
80. Linkhart TA, Mohan S, Baylink DJ (1996) Growth factors for bone growth and repair: IGF, TGF beta and BMP. Bone 19:1S-12S
81. Longobardi L, O'Rear L, Aakula S, Johnstone B, Shimer K, Chytil A, Horton WA, Moses HL, Spagnoli A (2006) Effect of IGF-I in the chondrogenesis of bone marrow mesenchymal stem cells in the presence or absence of TGF-beta signaling. J Bone Miner Res 21:626-636
82. Lu L, Yaszemski MJ, Mikos AG (2001) TGF-beta1 release from biodegradable polymer microparticles: its effects on marrow stromal osteoblast function. J Bone Joint Surg Am 83-A Suppl 1:S82-91
83. Lynch SE, Wisner-Lynch L, Nevins M, Nevins ML (2006) A new era in periodontal and periimplant regeneration: use of growth-factor enhanced matrices incorporating rhPDGF. Compend Contin Educ Dent 27:672-678; quiz 679-680
84. Marcopoulou CE, Vavouraki HN, Dereka XE, Vrotsos IA (2003) Proliferative effect of growth factors TGF-beta1, PDGF-BB and rhBMP-2 on human gingival fibroblasts and periodontal ligament cells. J Int Acad Periodontol 5:63-70
85. Martin I, Suetterlin R, Baschong W, Heberer M, Vunjak-Novakovic G, Freed LE (2001) Enhanced cartilage tissue engineering by sequential exposure of chondrocytes to FGF-2 during 2D expansion and BMP-2 during 3D cultivation. J Cell Biochem 83:121-128
86. Massague J (1990) The transforming growth factor-beta family. Annu Rev Cell Biol 6:597-641
87. Massague J (1998) TGF-beta signal transduction. Annu Rev Biochem 67:753-791
88. Mauck RL, Nicoll SB, Seyhan SL, Ateshian GA, Hung CT (2003) Synergistic action of growth factors and dynamic loading for articular cartilage tissue engineering. Tissue Eng 9:597-611
89. Meaney Murray M, Rice K, Wright RJ, Spector M (2003) The effect of selected growth factors on human anterior cruciate ligament cell interactions with a three-dimensional collagen-GAG scaffold. J Orthop Res 21:238-244
90. Meyer-Ingold W, Eichner W (1995) Platelet-derived growth factor. Cell Biol Int 19:389-398
91. Michalopoulos GK, Bowen WC, Zajac VF, Beer-Stolz D, Watkins S, Kostrubsky V, Strom SC (1999) Morphogenetic events in mixed cultures of rat hepatocytes and nonparenchymal cells maintained in biological matrices in the presence of hepatocyte growth factor and epidermal growth factor. Hepatology 29:90-100
92. Minamide A, Yoshida M, Kawakami M, Okada M, Enyo Y, Hashizume H, Boden SD (2007) The effects of bone morphogenetic protein and basic fibroblast growth factor on cultured mesenchymal stem cells for spine fusion. Spine 32:1067-1071
93. Miura Y, Parvizi J, Fitzsimmons JS, O'Driscoll SW (2002) Brief exposure to high-dose transforming growth factor-beta1 enhances periosteal chondrogenesis in vitro: a preliminary report. J Bone Joint Surg Am 84-A:793-799
94. Miyazono K (2000) TGF-beta signaling by Smad proteins. Cytokine Growth Factor Rev 11:15-22
95. Miyazono K, Kusanagi K, Inoue H (2001) Divergence and convergence of TGF-beta/BMP signaling. J Cell Physiol 187:265-276
96. Montesano R, Vassalli JD, Baird A, Guillemin R, Orci L (1986) Basic fibroblast growth factor induces angiogenesis in vitro. Proc Natl Acad Sci U S A 83:7297-7301
97. Moreau J, Chen J, Kaplan D, Altman G (2006) Sequential growth factor stimulation of bone marrow stromal cells in extended culture. Tissue Eng 12:2905-2912
98. Na K, Kim SW, Sun BK, Woo DG, Yang HN, Chung HM, Park KH (2007) Osteogenic differentiation of rabbit mesenchymal stem cells in thermo-reversible hydrogel constructs containing hydroxyapatite and bone morphogenic protein-2 (BMP-2). Biomaterials 28:2631-2637
99. Nakao A, Imamura T, Souchelnytskyi S, Kawabata M, Ishisaki A, Oeda E, Tamaki K, Hanai J, Heldin CH, Miyazono K, ten Dijke P (1997) TGF-beta receptor-mediated signalling through Smad2, Smad3 and Smad4. Embo J 16:5353-5362
100. Nevins M, Giannobile WV, McGuire MK, Kao RT, Mellonig JT, Hinrichs JE, McAllister BS, Murphy KS, McClain PK, Nevins ML, Paquette DW, Han TJ, Reddy MS, Lavin PT, Genco RJ, Lynch SE (2005) Platelet-derived growth factor stimulates bone fill and rate of attachment level gain: results of a large multicenter randomized controlled trial. J Periodontol 76:2205-2215
101. Nimni ME (1997) Polypeptide growth factors: targeted delivery systems. Biomaterials 18:1201-1225
102. Ornitz DM, Itoh N (2001) Fibroblast growth factors. Genome Biol 2:REVIEWS3005
103. Pandit A, Ashar R, Feldman D, Thompson A (1998) Investigation of acidic fibroblast growth factor delivered through a collagen scaffold for the treatment of full-thickness skin defects in a rabbit model. Plast Reconstr Surg 101:766-775
104. Papadopoulos CE, Dereka XE, Vavouraki EN, Vrotsos IA (2003) In vitro evaluation of the mitogenic effect of platelet-derived growth factor-BB on human periodontal ligament cells cultured with various bone allografts. J Periodontol 74:451-457
105. Parikh SN (2002) Bone graft substitutes: past, present, future. J Postgrad Med 48:142-148
106. Park H, Temenoff JS, Holland TA, Tabata Y, Mikos AG (2005) Delivery of TGF-beta1 and chondrocytes via injectable, biodegradable hydrogels for cartilage tissue engineering applications. Biomaterials 26:7095-7103
107. Piek E, Heldin CH, Ten Dijke P (1999) Specificity, diversity, and regulation in TGF-beta superfamily signaling. Faseb J 13:2105-2124
108. Reddi AH (2000) Morphogenesis and tissue engineering of bone and cartilage: inductive signals, stem cells, and biomimetic biomaterials. Tissue Eng 6:351-359

109. Reynolds BA, Weiss S (1996) Clonal and population analyses demonstrate that an EGF-responsive mammalian embryonic CNS precursor is a stem cell. Dev Biol 175:1-13
110. Ribatti D, Nico B, Vacca A, Roncali L, Presta M (1999) Endogenous and exogenous fibroblast growth factor-2 modulate wound healing in the chick embryo chorioallantoic membrane. Angiogenesis 3:89-95
111. Richard C, Liuzzo JP, Moscatelli D (1995) Fibroblast growth factor-2 can mediate cell attachment by linking receptors and heparan sulfate proteoglycans on neighboring cells. J Biol Chem 270:24188-24196
112. Rickard DJ, Sullivan TA, Shenker BJ, Leboy PS, Kazhdan I (1994) Induction of rapid osteoblast differentiation in rat bone marrow stromal cell cultures by dexamethasone and BMP-2. Dev Biol 161:218-228
113. Ripamonti U, Crooks J, Rueger DC (2001) Induction of bone formation by recombinant human osteogenic protein-1 and sintered porous hydroxyapatite in adult primates. Plast Reconstr Surg 107:977-988
114. Rose FR, Hou Q, Oreffo RO (2004) Delivery systems for bone growth factors - the new players in skeletal regeneration. J Pharm Pharmacol 56:415-427
115. Rotwein P (1991) Structure, evolution, expression and regulation of insulin-like growth factors I and II. Growth Factors 5:3-18
116. Rueger DC (2002) Biochemistry of bone morphogenetic proteins. In: Vukicevic S, Sampath KT (eds) Bone morphogenetic proteins: from laboratory to clinical practice. Birkhäuser, Basel, pp 1-18
117. Sales VL, Engelmayr GC Jr, Mettler BA, Johnson JA Jr, Sacks MS, Mayer JE, Jr. (2006) Transforming growth factor-beta1 modulates extracellular matrix production, proliferation, and apoptosis of endothelial progenitor cells in tissue-engineering scaffolds. Circulation 114:I193-199
118. Sampath TK, Reddi AH (1981) Dissociative extraction and reconstitution of extracellular matrix components involved in local bone differentiation. Proc Natl Acad Sci U S A 78:7599-7603
119. Sampath TK, Muthukumaran N, Reddi AH (1987) Isolation of osteogenin, an extracellular matrix-associated, bone-inductive protein, by heparin affinity chromatography. Proc Natl Acad Sci U S A 84:7109-7113
120. Satija NK, Gurudutta GU, Sharma S, Afrin F, Gupta P, Verma YK, Singh VK, Tripathi RP (2007) Mesenchymal stem cells: molecular targets for tissue engineering. Stem Cells Dev 16:7-23
121. Schmid C (1995) Insulin-like growth factors. Cell Biol Int 19:445-457
122. Seeherman H, Wozney JM (2005) Delivery of bone morphogenetic proteins for orthopedic tissue regeneration. Cytokine Growth Factor Rev 16:329-345
123. Slavin J (1995) Fibroblast growth factors: at the heart of angiogenesis. Cell Biol Int 19:431-444
124. Sleeman M, Fraser J, McDonald M, Yuan S, White D, Grandison P, Kumble K, Watson JD, Murison JG (2001) Identification of a new fibroblast growth factor receptor, FGFR5. Gene 271:171-182
125. Stewart CE, Rotwein P (1996) Growth, differentiation, and survival: multiple physiological functions for insulin-like growth factors. Physiol Rev 76:1005-1026
126. ten Dijke P, Yamashita H, Ichijo H, Franzen P, Laiho M, Miyazono K, Heldin CH (1994) Characterization of type I receptors for transforming growth factor-beta and activin. Science 264:101-104
127. Tropepe V, Sibilia M, Ciruna BG, Rossant J, Wagner EF, van der Kooy D (1999) Distinct neural stem cells proliferate in response to EGF and FGF in the developing mouse telencephalon. Dev Biol 208:166-188
128. Ullrich A, Berman CH, Dull TJ, Gray A, Lee JM (1984) Isolation of the human insulin-like growth factor I gene using a single synthetic DNA probe. Embo J 3:361-364
129. Urist MR (1965) Bone: formation by autoinduction. Science 150:893-899
130. Veilleux N, Spector M (2005) Effects of FGF-2 and IGF-1 on adult canine articular chondrocytes in type II collagen-glycosaminoglycan scaffolds in vitro. Osteoarthritis Cartilage 13:278-286
131. Wagner TU (2007) Bone morphogenetic protein signaling in stem cells–one signal, many consequences. Febs J 274:2968-2976
132. Warnke PH, Springer IN, Wiltfang J, Acil Y, Eufinger H, Wehmoller M, Russo PA, Bolte H, Sherry E, Behrens E, Terheyden H (2004) Growth and transplantation of a custom vascularised bone graft in a man. Lancet 364:766-770
133. Wei G, Jin Q, Giannobile WV, Ma PX (2006) Nanofibrous scaffold for controlled delivery of recombinant human PDGF-BB. J Control Release 112:103-110
134. Weisser J, Rahfoth B, Timmermann A, Aigner T, Brauer R, von der Mark K (2001) Role of growth factors in rabbit articular cartilage repair by chondrocytes in agarose. Osteoarthritis Cartilage 9 Suppl A:S48-S54
135. Westermark B, Heldin CH (1993) Platelet-derived growth factor. Structure, function and implications in normal and malignant cell growth. Acta Oncol 32:101-105
136. Westreich R, Kaufman M, Gannon P, Lawson W (2004) Validating the subcutaneous model of injectable autologous cartilage using a fibrin glue scaffold. Laryngoscope 114:2154-2160
137. Wiedlocha A, Sorensen V (2004) Signaling, internalization, and intracellular activity of fibroblast growth factor. Curr Top Microbiol Immunol 286:45-79
138. Woo YK, Kwon SY, Lee HS, Park YS (2007) Proliferation of anterior cruciate ligament cells in vitro by photo-immobilized epidermal growth factor. J Orthop Res 25:73-80
139. Wozney JM, Rosen V, Celeste AJ, Mitsock LM, Whitters MJ, Kriz RW, Hewick RM, Wang EA (1988) Novel regulators of bone formation: molecular clones and activities. Science 242:1528-1534
140. Wozney JM, Seeherman HJ (2004) Protein-based tissue engineering in bone and cartilage repair. Curr Opin Biotechnol 15:392-398
141. Wrana JL, Attisano L, Wieser R, Ventura F, Massague J (1994) Mechanism of activation of the TGF-beta receptor. Nature 370:341-347
142. Wrana JL, Attisano L (2000) The Smad pathway. Cytokine Growth Factor Rev 11:5-13

143. Würzler KK, Heisterkamp M, Böhm H, Kübler NR, Sebald W, Reuther JF (2004) Mandibular reconstruction with autologous bone and osseoinductive implant in the Gottingen minipig. Mund Kiefer Gesichtschir 8:75-82
144. Xue L, Tassiopoulos AK, Woloson SK, Stanton DL, Jr., Ms CS, Hampton B, Burgess WH, Greisler HP (2001) Construction and biological characterization of an HB-GAM/FGF-1 chimera for vascular tissue engineering. J Vasc Surg 33:554-560
145. Yamaoka H, Asato H, Ogasawara T, Nishizawa S, Takahashi T, Nakatsuka T, Koshima I, Nakamura K, Kawaguchi H, Chung UI, Takato T, Hoshi K (2006) Cartilage tissue engineering using human auricular chondrocytes embedded in different hydrogel materials. J Biomed Mater Res A 78:1-11
146. Yasuda A, Kojima K, Tinsley KW, Yoshioka H, Mori Y, Vacanti CA (2006) In vitro culture of chondrocytes in a novel thermoreversible gelation polymer scaffold containing growth factors. Tissue Eng 12:1237-1245
147. Yoon ST, Boden SD (2002) Osteoinductive molecules in orthopaedics: basic science and preclinical studies. Clin Orthop Relat Res 33-43
148. Zhang Y, Wang Y, Shi B, Cheng X (2007) A platelet-derived growth factor releasing chitosan/coral composite scaffold for periodontal tissue engineering. Biomaterials 28:1515-1522
149. Zhu Z, Lee CS, Tejeda KM, Giannobile WV (2001) Gene transfer and expression of platelet-derived growth factors modulate periodontal cellular activity. J Dent Res 80:892-897
150. Zofkova I (2003) Pathophysiological and clinical importance of insulin-like growth factor-I with respect to bone metabolism. Physiol Res 52:657-679
151. zur Nieden NI, Kempka G, Rancourt DE, Ahr HJ (2005) Induction of chondro-, osteo- and adipogenesis in embryonic stem cells by bone morphogenetic protein-2: effect of cofactors on differentiating lineages. BMC Dev Biol 5:1

IV   Engineering at the Cellular Level

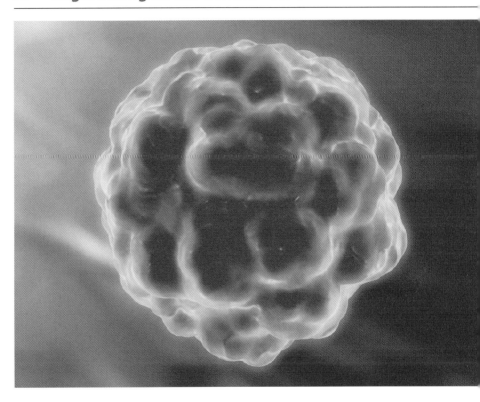

Fetal Tissue Engineering: Regenerative Capacity of Fetal Stem Cells

P. Wu, D. Moschidou, N. M. Fisk

Contents

12.1	Introduction	139
12.2	Advantages of Fetal Stem Cells	139
12.3	Fetal Stem Cell Sources	140
12.3.1	Bone Marrow, Liver, Blood	140
12.3.2	Amniotic Fluid, Placenta	141
12.3.3	Umbilical Cord Blood	141
12.3.4	Non-Haemopoietic Solid Organs	143
12.3.5	Harvesting Fetal Stem Cells and Their Characterisation	143
12.4	Potential for Tissue Repair	147
12.4.1	Microchimerism	147
12.4.2	Intrauterine Transplantation	149
12.4.3	Adult Tissue Repair	151
12.5	Conclusions	153
	References	153

12.1 Introduction

Considerable debate has focused on the contrasting merits of embryonic and adult stem cells. Fetal stem cells represent an intermediate cell type in this controversy. Adult stem cells have limited capacity to differentiate into fully functioning mature cell types of their tissue of origin (multipotent), whereas embryonic stem cells (ESCs) have the advantageous capacity to develop into all tissue types (pluripotent), including trophoblasts (totipotent). However, ESC research has been hampered by both safety concerns and ethical reservations due to requisite destruction of the blastocyst during harvesting. Transplantation of ESCs is almost invariably followed by the development of embryonal teratomas, precluding cell transplantation. On the other hand, adult stem cells have the advantage of greater accessibility, but the disadvantage of more limited proliferative capacity and restricted plasticity. Nevertheless, adult mesenchymal stem cells (MSCs) can differentiate into a range of mesoderm-derived tissues such as bone, fat and cartilage, while adult haemopoietic stem cells (HSCs) can reconstitute the haemopoietic system. The plasticity of ESC might seem to give them a therapeutic advantage over adult stem cells, whereas, in effect, this limits potential therapeutic application as a result of oncogenicity. So it is adult stem cells that hold promise for cell transplantation, while the eventual clinical use for ESCs is likely to be in tissue engineering applications.

12.2 Advantages of Fetal Stem Cells

Fetal stem cells can be found at various stages of human development with a declining gradient of potency as gestation advances, and teleologically lie betwixt and between ESCs and adult stem cells [41]. Compared with adult cells, they have greater differentiation potential and a competitive advantage in terms of engraftment [46, 95, 109], while, unlike ESCs, they do not form teratomas in vivo [1]. Early fetal MSCs, for instance, self renew faster in culture than adult MSC and senesce later, whilst retaining a stable phenotype. Therefore they are more readily suited for use on a therapeutic scale for either ex-vivo gene or cell therapy, fetal tissue engineering or intrauterine transplantation. First trimester human fetal MSCs are more primitive, have longer telomeres and greater telomerase activity compared to adult MSCs, and additionally express the allegedly embryonic-specific pluripotency markers [43]. Their multipotentiality is greater and they differentiate readily into muscle [22], oligodendrocytes [61] and haemopoietic cells [76]. They express a shared α2, α4, and α5β1 integrin profile with first trimester HSC, implicating them in homing and engraftment, and consistent with this, they have significantly greater binding to their respective extracellular matrix ligands than adult MSCs [30]. There seems no difference in transducibility with efficiencies of >95% using lentiviral vectors that have stable gene expression at both short and long time points, without affecting self renewal or multipotentiality [20]. Like adult MSCs, they express low levels of HLA I, but unlike them, they lack intracellular HLA II and take a longer time to express this on stimulation [40].

Umbilical cord blood (UCB) is a valuable source of HSCs, which are increasingly used as an alternative to bone marrow (BM) for transplantation as they are easily accessible and more readily available. On the other hand, although MSCs from cord blood have similar properties to first trimester fetal MSCs, they are difficult to isolate and are present in very low concentrations in term cord blood compared with UCB from preterm infants [32, 52]. Furthermore, Kogler et al. [65] demonstrated that only 35% of term UCB samples contain differentiating MSCs. Promisingly, MSCs have been obtained from term umbilical cord tissue (i.e. Wharton's jelly) at much higher efficiencies than term UCB [100]. Comparative studies between early and late fetal stem cells are awaited.

12.3
Fetal Stem Cell Sources

Fetal stem cells can be isolated from a variety of sources and have different properties depending on progenitor cell type. HSCs are found in abundance during embryonic and fetal life and originate from haemangioblasts in the yolk sac and the aorto-gonad-mesonephros (AGM) region [78]. More recently, Rhodes et al. [95] have shown that HSCs are not only present in murine placenta, but also that the placenta is an HSC generation site. These CD34$^+$ cells lack expression of lineage markers such as CD38 and HLA-DR. They have a slow doubling time, differentiate down the myeloid and lymphoid lineages in vitro, and can repopulate the entire haemopoietic system when transplanted into xenogeneic recipients [108].

Fetal MSCs can be isolated from fetal blood and haemopoietic organs in early pregnancy [14], but also from a variety of somatic organs including the liver, spleen, BM, femur, pancreas, and lung, as well as amniotic fluid and placenta across gestation and UCB after delivery.

12.3.1
Bone Marrow, Liver, Blood

HSC can easily be isolated from first trimester fetal blood and the major haemopoietic organs: the liver and BM. The relative abundance of HSCs in blood peaks in the mid second trimester, probably due to cells migrating from the liver to establish haemopoiesis in the fetal BM [25]. Fetal blood HSCs proliferate faster than those present in UCB or adult BM, and can give rise to all the haemopoietic lineages [13]. Fetal liver HSCs generate more progenitors than their adult BM counterparts and have a higher cloning efficiency [84, 97] as well as a tenfold competitive engraftment advantage over adult marrow cells when infused into immunodeficient fetal recipients [108].

In contrast, MSCs present in the fetal circulation from early gestation, and progressively decline in frequency in fetal blood during the late first trimester [14]. Their presence before initiation of BM haemopoiesis, their ability to support haemopoiesis in co-culture, their decline in frequency as haemopoiesis is established, along with comparable populations of MSCs being found in first trimester BM and liver is consistent with the hypothesis that fetal MSCs migrate to future definitive sites of haemopoiesis where they adhere to facilitate HSC engraftment and initiation of haemopoiesis [29].

In the presence of serum, fetal MSCs from blood, liver and bone marrow adhere to plastic and grow as spindle-shaped fibroblastic cells similar to adult BM MSCs. They are non-haemopoietic and non-endothelial, as they do not express CD45, CD34, CD14, CD31 or vWF, but express a number of adhesion molecules including HCAM-1 (CD44), VCAM-1 (CD106) and β1-integrin (CD29). In their undifferentiated state, MSCs show positive expression of stroma-associated markers CD73 (SH3 and SH4) and CD105 (SH2), the early bone marrow progenitor cell marker CD90 (thy-1), and the extracellular matrix proteins vimentin, laminin and fibronectin [42]. They demonstrate tremendous expansive capacity, and cycle faster than comparable adult BM-derived MSCs, having a doubling time of 24–30 h over 20 passages (50 population doublings) without differentiation [14, 39]. Like their adult counterparts, human fetal MSCs can differentiate into at least three different mesenchymal tissues: fat, bone and cartilage. More recent work shows that they may be able to differentiate into skeletal muscle [19], neurons and oligodendrocytes [60] as well as hepatocytes and blood [75]. Additionally, first trimester MSCs express the pluripotency stem cell markers Oct-4, Nanog, Rex-1, SSEA-3, SSEA-4, Tra-1-61 and Tra-1-81. MSCs also express more hTERT, are more readily expandable and senesce later in culture when compared with adult MSCs (Fig. 12.1) [43].

12.3.2
Amniotic Fluid, Placenta

The amniotic fluid (AF) has been regarded with great hope as a minimally-invasive source for fetal stem cells in utero, since isolation of cells from autologous AF would involve few, if any, ethical concerns. Various stem cells types have been isolated from AF but their actual origin yet to be determined due to the heterogeneous nature of AF cell content. A subpopulation of Oct-4 positive cells has been identified in second trimester AF, together with cells that express markers for neuronal stem cells. This contributes to the idea that pluripotent cells are present in AF [92]. Recently, De Coppi et al. [28] isolated cells from second trimester AF using c-kit, a tyrosine kinase receptor for the stem cell factor (SCF), as a selection marker. These stem cells have similar characteristics to fetal MSCs, can be clonally expanded and differentiated into cell types representative of all three embryonic germ layers both in vitro and in vivo. The stem cell properties of MSCs in second trimester AF have recently been confirmed and they appear similar to MSCs from other fetal sources [51, 59, 112].

Another accessible source of cells for autologous use is the placenta. Chorionic villus sampling (CVS) is used clinically to isolate placental cells with sufficiently minor invasion to allow diagnostic use. MSCs have been successfully isolated from second and third trimester placental samples, and represent of cells present in the placenta [5, 121]. Limited studies also suggest that MSCs are present in first trimester placenta [90]. Like AF MSCs, placental MSCs express typical markers such as CD166, CD105, CD73, and CD90, while they are negative for CD14, CD34 and CD45. Also, MSCs from the amnion membrane have recently been suggested to have some characteristics of cardiomyoblasts, and can differentiate into cells similar to cardiomyocytes and successfully integrate into cardiac tissues [122]. Placental MSCs can differentiate into cells of the adipogenic, chondrogenic, osteogenic, myogenic, endothelial and neurogenic lineages, as confirmed by several groups [5, 91, 121]. Additionally, placenta-derived MSCs express primitive markers such as FZD-9, SSEA-4, Oct-4, Nanog-3 and nestin when grown in medium used for ESC expansion [8].

12.3.3
Umbilical Cord Blood

Umbilical cord blood is a rich source of HSCs and progenitor cells, and can be used extensively as an

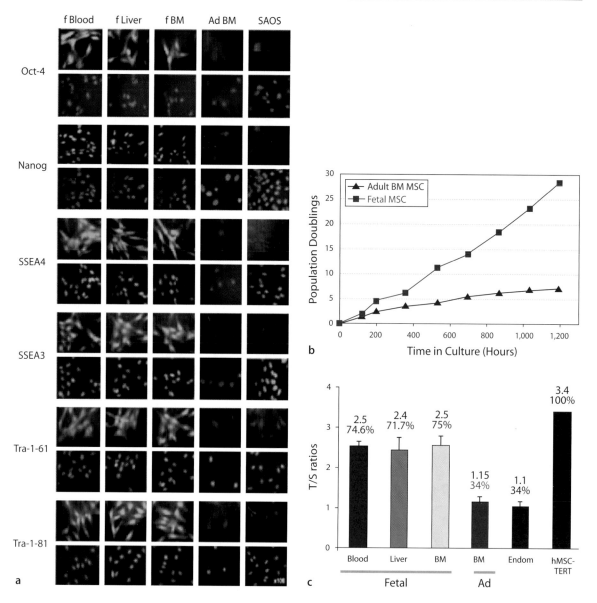

Fig. 12.1a–c Human first trimester fetal mesenchymal stem cell characteristics. **a** Expression of pluripotency markers Oct-4, Nanog, Rex-1, SSEA-3, SSEA-4, Tra-1-61 and Tra-1-81 by immunofluoerescence. **b** Growth kinetics of MSC from human first trimester fetal and adult bone marrow MSC seeded at 10,000 cells/cm² estimated by the cumulative population doublings over 50 days. **c** Telomerase activity showed greater activity in fetal samples compared with adult BM MSC and endometrial cells. Adapted from Guillot et al. [43]

alternative to BM and peripheral blood mobilised HSCs for transplantation in situations of marrow failure, malignancy and genetic disease [96]. The frequency of MSCs in UCB is far lower than that found in first trimester fetal blood or in BM [14]. However, even with sensitive culture techniques, UCB MSCs are isolated in only around a third of samples [12, 70]. This has considerable implications for private commercial blood banks offering directed autologous storage of UCB [33].

Wharton's jelly, the connective tissue of the human umbilical cord, is a rich source of MSCs, as shown by various studies [77, 83, 94]. One study has shown that these cells can be induced to differentiate into cells with neural characteristics, expressing neuron-specific enolase and other neural markers [80]. Kadner and co-workers derived autologous cells of myofibroblast origin by mincing cord blood vessels or whole cord [55, 54], while Takechi and co-workers showed that Wharton's jelly cells express vimentin, desmin, myosin and α-actin, as also demonstrated by Kadner et al. [54, 107].

Perhaps the richest postnatal source of mesenchymal progenitors is human umbilical cord perivascular cells. These cells have a doubling time of 20 hours at passage two, and produce over 10^{10} cells in one month in culture. They express SH2, SH3, Thy-1 and CD44, but are negative for CD34 and class I and II major histocompatibility antigens, which could be an advantage for allogeneic applications [89].

12.3.4
Non-Haemopoietic Solid Organs

MSCs have also been isolated from second trimester lung tissue, with similar properties to MSCs from other fetal tissues [50]. One difference is that fetal lung MSCs initially express CD34 at a high proportion, though they lose this expression later in culture [51]. Fetal pancreas and kidney are also a source of CD34$^-$ MSCs with a classic multipotent phenotype similar to that of other fetal MSCs [4, 48]. When fetal kidney MSCs were transplanted in a fetal lamb model, they showed site-specific engraftment and differentiation [3].

Neural stem cells (NSCs) have been found in many areas of the fetal brain [16] as well as the hippocampus and subventricular zone in the adult central nervous system (CNS) [53]. Fetal NSCs, like their adult counterparts, have the ability to self renew and differentiate into the three predominant cell types of the CNS: neurons, astrocytes and oligodendrocytes. When transplanted into immunodeficient newborn mice, they show engraftment, migration and site-specific neuronal differentiation up to 7 months later [113].

12.3.5
Harvesting Fetal Stem Cells and Their Characterisation

Collection of stem cells from cord blood as well as from fetuses in ongoing pregnancies for autologous use obviates the ethical concerns that have hampered ESC research. With the exception of cord blood and term placental tissues, fetal stem cells are more usually, however, collected from surplus fetal tissue after first or second trimester termination of pregnancy, subject to informed consent, institutional ethical approval, and compliance with national guidelines covering fetal tissue research.

MSC are isolated by removal of non-nucleated cells and standard selection for plastic adherence. Our group has applied ultrasound-guided cardiocentesis in consenting women undergoing surgical termination of pregnancy under general anaesthesia to characterise both HSCs and MSCs in first trimester blood [13, 14]. The harvest of fetal blood in continuing pregnancies is feasible using endoscopic and/ or ultrasound guided cord puncture, but it is technically challenging. Second trimester liver has been suggested as an alternative, although work in sheep suggests its collection carries a substantial fetal loss rate [105]. A more readily accessible source is AF or chorion villi, collected at clinically indicated diagnostic procedures.

Standard assays to identify human fetal MSCs (hfMSC) rely on their spindle-shaped morphology, selective adherence to a solid surface, proliferative potential, capacity to differentiate into lineages from the three germ layers and ability to repair tissues. Table 12.1 summarises the immunophenotype of various fetal stem cells. All fetal stem cells are CD45$^-$ and CD34$^-$ (except initially for fetal lung and spleen) and BM cells from first and second trimester exhibit similar characteristics. The majority of fetal stem cells are positive for SH2, CD29, CD44 and CD90; however, they are HLA-DR negative. Many fetal cells express pluripotency markers SSEA-4 and Oct-4, whereas other markers such as CD113 and CD133 are inconsistently expressed across different cell sources. Unfortunately, there is a lack of comparative studies between varying fetal cell sources and gestation, as most research have focused on

Table 12.1 Immunophenotype of the different types of fetal stem cells (+ positive, – negative, –/+ weakly positive or low expression). Adapted from Guillot et al. [41]

Antigen	CD No.	First trimester				Second trimester						Unknown	Postnatal				
		FB	FL	BM	Amnion	BM	Lung/Spleen	Pancreas	Kidney	Cartilage	AF	Muscle	UCB	Wharton's Jelly	HUCPV	Placenta	Amnion
	References	[14, 30, 43]			[90]	[9, 38]	[51]	[48]	[4]	[26]	[28, 59, 111]	[120]	[65, 92]	[58, 80, 115]	[7, 99]	[5, 9, 49, 121]	[79, 91, 102]
LCA	CD45	–	–	–	–	–	–	–	–	–	–	–	–	–	–	–	–
T10	CD38	–	–	–	–	–	–	–	–	–	–	–	–	–	–	–	–
LPS-R	CD14	–	–	–	–	–	–	–	–	–	–	–	–	–	–	–	–
Macrosialin	CD68	–	–	–	–	–	–	–	–	–	–	–	–	–	–	–	–
gp 105-120	CD34	–	–	–	+	–	+	–	–	–	–	–	–	–	–	–	–
PECAM	CD31	–	–	–	–	–	–	–	–	–	–	–	–	–	–	–	–
LFA-3	CD58					–/+	–/+						+				
ICAM-3	CD50	–	–	–	–	–	–	–	–	–	–	–	–	–	–	–	–
β1-integrin	CD29	+	+	+	+	–/+	+	+	+	+	+	–	+	+	+	+	+
HCAM-1	CD44	+	+	+	+	+	+	+	+	+	+	–	+	+	+	+	+
	CD51	–	–	–	–	–	–	–	–	–	–	–	–	–	–	–	–
T9	CD71					–	–/+	–	–				+				
Bp50	CD40	–	–	–	–	–	–	–	–	–	–	–	–	–	–	–	–
VCAM-1	CD106	+	+	+	–	–	–	–	–	–	–	–	–	–	–	–	–
B7.1	CD80	–	–	–	+	+	–	–	–	–	–	–	–	–	–	–	–
	CD166					+	–			+	–		+			–	+
B7.2	CD86	–	–	–	–	–	–	–	–	–	–	–	–	–	–	–	–
SH2	CD105	+	+	+	+	+	+	+	+	–	+	–	+	+	+	+	+
SH3	CD73	+	+	+	+	+	+	+	+	–	+	–	+	+	+	+	+

Chapter 12 Fetal Tissue Engineering: Regenerative Capacity of Fetal Stem Cells

Table 12.1 *(continued)* Immunophenotype of the different types of fetal stem cells (+ positive, –/+ weakly positive or low expression, – negative). Adapted from Guillot et al. [41]

Antigen	CD No.	First trimester				Second trimester						Unknown	Postnatal				
		FB	FL	BM	Amnion	BM	Lung/Spleen	Pancreas	Kidney	Cartilage	AF	Muscle	UCB	Wharton's Jelly	HUCPV	Placenta	Amnion
	References	[14, 30, 43]			[90]	[9, 38]	[51]	[48]	[4]	[26]	[28, 59, 111]	[120]	[65, 92]	[58, 80, 115]	[7, 99]	[5, 9, 49, 121]	[79, 91, 102]
SH4	CD73	+				+	+		+		+		+			+	
HLA-DR		–	–	–		–	–		–		–	–	–	–		–	–
VLA-4	CD49d	–					–		–		–		–	–			
VLA-5	CD49e	+	+		+		+		+		+		+	+		+	+
IL-2R	CD25		+				–		–				–				
IL-3R	CD123	–					–/+		–/+		–/+		–/+		–/+		
IL-7R	CD127						–		–		–						
TNF-R1,2	CD120a,b						–		–		–						
LFA-1	CD11a						–		–		–						
Aminopeptidase N	CD13	+			+			+			+		+	+		+	+
Neurothelin	CD147					–		–									
SCFR, c-kit	CD117	+	+			–		–	–		–/+	–	–	+	–/+		–/+
AC133	CD133	–				–			–		–		–/+	–			
Vimentin		+	+										+	+			
Laminin		+	+														
Fibronectin		+	+														
vWF		–	–														
ICAM-1	CD54	+	+				+		+		+		+			+	+

145

Table 12.1 (continued) Immunophenotype of the different types of fetal stem cells (+ positive, −/+ weakly positive or low expression, − negative). Adapted from Guillot et al. [41]

Antigen	CD No.	First trimester				Second trimester						Unknown	Postnatal				
		FB	FL	BM	Amnion	BM	Lung/Spleen	Pan-creas	Kidney	Carti-lage	AF	Muscle	UCB	Wharton's Jelly	HUCPV	Pla-centa	Amnion
References		[14, 30, 43]			[90]	[9, 38]	[51]	[48]	[4]	[26]	[28, 59, 111]	[120]	[65, 92]	[58, 80, 115]	[7, 99]	[5, 9, 49, 121]	[79, 91, 102]
Thy-1	CD90	+			+	+	+	+	+	+	+	+	+	+	+	+	−/+
Glyco-phorin A	CD235a												−				
CD146			+	+			+		+				+		+		
Nestin		+	+	+					+		+		+				
CK18		+	+	+					+				+				
Stro-1						+					+				−		
SSEA-4		+	+	+		+					+				−	+	+
Oct-4		+	+	+		+									−		+
Nanog		+	+	+		+											

comparing fetal stem cells with adult stem cells, particularly BM cells. Only one study compared MSC from first and third trimester placenta and showed little difference in lineage differentiation, despite the early gestation amnion epithelial cells expressing more markers (CD13, CD14, CD49e and CD105) than those from late gestation [90].

12.4
Potential for Tissue Repair

Pregnancy provides a naturally occurring endogenous paradigm, which attests to the ability of fetal stem cells to engraft tissues, and in some cases differentiate.

12.4.1
Microchimerism

Fetal stem cells trafficking into maternal blood during pregnancy sequester in maternal tissues, where they persist for decades. This phenomenon has been observed both in humans and in animals, but quantification has initially not been easy, as PCR used to identify Y-microchimerism did not allow for cell type identification [11]. In contrast, fluorescence-in-situ-hybridization (FISH) for the Y chromosome to indicate male and thus definitively fetal cells, allows for parallel cell immunophenotyping [87].

While persistent fetal cells were initially implicated in maternal autoimmune disease, more thorough studies with controls suggest that microchimeric fetal cells are found in controls without autoimmune disease, and in a wide range of organs, while animal studies have recently suggested that fetal microchimerism is both ubiquitous and may play a role in response to tissue injury [62].

The fetal cell type responsible for microchimerism is unknown and candidates include all cells in fetal blood or trophoblast with the potential to persist long-term, but evidence supports the involvement of stem cells such as MSCs or HSCs. Although the fetus and its cells are semi-allogeneic and should therefore be recognised by the maternal immune system, no reaction or rejection is apparent [72]. This may be due to the HLA status of hfMSC and their lack of ability to elicit alloreactive T-cell proliferative responses [40, 39]. Thus these immunomodulatory properties of fetal MSCs make them good candidates for long-term microchimerism. Additionally, in animal models infused allogeneic MSCs migrate to and engraft in host tissues [31]. Also, HSCs have been found in maternal blood up to 27 years after pregnancy. Finally, the presence of differentiated male cells in diseased post-reproductive female tissues implies that fetal stem cells entering the maternal blood can engraft, expand and differentiate within host tissues [86, 103].

12.4.1.1
Animals

Murine models of microchimerism have been developed in an attempt to identify the type and role of microchimeric cells, rather than their potential use in therapy [63, 64]. An early study by Liegeois et al. [73] demonstrated the presence of fetal cells from an earlier pregnancy in skin and lymphoid tissue. A more recent study showed that although frequency declined with increasing time postpartum, microchimeric fetal cells were still present in retired breeders [63]. Similarly, transgenic fetal cells were present in wild type murine maternal brains, with a peak frequency of 1.8% four weeks postpartum [107]. These animal studies have not elucidated the biological role of microchimeric fetal cells but it is increasingly suggested that microchimeric cells are involved in tissue repair rather than disease pathogenesis, since fetal cells are broadly recruited to sites of non-specific tissue injury [10, 64]. Intraperitoneal injection of vinyl-chloride in retired breeders showed a 48-fold increase in Y-genome equivalents in peripheral blood which was not seen in virgin controls, suggesting that fetal cells migrated to sites of injury in response to tissue damage [23]. Cell migration and tissue-specific differentiation was also observed in rats after chemical hepato-renal injury [114], while in a further model, chemical liver injury resulted in an increase in fetal microchimeric cells in the injured tissue [64]. In another model, fetal cells in postpartum maternal brain doubled after neural injury [107], while a more recent report demonstrated that GFP-labelled fetal cells recruited to sites of maternal skin inflammation during pregnancy expressed endothelial cell markers, implicating them in the tissue repair process [83].

12.4.1.2
Humans

It has long been known that fetal cells circulate in the blood of most pregnant women and can persist for decades postpartum [11]. In tissues, male cells have been identified in women with cervical cancer, thyroid tumours and skin lesions of polymorphic eruption of pregnancy (PEP). Although six out of eight women with cervical cancer had male cells, which in contrast were absent from the cervical tissue of healthy women and may thus suggest a role in the pathogenesis of cancer, the location of microchimeric cells within the cancerous tissues was not investigated [17]. Similarly, while male microchimeric cells have been found in a variety of benign and malignant thyroid tumours, their location has not been compared with adjacent healthy tissues [102]. In another study by Aractingi et al. [6], dermis and epidermis biopsies from pregnant women with PEP were examined for male DNA presence rather than cell phenotype. Male cells have also been identified in marrow-derived MSCs and rib sections from women with a history of at least one male pregnancy, but not in women without, which suggests that fetal stem cells circulating into the maternal blood engraft in marrow, where they remain for years (Fig. 12.2) [87]. Male cells have also been found clustered around tumour cells in lung or thymus tumour tissue and they were more likely to be located in diseased than healthy tissues [86].

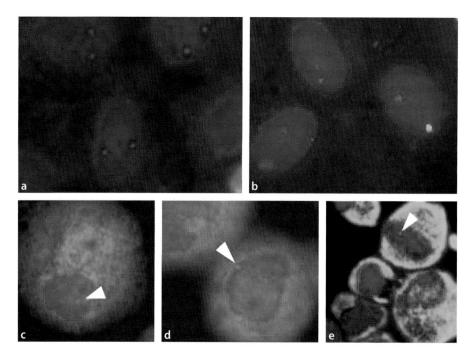

Fig. 12.2a–e Mesenchymal stem cells from adult female marrow are XX on FISH (**a**), controls and a male cell bearing a Y chromosome labelled with SpectrumGreen are shown in MSC cultures from women with sons (**b**). The Y chromosome in male MSC (*arrowhead*) was identified with alternative Y FISH probes, DYZ1 (**c**) and SRY (**d**), labelled with SpectrumOrange. Vimentin-FITC+ cells are shown in (**e**), where SRY (*arrowhead*) is also labelled with SpectrumOrange. Adapted from O'Donoghue et al. [87]

12.4.2 Intrauterine Transplantation

Fetal stem cells can be transplanted in utero to a fetal environment and this is a possible option in the treatment of hereditary haematological, metabolic and immunological diseases [34]. There are several advantages of intrauterine transplantation (IUT). As well as allowing pre-emptive treatment before the occurrence of irreversible end-organ damage in the fetus, there is an natural exponential expansion of cellular compartments and endogenous migration of stem cells in fetal life [68]. The stoichiometric benefit of the fetus being only about 30 g in size at 13 weeks is such that a much larger proportion of cells can be delivered prenatally compared with postnatally. Finally, it obviates the need for myeloablation, because of tolerance to foreign antigens in the first trimester fetus [39].

12.4.2.1 Allogeneic Versus Autologous

Although the immunological naiveté of the early gestation fetus should allow use of allogeneic stem cells for IUT, in utero HSC transplantation has so far only been successful in fetuses with immunodeficiency disorders. Indeed, allogeneic murine HSCs appear to induce an immune response compared to congenic donor cells [89].

The use of an autologous source should largely overcome the immunological barrier. This allows an additional cell therapy strategy, ex vivo gene therapy, which uses autologous stem cells transduced with an integrating vector before re-transplantation back into the patient (Fig. 12.3). While the harvest of fetal BM or brain is unlikely ever to be feasible in continuing pregnancies, the harvest of fetal liver has been reported in fetal sheep, albeit with significant

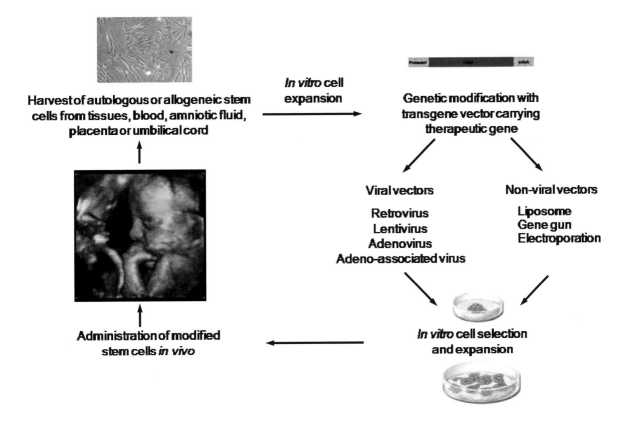

Fig. 12.3 Ex-vivo gene therapy strategy of fetal stem cells

morbidity and mortality [104]. Relatively safe techniques to access the fetal circulation have been developed over the last two decades for fetal diagnosis and therapy after 16 weeks, with a fetal loss rate of approximately one percent. For collection of hfMSC from earlier fetal blood, prior to 16 weeks gestation, procedures are limited by the small size of fetal vessels; the umbilical vein, for instance, measures only 2 mm in diameter at 12 weeks gestation. Nevertheless, one group has applied ultrasound-guided fetal blood samplings in ongoing pregnancies at risk of haemoglobinopathies with a loss rate of only 5% at gestations as early as 12 weeks [88]. Thin gauge embryo-fetoscopes also allow early fetal blood sampling from the umbilical vessels under direct visualisation (Fig. 12.4) [103]. Ultrasound-guided fetal blood sampling may be more appealing compared with the fetoscopic approach, as it is simpler, less equipment-intensive and experience of this technique for use at a later gestational age already exists in many specialist fetomaternal centres. We found that success rates of fetal blood sampling were comparable in fetoscopic (4/6) and ultrasound-guided (8/12) procedures; however procedural time was shorter with ultrasound-guidance [18]. To utilise hfMSCs as a target cell type for ex-vivo gene therapy, a sufficient volume of fetal blood is needed to isolate and expand hfMSCs. On the other hand, the risk of inducing hypovolaemic circulatory embarrassment in the fetus has to be considered. We showed that at least 50 µL are needed for hfMSC isolation, and when higher volumes of fetal blood are sampled in the first trimester, fetal bradycardia can occur post fetal blood sampling (1/3 using fetoscopy and 1/4 under ultrasound-guidance) [18].

12.4.2.2
Ex-Vivo Gene Therapy

Stem cells have considerable utility as targets for gene therapy because they can self renew, thus precluding the need for repeated administration of the gene vector. While this approach has already been validated in post-natal clinical trials with HSCs [2, 37, 45], the occurrence of leukaemia in three of 11 children treated in the French trial [67] and in one of ten children from the London trial [109] due to the insertion of the transgene near a proto-oncogene has raised serious concerns about the risk of insertional oncogenesis. Several lines of investigation are currently being pursued to reduce this risk, such as the use of insertional site screening prior to re-infusion, the use of safer and/or site specific vectors and the use of regulable vectors which could be switched off should an adverse event arise [67]. HfMSCs are readily transduced with integrating vectors without affecting their stem cell properties of self renewal and multilineage differentiation [20]. Transduced hfMSCs can be clonally expanded and work is now underway to select clones where integration occurred at "safe" regions of the human genome, in order to minimise the risk of any oncogenic event. The use of hfMSCs as vehicles for gene delivery would be applicable to a number of diseases such as osteogenesis imperfecta, the muscular dystrophies and various enzyme deficiency syndromes such as the mucopolysaccharidoses and lysosomal storage diseases.

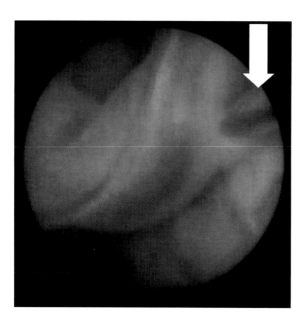

Fig. 12.4 Fetoscopic view showing needle (*arrow*) for fetal blood sampling via umbilical vein. Adapted from Chan et al. [18]

12.4.2.3
Animal Models

Intrauterine transplantation using HSCs has been well covered by a recent review [111], and in humans successful outcomes have been limited to immunodeficient fetuses, which can instead be transplanted more safely postnatally [35, 117,118]. In contrast, MSCs have immunomodulatory activities, and they engraft widely after intrauterine transplantation in animal models regardless of gestational age or immune competence at transplantation [71]. Taylor et al. [109] showed a tenfold competitive engraftment advantage of fetal liver relative to adult BM cells in fetal recipients. Another group also found long-term persistence of human cells with site-specific differentiation after transplanting human fetal kidney MSCs in a fetal lamb model [4]. Similarly, hfMSC engraft into multiple organ compartments after IUT in fetal sheep with evidence of site-specific differentiation [75, 100]. Our own experience is that first trimester human fetal blood MSCs transplanted in utero into immunocompetent MF1 wild type fetal mice can be detected postnatally by staining for human specific markers. When transplanted into immunodeficient damaged murine muscle, MSCs participated in regeneration of muscle by forming spectrin-positive muscle fibres, albeit at low levels (<1% fibres per muscle section) [22].

Muscular Dystrophy

Duchenne muscular dystrophy (DMD) is an X-linked recessive muscular dystrophy arising from a dystrophin gene mutation. Patients present with progressive myopathy, and often die by the third decade due to respiratory or cardiac failure. Previous work in our group on IUT in a DMD mouse model, the *mdx* mouse, achieved a low level of hfMSC engraftment (0.5–1.0%) in multiple organs, although the level was significantly higher in muscle than in non-muscle tissues [21]. This is presumably due to the lack of overt pathology in *mdx* model at the time of transplantation. Others have found that fetal liver cell transplants in *mdx* mice resulted in multicompartment engraftment and myogenic differentiation [74].

Osteogenesis Imperfecta

Osteogenesis imperfecta presents with in utero pathology, osteopenia and bone fragility and is due to abnormal collagen production by osteoblasts as a result of mutations in the α chains of collagen type I. The intermediate severity type III is progressively deforming with recurrent fractures from or before birth, short stature, and kyphoscoliosis predisposing to premature death. Using a mouse model of type III, *oim*, our group recently reported in utero transplantation of hfMSCs ameliorated the disease phenotype producing a clinically-relevant two-thirds reduction in fracture incidence, along with increased bone strength, length and thickness (Fig. 12.5) [44]. The marked improvement in skeletal phenotype from hfMSC transplantation was associated with engraftment levels in bone of only around 5%. This is in keeping with modest engraftment levels seen in both a paediatric BM transplantation trial and a single clinical case of hfMSC transplantation in a human fetus with osteogenesis imperfecta, resulting in long-term chimerism in the bone and BM, and a lack of alloreactivity to donor MSCs [47,69].

Osteopetrosis

Autosomal recessive osteopetrosis is a genetic disease where mineralised cartilage and bone cannot be degraded thus blocking BM formation. In its mouse model, IUT of unselected adult murine BM cells lead to improvements in 35% of mice transplanted, by rendering them either clinically asymptomatic or having a normal phenotype in terms of growth, ability to breed, and the presence of donor cells during the normal lifespan [36].

12.4.3
Adult Tissue Repair

The potential of fetal cells in postnatal tissue repair has been investigated using a number of animal models.

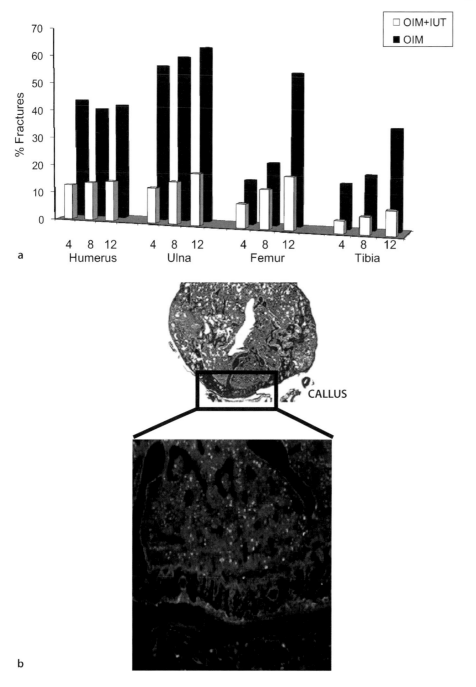

Fig. 12.5a,b IUT improves disease phenotype in *oim* mouse model of type III osteogenesis imperfecta. **a** IUT reduces long bone fractures. Incidence of fractured bones in humerus, ulna, femur and tibia of 4-, 8-, and 12-week-old OIM+IUT and OIM mice. $P<0.01$ in all groups. **b** Donor cells preferentially engraft at callus fractures demonstrated by visualisation of donor cells by FISH (H&E staining shown in *top panel*). Adapted from Guillot et al. [44]

12.4.3.1
Neural Disease

These show clustering of transplanted cells at sites of tissue injury with varying degrees of neural differentiation, and some functional improvement. A human MSC cell line derived from fetal marrow migrated to the lesion site in a mouse intracerebral haemorrhage stroke model, underwent neural differentiation and induced functional improvements up to 7 weeks post-transplantation [81]. Human umbilical cord blood cells transplanted into developing rats gave rise to cord-derived neural cells within the host brain [122]. Similar findings were shown using porcine umbilical cord matrix cells [115] and placental derived stem cells [66]. Using human umbilical cord matrix cells, Weiss et al. [116] demonstrated a phenotypic improvement in hemiparkinsonian rats after transplantation, while transplantation of amniotic epithelial cells led to partial amelioration in a similar fashion [57, 56]. Amniotic fluid MSCs transplanted into ischaemic brain differentiated into neural lineages, particularly astrocytes [24] and human amniotic stem cell clones pre-differentiated down the neural lineage can engraft in mouse brain [28].

12.4.3.2
Other Organ Systems

These similarly show site specific differentiation and suggest a degree of functional repair. In a rat myocardial infarction model, transplantation of fetal cardiac cells improved left ventricular ejection fraction, reduced left ventricular dilation and increased infarct wall thickness over 10 months [119]. Fetal liver has been shown to engraft into murine and rat models of liver failure [15, 27]. Human fetal liver cells transplanted into a mouse model of acute liver injury engrafted and repopulated the liver and decreased mortality compared with control animals [85]. Amniotic fluid derived stem cells on scaffold constructs transplanted into immunodeficient mice formed bony tissues at site of implantation which persisted for 18 weeks [28].

12.5
Conclusions

Fetal stem cells are an attractive class of stem cells for therapy because of their reduced immunogenicity and greater differentiation potential compared with adult cells, while not causing tumour formation. Gene therapy using transduced fetal stem cells utilises their self renewing properties to avoid repeated gene vector administration and, if combined with IUT, provides the exciting prospect of prenatally treating genetic diseases using autologous cells. Experimentally, adult tissue repair by fetal stem cells has been demonstrated in animal models, while in women fetal cells may repair sites of injury post pregnancy through microchimerism. Studies are now indicated into optimising engraftment and differentiation, and understanding the mechanisms behind therapeutic tissue repair to underpin future clinical applications for fetal stem cells.

References

1. Aguilar S, Nye E, Chan J et al (2007) Murine but not human mesenchymal stem cells generate osteosarcoma-like lesions in the lung. Stem Cells 25:1586–1594
2. Aiuti A, Slavin S, Aker M et al (2002) Correction of ADA-SCID by stem cell gene therapy combined with nonmyeloablative conditioning. Science 296:2410–2413
3. Almeida-Porada G, Porada C, Esmail D (2000) Differentiation potential of human metanephric stem cells: from mesenchyme to blood. Blood 96:494a
4. Almeida-Porada G, El Shabrawy D, Porada C et al (2002) Differentiative potential of human metanephric mesenchymal cells. Exp Hematol 30:1454–1462
5. Alviano F, Fossati V, Marchionni C et al (2007) Term amniotic membrane is a high throughput source for multipotent mesenchymal stem cells with the ability to differentiate into endothelial cells in vitro. BMC developmental biology 7:11
6. Aractingi S, Berkane N, Bertheau P et al (1998) Fetal DNA in skin of polymorphic eruptions of pregnancy. Lancet 352:1898–1901
7. Baksh D, Yao R, Tuan RS (2007) Comparison of proliferative and multilineage differentiation potential of human mesenchymal stem cells derived from umbilical cord and bone marrow. Stem Cells 25:1384–1392
8. Battula VL, Bareiss PM, Treml S et al (2007) Human placenta and bone marrow derived MSC cultured in serum-free, b-FGF-containing medium express cell surface

frizzled-9 and SSEA-4 and give rise to multilineage differentiation. Differentiation 75:279–291
9. Bernardo ME, Emons JA, Karperien M et al (2007) Human mesenchymal stem cells derived from bone marrow display a better chondrogenic differentiation compared with other sources. Connect Tissue Res 48:132–140
10. Bianchi DW (2007) Robert E. Gross Lecture. Fetomaternal cell trafficking: a story that begins with prenatal diagnosis and may end with stem cell therapy. J Pediatr Surg 42:12–18
11. Bianchi DW, Zickwolf GK, Weil GJ et al (1996) Male fetal progenitor cells persist in maternal blood for as long as 27 years postpartum. Proc Natl Acad Sci U S A 93:705–708
12. Bieback K, Kern S, Kluter H et al (2004) Critical parameters for the isolation of mesenchymal stem cells from umbilical cord blood. Stem Cells 22:625–634
13. Campagnoli C, Fisk N, Overton T et al (2000) Circulating hematopoietic progenitor cells in first trimester fetal blood. Blood 95:1967–1972
14. Campagnoli C, Roberts IA, Kumar S et al (2001) Identification of mesenchymal stem/progenitor cells in human first-trimester fetal blood, liver, and bone marrow. Blood 98:2396–2402
15. Cantz T, Zuckerman DM, Burda MR et al (2003) Quantitative gene expression analysis reveals transition of fetal liver progenitor cells to mature hepatocytes after transplantation in uPA/RAG-2 mice. Am J Pathol 162:37–45
16. Carpenter MK, Cui X, Hu ZY et al (1999) In vitro expansion of a multipotent population of human neural progenitor cells. Exp Neurol 158:265–278
17. Cha D, Khosrotehrani K, Kim Y et al (2003) Cervical cancer and microchimerism. Obstet Gynecol 102:774–781
18. Chan J, Kumar S, Fisk NM (2008) First trimester embryo-fetoscopic and ultrasound-guided fetal blood sampling for ex-vivo viral transduction of cultured human fetal mesenchymal stem cells. Hum Reprod 23:2427–37
19. Chan J, O'Donoghue K, Kennea N et al (2003) Myogenic potential of fetal mesenchymal stem cells. Ann Acad Med Singapore 32:S11–S13
20. Chan J, O'Donoghue K, de la Fuente J et al (2005) Human fetal mesenchymal stem cells as vehicles for gene delivery. Stem Cells 23:93–102
21. Chan J, Waddington SN, O'Donoghue K et al (2007) Widespread distribution and muscle differentiation of human fetal mesenchymal stem cells after intrauterine transplantation in dystrophic mdx mouse. Stem Cells 25:875–884
22. Chan J, O'Donoghue K, Gavina M et al (2006) Galectin-1 induces skeletal muscle differentiation in human fetal mesenchymal stem cells and increases muscle regeneration. Stem Cells 24:1879–1891
23. Christner PJ, Artlett CM, Conway RF et al (2000) Increased numbers of microchimeric cells of fetal origin are associated with dermal fibrosis in mice following injection of vinyl chloride. Arthritis Rheum 43:2598–2605
24. Cipriani S, Bonini D, Marchina E et al (2007) Mesenchymal cells from human amniotic fluid survive and migrate after transplantation into adult rat brain. Cell Biol Int 31:845–850
25. Clapp DW, Freie B, Lee WH et al (1995) Molecular evidence that in situ-transduced fetal liver hematopoietic stem/progenitor cells give rise to medullary hematopoiesis in adult rats. Blood 86:2113–2122
26. Cui Y, Wang H, Yu M et al (2006) Differentiation plasticity of human fetal articular chondrocytes. Otolaryngol Head Neck Surg 135:61–67
27. Dabeva MD, Petkov PM, Sandhu J et al (2000) Proliferation and differentiation of fetal liver epithelial progenitor cells after transplantation into adult rat liver. Am J Pathol 156:2017–2031
28. De Coppi P, Bartsch G Jr, Siddiqui MM et al (2007) Isolation of amniotic stem cell lines with potential for therapy. Nat Biotechnol 25:100–106
29. de la Fuente J, O'Donoghue K, Kumar S et al (2002) Ontogeny-related changes in integrin expression and cytokine production by fetal mesenchymal stem cells (MSC). Blood 100:526a
30. de la Fuente J, Fisk N, O'Donoghue K et al (2003) a2b1 and a4b1 integrins mediate the homing of mesenchymal stem/progenitor cells during fetal life. Haematol J 4:13
31. Devine SM, Bartholomew AM, Mahmud N et al (2001) Mesenchymal stem cells are capable of homing to the bone marrow of non-human primates following systemic infusion. Exp Hematol 29:244–255
32. Erices A, Conget P, Minguell JJ (2000) Mesenchymal progenitor cells in human umbilical cord blood. Br J Haematol 109:235–242
33. Fisk NM, Atun R (2008) Public-private partnership in cord blood banking. BMJ 336:642–644
34. Flake A, Zanjani E (1999) In utero hematopoietic stem cell transplantation: ontogenic opportunities and biologic barriers. Blood 94:2179–2191
35. Flake AW, Roncarolo MG, Puck JM et al (1996) Treatment of X-linked severe combined immunodeficiency by in utero transplantation of paternal bone marrow. N Engl J Med 335:1806–1810
36. Frattini A, Blair HC, Sacco MG et al (2005) Rescue of ATPa3-deficient murine malignant osteopetrosis by hematopoietic stem cell transplantation in utero. Proc Natl Acad Sci U S A 102:14629–14634
37. Gaspar HB, Parsley KL, Howe S et al (2004) Gene therapy of X-linked severe combined immunodeficiency by use of a pseudotyped gammaretroviral vector. Lancet 364:2181–2187
38. Gonzalez R, Maki CB, Pacchiarotti J et al (2007) Pluripotent marker expression and differentiation of human second trimester mesenchymal stem cells. Biochem Biophysical Res Commun 362:491–497
39. Gotherstrom C, Ringden O, Westgren M et al (2003) Immunomodulatory effects of human foetal liver-derived mesenchymal stem cells. Bone Marrow Transplant 32:265–272
40. Gotherstrom C, Ringden O, Tammik C et al (2004) Immunologic properties of human fetal mesenchymal stem cells. Am J Obstet Gynecol 190:239–245
41. Guillot PV, O'Donoghue K, Kurata H et al (2006) Fetal stem cells: betwixt and between. Semin Reprod Med 24:340–347

42. Guillot PV, Cui W, Fisk NM et al (2007) Stem cell differentiation and expansion for clinical applications of tissue engineering. J Cell Mol Med 11:935–944
43. Guillot PV, Gotherstrom C, Chan J et al (2007) Human first-trimester fetal MSC express pluripotency markers and grow faster and have longer telomeres than adult MSC. Stem Cells 25:646–654
44. Guillot PV, Abass O, Bassett JH et al (2007) Intrauterine transplantation of human fetal mesenchymal stem cells from first trimester blood repairs bone and reduces fractures in osteogenesis imperfecta mice. Blood 111:1717–1725
45. Hacein-Bey-Abina S, Le Deist F, Carlier F et al (2002) Sustained correction of X-linked severe combined immunodeficiency by ex vivo gene therapy. N Engl J Med 346:1185–1193
46. Harrison D, Zhong R, Jordan C et al (1997) Relative to adult marrow, fetal liver repopulates nearly five times more effectively long-term than short-term. Exp Hematol 25:293–297
47. Horwitz EM, Prockop DJ, Fitzpatrick LA et al (1999) Transplantability and therapeutic effects of bone marrow-derived mesenchymal cells in children with osteogenesis imperfecta. Nat Med 5:309–313
48. Huang H, Tang X (2003) Phenotypic determination and characterization of nestin-positive precursors derived from human fetal pancreas. Lab Invest 83:539–547
49. Igura K, Zhang X, Takahashi K et al (2004) Isolation and characterization of mesenchymal progenitor cells from chorionic villi of human placenta. Cytotherapy 6:543–553
50. in 't Anker PS, Noort WA, Kruisselbrink AB et al (2003) Nonexpanded primary lung and bone marrow-derived mesenchymal cells promote the engraftment of umbilical cord blood-derived CD34(+) cells in NOD/SCID mice. Exp Hematol 31:881–889
51. in 't Anker PS, Noort WA, Scherjon SA et al (2003) Mesenchymal stem cells in human second-trimester bone marrow, liver, lung, and spleen exhibit a similar immunophenotype but a heterogeneous multilineage differentiation potential. Haematologica 88:845–852
52. Javed MJ, Mead LE, Prater D et al (2008) Endothelial colony forming cells and mesenchymal stem cells are enriched at different gestational ages in human umbilical cord blood. Pediatr Res 64:68–73
53. Johansson CB, Momma S, Clarke DL et al (1999) Identification of a neural stem cell in the adult mammalian central nervous system. Cell 96:25–34
54. Kadner A, Zund G, Maurus C et al (2004) Human umbilical cord cells for cardiovascular tissue engineering: a comparative study. Eur J Cardiothorac Surg 25:635–641
55. Kadner A, Hoerstrup SP, Tracy J et al (2002) Human umbilical cord cells: a new cell source for cardiovascular tissue engineering. Ann Thorac Surg 74:S1422–S1428
56. Kakishita K, Nakao N, Sakuragawa N et al (2003) Implantation of human amniotic epithelial cells prevents the degeneration of nigral dopamine neurons in rats with 6-hydroxydopamine lesions. Brain Res 980:48–56
57. Kakishita K, Elwan MA, Nakao N et al (2000) Human amniotic epithelial cells produce dopamine and survive after implantation into the striatum of a rat model of Parkinson's disease: a potential source of donor for transplantation therapy. Exp Neurol 165:27–34
58. Karahuseyinoglu S, Cinar O, Kilic E et al (2007) Biology of stem cells in human umbilical cord stroma: in situ and in vitro surveys. Stem Cells 25:319–331
59. Kaviani A, Guleserian K, Perry TE et al (2003) Fetal tissue engineering from amniotic fluid. J Am Coll Surg 196:592–597
60. Kennea N, Fisk N, Edwards A et al (2003) Neural cell differentiation of fetal mesenchymal stem cells. Early Hum Dev 73:121–122
61. Kennea NL, Waddington S, O'Donoghue K et al (2008) Differentiation of human fetal mesenchymal stem cells into oligodendrocytes without cell fusion. (Submitted)
62. Khosrotehrani K, Johnson KL, Cha DH et al (2004) Transfer of fetal cells with multilineage potential to maternal tissue. JAMA 292:75–80
63. Khosrotehrani K, Johnson KL, Guegan S et al (2005) Natural history of fetal cell microchimerism during and following murine pregnancy. J Reprod Immunol 66:1–12
64. Khosrotehrani K, Reyes RR, Johnson KL et al (2007) Fetal cells participate over time in the response to specific types of murine maternal hepatic injury. Hum Reprod 22:654–661
65. Kogler G, Senksen S, Wernet P (2006) Comparative generation and characterization of pluripotent unrestricted somatic stem cells with mesenchymal stem cells from human cord blood. Exp Hematol 34:1589–1595
66. Kogler G, Senksen S, Airey JA et al (2004) A new human somatic stem cell from placental cord blood with intrinsic pluripotent differentiation potential. J Exp Med 200:123–135
67. Kohn DB, Sadelain M, Glorioso JC (2003) Occurrence of leukaemia following gene therapy of X-linked SCID. Nat Rev Cancer 3:477–488
68. Le Blanc K, Ringden O (2005) Immunobiology of human mesenchymal stem cells and future use in hematopoietic stem cell transplantation. Biol Blood Marrow Transplant 11:321–334
69. Le Blanc K, Gotherstrom C, Ringden O et al (2005) Fetal mesenchymal stem-cell engraftment in bone after in utero transplantation in a patient with severe osteogenesis imperfecta. Transplantation 79:1607–1614
70. Lee OK, Kuo TK, Chen WM et al (2004) Isolation of multipotent mesenchymal stem cells from umbilical cord blood. Blood 103:1669–1675
71. Liechty KW, MacKenzie TC, Shaaban AF et al (2000) Human mesenchymal stem cells engraft and demonstrate site-specific differentiation after in utero transplantation in sheep. Nat Med 6:1282–1286
72. Liegeois A, Escourrou J, Ouvre E et al (1977) Microchimerism: a stable state of low-ratio proliferation of allogeneic bone marrow. Transplant Proc 9:273–276
73. Liegeois A, Gaillard MC, Ouvre E et al (1981) Microchimerism in pregnant mice. Transplant Proc 13:1250–1252
74. Mackenzie TC, Shaaban AF, Radu A et al (2002) Engraftment of bone marrow and fetal liver cells after in utero transplantation in MDX mice. J Pediatr Surg 37:1058–1064

75. MacKenzie TC, Campagnoli C, Almeida-Porada G et al (2001) Circulating human fetal stromal cells engraft and differentiate in multiple tissues following transplantation into pre-immune fetal lambs. Blood 98:798
76. MacKenzie TS, Campagnoli C, Almcida-Porada G ct al (2001) Circulating human fetal stromal cells engraft and differentiate in multiple tissues following transplantation into pre-immune fetal lambs. Blood 98:328a
77. McElreavey KD, Irvine AI, Ennis KT et al (1991) Isolation, culture and characterisation of fibroblast-like cells derived from the Wharton's jelly portion of human umbilical cord. Biochem Soc Trans 19:29S
78. Medvinsky A, Dzierzak E (1996) Definitive hematopoiesis is autonomously initiated by the AGM region. Cell 86:897–906
79. Miki T, Lehmann T, Cai H et al (2005) Stem cell characteristics of amniotic epithelial cells. Stem Cells 23:1549–1559
80. Mitchell KE, Weiss ML, Mitchell BM et al (2003) Matrix cells from Wharton's jelly form neurons and glia. Stem Cells 21:50–60
81. Nagai A, Kim WK, Lee HJ et al. (2007) Multilineage potential of stable human mesenchymal stem cell line derived from fetal marrow. PLoS ONE 2:e1272
82. Naughton BA, San Roman J, Liu K (1997) Cells isolated from Wharton's jelly of the human umbilical cord develop a cartilage phenotype when treated with TGFβ in vitro. FASEB 11:19
83. Nguyen Huu S, Oster M, Uzan S et al (2007) Maternal neoangiogenesis during pregnancy partly derives from fetal endothelial progenitor cells. Proc Natl Acad Sci U S A 104:1871–1876
84. Nicolini U, Poblete A (1999) Single intrauterine death in monochorionic twin pregnancies. Ultrasound Obstet Gynecol 14:297–301
85. Nowak G, Ericzon BG, Nava S et al (2005) Identification of expandable human hepatic progenitors which differentiate into mature hepatic cells in vivo. Gut 54:972–979
86. O'Donoghue K, Sultan HA, Al-Allaf FA et al (2008) Microchimeric fetal cells cluster at sites of tissue injury in lung decades after pregnancy. Reprod Biomed Online 16:382–390
87. O'Donoghue K, Chan J, de la Fuente J et al (2004) Microchimerism in female bone marrow and bone decades after fetal mesenchymal stem-cell trafficking in pregnancy. Lancet 364:179–182
88. Orlandi F, Damiani G, Jakil C et al (1990) The risks of early cordocentesis (12–21 weeks): analysis of 500 procedures. Prenat Diagn 10:425–428
89. Peranteau WH, Endo M, Adibe OO et al (2007) Evidence for an immune barrier after in utero hematopoietic-cell transplantation. Blood 109:1331–1333
90. Portmann-Lanz CB, Schoeberlein A, Huber A et al (2006) Placental mesenchymal stem cells as potential autologous graft for pre- and perinatal neuroregeneration. Am J Obstet Gynecol 194:664–673
91. Prat-Vidal C, Roura S, Farre J et al (2007) Umbilical cord blood-derived stem cells spontaneously express cardiomyogenic traits. Transplant Proc 39:2434–2437
92. Prusa AR, Marton E, Rosner M et al (2003) Oct-4-expressing cells in human amniotic fluid: a new source for stem cell research? Hum Reprod 18:1489–1493
93. Purchio AF, Naughton BA, Roman JS (1999) Production of cartilage tissue using cells isolated from Wharton's Jelly. U.S. patent number 5,919,702
94. Rebel V, Miller C, Eaves C et al (1996) The repopulation potential of fetal liver hematopoietic stem cells in mice exceeds that of their liver adult bone marrow counterparts. Blood 87:3500–3507
95. Rhodes KE, Gekas C, Wang Y et al (2008) The emergence of hematopoietic stem cells is initiated in the placental vasculature in the absence of circulation. Cell Stem Cell 2:252–263
96. Rocha V, Sanz G, Gluckman E (2004) Umbilical cord blood transplantation. Curr Opin Hematol 11:375–385
97. Rollini P, Kaiser S, Faes-van't Hull E et al (2004) Long-term expansion of transplantable human fetal liver hematopoietic stem cells. Blood 103:1166–1170
98. Sarugaser R, Lickorish D, Baksh D et al (2005) Human umbilical cord perivascular (HUCPV) cells: a source of mesenchymal progenitors. Stem Cells 23:220–229
99. Schoeberlein A, Holzgreve W, Dudler L et al (2005) Tissue-specific engraftment after in utero transplantation of allogeneic mesenchymal stem cells into sheep fetuses. Am J Obstet Gynecol 192:1044–1052
100. Secco M, Zucconi E, Vieira NM et al (2008) Multipotent stem cells from umbilical cord: cord is richer than blood! Stem Cells 26:146–150
101. Soncini M, Vertua E, Gibelli L et al (2007) Isolation and characterization of mesenchymal cells from human fetal membranes. J Ttissue Eng Regen Med 1:296–305
102. Srivatsa B, Srivatsa S, Johnson KL et al (2001) Microchimerism of presumed fetal origin in thyroid specimens from women: a case-control study. Lancet 358:2034–2038
103. Surbek D, Tercanli S, Holzgreve W (2000) Transabdominal first trimester embryofetoscopy as a potential approach to early in utero stem cell transplantation and gene therapy. Ultrasound Obstet Gynecol 15:302–307
104. Surbek D, Schoeberlein A, Dudler L et al (2003) In utero transplantation of autologous and allogeneic fetal liver stem cells in fetal sheep. Am J Obstet Gynecol 189:S75
105. Surbek DV, Young A, Danzer E et al (2002) Ultrasound-guided stem cell sampling from the early ovine fetus for prenatal ex vivo gene therapy. Am J Obstet Gynecol 187:960–963
106. Takechi K, Kuwabara Y, Mizuno M (1993) Ultrastructural and immunohistochemical studies of Wharton's jelly umbilical cord cells. Placenta 14:235–245
107. Tan XW, Liao H, Sun L et al (2005) Fetal microchimerism in the maternal mouse brain: a novel population of fetal progenitor or stem cells able to cross the blood-brain barrier? Stem Cells 23:1443–1452
108. Taylor PA, McElmurry RT, Lees CJ et al (2002) Allogenic fetal liver cells have a distinct competitive engraftment advantage over adult bone marrow cells when infused into fetal as compared with adult severe combined immunodeficient recipients. Blood 99:1870–1872

109. Thrasher A (2007) Severe adverse event in clinical trial of gene therapy for X-SCID. http://www.esgct.org/upload/X-SCID_statement_AT.pdf
110. Tiblad E, Westgren M (2007) Fetal stem-cell transplantation. Best Pract Res Clin Obstet Gynaecol:22:189–201
111. Tsai MS, Lee JL, Chang YJ et al (2004) Isolation of human multipotent mesenchymal stem cells from second-trimester amniotic fluid using a novel two-stage culture protocol. Hum Reprod
112. Uchida N, Buck DW, He D et al (2000) Direct isolation of human central nervous system stem cells. Proc Natl Acad Sci U S A 97:14720–14725
113. Wang Y, Iwatani H, Ito T et al (2004) Fetal cells in mother rats contribute to the remodeling of liver and kidney after injury. Biochem Biophys Research Commun 325:961–967
114. Weiss ML, Mitchell KE, Hix JE et al (2003) Transplantation of porcine umbilical cord matrix cells into the rat brain. Exp Neurol 182:288–299
115. Weiss ML, Medicetty S, Bledsoe AR et al (2006) Human umbilical cord matrix stem cells: preliminary characterization and effect of transplantation in a rodent model of Parkinson's disease. Stem Cells 24:781–792
116. Wengler GS, Lanfranchi A, Frusca T et al (1996) In-utero transplantation of parental CD34 haematopoietic progenitor cells in a patient with X-linked severe combined immunodeficiency (SCIDXI). Lancet 348:1484–1487
117. Westgren M, Ringden O, Bartmann P et al (2002) Prenatal T-cell reconstitution after in utero transplantation with fetal liver cells in a patient with X-linked severe combined immunodeficiency. Am J Obstet Gynecol 187:475–482
118. Yao M, Dieterle T, Hale SL et al (2003) Long-term outcome of fetal cell transplantation on postinfarction ventricular remodeling and function. J Mol Cell Cardiol 35:661–670
119. Young HE, Steele TA, Bray RA et al (2001) Human reserve pluripotent mesenchymal stem cells are present in the connective tissues of skeletal muscle and dermis derived from fetal, adult, and geriatric donors. Anat Rec 264:51–62
120. Zhang Y, Li CD, Jiang XX et al (2004) Comparison of mesenchymal stem cells from human placenta and bone marrow. Chin Med J (Engl) 117:882–887
121. Zhao P, Ise H, Hongo M et al (2005) Human amniotic mesenchymal cells have some characteristics of cardiomyocytes. Transplantation 79:528–535
122. Zigova T, Song S, Willing AE et al (2002) Human umbilical cord blood cells express neural antigens after transplantation into the developing rat brain. Cell Ttransplant 11:265–274

Embryonic Stem Cell Use

J. Handschel, U. Meyer, H. P. Wiesmann

Contents

13.1	Introduction	159
13.2	Source and Culture Conditions	160
13.3	Cloning	161
13.4	Differentiation Capacity	161
13.5	Markers and Molecules	162
13.6	Clinical Considerations	162
13.7	Genetic Manipulations	163
13.8	Micromass Culture	163
13.9	Ethical Aspects	164
	References	164

13.1 Introduction

Over the last few decades reconstructive surgery and general medicine has shifted from a resection-oriented approach towards strategies focusing on the repair and regeneration of tissues. Artificial tissue substitutes containing metals, ceramics and polymers, to maintain skeletal function [6], and artificial devices, such as pacemakers and insulin pumps, have been used to reach this goal. These artificial materials and devices have significantly improved the possibility for clinicians to restore the form and, to some extent, the function of defective bones as well as to increased life expectancy of patients, e.g. with valvular heart disease or diabetes mellitus. Despite the fact that every artificial device has specific disadvantages, the use of biomaterials is currently a common treatment option in clinical practice. More detailed understanding exists concerning the transplantation of cells and tissues; thus, autografts are the second mainstay in clinical practice. The advantages of transplanting the body's own tissues ensure that autograft tissue transplantation can now be considered to be the "gold standard" in bone reconstruction. The reason for the primacy of tissue grafts over non-living biomaterials is that they contain living cells, thus possessing biological activity. The main disadvantages of using autografts are donor site morbidity and donor shortage [19]. Research is currently in progress into the use of cell-based approaches in reconstructive surgery, since cells are the driving elements for all repair and regeneration processes. As they synthesize and assemble the extracellular matrix, cells can be considered the basic unit needed for a biological regeneration strategy. Living cells can be used in a variety of ways to restore, maintain or enhance tissue functions [35, 38]. There are four principal ways in which cells are used to enhance bone formation [26]:

1. Transfer of cells as tissue blocks
2. In-situ cell activation
3. Implantation of isolated or cultured cells
4. Implantation of an extracorporally generated tissue construct.

In a broader sense, genetic engineering is another cell-based regeneration therapy, since the action of genes is directly related to the presence of living cells. The introduction of genetic information into cells, which is affected by different factors, is in its pre-clinical stage and may also become a therapeutic option in the near future.

Autologous cells are the first choice for tissue engineering, with allogenic and xenogenic cells providing secondary options. Each category can be subdivided according to whether the cells are in a more or less differentiated stage. Various mature cell lines as well as multipotential so-called mesenchymal progenitors have been successfully established in bone tissue engineering approaches [68]. Moreover, there are some reports using totipotent embryonic stem cells for tissue engineering of bone [28, 71]. Additionally, other bone cell lines such as genetically altered cell lines (sarcoma cells, immortalized cells, non-transformed clonal cell lines) have been developed and used to evaluate basic aspects of in-vitro cell behaviour in non-human settings. There are three groups of genetically unaltered natural cells that can be used for bone engineering. In contrast to completely undifferentiated cells such as embryonic stem cells [28, 71] or umbilical cord stem cells [34], there are terminally differentiated cells such as osteoblasts or chondroblasts. With regard to the differentiation stage, there is a third class of cells between these two groups of cells [26] (Fig. 13.1). This class, which is a focus of scientific and clinical studies today, is believed to contain multipotential stem cells, which are often called "mesenchymal stem cells" (MSCs) [25, 54] or "adult stem cells" [45]. However, in this chapter we will focus on embryonic stem cells (ESCs).

13.2
Source and Culture Conditions

ESCs are the major representative cell line of pluripotential cells. These cells were first isolated and grown in culture more than 20 years ago [40]. ESCs are routinely derived from the inner cell mass of blastocysts. ESCs exhibit two remarkable features in culture. First, they represent pluripotential embryonic precursor cells that give rise to any cell type in the embryo (Fig. 13.1). Pluripotency is defined as being the ability to generate all adult cell types. In contrast, totipotency is the ability to form all adult, germ line and extra-embryonic tissues. A second feature of ESCs is that they can proliferate indefinitely in culture and can be propagated indefinitely as a stable self-renewing cell population. ESCs have historically been maintained in co-culture with mitotically inactive fibroblasts [5, 10, 22]. This co-culture system is unnecessary for murine ESCs (mESCs) if the medium is supplemented with leukaemia inhibitory factor (LIF) [12, 61]. In the absence of LIF, ESCs differentiate into a morphologically mixed cell population, manifesting genes characteristic of endoderm and mesoderm [49]. Human ESCs (hESCs) have also been shown to preserve their pluripotent potential and karyotypic stability after multiple passages in culture [1]. After multiple passages, some authors observed genomic alterations which can also be detected in human cancer [39]. In contrast to mESCs, LIF is not able to support undifferentiated growth in feeder free hESC cultures [18]. One important factor for hESCs is basic fibroblast growth factor (bFGF), also known as FGF-2, which is normally added to the medium at a concentration of 4–8 ng/ml. Higher concentrations of bFGF have been shown to support hESC growth in the absence of murine embryonic fibroblasts [32, 66]. Despite many similarities between mESCs and hESCs, research findings with mESCs cannot be applied to hESCs one-to-one. Regarding the pathway for maintaining pluripotency, the differences between mESCs and hESCs are already elucidated [51].

Beside the various supplements of the medium, there is another factor affecting the karyotypic stability of hESCs. In general, cells can be passaged by mechanical dissociation (cut and paste) or enzymatic dissociation (e.g. collagenase or trypsin). It was demonstrated that for enzymatically passaged cells karyotypic abnormalities accumulate after several passages. In mechanically dissociated cells where there is massive cell-to-cell contact, significantly less divisions of the cells are expected and this might explain the more stable karyotype [31]. If enzymatic dissociation techniques are utilised, detached cells should be allowed to remain in clumps. Moreover, it is possible that the use of collagenase IV may be preferable over trypsin because it is less aggressive on the cells and enables them to remain in clusters [31]. Other authors suggest the use of trypsin without ethylenediaminetetraacetic acid (EDTA) [8].

By definition, ESCs have the potential to differentiate into every cell lineage.

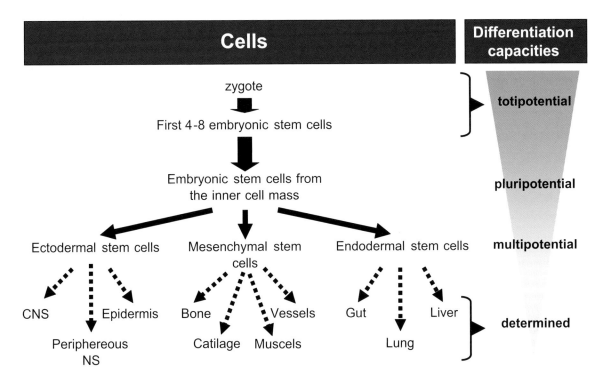

Fig. 13.1 Differentiation cascade

13.3 Cloning

Since the birth of Dolly in 1997, there has been tremendous interest in the field of cloning. However, cloned frogs were the first vertebrates derived from nuclear transfer, as reported in 1962 [24]. ESCs are also a source for cloning, of course. Two kinds of cloning have been described: reproductive and therapeutic cloning. Reproductive cloning means the generation of an embryo that has identical genetic material to its cell source. This embryo can be implanted into the uterus of a female to give rise to an infant that is a clone of the donor. This kind of cloning is banned in most countries for human applications. Therapeutic cloning is used to generate early-stage embryos that are explanted in culture to produce embryonic stem cell lines whose genetic material is identical to that of its source. These autologous stem cells have the potential to differentiate into almost any type of cell in the adult body. Thus, it could be used for tissue or organ regeneration or replacement [3, 29]. Currently, the efficiency of the overall cloning process is rather low, because the majority of embryos derived from animal cloning do not survive after implantation [62]. In addition, common abnormalities have been found in newborn clones if they survive to birth, including enlarged size, respiratory distress and defects of the kidney, liver, heart and brain [3, 14, 67, 69]. These may be related to the epigenetics of the cloned cells. Modifications of the chromatin structure or histones while the original DNA sequences remain intact may explain the above abnormalities [29].

13.4 Differentiation Capacity

Significant progress has been achieved in inducing murine and human cells to differentiate into particular types of cells, such as cardiomyocytes [33], neurons [37, 42] and smooth muscle cells [20]. Specifically, it has been shown by various investigators that ESCs can differentiate into osteogenic cells under selective culture conditions [13, 27, 28, 71]. The most common way to initiate osteogenic differentiation in ESCs is to supplement the medium with dexamethasone, ascorbic acid and ß-glycerol phosphate [5, 13, 27]. Moreover, vitamin D3 and BMP-2 have been found to promote osteogenic differentiation in ESCs [71]. ESCs are also used as an in-vitro model for trophoblast differentiation in the human placenta [59]. ESCs can be differentiated into primitive neuroectoderm after 8–10 days [53] or into mature granule neurons by sequential treatment with secreted factors (WNT1, FGF8, and RA) that initiate patterning in the cerebellar region of the neural tube, bone morphogenic proteins (BMP6/7 and GDF7) that induce early granule cell progenitor markers (MATH1, MEIS1, ZIC1), mitogens (SHH, JAG1) that control proliferation and induce additional granule cell markers (Cyclin D2, PAX2/6), and culture in glial-conditioned medium to induce markers of mature granule neurons [GABAalpha(6)r], including ZIC2, a unique marker for granule neurons [57].

13.5
Markers and Molecules

To determine the developmental potential of ESCs, functional analysis is the gold standard. However, molecular markers are often used to describe the actual stem cell state. The undifferentiated state is evident when observing the cells at the molecular level from the expression of various markers. Many of these markers are transcription factors expressed in ESCs or the inner cell mass and which have demonstrated roles in the maintenance of ESCs [51]. The best-characterised markers include the POU domain transcription factor Oct4 [48], the homeodomain transcription factor Nanog [11, 44] and the high-mobility group protein Sox2 [4]. Other markers include characteristic cell-surface antigens such as the glycomarkers SSEA-1, SSEA-3 and SSEA-4. Interestingly, in contrast to mESCs, hESCs do not exhibit high SSEA-1 reactivity, but instead are identified based on elevated SSEA-3 and SSEA-4 antigens [9]. In addition, hESCs present TRA-1-60 and TRA-1-81 antigens [56, 63], whereas only mESCs show the carbohydrate epitope N-acetylgalactosamine [46, 51]. The detailed transcription factor networks involved in ESC pluripotency have been reviewed elsewhere [7, 50].

During cell differentiation, markers of the undifferentiated cells are gradually turned off, whereas markers of differentiation are sequentially turned on. Early differentiation markers include LEFTYA, LEFTYB and NODAL [21]. In later differentiation stages, the cells exhibit specific markers of mature cells.

13.6
Clinical Considerations

An advantage of using ESCs instead of tissue-derived progenitor cells is that ESCs are immortal and could potentially provide an unlimited supply of differentiated osteoblast and osteoprogenitor cells for transplantation. In contrast, the proliferative, self-renewal and differentiation capacity of cells derived from adult tissues decreases with age [17, 55]. One major challenge in the use of ESCs for osteoregenerative therapies is overcoming immunological rejection from the transplant recipient. Interestingly, Burt and co-workers performed ESC transplantation in major histocompatibility complex (MHC)-mismatched mice without clinical or histological evidence of graft-versus-host disease (GVHD) [10]. In addition, recent data indicate the potential of ESCs to offer a possible solution for low-risk induction of tolerance not involving any immunosuppressions (reviewed by Zavazava [70]). Moreover, it might be possible to down-regulate the antigenicity of ESCs through suppression of MHC gene expression [28]. Another concern is that the cultivation and transplantation of such stem cells are accompanied by a tumourigenic differentiation. As it has been shown that undifferentiated ESCs give rise to teratomas and teratocarcinomas after implantation in animals, this potential misdevelopment constitutes a major problem for the clinical use of such cells [64].

13.7
Genetic Manipulations

Another important feature of hESCs is the capability of genetic manipulations [31]. This attribute enables scientists to perform various genetic changes in hESCs such as knocking-out a specific gene by homologous recombination [65], the knock-down of gene expression using siRNA [41] or the overexpression of genes either to induce tissue-specific gene expression or just to mark cells for monitoring [36]. One way to monitor the cells is to genetically modify them with a fluorescent tag. Specific cells transfected with, for example, enhanced green fluorescent protein (eGFP) can easily be localised, tracked during differentiation and even separated using a fluorescence-activated cell sorting machine (FACS). Genetic manipulation of hESCs can be achieved by viral infection or transfection of DNA [43]. Retroviruses, lentiviruses and adenoviruses have been used for viral infection to modify hESCs. Viral vectors demonstrate high transfection efficiencies and can provide good expression levels. However, viral transduction has safety concerns and requires viral packaging and extraction. Moreover, there are various non-viral transfection methods described in the literature (Table 13.1). However, the various transfection methods each have a different transfection efficiency. The range is between 65% for Nucleofector and below 30% for most lipid reagents [2, 31, 47, 60, 72].

The fusion of somatic cells with ESCs is another type of genetic manipulation. This means reprogramming a differentiated cell into one capable of giving rise to many different cell types, a pluripotent cell, which in turn could repopulate or repair diseased or damaged tissue. Because of the immunogenic advantages of these cells, this method could present beneficial applications in regenerative medicine [16, 52].

13.8
Micromass Culture

The in-vitro generation of tissues or organs usually requires the use of scaffold materials as well as cells mostly cultured in monolayers under appropriate conditions. When cultured in non-adherent Petri dishes, ESCs form spheroidal micromass bodies (or embryoid bodies) containing clustered cells and a non-artificial extracellular matrix which is produced by the cells. The cells in these micromass bodies gradually differentiate into various cells representing all three germ layers [30]. Various growth factors can be used to direct the differentiation toward a particular cell type [58]. After application of an osteogenic supplement murine ESCs form oval spheroidal bodies containing a calcified matrix in the centre. Over time, micromass bodies of ESCs show the development of a wide variety of morphologically and functionally different cell types, including neurons, cartilage, bone and cardiomyocytes [31], (for more details see Chapter 40, "Prospects of micromass culture technology in tissue engineering").

Table 13.1 Non-viral transfection methods

Transfection method	Transfection efficiency[a]	Reference
Electroporation	40%	[72]
FuGENE 6 transfection reagent	> 30%	[37, 47]
Lipofectamine 2000	> 30%	[37]
Nucleofector	65%	[60]
ExGen 500	> 20%	[2]
Effectene	> 20%	[2]
TransIT	> 40%	[47]

[a] The transfection efficiencies is described in the literature are for various cell types, not necessarily hESCs

13.9
Ethical Aspects

Opposition to the use of ESCs because of moral and religious concerns presents an obstacle to progress in cell-derived replacement treatment [15, 23]. Objections are mainly raised because experimentation with ESCs derived from human blastocysts requires the destruction of human embryos. The current debate on the ethical status of the human embryo has focused on its fragility and defencelessness, and the question as to what degree of respect a human embryo should merit in the light of recent medical progress. It is generally accepted that the blastocyst contains a complete set of genetic instructions and the capacity for the epigenetic determinations needed to develop into a human being. However, there is a dispute about the ethical and legal implications of using human primordial stem cells for tissue engineering, which includes a controversy as to whether absolute respect for individual human life already begins at conception [15, 23].

All in all, ESCs have been well-recognised as cells having some remarkable features. They have the potential to differentiate into all types of mature cells in the body and can be expanded in principle without limit. However, before these cells can be used in clinics many issues remain to be addressed. In addition, legal and ethical concerns might advocate the use of other pluripotent cells (e.g. umbilical cord blood cells) or the creation of ESCs by dedifferentiation.

Whereas up to now determined cells have been commonly used in clinical cell-based engineering strategies, further investigations might demonstrate the clinical feasibility and applicability of pluripotential cell lines such as ESCs in tissue engineering. It is predicted that the use of pluripotential cells will play a major role in the future of tissue engineering and regenerative medicine.

References

1. Amit M, Carpenter MK, Inokuma MS et al (2000) Clonally derived human embryonic stem cell lines maintain pluripotency and proliferative potential for prolonged periods of culture. Dev Biol 227:271-278
2. Arnold AS, Laporte V, Dumont S et al (2006) Comparing reagents for efficient transfection of human primary myoblasts: FuGENE 6, Effectene and ExGen 500. Fundam Clin Pharmacol 20:81-89
3. Atala A (2007) Engineering tissues, organs and cells. J Tissue Eng Regen Med 1:83-96
4. Avilion AA, Nicolis SK, Pevny LH et al (2003) Multipotent cell lineages in early mouse development depend on SOX2 function. Genes Dev 17:126-140
5. Bielby RC, Boccaccini AR, Polak JM et al (2004) In vitro differentiation and in vivo mineralization of osteogenic cells derived from human embryonic stem cells. Tissue Eng 10:1518-1525
6. Binderman I, Fin N (1990) Bone substitutesorganic, inorganic, and polymeric: Cell material interactions. In: Yamamuro T, Hench L and Wilson J eds) CRC Handbook of Bioactive Ceramics. CRC Press, Boca Raton, pp 45-51
7. Boiani M, Scholer HR (2005) Regulatory networks in embryo-derived pluripotent stem cells. Nat Rev Mol Cell Biol 6:872-884
8. Brimble SN, Zeng X, Weiler DA et al (2004) Karyotypic stability, genotyping, differentiation, feeder-free maintenance, and gene expression sampling in three human embryonic stem cell lines derived prior to August 9, 2001. Stem Cells Dev 13:585-597
9. Brimble SN, Sherrer ES, Uhl EW et al (2007) The cell surface glycosphingolipids SSEA-3 and SSEA-4 are not essential for human ESC pluripotency. Stem Cells 25:54-62
10. Burt RK, Verda L, Kim DA et al (2004) Embryonic stem cells as an alternate marrow donor source: engraftment without graft-versus-host disease. J Exp Med 199:895-904
11. Chambers I, Colby D, Robertson M et al (2003) Functional expression cloning of Nanog, a pluripotency sustaining factor in embryonic stem cells. Cell 113:643-655
12. Chambers I (2004) The molecular basis of pluripotency in mouse embryonic stem cells. Cloning Stem Cells 6:386-391
13. Chaudhry GR, Yao D, Smith A et al (2004) Osteogenic cells derived from embryonic stem cells produced bone nodules in three-dimensional scaffolds. J Biomed Biotechnol 2004:203-210
14. Cibelli JB, Campbell KH, Seidel GE et al (2002) The health profile of cloned animals. Nat Biotechnol 20:13-14
15. Cogle CR, Guthrie SM, Sanders RC et al (2003) An overview of stem cell research and regulatory issues. Mayo Clin Proc 78:993-1003
16. Collas P (2007) Dedifferentiation of cells: new approaches. Cytotherapy 9:236-244
17. D'Ippolito G, Schiller PC, Ricordi C et al (1999) Age-related osteogenic potential of mesenchymal stromal stem cells from human vertebral bone marrow. J Bone Miner Res 14:1115-1122
18. Daheron L, Opitz SL, Zaehres H et al (2004) LIF/STAT3 signaling fails to maintain self-renewal of human embryonic stem cells. Stem Cells 22:770-778
19. Damien CJ, Parsons JR (1991) Bone graft and bone graft substitutes: a review of current technology and applications. J Appl Biomater 2:187-208

20. Drab M, Haller H, Bychkov R et al (1997) From totipotent embryonic stem cells to spontaneously contracting smooth muscle cells: a retinoic acid and db-cAMP in vitro differentiation model. Faseb J 11:905-915
21. Dvash T, Mayshar Y, Darr H et al (2004) Temporal gene expression during differentiation of human embryonic stem cells and embryoid bodies. Hum Reprod 19:2875-2883
22. Evans MJ, Kaufman MH (1981) Establishment in culture of pluripotential cells from mouse embryos. Nature 292:154-156
23. Gilbert DM (2004) The future of human embryonic stem cell research: addressing ethical conflict with responsible scientific research. Med Sci Monit 10:RA99-103
24. Gurdon JB (1962) Adult frogs derived from the nuclei of single somatic cells. Dev Biol 4:256-273
25. Halleux C, Sottile V, Gasser JA et al (2001) Multi-lineage potential of human mesenchymal stem cells following clonal expansion. J Musculoskelet Neuronal Interact 2:71-76
26. Handschel J, Wiesmann HP, Depprich R et al (2006) Cell-based bone reconstruction therapies--cell sources. Int J Oral Maxillofac Implants 21:890-898
27. Handschel J, Berr K, Depprich R et al (2008) Osteogenic differentiation of embryonic stem cells. Head Face Med 4:10
28. Heng BC, Cao T, Stanton LW et al (2004) Strategies for directing the differentiation of stem cells into the osteogenic lineage in vitro. J Bone Miner Res 19:1379-1394
29. Hochedlinger K, Jaenisch R (2003) Nuclear transplantation, embryonic stem cells, and the potential for cell therapy. N Engl J Med 349:275-286
30. Itskovitz-Eldor J, Schuldiner M, Karsenti D et al (2000) Differentiation of human embryonic stem cells into embryoid bodies compromising the three embryonic germ layers. Mol Med 6:88-95
31. Kitsberg D (2007) Human embryonic stem cells for tissue engineering. Methods Mol Med 140:33-65
32. Klimanskaya I, Chung Y, Meisner L et al (2005) Human embryonic stem cells derived without feeder cells. Lancet 365:1636-1641
33. Klug MG, Soonpaa MH, Koh GY et al (1996) Genetically selected cardiomyocytes from differentiating embryonic stem cells form stable intracardiac grafts. J Clin Invest 98:216-224
34. Kogler G, Sensken S, Airey JA et al (2004) A new human somatic stem cell from placental cord blood with intrinsic pluripotent differentiation potential. J Exp Med 200:123-135
35. Langer R, Vacanti JP (1993) Tissue engineering. Science 260:920-926
36. Lavon N, Yanuka O, Benvenisty N (2004) Differentiation and isolation of hepatic-like cells from human embryonic stem cells. Differentiation 72:230-238
37. Lee SH, Lumelsky N, Studer L et al (2000) Efficient generation of midbrain and hindbrain neurons from mouse embryonic stem cells. Nat Biotechnol 18:675-679
38. Lysaght MJ, Reyes J (2001) The growth of tissue engineering. Tissue Eng 7:485-493
39. Maitra A, Arking DE, Shivapurkar N et al (2005) Genomic alterations in cultured human embryonic stem cells. Nat Genet 37:1099-1103
40. Martin GR (1981) Isolation of a pluripotent cell line from early mouse embryos cultured in medium conditioned by teratocarcinoma stem cells. Proc Natl Acad Sci U S A 78:7634-7638
41. Matin MM, Walsh JR, Gokhale PJ et al (2004) Specific knockdown of Oct4 and beta2-microglobulin expression by RNA interference in human embryonic stem cells and embryonic carcinoma cells. Stem Cells 22:659-668
42. McDonald JW, Liu XZ, Qu Y et al (1999) Transplanted embryonic stem cells survive, differentiate and promote recovery in injured rat spinal cord. Nat Med 5:1410-1412
43. Menendez P, Wang L, Bhatia M (2005) Genetic manipulation of human embryonic stem cells: a system to study early human development and potential therapeutic applications. Curr Gene Ther 5:375-385
44. Mitsui K, Tokuzawa Y, Itoh H et al (2003) The homeoprotein Nanog is required for maintenance of pluripotency in mouse epiblast and ES cells. Cell 113:631-642
45. Moosmann S, Hutter J, Moser C et al (2005) Milieu-adopted in vitro and in vivo differentiation of mesenchymal tissues derived from different adult human CD34-negative progenitor cell clones. Cells Tissues Organs 179:91-101
46. Nash R, Neves L, Faast R et al (2007) The lectin Dolichos biflorus agglutinin recognizes glycan epitopes on the surface of murine embryonic stem cells: a new tool for characterizing pluripotent cells and early differentiation. Stem Cells 25:974-982
47. Nguyen TH, Murakami A, Fujiki K et al (2002) Transferrin-polyethylenimine conjugate, FuGENE6 and TransIT-LT as nonviral vectors for gene transfer to the corneal endothelium. Jpn J Ophthalmol 46:140-146
48. Nichols J, Zevnik B, Anastassiadis K et al (1998) Formation of pluripotent stem cells in the mammalian embryo depends on the POU transcription factor Oct4. Cell 95:379-391
49. Niwa H, Miyazaki J, Smith AG (2000) Quantitative expression of Oct-3/4 defines differentiation, dedifferentiation or self-renewal of ES cells. Nat Genet 24:372-376
50. Niwa H (2007) How is pluripotency determined and maintained? Development 134:635-646
51. Ohtsuka S, Dalton S (2008) Molecular and biological properties of pluripotent embryonic stem cells. Gene Ther 15:74-81
52. Oliveri RS (2007) Epigenetic dedifferentiation of somatic cells into pluripotency: cellular alchemy in the age of regenerative medicine? Regen Med 2:795-816
53. Pankratz MT, Li XJ, Lavaute TM et al (2007) Directed neural differentiation of human embryonic stem cells via an obligated primitive anterior stage. Stem Cells 25:1511-1520
54. Pittenger MF, Mackay AM, Beck SC et al (1999) Multilineage potential of adult human mesenchymal stem cells. Science 284:143-147
55. Quarto R, Thomas D, Liang CT (1995) Bone progenitor cell deficits and the age-associated decline in bone repair capacity. Calcif Tissue Int 56:123-129
56. Reubinoff BE, Pera MF, Fong CY et al (2000) Embryonic stem cell lines from human blastocysts: somatic differentiation in vitro. Nat Biotechnol 18:399-404

57. Salero E, Hatten ME (2007) Differentiation of ES cells into cerebellar neurons. Proc Natl Acad Sci U S A 104:2997-3002
58. Schuldiner M, Yanuka O, Itskovitz-Eldor J et al (2000) Effects of eight growth factors on the differentiation of cells derived from human embryonic stem cells. Proc Natl Acad Sci U S A 97:11307-11312
59. Schulz LC, Ezashi T, Das P et al (2007) Human embryonic stem cells as models for trophoblast differentiation. Placenta 29(Suppl A):S10-S16
60. Siemen H, Nix M, Endl E et al (2005) Nucleofection of human embryonic stem cells. Stem Cells Dev 14:378-383
61. Smith AG, Heath JK, Donaldson DD et al (1988) Inhibition of pluripotential embryonic stem cell differentiation by purified polypeptides. Nature 336:688-690
62. Solter D (2000) Mammalian cloning: advances and limitations. Nat Rev Genet 1:199-207
63. Thomson JA, Itskovitz-Eldor J, Shapiro SS et al (1998) Embryonic stem cell lines derived from human blastocysts. Science 282:1145-1147
64. Trounson A (2002) Human embryonic stem cells: mother of all cell and tissue types. Reprod Biomed Online 4(Suppl 1):58-63
65. Urbach A, Schuldiner M, Benvenisty N (2004) Modeling for Lesch-Nyhan disease by gene targeting in human embryonic stem cells. Stem Cells 22:635-641
66. Wang G, Zhang H, Zhao Y et al (2005) Noggin and bFGF cooperate to maintain the pluripotency of human embryonic stem cells in the absence of feeder layers. Biochem Biophys Res Commun 330:934-942
67. Wilmut I, Young L, Campbell KH (1998) Embryonic and somatic cell cloning. Reprod Fertil Dev 10:639-643
68. Yamaguchi M, Hirayama F, Murahashi H et al (2002) Ex vivo expansion of human UC blood primitive hematopoietic progenitors and transplantable stem cells using human primary BM stromal cells and human AB serum. Cytotherapy 4:109-118
69. Young LE, Sinclair KD, Wilmut I (1998) Large offspring syndrome in cattle and sheep. Rev Reprod 3:155-163
70. Zavazava N (2003) Embryonic stem cells and potency to induce transplantation tolerance. Expert Opin Biol Ther 3:5-13
71. zur Nieden NI, Kempka G, Rancourt DE et al (2005) Induction of chondro-, osteo- and adipogenesis in embryonic stem cells by bone morphogenetic protein-2: effect of cofactors on differentiating lineages. BMC Dev Biol 5:1
72. Zwaka TP, Thomson JA (2003) Homologous recombination in human embryonic stem cells. Nat Biotechnol 21:319-321

The Unrestricted Somatic Stem Cell (USSC) From Cord Blood For Regenerative Medicine

G. Kögler

Contents

14.1 Biological Advantages of Cord Blood (CB) as a Stem Cell Resource 167
14.2 Unrestricted Somatic Stem Cells (USSCs) From CB 168
14.2.1 Isolation, Expansion and Characterization of USSCs from Fresh CB 168
14.2.2 Isolation, Expansion and Characterization of USSCs from Cryopreserved CB 171
14.2.3 Differentiation of USSCs Towards the Mesenchymal Lineage 171
14.2.4 Neural Differentiation of USSCs 172
14.2.5 Endodermal Differentiation of USSCs 172
14.3 Hematopoiesis Supporting Activity of USSCs 174
14.4 Conclusion: Future Efforts Towards the Regenerative Capacity of CB Non-hematopoietic Cells 174
References 175

14.1
Biological Advantages of Cord Blood (CB) as a Stem Cell Resource

In the 19 years since the first CB transplantation was reported in the *New England Journal of Medicine* [9], the fields of CB research, banking and transplantation have flourished [2]. Umbilical CB has made allogeneic hematopoietic stem cell transplantation available to patients who do not have an HLA-identical sibling or an unrelated donor [20]. CB transplantation has been used to successfully treat leukemia, lymphoma, myelodysplasia, aplastic anemia, hemoglobinopathies, metabolic storage diseases, and immunodeficiencies. There are more than 300,000 unrelated CB units stored in 40 banks, and more than 10,000 CB transplants have been performed [20].

Standardization of unrelated CB banking was initiated in Düsseldorf in 1992/1993; however, only due to the financial support of the German José Carreras Foundation in 1996 were we able to develop a program for high quality CB banking on an international basis [14]. As of April 2008, a total of 13,554 allogeneic CB products have been cryopreserved and 490 CB transplants were provided worldwide. At present the CB bank is the largest single standing CB bank in Europe (Fig. 14.1, www.netcord.org).

Compared with its allogeneic adult bone marrow counterpart, CB has substantial logistic and clinical advantages, such as: faster availability of banked units (if necessary within three days); extension of the donor pool because of the tolerance of one or two in six HLA-mismatches; lower severity and incidence of acute graft versus host disease, lower risk of transmission of infections by latent viruses, and absence of risks to donors [20]. In addition, the stem cell compartment in CB is less mature. This has been documented extensively for the hematopoietic system, including a higher proliferative potential in vitro as well as in vivo, which is associated with an extended life span of the stem cells and longer telomeres [23]. In addition, the frequency of these stem cells is also higher in CB than in adult bone marrow, probably due to the very rapid growth and

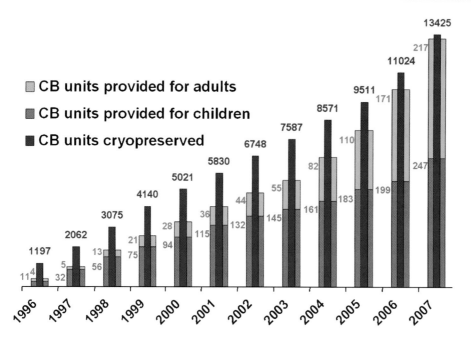

Fig. 14.1 Development of the José Carreras Cord Blood Bank

remodeling of tissue that is required during the fetal period. Hematopoietic stem cells, for example, proliferate and differentiate in a number of distinct sites such as in yolk sac, fetal liver, thymus, spleen and bone marrow, indicating migration through the circulation. The immaturity and increased frequency of hematopoietic stem cells suggest that CB might also be an attractive source for non-hematopoietic stem cells.

14.2
Unrestricted Somatic Stem Cells (USSCs) From CB

In 1999 our group described for the first time osteoblast precursor cells derived from human umbilical CB [25]. In 2000, adherently growing, fibroblast-like cells were identified in CB, which revealed an immunophenotype (CD45⁻, CD13⁺, CD29⁺, CD73⁺, CD105⁺) similar to bone marrow (BM)-derived mesenchymal stroma cells (MSCs) [8]. Soon thereafter, Campagnoli et al. [6] confirmed the MSC nature of such cells from fetal blood by showing that they have the potential to differentiate to osteocytes, adipocytes and chondrocytes. Goodwin et al. [10] described a cell population, which can give rise to cells with features of adipocytes, osteocytes or neural cells, thus, a mesodermal/ectodermal differentiation. Non-hematopoietic stem cells in CB appear not to be simply MSCs.

We were able to identify such a multipotent adherent cell (Fig. 14.2), which we have termed USSC [15, 16]. USSCs have the potential to differentiate into mesodermal osteoblasts, chondroblasts, adipocytes, hematopoietic cells and cardiomyocytes, into ectodermal cells of all three neural lineages as well as into endodermal hepatic cells [18, 21].

14.2.1
Isolation, Expansion and Characterization of USSCs from Fresh CB

Collection, processing and initial characterization of CB units obtained by the José Carreras CB bank in Düsseldorf from its 86 participating collection

Fig. 14.2 Characteristics of USSC. Spindle-shaped USSC plated at high density

sites/maternity hospitals (8,000 units per year) is performed routinely. Only about 30% of these units are suitable for hematopoietic banking, mainly due to cell number limitations [17]. Therefore, sufficient CB units are available for research purposes, and the donor mothers consent to the use for research if donations are not suitable for banking. In USSC cultures initiated, only in 43% were USSC colonies observed. On average there were four USSC colonies per CB unit. They grew into monolayers of fibroblastoid, spindle-shaped cells within 2–3 weeks (Fig. 14.2). Once established, when colonies were observed 32% reached passage seven or eight and beyond nine, respectively. If cells reached passage nine, frequently they could be expanded for more than 20 passages, yielding theoretically up to 10^{15} cells [15].

While long-time cultivation USSCs could be kept up to very high passage, the replicative lifespan of MSCs was shorter already ending up in P11 after 16 PD (Fig. 14.3). The better growth kinetic of USSCs was also reflected in shorter PD time (one doubling every 3.32 ± 0.88 days) if compared with MSCs (one doubling every 18.95 ± 10.73 days).

Culture conditions influenced the generation frequency and expansion potential. In the presence of Myelocult/10^{-7} M dexamethasone, USSC cultures were initiated from 57.1% of CB samples (52 out of 91), compared with 43% in DMEM/10^{-7}M dexamethasone/30% FCS (Fig. 14.4a). Although the generation frequency was higher in Myelocult medium, Myelocult grown cells showed a reduced expansion potential and a tendency to spontaneous osteogenic differentiation (Fig. 14.4b), associated with loss of

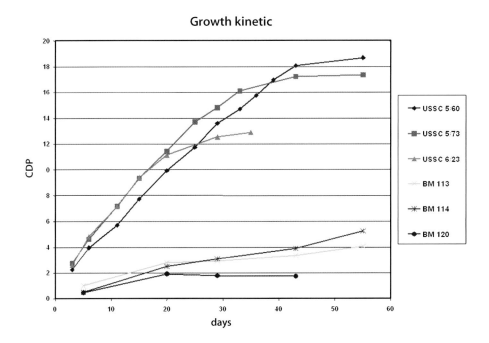

Fig. 14.3 Comparison of the growth kinetics of USSC and bone marrow (*BM*) MSC

multipotency (cells no longer differentiated along neural and endodermal pathways). Serum was identified as the most critical factor for generation and expansion, and FCS needs to be carefully pre-selected to allow for extensive amplification of multipotent USSCs. Currently, low glucose DMEM/30% FCS with (generation) or without (expansion) 10^{-7} M dexamethasone is the present standard in our group (Fig. 14.4a). Whether these effects of different culture conditions reflect a preferential outgrowth of distinct cell populations in the two media—multipotent USSCs in DMEM/10^{-7} M dexamethasone/30% FCS and more committed MSC-like cells triggered into differentiation by Myelocult/10^{-7} M dexamethasone—and, thus, a stem cell hierarchy, is currently unknown. Nevertheless, these results clearly indicate the importance of standardized conditions for generation and expansion of USSCs (Fig. 14.4c) and that multipotency of each USSC line has to be confirmed towards the mesodermal, ectodermal and endodermal lineage (Fig. 14.4d–f).

USSC lack expression of CD4, CD8, CD11a, CD11b, CD11c, CD14, CD15, CD16, CD18, CD25, CD27, CD31, CD33, CD34, CD40, CD45, CD49d, CD50, CD56, CD62E, CD62L, CD62P, CD80, CD86, CD106, CD117, cadherin V, glycophorin A and HLA-class II, but are positive for CD10, CD13, CD29, CD44, CD49e, CD58, CD71, CD73, CD90, CD105, CD146, CD166, vimentin, cytokeratin 8 and 18, von-Willebrand-factor as well as HLA-class I and show only limited expression of CD49b and CD123 [15, 16]. Due to their proliferative and differentiation capacity, USSCs are an interesting candidate for the future development of cellular therapies as well as support of hematopoietic reconstitution. Since generation and expansion under GMP conditions is

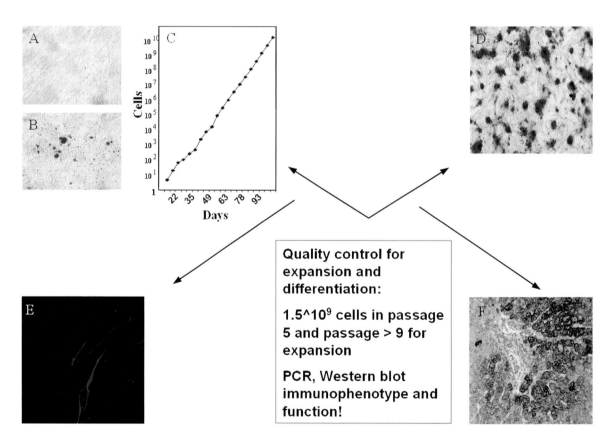

Fig. 14.4a–f Quality control of USSC. a Standard morphology in the presence of DMEM, 30% pretested FCS. b Nodule formation in the presence of corticoids (wrong FCS). c Growth kinetic for good USSC lines; d–f represent the differentiation potential towards the osteoblastic, neural and endodermal lineage

mandatory for use in clinical application, the automated cell processing system Sepax (BIOSAFE) with the CS900 separation kit can be used for mononuclear cell separation from CB in a similar way as described for bone marrow mononuclear cells (MNCs) [1]. The combination of this Sepax procedure together with the cell stack system (one, two, five and ten layers), results in cell numbers of 1×10^9 USSCs within four passages [19]. These USSC products could be cryopreserved, thawed and expanded further in clinical grade quality.

14.2.2
Isolation, Expansion and Characterization of USSCs from Cryopreserved CB

The generation of USSCs from cryopreserved and thawed CB samples was associated with difficulties as expected, since thawed CB contains many erythrocytes and erythroblasts as well as dead granulocytes, which might aggregate and the frequency of USSCs per MNC is extremely low [16]. Due to these difficulties to generate USSC from cryopreserved material, therapeutic application of USSCs probably will rely on the establishment of USSC cell banks, with the USSCs generated from fresh CB.

14.2.3
Differentiation of USSCs Towards the Mesenchymal Lineage

Differentiation of USSCs to *osteoblasts* can always (100% of all cultures) be induced by culturing the cells in the presence of dexamethasone, ascorbic acid and β-glycerol phosphate (Fig. 14.5a). Cells formed nodules, which stained with Alizarin red, an indication for osteoblast-typical calcification and functional

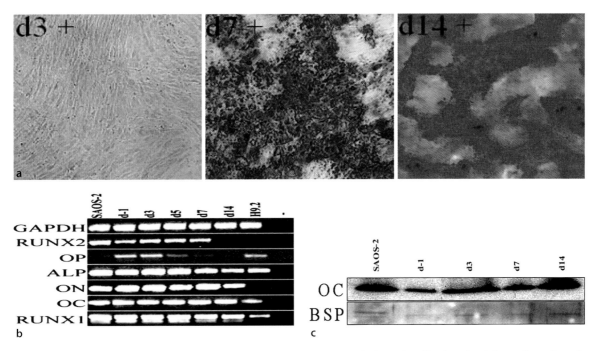

Fig. 14.5a–c For osteoblast differentiation, cells were cultured at 8,000 cells/cm² in 24-well plates in DMEM, 30% FCS, 10^{-7} M dexamethasone, 50 µM ascorbic acid-2 phosphate and 10 mM ß-glycerol phosphate (DAG). **a** Differentiation to osteoblasts is shown by Alizarin red staining to determine calcium deposition (×10 magnification). Osteoblastic nature was confirmed by RT-PCR (**b**) and Western blotting (**c**)

competency of the differentiated cells. Bone-specific alkaline phosphatase (ALP) activity was detected and continuous increase in Ca^{2+} release was documented. Osteogenic differentiation was further confirmed by increased expression of ALP, osteonectin, osteopontin, Runx1, bone sialo-protein and osteocalcin as well as Notch 1 and 4 detected by RT-PCR [15] (Fig. 14.5b) and confirmed on protein level (Fig. 14.5c). A pellet culture technique in the presence of dexamethasone, prolin, sodium pyruvat, ITS+ Premix and TGF-β1 was employed to trigger USSCs towards the *chondrogenic* lineage [15]. The chondrogenic nature of differentiated cells was assessed by Alcian blue staining and by expression analysis of the cartilage extracellular protein collagen type II.

For induction of *adipogenic* differentiation, USSCs were cultured with dexamethasone, insulin, IBMX and indometacin. For adipogenic differentiation we were able to identify two subsets of USSC lines: cells which could be differentiated towards adipocytes (which are more likely MSC) and USSC cell lines that could not be differentiated towards adipocytes but had otherwise a much better differentiation potential. Lack of adipogenic differentiation capacity of CB-derived MSC-like cells has also been reported by others [4, 13]. In contrast, bone marrow mesenchymal cells (BMMSCs) always showed adipogenic differentiation. Therefore, we postulate two functional subsets and the hypothesis that USSCs and MSCs can be easily distinguished in CB by simple adipogenic differentiation.

14.2.4
Neural Differentiation of USSCs

USSCs can be differentiated in the presence of NGF, bFGF, dibutyryl cAMP, IBMX and retinoic acid into neural cells—this is, at the moment, the primary quality control for multipotency (Fig. 14.6). These analyses have been performed in co-operation with Prof. H.W. Müller (Molecular Neurobiology, Department of Neurology, University of Düsseldorf Medical School, Düsseldorf) [15]. Neurofilament-positive (NF) cells were detected (70% of cells) and double immunostaining revealed co-localization of neurofilament and sodium-channel protein in a small proportion of cells. Expression of synaptophysin was detected after 4 weeks. USSC-derived neurons were identified that stained positive for TH (approximately 30% of the cells), the key enzyme of the dopaminergic pathway [11]. Choline acetyltransferase was detected in about 50% of the cells. Recent data demonstrated functional neural properties applying patch-clamp analysis and high performance liquid chromatography (HPLC) [11]. Although a rare event, voltage gated sodium channels could clearly be identified and the HPLC analysis confirmed synthesis and release of dopamine and serotonin by differentiated USSCs. The in-vitro data of USSC/MSC differentiation towards the neural lineage has in the mean time been confirmed by other groups [24]. Sun et al. [22] described voltage-sensitive and ligand-gated channels in differentiating fetal neural cells isolated from the non-hematopoietic fraction of human CB.

14.2.5
Endodermal Differentiation of USSCs

Previous experimental data showed that transplantation of USSCs in a non-injury model, the preimmune fetal sheep [15], and revealed more than 20% albumin-producing human parenchymal hepatic cells.

Meanwhile, we were able to establish endodermal differentiation of USSCs also in vitro [18, 21], applying protocols described for both embryonic and adult stem cells. USSCs were negative for the majority of endodermal markers tested by RT-PCR. Only expression of HGF and cytokeratins 8, 18 and 19 was observed on RT-PCR level. After induction of differentiation, USSCs never expressed NeuroD, HNF1, HNF3b, PDX-1, PAX4, insulin and α-fetoprotein, but did express, depending on the culture conditions, the common endodermal precursor markers GSC, Sox-17, HNF4α, GATA4 (but not HNF1 or HNF3ß) as well as albumin, Cyp2B6, Cyp3A4, Gys2 (liver development) and Nkx6.1 and ISL-1 (pancreatic development). Furthermore, glycogen storage (PAS staining) and albumin secretion were detected (Fig. 14.7). These results clearly indicate endodermal differentiation of USSCs in vitro [18, 21]. Based on RT-PCR data as well as functional data, an endodermal differentiation pathway could be delineated for USSCs [21].

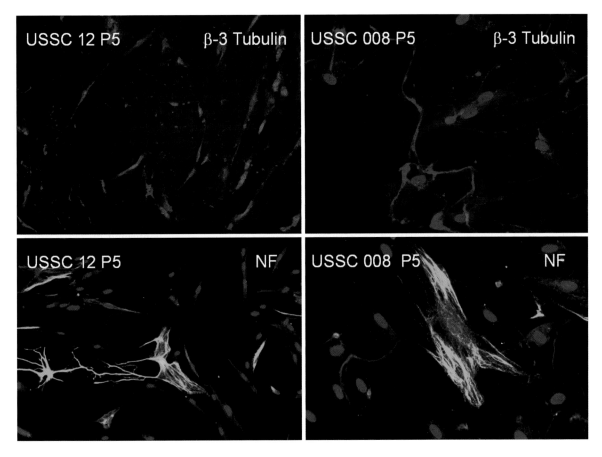

Fig. 14.6 In-vitro differentiation of USSC into neural cells. Differentiated cells showed positive immunoreactivity for the neuron-specific markers neurofilament and ß-3 tubulin

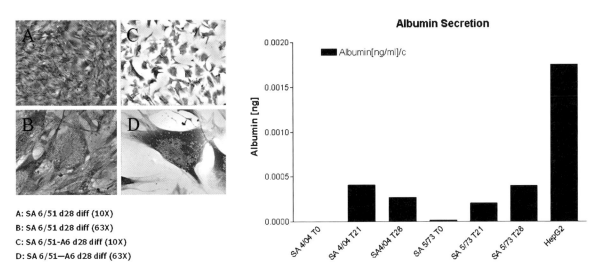

Fig. 14.7 *Left panel:* **a–d** PAS-staining of spindle-shaped undifferentiated USSC in DMEM A+B and angular hepatocyte-like cells in differentiation medium with HGF and OSM on Matrigel. *Right panel:* Secretion of human albumin. Quantitative determination of human albumin secretion in media during 28 days of differentiation measured by ELISA. Carcinoma cell line HepG2 was used as positive and negative control

14.3
Hematopoiesis Supporting Activity of USSCs

Recent experience shows that it is possible to also reconstitute adults after myeloablative or non-myeloablative conditioning by applying one large CB unit or by combining two CB units [3, 5]. Many adult patients among the provided transplants from the Düsseldorf CB bank received a double CB transplantation to increase hematopoietic progenitor and granulocyte numbers and to reduce the duration of posttransplant neutropenia. Recently, data presented by the EUROCORD registry in Paris showed excellent engraftment kinetics of the double CB transplants and high survival rates.

Other clinical approaches to improve reconstitution used the co-transplantation of a single CB unit together with highly purified CD34+ mobilized peripheral blood stem cells or bone marrow from a haploidentical related donor. The specialized BM stroma environment consisting of extracellular matrix and stroma cells has been shown to be crucial for hematopoietic regeneration after any stem cell transplantation. Therefore, co-transplantation of CB-USSCs may also improve engraftment. Our group has studied cytokine production and in vitro hematopoiesis-supporting stromal activity of USSC in vitro and in vivo in comparison with BM-derived MSCs [16]. USSC constitutively produced many cytokines, including SCF, LIF, TGF-1ß, M-CSF, GM-CSF, VEGF, IL-1ß, IL-6, IL-8, IL-11, IL-12, IL-15, SDF-1α and HGF. When USSCs were stimulated with IL-1ß, G-CSF was released. Production of SCF and LIF was significantly higher in USSCs compared with BMMSCs [16]. In order to determine the hematopoiesis supporting stromal activity of USSCs compared with BMMSCs, CB CD34+ cells were expanded in co-cultures. At 1, 2, 3 and 4 weeks, co-cultivation of CD34+ cells on the USSC layer resulted in a 14-fold, 110-fold, 151-fold and 183.6-fold amplification of total cells and in a 30-fold, 101-fold, 64-fold and 29-fold amplification of CFC, respectively. LTC-IC expansion at 1 and 2 weeks was with twofold and 2.5-fold significantly higher for USSCs than BMMSCs (single-fold and single-fold, respectively), but declined after day 21. In summary, USSCs produce functionally significant amounts of hematopoiesis supporting cytokines and are superior to BMMSCs in supporting the expansion of CD34+ cells from CB. USSCs are, therefore, suitable candidates for stroma-driven ex-vivo expansion of hematopoietic CB cells for short-term reconstitution (Fig. 14.8).

Recent in-vivo data of Chan et al. [7] have shown that USSCs induced a significant enhancement of CD34+ cell homing to both bone marrow and spleen. In the publication of Huang et al. [12], CB MSCs were used for the ex-vivo expansion of CB progenitors and resulted in a fast recovery after transplantation in the NOD-SCID mice.

14.4
Conclusion: Future Efforts Towards the Regenerative Capacity of CB Non-hematopoietic Cells

Ultimately, it would be of great scientific value to know whether the cells we call USSCs are multipotent or even pluripotent (germline differentiation?). Furthermore, it would be helpful to define antigens, which allow the enrichment of this cell population from each CB collection, and to define more precisely the distinct differentiation pathways with a read-out in selected animal models. In addition, the functionality of the USSC must be proven in comparison with the adult or fetal or even embryonic counterpart with regard to each differentiation pathway. On the basis of their multipotency and expansion under GMP conditions into large quantities, these USSCs or other non-hematopoietic cells from CB, when pretested for infectious agents and matched for the major transplantation antigens, may serve as a universal allogeneic stem cell source for the future development of cellular therapy for tissue repair and tissue regeneration.

Acknowledgments

The author would like to thank all the co-authors, and mainly Prof. P. Wernet, of our jointly published papers that were cited here and the Deutsche Forschungsgemeinschaft (DFG) for funding the research group FOR 717 including the project Ko2119/6-1. Studies reported from the author's laboratory con-

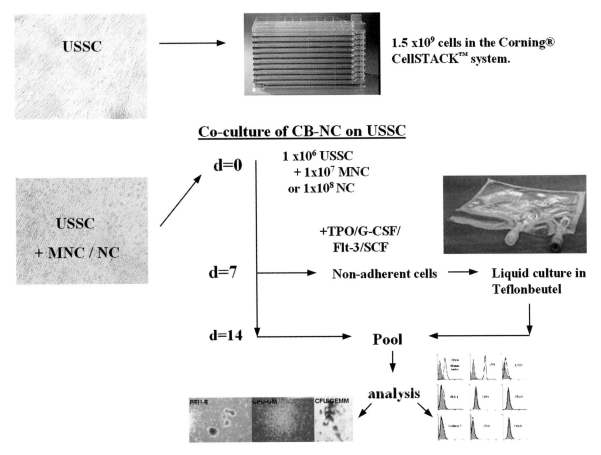

Fig. 14.8 Expansion of USSC can be performed in a closed system applying cell stacks (Costar Corning). In a ten-layer cell stack system, cell numbers of 1.5×10^9 USSC were obtained within 4 passages. These USSC can be used as a feeder layer to support hematopoietic cells in the presence of thrombopoietin (TPO), G-CSF, Flt-3 ligand and SCF. Readout after a subsequent culture of non adherent cells are the FACS analysis for the presence of immature CD34+ cells as well as the colony forming assay for BFU-E, CFU-GM and CFU-GEMM

cerning the hematopoiesis supporting activity, which were reviewed here, have been supported by the German José Carreras Leukemia Foundation grant DJCLS-R03/06; DJCLS-R07/05v and the EUROCORD III grant QLRT-2001-01918. Thanks to Teja F. Radke, Anja Buchheiser, PhD, Aurelie Lefort, and Simon Waclawczyk for their excellent technical support.

References

1. Aktas M, Radke TF, Strauer BE et al (2008) Separation of adult bone marrow mononuclear cells using the automated closed separation system Sepax. Cytotherapy 10:203–211
2. Ballen KK, Barker JN, Stewart SK et al (2008) Collection and preservation of cord blood for personal use. Biol Blood Marrow Transplant 14:356–363
3. Barker JN, Weisdorf DJ, DeFor TE et al (2005) Transplantation of 2 partially HLA-matched umbilical cord blood units to enhance engraftment in adults with hematologic malignancy. Blood 105:1343–1347
4. Bieback K, Kern S, Kluter H et al (2004) Critical parameters for the isolation of mesenchymal stem cells from umbilical cord blood. Stem Cells 22:625–634
5. Brunstein CG, Barker JN, Weisdorf DJ et al (2007) Umbilical cord blood transplantation after nonmyeloablative conditioning: impact on transplantation outcomes in 110 adults with hematologic disease. Blood 110:3064–3070
6. Campagnoli C, Roberts IA, Kumar S et al (2001) Identification of mesenchymal stem/progenitor cells in human first-trimester fetal blood, liver, and bone marrow. Blood 98:2396–2402

7. Chan SL, Choi M, Wnendt S et al (2007) Enhanced in vivo homing of uncultured and selectively amplified cord blood CD34+ cells by cotransplantation with cord blood-derived unrestricted somatic stem cells. Stem Cells 25:529–536
8. Erices A, Conget P, Minguell JJ (2000) Mesenchymal progenitor cells in human umbilical cord blood. Br J Haematol 109:235–242
9. Gluckman E, Broxmeyer HA, Auerbach AD et al (1989) Hematopoietic reconstitution in a patient with Fanconi's anemia by means of umbilical-cord blood from an HLA-identical sibling. N Engl J Med 321:1174–1178
10. Goodwin HS, Bicknese AR, Chien SN et al (2001) Multilineage differentiation activity by cells isolated from umbilical cord blood: expression of bone, fat, and neural markers. Biol Blood Marrow Transplant 7:581–588
11. Greschat S, Schira J, Küry P et al (2008) Unrestricted somatic stem cells from human umbilical cord blood can be differentiated into neurons with a dopamineric phenotype. Stem Cells and Development (in press)
12. Huang GP, Pan ZJ, Jia BB et al (2007) Ex vivo expansion and transplantation of hematopoietic stem/progenitor cells supported by mesenchymal stem cells from human umbilical cord blood. Cell Transplant 16:579–585
13. Kern S, Eichler H, Stoeve J et al (2006) Comparative analysis of mesenchymal stem cells from bone marrow, umbilical cord blood, or adipose tissue. Stem Cells 24:1294–1301
14. Kogler G, Callejas J, Hakenberg P et al (1996) Hematopoietic transplant potential of unrelated cord blood: critical issues. J Hematother 5:105–116
15. Kogler G, Senßken S, Airey JA et al (2004) A new human somatic stem cell from placental cord blood with intrinsic pluripotent differentiation potential. J Exp Med 200:123–135
16. Kogler G, Radke TF, Lefort A et al (2005) Cytokine production and hematopoiesis supporting activity of cord blood-derived unrestricted somatic stem cells. Exp Hematol 33:573–583
17. Kogler G, Tutschek B, Koerschgen L et al (2005) Die José Carreras Stammzellbank Düsseldorf im NETCORD/EUROCORD-Verbund: Entwicklung, klinische Ergebnisse, Perspektiven und Aufklärung werdender Eltern. Gynäkologe 38:836–846
18. Kogler G, Senßken S, Wernet P (2006) Comparative generation and characterization of pluripotent unrestricted somatic stem cells with mesenchymal stem cells from human cord blood. Exp Hematol 34:1589–1595
19. Radke TF, Buchheiser A, Lefort A et al (2007) GMP-conform generation and cultivation of USSC from cord blood using the SEPAX-separation method and a closed culture system applying cell stacks. Blood 110(11):367
20. Rocha V, Gluckman E (2007) Outcomes of transplantation in children with acute leukaemia. Lancet 369:1906–1908
21. Senßken S, Waclawczyk S, Knaupp AS et al (2007) In vitro differentiation of human cord blood-derived unrestricted somatic stem cells towards an endodermal pathway. Cytotherapy 9:362–378
22. Sun W, Buzanska L, Domanska-Janik K et al (2005) Voltage-sensitive and ligand-gated channels in differentiating neural stem-like cells derived from the nonhematopoietic fraction of human umbilical cord blood. Stem Cells 23:931–945
23. Vaziri H, Dragowska W, Allsopp RC et al (1994) Evidence for a mitotic clock in human hematopoietic stem cells: loss of telomeric DNA with age. Proc Natl Acad Sci U S A 91:9857–9860
24. Wang TT, Tio M, Lee W et al (2007) Neural differentiation of mesenchymal-like stem cells from cord blood is mediated by PKA. Biochem Biophys Res Commun 357:1021–1027
25. Wernet P, Callejas J, Enczmann J et al (1999) Osteoblast precursor cells derived from human umbilcal cord blood. Exp Hematol 27(Suppl 1):117

Mesenchymal Stem Cells: New Insights Into Tissue Engineering and Regenerative Medicine

F. Djouad, R. S. Tuan

Contents

15.1 Phenotype of Mesenchymal Stem Cells: Role Within the Bone Marrow Niche 177
15.2 Differentiation Potential of MSCs 179
15.3 Cartilage Tissue Engineering 181
15.3.1 Cells 181
15.3.2 Scaffold 182
15.3.3 Factors 184
15.4 Immunomodulatory Properties of MSCs .. 185
15.4.1 Immunophenotype of MSCs and Allogeneic Recognition 185
15.4.2 Immunosuppressive Properties of MSCs In Vitro 186
15.4.3 Immunosuppressive Properties of MSCs In Vivo 187
15.5 Conclusions 188
References 188

15.1 Phenotype of Mesenchymal Stem Cells: Role Within the Bone Marrow Niche

Mesenchymal stem cells (MSCs), a resident population of stem cells in many adult tissues, were first identified in the bone marrow and have been isolated from marrow aspirates on the basis of the strong capacity of these cells to adhere to plastic culture substrata [1]. In vitro, these cells are readily characterized, among the adherent hematopoietic cells (HSCs), by their ability to generate single-cell-derived colonies or fibroblast colony-forming units (CFU-Fs) [2–4]. Unlike HSCs, to date, MSCs do not have a profile of specific cell-surface markers, i.e., none of the markers known to be expressed by MSCs has been shown to enable isolation of pure MSCs [5]. However, combination of epitopes, in particular surface proteins, represents a promising strategy for distinct characterization of MSCs [6]. Due to the low frequency of MSCs (0.0001–0.01%) in nucleated cells derived from adult bone marrow [7, 8] and the lack of knowledge on specific cell surface markers and MSC location and origin, our current knowledge of this population of cells is largely based on analysis of the properties of culture-expanded cells, not on the primary colony-initiating cells. Although a body of studies argues for the existence of differences in antigen expression between unmanipulated MSCs and in-vitro-expanded MSCs, MSCs are commonly positive for homing-associated cell adhesion molecule (HCAM/CD44), a hyaluronic acid (HA) receptor that mediates cell attachment to HA [9], very late antigen-α1 (VLA-α1/CD49a), src homology 3 and 4 domains (SH3, SH4/CD73), Thy-1 (CD90), src homology 2 domain (SH2/endoglin/CD105), and activated leukocyte cell adhesion molecule (ALCAM/CD166). Contrary to HSCs, MSCs are negative for hematopoietic progenitor cell antigen (HPCA/CD34), leukocyte common antigen (LCA/CD45), or monocyte differentiation antigen (CD14). The antigen-labeling profiles are generally homogeneous from passage two. Interestingly, the surface markers reported to be highly expressed on culture-expanded MSCs are not equivalently discriminative with regard to MSCs. For example, almost all CD45⁻CD14⁻/CD73⁺ and CD45⁻CD14⁻/CD49a⁺ phenotype-identified subsets have been reported to have the ability to generate CFU-Fs, although only some cells express CD105 or CD90 [10]. However, comparisons of the various combinations by different laboratories show that the majority of subsets include either integrin-β1 (CD29), CD105 or both [11]. These contradictory observations may be explained by the alteration of surface antigen expression in vitro. Indeed, the mean of fluorescence, determined by flow cytometry, increased after 1–3 days of cell-substrate adherence, which favors an increase of some antigen sites on the cells. Furthermore, some of these discrepancies may be due to variations of sample origin, culture techniques, and media used among the different laboratories. Another surface marker characteristic for MSCs is the low-affinity nerve growth factor receptor (LNGFR/CD271). The recognized CD271⁺ populations were fractionated by fluorescence-activated cell sorting, and the clonogenic capacity of the sorted cells was analyzed for their ability to give rise to CFU-Fs. The results showed that only the CD271(bright) contained CFU-F cells. In this study, the authors proposed a novel panel of surface markers, including CD140b, CD340, and CD349, within the CD271(bright) population [12].

One of the first antibodies shown to enrich for CFU-Fs in fresh aspirates of human BM is STRO-1 [13]. STRO-1 is non-reactive with hematopoietic progenitors and a selection of STRO-1⁺ cells from human bone marrow results in a ten- to 20-fold enrichment of CFU-F relative to their incidence in unfractionated bone marrow [13, 14].

The biological activities and fate of MSCs are controlled by both intrinsic and regulatory mechanisms. The microenvironment of stem cells, also called "niche", is critical in regulating the balance between quiescence, self-renewal and the commitment of stem cells. Despite the complexity of the adult stem cell niche, some molecules have been identified as important niche components. Cells that make up the stem cell niche produce factors, which inhibit stem cell differentiation and maintain stemness potential. Among these factors, cell adhesion molecules play important roles in fixing the cells in specific niche within tissues, thereby regulating cell movement and coordinating the communication of cells with each other and with the extracellular matrix (ECM). Integrins are a large family of cell surface receptors, which mediate cell-matrix and cell-cell adhesion. An integrin receptor consists of two non-covalently bound subunits: α and β. These versatile receptors are crucial in mediating functional dialogue between the exterior and the interior of the cells to transmit inside-out and outside-in signals [15]. This bi-directional signaling mediates many functions, such as cell motility, proliferation, differentiation and death. MSCs display a positive profile for several members of the integrin family such as VLA-4 (integrin α4/β1), and integrins α6/β1, α8/β1 and α9/β1 [16]. Integrins CD49e and CD29 in HSC have been reported to participate in stem cell retention in the niche by mediating the adhesion of cells to the ECM molecules [17]. Elevated levels of integrins are often characteristic for stem cells. Recently, marrow-derived MSCs have been reported to represent a surrogate model for the cellular microenvironment of the hematopoietic niche, and affinity to the MSCs is significantly higher among hematopoietic progenitor cell (HPC) subsets with self-renewal capacity [18]. Indeed, the highly expressed adhesion molecules, such as fibronectin-1, cadherin-11, vascular cell adhesion molecule (VCAM-1), connexin 43, as well as integrins are involved in the specific interaction of HPC and MSCs. The primitive subsets of HPC associated with higher self-renewal capacity possess a higher affinity to MSCs than more committed progenitors [18]. Endoglin, an auxiliary component of the transforming growth factor β (TGF-β) receptor system, is also expressed on HSCs and has been shown to decrease cell migration by increasing cell-cell adhesion [19]. Endoglin is likely to be a good candidate to mediate stem cell homing. Indeed, stem cell proliferation requires mobilization from the niche, and restoration of quiescence is accompanied by a return to the niche. Similar to endoglin, activated leukocyte cell adhesion molecule, also involved in MSC-niche interactions, is expressed by undifferentiated MSCs and disappears once the cells enter the osteogenic pathway [20–22]. N-Cadherin is a member of the classical cadherins [23] which, like non-classical cadherins, contains several extracellular Ca^{2+} binding domains that participate in Ca^{2+}-mediated cell-cell adhesion. Cell-cell adhesion mediated by cadherins is also essential for the biology of MSCs. Changes in

N-cadherin expression in MSCs are suggested to be associated with parallel changes in cell-cell adhesion and differentiation into adipocytes, chondrocytes and osteoblasts. Different cadherin repertoires are required to provide cues for cell specification and commitment to a specific lineage [24–26]. Indeed, functional N-cadherin is required in a temporally and quantitatively specific manner for normal mesenchymal condensation and chondrogenesis during embryonic development. Interestingly, both overexpression of normal N-cadherin or a defective form of N-cadherin result in alterations in chondrogenic gene regulation [27, 28]. N-Cadherin is required for mesenchymal specification and condensation in the early developmental stages of chondrogenesis. However, persistence of N-cadherin expression prevents further progression from precartilage condensation to chondrocyte differentiation [28]. Interestingly, while constant high N-cadherin expression is seen during osteogenesis, low levels are required for adipocyte differentiation [29].

15.2
Differentiation Potential of MSCs

A key characteristic that defines MSCs is their multi-lineage differentiation potential. The osteogenic potential of fibroblast-like cells isolated from bone marrow was first described by Friedenstein et al. [1]. Subsequently, a body of studies showed that the differentiation potential of these fibroblastic-like cells was not limited to the osteogenic lineage but could involve multiple cell types of mesenchymal lineage (reviewed in [30]). Currently, MSCs are operationally defined as adherent fibroblastic-like cells capable of forming osteoblasts, adipocytes and, chondrocytes in vitro and in vivo [31, 32]. Under certain conditions, MSCs exhibit the ability to differentiate into ectoderm- and endoderm-like cells, but their differentiation into non-mesoderm lineages is still controversial. MSCs injected into the central nervous systems of newborn mice migrate throughout the brain, where they acquire morphological and phenotypic characteristics of astrocytes and neurons [33]. However, several studies have shown that MSC-derived neurons express an unbalanced number of neuronal proteins and a lack of the functional characteristics of neurons [34, 35]. Therefore, definitive evidence for the neural differentiation potential of MSCs is still lacking. The epithelial differentiation potential of MSCs was shown in a body of both in vitro and in vivo studies. In an animal model of lung injury induced by bleomycin exposure, MSC engraftment was enhanced and a small percentage of MSCs localized to areas of lung injury [36]. MSCs engrafted in lung differentiated into type I pneumocytes or acquired phenotypic characteristics of the main cell types in lung, such as fibroblasts, type I and II epithelial cells, and myofibroblasts [37]. In vitro, co-culture with heat-shocked small airway epithelial cells induced human MSCs to differentiate into epithelial-like cells [38]. In contrast, studies showed that MSCs do not contribute significantly to the structural regeneration of epithelial cells in the post-ischemic kidney or injured cornea [39, 40].

In contrast, the ability of MSCs to differentiate into multiple mesoderm-type cells, including osteoblasts, chondrocytes, and adipocytes, is clearly evident and, to date, is a key parameter for MSC characterization. However, definitive parameters to modulate biological pathways and to direct and maintain a specific differentiated state of MSCs is crucial for tissue engineering and regeneration approaches, and are still actively being investigated. For clinical applications, differentiation cascades that effect MSC commitment have to be understood in detail, and protocols must be established to guarantee controlled induction and guidance of cell differentiation toward the desired phenotype. In order to achieve these goals and to establish valid, sequential criteria to assess differentiation, one approach is to compare the molecular events accompanying MSC differentiation induction to those that occur during mesenchymal differentiation in vivo. The characteristics of these events are summarized below.

Endochondral ossification, the developmental process that leads to the formation of the long bones of the skeleton, begins with a chain of events that include mesenchymal condensation, chondrogenic differentiation, chondrocyte hypertrophy, and, matrix mineralization [41]. Proliferating chondroprogenitor mesenchymal cells are characterized by expression of collagen type I (Col I), whereas condensing pre-cartilage cells downregulate Col I and begin to express collagen type II (Col II), which is the predominant collagen type produced by the differentiated chondrocytes. Mature chondrocytes express Col II, Col IX, Col XI, and cartilage proteoglycan core protein (aggrecan) and link protein. Synthesis of Col X and a parallel decrease in the expression levels of Col II are characteristic of hypertrophic maturation of chondrocytes. Because of its restricted expression in hypertrophic cartilage and its capacity to bind calcium [42], Col X has been suggested to play a role in the transformation of cartilage into bone [43].

Molecular analysis of differentiation stages adopted by MSCs in the course of chondrogenesis has yielded variable results, and the patterns of gene expression observed seem to depend on the specific induction conditions [44–46]. In terms of the signaling pathways involved, TGF-β and BMP signals are mediated by Smad transcription factors that bind to TGF-β type I receptors that are phosphorylated following ligand binding to TGF-β type II receptors [47, 48]. Smads 1, 5, and 8 associate with BMP receptor and Smads 2 and 3 associate with TGF-β receptor are released into the cytoplasm upon phosphorylation, complex with Smad 4, and translocate into the nucleus where they regulate gene expression. Because TGF-β receptor- and BMP receptor-associated Smads compete for Smad 4 and other downstream signaling molecules, these pathways may antagonize each other, such that when one pathway is activated, the other is suppressed [49]. Overall, while it is clear that BMP and TGF-β mediated Smad signaling events are critical for induction of chondrogenesis, these pathways are only a part of the multiple signaling events that contribute to the regulation of chondrogenic commitment [50–52].

Osteogenesis of MSCs is critically dependent on BMPs, factors present in the demineralized bone matrix capable of inducing bone formation [53]. Various members, such as BMP-2, -4, -6, -7, and -9, are positive inducer of osteoblast differentiation of MSCs. Acting via the Smad pathway, BMPs stimulate the expression of the Id proteins (inhibitor of DNA binding/differentiation helix-loop-helix proteins), Msx2, and Dlx5 (for review see [54]). With Runx2 and Osterix, Msx2 and Dlx5 are the major transcription factors involved in osteogenesis. Msx2 acts on early committed progenitors and promotes their proliferation, whereas Dlx5 promotes osteoblast differentiation by inducing the expression of Runx2 and Osterix [55, 56]. Runx2 induces the early commitment of MSCs to osteochondrogenic progenitor, while the terminal differentiation event is affected by the action of Osterix. Among the osteogenic BMPs, BMP-6 is the most potent inducer of hMSC differentiation in vitro. Only BMP-6 expression is upregulated by dexamethasone treatment, while addition of exogenous BMP-6 consistently yields an osteoblastic phenotype via regulation of expression of the osteoblast lineage master regulatory transcription factors. Similarly, when BMPs are used in combination, only combinations containing BMP-6 promote robust mineralization. At the transcriptional level, the regulatory control mechanisms governing BMP-6-induced hMSC osteoblast differentiation appear to be similar to those involved in early skeletal development [57].

Adipogenesis, the process of differentiation of mesenchymal cells to mature adipocytes, is regulated by two important transcription factors: CCAAT/enhancer binding proteins (C/EBPs) and PPARγ. This differentiation process involves two major stages. The early stage involves the recruitment and proliferation of pre-adipocytes. The late stage is the differentiation of pre-adipocytes into adipocytes, including lipid accumulation [58]. Interestingly, in murine MSCs, it was shown that hyperglycemia accelerates the adipogenic induction of lipid accumulation, through an ERK1/2-mediated PI3K/Akt pathway, that results in an increase of PPARγ expression [59]. A number of studies strongly suggest that osteogenic and adipogenic differentiation of MSCs are reciprocally regulated. For example, it was shown that the increased differentiation of MSCs towards the adipocytic lineage observed in the bone marrow of osteoporotic patients occurs in detriment of the differentiation towards the osteogenic lineage [60, 61]. In agreement with these reports, Rickard et al. [62] have shown that intermittent exposure of MSCs to either parathyroid hormone (PTH) as well as a nonpeptide small molecule agonist of the PTH1 receptor, inhibited the expression of adipocyte phenotypic markers and adipocyte formation, while simultaneously stimulating the osteoblast-associated marker alkaline phosphatase. In this study, the authors suggest that the prevention of osteoprogenitor differentiation to the adipocyte lineage would be expected to contribute to the overall stimulatory effect of daily treatment with PTH on trabecular bone formation, mass and strength.

15.3
Cartilage Tissue Engineering

Since articular cartilage is structurally a relatively homogeneous tissue, consisting of sparsely distributed chondrocytes surrounded by abundant ECM with absence of vascularity and innervation, cartilage repair has been actively targeted for tissue engineering-based approaches. However, an immediate consideration is that, in vivo, articular cartilage, the load-bearing tissue of the joint, possesses a specialized architecture with poor spontaneous regeneration potential. Currently, despite various therapeutic options for the treatment of chondral and osteochondral defects, no approach offers the formation of a cartilage de novo with the structure, biochemical composition, mechanical properties and capacity of self-maintenance similar to healthy native articular cartilage. Subchondral bone penetration procedures based on the recruitment of stem cells from the marrow have been widely used to treat localized cartilage damages. These techniques include subchondral drilling, microfracture, and spongialization and offer good short- to intermediate-term results since they generate a fibrocartilage tissue characterized by its poor biomechanical properties compared with normal cartilage [63–65]. More recently, transplantation of cultured autologous chondrocytes has been proposed to repair deep cartilage defects in the femorotibial articular surface of the knee joint [66, 67], but in some patients has resulted in a poorly differentiated fibrocartilage [68]. Therefore, tissue engineering strategies utilizing combination of cells, biodegradable scaffold, and bioactive molecules to recapitulate natural processes of tissue regeneration and development appear as an attractive new approach.

15.3.1
Cells

The ideal source of cells for cartilage tissue engineering is still being determined. Currently, chondrocytes and MSCs are the most explored cells for their potential to repair cartilage. Since endogenous chondrocytes mediate the balance between synthesis and degradation of matrix components, i.e., cartilage homeostasis, chondrocytes are the most obvious choice. However, an important challenge is the production of chondrocytes in sufficient numbers to form tissue constructs of an appropriate size to fill a clinically relevant defect. Differentiated chondrocytes are round cells responsible for the production of ECM molecules such as Col II and glycosaminoglycans. Due to the limited supply of chondrocytes, their expansion is necessary, but when carried out in monolayer culture conditions, the cells become fibroblastic and lose their phenotype and characteristic pattern of matrix protein expression [69–71]. To prevent chondrocyte dedifferentiation in monolayer cultures, a variety of growth factors such as fibroblast growth factor-2 (FGF-2) has been used [72]. Monolayer expansion of articular chondrocytes on specific substrates such as aggrecan and Col I, or ceramic material (with osteogenic potential) modulates the dedifferentiated cell phenotype and stimulates their differentiation into either the chondrogenic or osteogenic lineage respectively [73, 74]. Another concern with the use of chondrocytes for cartilage repair deals with the need to surgically harvest cartilage from patients for cell isolation. Donor site morbidity as well as the quality of the isolated chondrocytes, influenced by trauma, age or disease processes, are major concerns [75, 76]. Recent years have witnessed an increased interest in using MSCs as an attractive alternative to chondrocytes for cartilage tissue engineering. MSCs are easy to isolate, expandable in culture and can be induced to undergo chondrogenesis under a variety of culture conditions, which usually include treatment with growth factors from the TGF-β superfamily and a three-dimensional, high-density culture environment, such as pellet and micromass cultures [77, 78]. Translating the micromass or pellet high cell density culture conditions to tissue engineering applications, MSCs have been seeded into biomaterial scaffolds such as fibrin, alginate hydrogels, agarose, hyaluronan, photopolymerizing gels, polyester foams and nanofibrous meshes, as well as in gel-fiber amalgams [79–84]. For in-vitro culture conditions, the efficiency of chondrogenesis was shown to be scaffold-dependent. Indeed, while alginate and agarose have similar macroscopic properties and are sometimes used interchangeably, recent evidence suggests that cell-scaffold interactions result in different cell behavior in response to biochemical and mechanical stimulation [85]. For example, the chondrogenic differentiation of rat MSCs in three-dimensional hydrogel systems induced by TGF-β1 is cell passage dependent, and MSC expansion in the presence of FGF-2, an enhancer of mesenchymal differentiation of stromal cells at late passage numbers, promotes cartilaginous matrix accumulation within alginate over time, whereas FGF-2 pretreatment decreases matrix accumulation by MSCs in agarose [86]. Interestingly, in a comparative study, we showed that while chondrogenic differentiation does occur in MSC-laden agarose hydrogels, the amount of the forming matrix and measures of its mechanical properties are lower than those produced by chondrocytes from the same group of donors. Furthermore, the fact that sulfated glycosaminoglycan content and mechanical properties plateau in MSC-laden constructs suggests that the results are not simply due to a delay while differentiation occurs. Therefore, while MSC-laden constructs can develop cartilage-like mechanical properties, further optimization must be done to achieve levels similar to those produced by differentiated chondrocytes [87].

Human liposuction aspirates contain multipotent cells also called adipose-derived stem cells (ADSCs) with at least a tri-lineage potential to differentiate into bone, cartilage and fat [88–90]. The chondrogenic differentiation potential of ADSCs has been determined in high density micromass cultures [91], in alginate [92], agarose [82] and collagen-based scaffolds [82, 93]. ADSCs supported in a fibrin glue matrix and implanted into a full-thickness chondral defects developed the early functional characteristics of native hyaline cartilage [94]. Currently, adipose tissue appears to be an attractive source because of its abundance, ease of harvesting, and limited donor-site morbidity. However, a lower ECM production and a lower Col II gene expression over other cell types suggest that the chondrogenic potential of ADSCs has to be optimized [46, 95].

Neocartilage tissue formation from synovium-derived stem cells cultured in micromasses, in alginate and collagen gels has also been achieved [96, 97]. When transferred to alginate gel cultures, synovium-derived stem cells assumed a rounded form. In a dose-dependent manner, BMP-2 stimulated the expression of Sox9, Col II and aggrecan. Interestingly, under optimal conditions, the expression levels of cartilage-specific genes were comparable with those of cultured articular cartilage-derived chondrocytes [97].

15.3.2
Scaffold

Scaffolds provide a three-dimensional environment that is desirable for the production of cartilaginous matrix. A variety of scaffolds have been assessed for MSC adhesion, proliferation, migration and differentiation. Scaffolds provide mechanical support while MSCs multiply and eventually differentiate into functional tissue-specific cells. It is thus desirable for the scaffold (1) to have a direct and controlled degradation profile that is coordinated with the production of ECM by the differentiating MSCs, (2) to promote cell viability, differentiation and proliferation, (3) to permit the diffusion of nutrients and waste products, (4) to adhere and integrate with the surrounding native tissue, (5) to fill the defect, and (6) to provide mechanical integrity congruent with the defect site.

The residence time of the scaffolds in the defect determined by its degradation rate is a key element in the design of tissue engineering scaffolds and in the application of particular scaffolds to specific repair models [98]. For example, in a model of osteochondral defects, the quick degradation of cross-linked polysaccharide polymer and the poly(DL-lactic-co-glycolic acid) scaffolds leads to rapid bone formation, while the slow degradation of HYAFF-11 and poly(L-lactic acid) prolongs the presence of cartilaginous tissue and delays endochondral bone formation [99]. In scaffold applications, since appropriate cellularity is critical to ensure adequate cell-cell interactions, cell seeding density has to be carefully considered. Enhanced cell-cell contact obtained at high cell-seeding density tends to improve ECM production and deposition [81]. Cell distribution and infiltration into the scaffold is dictated by the seeding method used and varies among different type of scaffolds, such as hydrogels, sponges, and fibrous meshes. Dynamic seeding in sponge and mesh scaffolds supports cell distribution and infiltration, whereas uniform cell distribution is obtained in hydrogels. Natural polymers, e.g., alginate, agarose, fibrin, chitosan, collagen, chondroitin sulfate, gelatin, hyaluronic acid and cellulose, usually display inferior mechanical properties and are often more readily susceptible to a variety of enzymatic degradations within the host tissues, compared with synthetic polymers such as poly(α-hydroxy esters), poly(ethylene glycol/oxide), poly(NiPAAn), poly(propylene fumarates), and polyurethanes, which are characterized by controllable and predictable chemical and physical properties [97, 100–108]. Hydrated cross-linked biomaterials, that mimic the natural structure of ECM with high water content in vivo and efficient exchange of nutrients and gases, offer an attractive alternative to high density cell pellet or synthetic polymer scaffold [109, 110].

Cell interactions with biomaterials play an important role in regulating cell function and ECM accumulation within tissue engineered constructs. For both mature and differentiating cell types, agarose and alginate hydrogels have been shown to support the chondrocytic phenotype, characterized by round cell morphologies and expression of cartilage ECM molecules [111–114]. Hydrogels are also capable of transducing mechanical loads to exert controlled forces on encapsulated cells, similar to physiological conditions. However, these chondrogenic hydrogels present limitations for cartilage engineering including slow degradation rates.

Hyaluronan or hyaluronic acid (HA), an important molecular component of articular cartilage, contributing to joint hydration, cell-matrix interactions, and mechanical integrity [115], is another polymer that has been used for cartilage repair. By varying properties of this scaffold, e.g., molecular weight and cross-link density, the mechanical, diffusional, and degradation properties of the HA hydrogel can be tailored to achieve a formulation optimized for cartilage repair. For example, Liu et al. [116] have shown that a co-cross-linked synthetic matrix scaffold composed of chemically modified HA and gelatin can be used as a cell delivery vehicle for osteochondral defect repair in a rabbit model, resulting in better delivery and retention of MSCs to support osteochondral differentiation during tissue repair than scaffold or MSCs alone.

Fibrous biomaterial meshes used for cell delivery influence cell behavior as a function of fiber diameter, alignment, and porosity/volume. Our laboratory has developed an electrospinning method to produce three-dimensional nanofibrous scaffolds using α-hydroxy polyesters [117]. The nanofibers structurally mimic the nanoscale fibrous components of the native ECM, and present advantages including high surface area to volume ratios and fully interconnected pores, as well as the possibility of fabrication of aligned fibers [117]. The level of chondrogenesis observed in MSCs seeded within the nanofibrous scaffolds was comparable with that observed for MSCs maintained as high-density cell aggregates or pellets [118].

The most commonly used synthetic polymers are the α-hydroxy polyesters, including polylactic acid (PLA), polyglycolic acid (PGA), or copolymers [poly(lactide-co-glycolide), PLGA], often in the form of fibers and sponges [119, 120]. PGA is the most hydrophilic and degrades into a natural metabolite. PLA is more hydrophobic with a slower degradation rate [121]. Interestingly, the molecular and cellular characterization of chondrocytes seeded in PGA and PLGA constructs implanted in a heterotopic animal model revealed that the faster bioresorbable PGA scaffold showed longer term cartilage phenotype [122]. The authors suggested that the cartilage-like tissue produced in PGA in vivo might be applicable for cartilage tissue engineering applications. The copolymer, PLGA, is frequently used as a cell carrier, but has unfavorable hydrophobicity characteristics [123, 124]. The problem of cell leakage from the scaffold during cell seeding in vitro can be overcome using alginate gel to seed cells in the scaffold to form an amalgam [125, 126]. While the polymer provided appropriate mechanical support and stability to the composite culture, the alginate hydrogel allows cell penetration as well as favoring cell rounding to promote mesenchymal chondrogenesis. These studies showed that stem cells seeded in the amalgam can undergo induction into chondrogenic phenotype both in vitro [125] and in vivo after being transduced with a replication deficient adenovirus carrying hTGF-β2 (Ad5-hTGF-β2) [126]. Such an approach may be a viable method to treat injuries of hyaline cartilage [126].

Finally, most efforts to tailor these scaffolds have focused on the chemical and mechanical properties of the biomaterial itself. Recently, Choi et al. [127] reported a strategy to micro-control the distribution of soluble materials within the scaffold via microfluidic networks embedded directly within the cell-seeded biomaterial. The authors described a functional microfluidic structure within a three-dimensional scaffold of calcium alginate seeded with bovine chondrocytes. These microfluidic channels support efficient exchange of solutes within the bulk of the scaffold with spatial and temporal control of the soluble environment experienced by the cells in their three-dimensional environment. Such an approach is promising in directing cellular activities towards the formation of spatially differentiated tissues.

15.3.3
Factors

Given the limited supply of differentiated cells with specific functions, current tissue engineering strategies principally involve inducing differentiation of MSCs or minimizing de-differentiation of chondrocytes. Growth factors have emerged as important tools in attempts to accomplish these goals. Transforming growth factor-β (TGF-βs) superfamily members, bone morphogenetic proteins (BMPs), and growth differentiation factors (GDFs) as well as insulin-like growth factors (IGFs), Wnt proteins and fibroblasts growth factors (FGFs), have all been functionally implicated in mesenchymal chondrogenesis [128].

BMPs were originally identified as proteins able to stimulate ectopic endochondral bone formation and are the most widely utilized inductive factor in cartilage and bone tissue engineering applications [129, 130]. Of all BMPs (BMP-2, -4, -5, -6, or -7) transfected in chondrocytes reassembled in three-dimensionally in alginate beads, only BMP-2 and BMP-7 showed regulation of the cartilage ECM genes, aggrecan and Col II. BMP-7 was able to promote and maintain chondrocyte phenotype expressing aggrecan and Col II and revert collagen expression profile from Col I to Col II during long-term culture. Moreover, BMP-7 stabilizes the artificial cartilage construct in vivo by protecting it from nonspecific infiltration and destruction. In contrast, BMP-2-expressing cells revealed no alteration in collagen type gene expression compared with control-transfected chondrocytes still displaying a de-differentiated phenotype [131]. In addition, BMP-2, -5, and -6 were shown to maintain and promote later stages of chondrocyte differentiation rather than initiation of maturation [132, 133], while BMP-7 promoted chondrocyte proliferation and inhibited terminal differentiation [134]. Similarly, in combination with IGF-1, BMP-7 was shown to increase proteoglycan production in both normal and OA chondrocyte cultures [135]. With regard to cartilage tissue engineering, only a subset of BMPs or sequentially administered amounts of particular BMPs may contribute to the achievement of a differentiated chondrocyte phenotype. BMPs were delivered into areas of partial or full-thickness defects by different vehicles, including Col I sponge, chitosan, autologous chondrocytes stimulated with BMPs, chondrocytes cultured in agarose in the presence of growth factors, BMPs mixed with perichondrium, as well as BMPs seeded on a fibrin matrix or encapsulated in the liposomes. Both BMP-2 and BMP-7 have been shown to accelerate and improve tissue repair by filling the defect areas with hyaline-like cartilage [136-140]. For example, Sellers et al. [138] have shown that implantation of BMP-2 impregnated collagen sponge into full-thickness defects in articular cartilage improved the histological appearance of the repair cartilage compared to that of empty defects and defects filled with a collagen sponge alone. This improvement was still evident as long as one year after implantation. Implantation of constructs obtained using mesenchymal cells infected with adenoviral constructs encoding BMP-2 or IGF-1 and suspended in fibrin glue into mechanically induced partial-thickness cartilage lesions in the patellar groove of the rat femur results in successful repair of the experimentally generated lesions [141].

TGF-β1, TGF-β2, and TGF-β3 have all been shown to induce chondrogenic differentiation of MSCs in vitro [45]. Although TGF-β1 was the factor initially reported to induce chondrogenesis of MSCs, stimulating Col II and proteoglycan expression in MSCs, TGF-β2 and TGF-β3 have been found to be superior to TGF-β1 in promoting chondrogenesis of hMSCs [45, 142]. It was reported that TGF-β1 was unable to induce the expression of the chondrocytic markers in monolayer cultures but had a chondrogenic effect on the chondrocytes in three-dimensional culture systems [143–145]. Therefore, TGF-β1 effects are dependent on the cells acquiring a spherical morphology, similar to cells in the lacunae of cartilage during development. Na et al. [144] have shown that chondrocytes encapsulated in hydrogels containing dexamethasone and TGF-β1 for differentiation accumulated abundant ECM rich in proteoglycans and polysaccharides, suggesting that the chondrocyte cells were encased in the hydrogel and then formed hyaline cartilage. In combination with

IGF-1 and FGF, TGF-β1 promotes chondrogenesis. For example, Worster et al. [146] demonstrated that MSCs cultured with IGF-I were significantly more chondrogenic when pre-treated with TGF-β1. The authors pointed out the importance of sequential release of IGF-I and TGF-β1 from modular designed PLGA scaffolds. In vivo, MSCs transduced with TGF-β1 seeded into chitosan completely repair the full-thickness defect with the formation of hyaline-like cartilage tissue [147].

Interestingly, dynamic deformational loading of chondrocyte laden agarose constructs applied concurrently with TGF-β3 supplementation resulted in significantly lower mechanical properties of the constructs, compared with constructs undergoing the same loading protocol applied after the discontinuation of the growth factor [148]. This study suggests that the optimal strategy using well-characterized conditions for the functional tissue engineering of articular cartilage, particularly to accelerate construct development, may incorporate sequential application of different growth factors and applied deformational loading [148, 149].

15.4
Immunomodulatory Properties of MSCs

The multilineage differentiation and transdifferentiation abilities of MSCs [4, 150–154], combined with their extensive proliferative capacity in vitro, make these adult stem cells excellent candidates for application in repair medicine such as in cardiac repair, bone disorders and metabolic diseases. Recent findings accentuate the superiority of these stem cells, revealing that they can escape immune recognition and modulate the immune function of major cell populations involved in alloantigen recognition and elimination, including antigen presenting cells, T cells and natural killer cells. This immunophenotype of MSCs, the low expression of human leukocyte antigen (HLA) major histocompatibility complex (MHC) class I, and the absence of co-stimulatory molecules, together with the observation that MSCs do not elicit a proliferative response from allogeneic lymphocytes, might open attractive possibilities to use universal donor MSCs for different therapeutic applications [155–158]. Despite current efforts focused on the mechanisms responsible for this immunoprivileged status, this phenomenon remains an intrigue.

15.4.1
Immunophenotype of MSCs and Allogeneic Recognition

As stated earlier, MSCs can be isolated from an increasing number of adult tissues, such as bone marrow, adipose tissue, fetal liver, cord blood or synovial membrane, although bone marrow is most often used [159–161], and can be expanded in vitro as monolayers of plastic-adherent cells typically with a fibroblast-like morphology. Identified and isolated as fibroblast colony-forming unit-fibroblasts (CFU-Fs), MSCs almost homogeneously lack hematopoietic surface markers including CD34 or CD45 [3, 4]. Currently, no specific marker for MSCs is available; however, their characterization relies on the expression of a specific pattern of adhesion molecules such as Thy-1 (CD90), endoglin (CD105), vascular cell adhesion molecule-1 (VCAM-1/CD106), SH2 and SH3 (for review, see [162]). In addition, MSCs express human leukocyte antigen (HLA) major histocompatibility complex (MHC) class I, negligible levels of both MHC class II and Fas ligand. MSCs do not express B7-1, B7-2, CD40, or CD40L co-stimulatory molecules [163]. Although MSCs have MHC molecules on the surface, their unique immunological properties have been suggested by the fact that engraftment of human MSCs occurred after intra-uterine transplantation into sheep, even when the transplant was performed after the fetuses became immunocompetent [164]. This finding has been supported by an emerging group of studies suggesting that MSCs escape recognition of alloreactive cells, or at least possess a hypoimmunogenic character [163, 165, 166]. Indeed, Potian et al. [167] showed that human MSCs express both HLA class I and II, display a lack of immunogenic potential, and are able to inhibit immune response, indicating that HLA expression on MSCs is not the major reason for immune escape and immune suppression. Similarly, MSCs failed to elicit a proliferative response when co-cultured with allogeneic lymphocytes, despite provision of a co-stimulatory signal delivered by an anti-CD28 antibody and pretreatment with IFN-γ to induce cell surface HLA class II expression in MSCs [167–170].

15.4.2
Immunosuppressive Properties of MSCs In Vitro

The presence of adult murine, baboon, and human MSCs in a mixed lymphocyte reaction (MLR) suppresses T-cell proliferation [156, 157, 163, 165, 169]. MSCs inhibit the proliferative activity of T lymphocytes when triggered by mitogens, such as concanavalin, protein A (SpA) and phytohemagglutinin, allogeneic lymphocytes, or professional antigen-presenting cells in a dose-dependent manner [156]. Indeed, the immunosuppressive properties of MSCs are most potent when they are present in excess (MSC to lymphocyte ratio >1:10), and are decreased or even reversed at low MSC concentration [156, 157]. This suppression induced by MSCs occurs when the cells are autologous to the stimulatory or the responder lymphocytes, or are derived from a third-party donor [156, 157, 165]. Although the reduction of lymphocyte proliferation can still be observed even when MSCs are added as late as day 5 in the MLR, the inhibitory effect is most potent when the cells are added at the beginning of the 6-day lymphocyte culture. Currently, although there is an emerging body of studies demonstrating immunosuppressive properties of MSCs in vitro, the mechanisms underlying the inhibitory activity remain poorly understood and, in fact, there are a number of controversies. Several studies have shown that the inhibition elicited by MSCs is mediated by soluble factors [156, 158]. However, supernatants from MSC cultures do not display a suppressive effect unless the MSCs have been co-cultured with lymphocytes. In this case, the conditioned supernatant obtained from a 4-day co-culture can suppress the proliferation of T cells, suggesting that activation of MSCs by lymphocytes is required for expression of any immunosuppressive factors [156]. TGF-β and hepatocyte growth factor (HGF) represent potential candidates that have been most studied, but it is now clear that the immunosuppressive activity of MSCs is not associated with the secretion of these factors [158, 169]. On the other hand, IL-10, a well-characterized anti-inflammatory cytokine, secreted by lymphocyte-activated MSCs, has been shown to play a role in the inhibitory effect of MSCs. Indeed, blocking IL-10 signaling partially restored T-cell proliferation and pro-inflammatory cytokine release inhibited by MSCs [171]. Another mechanism, based on the activity of indoleamine 2,3-dioxygenase (IDO), an enzyme which acts to impair protein synthesis by depletion of the essential amino acid tryptophan, has also been postulated [172]. IFN-γ produced by T lymphocytes or NK cells may promote the immune suppressive activity of MSCs through its ability to stimulate their IDO activity, which in turn suppresses T- or NK-cell proliferation [172, 173]. In agreement with these studies, using MSCs from placental tissue, immunosuppression was reversed when IDO activity was blocked, suggesting that the anti-proliferative effect of MSCs is specifically dependent on IDO production [174]. Interestingly, the emerging data on the mechanisms contributing to immune privilege in the pregnant uterus show striking similarity to the immunosuppressive effects of MSCs, including the production of IDO [172]. Therefore, the potential involvement of IDO activity in the anti-proliferative properties mediated by MSCs associated with the presence of MSCs in the placenta and in umbilical cord strongly supports the hypothesis that MSCs are involved in fetal tolerance [175, 176]. However, Tse et al. [169] have shown that the immunosuppressive properties of MSCs were not affected in the presence of tryptophan or an IDO inhibitor. Since IDO has been reported to induce apoptosis of thymocytes and T_{H1} cells, these results excluding a role for IDO are consistent with other studies demonstrating that MSCs do not induce apoptosis in the suppressed cultures [158, 169, 177, 178]. On the other hand, Plumas et al. [179] have recently shown that MSC inhibit T-cell proliferation by inducing apoptosis of activated T cells through the conversion of tryptophan into kynurenine.

Another potential mechanism for the observed MSC-induced immunosuppression is that they function through suppressive T cells. Currently, three subsets of regulatory T cells (Tregs) have been characterized which negatively regulate immune responses:

(1) naturally occurring CD4$^+$CD25^{high}foxP3$^+$ Tregs that act in an antigen-non-specific manner; (2) the induced CD4$^+$CD25$^{+/-}$ Tregs that act in an antigen-specific manner; and (3) CD8$^+$CD28$^-$ T suppressor subset recently identified in humans [180–182]. A possible involvement of Tregs has been suggested by several studies. Short-term co-cultures of MSCs and autologous or allogeneic PBMCs generate both powerful regulatory CD4$^+$ or CD8$^+$ T cells [183]. In addition, Batten et al. [184] showed that a cytokine profile lacking in pro-inflammatory cytokines is induced when T cells are cultured with MSCs and results in the generation of CD4$^+$CD25^{lo}CD69^{lo}FoxP3$^+$ Tregs. Aggarwal and Pittenger [185] also demonstrated an increase in the proportion of CD4$^+$CD25$^+$ in IL-2 stimulated lymphocytes co-cultured with MSCs, while Beyth et al. [171] showed that the suppressive properties of MSCs were not affected when CD25$^+$ cells were removed from the CD4$^+$ subpopulation. These studies suggest that MSCs do not generate new regulatory cells from naive T cells, but could stimulate the proliferation of Tregs.

Other explanations include the inhibition of the dendritic cell (DC) differentiation. Several studies have shown that MSCs also impaired the generation of DCs from monocytes, and induced a tolerogenic phenotype associated with IL-10 production, responsible for a failure in the antigen-presenting cell compartment [171, 185]. Indeed, during DC maturation, MSCs inhibit upregulation of antigen-presenting cell (APC)-related molecules, such as CD80 (B7-1) and CD86 (B7-2), and HLA-DR [186, 187]. More recently, Ramasamy et al. [188] have shown that monocytes, although they do not require the initiation of DNA synthesis or replication, nevertheless have to enter the cell cycle to become functional DCs. MSCs arrest monocytes in the G$_0$ phase of cell cycle and make them unable to stimulate allogeneic T cells. However, these convergent data on MSC effects on DC differentiation are insufficient to elucidate the mechanisms responsible for the suppressive properties of MSCs, since MSCs inhibit T-cell proliferation by mechanisms that do not require APCs, using direct stimulation with CD3 and CD28 antibodies, enriched T-cell populations, or T-cell clones [157, 163, 169, 189].

Clonal expansion or clonal anergy of T cells is determined by the presence or absence of a co-stimulatory signal, such as the interaction of CD28 on T cells with B7 on APCs. As MSCs lack surface expression of the co-stimulatory molecules, B7-1, B7-2 and CD40, it was assumed that MSCs can render T cells anergic. However, this possibility is not consistent with the study showing that retroviral transduction of MSCs with B7 did not result in increased T-cell proliferation [170].

15.4.3
Immunosuppressive Properties of MSCs In Vivo

The immunomodulatory effect of MSCs was first assessed in vivo in a preclinical baboon model of skin graft acceptance. In that study, Bartholomew et al. [165] observed a significant prolongation of skin graft survival with a single dose of intravenously administered MSCs, comparable with that obtained with the potent immunosuppressive drugs currently used in clinic.

Using genetically engineered MSCs, Djouad et al. [156] have shown in vivo that MSCs from various origins failed to elicit allogeneic T-cell proliferation in the host. Indeed, when implanted into allogeneic recipient, BMP-2-expressing murine MSCs were not rejected, but instead maintained their potential to differentiate toward cartilage and bone. Another important finding was the capacity of MSCs to allow the proliferation of allogeneic tumor cells that would otherwise be rejected by immunocompetent recipients when injected alone. When injected subcutaneously, the MSCs were seen localized in the stroma surrounding the tumor, whereas systemically infused MSCs could not be detected [156]. This potential side effect has to be considered in further clinical trials, and assessed and compared with the administration of various immunosuppressive drugs currently used in therapeutic protocols, such as total bone marrow transplantation.

Severe acute graft-versus-host disease (GVHD) is one of the major complications after hematopoietic stem-cell transplantation (HSCT). Treatment of severe GVHD is difficult and the condition is often associated with a high mortality. Le Blanc et al. [190] showed promising results with the intravenous administration of haplo-identical MSCs in a patient with severe treatment resistant grade IV acute GVHD of the gut and the liver. Administration of MSCs successfully treated the 9-year-old leukemia patient, who rapidly recovered with normalized bilirubin, a normal colon, and with no minimal residual disease in the blood or bone marrow one year after transplantation. Currently, this study offers the most striking effect of the use of MSCs to combat acute GVHD. However, 1.5 years after the transplantation, an attempt was made to discontinue immunosuppression, and the GVHD recurred. Immunosuppression was again initiated, but in the meantime the patient developed repeated episodes of pneumonia and died 19 months after the transplantation. In this case, although a profound immunomodulatory effect was seen, it is clear that tolerance was not induced by MSCs and that immunosuppression was needed [191].

Currently, there is clear evidence that MSCs modulate immune reactions in vitro and escape from immune surveillance in vivo. However, the mechanisms involved to permit better control of the immunosuppressive properties of MSCs in vivo, for both physiological and pathological conditions, and to develop optimal use of MSCs in future applications, such as the treatment of organ transplant rejection and autoimmune inflammatory disorders and tissue regeneration, still remain to be understood.

15.5
Conclusions

Tissue engineering is a promising approach in regenerative medicine, and in the case of cartilage repair, the goal is to restore function to the articular joint that has undergone pathological degeneration. A central issue in all tissue engineering-based therapies is finding an appropriate cell source. Currently, for cartilage repair, MSCs appear to be the best alternative to chondrocytes, which are limited in number in the native tissue and exhibit phenotype instability in culture. MSCs are in principle a potent cell source for regenerative medicine because, in addition to their multipotentiality, they are also immunosuppressive. However, the recent controversy concerning MSC function(s) underscores the critical need for standardization and thus for elucidation of the identity and phenotypic characteristics of MSCs. The future of cartilage tissue engineering depends on our knowledge of fundamental stem cell biology, our understanding of the molecular events during the chondrogenic process, and the development of technologies, e.g., applications of biomaterials and growth factors, for the long-term maintenance of the articular cartilage phenotype.

Acknowledgment

Supported by the NIH NIAMS Intramural Research Program (ZO1 AR41131).

References

1. Friedenstein AJ, Chailakhyan RK, Gerasimov UV (1987) Bone marrow osteogenic stem cells: in vitro cultivation and transplantation in diffusion chambers. Cell Tissue Kinet 20:263–272
2. Castro-Malaspina H, Gay RE, Resnick G, Kapoor N, Meyers P, Chiarieri D, McKenzie S, Broxmeyer HE, Moore MA (1980) Characterization of human bone marrow fibroblast colony-forming cells (CFU-F) and their progeny. Blood 56:289–301
3. Bruder SP, Jaiswal N, Haynesworth SE (1997) Growth kinetics, self-renewal, and the osteogenic potential of purified human mesenchymal stem cells during extensive subcultivation and following cryopreservation. J Cell Biochem 64:278–294
4. Pittenger MF, Mackay AM, Beck SC, Jaiswal RK, Douglas R, Mosca JD, Moorman MA, Simonetti DW, Craig S, Marshak DR (1999) Multilineage potential of adult human mesenchymal stem cells. Science 284:143–147
5. Kolf CM, Cho E, Tuan RS (2007) Mesenchymal stromal cells. Biology of adult mesenchymal stem cells: regulation of niche, self-renewal and differentiation. Arthritis Res Ther 9:204
6. Kemp KC, Hows J, Donaldson C (2005) Bone marrow-derived mesenchymal stem cells. Leuk Lymphoma 46:1531–1544
7. Sakaguchi Y, Sekiya I, Yagishita K, Muneta T (2005) Comparison of human stem cells derived from various mesenchymal tissues: superiority of synovium as a cell source. Arthritis Rheum 52:2521–2529
8. Dazzi F, Ramasamy R, Glennie S, Jones SP, Roberts I (2006) The role of mesenchymal stem cells in haemopoiesis. Blood Rev 20:161–171
9. Zhu H, Mitsuhashi N, Klein A, Barsky LW, Weinberg K, Barr M L, Demetriou A, Wu GD (2006) The role of

the hyaluronan receptor CD44 in mesenchymal stem cell migration in the extracellular matrix. Stem Cells 24:928–935
10. Boiret N, Rapatel C, Veyrat-Masson R, Guillouard L, Guerin JJ, Pigeon P, Descamps S, Boisgard S, Berger MG (2005) Characterization of nonexpanded mesenchymal progenitor cells from normal adult human bone marrow. Exp Hematol 33:219–225
11. Jackson L, Jones DR, Scotting P, Sottile V (2007) Adult mesenchymal stem cells: differentiation potential and therapeutic applications. J Postgrad Med 53:121–127
12. Buhring HJ, Battula VL, Treml S, Schewe B, Kanz L, Vogel W (2007) Novel markers for the prospective isolation of human MSC. Ann N Y Acad Sci 1106:262–271
13. Simmons PJ, Torok-Storb B (1991) Identification of stromal cell precursors in human bone marrow by a novel monoclonal antibody, STRO-1. Blood 78:55–62
14. Dennis JE, Carbillet JP, Caplan AI, Charbord P (2002) The STRO-1+ marrow cell population is multipotential. Cells Tissues Organs 170:73–82
15. Docheva D, Popov C, Mutschler W, Schieker M (2007) Human mesenchymal stem cells in contact with their environment: surface characteristics and the integrin system. J Cell Mol Med 11:21–38
16. Ip JE, Wu Y, Huang J, Zhang L, Pratt RE, Dzau VJ (2007) Mesenchymal stem cells utilize integrin beta1 not CX chemokine receptor 4 for myocardial migration and engraftment. Mol Biol Cell 18:2873–2882
17. Whetton AD, Graham GJ (1999) Homing and mobilization in the stem cell niche. Trends Cell Biol 9:233–238
18. Wagner W, Wein F, Roderburg C, Saffrich R, Faber A, Krause U, Schubert M, Benes V, Eckstein V, Maul H, Ho AD (2007) Adhesion of hematopoietic progenitor cells to human mesenchymal stem cells as a model for cell-cell interaction. Exp Hematol 35:314–325
19. Liu Y, Jovanovic B, Pins M, Lee C, Bergan RC (2002) Over expression of endoglin in human prostate cancer suppresses cell detachment, migration and invasion. Oncogene 21:8272–8281
20. Arai F, Ohneda O, Miyamoto T, Zhang XQ, Suda T (2002) Mesenchymal stem cells in perichondrium express activated leukocyte cell adhesion molecule and participate in bone marrow formation. J Exp Med 195:1549–1563
21. Bruder SP, Ricalton NS, Boynton RE, Connolly TJ, Jaiswal N, Zaia J, Barry FP (1998) Mesenchymal stem cell surface antigen SB-10 corresponds to activated leukocyte cell adhesion molecule and is involved in osteogenic differentiation. J Bone Miner Res 13:655–663
22. Schieker M, Pautke C, Haasters F, Schieker J, Docheva D, Bocker W, Guelkan H, Neth P, Jochum M, Mutschler W (2007) Human mesenchymal stem cells at the single-cell level: simultaneous seven-colour immunofluorescence. J Anat 210:592–599
23. Takeichi M, Inuzuka H, Shimamura K, Fujimori T, Nagafuchi A (1990) Cadherin subclasses: differential expression and their roles in neural morphogenesis. Cold Spring Harb Symp Quant Biol 55:319–325
24. Gumbiner BM (1996) Cell adhesion: the molecular basis of tissue architecture and morphogenesis. Cell 84:345–357
25. Vleminckx K, Kemler R (1999) Cadherins and tissue formation: integrating adhesion and signaling. Bioessays 21:211–220
26. Stains JP, Civitelli R (2005) Cell-cell interactions in regulating osteogenesis and osteoblast function. Birth Defects Res C Embryo Today 75:72–80
27. Modaresi R, Lafond T, Roman-Blas JA, Danielson KG, Tuan RS, Seghatoleslami MR (2005) N-cadherin mediated distribution of beta-catenin alters MAP kinase and BMP-2 signaling on chondrogenesis-related gene expression. J Cell Biochem 95:53–63
28. DeLise AM, Tuan RS (2002) Alterations in the spatiotemporal expression pattern and function of N-cadherin inhibit cellular condensation and chondrogenesis of limb mesenchymal cells in vitro. J Cell Biochem 87:342–359
29. Marie PJ (2002) Role of N-cadherin in bone formation. J Cell Physiol 190:297–305
30. Phinney DG, Prockop DJ (2007) Concise review: mesenchymal stem/multipotent stromal cells: the state of transdifferentiation and modes of tissue repair current views. Stem Cells 25:2896–2902
31. Aslan H, Zilberman Y, Kandel L, Liebergall M, Oskouian RJ, Gazit D, Gazit Z (2006) Osteogenic differentiation of noncultured immunoisolated bone marrow-derived CD105+ cells. Stem Cells 24:1728–1737
32. Muraglia A, Cancedda R, Quarto R (2000) Clonal mesenchymal progenitors from human bone marrow differentiate in vitro according to a hierarchical model. J Cell Sci 113:1161–1166
33. Kopen GC, Prockop DJ, Phinney DG (1999) Marrow stromal cells migrate throughout forebrain and cerebellum, and they differentiate into astrocytes after injection into neonatal mouse brains. Proc Natl Acad Sci U S A 96:10711–10716
34. Wislet-Gendebien S, Hans G, Leprince P, Rigo JM, Moonen G, Rogister B (2005) Plasticity of cultured mesenchymal stem cells: switch from nestin-positive to excitable neuron-like phenotype. Stem Cells 23:392–402
35. Cho KJ, Trzaska KA, Greco SJ, McArdle J, Wang FS, Ye JH, Rameshwar P (2005) Neurons derived from human mesenchymal stem cells show synaptic transmission and can be induced to produce the neurotransmitter substance P by interleukin-1 alpha. Stem Cells 23:383–391
36. Ortiz LA, Gambelli F, McBride C, Gaupp D, Baddoo M, Kaminski N, Phinney DG (2003) Mesenchymal stem cell engraftment in lung is enhanced in response to bleomycin exposure and ameliorates its fibrotic effects. Proc Natl Acad Sci U S A 100:8407–8411
37. Kotton DN, Ma BY, Cardoso WV, Sanderson EA, Summer RS, Williams MC, Fine A (2001) Bone marrow-derived cells as progenitors of lung alveolar epithelium. Development 128:5181–5188
38. Spees JL, Olson SD, Ylostalo J, Lynch PJ, Smith J, Perry A, Peister A, Wang MY, Prockop DJ (2003) Differentiation, cell fusion, and nuclear fusion during ex vivo repair of epithelium by human adult stem cells from bone marrow stroma. Proc Natl Acad Sci U S A 100:2397–2402
39. Duffield JS, Park KM, Hsiao LL, Kelley VR, Scadden DT, Ichimura T, Bonventre JV (2005) Restoration of tubular epithelial cells during repair of the postischemic

kidney occurs independently of bone marrow-derived stem cells. J Clin Invest 115:1743–1755
40. Ma Y, Xu Y, Xiao Z, Yang W, Zhang C, Song E, Du Y, Li L (2006) Reconstruction of chemically burned rat corneal surface by bone marrow-derived human mesenchymal stem cells. Stem Cells 24:315–321
41. DeLise AM, Fischer L, Tuan RS (2000) Cellular interactions and signaling in cartilage development. Osteoarthritis Cartilage 8:309–334
42. Kirsch T, von der Mark K (1991) Ca2+ binding properties of type X collagen. FEBS Lett 294:149–152
43. Kirsch T, Ishikawa Y, Mwale F, Wuthier RE (1994) Roles of the nucleational core complex and collagens (types II and X) in calcification of growth plate cartilage matrix vesicles. J Biol Chem 269:20103–20109
44. Sekiya I, Vuoristo JT, Larson BL, Prockop DJ (2002) In vitro cartilage formation by human adult stem cells from bone marrow stroma defines the sequence of cellular and molecular events during chondrogenesis. Proc Natl Acad Sci U S A 99:4397–4402
45. Barry F, Boynton R E, Liu B, Murphy JM (2001) Chondrogenic differentiation of mesenchymal stem cells from bone marrow: differentiation-dependent gene expression of matrix components. Exp Cell Res 268:189–200
46. Winter A, Breit S, Parsch D, Benz K, Steck E, Hauner H, Weber RM, Ewerbeck V, Richter W (2003) Cartilage-like gene expression in differentiated human stem cell spheroids: a comparison of bone marrow-derived and adipose tissue-derived stromal cells. Arthritis Rheum 48:418–429
47. Zou H, Choe KM, Lu Y, Massague J, Niswander L (1997) BMP signaling and vertebrate limb development. Cold Spring Harb Symp Quant Biol 62:269–272
48. Mehra A, Wrana JL (2002) TGF-beta and the Smad signal transduction pathway. Biochem Cell Biol 80:605–622
49. Candia AF, Watabe T, Hawley SH, Onichtchouk D, Zhang Y, Derynck R, Niehrs C, Cho KW (1997) Cellular interpretation of multiple TGF-beta signals: intracellular antagonism between activin/BVg1 and BMP-2/4 signaling mediated by Smads. Development 124:4467–4480
50. Xu D, Gechtman Z, Hughes A, Collins A, Dodds R, Cui X, Jolliffe L, Higgins L, Murphy A, Farrell F (2006) Potential involvement of BMP receptor type IB activation in a synergistic effect of chondrogenic promotion between rhTGFbeta3 and rhGDF5 or rhBMP7 in human mesenchymal stem cells. Growth Factors 24:268–278
51. Zhang W, Ge W, Li C, You S, Liao L, Han Q, Deng W, Zhao RC (2004) Effects of mesenchymal stem cells on differentiation, maturation, and function of human monocyte-derived dendritic cells. Stem Cells Dev 13:263–271
52. Goldring MB, Tsuchimochi K, Ijiri K (2006) The control of chondrogenesis. J Cell Biochem 97:33–44
53. Musgrave DS, Pruchnic R, Bosch P, Ziran BH, Whalen J, Huard J (2002) Human skeletal muscle cells in ex vivo gene therapy to deliver bone morphogenetic protein-2. J Bone Joint Surg Br 84:120–127
54. Satija NK, Gurudutta GU, Sharma S, Afrin F, Gupta P, Verma YK, Singh VK, Tripathi RP (2007) Mesenchymal stem cells: molecular targets for tissue engineering. Stem Cells Dev 16:7–23
55. Ryoo HM, Lee MH, Kim YJ (2006) Critical molecular switches involved in BMP-2-induced osteogenic differentiation of mesenchymal cells. Gene 366:51–57
56. Hu G, Lee H, Price SM, Shen MM, Abate-Shen C (2001) Msx homeobox genes inhibit differentiation through up-regulation of cyclin D1. Development 128:2373–2384
57. Friedman MS, Long MW, Hankenson KD (2006) Osteogenic differentiation of human mesenchymal stem cells is regulated by bone morphogenetic protein-6. J Cell Biochem 98:538–554
58. Avram MM, Avram AS, James WD (2007) Subcutaneous fat in normal and diseased states 3. Adipogenesis: from stem cell to fat cell. J Am Acad Dermatol 56:472–942
59. Chuang CC, Yang RS, Tsai KS, Ho FM, Liu SH (2007) Hyperglycemia enhances adipogenic induction of lipid accumulation: involvement of extracellular signal-regulated protein kinase 1/2, phosphoinositide 3-kinase/Akt, and peroxisome proliferator-activated receptor gamma signaling. Endocrinology 148:4267–4275
60. Rodriguez JP, Montecinos L, Rios S, Reyes P, Martinez J (2000) Mesenchymal stem cells from osteoporotic patients produce a type I collagen-deficient extracellular matrix favoring adipogenic differentiation. J Cell Biochem 79:557–565
61. Nuttall ME, Gimble JM (2000) Is there a therapeutic opportunity to either prevent or treat osteopenic disorders by inhibiting marrow adipogenesis? Bone 27:177–184
62. Rickard DJ, Wang FL, Rodriguez-Rojas AM, Wu Z, Trice WJ, Hoffman SJ, Votta B, Stroup GB, Kumar S, Nuttall ME (2006) Intermittent treatment with parathyroid hormone (PTH) as well as a non-peptide small molecule agonist of the PTH1 receptor inhibits adipocyte differentiation in human bone marrow stromal cells. Bone 39:1361–1372
63. Knutsen G, Engebretsen L, Ludvigsen TC, Drogset JO, Grontvedt T, Solheim E, Strand T, Roberts S, Isaksen V, Johansen O (2004) Autologous chondrocyte implantation compared with microfracture in the knee. A randomized trial. J Bone Joint Surg Am 86-A:455–464
64. Blevins FT, Steadman JR, Rodrigo JJ, Silliman J (1998) Treatment of articular cartilage defects in athletes: an analysis of functional outcome and lesion appearance. Orthopedics 21:761–767; discussion 767–768
65. Buckwalter JA, Mankin HJ (1998) Articular cartilage: degeneration and osteoarthritis, repair, regeneration, and transplantation. Instr Course Lect 47:487–504
66. Knutsen G, Drogset JO, Engebretsen L, Grontvedt T, Isaksen V, Ludvigsen TC, Roberts S, Solheim E, Strand T, Johansen O (2007) A randomized trial comparing autologous chondrocyte implantation with microfracture. Findings at five years. J Bone Joint Surg Am 89:2105–2112
67. Brittberg M, Lindahl A, Nilsson A, Ohlsson C, Isaksson O, Peterson L (1994) Treatment of deep cartilage defects in the knee with autologous chondrocyte transplantation. N Engl J Med 331:889–895
68. Roberts S, Hollander AP, Caterson B, Menage J, Richardson JB (2001) Matrix turnover in human cartilage repair tissue in autologous chondrocyte implantation. Arthritis Rheum 44:2586–2598

69. Thirion S, Berenbaum F (2004) Culture and phenotyping of chondrocytes in primary culture. Methods Mol Med 100:1–14
70. Stokes DG, Liu G, Coimbra IB, Piera-Velazquez S, Crowl RM, Jimenez SA (2002) Assessment of the gene expression profile of differentiated and dedifferentiated human fetal chondrocytes by microarray analysis. Arthritis Rheum 46:404–419
71. Darling EM, Athanasiou KA (2005) Rapid phenotypic changes in passaged articular chondrocyte subpopulations. J Orthop Res 23:425–432
72. Martin I, Vunjak-Novakovic G, Yang J, Langer R, Freed LE (1999) Mammalian chondrocytes expanded in the presence of fibroblast growth factor 2 maintain the ability to differentiate and regenerate three-dimensional cartilaginous tissue. Exp Cell Res 253:681–688
73. Darling EM, Athanasiou KA (2005) Retaining zonal chondrocyte phenotype by means of novel growth environments. Tissue Eng 11:395–403
74. Barbero A, Grogan SP, Mainil-Varlet P, Martin I (2006) Expansion on specific substrates regulates the phenotype and differentiation capacity of human articular chondrocytes. J Cell Biochem 98:1140–1149
75. Carver SE, Heath CA (1999) Influence of intermittent pressure, fluid flow, and mixing on the regenerative properties of articular chondrocytes. Biotechnol Bioeng 65:274–281
76. Lee CR, Grodzinsky AJ, Hsu HP, Martin SD, Spector M (2000) Effects of harvest and selected cartilage repair procedures on the physical and biochemical properties of articular cartilage in the canine knee. J Orthop Res 18:790–799
77. Majumdar MK, Wang E, Morris EA (2001) BMP-2 and BMP-9 promotes chondrogenic differentiation of human multipotential mesenchymal cells and overcomes the inhibitory effect of IL-1. J Cell Physiol 189:275–284
78. Awad HA, Halvorsen YD, Gimble JM, Guilak F (2003) Effects of transforming growth factor beta1 and dexamethasone on the growth and chondrogenic differentiation of adipose-derived stromal cells. Tissue Eng 9:1301–1312
79. Ma HL, Hung SC, Lin SY, Chen YL, Lo WH (2003) Chondrogenesis of human mesenchymal stem cells encapsulated in alginate beads. J Biomed Mater Res A 64:273–281
80. Erickson GR, Gimble JM, Franklin DM, Rice HE, Awad H, Guilak F (2002) Chondrogenic potential of adipose tissue-derived stromal cells in vitro and in vivo. Biochem Biophys Res Commun 290:763–769
81. Huang CY, Reuben PM, D'Ippolito G, Schiller PC, Cheung HS (2004) Chondrogenesis of human bone marrow-derived mesenchymal stem cells in agarose culture. Anat Rec A Discov Mol Cell Evol Biol 278:428–436
82. Awad HA, Wickham MQ, Leddy HA, Gimble JM, Guilak F (2004) Chondrogenic differentiation of adipose-derived adult stem cells in agarose, alginate, and gelatin scaffolds. Biomaterials 25:3211–3222
83. Caterson EJ, Li WJ, Nesti LJ, Albert T, Danielson K, Tuan RS (2002) Polymer/alginate amalgam for cartilage-tissue engineering. Ann N Y Acad Sci 961:134–138
84. Li WJ, Laurencin CT, Caterson EJ, Tuan RS, Ko FK (2002) Electrospun nanofibrous structure: a novel scaffold for tissue engineering. J Biomed Mater Res 60:613–621
85. Xu XL, Lou J, Tang T, Ng KW, Zhang J, Yu C, Dai K (2005) Evaluation of different scaffolds for BMP-2 genetic orthopedic tissue engineering. J Biomed Mater Res B Appl Biomater 75:289–303
86. Coleman RM, Case ND, Guldberg RE (2007) Hydrogel effects on bone marrow stromal cell response to chondrogenic growth factors. Biomaterials 28:2077–2086
87. Mauck RL, Yuan X, Tuan RS (2006) Chondrogenic differentiation and functional maturation of bovine mesenchymal stem cells in long-term agarose culture. Osteoarthritis Cartilage 14:179–189
88. Zuk PA, Zhu M, Ashjian P, De Ugarte DA, Huang JI, Mizuno H, Alfonso ZC, Fraser JK, Benhaim P, Hedrick MH (2002) Human adipose tissue is a source of multipotent stem cells. Mol Biol Cell 13:4279–4295
89. Zuk PA, Zhu M, Mizuno H, Huang J, Futrell JW, Katz AJ, Benhaim P, Lorenz HP, Hedrick MH (2001) Multilineage cells from human adipose tissue: implications for cell-based therapies. Tissue Eng 7:211–228
90. Dragoo JL, Samimi B, Zhu M, Hame SL, Thomas BJ, Lieberman JR, Hedrick MH, Benhaim P (2003) Tissue-engineered cartilage and bone using stem cells from human infrapatellar fat pads. J Bone Joint Surg Br 85:740–747
91. Chiou M, Xu Y, Longaker MT (2006) Mitogenic and chondrogenic effects of fibroblast growth factor-2 in adipose-derived mesenchymal cells. Biochem Biophys Res Commun 343:644–652
92. Lin Y, Luo E, Chen X, Liu L, Qiao J, Yan Z, Li Z, Tang W, Zheng X, Tian W (2005) Molecular and cellular characterization during chondrogenic differentiation of adipose tissue-derived stromal cells in vitro and cartilage formation in vivo. J Cell Mol Med 9:929–939
93. Masuoka K, Asazuma T, Hattori H, Yoshihara Y, Sato M, Matsumura K, Matsui T, Takase B, Nemoto K, Ishihara M (2006) Tissue engineering of articular cartilage with autologous cultured adipose tissue-derived stromal cells using atelocollagen honeycomb-shaped scaffold with a membrane sealing in rabbits. J Biomed Mater Res B Appl Biomater 79:25–34
94. Dragoo JL, Carlson G, McCormick F, Khan-Farooqi H, Zhu M, Zuk PA, Benhaim P (2007) Healing full-thickness cartilage defects using adipose-derived stem cells. Tissue Eng 13:1615–1621
95. Huang JI, Kazmi N, Durbhakula MM, Hering TM, Yoo JU, Johnstone B (2005) Chondrogenic potential of progenitor cells derived from human bone marrow and adipose tissue: a patient-matched comparison. J Orthop Res 23:1383–1389
96. Yokoyama A, Sekiya I, Miyazaki K, Ichinose S, Hata Y, Muneta T (2005) In vitro cartilage formation of composites of synovium-derived mesenchymal stem cells with collagen gel. Cell Tissue Res 322:289–298
97. Park Y, Sugimoto M, Watrin A, Chiquet M, Hunziker E. B (2005) BMP-2 induces the expression of chondrocyte-specific genes in bovine synovium-derived progenitor

cells cultured in three-dimensional alginate hydrogel. Osteoarthritis Cartilage 13:527–536
98. Solchaga LA, Yoo JU, Lundberg M, Dennis JE, Huibregtse BA, Goldberg VM, Caplan AI (2000) Hyaluronan-based polymers in the treatment of osteochondral defects. J Orthop Res 18:773–780
99. Solchaga LA, Temenoff JS, Gao J, Mikos AG, Caplan AI, Goldberg VM (2005) Repair of osteochondral defects with hyaluronan- and polyester-based scaffolds. Osteoarthritis Cartilage 13:297–309
100. Na K, Kim S, Woo DG, Sun BK, Yang HN, Chung HM, Park KH (2007) Synergistic effect of TGFbeta-3 on chondrogenic differentiation of rabbit chondrocytes in thermo-reversible hydrogel constructs blended with hyaluronic acid by in vivo test. J Biotechnol 128:412–22
101. Williams CG, Kim TK, Taboas A, Malik A, Manson P, Elisseeff J (2003) In vitro chondrogenesis of bone marrow-derived mesenchymal stem cells in a photopolymerizing hydrogel. Tissue Eng 9:679–88
102. Li WJ, Danielson KG, Alexander PG, Tuan RS (2003) Biological response of chondrocytes cultured in three-dimensional nanofibrous poly(epsilon-caprolactone) scaffolds. J Biomed Mater Res A 67:1105–1114
103. Aigner J, Tegeler J, Hutzler P, Campoccia D, Pavesio A, Hammer C, Kastenbauer E, Naumann A (1998) Cartilage tissue engineering with novel nonwoven structured biomaterial based on hyaluronic acid benzyl ester. J Biomed Mater Res 42:172–181
104. Hoshikawa A, Nakayama Y, Matsuda T, Oda H, Nakamura K, Mabuchi K (2006) Encapsulation of chondrocytes in photopolymerizable styrenated gelatin for cartilage tissue engineering. Tissue Eng 12:2333–2341
105. Kuo YC, Lin CY (2006) Effect of genipin-crosslinked chitin-chitosan scaffolds with hydroxyapatite modifications on the cultivation of bovine knee chondrocytes. Biotechnol Bioeng 95:132–144
106. De Franceschi L, Grigolo B, Roseti L, Facchini A, Fini M, Giavaresi G, Tschon M, Giardino R (2005) Transplantation of chondrocytes seeded on collagen-based scaffold in cartilage defects in rabbits. J Biomed Mater Res A 75:612–622
107. Vinatier C, Magne D, Moreau A, Gauthier O, Malard O, Vignes-Colombeix C, Daculsi G, Weiss P, Guicheux J (2007) Engineering cartilage with human nasal chondrocytes and a silanized hydroxypropyl methylcellulose hydrogel. J Biomed Mater Res A 80:66–74
108. Buschmann MD, Gluzband YA, Grodzinsky AJ, Kimura JH, Hunziker EB (1992) Chondrocytes in agarose culture synthesize a mechanically functional extracellular matrix. J Orthop Res 10:745–758
109. Balakrishnan B, Jayakrishnan A (2005) Self-cross-linking biopolymers as injectable in situ forming biodegradable scaffolds. Biomaterials 26:3941–3951
110. Shin H, Quinten Ruhe P, Mikos AG, Jansen JA (2003) In vivo bone and soft tissue response to injectable, biodegradable oligo(poly(ethylene glycol) fumarate) hydrogels. Biomaterials 24:3201–3211
111. Galois L, Hutasse S, Cortial D, Rousseau CF, Grossin L, Ronziere MC, Herbage D, Freyria AM (2006) Bovine chondrocyte behaviour in three-dimensional type I collagen gel in terms of gel contraction, proliferation and gene expression. Biomaterials 27:79–90
112. Giannoni P, Siegrist M, Hunziker EB, Wong M (2003) The mechanosensitivity of cartilage oligomeric matrix protein (COMP). Biorheology 40:101–109
113. Wong M, Siegrist M, Gaschen V, Park Y, Graber W, Studer D (2002) Collagen fibrillogenesis by chondrocytes in alginate. Tissue Eng 8 979–87.
114. Hung CT, Lima EG, Mauck RL, Takai E, LeRoux MA, Lu HH, Stark RG, Guo XE, Ateshian GA (2003) Anatomically shaped osteochondral constructs for articular cartilage repair. J Biomech 36:1853–1864
115. Knudson CB, Knudson W (2001) Cartilage proteoglycans. Semin Cell Dev Biol 12:69–78
116. Liu Y, Shu XZ, Prestwich GD (2006) Osteochondral defect repair with autologous bone marrow-derived mesenchymal stem cells in an injectable, in situ, cross-linked synthetic extracellular matrix. Tissue Eng 12:3405–3416
117. Li WJ, Cooper JA Jr, Mauck RL, Tuan RS (2006) Fabrication and characterization of six electrospun poly(alpha-hydroxy ester)-based fibrous scaffolds for tissue engineering applications. Acta Biomater 2:377–385
118. Li WJ, Tuli R, Okafor C, Derfoul A, Danielson KG, Hall DJ, Tuan RS (2005) A three-dimensional nanofibrous scaffold for cartilage tissue engineering using human mesenchymal stem cells. Biomaterials 26:599–609
119. Mooney DJ, Baldwin DF, Suh NP, Vacanti JP, Langer R (1996) Novel approach to fabricate porous sponges of poly(D,L-lactic-co-glycolic acid) without the use of organic solvents. Biomaterials 17:1417–1422
120. Freed LE, Marquis JC, Nohria A, Emmanual J, Mikos AG, Langer R (1993) Neocartilage formation in vitro and in vivo using cells cultured on synthetic biodegradable polymers. J Biomed Mater Res 27:11–23
121. Chung C, Burdick JA (2008) Engineering cartilage tissue. Adv Drug Deliv Rev 60:243–262
122. Zwingmann J, Mehlhorn, AT, Sudkamp N, Stark B, Dauner M, Schmal H (2007) Chondrogenic differentiation of human articular chondrocytes differs in biodegradable PGA/PLA scaffolds. Tissue Eng 13:2335–2343
123. Sittinger M, Reitzel D, Dauner M, Hierlemann H, Hammer C, Kastenbauer E, Planck H, Burmester GR, Bujia J (1996) Resorbable polyesters in cartilage engineering: affinity and biocompatibility of polymer fiber structures to chondrocytes. J Biomed Mater Res 33:57–63
124. Cao Y, Vacanti JP, Paige KT, Upton J, Vacanti CA (1997) Transplantation of chondrocytes utilizing a polymer-cell construct to produce tissue-engineered cartilage in the shape of a human ear. Plast Reconstr Surg 100:297–302; discussion 303–304
125. Caterson EJ, Nesti LJ, Li WJ, Danielson KG, Albert TJ, Vaccaro AR, Tuan RS (2001) Three-dimensional cartilage formation by bone marrow-derived cells seeded in polylactide/alginate amalgam. J Biomed Mater Res 57:394–403
126. Jin XB, Sun YS, Zhang K, Wang J, Shi TP, Ju XD, Lou SQ (2007) Tissue engineered cartilage from hTGF beta2 transduced human adipose derived stem cells seeded in PLGA/alginate compound in vitro and in vivo. J Biomed Mater Res A (in press)

127. Choi NW, Cabodi M, Held B, Gleghorn JP, Bonassar LJ, Stroock AD (2007) Microfluidic scaffolds for tissue engineering. Nat Mater 6 908–15.
128. Heng BC, Cao T, Lee EH (2004) Directing stem cell differentiation into the chondrogenic lineage in vitro. Stem Cells 22:1152–1167
129. Sampath TK, Reddi AH (1981) Dissociative extraction and reconstitution of extracellular matrix components involved in local bone differentiation. Proc Natl Acad Sci U S A 78:7599–7603
130. Urist MR, Mikulski A, Lietze A (1979) Solubilized and insolubilized bone morphogenetic protein. Proc Natl Acad Sci U S A 76:1828–1832
131. Kaps C, Bramlage C, Smolian H, Haisch A, Ungethum U, Burmester GR, Sittinger M, Gross G, Haupl T (2002) Bone morphogenetic proteins promote cartilage differentiation and protect engineered artificial cartilage from fibroblast invasion and destruction. Arthritis Rheum 46:149–162
132. Bailon-Plaza A, Lee AO, Veson EC, Farnum CE, van der Meulen MC (1999) BMP-5 deficiency alters chondrocytic activity in the mouse proximal tibial growth plate. Bone 24:211–216
133. Kameda T, Koike C, Saitoh K, Kuroiwa A, Iba H (2000) Analysis of cartilage maturation using micromass cultures of primary chondrocytes. Dev Growth Differ 42:229–236
134. Haaijman A, Burger EH, Goei SW, Nelles L, ten Dijke P, Huylebroeck D, Bronckers AL (2000) Correlation between ALK-6 (BMPR-IB) distribution and responsiveness to osteogenic protein-1 (BMP-7) in embryonic mouse bone rudiments. Growth Factors 17:177–192
135. Loeser RF, Chubinskaya S, Pacione C, Im HJ (2005) Basic fibroblast growth factor inhibits the anabolic activity of insulin-like growth factor 1 and osteogenic protein 1 in adult human articular chondrocytes. Arthritis Rheum 52:3910–3917
136. Frenkel SR, Saadeh PB, Mehrara BJ, Chin GS, Steinbrech DS, Brent B, Gittes GK, Longaker MT (2000) Transforming growth factor beta superfamily members: role in cartilage modeling. Plast Reconstr Surg 105:980–990
137. Sellers RS, Peluso D, Morris EA (1997) The effect of recombinant human bone morphogenetic protein-2 (rh-BMP-2) on the healing of full-thickness defects of articular cartilage. J Bone Joint Surg Am 79:1452–1463
138. Sellers RS, Zhang R, Glasson SS, Kim HD, Peluso D, D'Augusta DA, Beckwith K, Morris EA (2000) Repair of articular cartilage defects one year after treatment with recombinant human bone morphogenetic protein-2 (rh-BMP-2). J Bone Joint Surg Am 82:151–160
139. Grgic M, Jelic M, Basic V, Basic N, Pecina M, Vukicevic S (1997) Regeneration of articular cartilage defects in rabbits by osteogenic protein-1 (bone morphogenetic protein-7). Acta Med Croatica 51:23–27
140. Cook SD, Patron LP, Salkeld SL, Rueger DC (2003) Repair of articular cartilage defects with osteogenic protein-1 (BMP-7) in dogs. J Bone Joint Surg Am 85-A Suppl 3:116–123
141. Gelse K, von der Mark K, Aigner T, Park J, Schneider H (2003) Articular cartilage repair by gene therapy using growth factor-producing mesenchymal cells. Arthritis Rheum 48:430–441
142. Yamaguchi A (1995) Regulation of differentiation pathway of skeletal mesenchymal cells in cell lines by transforming growth factor-beta superfamily. Semin Cell Biol 6:165–173
143. Galera P, Redini F, Vivien D, Bonaventure J, Penfornis H, Loyau G, Pujol JP (1992) Effect of transforming growth factor-beta 1 (TGF-beta 1) on matrix synthesis by monolayer cultures of rabbit articular chondrocytes during the dedifferentiation process. Exp Cell Res 200:379–392
144. Na K, Kim S, Sun BK, Woo DG, Yang HN, Chung HM, Park KH (2008) Bioimaging of dexamethasone and TGF beta-1 and its biological activities of chondrogenic differentiation in hydrogel constructs. J Biomed Mater Res A (in press)
145. Yaeger PC, Masi TL, de Ortiz JL, Binette F, Tubo R, McPherson JM (1997) Synergistic action of transforming growth factor-beta and insulin-like growth factor-I induces expression of type II collagen and aggrecan genes in adult human articular chondrocytes. Exp Cell Res 237:318–325
146. Worster AA, Brower-Toland BD, Fortier LA, Bent SJ, Williams J, Nixon AJ (2001) Chondrocytic differentiation of mesenchymal stem cells sequentially exposed to transforming growth factor-beta1 in monolayer and insulin-like growth factor-I in a three-dimensional matrix. J Orthop Res 19:738–749
147. Guo CA, Liu XG, Huo JZ, Jiang C, Wen XJ, Chen ZR (2007) Novel gene-modified-tissue engineering of cartilage using stable transforming growth factor-beta1-transfected mesenchymal stem cells grown on chitosan scaffolds. J Biosci Bioeng 103:547–556
148. Lima EG, Bian L, Ng KW, Mauck RL, Byers BA, Tuan RS, Ateshian GA, Hung CT (2007) The beneficial effect of delayed compressive loading on tissue-engineered cartilage constructs cultured with TGF-beta3. Osteoarthritis Cartilage 15:1025–1033
149. Benjamin A, Byers RLM, Chiang IE, Tuan RS (2008) Transient exposure to TGF-β3 under serum-free conditions enhances the biomechanical and biochemical maturation of tissue-engineered cartilage. Tissue Engineering (in press)
150. Bianco P, Riminucci M, Kuznetsov S, Robey PG (1999) Multipotential cells in the bone marrow stroma: regulation in the context of organ physiology. Crit Rev Eukaryot Gene Expr 9:159–173
151. Bianco P, Gehron Robey P (2000) Marrow stromal stem cells. J Clin Invest 105:1663–1668
152. Friedenstein AJ, Latzinik NW, Grosheva AG, Gorskaya UF (1982) Marrow microenvironment transfer by heterotopic transplantation of freshly isolated and cultured cells in porous sponges. Exp Hematol 10:217–227
153. Song L, Webb NE, Song Y, Tuan RS (2006) Identification and functional analysis of candidate genes regulating mesenchymal stem cell self-renewal and multipotency. Stem Cells 24:1707–1718
154. Song L, Tuan RS (2004) Transdifferentiation potential of human mesenchymal stem cells derived from bone marrow. FASEB J 18:980–982

155. Glennie S, Soeiro I, Dyson PJ, Lam EW, Dazzi F (2005) Bone marrow mesenchymal stem cells induce division arrest anergy of activated T cells. Blood 105:2821–2827
156. Djouad F, Plence P, Bony C, Tropel P, Apparailly F, Sany J, Noel D, Jorgensen C (2003) Immunosuppressive effect of mesenchymal stem cells favors tumor growth in allogeneic animals. Blood 102:3837–3844
157. Krampera M, Glennie S, Dyson J, Scott D, Laylor R, Simpson E, Dazzi F (2003) Bone marrow mesenchymal stem cells inhibit the response of naive and memory antigen-specific T cells to their cognate peptide. Blood 101:3722–3729
158. Di Nicola M, Carlo-Stella C, Magni M, Milanesi M, Longoni PD, Matteucci P, Grisanti S, Gianni AM (2002) Human bone marrow stromal cells suppress T-lymphocyte proliferation induced by cellular or nonspecific mitogenic stimuli. Blood 99:3838–3843
159. Djouad F, Bony C, Haupl T, Uze G, Lahlou N, Louis-Plence P, Apparailly F, Canovas F, Reme T, Sany J, Jorgensen C, Noel D (2005) Transcriptional profiles discriminate bone marrow-derived and synovium-derived mesenchymal stem cells. Arthritis Res Ther 7:R1304–R1315
160. Jones EA, Kinsey SE, English A, Jones RA, Straszynski L, Meredith DM, Markham AF, Jack A, Emery P, McGonagle D (2002) Isolation and characterization of bone marrow multipotential mesenchymal progenitor cells. Arthritis Rheum 46:3349–3360
161. Nakahara H, Goldberg VM, Caplan AI (1991) Culture-expanded human periosteal-derived cells exhibit osteochondral potential in vivo. J Orthop Res 9 465–76.
162. Noel D, Djouad F, Jorgense C (2002) Regenerative medicine through mesenchymal stem cells for bone and cartilage repair. Curr Opin Investig Drugs 3:1000–1004
163. Le Blanc K, Tammik C, Rosendahl K, Zetterberg E, Ringden O (2003) HLA expression and immunologic properties of differentiated and undifferentiated mesenchymal stem cells. Exp Hematol 31:890–896
164. Liechty KW, MacKenzie TC, Shaaban AF, Radu A, Moseley AM, Deans R, Marshak DR, Flake AW (2000) Human mesenchymal stem cells engraft and demonstrate site-specific differentiation after in utero transplantation in sheep. Nat Med 6:1282–1286
165. Bartholomew A, Sturgeon C, Siatskas M, Ferrer K, McIntosh K, Patil S, Hardy W, Devine S, Ucker D, Deans R, Moseley A, Hoffman R (2002) Mesenchymal stem cells suppress lymphocyte proliferation in vitro and prolong skin graft survival in vivo. Exp Hematol 30:42–48
166. Maitra B, Szekely E, Gjini K, Laughlin MJ, Dennis J, Haynesworth SE, Koc ON (2004) Human mesenchymal stem cells support unrelated donor hematopoietic stem cells and suppress T-cell activation. Bone Marrow Transplant 33:597–604
167. Potian JA, Aviv H, Ponzio NM, Harrison JS, Rameshwar P (2003) Veto-like activity of mesenchymal stem cells: functional discrimination between cellular responses to alloantigens and recall antigens. J Immunol 171:3426–3434
168. Gotherstrom C, Ringden O, Westgren M, Tammik C, Le Blanc K (2003) Immunomodulatory effects of human foetal liver-derived mesenchymal stem cells. Bone Marrow Transplant 32:265–272
169. Tse WT, Pendleton JD, Beyer WM, Egalka MC, Guinan EC (2003) Suppression of allogeneic T-cell proliferation by human marrow stromal cells: implications in transplantation. Transplantation 75:389–397
170. Klyushnenkova E, Mosca JD, Zernetkina V, Majumdar MK, Beggs KJ, Simonetti DW, Deans RJ, McIntosh KR (2005) T cell responses to allogeneic human mesenchymal stem cells: immunogenicity, tolerance, and suppression. J Biomed Sci 12:47–57
171. Beyth S, Borovsky Z, Mevorach D, Liebergall M, Gazit Z, Aslan H, Galun E, Rachmilewitz J (2005) Human mesenchymal stem cells alter antigen-presenting cell maturation and induce T-cell unresponsiveness. Blood 105:2214–2219
172. Meisel R, Zibert A, Laryea M, Gobel U, Daubener W, Dilloo D (2004) Human bone marrow stromal cells inhibit allogeneic T-cell responses by indoleamine 2,3-dioxygenase-mediated tryptophan degradation. Blood 103:4619–4621
173. Krampera M, Cosmi L, Angeli R, Pasini A, Liotta F, Andreini A, Santarlasci V, Mazzinghi B, Pizzolo G, Vinante F, Romagnani P, Maggi E, Romagnani S, Annunziato F (2006) Role for interferon-gamma in the immunomodulatory activity of human bone marrow mesenchymal stem cells. Stem Cells 24:386-398
174. Jones BJ, Brooke G, Atkinson K, McTaggart SJ (2007) Immunosuppression by placental indoleamine 2,3-dioxygenase: a role for mesenchymal stem cells. Placenta 28:1174–1181
175. Baksh D, Yao R, Tuan RS (2007) Comparison of proliferative and multilineage differentiation potential of human mesenchymal stem cells derived from umbilical cord and bone marrow. Stem Cells 25:1384–1392
176. Nauta AJ, Fibbe WE (2007) Immunomodulatory properties of mesenchymal stromal cells. Blood 110:3499–3506
177. Fallarino F, Grohmann U, Vacca C, Orabona C, Spreca A, Fioretti MC, Puccetti P (2003) T cell apoptosis by kynurenines. Adv Exp Med Biol 527:183–190
178. Zappia E, Casazza S, Pedemonte E, Benvenuto F, Bonanni I, Gerdoni E, Giunti D, Ceravolo A, Cazzanti F, Frassoni F, Mancardi G, Uccelli A (2005) Mesenchymal stem cells ameliorate experimental autoimmune encephalomyelitis inducing T-cell anergy. Blood 106:1755–1761
179. Plumas J, Chaperot L, Richard MJ, Molens JP, Bensa JC, Favrot MC (2005) Mesenchymal stem cells induce apoptosis of activated T cells. Leukemia 19:1597–1604
180. Stephens LA, Barclay AN, Mason D (2004) Phenotypic characterization of regulatory CD4+CD25+ T cells in rats. Int Immunol 16:365–375
181. Filaci G, Fravega M, Negrini S, Procopio F, Fenoglio D, Rizzi M, Brenci S, Contini P, Olive D, Ghio M, Setti M, Accolla RS, Puppo F, Indiveri F (2004) Nonantigen specific CD8+ T suppressor lymphocytes originate from CD8+CD28– T cells and inhibit both T-cell proliferation and CTL function. Hum Immunol 65:142–156
182. Filaci G, Fravega M, Fenoglio D, Rizzi M, Negrini S, Viggiani R, Indiveri F (2004) Non-antigen specific CD8+ T suppressor lymphocytes. Clin Exp Med 4:86–92

183. Prevosto C, Zancolli M, Canevali P, Zocchi MR, Poggi A (2007) Generation of CD4+ or CD8+ regulatory T cells upon mesenchymal stem cell-lymphocyte interaction. Haematologica 92:881–888
184. Batten P, Sarathchandra P, Antoniw JW, Tay SS, Lowdell MW, Taylor PM, Yacoub MH (2006) Human mesenchymal stem cells induce T cell anergy and downregulate T cell allo-responses via the TH2 pathway: relevance to tissue engineering human heart valves. Tissue Eng 12:2263–2273
185. Aggarwal S, Pittenger MF (2005) Human mesenchymal stem cells modulate allogeneic immune cell responses. Blood 105:1815–1822
186. Djouad F, Charbonnier LM, Bouffi C, Louis-Plence P, Bony C, Apparailly F, Cantos C, Jorgensen C, Noel D (2007) Mesenchymal stem cells inhibit the differentiation of dendritic cells through an interleukin-6-dependent mechanism. Stem Cells 25:2025–2032
187. Maccario R, Podesta M, Moretta A, Cometa A, Comoli P, Montagna D, Daudt L, Ibatici A, Piaggio G, Pozzi S, Frassoni F, Locatelli F (2005) Interaction of human mesenchymal stem cells with cells involved in alloantigen-specific immune response favors the differentiation of CD4+ T-cell subsets expressing a regulatory/suppressive phenotype. Haematologica 90:516–525
188. Ramasamy R, Fazekasova H, Lam EW, Soeiro I, Lombardi G, Dazzi F (2007) Mesenchymal stem cells inhibit dendritic cell differentiation and function by preventing entry into the cell cycle. Transplantation 83:71–76
189. Le Blanc K, Tammik L, Sundberg B, Haynesworth SE, Ringden O (2003) Mesenchymal stem cells inhibit and stimulate mixed lymphocyte cultures and mitogenic responses independently of the major histocompatibility complex. Scand J Immunol 57:11–20
190. Le Blanc K, Rasmusson I, Sundberg B, Gotherstrom C, Hassan M, Uzunel M, Ringden O (2004) Treatment of severe acute graft-versus-host disease with third party haploidentical mesenchymal stem cells. Lancet 363:1439–1441
191. Le Blanc K, Ringden O (2005) Immunobiology of human mesenchymal stem cells and future use in hematopoietic stem cell transplantation. Biol Blood Marrow Transplant 11:321-334

Stem Cell Plasticity: Validation Versus Valedictory

N. D. Theise

Contents

16.1	Introduction	197
16.2	Cell Plasticity	198
16.3	Principles of Plasticity	198
16.3.1	Principle 1: Genomic Completeness	200
16.3.2	Principle 2: Cellular Uncertainty	201
16.3.3	Principle 3: Stochasticity of Cell Lineages	201
16.4	Issues of Experimental Design and Discourse	202
16.4.1	Nature and Severity of Injury	202
16.4.2	Methods of Detection	203
16.5	The Single-Cell Test	203
16.5.1	Some Cases for Plasticity	204
16.5.2	The Case Against...	204
16.6	Summary	206
	References	207

16.1 Introduction

The field of stem cell plasticity, by which most commentators and investigators mean plasticity of adult stem cells, was opened to great fanfare with the publication of three papers in *Science* in the last 2 years of the prior millennium. This trio, taken together, was trumpeted by *Science* as the "Breakthrough of the Year" for 1999 and showed that stem cells deriving from one organ were not only *not* restricted to producing cells of that organ (so, e.g., marrow-derived cells could become skeletal muscle) but were also not restricted to their embryonic germ layer of origin (neural stem cells produced blood and marrow-derived stem cells produced liver) [1–3].

These discoveries would simply have been exciting, with abundant scientific or therapeutic potential, but their entanglement in the extrascientific issues of the related field of embryonic stem cell plasticity led to extrascientific pressures and assessments which have often distorted the scientific findings themselves [4]. Thus, in the United States and Europe, where primarily some religious orthodoxies opposed the use of embryonic stem cells on religion-based, nonscientific grounds, the scientific establishment found it necessary to counter the "adult, so not embryonic" position of these polemicists with an overly simplified "not adult, so embryonic" response. This reply then formalized an investment on the part of scientists and scientific entities such as journals and granting bodies to downplay the positive findings regarding adult cell plasticity, as these were perceived as harming the cause of embryonic stem cell research. The result is widespread confusion for scientists, clinicians, and lay people alike.

In this review, the current consensus regarding adult stem cell plasticity will be reviewed along with scientific aspects of experimental design and how these may bias experimental results and interpretations. Hopefully, with this introduction to and summary of the field, the reader will understand some of the limitations of past experiments, some criteria with which to evaluate past and future reports, and join me in a degree of expectant optimism regarding the possibilities for future efforts. While many in the field are wearied by the political aspects of the fight to demonstrate the reality of adult stem cell plasticity, it is clear that the field continues apace, heading for vindication, rather than a valedictory that will mark its end.

16.2
Cell Plasticity

While the phrase most often encountered is "stem cell plasticity," this specificity for stem cells is actually not quite to the point. Cell plasticity in general is the theme, which may or may not involve a beginning with, or passage through, a functional stem cell phase of differentiation. Four basic pathways of cell plasticity have now been demonstrated [5, 6]. The first of these is the standard paradigm of both embryonic and fetal development and postnatal/adult tissue maintenance and reconstitution in response to injury. This pathway begins with a stem cell which undergoes asymmetric cell division that results in self-renewal as well as production of rapidly proliferative daughter cells giving rise to at least one, but usually several, differentiated cell types (Fig. 16.1a). Local microenvironmental effects including states of injury may influence which maturational lineages are followed in specific tissue locations. This pathway is commonly thought of as hierarchical and unidirectional reflecting progressive repression of unrelated genes, resulting in increasing lineage restriction until a "terminally differentiated" cell is produced.

The second pathway of cell plasticity is recognized primarily in nonmammalian vertebrates and in mammalian neoplasia. In this pathway, more fully differentiated cells "de-differentiate" toward a stem cell-like state and then re-differentiate along a different developmental lineage (Fig. 16.1b). The formation of a blastema from mature cells following amputation of a limb in amphibians is an example of this process [7]. Such de-differentiation is a commonly observed occurrence in clinical specimens from human malignancies [8]. It also seems to be maintained in mammalian fetuses: severing of a limb of a developing limb leads to regeneration through an amphibian-like process mediated in part by homeobox protein MSX1 [9].

That this pathway is at least potentially present in adult mammals, rather than having been lost during their evolutionary development, is demonstrated by recent findings from Ellen Heber-Katz (Wistar Institute) in the MRL mouse in which there is healing without scar and regeneration of organs (heart, brain, liver, and dermis/epidermis) similar to that seen in amphibians [10, 11]. At wound sites, a blastema is produced from differentiated cell populations that then re-differentiated into the cells of the regenerated limb or organ. Some of the mechanisms involved in this process include increased metalloproteinase production by circulating mononuclear cells that home to the site, thereby preventing both reorganization of subepithelial basement membrane and creation of dense scar.

The third and fourth pathways are those that are currently most typically referred to by the phrase "stem cell plasticity" and represent nuclear reprogramming either by microenvironmental influences to directly change a cell's differentiation state or by cytoplasmic influences when a cell of one type fuses with a cell of another type (Fig. 16.1c, d). While fusion was first brought forth as a critique of the earliest plasticity reports—i.e., that engraftment from transplanted cells in diverse organs arose not by relocation into the new site and then direct differentiation of the engrafted cell, but instead by fusion with a preexistent differentiated cell—we now know that both mechanisms occur. Which one predominates likely depends on the nature and severity of injury [12–16]. So, while fusion was first promoted as a critique of the previously published plasticity data, it turned out to be an important plasticity pathway in its own right.

16.3
Principles of Plasticity

In the first years of this research, my collaborator Diane Krause (Yale University) and I contemplated

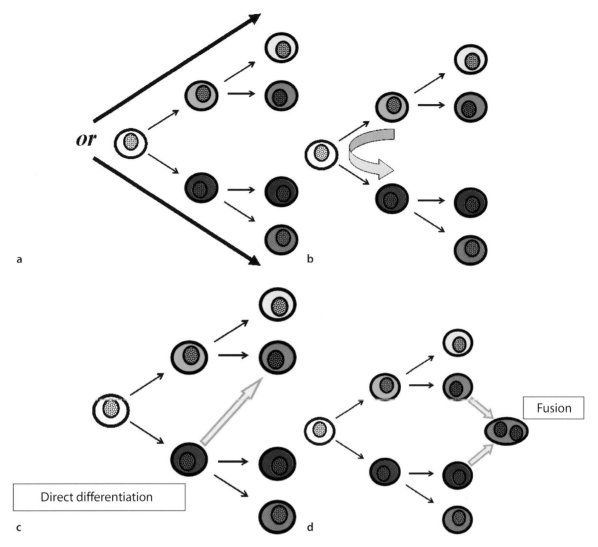

Fig. 16.1a–d Pathways of cell plasticity. **a** Cell lineages derive from "in tissue" stem/progenitor cells (in embryologic and fetal development and in maintenance and repair of adult tissues) which can shift to emphasize one lineage or another in response to stage of development or injury. **b** De-differentiation, in which differentiated cells of one lineage can revert to a blastema-like state and then re-differentiate along different lineages. This pathway is prominent in amphibian limb regeneration, but is present in mammals during fetal life and in neoplasia, at least. **c** Direct differentiation, in which cells transplanted directly into tissue or arriving through the circulation can engraft in a new site and directly take on site-appropriate differentiation, in response to microenvironmental cues. **d** Fusion, in which cells transplanted directly into tissue or arriving through the circulation fuse with pre-existing cells. The resulting cell can be binucleated or the nuclei can then fuse leading to a single, but tetraploid nucleus. This fusion event can be stable or the fused cell can undergo reduction division into two diploid cells

the reported findings of our own group and of others and formulated a "new paradigm" for cell plasticity comprising three principles (Table 16.1) [8, 17, 18]. These formulations were an attempt to organize the findings into a conceptual whole, offering guidelines for experimental design, evaluation of data, and thereby clarifying the nature, we hoped, of the long-standing dogmas that were being overturned.

Table 16.1 Principles of cell plasticity

1. Genomic Completeness	Any cell with the entire genome intact (i.e. without deletions, translocations, duplications, mutations) can potentially become any other cell type.
2. Cellular Uncertainty	Any attempt to analyze a cell necessarily alters the nature of the cell at the time of isolation, thereby altering outcomes of subsequent differentiation events.
3. Stochasticity of cell lineages	Descriptions of progenitors and progeny of any given cell must be expressed stochastically.

16.3.1
Principle 1: Genomic Completeness

This first principle states that any cell that contains the entire genome intact—without deletions, mutations, translocations, or multiplications—can potentially become any other cell type of the organism from which it has been isolated. This perhaps extreme statement of plasticity potential of all cells recognizes two concepts. The first is that the original experiments upon which the dogma of restricted differentiation was founded were not designed to demonstrate the full range of differentiative potential. These historic experiments involved the transfer of cells from one part of a developing embryo or fetus to another location. The well-known result is that up to a certain point in development, the transplanted cell took on the differentiation characteristics of the new location, but after a certain point such transfer did not result in reprogramming of the cell.

So the dogma developed that earlier in development there is a larger range of differentiation potential that is closed down systematically as development proceeds. Thus, the standard model of cell lineage development was one of increasing levels of irreversible gene restriction until a terminally differentiated cell was achieved. The first plasticity pathway described above was hence enshrined as the only possible normative cell lineage scheme.

The postulated molecular explanation for this, as noted, was progressive and irreversible gene repression. For several decades such mechanisms were indeed identified, generally grouped as epigenetic modifications and including methylation of the genome, particularly at CpG islands which would then, most often (though not always) block binding of transcription factors and prevent gene expression; methylation and deacetylation of histone proteins promoting the formation of heterochromatin and thereby making genes unavailable for transcription; and X inactivation, the random deletion of one X chromosome in all the cells of female organisms to prevent overexpression of the genes located on the X chromosome. With these mechanisms described, it seemed clear that irreversible gene restriction was indeed present, explaining the unidirectional, restricted nature of cell lineages.

However, in the same recent period of time in which the "new" plasticity was being identified in newly designed transplantation experiments, new data regarding mechanisms of gene repression began to accumulate, gradually demonstrating that gene restrictions were in fact not irreversible; rather they could be reversed, but circumstances had to trigger the necessary pathways which would allow that process. The original experiments in transplanting embryonic and fetal cells at different points in development were simply not providing the right signals to make such molecular reversals take place.

When we first published this concept of "genomic completeness" we had one kind of experiment which could unequivocally serve as proof of principle: the cloning of Dolly and, subsequently, many other animals [19]. Clearly, there were cytoplasmic factors in the ovum receiving the transplanted adult nucleus that could undo most if not all of the gene restrictions that had been laid down in that "terminally" differentiated donor cell. Even so, it was still something of a surprise when it was reported that even X inactivation in cloned animals had been reversed by the cloning process: all the cells of the cloned female did not have the same X inactivation as the original donor cell, but instead, the inactivated X chromosome had been reactivated, and random X inactivation was restored through development [20]. The mechanisms of gene de-repression are beyond the scope of this essay, but have been summarized elsewhere [21, 22]. Every few months, however, new such mechanisms are being identified, which are often tissue- and cell type-specific, meaning that these are coordinated and physiologically important processes.

The most recent entrants into this approach to exploring and developing the plasticity of adult cells have come through the synergy of parallel, not competing lines of embryonic and adult stem cell research. With the identification of the genes that underlie the pluripotentiality of embryonic stem cells, such as *nanog, oct4,* and *sox2,* several laboratories have recently published the effects of transfection of adult cells with these and other genes (e.g., c-*myc, klf4*) [23–25]. With these genes up-regulated, the adult cells, both murine and human, developed embryonic-like stem cell potential. Moreover, recent research with smaller molecules such as reversine suggests that exposure of cells to factors not requiring virally mediated gene transfers might be capable of similar transformation [26]. Thus, while the somewhat crude cloning process works, we discover ever-more-subtle mechanisms of inducing and manipulating the underlying molecular events. As we learn to exploit them we must admit that such plasticity is not the unique domain of cells from early in development.

cell meet. Formulating this principle, we recognized that while cell biologists often referred to separate steps of cell isolation and then cell conditioning, in fact cell isolation itself was at least potentially a conditioning step, in that isolation necessarily involves altering the microenvironment.

We included the word "potentially" in this principle, however, because we were not sure (indeed, we doubted) that it was an inherent aspect of cell behavior, believing that one day it might be possible, say with a perfected magnetic resonance imaging machine, to instantaneously identify all molecular and atomic components of a cell and thereby know everything about its state and how it will subsequently behave. This was a similar caveat to one issued by Heisenberg when he first pronounced the uncertainty of elementary particles: namely that this difficulty was perhaps merely a technological limitation. Ultimately, we now know that it is not based in our limited technology but is accepted as an inherent aspect of all material in the known universe. Dr. Krause and I were not (yet) making such claims. This topic, too, is beyond the range of this review, though it may have implications for how plasticity research is evaluated; detailed discussion suggesting that uncertainty is indeed "inherent" is available elsewhere [21, 22].

16.3.2
Principle 2: Cellular Uncertainty

This principle, echoing Heisenberg's uncertainty principle, states that "any attempt to observe a cell alters the state of that cell at the time of characterization and *potentially* alters the likelihood of subsequent differentiation events" [8, 17, 18, 27]. This principle reflects a dogma that still seems reasonable to hold, though cell biologists often somewhat blithely ignore it for reasons we will discuss below. As author Richard Lewontin has stated, "the internal and the external co-determine the cell" [28]. That is, the properties of a cell not only arise from its gene expression profile, but also from influences in the microenvironment. The nature of the cell arises dynamically where the inside and the outside of the

16.3.3
Principle 3: Stochasticity of Cell Lineages

The logical extension of the first two principles is that once a cell is in hand to be studied, one can never be absolutely certain about the cell lineage that produced that cell or about where it might go, in terms of differentiation, once it is manipulated. Instead, one must consider that cells are inherently stochastic and that the origin and fate of cells must be expressed stochastically [8, 17, 18]. While determinism versus stochasticity has been a long-standing argument in cell biology, the bulk of the evidence supports the less conservative, stochastic impression [29, 30].

Not only is stochasticity implied by the first two principles above, but can be demonstrated ex vivo in the varying progeny deriving from single, isolated hematopoietic progenitors for example, each plated cell producing a different assortment of hematopoietic progeny. On the molecular level, stochasticity of gene expression is also quite clearly demonstrated, so that, again, this characteristic is not due to limitations in our technology but in the behavior of cells.

An important example is the demonstration that euchromatin is highly mobile within the nucleoplasm, while the heterochromatin is quite fixed [31]. Fluorescent tagging of these two regions of the genome and observation in real time by fluorescence microscopy makes this clear. Moreover, the euchromatin's motion in the nucleoplasm, sometimes facing out of the chromatin domain, where it can freely interact with transcription factors in the nucleoplasm, and sometimes facing inward, where exposure to transcription factors is minimized, can best be described mathematically as a "random walk." Thus, gene expression itself has a stochastic element based, at least, on the free movement of the expressed portions of the genome.

16.4
Issues of Experimental Design and Discourse

It should be no surprise to any scientist reading this essay that experimental design affects the outcome of the experiment. This concept is oddly and dismayingly ignored by many of the investigators and commentators who critique adult stem cell research [32]. To some extent the ability to simultaneously pretend to scientific rigor while ignoring the implications of design choice as a conditioning factor stems from an increasingly outdated belief in cells as unitary objects with inherent properties independent of microenvironment. Thus, repeatedly, critics will quote negative results derived from one kind of injury to comment on the reliability of data from a different kind of injury, as though the body was not far more subtle in its properties than are the political influences on editorializing about stem cell research. We will consider examples of these kinds of comparative experiments and the deficiencies in the use of their data. Many of these examples will involve marrow to liver plasticity, as these are closest to my own area of expertise.

16.4.1
Nature and Severity of Injury

To explore how the nature of an injury can condition outcomes we may consider the derivation of intrahepatic stem/progenitor cells ("oval cells" in rodents, "ductular reactions" in humans) from bone marrow, a possibility with diverse findings in the papers addressing the topic. The first report by Petersen et al. of marrow to liver plasticity showed that some oval cells were marrow derived [3]. We confirmed that finding in humans, in the first marrow to liver plasticity paper [33]. Papers showing positive and negative results followed, with those of the Grompe and Dabeva laboratories coming down squarely on the side of the absence of such an effect [34, 35]. These negative-results papers have then been taken by most commentators as being the final word, even though the experimental models employed were not precise replications of the experimental methods of the earlier papers.

The recent paper of Oh et al., from Bryon Petersen's laboratory, clarifies the entire question by far more carefully exploring variations in methods [36]. Methena et al. had used three models for examining oval cell derivation from marrow cells [34]. One of these used the application of a typical and quite efficient oval cell generating protocol: the hepatotoxin retrorsine followed by partial hepatectomy. For the purposes of this experiment, experimental animals were first transplanted with bone marrow and oval cells generated after marrow engraftment was established. Petersen, recognizing that pyrrolizidine alkaloids such as retrorsine and monocrotaline affect rapidly proliferating cells, questioned whether the toxin itself might also inhibit the newly transplanted marrow, thus suppressing the marrow to liver pathway. Indeed, his intuition was correct: by reversing the order of injuries, first administering the hepatotoxin to injure the liver and then, during a time period in which the toxin has been eliminated, giving the marrow transplant, the partial hepatectomy produces equivalent numbers of oval cells—but now 20% of them are marrow derived. Thus, a widely understood "definitive" negative-results paper is seen to have conditioned the negative outcome.

The physiology is far more subtle than the discourse. What subtleties might be at play? We now know that some secreted factors are important for homing of marrow cells to sites of injury, including stromal derived factor-1 (SDF-1), matrix metalloproteinases 2 and 9, stem cell factor (SCF), and hepatocyte growth factor [37–40]. We have recently explored expression of SDF-1 and SCF in three models commonly used in marrow to liver plasticity research, the FAH-null mouse model of hereditary tyrosinemia type I, biliary injury due to DDC administration, and radiation injury at levels comparable to those used to achieve marrow transplant [41]. Unsurprisingly, the production of both serum factors is significantly different between all three models. Thus, to assume that plasticity results from one model provide a clear commentary on the findings of a completely different model is inappropriate if one is interested in scientific rigor.

Not only is the nature of the injury important, but its severity may play an important role. Despite this seemingly obvious concept, few publications have explored it. Herzog et al. did so with regard to radiation pneumonitis and found that extent of injury, assessed by measurements of cellular debris and protein in bronchoalveolar lavage fluid and apoptosis of pneumocytes in tissue sections, had a profound impact on marrow derivation of pneumocytes [42]. At radiation levels of 400 cGy and 600 cGy, despite greater than 80% and 90% marrow engraftment of donor cells, respectively, and significant lung injury, there was no marrow-to-lung engraftment. At 1,000 cGy, administered either as one dose or in split doses, there was a small increase of donor marrow engraftment to approximately 95%, but marked increases in lung injury by the above-described measurements. Only at these higher levels of pulmonary injury does pneumocyte engraftment appear.

16.4.2
Methods of Detection

Herzog et al. are also, in the same paper, among the first to very clearly delineate the ease with which engraftment can be overestimated without important controls [42]. By including CD45 staining in addition to Y-chromosome staining (the marker of cells of donor origin) and TTF-1 (marker of pneumocyte differentiation in the lung), some, though not all, seemingly $Y^+/TTF-1^+$ cells were found to actually be overlaps of circulating $CD45^+$ hematopoietic cells and pneumocytes even in 3-μm-thin tissue sections. In our original investigations we did not include this as a control, and thus we may well have overestimated the levels of engraftment [43].

Similarly, the choice of markers to indicate end organ cells of donor origin bears on detection. Two predominant methods have been used, sometimes in combination. One is the already mentioned male donor into female recipient with demonstration of the Y chromosome by in situ hybridization indicating donor derivation. The careful analysis of markers of hematopoietic origin and end organ differentiation are clearly important, as just described, because the Y chromosome may appear in one cell, but actually be in another that is overlapping.

The other common method of donor detection is expression of a protein in the donor that is not seen in the recipient, either in the case of a transgene in the donor, most commonly, or absence of a normative gene in the recipient. That transgene expression is variable is no surprise, though, again, it is not typically considered as an issue in plasticity experiments. However, while most green fluorescent protein (GFP) transgenic mice display GFP in all cell types, the expression is not truly ubiquitous as it is not present in every cell of every type. Thus, there may be high variabiliity of expression between strains and within individual tissues and cell types within each mouse. Swenson et al. have recently explored differences between three mice strains expressing green fluorescent protein and found profound and important differences among them [44].

16.5
The Single-Cell Test

From the earliest days of the plasticity field, it was clear that simple transplantation of large cell populations could not adequately address the question of whether true plasticity was observed. Could it not be possible, for example, that transplanted marrow-derived cells were not displaying plasticity, but that there were previously unidentified tissue-specific cells in the marrow: one cell type circulating to regenerate liver, another for brain, another for blood, another for muscle, etc, rather than a single cell whose progeny could do everything? To answer this question, plasticity had to be demonstrated at the single-cell level. The first paper to suggest such plasticity was that published in 2001 by my own research group [43]. A few others have followed in subsequent years, the relatively small number in part related to the difficulty of performing such experiments [4, 45–57]. The 15 relevant papers which have focused on single-cell plasticity of marrow-derived stem mesenchymal, hematopoietic or other stem cells that I have been able to identify are summarized in Table 16.2.

16.5.1
Some Cases for Plasticity

All but one of these papers demonstrate significant pluripentiality on a clonal basis, sometimes within the mesodermal lineage compartment (e.g., endothelium, cardiac, smooth and skeletal muscle, adipocytes, chondrocytes, glial cells, and osteoblasts), but also to endodermal (epithelia of lung, liver, gastrointestinal tract) and ectodermal (neuroectoderm, neurons, skin, and adnexa) lineages. Various techniques were used to induce and observe plasticity events: bone marrow transplantation, direct transplantation into sites of tissue injury, injection of retrovirally marked cells into blastocysts, coculture with cells of injured organs. Not all of these studies investigated whether differentiation occurred directly or by a stable cell fusion event, though when this was investigated, most experiments reported direct differentiation, though some identify fusion as well.

Regarding this last point, perhaps the most interesting and conceptually illustrative experiment is that reported by Spees et al, in which epithelial cells from injured lung tissue were cocultured with mesenchymal stem cells (MSCs) [55]. By observing and recording the behaviors of cells in real time, this team was able to observe not only the direct differentiation of MSCs into epithelial cells, but also the apparently active, pseudopod-mediated contact between MSCs and epithelial cells leading to fusion.

Thus, the latter is not only achieved from possible pressure effects pushing two cells into direct contact and forcing coalescence, but by some active, receptor/ligand-mediated process.

It would seem that, while the number of reports is relatively small, there is clearly solid evidence that some degree of adult stem cell plasticity takes place under the right circumstances. Why is there such confusion and controversy, then, in the field? The single "negative" result paper in this collection of studies is revealing and deserves a close look.

16.5.2
The Case Against...

This most famous negative-result paper is that of Wagers et al., from the laboratory of Irving Weissman, at Stanford University, titled "Little evidence for developmental plasticity of adult hematopoietic stem cells" [58]. The authors, purporting a desire to "rigorously test" the possibility that marrow-derived stem cells, performed the only other (to my knowledge) attempt to recreate our original single-cell, bone marrow transplant experiment, with investigation of multiorgan engraftment. They found seven hepatocytes and one Purkinje cell that were derived from a single c-kit$^+$thy1.1^{lo}lin$^-$Sca1$^+$ (KTLS) hematopoietic stem cell.

The lack of rigor in this paper, however, was pointed out by two Comment letters subsequently published in *Science,* one from the group of Helen Blau and one from our group [59, 60]. Points raised included

1. Wagers et al. used 6- to 12-week-old donor and 10- to 14-week-old recipient mice. Our donor and recipient mice were age-matched at 4 to 6 weeks old. In stem cell functioning, age matters, with younger stem cells out performing older cells in a variety of ways.
2. We analyzed engraftment 11 months following stem cell transplantation, while Wagers et al. analyzed mice from 4 to 9 months after transplantation. It remains unclear whether timing of sampling is important for such experiments.

Table 16.2 Single cell plasticity experiments with marrow-derived stem cells

Year	Citation	Cell type	Experiment	Differentiation	Fusion vs. Direct Differentiation
2001	Krause et al[29]	Two day homing, lin- HSC	BM Tx	Trilineage	Not investigated
2001	Halleux et al[14]	hMSC	In vitro	Osteocytes, adipocytes, chrondrocytes	Not investigated
2001	Jiang et al[24]	MAPC	Blastocyst Tx & BM Tx	Trilineage	Not investigated
2002	Grant et al[13]	Sca1+ ckit+ lin- HSC	BM Tx	Retinal endothelium	Not investigated
2002	Wagers et al[48]	ckit+thy1.1lo lin- Sca1+ HSC	BM Tx	Very rare hepatocytes, Purkinje cells	Not investigated
2003	Keene et al[27]	Murine MAPC	Blastocyst Tx	Neurons, glial cells	Not investigated
2003	Spees et al[43]	hMSC	In vitro	Bronchial epithelium	Both
2004	Jang et al[23]	Two day homing, lin- HSC	In vitro	Hepatocytes	Direct differentiation
2004	Kawada et al[26]	mMSC (neg results for HSC)	In tissue Tx	Cardiac myocytes	Not investigated
2005	Yoon et al[63]	hBMSC	In vivo/In vitro	Cardiac myocytes, smooth muscle, endothelium	Both
2006	De Bari et al[10]	hMSC	BM Tx and in tissue Tx	chondrocytes, osteoblasts, adipocytes, skeletal myocytes	Not investigated
2006	Lange et al[32]	rMSC	In vitro	Hepatocytes	Direct differentiation
2006	Tropel et al[57]	mMSC	In vitro	Neurons	Direct differentiation
2007	Cogle et al[9]	ckit+ thy1+ lin- Sca1+ HSC	BMTx	Colonic adenomas, Lung squamous cell cancer	Direct differentiation
2007	Chamberlain et al[7]	hMSC	In utero (sheep) BMTx	Hepatocytes	Direct differentiation

HSC hematopoietic stem cell, *BM* bone marrow, *Tx* transplant, *Trilineage* differentiation along several endodermal, mesodermal, and ectodermal lineages, *hMSC, mMSC, rMSC* human, mouse, and rat mesenchymal stem cells, *MAPC* multipotent adult progenitor cells

3. Wagers et al. reported that their mice had 0.03% to 71.6% peripheral blood engraftment, but engraftment levels in the four mice analyzed were not specified, nor was the level of blood engraftment stated for the time that the solid tissues were examined. Our levels of engraftment ranged from 30% to 91% in five mice, the engraftments in all of which were reported.
4. We isolated and purified our stem cells through functional isolation following negative selection for mature hematopoietic lineage markers ("lin negative selection"); Wagers et al. used purely phenotypic sorting to select for KTLS cells. Of possible import is that our lin negative selection included AA4.1 antibody, a step that removed hematopoietic progenitors and possibly allowed selection for an "earlier" stem cell with more plasticity potential, while that of Wagers et al. did not.
5. The parabiotic experiments performed by Wagers et al. would of course include the bone marrow cell population we tested, but the extensive acute

and continual chronic wound healing at the site of skin, subcutis, and vascular grafting, with expected production of extensive granulation tissue which we now know is extensively marrow-derived, may have acted as a significant sink for any circulating progenitors that might have engrafted elsewhere.

6. We have already discussed the difficulties inherent in using a transgene marker of donor cell origin, such as GFP. GFP was used by Wagers et al., but not only do they not demonstrate controls for native expression in the donor animals, this particular GFP$^+$ donor animal was developed by that laboratory and its native GFP expression is never, to my knowledge, described in any publication, let alone in this paper. Thus, the absence of engraftment could simply reflect poor GFP expression in the donor and is impossible to evaluate with certainty.

Why spend this much time critiquing this single negative result paper? Because a careful reader of the plasticity literature would be hard pressed to find a single discussion of plasticity that did not cite "Little evidence..." as a significant critique of the whole field, despite the comment from the Weissman research group, in a response to our Comment letter, that "our data are not directly comparable to those of Krause et al. and do not implicitly refute their observations" [61].

The power of the journal and prestige of the senior author guaranteed this paper a degree of polemical power that would haunt the field ever since its publication, despite this demurral published over 5 months later. A few months later, this exchange about the publishing of this paper between two noted stem cell researchers was recorded by journalist Cynthia Fox [62]:

Stem Cell Researcher 1: *"Very poor editorial judgement."*
Stem Cell Researcher 2: *"That paper has done damage, but I think it will all wash out."*

We wait for it to "wash out," still. I am confident, however, that we have begun to turn that corner. Nonetheless, one scientific fact about the relative capacity for plasticity between marrow-derived stem cells from Wagers et al. may be of prime significance. Perhaps its negative finding reflects that the KTLS cell is not indeed the "ultimate" hematopoietic cell the authors insist. Review of the cited single-cell plasticity studies certainly suggests that mesenchymal stem cells are quite readily, and perhaps much more readily, contributory to plasticity events and that the success of the 2-day homed cell of our experiments, developed by the Sharkis laboratory, reflects that it may be a more fundamental precursor of both hematopoietic and mesenchymal stem cells in the marrow. Direct comparison of these two cells to each other has not to my knowledge yet been published.

16.6
Summary

While there are many scientists and commentators with opinions regarding stem cell plasticity, and while many of these have decidedly negative opinions, it appears that this is not a result of balanced and unbiased review of the evidence. That adult cells have the potential under some circumstances to become pluripotent or even totipotent cannot be doubted given the proof of principle experiments of cloning in many species and by development of embryonic stem cell-like totipotency of mouse and adult somatic cells by retroviral gene transfer. To what extent plasticity happens physiologically probably depends on whether or not there is tissue injury and on the nature and severity of injury, if present.

The importance of these processes physiologically is still open to question, though even if it is of little physiologic importance, it certainly represents something that can be exploited for therapeutic or industrial purposes. While to date, the politics and other nonscientific conditioning factors that limit the potential of plasticity research drive many to say it is time for a valedictory for the field, the published data suggest that, rather, early claims of adult stem cell plasticity are vindicated and that ultimately only the limits of our experimental ingenuity will impede progress.

References

1. Bjornson CR, Rietze RL, Reynolds BA, et al (1999) Turning brain into blood: a hematopoietic fate adopted by adult neural stem cells in vivo. Science 283: 534–537
2. Ferrari G, Cusella-De Angelis G, Coletta M, et al. Muscle regeneration by bone marrow-derived myogenic progenitors. Science 279: 1528–1530 (1998)
3. Petersen BE, Bowen WC, Patrene KD, et al (1999) Bone marrow as a potential source of hepatic oval cells. Science 284: 1168–1170
4. Theise ND (2003) Stem cell research: elephants in the room. Mayo Clinic Proc 78: 1004.
5. Herzog EL, Chai L, Krause DS (2003) Plasticity of marrow-derived stem cells. Blood 102: 3483–93
6. Theise ND, Wilmut I (2003) Cell plasticity: flexible arrangement. Nature 425: 21
7. Brockes JP, Kumar A (2005) Appendage regeneration in adult vertebrates and implications for regenerative medicine. Science 310: 1919–23
8. Theise ND, Krause DS (2002) Toward a new paradigm of cell differentiation capacity. Leukemia 16: 542–548
9. Han M, Yang X, Farrington JE, et al (2003) Digit regeneration is regulated by Msx1 and BMP4 in fetal mice. Development 130: 5123–32
10. Hampton DW, Seitz A, Chen P, et al (2004) CNS response to injury in the MRL/MpJ mouse. Neuroscience 127: 821–32
11. Heber-Katz E, Chen P, Clark L, et al (2004) Regeneration in MRL mice: further genetic loci controlling the ear hole closure trait using MRL and M.m. Castaneus mice. Wound Repair Regen 12: 384–92
12. Alvarez-Dolado M, Pardal R, Garcia-Verdugo JM, et al (2003) Fusion of bone-marrow-derived cells with Purkinje neurons, cardiomyocytes and hepatocytes. Nature 425: 968–73
13. Harris RG, Herzog EL, Bruscia EM, et al (2004) Lack of a fusion requirement for development of bone marrow-derived epithelia. Science 305: 90–3
14. Newsome PN, Johannessen I, Boyle S, et al (2003) Human cord blood-derived cells can differentiate into hepatocytes in the mouse liver with no evidence of cellular fusion. Gastroenterology 124: 1891–900
15. Quintana-Bustamane O, Albarez-Barrientos A, Kofman AV, et al (2006) Hematopoietic mobilization in mice increases the presence of bone marrow-derived hepatocytes via in vivo cell fusion. Hepatology 43: 108–116
16. Willenbring H, Grompe M (2003) Embryonic versus adult stem cell pluripotency: in liver only fusion matters. J Assist Reprod Genet 20: 393–4
17. Theise ND (2003) New principles of cell plasticity. Comptes Rendus Biologies (Academie des Sciences, Paris) 325: 1039–43
18. Theise ND, Krause DS (2001) Suggestions for a new paradigm of cell differentiative potential. Blood Cells, Molecules, and Diseases 27: 625–631
19. Campbell KH, McWhir J, Ritchie WA, et al (1996) Sheep cloned by nuclear transfer from a cultured cell line. Nature 380: 64–66
20. Eggan K, Akutsu H, Hochedlinger K, Rideout W 3rd, Yanagimachi R, Jaenisch R X-chromosome inactivation in cloned mouse embryos. Science 290: 1578–1581 (2000)
21. Theise ND (2006) Implications of "post-modern biology" for pathology: the cell doctrine. Lab Invest 86: 335–44
22. Theise ND, Harris R (2006) Postmodern biology: (adult) (stem) cells are plastic, stochastic, complex, and uncertain. Handb Exp Pharmacol 174: 389–408
23. Okita K, Ichisaka T, Yamanaka S (2007) Generation of germline-competent induced pluripotent stem cells. Nature 448: 313–7
24. Takahashi K, Yamanaka S (2006) Induction of pluripotent stem cells from mouse embryonic and adult fibroblast cultures by defined factors. Cell 126: 663–76
25. Wernig M, Meissner A, Foreman R, et al (2007) In vitro reprogramming of fibroblasts into a pluripotent ES-cell-like state. Nature 448: 318–24
26. Chen S, Takanashi S, Zhang Q, et al (2007) Reversine increases the plasticity of lineage-committed mammalian cells. Proc Natl Acad Sci USA 104: 10482–7
27. Potten CS, Loeffler M (1990) Stem cells: attributes, cycles, spirals, pitfalls and uncertainties. Lessons for and from the crypt. Development 110: 1001–1020
28. Lewontin R (2000) It ain't necessarily so: The dream of the human genome and other illusions. New York Review of Books, New York
29. Kaern M, Elston TC, Blake WJ, et al (2005) Stochasticity in gene expression: from theories to phenotypes. Nat Rev Genet 6: 451–464
31. Bornfleth H, Edelmann P, Zink D, et al (1999) Quantitative motion analysis of subchromosomal foci in living cells using four-dimensional microscopy. Biophys J 77, 2871–2886 (1999)
30. Kurakin A (2005) Self-organization vs Watchmaker: stochastic gene expression and cell differentiation. Dev Genes Evol 215: 46–52
32. Theise ND (2005) Experimental design and discourse in adult stem cell plasticity. Stem Cell Reviews 1: 9–14
33. Theise ND, Nimmakayalu M, Gardner R, et al (2000) Liver from bone marrow in humans. Hepatology 32: 11–16
34. Menthena A, Deb N, Oertel M, et al (2004) Bone marrow progenitors are not the source of expanding oval cells in injured liver. Stem Cells 22: 1049–61
35. Wang X, Foster M, Al-Dhalimy M, et al (2003) The origin and liver repopulating capacity of murine oval cells. Proc Natl Acad Sci U S A. 100 Suppl 1: 11881–8
36. Oh SH, Witek RP, Bae SH, et al (2007) Bone marrow-derived hepatic oval cells differentiate into hepatocytes in 2-acetylaminofluorene/partial hepatectomy-induced liver regeneration. Gastroenterology 132: 1077–87
37. Hatch HM, Zheng D, Jorgensen ML, et al (2002) SDF-1alpha/CXCR4: a mechanism for hepatic oval cell activation and bone marrow stem cell recruitment to the injured liver of rats. Cloning Stem Cells 4: 339–51
38. Kollet O, Shivtiel S, Chen YQ, et al (2003) HGF, SDF-1, and MMP-9 are involved in stress-induced human CD34+ stem cell recruitment to the liver. J Clin Invest 112: 160–9

39. Kucia M, Wojakowski W, Reca R, et al (2006) The migration of bone marrow-derived non-hematopoietic tissue-committed stem cells is regulated in an SDF-1-, HGF-, and LIF-dependent manner. Arch Immunol Ther Exp (Warsz) 54: 121–35
40. Shyu WC, Lin SZ, Yen PS, et al (2007) Stromal cell-derived factor 1α promotes neuroprotection, angiogenesis and mobilization/homing of bone marrow-derived cells in stroke rats. J Pharmacol Exp Ther [Epub ahead of print]
41. Swenson ES, Kuwahara R, Krause DS, et al (2008) Physiological variations of stem cell factor and stromal-derived factor-1 in murine models of liver injury and regeneration. Liver International [In press]
42. Herzog EL, Van Arnam J, Hu B, et al (2006) Threshold of lung injury required for the appearance of marrow-derived lung epithelia. Stem Cells 24: 1986–92.
43. Krause DS, Theise ND, Collector MI, et al (2001) Multi-organ, multi-lineage engraftment by a single bone marrow-derived stem cell. Cell 105: 369–377.
44. Swenson ES, Price JG, Brazelton T, et al (2007) Limitations of green fluorescent protein as a cell lineage marker. Stem Cells 25: 2593–600
45. Chamberlain J, Yamagami T, Colletti E, et al (2007) Efficient generation of human hepatocytes by the intrahepatic delivery of clonal human mesenchymal stem cells in fetal sheep. Hepatology 46: 1935–45
46. Cogle CR, Theise ND, Fu DT, et al (2007) Bone marrow contributes to epithelial cancers in mice and humans as developmental mimicry. Stem Cells 25: 1881–1887
47. De Bari C, Dell'Accio F, Vanlauwe J, Eyckmans J, Khan IM, Archer CW, Jones EA, McGonagle D, Mitsiadis TA, Pitzalis C, Luyten FP. Mesenchymal multipotency of adult human periosteal cells demonstrated by single-cell lineage analysis. Arthritis Rheum. 2006 Apr;54(4):1209–21
48. Grant MB, May WS, Caballero S, et al (2002) Adult hematopoietic stem cells provide functional hemangioblast activity during retinal neovascularization. Nat Med 8: 607–12
49. Halleux C, Sottile V, Gasser JA, et al (2001) Multi-lineage potential of human mesenchymal stem cells following clonal expansion. J Musculoskelet Neuronal Interact 2: 71–6
50. Jang YY, Collector MI, Baylin SB, et al (2004) Hematopoietic stem cells convert into liver cells within days without fusion. Nat Cell Biol 6: 532–9
51. Jiang Y, Jahagirdar BN, Reinhardt RL, et al (2002) Pluripotency of mesenchymal stem cells derived from adult marrow. Nature 418: 41–9
52. Kawada H, Fujita J, Kinjo K, et al (2004) Nonhematopoietic mesenchymal stem cells can be mobilized and differentiate into cardiomyocytes after myocardial infarction. Blood 104: 3581–7
53. Keene CD, Ortiz-Gonzalez XR, Jiang Y, et al (2003) Neural differentiation and incorporation of bone marrow-derived multipotent adult progenitor cells after single cell transplantation into blastocyst stage mouse embryos. Cell Transplant 12: 201–13
54. Lange C, Bassler P, Lioznov MV (2005) Liver-specific gene expression in mesenchymal stem cells is induced by liver cells. World J Gastroenterol 11: 4497–504
55. Spees JL, Olson SD, Ylostalo J, et al (2003) Differentiation, cell fusion, and nuclear fusion during ex vivo repair of epithelium by human adult stem cells from bone marrow stroma. Proc Natl Acad Sci U S A 100: 2397–402
56. Tropel P, Platet N, Platel JC, et al (2006) Functional neuronal differentiation of bone marrow-derived mesenchymal stem cells. Stem Cells 24: 2868–76
57. Yoon YS, Wecker A, Heyd L, et al (2005) Clonally expanded novel multipotent stem cells from human bone marrow regenerate myocardium after myocardial infarction. J Clin Invest 115: 326–38
58. Wagers AJ, Sherwood RI, Christensen JL, et al (2002) Little evidence for developmental plasticity of adult hematopoietic stem cells. Science 297: 2256–9
59. Helen Blau, Tim Brazelton, Gilmor Keshet, et al (2002) Something in the Eye of the Beholder. Science 298: 361–363
60. Theise ND, Krause DS, Sharkis S (2003) Comments on: "Little evidence for stem cell plasticity." Science 299: 1317
61. Amy J. Wagers, Richard I. Sherwood, et al (2003) Response to Comment on "Little Evidence for Developmental Plasticity of Adult Hematopoietic Stem Cells" Science 299: 1318
62. Fox, C (2007) Cell of cells: the global race to capture and control the stem cell. W. W. Norton and Company, New York, p 31

V Engineering at the Tissue Level

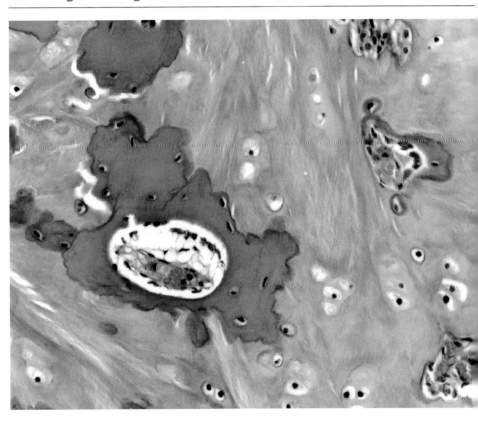

Bone Tissue Engineering

U. Meyer, H. P. Wiesmann, J. Handschel, N. R. Kübler

Contents

17.1	Introduction	211
17.2	Bone Repair Strategies	212
17.3	Membrane Techniques	212
17.4	Biophysical Effects	213
17.5	Distraction Osteogenesis	214
17.6	Biomolecules	215
17.7	Cell Transplantation Approaches	216
17.8	Flap Prefabrication	217
17.9	Extracorporal Strategies	218
17.10	Determined Bone Cells	220
17.11	Evaluation of Engineering Success	222
	References	229

17.1 Introduction

Bone repair is a subject of intensive investigation in reconstructive surgery (for review, see [1]). Current approaches in skeletal reconstructive surgery use biomaterials, autografts or allografts, although restrictions on all these techniques exist. These restrictions include donor site morbidity and donor shortage for autografts [2], immunological barriers for allografts, and the risk of transmitting infectious diseases. Numerous artificial tissue substitutes containing metals, ceramics, and polymers were introduced to maintain skeletal function [3]. However, each material has specific disadvantages, and none of these can perfectly substitute for autografts in current clinical practice. The use of biomaterials is a common treatment option in clinical practice. One important reason for the priority of tissue grafts over nonliving biomaterials is that they contain living cells and tissue-inducing substances, thereby possessing biological plasticity. Research is currently in progress to develop cell-containing hybrid materials and to create replacement tissues that remain interactive after implantation, imparting physiological functions as well as structure to the tissue or organ damaged by disease or trauma [4].

Bone tissue engineering, as in most other tissue engineering areas, exploits living cells in a variety of ways to restore, maintain, or enhance tissue functions [5, 6]. There are three principal therapeutic strategies for treating diseased or lost tissue in patients: (1) in situ tissue regeneration, (2) implantation of freshly isolated or cultured cells, and (3) implantation of a bone-like tissue assembled in vitro from cells and scaffolds. For in situ regeneration, new tissue formation is induced by specific scaffolds or external stimuli that are used to stimulate the body's own cells and promote local tissue repair. Cellular implantation means that individual cells or small cellular aggregates from the patient or a donor are injected directly into the damaged or lost region with or without a degradable scaffold. For tissue implantation, a complete three-dimensional tissue is grown in vitro using autologous or donor cells within a scaffold, which has to be implanted once it has reached "maturity" [7–9]. In this chapter we give some details of all of the above-mentioned strategies and present alternatives to extracorporal approaches that are important in clinical decision-making. Furthermore, it is mentioned that some of the techniques described here are clinically combined with extracorporal tissue-engineering methods.

17.2
Bone Repair Strategies

The use of autologous bone in bone defect reconstruction can be considered to be the gold standard. There are two classical ways to repair bone defects by use of autologous cells: one utilizes mechanisms of defect repair by enhancement of the local host cell population, while the other is based on transplantation of grafted bone (Fig. 17.1).

Augmentation of the host cell population can effectively be used to improve the healing of bone lesions. The ability to repair or reconstruct lost bone structure by this technique critically depends on the condition of the defect site. If the soft and hard tissues of the defect site are healthy, expansion of host cells is often successfully applied. In circumstances of impaired defect conditions such as wound infection and tissue necrosis or irritation, however, cellular augmentation of the local bone cell population will probably fail.

17.3
Membrane Techniques

Membrane techniques can be used to improve defect healing by ingrowth of local host cells in the defect site (Fig. 17.2). This kind of defect repair procedure (guided bone regeneration) is mainly used in maxillofacial surgery to repair bony defects in the maxilla and mandible. The principle of guided bone regeneration (GBR) is to effectively separate bone tissue from the soft tissue ingrowth by a barrier [10]. The application of the principle of GBR has proven to be successful in a number of controlled animal studies and clinical trials [11–13]. The healing pattern has been shown to involve all steps of de novo bone formation including blood clot formation, invasion by osteoprogenitor cells, and their terminal differentiation into osteoblasts. Under these circumstances the produced extracellular matrix finally mineralizes to form woven bone and later is remodeled into lamellar bone [14]. The success of GBR critically depends on the defect size and geometry.

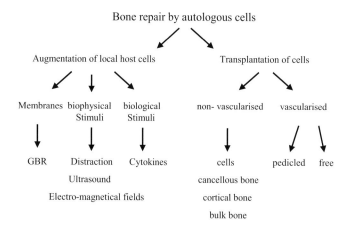

Fig. 17.1 Bone repair by autologous cells

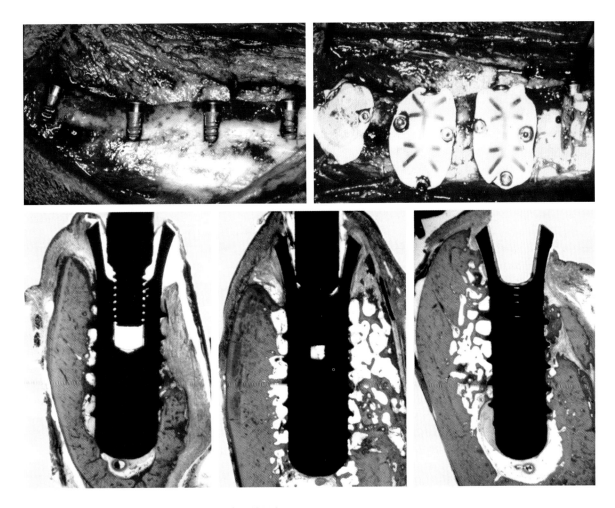

Fig. 17.2 Principles of guided bone regeneration (GBR)

Bone defects can be assumed to be repaired to a larger extent in cases where the defect is neighbored by more than two bone walls. The main limitation of GBR is therefore the inability to reconstruct geometrically complex or larger defect sites.

17.4 Biophysical Effects

Biophysical effects, in particular mechanical loading and electromagnetic signals, are clearly important regulators of bone formation. Indeed, the regenerative capacity of bone tissue is largely due to the bone's capacity to recognize the functional environment required for the emergence and maintenance of a structurally intact bone tissue. Biophysical stimulation methods have therefore been introduced and have proved successful in clinical practice. Ultrasound application and electromagnetic field exposure are, besides distraction osteogenesis, considered to be a special form of mechanical stimulation, classical examples of biophysically driven approaches to influencing bone formation. Ample evidence from various prospective double-blinded placebo-controlled clinical studies now documents the efficacy of mechanical and electrical stimulation (direct current, inductive coupling, capacitive coupling, and composite fields) and of mechanical stimulation (distraction osteogenesis, ultrasound stimulation, and fracture activation) in enhancing bone repair [15]. Because of its clinical efficacy, biophysical stimulation is now one of several methods to enhance bone formation in patients.

When applied to a patient, biophysical stimuli exert their effects at distinct skeletal sites. The response to ultrasound stimuli is altered by the anatomy of the application site. Ultrasound-related deformations through adjacent structures (joints, ligaments, and muscles) have a major impact on the resultant biophysical signal at the desired effector site. Both patient-specific and technique-specific factors play an important role. When biophysical treatment strategies are considered, the specific forms of biophysical stimulation and their related dose effect and application timing must be carefully determined and validated. Despite limited knowledge concerning the biological effects of biophysical stimulation, it has long been the goal of many scientists and clinicians to find applicable clinical tools for using biophysical stimuli in the context of bone healing. In recent decades, different treatment strategies have been reported to be successful in animal experiments as well as in clinical trials. This has led researchers and clinicians to conduct a multitude of defined experiments to assess the effects of different biophysical stimuli in facilitating new bone formation (for review, see [16]). Numerous investigations have been performed on electrical field exposure. When electrical stimuli are applied iatrogenically to a patient in order to enhance bone formation, one has to be aware that electrical stimuli exert their effects within the various hierarchical structures of the skeleton. The influence of external stimuli in skeletal regeneration can generally be addressed at organ, tissue, cellular, and molecular levels. Understanding the overall process of biophysical signaling requires an appreciation of these various levels of study and a knowledge of how one level relates to and influences the next. Electrical stimuli can be analyzed as they act on whole bones. Additionally, the contributions of electrical effects made on multicellular systems can be evaluated by examining the tissue reaction toward the biophysical microenvironment. Thirdly, cellular and molecular reactions critically sharpen the tissue response. The key requirement for using the effects of such biophysical forces is therefore to define the precise cellular response to the stimulation signal in an in vitro environment and to use well-established animal and clinical models to quantify and optimize the therapeutic stimulation approaches. Whereas this seems to be achievable through research collaboration among different disciplines using complementary approaches, electrical stimulation is not a standardized or common treatment option to regenerate skeletal defects.

17.5
Distraction Osteogenesis

Distraction osteogenesis frequently used to augment local bone in clinical settings [17] is nowadays an established bone reconstructive treatment option (Fig. 17.3). Histologically, distraction osteogenesis shares many of the features of fetal tissue growth and neonatal limb development, as well as normal fracture gap healing [18, 19]. Although the biomechanical, histological, and ultrastructural changes associated with distraction osteogenesis have been widely described, the molecular mechanisms governing the formation of new bone in the interfragmental gap of gradually distracted bone segments remain largely unclear. A growing body of evidence has emphasized the contribution of both local bone or bone precursor cells and neovascularity to bone formation during distraction [20]. Recent studies have implicated a growing number of cytokines that are intimately involved in the regulation of bone synthesis and turnover during distraction procedures [21]. The gene regulation of numerous cytokines (transforming growth factor-beta1 [TGF-β1], TGF-β2, TGF-β3, bone morphogenetic proteins, insulin-like growth factor-1, and fibroblast growth factor-2) and extracellular matrix proteins (osteonectin, osteopontin) during distraction osteogenesis have been best characterized and reviewed by Bouletreau et al. [22, 23]. Refinements in the surgical procedure of bone distraction have significantly improved clinical outcome [17]. Distraction osteogenesis has therefore become a mainstay in bone tissue engineering and has significantly improved our armamentarium for bone reconstructive procedures.

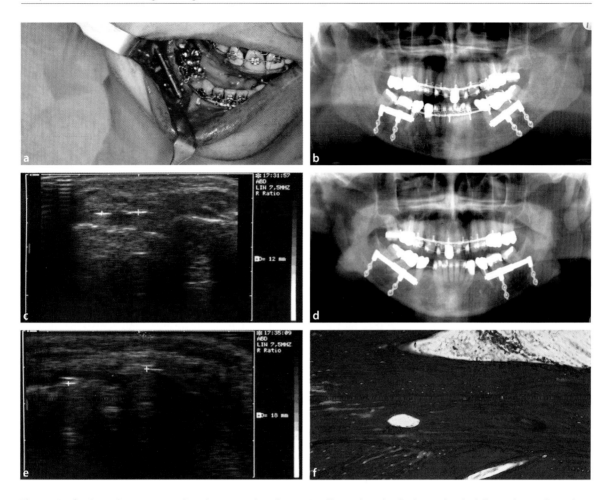

Fig. 17.3a–f Distraction osteogenesis. **a** intraoperative view, **b,d** radiographs at beginning and end of distraction, and **c,e** ultrasounds at the beginning and the end of distraction. **f** Histological appearance of the regenerated tissue

17.6
Biomolecules

Biomolecules with or without a carrier have been suggested to augment the local bone cell population (Table 17.1) [24]. It was suggested that bone reconstruction may be accomplished using various bioactive factors with varying potency and efficacy [25]. Members of the TGF-ß superfamily and other growth factors—notably, bone morphogenetic protein (BMP), insulin-like growth factor (IGF), platelet-derived growth factor (PDGF), and GDF—are currently being tested for their experimental use to engineer new bone. Evidence from a variety of studies indicate that exposure to osteoinductive growth factors can promote new bone formation [26]. Different healing environments require different types of bone tissue to be engineered. For example, a segmental long bone defect requires primarily cortical bone to be formed, whereas a maxillary alveolar crest augmentation requires primarily membranous (cancellous) bone. It is possible that different cascades of signaling molecules orchestrate these two healing processes. Alternatively, the same factors may be involved but have different outcomes because of the different mechanical environments or different host cell populations. It should be considered that

Table 17.1 Biomolecules used to augment bone cell reaction

Transforming growth factor-beta (TGF-ß)
Bone morphogenetic proteins (BMPs)
Insulin-like growth factors (IGFs)
Platelet-derived growth factors (PDGFs)
Fibroblast growth factors (FGFs)

different osteoinductive growth factors have a range of biological potency and with increased efficiency comes increased cost and often increased risk. There are currently three strategies for delivery of osteoinductive growth factors which are at various stages of development. The first strategy involves extracting and partially purifying a mixture of proteins from animal or human cortical bone, as described by Urist [27] and Urist and Strates for BMPs [28]. The second strategy involves the use of recombinant proteins (e.g., rhBMPs), which was enabled by the cloning and sequencing of many of these genes [29, 30]. The third and least-developed strategy is that of gene therapy, which involves delivery of the DNA encoding a growth factor rather than delivery of the protein itself [31–33]. It is also likely that there will not be one best strategy or one best growth factor and that success in the end may be determined by such factors as manufacturing cost and ease of use by the treating physicians or the flexibility to adapt to different bone tissue engineering environmental requirements. Regardless of the specific growth factor chosen, there are several common issues which when unsolved may present potential clinical limitations to all of the above strategies. These factors include dose, carrier, patient variability, and confounding clinical factors.

17.7
Cell Transplantation Approaches

Various donor sites allow the grafting of different types and amounts of bone cell-containing specimens. The use of autologous bone is accompanied by a grafting procedure and can be accompanied by donor site morbidity. Different surgical methods (Fig. 17.4) allow us to graft bone-forming tissue in the form of:
- periosteal flaps
- cancellous bone chips
- cortical bone chips
- bulk corticospongiosal grafts and
- vascularized bone.

Small bone defects can be treated with small bone grafts (periosteal flaps, cancellous, or cortical bone chips) provided local conditions are appropriate for bone healing. Endochondral bone from the ilium, tibia, or rib can be harvested, as well as membranous bone from the facial skeleton. Animal studies indicate that membranous bone is less prone to resorption compared with endochondral bone [34]. However, when the bone defect is large, bulk corticospongiosal grafts are necessary. Effective rigid fixation of the graft facilitates graft survival, and hematomas should be avoided. Problems with conventional grafts include their susceptibility to infection, the unknown survival of transplanted cells, and the unpredictable degree of resorption. The extent of resorption is dependent on the graft volume, the delay time before loading, and the condition of the transplant bed.

A major advance in bone reconstruction strategies was the use of vascularized bone tissue in combination with or without adjacent soft tissue (muscle, subdermal tissue, or skin). The soft tissue of the flap can be used to obliterate the defect space to a desired extent, to prevent infection, and to allow a better barrier against direct surface loads. Bone reconstruction by vascularized bone can be achieved through pedicled flaps or microvascular flaps. Several pedicled osteocutaneous or osteomuscular flaps have been established in clinical routine [35]: sternocleidomastoid muscle combined with the clavicle [36], trapezius muscle with parts of the scapula [37–39], pectoralis muscle with part of a rib or sternum [40–42], latissimus dorsi osteomyocutaneous flap [43], and the temporalis muscle with calvarian bone [44–46]. Free vascularized bone grafts provide recently the best possibility for reconstruction of large bone defects. Sufficient bone material with good quality corticocancellous bone can be harvested for use in reconstruction of large bone defects. A soft tissue paddle can be harvested together with the vascularized bone as for example an osteocutaneous flap. The iliac crest [47, 48], the fibula [49–52], and the scapula [53–56]

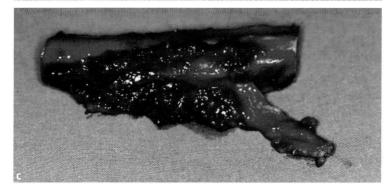

Fig. 17.4a–c Different methods for bone reconstruction. **a** cancellous bone, **b** cortical bone, and **c** microvascular fibular flap

are all well-established donor sites from which a large amount of bone can be grafted [57]. This type of bone reconstruction combines the advantage of enlarging bone transplants while maintaining the nutritional needs of cells. In current clinical practice, this technique overcomes limitations of other approaches (free nonvascularized bone graft transfer, tissue-engineered constructs) when larger defects need to be reconstructed. The main disadvantage of free vascularized grafts is donor site morbidity, especially when larger grafts are necessary.

17.8
Flap Prefabrication

Flap prefabrication is one of the most exciting areas in reconstructive surgery because of its bridging role between conventional reconstructive surgery and tissue engineering [58]. This new and exciting approach has gained major public interest since Vacanti and colleagues fabricated a human ear on a mouse back [59]. Using the prefabrication technique, tissues such as bone or cartilage can be preassembled to form composite tissues that can fit even major defect sites. Prefabrication can expand the versatility of a graft. Existing organs can be transformed into a transplantable state, or completely new organs can be created. In its simple form—e.g., pre-lamination—a new tissue can be created by burying cartilage, bone, skin, or mucosa underneath a fascia or in a muscle pouch, and it can later be harvested as a composite tissue free flap to replace lost tissue [58, 60–67]. Vascular induction is another approach in which new blood supply is generated in a present tissue to improve the survival of the transplanted construct. It was presented that bone chips wrapped in a vascular carrier such as muscular tissue became vascularized grafts [68]. The future of flap prefabrication can be envisioned to generate complex vascularized tissues in desired external shapes that mimic the defect to be replaced. The basis of such approaches may be the use of suitable vascular carriers, resorbable matrix components, attached cytokines, and incorporated cells. It was shown by various investigators that prefabrication can be started ex vivo and then be transferred to the in vivo situation [69], but can also be used solely in vivo. Warnke et al. [70] demonstrated, for example, the growth and transplantation of a custom vascularized bone graft in a man. The use of such complex tissue-engineering strategies demonstrates the general feasibility of applying these therapies in humans, but details of these therapy options have to be solved before they can be used as a routine treatment option.

17.9
Extracorporal Strategies

Extracorporal tissue engineering in the narrower (EU) definition of bone and cartilage engineering requires not only living cells, but additionally the interaction of three biological components: bone cells, the extracellular scaffolds, and in some instances growth factors. For engineering living tissues in vitro, cultured cells are commonly grown on two-dimensional or three-dimensional bioactive degradable biomaterials that provide the physical and chemical basis to guide their proliferation and differentiation. In bioreactors outside the body, the biomaterial is assembled to a complex three-dimensional construct. New approaches in extracorporal tissue-engineering strategies are aimed toward fabricating scaffold-free three-dimensional microtissue constructs. The technique of microtissue formation is described also in detail in this book. Additionally, the use of biomolecules seems not to be dispensable since the cells themselves may be stimulated to synthesize some of the needed biomolecules in a autocrine fashion. The assembly of cells into tissue substitutes is a highly orchestrated set of events that requires time scales ranging from seconds to weeks and dimensions ranging from 0.0001 to 10 cm. At the moment the techniques are moving from an experimental stage to the level of clinical application. Cell/scaffold-based approaches for tissue engineering of bone offer perspectives on future treatment concepts. The skeletal system contains a variety of different cell types: vascular cells, bone marrow cells, preosteoblasts, osteocytes, chondroblasts, chondrocytes, and osteoclasts, all executing distinct cellular functions to allow the skeleton to work as a highly dynamic, load-bearing organ. Whereas all these cells are necessary to build up a "real" skeleton, limited cell sources are regarded to be sufficient for engineering a "bone"—or "cartilage"-like construct in vitro.

Principal sources of cells for tissue engineering include those under xenogenic cells, autologous, and allogenic cells (Table 17.2). Each category can be subdivided according to whether the cells are stem cells (embryonic or adult) or whether the cells are in a more differentiated stage. Various mature cell lines as well as multipotent mesenchymal progenitors have been successfully established [71] in bone tissue engineering approaches. Additionally, other bone cell lines such as genetically altered cell lines (sarcoma cells, immortalized cells, and nontransformed clonal cell lines) were developed and used to evaluate basic aspects of in vitro cell behavior in nonhuman settings.

The stem cell of skeletal tissue is a hypothetical concept with only circumstantial evidence for its existence, and indeed, there seems to be a hierarchy of stem cells each with variable self-renewal potentials (for review, see [72]). Embryonic and adult stem cells can be distinguished. Embryonic stem cells reside in

Table 17.2 Principal cell sources used in tissue-engineering studies

Experimental studies	Preclinical studies	Clinical studies
Autologous	Autologous	Autologous
Allogenic	Allogenic	
Xenogenic	Xenogenic	
Immortalized		
Nontransformed clonal		
Sarcoma		

blastocysts. They were firstly isolated and grown in culture more than 20 years ago [73]. The primitive stem cells renewing skeletal structures have been given a variety of names including connective tissue stem cells, osteogenic cells [74], stromal stem cells [75], stromal fibroblastic cells [76], and mesenchymal stem cells [77]. Until today no nomenclature is entirely accurate based upon the developmental origins or differentiation capacities of these cells, but the last term, although defective, appears to be in favor at the moment. Stem cells have the capacity for extensive replication without differentiation, and they possess, as mentioned, a multilineage developmental potential allowing them to give rise to not only bone and cartilage, but tendon, muscle, fat, and marrow stroma. These expansion properties make stem cells, whether derived from the hematopoietic system (HSCs) or marrow (MSCs), a very interesting source of cells for tissue engineering of bone. MSCs are present in fetal tissue, postnatal bone marrow, and also in the bone marrow of adults [77]. A number of markers are expressed on MSCs. Some of these have been used not only to characterize these cells but also to enrich MSCs from populations of adherent bone marrow stromal cells. However, none of these markers seem to be specific for MSCs. Although the bone marrow serves as the primary reservoir for MSCs, their presence has also been reported in a variety of other tissues (periosteum and muscle connective tissue [78, 79] and fetal bone marrow, liver, and blood [80]). Transfer of these tissues (e.g., coverage of cartilage defects by periosteum) may act also through stem cell-induced repair mechanisms. Whereas MSCs have been identified in fetal blood [80] and infrequently observed in umbilical cord blood [81–83], it is not definitively resolved whether MSCs are present in steady-state peripheral blood of adults or not, but the number of stem cells in peripheral blood is probably extremely low [84].

A major problem in using adult stem cells in a given clinical situation is the difficulty in obtaining a significant number of stem cells to generate cell constructs of bigger size. Limitations of adult stem cell harvesting gave therefore rise to the use of fetal stem cells, particularly on fetal bone marrow (FBM). This cell source seems to be a promising source for tissue-engineering approaches, since they have a significantly reduced immunogenicity, compared with adult stem cells. In contrast to adult tissues, fetal tissues produce more abundant trophic substances and growth factors, which promote cell growth and differentiation to a greater extent. Research on fetal tissue transfer in animal models and in human experimental treatments has confirmed distinct advantages over adult tissues including lowered immunogenicity and higher percentage of primitive cells. Due to the lowered immunogenicity, fetus-derived stem cells seem to remove, to some extent, the problems of tissue typing. As such cells also have a high capacity to differentiate into the complete repertoire of mesenchymal cell lineages, combined with the ability for rapid cellular growth, differentiation, and reassembly, fetal cells may become a more important subject in tissue-engineering strategies. Fetal stem cells have, beneath their enormous potential for biomedical and tissue-engineering applications, some serious limitations. There is a lack of methods that enable tissue engineers to direct the differentiation of embryonic stem cells and to induce specific functions of the embryonic cell-generated tissue after transplantation [85]. Two other issues are of main concern [85]: (1) It must be assured that the cultivation and transplantation of such stem cells are not accompanied by a tumorigenic differentiation. As it was shown that undifferentiated ES cells give rise to teratomas and teratocarcinomas after implantation in animals, this potential misdevelopment constitutes a major problem for the clinical use of such cells. (2) It must be assured that the in vitro generation of cells does not lead to an immunological incompatibility toward the host tissue. Immunological incompatibilities can be avoided by using the somatic cell nuclear transfer methodology (SCNT) [86]. However, nuclear cloning methods are often criticized for this potential risk. The main problem in employing fetal cells for use in tissue engineering is not only to obtain such cells but perhaps more important the clinical, legal, and ethical issues involved in such treatment strategies, which are probably the most difficult barrier to overcome.

Many attempts have been undertaken to refine procedures for the propagation and differentiation of stem cells. Despite the various advantages of using tissue-derived adult stem cells over other sources of cells, there is some debate as to whether large enough populations of differentiated cells can be grown in vitro rapidly enough when needed clinically. At present, stem cells are, for example, not able to differentiate definitively into osteocyte-like cells which are competent to mineralize the pericellular region in a bone-like manner under in vitro conditions [87, 88]. This must be considered as one limitation for the use of stem cells in extracorporal skeletal tissue engineering. Much more, basic research is therefore necessary to assess the full potential of stem cell therapy to reconstitute skeletal mass. It is expected that many future studies will be directed toward the development of gene therapy protocols employing gene insertion strategies [89]. The concept that members of the BMP and TGF-β superfamily will be particularly useful in this regard has already been tested by many investigators [33, 90]. Genetic engineering to shape gene expression profiles may be, therefore, a future route for the use of stem cells in human tissue engineering, but this approach is at the moment far away from clinical applications.

17.10 Determined Bone Cells

Differentiated osteoblast-like cells serve as common cell lines in evaluating preclinical and clinical aspects of bone tissue engineering. Osteoblasts, derived from multiple sites of the skeleton, can be harvested in the form of precursor cells, lining cells, mature osteoblasts, or osteocytes. Cell separation is then performed by various techniques to multiplicate distinct cell lines in culture. A complementary approach in experimental settings can be distinguished from the use of primary cells: specifically modified cells. Whereas primary cells are commonly used in clinical cell-based engineering strategies, some of the genetic alterations of cells for use in bone reconstruction have until now been restricted to laboratory studies. Experiments using in vitro assay systems have yielded not only much basic information concerning the cultivation of these cell lines, but can also be used for tissue-engineering studies in basic and preclinical investigations. Each cell source has its inherent advantages and limitations.

As the use of autologous determined osteoblast-like cells does not raise, in contrast to the use of stem cells, either legal issues surrounding genetically altered cells or problems of immune rejection, autologous determined bone cells can be considered to be recently the most important cell source in bone tissue engineering [91]. Therefore, in current clinical practice, differentiated autologous osteoblast-like cells are the most desirable cell source. However, it is also true that these cells may be insufficient to rebuild damaged bone tissue in a reasonable time. A considerable number of cell divisions are needed to bulk the tissue to its correct size. Former studies have regarded the propagation of adult mature cells in culture as a serious problem, because it was thought that most adult tissues contained only a minority of cells capable of effective expansion. However, in numerous recent investigations it has been shown that bone cells proliferate in culture without losing their viability. Various sources of determined bone cells can be used for cultivation. Committed osteoprogenitors—i.e., progenitor cells restricted to osteoblast development and bone formation—can be identified by functional assays of their differentiation capacity in vitro, such as the colony-forming units assay (CFU-O assay). A wide variety of systemic, local, and positional factors regulate proliferation and differentiation of determined cells also, a sequence that is characterized by a series of cellular and molecular events distinguished by differential expression of osteoblast-associated genes including those for specific transcription factors, cell cycle-related proteins, adhesion molecules, and matrix proteins (for review, see [92]). The mature osteoblast phenotype is in itself heterogeneous with subpopulations of osteoblasts expressing only subsets of the known osteoblast markers, including those for cytokine, hormones, and growth factor receptors, raising the intriguing possibility that only certain osteoblasts are competent to respond to regulatory agents at particular points in time.

Cultures containing determined "osteoblastic" or "osteoblast-like" cells have been established from different cell populations in the lineage of osteogenic cells (osteoprogenitor cells, lining cells, osteoblasts, and osteocytes) (for review, see [93]). They can be derived from several anatomical sites, using different explant procedures. Bone cell populations may be derived from the cortical or cancellous bone, bone marrow, periosteum, and in some instances from other tissues. Isolation of cells can be performed by a variety of techniques, including mechanical disruption, explant outgrowth, and enzyme digestion [94]. Commonly used procedures to gain cells are digestive or outgrowth measures. Outgrowth of bone-like cells can be achieved through culturing of periosteum pieces or bone explants. Cells located within the periosteum and bone can differentiate into fibroblastic, osteogenic, or reticular cells [95–99]. Periosteal-derived mesenchymal precursor cells generate progenitor cells committed to one or more cell lines with an apparent degree of plasticity and interconversion [100–104]. In culture-expanded periosteum, cells were shown to retain the ability to heal a segmental bone defect after being reimplanted and induce osteogenic tissue when seeded into diffusion chambers [91, 105–107]. Outgrowth cultures of periosteum pieces favor the coculture of different cell types [108].

It remains controversial whether cortical or spongy bone, gained by different postsurgical processing techniques, is a better material of choice [109–111] to outgrow cells in culture. It has been suggested that particulate culturing is superior to bone chip culturing [112, 113], based on the assumption that when particle size decreases, the absolute square measure of the surface of the tissue specimens in the culture dish increases. An increased absolute square measure of the transplant surface would increase the amount of living cells released. Springer et al. [114] demonstrated in an experimental study, that bone chips obtained from trabecular bone provided a higher cell number than those raised from cortical bone. Surprisingly, they found that processing of spongy bone graft in the bone mill (leading to bone particulates) results in lower absolute amounts of osteoblast-like cells, whereas the use of the bone mill in cortical bone has much less impact on the number of cells counted. The authors speculated that the treatment of transplants with the bone mill or raising of transplants by rotating instruments should reduce the amount of bone cells supplied, suggesting that a decreased particle size is disadvantageous when cell outgrow methods are in use. Ecarot-Charrier and colleagues [115] were the first to present a method for isolating osteoblasts from newborn mouse calvaria using digestive enzymes in solution. As it was demonstrated that isolated osteoblasts gained through this method retained their unique properties in culture, tissue digestion became a common method of cell harvesting for in vitro purposes.

Once such matured cells have been isolated from the tissue, there are also in such culture strategies a number of parameters that influence the expression of the osteoblastic phenotype in cell culture, most important the culture medium, culture time, number of passages, and the presence of compounds. The presence of ascorbic acid, ß-glycerophosphate, and dexamethasone influence the expression of the osteoblastic phenotype in a differentiated manner. ß-Glycerophosphate for example induces phenotypic matrix maturation by enabling mineral formation in osteoblast-like cell cultures. Dexamethasone, as an additional factor, is described as inducing cell differentiation but imposes a negative effect on cell proliferation, indicative for a reciprocal and functionally coupled relationship between proliferation and differentiation. Therefore, it is convenient to select suitable experimental conditions for cultivation. As with stem cell cultivation, the culture conditions should be well defined in order to standardize the ex vivo grown product.

Because endochondral bone formation and frequently fracture repair proceed through a cartilaginous intermediate, some investigators have suggested that the transplantation of committed chondrocytes would also ameliorate bone regeneration [102]. Vacanti and colleagues [94] compared the ability of periosteal progenitors and articular chondrocytes to effect bone repair. They showed that periosteal cells from newborn calves seeded on a scaffold and implanted in critical-sized calvarial defects generated new bone. Specimens examined at early times contained material that grossly and histologically appeared to be cartilage. The scaffold seeded with chondrocytes also formed cartilage. However, no endochondral ossification was observed, since the transplanted specimens remained in a cartilaginous state. Therefore, chondrocytes proved ineffective as a cell-based therapy for tissue engineering of bone. Because mature cartilage is thought to produce factors that inhibit angiogenesis, implants seeded with committed chondrocytes may prevent the endochondral cascade by preventing vascular invasion. Cells derived from cartilage seem to be committed to retain their phenotype and, therefore, are unable to differentiate toward hypertrophic chondrocytes under the experimental conditions tested so far. In contrast, when precursor cells from the periosteum are provided, their primitive developmental state allows them to proceed through the entire chondrogenic lineage, ultimately becoming hypertrophic chondrocytes. The molecular basis for the difference in the phenotypic potential of these different cell types remains a mystery and is an area of intensive investigation.

17.11 Evaluation of Engineering Success

For the purpose of testing ex vivo-generated bone tissue substitutes, the experimental model should allow manipulation of the mode of fracture healing, the size and location of segmental bone defects, and the various forms of regeneration disturbances. A careful experimental design and an extended evaluation including the biomechanical characteristics of the newly formed tissue will improve our understanding of the biological basis and the clinical implications of bone tissue engineering (Table 17.3). A test of the conditions under which bone tissue engineering is advantageous over other techniques greatly relies on the experimental model selected [116]. Bone tissue engineering approaches are aimed at restoring large segments of the skeletal bone lost for various reasons, improving the healing of complicated fractures, and hopefully in the near future fully regenerating complex skeletal defects [117–119]. Animal models should effectively mimic the clinical situation in human bone healing. However, it is often desired that bone defects in animal models should fail to heal unless they are treated with a cell-based tissue

Table 17.3 Animal experimental and clinical evaluation methods

Method (in vivo)	Determination
Clinic	Form, function
X-ray	Hard tissue structure/overview
Microradiography	Hard tissue structure/details
Computed tomography	Hard and soft tissue structure/overview
µCT	Hard and soft tissue structure/details
NMR	Hard and soft tissue structure
Densitometry	Mineral content
Serum analysis	Bone formation and resorption marker, hormones, cytokines
Urine analysis	Bone formation and resorption marker, hormones, cytokines
Method (in vitro)	
Histology	Tissue structure

engineering strategy. Clinicians should be aware that bone tissue engineering approaches have unique features, as they differ in different fields of surgery. For example, bony regeneration is often observed after a sinus lift procedure in the maxillary sinus region, even in the absence of implanted biomaterial. In this case, elevation of the sinus mucosa seems to be the crucial procedure that induces bone growth. Bone regeneration in the maxillofacial area appears to take place in part by mechanisms distinct from those in the orthopedic field. In particular, the response to scaffold materials may be different depending on the nature of the surrounding intact bone and soft tissue. As an initial approach to determining the osteogenic potential of tissue-engineered bone constructs in vivo, heterotopic animal models are of special interest. They allow the evaluation of the biological performance of the hybrid material commonly implanted in ectopic sites. Implantation is performed for example in muscle pouches, fascial pouches, or subcutaneously. The investigation can be performed in immunocompetent as well as in immunocompromised animals. The next step of preclinical evaluation is the in vivo defect model. Schmitz and Hollinger [120] were the first to postulate a rationale for testing bone tissue engineering by using critical size defects in a hierarchy of animal models. It is known that the rate of bone repair varies inversely with order along the phylogenetic scale [121, 122]. Moreover, bone regenerative capacities differ significantly among animals of the same species [123, 124] and among loaded and nonloaded skeletal sites. Bone defects regenerate more actively in immature animals of a given species than in older ones. Therefore, from a clinical point of view it is advisable to test cellular hybrid materials for their ability to repair an osseous defect in a mature, adult animal [123]. Mature dogs, goats, pigs, and nonhuman primates have been successfully used to examine the regeneration of mandibular discontinuity defects [125]. These animal models allow us to follow the fate of the implanted material in a functional environment which is comparable to the human situation. This is particularly important in preparation for future US Food and Drug Administration trials [126], to establish these treatment options in daily clinical practice.

Cell based transplantation approaches are nowadays measures to improve tissue healing, since the legal and regulatory situation allows the transplantation of autologous cells. This kind of cell-based therapy enables us to accelerate bone formation. Figures 17.5–17.9 demonstrate the fundamental steps of an extracorporal tissue engineering strategy. Cells are harvested in a first step (Part I; Fig. 17.5). Autologous serum is gained from the patient to allow cells to multiplicate in a bioreactor system provided by a commercial tissue engineering company (co.don AG, Berlin; Part II; Fig. 17.6). Multiplicated cells are then coaxed with a scaffold (Part III; Fig. 17.7) and implanted in a critical size mandibular cyst (Part IV; Fig. 17.8). The radiological control (before cystectomy, directly afterwards, and 1 and 6 weeks after transplantation; Part V; Fig. 17.9) demonstrates a fast healing of the bony defect.

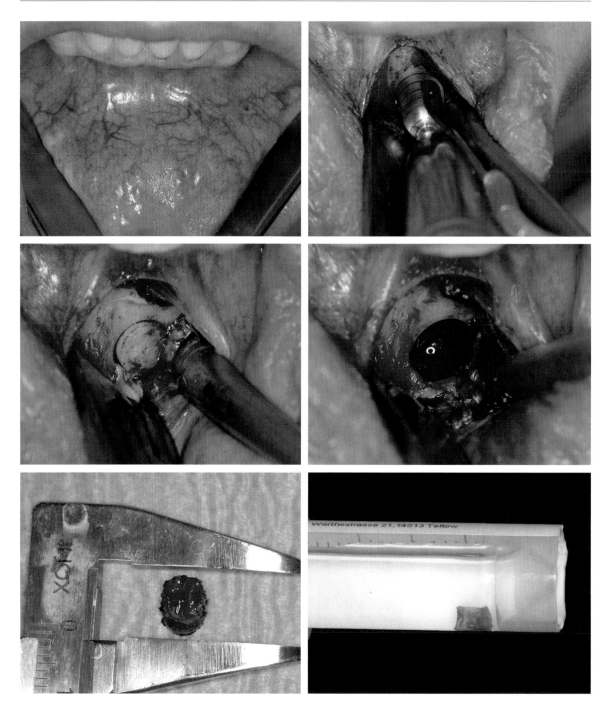

Fig. 17.5 Clinical proceeding of bone tissue engineering for bone reconstruction: part I: explantation

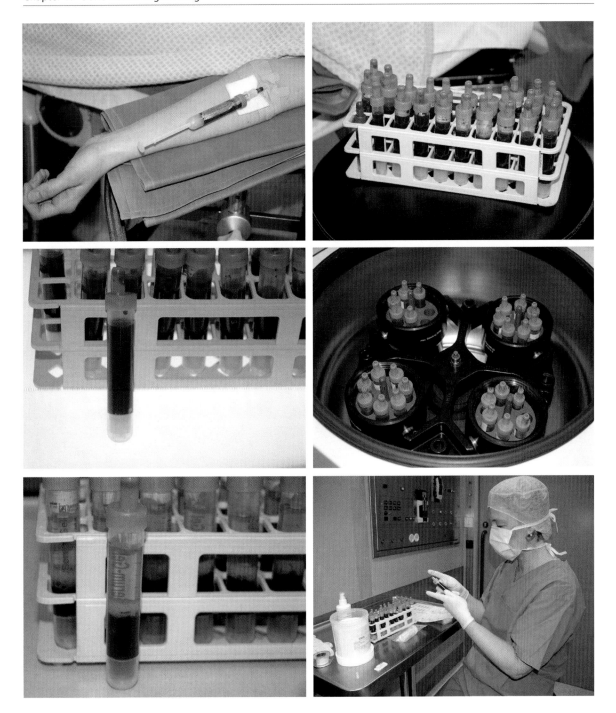

Fig. 17.6 Clinical proceeding of bone tissue engineering for bone reconstruction: part II: laboratory preparation

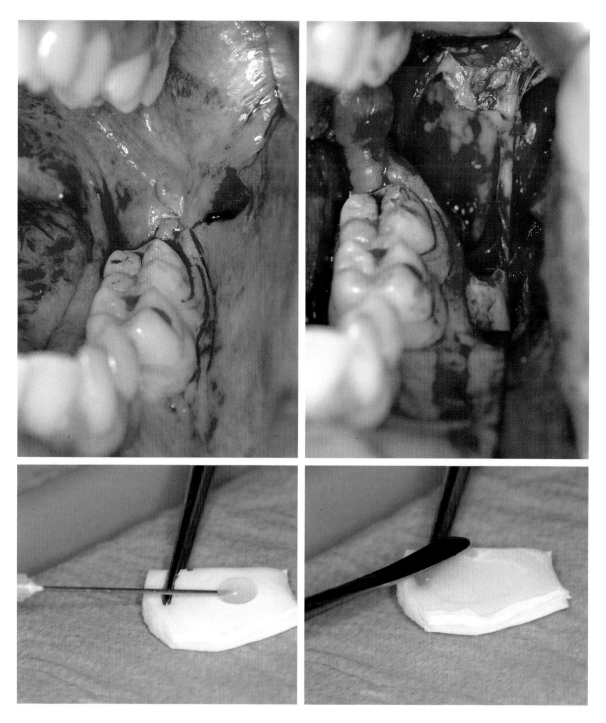

Fig. 17.7 Clinical proceeding of bone tissue engineering for bone reconstruction: part III: scaffold seeding

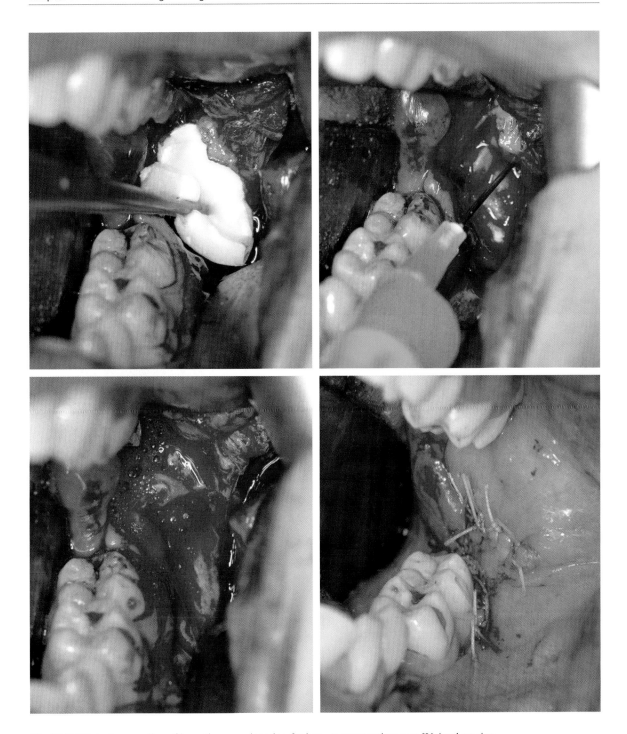

Fig. 17.8 Clinical proceeding of bone tissue engineering for bone reconstruction: part IV: implantation

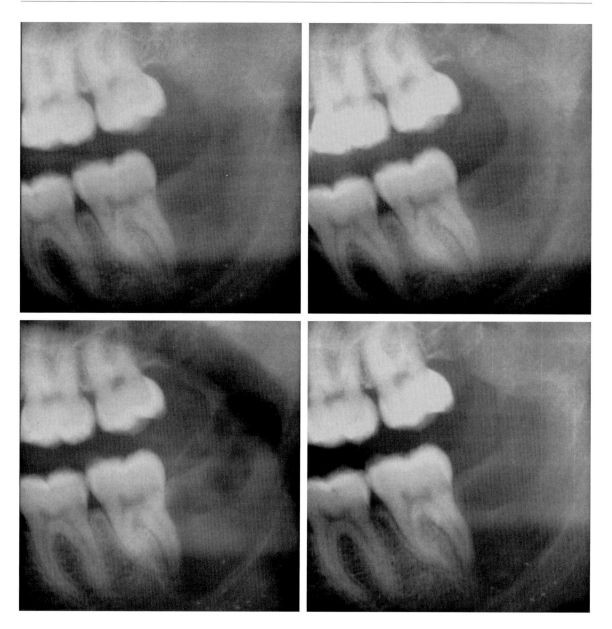

Fig. 17.9 Clinical proceeding of bone tissue engineering for bone reconstruction: part V: radiological control

References

1. Schultz O, Sittinger M, Haeupl T, et al (2000) Emerging strategies of bone and joint repair. Arthritis Res 2:433–436
2. Damien CJ, Parsons JR (1991) Bone graft and bone graft substitutes: a review of current technology and applications. J Appl Biomater 2:187–208
3. Binderman I, Fin N (1990) Bone substitutes organic, inorganic, and polymeric: Cell material interactions. CRC press Boca Raton
4. Alsberg E, Hill EE, Mooney DJ (2001) Craniofacial tissue engineering. Crit Rev Oral Biol Med 12:64–75
5. Langer R, Vacanti JP (1993) Tissue engineering. Science 260:920–926
6. Lysaght MJ, Reyes J (2001) The growth of tissue engineering. Tissue Eng 7:485–493
7. Loty C, Sautier JM, Boulekbache H, et al (2000) In vitro bone formation on a bone-like apatite layer prepared by a biomimetic process on a bioactive glass-ceramic. J Biomed Mater Res 49:423–434
8. Meyer U, Joos U, Wiesmann HP (2004a) Biological and biophysical principles in extracorporal bone tissue engineering. Part I. Int J Oral Maxillofac Surg 33:325–332
9. Schliephake H, Knebel JW, Aufderheide M, et al (2001) Use of cultivated osteoprogenitor cells to increase bone formation in segmental mandibular defects: an experimental pilot study in sheep. Int J Oral Maxillofac Surg 30:531–537
10. Lang NP, Hammerle CH, Bragger U, et al (1994) Guided tissue regeneration in jawbone defects prior to implant placement. Clin Oral Implants Res 5:92–97
11. Buser D, Dula K, Hirt HP, et al (1996) Lateral ridge augmentation using autografts and barrier membranes: a clinical study with 40 partially edentulous patients. J Oral Maxillofac Surg 54:420–432
12. Berglundh T, Lindhe J (1997) Healing around implants placed in bone defects treated with Bio-Oss. An experimental study in the dog. Clin Oral Implants Res 8:117–124
13. Fiorellini JP, Engebretson SP, Donath K, et al (1998) Guided bone regeneration utilizing expanded polytetrafluoroethylene membranes in combination with submerged and nonsubmerged dental implants in beagle dogs. J Periodontol 69:528–535
14. Hämmerle CH, Chiantella GC, Karring T, et al (1998) The effect of a deproteinized bovine bone mineral on bone regeneration around titanium dental implants. Clin Oral Implants Res 9:151–162
15. Brighton CT (1998) Breakout session. 4: Biophysical enhancement. Clin Orthop Relat Res (355 Suppl):S357–358
16. Chao EY, Inoue N (2003) Biophysical stimulation of bone fracture repair, regeneration and remodelling. Eur Cell Mater 6:72–84
17. Meyer U, Kleinheinz J, Joos U (2004b) Biomechanical and clinical implications of distraction osteogenesis in craniofacial surgery. J Craniomaxillofac Surg 32:140–149
18. Ilizarov GA (1992) The transosseous osteosynthesis. Theoretical and clinical aspects of the regeneration and growth of tissue. Springer New York
19. Sato M, Ochi T, Nakase T, et al (1999) Mechanical tension-stress induces expression of bone morphogenetic protein (BMP)-2 and BMP-4, but not BMP-6, BMP-7, and GDF-5 mRNA, during distraction osteogenesis. J Bone Miner Res 14:1084–1095
20. Choi IH, Ahn JH, Chung CY, et al (2000) Vascular proliferation and blood supply during distraction osteogenesis: a scanning electron microscopic observation. J Orthop Res 18:698–705
21. Meyer U, Meyer T, Schlegel W, et al (2001) Tissue differentiation and cytokine synthesis during strain-related bone formation in distraction osteogenesis. Br J Oral Maxillofac Surg 39:22–29
22. Bouletreau PJ, Warren SM, Longaker MT (2002a) The molecular biology of distraction osteogenesis. J Craniomaxillofac Surg 30:1–11
23. Bouletreau PJ, Warren SM, Spector JA, et al (2002b) Hypoxia and VEGF up-regulate BMP-2 mRNA and protein expression in microvascular endothelial cells: implications for fracture healing. Plast Reconstr Surg 109:2384–2397
24. Schliephake H (2002) Bone growth factors in maxillofacial skeletal reconstruction. Int J Oral Maxillofac Surg 31:469–469–484
25. Stevenson S, Horowitz M (1992) The response to bone allografts. J Bone Joint Surg Am 74:939–950
26. Hogan BL (1996) Bone morphogenetic proteins: multifunctional regulators of vertebrate development. Genes Dev 10:1580–1594
27. Urist MR (1965) Bone: formation by autoinduction. Science 150:893–899
28. Urist MR, Strates BS (1971) Bone morphogenetic protein. J Dent Res 50:1392–1406
29. Wozney JM, Rosen V, Celeste AJ, et al (1988) Novel regulators of bone formation: molecular clones and activities. Science 242:1528–1534
30. Ozkaynak E, Rueger DC, Drier EA, et al (1990) OP-1 cDNA encodes an osteogenic protein in the TGF-beta family. EMBO J 9:2085–2093
31. Fang J, Zhu YY, Smiley E, et al (1996) Stimulation of new bone formation by direct transfer of osteogenic plasmid genes. Proc Natl Acad Sci U S A 93:5753–5758
32. Boden SD, Titus L, Hair G, et al (1998) Lumbar spine fusion by local gene therapy with a cDNA encoding a novel osteoinductive protein (LMP-1). Spine 23:2486–2492
33. Lieberman JR, Le LQ, Wu L, et al (1998) Regional gene therapy with a BMP-2-producing murine stromal cell line induces heterotopic and orthotopic bone formation in rodents. J Orthop Res 16:330–339
34. Zins JE, Whitaker LA (1983) Membranous versus endochondral bone: implications for craniofacial reconstruction. Plast Reconstr Surg 72:778–785
35. Conley J (1972) Use of composite flaps containing bone for major repairs in the head and neck. Plast Reconstr Surg 49:522–526
36. Siemssen SO, Kirkby B, O'Connor TP (1978) Immediate reconstruction of a resected segment of the lower jaw,

using a compound flap of clavicle and sternomastoid muscle. Plast Reconstr Surg 61:724–735
37. Demergasso F, Piazza MV (1979) Trapezius myocutaneous flap in reconstructive surgery for head and neck cancer: an original technique. Am J Surg 138:533–536
38. Panje W, Cutting C (1980) Trapezius osteomyocutaneous island flap for reconstruction of the anterior floor of the mouth and the mandible. Head Neck Surg 3:66–71
39. Guillamondegui OM, Larson DL (1981) The lateral trapezius musculocutaneous flap: its use in head and neck reconstruction. Plast Reconstr Surg 67:143–150
40. Cuono CB, Ariyan S (1980) Immediate reconstruction of a composite mandibular defect with a regional osteomusculocutaneous flap. Plast Reconstr Surg 65:477–484
41. Green MF, Gibson JR, Bryson JR, et al (1981) A one-stage correction of mandibular defects using a split sternum pectoralis major osteo-musculocutaneous transfer. Br J Plast Surg 34:11–16
42. Lam KH, Wei WI, Siu KF (1984) The pectoralis major costomyocutaneous flap for mandibular reconstruction. Plast Reconstr Surg 73:904–910
43. Maruyama Y, Urita Y, Ohnishi K (1985) Rib-latissimus dorsi osteomyocutaneous flap in reconstruction of a mandibular defect. Br J Plast Surg 38:234–237
44. McCarthy JG, Zide BM (1984) The spectrum of calvarial bone grafting: introduction of the vascularized calvarial bone flap. Plast Reconstr Surg 74:10–18
45. McCarthy JG, Cutting CB, Shaw WW (1987) Vascularized calvarial flaps. Clin Plast Surg 14:37–47
46. Rose EH, Norris MS (1990) The versatile temporoparietal fascial flap: adaptability to a variety of composite defects. Plast Reconstr Surg 85:224–232
47. Taylor GI (1982) Reconstruction of the mandible with free composite iliac bone grafts. Ann Plast Surg 9:361–376
48. Shenaq SM (1988) Reconstruction of complex cranial and craniofacial defects utilizing iliac crest-internal oblique microsurgical free flap. Microsurgery 9:154–158
49. Taylor GI, Miller GD, Ham FJ (1975) The free vascularized bone graft. A clinical extension of microvascular techniques. Plast Reconstr Surg 55:533–544
50. Wei FC, Chen HC, Chuang CC, et al (1986) Fibular osteoseptocutaneous flap: anatomic study and clinical application. Plast Reconstr Surg 78:191–200
51. Wei FC, Seah CS, Tsai YC, et al (1994) Fibula osteoseptocutaneous flap for reconstruction of composite mandibular defects. Plast Reconstr Surg 93:294–304
52. Hidalgo DA (1989) Fibula free flap: a new method of mandible reconstruction. Plast Reconstr Surg 84:71–79
53. dos Santos LF (1984) The vascular anatomy and dissection of the free scapular flap. Plast Reconstr Surg 73:599–604
54. Swartz WM, Banis JC, Newton ED, et al (1986) The osteocutaneous scapular flap for mandibular and maxillary reconstruction. Plast Reconstr Surg 77:530–545
55. Baker SR (1989) Reconstruction of the Head and Neck. Churchill Livingston New York
56. Granick MS, Ramasastry SS, Newton ED, et al (1990) Reconstruction of complex maxillectomy defects with the scapular-free flap. Head Neck 12:377–385
57. Frodel JL Jr, Funk GF, Capper DT, et al (1993) Osseointegrated implants: a comparative study of bone thickness in four vascularized bone flaps. Plast Reconstr Surg 92:449–55
58. Tan BK, Chen HC, He TM, et al (2004) Flap prefabrication—the bridge between conventional flaps and tissue-engineered flaps. Ann Acad Med Singapore 33:662–666
59. Cao Y, Vacanti JP, Paige KT, et al (1997) Transplantation of chondrocytes utilizing a polymer-cell construct to produce tissue-engineered cartilage in the shape of a human ear. Plast Reconstr Surg 100:297–302
60. Alam MI, Asahina I, Seto I, et al (2003) Prefabrication of vascularized bone flap induced by recombinant human bone morphogenetic protein 2 (rhBMP-2). Int J Oral Maxillofac Surg 32:508–514
61. Jaquiery C, Rohner D, Kunz C, et al (2004) Reconstruction of maxillary and mandibular defects using prefabricated microvascular fibular grafts and osseointegrated dental implants—a prospective study. Clin Oral Implants Res 15:598–606
62. Keser A, Bozkurt M, Taner OF, et al (2004) Prefabrication of bone by vascular induction: an experimental study in rabbits. Scand J Plast Reconstr Surg Hand Surg 38:257–260
63. Schultze-Mosgau S, Lee BK, Ries J, et al (2004) In vitro cultured autologous pre-confluent oral keratinocytes for experimental prefabrication of oral mucosa. Int J Oral Maxillofac Surg 33:476–485
64. Staudenmaier R, Hoang TN, Kleinsasser N, et al (2004) Flap prefabrication and prelamination with tissue-engineered cartilage. J Reconstr Microsurg 20:555–564
65. Terheyden H, Menzel C, Wang H, et al (2004) Prefabrication of vascularized bone grafts using recombinant human osteogenic protein-1–part 3: dosage of rhOP-1, the use of external and internal scaffolds. Int J Oral Maxillofac Surg 33:164–172
66. The Hoang N, Kloeppel M, Staudenmaier R, et al (2005) Neovascularization in prefabricated flaps using a tissue expander and an implanted arteriovenous pedicle. Microsurgery 25:213–219
67. Top H, Aygit C, Sarikaya A, et al (2005) Bone flap prefabrication: an experimental study in rabbits. Ann Plast Surg 54:428–434
68. Fisher J, Wood MB (1987) Experimental comparison of bone revascularization by musculocutaneous and cutaneous flaps. Plast Reconstr Surg 79:81–90
69. Findlay M, Dolderer J, Cooper-White J, et al (2003) Creating large amounts of tissue for reconstructive surgery—a porcine model. Aust N Z J Surg 73:240
70. Warnke PH, Springer IN, Wiltfang J, et al (2004) Growth and transplantation of a custom vascularised bone graft in a man. Lancet 364:766–770
71. Yamaguchi M, Hirayama F, Murahashi H, et al (2002) Ex vivo expansion of human UC blood primitive hematopoietic progenitors and transplantable stem cells using human primary BM stromal cells and human AB serum. Cytotherapy 4:109–118
72. Triffitt JT (2002) Osteogenic stem cells and orthopedic engineering: summary and update. J Biomed Mater Res 63:384–389
73. Martin GR (1981) Isolation of a pluripotent cell line from early mouse embryos cultured in medium conditioned by

teratocarcinoma stem cells. Proc Natl Acad Sci U S A 78:7634–7638
74. Ham AW (1969) Histology. Lippincott Co. Philadelphia
75. Owen M (1988) Marrow stromal stem cells. J Cell Sci Suppl 10:63–76
76. Weinberg CB, Bell E (1986) A blood vessel model constructed from collagen and cultured vascular cells. Science 231:397–400
77. Caplan AI (1991) Mesenchymal stem cells. J Orthop Res 9:641–650
78. Nathanson MA (1985) Bone matrix-directed chondrogenesis of muscle in vitro. Clin Orthop Relat Res (200):142–158
79. Nakahara H, Dennis JE, Bruder SP, et al (1991) In vitro differentiation of bone and hypertrophic cartilage from periosteal-derived cells. Exp Cell Res 195:492–503
80. Campagnoli C, Roberts IA, Kumar S, et al (2001) Identification of mesenchymal stem/progenitor cells in human first-trimester fetal blood, liver, and bone marrow. Blood 98:2396–2402
81. Erices A, Conget P, Minguell JJ (2000) Mesenchymal progenitor cells in human umbilical cord blood. Br J Haematol 109:235–242
82. Gutierrez-Rodriguez M, Reyes-Maldonado E, Mayani H (2000) Characterization of the adherent cells developed in Dexter-type long-term cultures from human umbilical cord blood. Stem Cells 18:46–52
83. Mareschi K, Biasin E, Piacibello W, et al (2001) Isolation of human mesenchymal stem cells: bone marrow versus umbilical cord blood. Haematologica 86:1099–1100
84. Zvaifler NJ, Marinova Mutafchieva L, Adams G, et al (2000) Mesenchymal precursor cells in the blood of normal individuals. Arthritis Res 2:477–488
85. Wobus AM (2001) Potential of embryonic stem cells. Mol Aspects Med 22:149–164
86. Alison MR, Poulsom R, Forbes S, et al (2002) An introduction to stem cells. J Pathol 197:419–423
87. Jaiswal N, Haynesworth SE, Caplan AI, et al (1997) Osteogenic differentiation of purified, culture-expanded human mesenchymal stem cells in vitro. J Cell Biochem 64:295–312
88. Plate U, Arnold S, Stratmann U, et al (1998) General principle of ordered apatitic crystal formation in enamel and collagen rich hard tissues. Connect Tissue Res 38:149–57
89. Evans CH, Robbins PD (1995) Possible orthopaedic applications of gene therapy. J Bone Joint Surg Am 77:1103–1114
90. Oakes DA, Lieberman JR (2000) Osteoinductive applications of regional gene therapy: ex vivo gene transfer. Clin Orthop Relat Res (379 Suppl): 101–112
91. Ashton BA, Allen TD, Howlett CR, et al (1980) Formation of bone and cartilage by marrow stromal cells in diffusion chambers in vivo. Clin Orthop Relat Res (151):294–307
92. Yamaguchi A, Komori T, Suda T (2000) Regulation of osteoblast differentiation mediated by bone morphogenetic proteins, hedgehogs, and Cbfa1. Endocr Rev 21:393–411
93. Hutmacher DW, Sittinger M (2003) Periosteal cells in bone tissue engineering. Tissue Eng 9:S45–64
94. Vacanti CA, Kim W, Upton J, et al (1995) The efficacy of periosteal cells compared to chondrocytes in the tissue engineered repair of bone defects. Tissue Eng 1:301–301–308
95. Friedenstein AJ (1976) Precursor cells of mechanocytes. Int Rev Cytol 47:327–359
96. Nuttall ME, Patton AJ, Olivera DL, et al (1998) Human trabecular bone cells are able to express both osteoblastic and adipocytic phenotype: implications for osteopenic disorders. J Bone Miner Res 13:371–382
97. Triffitt JT, Oreffo ROC (1998) Osteoblast lineage. JAI Press, Inc. Connecticut
98. Dahir GA, Cui Q, Anderson P, et al (2000) Pluripotential mesenchymal cells repopulate bone marrow and retain osteogenic properties. Clin Orthop Relat Res (379 Suppl): 134–145
99. Bianco P, Riminucci M, Gronthos S, et al (2001) Bone marrow stromal stem cells: nature, biology, and potential applications. Stem Cells 19:180–192
100. Nakahara H, Goldberg VM, Caplan AI (1992) Culture-expanded periosteal-derived cells exhibit osteochondrogenic potential in porous calcium phosphate ceramics in vivo. Clin Orthop Relat Res (276):291–298
101. Park SR, Oreffo RO, Triffitt JT (1999) Interconversion potential of cloned human marrow adipocytes in vitro. Bone 24:549–554
102. Bahrami S, Stratmann U, Wiesmann HP, et al (2000) Periosteally derived osteoblast-like cells differentiate into chondrocytes in suspension culture in agarose. Anat Rec 259:124–130
103. Schantz JT, Hutmacher DW, Chim H, et al (2002) Induction of ectopic bone formation by using human periosteal cells in combination with a novel scaffold technology. Cell Transplant 11:125–138
104. Schantz JT, Hutmacher DW, Ng KW, et al (2002) Evaluation of a tissue-engineered membrane-cell construct for guided bone regeneration. Int J Oral Maxillofac Implants 17:161–174
105. Ohgushi H, Goldberg VM, Caplan AI (1989a) Repair of bone defects with marrow cells and porous ceramic. Experiments in rats. Acta Orthop Scand 60:334–339
106. Ohgushi H, Goldberg VM, Caplan AI (1989b) Heterotopic osteogenesis in porous ceramics induced by marrow cells. J Orthop Res 7:568–578
107. Nakahara H, Bruder SP, Goldberg VM, et al (1990) In vivo osteochondrogenic potential of cultured cells derived from the periosteum. Clin Orthop Relat Res (259):223–232
108. Meyer U, Szulczewski HD, Moller K, et al (1993) Attachment kinetics and differentiation of osteoblasts on different biomaterials. Cells Mater 3:129–129–140
109. Girdler NM, Hosseini M (1992) Orbital floor reconstruction with autogenous bone harvested from the mandibular lingual cortex. Br J Oral Maxillofac Surg 30:36–38
110. Chen NT, Glowacki J, Bucky LP, et al (1994) The roles of revascularization and resorption on endurance of craniofacial onlay bone grafts in the rabbit. Plast Reconstr Surg 93:714–22
111. Schwipper V, von Wild K, Tilkorn H (1997) Reconstruction of frontal bone, periorbital and calvarial defects with

111. autogenic bone. Mund Kiefer Gesichtschir 1 Suppl 1: 71–4
112. Marx RE, Miller RI, Ehler WJ, et al (1984) A comparison of particulate allogeneic and particulate autogenous bone grafts into maxillary alveolar clefts in dogs. J Oral Maxillofac Surg 42:3–9
113. Shirota T, Ohno K, Motohashi M, et al (1996) Histologic and microradiologic comparison of block and particulate cancellous bone and marrow grafts in reconstructed mandibles being considered for dental implant placement. J Oral Maxillofac Surg 54:15–20
114. Springer IN, Terheyden H, Geiss S, et al (2004) Particulated bone grafts—effectiveness of bone cell supply. Clin Oral Implants Res 15:205–212
115. Ecarot-Charrier B, Glorieux FH, van der Rest M, et al (1983) Osteoblasts isolated from mouse calvaria initiate matrix mineralization in culture. J Cell Biol 96:639–643
116. Einhorn TA (1999) Clinically applied models of bone regeneration in tissue engineering research. Clin Orthop Relat Res (367 Suppl):S59–67
117. Puelacher WC, Vacanti JP, Ferraro NF, et al (1996) Femoral shaft reconstruction using tissue-engineered growth of bone. Int J Oral Maxillofac Surg 25:223–228
118. Bruder SP, Kraus KH, Goldberg VM, et al (1998a) The effect of implants loaded with autologous mesenchymal stem cells on the healing of canine segmental bone defects. J Bone Joint Surg Am 80:985–996
119. Bruder SP, Kurth AA, Shea M, et al (1998b) Bone regeneration by implantation of purified, culture-expanded human mesenchymal stem cells. J Orthop Res 16:155-162
120. Schmitz JP, Hollinger JO (1986) The critical size defect as an experimental model for craniomandibulofacial nonunions. Clin Orthop Relat Res 205:299–308
121. Enneking WF, Morris JL (1972) Human autologous cortical bone transplants. Clin Orthop Relat Res 87:28–35
122. Prolo DJ, Pedrotti PW, Burres KP, et al (1982) Superior osteogenesis in transplanted allogeneic canine skull following chemical sterilization. Clin Orthop Relat Res 168:230–242
123. Harris WH, Lavorgna J, Hamblen DL, et al (1968) The inhibition of ossification in vivo. Clin Orthop Relat Res 61:52–60
124. Enneking WF, Burchardt H, Puhl JJ, et al (1975) Physical and biological aspects of repair in dog cortical-bone transplants. J Bone Joint Surg Am 57:237–252
125. Fennis JP, Stoelinga PJ, Jansen JA (2002) Mandibular reconstruction: a clinical and radiographic animal study on the use of autogenous scaffolds and platelet-rich plasma. Int J Oral Maxillofac Surg 31:281–286
126. Slavkin H (2000) Thoughts on the future of dental and craniofacial research. Compend Contin Educ Dent 21:927–930

Cartilage Engineering

J. Libera, K. Ruhnau, P. Baum, U. Lüthi,
T. Schreyer, U. Meyer, H. P. Wiesmann,
A. Herrmann, T. Korte, O. Pullig, V. Siodla

Contents

18.1	Introduction	233
18.2	Manufacturing and Clinical Application	234
18.3	Clinical Results	234
18.4	Preclinical Studies	237
18.5	In Vivo Integration and Differentiation Studies	238
18.6	In Vitro Differentiation and Maturation	239
	References	241

18.1 Introduction

Adult articular cartilage has a limited capacity for repair and cartilage defects often progress to osteoarthritis [7, 8]. The potential to repair hyaline cartilage is limited due to poor vascularisation, reduced migration and mitogenic characteristics of chondrocytes. In 1987 autologous chondrocytes transplantation was pioneered by Peterson and Brittberg to treat cartilage defects mainly in the knee joint. Peterson und Brittberg (1994) introduced a method where in vitro-propagated chondrocytes in suspension were transplanted under a periosteal flap to treat cartilage defects of human (ACT) [20, 21]. During the following years numerous in vitro and in vivo studies were performed to obtain insight into characteristics, function and behaviour of chondrocytes as well as to improve operational techniques, standardize clinical assessment and rehabilitation procedure. The first step for the operational improvement of the mentioned first generation of ACT was done by replacing the periosteal flap by a collagen membrane to cover the cartilage defect [23].

A second generation of cell-based transplants for cartilage repair was developed soon, combining propagated chondrocytes with scaffold materials. Chondrocytes were seeded on hyaluronic acid scaffolds, on collagen membranes or collagen gels and cultured, termed matrix-associated chondrocyte implantation (MACI) [2–4, 6, 19]. The cell-loaded membrane or scaffold is placed in the defect ground and fixed by fibrin glue. This technique avoids a close suturing of a periosteal flap.

As a third generation, chondrocytes were cultured three dimensionally without the use of any scaffold materials, offering a complete autologous three-dimensional (3D) autologous transplant that can be applied arthroscopically [1]. For this the cells were cultured in high density, where chondrocytes form so-called spheroids that only consist of the patient's own chondrocytes and a matrix that is synthesized by the chondrocytes themselves.

As for these techniques of the first-, second- and third-generation, cartilage biopsies were taken, chondrocytes isolated and transplanted back during a second operation, these techniques belong to the two-step procedures.

In parallel with these chondrocyte transplants, cartilage repair techniques were developed using cell free-scaffold materials, allowing a one-step procedure. This technique bases on microfracture of the subchondral bone, the in-bleeding, and migration of blood mesenchymal stem cells into a scaffold material that is placed into the cartilage defect [5, 13]. This technique is called autologous matrix-induced chondrogenesis (AMIC). Currently the improvement of this xenogenous one-step procedure is under development. For this, during the operation, chondrocytes are isolated out of a cartilage biopsy, mixed with bone marrow cells and seeded on a scaffold material [11]. A similar technique was induced by Lu et al. [15], seeding minced hyaline cartilage on a scaffold material. These cell- or tissue-loaded scaffolds are placed into the defect ground and fixed during the same operation.

This wide spectrum of techniques is indicative of the public and scientific interest to offer techniques that efficiently repair cartilage defects. The efficiency and safety of all these techniques need to be evaluated by clinical trials that will be base for the future market authorization under the new European legislation.

The following sections are focused on the clinical and preclinical results of the above-mentioned autologous chondrocyte spheroid technology.

18.2
Manufacturing and Clinical Application

Human chondrocytes are isolated out of a cartilage biopsy and after propagation transferred into a suspension culture system. Manufacturing is performed according to the German drug law and under the guidelines of GMP (good manufacturing practice).

Under the suspension culture conditions, chondrocytes aggregate and form spheroids (so-called chondrospheres®) initiated by cell-cell contact. Chondrocytes within spheroids are spherical in shape, allowing the expression of the hyaline-specific phenotype [1]. The synthesized hyaline-specific matrix components stay within the lumen of the cell aggregates, resulting in the formation of solid cell-matrix constructs. The size of chondrospheres® depends on the used cell number. Clinically, spheroids with a size of 600 to 800 µm are used (Fig. 18.1). This size allows an arthroscopic transplantation.

As the application and fixation of co.don chondrosphere® does not include any additional surgical procedure or additives such as a membrane, matrix or sealant, co.don chondrosphere® can be simply placed arthroscopically into the defects to treat (Fig. 18.1). The size of chondrospheres® of 500 up to 800 µm allows the use of 1-mm working channel to place the spheroids into the defect. After debriding the defect and removing the synovial fluid, spheroids were dropped into the defect e.g. using flexible cannulas. By using sterile arthroscopical instruments, spheroids can be distributed within the defect ground.

18.3
Clinical Results

Since 2004 co.don chondrosphere® has been clinically used to treat cartilage defects of the knee joint, talus, shoulder and recently of the hip. Pilot trials and

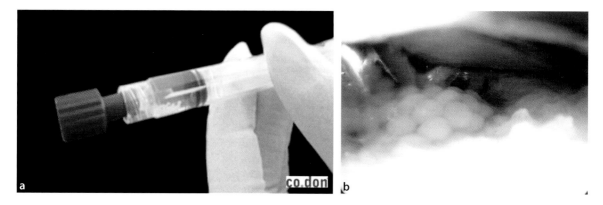

Fig. 18.1 **a** co.don® AG® chondrosphere® within a syringe. **b** Arthroscopic view after transplantation into an articular cartilage defect. (Fig. 1b courtesy of Dr. Schreyer, ev. Elisabethenstift Hospital, Darmstadt, Germany)

histological/immunohistochemical evaluations of the transplants as well as of regenerate biopsies indicated the efficiency of the autologous 3D chondrocyte transplants. Since 2004 more that 500 patients were treated.

Pilot trials were performed in three German clinical centres including 26 patients (Dr. Ruhnau, St. Marien-Hospital Buer), 10 patients (Dr. Baum, Gelenkklinik Gundelfingen) and 6 patients (Dr. Schreyer, ev. Hospital Elisabethenstift, Darmstadt). Patients included into the pilot trials at hospitals in the Buer and Gundelfingen were indicated for the treatment of full chondral defects, graded as Outerbridge III to IV, aged between 15 and 51 years. Defects had a size between 1.5 to 10.1 cm² and were located at the medial and lateral condyles, and patella trochleae. Patients included in the pilot trial in Darmstadt suffered full chondral, graded as Outerbridgew III to IV, with an average defect size of 3.9 cm², and average age of 36 years. Three patients were treated arthroscopically.

Cartilage biopsies were taken from a non-loaded area and sent together with an autologous blood sample to co.don® AG. Autologous chondrocytes were isolated and propagated in monolayer. After obtaining a sufficient cell number, chondrocytes were transferred into a chondrospheroid culture system. After additional 2 weeks, chondrosphere® processing was finished and the transplant shipped to the clinical site for immediate transplantation. The German society for cartilage and bone cell transplantation recommends a dose of 1 million chondrocytes per squared centimetre. Chondrosphere® is used in a concentration of about 3 million chondrocytes per squared centimetre.

The surgical treatment has been performed via mini-arthroscopy. Defect margins were sharply debrided back to the subchondral bone lamella without disturbing its integrity to avoid bleeding. Chondrospheres were dropped into the defect using a syringe. NaCl solution was removed and after a 20-min adhesion time the arthrotomies and incisions were closed in a standard fashion.

Postoperatively, the rehabilitation included 24-h leg resting, following a controlled passive- and later active-motion therapy. The loading depended on the defect location and size of the defects. Weight-bearing progression for a total of 8 weeks postoperatively was performed.

During the follow up of up to 24 months, at the centres in Buer and Gundelfingen, the Lysholm score, the IKDC (Ingvar Kamprad Design Centre; subjective knee evaluation form), and the WOMAC (Western Ontario and McMaster Universities osteoarthritis) index were evaluated. The statistical analysis was performed by Schicke (unaffiliated statistician in trauma surgery research, Helios hospital, Berlin-Buch). Second-look arthroscopies were performed 3 and 12 months after transplantation.

During the follow-up at the centres in Buer and Gundelfingen, all patients ($n=36$) reported an improved range of motion and loading as well as a pain decrease. The Lysholm score increased to 77.9 ± 2.9. For the IKDC a score of 61.1 ± 2.5 and for the WOMAC an index of 1.99 ± 0.3 was assessed. These results are comparable to other cartilage repair studies, where the Lysholm scores increased to 69 up to 86 [9, 18, 19], and the IKDC increased to 66 up to 81 [10, 22].

Nine patients of the group at the Gelenkklinik underwent a second-look arthroscopy (9 of 10). Already after 3 months defects were filled and regenerated cartilage was excellent integrated into the surrounding cartilage (Fig. 18.2). The surfaces of regenerates were smooth. Cartilage regenerates appeared white in colour, showing some round to oval, clear white areas, indicating crystallization points of cartilage differentiation originated from transplanted chondrospheres®. These spots disappeared over time and were not visible at second look arthroscopies after 11 months.

During the follow-up of 1 year, the patients at the ev. Elisabethenstift Hospital in Darmstadt were evaluated using the DGKKT, HSS, Lysholm, and Tegner scores (Fig. 18.3). All scores improved during the follow-up with an increase of the DGKKT score to about 77, the HS score to about 93 and the Lysholm score to about 83. The average Tegner score was 3.5. After treatment for all patients a full load bearing the treated joints was allowed.

co.don chondrosphere® offer advantages in two fields. The 3D chondrocyte transplant consists of strictly autologous in vitro-, de novo-synthesized cartilage matrix with cultured, low-differentiated and highly concentrated viable chondrocytes embedded in their in vitro-engineered autologous matrix. The surgical application is minimally invasive. The cell transplant adheres spontaneously on the base of the cartilage defect within a short time. There is no need for a shaping or for a fixation with a membrane. The cell migration of the surfacing chondrocytes of chondrosphere® explains the full integration between cartilage regenerate and surrounding articular cartilage tissue. A multicenter clinical trial needs to show the therapeutic benefit of co.don chondrosphere® for the treatment of focal cartilage defects within the knee joint.

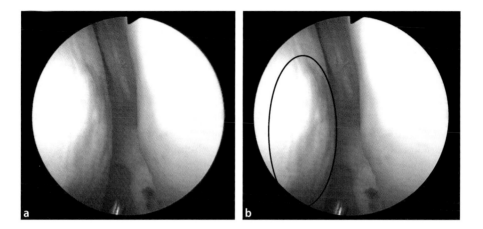

Fig. 18.2a,b Second-look arthroscopy, knee joint 3 months after co.don chondrosphere® treatment. Image on the *right* indicates original defect localization. (Images courtesy of Dr. Baum, Gelenkklinik, Gundelfingen, Germany)

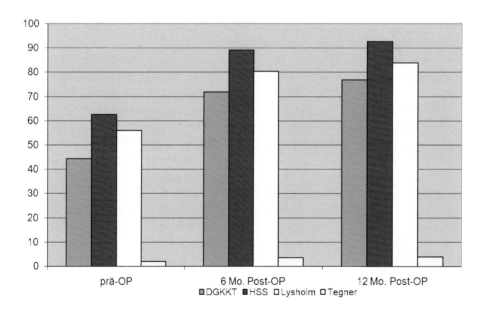

Fig. 18.3 Evaluation of patient outcome preoperatively, Prä-OP and 6 and 12 months after co.don chondrosphere® treatment

18.4 Preclinical Studies

To obtain first in vivo data on safety and efficacy of chondrosphere® a minipig pilot trial was performed (in cooperation with Dr. H. P. Wiesmann, University Hospital of Münster, and Dr. U. Meyer, University Hospital, Düsseldorf). It was tested whether autologous processed chondrospheres® can regenerate hyaline cartilage after transplantation into full-thickness chondral cartilage defects using the Göttinger mature minipig model (2 years old, around 35 kg; Ellegaard, Dalmose, Denmark).

Prior designing the trial, in vitro studies were performed revealing species-dependent proliferation, migration characteristics as well as a different capability to form in vitro 3D-tissue. Chondrocytes isolated from articular joints of human, sheep, horse, dog, cow and mini-pig were studied. Only for minipig and cow chondrocytes the formation of spheroids comparable to human chondrocytes were found. Based on this observation, the minipig model was used to analyze capacity of chondrocyte spheroids to heal cartilage defects. Additionally, the mature minipig model was chosen to minimize self-healing processes and to more approximate joint size, loading and cartilage time scale of healing to those of humans. This model had been successfully used in previous studies for skeletal reconstruction and cartilage repair [12, 16, 17]. The study was approved by the Animal Ethics Committee of the University of Münster.

For this purpose, 4-mm chondral cartilage defects were surgically created into the tibia-femoral joints of five minipigs. The created articular defects were then treated by transplanting autologous spheroid tissue constructs. chondrospheres® were allowed to adhere for 20 min. The untreated defects served as a control. After 1 week all animals were ambulating normally. No signs of infection were present during the experimental period. During preparation of the treated joints, there were no visible abnormalities of soft tissue or joint capsule in all animals, with no signs of synoviitis.

After 2 months in control, an overspreading of repairing scar tissue or an incomplete filling of the defect (Fig. 18.4c) and a deformation of the joint anatomy was observed (Fig. 18.4a). Within control defects, generated tissue revealed no expression of hyaline-specific proteoglycans and revealed a strong expression of collagen type I (Fig. 18.5). In one of the two regenerated control defects a moderate expression of collagen type II was found.

Hyaline cartilage generally was generated in chondrosphere® treated defects with chondrone appearance and a frequently columnar organization of chondrocytes in the near of subchondral bone. A positive safranin O-staining and collagen type II expression was found. Collagen type I expression was suppressed. Between the regenerated cartilage within chondrosphere® treated defects and the adjacent cartilage as well as the defect ground, no gaps or disturbances were visible either macroscopically or microscopically.

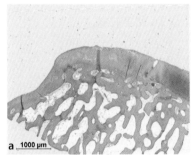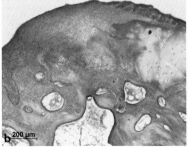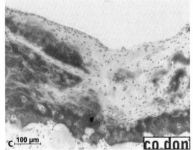

Fig. 18.4a–c Histological and immunohistochemical staining of repair tissue within nontreated cartilage defects using a minipig model. **a** Safranin O staining, **b** collagen type I staining, **c** collagen type II staining. Orange stain indicates expression of hyaline-specific proteoglycans (**a**); red stain indicates expression of collagen type I (**b**) and type II (**c**). For the treated defects, a defect filling height comparable to the adjacent cartilage was found (Fig. 18.5). Borders between the previously created defect and the surrounding cartilage were difficult to distinguish. All regenerated tissues were found to be smooth, homogenous and well integrated and no hypertrophy was observed

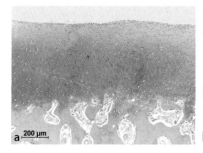 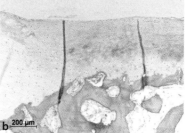 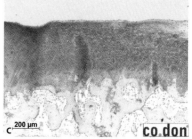

Fig. 18.5a–c Histological and immunohistochemical staining of repair tissue within co.don chondrosphere®-treated cartilage defects using a minipig model. **a** Safranin O staining, **b** collagen type I staining, **c** collagen type II staining. Orange stain indicates expression of hyaline-specific proteoglycans (**a**); red stain indicates expression of collagen type I (**b**) and type II (**c**)

18.5 In Vivo Integration and Differentiation Studies

The structural and functional integration is essential for long-term functionality of implants. The spheroid adhesion and subsequent remodelling are likely the crucial steps in structural and functional integration of engineered constructs into native cartilage. To assess the integrative capacity of chondrosphere®, studies were performed on co-cultures. chondrospheres were placed into created defects within human cartilage explants and were cultured in vitro or were placed subcutaneously into the severe combined immunodeficiency (SCID) mouse model (in cooperation with University Hospital Regensburg, Dr. T. Schubert, Dr. J. Schedel).

The results of the in vitro studies revealed that the adhesion and integration process of in vitro formed autologous chondrospheres consists of three phases: (1) initial fixation by cell mediated adhesion on host tissue; (2) widening and completion of adhesion area by migration of chondrosphere® surface chondrocytes at the irregular surface (e.g. fissures) of the host tissue; (3) shape adaptation of chondrosphere® transplant to the defect cavity by the synthesis and secretion of cartilage-specific structural and regulatory components, followed by the transplant remodelling, the filling of gaps between transplant and host tissue, as well as the biochemical integration into host tissue.

An excellent integration of chondrosphere® was found using the SCID mouse model, too. After 12 weeks, remodelled spheroids are characterized by a higher cell density than the native cartilage (Fig. 18.6).

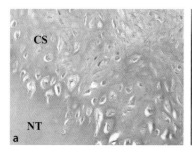 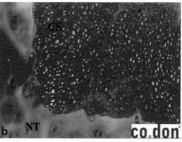

Fig. 18.6a,b Integration and maturation of chondrospheres in co-culture with native articular cartilage, transplanted subcutaneously in the SCID-mouse. **a** Four weeks, elastica van Giesson staining (×20), **b** collagen type II staining. *NT* native tissue, *CS* chondrospheres®. (In cooperation with Drs. Schubert and Schedel, University Hospital, Regensburg, Germany)

Between remodelled spheroids and native cartilage tissue, no gaps were visible. All adjacent surfaces between chondrospheres and native human tissue have shown no gaps. The de novo-secreted matrix by chondrosphere® chondrocytes is integrated into the native cartilage matrix.

18.6
In Vitro Differentiation and Maturation

Autologous chondrospheres were formed by 3D aggregation of patients own chondrocytes. Aggregation is followed by the synthesis and secretion of autologous matrix components [1]. For these aggregation and matrix formation processes no growth factors or other stimuli are used; the only supplement is the patient own serum. Chondrospheres are solid and elastic aggregates with a bright white surface. In dependence of the initial chondrocyte number per chondrosphere aggregate, the size of a single chondrospheres differs. For the treatment of focal knee cartilage defects within the knee joint, chondrospheres of a diameter of about 500–800 µm were used (Fig. 18.1).

As it was already described, during the ongoing spheroid culture, chondrocytes within chondrospheres are viable and produce high amounts of structural cartilage matrix components [1, 14]. The ongoing matrix secretion and maturation within the interior of chondrospheres is accompanied by a separation of chondrocytes. During maturation of chondrospheres few apoptotic and necrotic chondrocytes were observed, mainly found within areas of high matrix formation activity (in cooperation with Prof. A. Herrmann, Dr. T. Korte, Humboldt University of Berlin). This again confirms the regulated decrease in cell number during hyaline-matrix maturation processes. Raster electron microscopic studies revealed at the surface of chondrospheres a layer of elongated chondrocytes that are surrounded only by thin layer of matrix components. This layer is comparable to the surface of native hyaline cartilage. Additionally, a dense network of structural cartilage matrix components and matrix secretion processes were visible under the chondrocyte surface layer (Fig. 18.7) (in cooperation with Prof. A. Herrmann and Dr. Bleiss, Humboldt University of Berlin, Germany).

Fig. 18.7 Electron microscopy of co.don® AG chondrosphere®. Three weeks autologous 3D culture. (In cooperation with Prof. Herrmann and Dr. Bleiss, Humboldt University, Berlin, Germany)

Clinical success of autologous chondrosphere transplantation depends on the capability of chondrocytes to form hyaline cartilage after their transplantation into cartilage defects. Differentiation of monolayer propagated chondrocytes within chondrospheres is accompanied by the expression of collagen type II and the expression of further differentiation and hyaline-specific markers such as S100, aggrecan, hyaline-specific proteoglycans, SOX-9 and matrillin (part of analysis performed in cooperation with Dr. O. Pullig and Prof. Swoboda, University Hospital Erlangen). During the time course of the 3D-culture of chondrosphere®, the collagen type I expression decreases and is only found in the surface cell layers, comparable to surface chondrocytes within native hyaline cartilage. Additionally, chondrocyte differentiation and matrix maturation within chondrosphere® is regulated by the expression of endogenous chondrogenic growth factors. A differentiation dependent expression of the growth factors transforming growth factor beta (TGF-β), fibroblast growth factor 2 (FGF-2) and bone morphogenetic protein 2/4 (BMP2/4) (no immunohistochemical antibody for BMP2 available) was observed (Fig. 18.8, not shown for FGF-2). This underscores again the high intrinsic potential of isolated and cultured human chondrocytes to regulate their chondrogenesis. In parallel, the expression of interleukin 1 beta (IL-1β) and tumor necrosis factor (TNF-α), was not found (not shown). These factors are known to inhibit the chondrogenic differentiation.

Additionally, for chondrospheres a fusion could be demonstrated when several single spheroids were brought into close contact [1]. After transplantation of chondrospheres into focal cartilage defects, single chondrospheres need to merge with each other to form an integrated cartilage regenerate. The fusion process is accompanied by a reshaping and remodelling of the single spheroids. The gaps between single spheroids were filled with de novo-synthesized matrix and a compact construct is formed. Interestingly, during fusion a strong stimulation of hyaline-specific marker as well as chondrogenic growth factor expression as hyaline-specific proteoglycans, collagen type II, S100 antigen and aggrecan can be observed (Fig.

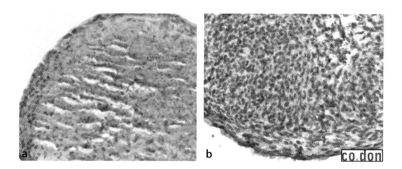

Fig. 18.8a,b Expression of growth factors by human chondrocytes cultured autologous as chondrospheres. Immunohistochemical staining for (a) TGF-β and (b) BMP2/4 after 8 weeks of culturing. Red staining indicates positive staining (×500)

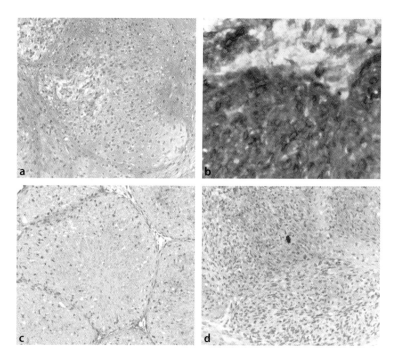

Fig. 18.9a–d Fused chondrosphere®. Histological and immunohistochemical staining. a Safranin O, b collagen type II, c aggrecan, d collagen type I. a–c 6 weeks (×250), d 2 weeks (×100) of culturing. Orange (safranin O) and red stains indicate positive staining

18.9). The density of chondrocytes within the interior of chondrospheres further decreased towards the density of native hyaline cartilage compared with the cell density within single chondrospheres. The increased expression of chondrogenic growth factors might indicate their influence of matrix formation and maturation within co.don chondrosphere®.

Our study indicates that the transplantation of in vitro-generated autologous cartilage microtissue by in vitro-expanded chondrocytes is a promising technique for the repair of full-thickness cartilage defects. The sole supplement to the culture medium was autologous serum and no foreign scaffold or matrix was used. The chondrosphere® technology seems to overcome an artificial scaffold use in cartilage tissue engineering strategies. Simplification of surgical technique and optimal integration of tissue engineered transplants are most advantages of co.don chondrosphere®. Clearly, additional studies are necessary to evaluate repair mechanism at different time points and to perform statistical analyses with regard to macroscopic and histological grading.

Acknowledgment

Parts of the in vitro studies were supported by Bundesministerium für Bildung und Forschung (Germany).

References

1. Anderer U, Libera J: In vitro engineering of human autogenous cartilage. *J Bone Miner Res* 17:1420–29, 2002
2. Andereya S, Maus U, Gavenis K, Gravius S, Stanzel S, Müller-Rath R, Miltner O, Mumme T, Schneider U: Treatment of patellofemoral cartilage defects utilizing a 3D collagen gel: two-year clinical results. *Z Orthop Unfall* 45(2):139–45, 2007
3. Bachmann G, Basad E, Lommel D, Steinmeyer J: MRI in the follow-up of matrix-supported autologous chondrocyte transplantation (MACI) and microfracture. *Radiologe* 44(8):773–82, 2004
4. Behrens P, Ehlers EM, Köchermann KU, Rohwedel J, Russlies M, Plötz W: New therapy procedure for localized cartilage defects. Encouraging results with autologous chondrocyte implantation. *MMW Fortschr Med* 141(45):49–51, 1999
5. Behrens P: Matrixgekoppelte Mikrofrakturierung. Ein neues Konzept zur Knorpeldefektbehandlung. *Arthroskopie* 8:193–197, 2005
6. Behrens P, Bitter T, Kurz B, Russlies M: Matrix-associated autologous chondrocyte Transplantation/implantation (MACT/MACI)—5-year follow-up. *Knee* 13(3):194–2 Epub, 2006
7. Davies-Tuck ML, Wluka AE, Wang Y, Teichtahl AJ, Jones G, Ding C, Cicuttini FM: The natural history of cartilage defects in people with knee osteoarthritis. *Osteoarthritis Cartilage*, 2007
8. Davis MA, Ettinger WH, Neuhaus JM, Cho SA, Hauck WW: The association of knee injury and obesity with unilateral and bilateral osteoarthritis of the knee. *Am J Epidemiol* 130:278–288, 1989
9. Dorotka R, Kotz R, Tratting S, Nehrer S: Mid-term results of autologous chondrocyte transplantation in knee and ankle. A one- to six-year follow-up study. *Z Rheumatol* 63(5):385–92, 2004
10. Flohe S, Schulz M: Prospektiver Vergleich zweier Matrix-gekoppelter Chondrozyten. Transplantationsverfahren zur Behandlung von Knorpelschäden im Kniegelenk. *Ger Med Sci* GM07dkou002, 2007
11. Hendriks J, de Bruijn E, Schotel R, van Blitterswijk CA, Riesle J: Cellular synergy for 1 step cartilage repair. Oral presentation at the ICRS, Warsaw, 2007
12. Hunziker EB: Biologic repair of articular cartilage. Defect models in experimental animals and matrix requirements. *Clin Orthop* 367:S135–S46, 1999
13. Kramer J, Böhrnsen F, Lindner U, Behrens P, Schlenke P, Rohwedel J: In vivo matrix-guided human mesenchymal stem cells. *Cell Mol Life Sci* 63(5):616–26, 2006
14. Libera J, Luethi U, Alasevic OJ: Autologous matrix-induced engineered cartilage transplantation. In: Zanasi S, Brittberg M, Marcacci M; Basic Science, clinical repair and reconstruction of articular cartilage defects: current status and prospects, Volume 1, p.591–600, Italy, 2006
15. Lu Y, Dhanaraj S, Wang Z, Bradley DM, Bowman SM, Cole BJ, Binette F: Minced cartilage without cell culture serves as an effective intraoperative cell source for cartilage repair. *J Orthop Res* 24(6):1261–70, 2006
16. Mainil-Varlet P, Riese F, Grogan S, Mueller W, Saager C, Jakob RP: Articular cartilage repair using a tissue engineered cartilage-like implant: an animal study. *Osteoarthritis Cartilage* 9:6–15, 2001
17. Meyer U, Runte C, Dirksen D, Stamm T, Fillies T, Joos U, Wiesmann HP: Image based biomimetric approach to design and fabrication of tissue engineered bone. *Comp Assisted Radiol Surg* 123:726–32, 2003
18. Nehrer S, Dorotka R, Schatz K, Bindreiter U, Kotz R: Klinische Ergebnisse nach matrixassitierter Knorpelzelltransplantation mit Hyaluronat Vlies. *Ger Med Sci* GM-04dgu0558, 2004
19. Ossendorf C, Kaps C, Kreuz PC, Burmester GR, Sittinger M, Erggelet C. Treatment of posttraumatic and focal osteoarthritic cartilage defects of the knee with autologous polymer-based three-dimensional chondrocyte grafts: 2-year clinical results. 9(2):R41, 2007
20. Peterson L, Minas T, Brittberg M, Nilsson A, Sjögren-Jansson E, Lindahl A: Two- to 9-year outcome after autologous chondrocyte transplantation of the knee. *Clin Orthop Relat Res* 374:212–34, 2000

21. Saris DB, Vanlauwe J, Victor J, Haspl M, Bohnsack M, Fortems Y, Vandekerckhove B, Almqvist KF, Claes T, Handelberg F, Lagae K, van der Bauwhede J, Vandenneucker H, Yang KG, Jelic M, Verdonk R, Veulemans N, Bellemans J, Luyten FP: Characterized chondrocyte implantation results in better structural repair when treating symptomatic cartilage defects of the knee in a randomized controlled trial versus microfracture. *Am J Sports Med* 36(2):235–46, 2008

22. Schmidt A, Johann K, Kunz M: Erste klinische Ergebnisse nach matrixgekoppelter autologer Chondrozytentransplantation am Kniegelenk. 35th congress of the DGRh and 21st annual conference of the ARO, Hamburg, 2007

23. Steinwachs M, Kreuz PC: Autologous chondrocyte implantation in chondral defects of the knee with a type I/III collagen membrane: a prospective study with a 3-year follow-up. *Arthroscopy* 23(4):381–7, 2007

Muscle Tissue Engineering

M. P. Lewis, V. Mudera, U. Cheema, R. Shah

Contents

19.1 Structure and Function 243
19.2 Myogenesis 244
19.3 Regeneration 245
19.4 Skeletal Muscle Engineering 245
19.5 Topographical Approaches 247
19.6 Material Scaffolds 247
19.7 Self-Organisation 249
19.8 Injectibles 249
19.9 Mechanical Approaches 249
19.10 Beyond the Basics: Vascularisation and Connection with Other Tissues 250
19.11 Conclusion 251
References 251

19.1 Structure and Function

Skeletal muscle is a classical example of "structure determining function". Successful strategies for clinical applications of tissue engineered skeletal muscle must recapitulate the processes that muscle undergoes during either embryonic development or adult regeneration. Current approaches to tissue engineering of skeletal muscle broadly utilize different aspects of these processes. To understand these approaches we must first begin with a consideration of these biological events.

Skeletal muscle is attached to the skeleton and is the most abundant type of muscle, accounting for 48% of body mass [1]. It is responsible for the voluntary control and active movement of the body and, in addition, protects the abdominal viscera and functions as an accessory to aid respiration. In the craniofacial region, skeletal muscle is responsible for facial expression and tongue, eye and jaw mobility [2].

Skeletal muscle consists of elongated, multinuclear muscle fibres encapsulated within connective tissue sheaths [3]. These "composites" of muscle fibres and connective tissue represent a contractile structure with high cell density and a high degree of cellular orientation and differentiation. Adult fibres have specialised endpoint attachments to tendons (myotendinous junctions) that, in conjunction with the cellular orientation, form an elaborate system for force transmission in a particular direction (vector) (Fig. 19.1).

The muscle sarcomere is the main contractile unit of skeletal muscle, where actin and myosin constitute the major structural components. These form thin and thick filaments that generate contraction by sliding alongside each other; this is the basis of muscle contraction (the "sliding filament" theory) whereby myosin heads attach to, "pull" (due to a conformational change in the protein) and release from the neighbouring actin filaments to affect the physical shortening of the muscle fibre. As the muscle contracts and shortens, there is naturally an increase in the cross-sectional area [4].

Skeletal muscles in different anatomical sites possess different contraction rates (e.g. eye muscles vs. spinal muscles) due to the presence of varying

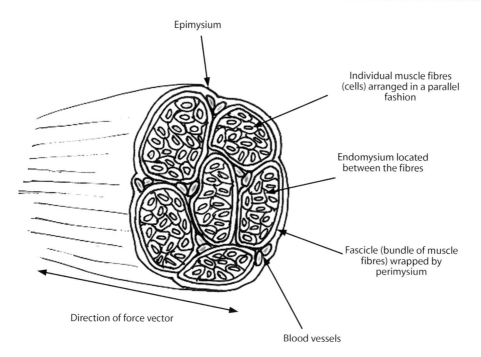

Fig. 19.1 Transverse section showing the macrostructure of skeletal muscle: the "composite" nature of the tissue (muscle fibres encased in connective tissue sheaths (endo-, epi- and perimysium) and unique parallel orientation of the muscle fibres is depicted

proportions of contractile protein isoforms that allow the tissue to meet individual physiological and functional requirements [5]. Determination of the contractile protein isoform profile is dependent upon both intrinsic and extrinsic factors; the former (genetic pre-programming) establishes a state of myogenic commitment in the early embryo. The extrinsic factors are able to modulate this baseline expression profile within certain limits, e.g. by neural, hormonal and physical signals [6–12].

The extracellular matrix (ECM) is necessary to act as a physical scaffold for the muscle fibres; it also has an essential role in the control and maintenance of cellular function. The ECM is composed of interstitial connective tissue and basal lamina that is in intimate contact with muscle fibres [13]. The basal lamina is composed of collagen IV, the glycoprotein laminin, ectactin and heparan sulphate proteoglycans (HSPGs) whilst the interstitial ECM (endo-, epi- and perimysium), is composed of collagen I, fibronectin and HSPGs [14]. As well as a structural role, many ECM components have a direct effect on the determination, movement, attachment, proliferation, alignment and fusion of cells committed to the myogenic lineage. For example, laminins present in the basal lamina act as major ligands for cell surface receptors involved in the transmission of force from the cell interior [15, 16]. Furthermore, proteoglycans are essential for binding growth factors to their receptors and proteolytic fragments of fibronectin and laminin act as chemotactic signals for myogenic cells [16–18]. Of vital importance to skeletal muscle regeneration are a population of mesenchymal fibroblasts that reside within this connective tissue milieu [19].

19.2
Myogenesis

The source of embryonic myogenic precursor cells (MPC; cells that are committed to the myogenic lineage) varies depending on the anatomical site of the muscle. The head, trunk and limb skeletal muscles develop as separate lineages in embryonic development. Skeletal muscle of the vertebrate body is derived from the somites, segmental blocks of paraxial mesoderm, which form either side of the neural tube [20]. The different myogenic precursor populations in the somite are first instructed to become myogenic by positive or negative signals emanating from neighbouring tissues, such as the neural tube, notochord, dorsal ectoderm and lateral mesoderm [21]. MPCs either proliferate in place or migrate to designated anatomical locations. Craniofacial muscles, such as the masseter muscle, are derived from unsegmented paraxial mesoderm and innervated by cranial nerves [21]. In vivo myogenesis is a complex process as myogenic cells progress through specific developmental stages: determination, differentiation and maturation. In mammals, myogenesis occurs in distinct waves; embryonic MPCs fuse during primary myogenesis and foetal/secondary MPCs, using primary myotubes as a scaffold, line up under the basement membrane of the primary myotubes and form secondary myotubes [16]. The secondary fibres form an independent basement membrane. Muscles split, become innervated, achieve their final pattern and grow. A further population of precursors, the satellite cells, persists in postnatal skeletal muscle [22]. Embryonic, foetal and postnatal cell populations have been shown to exhibit differences in myosin heavy chain isoforms [23], the expression of other cytoskeletal proteins such as desmin [24] and the expression of various cell adhesion molecules (e.g. M-cadherin) [25].

19.3
Regeneration

Although muscle is described as a "post-mitotic" tissue, it retains the capacity to regenerate; that is the restoration of lost nuclear material via fusion of new cellular bodies. This is accomplished by satellite cells, which fuse with adjacent fibres to provide a source of new myonuclei [26, 27]. Satellite cells are a normal constituent of all vertebrate skeletal muscles, regardless of age, fibre type or anatomical location [28] and they are located between individual muscle fibres and their associated basal lamina sheaths [29]. Most satellite cells appear to be quiescent and there is little migration in the adult under normal conditions [30].

Satellite cells can become active and migrate across the basal lamina and between groups of muscles, in response to mitogens and chemoattractants released locally when fibres are subjected to trauma, denervation or in response to stress induced by weight bearing exercise (micro trauma) in vivo, stretching (strain) or by explant and culture manipulations in vitro [31–33]. Some fractions of satellite cells are activated to re-enter the cell cycle and the progeny of activated cells (MPCs) undergo multiple rounds of division prior to fusing with existing or new muscle fibres [32, 34] (Fig. 19.2). After fusion, MPCs initiate the production of contractile proteins and restore the continuity and contractile function of the injured fibre [27].

19.4
Skeletal Muscle Engineering

Skeletal muscle may be deficient or lost as a consequence of primary or secondary causes: primary causes tend to be intrinsic and include congenital anomalies and diseases such as the muscular dystrophies, whereas secondary causes include surgery for the removal of cancer, trauma, endocrine and metabolic diseases, and the neuromuscular atrophies (e.g. poliomyelitis). Aesthetic and functional problems exist with the deficiency or loss of skeletal muscle, and often there is associated psychological distress resulting in a great need to provide replacement tissue.

There are a number of approaches to replace or reconstruct skeletal muscle of which muscle grafting from a local or distant donor site is the most commonly used. An example of this is the use of the pectoralis muscle to replace jaw muscle removed as part of the surgical procedure to treat invasive jaw cancers. Nevertheless, this technique has many limitations, which include donor site morbidity, the problem of transplanted muscle adapting to a new site with a different functional demand, and restoring aesthetics adequately.

Although tissue engineering of skeletal muscle has been slow to progress, it has great potential as a tool to study the formation and development of muscle as well as future applications in therapeutic

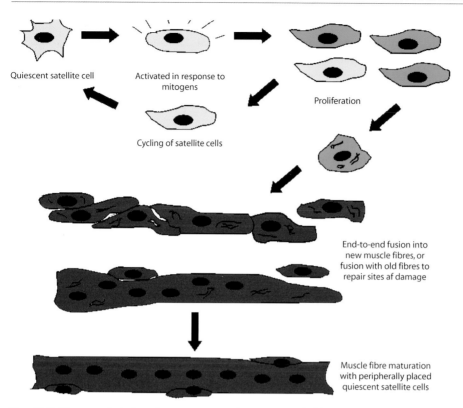

Fig. 19.2 Diagram depicting the events occurring during myogenesis: satellite cells become activated and proliferate with some of the satellite cells cycling and repopulating the pool. Other daughter cells will go on to fuse in an end-to-end fashion to produce new muscle fibres, or fuse to repair old fibres. Maturation of the muscle cells occurs with time and the satellite cells become quiescent and assume a position to the periphery of the fibre

replacement of tissues lost to damage due to disease and/or trauma. There is further need for the tissue engineering of skeletal muscle in terms of the potential in the pharmaceutical industry. Very often in drug development the benefits of using an in vitro testing methodology would far outweigh the more expensive in vivo testing, which also requires the additional ethical approval procedures required for such work. It is unlikely that this method of in vitro testing will completely remove the need for in vivo testing, but it will provide early testing alternatives.

The challenge is to fabricate a 3D viable tissue construct of sufficient mechanical strength to replace structures in vivo. Alongside this is the urgent need to develop monitoring tools to quantify functional outputs noninvasively and in a nondestructive manner. An ideal replacement construct would need to use an allogenic source of cells to make it commercially viable, but autologous cells have the current advantage of overcoming the host immune response. Critical choices on the source of cells for the ideal construct would have a combination of fibroblasts, myoblasts and satellite cells in the ideal proportion of the tissue to be replaced. The ability to use a single source of cells like mesenchymal stem cells or satellite cells would have a proven advantage if the right cues (signals) to drive them down specific lineages in a spatial 3D orientation were elucidated. Furthermore, cells from different anatomical sites and from different ages of muscle must have the ability to adapt to the mechanical and functional demands at the site of implantation. This is indeed a challenge that has been highlighted by a number of researchers [35–39]. Numerous investigations have been undertaken utilising transformed cell lines, which may provide standardised conditions for comparison in vitro; however, it is not possible to use such cells in the clinical situation, as they are essentially immortalised cells. Other studies have involved the use of primary cultures and have shown relative success over cell lines; for example, Dennis et al. [40] demonstrated engineered constructs from primary cells, which contracted regularly and vigorously as a syncytium compared with constructs engineered from transformed mouse skeletal muscle cell line (C2C12)/fibroblast co-cultures, which contracted singly and sporadically.

The choice of biomaterials used as substrates to support early cellular proliferation and differentiation should have tailored degradation rates such that the scaffold is eventually replaced by native tissue either in vitro or in vivo.

19.5
Topographical Approaches

As described earlier, one of the key events in myogenesis and regeneration is the "end-to-end" fusion of muscle precursor cells. A number of strategies have been used to define surface topography that will predicate cells into such an "end-to-end" formation using surface features such as grooves and channels. This can be achieved by adsorbing molecules onto a solid surface to generate "peaks and troughs" or by patterning solid surfaces with grooves. Self-assembled monolayers of, for example, alkylsiloxanes have been adsorbed onto a solid surface (silica) and the topography created supported C2C12 adhesion and proliferation, dependent upon the charge of the alkylsiloxane used [41]. Biological polymers have also been used with surface features introduced by laser ablation into fibronectin layers supporting C2C12 alignment [42], and by UV embossing of microchannels in polymeric films [43]. Grooving has been achieved in various polymers (such as polydimethylsiloxane [PDMS] and diethylenetriamine [DETA]) by plasma oxidation [44] and deep reactive ion etching [45]. These modified surfaces have supported the alignment of C2C12-derived cells and myotubes [44] and the myogenic differentiation of embryonic rat limb MPCs [45]. The spacing between the grooves is key for an optimal response with the ideal spacing being reported as varying from 6 to 100 µm.

19.6
Material Scaffolds

Biomaterials that are used for scaffold purposes may be constructed from synthetic or natural substances using a variety of techniques to impart desirable chemical, physical and biological characteristics to them. Biodegradable polyesters are frequently used and these include those derived from naturally occurring α-hydroxy acids (e.g. polyglycolic acid [PGA] and poly-l-lactic acid [PLLA]) and polycaprolactone (PCL). Other synthetic biomaterials that have been used include polyurethanes and polypropylene. Biological polymers have also been investigated and have focussed exclusively on collagen and alginates. Recently, there has also been interest in the use of inorganic polymers such as soluble phosphate glasses.

The format or morphology of the material presented to the cells is key and various processing techniques have been devised. These include extrusion, moulding, solvent casting or solid free form technology, utilised to construct the materials into fibres, porous sponges, tubular structures, hydrogel delivery systems and various other configurations [46, 47]. Although these morphologies have been used to enable the seeded cells to adhere, proliferate and differentiate into muscle fibres, the majority do not represent the anatomical arrangement of native skeletal muscle. Materials formed as fibres have the scope to provide a biomimetic scaffold with a parallel alignment to enable the seeded cells to assume the form of native skeletal muscle fibres both in vitro and in vivo.

In terms of biocompatibility, biological polymers are absolutely ideal but they often suffer from very poor mechanical properties and are not always conductive to fibre formation. Nevertheless, investigations have shown very positive results with both collagen and alginates. Cylindrical gels of the former have been implanted with both C2C12 and primary chick MPCs and maintained in bioreactors. Histology of these constructs indicated nascent myofibres (myotubes), which were metabolically active [48]. Both native and peptide modified alginate gels have been seeded with primary mouse and rat MPCs and investigated in both in vitro and in vivo environments. Myogenic differentiation was established with all configurations [49, 50].

For fibre-related morphologies, PGA has proved to be one of the first materials tested in tissue engineering applications due to its clinical acceptability as it is used in dissolvable sutures (US Food and Drug Administration approved). Primary rat MPC-loaded meshes consisting of 12-μm diameter fibres were implanted in vivo and allowed to integrate for up to 6 weeks with the resulting histology indicating the formation of multinucleated syncytia reminiscent of myofibres [1, 50, 51]. PLLA, a related polyester, can be formed into fibres by electrospinning [52] or extrusion [53] generating a wide range of fibre diameters (0.5–60 μm). Parallel arrays of fibres (mimicking the fibre arrangement of skeletal muscle) can be produced by tension alignment [52] or bundling [53]. Myotube formation and maturation was subsequently achieved after seeding of these scaffolds with C2C12 and primary human MPCs. Importantly, cellular response was only seen when the fibres were pre-coated with biological matrices such as Matrigel, laminin, fibronectin and collagen (or its denatured product, gelatin) [52, 53]. PCL is similarly well received clinically and C2C12 cell attachment and proliferation was supported on gravity-spun 80- to 150-μm diameter fibres in random arrays. Coating with a biopolymer (gelatin) was again a prerequisite [54].

The potential disadvantage of these "classical" degradable polyesters is that they are not particularly elastic whilst some newer materials can still be made into fibres but also possess highly improved elastomeric properties. Electrospun DegraPol® (a degradable block elastomeric polyesterurethane) meshes (10-μm diameter fibres) have been seeded with C2C12 and L6 cell lines as well as human MPCs and shown to support adhesion, proliferation and differentiation into multinucleate myotubes in vitro [55]. Once again, coating with Matrigel, collagen or fibronectin was required to elicit the optimum biological effect. Thus, although nonbiological organic polymers may be easy to manufacture to specification, a major drawback is the necessity to precoat the surfaces with a protein to enable cells to adhere. Interest has also focussed on tissue engineering skeletal muscles with scaffolds made from materials that are elastomeric if not degradable. Attachment, proliferation, migration and differentiation of the G8 myogenic cell line was supported on such a spun-cast material [56].

As can be appreciated, the fibre diameters are absolutely essential with the optimum diameters appearing to be in the 10- to 100-μm cross-sectional area. What is less widely investigated is the ideal spacing between such fibres although it has been suggested that 30- to 55-μm gaps between fibres was a requisite for nascent C2C12 muscle fibre (myotubes) formation on nondegradable polypropylene fibres; again laminin coating was vital [57].

Despite the very encouraging results with non-biological organic polymers, there are some problems emerging as investigations develop. One of the major issues appears to be the mode of degradation of the polymers in that they degrade by hydrolysis and the breakdown products are acidic in nature, which may interfere with the regenerative process. Furthermore, as the degradation response is not iterative to the biological response, if the biological response lags behind, then the constructs can fail without any control. Some attention has therefore been directed towards other materials that are biocompatible without the problems discussed above. Lin et al. [58] developed bioabsorbable glass fibres composed of calcium iron phosphate. The solubility rate of such glasses is correlated with composition and because it is highly linear with time, it is very predictable. These glasses are polymeric in nature and form fibres easily. Further modifications may be made by the addition of iron oxide, aluminium or magnesium, which have a strong effect on the glass network. Random meshes of heat drawn phosphate-based glass fibres (10-μm diameter) have been shown to support the attachment, proliferation and myogenic differentiation of both human and murine MPCs; however, in common with the organic polymeric fibres, coating of the glass fibres with gelatin or Matrigel was essential [59, 60].

19.7
Self-Organisation

An alternative strategy to using materials as either topographical signals or scaffolds is to allow myogenic cells to lie down and organise their own "native" scaffolds, i.e. self-organise. Dennis and Kosnik [61] seeded primary rat MPCs onto laminin-coated plates, which had previously been treated with Sylgard® to create a "non-stick" surface. In addition, artificial tendons were created using either laminin-coated silk sutures or acellularised muscle anchors to help establish the longitudinal axis of the engineered muscle construct ("myoid"). Over the period of 3–4 weeks, the cells fused into muscle fibres and formed a syncytium, which contracted and rolled from the edges to form a myoid with strong tendinous attachments. This technique has been successfully repeated with a variety of other cell lines and primary cultures [40, 62, 63]. In order to accelerate the process of self-organization, fibrin gels were used with sutures to provide the tendinous attachments. After the period of 7–10 days, the seeded rat MPCs had fused to form contracting muscle fibres that enabled the gels to roll from the edges to form myoids [64]. With maturation of these engineered constructs over time, it was anticipated that the MPCs would eventually synthesise their own ECM. The engineered myoids have produced forces only a fraction of those produced in vivo and thus these studies have provided the proof of concept, but are still far removed from providing implantable functional skeletal muscle tissue.

19.8
Injectibles

Substances that are a liquid when injected and then undergo a phase change to solidify or gel prove to be useful as this may be a way of encapsulating cells, which can prevent the immune response from destroying them if allogenic cell strategies are to be developed, which have a greater potential for commercialisation. Delivery of primary rat MPCs into muscle defects has been achieved using a double syringe system. Cells were mixed with thrombin and introduced simultaneously with fibrinogen into the defects. Polymerisation of the fibrin gels occurred within a minute and the resultant constructs contained donor nuclei within structures that stained positive for the muscle-specific protein desmin [65].

19.9
Mechanical Approaches

Mechanical forces generated in the growing embryo and in the active adult play a crucial role in organogenesis, as complex patterns of mechanical loading are applied to skeletal myoblasts and myofibres through the elongating skeleton and by foetal movements. It therefore follows that successful skeletal muscle tissue engineering requires the tissue to be mechanically sensitive.

Vandenburgh [66] reported the first 3D model of skeletal muscle, providing a 3D environment for MPC growth and application of mechanical strain to the muscle culture. Mechanical stimulation encouraged cell proliferation, myotube orientation and myotube longitudinal growth [66] with two- to fourfold longer myotubes being generated [36, 38].

Using a similar principle, 3D type I collagen gel seeded with C2C12 myoblasts was used by Okano et al. [67] to produce engineered muscle constructs. High density multinucleated myotubes orientated in one plane were developed using a process of cellular packing through centrifugation, to increase cell density [69]. Cyclic stretching resulted in a highly orientated hybrid muscular construct with both cells and collagen fibres aligned in the direction of stretch [69]. Constructs were inserted into subcutaneous spaces in the back of nude mice, and four weeks after implantation, a dense capillary network was formed in the vicinities and on the surface of the graft [68].

The platform technology developed by Vandenburgh and colleagues using 3D cell-seeded collagen/Matrigel constructs to generate endogenous tension has been further developed to generate parallel arrays of fused differentiated myotubes, by controlling the plane in which tension is generated [70, 71].

Through the application of mechanical load, initially ramp loading at 500 µm a day for 4 days, then after a period of static load application of varying cyclical loading regimens, Vandenburgh and colleagues were able to successfully increase the area and diameter of myofibres in their constructs by 40 and 12% respectively [71]. In a further development, other groups have found that mechanical stimulation of 3D collagen constructs of fused C2C12 myoblasts resulted in upregulation of *IGF-I* gene splicing and production, and it is likely that changes in myofibre size may be affected through this mechanism [70].

19.10
Beyond the Basics: Vascularisation and Connection with Other Tissues

Although progress has been made in the construction of muscle organoids by application of mechanical, topographical and chemical signals to ensure correct alignment and fusion, a fully functional skeletal muscle needs to be able to interact and respond to its environment. Perhaps the most important component of that is the establishment of a fully functional vascular system. The most prominent success in this area has been achieved using scaffolds composed of 50% PLLA and 50% polylactic-glycolic acid (PLGA) [72]. These scaffolds were coated with Matrigel, seeded with commercially available primary human endothelial cells and murine fibroblasts along with C2C12 cells. The C2C12 formed the basis of the engineered muscle tissue, the embryonic endothelial cells were able to induce an endothelial vessel network as well as induce the embryonic fibroblasts in the culture to differentiate into smooth muscle cells to help stabilize the vessel network [72]. An alternative approach is to, using microsurgical approaches, create an arteriovenous (AV) loop and to surround it with a polycarbonate chamber. The chamber can then be loaded with blocks of muscle tissue or primary MPCs. When muscle tissue was implanted (both rat and human), the majority of the tissue formed was adipose in nature although multinuclear, striated myofibres were also present. When primary rat MPCs were implanted into the chamber, myofibres were still formed and there was no adipose tissue present. In all cases, the tissues in the chamber were very well vascularised. Confirmation of the validity of such an approach has been made in studies where rat MPCs suspended within fibrinogen gels contained within cylindrical silicone chambers were placed around the femoral vessels of recipient animals. Histological examination revealed the features of developing skeletal muscle with evidence of myoblast fusion into multinucleated myotubes; in addition, von Willebrand staining for endothelial cells demonstrated the ingrowth of small vessels into the matrix [73]. Similar studies confirmed these results using molecular end-points to define myogenesis [74].

Functional connections to the nervous system are also required for effective function and there have been some efforts to recreate this. Rat foetal spinal cord explants have been introduced into the "self-assembly" models described above [75] leading to the fusion of some neural cells with myotubes and the formation of some neuromuscular-like junctions with clustering of acetylcholine receptors surrounded by neurofilament-stained neural extensions. The nerve-muscle constructs exhibited spontaneous baseline activity and in response to electrical field stimulation, demonstrated a twitch and tetanus response. In a similar experiment, rat MPCs have been incorporated into fibrin matrices along with spinal cord organotypic slice cultures [76]. Again, spontaneously contracting multinuclear and parallel-aligned myofibres were demonstrated with pharmacological tests suggesting the presence of neuromuscular contacts.

The final integrative connection needed is that between the muscle and bone i.e. tendon and hence the construction of myotendinous junctions (MTJ). This is a relatively unexplored area although a number of strategies have begun to emerge, based on technologies that have already proved successful for skeletal muscle. As with MPCs, rat tendon fibroblasts (tenocytes) can self-organize where confluent monolayers delaminate and "roll" into "self-organized tendons" (SOT). These can then be incorporated into "self-assembly" rat MPC constructs by allowing those constructs to form around the SOTs. Such constructs are functionally viable and are characteristic of neonatal MTJs [77]. Another approach has been to use the same glass fibres that support myogenic differentiation and to use them to support functional human tenocytes; the success of this approach means that constructs can be generated that have MTJ components interacting together [78].

19.11
Conclusion

There is no doubt that significant efforts are now going into generating functional skeletal muscle for implantation. The number of groups working on the area has increased considerably in the last decade and progress is being made. Successes to date have indicated that the ideal solution will probably entail a combination of approaches that recognize the system requires directional cues generated by non-permanent stimuli (e.g. topography, scaffolds), mechanical pre conditioning, the correct combinations of cells (e.g. myogenic, connective tissue, vascular), attachments to nerve and bone (via tendons) and a neural input. Finally, the milieu in which the muscle exists is also vital and the right "cocktail" of growth factors (e.g. insulin-like growth factors) and hormones (e.g. testosterone) is required. It is unclear how close we really are to transplanting functioning, tissue-engineered, skeletal muscle into patients but the signs are encouraging.

References

1. Saxena AK, Marler J, Benvenuto M et al (1999) Skeletal muscle tissue engineering using isolated myoblasts on synthetic biodegradable polymers: preliminary studies. Tissue Eng 5:525–32
2. Alsberg E, Hill EE, Mooney DJ (2001) Craniofacial tissue engineering. Crit Rev Oral Biol M 12:64–75
3. Purslow PP, Duance VC, Structure and function of intramuscular connective tissue. In: Hukins DWL (1990) Connective Tissue Matrix. CRC Press, Boca Raton, FL
4. Hanson J, Huxley HE (1954) Structural basis of the cross-striations in muscle. Nature 172: 530–2
5. Shuler CF, Dalrymple KR (2001) Molecular regulation of tongue and craniofacial muscle differentiation. Crit Rev Oral Biol Med 12:3–17
6. Miller JB, Crow MT, Stockdale FE (1985) Slow and fast myosin heavy chain content defines three types of myotubes in early muscle cell cultures. J Cell Biol 101:1643–50
7. Crow MT, Stockdale FE (1986) Myosin expression and specialization among the earliest muscle fibers of the developing avian limb. Dev Biol 113:238–54
8. Miller JB, Stockdale FE (1986) Developmental origins of skeletal muscle fibers: clonal analysis of myogenic cell lineages based on expression of fast and slow myosin heavy chains. P Natl Acad Sci USA 83:3860–4
9. Miller JB, Stockdale FE (1986) Developmental regulation of the multiple myogenic cell lineages of the avian embryo. J Cell Biol 103:2197–208
10. Schafer DA, Miller JB, Stockdale FE (1987) Cell diversification within the myogenic lineage: in vitro generation of two types of myoblasts from a single myogenic progenitor cell. Cell 48:659–70
11. Buonanno A, Apone L, Morasso MI et al (1992) The MyoD family of myogenic factors is regulated by electrical activity: isolation and characterization of a mouse *Myf-5* cDNA. Nucleic Acids Res 20:539–44
12. Hughes SM, Taylor JM, Tapscott SJ et al (1993) Selective accumulation of MyoD and myogenin mRNAs in fast and slow adult skeletal muscle is controlled by innervation and hormones. Development 118 1137–47
13. Sanes JR, Schachner M, Couvalt J (1986) Expression of several adhesive molecules (N-CAM, L1, J1, NILE, uvomorulin, laminin, fibronectin, and heparin sulfate proteoglycan) in embryonic, adult, and denervated adult skeletal muscles. J Cell Biol 102:420–431
14. Lewis MP, Machell JRA, Hunt NP et al (2001) The extracellular matrix of muscle-implications for manipulation of the craniofacial musculature. Eur J Oral Sci 109:209–221
15. Maley MAL, Davies MJ, Grounds MD (1995) Extracellular matrix, growth factors, genetics: their influence on cell proliferation and myotube formation in primary cultures of adult mouse skeletal muscle. Exp Cell Res 219:169–179
16. Gullberg D, Velling T, Lohikangas L et al (1998) Integrins during muscle development and in muscular dystrophies. Frontiers Biosci 3:d1039–1050
17. Adams JC, Watt FM (1993) Regulation of development and differentiation by the extracellular matrix. Development 117:1183–1198
18. Bischoff R (1997) Chemotaxis of skeletal muscle satellite cells. Dev Dyn 208:505–515
19. Fisher D, Rathgaber M (2006) An overview of muscle regeneration following acute injury. J Phys Ther Sci 18:57–66
20. Hawke TJ, Garry DJ (2001) Myogenic satellite cells: physiology to molecular biology. J Appl Physiol 91:534–551
21. Cossu G, Tajbakhsh S, Buckingham M (1996) How is myogenesis initiated in the embryo? Trends Genet 12:218–223
22. Stockdale FE, Miller JB (1987) The cellular basis of myosin heavy chain isoform expression during development of avian skeletal muscles. Dev Biol 123:1–9
23. Feldman JL, Stockdale FE (1991) Skeletal muscle satellite cell diversity: satellite cells form fibers of different types in cell culture. Dev Biol 143:320–334
24. Yablonka-Reuveni Z, Nameroff M (1990) Temporal differences in desmin expression between myoblasts from embryonic and adult chicken skeletal muscle. Differentiation 45:21–8
25. Rosen GD, Sanes JR, LaChance R et al (1992) Roles for the integrin VLA-4 and its counter receptor VCAM-1 in myogenesis. Cell 69:1107–1119
26. Moss FP, Leblond CP (1971) Satellite cells as the source of nuclei in muscles of growing rats. Anat Rec 170:421–435

27. Campion DR (1984) The muscle satellite cell: A review. Int Rev Cytol 87:225–251
28. Schultz E (1976) Fine structure of satellite cells in growing skeletal muscle. Am J Anat 147:49–70
29. Muir AR, Kanji AH, Allbrook D (1965) The structure of the satellite cells in skeletal muscle. J Anat 99:435–44
30. Bischoff R (1989) Analysis of muscle regeneration using single myofibers in culture. Med Sci Sports Exerc 21(5 Suppl):S164–72
31. Schultz E, Jaryszak DL, Valliere CR (1985) Response of satellite cells to focal skeletal muscle injury. Muscle Nerve 8:217–222
32. Bischoff R (1986) Proliferation of muscle satellite cells on intact myofibers in culture. Dev Biol 115: 129–139
33. Hughes SM, Blau HM (1990) Migration of myoblasts across basal lamina during skeletal muscle development. Nature 345:350–353
34. Grounds MD, Yablonka-Reuveni Z (1993) Molecular and cell biology of skeletal muscle regeneration. Mol Cell Biol Hum Dis Ser 3:210–56
35. Vandenburgh HH (1988) A computerized mechanical cell stimulator for tissue culture: Effects on skeletal muscle organogenesis. In Vitro Cell Dev Biol 24:609–619
36. Vandenburgh HH, Karlisch P (1989) Longitudinal growth of skeletal myotubes in vitro in a new horizontal mechanical cell stimulator. In Vitro Cell Dev Biol 25:607–616
37. Vandenburgh HH, Hatfaludy S, Sohar I et al (1990) Stretch induced prostaglandins and protein turnover in cultured skeletal muscle. Am J Physiol 259:C232–40
38. Vandenburgh HH, Swasdison S, Karlisch P (1991) Computer-aided mechanogenesis of skeletal muscle organs from single cells in vitro. FASEB J 5:2860–7
39. Vandenburgh HH, Karlisch P, Shansky J et al (1991) Insulin and IGF-I induce pronounced hypertrophy of skeletal myofibers in tissue culture. Am J Physiol 260:C475–84
40. Dennis RG, Kosnik PE, Gilbert ME et al (2001) Excitability and contractility of skeletal muscle engineered from primary cultures and cell lines. Am J Physiol Cell Physiol 280:C288–C295
41. Acarturk TO, Peel MM, Petrosko P et al (1999) Control of attachment, morphology, and proliferation of skeletal myoblasts on silanized glass. J Biomed Mater Res 44:355–70
42. Molnar P, Wang W, Natarajan A et al (2007) Photolithographic Patterning of C2C12 Myotubes using Vitronectin as Growth Substrate in Serum-Free Medium. Biotechnol Prog 23:265–8
43. Shen JY, Chan-Park MB, Feng ZQ et al (2006) UV-embossed microchannel in biocompatible polymeric film: application to control of cell shape and orientation of muscle cells. J Biomed Mater Res B Appl Biomater 77:423–30
44. Lam MT, Sim S, Zhu X et al (2006) The effect of continuous wavy micropatterns on silicone substrates on the alignment of skeletal muscle myoblasts and myotubes. Biomaterials 27:4340–7
45. Das M, Gregory CA, Molnar P et al (2006) A defined system to allow skeletal muscle differentiation and subsequent integration with silicon microstructures. Biomaterials 27:4374–80
46. Mooney DJ, Mikos AG (1999) Growing new organs. Sci Am 280:60–5
47. Hutmacher DW (2001) Scaffold design and fabrication technologies for engineering tissues-state of the art and future perspectives. J Biomater Sci Polymer Edn 12:107–124
48. Shansky J, Creswick B, Lee P et al (2006) Paracrine release of insulin-like growth factor 1 from a bioengineered tissue stimulates skeletal muscle growth in vitro. Tissue Eng 12:1833–41
49. Hill E, Boontheekul T, Mooney DJ (2006) Designing scaffolds to enhance transplanted myoblast survival and migration. Tissue Eng 12:1295–304
50. Kamelger FS, Marksteiner R, Margreiter E et al (2004) A comparative study of three different biomaterials in the engineering of skeletal muscle using a rat animal model. Biomaterials 25:1649–55
51. Saxena AK, Willital GH, Vacanti JP (2001) Vascularized three-dimensional skeletal muscle tissue-engineering. Biomed Mater Eng 11:275–81
52. Huang NF, Patel S, Thakar RG, Wu J, Hsiao BS, Chu B, Lee RJ, Li S (2006) Myotube assembly on nanofibrous and micropatterned polymers. Nano Lett 6:537–42
53. Cronin EM, Thurmond FA, Bassel-Duby R et al (2004) Protein-coated poly (l-lactic acid) fibers provide a substrate for differentiation of human skeletal muscle cells. J Biomed Mater Res A 69:373–81
54. Williamson MR, Adams EF, Coombes AG (2006) Gravity spun polycaprolactone fibres for soft tissue engineering: interaction with fibroblasts and myoblasts in cell culture. Biomaterials 27:1019–26
55. Riboldi SA, Sampaolesi M, Neuenschwander P et al (2005) Electrospun degradable polyesterurethane membranes: potential scaffolds for skeletal muscle tissue engineering. Biomaterials 26:4606–15
56. Mulder MM, Hitchcock RW, Tresco PA (1998) Skeletal myogenesis on elastomeric substrates: implications for tissue engineering. J Biomater Sci Polym Ed 9:731–48
57. Neumann T, Hauschka SD, Sanders JE (2003) Tissue engineering of skeletal muscle using polymer fiber arrays. Tissue Eng 9:995–1003
58. Lin ST, Krebs SL, Kadiyala S et al (1994) Development of bioabsorbable glass-fibers. Biomaterials 15:1057–1061
59. Ahmed I, Collins CA, Lewis MP et al (2004) Processing, characterisation and biocompatibility of iron-phosphate glass fibres for tissue engineering. Biomaterials 25:3223–3232
60. Shah R, Sinanan AC, Knowles JC et al (2005) Craniofacial muscle engineering using a 3-dimensional phosphate glass fibre construct. Biomaterials 26:1497–505
61. Dennis RG, Kosnik PE (2000) Excitability and isometric contractile properties of mammalian skeletal muscle constructs engineered in vitro. In Vitro Cell Dev Biol Anim 36:327–335
62. Kosnik PE, Faulkner JA, Dennis RG (2001) Functional development of engineered skeletal muscle from adult and neonatal rats. Tissue Eng 7:573–84
63. Baker EL, Dennis RG, Larkin LM (2003) Glucose transporter content and glucose uptake in skeletal muscle constructs engineered in vitro. In Vitro Cell Dev Biol Anim 39:434–9

64. Huang YC, Dennis RG, Larkin L et al (2005) Rapid formation of functional muscle in vitro using fibrin gels. J Appl Physiol 98:706–71
65. Beier JP, Stern-Straeter J, Foerster VT et al (2006) Tissue engineering of injectable muscle: three-dimensional myoblast-fibrin injection in the syngeneic rat animal model. Plast Reconstr Surg 118:1113–21
66. Vandenburgh HH (1983) Cell shape and growth regulation in skeletal muscle: exogenous versus endogenous factors. J Cell Physiol 116:363–71
67. Okano T, Satoh S, Oka T et al (1997) Tissue engineering of skeletal muscle. Highly dense, highly oriented hybrid muscular tissues biomimicking native tissues. Cell Transplant 6:109–118
68. Okano T, Matsuda T (1998) Muscular tissue engineering: capillary-incorporated hybrid muscular tissues in vivo tissue culture. Cell Transplant 7:435–42
69. Okano T, Matsuda T (1998) Tissue engineered skeletal muscle: preparation of highly dense, highly oriented hybrid muscular tissues. Cell Transplant 7:71–82
70. Cheema U, Brown RA, Yang S-Y et al (2005) Mechanical signals and *IGF-I* gene splicing involved in the development of skeletal muscle in vitro. J Cell Physiol 202:67–74
71. Powell CA, Smiley BL, Mills J et al (2002) Mechanical stimulation improves tissue-engineered human skeletal muscle. Am J Physiol Cell Physiol 283:C1557–65
72. Levenberg S, Rouwkema J, Macdonald M et al (2005) Engineering vascularized skeletal muscle tissue. Nat Biotechnol 23:879–84
73. Borschel GH, Dow DE, Dennis RG et al (2006) Tissue-engineered axially vascularized contractile skeletal muscle. Plast Reconstr Surg 117:2235–42
74. Bach AD, Arkudas A, Tjiawi J et al (2006) A new approach to tissue engineering of vascularized skeletal muscle. J Cell Mol Med 10:716–26
75. Larkin LM, Van der Meulen JH, Dennis RG et al (2006) Functional evaluation of nerve-skeletal muscle constructs engineered in vitro. In Vitro Cell Dev Biol Anim 42:75–82
76. Bach AD, Beier JP, Stark GB (2003) Expression of Trisk 51, agrin and nicotinic-acetycholine receptor epsilon-subunit during muscle development in a novel three-dimensional muscle-neuronal co-culture system. Cell Tissue Res 314:263–74
77. Larkin LM, Calve S, Kostrominova TY, Arruda EM (2006) Structure and functional evaluation of tendon-skeletal muscle constructs engineered in vitro. Tissue Eng 12:3149–58
78. Bitar M, Knowles JC, Lewis MP et al (2005) Soluble phosphate glass fibres for repair of bone-ligament interface. J Mater Sci Mater Med 16:1131–6

20 Tendon and Ligament Tissue Engineering: Restoring Tendon/Ligament and Its Interfaces

J. J. Lim, J. S. Temenoff

Contents

20.1 Tendon and Ligament 255
20.2 Physiology and Function 256
20.2.1 Structure 256
20.2.2 Cells 256
20.2.3 Extracellular Matrix 257
20.2.4 Muscle and Bone Insertion Points 258
20.2.5 Mechanical Properties 259
20.3 Injury and Healing 259
20.3.1 Injury 260
20.3.2 Healing 260
20.4 Current Reconstruction Procedures 260
20.5 Tissue Engineering 261
20.5.1 Scaffold Materials 262
20.5.2 Cell types 263
20.5.3 Differentiation of MSCs 264
20.5.4 Growth Factors 264
20.5.5 Interface Tissue Engineering for Bone Insertion Point Regeneration 265
20.6 Conclusions 265
References 266

20.1 Tendon and Ligament

Tendon and ligament injuries are very common. Over 800,000 people each year require medical attention for injuries to tendons, ligaments, or the joint capsule [14]. Unfortunately, tendon and ligament are relatively acellular and poorly vascularized tissues and have a poor capacity for healing [59, 65, 68, 119]. Suturing and grafts have had limited success in tendon and ligament repair, often resulting in poor healing, donor site morbidity, and insufficient mechanical properties [59, 65, 68, 124]. For this reason, there is currently a great deal of research on tendon and ligament tissue engineering.

To properly restore function to injured joints, it is important to first understand the normal physiology and function of tendon and ligament within these joints, including the structure-function relationship and mechanical properties of the tissue. These are critical considerations in the development of design parameters for engineered tissue. Successful tendon or ligament replacements must possess similar properties to the native tissue, because variations may result in lack of functionality or mechanical failure.

Tendon/ligament injury and healing are also important aspects of tissue engineering. A variety of tissue engineering approaches may be required depending on the type of injury sustained and the type of healing desired. This requires an understanding of the possible types of injury and how the body responds naturally to them. The body's natural healing response will be a careful consideration for graft biocompatibility, integration, and degradation.

Thus, details of native tendon/ligament structure, function, and response to injury have been included in this review, along with an overview of current reconstruction techniques and their associated limitations. It is hoped that many of the needs for tendon/ligament repair will be addressed by means of tissue engineering by developing replacement tissue to successfully return function to the injured region. This chapter concludes with a summary of the various biomaterial scaffolds, cell sources, and growth factors that are currently being investigated for such a use.

20.2
Physiology and Function

To properly engineer tendon/ligament tissue, it is important to understand how it is structured and regulated. Its strength is derived from its structure and is critical for proper function of replacement tissue. A thorough understanding of this structure-function relationship is important when determining the design parameters for tissue engineering.

Tendon is a tough band of fibrous connective tissue that connects muscle to bone; ligament is a similarly structured tissue connecting bone to bone. The major function of tendons is to transfer forces from the contraction of muscle to bone, resulting in locomotion or joint stability [119]. Tendons must withstand very large forces, which are mainly in the normal direction along the tendon, and the greatest forces are generally experienced during tension. During running, tensile forces can reach as high as 9 kN in the Achilles tendon, which is comparable to 12.5 times the body weight [61, 62].

Ligaments, however, mainly act to stabilize a joint where two bones come together in the joint capsule. Ligaments are necessary to limit and guide normal joint motion [58]. The anterior cruciate ligament (ACL), for example, is necessary to stabilize the knee and critical for normal joint kinematics [65]. Ligaments also provide support to many of the body's internal organs [27, 58]. Like tendons, ligaments must also withstand large tensile forces without injury or failure. The ACL regularly experiences forces ranging from 67 N when ascending stairs to 630 N while jogging [18] and can withstand tensile forces up 1,730 N before failure [89]. It is, however, important to note that both tendons and ligaments vary structurally depending on their locations in the body [75, 98, 127]. These structural differences are largely dependent on the magnitude and type of force that the tissue is designed to support [98]. To withstand these large forces, tendons and ligaments are structurally organized with collagen aligned in parallel to maximize tensile strength.

20.2.1
Structure

Tendons and ligaments are composed of a similar multi-unit hierarchical structure. From smallest to largest, tendon/ligament tissue is made up of collagen molecules, micro-fibrils, sub-fibrils, fibrils, collagen fibers, and fascicles [75, 119]. Fascicles then combine to form the gross structure of a tendon/ligament [65, 68, 75, 98, 119].

Tropocollagen is the basic structural unit of the collagen molecule. It is a right-handed triple helical domain, consisting of three left-handed polypeptide helices, forming a coiled-coil structure [58, 101]. Individual collagen chains are cross-linked to increase Young's modulus, reduce strain at failure, and increase tensile strength [27, 119]. The cross-linked tropocollagen molecules form collagen fibrils, the smallest functional unit of each tendon and ligament. Collagen molecules are assembled end-to-end in a quarter-staggered array to compose a fibril [119]. The hierarchical collagen structure of tendons and ligaments is aligned along the longitudinal axis, providing high tensile strength along this axis to withstand the large forces of the body [119].

20.2.2
Cells

Cells compose a relatively small proportion of tendon and ligament tissue, which is mostly extracellular matrix (ECM). Cells constitute only 20% of the total volume, while ECM and extracellular water compose the remaining 80% [66]. Tendons and ligaments contain small numbers of endothelial cells, synovial cells, and chondrocytes; fibroblasts, however, are the predominant cell type [27, 66, 68, 118, 119]. Fibroblasts are interspersed between collagen fiber bundles and are responsible for synthesizing and maintaining the ECM of the tissue [70, 119]. This is critical for the preservation of mechanical strength. Tendon and ligament fibroblasts also remodel the tissue during wound healing [119]. The fibroblasts attract reparative cells through chemotactic signals, release endogenous growth factors, and elicit a proper immune response [41]. Additionally, they are able to communicate with each other via a variety of means, including gap junctions [119]. This is especially important for the ability of tendon and ligament to respond and adapt to conditions of increased mechanical load with changes in gene expression and ECM protein synthesis [68, 69, 119].

20.2.3
Extracellular Matrix

Tendons and ligaments are composed of collagen, proteoglycans, glycoproteins, water, and cells. Water comprises 60–80% of the wet weight of tendon [68] and 55–65% of the wet weight of ligament [75].

20.2.3.1
Collagen

Other than water, the bulk of tendon and ligament tissue is composed of collagen, which imparts tensile strength to the tissues. For example, collagen contributes 75–85% of the dry weight of tendon, compared with 70–80% in the ligament [66, 75]. Collagen type I is the most abundant form of collagen within tendons and ligaments. It is responsible for 60% of the dry mass of the tendon and 95% of the total collagen in the tendon [75, 119]. Similarly, collagen type I makes up 90% of the total collagen in ligaments, slightly less than in tendon [66, 75].
The remaining collagen is mostly collagen types III and V. Type III forms smaller, less organized fibrils than type I, which may decrease mechanical strength, and is found at increased levels in aging tendons and at insertion sites of highly stressed tendons [29, 64, 119]. Collagen type V is mostly intercalated in the core of collagen type-I fibrils [119]. Collagen type V is believed to help regulate collagen type-I fibril growth [11]. Collagen types II, VI, IX, X, XI, and XII are also present in trace amounts [119, 124]. They are mainly localized at the fibrocartilaginous region of bone insertion sites and may strengthen the connection to bone by reducing stress concentration at the bone interface [119].

20.2.3.2
Proteoglycans

Although proteoglycans (PGs) occur in small quantities within tendons (1–2% dry weight) and ligaments (1–3% dry weight) [75], they appear to play a significant role in their mechanical properties. PG content has been found to vary with the site of tendon and depends on mechanical loading conditions, such as tension or compression. For example, in the bovine flexor digitorum profundus tendon, which acts mainly in compression, PGs contribute up to 3.5% of the dry weight; in bovine flexor tendon, which acts mainly in tension, PGs contribute to only 0.2–0.5% of the dry weight [115]. This suggests that PGs play a large role in imparting compressive properties in tendon.

PGs consist of a core protein with covalently linked glycosaminoglycan (GAG) chains [128]. Decorin is the most abundant PG in tendon and ligament [55, 128]. It is located on surface and middle portions of collagen fibrils [49] and thought to facilitate fibrillar slippage during deformation for improved tensile properties [94]. It also inhibits the formation of large collagen fibrils [114, 116, 128], allowing the tissue to adapt under tensile forces [128].

Aggrecan, another PG, is concentrated in regions of compression, such as the insertion site of tendon and ligament into bone [80, 85, 128]. It is negatively charged and holds large amounts of water within the fibrocartilage, allowing the tissue to resist compression [115, 128]. Aggrecan also acts as a lubricant, allowing the fibrils to slide over one another [128].

20.2.3.3
Glycoprotein

Tendons and ligaments also contain small amounts of glycoprotein. This is a protein that is covalently linked to a sugar molecule and does not contain the many GAG chains that are present in PGs. Glycoproteins present in tendons and ligaments include tenascin-C and fibronectin. The former appears to contribute to the mechanical stability of the ECM via interaction with collagen fibrils [28] and also influences growth factor activity [12] and inhibits β_1 integrin-dependent adhesion [95]. It also interacts with fibronectin [12] and upregulates expression under mechanical loading [118]. Fibronectin appears on the surface of collagens, and its expression is known to increase during wound healing [50, 56, 121]. It possesses cell-binding domains that bind cell-surface receptors such as integrins and syndecans. Fibronectin also binds ECM proteins such as collagens, heparin, fibrin, and fibronectin itself [101, 103].

20.2.3.4
Elastin

Tendons and ligaments also contain elastin, which accounts for 2–3% of the dry weight in tendons and up to 10–15% of the dry weight in ligament [75, 119]. Elastin is highly insoluble and has a very slow turnover rate [101]. The hydrophobic domains of elastin form large spirals with the side chains of hydrophobic amino acids held within the coil. Stretching of the coil opens up the hydrophobic center to water, and the thermodynamics of the system provides a driving force for the coils to return to their original length, resulting in high elasticity of the molecule [101]. Elastic fibers, composed of elastin and micro-fibrillar proteins, are believed to contribute to the elastic recovery of tendons and ligaments after stretching [119].

20.2.4
Muscle and Bone Insertion Points

Tendons and ligaments may connect to bone either directly or indirectly [75, 124, 126]. Direct insertions attach the tissue to bone at right angles via a transitional zone of fibrocartilage, while indirect insertions attach to bone at acute angles via Sharpey's fibers that extend through the periosteum into the bone, with no fibrocartilaginous zone [75, 124]. These insertion points with bone experience tensile, compressive, and shear forces, and the strains may be up to four times as much as those experienced in the tendon or ligament midsubstance [76]. Objects are generally the weakest at interfaces where two different materials meet. At this interface, stress concentrations tend to form as force is transferred from one material to another with differing properties, and this often will result in failure in this region [71]. Studies of the ACL, for example, have shown that regions at or near the bone insertion points experience the largest deformations and are most likely to fail [15, 40, 125].

To address this, direct insertion points consist of regions of tendon or ligament, fibrocartilage, mineralized fibrocartilage, and bone [71, 75, 124]. Collagen fibers are aligned in parallel in the tendon/ligament tissue, with fibroblasts embedded within the collagen matrix. In the fibrocartilage region, collagen fibrils are larger and no longer completely parallel [71]. The fibroblasts are also replaced with chondrocyte-like cells [71, 92]. Here, collagen type II is found in the pericellular matrix of the chondrocytes, although collagen type I is still the predominant collagen in the ECM [71, 92]. In the mineralized fibrocartilage region, chondrocytes become enlarged and more circular, and collagen type X is also present in this region [88, 92]. The mineralized fibrocartilage connects to bone through deep interdigitations, providing resistance to shear and tensile forces [71]. The bone is composed of osteoblasts, osteocytes, and osteoclasts among a highly calcified matrix. Collagen type I is the predominant collagen present; however, fibrocartilage-specific markers like collagen type II are no longer present [71]. This fibrocartilage region permits a gradual increase in stiffness, thus preventing the formation of stress concentrations in the tissue and decreasing the risk of failure [71]. It may resist bending forces in the tissue that would normally lead to fatigue failure [10, 99, 123]. Fibrocartilage plays a crucial role in the formation of a stable interface of tendon/ligament and bone.

On the opposite side of the tendon, there is a connection to the muscle at the myotendinous junction (MTJ), where collagen fibrils are inserted into deep recesses formed by myofibroblasts, allowing the large tensile forces of the muscle to be transmitted to the collagen fibers [119]. This junction also contributes to enhance muscle growth [9] and reduce tension on tendons during contraction; however, it is also the weakest point of the muscle-tendon unit [119]. For this reason, many indirect injuries, including sprains, strains, and rupture, are common at the MTJ [57, 68, 75, 111].

20.2.5
Mechanical Properties

Tendons and ligaments possess similar mechanical properties. The stress-strain curve has an initial nonlinear toe region where the tendon experiences up to 1.5–3.0% strain under low stress [75]. The toe region is a result of the stretching of the "crimp pattern" of collagen, a sinusoidal pattern that is thought to absorb shock and permit slight elongation without fibrous damage [58, 119]. The angle and length of the crimp pattern varies in different types of tendon and ligament tissue, and these differences affect the mechanical properties of the tendon [58, 119]. The toe region is followed by a linear region, where the tissue loses its crimp pattern and behaves as an elastic material. In general, Young's modulus ranges between 1.0 and 2.0 GPa [75]. The yield point of tendon and ligament is reached when strain ranges from 5 to 7% [75]. With a 5–7% strain, microscopic tearing of the fibers occurs, and then macroscopic failure occurs at 12–15% strain [75, 119]. Under further strain, the tendon ruptures. The tensile strength of tendons and ligaments ranges from 50 to 150 MPa [75].

Tendons and ligaments are viscoelastic tissues, meaning that they experience stress-relaxation, creep, and hysteresis in the stress-strain curve [75, 102, 119, 124]. Stress-relaxation and creep are time-dependent properties, for which mechanical properties vary over time, as stress or strain is applied. Stress-relaxation is defined as the decrease in stress over time when a constant strain is applied [75]. Creep is defined as the increase in strain over time that is observed under constant stress [75, 102]. Hysteresis in the stress-strain curve, where the unloading curve is lower than the loading curve, indicates a loss of energy in the loading process, and this is typical of viscoelastic materials [75]. Viscoelastic materials also are more deformable at low strain rates. At low strain rates, they absorb more energy but are less effective at transferring force. At high strain rates, they behave more stiffly and are more effective at transferring large forces [119]. The viscoelastic properties of tendons and ligaments are most likely due to collagen, water, and interactions between collagenous and noncollagenous proteins, including elastin and PGs [119, 124]. The movement of water within the collagen microstructure plays a large role in the time-dependent behavior of tendon/ligament tissue, because the water requires time to escape the tissue as load is applied [102].

To successfully tissue engineer a tendon or ligament, it is important to replicate these mechanical properties as closely as possible. Slight variations in both the time-independent and time-dependent properties may result in lack of functionality or mechanical failure.

20.3
Injury and Healing

A wide variety of tendon/ligament injuries are possible within the joint capsule. When designing tissue engineering approaches to treat these injuries, it is important to have a good understanding of the injuries and how the body responds to them. Different types of injury may require different tissue engineering solutions. An understanding of natural tendon/ligament healing is an important consideration for determining design parameters, such as biocompatibility, degradation properties, and rate of construct integration.

20.3.1
Injury

Tendon and ligament injuries may occur when a joint, such as the knee, experiences rapid twisting, excessive force, or direct trauma. Injury can be classified as either acute or chronic injury, and direct or indirect. An acute, direct injury may result from contusion, non-penetrating blunt injury, or laceration [68]. Acute, direct injuries may often result in rupture or tearing of the tendon or ligament.

Indirect injuries, however, often result from acute tensile overload or overuse injury caused by repetitive microtrauma [68]. In tendons, tensile overload and overuse injuries often result in injury to either the MTJ or the osteotendinous junction, because the healthy tendon midsubstance can generally withstand larger tensile loads than the muscle-tendon or bone-tendon interfaces [57, 68, 111, 112]. Chronic overuse injuries will generally result in inflammation, often without rupture or tearing.

20.3.2
Healing

Tendons and ligaments are largely acellular and poorly vascularized tissues and have a poor healing capacity [59, 65, 68, 119]. Their healing can be described in stages. The three steps involved are inflammation, cellular proliferation and ECM production, and remodeling [65, 68, 124, 127].

Inflammation involves serous fluid accumulation in the injured tendon/ligament and surrounding tissue almost immediately after injury and through the first 72 h after injury [59, 65]. Rupture of surrounding blood vessels causes a hematoma to form, which includes platelet aggregation and degranulation [59, 68]. Platelets release growth factors, vasodilators are activated, and mast cells release inflammatory chemicals [59, 65, 68]. The injured area becomes very sensitive and fragile as monocytes, leukocytes, and macrophages migrate to the injury [65]. Inflammatory cells phagocytose necrotic tissue and debris and break down the clot [68], and macrophages help recruit new fibroblasts to the area and promote angiogenesis [31, 43]. During this stage, a general increase in DNA, fibronectin, GAG, water, and collagen type III is experienced to stabilize the new ECM [68].

During the second stage—cellular proliferation and matrix repair—fibroblasts are present and disorganized vascular granulation tissue is formed [65, 68, 127]. Macrophages and mast cells decrease in number during this stage [68]. Studies have shown a higher ratio of collagen type III to type I production than normal levels and increased formation of ECM, as indicated by increased size of fibroblast endoplasmic reticulum [65, 68]. Collagen III and DNA are at their maximum amounts. These changes collectively act to optimize collagen production and gradually convert collagen type III to type I [46, 68]. Proliferation and ECM repair continue for approximately 6 weeks [59, 65].

During the third stage—remodeling—the new ECM matures into slightly disorganized hypercellular tissue; this lasts for several months [65]. Fibroblasts decrease in size and slow ECM synthesis, and collagen fibers orient longitudinally with the long axis of the tendon/ligament [68, 124]. The collagen type III to type I ratio returns to normal levels, and collagen cross-links, GAG, water, and DNA return to normal [68]. Optimal healing occurs when the continuity of collagen fibers is maintained; however, lack of organization and difference in crimp pattern in the newly formed tissue results in mechanical properties that are inferior to those observed pre-injury [65, 124, 127].

20.4
Current Reconstruction Procedures

While current procedures of tendon/ligament replacement aid in restoring locomotion, they do not replicate full joint function. Surgical techniques including sutures and grafts have experienced problems with mechanical strength, fatigue, wear, morbidity, and infection [59, 65, 68, 124]. Grafts also include issues regarding how to properly fix the tissue to the bone for a mechanically stable, biologically viable connection interface. These limitations demonstrate the need for tissue engineering alternatives for treatment of tendon/ligament injury.

When a tendon or ligament is unable to heal naturally, surgery may be necessary to repair it. For best results, tendon and ligament repair should be performed as close to the time of injury as possible [68]. Some minor tendon and ligament tears can be reattached by suturing the injured tissue back together [27]; however, the sutured tissue often heals poorly and is unable to replicate the strength required of native tendon, resulting in failure [68]. For this reason, tissue grafts are often required.

The most common graft procedure to replace an injured tendon or ligament involves an autograft, in which graft tissue is taken from the same patient. For reconstruction of an injured ACL, autografts are most often removed from the patellar tendon or hamstring tendon, and occasionally the quadriceps tendon [65, 124]. The patellar tendon is removed along with pieces of bone from the patella and tibia [65, 127]. This bone-patellar tendon-bone graft is anchored through a bone tunnel that is drilled through the tibia, drawn across the knee, and anchored into a tunnel drilled through the femur for ACL reconstruction [65, 71]. The bone or soft tissue of the graft is fed through the bone tunnel and may be secured into place using a variety of methods, including staples or interference screws [71].

These autografts have good initial mechanical strength, and they promote cell proliferation and new tissue growth [65, 75, 124, 127]; however, autografts still possess a number of disadvantages. Despite their good tensile strength, ACL autografts have been unable to properly stabilize the knee under rotatory loads [124, 127]. Furthermore, the long-term success of these grafts is largely dependent on revascularization of the graft tissue *in vivo*, which has proven difficult to achieve [66]. A limited supply of donor tissue is available for autografts, and additional surgery is required for tissue harvest, which often results in donor site morbidity, including pain, decreased motion, muscle atrophy, and tendonitis [65, 93, 118].

Another option for tendon/ligament replacement is an allograft, the graft material for which is removed from a cadaver and usually frozen [118]. ACL allografts most often involve cadaver patellar, hamstring, or Achilles tendons [65]. These allografts do not require secondary surgery for tissue harvest, and offer similar mechanical strength, cell proliferation, and new tissue growth to autografts [65, 75].

The disadvantages of allografts include potential disease transmission or bacterial infection [65, 93, 118]. They also carry the risk of unfavorable immunogenic response from the host [65, 93, 118].

A third option that has been investigated involves xenografts, which are removed from animals. Use of bovine tendons has been explored; however, only limited success has been shown [118]. In addition, use of xenografts has failed to receive approval from the Food and Drug Administration, due to cases of recurrent effusion, graft failure, and synovitis [118].

Another area that is currently being explored is the use of synthetic materials for tendon/ligament replacement. Synthetic graft materials include polytetrafluoroethylene, polypropylene, polyethylene terephthalate, polyethylene terephthalate-polypropylene, and carbon fibers [66, 93, 118]. Limited success has been demonstrated using these materials as tendon/ligament replacements. These synthetics are able to replace the initial function of ligament but tend to fail over time because they cannot fully replicate the mechanical behavior of native ligament [65, 66, 75, 93]. They are generally unable to withstand repeated elongation *in vivo*, often experiencing wear, fatigue, degradation, creep, permanent deformation, and stress-shielding [65, 75, 93, 118]. Additional problems involve fixation of the grafts, because the grafts must be securely anchored without physical damage and should support tissue in-growth [65, 118]. Even woven prostheses, which provide better mechanical properties, have experienced problems of axial splitting, low tissue infiltration, low extensibility, and abrasive wear [65].

20.5
Tissue Engineering

The limited ability of tendons and ligaments to heal on their own and the many limitations in surgical repair have resulted in the rapid development of the field of tendon and ligament tissue engineering. The goal of tendon and ligament tissue engineering is to produce tissue that initially possesses the mechanical strength to restore function to the injured joint, while promoting further tissue growth and remodeling over time. The new tissue should have similar mechanical properties to the native tendon and ligament to ensure proper functionality [118].

Clinically, tissue engineering could resolve many of the disadvantages associated with the current methods of tendon and ligament repair. Tissue-engineered tendons and ligaments would not require additional surgery for tissue harvest that is required for autografts, thus minimizing donor site morbidity. Also, tissue engineering would have a lower risk of immune rejection, infection, and disease transmission, which are common risks of allografts. Ideally, the surgical procedure of implantation and fixation of the tissue-engineered graft would be simple, reliable, and minimally invasive. A sturdy fixation of the graft material to the bone or muscle would be critical for proper functionality and would permit the patient to undergo immediate rehabilitation [118].

Tissue engineering applies a combination of biological, chemical, and engineering principles for the regeneration of new, functional tissue. The typical approach to tissue engineering utilizes a three-dimensional (3D) scaffold to promote tissue formation, cells that can be expanded and maintained *in vitro*, and the use of growth factors [65]. These components can also be combined with a bioreactor system and mechanical stimulation of the tissue, and success criteria can be assessed using a variety of quantitative methods *in vitro* [118]. Cell constructs are primarily assessed according to cell viability and proliferation, quantification of ECM production, overall tissue structure, and bulk mechanical properties [27]. Collectively, these measures indicate the degree of tissue formation and the mechanical functionality of the tissue.

A wide range of tissue engineering approaches are currently under investigation for tendon/ligament repair. Progress has been made in research using a variety of scaffold materials, cell sources, and growth factors.

20.5.1
Scaffold Materials

There are many design requirements that must be considered when developing a scaffold for tendon and ligament engineering. Of course, the material must be biocompatible, inducing a minimal inflammatory response when implanted into the body while promoting tissue formation. Also, the construct must properly permit the transport of nutrients and regulatory factors to the cells [54, 118]. Most cellular matrices are porous, promoting the growth of new tissue into the construct with cellular proliferation and ECM deposition [36, 54, 65]. Tissue in-growth helps to strengthen the scaffold and may help to integrate the construct at the fixation points [118]. Sufficient mechanical strength and stiffness are also necessary for proper tendon/ligament function without failure. The scaffold materials should have mechanical properties that are similar to the native tendon and ligament to immediately restore proper function and prevent stress-shielding of newly developing tissue [65, 118].

Many constructs for tendon and ligament tissue engineering are also designed to be biodegradable, so that new tissue can integrate with the scaffold and eventually replace it. This means that the construct should degrade at approximately the same rate that new tissue is formed [65, 118]. This provides adequate stabilization of the newly forming tissue while transmitting forces to the tissue for mechanical stimulation without stress-shielding [27, 65]. As the matrix gradually degrades, the tissue remodels and replaces the scaffold with functional tendon and ligament tissue [118].

As a result of these many design criteria, a wide variety of materials have been considered, including both natural and synthetic scaffold materials. Because tendon and ligament tissues are composed primarily of collagen type I, collagen type-I fibers and gels have been largely explored as a natural, 3D matrix for tendon and ligament tissue engineering [2, 6, 7, 13, 23, 24, 44, 45, 53, 63, 66, 77, 86, 87, 104, 107, 118]. Collagen is cross-linked in parallel to form a gel, which can act as a 3D environment for fibroblasts or mesenchymal stem cells (MSCs). Collagen constructs promote cell attachment, spreading, proliferation, and ECM production [2, 13, 44, 53, 63, 66, 77, 87, 104, 118]. Collagen also permits cell migration for potential tissue in-growth [23, 24, 86]. Collagen gels, however, have inferior mechanical properties to native tendon and ligament tissues, and, unless neotissue is formed concurrently, these properties tend to decrease over time [44, 45, 118].

Silk is another widely studied natural material for tendon and ligament tissue engineering [1, 3, 4, 19, 20, 82, 104]. Similar to collagen, silk scaffolds are also biocompatible and promote cell adhesion, proliferation, and ECM deposition [1, 3, 4, 19, 20, 82, 104]. Silk also has high linear stiffness, although when arranged in parallel its tensile strength and linear stiffness still fail to replicate those of native tendon and ligament [118]. Silk degrades slowly *in vivo*, losing its tensile strength after 1 year and completely degrading after 2 years [3]. By arranging the silk in a wire-rope structure, the silk scaffold is capable of achieving similar mechanical properties to native ligament by decreasing linear stiffness without affecting tensile strength [1, 3]. This silk scaffold was shown to support MSC attachment and spreading, proliferation, and tendon/ligament ECM deposition after 14 days [1].

Along with natural materials, there are also a multitude of synthetic materials under investigation as scaffolds for tendon and ligament tissue engineering. The most widely used synthetic materials are poly(L-lactic acid) (PLLA) [1, 2, 17, 22, 65, 73, 91, 118] and poly(lactic-co-glycolic acid) (PLGA) [21, 73, 96]. Others include polyurethane urea, poly(desaminotyrosyl-tyrosine ethyl carbonate) [poly (DTE carbonate)], and polydioxanone (PDS) [65]. Most of these polymeric materials are biodegradable by hydrolytic degradation of bonds. PLLA, for example, degrades by hydrolysis of its ester bonds, leaving lactic acid, which is biocompatible and easily removed from the body [47, 48, 73]. Polymer degradation varies with many factors, including pH, composition, cross-linking density, and hydrophilicity [48]. Control of these various factors permits modification of the degradation properties for a given function.

As polymeric fibers, arranged in parallel, the disadvantage with most of these materials is they do not possess the structural strength necessary for tendon and ligament. Also the parallel arrangement of the fibers may cause long-term failure due to fatigue, creep, and wear [65]. To strengthen these scaffolds, more complex designs have been developed involving woven or braided fibers, the mechanical properties of which more closely resemble tendons and ligaments [65].

20.5.2
Cell types

An obvious choice of cell type for tendon/ligament tissue engineering is tendon or ligament fibroblasts, since fibroblasts are the predominant cell type in tendon and ligament tissue. Fibroblasts primarily maintain the tendon and ligament ECM, including the collagen fibers which provide the tissue's mechanical strength. Studies have shown that tendon/ligament fibroblasts adhere and proliferate in a number of scaffolds [21, 39, 63, 67, 74, 96, 107]. They excrete typical tendon/ligament ECM—including collagen type I, collagen type III, and PGs [38, 39, 67, 74, 100]—and have been shown to upregulate ECM production under mechanical strain [67, 108, 120]. Studies have also verified that tendon/ligament fibroblasts are capable of migration and in-growth into the constructs [107, 109]. Ideally, an autologous cell source would be available for clinical purposes, but unfortunately, such a source of tendon and ligament fibroblasts is difficult to obtain and not readily available without invasive surgery [118].

Dermal fibroblasts have also been used for tendon/ligament tissue engineering. Unlike tendon/ligament fibroblasts, dermal fibroblasts are easily accessible from the skin, and may be a potential choice as an autologous cell source for tendon/ligament tissue engineering. Dermal fibroblasts have been shown to proliferate faster than ligament cells and did not have adverse effects when implanted *in vivo* [8]. Similarly, dermal fibroblasts also possess the ability to adhere and proliferate in a scaffold [30, 44, 45, 104], and they produce similar ECM to tendon/ligament fibroblasts, including collagen types I and III [30, 106]. Dermal fibroblasts are also capable of migration for potential tissue in-growth [23, 24]. It is still uncertain, however, whether dermal fibroblasts can be used as a successful cell replacement for tendon/ligament fibroblasts, due to structural differences between native skin and tendon/ligament tissues.

Another choice of cell type is mesenchymal stem cells (MSCs) from bone marrow. MSCs are adult stem cells that can selectively differentiate into any mesenchymal cell type, including osteoblasts, chondrocytes, adipocytes, and tendon/ligament fibroblasts. Unlike tendon/ligament fibroblasts, MSCs can be easily collected from adult bone marrow using a needle biopsy. Similar to fibroblasts, MSCs are capable of cell adhesion and proliferation in a variety of scaffolds [1, 13, 19, 26], and they excrete ECM components similar to those found in native tendons and ligaments [13, 19, 26, 91]. MSCs have been shown to differentiate into fibroblasts *in vitro*, under a variety of biochemical and mechanical stimuli [118], and are believed to improve wound healing *in vivo* [6, 13]. They may even have higher rates of proliferation and collagen excretion than tendon/ligament fibroblasts, making them good candidates for tendon and ligament tissue engineering [41]. MSCs could potentially serve as an autologous cell source for tendon/ligament engineering, without the risk of additional surgery or immune reaction [118].

20.5.3
Differentiation of MSCs

A wide variety of biochemical factors have been used to induce differentiation of MSCs down specific mesenchymal cell lineages. An assortment of growth factors have been shown to increase MSC proliferation and promote ECM production [84], and they will be discussed in section 20.5.4. Insulin is believed to promote protein expression and ECM production in soft connective tissue [84, 118], while ascorbate-2-phosphate has been shown to promote cell growth and is believed to play a role in the maintenance of connective tissues [32, 77, 84]. Oxygen has also been shown to affect ECM synthesis rates, proliferation rates, and *in vitro* development of tissues [16, 32, 90], although its role in the differentiation of MSCs into fibroblasts is not fully understood. Serum also contains many components, including amino acids, growth factors, vitamins, proteins, hormones, lipids, and minerals that are important in the stimulation of cell growth and differentiation [118].

Mechanical signals are known to play an important role in the growth and development of native ligaments *in vivo*, as well as *in vitro*. The application of loading can affect tissue development through both enhanced mass transport and the direct stimulation of cells [118]. Even in the absence of specific ligament growth factors, MSCs have been shown to differentiate into ligament-like cells under cyclic strain, with increased proliferation and tendon/ligament ECM excretion [1, 2, 4, 20]. The exact mechanism of mechanotransduction in cells is not fully understood. Loading may cause changes in the extracellular environment, cell shape, interfibrillar spacing, protein conformation, or a number of other factors to modify gene transcription and MSC differentiation [42, 117, 118].

20.5.4
Growth Factors

Tendon/ligament fibroblasts are known to respond to a wide range of growth factors. Generally, growth factors act in a dose-dependent manner and are believed to play a large role in tendon/ligament healing [81]. Growth factors such as transforming growth factor-β (TGF-β), basic fibroblast growth factor (bFGF), platelet-derived growth factor-BB (PDGF-BB), and insulin-like growth factor-1 (IGF-1) have been shown to upregulate fibroblast proliferation [25, 81, 105]. These four mentioned growth factors are believed to affect ECM production, namely collagen I and PGs [37, 81, 105]. They have also been shown to influence fibroblast migration [60, 81], and bFGF and vascular endothelial growth factor (VEGF) also play a role in angiogenesis [81].

Growth factors have also been suggested to play a significant role in MSC maintenance and differentiation. They can be combined or added sequentially in a large number of combinations for various effects on MSC differentiation. bFGF and FGF-2, TGF-β, EGF, and IGF-II were all shown to increase cell growth in MSCs and upregulate production of ECM characteristic of tendon/ligament [51, 83, 84, 110]. Additionally, when bFGF or EGF, then TGF were added sequentially, an increase in cell in-growth was observed [83].

Growth and differentiation factors (GDFs) 5, 6, and 7, which are also known as cartilage-derived morphogenetic proteins (CDMPs) 1, 2, and 3, respectively, are members of the bone morphogenetic protein (BMP) family and are believed to play a major role in the healing of tendon/ligament and differentiation of MSCs. Injection of GDF-5, -6, and -7 has improved healing capacity of tendon *in vivo* following injury in rats and rabbits [5, 33–35, 97, 113]. Also, studies of GDF-5 null mice suggest that GDF-5 plays a role in tendon/ligament development, collagen strength, and healing capacity [52, 78, 79]. Such mice developed weak tendons with less collagen [79], while adenovirus gene transfer of GDF-5 resulted in stronger, thicker tendons [97]. Additionally, ectopic implantation of GDF-5, -6, and -7 in rats, as well as intramuscular implantation in the quadriceps, promoted formation of new tendon/ligament tissue *in vivo* [122].

20.5.5
Interface Tissue Engineering for Bone Insertion Point Regeneration

Native tendons and ligaments, as well as tendon and ligament grafts, often fail at their interface with bones. For this reason, it is important not only to engineer tendon and ligament tissue successfully, but also to engineer the tissue's insertion point with bone. This interface can then be used in combination either with existing autograft and allograft materials or with current scaffolds for tendon and ligament tissue engineering to securely fix the materials to bone. A strong integration of graft material to the bone would significantly reduce the rate of failure of the grafts, while promoting proper healing.

As discussed in section 20.2.4, tendons and ligaments have a transition region of fibrocartilage where the tissue connects to bone. To engineer the bone insertion successfully, a scaffold would mimic the mechanical properties of the fibrocartilage, producing a gradual transition of mechanical properties from bone to tendon/ligament [71]. The scaffold would also facilitate the growth of and/or differentiation to multiple cells types, including fibroblasts and osteoblasts [71].

Multi-phase scaffolds have been developed with a gradient of mechanical, structural, and chemical properties aimed at regenerating the tendon-bone or ligament-bone insertion points [72]. This scaffold begins with fibrous PLGA, designed to facilitate tendon and ligament development; the addition of bioactive glass (BG) forms a composite material that is appropriate for bone formation. The PLGA-BG composite has been shown to support growth and differentiation of human osteoblast-like cells *in vitro* [72]. The interface, where the bone-forming composite and the ligament-forming PLGA meet, permits the direct interaction of osteoblasts and tendon/ligament fibroblasts. This interaction of the different cell types and different structural scaffolds may result in the formation of an interfacial fibrocartilage-like region to promote fixation of soft tissue to bone [71].

20.6
Conclusions

Tendons and ligaments are organized in a hierarchical manner with collagen aligned in parallel, in order to maximize tensile strength. This strength permits the tissue to withstand large, repetitive forces within the body without failure. For tissue repair to result in proper functionality and durability, it is important to replicate the mechanical properties of the native tendon/ligament. The structure-function relationship of native tendon/ligament tissues is an important consideration when determining design parameters for tissue engineering.

Occasionally, under direct trauma, excessive twisting, tensile overload, or overuse, tendon/ligament injury may occur, resulting in inflammation, tearing, or rupture of the tissue. These various injuries may require a range of tissue engineering treatments for proper tissue repair. The body's natural healing response to injury is also an important consideration when determining the local environment around a tissue engineered construct. This environment influences graft properties, such as biocompatibility, degradation properties, and rate of integration.

Current treatments for tendon/ligament reconstruction include autograft, allografts, and synthetic grafts; however, each of these has some major disadvantages, especially donor site morbidity, potential for disease transmission and immunorejection, and insufficient mechanical properties. These grafts also face concerns with secure fixation to the bone. Tissue engineering could potentially resolve many of these issues to achieve improved tendon/ligament repair.

There are many factors that must be considered when studying tendon/ligament tissue engineering and many questions left to be answered. The goal is to produce new tissue that restores function to the injured joint and provides the strength and durability to maintain function over time without failure. This requires engineering of both a viable tendon/ligament tissue and a means of stable integration of the tissue with bone. A wide variety of scaffolds, cell types, and culture conditions are being explored to address these needs. While no approach has yet completely replicated the outstanding mechanical properties of native tendon and ligament tissue, rapid and exciting progress is being made in the search for a tissue engineering solution for tendon and ligament injury.

References

1. Altman GH, Horan RL, Lu HH, et al (2002) Silk matrix for tissue engineered anterior cruciate ligaments. Biomaterials 23(20):4131–41.
2. Altman GH, Horan RL, Martin I, et al (2002) Cell differentiation by mechanical stress. Faseb J 16(2):270–2.
3. Altman GH, Diaz F, Jakuba C, et al (2003) Silk-based biomaterials. Biomaterials 24(3):401–16.
4. Altman GH, Lu HH, Horan RL, et al (2002) Advanced bioreactor with controlled application of multi-dimensional strain for tissue engineering. J Biomech Eng 124(6):742–9.
5. Aspenberg P, Forslund C (1999) Enhanced tendon healing with GDF 5 and 6. Acta Orthop Scand 70(1):51–4.
6. Awad HA, Boivin GP, Dressler MR, et al (2003) Repair of patellar tendon injuries using a cell-collagen composite. J Orthop Res 21(3):420–31.
7. Awad HA, Butler DL, Harris MT, et al (2000) In vitro characterization of mesenchymal stem cell-seeded collagen scaffolds for tendon repair: effects of initial seeding density on contraction kinetics. J Biomed Mater Res 51(2):233–40.
8. Bellincampi LD, Closkey RF, Prasad R, et al (1998) Viability of fibroblast-seeded ligament analogs after autogenous implantation. J Orthop Res 16(4):414–20.
9. Benjamin M, Ralphs JR (1996) Tendons in health and disease. Man Ther 1(4):186–191.
10. Benjamin M, Evans EJ, Rao RD, et al (1991) Quantitative differences in the histology of the attachment zones of the meniscal horns in the knee joint of man. J Anat 177:127–34.
11. Birk DE, Fitch JM, Babiarz JP, et al (1990) Collagen fibrillogenesis in vitro: interaction of types I and V collagen regulates fibril diameter. J Cell Sci 95 (Pt 4):649–57.
12. Bradshaw AD, Sage EH (2000) Regulation of cell behavior by matricellular proteins. In: Lanza RP, Langer R, Vananti J (eds) Principles of tissue engineering. Academic Press, San Diego, pp 119–27.
13. Butler DL, Awad HA (1999) Perspectives on cell and collagen composites for tendon repair. Clin Orthop Relat Res (367 Suppl):S324–32.
14. Butler DL, Dessler M, Awad H (2003) Functional tissue engineering: assessment of function in tendon and ligament repair. In: Guilak F, Butler DL, Goldstein SA, et al (eds) Functional tissue engineering. Springer, New York, pp 213–26.
15. Butler DL, Guan Y, Kay MD, et al (1992) Location-dependent variations in the material properties of the anterior cruciate ligament. J Biomech 25(5):511–8.
16. Carrier RL, Papadaki M, Rupnick M, et al (1999) Cardiac tissue engineering: cell seeding, cultivation parameters, and tissue construct characterization. Biotechnol Bioeng 64(5):580–9.
17. Charles-Harris M, Navarro M, Engel E, et al (2005) Surface characterization of completely degradable composite scaffolds. J Mater Sci Mater Med 16(12):1125–30.
18. Chen EH, Black J (1980) Materials design analysis of the prosthetic anterior cruciate ligament. J Biomed Mater Res 14(5):567–86.
19. Chen J, Altman GH, Karageorgiou V, et al (2003) Human bone marrow stromal cell and ligament fibroblast responses on RGD-modified silk fibers. J Biomed Mater Res A 67(2):559–70.
20. Chen J, Horan RL, Bramono D, et al (2006) Monitoring mesenchymal stromal cell developmental stage to apply on-time mechanical stimulation for ligament tissue engineering. Tissue Eng 12(11):3085–95.
21. Cooper JA, Lu HH, Ko FK, et al (2005) Fiber-based tissue-engineered scaffold for ligament replacement: design considerations and in vitro evaluation. Biomaterials 26(13):1523–32.
22. Cooper JA, Jr, Bailey LO, Carter JN, et al (2006) Evaluation of the anterior cruciate ligament, medial collateral ligament, achilles tendon and patellar tendon as cell sources for tissue-engineered ligament. Biomaterials 27(13):2747–54.
23. Cornwell KG, Downing BR, Pins GD (2004) Characterizing fibroblast migration on discrete collagen threads for applications in tissue regeneration. J Biomed Mater Res A 71(1):55–62.
24. Cornwell KG, Lei P, Andreadis ST, et al (2007) Crosslinking of discrete self-assembled collagen threads: Effects on mechanical strength and cell-matrix interactions. J Biomed Mater Res A 80(2):362–71.
25. Costa MA, Wu C, Pham BV, et al (2006) Tissue engineering of flexor tendons: optimization of tenocyte proliferation using growth factor supplementation. Tissue Eng 12(7):1937–43.
26. Cristino S, Grassi F, Toneguzzi S, et al (2005) Analysis of mesenchymal stem cells grown on a three-dimensional HYAFF 11-based prototype ligament scaffold. J Biomed Mater Res A 73(3):275–83.

27. Doroski DM, Brink KS, Temenoff JS (2007) Techniques for biological characterization of tissue-engineered tendon and ligament. Biomaterials 28(2):187–202.
28. Elefteriou F, Exposito JY, Garrone R, et al (2001) Binding of tenascin-X to decorin. FEBS Lett 495(1–2):44–7.
29. Fan L, Sarkar K, Franks DJ, et al (1997) Estimation of total collagen and types I and III collagen in canine rotator cuff tendons. Calcif Tissue Int 61(3):223–9.
30. Fawzi-Grancher S, De Isla N, Faure G, et al (2006) Optimisation of biochemical condition and substrates in vitro for tissue engineering of ligament. Ann Biomed Eng 34(11):1767–77.
31. Fenwick SA, Hazleman BL, Riley GP (2002) The vasculature and its role in the damaged and healing tendon. Arthritis Res 4(4):252–60.
32. Fermor B, Urban J, Murray D, et al (1998) Proliferation and collagen synthesis of human anterior cruciate ligament cells in vitro: effects of ascorbate-2-phosphate, dexamethasone and oxygen tension. Cell Biol Int 22(9–10):635–40.
33. Forslund C, Aspenberg P (2001) Tendon healing stimulated by injected CDMP-2. Med Sci Sports Exerc 33(5):685–7.
34. Forslund C, Aspenberg P (2003) Improved healing of transected rabbit Achilles tendon after a single injection of cartilage-derived morphogenetic protein-2. Am J Sports Med 31(4):555–9.
35. Forslund C, Rueger D, Aspenberg P (2003) A comparative dose-response study of cartilage-derived morphogenetic protein (CDMP)-1, -2 and -3 for tendon healing in rats. J Orthop Res 21(4):617–21.
36. Freed LE, Guilak F, Guo XE, et al (2006) Advanced tools for tissue engineering: scaffolds, bioreactors, and signaling. Tissue Eng 12(12):3285–305.
37. Fu SC, Wong YP, Cheuk YC, et al (2005) TGF-beta1 reverses the effects of matrix anchorage on the gene expression of decorin and procollagen type I in tendon fibroblasts. Clin Orthop Relat Res (431):226–32.
38. Funakoshi T, Majima T, Iwasaki N, et al (2005) Application of tissue engineering techniques for rotator cuff regeneration using a chitosan-based hyaluronan hybrid fiber scaffold. Am J Sports Med 33(8):1193–201.
39. Funakoshi T, Majima T, Iwasaki N, et al (2005) Novel chitosan-based hyaluronan hybrid polymer fibers as a scaffold in ligament tissue engineering. J Biomed Mater Res A 74(3):338–46.
40. Gao J, Rasanen T, Persliden J, et al (1996) The morphology of ligament insertions after failure at low strain velocity: an evaluation of ligament entheses in the rabbit knee. J Anat 189 (Pt 1):127–33.
41. Ge Z, Goh JC, Lee EH (2005) Selection of cell source for ligament tissue engineering. Cell Transplant 14(8):573–83.
42. Geiger B, Bershadsky A, Pankov R, et al (2001) Transmembrane crosstalk between the extracellular matrix--cytoskeleton crosstalk. Nat Rev Mol Cell Biol 2(11):793–805.
43. Gelberman RH, Chu CR, Williams CS, et al (1992) Angiogenesis in healing autogenous flexor-tendon grafts. J Bone Joint Surg Am 74(8):1207–16.
44. Gentleman E, Livesay GA, Dee KC, et al (2006) Development of ligament-like structural organization and properties in cell-seeded collagen scaffolds in vitro. Ann Biomed Eng 34(5):726–36.
45. Gentleman E, Lay AN, Dickerson DA, et al (2003) Mechanical characterization of collagen fibers and scaffolds for tissue engineering. Biomaterials 24(21):3805–13.
46. Gomez MA (1995) The physiology and biochemistry of soft tissue healing. In: Griffin LY (eds) Rehabilitation of the injured knee. Mosby Company, St. Louis, pp 34–44.
47. Gong Y, Zhou Q, Gao C, et al (2007) In vitro and in vivo degradability and cytocompatibility of poly(l-lactic acid) scaffold fabricated by a gelatin particle leaching method. Acta Biomater 3(4):531–40.
48. Gopferich A (1996) Mechanisms of polymer degradation and erosion. Biomaterials 17(2):103–14.
49. Graham HK, Holmes DF, Watson RB, et al (2000) Identification of collagen fibril fusion during vertebrate tendon morphogenesis. The process relies on unipolar fibrils and is regulated by collagen-proteoglycan interaction. J Mol Biol 295(4):891–902.
50. Grinnell F (1984) Fibronectin and wound healing. J Cell Biochem 26(2):107–16.
51. Hankemeier S, Keus M, Zeichen J, et al (2005) Modulation of proliferation and differentiation of human bone marrow stromal cells by fibroblast growth factor 2: potential implications for tissue engineering of tendons and ligaments. Tissue Eng 11(1–2):41–9.
52. Harada M, Takahara M, Zhe P, et al (2007) Developmental failure of the intra-articular ligaments in mice with absence of growth differentiation factor 5. Osteoarthritis Cartilage 15(4):468–74.
53. Henshaw DR, Attia E, Bhargava M, et al (2006) Canine ACL fibroblast integrin expression and cell alignment in response to cyclic tensile strain in three-dimensional collagen gels. J Orthop Res 24(3):481–90.
54. Hollister SJ (2005) Porous scaffold design for tissue engineering. Nat Mater 4(7):518–24.
55. Ilic MZ, Carter P, Tyndall A, et al (2005) Proteoglycans and catabolic products of proteoglycans present in ligament. Biochem J 385(Pt 2):381–8.
56. Jozsa L, Lehto M, Kannus P, et al (1989) Fibronectin and laminin in Achilles tendon. Acta Orthop Scand 60(4):469–71.
57. Kasemkijwattana C, Menetrey J, Bosch P, et al (2000) Use of growth factors to improve muscle healing after strain injury. Clin Orthop Relat Res (370):272–85.
58. Khatod M, Amiel D (2003) Ligament biochemistry and physiology. In: Pedowitz R, O'Connor JJ, Akeson WH (eds) Daniel's knee injuries. Lippincott Williams and Wilkens, Philadelphia, pp 31–42.
59. Khatod M, Akeson WH, Amiel D (2003) Ligament injury and repair. In: Pedowitz RA, O'Connor JJ, Akeson WH (eds) Daniel's knee injuries. Lippincott Williams and Wilkens, Philadelphia, pp 185–201.
60. Kobayashi K, Healey RM, Sah RL, et al (2000) Novel method for the quantitative assessment of cell migration: a study on the motility of rabbit anterior cruciate (ACL) and medial collateral ligament (MCL) cells. Tissue Eng 6(1):29–38.

61. Komi PV (1990) Relevance of in vivo force measurements to human biomechanics. J Biomech 23 Suppl 1:23–34.
62. Komi PV, Fukashiro S, Jarvinen M (1992) Biomechanical loading of Achilles tendon during normal locomotion. Clin Sports Med 11(3):521–31.
63. Koob TJ (2002) Biomimetic approaches to tendon repair. Comp Biochem Physiol A Mol Integr Physiol 133(4):1171–92.
64. Lapiere CM, Nusgens B, Pierard GE (1977) Interaction between collagen type I and type III in conditioning bundles organization. Connect Tissue Res 5(1):21–9.
65. Laurencin CT, Freeman JW (2005) Ligament tissue engineering: an evolutionary materials science approach. Biomaterials 26(36):7530–6.
66. Laurencin CT, Ambrosio AM, Borden MD, et al (1999) Tissue engineering: orthopedic applications. Annu Rev Biomed Eng 1:19–46.
67. Lee CH, Shin HJ, Cho IH, et al (2005) Nanofiber alignment and direction of mechanical strain affect the ECM production of human ACL fibroblast. Biomaterials 26(11):1261–70.
68. Lin TW, Cardenas L, Soslowsky LJ (2004) Biomechanics of tendon injury and repair. J Biomech 37(6):865–77.
69. Liu SH, Yang RS, al-Shaikh R, et al (1995) Collagen in tendon, ligament, and bone healing. A current review. Clin Orthop Relat Res (318):265–78.
70. Louie L, Yannas I, Spector M (1998) Tissue engineered tendon. In: Patrick CW, Mikos AG, McIntire LV (eds) Frontiers in tissue engineering. Elsevier Science Ltd., New York, pp 412–42.
71. Lu HH, Jiang J (2006) Interface tissue engineering and the formulation of multiple-tissue systems. Adv Biochem Eng Biotechnol 102:91–111.
72. Lu HH, El-Amin SF, Scott KD, et al (2003) Three-dimensional, bioactive, biodegradable, polymer-bioactive glass composite scaffolds with improved mechanical properties support collagen synthesis and mineralization of human osteoblast-like cells in vitro. J Biomed Mater Res A 64(3):465–74.
73. Lu HH, Cooper JA, Jr, Manuel S, et al (2005) Anterior cruciate ligament regeneration using braided biodegradable scaffolds: in vitro optimization studies. Biomaterials 26(23):4805–16.
74. Majima T, Funakosi T, Iwasaki N, et al (2005) Alginate and chitosan polyion complex hybrid fibers for scaffolds in ligament and tendon tissue engineering. J Orthop Sci 10(3):302–7.
75. Martin RB, Burr DB, Sharkey NA (1998) Mechanical properties of ligament and tendon. In: (eds) Skeletal tissue mechanics. Springer, New York, pp 309–46.
76. McGonagle D, Marzo-Ortega H, Benjamin M, et al (2003) Report on the Second international Enthesitis Workshop. Arthritis Rheum 48(4):896–905.
77. Meaney Murray M, Rice K, Wright RJ, et al (2003) The effect of selected growth factors on human anterior cruciate ligament cell interactions with a three-dimensional collagen-GAG scaffold. J Orthop Res 21(2):238–44.
78. Mikic B (2004) Multiple effects of GDF-5 deficiency on skeletal tissues: implications for therapeutic bioengineering. Ann Biomed Eng 32(3):466–76.
79. Mikic B, Schalet BJ, Clark RT, et al (2001) GDF-5 deficiency in mice alters the ultrastructure, mechanical properties and composition of the Achilles tendon. J Orthop Res 19(3):365–71.
80. Milz S, Tischer T, Buettner A, et al (2004) Molecular composition and pathology of entheses on the medial and lateral epicondyles of the humerus: a structural basis for epicondylitis. Ann Rheum Dis 63(9):1015–21.
81. Molloy T, Wang Y, Murrell G (2003) The roles of growth factors in tendon and ligament healing. Sports Med 33(5):381–94.
82. Moreau J, Chen J, Kaplan D, et al (2006) Sequential growth factor stimulation of bone marrow stromal cells in extended culture. Tissue Eng 12(10):2905–12.
83. Moreau JE, Chen J, Horan RL, et al (2005) Sequential growth factor application in bone marrow stromal cell ligament engineering. Tissue Eng 11(11–12):1887–97.
84. Moreau JE, Chen J, Bramono DS, et al (2005) Growth factor induced fibroblast differentiation from human bone marrow stromal cells in vitro. J Orthop Res 23(1):164–74.
85. Moriggl B, Jax P, Milz S, et al (2001) Fibrocartilage at the entheses of the suprascapular (superior transverse scapular) ligament of man-a ligament spanning two regions of a single bone. J Anat 199(Pt 5):539–45.
86. Murray MM, Spector M (2001) The migration of cells from the ruptured human anterior cruciate ligament into collagen-glycosaminoglycan regeneration templates in vitro. Biomaterials 22(17):2393–402.
87. Murray MM, Forsythe B, Chen F, et al (2006) The effect of thrombin on ACL fibroblast interactions with collagen hydrogels. J Orthop Res 24(3):508–15.
88. Niyibizi C, Sagarrigo Visconti C, Gibson G, et al (1996) Identification and immunolocalization of type X collagen at the ligament-bone interface. Biochem Biophys Res Commun 222(2):584–9.
89. Noyes FR, Grood ES (1976) The strength of the anterior cruciate ligament in humans and Rhesus monkeys. J Bone Joint Surg Am 58(8):1074–82.
90. Obradovic B, Carrier RL, Vunjak-Novakovic G, et al (1999) Gas exchange is essential for bioreactor cultivation of tissue engineered cartilage. Biotechnol Bioeng 63(2):197–205.
91. Ouyang HW, Toh SL, Goh J, et al (2005) Assembly of bone marrow stromal cell sheets with knitted poly (L-lactide) scaffold for engineering ligament analogs. J Biomed Mater Res B Appl Biomater 75(2):264–71.
92. Petersen W, Tillmann B (1999) Structure and vascularization of the cruciate ligaments of the human knee joint. Anat Embryol (Berl) 200(3):325–34.
93. Petrigliano FA, McAllister DR, Wu BM (2006) Tissue engineering for anterior cruciate ligament reconstruction: a review of current strategies. Arthroscopy 22(4):441–51.
94. Pins GD, Christiansen DL, Patel R, et al (1997) Self-assembly of collagen fibers. Influence of fibrillar alignment and decorin on mechanical properties. Biophys J 73(4):2164–72.
95. Probstmeier R, Pesheva P (1999) Tenascin-C inhibits beta1 integrin-dependent cell adhesion and neurite outgrowth on fibronectin by a disialoganglioside-mediated signaling mechanism. Glycobiology 9(2):101–14.

96. Qin TW, Yang ZM, Wu ZZ, et al (2005) Adhesion strength of human tenocytes to extracellular matrix component-modified poly(DL-lactide-co-glycolide) substrates. Biomaterials 26(33):6635–42.
97. Rickert M, Wang H, Wieloch P, et al (2005) Adenovirus-mediated gene transfer of growth and differentiation factor-5 into tenocytes and the healing rat Achilles tendon. Connect Tissue Res 46(4-5):175–83.
98. Rumian AP, Wallace AL, Birch HL (2007) Tendons and ligaments are anatomically distinct but overlap in molecular and morphological features-a comparative study in an ovine model. J Orthop Res 25(4):458–64.
99. Scapinelli R, Little K (1970) Observations on the mechanically induced differentiation of cartilage from fibrous connective tissue. J Pathol 101(2):85–91.
100. Schulze-Tanzil G, Mobasheri A, Clegg PD, et al (2004) Cultivation of human tenocytes in high-density culture. Histochem Cell Biol 122(3):219–28.
101. Scott-Burden T (1994) Extracellular Matrix: The Cellular Environment. NIPS 9:110–4.
102. Shrive NG, Thornton GM, Hart DA, et al (2003) Ligament mechanics. In: Pedowitz RA, O'Connor JJ, Akeson WH (eds) Daniel's knee injuries. Lippincott Williams and Wilkens, Philadelphia, pp 97–112.
103. Takahashi S, Leiss M, Moser M, et al (2007) The RGD motif in fibronectin is essential for development but dispensable for fibril assembly. J Cell Biol.
104. Takezawa T, Ozaki K, Takabayashi C (2007) Reconstruction of a hard connective tissue utilizing a pressed silk sheet and type-I collagen as the scaffold for fibroblasts. Tissue Eng 13(6):1357–66.
105. Thomopoulos S, Harwood FL, Silva MJ, et al (2005) Effect of several growth factors on canine flexor tendon fibroblast proliferation and collagen synthesis in vitro. J Hand Surg [Am] 30(3):441–7.
106. Tischer T, Vogt S, Aryee S, et al (2007) Tissue engineering of the anterior cruciate ligament: a new method using acellularized tendon allografts and autologous fibroblasts. Arch Orthop Trauma Surg.
107. Torres DS, Freyman TM, Yannas IV, et al (2000) Tendon cell contraction of collagen-GAG matrices in vitro: effect of cross-linking. Biomaterials 21(15):1607–19.
108. Toyoda T, Matsumoto H, Fujikawa K, et al (1998) Tensile load and the metabolism of anterior cruciate ligament cells. Clin Orthop Relat Res (353):247–55.
109. Trieb K, Blahovec H, Brand G, et al (2004) In vivo and in vitro cellular ingrowth into a new generation of artificial ligaments. Eur Surg Res 36(3):148–51.
110. Tsutsumi S, Shimazu A, Miyazaki K, et al (2001) Retention of multilineage differentiation potential of mesenchymal cells during proliferation in response to FGF. Biochem Biophys Res Commun 288(2):413–9.
111. Tuite DJ, Finegan PJ, Saliaris AP, et al (1998) Anatomy of the proximal musculotendinous junction of the adductor longus muscle. Knee Surg Sports Traumatol Arthrosc 6(2):134–7.
112. Vandervliet EJ, Vanhoenacker FM, Snoeckx A, et al (2007) Sports related acute and chronic avulsion injuries in children and adolescents with special emphasis on tennis. Br J Sports Med.
113. Virchenko O, Fahlgren A, Skoglund B, et al (2005) CDMP-2 injection improves early tendon healing in a rabbit model for surgical repair. Scand J Med Sci Sports 15(4):260–4.
114. Vogel KG (2004) What happens when tendons bend and twist? Proteoglycans. J Musculoskelet Neuronal Interact 4(2):202–3.
115. Vogel KG, Koob TJ (1989) Structural specialization in tendons under compression. Int Rev Cytol 115:267–93.
116. Vogel KG, Meyers AB (1999) Proteins in the tensile region of adult bovine deep flexor tendon. Clin Orthop Relat Res (367 Suppl):S344–55.
117. Vogel V, Sheetz M (2006) Local force and geometry sensing regulate cell functions. Nat Rev Mol Cell Biol 7(4):265–75.
118. Vunjak-Novakovic G, Altman G, Horan R, et al (2004) Tissue engineering of ligaments. Annu Rev Biomed Eng 6:131–56.
119. Wang JH (2006) Mechanobiology of tendon. J Biomech 39(9):1563–82.
120. Webb K, Hitchcock RW, Smeal RM, et al (2006) Cyclic strain increases fibroblast proliferation, matrix accumulation, and elastic modulus of fibroblast-seeded polyurethane constructs. J Biomech 39(6):1136–44.
121. Williams IF, McCullagh KG, Silver IA (1984) The distribution of types I and III collagen and fibronectin in the healing equine tendon. Connect Tissue Res 12(3–4):211–27.
122. Wolfman NM, Hattersley G, Cox K, et al (1997) Ectopic induction of tendon and ligament in rats by growth and differentiation factors 5, 6, and 7, members of the TGF-beta gene family. J Clin Invest 100(2):321–30.
123. Woo SL, Newton PO, MacKenna DA, et al (1992) A comparative evaluation of the mechanical properties of the rabbit medial collateral and anterior cruciate ligaments. J Biomech 25(4):377–86.
124. Woo SL, Abramowitch SD, Kilger R, et al (2006) Biomechanics of knee ligaments: injury, healing, and repair. J Biomech 39(1):1–20.
125. Woo SL, Gomez MA, Seguchi Y, et al (1983) Measurement of mechanical properties of ligament substance from a bone-ligament-bone preparation. J Orthop Res 1(1):22–9.
126. Woo SL, Gomez MA, Sites TJ, et al (1987) The biomechanical and morphological changes in the medial collateral ligament of the rabbit after immobilization and remobilization. J Bone Joint Surg Am 69(8):1200–11.
127. Woo SL, Debski RE, Zeminski J, et al (2000) Injury and repair of ligaments and tendons. Annu Rev Biomed Eng 2:83–118.
128. Yoon JH, Halper J (2005) Tendon proteoglycans: biochemistry and function. J Musculoskelet Neuronal Interact 5(1):22–34.

Neural Tissue Engineering and Regenerative Medicine

N. Zhang, X. Wen

Contents

21.1	Nervous System Injury and Regeneration	271
21.2	Tissue Engineering Strategies for Nervous System Repair	272
21.2.1	Peripheral Nervous System Repair (PNS-PNS)	273
21.2.2	Central Nervous System Repair (CNS-CNS)	275
21.2.3	Nerve Repair at the CNS-PNS Transition Zone	284
21.3	Conclusions	284
	References	285

21.1 Nervous System Injury and Regeneration

The nervous system in vertebrates is composed of two main divisions: the peripheral nervous system (PNS) and the central nervous system (CNS). The CNS includes the brain and the spinal cord; the PNS contains the cranial, spinal, and autonomic nerves, which (along with their branches) connect to the CNS. Neurons are the basic structural and functional elements in the nervous system. Each one contains a cell body, in which the cell nucleus is located; a number of dendrites, which transport electrical impulses to the cell body; and a number of axons, which transmit signals from the cell body to the peripheral regions to muscles and organs. In addition to neurons, the nervous system contains many types of neuroglial cells that support and protect neurons, including astrocytes, oligodendrocytes, and microglia in the CNS, and Schwann cells in the PNS.

Damage to the nervous system can occur due to ischemic, chemical, mechanical, or thermal factors. In addition to triggering a variety of cellular and molecular events, these insults may lead to transection of nerves, interruption of communications between nerve cell bodies and their supporting cells, disruption of the interrelations between neurons and their supporting cells, and the disruption of the blood-nerve barrier. A fundamental difference between PNS and the CNS neurons is the response to axotomy.

In the PNS, axonal sprouting and regeneration begin at the proximal end of a severed nerve following the removal of myelin and axonal debris by macrophages and Schwann cells. Supporting cells, including Schwann cells, macrophages, and monocytes, are able to form a synergistic combination in clearing cellular debris, releasing neurotrophins/cytokines, and promoting axonal outgrowth toward their synaptic targets [1]. Axonal regeneration proceeds until the outgrowing axons reach the distal stump of the severed nerve, where functional re-innervation with their distal targets occurs. PNS regeneration is most successful following crush injury where the damage is minor and the nerve fibers, rather than the endoneurial tubes, have been disrupted. The preservation of endoneurial tubes is critical to provide a guidance pathway lined up with growth-supportive extracellular matrix (ECM) molecules for axonal growth and regeneration.

In the CNS, axons do not regenerate in their native environment in response to injury. Myelin debris and other types of glycoproteins at the injury site are inhibitory for axonal regeneration. The presence of the blood-brain barrier retards the migration of macrophages to the injury site for debris clearance. Glial cells, such as astrocytes, do not provide trophic support for axonal regeneration. Oftentimes, they become reactive astrocytes that undergo hyperplasia (increase in number) and hypertrophy (increase in size) and participate in the formation of a dense scar tissue barrier that inhibits axonal regeneration. Studies have shown that a few neurons that had survived axotomy attempted regeneration and were eventually blocked from reaching their synaptic targets by the glial scar [2]. Further, the lack of an equivalent of an endoneurial tube in the CNS to provide an aligned and adhesive substrate as a guidance pathway for the regenerating axons has been considered as a major factor responsible for the failure of CNS regeneration.

The ability of nerves to regenerate is highly dependent on the location of the damage, whether it is on the nerve tract that connects PNS to PNS, CNS to CNS, or PNS to CNS. Current clinical treatment options for peripheral nerve injuries differ depending on the size of the lesion. For small size lesions (less than a couple of millimeters in length), surgical reconnection of the damaged nerve ends is performed by means of end-to-end suturing of individual fascicles within the nerve cable without introducing tension into the nerve. For large-size lesions that create gaps in the nerve, autologous nerve grafts or autografts that are harvested from another site in the body or another individual are used to fill in the lesion gaps. Problems associated with the use of grafts include risks of donor site shortage and morbidity, and the requirement of multiple surgeries. No effective treatments are available for CNS nerve injuries. In the presence of bone fragments in the vicinity of the injury site, surgeries are required for their removal to minimize secondary injury to the nerves. Management of the injury site generally includes the administration of anti-inflammatory drugs to reduce swelling of the damaged nerve stumps. Patients then enter a chronic phase of recovery, during which the remaining nerves are trained to compensate for the loss due to the injury.

21.2 Tissue Engineering Strategies for Nervous System Repair

Despite significant progress in the attempts to promote nerve repair and regeneration, structural and functional restoration of damaged nerves is far from satisfactory. For example, in the PNS, even with autologous nerve grafts, clinical functional recovery rates approach only 80%. To eliminate the need for autologous nerve grafts and multiple surgeries, alternatives are being developed in bioengineering strategies for PNS nerve regeneration that will facilitate the recovery and restoration of the damaged nerves at both cellular and functional levels. Use of nerve guidance channels is a promising nerve graft alternative. A variety of channels presenting physical, chemical, mechanical, and biological guidance cues to regenerating nerves have been developed with the potential for nerve repair following PNS and spinal cord injuries, and neurodegenerative pathologies [e.g., Parkinson's disease (PD)] [3–5].

When compared with PNS repair, CNS regeneration has posed greater challenges for therapies. Although it has been demonstrated that CNS axonal regeneration is possible, given permissive substrate and appropriate biological environment [6], the magnitude of the regeneration is too low for functional recovery. Both embryonic spinal cord grafts and peripheral nerve grafts have been shown to support axonal outgrowth in the CNS; however, the regenerating axons do not grow across the PNS-CNS transition zone (TZ) [7, 8]. Current bioengineering strategies have been focused on creating permissive substrates that combine biomaterials, cellular components, and neurotrophic factors to reconstruct the pathway for the regenerating axons inside the CNS [4] or across the PNS-CNS TZ [9, 10].

Regardless of the location of axonal regeneration, tissue engineering strategies for nervous system repair can be separated into four categories. These include axonal guidance devices, cell population recovery, drug delivery, and electrical stimulation. Devices of different designs and materials have been tested extensively for the creation of physical or chemical pathways for regenerating axons to cross the lesion site/scar zone to re-innervate with their appropriate targets. Axonal guidance approaches have been used for spinal cord and brain nerve tract regeneration. For the treatment of neurodegenerative diseases, strategies to replace or recover the neurons and the glial cells that are depleted or degenerated due to the pathologies are being developed. Table 21.1 lists the common types of neurodegenerative diseases and the affected cell populations. Attributing to advances made on the subject of stem cell biology, most of the neuronal and neural populations needed for nervous system repair can now be derived from stem cells in vitro for transplantation into the patient. However, due to the versatile nature of stem cells and their extreme sensitivity to local environment, functional consequences of stem cells or stem cell-derived phenotypes on the recipient remain unpredictable and need to be fully characterized. The goal is to produce specific functional differentiated neuronal and neural phenotypes in sufficient quantities and maintain in that differentiated state either in vitro for transplantation or in vivo in tissue environment as replacement cells for clinical uses. In addition to cell therapies, delivery of drugs or therapeutic agents to the site in need of repair is important to condition the local tissue environment to facilitate the regeneration. Drug delivery in a spatial and temporal fashion coincident with the regeneration process would allow axonal sprouting, extension along specific pathways, and synapse formation with appropriate targets. For example, agents that suppress the formation of scar tissue are delivered locally during the initial stage of the regeneration process to enhance the permissiveness of the environment for axonal sprouting and elongation. Subsequent administration of neurotrophic factors along the damaged pathways will provide guidance cues and trophic support necessary for directional axonal outgrowth and target re-innervations. In a strict sense, electrical stimulation is a bioengineering rather than a tissue engineering approach. Electrical stimulation techniques are often applied in combination with other strategies to further enhance axonal regeneration. Stimulating electrodes can be incorporated into neuronal bridging devices that also contain sustained agent delivery vehicles. Close apposition of the electrodes with regenerating neurons may improve the outgrowth rate and length of the regenerating axons. In the following section, we will discuss bioengineering strategies based on individual or a combination of these approaches to promote neural regeneration within the PNS, within the CNS, and across the PNS-CNS TZ.

21.2.1
Peripheral Nervous System Repair (PNS-PNS)

The nerve tracts of the PNS consist of thousands of axons and supporting cells. PNS axons are dependent on the neuronal cell bodies in the spinal cord or

Table 21.1 The common types of neurodegenerative diseases and the affected cell populations

Neurodegenerative diseases	The affected cell populations
Parkinson's disease	Degeneration of dopaminergic (DA) neurons
Huntington's disease	Degeneration of multiple sets of neuronal populations, including both cortical and striatal neurons
Alzheimer's disease	Brain-wide neuronal loss
Epilepsy	Cerebral cortex neuronal loss (e.g., pyramidal cells, dentate granule cells, and inhibitory interneurons); Neuronal damage (e.g., reduced arborization of dendritic tree, reduced number of GABA receptors, and reduced number of NMDA receptors
Multiple sclerosis	Degeneration of oligodendrocytes
Amyotrophic lateral sclerosis/Lou Gehrig's disease	Spinal cord anterior horn cell atrophy and the replacement of the large motor neurons by fibrous astrocytes, leading to myelin cell degeneration

the ganglia. Myelination of the axons by Schwann cells is critical for high-speed transduction of electrical impulses, resulting in transduction speeds of electrical signals along the axonal cable of about 100 m/s, compared with only 1 m/s in the absence of the myelin.

Axonal transection in the PNS generally triggers Wallerian degeneration, characterized by sequential axonal and myelin degeneration in the nerve segment that is distal to the site of transection, due to the loss of connections between the distal axons and the proximal axons. Accompanied by physiological and morphological changes in neurons and glial cells, this degeneration is also associated with macrophage-predominant inflammation, in which macrophages enter the area to remove the myelin and axonal debris. Immediately following transection (within hours), organelles and mitochondria start to accumulate in paranodal regions near the injury site, endoplasmic reticulum loses structure, and neurofilaments start to undergo degradation due to the influx of calcium ions and the activation of calpain. Axons become fragmented and phagocytosed. Degeneration of myelin occurs in a distal to proximal direction, and collapse of myelin results in the formation of voids. Fortunately, during the process, the basement membrane tube that encapsulates the Schwann cells and axons persists, allowing the line-up of Schwann cells and secretion of growth factors that attract axonal sprouting from the proximal terminal of the severed axons. The basement membrane tubes also serve as pathways for the regenerating axons. Subsequently, the Schwann cells re-myelinate the newly formed axons, which re-innervate with their appropriate targets in the destiny tissues or organs. The blood-nerve barrier loses integrity during the early stages of nerve degeneration and regeneration, and restores structures and functions over a chronic phase (months) of the regeneration.

As described previously, the repair of small lesions in the PNS nerve (gap size less than a couple of millimeters) can be done by surgical suturing. For the repair of lesions that result in large gaps in the nerve, entubulation approaches are necessary, in which a tubular construct is placed across the lesion gap to bridge the two stumps of the severed nerve to create a well-controlled regeneration environment. The entubulation sleeve serves as a barrier to prevent infiltration of scar tissue into the lumen where axons are regenerating. Cells or growth factors can be loaded into the sleeve to further enhance the regeneration. A number of studies have documented the effects of physical, chemical, biological, and electrical factors in the entubulation devices on PNS axonal regeneration. Physical factors include the morphology/biodegradability/porosity of the entubulation sleeve and the luminal material architecture (e.g., intraluminal channels or oriented nerve substratum). Chemical and biological factors are dictated by the types of biomolecular surface coatings on the device (sleeve and luminal matrices), the growth factors that are controlled released, and the types of cells that are pre-coated or incorporated into the device. Electrical factors depend on the spatial and temporal patterns of the electrical stimulations that are applied onto the regenerating axons. A wide variety of entubulation devices of materials of synthetic and/or biological origin have been tested to facilitate PNS nerve repair. These devices vary in shape (e.g., cylindrical or Y shaped), level of permeability (indiscriminately permeable, semi-permeable, or non-permeable), and luminal contents (exogenous agents or cells). In particular, the presence of distal nerve stump is important to attract axonal outgrowth and elongation in the bands of Bungner (arrays of Schwann cells and processes within basement membrane) over distances greater than 1 cm in adult rats. In the absence of distal nerve stump, axons can only grow 5 mm into the "unaided" entubulation tubes due to the lack of any target. In addition, Schwann cells and their basal laminae are also important elements in facilitating axonal re-growth in the PNS. In response to the injury, Schwann cells proliferate and form Bands of Bungner, providing substrate for axonal regeneration. Schwann cells are also able to secrete a wide variety of neurotrophic factors that support axonal sprouting and extension.

The PNS wound healing response in the presence of an entubulation tube is characterized by four phases, i.e., fluid phase, matrix phase, cellular phase, and axonal phase. Immediately following (within 24 h) tube grafting into the lesion gap to span the two stumps of the severed axons, the lumen of the tube is filled with wound fluid that is rich in neurotrophic molecules and matrix precursors. During the first week (2–6 days), the wound fluid in the tube lumen is replaced by an acellular, fibronectin-positive, laminin-negative fibrinous matrix. Within 7–14 days, the luminal area is invaded by various cells, including perineurial cells, fibroblasts, Schwann cells, and endothelial cells. Over a chronic period (15–21 days), axons gradually enter the tube as the last element, after the appearance of the non-neuronal cells and capillaries. Using the entubulation approach, guided regeneration of PNS axons can be monitored. For example, a circular mesh form of electrode of the same size as the cross-section of the served PNS nerve was placed between the two stumps to monitor axonal regeneration. Detection of electrical signals at the electrode indicated the arrival of the regenerating ascending axons onto the plane where the mesh resides.

21.2.2
Central Nervous System Repair (CNS-CNS)

Crucial differences in the adult PNS versus CNS environment following injury may contribute to the differences in the regeneration capability. It is well known that the PNS is capable of regeneration following injury. The regenerative responses may be mounted at multi-steps. First, there is an upregulation in the regeneration-associated genes (e.g., c-Jun, GAP-43, Tal-tubulin, etc.). Upregulation of some of these genes led to the alteration in the composition of the neuronal microtubules that directly participate in the formation of axonal branches [11]. For example, newly synthesized tubulins are delivered to the growth cones by slow axonal transport, where the increases in tubulin protein expression as a result of the upregulation in tubulin mRNA levels are reflected in the changes in the composition and properties of the axonal cytoskeleton that facilitate axonal re-growth [12]. Second, the unique structure called Bands of Bungner in the PNS is preserved following injury, which may aid in the elongation of regenerating axons. Further, there are Schwann cells in the PNS that provide nutrients for axon sprouting, guide axon outgrowth, as well as myelinate regenerated axons. In contrast, the CNS rarely recovers structurally and functionally after injury, suggesting the involvement of more than one of the steps in the regeneration failure. Since adult axons are capable of directional regeneration if directionally organized architecture and a permissive environment are provided, the lack of adult CNS regeneration is not due solely to the lack of intrinsic capability of injured neurons to regenerate. In the injured adult CNS, axotomized axons usually undergo "abortive sprouting", indicating that CNS axons can sprout after injury, but fail to elongate. Meanwhile, the inhibitory environment following CNS injury largely results from the chemorepulsive effect of the non-permissive ECM-rich glial scar tissue that forms at the lesion site. Multiple cell types contribute to the scar formation and exhibit inhibitory properties either through physical contact with regenerating axons or the secretion of inhibitory ECM molecules. These cells and their associated inhibitory molecules generally include oligodendrocytes and myelin-associated glycoproteins (MAGs), reactive astrocytes and chondroitin sulfate proteoglycans (CSPG), and reactive microglial cells.

CNS tissue biology varies between gray and white matter. In the spinal cord, gray matter is a butterfly-shaped region in which neuronal cell bodies are located. It also contains blood vessels, and glial cells such as astrocytes and microglia. White matter consists mostly of axons, as well as blood vessels and glial cells (including astrocytes and oligodendrocytes). Previous studies have indicated that CNS white matter, but not gray matter, is responsible for the regeneration failure in adult rats [13]. Further results indicated that the inhibitory effect of the white matter on neuronal cell adhesion and axonal elongation was largely mediated by the oligodendrocytes and myelin components in the adult white matter. In addition, recent data suggested that, in the absence of glial scarring, the disruption of fiber tract geometry that results in the loss of the spatial organization of cells and molecular factors is sufficient to pose a barrier for axonal regeneration in the adult spinal cord white matter [14]. In agreement with this view, the role of organized neural tissue architecture on axonal outgrowth was examined. In the normal nervous system with the existence of normal organized architecture, neurites were shown to be able to elongate and grow a long distance. However, when the architecture was destroyed by a lesion, irrespective of the type and extent of the injury, neurite outgrowth terminated at the spinal cord lesion site [15]. Organized tissue architecture is important in guiding cell migration, inducing appropriately aligned morphology, and directing axonal outgrowth. The only way to help the neurites to grow across the lesion is to create a bridge that restores the organized native tissue architecture and the biological environment lost at the lesion site [4, 15].

To this end, structural elements that present different types of guidance cues are engineered into the bridge. Physical guidance cues are nested in the unique structure and geometry of the scaffold that serves as the framework for regenerating axons to attach and orient for directional outgrowth. Chemical cues are delivered to the lesion site through biomolecular surface coatings of the scaffold or incorporation of sustained release schemes for therapeutic agents into the bridge. For example, the biological functions of the chemical cues include, but are not limited to:

1. Promoting axonal regeneration—(a) overcoming natural inhibitors of regeneration and inducing axonal growth, and (b) directing axonal path finding and synapse formation with their appropriate targets
2. Preventing expansion of the initial damage and minimizing the secondary injury—(a) blocking excitotoxic injury to the surviving cells and (b) preventing suicide of stressed cells due to the bolster defense mechanisms
3. Compensating for demyelination—(a) preventing dissipation of impulses at the demyelinated areas, (b) eliciting response of oligodendrocytes to remyelinate axons, and (c) replenishing lost oligodendrocytes
4. Inducing the directional differentiation of stem cells or precursor cells in the cord to replace the dead cells

Biological cues are incorporated into the bridge through cell seeding. The co-transplanted cells can be stem/precursor cells that are able to produce all the lost cell types in vivo under controlled local conditions or stem/precursor cell-derived phenotypes from in vitro engineering. These cells are expected to ameliorate local tissue environment following injury, replace the lost cells, provide permissive substrate for axonal outgrowth, and release functional biomolecules.

21.2.2.1
Spinal Cord Repair

Current bridging devices to restore a lost axonal pathway due to spinal cord injury use biological and synthetic materials in combination with cell and agent/biomolecule delivery. Biological bridges include peripheral nerve grafts (Schwann cells and basal lamina tubes), blood vessel grafts, acellular muscle grafts (basal lamina tubes), fetal tissue mini-grafts (a series of fetal grafts), decalcified bone channels, mesothelium, collagen tubes, pseudosynovial sheath, and amnion. In the synthetic bridges, biomaterials are fashioned into channels (with or without loadings of cells or intraluminal matrices) that create a well-controlled regeneration environment in the lumen. Cells can be co-transplanted into the bridges for enhanced regeneration. These include Schwann cells, immature astrocytes, macrophages, stem cells, olfactory bulb ensheathing cells, and genetically engineered cells. Biomolecules/neurotrophic factors or chemicals may either be incorporated into the bridges or sustain released to the lesion site to control the regeneration environment and promote desirable axonal responses (sprouting, elongation, and target re-innervation) in a temporal and spatial manner. Examples are the delivery of disinhibition agents to neutralize or block inhibitory molecules to suppress scar formation during the initial stage of the wound healing, sustained delivery of axonal growth-promoting factors along the regeneration pathway, and the administration of chemicals to remove scar-forming cells. In addition, X-irradiation to alter the cellular components at the lesion site and electrical/magnetic stimulation to accelerate unidirectional axonal outgrowth are applied in conjunction. These technologies are described in detail below.

Biological and Synthetic Bridges

Peripheral nerve grafts represent the most effective grafts so far in promoting axonal regeneration for CNS repair. Multi-nerve grafts have been used to reroute regenerating pathways from non-permissive white matter to permissive gray matter in the injured spinal cord [16].

Blood vessel grafts based on arteries and inside-out vein conduits have been used to bridge the spinal cord.

Acellular muscle grafts are another category of biological grafts that may facilitate spinal cord repair. Harvested muscle tissue was treated to preserve the basal lamina tubes while eliminating other components. When grafted into the lesion gap of the spinal cord, axons grew and oriented along the longitudinal axis of the tubes for a long distance and were closely associated with the tubes. The presence of the lamina tubes also helped in the myelination of the regenerating axons.

Series of fetal tissue minigrafts can be used individually or in bundle form to bridge the lesion gap in the spinal cord. These fetal spinal cord segments not only serve as a structural frame to restore the lost native tissue architectures in the lesioned spinal cord, but also provide trophic support for the regenerating axons by producing a spectrum of neurotrophic factors.

Hydrogel has also been used for spinal cord repair. The ease of tailoring the gelation temperature and the mechanical properties during the fabrication process has made it possible to produce hydrogels that mechanically match the native spinal cord and fit into the lesion site of irregular shapes. For example, alginate, a bioabsorbable long-chain polysaccharide, has been used to bridge the spinal cord after transection [17]. The alginate that was implanted into the lesion gap integrated well with the surrounding spinal tissue. Macrophages were seen infiltrating into the gel and at certain time points after the gel injection, myelinated axons were seen at the cross section of the lesion site that was filled with the gel.

Cell Transplantation in the Bridges

A variety of cells may be co-transplanted or loaded into the bridges for enhanced regeneration. Incorporation of supporting cells that are present in the native CNS tissue would provide biological cues similar to those in the native tissue to direct axonal outgrowth and perhaps myelination. In particular, Schwann cells have demonstrated the greatest efficacy in promoting axonal outgrowth. In an introductory DRG model, neurite outgrowth on Schwann cell-coated PLA surfaces was much more extensive than on non-coated controls. Schwann cell-loaded hollow fiber bridging channels represent the most promising strategy reported in the literature for CNS axonal regeneration [18].

Olfactory bulb ensheathing cells (OECs) are a unique cell type that share properties with both astrocytes in the CNS and Schwann cells in the PNS. Astrocyte-like features of OECs include anatomical location within the olfactory bulb, expression of glial fibrillary acidic protein (GFAP), and participation in the formation of glia limitans. Schwann cell-like features of OECs include (1) the ability to ensheath axons and support axonal re-growth, and (2) the myelinating capability. Attributing to the presence of OECs the failure of injured axons to regenerate within the mature CNS does not apply to the olfactory bulb. Normal and sectioned olfactory axons spontaneously grow within the bulb and establish synaptic contacts with targets. OECs offer several advantages over Schwann cells and astrocytes for CNS repair. Unlike Schwann cells, which are confined to the PNS, OECs accompany olfactory axons into the CNS, which confers the ability of these cells to survive and migrate within the CNS following transplantation, to secrete neurotrophic factors, and to promote the regeneration and re-myelination of damaged axons in a reactive astrocytic environment. The utility of OECs in PNS repair has been extensively demonstrated. When compared with conventional silicon tubes for PNS repair, which have limiting gap lengths of 10 mm for rats and 4 mm for mice, pre-filling the tubes with ECM (collagen or laminin gel) resulted in a greater nerve regeneration length of up to about 12 mm in rats. Further enhancement of the PNS nerve outgrowth length of 18 mm in rats was documented using tubes loaded with collagen gel + OECs. In addition, the myelination and fasciculation of the regenerated PNS axons to form axonal cables were also promoted relative to results obtained using tubes filled with collagen only. To bridge injured spinal cord, OECs have been used in combination with Schwann cells [19]. In this study, OECs and Schwann cells were loaded into a PAN/PVC hollow fiber membrane (HFM) channel that bridged the two stumps of the transected spinal cord. At different time points, the growth and formation of tissue cable inside the HFM conduit were examined and compared with those in the no-cells, channel-only controls. Two tissue cables were seen in the transection region—one inside and the other outside the HFM conduit. The inner tissue cable primarily consisted of regenerating axons along the loaded OECs and Schwann cells. The outside cable consisted of connective tissues.

Genetically engineered cells may be used as a source for sustained release of trophic factors. For example, genetically engineered cells to secrete neurotrophins had been transplanted to spinal cord injured animals to promote regeneration.

Stem cells represent abundant sources for cell therapy to repair the spinal cord. Different classes of stem cells have demonstrated potential in CNS repair, including neural stem cells (NSCs), hematopoietic stem cells (HSCs) and mesenchymal stem cells (MSCs), and embryonic stem cells (ESCs). NSCs that are resident in adult and developing CNS have been shown to give rise to neural and glial cells both in vitro [20, 21] and in vivo [20–22]. However, the in vivo differentiation of NSCs is dependent on the niche that they have been transplanted to. In adult CNS, there are neurogenic regions and non-neurogenic regions with different niche presentations. Transplantation to the neurogenic regions—such as dentate gyrus [22, 23] or subventricular zone (SVZ) [23]—leads to the differentiation of NSCs into neurons. In contrast, NSCs adopt glial fates in non-neurogenic regions, such as the spinal cord [20]. In addition, different fate determinations were observed in transplanted neural-restricted precursors (NRPs), a population committed to neural lineages at the time of isolation either from adult dentate gyrus [24] or through pre-differentiation of NSCs in vitro [25, 26], in normal versus injured spinal cord [27], suggesting the differences in stem cell niche under different tissue conditions. In particular, transplantation of a mixed population of NRPs and GRPs (glial-restricted precursors) that were isolated from fetal spinal cord into the injured adult spinal cord resulted in high survival, consistent migration of the transplanted cells out of the graft site, and robust differentiation into mature CNS phenotypes including neurons, astrocytes, and oligodendrocytes [25, 26]. Functional evaluation following transplantation of the NRP-GRP mixture into injured adult spinal cord indicated benefits in alleviating motor deficits, and enhancing the intraspinal circuitry [28]. These effects were perhaps mediated through a synergistic contribution of GRPs and NRPs in the mixed population, in which GRPs generated glial cells that supported the survival and differentiation of NRPs into functional neurons for cellular replacement in a non-neurogenic and non-permissive injured adult spinal cord environment [25].

HSCs and MSCs are stem cells that can be readily isolated from bone marrow, although HSCs have also been derived from umbilical cord blood [29] and fetal tissues [30]. HSCs have generated cells of neuronal [30, 31] and glial characteristics [29, 30] in vitro. However, the in vivo plasticity of HSCs has remained a matter of controversy. Some evidence from an in vivo trans-differentiation study indicates the trans-differentiation of HSCs into neurons and glial cells without cell fusion [32]; other studies have challenged the notion of plasticity of these stem cells by demonstrating the lack of neuronal differentiation of HSCs in the CNS [33–36]. Despite the disparate findings, locomotor improvement has been reported in mice receiving transplanted HSCs following contusion spinal cord injury [37]. In another study, functionally integrated neurons were identified from cross-species transplanted human HSCs into injured spinal cord in chicken embryos in the absence of fusion with the host cells [38]. More recently, our group has developed a biomaterial-based transplantation paradigm, which allows free exchange of soluble molecules but prohibits physical contact and, therefore, fusion of the transplanted cells with the host cells to examine the trans-differentiation potential of human umbilical cord blood-derived HSCs in adult brain tissue. We present (data in preparation for publication) solid definitive evidence for the trans-differentiation of HSCs into neural phenotypes specific to the CNS, including neurons, astrocytes, and oligodendrocytes in adult brain, validating a broader spectrum of phenotype specifications of HSCs that may qualify them for CNS repair. Although the ratio of neural trans-differentiation is very low (<5%), ongoing studies in our lab have demonstrated the promise of engineering approaches for high-yield production of desirable neuronal or neural phenotypes from HSCs in vitro.

Cell fusion is also a confounding issue in the evaluation of the trans-differentiation potential of MSCs, another subset of stem cells that can be derived from bone marrow. Neurogenic capacity of MSCs to generate neurons and glial cells has been shown both in vitro [39–41] and in vivo [42–44].

However, some studies failed to show this trans-differentiation [45, 46]. Nevertheless, transplantation of MSCs into lesioned spinal cord led to functional recovery [47, 48]. The only clinical trial to date involved intraspinal cord transplantation of ex vivo expanded autologous MSCs to human patients with amyotrophic lateral sclerosis, a neurodegenerative disease caused by the degeneration of motor neurons [49]. Besides the indications of the safety and feasibility of the procedure, no evaluation was performed to address the functional outcomes.

ESCs are pluripotent cells that are derived from the inner cell mass of the blastocytes during the early stage of development. They are able to give rise to all the cell types of the three germ layers [50] and, therefore, offer enormous potential in treating and curing a wide range of diseases. In particular, the spinal cord has been an attractive target for ESC-based therapeutic attempts. Due to the limitations with the acquisition of fetal or adult allogous cells for spinal cord repair, much research has been focused on deriving tissue-specific progenitor cells from human ESCs for neural transplantation. Myelination of the injured rat spinal cord was demonstrated using implanted mouse ESC-derived glial precursors for oligodendrocytes and astrocytes [51]. These results were correlated with a recent study, in which transplantation of oligodendrocytes derived from human ESCs led to the myelination of axons in a spinal cord contusion model in adult rats [52]. Transplantation of ESC-derived motor neurons into the spinal cord of adult paralyzed rat, in conjunction with delivery of glial cell-derived neurotrophic factors and agents to overcome myelin-mediated repulsion, was reported to mediate partial functional recovery from paralysis [53]. In a parallel study, transplantation of purified motoneuron precursors that were derived from ESCs transfected with MASH1 gene into the completely transected spinal cord of mice suppressed gliosis in the grafted spinal cord and promoted recovery in motor functions at a detectable level in electrophysiological assessment [54]. Regardless of the advance in utilizing ESCs or their derivatives for spinal cord repair, concerns are raised with regard to the potential generation of tetratomas or germinomas following transplantation due to the presence of any persistent undifferentiated ESCs in the donor pool. Protocols will have to be developed to purify the ESC-derived lineage-restricted precursors to deplete the populations of any undifferentiated ESCs.

Stimulated homologous macrophages are another type of cell that have been used for transplantation purposes in spinal cord repair. Different from CNS resident macrophages, which are inhibitory to axonal regeneration, macrophages that are pre-incubated ex vivo with segments of a PNS nerve (e.g., sciatic nerve) promote CNS axon regeneration, leading to partial recovery of paraplegic rats [55]. The pre-incubation with PNS nerve segments has enhanced the phagocytic activity of the macrophages, which is crucial for the axonal growth supportive properties of the stimulated macrophages. In a parallel study, implantation of macrophages that were pre-incubated with autologous skin was shown to improve motor functions and reduce spinal cyst formation in rats following spinal cord contusion injury [56]. Co-incubation with skin elevated the secretion of trophic factors [e.g., brain-derived neurotrophic factor (BDNF)] from the macrophages, whereas it reduced their production of tumor necrosis factor alpha (TNF-α). Thus, activated macrophages exhibit enhanced neuroprotective immune activity in the spinal cord, which was proposed to ameliorate the permissiveness and reduce the cytotoxicity of the injury environment [56].

Agent Delivery

Due to the extensive involvement of both cellular and molecular components in the CNS wound-healing response, and the presence of multiple molecules at the lesion site to inhibit axonal outgrowth, attempts have been made to overcome tissue reactivity around the lesion site through the delivery of agents/chemicals that remove the scar-forming cells, prevent the syntheses of inhibitory molecules, or block or degrade certain inhibitory molecules. For instance, in the disinhibition strategy, PD 168393, a factor known to block inhibitory molecules, was injected to the vicinity of the host tissue-bridge interface. Compared with the controls without agent delivery, a large number of axons were able to grow across the interface into the bridge. Some of the outgrowing axons were organized at the interface and exhibited aligned morphology [57].

A wide array of biomolecules to improve axonal regeneration and outgrowth has been delivered to the regeneration site. Among them are ECM components such as laminin, fibronectin, L1, heparin sulfate, etc. Neurotropins/neurotrophic factors are a class of biomolecules that have demonstrated effectiveness in promoting axonal regeneration.

Combinatorial Strategies

Currently, two types of biomaterial-based bridges have shown the most promising results. One uses entubulation tubes, and the other uses filament bundles. Using the entubulation approach, a tubular structure is placed across the lesion site to create a scar-free, well-controlled regeneration environment. The tubular sleeve prevents the infiltration of scar tissue into the lumen where regenerating axons grow. Cells and growth factors can be entubulated into the sleeve to facilitate axonal outgrowth. Semi-permeable HFMs have been used extensively as a preferable sleeve, attributing to its ease of fabrication and adjustments of properties, including the inner and outer surface morphologies, porosity, and permeability. Studies using HFM sleeves, by Xu and colleagues [18], demonstrated robust axonal outgrowth into the lumen of a HFM entubulation device at the spinal cord lesion site. However, the number of regenerating axons decreased dramatically as they approached the distal end of the device. Very few regenerating axons re-entered the host CNS environment; most turned back at the device-host interface, which limited the functional recovery. To facilitate axonal re-entry into the host CNS, Xu et al. [58] administered chondroitinase ABC, an enzyme known to block the synthesis of CSPG (an inhibitory proteoglycan deposited by reactive astrocytes), at this interface. As a result, numerous axons were able to grow across the HFM entubulation device-host CNS interface. Some were closely associated with and formed synapses with targets in the host CNS.

The idea of using filament bundles to guide directional axonal regeneration originated from the attempt to mimic the filamentous structures during nervous system development. During early embryonic stages, aligned radial glial cell processes form fan-like structures between the inner and the outer surfaces of developing neural tubes. Newly formed neurons migrate along the radial glial filaments and grow toward the outer surface of neural tubes to their proper targets. In addition, the guidance effect of filamentous structure on axonal outgrowth and regeneration during neuronal development is seen with the pioneer axons, also called the guiding axons, which are projected by the pioneer neurons to establish filamentous axonal tract for later-growing axons. Previous studies using filaments alone for spinal cord axonal regeneration evidenced the ability of unidirectional aligned filament bundles to guide the directional growth of both neurites and glial cells [59]. On bundles of 5 μm carbon filaments, both neurites and glial cells (e.g., Schwann cells) grew very well and were highly aligned along the longitudinal axis of the filament substrates. However, due to the lack of protective sleeve surrounding the filaments, inhibitory cells (e.g., reactive astrocytes) easily colonized the bundles and inhibited axonal regeneration and outgrowth. The other problem is that carbon filaments are non-degradable in vivo.

To combine the advantages offered by the entubulation sleeve and the aligned filamentous scaffold for nervous tissue repair, we have developed and tested a tissue-engineered HFM-based neuronal bridging device of a multi-filament entubulation configuration. Bundles of unidirectional aligned filaments are entubulated into the lumen of a semi-permeable biodegradable HFM. The aligned filament bundles provide guidance cues for the alignment of glial cells and regenerating axons. Selective ECM molecules are incorporated onto the surfaces of the filaments to enhance glial cell migration and axonal outgrowth. The biodegradable HFM serves as a sieve, preventing the penetration of the scar tissue into the lumen where regenerating axons grow along the filament bundles. The device can be easily connected to a syringe dura lock for the purpose of sterilization, cell/matrix loading, and ease of handling during surgery. Our preliminary evaluation of the device in vitro demonstrated its efficacy in promoting unidirectional alignment, directional outgrowth rate and length of neurites and glial cells, and close association between the neuritis and supportive glial cells [60]. Further testing of the device in vivo using a spinal cord contusion injury model in rats is underway in our laboratory.

21.2.2.2
Tissue Engineering Strategies for the Treatment of Neurodegenerative Diseases

Parkinson's Disease

PD is characterized by the destruction of the substantia nigrostriatal pathway due to the loss of dopaminergic (DA) neurons in the substantia nigra, which controls movement and balance. The ongoing death of dopamine-producing neurons in the substantia nigra leads to a shortage of dopamine release in the striatum. This triggers the acetylcholine producers in the striatum to over-stimulate their target neurons, and there is a subsequent chain reaction of abnormal signaling resulting in impaired mobility, loss of coordination, unstable posture, and tremor. The pathology of PD is associated with the presence of lewy bodies (microscopic protein deposits) in the cytoplasm of the dying neurons in the substantia nigra and the progressive deterioration of brain functioning. Several animal models have been developed for experimental PD. The most common PD model is the 6-hydroxydopamine (6-OHDA) lesion in rodents. By injecting 6-OHDA into the substantia nigra, the nigrostriatal pathway is disrupted to mimic PD. As an alternative, 1-methyl-4-phenyl-1,2,3,6-tetrahydropyridine (MPTP) is used in a non-human primate model of experimental PD.

Current treatment options for PD include deep brain stimulation, naked cell/tissue transplantation, cell/tissue encapsulation, controlled delivery of biomolecules, and axonal guidance strategies. Single or multi-electrode arrays can be implanted to stimulate the degenerating neurons in the mid-brain for enhanced function. However, due to the adverse tissue reactivity in the vicinity of the implanted electrodes, close apposition of the electrodes to the neurons is difficult to achieve; this has compromised the chronic effectiveness of deep brain stimulation in PD patients. Naked cells or tissues of desirable types can be transplanted into the brain to recover the cell loss in PD and also provide trophic support to preserve the functions of the remaining cells. However, problems with these strategies may rise due to the potential immunogenicity of the transplanted cells/tissue in the host tissue as a result of the physical contact between the transplant and the host tissue, as well as the migration of the transplanted cells in the host brain, and generation of undesired phenotypes with unpredictable consequences. To overcome these problems, cell/tissue encapsulation is performed prior to transplantation. Using the micro-encapsulation technique, cells/tissues are encapsulated into microparticles during the fabrication. Macro-encapsulation procedures usually use semi-permeable HFMs with a controllable level of permeability—sufficient to allow nutrients and growth factors to diffuse into, and the transport of secretory and waste products of the cells out of the membranes; but the entry of host cells, antibodies, and complement components into the membranes is prevented. For the anatomical and functional reconstruction of the disrupted nigrostriatal pathway, an axonal guidance device can be implanted along the pathway. Long-term controlled delivery schemes for biomolecules can be incorporated into all of these strategies.

The transplantation location of cells/tissues for the treatment of PD can also be varied. Due to the ease of transplantation, using the conventional heterotopic transplantation procedure, DA neurons/stem cells are transplanted to the striatum, a location where DA neurons do not normally reside. Although partial restoration of dopaminergic input to the striatum was documented, the problem with this location is that the native microenvironment at the striatum does not support the survival and functioning of DA neurons. The transplanted neurons/stem cells here lack the major afferent inputs that are crucial to regulate the function of nigrostriatal DA neurons. In the experimental PD model in rats, only some of the motor behaviors have been improved following this procedure, while more complex motor functions, such as skilled use of forelimb, have only been partially improved or have remained unaffected. In all, the heterotopic transplantation strategy does not result in normalization of the majority of PD symptoms or promotion of physiological recovery; therefore, it is not an ultimate cure for PD. Recently, homotopic transplantation has been performed—cells are transplanted into the substantia nigra where DA neurons do normally reside. The transplanted cells at the substantia nigra receive the appropriate afferent regulation from the microenvironment and, in conjunction with the axonal guidance strategy, may have the potential to restore the nigrostriatal pathway. Despite the current challenge with the transplantation procedure, homotopic transplantation may be a key element in the ultimate cure for PD. Our lab has designed a combinatorial strategy based on the homotopic transplantation of human embryonic stem cell (hES)-derived DA neurons and the implantation of a biodegradable HFM axonal guidance device that bridges the substantia nigra and the striatum for the reconstruction of the nigrostriatal pathway. Other investigators had created chemical bridges based on one-step or two-step injection of chemicals (such as kainic acid/ibotenic acid, which directly or indirectly induce the secretion of trophic elements from the fetal tissue) along the nigrostriatal pathway. Robust stream of TH-positive neurons was seen along the pathway following two-step injection of the chemicals [61].

Another strategy to treat neurodegenerative disorders is based on the mobilization and in situ phenotypic induction of endogenous progenitor cells. The persistence of NSCs and their committed neuronal progeny in the forebrain SVZ-olfactory bulb pathway and their neurogenic capacity throughout life suggest their potential utility in restoring the lost neuronal populations. A spectrum of humoral growth factors and chemotactic cytokines that are necessary to mobilize and site-specifically recruit endogenous NSCs to a local tissue site have been identified. We have focused on the idea that concurrent delivery of these factors in vivo may promote recruitment and accumulation of endogenous NSCs within a CNS defect site. Further delivery of neural inductive factors, such as BDNF, NT-3, GDNF, and CTNF, to the local tissue site may direct the in situ neural differentiation of the recruited NSCs into functional lineages for the restoration of the damaged neural pathway.

Hearing Repair

Repair or recovery of auditory functions in patients with impaired hearing or deafness has been centered on central auditory cortex stimulation. Loss of hearing function due to heredity, aging, or pathologies in the auditory system often results in disabilities in independence, communication, and lifestyle [62]. Statistical data from NIH/NIDCD show that hearing loss affects approximately 17 in 1,000 children under age 18 years, and the incidence increases with age: approximately 314 in 1,000 people over age 65 years and 40–50% of people aged 75 years or older have a hearing loss. Thus, this condition poses a major health care burden for our society, and there is a compelling need for effective interventional therapies for auditory disorders. Due to the size and anatomy of the inner ear, current therapies to treat a profoundly deaf or severely hearing-impaired patient are largely dependent on cochlear implants that have one or more electrodes to directly stimulate the surviving neurons in the auditory nerve while bypassing the damaged or missing sensory hair cells. These impulses are then sent to the brain, where a hearing person would recognize them as sound. Although cochlear implants have been used to partially restore hearing in more than 75,000 hearing-impaired people worldwide [63], their functionality is highly dependent on the remaining excitable auditory-nerve fibers and central-auditory pathways (i.e., the total number and the integrity of the auditory neurons available for stimulation) [64]. Degeneration or loss of these fibers or pathways severely compromises the effectiveness of conventional implants [65]. Studies clearly show a relationship between the total number of viable auditory neurons available for stimulation and the performance of subjects receiving cochlear implants [66]. Permanent sensorineural hearing loss is primarily associated with the loss of cochlear hair cells, which leads to a secondary wave of degeneration of the auditory neurons due to the loss of endogenous trophic support from the hair cells [67]. Subsequently, the afferent nerve fibers of the inner ear also undergo degeneration [68]. Previous data showed that the pattern of auditory nerve degeneration is closely associated with the loss of inner hair cells innervating those neurons [69, 70]. Thus, the loss of hair-cell function can result in degeneration of both the auditory and central neuron (brainstem) nuclei, which greatly aggravates the functional impairment. In many situations, most hair cells continue degenerating while diseases progress. Studies have indicated the efficacy of delivered neurotrophic agents in retarding disease progression, alleviating symptoms, and hastening functional recovery [71–74]. However, cessation of agent delivery often results in an exacerbation of the same disease relative to that in untreated conditions [68], suggesting a critical need for long-term sustained delivery with unlimited temporal profile.

Although Gillespie's result [68] is controversial, long-term delivery is necessary for deaf children with hair-cell loss.

In most of the available delivery strategies, repeated trans-tympanic blind injections or agent refills are necessary to maintain the local concentration of the agent in the diseased ear. These procedures significantly decrease the patients' compliance and increase the risks of infection and inflammation [75, 76]. More recently, the emergence of intra-ear perfusion delivery strategies, such as degradable carriers, microwicks, or miniosmotic pumps with catheters [75, 77–85] has offered new hope for the treatment and manipulation of auditory damages and diseases. A wide variety of pharmaceutical agents have been delivered locally using these techniques. However, these approaches are persistently problematic due to uneven delivery profiles, limited temporal delivery profiles, frequent agent refills, high cost for delivery of neurotrophins, and difficulty with retrieval when significant side effects occur. Furthermore, these approaches require additional implantations after the implantation of cochlear implant. Alternatively, using gene transfer techniques, long-term delivery of biological agents can be attained at the expense of compromised regulation of the level and site of agent expression [86]. The key parameters relevant to the efficacy of an agent delivery system remain poorly defined, and continuous, safe, uniform, cost-effective, retrievable, and refillable delivery to the inner-ear cavity for sensory function restoration and prevention of hearing loss has not been achieved.

To this end, we have further developed a "living" cochlear implant for long-term, continuous delivery of agents to targeted cochlear tissue based on tissue-engineering principles. The tissue engineered cochlear implant is a combination of a hollow-core cochlear implant with a retrievable cell-encapsulation insert consisting of a HFM and an encapsulated coil-based scaffold hosting genetically engineered neurotrophin-releasing cells. Orifices on the hollow-core cochlear implants allow for the release of the neurotrophins. The HFM is selectively permeable with regard to molecular weight for controlled release of therapeutic compounds. A re-sealable access port made of flexible medical grade polyurethane is attached to the proximal end of the cochlear implant and the port will be anchored onto the skull. The port will allow unlimited agent-loading, replacement, and retrieval through the hollow-core cochlear implant lumen without damage to the tissue or implant. Since only a small skin incision requiring local anesthesia is required to access the port, this very minor surgery can be done in a regular hospital procedure room. Our preliminary evaluation of the device in tissues has demonstrated effective, sustained delivery of agents at constant therapeutic levels by the encapsulated cells to the targeted tissue.

Vision Improvement

Electro-stimulating and recording electrodes have been used extensively as treatment for vision improvement. Utah array represents a prosthetic device for visual cortex stimulation. Utah array consists of an array of penetrating electrodes, which may be inserted directly into the visual cortex. The active tips of the implanted electrodes can reach nerve fibers at different levels and should allow much more focal stimulation. This technique may also be used for hearing repair as described above.

Cell transplantation is also a dominant approach to restore cell loss in visual system degeneration for functional protection and recovery. Multiple types of cells have been transplanted into animal models of vision impairment, including photoreceptors, retinal pigment epithelium cells, fetal tissue, and stem cells. In general, advantages of transplanted cells in vision improvement have been demonstrated. Bridging devices have been tested to direct the regeneration of retinal ganglion neurons. Delivery of neuroprotective agents and gene delivery using polymer carriers are applied to further promote axonal regeneration and functional restoration.

Pain Reliever

Development of pain relievers is an important area that has direct clinical relevance. Regular drugs have been used to relieve uncontrollable pains, including acute pains that result from widespread trauma and injury, and chronic pains due to inflammation, neuropathology, or cancer. In particular, local delivery of neuroactive substances, such as catecholamines, opioid peptides, and neurotrophic factors, is necessary. Depending on the properties of the substances to be delivered, biodegradable polymer-based delivery systems or genetically engineered cell delivery systems can be used. Biodegradable microspheres or nanoparticles are fabricated to incorporate agents, which are control released to a local tissue site for pain relief as a function of the degradation of the polymer carriers. Cells are genetically engineered to secrete agents of interest and delivered to the tissue either in the form of cell suspension (naked cells) or by encapsulating into microspheres or HFMs. A variety of animal models have been established to test the efficacy of different delivery approaches for pain relief. These models include acute injury, chronic injury, excitotoxic injury, ischemia, trauma, spinal cord injury, and brain trauma. Experimental data are available indicating the effectiveness of local delivery approaches in pain relieving. Cell encapsulation approaches are under clinical trial.

21.2.3
Nerve Repair at the CNS-PNS Transition Zone

Due to the significant difference between CNS and PNS environments, nerve regeneration at the CNS-PNS TZ poses great challenges. Immediately adjacent to the spinal cord, the dorsal root entry zone is the CNS-PNS transitional area. In anatomy, the TZ features a mosaic of CNS and PNS structures, yet the discontinuity of tissue types is obvious. The glial limitans that covers the CNS surface thickens at the TZ, and the tissue surrounding the myelinated or non-myelinated fibers changes from extracellular endoneurial connective tissue matrix containing collagen fibrils and fibroblasts on the peripheral side to astrocyte processes and extracellular tissue space on the central side. Axons are the only elements to completely traverse the TZ. The myelin sheaths surrounding the myelinated axons are formed by transitional Schwann cells peripherally and by an oligodendrocytic myelinating unit centrally. After traversing the TZ, sensory primary afferents terminate and synapse in the spinal cord (dorsal horn) in different areas corresponding to sensory modalities. Despite sprouting and forming growth cones, the inability of the interrupted dorsal root axons (PNS axons) to grow into the spinal cord (CNS) was first documented by Ramón y Cajal in 1928. Astrocyte processes of the TZ in the thickened glia limitans may constitute the major barrier for axon and Schwann cell migration across the TZ to access the cord [87]. Irradiation of the immature spinal cord to create deficiencies in the glia limitans has led to the invasion of large numbers of Schwann cells [88] and central growth of axons into the cord [89, 90].

Currently, efforts for axonal regeneration across the CNS-PNS TZ have been focused on two approaches: single or in combination. The first approach focuses on the alteration of the cellular and molecular components at the TZ to minimize the effects of inhibitory molecules within the gliotic tissue. In this regard, antibodies or enzymes directed against the molecules are administered at the TZ, in addition to localized delivery of axonal growth-promoting factors. A second approach involves the transplantation of embryonic or fetal tissue due to its plasticity and the ability to support axonal outgrowth in a glia scarring environment in adult CNS [9, 17, 87, 91]. The dramatic capacity of embryonic tissue to overcome the resistance to regenerate in mature CNS has been reported in numerous studies [9, 92]. Transplanted immature DRG tissue at the TZ was able to grow centrally along the root, deviate at the TZ to pass alongside, and penetrate the glia limitans to enter the cord [9]. The most promising results for axonal regeneration through the TZ have been achieved with OECs, which were transplanted to the TZ following the transaction of the rootlet [19]. Substantial regeneration of dorsal root axons that were closely associated with OECs was seen from the PNS across the TZ and deep into the CNS, where they terminated in the laminae in which they would have normally formed synapses. These findings suggest the potential of regenerating dorsal root axons in combination with OECs in traversing the gliotic tissue barrier in adult CNS [93, 94].

21.3
Conclusions

Tissue engineering approaches hold great promise for neural tissue repair/regeneration. Advances in developmental biology, biomaterials, cell and molecular biology, and neuroscience have furthered our understanding of the neural tissue formation process and environmental cues necessary for neural tissue regeneration. Mimicry of these cues by creating cell-scaffold constructs based on tissue engineering principles would direct and accelerate the tissue regeneration process, leading to enhanced anatomical reconstruction and functional restoration. Ongoing studies need to focus on combinatorial strategies that integrate individual strategies aimed at controlling local environment (suppressing local tissue reactivity and glia scar formation), guiding tissue regeneration (axonal guidance bridges), trophic support of regenerating axons (sustained delivery of neutrophic agents or agent-secreting cells), recovering cell loss (cell transplantation, or stem cell transplantation and directed differentiation in vivo), promoting functional integration of transplanted cells with the host tissue (scar-free transplant-host tissue interface), and developing neuro-prosthetic interface (electro-stimulation and recording). Neural repair/regeneration that leads to a decent degree of functional recovery or the ultimate cure of neurodegenerative diseases is only possible when the local tissue environment can be engineered to a level that induces normal behavior and functioning of cells similar to those exhibited by cells in normal tissues.

Acknowledgments

This work was made possible by the NIH/NINDS USA (R01 NS050243), NIH/NEI/NIDCD USA (R21 EY018467), Michael J. Fox Foundation for Parkinson's Research, Early Career Translational Research Awards in Biomedical Engineering from Wallace H. Coulter Foundation, South Carolina Spinal Cord Research Fund, National Science Foundation USA under Grant No. 0132573, NIH Grant Number P20 RR-016461 from the National Center for Research Resources (NCRR), USA, and URC Grant from the Medical University of South Carolina, USA.

References

1. Stoll G, Griffin JW, Li CY, Trapp BD. Wallerian degeneration in the peripheral nervous system: participation of both Schwann cells and macrophages in myelin degradation. J Neurocytol 1989;18(5):671–83.
2. Stichel CC, Muller HW. The CNS lesion scar: new vistas on an old regeneration barrier. Cell Tissue Res 1998;294(1):1–9.
3. Wen X, Tresco PA. Fabrication and characterization of permeable degradable poly(DL-lactide-co-glycolide) (PLGA) hollow fiber phase inversion membranes for use as nerve tract guidance channels. Biomaterials 2006; 27(20):3800–9.
4. Zhang N, Yan H, Wen X. Tissue-engineering approaches for axonal guidance. Brain Res Brain Res Rev 2005;49(1):48–64.
5. Zhang N, Zhang C, Wen X. Fabrication of semipermeable hollow fiber membranes with highly aligned texture for nerve guidance. J Biomed Mater Res A 2005;75(4):941–9.
6. Bray GM, Villegas-Perez MP, Vidal-Sanz M, Carter DA, Aguayo AJ. Neuronal and nonneuronal influences on retinal ganglion cell survival, axonal regrowth, and connectivity after axotomy. Ann N Y Acad Sci 1991;633:214–28.
7. Carlstedt T. Nerve fibre regeneration across the peripheral-central transitional zone. J Anat 1997;190 (Pt 1):51–6.
8. Stensaas LJ, Partlow LM, Burgess PR, Horch KW. Inhibition of regeneration: the ultrastructure of reactive astrocytes and abortive axon terminals in the transition zone of the dorsal root. Prog Brain Res 1987;71:457–68.
9. Fraher J, Dockery P, O'Leary D, Mobarak M, Ramer M, Bishop T, Kozlova E, Priestley E, McMahon S, Aldskogius H. The dorsal root transitional zone model of CNS axon regeneration: morphological findings. J Anat 2002;200(2):214.
10. Fraher JP. The transitional zone and CNS regeneration. J Anat 2000;196 (Pt 1):137–58.
11. Dent EW, Callaway JL, Szebenyi G, Baas PW, Kalil K. Reorganization and movement of microtubules in axonal growth cones and developing interstitial branches. J Neurosci 1999;19(20):8894–908.
12. Oblinger MM, Szumlas RA, Wong J, Liuzzi FJ. Changes in cytoskeletal gene expression affect the composition of regenerating axonal sprouts elaborated by dorsal root ganglion neurons in vivo. J Neurosci 1989;9(8):2645–53.
13. Savio T, Schwab ME. Rat CNS white matter, but not gray matter, is nonpermissive for neuronal cell adhesion and fiber outgrowth. J Neurosci 1989;9(4):1126–33.
14. Pettigrew DB, Shockley KP, Crutcher KA. Disruption of spinal cord white matter and sciatic nerve geometry inhibits axonal growth in vitro in the absence of glial scarring. BMC Neurosci 2001;2:8.
15. Davies SJ, Goucher DR, Doller C, Silver J. Robust regeneration of adult sensory axons in degenerating white matter of the adult rat spinal cord. J Neurosci 1999;19(14):5810–22.

16. Cheng H, Cao Y, Olson L. Spinal cord repair in adult paraplegic rats: partial restoration of hind limb function. Science 1996;273(5274):510–3.
17. Suzuki K, Suzuki Y, Ohnishi K, Endo K, Tanihara M, Nishimura Y. Regeneration of transected spinal cord in young adult rats using freeze-dried alginate gel. Neuroreport 1999;10(14):2891–4.
18. Xu XM, Guenard V, Kleitman N, Bunge MB. Axonal regeneration into Schwann cell-seeded guidance channels grafted into transected adult rat spinal cord. J Comp Neurol 1995;351(1):145–60.
19. Ramon-Cueto A, Plant GW, Avila J, Bunge MB. Long-distance axonal regeneration in the transected adult rat spinal cord is promoted by olfactory ensheathing glia transplants. J Neurosci 1998;18(10):3803–15.
20. Cao QL, Zhang YP, Howard RM, Walters WM, Tsoulfas P, Whittemore SR. Pluripotent stem cells engrafted into the normal or lesioned adult rat spinal cord are restricted to a glial lineage. Exp Neurol 2001;167(1):48–58.
21. Mokry J, Karbanova J, Filip S. Differentiation potential of murine neural stem cells in vitro and after transplantation. Transplant Proc 2005;37(1):268–72.
22. Shihabuddin LS, Horner PJ, Ray J, Gage FH. Adult spinal cord stem cells generate neurons after transplantation in the adult dentate gyrus. J Neurosci 2000;20(23):8727–35.
23. Fricker RA, Carpenter MK, Winkler C, Greco C, Gates MA, Bjorklund A. Site-specific migration and neuronal differentiation of human neural progenitor cells after transplantation in the adult rat brain. J Neurosci 1999;19(14):5990–6005.
24. Seaberg RM, van der Kooy D. Adult rodent neurogenic regions: the ventricular subependyma contains neural stem cells, but the dentate gyrus contains restricted progenitors. J Neurosci 2002;22(5):1784–93.
25. Lepore AC, Fischer I. Lineage-restricted neural precursors survive, migrate, and differentiate following transplantation into the injured adult spinal cord. Exp Neurol 2005;194(1):230–42.
26. Lepore AC, Han SS, Tyler-Polsz CJ, Cai J, Rao MS, Fischer I. Differential fate of multipotent and lineage-restricted neural precursors following transplantation into the adult CNS. Neuron Glia Biol 2004;1(2):113–126.
27. Cao QL, Howard RM, Dennison JB, Whittemore SR. Differentiation of engrafted neuronal-restricted precursor cells is inhibited in the traumatically injured spinal cord. Exp Neurol 2002;177(2):349–59.
28. Mitsui T, Shumsky JS, Lepore AC, Murray M, Fischer I. Transplantation of neuronal and glial restricted precursors into contused spinal cord improves bladder and motor functions, decreases thermal hypersensitivity, and modifies intraspinal circuitry. J Neurosci 2005;25(42):9624–36.
29. McGuckin CP, Forraz N, Allouard Q, Pettengell R. Umbilical cord blood stem cells can expand hematopoietic and neuroglial progenitors in vitro. Exp Cell Res 2004;295(2):350–9.
30. Hao HN, Zhao J, Thomas RL, Parker GC, Lyman WD. Fetal human hematopoietic stem cells can differentiate sequentially into neural stem cells and then astrocytes in vitro. J Hematother Stem Cell Res 2003;12(1):23–32.
31. Locatelli F, Corti S, Donadoni C, Guglieri M, Capra F, Strazzer S, Salani S, Del Bo R, Fortunato F, Bordoni A and others. Neuronal differentiation of murine bone marrow Thy-1- and Sca-1-positive cells. J Hematother Stem Cell Res 2003;12(6):727–34.
32. Cogle CR, Yachnis AT, Laywell ED, Zander DS, Wingard JR, Steindler DA, Scott EW. Bone marrow transdifferentiation in brain after transplantation: a retrospective study. Lancet 2004;363(9419):1432–7.
33. Castro RF, Jackson KA, Goodell MA, Robertson CS, Liu H, Shine HD. Failure of bone marrow cells to transdifferentiate into neural cells in vivo. Science 2002;297(5585):1299.
34. Koshizuka S, Okada S, Okawa A, Koda M, Murasawa M, Hashimoto M, Kamada T, Yoshinaga K, Murakami M, Moriya H and others. Transplanted hematopoietic stem cells from bone marrow differentiate into neural lineage cells and promote functional recovery after spinal cord injury in mice. J Neuropathol Exp Neurol 2004;63(1):64–72.
35. Roybon L, Ma Z, Asztely F, Fosum A, Jacobsen SE, Brundin P, Li JY. Failure of transdifferentiation of adult hematopoietic stem cells into neurons. Stem Cells 2006;24(6):1594–604.
36. Wagers AJ, Sherwood RI, Christensen JL, Weissman IL. Little evidence for developmental plasticity of adult hematopoietic stem cells. Science 2002;297(5590):2256–9.
37. Koda M, Okada S, Nakayama T, Koshizuka S, Kamada T, Nishio Y, Someya Y, Yoshinaga K, Okawa A, Moriya H and others. Hematopoietic stem cell and marrow stromal cell for spinal cord injury in mice. Neuroreport 2005;16(16):1763–7.
38. Sigurjonsson OE, Perreault MC, Egeland T, Glover JC. Adult human hematopoietic stem cells produce neurons efficiently in the regenerating chicken embryo spinal cord. Proc Natl Acad Sci U S A 2005;102(14):5227–32.
39. Bossolasco P, Cova L, Calzarossa C, Rimoldi SG, Borsotti C, Deliliers GL, Silani V, Soligo D, Polli E. Neuroglial differentiation of human bone marrow stem cells in vitro. Exp Neurol 2005;193(2):312–25.
40. Hung SC, Cheng H, Pan CY, Tsai MJ, Kao LS, Ma HL. In vitro differentiation of size-sieved stem cells into electrically active neural cells. Stem Cells 2002;20(6):522–9.
41. Woodbury D, Schwarz EJ, Prockop DJ, Black IB. Adult rat and human bone marrow stromal cells differentiate into neurons. J Neurosci Res 2000;61(4):364–70.
42. Deng YB, Yuan QT, Liu XG, Liu XL, Liu Y, Liu ZG, Zhang C. Functional recovery after rhesus monkey spinal cord injury by transplantation of bone marrow mesenchymal-stem cell-derived neurons. Chin Med J (Engl) 2005;118(18):1533–41.
43. Lee J, Kuroda S, Shichinohe H, Ikeda J, Seki T, Hida K, Tada M, Sawada K, Iwasaki Y. Migration and differentiation of nuclear fluorescence-labeled bone marrow stromal cells after transplantation into cerebral infarct and spinal cord injury in mice. Neuropathology 2003;23(3):169–80.
44. Zurita M, Vaquero J. Functional recovery in chronic paraplegia after bone marrow stromal cells transplantation. Neuroreport 2004;15(7):1105–8.

45. Neuhuber B, Timothy Himes B, Shumsky JS, Gallo G, Fischer I. Axon growth and recovery of function supported by human bone marrow stromal cells in the injured spinal cord exhibit donor variations. Brain Res 2005;1035(1):73–85.
46. Wu S, Suzuki Y, Ejiri Y, Noda T, Bai H, Kitada M, Kataoka K, Ohta M, Chou H, Ide C. Bone marrow stromal cells enhance differentiation of cocultured neurosphere cells and promote regeneration of injured spinal cord. J Neurosci Res 2003;72(3):343–51.
47. Vaquero J, Zurita M, Oya S, Santos M. Cell therapy using bone marrow stromal cells in chronic paraplegic rats: systemic or local administration? Neurosci Lett 2006;398(1–2):129–34.
48. Zurita M, Vaquero J. Bone marrow stromal cells can achieve cure of chronic paraplegic rats: functional and morphological outcome one year after transplantation. Neurosci Lett 2006;402(1–2):51–6.
49. Mazzini L, Fagioli F, Boccaletti R, Mareschi K, Oliveri G, Olivieri C, Pastore I, Marasso R, Madon E. Stem cell therapy in amyotrophic lateral sclerosis: a methodological approach in humans. Amyotroph Lateral Scler Other Motor Neuron Disord 2003;4(3):158–61.
50. Conley BJ, Young JC, Trounson AO, Mollard R. Derivation, propagation and differentiation of human embryonic stem cells. Int J Biochem Cell Biol 2004;36(4):555–67.
51. Brustle O, Jones KN, Learish RD, Karram K, Choudhary K, Wiestler OD, Duncan ID, McKay RD. Embryonic stem cell-derived glial precursors: a source of myelinating transplants. Science 1999;285(5428):754–6.
52. Nistor GI, Totoiu MO, Haque N, Carpenter MK, Keirstead HS. Human embryonic stem cells differentiate into oligodendrocytes in high purity and myelinate after spinal cord transplantation. Glia 2005;49(3):385–96.
53. Deshpande DM, Kim YS, Martinez T, Carmen J, Dike S, Shats I, Rubin LL, Drummond J, Krishnan C, Hoke A and others. Recovery from paralysis in adult rats using embryonic stem cells. Ann Neurol 2006;60(1):32–44.
54. Hamada M, Yoshikawa H, Ueda Y, Kurokawa MS, Watanabe K, Sakakibara M, Tadokoro M, Akashi K, Aoki H, Suzuki N. Introduction of the MASH1 gene into mouse embryonic stem cells leads to differentiation of motoneuron precursors lacking Nogo receptor expression that can be applicable for transplantation to spinal cord injury. Neurobiol Dis 2006;22(3):509–22.
55. Rapalino O, Lazarov-Spiegler O, Agranov E, Velan GJ, Yoles E, Fraidakis M, Solomon A, Gepstein R, Katz A, Belkin M and others. Implantation of stimulated homologous macrophages results in partial recovery of paraplegic rats. Nat Med 1998;4(7):814–21.
56. Bomstein Y, Marder JB, Vitner K, Smirnov I, Lisaey G, Butovsky O, Fulga V, Yoles E. Features of skin-coincubated macrophages that promote recovery from spinal cord injury. J Neuroimmunol 2003;142(1–2):10–6.
57. Koprivica V, Cho KS, Park JB, Yiu G, Atwal J, Gore B, Kim JA, Lin E, Tessier-Lavigne M, Chen DF and others. EGFR activation mediates inhibition of axon regeneration by myelin and chondroitin sulfate proteoglycans. Science 2005;310(5745):106–10.
58. Chau CH, Shum DK, Li H, Pei J, Lui YY, Wirthlin L, Chan YS, Xu XM. Chondroitinase ABC enhances axonal regrowth through Schwann cell-seeded guidance channels after spinal cord injury. Faseb J 2004;18(1):194–6.
59. Khan T, Dauzvardis M, Sayers S. Carbon filament implants promote axonal growth across the transected rat spinal cord. Brain Res 1991;541(1):139-45.
60. Wen X, Tresco PA. Effect of filament diameter and extracellular matrix molecule precoating on neurite outgrowth and Schwann cell behavior on multifilament entubulation bridging device in vitro. J Biomed Mater Res A 2006;76(3):626–37.
61. Lieberman DM, Corthesy ME, Cummins A, Oldfield EH. Reversal of experimental parkinsonism by using selective chemical ablation of the medial globus pallidus. J Neurosurg 1999;90(5):928–34.
62. Coles RR, Thompson AC, O'Donoghue GM. Intra-tympanic injections in the treatment of tinnitus. Clin Otolaryngol 1992;17(3):240–2.
63. Zeng FG. Auditory prostheses: past, present, and future. In: Zeng FG, Poper AN, Fay RR, editors. Cochlear implants: auditory prostheses and electric hearing. New York: Springer-Verlag; 2004. p 1–13.
64. Marzella PL, Clark GM. Growth factors, auditory neurones and cochlear implants: a review. Acta Otolaryngol 1999;119(4):407–12.
65. Shinohara T, Bredberg G, Ulfendahl M, Pyykko I, Olivius NP, Kaksonen R, Lindstrom B, Altschuler R, Miller JM. Neurotrophic factor intervention restores auditory function in deafened animals. Proc Natl Acad Sci U S A 2002;99(3):1657–60.
66. Gantz BJ, Woodworth GG, Knutson JF, Abbas PJ, Tyler RS. Multivariate predictors of audiological success with multichannel cochlear implants. Ann Otol Rhinol Laryngol 1993;102(12):909–16.
67. Webster M, Webster DB. Spiral ganglion neuron loss following organ of Corti loss: a quantitative study. Brain Res 1981;212(1):17–30.
68. Gillespie LN, Clark GM, Bartlett PF, Marzella PL. BDNF-induced survival of auditory neurons in vivo: Cessation of treatment leads to accelerated loss of survival effects. J Neurosci Res 2003;71(6):785–90.
69. Wang J, Powers NL, Hofstetter P, Trautwein P, Ding D, Salvi R. Effects of selective inner hair cell loss on auditory nerve fiber threshold, tuning and spontaneous and driven discharge rate. Hear Res 1997;107(1–2):67–82.
70. Wenthold RJ, McGarvey ML. Changes in rapidly transported proteins in the auditory nerve after hair cell loss. Brain Res 1982;253(1–2):263–9.
71. Staecker H, Kopke R, Malgrange B, Lefebvre P, Van de Water TR. NT-3 and/or BDNF therapy prevents loss of auditory neurons following loss of hair cells. Neuroreport 1996;7(4):889–94.
72. Ernfors P, Duan ML, ElShamy WM, Canlon B. Protection of auditory neurons from aminoglycoside toxicity by neurotrophin-3. Nat Med 1996;2(4):463–7.
73. Zheng JL, Stewart RR, Gao WQ. Neurotrophin-4/5, brain-derived neurotrophic factor, and neurotrophin-3 promote survival of cultured vestibular ganglion neurons and protect them against neurotoxicity of ototoxins. J Neurobiol 1995;28(3):330–40.
74. Lefebvre PP, Malgrange B, Staecker H, Moghadass M, Van de Water TR, Moonen G. Neurotrophins affect sur-

74. vival and neuritogenesis by adult injured auditory neurons in vitro. Neuroreport 1994;5(8):865–8.
75. Lehner R, Brugger H, Maassen MM, Zenner HP. A totally implantable drug delivery system for local therapy of the middle and inner ear. Ear Nose Throat J 1997;76(8):567–70.
76. Seidman MD. Glutamate Antagonists, Steroids, and Antioxidants as Therapeutic Options for Hearing Loss and Tinnitus and the Use of an Inner Ear Drug Delivery System. Int Tinnitus J 1998;4(2):148–154.
77. Silverstein H. Use of a new device, the MicroWick, to deliver medication to the inner ear. Ear Nose Throat J 1999;78(8):595–8, 600.
78. Light JP, Silverstein H. Transtympanic perfusion: indications and limitations. Curr Opin Otolaryngol Head Neck Surg 2004;12(5):378–383.
79. Light JP, Silverstein H, Jackson LE. Gentamicin perfusion vestibular response and hearing loss. Otol Neurotol 2003;24(2):294–8.
80. Silverstein H, Light JP, Jackson LE, Rosenberg SI, Thompson JH, Jr. Direct application of dexamethasone for the treatment of chronic eustachian tube dysfunction. Ear Nose Throat J 2003;82(1):28–32.
81. Silverstein H, Durand B, Jackson LE, Conlon WS, Rosenberg SI. Use of the malleus handle as a landmark for localizing the round window membrane. Ear Nose Throat J 2001;80(7):444–5, 448.
82. Schoendorf J, Neugebauer P, Michel O. Continuous intratympanic infusion of gentamicin via a microcatheter in Meniere's disease. Otolaryngol Head Neck Surg 2001;124(2):203–7.
83. Lefebvre PP, Staecker H. Steroid perfusion of the inner ear for sudden sensorineural hearing loss after failure of conventional therapy: a pilot study. Acta Otolaryngol 2002;122(7):698–702.
84. Kopke RD, Hoffer ME, Wester D, O'Leary MJ, Jackson RL. Targeted topical steroid therapy in sudden sensorineural hearing loss. Otol Neurotol 2001;22(4):475–9.
85. Heydt JL, Cunningham LL, Rubel EW, Coltrera MD. Round window gentamicin application: an inner ear hair cell damage protocol for the mouse. Hear Res 2004;192(1–2):65–74.
86. Bush RA, Lei B, Tao W, Raz D, Chan CC, Cox TA, Santos-Muffley M, Sieving PA. Encapsulated cell-based intraocular delivery of ciliary neurotrophic factor in normal rabbit: dose-dependent effects on ERG and retinal histology. Invest Ophthalmol Vis Sci 2004;45(7):2420–30.
87. Franklin RJ, Blakemore WF. Requirements for Schwann cell migration within CNS environments: a viewpoint. Int J Dev Neurosci 1993;11(5):641–9.
88. Gilmore SA, Sims TJ. Glial-glial and glial-neuronal interfaces in radiation-induced, glia-depleted spinal cord. J Anat 1997;190 (Pt 1):5–21.
89. Sims TJ, Gilmore SA. Regrowth of dorsal root axons into a radiation-induced glial-deficient environment in the spinal cord. Brain Res 1994;634(1):113–26.
90. Sims TJ, Gilmore SA. Regeneration of dorsal root axons into experimentally altered glial environments in the rat spinal cord. Exp Brain Res 1994;99(1):25–33.
91. Kawaguchi S, Iseda T, Nishio T. Effects of an embryonic repair graft on recovery from spinal cord injury. Prog Brain Res 2004;143:155–62.
92. Rosario CM, Aldskogius H, Carlstedt T, Sidman RL. Differentiation and axonal outgrowth pattern of fetal dorsal root ganglion cells orthotopically allografted into adult rats. Exp Neurol 1993;120(1):16–31.
93. Li Y, Field PM, Raisman G. Repair of adult rat corticospinal tract by transplants of olfactory ensheathing cells. Science 1997;277(5334):2000–2.
94. Raisman G. Use of Schwann cells to induce repair of adult CNS tracts. Rev Neurol (Paris) 1997;153(8–9):521–5.

Adipose Tissue Engineering

T. O. Acartürk

Contents

22.1 Introduction 289
22.1.1 Current Treatment Options and Limitations ... 290
22.1.2 Adipose Tissue Engineering as a Method to Overcome the Limitations of Conventional Therapies 290
22.2 Function, Anatomy and Physiology of Adipose Tissue 290
22.3 Isolation of ADSCs 292
22.4 ADSCs are Similar to Bone-Marrow-Derived Mesenchymal Stem Cells 292
22.5 Differentiation of ADSCs 294
22.5.1 Adipogenic 294
22.5.2 Osteogenic 294
22.5.3 Chondrogenic 296
22.5.4 Myogenic 296
22.5.5 Cardiac 296
22.5.6 Hepatic 297
22.5.7 Smooth Muscle Cells 297
22.5.8 Hematopoietic 297
22.5.9 Neurogenic 297
22.5.10 Vascular/Endothelial 297
22.5.11 Endocrine 297
22.6 Methods and Strategies in Adipose Tissue Engineering 298
22.6.1 Scaffold-Directed Tissue Regeneration 298
22.6.2 Injectable Cells or Cell-Carrier-Based Compositions 300
22.6.3 De Novo Adipogenesis 300
22.6.4 Cell-Based Therapy 300
22.6.5 Gene Delivery 300
22.6.6 Therapeutic Angiogenesis ... 301
22.6.7 Other 301
22.7 Clinical Implications 301
22.8 Future of Adipose Tissue Engineering 301
References 302

22.1 Introduction

Plastic, reconstructive and aesthetic surgeries aim to restore components of the body that have never formed, have been lost or are deformed through different mechanisms. Mostly, the affected body parts exist on visible areas, thus resulting in minor or major disfigurement. Often times, a large area of fat tissue is missing, resulting in contour abnormalities as well as exposure of vital organs. The etiology may be congenital (i.e., Romberg's disease, Poland syndrome), post-surgical (i.e., mastectomy, cancer ablation), traumatic (i.e., motor vehicle accident, burns) or due to aging and environmental factors [80]. Thus, the function of fat tissue to provide the body with normal contour cannot be overemphasized, as the loss would result in anxiety and major psychosocial impairment in addition to functional problems.

22.1.1
Current Treatment Options and Limitations

The current treatment options for such disease conditions are transfer of soft tissue from other parts of the body in the form of fat grafts or flaps, or use of implantable synthetic materials (i.e., silicone gel implants), each with their own limitations—such as donor site morbidity, flap loss, lack of available tissue, graft resorption or foreign body reaction.

Fat grafting has been used since 1910 to treat soft tissue loss due to aging or congenital reasons [91]. However, it often results in unpredictable outcomes due to resorptions ranging from 30% to 70% [12]. The main problem in fat graft loss is the ischemic insult on the adipocytes during the harvest due to trauma and after implantation due to lack of diffusion of oxygen and nutrients into the graft at the recipient site.

22.1.2
Adipose Tissue Engineering as a Method to Overcome the Limitations of Conventional Therapies

Recently, adipose tissue engineering has emerged as an alternative method to overcome the limitations of conventional treatments. The initial aim was to use the undifferentiated progenitor cells within the adipose tissue to generate mature fat tissue in the areas of the body that are needed. Later, it was found that the adipose-derived stem cells (ADSCs) are similar to other stem cells in their capacity to proliferate and differentiate into almost any cell in the body. The abundance of fat tissue and the ease of harvest has made ADSC-based therapies an area of growing interest and a clear alternative to other stem cell-based therapies.

22.2
Function, Anatomy and Physiology of Adipose Tissue

Adipose tissue functions as energy storage, heat production, metabolism and storage of hormones, regulation of appetite, food intake, fertility, reproduction and hematopoiesis, cushioning of bony prominences in contact with surfaces, and from a more reconstructive and aesthetic perspective formation contour throughout the body [43, 59]. It is found mainly within the subcutaneous areas of the body as well as surrounding the internal organs.

Brown and white fat are the two subtypes found in humans. Brown fat, which is more vascular and functions as a source of heat production mainly in the newborn, is gradually replaced by white fat as the organism ages [26]. In the mature adult, almost all the adipose tissue is white fat, which mainly functions as energy storage and release.

Adipose tissue is composed of lipid-filled adipocytes organized in a highly specialized 3D collagen matrix made up of stromal vascular cells with smooth muscle, endothelial cells, fibroblasts, blood cells and undifferentiated progenitor cells named as the pre-adipocytes [40]. Mature adipocytes, which are the primary cellular component, are composed of 90% lipid located within the cytoplasm of the cell [61]. The lipid content and metabolism are affected by hormonal changes, energy intake, drugs, infective agents, ischemia, trauma and environmental factors. The growth of adipose tissue is due to an increase in both the number and size of mature adipocytes. In addition, due to an abundant source of pre-adipocytes, there is always a capacity to form new fat cells throughout the life of the organism [68].

Pre-adipocytes are adipogenic precursor cells, which under proper stimuli will proliferate and differentiate into mature adipocytes (Fig. 22.1). The committed pre-adipocytes withdraw from the cell cycle and start accumulating lipid droplets, which gradually enlarge until terminally differentiated into mature adipocytes. The mature adipocytes usually contain a single large droplet of fat, which pushes the nucleus to the edge of the cell.

When compared with the lipid-filled, round and large adipocytes, the fusiform fibroblast-like

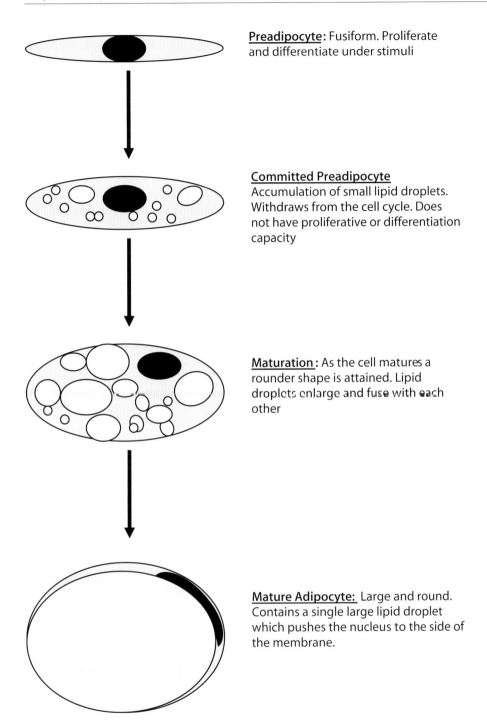

Fig. 22.1 Differentiation and maturation of a pre-adipocyte to a mature fat cell

pre-adipocytes are less likely to be injured during the harvest and processing of fat [3, 8, 61]. In addition, the mature adipocytes have decreased proliferative capacity in the culture environment and are not expandable due to their terminally differentiated state. In contrast, the pre-adipocytes can be easily obtained, cultured, expanded and can be induced to differentiate into cells of other lineage. Also pre-adipocytes are more resistant to traumatic effects of harvest and refinement, and can survive longer during ischemia and in nutrition-deprived environments than mature adipocytes [12, 96]. Thus, it is more advantageous to use pre-adipocytes in tissue engineering. These cells when cultured under conditions supporting stem cell morphology give rise to a more homogeneous population of cells, which are named as the ADSCs by many authors.

22.3
Isolation of ADSCs

The isolation of ADSCs starts with harvesting of adipose tissue from the organism, either in the form of lipoaspirate via liposuction or from an abdominoplasty specimen (Fig. 22.2). Such specimens should be microdissected to obtain small particles of fat lobules that are less than 1 cm^3. Following serial washings, collagenase is used to digest the stromal-vascular fraction within which the ADSCs reside. After filtration and centrifugation, the erythrocytes are lyzed and removed. The remaining cells are washed again before plating onto an expansion culture environment. Cells can be separated at 60% confluence before several passages. All steps should be performed under sterile conditions. Although slightly different modifications of the isolation techniques are used in different centers, the basic principles are more or less the same [11, 81]. However, it should be kept in mind that the specific technique of the harvesting and isolation procedure itself can affect the viability, proliferation and differentiation capacity of ADSCs. Other factors that might be an influence are donor age and gender, type and localization of adipose tissue, type and technique of surgical procedure, exposure to synthetic materials, plating density and media formulations [63, 81].

ADSCs obtained from human synovial adipose tissue of joints exhibited higher potential for chondrogenic differentiation than those from subcutaneous fat tissue [56]. In addition, the osteogenic potential of ADSCs obtained from visceral adipose tissue of rabbits was higher than that from subcutaneous adipose tissue [63]. Interestingly, in the clinical setting when fat tissue is transferred to a new location in the body in the form of a flap or graft, the size of the tissue in the new location changes according to the weight change of the patient [23].

Thus, each laboratory or center should characterize the molecular structure of the isolated cells at the end of the isolation procedure and adhere to strict guidelines for standardization of harvest techniques. Furthermore the isolated ADSC can be safely cryopreserved or expanded for additional use [69]. However, it should be remembered that although through multiple passages the ADSCs still retain capacity to differentiate, this declines with elevated number of passages [18, 75].

22.4
ADSCs are Similar to Bone-Marrow-Derived Mesenchymal Stem Cells

Adult stem cells or somatic stem cells by definition are a group of undifferentiated cells that can differentiate into specific cell types for either replacement or repair of tissues in the body [28, 92]. In contrast to other cell types, they can divide and replicate themselves and thus provide a constant source of stem cells [14]. Since they are not considered to be of any specific cell type, they do not perform functions executed by a differentiated cell. With the appropriate stimuli, they have the capacity to differentiate into a specific lineage. There are four types of human stem cells: embryonic, mesenchymal, hematopoietic and neural. The embryonic stem cells are derived from the inner lining of the embryo and are capable of differentiating to all kinds of tissue. However, due to ethical and legal considerations, these types of stem cells do not seem to be an ideal source for tissue engineering purposes in humans at present, or in the near future. Mesenchymal stem cells (MSCs) originate from the bone marrow and adipose tissue.

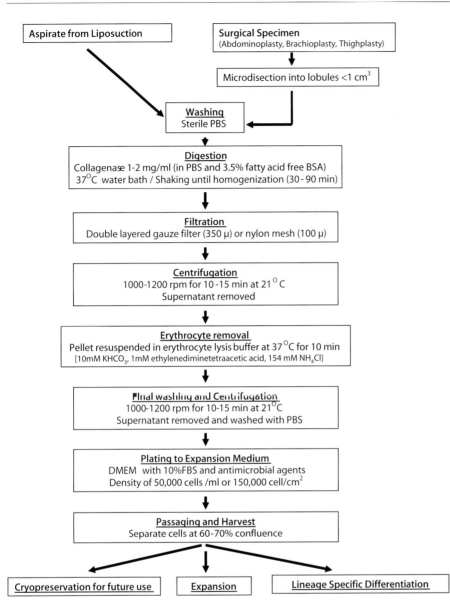

Fig. 22.2 Steps of harvesting adipose-derived stem cells. *PBS* phosphate-buffered saline, *DMEM* Dulbecco's modified Eagle Medium, *BSA* bovine serum albumin, *rpm* revolutions per minute, *FBS* fetal bovine serum

They have the capacity to differentiate into all mesenchymal tissues as well as neurogenic and endothelial lineage. Hematopoietic stem cells can differentiate into blood and immune cells as well as muscle and endothelium.

The most commonly used and most popular MSCs for regenerative medicine have been the multipotent stem cells derived from the bone marrow stroma.

In recent years, as more knowledge has been gained regarding the adipose tissue, ADSCs have emerged as a clear and possibly a better alternative.

Although not completely identical, ADSCs show striking similarities to MSCs derived from bone marrow stroma. In culture conditions, both cell types exhibit a fusiform fibroblast-like morphology [49]. The pattern of expression of genes and cell surface phenotypes are very similar between these cell types [88, 101]. CD surface marker antigens such as CD105, STRO-1, CD166(ALCAM) and CD 117, which have been shown to be expressed on totipotent and pluripotent stem cells, are almost always expressed on both of these cell types [27, 85]. Interestingly, ADSCs lack the expression of hematopoietic and endothelial surface markers. HLA-DR protein expression is seen in less than 1% of the ADSCs [7, 28].

Similar to bone-marrow-derived MSCs, the ADSCs have the capacity to differentiate into adipogenic, chondrogenic, myogenic, and osteogenic lineages [28]. In addition, these two cell types have a similar pattern of expression of genes and show the same surface markers once they are differentiated [25, 52].

Although there are many similarities, the proliferative and differentiating potential of these cell types may be different. ADSCs were found to show rapid growth, even when cultured at lower cell densities than MSCs obtained from bone marrow [16, 85]. In addition, the stem cell yield as measured by colony forming per gram of adipose tissue unit was observed to be five times higher than that of bone marrow [85]. There are many studies with contrasting results comparing the differentiation potential of these two cell types.

Another difference is the ease and amount of tissue that can be harvested from humans, with no complications. Bone marrow harvest done under local anesthesia, which is a painful procedure, yields approximately 50 ml of aspirate and 2.4×10^4 stem cells [5]; a liposuction of 200 ml, however, can easily be done yielding approximately 1×10^6 ADSCs [4].

22.5
Differentiation of ADSCs

ADSCs have the potential to differentiate into mesodermal as well as to non-mesodermal cells in vitro when exposed to various stimuli and conditions. The potential for multilineage differentiation makes the adipose tissue a valuable source for stem cells (Table 22.1).

22.5.1
Adipogenic

When ADSCs are grown in an adipogenic culture environment (DMEM and 10% FBS supplemented with insulin, dexamethasone) they withdraw from the cell cycle and differentiate into mature adipocytes [10, 28, 39, 100]. Other stimuli for adipogenic differentiation include IBMX (3-isobutyl-1-methyxanthine) and indomethacin. This differentiation can be observed by expression cell surface makers (lipoprotein lipase, leptin, and PPAR-γ-2), morphological changes (formation of lipid vacuoles staining positive for Oil red O stain) and reaction to circulating hormones (lipolysis with catecholamine stimulation) [18, 100, 101]. In addition, peroxisome proliferator-activated receptor gamma (PPAR-γ) is a key transcription factor for adipocyte differentiation [73]. This differentiation potential is retained under in vivo conditions when ADSCs are implanted within a scaffold with the pre-requisite that they are pre-differentiated into adipogenic lineage under culture conditions [29, 32, 94].

22.5.2
Osteogenic

ADSCs have the ability to differentiate into osteogenic lineage when grown in an osteogenic culture environment (DMEM and 10% FBS supplemented with dexamethasone, ascorbic acid and β-glycerophosphate) [28, 31, 101]. The osteoblastic differentiation can be evident by alkaline phosphatase activity and von Kossa staining for a calcified matrix, in addition to increased gene and protein expression. BMP-2, which is a strong stimulant for osteogenic differentiation has been shown to upregulate gene expression of Runx-2 which is one of the earliest transcription factors and forms the main regulator of bone formation by governing the on-off mechanism of cell growth and gene expression [20, 46, 50]. BMP-2 can also upregulate osteopontin gene expression, which is one of the most abundant non-collagenous proteins in bone

Table 22.1 Differentiation of adipose-derived stem cells into specific lineages; culture media; factors and stimulants inducing differentiation; Determination of differentiation; morphological and gene expression determinants of differentiation

Lineage of differentiation	Media	Factors and stimulants for differentiation	Determination of differentiation Morphological	Expression
Adipogenic	DMEM + 10% FBS	Insulin, dexamethasone, IBMX, indomethacin, rosiglitazone	Lipid vacuoles (oil red O stain)	Lipoprotein lipase, leptin, PPAR-γ-2
Osteogenic	DMEM + 10% FBS	Dexamethasone, ascorbic acid, β-glycerophosphate, bone morphogenetic protein (BMP)-2, vitamin D, valproic acid	Calcified matrix (von Kossa staining), mechanical loading	Alkaline phosphatase, Runx-1 and 2, osteopontin, BMP-2 and 4, osteonectin, osteocalcin, bone sialoprotein, BMP receptors I and II, PTH receptor
Chondrogenic	DMEM + 1% FBS	Ascorbic acid, insulin, transforming growth factor-β (1,2,3), insulin-like growth factor-1, BMP-4, -6, -7, fibroblast growth factor (FGF)-2	Proteoglycan-rich matrix (Alcian blue stain)	Aggrecan, α-1 chain collagen II, VI and IX, BMP-4,6,7
Myogenic	DMEM + 10% FBS + 5% HS	Dexamethasone, hydrocortisone, 5-azacytidine, direct contact with other muscle cells, experimental ischemia	Myotubes and multinucleation	MyoD1, Myf5, Myf6, myogenin, myosin heavy chain (MHC)
Cardiac	DMEM + 10% FBS + 5% HS	Interleukin-3, interleukin-6, stem cell factor	Pacemaker activity, response to adrenergic and cholinergic stimuli	Albumin, transthyretin, cytochrome 2E1, enhancer binding protein beta (C/EBPβ)
Hepatic	DMEM + 15% Human S	Hepatocyte growth factor, basic FGF, nicotinamide, dimethylsulfoxide, oncostatin	Cuboidal shape	
Smooth Muscle Cells	Medium MCDB 131 + 1% FBS	Heparin	Contract (with carbachol), relax (with atropine)	Smooth muscle cell-specific alpha actin (ASMA), calponin, caldesmon, SM22, MHC, smoothelin
Hematopoietic	DMEM + 5% HS (10% FBS)	Co-culture with hematopoietic cells? interleukins		CD34
Neurogenic	DMEM + 10-20% FBS	Azacytidine, valproic acid, hydroxyanisole, FGF, ethanol, epidermal growth factor (EGF), hydrocortisone	Nerve cell shape and morphology	Type III β-tubulin, glial fibrillary acidic protein (GFAP), nestin, neuN, intermediate filament M
Vascular/ Endothelial	DMEM/F12 ± 10% FBS	Leptin, vascular endothelial growth factor, monbutyril	Formation of vessel-like structures	CD1, von Willebrand factor
Endocrine	Serum-free DMEM/F12	Activin-A, exendin-4, HGF, pentagastrin, nicotinamide, high glucose concentration		Insulin, somatostatin, glucagon, Isl-1, Ipf-1, Ngn3

extracellular matrix [46]. Other proteins and genes specific to osteogenic lineage that are expressed after stimulation of ADSCs are osteonectin, osteocalcin, Runx-1, BMP-4, bone sialoprotein, BMP receptors I and II and the PTH receptor [30, 101]. Once differentiated, osteoblasts derived from ADSCs show morphological characteristics of regular bone cells, such as the response to mechanical loading and fluid shear stress [45, 90].

The potential for differentiation of ADSCs into osteogenic lineage is retained with increasing donor age [82]. Visceral ADSCs were found to have a higher osteogenic potential than subcutaneous ADSCs [63].

22.5.3
Chondrogenic

The chondrogenic differentiation of ADSCs is stimulated when in a chondrogenic media [DMEM and 1% FBS supplemented with insulin, transforming growth factor-β1 (TGF-β1) and ascorbic acid] [28]. The indicators for this differentiation are synthesis of sulfated proteoglycan-rich matrix detected with Alcian blue stain and synthesis of collagen II detected with collagen II-specific antibody immunostaining. BMP-4, BMP-6 and BMP-7 are the main regulators and strong stimulants of chondrogenic differentiation of ADSCs [22, 46]. After induction, aggrecan gene expression (chondroitin sulfate proteoglycan), and α-1 chain collagen II, VI and IX synthesis are increased [101]. Although fibroblast growth factor (FGF)-2 inhibits osteogenic differentiation of ADSCs, it is interesting to note that it has been shown to stimulate chondrogenic differentiation [9, 70].

There are contrasting findings whether ADSCs or MSCs have a better potential for osteogenic or chondrogenic differentiation. Several studies have indicated that the ADSCs have a lower osteogenic and chondrogenic differentiation potential than the bone-marrow-derived MSCs [34, 35]. In contrast, other studies have indicated that ADSCs have a higher potential for differentiation than MSCs derived from the same patient [16, 58].

22.5.4
Myogenic

ADSCs when grown in a myogenic environment (DMEM, 10% FBS and 5% horse serum supplemented with hydrocortisone, dexamethasone and 5-azacytidine) exhibit skeletal muscle cell phenotype characteristics such as formation of myotubes and multinucleation [17, 28, 48, 95]. They also start expressing early transcription factors such as MyoD1, myf5, myf6 and myogenin, followed by late transcription factors such as myosin heavy chain [42, 55, 101]. When transplanted into a murine model of Duchenne muscular dystrophy (*mdx* mice), ADSCs start regenerating muscle and expressing dystrophin, the protein missing in those with the disease [76]. It was found that direct contact with skeletal muscle cells was necessary for this transformation; this indicates an intrinsic potential of ADSCs for myogenic differentiation, especially after experimental ischemia [17, 48]. When ADSCs were transferred to injured skeletal muscle in a rabbit model, muscle weight, fiber cross sectional area and maximal contractile force were increased, indicating functional restoration [6]. When cultured in osteogenic, chondrogenic or adipogenic media, myogenically differentiated ADSCs did not differentiate into these cell types but retained their terminally differentiated state [42].

22.5.5
Cardiac

When the myogenic media is supplemented with IL-3, IL-6 and stem cell factor in a semi-solid environment, a cardiomyogenic differentiation pathway is preferred by the ADSCs [24, 65 71, 81]. The differentiated cells show pacemaker activity and respond to adrenergic and cholinergic stimuli similarly to cardiac myocytes [65]. In the in vivo setting, transplanted mouse or mice ADSCs have been shown to undergo differentiation with expression of cardiomyocyte-specific markers and reduce the infact area with improvement of left ventricular function [54, 85, 97]. The tissue transfer can occur by either transplanting monolayers grown in culture or cell injection into the coronary artery.

22.5.6
Hepatic

ADSCs were shown to differentiate into hepatic cells when grown in pro-hepatogenic medium [86]. This media contains 15% human serum, DMEM supplemented with hepatocyte growth factor (HGF), bFGF and nicotinamide [87]. Compared with bone-marrow-derived MSCs, ADSCs have a higher proliferative capacity and a longer culture period during hepatogenic differentiation. During differentiation, the ADSCs lose their fibroblastic morphology and attain a cuboidal shape similar to hepatocytes. Expression of hepatogenic markers such as albumin, transthyretin, cytochrome 2E1 and enhancer binding protein beta (C/EBPβ) increases during the culture period.

22.5.7
Smooth Muscle Cells

ADSCs cultured in smooth muscle differentiation medium (Medium MCDB 131 supplemented with 1% FBS plus 100 units/ml heparin) can express smooth muscle cell-specific alpha actin (ASMA), calponin, caldesmon, SM22, myosin heavy chain (MHC), and smoothelin [77]. In addition these newly differentiated smooth muscle cells, but not their precursors can contract and relax in direct response to carbachol and atropine, respectively.

22.5.8
Hematopoietic

Unlike bone-marrow-derived MSCs, complete hematopoietic differentiation of ADSCs has not been reported. However, ADSCs can support hematopoiesis in vitro by expanding numbers of myeloid and lymphoid progenitors when co-cultured with umbilical cord-blood-derived CD34(+) cells [41]. Exposure to lipopolysaccharides leads to secretion of both hematopoietic (granulocyte/monocyte, granulocyte, and macrophage colony stimulating factors, interleukin 7) and proinflammatory (interleukins 6, 8, and 11, tumor necrosis factor alpha) cytokines which, in turn, supports hematopoiesis similar to that for bone-marrow-derived MSCs. These myeloid and lymphoid elements are only short lived and do not have the capacity for self-renewal [15].

22.5.9
Neurogenic

ADSCs cultured in a neurogenic media (including azacytadine) can differentiate into neural cell lines by attaining nerve cell morphology. They express neuronal differentiation markers (type-III β-tubulin) and stain for glial fibrillary acidic protein, nestin, NeuN and intermediate filament M [78, 79].

When neuronal stem cells (NSCs) are co-cultured with human ADSCs, their proliferation decreases but differentiation into neuronal lineage increases relative to that when cultured alone [38]. Direct physical support rather than factors secreted by ADSCs is necessary for this effect.

22.5.10
Vascular/Endothelial

ADSCs can differentiate into endothelial cells under culture conditions indicated by expression of CD31 and other markers of mature endothelium [53]. In the mouse model, these cells can form vessel-like structures [66].

22.5.11
Endocrine

ADSCs in culture conditions when induced by activin-A, exendin-4, HGF and pentagastrin differentiated into cells having phenotypic characteristics of pancreatic endocrine cells expressing hormones such as insulin, somatostatin and glucagon [43, 89].

22.6
Methods and Strategies in Adipose Tissue Engineering

There are many strategies used for adipose tissue engineering. These are grouped under several categories depending on whether the ADSCs are used as a cell source to regenerate missing tissues or as carriers for various genes or proteins to alter other cell and tissue types (Table 22.2).

22.6.1
Scaffold-Directed Tissue Regeneration

A scaffold is a highly organized solid, semi-solid or gel-like three-dimensional structure that would allow cell attachment, migration, proliferation and differentiation. In both in vitro and in vivo settings, it should readily permit diffusion of nutrients, oxygen and metabolic byproducts for optimal cell function. Scaffolds are designed with the aim of carrying cells into the living organism in a certain architectural form. This is especially true for missing tissues needed for volume, stability and support (i.e., fat, bone and cartilage). An ideal scaffold is designed to resorb once the carried cells have reached a certain confluence and have replaced the structure or volume of the missing tissue. To discuss the characteristics of scaffolds is beyond the scope of this chapter, and only adipose tissue engineering-related literature will be discussed.

The basic strategy for adipose tissue engineering for volume replacement is to seed the scaffold with ADSCs in an in vitro setting supplemented with appropriate media and growth factors to permit attachment and proliferation of cells for a certain period of time. Later, the scaffold is transferred in vivo, where the cells continue to proliferate and eventually differentiate into mature fat cells as the scaffold material gradually remodels or absorbs [39, 62]. Although pre-differentiation of ADSCs into fat cells prior to in vivo implantation has been commonly used, post-implantation differentiation is possible using several growth factors [84, 99]. The former strategy of pre-differentiation prior to implantation is also preferred and is more effective if the ADSCs will be differentiated into another cell type [13, 21].

Several natural or synthetic types of polymers used as scaffolds in adipose tissue engineering are collagen,

Table 22.2 Strategies for adipose tissue engineering and clinical implications

Strategy	Clinical implication
1. Scaffold directed-tissue regeneration	Reconstruction of defects following trauma or cancer ablation and formation of 3D support structures (fat, cartilage, skeletal)
2. Injectable cells or cell-carrier-based compositions	Reconstruction of soft tissue (lipodystrophy, Romberg's disease, effects of aging) and skeletal defects
3. De novo adipogenesis	Reconstruction of soft tissue defects by inducing local stem cells
4. Cell-based therapy	Myocardial infarction, hepatic failure, diabetes mellitus, urinary incontinence, craniosynostosis, burns, vocal cord and pharyngeal augmentaion
5. Gene delivery	Treatment of genetic disorders (Duchenne muscular dystrophy, Osteogenesis imperfecta), induction of osteogenesis (BMP-2 delivery), neurological conditions (Parkinson's)
6. Therapeutic angiogenesis	Neovascularization, treatment of ischemic conditions (limb, cardiac, brain), burns
7. Other	Preclinical drug testing

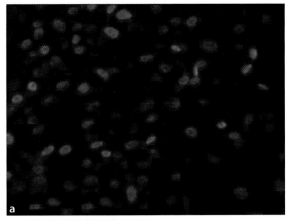
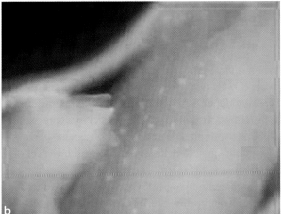

Fig. 22.3a–c Human ADSCs seeded on human demineralized bone matrix (DBM). **a** ADSCs in suspension. **b** Immediately after seeding onto demineralized bone matrix particles. The cells are round and have not attached to the surface. **c** After 2 weeks in culture. The cells are attached to the surface and have oriented and aligned following surface topography of DBM particles. (Fluorescent images of the DAPI nuclear stain.) Acartürk et al. Unpublished data

hyaluronic acid, alginate, polyethyleneglycol (PEG), polylactic acid (PLA), polyglycolic acid (PGA), polylactide-co-glycolic acid (PLGA) and fibrin glue [28, 32, 83, 93, 98]. In addition, many other synthetic or semi-synthetic materials are being tested [11]. The scaffold surface can be chemically modified by adding cell adhesion molecules which in turn bind fibronectin and laminin that will enhance cell attachment [8, 60, 67]. Optimal attachment of cells and cell spreading is an indication of cell survival and will also enhance growth and proliferation [1]. One other aim of cell surface modification is to increase biocompatibility. Cellular behavior can also be altered by changing the topography of the scaffold, including the pore size and shape. Large enough pore size is necessary to allow harboring fat cells filled with lipid droplets [32, 94]. However, porosity also affects the stability of the scaffold and, thus, there should be a balance between mechanical and cellular needs.

PEG scaffolds seeded with pre-differentiated adipocytes cultured from bone marrow-derived MSCs are able to form mature fat tissue and retain original shape and size of implant after implantation into subcutaneous pockets of nude mice [83]. Such scaffolds appear to be more stable than collagen sponges in long-term maintenance of shape and dimensions.

By pre-laminating the scaffold with certain materials or by adding growth factors, the ADSCs loaded onto the scaffold can be induced to differentiate into other cell types. Demineralized bone matrix (DBM), which contains quantities of bone morphogenic protein, can be used as a scaffold to harbor ADSCs as well as to induce them into an osteogenic pathway [47]. Our preliminary data also indicate that, under culture conditions, human ADSCs supplemented with osteogenic media implanted onto DBM particles attach to the surface of the material, align according to the surface topography and remain viable in culture for at least 2 weeks [2] (Fig. 22.3).

Various animal models of bone defects, together with a wide variety of resorbable scaffolds, have been used to investigate osteogenic differentiation and bone regeneration from ADSCs. Osteogenically pre-differentiated ADSCs seeded on PLA scaffolds placed in rat palatal defects and PLGA scaffolds placed on rat critical size calvarial defects lead to bone regeneration [13, 98]. Bone formation is also more robust if the length of the culture period is longer before implantation of the scaffold into the animal [98].

ADSCs when grown in a chondrogenic media and plated on fibrin glue scaffolds can regenerate full-thickness hyaline cartilage defects in articular surfaces of rabbits [19].

22.6.2
Injectable Cells or Cell-Carrier-Based Compositions

The same synthetic materials used as 3D scaffolds can be used as small spherical beads to be carriers for injectable forms of ADSCs. The cells survive and differentiate optimally when used in conjunction with a carrier substance—either a scaffold or beads. When injected into subcutaneous pockets of nude mice, human ADSCs lead to more tissue regeneration and adipogenic differentiation when combined with PLGA beads than when injected only as cell suspension [9]. ADSCs attached to PLGA spheres grown in myogenic media are able to form skeletal muscle when injected into subcutaneous pockets of nude mice [42]. These exogenously administered tissue engineered cells can fuse to repair and replace damaged fibers to treat individuals with chronic degenerative myopathies.

22.6.3
De Novo Adipogenesis

Exogenous growth factors can be administered to a local area to modulate the migration, proliferation and differentiation of endogenous ADSCs into mature adipocytes. In addition, de novo adipogenesis requires neovascularization, which is also controlled by these growth factors. bFGF released from gelatin microspheres that are incorporated into matrigel scaffolds can mobilize local ADSCs to the implantation site with infiltration of the matrigel [44]. New adipose tissue formation can be seen and is sustained up to 15 weeks. When bFGF is given in a controlled-release form, the adipogenic response is greater and sustained for a longer period of time [33]. Interestingly, addition of exogenous ADSCs did not enhance this adipogenic response, and bFGF alone was enough for in situ adipogenesis. Two or three different growth factors can be administered within the same delivery system, although with different predetermined release rates [51]. IGF-1 in addition to bFGF can control both the neovascularization and the physiological modulation of endogenous ADSCs.

22.6.4
Cell-Based Therapy

When injected into the urinary tract of nude mice, human ADSCs demonstrate morphological and phenotypic evidence of smooth muscle incorporation and differentiation with time [36]. In vivo expression of alpha-smooth muscle actin, which is an early marker of smooth muscle differentiation, was observed and cells remained viable for up to 12 weeks. Neurogenically induced human ADSCs, when transplanted into ischemic rat brains, were shown to improve motor and functional deficits [37]. These cells preferred to migrate into areas of ischemia when injected into the lateral ventricles. Cell-based therapy can also be used to treat hormone deficiencies, where pre-differentiated ADSCs can be given that act like organs secreting hormones [82].

22.6.5
Gene Delivery

ADSCs can be used for the purpose of gene delivery. Human ADSCs grown in culture and transfected with a BMP-2-carrying adenovirus can produce BMP-2 [64]. These cells, when transferred to critical size femoral defects in athymic rats, have been shown to regenerate bone. In addition, these cells can show in vitro osteogenic differentiation without the other osteogenic factors [20].

22.6.6
Therapeutic Angiogenesis

Therapeutic angiogenesis is a strategy that aims to treat various ischemic problems (ischemic cardiomyopathy, peripheral vascular disease, stroke, diabetes) or augment tissue survival by enhancing proliferation of collateral vessels (free tissue transfer, transplantation flap surgery). This is done by either using various angiogenic factors or transplantation of stem cells which in turn would either secrete these angiogenic growth factors or differentiate into endothelial cells. ADSCs are also equally angiogenic as bone-marrow-derived MSCs [57, 66]. ADSCs can secrete angiogenic growth factors [such as vascular endothelial growth factor (VEGF), HGF, placental growth factor (PGF), Ang-1 (angiopoietin) and TGF-β] [72]. The mechanism of angiogenesis is thought to be both by new vessel formation through differentiation into vascular elements and by upregulating the secretion of angiogenic cytokines via an autocrine or paracrine mechanism. This angiogenic potential of ADSCs can be further utilized in developing new vessel formation and ultimate perfusion to transferred or newly formed tissues in scaffold-base tissue engineering.

22.6.7
Other

ADSCs differentiated into various cell types in vitro can also potentially be used in preclinical drug testing.

22.7
Clinical Implications

Adipose tissue engineering has many clinical implications in plastic, reconstructive and aesthetic surgeries, and also in other areas of health and disease (Table 22.2). The most basic clinical implication is the soft-tissue reconstruction or augmentation. Fat grafting for soft tissue augmentation have been used for many years with variable success [12]. Mostly, the target areas are regions with only a minor to moderate deficiency of adipose tissue (lipoatrophy due to aging, Romberg's disease or HIV). Recently fat grafting has been explored for breast augmentation. The main downfall of autologous fat grafting is the loss of fat cells over time due to traumatic techniques of harvest and implantation. In addition, ischemia at the recipient site may augment death of implanted cells, and the regenerative capacity of mature fat cells is nil. In order to overcome this, purification of relatively trauma- and ischemia-resistant ADSCs from the lipoaspirates to yield a higher number of cells with regenerative capacity and the use of carrier materials (scaffold or beads) to provide a better architectural environment for optimal cell survival have been used; this forms the basis of adipose tissue engineering. In addition, some researchers believe that the effect of fat grafting comes from the survival of ADSCs in the lipoaspirate and, thus, application of purified ADSCs rather than whole lipoaspirates is more logical and feasible [74, 94]. In one clinical study, autologous ADSCs obtained from lipoaspirates that were expanded in culture were applied to patients with varying degrees of fibrosis, soft tissue deficiency, atrophy and ulcers due to radiation exposure [74]. After repeated treatments, dramatic improvement in symptoms and appearance was seen in over 90% of patients. The results were due to both adipogenesis and neovascularization of the ischemic tissues. When larger areas of filling or reconstruction are desired, the use of permanent of absorbable scaffolds can be used to deliver higher quantities of ADSCs deemed for adipogenic differentiation. More clinical trials are under way [80].

22.8
Future of Adipose Tissue Engineering

Since the initial research on adipose tissue engineering, many in vitro and in vivo studies have been done. Although there is still much to be learned from the lab application, the knowledge concerning use on human subjects is gradually emerging and will increase in the near future. There will be a time when a lab or facility will be an integral part of a plastic, reconstructive and aesthetic surgery department. Harvesting of autologous cells with a clinical application will become a "routine" surgical procedure.

One of the most important areas for future investigation and development will include vascularization of the engineered tissues to overcome the ischemic barriers for delivering larger volumes of the needed tissues with increased longevity. Also, much has to be learned in controlling the on-off mechanism of the proliferation and differentiation cycle of ADSCs, both to prevent early apoptosis and death of the delivered cells and to prevent them from becoming iatrogenically created tumors. Another area is to overcome the immunological barrier such that not only autologous ADSCs but also banked ADSCs from other donors can be used in health and disease conditions. This will lead to the development and spread of cell banks for the purpose of therapeutic as well as experimental purposes. Since fat is a readily abundant tissue easily harvested, ADSCs may fully take the place of bone marrow-derived MSCs for stem cell use in the clinical setting and for experimental purposes.

The future also holds optimization of harvest, isolation, preservation and delivery of ADSCs. One important area is the determination of optimal sites for harvesting ADSC with different differentiation and metabolic characteristics for various tissue engineering-based therapies. Since adipose tissue obtained from different locations (subcutaneous, visceral, synovial) have different characteristics, the fate of the grafts or cellular constructs may be affected accordingly. With the increase of efficiency of harvest and preservation methods, more cells per gram of tissue can be obtained and used. Creating and optimization of the ideal biomaterial as a carrier for ADSCs for various purposes will be another challenge. As better materials are developed, clinical applications will be easier, more efficient and longer lasting.

Considering the wide availability and ease of harvest, ADSCs (rather than other stem cells) may become the sole source of progenitor cells in the future for both research and therapeutic purposes.

References

1. Acartürk TO, Peel MM, Petrosko P et al. (1999) Control of attachment, morphology, and proliferation of skeletal myoblasts on silanized glass. J Biomed Mater Res 44:355–370.
2. Acartürk TO, Rubin PJ, Marra K, Bennet J. ADSCs grown on demineralized bone matrix. Unpublished Data.
3. Atala A, Lanza R (2002). Methods of tissue engineering. Academic Press, San Diego, CA.
4. Aust L, Devlin B, Foster SJ et al. (2004) Yield of human adipose derived adult stem cells from liposuction aspirates. Cytotherapy 6:7–14.
5. Bacigalupo A, Tong J, Podesta M et al. (1992) Bone marrow harvest for marrow transplantation: effect of multiple small (2 ml) or large (20 ml) aspirates. Bone Marrow Transplant 9:467–470.
6. Bacou F, el Andalousi RB, Daussin PA et al. (2004) Transplantation of adipose tissue-derived stromal cells increases mass and functional capacity of damaged skeletal muscle. Cell Transplant 13:103–111.
7. Barry FP, Murphy JM (2004) Mesenchymal stem cells: clinical applications and biological characterization. Int J Biochem Cell Biol 36:568–584.
8. Beahm EK, Walton RL, Patrick Jr CW (2003) Progress in adipose tissue construct development. Clin Plast Surg 30:547–558.
9. Chiou M, Xu Y, Longaker MT (2006) Mitogenic and chondrogenic effects of fibroblast growth factor-2 in adipose-derived mesenchymal cells. Biochem Biophys Res Commun 343:644–652.
10. Choi YS, Cha SM, Lee YY et al. (2006) Adipogenic differentiation of adipose tissue derived adult stem cells in nude mouse. Biochem Biophys Res Commun. 345:631–637.
11. Clavijo-Alvarez JA, Rubin JP, Bennett J et al. (2006) A novel perfluoroelastomer seeded with adipose-derived stem cells for soft-tissue repair. Plast Reconstr Surg. 118:1132–1142.
12. Coleman SR (2006) Structural fat grafting: more than a permanent filler. Plast Reconstr Surg 118:108S–120S.
13. Conejero JA, Lee JA, Parrett BM et al. (2006)Repair of palatal bone defects using osteogenically differentiated fat-derived stem cells. Plast Reconstr Surg. 117:857–863.
14. Conrad C, Huss R (2005) Adult stem cell lines in regenerative medicine and reconstructive surgery. J Surg Res 124:201–208.
15. Corre J, Barreau C, Cousin B et al. (2006) Human subcutaneous adipose cells support complete differentiation but not self-renewal of hematopoietic progenitors. J Cell Physiol 208:282–288.
16. De Ugarte DA, Morizono K, Elbarbary A et al. (2003) Comparison of multi-lineage cells from human adipose tissue and bone marrow. Cells Tissues Organs 174:101–109.
17. Di Rocco G, Iachininoto MG, Tritarelli A et al. (2006) Myogenic potential of adipose-tissue-derived cells. J Cell Sci 119:2945–2952.
18. Dicker A, Le Blanc K, Astrom G et al. (2005) Functional studies of mesenchymal stem cells derived from adult human adipose tissue. Exp Cell Res 308:283–290.
19. Dragoo JL, Carlson G, McCormick F et al. (2007) Healing Full-Thickness Cartilage Defects Using Adipose-Derived Stem Cells. Tissue Eng 13:1615–1621.

20. Dragoo JL, Choi JY (2003) Lieberman JR et al. Bone induction by BMP-2 transduced stem cells derived from human fat. J Orthop Res 21:622–629.
21. Dudas JR, Marra KG, Cooper GM et al. (2006) The osteogenic potential of adipose-derived stem cells for the repair of rabbit calvarial defects. Ann Plast Surg 56:543–548.
22. Estes BT, Wu AW, Guilak F (2006) Potent induction of chondrocytic differentiation of human adipose-derived adult stem cells by bone morphogenetic protein 6. Arthritis Rheum 54:1222–1232.
23. Flynn TC (2006) Does transferred fat retain properties of its site of origin? Dermatol Surg 32:405–6.
24. Gaustad KG, Boquest AC, Anderson BE et al. (2004) Gerdes AM, Collas P: Differentiation of human adipose tissue stem cells using extracts of rat cardiomyocytes. Biochem Biophys Res Commun 314: 420–427
25. Gimble JM (2003) Adipose tissue-derived therapeutics. Expert Opin Biol Ther 3:705–713.
26. Gregoire FM, Smas CM, Sul HS (1998) Understanding adipocyte differentiation. Physiol Rev 78:783–809.
27. Gronthos S, Franklin DM, Leddy HA et al. (2001) Surface protein characterization of human adipose tissue-derived stromal cells. J Cell Physiol 189:54–63.
28. Gomillion CT, Burg KJL (2006) Stem cells and adipose tissue engineering Biomaterials 27 (2006) 6052–6063.
29. Halbleib M, Skurk T, de Luca C et al. (2003) Tissue engineering of white adipose tissue using hyaluronic acid-based scaffolds. I: in vitro differentiation of human adipocyte precursor cells on scaffolds. Biomaterials 24:3125–3132.
30. Halvorsen YD, Franklin D, Bond AL et al. (2001) Extracellular matrix mineralization and osteoblast gene expression by human adipose tissue-derived stromal cells. Tissue Eng 7:729–741.
31. Hattori H, Sato M, Masuoka K et al. (2004) Osteogenic potential of human adipose tissue-derived stromal cells as an alternative stem cell source. Cells Tissues Organs 178:2–12.
32. Hemmrich K, von Heimburg D, Rendchen R et al. (2005) Implantation of preadipocyte-loaded hyaluronic acid based scaffolds into nude mice to evaluate potential for soft tissue engineering. Biomaterials 26:7025–7037.
33. Hiraoka Y, Yamashiro H, Yasuda K et al. (2006) In situ regeneration of adipose tissue in rat fat pad by combining a collagen scaffold with gelatin microspheres containing basic fibroblast growth factor. Tissue Eng. 12:1475–1487.
34. Hui JH, Li L, Teo YH et al. (2005) Comparative study of the ability of mesenchymal stem cells derived from bone marrow, periosteum, and adipose tissue in treatment of partial growth arrest in rabbit. Tissue Eng 11:904–912.
35. Im GI, Shin YW, Lee KB (2005) Do adipose tissue-derived mesenchymal stem cells have the same osteogenic and chondrogenic potential as bone marrow-derived cells. Osteoarthritis Cartilage 13:845–853.
36. Jack GS, Almeida FG, Zhang R et al. (2005) Processed lipoaspirate cells for tissue engineering of the lower urinary tract: implications for the treatment of stress urinary incontinence and bladder reconstruction. J Urol 174:2041–2045.

37. Kang SK, Lee DH, Bae YC et al. (2003) Improvement of neurological deficits by intracerebral transplantation of human adipose tissue-derived stromal cells after cerebral ischemia in rats. Exp Neurol 183:355–366.
38. Kang SK, Jun ES, Bae YC et al. (2003) Interactions between human adipose stromal cells and mouse neural stem cells in vitro. Brain Res Dev Brain Res 145:141–149.
39. Katz AJ, Llull R, Hedrick MH et al. (1999) Emerging approaches to the tissue engineering of fat. Clin Plast Surg 26:587–603.
40. Katz AJ, Mesenchymal cell culture: Adipose tissue. In: Atala A, Lanza R (2002) Methods of Tissue Engineering. Academic Press San Diego
41. Kilroy GE, Foster SJ, Wu X et al. (2007) Cytokine profile of human adipose-derived stem cells: Expression of angiogenic, hematopoietic, and pro-inflammatory factors. J Cell Physiol. 212:702–709.
42. Kim M, Choi YS, Yang SH et al. (2005) Muscle regeneration by adipose tissue-derived adult stem cells attached to injectable PLGA spheres. Biochem Biophys Res Commun 348:386–392.
43. Kim S, Moustaid-Moussa N (2000) Secretory, endocrine and autocrine/paracrine function of the adipocyte. J Nutr 130:3110S–3115S.
44. Kimura Y, Ozeki M, Inamoto T et al. (2002) Time course of de novo adipogenesis in matrigel by gelatin microspheres incorporating basic fibroblast growth factor. Tissue Eng 8:603–613.
45. Knippenberg M, Helder MN, Doulabi BZ et al. (2005) Adipose tissue-derived mesenchymal stem cells acquire bone cell-like responsiveness to fluid shear stress on osteogenic stimulation. Tissue Eng 11:1780–1788.
46. Knippenberg M, Helder MN, Zandieh Doulabi B et al. (2006) Osteogenesis versus chondrogenesis by BMP-2 and BMP-7 in adipose stem cells. Biochem Biophys Res Commun 342:902–908.
47. Koellensperger K, Markowicz M, Pallua N (2005) Demineralised bovine Spongiosa as three-dimensional matrix in adipose tissue engineering. 9th European Conference of Scientists and Plastic Surgeons 2005 Abstracts, S9.
48. Lee JH, Kemp DM (2006) Human adipose-derived stem cells display myogenic potential and perturbed function in hypoxic conditions. Biochem Biophys Res Commun 341:882–888.
49. Lee RH, Kim B, Choi I et al. (2004) Characterization and expression analysis of mesenchymal stem cells from human bone marrow and adipose tissue. Cell Physiol Biochem 14:311–324.
50. Lian JB, Javed A, Zaidi SK et al. (2004) Regulatory controls for osteoblast growth and differentiation: Role of Runx/Cbfa/AML factors. Crit Rev Eukaryot Gene Expr 14:1–41.
51. Masuda T, Furue M, Matsuda T (2004) Photocured, styrenated gelatin-based microspheres for de novo adipogenesis through corelease of basic fibroblast growth factor, insulin, and insulin-like growth factor I. Tissue Eng. 10:523–535.
52. Minguell JJ, Erices A, Conget P (2001) Mesenchymal stem cells. Exp Biol Med (Maywood) 226:507–520.

53. Miranville A, Heeschen C, Sengenes C et al. (2004) Improvement of postnatal neovascularization by human adipose tissue-derived stem cells. Circulation 110:349–355.
54. Miyahara Y, Nagaya N, Kataoka M et al. (2006). Monolayered mesenchymal stem cells repair scarred myocardium after myocardial infarction. Nat Med 12:459–465.
55. Mizuno H, Zuk PA, Zhu M et al. (2002) Myogenic differentiation by human processed lipoaspirate cells. Plast Reconstr Surg 109:199–209.
56. Mochizuki T, Muneta T, Sakaguchi Y et al. (2006) Higher chondrogenic potential of fibrous synovium- and adipose synovium-derived cells compared with subcutaneous fat-derived cells: distinguishing properties of mesenchymal stem cells in humans. Arthritis Rheum 54:843–853.
57. Nakagami H, Morishita R, Maeda K et al. (2005) Adipose Tissue-Derived Stromal Cells as a Novel Option for Regenerative Cell Therapy. J Atheroscler Thromb 13:77–81.
58. Nathan S, Das DS, Thambyah A et al. (2003) Cell based therapy in the repair of osteochondral defects: a novel use for adipose tissue. Tissue Eng 9:733–744.
59. Niemela SM, Miettinen S, Konttinen Y et al. (2007) Fat tissue: Views on reconstruction and exploitation. J Craniofac Surg 18:325–335.
60. Patel PN, Gobin AS, West JL et al. (2005) Poly(ethylene glycol) hydrogel system supports preadipocyte viability, adhesion, and proliferation. Tissue Eng 11:1498–1505.
61. Patrick CW (2004) Breast tissue engineering. Annu Rev Biomed Eng 6:109–130.
62. Patrick Jr CW, Zheng B, Johnston C et al. (2002) Long-term implantation of preadipocyte-seeded PLGA scaffolds. Tissue Eng 8:283–293.
63. Peptan IA, Hong L, Mao JJ (2006) Comparison of osteogenic potentials of visceral and subcutaneous adipose-derived cells of rabbits. Plast Reconstr Surg 117:1462–1470.
64. Peterson B, Zhang J, Iglesias R et al. (2005) Healing of critically sized femoral defects, using genetically modified mesenchymal stem cells from human adipose tissue. Tissue Eng 11:120–129.
65. Planat-Benard V, Menard C, Andre M et al. (2004) Spontaneous cardiomyocyte differentiation from adipose tissue stroma cells. Circ Res 94:223–229.
66. Planat-Benard V, Silvestre JS, Cousin B et al. (2004) Plasticity of human adipose lineage cells toward endothelial cells: physiology and therapeutic perspectives. Circulation 19:656–663.
67. Prichard HL, Reichert WM, Klitzman B (2007) Adult adipose-derived stem cell attachment to biomaterials. Biomaterials 28:936–946.
68. Prins JB, O'Rahilly S (1997) Regulation of adipose cell number in man. Clin Sci (Lond) 92:3–11.
69. Pu LL, Cui X, Fink BF et al. (2006) Adipose aspirates as a source for human processed lipoaspirate cells after optimal cryopreservation. Plast Reconstr Surg. 117:1845–1850.
70. Quarto N, Longaker MT (2006) FGF-2 inhibits osteogenesis in mouse adipose tissue-derived stromal cells and sustains their proliferative and osteogenic potential state. Tissue Eng 12:1405–1418.
71. Rangappa S, Fen C, Lee EH et al. (2003) Transformation of adult mesenchymal stem cells isolated from the fatty tissue into cardiomyocytes. Ann Thorac Surg 75:775–779.
72. Rehman J, Traktuev JE, Merfeld-Clauss S et al. (2004) Secretion of angiogenic and antiapoptotic factors by human adipose stromal cells. Circulation 109:1292–1298.
73. Reyes MR, Lazalde B (2007) Aortic preadipocyte differentiation into adipocytes induced by rosiglitazone in an in vitro model. In Vitro Cell Dev Biol Anim. 2007 Jun 13 [Epub ahead of print].
74. Rigotti G, Marchi A, Galie M et al. (2007) Clinical treatment of radiotherapy tissue damage by lipoaspirate transplant: a healing process mediated by adipose-derived adult stem cells. Plast Reconstr Surg 119:1409–1422.
75. Rodriguez AM, Elabd C, Delteil F et al. (2004) Adipocyte differentiation of multipotent cells established from human adipose tissue. Biochem Biophys Res Commun 315:255–263.
76. Rodriguez AM, Pisani D, Dechesne CA et al. (2005) Transplantation of a multipotent cell population from human adipose tissue induces dystrophin expression in the immunocompetent mdx mouse. J Exp Med 201:1397–1405.
77. Rodríguez LV, Alfonso Z, Zhang R et al. (2006) Clonogenic multipotent stem cells in human adipose tissue differentiate into functional smooth muscle cells. Proc Natl Acad Sci USA 103:12167–12172.
78. Romanov YA, Darevskaya AN, Merzlikina NV et al. (2005) Mesenchymal stem cells from human bone marrow and adipose tissue: Isolation, characterization, and differentiation potentialities. Bull Exp Biol Med 140:138–143.
79. Safford KM, Hicok KC, Safford SD et al. (2002) Neurogenic differentiation of murine and human adipose-derived stromal cells. Biochem Biophys Res Commun 294:371–379.
80. Sajjadian A, Magge KT (2007) Treating facial soft tissue deficiency: Fat grafting and adipose-derived stem cell tissue engineering. Aesthetic Surg J 27:100–104.
81. Schaffler A, Buchler C (2007) Concise review: adipose tissue-derived stromal cells-basic and clinical implications for novel cell-based therapies. Stem Cells. 25:818–827.
82. Shi YY, Nacamuli RP, Salim A et al. (2005) The osteogenic potential of adipose-derived mesenchymal cells is maintained with aging. Plast Reconstr Surg 116:1686–1696.
83. Stosich MS, Jeremy J et al. (2007). Adipose Tissue Engineering from Human Adult Stem Cells: Clinical Implications in Plastic and Reconstructive Surgery. Plast Reconstr Surg 119:71–83.
84. Strem BM, Hicok KC, Zhu M et al. (2005) Multipotential differentiation of adipose tissue-derived stem cells. Keio J Med 54:132–141.
85. Strem BM, Zhu M, Alfonso Z et al. (2005) Expression of cardiomyocytic markers on adipose tissue-derived cells in a murine model of acute myocardial injury. Cytotherapy 7:282–291.
86. Taléns-Visconti R, Bonora A, Jover R et al. (2006) Hepatogenic differentiation of human mesenchymal stem cells from adipose tissue in comparison with bone marrow mesenchymal stem cells. World J Gastroenterol 12:5834–5845.

87. Talens-Visconti R, Bonora A, Jover R et al. (2007) Human mesenchymal stem cells from adipose tissue: Differentiation into hepatic lineage. Toxicol In Vitro 21:324–329.
88. Tholpady SS, Llull R, Ogle RC et al. (2006) Adipose tissue: stem cells and beyond. Clin Plast Surg 33:55–62.
89. Timper K, Seboek D, Eberhardt M et al. (2006) Human adipose tissue-derived mesenchymal stem cells differentiate into insulin, somatostatin, and glucagon expressing cells. Biochem Biophys Res Commun 341:1135–1140.
90. Tjabringa GS, Vezeridis PS, Zandieh-Doulabi B et al. (2006) Polyamines modulate nitric oxide production and COX-2 gene expression in response to mechanical loading in human adipose tissue-derived mesenchymal stem cells. Stem Cells 24:2262–2269.
91. Toledo L, Mauad R (2002) Fat Injection: A 20- Year Revision. Clin Plast Surg 33:47–53.
92. Vats A, Tolley NS, Polak JM et al. (2002) Stem cells: sources and applications. Clin Otolaryngol Allied Sci 27:227–232.
93. von Heimburg D, Zachariah S, Low A et al. (2002) Influence of different biodegradable carriers on the in vivo behavior of human adipose precursor cells. Plast Reconstr Surg 108: 411–420.
94. von Heimburg D, Serov G, Oepen T et al. (2003) Fat tissue engineering. In: Ashammakhi N, Ferretti P. Topics in Tissue Engineering.
95. Wakitani S, Saito T, Caplan AI (1995) Myogenic cells derived from rat bone marrow mesenchymal stem cells exposed to 5-azacytidine. Muscle Nerve 18:1417–1426.
96. Wolter TP, Von Heimburg D, Stoffels I et al. (2005) Cryopreservation of mature human adipocytes: In vitro measurement of viability. *Ann Plast Surg* 55:408–413.
97. Yamada Y, Wang XD, Yokoyama S et al. (2006) Cardiac progenitor cells in brown adipose tissue repaired damaged myocardium. Biochem Biophys Res Commun 342:662–670.
98. Yoon E, Dhar S, Chun DE et al. (2007) In vivo osteogenic potential of human adipose-derived stem cells/poly lactide-co-glycolic acid constructs for bone regeneration in a rat critical-sized calvarial defect model. Tissue Eng 13:619–627.
99. Yuksel E, Weinfeld AB, Cleek R et al. (2000) De novo adipose tissue generation through long-term, local delivery of insulin and insulin like growth factor-1 by PLGA/PEG microspheres in an in vivo rat model: a novel concept and capability. Plast Reconstr Surg 105:1721–1729.
100. Zuk PA, Zhu M, Mizuno H et al. (2001) Multilineage cells from human adipose tissue: implications for cell-based therapies. Tissue Eng 7:211–228.
101. Zuk PA, Zhu M, Ashjian P et al. (2002) Human adipose tissue is a source of multipotent stem cells. Mol Biol Cell 13:4279–4295.

Intervertebral Disc Regeneration

J. Libera, Th. Hoell, H.-J. Holzhausen, T. Ganey,
B. E. Gerber, E. M. Tetzlaff, R. Bertagnoli,
H.-J. Meisel, V. Siodla

Contents

23.1 Background 307
23.2 Isolation, Expansion and 3D Culture of Disc-Derived Chondrocytes 308
23.3 Proliferation 308
23.4 Disc-Specific Matrix Formation and Maturation 309
23.5 Adhesion of Propagated Disc-Derived Chondrocytes to Native Disc Tissue 311
23.6 Migration of Propagated Disc-Derived Chondrocytes to Native Disc Tissue 311
23.7 Clinical Evaluation 313
23.8 Multicentric Clinical Trial 313
References 315

23.1 Background

Approximately 50–70% of patients who underwent sequestrectomy after experiencing a disc herniation still report of back pain [23, 25]. Persistent, severe lower back pains demand for additional surgical therapies in about 10% of these cases. Most of the described ailments are due to the fact that the surgically corrected disc, although reduced in size, is still exposed to full weight bearing. Decreases in disc height and intradiscal pressure as well as increases in radial disc bulge in relation to the mass of excised tissue were described by Brinckmann and Grootenboer using an in vitro model [3]. Increases in shear stress after partial denucleation in combination with the lack of a regenerative intrinsic healing process to substitute the removed disc tissue can lead to a progressive degeneration of the affected intervertebral disc [20, 21]. Later on, the instability of the disc results in degenerative changes in the adjacent levels, which may make surgical re-interventions and in the worst case a spinal-fusion surgery necessary. Therefore, disc restoration rather than sole discectomy seems to be the future for intervertebral disc treatments. As the metabolic activity of the disc chondrocytes was defined essential for the health of the disc [2, 7, 13, 18], options for the biological treatment and restoration of degenerated discs became the focus of novel treatment options.

One aim in developing novel therapy methods is the physiological reconstitution of the discectomy-derived defect to regain the functional state of the intervertebral disc and prevent further deleterious changes of adjoining vertebrae as well as extensive surgeries. The autologous disc chondrocyte transplantation (ADCT) is one promising method for the endogenous regeneration of the disc after discectomy and by this for the retardation of degenerative processes. This treatment involves the isolation and propagation of patient-specific disc chondrocytes according to the drug law regulations and transplantation of the amplified cells into the degenerated disc [8]. Given the value of disc chondrocytes to the metabolic health of the disc, the cell biological mode of function is of great interest.

Numerous studies have given insight into structure–function–failure relationships mostly attributable to a decreased disc cell density and a reduction in or alteration of the chondrocyte synthesis of cartilage-specific extracellular matrix proteins, such as certain collagen types and proteoglycans [2, 7, 13, 18]. Additionally, several in vitro studies on human nucleus pulposus and annulus fibrosus cells cultured in 3D alginate-, agarose-based, or pellet cultures or cell carriers were performed in terms of cell morphology, proliferation, and production of extracellular matrix components expanding the knowledge about the in vitro behavior of disc-derived cells [4, 5, 11, 12, 16, 24]. At least two phenotypically stable cell populations in the healthy adult human intervertebral disc were found producing either both collagen types I and II (annulus fibrosus cells) or only type II collagen (nucleus pulposus cells). These disc-derived chondrocytes remain viable and produce disc-specific matrix components in vitro while disc cell shape, gene expression and other cell functions are regulated by the extracellular environment [14]. In previous years, such studies could give insight into human and animal disc cell behavior after isolation and co-culturing using different co-culture and stimulation systems. But there is little known about cell biology aspects of cell cultures of human disc chondrocytes propagated and cultured without the use of any scaffolds and chemical stimuli.

The purpose of the current experimental study was the detailed characterization of monolayer as well as 3-dimensional (3D) spheroid cultures of human nucleus pulposus, annulus fibrosus or co-cultured cells to evaluate their differentiation and chondrogenic capability in view of their relevance for manufacturing and in vivo use as autologous transplants. Disc chondrocytes were isolated from sequestered tissue obtained during sequestrectomy. To show the chondrogenic potential of these cells, monolayer cell cultures were assessed for cell morphology, proliferation, and mRNA expression and 3D cultures were subjected to morphology, mRNA and protein expression analyses.

23.2
Isolation, Expansion and 3D Culture of Disc-Derived Chondrocytes

Intervertebral disc tissue was obtained from seven consenting volunteers undergoing sequestrectomy. Harvested discs were either separated into annulus fibrosus (AF) and nucleus pulposus (NP) or used non-separated (CC). Tissues were separated by the surgeon in the operating theatre using auto fluorescence effects of the intervertebral disc tissue [17]. Human serum for cell culturing was obtained from four additional donors. For all experiments and cell cultures, no growth factors, antibiotics or any other supplements were used. Cells were isolated by means of collagenase digestion and were cultured in medium supplemented with 1% glutamine and 10% human pooled serum in a humidified incubator (5% CO_2. 37°C). After reaching confluence, the cells were passaged by detaching using trypsin-EDTA and plating into new cell culture flasks. Cells were expanded up to passage 6 (P6). After reaching confluence of monolayer culture P2, cells were cultured as spheroids, as in our previous studies [1]. Briefly, detached cells were seeded at high density into cell non-adhesive cell culture dishes to initiate cell aggregation. Monolayer disc cells and disc cell aggregates were sampled for histological and immunohistochemical analyses. mRNA expression was analyzed from monolayers of passages 1 to 3 and from cell aggregates after 2, 4, and 6 weeks of cultivation by PCR. Immunohistochemical analyses were performed on paraffin-embedded and sectioned cell aggregates as well as on native disc tissue using rabbit antisera detected by fuchsin as a substrate for alkaline phosphatase-labeled secondary antibodies.

23.3
Proliferation

Monolayer cultures of AF, NP and CC cells were comparable with respect to cell morphology, and the rate and pattern of cell proliferation. All cells appeared flat and elongated isodiametric—a shape char-

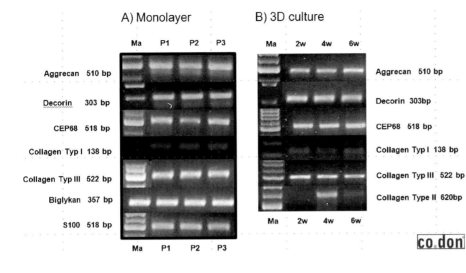

Fig. 23.1 mRNA expression of human disc-derived chondrocytes during propagation in monolayer and 3D cell culture system. Chondrocytes of annulus fibrosus and nucleus pulposus were cultured in co-culture. *P1, P2, P3* passages during monolayer culture. *2w, 4w, 6w* duration of cell culture in weeks

acteristic of monolayer cultures of undifferentiated mesenchymal cells, as also described for herniated disc chondrocytes by Lee et al. [19]. Proliferation analysis revealed minor donor-specific differences at low and high passages as well as a slightly stimulated, not significantly different, proliferation for co-culture cells (not shown). A stimulation of proliferation was also found for rabbit disc chondrocytes [22]. However, the exchange mechanism has not yet been characterized.

The chondrogenic phenotype of cultured cells was maintained during cell proliferation in monolayers (shown up to passage 3), as evidenced by mRNA analysis of molecular markers characteristic of AF and NP chondrocytes: type I collagen, type III collagen, biglycan, decorin, aggrecan, CEP68, and S100 (Fig. 23.1; only shows results for CC cultures). The expression of collagen type II during propagation in monolayer could not be detected in disc-derived chondrocytes (data not shown), as is the case for those isolated from knee joint cartilage.

23.4
Disc-Specific Matrix Formation and Maturation

Under high-density culture conditions, disc-derived chondrocytes aggregated and formed 3D cell-matrix structures—spheroids. The progression of aggregation was more rapid in cultures of CC cells than AF or NP cells. Morphological analysis revealed high cell density at early time points and, with increasing time, the strong and ongoing de novo synthesis of matrix in several discrete areas (Fig. 23.2; results only shown for CC cultures), resulting in an increase in matrix to cell ratio with time in culture. The distribution of cells and matrix were not spatially uniform, and the aggregates contained a surface layer of more elongated cells.

Analysis of mRNA expression during spheroid culturing confirmed the disc-specific differentiation of disc-derived chondrocytes. For annulus, nucleus and co-cultured chondrocytes, an expression of the proteoglycans aggrecan, biglycan (not shown), decorin, CEP68, S100, and collagens type I, II, and III was found (Fig. 23.1; results only shown for CC cultures). Most of the markers were expressed more in CC cultures than in separated AF or NP cultures. Under 3D CC culture conditions, the expression of collagen type II was detected. The expressions of collagens type II and III were higher than that of collagen type I for all three cell cultures. All of the analyzed proteoglycans were more strongly expressed than the different types of collagen.

Histological and immunohistochemical analyses confirmed the chondrogenic potential of isolated disc-derived chondrocytes. Collagen types I and III, and S100 (Fig. 23.2; data only shown for CC cultures) were detected in all tissue cultures at all time points.

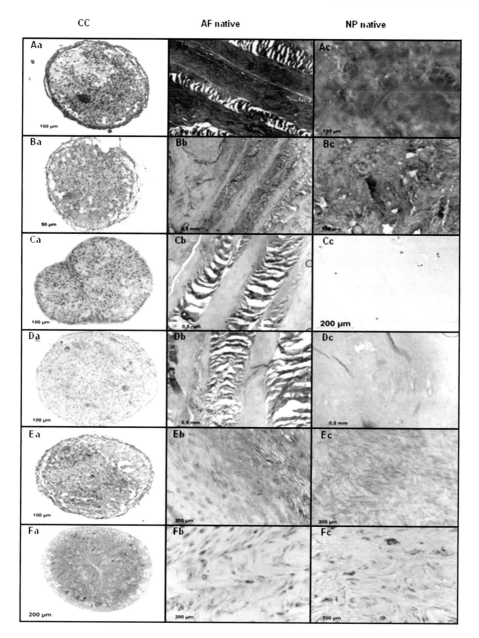

Fig. 23.2 Immunohistochemical and histological staining of human, 3D cultured, propagated disc-derived chondrocytes (*a*) and of human native annulus fibrosus (*b*) and nucleus pulposus (*c*). (*A*) SafraninO, (*B*) aggrecan, (*C*) S100, (*D*) collagen type I, (*E*) collagen type II and (*F*) type III. Disc-derived chondrocytes of AF and NP tissue were cultured in co-culture. *Orange staining* indicates expression of hyaline-specific proteoglycans using SafraninO staining. *Red staining* indicates positive expression of aggrecan, S100 and collagen types I, II, and III. Cell nuclei are counterstained with hematoxylin (*blue staining*)

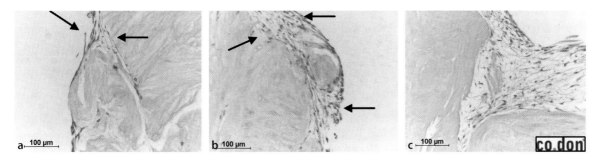

Fig. 23.3a–c Adhesion of propagated disc-derived chondrocytes on native human disc tissue (HE staining). Disc-derived chondrocytes adhere to the surface of native disc tissue, but also migrate into fissures of native disc tissue. (**a**) 2 weeks of co-culture, (**b,c**) 4 weeks of co-culture

However, the expression of collagen type II, aggrecan as well as a staining for SafraninO was only found in spheroids based on CC cells.

To compare in vitro differentiation of disc-derived chondrocytes with matrix secretion and maturation, native human intervertebral disc tissue was analyzed immunohistochemically. For both the AF and the NP matrix, an obvious staining for SafraninO, aggrecan, collagens type II and III, as well as a cellular staining for S100 were found. The expression of collagen type I was very low for the AF and the NP (Fig. 23.2).

These results indicated the constant phenotype of disc-derived chondrocytes during propagation, with the exception of collagen type-II expression. Under 3D culture conditions, disc-derived chondrocytes are able to form a matrix-cell construct and secrete disc-specific markers, as demonstrated using mRNA analysis. With the immunohistochemical protocol and the culture period and growth factor-free cell culture we used, not all the markers could be detected on the protein level. Additionally, as already observed during proliferation, the differentiation of disc-derived chondrocytes is stimulated when AF and NP cells were used as a co-culture. However, the mechanism is unknown.

23.5
Adhesion of Propagated Disc-Derived Chondrocytes to Native Disc Tissue

Human, propagated disc-derived chondrocytes adhere to native human disc tissue when cultured in co-culture with native tissue (Fig. 23.3). The suspended disc-derived chondrocytes directly adhered to extracellular tissue structures of the native tissue.

23.6
Migration of Propagated Disc-Derived Chondrocytes to Native Disc Tissue

Disc chondrocytes that were dropped and adhered to native disc tissue, started to migrate at the surface and into fissures of native tissue structures (Fig. 23.3). Additionally, disc-derived chondrocytes formed large cell-matrix networks that fused separated tissue fragments (Fig. 23.3, middle and right). This indicates not only a simple adhesion and migration of the chondrocytes but also a 3D formation of cell-matrix networks. The cell-matrix networks were observed at the surface as well as within fissures or cavities of the native tissue. Fissures and cavities were filled by these formed cell-matrix networks. The size of networks increased with ongoing culture time.

Secretion of disc-specific markers: Interestingly, already under static co-culture conditions, disc-derived chondrocytes express disc-specific markers. The formed cell-matrix networks stained positive for collagen types I, II, and III but also for S100 and SafraninO (Fig. 23.4, 23.5).

Collagens type I, II, and III are typical matrix components of disc tissue. As shown in Fig. 23.6, human annulus fibrosus and nucleus pulposus tissue showed weak staining for collagen type I, whereas a stronger expression of collagen types II and III was found.

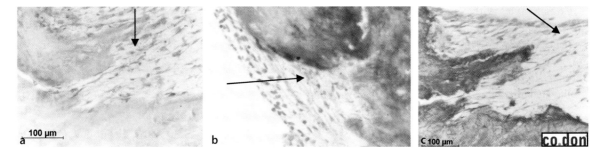

Fig. 23.4a–c Expression of collagens type I (**a**), II (**b**), and III (**c**) by propagated disc-derived chondrocytes in co-culture with native disc tissue. Cell-matrix networks stain positive for collagen type I, II and III. 2 weeks in co-culture. *Red staining* indicates positive expression. Cell nuclei are counterstained with hematoxylin (*blue staining*)

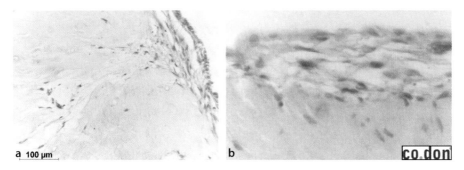

Fig. 23.5a,b Expression of hyaline-specific proteoglycans and S100 by propagated disc-derived chondrocytes in co-culture with native disc tissue. Cell-matrix networks stain positive SafraninO (**a**, **orange staining**) and for S100 (**b**, **red staining**). (**a**) 4 weeks, (**b**) 2 weeks in co-culture. Cell nuclei are counterstained with hematoxylin (blue staining)

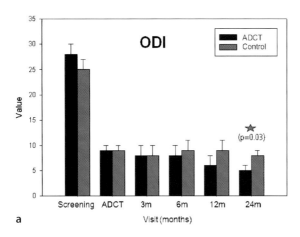
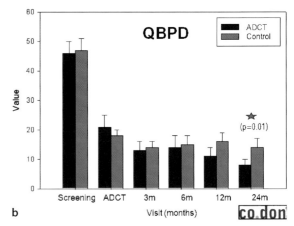

Fig. 23.6a,b Assessment of ODI and QBPD after sequestrectomy (control group) and sequestrectomy followed by ADCT (ADCT)

Chondrocytes within native human disc tissue express hyaline-specific proteoglycans that are stained orange by SafraninO. Additionally, it is native disc chondrocytes that express the hyaline-specific intracellular Ca-binding protein S100. Also, for human propagated disc-chondrocytes that were in co-culture with native human disc tissue, the expression of these specific proteoglycans and of S100 were observed (Fig. 23.5).

These results show the expression of disc-specific markers by propagated, human disc-derived chondrocytes cultured in a 3D co-culture system and support the phenotypic stability of these propagated disc-derived cells. Similar results were also found for human and animal annulus-fibrosus-derived chondrocytes cultured 3-dimensionally using an alginate culture system or porous silk scaffolds [4, 6, 11].

23.7
Clinical Evaluation

The clinical relevance of the repair of the intervertebral disc tissue by the ADCT has been evaluated in three pilot trials and is currently under evaluation in a prospective, randomized, multicenter, controlled clinical trial, EuroDISC (Bertagnoli et al., manuscript interims analysis in preparation). The first pilot trials were performed in Switzerland and Germany where patients were carefully selected to reflect the acute traumatic herniation. In Gerber, Switzerland, after 5–6 years, a full clinical and anatomical recovery period of patients with treated intervertebral discs was presented for a small pilot trial [9]. This trial included patients who suffered a massive lumbar disc herniation. After sequestrectomy, these patients were treated using autologous disc-derived chondrocyte transplants (SAS4, Vienna, 2004). A restructuring of the spine and a lack of Modic changes was shown using MRI. A 9-year follow-up of these patients was again presented as a good clinical and strong anatomical recovery period [first meeting of the Bone Research Society (BRS) and the British Orthopaedic Research Society (BORS), Southampton, 2006] [10]. A pilot trial of 14 patients performed by Meisel provided additional confidence that ADCT is a safe method which could provide a therapeutic benefit to patients with single level disc herniation [8]. In this trial, a sustained symptomatic pain relief and anatomical as well as functional recovery for the patients were observed.

23.8
Multicentric Clinical Trial

Sequestrectomy followed by ADCT was compared with sequestrectomy only. The main target of this study was to compare the amount of pain relief, and anatomical and functional recovery of patients who received ADCT with that of patients who were operated only by means of sequestrectomy.

Six German centers were included in the study, which fulfills all quality requirements for performing the study: Dr. Bertagnoli (Straubing), Dr. Meisel (Halle), Prof. Mayer (München), Dr. Curth (Potsdam), Prof. Weber (Saarbrücken), and Prof. Herdmann (Düsseldorf). The study was organized, managed and results evaluated by the contract research organization—IFE Europe GMBH (Essen, Germany). All patients included in this trial were treated by means of sequestrectomy, a standard method of treating lumbar disc herniation. Inclusion criteria were patients between 18 and 60 years of age, with a body mass index below 28 and a monosegmental disc herniation between levels L3 and S1. Exclusion criteria were sclerotic changes, edema, Modic changes grades II or III, spondylolisthesis, spinal stenosis, and other criteria, such as pregnancy. Patients included in the treatment group underwent sequestrectomy followed by ADCT. The control group underwent sequestrectomy and no further surgical procedure was performed. In the ADCT group, the sequester was obtained as biopsy material and collected under sterile operational conditions as well as a defined volume of patient blood for culturing of the chondrocytes. Each cell transplant was individually manufactured using the patients' own serum and disc chondrocytes, in accordance with Good Manufacturing Practice (GMP). A subgroup of 53 patients that reached a follow-up of 24 months at the time point of performing this subgroup analysis was evaluated: 27 ADCT-treated patients and 26 patients within the control group.

No placebo group was used; all patients received eligible treatment for their complaints. Double blinding of the trial was ethically not justifiable and not permitted by the ethics commission, because one treatment group would have been operated on without any benefit, but with the normal risks of a surgical procedure. The whole study was comprised of 104 evaluable patients.

Before transplantation, the integrity of the annulus ring was tested by volume pressure measurement. An in-line pressure gauge was specifically developed (Rehau, Rehau, Germany), allowing the measurement of pressure after injection of a bolus saline solution. In none of the patients was a leakage observed, indicating a closed annulus. After removal of the saline solution the cell suspension was injected through the same cannula.

Efficiency was assessed before sequestrectomy at the screening visit, at 3 months after sequestrectomy when the treatment group was treated by ADCT, and at 3, 6, 12, and 24 months after ADCT. As the primary target criterion after 12 months duration, the Oswestry Disability Index (ODI) was selected. As secondary criteria, the Quebec Back Pain Disability Scale (QBPD), Prolo scale for economic and functional status, and the Visual Analog Pain Scale (VAS) were used. The results of MRI were collected and evaluated, taking into consideration, for example, the intervertebral height, adjacent endplates, and deprivation of liquid.

After ADCT or control visit, 3 months after sequestrectomy, patients of both groups were compared with each other and were not relevantly different, based on the Oswestry disability total sumscore and index (Fig. 23.6). During the ongoing follow-up, a continuous reduction of pain and disability were observed, which were significantly improved for the ADCT-treated group. After 24 months, the ODI of the ADCT-treated group differed from the control group, showing a significance of $P=0.03$. Additionally, after 24 months, the QBPD differed significantly between the ADCT-treated and the control groups ($P=0.01$), confirming the significant clinical improvement in the ADCT patient group. For the VAS, there was a consistent trend toward decreased pain for both groups. This pain reduction was greater for the ADCT group, although there was no statistical significance relative to the control group. The Prolo score showed no difference for either study group (not shown).

MRI assessments were based on the investigation of three intervertebral discs. The height of the affected intervertebral disc was compared with the mean height of the two adjacent non-affected discs. Generally, the mean height of the affected discs of the control and treatment groups was lower than that of the non-affected discs. Comparison of mean intervertebral disc heights showed a slightly higher disc height for the ADCT-treated group than the control group after 2 years (not shown; not significant). Additionally, the content of liquid was analyzed using T2-weighted MRI images, visually assessing the signal intensity of the affected and the upper (1. non-affected segment) and lower (2. non-affected segment) non-affected discs (not shown). For each patient, the water content of the disc levels was evaluated as normal or decreased, and the percentage of normal and decreased levels was given. Within both study groups, the portion of affected discs showing a decreased liquid content was higher than the normal portion, as expected. As there is no validated method to quantitatively assess the water content of discs established, the retained data were difficult to interpret. Comparing the 1-year follow-up and the 2-year follow-up, an improvement of liquid content within the ADCT-treated group and an overall slight decrease of liquid content for the control group was found. In parallel, also the adjacent non-affected discs of the ADCT-treated group showed improved water content relative to that in the control group. Recurrent prolapses were observed. Only those occurring 3 months after sequestrectomy for the control group and after ADCT for the treated group are given here. Within the control group, recurrent prolapses were observed for four patients, within the ADCT-treated group for one patient.

In conclusion, these results clearly show the significant therapeutic benefit of this biological treatment method in comparison with the current standard surgical treatment of sequestrectomy only. The transplantation of autologous disc chondrocytes into discs pre-operated by sequestrectomy is assumed to delay or even inhibit the progression of degenerative disc disease by a regeneration of disc tissue—at least an improvement in general function and pain reduction is demonstrated. By implementing the ADCT, the range of non-fusion techniques can be enlarged and the number of fusion procedures reduced at the same time.

References

1. Anderer U, Libera J (2002) In vitro engineering of human autogenous cartilage. J Bone Miner Res 17(8):1420–9
2. Boos N, Weisbach S, Rohrbach H, et al. (2002) Classification of age related changes in lumbar intervertebral discs: 2002 Volvo Award in basic science. Spine 27:2631–44
3. Brinckmann P, Grootenboer H (1991) Change of disc height, radial disc bulge, and intradiscal pressure from discectomy. An in vitro investigation on human lumbar discs. Spine 16(6):641–6
4. Chang G, Kim HJ, Kaplan D, Vunjak-Novakovic G, Kandel RA (2007) Porous silk scaffolds can be used for tissue engineering annulus fibrosus. Eur Spine J 16(11):1848–57
5. Chelberg MK, Banks GM, Geiger DF, Oegema TR Jr. (1995) Identification of heterogeneous cell populations in normal human intervertebral disc. J Anat 186 (Pt 1):43–53
6. Chiba K, Andersson GB, Masuda K, Thonar EJ (1997) Metabolism of the extracellular matrix formed by intervertebral disc cells cultured in alginate. Spine 15;22(24):2885–93
7. Diwan AD, Parvataneni HK, Khan SN, Sandhu HS, Girardi FP, Cammisa FP Jr. (2000) Current concepts in intervertebral disc restoration. Orthop Clin North Am 31(3):453–64
8. Ganey TM, Meisel HJ (2002) A potential role for cell-based therapeutics in the treatment of intervertebral disc herniation. Eur Spine J 11 Suppl 2:206–14
9. Gerber BE, Siodla V, Josimovic-Alasevic O (2004) Five to six years follow up Results after biological disc repair by reimplantation of cultured autologous disc tissue. Poster presentation at SAS4, Vienna, Austria
10. Gerber BE, Biedermann M (2006) Nine Years Follow Up after the First Autologous Human Disc Regeneration and Replantation. 1st meeting of the Bone Research Society (BRS) and the British Orthopaedic Research Society (BORS) on July 5th–6th (2006), Southampton, UK
11. Gruber HE, Fisher EC Jr, Desai B, Stasky AA, Hoelscher G, Hanley EN Jr. (1997a) Human intervertebral disc cells from the annulus: three-dimensional culture in agarose or alginate and responsiveness to TGF-beta1. Exp Cell Res 235(1):13–21
12. Gruber HE, Stasky AA, Hanley EN Jr. (1997b) Characterization and phenotypic stability of human disc cells in vitro. Matrix Biol 16(5):285–8
13. Gruber HE, Hanley EN Jr. (1998) Analysis of aging and degeneration of the human intervertebral disc. Comparison of surgical specimens with normal controls. Spine 23(7):751–7
14. Gruber HE, Hanley EN Jr. (2000) Human disc cells in monolayer vs. 3D culture: cell shape, division and matrix formation. BMC Musculoskelet Disord 1:1
15. Gruber HE, Johnson TL, Leslie K, Ingram JA, Martin D, Hoelscher G, Banks D, Phieffer L, Coldham G, Hanley EN Jr. (2002) Autologous intervertebral disc cell implantation: a model using Psammomys obesus, the sand rat. Spine 27(15):1626–33
16. Gruber HE, Leslie K, Ingram J, Norton HJ, Hanley EN (2004) Cell-based tissue engineering for the intervertebral disc: in vitro studies of human disc cell gene expression and matrix production within selected cell carriers. Spine J 4(1):44–55
17. Hoell T, Huschak G, Beier A, Hüttmann G, Minkus Y, Holzhausen HJ, Meisel HJ (2006) Auto fluorescence of intervertebral disc tissue: a new diagnostic tool. Eur Spine J 15 Suppl 3:S345–53
18. Konttinen YT, Kääpä E, Hukkanen M, Gu XH, Takagi M, Santavirta S, Alaranta H, Li TF, Suda A (1999) Cathepsin G in degenerating and healthy discal tissue. Clin Exp Rheumatol 17(2):197–204
19. Lee JY, Hall R, Pelinkovic D, Cassinelli E, Usas A, Gilbertson L, Huard J, Kang J. (2001) New use of a three-dimensional pellet culture system for human intervertebral disc cells: initial characterization and potential use for tissue engineering. Spine 1;26(21):2316–22. Erratum in: Spine. 2007 Aug 1;32(17):1932
20. Lundon K, Bolton K (2001) Structure and function of the lumbar intervertebral disk in health, aging, and pathologic conditions. J Orthop Sports Phys Ther 31(6):291–303;304–6
21. Meakin JR, Redpath TW, Hukins DW (2001) The effect of partial removal of the nucleus pulposus from the intervertebral disc on the response of the human annulus fibrosus to compression. Clin Biomech (Bristol, Avon) 16(2):121–8
22. Okuma M, Mochida J, Nishimura K, Sakabe K, Seiki K (2000) Reinsertion of stimulated nucleus pulposus cells retards intervertebral disc degeneration: an in vitro and in vivo experimental study. J Orthop Res 18(6):988–97
23. Puolakka K, Ylinen J, Neva MH, Kautiainen H, Häkkinen A (2007) Risk factors for back pain-related loss of working time after surgery for lumbar disc herniation: a 5-year follow-up study. Eur Spine J Nov 23
24. Thonar E, An H, Masuda K (2002) Compartmentalization of the matrix formed by nucleus pulposus and annulus fibrosus cells in alginate gel. Biochem Soc Trans 30(Pt 6):874–8
25. Yorimitsu E, Chiba K, Toyama Y, Hirabayashi K (2001) Long-term outcomes of standard discectomy for lumbar disc herniation: a follow-up study of more than 10 years. Spine 15;26(6):652–7

Tissue Engineering of Ligaments and Tendons

P. Vavken

24

Contents

24.1	Introduction	317
24.2	The Need for Tissue Engineering in Ligament and Tendon Treatment	318
24.3	Lessons from Healthy and Healing Ligaments and Tendons	318
24.4	Tissue Engineering of the ACL	319
24.4.1	Choice of a Cell Source	320
24.4.2	The Biomaterial	320
24.4.3	Signaling	321
24.4.4	Approaches to Clinical ACL Tissue Engineering	322
24.5	Tissue Engineering of the Rotator Cuff	322
24.6	The Bone-Tendon Interface	323
24.7	Conclusion	325
	References	325

24.1 Introduction

Injuries to ligaments and tendons are important causes of pain and immobility in our society. In particular, the anterior cruciate ligament (ACL) and those tendons that constitute the rotator cuff do not heal after injury and, even after modern surgical treatments, long-term problems remain. Tissue engineering approaches hold much promise for both replacement and regeneration of damaged ligaments and tendons.

Numerous studies have been performed in this emerging field; however, thus far they point in various directions. In accordance with the tissue engineering triad "cells-biomaterials-signal" suggested by Bell, we try in this chapter to give a comprehensive, yet concise, overview of established cornerstones, important findings and current trends in the field of ligament and tissue engineering [1] (Fig. 24.1).

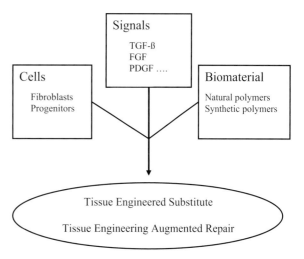

Fig. 24.1 Bell devised the triad of tissue engineering as a fundamental principle. The interactions among cells, biomaterials, and signals are used to create a biological substitute for lost or damaged tissue, or to enhance healing. It is important to note that there is significant overlap and reciprocity among the entities. For example, a biomaterial might act as a signal for cells by its composition and architecture. However, cells are able to synthesize their own biomaterial or remodel the biomaterial provided

24.2
The Need for Tissue Engineering in Ligament and Tendon Treatment

Current estimates present numbers of approximately 120,000 patients undergoing treatment for tendon or ligament defects annually in the US alone [33]. The most commonly torn ligament is the ACL, with an incidence of approximately 1 in 3,000 [20]. It is also a ligament with scarce to no healing capacity at all. Yet, it is quintessential for stability and, thus, function of the knee, in contrast to the posterior cruciate ligament, which shows many almost asymptomatic tears. Tears of the ACL cause pain and instability, and predispose patients to osteoarthritis in the long term. Hence, all treatment options in the management of the torn ACL need to be evaluated in the light of their short-term effectiveness as measured by pain, mechanical stability, and range of motion as well as their ability to prevent osteoarthritis in the long term. Historically, direct repair for the ACL was proposed as early as 1895, and this technique was further developed well into the 1970s and 1980s [39]. However, a number of studies showed poor outcome after primary repair, and this technique was subsequently abandoned [16, 54]. The current gold standard in ACL treatment is reconstruction using either the patella ligament, hamstring or quadriceps tendon. Allografts from cadavers and synthetic grafts are available, but their use is limited by availability and potential of disease transmission in the former and inflammatory reactions of foreign body type and poor outcome in the latter. Modern techniques in ACL reconstruction have consistently produced satisfactory results in joint stability, range of motion, and pain. However, recent studies have presented evidence of unabatedly high rates of osteoarthritis despite ACL reconstruction, even after controlling for other intra-articular damage caused by the initial trauma [17, 37, 59, 62].

Another important entity in musculoskeletal disease is the rotator cuff tear. The rotator cuff is a tendinous envelope consisting of four different muscles that encircle the humeral head and hold it in place during shoulder motion. Presently, the prevalence of shoulder pain attributable to problems of the rotator cuff is estimated between 7% and 30% [9]. Additionally, 4% of the population under the age of 40 years, and roughly 50% of the population above the age of 65 years have asymptomatic rotator cuff tears, 50% of which are expected to become symptomatic within 5 years [55, 64]. Generally, rotator cuff tears seem to be caused by gradual degeneration, repetitive micro-trauma, and overuse, an etiopathogenesis consistent with the insidious onset of symptoms and high proportion of elderly patients. However, a shift in affected populations toward younger patients has been noticed and is attributed to increased activity and sports in the present-day population. Furthermore, a number of risk factors other than age have been associated with rotator cuff injuries. Very much like the ACL tear, the torn rotator cuff causes pain, a loss of function, especially during over head motion, and predisposes patients to osteoarthritis of the shoulder. Despite the use of modern surgical techniques and careful perioperative management, high rates of treatment failures and recurrences have been reported [24–26, 29].

Summarized, ligament and tendon injuries are common causes of pain and loss of extremity function in both professional and private life. Hence, these diseases bring about individual suffering as well as societal burdens due to absenteeism and diminished productivity. Finally, the meaning of loss of ligament or tendon function is further aggravated by the fact that these diseases are independent risk factors of osteoarthritis, and very often remain so even after state-of-the-art surgical repair. Given this need for highly effective therapies for these increasingly common diseases, ligament and tendon tissue engineering might very well become an invaluable addition to the armamentarium of regenerative medicine in this field.

24.3
Lessons from Healthy and Healing Ligaments and Tendons

The purpose of a ligament or tendon is to withstand tensile forces. The healthy ACL supports loads of about 170 Newtons and will withstand up to 1,700 Newtons before failing [19]. These mechanical properties derive from the characteristics of collagen as a material as well as from structural properties.

Frank and Amiel were amongst the first to describe dense type-I collagen bundles surrounding fibroblasts with small additional contents of type-III collagen and glycosaminoglycans [3, 19]. On the structural level, both cells and fibers exhibit an undulating pattern, the so-called "crimp", which allows for stretching of the ligament [19]. This allows for a 6% elongation of the ligament before resulting in permanent damage (Fig. 24.2).

However, although it seems that the terms tendon and ligament are used somewhat interchangeably, and despite the use of tendons as ligament substitutes, it is important to note that there are distinct differences between these tissues. Ligaments have been reported to be more active metabolically, as suggested by higher cell numbers, higher DNA content, and more type-III collagen. Tendons, in turn, contain more total collagen but fewer glycosaminoglycans, which are important attractors of water. On the structural level, ligaments are arranged less regularly than tendons, but show the aforementioned, mechanically very important crimp. The significance of these differences for current and future clinical applications, however, is not fully understood.

Healing patterns of ligaments and tendons depend on a number of factors. It is a well-known conundrum that tears of the ACL will not heal, although those of the medial collateral ligament heal spontaneously. Tears of the rotator cuff also do not heal. Both these structures are intra-synovial, and it is likely that the lack of healing capacity is caused by factors of the intra-articular, or more precisely intra-synovial, environment. Wounded extra-synovial tissues produce and sustain a fibrin clot that serves as both a scaffold for inflammatory cell attachment and as a source of stimulatory cytokines from platelet activation. Within this clot, the damaged tissue is absorbed and new tissue is produced. In intra-synovial tissues, the formation of such a clot does not occur [40], a fact that is attributed to mechanical factors as well as biochemical factors such as tissue plasminogen activators. The resulting gap cannot be filled and the tissue stumps are covered by proliferating synovial cells and retract due to the production of smooth muscle actin-alpha in the matrix. Closing this gap and promoting cell migration into it is one of the most promising approaches in tissue-engineering-augmented repair of the ACL [40, 43, 44, 46] and the rotator cuff [12, 21-23, 53].

In summary, the main reason for insufficient healing of the ACL and rotator cuff is the lack of an adequate clot. This issue is targeted by tissue engineering applications and a solution to this problem would obviate the need to replace torn tissues, thus preserving the intricate architecture that will not be completely reproduced by a substitute. The success of such treatment in vitro should be evaluated by its mechanical strength, which is a function of both the quantitative and qualitative reproduction of cell-matrix interactions. Its clinical success has to be judged in the light of long-term effectiveness, especially since this is the area of weakness in current treatments.

24.4
Tissue Engineering of the ACL

Eugene Bell suggested the tissue engineering triad as a description of the bedrock of any tissue engineering approach [4]. Briefly, the deliberate choice of and the orchestrated interaction among cells, biomaterial, and signaling are the key to successful tissue engineering applications. Notably, the relationships among these three entities are marked by overlap, interdependence, and reciprocity. Cells may produce their own biomaterial, and biomaterials may act as signals in cell behavior.

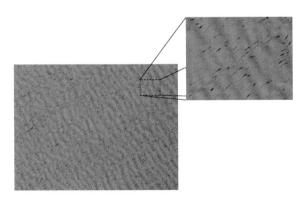

Fig. 24.2 A healthy ligament consists mostly of type-I collagen with interspersed fibroblasts. Both cells and fibrils show an undulating pattern that is responsible for many of the mechanical properties of the ligament. A tissue engineering application needs to mimic such architecture as closely as possible. A primary repair approach is more likely to do so than a complete replacement

24.4.1
Choice of a Cell Source

Various sources of cells have been studied for ACL tissue engineering. An optimal source would provide cells with a high proliferation rate and a high biosynthetic activity to build and remodel the ligament as fast and accurately as possible. Fibroblasts of different origins have been extensively studied, following the logic that highly differentiated cells possess all the phenotypic properties necessary to produce and maintain an adequately composed extracellular matrix. Cooper et al. compared rabbit fibroblasts from the ACL, the medial collateral ligament, the Achilles tendon, and the patellar tendon [10]. Of note, in this group, the ACL is the only intra-synovial tissue and the only one that does not heal spontaneously after injury. This study showed the highest rates of proliferation for cells obtained from tendons, but the highest gene expression for ACL cells. A similar study by Dunn and colleagues using rabbit ACL and patellar tendon fibroblasts showed similar attachment rates and proliferative and biosynthetic activity [15]. These differences are most likely due to random error, but should be kept in mind when translating results from animal studies into human applications. Brune et al. investigated fibroblasts from human healthy and ruptured ACL, considering the latter is the most likely and easily accessible source of fibroblasts for future clinical tissue engineering procedures [6]. This study found no difference in gene expression, taking into account the age of the donors. Yet, not only the origin, but also culture conditions have been shown to influence cellular behavior. When cultured on a collagenous ligament analog, patellar tendons and ACL fibroblasts showed a tenfold increase in collagen synthesis, and more rapid proliferation in tendon cells. Bellincampi focused on the in vivo survival of fibroblasts seeded on such a collagen analog and implanted into rabbit knees and subcutaneous pouches and showed viable cells for up to 6 weeks [5].

The major shortcomings of differentiated adult fibroblasts are their fairly low proliferation rates and relative scarcity. Another interesting source of cells are yet undifferentiated mesenchymal progenitor cells (MPCs). These cells can easily be obtained from the bone marrow or even isolated from adult tissues. They have a high proliferation rate and the potential to differentiate into multiple mesenchymal lineages. From early studies of MPCs, it is known that they differentiate in a site-specific manner after implantation into a sheep fetus [35]. They may also be cultured and differentiated into fibroblasts in vitro. Work by Altman et al. focusing on MPCs showed that these cells express ligament markers and form a proper ligament matrix when subjected to longitudinal and torsional strain in a bioreactor [2]. More recent work by the same group showed that a structural RGD-modified silk scaffold and the addition of growth factors further stimulated fibroblastic differentiation [8].

In a direct comparison of MPCs, ACL fibroblasts, and skin fibroblasts, presented by van Eijk, cells were seeded onto absorbable suture material and cultured for up to 12 days [61]. All cell types adhered to, proliferated and produced type-I collagen in culture. MPCs, however, showed the highest content of DNA and the highest rate of collagen synthesis in this experiment. There is still a small but persistent risk of faulty differentiation of MPCs that might lead to problems in a clinical application. Another important aspect is cell harvesting. Not unlike in currently employed cartilage repair procedures, fibroblasts might be obtained in an initial arthroscopic procedure, during which damages to other intra-articular tissue are assessed and addressed. It has been shown by Murray et al. that fibroblasts from human ACL are viable long after trauma and are able to migrate into a biomaterial used for tissue-engineering-augmented ACL repair [41, 42]. Another option would be a one-step technique, in which fibroblasts would be isolated in the theater and directly re-implanted, although low cell numbers might limit such a method. The classic way to obtain MPCs, in turn, is a bone marrow biopsy, which is a technically simple, yet considerably painful procedure.

24.4.2
The Biomaterial

After choosing a cell source, an appropriate biomaterial that satisfies a number of stipulations has to be selected. From the mechanical perspective, it has to resist tensile forces until sufficient ligament healing has been achieved. Furthermore, it should foster tissue remodeling by providing an environment that stimulates cellular attachment, growth, and biosynthetic activity. Biocompatibility and degradation rates that match and yield to tissue remodeling are also crucial. Additionally, safety is an important issue, since biomaterials might provoke inflammatory responses, cause arthrofibrosis, and lead to loss of joint function and other, even systematic, adverse reactions. Finally, the biomaterial has to be chosen according to the planned procedure. Tensile strength is less important than enhancement of cellular behavior in primary repair, while it is pivotal in replacement procedures.

Natural polymers have a long and successful history in tissue engineering, and collagen is an obvious choice for a tissue-engineered ligament. Collagen is and has been in clinical use for decades in suture materials and clotting agents, and its safety profile is well established. It is also used as biomaterial in clinically available tissue engineering methods, such as autologous chondrocyte implantation, and has been shown to enhance cellular phenotypes in this application [47]. Bovine collagen has been used in multiple studies to establish and sustain fibroblast cultures, yet with somewhat inconsistent results [5, 15]. However, the effect of collagen on cellular behavior depends not only on its mere presence, but also on material characteristics such as pore size, cross-linking, and fiber diameter [14, 47]. Other natural polymers that have been studied include hyaluronic acid, fibrin, and chitosan-alginate. Hyaluronic acid is a well-known biomaterial in tissue engineering and has been shown to beneficially influence cellular behavior. Using a rabbit model, Wiig and coworkers reported improved healing of a central ACL defect after injection of hyaluronic acid [63]. Cristino et al. showed MPC growth and differentiation in a modified hyaluronic acid-based scaffold [11]. It has also been shown to have a beneficial effect in the prevention of osteoarthritis in anterior-cruciate-deficient knees [56, 57]. Fibrin has the advantage of producing a biodegradable scaffold when mixed with thrombin; platelet-rich plasma, in turn, adds bioactive signals to these advantages [43-45].

Biomaterials such as poly(lactic acid), poly(glycolic acid), and other synthetic polymers have been used as suture materials, also with much success and few adverse events. The advantage of these polymers is that their composition can be controlled and adjusted for specific purposes. With modern processing techniques, these polymers can be spun into microfibers, which have been proven to enhance cell attachment by a high area-volume ratio and beneficial properties in mass transport of nutrients [34]. Additionally, these polymers have convincing mechanical properties. Perhaps of special interest in this group is silk, which holds a position at the intersection between naturally occurring and synthetically modified materials. Silk is an inert, biocompatible material with excellent mechanical properties, similar to the native ACL. Silk has also been used successfully as a scaffold for fibroblasts and shown to enhance fibroblastic differentiation of MPCs [8].

24.4.3
Signaling

Signaling is the third factor in the triad of tissue engineering. Its purpose is to direct cellular activity to achieve the desired outcome in cell growth or matrix production. For ligament tissue engineering, this is synthesis of an extracellular matrix with a high type I-type III collagen ratio. Numerous studies focused on signaling in ligament development and healing, yet it still remains unclear which factors are relevant and what the specifics of their mechanisms are. Hence, growth factors that have been associated with cell growth and differentiation were studied initially. Transforming growth factor beta (TGF-β), insulin-like growth factor (IGF), fibroblast growth factor (FGF), platelet-derived growth factor (PDGF), and the like have been shown to improve growth and bioactivity of fibroblasts [13]. In contrast, various growth factors have been shown to be produced by fibroblasts themselves and have been investigated for their use in tissue engineering of ligaments and tendons. However, the effects of many growth factors are still not completely understood and can hardly be adequately regulated. Rodeo et al., for example, used a growth factor cocktail in tendon repair and found a large increase in cell growth that eventually produced a mechanically inferior scar relative to the control group; thus was of detrimental effect [52].

Another very important source of cell stimuli is the chemical and mechanical environment of the cells. The structure of the used biomaterial, in terms of both chemical composition and structural properties, has been shown to stimulate cell attachment, growth, differentiation, and biosynthesis. Mechanical stimuli, such as tensile and torsional stress, also affect cellular behavior [2]. However, little is known about the details of and the interaction between these effects. Hence, the directed use of these stimuli in tissue engineering, beyond a general beneficial effect, is not yet possible.

24.4.4
Approaches to Clinical ACL Tissue Engineering

The information obtained in the aforementioned studies has been merged and translated into animal models. Generally, two philosophies exist in the management of the torn ACL: replacement using a tissue-engineered graft or tissue-engineering-augmented primary repair. In current clinical, non-tissue-engineering applications, replacement is preferred. This would mean creation of a complete ligament substitute. Advantages of such an approach are immediate mechanical function and a minimal change from currently used techniques; thus, a steep learning curve. However, the immediate mechanical strength also introduces two potential problems. On the one hand, it may cause stress shielding and thus deprive cells of important mechanical stimuli of bioactivity. On the other hand, the strength of the non-biological part of the graft must yield to the increasing strength of the remodeled tissue, otherwise the graft will fail eventually. A bio-artificial ACL has been presented by Goulet et al. [27]. In brief, this group suggested a complete substitute consisting of two bone plugs connected using a surgical thread. During culture, the cells attach to this thread and deposit a matrix rich in type-I collagen, thus building a ligament-like structure. This graft healed well in a goat model and showed tissue in-growth and vascularization in histology. After 13 months, it showed 36% of the mechanical strength of a normal goat ligament.

Another approach aims at supporting primary repair. The rationale of this approach is that the intricate nature of the ligament insertion, proprioceptive nerves, and the complex architecture of the ligament are preserved. Furthermore, the stumps serve as reservoirs of fibroblast. Murray and coworkers have described the specifics of such an approach in much detail. In summary, they demonstrated that human fibroblasts remain viable in the ACL stump and are able to migrate into a collagen scaffold, as could be used in an augmented primary repair procedure [41, 42]. Addition of platelet-rich plasma to this scaffold was shown to promote cellular migration and proliferation in a central defect model, thus stimulating healing [43]. Further examination revealed good defect filling in histology [45]. In another animal model, complete transections of the ACL in pigs were primarily repaired using the same technique and a significant improvement in mechanics was demonstrated [44] (Fig. 24.3).

24.5
Tissue Engineering of the Rotator Cuff

Very much like the ACL, tears of the rotator cuff have shown a low intrinsic capacity to heal and high recurrence rates, despite treatment, and many of the principles discussed to explain the ACL defect are also true for rotator cuff tears and will not be repeated. However, there are important differences between tendons and ligaments. Probably the most important difference is that tendons link muscles to bone rather than connecting bones to each other. For this reason, complete replacement by tissue-engineered substitutes is hardly possible, although tissue-engineering-enhanced repair is a very promising option in the management of rotator cuff tears. A number of biomaterials have been studied to support tendon repair in animal models. France et al. reported a primary repair that was augmented using PDS as an internal splint [18]. Koh and coworkers used PLLA patches to repair infraspinatus tears in a sheep model and showed a 1.25-fold stronger repair than when using normal sutures [32]. MacGillivray et al. used PGL patches in a similar model although failed to

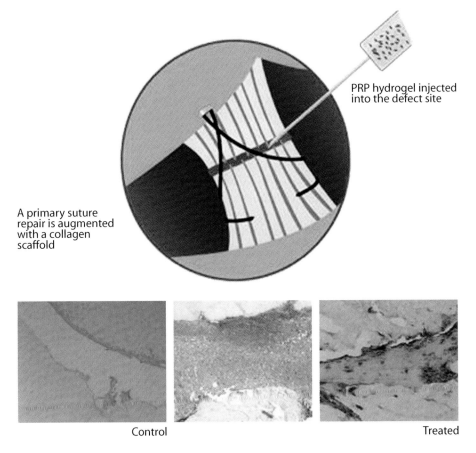

Fig. 24.3 Murray et al. developed a tissue-engineering-augmented primary repair technique for ACL tears using a collagen scaffold and platelet-rich plasma (PRP). Histological analysis showed excellent defect filling in a porcine model when compared with untreated defects (with permission)

show any differences between exposure and control group, a fact that might be attributed to a byproduct of PGA degradation—lactic acid—that decreases tissue pH and interferes with healing [38]. Adams reported the use of an acellular dermal matrix, with impressive mechanical results after 12 weeks in a canine model [1]. Improved healing of large rotator cuff tears was shown by Perry and Zalavras using porcine small intestinal submucosa in rats [48, 65]. Summarized, the results from these models have to be studied with caution. As many of the authors mentioned point out themselves, animals have different repair capacities than humans, which might influence the external validity of these findings. More importantly, however, the insertion of the rotator cuff tendons is intra-synovial in humans and extra-synovial in many animals [7]. Thus, animal models will show formation of a scar tissue, where a gap persists in human application.

24.6 The Bone-Tendon Interface

A particular problem in the surgical management of ACL and rotator cuff tears beyond tissue remodeling is the bone-ligament/tendon insertion. The insertion of a ligament or a tendon into the bone is highly complex [3, 19]. Specifically, it consists of four distinct tissues: tendon or ligament, non-mineralized fibrocartilage, mineralized fibrocartilage and, finally, bone. This complex structure avoids interfaces with high-stress gradients but provides a gradual transition between the properties of hard and soft tissues. It has been shown in various studies that such an interface is not remodeled in rotator cuff and ACL reconstruction [25, 52]. More specifically, in ACL reconstruction, micro-motion between the graft and bone stimulates the formation of a fibrous, mechanically inferior scar [49]. In the rotator cuff, a gap persists between the tendon and its bony insertion intra-synovially and not even a scar will be formed [52]. This gap has been documented with MRI refuting gross and histological appearance. These problems are considered one of the main reasons of treatment failure and are promising objects for tissue engineering (Fig. 24.4).

Most tissue engineering approaches in bone interface healing focus on the delivery of regulatory agents to the wound site. Rodeo et al. described the use of a growth factor mix on rotator cuff-tendon healing in a sheep model, but found only increased scar formation, and no true tissue regeneration [50, 52]. Rodeo also reported a beautiful description of the effect of micro-motion on ACL grafts in a rabbit model, presenting evidence that motion is greatest at tunnel apertures and stimulates osteoclast invasion [49]. In a subsequent study, Rodeo used an osteoprotegerin (OPG; an osteoclast inhibitor) and receptor activator of nuclear factor-kappa β ligand (RANKL; an osteoclast stimulator) delivery system to study osteoclastic activity in a rabbit model [51]. OPG led to decreased osteoclast numbers, increased bone formation, and smaller tunnel diameters than RANKL. Hays followed this approach and used liposomal clodronate in a rat ACL reconstruction model [30]. It was found to induce macrophage apoptosis and, thus, lead to decreased TGF-β secretion. In accordance with Rodeo's findings, this improved the results in morphology and biomechanics. Gulotta chose a biomaterial-based approach and augmented ACL reconstructions with a magnesium-based bone adhesive and found improved healing based on histological and biomechanical parameters [28]. For the interface between tendon and bone in the rotator cuff, fibrin-based scaffolds have been successfully employed in a rat model [60]. Currently, a clinical trial evaluating the effect of a platelet-rich fibrin matrix in rotator cuff repair is underway [Cascade PRFM Study: The Evaluation of Cascade Platelet-Rich Fibrin Matrix (PRFM) on Rotator Cuff Healing].

Chen used periosteum, a known source of both mesenchymal progenitors and cytokines, as an envelope for ACL grafts and showed maturing bone

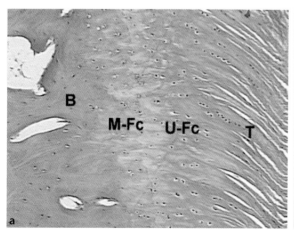
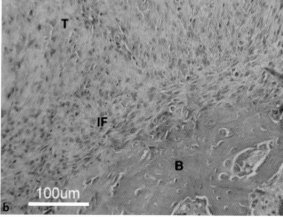

Fig. 24.4a,b The normal interface between the rotator cuff tendon and bone (a) shows four distinct layers: bone (B), mineralized fibrocartilage (M-FC), unmineralized fibrocartilage (U-Fc), and tendon (T). The treated shoulder (b) shows a fibrous scar with no further architecture at the interface (IF) between bone (B) and tendon (A). Thus, there is a high stress gradient across this interface, which might cause treatment failure (from Rodeo SA et al., 2007 [50], with permission)

over 12 weeks in a rabbit model [8]. Lim and Soon reported on ACL reconstruction in rabbit models using grafts coated with mesenchymal stromal cells in a fibrin glue carrier [36, 58]. In both studies, improved bone interface remodeling with formation of a mature fibrocartilaginous layer relative to a fibrous scar in control groups was found. Kanaya partially transected the ACL in rats and injected green-fluorescent-protein-labeled mesenchymal stromal cells into these knees [31]. After 4 weeks, he found covered, positively labeled defects in the treatment group and void defects in controls.

24.7 Conclusion

Tears of the ACL and the rotator cuff tendon are diseases characterized by both a high and increasing incidence and potentially serious and chronic long-term effects. Despite best efforts, the current standards of care are still associated with fairly high rates of recurrences and treatment failures, especially when considering the prevention of osteoarthritis as an outcome. Tissue engineering can augment and add to current treatments. Insufficient wound filling and persisting gaps have been identified as the most likely cause of lack of wound healing, and tissue-engineered defect filling using fibroblasts as well as progenitors has shown promising results. Furthermore, the recreation of a physiological bone-tendon/ligament interface has been identified as a main target in tissue engineering. Protocols using biomaterials as carrier systems for bioactive reagents and/or cells have been developed and successfully tested in animal models. The creation of a completely tissue-engineered ligament or tendon graft is possible, although its clinical application is complicated and relatively unlikely. Future studies will focus on the fusion of genetic engineering with tissue engineering in the field of ligament and tendon repair. Of further importance is the development of optimal animal models that mimic the mechanical, biochemical, and anatomical—especially in relation to intra/extra-synovial location—situation of the human more closely, in order to test potential tissue engineering applications.

Acknowledgements

The author is greatly indebted to Dr. Martha Meaney Murray for guidance and assistance in producing this text and to Ms. Shilpa Joshi for help with creating the figures.

References

1. Adams JE, Zobitz ME, Reach JS et al. (2006) Rotator cuff repair using an acellular dermal matrix graft: an in vivo study in a canine model. Arthroscopy: the journal of arthroscopic & related surgery: official publication of the Arthroscopy Association of North America and the International Arthroscopy Association 22:700–709
2. Altman GH, Horan RL, Martin I et al. (2002) Cell differentiation by mechanical stress. Faseb J 16:270–272
3. Amiel D, Frank C, Harwood F et al. (1984) Tendons and ligaments: a morphological and biochemical comparison. J Orthop Res 1:257–265
4. Bell E (2000) Tissue Engineering in Perspective. In: Lanza RP and Vacanti J eds) Principles of Tissue Engineering (Second Edition). Academic Press, pp xxxv–xli
5. Bellincampi LD, Closkey RF, Prasad R et al (1998) Viability of fibroblast-seeded ligament analogs after autogenous implantation. J Orthop Res 16:414–420
6. Brune T, Borel A, Gilbert TW et al. (2007) In vitro comparison of human fibroblasts from intact and ruptured ACL for use in tissue engineering. European cells & materials 14:78-90; discussion 90–71
7. Carpenter JE, Thomopoulos S, Soslowsky LJ (1999) Animal models of tendon and ligament injuries for tissue engineering applications. CORR S296–311
8. Chen J, Altman GH, Karageorgiou V et al. (2003) Human bone marrow stromal cell and ligament fibroblast responses on RGD-modified silk fibers. J Biomed Mater Res A 67:559–570
9. Coghlan JA, Buchbinder R, Green S et al. (2008) Surgery for rotator cuff disease. Cochrane database of systematic reviews (Online) CD005619
10. Cooper JA, Bailey LO, Carter JN et al. (2006) Evaluation of the anterior cruciate ligament, medial collateral ligament, achilles tendon and patellar tendon as cell sources for tissue-engineered ligament. Biomaterials 27:2747–2754
11. Cristino S, Grassi F, Toneguzzi S et al. (2005) Analysis of mesenchymal stem cells grown on a three-dimensional HYAFF 11-based prototype ligament scaffold. J Biomed Mater Res A 73:275–283
12. Derwin KA, Baker AR, Codsi MJ et al. (2007) Assessment of the canine model of rotator cuff injury and repair. J Shoulder Elbow Surg 16:S140–148
13. Deuel T, Chang Y (2007). Growth factors, Third Edition edn. Academic Press
14. Dorotka R, Toma CD, Bindreiter U et al. (2005) Characteristics of ovine articular chondrocytes in a three-di-

mensional matrix consisting of different crosslinked collagen. J Biomed Mater Res B Appl Biomater 72:27–36
15. Dunn MG, Liesch JB, Tiku ML et al. (1995) Development of fibroblast-seeded ligament analogs for ACL reconstruction. J Biomed Mater Res 29:1363–1371
16. Feagin JA, Jr., Curl WW (1976) Isolated tear of the anterior cruciate ligament: 5-year follow-up study. Am J Sports Med 4:95–100
17. Ferretti A, Conteduca F, De Carli A et al. (1991) Osteoarthritis of the knee after ACL reconstruction. International orthopaedics 15:367–371
18. France EP, Paulos LE, Harner CD et al. (1989) Biomechanical evaluation of rotator cuff fixation methods. Am J Sports Med 17:176–181
19. Frank C, Amiel D, Woo SL et al. (1985) Normal ligament properties and ligament healing. CORR 15–25
20. Frank CB, Jackson DW (1997) Current Concepts Review—The Science of Reconstruction of the Anterior Cruciate Ligament. J Bone Joint Surg Am 79:1556–1576
21. Galatz L, Rothermich S, VanderPloeg K et al. (2007) Development of the supraspinatus tendon-to-bone insertion: localized expression of extracellular matrix and growth factor genes. J Orthop Res 25:1621–1628
22. Galatz LM, Ball CM, Teefey SA et al. (2004) The outcome and repair integrity of completely arthroscopically repaired large and massive rotator cuff tears. J Bone Joint Surg Am 86-A:219–224
23. Galatz LM, Rothermich SY, Zaegel M et al. (2005) Delayed repair of tendon to bone injuries leads to decreased biomechanical properties and bone loss. J Orthop Res 23:1441–1447
24. Gerber C, Schneeberger AG, Beck M et al. (1994) Mechanical strength of repairs of the rotator cuff. J Bone Joint Surg Br 76:371–380
25. Gerber C, Schneeberger AG, Perren SM et al. (1999) Experimental rotator cuff repair. A preliminary study. J Bone Joint Surg Am 81:1281–1290
26. Gerber C, Fuchs B, Hodler J (2000) The results of repair of massive tears of the rotator cuff. JBJS 82:505–515
27. Goulet F, Germaine L, Rancourt D et al. (2007). Tendons and ligaments, 3rd ed. edn. Elsevier
28. Gulotta LV, MD, Kovacevic D et al. (2008) Augmentation of Tendon-to-Bone Healing With a Magnesium-Based Bone Adhesive. The American journal of sports medicine
29. Harryman DT, Mack LA, Wang KY et al. (1991) Repairs of the rotator cuff. Correlation of functional results with integrity of the cuff. JBJS 73:982–989
30. Hays PL, Kawamura S, Deng XH et al. (2008) The role of macrophages in early healing of a tendon graft in a bone tunnel. The Journal of bone and joint surgery American volume 90:565–579
31. Kanaya A, Deie M, Adachi N et al. (2007) Intra-articular injection of mesenchymal stromal cells in partially torn anterior cruciate ligaments in a rat model. Arthroscopy: the journal of arthroscopic & related surgery: official publication of the Arthroscopy Association of North America and the International Arthroscopy Association 23:610–617
32. Koh JL, Szomor Z, Murrell GA et al. (2002) Supplementation of rotator cuff repair with a bioresorbable scaffold. The American journal of sports medicine 30:410–413
33. Langer R, Vacanti JP (1993) Tissue engineering. Science 260:920–926
34. Lee CH, Shin HJ, Cho IH et al. (2005) Nanofiber alignment and direction of mechanical strain affect the ECM production of human ACL fibroblast. Biomaterials 26:1261–1270
35. Liechty KW, MacKenzie TC, Shaaban AF et al. (2000) Human mesenchymal stem cells engraft and demonstrate site-specific differentiation after in utero transplantation in sheep. Nat Med 6:1282–1286
36. Lim JK, Hui J, Li L et al. (2004) Enhancement of tendon graft osteointegration using mesenchymal stem cells in a rabbit model of anterior cruciate ligament reconstruction. Arthroscopy: the journal of arthroscopic & related surgery: official publication of the Arthroscopy Association of North America and the International Arthroscopy Association 20:899–910
37. Lohmander LS, Ostenberg A, Englund M et al. (2004) High prevalence of knee osteoarthritis, pain, and functional limitations in female soccer players twelve years after anterior cruciate ligament injury. Arthritis Rheum 50:3145–3152
38. MacGillivray JD, Fealy S, Terry MA et al. (2006) Biomechanical evaluation of a rotator cuff defect model augmented with a bioresorbable scaffold in goats. Journal of shoulder and elbow surgery/American Shoulder and Elbow Surgeons [et al] 15:639–644
39. Mayo Robson A (1903) Ruptured cruciate ligaments and their repair by operation. Ann Surg 37:716–718
40. Murray MM, Martin SD, Martin TL et al. (2000) Histological Changes in the Human Anterior Cruciate Ligament After Rupture. J Bone Joint Surg Am 82:1387–1397
41. Murray MM, Spector M (2001) The migration of cells from the ruptured human anterior cruciate ligament into collagen-glycosaminoglycan regeneration templates in vitro. Biomaterials 22:2393–2402
42. Murray MM, Bennett R, Zhang X et al. (2002) Cell outgrowth from the human ACL in vitro: regional variation and response to TGF-beta1. J Orthop Res 20:875–880
43. Murray MM, Spindler KP, Devin C et al. (2006) Use of a collagen-platelet rich plasma scaffold to stimulate healing of a central defect in the canine ACL. J Orthop Res 24:820–830
44. Murray MM, Spindler KP, Abreu E et al. (2007) Collagen-platelet rich plasma hydrogel enhances primary repair of the porcine anterior cruciate ligament. J Orthop Res 25:81–91
45. Murray MM, Spindler KP, Ballard P et al. (2007) Enhanced histologic repair in a central wound in the anterior cruciate ligament with a collagen-platelet-rich plasma scaffold. J Orthop Res 25:1007–1017
46. Murray MM, Spindler KP, Ballard P et al. (2007) Enhanced histologic repair in a central wound in the anterior cruciate ligament with a collagen-platelet-rich plasma scaffold. J Orthop Res
47. Nehrer S, Breinan HA, Ramappa A et al. (1997) Canine chondrocytes seeded in type I and type II collagen implants investigated in vitro. J Biomed Mater Res 38:95–104

48. Perry SM, Gupta RR, Van Kleunen J et al. (2007) Use of small intestine submucosa in a rat model of acute and chronic rotator cuff tear. Journal of shoulder and elbow surgery/American Shoulder and Elbow Surgeons [et al] 16:S179–183
49. Rodeo SA, Kawamura S, Kim HJ et al. (2006) Tendon healing in a bone tunnel differs at the tunnel entrance versus the tunnel exit: an effect of graft-tunnel motion? The American journal of sports medicine 34:1790–1800
50. Rodeo SA (2007) Biologic augmentation of rotator cuff tendon repair. Journal of shoulder and elbow surgery / American Shoulder and Elbow Surgeons [et al] 16:S191–197
51. Rodeo SA, Kawamura S, Ma CB et al. (2007) The effect of osteoclastic activity on tendon-to-bone healing: an experimental study in rabbits. JBJS 89:2250–2259
52. Rodeo SA, Potter HG, Kawamura S et al. (2007) Biologic augmentation of rotator cuff tendon-healing with use of a mixture of osteoinductive growth factors. The Journal of bone and joint surgery American volume 89:2485–2497
53. Safran O, Derwin KA, Powell K et al. (2005) Changes in rotator cuff muscle volume, fat content, and passive mechanics after chronic detachment in a canine model. J Bone Joint Surg Am 87:2662–2670
54. Sandberg R, Balkfors B, Nilsson B et al. (1987) Operative versus non-operative treatment of recent injuries to the ligaments of the knee. A prospective randomized study. J Bone Joint Surg Am 69:1120–1126
55. Sher JS, Uribe JW, Posada A et al. (1995) Abnormal findings on magnetic resonance images of asymptomatic shoulders. J Bone Joint Surg Am 77:10–15
56. Smith GN, Jr., Mickler EA, Myers SL et al. (2001) Effect of intraarticular hyaluronan injection on synovial fluid hyaluronan in the early stage of canine post-traumatic osteoarthritis. J Rheumatol 28:1341–1346
57. Sonoda M, Harwood FL, Amiel ME et al. (2000) The effects of hyaluronan on the meniscus in the anterior cruciate ligament-deficient knee. J Orthop Sci 5:157–164
58. Soon MY, Hassan A, Hui JH et al. (2007) An analysis of soft tissue allograft anterior cruciate ligament reconstruction in a rabbit model: a short-term study of the use of mesenchymal stem cells to enhance tendon osteointegration. The American journal of sports medicine 35:962–971
59. Spindler KP, Warren TA, Callison JC, Jr. et al. (2005) Clinical outcome at a minimum of five years after reconstruction of the anterior cruciate ligament. J Bone Joint Surg Am 87:1673–1679
60. Thomopoulos S, Soslowsky LJ, Flanagan CL et al. (2002) The effect of fibrin clot on healing rat supraspinatus tendon defects. Journal of shoulder and elbow surgery/American Shoulder and Elbow Surgeons [et al] 11:239–247
61. Van Eijk F, Saris DB, Riesle J et al. (2004) Tissue engineering of ligaments: a comparison of bone marrow stromal cells, anterior cruciate ligament, and skin fibroblasts as cell source. Tissue Eng 10:893–903
62. von Porat A, Roos EM, Roos H (2004) High prevalence of osteoarthritis 14 years after an anterior cruciate ligament tear in male soccer players: a study of radiographic and patient relevant outcomes. Ann Rheum Dis 63:269–273
63. Wiig ME, Amiel D, VandeBerg J et al. (1990) The early effect of high molecular weight hyaluronan (hyaluronic acid) on anterior cruciate ligament healing: an experimental study in rabbits. J Orthop Res 8:425–434
64. Yamaguchi K, Tetro AM, Blam O et al. (2001) Natural history of asymptomatic rotator cuff tears: a longitudinal analysis of asymptomatic tears detected sonographically. J Shoulder Elbow Surg 10:199–203
65. Zalavras CG, Gardocki R, Huang E et al. (2006) Reconstruction of large rotator cuff tendon defects with porcine small intestinal submucosa in an animal model. Journal of shoulder and elbow surgery / American Shoulder and Elbow Surgeons [et al] 15:224–231

25 Tissue Engineering of Cultured Skin Substitutes

R. E. Horch

Contents

25.1	Introduction	329
25.2	Currently Available Skin Substitutes According to Their Fundamental Features	330
25.3	Problems of Skin Substitution in Major Burns	330
25.4	Cultured Skin Developments	332
25.5	Cultured "Sheet Grafts" (Cultured Epithelial Autografts)	333
25.6	Cultured Human Cell Suspension Grafting	334
25.6.1	Membrane Cell Delivery Systems	335
25.6.2	Alloplastic or Mixed Synthetic-Biological Cell Carriers	335
25.6.3	Cultured Cells and Biological Carriers	336
25.7	Outlook	337
	References	338

25.1 Introduction

Wound closure and repair of skin defects belong to the elementary processes maintaining structural integrity of the human body, which relies on an intact skin barrier to protect against environmental influences. Skin, also known as the integument, not only is the body's largest laminar organ, but also fulfils many complex duties, such as being part of the immune response. Within this process, the primary aim of restitution is immediate re-epithelialization of any wounded surface [152].

Following Barronio's sheep-skin transplant experiments (1804), it was the ground-breaking report of human skin grafting by Reverdin in 1871 that ever since has set the gold standard for skin replacement by means of transplantation of the patient's own skin. Inspired by Theodor Billroth's observation that small islands of epithelium sometimes appeared in the granulation tissue in cases of severe burns, then extended rapidly, thus contributing to the epithelialization of a wound, Reverdin investigated what would happen if small pieces of skin were placed directly on the granulating tissue. His successful experiments in a 53-year-old patient convinced the surgeons of his time and set the pace for further developments. This procedure may be performed either as a full-thickness or as any form of split-thickness skin grafting. Hence, any tissue-engineered skin substitute will have to compare to the performance of standard autologous skin grafts over time.

Cultured human skin substitutes and living skin equivalents were initially developed to overcome the shortage of skin donor sites in those with extensive skin wounds, such as major burns. The intention was to prevent infection and desiccation and deliver cell guidance by dermal elements [82]. With such extensive burns, the wounded surfaces and loss of skin demanded the invention of various temporary or permanent skin substitutes. One simple reason is that there are too few skin resources on the patient's own body that would allow for recovery.

However, to achieve a long-term recovery, properties of both the dermal and the epidermal layers of skin need to be incorporated. When compared with conventional skin grafts, the use of cultured skin substitutes to close large wounds has reduced the amount of donor skin required more than tenfold. It also has reduced the number of surgeries to harvest donor skin, at the same time decreasing the time to recovery of severely burn-injured patients.

The life-saving effects of cultured human skin substitutes in treating major burns have propagated their use in treating chronic wounds that are otherwise difficult to heal. However, the results have not been as expected and, thus, such treatment is therefore still a matter of controversy. Various technical advances are under investigation to optimize tissue-engineered skin substitute performance in the latter indications [58, 72, 74, 76, 78, 94].

Nevertheless, tissue engineering and skin cell culture research has—besides the immediate clinical effects—significantly contributed to the basic understanding of skin regeneration and biology [88, 92, 108, 110, 127, 145].

Table 25.1 Systematic classification of principally currently available biological and synthetic skin substitutes

Biologicals (Naturally occurring tissues)
– Cutaneous allografts (γ-irradiated, deep frozen, glycerolized)
– Cutaneous xenografts
– Amniotic membranes
Skin substitutes
– Synthetic bilaminates
– Collagen-based composites
Collagen-based dermal analogues
Deepithelized allografts
Culture-derived tissue substitutes (see also Tab.2)
– Cultured autologous keratinocytes (sheet grafts, cell suspensions)
– Bilayer human tissue
– Polyglycolic or acid mesh
– Fibroblast-seeded dermal analogs
– Collagen-glycosaminoglycan matrix
– Epithelial seeded dermal analogs

25.2
Currently Available Skin Substitutes According to Their Fundamental Features

The biological process of wound closure requires a material that restores the epidermal barrier function and becomes integrated into the healing wound. In contrast, materials designed for wound coverage rely mainly on the ingrowth of granulation tissue for adhesion [95]. The latter are best suited for superficial burns and help to improve the healing environment for epithelial regeneration [18, 102].

Conceptually, skin substitutes may be classified as either permanent or temporary; epidermal, dermal or composite; and biological or alloplastic (synthetic). Biological components can be autogenic, allogenic, or xenogenic. Thus, research efforts of different groups are focused on several possible permutations of these traits, whereas practically most designs rely on a permanent or temporary engraftment of the material.

Following the principal action of the materials popular and currently available, one can discern biological and synthetic skin substitutes to be categorized into three different groups of clinical application purposes (Tables 25.1 and 25.2).

A) **Temporary:** material designed to be placed on a fresh wound (partial thickness) and left until healed
B) **Semi-permanent:** material remaining attached to the excised wound and eventually replaced by autogenous skin grafts
C) **Permanent:** incorporation of an epidermal analogue, dermal analogue, or both as a permanent replacement

25.3
Problems of Skin Substitution in Major Burns

There is no doubt that the driving force to develop cultured skin substitutes arose from the clinical need of large amounts of autologous skin grafts to treat extensive burns. Rapid and effective burn wound

Table 25.2 Overview over currently commercially available or marketed matrices and products for tissue-engineered skin substitutes

Material	Brand name	Manufacturer
Collagen gel + cult. Allog. HuK + allog. HuFi	Apligraf™ (earlier name: Graftskin™)	Organogenesis, Inc., Canton, MA
cult. Autol HuK	Epicell™	Genzyme Biosurgery, Cambridge, MA
PGA/PLA + ECMP DAHF	Transcyte™	Advanced Tissue Sciences, LaJolla, CA
Collagen GAG-polymer + silicone foil	Integra™	Integra LifeScience, Plainsborough, NJ
Acellular dermis	AlloDerm™	Lifecell Corporation, Branchberg, NJ
HAM + cult. HuK	Laserskin™	Fidia Advanced Biopolymers, Padua, Italy
PGA/PLA + allog. HuFi	Dermagraft™	Advanced Tissue Sciences, LaJolla, CA
Collagen + allog HuFi + allog HuK	Orcel™	Ortec International, Inc., New York, NY
Fibrin sealant + cult. autol HuK	Bioseed™	BioTissue Technologies, Freiburg, Germany
PEO/PBT + autol. HuFi + cult autol HuK	Polyactive™	HC Implants
HAM + HuFi	Hyalograft 3D™	Fidia Advanced Biopolymers, Padua, Italy
Silicone + nylon mesh + collagen	Biobrane™	Dow Hickham/Bertek Pharmac., Sugar Land, Tx

ECMP extracellular matrix proteins, *DAHF* derived from allog. HuFi, *GAG* glycosaminoglycan, *PGA* polyglycolic acid (Dexon™), *PLA* polylactic acid (Vicryl™), *PEO* polyethylene oxide, *PBT* poly(butylene)terephthalate, *cult.* cultured, *autol.* autologous, *allog.* allogeneic, *HuFi* human fibroblasts, *HuK* human keratinocytes, *HAM* microperforated hyaluronic acid membrane (benzilic esters of hyaluronic acid *HYAFF-11*®)

closure is one of the most important aspects in the treatment of burn patients, because the patient remains in a sub-septic condition until all skin defects are closed. Hence, modern treatment algorithms in burn surgery are based on the perception that early surgical removal of heat-denatured proteins and devitalized tissue from a wound (staged serial debridement between the second and tenth day after the injury = early necrectomy) turns the burn into an excisional (primary) wound that can heal faster than a secondary wound [1, 41, 46, 57, 134].

This strategy of (post)-primary excision approaches has reduced mortality, morbidity and later reconstructive measures by a factor of 50% when compared with prior results obtained by awaiting spontaneous separation of the burn eschar with later grafting. It is generally accepted that excision and the reconstitution of a functional skin as early as possible—once the patient is stabilized—is crucial for the further course and for the patient's survival [69, 77, 86, 121] and leads to a better wound healing with improved functional results [4, 112, 114].

One tool to overcome the initial lack of autografts is the use of homografts or synthetic replacements. Such allogenic or alloplastic skin substitutes provide a temporary solution until definitive skin cover can be achieved [1, 52, 69, 77, 113, 120, 133, 143, 144, 147, 161, 174]. It has been reported that allogenic skin grafts may be partly or completely integrated into the healing wound initially [167]. This helps to bridge the critical time gap in the early phase of burn treatment, but in the further course all allografts irrevocably undergo immunogenic rejection when no life-long immunosuppression is administered [5, 63, 69, 70, 77, 87, 98, 144, 159, 167, 174]. Theoretically, the application of in vitro cultivated autologous skin substitutes is therefore able to overcome this specific deficit of today's burn treatment methods.

25.4
Cultured Skin Developments

Since the development of the method for growing epithelial sheets with the support of lethally irradiated 3T3-feeder layer cells in submerged culture conditions by Rheinwald and Green in 1975 [138], cultured epithelium has been used as grafting material in different clinical situations, such as the treatment of burn wounds, chronic skin ulcers and oral mucosal defects. This technique allows growth of the epidermis from a single small skin biopsy to the quantity of the complete body's surface within 3–4 weeks. Isolated colonies of epithelial cells expand into broad sheets of undifferentiated epithelial cells and are then taken to secondary culture. Tremendous research efforts of many groups all over the world led to the formulation of perfectly defined commercially available media, which enable keratinocytes to be cultured without a feeder layer and serum free. It was the combination of the modern concepts of early staged burn debridement together with temporary skin cover and the application of expanded keratinocyte grafts multiplied in culture that gave rise to the hope that, in the future, each burn wound could be covered within 3–4 weeks [35, 40, 56, 63, 65, 66, 69, 74, 76, 169].

From the author's point of view, the development of in-vitro cultured skin substitutes since the very first clinical applications may be characterized following two principally different skin culture approaches:

– Cell culture of **multilayered epithelial transplants** (commonly termed "sheet grafts") [17, 18, 48, 55, 59, 66, 72, 74, 76, 111, 129, 138, 140, 141, 160, 166]
– Growth and construction of **composite multilayered dermal-epidermal analogues** [10, 14, 17, 19–23, 31, 32, 35, 36, 40, 43, 48, 52–54, 62, 65, 66, 74, 76, 78, 97, 100, 102, 129, 146, 148, 150, 153, 155–157, 163, 164, 168–170]
– Growth and transplantation of **pre-confluent cell grafts** (cultured or non-cultured) [15, 17, 34, 44, 50, 61, 73–78, 81, 84, 85, 95, 101, 102, 104, 106, 119, 154, 160, 163, 166]

The following summary highlights the most familiar techniques to produce and apply cultured skin substitutes, including currently available methods and research directions (Table 25.3). Regarding these different approaches—mainly based on clinical problems with cultured skin products—some implications for the further development and research can be concluded.

Table 25.3 Summary of possible skin substitute techniques utilizing cultured human keratinocytes with regard to the various possible designs that are currently used or experimentally developed

I. Autologous Cultured Human Keratinocytes
1. Autologous epidermal sheet transplants ("sheet grafts" = gold standard)
2. In-vitro cultured and constructed dermo-epidermal autologous transplants:
2.1. Keratinocytes on a collagen gel+fibroblasts
2.2. Keratinocyte sheets+collagen-glycosaminoglycan membrane+Fibroblasts
2.3. Keratinocyte sheets on a layer of fibrin-gel
2.4. Keratinocyte sheets on cell free pig dermis
2.5. Keratinocyte sheets on cell free human dermis
2.6. Keratinocytes on bovine or equine collagen matrices
2.7. Keratinocyte sheets on micro-perforated hyaluronic acid membranes
2.8. Keratinocyte sheets on collagen+chondroitin-6-sulfate with silicon membrane coverage (living skin equivalent)
3. Combination of allogeneic dermis (in vivo) with epidermal sheets
4. non-confluent keratinocyte suspensions
as a spray suspended in saline solutions
as spray or clots suspended in a fibrin matrix
4.1. exclusively
4.2. in clinical combination with fresh or preserved allogeneic skin
4.4. as non confluent keratinocyte monolayers on equine or bovine collagen matrices or on top of hyaluronic acid membranes
4.6. in combination with collagen-coated nylon on silicone backing
4.7. dissociated keratinocytes without cell culture
4.8. Outer root sheath cells (from plucked hair follicles) cultured or without culture
5. Three dimensional cell cluster cultures (spherocytes)
5.1. Cultured on microspheres as carrier systems (experimentally on: dextrane, collagen, hyaluronic acid)
5.2. Cell seeded microspheres+allografts/biomaterials

Table 25.3 *(continued)* Summary of possible skin substitute techniques utilizing cultured human keratinocytes with regard to the various possible designs that are currently used or experimentally developed

II. Allogeneic Keratinocytes
Allogeneic keratinocytes
6.1. Keratinocyte -sheets—(as a temporary wound cover)
6.2. Allogeneic keratinocyte suspensions (experimentally)
6.3. Syngenic-allogeneic keratinocytes
In-vitro constructed dermo-epidermal composites/analogues
6.4. Keratinocytes and fibroblasts (collagen matrices)

25.5
Cultured "Sheet Grafts" (Cultured Epithelial Autografts)

The growth of multilayered cultured epithelial sheets of human autologous keratinocytes (CEA = cultured epithelial autografts, so-called "sheet grafts") was the primary attempt to permanently substitute the patient's lost skin. Nevertheless, various factors have tempered a more widespread use of these grafts. Therefore, there is scientific controversial discussion about the best indications and advantages or disadvantages to using this technique [54, 115, 120, 141]. In particular, the extremely high costs have been considered as a practical problem. In a clinical trial, conventional meshed autografts were found to be superior to CEA for containing hospital cost and diminishing the length of stay in hospital, while decreasing the number of readmissions for reconstruction of contractures. In our own experience, a survey of the costs of survival of a patient with 88% TBSA (total body surface area) burns who received CEA were estimated to sum up to Euro 425,000 (currently US $673,497) [67, 83]. A review of relevant data from the literature revealed the cost of successful treatment of 1% of TBSA with CEA in 1995 to account for some Euro 148,131.70 (US $234,951.00 at today's exchange rates). In addition it has to be taken into account that the reported "take" (engraftment) rates of CEA and the necessity to regraft the treated areas are extremely variable and hard to figure out [100, 172].

One of the main causes of cultured graft failure is believed to be wound infection, i.e., clinically significant bacterial contamination. During the very first days in particular after CEA grafting, the fragile epithelium and the not yet fully developed dermoepidermal junction are much more likely to be damaged by the effects of bacterial infection than a meshed skin graft, which is more robust against bacterial colonization [85].

In the literature, there is a discrepancy between data from multicenter trials from the beginning era of CEA grafting, when high engraftment rates of up to 100% of cultured CEA were reported, and later reports from others and especially from single-centre experiences with a larger number of CEA-grafted patients. The latter reported remarkably lower take rates that ranked between 15% and 65% [35, 36, 40, 66, 100, 112, 118, 120, 137, 148, 149]. The true "take rate" derived from cumulative literature meta-analysis may therefore reach an average value of between 50% and 60% or less [51]. This is consistent with the original data of the pioneering works. The engraftment rate was initially reported to vary considerably between 0% and more than 80% take in adults and 50% take in children [36, 54].

Secondary devices are necessary to manipulate CEA, because of the few cell layers they are fragile and very delicate to handle [54]. The lack of adherence and the tendency to form blisters even months after the engraftment when exhibited to shearing forces are unsolved problems with this technique. Among other possible explanations, it is believed that the abnormal structure of the anchoring fibrils under culture conditions that are also affected by the enzymatic treatment of sheets to detach them from the culture flasks immediately before the grafting process may be one reason for CEA failing to adhere to the wound [31, 35, 77, 107, 126, 155–157]. Until now, no whole body surface but rather parts of the body have been successfully treated with CEA sheets in major burns.

CEA grafting alone does not solve the problem of the lack of dermis that is associated with third-degree burns or with full-thickness chronic wounds. Therefore, several attempts have been made to develop either dermal analogues or combinations of both epithelial cells and dermal substitutes, as well as other matrix cells such as fibroblasts [5, 7, 45, 78, 100, 116, 120, 153, 163, 168].

Because of the unexpected clinical performance problems with CEA, the temporary coverage of debrided wounds with other materials has been favoured. This procedure relies on the engraftment of at least parts of allografted dermis that remain after the immunogenic rejection process or after surgical removal of the allogenic epidermis [80]. In major burns, it has been shown that early excision and covering of wounds with allografts can keep the wound bed clean and well-vascularized, enhancing the likelihood of sheet graft take [69, 70, 77, 80, 103, 116, 141, 173]. Such temporary allogenic skin covers may serve as a biological barrier and prevent infection. It renders an in-vivo-culture wound healing environment after the surgical debridement. Subsequent autografting may be performed on the integrated parts while the allogenic epidermal part undergoes rejection. Donor shortage of such grafts is a common problem in most countries. When readily available xenografts (tissue from another species) are used, the problem of immediate rejection is not solved, and they cannot be permanently re-vascularized from the wound. This process leads to tissue break down and xenograft sloughing [75].

Observations of human allografts to be incorporated with their dermal parts have facilitated the use of extremely meshed or minced autografts under the protective layer of such biological dressings [79]. Even in patients with very limited donor sites, repeated grafting of considerably expanded autoskin sequential cover of major burn wounds can be achieved with a permanent result [4]. Limited socio-economic resources in various health systems over the world have led to further progress in using such conventional techniques and help circumvent the use of cultured skin substitutes whenever possible for reasons of cost. One of these developments is the advent of "microskin grafting" [69, 70, 77, 103, 116, 141, 173].

25.6
Cultured Human Cell Suspension Grafting

In as early as 1895, the German surgeon von Mangoldt published his observations on "epithelial cell seeding" to treat chronic wounds and wound cavities. He noted surprisingly good clinical results when using this approach [109]. Epithelial cells or cell clusters harvested by scraping off superficial epithelium from a patient's forearm with a surgical blade "until fibrin was exuded from the wound" were grafted to various wounds. A reduced donor site morbidity and a more regular aspect of the resurfaced wounds was claimed relative to the method of Reverdin. He also described his impression that single cells or cell clusters would better attach to a wound bed than would conventional pieces of skin.

Modifications of this method were published later but for the fear of epithelial cell cyst formation were not widely adopted [131]. Experimentally, epidermal cell suspensions were transplanted in saline solutions to full-thickness pig wounds, in 1952, but did not yield consistent results [24]. What stimulated further research was Hunyadi's report of successful transplantation of non-cultured keratinocytes, gained by trypsinization from biopsies and suspended in a fibrin matrix to heal chronic venous leg ulcers when suspending the cells in a fibrin sealant versus saline solution [74, 89, 90]. Fibrin is a naturally occurring substrate that plays a key role in all wound healing processes. Positive results with the use of fibrin sealant to fix skin grafts on burns and other wounds have been published [6, 9, 16, 60, 71, 74, 76, 81, 89–91, 129, 162]. It has been shown in cell culture studies that the relative percentage of holoclones, meroclones and paraclones of basal keratinocytes can be maintained in keratinocytes that are cultivated on fibrin. Also, fibrin has been shown not to induce clonal conversion, and this means there is not a loss of epidermal stem cells. The clonogenic ability of keratinocytes, their growth rate, and the long-term proliferative potential are not adversely affected by such culture systems. When fibrin-cultured autografts bearing stem cells are applied on massive full-thickness burns, the "take rate" of such suspended keratinocytes is reproducible and permanent. In addition, the fibrin adhesive allows a significant reduction in cost of cultured autografts and obviously eliminates some of the problems related to their handling and to the difficult transportation [129].

We have described repeatedly in experimental and clinical trials, with 6 patients and 14 transplantations, that extensive burned areas up to 88% TBSA can be successfully covered with a cultured keratinocyte-fibrin-sealant suspension using a commercially available, two-component fibrin sealant [21, 67, 72, 74, 76, 84, 101, 154]. Such KFSs (=keratinocyte-fibrin suspensions) were available to be grafted within only 10 days after seeding. This is due to the fact that no epidermal differentiation is necessary for single cell suspensions, whereas standard CEAs are available only after 3 weeks, which is critical for the survival of severely burned patients [73, 112].

Clinical findings of a lack of dermal structures when utilizing keratinocytes alone have led to the combination of a preliminary wound bed preparation by allografts, which is then followed by subsequent KFS transplantation together with meshed split-thickness allograft skin as an overlay. This additionally helps to secure the grafted cells and protects them from the impending desiccation during the healing process. While the allografts heal initially with clearly visible signs of a revascularization similar to autologous skin grafts, a mild and progressive immunogenic rejection period is noted after 12–14 days [86]. We observed a stable wound coverage within two more weeks, when the epithelial allograft parts had been rejected. This multimodal approach induced resilient wound closure, even over stress-prone areas such as the knees and elbow joints, and good mechanical stability in contrast to the effects noted after simple epithelial grafting without allografts [21, 67, 72, 74, 76, 154].

Various authors have fostered the use of allogenic keratinocytes as a source of readily available grafts of cultured epithelium [49, 136]. Superficial second-degree wounds in those severely injured by burns are thought to be immediately covered by such cultured allografts. Also, third-degree wounds may be covered with a biological skin substitute until sufficient amounts of autografts are available again. Skin graft donor sites have been reported to heal faster and allow for repeated harvest of thin skin grafts. However, the time for such allogenic cells or substitutes to persist in the wound and the timely course of rejection remains unclear and is subject to controversial discussions [35, 40, 56, 63, 65, 66, 69, 76–78, 112, 115, 117, 118, 124, 141, 143, 169]. Commercially available bilayer skin substitutes have been reported also to support keratinocyte cell suspensions until their engraftment in full-thickness animal wounds. It must be shown to what extent human keratinocytes will behave similarly when applied to human burn wounds under clinical conditions [95].

25.6.1
Membrane Cell Delivery Systems

A cell membrane delivery system is a means of basic mechanical support to culture keratinocytes and support the transportation of such cells to wound beds while protecting such cell grafts in the early post-transplantation period [100, 132]. Various systems have been published that are based on either biological or alloplastic carrier materials. At the moment they have not been widely adopted in clinical practice [106, 166].

25.6.2
Alloplastic or Mixed Synthetic-Biological Cell Carriers

Problems with purely epithelial cell grafts have induced the combination of cultured autologous keratinocytes with various alloplastic dermal regeneration templates [95, 171]. Yannas and coworkers reported in 1989 that trypsinized whole-skin-cell suspensions (uncultured keratinocytes and fibroblasts) centrifuged into a collagen-glycosaminoglycan (C-GAG) matrix and grafted onto guinea pigs were able to facilitate the regeneration of a healthy epidermis [171]. Clinically, the centrifugation of cells into the membrane and the combination of cell-seeding techniques together with commercially available off-the-shelf products with a number of regulatory questions have hindered a more widespread use of these methods when compared with those that rely only on the material from individual laboratories [81, 100].

For many years, a well-known example of compound materials has been propagated for reconstructive and burn surgery [95]. It consists of a bilayer membrane product with (1) a well-characterized, so-called "dermal portion" that consists of a porous lattice of fibres of a cross-linked bovine collagen and GAG that is supposed to be replaced by new collagen synthesized by fibroblasts that grow in on full-thickness wounds [27–30, 122] and (2) a so-called "epidermal" cover of synthetic polysiloxane polymer (silicone). It is propagated so that once the collagen layer has been integrated into the wound bed, the silicone layer may either start to separate spontaneously or has to be peeled away and replaced by an ultra-thin, split-thickness skin graft (SSG) of approximately 0.1 mm. Cultured autologous keratinocyte sheets have been proposed as an alternative form of definitive wound cover for this manoeuvre, but results have not yet been consistent [95]. In experiments from our group, cultured keratinocytes seeded directly into such a bilayer matrix performed well in the laboratory. However, they did not succeed in nude mice full-thickness wounds, whereas others have reported on the reformation of epithelium in full-thickness mouse wounds [32, 95, 151].

To further improve the handling of cultured non-confluent grafts and accelerate the availability, we have investigated pre-confluent cultured monolayers of human keratinocytes on synthetic polymeric membranes or on polyurethane. Monolayer grafting would allow for early coverage of excised burn wounds within only a few days of receiving the biopsy. When such films are inverted onto the wound, it is thought that the keratinocytes migrate to the recipient wound bed to form an epithelium [106]. In order to maintain a single layer of keratinocytes with basal characteristics, serum-free culture is recommended for this type of cell transfer.

25.6.3
Cultured Cells and Biological Carriers

Pre-confluent human keratinocyte monolayer grafts have also been used in combination with biological carrier materials. In contrast to the most published approach to mimic human skin using bilayered skin substitutes, proliferating basal cells which are responsible for the initial reformation of an epithelium can be transplanted upside down with the carrier on top as a bio-dressing. From our previous studies [76, 101] and clinical trials [94], we concluded the combination of different surgical approaches and culture techniques (such as allografting together with simultaneously delivering cultured human keratinocytes) to be of potential benefit [17, 72, 74, 76–78, 86, 101, 104] (Fig. 25.1). Using fibrin sealant as a biological cell carrier, it has been published that spray techniques optimize cell delivery and do not harm cell viability [50]. This enables the dispersion and distribution of cultured cells to a maximum surface unlike our initial traditional approach without spray systems. We have gained similar insights experimentally [78] and clinically. Ronfard and co-workers cultured keratinocytes on a stabilized fibrin sealant in the gel phase to optimize keratinocyte growth and delivery [140]. This technique offers a reliable and simple method to deliver keratinocytes and has been demonstrated both experimentally and clinically [78]. Because this method was developed mainly for burn victims, it has not been widely adopted by others for wound treatment [175].

To avoid the enzymatic detachment from the culture dishes before the grafting procedure as a potentially harmful action to the anchoring fibrils of cultured epithelial cells [35, 37], growing epithelia on dermal matrices has been attempted to facilitate the handling procedure, and at the same time to deliver a dermal analogue. One of the most common materials has been C-GAG, with or without the cover of a gas-permeable silastic membrane that serves as a barrier to fluid loss [124]. The question whether dermal fibroblasts seeded into such composites are necessary has not yet been adequately answered [5, 14, 25, 43, 64, 123–125, 133, 135, 137, 150, 158, 165, 167, 173, 175].

One of the problems with such types of composite grafts is the distance between the matrix material and the keratinocyte layer; the graft acts as a barrier towards nutrients necessary for keratinocyte survival on top of such composites. The short- and long-term clinical efficacy of these grafts remains to be shown (Fig. 25.2).

Fig. 25.1 Secondary culture of confluent human keratinocytes on a collagen type-I carrier in a non-stratified monolayer

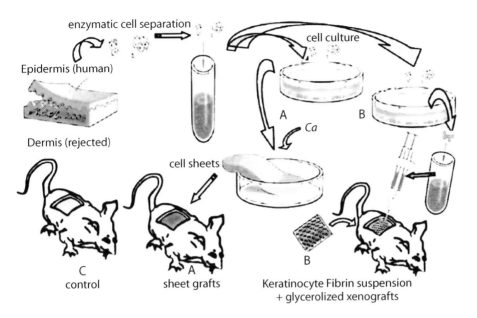

Fig. 25.2 Experimental setup of testing the performance of tissue-engineered skin substitutes in full-thickness nude mouse skin wounds. Here, comparison of cultured and stratified epithelial sheet grafts after calcium addition (**A**) versus sub-confluently cultured non-stratified keratinocyte single cell suspensions in fibrin sealant (**B**) with xenogenic meshed allograft skin as an overlay versus control wounds (**C**); model mimicking the clinical reality in large burn wounds (modelled after Horch et al. 2005; [81])

"Upside-down" transplantation with the keratinocyte layer directed towards the wound bed and underneath the matrix, with the collagen component serving as a carrier and as a biological dressing on top utilizing sub-confluent monolayers of cultured human keratinocytes instead of multilayered sheet grafts has been shown to function experimentally [76]. Keratinocytes are thereby brought into the immediate neighbourhood of the recipient wound bed and are not lost during the engraftment process. In full-thickness nude mice wounds, we were able to show the feasibility of "upside-down" monolayer grafting procedures [78]. The advantage may be a simplified handling and a potentially higher number of transplanted cells that are more likely to survive in a more natural wound environment. Similar findings are known from classical surgical manoeuvres, such as buried chip skin grafting [68, 70, 99]. As with this method, with the combination of various approaches with the in vitro expansion of graftable cells together with the advantages of appropriate materials and grafting techniques, other strategies may well become feasible alternatives to the methods applied today in treating chronic problem wounds or major burns.

25.7
Outlook

Without doubt, the discoveries that multipotent cells can be isolated from many tissues of the body has led to the concept of potential organ-specific regeneration [33, 38, 39, 42, 47, 88, 92, 93, 127]. Loosely referred to as stem cells, these cells in the adult body resemble pluripotent cell populations, such as those derived from the early embryo, which may contribute to virtually any type of tissue under appropriate experimental or biological conditions [142]. In contrast to the situation for embryonic stem cells from undetermined early organisms without any history of differentiation, it is not yet understood why and how a population of "reserve stem cells" can retain their capability in the adult human organism [142].

It is likely that application of multipotent adult stem cells may become a feasible tool in skin regeneration when ways are discovered to influence the lineage of such cells and guide their growth and differentiation [2, 8, 11, 13, 105]. Since the source of epithelial regeneration stems from basal keratinocytes, these cells have therefore been recognized to behave similarly to adult stem cells [129]. The group of DeLuca and co-workers [142] has performed tremendous work to characterize the potential stem-cell-like behaviour of basal keratinocytes into holoclones and meroclones [139]. From these observations, it was concluded that the proliferative compartment of stratified squamous epithelia consists of stem and transient amplifying (TA) keratinocytes [7, 12, 72, 104]. Furthermore, this group also demonstrated that epithelial cells from the corneal limbus have the capacity to reform a corneal epithelium with translucency after a culture period and re-transplantation into the recipient eyes of patients [130]. By their description of the cell doubling rates observed under culture conditions, an intrinsic "cell doubling clock" has been postulated. Because of the clinical need for surgery of defects of the anterior ocular surface, new perspectives for autologous cultured conjunctival epithelium therapies have been suggested [14, 128, 129].

Within the last decade, many hopes have been raised by the technique of gene therapy when combined with keratinocyte culture as a tool to improve cultured skin substitute performance [26]. It is believed to accelerate cell growth during the in vitro and in vivo process and has been addressed to modulate difficult-to-heal wounds [3, 7, 72]. Currently, various practical approaches and scientific principles to replace skin temporarily or permanently are advancing at a rapid rate. Although both research in this area and the clinical application of cultured human skin substitutes are expensive, further progress to optimize skin substitute performance using tissue engineering procedures is necessary. Improved cosmesis and the ultimate regaining of lost skin functions (such as sensitivity, elasticity, normal physiological sweat gland and dermal appendage function), as well as regaining a normal pigmentation with invisible scars are the ultimate goals of future endeavours in this field.

The evolving and steadily growing insights to and the concepts of modern tissue engineering applied to dermal replacement following burn injury or dealing with chronic wounds may well overcome many of today's limits to substitute skin [96]. It is our true belief that through both basic and clinical research, major improvements will be made to understand and effectively deal with the hitherto unsolved problems of wound healing. The aim should be to replace with a truly functional skin inclusive of all dermal appendages that are of permanent skin-like quality. As we stated earlier, perhaps the definitive breakthrough will come through the ability to replace worn out, defective or damaged body parts using technologies that resemble natural regeneration.

References

1. Achauer BM, Martinez SE. Burn wound pathophysiology and care. Crit Care Clin 1 (1): 47–58, 1985.
2. Adams SW, Wang L, Fortney J, Gibson LF. Etoposide differentially affects bone marrow and dermal derived endothelial cells. J Cell Mol Med 8 (3): 338–48, 2004.
3. Agrawal N, You H, Liu Y, Chiriva-Internati M, Bremner J, Garg T, Grizzi F, Krishna Prasad C, Mehta JL, Hermonat PL. Generation of recombinant skin in vitro by adeno-associated virus type 2 vector transduction. Tissue Eng 10 (11–12): 1707–15, 2004.
4. Alexander J, MacMillan B, Law E, Kittur D. Treatment of severe burns with widely meshed skin autograft and meshed skin allograft overlay. J Trauma 21: 433–438, 1981.
5. Alsbjorn B. In search of an ideal skin substitute. Scand J Plast Reconstr Surg 18 (1): 127–33, 1984.
6. Altmeppen J, Hansen E, Bonnlander GL, Horch RE, Jeschke MG. Composition and characteristics of an autologous thrombocyte gel. J Surg Res 117 (2): 202–7, 2004.
7. Andree C, Voigt M, Wenger A, Erichsen T, Bittner K, Schaefer D, Walgenbach KJ, Borges J, Horch RE, Eriksson E, Stark GB. Plasmid gene delivery to human keratinocytes through a fibrin-mediated transfection system. Tissue Eng 7 (6): 757–66, 2001.
8. Anglani F, Forino M, Del Prete D, Tosetto E, Torregrossa R, D'Angelo A. In search of adult renal stem cells. J Cell Mol Med 8 (4): 474–87, 2004.
9. Archambault M, Yaar M, Gilchrest BA. Keratinocytes and fibroblasts in a human skin equivalent model enhance melanocyte survival and melanin synthesis after ultraviolet irradiation. J Invest Dermatol 104 (5): 859–67, 1995.
10. Auger FA, Rouabhia M, Goulet F, Berthod F, Moulin V, Germain L. Tissue-engineered human skin substitutes developed from collagen-populated hydrated gels: clinical and fundamental applications. Med Biol Eng Comput 36 (6): 801–12, 1998.
11. Badiu C. Genetic clock of biologic rhythms. J Cell Mol Med 7 (4): 408–16, 2003.

12. Bajaj B, Behshad S, Andreadis ST. Retroviral gene transfer to human epidermal keratinocytes correlates with integrin expression and is significantly enhanced on fibronectin. Hum Gene Ther 13 (15): 1821–31, 2002.
13. Baksh D, Song L, Tuan RS. Adult mesenchymal stem cells: characterization, differentiation, and application in cell and gene therapy. J Cell Mol Med 8 (3): 301–16, 2004.
14. Balasubramani M, Kumar TR, Babu M. Skin substitutes: a review. Burns 27 (5): 534–44, 2001.
15. Bannasch H, Fohn M, Unterberg T, Bach AD, Weyand B, Stark GB. Skin tissue engineering. Clin Plast Surg 30 (4): 573–9, 2003.
16. Bannasch H, Fohn M, Unterberg T, Knam F, Weyand B, Stark GB. Skin tissue engineering. Chirurg 74 (9): 802–7, 2003.
17. Bannasch H, Horch RE, Tanczos E, Stark GB. Treatment of chronic wounds with cultured autologous keratinocytes as suspension in fibrin glue. Zentralbl Chir 125 Suppl 1: 79–81, 2000.
18. Bannasch H, Kontny U, Kruger M, Stark GB, Niemeyer CM, Brandis M, Horch RE. A semisynthetic bilaminar skin substitute used to treat pediatric full-body toxic epidermal necrolysis: wraparound technique in a 17-month-old girl. Arch Dermatol 140 (2): 160–2, 2004.
19. Beele H. Artificial skin: past, present and future. Int J Artif Organs 25 (3): 163–73, 2002.
20. Bell E, Ehrlich HP, Buttle DJ, Nakatsuji T. Living tissue formed in vitro and accepted as skin-equivalent tissue of full thickness. Science 211 (4486): 1052–4, 1981.
21. Bell E, Sher S, Hull B. The living skin-equivalent as a structural and immunological model in skin grafting. Scan Electron Microsc (Pt 4): 1957–62, 1984.
22. Bell E, Sher SE, Hull BE, Sarber RL, Rosen S. Long-term persistance in experimental animals of components of skin-equivalent grafts fabricated in the laboratory. Adv Exp Med Biol 172: 419–33, 1984.
23. Benathan M, Labidi-Ubaldi F. Living epidermal and dermal substitutes for treatment of severely burned patients. Rev Med Suisse Romande 118 (2): 149–53, 1998.
24. Billingham R, Reynolds J. Transplantation studies on sheet of pure epidermal epithelium and of epidermal cell suspensions. Br J Plast Surg 23: 25–32, 1952.
25. Black AF, Berthod F, L'Heureux N, Germain L, Auger FA. In vitro reconstruction of a human capillary-like network in a tissue-engineered skin equivalent. Faseb J 12 (13): 1331–40, 1998.
26. Bleiziffer O, Eriksson E, Yao F, Horch RE, Kneser U. Gene transfer strategies in tissue engineering. J Cell Mol Med 11 (2): 206–23, 2007.
27. Boyce ST. Design principles for composition and performance of cultured skin substitutes. Burns 27 (5): 523–33, 2001.
28. Boyce ST. Skin substitutes from cultured cells and collagen-GAG polymers. Med Biol Eng Comput 36 (6): 791–800, 1998.
29. Boyce ST, Foreman TJ, English KB, Stayner N, Cooper ML, Sakabu S, Hansbrough JF. Skin wound closure in athymic mice with cultured human cells, biopolymers, and growth factors. Surgery 110 (5): 866–76, 1991.
30. Boyce ST, Glatter R, Kitzmiller WJ. Case studies: treatment of chronic wounds with cultured skin substitutes. Ostomy Wound Manage 41 (2): 26–8, 30, 32 passim, 1995.
31. Breitkreutz D, Bohnert A, Herzmann E, Bowden PE, Boukamp P, Fusenig NE. Differentiation specific functions in cultured and transplanted mouse keratinocytes: environmental influences on ultrastructure and keratin expression. Differentiation 26 (2): 154–69, 1984.
32. Burke JF, Yannas IV, Quinby WC, Jr, Bondoc CC, Jung WK. Successful use of a physiologically acceptable artificial skin in the treatment of extensive burn injury. Ann Surg 194 (4): 413–28, 1981.
33. Chivu M, Diaconu CC, Brasoveanu L, Alexiu I, Bleotu C, Banceanu G, Miscalencu D, Cernescu C. Ex vivo differentiation of umbilical cord blood progenitor cells in the presence of placental conditioned medium. J Cell Mol Med 6 (4): 609–20, 2002.
34. Cohen M, Bahoric A, Clarke HM. Aerosolization of epidermal cells with fibrin glue for the epithelialization of porcine wounds with unfavorable topography. Plast Reconstr Surg 107 (5): 1208–15, 2001.
35. Compton CC, Gill JM, Bradford DA, Regauer S, Gallico GG, O'Connor NE. Skin regenerated from cultured epithelial autografts on full-thickness burn wounds from 6 days to 5 years after grafting. A light, electron microscopic and immunohistochemical study. Lab Invest 60 (5): 600–12, 1989.
36. Compton CC, Nadire KB, Regauer S, Simon M, Warland G, O'Connor NE, Gallico GG, Landry DD. Cultured human sole-derived keratinocyte grafts re-express site-specific differentiation after transplantation. Differentiation 64 (1): 45–53, 1998.
37. Compton CC, Press W, Gill JM, Bantick G, Nadire KB, Warland G, Fallon JT, 3rd, Vamvakas EC. The generation of anchoring fibrils by epidermal keratinocytes: a quantitative long-term study. Epithelial Cell Biol 4 (3): 93–103, 1995.
38. Constantinescu S. Stemness, fusion and renewal of hematopoietic and embryonic stem cells. J Cell Mol Med 7 (2): 103–12, 2003.
39. Constantinescu SN. Stem cell generation and choice of fate: role of cytokines and cellular microenvironment. J Cell Mol Med 4 (4): 233–248, 2000.
40. De Luca M, Albanese E, Bondanza S, Megna M, Ugozzoli L, Molina F, Cancedda R, Santi PL, Bormioli M, Stella M, et al. Multicentre experience in the treatment of burns with autologous and allogenic cultured epithelium, fresh or preserved in a frozen state. Burns 15 (5): 303–9, 1989.
41. Demling RH, DeSanti L. Management of partial thickness facial burns (comparison of topical antibiotics and bio-engineered skin substitutes). Burns 25 (3): 256–61, 1999.
42. Doss MX, Koehler CI, Gissel C, Hescheler J, Sachinidis A. Embryonic stem cells: a promising tool for cell replacement therapy. J Cell Mol Med 8 (4): 465–73, 2004.
43. Dubertret L, Coulomb B. Reconstruction of human skin in culture. C R Seances Soc Biol Fil 188 (3): 235–44, 1994.

44. Dubertret L, Coulomb B, Saiag P, Bertaux B, Lebreton C, Heslan M, Breitburd F, Bell E, Baruch J, Guilbaud J. Reconstruction in vitro of a human living skin equivalent. Life Support Syst 3 Suppl 1: 380–7, 1985.
45. Eaglstein WH, Iriondo M, Laszlo K. A composite skin substitute (graftskin) for surgical wounds. A clinical experience. Dermatol Surg 21 (10): 839–43, 1995.
46. Ehrlich HP. Control of wound healing from connective tissue aspect. Chirurg 66 (3): 165–73, 1995.
47. Filip S, English D, Mokry J. Issues in stem cell plasticity. J Cell Mol Med 8 (4): 572–7, 2004.
48. Foyatier JL, Faure M, Hezez G, Masson C, Paulus C, Chomel P, Latarjet J, Delay E, Thomas L, Adam C, et al. Clinical application of grafts of cultured epidermis in burn patients. Apropos of 16 patients. Ann Chir Plast Esthet 35 (1): 39–46, 1990.
49. Fratianne R, Schafer IA. Keratinocyte Allografts Act as a Biological Bandage to Accelerate Healing of Split Thickness Donor Sites. In: Horch RE, Munster AM, Achauer B, eds. Cultured Human Keratinocytes and Tissue Engineered Skin Substitutes. Stuttgart: Thieme, pp. 316–325, 2001.
50. Fraulin FO, Bahoric A, Harrop AR, Hiruki T, Clarke HM. Autotransplantation of epithelial cells in the pig via an aerosol vehicle. J Burn Care Rehabil 19 (4): 337–45, 1998.
51. Freising C, Horch RE. Clinical results of cultivated keratinocyzes to treat burn injuries—a metaanalysis. In: Horch RE, Munster AM, Achauer B, eds. Cultured Human Keratinocytes and Tissue Engineered Skin Substitutes. Stuttgart: Thieme, pp. 220–226, 2001.
52. Gallico GG, 3rd. Biologic skin substitutes. Clin Plast Surg 17 (3): 519–26, 1990.
53. Gallico GG, 3rd, O'Connor NE. Cultured epithelium as a skin substitute. Clin Plast Surg 12 (2): 149–57, 1985.
54. Gallico GG, 3rd, O'Connor NE, Compton CC, Kehinde O, Green H. Permanent coverage of large burn wounds with autologous cultured human epithelium. N Engl J Med 311 (7): 448–51, 1984.
55. Gallico Gr, O'Connor N, Compton C, Kehinde O, H G. Permanent coverage of large burn wounds with autologous cultured human epithelium. N Engl J Med 311 (7): 448–451, 1984.
56. Green H, Kehinde O, Thomas J. Growth of cultured human epidermal cells into multiple epithelia suitable for grafting. Proc Natl Acad Sci U S A 76 (11): 5665–8, 1979.
57. Greenfield E, Jordan B. Advances in burn wound care. Crit Care Nurs Clin North Am 8 (2): 203–15, 1996.
58. Griffiths M, Ojeh N, Livingstone R, Price R, Navsaria H. Survival of Apligraf in acute human wounds. Tissue Eng 10 (7–8): 1180–95, 2004.
59. Grossman N, Slovik Y, Bodner L. Effect of donor age on cultivation of human oral mucosal keratinocytes. Arch Gerontol Geriatr 38 (2): 114–22, 2004.
60. Gyulai R, Hunyadi J, Kenderessy-Szabo A, Kemeny L, Dobozy A. Chemotaxis of freshly separated and cultured human keratinocytes. Clin Exp Dermatol 19 (4): 309–11, 1994.
61. Hafez AT, Bagli DJ, Bahoric A, Aitken K, Smith CR, Herz D, Khoury AE. Aerosol transfer of bladder urothelial and smooth muscle cells onto demucosalized colonic segments: a pilot study. J Urol 169 (6): 2316–9; discussion 2320, 2003.
62. Harriger MD, Supp AP, Swope VB, Boyce ST. Reduced engraftment and wound closure of cryopreserved cultured skin substitutes grafted to athymic mice. Cryobiology 35 (2): 132–42, 1997.
63. Heimbach D, Luterman A, Burke J, Cram A, Herndon D, Hunt J, Jordan M, McManus W, Solem L, Warden G, et al. Artificial dermis for major burns. A multi-center randomized clinical trial. Ann Surg 208 (3): 313–20, 1988.
64. Hergrueter CA, O'Connor NE. Skin substitutes in upper extremity burns. Hand Clin 6 (2): 239–42, 1990.
65. Herndon DN, Rutan RL. Comparison of cultured epidermal autograft and massive excision with serial autografting plus homograft overlay. J Burn Care Rehabil 13 (1): 154–7, 1992.
66. Hickerson WL, Compton C, Fletchall S, Smith LR. Cultured epidermal autografts and allodermis combination for permanent burn wound coverage. Burns 20 Suppl 1: S52–5; discussion S55–6, 1994.
67. Horch R, GB. Economy of skin grafting in burns. Hospital—J Eur Assoc Hosp Man 3: 6–9, 2001.
68. Horch R, Roosen J. Anal cancer: current concepts and treatment results. Nippon Geka Hokan 58 (5): 391–7, 1989.
69. Horch R, Stark GB, Kopp J, Spilker G. Cologne Burn Centre experiences with glycerol-preserved allogeneic skin: Part I: Clinical experiences and histological findings (overgraft and sandwich technique). Burns 20 Suppl 1: S23–6, 1994.
70. Horch R, Stark GB, Spilker G. [Treatment of perianal burns with submerged skin particles]. Zentralbl Chir 119 (10): 722–5, 1994.
71. Horch RE. Future perspectives in tissue engineering. J Cell Mol Med 10 (1): 4–6, 2006.
72. Horch RE, Andree C, Kopp J, Tanczos E, Voigt M, Bannasch H, Walgenbach KJ, Dai FP, Bittner K, Galla TJ, Stark GB. Gene therapy perspectives in modulation of wound healing. Zentralbl Chir 125 Suppl 1: 74–8, 2000.
73. Horch RE, Bannasch H, Kopp J, Andree C, Ihling C, Stark GB. Keratinocytes suspended in fibrin glue (KFGS) restore dermo-epidermal junction better than conventional sheet grafts (CEG). Plast Surg Forum 19: 23–25, 1996.
74. Horch RE, Bannasch H, Kopp J, Andree C, Stark GB. Single-cell suspensions of cultured human keratinocytes in fibrin-glue reconstitute the epidermis. Cell Transplant 7 (3): 309–17, 1998.
75. Horch RE, Bannasch H, Stark GB. Combined grafting of cultured human keratinocytes as a single cell suspension in fibrin glue and preserved dermal grafts enhances skin reconstitution in athymic mice full-thickness wounds. Eur J Plast Surg 22: 237–243, 1999.
76. Horch RE, Bannasch H, Stark GB. Transplantation of cultured autologous keratinocytes in fibrin sealant biomatrix to resurface chronic wounds. Transplant Proc 33 (1–2): 642–4, 2001.

77. Horch RE, Corbei O, Formanek-Corbei B, Brand-Saberi B, Vanscheidt W, Stark GB. Reconstitution of basement membrane after "sandwich-technique" skin grafting for severe burns demonstrated by immunohistochemistry. J Burn Care Rehabil 19 (3): 189–202, 1998.
78. Horch RE, Debus M, Wagner G, Stark GB. Cultured human keratinocytes on type I collagen membranes to reconstitute the epidermis. Tissue Eng 6 (1): 53–67, 2000.
79. Horch RE, Jeschke MG, Spilker G, Herndon DN, Kopp J. Treatment of second degree facial burns with allografts—preliminary results. Burns 31 (5): 597–602, 2005.
80. Horch RE, Jeschke MG, Spilker G, Herndon DN, Kopp J. Treatment of second degree facial burns with allografts—preliminary results. Burns 31: 225–231, 2005.
81. Horch RE, Kopp J, Kneser U, Beier J, Bach AD. Tissue engineering of cultured skin substitutes. J Cell Mol Med 9 (3): 592–608, 2005.
82. Horch RE, Munster AM, Achauer B. Cultured human keratinocytes and tissue engineered skin substitutes. Stuttgart: Thieme, 2001.
83. Horch RE, Stark GB. Economy of skin grafting in burns. Hospital—J Eur Assoc Hosp Man 3: 6–9, 2001.
84. Horch RE, Stark GB, Kopp J. Histology after grafting of cultured keratinocyte-fibrin-glue-suspension (KFGS) with allogenic split-thickness-skin (STS) overlay. ellipse 11: 21–26, 1994.
85. Horch RE, Stark GB, Kopp J, Andree C. Dermisersatz nach drittgradigen Verbrennungen und bei chronischen Wunden—Neue Erkenntnisse zur Morphologie nach Fremdhauttransplantation in Kombination mit kultivierten Keratinozyten. Transplantationsmedizin 7: 99–103, 1995.
86. Horch RE, Stark GB, Kopp J, Spilker G. Cologne Burn Centre experiences with glycerol-preserved allogeneic skin: Part I: Clinical experiences and histological findings (overgraft and sandwich technique). Burns 20 Suppl 1: S23–6, 1994.
87. Horch RE, Stark GB, Spilker G. Treatment of perianal burns with submerged skin particles. Zentralbl Chir 119 (10): 722–5, 1994.
88. Hristov M, Weber C. Endothelial progenitor cells: characterization, pathophysiology, and possible clinical relevance. J Cell Mol Med 8 (4): 498–508, 2004.
89. Hunyadi J, Farkas B, Bertenyi C, Olah J, Dobozy A. Keratinocyte grafting: a new means of transplantation for full-thickness wounds. J Dermatol Surg Oncol 14 (1): 75–8, 1988.
90. Hunyadi J, Farkas B, Bertenyi C, Olah J, Dobozy A. Keratinocyte grafting: covering of skin defects by separated autologous keratinocytes in a fibrin net. J Invest Dermatol 89 (1): 119–20, 1987.
91. Hunyadi J, Farkas B, Olah J, Bertenyi C, Dobozy A. Keratinocyte transplantation: covering of skin defects with autologous keratinocytes. Orv Hetil 128 (46): 2409–11, 1987.
92. Iwami Y, Masuda H, Asahara T. Endothelial progenitor cells: past, state of the art, and future. J Cell Mol Med 8 (4): 488–97, 2004.
93. Jansen J, Hanks S, Thompson JM, Dugan MJ, Akard LP. Transplantation of hematopoietic stem cells from the peripheral blood. J Cell Mol Med 9 (1): 37–50, 2005.
94. Johnsen S, Ermuth T, Tanczos E, Bannasch H, Horch RE, Zschocke I, Peschen M, Schopf E, Vanscheidt W, Augustin M. Treatment of therapy-refractive ulcera cruris of various origins with autologous keratinocytes in fibrin sealant. Vasa 34 (1): 25–9, 2005.
95. Jones I, Currie L, Martin R. A guide to biological skin substitutes. Br J Plast Surg 55 (3): 185–93, 2002.
96. Kneser U, Arkudas A, Ohnolz J, Heidner K, Bach AD, Kopp J, Horch RE. Vascularized Bone Replacement for the Treatment of Chronic Bone Defects—Initial Results of Microsurgical Solid Matrix Vascularization. EWMA Journal 5: 14–19, 2005.
97. Kogan L, Govrin-Yehudain J. Vertical (two-layer) skin grafting: new reserves for autologic skin. Ann Plast Surg 50 (5): 514–6, 2003.
98. Kohnlein HE. Skin transplantation and skin substitutes. Langenbecks Arch Chir 327: 1090–106, 1970.
99. Kopp J, Bach AD, Kneser U, Polykandriotis E, Loos B, Krickhahn M, Jeschke MG, Horch RE. Indication and clinical results of buried skin grafting to treat problematic wounds. Zentralbl Chir 129 Suppl 1: S129–32, 2004.
100. Kopp J, Jeschke M, Bach A, Kneser U, Horch R. Applied Tissue Engineering in the Closure of severe Burns and Chronic wounds using cultured human autologous keratinocytes in a natural Fibrin Matrix. Cells Tissue Banking 5: 212–217, 2004.
101. Kopp J, Jeschke MG, Bach AD, Kneser U, Horch RE. Applied tissue engineering in the closure of severe burns and chronic wounds using cultured human autologous keratinocytes in a natural fibrin matrix. Cell Tissue Bank 5 (2): 89–96, 2004.
102. Kopp J, Jiao XY, Bannasch H, Horch RE, Nagursky H, Voigt M, Stark GB. Membrane Cell Grafts (MCG), fresh and frozen, to cover full thickness wounds in athymic nude mice. Eur J Plast Surg 22: 213–219, 1999.
103. Kopp J, Wang GY, Horch RE, Pallua N, Ge SD. Ancient traditional Chinese medicine in burn treatment: a historical review. Burns 29 (5): 473–8, 2003.
104. Kopp J, Wang GY, Kulmburg P, Schultze-Mosgau S, Huan JN, Ying K, Seyhan H, Jeschke MD, Kneser U, Bach AD, Ge SD, Dooley S, Horch RE. Accelerated wound healing by in vivo application of keratinocytes overexpressing KGF. Mol Ther 10 (1): 86–96, 2004.
105. Kuehnle I, Goodell MA. The therapeutic potential of stem cells from adults. Bmj 325 (7360): 372–6, 2002.
106. Levi M, Friederich PW, Middleton S, de Groot PG, Wu YP, Harris R, Biemond BJ, Heijnen HF, Levin J, ten Cate JW. Fibrinogen-coated albumin microcapsules reduce bleeding in severely thrombocytopenic rabbits. Nat Med 5 (1): 107–11, 1999.
107. Lin S, Lai C, Chou C, Tsai C, Wu K, Chang C. Microskin autograft with pigskin xenograft overlay: a preliminary report of studies on patients. Burns 18: 321–325, 1992.
108. Manea A, Constantinescu E, Popov D, Raicu M. Changes in oxidative balance in rat pericytes exposed to diabetic conditions. J Cell Mol Med 8 (1): 117–26, 2004.
109. Mangoldt F. Die Überhäutung von Wundflächen und Wundhöhlen durch Epithelaussaat, eine neue Methode der Transplantation. Deut Med Wschr 21: 798–799, 1895.

110. McNulty JM, Kambour MJ, Smith AA. Use of an improved zirconyl hematoxylin stain in the diagnosis of Barrett's esophagus. J Cell Mol Med 8 (3): 382–7, 2004.
111. Meana A, Iglesias J, Del Rio M, Larcher F, Madrigal B, Fresno MF, Martin C, San Roman F, Tevar F. Large surface of cultured human epithelium obtained on a dermal matrix based on live fibroblast-containing fibrin gels. Burns 24 (7): 621–30, 1998.
112. Munster AM. Cultured skin for massive burns. A prospective, controlled trial. Ann Surg 224 (3): 372–5; discussion 375–7, 1996.
113. Munster AM. New horizons in surgical immunobiology. Host defence mechanisms in burns. Ann R Coll Surg Engl 51 (2): 69–80, 1972.
114. Munster AM. Use of cultured epidermal autograft in ten patients. J Burn Care Rehabil 13 (1): 124–6, 1992.
115. Munster AM. Whither [corrected] skin replacement? Burns 23 (1): v, 1997.
116. Munster AM, Smith-Meek M, Shalom A. Acellular allograft dermal matrix: immediate or delayed epidermal coverage? Burns 27 (2): 150–3, 2001.
117. Munster AM, Smith-Meek M, Sharkey P. The effect of early surgical intervention on mortality and cost-effectiveness in burn care, 1978–91. Burns 20 (1): 61–4, 1994.
118. Munster AM, Weiner SH, Spence RJ. Cultured epidermis for the coverage of massive burn wounds. A single center experience. Ann Surg 211 (6): 676–9; discussion 679–80, 1990.
119. Myers SR, Grady J, Soranzo C, Sanders R, Green C, Leigh IM, Navsaria HA. A hyaluronic acid membrane delivery system for cultured keratinocytes: clinical "take" rates in the porcine kerato-dermal model. J Burn Care Rehabil 18 (3): 214–22, 1997.
120. Nanchahal J, Dover R, Otto WR. Allogeneic skin substitutes applied to burns patients. Burns 28 (3): 254–7, 2002.
121. Nanchahal J, Ward CM. New grafts for old? A review of alternatives to autologous skin. Br J Plast Surg 45 (5): 354–63, 1992.
122. O'Brien FJ, Harley BA, Yannas IV, Gibson LJ. The effect of pore size on cell adhesion in collagen-GAG scaffolds. Biomaterials 26 (4): 433–41, 2005.
123. Ojeh NO, Frame JD, Navsaria HA. In vitro characterization of an artificial dermal scaffold. Tissue Eng 7 (4): 457–72, 2001.
124. Orgill DP, Straus FH, 2nd, Lee RC. The use of collagen-GAG membranes in reconstructive surgery. Ann N Y Acad Sci 888: 233–48, 1999.
125. Ozerdem OR, Wolfe SA, Marshall D. Use of skin substitutes in pediatric patients. J Craniofac Surg 14 (4): 517–20, 2003.
126. Parisel C, Saffar L, Gattegno L, Andre V, Abdul-Malak N, Perrier E, Letourneur D. Interactions of heparin with human skin cells: binding, location, and transdermal penetration. J Biomed Mater Res 67A (2): 517–23, 2003.
127. Paunescu V, Suciu E, Tatu C, Plesa A, Herman D, Siska IR, Suciu C, Crisnic D, Nistor D, Tanasie G, Bunu C, Raica M. Endothelial cells from hematopoietic stem cells are functionally different from those of human umbilical vein. J Cell Mol Med 7 (4): 455–60, 2003.
128. Pellegrini G, Dellambra E, Paterna P, Golisano O, Traverso CE, Rama P, Lacal P, De Luca M. Telomerase activity is sufficient to bypass replicative senescence in human limbal and conjunctival but not corneal keratinocytes. Eur J Cell Biol 83 (11–12): 691–700, 2004.
129. Pellegrini G, Ranno R, Stracuzzi G, Bondanza S, Guerra L, Zambruno G, Micali G, De Luca M. The control of epidermal stem cells (holoclones) in the treatment of massive full-thickness burns with autologous keratinocytes cultured on fibrin. Transplantation 68 (6): 868–79, 1999.
130. Pellegrini G, Traverso CE, Franzi AT, Zingirian M, Cancedda R, De Luca M. Long-term restoration of damaged corneal surfaces with autologous cultivated corneal epithelium. Lancet 349 (9057): 990–3, 1997.
131. Pels-Leusden F. Die Anwendung des Spalthautlappens in der Chirurgie. Deut Med Wschr 31: 99–102, 1905.
132. Petzoldt JL, Leigh IM, Duffy PG, Masters JR. Culture and characterisation of human urothelium in vivo and in vitro. Urol Res 22 (2): 67–74, 1994.
133. Phillips TJ. Biologic skin substitutes. J Dermatol Surg Oncol 19 (8): 794–800, 1993.
134. Prasanna M, Singh K, Kumar P. Early tangential excision and skin grafting as a routine method of burn wound management: an experience from a developing country. Burns 20 (5): 446–50, 1994.
135. Pruitt BA, Jr. The evolutionary development of biologic dressings and skin substitutes. J Burn Care Rehabil 18 (1 Pt 2): S2–5, 1997.
136. Rab M, Koller R, Ruzicka M, Burda G, Kamolz LP, Bierochs B, Meissl G, Frey M. Should dermal scald burns in children be covered with autologous skin grafts or with allogeneic cultivated keratinocytes?—"The Viennese concept". Burns 31 (5): 578–86, 2005.
137. Raghunath M, Meuli M. Cultured epithelial autografts: diving from surgery into matrix biology. Pediatr Surg Int 12 (7): 478–83, 1997.
138. Rheinwald JG, Green H. Serial cultivation of strains of human epidermal keratinocytes: the formation of keratinizing colonies from single cells. Cell 6 (3): 331–43, 1975.
139. Rheinwald JG, Hahn WC, Ramsey MR, Wu JY, Guo Z, Tsao H, De Luca M, Catricala C, O'Toole KM. A two-stage, p16(INK4A)- and p53-dependent keratinocyte senescence mechanism that limits replicative potential independent of telomere status. Mol Cell Biol 22 (14): 5157–72, 2002.
140. Ronfard V, Broly H, Mitchell V, Galizia JP, Hochart D, Chambon E, Pellerin P, Huart JJ. Use of human keratinocytes cultured on fibrin glue in the treatment of burn wounds. Burns 17 (3): 181–4, 1991.
141. Ronfard V, Rives JM, Neveux Y, Carsin H, Barrandon Y. Long-term regeneration of human epidermis on third degree burns transplanted with autologous cultured epithelium grown on a fibrin matrix. Transplantation 70 (11): 1588–98, 2000.
142. Rosenthal N. Prometheus's vulture and the stem-cell promise. N Engl J Med 349 (3): 267–74, 2003.
143. Rouabhia M. In vitro production and transplantation of immunologically active skin equivalents. Lab Invest 75 (4): 503–17, 1996.

144. Rouabhia M, Germain L, Bergeron J, Auger FA. Allogeneic-syngeneic cultured epithelia. A successful therapeutic option for skin regeneration. Transplantation 59 (9): 1229–35, 1995.
145. Schiera G, Bono E, Raffa MP, Gallo A, Pitarresi GL, Di Liegro I, Savettieri G. Synergistic effects of neurons and astrocytes on the differentiation of brain capillary endothelial cells in culture. J Cell Mol Med 7 (2): 165–70, 2003.
146. Seah CS. Skin graft and skin equivalent in burns. Ann Acad Med Singapore 21 (5): 685–8, 1992.
147. Shakespeare P. Burn wound healing and skin substitutes. Burns 27 (5): 517–22, 2001.
148. Shakespeare P. Skin substitutes—benefits and costs. Burns 27 (5): vii–viii, 2001.
149. Shakespeare PG. Cost effectiveness of skin substitutes. A commentary on the debate at the 10th ISBI Congress, Jerusalem 1998. International Society for Burn Injuries. Burns 25 (2): 179–81, 1999.
150. Sheridan RL, Moreno C. Skin substitutes in burns. Burns 27 (1): 92, 2001.
151. Sheridan RL, Morgan JR, Cusick JL, Petras LM, Lydon MM, Tompkins RG. Initial experience with a composite autologous skin substitute. Burns 27 (5): 421–4, 2001.
152. Slavin J. The role of cytokines in wound healing. J Pathol 178 (1): 5–10, 1996.
153. Smola H, Stark HJ, Thiekotter G, Mirancea N, Krieg T, Fusenig NE. Dynamics of basement membrane formation by keratinocyte-fibroblast interactions in organotypic skin culture. Exp Cell Res 239 (2): 399–410, 1998.
154. Stark GB, Horch RE, Voigt M, Tanczos E. Biological wound tissue glue systems in wound healing. Langenbecks Arch Chir Suppl Kongressbd 115: 683–8, 1998.
155. Stark HJ, Baur M, Breitkreutz D, Mirancea N, Fusenig NE. Organotypic keratinocyte cocultures in defined medium with regular epidermal morphogenesis and differentiation. J Invest Dermatol 112 (5): 681–91, 1999.
156. Stark HJ, Szabowski A, Fusenig NE, Maas-Szabowski N. Organotypic cocultures as skin equivalents: A complex and sophisticated in vitro system. Biol Proced Online 6: 55–60, 2004.
157. Stark HJ, Willhauck MJ, Mirancea N, Boehnke K, Nord I, Breitkreutz D, Pavesio A, Boukamp P, Fusenig NE. Authentic fibroblast matrix in dermal equivalents normalises epidermal histogenesis and dermoepidermal junction in organotypic co-culture. Eur J Cell Biol 83 (11–12): 631–45, 2004.
158. Supp DM, Supp AP, Bell SM, Boyce ST. Enhanced vascularization of cultured skin substitutes genetically modified to overexpress vascular endothelial growth factor. J Invest Dermatol 114 (1): 5–13, 2000.
159. Suzuki T, Ui K, Shioya N, Ihara S. Mixed cultures comprising syngeneic and allogeneic mouse keratinocytes as a graftable skin substitute. Transplantation 59 (9): 1236–41, 1995.
160. Tanczos E, Horch RE, Bannasch H, Andree C, Walgenbach KJ, Voigt M, Stark GB. Keratinocyte transplantation and tissue engineering. New approaches in treatment of chronic wounds. Zentralbl Chir 124 Suppl 1: 81–6, 1999.
161. van Luyn MJ, Verheul J, van Wachem PB. Regeneration of full-thickness wounds using collagen split grafts. J Biomed Mater Res 29 (11): 1425–36, 1995.
162. Vanscheidt W, Ukat A, Horak V, Bruning H, Hunyadi J, Pavlicek R, Emter M, Hartmann A, Bende J, Zwingers T, Ermuth T, Eberhardt R. Treatment of recalcitrant venous leg ulcers with autologous keratinocytes in fibrin sealant: a multinational randomized controlled clinical trial. Wound Repair Regen 15 (3): 308–15, 2007.
163. Voigt M, Schauer M, Schaefer DJ, Andree C, Horch R, Stark GB. Cultured epidermal keratinocytes on a microspherical transport system are feasible to reconstitute the epidermis in full-thickness wounds. Tissue Eng 5 (6): 563–72, 1999.
164. Wang HJ, Bertrand-de Haas M, van Blitterswijk CA, Lamme EN. Engineering of a dermal equivalent: seeding and culturing fibroblasts in PEGT/PBT copolymer scaffolds. Tissue Eng 9 (5): 909–17, 2003.
165. Wang TW, Huang YC, Sun JS, Lin FH. Organotypic keratinocyte-fibroblast cocultures on a bilayer gelatin scaffold as a model of skin equivalent. Biomed Sci Instrum 39: 523–8, 2003.
166. Wright KA, Nadire KB, Busto P, Tubo R, McPherson JM, Wentworth BM. Alternative delivery of keratinocytes using a polyurethane membrane and the implications for its use in the treatment of full-thickness burn injury. Burns 24 (1): 7–17, 1998.
167. Wu J, Barisoni D, Armato U. An investigation into the mechanisms by which human dermis does not significantly contribute to the rejection of allo-skin grafts. Burns 21 (1): 11–6, 1995.
168. Xu W, Germain L, Goulet F, Auger FA. Permanent grafting of living skin substitutes: surgical parameters to control for successful results. J Burn Care Rehabil 17 (1): 7–13, 1996.
169. Yannas IV, Burke JF, Orgill DP, Skrabut EM. Wound tissue can utilize a polymeric template to synthesize a functional extension of skin. Science 215 (4529): 174–6, 1982.
170. Yannas IV, Burke JF, Warpehoski M, Stasikelis P, Skrabut EM, Orgill D, Giard DJ. Prompt, long-term functional replacement of skin. Trans Am Soc Artif Intern Organs 27: 19–23, 1981.
171. Yannas IV, Lee E, Orgill DP, Skrabut EM, Murphy GF. Synthesis and characterization of a model extracellular matrix that induces partial regeneration of adult mammalian skin. Proc Natl Acad Sci U S A 86 (3): 933–7, 1989.
172. Yasushi F, Koichi U, Yuka O, Kentaro K, Hiromichi M, Yoshimitsu K. 026 Treatment with Autologous Cultured Dermal Substitutes (CDS) for Burn Scar Contracture in Children. Wound Repair Regen 12 (1): A11, 2004.
173. Zhao Y, Wang X, Lu S. Identifying the existence of cultured human epidermal allografts with PCR techniques. Zhonghua Wai Ke Za Zhi 33 (7): 387–9, 1995.
174. Zhao YB. Primary observation of prolonged survival of cultured epidermal allografts. Zhonghua Wai Ke Za Zhi 30 (2): 104–6, 125–6, 1992.
175. Ziegler UE, Debus ES, Keller HP, Thiede A. Skin substitutes in chronic wounds. Zentralbl Chir 126 Suppl 1: 71–4, 2001.

Dental Hard Tissue Engineering
J. M. Mason, P. C. Edwards

Contents

26.1	Introduction	345
26.1.1	Bone Graft Gold Standard	346
26.1.2	Materials Science	346
26.1.3	Use of Biologics	346
26.1.4	Role of Inflammation and Fibrosis	347
26.1.5	Regeneration Versus Scar Formation	347
26.1.6	Practical Considerations for Bone Regeneration Technology	348
26.2	Basic Tissue Engineering Principles	348
26.3	Approaches to Apply Tissue Engineering Techniques	349
26.3.1	Group 1: Scaffold/Matrix Material	350
26.3.2	Group 2: Protein or Gene Delivery	350
26.3.3	Group 3: Cells/"Stem Cells" Delivery	351
26.3.4	Group 4: Scaffold Plus Protein or Gene Delivery	352
26.3.5	Group 5: Scaffold+Cells Delivery	353
26.3.6	Group 6: Protein (or Gene)+Cells	354
26.3.7	Group 7: Scaffold+Protein (or Gene)+Cells	354
26.3.8	Practical Issues Involving Gene and Cell Delivery	355
26.4	Clinical Application of Biologics for the Regeneration of Craniofacial Hard Tissue	356
26.4.1	Osseointegration of Dental Implants	356
26.4.2	Grafting of Dentoalveolar Bone Defects	358
26.4.3	Periodontal Regeneration	358
26.4.4	Dentin/Pulp Regeneration	360
26.4.5	Articular Cartilage Repair	361
26.4.6	Complete Regeneration of Teeth and Jaws	362
26.5	Conclusions	362
	References	363

26.1 Introduction

Improved methods of bone regeneration are needed to greatly improve the medical treatment options to vast numbers of patients suffering from different types of bone injury. Bone can be injured in many ways: as a result of trauma or aging, and from diseases such as cancer, osteoporosis, and periodontal disease. Drug side effects, as seen when using bisphosphonate for treatment of osteoporosis and metastatic bone cancer, can also destroy bone [1, 2]. Specific indications for bone repair in dental and craniofacial reconstruction include bone augmentation prior to prosthetic reconstruction, fracture repair, and repair of bone defects secondary to trauma, tumor resection, and congenital deformities. In roughly half of all individuals in the United States, periodontal disease alone is projected to result in the loss of six teeth per individual by age

65 years [3, 4]. Clearly, bone loss remains a major health care concern.

26.1.1
Bone Graft Gold Standard

Autologous iliac crest bone is currently the gold standard in bone graft material. However, there are some significant disadvantages regarding this source of graft material, including the limited amount of graft material available and the requirement for another surgical site, which has the potential for significant donor site morbidity. Other graft materials, including allograft- and xenograft-derived bone, although eliminating the need for a second surgical site, introduce the possibility of disease transmission from the donor or have the potential to elicit an immune reaction and delayed or insufficient bone repair. In addition, in some clinical applications, allografts fail in one-third of cases [5]. Development of novel bone graft substitutes is clearly desirable. We share the view of many others that an ideal bone graft material should mimic as closely as possible the positive attributes of iliac crest bone [6, 7]. Three important components of an ideal bone graft material include: (1) the presence of sufficient numbers of cells having bone regenerative capacity; (2) the presence of regulatory proteins important in bone formation to orchestrate the regenerative process and serve as the osteoinductive component of the graft; and (3) a biocompatible scaffold with osteoconductive properties [human bone is a natural composite ~40% volume (70% mass) of the inorganic bone mineral $Ca_{10}(PO_4)_6(OH)_2$ (hydroxyapatite) in a matrix of organic collagen]. As defined by Cornell et al. [8], an osteoconductive material facilitates the attachment, migration, and differentiation of osteoblastic progenitors to allow the bone-healing response to progress throughout the site. Collagen is listed as an osteoconductive material due to its structure, which promotes mineral deposition. It also binds matrix proteins that initiate and control mineralization. In-depth reviews of the vast body of research exploring the use of these components in bone repair have been published [8, 9].

26.1.2
Materials Science

Novel approaches are sorely needed to improve bone regeneration for the aforementioned conditions; consequently, many varied approaches to dental hard tissue engineering have been developed in recent years. These approaches can be separated into two basic types—those that employ biologics and those that do not. First, using the traditional approach, defects are filled with a variety of foreign scaffold materials (such as hydroxyapatite, bioactive glass, tricalcium phosphate ceramics, and osteoactive polymers) to foster bone regeneration. These materials generally provide structural support, and most provide osteoconductive properties to augment the bone healing process. There is a large body of published work from materials scientists that has focused on these non-biologics approaches to bone regeneration. This approach has been in development for many years, it continues to evolve, and a number of excellent reviews have been written summarizing it [10, 11, 12]. While these osteoconductive materials have been shown to augment bone regeneration to varying effect, there is definitely room for improvement in the regeneration rate, and consistency and quality of repair tissue. While the development of novel scaffold materials is the most direct approach to regenerate bone and likely to be a critical component of any bone regeneration product in the future, supplemental biological components are required for optimal bone regeneration. Consequently, the focus of this review will be on tissue engineering approaches that employ potent biologics components alone or in combination with matrix materials to promote bone regeneration.

26.1.3
Use of Biologics

A more recently developed approach delivers biologics (any combination of proteins, genes, or cells with or without a scaffold) to promote bone regeneration. As reviewed below, delivery of proteins, genes, or cells individually as "biological drugs" has been shown to improve the kinetics and quality of bone regeneration. However, based on our own research and that of many others, we hold the view that the most effective approach is one that combines delivery of cells having osteogenic potential with potent morphogens that direct both the implanted cells (as well as certain resident cells) to regenerate bone. Given the problems inherent with protein delivery (discussed below), we favor an approach in which cells that have been genetically enhanced with morphogen genes are delivered to bone injuries in the presence of a suitable scaffold (matrix) material. This approach (reviewed below) has been given many names by different investigators, but in this review it will be referred to by its most descriptive name—gene-enhanced tissue engineering (GETE). GETE is developed to mimic the three-component system of an ideal bone graft by combining the technologies of gene therapy and tissue engineering. In effect, GETE mimics the types of components used by nature for bone formation during skeletal development by substituting comparable components to an injury site when bone regeneration is needed.

26.1.4
Role of Inflammation and Fibrosis

Another important issue to be considered in a technology development plan that models initial tissue formation to the regeneration of injured tissue is the role played by inflammation and fibrosis. These events are problematic following tissue injury but not during initial stages of tissue formation. The following description is generally applicable to all tissues: tissue injury is characterized by three phases—inflammatory, fibroblastic, and remodeling [13]. Injury to the tissue triggers a cascade of events that results in increased local tissue damage beyond the level of the initial trauma. Capillaries engorge and increase their permeability, resulting in an influx of exudate (fluid, platelets, and cells) into the surrounding tissue. Accumulation of exudate results in swelling that causes additional local tissue damage [14]. The inflammatory phase is further distinguished by infiltration from the surrounding tissues of inflammatory cells (neutrophils, macrophages) that are attracted by chemokines released by degranulating platelets, damaged cells, and the first patrolling leukocytes arriving at the injury site [15]. The inflammatory cells phagocytose necrotic tissue and clot material [16] and are known to release a variety of growth factors and cytokines that may also be involved in the fibroblastic response [17]. The infiltrating platelets and cells also release factors such as transforming growth factor beta 1 (TGF-β1), platelet-derived growth factors AA/AB (PDGF), and epidermal growth factor (EGF) that promote local fibroblast proliferation [18, 19]. During the fibroblastic phase, the fibroblasts proliferate and synthesize collagen and extracellular matrix. Finally, during remodeling, the newly made collagen fibers are reorganized to a final structure. Thus, fibroblasts are essential for healing but they can also over respond to injury with excessive growth and collagen deposition, resulting in scar formation. This repair tissue often has greatly reduced function when compared with the original uninjured tissue.

26.1.5
Regeneration Versus Scar Formation

In general, the body is capable of repairing minor injuries to different tissues often to full or near-full function. However, this capacity is limited for most tissues in post-natal situations, and bone is no exception. Following larger injuries, many tissues in the body are capable only of repair, often characterized by fibroblast overgrowth to "fill" the void left by the injury with scar tissue, rather than true regeneration of tissue to its previous fully functional state. Such scar tissue formation is problematic due to reduced function and because once this filler tissue is generated, it interferes with regeneration of the fully functional tissues. This occurs in many of the body's tissues and is plainly exemplified in nerve regeneration where the body is often quite capable of slow re-growth of neural networks, but the more rapid formation of scar tissue ultimately prevents proper innervation of the distal organs. In effect, the problem to be overcome is one of evolution. To optimize short-term survival, nature has evolved to select repair mechanisms resulting in a "quick fix" to rapidly patch injuries; unfortunately this patch is mainly scar tissue. Such short-term fixes have a selective advantage over the regenerative process because true tissue regeneration is an incredibly complex process, requiring differential gene expression in various cells orchestrating involvement of multiple cell types and, therefore, it is by nature a slower process and loses the race to the rapid fibroblastic response. Regeneration is disfavored because mammals in the wild do not survive if they remain incapacitated for long periods of time. From an evolutionary stand point, it is better to heal quickly with less than full function than to remain incapacitated for much longer periods of time required for regeneration to restore full tissue function. Simply put, there is a competition to repair injury through fibrosis versus regeneration; fibrosis is a faster process and therefore wins the race. Thus, a dilemma exists. How can we overcome nature's predilection to rapidly fill injuries with scar tissue while simultaneously optimizing the speed and quality of regeneration? We believe that the answer is that tissue regeneration approaches must both promote regenerative pathways and suppress fibrotic ones.

Thus, controlling the response of the resident fibroblasts to injury is the key to a process whereby tissues are regenerated versus where fibroblasts overcompensate to injury resulting in additional scarring. While much has been learned about this process through studies that merely observed repair, knowledge is still lacking on the challenging problem of how to actively control the response of fibroblasts such that healing is promoted and scarring is suppressed. Other issues that come into play for optimal tissue regeneration include: assuring presence of an adequate blood supply to deliver nutrients to the healing tissue and proper remodeling of the repair tissue in response to biomechanical force [20]. All of these issues should be considered in novel approaches to hard tissue engineering.

26.1.6
Practical Considerations for Bone Regeneration Technology

It is evident that delivery of powerful drugs with pluripotent effects or multiple drugs each with lesser effects will be needed for optimal tissue regeneration. From a practical viewpoint, it is better to identify and deliver pluripotent factors that concurrently favor regeneration and suppress scarring rather than attempt to develop approaches that require delivery of many drugs to achieve adequate regeneration. This leads to a critical point to consider in the review of new technologies for clinical application to engineering of dental hard tissues (or any tissues for that matter); that is, the practicality and economic feasibility of the technology being developed. After all, it will be companies, not academics that will ultimately bring new drugs to the masses, and the technology must make sense from an economic and regulatory viewpoint in addition to being scientifically sound. This topic is usually completely neglected in academic circles, yet it is an issue of fundamental importance best considered at the earliest planning stages of technology development. Many novel approaches to bone regeneration may eventually suffice for treatment of small numbers of patients with specific hard tissue injuries, but they may not be practical or economically feasible for widespread application to large numbers of patients. Consequently, in addition to a scientific critique of various technologies potentially useful for dental hard tissue engineering, this review will bear a critical eye to the practical issues related to the technologies being developed. A review focusing on the commercial impact of stem-cell-related approaches to dental regeneration was recently published [21].

26.2
Basic Tissue Engineering Principles

As mentioned, it is our view that an ideal bone graft substitute should mimic the complex mixture of components present in iliac crest bone. Bone graft materials should provide cells of bone regenerative capacity (osteogenic cells), contain protein or other factors that signal the induction of new bone formation (an osteoinductive environment) and supply a scaffold conducive to new bone formation (an osteoconductive environment). Some or all of these three components are used in dental hard tissue engineering. Many interdisciplinary groups of materials scientists, cell biologists, molecular biologists, animal surgeons, and clinicians (often collectively known as tissue en-

Table 26.1 Seven approaches to dental hard tissue engineering

Delivered agent	Components of an ideal bone graft substitute		
	Osteoconductive agent (scaffold)	Osteoinductive agent (protein or gene)	Cellular component
1. Scaffold/matrix materials	√		
2. Protein (or gene)		√	
3. Cells/"stem cells"			√
4. Scaffold+protein (or gene)	√	√	
5. Scaffold+cells	√		√
6. Protein (or gene)+cells		√	√
7. Scaffold+protein (or gene)+cells	√	√	√

gineers and regenerative medicine specialists) are directing major efforts at identifying the optimal single component or combination of these components in an attempt to develop the best performing tissue engineered bone. Table 26.1 depicts all seven of the theoretical combinations of these components given a three-variable system. Essentially all tissue engineering approaches to bone regeneration fit into one of these seven groupings. We present these with the understanding that the scaffold material generally supplies the osteoconductive component, the osteoinductive agent is usually supplied as protein through either direct protein delivery or delivery of a gene that then causes the genetically enhanced cells to produce the protein (other osteoinductive agents such as chemicals are uncommonly used), and the cellular component is either supplied exogenously or relies on intrinsic cells present at the injury site. Tissue-engineered bone or teeth grown ex vivo and then implanted as grafts also would have been initially formed using one of these seven groupings [22]. As stated above, this review will focus on six of the approaches (groups 2–7 in Table 26.1) that employ biologics components in dental hard tissue engineering.

26.3
Approaches to Apply Tissue Engineering Techniques

All seven approaches work to varying degrees to regenerate bone and other hard tissues. A brief synopsis of each is presented here prior to a more detailed review of the literature in the sections that follow. Clearly, the most economically feasible approach is generally the simplest, represented by group 1—scaffold material delivery. This approach has been used for many years [23]. Unfortunately, however, even modern-day scientists focused on developing novel scaffold materials have not yet found an ideal material [11, 13]. This is likely because this group lacks a suitable osteoinductive agent and sufficient osteogenic cells at the repair site. Group 2—protein delivery [often of proteins in the bone morphogenetic protein (BMP) family]—suffers mainly from short duration of effective protein half-life [24, 25] and the fact that supraphysiological amounts of protein are used in bolus, which can result in suboptimal bone quality [26]. Because no supplemental cells of osteogenic potential are supplied, this approach can also suffer from a lack of intrinsic osteogenic cells in the repair site, particularly in aged or diseased tissues. It has been reported in an in vitro study that the number of osteoprogenitor cells in bone marrow of rats is reduced as a function of aging [27]. However, other in vitro studies have reported that the absolute number of human osteoprogenitor cells is not reduced with aging, but rather there is a reduction in the proliferative capacity of osteoprogenitor cells or their responsiveness to biological factors, leading to an alteration in subsequent differentiation that is responsible for age-related decline in bone healing [28]. Yet another group reported that the osteogenic potential (ability to differentiate into osteoblasts) of human marrow-derived cells does not decrease as a function of aging in late adulthood, suggesting it is a reduction in cell number [29]. Whatever the case, the literature suggests that intrinsic cells of full osteogenic potential may be limiting in certain patient populations. Group 2—gene delivery alone (often of morphogen genes)—can similarly suffer from lack of sufficient osteogenic cells in the repair site. In group 3, cells delivered to injury sites generally perform better when implanted in a scaffold material. In group 4, scaffold plus protein supplies both the osteoconductive and osteoinductive components, but may suffer from protein half-life and lack of osteogenic cells, as described. Similarly, also in group 4, scaffold plus gene delivery requires sufficient numbers of osteogenic cells to be present for genetic enhancement and bone regeneration. In group 5, scaffold plus cells can regenerate bone, but generally at a slower rate than when osteoinductive agents are co-delivered. Protein plus cells has been used very effectively (group 6) to regenerate bone, particularly when stem cells from mesenchymal tissues are used, although inclusion of a scaffold material is preferred. Similarly, group-6 genetically enhanced cells have been used to regenerate bone effectively, but a scaffold material for cell delivery is also preferred when using this approach. Finally, group 7 combines all three components and can very effectively regenerate bone. However, this approach suffers from a business perspective because it is the most complex approach and it triggers government regulatory review as both a device and a biological drug.

The following literature review is not intended to be all inclusive. Rather, it is presented as a topical review of current approaches to dental hard tissue engineering, not a historical review of this subject. There exists an enormous amount of relevant published material, including both primary data and review articles, the vast majority of which supports the conclusions summarized above. We have focused on the primary data reports of in vivo studies but also included references to other review articles for those interested in more detail. We apologize to the many authors whose primary data could not be cited in the present review which generally focuses on the most recent publications in each of the seven groups.

26.3.1
Group 1: Scaffold/Matrix Material

Although we will not review the large number of scaffold materials that are continually being developed, excellent reviews of the latest in scaffold [30] and nanotechnology [31] development have recently been published elsewhere. The latter review describes cutting-edge research into how novel nanoscale molecules mimic natural extracellular matrix and how nanotechnology meshes well with biologics delivery systems for enhanced tissue regeneration. Another recent review focuses on the increasingly appreciated but less thoroughly examined impact of biomechanical forces on synthetic implants and surrounding bone following dental applications of bone tissue engineering [32].

26.3.2
Group 2: Protein or Gene Delivery

The direct delivery of BMP is a well-established and validated procedure for improving surgical success in a number of defined orthopedic situations. The United States Food and Drug Administration (FDA) recently approved rhBMP-2 as an alternative to autogenous bone grafting for dental surgical procedures involving maxillary sinus augmentation and localized ridge augmentation. This approach is discussed in more detail below.

Another method of achieving localized presence of supplemental proteins at the defect site is through direct gene transfer (delivery of plasmid DNA or viral vectors directly to the site) followed by in vivo transfection/transduction of the resident cells. The advantage of this is that proteins can be produced locally by the resident cells at the defect site. Proteins can be expressed for an extended period of time in supraphysiological amounts or at physiological levels, depending on the vector system chosen. The expressed proteins can have both autocrine and paracrine effects depending on whether the genetically enhanced cells become altered to serve as active constituents of the repair tissue or whether their role is limited to merely serving as "protein factories" to aid the non-genetically enhanced resident cells to elicit repair. One benefit of gene delivery is that the expansive knowledge base of developmental biology can be brought to bear in identifying genes of potential importance in bone formation. The natural process of bone formation can be mimicked and an empirical approach used to determine which genes perform optimally in bone regeneration technology.

Many of the early reports involving gene delivery for bone regeneration used the simplest and most direct delivery approach. Plasmid DNA or viral vectors were directly delivered to defect sites without scaffolds or supplemental cell components, often with good bone regeneration results. Due to its simplicity, this approach is still being actively pursued. In a recent report, Park et al. described use of a BMP2 expression plasmid that was delivered as a liposome complex to the peri-implant area of pig calvarial defects, with or without autologous bone graft [33]. By 4 weeks, bone regeneration was improved in the BMP2 groups relative to controls. In another report, an adeno associated vector (AAV) expressing the vascular endothelial growth factor ($VegF_{164}$) gene was directly delivered to the mandibular condyle of rats, resulting in a significantly increased length and width of condylar head compared with controls [34]. In contrast to the plasmid-BMP2 results, which showed effects within 30 days of treatment, the significance of the AAV-VegF effect was only observed from 30 to 60 days post-treatment and not at earlier time points. This is consistent with the fact that plasmid transfection is typically used as a transient overexpression system with maximal (often supraphysiological levels of) protein expression in the first 2–4 days post-transfection, while AAV transduction gives lower initial levels (more in line with physiological levels) of gene expression but for much longer periods of time (weeks to months). These studies confirm numerous previous reports that various members of the BMP superfamily are beneficial for bone regeneration and that neovascularization is beneficial to mandibular condylar growth [35].

The main attractiveness of this approach is its simplicity and the commercial appeal of an "off the shelf" product that can merely be injected into the defect site to foster bone regeneration. This approach may prove clinically useful for varied bone regeneration applications, particularly for regeneration of smaller sized defects. However, the drawback of targeting only resident cells that may be limited in number or functionality for bone regenerative applications has encouraged development of other approaches that include delivery of a scaffold and/or cellular component.

26.3.3
Group 3: Cells/"Stem Cells" Delivery

Various cell types have been delivered to defects without scaffold material or osteoinductive agent and have proven useful for regeneration of various tissues. Many groups describe the cells used as "stem cells;" however, this term is often misused. Many of the cell types used are in truth not stem cells per se, but rather cells that have some degree of plasticity when exposed to appropriate factors. These cells have the ability to differentiate into different cell types in response to certain signals. The issue of cellular plasticity, transdifferentiation, and dedifferentiation is a hot topic that remains controversial [36]. It is our view that populations of fibroblast-like cells present in many tissues in the body are differentially plastic and that, when given the appropriate signals, these cells can and do transdifferentiate into other cell types and become incorporated into the repair tissue. Data supporting this view will be presented in section 26.3.7.

In an interesting study toward development of tissue-engineered teeth, various murine cell types (embryonic stem cells, neural stem cells, and adult bone-marrow-derived cells) developed tooth structures and bone in adult renal capsules when combined with embryonic oral epithelium [37]. Transfer of these cell combinations into adult jaw resulted in development of tooth structures. This suggests that multiple cell types can be used for tooth engineering. A later study in a rat model used apical bud cells instead of embryonic oral epithelium to supply inductive factors and demonstrated that dental pulp stem cells formed typical tooth-shaped tissues with amelogenesis and dentinogenesis in renal capsules while the bone-marrow-derived cells lacked enamel formation [38]. These investigators suggested that dental cells outperform non-dental cell populations for tooth regeneration applications, but clearly much more work needs to be done before coming to such a conclusion. Similar tooth regeneration studies and results have been recently reported using dental epithelial cell lines rather than primary epithelial cells as inducers [39]. A recent review focusing on tooth regeneration was recently published [40].

Although delivery of just cellular components to defects can improve tissue regeneration, when direct comparisons are made, it is generally observed that tissue regeneration is better when two or more of the three components of an ideal bone graft are supplied to defects. One study that highlights this is a recent paper by Hsu et al. in which a lentiviral-BMP2 vector was used to transduce rat bone marrow stromal cells followed by implant into rat segmental femoral defects [41]. All ten of ten defects healed in 8 weeks, while none of the defects healed that had been treated with either non-transduced cells or green fluorescent protein (GFP) control transduced cells. Clearly, delivery of cells alone (or control transduced cells) was not sufficient for quality bone regeneration. Supplementation with an osteoinductive factor was required for effective bone regeneration, as was delivery of sufficient numbers of cells, because only one of ten defects healed when fivefold fewer lenti-BMP2 transduced cells were implanted. This is only one recent of many studies that have repeatedly shown that both osteoinductive factors and a supplemental cell component (with a sufficient number of cells delivered) are necessary for superior bone regeneration.

26.3.4
Group 4: Scaffold Plus Protein or Gene Delivery

Group 4 involves two of the three components of an ideal bone graft material—the osteoconductive scaffold and the osteoinductive signal, the latter of which can be provided by delivered protein or through gene delivery. The two approaches have different strengths and weaknesses and will be discussed separately. The scaffold plus protein delivery approach will be discussed first. Identifying the best pairings of different osteoconductive scaffolds with different osteoinductive factors for optimal bone regeneration has been a long standing goal for many investigators. A typical finding was recently reported in which recombinant human BMP2 was delivered to osseous defects along with implants 5 months after bilateral extraction of mandibular four premolars and first molar in canine mandibles [42]. The rhBMP2 was delivered in polylactide/glycolide polymer scaffolds. After 4 and 12 weeks, histomorphometric analyses were performed. The presence of rhBMP2 resulted in better percentage contact, new bone area, and percentage defect fill than controls lacking rhBMP2 at 4 weeks, but not at 12 weeks. In another report using an essentially analogous model system employing rhBMP2 delivery, no improvement in histomorphometric measurements at either 4 or 12 weeks post-implant was observed in the canine mandibles [43]. These studies help illustrate the problem with use of protein delivery. Possibly due to the short physiological half-life of these proteins, short-term improvements in bone regeneration are observed, although not consistently, and later time points may show no improvements at all. This problem is not limited to delivery of rhBMP2. A synthetic peptide (P-15) analog of collagen delivered in an organic bovine bone mineral failed to enhance bone regeneration after 4 months in two experimental membrane-protected defects (periodontal and intra-bony) in a canine model [44]. We and others [45] contend that gene delivery approaches may be superior to protein delivery approaches for craniofacial applications.

The scaffold plus gene delivery approach has shown promise in craniofacial applications. Recently, a novel mixture of polyplex nanomicelle (a complex of polyethyleneglycol with a catiomer of P[Asp-DET] and plasmid DNA) was developed [46]. This nanomicelle/DNA mixture was delivered to mouse calvarial bone defects in a calcium phosphate cement scaffold. The constitutively active form of activin receptor-like kinase 6 (caALK6) and runt-related transcription factor 2 (Runx2) expression plasmids were delivered resulting in substantial bone formation without inflammation after 4 weeks. Although from a preliminary study, this reflects the fact that investigators continue actively pursuing novel mixtures of transfection agents, genes, and scaffolds for optimal bone regeneration. However, plasmid delivery continues to suffer from generally poor transfection efficiency, resulting in an overall transient and low level of transgene expression. A review focusing on scaffolds for DNA delivery was recently published [47]. In comparison, viral vectors generally give higher levels of protein expression for longer durations, often resulting in excellent bone regeneration. As an example, an adenoviral vector bearing the BMP7 gene was delivered in a collagen scaffold to osseous defects with titanium implant fixtures that were placed 1 month following extraction of rat maxillary first molars [48]. Good BMP7 expression was observed for 10 days at osteotomy sites resulting in enhanced alveolar bone fill, new coronal bone formation, and new bone-to-implant contact. In an interesting recent study, adenoviral vectors bearing the BMP2 gene were lyophilized and delivered to rat calvarial defects with bone regeneration assessed 5 weeks later [49]. This study demonstrates that viral vectors can be lyophilized and retain biological activity following in vivo delivery to calvarial defects. In addition, the controls for this study allowed direct comparison of effectiveness of several groups in the same model of bone regeneration. The study demonstrated that delivery of gelatin scaffold alone (group 1 of this review) was not effective in bone regeneration, and delivery of "free-form" adeno-BMP2 vector (group 2 of this review) had only modest effects. In comparison, the adeno-BMP2 vector lyophilized on gelatin sponges (group 4 of this review) resulted in 80% regeneration of the calvarial defects. These findings are reflective of many studies that generally support the concept that better bone regeneration is observed when more of the three components of an ideal bone graft are delivered to craniofacial defects.

26.3.5
Group 5: Scaffold + Cells Delivery

Another active area of research has investigators pairing different cell sources (cells that have bone regenerative potential) with different scaffolds for hard tissue regeneration. Many pairings show promise for regeneration of different hard tissues. A recent study of surface properties of poly(lactic-co-glycolic acid) (PLGA) scaffolds found that seeded human mesenchymal stem cells from dental pulp adhered better to microcavity-rich scaffolds than to smooth-surface scaffolds, resulting in better bone formation in a rat model [50]. Similarly, another group used mesenchymal cells from the root apical papilla of human teeth and periodontal ligament stem cells to form a root/periodontal complex that supports a porcelain crown in a mini-pig model [51]. Two recent studies demonstrated that supplemental delivery of cells with scaffolds gives superior results in regeneration over delivery of scaffold alone. Autologous cultured canine bone-marrow-derived stromal cells delivered with collagen-coated β-tricalcium phosphate (TCP) were superior at regenerating cranial bone, after 3 and 6 months, than collagen coated β-TCP, autologous bone, or fibrin glue delivered alone without the stromal cells [52]. The second study involved extraction of a single rabbit incisor followed by immediate implantation into the empty extraction sockets of bone marrow mesenchymal stem cells in a scaffold composed of poly-L-lactic acid:polyglycolic acid composite (PLG) [53]. Bone regeneration was assessed 4 weeks later, both histologically and radiographically. Results indicated preservation of alveolar bone wall in the sockets treated with cells and scaffold but not in those treated with only scaffold material; again demonstrating the value of a supplemental cell component for bone regeneration. Unfortunately, cell and scaffold delivery is sometimes insufficient for regeneration of more complex hard tissues. A heterogeneous cell population from canine tooth buds delivered with scaffold was unable to form enamel or dental roots 24 weeks following implant into extraction sockets (although tubular dentin and bone were regenerated) [54]. This same group then attempted to solve the tooth morphology problem by sequentially seeding porcine mesenchymal cells at the base of the scaffold with epithelial cells layered on top followed by implant into immunocompromised rats [55]. Tooth morphology was improved using the sequential seeding approach. However, even with these advances, delivery of cells plus scaffold may still not match the performance of autologous iliac crest bone grafts because of the lack of sufficient osteoinductive agent.

In a clinical sinus augmentation study, it was recently reported that the autologous bone grafts outperformed implants consisting of PLGA scaffolds seeded with cultured human osteoblasts [56]. A review focusing on craniofacial tissue engineering using cell and scaffold delivery was recently published [57].

26.3.6
Group 6: Protein (or Gene) + Cells

There are few reports of investigators delivering only these two components for hard tissue engineering. One approach that has been used with some success is the delivery of cells with platelet-rich plasma (PRP). In earlier work, delivery of canine mesenchymal stem cells and PRP to mandibular defects in dogs outperformed delivery of PRP or cells alone when bone regeneration was assessed histologically and histomorphometrically after 2 months [58]. This study also reported that the "injectable tissue-engineered bone" provided stable results when used simultaneously with implant placement in the maxilla or mandible of three human patients. A more recent study from a separate group of investigators confirmed the utility of the PRP and concluded that regeneration of mandibular bone requires the presence of osteoinductive factors supplied by the PRP [59]. The reason that few reports exist for this group is because most investigators find that cell delivery requires some form of scaffold for optimal performance.

26.3.7
Group 7: Scaffold + Protein (or Gene) + Cells

Many investigators have reported excellent results in bone regeneration following administration of scaffold, morphogen gene, and cells [60, 61]. Human adipose-derived mesenchymal stem cells (fibroblast-like cells from adipose) were transduced with adeno-BMP2 in culture, applied to a collagen-ceramic scaffold, and implanted into femoral defects in nude rats resulting in healing in 11 of 12 cases by 8 weeks [62].

The scaffold alone and scaffold plus non-transduced cells did not heal, indicating the requirement of morphogen presence. Similar results were also recently reported using BMP-2 gene-enhanced fibroblast-like cells from adipose in a β-TCP carrier delivered into canine ulnar bone defects [63]. These results agree with our previously published report in which rabbit gingival fibroblasts, periosteal-derived mesenchymal cells, and fibroblast-like cells from adipose where retrovirally transduced with the sonic hedgehog (SHH) gene and selected in culture prior to implant in collagen/alginate scaffolds into rabbit calvarial defects [64]. The mesenchymal cells from all three donor tissues effectively regenerated bone, indicating that mesenchymal fibroblasts from various tissue sources have bone regenerative capacity. However, bone regeneration was only observed when the cells were genetically enhanced with the SHH gene indicating that these cells are not "stem" cells per se. Another group recently reported that fibroblasts transduced with osteogenic differentiation factors (constitutively active activin receptor-like kinase 6 and runt-related transcription factor 2) healed murine calvarial bone defects in 4 weeks [65]. Delivery of protein with cells and scaffold has also been successful. Cultured dog mesenchymal cells from iliac crest were delivered with dental implants into mandible defects in fibrin scaffold supplemented with proteins present in PRP [66]. Bone-implant contact was greatest (53%) with cells/fibrin/PRP when compared with cells/fibrin and fibrin alone after 2, 4, and 8 weeks. Taken together, these studies suggest that mesenchymal cells from many tissues can effectively regenerate bone when supplemented with morphogens or differentiation factors as genes or proteins, and that delivery of scaffold + protein (or gene) + cells outperforms delivery of only two components.

The mechanism by which morphogens and differentiation factors improve bone regeneration remains a contentious issue. We and others support the concept that the implanted mesenchymal cells are differentially plastic and that these cells transdifferentiate in response to the transgene signaling, resulting in improved bone regeneration. In addition, we and many others have provided direct evidence that the implanted cells become incorporated into the developing bone, suggesting that the cell component plays an important role in the process. However, others reported that the implanted cells function only as factories to produce the protein factors locally that induce osteogenesis by the native cells present locally, and they do not contribute to the newly formed tissue [67]. As is the case many times when groups report contrasting findings, we believe it is likely that the implanted cells may or may not transdifferentiate into bone, depending on the type of implanted cells and transgene chosen for study as well as the details of the particular study. Different cell sources may have different capacities to transdifferentiate in response to different transgenes, and clearance from tissues may be related to scaffold used, transgene toxicity, and vector used for gene enhancement. As an example, we have found that overexpression of morphogen genes from strong enhancer/promoters such as cytomegalovirus is toxic to different target cells; consequently, the target cells express and secrete the morphogen for a few days until the toxicity kills the implanted cells, which is why they do not appear in regenerated tissues. However, the implanted cells did function successfully as protein factories to impact on native non-transduced cells in the local environment. If plenty of cells with bone regenerative capacity are available locally, there may not be a need for the implanted cells to contribute to the repair tissue. But this is not always the case. Therefore, we designed our experiments to express these potent morphogen genes off of the relatively weak β-actin promoter and observed that the implanted cells do indeed become incorporated into the repair tissue.

26.3.8
Practical Issues Involving Gene and Cell Delivery

Review of the literature clearly indicates that all of the aforementioned groups have shown some degree of success in regeneration of various dental hard tissues. Some approaches will work fine for regeneration of small defects or just for bone. Regeneration of large defects or more complex hard tissues will likely require the more complex approaches. This review has been structured to present the technologically simpler groups first followed by the more complex approaches. Clearly, delivery of one component is less complicated than delivery of two, with three component delivery being the most complex. This complexity is not just scientific. Regulatory agencies (at least in the U.S.) consider the delivery of biologics differently (read more rigorously) than that of chemicals, with which they have much more experience and a higher comfort level. Group 1 is the simplest and most conventional pharmaceutical and will have the easiest regulatory path to a commercial product. In general, this group is considered a device or chemical drug; groups 2, 3, and 6 are considered biological drugs; and groups 4, 5, and 7 are considered "combination products", being part device and part biologics. From the point of view of development of a commercial regeneration product, the path of least resistance comes into play. Consequently, the simplest approach that works well will be the best to use for a particular application. In our view, it is likely that many of these groups already have or will in the future produce viable commercial products for dental hard tissue regeneration [obviously, we focus on product commercialization as the ultimate mark of success, because if it is not commercialized, the regeneration technology (product) cannot be brought to the masses, which is the ultimate goal of translational biomedical research].

Regarding the groups that involve cell delivery, a "universal donor" approach is favored by commercial entities. This approach, in which a cell line is produced en masse and used to treat multiple patients, is a process that conventional drug manufacturers are familiar and comfortable with, but, in our view, may not function well scientifically for many applications. We believe that patients will have the highest comfort level with use of their own cells in customized therapy given the risks of infectious disease transmission, immune rejection, and risk of tumor formation with use of cultured foreign donor cells (including embryonic stem cells). Consequently, identification of an abundant autologous cell source that effectively regenerates tissues (or cells that naturally are or can be made to be differentially plastic and therefore regenerative) is of practical importance. Readily available populations of autologous cells that do not require extensive purification and that could be rapidly processed and returned to the body without expensive and time-consuming expansion in culture would be ideal. Many investigators, including ourselves, have focused efforts in identifying such cells.

26.3.8.1
Protein Delivery Issues

The main problems to be overcome with protein delivery are increasing the effective half-life of the protein, reducing the amount of protein required for physiological effect, and having the appropriate amount of protein present in the proper location for the minimally required period of time. These are difficult problems to overcome, but a large effort continues to seek solutions.

26.3.8.2
Gene Delivery Issues

Some investigators propose delivery of multiple genes for hard tissue regeneration. Delivery of two or more genesis of academic interest but may not be the most practical given that the FDA views each vector as an individual drug that will require extensive and costly testing. It is more practical to identify "master control genes" that coordinate expression of several genes important in bone regeneration and to use these individual genes for initial clinical trials. At later dates, multiple vectors bearing different master control genes could be employed to improve upon results of single gene therapy should it be deemed necessary. It is also important to understand that there is not one ideal gene delivery system for all applications. Each system has its own strengths and weaknesses—all are flawed—but they are the best that current technology has to offer for obtaining transient or stable production of proteins from cells in a physiologically relevant manner. Another issue is that technology is not currently advanced to the point where whole organs or complex multiple tissue systems can be regenerated. Although this is a noble theoretical goal to pursue, this review has focused on what is possible in the near term; what is feasible in the next decade. Along similar lines, in vivo regulateable gene expression systems are not yet very well developed or reliable [68]. Although promising for the future, the current systems (i.e., the tet on/off, rapamycin, light inducible systems) continue to give variable results in in vitro and in vivo systems. At present, it is best to use established technologies to express genes for transient or longer term time frames. Different vector systems enable different durations of gene expression. Another issue of practical importance is that, in general, delivery of genes that express secreted factors may work better than those expressing non-secreted factors because secreted factors can impact on neighboring and distant cells, not just on the relatively small number of cells that are transduced. For this reason, we believe that delivery of non-secreted proteins (such as transcription factors) will not be as efficacious as secreted factors, although more work must be done in this area.

26.4
Clinical Application of Biologics for the Regeneration of Craniofacial Hard Tissue

Practical applications for biologics-based tissue-regenerative therapies in the oral and maxillofacial complex are wide ranging. Objectives include improving the local environment into which dental implants are placed, grafting of dentoalveolar bone defects (e.g., fracture repair, dentoalveolar augmentation prior to prosthetic reconstruction, and repair of facial bone defects secondary to trauma, tumor resection, or congenital deformities), periodontal regeneration, pulp capping/dentin regeneration and articular cartilage repair [69, 70]. Long-term prospects could ultimately include the capacity to regenerate entire teeth and jaws. These approaches are discussed below.

26.4.1
Osseointegration of Dental Implants

Osseointegrated dental implants are titanium-based tooth root substitutes that are surgically placed into the alveolar bone, forming a permanent connection with adjacent bone following osteoblast-induced bone ingrowth. They have gained widespread acceptance in the treatment of edentuous areas of the jaws. The rapid increase in the use of dental implants is based in part on their relative ease of placement and the very high overall clinical success rate, particularly when used in an ideal setting. However, in many patients, optimal placement of dental implants is compromised by previous alveolar bone loss. Autologous bone, demineralized freeze-dried bone, and hydroxyapatite are all widely used to improve the local environment into which dental implants are placed by increasing the volume of surrounding alveolar bone.

Considerable effort has been devoted toward identifying surface treatment approaches to increase the success of osseointegration, particularly in less than ideal clinical situations. Until recently, this primarily involved modifications of surface geometry, such as porosity, or the addition of thin layers of calcium phosphate and/or hydroxyapatite to the implant surface [71]. Newer approaches currently under investigation involve applying osteoinductive protein factors or other biomimetic compounds, either into the newly created surgical site or directly bound to the implant surface. For example, a recently described approach employing an RGD (arginine–glycine–aspartic acid) cell adhesion peptide-modified matrix containing covalently bound peptides of parathyroid hormone afforded bone regeneration similar to autogenous bone in a titanium dental implant model in the dog [72]. The reader is referred to a recent review by Moioli et al. [73] which details the varied approaches already explored.

Much of the work in this area has revolved around the use of BMP delivery. The FDA recently approved rhBMP-2, produced from an engineered Chinese hamster ovary cell line, in a bovine collagen carrier (trade name INFUSE® Bone Graft, Medtronic Inc.) delivering 1.5 mg/ml rhBMP-2 as an alternative to autogenous bone grafting for dental surgical procedures involving maxillary sinus augmentation and localized ridge augmentation. Even prior to FDA approval, recombinant BMP, primarily BMP-2 and to a lesser extent BMP-7, was being used by dentists in an attempt to increase the success rate of implant integration in compromised bone sites.

In some studies [74], no significant difference was noted with rhBMP-7-treatment when measuring bone apposition around implants compared with untreated implants in the dog. In a more recent dog model [75], the authors of the study stated that the addition of autogenous bone graft or BMP-2 to the surgical site had no discernible beneficial effect on implant osseointegration, as judged by both bone histomorphometry and implant stability. The source of the BMP-2 in this study was actually a demineralized bone matrix (DBM) in a gelatin carrier (Regenafil; Regeneration Technologies Inc.). No information was provided as to the protein dose used. Although preparations of DBM contain numerous osteogenic protein factors, the concentration of BMPs in DBM is in the nanogram/cm^3 or less range, several orders of magnitude lower than the concentration of FDA-approved formulations of rhBMP-2. The outcomes from another study suggest that the addition of BMP-2 may not offer much benefit over the use of collagen 1 alone [76].

However, these findings are in contrast to numerous other studies that have shown a significant benefit with BMP-2 use. The early studies by Bessho et al. [77] demonstrated increased bond strength at the bone–titanium implant interface with the use of purified BMP. Improved outcomes were noted in a recent rat study in which implants coated with low doses (5–20 µg/site) of rhBMP-2 were placed subcutaneously into the ventral thoracic region. Interestingly, in this model, the greatest amount of ectopic bone formation was observed at the lowest dose [78].

Another common clinical scenario in implant dentistry is that of the patient with inadequate bone height in the posterior maxilla, resulting in proximity of the maxillary sinus to the planned implant placement site. Although numerous "sinus lift" techniques are employed in clinical practice to increase bone height prior to implant placement in the compromised maxilla, most require the use autologous bone or DBM. Boyne et al. [79] demonstrated that rhBMP-2 in a collagen matrix carrier resulted in a mean increase in bone height of 8.5 mm when used for maxillary sinus floor augmentation prior to implant placement in a group of 12 patients.

Recently, BMP-7 gene delivery to dental implants has been shown to increase alveolar bone defect fill, coronal bone formation, and bone-to-implant contact [80].

26.4.2
Grafting of Dentoalveolar Bone Defects

Conventional approaches to the management of dentoalveolar bone defects are based to a large extent on the size and location of the defect as well as the desired long-term prosthetic treatment plan. While smaller defects can often be treated with autologous bone grafting for neighboring dentoalveolar sites, medium-sized defects are usually reconstructed with non-vascularized "free" bone grafts harvested from extraoral donor sites, usually the iliac crest or tibia. Failure rates are increased in larger defects that involve oral mucosa, bone and overlying facial skin (termed composite defects), as well as in sites that have been compromised by exposure to therapeutic doses of radiation. In these situations, vascularized free flaps are preferred, having success rates approximating 90%. This microvascular surgical technique—in which bone is harvested along with the overlying soft tissue and vascular component, often from the fibula, scapula or radius—permits immediate restoration of blood flow to the graft, thereby enhancing wound healing. The principal disadvantage with all of these approaches is the need for a second surgical (donor) site.

A number of exciting new techniques are being explored to enhance bone regeneration in the oral cavity. It should be pointed out that these techniques, when used in the oral cavity, offer little in the way of differences relative to bone regeneration at other sites. Hence, the reader is referred to previous sections of this chapter dealing with bone regeneration.

26.4.3
Periodontal Regeneration

The periodontal attachment comprises the gingiva, periodontal ligament (PDL), tooth root cementum, surrounding alveolar bone, and a heterogeneous population of additional tissues that function together to attach the tooth to the supporting alveolar bone. Microbial colonization of these structures leads to periodontal disease, a complex process mediated by the host immune response to microbial agents [81] and characterized by destruction of the attachment fibers and supporting alveolar bone leading to tooth mobility and eventual tooth loss. Mild periodontal disease affects an estimated 90% of the adult population [82], with more advanced disease affecting approximately 15% of adults over the age of 30 years [83]. The economic cost of treating and preventing periodontal disease is high, estimated at US $14 billion in the United States of America alone in 1999 [84].

Conventional treatment of periodontal disease has focused on removing diseased tissue to promote an ideal environment for periodontal repair. These approaches generally result in repair instead of regeneration, characterized by formation of a non-physiological epithelial reattachment, known as a long junctional attachment, formed by crevicular keratinocytes that migrate into the pocket. Many periodontal regenerative protocols also involve the use of a guided tissue regeneration (GTR) approach, in which a cell-impermeable barrier is placed between the root and the crevicular epithelium in an effort to retard the migration of crevicular epithelium into this space thereby promoting periodontal regeneration over repair. GTR is also employed as a sole modality treatment in conjunction with conventional open surgical root debridement, although the overall response rate is unpredictable [84].

True regeneration of lost attachment has been hampered by the necessity to regenerate several tissue types, including root cementum, alveolar bone, and intervening periodontal ligament fibers, in a coordinated fashion. Recently, several promising approaches to periodontal tissue regeneration have been developed. Unfortunately, proper clinical evaluation of the success of these techniques has been hampered by a lack of consistency in treatment protocols, disparities in the methods used to quantify treatment outcomes, and the observation that certain as of yet unidentified subpopulations of patients appear to respond to treatment better than others.

Although autogenously harvested cancellous bone from the iliac crest, maxillary tuberosity area, and tooth extraction sockets all produce statistically significant bone fill, the ability of these graft materials to regenerate periodontal attachment has not been as notable. Drawbacks with the use of fresh iliac crest bone include an increased incidence of root resorption at the regeneration site and the necessity for a second surgical site. Moreover, histological evidence of true periodontal regeneration has been limited [85]; in many instances bone regeneration around the tooth root is seen in association with the formation of a long junctional epithelium.

Although decalcified freeze-dried allogeneic bone (DFDB) obtained from commercial tissue banks [86] has been shown in human clinical studies to regenerate periodontal attachment, this effect is concentrated to the base of the defect. The potential of DFDB to elicit a host immune response, the risk of disease transmission, and the variability in biological activity between different preparations has further dampened enthusiasm for this approach.

Consequently, significant effort has been placed on developing new methods for periodontal regeneration that involve direct delivery of secreted growth factors. The success of these approaches rests in part on the presence of putative precursor cells within the vicinity of the periodontal attachment that are capable of differentiating into the more specialized cell types required for the reconstruction of a functional periodontal attachment apparatus when stimulated with appropriate growth factors. Putative protein factors common to both cementum and bone include those that stimulate osteogenesis (e.g., bone morphogenetic protein-2 and -7, IGF-I and IGF-II), those that promote cellular differentiation (e.g. PDGF) and angiogenesis (e.g. VegF) [87].

Emdogain (Strauman AG, Basel, Switzerland), a mixture of enamel matrix proteins isolated from developing porcine teeth, was recently approved by the FDA for regeneration of angular intra-bony periodontal defects. Its mechanism of action is not known. It also contains numerous osteogenic proteins, including both TGF-β and BMPs [88]. Overall, Embdogain's efficacy at promoting periodontal regeneration [89] is similar to that of GTR alone. Sporadic cases of external root resorption following the use of Embdogain have been reported [90].

The usefulness of PRP, a concentrate of autologous whole blood obtained following the centrifugation of the patient's own plasma, at promoting periodontal regeneration is also most likely related to it functioning as a reservoir of growth factors such as PDGF and TGF-β [91, 92].

While local delivery of recombinant human BMP-2 (rhBMP-2) and rhBMP-7 to sites of periodontal attachment loss may prove beneficial, one potential concern has been the risk of ankylosis, the development of a direct non-physiological connection between the tooth root and the neighboring alveolar bone without intervening periodontal ligament fibers. This was evident in a dog model when using BMP-2 [93]; whereas, in baboons, the use of BMP-7 was not associated with ankylosis [94]. It is evident from these studies that treatment outcomes are heavily influenced by variables such as the animal model and type of defect created, whether the treated teeth are in occlusion, in addition to the specific protein carrier used [112]. Other growth factors employed with varying success have included PDGF±IGF-I [95, 96], FGF-2 [97], TGF-β1 [98], and brain-derived neurotrophic factor [99]. Several reviews detailing the strengths and weaknesses of these different growth factors for periodontal regeneration are available [100, 101, 102].

One potential shortcoming of all of these protein delivery techniques is that these factors, which have a short *in vivo* half-life once released, are delivered as a single non-physiological bolus. Development of controlled-release delivery approaches has the potential to significantly increase their clinical effectiveness [103]

Another promising potential approach to periodontal regeneration is the use of a cell-delivery approach [104]. The exact source of periodontal precursor cells has yet to be determined, although a population of multipotent postnatal stem cells have been isolated from human PDL (PDLSCs) that are capable of generating cementum/PDL-like structures when transplanted into immunodeficient rats [105, 106].

The use of gene-enhanced tissue engineering to express growth factors that are involved in the initial formation of both dental and periodontal attachment tissues in conjunction with cell delivery has definite promise if this approach could be adapted for use with easily harvestable fully mature cells (e.g., gingival fibroblasts, periodontal ligament fibroblasts). In short, this approach is intended to mimic the normal biological process that occurs as these tissues are formed early in development.

While still in the very early stages of development, several "proof of concept" studies have established that GETE may offer potential in periodontal regeneration. Using syngeneic dermal fibroblasts transduced *ex vivo* with an adenoviral vector expressing BMP-7 [107], significant bridging of surgically created periodontal–alveolar bone defects was seen in the rat. Interestingly, new bone formation occurred through a process of endochondral ossification. In another approach, direct *in vivo* transfer of PDGF-B was shown to stimulate both alveolar bone and cementum regeneration in a rat acute periodontitis model [108].

26.4.4
Dentin/Pulp Regeneration

The goal of modern restorative dentistry is to functionally and, where possible, cosmetically restore tooth structure destroyed from dental caries. The currently used restorative materials include metal and polymer-based materials; primarily silver amalgam, resin-based composites and metal or porcelain crowns. While highly effective at preserving teeth, these materials have a limited life span. It has been estimated that in the United States alone, two-thirds of the 300 million restorations placed by dentists each year involve the replacement of failed restorations [109]. In addition, a significant number of these restored teeth eventually undergo pulpal necrosis, requiring either tooth extraction or endodontic treatment. Therefore, development of novel techniques to regenerate specific components of an otherwise viable tooth, as opposed to repairing lost tooth structure, would have significant societal benefits. The lack of enamel-forming cells in fully developed erupted teeth precludes using cell-based approaches for enamel regeneration. In contrast, the regeneration of dentin is feasible because it is in intimate contact with pulpal tissue, forming a "dentin–pulp complex".

During primary tooth formation, dentin is produced by odontoblastic cells located within the pulp. Following tooth eruption, pulpal tissue retains a limited potential to repair itself. Specifically, perivascular progenitor cells in the pulp [110] can be triggered to differentiate into odontoblastic-like cells under the influence of specific growth factors, including BMP-2, BMP-7, and insulin-like growth factor-1 (IGF-1) [111, 112, 113, 114, 115, 116, 117].

Attempts at regenerating dental hard tissue using these growth factors have shown mixed results. Intrapulpal application of TGF-β1 induces differentiation of odontoblast-like cells and reparative dentin formation in the immediate vicinity of the application site [118], and dentin matrix extract, a complex mixture of bioactive molecules, induces differentiation of pulp progenitor cells into odontoblast-like cells *in vitro* [119]. When translated to in vivo models, results with TGF-β 1 [120], dentin matrix extract [121, 122, 123], supraphysiological doses of recombinant BMPs [124], bone sialoprotein [125], and amelogenin gene splice products [126] have, for the most part, been less than impressive, affording either minimal dentin formation or excessive amounts of ectopic bone-like product. In fact, pulp capping with MTA resulted in improved dentinoblast differentiation when compared with recombinant BMP-7 [125]. Presumably, delivery of a single dose of protein does not provide the sustained physiological delivery required for ideal tissue regeneration. In addition, supraphysiological doses of some factors appears to have an inhibitory effect on hard tissue regeneration [122].

Human adult dental pulp contains a population of cells ("dentin pulp stem cells") with stem cell-like properties, including self-renewal and the ability to differentiate into adipocytes and neural-like cells [127, 128, 129]. These dentin pulp stem cells appear to hold promise for dentin/pulp regeneration, especially when supplemented with appropriate growth factors [130]. Likewise, deciduous teeth [131] contain a population of immature multipotent cells ("stem cells from human exfoliated deciduous teeth") that can be induced to form dentin-like structures.

The use of GETE in dentin/pulp regeneration is still in the earliest phase of development. Ex vivo gene transfer of BMP-7 using dermal fibroblasts induces reparative dentinogenesis and regeneration of the dentin–pulp complex in a ferret model of reversible pulpitis [132]. However, in the same model, *in vivo* transduction with BMP-7 recombinant adenovirus was ineffective. Ultrasound-mediated delivery of BMP-11 cDNA to mechanically exposed canine pulp tissue resulted in reparative dentin formation [133, 134]. Similarly, *ex vivo* transplantation of BMP-11-transfected dental pulp stem cells stimulates reparative dentin formation in the dog model [135].

There are significant local issues that will need to be overcome before these approaches will be clinically feasible, not the least of which are the hemodynamic changes in the infected pulp that compromise the overall vitality of the dentin–pulp complex. Absent an adequate supply of viable pulpal tissue and pulp progenitor stem cells, *in vivo* gene therapy techniques will not be successful. In those situations, *ex vivo* approaches, in which cells are isolated outside the tooth and enhanced with growth factors prior to transplantation into the tooth, will be required. With either approach, the cells would require a source of oxygen and nutrients to sustain viability. In addition to neovascularization, complete restoration of the dentin–pulp complex will also require regeneration of the pulpal nerve supply. Key questions regarding the signal transduction mechanisms regulating the dentin–pulp complex remain unanswered. Until our understanding of these mechanisms increases, further exploration of these approaches will be slow.

26.4.5
Articular Cartilage Repair

Osteoarthritis-induced loss of condylar head articular cartilage is a significant public health problem, contributing to temporomandibular joint (TMJ) dysfunction in a sizeable number of patients. Although reconstruction of the mandibular condyle has been performed with various autogenous graft materials, the costochondral graft remains one of the most widely employed techniques. However, as with all surgical bone augmentation approaches, these techniques result in the creation of secondary bone defects at the donor site.

In contrast to efforts at regenerating articular cartilage in the knee, similar efforts are at a very preliminary stage in the jaws. It can be argued that the regenerative techniques developed to treat cartilage defects in the knees, including autologous chondrocyte implantation and periosteal transfer, should be generally transferable to the TMJ. Interesting approaches being developed for the knee include the use minimally invasive techniques in which cells and/or drugs in a gel-based carrier are injected into the defect site.

One readily adaptable protocol for articular cartilage regeneration involves the isolation of fully differentiated autologous chondrocytes followed by cell expansion in culture, and placement into the defect site in a suitable carrier (especially alginate, as this seems to assist in maintaing chondrocytes in the differentiated state), potentially coupled with the addition of specific growth factors. However, this approach is limited by the difficulty in isolating sufficient numbers of donor cells, since harvested chondrocytes lose their chondrocyte phenotype during two-dimensional serial expansion. Moreover, chondrocytes compose only a small percentage of the overall structure of articular cartilage. For a recent comprehensive review of the various parameters, the reader is referred to Wang and Detamore [136].

The use of micromass tissue-culture techniques, in which cells are induced to remain in the chondrogenic phenotype in the presence of a chondrogenic media and appropriate environmental cues (e.g., centrifugal force to simulate weight bearing) without the need for foreign matrices, offers significant promise in this area [137].

Few studies have critically evaluated the potential of these modalities to regenerate articular cartilage in the TMJ *in vivo*. One of the problems in assessing the outcome of any regenerative intervention in the TMJ is the wide range of regenerative ability of the condylar cartilage; which is not only species and age dependent, but also varies depending on the type of defect created. For example, surgically created full-thickness defects placed in the mandibular cartilage completely heal without intervention in young sheep [138]. In the adult dog, unilateral condylectomy is associated with partial regeneration of condylar structure at 3 months, with irregular cartilage formation [139], while the immature mini-pig retains the ability to regenerate a condyle-like structure if periosteum, capsule, and disc are preserved [140].

BMP delivery for cartilage regeneration in the TMJ has only recently been investigated. In one study in a rabbit model [141], rhBMP-2 at doses of 3 and 15 µg per defect induced proliferation of cartilage, whereas fibrous tissue proliferation predominated at the 0.6-µg rhBMP-2 dose. Very exciting is the potential use of gene-enhanced tissue engineering for cartilage reconstruction.

26.4.6
Complete Regeneration of Teeth and Jaws

One of the long-term goals espoused by researchers in the field of dentofacial tissue regeneration is the regeneration of entire teeth and jaws. Although tooth-like tissues have been engineered in the omenta of athymic rats by implanting single cell suspensions isolated from porcine third molar tooth follicles [142], the resulting product resembles more of a hamartomatous tooth-like tissue than a fully developed mature tooth [143]. Similarly, explants of adult bone marrow stem cells and oral epithelium from E10.0 mouse embryos form crude tooth-like tissues when grown in kidney capsules [144].

The possibility of regenerating a section of mandible, in this case removed secondary to a large odontogenic tumor, was demonstrated by Moghadam et al. [145]. A bioimplant of demineralized bone matrix in a malleable poloxamer gel (DynaGraft Gel, GenSci Regenerative Sciences, Irvine, Ca), supplemented with 200 mg of "native human BMP", was used. Unfortunately, the exact source and type of BMP was not described; and it is not possible to determine the relative contributions of the DBM and the gel versus the added BMP to the overall amount of bone regeneration.

Warnke et al. [146] recently reported a successful proof of principal study in which a patient's body, using a surgically created pouch in the latissimus dorsus muscle, was used as an incubator to produce an osseous implant approximating the size of a human mandible. Specifically, a titanium mesh cage was formed around a Teflon model constructed using computer-aided design from data obtained by 3-dimensional computed tomography of the patient's jaws. The supporting cage was filled with blocks and granules of bovine bone (Bio-Oss, Geistlich Biomaterials), and supraphysiological doses of recombinant BMP-7 and bovine type-I collagen. Unenriched whole bone marrow aspirate from the iliac crest was added as a presumed source of stromally derived mesenchymal stem cells.

26.5
Conclusions

The subject of tissue regeneration is still in its infancy. To date, no single approach to tissue regeneration has proven superior to others. In fact, the availability of numerous approaches, coupled with our limited knowledge of how the myriad of different potential growth factors and matrices interact, makes identification of a single "ideal" approach to tissue regeneration impractical. Depending on the particular clinical situation, each approach has its own merits and its own set of unfavorable features. It is our contention that, as understanding of the molecular interactions between the different growth factors involved in development of the dentoalveolar structures increases, the number of potential therapeutic agents will likewise grow.

It can be argued that the oral cavity is an ideal environment in which to investigate the different parameters that contribute to successful hard tissue regeneration. Compared with other sites in the body, the oral cavity offers easy access and observability, the ability to limit the potentially negative effects of immediate graft loading, minimal tendency for scar formation, and a rich vascular supply. This excitement must, however, be tempered to some extent by one of the unique features of the oral cavity, namely the relatively easy access of the abundant mixed flora of the oral cavity to the underlying alveolar bone as a result of the direct connection between the roots of the teeth and the surrounding alveolar bone.

The presence of many tissue types that must be intimately integrated to function properly complicates development of such approaches. Moreover, successful regeneration will require the sequential coordination of a number of tightly related processes, many of which will need to be elucidated before ideal conditions can be developed. Ultimately, the key to developing successful approaches to the regeneration of dentofacial hard tissues will be the ability to devise techniques that produce the best quality hard tissue and that are both clinically and economically feasible.

References

1. Ruggiero SL, Mehrotra B, Rosenberg TJ, Engroff SL. Osteonecrosis of the jaws associated with the use of bisphosphonates: a review of 63 cases. J Oral Maxillofac Surg. 2004;62(5):527–34.
2. Pickett FA. Bisphosphonate-associated osteonecrosis of the jaw: a literature review and clinical practice guidelines. J Dent Hyg. 2006;80(3):10.
3. Oliver RC, Brown LJ, Loe H. Periodontal diseases in the United States population. J. Periodontol 1998;69:269–278.
4. Albandar JM, Brunelle JA, Kingman A. Destructive periodontal disease in adults 30 years of age and older in the United States, 1988–1994. J Periodontol. 1999 Jan;70(1):13–29.
5. Kwong LM, Jasty M, Harris WH. High failure rate of bulk femoral head allografts in total hip acetabular reconstructions at 10 years. J Arthroplasty 1993: 8;341–346.
6. Khan SN, Tomin E, Lane JM. Clinical applications of bone graft substitutes. Orthop Clin North Am 2000: 31(3); 389–398.
7. Vaccaro AR. The role of the osteoconductive scaffold in synthetic bone graft. Orthopedics 2002: 25(5 Suppl); S571–578.
8. Fleming JE, Cornell CN, Muschler GF. Bone cells and matrices in orthopedic tissue engineering. Orthop Clin North Am 2000: 31; 357–374.
9. Lieberman JR, Ghivizzani SC, Evans CH. Gene transfer approaches to the healing of bone and cartilage. Molecular Therapy 2002: 6; 141–147.
10. Gross JS. Bone grafting materials for dental applications: a practical guide. Compend Contin Educ Dent. 1997 Oct;18(10):1013–8, 1020–2, 1024.
11. LeGeros, RZ. Properties of Osteoconductive Biomaterials: Calcium Phosphates. Clinical Orthopaedics & Related Research. 2002;395:81–98.
12. Damien CJ, Parsons JR. Bone graft and bone graft substitutes: A review of current technology and applications. Journal of Applied Biomaterials 2004;2(3):187–208.
13. Gelberman RH, Manske PR, Akeson WH, Woo SL, Lundborg G, Amiel D. Flexor tendon repair. J Orthop Res 1986;4;119–128.
14. Leukocyte migration and inflammation. In: Goldsby RA, Kindt TJ, Osborne BA, editors. Kuby Immunology, 4th Edition. New York. W.H. Freeman & Co Ltd: 2000;379–387.
15. Martin P, Parkhurst SM. Parallels between tissue repair and embryo morphogenesis. Development 2004:3021–3034.
16. Gelberman RH, Bandeberg JS, Manske PR, Akeson WH. The early stages of flexor tendon healing: a morphologic study of the first fourteen days. J Hand Surg Am 1985:10;776–784.
17. Rappolee DA, Mark D, Banda MJ, Werb Z. Wound macrophages express TGF-alpha and other growth factors in vivo: analysis by mRNA phenotyping. Science 1988:241;708–712.
18. Zagai U, Fredriksson K, Rennard SI, Lundahl J, Skold CM. Platelets stimulate fibroblast-mediated contraction of collagen gels. Respiratory Research 2003:4;13.
19. Hasleton PS, Roberts TE. Adult respiratory distress syndrome—an update. Histopathology 1999:34;285–294.
20. Beredjiklian PK. Biologic aspects of flexor tendon laceration and repair. J Bone Joint Surg 2003:85A;539–550.
21. Garcia-Godoy F, Murray PE. Status and potential commercial impact of stem cell-based treatments on dental and craniofacial regeneration. Stem Cells Dev. 2006;15(6):881–887.
22. Modino SA, Sharpe PT. Tissue engineering of teeth using adult stem cells. Arch Oral Biol. 2005;50(2):255–258.
23. Ring ME. A thousand years of dental implants: a definitive history—part 1. Compend Contin Educ Dent. 1995;16(10):1060–1064.
24. Pepinsky RB, Shapiro RI, Wang S, Chakraborty A, Gill A, Lepage DJ, Wen D, Rayhorn P, Horan GS, Taylor FR, Garber EA, Galdes A, Engber TM. Long-acting forms of Sonic hedgehog with improved pharmacokinetic and pharmacodynamic properties are efficacious in a nerve injury model. Journal of Pharmaceutical Sciences 2002: 91(2); 371–387.
25. Wozney JM. Overview of Bone Morphogenic Proteins. SPINE 2002: 27; S2–S8.
26. Groeneveld EHJ, Burger EH. Bone Morphogenic Proteins in human bone regeneration. European Journal of Endocrinology 2000: 142; 9–21.
27. Quarto R, Thomas D, Liang CT. Bone progenitor cell deficits and the age-associated decline in bone repair capacity. Calcif Tissue Int. 1995: 56;123–129.
28. Oreffo RO, Bord S, Triffitt JT. Skeletal progenitor cells and ageing human populations. Clin Sci (Lond) 1998: 94;549–555.
29. Leskela HV, Risteli J, Niskanen S, Koivunen J, Ivaska KK, Lehenkari P. Osteoblast recruitment from stem cells does not decrease by age at late adulthood. Biochem Biophys Res Commun. 2003: 311; 1008–1013.
30. Moioli EK, Clark PA, Xin X, Lal S, Mao JJ. Matrices and scaffolds for drug delivery in dental, oral and craniofacial tissue engineering. Adv Drug Deliv Rev. 2007;59(4–5):308–324.
31. Goldberg M, Langer R, Jia X. Nanostructured materials for applications in drug delivery and tissue engineering. J Biomater Sci Polym Ed. 2007;18(3):241–268.
32. Earthman JC, Li Y, VanSchoiack LR, Sheets CG, Wu JC. Reconstructive materials and bone tissue engineering in implant dentistry. Dent Clin North Am. 2006;50(2):229–244.
33. Park J, Lutz R, Felszeghy E, Wiltfang J, Nkenke E, Neukam FW, Schlegel KA. The effect on bone regeneration of a liposomal vector to deliver BMP-2 gene to bone grafts in peri-implant bone defects. Biomaterials. 2007;28(17):2772–2782.
34. Rabie AB, Dai J, Xu R. Recombinant AAV-mediated VEGF gene therapy induces mandibular condylar growth. Gene Ther. 2007;14(12):972–980.
35. Gerber HP, Vu TH, Ryan AM, Kowalski J, Werb Z, Ferrara N. VegF couples hypertrophic cartilage remodeling,

36. Kashofer K, Bonnet D. Gene therapy progress and prospects: stem cell plasticity. Gene Ther. 2005;12(16):1229–1234.
37. Ohazama A, Modino SA, Miletich I, Sharpe PT. Stem-cell-based tissue engineering of murine teeth. J Dent Res. 2004;83(7):518–522.
38. Yu J, Wang Y, Deng Z, Tang L, Li Y, Shi J, Jin Y. Odontogenic capability: bone marrow stromal stem cells versus dental pulp stem cells. Biol Cell. 2007; [Epub ahead of print]
39. Komine A, Suenaga M, Nakao K, Tsuji T, Tomooka Y. Tooth regeneration from newly established cell lines from a molar tooth germ epithelium. Biochem Biophys Res Commun. 2007;355(3):758–763.
40. Onyekwelu O, Seppala M, Zoupa M, Cobourne MT. Tooth development: 2. Regenerating teeth in the laboratory. Dent Update. 2007;34(1):20–29.
41. Hsu WK, Sugiyama O, Park SH, Conduah A, Feeley BT, Liu NQ, Krenek L, Virk MS, An DS, Chen IS, Lieberman JR. Lentiviral-mediated BMP-2 gene transfer enhances healing of segmental femoral defects in rats. Bone 2007;40(4):931–938.
42. Jones AA, Buser D, Schenk R, Wozney J, Cochran DL. The effect of rhBMP-2 around endosseous implants with and without membranes in the canine model. J Periodontol. 2006;77(7):1184–1193.
43. Salata LA, Burgos PM, Rasmusson L, Novaes AB, Papalexiou V, Dahlin C, Sennerby L. Osseointegration of oxidized and turned implants in circumferential bone defects with and without adjunctive therapies: an experimental study on BMP-2 and autogenous bone graft in the dog mandible. Int J Oral Maxillofac Surg. 2007;36(1):62–71.
44. Artzi Z, Weinreb M, Tal H, Nemcovsky CE, Rohrer MD, Prasad HS, Kozlovsky A. Experimental intrabony and periodontal defects treated with natural mineral combined with a synthetic cell-binding Peptide in the canine: morphometric evaluations. J Periodontol. 2006;77(10):1658–1664.
45. Nussenbaum B, Krebsbach PH. The role of gene therapy for craniofacial and dental tissue engineering. Adv Drug Deliv Rev. 2006;58(4):577–591.
46. Itaka K, Ohba S, Miyata K, Kawaguchi H, Nakamura K, Takato T, Chung UI, Kataoka K. Bone Regeneration by Regulated In Vivo Gene Transfer Using Biocompatible Polyplex Nanomicelles. Mol Ther. 2007; [Epub ahead of print]
47. De Laporte L, Shea LD. Matrices and scaffolds for DNA delivery in tissue engineering. Adv Drug Deliv Rev. 2007;59(4–5):292–307.
48. Dunn CA, Jin Q, Taba M, Franceschi RT, Bruce Rutherford R, Giannobile WV. BMP gene delivery for alveolar bone engineering at dental implant defects. Mol Ther. 2005;11(2):294–299.
49. Hu WW, Wang Z, Hollister SJ, Krebsbach PH. Localized viral vector delivery to enhance in situ regenerative gene therapy. Gene Ther. 2007;14(11):891–901.
50. Graziano A, d'Aquino R, Cusella-De Angelis MG, Laino G, Piattelli A, Pacifici M, De Rosa A, Papaccio G. Concave pit-containing scaffold surfaces improve stem cell-derived osteoblast performance and lead to significant bone tissue formation. PLoS ONE. 2007;2:e496.
51. Sonoyama W, Liu Y, Fang D, Yamaza T, Seo BM, Zhang C, Liu H, Gronthos S, Wang CY, Shi S, Wang S. Mesenchymal stem cell-mediated functional tooth regeneration in Swine. PLoS ONE. 2006;1:e79.
52. Umeda H, Kanemaru S, Yamashita M, Kishimoto M, Tamura Y, Nakamura T, Omori K, Hirano S, Ito J. Bone regeneration of canine skull using bone marrow-derived stromal cells and beta-tricalcium phosphate. Laryngoscope. 2007;117(6):997–1003.
53. Marei MK, Nouh SR, Saad MM, Ismail NS. Preservation and regeneration of alveolar bone by tissue-engineered implants. Tissue Eng. 2005;11(5–6):751–767.
54. Honda MJ, Ohara T, Sumita Y, Ogaeri T, Kagami H, Ueda M. Preliminary study of tissue-engineered odontogenesis in the canine jaw. J Oral Maxillofac Surg. 2006;64(2):283–289.
55. Honda MJ, Tsuchiya S, Sumita Y, Sagara H, Ueda M. The sequential seeding of epithelial and mesenchymal cells for tissue-engineered tooth regeneration. Biomaterials. 2007;28(4):680–689.
56. Zizelmann C, Schoen R, Metzger MC, Schmelzeisen R, Schramm A, Dott B, Bormann KH, Gellrich NC. Bone formation after sinus augmentation with engineered bone. Clin Oral Implants Res. 2007;18(1):69–73.
57. Mao JJ, Giannobile WV, Helms JA, Hollister SJ, Krebsbach PH, Longaker MT, Shi S. Craniofacial tissue engineering by stem cells. J Dent Res. 2006;85(11):966–979.
58. Yamada Y, Ueda M, Hibi H, Nagasaka T. Translational research for injectable tissue-engineered bone regeneration using mesenchymal stem cells and platelet-rich plasma: from basic research to clinical case study. Cell Transplant. 2004;13(4):343–355.
59. Wojtowicz A, Chaberek S, Urbanowska E, Ostrowski K. Comparison of efficiency of platelet rich plasma, hematopoieic stem cells and bone marrow in augmentation of mandibular bone defects. N Y State Dent J. 2007;73(2):41–45.
60. Kimelman N, Pelled G, Helm GA, Huard J, Schwarz EM, Gazit D. Gene- and stem cell-based therapeutics for bone regeneration and repair. Tissue Eng. 2007;13:1135–1150.
61. Phillips JE, Gersbach CA, García AJ. Virus-based gene therapy strategies for bone regeneration. Biomaterials 2007;28(2):211–229.
62. Peterson B, Zhang J, Iglesias R, Kabo M, Hedrick M, Benhaim P, Lieberman JR. Healing of critically sized femoral defects, using genetically modified mesenchymal stem cells from human adipose tissue. Tissue Eng. 2005;11:120–129.
63. Li H, Dai K, Tang T, Zhang X, Yan M, Lou J. Bone regeneration by implantation of adipose-derived stromal cells expressing BMP-2. Biochem Biophys Res Commun. 2007;356:836–842.
64. Edwards P, Ruggiero S, Fantasia J, Burakoff R, Moorji S, Paric E, Razzano P, Grande DA, Mason JM. Sonic Hedgehog Gene Enhanced Tissue Engineering for Bone Regeneration. Gene Therapy 12:75–86, 2005.

65. Ohba S, Ikeda T, Kugimiya F, Yano F, Lichtler AC, Nakamura K, Takato T, Kawaguchi H, Chung UI. Identification of a potent combination of osteogenic genes for bone regeneration using embryonic stem (ES) cell-based sensor. FASEB J. 2007;21:1777–1787.
66. Ito K, Yamada Y, Naiki T, Ueda M. Simultaneous implant placement and bone regeneration around dental implants using tissue-engineered bone with fibrin blue, mesenchymal stem cells and platelet-rich plasma. Clin Oral Implants Res. 2006;17:579–586.
67. Fouletier-Dilling CM, Gannon FH, Olmsted-Davis EA, Lazard Z, Heggeness MH, Shafer JA, Hipp JA, Davis AR. Efficient and rapid osteoinduction in an immune-competent host. Hum Gene Therapy 2007;18:733–745.
68. Pestell R. Light-activated gene therapy, new selective therapies for disease. J Nuc Med 2006;47:21–22N.
69. Grande DA, Breitbart AS, Mason JM, Paulino C, Laser J, Schwartz RE. Cartilage tissue engineering: current limitations and solutions. Clin Orthop Rel Res 1999, 367: S176–185.
70. Grande DA, Mason J, Light E, Dines D. Stem cells as platforms for delivery of genes to enhance cartilage repair. J Bone Joint Surg 2003, 85 Supp 2:111–116.
71. Liu Y, de Groot K, Hunziker EB. Osteoinductive implants: the mise-en-scene for drug-bearing biomimetic coatings. Ann Biomed Eng. 2004 Mar;32:398–406.
72. Jung RE, Cochran DL, Domken O, Seibl R, Jones AA, Buser D, Hammerle CH. The effect of matrix bound parathyroid hormone on bone regeneration. Clin Oral Implants Res. 2007 Jun;18:319–25.
73. Moioli EK, Clark PA, Xin X, Lal S, Mao JJ. Matrices and scaffolds for drug delivery in dental, oral and craniofacial tissue engineering. Adv Drug Deliv Rev. 2007;59:308–24.
74. Cook SD, Salkeld SL, Rueger DC. Evaluation of recombinant human osteogenic protein-1 (rhOP-1) placed with dental implants in fresh extraction sockets. J Oral Implantol 1995: 21: 281–289.
75. Salata LA, Burgos PM, Rasmusson L, Novaes AB, Papalexiou V, Dahlin C, Sennerby L. Osseointegration of oxidized and turned implants in circumferential bone defects with and without adjunctive therapies: an experimental study on BMP-2 and autogenous bone graft in the dog mandible. Int J Oral Maxillofac Surg. 2007 Jan;36:62–71.
76. Schliephake H, Aref A, Scharnweber D, Bierbaum S, Roessler S, Sewing A. Effect of immobilized bone morphogenic protein 2 coating of titanium implants on peri-implant bone formation. Clin Oral Implants Res. 2005;16:563–9.
77. Bessho K, Carnes DL, Cavin R, Chen HY, Ong JL. BMP stimulation of bone response adjacent to titanium implants in vivo. Clin Oral Implants Res. 1999;10:212–8.
78. Hall J, Sorensen RG, Wozney JM, Wikesjö UM. Bone formation at rhBMP-2-coated titanium implants in the rat ectopic model. J Clin Periodontol. 2007;34:444–51.
79. Boyne PJ, Marx RE, Nevins M, Triplett G, Lazaro E, Lilly LC, Alder M, Nummikoski P. A feasibility study evaluating rhBMP-2/absorbable collagen sponge for maxillary sinus floor augmentation. Int J Periodontics Restorative Dent. 1997;17:11–25.
80. Dunn CA, Jin Q, Taba M, Franceschi RT, Rutherford BR, Giannobile WV. BMP gene delivery for alveolar bone engineering at dental implant defects. Mol Ther. 2005;11:294–9.
81. Pihlstrom BL, Michalowicz BS, Johnson NW. Periodontal diseases. Lancet 2005, 366:1809–1820.
82. American Academy of Periodontology. Position paper: Epidemiology of periodontal diseases. J Periodontol 2005, 76:1406–1419.
83. Brown LJ, Johns BA, Wall TP. The economics of periodontal diseases. Periodontol 2000:2002, 29:223–234.
84. Needleman I, Tucker R, Giedrys-Leeper E, Worthington H. Guided tissue regeneration for periodontal intrabony defects-a Cochrane Systematic Review. Periodontol 2000 2005, 37:106–123.
85. American Academy of Periodontology. Position paper. Periodontal regeneration. J Periodontol 2005, 76:1601–1622.
86. Reynolds MA, Aichelmann-Reidy ME, Branch-Mays GL, Gunsolley JC. The efficacy of bone replacement grafts in the treatment of periodontal osseous defects. A systematic review. Ann Periodontol 2003, 8:227–265.
87. Murphy WL, Simmons CA, Kaigler D, Mooney DJ. Bone regeneration via a mineral substrate and induced angiogenesis. J Dent Res 2004, 83:204–210.
88. Suzuki S, Nagano T, Yamakoshi Y, Gomi K, Arai T, Fukae M, Katagiri T, Oida S. Enamel matrix derivative gel stimulates signal transduction of BMP and TGF-beta. J Dent Res 2005, 84:510–514.
89. Esposito M, Grusovin MG, Coulthard P, Worthington HV. Enamel matrix derivative (Emdogain™) for periodontal tissue regeneration in intrabony defects. The Cochrane Database of Systemic Reviews 2005, 4: Art. No.;CD003875.pub2.
90. St George G, Darbar U, Thomas G. Inflammatory external root resorption following surgical treatment for intra-bony defects: a report of two cases involving Emdogain and a review of the literature. J Clin Periodontol. 2006;33:449–54.
91. Okuda K, Tai H, Tanabe K, Suzuki H, Sato T, Kawase T, Saito Y, Wolff LF, Yoshiex H. Platelet-rich plasma combined with a porous hydroxyapatite graft for the treatment of intrabony defects in humans: a comparative controlled clinical study. J Periodontol 2005, 76:890–898.
92. Sammartino G, Tia M, Marenzi G, diLauro AE, D'Agostino E, Claudio PP. Use of autologous platelet-rich plasma (PRP) in periodontal defect treatment after extraction of impacted third molars. J Oral Maxillofac Surg 2005, 63:766–770.
93. Sigurdsson TJ, Nygaard L, Tatakis DN, Fu E, Turek TJ, Jin L, Wozney JM, Wikesjo UM. Periodontal repair in dogs: evaluation of rhBMP-2 carriers. Int J Periodont Res Dent 1996, 16:524–537.
94. Ripamonti U, Heliotis M, Rueger DC, Sampath TK. Induction of cementogenesis by recombinant human osteogenic protein-1 (hOP-1/BMP-7) in the baboon (Papio ursinus). Arch Oral Biol 1996, 41:121–126.
95. Giannobile WV, Finkelman RD, Lynch SE. Comparison of canine and non-human primate animal models for periodontal regenerative therapy: results following a sin-

gle administration of PDGF/IGF-I. J Periodontol 1994, 65:1158–1168.
96. Howell TH, Fiorellini JP, Paquette DW, Offenbacher S, Giannobile WV, Lynch SE. A phase I/II clinical trial to evaluate a combination of recombinant human platelet-derived growth factor-BB and recombinant human insulin-like growth factor-I in patients with periodontal disease. J Periodontol 1997, 68:1186–1193.
97. Murakami S, Takayama S, Kitamura M, Shimabukuro Y, Yanagi K, Ikezawa K, Saho T, Nozaki T, Okada H. Recombinant human fibroblast growth factor (bFGF) stimulates periodontal regeneration in class II furcation defects created in beagle dogs. J Periodont Res 2003, 38:97–103.
98. Wikesjo UM, Razi SS, Sigurdsson TJ, Tatakis DN, Lee MB, Ongpipattanakul B, Nguyen T, Hardwick R. Periodontal repair in dogs: effect of recombinant human transforming growth factor-beta1 on guided tissue regeneration. J Clin Periodontol 1998, 25:475–481.
99. Takeda K, Shiba H, Mizuno N, Hasegawa N, Mouri Y, Hirachi A, Yoshino H, Kawaguchi H, Kurihara H. Brain-derived neurotrophic factor enhances periodontal tissue regeneration. Tissue Eng 2005, 11:1618–1629.
100. Anusaksathien O, Giannobile WV. Growth factor delivery to re-engineer periodontal tissues. Curr Pharm Biotech 2002, 3:129–139.
101. King GN, Cochran DL. Factors that modulate the effects of bone morphogenetic protein-induced periodontal regeneration: a critical review. J Periodontol 2002, 73:925–936.
102. Nakashima M, Reddi H. The application of bone morphogenetic proteins to dental tissue engineering. Nat Biotech 2003, 21:1025–1032.
103. King GN. The importance of drug delivery to optimize the effects of bone morphogenetic proteins during periodontal regeneration. Curr Pharm Biotech 2001, 2:131–142.
104. Kawaguchi H, Hayashi H, Mizuno N, Fujita T, Hasegawa N, Shiba H, Nakamura S, Hino T, Yoshino H, Kurihara H, Tanaka H, Kimura A, Tsuji K, Kato Y. Cell transplantation for periodontal disease. A novel periodontal tissue regenerative therapy using bone marrow mesenchymal stem cells (In Japanese). Clin Calcium 2005, 15:99–104.
105. Seo BM, Miura M, Gronthos S, Bartold PM, Batouli S, Brahim J. Investigation of multipotent stem cells from human periodontal ligament. Lancet 2004, 364:149–155.
106. Seo BM, Miura M, Sonoyama W, Coppe C, Stanyon R, Shi S. Recovery of stem cells from cryopreserved periodontal ligament. J Dent Res 2005, 84:907–912.
107. Jin QM, Anusaksathien O, Webb SA, Rutherford RB, Giannobile WV. Gene therapy of bone morphogenetic protein for periodontal tissue engineering. J Periodontol 2003, 74:202–213.
108. Jin QM, Anusaksathien O, Webb SA, Printz MA, Giannobile WV. Engineering of tooth-supporting structure by delivery of PDGF gene therapy vectors. Mol Ther 2004, 9:519–526.
109. Arnst C, Carey J. Biotech bodies. Business Week 1998, July:42–49.
110. Shi S, Gronthos S. Perivascular niche of postnatal mesenchymal stem cells in bone marrow and dental pulp. J Bone Mineral Res 2003, 18:696–704.
111. Saito T, Ogawa M, Hata Y, Bessho K. Acceleration effect of human recombinant bone morphogenetic protein-2 on differentiation of human pulp cells into odontoblasts. J Endodont 2004, 30:205–208.
112. AS, Camps J, Dejou J, About I. Activation of human dental pulp progenitor/stem cells in response to odontoblast injury. Arch Oral Biol 2005, 50:103–108.
113. Zhang YD, Chen Z, Song YO, Lin C, Chen YP. Making a tooth: growth factors, transcription factors, and stem cells. Cell Res 2005, 15:301–316.
114. Cheifetz S. BMP receptors in limb and tooth formation. Crit Rev Oral Biol Med 1999, 10:182–198.
115. Helder MN, Kasg H, Bervoets TJ, Vukicevic S, Burger EH, D'Souza RN, Wöltgens JH, Karsenty G, Bronckers AL. Bone morphogenetic protein-7 (osteogenic protein-1, OP-1) and tooth development. J Dent Res 1998, 77:545–554.
116. Mitsiadis TA, Rahiotis C. Parallels between tooth development and repair: conserved molecular mechanisms following carious and dental injury. J Dent Res 2004, 83:896–902.
117. Yamashiro T, Tummers M, Thesleff I. Expression of bone morphogenetic proteins and Msx genes during root formation. J Dent Res 2003, 82:172–176.
118. Tziafas D, Alvanou A, Komnenou A, Gasic J, Papadimitriou S. Effects of basic fibroblast growth factor, insulin-like growth factor-II and transforming growth factor beta1 on dental pulp cells after implantation in dog teeth. Arch Oral Biol 1998, 43:431–444.
119. Jin LJ, Ritchie HH, Smith AJ, Clarkson BH. In vitro differentiation and mineralization of human dental pulp cells induced by dentin extract. In vitro Cell Dev Biol Anim 2005, 41:232–238.
120. Tziafas D, Belibasakis G, Veis A, Papadimitriou S. Dentin regeneration in vital pulp therapy: design principles. Adv Dent Res 2001, 15:96–100.
121. Smith AJ, Patel M, Graham L, Sloan AJ, Cooper PR. Dentine regeneration: key roles for stem cells and molecular signalling. Oral Biosci Med 2005, 2:127–132.
122. Ishizaki NT, Matsumoto K, Kimura Y, Wang X, Yamashita A. Histopathological study of dental pulp tissue capped with enamel matrix derivative. J Endodon 2003, 29:176–179.
123. Tziafas D, Kalyva M, Papadimitriou S. Experimental dentin-based approaches to tissue regeneration in vital pulp therapy. Connec Tis Res 2002, 43:391–395.
124. Andelin WE, Shbahang S, Wright K, Torabinejad M. Identification of hard tissue after experimental pulp capping using dentin sialoprotein (DSP) as a marker. J Endodont 2003, 29:646–650.
125. Decup F, Six N, Palmier B, Buch D, Lasfargues JJ, Salih E, Goldberg M. Bone sialoprotein-induced reparative dentinogenesis in the pulp of rat's molar. Clin Oral Investig 2000, 4:110–119.
126. Goldberg M, Six N, Decup F, Lasfargues JJ, Salih E, Tompkins K, Veis A. Bioactive molecules and the future of pulp therapy. Am J Dent 2003, 16:66–76.
127. Shi S, Bartold P, Miura M, Seo B, Robey P, Gronthos S. The efficacy of mesenchymal stem cells to regenerate and repair dental structures. Orthod Craniofac Res 2005, 8:191–199.

128. Gronthos S, Brahim J, Li W, Fisher LW, Cherman N, Boyde A, DenBesten P, Robey PG, Shi S. Stem cell properties of human dental pulp stem cells. J Dent Res 2002, 81:531–535.
129. Pierdomenico L, Bonsi L, Calvitti M, Rondelli D, Arpinati M, Chirumbolo G, Becchetti E, Marchionni C, Alviano F, Fossati V, Staffolani N, Franchina M, Grossi A, Bagnara GP. Multipotent mesenchymal stem cells with immunosuppressive activity can be easily isolated from dental pulp. Transplant 2005, 80:836–842.
130. Iohara K, Makashima M, Ito M, Ishikawa M, Nakasima A, Akamine A. Dentin regeneration by dental pulp stem cell therapy with recombinant human bone morphogenetic protein 2. J Dent Res 2004, 83:590–595.
131. Miura M, Gronthos S, Zhao M, Lu B, Fisher LW, Robey PG, Shi S. SHED. stem cells from human exfoliated deciduous teeth. PNAS 2003, 100:5807–5812.
132. Rutherford RB. BMP-7 gene transfer to inflamed ferret dental pulps. Eur J Oral Sci 2001, 109:422–424.
133. Nakashima M, Tachibana K, Iohara K, Ito M, Ishikawa M, Akamine A. Induction of reparative dentin formation by ultrasound-mediated gene delivery of growth/differentiation factor 11. Human Gene Ther 2003, 14:591–597.
134. Nakashima M, Mizunuma K, Murakami T, Akamine A. Induction of dental pulp stem cell differentiation into odontoblasts by electroporation-mediated gene delivery of growth/differentiation factor 11 (Gdf11). Gene Ther 2002, 9:814–818.
135. Nakashima M, Iohara K, Ishikawa M, Ito M, Tomoklyo A, Tanaka T, Akamine A. Stimulation of reparative dentin formation by ex vivo gene therapy using dental pulp stem cells electrotransfected with growth/differentiation factor 11 (Gdf11). Human Gene Ther 2004, 15:1045–1053.
136. Wang L, Detamore MS. Tissue Engineering the Mandibular Condyle. Online 2007.
137. Handschel JG, Depprich RA, Kubler NR, Wiesmann HP, Ommerborn M, Meyer U. Prospects of micromass culture technology in tissue engineering. Head Face Med. 2007; 3:4.
138. Güvena O, Metinb M, Keskina A. Remodelling in young sheep: a histological study of experimentally produced defects of the TMJ. Swiss Med Wkly 2003;133:423–426.
139. Miyamoto H, Shigematsu H, Suzuki S, Sakashita H. Regeneration of mandibular condyle following unilateral condylectomy in canines Journal of Cranio-Maxillofacial Surgery (2004) 32, 296–302.
140. Zouhary KJ, Feinberg SE. Condylar regeneration in adolescent minipigs. J Oral Maxillofac Surg 60;52 2002 (suppl 1).
141. Suzuki T, Bessho K, Fujimura K, Okubo Y, Segami N, Iizuka T. Regeneration of defects in the articular cartilage in rabbit temporomandibular joints by bone morphogenetic protein-2 British Journal of Oral and Maxillofacial Surgery (2002) 40, 201–206.
142. Young CS, Terada S, Vacanti JP, Honda M, Bartlett JD, Yelick PC. Tissue engineering of complex tooth structures on biodegradable polymer scaffolds. J Dent Res 2002, 81:695–700.
143. Honda J, Sumita Y, Kagami H, Ueda M. Histologic and immunohistochemical studies of tissue engineered odontogenesis. Arch Histol Cytol 2005, 68:89–101.
144. Modino SA, Sharpe PT. Tissue engineering of teeth using adult stem cells. Arch Oral Biol 2005, 50:255–258,
145. Moghadam HG, Urist MR, Sandor GK, Clokie CM. Successful mandibular reconstruction using a BMP bioimplant. J Craniofa Surg 2001;12:119–27.
146. Warnke PH, Springer IN, Wiltfang J, Acil Y, Eufinger H, Wehmöller M, Russo PA, Bolte H, Sherry E, Behrens E, Terheyden H. Growth and transplantation of a custom vascularised bone graft in a man Lancet 2004; 364: 766–70.

Mucosa Tissue Engineering

27

G. Lauer

Contents

27.1	Introduction	369
27.2	Optimizing Culture Conditions	369
27.3	Tissue Engineering of Epithelial Mucosa Graft	371
27.4	Clinical Application of the Tissue Engineered Epithelial Mucosa Graft and Histological Assessment	372
27.5	Histological and Immunohistological Assessment	373
27.6	Optimizing the Mucosa Graft: Developing Gingival Keratinocyte—Fibroblast Construct	375
27.7	Gingival Keratinocyte—Gingival Fibroblast Co-cultures	376
27.7.1	Sandwich Constructs	376
27.7.2	Composite Constructs	377
27.8	Cryopreservation	377
27.9	Epithelial Stem Cells	378
	References	379

27.1 Introduction

In oral and maxillofacial surgery as well as in other plastic reconstructive procedures there is the need for mucosa or gingiva transplants. Free mucosa transplants from the inner cheek or the palate are the main sources [3, 8, 30, 53] but they are limited in size. Therefore, either skin transplants or intestinal mucosa are used to cover extensive defects. A great disadvantage beside the donor site morbidity is that epithelium from a different location, and consequently of different characteristics and qualities, is transferred into the mouth and shows negligible assimilation even years after transplantation [5, 48, 56]. In particular, in the oral cavity where split skin grafts are pierced by dental implants there is chronic inflammation [32, 51]. These facts clearly underline the need to develop tissue engineered mucosa transplants.

Transplantation of cultured skin was established in humans for the treatment of burns a quarter of a century ago [9, 38]. A prerequisite, therefore, was the culturing of epithelial cells in bulk. The cultivation of adult mammalian skin epithelium in vitro has been reported by Medawar [29], and the findings that mouse fibroblasts and the use of epidermal growth factor have a permissive effect on keratinocyte proliferation [45, 49] allowed for culturing of epithelial sheets of sufficient size. However, these conditions for culturing primary epithelial cells were not really suitable to grow transplants to be used in elective surgery. Nevertheless, the first clinical applications of oral mucosa cultured with 3T3 mouse feeder cells and fetal calf serum were reported in tumor-related surgery and periodontal surgery [6, 20].

27.2 Optimizing Culture Conditions

For translational research on applying mucosa/gingival grafts clinically, small gingival biopsies (max. 6 mm diameter) are required. From patients undergoing dental surgery (implants or tooth extractions) or especially for the purpose of grafting, tissue was harvested and stored in culture medium. For preparing autogenous serum, 50 ml of venous blood was taken additionally. The studies were approved by the Ethic Committees of Dresden University and Freiburg University (15022002, 94/99). For the application in elective surgery a series of experiments were undertaken to establish a reliable culture/tissue engineering protocol.

The efficacy of the explant culture technique was demonstrated by comparing it to establishing cultures from single cell suspension. Therefore, after removing the connective tissue the biopsies were divided into 1-mm explants, seeded on plastic petri dishes (3.5 cm, Costar, Cambridge, MA, USA), and covered with a little culture medium. For single cell suspensions the explants were immersed in 0.25% trypsin in Hanks balanced salt solution for 3×30 min. The disaggregated cells were collected in culture medium and centrifuged at $400 \times g$. The pellets were resuspended and cells were seeded either on a feeder layer of 3T3 mouse fibroblasts or per droplet of 200 μl (approx. 5 mm diameter) [21, 25, 28]. Cultures were kept in Dulbecco's Modified Eagle's Medium (DMEM) and Ham's Nutrient Factor 12 ratio 3:1 (Gibco, Eggenstein, Germany) and a humidified atmosphere of 95% air and 5% CO_2 at 37°C for at least 24 h. Medium additives were 10% fetal calf serum (Gibco, Eggenstein, Germany) or 10% human autogenous serum; further, insulin 5 mg/l, hydrocortisone 0.2 mg/l, cholera toxin 0.0085 mg/l, adenine 24 mg/l (all Sigma, Munich, Germany), epidermal growth factor 10 μg/l (Boehringer, Mannheim, Germany), penicillin G 100,000 U/l, and streptomycin 100,000 μg/l (both Seromed Biochrom, Berlin, Germany). The explant culture method proved to be most reliable to instigate gingival keratinocyte primary cultures (Table 27.1).

The extracellular matrix proteins fibronectin and laminin were tested as the coating of culture dishes to support explant adherence and cell outgrowths (Table 27.2). After 10 days an average 98.4 mm² of the fibronectin coated, 98.4 mm² of the laminin coated, and 98.7 mm² of the uncoated culture dishes were covered by keratinocytes. From day 10 to day 20 of culture there was an increase of the covered areas by 56.2 mm² for fibronectin, 52.1 mm² for laminin, and 48.9 mm² for plastic dishes. The differences were statistically not significant (Friedman test, paired student t-test) [28].

To assess the efficacy of autogenous serum supporting cell growth, an intraindividual comparison on gingival keratinocyte cultures was carried out. After culturing gingival keratinocytes for 24 days in medium either with autogenous or fetal calf serum, cell covered surfaces planimetrically revealed that

Table 27.1 Success rate to establish primary gingival keratinocyte cultures

Patient age	Explant culture	Single cell suspension feeder cell culture	Single cell suspension droplet culture
< 40 years	90%	66%	30%
> 40 years	75%	50%	0%

Table 27.2 Influence of extracellular matrix proteins on the outgrowths of primary gingival keratinocytes from explants

	Keratinocyte covered surface (mm²)		
Culture period	No coating	Fibronectin	Laminin
10 days	98.7	98.4	98.4
20 days	147.6	154.6	150.5
Difference day 20/day 10	48.9	56.2	52.1

growth promotion of both sera was equivalent within each patient [22].

To evaluate the influence of the donor age on growth and proliferation of gingival keratinocytes in explant cultures, biopsies of five patients younger than 40 years of age and of five older than 40 years were used. Every fourth day of culture up to 60 days, the DNA synthesis rate (labeling index) and DNA content (parameter for epithelial growth) were determined. Gingival keratinocytes migrated from the explants onto the floor developing epithelial islets. During the early culture period migrating single cells were observed on these islets (Fig. 27.1). The island expanded radially forming confluent gingival epithelial layers. After a period of 40 to 60 days in primary culture the epithelial layers started to disintegrate. Keratinocytes detached from the culture dish leaving gaps in the monolayer. The remaining cells changed their cobblestone shape to a stellate morphology.

The DNA synthesis rate revealed that in cultures of younger patients proliferation is significantly higher between day 8 and day 12, whereas in cultures of older patients there was a significantly higher proliferation peak between day 24 and day 28. However, the DNA content as a parameter of cell growth measured over the whole culture period was two thirds higher in cultures of younger patients. Thus, there is a clear age dependency, although culturing of gingival keratinocytes from old patients is possible in sufficient amounts [28].

27.3 Tissue Engineering of Epithelial Mucosa Graft

To simplify the handling of cultured gingival keratinocyte grafts for the surgeon, biomaterials generally used in dental surgery were tested to act as a carrier material instead of the Vaseline gauze which was originally used for keratinocyte grafts [6, 38]. The membranes/foils tested were made of polylactide, collagen I (Tissue Foil, Baxter Immuno Inc., Heidelberg, Germany). After establishing gingival keratinocyte primary cultures using the explant technique (see above), the cells were trypsinized, seeded in a density of 20,000–50,000 cells per cm² on the carrier materials, and cultured in DMEM/KCSFM 1:1 (Gibco Inc., Eggenstein, Germany) supplemented with either fetal calf serum or human autogenous serum 10% (for grafting), and penicillin G 100,000 IU/l (Seromed Inc., Berlin, Germany). The cells adhered well to both the collagen and the polylactide foils (Fig. 27.2) forming an epithelial layer [10].

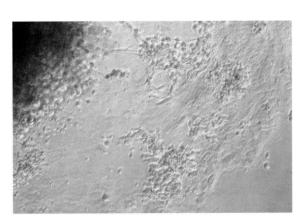

Fig. 27.1 Explant (*E*) surrounded by gingival keratinocyte island (border—*arrow*). Migrating cells (*MC*) on the epithelial layer. (×200)

Fig. 27.2 Keratinocyte layer on polylactide foil. (×1000)

27.4 Clinical Application of the Tissue Engineered Epithelial Mucosa Graft and Histological Assessment

In the field of maxillofacial and plastic reconstructive surgery, there are different pathologies and indications requiring coverage with a mucosal lining. Open wound surfaces are created during different pre-prosthetic procedures, e.g., vestibuloplasty, freeing of the tongue after local defect closure due to tumor resection, or when mucosal lining is used to reconstruct the urethra. Further, in a preliminary study tissue engineered mucosa was used for prelaminating the radial forearm flap.

Since 1991 the grafting of cultured gingival keratinocytes, and consecutively from 1995 onwards the grafting of tissue engineered gingival mucosa, has been performed in nearly 100 patients. The healing of these grafts has been assessed clinically, histologically, and immunohistologically. Since 2002 the tissue engineered graft has been further developed using the perfusion culture technique as well as creating a fibroblast layer (see below) underneath the epithelial sheet.

The aim of the vestibuloplasty was to extend loco stabile mucosa either in the denture-bearing area of cover dentures or around dental implants. After epiperiosteal dissection the mucosal defect was covered by the tissue engineered epithelial mucosa graft.

In a clinical follow-up study on 25 patients the healing of the tissue engineered mucosa graft (gingival keratinocytes on collagen foils) after vestibuloplasty was investigated. Ten days after grafting there was a pale pink vulnerable surface due to a lack of

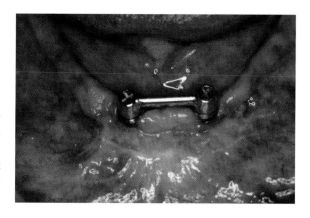

Fig. 27.3 Vestibulum of the lower jaw 10 years after grafting a tissue engineered mucosa graft, which consisted of a gingival keratinocyte layer only. Around the dental implants there is hardly any keratinized gingiva. However, there is loco-stabile mucosa between the white scar line and the implants. This distance (depth of the vestibule) is listed in Table 3

keratinization and submucosal tissue. Twenty days after grafting the surface was stable showing some keratinization. After 6 months to 1 year the texture of the tissue changed. There was a rim of keratinized tissue close to the alveolar crest, whereas there was a nonkeratinized mucosa towards the scar line in the vestibule (Fig. 27.3). The scar line determined the former caudal border of the tissue dissection. The depth of the vestibule, i.e., the distance between this white scar line and the alveolar crest, was used as a parameter of wound shrinkage in the follow-up for up to 10 years. After the initial wound shrinkage within the first 6 months post operation, further reduction of the vestibule depth was small as seen within the observation period of up to 12 years (Table 27.3).

Table 27.3 Depth of the vestibule after vestibuloplasty and grafting of tissue engineered mucosa

Period postoperative	Region in dental arch of the mandible					
	45	43	41	31	33	35
One week	6	10	11	11	10	6
Four weeks	5	7.5	8	8	7.5	5
Six months	5	6	7	7	6	5
One year	2	5	6	6	5	1
Up to ten years	2	5	6	6	5	1

The purpose of the freeing of the tongue operation and mucosa grafting of the floor of the mouth was to improve movement of the tongue and subsequently food intake, swallowing, and speech as well as to allow for prosthodontic rehabilitation in patients suffering previously from a squamous cell carcinoma (anterior or lateral floor of the mouth, T stage: T_1 and T_2). Prior to freeing of the tongue, patients had a disease-free interval of between 6 and 18 months after tumor resection and local defect coverage with a tongue flap. In a clinical study on ten patients the wound healing and graft take was as described above and the mobility of the tongue had improved. It was excellent in three and good in six cases. There was a good improvement of speech in seven patients and prosthodontic restorations were possible in eight patients [24].

27.5
Histological and Immunohistological Assessment

From patients who had undergone vestibuloplasty or freeing of the tongue, biopsies were taken during the follow-up period of up to 10 years. The histological findings changed characteristically within the first 6 months after grafting, but later on morphology was stable: on day 14 a multilayered epithelium had formed expressing CK 1, 2, 10, and 11 in the suprabasal and CK 5, 6, and 17 in all layers. There was a continuous basal membrane. Four to 8 weeks postoperatively an epithelium of 20 to 40 layers without differentiation into strata had appeared. The CK staining pattern had not yet changed. In the postoperative observation period, the histologic configuration of the grafted tissue engineered mucosa changed to a normal appearance. From 6 months onwards a differentiated mucosal epithelium was observed and the different strata could be distinguished. In biopsies taken close to the alveolar ridge, a more prominent formation of the rete ridges was observed (Fig. 27.4), as well as a well-developed stratum corneum. The cell differentiation, as judged by the cytokeratin staining, resembled that of the nongrafted mucosa. CK 5 and 6 were mainly expressed in the basal and adjacent suprabasal layers, whereas the CK 1, 2, 10,

Fig. 27.4 Immunohistology of keratinized gingiva from an area of transplanted tissue engineered gingival keratinocytes. CK 1, 2, 10, 11 is expressed in all upper layers but not in the basal layer. ($\times 500$)

11 reaction was restricted to all upper strata except the basal cells. These findings were noted in biopsies taken at 10 years post grafting [23].

Tissue engineered mucosa was used to cover the harvesting site of the full mucosa graft for urethra reconstruction. Today, urethra reconstruction profits from buccal mucosa grafts as standard procedure [8, 18, 35]. However, there are local sequels at the donor site, like submucosal scarring with contracture and subsequent web formation and limitation of movement of the mandible [54]. In order to avoid these complications and to mitigate donor site morbidity, the application of tissue engineered oral mucosa on the intraoral mucosa donor side was initiated [26].

Six weeks prior to urethra reconstruction tissue engineering of autogenous mucosa grafts was started.

Full buccal mucosa grafts were taken up to sizes of 11 × 3 cm. The defects were covered by stitching the tissue engineered mucosa graft to the wound bed and a dressing of Vaseline gauze was packed on top. At the time of intraoral wound dressing removal, usually 8–10 days postoperatively, a still vulnerable epithelial wound surface covered with little fibrin was observed in all patients. The wound surfaces stabilized during the next few days. After 1 month there was minimal superficial scar formation, which did not deteriorate at any later follow-up.

Early post operation patients suffered only from moderate discomfort intraorally. Preoperative maximal incisal opening was reached after an average time of 2 to 6 weeks postoperatively. Accompanying initial limitation of mouth opening was a deviation to the wound site which settled with increase of mouth opening.

During the healing period the histological and immunohistological features resembled those described above. In biopsies taken from 6 months postoperatively onwards, a differentiated mucosal epithelium was observed. The different strata could be distinguished (basal stratum, spinous stratum, distentum stratum). The cell differentiation as judged by the CK staining resembled that of nongrafted alveolar mucosa. The CK 5/6 were mainly expressed in the basal and adjacent suprabasal layer, whereas the CK 13 reaction was predominantly restricted to all upper strata.

The prelamination of the radial forearm flap (RFF) with tissue engineered mucosa was tested to omit the disadvantages of the conventional flap. The RFF [58] is widely used in reconstructive surgery. For intraoral application [50] its pliability is a particular feature but the transplanted skin can maintain keratinization and hair growths, a distinct disadvantage for prosthodontic rehabilitation [32]. Therefore, the prelamination of the RFF with mucosa or split-skin grafts has been introduced [30, 57]. However, both techniques require additional graft harvesting and the amount of mucosa harvested from the oral cavity is limited. As alternative the prelamination procedure of the lower forearm flap with tissue engineered mucosa has been tested in six male patients (age from 53 to 72 years) suffering from squamous cell carcinoma of the oral cavity (size T2 to T4) [27]. Biopsies of the suspected area for histopathological confirmation and of clinically healthy mucosa from the contralateral buccal plane were taken. From the latter biopsy, primary cultures of mucosa keratinocytes were established and additionally a small sample was tested histopathologically to exclude metaplasia. Tissue engineering of the mucosa graft and the prelamination of the RFF was performed, while standard screening procedures were performed to exclude metastasis and fitness for general anesthesia.

Under local infiltration anesthesia (Articain®, Aventis, Frankfurt, Germany) prelamination was performed via a 9- to 10-cm longitudinal skin incision on the palmar site of the lower arm preparing a subcutaneous pouch above the superficial veins. After placing the collagen film seeded with mucosal keratinocytes facing vein and artery, the wound was closed.

After approximately 7 days tumor resection creating defects of 6 × 8 cm and lymph node surgery were carried out under general anesthesia. Simultaneously to tumor surgery, the prelaminated flap was raised via the previous incision dissecting first the island part of the flap with the newly integrated film on top and then the two pedicles—radial artery with concomitant veins and cephalic vein. After ligating these two vessels, the flap was transferred into the oral cavity and the wound at the lower arm closed primarily. The tissue engineered mucosa forearm flap was spread into the defect and fixed to the resection borders by interrupted sutures 2-0 and a second continuous suture (Vicryl®, Ethicon, Hamburg, Germany). Compared to the conventional forearm flap, greater care had to be taken not to rupture the soft edges of the flap. The recipient vessels were anastomosed to the upper thyroid artery and thyroid vein. Redon suction drains were placed and the wounds at the neck were closed layer by layer.

Routine postoperative care was given concerning elective extubation, postoperative patient mobilization, intra- and extraoral wound care, and suture removal. The patients were fed via a stomach tube for at least 7 days before subsequent adaptation to oral feeding.

During the postoperative healing period the flap appeared a pale-pink color. Within 2 to 3 weeks postoperatively the collagen film disintegrated revealing a firm but still pliable, vulnerable, easily bleeding, partly fibrin slush-covered mucosa-like surface, and over the next few weeks it developed into that of clinically normal appearing mucosa. However, there was shrinkage of the wound surface but the mobility of the tongue and the mouth opening stayed good. After 3 to 6 months the borders of the flap could no longer be distinguished.

Microvascular anastomosis worked well in all patients but in one case, due to retention of saliva, consecutive infection at the right buccal cheek occurred. The infection settled after regular wound drainage without jeopardizing the flap.

Immunohistology of samples taken at the time of grafting revealed at the prelaminated RFF an interrupted keratinocyte layer. Three weeks after grafting, in areas with clinically firmer tissue, a thicker epithelial layer had formed, with a well-developed basement membrane similar to histology done after 3 to 6 months.

In all patients the donor region healed uneventfully during the 6 to 8 day period of the prelamination. The collagen film of the tissue engineered mucosa had adhered firmly to the underlying tissue, although in one case a hematoma had formed between the skin and collagen film. After harvesting the graft, the wounds at the lower forearm healed primarily resulting in acceptable scar tissue without impairment in hand movement or finger force.

27.6
Optimizing the Mucosa Graft: Developing Gingival Keratinocyte—Fibroblast Construct

From the clinical application it is known that there is considerable wound shrinkage, which may be due to lack of differentiation in the graft. The environment of epithelial cells may influence their differentiation as, for example, restoration of differentiation was described for keratinocytes in vitro, e.g., when using an air-liquid culture technique [2] or when combining keratinocytes with a submucosal layer of fibroblasts [55]. Further, perfusion culture systems help to maintain or promote a high level of cell differentiation in epithelial cells in vitro [1, 17 31]. However, whether such an effect on cell differentiation is also observed on human oral keratinocytes has only been poorly studied. Therefore, two approaches were made:

1) Gingival keratinocytes were cultured in a perfusion culture system.
2) Gingival keratinocytes were grown in as co-cultures together with gingival fibroblasts.

For perfusion cultures, primary gingival keratinocytes cultures were trypsinized and seeded in a concentration of 200,000 cells on polycarbonate membranes (Corning-Costar, Bodenheim, Germany; diameter 1.3 cm, 3 µm pore size), mounted in sterile carrier rings (Minucells and Minutissue, Bad Abbach, Germany). These secondary cultures were kept in Keratinocyte-SFM (Gibco, Eggenstein, Germany) and DF-Medium (see above) in a ratio of 1:1 and additives 2.5% (v/v) inactivated fetal calf serum, 1% (v/v) penicillin/streptomycin, and 7% (v/v) HEPES buffer as:

1) *Standard cultures*, cell-seeded membranes mounted on tissue carriers in six-well plates, were left in the cell incubator for 16 days (Heraeus Instruments, Osterode, Germany) at 37°C, 95% air and 5% CO_2 humidity atmosphere, medium being changed every third day.
2) *Perfusion cultures*, cell-seeded membranes mounted on tissue carriers, were transferred into a perfusion culture container (Minucells and Minutissue, Bad Abbach, Germany) after 48 h of cell adherence. Cells were perfused for another 14 days continuously with KD-Medium at a rate of 0.7 ml/h (IPC-N8 peristaltic pump, Ismatec, Zürich, Switzerland).

Cultures were processed for light microscopy by embedding the specimens in methyl methacrylate (Technovit 8100, Kulzer, Wehr, Germany) at 4°C after fixation in 3% buffered paraformaldehyde and dehydration. Semithin sections were cut and prepared for histology or immunohistochemistry. Antibodies against cytokeratins 1, 2, 10, 11; 5, 6; etc. were applied by indirect immunoincubation using the avidin-biotin technique. Slides were examined using a BX-61 apparatus (Olympus, Japan) and photographs were taken by using a digital camera in combination with analysis software (Soft Imaging Systems, Münster, Germany).

Morphology after perfusion culture showed continuous dense cell growth with a mean of 3.4 cell layers (Fig. 27.5) and a standard deviation of 0.4 cell layers only. After standard culture the continuous cell layers were interrupted by areas with little growth,

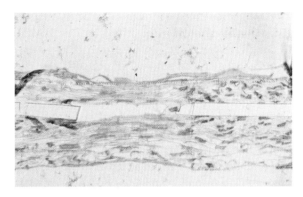

Fig. 27.5 Perfusion culture of gingival keratinocytes. Gingival keratinocytes cultured on polycarbonate membrane (PCM) under perfusion conditions show an increase in cell layers. ($\times 300$)

and histologically there was an average of 2.4 layers and a standard deviation of one layer. The increase in cell layers had been reported previously [36] for perfusion cultures of a human oral mucosal keratinocyte cell line.

After perfusion, cells attached to the polycarbonate membrane had more a cuboid shape with round nuclei, whereas the cells forming the top layers were flat without a smooth surface. After standard culture the cells and their nuclei showed flat shapes and the cytoplasms contained numerous vacuoles. The fewer vacuoles in the cytoplasm of the perfusion cultures may be signs of a less impaired metabolism of the fatty acids of keratinocytes [52], as vacuoles are interpreted as the physiological answer to cellular damage [11]; so in this respect, the cells in perfusion culture seem to suffer less cellular stress.

With respect to the expression of cytokeratins as markers of differentiation, for CK 13, CK 14, and CK 1, 2, 10, 11 differences were found. CK 13, a marker for suprabasal cells in nonkeratinizing epithelia, was very strongly expressed in all cell layers during the adherence phase and under standard culture conditions. After perfusion culture the CK 13 expression was limited to a few cells only at the basal aspect of the epithelium.

After standard culture anti CK 14 reacted mainly with the cells close to the membrane. After perfusion culture, CK 14 was only seen in cells that were close to the pores or filling the pores.

After perfusion culture CK 1, 2, 10, 11, markers of terminal differentiation of cornified epithelium, showed a positive staining reaction in the cytoplasm of all cells within the whole epithelium. After standard culture fewer cells were binding to the antibody. Only cells in close relation to the carrier membrane were expressing these cytokeratins.

Consequently, the culture conditions influence the differentiation pathway of the oral mucosa cells. Perfusion culture enhances the expression of the terminal differentiation markers CK 1, 2, 10, 11, indicating a differentiation as gingival keratinocytes [33], whereas after standard culture cells support differentiation as alveolar mucosa cells expressing CK 13 [41]. Hence, these morphological and cell biological changes clearly indicate a higher differentiation of oral keratinocytes cultured under perfusion conditions.

27.7
Gingival Keratinocyte—Gingival Fibroblast Co-cultures

To create mucosa looking like gingival epithelium after transplantation, the tissue engineered mucosa graft needs a fibrous connective tissue. Beside perfusion culture conditions, the importance of the submucosa connective tissue/fibroblast component has been demonstrated in vivo in transplantation studies as well as for in vitro investigations [7, 16, 42, 55]. Therefore, gingival biopsies were separated in epithelial cells and in fibroblasts. Primary gingival epithelial cultures were established using the explant technique and for fibroblast cultures the single cell suspension technique.

Different approaches were made to create complex keratinocyte fibroblast constructs, namely: (1) sandwich constructs, consisting of a fibroblast and a keratinocyte in two different biomaterials put on top of each other and (2) composite constructs, consisting of keratinocytes and fibroblasts in one biomaterial.

27.7.1
Sandwich Constructs

Sandwich constructs were prepared by culturing fibroblasts on one biomaterial and the gingival keratinocytes on another biomaterial. Both cell types were seeded on different materials like tissue fascie, Vicryl net, Xenoderm (Baxter, Heidelberg; Ethicon, Hamburg; MLP, Ludwigslust). It was possible to culture the fibroblasts on the Vicryl net and the keratinocytes on the collagen foil. To receive such a sandwich construct the connective tissue component (fibroblasts in a Vicryl net) and the epithelial component (keratinocytes on collagen foil) were put on top of each other for transplantation. Using this approach the clinical application was feasible and tested on several patients as mucosa cover after vestibuloplasty [47]. First the fibroblasts on Vicryl nets and then the gingival keratinocytes on tissue foil collagen sheets were placed onto the periosteal wound surface. The sandwich graft was fixed with an acrylic splint for 14 days. When removing the splint a vulnerable but mostly epithelialized wound surface was visible. The collagen foil had completely disappeared; the Vicryl net was lying on top of the wound surface and could be easily lifted off; 21 days postoperative a complete epithelization was seen. Six months after vestibuloplasty in the grafted area a mainly keratinized mucosa was found (Fig. 27.6). The clinical feature was confirmed in biopsies. Histological controls showed a cornified epithelium with rete ridges as well as the expression of cytokeratins 1, 2, 10, and 11 immunohistologically.

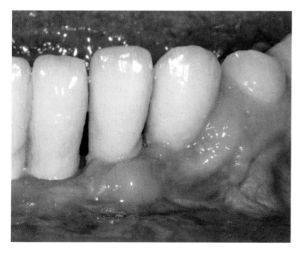

Fig. 27.6 Vestibulum of the lower jaw 2 years after grafting a tissue engineered sandwich graft, which consisted of gingival keratinocyte on a collagen membrane and gingival fibroblasts on Vicryl net. Around the dental implants (*I*) there is a considerable rim of keratinized gingival keratinocytes (*GK*)

27.7.2
Composite Constructs

Composite constructs or organotypical cultures of oral mucosa have been reported in several ways. One type of construct was manufactured by seeding the epithelial cell on top of type I collagen matrix containing the fibroblast. Choosing fibroblasts of different origin influences the differentiation characteristics of the epithelial cells. For example, oral keratinocytes seeded on top of dermal fibroblasts are more differentiated expressing CK 10 [39]. In long-term organotypic cultures of oral mucosa, alveolar fibroblasts influence skin keratinocytes to express cytokeratins 19, 4, 13 which are characteristic for keratinocytes of alveolar mucosa [4]. The group of Mackenzie developed a technique to construct large organotypical cultures of oral mucosa which can be used clinically as graft by adjusting the concentration of fibroblasts in the collagen I gel and plating the keratinocytes on top prior to contraction [13]. Similar mucosa composite constructs, however, of smaller size have been used in our lab for studies on cryopreservation (see below). The development and the clinical application of a complex tissue engineered graft from cultured oral mucosa and a skin-derived cadaveric dermal equivalent (AlloDerm®) have been reported [14, 40]. However, this construct lacks living fibroblasts.

27.8
Cryopreservation

Cryopreservation of tissue engineered mucosa may become an issue of the increasing clinical application that needs to be addressed. At present, tissue engineered transplants are prepared for each patient individually and cannot be stored because cell proliferation and cell viability is not unlimited, e.g., after a culture period of approximately 60 days proliferation of the cells ceases [28]. However, what happens when the transplantation of an individually prepared graft has to be postponed as the patient becomes sick? Knowing that there is only a certain time frame for the cells in the transplant to be proliferating, viable, and differentiated, the transplant must be preserved. The only way to achieve this goal is cryopreservation.

Cryopreservation is well established for single cell suspensions [44], but protocols and procedures to cryopreserve complete tissues are not available. Therefore, the aim was to see whether a cryopreserved tissue engineered human mucosa transplant is viable and leads after healing to a specific tissue formation at the recipient site when grafted in an animal model.

After primary culture for approximately 3 weeks, fibroblasts and gingival keratinocytes were used to tissue engineer composite mucosa constructs. Prior to cryopreservation the culture medium DMEM and Ham F 12 was changed to a custom-made cryoprotective medium and incubated for 1 h. Then, a stepwise cryoprocedure down to –70°C was performed. After rapid thawing the transplants were washed in normal culture medium and implanted subcutaneously in dissected pouches on the back of nude rats. As host and donor tissue originate from different species, the transplanted cells could be identified using human tissue specific antibodies. After wound closure there were healing periods of either 1 week or 3 weeks.

After cryopreservation tissue engineered mucosa maintained some tissue formation with fibroblasts and keratinocytes in the collagen matrix. Immunohistology of cryopreserved tissue 1 week after transplantation revealed no proper tissue formation, only some clusters of keratinocytes. In the transplants, cells labeled with BrdU were identified providing evidence that cells labeled in culture prior to cryopreservation and grafting survive both procedures and integrate at the recipient site. Three weeks after transplantation an epithelial fibroblast tissue had formed (Fig. 27.7) with a keratinocyte layer which, however, is not as mature as that seen when tissue engineered grafts are used to cover epithelial defects in the normal clinical setting. There are only a few reports about cryopreservation of keratinocytes cultured as cell layers or on starch or collagen sheets [12, 43]. Other cryopreserved tissue engineered constructs are pancreatic and dermal substitutes [19, 34]. However, for all these cells viability and function, which is around 60%, were assessed only in vitro after direct thawing. The transplantation of cryopreserved tissue is only described for heart valves either seeded with endothelial cells or without [37, 59]. However, between the two groups there were no differences with respect to change in formation of the valve collagen matrix after a healing period of up to 42 days. To our knowledge this is the first description of cryopreserved tissue engineered mucosa transplants consisting of gingival keratinocytes and fibroblasts in a matrix, and therefore the in vivo proof of function by forming a human epithelial fibroblast tissue after grafting.

27.9
Epithelial Stem Cells

Epithelial stem cells will be of interest as an approach to master shortcomings encountered in previous studies. Although in clinical studies the usefulness of cultured oral keratinocytes was demonstrated (see above), there are still limitations of these grafts when used in oral surgery and plastic reconstructive surgery. The lack of sufficient size and stratification are still apparent. However, to further pursue cell/tissue based therapeutic approaches these issues need to be addressed and further developments of oral epithelial tissue are required. In this respect the use of stem cells, an important topic also in other areas of regenerative medicine, has to be considered. Epithelial stem cells alone or in combination with a connective tissue base are promising. The isolation of epithelial stem cells from epidermis or oral mucosa

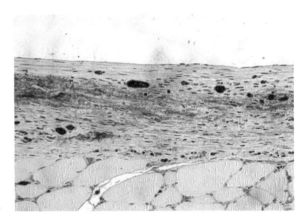

Fig. 27.7 Human gingival keratinocyte-gingival fibroblast construct after cryopreservation and 3 weeks implantation in the nude rat. A keratinocyte layer (*KL*) has formed at the subcutaneous implantation side. (×200)

has been reported. These cells are characterized by a complete take in the wound bed and a higher rate of forming a complete, highly stratified epithelial layer on the dermal matrix [15, 46].

References

1. Aigner J, Kloth S, Jennings ML, Minuth WW (1995) Transitional differentiation patterns of principal and intercalated cells during renal collecting duct development. Epithelial Cell Biol 4:121–130
2. Asselineau D, Bernard BA, Bailly C, Darmon M (1985) Epidermal morphogenesis and induction of the 67 kD keratin polypeptide by culture of human keratinocytes at the liquid-air interface. Exp Cell Res 159:536–539
3. Björn H (1963) Free transplantation of gingiva propria. Odontol Revy 14:323–331
4. Chinnathambi S, Tomanek-Chalkley A, Ludwig N, King E, DeWaard R, Johnson G, Wertz PW, Bickenbach JR (2003) Recapitulation of oral mucosal tissues in long-term organotypic culture. Anatl Rec Part A 270A:162–174
5. Dellon AL, Tarpley TM, Chretien PB (1976) Histologic evaluation of intraoral skin grafts and pedicle flaps in human. J Oral Maxillofac Surg 34:789–795
6. DeLuca M, Albanese E, Megna M, Cancedda R, Mangiante PE, Cadon, A, Franzi AT (1990) Evidence that human oral epithelium reconstituted in vitro and transplanted onto patients with defects in the oral mucosa retains properties of the original donor site. Transplantation 50:454–459
7. El Ghalbzouri A, Lamme E, Ponec M (2002) Crucial role of fibroblasts in regulating epidermal morphogenesis. Cell Tissue Res 310:189–199
8. Fichtner J, Fisch M, Filipas D, Thuroff JW, Hohenfellner R (1998) Refinements in buccal mucosal grafts urethroplasty for hypospadias repair. World J Urol 16:192–194
9. Gallico GG, O'Connor NE, Compton CC, Kehinde O, Green H (1984) Permanent coverage of large burn wounds with autologous cultured human epithelium. N Engl J Med 311:448–451
10. Gutwald R, Lauer G, Otten JE, Schilli W (1994) Epithelzellen und Fibroblasten der Gingiva auf resorbierbaren Membranen—Gewebetransfer zur Wundheilung? Dtsch Zahnärztl Z 49:1015–1018
11. Henics T, Wheatley DN (1999) Cytoplasmic vacuolation, adaptation and cell death: a view on new perspectives and features. Biol Cell Sep 91:485–98
12. Hibino Y, Hata K, Horie K, Torii S, Ueda M (1996) Structural changes and cell viability of cultured epithelium after freezing storage. J Craniomaxillofac Surg 24:346–351
13. Igarashi M, Irwin CR, Locke M, Mackenzie IC (2003) Construction of large area organotypical cultures of oral mucosa and skin. J Oral Pathol Med 32:422–430
14. Izumi K, Feinberg SE, Iida A, Yoshizawa M (2003) Intraoral grafting of an ex vivo produced oral mucosa equivalent: a preliminary report. Int J Oral Maxillofac Surg 32:188–197
15. Izumi K, Tobita T, Feinberg SE (2007) Isolation of human oral keratinocyte progenitor/stem cells. J Dent Res 86:341–346
16. Karring T, Lang NP, Löe H (1975) The role of gingival connective tissue in determining epithelial differentiation. J Periodont Res 10:1–11
17. Kloth S, Eckert E, Klein SJ, Monzer J, Wanke C, Minuth WW (1998) Letter to the Editor—Gastric epithelium under organotypic perfusion culture. In Vitro Cell Dev Biol Anim 34:515–7
18. Kropfl D, Tucak A, Prlic D, Verweyen A (1998) Using buccal mucosa for urethral reconstruction in primary and re-operative surgery. Eur Urol 34:216–220
19. Kubo K, Kuroyanagi Y (2005) The possibility of long-term cryopreservation of cultured dermal substitute. Artif Organs 29:800–805
20. Langdon JD, Leigh IM, Navsaria HA, Williams DM (1990) Autologous oral keratinocyte grafts in the mouth. Lancet 335:1472–1473
21. Lauer G (1994) Autografting of feeder-cell free cultured gingival epithelium—method and clinical application. J Craniomaxillofac Surg 22:18–22
22. Lauer G (1997) Autogenous serum for culturing keratinocyte autografts. In: Phillips GO, von Versen R, Strong DM, Nather A (eds) Advances in tissue banking, vol 1. World Scientific, Singapore, pp 183–187
23. Lauer G (2002) Tissue Engineering autologer Mundschleimhaut—Perspektive für das periimplantäre Welchgewebemanagement. Implantologie 10:159–174
24. Lauer G, Schimming R (2001) Tissue engineered mucosa graft for reconstruction of the intraoral lining after freeing of the tongue. A clinical and immunohistological study. J Oral Maxillofac Surg 59:169–175
25. Lauer G, Otten JE, von Specht BU, Schilli W (1991) Cultured gingival epithelium. A possible suitable material for pre-prosthetic surgery. J Craniomaxillofac Surg 19:21–26
26. Lauer G, Schimming R, Frankenschmidt A (2001a) Intraoral wound closure with tissue engineered mucosa—new perspectives for urethra reconstruction. Plast Reconstr Surg 107:25–33
27. Lauer G, Schimming R, Gellrich NC, Schmelzeisen R (2001b) Prelaminating the fascial radial forearm flap by tissue engineered mucosa—improvement of donor and recipient site. Plast Reconstr Surg 108:1564–1572
28. Lauer G, Siegmund C, Hübner U (2003) Influence of donor age and culture conditions on tissue engineering of mucosa autografts. Int J Oral Maxillofac Surg 32:305–312
29. Medawar PB (1948) The cultivation of adult mammalian skin epithelium in vitro. Quart J Microsc Sci 89:187–190
30. Millesi W, Millesi-Schobel G, Glaser C (1998) Reconstruction of the floor of the mouth with facial radial forearm flap prelaminated with autologous mucosa. Int J Oral Maxillofac Surg 27:106–110
31. Minuth WW, Kloth S, Aigner J, Sittinger M, Röckl W (1996) Approach to an organo-typical environment for cultured cells and tissues. BioTechniques 20: 498–501

32. Mitchell DL, Synnott SA, van Dercreek JA (1990) Tissue reaction involving an intraoral skin graft and cp titanium abutments: a clinical report. Int J Oral Maxillofac Implants 5:79–84
33. Moll R, Franke WW, Schiller DL, Geiger B, Krepler R (1982) The catalog of human cytokeratins: patterns of expression in normal epithelia, tumors and cultured cells. Cell 31:11–24
34. Mukherjee N, Chen Z, Sambanis A, Song Y (2005) Effects of cryopreservation on cell viability and insulin secretion in a model tissue-engineered pancreatic substitute (TEPS). Cell Transplant 14:449–456
35. Naude JH (1999) Buccal mucosal grafts in the treatment of ureteric lesions. BJU Int 83:751–754
36. Navarro FA, Mizuno S, Huertas JC, Glowacki J, Orgill DP (2001) Perfusion of medium improves growth of human oral neomucosal tissue constructs. Wound Repair Regen 9:507–512
37. Numata S, Fujisato T, Niwaya K, Ishibashi-Ueda H, Nakatani T, Kitamura S (2004) Immunological and histological evaluation of decellularized allograft in a pig model: comparison with cryopreserved allograft. J Heart Valve Dis 13:984–990
38. O'Connor NE, Mulliken JB, Banks-Schlegel S, Kehinde O, Green H (1981) Grafting of burns with cultured epithelium prepared from autologous epidermal cells. Lancet 10:75–79
39. Okazaki M, Yoshimura K, Suzuki Y, Harii K (2003) Effects of subepithelial fibroblasts on epithelial differentiation in human skin and oral mucosa: heterotypically recombined organotypic culture model. Plast Reconstr Surg 112:784–792
40. Ophof R, van Rheden REM, Von den Hoff JW, Schalkwijk J, Kuijpers-Jagtman AM (2002) Oral keratinocytes cultured on dermal matrices form a mucosa-like tissue. Biomaterials 23:3741–3748
41. Ouhayoun JP, Gosselin F, Forest N, Winter S, Franke WW (1985) Cytokeratin patterns of human oral epithelia: differences in cytokeratin synthesis in gingival epithelium and adjacent alveolar mucosa. Differentiation 30:123–129
42. Ouhayoun JP, Sawaf MH, Goffaux JC, Etienne D, Forest N (1988) Reepithelization of a palatal connective tissue graft transplanted in a non-keratinized alveolar mucosa: a histological and biochemical study in humans. J Periodont Res 23:127–133
43. Pasch J, Schiefer A, Heschel I, Rau G (1999) Cryopreservation of keratinocytes in a monolayer. Cryobiology 39:158–168
44. Pasch J, Schiefer A, Heschel I, Dimoudis N, Rau G (2000) Variation of the HES concentration for the cryopreservation of keratinocytes in suspensions and in monolayers. Cryobiology 41:89–96
45. Peehl DM, Ham RG (1980) Clonal growth of human keratinocytes with small amounts of dialyzed serum. In Vitro 16:526–540
46. Pellegrini G, Ranno R, Stracuzzi G, Bondanza S, Guerra L, Zambruno G, Micali G, De Luca M (1999) The control of epidermal stem cells (Holoclones) in the treatment of massive full-thickness burns with autologous keratinocytes cultured on fibrin. Transplantation 68:868–879
47. Pradel W, Blank A, Lauer G (2002) Klinischer Einsatz von im Tissue Engineering hergestellten Gingivakeratinozyten-Gingivafibroblasten-Konstrukten als Weichgewebsersatz. Dtsch Zahnärztl Z 57:709–712
48. Reuther JF, Steinau HU (1980) Mikrochirurgische Dünndarmtransplantation zur Rekonstruktion großer Defekte in der Mundhöhle. Dtsch Z Mund Kiefer Gesichtschir 4:131–136
49. Rheinwald JG, Green H (1975) Serial cultivation of strains of human epidermal keratinocytes: the formation of keratinizing colonies from single cells. Cell 6:331–344
50. Soutar DS, Scheker LR, Tanner NSB, McGregor IA (1983) The radial forearm flap: a versatile method for intraoral reconstruction. Br J Plast Surg 36:1–8
51. Spitzer WJ, Steinhäuser EW (1989) Versorgung des zahnlosen Unterkiefers mit enossalen Implantaten in Kombination mit konventionellen präprothetischen Operationen. Z Zahnärztl Implantol 5:3–6
52. Stark HJ, Baur M, Breitkreutz D, Mirancea N, Fusenig NE (1999) Organotypic keratinocyte cocultures in defined medium with regular epidermal morphogenesis and differentiation. J Invest Dermatol 112: 681–91
53. Sullivan HC, Atkins JH (1968) Free autogenous gingival grafts. III Utilization of grafts in the treatment of gingival recessions. Periodontics 6:152–158
54. Tolstunov L, Pogrel MA, McAninch JW (1997) Intraoral morbidity following free buccal mucosa graft harvesting for urethroplasty. Oral Surg Oral Med Oral Pathol Oral Radiol Endod 84:480–482
55. Tomakidi P, Breitkreuz D, Fusenig NE, Zöller J, Kohl A, Komposch G (1998) Establishment of oral mucosa phenotype in vitro in correlation to epithelial anchorage. Cell Tissue Res 292:355–66
56. Umeda T (1969) Experimental autotransplantation of full thickness skin into the mouth. Oral Surg 23:709–715
57. Wolff KD, Ervens J, Hoffmeister B (1995) Improvement of the radial forearm donor site by prefabrication of fascial-split-thickness skin grafts. Plast Reconstr Surg 98:358–362
58. Yang G, Chen B, Gao Y (1981) Forearm free skin flap transplantation. Natl Med J China 61:139
59. Yokose S, Fukunaga S, Tayama E, Kato S, Aoyagi S (2002) Histological and immunohistological study of cryopreserved aortic valve grafts: the possibility of a clinical application for cryopreserved aortic valve xenograft. Artif Organs 26:407–415

Tissue Engineering of Heart Valves

C. Lüders, C. Stamm, R. Hetzer

Contents

28.1 Introduction 381
28.2 Suitable Cell Sources 381
28.3 Scaffold Materials 382
28.3.1 Natural Matrices 383
28.3.2 Biological Matrices 383
28.3.3 Synthetic Matrices 383
28.4 Culture Conditions 383
28.5 Conclusion and Future Perspectives ... 384
References 384

28.1 Introduction

Valvular heart disease is an important cause of morbidity and mortality worldwide. Valve replacement represents the most common surgical therapy for end-stage valvular heart disease. Currently, 300,000 procedures are performed annually worldwide. Furthermore, eight of 1000 children are born with congenital cardiac defects. Every fifth of these needs a heart valve replacement. Currently, clinically available cardiovascular prosthetic substitutes, including xenografts, mechanical prostheses, and homografts, function well but have some disadvantages in common [1–3]. They consist of foreign, nonviable materials which entail the risk of thromboembolism and the lack of ability to repair, remodel, and grow, which leads to multiple reoperations. Pediatric patients are of particular interest in this context because they "outgrow" the prostheses so that multiple reoperations and considerable suffering for the patients and their families are the consequence [4]. Tissue engineering could be an alternative in overcoming these disadvantages. The interdisciplinary approach of tissue engineering combines principles of engineering and materials science with biology and vascular surgery to fabricate viable and functional prostheses from autologous, living cells with the aim of long-lasting replacement or reconstruction of the dysfunctional native tissue. Using autologous cells, these viable prostheses should have the potential to integrate, grow, remodel, and repair and therefore to conceivably make reoperations unnecessary. There are three major requirements for a successful tissue engineered cardiovascular substitute: (a) isolation of a suitable cell source, if possible of an autologous origin; (b) a suitable 3-D scaffold; and (c) in vitro culture conditions conducive to fabrication of the construct before implantation.

28.2 Suitable Cell Sources

Several potential cell sources for the tissue engineering of cardiovascular structures have been evaluated in recent years. The ideal cell source should be autologous and cells should be available in large quantities. With regard to heart valve substitutes,

 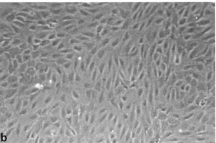

Fig. 28.1 a Myofibroblast culture, 10× magnification; b human umbilical vein endothelial cell (HUVEC) culture, 10× magnification

two cell types are required for the tissue engineered constructs. Firstly, myofibroblast-like cells which have the potential to differentiate into fibroblasts and smooth muscle cells. These cells produce the extracellular matrix (ECM) and are responsible for the development of tissue with the mechanical properties of native heart valves. Secondly, endothelial cells are necessary to cover the surface of the tissue engineered valves. These cells represent a blood-compatible layer, thus enabling thrombotic complications to be avoided (Fig. 28.1).

There are a variety of approaches to harvesting autologous cells such as biopsies from organs or segments from vascular vessels. During the past decade, venous and aortic human fibroblasts [5], human marrow stromal cells [6], and human umbilical cord cells [7, 8] have been investigated in intensive experiments. Umbilical cord vessels are a rich source of vascular cells and their progenitors for several potential clinical applications. These autologous cells could be used as a potential cell source, particularly for pediatric patients and young adults with congenital heart defects. Naturally, harvesting autologous cells requires an invasive operation, e.g., to remove a small segment of an artery or vein. Using vascular cells from umbilical cords as an autologous cell source avoids the excision of pieces of vessels and therefore the risk of infectious complications. In recent years, interest in the use of stem cells has rapidly grown. Human stem cells and progenitor cells have been isolated from a variety of sources, i.e., human bone marrow, adipose tissue, umbilical cord blood, and amniotic-derived cells. Their autologous origin, high proliferation capacity, and potential for differentiation into vascular phenotypes make this cell type suitable for the tissue engineering of cardiovascular structures such as heart valves. In comparison to embryonic stem cells, adult stem cells are less controversial and are associated with fewer ethical concerns.

Although the feasibility of culturing autologous cells from fresh material in vitro and retransplanting them has been demonstrated in several animal models, cryopreservation of autologous cells for the tissue engineering of cardiovascular structures is not well established. Cryopreservation is a procedure that permits the long-term preservation of functional, living cells and tissues for scientific research as well as for many medical applications, such as bone marrow transplantation or in vitro fertilization. As tissue engineering becomes more important, cryopreservation will play a vital role in establishing individual cell banks. Cryopreserved autologous cells bear one major advantage, which is the possibility of banking of a large quantity of cells so that they are available when the patient is ready to receive the transplantation (Fig. 28.2).

28.3
Scaffold Materials

Several research experiments have demonstrated the possibility of fabricating a tissue-like structure in vitro, which remodels into functional tissue in vivo. The major problem is the composition of the scaffold. The scaffold serves as a physical and structural support and template for cell adhesion and tissue development. In recent years much effort has been

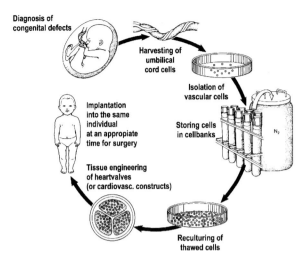

Fig. 28.2 Principle of heart valve tissue engineering using cryopreserved vascular human umbilical cord cells

expended on analyzing the optimal chemical and physical configuration of different biomaterials and their interaction with living cells to produce a functional tissue engineered construct. Therefore, the selection of an optimal scaffold is of particular interest. The biomaterials require a 3-D structure and can be permanent or biodegradable. They can be of natural or biological origin, synthetic materials, or hybrids. All these materials should be compatible to living cells in vitro and in vivo, without cytotoxic degradation products or immune response reactions, and should have adequate hemodynamic and mechanical functions and therefore life-long durability. Biomaterials can also incorporate signaling factors that the materials may offer. These include the release of growth and differentiation factors, cytokines, or specific receptors.

28.3.1
Natural Matrices

Collagen-based matrices have been commonly used for tissue engineered constructs for blood vessels. Furthermore, the use of fibrin gel isolated from patients' own blood as an autologous matrix has been investigated [9]. One advantage is the homogeneous distribution of cells within the gel from the beginning caused by the fast immobilization of the cells during the polymerization of the gel.

28.3.2
Biological Matrices

Biological scaffolds include porcine or human material that are decellularized by enzymatic or detergent reactions. The decellularized matrices offer the ECM, a mixture of functional proteins such as collagens, fibronectin, elastin, and others, glycoproteins and proteoglycans [10]. Theoretically, decellularized matrices are biodegradable but they will not work for all applications. They all have a heterologous origin, which makes immunosuppression therapy necessary.

28.3.3
Synthetic Matrices

Alternatively, synthetic biodegradable polymers, such as polyglycolic acid (PGA), polylactide acid (PLA), or poly(4-hydroxybutyrate) (P4HB), have been demonstrated to be applicable for cardiovascular tissue engineering [11–14]. PGA was the first polymer used for the successful fabrication of a tissue engineered tissue. Recent studies have demonstrated the feasibility of combinations of different polymers resulting in copolymers that provide thermoplasticity and better mechanical properties. Major advantages of synthetic matrices are their controllable biodegradation properties and their high elasticity.

28.4
Culture Conditions

Previous studies have demonstrated that cells seeded onto biomaterials need specific cell culture conditions, the so-called "in vitro conditioning" [13]. Before implantation the cells have to be "trained"

 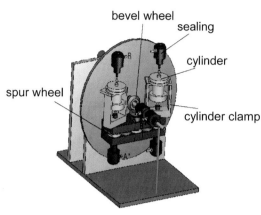

Fig. 28.3 *Left:* cell seeding device; *right:* scheme of the modified cell seeding device in the bioreactor (both: Laboratory for Tissue Engineering, German Heart Institute Berlin)

for their intended function. Therefore, the cell culture conditions have to be optimized to guarantee a functional in vitro fabricated cardiovascular substitute. In contrast to static cell culture conditions the distribution of cells seeded onto biomaterials, especially onto polymers, is more uniform under dynamic conditions. Further, in a pulsatile flow environment the cells orientate in one direction, express a higher ECM protein concentration, and demonstrate comparable mechanical properties to native tissue. Therefore, much effort has been dedicated to the development of new cell seeding devices and bioreactors (Fig. 28.3) [15, 16].

28.5
Conclusion and Future Perspectives

Recently used commercial cardiovascular prostheses, i.e., heart valve replacements, function well and improve the survival time and the quality of life of patients. However, they have several disadvantages in common: they all represent foreign, nonliving substitutes which are associated with the risk of rejection and inflammation, and the need of anticoagulation or immunosuppressive therapy. Further, they do not have the potential to grow, integrate, or remodel. To overcome these shortcomings, tissue engineering represents an alternative therapy. Progress has been made in engineering the various components of the cardiovascular system, i.e., blood vessels and heart valves. Recently, cells from several sources have been characterized and analyzed, and knowledge of biocompatible scaffolds and the in vitro conditioning process has rapidly increased. With the use of autologous cells the tissue engineered constructs are very similar to native tissue and therefore the risk of inflammation and rejection is minimized. In summary, the tissue engineering of cardiovascular structures represents an alternative therapy and is expected to play an important role in future human clinical applications.

References

1. Braunwald E. Valvular heart disease. In: Heart disease, 5th edn, 1997, Braunwald E ed., Saunders, Philadelphia
2. Hammermeister KE, Sethi GK, Henderson WG et al. A comparison of outcomes in men 11 years after heart-valve replacement with mechanical valve or bioprosthesis. N Engl J Med 1993; 328:1289–1296
3. Vongpatanasin W, Hillis D, Lange RA. Prosthetic heart valves. N Engl J Med 1996; 335:407–416
4. Cannegieter SC, Rosendaal FR, Briet E. Thromboembolic and bleeding complications in patients with mechanical heart valve prostheses. Circulation 1994; 89:635–641
5. Schnell AM, Hoerstrup SP, Zund G, Kolb S, Sodian R, Visjager JF, Grunenfelder J, Suter A, Turina M. Optimal

cell source for cardiovascular tissue engineering: venous vs. aortic human myofibroblasts. Thorac Cardiovasc Surg 2001; 49(4):221–225
6. Hoerstrup SP, Kadner A, Melnitchouk S, Trojan A, Eid K, Tracy J, Sodian R, Visjager JF, Kolb SA, Grunenfelder J, Zund G Turina M. Tissue engineering of functional trileaflet heart valves from human marrow stromal cells. Circulation 2002; 106 [suppl I]:I143–I150
7. Sodian R, Lüders C. Krämer L, Kübler WM, Shakibaei M, Reichart B, Däbritz S, Hetzer R. Tissue engineering of autologous human heart valves using cryopreserved vascular umbilical cord cells. Ann Thorac Surg 2006; 81(6):2207–2216
8. Kadner A, Hoerstrup SP, Breymann C, Maurus CF, Melnitchouk S, Kadner G, Turina M. Human umbilical cord cells: a new cell source for cardiovascular tissue engineering. Ann Thorac Surg 2002; 74(4):S1422–1428
9. Jockenhoevel S, Zünd G, Hoerstrup S, Chalabi K, Sachweh J, Demircan B, Turina M. Fibrin gel—advantages of a new scaffold in cardiovascular tissue engineering. Eur J Cardiothorac Surg 2001; 19:424–430
10. Badylak SF. The extracellular matrix as a scaffold for tissue reconstruction. Semin Cell Dev Biol 2002; 13(5):377–383
11. Shinoka T, Breuer CK, Tanel RE, Zund G, Miura T, Ma PX, Langer R, Vacanti JP, Mayer JE Jr. Tissue engineering heart valves: valve leaflet replacement study in a lamb model. Ann Thorac Surg 1995; 60(6 Suppl):S513–516
12. Ye Q, Zünd G, Jockenhoevel S, Schoeberlein A, Hoerstrup S, Grunenfelder J, Benedikt P, Turina M. Scaffold precoating with human autologous extracellular matrix for improved cell attachment in cardiovascular tissue engineering. ASAIO J 2000; 46:730–733
13. Hoerstrup SP, Sodian R, Daebritz S, Wang J, Bacha EA, Martin DP, Moran AM, Guleserian KJ, Sperling JS, Kaushal S, Vacanti JP, Schoen FJ, Mayer JE Jr. Functional living trileaflet heart valves grown in vitro. Circulation 2000; 102(19 Suppl 3):III44–449
14. Sodian R, Lüders C. Krämer L, Kübler WM, Shakibaei M, Reichart B, Däbritz S, Hetzer R. Tissue engineering of autologous human heart valves using cryopreserved vascular umbilical cord cells. Ann Thorac Surg 2006; 81(6):2207–2216
15. Sodian R, Lemke T, Fritsche C, Hoerstrup SP, Fu P, Potapov EV, Hausmann H, Hetzer R. Tissue-engineering bioreactors: a new combined cell-seeding and perfusion system for vascular tissue engineering. Tissue Eng 2002; 8(5):863–870
16. Lüders C, Sodian R, Shakibaei M, Hetzer R. Short-term culture of human neonatal myofibroblasts seeded using a novel three-dimensional rotary seeding device. ASAIO J 2006; 52:310–314

VI Engineering at the Organ Level

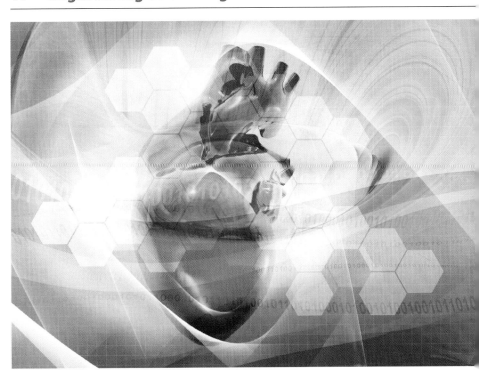

Breast Tissue Engineering

E. Geddes, X. Wu, C. W. Patrick Jr.

Contents

29.1	Introduction	389
29.2	Standard of Care	390
29.3	Overall Strategies	390
29.4	Adipose Tissue Constructs	391
29.5	Naturally Derived Materials	392
29.6	Synthetic Materials	393
29.7	Future Grand Challenges	394
	References	395

29.1 Introduction

From repairing small soft tissue defects of the face to rebuilding the entire breast mound subsequent to radical mastectomy, reconstructive surgeons face a multitude of challenges. Although much has been achieved, the inadequacies of current restorative and reparative techniques have served as the clinical impetus for developing rational and translatable adipose tissue engineering strategies. This chapter focuses on these state-of-the-art tissue engineering strategies as applied to breast cancer rehabilitation, specifically post-oncologic breast tissue engineering.

Adipose tissue engineering can be defined as the scaffold- and bioactive factor-guided generation of new autologous fat for the correction of soft tissue deficits. These deficits and the resulting contour deformities caused by trauma, tumor resection, congenital abnormalities, and aging run the gamut in terms of volume and impact the form, function, and psychological well-being of the patient. With sophisticated biodegradable scaffolds and autologous cells as engineering materials, a patient-specific construct can ideally be designed in the laboratory to grow adipose tissue with precision to the desired shape and size. The ultimate application of adipose tissue engineering is breast reconstruction subsequent to tumor resection.

Breast cancer is the second most occurring cancer in women, after skin cancer, affecting one in every eight women in America and one in every three women with cancer [1]. Sadly, studies show that many women who have undergone a mastectomy tend to suffer from a syndrome "marked by anxiety, insomnia, depressive attitudes, occasional ideas of suicide, and feelings of shame and worthlessness" [2]. Restoring the breast mound following a mastectomy has been proven to alleviate the sense of mutilation and suffering that women experience post-surgery [3]. This underscores the necessity of breast reconstruction for the maintenance of self-esteem and a healthy body image. Understandably, breast reconstruction is now an option offered to any woman undergoing surgery in the management of breast cancer.

Due to the large number of clinical occurrences, breast reconstruction following lumpectomy or radical mastectomy has become the sixth most common reconstructive procedure performed in the United States [4]. It involves replacing missing skin and soft tissue volume to recreate the appearance of the breast, which can be particularly challenging depending on the volume involved and relative paucity of adequate fill. Breast tissue engineering is the promising enterprise that may soon abrogate these limitations and offer an innovative alternative to traditional reconstructive procedures. Instead of fabricating a "one size fits all" construct, it is feasible to use imaging modalities to engineer a patient-specific adipose construct, allowing for correction of the tissue deficit and restoration of the breast form. Generating replacement breast tissue is the best option for reconstructing the breast mound while maintaining tactile sensation. For all women who must undergo mastectomy, the precisely engineered adipose tissue offers the patient the opportunity to improve their quality of life, the function of the breast, and psychosocial outcome without the risks and drawbacks of current surgical strategies and alternatives.

29.2
Standard of Care

Prior to the development of contemporary tissue transfer techniques, women with breast cancer faced a difficult choice between no breast restoration, wearing an external prosthesis, or the implantation of prosthetic devices. Today, when planning a reconstruction as part of the sequela of cancer care, a woman is presented with a number of alternatives. One option is to use autologous tissue to reconstruct the breast. This is the procedure most commonly performed for patients following a mastectomy and remains the gold standard for comparing adipose tissue engineering strategies. The use of autologous tissues tends to produce a breast mound that better recreates the shape, contour, softness, and fullness of the natural breast than the use of implants [5]. Initially, the transfer of autologous fat alone was investigated as a virtually limitless source of material for soft tissue repair [6]. It is readily available and most patients possess a generous supply that can be easily harvested without producing significant contour defects. Despite the theoretical advantages, free fat grafting has thus far demonstrated poor results [7], with a reduction in over half of the graft volume due to progressive resorption [8–11].

For defects characterized by a large loss of tissue volume, such as in a mastectomy, reconstruction commonly involves the transfer of variable amounts of skin, fat, and muscle from the lower abdominal wall (transverse rectus abdominis musculocutaneous (TRAM) flap) or the back (latissimus dorsi musculocutaneous flap). Tissue availability must be considered for each donor site based on the patient [12]. The surgery, however, is quite extensive and requires additional procedures to reconstruct the nipple and areola and to refine the breast mound. More so, contour deformities from the donor site may result. Finally, complications such as prolonged pain, abdominal weakness, an extended recovery, and occasional areas of necrosis in the TRAM flap may occur following the surgery [12].

A second option is to use autologous tissue in combination with a prosthetic device, especially in the case where a patient's own tissue does not contribute an adequate volume for reconstruction [5, 12]. The advantage of using an implant is that it can be manufactured in a broad range of sizes, contours, profiles, and textures. The most common complication, however, is capsular contraction [5, 12, 13]. Inserting an implant inevitably leads to the formation of a capsule of fibrous scar tissue due to a foreign body immunologic response [14–17]. Over time the capsule constricts, making the augmented breast feel harder and firmer than desired [5, 18, 19]. Much to the patient's dismay, this often results in a spherical breast appearance, chronic chest wall discomfort, asymmetry, and restricted shoulder or arm movement [5, 19]. The potential for calcium deposits accumulating in the capsule is also problematic and can interfere with future tumor detection.

29.3
Overall Strategies

Although the aforementioned reconstructive procedures represent great strides in the rehabilitation of breast cancer, the limitations underscore the need for engineered soft tissue alternatives. Various adipose tissue engineering strategies are currently being investigated as a means to repair and restore the breast. In general, tissue engineering involves seeding cells on a three-dimensional natural, synthetic, or hybrid scaffold and introducing the appropriate tissue induction and differentiation growth factors to promote proliferation and tissue maturation. The ultimate goal is to develop "biological substitutes that restore, maintain, or improve tissue function" [20, 21]. The properties of the material composing the scaffold are critical. Materials must be able to mechanically support and guide tissue formation, such as adipogenesis. They must also be biocompatible, biodegradable, easily processed, resistant to mechanical strain, and easily shaped to the surgeon's specifications [22]. Numerous scaffolds, ranging from nonwoven fiber and hydrogel extracellular matrix-derived materials to synthetic polymers in the form of foams, nonwoven fibers, and hydrogels, are being investigated. Ideally, the scaffold should recapitulate the endogenous extracellular matrix as much as possible. Acellular tissue engineering constructs are implanted in patients and rely on recruitment of surrounding cells or are tailored to remain acellular. Cell-seeded tissue engineering constructs either are grown ex vivo in sophisticated bioreactors and then implanted in the patient, or are placed directly in vivo with the patient serving as a bioreactor. Preadipocytes, precursor cells that differentiate into mature adipocytes, can be seeded onto a scaffold and allowed to proliferate and differentiate to promote the formation of adipose tissue.

29.4 Adipose Tissue Constructs

There are numerous methods with which to categorize and discuss adipose tissue engineered constructs, including delineating animal models, source of cells, and level of bioactive factor incorporation. We have elected to discuss the constructs in terms of whether the scaffold involves natural or synthetic biomaterials (Fig. 29.1). Space limitations prevent an exhaustive presentation of the constructs that have been employed. Consequently, the examples presented are meant to be representative rather than a detailed review of all investigative studies.

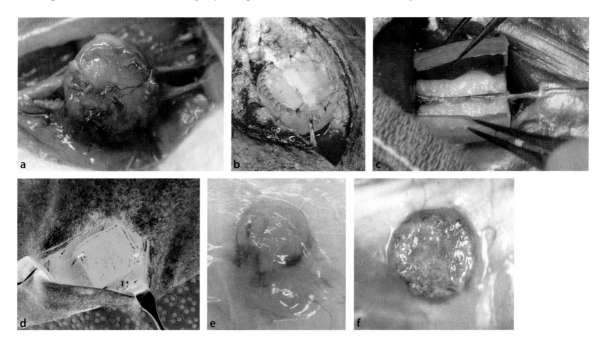

Fig. 29.1a–f Examples of constructs in various animal models. **a** Rat with dome-shaped chamber filled with preadipocyte-seeded, crosslinked collagen-chitosan and vascular pedicle at 1 month. **b** Pig with dome-shaped chamber filled with preadipocyte-seeded Matrigel and vascular pedicle at $t=0$. **c** Pig with tube-shaped chamber filled with crosslinked collagen-chitosan and vascular pedicle at $t=0$. **d** Preadipocyte-seeded nonwoven fiber polymer (proprietary Johnson & Johnson material) in a rat subcutaneous pocket model at $t=0$. **e** Preadipocyte-seeded crosslinked collagen-chitosan in a rat subcutaneous pocket model at 2 weeks [32]. **f** Preadipocyte-seeded poly(l-lactic-co-glycolic acid) scaffold in a rat subcutaneous pocket model at 2 weeks [35, 36]

29.5
Naturally Derived Materials

Several naturally derived materials have been employed as candidate scaffolds in adipose tissue engineering (Fig. 29.2). Fibrin glue (containing fibronectin as a key component) has been utilized as a delivery vehicle for preadipocytes placed into a vascularized capsule (formed by previous implantation of a silicone block) [23]. In this preparation, adipose tissue was noted for up to 12 months in a rat model. Moreover, in vitro adipogenesis has been demonstrated in porous gelatin sponges (Gelfoam) [24]. In addition, Von Heimburg and colleagues have studied two types of hyaluronan-based devices for adipose tissue engineering [25, 26]. The hyaluronic acid-based devices were manufactured into sponges or scaffolds composed of nonwoven fibers. The sponges demonstrated open, interconnecting pores ranging from 50 to 340 microns while the nonwoven mesh had an interfiber distance of 100–300 microns. In vivo studies conducted in mice showed that the hyaluronan sponge proved to be a better scaffold than the hyaluronan nonwoven carrier due to the larger, interconnected pores. After 8 months, a greater number of adipocytes were found in the sponge compared to the nonwoven mesh. The large pore size is important for the preadipocytes to incorporate lipids and enlarge during differentiation. Pore size was also found to be a factor in adipogenesis in several proprietary Johnson & Johnson biodegradable polymer sponges and nonwoven fibers [27]. In another study, human preadipocytes seeded in mice demonstrated both improved expansion and differentiation on hyaluronic acid modified carriers (HYAFF 11) as compared to collagen scaffolds examined 3 and 8 weeks after seeding [28]. The increased penetration of cells into the matrix in the HYAFF 11 preparations was felt likely to be due not only to a larger pore size (120 µm), affording greater vascular ingrowth, but also to a positive stimulatory effect of HYAFF 11 on adipogenesis [28].

Porous alginate, a naturally derived hydrogel, has also been investigated as a construct for soft tissue engineering. Halberstadt et al. modified alginate with RGD peptide sequences to allow cells to adhere to the construct [29]. In vitro studies demonstrated that the porous alginate-RGD material supported cell attachment, adhesion, and proliferation. Small-animal studies performed over 6 months showed that the implanted material

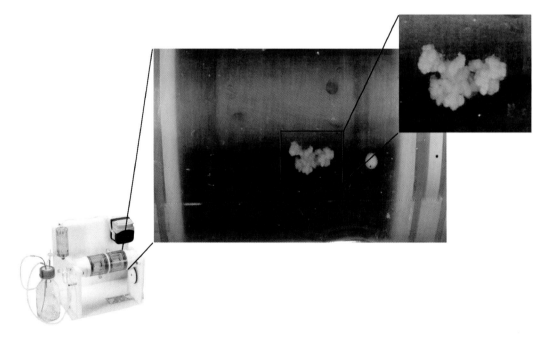

Fig. 29.2 Ex vivo adipogenesis. Preadipocytes have been placed in perfused, low shear bioreactors in the absence and presence of synthetic microcarriers [47]. Adipose tissue readily forms (see inset)

was also conducive to tissue ingrowth and did not elicit major inflammatory responses [30]. In a large-animal model (sheep), the material was seeded with preadipocytes and injected into the nape of the neck. Well-defined adipose tissue was identified within the hydrogel at 1 and 3 months. Unfortunately, it cannot be determined whether the adipose tissue growth resulted from the previously seeded preadipocytes in the material or from resident preadipocytes [29].

Collagen hydrogels have also been investigated as a three-dimensional biological matrix upon which preadipocytes are cocultured with human mammary epithelial cells in order to form tissue that closely resembles the normal human breast. Histologic analysis of the collagen gels from in vitro studies indicates a pattern of ductal structures of human mammary epithelial cells within clusters of adipocytes similar to the architecture of breast tissue [31]. In addition, porous collagen scaffolds have been fabricated and assessed in vivo. The collagen scaffolds are constructed based on a directional solidification method followed by freeze-drying to obtain a uniformly porous structure. Recently, glutaraldehyde-crosslinked collagen-chitosan hybrid materials were shown to support adipogenesis and concomitantly angiogenesis within a rat model [32].

29.6 Synthetic Materials

Traditional and innovative polymer chemistry strategies have been employed to develop synthetic biomaterials for adipose tissue engineering. Poly(glycolic acid) (PGA) scaffolds reinforced with poly(l-lactic acid) (PLLA) and infused with human preadipocyte-seeded fibrin demonstrated in vivo adipogenesis for up to 6 weeks in mice [33]. In addition, PGA polymers seeded with 3T3-L1 cells demonstrated fatlike tissues up to 1 month in mice [34]. In earlier studies, rigid poly(l-lactic-co-glycolic acid) (PLGA) polymer foams seeded with rat preadipocytes were implanted subcutaneously into male Lewis rats in an effort to generate de novo adipose tissue. Results from short-term (2–5 weeks) and long-term (1–12 months) studies demonstrated the successful formation of adipose tissue for up to 2 months [35, 36]. The volume of generated adipose tissue then began to decrease and was completely resorbed by 5 months. The resorption of adipose tissue after 2 months may be due to factors such as a lack of adequate vascularization, lack of support structure after the PLGA degraded, the anatomical site of implantation not being conducive to long-term maintenance of adipose tissue, or a limitation related to the small-animal model used. The long-term maintenance of generated tissue is a challenge for all tissue engineering applications.

Fluortex monofilament-expanded polytetrafluoroethylene (52 micron pore size) has also been studied in vitro as a potential scaffold for adipose tissue engineering [37]. Preadipocytes do not attach to uncoated polytetrafluoroethylene and, hence, human collagen, albumin, and fibronectin coatings were studied to optimize seeding efficiency. Fibronectin coating resulted in a significantly higher number of attached human preadipocytes than the collagen or albumin coatings. Human preadipocytes were able to proliferate and differentiate into adipocytes on the fibronectin-coated expanded polytetrafluoroethylene in vitro over a period of 120 h. In vivo studies have not been conducted. In another study, 3T3-L1 cells were shown to adhere to and lipid load on nondegradable fibrous polyethylene terephthalate scaffolds [38].

Tissue engineering requires precise control of cellular responses. As a result, several synthetic materials have been derivatized with specific peptide sequences known to mediate cell adhesion. A composite scaffold of fibronectin-coated alloplast (expanded polytetrafluoroethylene mesh) was noted to enhance adipocyte differentiation [37]. RGD-derivatized alginate was previously discussed.

Proteolytically degradable peptides can be incorporated into the backbone of the polymer to form hydrogels that are degraded by cell-secreted enzymes. For instance, polyethylene glycol (PEG) can be modified with the LGPA peptide sequence to form a polymer degradable by collagenase [39] (Fig. 29.3). Preadipocyte adhesion sites can also be coupled to PEG using the YIGSR peptide sequence [39]. YIGSR is a cell binding peptide found on laminin. Wu and Patrick have shown that preadipocytes bind preferentially to laminin-1 and that cell adhesion to and migration on laminin-1 is mediated by the $\alpha_1\beta_1$ integrin [39, 40]. Santiago et al. have similarly shown increased adherence of adipose-derived stem cells to polycaprolactone surfaces derivatized with the laminin

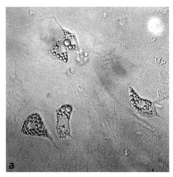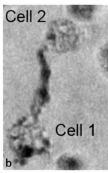

Fig. 29.3 Preadipocytes within **a** a natural fibrin hydrogel and **b** synthetic YIGSR- and LGPA-derived PEG hydrogel [39, 42]

cell-binding peptide sequences YIGSR, IKVAV, and RGD [41]. The preadipocytes can be mixed into a PEG solution and then photopolymerized into a hydrogel. Moreover, the PEG hydrogels were shown to possess material and rheological properties similar or superior to human subcutaneous adipose tissue [42]. Combining the degradable PEG with the polymer coupled with cell adhesion sites produces a synthetic hydrogel that has been shown to be adequate scaffold material in vitro. Alhadlaq and colleagues implanted PEG diacrylate hydrogels seeded with mesenchymal stem cells and demonstrated in vivo adipogenesis for up to 4 weeks in mice [43]. The clinical benefit of having a polymerizable hydrogel such as PEG is that the cell–polymer solution can easily be injected into the defect to be corrected and then photopolymerized in situ into a hydrogel. The need for complex surgical intervention would thus be eliminated.

29.7
Future Grand Challenges

Current tissue engineering studies, including adipose tissue engineering studies, while full of promise, possess several profound challenges. First, large-animal models need to be employed in preclinical investigation. Long-term maintenance of tissue engineered constructs has remained an elusive goal for all tissue types, including adipose tissue. To a large extent, small-animal models have been relied upon to demonstrate proof of concept. Numerous investigations in small-animal models attest to the feasibility of adipose tissue engineering. However, these tissue engineered constructs are small and fail to approach a size adequate enough to approximate clinical relevance. Neither of these limitations can be overcome by using small-animal models. Hence, researchers need to develop and characterize an appropriate large-animal model with quantitative metrics comparable to human dimensions in order to achieve the goal of restoring the human breast mound subsequent to mastectomy. Our laboratory is using porcine models to assess long-term maintenance of clinically sized adipose constructs (Fig. 29.1b, c).

Second, an adequate blood supply must be codeveloped with the generated adipose tissue. Many current scaffolding strategies show promise for adipose tissue growth, but are limited by the level of vascular support of the scaffold. Adipose tissue is highly vascular, possessing resting values of blood flow and capillary filtration coefficients two to three times higher than those in skeletal muscle. This suggests that a rich vascular network is crucial to support the demands of an adipose construct. It is the most critical factor limiting the size, maintenance, and quality of the construct. While angiogenesis and capillary ingrowth accompany adipogenesis, in the latter stages of wound healing vascular ingrowth regresses. An in vivo application of a defined vascular pedicle, analogous to the axial blood vessel of a flap, may be one of the ingredients necessary to ensure long-term support of the construct [44–46]. Complicating the scenario in terms of breast reconstruction is the fact many of the desired implantation areas will be irradiated as part of breast cancer treatment, severely impeding vascularization and wound healing. This fact will need to be addressed in future adipose tissue engineering strategies.

In addition, it is currently unclear what cell source is the best for clinical applications. Investigators are studying adult and embryonic stem cells and preadipocytes. Preadipocytes may be the optimum choice given their relative number and ease of harvest and isolation from lipoaspirates. However, do we need to inject or implant cells? It remains to be determined how large a role local recruitment of resident stem cells and preadipocytes plays in adipogenesis.

Finally, funding entities need to continue to support adipose tissue engineering. Too many funding agencies view adipose tissue engineering as largely an aesthetic or cosmetic application. However, the reconstructive need for adipose tissue engineering is real and we will do patients a disservice if we ignore this research endeavor.

References

1. American Cancer Society. Breast cancer statistics for 2002, vol. 2002: American Cancer Society; 2002.
2. Renneker R, Cutler M. Psychological problems of adjustment to cancer of the breast. JAMA 1952;148:834.
3. Jacobson N. The socially constructed breast: breast implants and the medical construction of need. Am J Public Health 1998;88:1254–1261.
4. American Society of Plastic Surgeons. 2004 National clearinghouse of plastic surgery statistics, vol. 2005: American Society of Plastic Surgeons; 2005.
5. Robb GL. Reconstructive surgery. In: Hunt KK, Robb GL, Strom EA, Ueno NT, eds., Breast Cancer. New York: Springer; 2001:223–253.
6. Coleman WP III. Fat transplantation. Dermatol Clin 1999;17:891–898.
7. Ellenbogen R. Invited Comment. Aesthetic Plast Surg 1990;24:197.
8. Niechajev I, Sevcuk O. Long term results of fat transplantation: clinical and histologic studies. Plast Reconstr Surg 1994;94:496–506.
9. Matsudo P, Toledo L. Experience of injected fat grafting. Aesthetic Plast Surg 1988;12:35–38.
10. Chajchir A, Benzaquen I. Liposuction fat grafts in face wrinkles and hemifacial atrophy. Aesthetic Plast Surg 1986;10:115–117.
11. de la Fuente A, Tavora T. Fat injection for the correction of facial lipodistrophies: a preliminary report. Aesthetic Plast Surg 1988;12:39–43.
12. Bostwick J III. Plastic and reconstructive surgery, vols. 1&2 (2nd edn). St. Louis: Quality Medical Publishing; 2000.
13. Robb G, Miller M, Patrick CW Jr. Breast reconstruction. In: Atala A, Lanza R, eds. Methods in tissue engineering. San Diego: Academic 2001; 881–889.
14. Baran CN, Peker F, Ortak T, Sensoz O, Baran NK. A different strategy in the surgical treatment of capsular contracture: leave capsule intact. Aesthetic Plast Surg 2001;25:427–431.
15. Coleman DJ, Foo ITH, Sharpe DT. Texture of smooth implants for breast augmentation? A prospective controlled trial. Br J Plast Surg 1991;44:444–448.
16. Peters W, Pritzker K, Smith D, et al. Capsular calcification associated with silicone breast implants: incidence, derminants, and characterization. Ann Plast Surg 1998;41:348–360.
17. Pollock H. Breast capsular contracture: a retrospective study of textured versus smooth silicone implants. Plast Reconstr Surg 1993;91:404–407.
18. Bosetti M, Navone R, Rizzo E, Cannas M. Histochemical and morphometric observations on the new tissue formed around mammary expanders coated with pyrolytic carbon. J Biomed Mater Res 1998;40:307–313.
19. Gerszten PC. A formal risk assessment of silicone breast implants. Biomaterials 1999;20:1063–1069.
20. Patrick CW Jr. Tissue engineering of fat. Surg Oncol 2000;19:302–311.
21. Patrick CW Jr. Tissue engineering strategies for soft tissue repair. Anat Rec 2001;263:361–366.
22. Patel PN, Patrick CW Jr. Materials employed for breast augmentation and reconstruction. In: Ma PX, Elisseeff J, eds. Scaffolding in tissue engineering. New York: Marcel Dekker; 2005.
23. Wechselberger G, Russell RC, Neumeister MW, Schoeller T, Piza-Katzer H, Rainer C. Successful transplantation of three tissue-engineered cell types using capsule induction technique and fibrin glue as a delivery vehicle. Plast Reconstr Surg 2002;110:123–129.
24. Hong L, Peptan I, Clark P, Mao JJ. Ex vivo adipose tissue engineering by human marrow stromal cell seeded gelatin sponge. Ann Biomed Eng 2005;33:511–517.
25. von Heimburg D, Zachariah, S, Heschel I, Kuhling H, Schoof H, Hafemann B, Pallua N. Human preadipocytes seeded on freeze-dried collagen scaffolds investigated in vitro and in vivo. Biomaterials 2001;22:429–438.
26. von Heimburg D, Zachariah S, Low A, Pallua N. Influence of different biodegradable carriers on the in vivo behavior of human adipose precursor cells. Plast Reconstr Surg 2001;108:411–420.
27. Roweton S, Freeman L, Patrick CW Jr, Zimmerman M. Preadipocyte-seeded absorbable matrices. Johnson & Johnson Excellence in Science Symposium. New Jersey; 2000.
28. Heimburg D, Zachariah S, Low A, Pallua N. Influence of different biodegradable carriers on the in vivo behavior of human adipose precursor cells (discussion). Plast Reconstr Surg 2001;108:421–422.
29. Halberstadt C, Austin C, Rowley J, Culberson C, Loebsack A, Wyatt S, Coleman S, Blacksten L, Burg K, Mooney D, Holder W Jr. A hydrogel material for plastic and reconstructive applications injected into the subcutaneous space of a sheep. Tissue Eng 2002;8:309–319.
30. Halberstadt CR, Mooney DJ, Burg KJL, et al. The design and implentation of an alginate material for soft tissue engineering. Sixth World Biomaterial Congress. Kamuela, Hawaii; 2000.
31. Huss FRM, Kratz G. Mammary epithelial cell and adipocytes co-culture in a 3-D matrix: the first step towards tissue-engineered human breast tissue. Cells Tissues Organs 2001;169:361–367.
32. Wu X, Black L, Santacana-Laffitte G, Patrick CW Jr. Preparation and assessment of glutaraldehyde cross-linked collagen–chitosan hydrogels for adipose tissue engineering. J Biomed Mater Res 2007;81A:59–65.
33. Cho SW, Kim SS, Rhie JW, Cho HM, Cha YC, Kim BS. Engineering of volume-stable adipose tissues. Biomaterials 2005;26:3577–3585.
34. Fischbach C, Spruss T, Weiser B, et al. Generation of mature fat pads in vitro and in vivo utilizing 3-D long-

35. Patrick CW Jr, Chauvin PB, Reece GP. Preadipocyte seeded PLGA scaffolds for adipose tissue engineering. Tissue Eng 1999;5:139–151.
36. Patrick CW Jr, Zheng B, Johnston C, Reece GP. Long-term implantation of preadipocyte seeded PLGA scaffolds. Tissue Eng 2002;8:283–293.
37. Kral JG, Crandall DL. Development of a human adipocyte synthetic polymer scaffold. Plast Reconstr Surg 1999;104:1732–1738.
38. Kang X, Xie Y, Kniss DA. Adipose tissue model using three-dimensional cultivation of preadipocytes seeded onto fibrous polymer scaffolds. Tissue Eng 2005;11:458–468.
39. Patel PN, Gobin AS, West JL, Patrick CW Jr. Poly(ethylene glycol) hydrogel system supports preadipocyte viability, adhesion, and proliferation. Tissue Eng 2005;11:1498–1505.
40. Patrick CW Jr, Wu X. Integrin-mediated preadipocyte adhesion and migration on laminin. Ann Biomed Eng 2003;31:505–515.
41. Santiago L, Nowak R, Rubin J, Marra K. Peptide-surface modification of poly(caprolactone) with laminin-derived sequences for adipose-derived stem cell applications. Biomaterials 2006;27:2962–2969.
42. Patel PN, Smith CK, Patrick CW Jr. Rheological and recovery properties of poly(ethylene glycol) diacrylate hydrogels and human adipose tissue. J Biomed Mater Res 2005;73A:313–319.
43. Alhadlaq A, Tang M, Mao JJ. Engineered adipose tissue from human mesenchymal stem cells maintains predefined shape and dimension: implications in soft tissue augmentation and reconstruction. Tissue Eng 2005;11:556–566.
44. Beahm E, Walton R, Patrick CW Jr. Progress in adipose tissue construct development. Clin Plast Surg 2003;30:547–558.
45. Beahm E, Wu L, Walton RL. Lipogenesis in a vascularized engineered construct. American Society of Reconstructive Microsurgeons. San Diego; 2000.
46. Walton RL, Beahm EK, Wu L. De novo adipose formation in a vascularized engineered construct. Microsurgery 2004;24:378–384.
47. Frye C, Patrick CW Jr. Three-dimensional adipose tissue model using low shear bioreactors. In Vitro Cell Dev Biol Anim 2006;42:109–114.

Bioartificial Liver

J.-K. Park, S.-K. Lee, D.-H. Lee, Y.-J. Kim

Contents

30.1	Introduction	397
30.2	Clinically Studied BAL Systems	398
30.3	New BAL Systems Currently in Preclinical or In Vitro Testing	400
30.4	Recent Trends in the Development of BAL Systems	402
30.5	The Search for an Alternative Source of Functional Hepatocytes: a Major Issue in the Development of BAL Systems	404
30.6	Practical Considerations for BAL Development	406
30.6.1	Bioreactor Design and Culture Methods	406
30.6.2	Source Pigs and Hepatocyte Isolation	406
30.6.3	Preclinical Studies	406
30.6.4	Clinical Studies	406
30.7	Conclusions	407
	References	407

30.1 Introduction

Despite the recent advances in supportive therapies, there is still no direct, satisfactory treatment for end-stage liver failure. Orthotopic liver transplantation is the only life-saving treatment currently available for acute liver failure and acute-on-chronic liver disease [4, 12]. However, due to a severe shortage of donor organs, only 25% of patients on waiting lists actually receive a liver transplant and almost 2,000 patients on the waiting lists die each year in the United States (liver waiting list, United Network for Organ Sharing, 2003, available at www.unos.org). Furthermore, nearly 90% of patients with fulminant hepatic failure (FHF) will die unless they receive the transplantation [72]. FHF is a severe form of hyperacute liver failure; it is defined by the appearance of encephalopathy within 8 weeks after the onset of jaundice in a patient without previous known liver diseases. If essential liver functions can be restored during the critical phase of liver failure, by either artificial or auxiliary methods, it might be possible to improve the survival rate of these patients without transplantation. Therefore, more donor livers will be available and costly transplantation procedures can be avoided [24]. Various nonbiological approaches, such as hemodialysis, hemoperfusion, plasmapheresis, and plasma exchange, have had a limited success because of the insufficient replacement of the synthetic and metabolic functions of the liver in these systems [16, 38, 60, 92]. On the other hand, extracorporeal biological treatments including whole-liver perfusion, liver-slice perfusions, and cross hemodialysis, have shown some beneficial results, but they are difficult to implement in a clinical setting [5, 62, 77].

For these reasons, many investigators have attempted to develop various extracorporeal bioartificial liver (BAL) systems. Typically, a BAL system incorporates hepatocytes into a specially designed bioreactor in which the cells are immobilized, cultured, and induced to perform the hepatic functions by processing the blood or plasma of liver failure patients [7, 50]. The BAL system is expected to provide whole-liver functions, including synthesis of plasma proteins and drug-metabolizing capacities, which are not supported by artificial liver systems.

There are several reviews that describe the overall status [1, 68], history [10, 57], critical issues [6, 9], cell sources [8, 53], general bioreactor and system designs [17, 61], preclinical and clinical results of current BAL systems [31, 37, 87], and future perspectives [33, 61]. In this chapter, first we summarize the characteristics of current BAL systems in clinical and preclinical studies; after that we look into the recent trends in the development of BAL systems, efforts for alternative cell sources, and practical considerations for the development of a BAL system.

30.2
Clinically Studied BAL Systems

To date, nine types of BAL system have been studied clinically, and the safety and efficacy of each of these BAL systems has been reported. None of the BAL systems studied thus far were proven to provide sufficient liver support or to result in increased survival rates [37, 87]. Safety and performance evaluations of the new BAL systems as well as more controlled, large-scale clinical trials of the early BAL systems are currently underway in several countries. As a brief review, the characteristics and major clinical findings of six representative BAL systems are listed in Table 30.1.

The Extracorporeal Liver Assist Device (ELAD) is the only BAL system that uses the human hepatocyte cell line C3A, which is derived from the hepatoma cell line HepG2 [21, 78, 79]. C3A cells express normal liver-specific pathways such as ureogenesis, gluconeogenesis, and cytochrome P-450 activities, and produce albumin and α-fetoprotein [89]. The cells are grown in the extracapillary space of hollow-fiber cartridges, and blood flows through the lumen of the hollow fibers. A portion of the patient's plasma is ultrafiltrated through a membrane (70 kD) and is in direct contact with the C3A cells. In a pilot, controlled trial on 24 acute liver-failure patients, there was no significant difference in survival rate, renal function, and biochemical parameters between the ELAD-treated patients and the controls [21]. In contrast, the plasma ammonia and bilirubin levels were increased after the ELAD treatment, which may have resulted from the use of a functionally deficient cell line.

The HepatAssist system developed by Demetriou and coworkers was examined in the largest controlled clinical trial thus far [2, 13, 14, 58, 69, 90]. In this system, microcarrier-attached cryopreserved porcine hepatocytes are placed in the extracapillary space [13]; these authors used 0.2-μm-pore hollow-fiber membranes, a size that is sufficiently small to block the passage of whole cells. Separated plasma passes through an activated charcoal column and then flows through the lumen of the hollow fibers. The concentrations of both total bilirubin and ammonia decreased by 18% of the initial concentrations.

The Liver Support System (LSS), which was developed by Gerlach and coworkers in Berlin, consists of a unique bioreactor with four different capillary membranes that are woven into a three-dimensional lattice [28, 70]. These capillaries independently and locally provide oxygen, nutrients, and plasma perfusate inflow and outflow to hepatocytes. The bioreactor enables the spontaneous aggregation of parenchymal cells in a coculture with nonparenchymal cells, forming tissue-like structures. The LSS is the only system that uses primary human hepatocytes isolated from discarded donor livers as well as porcine hepatocytes [71]. In phase I of a study using human hepatocytes, the LSS was combined with a single-pass albumin dialysis device called the Modular Extracorporeal Liver Support (MELS) [71].

The Bioartificial Liver Support System, which was developed at Pittsburgh University and commercialized by Excorp Medical (Minneapolis, USA) uses cellulose acetate hollow fibers with a 100-kD-molecular-weight cut-off (MWCO) [44, 63]. Primary porcine hepatocytes (70–120 g) are mixed with 20% vol/wt of a 3.1% collagen solution and the mixture is infused into the extracapillary space. The blood ammonia levels decreased by 33% compared with the initial levels. However, the patient's neurological state is not improved significantly [44].

The radial flow bioreactor (RFB)-BAL system was developed at the University of Ferrara, Italy [52, 54]. In this system, the patient's plasma passes from the

Chapter 30 Bioartificial Liver

Table 30.1 Characteristics and clinical results of bioartificial liver (BAL) systems. Reproduced with permission from Park and Lee [61]. *ELAD* Extracorporeal Liver Assist Device, *LSS* Liver Support System, *BLSS* Bioartificial Liver Support System, *RFB-BAL* radial flow bioreactor, *AMC-BAL* Academic Medical Center BAL, *Univ.* University

BAL system	Characteristics of six BAL systems						Clinical results		
BAL system (former/current company or institute)	Bioreactor configuration	Shape of cells	Hepatocyte source	Immunological barrier	Perfusion (plasma separation rate, ml/min)	Reactor flow rate (ml/min)	Neurological improvement	Ammonia removal	Bilirubin elimination
Hollow-fiber systems									
ELAD (VitaGen/Vital Therapies)	Hepatocyte cell line (C3A), Plasma, Blood	Tightly packed aggregates	Human cell line (C3A) (60 g)	Yes (70 kD)	Blood	200	Probably	−8% (increased)	−20% (increased)
HepatAssist (Circe Biomedical/Arbios Systems)	Microcarrier-attached hepatocytes, Plasma	Microcarrier-attached irregular aggregates	Cryopreserved porcine (5–7×10^9)	None	Plasma (50)	400	Yes	18%	18%
LSS (Charite, Humbolt Univ., Germany)	Gas, Plasma in, Hepatocytes, Plasma out	Tissue-like organoids	Porcine (200–600 g)	Yes (300 kD)	Plasma (31)	100–200	Yes	Uncertain	Some improvement
BLSS (Excorp Medical)	Hepatocytes in collagen gel, Blood	Collagen gel entrapped	Porcine (70–120 g)	Yes (100 kD)	Blood	100–250	No	33%	6%
Porous matrix systems									
RFB-BAL (Univ. of Ferrara, Italy)	Plasma, Hepatocytes	Aggregates	Porcine (200 g)	None	Plasma (22)	200–300	Yes	33%	11%
AMC-BAL (Hep-Art)	Hepatocytes, Gas, Plasma	Small aggregates	Porcine (1.2×10^{10})	None	Plasma (40–50)	150	Yes	44%	31%

center to the periphery of the module. The rate of oxygen consumption is measured to evaluate the function of the bioreactor; when the rate decreases during the treatment of a patient, it indicates the necrosis of hepatocytes. The RFB-BAL system decreased the mean ammonia and bilirubin levels by 33% and 11%, respectively [54].

The Academic Medical Center (AMC)-BAL system developed by Chamuleau and coworkers at the University of Amsterdam, The Netherlands, consists of a nonwoven hydrophilic polyester matrix for cell attachment, hollow fibers for oxygen supply, and a spacer between the matrix [26, 27, 75, 86]. The matrix, with a thickness of 4 mm and a total surface area of 5610 cm^2, is spirally wound with longitudinal hollow fibers. In the AMC-BAL reactor, the patient's plasma and small hepatocyte aggregates are in direct contact, resulting in an optimal mass transfer, and thus optimal oxygen supply is achieved by hollow fibers in the bioreactor. After treatment with the AMC-BAL, the plasma ammonia and bilirubin concentrations decreased by 44% and 31%, respectively [86].

With regard to hepatic functional performance, the HepatAssist system has three important disadvantages. First, it uses a relatively small amount of cryopreserved hepatocytes, only 5% of the liver mass. Second, treatment time is very short, 6–8 h. In most other BAL systems, patients are supported for at least 12–24 h. Third, the mass transfer between hepatocytes and plasma occurs only by diffusion and back filtration through the hollow fiber membrane in the bioreactor. The other five systems use more than 10 billion hepatocytes (10% of the liver mass), and in the RFB-BAL, MELS, and AMC-BAL systems, the patient's plasma is in direct contact with hepatocytes. The major factor that influences mass transfer efficiency between the BAL systems and liver-failure patients is plasma or blood exchange rates, ranging from 22 to 50 ml/min and 150 to 250 ml/min for plasma and blood, respectively. The plasma exchange rates of the RFB-BAL (22 ml/min) and MELS (31 ml/min) systems are low when their cell mass is considered.

It is very difficult to compare their performance or choose the best one among the six BAL systems due to the complexity of patient groups, the different perfusion types, and other conditions in a clinical setting. There is only one recent report of a carefully controlled comparison study of the AMC-BAL and MELS systems performed in an in vitro set-up [66]. However, considering only the performance-specific parameters, such as cell mass and mass transfer rate, the RFB-BAL, MELS, and AMC-BAL systems are expected to have a higher performance than the others.

30.3
New BAL Systems Currently in Preclinical or In Vitro Testing

In addition to the aforementioned BAL systems in clinical studies, various other BAL systems have been investigated and examined for their in vitro and in vivo performance. Here, eight noteworthy BAL systems are considered. Their unique bioreactor configurations are demonstrated schematically in Table 30.2. The LIVERx2000 systems have already entered the clinical study phase. However, because their clinical results have not yet been reported, the BAL system is described in this section.

The LIVERx2000 system was designed and developed by Hu et al. at the University of Minnesota and commercialized by Algenix (St. Paul, MN, USA) [30, 59]. The hepatocytes suspended in a collagen gel are injected into the lumen of a hollow fiber with a MWCO of 100 kD and the extracapillary compartment is perfused with a recirculating medium. After 24 h, the gel contracts, creating a third space, is perfused with the medium, and the extrafiber space is perfused with whole blood.

The LIVERAID-BAL system was developed by Arbios Systems (Los Angeles, CA, USA), and was founded by the principal investigator of the HepatAssist system (information available at the Arbios Systems Website, http://www.arbios.com). This BAL system utilizes a multicompartment hollow-fiber module with a unique fiber-within-fiber geometry. This geometry allows for the integration of liver-cell therapy and blood detoxification within a single module.

The oxygenating hollow-fiber bioreactor (OXY-HFB)-BAL system was developed by Jasmund et al. at Eberhard Karls University, Germany [35]. This system consists exclusively of oxygenating and

Table 30.2 BAL systems in in vitro and preclinical tests. Reproduced with permission from Park and Lee [61]. *UCLA* University of California, Los Angeles, *OXY-HFB* oxygenating hollow-fiber bioreactor, *HALSS* hybrid artificial liver support system, *PUF* polyurethane foam, *FMB-BAL* flat-membrane bioreactor BAL

	System (former/current company or institute) [Ref.]	Bioreactor configuration	System (former/current company or institute)	Bioreactor configuration
Hollow-fiber systems	LIVERx2000 (Algenix) [30]	Hepatocytes in collagen gel		
	LIVERAID (Arbios Systems) [www.arbios.com]	Charcoal and resin / Blood / Hepatocytes	Encapsulation systems	UCLA-BAL (UCLA, USA) [18] — Encapsulated hepatocytes / Blood
	OXY-HFB (Eberhard Karls Univ., Germany) [35]	Plasma / Hollow fibers / Gas or water / Hepatocyte		LifeLiver (LifeCord) [42] — Optimum size hepatocyte spheroids / Plasma
	LLS-HALSS (Kyushu Univ., Japan) [51]	Hollow fiber / Blood / Hepatocyte organoids	Porous matrix system	PUF-HALSS (Kyushu Univ., Japan) [56] — Hepatocyte spheroids in polyurethane foam / Plasma
			Flat-membrane system	FMB-BAL (Eberhard Karls Univ., Germany) [11] — Plasma / Collagen / Hepatocytes / Gas permeable membrane / O_2

integral heat-exchange fibers, and has a simple and effective design. Primary liver cells are seeded onto the surface of the fibers in the extrafiber space. Oxygen requirements are supplied and temperature is controlled via the fibers. The patient's plasma is perfused through an extrafiber space and brought into direct hepatocellular contact.

The liver lobule-like structure module (LLS)-BAL system, which was designed by Mizumoto and Funatsu at Kyushu University, Japan, has many hollow fibers that act as a blood capillary system and are regularly arranged, close to each other [51]. Hepatocytes are inoculated by a centrifugal force, at a high density in the outer space of the hollow fibers. They form an organoid and perform many liver-specific functions.

The University of California, Los Angeles (UCLA)-BAL system developed by Dixit and Gitnick, at UCLA, USA, involves the direct hemoperfusion of microencapsulated hepatocytes in an extracorporeal chamber [18]. The hepatocytes are located in biocompatible and semipermeable calcium alginate-poly-l-lysine-sodium alginate composite membranes, where they can freely interact in various biological reactions. The microencapsulated hepatocytes are simultaneously immunoisolated from the components of blood elements, thereby preventing any adverse immunological reactions. Microcapsules have a high surface area, a high capacity for hepatocytes, and high membrane permeability. All of these characteristics should result in a more effective and efficient metabolic exchange of metabolites with hepatocytes.

We have developed an alginate-entrapped hepatocyte spheroid BAL system that has a configuration similar to that of the UCLA-BAL system, and it has now been commercialized by LifeCord (Seoul, Korea), under the name LifeLiver. In this BAL system, the porcine hepatocytes are suspension cultured for 9–18 h to form spherical aggregates, called spheroids, before entrapment. In previous in vitro studies, hepatocyte spheroids have a higher performance of liver-specific functions than dispersed single hepatocytes, particularly detoxification in plasma or serum [42]. These hepatocyte spheroids are immobilized in calcium alginate beads by a high-capacity solution dropping apparatus within 10 min. The bioreactor packed with the gel-bead-containing hepatocyte spheroids is perfused with a downward flow of medium, which is generated by creating a difference in the medium level between the reservoir and outlet chamber placed before and after the bioreactor, respectively. The gravity perfusion system completely prevents any damage to the gel beads. The efficiency and safety of the LifeLiver system has been actively evaluated at anhepatic pigs and healthy beagle dogs, respectively (data not shown).

The multicapillary polyurethane foam module (PUF)-BAL system developed at Kyushu University consists of a cylindrical PUF block with many capillaries in a triangular arrangement to form a flow channel [56]. The diameter of each capillary is 1.5 mm, and the capillaries are arranged at a 3.0-mm pitch. The inoculated hepatocytes located in the pores form 100- to 150-μm-diameter spheroids within 24 h and are in direct contact with patient's plasma.

The flat-membrane bioreactor (FMB)-BAL system developed by de Bartolo et al. at Eberhard Karls University, Germany [11], and Shito et al. [73] at Masachusetts General Hospital (USA) comprises a multitude of stackable flat-membrane modules, each having an oxygenating surface area. The bioreactor is prepared for cell culture by coating the oxygen-permeable membrane with type I collagen. After the seeding and attachment of hepatocytes, a second matrix layer of collagen is placed on the top of the cell layer. This unique culture configuration is called a "sandwich culture," which is the most stable culture method for hepatocytes [19].

In observing the aforementioned eight new BAL systems, two notable trends of the bioreactor design for the BAL systems are recognized. One is that an additional channel for nutrients or oxygen supply is integrated into the bioreactor (LIVERx2000, LIVERAID-BAL, OXY-HFB-BAL, and FMB-BAL). The other is that more improved hepatocyte culture techniques, such as the use of spheroids, cylindrical organoids, and sandwich culture, are applied for long-term and enhanced liver functions (LLS-BAL, LifeLiver, PUF-BAL, and FMB-BAL).

30.4
Recent Trends in the Development of BAL Systems

Recently, the first prospective, multicenter, randomized, controlled clinical trial of a HepatAssist system has been conducted on 171 patients in 11 United States and 9 European medical centers [15]. This trial did not achieve its primary endpoint (30-day survival rate) in the overall study population. For the entire patient population, the 30-day survival rate was 71% for the BAL treatment versus 62% for the control ($P=0.26$). However, when adjusting for the impact of the disease etiology and liver transplantation on their survival, patients with FHF and subfulminant hepatic failure treated with the BAL had a statistically significant ($P=0.048$) survival advantage compared to the controls receiving standard medical care. The concept of a BAL (i.e., the use of isolated hepatocytes cultured in a bioreactor for extracorporeal support in a liver-failure patient) was partially validated by these results. This clinical trial was a milestone in the history of BAL technology and provided a chance to carefully consider the requirements of a BAL system to support liver functions.

Recently, the membrane of the bioreactor and the reactor flow rate of the ELAD system has been modified from 70 kD to 120 kD and 200 ml/min to 500 ml/min, respectively [48]. More recently, the principal developers of the HepatAssist system fabricated an improved version, the HepatAssist-2 system, which has specifications similar to those of their early BAL system except for the increased cell mass [68]. The HepatAssist-2 system can accommodate 15 billion hepatocytes; 3 times that of the former BAL system. Moreover, the AMC-BAL system developed at the AMC, which is the most novel noteworthy BAL system that has shown impressive phase I clinical results [86], uses 12 billion fresh (not cryopreserved) hepatocytes (twofold that used in the HepatAssist system), and the hepatocytes loaded in the bioreactor form small aggregates in the porous matrix. The hepatocyte aggregates are directly perfused with plasma without any immunological barriers. The bioreactor used in the AMC-BAL system was recently upgraded for optimal cell distribution and system functionality [88]. The recent modifications of the BAL systems mentioned herein are summarized in Table 30.3.

From the characteristics of the modified ELAD, HepatAssist-2, AMC-BAL, MELS, and other BAL systems, a large animal study of which has recently been completed, the following three trends can be identified. First, the cell mass employed is increasing and most BAL systems use approximately 10–20 billion fresh hepatocytes. Second, immunological barriers are not a requirement for a BAL system. This is because the short-term contact with xenogenic cells does not induce any immunological complications, and the immunological barriers are frequently removed for the maximum mass transfer. Third, hepatocytes are cultured as an organoid with appropriate cell-cell interactions, which can enhance and prolong the hepatic functions. The characteristics of the bioreactor used in the early and more recent BAL systems, including these three technological changes, are compared in Table 30.4.

If methods of long-term preservation of hepatocytes without a loss of cellular activity were developed, the availability and logistics of BAL systems will be significantly improved. Therefore, until a preservation method is fully established, centralized regional liver support centers such as the Charite Center in Germany [71] are needed to increase the frequency of supporting FHF and subfulminant hepatic failure patients, up to two to three cases per a week. At this frequency, the continuous maintenance of ready-to-use BAL systems may be reasonable and economical.

Table 30.3 Modifications of clinically evaluated BAL systems

BAL system	Modifications of later version	Aim of change
ELAD	– Increased cell mass (60 → 400 g)	Performance enhancement
	– Increased membrane pore size (70 → 120 kDa)	Performance enhancement
	– Increased reactor flow rate (200 → 500 ml/min/cartridge)	Performance enhancement
	– Perfusion medium (blood → plasma)	Safety enhancement
HepatAssist	– Increased cell number ($7 \to 15 \times 10^9$ cells)	Performance enhancement
	– Increased reactor flow rate (400 → 800 ml/min)	Performance enhancement
AMC-BAL	– Improved reactor features for even cell distribution and perfusion	Performance enhancement
	– Number of cell loading ports (2 → 3)	Performance enhancement
	– Gas exchange surface (0.75 → 1.08 m^2)	Performance enhancement
	– Number of matrix windings with hollow fibers (15 → 35)	Performance enhancement
	– Optimized reactor design parameters (e.g., matrix thickness)	

Table 30.4 Overall comparison of bioreactor characteristics of early and recent BAL systems

Bioreactor characteristics	Early BAL systems (HepatAssist, ELAD, BLSS)	Recent BAL systems in clinical and preclinical studies (LSS, AMC-BAL, and BAL systems list in Table 30.2)
Cell sources	– Cryopreserved or fresh porcine hepatocytes – Hepatoma cell line	– Freshly isolated porcine hepatocytes
Cell number	– $5\sim10 \times 10^9$ cells	– $1.2\sim4.0 \times 10^{10}$ cells
Culture method	– Microcarrier attached – Single cell immobilized	– Aggregates or organoids
Mass transfer	– Limited mass transfer	– Direct contact with hepatocytes
Treatment time	– 6~12 h for porcine cells – Up to 168 h for hepatoma cell line	– Over 12 h
Major concerns of reactor design	**Safety**>Performance	Safety<**Performance**

30.5 The Search for an Alternative Source of Functional Hepatocytes: a Major Issue in the Development of BAL Systems

The cellular component of most BAL systems in clinical and preclinical studies relies on primary porcine hepatocytes, in spite of the finding that human primary hepatocytes are ideal for a BAL system [8, 53, 65]. This is because the high demand for donor organs makes it unlikely that sufficient excess tissues would be available for nontransplantation application. Due to their easy availability and high functional activities, porcine hepatocytes are the alternative choice until other human-origin hepatocytes are available.

Promising sources of human hepatocytes for BAL systems in the future are stem cells; these are classified into two types according to the site of origin:
1. Intrahepatic stem cells. These include mature hepatocytes [47, 67], oval cells [64], small hepatocytes [49, 55], hepatoblasts [83], fetal hepatocytes [20], and hepatocyte-colony-forming units in culture [80, 81].
2. Extrahepatic stem cells. These include hematopoietic stem cells [3, 34], embryonic stem cells [32, 76, 82], and mesenchymal stem cells [36, 43, 74].

These human-origin stem cells will be actively utilized for cell transplantation for metabolic liver disease [25] and hepatic tissue-engineering fields [17, 40].

As another source of hepatocytes, established cell lines that can be sufficiently expanded have been approached via the development of tumor-derived cell lines [89], immortalized cell lines [91], and their gene manipulations [22, 39, 84]. The C3A subclone of HepG2 has already been used in clinical trials despite its poor liver functions [79]. To enhance their weak or lost liver functions that are important for liver support such as ammonia removal activity, the glutamine synthetase gene was transfected and transfected cells demonstrated a high ammonia removal activity in a bioreactor [22, 23].

In the field of xenotransplantation, alpha-1,3-galactosyltransferase-knockout pigs [85] and transgenic pigs expressing the human complement activation-control protein were developed in an effort to reduce severe hyperacute rejection in baboons [45]. The livers from these transgenic pigs are better cell sources for the improved BAL systems. In addition to these efforts to reduce immune responses, mature human hepatocytes can be acquired from humanized "animal" livers [41, 46]. The humanized liver was formed by transplanting human hepatocytes into the liver of severe combined immunodeficient mice expressing urokinase plasminogen activator and then repopulating the liver with the transplanted human cells [46]. However, this technology needs to be expanded to larger species before it can be applied to BAL systems. These various approaches to providing a sufficient alternative cell source for BAL systems and their characteristics are listed in Table 30.5.

A BAL system should be custom-made for emergency situations. Therefore, to use hepatic stem cells as a hepatocyte source it will be necessary to cultivate them in large quantities and differentiate them into functional mature hepatocytes in vitro. Although the aforementioned hepatic stem cells may have differentiation potential from a basic scientific point of view, they must satisfy two major requirements for applications. First, they must have a proliferative capacity of up to at least 10 billion cells (the minimal cell number for a BAL system) without a loss of differentiation potential. For this, it is believed that the special bioreactor technology will be involved [20, 32]. Second, they must differentiate to functional mature hepatocytes with a maximal yield based on the precise lineage-specific differentiation protocol. These major obstacles as well as several other technical challenges [29] need to be overcome in order to develop clinically applicable BAL systems.

For these reasons, we are unable to estimate how many years it will take to develop low-immunogenic, high-performance, and reasonable-cost BAL systems comprising human mature hepatocytes with unlimited supply. Until that time, the current BAL systems should be continually improved and optimized for patients currently suffering from life-threatening liver failure.

Table 30.5 Characteristics of current and future hepatocyte source for BAL systems

	Cell source [Ref.]		Advantages	Disadvantages
Current source (clinically applied)	Human primary hepatocytes [71]		– High functional activity – Good compatibility	– Limited supply
	Hepatoma derived cell line [21]		– Good availability – Low immune response	– Low hepatic functions – Tumorigenic risk
	Porcine primary hepatocytes [13, 90]		– Good availability – High functional activity	– Immune response – Zoonosis – Physiological incompatibility
Cell source candidates for future BAL system	Stem cells (human origin)	– Intrahepatic stem cells (oval cells, small hepatocytes etc.) [49, 64] – Hematopoietic stem cells [34] – Mesenchymal stem cells [36, 43, 74] – Fetal liver cells [20] – Embryonic stem cells [32, 76, 82]	– Good compatibility – Low immune response	– Low hepatic functions – Technical difficulties – High cost predicted
	Hepatoma or immortalized cell line (human origin)	– High functional immortalized cell line [89, 91] – Reversibly immortalized cell line [39] – Recombinant hepatoma cell line [22, 23, 84]	– Enhanced hepatic functions or reduced malignancies than former cell lines – Good availability – Low immune response	– Low hepatic functions – Tumorigenic risk
	Porcine (xenogenic)	– Transgenic pig primary hepatocytes [45, 85] – Humanized pig liver [41, 46]	– Reduced immune response – Enhanced compatibility – Good availability – High functional activity	– Immune response – Technical difficulties – Zoonosis

30.6
Practical Considerations for BAL Development

The initial step of designing and evaluating a hepatocyte bioreactor is the most important and difficult procedure in the development of a new BAL system. However, there are several obstacles awaiting the inventor at each stage in development. To overcome these difficulties, serious and substantial discussions need to take place between researchers with a variety of scientific backgrounds, including cell culture engineers, bioreactor engineers, hepatologists, transplantation surgeons, medical device engineers, veterinarians, and virologists. The organization of the interdisciplinary development team is the most important factor in the successful development of these systems.

Since the livers from pathogen-free pigs are the only source of functional hepatocytes currently available, most of the BAL systems used in clinical and preclinical studies employed porcine hepatocytes. The major issues and miscellaneous practical considerations for BAL systems that use porcine hepatocytes are listed below.

30.6.1
Bioreactor Design and Culture Methods

1. Sufficient cell capacity (up to 500 g of hepatocytes).
2. Culture media formulation; devoid of animal-derived products, including fetal bovine serum.
3. Integral oxygenation (if not, perfusion flow rates should exceed $150\,ml/min/10^{10}$ cells to prevent oxygen limitation).
4. Immune barrier (not a prerequisite, but offers several advantages for safety and efficacy and may act as a mass transfer barrier).
5. Perfusion medium (blood or plasma, in the case of plasma, the type of separation method, membrane or centrifugation).
6. Proper cell–cell interactions among cultured hepatocytes for enhanced and extended functionality.

30.6.2
Source Pigs and Hepatocyte Isolation

1. Specified pathogen-free pigs satisfying general xenotransplantation product guidelines (guidance for industry: source animal, product, preclinical, and clinical issues concerning the use of xenotransplantation products in humans. Final guidance, United States Food and Drug Administration, April 2003).
2. Narrowing the ranges of weight and age of source pigs for consistent isolation results.
3. Type, grade, and concentration of collagenase used for hepatocyte isolation.
4. Precise optimization of the isolation conditions for highly reproducible hepatocyte quality and quantity.

30.6.3
Preclinical Studies

1. In vitro confirmation of the broad spectrum of hepatic functions including, for example, detoxification, synthesis, drug metabolism.
2. Defining the criteria of cell quantity (viable cell number, mass, or downed cell volume).
3. Liver-failure animal model (e.g., small and/or large animal, anhepatic or ischemic model, surgical or drug-induction model, recoverable or not, full care or minimal care).
4. Combination with artificial detoxifying elements such as, for example, charcoal column, ion-exchange resin, and albumin dialysis unit.

30.6.4
Clinical Studies

1. Strict regulation of xenotransplantation to protect against infectious agents.
2. Preculture period and standby conditions.

3. Set-up of fast sterility test methods such as polymerase chain reaction (PCR) and reverse transcriptase-PCR.
4. Manufacturing facilities to guarantee aseptic and reproducible processing (including separate animal surgery facility for liver extraction).
5. Functional activity monitoring of the bioreactor during treatment (e.g., oxygen uptake rate) that offers proper information for gauging when the treatment should stop.

30.7 Conclusions

Although there are still many obstacles to overcome, it is expected that highly functional human hepatocytes will become available within a decade with the development of stem-cell technology. Hepatocytes derived from stem cells are first utilized for cell transplantation therapy. However, BAL systems also require human hepatocytes generated from stem cells. According to recent clinical results, BAL systems using porcine hepatocytes have proven to be safe, but immunological rejection and zoonosis remain major problems. If a sufficient number of human hepatocytes can be used as a cell source for BAL systems, the corresponding progress in bioreactor development and the better understanding of cell biology and the physiology of liver failure will allow the development of next-generation BAL systems. The next generation of BAL systems might have the following characteristics.

1. Human mature hepatocytes from stem cells cocultured as three-dimensional liver-like structures.
2. Entire liver functions, including bile excretion, can be supported for weeks to months.
3. The ability to treat both chronic and acute liver-failure patients.

If development of the BAL system were to progress and the system were to reach the level of hemodialysis for renal-failure patients, it is expected that the number of patients suffering from hepatic failure would decrease, regardless of a shortage of donor livers in the near future. Moreover, a diseased liver can be supported or substituted with new tissues generated by liver tissue engineering. As described above, these developing BAL systems could lead to an improved survival rate of patients suffering from acute or chronic hepatic failure in the near future.

Acknowledgments

We would like to acknowledge the contributions of our colleagues at Dongguk University, Samsung Medical Center, and LifeCord. We thank the Society for Biotechnology, Japan (SBJ) for granting us permission to reproduce the tables and descriptions used in our previous review articles published in Journal of Bioscience and Bioengineering.

References

1. Allen JW, Hassanein T, Bhatia SN (2001) Advances in bioartificial liver devices. Hepatology 34:447–455
2. Arkadopoulos N, Detry O, Rozga J, Demetriou AA (1998) Liver assist systems: state of the art. Int J Artif Organs 21:781–787
3. Austin TW, Lagasse E (2003) Hepatic regeneration from hematopoietic stem cells. Mech Dev 120:131–135
4. Bernuau J, Rueff B, Benhamou JP (1986) Fulminant and subfulminant liver failure: definitions and causes. Semin Liver Dis 6:97–106
5. Burnell JM, Dawborn JK, Epstein RB, Gutman RA, et al (1967) Acute hepatic coma treated by cross-circulation or exchange transfusion. N Engl J Med 276:935–943
6. Busse B, Smith MD, Gerlach JC (1999) Treatment of acute liver failure: hybrid liver support. A critical overview. Langenbecks Arch Surg 384:588–599
7. Cao S, Esquivel CO, Keeffe EB (1998) New approaches to supporting the failing liver. Annu Rev Med 49:85–94
8. Chamuleau RA, Deurholt T, Hoekstra R (2005) Which are the right cells to be used in a bioartificial liver? Metab Brain Dis 20:327–335
9. Chan C, Berthiaume F, Nath BD, Tilles AW, et al (2004) Hepatic tissue engineering for adjunct and temporary liver support: critical technologies. Liver Transpl 10:1331–1342
10. Court FG, Wemyss-Holden SA, Dennison AR, Maddern GJ (2003) Bioartificial liver support devices: historical perspectives. ANZ J Surg 73:739–748
11. De Bartolo L, Jarosch-Von Schweder G, Haverich A, Bader A (2000) A novel full-scale flat membrane bio-

reactor utilizing porcine hepatocytes: cell viability and tissue–specific functions. Biotechnol Prog 16:102–108
12. de Rave S, Tilanus HW, van der Linden J, de Man RA, et al (2002) The importance of orthotopic liver transplantation in acute hepatic failure. Transpl Int 15:29–33
13. Demetriou AA, Rozga J, Podesta L, Lepage E, et al (1995) Early clinical experience with a hybrid bioartificial liver. Scand J Gastroenterol Suppl 208:111–117
14. Demetriou AA, Watanabe F, Rozga J (1995) Artificial hepatic support systems. Prog Liver Dis 13:331–348
15. Demetriou AA, Brown RS Jr, Busuttil RW, Fair J, et al (2004) Prospective, randomized, multicenter, controlled trial of a bioartificial liver in treating acute liver failure. Ann Surg 239:660–667; discussion 667–670
16. Denis J, Opolon P, Delorme ML, Granger A, et al (1979) Long-term extra-corporeal assistance by continuous haemofiltration during fulminant hepatic failure. Gastroenterol Clin Biol 3:337–347
17. Diekmann S, Bader A, Schmitmeier S (2006) Present and future developments in hepatic tissue engineering for liver support systems. Cytotechnology 50:163–179
18. Dixit V, Gitnick G (1998) The bioartificial liver: state-of-the-art. Eur J Surg Suppl:71–76
19. Dunn JC, Tompkins RG, Yarmush ML (1991) Long-term in vitro function of adult hepatocytes in a collagen sandwich configuration. Biotechnol Prog 7:237–245
20. Ehashi T, Ohshima N, Miyoshi H (2006) Three-dimensional culture of porcine fetal liver cells for a bioartificial liver. J Biomed Mater Res A 77:90–96
21. Ellis AJ, Hughes RD, Wendon JA, Dunne J, et al (1996) Pilot-controlled trial of the extracorporeal liver assist device in acute liver failure. Hepatology 24:1446–1451
22. Enosawa S, Miyashita T, Fujita Y, Suzuki S, et al (2001) In vivo estimation of bioartificial liver with recombinant HepG2 cells using pigs with ischemic liver failure. Cell Transplant 10:429–433
23. Enosawa S, Miyashita T, Saito T, Omasa T, Matsumura T (2006) The significant improvement of survival times and pathological parameters by bioartificial liver with recombinant HepG2 in porcine liver failure model. Cell Transplant 15:873–880
24. Farmer DG, Anselmo DM, Ghobrial RM, Yersiz H, et al (2003) Liver transplantation for fulminant hepatic failure: experience with more than 200 patients over a 17-year period. Ann Surg 237:666–675; discussion 675–666
25. Fisher RA, Strom SC (2006) Human hepatocyte transplantation: worldwide results. Transplantation 82:441–449
26. Flendrig LM, la Soe JW, Jorning GG, Steenbeek A, et al (1997) In vitro evaluation of a novel bioreactor based on an integral oxygenator and a spirally wound nonwoven polyester matrix for hepatocyte culture as small aggregates. J Hepatol 26:1379–1392
27. Flendrig LM, Maas MA, Daalhuisen J, Ladiges NC, et al (1998) Does the extend of the culture time of primary hepatocytes in a bioreactor affect the treatment efficacy of a bioartificial liver? Int J Artif Organs 21:542–547
28. Gerlach JC (1996) Development of a hybrid liver support system: a review. Int J Artif Organs 19:645–654
29. Gerlach JC (2006) Bioreactors for extracorporeal liver support. Cell Transplant 15 Suppl 1:S91–103
30. Hu W-S, Friend JR, Wu FJ, Sielaff T, et al (1997) Development of a bioartificial liver employing xenogeneic hepatocytes. Cytotechnology 23:29–38
31. Hui T, Rozga J, Demetriou AA (2001) Bioartificial liver support. J Hepatobiliary Pancreat Surg 8:1–15
32. Imamura T, Cui L, Teng R, Johkura K, et al (2004) Embryonic stem cell-derived embryoid bodies in three-dimensional culture system form hepatocyte-like cells in vitro and in vivo. Tissue Eng 10:1716–1724
33. Jalan R, Sen S, Williams R (2004) Prospects for extracorporeal liver support. Gut 53:890–898
34. Jang YY, Collector MI, Baylin SB, Diehl AM, et al (2004) Hematopoietic stem cells convert into liver cells within days without fusion. Nat Cell Biol 6:532–539
35. Jasmund I, Langsch A, Simmoteit R, Bader A (2002) Cultivation of primary porcine hepatocytes in an OXY-HFB for use as a bioartificial liver device. Biotechnol Prog 18:839–846
36. Jiang Y, Jahagirdar BN, Reinhardt RL, Schwartz RE, et al (2002) Pluripotency of mesenchymal stem cells derived from adult marrow. Nature 418:41–49
37. Kjaergard LL, Liu J, Als-Nielsen B, Gluud C (2003) Artificial and bioartificial support systems for acute and acute-on-chronic liver failure: a systematic review. JAMA 289:217–222
38. Knell AJ, Dukes DC (1976) Dialysis procedures in acute liver coma. Lancet 2:402–403
39. Kobayashi N, Okitsu T, Nakaji S, Tanaka N (2003) Hybrid bioartificial liver: establishing a reversibly immortalized human hepatocyte line and developing a bioartificial liver for practical use. J Artif Organs 6:236–244
40. Kulig KM, Vacanti JP (2004) Hepatic tissue engineering. Transpl Immunol 12:303–310
41. Laconi E (2006) The past, present and future of xeno-derived liver cells. Curr Opin Organ Transplant 11:654–658
42. Lee DH, Lee JH, Choi JE, Kim YJ, Kim SK, Park JK (2002) Determination of optimum aggregates of porcine hepatocytes as a cell source of a bioartificial liver. J Microbiol Biotechnol 12:735–739
43. Lee KD, Kuo TK, Whang-Peng J, Chung YF, et al (2004) In vitro hepatic differentiation of human mesenchymal stem cells. Hepatology 40:1275–1284
44. Mazariegos GV, Kramer DJ, Lopez RC, Shakil AO, et al (2001) Safety observations in phase I clinical evaluation of the Excorp Medical Bioartificial Liver Support System after the first four patients. ASAIO J 47:471–475
45. McCurry KR, Kooyman DL, Alvarado CG, Cotterell AH, et al (1995) Human complement regulatory proteins protect swine-to-primate cardiac xenografts from humoral injury. Nat Med 1:423–427
46. Meuleman P, Libbrecht L, De Vos R, de Hemptinne B, et al (2005) Morphological and biochemical characterization of a human liver in a uPA-SCID mouse chimera. Hepatology 41:847–856
47. Michalopoulos GK, DeFrances MC (1997) Liver regeneration. Science 276:60–66
48. Millis JM, Cronin DC, Johnson R, Conjeevaram H, et al (2002) Initial experience with the modified extracorporeal liver-assist device for patients with fulminant he-

patic failure: system modifications and clinical impact. Transplantation 74:1735–1746
49. Mitaka T, Mizuguchi T, Sato F, Mochizuki C, et al (1998) Growth and maturation of small hepatocytes. J Gastroenterol Hepatol 13 Suppl:S70–77
50. Mito M (1986) Hepatic assist: present and future. Artif Organs 10:214–218
51. Mizumoto H, Funatsu K (2004) Liver regeneration using a hybrid artificial liver support system. Artif Organs 28:53–57
52. Morsiani E, Brogli M, Galavotti D, Bellini T, et al (2001) Long-term expression of highly differentiated functions by isolated porcine hepatocytes perfused in a radial-flow bioreactor. Artif Organs 25:740–748
53. Morsiani E, Brogli M, Galavotti D, Pazzi P, et al (2002) Biologic liver support: optimal cell source and mass. Int J Artif Organs 25:985–993
54. Morsiani E, Pazzi P, Puviani AC, Brogli M, et al (2002) Early experiences with a porcine hepatocyte-based bioartificial liver in acute hepatic failure patients. Int J Artif Organs 25:192–202
55. Nakajima-Nagata N, Sakurai T, Mitaka T, Katakai T, et al (2004) In vitro induction of adult hepatic progenitor cells into insulin-producing cells. Biochem Biophys Res Commun 318:625–630
56. Nakazawa K, Ijima H, Fukuda J, Sakiyama R, et al (2002) Development of a hybrid artificial liver using polyurethane foam/hepatocyte spheroid culture in a preclinical pig experiment. Int J Artif Organs 25:51–60
57. Naruse K, Nagashima H, Sakai Y, Kokudo N, et al (2005) Development and perspectives of perfusion treatment for liver failure. Surg Today 35:507–517
58. Neuzil DF, Rozga J, Moscioni AD, Ro MS, et al (1993) Use of a novel bioartificial liver in a patient with acute liver insufficiency. Surgery 113:340–343
59. Nyberg SL, Shatford RA, Payne WD, Hu WS, et al (1992) Primary culture of rat hepatocytes entrapped in cylindrical collagen gels: an in vitro system with application to the bioartificial liver. Rat hepatocytes cultured in cylindrical collagen gels. Cytotechnology 10:205–215
60. Opolon P (1979) High-permeability membrane hemodialysis and hemofiltration in acute hepatic coma: experimental and clinical results. Artif Organs 3:354–360
61. Park JK, Lee DH (2005) Bioartificial liver systems: current status and future perspective. J Biosci Bioeng 99:311–319
62. Pascher A, Sauer IM, Hammer C, Gerlach JC, et al (2002) Extracorporeal liver perfusion as hepatic assist in acute liver failure: a review of world experience. Xenotransplantation 9:309–324
63. Patzer JF 2nd, Mazariegos GV, Lopez R (2002) Preclinical evaluation of the Excorp Medical, Inc, Bioartificial Liver Support System. J Am Coll Surg 195:299–310
64. Petersen BE, Bowen WC, Patrene KD, Mars WM, et al (1999) Bone marrow as a potential source of hepatic oval cells. Science 284:1168–1170
65. Poyck PP, Hoekstra R, van Wijk AC, Attanasio C, et al (2007) Functional and morphological comparison of three primary liver cell types cultured in the AMC bioartificial liver. Liver Transpl 13:589–598

66. Poyck PP, Pless G, Hoekstra R, Roth S, et al (2007) In vitro comparison of two bioartificial liver support systems: MELS Cell Module and AMC-BAL. Int J Artif Organs 30:183–191
67. Rhim JA, Sandgren EP, Degen JL, Palmiter RD, et al (1994) Replacement of diseased mouse liver by hepatic cell transplantation. Science 263:1149–1152
68. Rozga J (2006) Liver support technology—an update. Xenotransplantation 13:380–389
69. Rozga J, Holzman MD, Ro MS, Griffin DW, et al (1993) Development of a hybrid bioartificial liver. Ann Surg 217:502–509; discussion 509–511
70. Sauer IM, Obermeyer N, Kardassis D, Theruvath T, et al (2001) Development of a hybrid liver support system. Ann N Y Acad Sci 944:308–319
71. Sauer IM, Zeilinger K, Obermayer N, Pless G, et al (2002) Primary human liver cells as source for modular extracorporeal liver support—a preliminary report. Int J Artif Organs 25:1001–1005
72. Schiodt FV, Atillasoy E, Shakil AO, Schiff ER, et al (1999) Etiology and outcome for 295 patients with acute liver failure in the United States. Liver Transpl Surg 5:29–34
73. Shito M, Tilles AW, Tompkins RG, Yarmush ML, et al (2003) Efficacy of an extracorporeal flat-plate bioartificial liver in treating fulminant hepatic failure. J Surg Res 111:53–62
74. Snykers S, Vanhaecke T, Papeleu P, Luttun A, et al (2006) Sequential exposure to cytokines reflecting embryogenesis: the key for in vitro differentiation of adult bone marrow stem cells into functional hepatocyte-like cells. Toxicol Sci 94:330–341; discussion 235–339
75. Sosef MN, Van De Kerkhove MP, Abrahamse SL, Levi MM, et al (2003) Blood coagulation in anhepatic pigs: effects of treatment with the AMC-bioartificial liver. J Thromb Haemost 1:511–515
76. Soto-Gutierrez A, Navarro-Alvarez N, Rivas-Carrillo JD, Chen Y, et al (2006) Differentiation of human embryonic stem cells to hepatocytes using deleted variant of HGF and poly-amino-urethane-coated nonwoven polytetrafluoroethylene fabric. Cell Transplant 15:335–341
77. Stockmann HB, Hiemstra CA, Marquet RL, Yzermans JN (2000) Extracorporeal perfusion for the treatment of acute liver failure. Ann Surg 231:460–470
78. Sussman NL, Kelly JH (1993) Improved liver function following treatment with an extracorporeal liver assist device. Artif Organs 17:27–30
79. Sussman NL, Gislason GT, Conlin CA, Kelly JH (1994) The Hepatix extracorporeal liver assist device: initial clinical experience. Artif Organs 18:390–396
80. Suzuki A, Nakauchi H, Taniguchi H (2003) In vitro production of functionally mature hepatocytes from prospectively isolated hepatic stem cells. Cell Transplant 12:469–473
81. Suzuki A, Zheng YW, Kaneko S, Onodera M, et al (2002) Clonal identification and characterization of self-renewing pluripotent stem cells in the developing liver. J Cell Biol 156:173–184
82. Tabei I, Hashimoto H, Ishiwata I, Tachibana T, et al (2005) Characteristics of hepatocytes derived from early

ES cells and treatment of surgically induced liver failure rats by transplantation. Transplant Proc 37:262–264
83. Tanimizu N, Miyajima A (2004) Notch signaling controls hepatoblast differentiation by altering the expression of liver-enriched transcription factors. J Cell Sci 117:3165–3174
84. Terada S, Kumagai T, Yamamoto N, Ogawa A, et al (2003) Generation of a novel apoptosis—resistant hepatoma cell line. J Biosci Bioeng 95:146–151
85. Tseng YL, Kuwaki K, Dor FJ, Shimizu A, et al (2005) alpha1,3-Galactosyltransferase gene-knockout pig heart transplantation in baboons with survival approaching 6 months. Transplantation 80:1493–1500
86. van de Kerkhove MP, Di Florio E, Scuderi V, Mancini A, et al (2002) Phase I clinical trial with the AMC-bioartificial liver. Int J Artif Organs 25:950–959
87. van de Kerkhove MP, Hoekstra R, Chamuleau RA, van Gulik TM (2004) Clinical application of bioartificial liver support systems. Ann Surg 240:216–230
88. van de Kerkhove MP, Poyck PP, van Wijk AC, Galavotti D, et al (2005) Assessment and improvement of liver specific function of the AMC-bioartificial liver. Int J Artif Organs 28:617–630
89. Wang L, Sun J, Li L, Mears D, et al (1998) Comparison of porcine hepatocytes with human hepatoma (C3A) cells for use in a bioartificial liver support system. Cell Transplant 7:459–468
90. Watanabe FD, Mullon CJ, Hewitt WR, Arkadopoulos N, et al (1997) Clinical experience with a bioartificial liver in the treatment of severe liver failure. A phase I clinical trial. Ann Surg 225:484–491; discussion 491–494
91. Yoon JH, Lee HV, Lee JS, Park JB, et al (1999) Development of a non-transformed human liver cell line with differentiated-hepatocyte and urea-synthetic functions: applicable for bioartificial liver. Int J Artif Organs 22:769–777
92. Yoshiba M, Sekiyama K, Iwamura Y, Sugata F (1993) Development of reliable artificial liver support (ALS)—plasma exchange in combination with hemodiafiltration using high-performance membranes. Dig Dis Sci 38:469–476

31 Pancreas Engineering

R. Cortesini, R. Calafiore

Contents

31.1	Introduction	411
31.2	Islet Transplantation in Humans: State of the Art	412
31.3	Immunosuppression of Allograft Rejection and Autoimmunity	414
31.4	Immunomodulation by Gene Transfer in Transplanted Islets	415
31.5	Cell Sources for Tissue Engineering	416
31.5.1	Embryonic Stem Cells	416
31.5.2	Adult Stem Cells	418
31.6	Microencapsulation	419
31.7	Organogenesis of the Pancreas	423
31.8	Conclusions	423
	References	424

31.1 Introduction

Diabetes mellitus is one of the most diffuse and destructive diseases. It affects more than 190 million individuals in the world and is the fifth leading killer of Americans, with 73,000 deaths every year. Diabetes is a disease in which the body's failure to regulate glucose can lead to serious and even fatal complications. The incidence of diabetes has increased by 61% since 1991. The increasing prevalence and huge cost, exceeding 100 billion dollars annually in the United States, create the need for innovative research in this field [1].

There are two forms of diabetes, type I and type II, which differ in their pathogenesis. Type I is an autoimmune disease, presenting primarily during the first two decades of life and resulting from the destruction of insulin-producing β cells in the pancreatic islets of Langerhans. Type II generally presents in adults, is characterized by insulin resistance in the tissues and eventually β-cell failure in about 50% of the cases. In both types of diabetes the islets of Langerhans are the targets of destruction [2, 3].

Islets of Langerhans are discrete clusters of endocrine cells scattered throughout the pancreas. Islet numbers vary between species, but in general range from 100,000 to 2.5 million per pancreas and vary in size from about 50 to 300 μm in diameter. Each islet contains several thousand hormone-secreting cells, comprising insulin-secreting β cells, glucagon-secreting α cells, somatostatin-secreting δ cells and pancreatic polypeptide-secreting PP cells. The β cells, which comprise about 70% of the endocrine cells in the islets, are unique in their ability to express the preproinsulin gene by mechanisms that are fairly well, if not completely, understood. Transplantation of the pancreas, pancreatic islets or genetically engineered producing cells aims to provide the recipient with a continuous source of insulin.

The cornerstone of diabetes treatment is the administration of exogenous insulin, but adequate control of glucose level cannot always be achieved in patients with type I diabetes mellitus, who display frequent episodes of hypoglycemia. The rate of complications due to chronic hyperglycemia is also increased in patients with poorly controlled diabetes. The more serious complications occur at the microvascular level and include retinopathy and blindness, nephropathy, neuropathy, foot ulcers, and cardiovascular disease.

Although strict control of hyperglycemia through adequate diet and well-planned insulin therapy can reduce the intensity and progression of diabetic complications, a stable state of euglycemia in rarely achieved.

Over the last 40 years, there has been an intensive search for an effective treatment of diabetes. Pancreas transplantation has been the first attempt to cure this disease surgically, implanting a new pancreas. The first pancreas transplant in a human was performed by Kelly and Lillehey on December 1966 at the University of Minnesota [4]. The pancreas was transplanted simultaneously with a kidney in a uremic diabetic patient. Endocrine function was sustained for several weeks before rejection of both grafts occurred. Little progress has been made over the following years, and by 1980 only 105 pancreas transplants had been reported worldwide. In 1978 the transplantation group from Lyon, led by J-M Dubernard, reported a new technique in which the pancreatic duct of the segmental pancreas graft was obstructed by neoprene injection, which blocked the supply of pancreatic enzymes. This technique yielded encouraging results, especially when the kidney and pancreas were transplanted together in diabetic patients with renal failure. With the advent of cyclosporine, quadruple immunosuppression (induction with anti-T-cell antibodies plus three-drug maintenance therapy with calcineurin inhibitor, antimetabolite, and steroids) quickly became accepted worldwide. The number of patients with functioning grafts increased and some of them remained insulin-free for more than 20 years (Registry 2007, [5]). It is now accepted that vascularized, whole-pancreas transplants restore normoglycemia reliably and maintain normal, long-term glucose levels. More than 25,000 pancreas transplants have been performed worldwide by 2007, securing improvement of the quality of life and reversal of some secondary complications [5]. However, despite its clinical success, pancreas transplantation is a major operation with associated morbidity and mortality. The shortage of cadaveric donors and occurrence of serious side effects from chronic immunosuppression remain major problems.

Another important approach that has the advantage of eliminating major surgery is the implantation of human pancreatic islets. Islets transplantation involves the transfer of healthy islets from a deceased donor to a diabetic patient. Beta cells can be isolated, their function can be tested before implantation, and eventually they may be cryopreserved. Typically islets are implanted via the portal vein into the liver, where they produce insulin. In some patients the transplanted islets can control glucose levels for more than 5 years. However, several problems remain unresolved. First, the availability of islets and probability of maintaining them in a functional state and capable of growing is still unpredictable. Second, powerful immunosuppression based on sirolimus, tacrolimus, and blocking antibodies must be administered. Third, patients who reject the graft develop anti-donor human leukocyte antigen (HLA) antibodies, which preclude a secondary transplant. These problems raise the question: are islet transplants preferable to continuous insulin treatment?

New technologies that can overcome the difficulties encountered so far in the treatment of diabetes are desperately needed. New sources of islet cells, new techniques permitting pancreas organogenesis, islet neogenesis, or microencapsulation of pancreatic islet transplants are being considered. We will discuss some of these possibilities in this chapter.

31.2
Islet Transplantation in Humans: State of the Art

The idea of transplanting isolated pancreatic islets germinated nearly simultaneously at the University of Minnesota, University of Pennsylvania, and University of Giessen. The first clinical transplant of adult allogeneic pancreatic islets was performed in 1974 at the University of Minnesota [6]. The human islet allograft functioned poorly. In contrast, clinical islet autografts, performed after pancreatectomy for chronic pancreatitis at the University of Minnesota and at University of Giessen in 1977 and 1978 displayed good function. The results showed that viable human islets can be isolated and that they produce insulin when transplanted via the portal vein in the liver [6]. By the end of the 20th century, a few hundred islet allografts had been transplanted in diabetic patients. Most of these transplants were performed at four institutions in the United States (Minneapolis, Pittsburgh, St. Louis, and Miami). Other centers from Canada (Edmonton), and Europe (Giessen, Geneva, and Milan) also got involved. A detailed review of the cumulative world experience in clinical islet transplantation showed that the major factors responsible for the failure of most transplants performed until 1999 were: (1) inadequate transplant mass, (2) inadequate islet function, (3) inadequate immunosuppression, and (4) diabetogenic effects of immunosuppressive drugs [7]. To overcome these problems, a new protocol was studied at the University of Alberta in Edmonton, Canada, by a group lead by James Shapiro. Their research proved to be a real turning point in clinical islet transplantation. The approach was based on the use of an adequate mass of islets, multiple islet transplants, and immunosuppressive treatment with daclizumab, and low doses of sirolimus and tacrolimus (without corticosteroids). Follow-up studies of more than 50 patients treated with the Edmonton Protocol, confirmed insulin independence rates of 80% after 1 year [8]. Thereafter, this protocol was adopted worldwide.

Between January 1990 and December 2004, 458 pancreatic islet transplants have been reported to the International Islet Transplant Registry (ITR) based at the University of Giessen, Germany. The collaborating centers were from North America and Europe. Data analysis of the ITR showed a 1-year patient survival rate of 97% with a functioning islet graft in 82% of the cases. However, insulin independence was achieved in only in 43% of the cases [7].

The National Institute of Diabetes and Digestive and Kidney Disease founded the Collaborative Islets Transplant Registry (CITR) for data collection from North American programs between 1999 and 2006. All 45 North American centers with an islet transplantation program submitted the results obtained during 1999–2006. The reported data included 292 allograft recipients with complete follow-up and 579 infusions of islets obtained from a total of 634 donors [5].

The majority of the patients with an islet-alone transplant (60%) received at the time of the first islet infusion daclizumab, sirolimus, and tacrolimus. Antithymocyte globulin was given alone or in combination with the other drugs to 11% of the patients.

An increasing percentage of islet-alone recipients were reinfused: 12% by day 30, 37% by day 75, 57% by month 6, and 69% by the end of year 1. The percentage of patients who were insulin independent without reinfusion remained fairly constant (at 9–13%) throughout the 1st year. However, an increasing prevalence of graft loss and decreasing prevalence of insulin independence over time since the last infusion was found regardless of the total number of infusions given. The prevalence of insulin independence declined from about 60% after 4 months to about 24% after 3 years (1,100 days) after the last infusion. The trend was similar in patients with two or three infusions. Overall, at 5 years after transplantation, 85% of islet recipients had measurable plasma C-peptide and well-controlled HBA1c levels. They required a significantly lower amount of daily insulin, had virtually no clinical hypoglycemia and experienced freedom from use of exogenous insulin in 10% of the cases [9].

There are still major problems in the field of islet transplantation, which can be summarized as follows [9]:

1. Approximately 50% of the islets transplanted do not engraft and/or display low functional capacity because of isolation procedures, local inflammatory and apoptotic events mediated by cytokines, or by local factors (hypoxia) within the hepatic microenvironment.

2. Isolated islet grafts were more prone to rejection and recurrence of autoimmunity than whole pancreases allotransplanted in diabetic patients. Furthermore, the development of alloantibodies in a large percentage of islet recipients diminished their chance of receiving future islet, pancreas, or kidney transplants.

3. Because there is no marker for monitoring islet graft rejection, and biopsy of the graft cannot be performed, relatively high doses of immunosuppressive drugs are being administered. As a consequence there is a high incidence of side effects, such as nephrotoxicity, leucopenia, infection, gastrointestinal intolerance, and anemia. Such adverse events impose prompt reduction of immunosuppression, which in turn results in allosensitization, production of anti-HLA antibodies, and accelerated destruction of the graft.

4. The liver, which appears to be an immunologically privileged site, requiring less immunosuppression

when transplanted, does not protect allogeneic islets dispersed through the portal vein, facilitating instead allosensitization and production of alloantibodies.

In conclusion, islet transplantation continues to show short-term benefits in terms of insulin independence, normal or near normal HbA1c levels, and decreased frequency of hypoglycemic episodes. However, long-term primary efficacy, safety of immunosuppression, and effects on secondary complications require further studies.

31.3
Immunosuppression of Allograft Rejection and Autoimmunity

The long-term success of organ, tissue, and cell transplantation is frequently limited by rejection and/or recurrence of the original disease. T-cell recognition of foreign major histocompatibility complex (MHC) antigens occurs by two mechanisms: direct allorecognition of intact donor MHC molecules expressed on the membrane of graft cells, and indirect allorecognition of donor MHC molecules that are processed and presented as peptides bound to the MHC class II molecules displayed on the surface of host antigen-presenting cells (APC). The mechanism of indirect allorecognition is similar to that involved in the development of autoimmunity. In both cases T cells first recognize a dominant (self or nonself) epitope. With progression of the disease the response spreads to other cryptic epitopes, possibly due to cytokine-induced alterations of endosomal enzymes with consequent changes in the cleavage of proteins during antigen processing [10].

Effective priming of CD4 T lymphocytes requires cell-to-cell interactions between APC and T cells. This interaction allows bidirectional costimulatory signals, which activate the APC presenting the processed antigen and those T cells with cognate-specific T-cell receptors. T-cell receptor triggering results in upregulation of CD40L (CD154) expression on activated T cells. CD40L interacts with CD40 on APC, inducing them to upregulate B7 (CD80 and CD86) costimulatory molecules. These B7 molecules interact with their counter-receptor (CD28), augmenting CD40L expression and inducing CD4 T-helper (Th) cell proliferation. Signaling along the B7 pathway is bidirectional and enables the conditioning of dendritic cells (DC), as evidenced by induction of interleukin (IL)-12, IL-6, and interferon production, expression of indoleamine 2,3-dioxygenase (IDO), and activation of the mitogen-activated protein kinase pathway. T-cell activation in the absence of B7 costimulation results in T-cell paralysis or anergy [11].

The differentiation of $CD8^+$ cytotoxic T cells (CTL) is contingent upon recognition of MHC class I/peptide complexes on APC that have been "licensed" or conditioned by $CD4^+$ Th cells, via the CD40-signaling pathway. For effective CTL priming, different peptides of the same protein must be recognized on the same APC by $CD4^+$ and $CD8^+$ T cells. Emerging evidence indicates that DC have a specialized capacity to process exogenous antigens into the MHC class I pathway, being able to cross-prime $CD8^+$ T cells. Stimulatory pathogens and inflammatory cytokines can also license DC to stimulate T cell responses [11].

Alternatively DC can be "licensed by certain cytokines" to become a tolerogenic, via the upregulation of inhibitory receptors (such as ILT3 and ILT4) and downregulation of costimulatory molecule expression on their surface. Tolerogenic DC induce anergy in Th cells and elicit the differentiation of CD8 and CD4 regulatory T cells (Treg). There is a bidirectional interaction between Treg and APC, which resides in the capacity of Treg to induce tolerogenic APC and that of tolerogenic APC to induce Treg. Antigen-specific T-suppressor cells (Ts or Treg) block the differentiation of $CD8^+$ CTLs and cytokine-producing CD4 Th cells. The potential of Treg/Ts to mediate tolerance to self and nonself antigens is being studied extensively [11].

Suppression of the immune response to allogeneic transplants is currently accomplished by use of several drugs that inhibit T-cell activation and differentiation. Certain immunosuppressive drugs such as cyclosporin and steroids cannot be used in pancreas/islet transplantation because of their diabetogenic effect. Tacrolimus, sirolimus, and daclizumab are the basic components of the Edmonton protocol, which, however, does not secure long-term survival of islet allografts.

Numerous studies are ongoing in both human and nonhuman primates to discriminate between rejection, recurrence of autoimmunity, or islet exhaustion as primary causes of islet transplant failure [12]. Major efforts are also being made to develop corticoid-sparing immunosuppressive therapy, ultimately aiming to induce allospecific tolerance. Such attempts include the blockade of the CD40-CD40L costimulatory pathway, which is deemed to be crucial to the activation and differentiation of T-effector cells [11]. Blockade of this pathway is particularly relevant to islet cell transplantation in view of the recent finding that pancreatic islet cells express CD40, which is upregulated by proinflammatory cytokines [13].

CD154 (CD40 ligand) blockade by use of monoclonal antibodies (mAb) has been shown to prolong islet allograft survival in chemically induced and spontaneously diabetic (nonobese diabetic, NOD) mice as well as in nonhuman primates, yet long-term allograft survival has not been achieved [14, 15]. A synergistic effect on prolonging islet cell allograft survival has been accomplished by simultaneously targeting CD154 and lymphocyte function-associated antigen-1 (LFA-1) [16].

Extensive studies were also performed in NOD mice transplanted with allogeneic islets under the cover of "Edmonton-like" immunosuppression supplemented with anti-CD154, anti-LFA and/or anti-CD45RO/RB mAb [14, 16–18].

More recently, the NOD/severe combined immune deficiency (NOD/SCID) mouse system has been adopted to study the survival of human pancreatic islet cells transplanted into mice that have been reconstituted with human peripheral blood mononuclear cells (PBMC) and treated with immunomodulatory agents [18]. Treatment with CD45RO/RB mAb and rapamycin induced long-term tolerance, possibly through apoptosis of activated CD4+ T cells, induction of both CD4+ and CD8+ T-regulatory cells, and modulation of DC [18]. Long-term survival of islets transplanted under the kidney capsule of humanized NOD/SCID mice has also been achieved by treatment with ILT3-Fc protein (Cortesini and Suciu-Foca, recently in Diabetes 2008).

Animal models have been instrumental in the advancement of our understanding of the mechanisms involved in the immune response against self and nonself antigens. Multiple transgenic, knockout and reconstituted models of autoimmune diseases have been developed over the past two decades [19, 20]. Although many biological mechanisms are similar in rodents and humans, there are several structural and functional differences, which render the extrapolation of experimental results to clinical practice quite difficult [20]. Thus, the creation of humanized mice, defined as immunodeficient mice engrafted with human hematopoietic stem cells or PBMC, provided a powerful tool for preclinical testing of new immunomodulatory agents and study of human immune responses [20]. This is particularly true for certain human genes that have no ortholog in rodents, such as ILT3.

31.4
Immunomodulation by Gene Transfer in Transplanted Islets

Tissue-specific gene therapy may provide a strategy for protecting islet grafts from the host environment without systemic immunosuppression. Genes can be transfected ex vivo to the donor islets before transplantation to provide protection from the host's autoimmune and alloimmune response in the microenvironment of the graft.

Bioengineered insulin-positive aggregates can be transplanted in the spleen, portal vein or under the renal capsule [21]. The latter has the advantage of permitting easy access to the graft for studying changes that occur within and around the implant.

Prolongation of graft survival by gene transfer of immunoregulatory molecules has been accomplished in various experimental models. Adenoviral gene transfer of IDO to pancreatic islets from prediabetic NOD mouse donors conferred protection of the islets from the attack of adoptively transferred NOD diabetogenic T cells and prolonged survival in NOD/SCID recipient mice [22]. Similarly, mice that received islet transplants overexpressing manganese superoxide dismutase (MnSOD; a nuclear-encoded mitochondrial antioxidant enzyme expressed in response to inflammatory cytokines), were euglycemic 50% longer than mice transplanted with untreated islets [23]. Cotransfection of IDO and MnSOD showed no synergistic relationship, yet demonstrated that gene combinations can be used to more efficiently protect islet grafts from diabetogenic T cells [24].

Recently, it has been shown that transplantation of islets infected with an adenoviral vector expressing sCD40-Ig prolonged significantly their survival in allogeneic recipient mice with clinically induced diabetes. Local production of sCD40 shielded these genetically modified islets from mononuclear cell infiltration, activation of T cells, and subsequent graft rejection [25].

Taken together, these data suggest that genetic modification of islet cells facilitate the induction of long-term tolerance.

31.5
Cell Sources for Tissue Engineering

There are several different sources of cells that could be used for tissue engineering. These include embryonic stem cells (ESC) or germ cells, adult stem cells such as bone marrow stromal (mesenchymal) stem cells, and mature (non-stem) cells from the patient.

31.5.1
Embryonic Stem Cells

ESC exhibit two important properties: the ability to proliferate in an undifferentiated but pluripotent state and the ability to differentiate into many specialized cell types. ESC can be isolated from the inner cell mass of the embryo during the blastocyst stage and are usually grown on human feeder cells.

Human ESC have been shown to differentiate in vitro into cells from all three embryonic germ layers. Under appropriate conditions, ESC grow, generating ectodermal, mesodermal, and endodermal cell lineages. At this stage ESC are forming the typical cell aggregates called embryoid bodies (EB). Skin and neurons have been formed, indicating ectodermal differentiation. Blood, cardiac cells, cartilage, endothelial cells, and muscle have been differentiated from mesodermal tissue and pancreatic cells have been generated, indicating endodermal differentiation [26].

Although ESC have the highest potential to differentiate into various tissues, their use raises significant ethical and political concerns in the United States and several European nations because harvesting of human ESC requires the destruction of the human embryo. These obstacles to ESC research have accelerated the search for alternative sources of stem cells.

Therapeutic cloning may generate a viable source of ESC that avoids the ethical and political controversies [27]. Recently at Harvard University researchers have fused human embryonic stem cells (hESC) with human fibroblasts, generating hybrid cells that maintain a stable tetraploid DNA content and have morphologized growth and antigen expression patterns characteristic of hESC. Differentiation of hybrid cells in vivo and in vitro evolved in cell types representative of each embryonic germ layer. Analysis of these cells showed that the somatic genome was reprogrammed to an embryonic state. It is hoped that these cells will provide the material from which the elusive reprogramming factors can be identified, synthesized, and used to redirect cytogenesis. The embryonic cell takeover is so complete that the fused cell behaves just like one. Once the two nuclei of the "combined" cell fuse together, it is virtually impossible to disentangle their chromosomes. However, after the membrane surrounding each nucleus has dissolved, there might be a short interval during which the embryonic cell's chromosomes could somehow be evicted from the cell. The resulting diploid cell should contain only the patient's genes, representing an excellent source of material for a therapeutic approach.

31.5.1.1
Nuclear Transfer

Nuclear cloning, also called nuclear transfer or nuclear transplantation, is the technique that involves the introduction of a nucleus from a donor cell into an enucleated ovocyte to generate an embryo with a genome identical to that of the donor [28]. When transferred to the uterus of a female recipient the embryo develops into an adult organism. This technique has been used for the first time for cloning frogs in 1962 and later sheep (100 cattle, goats, mice, pigs, and dogs) [29]. The cloning of Dolly demonstrated that nuclei from mammalian differentiated cells can be reprogrammed to an undifferentiated state by transacting factors present in ovocytes.

Nuclear cloning can be reproductive or therapeutic. In humans, reproductive cloning is banned. Therapeutic cloning is used to create embryos that are genetically identical to the stem cell donor. These autologous stem cells have the potential to differentiate into any type of tissues and be of use for tissue and organ replacement therapy. Therefore, theoretically, therapeutic cloning, also called somatic cell nuclear transfer, provides a limitless alternative source of transplantable cells [28].

31.5.1.2
Induction of Pluripotent Stem Cells

The discovery of transacting factors, which are present in ovocytes, generated the search for similar factors that can induce reprogramming without somatic cell nuclear transfer. Recently, two revolutionary studies described concomitantly the generation of induced pluripotent stem cells (iPSC) from somatic cells by the transduction of defined transcription factors.

One of these studies demonstrated the generation of iPSC from adult human dermal fibroblasts with four factors: OCT 3/4, SOX2, Klf4, and c-Myc. These cells were similar in morphology and pluripotency to ESC, differentiating into cell types of the three germ layers in vitro and in teratomas [30].

The second study used a combination of factors lacking c-Myc, which causes death and differentiation of hESC. They showed that four factors (OCT4, SOX2, NANOG, and LIN 28) are sufficient to reprogram human somatic cells (newborn foreskin fibroblasts) to pluripotent stem cells that have the essential characteristics of ESC.

These iPSC have normal karyotypes, telomerase activity, and cell surface markers that are similar to ESC and should be useful for generating tissues for autologous transplantation. Further studies were cautionately suggested to avoid mutation through viral integration [31].

The importance of these studies resides in the fact that they do not involve the destruction of human preimplantation embryos.

31.5.1.3
Differentiation of Insulin-Producing Cells

Pancreatic cell lines have been widely used as in vitro models for research on β-cell physiology [32]. This is in part because cell propagation to large numbers is relatively straightforward, the cells are characterized as clonal, and they exhibit many of the functions found in native β cells. Pancreatic β-cell lines are derived from insulinoma cells via the introduction and expression of tumor genes. The use of cell lines for cellular therapies is problematic because of serious risks for tumorigenesis [33].

ESC are derived from the inner cell mass of a blastocyst. When ESC are cultivated on nonadherent surfaces, they form EB in which spontaneous differentiation occurs toward cell types of all three embryonic layers. Treatment of dispersed ESC or EB with different compounds has also been attempted in combination with a selection of cells positive for the neurofilament protein nestin. Active formation of insulin in differentiated cells should also be analyzed via several methods, such as C-peptide staining, metabolic labeling, demonstration of biphasic insulin release upon glucose challenge, and transplantation assays.

Differentiated ESC can be enriched with a desired cell type through the introduction of transcription factor genes. Constitutive expression of the Pdx-129, Pax4,29, or Nkx2.230 transcription factor leads to the differentiation of ESC expressing pancreatic endocrine markers [32].

Cell trapping is another approach using gene insertion for selection of insulin-producing cells. ESC are stably transfected with a construct of a drug-resistance gene downstream of a promoter that is active in β cells. However, promoters may be active in multiple, distinct cells types, thus confounding the selection outcome.

Further advances in the generation of ESC-derived pancreatic β cells will require the development of methodologies for multistep exposure to physiologically relevant stimuli. Mouse and human ESC have been successfully induced to definitive endoderm [34, 35], which is a first step toward pancreatogenesis. D'Amour et al. [36] reported the conversion of hESC to endocrine cells capable of synthesizing insulin, glucagon, somatostatin, pancreatic polypeptide, and ghrelin. However, C-peptide secretion in response to glucose is minimal and many of the differentiated cells express multiple hormones, pointing to an immature phenotype [32].

31.5.2
Adult Stem Cells

There are two major challenges that must be addressed in order to use pancreatic β-cell replacement therapy on a large scale. The first is to shield the insulin-producing cells from autoimmunity and alloimmunity, as discussed earlier. The second hurdle is to provide an adequate mass of cells for transplantation. Successful islet transplantation in humans requires four or more pancreas donors for one single diabetic recipient. Given the fact that there are millions of patients with type 1 diabetes, but only a few thousand pancreases available each year, the impracticability of this approach is obvious.

Three alternative approaches are being intensively investigated to circumvent the need for insulin-producing islet cells: (1) the production of surrogate cells by genetically modifying nonendocrine cells to secrete insulin in response to a glucose challenge, (2) the transdifferentiation of nonendocrine stem/progenitor cells or mature cells to glucose-responsive adult tissue, and (3) the regulated differentiation of islet stem/progenitor cells to produce large numbers of mature, functional islets [3].

Various groups have focused on the possibility of generating glucose-sensing, insulin-secreting non-islet cells (referred to as surrogate cells) as replacement tissue for β cells. Surrogate cells may escape host autoimmune and alloimmune responses [40].

Falqui and coworkers transfected human fibroblasts and myoblasts, and rat hepatocytes with a genetically modified human proinsulin gene. Transplantation of insulin-producing primary human fibroblasts into laboratory animals resulted in a decrease in blood glucose levels, thus demonstrating the feasibility of this approach [37].

An alternative approach was used by Cheung and coworkers, who transfected gut-associate K cells with the human preproinsulin gene. The transfected cells were then injected into mouse embryos. These implanted mice produced human insulin within the gut, and were protected from diabetes following deliberate destruction of the endogenous β-cell mass [38].

A third approach has been to promote and control the (trans)-differentiation of non-islet (usually non-pancreatic) cells with stem-cell characteristics. This has been especially successful using hepatic oval stem cells. Yang et al. used hepatic stem cells that are known to be capable of differentiating to either hepatocytes or bile-duct epithelium. They showed that such stem cells differentiate into endocrine hormone-producing cells when stimulated with high glucose concentrations [39].

Hepatic "oval" stem cells also form islet-like aggregates in culture and express pancreatic islet cell differentiation markers such as Pdx-1, Pax4, Pax6, Nkx2.2, Nkx6.1, and islet hormones. These aggregates secrete insulin when challenged with glucose and reverse diabetes when transplanted in streptozotocin-treated NOD/SCID mice [39].

Transdifferentiation offers an interesting and potentially important approach to generating surrogate islet cells. Several investigators have reported the isolation and differentiation of stem cells derived from adult pancreatic ductal structures expressing endocrine hormones. All three major cellular compartments of the adult pancreas (ducts, acinar, and islets) are presumed to contain stem cells or precursors capable of differentiating, transdifferentiating or dedifferentiating into cells that have the potential to become endocrine cells [3].

Peck and coworkers first demonstrated the successful growth of functional islets from partially digested pancreatic tissue in long-term cultures [34–36]. A series of sequential steps permitted successive growth, differentiation, and maturation of the cultured cell populations. According to their protocol, ductal epithelial cells and islets are first isolated from digested pancreas then cultured in a growth restrictive medium that enriches for an epithelial cell subpopulation capable of forming monolayers with a neuroendocrine-cell-like phenotype. Next, islet-like structures bud from the epithelial-like monolayers and proliferate. Cultures are stimulated with high concentrations of glucose and various growth factors to force the maturation of the cells. Finally, these islet-like structures are transplanted to promote final maturation of the β cells exhibiting regulated glucose-responsiveness [3, 40–43].

The functional capacity of in-vitro-grown mouse islets has been extensively investigated in NOD mice [42]. Evidence has been provided that such islets restored euglycemia successfully.

Stem cells derived from the bone marrow are also considered a prospective source for pancreatic β cells [44, 45]. Umbilical cord blood (UCB) is also known to contain progenitor cells. Transplantation of human UCB cells to diabetic mice led to correction of hyperglycemia, presumably because of in vivo differentiation of the cord cells to insulin-producing cells [46]. Other tissues that have also been investigated for the presence of progenitor cells that can transdifferentiate to insulin-producing cells include cells isolated from the spleen [47], adipose tissue [48], salivary glands [49], and central nervous system [50].

The risk of tumorigenicity is of major concern regarding the use of ESC in therapies. In addition to oncogenicity, the immunocompatibility of hESC-derived cells should be scrutinized.

The possibility that adult stem-cell-derived, in-vitro generated islets may soon be an alternative to cadaver-derived islets for treating diabetic patients is rapidly gaining credibility.

Although it has been known for many years that stem cells associated with the pancreatic ducts are present in both healthy and diabetic individuals, the ability to stimulate these stem cells in vitro to differentiate into functional islets represents a breakthrough that could have limitless potential for therapeutic intervention in type 1 diabetes. However, results from a growing number of laboratories indicate that although one can initiate the expansion and differentiation of such stem cells, a basic understanding of the mechanisms that control this process is still lacking. Identification of genes that are expressed during various ontogenetic stages of differentiation along the endocrine pathway may ultimately provide the information necessary to sequentially stimulate stem cells to mature to end-stage islets using mixtures of appropriate growth factors. Alternatively, the immature islets may be ideally suited for implantation since they appear to have the potential to respond to the in vivo environment, thereby establishing homeostasis with the recipient through vascularization, expansion of the β-cell mass, and differentiation to insulin-secreting cells.

31.6
Microencapsulation

The bioartificial pancreas is a system powered by live, insulin-secreting human or animal pancreatic β cells or genetically engineered pancreatic or extrapancreatic stem cells. Such a device, named the "biohybrid artificial pancreas" (BHAP), contains artificial micro- or macromembranes that protect live insulin-secreting cells from the immune attack of host cells [51]. The BHAP may be useful for treatment of both type 1 (insulin-dependent) and type 2 (insulin-requiring) diabetes mellitus, since insulin-secreting cells, living within selective permeable and biocompatible artificial membranes could provide a continuous supply of insulin. As discussed earlier, transplantation of an endocrine pancreas as either a whole organ or isolated human islet cells have enabled the reversal of hyperglycemia for variable periods of time in chronically immunosuppressed patients with type 1 diabetes mellitus. However, the requirement of 2–3 pancreases per recipient to correct hyperglycemia poses a serious problem, given the shortage of cadaver donors [52, 53]. For this reason, sources of nonhuman insulin-secreting cells/tissues have been considered. One such example is adult or neonatal porcine islets, since porcine insulin is well tolerated by humans. Evidence has been provided that long-term cell survival with concurrent positive effects on metabolic control can be achieved by xenotransplanting type 1 diabetic patients with neonatal porcine islets of Langerhans protected with Sertoli cells inside an autologous collagen-generating device implanted subcutaneously [54].

Unlike macrodevices, microcapsules hold the advantage of offering a far better volume:surface ratio, which implies faster diffusion kinetics of hormone and nutrients. Microcapsules are easy to handle and graft under local anesthesia; they represent the only immunoprotective devices that have been evaluated extensively and shown to function, especially when made of alginate-based biopolymers complexed with polyaminoacids (i.e., poly-L-lysine, polyornithine) [55–59]. A method for engineering microcapsules from (lypopolysaccharide and endotoxin free) highly purified alginate/poly-L-ornithine has been developed at the University of Perugia and used for immunoprotection of islet allo- or xenografts.

The latset generation of microcapsules, based on the use of alginate-poly-L-ornithine-alginate, measures an average 500 μm in diameter and are named "medium-sized" microcapsules (Fig. 31.1). They have an excellent volume:surface ratio and a reasonable volume for experimental or clinical transplantation. These capsules maintain a highly uniform size and shape both when are they empty (Fig. 31.2) and filled with islet cells (Fig. 31.3). They fulfill the requirements of United States Pharmacopenia, being made of highly purified pyrogen and endotoxin-free alginate and having a size and geometry that is well tolerated without eliciting an inflammatory reaction.

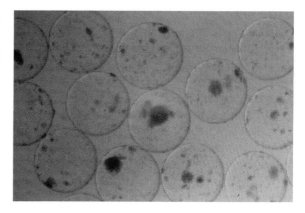

Fig. 31.3 Neonatal porcine islet containing alginate/poly-L-ornithine microcapsules. Staining with diphenylthiocarbazone (inverted phase microscopy 10×). University of Perugia

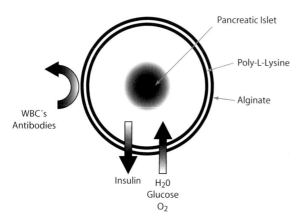

Fig. 31.1 Schematic representation of islet containing alginate/poly-l-ornithine microcapsules (University of Perugia). *WBCs* White blood cells

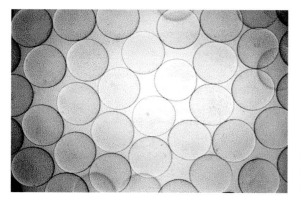

Fig. 31.2 Alginate/poly-L-ornithine empty microcapsules (inverted phase microscopy, 10×). University of Perugia

This efficient microencapsulation system has been gradually improved by means of multiple sequential filtration, dialysis, and use of pharmaceutical-grade raw powder of immunoglobulin-M-enriched sodium alginate. Empty microcapsules fabricated with this alginate and overlayed with poly-L-ornithine and an outer alginate film have been proven to be biocompatible, avoiding a host inflammatory reaction in rodents, dogs, primates, and ultimately in humans [59, 60].

Luca et al. have demonstrated that adult and neonatal porcine islets enable reversal of hyperglycemia in nonimmunosuppressed diabetic rodents and large-sized animals [61]. Although porcine endogenous retrovirus (PERV) transmission to humans has been viewed as a potential risk to xenotransplantation [61, 62], no signs of PERV transmission has been seen in patients who had been grafted with porcine islets [63]. Microcapsules per se might represent a diffusion barrier for PERV [64], while macrodevices do not have a similar advantage [65].

Because microcapsules represent immunoisolatory and biocompatible barriers, they are expected to prevent allo- or xenospecific immune responses. It seems less difficult to achieve this goal with allografts than with xenografts. Some investigators claim that alginate-based microcapsules cannot protect islet xenografts and that general immunosuppression will still be needed. We believe that this strategy, although possibly useful, would considerably weaken the concept that microcapsules confer physical immunoprotection to pancreatic islets. It may be preferable to improve and eventually potentiate their immunoprotective capacity using composite microcapsules. Calafiore et al. have explored the possibility of creating multiple compartments (Fig. 31.4) within conventional microcapsules, where islets are placed together with agents that have anti-inflammatory properties [66] and antioxidizing effects [67]. This strategy improved significantly both acceptance of the encapsulated islet grafts and the viability of the enveloped islet cells (Fig. 31.5). These authors initiated a pilot clinical trial in which microencapsulated human islets were implanted intraperitoneally into patients with type 1 diabetes who were not subjected to immunosuppression. The preliminary data showed that the transplant procedure was simple, noninvasive, painless, and had no side effects. Although exogenous insulin supplementation could not be withdrawn, the patients displayed a decline in glycohemoglobin, disappearance or less frequent hypoglycemic episodes, and responsiveness to an oral glucose tolerance test, indicating that the grafts had a significant metabolic impact (Fig. 31.6) [68].

Most allogeneic or xenogeneic islet-containing-arginate-poly-L-lysine/poly-L-ornithine microcapsules have been implanted intraperitoneally into diabetic patients who were not immunosuppressed. Because of the lack of standardized materials and procedures it is difficult to assess impartially the results reported by different laboratories. In general, studies on rodents with chemically induced diabetes yielded better results than large-animal trials in terms of full remission of hyperglycemia following transplantation of microencapsulated islets [69]. A trial of arginate-poly-L-lysine-encapsulated porcine islet xenografts into spontaneously diabetic, nonimmunosuppressed monkeys revealed full remission of hyperglycemia and successful withdrawal of exogenous insulin for up to 3 years of posttransplantation follow-up [70]. This striking, still unmatched result suggests that under ideal conditions microcapsules represent an effective, biocompatible and immunoselective physical barrier that enables immunoprotection of islet xenografts without immunosuppression of the host.

MULTICOMPARTMENTAL MICROCAPSULES

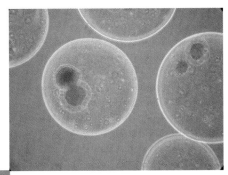

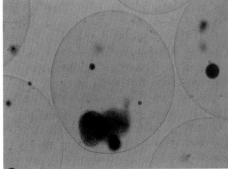

-neonatal porcine islets
-anti-oxidizing agents
(Vit D, E)

Fig. 31.4 Multicompartment alginate/poly-L-lysine microcapsules containing neonatal pig islets and nanoencapsulated, antioxidizing vitamins (*Vit.*) D and E. Inverted phase microscopy (20×). University of Perugia

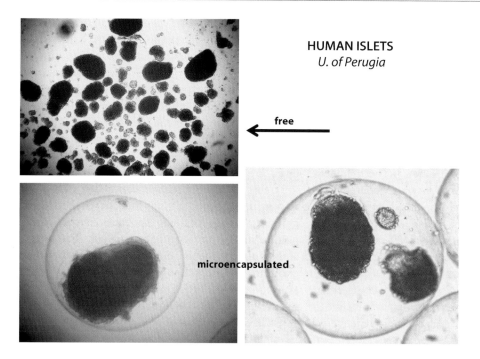

Fig. 31.5 Human pancreatic islets freshly stained with diphenylthiocarbazone before (upper left, 10×) and after (lower left and right, 40×) microencapsulation, under inverted phase microscopy

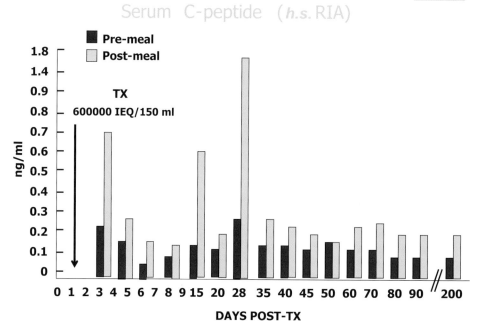

Fig. 31.6 Serum C-peptide levels before and after peritoneal transplant of microencapsulated human islets (*Tx*; patient no. 2) with no immunosuppression

It is obvious that the search for new biopolymers from which to engineer high-performance microcapsules should be continued. In particular, the membrane's durability, immunoselectiveness, and biocompatibility are issues that deserve the highest attention. In light of the increasing success of an artificial scaffold whereby the islets are deposited under much better conditions of nutrition supply [70], the question arises: should microencapsulated islets be implanted as "loose" microspheres or rather incorporated within laminar prevascularized membranes (to provide nutrient/oxygen supply and improve metabolic insulin kinetics)? Also, are the capsules themselves useful or should they be replaced by artificial scaffolds? [71] These questions need to be addressed in the future, to achieve further progress. Most importantly, microcapsules need to be made of highly purified biopolymers [72] in order to guarantee an excellent product.

31.7
Organogenesis of the Pancreas

Organogenesis consists of growing in situ a new organ such as a kidney, for example, from the metanephros transplanted into an adult animal. To increase its functionality, the tissue is obtained from an embryo at an appropriately early stage and can be incubated with specific growth factors prior to transplantation. When transplanted into the omentum of another animal, this structure grows, differentiates, and matures into a functioning organ [73, 74].

Organogenesis of an endocrine pancreas from a transplanted embryonic anlagen was successfully achieved by Marc Hammerman and collaborators [74]. Hammerman et al. have developed a method that permits the transplantation of developing fetal or neonatal pancreatic tissue (pancreatic anlagen), which has an increased capacity for pancreatic β-cell expansion postimplantation [73, 74]. Because the pancreas originates from a single embryonic precursor tissue, it is possible to expand pancreatic anlagen to amounts that are sufficient to render a diabetic host euglycemic. Experimentally induced diabetes has been treated successfully in rodents using embryonic pancreas transplants. Embryonic rat pancreas anlagen transplanted into the omentum of adult syngeneic rats generated a novel pancreas (consisting of islet tissue within stroma surrounded by peritoneal fat) 15 weeks after transplantation. No acinar tissue was present. Electron microscopy studies showed pancreatic β cells containing neurosecretory granules with eccentric dense cores, representing crystallized insulin. Pancreatic anlagen transplanted into streptozotocin-treated diabetic rats established normoglycemia that persisted for 4 months. In these transplanted animals, the glucose tolerance test was normal [74].

Rat pancreatic anlagen were also transplanted into the omentum of mice in a concordant xenotransplantation model. The recipient animals were treated with costimulatory blocking agents. The novel pancreas developed and functioned perfectly. This technique is able to generate a novel bioengineered organ, which, although not consisting of a pure population of islets, is islet-enriched and can be safely transplanted into the omentum, a site that is drained by the portal vein [74].

More recently, this group of investigators transplanted pig pancreatic primordia into the mesentery of diabetic rhesus macaque momkeys; insulin requirements were reduced over a 22-month period in one animal [75].

The experience gained from these experiments suggests that it may become possible to transplant an embryonic pig pancreas, obtained early during gestation, to an intraperitoneal site that would permit secretion of insulin for controlling glycemia in diabetic patients.

31.8
Conclusions

The field of transplantation medicine is in a new era of expansion that encompasses tissue engineering and organogenesis. New strategies include: (1) the use of living cells to restore, maintain or enhance tissue function; (2) the generation of new organs from xenogeneic embryos or from embryos cloned using nuclear transfer techniques; (3) the ex vivo production of vessels, cardiac muscle, and heart valves that have all of the characteristic of living tissues; (4) the generation of new organs, especially gastrointestinal, from tissue-specific organoids seeded on biodegradable scaffolds.

These exciting new developments are emerging as a result of contemporary advances in biomaterial technology and developmental biology. Research is directed toward the development of new ways to synthesize biomaterials and control the response of the human body to such materials. The tissue-engineering approach to the repair and regeneration of cardiovascular, liver, and pancreatic tissues is very imaginative and holds great potential, although to date efforts have been largely empirical. Key scientific inputs and major technological advances are necessary to bring these approaches to clinical reality.

Organogenesis is still at a very early stage of development. To further expand this field, advances in developmental biology, genetics, and immunology will be required. Ethical problems raised as a result of the use of embryonic tissue must be discussed and resolved. Although the advantage of using pig embryos for tissue engineering and xenotransplantation is obvious, the possibility of retrovirus transmission from pig to human recipients must first be excluded. Studies on nuclear transfer have demonstrated that cells obtained by therapeutic cloning can be successfully harvested, expanded in culture, and transplanted onto biodegradable scaffolds and into the autologous host [76]. Therapeutic cloning offers a potential limitless source of cells for the regeneration of tissue and organs in vivo, yet it poses important ethical problems.

Transplantation medicine will include cell and tissue replacement, bioartificial organs, and cloned organs. Although today it may sound like futuristic science, it may become a clinical reality in the near future.

Acknowledgement

This work was supported by grants from the Interuniversitary Organ Transplantation Consortium, Rome, Italy.

References

1. Hogan, P., T. Dall, and P. Nikolov. 2003. Economic costs of diabetes in the US in 2002. Diabetes Care 26:917–932
2. Calne, R. 2005. Cell transplantation for diabetes. Philos Trans R Soc Lond B Biol Sci 360:1769–1774
3. Peck, A.B., and V. Ramiya. 2004. In vitro-generation of surrogate islets from adult stem cells. Transpl Immunol 12:259–272
4. Kelly, W.D., R.C. Lillehei, F.K. Merkel, Y. Idezuki, and F.C. Goetz. 1967. Allotransplantation of the pancreas and duodenum along with the kidney in diabetic nephropathy. Surgery 61:827–837
5. CITR Coordinating Center CITR Annual Report 2007 NIH Bethesda, MD 2007
6. Najarian J, DER Sutherland and M. Steffes. 1975 Isolation of Human islets of Langerhans for Transplantation, Transplant Proc 7:611–13
7. Bretzel, R.G., H. Jahr, M. Eckhard, I. Martin, D. Winter, and M.D. Brendel. 2007. Islet cell transplantation today. Langenbecks Arch Surg 392:239–253
8. Ryan, E.A., J.R. Lakey, B.W. Paty, S. Imes, G.S. Korbutt, N.M. Kneteman, D. Bigam, R.V. Rajotte, and A.M. Shapiro. 2002. Successful islet transplantation: continued insulin reserve provides long-term glycemic control. Diabetes 51:2148–2157
9. Bromberg, J.S., B. Kaplan, P.F. Halloran, and R.P. Robertson. 2007. The islet transplant experiment: time for a reassessment. Am J Transplant 7:2217–2218
10. Ciubotariu, R., Z. Liu, A.I. Colovai, E. Ho, S. Itescu, S. Ravalli, M.A. Hardy, R. Cortesini, E.A. Rose, and N. Suciu-Foca. 1998. Persistent allopeptide reactivity and epitope spreading in chronic rejection of organ allografts. J Clin Invest 101:398–405
11. Vlad, G., R. Cortesini, and N. Suciu-Foca. 2005. License to heal: bidirectional interaction of antigen-specific regulatory T cells and tolerogenic APC. J Immunol 174:5907–5914
12. Inverardi, L., N.S. Kenyon, and C. Ricordi. 2003. Islet transplantation: immunological perspectives. Curr Opin Immunol 15:507–511
13. Klein, D., F. Barbe-Tuana, A. Pugliese, H. Ichii, D. Garza, M. Gonzalez, R.D. Molano, C. Ricordi, and R.L. Pastori. 2005. A functional CD40 receptor is expressed in pancreatic beta cells. Diabetologia 48:268–276
14. Molano, R.D., A. Pileggi, T. Berney, R. Poggioli, E. Zahr, R. Oliver, C. Ricordi, D.M. Rothstein, G.P. Basadonna, and L. Inverardi. 2003. Prolonged islet allograft survival in diabetic NOD mice by targeting CD45RB and CD154. Diabetes 52:957–964
15. Kenyon, N.S., M. Chatzipetrou, M. Masetti, A. Ranuncoli, M. Oliveira, J.L. Wagner, A.D. Kirk, D.M. Harlan, L.C. Burkly, and C. Ricordi. 1999. Long-term survival and function of intrahepatic islet allografts in rhesus monkeys treated with humanized anti-CD154. Proc Natl Acad Sci U S A 96:8132–8137
16. Berney, T., A. Pileggi, R.D. Molano, R. Poggioli, E. Zahr, C. Ricordi, and L. Inverardi. 2003. The effect of simul-

taneous CD154 and LFA-1 blockade on the survival of allogeneic islet grafts in nonobese diabetic mice. Transplantation 76:1669–1674
17. Molano, R.D., A. Pileggi, T. Berney, R. Poggioli, E. Zahr, R. Oliver, T.R. Malek, C. Ricordi, and L. Inverardi. 2003. Long-term islet allograft survival in nonobese diabetic mice treated with tacrolimus, rapamycin, and anti-interleukin-2 antibody. Transplantation 75:1812–1819
18. Gregori, S., P. Mangia, R. Bacchetta, E. Tresoldi, F. Kolbinger, C. Traversari, J.M. Carballido, J.E. de Vries, U. Korthauer, and M.G. Roncarolo. 2005. An anti-CD45RO/RB monoclonal antibody modulates T cell responses via induction of apoptosis and generation of regulatory T cells. J Exp Med 201:1293–1305
19. Gudjonsson, J.E., A. Johnston, M. Dyson, H. Valdimarsson, and J.T. Elder. 2007. Mouse models of psoriasis. J Invest Dermatol 127:1292–1308
20. Shultz, L.D., F. Ishikawa, and D.L. Greiner. 2007. Humanized mice in translational biomedical research. Nat Rev Immunol 7:118–130
21. Soria, B., E. Roche, G. Berna, T. Leon-Quinto, J.A. Reig, and F. Martin. 2000. Insulin-secreting cells derived from embryonic stem cells normalize glycemia in streptozotocin-induced diabetic mice. Diabetes 49:157–162
22. Alexander, A.M., M. Crawford, S. Bertera, W.A. Rudert, O. Takikawa, P.D. Robbins, and M. Trucco. 2002. Indoleamine 2,3-dioxygenase expression in transplanted NOD Islets prolongs graft survival after adoptive transfer of diabetogenic splenocytes. Diabetes 51:356–365
23. Bertera, S., A.M. Alexander, M.L. Crawford, G. Papworth, S.C. Watkins, P.D. Robbins, and M. Trucco. 2004. Gene combination transfer to block autoimmune damage in transplanted islets of Langerhans. Exp Diabesity Res 5:201–210
24. Bertera, S., X. Gene, Z. Tawadrous, R. Bottino, A.N. Balanumrugan, W.A. Rudert, P. Drain, S.C. Watkins, M. Trucco. 2003. Body window-enabled in vivo multicolor imaging of transplanted mouse islets expressing an insulin-Timer fusion protein. Biotechniques 35:718–722
25. Rehman, K.K., S. Bertera, M. Trucco, A. Gambotto, and P.D. Robbins. 2007. Immunomodulation by adenoviral-mediated SCD40-Ig gene therapy for mouse allogeneic islet transplantation. Transplantation 84:301–307
26. Roche, E., J.A. Reig, A. Campos, B. Paredes, J.R. Isaac, S. Lim, R.Y. Calne, and B. Soria. 2005. Insulin-secreting cells derived from stem cells: clinical perspectives, hypes and hopes. Transpl Immunol 15:113–129
27. Shufaro, Y., and B.E. Reubinoff. 2004. Therapeutic applications of embryonic stem cells. Best Pract Res Clin Obstet Gynaecol 18:909–927
28. Koh, C.J., and A. Atala. 2004. Therapeutic cloning applications for organ transplantation. Transpl Immunol 12:193–201
29. Armstrong, L., and M. Lako. 2006. The future of human nuclear transfer? Stem Cell Rev 2:351–358
30. Takahashi, K., K. Tanabe, M. Ohnuki, M. Narita, T. Ichisaka, K. Tomoda, and S. Yamanaka. 2007. Induction of pluripotent stem cells from adult human fibroblasts by defined factors. Cell 131:861–872
31. Yu, J. M.A. Vodyanik, K. Smuga-Otto, J. Antosiewicz-Bourget, J.L. Frane, S. Tian, J. Nie, G.A. Jonsdottir, V. Ruotti, R. Stewart, I.I. Slukvin, and J.A. Thomson. 2007. Induced pluripotent stem cell lines derived from human somatic cells. Science 318:1917–1920
32. Lock, L.T., and E.S. Tzanakakis. 2007. Stem/Progenitor cell sources of insulin-producing cells for the treatment of diabetes. Tissue Eng 13:1399–1412
33. Radvanyi, F., S. Christgau, S. Baekkeskov, C. Jolicoeur, and D. Hanahan. 1993. Pancreatic beta cells cultured from individual preneoplastic foci in a multistage tumorigenesis pathway: a potentially general technique for isolating physiologically representative cell lines. Mol Cell Biol 13:4223–4232
34. Kubo, S., H.K. Takahashi, M. Takei, H. Iwagaki, T. Yoshino, N. Tanaka, S. Mori, and M. Nishibori. 2004. E-prostanoid (EP)2/EP4 receptor-dependent maturation of human monocyte-derived dendritic cells and induction of helper T2 polarization. J Pharmacol Exp Ther 309:1213–1220
35. Yasunaga, M., S. Tada, S. Torikai-Nishikawa, Y. Nakano, M. Okada, L.M. Jakt, S. Nishikawa, T. Chiba, and T. Era. 2005. Induction and monitoring of definitive and visceral endoderm differentiation of mouse ES cells. Nat Biotechnol 23:1542–1550
36. D'Amour, K.A., A.D. Agulnick, S. Eliazer, O.G. Kelly, E. Kroon, and E.E. Baetge. 2005. Efficient differentiation of human embryonic stem cells to definitive endoderm. Nat Biotechnol 23:1534–1541
37. Falqui, L., S. Martinenghi, G.M. Severini, P. Corbella, M.V. Taglietti, C. Arcelloni, E. Sarugeri, L.D. Monti, R. Paroni, N. Dozio, G. Pozza, and C. Bordignon. 1999. Reversal of diabetes in mice by implantation of human fibroblasts genetically engineered to release mature human insulin. Hum Gene Ther 10:1753–1762
38. Cheung, A.T., B. Dayanandan, J.T. Lewis, G.S. Korbutt, R.V. Rajotte, M. Bryer-Ash, M.O. Boylan, M.M. Wolfe, and T.J. Kieffer. 2000. Glucose-dependent insulin release from genetically engineered K cells. Science 290:1959–1962
39. Yang, L., S. Li, H. Hatch, K. Ahrens, J.G. Cornelius, B.E. Petersen, and A.B. Peck. 2002. In vitro trans-differentiation of adult hepatic stem cells into pancreatic endocrine hormone-producing cells. Proc Natl Acad Sci U S A 99:8078–8083
40. Peck, A.B., J.G. Cornelius, D. Schatz, and V.K. Ramiya. 2002. Generation of islets of Langerhans from adult pancreatic stem cells. J Hepatobiliary Pancreat Surg 9:704–709
41. Cornelius, J.G., V. Tchernev, K.J. Kao, and A.B. Peck. 1997. In vitro-generation of islets in long-term cultures of pluripotent stem cells from adult mouse pancreas. Horm Metab Res 29:271–277
42. Ramiya, V.K., M. Maraist, K.E. Arfors, D.A. Schatz, A.B. Peck, and J.G. Cornelius. 2000. Reversal of insulin-dependent diabetes using islets generated in vitro from pancreatic stem cells. Nat Med 6:278–282
43. Bonner-Weir, S., M. Taneja, G.C. Weir, K. Tatarkiewicz, K.H. Song, A. Sharma, and J.J. O'Neil. 2000. In vitro

cultivation of human islets from expanded ductal tissue. Proc Natl Acad Sci U S A 97:7999–8004
44. Hess, D., L. Li, M. Martin, S. Sakano, D. Hill, B. Strutt, S. Thyssen, D.A. Gray, and M. Bhatia. 2003. Bone marrow-derived stem cells initiate pancreatic regeneration. Nat Biotechnol 21:763–770
45. Mathews, V., P.T. Hanson, E. Ford, J. Fujita, K.S. Polonsky, and T.A. Graubert. 2004. Recruitment of bone marrow-derived endothelial cells to sites of pancreatic beta-cell injury. Diabetes 53:91–98
46. Zhao, Y., H. Wang, and T. Mazzone. 2006. Identification of stem cells from human umbilical cord blood with embryonic and hematopoietic characteristics. Exp Cell Res 312:2454–2464
47. Kodama, S., W. Kuhtreiber, S. Fujimura, E.A. Dale, and D.L. Faustman. 2003. Islet regeneration during the reversal of autoimmune diabetes in NOD mice. Science 302:1223–1227
48. Timper, K., D. Seboek, M. Eberhardt, P. Linscheid, M. Christ-Crain, U. Keller, B. Muller, and H. Zulewski. 2006. Human adipose tissue-derived mesenchymal stem cells differentiate into insulin, somatostatin, and glucagon expressing cells. Biochem Biophys Res Commun 341:1135–1140
49. Hisatomi, Y., K. Okumura, K. Nakamura, S. Matsumoto, A. Satoh, K. Nagano, T. Yamamoto, and F. Endo. 2004. Flow cytometric isolation of endodermal progenitors from mouse salivary gland differentiate into hepatic and pancreatic lineages. Hepatology 39:667–675
50. Hori, Y., X. Gu, X. Xie, and S.K. Kim. 2005. Differentiation of insulin-producing cells from human neural progenitor cells. PLoS Med 2:e103
51. Colton, C.K. 1995. Implantable biohybrid artificial organs. Cell Transplant 4:415–436
52. Ryan, E.A., D. Bigam, and A.M. Shapiro. 2006. Current indications for pancreas or islet transplant. Diabetes Obes Metab 8:1–7
53. Calafiore, R. 1997. Perspectives in pancreatic and islet cell transplantation for the therapy of IDDM. Diabetes Care 20:889–896
54. Valdes-Gonzalez, R.A., L.M. Dorantes, G.N. Garibay, E. Bracho-Blanchet, A.J. Mendez, R. Davila-Perez, R.B. Elliott, L. Teran, and D.J. White. 2005. Xenotransplantation of porcine neonatal islets of Langerhans and Sertoli cells: a 4-year study. Eur J Endocrinol 153:419–427
55. Mendez, A.J., L. Inverardi, C. Ricordi, and N.S. Kenyon. 2007. The Miami results on porcine islet-Sertoli cell xenotransplantation. Xenotransplantation 14:89
56. Toso, C., Z. Mathe, P. Morel, J. Oberholzer, D. Bosco, D. Sainz-Vidal, D. Hunkeler, L.H. Buhler, C. Wandrey, and T. Berney. 2005. Effect of microcapsule composition and short-term immunosuppression on intraportal biocompatibility. Cell Transplant 14:159–167
57. Basta, G., P. Sarchielli, G. Luca, L. Racanicchi, C. Nastruzzi, L. Guido, F. Mancuso, G. Macchiarulo, G. Calabrese, P. Brunetti, and R. Calafiore. 2004. Optimized parameters for microencapsulation of pancreatic islet cells: an in vitro study clueing on islet graft immunoprotection in type 1 diabetes mellitus. Transpl Immunol 13:289–296
58. Orive, G., R.M. Hernandez, A.R. Gascon, R. Calafiore, T.M. Chang, P. De Vos, G. Hortelano, D. Hunkeler, I. Lacik, A.M. Shapiro, and J.L. Pedraz. 2003. Cell encapsulation: promise and progress. Nat Med 9:104–107
59. Desai, T.A., W.H. Chu, G. Rasi, P. Sinibaldi-Vallebona, E. Guarino, and M. Ferrari. 1999. Microfabricated biocapsules provide short-term immunoisolation of insulinoma xenografts. Biomed Microdevices 1:131–138
60. Duvivier-Kali, V.F., A. Omer, R.J. Parent, J.J. O'Neil, and G.C. Weir. 2001. Complete protection of islets against allorejection and autoimmunity by a simple barium-alginate membrane. Diabetes 50:1698–1705
61. Luca, G., C. Nastruzzi, M. Calvitti, E. Becchetti, T. Baroni, L.M. Neri, S. Capitani, G. Basta, P. Brunetti, and R. Calafiore. 2005. Accelerated functional maturation of isolated neonatal porcine cell clusters: in vitro and in vivo results in NOD mice. Cell Transplant 14:249–261
62. Patience, C., Y. Takeuchi, and R.A. Weiss. 1997. Infection of human cells by an endogenous retrovirus of pigs. Nat Med 3:282–286
63. Fishman, J.A., and C. Patience. 2004. Xenotransplantation: infectious risk revisited. Am J Transplant 4:1383–1390
64. Elliott, R.B., L. Escobar, O. Garkavenko, M.C. Croxson, B.A. Schroeder, M. McGregor, G. Ferguson, N. Beckman, and S. Ferguson. 2000. No evidence of infection with porcine endogenous retrovirus in recipients of encapsulated porcine islet xenografts. Cell Transplant 9:895–901
65. Pakhomov, O., L. Martignat, J. Honiger, B. Clemenceau, P. Sai, and S. Darquy. 2005. AN69 hollow fiber membrane will reduce but not abolish the risk of transmission of porcine endogenous retroviruses. Cell Transplant 14:749–756
66. Ricci, M., P. Blasi, S. Giovagnoli, C. Rossi, G. Macchiarulo, G. Luca, G. Basta, and R. Calafiore. 2005. Ketoprofen controlled release from composite microcapsules for cell encapsulation: effect on post-transplant acute inflammation. J Control Release 107:395–407
67. Luca, G., C. Nastruzzi, G. Basta, A. Brozzetti, A. Saturni, D. Mughetti, M. Ricci, C. Rossi, P. Brunetti, and R. Calafiore. 2000. Effects of anti-oxidizing vitamins on in vitro cultured porcine neonatal pancreatic islet cells. Diabetes Nutr Metab 13:301–307
68. Calafiore, R., G. Basta, G. Luca, A. Lemmi, M.P. Montanucci, G. Calabrese, L. Racanicchi, F. Mancuso, and P. Brunetti. 2006. Microencapsulated pancreatic islet allografts into nonimmunosuppressed patients with type 1 diabetes: first two cases. Diabetes Care 29:137–138
69. Lum, Z.P., I.T. Tai, M. Krestow, J. Norton, I. Vacek, and A.M. Sun. 1991. Prolonged reversal of diabetic state in NOD mice by xenografts of microencapsulated rat islets. Diabetes 40:1511–1516
70. Sun, Y., X. Ma, D. Zhou, I. Vacek, and A.M. Sun. 1996. Normalization of diabetes in spontaneously diabetic cynomologus monkeys by xenografts of microencapsulated

porcine islets without immunosuppression. J Clin Invest 98:1417–1422

71. Dufour, J.M., R.V. Rajotte, M. Zimmerman, A. Rezania, T. Kin, D.E. Dixon, and G.S. Korbutt. 2005. Development of an ectopic site for islet transplantation, using biodegradable scaffolds. Tissue Eng 11:1323–1331

72. Thanos, C.G., R. Calafiore, G. Basta, B.E. Bintz, W.J. Bell, J. Hudak, A. Vasconcellos, P. Schneider, S.J. Skinner, M. Geaney, P. Tan, R.B. Elliot, M. Tatnell, L. Escobar, H. Qian, E. Mathiowitz, and D.F. Emerich. 2007. Formulating the alginate-polyornithine biocapsule for prolonged stability: evaluation of composition and manufacturing technique. J Biomed Mater Res A 83:216–224

73. Hammerman, M.R. 2004. Organogenesis of endocrine pancreas from transplanted embryonic anlagen. Transpl Immunol 12:249–258

74. Rogers, S.A., H. Liapis, and M.R. Hammerman. 2003. Intraperitoneal transplantation of pancreatic anlagen. ASAIO J 49:527–532

75. Rogers, S.A., F. Chen, M.R. Talcott, C. Faulkner, J.M. Thomas, M. Thevis, and M.R. Hammerman. 2007. Long-term engraftment following transplantation of pig pancreatic primordia into non-immunosuppressed diabetic rhesus macaques. Xenotransplantation 14:591–602

76. Auchincloss, H., and J.V. Bonventre. 2002. Transplanting cloned cells into therapeutic promise. Nat Biotechnol 20:665–666

Tissue-Engineered Urinary Bladder

A. M. Turner, J. Southgate

Contents

32.1	The Urinary Bladder	429
32.1.1	Urothelium	429
32.1.2	Bladder Development	430
32.2	Bladder Reconstruction	432
32.2.1	Clinical Need for Reconstruction	432
32.2.2	Current Reconstructive Options	432
32.3	Basic Tissue-Engineering Principles	433
32.3.1	Tissue Engineering	433
32.3.2	Approaches to Tissue Engineering	433
32.3.3	Applied Tissue Engineering	433
32.3.4	Developmental Tissue Engineering	436
32.4	Conclusions	437
	References	437

32.1 The Urinary Bladder

Once thought of as a passive urine storage vessel, the urinary bladder is now recognised as a dynamic, compliant mechanosensory organ, dependent upon specialised structure-function relationships to preserve renal function and urinary continence. Understanding these relationships is central to the development of successful reconstruction approaches to restore urinary capacity and continence in patients with dysfunctional or diseased bladders. In recent years, the refinement of cell culture techniques to generate clinically useful quantities of normal urothelial and bladder smooth muscle cells in the laboratory has progressed alongside advances in biomaterial technologies. The combining of these fields has produced a wide variety of potential bladder tissue engineering and regenerative reconstruction strategies for the future.

The bladder functions to store urine at safe, physiological pressures, and in so doing protects the upper renal tracts from damage [1]. Its specialised structure and biomechanical properties confer high distensability, capacity and compliance, and the ability to undergo repeated voiding cycles. The adult mammalian bladder is composed of four distinct layers. The innermost layer is the urothelium, which abuts the intravesicular space and is separated by a basement membrane from the lamina propria. The lamina propria contains nervous and vascular structures embedded within a collagenous matrix. Outside the lamina propria are three loosely arranged layers of smooth muscle, and the bladder is covered externally by a single serosal layer.

32.1.1 Urothelium

The urothelium is a transitional epithelium with a single row of basal cells attached to the basal lamina, an intermediate zone of variable cell layers and a single row of superficial cells. The superficial or "umbrella" cells are highly specialised and are primarily responsible for the urinary barrier that enables the bladder to store urine in an unaltered state from the kidneys. The urinary barrier has two major components: the paracellular barrier is formed by intercellular tight junctions, located at the level of the superficial cells [2, 3], whereas the transcellular barrier is constituted by unique plaques of asymmetric unit membrane (AUM) present in the apical superficial cell membranes [4, 5]. The AUM provides an unequivocal marker of terminally differentiated urothelial cells and is formed by the interactions between four urothelium plaque proteins (uroplakins) [6, 7]. Manufactured in the Golgi apparatus of the superficial urothelial cell, the AUM plaques are transported to the apical membrane in fusiform vesicles that unfold onto the luminal surface. During bladder filling, the rate of fusiform vesicle export onto the surface membrane exceeds membrane resorption, thereby maintaining an effective urinary barrier of increasing surface area [8]. This has the effect of protecting the kidneys during bladder filling, by maintaining the bladder as a low-pressure reservoir.

The urothelium is an extremely stable epithelium with a very slow constitutive rate of cell turnover, but is able to regenerate and redifferentiate rapidly after acute injury [4, 9]. This regenerative capacity can be exploited to advantage when developing tissue-engineering strategies to replace diseased or damaged bladder tissue using urothelial cells propagated in vitro.

32.1.2
Bladder Development

The gross embryological development of the bladder and associated structures is complex and outside the scope of this review [10]. However, a critical part of the developmental process is the reciprocal interactions between epithelial and mesenchymal precursors that lead to the correct formation of the differentiated urothelium and smooth muscle components of the post-natal bladder tissue. In vivo, foetal epithelial and smooth muscle development is controlled by both permissive and instructive cell-signalling interactions. Whereas permissive signals allow the recipient cell to differentiate along a predetermined programme, instructive signals provide additional information that influences the ultimate fate of the target tissue. Largely overlooked, but potentially central to successful bladder tissue engineering, is the concept that developmental processes might be exploited to promote the development of functional engineered bladder tissues. Relevant strategies might involve either the use of therapeutic stem cells, or incorporation into "smart" biomaterials of peptide growth factors involved in urothelial:mesenchymal cell interactions during foetal bladder development.

32.1.2.1
Lessons from Bladder Smooth Muscle Development

In the bladder, mesenchymal differentiation to smooth muscle is characterised by the formation of organised smooth muscle bundles and the expression of specific protein markers, such as smooth muscle α-actin, γ-actin, smooth muscle myosin, vinculin, desmin and laminin [11]. The role of the urothelium in bladder smooth muscle development has been studied by Baskin and colleagues, who demonstrated that differentiation of isolated rat foetal bladder mesenchymal tissue into smooth muscle was dependent upon an inductive interaction with urothelium [11]. The age of the urothelium (embryonic, newborn or adult) was shown to be unaffected by the extent to which bladder smooth muscle differentiation occurred [12, 13] and, interestingly, smooth muscle differentiation was also induced, albeit to different degrees, by implanting other epithelia onto the mesenchyme. This shows that this particular signal was not specific to urothelium and thus was a permissive interaction [13]. The effect was also shown to traverse species barriers, with mouse urothelium instructing rat bladder mesenchyme to differentiate into rat smooth muscle [14].

Although, not fully elucidated, such interactions are thought to be mediated by peptide growth factors, with keratinocyte growth factor (KGF) and transforming growth factors (TGF)-α and -β amongst the likely mediators for urothelial and bladder smooth muscle development (reviewed in [15]). Vascular endothelial growth factor (VEGF), which is expressed by both urothelial and mesenchymal compartments of the developing bladder, has also been suggested to be a key factor for angiogenesis, a hypothesis reinforced by studies in which exogenous VEGF was added to murine embryonic bladder organ culture [16]. Not only was endothelial cell development promoted, but there was also a significant positive growth and differentiation effect on the urothelial and smooth muscle components [16].

Whereas foetal mesenchyme matured when implanted under the renal capsule of adult syngeneic hosts in the presence of urothelium [11, 12], smooth muscle expression was respectively decreased or absent when foetal urothelial/mesenchymal constructs or foetal mesenchymal tissue alone were placed orthotopically in the suburothelial space of the adult rat bladder [12]. It was suggested that inhibitory factors were at play, and indeed it was shown that the peptide growth factors responsible for stimulating smooth muscle differentiation and inhibition of urothelial proliferation (e.g. TGF-β2/3) decreased markedly in adulthood; simultaneously, growth factors such as KGF and TGF-α increased, resulting in preferential epithelial proliferation over further smooth muscle differentiation [12, 15]. More recent work has confirmed the timing of maturation of foetal mesenchyme in the mouse bladder using immunohistology combined with DNA microarray analyses [17]. As expected, a plethora of genes were found to be up- and down-regulated in both epithelial and mesenchymal compartments. Most interesting among them were well-characterised regulators of vascular smooth muscle differentiation [17], including serum-response factor (SRF) and SRF-regulated angiotensin receptor II, and TGF-β2; the latter confirming previous findings [15]. The finding that TGF-β2 was also up-regulated in the bladder epithelium around the time of mesenchymal maturation was suggested to imply inductive and regulatory roles [17].

32.1.2.2
Lessons from Urothelial Development

It is accepted that inductive signalling between epithelia and mesenchyme is a reciprocal arrangement, leading to functional differentiation of the urothelium. Oottamasathien and colleagues showed that cultured rat urothelial cells only differentiated to express uroplakins when recombined with (murine) embryonic bladder mesenchyme [18]. As expected, mesenchyme also differentiated into bladder smooth muscle in the presence of urothelium; alone, however, neither component differentiated into mature tissue. Whereas the evidence suggests that the urothelial influence on smooth muscle development is permissive rather than instructive, a large body of work has demonstrated the ability of mesenchyme to influence the epithelial differentiation programme and to elicit the transdifferentiation of adult epithelia in recombination experiments (reviewed in [19]). The demonstration that exogenous growth factors influence mesenchymal:epithelial interactions in vitro indicates that this is a useful platform for translating this work to a bladder engineering focus [15, 16].

Other studies have supported the suggestion that the inductive influence of foetal mesenchyme on epithelium is attenuated in maturity. In unpublished work from our laboratory, Fraser and colleagues showed that the combination of a cultured porcine urothelial cell sheet with de-epithelialised adult porcine gastric stroma in organ culture resulted in maturation of the urothelium with expression of markers of terminal differentiation. This suggested that differentiation and urothelium-specific gene expression occurred in response to permissive, not instructive, stromal signals. As yet there is no evidence to suggest that mature stromal or smooth muscle tissues can induce epithelial transdifferentiation. Thus, there seems to be a demarcation between the properties of immature and adult stromal tissues, with switches in growth factor profiles suggested to be responsible for the different responses. The critical, but to date unanswered question is whether these responses could be harnessed to develop appropriate constructs to support in vitro or in vivo development (i.e. growth and differentiation) of post-natal bladder tissue.

Consideration of these processes raises profound questions as to the direction tissue engineering and regenerative medicine should take; is it enough just to harness the repair mechanisms of the recipient or should we look towards more developmental approaches in order to produce ready-made tissues with complex functions?

32.2
Bladder Reconstruction

32.2.1
Clinical Need for Reconstruction

The need for surgical reconstruction of the bladder results from several convergent disease processes that lead to intractable incontinence that cannot be controlled by conservative or medical therapies and/or the presence of rising urinary pressures, risking damage to the kidneys. The causes of these problems are varied, but commonly result from reflex, unstable or non-compliant bladder activity, or a low-volume bladder that cannot hold normal volumes of urine. Examples include the neuropathic bladder (e.g. secondary to myelomeningocele or spinal cord injury), which results in a non-compliant and reflex bladder, severe detrusor instability, painful bladder syndromes such as interstitial cystitis and the low-volume bladder resulting from radical resection for cancer or trauma.

32.2.2
Current Reconstructive Options

The ultimate aim of bladder reconstruction is to increase the capacity and compliance so that continence is improved and intravesical pressures are reduced, protecting the upper tracts from damage. Currently, the most commonly performed surgical procedure to augment or replace the bladder involves the incorporation of a vascularised tissue graft into the bivalved bladder. A segment of bowel, most commonly the ileum, is isolated on its vascular pedicle, detubularised and sutured as a patch into the bladder, thus increasing its volume. So-called "enterocystoplasty", often combined with clean intermittent self catheterisation via the urethra or iatrogenic vesicostomy (e.g. as described by Mirtrofanoff [20]), has enhanced the quality of life for many patients by improving continence and preventing further damage to the kidneys.

Although enterocystoplasty is successful in a large proportion of patients [21, 22], it is associated with several potentially serious complications; these are largely attributable to the fact that the bowel mucosa is structurally and physiologically incompatible with exposure to urine. These complications include mucus production by the bowel, which acts as a nidus for bladder stone formation, bacteriuria, metabolic disturbances and malignancy (reviewed in [1, 22]). Although a late and rare complication of enterocystoplasty, the development of carcinoma is a serious event with a significant mortality of about 30% [23]. Most are adenocarcinomas located at the enterovesical anastomosis [24]. Many of these cases have appeared in patients with multiple risk factors (including genitourinary tuberculosis) and so it has been unclear whether the indication for the procedure, the procedure itself or another factor led to neoplasia. Recently, however, it has been suggested that enterocystoplasty itself may be an independent risk factor for urothelial cell carcinoma of the bladder [25].

The ideal tissue for bladder augmentation is a smooth muscle combined with a urothelial lining. Unfortunately, the nature of the underlying pathology in bladders that require augmentation generally means that there is a lack of adequate urothelial tissue for reconstruction. Currently, only two specific procedures can address this problem directly. Autoaugmentation involves performing a detrusor myectomy to produce an iatrogenic bladder diverticulum that can increase capacity and compliance by absorbing increased intravesical pressures associated with the poorly compliant bladder. Studies have shown that detrusor myectomy is successful in 80% of patients with idiopathic detrusor instability, but is largely unsuccessful in those with neuropathic bladder secondary to myelomeningocele [26, 27]. Ureterocystoplasty utilises the otherwise ineffective megaureter in patients with chronic unilateral hydroureteronephrosis in association with neuropathic bladder, by reconfiguring it for incorporation into the bladder. Although this procedure has demonstrated promising results [28–32], it is limited to those patients with a megaureter. Therefore, although autoaugmentation and ureterocystoplasty ensure urothelial continuity within the neobladder, most surgeons use enterocystoplasty as the principal method of augmentation.

Given that the fundamental problem of bladder augmentation is the lack of sufficient quantities of native tissue, current research has turned to the use of biomaterials and tissue-engineering techniques as alternative approaches.

32.3
Basic Tissue-Engineering Principles

32.3.1
Tissue Engineering

Tissue engineering seeks to combine cells and tissues, propagated in vitro, with suitable biomaterials to provide constructs for the replacement or reconstruction of absent or dysfunctional native tissue. Methods to isolate and propagate urothelial and smooth muscle cells to clinically useful quantities are well established and are described elsewhere (e.g. [33, 34]). Biomaterials may be natural and either native or foreign to the recipient, or synthetic. Historically, the use of hard or non-compliant materials to incorporate into the bladder has met with failure due to biomechanical failure or biological incompatibility (described in section 32.3.3.2 below) and so the development of modern bladder tissue-engineering technologies has arisen as a direct result of the need to improve current reconstructive options and build upon lessons learned in the past.

32.3.2
Approaches to Tissue Engineering

There are two main approaches to bladder tissue engineering: passive and active. Passive tissue engineering describes techniques whereby biomaterials are incorporated alone into the bladder wall, whereupon they are expected to become recellularised by ingrowth from the remaining native tissue. Alternatively, active tissue engineering encompasses the propagation of clinically useful quantities of autologous cells in the laboratory and combination with a biomaterial, prior to incorporation into the reconstructed bladder. In the latter case, it is important to consider whether the cells seeded onto the biomaterial will be capable of forming a differentiated, functional tissue as a result of the body acting as an "internal bioreactor", or whether alternative "external bioreactor" approaches are required to generate a functional tissue ex vivo, prior to incorporation into the reconstructed bladder. The advantage of the latter approach is that it would allow constructs with the morphological and functional properties of the native tissue to be implanted into the bladder, hence providing an effective urinary barrier from the outset.

Urothelium-associated differentiation markers have long been used not only to indicate molecular differentiation, but also to imply functional maturity. However, since Cross et al. [35] developed a method to promote functional differentiation of in-vitro-propagated normal human urothelial cells, we have been able to compare directly the relationship between morphological differentiation and barrier function. When investigated in porcine urothelial cell sheets, it was found that expression of urothelial differentiation markers did not necessarily correlate with barrier function as assessed by transepithelial resistance and permeability studies [36]. This suggests that although indicative, immunohistochemical studies cannot be used as an unequivocal surrogate marker of tissue function.

32.3.3
Applied Tissue Engineering

32.3.3.1
Vascularised Native Tissue

As the complications associated with enterocystoplasty rest with exposure of the bowel mucosa to urine, it would seem logical that removal of this epithelial layer prior to incorporation into the bladder should eliminate them. Unfortunately, as demonstrated in animal models, the augment undergoes significant fibrosis and shrinkage, secondary to chemical irritation, infection, vascular insufficiency or damage to the graft during dissection [37–41]. It has been shown that such fibrosis can be prevented by the presence of an inflated silicon balloon within the bladder for 2 weeks after reconstruction [42]. Other vascularised tissue flaps have been used for bladder augmentation and have shown mixed results; segments of abdominal wall and gracilis muscle underwent severe fibrosis [43], whereas omental flaps (omentocystoplasty) have shown promising results in both animal models and in the clinical setting [44].

It could be argued that fibrosis of de-epithelialised bowel also occurred because the epithelium had been removed and not replaced, resulting in smooth muscle de-differentiation, similar to that experienced by Moriya et al. [45]. This suggestion is supported by the observation that when urothelium was placed in contact with a stromal patch prior to augmentation, shrinkage of the graft was minimal or absent [37, 39, 43]. These studies used fresh urothelial strips and both non-cultured and cultured urothelial +/− smooth muscle cell suspensions with which to line the stromal segment.

The work of Fraser et al. [38] also addressed this issue and showed that a successful bladder augment could be achieved in a minipig model using so-called "composite cystoplasty", where in-vitro-propagated autologous urothelial cells were laid as sheets onto a vascularised, de-epithelialised uterine tissue and then incorporated into the bladder [38]. Although the study achieved a functionally augmented bladder and some expression of AUM was observed in the urothelial lining of the composite uterocystoplasty, the presence of inflammation within both augmenting and native segments raised the question as to the effectiveness of the cell sheets as a urinary barrier.

The results of this study support the idea of using the external bioreactor concept to produce and implant a functional barrier urothelium in vitro [35, 36]. In the context of composite cystoplasty, this approach should address the hypothesis that implanting a functional tissue would encourage the rapid establishment of an effective urinary barrier and limit inflammation, by providing the stroma with immediate protection from urine-mediated damage.

32.3.3.2
Natural Materials: Collagen

Collagen is the most abundant protein within the extracellular matrix (ECM) and, as the major structural protein in the body, imparts strength and stability to natural materials. It is an ideal material for tissue-engineering purposes, as it lacks significant immunogenicity, supports cell growth and can be physically manipulated to suit the application. However, it falls short in its purified form, as it loses tensile strength and succumbs to tearing during suturing [46, 47]. For this reason, it is often combined with synthetic materials to improve strength.

32.3.3.3
Natural Matrices: Small Intestinal Submucosa and Bladder Acellular Matrix Graft

Small intestinal submucosa (SIS) and bladder acellular matrix graft (BAMG) are produced from the submucosa of porcine small intestine and bladder, respectively. As they are heavily laden with immunogenic cellular material, these collagen- and elastin-rich materials are subjected to extensive decellularisation procedures. Once implanted into the body, they slowly degrade, acting as temporary scaffolds to support the ingrowth of cells and tissue development. BAMG may be generated by the decellularisation of split- or full-thickness bladder tissue [48–57]. Both SIS and BAMG have shown promising results when incorporated into the bladders of rats, dogs and pigs, with regeneration of both urothelial and smooth muscle compartments and evidence of revascularisation and reinnervation [51–55, 58–60]. In addition, appropriate contractile and relaxation responses of both types of regenerated patch have been reported when chemically stimulated, albeit of lower magnitude than that exhibited by the respective native tissues [51, 60]. SIS and BAMG have been shown to differ in their compliance properties, with non-regenerated SIS showing much lower values than non-regenerated split- and full-thickness BAMG [49, 56, 57, 59].

Lithogenesis, graft shrinkage and incomplete or disorganised smooth muscle structure are problems encountered with bladder-reconstructive techniques using SIS or BAMG. Stone formation is a particular, but not exclusive, problem in experimental rodent models, with up to 75–80% of animals being affected [51, 60]. Fibroproliferative change, which may result in graft shrinkage of up to 48% [48], is progressive and defeats the primary purposes of reconstruction to increase capacity and compliance.

Original graft size is a key feature when considering the extent to which smooth muscle cells can infiltrate and organise within the material. In experimental rat models, where graft sizes are often ≤0.5 cm², infiltration is more efficient than in larger animals, where smooth muscle bundles in the central areas of grafts as large as 46 cm² have been reported as "scanty" [48, 51]. Clearly, this would be a problem with grafts for use in man, which may be of an equivalent or larger size.

Because decellularised materials appear to retain some biological activity and ECM structure is unique to each tissue, in the context of bladder-tissue engineering it may be expected that BAMG should be more efficient at encouraging ingrowth of tissue, due to a more appropriate histioarchitecture and the presence of relevant growth factors [56, 57]. However, there must be a balance maintained between adequate processing to limit immunogenicity and retaining factors important to cell growth. It is worth noting that some commercially available SIS has been shown to contain porcine nuclear residues and to be cytotoxic to urothelial cells [61].

Two groups have tested the effects of seeding urothelial and smooth muscle cells onto BAMG and SIS in vitro, respectively [62, 63]. Both groups found that smooth muscle cell infiltration into the matrices was enhanced when urothelial/smooth muscle cell co-culture was performed, rather than seeding smooth muscle cells alone. Ram-Liebig and colleagues demonstrated enhanced proliferation and migration of de-differentiated smooth muscle cells into BAMG when incubated with medium previously conditioned with urothelium [64]. They noted that proliferation of smooth muscle cells was more marked when medium from proliferative urothelium was used, compared to terminally differentiated urothelium. These interesting observations add to the evidence for reciprocal interactions between urothelium and stromal compartments being mediated via peptide growth factors, the expression of which is modulated by the tissue status (i.e. foetal versus mature versus regenerative). This group also observed that proliferation and invasion was further enhanced when mechanical stimulation of the matrices was employed. It has long been recognised that mechanical stress plays a part in cell processes such as proliferation and biosynthetic activity, as described in myofibroblasts by Grinnell [65]. In the case of bladder-tissue engineering, such biomechanical conditioning simulates the normal fill:void cycles of the bladder, with the aim of encouraging tissue functionalisation and improved biomimetic properties [66].

Going one step further, the permissive induction of foetal mesenchyme to a smooth muscle phenotype by foetal urothelium was assessed on BAMG in vivo [67]. BAMG, seeded with foetal urothelium, was placed subcutaneously or under the renal capsule of athymic mice, replacing the mesenchymal component of previous experiments. Fibroblasts were recruited from surrounding tissue, which infiltrated the acellular matrix and then underwent differentiation to a smooth muscle phenotype [67].

This experiment showed the importance of inductive signalling even within modified natural matrices, and also suggested that smooth muscle regeneration is not dependent upon the origin of infiltrating mesenchymal cells. Further work by Moriya et al. seems to support the theory of maturity-related growth-factor switching [45]. Cultured adult rat urothelial cells were seeded onto BAMG and implanted into rat small-bowel mesentery. The infiltrating mesenchymal cells duly differentiated into smooth muscle cells, but the phenotype did not persist over time, with loss of smooth muscle histological markers and the presence of fibrocytes [45]. The authors suggested that this may have occurred either because the mesenchymal cells were of non-bladder origin, or because the urothelial covering was insufficient to maintain inductive signalling, the latter serving as a reminder that mesenchyme does not readily differentiate in the absence of epithelium. It is also worth considering that the growth factor milieu in the context of adult urothelial cells may have limited capacity to support smooth muscle differentiation.

32.3.3.4
Synthetic Materials

Synthetic biomaterials, which can be processed to produce the desired strength, microstructure, permeability and biodegradable properties are, for many reasons, more desirable than natural materials, but may lack properties compatible with cell attachment and growth [68, 69]. The early use of synthetic materials to reconstruct the bladder, such as plastics, polyvinyl sponge, polytetrafluoroethylene (Teflon) and Japanese paper, was characterised by failure, mainly due to biological and mechanical incompatibility [70–73].

The properties of synthetic biomaterials that influence cell behaviour are generally poorly understood. The proliferation, migration and differentiation of different cell types vary with the structure of the material. For example, a more natural arrangement of smooth muscle cells was observed when grown on electrospun polystyrene scaffolds whose fibres were aligned as collagen fibres in vivo, as opposed to a random arrangement [74], and materials with an elastic modulus most similar to the bladder provided a better environment for the propagation of urothelial and bladder smooth muscle cells in vitro [75]. In another study, a natural:synthetic hybrid material supported urothelial stratification as a sponge, but not as a gel, whereas smooth muscle differentiation was maintained on a gel, but not on a sponge [76].

Polylactic-co-Glycolic Acid

Polylactic-co-glycolic acid (PLGA) is an established biomaterial that is used in the form of sutures and meshes in modern surgical practice (Vicryl). It is non-toxic, biodegradable and biocompatible with both normal human urothelial and bladder smooth muscle cells [77–79]. It has been used alone or in combination with other materials in the development of bladder-reconstruction procedures. Oberpenning et al. coated a polyglycolic acid (PGA) mesh with PLGA and moulded the material into the shape of a bladder [80]. Urothelium and smooth muscle cells from beagle dogs were propagated in vitro then seeded onto the inner and outer surfaces of the biomaterial, respectively. The autologous cell constructs were implanted in vivo onto a dog's bladder base after trigone-sparing cystectomy. The neobladders were coated with fibrin glue then given an omental wrap, to encourage angiogenesis. The animals were monitored for up to 11 months, but after 3 months the polymer had degraded and the neobladder was reported to be vascularised, innervated and to contain organised smooth muscle bundles and a stratified urothelium, which was positive with antibodies against AUM [80].

Hybrid Materials

Hybrid materials consisting of both natural and synthetic components have the advantage of conformability, malleability and the benefit of a temporary scaffold being present during cell infiltration and organisation. Such materials have been used with some success, with Atala and colleagues making the transition from canine model to clinical trials [81]. Autologous smooth muscle and urothelial cells were seeded onto collagen-PGA hybrid scaffolds and implanted into patients with severely neuropathic bladders, where the constructs were wrapped in omentum. Although some patient benefit was reported [71], further work is required to critically assess the outcomes of the procedure in terms of the histioarchitecture and differentiation status of the engineered bladder, and in terms of continent capacity and patient health/quality of life measures.

32.3.4
Developmental Tissue Engineering

32.3.4.1
Stem Cell Technology

Dedicated bladder-specific precursor cells have yet to be identified, whether from resident or peripheral (i.e. recruited) stem-cell pools. Given the self-renewing properties of the urothelium, it may even be the case that there is no such population of urothelial stem cells. However, given correct developmental stimuli, pluripotent cells have the potential to be of use in cases where an autologous biopsy would not provide sufficient healthy cells to initiate an effective cell culture. For example, Oottamasathien et al. showed that mouse embryonic stem cells, when recombined with embryonic rat bladder mesenchyme, developed into mature, terminally differentiated urothelium [82]. Smooth muscle and neuronal tissue also developed in an organised fashion. Furthermore, passive infiltration of urothelium and smooth muscle into SIS bladder patches in animal models has been shown to be enhanced by the presence of embryonic stem cells [83, 84]. The possibility that stem cells could release positive growth or chemotactic factors was supported by in vitro observations that stem-cell-conditioned medium elicited migratory and proliferative responses in urothelial and smooth muscle cells in co-culture [85].

32.3.4.2
Smart Biomaterials

Smart biomaterials incorporate growth factors designed to give them enhanced biological properties to promote a particular outcome [68]. The lessons from bladder development suggest that incorporation of specific factors may enhance the regenerative/developmental capacity of bladder-derived cells. An early example of biomaterial modification was described by Danielsson et al., whose polyester-urethane-serum-pretreated foam was designed to encourage smooth muscle attachment and growth [68]. Considerable attention has focused on incorporating VEGF into materials with the aim of inducing vascularisation and rapid integration of graft constructs. For example, synthetic hydrogel matrices were integrated with VEGF, which was activated in response to the release of matrix metalloproteinase-2 by infiltrating human umbilical vein endothelial cells, producing local angiogenesis [86].

32.4
Conclusions

Enterocystoplasty has for many years been the mainstay of treatment for the small, non-compliant bladder that is refractory to conservative and medical therapies. However, the complications of this procedure, which result primarily from urinary interactions with the bowel mucosa, may be at best inconvenient and at worst life-threatening. Tissue-engineering strategies have been developed to construct augmenting segments of bladder to increase capacity and compliance. Most strategies have relied upon the regenerative capacity of the body to repopulate natural or synthetic matrices, or to condition immature cells grown in the laboratory and implanted onto biomaterial scaffolds. The evidence suggests, however, that those tissue-engineering strategies that more closely mimic native structural and physiological processes are the most likely to succeed. Composite cystoplasty, combining an autologous in-vitro-generated differentiated and functional urothelium with a fully vascularised surrogate smooth muscle segment, is a promising approach to overcoming the problems associated with conventional enterocystoplasty. In the future, we suggest that in-depth understanding of epithelial:mesenchymal signalling during embryological development may shift attention from a regenerative approach towards harnessing developmental processes to influence the outcome of cells grown in culture.

References

1. Thomas DF: Surgical treatment of urinary incontinence. Arch Dis Child, 76: 377, 1997
2. Acharya P, Beckel J, Ruiz WG. et al: Distribution of the tight junction proteins ZO-1, occludin, and claudin-4, -8, and -12 in bladder epithelium. Am J Physiol Renal Physiol, 287: F305, 2004
3. Varley CL, Garthwaite MA, Cross W et al: PPARgamma-regulated tight junction development during human urothelial cytodifferentiation. J Cell Physiol, 208: 407, 2006
4. Hicks RM: The mammalian urinary bladder: an accommodating organ. Biol Rev Camb Philos Soc, 50: 215, 1975
5. Hicks RM, Ketterer B, Warren RC: The ultrastructure and chemistry of the luminal plasma membrane of the mammalian urinary bladder: a structure with low permeability to water and ions. Philos Trans R Soc Lond B Biol Sci, 268: 23, 1974
6. Wu XR, Manabe M, Yu J et al: Large scale purification and immunolocalization of bovine uroplakins I, II, and III. Molecular markers of urothelial differentiation. J Biol Chem, 265: 19170, 1990
7. Yu J, Lin JH, Wu XR et al: Uroplakins Ia and Ib, two major differentiation products of bladder epithelium, be-

long to a family of four transmembrane domain (4TM) proteins. J Cell Biol, 125: 171, 1994
8. Truschel ST, Wang E, Ruiz WG et al: Stretch-regulated exocytosis/endocytosis in bladder umbrella cells. Mol Biol Cell, 13: 830, 2002
9. Lavelle J, Meyers S, Ramage R et al: Bladder permeability barrier: recovery from selective injury of surface epithelial cells. Am J Physiol Renal Physiol, 283: F242, 2002
10. Staack A, Hayward SW, Baskin LS et al: Molecular, cellular and developmental biology of urothelium as a basis of bladder regeneration. Differentiation, 73: 121, 2005
11. Baskin LS, Hayward SW, Young P et al: Role of mesenchymal-epithelial interactions in normal bladder development. J Urol, 156: 1820, 1996
12. Baskin L, DiSandro M, Li Y et al: Mesenchymal-epithelial interactions in bladder smooth muscle development: effects of the local tissue environment. J Urol, 165: 1283, 2001
13. DiSandro MJ, Li Y, Baskin LS et al: Mesenchymal-epithelial interactions in bladder smooth muscle development: epithelial specificity. J Urol, 160: 1040, 1998
14. Baskin LS, Hayward SW, Sutherland RA. et al: Cellular signaling in the bladder. Front Biosci, 2: d592, 1997
15. Baskin LS, Hayward SW, Sutherland RA et al: Mesenchymal-epithelial interactions in the bladder. World J Urol, 14: 301, 1996
16. Burgu B, McCarthy LS, Shah V et al: Vascular endothelial growth factor stimulates embryonic urinary bladder development in organ culture. BJU Int, 98: 217, 2006
17. Li J, Shiroyanagi Y, Lin G et al: Serum response factor, its cofactors, and epithelial-mesenchymal signaling in urinary bladder smooth muscle formation. Differentiation, 74: 30, 2006
18. Oottamasathien S, Williams K, Franco OE et al: Bladder tissue formation from cultured bladder urothelium. Dev Dyn, 235: 2795, 2006
19. Cunha GR, Hayashi N, Wong YC: Regulation of differentiation and growth of normal adult and neoplastic epithelia by inductive mesenchyme. Cancer Surv, 11: 73, 1991
20. Mitrofanoff P: Trans-appendicular continent cystostomy in the management of the neurogenic bladder. Chir Pediatr, 21: 297, 1980
21. Beier-Holgersen R, Kirkeby LT, Nordling J: 'Clam' ileocystoplasty. Scand J Urol Nephrol, 28: 55, 1994
22. Greenwell TJ, Venn SN, Mundy AR: Augmentation cystoplasty. BJU Int, 88: 511, 2001
23. Gough D. C: Enterocystoplasty. BJU Int, 88: 739, 2001
24. Woodhams SD, Greenwell TJ, Smalley T et al: Factors causing variation in urinary N-nitrosamine levels in enterocystoplasties. BJU Int, 88: 187, 2001
25. Soergel TM, Cain MP, Misseri R et al: Transitional cell carcinoma of the bladder following augmentation cystoplasty for the neuropathic bladder. J Urol, 172: 1649, 2004
26. Kumar SP, Abrams PH: Detrusor myectomy: long-term results with a minimum follow-up of 2 years. BJU Int, 96: 341, 2005
27. Marte A, Di Meglio D, Cotrufo AM et al: A long-term follow-up of autoaugmentation in myelodysplastic children. BJU Int, 89: 928, 2002
28. Churchill BM, Aliabadi H, Landau EH. et al: Ureteral bladder augmentation. J Urol, 150: 716, 1993
29. Hitchcock RJ, Duffy PG, Malone PS: Ureterocystoplasty: the "bladder" augmentation of choice. Br J Urol, 73: 575, 1994
30. Landau EH, Jayanthi VR, Khoury AE. et al: Bladder augmentation: ureterocystoplasty versus ileocystoplasty. J Urol, 152: 716, 1994
31. Nahas WC, Lucon M, Mazzucchi E et al: Clinical and urodynamic evaluation after ureterocystoplasty and kidney transplantation. J Urol, 171: 1428, 2004
32. Tekgul S, Oge O, Bal K. et al: Ureterocystoplasty: an alternative reconstructive procedure to enterocystoplasty in suitable cases. J Pediatr Surg, 35: 577, 2000
33. Kimuli M, Eardley I, Southgate, J: In vitro assessment of decellularized porcine dermis as a matrix for urinary tract reconstruction. BJU Int, 94: 859, 2004
34. Southgate J, Hutton KA, Thomas DF. et al: Normal human urothelial cells in vitro: proliferation and induction of stratification. Lab Invest, 71: 583, 1994
35. Cross WR, Eardley I, Leese HJ et al: A biomimetic tissue from cultured normal human urothelial cells: analysis of physiological function. Am J Physiol Renal Physiol, 289: F459, 2005
36. Turner AM, Subramaniam R, Thomas DF, Southgate J: Generation of a functional, differentiated porcine urothelial tissue in vitro. Eur Urol, 54: 1423, 2008
37. Aktug T, Ozdemir T, Agartan C. et al: Experimentally prefabricated bladder. J Urol, 165: 2055, 2001
38. Fraser M, Thomas DF, Pitt E et al: A surgical model of composite cystoplasty with cultured urothelial cells: a controlled study of gross outcome and urothelial phenotype. BJU Int, 93: 609, 2004
39. Hafez AT, Afshar K, Bagli DJ. et al: Aerosol transfer of bladder urothelial and smooth muscle cells onto demucosalized colonic segments for porcine bladder augmentation in vivo: a 6-week experimental study. J Urol, 174: 1663, 2005
40. Motley RC, Montgomery, BT, Zollman, PE. et al: Augmentation cystoplasty utilizing de-epithelialized sigmoid colon: a preliminary study. J Urol, 143: 1257, 1990
41. Salle JL, Fraga JC, Lucib, A et al: Seromuscular enterocystoplasty in dogs. J Urol, 144: 454, 1990
42. Lima SV, Araujo LA, Vilar FO: Nonsecretory intestinocystoplasty: a 10-year experience. J Urol, 171: 2636, 2004
43. Schoeller T, Neumeister MW, Huemer GM et al: Capsule induction technique in a rat model for bladder wall replacement: an overview. Biomaterials, 25: 1663, 2004
44. Kiricuta I, Goldstein AM: The repair of extensive vesicovaginal fistulas with pedicled omentum: a review of 27 cases. J Urol, 108: 724, 1972
45. Moriya K, Kakizaki H, Watanabe S et al: Mesenchymal cells infiltrating a bladder acellular matrix gradually lose smooth muscle characteristics in intraperitoneally regenerated urothelial lining tissue in rats. BJU Int, 96: 152, 2005

46. Elbahnasy AM, Shalhav A, Hoenig DM. et al: Bladder wall substitution with synthetic and non-intestinal organic materials. J Urol, 159: 628, 1998
47. Hattori K, Joraku A, Miyagawa T et al: Bladder reconstruction using a collagen patch prefabricated within the omentum. Int J Urol, 13: 529, 2006
48. Brown AL, Farhat W, Merguerian PA. et al: 22 week assessment of bladder acellular matrix as a bladder augmentation material in a porcine model. Biomaterials, 23: 2179, 2002
49. Dahms SE, Piechota HJ, Dahiya R et al: Composition and biomechanical properties of the bladder acellular matrix graft: comparative analysis in rat, pig and human. Br J Urol, 82: 411, 1998
50. Merguerian PA, Reddy PP, Barrieras DJ. et al: Acellular bladder matrix allografts in the regeneration of functional bladders: evaluation of large-segment (>24 cm) substitution in a porcine model. BJU Int, 85: 894, 2000
51. Piechota HJ, Dahms SE, Nunes LS. et al: In vitro functional properties of the rat bladder regenerated by the bladder acellular matrix graft. J Urol, 159: 1717, 1998
52. Probst M, Dahiya R, Carrier S et al: Reproduction of functional smooth muscle tissue and partial bladder replacement. Br J Urol, 79: 505, 1997
53. Probst M, Piechota HJ, Dahiya R et al: Homologous bladder augmentation in dog with the bladder acellular matrix graft. BJU Int, 85: 362, 2000
54. Reddy PP, Barrieras DJ, Wilson G et al: Regeneration of functional bladder substitutes using large segment acellular matrix allografts in a porcine model. J Urol, 164: 936, 2000
55. Sutherland RS, Baskin LS, Hayward SW. et al: Regeneration of bladder urothelium, smooth muscle, blood vessels and nerves into an acellular tissue matrix. J Urol, 156: 571, 1996
56. Badylak SF: Xenogeneic extracellular matrix as a scaffold for tissue reconstruction. Transpl Immunol, 12: 367, 2004
57. Bolland F, Korossis S, Wilshaw SP et al: Development and characterisation of a full-thickness acellular porcine bladder matrix for tissue engineering. Biomaterials, 28: 1061, 2007
58. Kropp BP, Rippy MK, Badylak SF et al: Regenerative urinary bladder augmentation using small intestinal submucosa: urodynamic and histopathologic assessment in long-term canine bladder augmentations. J Urol, 155: 2098, 1996
59. Kropp BP, Sawyer BD, Shannon HE. et al: Characterization of small intestinal submucosa regenerated canine detrusor: assessment of reinnervation, in vitro compliance and contractility. J Urol, 156: 599, 1996
60. Vaught JD, Kropp BP, Sawyer BD. et al: Detrusor regeneration in the rat using porcine small intestinal submucosal grafts: functional innervation and receptor expression. J Urol, 155: 374, 1996
61. Feil G, Christ-Adler M, Maurer S et al: Investigations of urothelial cells seeded on commercially available small intestine submucosa. Eur Urol, 50: 1330, 2006
62. Brown AL, Brook-Allred TT, Waddell JE. et al: Bladder acellular matrix as a substrate for studying in vitro bladder smooth muscle-urothelial cell interactions. Biomaterials, 26: 529, 2005
63. Zhang Y, Kropp BP, Moore P et al: Coculture of bladder urothelial and smooth muscle cells on small intestinal submucosa: potential applications for tissue engineering technology. J Urol, 164: 928, 2000
64. Ram-Liebig G, Ravens U, Balana B et al: New approaches in the modulation of bladder smooth muscle cells on viable detrusor constructs. World J Urol, 24: 429, 2006
65. Grinnell,F: Fibroblasts, myofibroblasts, and wound contraction. J Cell Biol, 124: 401, 1994
66. Korossis S, Bolland F, Ingham E et al: Review: tissue engineering of the urinary bladder: considering structure-function relationships and the role of mechanotransduction. Tissue Eng, 12: 635, 2006
67. Master VA, Wei G, Liu W et al: Urothlelium facilitates the recruitment and trans-differentiation of fibroblasts into smooth muscle in acellular matrix. J Urol, 170: 1628, 2003
68. Danielsson C, Ruault S, Simonet M. et al: Polyesterurethane foam scaffold for smooth muscle cell tissue engineering. Biomaterials, 27: 1410, 2006
69. Vacanti JP, Langer R: Tissue engineering: the design and fabrication of living replacement devices for surgical reconstruction and transplantation. Lancet, 354 Suppl 1: SI32, 1999
70. Bohne AW, Urwiller KL: Experience with urinary bladder regeneration. J Urol, 77: 725, 1957
71. Fujita K: The use of resin-sprayed thin paper for urinary bladder regeneration. Invest Urol, 15: 355, 1978
72. Kelami A, Dustmann HO, Ludtke-Handjery A et al.: Experimental investigations of bladder regeneration using Teflon-felt as a bladder wall substitute. J Urol, 104: 693, 1970
73. Kudish HG: The use of polyvinyl sponge for experimental cystoplasty. J Urol, 78: 232, 1957
74. Baker SC, Atkin N, Gunning PA et al: Characterisation of electrospun polystyrene scaffolds for three-dimensional in vitro biological studies. Biomaterials, 27: 3136, 2006
75. Rohman G, Pettit JJ, Isaure F et al: Influence of the physical properties of two-dimensional polyester substrates on the growth of normal human urothelial and urinary smooth muscle cells in vitro. Biomaterials, 28: 2264, 2007
76. Nakanishi Y, Chen G, Komuro H et al: Tissue-engineered urinary bladder wall using PLGA mesh-collagen hybrid scaffolds: a comparison study of collagen sponge and gel as a scaffold. J Pediatr Surg, 38: 1781, 2003
77. Pariente JL, Kim BS, Atala A: In vitro biocompatibility assessment of naturally derived and synthetic biomaterials using normal human urothelial cells. J Biomed Mater Res, 55: 33, 2001
78. Pariente JL, Kim BS, Atala A: In vitro biocompatibility evaluation of naturally derived and synthetic biomaterials using normal human bladder smooth muscle cells. J Urol, 167: 1867, 2002
79. Scriven SD, Trejdosiewicz LK, Thomas DFM. et al: Urothelial cell transplantation using biodegradable synthetic scaffolds. J Mater Sci Mater Med, 12: 991, 2001

80. Oberpenning F, Meng J, Yoo JJ et al: De novo reconstitution of a functional mammalian urinary bladder by tissue engineering. Nat Biotechnol, 17: 149, 1999
81. Atala A, Bauer SB, Soker S et al: Tissue-engineered autologous bladders for patients needing cystoplasty. Lancet, 367: 1241, 2006
82. Oottamasathien S, Wang Y, Williams K. et al: Directed differentiation of embryonic stem cells into bladder tissue. Dev Biol, 304: 556, 2007
83. Chung SY, Krivorov NP, Rausei V et al: Bladder reconstitution with bone marrow derived stem cells seeded on small intestinal submucosa improves morphological and molecular composition. J Urol, 174: 353, 2005
84. Frimberger D, Morales N, Shamblott M et al: Human embryoid body-derived stem cells in bladder regeneration using rodent model. Urology, 65: 827, 2005
85. Frimberger D, Morales N, Gearhart JD. et al: Human embryoid body-derived stem cells in tissue engineering-enhanced migration in co-culture with bladder smooth muscle and urothelium. Urology, 67: 1298, 2006
86. Zisch AH, Lutolf MP, Ehrbar M et al: Cell-demanded release of VEGF from synthetic, biointeractive cell ingrowth matrices for vascularized tissue growth. FASEB J, 17: 2260, 2003

Cell-Based Regenerative Medicine for Heart Disease

C. Stamm, C. Lüders, B. Nasseri, R. Hetzer

Contents

33.1	Introduction	441
33.2	Myocardial Regeneration	441
33.3	Cardiac Cell Therapy	443
33.3.1	Embryonic Stem Cells	444
33.4	Bone Marrow Cells and Heart Disease	444
33.4.1	Non-haematopoietic Stem Cells	445
33.5	Clinical Cell Therapy	446
33.5.1	Bone Marrow Mononuclear Cells in Acute Infarction	446
33.5.2	Bone Marrow MNCs in Chronic Ischaemia	446
33.5.3	Enriched Stem-Cell Products	446
33.5.4	Mesenchymal Stem Cells	447
33.5.5	Cytokine-Induced Bone Marrow Cell Mobilisation	447
33.6	Unresolved Problems	447
33.6.1	Patient Selection	447
33.6.2	Concomitant Procedures	448
33.6.3	Cell Delivery	448
33.6.4	Timing	449
33.6.5	Cell Survival	449
33.6.6	Dosage	449
33.6.7	Age	449
33.7	Summary	450
	References	450

33.1 Introduction

Many approaches to improve health by stimulating the body's own capacity for healing have been around for a long time. Some components of today's regenerative medicine, however, are indeed fundamentally new developments. One of those is the concept of increasing the number of contractile cells in the heart to cure heart failure, either by stimulating intrinsic regeneration processes or by adding exogenous cells. This chapter will briefly summarise the background of cardiac regenerative medicine, and attempt a critical appraisal of the current efforts to translate experimental cell therapy approaches into the clinical setting.

33.2 Myocardial Regeneration

One of the traditional paradigms in cardiovascular medicine is that myocardial hypertrophy in response to increased workload is not associated with an increase in myocyte number, because cardiomyocytes permanently and irreversibly rest in the G_1/G_0 phase of the cell cycle (Fig. 33.1). This concept of the heart as a postmitotic organ is also based on the lack of observable regeneration after myocardial infarction. On the other hand, it has been difficult to reconcile this notion with quantitative data on apoptotic cell death in various diseases. Even the most conservative estimates imply that the entirety of cardiomyocytes would have disappeared after some time, unless myocytes are constantly replaced [1]. The question is whether myocytes are replaced—if they are replaced—by mitotic division of pre-existing cardiomyocytes, by proliferation and differentiation of resident cardiomyocyte progenitor cells, or by cells originating from the bone marrow that have migrated to the heart.

Solid evidence of a permanent cellular turnover in adult human myocardium comes from gender-mismatched organ transplantation. When a female has received a male bone marrow transplant, donor cells can be detected in the heart after many years, predominately in the coronary vasculature, but also in the interstitium and the myocardium [2–4]. Similar observations have been made in donor hearts from females that were transplanted into a male recipient [5, 6]. There is little doubt that the coronary microvasculature is subject to significant endothelial cell turnover; with respect to cardiomyocytes, however, controversy remains. Convincing morphologic evidence of mitotic cardiomyocyte division has not yet been provided, and it is indeed difficult to imagine how such highly complex mature cells consisting mainly of the contractile apparatus and its supporting structures, should be able to regress to a state that allows for cell division. An intuitively more appealing concept is that putative neo-cardiomyocytes originate from resident cardiac stem cells. The presence of myogenic progenitor cells in skeletal muscle has long been known, but the existence of similar progenitor cells in the post-natal myocardium was considered impossible. However, recent experimental data indicate that several types of cardiac muscle stem cells might be involved in physiologic regeneration attempts [7]. In rodents, putative cardiac stem cell populations have been described based on expression of c-kit, Sca-1, and Isl-1, but also CD34, FLK-1, CD31 and GATA-4, in varying combinations with the pres-

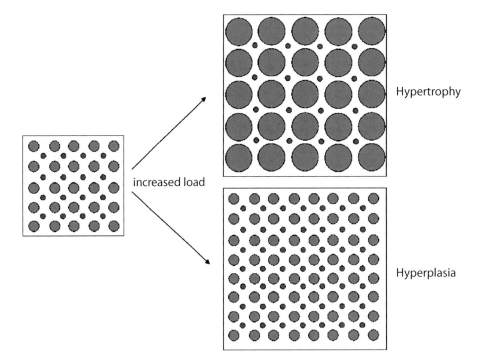

Fig. 33.1 *Top* If faced with increased work load, the myocardium responds by hypertrophy of cardiomyocytes. The diameter and mass of cardiomyocytes (*grey*) increase, but not their number. Note that there is also a tendency to develop a perfusion mismatch, since the hypertrophic response is not always accompanied by an increased growth of blood vessels (*red*). *Bottom* Hyperplasia, and increase in cardiomyocyte (and blood vessel) number by mitotic cell division appears to occur only in foetal and neonatal myocardium. The main goal of cardiac regenerative medicine is to elicit a hyperplasia response in adult diseased myocardium, either by adding exogenous cells or by facilitating proliferation of intrinsic myocytes

ence or absence of other surface markers [8–10]. As Torella et al. have pointed out [7], those are probably phenotypic variations of a unique cell type, with the exception of Isl-1+ cells in the right heart that may be remnants of the cardiac primordia. Rodent cardiac stem cells have been expanded ex vivo and were used successfully for heart muscle regeneration in syngenic and allogenic models of myocardial infarction. Other groups isolated c-kit+ cardiac stem cells from human endomyocardial biopsies or surgical samples, confirmed their cardiomyogenic differentiation potential in vitro, and applied them to xenogenic experimental models of myocardial infarction [11–14]. Nevertheless, some conceptual issues still need to be addressed: Why does the heart not regenerate in many patients if it really contains substantial numbers of powerful tissue-specific stem cells? Is cardiac stem cell function impaired in patients with severe heart disease? If so, can autologous circulating stem cells be expected to be of therapeutic use? To answer these questions, more detailed basic research and preclinical studies will have to be carried out prior to serious clinical translation attempts.

33.3 Cardiac Cell Therapy

The primary goal of cardiac cell therapy is to increase the number of contractile cells in the ventricular myocardium so that heart function can return to normal. However, this concept has faded from the spotlight in favour of surrogate theories, including paracrine effects supporting angiogenesis, modulation of extracellular matrix components, supportive effects on cardiomyocytes suffering from ischaemic stress, and even stimulating interactions with resident cardiac progenitor cells. Originally, immortalised myocyte lines and neonatal cardiomyocytes were used for transplantation experiments [15–17], and the notion that exogenous contractile cells may be able to incorporate in post-natal myocardium was revolutionary. At the same time, it became clear that transplanted cardiomyocytes will not survive in terminally ischaemic tissue. A solution to this problem seemed to be the use of skeletal muscle progenitor cells [15, 18]. These have a high tolerance to ischaemia and maintain contractile work even through prolonged periods of anaerobic metabolism. Myoblast transplantation as part of a surgical procedure was introduced into the clinical arena in 2001 [19]. Initial feasibility studies were promising and laid the foundation for an avalanche of cell therapy studies. However, it soon became clear that skeletal myoblasts lack the capacity to electrically couple with surrounding cardiomyocytes because they do not express the intercellular communication protein connexin 43. Thus, they do not form "connexon" ion channels that are typical of cardiomyocytes [20], which maintain their single-cell integrity but connect with their neighbours via gap junction connexons to form a functional syncytium for propagation of excitation from cell to cell. In contrast, skeletal myoblasts and their progeny fuse to form multinucleated myofibres and connect with one specific motoneuron (Fig. 33.2). This is a prerequisite for the rapid and fine-controlled adjustment of skeletal muscle contractile force. Therefore, skeletal myofibres in the heart, although they readily survive, remain isolated from the surrounding myocardium and

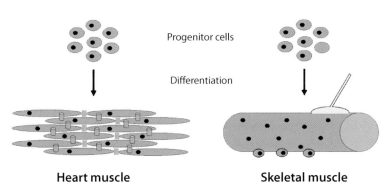

Fig. 33.2 Schematic depiction of the inter-cell behaviour of cardiomyocyte and skeletal myocyte progenitors upon differentiation. Cardiomyocytes retain their cellular integrity but connect electrically via low-resistance ion channels (connexons, *green*) and mechanically via intercalated disks. In contrast, skeletal myocytes fuse to form multinucleated myofibers, which are electrically isolated from their surroundings and coupled with one motoneuron (*yellow*)

may act as arrhythmogenic foci. Some investigators report that they have never encountered arrhythmia problems [21]. However, given that the observed improvement in contractility is, at best, very mild, the majority of clinicians have abandoned skeletal myoblasts for the treatment of heart failure.

33.3.1
Embryonic Stem Cells

For more than one decade, experimental work on embryonic stem cells (ESCs) for myocyte reproduction has progressed steadily [22]. There is evidence that the host myocardium can control the specific cardiomyocyte differentiation of a limited number of implanted ESCs, but once a certain threshold has been exceeded, uncontrolled proliferation and differentiation with teratoma formation occurs. Researchers and industry have therefore focused on predifferentiated cardiomyocytes from ESCs, which can—theoretically—be produced in large quantities in vitro, prior to implantation into the diseased heart [23–25]. Clinical translation of this technology, however, is still hampered by several fundamental biological and biotechnological problems:
1. Theoretically, even a single naïve ESC can give rise to a teratoma in the heart. Therefore, ESC-derived myocytes or myocyte progenitor cell products must have 100% purity, requiring complex and reliable cell-processing techniques.
2. The immunogenicity of ESCs and their in vitro progeny is incompletely understood, and it is unlikely that ESC products can be transplanted in allogenic fashion.
3. The ethical debate surrounding ESC procurement from viable human embryos severely hampers further development of ESC technology.

ESC-like cells can also be produced without the need to destroy an embryo, by therapeutic cloning, reprogramming of somatic cells, or from germ-cell progenitors. However, therapeutic cloning by nuclear transfer results in a hybrid cell with nuclear DNA from one organism and mitochondrial DNA from another [26], and the long-term consequences are not known. ESC-like cell production from germ-cell progenitors has so far only been successful in rodents [27]. Very recently, successful genetic reprogramming of skin fibroblasts has been reported, resulting in dedifferentiation into cells with ESC characteristics ("induced pluripotent stem cells") [28–30]. This technique could solve the ethical problems surrounding ESCs and the unclear situation regarding the immunogenicity of allogenic ESCs. On the other hand, it requires virus-mediated transfection of cells with several stemness-encoding genes, and the in vivo consequences are unknown. Nevertheless, a door has been opened that may lead to completely new options for obtaining pluripotent cells that can be processed into autologous cardiomyocyte cell products, and this will surely stimulate a tremendous amount of research and development work.

33.4
Bone Marrow Cells and Heart Disease

Blood and blood vessel cells have common ancestors, and vestiges of this unity persist throughout life. There is a constant turnover of endothelial cells, and bone marrow cells are recruited to participate in vascular regeneration processes. Cardiovascular disease and endothelial progenitor cell number and function correlate closely, and marrow-derived cells of haematopoietic and/or endothelial lineage participate in angiogenesis processes in ischaemic tissue. Kocher and colleagues were among the first to successfully use human CD34+ cells in a rat model of myocardial infarction [31], and the increased growth of small blood vessels was associated with a marked improvement of contractility. In large animals, however, the impact of neoangiogenesis on contractility is less pronounced, while in humans it is often negligible. Another apparent breakthrough was also reported in 2001 [32]. C-kit+ lin– cells were isolated from the bone marrow of green fluorescent protein (GFP)-expressing transgenic mice and implanted in the infarcted myocardium of non-GFP expressing animals, and indeed both GFP+ blood vessels and GFP+ contractile cells were visualised later. The revolutionary implication was that adult bone marrow stem cells can differentiate into both endothelial cells and cardiomyocytes, driven by factors present in the host myocardium. While this report led to clinical pilot studies all over the world, other groups doubted the

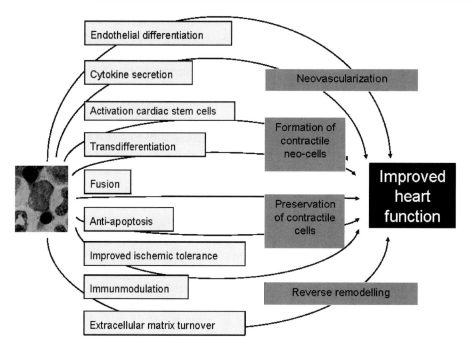

Fig. 33.3 Possible effects of marrow-derived stem cells on ischaemic myocardium. Not every mechanism has been directly observed with every subtype of bone marrow stem cell. Myogenic transdifferentiation of haematopoietic stem cells in particular is a highly controversial topic. Many proponents of cardiac cell therapy believe that in summary, the different cell-therapy-induced changes in the myocardium result in a net improvement of function. On the other hand, the opponents argue that one must first identify an exact mechanism-of-action in experimental studies, before clinical trials are justified

unlimited plasticity of adult bone marrow cells and failed to detect relevant cardiomyocyte differentiation of murine bone marrow stem cells in vivo [33]. Controversy regarding this issue remains to date; variations in technical details may explain the differing results. However, even if some haematopoietic stem cells can indeed be driven to express myocyte-specific markers, the frequency of such events in humans is probably too small to guarantee a relevant clinical effect (Fig. 33.3) [34].

33.4.1
Non-haematopoietic Stem Cells

Bone marrow contains an extensive cell-rich stroma that was long believed to solely support the proliferation of haematopoietic cells [35]. In the 1990s, however, it became clear that many stroma cells have the capacity to self-renew and to differentiate into lineages that normally originate from the embryonic mesenchyme [36, 37]. Marrow-derived mesenchymal stromal/stem cells (MSCs) can be easily induced to differentiate into bone, cartilage and fat cells, and MSC-based cell products have already found clinical applications in the regeneration of cartilage and bone defects. Moreover, MSCs have strong immunosuppressive effects both in vitro and in vivo [38]. In contrast to haematopoietic stem cells, MSCs may be able to obtain a cardiomyocyte phenotype, provided they are stimulated accordingly [39]. It is important to note, however, that usually epigenetic modulation of the transcriptional profile by DNA-demethylation with 5-azacytidine needs to be employed. Demethylation reduces the stability of DNA-silencing signals and thus confers non-specific gene activation. Under the influence of 5-azacytidine, MSCs obtain a wide range of different phenotypes, and spontaneously beating myocyte-like cells are isolated and selectively expanded. The resulting cell population shows many of the morphologic, proteomic, and functional characteristics of true cardiomyocytes [40]. Whether physiologic in vivo signals are sufficient to drive naïve MSCs into a myogenic lineage without significant exogenous triggers, however, remains unclear [41, 42].

33.5 Clinical Cell Therapy

33.5.1 Bone Marrow Mononuclear Cells in Acute Infarction

Bone marrow mononuclear cells (MNCs) are the commonest cell products used in patients with acute myocardial infarction, where they are injected into the infarct vessel after it has been reopened by balloon dilation and stent placement. This straightforward cell therapy approach has been performed in thousands of patients world-wide. Following small-scale pilot trails, the first randomised, placebo-controlled study comparing intracoronary MNC injection with standard treatment of acute myocardial infarction was the BOOST (BOne marrOw transfer to enhance ST-elevation infarct regeneration) trial [43]. At 6 months follow-up, cell-treated patients had a significantly higher left-ventricular ejection fraction (LVEF) than control patients. Subsequently, several similar studies were conducted by other groups [44, 45]. Some of those trials clearly produced a negative result, with no difference in outcome between cell-treated and control patients [46, 47]. In the multicentre trial coordinated by the Frankfurt group, LVEF rose by 5.5% in cell-treated patients, and by 3.0% in the control group [48]. The difference proved statistically significant, but it remains to be seen if such a modest effect translates into a relevant clinical benefit. Other reports focused on clinical exercise tolerance and quality-of-life data, and there seems to be a slight advantage for patients who have received cell therapy [49]. Ultimately, a significant reduction of MACE (major cardiac adverse events—death, re-infarction, and need for re-intervention) needs be proven.

33.5.2 Bone Marrow MNCs in Chronic Ischaemia

Patients with chronic myocardial ischaemia have also been treated with bone marrow MNC products. Here, direct intramyocardial injection during cardiac surgery dominates, but intracoronary cell delivery has also been performed. Moreover, there are reports on catheter-based transendocardial intramuscular injection of bone marrow MNCs [50]. Again, some trials have shown a modest benefit, while others have produced an essentially negative result [51]. The same must be said regarding surgical injection of MNCs in conjunction with bypass surgery. In early pilot studies, an improvement of regional ventricular wall motion in cell-treated areas was observed, but this did not result in better global heart function as compared with routine bypass surgery [52]. Our own experience was very similar. We treated a series of patients undergoing bypass surgery for chronic ischaemic heart disease, and although we were able to detect improved myocardial function in cell-treated segments, this did not result in better global ventricular function as assessed by LVEF (unpublished data).

33.5.3 Enriched Stem-Cell Products

Specific progenitor cell products can be prepared using clinical-grade immunomagnetic selection for either CD34 or CD133; negative selection for CD45 is also possible. Another strategy is the in vitro expansion of bone marrow MNCs, with or without addition of differentiation-inducing or -suppressing substances. The clinician must choose between freshly isolated CD34- or CD133-enriched cell products with high purity, well-defined characteristics, but rather low cell dose, and high-dose, in vitro expanded cell products that contain incompletely characterised cells and take several weeks to be prepared [53]. In conjunction with coronary artery bypass graft (CABG) surgery, intramyocardial injection of CD34+ bone marrow cells can result in a nearly 10% higher LVEF than CABG surgery alone [54]. Our group has focused on CD133+ cells because they are believed to contain a subpopulation of cells that are even more immature than CD34+ cells. In 2001, we started a feasibility and safety study in 10 patients where no procedure-related adverse events were observed [55, 56], and subsequently conducted a controlled study in 40 patients. Here, CABG and CD133+ cell injection led to a significantly higher LVEF at 6 months follow-up than CABG surgery alone [57]. Whether this benefit is maintained in the long run remains to be determined. Other groups have isolated CD133 cells for surgical delivery from peripheral blood, following mobilisation from the marrow with colony-stimulating factor (G-CSF) [58]. This procedure yields a substantially higher cell dose, but requires several days for cell preparation. Enriched bone marrow stem-cell products have also been injected using catheter-based systems, and preliminary data indicate an improvement of left-ventricular function over placebo-treated patients [59]. Taken together, the intracardiac transplantation of purified progenitor cell products seems to yield a greater benefit than the use of mononuclear cells, but definitive proof is still lacking.

33.5.4
Mesenchymal Stem Cells

Mesenchymal stem cells have reached the clinical cardiovascular arena later than their haematopoietic counterparts, partly because they require to be cultivated and expanded in vivo over several weeks when used in autologous fashion, and partly because there were reports on microinfarction after intracoronary injection of MSC products in animal models, and on a potential risk of bone, cartilage and adipose tissue formation in the heart. A unique characteristic of MSCs is their immunomodulatory potency; allogenic MSCs may escape detection and elimination by the host immune system [60, 61] because they express only very low levels of major histocompatibility complex (MHC) class II. Moreover, they can actively suppress immunologic processes and reduce inflammation by inhibiting T-cell proliferation, and it is believed that MSCs beneficially influence solid organ injury by shifting the local balance of pro- and anti-inflammatory cytokines. To date, there is only anecdotal information on the use of autologous, bone-marrow-derived MSCs for cardiac repair. A clinical pilot study testing an allogenic off-the-shelf a product in patients with myocardial infarction by peripheral intravenous injection is currently being performed in North America, and initial results are positive.

33.5.5
Cytokine-Induced Bone Marrow Cell Mobilisation

Here, the idea is that stem/progenitor cells mobilised from marrow using G-CSF are attracted to the ischaemic heart and initiate the regeneration processes [62]. However, the number of mature leukocytes also rises markedly, and this has raised concerns regarding the safety of G-CSF treatment. Because bone marrow cell mobilisation did not result in a favourable outcome in a nonhuman primate model of myocardial infarction, this pharmacological approach has been viewed with some restraint. In clinical trials, an unexpectedly high rate of in-stent restenosis was observed, and incidences such as acute re-infarction and sudden death have occurred in other G-CSF studies [63]. Nevertheless, several controlled clinical efficacy studies have subsequently been performed, but the outcome data in terms of heart function improvement are equivocal.

33.6
Unresolved Problems

Although the experimental basis of myocardial cell therapy is incomplete, numerous clinical trials and "routine" applications have already been initiated. Attention focuses mainly on the cell product to be used, but there are numerous other factors that need to be considered to maximise the likelihood of successful cell-based myocardial regeneration. Several of these will now be discussed.

33.6.1
Patient Selection

Obviously, a novel therapeutic approach is best evaluated in a uniform cohort of patients with well-defined patient- and disease-related characteristics. Regarding cell therapy for heart disease, however, this is more difficult than it seems. In the majority of patients, myocardial cell therapy has been performed within several days after the onset of myocardial infarction symptoms, when patients cannot be selected according to their pretreatment heart function. Consequently, such patient cohorts cover a wide range of left ventricular contractility at baseline, rendering the analysis and interpretation of the data more difficult than in a uniform group of patients. In trials of patients with chronic myocardial ischaemia, the degree of contractile dysfunction is usually better defined. Finally, patients with non-ischaemic heart disease have also been subject to cell-therapy approaches, although the body of preclinical data is much smaller than for ischaemic heart disease. Here, one problem is the diversity of underlying aetiologies for non-ischaemic heart failure, including inflammatory processes, genetic diseases, and a large group of "idiopathic" cases where no cause can be established.

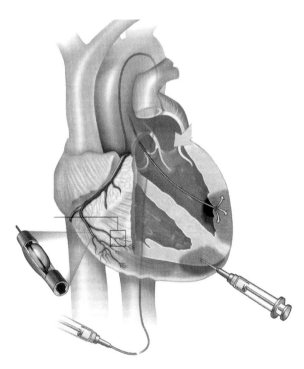

Fig. 33.4 Techniques to deliver cell products to the heart in the clinical setting. From left to right: intracoronary injection with an inflated percutanueous transluminal coronary angioplasty balloon; catheter-based transendocardial injection in the heart muscle; transepicardial injection requiring surgical access to the heart

33.6.2
Concomitant Procedures

Cell therapy has been performed as a true stand-alone procedure in only very few studies. In acute infarction, it is usually combined with catheter-based reopening of the infarct vessel. In patients with chronic myocardial ischaemia, cells are often delivered at the time of a coronary artery bypass operation. There are also reports on studies where intramyocardial cell therapy was performed with heart valve surgery [64]. Another option is the intramyocardial delivery of cells at the time of implantation of a ventricular assist device in end-stage heart failure patients with the most urgent need for innovative therapeutic strategies. A further possibility is the combination of myocardial cell therapy with transmural myocardial laser revascularisation, with or without additional bypass surgery [65]. The rationale is that the laser injury induces a local inflammatory response that may stimulate transplanted bone marrow cells and enhance their regenerative potential. However, little experimental evidence supports this hypothesis.

33.6.3
Cell Delivery

The most frequent way to delivery cells to the heart is catheter-based injection into the coronary arteries, requiring transmigration of cells through the vascular wall into the myocardium (Fig. 33.4). Using catheter-based systems, direct intramyocardial cell delivery is also possible, often combined with intracardiac electrical mapping of the left ventricle to identify the ischaemic area-of-interest. There is also a catheter system that is guided into the epicardial veins and allows for injection into the heart muscle. For cell delivery under direct vision using surgical techniques, any commercially available syringe and needle system can be used, but industry has also developed special cell-injection needles with side-holes. In some trials, cells are being delivered by peripheral venous injection, relying on chemokines produced in the heart that may attract stem cells from the periphery. A fraction of those cells indeed ends up in the heart, but the majority undergoes first-pass trapping in the lung, or is eliminated by the reticuloendothelial system in liver and spleen. Each of these cell-delivery techniques involves fundamentally different biologic processes, which clearly have a major impact on the therapeutic efficacy.

33.6.4
Timing

It is not known whether there is an ideal time point for cell therapy in patients with ischaemic heart disease. Emergency treatment of acute infarction is usually performed by the interventional cardiologist, who may also decide to perform intracoronary injection of a rapidly available cell product. When patients with acute infarction need emergency surgery, it is usually not feasible to arrange for concomitant cell therapy. The situation in chronic ischaemic heart disease is very different. In the post-infarct or chronically ischaemic myocardium, a substantial net loss of contractile tissue mass has occurred, and there is diffuse or localised scar formation. Intuitively, the longer the interval between myocardial infarction and cell treatment, the smaller the chance to achieve a beneficial effect. This notion, however, is presently not supported by clear-cut data.

33.6.5
Cell Survival

When cells are injected into diseased myocardium, most of them do not survive. The magnitude of cell death upon intramyocardial transplantation is difficult to measure, but the suggested survival rate ranges between 0.1% and 10% [66, 67]. The ischaemic myocardium is a hostile environment due to local hypoxia, acidosis, lack of substrates and accumulation of metabolites. Moreover, the infarct area is infiltrated by phagocytic cells that remove cell debris, and many transplanted cells are probably lost in this "clean-up" process. Third, the mechanical forces in the myocardium are substantial. Transmural pressure is high during systole, there are shear forces between contracting myofibres and layers, and a marrow cell is not well equipped to withstand such stress. However, the rate of cell death can be slowed. Transfection of marrow stromal cells with genes encoding for the anti-apoptotic proteins AKT or Bcl-2 has been shown to greatly improve cell survival and regenerative capacity [68, 69]. Pretreatment of endothelial progenitor cells with endothelial-nitric-oxide-synthase-enhancing substances also appears to have a beneficial effect [70], similar to those observed with statins [71]. Hypoxic preconditioning or heat shock prior to cell injection might also help, since both of these activate anti-apoptotic and nitric-oxide-related signalling pathways [72].

33.6.6
Dosage

The normal adult heart weighs between 250 and 350 g, and around 80% of the myocardial mass consists of approximately 1×10^9 cardiomyocytes. Assuming that 20% of the cardiomyocytes are lost following myocardial infarction, one would ultimately need 20×10^8 surviving neomyocytes weighing nearly 50 g to completely reconstitute the myocardium. Given the high rate of cell death upon transplantation into the heart and the presumably very low number of adult stem cells that actually differentiate into myocytes, it becomes clear that with currently available cell products, we are far from being able to replace all lost heart muscle tissue. Hypothetically, one would either need to transplant a very large volume of cardiomyocyte suspension or progenitor cells with a high proliferative capacity in vivo to achieve *restitutio ad integrum*—complete restitution.

33.6.7
Age

Two questions need to be addressed regarding cardiac cell therapy in elderly patients:
1. Can the aged recipient heart be repaired at all?
2. Are autologous cell products derived from elderly patients suitable for myocardial regeneration?

Theoretically, stem cells constantly renew themselves, but it has become clear that bone marrow stem cells undergo ageing processes and are affected by remote diseases. Ageing may not just be imposed on marrow stem cells by external and internal stressors, but seems to be an active process that helps protect the organism. In essence, the organism sacrifices its regenerative capacity in order to counteract the increasing tendency to develop cancer, but the functional consequences vary between the different types of stem and progenitor cells. Although this issue has been addressed in several experimental studies, the impact of patient age on the outcome of clinical cell therapy for heart disease has not yet been studied systematically.

33.7
Summary

While cardiac cell therapy has initially caused tremendous excitement, it is currently viewed with reluctance. There have been—and will be more—substantial obstacles and setbacks along the road to success, but only about 15 years have so far been spent trying to develop therapies involving cell-based cardiac regeneration. The early clinical use of bone-marrow-derived cells for heart disease has been much criticised, but it is understandable that physicians began by testing the clinical efficacy of bone marrow cells, before moving on to do further research on more powerful, but also more complex cell products. Although the systematic evaluation of marrow-derived cell products is by no means complete, there is little hope that truly curative therapeutic strategies will evolve using simple autologous cell products. On the other hand, ESCs and induced pluripotent stem cells technology has steadily progressed, and there is little doubt that it will eventually yield effective clinical cell replacement therapies. Given its biological, medical and regulatory complexity, clinical translational of those therapies will require an unprecedented degree of interdisciplinary collaboration. Moreover, novel modes of interaction between researchers, health-care providers and industry are required, because the traditional concept of intellectual property and its commercialisation barely cover the entire cell therapy value chain. If this can be achieved, cell-based regeneration therapies will sooner or later transform not only cardiovascular medicine.

References

1. Rota M, Hosoda T, De Angelis A, et al. The young mouse heart is composed of myocytes heterogeneous in age and function. Circ Res; 2007:387–99
2. Bittmann I, Hentrich M, Bise K, et al. Endothelial cells but not epithelial cells or cardiomyocytes are partially replaced by donor cells after allogeneic bone marrow and stem cell transplantation. J Hemather Stem Cell Res; 2003;12:359–66
3. Deb A, Wang S, Skelding KA, et al. Bone marrow-derived cardiomyocytes are present in adult human heart: a study of gender-mismatched bone marrow transplantation patients. Circulation 2003;107:1247–9
4. Jiang S, Walker L, Afentoulis M, et al. Transplanted human bone marrow contributes to vascular endothelium. Proc Natl Acad Sci U S A 2004;101:16891–6
5. Minami E, Laflamme MA, Saffitz JE, et al. Extracardiac progenitor cells repopulate most major cell types in the transplanted human heart. Circulation 2005;112:2951–8
6. Laflamme MA, Myerson D, Saffitz JE, et al. Evidence for cardiomyocyte repopulation by extracardiac progenitors in transplanted human hearts. Circ Res 2002;90:634–40
7. Torella D, Ellison GM, Karakikes I, et al. Resident cardiac stem cells. Cell Mol Life Sci 2007;64:661–73
8. Beltrami AP, Barlucchi L, Torella D, et al. Adult cardiac stem cells are multipotent and support myocardial regeneration. Cell 2003;114:763–76
9. Oh H, Bradfute SB, Gallardo TD, et al. Cardiac progenitor cells from adult myocardium: homing, differentiation, and fusion after infarction. Proc Natl Acad Sci USA 2003;100:12313–8
10. Oyama T, Nagai T, Wada H, et al. Cardiac side population cells have a potential to migrate and differentiate into cardiomyocytes in vitro and in vivo. J Cell Biol 2007;176:329–41
11. Bearzi C, Rota M, Hosoda T, et al. Human cardiac stem cells. Proc Natl Acad Sci U S A 2007;104:14068–73
12. Messina E, De Angelis L, Frati G, et al. Isolation and expansion of adult cardiac stem cells from human and murine heart. Circ Res 2004;95:911–21
13. Smith RR, Barile L, Cho HC, et al. Regenerative potential of cardiosphere-derived cells expanded from percutaneous endomyocardial biopsy specimens. Circulation 2007;115:896–908
14. Laugwitz KL, Moretti A, Lam J, et al. Postnatal isl1+ cardioblasts enter fully differentiated cardiomyocyte lineages. Nature 2005;433:647–53
15. Chiu RC, Zibaitis A, Kao RL. Cellular cardiomyoplasty: myocardial regeneration with satellite cell implantation. Ann Thorac Surg 1995;60:12–8

16. Marelli D, Desrosiers C, el-Alfy M, et al. Cell transplantation for myocardial repair: an experimental approach. Cell Transplant 1992;1:383–90
17. Leor J, Patterson M, Quinones MJ, et al. Transplantation of fetal myocardial tissue into the infarcted myocardium of rat. A potential method for repair of infarcted myocardium? Circulation 1996;94:II332–6
18. Taylor DA, Atkins BZ, Hungspreugs P, et al. Regenerating functional myocardium: improved performance after skeletal myoblast transplantation. Nat Med 1998;4:929–33
19. Menasche P, Hagege AA, Scorsin M, et al. Myoblast transplantation for heart failure. Lancet 2001;357:279–80
20. Rubart M, Soonpaa MH, Nakajima H, et al. Spontaneous and evoked intracellular calcium transients in donor-derived myocytes following intracardiac myoblast transplantation. J Clin Invest 2004;114:775–83
21. Dib N, Michler RE, Pagani FD, et al. Safety and feasibility of autologous myoblast transplantation in patients with ischemic cardiomyopathy: four-year follow-up. Circulation 2005;112:1748–55
22. Behfar A, Zingman LV, Hodgson DM, et al. Stem cell differentiation requires a paracrine pathway in the heart. FASEB J 2002;16:1558–66
23. Dai W, Field LJ, Rubart M, et al. Survival and maturation of human embryonic stem cell-derived cardiomyocytes in rat hearts. J Mol Cell Cardiol 2007;43:504–16
24. Menard C, Hagege AA, Agbulut O, et al. Transplantation of cardiac-committed mouse embryonic stem cells to infarcted sheep myocardium: a preclinical study. Lancet 2005;366:1005–12
25. Behfar A, Perez-Terzic C, Faustino RS, et al. Cardiopoietic programming of embryonic stem cells for tumor-free heart repair. J Exp Med 2007;204:405–20
26. Byrne JA, Pedersen DA, Clepper LL, et al. Producing primate embryonic stem cells by somatic cell nuclear transfer. Nature 2007;450:497–502
27. Guan K, Wagner S, Unsold B, et al. Generation of functional cardiomyocytes from adult mouse spermatogonial stem cells. Circ Res 2007;100:1615–25
28. Yu J, Vodyanik MA, Smuga-Otto K, et al. Induced pluripotent stem cell lines derived from human somatic cells. Science 318:1917–20
29. Takahashi K, Tanabe K, Ohnuki M, et al. Induction of pluripotent stem cells from adult human fibroblasts by defined factors. Cell 2007;131:861–72
30. Takahashi K, Yamanaka S. Induction of pluripotent stem cells from mouse embryonic and adult fibroblast cultures by defined factors. Cell 2006;126:663–76
31. Kocher AA, Schuster MD, Szabolcs MJ, et al. Neovascularization of ischemic myocardium by human bone-marrow-derived angioblasts prevents cardiomyocyte apoptosis, reduces remodeling and improves cardiac function. Nat Med 2001;7:430–6
32. Orlic D, Kajstura J, Chimenti S, et al. Bone marrow cells regenerate infarcted myocardium. Nature 2001;410:701–5
33. Murry CE, Soonpaa MH, Reinecke H, et al. Haematopoietic stem cells do not transdifferentiate into cardiac myocytes in myocardial infarcts. Nature 2004;428:664–8
34. Rota M, Kajstura J, Hosoda T, et al. Bone marrow cells adopt the cardiomyogenic fate in vivo. Proc Natl Acad Sci U S A 2007;104:17783–8
35. Friedenstein AJ, Petrakova KV, Kurolesova AI, et al. Heterotopic of bone marrow. analysis of precursor cells for osteogenic and hematopoietic tissues. Transplantation 1968;6:230–47
36. Prockop DJ. Marrow stromal cells as stem cells for non-hematopoietic tissues. Science 1997;276:71–4
37. Caplan AI. Mesenchymal stem cells. J Orthop Res 1991;9:641–50
38. Le Blanc K, Ringden O. Immunomodulation by mesenchymal stem cells and clinical experience. J Int Med 2007;262:509–25
39. Makino S, Fukuda K, Miyoshi S, et al. Cardiomyocytes can be generated from marrow stromal cells in vitro. J Clin Invest 1999;103:697–705
40. Hakuno D, Fukuda K, Makino S, et al. Bone marrow-derived regenerated cardiomyocytes (CMG cells) express functional adrenergic and muscarinic receptors. Circulation 2002;105:380–6
41. Hattan N, Kawaguchi H, Ando K, et al. Purified cardiomyocytes from bone marrow mesenchymal stem cells produce stable intracardiac grafts in mice. Cardiovasc Res 2005;65:334–44
42. Toma C, Pittenger MF, Cahill KS, et al. Human mesenchymal stem cells differentiate to a cardiomyocyte phenotype in the adult murine heart. Circulation 2002;105:93–8
43. Wollert KC, Meyer GP, Lotz J, et al. Intracoronary autologous bone-marrow cell transfer after myocardial infarction: the BOOST randomised controlled clinical trial. Lancet 2004;364:141–8
44. Assmus B, Fischer-Rasokat U, Honold J, et al. Transcoronary transplantation of functionally competent BMCs is associated with a decrease in natriuretic peptide serum levels and improved survival of patients with chronic postinfarction heart failure: results of the TOPCARE-CHD Registry. Circ Res 2007;100:1234–41
45. Assmus B, Honold J, Schachinger V, et al. Transcoronary transplantation of progenitor cells after myocardial infarction. N Engl J Med 2006;355:1222–32
46. Janssens S, Dubois C, Bogaert J, et al. Autologous bone marrow-derived stem-cell transfer in patients with ST-segment elevation myocardial infarction: double-blind, randomised controlled trial. Lancet 2006;367:113–21
47. Lunde K, Solheim S, Aakhus S, et al. Intracoronary injection of mononuclear bone marrow cells in acute myocardial infarction. N Engl J Med 2006;355:1199–209
48. Schachinger V, Erbs S, Elsasser A, et al. Intracoronary bone marrow-derived progenitor cells in acute myocardial infarction. N Engl J Med 2006;355:1210–21
49. Lunde K, Solheim S, Aakhus S, et al. Exercise capacity and quality of life after intracoronary injection of autologous mononuclear bone marrow cells in acute myocardial infarction: results from the Autologous Stem cell Transplantation in Acute Myocardial Infarction (ASTAMI) randomized controlled trial. Am Heart J 2007;154:710 e1–8

50. Perin EC, Dohmann HF, Borojevic R, et al. Improved exercise capacity and ischemia 6 and 12 months after transendocardial injection of autologous bone marrow mononuclear cells for ischemic cardiomyopathy. Circulation 2004;110:II213–8
51. Brehm M, Strauer BE. Stem cell therapy in postinfarction chronic coronary heart disease. Nat Clin Pract Cardiovasc Med 2006;3:S101–4
52. Galinanes M, Loubani M, Davies J, Chin D, Pasi J, Bell PR. Autotransplantation of unmanipulated bone marrow into scarred myocardium is safe and enhances cardiac function in humans. Cell Transpl 2004;13:7–13
53. Erbs S, Linke A, Adams V, et al. Transplantation of blood-derived progenitor cells after recanalization of chronic coronary artery occlusion: first randomized and placebo-controlled study. Circ Res 2005;97:756–62
54. Patel AN, Geffner L, Vina RF, et al. Surgical treatment for congestive heart failure with autologous adult stem cell transplantation: a prospective randomized study. J Thorac Cardiovasc Surg 2005;130:1631–8
55. Stamm C, Kleine HD, Westphal B, et al. CABG and bone marrow stem cell transplantation after myocardial infarction. Thorac Cardiovasc Surg 2004;52:152–8
56. Stamm C, Westphal B, Kleine HD, et al. Autologous bone-marrow stem-cell transplantation for myocardial regeneration. Lancet 2003;361:45–6
57. Stamm C, Kleine HD, Choi YH, et al. Intramyocardial delivery of CD133+ bone marrow cells and coronary artery bypass grafting for chronic ischemic heart disease: safety and efficacy studies. J Thorac Cardiovasc Surg 2007;133:717–25
58. Pompilio G, Cannata A, Peccatori F, et al. Autologous peripheral blood stem cell transplantation for myocardial regeneration: a novel strategy for cell collection and surgical injection. Ann Thorac Surg 2004;78:1808–12
59. Losordo DW, Schatz RA, White CJ, et al. Intramyocardial transplantation of autologous CD34+ stem cells for intractable angina: a phase I/IIa double-blind, randomized controlled trial. Circulation 2007;115:3165–72
60. Amado LC, Saliaris AP, Schuleri KH, et al. Cardiac repair with intramyocardial injection of allogeneic mesenchymal stem cells after myocardial infarction. Proc Natl Acad Sci U S A 2005;102:11474–9
61. Grinnemo KH, Mansson A, Dellgren G, et al. Xenoreactivity and engraftment of human mesenchymal stem cells transplanted into infarcted rat myocardium. J Thorac Cardiovasc Surg 2004;127:1293–300
62. Orlic D, Kajstura J, Chimenti S, et al. Mobilized bone marrow cells repair the infarcted heart, improving function and survival. Proc Natl Acad Sci U S A 2001;98:10344–9
63. Ince H, Stamm C, Nienaber CA. Cell-based therapies after myocardial injury. Curr Treat Options Cardiovasc Med 2006;8:484–95
64. Messas E, Bel A, Morichetti MC, et al. Autologous myoblast transplantation for chronic ischemic mitral regurgitation. J Am Coll Cardiol 2006;47:2086–93
65. Klein HM, Ghodsizad A, Borowski A, et al. Autologous bone marrow-derived stem cell therapy in combination with TMLR. A novel therapeutic option for endstage coronary heart disease: report on 2 cases. Heart Surg Forum 2004;7:E416–9
66. Yau TM, Kim C, Ng D, et al. Increasing transplanted cell survival with cell-based angiogenic gene therapy. Ann Thorac Surg 2005;80:1779–86
67. Zhang M, Methot D, Poppa V, et al. Cardiomyocyte grafting for cardiac repair: graft cell death and anti-death strategies. J Mol Cell Cardiol 2001;33:907–21
68. Li W, Ma N, Ong LL, et al. Bcl-2 engineered MSCs inhibited apoptosis and improved heart function. Stem Cells 2007;25:2118–27
69. Mangi AA, Noiseux N, Kong D, et al. Mesenchymal stem cells modified with Akt prevent remodeling and restore performance of infarcted hearts. Nat Med 2003;9:1195–201
70. Sasaki K, Heeschen C, Aicher A, et al. Ex vivo pretreatment of bone marrow mononuclear cells with endothelial NO synthase enhancer AVE9488 enhances their functional activity for cell therapy. Proc Natl Acad Sci U S A 2006;103:14537–41
71. Spyridopoulos I, Haendeler J, Urbich C, et al. Statins enhance migratory capacity by upregulation of the telomere repeat-binding factor TRF2 in endothelial progenitor cells. Circulation 2004;110:3136–42
72. Maurel A, Azarnoush K, Sabbah L, et al. Can cold or heat shock improve skeletal myoblast engraftment in infarcted myocardium? Transplantation 2005;80:660–5

Part D
Technical Aspects

VII Biomaterial Related Aspects

Biomaterials

H. P. Wiesmann, U. Meyer

Contents

34.1 Introduction 457
34.2 Synthetic Organic Materials 458
34.3 Synthetic Inorganic Materials 460
34.4 Natural Organic Materials 461
34.5 Natural Inorganic Materials 464
     References 465

34.1 Introduction

Application of non-living biomaterials can be conceptualised as the use of materials to replace lost structures, augment existing structures or promote new tissue formation [4, 5]. Common degradable and non-degradable implant materials can be divided into synthetically produced metals and metallic alloys, ceramics, polymers, and composites or modified natural materials [53]. Whereas non-resorbable materials like steel or titanium alloys are commonly used for prosthetic devices, resorbable substitute materials are currently being investigated for their utility in bone and cartilage replacement therapies. Whether or not a material is biodegradable, its surface properties will influence the initial cellular events at the cell–material interface. A major difference between degradable and non-degradable implants is that the surface adhesion towards osteoblasts or chondrocytes is changing in degradable materials, while it remains constant in non-degradable implants. The clinical fate of implants, substitute materials and scaffolds used in tissue engineering strategies depends critically upon the underlying material [54] and the mechanical properties of the material-based scaffold (Table 34.1). In the design process of cell-based implants and engineered bone and cartilage substitutes, it is important to consider the cellular behaviour of the desired cell source towards the material (for review see [66]).

Four types of material have been studied experimentally and/or clinically as bone and cartilage substitute materials or scaffold materials for applications in tissue engineering. These include various groups of synthetic organic materials:

1. Biodegradable and bioresorbable polymers that have been used for clinically established products, such as polyglycolide, optically active and racemic polylactides, polydioxanone and polycaprolactone.
2. Polymers currently under clinical investigation, such as polyorthoester, polyanhydrides and polyhydroxyalkanoate.
3. Entrepreneurial polymeric biomaterials, such as for example poly (lactic acid-co-lysine), synthetic inorganic materials (e.g. hydroxyapatite—HA, calcium/phosphate composites, glass ceramics), organic materials of natural origin (e.g. collagen, fibrin, hyaluronic acid) and inorganic material of natural origin (e.g. coralline HA).

Table 34.1 Mechanical properties of biomaterials. *A-W* Apatite-wollastonite, *HA* hydroxyapatite, *PLGA* poly(lactic-co-glycolic acid)

Material	Young's modulus (GPa)	Strength (MPa)	
		Compressive	Bending
Cortical bone	7–30	100–230	50–150
Cancellous bone	0.05–0.5	2–12	–
Cartilage	0.2–0.3	0.01–3	–
Synthetic HA	80–110	500–1000	115–200
PLGA 85/15	2	0.34	–
PLGA 75/25	2	0.34	–
A-W glass-ceramic	118	1080	220
Collagen tendon	1.5	0.14	–

Several reports have been published on the physicochemical properties and surface characteristics of these biodegradable and bioresorbable polymers [29]. It was found that the features of the material surface are important for the correct implantation and coverage by autochthonic cells.

A comprehensive overview of the materials used in bone and cartilage tissue engineering is given in this chapter. The surface properties of materials will also be discussed, since the material characteristics and surface properties of materials are closely related. The list of materials used for bone tissue engineering differs from that of cartilage engineering, in that organic materials are mainly used for bone tissue engineering.

34.2 Synthetic Organic Materials

Various polymers are used extensively for the preparation of bone and cartilage scaffolds (Fig. 34.1). Most of them are fabricated on the basis of polyhydroxyacids such as polylactides, polyglycolides and their copolymers. Other synthetic materials include polyethylene oxide, polyvinyl alcohol, polyacrylic acid and polypropylene fumarate-co-ethylene glycol. Scaffolds based on biodegradable polyhydroxyacids, such as poly-lactide-glycolide copolymers, have been widely used as materials for bone and cartilage tissue engineering, since these polymers support the attachment and proliferation of both cell types [7, 9, 23, 24, 34, 68, 72]. Polylactic acid (PLA) and polyglycolic acid (PGA) copolymers degrade prevailingly via chemical hydrolysis of the hydrolytically unstable ester bonds into lactic acid and glycolic acid, which are non-toxic and can be removed from the body by normal metabolic pathways [1]. In addition, enzymatic degradation has been reported for those copolymers [9]. It was found that the degradation behaviour of polyhydroxyacids depends on the polymer structure and surface properties (molecular weight, copolymer composition, crystallinity, overall material surface, etc.) and environmental conditions (medium exchange, temperature, polymer/host interaction and pH; Table 34.2). The degradation rate of polyhydroxyacids can be adjusted by changes in the PLA:PGA ratio, the molecular weight of each component, the crystallinity and other factors in order to support the slow degradation rate. This degradation rate may be similar or slightly lower than the rate of tissue formation in the defect site [6]. It is important to note in this respect that the repair or remodelling turnover time of cartilage and bone differs significantly. Limited resistance of polyhydroxyacid materials to loading, as well as their limited deformability and elasticity are disadvantageous in their use for bone and cartilage tissue engineering strategies, especially if mechanical competence is required [22]. Elastomeric polyurethanes are materials with enhanced molecular stability in vivo. Newly designed biodegradable polyurethanes were shown to degrade in vivo to non-toxic by-products [8, 12, 14–19, 35]. Based on these findings, biodegradable polyurethanes have been used successfully as cancellous bone graft substitutes in animals [20, 21]. However, very little attention has been paid to biodegradable polyurethanes as scaffolding materials for chondrocytes. To date, only hydroxybutyrate-co-valerate polyol-based polyurethane has been used to culture rat chondrocytes [64].

Table 34.2 Polymer properties

Polymer	Strength (GPa)	Degradation time (months)	Degradation products
Polyester (polyglycolic acid)	7.0	6–12	Glycolic acid
Polyester (poly(L)-lactic acid)	2.7	>24	l-Lactic acid
Polyester (poly(D,L-lactic-co-glycolic) acid (85/15))	2.0	1–6	d,l-Lactic acid, glycolic acid
Polyester (polycaprolactone)	0.4	>24	Caproic acid
Polyester (polydioxanone)	–		
Polyester (polypropylenefumarat)	2–30	–	Fumaric acid, propylene glycol
Polyanhydrides	0.0001	12	Dicarboxylic acid
Polycarbonates	–	Slow	Tyrosine, carbon dioxide
Polyurethanes	0.008–0.004	1–2	Lysine, glycolic acid and caproic acid
Polyphosphazenes	–	>1	Phosphates, ammonia

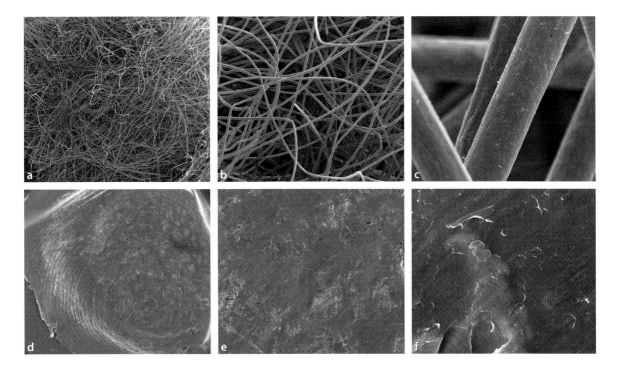

Fig. 34.1a–f Scanning electron microscope view of hyaluronic-acid-based biomaterials. The biomaterial can be processed as a fibre-based scaffold (**a–c**) or as a bulk material (**d–f**). From *left* to *right*: detailed aspects of material structure and surface

34.3
Synthetic Inorganic Materials

Various types of synthetic inorganic material have been developed for skeletal replacement therapies (Table 34.3). These synthetic substrates, capable of supporting the natural process of bone remodelling, were used mainly in bone tissue engineering applications that include the ex vivo generation of cell-scaffold complexes, in vivo resorbable bone cements, implantable coatings that enhance the bonding of natural bone to the implant, various forms of implantable prostheses and bone-repair agents [10, 41, 59, 71]. Synthetic inorganic materials are seldom used in cartilage tissue engineering strategies, since the matrix of cartilage does not contain crystalline inorganic materials (except for the calcified cartilage layer). Among the materials used as scaffolds for bone tissue engineering, calcium-phosphate-containing ceramics are most frequently exploited in bone-replacement strategies (Fig. 34.2). These are: HA [$Ca_{10}(PO_4)_6(OH)_2$], β-tricalcium phosphate [TCP; $Ca_3(PO_4)_2$] and HA/β-tricalcium phosphate bi-phase ceramics. Calcium phosphate materials have differing degrees of stoichiometry. HA is the most frequently used of several calcium phosphorous (CaP) compounds that are near the primary ionic component of natural bone. Pure HA has the stoichiometric Ca:P ratio of 1.67, lattice parameters: a-axis=0.94 nm, and c-axis=0.69 nm, and the presence of only the OH and PO_4 absorption bands in their infrared spectra. Other CaP materials also recommended for scaffolds in bone tissue engineering include octacalcium phosphate [31], whitlockite, or magnesium-substituted tricalcium phosphates [13], zinc-substituted tricalcium phosphate [32], carbonate-substituted apatites [61] and fluoride-substituted apatites [11, 13, 33, 56]. Substitution of single elements in the calcium phosphate or apatite structure affects the crystal and dissociation properties of CaP. Carbonate substitution, for example, causes not only the formation of smaller and more soluble apatite particles, but also a better pH stability, whereas fluoride incorporation has the opposite effects upon material degradation, while having no effect on the pH stability [37, 39, 42–44, 50, 57, 76]. Magnesium incorporation in apatite is limited but causes a reduction in crystallinity

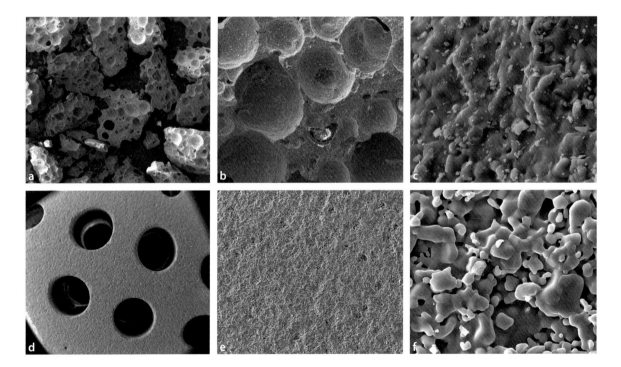

Fig. 34.2a–f Scanning electron microscope view of two different tricalcium-phosphate-based biomaterials (**a–c** vs. **d–f**), showing different gross morphologies. From *left* to *right*: detailed aspects of material structure and surface

Table 34.3 Naturally occurring and artificial polymeric scaffold materials

	Protein based	Carbohydrate based	Synthetic
Natural	Collagen Fibrin	Hyaluronan	
Artificial	Gelatine	Polylactic acid Polyglycolic acid Chitosan Agarose Alginate	Polymethlymethacrylate Polyethylene Polytetrafluoroethylene Polybutyric acid Polycarbonate Polyesterurethane

(smaller crystal size) and increases the extent of dissolution [37, 39, 45, 47–50]. Magnesium or zinc substitution in different calcium materials also affects the properties [39, 51, 60].

Properties of ceramic materials can be influenced by the fabrication process. Parameters like porosity, crystal size, composition and dissolution properties determine to a great extent the fate of the material in the in vivo and in vitro situation [38, 39, 46]. Due to their ionic, hydrophilic composition, ceramic materials have a particular affinity to bind proteins. They may therefore be suitable carriers for bioactive peptides or bone growth factors [58, 63, 70]. However, it is important to note that although CaP biomaterials are osteoconductive, they do not have osteoinductive properties, meaning that they are unable to support de novo bone tissue generation at non-bony sites.

34.4
Natural Organic Materials

Naturally derived materials have frequently been exploited in tissue engineering applications because they are either components of or have macromolecular properties similar to the natural extracellular matrix (ECM; Table 34.4). For example, collagens, fibrin, hyaluronan or some proteoglycans are the main proteins of hard-tissue ECM of vertebrates (Fig. 34.3). Non-mammalian molecules like alginate and chitosan are candidates for scaffold materials. Despite the fact that these are not present in human tissues, they have also been shown to interact in a favourable manner with the surface of implant devices and have been utilised as scaffold materials for bone and cartilage tissue engineering.

Collagens are attractive materials for bone and cartilage tissue engineering, as this group of secreted proteins is present in skeletal tissues, where they constitute the main substrate of the ECM [26, 65, 74]. Whereas collagen type I is the predominantly expressed collagen found in bone tissue, collagen type II is present in the ECM of cartilage. As an ideal scaffold material mimicking the real situation, collagen-based scaffolds for bone or cartilage should differ in their composition. The basic structure of all collagen is composed of three polypeptide chains, building up a three-stranded rope structure. The various types of collagen are naturally degraded by secreted collagenases. Thus, the degradation of collagen fibres is

Table 34.4 Calcium phosphate biomaterials. *TCP* β-tricalcium phosphate

Natural	Freeze dried/banked human bones Bovine bone derived (BioOss, Endobone, Trubone) Coral derived (Interpore, Pro-osteon)
Artificial—bulk	Calcium hydroxyapatite (Calcitite, Durapatite, Osteograf, Alveograf) Beta-tricalcium phosphate (Synthograf, Augmen) Biphasic HA/TCP materials (Osteosynt)
Artificial—cement	Calcium phosphate cement (Endobone)

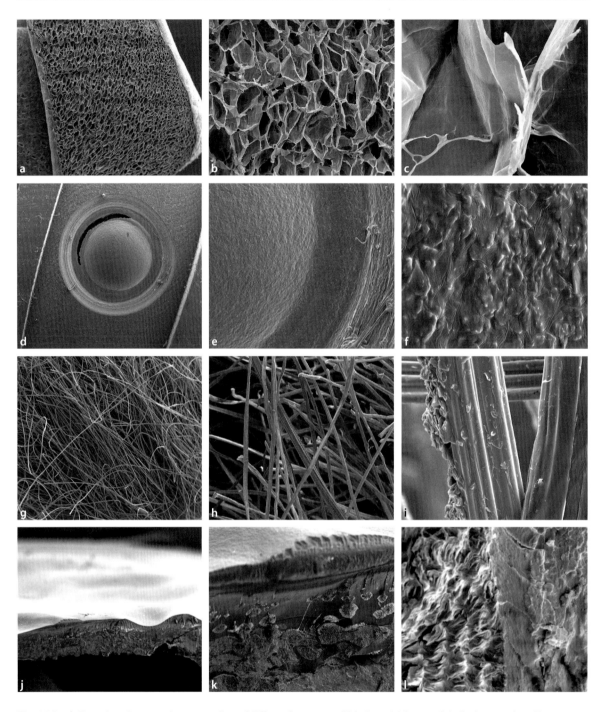

Fig. 34.3a–l Scanning electron microscope view of differently structured biodegradable materials for bone and cartilage reconstruction. Materials can have a spongiosa-like (**a–c**), bulk (**d–f**), fibre-like (**g–i**) or compound (**j–l**) structure. From *left* to *right*: detailed aspects of material structure and surface

controlled locally by cells present in the engineered tissue, and the balance between collagen synthesis and the activities of collagenases is expected to determine the turnover rate of the carrier. Moreover, it is important that collagen-based scaffolds do not usually provoke a foreign-tissue response. From a biomaterials perspective, it should be noted that the features of collagen scaffolds can be modified according to the experimental or clinical needs. Collagen-based scaffolds can be created artificially and their mechanical properties altered to some extent by various chemical and physical techniques. Modification of collagen can be performed by incubation of collagen with chemical cross-linkers (i.e. glutaraldehyde, formaldehyde, carbodiimide), by physical treatments (i.e. ultraviolet irradiation, freeze-drying, heating) and copolymerisation with other polymers (e.g. polyhydroxyacids) [36]. Importantly, the mechanical strength of all collagen scaffolds is limited in clinical terms (since the elasticity modulus of collagen is much lower than that of bone), even when molecules are cross-linked. By using defined collagen types for scaffold production, cellular differentiation can be promoted (e.g. collagen type I for osteoblastic differentiation and collagen type II for chondrocytic differentiation).

During the last decade, hyaluronan has gained increased attention as a scaffold material for cartilage tissue engineering [62]. Hyaluronan is the simplest glycosaminolglycan and is found especially in cartilage tissue. Hydrogels of hyaluronan are formed by various chemical treatments [28]. Hyaluronan has also been combined with both collagen and alginate to form composite hydrogels. Similarly to collagens, hyaluronan is naturally degraded by secreted proteases, termed hyaluronidases, allowing tissue turnover by cells in the skeletal defect site.

Fibrin, associated with fibronectin, has been shown to support keratinocyte and fibroblast growth both in vitro and in vivo, and appears to enhance cellular motility in the wound [52]. When used as a scaffold system for cultured mesenchymal cells, fibrin may provide some advantages over other natural materials [30]. Fibrin glue works as an adhesive by emulating the exudative phase of wound healing. Early products were made with human fibrin concentrate and thrombin. When the two substances are mixed, the thrombin, in the presence of calcium, converts fibrinogen to fibrin. The resulting fibrin polymer has a stable structure that facilitates the growth of collagen-producing fibroblasts [55]. Further development has led to the addition of factor XIII, a fibrin-stabilising factor present in blood, or aprotinin, which is an antiplasmin that protects the fibrin polymer clot from premature fibrinolysis. Fibrin deposition depends on the relative rates of formation, degradation and dissolution. Fibrin glue has also been shown to be a suitable delivery vehicle for exogenous growth factors that may in the future be used to accelerate wound healing.

Alginate has been used in a variety of tissue engineering applications, because it can be processed as a gel and has a low toxicity [25]. Alginate as a polymer is composed of polymerised G and M monomers. Gels are formed when divalent cations cooperatively interact with blocks of monomers to form ionic bridges between different polymer chains. The cross-linking density, and thus the mechanical competence and pore size of the ionically cross-linked gels, can be adjusted by different techniques: by varying the M:G ratio or the molecular weight of the polymer chain, as well as by covalently cross-linking alginate with adipic hydrazide or polyethylene glycol [69]. Ionically cross-linked alginate hydrogels do not specifically degrade, but undergo slow, uncontrolled dissolution.

Chitosan is sometimes used as a scaffold material in tissue engineering applications because it is structurally similar to naturally occurring glycosaminoglycans and is degradable by human enzymes [75]. It is a linear polysaccharide of (1–4)-linked D-glucosamine and N-acetyl-D-glucosamine residues. Dissolved chitosan can be fabricated as a gel by various methods (increasing the pH, extruding the solution into a non-solvent, by glutaraldehyde cross-linking, ultraviolet irradiation and thermal variations) [2, 3]. The crystallinity of chitosan, depending on the degree of N-deacetylation, influences the kinetics of degradation. Degradation of chitosan is performed experimentally through lysozyme action. As the degradation rate of chitosan is inversely related to the degree of crystallinity [2, 3], degradation kinetics can be adjusted by producing gels with different degrees of N-deacetylisation.

34.5
Natural Inorganic Materials

HA of natural origin can be gained from bone of various sources [40, 47–49, 73] or special species of marine corals (Table 34.4) [27, 38–40, 67]. Naturally occurring HAs are not pure, but contain some of the minor and trace elements originally present in bone mineral or coral mineral. Bone-derived apatite contains different elements and ions (magnesium, sodium, carbonate, trace elements) in varying concentrations that are important in bone physiology [37, 39]. Coral-derived apatite or coralline HA is similar in composition with magnesium and carbonate [38, 45, 47–49]. HA is used as scaffold material in various forms. The composition and structure of HA depends on the method of preparation: (1) with the organic matrix, unsintered; (2) without organic matrix, unsintered or (3) without the organic matrix and sintered. The difference between sintered and unsintered HA is that unsintered bone mineral consists of small crystals of bone apatite (carbonate HA), whereas sintered bone mineral consists of much larger apatite crystals without carbonate when heated above 1000°C. The unsintered material containing organic matrix possesses the unique bone-like structure in the context of the three-dimensional macro- and microstructure. The naturally present HAs have an interconnecting macroporosity that enables the ingrowth of cells into the scaffold structure both in vivo and in vitro. The degradation characteristics of HA as well as the cell reactions towards the material depend critically upon the source of the material (Fig. 34.4).

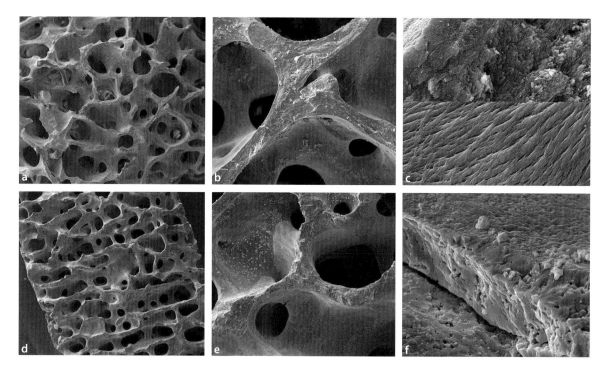

Fig. 34.4a–l Scanning electron microscope view of different hydroxyapatite-based materials. Materials differ in particle size and shape, porosity and surface microstructure. Hydroxyapatite-based biomaterials are commonly used for bone reconstruction approaches

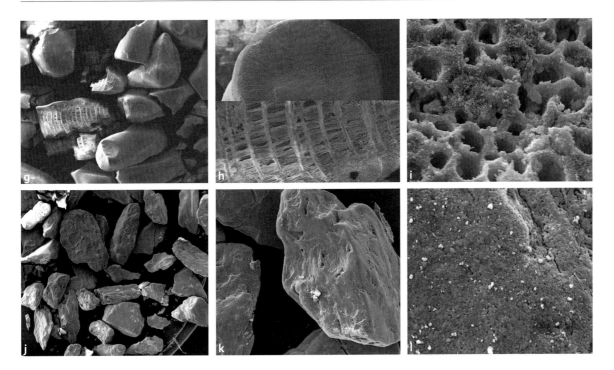

Fig. 34.4a–l (*continued*) Scanning electron microscope view of different hydroxyapatite-based materials. Materials differ in particle size and shape, porosity and surface microstructure. Hydroxyapatite-based biomaterials are commonly used for bone reconstruction approaches

References

1. Athanasiou KA, Shah AR, Hernandez RJ, LeBaron RG (2001) Basic science of articular cartilage repair. Clin Sports Med 20:223–247
2. Berger J, Reist M, Mayer JM, Felt O, Peppas NA, Gurny R (2004a) Structure and interactions in covalently and ionically crosslinked chitosan hydrogels for biomedical applications. Eur J Pharm Biopharm 57:19–34
3. Berger J, Reist M, Mayer JM, Felt O, Gurny R (2004b) Structure and interactions in chitosan hydrogels formed by complexation or aggregation for biomedical applications. Eur J Pharm Biopharm 57:35–52
4. Burg KJ, Holder WD Jr, Culberson CR, Beiler RJ, Greene KG, Loebsack AB, Roland WD, Eiselt P, Mooney DJ, Halberstadt CR (2000a) Comparative study of seeding methods for three-dimensional polymeric scaffolds. J Biomed Mater Res 51:642–649
5. Burg KJ, Porter S, Kellam JF (2000b) Biomaterial developments for bone tissue engineering. Biomaterials 21:2347–2359
6. Burkart AC, Schoettle PB, Imhoff AB (2001) [Surgical therapeutic possibilities of cartilage damage]. Unfallchirurg 104:798–807
7. Chu CR, Coutts RD, Yoshioka M, Harwood FL, Monosov AZ, Amiel D (1995) Articular cartilage repair using allogeneic perichondrocyte-seeded biodegradable porous polylactic acid (PLA): a tissue-engineering study. J Biomed Mater Res 29:1147–1154
8. Elema H, de Groot JH, Nijenhuis AJ, Pennings AJ, Veth RP, Klompmaker J, Jansen HW (1990) Use of biodegradable polymer implants in meniscus reconstruction. 2. Biological evaluation of porous biodegradable implants in menisci. Colloid Polym Sci 268:1082–1088
9. Freed LE, Marquis JC, Nohria A, Emmanual J, Mikos AG, Langer R (1993) Neocartilage formation in vitro and in vivo using cells cultured on synthetic biodegradable polymers. J Biomed Mater Res 27:11–23
10. Friedman CD, Costantino PD, Takagi S, Chow LC (1998) Bone Source hydroxyapatite cement: a novel biomaterial for craniofacial skeletal tissue engineering and reconstruction. J Biomed Mater Res 43:428–432
11. Frondoza CG, LeGeros RZ, Hungerford DS (1998) Effect of bovine bone derived materials on human osteoblast-like cells in vitro. In: LeGeros RZ, LeGeros JP (eds) Bioceramics 11. World Scientific, Singapore, pp 289–291
12. Galletti G, Gogolewski S, Ussia G, Farruggia F (1989) Long-term patency of regenerated neoaortic wall following the implant of a fully biodegradable polyurethane

prosthesis: experimental lipid diet model in pigs. Ann Vasc Surg 3:236–243
13. Gatti A, LeGeros RZ, Monari E, Tanza D (1998) Preliminary in vivo evaluation of synthetic calcium phosphate materials. In: LeGeros RZ, LeGeros JP (eds) Bioceramics 11, World Scientific, Singapore, pp 399–402
14. Gogolewski S, Galletti G (1986) Degradable vascular prosthesis from segmented polyurethanes. Colloid Polym Sci 264:854–858
15. Gogolewski S, Pennings AJ (1982) Biodegradable materials of polylactides. IV. Porous biomedical materials based on mixtures of polylactides and polyurethanes. Macromol Chem Rapid Commun 3:839–845
16. Gogolewski S, Pennings AJ (1983) An artificial skin based on biodegradable mixtures of polylactides and polyurethanes for full thickness wound covering. Macromol Chem Rapid Commun 4:675–680
17. Gogolewski S, Pennings AJ, Lommen E, Nieuwenhuis P, Wildevuur CRH (1983) Growth of a neo-artery induced by a biodegradable polymeric vascular prosthesis. Macromol Chem Rapid Commun 4:213–219
18. Gogolewski S, Galletti G, Ussia G (1987a) Polyurethane vascular prosthesis in pigs. Colloid Polym Sci 265:774–778
19. Gogolewski S, Walpoth B, Rheiner P (1987b) Polyurethane microporous membranes as pericardial substitutes. Colloid Polym Sci 265:971–977
20. Gogolewski S, Gorna K, Rahn B, Wieling R (2001) Biodegradable polyurethane cancellous bone graft substitute promotes bone regeneration in the iliac crest defects. Transactions 24, 573, 27th Annual Meeting, Society for Biomaterials, April 24–29, Saint Paul, MN, USA
21. Gogolewski S, Gorna K, Turner AS (2002) Regeneration of bicortical defects in the iliac crest of estrogen deficient sheep using new biodegradable polyurethane cancellous bone graft substitutes. A pilot study 48th Annual Meeting, Orthopaedic Research Society, February 10–13, Dallas, TX, USA
22. Grad S, Zhou L, Gogolewski S, Alini M (2003) Chondrocytes seeded onto poly (L/DL-lactide) 80%/20% porous scaffolds: a biochemical evaluation. J Biomed Mater Res A 66:571–579
23. Grande DA, Halberstadt C, Naughton G, Schwartz R, Manji R (1997) Evaluation of matrix scaffolds for tissue engineering of articular cartilage grafts. J Biomed Mater Res 34:211–220
24. Gugala Z, Gogolewski S (2000) In vitro growth and activity of primary chondrocytes on a resorbable polylactide three-dimensional scaffold. J Biomed Mater Res 49:183–191
25. Gutowska A, Jeong B, Jasionowski M (2001) Injectable gels for tissue engineering. Anat Rec 263:342–349
26. Höhling, HJ, Arnold S, Barckhaus RH, Plate U, Wiesmann HP (1995) Structural relationship between the primary crystal formations and the matrix macromolecules in different hard tissues. Discussion of a general principle. Connect Tissue Res 33:171–178
27. Holmes RE (1979) Bone regeneration within a coralline hydroxyapatite implant. Plast Reconstr Surg 63:626–633
28. Hubbell JA (2003) Materials as morphogenetic guides in tissue engineering. Curr Opin Biotechnol 14:551–558
29. Hutmacher DW (2000) Scaffolds in tissue engineering bone and cartilage. Biomaterials 21:2529–2543
30. Hutmacher DW, Goh JC, Teoh SH (2001a) An introduction to biodegradable materials for tissue engineering applications. Ann Acad Med Singapore 30:183–191
31. Kamakura S, Sasano Y, Homma H, Suzuki O, Kagayama M, Motegi K (1999) Implantation of octacalcium phosphate (OCP) in rat skull defects enhances bone repair. J Dent Res 78:1682–1687
32. Kawamura H, Ito A, Miyakawa S, Layrolle P, Ojima K, Ichinose N, Tateishi T (2000) Stimulatory effect of zinc-releasing calcium phosphate implant on bone formation in rabbit femora. J Biomed Mater Res 50:184–190
33. Kazimiroff J, Frankel SR, LeGeros RZ (1996) Bone/biomaterial interface: autoradiographic assessment. In: Kokubo T, Nakamura T, Mijaji F (eds) Bioceramics 9, Elsevier, London, pp 169–172
34. Kim WS, Vacanti JP, Cima L, Mooney D, Upton J, Puelacher WC, Vacanti CA (1994) Cartilage engineered in predetermined shapes employing cell transplantation on synthetic biodegradable polymers. Plast Reconstr Surg 94:233–237; discussion 238–240
35. Klompmaker J, Jansen HW, Veth RP, de Groot JH, Nijenhuis AJ, Pennings AJ (1991) Porous polymer implant for repair of meniscal lesions: a preliminary study in dogs. Biomaterials 12:810–816
36. Knott L, Bailey AJ (1998) Collagen cross-links in mineralizing tissues: a review of their chemistry, function, and clinical relevance. Bone 22:181–187
37. LeGeros RZ (1981) Apatites in biological systems. Prog Crystal Growth Charact 4:1–45
38. LeGeros RZ (1988) Calcium phosphate materials in restorative dentistry: a review. Adv Dent Res 2:164–180
39. LeGeros RZ (1991) Calcium phosphates in oral biology and medicine. Monogr Oral Sci 15:1–201
40. LeGeros RZ (1992) Materials for Bone Repair, Augmentation and implant coatings. In: Niwa S (ed) Proceedings of the International Seminar of Orthopedic Research, Nagoya, 1990. Springer-Verlag, Tokyo, pp 147–174
41. LeGeros RZ (2002) Properties of osteoconductive biomaterials: calcium phosphates. Clin Orthop Relat Res (395):81–98
42. LeGeros RZ, Tung MS (1983) Chemical stability of carbonate- and fluoride-containing apatites. Caries Res 17:419–429
43. LeGeros RZ, Trautz OR, LeGeros JP, Klein E, Shirra WP (1967) Apatite crystallites: effect of carbonate on morphology. Science 155(1409–11)
44. LeGeros RZ, LeGeros JP, Trautz OR, Shirra WP (1971) Conversion of monetite, $CaHPO_4$, to apatites: effect of carbonate on the crystallinity and the morphology of the apatite crystallites. Adv Xray Anal 14:57–66
45. LeGeros RZ, Daculsi G, Kijkowska R (1989) The effect of magnesium on the formation of apatites whitlockites. In: Itokawa Y, Durlach J (eds) Magnesium in Health and Disease, J Libbey, New York, pp 11–19
46. Legeros RZ, Orly I, Gregoire M, Daculsi G (1991) Substrate surface dissolution and interfacial biological mineralization. In: Davies JED (ed) The Bone Bio-

material Interface. University of Toronto Press, Toronto, pp 76–88
47. LeGeros RZ, Bautista C, LeGeros JP (1995a) Comparative properties of bioactive bone graft materials. In: Hench L, Wilson-Hench J (eds) Bioceramics 8. Pergamon, New York, pp 81–87
48. LeGeros RZ, Kijkowska R, Bautista C, LeGeros JP (1995b) Synergistic effects of magnesium and carbonate on properties of biological and synthetic apatites. Connect Tissue Res 33:203–209
49. LeGeros RZ, LeGeros JP, Daculsi G, Kijkowska R (1995c) Calcium phosphate biomaterials: preparation, properties and biodegradation. In: Wise DL, Trantolo DJ, Altobelli DE (eds) Encyclopedic Handbook of Biomaterials and Bioengineering. Part 1. Marcel Dekker, New York, pp 1429–1463
50. LeGeros RZ, Sakae T, Bautista C, Retino M, LeGeros JP (1996) Magnesium and carbonate in enamel and synthetic apatites. Adv Dent Res 10:225–231
51. LeGeros RZ, Bleiwas CB, Retino M, Rohanizadeh R, LeGeros JP (1999) Zinc effect on the in vitro formation of calcium phosphates: relevance to clinical inhibition of calculus formation. Am J Dent 12:65–71
52. Marx G (2003) Evolution of fibrin glue applicators. Transfus Med Rev 17:287–298
53. Meyer U, Joos U, Jayaraman M, Stamm T, Hohoff A, Fillies T, Stratmann U, Wiesmann HP (2004) Ultrastructural characterization of the implant/bone interface of immediately loaded dental implants. Biomaterials 25:1959–1967
54. Meyer U, Joos U, Wiesmann HP (2004) Biological and biophysical principles in extracorporal bone tissue engineering. Part I. Int J Oral Maxillofac Surg 33:325–332
55. Michel D, Harmand MF (1990) Fibrin seal in wound healing: effect of thrombin and [Ca2+] on human skin fibroblast growth and collagen production. J Dermatol Sci 1:325–333
56. Monroe EA, Votava W, Bass DB, McMullen J (1971) New calcium phosphate ceramic material for bone and tooth implants. J Dent Res 50:860–861
57. Moreno EC, Kresak M, Zahradnik RT (1977) Physicochemical aspects of fluoride-apatite systems relevant to the study of dental caries. Caries Res 11:142–171
58. Ohgushi H, Caplan AI (1999) Stem cell technology and bioceramics: from cell to gene engineering. J Biomed Mater Res 48:913–927
59. Ohgushi H, Miyake J, Tateishi T (2003) Mesenchymal stem cells and bioceramics: strategies to regenerate the skeleton. Novartis Found Symp 249:118–127; discussion 127–132, 170–174, 239–241
60. Okazaki M, LeGeros RZ (1992) Crystallographic and chemical properties of Mg-containing apatites before and after suspension in solutions. Magnes Res 5:103–108
61. Okazaki M, Matsumoto T, Taira M, et al (1998) CO_3apatite preparations with solubility gradient: potential degradable biomaterials. In: LeGeros RZ, LeGeros JP (eds), Bioceramics 11. World Scientific, Singapore, pp 85–88
62. Pavesio A, Abatangelo G, Borrione A, Brocchetta D, Hollander AP, Kon E, Torasso F, Zanasi S, Marcacci M (2003) Hyaluronan-based scaffolds (hyalograft C) in the treatment of knee cartilage defects: preliminary clinical findings. Novartis Found Symp 249:203–217; discussion 229–233, 234–238, 239–241
63. Reddi AH (2000) Morphogenesis and tissue engineering of bone and cartilage: inductive signals, stem cells, and biomimetic biomaterials. Tissue Eng 6:351–359
64. Saad B, Neuenschwander P, Uhlschmid GK, Suter UW (1999) New versatile, elastomeric, degradable polymeric materials for medicine. Int J Biol Macromol 25:293–301
65. Sanchez C, Arribart H, Guille MM (2005) Biomimetism and bioinspiration as tools for the design of innovative materials and systems. Nat Mater 4:277–288
66. Shin H, Jo S, Mikos AG (2003) Biomimetic materials for tissue engineering. Biomaterials 24:4353–4364
67. Shors EC, Holmes RE (1993) Porous Hydroxyapatite. In: Hench LL, Wilson Hench J (eds) An Introduction to Bioceramics. World Scientific, London, pp 181–193
68. Sittinger M, Bujia J, Rotter N, Reitzel D, Minuth WW, Burmester GR (1996) Tissue engineering and autologous transplant formation: practical approaches with resorbable biomaterials and new cell culture techniques. Biomaterials 17:237–242
69. Thomas S (2000) Alginate dressings in surgery and wound management—Part 1. J Wound Care 9:56–60
70. Toquet J, Rohanizadeh R, Guicheux J, Couillaud S, Passuti N, Daculsi G, Heymann D (1999) Osteogenic potential in vitro of human bone marrow cells cultured on macroporous biphasic calcium phosphate ceramic. J Biomed Mater Res 44:98–108
71. Uemura T, Dong J, Wang Y, Kojima H, Saito T, Iejima D, Kikuchi M, Tanaka J, Tateishi T (2003) Transplantation of cultured bone cells using combinations of scaffolds and culture techniques. Biomaterials 24:2277–2286
72. Vacanti CA, Langer R, Schloo B, Vacanti JP (1991) Synthetic polymers seeded with chondrocytes provide a template for new cartilage formation. Plast Reconstr Surg 88:753–759
73. Valentini P, Abensur D, Wenz B, Peetz M, Schenk R (2000) Sinus grafting with porous bone mineral (Bio-Oss) for implant placement: a 5-year study on 15 patients. Int J Periodontics Restorative Dent 20:245–253
74. Wiesmann HP, Meyer U, Plate U, Höhling HJ (2005) Aspects of collagen mineralization in hard tissue formation. Int Rev Cytol 242:121–156
75. Yilmaz E (2004) Chitosan: a versatile biomaterial. Adv Exp Med Biol 553:59–68
76. Zapanta-LeGeros R (1965) Effect of carbonate on the lattice parameters of apatite. Nature 206:403–404

Biomaterial-Related Approaches: Surface Structuring

G. Jell, C. Minelli, M. M. Stevens

Contents

35.1	Introduction	469
35.1.1	Mimicking the Extracellular Matrix	469
35.1.2	Micromechanical Surface Properties	470
35.2	Chemical Modification of Material Surfaces	471
35.2.1	Influence of Chemistry on Protein Adsorption	471
35.2.2	Influence of Chemistry on Cell Behaviour	473
35.2.3	Engineering Surface Chemistry	474
35.3	Biofunctionalisation of Material Surfaces	476
35.3.1	Chemical and Biochemical Patterning of Surfaces	477
35.4	Topographical Modification of Material Surfaces	478
35.4.1	Influence of Topography on Protein Adsorption and Cell Behaviour	478
35.4.2	Engineering Surface Topography	481
35.5	Conclusions	482
	References	482

35.1 Introduction

Engineering the surface of materials to promote cell adhesion and direct cellular behaviour is vital for the development of materials capable of restoring, replacing and/or enhancing tissue function. Cells are inherently sensitive to physical, biochemical and chemical stimuli from their surroundings. In vivo, the local cell environment or "niche" provides specific environmental cues that determine cell-specific recruitment, migration, proliferation, differentiation and the production of the numerous proteins needed for hierarchical tissue organisation. However, when cells are cultured in vitro or when materials are implanted into the body, cells encounter very different, unfamiliar surfaces and environments (Figs. 35.1 and 35.2). This chapter will introduce various approaches to modifying these unfamiliar material surfaces to promote desirable cell responses.

35.1.1 Mimicking the Extracellular Matrix

In vivo, cells are surrounded by a biological matrix comprising of tissue-specific combinations of insoluble proteins (e.g. collagen and elastin), glycosaminoglycans (e.g. hyaluronan) and inorganic crystals (in bone) that are collectively referred to as the extracellular matrix (ECM). The varied composition of the ECM components not only provides not the physical architecture and mechanical strength to the tissue, but also contains a reservoir of cell-signalling motifs (ligands) and growth factors that guide cellular anchorage and behaviour. The spatial distribution

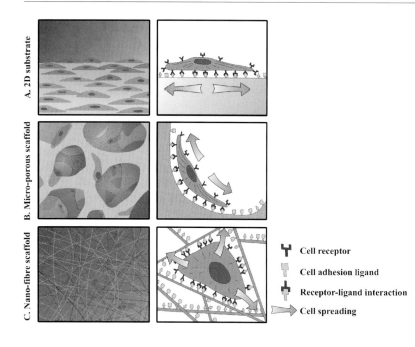

Fig. 35.1a–c The influence of two-dimensional (2D) and three-dimensional (3D) environments on cell behaviour. 2D substrates (e.g. cell culture plastic; (**a**)) and 3D microporous tissue scaffolds (**b**), have very different influences on cell behaviour when compared to nanostructured environments, such as extracellular matrix (ECM) or nanofibrous scaffolds (**c**). Nanostructured environments provide, topographical, biological and mechanical inputs in three dimensions, which can dramatically affect cell spreading/contraction and associated intracellular signalling. Most cells in vivo require cues from a surrounding 3D environment to form functional tissue. This figure was adapted and reprinted with permission from Stevens and George (2005) [71]. Full-stops missing from all figure legends

and concentration of ECM ligands, together with the tissue-specific topography and mechanical properties (in addition to signals from adjacent cells—juxtacrine signalling—and from the surrounding fluid), provide signalling gradients that direct cell migration and cellular production of ECM constituents. In this dynamic environment, the bidirectional flow of information between the ECM and the cells mediates gene expression, ECM remodelling and ultimately tissue/organ function.

Native ECM exhibits macroscale to nanoscale patterns of chemistry and topography [71], and it is therefore somewhat unsurprising that cells respond to these various scales of chemically and/or topographically patterned features [2, 29, 73]. Most in vitro studies, however, have explored cell responses to topographical or chemically patterned features on flat, two-dimensional (2D) surfaces and concentrated less on how these cellular responses translate to the three-dimensional (3D) environments found in native tissues [35, 42, 52, 77]. Intimate contact between cells and the surrounding 2D or 3D environment alters the mechanical stresses placed on the cell and leads to changes in cell morphology and function (Fig. 35.1) [38, 52, 71].

35.1.2 Micromechanical Surface Properties

The microscale mechanical properties of a substrate (in addition to the bulk mechanical properties) are critical for determining cell behaviour, including directing stem cell differentiation, cell migration and tissue growth [30, 32, 66, 72]. Cells not only adhere to surfaces, but also "pull" on the surface substrate (and adjacent cells). The relative substrate resistance or "give" encountered activates various mechanotransduction and cellular pathways, which in turn trigger gene expression. In vivo micromechanical stimuli are important environmental cues that enable cell attachment, migration and organogenesis. Matching the mechanical surface properties of the tissue scaffold to the specific mechanical characteristics of the specific tissue site is therefore vital for controlling cell behaviour.

Modifying material surface structures to mimic aspects of the ECM in order to provide the tissue-specific cues to direct cell behaviour and trigger tissue regeneration, is an important aspect of many tissue-engineering strategies. Material surface prop-

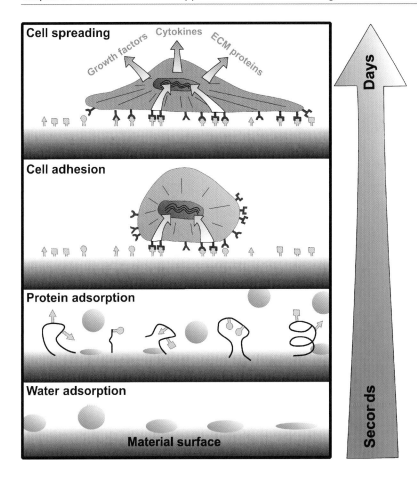

Fig. 35.2 A time line of the biological response to material surfaces. Upon implantation into a biological system, water molecules interact with the material's surface, closely followed by protein adsorption. Material surface-bound proteins provide the recognition sites that enable cell adhesion via specific cell receptors (e.g. integrins). The chemistry and topography of the surface affects the abundance, conformation, orientation and distribution of the surface-bound protein. These factors are critical for cellular recognition of protein binding sites, attachment and subsequent behaviour [5, 12, 27]. For example, the adsorbed protein may be orientated in such a manner, or undergo conformational changes that prevent cell receptor recognition [64]. Anchorage to a substrate is vital for cell survival for most cell types. Surface-bound proteins, cell-membrane bound receptors and cytoplasmic proteins form focal adhesion points. These focal adhesion points interact with the cytoskeleton to induce signal transduction, gene expression, and subsequently protein production and ECM remodelling. Cell adhesion and behavioural adaptation to a material surface can occur in a matter of minutes, hours or days depending on the cell type, substrate and environment

erties can be modified by physicochemical modification, biofunctionalisation and/or topographical surface structuring. Methods for implementing these modifications and their impact on protein adsorption and cell behaviour are discussed in detail within this chapter.

of biofunctional specificity exhibited on the material surface depends on the level and complexity of the surface-modification approach utilised (Fig. 35.3) [10]. The mechanisms underlying protein and cell response to various surface chemistries, and the technologies used to engineer these chemistries are discussed in the section below.

35.2 Chemical Modification of Material Surfaces

35.2.1 Influence of Chemistry on Protein Adsorption

Controlling the surface chemistry of materials enables us, to some extent, to dictate the rate of protein adsorption, the functionality of adsorbed proteins and subsequent cell adhesion (Fig. 35.2). The degree

Protein adsorption occurs through a complex series of events that depends upon both the material's surface properties and the nature of the individual protein.

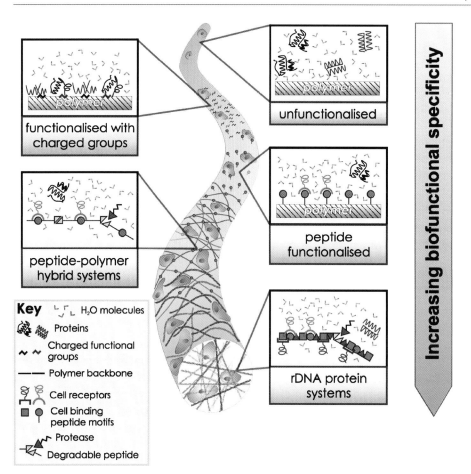

Fig. 35.3 Enhancing material biofunctionality. Several different examples of polymer surface engineering are presented here, ordered by increasing biofunctional specificity. Unmodified polymer surfaces non-specifically adsorb proteins through weak interactions between the protein–water and water–surface interfaces. Chemical modification of the polymer scaffold conferring charged end-groups (e.g. –OH$^-$, –S$^-$, –COO$^-$, NH^{3+}) to its surface may lead to protein adsorption and structural rearrangements via electrostatic interactions. Increased biofunctionality can be achieved by attaching specific peptide motifs, such as RGD (arginine-glycine-aspartic acid), which can bind to cell receptors (e.g. integrins such as αvβ3) and induce "firm" cell anchorage. The scaffold can be further enhanced by including biologically recognised and responsive peptide sequences (e.g. protease-sensitive degradation sites) that control scaffold degradation and cell migration. Finally, larger biologically relevant proteins can be produced using recombinant DNA technology and incorporated into the scaffold. This figure was reprinted with permission from Bonzani et al. (2006) [10]

When a protein binds to a surface, its structure can become distorted due to some regions of the protein possessing a higher affinity to the surface than others. Proteins bound to surfaces can therefore exhibit changes in their biological activity.

A protein approaching a material's surface is subjected to electrostatic and Van der Waals forces (e.g. forces involving the polarisation of uncharged molecules), both of which can affect protein adsorption. Nonetheless, protein adsorption depends primarily on the interplay between water molecules with the material's surface. It is generally accepted that hydrophobic surfaces can tightly adsorb, but also distort, proteins, rendering them inactive. Conversely, hydrophilic surfaces typically resist the adsorption of proteins but preserve their original conformation and thereby bioactivity. This behaviour originates from the polar nature of water molecules and their tendency, in bulk, to self-associate via transient hydrogen bonds forming an interconnected 3D network [75]. The self-association between water molecules is promoted close to purely non-polar surfaces because of the limited number of neighbouring molecules and polar sites available to bind via hydrogen bonds.

Water molecules in proximity to a polar substrate, however, bind strongly onto its surface and water self-association is disrupted (Fig. 35.4a). The energy required for the displacement of such molecules is prohibitive to proteins approaching hydrophilic surfaces. Hence, proteins in the proximity of hydrophilic surfaces are generally only trapped within bound water layers. On non-polar substrates, however, proteins can easily displace water molecules and adsorb onto the surface (Fig. 35.4b). Here, the non-polar sites in the protein associate with the hydrophobic surface, while the water molecules tend to maximise the number of their hydrogen bonds and therefore reduce their contact with non-polar interfaces. As a consequence, proteins remain in intimate contact with non-polar surfaces. The chemistry of a material's surface can be such as to allow the formation of further bonds between a protein and a surface, inducing unfolding and conformational changes of the protein. Protein unfolding can cause the exposure of new cell-binding sites or conversely distort cell-binding sites, thereby inhibiting biofunctionality.

35.2.2
Influence of Chemistry on Cell Behaviour

A cell approaching a surface experiences electrostatic, Van der Waals and steric stabilisation forces, the magnitude of which varies with the distance between the cell and the surface. The electrostatic forces can be either attractive or repulsive depending on the charge associated with the substrate surface, the cell surface carrying a net negative charge [31]. The Van der Waals forces are always attractive, whilst the steric stabilisation forces are always repulsive. The steric stabilisation forces originate from the osmotic imbalance in the gap between the cell and the surface, where the water molecules have been forced out. However, cell adhesion that relies only on these unspecific interactions is fairly weak.

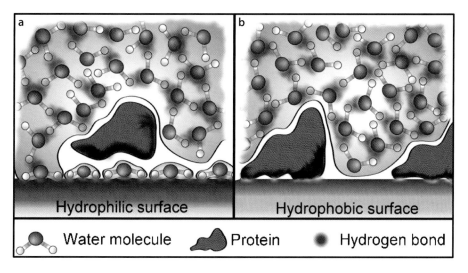

Fig. 35.4a,b Schematic of the interaction between water molecules and surfaces. **a** Water molecules tightly bind onto polar (hydrophilic) surfaces, disrupting water self-association. The proteins in proximity of these surfaces are trapped within bound water layers. **b** Water molecules tend to reduce their contact with non-polar (hydrophobic) interfaces, compelling proteins to associate with non-polar surfaces via hydrophobic interactions

Receptor-ligand binding is vital for the specific adhesion of the cell to a material. Ligand binding sites for the cell can be provided by proteins on the substrate surface, or by specific functional groups on the material surface.

35.2.3
Engineering Surface Chemistry

Numerous materials have been used in tissue engineering and regenerative medicine applications and a tremendous variety of surface structuring techniques have been developed. In general, the techniques used to chemically or biochemically structure the surface of materials can be divided in three groups (Table 35.1): (1) techniques that lead to the modification of the original material's surface, for example via implantation of new atoms, or via oxidation of the outermost atomic layers; (2) techniques that involve the deposition and adsorption of non-covalently linked molecules or biomolecules onto the materials' surface, producing a coating layer; (3) techniques that lead to the formation of a covalently linked coating layer formed by molecules (e.g. silanes, thiols) or biomolecules. Ideally, the surface modification of a material should not affect its bulk or functional properties. For example, surface coatings should be thin enough to provide the cells with the proper mechanical support granted by the underlying substrate, but not so thin as to affect the uniformity, durability or functionality of the interface.

Table 35.1 Physicochemical techniques commonly utilised to modify material surfaces (adapted from Ratner et al. (2004) [63])

Strategy	Technique	Materials
Modification of the original surface	Ion-beam implantation	Metals, ceramics, glasses
	Plasma etching	Polymers, metals, ceramics, glasses
	Corona discharge	Polymers, metals, ceramics, glasses
	Ion exchange	Polymers, metals, ceramics, glasses
	Ultraviolet irradiation	Polymers, metals, ceramics, glasses
	Oxidation	Polymers, metals, ceramics, glasses
	Surface chemical reaction	Polymers
	Conversion coatings	Metals
Non-covalent coatings	Solvent coatings	Polymers, metals, ceramics, glasses
	Langmuir-Blodgett film deposition	Polymers, metals, ceramics, glasses
	Surface-active additives	Polymers, metals, ceramics, glasses
	Vapour deposition	Polymers, metals, ceramics, glasses
Covalently attached coatings	Radiation grafting	Polymers
	Photografting	Polymers, glasses
	Plasma-assisted techniques	Polymers, metals, ceramics, glasses
	Ion-beam sputtering	Polymers, metals, ceramics, glasses
	Chemical vapour deposition	Metals, ceramics, glasses
	Chemical grafting	Polymers, metals, ceramics, glasses
	Silanisation	Polymers, metals, ceramics, glasses
	Biomolecule immobilisation	Polymers, metals, ceramics, glasses

A critical aspect of a material's functional surface is its behaviour when immersed in water or physiological solutions. Ionisable functional groups such as acids and amines are affected by environmental pH and ionic concentration. The net charge at the material's surface will depend on these two parameters and in turn influence adhering proteins. Furthermore, surface coatings can be prone to swelling and delamination. The covalent binding of the surface coating, direct or via a chemical interlayer, reduces the risk of delamination. Cell and protein interactions with a material surface must therefore be studied in the context of an appropriate simulated in vivo environment. Such considerations are of critical importance for the engineering and validation of materials for tissue engineering.

35.2.3.1
Ion-Beam Implantation

Ion beam implantation can be used to modify the original surface of polymers, metals, ceramics and glasses. Accelerated ions are injected into a material to modify its surface properties and confer a higher resistance to wear and corrosion, as well as lubricity, toughness, conductivity and bioactivity. From the point of view of cell-material interface structuring, the controlled implantation of cations or anions into a surface can reproducibly alter surface charge and chemistry. The technique has been reported to enhance the clinical performance of titanium orthopaedic implants, improving for example their biocompatibility and enhancing growth of bone tissue in vivo [37, 53].

35.2.3.2
Chemical Reactions

The surface of biomaterials can be modified via chemical reactions, to confer upon the materials precise surface energy, charge or functional groups in order to tailor their levels of interaction with proteins and cells. The surfaces can be pretreated in to display reactive chemical groups and guarantee the covalent binding of a functional coating. For example, the radiation grafting of chemical groups such as –OH, –COOH and –NH_2 onto the surface of relatively inert polymers has been used extensively to modify the surface of biomaterials [40, 62]. Energy sources like ionising radiation sources, ultraviolet radiation and high-energy electron beams are used to break the chemical bonds at the surface of the material to be grafted, allowing the formation of free radicals and other reactive species. The surface is then exposed to, and reacts with, molecules that will form the surface functional coating of the material.

35.2.3.3
Silanisation

A typical liquid-phase chemical modification is silanisation, which involves reactions of silane molecules with hydroxylated glass, silicon, alumina or quartz surfaces. Silanisation can produce stable covalently linked coatings, although silane polymerisation in the presence of water molecules can lead to the formation of thick irregular layers. Silane molecules can be synthesised with precise chemical groups at their exposed free ends. The silanisation of a material's surface is therefore a relatively fast and simple method by which to tune the chemical properties of a surface.

35.2.3.4
Self-Assembled Monolayers

Self-assembled monolayers (SAMs) are molecular assemblies that form spontaneously by the adsorption of an active surfactant onto a solid [74]. They are characterised by a highly ordered and stable structure due to the firm anchoring of their molecules to the substrate (e.g. via thiol groups on gold, silanes on glass, etc.) and to the Van der Waals interactions among the molecular chains. SAMs are inherently simple to manufacture and thus technologically attractive for surface engineering. Alkanethiols on gold are among the most well known SAMs used to modify the surfaces of materials and have been used as model systems for studying protein adsorption [51, 57]. While the thiols provide firm binding of the molecules to the gold, the free-end groups of the alkyl chains can be functionalised in order to exhibit, for example, hydrophobic or hydrophilic terminations (typically CH_3 and OH respectively), non-fouling polymer short chains (e.g. ethylene glycol) or polysaccharide groups. Other examples of SAMs are alkyl-silanes on glass, alkanethiols on metals [55] and fatty acid molecules on aluminium oxide [1]. Moreover, some classes of molecules such as proteins, porphyrins, nucleotide bases and aromatic ring hydrocarbons can also form SAMs without the presence of an alkyl chain [4].

35.2.3.5
Plasma-Assisted Techniques

Radiofrequency glow discharge plasma-induced surface modification processes use low-pressure ionised gases for ablation, etching or coating of a material's surface. Controlling the plasma deposition conditions allows tuning of the functional group density at the material surface and therefore the level of interaction between the material and the biological tissue. These techniques offer the possibility of applying a well-adhered and uniform coating to flat or complex 3D surfaces. Among the plasma-assisted techniques, plasma polymerisation [34] is generating interest in the biomaterials community due to the wide range of materials that can be used as substrates (e.g. polyethylenes, polystyrene, polyurethanes, ceramics, cellulose) and the ability to generate uniform polymer coatings on (flat surface) fabrics and porous materials. The polymer coatings created with this technique have unique properties and are difficult to synthesise using conventional chemical methods. Recently, silver-containing polymer films have been engineered with plasma-assisted deposition processes and proposed as coatings with inherent long-term resistance to bacterial adhesion and colonisation [6, 43].

35.3
Biofunctionalisation of Material Surfaces

Artificial materials can be endowed with precise biological functionalities by immobilising bioactive molecules such as enzymes, peptides and proteins, drugs and cell receptor ligands onto their surfaces. These biomolecules can be simply adsorbed onto the material's surface, or covalently linked via chemical groups previously created on the surface. The biological response following the surface biomodification of a material depends on structural parameters, such as the density of the ligands, their spatial distribution, their steric hindrance and their colocalisation with synergistic ligands [69]. A common strategy used to confer the required flexibility to cell binding sites is the insertion of bioinert spacers (e.g. low-molecular-weight polyethylene glycol [39, 68] and a GGGG peptide sequence [49]) between bioactive peptide ligands and the surface.

The functionalisation of a material surface via adsorption or chemical binding of ECM elements is a common approach utilised to promote cell adhesion. For example, a popular research strategy employed to improve the blood-compatibility properties of vascular grafts comprises seeding endothelial cells onto their surfaces [59]. The complete coverage of the biomaterial's surface by these cells (endothelialisation) inhibits thrombosis, prevents intimal hyperplasia, and thereby increases the patency of vascular grafts. Several bioresponsive peptides, proteins and growth factors can be incorporated into biomaterial surfaces to alter protein adhesion, endothelial or endothelial progenitor cell recruitment and endothelialisation. For example arginine-glycine-aspartate (RGD) motifs have been immobilised onto the material surface in an attempt to mimic the cell-binding domains of the ECM. The RGD peptide sequence has been reported to increase the adhesion and spreading of human endothelial cells relative to smooth muscle cells, fibroblasts or blood platelets [41], whilst the peptide surface density controls cellular response and behaviour. Surface functionalisation via immobilisation of bioactive molecules thus enables the creation of materials that have selective cell adhesion and can spatially organise cells.

35.3.1
Chemical and Biochemical Patterning of Surfaces

The chemical and biochemical patterning of surfaces provides a means of controlling cellular spatial organisation. This is shown elegantly in the example of Fig. 35.5, where mouse melanoma cells have been cultured onto patterned fibronectin (a protein responsible for cell binding) substrata [46]. Microcontact printing [51] was used to create spots of $-CH_3$-terminated alkanethiols on gold-coated glass, while the rest of the surface was covered by ethylene-glycol-terminated alkanethiols. The surfaces were then incubated in a protein solution containing fibronectin (red), which readily adsorbed on the hydrophobic spots. Fluorescence microscopy performed after cell culture revealed cell actin filaments (green) distributed throughout the cell periphery on a homogeneous substratum (Fig. 35.5a) and patterned surfaces exhibiting fibronectin spots less that 2 µm apart (Fig. 35.5b and c). In contrast, the actin cytoskeleton of the spreading cells forms clearly defined stress fibres between adjacent dots when their spacing is between 10 and 20 µm (Fig. 35.5d–f). Cell spreading is limited

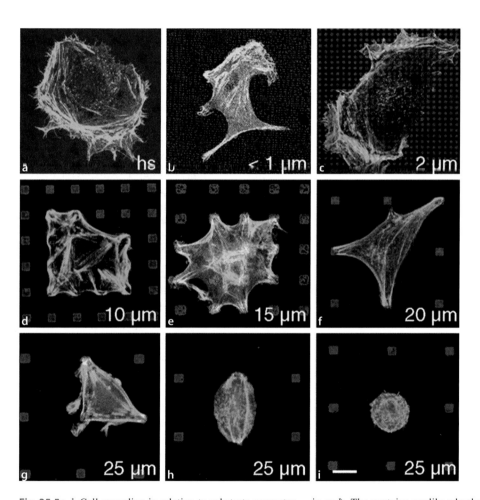

Fig. 35.5a–i Cell spreading in relation to substrate geometry. Mouse melanoma cells cultured on fibronectin patterned substrata. Micro-contact-printing was used to create squares of self-assembled monolayers (SAMs) of hydrophobic alkanethiols on gold, while the uncoated gold surface was covered with SAMs of hydrophilic alkanethiols. The patterned surface was exposed to a protein solution containing fibronectin (labelled in *red*). The proteins readily adsorbed onto the hydrophobic patterns. Cell actin filaments were labelled with a green fluorescent dye. Cells cultured on such substrates were influenced by the pattern geometry at the surface. The spacing between hydrophobic spots is noted in the bottom right corner of **b–i**. Scale bar: 10 µm. *hs* Homogeneous substrate. This figure was reprinted with permission from Lehnert et al. (2004) [46]

once the distances between spots are of the order of 25 μm, and the cell assumes a triangular, ellipsoid or round shape (Fig. 35.5g–i).

In the previous example, patterns of alkanethiol SAMs with different surface end-groups were shown to direct protein adsorption via hydrophobic interactions. Other techniques can also be used to create specific surface chemical patterns with similar end goals. For example, patterned bimetal surfaces comprising a combination of titanium, aluminium, vanadium and niobium have been used to study protein adsorption and the adhesion of primary human osteoblasts [67]. Single metal substrates with a homogeneous surface were used as a control. The patterned substrates were fabricated by photolithographic techniques and exhibited at their surface six spatially patterned regions, each characterised by an array of dots or stripes with 50, 100 and 150 μm diameter/width. Both serum protein and cell adhesion depended on the material (with a marked preference for the non-aluminium surfaces) and the cell alignment relative to pattern geometry was detectable after 2 h and fully developed after 18 h of incubation. In another example, the spatial organisation of neuronal cells, together with their differentiation, was controlled by pattern-specific cell binding receptors on a fluorinated ethylene propylene (FEP) surface [61]. A pattern of reactive hydroxyl groups was first created on a homogeneous FEP surface utilising a radiofrequency glow discharge process. Secondly, the FEP surface was exposed to oligopeptide sequences derived from laminin (an ECM molecule that promotes neural cell attachment and differentiation), which covalently bound to the hydroxyl groups. Finally, neuronal cells were cultured on such surfaces in vitro. Cell attachment and neurite extension was localised to the oligopeptide pattern, whereas the surrounding unmodified FEP inhibited the adsorption of proteins from the culture media and, therefore, cell attachment.

35.4
Topographical Modification of Material Surfaces

For over 40 years it has been known that cell behaviour is influenced by microscale surface topographical features [13, 16]. More recently, it has become clear through technological advances in material processing and sensitive analytical techniques, that nanotopography also influences protein adsorption and cell behaviour [7, 24, 29, 70, 79]. Indeed, cellular behaviour is determined not only by the size of feature [7], but also feature shape [17, 24] and the geometric/spatial arrangement of the features (i.e. the frequency and randomness of the structures) [19, 25, 26].

35.4.1
Influence of Topography on Protein Adsorption and Cell Behaviour

Cellular adhesion, orientation and ECM production have been investigated in response to a wide variety of surface features, such as ridges, grooves, pits, ribbons, islands, spikes and single cliffs [8, 11, 15, 28, 29, 33, 50, 76]. An important outcome of these studies is that the scale and spatial arrangement of the topographical features (see section 35.3.1) influences cell behaviour (Table 35.2).

The topographical scale of a scaffold designed for tissue regeneration will influence protein adsorption and cell behaviour. Nanoscale topographical features, for example have relatively large surface areas compared to microscale features, thereby allowing greater protein adsorption. Protein adsorption and conformation, and consequently cell behaviour may also be influenced by an asymmetric distribution of surface chemistries and charges, which are intrinsically associated with nanostructures (e.g. the high charge density associated with sharp projections). The importance of topographical scale on protein adsorption can be demonstrated by observing albumin and fibrinogen adsorption onto nanosized spheres of various diameters (and thereby curvatures) [65]. In this study, different-shaped proteins, albumin (a small globular protein) and fibrinogen (a rod-like protein), were found to have differing adsorption profiles and conformations depending upon sphere size [65]. Cell filopodia (nanosized cellular extensions that appear to form an integral part of the system through which cells "sense" topographical features; Fig. 35.6) also appear to increase in number when cells encounter a nanometrically featured surface (compared to a microsized or flat surface) [26, 56].

Table 35.2 Techniques commonly utilised to modify material surface topography. *IL* Interleukin

Method	Disordered/ordered spatial pattern	Type of feature	Example of biological response
Polymer demixing	Disordered	Pits, bands or ribbons	Generally decrease cell adhesion and spreading [20, 21, 23]
Photolithography	Both	Pits, grooves and ridges	Reduced cell spreading [2]
Electron-beam lithography	Both	Pits, grooves and ridges	Reduced cell spreading [9], enhanced osteoblast differentiation depending upon disorder pattern [26]
Colloidal lithography	Disordered	Pillars and pits	Reduced cell spreading, decreased IL-6 and IL-8 production [2, 36]
Nanoprinting	Both	Pits, grooves and ridges	Anisotropic morphological behaviour [47]
Laser machined	Both	Pits, grooves and ridges	Nanoscale no effect, microscale caused cytoskeletal changes [7]
Dip coating	Both	Ridges, bands and pillars	Unpublished data
Embossing	Both	Pits and grooves	Increased inflammatory response (IL-6 and IL-1 production) [36]
Sandblasting	Disordered	Various roughness	Increased osteoblast adhesion and decreased fibroblast adhesion [44]
Grinding	Disordered	Various roughness	Enhanced osteoblast differentiation on rough surfaces [48]
Acid etching	Disordered	Various roughness	Enhanced bone nodule formation in vitro [44]

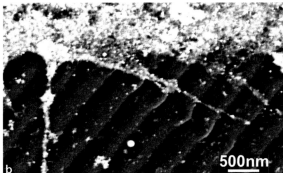

Fig. 35.6a,b Scanning electron microscopy images of osteoblast fillipodia "exploring" local environments when cultured on tissue culture plastic (a) and laser cut nanogrooved Kapton (polyimide polymer; b). Fillipodia interact with other cells (a) and substrate features (b) Fig.35.6a,b was kindly obtained from M. Ball and R. Sherlock (National Centre for Laser Applications, National University of Ireland, Galloway)

Cellular response to various topographical surface features may be governed not only by the differing protein adsorption profiles (caused by the scale, shape and spatial location of the topographical features), but also by physical interactions with those topographical features. Different surface structured features are likely to cause complex cellular physical stresses, resulting in differential cytoskeletal tensions, which in turn activate various mechanotransductive processes and gene expression. Indeed, the elastic modulus of osteoblast cells (as determined by atomic force microscopy) has been shown to increase when cultured on polymer demixed nanotopographic surfaces (11- to 38-nm-high islands) compared to flat, control surfaces. The impact of some common surface structures and their spatial arrangement on cell behaviour is discussed below.

35.4.1.1
Roughness

Surface roughness is often quantified by R_a, which describes the average distance between surface peaks and valleys. The term "roughness" does not, however, define the surface feature shape, and two surfaces with similar R_a can appear very different. Surface roughness can be created by acid etching, grit blasting and grinding (Table 35.2) [3, 27, 45, 50]. These methods not only generate distinct surface topographies, but also induce changes in surface chemistry [58, 80]. The various feature shapes and surface chemistries of materials with similar R_a values may explain the often conflicting reports on the effect of roughness on cell behaviour [8]. In the area of bone research, for example, whilst the majority of reports have shown that surface roughness increases desirable osteoblast behaviour and osseointegration of implants [27, 80], others have reported reduced osteoblast adhesion and undesirable behaviour [3]. The majority of features generated by machining techniques are on the microscale, but nanometrically deep features have also been produced (by reducing the dry etch time) [76] and been shown to increase the adhesion and orientation of fibroblastic, endothelial, epithelial and macrophage cell types [76].

35.4.1.2
Grooves, Steps and Ridges

The alignment of many different cell types along grooves, steps and ridges has been reported [11, 14, 22, 47, 60, 79]. The phenomenon of cell alignment, called contact guidance, depends upon several parameters including groove depth, groove and ridge width, and cell type [17]. The sensitivity of cells to groove dimensions is illustrated by how two different types of neuronal cells (embryonic *Xenopus* spinal cord neurons and rat hippocampal neurons) grown on quartz etched with a series of parallel grooves, orientated themselves depending upon groove depth. *Xenopus* neurites grew parallel to grooves as shallow as 14 nm, whilst hippocampal neurites grew parallel to deep, wide grooves but perpendicular to shallow, narrow ones [60]. Furthermore, the authors reported that topography not only provided important neurite guidance information, but that substratum topography is also a potent morphogenetic factor for developing central nervous system neurons [60]. Macrophage behaviour has also been shown to be guided by grooves of various scale; 30 nm grooves caused parallel macrophage alignment, but only deeper grooves (71 nm) modified phagocytic ability [76].

35.4.1.3
Random Versus Ordered Geometric Arrangement

In vivo, the ECM of the different tissues displays various degrees of spatial topographical order (e.g. regular-patterned inorganic crystal distribution between collagen fibres in bone). Biomaterials, however, often present a more random, chaotic topography (and surface chemistry). For example, the average plastic culture dish has a random spatial arrangement of ridges approximately 10 nm high on the culture surface [17]. Similarly polished metal surfaces, such as the "ball" on a ball-and-socket hip joint has a scratched surface with grooves and ridges between 20 and 50 nm deep [17]. The degree of spatial disorder has been shown to affect cellular behaviour [8, 19, 25, 26]. Cells respond differently to surfaces with ordered topography compared to surfaces with disordered topography of a similar R_a value. For example, ordered surface arrays (orthogonal or hexagonal) of nanopits in polycaprolactone or polymethylmethacrylate reduce cell adhesion compared with disordered arrays or planar surfaces [19]. Similarly, a certain degree of nanoscale disorder has been shown to stimulate human mesenchymal stem cells to produce bone mineral in vitro, in the absence of osteogenic supplements (Fig. 35.7) [9].

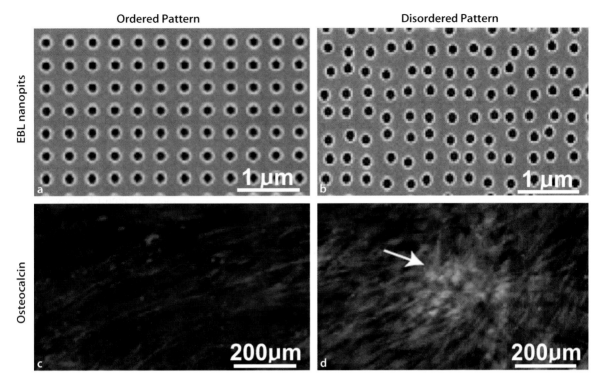

Fig. 35.7a–d Random versus ordered symmetry. Nanotopographies of increasing disorder were fabricated by electron beam lithography (**a** and **b**). The pits generated (120 nm in diameter and 100 nm deep) were in a square arrangement (**a**) with increasing disorder (**b**), displaced square ± 50 nm from true centre). Nanoscale disorder stimulated human mesenchymal stem cells to increase the expression of the bone-specific ECM protein osteopontin (**d**, *arrow*) compared to the ordered structure (**c**). Adapted with permission from Dalby et al. (2007) [26]

35.4.2
Engineering Surface Topography

Numerous topographical surface-structuring techniques (used either singly or in combination with other surface-structuring techniques) and their interactions with a variety of cellular processes have been studied (Table 35.2). Technological developments that have enabled the manufacture of surfaces with controllable topographical features ranging from the micro- to nanoscale have enabled a greater systemic investigation and understanding of their interaction with proteins and cells. Some of the current and commonly used topographical surface-structuring techniques (lithography and polymer demixing) are discussed below.

Lithography encompasses a variety of techniques that are widely used to create various surface topographies [54]. Photolithography is a technique commonly used for the fabrication of microscale topographies. This process involves coating a material (often a silicon wafer) with a photoresist, over which a mask with a desired pattern is placed, followed by exposure to a near-ultraviolet light source. The mask selectively allows light through to the wafer, thereby recreating the pattern in the photoresist. At present, photolithography can reliably produce features of 1–2 μm, but is not capable of nanoscale structures. In contrast, electron-beam lithography (EBL) has high-resolution capabilities with single 4-nm beam diameter features possible (although the feature diameter increases to typically over 40 nm for the large surface arrays necessary for biological research). EBL involves the use of high-energy electrons to expose an electron-sensitive resist and, unlike photolithography, a physical mask is not needed to pattern the surface. Fabrication using EBL can be time consuming and costly. EBL is therefore often used to produce a nanofeatured patterned master that can then be replicated with a curable polymer, such as polydimethylsiloxane. Colloidal lithography provides an alternative and inexpensive method for creating controlled nanoscale features and involves the use of nanocolloids as an etch mask. These nanocolloids are dispersed as a monolayer and are electrostatically self-assembled over a surface. Directed reactive ion-beam bombardment or film evaporation can then be used to etch away the area surrounding the nanocolloids, as well as the nanocolloids themselves [54, 78].

Polymer demixing is another widely employed method for creating topographical structures and involves the spontaneous phase separation of polymer blends (e.g. polystyrene-poly(bromostyrene) demixing), cast onto material surfaces. Polymer demixing is a relatively quick and cheap manufacturing method, where careful adjustment of polymer ratio and concentration can create nanoscale or microscale pits, islands and ribbons [21, 23]. Features created using polymer demixing have a somewhat disordered spatial arrangement, yet very precise control can be achieved in the vertical scale. Nanometric features created by polymer demixing, however, often tend to exhibit larger micrometric structures in one or more planes and can exhibit different chemistries in addition to topographies.

When designing topographical experiments it is also worth noting that truly flat surfaces, for control purposes, are rare in nature and are fairly difficult to manufacture. Crystal cleavage surfaces, such as silicon, mica and glass coverslips are relatively flat (R_a values of a few nanometres), but polishing or similar processes induce various chaotic topography and surface chemistry [17].

35.5
Conclusions

This chapter has shown that both the chemistry and topography of the biomaterial surface influences protein adsorption and cell behaviour. Different topographical features are, however, invariably accompanied by chemical heterogeneities, and thus elucidating the topographical from chemical effects on biological behaviour is complex [15, 18]. The cellular response to structured surfaces is influenced by both the differential adsorption of various extracellular macromolecules onto the substrate (determined largely by the surface chemistry) and the cytoskeletal reorganisation caused by feature shape.

For many tissue-engineering strategies, topographical, chemical and biofunctional surface structuring could enable dynamic interactions of the biomaterial with cells and other biological components to ensure hierarchical tissue formation concurrent with scaffold degradation. Technological advances have enabled the creation of surfaces with well-defined surface properties, which combined with sensitive surface-characterisation techniques have unquestionably deepened our understanding of surface chemical and topographical effects on cell behaviour. This understanding could lead to more rational approach to the engineering of synthetic biomolecular materials capable of mimicking important aspects of the complexity of native ECM.

References

1. Allara, D. L., R. G. Nuzzo, Langmuir, 1985, 1, 45–52
2. Andersson A. S., F. Backhed, A. von Euler, A. Richter-Dahlfors, D. Sutherland, B. Kasemo, Biomaterials, 2003, 24, 3427–36
3. Anselme, K., P. Linez, M. Bigerelle, D. Le Maguer, A. Le Maguer, P. Hardouin, H. F. Hildebrand, A. Iost, J. M. Leroy, Biomaterials, 2000, 21, 1567–77
4. Ariga, K., T. Nakanishi, T. Michinobu, J Nanosci Nanotechnol, 2006, 6, 2278–301
5. Arima, Y., H. Iwata, Biomaterials, 2007, 28, 3074–82
6. Balazs, D. J., K. Triandafillu, P. Wood, Y. Chevolot, D. C. van, H. Harms, C. Hollenstein, H. J. Mathieu, Biomaterials, 2004, 25, 2139–51
7. Ball, M. D., U. Prendergast, C. O'Connell, R. Sherlock, Exp Mol Pathol, 2007, 82, 130–4
8. Ball, M., D. M. Grant, W. J. Lo, C. A. Scotchford, J Biomed Mater Res A, 2007 (in press)
9. Biggs, M. J. P., R. G. Richards, N. Gadegaard, C. D. W. Wilkinson, M. J. Dalby, J Mater Sci Mater Med, 2007, 18, 399–404
10. Bonzani, I. C., J. H. George, M. M. Stevens, Curr Opin Chem Biol, 2006, 10, 568–75
11. Bruinink, A., E. Wintermantel, Biomaterials, 2001, 22, 2465–73
12. Cai, K. Y., J. Bossert, d K. D. Jandt, Colloids Surf B, 2006, 49, 136–44

13. Clark, P., P. Connolly, A. S. G. Curtis, J. A. T. Dow, C. D. W. Wilkinson, Development, 1987, 99, 439–48
14. Clark, P., P. Connolly, A. S. G. Curtis, J. A. T. Dow, C. D. W. Wilkinson, Development, 1990, 108, 635–44
15. Curtis, A., C. Wilkinson, Biomaterials, 1997, 18, 1573–83
16. Curtis, A. S., M. Varde, J Natl Cancer Inst, 1964, 33, 15–26
17. Curtis, A. S., C. D. Wilkinson, J Biomater Sci Polym Ed, 1998, 9, 1313–29
18. Curtis, A. S., C. W. D. Wilkinson, B. Wojciak, Abstracts of Papers of the American Chemical Society, 1994, 207, 269–COLL
19. Curtis, A. S., N. Gadegaard, M. J. Dalby, M. O. Riehle, C. D. W. Wilkinson, G. Aitchison, IEEE Trans Nanobiosci, 2004, 3, 61–5
20. Dalby, M. J., M. O. Riehle, H. Johnstone, S. Affrossman, A. S. G. Curtis, Biomaterials, 2002, 23, 2945–54
21. Dalby, M. J., S. J. Yarwood, M. O. Riehle, H. J. H. Johnstone, S. Affrossman, A. S. G. Curtis, Exp Cell Res, 2002, 276, 1–9
22. Dalby, M. J., M. O. Riehle, S. J. Yarwood, C. D. W. Wilkinson, A. S. G. Curtis, Exp Cell Res, 2003, 284, 274–82
23. Dalby, M. J., D. Giannaras, M. O. Riehle, N. Gadegaard, S. Affrossman, A. S. G. Curtis, Biomaterials, 2004, 25, 77–83
24. Dalby, M. J., M. O. Riehle, H. Johnstone, S. Affrossman, A. S. G. Curtis, Cell Biol Int, 2004, 28, 229–36
25. Dalby, M. J., D. McCloy, M. Robertson, H. Agheli, D. Sutherland, S. Affrossman, R. O. C. Oreffo, Biomaterials, 2006, 27, 2980–7
26. Dalby, M. J., N. Gadegaard, R. Tare, A. Andar, M. O. Riehle, P. Herzyk, C. D. Wilkinson, R. O. Oreffo, Nat Mater, 2007, 6, 997–1003
27. Deligianni, D. D., N. Katsala, S. Ladas, D. Sotiropoulou, J. Amedee, Y. F. Missirlis, Biomaterials, 2001, 22, 1241–51
28. denBraber E. T., J. E. deRuijter, H. T. J. Smits, L. A. Ginsel, A. F. vonRecum, J. A. Jansen, Biomaterials, 1996, 17, 1093–9
29. Desai, T. A., Med Eng Phys, 2000, 22, 595–606
30. Discher, D. E., P. Janmey, Y. L. Wang, Science, 2005, 310, 1139–43
31. Elul, R., J Physiol, 1967, 189, 351–65
32. Engler, A. J., F. Rehfeldt, S. Sen, D. E. Discher, 2007, 83, 521–545
33. Flemming, R. G., C. J. Murphy, G. A. Abrams, S. L. Goodman, P. F. Nealey, Biomaterials, 1999, 20, 573–88
34. Forch, R., A. N. Chifen, A. Bousquet, H. L. Khor, M. Jungblut, L. Q. Chu, Z. Zhang, I. Osey-Mensah, E. K. Sinner, W. Knoll, Chem Vap Deposition, 2007, 13, 280–94
35. George, J. H., M. S. Shaffer, M. M. Stevens, J Exp Nanosci, 2006, 1, 1–12
36. Giavaresi, G., M. Tschon, J. H. Daly, J. J. Liggat, D. S. Sutherland, H. Agheli, M. Fini, P. Torricelli, R. Giardino, J Biomater Sci Polym Ed, 2006, 17, 1405–23
37. Hanawa, T., Y. Kamiura, S. Yamamoto, T. Kohgo, A. Amemiya, H. Ukai, K. Murakami, K. Asaoka, J Biomed Mater Res, 1997, 36, 131–6
38. Hansen, J. C., J. Y. Lim, L. C. Xu, C. A. Siedlecki, D. T. Mauger, H. J. Donahue, J Biomech, 2007, 40, 2865–71
39. Hern, D. L., J. A. Hubbell, J Biomed Mater Res, 1998, 39, 266–76
40. Hoffmann, A. S., Radiat Phys Chem, 1981, 18, 323–42
41. Hubbell, J. A., Biotechnology (N Y), 1995, 13, 565–76
42. Khor, H. L., Y. Kuan, H. Kukula, K. Tamada, W. Knoll, M. Moeller, D. W. Hutmacher, Biomacromolecules, 2007, 8, 1530–40
43. Kumar, R., S. Howdle, H. Munstedt, J Biomed Mater Res B Appl Biomater, 2005, 75, 311–9
44. Kunzler, T. P., T. Drobek, M. Schuler, N. D. Spencer, Biomaterials, 2007, 28, 2175–82
45. Lauer, G., M. Wiedmann-Al-Ahmad, J. E. Otten, U. Hubner, R. Schmelzeisen, W. Schilli, Biomaterials, 2001, 22, 2799–809
46. Lehnert, D., B. Wehrle-Haller, C. David, U. Weiland, C. Ballestrem, B. A. Imhof, M. Bastmeyer, J Cell Sci, 2004, 117, 41–52
47. Lehnert, S., M. B. Meier, U. Meyer, L. F. Chi, H. P. Wiesmann, Biomaterials, 2005, 26, 563–70
48. Lincks, J., B. D. Boyan, C. R. Blanchard, C. H. Lohmann, Y. Liu, D. L. Cochran, D. D. Dean, Z. Schwartz, Biomaterials, 1998, 19, 2219–32
49. Loebsack, A., K. Greene, S. Wyatt, C. Culberson, C. Austin, R. Beiler, W. Roland, P. Eiselt, J. Rowley, K. Burg, Mooney D, Holder W, Halberstadt C, J Biomed Mater Res, 2001, 57, 575–81
50. Martin, J. Y., Z. Schwartz, T. W. Hummert, D. M. Schraub, J. Simpson, J. Lankford, D. D. Dean, D. L. Cochran, B. D. Boyan, J Biomed Mater Res, 1995, 29, 389–401
51. Mrksich, M., C. S. Chen, Y. Xia, L. E. Dike, D. E. Ingber, G. M. Whitesides, Proc Natl Acad Sci U S A, 1996, 93, 10775–8
52. Mwenifumbo, S., M. S. Shaffer, M. M. Stevens, J Mater Chem, 2007, 17, 1894–902
53. Nayab, S. N., F. H. Jones, I. Olsen, Biomaterials, 2005, 26, 4717–27
54. Norman, J., T. Desai, Ann Biomed Eng, 2006, 34, 89–101
55. Parviz, B. A., D. Ryan, G. M. Whitesides, IEEE Trans Adv Packag, 2003, 26, 233–41
56. Popat, K. C., L. Leoni, C. A. Grimes, T. A. Desai, Biomaterials, 2007, 28, 3188–97
57. Prime, K. L., G. M. Whitesides, Science, 1991, 252, 1164–7
58. Pypen, C. M. J. M., H. Plenk, M. F. Ebel, R. Svagera, J. Wernisch, J Mater Sci Mater Med, 1997, 8, 781–4
59. Rabkin, E., F. J. Schoen, Cardiovasc Pathol, 2002, 11, 305–17
60. Rajnicek, A. M., S. Britland, C. D. Mccaig, J Cell Sci, 1997, 110, 2905–13
61. Ranieri, J. P., R. Bellamkonda, E. J. Bekos, J. A. Gardella, Jr, H. J. Mathieu, L. Ruiz, P. Aebischer, Int J Dev Neurosci, 1994, 12, 725–35
62. Ratner, B. D., J Biomed Mater Res, 1980, 14, 665–87
63. Ratner B. D., Biomaterial Science—an Introduction to Materials in Medicine. Elsevier, New York, 2004
64. Roach, P., D. Farrar, C. C. Perry, J Am Chem Soc, 2005, 127, 8168–73

65. Roach, P., D. Farrar, C. C. Perry, J Am Chem Soc, 2006, 128, 3939–45
66. Saez, A., M. Ghibaudo, A. Buguin, P. Silberzan, B. Ladoux, Proc Natl Acad Sci U S A, 2007, 104, 8281–6
67. Scotchford, C. A., M. Ball, M. Winkelmann, J. Voros, C. Csucs, D. M. Brunette, G. Danuser, M. Textor, Biomaterials, 2003, 24, 1147–58
68. Shin, H., S. Jo, A. G. Mikos, J Biomed Mater Res, 2002, 61, 169–79
69. Shin, H., S. Jo, A. G. Mikos, Biomaterials, 2003, 24, 4353–64
70. Simon, K. A., E. A. Burton, Y. B. Han, J. Li, A. Huang, Y. Y. Luk, J Am Chem Soc, 2007, 129, 4892–3
71. Stevens, M. M., J. H. George, Science, 2005, 310, 1135–8
72. Taqvi, S., K. Roy, Biomaterials, 2006, 27, 6024–31
73. Teixeira, A. I., G. A. Abrams, P. J. Bertics, C. J. Murphy, P. F. Nealey, J Cell Sci, 2003, 116, 1881–92
74. Ulman, A., Chem Rev, 1996, 96, 1533–54
75. Vogler, E. A., Adv Colloid Interface Sci, 1998, 74, 69–117
76. Wojciak-Stothard, B., A. Curtis, W. Monaghan, K. Macdonald, C. Wilkinson, Exp Cell Res, 1996, 223, 426–35
77. Woo, K. M., V. J. Chen, P. X. Ma, J Biomed Mater Res A, 2003, 67A, 531–7
78. Wood, M. A., J R Soc Interface, 2007, 4, 1–17
79. Yim, E. K. F., S. W. Pang, K. W. Leong, Exp Cell Res, 2007, 313, 1820–9
80. Zhou, W., X. X. Zhong, X. C. Wu, L. Q. Yuan, Z. C. Zhao, H. Wang, Y. X. Xia, Y. Y. Feng, J. He, W. T. Chen, Surf Coat Technol, 2006, 200, 6155–60

36 Mineralised Collagen as Biomaterial and Matrix for Bone Tissue Engineering

M. Gelinsky

Contents

36.1 Introduction 485
36.2 Extracellular Matrix of Bone Tissue 485
36.3 Synthetically Mineralised Collagen 486
36.4 Scaffolds Made of Mineralised Collagen .. 487
36.5 Cell and Tissue Response
  to Mineralised Collagen 489
36.6 Establishing an In Vitro Model for Bone
  Remodelling 491
36.7 Outlook 492
  References 493

a highly organised nanocomposite of collagen type I fibrils and mineral phase hydroxyapatite (HAP). Many attempts have been made in the last decades at developing artificial materials which mimic this matrix. This chapter discusses one particular method, in which collagen fibril reassembly and HAP nanocrystal formation take place simultaneously, leading to a real nanocomposite of both phases—mineralised collagen. The formation process, several scaffold types, their properties and possible applications are described. In addition, first approaches to setting up an in vitro model for bone remodelling based on this artificial ECM are presented.

36.1 Introduction

An optimal scaffold for tissue engineering applications would mimic the properties of the extracellular matrix (ECM) of those tissues to be regenerated perfectly and completely. In the case of bone, this is simply not possible, because the ECM of bone is adapted exactly to the local requirements, especially with regard to the mechanical situation. Nevertheless, a "biomimetic" scaffold material which resembles the composition and (micro)structure of bone ECM could be a suitable matrix for bone cell cultivation, allowing the generation of mineralised tissue in vitro and regenerative therapy of bone defects in vivo. On the lowest level, bone ECM consists of

36.2 Extracellular Matrix of Bone Tissue

In mammals, several types of bone are present or are formed temporarily during development or healing, mainly compact (cortical), trabecular (spongy) and woven bone [21]. They are organised in a hierarchical manner, but at the lowest level, all consist of a highly organised nanocomposite, made of fibrillar collagen type I and HAP nanocrystals. Collagen is synthesised by osteoblasts, which also express alkaline phosphatase (ALP), responsible for calcium phosphate mineral formation. A variety of non-collagenous proteins like osteocalcin and osteopontin, also expressed by osteoblasts, are responsible for control of the matrix formation and mineralisation processes, but the molecular mechanisms are not completely understood [4]. With the exception of woven bone, collagen fibrils are deposited in a sheet-like manner and with a parallel fibre alignment (called *"lamellae"*) into the free space, created by resorbing osteoclasts during bone remodelling. *Lamellae* form osteons in compact and trabecules in spongy bone. These structures are responsible for the outstanding mechanical properties of bone tissue and its perfect adaptation to the local force distribution [17]. Compact bone has only pores with diameters in the micrometer range, filled either with blood capillaries (Haversian channels, located in the centre of the osteons) or osteocytes (*lacunae*—interconnected by the *canaliculi* pore system). In contrast, the trabecules in spongy bone form a highly open porous structure with pore widths of up to a few millimetres. This is the location of the bone marrow, responsible for blood cell development and one of the main sources of mesenchymal stem cells in mammals.

Whereas the macro- and microstructure of the several types of bone tissue seem to be understood in the main, the interaction between the collagen fibrils and nanoscopical HAP crystals at the nanometer level is still a subject of intensive investigation. Obviously, there are big differences between the bone nanostructure depending on species, location and age. The most prominent distinction is between adult cortical or trabecular bone on the one hand and foetal woven bone on the other hand. Whereas in the former, platelet-like HAP nanocrystals are mainly located inside the collagen fibrils, in the latter, bigger and more needle-shaped crystals are also found in between the fibrils [19].

method, known as synchronous biomineralisation of collagen, developed by Bradt and co-workers [5]. The principles of this process are shown in Fig. 36.1. Briefly, a solution of (acid-soluble) collagen I in diluted hydrochloric acid is mixed with a calcium chloride solution. The ionic strength is adjusted by addition of NaCl, and the pH is raised to a value of 7 by adding TRIS (tris(hydroxymethyl)aminomethane) and phosphate buffer. Warming up the mixture to 37°C initiates collagen fibril reassembly—and the presence of calcium as well as phosphate ions in the solution leads to precipitation of calcium phosphate. Initially, amorphous calcium phosphate phases are formed (causing a milky turbidity) which are then slowly transformed into nanocrystalline HAP as the most stable phase at neutral pH. The conditions are adjusted in such a way that collagen fibril reconstitution and HAP formation take place simultaneously.

The growing collagen fibrils act as a template for calcium phosphate crystallisation with the consequence that a homogenous nanocomposite is formed, consisting of about 30 wt% collagen and 70 wt% nanocrystalline HAP. Besides the mineralised collagen fibrils appearing as a cloudy white precipitate floating in the reaction mixture, no unbound crystals can be found at the bottom of the flask, and the surrounding solution is clear. The latter verifies the complete transformation of the initially formed amorphous calcium phosphate phase into crystalline HAP, bound exclusively to the collagen matrix. Finally, the product, which can be described as mineralised collagen fibrils, can be isolated by centrifugation. The nanocomposite has been fully characterised,

36.3 Synthetically Mineralised Collagen

Many different attempts have been made up to now to synthesise collagen type I-HAP nanocomposites, mimicking the ECM of bone. Whereas some are not much more than simple mixtures between both phases, others try to use optimised conditions by controlling several parameters such as concentration, temperature and pH, leading to the formation of defined composite materials. Some of the collagen-HAP composites were reviewed recently by Wahl and Czernuszka [20]. Here we will focus on one

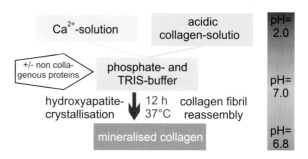

Fig. 36.1 Schematic representation of the synchronous biomineralisation process, leading to the formation of a homogenous collagen type I–HAP nanocomposite. By addition of non-collagenous proteins or other substances together with the buffer solution, the influence thereof on collagen fibril reassembly and mineralisation can be investigated

including Fourier-transform infrared spectroscopy (FT-IR), X-ray diffraction (XRD) analyses [5] and transmission electron microscopy (TEM), as well as high-resolution TEM (HRTEM) [5, 11], and it can clearly be demonstrated that the calcium phosphate phase is HAP.

In Fig. 36.2, TEM images of the nanocomposite are presented. The samples were prepared without an additional staining, which is why the collagen fibrils are not clearly visible due to their low material contrast. The HAP nanocrystals can be seen as dark, needle- or platelet-like objects with a maximum size of about 80 nm. It can be assumed that the collagen fibrils, covered with the HAP nanocrystals, are located in the light, elongated areas seen in the micrographs (along the arrows). In some regions (marked with circles), the HAP crystals seem to be partly oriented with their long axis (proven by high-resolution TEM to be the crystallographic c axis of the hexagonal HAP lattice [11]) parallel to the collagen fibres. Concerning the size, orientation and position of the HAP crystals with regard to the collagen fibrils, the TEM images have a great similarity to those taken of foetal human woven bone [19].

The process described above leads to a homogenous nanocomposite which mimics the ECM of healthy bone tissue concerning its basic chemical composition—and which shows a good correspondence to woven bone regarding its nanostructures. In addition, the process can easily be used to investigate the effects of non-collagenous proteins on collagen fibril reassembly and mineralisation. For this, the substance to be studied is added to the reaction mixture together with the buffer solution (Fig. 36.1). Utilising this approach, we have started to investigate non-collagenous proteins like osteocalcin and osteopontin and model substances thereof, like poly(L-aspartic acid) [5].

36.4
Scaffolds Made of Mineralised Collagen

Using the mineralised collagen fibrils as a starting material, several types of 2D and 3D scaffolds have been developed over the past years. An overview is

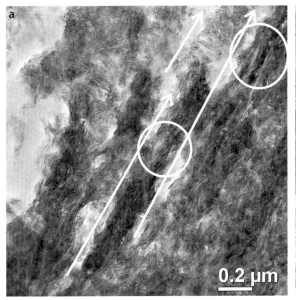 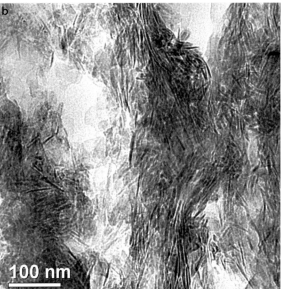

Fig. 36.2a,b TEM micrographs of mineralised collagen. The samples were prepared without contrasting for collagen. The assumed positions of the collagen fibrils are indicated by *arrows*, and the regions with an assumed parallel orientation of HAP crystals and collagen fibres are shown by *circles*. TEM images: Dr. Paul Simon, MPI CPfS, and Dept. of Structural Physics, TUD (both Dresden, Germany)

given in Table 36.1, which also contains the related references. By densification of the suspended nanocomposite by means of a vacuum filtration process, using a glass filter frit, a flat, membrane-like material ("tape") can be achieved. After cross-linking with the carbodiimide derivative EDC (N-(3-dimethylaminopropyl)-N'-ethylcarbodiimide), the tape is stable against water and cell culture medium. The freeze-dried material can be stored and also sterilised by gamma irradiation for use in cell culture and animal experiments. The material exhibits pores only in the micrometer range, which can be seen in Fig. 36.3a. Pores with such diameters are too small for cell ingrowth, which is why the tapes can be used as a 2D cell culture substrate. In wet state, the membranes are flexible and elastic; tensile strength was determined to be about 1 MPa at a fracture strain of approximately 20%.

Storing of the tapes in acidic TRIS buffer leads to dissolution of the HAP mineral phase. The resulting product is a pure, cross-linked collagen membrane which cannot be achieved directly, as non-mineralised collagen fibres cannot be densified by means of a vacuum filtration process (because they clog the pores of the glass filter frit immediately). This material was developed and tested as a scaffold for tissue engineering of skin and oral mucosa.

By freeze-drying concentrated suspensions of mineralised collagen, filled in appropriate moulds, 3D scaffolds with an interconnecting pore system can be produced (Fig. 36.3b). Cross-linking with an EDC solution in 80% (v/v) ethanol makes the porous structure stable under cell culture conditions, whereas some commercially available non-mineralised collagen type I sponges collapse spontaneously when wetted. The mean pore diameter can be controlled by the freezing conditions. Pore sizes of about 180 µm were demonstrated to be suitable in respect to homogenous cell seeding and mechanical stability. For samples with a mean pore size of 180 µm, a total porosity of 72% was measured. The scaffolds are highly elastic in the wet state, which was shown in cyclic loading experiments. Due to their flexibility, a compressive strength cannot be determined, but at 50% uniaxial compression, the stress was found to be about 28 kPa, compared to only 2–3 kPa, which was measured for non-mineralised collagen sponges produced under similar conditions.

Table 36.1 Different types of scaffolds, developed using the mineralised collagen nanocomposite as starting material

Description	References
Membrane ("tape")	[3, 6, 7]
Demineralised membrane	[12]
Porous 3D scaffold	[2, 9, 11, 18, 23]
Biphasic, but monolithic, 3D scaffold	[10]

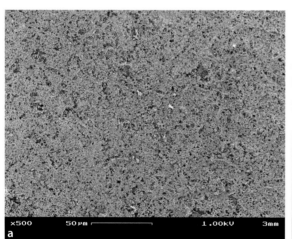 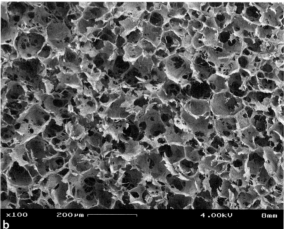

Fig. 36.3a,b SEM images of two different types of scaffolds, developed from mineralised collagen. a "Tape"—a membrane-like material, exhibiting pores only in the micrometer range (×500). b Porous 3D scaffold (section, ×100)

Regarding the mechanical requirements for clinical use, one has to distinguish between implantation in a load-bearing and non-load-bearing situation. For the first, the scaffolds must at least be stable enough to withstand the internal body pressure by preserving the openness of the pores. Real load-bearing applications are seldom for degradable bone replacement materials, because additional stabilisation by metal devices like plates, nails or *fixateur externe* are mostly necessary to guarantee axial stability of the defect site [8]. This is why for fully load-bearing applications, normally non-resorbable implant materials are used.

By covering a layer of suspended mineralised collagen with a layer of a non-mineralised collagen/hyaluronic acid composite, biphasic, but monolithic, scaffolds can be achieved after joint freeze-drying and chemical cross-linking. These devices were developed for the therapy of osteochondral defects—deep lesions of the articular cartilage, in which the underlying bone is also already damaged. The mineralised, HAP-containing layer of the scaffold is designed to fill the bony part of the defect, whereas the non-mineralised one fits to the chondral tissue. In contrast to many solutions for osteochondral defects, suggested by other researchers [16], this type of biphasic scaffold overcomes the problem of joining the two different parts together: by combining them in the liquid state, followed by joint lyophilisation and cross-linking, they are fused to a unified whole.

With this set of 2D and 3D scaffolds, consisting of mineralised collagen or composites, containing non-mineralised collagen, many possibilities for research on tissue formation and regeneration are available. Some approaches are described below.

36.5
Cell and Tissue Response to Mineralised Collagen

An optimal scaffold material for healing of bone defects or for use as a matrix in tissue engineering should fulfil the following requirements: it should support cell adhesion and proliferation, stimulate cell differentiation (including synthesis and mineralisation of new ECM) and it should be degradable to allow complete remodelling to healthy tissue in time. Three-dimensional materials, in addition, must provide an interconnecting porosity, suitable either for facile and homogenous cell seeding in vitro or fast and deep cell and blood capillary ingrowth in vivo. The optimal pore size is a matter of controversial discussion in the literature [14], but we found pore diameters in the range of 180 μm suitable for both the requirements in vitro and in vivo [11, 23]. Furthermore, the diameters of the pore interconnections seem to be of more importance than the pore size itself.

The aim of mimicking structure and composition of the ECM of bone by artificial scaffold materials does not include the goal of imitating the macroporosity of the tissue. On the one hand, after implantation of a porous implant material, the pore space is invaded by cells and later filled with tissue—which means that to mimic, for example, the structure of spongy bone, the generation of an inverse porosity would be necessary. But pores with the dimensions of *trabeculae* would be too small for fast tissue ingrowth, which is why such an approach would not lead to proper results. On the other hand, the trabecular structure is, as already mentioned above, perfectly adapted to the local mechanical situation—which can hardly be copied by scaffold fabrication methods. Due to the fact that collagen-based materials are usually rapidly degraded in vivo, the organism will remodel the scaffold irrespective of its macroporous structure depending on the local demands if the pores are suitable for cell and blood capillary invasion.

Osteoblast-like cell lines (ST-2, 7F2), as well as primary osteoblasts derived from rats and humans, were shown to adhere and proliferate well on the 2D and 3D matrices of mineralised collagen, mentioned above. The materials were further tested using human mesenchymal stromal cells, derived from bone marrow (hBMSC), because these cells play a major role in regenerative processes in vivo and are mostly used for tissue engineering purposes in vitro. In several studies, we could demonstrate that hBMSC can be cultivated on the tapes as well as the porous 3D scaffolds for several weeks, applying static as well as perfusion cell culture conditions [2, 3, 15]. On the membrane-like materials, cell adhesion is slightly reduced in comparison to cell culture polystyrene, but hBMSC proliferate well, forming a dense layer on the tape surface after a few weeks. Mineralised collagen, mimicking bone ECM, seems to stimulate osteogenic differentiation of hBMSC: such cells showed a significantly higher specific activity of the osteoblastic marker alkaline phosphatase (ALP) when cultivated on the tapes in comparison to cell culture polystyrene (Fig. 36.4) [3]. Cell culture was carried out using standard (DMEM low glucose plus 10% foetal calf serum) as well as osteogenic cell culture medium, additionally containing dexamethasone, β-glycerophosphate and ascorbic acid 2-phosphate. The experimental conditions have been described in detail elsewhere [3].

For porous 3D scaffolds, a homogenous cell distribution inside the matrix and good seeding efficiency are of great importance. Materials exhibiting too small pores or an insufficient interconnectivity can only be seeded with cells on the outer surface. By contrast, too big pore diameters led to difficult seeding procedures and often to a low seeding efficiency.

The 3D scaffolds of mineralised collagen, described above, can be seeded rather homogenously and with a high efficiency. In an extensive assay, it could be demonstrated that on 94% of all samples tested ($n=60$), more than 90% of the cells became adherent on or in the porous matrices [15]. After seeding of cylindrical scaffolds with a diameter of 10 mm and a height of 7 mm with 2×10^5 hBMSC each, cells could be found up to a depth of 4–5 mm [11]. In Fig. 36.5 a SEM image of a section through a hBMSC-seeded scaffold after 2 days of cultivation is presented. The horizontal section was made about 2–3 mm below the surface onto which the cells were seeded. As in the case of the 2D membranes, hBMSC show good proliferation rates on the porous 3D scaffolds and can easily be differentiated towards the osteogenic lineage [2].

The three-dimensional material was tested successfully in animal experiments. In a rat model,

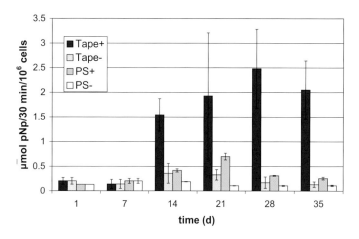

Fig. 36.4 Specific ALP activity of induced (OS+) and non-induced (OS−) hBMSC seeded on tapes of mineralised collagen in comparison to those cultivated on cell culture polystyrene (PS) over a period of 35 days. Dr. A. Bernhardt and Dr. A. Lode, both MBC, TUD

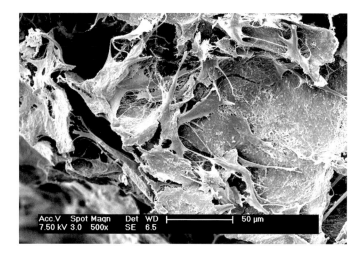

Fig. 36.5 SEM image of a section through a hBMSC-seeded scaffold after 2 days of cultivation. The horizontal section was made about 2–3 mm below the surface onto which the cells were seeded. ×500. Image: Dr. A. Bernhardt, MBC, TUD

implantation in subcutaneous tissue led to a fast resorption by phagocytosis, and the samples had nearly completely disappeared after 8 weeks. Implantation of the porous scaffolds in a 2×3 mm bone defect, made in the femur of rats, resulted in total remodelling, which was completed after about 12 weeks [23]. Interestingly, it could be observed that invaded osteoblasts deposited new mineralised matrix directly onto the artificial material—and that the scaffold was actively degraded by osteoclasts, which was proven by positive staining for tartrate-resistant acid phosphatase (TRAP). The fast tissue and also blood capillary ingrowth, detected as soon as 1 week after implantation, confirmed the suitability of the pore system for fast-scaffold remodelling in vivo. Fast vascularisation of the interconnecting pores could also be demonstrated by testing the 3D scaffolds in the chorioallantois membrane assay [1]. In another animal experiment, the porous 3D scaffolds were used in combination with (load-bearing) titanium cages for spinal fusion in sheep. Additionally, two modifications of the three-dimensional material were investigated, which were prepared by functionalisation of the mineralised collagen matrix with either rhBMP-2 (bone morphogenic protein 2) or a cyclic RGD peptide, known to enhance cellular adhesion [13]. It could be shown that the two functionalised matrices were as effective in bony fusion as autologous spongiosa, still seen as the gold standard for this procedure [18]. Due to the limited amount of autologous bone graft, which can be harvested in one patient, and the risks and morbidity of the surgical procedure, a suitable artificial material for use in spinal fusion is highly demanded. The porous 3D scaffolds, functionalised with RGD, might be an optimal candidate for this because that modification is much cheaper than those with BMP, more storable and also less sensitive with regard to denaturation.

Wiesmann and co-workers investigated the porous 3D scaffolds successfully in a tissue-engineering approach. The material was pre-seeded with autologous cells, outgrown from periosteum of minipigs. Prior to implantation in a defect at the lower jaw bone, cells were stained with a non-toxic fluorescence dye. Two weeks later, the implants were harvested, and fluorescence microscopy revealed that the cells, seeded onto the scaffolds in vitro, were still alive and had started to remodel the material. Some were found to have been transformed to osteocytes, embedded in new mineralised ECM [22].

36.6
Establishing an In Vitro Model for Bone Remodelling

The similarity of the artificial mineralised collagen matrices to ECM of bone and the observation that the materials are readily resorbed by osteoclasts in vivo led to the consideration of establishing an in vitro model for the complex processes of bone remodelling using those scaffolds. Having already investigated growth and osteogenic differentiation of osteoblasts and hBMSC on the 2D and 3D matrices (see 36.5), the next step was to achieve osteoclast formation and resorption of the mineralised collagen in cell culture. For this, primary human monocytes isolated from buffy coats were seeded onto the tapes with a density of $5 \times 10^5/\mathrm{cm}^2$ and cultivated using a cell culture medium containing MCSF (macrophage colony stimulating factor) and RANK ligand (RANKL) at concentrations of 50 ng/ml each [7]. Figure 36.6 shows a confocal laser scanning microscope (cLSM) image of multinucleated osteoclast-like cells 3 weeks after seeding of the monocytes onto a tape made of

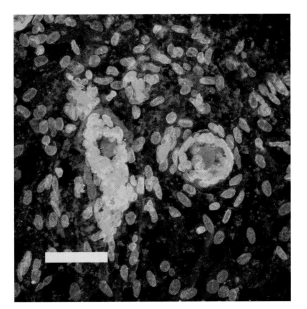

Fig. 36.6 Confocal LSM image of multinucleated osteoclast-like cells formed from human primary monocytes, surrounded by mononuclear cells on the surface of a mineralised collagen tape, 3 weeks after seeding; phalloidine-rhodamine staining of actin (*red*) and DAPI staining of cell nuclei (*blue*). The yellow bar represents 50 µm. Imaging: Dr. Th. Hanke (MBC, TUD)

mineralised collagen. In the centre of the micrograph, two large cells are visible, containing more than 20 nuclei (blue) each, surrounded by mononuclear cells. The large cells possess organised, ring-shaped actin patterns (red), typical for osteoclasts.

For approaching closer to the situation in living bone, a co-culture consisting of mouse osteoblasts (cell line ST-2) and osteoclast-like cells (derived from human monocytes as described) was established. Cells from different species were chosen to be able to distinguish between both in RT-PCR analyses. The cell culture medium necessary for the co-culture contained osteogenic supplements (see above) as well as MCSF and RANKL. It could be shown by RT-PCR that the osteoblasts transcribed the typical markers collagen type I, osteocalcin and ALP—and the osteoclast-like cells tartrate-resistant acid phosphatase (TRAP), cathepsin K and chloride channel 7 (CLC 7) [7]. TEM micrographs of thin sections of embedded samples revealed active resorption of the mineralised collagen matrix by the osteoclast-like cells in an endocytosis-like mechanism (Fig. 36.7). This might be due to the microporosity of the tapes (Fig. 36.3a), preventing the formation of a really tight sealing zone as osteoclasts form normally when degrading bone matrix in vivo. Therefore, lowering of the pH beneath the osteoclasts, necessary for HAP dissolution, seems to be difficult to achieve—which is why the cells might begin to phagocytose the artificial mineralised collagen matrix in a more macrophage-like manner. Additional experiments using densified, non-porous matrices shall bring to light the underlying mechanism.

For coming still closer to the situation in living bone, a solely human co-culture system was established next. Osteoclast-like cells derived from primary monocytes were co-cultivated with osteogenically induced hBMSC on membranes of mineralised collagen. For gene expression analyses, both cell types were cultured and separated from each other using transwell inserts. Beside investigations on the transcript level, some typical protein markers were quantified, and all samples were studied with SEM and TEM. The results confirmed those of the murine/human co-culture described above: again the formation of actively resorbing osteoclasts could be observed, and both cell types expressed their typical markers. Details will be reported elsewhere.

36.7 Outlook

After having established an in vitro model for bone remodelling by co-cultivating human osteoclast-like cells together with osteogenically differentiated hBMSC on membranes of mineralised collagen as an artificial bone ECM, this system might be useful in investigating further the complex mechanisms of cellular crosstalk as well as matrix synthesis and degradation. In addition, the co-culture should also be established with the porous 3D scaffolds for a better representation of the situation in living bone tissue.

The porous 3D scaffold material, functionalised with a cyclic RDG peptide for enhancement of cell attachment, shall be further optimised for use in spinal fusion, where artificial materials are highly demanded as an alternative for autologous bone graft.

In conclusion, we have started to utilise the elastic properties of the porous 3D matrices to study effects of cyclic mechanical loading on osteoblast and hBMSC proliferation and differentiation. A long-term objective of this research might be to combine mechanical stimulation and co-cultivation of osteoblasts and osteoclasts to mimic the complex situation of the remodelling processes occurring in living bone tissue.

Fig. 36.7 TEM image showing osteoclast-like cells derived from primary human monocytes in co-culture with ST-2 mouse osteoblasts on the surface of a tape, degrading and phagocytosing the mineralised collagen. Sate 3 weeks after seeding of the cells; ×5,000. Cell cultivation by H. Domaschke and TEM image kindly provided by Dr. habil. R. Fleig (both TU Dresden)

Acknowledgements

Most of the work described here was carried out at the research group Tissue Engineering and Biomineralisation at the Max Bergmann Center of Biomaterials, Dresden University of Technology, Germany. The author in particular wants to acknowledge Dr. Anne Bernhardt, Dr. Rainer Burth, Dr. Ulla König and Dr. Anja Lode for their contributions. The co-culture studies were done in close and amicable cooperation with Prof. Angela Rösen-Wolff, University Hospital Dresden, and her co-workers Hagen Domaschke and Dr. Sebastian Thieme. Our work was funded by the German Federal Ministry of Education and Research (BMBF), the Saxonian Ministry of Science and Arts (SMWK) and the DFG Research Center for Regenerative Therapies Dresden (CRTD), of which the author is a founding member. We are grateful to the companies Biomet Europe (Berlin) and Syntacoll (Saal/Donau, Germany) for their support. Without the continuous encouragement by and stimulating discussions with Professor Wolfgang Pompe, this work would not have been possible.

References

1. Auerbach R, Lewis R, Shinners B et al (2003) Angiogenesis assays: a critical overview. Clin Chem 49:32–40
2. Bernhardt A, Lode A, Mietrach C et al. (2008) In vitro osteogenic potential of human bone marrow stromal cells cultivated in porous scaffolds from mineralised collagen. J Biomed Mater Res A (in press, DOI 10.1002/jbm.a.32144)
3. Bernhardt A, Lode A, Boxberger S et al (2008) Mineralised collagen—an artificial, extracellular bone matrix—improves osteogenic differentiation of mesenchymal stem cells. J Mater Sci Mater Med 19:269–275
4. Bilezikian JP, Raisz LG, Rodan GA (eds) (2002) Principles of bone biology. Academic Press, London
5. Bradt JH, Mertig M, Teresiak A et al (1999) Biomimetic mineralization of collagen by combined fibril assembly and calcium phosphate formation. Chem Mater 11:2694–2701
6. Burth R, Gelinsky M, Pompe W (1999) Collagen-hydroxyapatite tapes—a new implant material. Tech Textile 8:20–21
7. Domaschke H, Gelinsky M, Burmeister B et al (2006) In vitro ossification and remodeling of mineralized collagen I scaffolds. Tissue Eng 12:949–958
8. Epari DR, Kassi JP, Schell H et al (2007) Timely fracture-healing requires optimization of axial fixation stability. J Bone Joint Surg Am 89:1575–1585
9. Gelinsky M, König U, Sewing A et al (2004) Porous scaffolds made of mineralised collagen—a biomimetic bone graft material. Mat-wiss Werkstofftech 35:229–233 (in German)
10. Gelinsky M, Eckert M, Despang F (2007) Biphasic, but monolithic scaffolds for the therapy of osteochondral defects. Int J Mater Res 98:749–755
11. Gelinsky M, Welzel PB, Simon P et al (2008) Porous three dimensional scaffolds made of mineralised collagen: preparation and properties of a biomimetic nanocomposite material for tissue engineering of bone. Chem Eng J 137:84–96
12. Gelinsky M, Bernhardt A, Eckert M et al (2008) Biomaterials based on mineralised collagen: an artificial extracellular matrix. In: Watanabe M, Okuno O (eds) Interface oral health science 2007. Springer, Tokyo, pp 323–328
13. Kantlehner M, Schaffner P, Finsinger D et al (2000) Surface coating with cyclic RGD peptides stimulates osteoblast adhesion and proliferation as well as bone formation. Chembiochem 1:107–14
14. Karageorgiou V, Kaplan D (2005) Porosity of 3D biomaterial scaffolds and osteogenesis. Biomaterials 26:5474–5491
15. Lode A, Bernhardt A, Boxberger S et al (2006) Cultivation of mesenchymal stem cells on a three-dimensional artificial extracellular bone matrix. In: Nadolny AJ (ed) Biomaterials in regenerative medicine. Vienna Scientific Centre of the Polish Academy of Sciences, Vienna, pp 167–172
16. Martin I, Miot S, Barbero A et al (2007) Osteochondral tissue engineering. J Biomech 40:750–765
17. Rho JY, Kuhn-Spearing L, Zioupos P (1998) Mechanical properties and the hierarchical structure of bone. Med Eng Phys 20:92–102
18. Scholz M, Schleicher P, Koch C et al (2007) Cyclic-RGD is as effective as BMP-2 in anterior interbody fusion of the sheep cervical spine. Eur Spine J (under review)
19. Su X, Sun K, Cui FZ et al (2003) Organization of apatite crystals in human woven bone. Bone 32:150–162
20. Wahl D, Czernuszka JT (2006) Collagen–hydroxyapatite composites for hard tissue repair. Eur Cells Mater 11:43–56
21. Weiner S, Wagner HD (1998) The material bone: structure-mechanical function relations. Annu Rev Mater Sci 28:271–298
22. Wiesmann HP et al. Manuscript in preparation
23. Yokoyama A, Gelinsky M, Kawasaki T et al (2005) Biomimetic porous scaffolds with high elasticity made from mineralised collagen—an animal study. J Biomed Mater Res B Appl Biomater 75B:464–472

Hydrogels for Tissue Engineering

J. Teßmar, F. Brandl, A. Göpferich

Contents

37.1	Introduction	495
37.2	Hydrogels as Biomaterials	496
37.3	Hydrogels and Their Functions in Tissue Engineering	498
37.4	Chemical Structures of Gel-Forming Polymers	498
37.4.1	Natural Polymers	499
37.4.2	Synthetic Polymers	501
37.5	Mechanisms of Cross-Linking	504
37.5.1	Physical Cross-Linking	504
37.5.2	Chemical Cross-Linking	505
37.6	Design Criteria and Desirable Properties for Hydrogels	507
37.6.1	Mechanical Properties and Mesh Size	507
37.6.2	Permeability for Drug Substances and Nutrients	507
37.6.3	Biomimetic Modifications—Interaction with Cells	508
37.6.4	Degradation and Elimination from the Patient	509
37.7	Tools for the Characterization of Hydrogels	509
37.7.1	Nuclear Magnetic Resonance	509
37.7.2	Swelling Studies for the Determination of Network Parameters	510
37.7.3	Fluorescence Recovery After Photobleaching	511
37.7.4	Rheological Characterization of Hydrogels	511
37.8	Conclusions and Outlook	513
	References	513

37.1 Introduction

Today's tissue engineering approaches rely on two different types of polymeric cell carriers. First, there are the well-established solid scaffolds, such as poly(α-hydroxy esters), which are generally based on lipophilic but hydrolytically degradable polymers that were originally designed as degradable sutures or drug-releasing matrix materials. Alternatively, new and promising strategies rely on hydrophilic polymer networks; these are based on hydrogels, which are also degradable polymers. Hydrophilic polymer network systems are suitable for many of the soft tissue engineering applications that do not require the strong mechanical support of solid scaffolds, but rather a flexible material that mimics the extracellular matrix (ECM). Highly hydrated hydrophilic polymer networks contain pores and void space between the polymer chains (Fig. 37.1); this provides many advantages over the common solid scaffold materials, including an enhanced supply of nutrients and oxygen for the cells. Pores within the network provide room for cells, and after proliferation and expansion, for the newly formed tissue. Their formation can be controlled using chemical modifications of the hydrogel network.

Despite the fact that all hydrogels contain approximately 90% water and most appear to behave similarly, there are very distinct differences between various gels, and some gel-forming polymers exhibit very unique properties that will be highlighted and further explained in this chapter. The observed differences between individual gels are dependent on the chemical structure of the materials that are used. Chemical structure can significantly affect the behavior of cells seeded on the materials, such as in the cases of altering differentiation or guiding cellular movement. For the latter example, material characteristics such as biomechanical properties and degradability become most important. These features of the material must be carefully considered and controlled prior to application of the hydrogel in vivo [15a].

37.2
Hydrogels as Biomaterials

Hydrogels are considered mechanically stable because they require an applied force to exhibit flow. They are easily deformable systems that contain at least two components: a colloidally dispersed solid with long and branched parts and a liquid, in most cases water, as a dispersant [33]. The solid component usually forms a three-dimensional network, which is stabilized via chemical or physical interactions of the individual molecules. Gel-forming molecules can either be high molecular weight polymers or, in rare cases, small molecules with very strong interactions [79].

Typical hydrogels for tissue engineering are able to trap large volumes of water in the aforementioned void spaces and pores. The enmeshed water forms a continuous maze of canals through the polymer-rich hydrogel network (Fig. 37.1). The pores in the polymer network are originally formed around encapsulated cells or other porogens such as particles [16] or gas bubbles [4]. They grow upon degradation of the hydrogel, and this is the most crucial consideration for the invasion of cells into the hydrogels, since cells are significantly larger than usual void spaces in cross-linked polymers. In preparing the hydrogel matrix, the dispersed materials are various hydrophilic polymers that are sufficiently long or branched to maintain the structural integrity of an entire viscoelastic system [81].

Based on the method of cross-linking present in the hydrogel, two different types of gel systems can be distinguished. These types only differ in the cross-link stability, and the overall behavior of the systems is similar, if drug release kinetics or mechanical properties are considered [79]. The first type of gel systems are *physically cross-linked* systems, where the polymer chains are linked by weak chain entanglements or physical interactions caused by temperature-induced dehydration of lipophilic monomers in the polymer chains or by certain structures, which are already present in the polymer chains (e.g., charged groups, lipophilic amino acids or monomers). These interactions are non-permanent and often in a constant exchange of the polymer partners, and they can be broken apart by temperature changes or simple dilution. Physical interactions are based on van der Waals forces between lipophilic amino acids, dipole–dipole interactions or electrostatic interactions between divalent ions and the

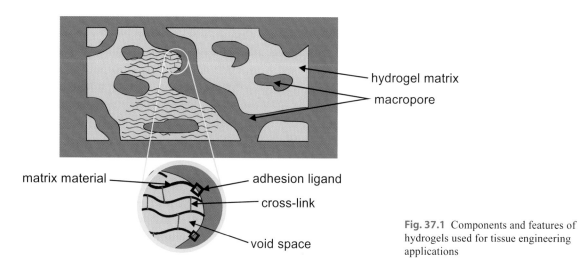

Fig. 37.1 Components and features of hydrogels used for tissue engineering applications

polymers. The last case leads to the formation of larger networks of several polymer strands in solution, resulting in the overall formation of a viscoelastic gel system. Physically stabilized gels tend to exhibit concentration-dependent strength, becoming dramatically weaker upon dilution of the gel, which leads to separation of the entangled polymer chains and a decrease in the viscosity of the system until only a polymer solution remains (Fig. 37.2a).

The second type of system, *chemically cross-linked* gels, consists of stable chemical cross-links between single long polymer chains or a network of smaller polymers forming multivalent cross-links at their ends. Consequently, these gels show limited swelling upon exposure to increasing volumes of solvent under physiological conditions. Swelling stops completely when the polymer chains in the hydrogel are fully elongated (Fig. 37.2b) [26]. If the polymer

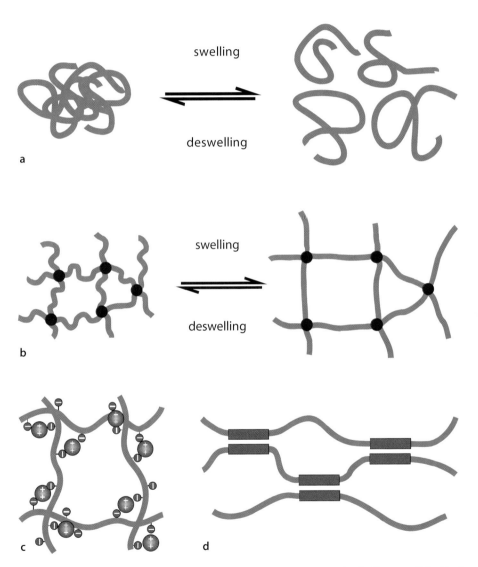

Fig. 37.2 a Physically cross-linked hydrogels with loose entanglements; upon swelling they become loose networks and eventually polymer solutions upon further dilution. **b** Chemically cross-linked hydrogels with defined cross-linking spots and an equilibrium swelling upon exposure to excess of water. **c** Physically cross-linked hydrogels mediated by opposite charges of polymer and divalent cations. **d** Physically cross-linked hydrogels mediated by lipophilic regions in the polymer caused by thermal removal of hydrate shells

chains and the cross-linking points are chemically stable for an extended period of time, the hydrogels are suitable for applications where they would be in contact with large amounts of water, such as during implantation into a patient's body, and also for long-term applications, for example, hydrophilic gel-like contact lenses, which also consist of highly hydrated acrylate or silicone gels shaped according to the visual impairment of the patient [75]. For tissue engineering, chemically cross-linked hydrogels offer the possibility to change the type of cross-linker—degradable or not degradable—and the cross-link density, which allows fine-tuning of the systems to obtain a certain mechanical strength and, additionally, degradation kinetics to control the fate of the implanted cell carriers [63, 78, 106].

37.3
Hydrogels and Their Functions in Tissue Engineering

Based on their chemical structure and their physical properties, hydrogels are able to perform many functions for tissue engineering [56, 59, 74]. Firstly, hydrogels can function just as cell carriers, supporting the three-dimensional growth of cells. In this case, hydrogels must be biocompatible in order to prevent interference with the proliferation and growth of adhering cells. Moreover, the hydrophilic material has to provide a sufficient number of adhesion sites to allow close contact between the cells and the material. Most hydrogels must be chemically modified through the introduction of cell adhesion ligands (Fig. 37.1) [38, 109], which can be small peptides that associate with cell surface receptors or lipophilic regions along the polymer chains that mediate the adsorption of cell adhesive proteins from the culture media [39, 55].

Another possible function of hydrogels in tissue engineering involves their application as a carrier for drugs, a technique which is frequently used in conventional drug-delivery applications [81]. For purposes relating to tissue engineering, however, growth factors, which can act directly on the locally present cells and support their development and differentiation to the newly formed tissue, are released [96, 97, 107]. The localized delivery of the growth factors is also useful in preventing side effects of these potent molecules at other sites. To ensure localized delivery, further immobilization of the factors within the hydrogel network is mandatory. Generally, hydrogel delivery systems for tissue engineering can be designed based on previously applied principles for other drug-delivery applications [79, 81]. The unique features of hydrogels that are exploited for delivery applications are the free diffusivity of incorporated substances, the general suitability for parenteral application of most materials and the fact that due to the hydrophilicity, less denaturation and irreversible adsorption of lipophilic and sensitive drug molecules can occur.

As a final application, hydrogels can also be used as "space fillers" for various tissues that are not capable of being restored or replaced in a short time. Here, the hydrogels offer the ability to act as a template to guide the ingrowth of repair cells, and with the support of incorporated drug substances, they are able to control the differentiation and overall behavior of these cells [40, 41]. The use of water-soluble hydrogels presents the possibility of introducing an aqueous solution of the polymer at the defect site using minimally invasive techniques. After this step, necessary stimuli are provided to induce cross-linking in order to obtain the solidified polymer network, which then remains at the site of application. With this method, it is possible to inject cell-loaded polymer solutions in irregularly shaped defects and cross-link the polymers in the presence of the living cells [99, 100].

37.4
Chemical Structures of Gel-Forming Polymers

Based on the intended application, many different polymers can be chosen to provide suitable hydrogels for tissue engineering. Accordingly, there are many different reviews that highlight advantages and disadvantages of different polymers for various applications, which are recommended for further reading [90]. Here, some of the important polymers and their material-dependent characteristics will be discussed.

37.4.1
Natural Polymers

Initial applications of hydrogel-forming polymers focused on the use of natural materials. These spanned a great deal of materials, from mammalian extracellular matrices to polymers derived from selected plants, like algae (Table 37.1). All of these polymers have weak intrinsic gelling capacity, but tissue-engineering applications require chemical modifications to increase mechanical and chemical stability. The primary concerns associated with natural polymers are significant batch-to-batch variations and the potential risk of infection or immunogenic reactions for polymers derived, for example, from animals.

37.4.1.1.
Collagen and Gelatin

Collagen is one of the main components of the extracellular matrix in many mammalian tissues. It is composed of triple-helical peptide strands that arrange in several tissue-specific combinations, which allow the identification of different collagen types with respect to the differentiation state of the producing cells [104]. Collagen is known to form thermo-reversible gels, because it contains hydrophilic and lipophilic amino acids that exhibit changes in solubility that are temperature dependent. Due to the fact that collagen is derived from natural sources that include animals, there are always concerns of immunogenicity and

Table 37.1 Examples of hydrogel-forming polymers of natural origin

	Chemical structure	Modifications/cross-linking
Collagen		Physical cross-linking induced by temperature; chemical cross-linking with glutaraldehyde or carbodiimides.
Gelatin		Triple helical polypeptide, main structural component of extracellular matrix, various types and derivatives. Partially hydrolyzed protein derived from collagen
Fibrin	Protein involved in blood clotting	Physical cross-linking after enzymatic activation
Hyaluronic acid	(chemical structure)	Physical cross-linking and chemical modifications for chemical cross-linking
Alginate	(chemical structure)	Physical cross-linking mediated by divalent cations; chemical cross-linking with amines
Chitosan	(chemical structure)	Physical cross-linking mediated via anions; chemical cross-linking with glutaraldehyde

contamination with viruses. Furthermore, collagen only forms very weak gels that need further chemical cross-linking through temperature-induced cross-linking by dehydration or other methods that will be discussed later [98].

Gelatin is a hydrogel-forming polymer that is directly derived from collagen. It can be obtained via basic or acidic hydrolysis of collagen from different tissues of various mammalian species or fish. Depending on whether the hydrolysis is basic or acidic, gelatin can change its isoelectric point from 4.8 to 9.4, and consequently its gelation behavior varies at different solution pH. With regards to biocompatibility, it is similar to collagen. And it also forms only very weak gels without chemical modification. Both collagen and gelatin have the advantage of already containing a sufficient number of adhesion sites for cells; these include the amino acid sequences RGD (arginine-glycine-aspartic acid) or GER (glycine-glutamic acid-arginine). Because of this, further functionalization is not necessary to promote cellular adhesion.

37.4.1.2
Fibrin and Silk Hydrogels

Fibrin is another protein-based, hydrogel-forming polymer. Fibrin gel formation naturally occurs as part of the blood coagulation process in damaged blood vessels and wounds. Fibrin is enzymatically obtained by cleavage of fibrinogen in the presence of thrombin. The liberated fibrin then forms distinct aggregates that lead to coagulation and the formation of gels with material properties that strongly depend on the chosen reaction conditions, Ca^{2+} concentration and thrombin concentration [11, 60]. The obtained hydrogels show excellent biocompatibility for many tissue-engineering applications [32], and the materials are commonly used as wound sealant in surgical procedures. However, their long-term stability is very limited, as fibrin is readily degraded by fibrinolysis in the patient, which is the naturally occurring elimination mechanism during wound healing. Nevertheless, fibrin is a frequently used polymer due to the extensive data available for this material based on previous surgical and wound treatment procedures.

Similar to fibrin, hydrogel-forming silk proteins have also been investigated as possible scaffold materials. Gelation time and mechanical strength of silk hydrogels are strongly dependent on the protein concentration as well as the presence and addition of metal ions that affect protein folding [46]. However, for all applications involving silk, the solvents (LiBr solutions or hexafluoroisopropanol) used for preparation of the gels are very toxic and must be completely removed prior to contact with cells. Therefore, silk hydrogel preparation can only be performed prior to the addition of cells to the tissue-engineering material.

37.4.1.3
Hyaluronan/Hyaluronic Acid

As an alternative to protein-based, hydrogel-forming polymers, several naturally occurring polysaccharides were used to prepare tissue-engineering scaffolds and cell substrates. Various salts of hyaluronic acid, which are found all over the human natural extracellular matrix, are an example of such a polysaccharide. Hyaluronic acid is naturally involved in tissue repair and is also the main component of the ECM of cartilage, making it an ideal material for cartilage tissue engineering [94]. Because of its hydrophilic nature, it requires further modification with adhesion-mediating peptides to allow sufficient cell attachment. Additional methods of chemical cross-linking using different linkers have been investigated to improve the mechanical properties of the obtained hydrogels for long-term applications [23, 54, 91]. Due to its natural origin and the fact that hyaluronic acid is a component of the natural ECM, it is widely accepted as a hydrogel-forming polymer and is already used clinically as a wound dressing and an antiadhesive barrier.

37.4.1.4
Alginate and Chitosan

Alginate is also a hydrophilic and negatively charged polysaccharide that can be used to form hydrogels.

It is derived from brown algae and is obtained after several extraction and hydrolysis steps [2]. The most notable characteristic of this gel-forming polymer is the physical gelation mechanism that is mediated by divalent cations such as calcium (Ca^{2+}) or magnesium (Mg^{2+}). This physical gelation only occurs with consecutive blocks of guluronic acid (G-blocks), which is one of the two components of alginate, and works through the formation of egg-carton-like structures that are able to complex the calcium ions between neighboring polymer chains. The other component of alginate, isomeric mannuronic acid blocks (M-blocks), does not take part in the cross-linking step. As a result, the properties of the obtained gels can be manipulated in many ways, including the alteration of the hydrolysis time to give control of the chain length or selection of the plant origin of the alginate to change the statistical occurrence of the G-blocks. Because their properties are tailorable, alginate hydrogels are commonly used for cell encapsulation and embedding. Alginate hydrogels also present a further advantage, because gelation can be initiated by simply dripping cell-containing alginate solutions into Ca^{2+} containing buffers, eventually leading to the obtainment of single encapsulated cells [50, 73].

Other applications of alginate as a hydrogel-forming polymer make use of chemically cross-linked or modified gels, which allow further control over gel stability and degradation. These properties depend primarily on the removal of divalent cations from the physically cross-linked gels. As an example of how modified gels function, incorporation of oxidized species enhances degradation of the polymer chains, while additional chemical cross-links can slow down the removal of the hydrogel from the application site [2].

Chitosan, which is derived from arthropod exoskeletons, is another polysaccharide that plays an important role as a hydrogel-forming polymer for tissue engineering. Because of its chemical structure, it shares some characteristics with glycosaminoglycans from articular cartilage of mammals, and it is therefore used frequently as a scaffold material for cartilage and bone tissue engineering [29, 34]. Depending on preparation conditions, chitosan possesses varying amounts of deacetylated amine groups, and this has a pronounced effect on its solubility and the ability to control its gelation properties. Solubility and gelation are mediated via the positive charge of the amine group present in the glucosamine residues. These groups are ideal anchors for chemical modifications, as the amines are much more reactive than surrounding hydroxyl groups. Both alginate and chitosan are readily available and exhibit several potential sites for functionalization, making them very interesting for further study and highlighting their potential for future applications.

37.4.2
Synthetic Polymers

Synthetic polymers have been developed for the preparation of suitable hydrogels for tissue engineering as an alternative to the aforementioned natural polymers. Synthetic polymers offer a more diverse set of material properties along with an enhanced ability to adjust the properties of the resulting hydrogels, since copolymerization with other monomers and alteration of the polymer block structure can result in dramatic changes of the overall hydrogel properties (Table 37.2). Additionally, synthetic polymers are less prone to undesirable issues such as remaining byproducts (allergenic or pathogenic) and batch-to-batch variations, which are common problems associated with natural polymers. Finally, synthetic polymers can be specifically designed to eliminate the risk of immunogenicity and pathogenicity that can come along with some natural polymers.

37.4.2.1
Poly(lactic acid) and Poly(propylene fumarate) Derived Copolymers

One of the most prevalent examples of synthetic polymers, and one that exemplifies how easily synthetic polymer properties can be adjusted, is poly(lactic acid) (PLA). PLA polymers are generally considered to be lipophilic polymers that only take up about 5–10% water in aqueous surrounding, meaning that they would not traditionally be classified as hydrogel-forming polymers. However, through copolymerization with more hydrophilic monomers or the incorporation of short poly(ethylene glycol) (PEG) chains, PLA polymers can even be rendered water soluble.

Table 37.2 Examples of synthetic hydrogel-forming polymers

	Structure	Modifications/cross-linking
Derivatives of lipophilic polymers	Poly(lactic acid) copolymer	Chemical cross-linking of the unsaturated double bonds with radical initiation (temperature or UV light)
	Poly(propylene fumarate) copolymer	
PEG derivatives	Poly(ethylene glycol) acrylate	Chemical cross-linking of the unsaturated double bonds with radical initiation (temperature or UV light)
	Oligo(poly(ethylene glycol)fumarate)	
Vinyl- and acrylate-derived polymers	Poly(vinyl alcohol)	Physical cross-linking; chemical cross-linking with glutaraldehyde and other divalent linkers
	Poly(hydroxy ethyl methacrylate)	
Synthetic peptides	Different compositions of oligo- and polypeptides, eventually in combination with other polymers	Physical cross-linking with subsequent chemical fixation

In this case, an additional incorporation of cross-linkable functional groups is required to achieve a suitable cross-linked network with sufficient stability during application [88, 103]. Even with the extensive modifications, the still-present PLA segments exhibit two important advantages. First, the segments support cell adhesion to a certain extent, and second, the segments are degradable, making the entire cross-linked gel network biodegradable and facilitative to elimination from the patient [18, 20].

Similar modifications have now also been used on other lipophilic polymers that were already being used for tissue engineering. By introducing PEG chains into the extremely lipophilic (i.e., water uptake ~ 0%) poly(propylene fumarate) (PPF), copolymers have been obtained that are capable of forming thermo-reversible physically cross-linked hydrogels due to their alternating three-block structure [4, 6]. In contrast to the hydrogels formed with PLA, PPF copolymers offer the possibility for permanent chemical cross-links due to the unsaturated double bonds in the lipophilic segments of the polymer [5, 92]. Radical cross-linking using temperature or UV-light-induced decomposition of radical initiators, therefore, leads to the formation of links between the individual polymer chains, while the still-present ester bonds render the whole hydrogel biodegradable [7]. PLA and PPF present unique possibilities through the adjustment of the hydrophilic–lipophilic balance, allowing for the control of cellular differentiation and behavior. Inclusion of degradable links enhances biodegradability and the control of the material's fate once it is in the patient.

37.4.2.2
Poly(ethylene glycol) Derivatives

To obtain hydrogels from PEG, which is a readily water-soluble polymer, several chemical modifications are established with different cross-linking strategies and resulting hydrogel structures [52]. PEG derived hydrogels make up the majority of past and present hydrogel systems, and this is due to the excellent biocompatibility and the versatility of this material. Through chemical modification, PEG allows tremendous variations in obtained hydrogels with respect to mesh size, swelling extent and cross-linking chemistry, as well as degradability.

Since PEG polymer chains of all lengths are readily water soluble, a suitable terminal functional group has to be introduced to enable cross-linking and the formation of a stable network. In addition to the already-mentioned acrylate groups [15, 19, 108], several other systems have been used, including double-bond-containing fumarate esters [43, 44, 99, 100], vinyl sulfon derivatives cross-linkable with thiol groups [61, 64] and N-hydroxy succinimide esters, which covalently bind amine groups to corresponding amides [14]. The latter two functionalization methods require two different polymer components in two hydrogel-forming polymer solutions. Specifically, the active polymer and a second component that contains either the amine or the thiol functionalities must be present. For the second component, either suitable PEG derivatives or custom-synthesized peptides bearing the relevant functionality for cross-linking can be used. In all cases, the resulting hydrogels must be obtained with suitable peptide adhesion sites, or with enzymatically degradable cross-links, which favor the controlled elimination from the patient [86, 87]. Adjusting the cross-linking and polymer parameters enables many variation possibilities, and this is what makes PEG-derived hydrogels a prime material for tissue engineering, especially in situations where new adhesion ligands must be used or strategies for the controlled degradation of hydrogel systems need to be investigated [28].

37.4.2.3
Vinyl- and Acrylate-Derived Polymers

Another frequently used polymer class is based on hydrophilic derivatives of vinyl- and acrylate-derived polymers. Prominent examples of these polymers include poly(vinyl alcohol) (PVA) and poly(hydroxy ethyl methacrylate) (PHEMA). These polymers tend to form weak physically cross-linked hydrogels on their own. For long-term applications, they require additional chemical cross-linking [17, 68, 89]. Due to the fact that the main polymer chains lack ester bonds and are not degradable in vivo, the resulting hydrogels are either not degraded at all in the patient or copolymerization with degradable PEG derivatives or degradable cross-links yields hydrogels that are degraded at the cross-link and then are sufficiently small to be excreted renally. Hydrogels derived from these polymers are usually used for in vitro investigations and studies of cellular adhesion, where their long-term stability is of great advantage. One exception involves the cultivation of cells without the application of proteolytic enzymes for cell harvest. Methods have been described in which cells are cultured on a gel substratum, altering its hydrophilicity upon temperature change. This eventually leads to the detachment of whole cell sheets without the detrimental effects of trypsinisation on cell surface receptors [69, 76].

stability that is needed for a given application, it is important to determine if the cross-linking can be performed before the application or inside the defect with implanted cells or adjacent to the healthy tissue. The most important consideration for the cross-linking process is the biocompatibility of the chosen chemicals and of the cross-linking reaction, which determines the survival of cells in and next to the hydrogel. If a certain cross-linking process is biologically safe, the immediate in situ gelation of the hydrogel system inside the defect offers unique advantages, such as a perfect fit in the defect and the elimination of void spaces, in addition to the significantly decreased size of scars and trauma to the patient.

37.4.2.4
Synthetic Peptides

Beyond the synthetic polymers described herein, recent developments are aimed at the application of synthetic self-assembling peptides for the creation of hydrogel biomaterials [9, 12, 110]. These specially designed peptides consist of short oligomers that aggregate in aqueous solution and form distinct beta sheet structures with charged amino acids on the outside, allowing for further interactions with ions contained in physiological fluids. The observed hydrogel formation can be compared with the fibrin gel formation. Further modifications of the short peptides and the architecture of the formed hydrogels will certainly lead to improved hydrogels suitable for tissue engineering. Development of new materials based on this strategy is destined to lead to new and promising materials that have tailorable gelling properties and long-term stability in vivo.

37.5
Mechanisms of Cross-Linking

It has been pointed out that many of the aforementioned polymers need additional cross-linking and chemical modification to obtain appropriate hydrogel properties. In addition to deciding on the gel

37.5.1
Physical Cross-Linking

In general, all physical cross-linking processes provide weak and non-permanent bonds between single polymer chains. The simplest gel involves the formation of physical entanglements between different polymer chains (Fig. 37.2b). The length or the branching of the polymer controls the strength of such gels to a certain extent; however, the gels will always lose mechanical integrity upon dilution, making it extremely challenging to apply them in vivo, especially if the defect site is not properly confined.

A second possibility for the formation of physically cross-linked gels exists through the formation of lipophilic interactions between individual polymer molecules. To achieve this, the polymers must contain segments that tend to alter their hydration shells in response to increased solution temperature or other environmental changes. Increased van der Waals interactions then lead to the aggregation of the chains and subsequent gel formation (Fig. 37.2d). Typically, these thermo-reversible systems are liquids at room temperature, and they allow injection of cells and handling of the system. After initial steps are completed, they are heated to near body temperature, causing the gelation of the system. As indicated by the name, the process can be reversed by lowering the temperature; this is why physical cross-linking of this nature is often combined with a permanent chemical cross-linking step.

A third type of physical cross-linking is mediated by opposite charges of the polymers and added divalent cations or anions. This effect is primarily observed with anionic polysaccharide-derived polymers that can be easily solidified using calcium ions. Alginate is an example for a gel that uses this type of physical cross-linking; a simple method can be used to encapsulate single cells by dropping an alginate-containing cell suspension in a solution containing calcium ions (Fig. 37.2c). This ionic interaction applies also to chitosan-derived hydrogels, which can be solidified via the addition of oppositely charged polymers, such as glycosaminoglycans or even alginates. Despite the fact that a more robust hydrogel is formed by the addition of the counterions, the obtained gels are sensitive to changes in the ionic milieu, which can easily lead to a washout of the calcium ions in the case of alginate. Consequently, physically linked gels of this nature are often additionally stabilized using permanent chemical cross-links together with the ionic cross-linkers.

37.5.2
Chemical Cross-Linking

37.5.2.1
Radical Cross-Linking

Probably the most frequently utilized cross-linking scheme for chemically stabilized hydrogels is radical cross-linking. This is because of the easy initiation of the cross-linking step and the versatile preparation of suitable hydrogel-forming polymers, which need only to be modified with acrylic acid to introduce the necessary double bonds for radical cross-linking. This is generally achieved by attaching acryloyl chloride to terminal hydroxyl or amine functionalities; alternatively, monomers can be introduced into the polymer that already carry the double bond within the chain. One possibility for the latter procedure involves the integration of fumaric acid into the polymer. Fumaric acid provides two possible acid groups for further polymerization and also carries a double bond that can be exploited for later cross-linking.

After incorporating the double bond, the linkage between two adjacent chains is formed by the activation of radical starters that produce the more stable starter radicals (Fig. 37.3a). Two different types of cross-linking exist: in one case, the cross-linking is initiated by increased temperature, while in the other, cross-linking begins when radical starters decompose upon irradiation by UV light [74]. Especially with the second procedure, the biocompatibility of the cross-linking process must be carefully evaluated, since cells present during the cross-linking step can be harmed by either the formed radicals or the UV light. If the hydrogels come into contact with cells shortly after the cross-linking procedure, they should be carefully rinsed to remove residual amounts of toxic radicals and monomers.

37.5.2.2
Aldehyde Linkers

Divalent aldehydes can be used to form linkages between polymer chains as an alternative to the radical procedure. However, this procedure involves even harsher conditions, since first the amines of the used polymers are reacted with the aldehyde, yielding the Schiff's base, which in a second step often has to be reduced to obtain a stable amine linkage (Fig. 37.3b). This process also harms living cells; therefore, aldehyde cross-links are usually applied for prefabricated hydrogels in the absence of cells.

37.5.2.3
Active Ester Linkers

A more gentle method for the cross-linking of hydrogels involves the application of active esters of carboxylic acids. To obtain these molecules, carboxylic acids contained in the polymers are converted into esters of N-hydroxy succinimide (NHS), which yields an ester that is readily attacked by free amine groups and subsequently converted into stable amide linkages (Fig. 37.3c). Storage of the active ester requires the absence of water to eliminate hydrolysis of the ester form, which reduces the reactivity of the polymer. One drawback of this cross-linking procedure is the fact that two different components have to be mixed thoroughly shortly before application to obtain homogenous gels. For successful formation of the network structure, an amine-containing compound must be included to react with the NHS esters. This can be done, for example, by addition of amine derivatives of PEG or by adding peptides that bear two amine functionalities or other bifunctional amine components [14].

Fig. 37.3 **a** Radical cross-linking of unsaturated acrylate groups by UV light or thermally decomposing radical initiators. **b** Cross-linking of amine groups using bivalent glutaraldehyde and reductive amination. **c** Formation of amide bonds between succinimidyl esters and amine groups. **d** Michael-type addition of thiols to vinyl sulfon groups

37.5.2.4
Thiol-Reactive Linkers

A final cross-linking process makes use of thiol-reactive compounds such as vinyl sulfones or vinyl sulfoxides, which react with thiol-containing compounds in a comparable fashion to active esters (Fig. 37.3d) [62, 87]. Possible advantages of these systems include the reduced extent of hydrolysis in aqueous polymer solutions and a reduced frequency of free thiol groups in the natural environment, which reduces possible reactions with cell surface proteins and adjacent tissues. However, it must be mentioned that the formed thioether linkage with cysteine groups does not occur for common proteins and peptides, and the gels will therefore release modified amino acids to the surrounding tissues. Additionally, the cross-linking has to be conducted at a slightly higher pH to activate the free thiol groups sufficiently, and this altered environment can be harmful to some cell types. Concurrently, care must be taken to prevent oxidation of the free thiol groups to disulfides.

37.6
Design Criteria and Desirable Properties for Hydrogels

37.6.1
Mechanical Properties and Mesh Size

Mechanical properties, such as elasticity and stiffness, are the most important material characteristics of hydrogels, as they determine the capability of the hydrogels to withstand mechanical forces exerted on them by attached cells and surrounding tissues [15]. The mechanical stability of a hydrogel can be influenced in multiple ways, including alteration of the polymer content and cross-linking density. Increasing either parameter leads to increased gel strength [49]. However, changing the initial hydrogel composition can also lead to significantly decreased mesh size within the polymer network, and this prevents or hinders the ingrowth of cells that is needed for the formation of the replacement tissue. This is a result of the fact that many more links must be cleaved in order to provide enough space for cell growth. Alternative approaches to increase the mechanical stability of hydrogels rely on designing copolymers with mechanically stronger monomers or creating closer aggregates of single polymer strands to increase the overall polymer strength.

Beyond the original composition of the hydrogel, material properties are also affected during the application or degradation of the hydrogel. The extent of water uptake in the gels determines the swelling equilibrium, which is dependent on the balance of lipophilic and hydrophilic regions in the polymer. High water uptake in the hydrogels can therefore lead to an increased mesh size and a corresponding decrease in mechanical stability. Additional changes can occur, if the polymer chains or the incorporated cross-links are sensitive to enzymatic activity or are susceptible to hydrolysis. Both of these effects lead to a breakdown of the hydrogel network structure and a subsequent loss of mechanical stability. In the case of controlled cellular ingrowth, these effects are desirable and actually intended. In all situations, a meticulous mechanical and structural evaluation of the hydrogels in combination with cells or physiological media must be undertaken to insure their successful application in vitro or in vivo.

37.6.2
Permeability for Drug Substances and Nutrients

For many applications, hydrogel systems must also be able to provide sufficient mobility of incorporated substances, such as nutrients or oxygen in the case of cellular ingrowth. Controlled release is also possible for the delivery of growth factors and other biologically active molecules, which are used to locally influence cell behavior. In both of these cases, beyond mesh and pore size, the charge or the adhesiveness (lipophilicity) of the network polymers is an important consideration as it can influence the diffusion of factors that influence cell growth and differentiation [105]. Networks with a net charge similar to the factor of interest can prevent these molecules from entering the void space, while opposite charges can lead to strong attractive interactions between the molecules and the polymer network, restricting further diffusion and movement of molecules. This restriction is especially useful for long-term drug release [40, 41]. Attractive interactions can furthermore be used to immobilize active substances such as growth factors on the hydrogel network, e.g. by forming complexes with chemically bound heparin. This approach can then be used to locally store proteins in the hydrogel in a way that mimics the physiological processes in natural ECM. Evaluation of the mobility of the applied substances can be performed using techniques like NMR and FRAP, which will be described later in this chapter.

37.6.3
Biomimetic Modifications—Interaction with Cells

To provide sufficient interactions with cells, most hydrogels must be modified to support the adhesion and the growth of cells. Most approaches mimic mechanisms of the natural ECM, and therefore they are considered as biomimetic [30]. Because hydrogel-forming polymers are very hydrophilic, they suppress the adsorption of most proteins. This is an advantage for applications such as controlled drug delivery, where loss of proteins due to adsorption is undesirable. However, for processes like cell adhesion, adsorbed proteins are necessary to mediate the interactions of the biomaterials with the cellular adhesion receptors, e.g., integrins. Consequently, unmodified hydrogels exhibit a very low cellular adhesion capacity and must be altered in order to adsorb an appreciable amount of serum proteins, or they must contain the relevant receptor ligands for the attachment of cells [109]. For this purpose, short peptides derived from natural extracellular proteins like fibronectin, osteopontin, collagen or laminin are chosen to provide sufficient interactions.

It has been shown that for many lipophilic biomaterials, peptide sequences like RGD (derived from fibronectin), GER (derived from collagen) or their combinations with other short peptides sequences provide a sufficient number of interactions to mediate cellular attachment [39, 47, 58, 111]. Further modifications of this sequences, such as by steric fixation of the amino acids by cyclization, are able to address certain integrin receptor subtypes, enabling control of the adhering cell types [39]. To attach these sequences to hydrogels for tissue engineering, the peptides can be linked to the backbone of the hydrogel-forming polymer, in which case they are difficult to access for cells, or they can be added to the hydrogel by incorporation of a spacer molecule, which provides enough flexibility for the interaction with the cellular receptors [27]. For the latter, PEG derivatives of the peptide can be used. For example, the PEG derivative can be linked to the hydrogel via terminal acrylate bonds during the polymerization step of the whole hydrogel [27, 67].

Further modifications of hydrogels can be used to stimulate the production of an improved extracellular matrix in newly formed tissue. This is for example needed in cases such as during the mineralization of engineered bone. In this process, carboxylic or phosphoric acid groups included in the hydrogel structure can support the formation of nucleation sites for the biomineralization of calcium phosphate or calcium carbonate. The mineralization is needed to provide stability for bone tissue [35, 95].

One other improvement to hydrogels that can be made involves using them as a combination of cell carrier and drug-releasing system. This is achieved by incorporating drug substances like growth factors or small chemical entities in the hydrogel or a separate carrier [96, 97, 107]. These additionally released or incorporated substances can further improve the growth and differentiation of the adhering cells. For these applications, hydrogels offer significant advantages, as they do not interact unspecifically with most of the applied proteins or drug substances. This reduces the likelihood of denaturation and additionally allows free diffusion of the applied molecules to the cells or their site of action. The strategies that are applied to attain these combination systems include the addition of soluble bioactive substances to the polymer solvent, the addition of protein-loaded microparticles for prolonged release [40, 41, 42] or, the most sophisticated strategy, the covalent attachment of the growth factors to the hydrogel network or other biomaterials. These last systems allow increased control over stimulation of exposed receptors on the cell surfaces. Longer application of factors is also possible with this technique, and their local activity can be better controlled, since enzymatic inactivation of the proteins is reduced and their diffusion away from the defect site is hampered due to the covalent immobilization [28, 53]. Covalent immobilization is even more important for the attachment of small chemical entities, which would otherwise readily diffuse away from the application site. Degradable attachment sites for low molecular weight drugs even allow retarded release rates to affect cells in long-term applications [8].

37.6.4
Degradation and Elimination from the Patient

To control the last important property of in vivo applied hydrogels, their elimination from the patient, several strategies have been developed to create usable materials. An essential prerequisite for elimination is the breakdown of the hydrogel into its separate water-soluble polymer chains, which must be small enough to pass the renal barrier for elimination. In the case of PEG derivatives, the upper size limit is about 40,000 Da [102]. The breakdown can be mediated by hydrolytically cleavable bonds, like esters, or more specifically by enzymatic cleavage if protease-sensitive peptide cross-links are used. For natural polymers and also PLA copolymers, the physiological metabolism of the monomers allows the elimination through other pathways such as the digestion of the implanted hydrogel.

Since degradation plays a major role, possible degradation mechanisms and kinetics need to be thoroughly characterized before application of a new hydrogel system in vivo. Important considerations include the swelling extent during degradation and the toxicity of the released breakdown product. Several techniques have been applied to elucidate the degradation kinetics, and it is actually possible to tailor materials according to the development of the implanted cells and the subsequently growing tissues.

37.7
Tools for the Characterization of Hydrogels

Several of the described properties of hydrogels, along with the sophisticated variations that have been shown, require analytical techniques to prove that the correct structure of the system has been attained. Some of the relevant methods are described in the following sections.

37.7.1
Nuclear Magnetic Resonance

Nuclear magnetic resonance (NMR) spectroscopy is a powerful analytical method that exploits the magnetic properties of certain nuclei. The most common techniques, ^{1}H NMR and ^{13}C NMR spectroscopy, have been widely applied to study structural parameters of polymers such as composition, tacticity and sequence distribution. NMR spectroscopy and related technologies are also useful for the characterization of hydrogels [70]. Since the spin-lattice (T_1) and spin-spin relaxation times (T_2) depend on the mobility of the molecules, NMR relaxation experiments may be used to distinguish between chemical species of different mobility (e.g., monomers and polymers). For example, measuring T_2 allows the monitoring of cross-linking reactions. High-resolution ^{13}C NMR spectra can be used to estimate the cross-linking density of hydrogels, as the line-width of NMR signals increases with decreasing T_2. NMR relaxation studies may be used for the quantification of bound, intermediate and free water within swollen hydrogels.

Magnetic resonance imaging (MRI) is another NMR-based technology that has been successfully applied to the characterization of hydrogels. MRI has been used to monitor changes in water uptake, thus allowing studies on the swelling processes with high spatial and temporal resolution [48, 59, 84]. Ramaswamy et al. used MRI as a noninvasive technique to investigate the integration of hydrogels into articular cartilage [85]. Modified MRI systems further enable nondestructive and localized characterization of mechanical properties of various materials including hydrogels, living tissues and tissue-engineered constructs [65, 66, 77].

37.7.2
Swelling Studies for the Determination of Network Parameters

Besides a structural characterization, which includes evaluating molecular weight and extent of cross-linking, swelling studies of hydrogels are used to determine the characteristics of the polymer networks. The nanostructure of cross-linked gels can be described by three different parameters: (1) the polymer volume fraction in the swollen state, v_{2s}, (2) the number average molecular weight between cross-links, $\overline{M}_c$, and (3) the theoretical network mesh size or the distance between two adjacent cross-links, ξ [59, 80] (Fig. 37.4). The polymer volume fraction in the swollen state (v_{2s}) describes the amount of water that can be imbibed by the hydrogel. It is the reciprocal of the volumetric swelling ratio (Q), and it is defined as the ratio of the polymer volume (V_p) to the volume of the swollen gel (V_g) [59].

The molecular weight between two adjacent cross-links ($\overline{M}_c$) characterizes the degree of cross-linking. In neutral networks, $\overline{M}_c$ can be expressed by the Flory-Rehner equation as a function of v_{2s}, the average molecular weight of the linear polymer chains ($\overline{M}_n$), the specific volume of the polymer ($\overline{v}$), the molar volume of water (V_1), and the polymer-water interaction parameter (χ_{12}) [59]. More complex versions of the Flory-Rehner expression can be used to describe the swelling of ionic gels or other hydrogels. These have been developed by Peppas et al. and can be found in the literature [80].

The network mesh size, ξ, is an indicator of the distance between consecutive cross-links. It is affected by $\overline{M}_c$, the chemical structure of the monomer, and further external stimuli (e.g., temperature, pH and ionic strength) [22, 59]. In the swollen state, the mesh size of hydrogels typically ranges from 5 to 100 nm. Comparing these values with the hydrodynamic radii of entrapped molecules allows an estimation of their rates of diffusion. Most small-molecule drugs, for example, will not be retarded in gel matrices; sustained release of macromolecules such as proteins, however, is possible for swollen hydrogels [59, 101]. Cross-linked hydrogels can also serve as carriers for particulate drug-delivery systems [101] or as scaffolds for living cells [31], where there is no mobility with the cross-links still in place.

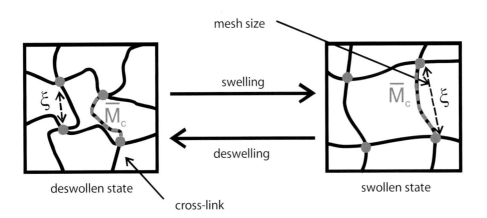

Fig. 37.4 Schematic of mesh size in hydrogels at shrunken or swollen states

37.7.3
Fluorescence Recovery After Photobleaching

Fluorescence recovery after photobleaching (FRAP) is a versatile tool to study the mobility of fluorescent molecules at the microscopic level. The basic instrumentation for FRAP experiments is comprised of an optical microscope equipped with a light source to bleach arbitrary regions within the sample and some fluorescent probe molecules. Most confocal laser scanning microscopes (CLSMs) provide these features, making them an excellent standard tool to perform FRAP measurements. The experiment begins by recording a background image of the sample. This is called the prebleach image. Next, an intense laser beam bleaches the molecules in the region of interest, causing a drop in fluorescence compared to the prebleach image. Immediately after the bleaching process, the recovery of fluorescence within the bleached spot due to diffusion of fluorescent molecules from the surrounding unbleached areas into the bleached area is measured by an attenuated laser beam. If all fluorescent molecules in the sample are mobile, the fluorescence intensity in the bleached area will recover, and it will be similar to the prebleach intensity. If some of the fluorescent molecules are immobilized, the final intensity will be less than the prebleach intensity. The diffusion coefficient of the fluorescent molecules (D) can be calculated from the obtained recovery profile [13, 72, 93]. Typical diffusion coefficients for molecules in phosphate-buffered saline are 3.83×10^{-6} cm²/s for a low molecular weight fluorescent dye and 3.64×10^{-7} cm²/s for polysaccharides with molecular weights of 100 kDa. Once these molecules are included in agarose hydrogels, the diffusion of the dye is reduced to 50%. The larger polysaccharide, however, has only one-tenth of its original diffusivity [51].

In addition to characterization of the diffusivity of molecules within hydrogels [10, 24, 82], FRAP experiments also provide deeper insights into the hydrogel architecture itself. De Smedt et al. investigated the mobility of fluorescein isothiocyanate (FITC)-labeled dextrans in dextran methacrylate gels and correlated the results with rheological measurements and release data [25]. Gribbon et al. used FRAP techniques to study the macromolecular organization of aggrecan, a high molecular weight proteoglycan that forms networks at concentrations close to those found physiologically [36]. The network architecture of hydrogels was also studied by Pluen et al. They measured the diffusion coefficients (D) of different types of macromolecules such as DNA, dextrans, polymer beads and proteins in 2% agarose gels. Fitting the data to different theoretical models allowed for the estimation of various gel parameters, including mean pore size, agarose fiber radius, and hydraulic permeability [83].

Leone et al. prepared microporous hydrogels from different polysaccharides, including hyaluronate, alginate, and carboxymethylcellulose by forcing CO_2 bubbles through the cross-linked matrices. Subsequent FRAP experiments showed a clear relationship between the pore size and the diffusivity of bovine serum albumin (BSA) [57]. Similarly, FRAP techniques were used by Guarnieri et al. to study the effects of fibronectin and laminin on the structural, mechanical and transport characteristics of collagen gels [37]. A model that allows for the prediction of diffusivities in complex, multicomponent hydrogels, such as the natural ECM, was developed by Kosto et al. [51]. FRAP techniques have also been applied to degrading networks [21, 71]. Burke et al. examined the influence of enzyme treatment on the diffusion of FITC-dextrans in guar hydrogels. Depending on the used enzyme, β-mannanase or α-galactosidase, they found an increase or decrease in the diffusivities of FITC-dextrans, respectively. This was due to the fact that β-mannanase cleaves glycosidic bonds in the mannan backbone of guar, causing a drop in viscosity, while α-galactosidase removes galactose branches, inducing syneresis of the guar network [21].

37.7.4
Rheological Characterization of Hydrogels

Among other methods, the mechanical properties of hydrogels can be characterized by tensile tests, compression tests and dynamic mechanical analysis (DMA) [1]. For uniaxial tensile testing, dog-bone-shaped samples are placed between two clamps and stretched at constant extension rates. From these experiments, the Young's modulus of the material can be determined. It is defined as the ratio of tensile stress to tensile strain, whereas the maximal tensile stress carried by a material is defined as the tensile strength. Similarly, the compressive modulus is defined as the ratio of compressive stress to compressive strain. Compressive testing is generally performed by uniaxially compressing cylindrical specimens between two smooth and impermeable platens. This is known as unconfined compressive testing. The compressive strength is defined as the maximal compressive stress that a sample can withstand. Both Young's modulus and compressive modulus are measures of the stiffness of a given material, which reflects the resistance of an elastic body against the deflection of an applied force.

DMA is typically performed to assess the viscoelastic behavior of materials. In rheological terms, "viscoelastic" means the concomitance of viscous or "liquid-like" and elastic or "solid-like" behavior. For DMA assessments, an oscillating rheometer applies a sinusoidal shear load to the sample. A stress transducer measures the applied shear stress (σ^*), while the strain induced in the sample (γ^*) is measured using a strain gauge. The complex shear modulus G^* is defined as follows:

$$G^* = G' + iG'' = \frac{\sigma^*}{\gamma^*} \tag{37.1}$$

G' is referred to as the real part of G^*, or the elastic or storage modulus, and it represents the relative degree of a material to recover or exhibit an "elastic response." G'' is known as the imaginary part of G^*, or the viscous or loss modulus, and it represents the relative degree of a material to flow or display a "viscous response." Measuring G^* against the shear stress or shear strain, respectively, permits the determination of the stiffness and strength of a given hydrogel [3, 45].

Oscillatory shear experiments can be used to monitor gelation kinetics and to distinguish between different types of network architectures [45]. In situ gelling systems, for example, can be characterized by measuring the evolution of G' and G'' over time (Fig. 37.5). Typically, the uncross-linked sample will behave like a viscous fluid ($G'' > G'$). After an initial lag time, both moduli increase, but G' increases faster than G''. The crossover of G' and G''

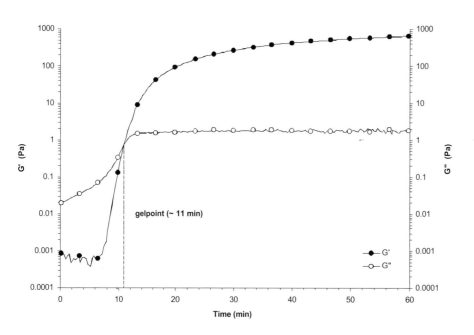

Fig. 37.5 Rheogram of the gel formation of a chemically cross-linking polymer based on NHS esters and amine functionalized poly(ethylene glycols). For further details on the polymer and measurement conditions, see [14]

at a certain time point of the experiment is regarded as the gelpoint. Thereafter, the structure will behave predominantly as an elastic solid ($G'>G''$). Both moduli will reach a plateau when the cross-linking reaction is complete. Similar experiments can be performed for thermosensitive systems by measuring G' and G'' against temperature.

Recording G' and G'' against the oscillatory frequency allows the distinguishability of covalently cross-linked hydrogels from weaker entangled networks [45]. While in the former case, both moduli are frequency insensitive, there is a "crossover" of G' and G'' at a characteristic frequency for entanglement networks. These gels tend to flow as high viscous liquids at low frequencies during storage, but they behave like elastic solids under high-frequency loads such as those encountered during injection or spraying processes. Thus, gel type can be identified through the use of a simple frequency sweep experiment.

37.8
Conclusions and Outlook

Hydrogel materials that have been developed thus far have excellent biocompatibility due to the enormous amount of incorporated water and the unhindered diffusion of oxygen and small molecular weight nutrients. When properly chosen, cross-linking techniques can be applied to a vast variety of different tissues ranging from cartilage to soft tissues. Cell growth can also be maintained and enhanced in these materials using appropriate cross-linking.

Important improvements have been described, including hydrogels that can be remodeled and degraded at different application sites in the patient via enzymatic activity on biomimetic peptide linkages. These altered cross-link structures allow the production of hydrogels that more closely match natural extracellular matrices and present the promise of improved replacement tissues, which are rebuilt and restructured in a way similar to natural wound healing.

Further improvements are still needed for hydrogels that are intended for replacement of load-bearing tissues such as bone. Thus far, the limited mechanical stability of hydrogels has been a major drawback of hydrogels, as they are not comparable to standard replacement ceramics in this regard. For applications like bone, material combinations of drug-releasing hydrogels with solid scaffolds may provide the necessary support for the growth of cells and tissues in vivo, diminishing the need for further external fixation.

References

1. Anseth KS, Bowman CN, Brannon-Peppas L (1996) Mechanical properties of hydrogels and their experimental determination. Biomaterials 17 (17): 1647–1657
2. Augst AD, Kong HJ, Mooney DJ (2006) Alginate hydrogels as biomaterials. Macromolecular Bioscience 6 (8): 623–633
3. Barnes HA, Hutton JF, Walters K (1989) An introduction to rheology. Elsevier, Amsterdam
4. Behravesh E, Jo S, Zygourakis K, Mikos AG (2002) Synthesis of in situ cross-linkable macroporous biodegradable poly(propylene fumarate-co-ethylene glycol) hydrogels. Biomacromolecules 3 (2): 374–381
5. Behravesh E, Mikos AG (2003) Three-dimensional culture of differentiating marrow stromal osteoblasts in biomimetic poly(propylene fumarate-co-ethylene glycol)-based macroporous hydrogels. Journal of Biomedical Materials Research, Part A 66A (3): 698–706
6. Behravesh E, Shung AK, Jo S, Mikos AG (2002) Synthesis and characterization of triblock copolymers of methoxy poly(ethylene glycol) and poly(propylene fumarate). Biomacromolecules 3 (1): 153–158
7. Behravesh E, Timmer MD, Lemoine JJ, Liebschner MAK, Mikos AG (2002) Evaluation of the in vitro degradation of macroporous hydrogels using gravimetry, confined compression testing, and microcomputed tomography. Biomacromolecules 3 (6): 1263–1270
8. Benoit DSW, Nuttelman CR, Collins SD, Anseth KS (2006) Synthesis and characterization of a fluvastatin-releasing hydrogel delivery system to modulate hMSC differentiation and function for bone regeneration. Biomaterials 27 (36): 6102–6110
9. Bergmann A, Teßmar J, Owen A (2007) Influence of electron irradiation on the crystallisation, molecular weight and mechanical properties of poly-(R)-3-hydroxybutyrate. Journal of Materials Science 42(11): 3732–3738
10. Berk DA, Yuan F, Leunig M, Jain RK (1993) Fluorescence photobleaching with spatial Fourier analysis: measurement of diffusion in light-scattering media. Biophys J 65 (6): 2428–2436
11. Blomback B, Bark N (2004) Fibrinopeptides and fibrin gel structure. Biophysical Chemistry 112 (2–3): 147–151
12. Blunk T, Sieminski AL, Gooch KJ, Courter DL, Hollander AP, Nahir M, Langer R, Vunjak-Novakovic G, Freed LE (2002) Differential effects of growth factors

on tissue-engineered cartilage. Tissue Engineering 8 (1): 73–84
13. Braeckmans K, Peeters L, Sanders NN, De Smedt SC, Demeester J (2003) Three-dimensional fluorescence recovery after photobleaching with the confocal scanning laser microscope. Biophys J 85 (4): 2240–2252
14. Brandl F, Henke M, Rothschenk S, Gschwind R, Breunig M, Blunk T, Tessmar J, Göpferich A (2007) Poly(ethylene glycol) based hydrogels for intraoccular applications. Advanced Engineering Materials 9 (12): 1141–1149
15. Bryant SJ, Bender RJ, Durand KL, Anseth KS (2004) Encapsulating chondrocytes in degrading PEG hydrogels with high modulus: engineering gel structural changes to facilitate cartilaginous tissue production. Biotechnology and Bioengineering 86 (7): 747–755
15a. Brandl F, Sommer F, Goepferich A (2007) Rational design of hydrogels for tissue engineering: Impact of physical factors on cell behavior. Biomaterials 28 (2), 134–146
16. Bryant SJ, Cuy JL, Hauch KD, Ratner BD (2007) Photopatterning of porous hydrogels for tissue engineering. Biomaterials 28 (19): 2978–2986
17. Bryant SJ, Davis-Arehart KA, Luo N, Shoemaker RK, Arthur JA, Anseth KS (2004) Synthesis and characterization of photopolymerized multifunctional hydrogels: water-soluble poly(vinyl alcohol) and chondroitin sulfate macromers for chondrocyte encapsulation. Macromolecules 37 (18): 6726–6733
18. Bryant SJ, Durand KL, Anseth KS (2003) Manipulations in hydrogel chemistry control photoencapsulated chondrocyte behavior and their extracellular matrix production. Journal of Biomedical Materials Research, Part A 67A (4): 1430–1436
19. Burdick JA, Anseth KS (2002) Photoencapsulation of osteoblasts in injectable RGD-modified PEG hydrogels for bone tissue engineering. Biomaterials 23 (22): 4315–4323
20. Burdick JA, Philpott LM, Anseth KS (2001) Synthesis and characterization of tetrafunctional lactic acid oligomers: a potential in situ forming degradable orthopaedic biomaterial. Journal of Polymer Science, Part A: Polymer Chemistry 39 (5): 683–692
21. Burke MD, Park JO, Srinivasarao M, Khan SA (2000) Diffusion of macromolecules in polymer solutions and gels: a laser scanning confocal microscopy study. Macromolecules 33 (20): 7500–7507
22. Canal T, Peppas NA (1989) Correlation between mesh size and equilibrium degree of swelling of polymeric networks. J Biomed Mater Res 23 (10): 1183–1193
23. Crescenzi V, Cornelio L, DiMeo C, Nardecchia S, Lamanna R (2007) Novel hydrogels via click chemistry: synthesis and potential biomedical applications. Biomacromolecules 8 (6): 1844–1850
24. Cutts LS, Roberts PA, Adler J, Davies MC, Melia CD (1995) Determination of localized diffusion coefficients in gels using confocal scanning laser microscopy. J Microsc 180 (2): 131–139
25. De Smedt SC, Meyvis TKL, Demeester J, Van Oostveldt P, Blonk JCG, Hennink WE (1997) Diffusion of macromolecules in dextran methacrylate solutions and gels as studied by confocal scanning laser microscopy. Macromolecules 30 (17): 4863–4870
26. De SK, Aluru NR, Johnson B, Crone WC, Beebe DJ, Moore J (2002) Equilibrium swelling and kinetics of pH-responsive hydrogels: models, experiments, and simulations. Journal of Microelectromechanical Systems 11 (5): 544–555
27. DeLong SA, Gobin AS, West JL (2005) Covalent immobilization of RGDS on hydrogel surfaces to direct cell alignment and migration. Journal of Controlled Release 109 (1–3): 139–148
28. DeLong SA, Moon JJ, West JL (2005) Covalently immobilized gradients of bFGF on hydrogel scaffolds for directed cell migration. Biomaterials 26 (16): 3227–3234
29. Di Martino A, Sittinger M, Risbud MV (2005) Chitosan: A versatile biopolymer for orthopaedic tissue-engineering. Biomaterials 26 (30): 5983–5990
30. Drotleff S, Lungwitz U, Breunig M, Dennis A, Blunk T, Tessmar J, Gopferich A (2004) Biomimetic polymers in pharmaceutical and biomedical sciences. European Journal of Pharmaceutics and Biopharmaceutics 58 (2): 385–407
31. Drury JL, Mooney DJ (2003) Hydrogels for tissue engineering: scaffold design variables and applications. Biomaterials 24 (24): 4337–4351
32. Eyrich D, Brandl F, Appel B, Wiese H, Maier G, Wenzel M, Staudenmaier R, Goepferich A, Blunk T (2007) Long-term stable fibrin gels for cartilage engineering. Biomaterials 28 (1): 55–65
33. Falbe J, Regitz M (2007) Roempp Chemie Lexikon, 9th edn. Georg Thieme Verlag, Stuttgart
34. Francis Suh J-K, Matthew HWT (2000) Application of chitosan-based polysaccharide biomaterials in cartilage tissue engineering: a review. Biomaterials 21 (24): 2589–2598
35. Grassmann O, Lobmann P (2003) Biomimetic nucleation and growth of $CaCO_3$ in hydrogels incorporating carboxylate groups. Biomaterials 25 (2): 277–282
36. Gribbon P, Hardingham TE (1998) Macromolecular diffusion of biological polymers measured by confocal fluorescence recovery after photobleaching. Biophys J 75 (2): 1032–1039
37. Guarnieri D, Battista S, Borzacchiello A, Mayol L, De Rosa E, Keene DR, Muscariello L, Barbarisi A, Netti PA (2007) Effects of fibronectin and laminin on structural, mechanical and transport properties of 3D collagenous network. J Mater Sci Mater Med 18 (2): 245–253
38. Hern DL, Hubbell JA (1998) Incorporation of adhesion peptides into nonadhesive hydrogels useful for tissue resurfacing. Journal of Biomedical Materials Research 39 (2): 266–276
39. Hersel U, Dahmen C, Kessler H (2003) RGD modified polymers: biomaterials for stimulated cell adhesion and beyond. Biomaterials 24 (24): 4385–4415
40. Holland TA, Tabata Y, Mikos AG (2003) In vitro release of transforming growth factor-beta 1 from gelatin microparticles encapsulated in biodegradable, injectable oligo(poly(ethylene glycol) fumarate) hydrogels. Journal of Controlled Release 91 (3): 299–313
41. Holland TA, Tabata Y, Mikos AG (2005) Dual growth factor delivery from degradable oligo(poly(ethylene

glycol) fumarate) hydrogel scaffolds for cartilage tissue engineering. Journal of Controlled Release 101 (1–3): 111–125
42. Holland TA, Tessmar JKV, Tabata Y, Mikos AG (2004) Transforming growth factor-[beta]1 release from oligo(poly(ethylene glycol) fumarate) hydrogels in conditions that model the cartilage wound healing environment. Journal of Controlled Release 94 (1): 101–114
43. Jo S, Shin H, Mikos AG (2001) Modification of oligo(poly(ethylene glycol) fumarate) macromer with a GRGD peptide for the preparation of functionalized polymer networks. Biomacromolecules 2 (1): 255–261
44. Jo S, Shin H, Shung AK, Fisher JP, Mikos AG (2001) Synthesis and characterization of oligo(poly(ethylene glycol) fumarate) macromer. Macromolecules 34 (9): 2839–2844
45. Kavanagh GM, Ross-Murphy SB (1998) Rheological characterisation of polymer gels. Prog Polym Sci 23 (3): 533–562
46. Kim UJ, Park J, Li C, Jin HJ, Valluzzi R, Kaplan DL (2004) Structure and properties of silk hydrogels. Biomacromolecules 5 (3): 786–792
47. Knight CG, Morton LF, Peachey AR, Tuckwell DS, Farndale RW, Barnes MJ (2000) The collagen-binding A-domains of integrins alpha 1beta 1 and alpha 2beta 1 recognize the same specific amino acid sequence, GFOGER, in native (triple-helical) collagens. Journal of Biological Chemistry 275 (1): 35–40
48. Knoergen M, Arndt KF, Richter S, Kuckling D, Schneider H (2000) Investigation of swelling and diffusion in polymers by ^{1}H NMR imaging: LCP networks and hydrogels. J Mol Struct 554 (1): 69–79
49. Kong HJ, Lee KY, Mooney DJ (2003) Nondestructively probing the cross-linking density of polymeric hydrogels. Macromolecules 36 (20): 7887–7890
50. Kong HJ, Smith MK, Mooney DJ (2003) Designing alginate hydrogels to maintain viability of immobilized cells. Biomaterials 24 (22): 4023–4029
51. Kosto KB, Deen WM (2004) Diffusivities of macromolecules in composite hydrogels. AIChE Journal 50 (11): 2648–2658
52. Krsko P, Libera M (2005) Biointeractive hydrogels. Materials Today (Oxford, United Kingdom) 8 (12): 36–44
53. Kuhl PR, Griffith-Cima LG (1996) Tethered epidermal growth factor as a paradigm for growth factor-induced stimulation from the solid phase [published erratum appears in Nat Med 1997 Jan; 3 (1): 93]. Nat. Med. 2 (9): 1022–1027
54. Leach JB, Schmidt CE (2005) Characterization of protein release from photocrosslinkable hyaluronic acid-polyethylene glycol hydrogel tissue engineering scaffolds. Biomaterials 26 (2): 125–135
55. LeBaron RG, Athanasiou KA (2000) Extracellular matrix cell adhesion peptides: functional applications in orthopedic materials. Tissue Engineering 6 (2): 85–103
56. Lee KY, Mooney DJ (2001) Hydrogels for tissue engineering. Chem. Rev. (Washington, D.C.) 101 (7): 1869–1879
57. Leone G, Barbucci R, Borzacchiello A, Ambrosio L, Netti PA, Migliaresi C (2004) Preparation and physicochemical characterization of microporous polysaccharidic hydrogels. J Mater Sci Mater Med 15 (4): 463–467
58. Lieb E, Hacker M, Tessmar J, Kunz-Schughart LA, Fiedler J, Dahmen C, Hersel U, Kessler H, Schulz MB, Gopferich A (2005) Mediating specific cell adhesion to low-adhesive diblock copolymers by instant modification with cyclic RGD peptides. Biomaterials 26 (15): 2333–2341
59. Lin CC, Metters AT (2006) Hydrogels in controlled release formulations: network design and mathematical modeling. Advanced Drug Delivery Reviews 58 (12–13): 1379–1408
60. Linnes MP, Ratner BD, Giachelli CM (2007) A fibrinogen-based precision microporous scaffold for tissue engineering. Biomaterials 28 (35): 5298–5306
61. Lutolf MP, Hubbell JA (2003) Synthesis and physicochemical characterization of end-linked poly(ethylene glycol)-co-peptide hydrogels formed by Michael-type addition. Biomacromolecules 4 (3): 713–722
62. Lutolf MP, Hubbell JA (2005) Synthetic biomaterials as instructive extracellular microenvironments for morphogenesis in tissue engineering. Nature Biotechnology 23 (1): 47–55
63. Lutolf MP, Lauer-Fields JL, Schmoekel HG, Metters AT, Weber FE, Fields GB, Hubbell JA (2003) Synthetic matrix metalloproteinase-sensitive hydrogels for the conduction of tissue regeneration: engineering cell-invasion characteristics. Proceedings of the National Academy of Sciences of the United States of America 100 (9): 5413–5418
64. Lutolf MP, Weber FE, Schmoekel HG, Schense JC, Kohler T, Mueller R, Hubbell JA (2003) Repair of bone defects using synthetic mimetics of collagenous extracellular matrices. Nature Biotechnology 21 (5): 513–518
65. Madelin G, Baril N, De Certaines JD, Franconi JM, ThiaudiŠre E (2004) NMR characterization of mechanical waves. In: Webb GA (ed) Annual reports on NMR spectroscopy. Academic Press, Amsterdam
66. Manduca A, Oliphant TE, Dresner MA, Mahowald JL, Kruse SA, Amromin E, Felmlee JP, Greenleaf JF, Ehman RL (2001) Magnetic resonance elastography: non-invasive mapping of tissue elasticity. Med Image Anal 5 (4): 237–254
67. Mann BK, Tsai AT, Scott BT, West JL (1999) Modification of surfaces with cell adhesion peptides alters extracellular matrix deposition. Biomaterials 20 (23–24): 2281–2286
68. Martens PJ, Bryant SJ, Anseth KS (2003) Tailoring the degradation of hydrogels formed from multivinyl poly(ethylene glycol) and poly(vinyl alcohol) macromers for cartilage tissue engineering. Biomacromolecules 4 (2): 283–292
69. Masuda S, Shimizu T, Yamato M, Okano T (2008) Cell sheet engineering for heart tissue repair. Advanced Drug Delivery Reviews 60 (2): 277–285
70. Mathur AM, Scranton AB (1996) Characterization of hydrogels using nuclear magnetic resonance spectroscopy. Biomaterials 17 (6): 547–557
71. Meyvis TKL, Van Oostveldt P, Hennink WE, Demeester J (1998) Correlation between rheological characteristics of enzymatically degrading dextran glycidyl methacrylate

gels and the mobility of macromolecules inside these gels. In te Nijenhuis K, Mijs WJ (eds) The Wiley polymer networks group review series: chemical and physical networks. Wiley, Weinheim

72. Meyvis TKM, De Smedt SC, Van Oostveldt P, Demeester J (1999) Fluorescence recovery after photobleaching: a versatile tool for mobility and interaction measurements in pharmaceutical research. Pharm Res 16 (8): 1153–1162

73. Murphy CL, Sambanis A (2001) Effect of oxygen tension and alginate encapsulation on restoration of the differentiated phenotype of passaged chondrocytes. Tissue Engineering 7 (6): 791–803

74. Nguyen KT, West JL (2002) Photopolymerizable hydrogels for tissue engineering applications. Biomaterials 23 (22): 4307–4314

75. Nicolson PC, Vogt J (2001) Soft contact lens polymers: an evolution. Biomaterials 22 (24): 3273–3283

76. Okano T, Yamada N, Sakai H, Sakurai Y (1993) A novel recovery system for cultured cells using plasma-treated polystyrene dishes grafted with poly(N-isopropylacrylamide). Journal of Biomedical Materials Research 27 (10): 1243–1251

77. Othman SF, Xu H, Royston TJ, Magin RL (2005) Microscopic magnetic resonance elastography (microMRE). Magn Reson Med 54 (3): 605–615

78. Park Y, Lutolf MP, Hubbell JA, Hunziker EB, Wong M (2004) Bovine primary chondrocyte culture in synthetic matrix metalloproteinase-sensitive poly(ethylene glycol)-based hydrogels as a scaffold for cartilage repair. Tissue Engineering 10 (3/4): 515–522

79. Peppas NA, Bures P, Leobandung W, Ichikawa H (2000) Hydrogels in pharmaceutical formulations. European Journal of Pharmaceutics and Biopharmaceutics 50 (1): 27–46

80. Peppas NA, Huang Y, Torres-Lugo M, Ward JH, Zhang J (2000) Physicochemical foundations and structural design of hydrogels in medicine and biology. Annu Rev Biomed Eng 2 (1): 9–29

81. Peppas N, Khare AR (1993) Preparation, structure and diffusional behavior of hydrogels in controlled release. Advanced Drug Delivery Reviews 11 (1–2): 1–35

82. Perry PA, Fitzgerald MA, Gilbert RG (2006) Fluorescence recovery after photobleaching as a probe of diffusion in starch systems. Biomacromolecules 7 (2): 521–530

83. Pluen A, Netti PA, Jain RK, Berk DA (1999) Diffusion of macromolecules in agarose gels: comparison of linear and globular configurations. Biophysical Journal 77 (1): 542–552

84. Prior-Cabanillas A, Barrales-Rienda JM, Frutos G, Quijada-Garrido I (2007) Swelling behaviour of hydrogels from methacrylic acid and poly(ethylene glycol) side chains by magnetic resonance imaging. Polym Int 56 (4): 506–511

85. Ramaswamy S, Wang DA, Fishbein KW, Elisseeff JH, Spencer RG (2006) An analysis of the integration between articular cartilage and nondegradable hydrogel using magnetic resonance imaging. J Biomed Mater Res B 77 (1): 144–148

86. Rizzi SC, Ehrbar M, Halstenberg S, Raeber GP, Schmoekel HG, Hagenmueller H, Mueller R, Weber FE, Hubbell JA (2006) Recombinant protein-co-PEG networks as cell-adhesive and proteolytically degradable hydrogel matrixes. Part II: biofunctional characteristics. Biomacromolecules 7 (11): 3019–3029

87. Rizzi SC, Hubbell JA (2005) Recombinant protein-co-PEG networks as cell-adhesive and proteolytically degradable hydrogel matrixes. Part I: development and physicochemical characteristics. Biomacromolecules 6 (3): 1226–1238

88. Sanabria-DeLong N, Agrawal SK, Bhatia SR, Tew GN (2007) Impact of synthetic technique on PLA-PEO-PLA physical hydrogel properties. Macromolecules 40 (22): 7864–7873

89. Schmedlen RH, Masters KS, West JL (2002) Photocrosslinkable polyvinyl alcohol hydrogels that can be modified with cell adhesion peptides for use in tissue engineering. Biomaterials 23 (22): 4325–4332

90. Seal BL, Otero TC, Panitch A (2001) Polymeric biomaterials for tissue and organ regeneration. Mater. Sci. Eng. R 34 (4–5): 147–230

91. Shu XZ, Liu Y, Palumbo F, Prestwich GD (2003) Disulfide-crosslinked hyaluronan-gelatin hydrogel films: a covalent mimic of the extracellular matrix for in vitro cell growth. Biomaterials 24 (21): 3825–3834

92. Shung AK, Behravesh E, Jo S, Mikos AG (2003) Crosslinking characteristics of and cell adhesion to an injectable poly(propylene fumarate-co-ethylene glycol) hydrogel using a water-soluble crosslinking system. Tissue Engineering 9 (2): 243–254

93. Sniekers YH, van Donkelaar CC (2005) Determining diffusion coefficients in inhomogeneous tissues using fluorescence recovery after photobleaching. Biophys J 89 (2): 1302–1307

94. Solchaga LA, Gao J, Dennis JE, Awadallah A, Lundberg M, Caplan AI, Goldberg VM (2002) Treatment of osteochondral defects with autologous bone marrow in a hyaluronan-based delivery vehicle. Tissue Engineering 8 (2): 333–347

95. Song J, Saiz E, Bertozzi CR (2003) A new approach to mineralization of biocompatible hydrogel scaffolds: an efficient process toward 3-dimensional bonelike composites. Journal of the American Chemical Society 125 (5): 1236–1243

96. Tabata Y (2003) Tissue regeneration based on growth factor release. Tissue Engineering 9 (4): 5–15

97. Tabata Y (2005) Significance of release technology in tissue engineering. Drug Discovery Today 10 (23/24): 1639–1646

98. Tabata Y, Miyao M, Ozeki M, Ikada Y (2000) Controlled release of vascular endothelial growth factor by use of collagen hydrogels. Journal of Biomaterials Science, Polymer Edition 11 (9): 915–930

99. Temenoff JS, Park H, Jabbari E, Conway DE, Sheffield TL, Ambrose CG, Mikos AG (2004) Thermally crosslinked oligo(polyethylene glycol) fumarate) hydrogels support osteogenic differentiation of encapsulated marrow stromal cells in vitro. Biomacromolecules 5 (1): 5–10

100. Temenoff JS, Park H, Jabbari E, Sheffield TL, LeBaron RG, Ambrose CG, Mikos AG (2004) In vitro osteogenic differentiation of marrow stromal cells encapsulated in biodegradable hydrogels. Journal of Biomedical Materials Research, Part A 70A (2): 235–244

101. Tessmar J, Goepferich A (2007) Matrices and scaffolds for protein delivery in tissue engineering. Adv Drug Deliv Rev 59 (4–5): 274–291

102. Tessmar JK, Goepferich AM (2007) Customized PEG-derived copolymers for tissue-engineering applications. Macromolecular Bioscience 7 (1): 23–39

103. Tew GN, Aamer KA (2003) Triblock PLA-PEO-PLA hydrogels: structure and mechanical properties. Polymeric Materials Science and Engineering 89: 236–237

104. Wallace DG, Rosenblatt J (2003) Collagen gel systems for sustained delivery and tissue engineering. Advanced Drug Delivery Reviews 55 (12): 1631–1649

105. Watkins AW, Anseth KS (2005) Investigation of molecular transport and distributions in poly(ethylene glycol) hydrogels with confocal laser scanning microscopy. Macromolecules 38 (4): 1326–1334

106. West JL, Hubbell JA (1999) Polymeric biomaterials with degradation sites for proteases involved in cell migration. Macromolecules 32 (1): 241–244

107. Whitaker MJ, Quirk RA, Howdle SM, Shakesheff KM (2001) Growth factor release from tissue engineering scaffolds. Journal of Pharmacy and Pharmacology 53 (11): 1427–1437

108. Williams CG, Kim TK, Taboas A, Malik A, Manson P, Elisseeff J (2003) In vitro chondrogenesis of bone marrow-derived mesenchymal stem cells in a photopolymerizing hydrogel. Tissue Engineering 9 (4): 679–688

109. Zajaczkowski MB, Cukierman E, Galbraith CG, Yamada KM (2003) Cell-matrix adhesions on poly(vinyl alcohol) hydrogels. Tissue Engineering 9 (3): 525–533

110. Zhang S (2003) Fabrication of novel biomaterials through molecular self-assembly. Nat Biotech 21 (10): 1171–1178

111. Zhang WM, Kapyla J, Puranen JS, Knight CG, Tiger CF, Pentikainen OT, Johnson MS, Farndale RW, Heino J, Gullberg D (2003) Alpha 11 beta 1 integrin recognizes the GFOGER sequence in interstitial collagens. Journal of Biological Chemistry 278 (9): 7270–7277

VIII Scaffold Related Aspects

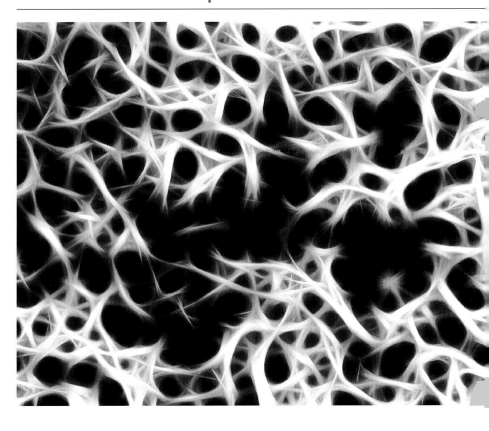

Defining Design Targets for Tissue Engineering Scaffolds

S. J. Hollister, E. E. Liao, E. N. Moffitt, C. G. Jeong, J. M. Kemppainen

Contents

38.1	Introduction	521
38.2	Tissue Mechanical and Mass Transport Properties	522
38.3	Influence of Mechanical and Mass Transport Microenvironments on Cell Response	524
38.4	Proposed Design Targets for Tissue Engineering Scaffolds	526
38.5	Progress in Designing and Fabricating Scaffolds to Meet Design Targets	527
38.6	Experiments Relating Scaffold Design Targets to Tissue Regeneration	532
38.7	Conclusions	534
	References	535

38.1 Introduction

Biological tissues are extremely complex three-dimensional (3D) structures with concomitant complicated mechanical function and mass transport characteristics. Tissue engineering seeks to recapitulate this complex structure and function using biomaterial scaffolds delivering therapeutic biologics such as cells, proteins, and genes for tissue reconstruction. It is clear that the biomaterial/biologic construct cannot replicate the complex tissue milieu, including multiple cell types interacting with numerous cytokines to produce extracellular matrices having hierarchical features exhibiting highly nonlinear, biphasic mechanical function. The biomaterial/biologic construct is, at best, a crude approximation to the normal tissue milieu. To improve the clinical potential of tissue engineering/regenerative medicine, we must be able to relate the goodness of this approximation to the success of tissue regeneration. In essence, we must be able to define relevant design criteria for tissue engineering therapies. For the scaffold, the focus of this chapter, the pertinent question becomes:

"How closely does a biomaterial scaffold have to approximate the normal tissue structure, mechanical function, mass transport, and cell-matrix interaction as a function of time to achieve desired tissue reconstruction?"

Answering this question is an arduous task, requiring that we:

1. Characterize normal tissue structure, mechanical function, mass transport, and cell-matrix interactions
2. Engineer scaffolds with well-controlled structure, mechanical function, mass transport, and cell matrix that can be quantitatively compared to normal tissue characteristics accounting for degradation
3. Perform in vivo experiments where tissue regeneration success (defined as the relationship of engineered tissue characteristics to normal tissue characteristics) is correlated to the engineered scaffold characteristics

This chapter will not answer the proposed question, but rather propose potential mechanical and mass transport design targets for tissue engineering scaffolds based on reviews of tissue properties, cell responses to microenvironments, and what is known about scaffold properties and how these properties affect tissue regeneration. We first review tissue mechanical and mass transport properties as a first approximation for scaffold design targets. We then discuss how mechanical and mass transport properties may affect cell behavior and phenotype as additional considerations for defining scaffold design targets. Finally, we review mechanical and mass transport properties of currently engineered scaffolds and how they relate to tissue regeneration.

Tissue function will be characterized by mechanical constitutive models (linear elastic, nonlinear elastic, viscoelastic, and biphasic), and mass transport will be characterized by diffusion, permeability, and porosity. We have chosen to focus on scaffold mechanical function and mass transport for two reasons. First, scaffolds must provide temporary mechanical function within a tissue defect during tissue regeneration. If the scaffold does not have adequate mechanical properties, then function cannot be maintained, and the reconstruction will fail. Second, both mass transport and mechanical properties influence tissue regeneration by shaping the microenvironment influencing tissue regeneration. Mass transport obviously shapes the cell microenvironment by controlling cell nutrient access and waste removal. Mechanical properties affect tissue regeneration by affecting cell differentiation through mechanotransduction. Based on the review, we will postulate design targets based on three tissue types:

1. Vascularized hard tissue (cortical and trabecular bone)
2. Vascularized soft tissues (cardiovascular, muscle, and neural tissues)
3. Avascular soft tissues (articular cartilage, fibrocartilaginous tissues, ligaments/tendons)

38.2
Tissue Mechanical and Mass Transport Properties

Characterization of tissue mechanical and mass transport properties is one of the largest and most active research areas in biomedical engineering. Numerous texts [16, 25, 26, 43, 48] and thousands of articles have been written concerning these topics. Although ideally one would seek to design scaffolds based on mechanical loads experienced by a specific tissue, these loads are notoriously difficult to accurately quantify. In tissue engineering, an alternative target is to design scaffolds whose mechanical properties are as close as possible to native tissues [7, 26, 68]. It is assumed that any scaffold matching native tissue properties will be able to withstand the mechanical loads that a normal tissue withstands. Furthermore, the mechanical microenvironment experienced by cells within the scaffold will be similar to the microenvironment experienced by cells of the desired phenotype normal tissue, thereby helping to maintain desired phenotype, or in the case of progenitor cells, differentiate into the desired phenotype. It is not known, of course, whether matching specific tissue properties exactly or some percentage of these properties is a valid design goal; this can only be determined through experiment.

The relevant issue for scaffold mechanical design becomes what is the most appropriate constitutive model to use as a target. While bone may adequately be characterized using linear elastic constitutive models, soft tissues are more appropriately characterized using nonlinear elastic or viscoelastic constitutive models for tissue classically considered as single phase (ligaments, tendons, skin, neural and cardiovascular tissues) or using biphasic/poroelastic constitutive models for tissues considered as fluid/solid mixtures (articular cartilage and fibrocartilaginous tissues such as the meniscus and intervertebral disc). For biphasic materials, the solid phase is typically considered to be an incompressible linear or nonlinear elastic material, and the fluid is considered to be an incompressible Newtonian fluid, whose flow behavior is governed by Darcy's Law and characterized by permeability. Since describing the various constitutive models is beyond the scope of this paper, we have chosen a very simplified 1D nonlinear elastic model widely used to characterize tissue solid matrix behavior [48]:

$$\sigma = A(e^{B\varepsilon} - 1) \quad (38.1)$$

Where ε is the strain (small or large strain), σ is the stress (Cauchy for small strain; 1st Piola–Kirchoff stress for large strain), and A and B are coefficients fit to experimental data. Using this model, A corresponds

Table 38.1 Nonlinear and linear elastic mechanical properties for tissues fit to the model $\sigma = A(e^{B\epsilon} - 1)$. Results also give the tangent modulus, calculated as $E^{tangent} = ABe^{B\epsilon}$, for 1 and 10% strain. Different values at 1 and 10% strain indicate a nonlinear response. Results show the tremendous variation in tissue mechanical properties ranging from stiff linear elastic (bone) to very compliant nonlinear elastic for smooth muscle, fat, cardiovascular, and skin tissues

Tissue (species)	A coefficient range in MPa	B coefficient range	Tangent modulus at 1% strain (MPa) = $ABe^{B\epsilon}$	Tangent modulus at 10% strain (MPa) = $ABe^{B\epsilon}$	Reference for data fit
Trabecular bone (human)	100.0–100,000.0	0.06–0.0152	6.0–1,520.0	6.0–1,520.0	21
Meniscus	1.6–3.2	27.5–31.9			48
Articular cartilage	0.3–2.1	1.3–5.0	0.4–11.0	0.444–17.3	48
Medial collateral ligament (human)	0.3	12	4.06	11.95	56
Intervertebral disc—fibrocartilage	0.05–0.07	4.95–11.9	0.26–0.94	0.41–2.74	34
Spinal cord grey matter (cow)	0.0066	9.06	0.066	0.148	28
Spinal cord white matter (cow)	0.0041	6.54	0.029	0.052	28
Bladder smooth muscle (rat)	0.0022	25.7	0.073	0.739	20
Fat (human)	0.002	9.64	0.002	0.005	45
Heart valve	0.00153	28.81	0.06	0.78	71
Myocardium	0.0013	6.0	0.008	0.014	62
Skin/subcutaneous tissue	0.000057	21.52	0.0015	0.011	76

to an initial tangent modulus at low strains, while B characterizes the degree of nonlinearity, with higher B indicating greater nonlinear behavior. Values of A and B for tissues vary widely (Table 38.1), showing stiff, linear results for bone to extremely compliant very nonlinear behavior for skin and myocardium. The degree of nonlinearity is also reflected in tangent modulus variation at 1% and 10% strain. If the tangent modulus doesn't vary, the material is linear. Greater nonlinearity is given by large changes in tangent modulus. Bone is linear with tangent modulus in a 10–1,000 MPa range that does not change with strain. Intervertebral discs and the medial collateral ligaments are moderately nonlinear, while the heart valve, myocardium, smooth muscle, and skin are highly nonlinear. Thus, matching the complete (yet still highly simplified) mechanical behavior of soft tissues will require scaffolds exhibiting nonlinear behavior with tangent moduli ranging from 0.05 to 5 MPa.

The second key tissue characteristic design target is mass transport. Mass transport may be characterized by both diffusivity (Brownian motion of solutes through a porous matrix) and permeability (fluid flow through a porous matrix under applied pressure gradient). Both these measures are weakly correlated to porosity but do not necessarily increase as porosity increases. Obviously, replacing large volumes of vascularized tissue requires the ability to fully perfuse the scaffold. Given extreme difficulties in fabricating a vascular system within a scaffold, increasing scaffold diffusivity and permeability is the next best thing to both increase scaffold perfusion and increase vascular ingrowth when delivering angiogenic biologics like vascular endothelial growth factor (VEGF).

Tissue values of diffusivity and permeability roughly reflect cell metabolic needs within the tissue. Thus, it comes as no surprise that avascular cartilaginous tissues and ligaments have low diffusivity and permeability compared to vascularized tissues like bone and cardiovascular tissues (Table 38.2).

It is important to note that diffusivity coefficients depend on the size of the molecule used in transport studies, with larger molecular weight leading to lower diffusivity coefficients as demonstrated for hy-

Table 38.2 Diffusion and permeability properties of selected tissues. Note that permeability data are taken from direct fluid permeation measurements for cortical bone, trabecular bone, medial collateral ligament, muscle, and valve while permeability measurements for the intervertebral disc, meniscus, and articular cartilage are obtained by fitting confined or unconfined compression to biphasic theory solutions

Tissue (species)	Diffusion coefficient (mm²/s)	Permeability (m⁴/Ns)	Reference
Trabecular bone (human)		$0.003-11e^{-6}$	63
Cortical bone (bovine and canine)	$3.3e^{-4}$	$0.9-7.8e^{-11}$	70, 63
Meniscus (human)		$1-4e^{-15}$	30
Articular cartilage (human)	$5-70e^{-6}$ (3–40 kDa Dextran) Leddy	$0.01-19.5e^{-15}$	35, 57
Medial collateral ligament (human)		$0.04-0.85e^{-15}$	72
Intervertebral disc—fibrocartilage		$1.4-2.1e^{-15}$	74
Abdominal muscle (rat)		$5.1e^{-11}$	79
Heart valve (rat)	$5.9e^{-7}$ (middle layer) $1.0e^{-5}$ (intima layer)	$2.3e^{-17}$ (middle layer) $1.1e^{-13}$ (intima layer)	81, 82

drogels and articular cartilage [34]. It is also important to note that the bone and ligament permeability were measured using direct permeation experiments where a fluid flows through the material under a hydraulic head while the cartilage, meniscus, and intervertebral disc permeability were measured by fitting biphasic theory coefficients (aggregate modulus and permeability) to either creep or stress relaxation results from confined compression tests. As with mechanical properties, there is a wide range of tissue permeability and diffusivity coefficients to target for mass transport design. However, unlike mechanical properties, mass transport is constrained by the need to maintain pore connectivity to allow for contiguous cell seeding and/or migration.

38.3
Influence of Mechanical and Mass Transport Microenvironments on Cell Response

Effective tissue mechanical (Table 38.1) and mass transport properties (Table 38.2) are local properties averaged over a tissue volume at least 2–3 mm on a side. This volume is much larger than a single cell; thus, the effective mechanical and mass properties represent the average local cell mechanical and mass transport microenvironment. In effect, tissue mechanical and mass transport properties are measures of the tissue architecture that acts to filter mechanical and mass transport signals to cells. The local microenvironment formed by the matrix, including chemical (resulting from mass transport) and mechanical signals influences cell phenotype and matrix production.

It has been amply documented that these mass transport and mechanical signals affect cell differentiation and matrix production. Partial oxygen pressure (PO_2) is a prime example of a mass transport signal affecting cell differentiation. For example, low oxygen tensions have been associated with support of cartilage regeneration and inhibition of bone regeneration. Domm et al. [13] reported that dedifferentiated bovine chondrocytes cultured in 5% partial oxygen pressure (PO_2) showed cartilage matrix production while those cultured in 21% PO_2 did not produce cartilage. Likewise, chondrocytes cultured in monolayer did not produce cartilage matrix under any level of PO_2. Malda et al. [41] also found that low oxygen tension promoted redifferentiation of dedifferentiated human nasal chondrocytes. These results suggest that both low diffusivity (restricting oxygen diffusion) and cell shape are important for enhancing cartilage regeneration.

In contrast to chondrocyte behavior, osteoblasts are adversely affected by low PO_2. Utting et al. [69] found that reducing PO_2 from 21% to 2% significantly reduced mineralized nodule formation and expression of mRNA for alkaline phosphatase (ALP) and osteocalcin (OC) for rat osteoblasts, indicating inhibition of osteogenic differentiation.

Bone marrow stromal cell differentiation is also significantly influenced by PO_2. Robins et al. [60] found that low PO_2 increased expression of the Sox9 promoter and upregulated expression of chondrogenic genes for aggregan and collagen II. D'Ippolito et al. [12] found that 3% PO_2 inhibited expression of osteoblastic markers, including OC, Runx2, bone sialoprotein, and ALP by bone marrow stromal cells cultured in osteogenic media. They also found that the low PO_2 supported maintenance of stemness characteristics, in addition to inhibiting osteogenic differentiation. Thus, low PO_2 is associated with maintenance of a chondrogenic phenotype and inhibition of osteogenic phenotype while high PO_2 is associated with the reverse.

Mechanical stimulus is another microenvironmental factor shown to influence cell differentiation and matrix deposition. The hypothesis that tissue form and function is related to mechanical stimulus and that mechanical stimulus could directly affect cell activity can be traced back to Julius Wolff in 1892 and Wilhelm Roux in 1895. Perren [53] proposed a modern adaptation specifying that strain magnitude was related to tissue differentiation in fracture gaps, with 2% strain leading to bone formation, 10% strain leading to cartilage formation at the beginning of endochondral ossification, and 100% strain leading to fibrous tissue formation initiating a non-union and pseudoarthrosis. This theory clearly suggests that mechanical stimulus affects differentiation of pluripotent cells present from bone marrow and periosteum in a fracture gap.

There have been numerous computational implementations of mechanobiology theories since the 1980s. Carter and colleagues [9, 40] developed a theory stating that low distortional and hydrostatic stress is conducive to bone growth, moderate compressive hydrostatic stress is conducive to hyaline cartilage growth, high compressive hydrostatic stress is conducive to fibrocartilage growth, and high tensile and distortional strain is conducive to fibrous tissue growth. Prendergast and colleagues [8, 33, 54] also proposed a mechanobiologic tissue growth theory based on fluid flow and octahedral shear strain, modeling tissues as poroelastic or biphasic materials. Low flow and shear strain is associated with bone growth from mesenchymal cells; moderate flow and shear strain is associated with cartilage formation; and high flow and shear strain is associated with fibrous tissue formation. Despite differences in specifics of these mechanobiologic theories, they all postulate low levels of strain are associated with bone formation, moderate levels of strain with cartilage formation, and high levels of strain with fibrous tissue formation.

Recent experiments on stem cell differentiation have confirmed aspects of mechanobiologic theories. Simmons et al. [65] reported that mesenchymal stem cells (MSC) subject to 3% strain on a silastic flex cell produced 2.3 times the mineral of MSC not subject to strain. Ng et al. [50] applied 10% dynamic strain to chondrocytes in agarose and found that this mechanical stimulus increased both the functional and biochemical properties of cartilage matrix. Altman et al. [2] reported that application of longitudinal strain to MSC in a silk scaffold led to upregulation of collagen I, collagen III, and tenascin-C, all markers of ligament tissue. These markers were not expressed in cells that were not mechanically stimulated. Altman et al. applied a rotational displacement but did not report a strain level. They also noted that the effect of mechanical stimulus was dependent on MSC maturation, with mechanical stimulation up to 3 days being detrimental, but at day 9, mechanical stimuli had a beneficial effect. Riha et al. [59] applied cyclical strain of 10% to mouse embryonic progenitor cells. They found that strain promoted differentiation of progenitor cells into a smooth muscle cell lineage, with cells expressing smooth muscle markers including smooth muscle specific-actin (α-SMA) and smooth muscle myosin heavy chain (SMMHC). The cited studies are only a small sampling of numerous reports indicating that mechanical and mass transport microenvironments affect cell differentiation and matrix deposition.

38.4 Proposed Design Targets for Tissue Engineering Scaffolds

Although scaffold design to directly produce a desired cell microenvironment is desirable, it is problematic for two reasons. First, designing directly for in vivo cell microenvironments is sensitive to assumed boundary conditions, both loading and mass transport, which are rarely known to any reasonable degree of accuracy. Second, even if boundary conditions were known, computationally designing scaffolds at the cell level for a large anatomic region would not be numerically feasible. Considering scaffold design, however, it is the scaffold architecture that filters the boundary conditions input to provide the cell microenvironment. Thus, without knowing detailed boundary condition information, designing architectures to achieve desired effective mechanical and mass transport properties is likely the next best solution. If one assumes that the mechanical and mass transport boundary conditions are appropriate for the tissue type (given, this is a significant assumption), then an appropriately designed architecture will filter the applied boundary conditions to deliver the correct microenvironment signals for cell differentiation and matrix deposition in an average sense over a 1–3-mm volume, illustrated schematically in Fig. 38.1.

Brand [6] first proposed the hypothesis that ECM serves as both a mass transport and mechanical filter to cells. Thus, it seems a reasonable starting point that scaffolds should also serve as a filter to both implanted and migrating cells, albeit a much cruder filter than ECM.

Under this assumption, it makes sense as a first approximation to design scaffold architecture to achieve appropriate mechanical and mass transport properties. Another microenvironmental feature affecting cell differentiation is of course the base scaffold material and cell interactions with this material. This is seen most predominantly with bone, where calcium phosphate ceramics are known to be osteoconductive, enhancing bone formation based on material effects alone. Thus, scaffold design targets would include mechanical properties, mass transport properties, and cell-material interaction. The available design parameters are material and material/pore arrangement in 3D space, i.e., scaffold architecture.

A complicating factor in achieving desired design targets is that mechanical properties, mass transport properties, and cell-material interaction are coupled, often inversely. Obviously, choice of material will fix cell-material interaction characteristics (unless surface modification is performed), as well as the range of mechanical properties that may be achieved. The bulk material will set the maximum mechanical properties that can be achieved; any designed porous architecture will then produce effective mechanical properties ranging from the bulk material to zero, which is the theoretical lower bound. The bulk material may also have some intrinsic minimum mass transport characteristics due to inherent microporosity (highly dependent on processing) that will set the minimum mass transport properties. By implementing a designed porosity, it is intuitive that mass transport properties will increase while mechanical properties will decrease. Simply put, choice of scaffold material will fix the cell-interaction characteristics as well as the maximum mechanical properties

Fig. 38.1 Schematic illustration of the scaffold architecture filter hypothesis. Applied mechanical and mass transport boundary conditions will be filtered by the effective mechanical and mass transport properties of the scaffold to cells that are attached to and are within the scaffold. Based on a hypothesis proposed by Brand [6] that tissue ECM filters signals to cells in the ECM

and minimum mass transport properties. Scaffold architecture design will determine where the actual effective mechanical and mass transport properties fall within minimum and maximum bounds.

Mathematical bounds have been derived specifying attainable values of a normalized bulk modulus (defined as $K = 0.33 \times (C_{11} + 2 \times C_{12})$, where C_{11} and C_{12} are elastic constants) and normalized diffusivity coefficient as a function of porosity (Fig. 38.2) [18].

There are two important aspects to this bounds graph. First, since these are absolute bounds, any designed scaffold architecture must produce effective bulk moduli and diffusivity within the outline. Thus, given a porosity and cubic symmetry, we can know a priori what effective mechanical and mass transport properties can be achieved for any material and porosity. These bounds thus become a gauge of whether desired properties are even achievable. Second, these bounds clearly show the inverse trade-off between modulus and mass transport. If we desire increased mechanical properties, they come at the expense of mass transport and vice versa.

These mathematical bounds can be used to define relative design targets for specific tissue applications. From Tables 38.1 and 38.2, we can roughly characterize the following targets based on effective tissue properties:

1. Vascularized hard tissue: high bulk scaffold modulus, high scaffold diffusivity
2. Vascularized soft tissues: low bulk scaffold modulus, high scaffold diffusivity
3. Avascular soft tissues: low bulk scaffold modulus, low scaffold diffusivity

An issue that may be raised is the use of a bulk modulus as a measure of mechanical properties, since most soft tissues are assumed to be incompressible with an infinite bulk modulus. However, Bahhvalov et al. [3] demonstrated that any solid with pores will have effective compressible elastic properties, even if the base material is incompressible. Thus, scaffolds, which are solids incorporating pores, will be compressible, and thus the bulk modulus is a fair way to characterize scaffold mechanical properties.

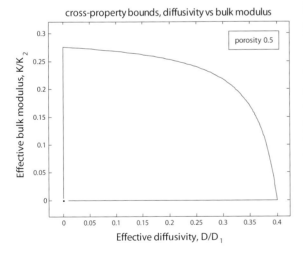

Fig. 38.2 An example of dual bulk modulus and diffusivity bounds for a porosity of 50%. The tradeoff between mechanical and mass transport properties is readily apparent. For example, achieving the highest bulk modulus can only be achieved at zero diffusivity while the highest diffusivity can only be achieved at zero bulk modulus. In-between values of bulk modulus and diffusivity represent a trade-off between the two properties

38.5
Progress in Designing and Fabricating Scaffolds to Meet Design Targets

The ability to meet proposed designed targets relies on material choice and 3D porous architecture. Material choice for tissue engineering includes calcium phosphate ceramics, polymers, and metals. We can further categorize polymers into whether they are used below their glass transition temperature (Tg) and are relatively stiff undergoing small deformation, or are used above their Tg and are relatively elastic undergoing large deformations, thereby termed "elastomers." Widely used, more rigid polymers include polylactic acid (PLA; Tg = 50–65°C), polyglycolic acid (PGA; Tg = 35–40°C), and their copolymers. Polycaprolactone (PCL; Tg = −60°C) is widely used above its Tg and can undergo relatively large strains but is much more rigid than other elastomeric polymers, including poly(glycerol) sebacate (PGS; Tg = −26°C) and poly (1,8) octane diol citrate (POC; Tg = −5–10°C), that have very compliant properties in addition to undergoing large strains.

Material choice will determine the bounds of attainable mechanical properties as well as cell-material interaction characteristics. Cell-material interactions are extremely complex, depending on many factors, including surface roughness and protein absorption. We restrict our attention to how scaffold materials should be chosen for tissue applications in terms of maximum achievable mechanical property bounds versus native tissue properties. In terms of maximum achievable mechanical modulus, titanium, CaP ceramics, PLA, PGA, and PCL are the only materials that can achieve a modulus at least in the low range of trabecular bone. Unless made into composites with CaP [55], truly elastomeric polymers like POC and PGS are likely not appropriate for bone based purely on mechanical considerations. The question of material choice becomes less clear when looking at minimum bounds, as theoretically any bulk material with pores can achieve a stiffness approaching the stiffness of the second phase, simply by making isolating particles of the stiffer phase. However, it is not practical in most cases to make such structures; thus, only polymers have been seen as suitable soft tissue replacements based on mechanical properties. Based solely on bulk material characteristics, we can make the following material design choices:

1. Vascularized hard tissue: titanium, CaP, PGA, PLA, PCL, polymer/CaP composites
2. Vascularized soft tissues: PGA, PLA, PCL, POC, PGS
3. Avascular soft tissues: PGA, PLA, PCL, POC, PGS

Material choices give broad ranges of maximum mechanical properties and theoretically minimum mass transport properties. The key question is how to design the 3D porous architecture to achieve desired effective properties within the minimum and maximum bounds. Traditionally, the 3D porous architecture has been largely controlled by the processing method, such as porogen leaching, emulsion, or gas foaming, with design input being limited to porogen size, temperature, or pressures that have limited correlation with final architecture [66]. The methods must generally create high porosity structures to satisfy pore connectivity constraints needed to ensure contiguous cell seeding/migration.

Using a variation of porogen leaching called the vibrating particle method, Agrawal et al. [1] measured both modulus and permeability of 50/50 PLA/PGA copolymer (PLGA) scaffolds (Fig. 38.3a) of 80%, 87%, and 92% porosity at 0, 2, 4, and 6 weeks in vitro degradation. Initial moduli ranged from 0.1 MPa (92% porosity) to 0.26 MPa (80% porosity), increased to 1.79–1.88 MPa at 2 weeks, and then decreased again. Interestingly, lower porosity scaffolds showed faster degradation, likely due to autocatalysis. Autocatalysis is the process by which acidic degradation products will catalyze and increase the rate of the hydrolytic degradation if not removed. Permeability, measured by permeation experiments in which fluid flows under pressure through the scaffold and flow rate is measured, initially ranged from 2.1×10^{-9} m^4/Ns (80% porosity) to 16.1×10^{-9} m^4/Ns (92% porosity). The permeability for most specimens decreased over time, reflecting compacting of specimens after mass loss. Effective properties were highly variable, with coefficients of variation (standard deviation/mean, abbreviated as COV) ranging from 40 to 62% for modulus and 40 to 133% for permeability. Zhang et al. [83] used a combined porogen leaching and compression molding technique with spherical and cubic shaped porogens to create 85:15 PLGA scaffolds with 78–97% porosity. The resulting moduli ranged from 0.5 to 13 MPa, but permeability and COV were not presented. Wu and Ding [77] also used a compression molding/porogen leaching technique for PLGA scaffolds of different PLA/PGA ratios with 87% porosity. They measured moduli ranging from 6.6 to 12.9 MPa. COV ranged from 4 to 14%. Interestingly, they found that during degradation up to 12–24 weeks, moduli initially increased before eventually decreasing despite a constant decrease in polymer molecular weight. They attributed the increase to pore compaction resulting from collapse of the architecture. Riddle and Mooney [58] used a gas foaming/particulate leaching (GF/PL) to make 95% porous poly(lactide-co-glycolide) scaffolds with connected porosity. In GF/PL, polymer particles and porogen are placed in a high pressure gas chamber, which causes the polymer to foam around the porogen. Riddle and Mooney found that polymer particle size significantly affected mechanical modulus, with particles less than 250 μm consistently producing the highest modulus between 0.4 and 0.5 MPa, with COV ranging from 2 to 12%.

A second category of scaffold architectures are woven (Fig. 38.3b) and nonwoven fibrous architec-

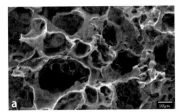
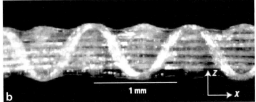
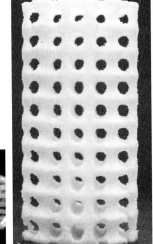

Fig. 38.3a–c Examples of three types of scaffold architectures. **a** Porogen leached sponge architecture from PLGA (reprinted with permission from Biomaterials 24:181–194, "Indirect solid free form fabrication of local and global porous, biomimetic and composite 3D polymer-ceramic scaffolds"). **b** Woven fiber PGA scaffold architecture (reprinted with permission from Nature Materials 6:162–167, 2007, "A biomimetic three-dimensional woven composite scaffold for functional tissue engineering of cartilage.") **c** An SLS-fabricated 3D PCL scaffold with spherical pore architecture

tures. Electrospinning is a commonly used technique to create fibers from PLGA, PCL, collagen, chitosan, and other materials ranging in size from nanometers to millimeters for nonwoven fibrous scaffold architectures [5]. Fibers are created by applying a strong electric field to the nozzle from which a polymer melt or droplet in solution is extruded. The extruded fibers may be amorphous or aligned by spinning along a rotating drum. Amorphous fibers will lead to isotropic effective scaffold properties, while aligned fibers will produce anisotropic effective scaffold properties. Nerurkar et al. [49] created aligned 85% porous PCL fibrous scaffolds by spinning PCL in solution at different alignment angles. They found a tensile modulus of 20 MPa along the fiber direction dropping to 1 MPa at 90° to the fiber, demonstrating the significant anisotropy of aligned fibers. Thomas et al. [67] examined the effect of rotating drum speeds of 0, 3,000 revolutions per minute (rpm), and 6,000 rpm on effective PCL nanofiber properties. Increasing rotation speed increased effective tensile modulus 7.1 MPa at 0 rpm to 33.2 MPa at 6,000 rpm. They attributed the increased effective modulus to increased fiber alignment and decreased fiber spacing at higher rpm. The coefficients of variation ranged from 6% to 11%.

Li and Mak [36] measured permeability of 95% porous nonwoven fibrous PGA scaffolds. They found initial permeability of $2.6e^{-6}$ m^4/Ns, which increased over 14 days to $4.7e^{-6}$ m^4/Ns. The coefficient of variation was approximately 8%. Moutos et al. [47] created scaffolds by weaving 104 μm diameter PGA filaments in a layered structure and then combining this with agarose hydrogel. Two different scaffold architectures were created, one with 390 μm × 320 μm × 104 μm pores having 70% porosity and a second with 390 μm × 320 μm × 104 μm pores having 74% porosity. Aggregate modulus, Young's modulus, and permeability were determined from confined compression, unconfined compression, and tensile tests. Results demonstrated that small pore scaffolds had higher aggregate (199 vs. 138 KPa) and Young's moduli (77 vs. 68 KPa) than large pore scaffolds, but that large pore scaffolds had higher permeability ($0.9e^{-15}$ vs. $0.4e^{-15}$ m^4/Ns). Tensile moduli ranged from 300 to 500 MPa, with changes in tangent moduli at 0 and 10% strain, indicating nonlinear behavior in tension. Coefficients of variations ranged from 8 to 31% for modulus and 15 to 40% for permeability. Moutos et al. noted that the woven fiber scaffolds exhibited moduli, tension-compression nonlinearity, and permeability in the range of articular cartilage. However, it is important to note that permeability measured by confined compression tests [47] differs by 6–9 orders of magnitude from permeability determined by fluid permeation experiments for scaffolds with similar porosity [1, 36].

Although both fiber and porogen-leaching scaffolds have been widely used for scaffold architectures, it is difficult to incorporate these architectures into complex anatomic defect shapes. Furthermore, although fiber-based scaffolds allow more control over architecture than porogen leaching, neither technique allows a level of control that allows variation of pore shapes or controllable gradients in pore architecture. A much higher level of control over architecture is possible through the use of solid free-form fabrication (SFF) techniques [23, 27]. These techniques lay down material on a layer-by-layer basis directed by a computer file detailing both the internal architecture and external shape of a structure. SFF can be used to either build a scaffold directly from a biomaterial (direct SFF), or build a scaffold mold into which a biomaterial can be cast (indirect SFF). Direct SFF is more difficult to implement since the biomaterial must be processed such that it is compatible with the physics of a particular SFF system. In general, the most commonly used direct SFF approaches have been nozzle deposition systems that directly deposit materials in patterns or systems in which material beds, solid or liquid, are bonded in a pattern by an overhead laser (as in selective laser sintering or stereolithography) or a chemical spray system (as in 3D printing).

Nozzle-based systems have been used with polymers, ceramics, and polymer/ceramic composites. Hutmacher and colleagues [26, 28, 80] developed a technique to build PCL scaffolds using a fused deposition modeling (FDM) system. For FDM, a PCL fiber was extruded that could then be heated and deposited using the FDM nozzle system. Using this technique, Zein et al. [80] created PCL scaffolds with different material layup patterns with porosity ranging from 48 to 77%. These scaffolds had moduli ranging from 4 to 77 MPa and yield strength ranging from 0.4 to 3.6 MPa, both properties in the low range of trabecular bone. Coefficients of variation ranged from 4 to 12% for modulus and 4 to 14% for yield strength. Liu et al. [39] developed a low temperature nozzle deposition technique to build PLLA/TCP and PLGA/TCP composite scaffolds using a ratio of 70% polymer to 30% TCP. The scaffolds had porosity ranging from 74 to 81% and exhibited linear elastic behavior with elastic moduli ranging from 17 to 23 MPa and yield strength ranging from 0.75 to 1.4 MPa. Coefficients of variation ranged from 5 to 17% for modulus and 8 to 33% for yield strength. Woodard et al. [75] created 41% macroporous hydroxyapatite scaffolds using a robotic nozzle deposition system. Mixtures of poly(methyl-methacrylate) (PMMA) microspheres were also mixed with the HA suspension to produce microporosity upon burnout of the PMMA. Both scaffolds exhibited linear elastic behavior with macroporous scaffolds having a modulus/ultimate strength of 1,110/34.4 MPa and macro/microporous scaffolds having a modulus/ultimate strength of 1,240/27.4 MPa. Coefficients of variation ranged from 7 to 23% for modulus and 6 to 15% for ultimate strength. These values were in the high range of trabecular bone for compressive properties. Moroni et al. [46] used a Bioplotter nozzle deposition system to fabricate copolymer polyethyleneoxide-terephtalate (PEOT) and polybutylene-terephtalate (PBT) scaffolds with porosity ranging from 29 to 91%. Viscoelastic properties of storage modulus and the damping factor (tangent of the lag angle between time-dependent stress and strain) were determined as a function of porosity and compared to bovine cartilage. In general, with increasing porosity, storage modulus increased from 0.19 to 13.7 MPa, while the damping factor decreased from 0.20 to 0.08. Coefficients of variation ranged from 3 to 19% for storage modulus and from 7 to 16% for damping factor. In comparison, articular cartilage had a storage modulus of 9.6 MPa (19% coefficient of variation) and a damping factor of 0.18 (17% coefficient of variation).

Material bed systems have been used with polymer or polymer/ceramics composites. Giordano et al. [19] were one of the first groups to build biomaterials directly using a 3D printing system. They built dense PLLA structures with elastic moduli ranging from 187 to 601 MPa and tensile strength ranging from 4.8 to 14.2 MPa. COV were 0.7–11% for modulus and 1–15% for strength. Our own group [73] fabricated PCL scaffolds using selective laser sintering (SLS) (Fig. 38.3c) with porosity ranging from 37.5 to 55% that exhibited moduli ranging from 54 to 65 MPa and yield strength ranging from 2.0 to 3.2 MPa. Coefficients of variation ranged from 3.5 to 4.5% for modulus and 3.5 to 15.5% for yield strength.

Although direct SFF provides highly controllable and automatic fabrication, it is limited by the need to process a given biomaterial such that it can be fabricated by a specific commercial SFF machine. For example, a material must be photopolymerizable for stereolithography or have a sub-100-μm particle size and not agglomerate for SLS. Therefore, many research groups have utilized indirect SFF in which an inverse mold is made by SFF and the biomaterials (including HA, PCL, PLGA, and collagen) are cast into the mold [11, 66]. Our own research group has done extensive work creating scaffolds from HA, PCL, PLGA, and POC. Chu et al. [11] were one of the first to utilize this technique using HA, creating 40% porous scaffolds. These scaffolds had a compressive modulus of 1,400 MPa and compressive strength of 30 MPa. Coefficients of variation were 28.5% and 26% for modulus and strength, respectively. Taboas et al. [66] extended the inverse SFF technique to create polymer and polymer ceramics composites. They also combined inverse SFF with traditional porogen leaching and emulsion techniques to create hierarchical pores structures but did not present mechanical or mass transport properties for these scaffolds. We have further extended the inverse SFF techniques reported in Taboas et al. [66] to fabricate PCL and POC scaffolds, measuring effective nonlinear elastic, viscoelastic, and permeability properties. We found that 30%, 50%, and 70% porous POC exhibited nonlinear elastic properties when tested dry, similar to Kang et al. However, an interesting result was that, in addition to decreasing tangent moduli with increasing porosity, the mechanical behavior also became more linear. This is likely due to less material available to provide a stiffening response as the scaffold deforms. We found that PCL with designed porosity of 53%, 63%, and 70% had permeability ranging from $1.4e^{-7}$ to $12.8e^{-7}$ m^4/Ns when dry (COV 33%) and permeability ranging from $2.9e^{-7}$ to $15.4e^{-7}$ m^4/Ns (COV 15–24%) when wet. Reduction in wet state permeability is likely due to PCL contraction. Liu et al. [38] developed an inverse SFF technique to create collagen I scaffolds. All scaffolds were dehydrated under vacuum pressure with some scaffolds subsequently cross-linked using lysine. Viscoelastic storage moduli and damping factors were measured, with storage moduli ranging from 0.1 to 1 MPa, with higher moduli found for cross-linked collagen. Moduli decreased with increasing strain, indicating a nonlinear response. Damping factors ranged from 0.3 to 1.2, higher than values for PEOT/PBT and articular cartilage measured by Moroni et al. [46].

In summary, a great deal of progress has been made in scaffold fabrication from a variety of polymeric, ceramic, polymer-ceramic composites, and even titanium. This control over mechanical and mass transport properties is critical for determining how these parameters affect in vivo performance and tissue regeneration. It becomes very difficult to determine mechanical and mass transport design targets based on in vivo experiments if these properties cannot be controlled with small variations. Furthermore, developing scaffolds for clinical use will require small variations to satisfy regulatory requirements.

Another important issue besides property control is the range and type of mechanical and mass transport properties that can be achieved through the coupling of material synthesis and 3D architecture design. To date, the vast majority of studies reported linear elastic properties and yield strength. This is appropriate for bone tissue engineering, as bone exhibits linear elastic behavior. However, as summarized in Sect. 38.2, soft tissue often exhibit nonlinear elastic and viscoelastic behavior. Therefore, it is important to characterize the nonlinear and viscoelastic behavior of scaffolds as well when used for soft tissue engineering. Besides the works of Moroni et al. [46] and Liu et al. [38] detailing viscoelasticity, Moutous et al. [47] detailing biphasic properties, and our own group detailing nonlinear elastic properties (this chapter), little has been reported on such properties. For nonlinear elasticity and viscoelasticity, material choice becomes crucial, as the base material will determine whether nonlinear elastic or viscoelastic behavior can be achieved at all, with architecture design modulating the range of such properties. In this sense, the choice of elastomeric polymers like PGS or POC may be better for engineering very nonlinearly compliant soft tissues.

The majority of scaffold mechanical characterization has been done in a dry environment at room temperature (21–23°C), not considering the effect of sterilization. However, most polymeric scaffold materials will be significantly affected by all factors, and of course clinical use will require sterilization followed by implantation in a wet environment at body temperature (37°C). For example, Wu et al. [78] found that PDLLA scaffolds tested in a wet state experienced a 20% reduction in modulus and a 10% reduction in strength. Likewise, testing at body temperature reduced the modulus between 20 and 30%. Filipczak et al. [15] studied the effect of sterilization by irradiation on PCL mechanical properties for doses ranging from 25 to 150 kGy. They found that increasing radiation doses increased the elastic modulus by up to 30%, but decreased ductility as strain at yield from 9.4 to 8.4%. Since both environmental and sterilization procedures will be combined for in vivo implantation, it is important to study their effect in combination on effective mechanical and mass transport properties.

The issue of controlling and designing mass transport properties is almost completely determined by pore architecture design. Base materials will have some microporosity (defined here as pores < 10 μm in size), resulting from the fabrication method, but this microporosity is likely to be random and tortuous, leading to low values of diffusivity and permeability. Thus, control over mass transport properties necessitates design-based diffusivity and fluid flow analysis, coupled with fabrication methods that can reproduce pore designs with high fidelity. In this sense, fiber spinning and weaving provide moderate control, but again SFF techniques provide the highest level of fabrication control. The second issue with mass transport design is the definition of permeability. Fluid permeation experiments provide a direct measure of permeability by pressurized flow through the scaffold while confined compression tests of the scaffold fit permeability and aggregate modulus to stress relaxation data based on biphasic theory. However, these two methods have produced vastly different permeability data (10^{-6}–10^{-9} m^4/Ns from permeation experiments versus 10^{-15} m^4/Ns from confined compression tests). These differences can be traced to the analytical isotropic biphasic theory solution. This solution will not exhibit stress relaxation unless the permeability is below 10^{-13} m^4/Ns. While the theory works well for cartilage having extremely small pores and large proteoglycan molecules to impede fluid flow, it may not work well for fabricated scaffolds that have pores on the order of hundreds of microns. Thus, fluid permeation experiments may provide a more consistent way to characterize scaffold permeability. A summary of scaffold mechanical and mass transport properties organized by fabrication method is given in Table 38.3.

38.6
Experiments Relating Scaffold Design Targets to Tissue Regeneration

Engineering scaffolds consistently with small variations to match the extreme range of tissue mechanical and mass transport properties is the first necessary (and very difficult!) step in defining scaffold design targets. However, it is not sufficient to determine whether matching these ranges is important for improving tissue regeneration. Obviously, the scaffold must not mechanically fail before tissue regeneration is complete. Beyond this, not much is known as to how the level of mechanical and mass transport properties influence tissue regeneration. Will a scaffold with 30% of native trabecular bone linear elastic modulus and permeability support sufficient bone regeneration? Will a scaffold that has twice the stiffness and four times the permeability of native articular cartilage still support cartilage regeneration? Is it necessary to match the complete nonlinear stress strain curve of a soft tissue or just the initial compliant portion? To answer these questions, the importance of testing scaffolds with small mechanical and mass transport variations resulting from controlled architectures (regardless of fabrication method) becomes readily apparent as it is difficult to interpret experimental results when property and architecture variations are large.

Use of scaffold with well-controlled pore architecture has already challenged accepted conclusions made using scaffolds with large architecture variations. For example, despite more than 30 years of study, no definitive conclusion has been made regarding the effect of pore size on bone regeneration [32]. Some studies have suggested optimal pore sizes of 100–500 μm, while others have suggested only that a minimum pore size of 75 μm or 300 μm is necessary to enhance bone growth [4, 10, 14, 17, 32]. Still, other studies have stated that bone growth is not sensitive to pore size, but rather to pore *interconnectivity* size (which is closely correlated to permeability) [22, 29, 37, 44, 52]. Work by our group [24, 64] using scaffolds with controlled and homogeneous pore distributions have demonstrated that there was no significant difference in bone regeneration for pore sizes in the range of 400–1,200 μm using both a mouse subcutaneous and a minipig mandibular de-

Table 38.3 Elastic and permeability properties of selected scaffolds, organized by fabrication technique. Most techniques show similar variation in properties, although the SFF techniques have created a much broader range of architectures as reflected by porosity ranges

Fabrication method	Material	Porosity	Tangent modulus at 1% strain (MPa) = ABe^{BE}	Tangent modulus at 10% strain (MPa) = ABe^{BE}	COV	Permeability (m^4/Ns)	COV	Ref
Porogen leaching and compression molding								
PL	PLGA	80–92%	0.1–1.88*	0.1–1.88*	40–62%	$2.1e^{-9}$–$16.1e^{-9}$*	40–133%	1
CM/PL	PLLA	78–97%	0.5–13	0.5–13	–	–	–	83
CM/PL	PLGA	87%	6.6–12.9	6.6–12.9	4–14%	–	–	77
GF/PL	PLGA	95%	0.4–0.5	0.4–0.5	2–12%	–	–	58
Fiber architectures								
Electrospun	PCL	85%	1–20**	1–20**	–	–	–	49
Electrospun	PCL	–	7.1–33.2**	7.1–33.2**	6–11%	–	–	67
Fiber	PGA	95%	–	–	–	$2.6e^{-6}$–$4.7e^{-6}$*	8%	36
Woven fiber	PGA	70–74%	0.14–0.2 300–500^	0.14–0.2 300–500^	8–31%	$0.4e^{-15}$–$0.9e^{-15}$	15–40%	47
Solid free-form fabrication methods								
FDM	PCL	48–77%	4–77	4–77	4–12%	–	–	80
ND	PLGA/PLLA/TCP	74–81%	17–23	17–23	5–17%	–	–	39
ND	HA	41%	1,110–1,240	1,110–1,240	7–23%	–	–	75
ND	PEOT/PBT	29–91%	0.2–13.7	0.2–13.7	3–19%	–	–	46
3DP	PLLA	0%	187–601	187–601	0.7–11%	–	–	19
SLS	PCL	37–55%	54–65	54–65	4–5%	–	–	73
Inverse SFF	HA	40%	1,400	1,400	28%	–	–	11
Inverse SFF	Col I	–	0.1–1	0.1–1	–	–	–	38
Inverse SFF	POC	30/50/70%	0.35–1.05	0.35–1.5	14–53%			CG
Inverse SFF	PCL—dry	–	–	–	–	$6.9e^{-6}$–$40e^{-6}$		CG
Inverse SFF	PCL—wet	–	–	–	–	$2.9e^{-7}$–$15.4e^{-7}$		CG

*Ranges reflect material degradation over time
**Ranges reflect anisotropy of fiber matrices, modulus measured in different directions
^Ranges reflect difference in compression and tension modulus

CG current chapter, *PL* porogen leaching, *CM* compression molding, *GF* gas foaming, *FDM* fused deposition molding, *ND* nozzle deposition, *3DP* 3D printing, *SLS* selective laser sintering, *Col I* collagen I

fect model. These scaffolds had complete designed pore interconnectivity, supporting the postulate that pore interconnectivity is more critical to bone growth in the scaffold than pore size.

Similar confounding results of architecture on bone regeneration can be seen when investigating porosity. Conventional wisdom suggests that increasing porosity should lead to increased bone regeneration due to enhanced mass transport. Indeed, studies such as that by Okamoto et al. [51] demonstrated increasing bone growth in bone marrow stem-cell-seeded HA scaffolds with porosity increasing from 30 to 50 to 70%. However, Okamoto et al. noted all scaffolds had pore sizes ranging from 1 to 1,000 μm with restricted pore connections, especially the 30% porosity scaffolds. Our research group also studied the effect of 30, 50, and 70% porous PPF/TCP scaffolds seeded with BMP-7 transduced human gingival fibroblasts on bone regeneration. These results showed no statistically significant difference in regenerate bone volume between scaffolds of different porosity. When normalized by pore volume, the lower porosity scaffolds actually had more bone ingrowth. These results support the theory the pre interconnectivity, or equivalently permeability, is a more relevant architectural design variable than pore size or porosity.

Controlled scaffold architecture has also been shown to influence cartilage regeneration. Malda et al. [42] studied cartilage regeneration in PEGT/PBT scaffolds made by 3D nozzle deposition (3DND) and compression molding/porogen leaching (3DND). Both scaffolds were between 75 and 80% porous, both the 3D-deposited scaffolds had interconnected pores with uniform 525 μm pore size while the CMPL scaffold had tortuous pore architecture with a wide range of pore sizes having a mean of 182 μm. Dynamic stiffness of the 3DND scaffolds averaged 4.33 MPa (COV of 12%), within the range of both human (4.5 MPa) and bovine (4.1 MPa) articular cartilage, compared to the CMPL scaffolds having 1.72 MPa dynamic stiffness (COV of 19%). Subcutaneous in vivo implantation with chondrocytes demonstrated significantly more cartilage matrix in the interconnected 3DND scaffold architecture.

Our group has studied the effect of ellipsoidal versus cubic-shaped pores on cartilage regeneration from both chondrocytes and bone marrow stromal cells (BMSC) in vitro and subcutaneously in vivo. Scaffolds were 50% porous (cubic pore) and 52% porous (ellipsoidal pore) with both having completely interconnected porosity. The ellipsoidal pore scaffold was computed to have only 56% of the diffusivity and 17% of the permeability of the cubic pore scaffold. Results with both chondrocytes and BMSC demonstrated increased cartilage matrix production in ellipsoidal pores compared to cubic pores. This result was attributed to the increased cell aggregation in ellipsoidal pore structures determined by Rhodamine-conjugated peanut agglutinin (PNA) staining.

38.7
Conclusions

In conclusion, significant progress has been made in biomaterial scaffold fabrication, including the ability to fabricate scaffolds from a wide range of materials and, with the advent of SFF, to control architecture designs. These fabricated scaffolds cover a wide range of mechanical and mass transport properties and now offer the possibility of creating these properties to match the tremendous variations in tissue mechanical and mass transport properties. Remarkably, the COV of different scaffold fabrication methods are similar, although SFF methods have provided a wider variation in architecture designs and resulting mechanical and mass transport properties. Tissue properties, however, run the entire gamut from linear elastic to nonlinear elastic and viscoelastic.

Despite the significant progress made in scaffold design and fabrication, we have surprisingly little guidance on how to best engineer scaffolds for a specific application. At best, it has been suggested that scaffold mechanical and mass transport properties should be in the range of native tissue properties. This lack of guidance can be attributed to two factors. First, without a broad range of available materials and controlled design/fabrication, it has been difficult to create scaffolds that can match tissue properties. Indeed, there is still a great need for both more sophisticated characterizations of scaffold properties to include nonlinear elasticity and viscoelasticity to determine just how well scaffolds can match tissue behavior. Second, however, it is not known how precisely we need to match these properties to enhance tissue regeneration. This can only be accomplished by first fabricating a wide range of scaffolds that match tissue properties in varying degrees, and then testing these scaffolds in appropriate animal models to experimentally assess the effect of scaffold architecture and material on tissue regeneration. Specifically, we will need to answer questions on how permeability and diffusivity affect tissue regeneration, as well as stiffness, and for nonlinear elastic materials, if all or only part of the nonlinear stress strain curve needs to be matched to enable load bearing and enhance tissue regeneration for soft tissues. It is only through such scaffold fabrication, scaffold characterization, and in vivo experiments that we can develop guidelines for scaffold design.

Acknowledgments

Portions of the authors' work was supported by the NIH, including R01 DE13608, R01 DE 016129, and R01 AR 053379. We would also like to acknowledge the contributions of Rachel Schek, Juan Taboas, Colleen Flanagan, Eiji Saito, Heesuk Kang, Dr. Stephen Feinberg, and Dr. Paul Krebsbach to the work reported in this chapter.

References

1. Agawam CM, McKinney JS, Lanctot D, Athanasiou KA (2000) Effects of fluid flow on the in vitro degradation kinetics of biodegradable scaffolds for tissue engineering. Biomaterials 21:2443–2452
2. Altman GH, Horan RL, Martin I, Farhadi J, Stark PR, Volloch V, Richmond JC, Vunjak-Novakovic G, Kaplan DL (2002) Cell differentiation by mechanical stress. FASEB J 16:270–272
3. Bakhvalov NS, Bogachev KY, Eglit ME (1996) Numerical calculation of effective elastic moduli for incompressible porous material. Mech Comp Mat 32:399–405
4. Bobyn JD, Pilliar RM, Cameron HU, Weatherly GC (1980) The optimum pore size for the fixation of porous-surfaced metal implants by the ingrowth of bone. Clin Orthop Relat Res 150:263–270
5. Boudriot U, Dersch R, Greiner A, Wendorff JH (2006) Electrospinning approaches toward scaffold engineering—a brief overview. Artif Organs 30:785–792
6. Brand RA (1992) Autonomous informational stability in connective tissues. Med Hypotheses 37:107–114
7. Brekke JH, Toth JM (1998) Principles of tissue engineering applied to programmable osteogenesis. J Biomed Mater Res 43:380–398
8. Byrne DP, Lacroix D, Planell JA, Kelly DJ, Prendergast PJ (2007) Simulation of tissue differentiation in a scaffold as a function of porosity, Young's modulus and dissolution rate: application of mechanobiological models in tissue engineering. Biomaterials 28:5544–5554
9. Carter DR, Beaupre GS (2001) Skeletal function and form: mechanobiology of skeletal development, aging and regeneration, 1st edn. Cambridge University Press, Cambridge
10. Cheung HY, Lau KT, Lu TP, Hui D (2007) A critical review on polymer-based bio-engineered materials for scaffold development. Composites: Part B 38:291–300
11. Chu TM, Orton DG, Hollister SJ, Feinberg SE, Halloran JW (2002) Mechanical and in vivo performance of hydroxyapatite implants with controlled architectures. Biomaterials 23:1283–1293
12. D'Ippolito G, Diabira S, Howard GA, Roos BA, Schiller PC (2006) Low oxygen tension inhibits osteogenic differentiation and enhances stemness of human MIAMI cells. Bone 39:513–522
13. Domm C, Schunke M, Christesen K, Kurz B (2002) Redifferentiation of dedifferentiated bovine articular chondrocytes in alginate culture under low oxygen tension. Osteoarthritis Cartilage 10:13–22
14. Eggli PS, Muller W, Schenk RK (1988) Porous hydroxyapatite and tricalcium phosphate cylinders with two different pore size ranges implanted in the cancellous bone of rabbits. A comparative histomorphometric and histologic study of bony ingrowth and implant substitution. Clin Orthop Relat Res (232):127–138
15. Filipczak K, Wozniak M, Ulanski P, Olah L, Przybytniak G, Olkowski RM, Lewandowska-Szumiel M, Rosiak JM (2006) Poly(epsilon-caprolactone) biomaterial sterilized by E-beam irradiation. Macromol Biosci 6:261–273
16. Fung YC (1993) Biomechanics: mechanical properties of living tissues, 2nd edn. Springer, Berlin
17. Galois L, Mainard D (2004) Bone ingrowth into two porous ceramics with different pore sizes: an experimental study. Acta Orthop Belg 70:598–603
18. Gibiansky LV, Torquato S (1996) Connection between the conductivity and bulk modulus of isotropic composite materials. Proc R Soc Lond A 452:253–283
19. Giordano RA, Wu BM, Borland SW, Cima LG, Sachs EM, Cima MJ (1996) Mechanical properties of dense polylactic acid structures fabricated by three dimensional printing. J Biomater Sci Polym Ed 8:63–75
20. Gloeckner DC, Sacks MS, Fraser MO, Somogyi GT, de Groat WC, Chancellor MB (2002) Passive biaxial mechanical properties of the rat bladder wall after spinal cord injury. J Urol 167:2247–2252
21. Goulet RW, Goldstein SA, Ciarelli MJ, Kuhn JL, Brown MB, Feldkamp LA (1994) The relationship between the structural and orthogonal compressive properties of trabecular bone. J Biomech 27:375–389
22. Hing KA, Best SM, Tanner KE, Bonfield W, Revell PA (2004) Mediation of bone ingrowth in porous hydroxyapatite bone graft substitutes. J Biomed Mater Res A 68:187–200

23. Hollister SJ (2005) Porous scaffold design for tissue engineering. Nat Mater 4:518–524
24. Hollister SJ, Lin CY, Saito E, Lin CY, Schek RD, Taboas JM, Williams JM, Partee B, Flanagan CL, Diggs A, Wilke EN, Van Lenthe GH, Muller R, Wirtz T, Das S, Feinberg SE, Krebsbach PH (2005) Engineering craniofacial scaffolds. Orthod Craniofac Res 8:162–173
25. Holzapfel GA, Ogden RW (2006) Mechanics of biological tissues. Springer, Berlin
26. Humphrey JD (2002) Cardiovascular solid mechanics: cells, tissues and organs, 1st edn. Springer, Berlin
27. Hutmacher DW (2001) Scaffold design and fabrication technologies for engineering tissues—state of the art and future perspectives. J Biomater Sci Polym Ed 12:107–124
28. Ichihara K, Taguchi T, Sakuramoto I, Kawano S, Kawai S (2003) Mechanism of the spinal cord injury and the cervical spondylotic myelopathy: new approach based on the mechanical features of the spinal cord white and gray matter. J Neurosurg 99:278–285
29. Jones AC, Arns CH, Sheppard AP, Hutmacher DW, Milthorpe BK, Knackstedt MA (2007) Assessment of bone ingrowth into porous biomaterials using MICRO-CT. Biomaterials 28:2491–2504
30. Joshi MD, Suh JK, Marui T, Woo SL (1995) Interspecies variation of compressive biomechanical properties of the meniscus. J Biomed Mater Res 29:823–828
31. Kang Y, Yang J, Khan S, Anissian L, Ameer GA (2006) A new biodegradable polyester elastomer for cartilage tissue engineering. J Biomed Mater Res A 77:331–339
32. Karageorgiou V, Kaplan D (2005) Porosity of 3D biomaterial scaffolds and osteogenesis. Biomaterials 26:5474–5491
33. Kelly DJ, Prendergast PJ (2006) Prediction of the optimal mechanical properties for a scaffold used in osteochondral defect repair. Tissue Eng 12:2509–2519
34. Klisch SM, Lotz JC (1999) Application of a fiber-reinforced continuum theory to multiple deformations of the annulus fibrosus. J Biomech 32:1027–1036
35. Leddy HA, Guilak F (2003) Site-specific molecular diffusion in articular cartilage measured using fluorescence recovery after photobleaching. Ann Biomed Eng 31:753–760
36. Li J, Mak AF (2005) Hydraulic permeability of polyglycolic acid scaffolds as a function of biomaterial degradation. J Biomater Appl 19:253–266
37. Li JP, Habibovic P, van den Doel M, Wilson CE, de Wijn JR, van Blitterswijk CA, de Groot K (2007) Bone ingrowth in titanium implants produced by 3D fiber deposition. Biomaterials 28:2810–2820
38. Liu CZ, Xia ZD, Han ZW, Hulley PA, Triffitt JT, Czernuszka JT (2007) Novel 3D collagen scaffolds fabricated by indirect printing technique for tissue engineering. J Biomed Mater Res B Appl Biomater
39. Liu L, Xiong Z, Yan Y, Hu Y, Zhang R, Wang S (2007) Porous morphology, porosity, mechanical properties of poly(alpha-hydroxy acid)-tricalcium phosphate composite scaffolds fabricated by low-temperature deposition. J Biomed Mater Res A 82:618–629
40. Loboa EG, Beaupre GS, Carter DR (2001) Mechanobiology of initial pseudarthrosis formation with oblique fractures. J Orthop Res 19:1067–1072
41. Malda J, van Blitterswijk CA, van Geffen M, Martens DE, Tramper J, Riesle J (2004) Low oxygen tension stimulates the redifferentiation of dedifferentiated adult human nasal chondrocytes. Osteoarthritis Cartilage 12:306–313
42. Malda J, Woodfield TB, van der Vloodt F, Kooy FK, Martens DE, Tramper J, van Blitterswijk CA, Riesle J (2004) The effect of PEGT/PBT scaffold architecture on oxygen gradients in tissue engineered cartilaginous constructs. Biomaterials 25:5773–5780
43. Martin RB, Burr DB, Sharkey NA (1998) Skeletal tissue mechanics, 1st edn. Springer, Berlin
44. Mastrogiacomo M, Scaglione S, Martinetti R, Dolcini L, Beltrame F, Cancedda R et al (2006) Role of scaffold internal structure on in vivo bone formation in macroporous calcium phosphate bioceramics. Biomaterials 27:3230–3237
45. Miller-Young JE, Duncan NA, Baroud G (2002) Material properties of the human calcaneal fat pad in compression: experiment and theory. J Biomech 35:1523–1531
46. Moroni L, de Wijn JR, van Blitterswijk CA (2006) 3D fiber-deposited scaffolds for tissue engineering: influence of pores geometry and architecture on dynamic mechanical properties. Biomaterials 27:974–985
47. Moutos FT, Freed LE, Guilak F (2007) A biomimetic three-dimensional woven composite scaffold for functional tissue engineering of cartilage. Nat Mater 6:162–167
48. Mow VC, Huiskes R (2005) Basic orthopaedic biomechanics and mechano-biology, 3rd edn. Lippincott Williams & Wilkins, Philadelphia
49. Nerurkar NL, Elliott DM, Mauck RL (2007) Mechanics of oriented electrospun nanofibrous scaffolds for annulus fibrosus tissue engineering. J Orthop Res 25:1018–1028
50. Ng KW, Mauck RL, Statman LY, Lin EY, Ateshian GA, Hung CT (2006) Dynamic deformational loading results in selective application of mechanical stimulation in a layered, tissue-engineered cartilage construct. Biorheology 43:497–507
51. Okamoto M, Dohi Y, Ohgushi H, Shimaoka H, Ikeuchi M, Matsushima A, Yonemasu K, Hosoi H (2006) Influence of the porosity of hydroxyapatite ceramics on in vitro and in vivo bone formation by cultured rat bone marrow stromal cells. J Mater Sci Mater Med 17:327–336
52. Otsuki B, Takemoto M, Fujibayashi S, Neo M, Kokubo T, Nakamura T (2006) Pore throat size and connectivity determine bone and tissue ingrowth into porous implants: three-dimensional micro-CT based structural analyses of porous bioactive titanium implants. Biomaterials 27:5892–5900
53. Perren SM (1979) Physical and biological aspects of fracture healing with special reference to internal fixation. Clin Orthop Relat Res 138:175–196
54. Prendergast PJ, Huiskes R, Soballe K (1997) ESB Research Award 1996. Biophysical stimuli on cells during tissue differentiation at implant interfaces. J Biomech 30:539–548

55. Qiu H, Yang J, Kodali P, Koh J, Ameer GA (2006) A citric acid-based hydroxyapatite composite for orthopedic implants. Biomaterials 27:5845–5854
56. Quapp KM, Weiss JA (1998) Material characterization of human medial collateral ligament. J Biomech Eng 120:757–763
57. Reynaud B, Quinn TM (2006) Anisotropic hydraulic permeability in compressed articular cartilage. J Biomech 39:131–137
58. Riddle KW, Mooney DJ (2004) Role of poly(lactide-co-glycolide) particle size on gas-foamed scaffolds. J Biomater Sci Polym Ed 15:1561–1570
59. Riha GM, Wang X, Wang H, Chai H, Mu H, Lin PH, Lumsden AB, Yao Q, Chen C (2007) Cyclic strain induces vascular smooth muscle cell differentiation from murine embryonic mesenchymal progenitor cells. Surgery 141:394–402
60. Robins JC, Akeno N, Mukherjee A, Dalal RR, Aronow BJ, Koopman P, Clemens TL (2005) Hypoxia induces chondrocyte-specific gene expression in mesenchymal cells in association with transcriptional activation of Sox9. Bone 37:313–322
61. Roy TD, Simon JL, Ricci JL, Rekow ED, Thompson VP, Parsons JR (2003) Performance of degradable composite bone repair products made via three-dimensional fabrication techniques. J Biomed Mater Res A 66:283–291
62. Sacks MS, Chuong CJ (1993) Biaxial mechanical properties of passive right ventricular free wall myocardium. J Biomech Eng 115:202–205
63. Sander EA, Nauman EA (2003) Permeability of musculoskeletal tissues and scaffolding materials: experimental results and theoretical predictions. Crit Rev Biomed Eng 31:1–26
64. Schek RM, Wilke EN, Hollister SJ, Krebsbach PH (2006) Combined use of designed scaffolds and adenoviral gene therapy for skeletal tissue engineering. Biomaterials 27:1160–1166
65. Simmons CA, Matlis S, Thornton AJ, Chen S, Wang CY, Mooney DJ (2003) Cyclic strain enhances matrix mineralization by adult human mesenchymal stem cells via the extracellular signal-regulated kinase (ERK1/2) signaling pathway. J Biomech 36:1087–1096
66. Taboas JM, Maddox RD, Krebsbach PH, Hollister SJ (2003) Indirect solid free form fabrication of local and global porous, biomimetic and composite 3D polymer-ceramic scaffolds. Biomaterials 24:181–194
67. Thomas V, Jose MV, Chowdhury S, Sullivan JF, Dean DR, Vohra YK (2006) Mechano-morphological studies of aligned nanofibrous scaffolds of polycaprolactone fabricated by electrospinning. J Biomater Sci Polym Ed 17:969–984
68. Thomson RC, Yaszemski MJ, Powers JM, Mikos AG (1995) Fabrication of biodegradable polymer scaffolds to engineer trabecular bone. J Biomater Sci Polym Ed 7:23–38
69. Utting JC, Robins SP, Brandao-Burch A, Orriss IR, Behar J, Arnett TR (2006) Hypoxia inhibits the growth, differentiation and bone-forming capacity of rat osteoblasts. Exp Cell Res 312:1693–1702
70. Wang L, Wang Y, Han Y, Henderson SC, Majeska RJ, Weinbaum S, Schaffler MB (2005) In situ measurement of solute transport in the bone lacunar-canalicular system. Proc Natl Acad Sci U S A 102:11911–11916
71. Weinberg EJ, Kaazempur-Mofrad MR (2006) A large-strain finite element formulation for biological tissues with application to mitral valve leaflet tissue mechanics. J Biomech 39:1557–1561
72. Weiss JA, Maakestad BJ (2006) Permeability of human medial collateral ligament in compression transverse to the collagen fiber direction. J Biomech 39:276–283
73. Williams JM, Adewunmi A, Schek RM, Flanagan CL, Krebsbach PH, Feinberg SE, Hollister SJ, Das S (2005) Bone tissue engineering using polycaprolactone scaffolds fabricated via selective laser sintering. Biomaterials 26:4817–4827
74. Williams JR, Natarajan RN, Andersson GB (2007) Inclusion of regional poroelastic material properties better predicts biomechanical behavior of lumbar discs subjected to dynamic loading. J Biomech 40:1981–1987
75. Woodard JR, Hilldore AJ, Lan SK, Park CJ, Morgan AW, Eurell JAC, Clark SG, Wheeler MB, Jamison RD, Wagoner Johnson AJ (2007) The mechanical properties and osteoconductivity of hydroxyapatite bone scaffolds with multi-scale porosity. Biomaterials 28:45–54
76. Wu JZ, Cutlip RG, Andrew ME, Dong RG (2007) Simultaneous determination of the nonlinear-elastic properties of skin and subcutaneous tissue in unconfined compression tests. Skin Res Technol 13:34–42
77. Wu L, Ding J (2004) In vitro degradation of three-dimensional porous poly(d,l-lactide-co-glycolide) scaffolds for tissue engineering. Biomaterials 25:5821–5830
78. Wu L, Zhang J, Jing D, Ding J (2006) "Wet-state" mechanical properties of three-dimensional polyester porous scaffolds. J Biomed Mater Res A 76:264–271
79. Zakaria E, Lofthouse J, Flessner MF (1997) In vivo hydraulic conductivity of muscle: effects of hydrostatic pressure. Am J Physiol 273:H2774–2782
80. Zein I, Hutmacher DW, Tan KC, Teoh SH (2002) Fused deposition modeling of novel scaffold architectures for tissue engineering applications. Biomaterials 23:1169–1185
81. Zeng Z, Yin Y, Huang AL, Jan KM, Rumschitzki DS (2007) Macromolecular transport in heart valves. I. Studies of rat valves with horseradish peroxidase. Am J Physiol Heart Circ Physiol 292:H2664–2670
82. Zeng Z, Yin Y, Jan KM, Rumschitzki DS (2007) Macromolecular transport in heart valves. II. Theoretical models. Am J Physiol Heart Circ Physiol 292:H2671–2686
83. Zhang J, Wu L, Dianying J, Ding J (2005) A comparative study of porous scaffolds with cubic and spherical macropores. Polymer 46:4979–4985

Scaffold Structure and Fabrication

H. P. Wiesmann, L. Lammers

Whereas the properties of a material's surface directly influence single cell behaviour, the three-dimensional scaffold structure plays a critical role in the orchestration of tissue formation both in vitro and in vivo. Whereas the microstructure of a material refers to the material at the nanoscale or microscale level (mainly used to characterise material surfaces), scaffold architecture defines the structure of the material in space at a tissue-length scale. Scaffolds not only provide the structural basis for cells to form a three-dimensional tissue-like construct in vitro, but they also determine the features of mass transport (diffusion and convection). The scaffold architecture affects both single-cell parameters (e.g. cell viability, cell migration, cell differentiation) and the composition of the generated tissue substitute.

The following scaffold characteristics can be distinguished:
– Scaffold composition
– External geometry
– Macrostructure
– Microstructure
– Interconnectivity
– Surface/volume ratio
– Mechanical competence
– Degradation characteristics
– Chemical properties

Scaffolds can be grouped into different categories: underlying material, macrostructure (spatial geometry), microstructure (bulk versus porous), mechanical properties and degradation characteristics [13].

Whereas the micro- and nano-properties are related to a great extent to the material surface, the micro- and macro-properties are also critical for tissue development. Scaffolds serve as space holders for cells and allow ingrowth of surrounding tissues into the reconstruction site after transplantation. Thus, they provide structures that facilitate the three-dimensional proliferation, differentiation and orientation of cells in order to enable a tissue-like construct growth ex vivo (Fig. 39.1). Scaffolds facilitate the transfer of loads to surrounding tissues and ideally allow the reconstruction site to be mechanically competent directly after insertion.

Scaffolds also provide a space in which tissue development and maturation towards complex multicellular systems can occur [45]. Two classical ways involve fabricating cell-containing scaffolds. A commonly used method is to coax cells with the scaffold by seeding cells on the scaffold (Fig. 39.2a, b). This approach is simple and fast, since cells attach to the scaffold within the first 24 h. In contrast to the second method (to cultivate and maturate cells within a scaffold in a bioreactor), this approach is limited by an inhomogenous distribution of cells. Various techniques to coax cells with scaffolds are in use. The most commonly used method, static cell seeding, involves the placement of the scaffold in a bioreactor followed by adding a cell suspension to allow the adhesion of cells. However, the resulting cell distribution in the scaffold is random and dependent on the gross scaffold structures, but in general, the majority of the cells attach only to the outer surfaces. Wetting techniques of scaffolds prior to cell seeding allow for displacement of air-filled pores with water and facilitate penetration of the cell suspension into these pores.

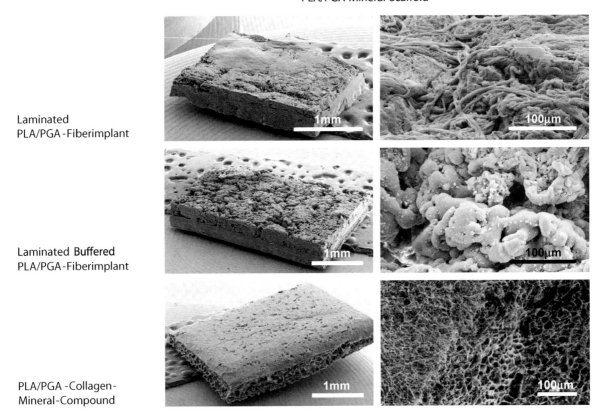

Fig. 39.1 Scanning electron microscopical view of different scaffolds (*left*: overview, *right*: detail) fabricated by the conventional polymer fabrication technique. Polymer scaffolds can be altered by inclusion of buffer systems or mineral compounds in order to optimise their properties

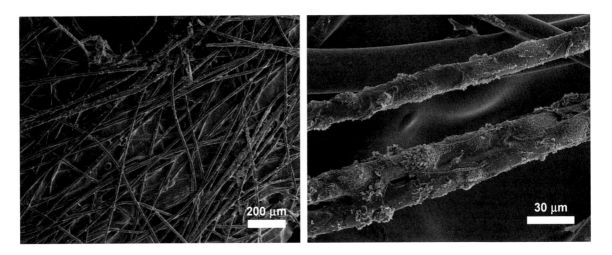

Fig. 39.2 Scanning electron microscopical view of cell/scaffold construct. Seeding of osteoblasts to a fibre scaffold leads to an adhesion of cells at the fibre surface without filling the space between fibres

In contrast to simple seeding techniques, cells can also be coaxed with scaffolds by injection or by applying a vacuum to ensure penetration of the cell suspension. By using this technique, cells can be placed more homogenously throughout the three-dimensional space. However, the uniformity is lost under static culture conditions because of the nutrient and oxygen diffusion limitations within the scaffold. When cells are seeded on scaffolds, high cell densities are necessary to achieve uniform tissue regeneration. The effects of cell seeding density on cartilage formation indicate that high numbers of cells are required to gain a more mature tissue formation [47, 48, 4, 11].

The second approach of coaxing cells with scaffolds is the long-term culture of cells with scaffolds. By using this technique, cells have to proliferate, migrate and differentiate during culture with the scaffolds (Fig. 39.3a–f). This longer-lasting method has the main advantage that cells are not restricted to the outer surface of the scaffold but may grow into deeper scaffold structures. As bone and cartilage differs significantly in the tissue composition at the different levels (cellular level, supracellular level, tissue level), the "optimal" scaffold properties can be assumed to differ between scaffolds used for bone and cartilage tissue engineering [49]. Reviewing the literature reveals that bioorganic scaffold materials (e.g. collagen, fibrin, hyaluronan, alginate, chitosan) and polymer materials (e.g. polyhydroxy acids) are commonly used for both (cartilage and bone) tissue engineering strategies [1], whereas organic materials (hydroxyapatite, tricalcium phosphates) are preferably used in bone tissue engineering. Independent from the underlying material, various basic scaffold characteristics can be distinguished:

- **Scaffold composition** Materials that constitute the scaffold can be distinguished by the chemical composition. Pure, non-organic materials can be distinguished from composite materials (also containing organic materials) and sole organic materials. Materials can also be grouped by whether they are in a solid or gel-like condition (hydrogels, for example, contain a water content >30% by weight).
- **Macrostructure** The macrostructure reflects the external geometry and gross internal structure of the scaffold. A three-dimensional scaffold that is congruent to the external geometry of the tissue to be replaced is desired for scaffold placement and fixation in the clinical situation. A direct contact between the defect borders and the scaffold will enhance the interaction between the ex-vivo-generated construct and the host site. An optimal internal scaffold structure will promote tissue formation by mimicking the native environment.
- **Porosity and pore interconnectivity** Scaffolds are constituted of either bulk materials or they have a pore or tube geometry. Pores or tubes can be introduced in scaffolds in an isolated fashion or they can be interconnected. An advantage of an interconnected porous or tubular system is the improved nutritional supply (by diffusion, convection or directed fluid flow) in deeper scaffold areas, thereby enabling cells to survive in these regions. A high interconnectivity of pores or tubes facilitates hydrodynamic microenvironments with minimal diffusion constraints, but at the same time reduces the mechanical strength. As spongiosal bone tissue has a porous structure, these scaffold features seem to be advantageous, especially in the context of ex vivo bone tissue engineering.
- **Pore size/tube diameter** Tissue generation ex vivo and regeneration in vivo may be achieved by using scaffolds with optimal hollow structures (tube/pore combinations). The size of these hollow structures not only improves the fluid flow through the scaffold but may also allow vascular ingrowth (in vitro, as well as after transplantation in vivo) and therefore improve the outcome of tissue engineering treatments. From a conceptual point of view, cartilage tissue as a non-vascular tissue does not benefit from porous scaffolds, whereas bone formation critically depends on vascularisation. As researchers indicated the need for pore sizes ranging from 200–500 µm for vascular ingrowth, scaffolds containing tubular structures of such diameters seem to be beneficial in bone tissue engineering applications.
- **Surface/volume ratio** A high overall material surface area to volume ratio is beneficial in respect to allowing large numbers of cells to attach and migrate into porous scaffolds. Since pore diameter and internal surface area are related linearly, a compromise between surface gain and mechanical stability has to be found depending on the application of the scaffold.

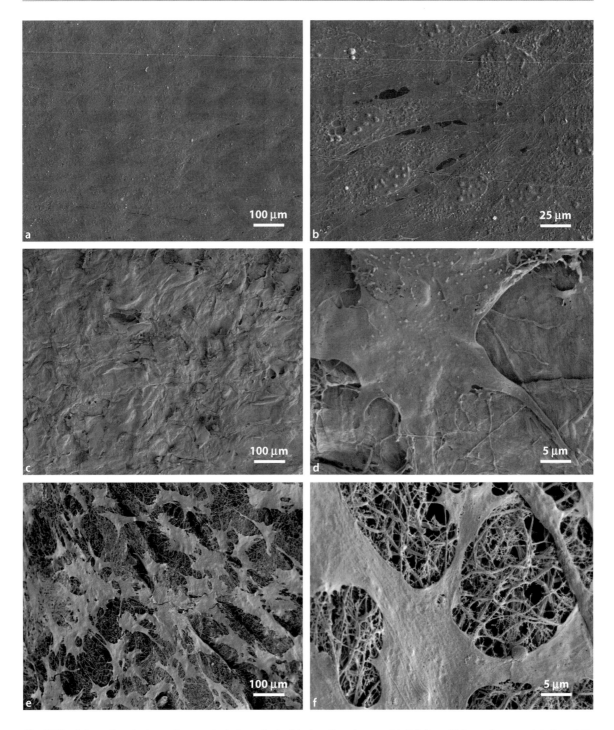

Fig. 39.3a–f Scanning electron microscopical view of osteoblasts cultured on a petri dish (**a** and **b**), a collagen matrix (**c** and **d**), and a fibrin matrix (**e** and **f**). Cells with an osteoblastic appearance cover the surface of scaffolds

- **Mechanical properties** Scaffolds should ideally have sufficient mechanical strength during in vitro culturing as well as the strength to resist the physiological mechanical environment in regenerating load-bearing tissues (cartilage, bone) at the desired implantation site.
- **Degradation characteristics** The ideal scaffold degradation must be adjusted appropriately such that it parallels the rate of new tissue formation and at the same time retains sufficient structural integrity until the newly grown tissue has replaced the scaffold's supporting function (for review, see [24]). Additionally, degradation of the scaffold should not be accompanied by lowering the physiological pH in the in vitro environment or at the defect site (Fig. 39.4a, b). Moreover, during the degradation process, there should be no release of toxic degradation products. Multiple parameters determine the degradation of a scaffold in the in vitro and in vivo situation (Table 39.1).
- **Chemical properties** A scaffold should ideally be used as a carrier for proteins. The proteins, bound to the scaffold material, should be biologically active, and they should be released in a predetermined fashion.

A major technical challenge in scaffold production is maintaining high levels of accurate control over the described macrostructural (e.g. external geometry, spatial composition, mechanical strength, density, porosity), microstructural (e.g. pore size, pore interconnectivity, degradation) and nanostructural

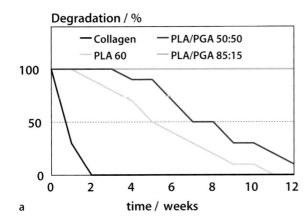
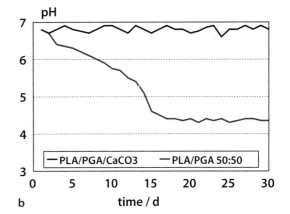

Fig. 39.4a,b Scaffold characteristics can be adjusted by chemical modification techniques. The degradation kinetics of polyesters can be adjusted by a defined co-polymerisation of different polymers (e.g. PLA and PGA) (**a**). Additionally, pH stability during scaffold degradation can be reached through an inclusion of a carbonate buffer system in the scaffold (**b**)

Table 39.1 Parameters affecting scaffold degradation

Parameters affecting scaffold degradation			
Material	Scaffold	In vitro	In vivo
– Composition	– Size	– Medium	– Defect site
– Structure	– Mass	– pH	– Species
– Configuration	– Porosity	– Fluid flow	– Inflammation
– Molecular weight	– Surface/volume ratio	– Biomechanics	– Immunology
– Molecular orientation	– Shape	– Temperature	– Biomechanics
– Cross links	– Surface topography	– Cell sources	– Tissue turnover
– Chain motility	– Surface chemistry	– Cell number	– Enzymes
		– Cell activity	– Vascularisation
		– Ionic strength	

(e.g. surface topography or surface physicochemistry) properties at the same time. Ideally, the processing technique used to fabricate the scaffolds should also allow use of various underlying materials. Besides the fabrication techniques that largely define the scaffolds' macro- and microstructure, there is the variety of natural or synthetic scaffolding materials to consider. It is important to note that each underlying material possesses different processing requirements and varying degrees of processability to form scaffolds. The design of complex scaffolds composed of different materials reaches a further step of complexity in scaffold fabrication. Various engineering parameters (technical procedure, fabrication accuracy, automation) determine the result of the scaffold production.

- **Technical procedure** The processing technique should not change the chemical and biophysical properties of the scaffold nor cause any damage to the material structure.
- **Fabrication accuracy** The fabrication process should create spatially accurate three-dimensional scaffolds at the macro- and microscale level.
- **Automation** The applied technique should be automated in order to ease the production process, reduce the inter-batch inconsistency, ideally reduce the fabrication time and at the same time be cost effective.

As mentioned, the applied scaffold fabrication technique is to a main extent determined by the underlying material. It is obvious that natural-containing, non-organic materials are commonly processed by methods other than, for example, synthetic organic materials. Additionally, naturally occurring non-organic materials are generated by arrays of techniques different from those for synthetic organic materials. Fabrication of non-organic materials is performed by techniques commonly used in ceramic technology, whereas a wide range of techniques have been developed to process synthetic and natural (organic) polymer materials into scaffold structures [35, 36].

Conventional polymer fabrication techniques can be distinguished from the newly introduced solid-free form fabrication techniques (SFF) [38] (Table 39.2). Conventional fabrication techniques are defined as processes that create scaffolds having a bulk or porous (interconnected or non-interconnected) structure which lacks any long-range channelling microstructure [51, 52]. Solid-free form fabrication, in contrast, uses layer manufacturing processes to form scaffolds directly from computer-generated models, thereby enabling the introduction of hollow or tubular structures in scaffolds. Additionally, SFF enables the creation of the external geometry of the scaffold with high precision [50]. Conventional scaffold fabrication is often used in scaffold fabrication for bone and cartilage tissue engineering. Commonly used techniques are: solvent casting/particulate leaching, phase inversion/particulate leaching, fibre meshing/bonding, melt moulding, gas foaming/high pressure processing, membrane lamination, hydrocarbon templating, freeze drying, emulsion freeze drying, solution casting and combinations of these techniques. The principles and technical procedures of these techniques can be found in various reviews

Table 39.2 Scaffold fabrication techniques

Scaffold fabrication techniques	
Conventional	Rapid prototyping
- Solvent casting/particulate leaching	- Fused deposition modeling
- Phase inversion/particulate leaching	- Three dimensional printing
- Fibre meshing/bonding	- Three dimensional plotting
- Melt moulding	- Selective laser sintering
- Gas foaming	- Laminated object manufact.
- Membrane lamination	- Stereolitographie
- Hydrocarbon templating	- Multiphase jet solidification
- Freeze drying	
- Emulsion freeze drying	
- Solution casting	

[13, 32, 22]. A variety of produced scaffolds has been successfully applied to engineer bone and cartilage tissue ex vivo and was investigated for success in animal experimental and clinical situations. In order to understand the advantages and limitations of each processing method, a description of the technical aspects is given in brief.

Solvent casting/particulate leaching is a commonly used and investigated method for the preparation of tissue engineering scaffolds. Scaffolds produced by this technique have been used in a multitude of studies for bone and cartilage tissue engineering with favourable results [33]. This method, first described by Mikos et al. in 1994, is based on dispersing mineral (e.g. sodium chloride, sodium tartrate and sodium citrate) or organic (e.g. saccharose) particles in a polymer solution. The dispersion process is then performed by casting or by freeze-drying in order to produce porous scaffolds. By modulation of the basic process, different researchers [28, 2, 53, 31] aimed to improve the scaffold structure. They introduced variations of the basic processing principle to alter the polymer structure to meet desired needs: to increase the porogen/polymer ratio, avoid crystal deposition, increase the pore interconnectivity, and obtain fully interconnected scaffolds with a high degree of porosity [2, 31]. Solvent casting/particulate leaching has especially been used to fabricate poly(L-lactic acid) (PLLA) or poly(lactide-co-glycolide) (PLGA) scaffolds, but the method has also been applied to process other polymers [8]. Despite the relative ease of fabricating scaffolds using this approach, this method has some inherent disadvantages, such as the possible use of highly toxic solvents [12, 50, 10], the limitation of producing only thin scaffolds [50, 8] and impaired mechanical properties [10].

Phase inversion/particulate leaching is a closely related technique to solvent casting. The difference is that instead of allowing the solvent to evaporate, in phase inversion/particulate leaching, the solution film is placed in water. The subsequent phase inversion causes the polymer to precipitate [11]. The main advantage for tissue engineering purposes when compared to the basic method of solvent casting is that crystal deposition is avoided, and therefore this approach enables the production of thicker scaffolds. It was shown from a technical point of view that when using phase inversion, scaffolds with improved interconnectivity and morphologies mimicking trabecular bone could be fabricated [11]. Based on experimental biological studies, these scaffolds have shown on a number of occasions that they can support the growth of osteoblast-like cells and facilitate the consequent bone matrix formation [11, 16].

Fibre bonding is based on the presence of fibre-like polymer structures. In this process, individual fibres are woven or knitted into three-dimensional scaffolds with variable pore size. By using fibres with different diameters and by connecting the fibres in different special configurations, a high degree of variability in the scaffold structure can be achieved. Whereas this technique enables the creation of large surface areas and thereby improves nutrient diffusion [29, 17], the fibre bonding technique is difficult in respect to the adjustment of the desired porosity and pore size. Additionally, there is the possibility of solvent residues in the scaffold which may be harmful to cells or cells at the host site [23, 12, 50, 21].

Melt-based technologies are often used when porous scaffolds are desired. Melt moulding/particulate leaching is of all melt-based techniques the most commonly used. Through this technique, a polymer is mixed with a porogen prior to loading it into a mould [44]. The mould, shaped in the form of the desired defect geometry, is then heated above the glass transition temperature of the used polymer. After reaching the glass transition temperature, the composite material is immersed in a solvent for the selective dissolution of the porogen [14]. As this technique enables the fabrication of porous scaffolds of defined shapes by creating a defined (external) mould geometry [14, 23, 21], this method has gained wider attraction in clinical tissue engineering strategies [27]. Furthermore, this method also offers independent control of the pore size and porosity by varying the amount and size of the porogen crystals [44, 23, 21].

Extrusion and injection moulding are other melt-based techniques used for the processing of scaffolds [8, 9, 10]. As with particulate leaching, generation of a scaffold [10] with a high degree of porosity and interconnectivity is possible. By resembling the porous structure of trabecular bone, the mechanical properties of such scaffolds can be adjusted in order to reach the compressive modulus and compressive strength of trabecular bone. As the control of pore distribution is impaired by this method, nutritional demands of cells are difficult to meet.

Gas foaming/high pressure as another processing technique is based on the CO_2 saturation of polymer disks through their exposure to high-pressure CO_2 [30]. An advantage of this method is the possibility of obtaining scaffolds with a high degree of porosity and at the same time producing pores with a pore size in the range of 100 μm [30]. However, a low mechanical strength and a poorly defined pore structure limit the widespread use of this technique [40].

A further method is based on a thermally induced phase separation (freeze-drying), which occurs when the temperature of a homogeneous polymer solution, previously poured into a mould, is decreased. After the phase-separated system is stabilised, the solvent-rich phase is removed by vacuum sublimation, leaving behind the polymeric foam. This technique allows the production of scaffolds consisting of natural and synthetic polymers [25, 26]. As the processing conditions are technically challenging and the obtained scaffolds have a low mechanical competence accompanied by a reduced pore size, the application of this technique is uncommon [12, 50].

The varying success rates of scaffold fabrication by the above-mentioned methods may be to some extent technique inherent. Additional limitations of conventional techniques are the need for manual intervention and the inconsistent and inflexible processing procedures. As the ideal scaffold material for bone and cartilage tissue engineering is not known, it is speculative to comment on the feasibility of each processing technique. Much experimental work has to be performed to gain a more detailed insight into the success of the various conventionally fabricated scaffolds, concerning the biological and clinical outcome in bone and tissue engineering strategies.

As new processing technologies were developed in the last decades, they were applied in tissue engineering for scaffold fabrication to overcome the limitations of conventional processing techniques. Rapid prototyping (RP) [12], a synonym for solid free-form fabrication [21], is the most popular one. Rapid prototyping is a computer-based design and fabrication technique that enables the production of ordered external and internal scaffold structures. The main advantages of the SFF technique over conventional techniques are: the possibility of a defined external and internal scaffold structure fabrication, the computer-controlled fabrication process, the improved processing accuracy and eventual coaxing of scaffolds with cells. As with other computer-based techniques, automation and standardisation of processing protocols can be more easily achieved. The data sets used to design and fabricate scaffolds can be based on virtual data sets or computer-based medical imaging technologies (e.g. CT scans) and other technologies [12, 21]. Rapid prototyping is a method that can not only individualise the external and internal defect geometry but can be envisioned in the future also in the control of anisotropic scaffold nanostructures.

The solid free-form technique can be considered for layered manufacturing. In the case of scaffold fabrication, the three-dimensional geometry is created layer by layer via defined processing techniques. Through the possibility of implementing macroscopic, microscopic and possibly future nanoscopic structural features, even in different areas of the scaffold, tissue-like structures may be reached. A complex scaffold micro- and nanostructure closely relating to the desired biological basis tissue can be assumed to be advantageous when cell/scaffold constructs with multiple cell types (e.g. bone and endothelial cells), arranged in hierarchical orders, are desirable [37, 46]. Solid free-form fabrication techniques used for tissue engineering purposes are fused deposition modelling (FDM) [14, 15, 54], three-dimensional printing [6, 34, 41, 18, 42], three-dimensional plotting [19, 3, 20] and a variety of other approaches [7, 5, 43]. Whereas extensive research is done on a material-based or cell-biology-based level, only a small number of SFF techniques have recently been exploited for the clinical use of extracorporally grown tissues. As the advantages of these techniques are impressive and the logistic and technical problems are likely to be solved in the near future, it can be assumed that these techniques will become standard measures to create scaffolds.

Through fused deposition modelling, scaffolds with a highly ordered microstructure can be fabricated. The avoidance of organic solvents is the main advantage of this technique from a technical point of view. In contrast, incorporation of growth factors is difficult to achieve, because elevated temperatures during processing will destroy the structure and activity of such growth factors [21]. Fused deposition modelling was shown to generate scaffolds with adequate porosity supporting the ingrowth and survival of cells and tissues [14, 54]. It was demonstrated by Schantz et al. [39] that osteoid formation could be obtained when these scaffolds, coaxed with periosteal cells under osteogenic conditions in vitro, were implanted in vivo in a subcutaneous animal model. As the processing techniques leads to elevated material temperature, cells can be coaxed with the material only after the scaffold is produced.

From the view of tissue engineering, three-dimensional printing is a promising prototyping technique [6, 41]. The main advantage of three-dimensional printing is the fact that scaffolds can be created at room temperature. This method thereby allows the seeding of cells or the incorporation of growth factors during the fabrication process [12]. The technique, based on the printing of a liquid binder onto thin layers of powder, enables the coaxing of cells with the polymer material during layered manufacturing and therefore locates cells in deeper structures of the scaffold [34, 12, 21]. By stacking of multiple cell-containing material layers in a scaffold with a defined tunnel structure, the viability of cells in central parts of the scaffold can be envisioned. This promising technique was successfully exploited in bone tissue engineering [34, 12, 21]. In addition to the cell biological advantages of processing at room temperature, as well as the deposition of cells in central areas of the scaffolds and an improved tube and pore geometry, a further advantage of the technique is the possibility of fabricating complex scaffolds containing different materials with different substructures [6, 34, 18, 42]. A main disadvantage of the three-dimensional printing technique is the fact that the microstructure of the scaffolds is dependent on the powder size of the stock material. An additional technical problem involves closure of the pores by the underlying material.

Three-dimensional plotting is a commonly used method of SFF fabrication for processing hydrogels [19, 3, 20]. A hydrogel-based scaffold formation of special relevance in cartilage tissue engineering is achieved by chemical reactions of co-reactive components that are placed in two component dispensers, or one component can be plotted in a liquid medium containing a co-reactive component [20]. This technique has, like three-dimensional printing, the advantage of operating at physiological conditions (e.g. physiological temperature). Scaffolds can also be enriched homogenously throughout the scaffold volume with cells or growth factors during the fabrication process.

References

1. Agrawal CM, Ray RB (2001) Biodegradable polymeric scaffolds for musculoskeletal tissue engineering. J Biomed Mater Res 55:141–150
2. Agrawal CM, Mckinney JS, Huang D, Athanasiou KA (2000) Synthetic bioabsorbable polymers for implants, 1st edn. ASTM, Philadelphia
3. Ang TH, Sultana FSA, Hutmacher DW, Wong YS, Fuh JYH, Mo XM, Loh HT, Burdet E, Teoh SH (2002) Fabrication of 3D chitosan-hydroxyapatite scaffolds using a robotic dispensing system. Mater Sci Eng C 20(1–2):35–42
4. Burg KJ, Holder WD Jr, Culberson CR, Beiler RJ, Greene KG, Loebsack AB, Roland WD, Eiselt P, Mooney DJ, Halberstadt CR (2000) Comparative study of seeding methods for three-dimensional polymeric scaffolds. J Biomed Mater Res 51(4):642–649
5. Chu TM, Halloran JW, Hollister SJ, Feinberg SE (2001) Hydroxyapatite implants with designed internal architecture. J Mater Sci Mater Med 12(6):471-478
6. Cima LG, Vacanti JP, Vacanti C, Ingber D, Mooney D, Langer R (1991) Tissue engineering by cell transplantation using degradable polymer substrates. J Biomech Eng 113(2):143–151
7. Fedchenko F (1996) Stereolithography and other RP&M technologies. ASME Press, Dearborn, p 2
8. Gomes ME, Salgado AJ, Reis RL (2002b) Polymer based systems on tissue engineering: replacement and regeneration, 1st edn. Kluwer, Dordrecht, p 221
9. Gomes ME, Ribeiro AS, Malafaya PB, Reis RL, Cunha AM (2001) A new approach based on injection moulding to produce biodegradable starch-based polymeric scaffolds: morphology, mechanical and degradation behaviour. Biomaterials 22(9):883–889
10. Gomes ME, Godinho JS, Tchalamov D, Cunha AM, Reis RL (2002a) Alternative tissue engineering scaffolds based on starch: processing methodologies, morphology, degradation and mechanical properties. Mater Sci Eng C 20(1–2):19–26
11. Holy CE, Shoichet MS, Davies JE (2000) Engineering three-dimensional bone tissue in vitro using biodegradable scaffolds: investigating initial cell-seeding density and culture period. J Biomed Mater Res 51(3):376–382
12. Hutmacher DW (2000) Scaffolds in tissue engineering bone and cartilage. Biomaterials 21(24):2529–2543
13. Hutmacher DW (2001) Scaffold design and fabrication technologies for engineering tissues—state of the art and future perspectives. J Biomater Sci Polym Ed 12:107–124
14. Hutmacher DW, Kirsch A, Ackermann KL, et al (2001a) A tissue engineered cell-occlusive device for hard tissue regeneration—a preliminary report. Int J Periodontics Restorative Dent 21:49–59
15. Hutmacher DW, Schantz T, Zein I, Ng KW, Teoh SH, Tan KC (2001b) Mechanical properties and cell cultural response of polycaprolactone scaffolds designed and fabricated via fused deposition modeling. J Biomed Mater Res 55(2):203–216

16. Karp JM, Shoichet MS, Davies JE (2003) Bone formation on two-dimensional poly(DL-lactide-co-glycolide) (PLGA) films and three-dimensional PLGA tissue engineering scaffolds in vitro. J Biomed Mater Res A 64(2):388–396
17. Kim BS, Mooney DJ (1998) Engineering smooth muscle tissue with a predefined structure. J Biomed Mater Res 41(2):322–332
18. Lam CXF, Mo XM, Teoh SH, Hutmacher DW (2002) Scaffold development using 3D printing with a starch-based polymer. Mater Sci Eng C 20(1–2):49–56
19. Landers R, Mulhaupt R (2000) Desktop manufacturing of complex objects, prototypes and biomedical scaffolds by means of computer-assisted design combined with computer-guided 3D plotting of polymers and reactive oligomers. Macromol Mater Eng 282:17–21
20. Landers R, Hubner U, Schmelzeisen R, Mülhaupt R (2002) Rapid prototyping of scaffolds derived from thermoreversible hydrogels and tailored for applications in tissue engineering. Biomaterials 23(23):4437–4447
21. Leong KF, Cheah CM, Chua CK (2003) Solid freeform fabrication of three-dimensional scaffolds for engineering replacement tissues and organs. Biomaterials 24(13):2363–2378
22. Liu X, Ma PX (2004) Polymeric scaffolds for bone tissue engineering. Ann Biomed Eng 32(3):477–486
23. Lu L, Mikos AG (1996) The importance of new processing techniques in tissue engineering. MRS Bull 21(11):28–32
24. Lu L, Zhu X, Valenzuela RG, Currier BL, Yaszemski MJ (2001) Biodegradable polymer scaffolds for cartilage tissue engineering. Clin Orthop Relat Res (391 Suppl):S251–S270
25. Malafaya PB, Reis RL (2003) Key engineering materials. Trans Tech Pub, Zurich 240–242:39
26. Mao JS, Zhao LG, Yin YJ, Yao KD (2003) Structure and properties of bilayer chitosan-gelatin scaffolds. Biomaterials 24(6):1067–1074
27. Meyer U, Szuwart T, Runte C, Dierksen D, Büchter A, Wiesmann HP (2005) Computer-aided bone tissue engineering. Int J Oral Maxillofac Implants in press
28. Mikos AG, Thorsen AJ, Czerwonka LA, Bao Y, Langer R, Winslow DN, Vacanti JP (1994) Preparation and characterization of poly(L-lactic acid) foams. Polymer 35:1068
29. Mikos AG, Bao Y, Cima LG, Ingber DE, Vacanti JP, Langer R (1993) Preparation of poly(glycolic acid) bonded fiber structures for cell attachment and transplantation. J Biomed Mater Res 27(2):183–189
30. Mooney DJ, Baldwin DF, Suh NP, Vacanti JP, Langer R (1996) Novel approach to fabricate porous sponges of poly(D,L-lactic-co-glycolic acid) without the use of organic solvents. Biomaterials 17(14):1417–1422
31. Murphy WL, Dennis RG, Kileny JL, Mooney DJ (2002) Salt fusion: an approach to improve pore interconnectivity within tissue engineering scaffolds. Tissue Eng 8(1):43–52
32. Nishimura I, Garrell RL, Hedrick M, Iida K, Osher S, Wu B (2003) Precursor tissue analogs as a tissue-engineering strategy. Tissue Eng 9 Suppl 1:S77–S89
33. Ochi K, Chen G, Ushida T, Gojo S, Segawa K, Tai H, Ueno K, Ohkawa H, Mori T, Yamaguchi A, Toyama Y, Hata J, Umezawa A (2003) Use of isolated mature osteoblasts in abundance acts as desired-shaped bone regeneration in combination with a modified poly-DL-lactic-co-glycolic acid (PLGA)-collagen sponge. J Cell Physiol 194(1):45–53
34. Park A, Wu B, Griffith LG (1998) Integration of surface modification and 3D fabrication techniques to prepare patterned poly(L-lactide) substrates allowing regionally selective cell adhesion. J Biomater Sci Polym Ed 9(2):89–110
35. Peter SJ, Miller MJ, Yasko AW, Yaszemski MJ, Mikos AG (1998a) Polymer concepts in tissue engineering. J Biomed Mater Res 43(4):422–427
36. Peter SJ, Miller ST, Zhu G, Yasko AW, Mikos AG (1998b) In vivo degradation of a poly(propylene fumarate)/beta-tricalcium phosphate injectable composite scaffold. J Biomed Mater Res 41(1):1–7
37. Quarto R, Mastrogiacomo M, Cancedda R, Kutepov SM, Mukhachev V, Lavroukov A, Kon E, Marcacci M (2001) Repair of large bone defects with the use of autologous bone marrow stromal cells. N Engl J Med 344(5):385–386
38. Sachlos E, Czernuszka JT (2003) Making tissue engineering scaffolds work. Review: the application of solid freeform fabrication technology to the production of tissue engineering scaffolds. Eur Cell Mater 5:29–39; discussion 39–40
39. Schantz JT, Hutmacher DW, Chim H, Ng KW, Lim TC, Teoh SH (2002) Induction of ectopic bone formation by using human periosteal cells in combination with a novel scaffold technology. Cell Transplant 11(2):125–138
40. Shea LD, Wang D, Franceschi RT, Mooney DJ (2000) Engineered bone development from a pre-osteoblast cell line on three-dimensional scaffolds. Tissue Eng 6(6):605–617
41. Shen F, Cui YL, Yang LF, Yao KD, Dong XH, Jia WY, Shi HD (2000) A study on the fabrication of porous chitosan/gelatin network scaffold for tissue engineering. Polym Int 49:1596
42. Sherwood JK, Riley SL, Palazzolo R, Brown SC, Monkhouse DC, Coates M, Griffith LG, Landeen LK, Ratcliffe A (2002) A three-dimensional osteochondral composite scaffold for articular cartilage repair. Biomaterials 23(24):4739–4751
43. Taboas JM, Maddox RD, Krebsbach PH, Hollister SJ (2003) Indirect solid free form fabrication of local and global porous, biomimetic and composite 3D polymer-ceramic scaffolds. Biomaterials 24(1):181–194
44. Thomson RC, Yaszemski MJ, Powers JM, Mikos AG (1995) Fabrication of biodegradable polymer scaffolds to engineer trabecular bone. J Biomater Sci Polym Ed 7(1):23–38
45. Tsang VL, Bhatia SN (2004) Three-dimensional tissue fabrication. Adv Drug Deliv Rev 56(11):1635–1647
46. Vacanti CA, Bonassar LJ, Vacanti MP, Shufflebarger J (2001) Replacement of an avulsed phalanx with tissue-engineered bone. N Engl J Med 344(20):1511–1514
47. Vunjak-Novakovic G, Obradovic B, Martin I, Bursac PM, Langer R, Freed LE (1998) Dynamic cell seeding of

polymer scaffolds for cartilage tissue engineering. Biotechnol Prog 14(2):193–202
48. Vunjak-Novakovic G, Martin I, Obradovic B, Treppo S, Grodzinsky AJ, Langer R, Freed LE (1999) Bioreactor cultivation conditions modulate the composition and mechanical properties of tissue-engineered cartilage. J Orthop Res 17(1):130–138
49. Wiesmann HP, Joos U, Meyer U (2004) Biological and biophysical principles in extracorporal bone tissue engineering. Part II. Int J Oral Maxillofac Surg 33(6):523–530
50. Yang S, Leong KF, Du Z, Chua CK (2001) The design of scaffolds for use in tissue engineering. Part I: traditional factors. Tissue Eng 7(6):679–689
51. Yang S, Leong KF, Du Z, Chua CK (2002a) The design of scaffolds for use in tissue engineering. Part II: rapid prototyping techniques. Tissue Eng 8(1):1–11
52. Yang Y, Magnay JL, Cooling L, El HA (2002b) Development of a 'mechano-active' scaffold for tissue engineering. Biomaterials 23(10):2119–2126
53. Yoon JJ, Park TG (2001) Degradation behaviors of biodegradable macroporous scaffolds prepared by gas foaming of effervescent salts. J Biomed Mater Res 55(3):401–408
54. Zein I, Hutmacher DW, Tan KC, Teoh SH (2002) Fused deposition modeling of novel scaffold architectures for tissue engineering applications. Biomaterials 23(4):1169–1185

Prospects of Micromass Culture Technology in Tissue Engineering

J. Handschel, H. P. Wiesmann, U. Meyer

Contents

40.1 Background 551
40.2 What Is the Theory of Micromass Technique? 551
40.3 Technical Aspects of the Micromass Technology 552
40.4 Cell Sources for Micromass Technology .. 553
40.5 Future Prospects and Challenges 554
References 555

40.1 Background

The in vitro formation of bone- or cartilaginous-like tissue for subsequent implantation [1, 13, 15] or other organs is, as described, commonly performed using scaffolds. Various scaffold materials have been introduced, each showing specific advantages and disadvantages in vitro and in vivo. Recently, there has been a controversy (e.g., biocompatibility, biodegradability) concerning the use of artificial scaffolds compared to the use of a natural matrix [24]. Skeletal defect regeneration by extracorporally created tissues commonly exploits a three-dimensional cell-containing artificial scaffold. As indicated before, a number of in vitro studies have been performed to evaluate the cell behaviour in various three-dimensional artificial scaffold materials [10, 18, 19, 25]. Whereas most of these materials were generally shown to allow spacing of skeletal cells in a three-dimensional space, not all materials promote the ingrowth of cells within the scaffolds [7]. Rather, supporting cellular function depends, as described, on multiple parameters, such as the chosen cell line, the underlying material, the surface properties and the scaffold structure.

Some in vitro studies indicate that a material itself may impair the outcome of ex vivo tissue formation when compared to a natural-tissue-containing matrix. Additionally, in the in vivo situation, defect regeneration can be critically impaired by the immunogenity of the material, the unpredictable degradation time, and by side effects caused by degradation products [24]. Based on these considerations, matrices close to the natural extracellular matrix are regarded as most promising in skeletal tissue engineering by some researchers. A recently elaborated approach in extracorporeal tissue engineering is therefore the avoidance of non-degradable scaffolds that are resorbed at a different time rate than the skeletal tissue regeneration by itself proceeds. Therefore, new approaches have been invented to overcome these problems by renouncing scaffolds.

40.2 What Is the Theory of Micromass Technique?

It is well known that tissue explants can regenerate complete organisms [14]. Basic research has indicated that regeneration of simple animals and microtissues can be achieved by reaggregation approaches using the micromass technique [22]. Investigations on skeletal development gave first insight into this micromass biology [5, 6, 17]. The micromass technology relies to a great extent on the presence of the proteinacious extracellular matrix. As described before, the extracellular matrix may exert both direct and indirect influences on cells and consequently modulate their behaviour. At the same time, these cells alter the composition of the extracellular matrix. This may be accomplished in a variety of fashions, including differential expression of particular extracellular matrix components and/or proteases such as metalloproteinases by cells in the local microenvironment.

Whereas most investigations concerning micromass technology were performed in developmental studies, only limited literature is available concerning the use of this technique in tissue engineering [20]. A large body of evidence has confirmed that a minimal cell number is required in three-dimensional, tissue-like constructs to induce the differentiation of mesenchymal precursors along the chondrogenic and osteogenic pathways (reviewed in [9]). In contrast, mesenchymal precursors seeded in low-density micromasses adopt features of a fibroblastic phenotype and abolish cell differentiation when mimicking a low-density condensation [4, 11, 12]. These findings indicate that a "critical" cell mass is necessary to proceed with a specific extracellular matrix formation. A threshold amount of precursor cells is necessary to form a three-dimensional extracellular matrix structure around these cell masses and promote their differentiation. The extracellular matrix in the microenvironment then interacts with cells to further develop towards a specific tissue. The absence of the requisite extracellular matrix components would lead to decreased recruitment of precursors to the condensations, causing a subsequent deficiency in chondrocyte or osteoblast differentiation. In vitro studies with chondrocytes confirmed these findings, showing that the ability of mesenchymal precursors to initiate chondrogenic differentiation is dependent upon cell configuration within a condensation process, which varies by the density of the condensation [2].

40.3
Technical Aspects of the Micromass Technology

In the context of tissue engineering, ex vivo tissue generation may be optimised by the use of cell reaggregation technology. The reaggregate approach is a method to generate, in an attempt to mimic the in vivo situation, a tissue-like construct from dispersed cells, under special culture conditions. Therefore, the self-renewal (cell amplification), spatial sorting and self-organisation of multipotential stem cells, in combination with the self-assembly of determined cells, form the basis for such an engineering design option. Technically, cells are dissociated, and the dispersed cells are then reaggregated into cellular spheres [20]. In order to technically refine scaffold-free spheres, cells are kept either in regular culture dishes (as gravitory cultures), in spinner flasks or in more sophisticated bioreactors. In contrast to conventional monolayer cell cultures, in which cells grow in only two dimensions on the flat surface of a plastic dish, suspension cultures allow tissue growth in all three dimensions.

It has been observed that cells in spheres exert higher proliferation rates than cells in monolayer cultures, and their differentiation more closely resembles that seen in situ. This finding may be based on the spatial configuration in a three-dimensional matrix network. Different culture parameters (sizes of the culture plate, movement in a bioreactor, coating of culture walls) are all crucial to the process. Roller tube culture systems have been shown to be suitable for cultivation of tissue explants in suspension. The cultivated and fabricated tissues may be used for studying the primary mixing of cells and the patterns of cell differentiation and growth within growing spheres in order to improve the outcome of microsphere cultivation. In addition, some culture conditions could aid the development of high-throughput systems and allow manipulation of individual spheres.

It seems worthwhile elaborating new bioreactor technologies and culture techniques to improve the ex vivo growth of scaffold-free tissues. Technically, short-term reaggregation experiments, which last from minutes to a few hours, can be distinguished from long-term studies. Short-term reaggregation has been used widely to evaluate basic principles of cell–cell interactions and cell–matrix interactions, whereas long-term cultivation (days to several weeks) is suitable in ex vivo tissue engineering strategies. Recent studies on the reaggregation approach aim to solve two aspects: to fabricate scaffold-free, three-dimensional tissue formation and at the same time to investigate basic principles of cellular self-assembly [3, 23]. As in monolayer cultures, which facilitate the study of cell–material interactions, suspension cultures allow the evaluation of cell action towards a three-dimensional space. The reaggregate approach enables the following of tissue formation from single-cell sources to organised spheres in a controlled environment. Thus, the inherent fundamentals of tissue engineering are better revealed. Additionally, as the newly formed tissue is void of an artificial material, it more closely resembles the in vivo situation.

40.4 Cell Sources for Micromass Technology

Cells from cartilage and/or bone have been found to be suitable cell sources for such ex vivo reaggregate approaches. Anderer and Libera [1] developed an autologous spheroid system to culture chondrocytes and osteoblasts without adding xenogenous serum, growth factors or scaffolds, considering that several growth factors and scaffolds are not permitted for use in clinical applications. It was demonstrated by such an approach that autologous chondrocytes and osteoblasts cultured in the presence of autologous serum form a three-dimensional micro-tissue that generates its own extracellular matrix. Chondrocyte-based microtissue had a characteristic extracellular space that was similar to the natural matrix of hyaline cartilage. Osteoblasts were also able to build up a microtissue similar to that of bone repair tissue without collagen-associated mineral formation.

Currently, our own results show that murine embryonic stem cells are able to build spheroid micromass bodies, too. Cultured with an osteogenic supplement (dexamethasone, ascorbic acid, ß-glycerolphosphate), embryonic stem cells formed a mineralized matrix in the centre of the spheroid body (Fig. 40.1). The fabrication of a self-assembled skeletal tissue seems not to be limited towards certain species, as results from bovine and porcine chondrocyte and osteoblast cultivation led to the formation of species-related, cartilage-like or bone-like tissue. However, conditions allowing cartilage formation in one species are not necessarily transposable to other species. Therefore, results with animal models should be cautiously applied to humans. In addition, for tissue-engineering purposes, the number of cell duplications must be, for each species, carefully monitored to remain in the range of amplification allowing redifferentiation and chondrogenesis [8]. Recently, micromass cultures were established using human umbilical cord blood cells (Fig. 40.2). The shape of these bodies was more round than the micromass bodies of the embryonic stem cells, and the spreading of the cells within these bodies was different.

It was recently observed that even complex cellular systems can be generated ex vivo without the use of scaffolds. Co-cultures of osteoblasts and endothelial cells, for example, resulted in the formation of a bi-cellular micromass tissue renouncing

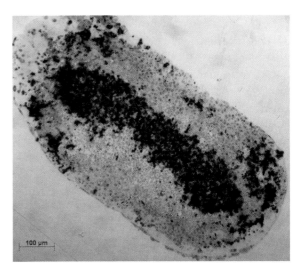

Fig. 40.1 Micromass culture of murine embryonic stem cells after 21 days of culture with an osteogenic medium. The Alizarin staining (*red*) shows the mineralized extracellular matrix in the centre of the spheroid body. (Alizarin red staining)

Fig. 40.2 Micromass culture of human umbilical cord blood cells

any other materials. Other organotypic cultures, used to develop engineered tissues other than of skeletal origin, confirm that it is feasible to create tissue substitutes based on reaggregated sphere technology. Examples of these strategies include liver reconstruction, synthesis of an artificial pancreas, restoration of heart valve tissue and cardiac organogenesis in vitro [16].

40.5
Future Prospects and Challenges

Several investigations have suggested that after in vivo transfer of such reaggregates, tissue healing is improved in the sense of a repair tissue that mimics the features of the original skeletal tissue [1, 21]. Especially, preclinical and clinical cartilage repair studies demonstrated that tissue formation resembled more closely the natural situation. The transplantation of reassembled chondrogenic microtissues is able to impair the formation of fibrocartilage by suppression of type I collagen expression while promoting the formation of proteoglycan accompanied by a distinct expression of type II collagen. It can be assumed that the volume of the observed repair tissue was formed by the implanted chondrosphere itself as well as by host cells located in the superficial cartilage defect. The mechanisms by which chondrospheres promote defect healing are complex and not completely understood. Van der Kraan et al. [24] reviewed the role of the extracellular matrix in the regulation of chondrocyte function in the defect site and the relevance for cartilage tissue engineering. Numerous other studies have confirmed that the extracellular matrix of articular cartilage can be maintained by a distinct number of chondrocytes and that the extracellular matrix plays an important role in the regulation of chondrocyte function. In in-vitro-generated, cartilage-like tissue, a time-dependent increase in the expression of collagen type II, S-100 and cartilage-specific proteoglycans, paralleled by a reduced cell–matrix ratio, was observed in the microspheres [21]. The transplanted cell–matrix complex was determined to be responsible for the observed chondrocyte proliferation, differentiation and hyaline cartilage-like matrix maturation in vivo.

The inductive properties of the implantation site may also be beneficial when a stem-cell-based microtissue strategy is chosen. Stem cell tissue engineering using foetal or adult stem cells in combination with sphere technologies leads to implantable stem-cell-driven tissues (unpublished data). Typically, stem cells must be amplified to large quantities in suspension cultures and have access to appropriate growth factors to establish specially organised histotypical spheres. These spheres can then be implanted into the lesioned skeletal or organ site. Cells in micromass spheres sprout into the surrounding areas when transferred to adherent surfaces. The cells show a centrifugal outgrowth of the bodies as is shown by human umbilical cord blood cells (Fig. 40.3).

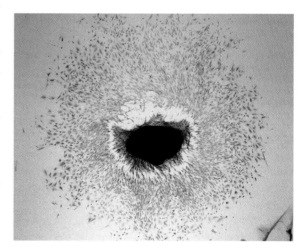

Fig. 40.3 Centrifugal outgrowth of umbilical cord blood cells (HE staining)

Therefore, using cultured spheres, it is possible to transfer large amounts of cells into scaffolds or directly into tissue defects. A high number of cells are probably necessary to promote tissue regeneration at lesioned skeletal or organ sites.

Actually, most reports regarding micromass technology use pluripotent cells (embryonic stem cells, umbilical cord blood cells) or mature cells (e.g., osteoblasts, chondroblasts).

Although adult stem cells of various origins can also transdifferentiate into distinct cell types, the transformation of these cell types into functioning tissues and their successful implantation by reaggregation technology need further elaboration.

Taken together, the micromass culture is a promising technology which can already transfer high amounts of cells (e.g., on scaffold materials). In addition, micromass cultures with embryonic stem cells may resemble organ development in a very early stage and may contribute to a better understanding of growth and development. However, the engineering of complex organs with micromass technology is currently far off and needs much more research.

References

1. Anderer U, Libera J (2002) In vitro engineering of human autogenous cartilage. J Bone Miner Res 17:1420–1429
2. Archer CW, Rooney P, Cottrill CP (1985) Cartilage morphogenesis in vitro. J Embryol Exp Morphol 90:33–48
3. Battistelli M, Borzi RM, Olivotto E et al. (2005) Cell and matrix morpho-functional analysis in chondrocyte micromasses. Microsc Res Tech 67:286–295
4. Cottrill CP, Archer CW, Wolpert L (1987) Cell sorting and chondrogenic aggregate formation in micromass culture. Dev Biol 122:503–515
5. DeLise AM, Stringa E, Woodward WA et al. (2000) Embryonic limb mesenchyme micromass culture as an in vitro model for chondrogenesis and cartilage maturation. Methods Mol Biol 137:359–375
6. Edwall-Arvidsson C, Wroblewski J (1996) Characterization of chondrogenesis in cells isolated from limb buds in mouse. Anat Embryol (Berl) 193:453–461
7. Freed LE, Hollander AP, Martin I et al. (1998) Chondrogenesis in a cell–polymer-bioreactor system. Exp Cell Res 240:58–65
8. Giannoni P, Crovace A, Malpeli M et al. (2005) Species variability in the differentiation potential of in vitro-expanded articular chondrocytes restricts predictive studies on cartilage repair using animal models. Tissue Eng 11:237–248
9. Hall BK, Miyake T (1992) The membranous skeleton: the role of cell condensations in vertebrate skeletogenesis. Anat Embryol (Berl) 186:107–124
10. Handschel J, Berr K, Depprich R et al. (2007) Compatibility of embryonic stem cells with biomaterials. J Biomat Appl accepted 2008
11. Hattori T, Ide H (1984) Limb bud chondrogenesis in cell culture, with particular reference to serum concentration in the culture medium. Exp Cell Res 150:338–346
12. Hurle JM, Hinchliffe JR, Ros MA et al. (1989) The extracellular matrix architecture relating to myotendinous pattern formation in the distal part of the developing chick limb: an ultrastructural, histochemical and immunocytochemical analysis. Cell Differ Dev 27:103–120
13. Hutmacher DW (2000) Scaffolds in tissue engineering bone and cartilage. Biomaterials 21:2529–2543
14. Kelm JM, Fussenegger M (2004) Microscale tissue engineering using gravity-enforced cell assembly. Trends Biotechnol 22:195–202
15. Lindenhayn K, Perka C, Spitzer R et al. (1999) Retention of hyaluronic acid in alginate beads: aspects for in vitro cartilage engineering. J Biomed Mater Res 44:149–155
16. Lu HF, Chua KN, Zhang PC et al. (2005) Three-dimensional co-culture of rat hepatocyte spheroids and NIH/3T3 fibroblasts enhances hepatocyte functional maintenance. Acta Biomater 1:399–410
17. Mello MA, Tuan RS (1999) High density micromass cultures of embryonic limb bud mesenchymal cells: an in vitro model of endochondral skeletal development. In Vitro Cell Dev Biol Anim 35:262–269
18. Meyer U, Joos U, Wiesmann HP (2004) Biological and biophysical principles in extracorporal bone tissue engineering. Part I. Int J Oral Maxillofac Surg 33:325–332
19. Meyer U, Joos U, Wiesmann HP (2004) Biological and biophysical principles in extracorporal bone tissue engineering. Part III. Int J Oral Maxillofac Surg 33:635–641
20. Meyer U, Wiesmann HP (2005) Bone and cartilage tissue engineering. Springer, Heidelberg
21. Meyer U, Wiesmann HP, Büchter A et al. (2006) Cartilage defect regeneration by ex-vivo engineered autologous mikro-tissue. Osteoarthr Cartil submitted
22. Sanchez Alvarado A (2004) Regeneration and the need for simpler model organisms. Philos Trans R Soc Lond B Biol Sci 359:759–763
23. Tare RS, Howard D, Pound JC et al. (2005) Tissue engineering strategies for cartilage generation—micromass and three dimensional cultures using human chondrocytes and a continuous cell line. Biochem Biophys Res Commun 333:609–621
24. van der Kraan PM, Buma P, van Kuppevelt T et al. (2002) Interaction of chondrocytes, extracellular matrix and growth factors: relevance for articular cartilage tissue engineering. Osteoarthr Cartil 10:631–637
25. Wiesmann HP, Joos U, Meyer U (2004) Biological and biophysical principles in extracorporal bone tissue engineering. Part II. Int J Oral Maxillofac Surg 33:523–530

IX Laboratory Aspects and Bioreactor Use

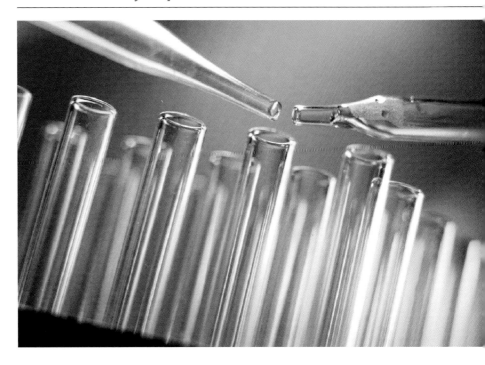

41 Laboratory Procedures – Culture of Cells and Tissues

C. Naujoks, K. Berr, U. Meyer

Contents

41.1	Introduction	559
41.2	Cell Sources	559
41.2.1	Genetically Modified Cell Lines	564
41.3	Cell Culture	565
41.3.1	Monolayer	565
41.3.2	Micromass Culture	565
41.3.3	Agarose and Alginate Culture	565
41.3.4	Coculture	566
41.3.5	Marrow-Derived Stem Cells	567
41.3.6	Determined Bone Cells	570
41.3.7	Determined Chondrogenic Cells	575
41.3.8	Osteoclasts	575
41.3.9	Endothelial Cells	576
41.4	Cell Culture Evaluation Techniques	577
41.4.1	Immunostaining	586
41.4.2	Cryopreservation	587
	References	588

are necessary for the construction and design of artificial living tissues. The potential to improve, maintain, and restore tissue function may enhance quality of life and health care, for instance by overcoming the shortages of donor tissues and organs. Tissue engineering and cell cultivation include cell sourcing, manipulation of cell function, and the construction of living tissues.

In general there are four steps that need to be considered for tissue design. The first step in any attempt to engineer a tissue is identification of the cells to be employed and how these cells can be harvested, isolated, and enriched. Subsequently a method may need to be established that alters the functional characteristics of the cells in order to achieve a desired behavior. The third step will be to achieve a three-dimensional architecture that reflects the functional characteristics of the specific tissue that is mimicked. The final step is to establish a cost-effective manufacturing process for the tissue (e.g., bioreactor technology). In this chapter, a review of different cell-cultivation methods, tissues, and laboratory procedures is given. The description of cell handling and culture will focus mainly on bone- and cartilage-derived cells. The general principles of cell culture will also be valid for many other cell types.

41.1 Introduction

Tissue engineering and cell cultivation are common methods of basic research that help to provide an understanding of the biology of cells and tissues. Even today there remain many challenges to face. These methods may support the control of processes that

41.2 Cell Sources

The availability of well-defined cells or progenitor cells that are able to differentiate into the desired cell types is crucial for tissue engineering processes. In addition to their homogeneity, the cells have to be free from all pathogens and contaminants.

The principal sources of cells for tissue engineering include xenogenic cells as well as autologous and allogenic cells (Table 41.1). According to the source of the cells the immune acceptance of the host can vary. While autologous cells do not usually cause immune responses, the use of cells from allogenic and xenogenic sources requires the engineering of immune acceptance.

Each category can be subdivided according to whether the cells are stem cells (embryonic or adult) or whether the cells are in a more differentiated stage (Table 41.2, Fig 41.1). Unlimited self-renewal and the potential to differentiate are the major capabilities of stem cells. The stem cells can be divided according to their origin, first into embryonic stem cells derived from the inner cell mass of preimplantation embryos, and second into adult stem cells residing in mature tissues and organs (e.g., mesenchymal stem cells, MSCs). Various mature cell lines as well as multipotent mesenchymal progenitors have been successfully established using tissue-engineering approaches [149]. Other cell lines, like genetically altered cell lines (sarcoma cells, immortalized cells, nontransformed clonal cell lines) have also been developed and used to evaluate basic aspects of in vitro cell behavior in nonhuman settings.

To date a variety of different stem cells have been isolated, which can be assigned to the different types and sources as shown in Fig 41.1. Embryonic stem cells are regarded as totipotent cells that are capable of differentiating into all cell and tissue types.

Embryonic stem cells reside in blastocysts and are derived from the inner cell mass, as mentioned earlier. They were isolated for the first time more than two decades ago and have successfully been grown in culture [75]. These cells have maintained the capability of unlimited self-renewal and pluripotent development potency and therefore can develop into cells and tissues of all three germ layers. Human embryonic stem cells (hESC) open up new opportunities for regenerative medicine. In order to make hESCs available for clinical applications it is necessary to define processes to differentiate these cells into specific lineages or cell types. So far different protocols to harvest embryonic stem cells from a variety of species have been used. In 1998 the first hESCs were derived [131] with the aid of mitotically inactivated mouse embryonic fibroblasts as the feeder layer.

For the proliferation of embryonic stem cells it is important to inhibit spontaneous differentiation, for example with the aid of basic fibroblast growth factor (bFGF) [147]. These culture conditions can not maintain all cells in an undifferentiated state and will finally lead to heterologous populations of differentiated and undifferentiated cells.

Table 41.1 Cell sources

Type	Origin	Advantages
Autologous	Patient himself	Immune acceptable
Allogenic	Other human source	May induce immune rejection
Xenogenic	Other different species	No immune acceptance

Table 41.2 Stem cells

Type	Origin	Differentiation
Embryonic	Embryo or fetal tissue	All types of tissue
Hematopoietic	Adult bone marrow	Blood cells
Neuronal	Fetal brain	Neurons, glia
Mesenchymal	Adult bone marrow	Muscle, bone, cartilage, tendon

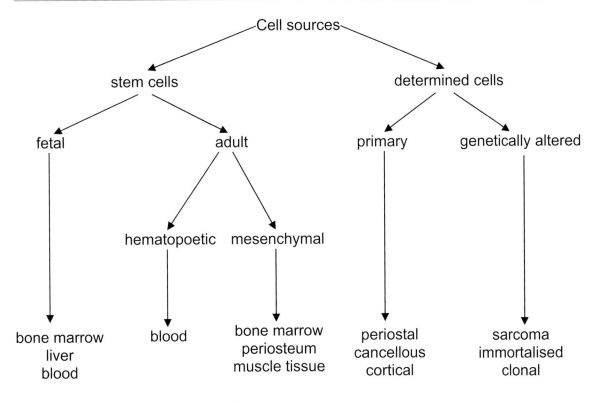

Fig. 41.1 Different types and sources of stem cells

The "embryoid body" formation is often used to differentiate embryonic stem cells. These aggregates form spontaneously during the cultivation of embryonic stem cells on low-attachment culture plastics, and are known as embryoid bodies (EBs). These EBs imitate an early embryo development in vivo and probably contain all three primitive germ layers. By modifying culture conditions, the differentiation to a variety of lineages can be triggered, but it is still difficult to achieve homogenous populations within EBs.

Another approach for the isolation of homogenous cell lines works through labeling of specific cell lineages with fluorescent antibodies and the subsequent purification of these cells by fluorescence-assisted cell sorting. It is assumed that cells of the same lineage are carrying the same markers on their surface that allow for binding of specific antibodies [12].

Skeletal tissue stem cells are a hypothetical concept, there being only circumstantial evidence for their existence, and indeed, there seems to be a hierarchy of stem cells each with variable self-renewal potentials [133]. The primitive stem cells that renew skeletal structures have been given a variety of names including connective tissue stem cells, osteogenic cells [48], stromal stem cells [99, 100], stromal fibroblastic cells [139], and MSCs [18]. There is as yet no entirely accurate nomenclature based upon the developmental origins or differentiation capacities of these cells, but the latter term, although defective, appears to be in favor at the moment. Stem cells have the capacity for extensive replication without differentiation, and they possess, as mentioned, a multilineage developmental potential that allows them to give rise to not only one single tissue. MSCs may develop into bone, cartilage, tendon, muscle, fat, or marrow stroma depending on the cultivation condition applied. Their vast expansion potential makes stem cells a very interesting source of cells for tissue engineering of bone and cartilage independently from their source. MSCs are present in fetal tissue, postnatal bone marrow, and in the bone marrow of adults [18].

MSCs are characterized by a profile of surface markers and the expression of extracellular matrix proteins. While hematopoietic antigens (CD45-, CD34-, CD14-) are absent, the stroma-associated markers CD29 (β1-integrin), CD73, CD105

(SH2), CD44 (HCAM1), and CD90 (thy-1), and extracellular matrix proteins vimentin, laminin, and fibronectin are expressed. Some of these have been used not only to characterize these cells, but also to enrich MSCs from populations of adherent bone marrow stromal cells. However, none of these markers taken by itself seems to be specific for MSCs. Although in adults bone marrow serves as the primary reservoir for MSCs, their presence has also been reported in a variety of other tissues like periosteum and muscle connective tissue [89, 91]. Fetal MSCs can be isolated from bone marrow, liver, and blood [17]. Transfer of these tissues may also act through stem-cell-induced repair mechanisms as it is assumed in the coverage of cartilage defects by periosteum. MSCs that are residing in these tissues can be used to treat patients from whom they were isolated as an autologous transplant.

Another source of stem cells is the umbilical cord blood present in the cord at delivery. Some populations of these cord blood cells are able to differentiate into multiple cell types and may be used in regenerative medical therapies (e.g., to treat diabetes mellitus). Fetal MSCs are less lineage-committed than adult MSCs in humans and primates [43, 68, 83]. While MSCs have been identified in fetal blood [17] and infrequently observed in umbilical cord blood [31, 47, 74], it remains to be established whether MSCs are present in the steady-state peripheral blood of adults or not, but the number of stem cells in peripheral blood is probably extremely low [153].

Adult stem cells, which can be found in tissues, organs and blood, show limited replicative capacity, but at the same time do not evoke the immunogenity found with embryonic stem cells. These characteristics of adult stem cells could bypass the need for immunosuppression, while the low abundance (1 stem cell per 100,000 bone marrow cells), restricted differentiation potential, and poor growth of these cells make them ineligible for transplantation. In 1974 Friedenstein et al. discovered a cell type in bone marrow explants that is now referred to as marrow stromal cells [35].

A major problem in using adult stem cells in a given clinical situation is the difficulty of obtaining a significant number for the generation of larger cell constructs. The limitations of adult stem-cell harvesting lead to the concept of employing fetal stem cells, particularly from fetal bone marrow. These cells seem to be a more appropriate source for tissue engineering approaches, since they show significantly reduced immunogenity compared with adult stem cells. Moreover, fetal tissues produce more abundant trophic substances and growth factors, thereby promoting cell growth and differentiation at the site of implantation. Research on fetal tissue transfer in animal models and in human experimental treatments has confirmed distinct advantages over adult tissues including lowered immunogenicity and a higher percentage of primitive cells. Due to their lower immunogenic potential, fetus-derived stem cells seem to alleviate to some extent the problems of tissue typing. As such cells also have a high capacity to differentiate into the complete repertoire of mesenchymal cell lineages, combined with their ability for rapid cellular growth, differentiation, and reassembly, fetal cells may become a more important subject in tissue engineering strategies (Fig. 41.2). Fetal stem cells have some serious limitations, despite their enormous potential for biomedical and tissue engineering applications. There is a lack of methods that enable tissue engineers to direct the differentiation of embryonic stem cells and to induce specific functions of the embryonic-cell-generated tissue after transplantation [141]. Two other issues are of main concern:

1. It must be assured that the cultivation and transplantation of such stem cells are not accompanied by a tumorigenic differentiation. Since it was shown that undifferentiated embryonic stem cells give rise to teratomas and teratocarcinomas after implantation in animals, the risk of misdevelopment constitutes a major problem for the clinical use of such cells.
2. It must be assured that the in vitro generation of cells does not lead to an immunological incompatibility toward the host tissue. Immunological incompatibilities can be avoided by using somatic cell nuclear transfer methodology [3].

However, nuclear cloning methods are often criticized for this potential risk. The main problem in employing fetal cells for use in tissue engineering is not only to obtain such cells, but perhaps more importantly the clinical, legal, and ethical issues involved in such treatment strategies.

Many attempts have been undertaken to refine procedures for the propagation and differentiation of stem cells. Despite the various advantages of using

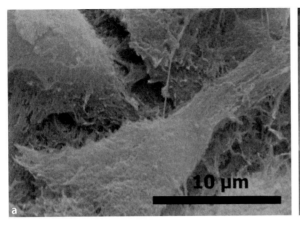 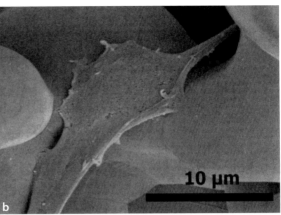

Fig. 41.2a,b Fetal stem-cell-based tissue engineering. Fetal stem cell cultured under defined conditions on an artificial material, displaying the phenotypic appearance of mature bone cells. **a** Scanning electron microscope (SEM) picture of osteoblasts at the surface of a bone trabecula. **b** SEM view of a murine fetal stem cell cultured on a polylactic acid/polyglycolic acid scaffold

tissue-derived adult stem cells over other sources of cells, there is some debate as to whether large enough populations of differentiated cells can be grown in vitro rapidly enough when needed clinically. At present, stem cells are, for example, not able to differentiate definitively into osteocyte-like cells that are competent to mineralize the pericellular region in a bone-like manner under in vitro conditions [56, 107]. This must be considered as one limitation for the use of stem cells in extracorporeal skeletal tissue engineering. Much more basic research is therefore necessary to assess the full potential of stem-cell therapy to reconstitute skeletal mass or other tissues and organs. It is expected that many future studies will be directed toward the development of gene therapy protocols employing gene-insertion strategies [32]. The concept that members of the bone morphogenetic proteins (BMP) and the transforming growth factor-beta (TGF-βa) superfamily will be particularly useful in this regard has already been tested by many investigators [71, 93]. Genetic engineering to shape gene expression profiles may thus open a future route for the use of stem cells in human tissue engineering, but this approach is at the moment even further removed from clinical application.

There are controversial opinions about the ideal stage of differentiation of cells used for transplantation. If cells are transplanted before they have reached their determined state, they may not completely transform to the required cell type. Amongst other reasons, the implantation site may lack some signals and extracellular components compared to the normal developmental setting, which could be compulsory for the maturation of these cells. In addition, such dysfunctional conditions may result in a higher risk of tumor formation. On the other hand, if predetermined cells that have already reached their full phenotype are transplanted, they may no longer be capable of adapting to the in vivo environment.

Differentiated osteoblast- or chondrocyte-like cells serve as common cell lines in evaluating preclinical and clinical aspects of bone and cartilage tissue engineering. Osteoblasts, derived from multiple sites of the skeleton, can be harvested as precursor cells, lining cells, mature osteoblasts, or osteocytes. Cell separation is then performed by various techniques in order to expand distinct cell lines in culture. Chondrocytes can also be gained from the various cartilage-containing tissue sites; they are easy to harvest from these sources and are already present in a mature differentiation stage.

A complementary approach in experimental settings with primary cells is the use of specifically modified cells. While primary cells are commonly used in clinical cell-based engineering strategies, so far the application of genetically altered cells in bone or cartilage reconstruction is still restricted to laboratory studies. Experiments using in vitro assay systems have yielded not only much basic information concerning the cultivation of these cell lines, but can also be used for tissue engineering studies in basic and preclinical investigations.

41.2.1
Genetically Modified Cell Lines

Cell lines that are currently used for evaluating aspects of extracorporeal bone tissue engineering include, in addition to primary cells: (1) osteosarcoma cell lines, (2) intentionally immortalized cell lines, and (3) nontransformed clonal cell lines [59]. Relevant characteristic features of each cell type will now be described.

Osteosarcoma cell lines are known to display patterns of gene expression, modes of adhesion, and signal transduction pathways that in certain aspects resemble those of normal, nontransformed bone cells. Most of the osteosarcoma cell lines used, however, do not display a complete pattern of in vitro differentiation. The development of established clonal osteoblast-like cells from rat osteosarcomas (MG-63, UMR, and ROS series) provided cell lines that were homogeneous, phenotypically stable, and easy to propagate and maintain in culture [137]. They share many of the properties of nontransformed osteoblasts. But, as with cancer cells, these cells are transformed and display an aberrant genotype, have an uncoupled proliferation/differentiation relationship, and exhibit phenotypic instability in long-term culture. As these osteoblast-like cells do not reflect the normal phenotype of primary osteoblast-like cells, they are not suitable for the evaluation of biomaterial-related aspects in tissue engineering.

Other approaches have been to use clonally derived immortalized or spontaneously immortalized cell lines (neonatal mouse MC3T3E1 and fetal rat RCJ cell lines; see [30]). Although none of these cell lines behaves exactly alike and their behavior in cell culture differ considerably [9], they share some common features like alkaline phosphatase activity, collagen type I production, and bone-like nodule formation. Due to their different backgrounds these cells can be in different stages of growth and development under cell culture conditions. The variability of the resulting phenotypic features depends on the chosen cell culture situation. Conditionally transformed immortalized human osteoblast cell lines were developed by various researchers and aimed to investigate the behavior of osteoblasts under the influence of external stimuli. Xiaoxue et al. [146], for example, assessed the generation of an immortalized human stromal cell line that contains cells that are able to differentiate into osteoblastic cells. Concerning the use of immortalized cells in ex vivo tissue-engineering approaches, it is important to recognize that all cell lines impose the disadvantage of having unique phenotypes, so that the morphological sensitivity toward a changing environment, like material surface or external stimuli, is impaired.

Considering the features of genetically altered cells, it becomes obvious that nontransformed and primary cultured osteoblasts are advantageous in extracorporeal tissue engineering, since these cells display a well-defined inverse relationship between proliferation and differentiation [101]. Measures of osteoblast-specific matrix protein expression define valuable reference points for the study of regulated osteoblast physiology, especially when a substratum-dependent reaction is under investigation. Oreffo and Triffitt [97] demonstrated that primary cells are able to react sensitively to minor structural alterations in their surroundings, a key feature that is advantageous in tissue engineering concepts. The use of primary and nontransformed cells is advisable for assessing cellular reactions toward scaffolds. It should be noticed that the reaction of cells toward the material is also dependent on the cellular maturation stage [13]. It should also be emphasized that the behavior of osteoblasts on artificial surfaces is dependent on the experimental cell culture conditions [24].

Clonal sarcoma cell lines with cartilage phenotypes have been established from various sources [65]. The human chondrosarcoma cell line HCS-2/8 exhibits a polygonal to spherical morphology as the cells become confluent. After reaching confluence, the cells continue to proliferate slowly and to form nodules, which show metachromatic features [136]. Observation of the cells with the aid of an electron microscope reveals that the cultures display features of chondrocytes and produce an extracellular matrix consisting of thin collagen-like fibrils. Immortalized chondrocytes have been generated that serve as reproducible models for studying chondrocyte function. Immortalization of chondrocytes increases the life span and proliferative capacity, but does not necessarily stabilize the differentiated phenotype. Primary human chondrocytes have been subjected to treatments with viral sequences like SV40-Tag, HPV-16 E6/E7, or the enzyme telomerase by retrovirally mediated transduction, and transformed cells have been selected with appropriate substances like neomycin. It has been observed that stable integration of an immortalizing gene stabilizes proliferative capacity, whereas it does not affect the differentiated chondrocyte phenotype.

41.3
Cell Culture

After a certain cell type has been chosen, the next steps are to isolate and enrich these cells, cultivate them, and subsequently manipulate the cell function in order to obtain the desired behavior. There is a range of techniques available for these processes offering choices of different parameters concerning the cell source used for the tissue engineering approach.

There are different models with which to cultivate cells and tissues. It is very important to select the model that mimics the in vivo conditions in most aspects and meets the demands of cells. It is known that chondrocytes do not have cell junctions to other cells because they were inserted into extracellular matrix, whereas fat cells many cell junctions keeping them in direct contact. Requirements like these can influence the phenotype and the proliferation rate of cell cultures. Another point that must be accommodated is the turnover of the extracellular matrix and the feeding situation. This can be achieved by the use of perfusion systems that allow the regulation of the turnover.

41.3.1
Monolayer

The monolayer is a two-dimensional model that is used to cultivate cells with the major underlying assumption that the cells must be isolated from their respective extracellular matrices. As early as 1952, Moscona and Moscona that cells of the early chick embryo can be isolated by the use of trypsin [84], and Rinaldini isolated cells from connective tissue with the aid of a collagenase [114]. In principal, cells were isolated with the aid of special enzymes and suspended into a culture medium. As soon as these cells come into contact with a suitable surface, like culture vessels, they build connections with the material and adhere more or less tightly to it. By using trypsin, the cells can be separated from the material and can be seeded again, a procedure that is designated as "passage." In order to examine differentiation rates and different cellular parameters under defined conditions, this technique is used for controlled manipulation and investigation of cell layers. The advantages of this method are the very simple handling and the control of proliferation rates. In most primary cultures of these types, phenotype alteration may occur after few passages and accounts for the major disadvantage of this technique.

41.3.2
Micromass Culture

The extracellular matrix has direct and indirect influences on cells and their behavior. In natural tissues the matrix is secreted by the cells themselves, which are thereby controlling the composition of the extracellular matrix. With this technique, cells are dissociated and the dispersed cells are reaggregated into cellular spheres [79]. In contrast to the monolayer, a growth in all three dimensions is allowed. By using new bioreactor technologies, it is possible to improve the ex vivo growth of scaffold-free tissues.

41.3.3
Agarose and Alginate Culture

Already in 1970 Horwitz and Dorfman [52] cultivated the chondrocytes of an embryonic chick on agarose. In this setting the bottom of the culture scale is coated with 80°C autoclaved fluid agarose that solidifies below its gelling point. This treatment of the plastic material prevents the adhesion of cells on the bottom of the culture dishes. A suspension of cells in culture medium is seeded onto the coated culture vessel and supplied with medium appropriate for the culture. At the end of the experiment the agarose is removed by the application of pronase and subsequent centrifugation of the suspension. The major disadvantage of the technique is its relatively complex setup.

Alginate, a seaweed polysaccharide composed of 1–4 β-D-mannuronic acid and α-L-glucuronic acid, can be used instead of agarose. Kupchik et al. [66] first applied this technique to cultivate human cells of human pancreatic adenocarcinoma.

41.3.4
Coculture

Cocultures of various cell lines (osteoblasts and osteoclasts, osteoblasts and endothelial cells as well as the coculture of differentiated and stem cells) can be used to improve extracorporeal tissue engineering strategies [113, 149]. The major advantage of coculture of differentiated cells and stem cells for use in extracorporeal tissue engineering strategies is that the pool of stem cells allows the renewing of cells and, via efficient signal transduction pathways, the induction of a differentiated cell population. As with pure differentiated cell coculture systems, autocrine and paracrine factors secreted by one cell type readily interact with other cell types when different cell lines are in use. Hence, cocultures provide a more physiological environment for cell regeneration in vitro that more closely resembles the in vivo conditions, especially in bone tissue engineering. However, it has recently been reported that under certain circumstances intimate physical contact of different cells during coculture may lead to the fusion of different cell types, resulting in the formation of heterokaryons [130, 151].

Three different approaches for a coculture of cells are possible:

1. The culture of two or more cell lines in one chamber (Fig. 41.3).
2. The culture of two cells in one chamber, separated by a selectively permeable membrane (Fig. 41.4).
3. The exchange of medium secreted by cells (when cells are cultivated in different chambers; Fig. 41.5).

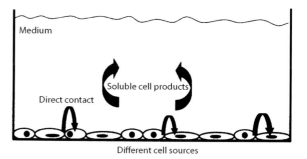

Fig. 41.3 Coculture: two cell types in one chamber

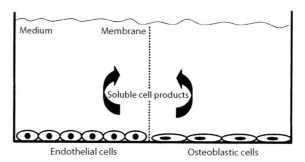

Fig. 41.4 Coculture: two cell types in one chamber, separated by a permeable membrane

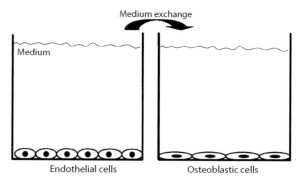

Fig. 41.5 Coculture: different cell types are cultivated in different chapters followed by a medium exchange

It is known that the last two approaches work through a culture medium conditioned by one type of cell that contains various soluble factors capable of inducing cell differentiation. As such factors are easy to transfer, a conditioned medium seems to be desirable for cell scaffold maturation in vitro. The advantage of using cell-conditioned media or the separation of two different cell lines by a membrane over pure coculture techniques is that they are not accompanied by the risk of heterokaryon cell function. In addition, filtering of conditioned media can significantly reduce the risk of contamination with a different cell type, and decreases the risk of disease transmission. The major disadvantage of using membranes or exchange medium, as opposed to conventional coculture, is that there is no intimate physical contact and regulatory crosstalk between the different cell types. The subsequent limitation of signal transduction between the various cells is also not close to the in vivo situation. Indeed, several studies have shown that in some cases conditioned media failed to elicit cell proliferation and differentiation, when compared to a "pure" coculture [110].

41.3.5
Marrow-Derived Stem Cells

Bone-marrow-derived stem cells are commonly gained through a marrow aspiration procedure. Common locations to harvest bone marrow are the iliac crest and the sternum [11, 15, 56, 60, 81]. Unfractioned fresh autologous or syngeneic bone marrow was used in the first attempts to create tissue-engineered bone [45, 94, 143]. Because bone marrow is known to contain osteogenic and chondrogenic precursor cells, its use was perceived to potentially facilitate bone and cartilage regeneration. Principally, MSCs were harvested for these investigations by marrow aspiration, expanded in culture, and then reimplanted. Studies using this protocol have indicated that mouse marrow fibroblastic cells, gained through the aspiration procedure, implanted locally or injected systemically homed to bony sites and persisted there, participating in the regenerative processes [104, 140].

Various investigations used different methods to selectively isolate and enrich marrow stromal cells [55, 95, 96]. Isolation of MSCs is based on density gradient centrifugation and cell culturing techniques. Various cell culture techniques have been developed to allow MSCs to be cultured and expanded without undergoing differentiation [15, 88, 89]. The phenotype of the cells appeared to be stable throughout the culture period without loss in osteogenic, chondrogenic, or adipogenic potential [15, 106]. It is important to note that presently no unique phenotype has been identified that permits the reproducible isolation of MSC precursors with predictable developmental potential. The isolation of stromal cells is mainly dependent on their ability to adhere to plastic as well as their "selective" expansion potential. Human bone marrow progenitor cells were shown to be isolated and enriched by using selective markers [126, 128]. As several markers are expressed on MSCs, some of these have been used to selectively gain subpopulations of more determined cells. Phenotypic markers of MSCs (as well as hematopoietic stem cells) belong to different classes [98]: surface antigens, secreted proteins, surface receptors, cytoskeletal proteins, and extracellular matrix proteins. Phenotypic markers of osteogenic MSCs are:

1. Surface antigens: CD13, CD29, CD44, CD49a, CD71, CD90, CD105, CD114, CD166, human leukocyte antigen (HLA)-ABC, glycophorin A, gp130, intercellular adhesion molecule-1/2, Mab 1740, p75 nerve growth factor-R, SH3, SH4, Stro-1, and HLA class II.
2. Secreted proteins: interleukins (IL) IL1α, IL6, IL7, IL8, IL11, IL12, IL14, IL15, LIF, stem cell factor (SCF), FMS-like tyrosine kinase 3 ligand, granulocyte-monocyte-colony stimulating factor (GM-CSF), granulocyte-colony stimulating factor (G-CSF), and monocyte-colony stimulating factor (M-CSF).
3. Surface receptors: Il1-R, Il3-R, Il4-R, Il6-R, Il7-R, leukemia inhibitory factor receptor, SCF receptor, G-CSF receptor, vascular cell adhesion molecule-1, activated leukocyte cell adhesion molecule-1, lymphocyte-function-associated antigen-3, interferon-γ receptor, tumor necrosis factor (TNF)1R, TNF2R, TGFβ1R, TGFβ2R, bFGF receptor, platelet-derived growth factor receptor, epidermal growth factor receptor.

4. Cytoskeletal proteins: α-smooth muscle actin, glial fibrillary acidic protein.
5. Extracellular matrix components: collagen types I, III, IV, V, and VI, fibronectin, laminin, hyaluronan, and proteoglycans.

Hematopoietic stem cells express the following surface antigens: CD11a, CD11b, CD14, CD34, CD45, CD133, ABCG2, cKit, and Sca-1.

Human MSCs derived from bone marrow have not only been reported to maintain their differentiation capacity into osteogenic lineages for over 40 cell doublings [112], but also when cultured in the presence of dexamethasone and ascorbic acid, or "selectively" enriched through phenotypic markers, purified MSCs undergo a development characterized by the transient induction of alkaline phosphatase, expression of bone matrix protein mRNAs, and deposition of calcium [106]. Cells were suggested to be able to form mineral-like foci indicative of an osteoprogenitor phenotype [8, 9], but it remains to be established whether this mineral formation resembles the mineral formation present in bone tissue. Phenotypic markers of chondrogenic MSCs also include surface antigens, secreted proteins, surface receptors, cytoskeletal proteins, and extracellular matrix proteins:

1. Surface antigens: CD44, CD59, CDC42, and HLA class IC.
2. Secreted proteins: TGFβ, insulin-like growth factor (IGF), and BMP6.
3. Surface receptors: TGF-β3-R and BMP6-R.
4. Cytoskeletal proteins: β-actin.
5. Extracellular matrix components: collagen types I, II, III, IX, and XI, aggrecan, biglykan, chondromodulin, and fibromodulin.

As with osteogenic cells, many investigators have used enrichment techniques for MSCs for cartilage engineering because of the ability of this cell type to differentiate into chondrocyte-like cells. Stem cells that are the target cells for cartilage-inducing factors retain thereby the ability to differentiate to become more determined, functional chondrocytes. It was suggested that it is possible by various measures to induce MSCs to differentiate into a chondrogenic lineage both in vivo and in vitro [19, 58, 82]. Various growth factors such as TGF-β3, BMP-6, and IGF-1 have been used to realize this goal [33, 37, 64, 76, 85, 116, 118, 123, 124, 145].

The in vitro chondrogenic differentiation of postnatal mammalian bone-marrow-derived progenitor cells has been described [58, 152]. This in vitro system is applicable for evaluating the effects of particular factors on the chondrogenic process. The various growth factors can be used separately or together. It is assumed that the so far only limited success of chondrogenic differentiation may be attributable to the fact that the in vivo environment contains a combination of various factors and matrices, with their content changing according to the cell differentiation process. The chondrogenic potential of these progenitor cells is therefore orchestrated by the combination of numerous factors. In vitro systems indicate that it is not only marrow-derived progenitor cells that have the ability to differentiate into chondrocytes. The presence of hypertrophic chondrocytes, as indicated by the presence of type X collagen mRNA and protein, is of special relevance [58, 152].

The presence of a metachromatic-staining matrix, the chondrocytic appearance of cells, and the immunochemical detection of type II collagen can be used in cell culture assays as indicators demonstrating that the tissue generated by marrow-derived cells is cartilage. In addition, the characterization of immunological features is usually applied on culture-expanded cells and not on primary cells. However, none of the aforementioned markers are specific for MSCs, thus hampering the isolation of pure populations of MSCs. After culture expansion, various stromal cell colonies are commonly present. It seems that a proportion of the cultured cells remain multipotent and maintain multilineage potential into osteogenic, chondrogenic, myogenic, endothelogenic, thymogenic, and adipogenic lineages [106]. These results indicate that even after isolation and enrichment strategies, some cells in culture retain multipotentiality.

As with differentiated cells, it is important also to consider the basic cell culture conditions for stem cell cultivation. In general, cell adhesion and cell differentiation are regulated by the conditions of the culture milieu. Cell adhesion is best when cells are cultivated in a serum-containing media and when the surface of the culture dish is coated with proteins. In the absence of proteins, cells can hardly adhere to the surface as they have to synthesize the proteins required to build up the proteinaceous interface (Fig. 41.6). Standard conditions for expansion of stem cells have been established by several studies. These conditions include cell density as well as the presence of serum (in most instances fetal bovine serum). As cell density is known to be a critical factor affecting the growth of cells, attempts to cultivate cells were performed to enrich cells above a critical cell density. Cells can also be grown directly – that is, unmanipulated after collection – or more often after density gradient separation. For clinical applications of stem cells used in extracorporeal tissue engineering, it is imperative that in vitro culture protocols should be devoid of animal or human products, to avoid potential contamination with pathogens. By using defined standard conditions, it is possible to reduce the variability within the culture milieu, therefore also providing a more stringent level of quality control. A special problem of supplemented allogenic animal or human proteins is the possibility that such proteins may enhance the antigenicity of cells upon transplantation. Ideally, the ideal culture milieu for stem cell differentiation in vitro should therefore be chemically defined, and either be serum free, utilize synthetic serum replacements, or use autologous serum [42, 144]. The possible supplementation of specific recombinant cytokines and growth factors may enhance the growth and differentiation of cells, especially when serum-free medium is in use. The major problems with culturing stem cells or more differentiated cells under serum-free conditions are that cells generally tend to have a lower mitotic index, become apoptotic, and display poor adhesion in the absence of serum compounds [144]. Serum supplementation is therefore usually required for in vitro culture of stem cells.

Autologous serum can be considered to be the ideal source of nutritional and stimulatory support of cells when they are used in immunocompetent animals or in clinical trials. For clinical applications, there are several reasons that serum should in future applications be completely eliminated from the in vitro culture milieu. The use of serum is accompanied by some disadvantages:

1. The composition of serum is generally poorly defined. Serum batches possess a considerable degree of individual variation, even when obtained from the same patient.
2. It is important to note that even serum is not completely physiological, since it is essentially a pathological fluid formed in response to blood clotting.
3. Based on the individual patient situation, serum contains variable levels of growth and differentiation factors. The presence of these factors may therefore influence to an unknown extent the desired differentiation of stem cells into specific and well-defined lineages.

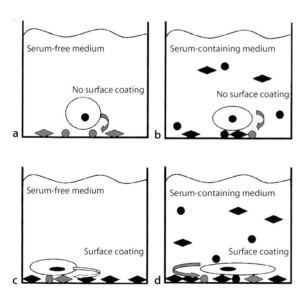

Fig. 41.6a–d Culture-condition-dependent cell attachment. a Cell adhesion is impaired under serum-free culture conditions, especially when the substrate is not coated by adhesion proteins. Cell adhesion improves when a serum-containing medium is used (b), or when the surface is coated prior to cell seeding, even in the absence of a serum-containing medium (c). Cell adhesion is best under conditions of surface coating in combination with the use of serum-containing medium (d)

The development of chemically defined synthetic serum substitutes is therefore of special relevance. Commercially available synthetic serum substitutes [42, 144] containing protein-based cytokines and growth factors or synthetic chemical compounds are currently under intensive investigation [2, 16, 36, 46, 63, 72, 105, 111, 115, 121, 142]. Such synthetic chemicals, possessing the advantage of being more stable compared to protein-based cytokines and growth factors, may improve not only the outcome of stem-cell based tissue engineering approaches, but also strategies that use mature cells.

Despite the success that has been obtained using bone marrow-derived stem cells, one biologic consideration limits its widespread application. Frequently, it is not feasible to obtain sufficient amounts of bone marrow with the requisite number of osteoprogenitor cells by marrow aspiration. In addition, the age-related decrease in bone marrow components, which is accompanied by a partial loss of precursor cells [29, 109], is a frequent clinical limitation to the achievement of sufficient numbers of stem cells. As mentioned before, the outcome of the in vitro use of bone marrow explants is critically dependent upon the transfer of sufficient numbers of these progenitors. Therefore, the use of marrow-derived stem cells may be least applicable in those situations where it is most needed. It was shown that osteoprogenitors represent approximately 0.001% of the nucleated cells in healthy adult marrow [15, 51], which is an indication of the practical problems surrounding the collection of "pure" stem cell sources by the aforementioned cell culture strategies. Therefore, improvements in all aspects of stem cell culture are necessary to further select, expand, and administer the progenitor marrow cell fraction in order to achieve clinically relevant numbers of osteogenic or chondrogenic stem cells.

41.3.6
Determined Bone Cells

It has long been known that the vast capacity for regeneration of bone is based on the presence of differentiated osteoblasts [7]. As the use of determined osteoblast-like cells does not raise legal issues or problems of immune rejection, in contrast to the use of stem cells, determined bone cells have recently been considered to be the most important cell source in bone tissue engineering. Therefore, in current clinical practice, differentiated autologous osteoblast-like cells are the most desirable cell source (Fig. 41.7). However, these cells may be insufficient to rebuild damaged bone tissue within a reasonable time. A considerable number of cell divisions is needed to bulk the tissue to its correct size. Former studies have regarded the propagation of adult mature cells in culture as a serious problem, because it was thought that most adult tissues contained only a minority of cells capable of effective expansion. However, in numerous recent investigations it has been shown that bone cells proliferate in culture without losing their viability. Various sources of determined bone cells can be used for cultivation. Committed osteoprogenitors (i.e., progenitor cells restricted to osteoblast development and bone formation) can be identified by functional assays of their differentiation capacity in vitro, such as the colony forming units assay. A wide variety of systemic, local, and positional factors also regulate the proliferation and differentiation of determined cells, a sequence that is characterized by a series of cellular and molecular events distinguished by the differential expression of osteoblast-associated genes, including those for specific transcription factors, cell-cycle-related proteins, adhesion molecules, and matrix proteins [148]. The mature osteoblast phenotype is in itself heterogeneous, with subpopulations of osteoblasts expressing only subsets of the known osteoblast markers, including those for cytokines, hormones, and growth factor receptors, raising the intriguing possibility that only certain osteoblasts are competent to respond to regulatory agents at particular points in time.

Cultures containing determined "osteoblastic" or osteoblast-like" cells have been established from different cell populations in the lineage of osteogenic cells (osteoprogenitor cells, lining cells, osteoblasts, and osteocytes) [54]. They can be derived from several anatomical sites, using different explant procedures. Bone cell populations may be derived from cortical or cancellous bone, bone marrow, periosteum, and in some instances from other tissues. Isolation of cells can be performed by a variety of techniques, including mechanical disruption, explant outgrowth, and enzyme digestion [135]. Commonly used procedures to gain cells are digestive or outgrowth measures. Outgrowth of bone-like cells can be achieved through culturing of periosteum pieces or bone explants. Cells located within the periosteum and bone can differentiate into fibroblastic, osteogenic, or reticular cells (Fig. 41.8) [11, 25, 34, 92, 134]. Periosteal-derived mesenchymal precursor cells generate progenitor cells committed to one or more

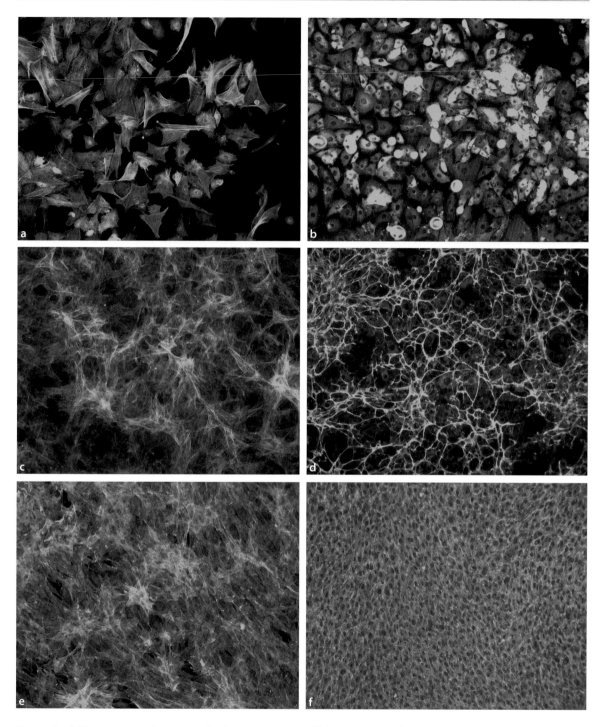

Fig. 41.7a–f Fluorescence microscopy of cell behavior and extracellular matrix formation in osteoblast culture. During culture, subconfluently located cells (**a**) proliferate over time (**b**) in an attempt to form a confluent cell layer (**f**). At the same time, a dense extracellular matrix formation can be observed between cells (**c–e**)

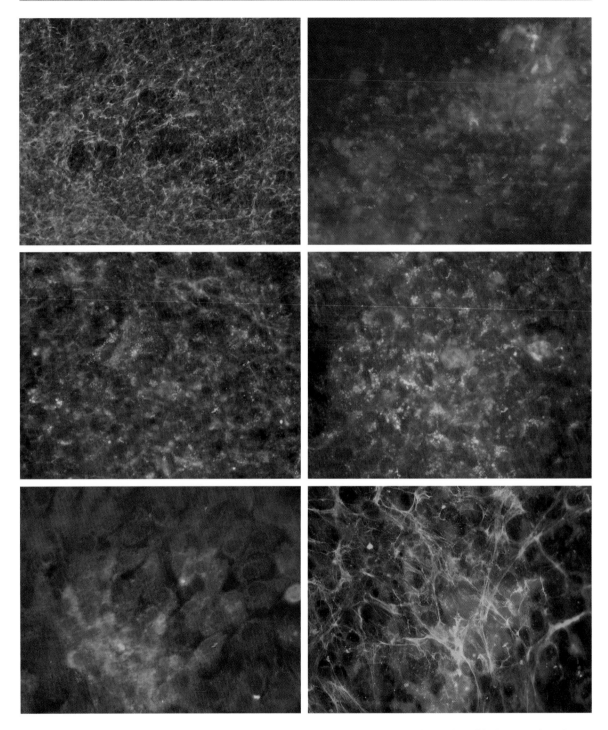

Fig. 41.8 Fluorescence microscopy of the mineral-matrix relationship in osteoblast culture. Mineral is deposited in direct contact to the extracelular matrix network. The green dye visualizes the presence of immunostained proteins, blue coloring is indicative of the formation of hydroxyapatite. The colocalization approach gives insight into aspects of biomineral formation in osteoblast culture

cell lines with an apparent degree of plasticity and interconversion [10, 90, 102, 119, 120]. In culture, expanded periosteum cells were shown to retain the ability to heal a segmental bone defect after being reimplanted, and to induce osteogenic tissue when seeded into diffusion chambers [7, 87, 95, 96]. Outgrowth cultures of periosteum pieces favor the coculture of different cell types [80].

It remains controversial whether cortical or spongy bone, gained by different postsurgical processing techniques, is a better material of choice [22, 41, 122] to outgrow cells in culture. It has been suggested that particulate culturing is superior to bone-chip culturing [77, 125], based on the assumption that when particle size decreases, the absolute square measure of the surface of the tissue specimens in the culture dish increases. An increased absolute square measure of the transplant surface would increase the amount of living cells released. Springer et al. demonstrated in an experimental study that bone chips obtained from trabecular bone provided a higher cell number than those raised from cortical bone [127]. Surprisingly, they found that the processing of spongy bone graft in the bone mill (leading to bone particulates) results in lower absolute amounts of osteoblast-like cells, whereas the use of the bone mill for cortical bone has much less impact on the number of cells counted. The authors speculated that the treatment of transplants with the bone mill or raising of transplants by rotating instruments should reduce the amount of bone cells supplied, suggesting that a decreased particle size is disadvantageous when cell-outgrow methods are in use. Ecarot-Charrier and colleagues were the first to present a method for isolating osteoblasts from newborn mouse calvaria using digestive enzymes in solution [28]. As it was demonstrated that isolated osteoblasts gained through this method retained their unique properties in culture, tissue digestion became a common method of cell harvesting for in vitro purposes.

Once such matured cells have been isolated from the tissue, there are also in such culture strategies several parameters that influence the expression of the osteoblastic phenotype in cell culture. The most important of these are the culture medium, culture time, number of passages, and the presence of compounds. The presence of ascorbic acid, β-glycerophosphate, and dexamethasone influence the expression of the osteoblastic phenotype in a differentiated manner. β-Glycerophosphate, for example, induces phenotypic matrix maturation by enabling mineral formation in osteoblast-like cell cultures (Fig. 41.9). Dexamethasone, as an additional factor, is described as inducing cell differentiation, but imposes a negative effect on cell proliferation, indicative of a reciprocal and functionally coupled

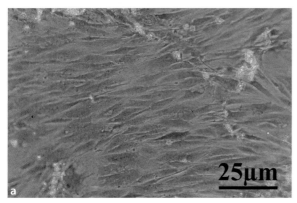
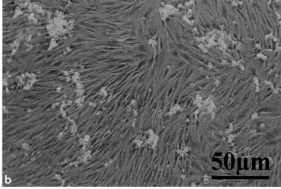

Fig. 41.9a–h Mineral formation and alkaline phosphatase (ALP) expression in osteoblast culture. **a–d** The amount of biomineral formation can be observed through the deposition of calcium-phosphate-containing minerals (*bright areas*) after 1, 2, 3, and 4 weeks. **e–h** The extent of ALP expression (*dark areas*) correlates with the biomineral deposition after 1, 2, 3 and 4 weeks of culture

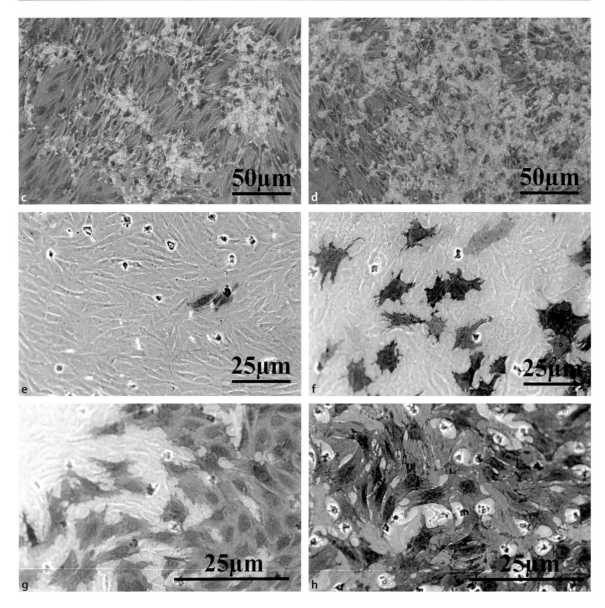

Fig. 41.9a–h *(continued)* Mineral formation and alkaline phosphatase (ALP) expression in osteoblast culture. **a–d** The amount of biomineral formation can be observed through the deposition of calcium-phosphate-containing minerals (*bright areas*) after 1, 2, 3, and 4 weeks. **e–h** The extent of ALP expression (*dark areas*) correlates with the biomineral deposition after 1, 2, 3 and 4 weeks of culture

relationship between proliferation and differentiation. Therefore, it is convenient to select suitable experimental conditions for cultivation. As with stem-cell cultivation, the culture conditions should be well defined in order to standardize the ex-vivo-grown product.

Because endochondral bone formation, and frequently fracture repair, proceed through a cartilaginous intermediate, some investigators have suggested that the transplantation of committed chondrocytes would also ameliorate bone regeneration [10]. Vacanti and colleagues compared the ability of periosteal progenitors and articular chondrocytes to effect bone repair [135]. They showed that periosteal cells from newborn calves seeded on a scaffold and implanted in critical-sized calvarial defects generated new bone. Specimens examined at early times contained material that grossly and histologically appeared to be cartilage. The scaffold seeded with chondrocytes also formed cartilage. However, no endochondral ossification was observed, since the transplanted specimens remained in a cartilaginous state. Therefore, chondrocytes proved ineffective as a cell-based therapy for tissue engineering of bone. Because mature cartilage is thought to produce factors that inhibit angiogenesis, implants seeded with committed chondrocytes may prevent the endochondral cascade by preventing vascular invasion. Cells derived from cartilage seem to be committed to retain their phenotype, and are therefore unable to differentiate toward hypertrophic chondrocytes under the experimental conditions tested so far. In contrast, when precursor cells from the periosteum are provided, their primitive developmental state allows them to proceed through the entire chondrogenic lineage, ultimately becoming hypertrophic chondrocytes. The molecular basis for the difference in the phenotypic potential of these different cell types remains mysterious and is an area of intensive investigation.

of immature bovine articular cartilage [27]. These cells, characterized as determined chondrogenic cells, were shown to allow appositional growth of the articular cartilage from the articular surface [50]. Therefore, when chondrocytes are aimed to generate a cartilage-like structure ex vivo, it seems to be reasonable not to gain full-thickness cartilage implants, but to use subpopulations of chondrocytes. Separation of cartilage zones after the explantation and before cultivation with a selective subpopulation may therefore provide a tool with which to improve tissue engineering strategies using determined cells. Phenotypic plasticity was tested by a series of in ovo injections, where colony-derived populations of these chondroprogenitors were engrafted into a variety of connective tissue lineages, thus confirming that this population of cells has properties akin to those of a progenitor cell. The high colony-forming ability and the capacity to successfully expand these progenitor populations in vitro [27] may further aid our knowledge of cartilage development and growth and may provide novel solutions in ex vivo cartilage tissue engineering strategies.

41.3.7
Determined Chondrogenic Cells

Clinical cartilage engineering strategies have, to date, predominantly focused on the use of an unselected source of chondrocytes [14]. In the ongoing search to improve chondrocyte cell lines, the use of specific chondrocyte populations are now being considered to investigate whether an improved cartilaginous structure would be generated in vivo and in vitro by these specifically selected populations of determined chondrocytes [111]. As distinct phenotypic and functional properties of chondrocytes across the zones of articular cartilage are present, it seemed reasonable to search for the best source of chondrocyte subpopulations [138]. It was reported in this respect that a combination of mid- and deep-zone chondrocytes seems to be more suitable for the ex vivo generation of a hyaline-like cartilage tissue. Dowthwaite et al. have recently reported on an isolation technique for chondrocytes that reside in the superficial zone

41.3.8
Osteoclasts

Osteoclasts may be used in bone tissue engineering strategies, since these cells play an important role in bone physiology, and since a culture containing both cell lines would be more closely mimicking the in vivo bone situation. Osteoclasts are derived from mononuclear cells of hemopoietic bone marrow and peripheral blood. The potential to differentiate into mature osteoclasts in culture is maintained by a variety of cells [21, 49], including hematopoietic stem cells and colony-forming precursors cells, elicited by hematopoietic colony-stimulating factors, and their progeny in the colonies [150]. Moreover, well-differentiated cells, such as monocytes and even mature macrophages, are capable of differentiating into osteoclasts [1]. As the differentiation pathway is common to that of macrophages and dendritic cells, factors can be used to selectively induce differentiation. A promyeloid precursor can differentiate into an osteoclast, a macrophage, or a dendritic cell, depending on whether it is exposed to some osteoclast-inducing factors, such as receptor activator of nuclear factor kappa B ligand (RANKL), osteoprotegerin ligand (OPGL), osteoclast differentiation factor (ODF), M-CSF, and GM-CSF. As some groups demonstrated that bone marrow stromal cells and osteoblasts produce membrane-bound and soluble RANKL/TNF-related activation-induced cytokine/OPGL/ODF factors, an important positive regulator of osteoclast formation [67] is a coculture of osteoclasts with osteoblasts. Alternatively, culturing of osteoclasts can be performed by enrichment of osteoclast precursor cells from the peripheral blood or from bone marrow. Periosteum outgrowth techniques allow also the propagation of osteoclastic cells from monocytes located in the periosteum [129]. Recent research suggests that multinucleated osteoclasts can be cultivated by adding alveolar mononuclear cells to newborn rat calvaria osteoblasts in vitro.

41.3.9
Endothelial Cells

The additional use of vascular cells offers several theoretical advantages over approaches of extracorporeal bone tissue engineering, exploiting only bone cells as a single cell type. As a cell-based strategy, endothelial cell therapy promises to deliver at the same time the substrate (endothelial cells) and cytokines and growth factors. Endothelial progenitor cells as well as mature endothelial cells are capable of settling in areas of bone neovascularization, thus exerting their effects at sites in need of new blood vessel growth [5, 6]. As these cells are present in nearly all sites of the body, they exhibit no unfavorable side effects when transplanted autologously. The induction of angiogenesis as well as the participation in new vessel formation makes them appealing components for bone tissue engineering. Endothelial cells promote synergistic vasculogenesis and bone formation. The action of endothelial cells on osteoblast function was demonstrated recently through various cell culture studies.

Endothelial progenitor cells were first described by Asahara and colleagues in 1997 [5, 6]; since then, significant progress has been made in defining the origin and lineage of these cells. Several studies provide evidence that endothelial progenitors originate in the bone marrow and are then selectively recruited to sites of neovascularization. Therefore, bone marrow cells as well as cells gained from peripheral vessels are used for tissue engineering approaches.

The relative ease of isolating and expanding mature endothelial cells (from explanted blood vessels) or endothelial progenitor cells (from bone marrow) makes them an attractive source of autologous vascular cells for the generation of a vascularized scaffold complex in vitro. Studies have demonstrated that endothelial cells form ring-like structures in two-dimensional cultures and tubules in three-dimensional extracellular matrices in vitro. It is assured that they are able to induce vascular invasion in the host tissue if implanted [4, 86]. Studies with various cell lines indicated that patency rates are strongly correlated with the amount of host cells (smooth muscle cells and endothelial cells) incorporated into the graft [26, 57, 70, 139]. Whereas most of these investigations were not intended to induce the formation of vascular matrices for tissue engineering, the findings are encouraging in light of recent work showing the potential of axial vessels to vascularize cellular scaffolds in vitro and in vivo [23].

By using osteoblast-like cells and endothelial cells it seems to be possible to create complex tissue-engineered vascularized bone constructs. In recent studies tissue-engineered bone constructs have been fabricated by combining autologous vascular cells and bone cells in a porous scaffold structure. Endothelial cells may accelerate the defect healing process by improving the cell nutrition in the scaffolds as well as in the defect site. It was shown that the success of grafts seeded with endothelial cells was significantly greater than that of nonseeded grafts [62]. Vacanti and coauthors demonstrated, also in in vivo studies, that bone replacement materials could be effectively combined with autologous endothelial cells [135]. An additional advantage of using endothelial cells is their potential of thrombus regression. As thrombosis is a serious problem in tissue engineering, prevention of microvascular failures is of major importance after a tissue-engineered bone construct is implanted in vivo. In the future, endothelial cells offer new perspectives to improve the ex vivo formation of a more "mature" bone construct, thereby accelerating the process of new blood vessel formation in vivo. With our recent understanding of the physiological roles of endothelial cells, the importance of vasculogenesis in extracorporeal bone tissue engineering has come into focus, especially in light of the fact that cell survival balanced by nutrition is one of the main limiting steps in scaling up bone constructs for clinical use.

41.4 Cell Culture Evaluation Techniques

The success of cell culturing is monitored by both biochemical and morphological investigations (Figs. 41.10–41.17). Biochemical criteria include all aspects from the molecular features of the cell to the composition of the extracellular matrix (Table 41.3). Morphological aspects range from the ultrastructural analysis of cells and matrix components to the spatial relationships of cells in a three-dimensional tissue construct (Table 41.4). Initially, biochemical assays used traditional methods of protein chemistry (e.g., immunostaining, electrophoresis, and blot techniques; Figs. 41.10 and 41.11). More recently, the availability of genetic probes for many of the bone and cartilage proteins has made it easy to assess gene expression of multiple phenotypic markers by cells gene chips. Advancements in imaging and analytical techniques allow morphological insight into the creation of engineered tissue by using high-resolution pictures (transmission electron microscopy, scanning electron microscopy, X-ray diffraction, atomic force microscopy) [73]. In addition, some of the new techniques enable the combined biochemical and morphological analysis of probes (time-of-flight secondary ion mass spectroscopy).

Biomechanical features of cells in culture can be determined by special investigations. Cell adhesion strength, for example, can be assessed indirectly by cell detachment evaluations. The relative strength of cell adhesion can be determined by trypsination experiments (Fig. 41.13). The amount of adhering cells is an indicator of the number and strength of focal adhesions and can provide information concerning the biology of cell–material interactions.

Table 41.3 Biochemical evaluation methods. *HPLC* High-performance liquid chromatography, *ELISA* enzyme-linked immunosorbent assay, *IRMA* immunoradiometric assay, *PCR* polymerase chain reaction

Method	Determination
Cell counter	Cell number, cell volume
Colorimetric assays	Cell viability, calcium phosphate content
Fluorescence staining	Cytoskeletal proteins, minerals
Immunostaining	Cell and matrix components
Gel electrophoresis	Proteins, nonproteinaceous components
HPLC	Proteins, nonproteinaceous components
Capillary electrophoresis	Proteins, nonproteinaceous components
Western blot	Proteins
ELISA	Proteins, nonproteinaceous components
IRMA	Proteins, nonproteinaceous components
Radioimmunassay	Proteins, nonproteinaceous components
DNA analysis (PCR)	DNA
Gene chips	Gene composition

Table 41.4 Microscopical methods used in tissue engineering. *SMS* Scanning mass spectrometry, *ToF SIMS* time-of-flight secondary low-mass spectrometry, *EDX* energy-dispersive X-ray analysis

Method	Resolution (nm)	Immunostaining	Element analysis	Molecule analysis
Phase-contrast microscopy	100	−	−	−
Conventional microscopy	100	+	−	−
Fluorescence microscopy	100	+	−	−
Scanning electron microscopy	10	+	+ (EDX)	−
Transmission electron microscopy	0.1	+	+ (EDX)	−
Atomic-force microscopy	0.01	−	−	−
Laser SMS	10	−	+	−
ToF SIMS	10	−	+	−

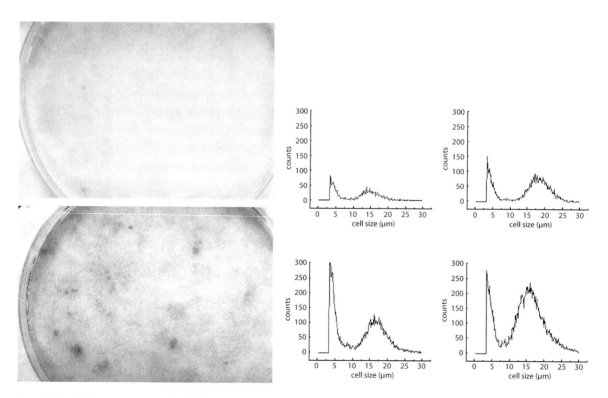

Fig. 41.10 ALP expression in a Petri dish, and the size distribution of osteoblasts in vitro. ALP activity increases over time in culture. In addition, the cell size increases during cell growth in the culture dish (Coulter counter measurements, shown in graphic form on the right side). Both parameters are indicative of viable osteoblast-like cells

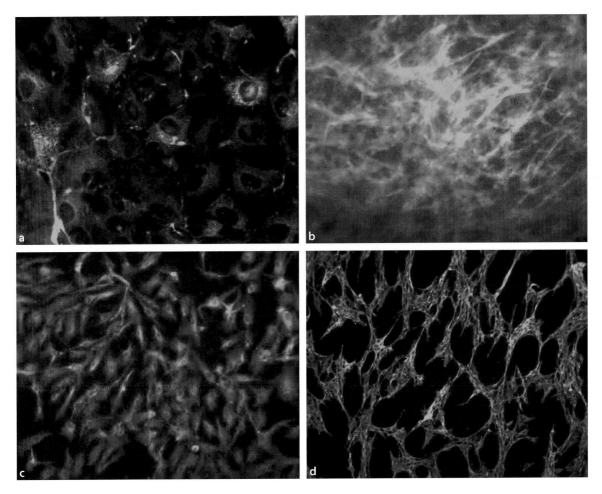

Fig. 41.11a–d Coculture of osteoblasts and endothelial cells. Cells have no direct contact but can interact over the media. **a,b** Collagen type I expression of osteoblasts in single (**a**) and in coculture (**b**) after 1 week. **c,d** Actin expression of endothelial cells in single (**c**) and in coculture (**d**) – the ring format occurs only in coculture

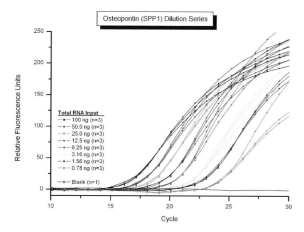

Fig. 41.12 Quantitative osteopontin detection by polymerase chain reaction (PCR). The quantitative analysis of gene expression by PCR technology is a valuable tool in the determination of cell activity states in culture

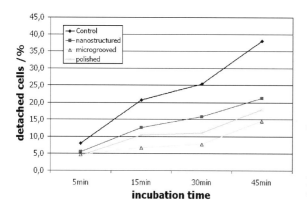

Fig. 41.13 Relative strength of cell adhesion on different surfaces can be determined by trypsinization with diluted trypsin at different time points, plotting the percentage of detached cells versus trypsinization time. The example shows cell attachment on a control surface compared with attachment on different surface nanostructures of titanium

It is generally important to consider that the outcome of individual cell culture determinations is dependent upon the chosen cell culture parameters. Characterization of cells can be adequately performed by determination of the scope of synthesized extracellular matrix proteins (Fig. 41.14). The results of such immunostaining determinations are dependent not only on the preferred cell source, but also on the chemicals used in the cell culture. The expression and visualization of synthesized matrix proteins, for example, differ according to the preferred blocking system (Fig. 41.15) and serum batches (Fig. 41.16). One of the most influencing factors in cell differentiations, as mentioned before, is the special placement of cells. The phenotypic appearance as well as the differentiation capacity of chondrocytes differ significantly between cultures of cells in a two- or three-dimensional cell culture system (Fig. 41.17), indicative for the complex reactivity of cells toward their cell culture microenvironment. Advancements in genetic analysis techniques allow the determination of cell differentiation not only at the protein level, but also at the level of gene expression (Fig. 41.13). Nowadays, the use of gene chips enables a comprehensive insight into the gene activity of cells (Fig. 41.18, Table 41.5). The application of gene chip technology can therefore be accepted to significantly improve the possibility of achieving a more detailed insight into cell behavior, especially when the screening results of gene chip technology can be confirmed by polymerase chain reaction. Evaluation of cell-based tissue engineering strategies in vitro and in vivo can now be determined more precisely by gene analysis technology. It is generally mandatory, in order to compare results of different in vitro experiments, to consider the cell culture outcome in light of the individual experimental setting and the chosen analytical techniques.

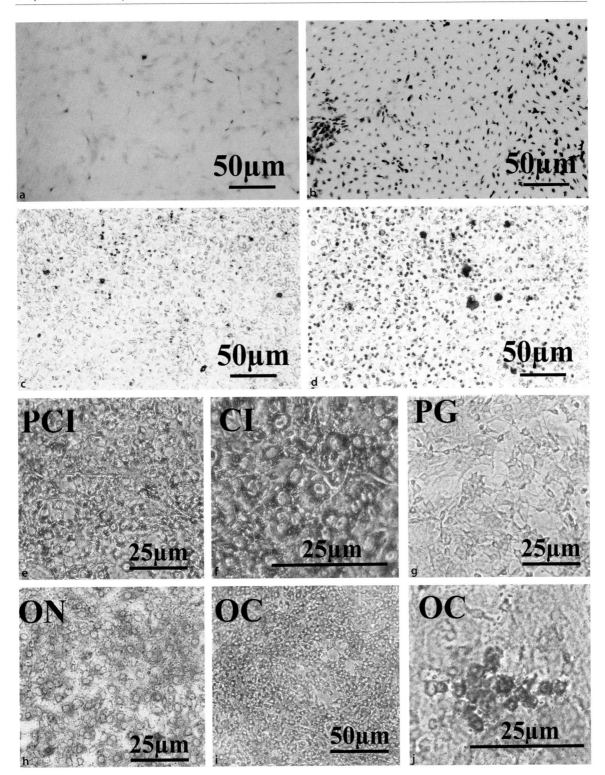

Fig. 41.14a–j Cell cultures can be well characterized by various staining procedures (**a–d**) and a comprehensive analysis of the array of proteins synthesized by the cells (**e–j**, immunostaining). The presence of procollagen type I (*PCI*; **e**), collagen type I (*CI*; **f**), proteoglykanes (*PG*; **g**), osteonectin (*ON*; **h**), and osteocalcin (*OC*; **i, j**) is indicative of a mature osteoblast cell culture

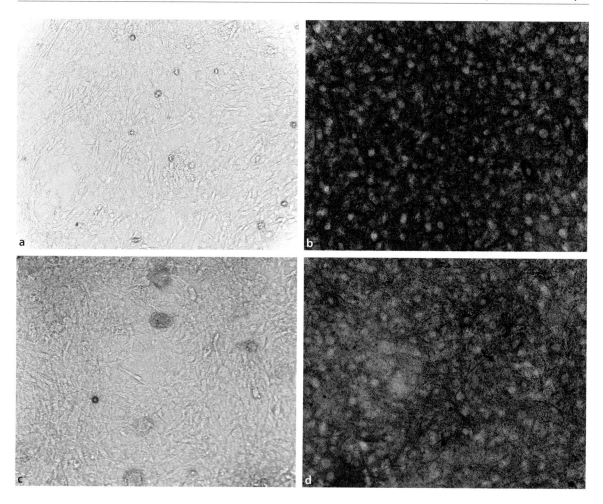

Fig. 41.15a–d Osteocalcin expression of osteoblasts. The staining intensity depends on the blocking system used, which is indicative of the difficulty surrounding the quantitative comparison of the results of different cell culture investigations. Defining the culture condition is a main requirement in cell biology research. **a–d** Candor blocking solution provides the correct osteocalcin expression in 1 day (**a**), 7 days (**c**) and 14 days (**d**) in comparison to the overstaining by using a bovine serum albumin blocking solution

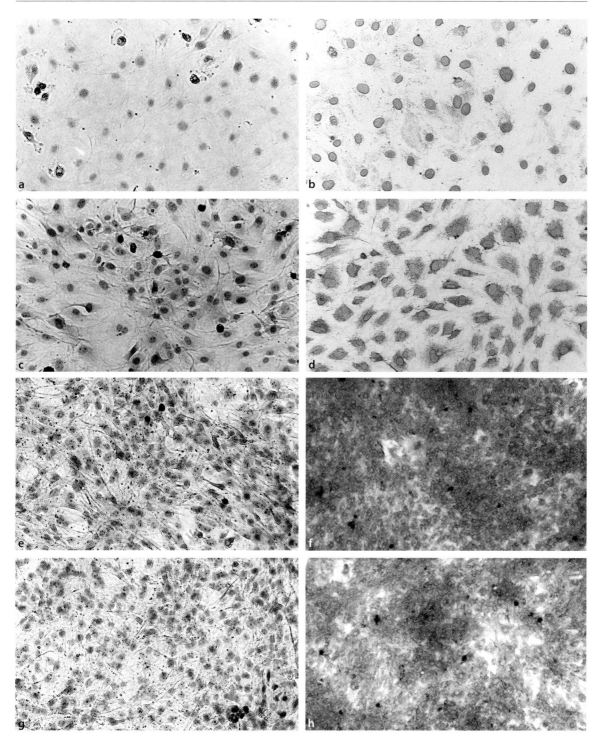

Fig. 41.16a–h Osteoblast monolayer culture. The outcome of the cell cultures is dependent on the serum concentration; different serum concentrations lead to a significant difference in the growth and differentiation characteristics of cells during cell culture. These micrographs show cells cultured in serum-free conditions (**a,b**), 2% serum (**c,d**), 10% serum (**e,f**), and 15% serum (**g,h**), and stained with Richardson's stain (**a,c,e,g**) or for osteonectin expression (**b,d,f,h**). No clear difference can be seen between cells cultured in 10% serum and 15% serum

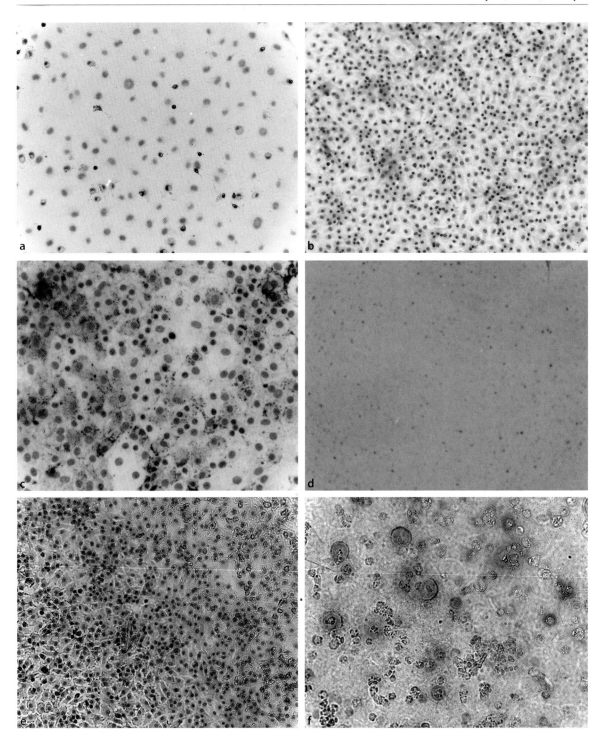

Fig. 41.17a–f Staining of chondrocytes in culture. The appearance of chondrocytes differs significantly between cells cultured in a monolayer and in a three-dimensional gel system. **a** Monolayer culture with low cell density; **b** monolayer culture with high cell density; **c** expression of collagen type I in a monolayer culture; **d** expression of collagen type II in a monolayer culture; **e** expression of collagen type I in a three-dimensional gel culture; **f** expression of collagen type II in a three-dimensional gel culture. Chondrocyte differentiation is sensitive to the special environment. The different collagen expressions of monolayer and gel culture represents the unequal cell differentiation

Chapter 41 Laboratory Procedures – Culture of Cells and Tissues

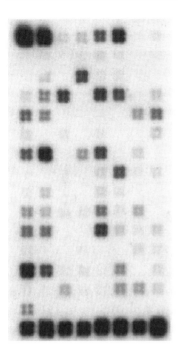

Fig. 41.18 The staining pattern on an osteogenesis array gene chip (Super Array, Bioscience, USA). Expression is measured on this array as the intensity of the spots. Genes with higher expression appear darker on the image

Table 41.5 Location of the different genes on the osteogenesis array gene chip (see Fig. 41.18, Super Array Bioscience, USA)

Array layout							
Gapdh	Ahsg	Akp2	Ambn	Anxa5	Bgn	Bmp1	Bmp2
1	2	3	4	5	6	7	8
Bmp3	Bmp4	Bmp5	Bmp6	Bmp7	Bmp8a	Bmp8b	Bmpr1a
9	10	11	12	13	14	15	16
Bmpr1b	Bmpr2	Calcr	Calcr	Cdh11	Col10a1	Col11a1	Col11a1
17	18	19	20	21	22	23	24
Col14a1	Col15a1	Col18a1	Col19a1	Col1a1	Col1a2	Col2a1	Col3a1
25	26	27	28	29	30	31	32
Col4a1	Col4a2	Col4a3	Col4a4	Col4a5	Col4a6	Col5a1	Col6a1
33	34	35	36	37	38	39	40
Col6a2	Col7a1	Col8a1	Col9a1	Col9a3	Comp	Csf2	Csf3
41	42	43	44	45	46	47	48
Ctsk	Dcn	Dmp1	Dspp	Egf	Enam	Fgf1	Fgf2
49	50	51	52	53	54	55	56
Fgf3	Fgfr1	Fgfr2	Fgfr3	Flt1	Fn1	Gdf10	Ibsp
57	58	59	60	61	62	63	64
Icam1	Igf1	Igf1r	Itga2	Itga2b	Itga3	Itgam	Itgav

Table 41.5 (continued) Location of the different genes on the osteogenesis array gene chip (see Fig. 41.18, Super Array Bioscience, USA)

Array layout							
65	66	67	68	69	70	71	72
Itgb1	Mglap	Mmp10	Mmp13	Mmp2	Mmp8	Mmp9	Msx1
73	74	75	76	77	78	79	80
Nfkb1	Pdgfa	Phex	Runx2	Scarb1	Serpinh	Smad1	Smad2
81	82	83	84	85	86	87	88
Smad3	Smad4	Smad5	Smad6	Smad7	Smad9	Sost	Sox9
89	90	91	92	93	94	95	96
Sparc	Spp1	Tfip11	Tgfb1	Tgfb2	Tgfb3	Tgfbr1	Tgfbr2
97	98	99	100	101	102	103	104
Tgfbr3	Tnf	Tuft1	Twist1	Twist2	Vcam1	Vdr	Vegfa
105	106	107	108	109	110	111	112
Vegfb	Vegfc	PUC18	Blank	Blank	AS1R2	AS1R1	AS1
113	114	115	116	117	118	119	120
Rps27a	B2m	Hspcb	Hspcb	Ppia	Ppia	BAS2C	Bas2C
121	122	123	124	125	126	127	128

41.4.1
Immunostaining

Immunohistology can be used to detect the presence of special cell markers (antigens) of the cultivated cells to match a special tissue as an adjunct to morphological diagnosis. In 1982 the use of monoclonal antibodies for the histopathological diagnosis of human malignancy was published by Gatter and colleagues [40].

The immunocytological investigation of cells is based on the specific affinity of antibodies against epitopes present on the surface of the examined cell populations. If these antibodies have been linked to fluorescent dyes, unfixed cells in suspension carrying these epitopes can be detected through excitation with light of the appropriate wavelength and then measurement of the corresponding emission wavelength. The cells may also haven been immobilized on glass slides or other surfaces and be distinguished with the aid of a fluorescence microscope through the fluorescence properties of the antibodies used.

Another common method of visualizing the presence of specific epitopes is through the application of enzyme-linked antibodies, where an insoluble dye is formed by the enzyme at sites of specific binding. A broad range of antibodies against known epitopes can be obtained from different suppliers. Advanced genetic engineering of antibody libraries has lead to the fast accessibility of specific antibodies against peptides of different origins. So far, the nomenclature of all available antibodies has not been subjected to a systematic approach, which may hamper the search for antibodies against peptides or proteins other than clusters of designation (CDs). For example, to control the cell cycle and apoptosis it has been demonstrated in several tumor studies that the expression of cyclin-dependent kinase proteins ($p16^{INK4a}$, $p21^{WAF1}$, $p27^{Kip1}$, $p57^{Kip2}$) are checkpoint proteins that represent prognostic indicators. Proteins that are related to cell proliferation and cell death are very interesting because they may provide information about the prognosis, and further, may serve as a potential therapeutic marker. As a consequence, a plethora of antibodies against such proteins is now available [69].

Immunohistochemistry can be used for more than merely demonstrate the presence of proteins as markers of a particular cell or tissue type. Due to the increased sensitivity and specificity of antibodies, even genetic abnormalities like mutations or specific chromosomal translocations can be detected. The prerequisite for such an assay is that these genetic abnormalities result in novel chimeric proteins or gene amplification. In the future, this "genogenic immunohistochemistry" [44] may be a new molecular therapeutic target for treatment and an effective substitute for the more expensive and time-consuming genetic studies that are used today.

There has been major progress in the treatment of cancers involving the use of antibody technology. Humanized and chimeric monoclonal antibodies have been produced to target molecules that are expressed on the cell surface of cancer cells and participate in the regulation of growth and proliferation. The best response to this treatment generally occurs in tumors expressing large amounts of these target molecules.

41.4.2
Cryopreservation

In most applications it is important to store cultivated and functionalized cells or tissues for a certain period of time for future scientific use or analysis. A suitable way of storing live cells is the method of cryopreservation. The advantages of cryopreservation are its cost effectiveness, genetic stability, and no risk of microbial contamination, if appropriate measures are taken.

Due to the strong thermal, chemical, and physical changes of the surrounding fluid and the cells themselves during the process of cryopreservation there is a high risk of biological damage for cells. In order to prevent cells and tissues from damage during the potential destructive processes of freezing and thawing, the process of cryopreservation has to be optimized.

The extreme temperature changes during the heat transfer between cells and the surrounding liquid during freezing and thawing are responsible for the thermal damage of cells. In addition, the chemical changes bear the risk of cell destruction during the process of cryopreservation. Alteration of the chemical environment can be reduced by the addition of cryoprotectants before freezing and their removal after thawing. There is also the risk that cryoprotectants not only reduce some of the chemical changes but even create chemical changes through their own chemical behavior. It is important that the chosen cryoprotectant and the applied protocol for adding and removing the additive exert a minimum of deleterious effects to the tissue. The composition of the medium, the temperature, and the period of time taken for each step of the process are the three parameters of the protocol that can be changed to minimize the damaging effects of the cryopreservation process. Further chemical changes are caused by the formation of ice during thawing and freezing. The cooling rate is crucial in order to minimize chemical changes caused by the formation of ice. The depletion of water by extracellularly forming ice crystals leads to a chemical potential difference across the cell membrane and results in driving water out of the cells by osmosis. If a slow cooling rate is applied and the intracellular solution can equilibrate with the external environment by expressing water through the cell membrane, the cell will be massively dehydrated with decreasing temperature. On the other hand, there will be an intracellular formation of ice crystals when the selected cooling rate is too fast and low temperatures are reached before a certain dehydration of the cells has occurred. Mazur et al. discovered in 1972 that the post-thaw viability decreases with increasing rate of cooling [78], and Karlsson detected a mechanical disruption of plasma membrane and other cell structures in 1993 [61]. The presence of ice and the mechanical forces caused by thermal changes can damage the cells. For the correct choice of the most appropriate cooling rate there are two forces affecting cells or tissues during the process of cryopreservation that have to be taken into account: first, the formation of intracellular ice crystals, which takes place foremost at high cooling rates, and second, the solution effects dominant at low cooling rates. In fact, the thermal, chemical, and physical damaging mechanisms are linked and interact as described earlier. It is important to determine the optimal cooling rate at which these two mechanisms of destruction are balanced and the probability of cell survival reaches maximum. The optimal protocol differs from cell type to cell type and from tissue to tissue due to the biophysical properties, like for example, water permeability of the cell membrane [132].

There are some cryoprotectant additives that have cryoprotective properties, like glycerol and dimethyl sulfoxide (DMSO). Polge and colleagues first discovered the cryoprotectant effect of glycerol in 1949 [108]. Cryoprotectants can be subdivided into permeating compounds like glycerol, DMSO and ethylene glycol, or nonpermeating compounds like polyvinyl pyrrolidone and various sugars. So far the detailed underlying mechanisms for each group are not completely understood, but compounds of both groups reduce the concentration of intracellular water, thereby leading to a reduction in ice crystal formation. The permeating cryoprotectants also act as solvent diluents when water is removed during freezing so that less solution effects occur. Nevertheless, the proteins of the cell membrane become stabilized during this process. It is important to be aware that the cryoprotectants themselves can be cytotoxic if the cells have been exposed to them for a longer period of time and in high concentrations. As the cell membrane is more permeable to water than to cryoprotectants, the hypertonic environment of extracellular cryoprotectants leads to a shrinking of the cells. The rate of this loss of water is higher than the diffusion rate of the cryoprotectant into the cells. After a while the osmotic potential across the membrane decreases and the cryoprotectant permeation will dominate. This leads to an increase in cell size due to water reentry, while the removal of cryoprotectant shows opposite dynamics. Such large excursions of cell volume can cause deleterious effects [39].

When we focus on the cryobiology of tissue there are several size effects that must be considered in addition to the effects mentioned above. The presupposition of spatial uniformity of cells can not be used when we look at tissues. Due to the macroscopic dimension of tissues, any gradients in temperature, pressure, and solute concentration can not be neglected because of their spatial variations. The boundary conditions of the system are the only ones that can be controlled and changed to influence the condition of the tissue during the process of cryopreservation. Depending on the type of tissue, there are different transport kinetics of the cryoprotectant during the addition and removal after thawing. For DMSO it has been shown that it penetrates a tissue to a depth of approximately 1 mm after 15 min and 2 mm after 2 h. In order to avoid the damage caused by excessive exposure to cryoprotectants, incubation times of tissues were often minimized, which lead to incomplete equilibration with low cryoprotectant concentrations in the tissue interior [20, 38, 53].

As described earlier, the cooling rate is the most important parameter for the outcome of cryopreservation. Depending on the tissue size and any variations in thermal conductivity, the temperature change to a cell will largely depend on its location within the tissue. In those parts of the specimen where optimum rates of cooling and thawing differ from the optimum treatment, extensive cell damage should be expected. In addition, different cell types show individual optimum rates of cooling and thawing, which leads to difficulties in the cryopreservation of tissues consisting of multiple cell species [61].

Due to the volumetric expansion resulting from the water–ice phase transition mechanical stress can exceed the strength of the frozen material, causing fractures in the tissue [103]. This is of major concern for tissues with significant luminal spaces like blood vessels or liver tissue. The formation of ice in luminal spaces may cause dehydration of the surrounding tissue and exert pressure onto the tissue structures [117], which may lead to extensive damage.

References

1. Akagawa, K. S, N. Takasuka, Y. Nozaki, I. Komuro, M. Azuma, M. Ueda, M. Naito, K. Takahashi (1996) Generation of CD1 + RelB + dendritic cells and tartrate-resistant acid phosphatase-positive osteoclast-like multinucleated giant cells from human monocytes. Blood 88:4029–39
2. Akita, J, M. Takahashi, M. Hojo, A. Nishida, M. Haruta, Y. Honda (2002) Neuronal differentiation of adult rat hippocampus-derived neural stem cells transplanted into embryonic rat explanted retinas with retinoic acid pretreatment. Brain Res 954:286–93
3. Alison, M. R, R. Poulsom, S. Forbes, N. A. Wright (2002) An introduction to stem cells. J Pathol 197:419–23
4. Al-Khaldi, A, N. Eliopoulos, K. Lachapelle, J. Galipeau (2001) EGF-dependent angiogenic response induced bx in situ cultured marrow stromal cells. Anaheim, Calif: American Heart Association Scientific Session, November, pp 11–4
5. Asahara, T, T. Murohara, A. Sullivan, M. Silver, R. van der Zee, T. Li, B. Witzenbichler, G. Schatteman, J. M. Isner (1997) Isolation of putative progenitor endothelial cells for angiogenesis. Science 275:964–7
6. Asahara, T, H. Masuda, T. Takahashi, C. Kalka, C. Pastore, M. Silver, M. Kearne, M. Magner, J. M. Isner (1999) Bone marrow origin of endothelial progenitor cells re-

sponsible for postnatal vasculogenesis in physiological and pathological neovascularization. Circ Res 85:221–8
7. Ashton, B. A, T. D. Allen, C. R. Howlett, C. C. Eaglesom, A. Hattori, M. Owen (1980) Formation of bone and cartilage by marrow stromal cells in diffusion chambers in vivo. Clin Orthop Relat Res 151:294–307
8. Aubin, J. E (1998) Bone stem cells. J Cell Biochem Suppl 30–31:73–82
9. Aubin, J. E (1998) Advances in the osteoblast lineage. Biochem Cell Biol 76:899–910
10. Bahrami, S, U. Stratmann, H. P. Wiesmann, K. Mokrys, P. Bruckner, T. Szuwart (2000) Periosteally derived osteoblast-like cells differentiate into chondrocytes in suspension culture in agarose. Anat Rec 259:124–30
11. Bianco, P, M. Riminucci, S. Gronthos, P. G. Robey (2001) Bone marrow stromal stem cells: nature, biology, and potential applications. Stem Cells 19:180–92
12. Bourne S, Polak JM, Hughes SPF, Buttery LDK (2004) Osteogenic differentiation of mouse embryonic stem cells: differential gene expression analysis by cDNA microarray and purification of osteoblasts by cadherin-11 magnetically activated cell sorting. Tissue Eng 10:796–806
13. Boyan, B. D, T. W. Hummert, D. D. Dean, Z. Schwartz (1996) Role of material surfaces in regulating bone and cartilage cell response. Biomaterials 17:137–46
14. Brittberg, M, A. Lindahl, A. Nilsson, C. Ohlsson, O. Isaksson, L. Peterson (1994) Treatment of deep cartilage defects in the knee with autologous chondrocyte transplantation. N Engl J Med 331:889–95
15. Bruder, S. P, N. Jaiswal, S. E. Haynesworth (1997) Growth kinetics, self-renewal, and the osteogenic potential of purified human mesenchymal stem cells during extensive subcultivation and following cryopreservation. J Cell Biochem 64:278–94
16. Buttery, L. D, S. Bourne, J. D. Xynos, H. Wood, F. J. Hughes, S. P. Hughes, V. Episkopou, J. M. Polak (2001) Differentiation of osteoblasts and in vitro bone formation from murine embryonic stem cells. Tissue Eng 7:89–99
17. Campagnoli, C, I. A. Roberts, S. Kumar, P. R. Bennett, I. Bellantuono, N. M. Fisk (2001) Identification of mesenchymal stem/progenitor cells in human first-trimester fetal blood, liver, and bone marrow. Blood 98:2396–402
18. Caplan, A. I (1991) Mesenchymal stem cells. J Orthop Res 9:641–50
19. Caplan, A. I (2000) Mesenchymal stem cells and gene therapy. Clin Orthop Relat Res (379 Suppl)S67–70
20. Carpenter, J. F, Dawson P. E (1991) Quantitation of dimethyl sulfoxide in solutions and tissues by high-performance liquid chromatography. Cryobiology 28:210–15
21. Chambers, T. J (2000) Regulation of the differentiation and function of osteoclasts. J Pathol 192:4–13
22. Chen, N. T, J. Glowacki, L. P. Bucky, H. Z. Hong, W. K. Kim, M. J. Yaremchuk (1994) The roles of revascularization and resorption on endurance of craniofacial onlay bone grafts in the rabbit. Plast Reconstr Surg 93:714–22; discussion 723–4
23. Chung, S, A. Hazen, J. P. Levine, G. Baux, W. A. Olivier, H. T. Yee, M. S. Margiotta, N. S. Karp, G. C. Gurtner (2003) Vascularized acellular dermal matrix island flaps for the repair of abdominal muscle defects. Plast Reconstr Surg 111:225–32
24. Coelho, M. J, M. H. Fernandes (2000) Human bone cell cultures in biocompatibility testing. Part II: effect of ascorbic acid, beta-glycerophosphate and dexamethasone on osteoblastic differentiation. Biomaterials 21:1095–102
25. Dahir, G. A, Q. Cui, P. Anderson, C. Simon, C. Joyner, J. T. Triffitt, G. Balian (2000) Pluripotential mesenchymal cells repopulate bone marrow and retain osteogenic properties. Clin Orthop Relat Res (379 Suppl) S134–45
26. Deutsch, M, J. Meinhart, T. Fischlein, P. Preiss, P. Zilla (1999) Clinical autologous in vitro endothelialization of infrainguinal ePTFE grafts in 100 patients: a 9-year experience. Surgery 126:847–55
27. Dowthwaite, G. P, J. C. Bishop, S. N. Redman, I. M. Khan, P. Rooney, D. J. Evans, L. Haughton, Z. Bayram, S. Boyer, B. Thomson, M. S. Wolfe, C. W. Archer (2004) The surface of articular cartilage contains a progenitor cell population. J Cell Sci 117:889–97
28. Ecarot-Charrier, B, F. H. Glorieux, M. van der Rest, G. Pereira (1983) Osteoblasts isolated from mouse calvaria initiate matrix mineralization in culture. J Cell Biol 96:639–43
29. Egrise, D, D. Martin, A. Vienne, P. Neve, A. Schoutens (1992) The number of fibroblastic colonies formed from bone marrow is decreased and the in vitro proliferation rate of trabecular bone cells increased in aged rats. Bone 13:355–61
30. Elgendy, H. M, M. E. Norman, A. R. Keaton, C. T. Laurencin (1993) Osteoblast-like cell (MC3T3-E1) proliferation on bioerodible polymers: an approach towards the development of a bone-bioerodible polymer composite material. Biomaterials 14:263–9
31. Erices, A, P. Conget, J. J. Minguell (2000) Mesenchymal progenitor cells in human umbilical cord blood. Br J Haematol 109:235–42
32. Evans, C. H, P. D. Robbins (1995) Possible orthopaedic applications of gene therapy. J Bone Joint Surg Am 77:1103–14
33. Frenz, D. A, W. Liu, J. D. Williams, V. Hatcher, V. Galinovic-Schwartz, K. C. Flanders, T. R. Van de Water (1994) Induction of chondrogenesis: requirement for synergistic interaction of basic fibroblast growth factor and transforming growth factor-beta. Development 120:415–24
34. Friedenstein, A. J (1976) Precursor cells of mechanocytes. Int Rev Cytol 47:327–59
35. Friedenstein, A. J, U. F. Deriglasova, N. N. Kulagina, A. F. Panasuk, S. F. Rudakowa, E. A. Luria, I. A. Rudakow (1974) Precursors for fibroblasts in different populations of hematopoietic cells as detected by in vitro colony assay method. Exp Hematol 2:83–92
36. Fukuda, K (2002) Molecular characterization of regenerated cardiomyocytes derived from adult mesenchymal stem cells. Congenit Anom (Kyoto) 42(1)1–9
37. Fukumoto, T, J. W. Sperling, A. Sanyal, J. S. Fitzsimmons, G. G. Reinholz, C. A. Conover, S. W. O'Driscoll

(2003) Combined effects of insulin-like growth factor-1 and transforming growth factor-beta1 on periosteal mesenchymal cells during chondrogenesis in vitro. Osteoarthritis Cartilage 11:55–64

38. Fuller, B. J, A. L. Busza, E. Proctor (1989) Studies on cryoprotectant equilibration in the intact rat liver using nuclear magnetic resonance spectroscopy: a noninvasive method to asses distribution of dimethyl sulfoxide in tissues. Cryobiology 26:112–8

39. Gao, D. Y, J. Liu, C. Liu, L. E. McGann, P. F. Watson, F. W. Kleinhans, P. E. Mazur, E. S. Critser, J. K. Critser (1995) Preventing of osmotic injury to human spermatozoa during addition and removal of glycerol. Hum Reprod 10:1109–22

40. Gatter, K. C, Z. Abdulaziz, P. Beverley (1982) Use of monoclonal antibodies for the histopathological diagnosis of human malignancy. J Clin Pathol 35:1253–67

41. Girdler, N. M, M. Hosseini (1992) Orbital floor reconstruction with autogenous bone harvested from the mandibular lingual cortex. Br J Oral Maxillofac Surg 30:36–8

42. Goldsborough, M. D, M. L. Tilkins, P. J. Price, J. Lobo-Alfonso, J. R. Morrison, M. E. Stevens (1998) Serum-free culture of murine embryonic stem (ES) cells. Focus 20:8–12

43. Gotherstrom, C, A. West, J. Liden, M. Uzunel, R. Lahesmaa, K. Le Blanc (2005) Difference in gene expression between human fetal liver and adult bone marrow mesenchymal stem cells. Hematol J 90:1017–26

44. Gown A. M (2002) Genogenic immunohistochemistry: a new era in diagnostic immunohistochemistry. Curr Diagn pathol 8:193–200

45. Grundel, R. E, M. W. Chapman, T. Yee, D. C. Moore (1991) Autogeneic bone marrow and porous biphasic calcium phosphate ceramic for segmental bone defects in the canine ulna. Clin Orthop Relat Res (266):244–58

46. Guan, K, H. Chang, A. Rolletschek, A. M. Wobus (2001) Embryonic stem cell-derived neurogenesis. Retinoic acid induction and lineage selection of neuronal cells. Cell Tissue Res 305:171–6

47. Gutierrez-Rodriguez, M, E. Reyes-Maldonado, H. Mayani (2000) Characterization of the adherent cells developed in Dexter-type long-term cultures from human umbilical cord blood. Stem Cells 18:46–52

48. Ham, A. W (1969) Histology. Lippincott, Philadelphia, p 247

49. Hayashi, S, T. Yamane, A. Miyamoto, H. Hemmi, H. Tagaya, Y. Tanio, H. Kanda, H. Yamazaki, T. Kunisada (1998) Commitment and differentiation of stem cells to the osteoclast lineage. Biochem Cell Biol 76:911–22

50. Hayes, D. W. Jr, R. K. Averett (2001) Articular cartilage transplantation. Current and future limitations and solutions. Clin Podiatr Med Surg 18:161–76

51. Haynesworth, S. E, J. Goshima, V. M. Goldberg, A. I. Caplan (1992) Characterization of cells with osteogenic potential from human marrow. Bone 13:81–8

52. Horwitz, A. L, A. Dorfman (1970) The growth of cartilage cells in soft agar and liquid suspension. J Cell Biol 45:434–8

53. Hu, J. F, L. Wolfinbarger (1994) Dimethyl sulfoide concentration in fresh and cryopreserved porcine valved conduit tissue. Cryobiology 31:461–7

54. Hutmacher, D. W, M. Sittinger (2003) Periosteal cells in bone tissue engineering. Tissue Eng 9:S45–64

55. Jackson, I. T, L. R. Scheker, J. G. Vandervord, J. G. McLennan (1981) Bone marrow grafting in the secondary closure of alveolar-palatal defects in children. Br J Plast Surg 34:422–5

56. Jaiswal, N, S. E. Haynesworth, A. I. Caplan, S. P. Bruder (1997) Osteogenic differentiation of purified, culture-expanded human mesenchymal stem cells in vitro. J Cell Biochem 64:295–312

57. Jarrell, B. E, S. K. Williams, G. Stokes, F. A. Hubbard, R. A. Carabasi, E. Koolpe, D. Greener, K. Pratt, M. J. Moritz, J. Radomski, et al (1986) Use of freshly isolated capillary endothelial cells for the immediate establishment of a monolayer on a vascular graft at surgery. Surgery 100:392–9

58. Johnstone, B, T. M. Hering, A. I. Caplan, V. M. Goldberg, J. U. Yoo (1998) In vitro chondrogenesis of bone marrow-derived mesenchymal progenitor cells. Exp Cell Res 238:265–72

59. Jones, D. B, H. Nolte, J. G. Scholubbers, E. Turner, D. Veltel (1991) Biochemical signal transduction of mechanical strain in osteoblast-like cells. Biomaterials 12:101–10

60. Joyner, C. J, A. Bennett, J. T. Triffitt (1997) Identification and enrichment of human osteoprogenitor cells by using differentiation stage-specific monoclonal antibodies. Bone 21:1–6

61. Karlsson, J. O. M, E. G. Carvalho, I. H. M. Borel Rinkes, R. G. Tompkins, M. L. Yarmush, M. Toner (1993) Nucleation and growth of ice crystals inside cultured hepatocytes during freezing in the presence of dimethyl sulfoxide. Biophysical Journal 65:2524–36

62. Kaushal, S, G. E. Amiel, K. J. Guleserian, O. M. Shapira, T. Perry, F. W. Sutherland, E. Rabkin, A. M. Moran, F. J. Schoen, A. Atala, S. Soker, J. Bischoff, J. E. Mayer, Jr (2001) Functional small-diameter neovessels created using endothelial progenitor cells expanded ex vivo. Nat Med 7:s1035–40

63. Kim, B. J, J. H. Seo, J. K. Bubien, Y. S. Oh (2002) Differentiation of adult bone marrow stem cells into neuroprogenitor cells in vitro. Neuroreport 13:1185–8

64. Kucich, U, J. C. Rosenbloom, D. J. Herrick, W. R. Abrams, A. D. Hamilton, S. M. Sebti, J. Rosenbloom (2001) Signaling events required for transforming growth factor-beta stimulation of connective tissue growth factor expression by cultured human lung fibroblasts. Arch Biochem Biophys 395:103–12

65. Kudawara, I, N. Araki, A. Myoui, Y. Kato, A. Uchida, H. Yoshikawa (2004) New cell lines with chondrocytic phenotypes from human chondrosarcoma. Virchows Arch 444:577–86

66. Kupchik, H. Z, R. S. Langer, C. Haberern, S. El-Deriny, M. O'Brien (1983) A new method for the three-dimensional in vitro growth of human cancer cells. Exp Cell Res 147:454–60

67. Lacey, D. L, E. Timms, H. L. Tan, M. J. Kelley, C. R. Dunstan, T. Burgess, R. Elliott, A. Colombero, G. Elliott, S. Scully, H. Hsu, J. Sullivan, N. Hawkins, E. Davy, C. Capparelli, A. Eli, Y. X. Qian, S. Kaufman, I. Sarosi, V. Shalhoub, G. Senaldi, J. Guo, J. Delaney, W. J. Boyle (1998) Osteoprotegerin ligand is a cytokine that regulates osteoclast differentiation and activation. Cell 93:165–76
68. Lee, C. C. I, F. Ye, A. F. Tarantal (2006) Comparison of growth and differentiation of fetal and adult rhesus monkey mesenchymal stem cells. Stem Cells Dev 15:209–20
69. Leong, A. S. Y, T. Y. M. Leong (2006) Newer developments in immunohistology. J Clin Pathol 59:1117–26
70. L'Heureux, N, S. Paquet, R. Labbe, L. Germain, F. A. Auger (1998) A completely biological tissue-engineered human blood vessel. FASEB J 12:47–56
71. Lieberman, J. R, L. Q. Le, L. Wu, G. A. Finerman, A. Berk, O. N. Witte, S. Stevenson (1998) Regional gene therapy with a BMP-2-producing murine stromal cell line induces heterotopic and orthotopic bone formation in rodents. J Orthop Res 16:330–9
72. Liu, S, Y. Qu, T. J. Stewart, M. J. Howard, S. Chakrabortty, T. F. Holekamp, J. W. McDonald (2000) Embryonic stem cells differentiate into oligodendrocytes and myelinate in culture and after spinal cord transplantation. Proc Natl Acad Sci U S A 97:6126–31
73. Manso, M, S. Ogueta, J. Perez-Rigueiro, J. P. Garcia and J. M. Martinez-Duart (2002) Testing biomaterials by the in-situ evaluation of cell response. Biomol Eng 19:239–42
74. Mareschi, K, E. Biasin, W. Piacibello, M. Aglietta, E. Madon, F. Fagioli (2001) Isolation of human mesenchymal stem cells: bone marrow versus umbilical cord blood. Haematologica 86:1099–100
75. Martin, G. R (1981) Isolation of a pluripotent cell line from early mouse embryos cultured in medium conditioned by teratocarcinoma stem cells. Proc Natl Acad Sci U S A 78:7634–8
76. Martin, I, M. Jakob, D. Schafer, W. Dick, G. Spagnoli, M. Heberer (2001) Quantitative analysis of gene expression in human articular cartilage from normal and osteoarthritic joints. Osteoarthritis Cartilage 9:112–8
77. Marx, R. E, R. I. Miller, W. J. Ehler, G. Hubbard, T. I. Malinin (1984) A comparison of particulate allogeneic and particulate autogenous bone grafts into maxillary alveolar clefts in dogs. J Oral Maxillofac Surg 42:3–9
78. Mazur, P, S. P. Leibo, E. H. Y. Chu (1972) A two-factor hypothesis of freezing injury. Exp Cell Res 71:345–55
79. Meyer, U, H. P. Wiesman (2005) Bone and Cartilage Tissue Engineering. Springer, Heidelberg, Berlin, Tokyo, New York
80. Meyer, U, H. D. Szulczewski, K. Möller, H. Heide, D. B. Jones (1993) Attachment kinetics and differentiation of osteoblasts on different biomaterials. Cells Mater 3:129–40
81. Meyer, U, T. Meyer, D. B. Jones (1998) Attachment kinetics, proliferation rates and vinculin assembly of bovine osteoblasts cultured on different pre-coated artificial substrates. J Mater Sci Mater Med 9:301–7
82. Minguell, J. J, A. Erices, P. Conget (2001) Mesenchymal stem cells. Exp Biol Med (Maywood) 226:507–20
83. Mirmalek-Sani, S. H, R. S. Tare, S. M. Morgan, H. I. Roach, D. I. Wilson, N. A. Hanley, R. O. C. Oreffo (2006) Characterization and multipotentiality of human fetal femur-derived cells: implications for skeletal tissue regeneration. Stem Cell 24:1042–53
84. Moscona, H, A. Moscona (1952) The dissociation and aggregation of cells from organ rudiments of early chick embryo. J Anat 86:287–301
85. Muraglia, A, R. Cancedda, R. Quarto (2000) Clonal mesenchymal progenitors from human bone marrow differentiate in vitro according to a hierarchical model. J Cell Sci 113:1161–6
86. Murayama, T, O. M. Tepper, M. Silver, H. Ma, D. W. Losordo, J. M. Isner, T. Asahara, C. Kalka (2002) Determination of bone marrow-derived endothelial progenitor cell significance in angiogenic growth factor-induced neovascularization in vivo. Exp Hematol 30:967–72
87. Nakahara, H, S. P. Bruder, V. M. Goldberg, A. I. Caplan (1990) In vivo osteochondrogenic potential of cultured cells derived from the periosteum. Clin Orthop Relat Res (259):223–32
88. Nakahara, H, S. P. Bruder, S. E. Haynesworth, J. J. Holecek, M. A. Baber, V. M. Goldberg, A. I. Caplan (1990) Bone and cartilage formation in diffusion chambers by subcultured cells derived from the periosteum. Bone 11:181–8
89. Nakahara, H, J. E. Dennis, S. P. Bruder, S. E. Haynesworth, D. P. Lennon, A. I. Caplan (1991) In vitro differentiation of bone and hypertrophic cartilage from periosteal-derived cells. Exp Cell Res 195:492–503
90. Nakahara, H, V. M. Goldberg, A. I. Caplan (1992) Culture-expanded periosteal-derived cells exhibit osteochondrogenic potential in porous calcium phosphate ceramics in vivo. Clin Orthop Relat Res (276):291–8
91. Nathanson, M. A (1985) Bone matrix-directed chondrogenesis of muscle in vitro. Clin Orthop Relat Res (200):142–58
92. Nuttall, M. E, A. J. Patton, D. L. Olivera, D. P. Nadeau, M. Gowen (1998) Human trabecular bone cells are able to express both osteoblastic and adipocytic phenotype: implications for osteopenic disorders. J Bone Miner Res 13:371–82
93. Oakes, D. A, J. R. Lieberman (2000) Osteoinductive applications of regional gene therapy: ex vivo gene transfer. Clin Orthop Relat Res (379 Suppl):S101–12
94. Ohgushi, H, A. I. Caplan (1999) Stem cell technology and bioceramics: from cell to gene engineering. J Biomed Mater Res 48:913–27
95. Ohgushi, H, V. M. Goldberg, A. I. Caplan (1989) Repair of bone defects with marrow cells and porous ceramic. Experiments in rats. Acta Orthop Scand 60:334–9
96. Ohgushi, H, V. M. Goldberg, A. I. Caplan (1989) Heterotopic osteogenesis in porous ceramics induced by marrow cells. J Orthop Res 7:568–78
97. Oreffo, R. O, J. T. Triffitt (1999) Future potentials for using osteogenic stem cells and biomaterials in orthopedics. Bone 25:5S–9S

98. Otto, W. R, J. Rao (2004) Tomorrow's skeleton staff: mesenchymal stem cells and the repair of bone and cartilage. Cell Prolif 37:97–110
99. Owen, M (1988) Marrow stromal stem cells. J Cell Sci Suppl 10:63–76
100. Owen, M, A. J. Friedenstein (1988) Stromal stem cells: marrow-derived osteogenic precursors. Ciba Found Symp 136:42–60
101. Owen, T. A, M. Aronow, V. Shalhoub, L. M. Barone, L. Wilming, M. S. Tassinari, M. B. Kennedy, S. Pockwinse, J. B. Lian, G. S. Stein (1990) Progressive development of the rat osteoblast phenotype in vitro: reciprocal relationships in expression of genes associated with osteoblast proliferation and differentiation during formation of the bone extracellular matrix. J Cell Physiol 143:420–30
102. Park, S. R, R. O. Oreffo, J. T. Triffitt (1999) Interconversion potential of cloned human marrow adipocytes in vitro. Bone 24:549–54
103. Peeg, D.E, M. C. Wusteman, S. Boylan (1997) Fractures in cryopreserved elastic arteries. Cryobiology 34:183–92
104. Perka, C, O. Schultz, R. S. Spitzer, K. Lindenhayn, G. R. Burmester, M. Sittinger (2000) Segmental bone repair by tissue-engineered periosteal cell transplants with bioresorbable fleece and fibrin scaffolds in rabbits. Biomaterials 21:1145–53
105. Phillips, B. W, C. Vernochet, C. Dani (2003) Differentiation of embryonic stem cells for pharmacological studies on adipose cells. Pharmacol Res 47:263–8
106. Pittenger, M. F, A. M. Mackay, S. C. Beck, R. K. Jaiswal, R. Douglas, J. D. Mosca, M. A. Moorman, D. W. Simonetti, S. Craig, D. R. Marshak (1999) Multilineage potential of adult human mesenchymal stem cells. Science 284:143–7
107. Plate, U, S. Arnold, U. Stratmann, H. P. Wiesmann, H. J. Höhling (1998) General principle of ordered apatitic crystal formation in enamel and collagen rich hard tissues. Connect Tissue Res 38:149–57; discussion 201–5
108. Polge, C, A. U. Smith, A. S. Parkes (1949) Revival of spermatozoa after vitrification and dehydratation at low temperatures. Natrure 164:666–76
109. Quarto, R, D. Thomas, C. T. Liang (1995) Bone progenitor cell deficits and the age-associated decline in bone repair capacity. Calcif Tissue Int 56:123–9
110. Rangappa, S, J. W. Entwistle, A. S. Wechsler, J. Y. Kresh (2003) Cardiomyocyte-mediated contact programs human mesenchymal stem cells to express cardiogenic phenotype. J Thorac Cardiovasc Surg 126:124–32
111. Redman, S. N, S. F. Oldfield, C. W. Archer (2005) Current strategies for articular cartilage repair. Eur Cell Mater 9:23–32; discussion 23–32
112. Reyes, M, T. Lund, T. Lenvik, D. Aguiar, L. Koodie, C. M. Verfaillie (2001) Purification and ex vivo expansion of postnatal human marrow mesodermal progenitor cells. Blood 98:2615–25
113. Richards, M, C. Y. Fong, W. K. Chan, P. C. Wong, A. Bongso (2002) Human feeders support prolonged undifferentiated growth of human inner cell masses and embryonic stem cells. Nat Biotechnol 20:933–6
114. Rinaldini, L. M. J (1959) The isolation of living cells from animal tissues. Int Rev Cytol 7:587–647
115. Rogers, J. J, H. E. Young, L. R. Adkison, P. A. Lucas, A. C. Black, Jr (1995) Differentiation factors induce expression of muscle, fat, cartilage, and bone in a clone of mouse pluripotent mesenchymal stem cells. Am Surg 61:231–6
116. Rosado, E, Z. Schwartz, V. L. Sylvia, D. D. Dean, B. D. Boyan (2002) Transforming growth factor-beta1 regulation of growth zone chondrocytes is mediated by multiple interacting pathways. Biochim Biophys Acta 1590:1–15
117. Rubinsky, B, C. Y. Lee, J. Bastacky, G. Onik (1990) The process of freezing and the mechanism of damage during hepatic cryosurgery. Cryobiol 27:85–97
118. Schaefer, J. F, M. L. Millham, B. de Crombrugghe, L. Buckbinder (2003) FGF signaling antagonizes cytokine-mediated repression of Sox9 in SW1353 chondrosarcoma cells. Osteoarthritis Cartilage 11:233–41
119. Schantz, J. T, D. W. Hutmacher, H. Chim, K. W. Ng, T. C. Lim, S. H. Teoh (2002) Induction of ectopic bone formation by using human periosteal cells in combination with a novel scaffold technology. Cell Transpl 11:125–38
120. Schantz, J. T, D. W. Hutmacher, K. W. Ng, H. L. Khor, M. T. Lim, S. H. Teoh (2002) Evaluation of a tissue-engineered membrane-cell construct for guided bone regeneration. Int J Oral Maxillofac Implants 17:161–74
121. Schinstine, M, L. Iacovitti (1997) 5-Azacytidine and BDNF enhance the maturation of neurons derived from EGF-generated neural stem cells. Exp Neurol 144:315–25
122. Schwipper, V, K. von Wild, H. Tilkorn (1997) [Reconstruction of frontal bone, periorbital and calvarial defects with autogenic bone]. Mund Kiefer Gesichtschir 1:S71–4
123. Sekiya, I, D. C. Colter, D. J. Prockop (2001) BMP-6 enhances chondrogenesis in a subpopulation of human marrow stromal cells. Biochem Biophys Res Commun 284:411–8
124. Sekiya, I, J. T. Vuoristo, B. L. Larson, D. J. Prockop (2002) In vitro cartilage formation by human adult stem cells from bone marrow stroma defines the sequence of cellular and molecular events during chondrogenesis. Proc Natl Acad Sci U S A 99:4397–402
125. Shirota, T, K. Ohno, M. Motohashi, K. Michi (1996) Histologic and microradiologic comparison of block and particulate cancellous bone and marrow grafts in reconstructed mandibles being considered for dental implant placement. J Oral Maxillofac Surg 54:15–20
126. Simmons, P. J, B. Torok-Storb (1991) Identification of stromal cell precursors in human bone marrow by a novel monoclonal antibody, STRO-1. Blood 78:55–62
127. Springer, I. N, H. Terheyden, S. Geiss, F. Harle, J. Hedderich, Y. Acil (2004) Particulated bone grafts – effectiveness of bone cell supply. Clin Oral Implants Res 15:205–12
128. Stewart, K, S. Walsh, J. Screen, C. M. Jefferiss, J. Chainey, G. R. Jordan, J. N. Beresford (1999) Further characterization of cells expressing STRO-1 in cultures of adult human bone marrow stromal cells. J Bone Miner Res 14:1345–56
129. Szulczewski, D. H, U. Meyer, K. Möller, U. Stratmann, S. B. Doty, D. B. Jones (1993) Characterisation of bovine

osteoclasts on an ionomeric cement in vitro. Cells Mater 3:83–92
130. Terada, N, T. Hamazaki, M. Oka, M. Hoki, D. M. Mastalerz, Y. Nakano, E. M. Meyer, L. Morel, B. E. Petersen, E. W. Scott (2002) Bone marrow cells adopt the phenotype of other cells by spontaneous cell fusion. Nature 416:542–5
131. Thomson, J. A, J. Itskovitz-Eldor, S. S. Shapiro, M. A. Waknitz, J. J. Swiergiel, V. S. Marschall, J. M. Jones (1998) Embryonic stem cell lines derived from human blastocysts. Science 282:1145–7
132. Tonner, M, E. G. Cravalho, J. Stachecki, T. Fitzgerald, R. G. Tompkins, M. L. Yarmush, D. R. Armant (1993) Nonequilibrium freezing of one-cell mouse embryos. Biophys J 64:1908–21
133. Triffitt, J. T (2002) Osteogenic stem cells and orthopedic engineering: summary and update. J Biomed Mater Res 63:384–9
134. Triffitt, J. T, R. O. C. Oreffo (1998) Osteoblast lineage. In: Zaidi, M (ed) Advances in Organ Biology. Molecular and Cellular Biology of Bone, Advances in Organ Biology Series. JAI Press, Connecticut, 5B, pp 429–51
135. Vacanti, C. A, W. Kim, J. Upton, D. Mooney, J. P. Vacanti (1995) The efficacy of periosteal cells compared to chondrocytes in the tissue engineered repair of bone defects. Tissue Eng 1:301–8
136. Vautier, D, J. Hemmerle, C. Vodouhe, G. Koenig, L. Richert, C. Picart, J. C. Voegel, C. Debry, J. Chluba, J. Ogier (2003) 3-D surface charges modulate protrusive and contractile contacts of chondrosarcoma cells. Cell Motil Cytoskeleton 56:147–58
137. Wada, Y, H. Kataoka, S. Yokose, T. Ishizuya, K. Miyazono, Y. H. Gao, Y. Shibasaki, A. Yamaguchi (1998) Changes in osteoblast phenotype during differentiation of enzymatically isolated rat calvaria cells. Bone 22:479–85
138. Waldman, S. D, M. D. Grynpas, R. M. Pilliar, R. A. Kandel (2003) The use of specific chondrocyte populations to modulate the properties of tissue-engineered cartilage. J Orthop Res 21:132–8
139. Weinberg, C. B, E. Bell (1986) A blood vessel model constructed from collagen and cultured vascular cells. Science 231:397–400
140. Wiesmann, H. P, N. Nazer, C. Klatt, T. Szuwart, U. Meyer (2003) Bone tissue engineering by primary osteoblast-like cells in a monolayer system and 3-dimensional collagen gel. J Oral Maxillofac Surg 61:1455–62
141. Wobus, A. M (2001) Potential of embryonic stem cells. Mol Aspects Med 22:149–64
142. Wobus, A. M, G. Kaomei, J. Shan, M. C. Wellner, J. Rohwedel, G. Ji, B. Fleischmann, H. A. Katus, J. Hescheler, W. M. Franz (1997) Retinoic acid accelerates embryonic stem cell-derived cardiac differentiation and enhances development of ventricular cardiomyocytes. J Mol Cell Cardiol 29:1525–39
143. Wolff, D, V. M. Goldberg, S. Stevenson (1994) Histomorphometric analysis of the repair of a segmental diaphyseal defect with ceramic and titanium fibermetal implants: effects of bone marrow. J Orthop Res 12:439–46
144. Wong, M, R. S. Tuan (1993) Nuserum, a synthetic serum replacement, supports chondrogenesis of embryonic chick limb bud mesenchymal cells in micromass culture. In Vitro Cell Dev Biol Anim 29A:917–22
145. Worster, A. A, B. D. Brower-Toland, L. A. Fortier, S. J. Bent, J. Williams, A. J. Nixon (2001) Chondrocytic differentiation of mesenchymal stem cells sequentially exposed to transforming growth factor-beta1 in monolayer and insulin-like growth factor-I in a three-dimensional matrix. J Orthop Res 19:738–49
146. Xiaoxue, Y, C. Zhongqiang, G. Zhaoqing, D. Gengting, M. Qingjun, W. Shenwu (2004) Immortalization of human osteoblasts by transferring human telomerase reverse transcriptase gene. Biochem Biophys Res Commun 315:643–51
147. Xu, C. H, M. S. Inokuma, J. Denham, K. Golds, P. Kundu, J. D. Gold, M. K. Carpenter (2001) Feeder-free growth of undifferentiated human embryonic stem cells. Nat Biotechnol 19:971–4
148. Yamaguchi, A, T. Komori, T. Suda (2000) Regulation of osteoblast differentiation mediated by bone proteins, hedgehogs, and Cbfa1. Endocr Rev 21:393–411
149. Yamaguchi, M, F. Hirayama, H. Murahashi, H. Azuma, N. Sato, H. Miyazaki, K. Fukazawa, K. Sawada, T. Koike, M. Kuwabara, H. Ikeda, K. Ikebuchi (2002) Ex vivo expansion of human UC blood primitive hematopoietic progenitors and transplantable stem cells using human primary BM stromal cells and human AB serum. Cytotherapy 4:109–18
150. Yamazaki, H, T. Kunisada, T. Yamane, S. I. Hayashi (2001) Presence of osteoclast precursors in colonies cloned in the presence of hematopoietic colony-stimulating factors. Exp Hematol 29:68–76
151. Ying, Q. L, J. Nichols, E. P. Evans, A. G. Smith (2002) Changing potency by spontaneous fusion. Nature 416:545–8
152. Yoo, J. U, T. S. Barthel, K. Nishimura, L. Solchaga, A. I. Caplan, V. M. Goldberg, B. Johnstone (1998) The chondrogenic potential of human bone-marrow-derived mesenchymal progenitor cells. J Bone Joint Surg Am 80:1745–57
153. Zvaifler, N. J, L. Marinova-Mutafchieva, G. Adams, C. J. Edwards, J. Moss, J. A. Burger, R. N. Maini (2000) Mesenchymal precursor cells in the blood of normal individuals. Arthritis Res 2:477–88

42. Bioreactors in Tissue Engineering: From Basic Research to Automated Product Manufacturing

D. Wendt, S. A. Riboldi

Contents

42.1	Introduction	595
42.2	Bioreactors in Tissue Engineering: Key Features	596
42.2.1	Cell Seeding on Three-Dimensional Matrices	596
42.2.2	Maintenance of a Controlled Culture Environment	598
42.2.3	Physical Conditioning of Developing Tissues	599
42.3	Sensing in Tissue-Engineering Bioreactors	599
42.3.1	Monitoring of the Milieu	600
42.3.2	Monitoring of the Construct	602
42.4	Bioreactor-Based Manufacturing of Tissue-Engineering Products	603
42.4.1	Automating Conventional Cell-Culture Techniques	604
42.4.2	Automating Tissue-Culture Processes	606
42.4.3	Production Facilities: Centralized Versus Decentralized	607
42.4.4	Intraoperative "Manufacturing" Approaches	608
42.5	Conclusions and Future Perspectives	608
	References	609

42.1 Introduction

"Bioreactors," a term generally associated with classical industrial bioprocesses such as fermentation, was initially used in tissue engineering applications to describe little more than simple mixing of a Petri dish. Over the last two decades, bioreactors used in tissue engineering research evolved, not only for the function of engineering in vitro various types of biological tissues (e.g., skin, tendons, blood vessels, cartilage, and bone), but also to serve as defined model systems supporting investigations on cell function and tissue development. In recent years, as bioreactors continued to progress in sophistication, the term has gradually become synonymous with sophisticated devices enabling semiautomated, closely monitored, and tightly controlled cell and tissue culture. In particular, by controlling specific physicochemical culture parameters at defined levels, bioreactors provide the technological means with which to perform controlled studies aimed at understanding the effects of specific biological, chemical, or physical cues on basic cell functions in a three-dimensional spatial arrangement. Moreover, bioreactors successfully make up for limitations of conventional manual methods when driving the development of structurally uniform and functionally effective three-dimensionally engineered constructs.

Despite the impressive progress achieved by researchers in the field of bioreactor-based tissue engineering, it is evident that the need for safe and clinically effective autologous tissue substitutes remains unsatisfied. In order to successfully translate tissue-engineering technologies from bench to bedside, numerous challenges remain to be addressed. To this end, of prime consideration is the fact that the clinical efficacy of a tissue-engineered product will need to be accompanied by a *cost-effective manufacturing process* and *compliance to the evolving regulatory framework* in terms of quality control and good manufacturing practice (GMP) requirements. Tissue-engineering products manufactured by labor-intensive, manual, benchtop cell- and tissue-culture protocols may find difficulty in competing with alternative therapeutic options, concerning safety and cost:benefit ratio. On the contrary, *bioreactors* as a means to generate and maintain a controlled culture environment and enable directed tissue growth, could represent the key element for the development of automated, standardized, traceable, cost-effective, and safe manufacturing of engineered tissues for clinical applications.

In this chapter we discuss the role of bioreactors in the translational paradigm of tissue-engineering approaches from basic research to streamlined tissue manufacturing. We first review the key functions of bioreactors traditionally employed in research applications (Sect. 42.2). Subsequently, having identified current sensor and monitoring techniques as a significant bottleneck toward the implementation of successful feedback-controlled strategies to optimize the culture progression, we give a brief overview of the basic principles of sensing in tissue engineering bioreactors (Sect. 42.3). Finally, we describe and critically discuss examples, potentials, and challenges for bioreactor-based manufacturing of tissue-engineered products (Sects. 42.4 and 42.5).

42.2
Bioreactors in Tissue Engineering: Key Features

The use of bioreactors in scientific research is gaining increasing importance, both as a means to direct the in vitro development of living, functional substitutes of biological tissues and as dynamic culture model systems, when studying the fundamental mechanisms of cell function. In both cases, general functions of bioreactors are to initiate, maintain, and direct cell cultures and tissue development in a three-dimensional, physicochemically defined, tightly controlled, aseptic environment. In the following, the key features of bioreactors commonly used for research purposes in the field of tissue engineering are described (Fig. 42.1) [9, 33, 47].

42.2.1
Cell Seeding on Three-Dimensional Matrices

Traditionally, the delivery of a cell suspension within a three-dimensional scaffold is performed manually by means of pipettes, and relying on gravity as a leading principle for cell settlement and subsequent adhesion to the scaffold pores. Such a seeding method, besides being scarcely reproducible due to marked intra- and interoperator variability, is inevitably characterized by poor efficiency and nonuniformity of the resulting cell distribution within the scaffold [68]. The usual "static" seeding method may yield particularly inhomogeneous results when *thick and/or low-porosity scaffolds* are used, since gravity may not suffice for the cells to penetrate throughout the scaffold pores. When dealing with *human cell sources*, optimizing the efficiency of seeding will be crucial in order to maximize the utilization of cells that can be obtained from the rather limited tissue biopsy samples.

Hence a variety of "dynamic" cell-seeding techniques, relying on the use of bioreactors, have been recently developed with the aim of increasing the quality, reproducibility, efficiency, and uniformity of the seeding process as compared to conventional static methods. Spinner flasks [67], wavy-walled reactors [8], and rotating wall vessels [19] are only examples of the numerous devices found in the literature. However, the most promising approach, enabling efficient and uniform seeding of different cell types in scaffolds of various morphologies and porosities, proved to be "perfusion seeding," comprising direct perfusion of a cell suspension through the pores of a three-dimensional scaffold [6, 10, 24,

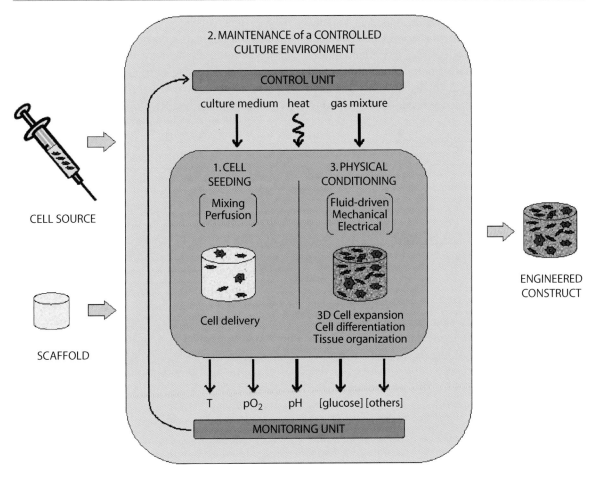

Fig. 42.1 Schematic representation of the key functions of bioreactors used in research applications for tissue engineering, as described in Sect. 42.2.1. Cell seeding of three-dimensional (3D) matrices: bioreactors can maximize cell utilization, control cell distribution, and improve the reproducibility of the cell-seeding process. 2. Maintenance of a controlled culture environment: bioreactors that monitor and control culture parameters can provide well-defined model systems to investigate the fundamental aspects of cell function and can be used to enhance the reproducibility and overall quality of engineered tissues. 3. Physical conditioning of the cell/scaffold constructs: bioreactors that apply physiological regimes of physical stimulation can improve the structural and functional properties of engineered tissues. T Temperature, pO_2 oxygen partial pressure, [ ] concentrations

25, 28, 63, 65, 68, 69, 72]. Such an efficient method, relying on active driving forces rather than on gravity for the fluid to penetrate the scaffold pores, proved to be particularly suitable when seeding cells into thick scaffolds of low porosity [68]. Interestingly, the principle of perfusion has been used in the field of heart-valve tissue engineering and for in vitro transformation of porcine valves into human valves, enabling decellularization of valve grafts of xenogenic origin and subsequent recellularization with human cells [26].

When defining and optimizing seeding protocols (i.e., the selection of parameters such as cell concentration in the seeding suspension, medium flow rate, flow directions, and timing of the perfusion pattern), most of the studies found in the literature have relied upon experimental, application-specific, trial-and-error investigations, without the support of theoretical models. However, the inherent complexity of a dynamic seeding system represents a major challenge for modeling, due to high dependence on the specific cell type and scaffold implemented (e.g., complex pore architecture and related fluid dynamics, kinetics of cell adhesion, molecular mechanics, and biomaterial properties). A notable effort in this direction was described by Li and coauthors, who developed and validated a mathematical model allowing the predictive evaluation of the maximum seeding density achievable within matrices of different porosities, in a system enabling filtration seeding at controlled flow rates [30].

42.2.2
Maintenance of a Controlled Culture Environment

Early bioreactors, developed for research purposes in the 1980s and 1990s, were generally meant to be positioned inside cell culture incubators while in use. In such configurations, monitoring and control of key environmental parameters for homeostatic maintenance of cell cultures (such as temperature, atmosphere composition, and relative humidity) were supplied by the incubators themselves. More recently, a spreading demand for automated, user-friendly, and operationally simple bioreactor systems for cell and tissue culture catalyzed research toward the development of stand-alone devices integrating the key function of traditional cell-culture incubators, namely environmental control. Implications at a "design" level of the new bioreactor concept will be analyzed in details in Sects. 42.3 and 42.4, where sensing and automation will be discussed extensively.

Here we will rather focus on the crucial role that bioreactors are known to play in the maintenance of local homeostasis, at the level of the engineered construct, specifically via oxygen and metabolite supply and waste product removal. The high degree of structure heterogeneity of three-dimensionally engineered constructs cultured in static conditions (i.e., the presence of a necrotic central region surrounded by a dense layer of viable cells) suggests that diffusional transport does not properly assure uniform and efficient mass transfer within the constructs [16]. On the contrary, convective media flow around the construct and, to an even greater extent, direct medium perfusion through its pores, can aid overcoming diffusional transport limitations. Bioreactors that perfuse the culture medium directly through the pores of a scaffold have therefore been employed in the engineering of various tissues, demonstrating that perfusion enhances calcified matrix deposition by marrow-derived osteoblasts [3, 24, 25, 57], viability, proliferative capacities and expression of cardiac-specific markers of cardiomyocytes [15, 52], cell proliferation in engineered blood vessels [28], and extracellular matrix deposition, accumulation, and uniform distribution by chondrocytes [13, 69]. In this context, a deep understanding of the basic mechanisms underlying perfusion-associated cell proliferation/differentiation and matrix production will be challenging to achieve, since the relative effects of perfusion-induced mechanical stresses acting on cells and enhanced mass transfer of chemical species, or a combination of the factors, cannot be easily discerned. As a result, similar to perfusion-seeding parameters, optimization of perfusion culture conditions is commonly achieved from an experimental, trial-and-error approach. In future applications, both the design of new perfusion bioreactors and the optimization of their operating conditions will derive significant benefits from computational fluid dynamics modeling aimed at estimating fluid velocity and shear profiles [11, 20, 45], as well as biochemical species concentrations within the pores of three-dimensional scaffolds. A more comprehensive strategy that could help to elucidate and decouple the effects of mechanical stimuli and specific species should: (1) combine theoretical and experimental approaches (i.e., validate simplified models with experimental data) [53] and (2) make use of sensing and control technologies to monitor culture progression and adapt the culture conditions in a feedback-controlled loop, aimed at reestablishing homeostatic parameters. Technological platforms enabling the latter approach in particular will be discussed in Sect. 42.3.

Another noteworthy factor heavily hindering homeostatic control in cell-culture systems is the abrupt change in the concentration of metabolites/catabolites, signal molecules, and pH when culture medium is exchanged in periodic batches. In traditional static culture procedures, the smoothening of these step-shaped variations can be achieved by performing partial medium changes; however this requires additional repeated manpower involvement. Bioreactor technology offers a better solution by enabling either semicontinuous automatic replenishment of exhausted media at defined time-points, or feedback-controlled addition of fresh media, aimed at reestablishing a homeostatic parameter to a predefined set point (e.g., pH) [27, 49, 63].

42.2.3 Physical Conditioning of Developing Tissues

Over the centuries, several in vivo and ex vivo studies have contributed to demonstrate that physical forces (i.e., hydrodynamic/hydrostatic, mechanical, and electrical) play a key role in the development of tissues and organs during embryogenesis, as well as their remodeling and growth in postnatal life. Based on these findings, and in an attempt to induce the development of biological constructs that resemble the structure and function of native tissues, tissue engineers have aimed to recreate in vitro a physical environment similar to that experienced by tissues in vivo. For this purpose, numerous bioreactors have been developed, enabling controlled and reproducible dynamic conditioning of three-dimensional constructs for the generation of functional tissues.

Bioreactors applying fluid-driven mechanical stimulation, for example, were employed for the investigation of developmental mechanisms via the establishment of shear stresses acting directly on cells (e.g., in the case of cartilage [53], bone [3], and cardiac tissue [50]), via the creation of a differential pressure (e.g., for blood vessels [64] and heart valves [40]) or combining these two mechanisms (again with vessels [22, 23, 42, 58] and heart valves [18]). The in vitro engineering of tissues, wich are natively exposed to mechanical cues in vivo, has frequently been reported to be enhanced by means of bioreactors enabling *mechanical conditioning*, namely direct tension (e.g., tendons, ligaments, skeletal muscle tissue [1, 31, 48], and cardiac tissue [17]), compression (e.g., cartilage [12, 14]) and bending (e.g., bone [38]). Similarly, interesting findings on the effect of electrical stimulation on the development of excitable tissues were derived by conditioning skeletal muscle [44, 48] and cardiac constructs [51]. Moreover, a promotion of neural gene expression by activation of calcium channels was observed as a result of the application of physiological electrical patterns to primary sensory neurons [7].

Consistent with the tight correlation that exists in nature between the structure and function of biological tissues (the spatial arrangement of load-bearing structures in long bones and the presence of tightly parallel arrays of fibers in skeletal muscles being just two examples of this principle), appropriate tissue structural arrangements have been induced in vitro via the dynamic conditioning of engineered tissues. Physical conditioning was shown to be an effective means of improving cell/tissue structural organization, mainly through the mutual influence that cells and extracellular proteins exert reciprocally via integrin binding [10, 21, 56, 70].

As previously discussed with respect to flow-associated effects in perfusion bioreactors, it is imperative to emphasize the fact that current scientific knowledge is far from allowing a deep understanding of the mechanoresponsive dynamics of cell function. As a consequence, the idea of precisely directing tissue development in vitro by means of specific physical cues remains an immense challenge. In evidence of this fact, one only needs to examine the vast array of model systems that can be found in the literature, and moreover, the various magnitudes, frequencies, and durations of the applied physical stimuli, even with reference to one single tissue type. Deriving conclusions from different model systems becomes more challenging when considering that the same physical cue (e.g., compression) may result in the alteration of many secondary variables (e.g., tension, hydrostatic pressure, flow-shear, streaming potentials, and mass transfer) according to the distinguishing features of the model system and, in particular, to the specific scaffold employed.

42.3 Sensing in Tissue-Engineering Bioreactors

The earliest form of "bioreactors" used for tissue-engineering applications were rather naive devices, often built out of pre-existing laboratory equipment (e.g., Petri dishes, culture flasks, and multiwell plates). With time, bioreactor systems underwent progressive evolution, increasing in their sophistication and complexity through the addition of various types of tubing, connectors, and fittings, and mixing of the culture media. However, these devices suffered from several drawbacks, requiring more and more laborious assembly and operations in laminar-flow cabinets before initiating cell culture, and the repeated need for intervention of skilled operators for medium-exchange procedures, ultimately leading to marked user-related variability of the reactor performance. Furthermore, due to the still-immature integration of sensing technologies in the field of cell and tissue culture, monitoring of the culture process was traditionally based on qualitative observations (e.g., the color of the culture medium) and upon end-point destructive investigations, undermining the opportunity to adapt and optimize the culture protocols to the actual developmental state of the engineered constructs.

More recently, the increasing use of bioreactors in research, and their foreseen introduction into clinically related applications, are increasing the demand for automated, user-friendly, and ergonomic devices, integrating complex functions (e.g., environmental conditioning, medium exchange, and monitoring of key parameters) not necessarily conditional on the presence of skilled operators. As a result, bioreactors are gradually turning into automated devices of increased "internal" sophistication that are simple to operate and have a higher degree of reliability and standardization.

A fundamental feature of these new-generation bioreactors will be the monitoring of key parameters indicative of the culture progression (e.g., concentration of metabolites/catabolites in the medium, construct permeability, scaffold mechanical properties). Sensing in tissue-culture bioreactors represents a major premise in order to achieve essential objectives, such as:
1. Elucidating the physical and biochemical mechanisms underlying tissue development.
2. Implementation of feedback-controlled strategies aimed at optimizing the culture progression.
3. Tracing of the culture process.
4. Automation and scale-up of tissue manufacturing.

Monitoring the partial pressure of oxygen and carbon dioxide in the culture medium, and detecting the concentrations of glucose and lactate, for instance, allow the quantitative evaluation of the metabolic behavior of cultured cells, thus supporting/substituting subjective and qualitative conclusions traditionally derived by simply observing the color of the medium. On the basis of such information, it is then possible to enact strategies (e.g., injection of fresh medium or supplements, modification of the medium flow rate in perfusion bioreactors) aimed at adapting the dynamic culture environment to the actual developmental state of the engineered construct.

Given the crucial role that monitoring and feedback-control techniques can play in bioreactors for cell- and tissue-culture applications, we will present an overview of sensing strategies and issues in the following sections. Borrowing from the nomenclature of Mason et al. [36] and Starly and Choubey [59], the noteworthy parameters that should be adequately monitored and controlled during in vitro organogenesis can be classified into two main categories, namely the *milieu parameters* and the *construct parameters*. Milieu parameters are then further subdivided into the physical (e.g., temperature, pressure, and flow rate), chemical (e.g., pH, dissolved oxygen and carbon dioxide, chemical contaminants, and concentration of significant metabolites/catabolites such as glucose, lactate, or secreted proteins), and biological (e.g., sterility). Similarly, the construct parameters can be different in nature: physical (e.g., stiffness, strength, and permeability), chemical (e.g., composition of the scaffold and of the developing extracellular matrix), and biological (e.g., cell number and proliferation rate, concentration of intracellular proteins, and cell viability).

42.3.1
Monitoring of the Milieu

Several technological solutions were developed in recent years to monitor the milieu parameters. In addition to the general requirements that sensors need to meet in common practice (i.e., accuracy, sensitivity, and specificity), the probes employed in cell and tissue culture are required to fulfill peculiar specifi-

	ADVANTAGES	DRAWBACKS
INDIRECT SENSING	• high specificity • high sensitivity	• lag time → artifacts • impaired feedback
INVASIVE SENSING	• high specificity • high sensitivity	• need for sterility • difficult calibration
NONINVASIVE SENSING	• no need for sterility • ease of calibration	• need of transparency • low specificity • low sensitivity

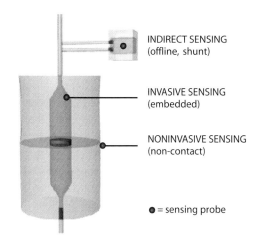

Fig. 42.2 The three main modalities used to monitor the milieu in bioreactors, as described in Sect. 42.3.1. "Indirect sensing" is performed directly on the culture media, but via sampling means, either by offline analysis or shunt sensing. "Invasive sensing" implies that the sensor probe is placed directly inside the culture chamber of the bioreactor, either immersed in the culture fluid or in direct contact with the engineered construct. "Noninvasive sensing" involves sensors that do not come into contact with the interior of the culture chamber, but are capable of measuring via interrogation through the bioreactor wall. This figure was adapted from Rolfe 2006 [54] by Rosaria Santoro

cations. In particular, they might need to be rather small in size compared to many commercially available sensors, have a lifetime of several weeks, unless they are sufficiently low cost to be disposed of and replaced during culture, and must ensure a stable response over time, since repeated calibration might be difficult to carry out due to the accessibility of the sensors during culture. Regardless of the technical details related to the specific parameter under investigation (reviewed in [54]), sensors for milieu monitoring can be generally classified in different categories according to the position of the sensing probe relative to the culture chamber of the reactor (Fig. 42.2) [36].

disposable or capable of withstanding repeated sterilization protocols. Integrating these sensors within a bioreactor for clinical applications may have limitations with regard to the potential high cost of single-use disposable sensors. Clear advantages of invasive sensing are the high precision and accuracy achieved with the measurement. The most common invasive sensors currently in use are based on optical and electrochemical principles: fiber-optic fluorescence-based sensors, for example, have been manufactured for the measurement of dissolved oxygen and of pH, while typical electrochemical sensors are membranes functionalized with appropriate enzymes for the detection of glucose or urea.

42.3.1.1
Invasive

Invasive (embedded) sensors are those whose probes are placed directly inside the culture chamber of the bioreactor, either immersed in the culture fluid or in direct contact with the engineered construct. Clearly, invasive sensors must be sterile and therefore either

42.3.1.2
Noninvasive

Noninvasive (noncontact) sensors are sensors that do not come in contact with the interior of the culture chamber, and are capable of measuring via interrogation through the bioreactor wall, for example by using ultrasound or optical methods such as spectrophotometry or fluorimetry. This approach, while avoiding the sterility issues associated with invasive sensors, implies that the bioreactor wall must be either entirely or locally transparent to the investigating wave. Moreover, due to the presence of an intermediate material between the probe and the object of interest, it is more challenging to achieve high specificity and high sensitivity of the measurement with these sensors. Typically, the flow rate of the medium in perfusion bioreactors is detected via noninvasive techniques, mainly based on Doppler velocimetry. Doppler optical coherence tomography (DOCT), for example, is a novel technique that allows noninvasive imaging of the fluid flow at the micron level, in highly light-scattering media or biological tissues. Derived from clinical applications, DOCT has been adapted to characterize the flow of culture medium through a developing engineered vascular construct within a bioreactor chamber [36].

42.3.1.3
Indirect

Indirect sensing is performed directly on the culture media, but via sampling means. The two main options included in this category are "offline analysis" and "shunt sensing." In the first case, manual or automated online medium sampling is performed (with possible negative implications for the sterility of the closed system) and analyses are conducted with common instruments for bioanalytics (e.g., blood-gas analyzers). In the latter, the measurement is carried out directly within the fluid, driven through a sensorized shunting loop and later either returned to the body of the bioreactor or discarded. Since probes do not need to be placed inside the culture chamber, indirect sensing can be performed by means of advanced, accurate instruments, with clear advantages in terms of specificity and sensitivity with respect to the invasive method. On the other hand, the lag-time introduced by sampling can heavily hinder the significance of the measurement itself (with possible introduction of artifacts) and impair the efficacy of feedback-control strategies. With this method, the partial pressures of oxygen and carbon dioxide, pH, and glucose concentration in the culture medium are typically measured [62]; protein and peptide analysis can be also conducted via spectrophotometric and fluorimetric assays within shunting chambers.

42.3.2
Monitoring of the Construct

While monitoring of the milieu is gradually entering the practice of bioreactor-based tissue engineering, monitoring the function and structure of developing engineered constructs remains a relatively uncharted area and a highly challenging field of research [43]. In this application, it would be limiting to use the term "sensor" in the traditional sense, since the techniques currently under study are based on highly sophisticated cutting-edge technology, often inherited from rather unrelated fields (e.g., clinics, telecommunications). Systems for the nondestructive online monitoring of the construct developmental state would allow continuous and immediate optimization of the culture protocol to the actual needs of the construct itself, thus overcoming the drawbacks traditionally related to the use of endpoint detection methods or fixed time-point analyses. Typically, research is being driven by the need for real-time characterization of the functional and morphological properties of engineered constructs, at both the micro- and at the macro-scale. The following paragraphs give a brief overview of the most recent techniques applied in bioreactors for these cited purposes.

42.3.2.1
Monitoring of the Functional Properties

This comprises characterization of the construct's overall physical properties (e.g., strength, elastic modulus, and permeability), but also monitoring of cell function within the engineered construct itself (e.g., in terms of proliferation, viability, metabolism, phenotype, biosynthetic activity, and adhesive forces). In this context, Stephens and collaborators [60] recently proposed a method to image real-time cell/material interactions in a perfusion bioreactor, based on the use of an upright microscope. The kinetics of cell aggregation and organoid assembly in rotating wall vessel bioreactors, instead, could be performed according to the method developed by Botta et al. [4], relying on a diode pumped solid-state laser and on a CCD video camera. Boubriak and coauthors [5] recently proposed the use of microdialysis for detecting local changes in cellular metabolism (i.e., glucose and lactate concentrations) within a tissue-engineered construct. By means of this method, concentration gradients could be monitored within the construct, with the highest lactate concentrations in the construct center, thus allowing early detection of inappropriate local metabolic changes.

of the number, size, and distribution of mineralized particles within the construct [46].

In conclusion, there is evidence in the field that novel techniques are being developed for nondestructive, continuous, online monitoring of cell- and tissue-culture processes in bioreactors. Such advanced monitoring systems could represent useful tools with which to potentially clarify still-unknown aspects of the cellular response in dynamic culture conditions (e.g., the previously mentioned relative contributions of increased mass transport and shear stress in improving the quality of engineered tissues), as well as a step toward the automation and in-process control of the tissue-manufacturing process, which will be the subject of the following sections.

42.3.2.2
Monitoring of the Morphological Properties

Monitoring of the morphological properties of engineered constructs essentially encompasses assessing the amount, composition, and distribution of the extracellular matrix that is being deposited throughout the scaffold during bioreactor culture. Monitoring morphological properties is particularly pertinent when engineering tissues whose function is strictly dependent on the structural organization of their extracellular matrix (e.g., bone and tendons). In this context, Optical coherence tomography (OCT) has been successfully employed as a real-time, nondestructive, noninvasive tool with which to monitor the production of extracellular matrix within engineered tendinous constructs in a perfusion bioreactor [2]. OCT is analogous to conventional clinical ultrasound scanning, but by using near-infrared light sources instead of sound, it enables higher resolution images (1–15 µm vs 100–200 µm). The technique is compact and flexible in nature, as well as relatively low cost since it can be implemented by commercially available optic fibers [37]. However, the most promising technique in the field of real-time imaging is undoubtedly micro-computed tomography. Using this technique, the mineralization within a three-dimensional construct cultured in a perfusion bioreactor was monitored over time, allowing quantification

42.4
Bioreactor-Based Manufacturing of Tissue-Engineering Products

During the initial phase of the emerging tissue-engineering field, we have been mainly consumed by the new biological and engineering challenges posed in establishing and maintaining three-dimensional cell and tissue cultures. After nearly two decades, with exciting and promising research advancements, tissue engineering is now at the stage where it must begin to translate this research-based technology into large-scale and commercially successful products. However, just as other biotechnological and pharmaceutical industries came to realize in the past, we are ultimately faced with the fact that even the most clinically successful products will need to demonstrate: (1) cost-effectiveness and cost-benefits over existing therapies, (2) absolute safety for patients, manufacturers, and the environment, and (3) compliance to the current regulations. But what has been hindering cell-based engineered products from reaching the market and what can be done to increase their potential for clinical and commercial success? In this section, we describe and discuss several commercial manufacturing strategies that allowed the first cell-based products to enter the clinical practice. Moreover, we comment on the potential of bioreactor-based manufacturing approaches to improve the clinical and commercial success of engineered products by controlling, standardizing, and automating cell- and tissue-culture procedures in a cost-effective and regulatory-compliant manner. In particular, a brief insight will be given in regards to: (1) techniques aimed at automating conventional cell culture techniques (i.e., two-dimensional monolayer cell expansion), (2) strategies to automate and streamline tissue-culture processes (i.e., comprising automation of the three-dimensional culture phase), (3) centralized and decentralized manufacturing approaches, and (4) the rising interest in "intraoperative strategies."

42.4.1
Automating Conventional Cell-Culture Techniques

As described in Sect. 42.2, the basic procedures for generating engineered tissues have traditionally been based around conventional manual benchtop cell- and tissue-culture techniques. It is therefore quite natural that these manual techniques, due to their simplicity and widespread use, were included in the initial phases of product development, and ultimately in the final manufacturing processes of early cell-based products. Manual techniques remain particularly appealing for start-up companies since the simple level of technology minimizes the initial development time and investment costs, allowing for more rapid entry to clinical trials and into the market. An example of the straightforward benchtop-based manufacturing process is that employed by Genzyme Tissue Repair (Cambridge, MA, USA) for the production of Carticel®, an *autologous* cell transplantation product for the repair of articular cartilage defects that is currently used in the clinic. To manufacture the Carticel product, a cartilage biopsy is harvested upon surgical intervention and sent to a central facility where the chondrocytes are isolated and expanded in monolayer culture to generate a sufficient number of cells [39] using routine culture systems (i.e., manually by a laboratory technician, inside a biological safety cabinet, housed in a Class 10,000 clean room). Hyalograft C™, marketed by Fidia Advanced Biopolymers (Abano Terme, Italy), is an alternative autologous cell-based product for the treatment of articular cartilage defects, also manufactured through conventional benchtop techniques. Similar to the production of Carticel, chondrocytes are first isolated from a biopsy and expanded in tissue-culture flasks by highly trained technicians at a central manufacturing facility. To generate the Hyalograft C cartilage graft, the expanded chondrocytes are then manually seeded onto a three-dimensional scaffold and cultured for 14 days by specialized technicians using routine tissue-culture techniques [55]. For both of these products, the simple production systems kept initial product development costs down as the products were established within the marketplace. However, these manufacturing processes require a large number of manual and labor-intensive manipulations and, moreover, due to the autologous nature of the products, each cell preparation must be treated individually. For instance, whereas in general laboratory practice a large quantity of flasks is routinely processed simultaneously in a sterile hood in parallel, for manufacturing autologous products, cells/flasks derived from a single donor would be removed from the incubator, introduced into the sterile hood, and processed individually (e.g., media exchanged, cells trypsinized and passaged, and so on), and only following a detailed and validated cleaning/decontamination procedure would the entire procedure be repeated with cells/flasks derived from a different donor. As a result, the production costs of these products are rather high, they possess inherent risks for contamination and intra- and interoperator variability, and would be difficult to scale as product demands increase. It is therefore becoming more and more evident that tissue-engineering firms will inevitably have to follow in the footsteps of other biotechnology fields and begin to introduce process engineering into their manufacturing strategies.

Robotic systems have proven highly effective at automating and controlling sophisticated manufacturing process for a wide variety of industries such as the computer and automotive fields. Recently, robotics have also been developed for use in biotechnology applications (e.g., Cell^{host} system, from Hamilton, Bonaduz, Switzerland; and SelecT™ from The Automation Partnership, Royston, UK), capable of performing several routine but laborious cell-culture processes such as the maintenance and expansion of multiple simultaneous cell lines and the automated culture of embryonic stem cells. Robotics could also be an attractive option in the tissue-engineering field [29, 35]. For instance, in an attempt at automating the labor-intensive phase of expanding epithelial cells for dermal tissue-engineering purposes, a closed bioreactor system with integrated robotics technology was designed to perform both automated medium exchange and cell passaging [27]. Online measurements of medium components and simulations of cell-growth kinetics could be used to determine the timing for medium exchanges and to predict cell confluence and scheduling cell passages. Moreover, the automatic and closed environment of the system minimizes the number of potential aseptic handling steps, reducing the number of contaminations as compared to manual performance of the same process (Fig. 42.3). This study demonstrates the significance of monitoring and control for bioreactors discussed earlier in this chapter, as well as the value of implementing innovative technological concepts more conducive to automation. However, it remains questionable whether complex and rather costly robotic systems could actually demonstrate a real cost-benefit by replacing manual cell culture techniques in a manufacturing process.

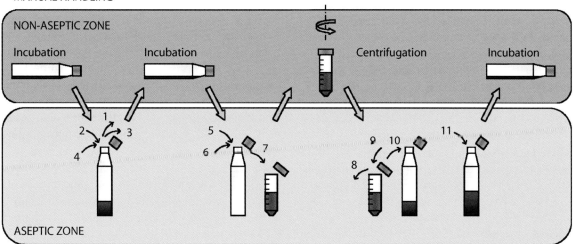

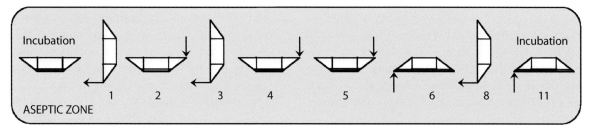

Fig. 42.3 Schematic description of automated and manual operations for cell passaging. A closed bioreactor system with integrated robotics technology was developed to automate and streamline the conventional manual techniques of epithelial cell expansion and cell passaging [27]. Automated cell passaging in the bioreactor was performed through a series of robotic procedures that rotated the growth chamber 90° to various positions and removed and filled the chamber with the various liquid components. The growth chamber had two culture surfaces with different surfaces areas – $S_1 = 6.25$ cm² and $S_2 = 25.0$ cm² for the first and second passages, respectively. A schematic representation of automated and manual operations for cell passaging follows: *1* removal of spent medium; *2* rinse of culture surface with phosphate-buffered saline (PBS); *3* removal of PBS; *4* addition of trypsin solution; *5* addition of trypsin inhibitor solution; *6* addition of fresh medium; *7* transfer of cells; *8* removal of trypsin and trypsin inhibitor solutions; *9* resuspension of cells; *10* inoculation into flask; *11* addition of fresh medium. *Thick grey arrows* indicate vessel transfer between aseptic and nonaseptic environments, and *thin black arrows* indicate liquid handlings for cell passage. This figure was adapted from Kino-Oka et al. 2005 [27]

As an alternative to essentially mimicking established manual procedures, bioreactor systems that implement novel concepts and techniques that streamline the conventional engineering processes will likely have the greatest impact on manufacturing. As opposed to the standard process for cell expansion, in which cells are cultured in a flask until reaching confluence, trypsinized, and replated in multiple flasks, culturing cells on an expandable membrane could minimize or bypass the need for cell passages. Using a bioreactor developed by Cytomec (Spiez, Switzerland), cells can be seeded onto a small circular membrane that can be gradually stretched radially by the bioreactor (similar to the iris of the eye) to continually provide cells with space to grow, without the need for serial passaging [66]. Technology used to engineer cell sheets [71] could also be implemented within a bioreactor design to facilitate the streamlining and automation of the cell-expansion process. Culture surfaces in the bioreactor could be coated with the temperature-responsive polymer poly(N-isopropylacrylamide), which allows for the detachment of cells by simply lowering the temperature, thereby eliminating the need for trypsin and the associated time-consuming processing steps.

42.4.2
Automating Tissue-Culture Processes

While the systems described in Sect. 42.4.1 illustrate various approaches to the automation and streamlining of traditional cell culture processes, these "two-dimensional phases" are typically only the starting point for most tissue-engineering strategies, and represent only one component of the numerous key bioprocesses required to generate three-dimensional tissue grafts. As discussed in Sect. 42.2, the structure, function, and reproducibility of engineered constructs can be dramatically enhanced by employing bioreactor-based strategies to establish, maintain, and possibly physically condition cells within the three-dimensional environment (the "three-dimensional phases"). Therefore, ideally, bioreactor systems would be employed to automate and control the entire manufacturing process (both two- and three-dimensional phases), from cell isolation through to the generation of a suitable graft. The advantages of this comprehensive approach would be manifold. A closed, standardized, and operator-independent system would possess great benefits in terms of safety and regulatory compliance, and despite incurring high product-development costs initially, these systems would have great potential to improve the cost-effectiveness of a manufacturing process, maximizing the potential for large-scale production in the long-term.

Advanced Tissue Sciences (La Jolla, California, USA) was the first tissue-engineering firm to address the issues of automation and scale-up for their production of Dermagraft®, an *allogenic* product manufactured with dermal fibroblasts grown on a scaffold for the treatment of chronic wounds such as diabetic foot ulcers (currently manufactured by Smith and Nephew, London, UK) [32]. Skin grafts were generated in a closed manufacturing system within bioreactor bags inside which cells were seeded onto a scaffold, cell-scaffold constructs were cultivated, cryopreservation was performed, and finally that also served as the transport container in which the generated grafts were shipped to the clinic [41]. Eight grafts could be manufactured within compartments of a single bioreactor bag, and up to 12 bags could be cultured together with automated medium perfusion using a manifold, allowing the scaling of a single production run to 96 tissue grafts. Nevertheless, despite this early effort to automate the tissue engineering process, the production system was not highly controlled and resulted in many batches that were defective, ultimately contributing to the overall high production costs [33]. Considering that significant problems were encountered in the manufacturing of this *allogenic* product, tremendous challenges clearly lie ahead in order to automate and scale the production of *autologous* grafts (technically, biologically, and in terms of regulatory issues), particularly since cells from each patient may be highly variable and cells must be processed as completely independent batches.

A particularly appealing approach to the automation of autologous cell-based product production would be based on a modular design, where the bioprocesses for each single cell source are performed in individual, dedicated, closed-system subunits. In this strategy, a manufacturing process can be scaled-up, or perhaps more appropriately considered "scaled-out" [34], simply by adding more units to the production as product demand increases. This strategy is exemplified by the concept of the Autologous Clinical Tissue Engineering System (ACTES™), previously under development by Millenium Biologix (then in Kingston, Ontario, Canada). As a compact, modular, fully automated, and closed bioreactor system, ACTES would digest a patient's cartilage biopsy specimen, expand the chondrocytes, and provide either an autologous cell suspension, or an osteochondral graft (CartiGraft™) generated by seeding and culturing the cells onto the surface of an osteoconductive porous scaffold. Clearly, full automation of an entire tissue-engineering process possesses the greatest risks upfront, requiring considerable investment costs and significant time to develop a highly technical and complex bioreactor system such as ACTES. In fact, Millenium Biologix was forced to file for bankruptcy in late 2006, and the ACTES system never reached the production stage.

Bioreactor designs could be dramatically simplified, and related development costs significantly reduced, if we reevaluate the conventional tissue-engineering paradigms and could streamline the numerous individual processing steps. A bioreactor-based concept was recently described by Braccini et al. for the engineering of osteoinductive bone grafts [6], which would be particularly appealing to implement in a simple and streamlined manufacturing process. In this approach, cells from a bone marrow aspirate were introduced directly into a perfusion bioreactor, without the conventional phase of selection and cell expansion on plastic dishes; bone marrow stem cells could be seeded and expanded directly within the three-dimensional ceramic scaffold, ultimately producing a highly osteoinductive graft, in a single perfusion bioreactor system. Simplified tissue-engineering processes could be the key to future manufacturing strategies by requiring a minimal number of bioprocesses and unit operations, facilitating simplified bioreactor designs with reduced automation requirements, and permitting compact designs, with the likely result of reduced product development and operating costs.

For the enthusiastic engineer, developing a fully automated and controlled system would probably necessitate state-of-the-art systems to monitor and control a full range of culture parameters, and when possible, to monitor cell behavior and tissue development throughout the production process [33]. Significant benefits would derive from implementing sensing and monitoring devices within the manufacturing system in terms of the traceability and safety of the process itself, features that are crucial to compliantly face current GMP guidelines. However, sensors and control systems will add significant costs to the bioreactor system. Keeping in mind that low-cost bioreactor systems will be required for a cost-effective manufacturing process, it will be imperative to identify the essential process and construct parameters to monitor and control to standardize production and which can provide meaningful quality control and traceability data. In this context, the monitoring and control of bioreactor systems as discussed in Sect. 42.3 will be crucial at the research stage of product development in order to identify these key parameters and to establish standardized production methods.

42.4.3
Production Facilities: Centralized Versus Decentralized

To date, all tissue-engineering products currently on the market have been and continue to be manufactured within centralized production facilities. While manufacturing a product at central locations has the clear advantage of enabling close supervision over the entire production process, this requires establishing and maintaining large and expensive GMP facilities. Unlike the production of other biotech products such as pharmaceuticals, however, critical processes and complicated logistical issues (e.g., packaging, shipping, and tracking of living biopsies and engineered grafts), and the considerable associated expenses, must be considered for the centralized production of engineered tissue grafts.

As an alternative to manufacturing engineered products within main centralized production facilities, a decentralized production system, such as a fully automated closed-bioreactor system (e.g., ACTES), could be located on-site within the confines of a hospital. This would eliminate the complex logistical issues of transferring biopsy samples and engineered products between locations, eliminate the need for large and expensive GMP tissue-engineering facilities, facilitate scale-up, and minimize labor-intensive operator handling. On the other hand, as previously mentioned in the context of fully automated closed bioreactors systems, a decentralized manufacturing strategy will clearly involve the greatest up-front risks in terms of development time and costs.

42.4.4
Intraoperative "Manufacturing" Approaches

During the first two decades of the tissue-engineering field, most research was aimed at the in vitro generation of tissue grafts that resemble the composition and function of native tissues. Trends may be changing. Perhaps due in part to the realization of the current high costs of engineering mature tissue grafts, there is now great emphasis on determining the minimal maturation stage of the graft (i.e., only cells seeded onto a scaffold, cells primed for (re-)differentiation within a scaffold, or a functional graft) that will promote defect repair in vivo (capitalizing on the in vivo "bioreactor"), with the ultimate goal of developing intraoperative therapies. In spite of a potential future paradigm shift, bioprocess engineering will continue to serve numerous vital roles in the tissue-engineering/-regenerative medicine field. Bioreactors could be used to automate the isolation of cells from a biopsy sample for intraoperative cell therapies (e.g., Biosafe from Sepax, Eysins, Switzerland), or to rapidly seed the isolated cells into a three-dimensional scaffold for immediate implantation. Moreover, bioreactors will continue to be critical for in vitro research applications to identify the requirements for the "in vivo bioreactors" [61], and supporting the shift from tissue-engineering approaches to the more challenging field of regenerative medicine.

42.5
Conclusions and Future Perspectives

The ex vivo generation of living tissue grafts has presented new biological and engineering challenges for establishing and maintaining cells in three-dimensional cultures, therefore necessitating the development of new biological models as compared to those long-established for traditional cell culture. In this context, bioreactors represent a key tool in the tissue-engineering field, from the initial phases of basic research through to the final manufacturing of a product for clinical applications.

As we have seen from past and present tissue-engineering manufacturing strategies, manual benchtop-based production systems allowed engineered products to reach the clinic, despite their rather high cost and limitations for potential scale-up. Higher-level technology involves longer development time, increased costs, and the risk of technical difficulties, but on the other hand, maximizes the potential for a safe, standardized, scaleable, and cost-effective manufacturing process. Therefore, fundamental knowledge gained through the use of well-defined and controlled bioreactor systems at the research level will be essential to define, optimize, and moreover, streamline the key processes required for efficient manufacturing models.

The translation of bioreactors initially developed for research applications into controlled and cost-effective commercial manufacturing systems would benefit from collaborations between tissue-engineering firms, academic institutions, and industrial partners with expertise in commercial bioreactor and automation systems. Academic partners would be key to provide the fundamental aspects of the system, while industrial partners could provide essential elements of automation, as well as making the system user-friendly and compliant with regulatory criteria. Working toward this ambitious goal, several multidisciplinary consortia have already been established within Europe (e.g., REMEDI, AUTOBONE, STEPS) to develop automated and scaleable systems and processes to streamline and control the engineering of autologous cell-based grafts, such that the resulting products meet specific regulations and criteria regarding efficacy, safety, and quality, in addition to being cost-effective. Efforts in this direction will help to make tissue-engineered products more clinically accessible and will help to translate tissue-engineering from research-based technology to a competitive commercial field.

References

1. Altman GH, Lu HH, Horan RL, et al (2002) Advanced bioreactor with controlled application of multi-dimensional strain for tissue engineering. J Biomech Eng 124:742–749
2. Bagnaninchi PO, Yang Y, Zghoul N, et al (2007) Chitosan microchannel scaffolds for tendon tissue engineering characterized using optical coherence tomography. Tissue Eng 13:323–331
3. Bancroft GN, Sikavitsas VI, van den DJ, et al (2002) Fluid flow increases mineralized matrix deposition in 3D perfusion culture of marrow stromal osteoblasts in a dose-dependent manner. Proc Natl Acad Sci U S A 99:12600–12605
4. Botta GP, Manley P, Miller S, et al (2006) Real-time assessment of three-dimensional cell aggregation in rotating wall vessel bioreactors in vitro. Nat Protoc 1:2116–2127
5. Boubriak OA, Urban JP, Cui Z (2006) Monitoring of metabolite gradients in tissue-engineered constructs. J R Soc Interface 3:637–648
6. Braccini A, Wendt D, Jaquiery C, et al (2005) Three-dimensional perfusion culture of human bone marrow cells and generation of osteoinductive grafts. Stem Cells 23:1066–1072
7. Brosenitsch TA, Katz DM (2001) Physiological patterns of electrical stimulation can induce neuronal gene expression by activating N-type calcium channels. J Neurosci 21:2571–2579
8. Bueno EM, Laevsky G, Barabino GA (2007) Enhancing cell seeding of scaffolds in tissue engineering through manipulation of hydrodynamic parameters. J Biotechnol 129:516–531
9. Chen HC, Hu YC (2006) Bioreactors for tissue engineering. Biotechnol Lett 28:1415–1423
10. Chen JP, Lin CT (2006) Dynamic seeding and perfusion culture of hepatocytes with galactosylated vegetable sponge in packed-bed bioreactor. J Biosci Bioeng 102:41–45
11. Cioffi M, Boschetti F, Raimondi MT, et al (2006) Modeling evaluation of the fluid-dynamic microenvironment in tissue-engineered constructs: a micro-CT based model. Biotechnol Bioeng 93:500–510
12. Davisson T, Kunig S, Chen A, et al (2002) Static and dynamic compression modulate matrix metabolism in tissue engineered cartilage. J Orthop Res 20:842–848
13. Davisson T, Sah RL, Ratcliffe A (2002) Perfusion increases cell content and matrix synthesis in chondrocyte three-dimensional cultures. Tissue Eng 8:807–816
14. Demarteau O, Wendt D, Braccini A, et al (2003) Dynamic compression of cartilage constructs engineered from expanded human articular chondrocytes. Biochem Biophys Res Commun 310:580–588
15. Dvir T, Benishti N, Shachar M, et al (2006) A novel perfusion bioreactor providing a homogenous milieu for tissue regeneration. Tissue Eng 12:2843–2852
16. Fassnacht D, Portner R (1999) Experimental and theoretical considerations on oxygen supply for animal cell growth in fixed-bed reactors. J Biotechnol 72:169–184
17. Fink C, Ergun S, Kralisch D, et al (2000) Chronic stretch of engineered heart tissue induces hypertrophy and functional improvement. FASEB J 14:669–679
18. Flanagan TC, Cornelissen C, Koch S, et al (2007) The in vitro development of autologous fibrin-based tissue-engineered heart valves through optimised dynamic conditioning. Biomaterials 28:3388–3397
19. Freed LE, Vunjak-Novakovic G (1997) Microgravity tissue engineering. In Vitro Cell Dev Biol Anim 33:381–385
20. Galbusera F, Cioffi M, Raimondi MT, et al (2007) Computational modeling of combined cell population dynamics and oxygen transport in engineered tissue subject to interstitial perfusion. Comput Methods Biomech Biomed Eng 10:279–287
21. Grad S, Lee CR, Gorna K, et al (2005) Surface motion upregulates superficial zone protein and hyaluronan production in chondrocyte-seeded three-dimensional scaffolds. Tissue Eng 11:249–256
22. Hahn MS, McHale MK, Wang E, et al (2007) Physiologic pulsatile flow bioreactor conditioning of poly(ethylene glycol)-based tissue engineered vascular grafts. Ann Biomed Eng 35:190–200
23. Hoerstrup SP, Zund G, Sodian R, et al (2001) Tissue engineering of small caliber vascular grafts. Eur J Cardiothorac Surg 20:164–169
24. Janssen FW, Hofland I, van OA, et al (2006) Online measurement of oxygen consumption by goat bone marrow stromal cells in a combined cell-seeding and proliferation perfusion bioreactor. J Biomed Mater Res A 79:338–348
25. Janssen FW, Oostra J, Oorschot A, et al (2006) A perfusion bioreactor system capable of producing clinically relevant volumes of tissue-engineered bone: in vivo bone formation showing proof of concept. Biomaterials 27:315–323
26. Karim N, Golz K, Bader A (2006) The cardiovascular tissue-reactor: a novel device for the engineering of heart valves. Artif Organs 30:809–814
27. Kino-Oka M, Ogawa N, Umegaki R, et al (2005) Bioreactor design for successive culture of anchorage-dependent cells operated in an automated manner. Tissue Eng 11:535–545
28. Kitagawa T, Yamaoka T, Iwase R, et al (2006) Three-dimensional cell seeding and growth in radial-flow perfusion bioreactor for in vitro tissue reconstruction. Biotechnol Bioeng 93:947–954
29. Knoll A, Scherer T, Poggendorf I, et al (2004) Flexible automation of cell culture and tissue engineering tasks. Biotechnol Prog 20:1825–1835
30. Li Y, Ma T, Kniss DA, et al (2001) Effects of filtration seeding on cell density, spatial distribution, and proliferation in nonwoven fibrous matrices. Biotechnol Prog 17:935–944

31. Mantero S, Sadr N, Riboldi SA, et al (2007) A new electro-mechanical bioreactor for soft tissue engineering. JABB 5:107–116
32. Marston WA, Hanft J, Norwood P, et al (2003) The efficacy and safety of Dermagraft in improving the healing of chronic diabetic foot ulcers: results of a prospective randomized trial. Diabetes Care 26:1701–1705
33. Martin I, Wendt D, Heberer M (2004) The role of bioreactors in tissue engineering. Trends Biotechnol 22:80–86
34. Mason C, Hoare M (2006) Regenerative medicine bioprocessing: the need to learn from the experience of other fields. Regen Med 1:615–623
35. Mason C, Hoare M (2007) Regenerative medicine bioprocessing: building a conceptual framework based on early studies. Tissue Eng 13:301–311
36. Mason C, Markusen JF, Town MA, et al (2004) Doppler optical coherence tomography for measuring flow in engineered tissue. Biosens Bioelectron 20:414–423
37. Mason C, Markusen JF, Town MA, et al (2004) The potential of optical coherence tomography in the engineering of living tissue. Phys Med Biol 49:1097–1115
38. Mauney JR, Sjostorm S, Blumberg J, et al (2004) Mechanical stimulation promotes osteogenic differentiation of human bone marrow stromal cells on 3-D partially demineralized bone scaffolds in vitro. Calcif Tissue Int 74:458–468
39. Mayhew TA, Williams GR, Senica MA, et al (1998) Validation of a quality assurance program for autologous cultured chondrocyte implantation. Tissue Eng 4:325–334
40. Mol A, Driessen NJ, Rutten MC, et al (2005) Tissue engineering of human heart valve leaflets: a novel bioreactor for a strain-based conditioning approach. Ann Biomed Eng 33:1778–1788
41. Naughton GK (2002) From lab bench to market: critical issues in tissue engineering. Ann N Y Acad Sci 961:372–385
42. Niklason LE, Gao J, Abbott WM, et al (1999) Functional arteries grown in vitro. Science 284:489–493
43. Pancrazio JJ, Wang F, Kelley CA (2007) Enabling tools for tissue engineering. Biosens Bioelectron 22:2803–2811
44. Pedrotty DM, Koh J, Davis BH, et al (2005) Engineering skeletal myoblasts: roles of three-dimensional culture and electrical stimulation. Am J Physiol Heart Circ Physiol 288:H1620–H1626
45. Porter B, Zauel R, Stockman H, et al (2005) 3-D computational modeling of media flow through scaffolds in a perfusion bioreactor. J Biomech 38:543–549
46. Porter BD, Lin AS, Peister A, et al (2007) Noninvasive image analysis of 3D construct mineralization in a perfusion bioreactor. Biomaterials 28:2525–2533
47. Portner R, Nagel-Heyer S, Goepfert C, et al (2005) Bioreactor design for tissue engineering. J Biosci Bioeng 100:235–245
48. Powell CA, Smiley BL, Mills J, et al (2002) Mechanical stimulation improves tissue-engineered human skeletal muscle. Am J Physiol Cell Physiol 283:C1557–C1565
49. Prenosil JE, Kino-Oka M (1999) Computer controlled bioreactor for large-scale production of cultured skin grafts. Ann N Y Acad Sci 875:386–397
50. Radisic M, Euloth M, Yang L, et al (2003) High-density seeding of myocyte cells for cardiac tissue engineering. Biotechnol Bioeng 82:403–414
51. Radisic M, Park H, Shing H, et al (2004) Functional assembly of engineered myocardium by electrical stimulation of cardiac myocytes cultured on scaffolds. Proc Natl Acad Sci U S A 101:18129–18134
52. Radisic M, Yang L, Boublik J, et al (2004) Medium perfusion enables engineering of compact and contractile cardiac tissue. Am J Physiol Heart Circ Physiol 286:H507–H516
53. Raimondi MT, Moretti M, Cioffi M, et al (2006) The effect of hydrodynamic shear on 3D engineered chondrocyte systems subject to direct perfusion. Biorheology 43:215–222
54. Rolfe P (2006) Sensing in tissue bioreactors. Meas Sci Technol 17:578–583
55. Scapinelli R, Aglietti P, Baldovin M, et al (2002) Biologic resurfacing of the patella: current status. Clin Sports Med 21:547–573
56. Shangkai C, Naohide T, Koji Y, et al (2007) Transplantation of allogeneic chondrocytes cultured in fibroin sponge and stirring chamber to promote cartilage regeneration. Tissue Eng 13:483–492
57. Sikavitsas VI, Bancroft GN, Lemoine JJ, et al (2005) Flow perfusion enhances the calcified matrix deposition of marrow stromal cells in biodegradable nonwoven fiber mesh scaffolds. Ann Biomed Eng 33:63–70
58. Sodian R, Lemke T, Fritsche C, et al (2002) Tissue-engineering bioreactors: a new combined cell-seeding and perfusion system for vascular tissue engineering. Tissue Eng 8:863–870
59. Starly B, Choubey A (2007) Enabling sensor technologies for the quantitative evaluation of engineered tissue. Ann Biomed Eng 36:30–40
60. Stephens JS, Cooper JA, Phelan FR Jr, et al (2007) Perfusion flow bioreactor for 3D in situ imaging: investigating cell/biomaterials interactions. Biotechnol Bioeng 97:952–961
61. Stevens MM, Marini RP, Schaefer D, et al (2005) In vivo engineering of organs: the bone bioreactor. Proc Natl Acad Sci U S A 102:11450–11455
62. Sud D, Mehta G, Mehta K, et al (2006) Optical imaging in microfluidic bioreactors enables oxygen monitoring for continuous cell culture. J Biomed Opt 11:050504
63. Sun T, Norton D, Haycock JW, et al (2005) Development of a closed bioreactor system for culture of tissue-engineered skin at an air–liquid interface. Tissue Eng 11:1824–1831
64. Thompson CA, Colon-Hernandez P, Pomerantseva I, et al (2002) A novel pulsatile, laminar flow bioreactor for the development of tissue-engineered vascular structures. Tissue Eng 8:1083–1088
65. Timmins NE, Scherberich A, Fruh JA, et al (2007) Three-dimensional cell culture and tissue engineering in a T-CUP (tissue culture under perfusion). Tissue Eng 13:2021–2028
66. Vonwil D, Barbero A, Quinn T, et al (2007) Expansion of adult human chondrocytes on an extendable surface:

a strategy to reduce passageing-related dedifferentiation. Eur Cell Mater 13:17
67. Vunjak-Novakovic G, Martin I, Obradovic B, et al (1999) Bioreactor cultivation conditions modulate the composition and mechanical properties of tissue-engineered cartilage. J Orthop Res 17:130–138
68. Wendt D, Marsano A, Jakob M, et al (2003) Oscillating perfusion of cell suspensions through three-dimensional scaffolds enhances cell seeding efficiency and uniformity. Biotechnol Bioeng 84:205–214
69. Wendt D, Stroebel S, Jakob M, et al (2006) Uniform tissues engineered by seeding and culturing cells in 3D scaffolds under perfusion at defined oxygen tensions. Biorheology 43:481–488
70. Wernike E, Li Z, Alini M, et al (2007) Effect of reduced oxygen tension and long-term mechanical stimulation on chondrocyte-polymer constructs. Cell Tissue Res 331:473–483
71. Yang J, Yamato M, Shimizu T, et al (2007) Reconstruction of functional tissues with cell sheet engineering. Biomaterials 28:5033–5043
72. Zhao F, Ma T (2005) Perfusion bioreactor system for human mesenchymal stem cell tissue engineering: dynamic cell seeding and construct development. Biotechnol Bioeng 91:482–493

The Evolution of Cell Printing

B. R. Ringeisen, C. M. Othon, J. A. Barron, P. K. Wu, B. J. Spargo

Contents

43.1	Introduction	613
43.2	Approaches to Cell Printing	614
43.2.1	Jet-Based Methods	615
43.2.2	Extrusion Methods	624
43.3	Compare and Contrast Cell-Printing Methodologies	625
43.4	Conclusion	628
	References	629

43.1 Introduction

Tissue and organs are highly complex systems with innate heterogeneous components, each with their own structure and function. Many facets of this structure have micron-scale features (e.g., capillaries, sinuses, cell–cell contacts, extracellular matrix). Traditional approaches in tissue engineering and regenerative medicine attempt to recreate this structure and function in vitro by randomly seeding cells onto 3D scaffolds. These 3D scaffolds provide the structure and an environment for seeded cells to differentiate into tissue-like materials [17, 67, 82]. The structure and function of natural tissue is replicated by using either multifunction stem cells (e.g., pluripotent, mesenchymal, embryonic) or highly sophisticated scaffolds (e.g., micro-/nanostructured, chemically/biologically functionalized), or both [14, 24, 36, 37, 41–43, 49, 52, 71, 78, 89]. There are tissue-engineering success stories, but most are for simple, homogeneous systems such as skin and thin membrane (bladder) replacements. More complex tissues and organs have eluded researchers to date, mainly due to three problems with current tissue-engineering approaches:

1. The inability to achieve adequate nutrient and waste diffusion (inadequate vascularization of 3D constructs).
2. The inability to replicate the heterogeneous (multicomponent with different molecules and cells) nature of tissues/organs on a relevant scale (10's to 100's of micrometers).
3. Achieving proper cell differentiation in vitro.

Cell printing, or organ printing[1] as it is sometimes referred to in both popularized and research literature [11, 46, 47, 83], is the ability to "drop-and-place" viable mammalian cells into organized 2D and 3D patterns [5–9, 13, 22, 23, 53, 54, 63]. If cell printing could successfully create 3D, heterogeneous patterns of cells and proteins, it could provide a unique solution to many of the roadblocks that are preventing significant progress in tissue engineering. At the turn

---

[1] We will refer to these approaches as "cell printing" rather than "organ printing" because in our opinion it more accurately represents the actual scientific methodology. Not surprisingly, a simple Web search (google.com) reveals that "organ printing" is referenced nearly 20 times more often than "cell printing" (as of 7.30.2007, 11,100 vs. 610 hits), indicating a great desire by the public and media to possess this science-fiction-sounding capability.

of the millennium, cell printing was in its infancy, with only one manuscript detailing the ability to "direct write" mammalian cells. Even at this point in time, the authors of this first manuscript had the foresight to imagine the power of cell printing if mastered: "Potentially, multiple cell types can be placed at arbitrary positions with micrometer precision in an attempt to recapitulate the complex 3D cellular organization of native tissues" [53].

A few short years later, these small steps had provided the foundation for a vibrant research community with diverse approaches and experiments being performed all over the world [64]. The field of cell printing is still quite young, but progress is quickly being made. However, news outlets have begun to tout cell-printing achievements as revolutionary advances in medicine. For example, *The New Scientist* elaborately described how laser-based cell printers may one day be used to "create joints and organs" [91]. Ink-jet cell printers have been glorified in the media, even being highlighted on the Discovery Channel in its "2057" program focusing on what life may be like 50 years from now. The television program advertises that cell printers could be used to "custom-build organs from scratch" (discovery.com). ABC News has reported that organ printing "could have the same impact as Guttenberg's press" (ABCNews.go.com). Thus, there is now quite a large, worldwide expectation that cell printers will deliver organs to you when yours fail. This "buzz" is creating unrealistic expectations that are not yet backed by adequate scientific findings.

First of all, there has not been a single in vivo experiment (animal implantation study) published in the literature using a cell-printed scaffold. What is available is a significant base of scientific proof, or even better stated, "proof-of-principal" that cell printing works, AND that it works in more than one way [64]. It works with laser pulses, with ink jets, with extrusion micropens, and with electrohydrodynamic jetting (EHDJ) nozzles. We have an elaborate amount of experimentation detailed in the literature that backs the premise of printing patterns of living, functional cells.

This chapter will summarize the cell-printing literature that has been published over the past 8 years. We will then compare and contrast each technique in terms of its proven print specifications (e.g., speed, resolution) and the ability to print cells with high viability and functionality. The conclusion will present a realistic future of cell printing with a potential pathway that could achieve true scientific and medical breakthroughs.

43.2 Approaches to Cell Printing

The basic concept of "building" tissue or organs layer-by-layer from individual components has not changed from the first cell-printing publications [53, 54]. However, not all cell printers can go about this process in the same fashion. Specifically, there are two general approaches, termed "structural" and "conformal" cell printing, that can be used to print tissue constructs [64]. Both approaches result in constructs that are fabricated from the bottom up (i.e., layer-by-layer or cell-by-cell) and have heterogeneous cell and biomolecular structure in three dimensions. Structural cell printing requires that the same tool print the scaffolding, cells, and biomolecules simultaneously or sequentially. Conformal cell printing is a hybrid approach that prints cells and biomolecules into thin layers of prefabricated scaffolding (Fig. 43.1).

The concept of structural cell printing is enticing because the printer is the only tool necessary to build the entire 3D cell construct. However, there are two potential problems with this concept. The most daunting limitation of structural cell printing is in the choice of scaffolding materials available to be printed. Many of the most sophisticated tissue scaffold materials require harsh solvents (e.g., acids, organics) and complex preparations, and are therefore unfit to be used simultaneously as a matrix for living cells. Therefore, a structural cell printer will most likely use thermoreversible gels or hydrogels that can be printed through nozzles or print-heads in liquid form and then gel into more structured materials post-printing. Unfortunately, many of these materials lack rigidity and structural integrity, limiting the types of applications the printed cell constructs could be used in (e.g., soft tissues) [1, 29, 35]. Scaffold materials for structural printers should also be chosen that enable melding of individual printed tubes or droplets into a contiguous, uniform, and adherent 3D structure.

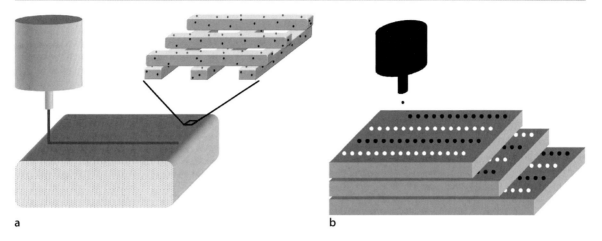

Fig. 43.1a,b Schematic representations of "structural" (**a**) and "conformal" (**b**) cell printing (Copyright Wiley-VCH Verlag GmbH & Co. KGaA. Reproduced with permission)

The second issue that needs to be resolved for structural cell printing is the competition between deposition speed and resolution. Structural cell printers would most likely require relatively large pixel (or droplet) resolutions on the scale of 0.3–1 mm to enable deposition of enough cells and materials to create a freestanding, macroscopic scaffold. For some simple, homogeneous tissues such as the bladder or cartilage, this resolution may be adequate. However, for more sophisticated tissues and organs such as the liver and kidney, cell and structural heterogeneity is present on the 0.01-mm scale, making it difficult to envision structural cell printing being used for these tissues.

Conformal cell printing is a hybrid cell-printing approach that builds upon previously developed scaffolds (Fig. 43.1b) by printing cells to thin layers of prefabricated scaffolds [64]. Even though conformal cell printing requires more equipment than the cell printer alone, there is the potential to use many more types of scaffolding materials. Specifically, because the scaffold does not need to be printed, the pre- and postprocessing steps to form the material would not need to be conducive to printing or cell viability (prefabricated layers of scaffolding). It would therefore be possible to use the highly sophisticated scaffolds currently being used by traditional tissue-engineering approaches [14, 36, 37, 41–43, 71, 78]. Conformal cell printing may also have the advantage of higher resolution, at least initially, based on published reports of laser-based and piezo-tip inkjet conformal printers that have achieved single-cell resolution [9, 51, 68]. Conversely, these approaches require more than just a cell printer in the experimental design (e.g., scaffold slicer) and face potential problems with overall structural integrity. However, there is a literature example of scaffolds that can be sliced into discrete layers (similar to tissue sections) [57], and surgical or natural adhesives could be externally added or printed to help interconnect each layer, enhancing the structural integrity of the finished construct. Overall, both structural and conformal cell printing could theoretically be used to form a 3D tissue construct with cellular and biomolecular heterogeneity (cell/molecular patterns, cell/molecular diversity), mimicking the structure of natural tissue or organs.

43.2.1
Jet-Based Methods

Jet-based cell printers utilize an energy source (e.g., laser, thermal, vibration, voltage) to create a fluid jet that carries cells from a liquid reservoir to a piece of "biopaper." The biopaper usually consists of a moist environment such as a hydrogel or wetted polymer scaffold so that the droplet of printed cell solution does not dry and immediately damage the printed cells. When the jetting apparatus and the biopaper are on high-speed positioning stages, cell patterns can be printed just as a desktop printer would print ink patterns on a sheet of regular paper. Jet-based methods are distinct from extrusion methods (Sect. 43.2.2) in that the velocity of the cells during transit from the reservoir to the biopaper is significantly higher (≥ 10 m/s vs. < 1 m/s). There are currently four jet-based cell printers that have successfully printed viable mammalian cells: laser-guided direct write (LGDW), modified laser-induced forward transfer (LIFT), modified ink jet, and EHDJ.

43.2.1.1
Laser-Guided Direct Write

LGDW was the first method used to print viable mammalian cells, and it did so by utilizing radiation pressure to direct cell deposition with micrometer resolution [53]. The radiation pressure arises from the scattering of energetic photons off the surface of the cell in a laser beam. If the index of refraction of the cell is larger than that of the surrounding medium, it experiences both a radial force pulling the cell toward the center of the laser beam and an axial force pushing the cell in the direction of the propagation of the light [2]. LGDW uses a low-numerical-aperture lens to weakly focus the laser light, resulting in cell passage over large distances, up to 7 mm [54].

For cell printing, LGDW consists of a weakly focused laser beam, cell suspension (either liquid or aerosol) to be deposited, and a computer-controlled (computer-aided design/computer-aided manufacturing, CAD/CAM) substrate (biopaper), as illustrated in Fig. 43.2. For biological suspensions, a laser with a wavelength in the near-infrared part of the spectrum is chosen because it has been found to cause the least damage to living cells [34, 39]. The incident light focuses and directs the particles onto the surface of the receiving substrate. To minimize the effects of the fluid motion of the suspension medium, a hollow optical fiber may be employed to separate the suspension medium from the deposited target material [53]. In this configuration, cells are forced into a hollow optical fiber and directed toward the receiving substrate.

The first LGDW cell-printing experiments were performed on embryonic-chick spinal-cord cells, which were jetted out of the narrow fiber's orifice (Fig. 43.2b, 30–100 μm inside diameter of fiber) [53, 54]. This experiment resulted in continuous direct writing of tens of cells deposited over a ~ 30-min period (Fig. 43.3). A total of 26 clusters of cells were

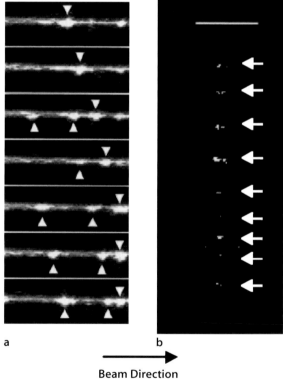

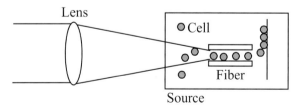

Fig. 43.2 Laser-guided direct write apparatus when configured to perform cell printing experiments (Copyright Wiley-VCH Verlag GmbH & Co. KGaA. Reproduced with permission)

Fig. 43.3a,b The first example of cell printing. **a** Embryonic chick spinal cord cells being transported down a hollow optical fiber. **b** Clusters of cells residing on a bare glass slide after being printed through the hollow fiber (Copyright Wiley-VCH Verlag GmbH & Co. KGaA. Reproduced with permission)

printed onto a glass cover slip with an average spacing of ~40 μm between the spots. Many of the areas had multiple cells, up to a maximum of 5 cells/spot. However, some of the cell clusters did not adhere to the glass slide, suggesting that cell–surface adhesion did not regularly occur. This result is not surprising because both liquid media and cells were printed to bare glass with no functional coating to enhance cell adhesion to the surface. Without such preconditioning, it is unlikely that the printed cells would remain localized in the printed regions, resulting in the observed pattern loss. However, 60% of the printed clusters were visible after the experiment, and limited 3D capabilities were demonstrated by stacking two cells on top of one another. Limited additional results examined cell viability and functionality (neurite outgrowth). However, no specific viability or functionality assays were performed, so quantitative percentages were not reported.

One possible limitation to LGDW cell printing is low cell throughput (2.5 cells/min). Several methods to potentially enhance print speeds and cell throughput are presented in these articles, ranging from using higher cell density to using a more intense laser. However, these solutions may result in a higher occurrence of clogging or an increased potential for cell damage. More recent work has reported utilizing LGDW to print viable embryonic chick forebrain neurons, but it is yet to be seen whether this technique can print cells fast enough to become relevant to the tissue-engineering community [21].

43.2.1.2
Modified Laser-Induced Forward Transfer

Successful biological printing experiments using modified LIFT techniques have been reported by several different research groups [6, 8, 15, 16, 19, 22, 23, 63]. All modified LIFT cell printers are unique because they rely upon a focused laser beam to achieve micron-scale resolution, rather than a capillary or small-diameter orifice (e.g., print-head, syringe). This characteristic enables cells to be printed without the clogging concerns or viscosity/material constraints that may be placed on orifice-based techniques. The cell reservoir for modified LIFT printers is a thin layer of cells or cell solution on a transparent support, usually quartz or glass. This reservoir is inverted, loaded into a CAD/CAM controlled system, and placed in close proximity (sandwich) to a section of biopaper that will receive the printed cells (Fig. 43.4). The cell-printing process is

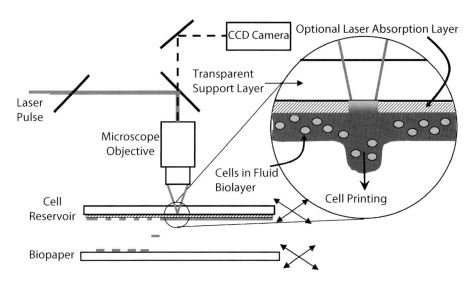

Fig. 43.4 Schematic of modified laser-induced forward transfer (LIFT) cell printers with an optional laser absorption layer between the transparent support and the cell suspension

photo-initiated, with a laser pulse focused through the transparent backing of the cell reservoir. Each laser pulse results in a droplet of cells being jetted away from the cell reservoir and then deposited onto the biopaper. There are two mechanisms reported in the literature: (1) light absorption by and subsequent ablation of a matrix material (e.g., hydrogel, aqueous solution) that surrounds the cells in the reservoir, or (2) light absorption by an energy conversion layer that is located between the transparent backing and the cell solution in the reservoir. The first method is commonly referred to as matrix-assisted pulsed laser evaporation direct write, or matrix assisted pulsed laser direct write (MAPLE DW). The second method is referred to as biological laser printing (BioLP) or absorbing film-assisted LIFT. Each method has a significant publication record that details successful printing experiments with high percentages of viable and functional cells.

The first biological laser-based printing experiments were performed using MAPLE DW to deposit micron-scale patterns of active proteins and viable bacteria [60, 61]. Shortly after this, additional experiments demonstrated that mammalian cells could be printed to hydrogel surfaces using the same technique [63]. In all cases, a matrix material, such as cell media or a hydrogel, was used in the cell reservoir (with the cells) to absorb the incident laser pulse and initiate the printing. It is important to note that even though the printing mechanism is initiated by ablation (matrix evaporation), bulk *fluid* is transferred from the cell reservoir to the biopaper in all cases [4, 23, 90]. Several years later, researchers determined that a significant portion of the incident ultraviolet laser light is not absorbed by the matrix materials, thereby exposing the active biological elements to potentially damaging elements [8]. To prohibit ultraviolet exposure to the cells and to enhance the reproducibility of the laser–material interaction, a laser absorption layer was added between the transparent support and the cell suspension. This new approach was referred to as BioLP, and much larger and more reproducible arrays of proteins and living mammalian cells have been reported when using this technique [4, 7–9, 13].

Examples of modified LIFT (MAPLE DW and BioLP) cell printing are shown in Fig. 43.5. Panels a–c are fluorescence micrographs of a live/dead assay performed on MAPLE-DW-printed mouse pluripotent embryonal carcinoma cells [63]. Panel a shows a micrograph of cells printed to a dry, bare, glass surface. The orange fluorescence in the micrograph resulted from exposure to a membrane-exclusion dye and indicates that essentially all cell membranes were compromised under these conditions. The lysing was most likely due to shear stress and/or drying during the printing process. When the biopaper was modified with a thin layer (20–50 μm thick) of basement membrane hydrogel (Matrigel), cells survived (green fluorescence is due to the enzymatic hydrolysis of calcein AM to calcein and indicates live cells) and proliferated postprinting (Fig. 43.5b, c). By placing a hydrogel layer on the biopaper, both the drying effects and shear forces were counteracted. The percentage of viable cells is shown to be proportional to the thickness of the hydrogel layer, with approximately 50% viability for a 20-μm-thick hydrogel and near-100% viability for layers greater than 40 μm thick.

Figure 43.5d is a fluorescence micrograph of an array of human osteosarcoma cells printed by BioLP [8]. A 7×4 array was printed to a 200-μm-thick hydrogel scaffolding layer from a Dulbecco's modified essential medium-based "bioink" with approximately 1.0×10^8 cells/ml [8]. A live/dead stain was exposed to the biopaper 24 h postprinting to demonstrate near-100% viability of the osteosarcoma cells. There is a slight distribution in terms of number of cells per spot in the cell array (3–10 cells/spot). This distribution arises from the statistical sampling that occurs during the printing process and does not reflect the much lower spot-to-spot reproducibility of the printer (±3%). It is important to note that almost all cell printers that work with low cell counts per spot will battle this statistical sampling error unless the cells can be selected during the printing process. Because modified LIFT techniques use optical means to print cells, it is possible that this approach could be used to preselect the same number of cells to be printed with each laser pulse. However, such sampling would greatly slow the printing process and has yet to be demonstrated. Several other cell-printing experiments have been demonstrated by BioLP. These experiments include showing near-100% viability postprinting for many cell lines (both normal and carcinoma), single-cell printing, the use of several different laser absorption layers, and achieving adjacent patterns of different cell types with resolution similar to the heterogeneous structure of natural tissue (10's of micrometers) [6–9, 13, 62].

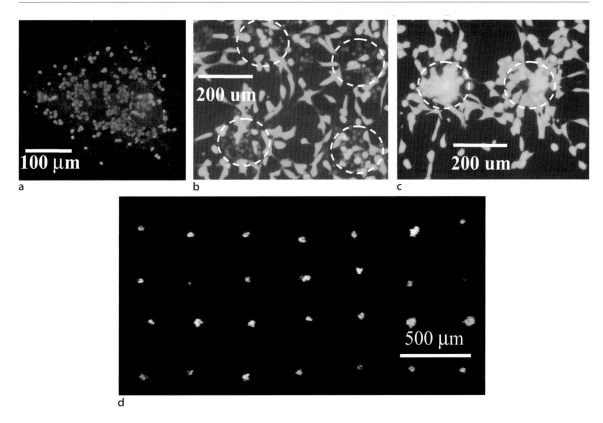

Fig. 43.5a–d Cell viability via modified LIFT printing approaches. Each fluorescence micrograph shows a live/dead assay 24 h post-printing. **a** Matrix-assisted pulsed laser direct write (MAPLE DW) cells printed to bare glass slide do not survive (*green fluorescence* = viable or alive; *orange fluorescence* = no viability or dead). **b** Viability of 50% when cells were printed to 20-μm-thick hydrogel. **c** Near-100% viability when cells were printed to 40-μm-thick hydrogel. **d** Biological laser printing (BioLP) enables large and reproducible arrays of viable cells to be printed with sub-100-μm-spot resolution (reproduced with kind permission of Springer Science and Business Media)

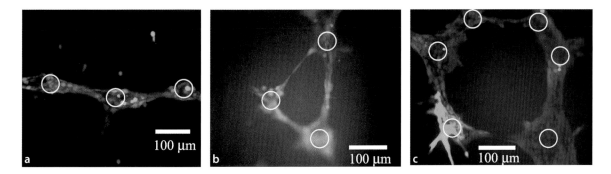

Fig. 43.6a–c Fluorescence micrographs (live/dead) of bovine aortic endothelial cells printed viably by BioLP into a hydrogel layer. The pictures show sub-100-μm resolution and directed growth and partial differentiation of the endothelial cells

BioLP has also printed bovine aortic endothelial cells (BAECs) onto a hydrogel surface where apparent cell–cell signaling and cell differentiation transformed the initially unconnected pattern (individual, separated spots highlighted by dashed circles) into an interconnected line (Fig. 43.6a), triangle (Fig. 43.6b), and circle (Fig. 43.6c) [13]. This asymmetric growth is an indication that the cells retained their functionality through the printing process and demonstrates how cell printing can be used to begin forming vessel-like structures. The cell–cell signaling, surface migration, and cell differentiation that is shown in Fig. 43.6 is essential if cell-printed scaffolds are going to grow into functional tissues.

BioLP was also the first cell-printing technique to demonstrate conformal cell printing. As shown in Fig. 43.7, BioLP was used to print viable osteosarcoma cells throughout a prefabricated 3D hydrogel via a layer-by-layer approach. Each panel shows a fluorescence micrograph (live/dead assay) of a different layer of cell-printed biopaper (hydrogel) [8]. After each layer of biopaper was spread, BioLP was then used to print high-resolution cell spots, where each layer was ~75 µm thick with a cell pattern resolution of ~100 µm. These images demonstrate one of the ultimate goals of cell printing, precise and controlled cell seeding throughout a 3D hydrogel scaffold.

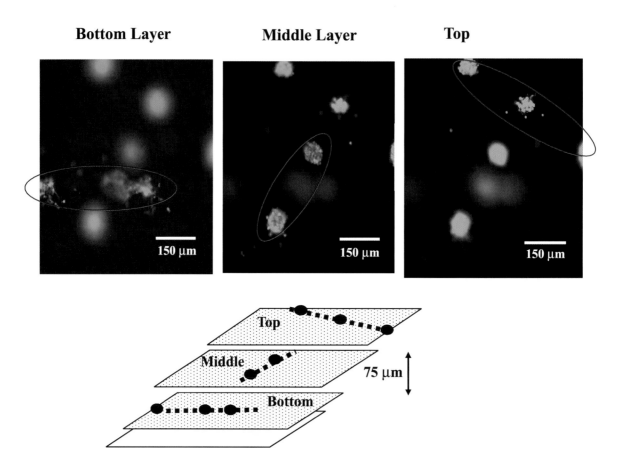

Fig. 43.7 An example of using BioLP to achieve high-resolution, 3D cell printing. A three-layer pattern of viable osteosarcoma cells was printed into a hydrogel scaffold

43.2.1.3
Modified Ink Jet

Both thermal and piezoelectric ink-jet printers have been modified to enable cell printing [30, 45, 47, 48, 50, 51, 64, 65, 68–70, 81, 85]. In order to be used for cell printing, the thermal or piezo-tip print-heads and ink cartridges are modified to allow bioinks to be printed. Examples of cell bioinks are thermoreversible polymers, various hydrogels, or collagen [47, 58, 69, 85]. The approach to printing cells is similar to what is shown schematically in Fig. 43.1, where the apparatus consists of a print-head or nozzle, an ink cartridge/reservoir, and a computer-controlled receiving substrate. The ink-jet print-heads energize the transfer of bioink via thermal (heating/evaporative) or piezoelectric (vibration) mechanisms and are used for both parallel (multiple heads) and serial (single head) deposition.

Thermal Ink Jet

The first viable cells printed by an ink-jet device used a modified thermal print-head [11, 46, 83]. Ink cartridges were washed and filled with cell media and suspended BAECs (1×10^5 cells/ml). Live/dead assays were performed on droplets of cell solution (150 μm diameter, 15 nl, ~1.5 cells/drop) printed to Matrigel, and lines of printed cells were shown to be viable and adherent to the hydrogel surface up to 3 days postprinting. Additional experiments utilized a thermal ink-jet printer to form multiple layers of BAEC aggregates in both a thermoreversible gel and collagen type I gel [11, 46]. Layers of gel and cells were printed serially, yielding cellular structures that consisted of fused cell aggregates. Aggregates were approximately 670 μm in diameter. After 24 h the cells appeared to fuse with other aggregates printed in close proximity. The thermoreversible gel was chosen to pursue the concept of "structural" cell printing, as both the gel and cells were printed simultaneously and the cells were compatible with the gelling process. Specifically, when held at room temperature, the gel precursor is fluid-like and allows functional printing, while gelling is initiated on the biopaper by warming to 37°C.

Thermal ink-jet printing has since been used for the successful deposition of several other cell types including primary cortical neurons [85, 86]. Figure 43.8a is an optical micrograph showing a printed ring pattern of primary cortical neurons onto a collagen hydrogel biopaper [12, 85]. The line-width of the printed ring was approximately 0.6 mm with an inside diameter of ~1.5 mm (Fig. 43.8a). Many dead cells were observed in the printed neuronal ring, but a few neurons had extended their processes at day 3 postprinting (Fig. 43.8b). By day 7 (Fig. 43.8c), some processes had extended to 40 μm in length. All

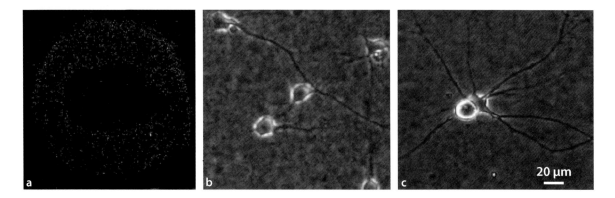

Fig. 43.8 a Thermal ink jet printing of embryonic motoneuron cells onto a pre-fabricated collagen hydrogel layer. **b** Optical micrograph showing extension of processes 13 days postprinting. **c** Higher-magnification of a thermal ink-jet-printed embryonic motoneuron cell at 7 days postprinting

results to date using a thermal ink-jet printer to deposit cell patterns have yielded a wide line resolution of between 300 and 600 µm. This width seems to be characteristic of thermal ink jets (many similar ring structures published in the literature with almost identical dimensions and numbers of cells), but may present the advantage of yielding high-throughput cell printing. Published estimates indicate that 100,000 cells could potentially be printed per second by using multiple print-heads [86].

Printed cortical neurons were then analyzed for retained functionality by utilizing both immunocytochemical staining and patch-clamp measurements. Figure 43.9a shows the measured electrical firings of printed cortical neurons from patch-clamp experiments, demonstrating a similar firing pattern as nonprinted cells [12]. Figure 43.9b is a confocal fluorescence micrograph of a printed neuron after exposure to fluorescence-labeled antibodies for microtubular-associated-protein-2, which is often present in extended processes as a signal of successful neuron differentiation. These experiments are a good indication that the cortical cells that survive the printing process (~75% viability) retain their ability to form active and functional processes.

A statistically relevant number of cells are found to be lysed during the thermal ink-jet printing process (~3–10%), and the cytotoxicity of the bioink, usually a modified Dulbecco's phosphate buffer saline solution, is also significant at approximately 15% [85]. The cytotoxicity of the printing solution is a difficult problem to overcome in all cell-printing experiments. It is generally accepted that utilizing suspended cells in printing experiments enhances a cell's ability to resist lysing during printing (i.e., the spherical nature of cells, as compared to adherent cells with pseudopodia extension, enhances resiliency to various printing stressors such as, for example, shear and heat). However, maintaining adherent cells in a spherical state for long periods of time (i.e., in the cell reservoir) is often difficult, making rapid deposition or convenient bioink replenishment a priority to realize successful cell printing.

Recent thermal ink-jet cell-printing experiments have moved away from printing cells and scaffolding together in lieu of a two-step approach: depositing scaffold first followed by cell printing. NT2 neural cell structures have been printed into 3D scaffolds via this two-step thermal ink-jet printing process [85]. The first step was to plate a thin layer of fibrinogen onto the receiving substrate. Thrombin was then printed on top of the fibrinogen to make a fibrin gel. The gelling process required a few minutes, and neurons were then printed onto the gelled surface. This procedure was repeated to form a 3D neural cell construct.

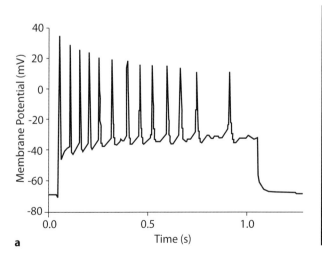 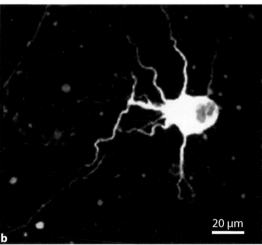

Fig. 43.9 a Patch-clamp measurements on thermal-ink-jet-printed motoneuron cells. **b** Confocal microscopy image of thermal-ink-jet-printed motoneurons stained with anti- microtubular associated-protein-2 monoclonal antibodies 15 days postprinting

Piezoelectric Ink Jet

Viable cells have also been printed using piezo-tip ink-jet printers, although less information is available in the literature [51, 68–70]. Several variations of piezoelectric jetting nozzles have been used to print cell suspensions, making the summary of cell-printing accomplishments difficult. In order to prevent clogging of these small capillaries, the cell concentration used in these experiments is often lower (10^4–10^6 cells/ml) than for thermal ink jet (10^7 cells/ml) and laser-based techniques (up to 10^8 cells/ml). This lower cell concentration results in low overall cell throughput. In general, this approach to cell printing relies upon high droplet print rates of between 1000 and 10,000 drops per second to enable reasonable numbers of cells to be deposited. In extreme cases where small volume drops are printed (~10 pl), as few as 1 drop in 10,000 contains a cell [69, 70]. Other researchers have demonstrated one to four cells per printed drop (~1–10 nl/drop), but cell viability was not adequately addressed [51]. Also, in order to achieve this increased cell throughput, larger volumes per drop were printed, resulting in a compromise in spot resolution (>200 μm diameters compared to ~10 μm for regular ink). Some cell viability postprinting has been demonstrated by observing cell adherence and proliferation, but no quantitative measurement of cell viability has been published.

More recent publications indicate a renewed interest in piezoelectric ink-jet printing for depositing patterns of bacteria [20, 44, 73]. Because the cell diameters of bacteria are almost ten times smaller than eukaryotic cells, there would be a much lower risk of clogging, enabling higher cell concentrations to be used. It is possible that piezoelectric ink-jet printing may not play a pivotal role in mammalian cell printing and tissue engineering based purely on these cell throughput issues.

43.2.1.4 Electrohydrodynamic Jetting or Bio-Electrospray

There has been a recent explosion in publications that use EHDJ or bio-electrospray nozzles to deposit living cells onto surfaces [18, 25, 27, 28, 55, 77]. However, many of these publications present very similar data and will therefore be briefly summarized together here. The basic components of this cell-printing approach are a syringe pump that supplies cell solution to a ~500 μm internal diameter needle/nozzle, which is connected to a power supply. One example of the nozzles used for EHDJ cell deposition is shown in Fig. 43.10a. Jetting of cell solution is achieved by raising the electric field strength to approximately 0.5 kV/mm between the nozzle and ground electrode attached to the receiving substrate several millimeters from the nozzle. Viable cells have been printed to various culture vessels including glass slides and nylon membranes.

The best example of EHDJ cell printing was performed on Jurkat cells [28], although mouse neuronal cells and human glial cells have also been deposited [18, 26]. Fig. 43.10b and c show two optical micrographs: control Jurkat cells that were deposited through the nozzle with no applied voltage, and printed Jurkat cells passed through the needle with applied voltage and deposited onto the receiving substrate, respectively. The jetted cells appear undamaged postprinting. Cell growth was observed postprinting and no change in the growth curve was observed when compared to control cells.

There has been no demonstration in the literature to suggest that EHDJ can be used to create a controlled cell pattern. This technique is known to produce a wide array of spot sizes ranging from hundreds of microns to over millimeters in diameter for the same experiment [18]. Jayasinghe et al. report that stable jetting can increase the resolution and reproducibility of EHDJ [26]; however, the results presented in this manuscript contradict the claim that sub-300-μm spots can be deposited as there is no coherent or reproducible pattern demonstrated. It is difficult to imagine that this approach would ever be useful for high-resolution cell printing, as reproducibility and reduced spot size are an absolute requirement to create cell scaffolds that contain heterogeneous components on the scale of living tissue (<100 μm).

EHDJ does present the advantage of a continuous stream of cell solution, and researchers are attempting to take advantage of this by "electrospinning" cells into polymer threads [77]. This aspect of EHDJ is unique and could present interesting advantages in the field of tissue engineering, especially the concept of creating encapsulated multicomponent threads.

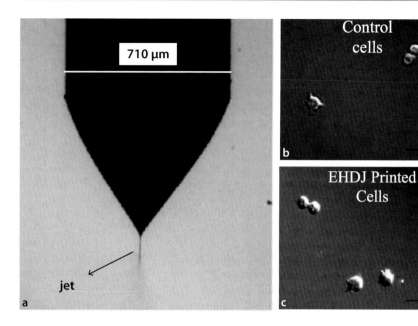

Fig. 43.10 a Electrohydrodynamic jetting nozzle used for cell deposition/printing. b Control Jurkat cells that have not been printed via electrohydrodynamic jetting (EHDJ). c Jurkat cells printed via EHDJ 250 min postprinting (Copyright Wiley-VCH Verlag GmbH & Co. KGaA. Reproduced with permission)

43.2.2
Extrusion Methods

Two recent articles have reviewed developments in rapid prototyping tools for printing 3D tissue scaffolds [74, 88]. Some of these approaches are also referred to as solid freeform fabrication tools. In all cases, these techniques utilize large diameter (>300 μm) orifices (micropen, syringe, or plotter head) in conjunction with a fluid pump to push polymer or precursor solutions into 3D assemblies, with microscopic structure throughout the interior. Because these methods rely upon pressurizing fluid through a needle or micropen, we will classify these cell printing approaches as extrusion-based methods. These technologies are most often used to fabricate acellular scaffolding, so in order to perform tissue-engineering studies, the cells would need to be seeded randomly onto the material after completion of the scaffold printing [32, 33, 35, 38, 75, 79, 84]. However, there are a few recent examples in the literature demonstrating that these technologies can also deposit cells simultaneously with scaffolding materials and therefore qualify as cell printers [31, 72, 80, 87].

One of the best examples of 3D structural cell printing was performed by the extrusion of rat hepatocytes and a gelatin-based precursor through a syringe needle rapid-prototyping apparatus [80, 87]. An approximately 3 mm × 3 mm × 2 mm (height) gelatin/hepatocyte structure was formed with a periodic channel structure repeated throughout the entire printed scaffold. The channel diameter ranged between 100 and 300 μm and was designed to enhance nutrient delivery and waste removal. A cell density of 1.5×10^6 cells/ml was used during the deposition. The gelatin structure was cross-linked postprinting by a 5-s exposure to 2.5% glutaraldehyde solution. The authors show that cross-linking is necessary to maintain the printed structure for relevant culture periods (>1 month). Five minutes of cross-linking resulted in much firmer scaffolds, but no cells survived this more damaging exposure. Even 5 s exposure to a concentrated glutaraldehyde solution (often used for cell/tissue fixation), such as the one used for these experiments, could result in genetic or phenotype damage to the hepatocytes embedded in the gelatin structure [59, 66]. Enzyme activity data are presented that suggest some hepatocyte necrosis throughout the experiment, but the authors believe this cell death was induced by limited nutrient supply and waste removal rather than exposure to glutaraldehyde. The gelatin appears to protect the hepatocytes embedded in the structure, while cells on the outside of the walls died. Further genetic and phenotype analysis of the printed cells needs to be performed to determine what level of damage is incurred during this cross-linking procedure. As discussed earlier, the gelling process and its compatibility with living cells is a problem that must be faced for all structural cell printers.

Significant percentages of cells remained viable (~90%) throughout the printed scaffold over the 45-day culture experiment [80, 87]. Figure 43.11a and b shows two micrographs (at different magnification) of the gelatin/hepatocyte structure after 6 days of culture. Figure 43.11c shows a histopathological section of the scaffold after 45 days of culture. These images demonstrate that cells are embedded throughout the scaffold walls and that there are well-defined channels throughout the structure. Higher magnification shows healthy cell morphology throughout the scaffold. This is a significant accomplishment because the scaffold thickness (~2 mm) would suggest that cell growth and viability in the inner portions would be reduced. The printed microchannels appear to successfully aid in nutrient diffusion and waste removal.

The BioAssembly Tool (BAT) has also been used for structural cell printing. A 15-layer fibroblast/polyoxyethylene-polyoxypropylene structure was printed with a pneumatic pen (inner diameter ~450 μm) attached to a rapid prototyping apparatus [72]. The system enables a wide range of deposition speeds (12 nl/s–1 ml/s) and is configured to enable inline curing of polymer precursors.

The BAT has been used to print polymer/fibroblast structures [72]. The printed tubular scaffolds have wall thickness of ~500 μm and an inner diameter of ~1 mm. Approximately 60% of the fibroblasts remained viable, as assayed immediately postprinting. No explanation was given as to why such a large percentage of cells died during the extrusion process, but it was indicated that larger pen diameters resulted in higher percentages of viable cells (up to 86% and down to 46% viable). These data suggest that there are shear forces present during the deposition process that may damage a significant fraction of the cells. The size and shape of the printed constructs are impressive, and cell-embedded structures were made from both polymers and collagen I. However, the extent of cell damage that occurs during extrusion needs to be investigated.

43.3 Compare and Contrast Cell-Printing Methodologies

After surveying the literature, nearly all cell printers can be categorized as either a "structural" or "conformal" approach. This categorization is dependent mainly upon two factors: (1) whether the printer operates as "drop-on-demand" or has "continuous" depo-

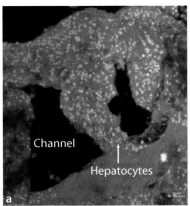
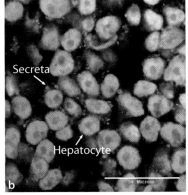
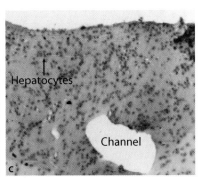

Fig. 43.11a–c Hepatocytes printed simultaneously with a gelantin/chitosan scaffold via a 3D micropositioning syringe-based system. **a** Laser scanning confocal micrograph showing hepatocytes imbedded in gelatin/chitosan matrix with channels. **b** Higher magnification. **c** Hematoxylin-eosin staining of a cross-section of the printed hepatocyte structure

sition, and (2) the achievable line or spot resolution. The only techniques to successfully deposit structurally defined, macroscopic scaffolds that endure in culture for several weeks are extrusion cell printers. These techniques have the ability to continuously deposit a curable solution of precursor and cells. Various polymerization methods (e.g., temperature, ultraviolet, chemical) are then used postdeposition to increase the structural rigidity of the printed structure. Care must be taken during the polymerization to not damage the cells present throughout the printed scaffold. Extrusion cell printers require large orifices (>300 μm in diameter) so that sufficient material can be deposited to form macroscopic, contiguous tissue scaffolds.

AT the other end of the spectrum, there are several cell printers that have higher-resolution drop-on-demand capabilities. These technologies would not realistically be used for printing entire free-standing scaffolds because the very nature of high-resolution printing requires less material be deposited. Less material usually translates to slower printing, especially when compared with extrusion-based printing, making it more difficult to entirely print millimeter-to-centimeter-scale structures. It is our opinion that a conformal printing approach would be most useful for drop-on-demand cell printing (e.g., modified LIFT, ink jet). This type of approach would enable heterogeneous cell patterns to be formed in a layer-by-layer fashion, perhaps with resolution on the scale of the heterogeneity present in natural tissues and organs. By adding (not printing) entire layers of scaffolding, conformal printing could then be used to build macroscopic tissue scaffolds with inherent cell and material diversity.

Table 43.1 compares the demonstrated capabilities (resolution, print speed, cell throughput, load volume, and cell viability) of all cell printers as reported in the peer-reviewed literature [64]. Perhaps the most important values are the percentage of viable cells achieved postprinting. Thermal ink-jet (70–95%), modified-LIFT (95–100%), and extrusion-based techniques (~50–90%) have reported a significant percentage of viable cells postprinting as assayed through fluorescence microscopy and live/dead cytochemical staining experiments. Cell viability in the thermal ink-jet process is reduced mainly by the cytotoxicity of the bioink, with longer print times resulting in lower percentages of viable cells (<70%), while small-diameter micropens may result in lower viability due to shear forces. As shown in Table 43.1, many cell printers do not have live/dead assay data to quantitatively determine the percentage of viable cells postprinting.

Cell-printer resolution, print speed, cell throughput, and load volume will all have varying levels of relevancy to the tissue-engineering community depending upon the application (e.g., organ, tissue, injury site). If the goal is to repair a small injury (e.g., bone, spinal cord), then a relatively small bridging scaffold could be fabricated to fill the injury site. This type of repair would require fewer cells to be deposited accurately (micrometer resolution) over small length scales (mm–cm). If the objective is to print entire replacement tissues (or organs), faster print speeds (>10^5 cells/s) with less stringent resolution would be required. Conformal cell printing may enable larger structures to be printed with much higher resolution (<100 μm) than current demonstrations of structural cell printing (>300 μm). The best cell printers will be those that combine speed with high resolution (<100 μm pixel size), preferably fast enough to print large structures with pixel sizes on the order of a single cell (~10 μm). Reproducibility is already at acceptable levels (~3%) for almost all cell printers (with the exception of EHDJ), and most printers have reached their fundamental limit in pixel size (i.e., matching size of orifice).

Load volume may also play a key role in cell printing. Often cell types relevant to tissue engineering (stem cells, harvested primary cells) are rare, difficult to harvest, grow slowly, or lose functionality over time in vitro. In these cases a cell printer like modified-LIFT (down to 500-nl load volume) would be preferable, because these approaches would enable small numbers of cells to be concentrated into ultrasmall volumes [4, 8, 13]. Because there are no capillaries or orifices used during modified LIFT printing, more concentrated cell solutions could be used (10^8 cells/ml) when compared to other technologies.

Another important consideration in cell printing is the ability to perform experiments in parallel and/or achieve cell throughputs of 10^3–10^5 cells/s. Some cell printers have already demonstrated these cell throughputs with a single print-head, while others have made theoretical arguments (Table 43.1). Ink-jet printers are often configured to print multiple types of "inks"

Table 43.1 Comparison between demonstrated capabilities of different cell printers. *LGDW* Laser-guided direct write, *LIFT* laser-induced forward transfer, *EHDJ* electrohydrodynamic jetting, *BAT* Bio-Assembly Tool (Copyright Wiley-VCH Verlag GmbH & Co. KGaA. Reproduced with permission)

	Spot size or resolution (µm)	Print speed	Maximum cell throughput (cells/s)†	Load volume (ml)	Cell viability (%)*
LGDW	10–30	Continuous (9×10^{-8} ml/s)	0.04	Not reported	–
Modified LIFT	30–100	10^2 drops/s	10^4	$0.5–20 \times 10^{-3}$	95–100
Ink jet -thermal	>300	5×10^3 drops/s	850	0.3–0.5	75–90
-Piezo-tip	Not reported	1×10^4 drops/s	2	0.3–0.5	–
EHDJ	50–1000	Continuous (0.01 ml/s)	2×10^4	2–5	–
Syringe extrusion	>300	Continuous (not reported)	Not reported	0.6	>90
BAT	>300	Continuous (12 nl/s – 1 ml/s)	Not reported	3	46–86

†Demonstrated cell throughput per orifice or cell reservoir. Often print-heads can contain multiple orifices
*Reported % viability through live/dead assay. LGDW, Piezo-tip and EHDJ have reported cell viability without giving specific %

in parallel. A thermal ink-jet cell printer has deposited cells from 50 independent chambers, increasing print speeds up to 250,000 drops/s (~4×10^4 cells/s) [86]. This speed was not demonstrated, but is a theoretical limit. EHDJ is unique in that it can be configured to operate with multiple needles, enabling parallel or unique simultaneous deposition (e.g., concentric tubes, fibers) [10, 40]. Modified LIFT has demonstrated cell throughput of greater than 10^3 cells/s (100 drops/s and 10 cells/drop) and has the ability to print cells from multiple adjacent cell reservoirs [4, 8, 63]. The modified LIFT technique can also be performed in parallel by utilizing different laser sources or beam-splitters for simultaneous deposition from multiple cell reservoirs. Overall, it appears that there is good reason to believe that cell printers will report more parallel deposition (e.g., different cell types, high throughput) and achieve higher cell throughputs in coming years.

The biggest weakness in the cell-printing literature is the lack of reported damage assays performed on printed cells. It is obvious that no matter how cells are printed, they undergo some level of stress during the deposition process. There is shear stress as cells are pushed through a capillary or orifice either via pressure, heat, or vibrations. There is the potential for heat stress as a laser or high-voltage source is used to actuate cell transfer. In all cases, cells impact the biopaper at some incident velocity, which would induce extreme deceleration (potentially greater than $10^6 \times g$) and shear stress. Often it is argued in the literature that cells endure these forces for such short times (ns–µs) that little to no damage is incurred. However, only one cell printer has studied the potential expression of heat-shock protein (HSP) by printed cells as a marker of stress endured during the deposition process. HSP is known to be expressed by cells that have undergone stress ranging from moderate temperature rises to shear stress [3, 56, 76].

Figure 43.12 is a series of micrographs that show a modified LIFT experiment probing whether BAECs printed by BioLP expressed HSP postprinting [9, 13]. Figure 43.12a is a fluorescence micrograph of positive control BAECs that were not printed, but incubated at 45°C for 1 h and then assayed for HSP60/70 expression. Figure 43.12b is the same experiment performed on nonprinted BAECs that had been incubated at 37°C for 1 h (negative control). Figure 43.12c–f shows micrographs of BioLP-printed BAECs at four different laser energies, ranging from 0.15 J/cm² to 1.5 J/cm². The low range is the laser fluence used under normal cell-printing conditions, so this experiment investigated typical to extreme conditions. Cell viability was not found to decrease until 1.5 J/cm², at which point it was reduced from 100% to 90%. Statistical analysis of the micrographs found insignificant changes in HSP60/70 expression for the printed cells when compared to the negative control cells, although a slight increase (within error) was observed for the highest energy. These results indicate that the heat and shear stress endured by cells during BioLP deposition was not enough to trigger expression of HSP. All cell printers should perform similar experiments so that damage levels can be determined. Other good measures of cell damage during printing are assays that probe cell differentiation, cell–cell communication, cell migration, and retained genotype.

43.4
Conclusion

Cell printing is still in its infancy. Research needs to focus less on the science-fiction concept of organ printing and more on the fundamental science and engineering of cell printing. Specifically, more experiments need to be performed to demonstrate spot or line resolution and cell throughput. In addition, researchers need to determine quantitatively the percentage of cells that remain viable postprinting. Damage assays, such as the HSP experiments described above, also need to be performed to ascertain whether cells undergo relevant amounts of heat and/or shear stress during deposition. Other assays can also be performed to ensure that the genotype and phenotype of the cells are retained postprinting. Only after these types of experiments are completed should researchers move on to print 3D tissue structures.

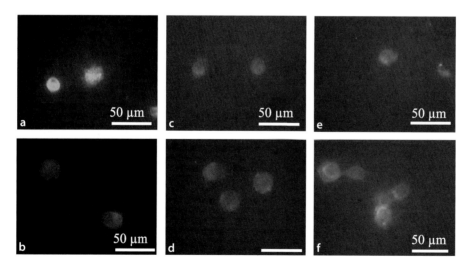

Fig. 43.12a–f Heat-shock protein (HSP) expression in control and BioLP-printed bovine aortic endothelial cells. **a** Fluorescence micrograph of positive control cells exposed to anti-HSP60/70 after incubation for 1 h at 45°C. **b** Negative control cells incubated for 1 h at 37°C. **c** BioLP-printed cells at 0.15 J/cm², incubated for 1 h at 37°C, then immunoassayed. **d** 0.30 J/cm². **e** 0.75 J/cm². **f** 1.5 J/cm²

Ultimately, printed scaffolds should be compared with traditional random cell-seeding approaches through in vivo testing followed by histopathological examination of the tissue implants. Through this methodical in vivo testing we will truly determine whether cell printing can contribute to the field of regenerative medicine. The potential is there for cell printers to create unique heterogeneous, multicell, channeled, even vascularized cell scaffolds that mimic the natural structure of tissue and organs. However, it is yet to be seen whether this revolutionary concept will be fully realized through laboratory experimentation.

References

1. An YH (2001) Regaining chondrocyte phenotype in thermo-reversible gel culture. Anat Rec 263:336
2. Ashkin A (1970) Acceleration and trapping of particles by radiation pressure. PRL 24:156
3. Azuma N, Akasaka N, Kito H, et al (2001) Role of p38 MAP kinase in endothelial cell alignment induced by fluid shear stress. Am J Physiol Heart Circ Physiol 280:H189
4. Barron J, Young H, Dlott D, et al (2005) Printing of protein microarrays via a capillary-free fluid jetting mechanism. Proteomics 5:4138
5. Barron JA, Ringeisen BR, Kim H, et al (2004) Application of laser printing to mammalian cells. Thin Solid Films 453–454:383
6. Barron JA, Rosen R, Jones-Meehan J, et al (2004) Biological laser printing of genetically modified Escherichia coli for biosensor applications. Biosens Bioelectron 20:246
7. Barron JA, Spargo BJ, Ringeisen BR (2004) Biological laser printing of three dimensional cellular structures. Appl Phys A Mater Sci Process 79:1027
8. Barron JA, Wu P, Ladouceur HD, et al (2004) Biological laser printing: a novel technique for creating heterogeneous 3-dimensional cell patterns. Biomed Microdevices 6:139
9. Barron JA, Krizman David B, Ringeisen BR (2005) Laser printing of single cells: statistical analysis, cell viability, and stress. Ann Biomed Eng 33:121
10. Bocanegra R, Galan D, Marquez M, et al (2005) Multiple electrosprays emitted from an array of holes. J Aerosol Sci 36:1387
11. Boland T, Mironov V, Gutowska A, et al (2003) Cell and organ printing 2: fusion of cell aggregates in three-dimensional gels. Anat Rec A Discov Mol Cell Evol Biol 272:497
12. Boland T, Xu T, Damon B, et al (2006) Application of inkjet printing to tissue engineering. Biotechnol J 1:1910
13. Chen CY, Barron JA, Ringeisen BR (2006) Cell patterning without chemical surface modification: cell-cell interacftions between bovine aortic endothelial cells (BAEC) on a homogeneous cell-adherent hydrogel. Appl Surf Sci 252:8641
14. Chen VJ, Ma PX (2004) Nano-fibrous poly(l-lactic acid) scaffolds with interconnected spherical macropores. Biomaterials 25:2065
15. Chrisey D, Pique A, McGill R, et al (2003) Laser deposition of polymer and biomaterial films. Chem Rev 103:553
16. Colina M, Serra P, Fernandez-Pradas JM, et al (2005) DNA deposition through laser induced forward transfer. Biosens Bioelectron 20:1638
17. Cortesini R (2005) Stem cells, tissue engineering and organogenesis in transplantation. Transpl Immunol 15:81
18. Eagles PAM, Qureshi AN, Jayasinghe SN (2006) Electrohydrodynamic jetting of mouse neuronal cells. Biochem J 394:375
19. Fernandez-Pradas JM, Colina M, Serra P, et al (2004) Laser-induced forward transfer of biomolecules. Thin Solid Films 453–454:27
20. Flickinger M, Schottel J, Bond D, et al (2007) Painting and printing living bacteria. Biotechnol Prog 23:2
21. Guduru S, Narasimhan S, Birchfield S, et al (2005) Analysis of neurite outgrowth for a laser patterned neuronal culture. Proceedings of the 2nd International IEEE EMBS Special Topic Conference on Neural Engineering, 2005, Arlington, VA
22. Hood BL, Darfler MM, Guiel TG, et al (2005) Proteomic analysis of formalin-fixed prostate cancer tissue. Mol Cell Proteomics 4:1741
23. Hopp B, Smausz T, Kresz N, et al (2005) Survival and proliferative ability of various living cell types after laser-induced forward transfer. Tissue Eng 11:1817
24. Humes H (2005) Stem cells: the next therapeutic frontier. Trans Am Clin Climatol Assoc 116:167
25. Jayasinghe S, Townsend-Nicholson A (2006) Bio-electrosprays: the next generation of electrified jets. Biotechnol J 1:1018
26. Jayasinghe S, Townsend-Nicholson A (2006) Stable electric-field driven cone-jetting of concentrated biosuspensions. Lab Chip 6:1086
27. Jayasinghe SN, Eagles PAM, Qureshi AN (2006) Electric field driven jetting: an emerging approach for processing living cells. Biotechnol J 1:86
28. Jayasinghe SN, Qureshi AN, Eagles PAM (2006) Electrohydrodynamic jet processing: an advanced electric-field-driven jetting phenomenon for processing living cells. Small 2:216
29. Jeong B, Gutowska A (2002) Lessons from nature: stimuli-responsive polymers and their biomedical applications. Trends Biotechnol 20:305
30. Kesari P, Xu T, Boland T (2005) Layer-by-layer printing of cells and its application to tissue engineering. Mater Res Soc Symp Proc 845:111
31. Khalil S, Nam J, Sun W (2005) Multi-nozzle deposition for construction of 3D biopolymer tissue scaffolds. Rapid Prototyping J 11:9
32. Kim S, Utsunomiya H, Koski J, et al (1998) Survival and function of hepatocytes on a novel three-dimensional

synthetic biodegradable polymer scaffold with an intrinsic network of channels. Ann Surg 228:8
33. Koegler W, Griffith L (2004) Osteoblast response to PLGA tissue engineering scaffolds with PEO modified surface chemistries and demonstration of patterned cell response. Biomaterials 25:2819
34. Konig K, Tadir Y, Patrizio P, et al (1996) Effects of ultraviolet exposure and near infrared laser tweezers on human spermatozoa. Hum Reprod 11:2162
35. Landers R, Hubner U, Schmelzeisen R, et al (2002) Rapid prototyping of scaffolds derived from thermoreversible hydrogels and tailored for applications in tissue engineering. Biomaterials 23:4437
36. Lee KY, Mooney DJ (2001) Hydrogels for tissue engineering. Chem Rev 101:1869
37. Lee KY, Peters MC, Mooney DJ (2001) Controlled drug delivery from polymers by mechanical signals. Adv Mater (Weinheim, Germany) 13:837
38. Leong K, Cheah C, Chua C (2003) Solid freeform fabrication of three-dimensional scaffolds for engineering replacement tissues and organs. Biomaterials 13:2363
39. Liang H, Vu K, Krishnan P, et al (1996) Wavelength dependence of cell cloning efficiency after optical trapping. Biophys J 70:1529
40. Loscertales G (2002) Micro/nano encapsutation via electrified coaxial liquid jets. Science 295:1695
41. Ma PX (2004) Scaffolds for tissue fabrication. Mater Today 7:30
42. Ma PX, Choi J-W (2001) Biodegradable polymer scaffolds with well-defined interconnected spherical pore network. Tissue Eng 7:23
43. Ma PX, Zhang R (1999) Synthetic nanoscale fibrous extracellular matrix. J Biomed Mater Res 46:60
44. Merrin J, Leibler S, Chuang J (2007) Printing multistrain bacterial patterns with a piezoelectric inkjet printer. PLoS ONE 2:e663
45. Miller ED, Fisher GW, Weiss LE, et al (2006) Dose-dependent cell growth in response to concentration modulated patterns of FGF-2 printed on fibrin. Biomaterials 27:2213
46. Mironov V, Boland T, Trusk T, et al (2003) Organ printing: computer-aided jet-based 3D tissue engineering. Trends Biotechnol 21:157
47. Mironov V, Markwald RR, Forgacs G (2003) Organ printing: self-assembling cell aggregates as "bioink". Sci Med 9:69
48. Mironov V, Boland T, Trusk T, et al (2004) Organ printing: computer-aided jet-based 3D tissue engineering. [Erratum to document cited in CA140:038057]. Trends Biotechnol 22:265
49. Murugan R, Ramakrishna S (2006) Review article: nanofeatured scaffolds for tissue engineering: a review of spinning methodologies. Tissue Eng 12:435
50. Nakamura M (2004) Application of inkjet technology for tissue engineering. Bio Industry 21:68
51. Nakamura M, Kobayashi A, Takagi F, et al (2005) Biocompatible inkjet printing technique for designed seeding of individual living cells. Tissue Eng 11:1658
52. Norman J, Desai T (2006) Methods for fabrication of nanoscale topography for tissue engineering scaffolds. Ann Biom Eng 34:89
53. Odde DJ, Renn MJ (1999) Laser-guided direct writing for applications in biotechnology. Trends Biotechnol 17:385
54. Odde DJ, Renn MJ (2000) Laser-guided direct writing of living cells. Biotechnol Bioeng 67:312
55. Odenwälder P, Irvine S, McEwan J, et al (2007) Bio-electrosprays: a novel electrified jetting methodology for the safe handling and deployment of primary living organisms. Biotechnol J 2:622
56. Oishi Y, Taniguchi K, Matsumoto H, et al (2003) Differential responses of HSPs to heat stress in slow and fast regions of rat gastrocnemius muscle. Muscle Nerve 28:587
57. Othman S, Xu H, Royston T, et al (2005) Microscopic magnetic resonance elastography (microMRE). Magn Reson Med 54:605
58. Pardo L, Boland T (2003) Characterization of patterned self-assembled monolayers and protein arrays generated by teh ink-jet method. Langmuir 19:1462
59. Paull T, Fleming J (1990) Upregulation of E. coli 39kDa proteins induced by glutaraldehyde and formaldehyde. Curr Microbiol 21:117
60. Ringeisen BR, Chrisey DB, Pique A, et al (2001) Generation of mesoscopic patterns of viable Escherichia coli by ambient laser transfer. Biomaterials 23:161
61. Ringeisen B, Wu P, Kim H, et al (2002) Picoliter-scale protein microarrays by laser direct write. Biotechnol Prog 18:1126
62. Ringeisen BR, Barron JA, Spargo BJ (2004) Novel seeding mechanisms to form multilayer heterogeneous cell constructs. Mater Res Soc Symp Proc EXS-1:105
63. Ringeisen BR, Kim H, Barron JA, et al (2004) Laser printing of pluripotent embryonal carcinoma cells. Tissue Eng 10:483
64. Ringeisen B, Othon C, Barron J, et al (2006) Jet-based methods to print living cells. Biotechnol J 1:930
65. Roth EA, Xu T, Das M, et al (2004) Ink-jet printing for high-throughput cell patterning. Biomaterials 25:3707
66. Sabatini D, Bensch K, Barrnett R (1963) Cytochemistry and electron microscopy. J Cell Biol 17:19
67. Saltzman WM (2004) Tissue Engineering : Engineering Principles for the Design of Replacement Organs and Tissues (1st edn). Oxford University Press New York, USA
68. Saunders R, Bosworth L, Gough J, et al (2004) Selective cell delivery for 3D tissue culture and engineering. Eur Cells Mater 7:84
69. Saunders R, Derby B, Gough J, et al (2004) Ink-jet printing of human cells. Mater Res Soc Symp Proc EXS-1:95
70. Saunders R, Gough J, Derby B (2005) Ink jet printing of mammalian primary cells for tissue engineering applications. Mater Res Soc Symp Proc 845:57
71. Shea L, Smiley E, Bonadio J, et al (1999) DNA delivery from polymer matrices for tissue engineering. Nature Biotechnology 17:551
72. Smith CM, Stone AL, Parkhill RL, et al (2004) Three-dimensional bioassembly tool for generating viable tissue-engineered constructs. Tissue Eng 10:1566
73. Sumerel J, Lewis J, Doraiswamy A, et al (2006) Pizoelectric ink jet processing of materials for medical and biological applications. Biotechnol J 1:976

74. Sun W, Lal P (2002) Recent developments on computer aided tissue engineering, a review. Comput Methods Programs Biomed 67:85
75. Taylor P, Sachlos E, Dreger S, et al (2006) Interaction of human valve interstitial cells with collagen matrices manufactured using rapid prototyping. Biomaterials 27:2733
76. Tazi K, Barriere E, Moreau R, et al (2002) Role of shear stress in aortic eNOS up-regulation in rats with biliary cirrhosis. Gastroenterology 122:1869
77. Townsend-Nicholson A, Jayasinghe S (2006) Cell electrospinning: a unique biotechnique for encapsulating living organisms for generating active biological microthreads/scaffolds. Biomacromolecules 7:3364
78. Vasita R, Katti D (2006) Growth factor-delivery systems for tissue engineering: a materials perspective. Expert Rev Med Devices 3:29
79. Vozzi G, Flaim C, Ahluwalia A, et al (2003) Fabrication of PLGA scaffolds using soft lithography and microsyringe deposition. Biomaterials 24:2533
80. Wang X, Yan Y, Pan Y, et al (2006) Generation of three-dimensional hepatocyte/gelatin structures with rapid prototyping system Tissue Eng 12:83
81. Watanabe K, Miyazaki T, Matsuda R (2003) Growth factor array fabrication using a color ink jet printer. Zoolog Sci 20:429
82. Williams D, Sebastine I (2005) Tissue engineering and regenerative medicine: manufacturing challenges. IEE Proc Nanobiotechnol 152:207
83. Wilson WC Jr, Boland T (2003) Cell and organ printing 1: protein and cell printers. Anat Rec A Discov Mol Cell Evol Biol 272:491
84. Woodfield T, Malda J, de Wijn J, et al (2004) Design of porous scaffolds for cartilage tissue engineering using a three-dimensional fiber-deposition technique. Biomaterials 25:4149
85. Xu T, Gregory CA, Molnar P, et al (2006) Viability and electrophysiology of neural cell structures generated by the inkjet printing method. Biomaterials 27:3580
86. Xu T, Jin J, Gregory C, et al (2005) Inkjet printing of viable mammalian cells. Biomaterials 26:93
87. Yan Y, Wang X, Pan Y, et al (2005) Fabrication of viable tissue-engineered constructs with 3D cell-assembly technique. Biomaterials 26:5864
88. Yang S, Leong K, Du Z, et al (2002) The design of scaffolds for use in tissue engineering. Part II. Rapid prototyping techniques. Tissue Eng 8:1
89. Yang SF, Leong KF, Du ZH, et al (2002) The design of scaffolds for use in tissue engineering. Part 1. Traditional factors. Tissue Eng 7:679
90. Young D, Auyeung R, Piqué A, et al (2001) Time resolved optical microscopy of a laser-based forward transfer process. Appl Phys Lett 78:3139
91. Zandonella C (2001) New body parts at the stroke of a laser pen. New Sci 170:24

Biophysical Stimulation of Cells and Tissues in Bioreactors

H. P. Wiesmann, J. Neunzehn, B. Kruse-Lösler, U. Meyer

Contents

44.1	Introduction	633
44.2	Biophysical Parameters of Bone Physiology	634
44.2.1	Organ Level	634
44.2.2	Tissue Level	635
44.2.3	Cellular Level	636
44.3	Basic Bioreactors	637
44.4	Complex Bioreactors	638
44.5	Bioreactor Systems	638
44.6	Theoretical Concepts of Bioreactor Features	639
44.7	Experimental Data	640
	References	644

44.1 Introduction

Although the regeneration of complex tissue structures in all sites of the body is still a vision of the future, the first steps towards introducing tissue-engineering-based therapies into surgery have already been made [1, 38, 58]. Bioreactors can be defined as devices in which biological or biochemical processes, or both, are re-enacted under controlled conditions (for example, pH, temperature, pressure, oxygen supply, nutrient supply and waste removal). Most bioreactors were originally developed to test biomaterials, but some of them were also invented with the objective of extracorporeal tissue growth [39]. Various types of bioreactor have therefore been developed and tested for their use in tissue engineering. The ex vivo generation of larger three-dimensional tissues in particular requires the development of enlarged devices and systems that allow cells to locate themselves in a three-dimensional space, and at the same time satisfy the physicochemical demands of large cell masses that are commonly connected to a scaffold [22, 61]. The design and performance of bioreactors is therefore of special relevance in recent and future tissue-engineering strategies. It is important to recognise that most tissues are stimulated by mechanical forces, but react very differently towards the microenvironment (for example, oxygen tension). Bioreactors with defined mechanical properties are of special concern in order to optimise the ex vivo growth of the desired cell or scaffold constructs [34, 40, 41].

As nutrition and stimulation of cells or cell or scaffold constructs, control over environmental conditions, duration of the culture, online evaluation of biological variables and automation of bioprocesses have to be considered for the development and use of bioreactors, this chapter discusses the various aspects in the context of hard tissue engineering. Special attention is paid on tissue engineering of connective tissue like bone, since this kind of regeneration is currently one of the most clinically applicable.

44.2 Biophysical Parameters of Bone Physiology

In order to use these bioreactors for bone tissue engineering it is important to understand the individual features of bone physiology. Bone is a complex, highly organised tissue. It comprises a structured extracellular matrix composed of inorganic and organic elements that contain several types of cell, which are responsible for its metabolism and upkeep and are responsive to various signals [7]. The formation of bone is a multistep process that is characterised by interactions between the various cells of the bony tissue, components of the extracellular matrix, and inorganic minerals [33]. As bone has to be structurally adapted to various functions, the activity of cells and the subsequent reaction in the tissue varies with the type, the anatomical site, and the loading conditions [8]. Craniofacial bones are specialised structures in which muscles, joints, and teeth are adapted to operate in complex synergy within the highly developed masticatory system. Biomechanical transfer of load is a key effector in most craniofacial bones, so it has an important role in humans because load-related deformation of tissue can both increase bone formation and decrease bone resorption [56]. Indeed, the absence of load can lead to a reduction in the production of bone-matrix protein and loss of mineral content, as occurs in the lower limbs of astronauts who have long flights in space. Loading leads to a complex sequence of biophysical events. Deformation of cells in the microenvironment of the tissue and fluid-flow-related generation of electrical potentials are the main consequences of loading in bone [17].

When mechanical or electrical stimuli are applied therapeutically to improve the formation of bone, one has to be aware that they exert their effects within the various structures of the skeleton. The influence of physical factors in skeletal regeneration can generally be considered at the organ, tissue, cellular and molecular levels. Understanding the overall process of biophysical signalling requires an appreciation of these various branches of study and knowledge of how one process relates to and influences the next. Firstly, biophysical stimuli can be analysed for their effect on whole bones. Secondly, the effects on multicellular systems can be evaluated by examining the reaction of tissue to the biophysical microenvironment. Thirdly, cellular and molecular reactions can be detected by measuring effects at the microscale level.

Biophysical stimuli exert their effects in humans at distinct skeletal sites. The relationship between the stimulus that is applied clinically and the outcome in terms of histological and clinical success is complex. Physical stimuli are altered by the structure of the site where the stimulus is applied. The transfer of load through adjacent structures (joints, ligaments and muscles) and the absorption of load by exposure to an electrical field have an important bearing on the resultant biophysical signal at the desired effector site. Both patient-specific and technique-specific factors play an important part. The specific forms of therapeutic biophysical stimulation and the effect of dose and timing of the application must be calculated and validated carefully for the different types of bony tissue. Key requirements of using biophysical forces are therefore to define the precise cellular response to the stimulation signal in an in-vitro environment and to use well-established animal and clinical models to quantify and optimise the type of therapeutic stimulation. This seems to be achievable through collaboration among different disciplines, using defined scientific methods.

44.2.1 Organ Level

Some skeletal characteristics have to be taken into account when considering biophysical effects on bones. The main purpose of bone is to provide structural strength commensurate with its mechanical use. This means that bones provide enough strength to prevent normal physical loads from damaging the integrity of the tissue. Recent research has indicated that bones can adapt amazingly well to a changing functional environment; this is often referred to as phenotype plasticity [20]. Specific biophysical signals are assumed to be largely responsible for controlling this adaptive mode of modelling of bony tissue. When loads act on whole bones, the tissue begins to deform, causing local strains (typically reported in units of microstrain, μstrain; $10{,}000\,\mu$strain$=1\%$ change in length). Various investigators have shown that reactive loads give rise to strains at fundamental frequencies ranging from 1 to 10 Hz. Peak magnitudes of strain measured in various species were found to have a remarkable similarity, ranging from 2,000 to 3,500 μstrain, with no great difference between long bones and craniofacial bones. Lanyon et al. [31] showed that within a single period of loading, the remodelling process was saturated after only a few (<50) loading cycles. Repeated applications of load then produced no extra effect. Frost offered a unique theory to explain the load-related modelling and remodelling processes in bone [20]. He showed that the adaptive mechanisms included basic multicellular units. Effector cells within these units function in an interdependent manner. While hormones may bring about as much as 10% of the postnatal changes in bone strength and mass, 40% are established by mechanical factors. This effect, which is reflected in the loss of more than 40% of bone mass in the lower limbs of patients with paraplegia, stresses the effect of biophysical signals on bone modelling and remodelling [17].

The functional biological environment of any bony tissue is thus derived from a dynamic interaction between various active basic multicellular units exposed to a biophysical microenvironment that undergoes continuous load-related changes. Formation of bone occurs through drifts of formation and resorption to reshape, thicken, and strengthen a bone or trabecula by moving its surfaces around in tissue space. Remodelling of bone also involves both resorption and formation. Basic multicellular units turn bone over in small packets through a process in which an activating event causes some resorption followed by formation. This basic multicellular unit-based remodelling operates in two modes: "conservation mode" and "disuse mode". Specific ranges of threshold of strain seem to control which of these two modes is active at any given time. While the strain-related bone-modelling theory was long assumed to be applicable only to intact bones, recent research has shown that this adaptive behaviour is also present in the healing processes in bone.

44.2.2
Tissue Level

It is important to recognise that mechanical and electrical effects are exerted simultaneously in bone [53]. Complex mechanical and electrical interactions during the modelling and remodelling processes integrate the action of several signals to form the final response. Deformation of tissue is a key stimulus in bone physiology and leads to a complex non-uniform biophysical environment within bony tissue, consisting of fluid flow, direct mechanical strain, and electrokinetic effects on bone cells. At the tissue level, some researchers assumed that the differentiating tissue was a continuous material and evaluated biophysical signals by characterising the stimulus in terms of mechanical engineering quantities, such as stress and strain. Based on the material properties of tissue and approximations of tissue loading in the different experimental and clinical conditions, these quantities were calculated throughout the tissue and were related to various patterns of differentiation in the tissue. The biophysical mechanisms underlying the tissue response were directly related to mechanical effects, electromechanical effects, or increase in molecular transport mechanisms [53]. Pressure, distortion, pressure gradients, and dissipation of energy are additional mechanical quantities at the tissue level that were measured and related to tissue responses by other authors [17].

Load-induced flow of interstitial fluid may provide a convergence feature between electrical and mechanical signals and so trigger the formation of hard tissue in adaptation [10]. Loading of bone leads to a deformation of cells in the microenvironment of the tissue and to a fluid-flow-related generation of electrical potentials that exert effects on adjacent cells. Pressure gradients from the mechanical loading of bony tissue elongate cells and the mineralised matrix and move extracellular fluid radially outward. Electric fields of a magnitude and frequency that would occur naturally as a result of physiological loads on bone through piezoelectric effects, streaming potentials, or a combination of these, have been shown to modulate the formation of bone in the craniofacial skeleton.

The sensitivity of the response of the osteoblastic network was found to be much greater when the cells were part of a tissue than when they were evaluated as single entities. The bony tissue integrates and amplifies a complex physical signal, such as a mechanical strain or functionally induced fluid flow by transmitting the signal from the signal-detecting cells to the change-effecting cells. Bone cells are coupled functionally by gap junctions. Gap junctions, which are well suited to the integration and amplification of the response of bone cell networks to biophysical signals, are therefore important mediators of the effector response. The role of gap junctions in signal transduction at the tissue level is exemplified in the response of bone cells to electromagnetic fields. In contrast to single cells, the intercellular network was found to contribute to the ability of electromagnetic fields to stimulate the activity of alkaline phosphatase, a marker for osteoblastic differentiation, which suggests that gap junctions are involved in the amplification of electromagnetic-field-stimulated osteoblastic differentiation. Gap junctions also contribute to the responsiveness of bone cells to other biophysical signals, such as cell deformation and fluid flow, indicating their special role in transmitting signals from single cells into the tissue.

The final development of bone at the tissue level is the provision of a collagenous matrix that is mineralised to give the bony tissue the desired mechanical properties [46]. The phenomenon of mineralisation is therefore an important regulator in the formation of hard tissue. Numerous studies have shown that both the collagen microarchitecture and the mineralisation process of bone are influenced by biophysical forces. Biophysical stimuli exert their effects not only on the proliferation and differentiation of bone cells, but also on the collagen-associated emergence and maintenance of mineralisation of the matrix [59]. Collagens are important stress-carrying proteins that are sensitive to the application of mechanical strains. Low strains lead to a straightening of collagen fibres, whereas higher strains cause molecular gliding within the fibrils, resulting in disruption of the fibrillar organisation and ultimately impeding the formation of hard tissue. Ultrastructural data support the hypothesis of a structural alteration of collagen under tension stress. Recent research indicates that the tension-related assembly of collagen fibres may be associated with different methods of mineralisation. At low strains, mature crystals form in tension-related bone healing. In contrast, even moderately high strains are associated with the formation of immature, predominantly developing apatite crystals, as shown by electron microscopy, diffraction analysis, and measurement of elements by radiographic analysis. Biophysical signals also seem to alter the chemical composition and properties of the cellular microenvironment at the time of nucleation of crystals and during their subsequent growth. Changes in the concentration and chemical composition of the extracellular fluids that occur during the mineralisation process are controlled by external biophysical forces and have an influence on the mineralisation process. While it is not known which factors control the initiation of mineral deposition and that are responsible for the collagen-associated mineralisation of the matrix, mechanical tension and field exposure have been identified as potent regulators of the composition of elements in the extracellular matrix and of the pattern of formation of minerals [59].

44.2.3
Cellular Level

Osteoblasts and osteocytes that are connected with each other through gap junctions act as the sensors of local bone strains and are appropriately located in the bone for this function. Because the formation of bony tissue is based mainly on the action of osteoblasts, these cells are of special relevance. Formation of bone involves a complex pattern of cellular events that are initiated by proliferation and differentiation of mesenchymal cells into bone-forming cells, finally resulting in mineralisation of the extracellular matrix. Conversion of a biophysical force into a cellular response by signal transduction is an essential mechanism, allowing single bone cells to respond to a changing biophysical environment.

Many different mechanisms of biophysical transduction have been suggested at the cellular level [27]. In-vitro studies have related signals at the cellular level, from changes of cell shape, cell pressure, and local oxygen tension, to patterns of the composition of components of the extracellular matrix [8]. Other signals that have been assumed to transform biophysical signals include bending of cilia, changes of temperature, alteration in the ion constitution, and localised electrical potentials [27]. Signal transduction itself can be categorised in an idealised manner into:
1. Signal coupling, the transduction of a biophysical force applied to the tissue into a local mechanical signal perceived by a bone cell.
2. Biochemical coupling, the transduction of the local biophysical signal into biochemical signal cascades that alter expression of genes or activation of proteins.

3. Transmission of signals from the sensor cells to effector cells; and ultimately, the effector cell response, which forms or removes bone.

The molecular level is the most specific level for the study of biophysical signalling. Signals at the molecular level may include cytoskeletal damage or disruption, receptor binding, growth factors, and stretch-activated activity of the ion channel [37]. Many of these signals at molecular level have also been correlated with changes in specific activities of cells [26]. Most authors have suggested that the biochemical coupling process of biophysical forces might be related directly to deformations of ultrastructural organelles or proteins, thereby converting the biophysical information into a biochemical signal. The strain sensor is assumed to be linked to the cytoskeleton, although other hypotheses have been suggested [26]. If the strain sensor is located in the cytoskeleton, then deformations of this structure would not be homogeneous because different compartments would have different mechanical properties and the weakest link would be subject to most deformation. In this model, elongation of bone cells influences subsequent transcriptional events. As the oestrogen receptor is involved in the adaptive response of osteoblasts to mechanical strain, transduction of biophysical force seems also to be connected closely with the hormonal regulation of turnover in bone, which indicates the complexity of the behaviour of bone cells towards signals at the molecular level. In summary, we accept the evidence supporting the concept that biophysical signals are able to regulate the behaviour of bone cells, but the relationship between defined biophysical forces and resultant effects on the behaviour of bone cells is to some extent controversial. Whereas some investigations have found that biophysical forces, including deformation of substrate, hydrostatic pressure, fluid flow, or hypergravity, increase the expression or synthesis of markers of differentiation of bone cells, others have failed to do so or have even reported that biophysical stimuli inhibit cell proliferation and differentiation. This contradictory evidence may be related to the fact that various biophysical stimuli (physiological or non-physiological) were exerted on different bone cells (primary cells, altered cells) in the various investigations.

44.3 Basic Bioreactors

The culture dish is the simplest and most widely used bioreactor in most of the recent tissue-engineering techniques. Its main advantage is that it is easy to handle and economical to manufacture. Flasks are also commonly used bioreactors for most of the current cell multiplication strategies [30]. However, both types of "bioreactor" are of limited value, particularly when considering that a three-dimensional bone construct should ideally be fabricated for the reconstruction of defects in the "real" clinical situation. The major advantage of cell cultivation in monolayers is that cells in monolayer cultures are not generally nutrient-limited, and rapid multiplication of cells can be achieved as passive diffusion is more than adequate to supply the cell layers. As the distance of diffusion in a monolayer is lower than 100–200 µm, supply of oxygen and soluble nutrients is not critical. Whenever cells are supposed to grow in multiple layers or cells are located in scaffolds, access to substrate as well as signalling molecules, growth factors and nutrients (oxygen, glucose, amino acids, and proteins) and clearance from metabolic products of metabolism (CO_2, lactate, and urea) are critical to cell survival [57].

The diffusion of these substrates into and out of the cell-containing microenvironment is a prerequisite to scale up cell or material devices towards tissue-like products. The transport of the various substrates can be mediated by inducing fluid flow in the cell-containing culture dishes or flasks, thereby creating more complex (for example, multilayer cell-scaffold) constructs. Cell growth in a spinner flask as a next step therefore provides continuous exposure of the cells to various nutrients by convection. Convection is of special relevance in tissue engineering since cells are naturally embedded in a dense extracellular matrix, which plays a regulatory role in enhancing the transport of large molecules such as proteins and growth factors. Another mechanism for mass transport (particularly for small molecules) in tissues like cartilage or bone is passive diffusion along concentration gradients. Various techniques have been introduced in bioreactors to create pressure gradients driving the fluid flow. Stirring the culture medium can improve the nutrient supply of cells. As one of the most basic bioreactors, the stirred flask induces mixing of oxygen and nutrients throughout the medium and reduces the concentration boundary layer at the construct surface. If the turbulent flow generated within stirred-flask bioreactors is unfavourable and leads to unpredictable flow effects in different regions of the multilayered cell or scaffold complex, it complicates homogeneous tissue formation.

44.4
Complex Bioreactors

When cells are located in a more complex, scaled-up scaffold, the nutritious medium, even when stirred, is usually limited to the outer surfaces or the pores of the scaffold. Particularly when multicellular constructs are fabricated (for example, by co-culturing osteoblasts and endothelial cells), cells compete with the various cell sources that are located within a scaffold. Newly developed bioreactor systems, fabricated to improve cell survival in scaled-up scaffolds, reach a new step of complexity in tissue-engineering techniques.

Oxygen and nutrient concentration, medium or cell volume ratio, oxygen requirement and consumption of cells, metabolite removal and fluid flow within the site have a dramatic impact on the growth and survival of the tissue substitute in enlarged constructs, and when controlled can be used as major positive modulators in elaborate bioreactors. A dynamic laminar flow environment was shown to be a desirable and efficient way to reduce the diffusional limitations of nutrients and metabolites, and therefore permits the fabrication of tissue equivalents, as has been demonstrated using chondrocytes and osteoblasts [49]. Cartilage-like matrix synthesis by chondrocytes and growth, differentiation, and deposition of mineralised matrix by bone cells are enhanced by direct perfusion bioreactors [40]. It is obvious that the flow rate in the microenvironment of cells is responsible to a great extent for the effects of medium perfusion. Therefore, in optimising a perfusion bioreactor for tissue engineering, one must carefully address the balance between the extent of nutrient supply, the transport of metabolites to and away from cells, and the fluid-induced shear stress effects on cells located at the surface and in the porous structures of the scaffold [48]. Biochemical and biomechanical properties superior to those of static or stirred-flask cultures and approaching those of native tissue were achieved after a longer period of cultivation under laminar flow in the case of cartilage engineering [14]. When considering cell nutrition, it is important to note that the structure and porosity of scaffold materials, the overall cell or scaffold construct size and the diffusion through the biomaterial also influence and regulate cell viability in bioreactors. An adequate supply of nutrients is of major importance in manufacturing scaled-up, three-dimensional tissues, since cells are not only supposed to colonise the surface of scaffolds, but should also be located within the scaffold in a viable state. The concept of improving the scaffold design with the aim of increasing the nutrition of cells in complex constructs using a combination of laminar flow with tubular structures located in the scaffold is considered to be crucial for the development of advanced bioreactor systems. Mass transport of nutrients and oxygen in central parts of cell scaffold complexes, and therefore improvement of metabolic function of cells could be achieved by medium flow either through or around semipermeable tubes. Further development of this concept is perfusion directly from bioreactor-inherent tubes through prefabricated interconnecting tubes and pores of scaffolds, with the advantage of reducing mass transfer limitations, particularly in the central parts of the scaffolds. The possibility of direct perfusion allows active delivery of nutrients in complex systems, as shown by Curtis and Riehle [13].

44.5
Bioreactor Systems

From a technical point of view, bioreactors can be classified according to their environmental design properties. Bioreactors, connected to ports and filters for gas exchange, can be regarded as more "closed systems" compared with conventional dishes and flasks. "Open systems", such as culture-dish systems, require individual manual handling for medium exchange, cell seeding, and so on, which ultimately limits their usefulness when high manufacturing standards (such as approaches for clinical use) are required.

Major advantages are offered by closed bioreactor systems, since sterility can be assured and the viability of the tissue product maintained. Closed and automated bioreactors require state-of-the-art systems for monitoring and adjusting physicochemical as well as mechanical variables. For reproducibility and standardisation, the culture conditions, such as temperature, pH, nutrients and oxygen supply, should be adjustable. An improvement in bioreactor design would be the addition of devices that allow the on-line evaluation of the cell or scaffold maturation. The introduction of non-destructive analysis tools to monitor material variables (e-modulus, degradation), biological variables (cell number, cell differentiation) and metabolic variables (pH, oxygen concentration, nutrient concentration) is helpful in adjusting culture conditions towards tissue needs. Advanced techniques (oxygen measurements, substrate monitoring, flow determination, fluorescence microscopy, micro-computerised tomography) that aim to predict the development of the tissue construct over time, have now been initially introduced in the fabrication of various tissue-engineered products.

44.6
Theoretical Concepts of Bioreactor Features

Theoretical concepts of bioreactor features are helpful in the design and fabrication processes of new bioreactor generations. Mathematical modelling, for example, offers the possibility of gaining insight into the nutrient supply in cell or scaffold constructs as well as in calculating the diffusion-related flow conditions in the deeper structures of bulk materials. With the aid of recently available computational tools, variables such as flow fields, shear stresses and mass transport in scaffold-containing bioreactors can be calculated. Mathematical modelling enables researchers to determine the relationship between mass transport and cell viability or to calculate the momentum and oxygen transport within concentric cylinder bioreactors. For example, theoretical modelling was applied to investigation of the relationships between cell density, diffusion distance and cell viability within a cell or scaffold construct under a given nutrient supply. Muschler et al. used a system of differential equations to calculate the relationship between cell density, diffusion distance and cell viability in materials [43]. Modelling of these variables was performed to estimate when conditions of hypoxia will appear. It was calculated that an increase in scaffold dimensions by a factor of 5 decreases the maximum concentration of cells that can be delivered by oxygen by a factor of 25. This theoretical assumption confirms the findings that flat or porous scaffold structures work better than bulk materials. A combined approach of two mathematical models (combining the calculation of both flow and diffusion conditions) seems to be promising in order to most closely reflect the "real" situation for oxygen supply and consumption in complex scaffold systems.

The complexity of cell nutrition modelling is becoming even more obvious when considering that cell survival depends not only on oxygen tension, but also on the delivery, consumption and removal of nutrients and metabolic products [47]. Since the diffusion of oxygen is relatively slow in some of the biomaterials, while oxygen consumption is high, and the transport of other nutrients such as glucose and amino acids to and away from cells is more favourable in some biomaterials than in others, modelling of complex systems is difficult. Therefore, mathematical modelling is one way of delivering data about nutrition-related cell reactions, but collection of experimental data and comparison between theoretical and real data is of special relevance.

A recent field in which mathematical modelling plays an important role is in the assessment of the micromechanical environment of cells. Theoretical investigations at different structural levels (monolayers, multiple layers, three-dimensional constructs) help to improve the comprehensive understanding of micromechanical effects. As the design and development of load devices in bioreactors must incorporate the fact that small deformations generated by load transfer through the scaffold and the substrate surface to cells, it is important to realise that they will have profound effects on cell behaviour [50] and assessment of deformations at the various levels of hierarchy.

Finite element analysis (FEA) is commonly used to define the stress and strain fields, because other ex-

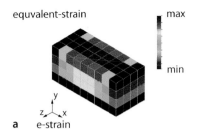

Fig. 44.1a,b Finite-element method used to calculate the strain environment in the mechanically loaded probes. Boundary conditions are modelled according to the stamp-induced deformation of the probes in the tissue chamber. The finite-element model predicts a nearly uniform and isotropic distribution of strain in the central portion of the collagen gel. Effects of activation of the stamp on the distribution of surface strain can be seen as the various colours on this graph. The range of total strains can be adjusted by the amplifier

perimental measures are not able to determine load-related deformations in the microenvironment of a cell-containing scaffold (Fig. 44.1). The accuracy of FEA is limited, as in most approaches the matrix structure is considered a homogeneous incompressible structure, whereas in reality the cell-containing scaffold is generally more complex. Another limitation is that bioreactor and scaffold variables are implemented on a simplified database. Despite these simplifications, calculated strain fields seem to reflect the real deformations in the tissue chambers. At the very least, the calculated magnitude that is biologically relevant allows discrimination between no, physiological, and hyperphysiological loads [60].

An advanced loading device for the application of specific compressive strains to tissue cylinders and measurement of resulting deformations was developed by Jones et al. [28]. The device allows it to load the specimen at frequencies between 0.1 and 50 Hz and amplitudes over 7,000 µstrain.

44.7
Experimental Data

As oxygen presents the main limiting factor in cell survival in bioreactors containing scaled-up artificial tissues, but also has a modulatory effect on cell differentiation, most experiments focus on the effects of oxygen on cell growth and differentiation. A higher oxygen tension was demonstrated, for example, for osteoblastic differentiation, whereas prolonged hypoxia favours the formation of cartilage or fibrous tissue in mesenchymal cell cultures [34]. The experimental findings also showed that at least some of the connective tissue progenitors in bone and bone marrow tended to survive under hypoxic conditions. Many stem and progenitor cells, including connective tissue progenitors in bone, exhibited a remarkable tolerance to, and are even stimulated by hypoxia, not unlike endothelial cells. In metabolically active tissues such as trabecular bone and bone marrow, for example, the distance that oxygen must diffuse between a capillary lumen and a cell membrane is almost never more than 40–200 µm. This diffusion distance is critical in maintaining the balance between oxygen delivery to a bone cell and consumption of oxygen by cells, both in native tissues and in extracorporeally engineered tissue. The adjustment of oxygen tension in bioreactors is therefore a critical aspect in bioreactor design.

As few cells tolerate diffusion distances of >200 µm, materials should be limited in their bulk size (particularly in the case of bone tissue engineering) [16]. Osteoblasts seeded on to porous scaffolds in vitro form a viable tissue that is no thicker than 0.2 mm. Similar observations have been reported for different cell types cultured in three-dimensional scaffolds under diffusion conditions. Deposition of mineralised matrix by stromal osteoblasts cultured into poly (DL-lactic-co-glycolic acid) foams reached a maximum penetration depth of 240 μm from the top surface.

Cartilage tissue is exceptional, in contrast to bone and most other tissues, because chondrocytes retain viability or may even be stimulated by a low oxygen concentration. Not unlike osteoblasts, chondrocyte function is dependent not only on the oxygen concentration present, but also to a large degree on various other nutrients. It has been shown that chondrocyte function is disturbed below a distinct level of nutritional supply. For example, glycosaminoglycan deposition by chondrocytes cultured on poly(glycolic acid) meshes was found to be poor in central parts of three-dimensional constructs. Stirring-induced fluid flow in the flask was accompanied by an increase in the synthesis of glucosaminoglycans and their fragments, which seems indicative of improved chondrocyte function through an improved delivery of metabolites [14].

As the mechanical forces imposed by muscular contractions, body movement and various other external loadings are continuously acting on the skeletal system, externally applied mechanical forces may improve the ex vivo formation of skeletal tissues. Proliferation of osteoblasts and chondrocytes, cell orientation, gene activity and other features of cellular activity can be modulated mechanically in ex-vivo-generated tissues [12]. Mechanical forces applied from outside or organised from within a scaffold are beneficial in bone engineering [51]. Engineering bone and cartilage substitutes in vitro has been developed towards advanced bioreactor systems that mimic not only the three-dimensional morphology [32], but also the mechanical situation of bones or joints [11]. The validity of this principle, particularly in the context of musculoskeletal tissue engineering, has been demonstrated by various studies [25]. Although improvement of the structural and functional properties of engineered bone and cartilage tissue by mechanical conditioning has been demonstrated by numerous proof-of-principle studies, the specific mechanical variables or regimens of application (magnitude, frequency, duration and mode of load application) for an optimal stimulation of bone or cartilage tissue remain mostly unknown [6].

Whereas most of the studies that have investigated the effects of load on osteoblasts were performed in monolayer cultures, and investigations of load-related chondrocyte reactions were done in explant cultures, new experimental data have confirmed that defined mechanical stimulation improves tissue formation in complex bioreactor systems (Figs. 44.2 and 44.3) [40].

Although measurement of strain and stresses in culture dishes and bioreactors can be done experimentally as well as theoretically, limitations of both approaches exist. Custom-designed bioreactors for the mechanical stimulation of skeletal tissues by defined deformations and experimental systems for the application of defined mechanical loads towards cells in culture have been developed [21, 25]. Some techniques use biaxial strain devices in an effort to apply homogeneous, isotropic strain to cells cultured in a monolayer, and use FEA to predict the appropriate geometry for a uniform strain field. The accuracy of the strain field calculations was confirmed experimentally [3]. A recently elaborated method to apply defined loads to three-dimensional tissue specimens is pressure application through lever-arm-connected stamps in a deformable tissue chamber. This method allows a precise application of stamp movement with the subsequent deformation of the scaffold, as revealed by the mathematical calculation of measurement of strain and stress fields within the specimen (Fig. 44.4) [40].

In the context of bone and cartilage fabrication, it is important to note that tissues at different stages of development or skeletal sites might require different management of mechanical conditioning. It also has to be considered that cell deformations resulting from mechanical loading inside a bulk material may differ from deformations of surface-bound cells [9]. Bone cells are far more sensitive to mechanical deformations than most other cell types (such as chondrocytes) because their physiological strain environment is much lower than that of other types of cell.

Several experimental studies have investigated the effects of load on cells in different types of bioreactor. As load plays a predominant role for the skeletal system, most studies have been performed with bony and cartilaginous cells. Different investigators embedded osteoblast-like cells in gels that were subjected to cyclical tension forces and found that cell orientation and gene activity were altered as a result of cyclical mechanical loading. The most effective frequency of this loading lay at around 1 Hz. Most in-vitro studies referred to mechanical stress as

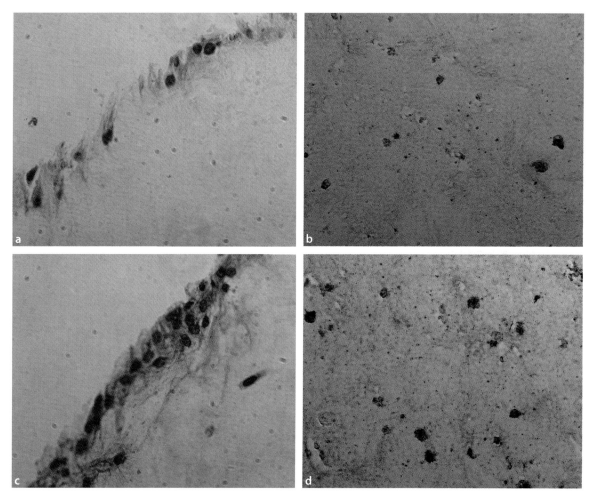

Fig. 44.2a–d Histological examination of the collagen gel and osteoblasts. Non-loading of osteoblasts (2000 µstrain, 200 cycles/day for 21 days) leads, in comparison to stimulated cells, to a reduced proliferation of osteoblasts (**a**) accompanied by a low synthesis of osteocalcin (**b**). Stimulated cells revealed a multilayer growth (**c**) and a more intense stain for osteocalcin (**d**). **a** and **c** Haematoxylin and eosin stain, original magnification ×40. **b** and **d** Immunohistological detection of osteocalcin synthesis in specimens at higher magnification (×40)

a stimulant of the proliferation of osteoblasts [4, 5, 26, 54]. The results of various mechanical stimulation studies revealed an altered expression of bone-specific proteins (alkaline phosphatase, osteopontin and osteocalcin) depending on the techniques used for loading [29, 36]. Effects ascribed to forces that exposed cells to physiological strains were also demonstrated on bone cells when they were cultured under fluid-flow-induced mechanical loads in spinner flasks [44, 45]. The sensitivity of osteoblasts to fluid shear stress is well established for cell cultures within flow chambers [24, 55]. Rotating-wall vessel reactors, which were originally designed to simulate a microgravity environment, have been tested mainly in cartilage and bone tissue engineering [2]. The centrifugal forces generated by definable and controlled rotation rates of the bioreactor also stimulate cells mechanically [52].

Recent investigations have shown that bone cells seeded onto polymer constructs and grown in rotating bioreactors displayed minimal differentiation towards the osteoblastic phenotype. A low alkaline phosphatase activity compared with static controls, a reduced extracellular matrix protein synthesis in the medium during long-term cultivation, and a reduced calcium deposition were found. These findings contradict earlier reports on the beneficial effects of culture in rotating-wall vessels on osteoblastic cells

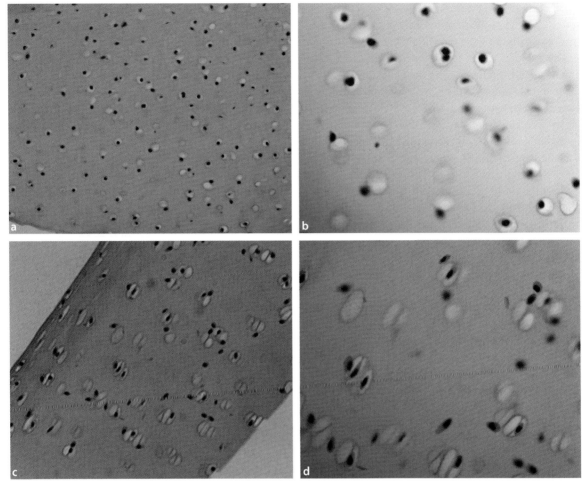

Fig. 44.3a–d Histological examination of cartilage construct (haematoxylin and eosin stain). **c** Chondrocytes in the specimens of cartilage activated for 21 days at 2000 µstrain proliferated more than unstimulated controls (**a**; original magnification ×10). **d** Cells had more mitotic activity under physiological loading as seen at a higher magnification (×40) compared with unstimulated specimens (**b**)

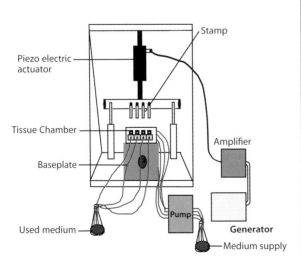

Fig. 44.4 Diagram and photograph of the bioreactor showing the mechanism for converting the motion of the piezoelectric-driven stamp into deformation of the probes in the tissue chamber. The frequency and type of load can be modulated by the frequency generator

[2]. The underlying causes of the low inductive properties of some of the rotating-wall vessel bioreactors in skeletal tissue formation are not understood, but unphysiological forces may be responsible for the experimental findings. The mechanisms whereby mechanical stimulation leads to cell proliferation and expression of tissue-specific genes reflect the in-vivo situation of load-induced osteoblast behaviour [16]. In general, all bioreactor systems that expose cell or polymer constructs to biophysical stimuli not exceeding a physiologically tolerable range may support the differentiation of precursor cells towards the mature phenotype and increase the function of mature cell types. The sum of experimental data confirms that engineered skeletal tissue provides a good example of how mimicking the native mechanical environment of cells can be beneficial for the ex vivo fabrication of bone or cartilage tissues [40]. Under conditions of optimised periodic strains, the mechanical properties of engineered tissues seem to improve significantly. In addition, the synthesis of matrix proteins and collagen, both being components of the secreted cell environment, can be improved by dynamic loading. Scaffold deformation on a physiological scale seems therefore to improve extracorporeal cell or scaffold maturation in bioreactors.

In contrast to the well-understood physiological microstrain environment for bone [19], less is known about the physiological load environment of cartilage [18]. As a mechanically favourable load environment also seems to be important for cartilage bioprocessing, many bioreactors were introduced to load cartilage explants by external forces, but only a few bioreactors were developed to provide a uniform and quantifiable mechanical environment in which to promote chondrocyte proliferation and matrix growth under well-defined and controllable conditions. Basically, spinner flasks, rotating-wall vessel bioreactors, and perfusion flow bioreactors were used to generate load-related cartilage tissue maturation. It was shown that chondrocyte proliferation and differentiation could be best achieved by dynamical load [5, 18, 35]. Bulk scaffold materials (gels) are, in contrast to bone tissue engineering, the preferred scaffold structure for cartilage tissue fabrication. An even more uniform cartilage tissue production was achieved in mechanically stimulated tissue cylinders, an effect attributed to the homogeneous hydrodynamic and favourable mass-transport kinetics in the bioreactor [35].

Based on the discovery of piezoelectric potentials in bone tissue in the late sixties, the idea of influencing bone cell behaviour in the in-vitro environment by the use of electrical fields was postulated [15]. Although many studies have investigated the effects of electromagnetic fields on bone cells and demonstrated modification of such cell behaviour, cellular mechanisms of action are not exactly understood. Electromagnetic effects at biomaterial surfaces were attributed as an influencing factor in cell behaviour, as material-specific effects work through electrical changes at the surface. Several studies were performed in engineering strategies for a precise analysis of electromagnetic effects in osteoblast-like primary cultures on cell proliferation or differentiation and initialisation of mineral formation, and effects at material interfaces. Recent investigations have shown that the presence of electric fields in bioreactors has distinct effects on osteoblasts in vitro [23] and increases mineral formation in cultured osteoblasts. Long-term electrical stimulation of osteoblasts in bioreactors alter the pattern of gene expression and biochemical processes at artificial surfaces, resulting in increased extracellular matrix synthesis, which leads to improved biomineral formation [59]. In this respect, the application of electrical fields in bioreactors seems to be a promising approach in extracorporeal bone tissue engineering. Piezoelectric potentials are also detected in loaded cartilage, but knowledge of the effects on cartilage tissue is limited. Approaches to stimulate chondrocyte-like cells with electrical fields in bioreactors are therefore seldom made.

References

1. Abukawa H., M. Shin, W.B. Williams, J.P. Vacanti, L.B. Kaban, M.J. Troulis (2004) Reconstruction of mandibular defects with autologous tissue-engineered bone. J Oral Maxillofac Surg 62:601–606
2. Botchwey E.A., S.R. Pollack, E.M. Levine, C.T. Laurencin (2001) Bone tissue engineering in a rotating bioreactor using a microcarrier matrix system. J Biomed Mater Res 55:242–253
3. Bottlang M., M. Simnacher, H. Schmitt, R.A. Brand, L. Claes (1997) A cell strain system for small homogeneous strain applications. Biomed Tech (Berl) 42:305–309
4. Brown T.D. (2000) Techniques for mechanical stimulation of cells in vitro: a review. J Biomech 33:3–14
5. Buschmann M.D., Y.A. Gluzband, A.J. Grodzinsky, E.B. Hunziker (1995) Mechanical compression modulates

matrix biosynthesis in chondrocyte/agarose culture. J Cell Sci 108:1497–1508
6. Butler D.L., S.A. Goldstein, F. Guilak (2000) Functional tissue engineering: the role of biomechanics. J Biomech Eng 122:570–575
7. Carter D.R., T.E. Orr (1992) Skeletal development and bone functional adaption. J Bone Miner Res 7:389S–395S
8. Carter D.R., G.S. Beaupré, N.J. Giori, J. Helms (1998) Mechanobiology of skeletal regeneration. Clin Orthop Relat Res 1:41S–55S
9. Casser-Bette M., A.B. Murray, E.I. Closs, V. Erfle, J. Schmidt (1990) Bone formation by osteoblast-like cells in a three-dimensional cell culture. Calcif Tissue Int 46:46–56
10. Chao E.Y., N. Inoue (2003) Biophysical stimulation of bone fracture repair, regeneration and remodeling. Eur Cell Mater 6:72–84
11. Cheng G.C., W.H. Briggs, D.S. Gerson, P. Libby, A.J. Grodzinsky, M.L. Gray, R.T. Lee (1997) Mechanical strain tightly controls fibroblast growth factor-2 release from cultured human vascular smooth muscle cells. Circ Res 80:28–36
12. Cheng M.Z., G. Zaman, S.C. Rawlinson, R.F. Suswillo, L.E. Lanyon (1996) Mechanical loading and sex hormone interactions in organ cultures of rat ulna. J Bone Miner Res 11:502–511
13. Curtis A., M. Riehle (2001) Tissue engineering: the biophysical background. Phys Med Biol 46:R47–R65
14. Darling E.M., K.A. Athanasiou (2003) Articular cartilage bioreactors and bioprocesses. Tissue Eng 9:9–26
15. Domen J. (2000) The role of apoptosis in regulating hematopoiesis and hematopoietic stem cells. Immunol Res 22:83–94
16. Duncan R.L., C.H. Turner (1995) Mechanotransduction and the functional response of bone to mechanical strain. Calcif Tissue Int 57:344–358
17. Einhorn T.A. (1996) Biomechanics of Bone, Principles of Bone Biology. Academic Press, New York, pp 25–37
18. Freed L.E., G. Vunjak-Novakovic (1997) Microgravity tissue engineering. In Vitro Cell Dev Biol Anim 33:381–385
19. Frost H.M. (2000) Why the ISMNI and the Utah paradigm? Their role in skeletal and extraskeletal disorders. J Musculoskelet Neuronal Interact 1:5–9
20. Frost H.M. (2000) The Utah paradigm of skeletal physiology: an overview of its insights for bone, cartilage and collagenous tissue organs. J Bone Miner Metab 18:305–316
21. Granet C., N. Laroche, L. Vico, C. Alexandre, M.H. Lafage-Proust (1998) Rotating-wall vessels, promising bioreactors for osteoblastic cell culture: comparison with other 3D conditions. Med Biol Eng Comput 36:513–519
22. Griffith L.G., G. Naughton (2002) Tissue engineering – current challenges and expanding opportunities. Science 295:1009–1014
23. Hartig M., U. Joos, H.P. Wiesmann (2000) Capacitively coupled electric fields accelerate proliferation of osteoblast-like primary cells and increase bone extracellular matrix formation in vitro. Eur Biophys J 29:499–506
24. Hillsley M.V., J.A. Frangos (1994) Bone tissue engineering: the role of interstitial fluid flow. Biotechnol Bioeng 43:573–581
25. Hung C.T., R.L. Mauck, C.C. Wang, E.G. Lima, G.A. Ateshian (2004) A paradigm for functional tissue engineering of articular cartilage via applied physiologic deformational loading. Ann Biomed Eng 32:35–49
26. Jones D., G. Leivseth, J. Tenbosch (1995) Mechanoreception in osteoblast-like cells. Biochem Cell Biol 73:525–534
27. Jones D.B., H. Nolte, J.G. Scholubbers, E. Turner, D. Veltel (1991) Biochemical signal transduction of mechanical strain in osteoblast-like cells. Biomaterials 12:101–110
28. Jones D.B., E. Broeckmann, T. Pohl, E.L. Smith (2003) Development of a mechanical testing and loading system for trabecular bone studies for long term culture. Eur Cell Mater 5:48–59 discussion 59–60
29. Krishnan L., J.A. Weiss, M.D. Wessman, J.B. Hoying (2004) Design and application of a test system for viscoelastic characterization of collagen gels. Tissue Eng 10:241–252
30. Langer R., J.P. Vacanti (1993) Tissue engineering. Science 260:920–926
31. Lanyon L.E., W.G.J. Hampson, A.E. Goodship, J.S. Shah (1975) Bone deformation recorded in vivo from strain gauges attached to the human tibial shaft. Acta Orthop Scand 46:256–268
32. Lee A.A., T. Delhaas, L.K. Waldman, D.A MacKenna, F.J. Villarreal, A.D. McCulloch (1996) An equibiaxial strain system for cultured cells. Am J Physiol 271:C1400–C1408
33. Lian J.B., G.S. Stein (1992) Concepts of osteoblast growth and differentiation: basis for modulation of bone cell development and tissue formation. Crit Rev Oral Biol Med 3:269–305
34. Malda J., D.E. Martens, J. Tramper, C.A. van Blitterswijk, J. Riesle (2003) Cartilage tissue engineering: controversy in the effect of oxygen. Crit Rev Biotechnol 23:175–194
35. Martin I., B. Obradovic, S. Treppo, A.J. Grodzinsky, R. Langer, L.E. Freed, G. Vunjak-Novakovic (2000) Modulation of the mechanical properties of tissue engineered cartilage. Biorheology 37:141–147
36. Masi L., A. Franchi, M. Santucci, D. Danielli, L. Arganini, V. Giannone, L. Formigli, S. Benvenuti, A. Tanini, F. Beghè, et al (1992) Adhesion, growth, and matrix production by osteoblasts on collagen substrata. Calcif Tissue Int 51:202–212
37. Meyer U., H.P. Wiesmann (2005) Tissue engineering: a challenge of today's medicine. Head Face Med 1:2
38. Meyer U., H.D. Szulczewski, K. Moller, H. Heide, D.B. Jones (1993) Attachment kinetics and differentiation of osteoblasts on different biomaterials. Cells Mater 3:129–140
39. Meyer U., T. Meyer, D.B. Jones (1997) No mechanical role for vinculin in strain transduction in primary bovine osteoblasts. Biochem Cell Biol 75:81–86
40. Meyer U., U. Joos, H.P. Wiesmann (2004) Biological and biophysical principles in extracorporal bone tissue engineering. Part III. Int J Oral Maxillofac Surg 33:635–641
41. Meyer U., A. Buchter, N. Nazer, H.P. Wiesmann (2006) Design and performance of a bioreactor system for me-

chanically promoted three-dimensional tissue engineering. Br J Oral Maxillofac Surg 44:134–140
42. Meyer U., B. Kruse-Losler, H.P. Wiesmann (2006) Principles of bone formation driven by biophysical forces in craniofacial surgery. Br J Oral Maxillofac Surg 44:289–295
43. Muschler G.F., C. Nakamoto, L.G. Griffith (2004) Engineering principles of clinical cell-based tissue engineering. J Bone Joint Surg Am 86-A:1541–1558
44. Ogata T. (2000) Fluid flow-induced tyrosine phosphorylation and participation of growth factor signaling pathway in osteoblast-like cells. J Cell Biochem 76:529–538
45. Pavalko F.M., N.X. Chen, C.H. Turner et al (1998) Fluid shear-induced mechanical signaling in MC3T3-E1 osteoblasts requires cytoskeleton-integrin interactions. Am J Physiol 275:C1591–C1601
46. Plate U., S. Arnold, U. Stratmann, H.P. Wiesmann, H.J. Hohling (1998) General principle of ordered apatitic crystal formation in enamel and collagen rich hard tissues. Connect Tissue Res 38:149–157
47. Plate U., T. Polifke, D. Sommer, J. Wunnenberg, H.P. Wiesmann (2006) Kinetic oxygen measurements by CVC96 in L-929 cell cultures. Head Face Med 2:6
48. Ratcliffe A., L.E. Niklason (2002) Bioreactors and bioprocessing for tissue engineering. Ann N Y Acad Sci 961:210–215
49. Risbud M.V., M. Sittinger (2002) Tissue engineering: advances in in vitro cartilage generation. Trends Biotechnol 20:351–356
50. Rubin C.T., L.E. Lanyon (1987) Kappa Delta Award paper. Osteoregulatory nature of mechanical stimuli: function as a determinant for adaptive remodeling in bone. J Orthop Res 5:300–310
51. Schmelzeisen R., R. Schimming, M. Sittinger (2003) Making bone: implant insertion into tissue-engineered bone for maxillary sinus floor augmentation – a preliminary report. J Craniomaxillofac Surg 31:34–39
52. Schwarz R.P., T.J. Goodwin, D.A. Wolf (1992) Cell culture for three-dimensional modeling in rotating-wall vessels: an application of simulated microgravity. J Tissue Cult Methods 14:51–57
53. Spadaro J.A. (1997) Mechanical and electrical interactions in bone remodeling. Bioelectromagnetics 18:193–202
54. Stein G.S., J.B. Lian (1993) Molecular mechanisms mediating proliferation/differentiation interrelationships during progressive development of the osteoblast phenotype. Endocr Rev 14:424–442
55. Toma C.D., S. Ashkar, M.L. Gray, J.L. Schaffer, L.C. Gerstenfeld (1997) Signal transduction of mechanical stimuli is dependent on microfilament integrity: identification of osteopontin as a mechanically induced gene in osteoblasts, J Bone Miner Res 12:1626–1636
56. Van Eijden T.M. (2000) Biomechanics of the mandible. Crit Rev Oral Biol Med 11:123–136
57. Vunjak-Novakovic G. (2003) The fundamentals of tissue engineering: scaffolds and bioreactors. Novartis Found Symp 249:34–46 discussion 46–51, 170–4, 239–41
58. Warnke P.H., I.N. Springer, J. Wiltfang, Y. Acil, H. Eufinger, M. Wehmöller, P.A. Russo, H. Bolte, E. Sherry, E. Behrens, H. Terheyden (2004) Growth and transplantation of a custom vascularised bone graft in a man. Lancet 364:766–770
59. Wiesmann H., M. Hartig, U. Stratmann, U. Meyer, U. Joos (2001) Electrical stimulation influences mineral formation of osteoblast-like cells in vitro. Biochim Biophys Acta 1538:28–37
60. Winston F.K., E.J. Macarak, S.F. Gorfien, L.E. Thibault (1989) A system to reproduce and quantify the biomechanical environment of the cell. J Appl Physiol 67:397–405
61. Wu W., X. Feng, T. Mao et al (2007) Engineering of human tracheal tissue with collagen-enforced poly-lactic-glycolic acid non-woven mesh: a preliminary study in nude mice. Br J Oral Maxillofac Surg 45:272–278

Microenvironmental Determinants of Stem Cell Fate

R. L. Mauck, W-J. Li, R. S. Tuan

Contents

45.1	Introduction	647
45.1.1	Developmental Determinants of Cell Fate	647
45.1.2	Mechanical and Topographic Control of Musculoskeletal Development	648
45.1.3	Developmental Signals in Regenerative Medicine	648
45.2	Stem Cells for Regenerative Medicine	648
45.3	Engineered Scaffolds for Regenerative Medicine	649
45.4	In Vitro Modulation of the Stem Cell Fate Through Alterations in the Microenvironment	650
45.4.1	Passive Modulation of Stem Cell Fate	650
45.4.2	Active Modulation of Stem Cell Fate	656
45.5	Summary and Future Directions	659
	References	659

45.1 Introduction

45.1.1 Developmental Determinants of Cell Fate

Within the developing embryo, undifferentiated cells make fate decisions en route to their assumption of an adult phenotype and biologic function. In simple model systems, such as *Caenorhabditis elegans*, the fate of each cell has been mapped to its final disposition, and as such, the decision tree can be investigated in detail for factors that modulate each lineage specification [70]. In more complex systems, including mammals, no such cell-specific decision tree exists. Nevertheless, much insight has been gained by studying the formation of specific organs and musculoskeletal units and their fate determinants.

Fate decisions toward musculoskeletal phenotypes can be driven by several factors, with differentiation representing a decision based on the sum of these inputs, including what the cells are acting upon, and what is acting upon them. It has long been appreciated that soluble factors are primary determinants in cells transitioning from one phenotype to another. Early in embryonic development, with the formation of the Spemann organizer, local gradients of soluble factors are generated that initiate partitioning of the embryo into substructures (i.e., endoderm, ectoderm) [49]. In the developing limb bud, fibroblast growth factor signaling from the apical ectodermal ridge plays a role in defining the extent of limb progression and its eventual bifurcations into discrete elements [49]. Alterations in the origin or amount of these soluble cues can have severe consequences on the shape of the formed appendage [116].

45.1.2
Mechanical and Topographic Control of Musculoskeletal Development

In addition to soluble cues, the mechanics and topography of the microenvironment in which cells reside has been implicated in fate decisions and tissue differentiation. This microenvironment includes several factors such as cell density, adhesion, organization and spatial cues, matrix stiffness, and intrinsic and extrinsic mechanical forces. For example, in the developing limb, the condensation of a large number of cells into a small volume is required for the chondrogenic differentiation of limb bud cells [32]. These aggregates form the cartilage anlagen (or template) that defines the position and extent of bone in the adult organism. In developing tendon, meniscus, and annulus fibrosus, cell alignment is prescribed such that matrix deposition occurs in a prevailing direction [25, 54, 55]. This ordered matrix deposition engenders the direction-dependent mechanical properties and function of these tissues in the adult. Mechanical forces may arise in developing tissues by virtue of differential growth at two conjoined surfaces—a concept termed growth-generated stresses and strains [56]. For example, the swelling cartilage rudiment engages the surrounding perichondrium as it increases in size and imbibes water due to charged proteoglycan accumulation [57]. If the perichondrium is breached, this growth-induced tensile stress fails to develop, and chondrogenesis in the entire limb bud is altered [3]. More explicitly, after muscles develop, active force generation influences the differentiation of structures within their sphere of influence [99, 100]. For example, blocking muscle contraction in ovo in a developing chick results in misshapen limbs that lack functional joints. Indeed, in the tibiofemoral joint, regression of soft-tissue elements (meniscus, cartilage, ligaments) occurs and the bony elements of the joint fuse in the absence of mechanical loading [91, 92, 94]. At ligament-to-bone insertion sites, mature function is only achieved with joint motion and load-bearing use [134]. Clearly, these microenvironmental topographical and mechanical factors are important in developmental processes, and as such warrant further study in controlled experimental settings.

45.1.3
Developmental Signals in Regenerative Medicine

Tissue-engineering strategies have been developed for the de novo formation of adult tissues with functional properties. Investigation of the microenvironmental determinants of cell fate and tissue development during skeletogenesis may provide helpful insights for optimization of these regenerative strategies. Indeed, considerable efforts have focused on controlling the micro- and macromechanical environment to promote maturation of tissue-engineered constructs [52, 61, 66]. We posit that not only are these adult post-natal signals important, but that this focus should be extended to encompass developmentally relevant signals, including loading durations, duty cycles, and magnitudes. For example, while the adult heart beats ~60 times per minute, the fetal heart beats at more than twice that rate. In ovo, the motile activity of the developing chick embryo peaks around incubation day 14, and then diminishes as further growth constrains movement [93]. While not measured, muscle contraction changes from isotonic to isometric under these circumstances [94]. Thus, mechanical conditioning aimed at rapid maturation of neo-tissue may benefit from a developmentally inspired approach. Comprehending the formation, maturation, and repair mechanisms at work in these developing tissues may provide insights into adult tissue repair and the most appropriate culture conditions for tissue regeneration.

45.2
Stem Cells for Regenerative Medicine

In order to adequately test developmental paradigms for driving neo-tissue formation in regenerative medicine, a versatile and plastic cell type is required.

To date, the most common practice in tissue engineering is to employ "primary, fully differentiated," cells isolated from the adult tissue of interest. These cells by definition possess the appropriate phenotype for producing differentiated tissues; however, their clinical applicability is limited. This is due to the scarcity of healthy cells, the potentially diseased nature of the source tissue, complications with tissue harvest, and the widely noted decrease in the capacity of primary cells due to aging or senescence [20, 68, 129].

Attention has therefore focused on available cells that possess a multilineage differentiation potential. Embryonic stem (ES) cells are the most primordial of cell types and are totipotent (i.e., they are able to differentiate along all lineages required to establish the adult form). ES cells have increasingly been used in regenerative medicine therapies [39, 65], although concerns related to the stability and homogeneity of the differentiated phenotype and the potential for teratoma formation currently limit their clinical application. The technique of controlling the fate of ES cells in vitro is complicated and far from mature. Recent reports have also shown that adult cells can be modified to take on pluripotent characteristics [106, 126]. Specifically, adult dermal fibroblasts can be reprogrammed toward an ES-like state with manipulation of several critical transcription factors responsible for maintenance of cell "stemness." These induced pluripotent stem cells are promising, offering as they do the potential for patient-specific cell therapies, although they, like ES cells, remain far from clinical application.

More widely studied for musculoskeletal applications are adult tissue-derived mesenchymal stem cells (MSCs). These cells are a versatile cell type that can progress through multiple lineage-specific pathways when provided with the appropriate soluble and microenvironmental cues [10, 18, 123]. MSCs are easily isolated from bone marrow aspirates via plastic adherence [46] or by selecting for surface markers (i.e., CD105 and STRO-1 [78, 79, 118]). In addition to bone marrow, comparable cell types have also been isolated from adipose tissue, periosteum, trabecular bone chips, cartilage, and meniscus [41, 85, 98, 104, 127]. MSCs are characterized by their ready expansion capacity [15] and their ability to differentiate along numerous mesenchymal pathways, including fat, muscle, bone, and cartilage [17, 110, 111]. MSC chondrogenic differentiation is initiated in high-density pellet cultures with growth factors from the transforming growth factor-β (TGF-β) superfamily, including TGF-β1 [6, 34, 63] and TGF-β3 [77], and bone morphogenetic protein-2 [33, 79] in the presence of ascorbate and dexamethasone. Osteogenic differentiation occurs in monolayer culture in serum containing medium supplemented with β-glycerol phosphate and 1,25-dihydroxy vitamin D_3 [9, 12]. Likewise, adipogenic differentiation can be achieved in monolayer culture in serum containing medium with the addition of insulin and dexamethasone [124]. Most strikingly, these cells can be induced to transdifferentiate directly from one mesenchymal lineage to another with changes in microenvironmental factors [124]. These cells thus provide a useful framework for studying musculoskeletal differentiation in vitro.

45.3 Engineered Scaffolds for Regenerative Medicine

Tissue-engineering strategies combine differentiated or undifferentiated target cells expanded in vitro with 3D, natural or synthetic biomaterial scaffolds. The cellular scaffold in the presence of growth-supporting stimuli gradually turns into a natural tissue-like cellular implant. Biologically functional scaffolds play a critical role in the process of tissue regeneration, as they serve as a vehicle for delivering cells to the implant site as well as providing a 3D structure for cell attachment, migration, proliferation, and differentiation. In a sense, the scaffold is an artificial extracellular matrix (ECM). Cells are in contact with ECM that is synthesized, organized, and maintained by the cells themselves, while the ECM functions as a physical protector to the resident cells and acts to regulate cellular activities [115]. In native tissues, the ECM continuously undergoes dynamic turnover in response to microenvironmental signals transmitted to the cells within [50]. Extrapolating from these native interactions, a scaffold that serves as a functional, temporary ECM must involve optimal cell-matrix interactions, as well as provide biologically favorable cell–cell interactions. The success of the tissue-engineering process is highly dependent on the chemical and physical properties of the biomaterial scaffold. Whether the biomaterial scaffold serves as a 3D matrix for in vitro culture or functions as a template to recruit surrounding host cells to conduct the repair process, a principal objective of scaffolds is to simulate native ECM until cells seeded within the scaffold and/or derived from the host tissue can synthesize a new, natural matrix.

3D cell culture in engineered scaffolds, compared to 2D culture on culture plastic surfaces, more closely resembles the in vivo cellular environment and creates biologically favorable cell–cell and cell–matrix interactions. The biological relevance of 3D culture is most likely a consequence of cell–matrix interactions that proceed in a complicated, bidirectional (outside-in and inside-out) manner [28], and are likely to be critical for effective tissue regeneration. The architecture of the 3D biomaterial scaffold has a macroarchitecture that is defined by the gross structure of the entire scaffold and can be fabricated into a desired dimension and shape, depending on the type of tissue being engineered. On the other hand, the microarchitecture is defined by the building-component structures, such as spheres, fibers, or gels, and their surface topographies and adhesive/interactive characteristics. The mechanical properties of the scaffold and transport efficiency of nutrient/waste between the scaffold and the environment are determined by both the macro- and microarchitecture, whereas the biological response to the scaffold is regulated primarily by the microarchitecture of the cell–material interface.

For studies of cell biology and tissue-engineering applications, the selection of scaffold materials and architectures should be properly considered to ensure biocompatibility with the seeded cells and to promote specific cellular function. An ideal scaffold should be biocompatible before and after degradation of the scaffolding material, promote cellular activities, and possess favorable mechanical properties [62]. It is known that both the chemical [69] and physical properties [11] of a scaffold affect cell behavior. The mechanism of cell regulation is controlled by the combination of identity, density, and presentation of the ligand present on the surface of surrounding scaffolds or cells. Numerous attempts, through the use of natural polymers [140] or synthetic polymers incorporating bioactive peptides [58], have added, increased, or modified biologically favorable ligands to improve the chemical biocompatibility of scaffolds. On the other hand, scaffold architecture and mechanical properties are also known to regulate cell response [45]. As is described through the remainder of the chapter, scaffolds and material surfaces, in combination with defined external mechanical environments, can provide a complex interactive environment at numerous length scales to control progenitor cell fate decisions.

45.4
In Vitro Modulation of the Stem Cell Fate Through Alterations in the Microenvironment

Various in vitro systems have been developed to investigate microenvironmental control of stem cell fate. We separate these into two domains: passive systems and active systems. Passive systems are so noted as the interaction is mediated by the stem cell's outward perception and response to its microenvironment. Active systems are those in vitro methods in which a defined mechanical signal is superimposed onto the existing passive microenvironment to modulate differentiation. Collectively, these in vitro stimulation systems can recapitulate numerous features of embryonic development and adult function.

45.4.1
Passive Modulation of Stem Cell Fate

Adhesion is the first event to take place when cells are seeded onto a biomaterial surface or come into contact with one another, and passive systems exert their influence through this mechanism. Cell migration, proliferation, and/or differentiation can take place only after cells are securely adhered [51]. In the native biologic milieu, cells bind to ECM ligands in their microenvironment, consisting of specific peptide sequences, via their integrin receptors [48]. In contrast, cell adhesion to a polymeric material or scaffold is necessarily mediated by plasma/serum proteins adsorbed onto the polymer surface [97]. Protein adsorption to the polymer surface is affected by the hydrophilicity [120], the surface energy of the polymer [13], and/or the scaffolding architecture [136].

45.4.1.1 Cell–Cell Contact, Extracellular Adhesion, and Cell Fate

One of the first cell types to be investigated for musculoskeletal differentiation was limb-bud mesenchymal cells. When placed in high-density micromass cultures, these cells first transition through a chondrocyte-like phenotype to a bone-like phenotype [32]. Control of early chondrogenic fate decisions by limb-bud cells is dependent on a close cellular interaction that can be modulated by cationic molecules such as poly-L-lysine (allowing tighter cell packing and enhancing proteoglycan retention) [113]. In addition, early expression and local deposition of ECM elements that populate the microenvironment, including exon IIIA of fibronectin, are critical for condensation events to proceed [47]. When first established, culture systems for adult MSC chondrogenesis used high-density pellets to mimic limb bud and micromass cultures [53, 63], with many of the same signaling pathways observed [26, 43, 44, 116]. As with embryonic limb-bud cells, cell–cell contact plays a significant role in MSC differentiation. Incubation of N-cadherin blocking antibodies abrogates MSC chondrogenesis in pellet cultures [130]. More recent studies have shown that cell-to-cell contact is not an absolute necessity, as MSC chondrogenesis and osteogenesis can occur in isolated cells seeded in 3D scaffolds such as fibrin [137], agarose [7, 88], hyaluronan (HA)/gelatin [4,7], polyethylene glycol (PEG) [1, 135], and alginate hydrogels [7, 41, 76, 79], as well as in polyester foams [82] and polyester/alginate amalgams [21, 22]. In 3D hydrogels, MSC viability and differentiation is dependent on the adhesive microenvironment. For example, inclusion of RGD-moieties enhances MSC interaction with non-adhesive 3D hydrogels, and promotes cell viability [103, 112]. However, in this same context, too much adhesion can limit chondrogenesis, as evidenced by recent studies in RGD-modified alginate showing decreased levels of ECM production by MSCs with increasing adhesivity [27]. In one recent study [24] comparing photocrosslinkable PEG gels with photocrosslinkable HA gels, human MSCs underwent chondrogenesis to a greater extent in the more natural, biologically inspired, 3D HA environment (Fig. 45.1).

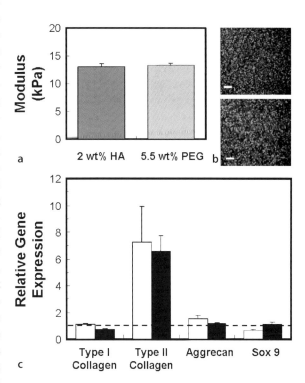

Fig. 45.1a–c Biomaterial control of mesenchymal stem cell (MSC) differentiation. MSC chondrogenesis is enhanced in biologic hyaluronan (HA) hydrogels compared to synthetic polyethylene glycol (PEG) hydrogels. MSCs were cultured in gels of similar properties (**a**, compressive modulus) yielding constructs with similar MSC viability (**b**) 24 h after polymerization (scale bar = 200 μm; top: HA, bottom: PEG). **c** Chondrogenic gene expression after 7 (*white*) and 14 days (*black*) was enhanced in HA gels compared to PEG gels. *Significant difference ($p < 0.05$) between HA and PEG hydrogels. Adapted from Chung and Burdick (2008) [24] with permission from Mary Ann Liebert

45.4.1.2
Surface and Matrix Elasticity and Cell Fate

In addition to specific cell–cell and cell–ECM interactions, the stiffness of the external milieu can influence MSC differentiation. This finding has its antecedents in cancer biology, where transformation from quiescent to invasive phenotypes is determined at least in part by the surrounding matrix stiffness [109]. For MSCs, external stiffness also controls lineage specification. For example, MSCs seeded onto collagen-coated acrylamide gels exhibited lineage-specific expression profiles based on the substrate stiffness. Neurogenic differentiation was observed on soft substrates mimicking brain, myogenic differentiation was observed on intermediate stiffness matrices typical of muscle, and osteogenic differentiation was observed on harder matrices representing newly formed osteoids [40]. At present it is unclear how these stiffness changes might influence MSC differentiation when the cells are completely surrounded by a 3D matrix, as changing the mechanical properties of the 3D environment can alter several other factors, including permeability and diffusivity. Nevertheless, further study of this intriguing phenomenon is warranted.

45.4.1.3
Shape Effects on Cell Fate

As described above, cell shape and substrate topography play an important role in fate decisions during development. Several in vitro studies have explored this concept in monolayer cultures at varying seeding densities and on patterned surfaces. Most interestingly, the ability to spread on a substrate in low-density cultures promoted MSC osteogenesis, while maintenance of a rounded shape promoted adipogenesis [89]. In this same study, the degree of spreading was controlled with micropatterned islands, with the finding that a smaller cell footprint promoted adipogenesis at the expense of osteogenesis (Fig. 45.2). Moreover, disruption of the actin cytoskeleton (a determinant of cell shape and active tension within the cell) prompted this same fate decision. Actin cytoskeleton organization is also important in chondrogenic fate decisions. It has long been appreciated that disruption of actin polymerization spurs the redifferentiation of culture-expanded chondrocytes [14], and it has recently been reported that this same treatment can promote chondrogenesis in embryoid-body-derived ES cells [141].

45.4.1.4
Two- and Three-Dimensional Topography Effects on Cell Fate

Cells behave differently when cultured on substrates with varying topography, suggesting that they are capable of distinguishing geometric properties of substrates, such as shape and/or roughness. For example, substrate surface roughness has a regulatory effect on cell morphology, cytoskeletal organization, and proliferation [30, 31, 119]. When osteoblasts are cultured in a nanophase ceramic, compared to micro-grain-size ceramics, adhesion, proliferation, synthesis of alkaline phosphatase, and deposition of a calcium-containing mineral were all enhanced [132, 133]. More recently, these same surface modifications producing semiordered and random nanotopographies have been shown to alter the degree of MSC osteogenesis [29]. Favorable biological functions of osteoblasts are increased when cultured on carbon nanofibers of decreasing diameter [38]. Nanoscale parallel ridges of increasing heights progressively promote cell alignment on patterned surfaces [81], and nanogratings (0.35–10 μm length scale) promote MSC alignment and differentiation toward a neuronal lineage [139]. These findings suggest that a scaffold composed of nanometer-scale components and/or topographical features is biologically favorable and that nanometer-scale structural components should be preferred for the fabrication of functional tissue-engineered scaffolds.

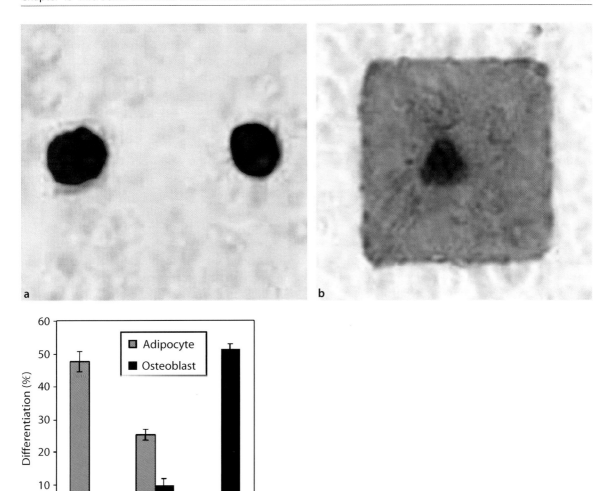

Fig. 45.2a–c Substrate topography and cell size influence MSC differentiation. MSCs were plated onto adhesive islands of differing sizes and cultured in the presence of media supportive of both osteogenic and adipogenic differentiation. Smaller islands (**a**) induced adipogenic differentiation, while large islands (**b**) induced osteogenic differentiation, as evidenced by staining of lipid deposition and alkaline phosphatase activity, respectively. **c** Quantification of adipogenic and osteogenic differentiation over a range of island sizes supports this finding. Adapted from McBeath et al. (2004) [89] with permission from Elsevier

45.4.1.5
Nanotopography in 3D Culture: Electrospun Nanofibrous Scaffolds

Given the potential of nanofeatured extracellular microenvironments, our group has focused extensively on the production and characterization of nanofibrous scaffolds created through electrospinning. Among nanostructures, nanofibers are particularly suitable for use as scaffolding components. The unique features of nanosized fibers and natural ECM-like fiber morphology [72] improve scaffolding physical properties, including high porosity, variable pore-size distribution, and a high surface-to-volume ratio, and most importantly, enhance favorable biological activities. Woo et al. recently reported that nanofibrous scaffolds show enhanced adsorption of cell-adhesion ECM molecules, which may enhance cell adhesion [136]. The high surface-area-to-volume ratio of nanofibrous scaffolds should result in greater adsorption of adhesion molecules than other scaffolds. In fact, approximately four times as much serum proteins adsorbed to nanofibrous scaffolds compared to scaffolds with solid pore walls [136]. Although the mechanisms by which a nanofibrous scaffold acts as a selective substrate are not yet known, it is clear that the enhanced adsorption of adhesion molecules enhances cell adhesion on nanofibrous scaffolds.

Previous reports have demonstrated that nanofibers as scaffolds are more biologically favorable when compared to microfibers, suggesting that cells can distinguish between fiber diameters and respond differently on the same material [71, 138]. One such study showed that chondrocytes seeded into microfibrous poly-L-lactic acid scaffolds spread and displayed a dedifferentiated, fibroblast-like morphology, whereas chondrocytes on nanofibrous scaffolds maintained a chondrocyte-like morphology (Fig. 45.3) [71]. Cell proliferation and the production of cartilaginous ECM were enhanced in nanofibrous cultures compared to microfibrous cultures. Another study reported a greater percentage of neural stem cells exhibiting a neuron-like morphology on nanofibrous scaffolds compared to microfibrous scaffolds [138]. Nanofibrous structures physically resemble basement membrane, and have been used to culture ES cells to maintain their "stemness" [102]. ES-cell-seeded nanofibrous cultures maintained significantly larger colonies of undifferentiated cells and enhanced proliferation compared to controls. It has also been reported that MSCs maintained in nanofibrous scaffolds were capable of differentiating along adipogenic, chondrogenic, or osteogenic lineages [74].

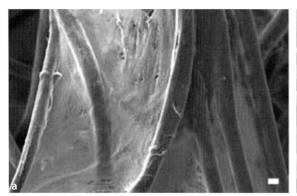
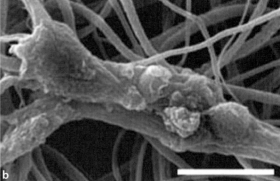
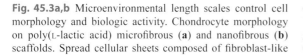

Fig. 45.3a,b Microenvironmental length scales control cell morphology and biologic activity. Chondrocyte morphology on poly(L-lactic acid) microfibrous (**a**) and nanofibrous (**b**) scaffolds. Spread cellular sheets composed of fibroblast-like cells spanned between microfibers after 28 days, while cellular aggregates composed of globular, chondrocyte-like cells grew on nanofibers (scale bar = 10 µm). Adapted from Li et al. (2006) [71] with permission from Mary Ann Liebert

The unique architectural features of nanofibrous scaffolds, especially their ultrafine structure, are generally believed to play a direct and/or indirect role in the regulation of cellular activities. However, studies on cell-nanofiber interactions are relatively new, and the specific mechanisms by which nanofiber properties regulate cellular activity are largely unknown. From recent studies, it is increasingly clear that integrin receptors, cytoskeleton, and signaling pathways involving focal adhesion kinase (FAK), Rho, Rac, and Cdc42 GTPases play a significant role [101, 114]. Schindler et al. demonstrated that cells cultured on a nanofiber surface have less-defined, punctate patterns of vinculin and FAK (molecules mediating cell adhesion) at the edge of lamellipodia [114], while these same cells on flat glass surfaces accumulated vinculin and FAK in a streaky pattern and promoted actin filament formation. The decrease of FAK at the adhesion sites and the formation of cortical actin are characteristics of "3D-matrix adhesion," which is different from the focal adhesion and fibrillar adhesion commonly formed on a flat surface [28]. Thus, cells establish 3D-matrix adhesion in nanofibrous scaffolds, and such an in vitro culture may bring the cells closer to the in vivo microenvironment.

In addition to the formation of 3D-matrix adhesion, the family of Rho GTPases, Rho, Rac, and Cdc42, each of which controls distinct downstream signaling pathways, is regulated by the nanofibrous structure [101]. Cells cultured on a nanofibrous surface extensively activate Rac but not Rho or Cdc42. This finding that nanofibrous scaffolds can induce preferential Rac activation suggests that nanofibrous scaffolds provide physical as well as spatial cues to activate intracellular signaling pathways that are essential to mimic in-vivo-like tissue growth. Indeed, we have shown that cell organization can be tuned on these nanofibrous scaffolds by controlling their organization. Altering the collection of nanofibers by introducing a rotating mandrel produced an array of scaffold organizations, ranging from randomly distributed fibers to those that were fully aligned [73]. This alignment dictated immediate cellular alignment and actin organization (Fig. 45.4) [73], as well as the orientation and mechanical properties of ECM deposited by MSCs after 10 weeks in culture [8]. Although not yet established, it is likely that this novel 3D topographical network will further influence stem cell fate decisions, just as ordered topography controls fate decisions in the developing embryo.

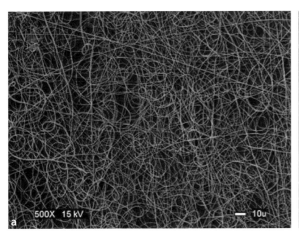
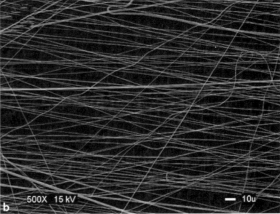

Fig. 45.4a–d Extracellular scaffold topography controls MSC orientation and cytoskeletal architecture. Human MSCs seeded on random (**a**) and aligned (**b**) nanofibrous scaffolds adopt different morphologies after 24 h in culture. Similarly, the actin cytoskeleton of these cells is differentially regulated by the three-dimensional (3D) topography, with a fine actin network observed on random (**c**) scaffolds, and a dense organized array of actin stress fibers observed on aligned (**d**) scaffolds (scale bar = 10 μm). Adapted from Li et al. (2007) [73] with permission from Elsevier

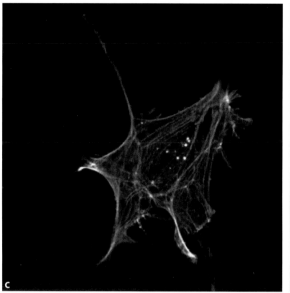
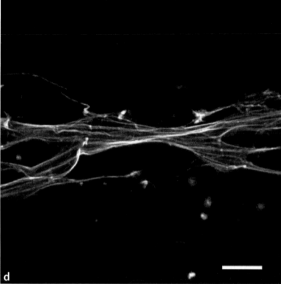

Fig. 45.4a–d *(continued)* Extracellular scaffold topography controls MSC orientation and cytoskeletal architecture. Human MSCs seeded on random (**a**) and aligned (**b**) nanofibrous scaffolds adopt different morphologies after 24 h in culture. Similarly, the actin cytoskeleton of these cells is differentially regulated by the three-dimensional (3D) topography, with a fine actin network observed on random (**c**) scaffolds, and a dense organized array of actin stress fibers observed on aligned (**d**) scaffolds (scale bar = 10 μm). Adapted from Li et al. (2007) [73] with permission from Elsevier

45.4.2
Active Modulation of Stem Cell Fate

Active systems apply defined mechanical loading to cells in culture. Cells can be mechanically stimulated in 2D (on flat or textured surfaces) or 3D (embedded or seeded within an engineered network). Ideally, the applied mechanical loading environment would in some way replicate an in vivo loading condition, or at least a portion thereof, that arises during development or with adult load-bearing use. Recent studies have shown that even fully differentiated tissues alter their phenotype with changes in the mechanical environment [80, 83]. This appreciation of physical forces in the development and maintenance of cellular phenotype is clearly related to tissue-engineering approaches based on stem cells. Indeed, some of the very signals that cause differentiation may be used to optimize construct growth [86, 87]. A growing body of literature has begun to explore these mechanically induced microenvironmental effects on MSCs differentiation and subsequent tissue maturation.

45.4.2.1
Static and Dynamic Stretch

One of the most common physical stimulation methods is the application of tensile strain. This is a developmentally relevant signal, with static stretch arising from differential growth of two conjoined layers, and dynamic stretch occurring when muscles in the developing embryo begin regular contraction. Static stretch applied to the substrate on which limb-bud micromasses are cultured causes a decrease in chondrogenic differentiation in an RGD-dependent fashion, implicating integrin binding in this developmental event [105]. Similar studies using adult MSCs have shown that static and dynamic stretch can also regulate fate decisions. For example, cyclic substrate stretch enhances MSC osteogenesis via ERK1/2 signaling [117]. More recently, it has been demonstrated that dynamic stretch regimens not only enhance osteogenesis, but actively downregulates other mesenchymal lineages (such as adipogenesis and chondrogenesis) [131].

The mode of stretch may also influence the response of MSCs to mechanical signals. For example, one recent study compared the effect of equiaxial strain (equal in perpendicular directions) versus uniaxial strain (in only one direction) of the substrate on MSC differentiation [107]. Results from this study showed that dynamic uniaxial, but not equiaxial, strain yielded a transient increase in collagen type I production by MSCs, suggesting that the cells may differentially perceive the directionality of the applied tensile deformation. This concept of anisotropic (different in different directions) mechanosensing has been further investigated by seeding MSCs on parallel microgrooves created by soft lithography. Normally, MSCs will reorient to minimize the distance over which strain is applied such that cells align perpendicular to the direction of applied stretch. By forcing cells to take on a specific orientation along microgrooves, and applying dynamic substrate strain in that direction, a greater molecular response was observed [67]. These findings demonstrate that the combination of active (stretch) and passive (organized microgrooves) determinants of cell fate can produce nonintuitive findings that have developmental implication.

In addition to these 2D-substrate-stretching studies, an increasing number of investigators have examined the effect(s) of static and dynamic stretch on MSC-seeded 3D constructs. This focus is motivated by the need for tissue-engineered constructs (such as tendons and ligaments, for example) that are exposed to such physical forces in vivo. Several custom-built and commercially available systems are now available for this application. One of the earliest studies of MSC differentiation in 3D culture in response to stretch was for ligament applications [2]. In that study, dynamic stretching and torsion applied to MSC-seeded collagen gels increased the expression of fibrous tissue markers compared to nonstimulated controls. Additional studies using MSC-seeded collagen sponges have shown that dynamic loading increases fibrous markers (expression of collagen types I and III) as well as the mechanical properties of the constructs after 2 weeks of stimulation [64]. In another study, MSCs seeded on a collagen/glycosaminoglycan scaffold and dynamically loaded for 1 week in chondrogenic medium conditions showed enhanced cartilage-specific ECM production with dynamic stretch [90]. One interesting observation in studies of this kind is that the "developmental stage" of an MSC might influence the response to mechanical stimuli. For example, MSCs seeded on silk fibroin scaffolds were most sensitive to mechanical stimulation after they had developed a self-generated biologic microenvironment in the context of their 3D scaffold [23].

45.4.2.2
Static and Dynamic Compression

Just as one rapidly growing layer can exert tensile force on a slowly growing conjoined layer, so too can compressive forces be exerted on a fast-growing layer. In addition, with active muscle contraction, compressive forces across developing joints produce compressive forces on the intervening structures. To apply direct compression to developmentally relevant cells, a 3D environment is required in which cells may be distributed and which can transduce mechanical signals. Both limb-bud cells and MSCs have been seeded in a variety of the 3D scaffolds as described above. Early work using compressive loading showed that static compression promoted limb-bud chondrogenesis in collagen gels, as evidenced by increased cartilage-specific gene expression [125]. In later studies using limb-bud cells in agarose, dynamic compressive loading increased chondrogenesis to varying levels, depending on the duty cycle and magnitude of the applied compression [36, 37]. Interestingly, this mechanical loading altered soluble factor concentration in the environment, with conditioned media from loaded samples inducing higher levels of chondrogenesis in free-swelling controls [35]. Recent studies have extended this work to examine both adult MSCs and ES cells in 3D culture. Dynamic compression of rabbit and human MSCs in agarose and alginate improved chondrogenesis [16, 59, 60]. Bovine MSCs in agarose increased chondrogenic differentiation depending on the duration and frequency of loading (Fig. 45.5) [84]. Interestingly, a short duration of loading (5 days, 3 hours/day) was found to improve cartilage-specific ECM deposition over a 4-week time course. Comparable studies in HA sponges [4] and PEG hydrogels [128] show similar improvements in MSC chondrogenesis

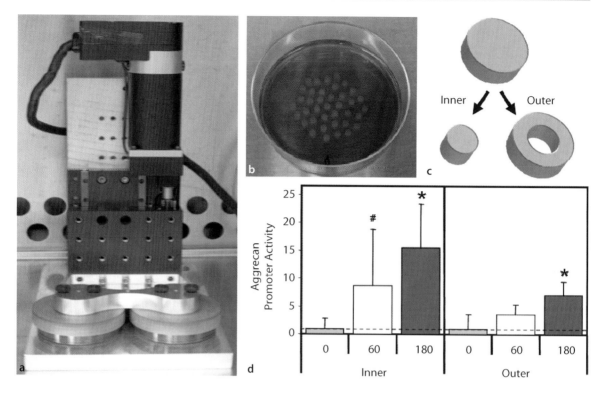

Fig. 45.5a–d Dynamic compressive loading enhances MSC chondrogenesis in 3D culture. A compressive loading bioreactor (**a**) was used to apply a defined cyclic compressive deformation to MSC-seeded cylindrical agarose constructs (**b**). Dynamic loading at 1 Hz for increasing durations enhanced aggrecan promoter activity in both the central core and outer annulus (**c**). In each region, promoter activity (**d**) was normalized to the corresponding free swelling (0 min of loading) control in that region. #$p<0.10$, *$p<0.05$ vs. 0 min (no load) control, $n=3-4$. Adapted from Mauck et al. (2007) [84] with permission from Springer-Verlag

with dynamic loading. Comparing adult stem cells to ES cells, it was recently noted that dynamic loading improved chondrogenesis of MSCs even in the absence of soluble chondrogenic factors, while ES cells required prior chondrogenesis for dynamic loading to have an anabolic effect [128].

45.4.2.3
Hydrostatic Pressurization

In addition to mechanical forces that elicit changes in the shape of developing structures, hydrostatic pressurization may also play a role in stem cell fate decisions. It has long been postulated, based on poroelastic model simulations of bone development, that hydrostatic stress elicits enhanced chondrogenesis, while shear stresses cause bone formation [19]. It has also been appreciated that in adult cartilage, hydrostatic pressurization arises from mechanical use, and transmits a significant portion of the stresses transferred between articulating surfaces, sparing the solid component of the ECM [122]. In fully differentiated chondrocytes and cartilage tissue, static and dynamic hydrostatic pressure modulates ECM biosynthesis [108, 121]. Translating these concepts to differentiating cells, several studies have investigated the effect of fluid pressurization on MSC chondrogenesis, using a range of pressure magnitudes. Most of these studies apply hydrostatic pressure through a fluid phase to MSCs in pellet culture. These studies show that the magnitude and duration of pressurization, as well as inclusion of soluble prochondrogenic factors, can enhance chondrogenesis [5, 95, 96]. More recently, it was demonstrated that steady versus graded levels of dynamic pressurization differentially influence chondrogenesis of MSCs in agarose gels, as evidenced by changes in Sox9 gene expression [42]. Even very low magnitudes of hydrostatic pressure (0.1 MPa) applied at low frequencies (0.25 Hz) for long durations can elicit a positive chondrogenic response by MSCs seeded on fibrous scaffolds [75].

45.5 Summary and Future Directions

This chapter identifies the developmental and microenvironmental concepts that have a bearing on tissue engineering with adult stem cells. We posit that a multiscale approach should be taken in each subdiscipline intrinsic to this endeavor (Fig. 45.6). Scaffolds must encompass signals originating at the cell material interface, including surface roughness, dimensionality, and stiffness, to the tissue-scale level of matrix organization and anisotropy. Mechanical loading of both spatial and temporal relevance should be derived from development, and applied to promote progenitor cell differentiation at the appropriate time and in the appropriate space. Normal developmental processes result in functional tissues, and may serve as a source for successful, developmentally inspired engineered replacement tissues. Clearly there remains a considerable amount of work to better understand the developmental microenvironment in which these diverse signals are interpreted and consolidated as progenitor cells make fate decisions. However, this approach represents a worthy pursuit in regenerative medicine with adult stem cells, as these cells offer hope for the successful replacement or repair of all musculoskeletal tissues.

Acknowledgment

Supported in part by the Intramural Research Program, NIAMS, NIH (ZO1 AR 41131).

References

1. Alhadlaq, A., J.H. Elisseeff, L. Hong, et al (2004) Adult stem cell driven genesis of human-shaped articular condyle. Ann Biomed Eng, 32:911–23
2. Altman, G.H, R.L. Horan, I. Martin, et al (2002) Cell differentiation by mechanical stress. FASEB J, 16:270–2
3. Amprino, R. (1985) The influence of stress and strain in the early development of shaft bones. An experimental study on the chick embryo tibia. Anat Embryol (Berl), 172:49–60
4. Angele, P, D. Schumann, M. Nerlich, et al (2004) Enhanced chondrogenesis of mesenchymal progenitor cells loaded in tissue engineering scaffolds by cyclic, mechanical compression. Trans Orthop Res Soc, 29:835
5. Angele, P, J.U. Yoo, C. Smith, et al (2003) Cyclic hydrostatic pressure enhances the chondrogenic phenotype of human mesenchymal progenitor cells differentiated in vitro. J Orthop Res, 21:451–7
6. Awad, H, Y.-D.C. Halvorsen, J.M. Gimble, et al (2003) Effects of transforming growth factor beta1 and dexamethasone on the growth and chondrogenic differentiation of adipose-derived stromal cells. Tissue Eng, 9:1301–1312
7. Awad, H.A, M.Q. Wickham, H.A. Leddy, et al (2004) Chondrogenic differentiation of adipose-derived adult stem cells in agarose, alginate, and gelatin scaffolds. Biomaterials, 25:3211–22
8. Baker, B.M, R.L. Mauck (2007) The effect of nanofiber alignment on the maturation of engineered meniscus constructs. Biomaterials, 28:1967–77
9. Baksh, D, G.M. Boland, R.S. Tuan (2007) Cross-talk between Wnt signaling pathways in human mesenchymal

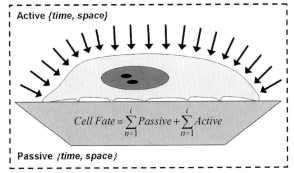

Fig. 45.6 Schematic of microenvironmental signals influencing stem cell fate. Stem cells receive and integrate temporally and spatially varying passive and active cues from their evolving microenvironment during differentiation. Quantification of the mechanisms that underlie the relationship between these developmentally relevant signals may be used to promote lineage-specific differentiation of adult stem cells for regenerative medicine applications

stem cells leads to functional antagonism during osteogenic differentiation. J Cell Biochem, 101:1109–24
10. Baksh, D, L. Song, R.S. Tuan (2004) Adult mesenchymal stem cells: characterization, differentiation, and application in cell and gene therapy. J Cell Mol Med, 8:301–16
11. Bhardwaj, T, R.M. Pilliar, M.D. Grynpas, et al (2001) Effect of material geometry on cartilagenous tissue formation in vitro. J Biomed Mater Res, 57:190–9
12. Boland, G.M, G. Perkins, D.J. Hall, et al (2004) Wnt 3a promotes proliferation and suppresses osteogenic differentiation of adult human mesenchymal stem cells. J Cell Biochem, 93:1210–30
13. Boyan, B.D, T.W. Hummert, D.D. Dean, et al (1996) Role of material surfaces in regulating bone and cartilage cell response. Biomaterials, 17:137–46
14. Brown, P.D, P.D. Benya (1988) Alterations in chondrocyte cytoskeletal architecture during phenotypic modulation by retinoic acid and dihydrocytochalasin B-induced reexpression. J Cell Biol, 106:171–9
15. Bruder, S.P, N. Jaiswal, S.E. Haynesworth (1997) Growth kinetics, self-renewal, and the osteogenic potential of purified human mesenchymal stem cells during extensive sub-cultivation and following cryopreservation. J Cell Biochem, 64:278–94
16. Campbell, J.J, D.A. Lee, D.L. Bader (2006) Dynamic compressive strain influences chondrogenic gene expression in human mesenchymal stem cells. Biorheology, 43:455–70
17. Caplan, A.I (1991) Mesenchymal stem cells. J Orthop Res, 9:641–50
18. Caplan, A.I, S.P. Bruder (2001) Mesenchymal stem cells: building blocks for molecular medicine in the 21st century. Trends Mol Med, 7:259–64
19. Carter, D.R, M. Wong (1988) The role of mechanical loading histories in the development of diarthrodial joints. J Orthop Res, 6:804–16
20. Carver, S.E, C.A. Heath (1999) Influence of intermittent pressure, fluid flow, and mixing on the regenerative properties of articular chondrocytes. Biotechnol Bioeng, 65:274–81
21. Caterson, E.J, W.J. Li, L.J. Nesti, et al (2002) Polymer/alginate amalgam for cartilage-tissue engineering. Ann N Y Acad Sci, 961:134–8
22. Caterson, E.J, L.J. Nesti, W.J. Li, et al (2001) Three-dimensional cartilage formation by bone marrow-derived cells seeded in polylactide/alginate amalgam. J Biomed Mater Res, 57:394–403
23. Chen, J, R.L. Horan, D. Bramono, et al (2006) Monitoring mesenchymal stromal cell developmental stage to apply on-time mechanical stimulation for ligament tissue engineering. Tissue Eng, 12:3085–95
24. Chung, C, J.A. Burdick (2008) Enhanced chondrogenic differentiation of mesenchymal stem cells in hyaluronan hydrogels. Tissue Eng (in press)
25. Clark, C.R, J.A. Ogden (1983) Development of the menisci of the human knee joint. Morphological changes and their potential role in childhood meniscal injury. J Bone Joint Surg Am, 65:538–47
26. Coleman, C.M, R.S. Tuan (2003) Functional role of growth/differentiation factor 5 in chondrogenesis of limb mesenchymal cells. Mech Dev, 120:823–836
27. Connelly, J.T, A.J. Garcia, M.E. Levenston (2007) Inhibition of in vitro chondrogenesis in RGD-modified three-dimensional alginate gels. Biomaterials, 28:1071–83
28. Cukierman, E, R. Pankov, K.M. Yamada (2002) Cell interactions with three-dimensional matrices. Curr Opin Cell Biol, 14:633–9
29. Dalby, M.J, N. Gadegaard, A.S. Curtis, et al (2007) Nanotopographical control of human osteoprogenitor differentiation. Curr Stem Cell Res Ther, 2:129–38
30. Dalby, M.J, M.O. Riehle, H. Johnstone, et al (2002) In vitro reaction of endothelial cells to polymer demixed nanotopography. Biomatcrials, 23:2945–54
31. Dalby, M.J, M.O. Riehle, H.J. Johnstone, et al (2002) Polymer-demixed nanotopography: control of fibroblast spreading and proliferation. Tissue Eng, 8:1099–108
32. DeLise, A.M, L. Fischer, R.S. Tuan (2000) Cellular interactions and signaling in cartilage development. Osteoarthritis Cartilage, 8:309–34
33. Denker, A.E, A.R. Haas, S.B. Nicoll, et al (1999) Chondrogenic differentiation of murine C3H10T1/2 multipotential mesenchymal cells: I. Stimulation by bone morphogenetic protein-2 in high-density micromass cultures. Differentiation, 64:67–76
34. Denker, A.E, S.B. Nicoll, R.S. Tuan (1995) Formation of cartilage-like spheroids by micromass cultures of murine C3H10T1/2 cells upon treatment with transforming growth factor-beta 1. Differentiation, 59:25–34
35. Elder, S.H (2002) Conditioned medium of mechanically compressed chick limb bud cells promotes chondrocyte differentiation. J Orthop Sci, 7:538–43
36. Elder, S.H, J.H. Kimura, L.J. Soslowsky, et al (2000) Effect of compressive loading on chondrocyte differentiation in agarose cultures of chick limb–bud cells. J Orthop Res, 18:78–86
37. Elder, S.H, S.A. Goldstein, J.H. Kimura, et al (2001) Chondrocyte differentiation is modulated by frequency and duration of cyclic compressive loading. Ann Biomed Eng, 29:476–82
38. Elias, K.L, R.L. Price, T.J. Webster (2002) Enhanced functions of osteoblasts on nanometer diameter carbon fibers. Biomaterials, 23:3279–87
39. Elisseeff, J.H (2004) Embryonic stem cells: potential for more impact. Trends Biotechnol, 22:155–6
40. Engler, A.J, S. Sen, H.L. Sweeney, et al (2006) Matrix elasticity directs stem cell lineage specification. Cell, 126:677–89
41. Erickson, G.R, J.M. Gimble, D.M. Franklin, et al (2002) Chondrogenic potential of adipose tissue-derived stromal cells in vitro and in vivo. Biochem Biophys Res Commun, 290:763–9
42. Finger, A.R, C.Y. Sargent, K.O. Dulaney, et al (2007) Differential effects on messenger ribonucleic acid expression by bone marrow-derived human mesenchymal stem cells seeded in agarose constructs due to ramped and steady applications of cyclic hydrostatic pressure. Tissue Eng, 13:1151–8
43. Fischer, L, G. Boland, R.S. Tuan (2002) Wnt-3A enhances bone morphogenetic protein-2-mediated chondrogenesis of murine C3H10T1/2 mesenchymal cells. J Biol Chem, 277:30870–8

44. Fischer, L, G. Boland, R.S. Tuan (2002) Wnt signaling during BMP-2 stimulation of mesenchymal chondrogenesis. J Cell Biochem, 84:816–31
45. Flemming, R.G, C.J. Murphy, G.A. Abrams, et al (1999) Effects of synthetic micro- and nano-structured surfaces on cell behavior. Biomaterials, 20:573–88
46. Friedenstein, A.J, J.F. Gorskaja, N.N. Kulagina (1976) Fibroblast precursors in normal and irradiated mouse hematopoietic organs. Exp Hematol, 4:267–74
47. Gehris, A.L, E. Stringa, J. Spina, et al (1997) The region encoded by the alternatively spliced exon IIIA in mesenchymal fibronectin appears essential for chondrogenesis at the level of cellular condensation. Dev Biol, 190:191–205
48. Giancotti, F.G, E. Ruoslahti (1999) Integrin signaling. Science, 285:1028–32
49. Gilbert, S.J (2000) Developmental Biology (6th edn). Sinauer, New York
50. Gray, M.L, A.M. Pizzanelli, A.J. Grodzinsky, et al (1988) Mechanical and physiochemical determinants of the chondrocyte biosynthetic response. J Orthop Res, 6:777–92
51. Grinnell, F (1978) Cellular adhesiveness and extracellular substrata. Int Rev Cytol, 53:65–144
52. Guilak, F, D.L. Butler, S.A. Goldstein (2001) Functional tissue engineering: the role of biomechanics in articular cartilage repair. Clin Orthop, S295–305
53. Haas, A.R, R.S. Tuan (1999) Chondrogenic differentiation of murine C3H10T1/2 multipotential mesenchymal cells: II. Stimulation by bone morphogenetic protein-2 requires modulation of N-cadherin expression and function. Differentiation, 64:277–89
54. Hayes, A.J, M. Benjamin, J.R. Ralphs (1999) Role of actin stress fibres in the development of the intervertebral disc: cytoskeletal control of extracellular matrix assembly. Dev Dyn, 215:179–89
55. Hayes, A.J, M. Benjamin, J.R. Ralphs (2001) Extracellular matrix in development of the intervertebral disc. Matrix Biol, 20:107–21
56. Henderson, J.H, D.R. Carter (2002) Mechanical induction in limb morphogenesis: the role of growth-generated strains and pressures. Bone, 31:645–53
57. Henderson, J.H, L. de la Fuente, D. Romero, et al (2007) Rapid growth of cartilage rudiments may generate perichondrial structures by mechanical induction. Biomech Model Mechanobiol, 6:127–37
58. Hersel, U, C. Dahmen, H. Kessler (2003) RGD modified polymers: biomaterials for stimulated cell adhesion and beyond. Biomaterials, 24:4385–415
59. Huang, C.-Y, K. Hagar, L.E. Frost, et al (2004) Effects of cyclic compressive loading on chondrogenesis of rabbit bone marrow-derived mesenchymal stem cells. Trans Orthop Res Soc, 29:161
60. Huang, C.Y, P.M. Reuben, H.S. Cheung (2005) Temporal expression patterns and corresponding protein inductions of early responsive genes in rabbit bone marrow-derived mesenchymal stem cells under cyclic compressive loading. Stem Cells, 23:1113–21
61. Hung, C.T, R.L. Mauck, C.C. Wang, et al (2004) A paradigm for functional tissue engineering of articular cartilage via applied physiologic deformational loading. Ann Biomed Eng, 32:35–49
62. Hutmacher, D.W (2001) Scaffold design and fabrication technologies for engineering tissues—state of the art and future perspectives. J Biomater Sci Polymer Edn, 12:107–124
63. Johnstone, B, T.M. Hering, A.I. Caplan, et al (1998) In vitro chondrogenesis of bone marrow-derived mesenchymal progenitor cells. Exp Cell Res, 238:265–72
64. Juncosa-Melvin, N, K.S. Matlin, R.W. Holdcraft, et al (2007) Mechanical stimulation increases collagen type I and collagen type III gene expression of stem cell-collagen sponge constructs for patellar tendon repair. Tissue Eng, 13:1219–26
65. Kim, M.S, N.S. Hwang, J. Lee, et al (2005) Musculoskeletal differentiation of cells derived from human embryonic germ cells. Stem Cells, 23:113–23
66. Kuo, C.K, W.J. Li, R.L. Mauck, et al (2006) Cartilage tissue engineering: its potential and uses. Curr Opin Rheumatol, 18:64–73
67. Kurpinski, K, J. Chu, C. Hashi, et al (2006) Anisotropic mechanosensing by mesenchymal stem cells. Proc Natl Acad Sci U S A, 103:16095–100
68. Lee, C.R, A.J. Grodzinsky, H.P. Hsu, et al (2000) Effects of harvest and selected cartilage repair procedures on the physical and biochemical properties of articular cartilage in the canine knee. J Orthop Res, 18:790–9
69. Lee, J.H, H.W. Jung, I.K. Kang, et al (1994) Cell behaviour on polymer surfaces with different functional groups. Biomaterials, 15:705–11
70. Lee, J.Y, B. Goldstein (2003) Mechanisms of cell positioning during C. elegans gastrulation. Development, 130:307–20
71. Li, W.J, Y.J. Jiang, R.S. Tuan (2006) Chondrocyte phenotype in engineered fibrous matrix is regulated by fiber size. Tissue Eng, 12:1775–85
72. Li, W.J, C.T. Laurencin, E.J. Caterson, et al (2002) Electrospun nanofibrous structure: a novel scaffold for tissue engineering. J Biomed Mater Res, 60:613–21
73. Li, W.J, R.L. Mauck, J.A. Cooper, et al (2007) Engineering controllable anisotropy in electrospun biodegradable nanofibrous scaffolds for musculoskeletal tissue engineering. J Biomech, 40:1686–93
74. Li, W.J, R. Tuli, X. Huang, et al (2005) Multilineage differentiation of human mesenchymal stem cells in a three-dimensional nanofibrous scaffold. Biomaterials, 26:5158–66
75. Luo, Z.J, B.B. Seedhom (2007) Light and low-frequency pulsatile hydrostatic pressure enhances extracellular matrix formation by bone marrow mesenchymal cells: an in-vitro study with special reference to cartilage repair. Proc Inst Mech Eng [H], 221:499–507
76. Ma, H.-L, S.-C. Hung, S.-Y. Lin, et al (2003) Chondrogenesis of human mesenchymal stem cells encapsulated in alginate beads. J Biomed Mat Res, 64A:273–81
77. Majumdar, M.K, V. Banks, D.P. Peluso, et al (2000) Isolation, characterization, and chondrogenic potential of human bone marrow-derived multipotential stromal cells. J Cell Physiol, 185:98–106
78. Majumdar, M.K, M. Keane-Moore, D. Buyaner, et al (2003) Characterization and functionality of cell surface molecules on human mesenchymal stem cells. J Biomed Sci, 10:228–41

79. Majumdar, M.K, E. Wang, E.A. Morris (2001) BMP-2 and BMP-9 promotes chondrogenic differentiation of human multipotential mesenchymal cells and overcomes the inhibitory effect of IL-1. J Cell Physiol, 189:275–84
80. Malaviya, P, D.L. Butler, G.P. Boivin, et al (2000) An in vivo model for load-modulated remodeling in the rabbit flexor tendon. J. Orthop. Res, 18:116–25
81. Manwaring, M.E, J.F. Walsh, P.A. Tresco (2004) Contact guidance induced organization of extracellular matrix. Biomaterials, 25:3631–8
82. Martin, I, R.F. Padera, G. Vunjak-Novakovic, et al (1998) In vitro differentiation of chick embryo bone marrow stromal cells into cartilaginous and bone-like tissues. J Orthop Res, 16:181–9
83. Matayas, J.R, M.G. Anton, N.G. Shrive, et al (1995) Stress governs tissue phenotype at the femoral insertion of the rabbit MCL. J. Biomech, 28:147–57
84. Mauck, R.L, B.A. Byers, X. Yuan, et al (2007) Regulation of Cartilaginous ECM Gene Transcription by chondrocytes and MSCs in 3D culture in response to dynamic loading. Biomech Model Mechanobiol, 6:113–25
85. Mauck, R.L, G.J. Martinez-Diaz, X. Yuan, et al (2007) Regional variation in meniscal fibrochondrocyte multilineage differentiation potential: implications for meniscal repair. Anat Rec, 290:48–58
86. Mauck, R.L, S.L. Seyhan, G.A. Ateshian, et al (2002) Influence of seeding density and dynamic deformational loading on the developing structure/function relationships of chondrocyte-seeded agarose hydrogels. Ann Biomed Eng, 30:1046–56
87. Mauck, R.L, M.A. Soltz, C.C. Wang, et al (2000) Functional tissue engineering of articular cartilage through dynamic loading of chondrocyte-seeded agarose gels. J Biomech Eng, 122:252–60
88. Mauck, R.L, X. Yuan, R.S. Tuan (2006) Chondrogenic differentiation and functional maturation of bovine mesenchymal stem cells in long-term agarose culture. Osteoarthritis Cartilage, 14:179–89
89. McBeath, R, D.M. Pirone, C.M. Nelson, et al (2004) Cell shape, cytoskeletal tension, and RhoA regulate stem cell lineage commitment. Dev Cell, 6:483–95
90. McMahon, L.A, A.J. Reid, V.A. Campbell, et al (2008) Regulatory effects of mechanical strain on the chondrogenic differentiation of MSCs in a collagen-GAG scaffold: experimental and computational analysis. Ann Biomed Eng, 36:185–94
91. Mikic, B, A.L. Isenstein, A.B. Chhabra (2004) Mechanical modulation of cartilage structure and function during embryogenesis of the chick. Ann Biomed Eng, 32:18–25
92. Mikic, B, T. Johnson, E.B. Hunziker (2000) Differential effects of embryonic immobilization on the development of fibrocartilaginous skeletal elements. Trans Orthop Res Soc, 25:969
93. Mikic, B, T.L. Johnson, A.B. Chhabra, et al (2000) Differential effects of embryonic immobilization on the development of fibrocartilaginous skeletal elements. J Rehabil Res Dev, 37:127–33
94. Mitrovic, D (1982) Development of the articular cavity in paralyzed chick embryos and in chick limb buds cultured on chorioallantoic membranes. Acta Anat, 112:313–24
95. Miyanishi, K, M.C. Trindade, D.P. Lindsey, et al (2006) Dose- and time-dependent effects of cyclic hydrostatic pressure on transforming growth factor-beta3-induced chondrogenesis by adult human mesenchymal stem cells in vitro. Tissue Eng, 12:2253–62
96. Miyanishi, K, M.C. Trindade, D.P. Lindsey, et al (2006) Effects of hydrostatic pressure and transforming growth factor-beta 3 on adult human mesenchymal stem cell chondrogenesis in vitro. Tissue Eng, 12:1419–28
97. Nikolovski, J, D.J. Mooney (2000) Smooth muscle cell adhesion to tissue engineering scaffolds. Biomaterials, 21:2025–32
98. Noth, U, A.M. Osyczka, R. Tuli, et al (2002) Multilineage mesenchymal differentiation potential of human trabecular bone derived cells. J Orthop Res, 20:1060–9
99. Nowlan, N.C, P. Murphy, P.J. Prendergast (2007) Mechanobiology of embryonic limb development. Ann N Y Acad Sci, 1101:389–411
100. Nowlan, N.C, P. Murphy, P.J. Prendergast (2008) A dynamic pattern of mechanical stimulation promotes ossification in avian embryonic long bones. J Biomech, 41:249–58
101. Nur, E.K.A, I. Ahmed, J. Kamal, et al (2005) Three dimensional nanofibrillar surfaces induce activation of Rac. Biochem Biophys Res Commun, 331:428–34
102. Nur, E.K.A, I. Ahmed, J. Kamal, et al (2006) Three-dimensional nanofibrillar surfaces promote self-renewal in mouse embryonic stem cells. Stem Cells, 24:426–33
103. Nuttelman, C.R, M.C. Tripodi, K.S. Anseth (2005) Synthetic hydrogel niches that promote hMSC viability. Matrix Biol, 24:208–18
104. O'Driscoll, S.W, F.W. Keeley, R.B. Salter (1986) The chondrogenic potential of free autogenous periosteal grafts for biological resurfacing of major full-thickness defects in joint surfaces under the influence of continuous passive motion. An experimental investigation in the rabbit. J Bone Joint Surg Am, 68:1017–35
105. Onodera, K, I. Takahashi, Y. Sasano, et al (2005) Stepwise mechanical stretching inhibits chondrogenesis through cell-matrix adhesion mediated by integrins in embryonic rat limb-bud mesenchymal cells. Eur J Cell Biol, 84:45–58
106. Park, I.H, R. Zhao, J.A. West, et al (2008) Reprogramming of human somatic cells to pluripotency with defined factors. Nature, 451:141–6
107. Park, J.S, J.S. Chu, C. Cheng, et al (2004) Differential effects of equiaxial and uniaxial strain on mesenchymal stem cells. Biotechnol Bioeng, 88:359–68
108. Parkkinen, J.J, J. Ikonen, M.J. Lammi, et al (1993) Effects of cyclic hydrostatic pressure on proteoglycan synthesis in cultured chondrocytes and articular cartilage explants. Arch Biochem Biophys, 300:458–65
109. Paszek, M.J, N. Zahir, K.R. Johnson, et al (2005) Tensional homeostasis and the malignant phenotype. Cancer Cell, 8:241–54
110. Pittenger, M.F, A.M. Mackay, S.C. Beck, et al (1999) Multilineage potential of adult human mesenchymal stem cells. Science, 284:143–47
111. Prockop, D.J (1997) Marrow stromal cells as stem cells for nonhematopoietic tissues. Science, 276:71–4

112. Salinas, C.N, B.B. Cole, A.M. Kasko, et al (2007) Chondrogenic differentiation potential of human mesenchymal stem cells photoencapsulated within poly(ethylene glycol)-arginine-glycine-aspartic acid-serine thiol-methacrylate mixed-mode networks. Tissue Eng, 13:1025–34

113. San Antonio, J.D, O. Jacenko, M. Yagami, et al (1992) Polyionic regulation of cartilage development: promotion of chondrogenesis in vitro by polylysine is associated with altered glycosaminoglycan biosynthesis and distribution. Dev Biol, 152:323–35

114. Schindler, M, I. Ahmed, J. Kamal, et al (2005) A synthetic nanofibrillar matrix promotes in vivo-like organization and morphogenesis for cells in culture. Biomaterials, 26:5624–31

115. Scully, S.P, J.W. Lee, P.M.A. Ghert, et al (2001) The role of the extracellular matrix in articular chondrocyte regulation. Clin Orthop Relat Res, S72–89

116. Shum, L, C.M. Coleman, Y. Hatakeyama, et al (2003) Morphogenesis and dismorphogenesis of the appendicular skeleton. Birth Def Res (Part C) Embryo Today, 69:102–22

117. Simmons, C.A, S. Matlis, D.J. Thornton, et al (2003) Cyclic strain enhances matrix mineralization by adult human mesenchymal stem cells via the extracellular signal-regulated kinase (ERK1/2) signaling pathway. J Biomech, 36:1087–96

118. Simmons, P.J. and B. Torok-Storb (1991) Identification of stromal cell precursors in human bone marrow by a novel monoclonal antibody, STRO-1. Blood, 78:55–62

119. Sinha, R.K, F. Morris, S.A. Shah, et al (1994) Surface composition of orthopaedic implant metals regulates cell attachment, spreading, and cytoskeletal organization of primary human osteoblasts in vitro. Clin Orthop Relat Res, 258–72

120. Smetana, K Jr (1993) Cell biology of hydrogels. Biomaterials, 14:1046–50

121. Smith, R.L, S.F. Rusk, B.E. Ellison, et al (1996) In vitro stimulation of articular chondrocyte mRNA and extracellular matrix synthesis by hydrostatic pressure. J Orthop Res, 14:53–60

122. Soltz, M.A, G.A. Ateshian (1998) Experimental verification and theoretical prediction of cartilage interstitial fluid pressurization at an impermeable contact interface in confined compression. J Biomech, 31:927–34

123. Song, L, D. Baksh, R.S. Tuan (2004) Mesenchymal stem cell-based cartilage tissue engineering: cells, scaffold and biology. Cytotherapy, 6:596–601

124. Song, L, R.S. Tuan (2004) Transdifferentiation potential of human mesenchymal stem cells derived from bone marrow. FASEB J, 18:980–2

125. Takahashi, I, G.H. Nuckolls, K. Takahashi, et al (1998) Compressive force promotes sox9, type II collagen and aggrecan and inhibits IL-1beta expression resulting in chondrogenesis in mouse embryonic limb bud mesenchymal cells. J Cell Sci, 111 (Pt 14):2067–76

126. Takahashi, K, K. Tanabe, M. Ohnuki, et al (2007) Induction of pluripotent stem cells from adult human fibroblasts by defined factors. Cell, 131:861–72

127. Tallheden, T, J.E. Dennis, D.P. Lennon, et al (2003) Phenotypic plasticity of human articular chondrocytes. J Bone Joint Surg, 85A:93–100

128. Terraciano, V, N. Hwang, L. Moroni, et al (2007) Differential Response of adult and embryonic mesenchymal progenitor cells to mechanical compression in hydrogels. stem cells. Stem Cells, 25:2730–8

129. Tran-Khanh, N, C.D. Hoemann, M.D. McKee, et al (2005) Aged bovine chondrocytes display a diminished capacity to produce a collagen-rich, mechanically functional cartilage extracellular matrix. J Orthop Res, 23:1354–62

130. Tuli, R, S. Tuli, S. Nandi, et al (2003) Transforming growth factor-beta-mediated chondrogenesis of human mesenchymal progenitor cells involves N-cadherin and mitogen-activated protein kinase and Wnt signaling cross-talk. J Biol Chem, 278:41227–36

131. Ward, D.F, Jr, R.M. Salasznyk, R.F. Klees, et al (2007) Mechanical strain enhances extracellular matrix-induced gene focusing and promotes osteogenic differentiation of human mesenchymal stem cells through an extracellular-related kinase-dependent pathway. Stem Cells Dev, 16:467–80

132. Webster, T.J, C. Ergun, R.H. Doremus, et al (2000) Enhanced functions of osteoblasts on nanophase ceramics. Biomaterials, 21:1803–10

133. Webster, T.J, R.W. Siegel, R. Bizios (1999) Osteoblast adhesion on nanophase ceramics. Biomaterials, 20:1221–7

134. Wei, X, K. Messner (1996) The postnatal development of the insertions of the medial collateral ligament in the rat knee. Anat Embryol (Berl), 193:53–9

135. Williams, C.G, T.K. Kim, A. Taboas, et al (2003) In vitro chondrogenesis of bone marrow-derived mesenchymal stem cells in a photopolymerizing hydrogel. Tissue Eng, 9:679–88

136. Woo, K.M, V.J. Chen, P.X. Ma (2003) Nano-fibrous scaffolding architecture selectively enhances protein adsorption contributing to cell attachment. J Biomed Mater Res A, 67:531–7

137. Worster, A.A, B.D. Brower-Toland, L.A. Fortier, et al (2001) Chondrocytic differentiation of mesenchymal stem cells sequentially exposed to transforming growth factor-beta1 in monolayer and insulin-like growth factor-1 in a three-dimensional matrix. J Orthop Res, 19:738–49

138. Yang, F, R. Murugan, S. Wang, et al (2005) Electrospinning of nano/micro scale poly(L-lactic acid) aligned fibers and their potential in neural tissue engineering. Biomaterials, 26:2603–10

139. Yim, E.K, S.W. Pang, K.W. Leong (2007) Synthetic nanostructures inducing differentiation of human mesenchymal stem cells into neuronal lineage. Exp Cell Res, 313:1820–9

140. Zhang, S (2003) Fabrication of novel biomaterials through molecular self-assembly. Nat Biotechnol, 21:1171–8

141. Zhang, Z, J. Messana, N.S. Hwang, et al (2006) Reorganization of actin filaments enhances chondrogenic differentiation of cells derived from murine embryonic stem cells. Biochem Biophys Res Commun, 348:421–7

Part E

Transplantation Issues

X Functional Aspects in Biological Engineering

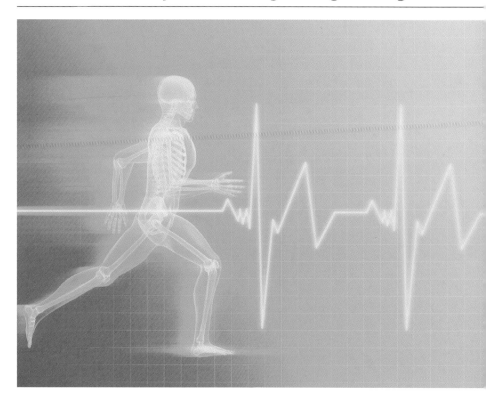

Perfusion Effects and Hydrodynamics

R. A. Peattie, R. J. Fisher

Contents

46.1	Introduction and Motivation	669
46.2	Perfusion	671
46.3	Elements of Theoretical Hydrodynamics	672
46.3.1	Elements of Continuum Mechanics	672
46.3.2	Flow in Tubes	674
46.4	Pulsatile Flow	678
46.4.1	Hemodynamics in Rigid Tubes: Womersley's Theory	678
46.4.2	Hemodynamics in Elastic Tubes	680
46.4.3	Turbulence in Pulsatile Flow	682
46.5	Models and Computational Techniques	683
46.5.1	Approximations to the Navier–Stokes Equations	683
46.5.2	Computational Fluid Dynamics	683
	References	684

46.1 Introduction and Motivation

Biological processes within living systems are significantly influenced by the flow of liquids and gases. Biomedical engineers must therefore understand hydrodynamic phenomena [1] and their vital role in the biological processes that occur within the body [2]. In particular, tissue engineers are concerned with perfusion effects in the cellular microenvironment, and the ability of the circulatory and respiratory systems to provide a whole-body communication network with dynamic response capabilities. Understanding the fundamental principles of the fluid flow involved in these processes is essential to accomplishing the primary objective of mimicking tissue behavior, whether in extracorporeal devices or for in vivo cellular therapy (i.e., to know how tissue function can be built, reconstructed and/or modified to be clinically applicable).

From a geometric and flow standpoint, the body may be considered a network of highly specialized and interconnected organ systems. The key elements of this network for transport and communication are its pathway (the circulatory system) and its fluid (blood). Of most interest for engineering purposes is the ability of the circulatory system to transport oxygen and carbon dioxide, glucose, other nutrients, and metabolites, and signal molecules to and from the tissues, as well as to provide an avenue for stress-response agents from the immune system, including cytokines, antibodies, leukocytes, and macrophages, and system repair agents such as stem cells and platelets. The bulk transport capability provided by convective flow helps to overcome the large diffusional resistance that would otherwise be offered by such a large entity as the human body. At rest, the mean blood circulation time is of the order of 1 min. Therefore, given that the total amount of blood circulating is about 76–80 ml/kg (5.3–5.6 l for a 70-kg "standard male"), the flow from the heart to this branching network is about 95 ml/s. This and other order of magnitude estimates for the human body are available elsewhere (e.g., [2–4]).

Although the main fluids considered in this chapter are blood and air, other fluids such as urine, perspiration, tears, ocular aqueous and vitreous fluids, and the synovial fluid in the joints can also be important in evaluating tissue system behavioral responses to induced chemical and physical stresses. For purposes of analysis, these fluids are often assumed to exhibit Newtonian behavior, although the synovial fluid and blood under certain conditions can be non-Newtonian. Since blood is a suspension, it has interesting properties; it behaves as a Newtonian fluid for large shear rates, but is highly non-Newtonian for low shear rates. The synovial fluid exhibits viscoelastic characteristics that are particularly suited to its function of joint lubrication, for which elasticity is beneficial. These viscoelastic characteristics must be accounted for when considering tissue therapy for joint injuries.

Further complicating analysis is the fact that these fluids travel through three-dimensional passageways that are often branched and usually distensible. In these pathways, disturbed or turbulent flow regimes may be mixed with stable, laminar regions. For example, pulsatile blood flow is laminar in many parts of a healthy circulatory system in spite of the potential for peak Reynolds numbers (defined below) of the order of 10,000. However, "bursts" of turbulence are detected in the aorta during a fraction of each cardiac cycle. An occlusion or stenosis in the circulatory system, such as the stenoses created by atherosclerotic deposits, will promote the development of turbulence. Airflow in the lung is normally stable and laminar during inspiration, but less so during expiration, and heavy breathing, coughing or an obstruction can result in turbulent flow, with Reynolds numbers of 50,000 a possibility.

Although the elasticity of vessel walls can significantly complicate fluid flow analysis, biologically it provides important advantages. For example, pulsatile blood flow induces accompanying expansions and contractions in healthy elastic-wall vessels. These wall displacements then influence flow fields within the vessels. Elastic behavior maintains the norm of laminar flow that minimizes wall stress, lowers flow resistance and thus energy dissipation, and fosters maximum life of the vessel. In combination with pulsatile flow, distensibility permits strain-relaxation of the wall tissue with each cardiac cycle, which provides an exercise routine promoting extended "on-line" use.

The term *perfusion* is used in engineering biosciences to identify the rate of blood supplied to a tissue or organ. Clearly, perfusion of in vitro tissue systems is necessary to maintain cell viability along with functionality to mimic in vivo behavior. Furthermore, it is highly likely that cell viability and normoperative metabolism are dependent on the three-dimensional structure of the microvessels distributed through any tissue bed, which establishes an appropriate microenvironment through both biochemical and biophysical mechanisms [5]. This includes transmitting intracellular signals through the extracellular matrix.

The primary objective of this chapter is to develop an appreciation for the most important ideas of fluid dynamics. Hydrodynamic and hemodynamic principles have many important applications to tissue engineering. In fact, the interaction between fluids and tissue is of paramount importance to the successful development of viable tissues, both in vivo and in vitro. The strength of adhesion and dynamics of detachment of mammalian cells from engineered biomaterials and scaffolds are important subjects of ongoing research [6], as are the effects of shear on receptor-ligand binding at the cell-fluid interface. Flow-induced shear stress has numerous critical consequences for cells, altering transport across the cell membrane, receptor density and distribution, binding affinity and signal generation, as well as subsequent trafficking within the cell [7]. In addition, the design and use of perfusion systems such as membrane biomimetic reactors and hollow fibers is most effective when careful attention is given to issues of hydrodynamic similitude. Similarly, understanding the role of fluid mechanical phenomena in arterial disease and subsequent therapeutic applications is clearly dependent on an appreciation of hemodynamics. Working with team members with expertise in fluid mechanics can therefore often be of substantial benefit toward the completion of tissue-engineering projects. However, understanding their approach and having a fundamental grasp of the technology and its terminology is a prerequisite for effective communication.

A thorough treatment of the mathematics needed for model development and analysis is beyond the scope of this volume, and is presented in numerous sources [1, 2]. Herein, the significance of issues surrounding perfusion in tissue-engineering settings is discussed. Because of the importance of those issues, major principles of hemodynamic flow and transport are then briefly described, with the goal of providing physical understanding of the important topics. Solution methods are summarized, and the benefits associated with use of computational fluid dynamics (CFD) packages are described. Tissue engineers often find CFD a very powerful and versatile tool for analysis and solution of many complex but crucial problems. For example, the analysis and design of bioreactors is highly dependent on understanding the nonideal flow patterns that exist within the reactor vessel. In principle, if the complete velocity distribution of the reactor is known, mixing and transport characteristics and the pressure and stress profiles affecting cellular processes can be fully predicted. Once considered an impractical approach, this is now obtainable by computing the velocity distribution using the CFD-based procedures.

46.2
Perfusion

Maintenance of the appropriate microenvironment in tissue culture systems, whether for in vivo, ex vivo, or in vitro applications, is crucial for successful establishment of viable tissue. Dynamic similitude between the cultured tissue and its native in vivo environment, with respect to both chemical and physical phenomena, is essential for optimum culture performance and is of particular concern in extracorporeal devices. Perfusion, either with actual body fluids or with engineered surrogates, establishes the culture microenvironment, although the need for rapid delivery must be balanced with possible detrimental efforts from high shear forces and inappropriate signal generation. Moreover, perfusion difficulties in vivo can lead to tissue necrosis and transplant failure, whether of a whole organ or an ex vivo system [8–10].

A properly designed and constructed perfusion system can eliminate many of the current problems associated with tissue culture systems. For example, although the production and longevity of hematopoietic cultures can be improved through optimal manual feeding protocols, such protocols are too labor intensive for large-scale use. Furthermore, subjecting the cultures to the frequent physical disruptions associated with manual feeding leads to large changes in culture conditions and possible contamination at each feeding. These complications restrict the optimization of the culture environment and provide incentives to develop continuously perfused bioreactors. By delivery of appropriate substrates, perfusion bioreactors support the development and maintenance of accessory cell populations, resulting in significant endogenous growth factor production and enhanced culture success.

Design of an effective perfusion system for a given application is highly dependent upon knowledge of the specific requirements of that tissue system [11]. Specifications describing cellular function in vivo must be identified to characterize the tissue microenvironment and identify communication requirements [5]. Key elements to consider in determining how tissue can best be built, reconstructed, and/or modified are based on the following axioms [12]:

1. In organogenesis and wound healing, proper intercellular communications are of paramount concern since a systematic and regulated response is required from all participating cells.
2. The function of fully formed organs is strongly dependent on the coordinated function of multiple cell types, with tissue function based on multicellular aggregates.
3. The functionality of an individual cell is strongly affected by its microenvironment (typically within 100 µm of the cell).
4. This microenvironment is further characterized by:
    a. Neighboring cells via cell–cell contact and/or the presence of molecular signals such as soluble growth factors.
    b. Transport processes and physical interactions with the extracellular matrix.
    c. The local geometry, in particular its effects on microcirculation.

More specific details related to particular reactor and perfusion systems may be found elsewhere [9–13].

46.3 Elements of Theoretical Hydrodynamics

It is essential that tissue engineers understand both the advantages and the limitations of mathematical theories and models of biological phenomena, since those models often involve approximations that are not always fully justified in biological systems. Avoiding erroneous conclusions at a minimum requires understanding the assumptions underlying the models. Mechanical theories often begin with Newton's second law ($F = ma$). When applied to continuous distributions of Newtonian fluids; Newton's second law gives rise to the Navier-Stokes equations. In brief, these equations provide an expression governing the motion of fluids such as air and water, for which the rate of motion is linearly proportional to the applied stress producing the motion. Below, the basic concepts from which the Navier-Stokes equations have been developed are summarized along with a few general ideas about boundary layers and turbulence. Applications to the vascular system are then treated in the context of pulsatile flow, since pulsatile flows underlie vascular hemodynamics and are crucial for many aspects of clinical evaluations. It is hoped that this generalized approach will facilitate understanding of the major principles of fluid flow, as well as foster appreciation of the complexities involved in analytic solutions relevant to pulsatile phenomena.

46.3.1 Elements of Continuum Mechanics

The theory of fluid flow, together with the theory of elasticity, make up the field of continuum mechanics, the study of the mechanics of continuously distributed materials. Such materials may be either solid or fluid, or may have intermediate viscoelastic properties. Since the concept of a continuous medium, or continuum, does not take into consideration the molecular structure of matter, it is inherently an idealization. However, as long as the smallest length scale in any problem under consideration is very much larger than the size of the molecules making up the medium and the mean free path within the medium, for mechanical purposes all mass may safely be assumed to be continuously distributed in space. As a result, the density of materials can be considered to be a continuous function of spatial position and time.

Continuum mechanics principles are stated through two types of relationships: *constitutive equations*, which express the properties of particular materials, and *field equations*, which express the general mechanical principles that apply to all materials. For example, general conservation principles of mass, momentum, and energy are field equations valid for all types of continuous media. In contrast, stress-strain relationships, which predict the response of particular materials to applied loads and stresses, are constitutive equations and pertain only to the specific material in question.

46.3.1.1 Constitutive Equations

The response of any fluid to applied forces and temperature disturbances can be used to characterize the material. For this purpose, functional relationships between applied stresses and the resulting rate of strain field of the fluid are needed. Fluids that are homogeneous and isotropic, and for which there is a linear relationship between the state of stress within the fluid s_{ij} and the rate of strain tensor ξ_{ij}, where i and j denote the Cartesian coordinates x, y, and z, are called *Newtonian*. In physiologic settings, Newtonian fluids normally behave as if incompressible. For such fluids, it can be shown that:

$$s_{ij} = -P\delta_{ij} + 2\mu\xi_{ij} \tag{46.1}$$

where μ represents the dynamic viscosity of the fluid and $P = P(x, y, z)$ is the fluid pressure.

46.3.1.2 Conservation (Field) Equations

In vector notation, conservation of mass for a continuous fluid is expressed through:

$$\frac{\partial \rho}{\partial t} + \nabla \cdot \rho \mathbf{u} = 0 \tag{46.2}$$

where t denotes time, ρ is the fluid density and $\mathbf{u} = \mathbf{u}(x, y, z)$ is the vector velocity field. When the fluid is incompressible, density is constant and Eq. 46.2 reduces to the well-known *continuity condition*, $\nabla \cdot \mathbf{u} = 0$. Alternatively, the continuity condition can be expressed in terms of Cartesian velocity components (u, v, w) as:

$$\frac{\partial u}{\partial x} + \frac{\partial v}{\partial y} + \frac{\partial w}{\partial z} = 0 \quad (46.3)$$

The basic equation of Newtonian fluid motion, the Navier-Stokes equations, can be developed by substitution of the constitutive relationship for a Newtonian fluid (Eq. 46.1) into the Cauchy principle of momentum balance for a continuous material [14]. In stating momentum balance for a continuously distributed fluid, the acceleration of a fluid particle to which forces are being applied must be expressed through the *material derivative* $D\mathbf{u}/Dt$, where $D\mathbf{u}/Dt = \partial \mathbf{u}/\partial t + (\mathbf{u} \cdot \nabla)\mathbf{u}$. That is, the velocity of a fluid particle may change for either of two reasons: because the particle accelerates or decelerates with time (*temporal acceleration*) or because the particle moves to a new position, at which the velocity has different magnitude and/or direction (*convective acceleration*). The total time derivative of the velocity is a sum of both effects.

A flow field for which $\partial/\partial t = 0$ for all possible properties of the fluid and its flow is described as *steady*, to indicate that it is independent of time. However, the statement $\partial \mathbf{u}/\partial t = 0$ does not imply $D\mathbf{u}/Dt = 0$, and similarly $D\mathbf{u}/Dt = 0$ does not imply that $\partial \mathbf{u}/\partial t = 0$. Using the material derivative, the Navier-Stokes equations for an incompressible fluid can be written in vector form as:

$$\frac{D\mathbf{u}}{Dt} = \mathbf{B} - \frac{1}{\rho}\nabla P + \nu \nabla^2 \mathbf{u} \quad (46.4)$$

where $\mathbf{B}$ represents the body force field experienced by the fluid and ν is the fluid kinematic viscosity, μ/ρ.

Expanded in full, the Navier-Stokes equations are three simultaneous, nonlinear scalar equations, one for each component of the velocity field. In Cartesian coordinates, they take the form:

$$\frac{\partial u}{\partial t} + u\frac{\partial u}{\partial x} + v\frac{\partial u}{\partial y} + w\frac{\partial u}{\partial z} =$$
$$B_x - \frac{1}{\rho}\frac{\partial P}{\partial x} + \nu\left(\frac{\partial^2 u}{\partial x^2} + \frac{\partial^2 u}{\partial y^2} + \frac{\partial^2 u}{\partial z^2}\right) \quad (46.5)$$

$$\frac{\partial v}{\partial t} + u\frac{\partial v}{\partial x} + v\frac{\partial v}{\partial y} + w\frac{\partial v}{\partial z} =$$
$$B_y - \frac{1}{\rho}\frac{\partial P}{\partial y} + \nu\left(\frac{\partial^2 v}{\partial x^2} + \frac{\partial^2 v}{\partial y^2} + \frac{\partial^2 v}{\partial z^2}\right) \quad (46.6)$$

$$\frac{\partial w}{\partial t} + u\frac{\partial w}{\partial x} + v\frac{\partial w}{\partial y} + w\frac{\partial w}{\partial z} =$$
$$B_z - \frac{1}{\rho}\frac{\partial P}{\partial z} + \nu\left(\frac{\partial^2 w}{\partial x^2} + \frac{\partial^2 w}{\partial y^2} + \frac{\partial^2 w}{\partial z^2}\right) \quad (46.7)$$

Flow fields may be determined by solution of the Navier-Stokes equations, provided $\mathbf{B}$ is known. This is generally not a difficulty, since the only body force normally significant in tissue-engineering applications is gravity. For an incompressible flow, there are then four unknown dependent variables, the three components of velocity and the pressure P, and four governing equations, the three components of the Navier-Stokes equations and the continuity condition. It is important to emphasize that this set of equations is *not* sufficient to calculate the flow field when the flow is compressible or involves temperature changes, since pressure, density, and temperature are then interrelated, which introduces new dependent variables to the problem.

Solution of the Navier-Stokes equations also requires that boundary conditions, and sometimes also initial conditions, be specified for the flow field of interest. By far the most common boundary condition in physiologic flows and other tissue-engineering settings is the so-called no-slip condition, which requires that the layer of fluid elements in contact with a boundary have the same velocity as the boundary itself. For an unmoving, rigid wall, as in a pipe, this velocity is zero. However, in the vasculature, vessel walls expand and contract during the cardiac cycle.

Flow patterns and accompanying flow field characteristics depend largely on the values of governing dimensionless parameters. There are many such parameters, each relevant to specific types of flow settings, but the principle parameter of steady flows is the Reynolds number, Re, defined as:

$$\text{Re} = \frac{\rho UL}{\mu} \quad (46.8)$$

where U is a characteristic velocity of the flow field and L is a characteristic length. Both U and L must be selected for the specific problem under study, and in general both will have different values in different problems. For pipe flow, U is most commonly selected to be the mean velocity of the flow with L the pipe diameter.

It can be shown that the Reynolds number represents the ratio of inertial forces to viscous forces in the flow field. Flows at sufficiently low Re therefore behave as if highly viscous, with little to no fluid acceleration possible. At the opposite extreme, high Re flows behave as if lacking viscosity. One consequence of this distinction is that very high Reynolds number flow fields may at first thought seem to contradict the no-slip condition, in that they seem to "slip" along a solid boundary, exerting no shear stress. This dilemma was first resolved in 1905 with Prandtl's introduction of the *boundary layer*, a thin region of the flow field adjacent to the boundary in which viscous effects are important and the no-slip condition is obeyed [15–17].

46.3.1.3
Turbulence and Instabilities

Flow fields are broadly classified as either laminar or turbulent to distinguish between smooth and irregular motion, respectively. Fluid elements in laminar flow fields follow well-defined paths indicating smooth flow in discrete layers or "laminae," with minimal exchange of material between layers due to the lack of macroscopic mixing. The transport of momentum between system boundaries is thus controlled by molecular action, and is dependent on the fluid viscosity.

In contrast, many flows in nature as well as engineered applications are found to fluctuate randomly and continuously, rather than streaming smoothly, and are classified as turbulent. Turbulent flows are characterized by a vigorous mixing action throughout the flow field, which is caused by *eddies* of varying size within the flow. Physically, the two flow states are linked, in the sense that any flow can be stable and laminar if the ratio of inertial to viscous forces is sufficiently small. Turbulence results when this ratio exceeds a critical value, above which the flow becomes unstable to perturbations and breaks down into fluctuations. Because of these fluctuations, the velocity field **u** in a turbulent flow field is not constant in time. Although turbulent flows therefore do not meet the aforementioned definition for steady, the velocity at any point presents a statistically distinct time-average value that is constant. Turbulent flows are described as *stationary*, rather than truly unsteady.

Fully turbulent flow fields have four defining characteristics [17, 18]: they fluctuate randomly, they are three-dimensional, they are dissipative and they are dispersive. The turbulence intensity, I, of any flow field is defined as the ratio of velocity fluctuations, u', to time-average velocity $\bar{u}$, $I = u'/\bar{u}$.

Steady flow in straight, rigid pipes is characterized by only one dimensionless parameter, the Reynolds number. It was shown by Osborne Reynolds that for $Re < 2000$, incidental disturbances in the flow field are damped out and the flow remains stable and laminar. For $Re > 2000$, brief bursts of fluctuations appear in the velocity separated by periods of laminar flow. As Re increases, the duration and intensity of these bursts increases until they merge together into full turbulence. Laminar flow may be achieved with Re values as large as 20,000 or greater in extremely smooth pipes, but it is unstable to flow disturbances and rapidly becomes turbulent if perturbed.

Since the Navier-Stokes equations govern all the behavior of any Newtonian fluid flow, it follows that turbulent flow patterns should be predictable through analysis based on those equations. However, although turbulent flows have been investigated for more than a century and the equations of motion analyzed in great detail, no general approach to the solution of problems in turbulent flow has been found. Statistical studies invariably lead to a situation in which there are more unknown variables than equations, which is called the closure problem of turbulence. Efforts to circumvent this difficulty have included phenomenologic concepts such as eddy viscosity and mixing length, as well as analytical methods including dimensional analysis and asymptotic invariance studies. Even formal statistical analyses, however, have been unable to produce a generally valid solution.

46.3.2
Flow in Tubes

Flow in a tube is the most common fluid dynamic phenomenon in the physiology of living organisms, and is the basis for transport of nutrient molecules, respiratory gases, hormones, and a variety of other important solutes throughout the bodies of all complex living plants and animals. Only single-celled organisms, and multicelled organisms with small numbers of cells, can survive without a mechanism for transporting such molecules, although even these organisms exchange materials with their external environment through fluid-filled spaces. Higher organisms, which need to transport molecules and materials over larger distances, rely on organized systems of directed flows through networks of tubes to carry fluids and solutes. In human physiology, the circulatory system, which consists of the heart, the blood vessels of the vascular tree and the fluid, blood, and which serves to transport blood throughout the body tissues, is perhaps the most obvious example of an organ system dedicated to creating and sustaining flow in a network of tubes. However, flow in tubes is also a central characteristic of the respiratory, digestive, and urinary systems. Furthermore, the immune system utilizes systemic circulatory mechanisms to facilitate transport of antibodies, white blood cells, and lymph throughout the body, while the endocrine system is critically dependent on blood flow for delivery of its secreted hormones. In addition, reproductive functions are also based on fluid flow in tubes. Thus, seven of the ten major organ systems depend on flow in tubes to fulfill their functions.

46.3.2.1
Steady Poiseuille Flow

The most basic state of motion for fluid in a pipe is one in which the motion occurs at a constant rate, independent of time. The pressure-flow relationship for laminar, steady flow in round tubes is called Poiseuille's Law, after J.L.M. Poiseuille, the French physiologist who first derived the relationship in 1840 [19]. Accordingly, steady flow through a pipe or channel that is driven by a pressure difference between the pipe ends of just sufficient magnitude to overcome the tendency of the fluid to dissipate energy through the action of viscosity is called Poiseuille flow.

Strictly speaking, Poiseuille's Law applies only to steady, laminar flow through pipes that are straight, rigid, and infinitely long, with uniform diameter, so that effects at the pipe ends may be neglected without loss of accuracy. However, although neither physiologic vessels nor industrial tubes fulfill all of these conditions exactly, Poiseuille relationships have proven to be of such widespread usefulness that they are often applied even when the underlying assumptions are not met. As such, Poiseuille flow can be taken as the starting point for analysis of cardiovascular, respiratory and other physiologic flows of interest.

A straight, rigid, round pipe is shown in Fig. 46.1, with x denoting the pipe axis and a the pipe radius. Flow in the pipe is governed by the Navier Stokes equations (Eqs. 46.4–46.7). It can be shown from Eqs. 46.5–46.7 that in this flow field $\partial P/\partial x$ can be at most a constant. If the constant is designated $-\kappa$, the full equations reduce to $d^2u/dr^2 + (1/r)(du/dr) = -\kappa/\mu$, with the conditions that the flow field must be symmetric about the pipe center line (i.e., $du/dr|_{r=0} = 0$), and the no-slip boundary condition applies at the wall, $u = 0$ at $r = a$. Under these conditions, the velocity field solution is $u(r) = (\kappa/4\mu)(a^2 - r^2)$.

The velocity profile described by this solution has the familiar parabolic form known as Poiseuille flow (Fig. 46.1). The velocity at the wall ($r = a$) is clearly zero, as required by the no-slip condition, while as expected for physical reasons, the maximum velocity occurs on the axis of the tube ($r = 0$), where

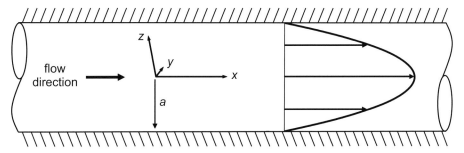

Fig. 46.1 Parabolic velocity profile characteristic of Poiseuille flow in a round pipe of radius a. x, y, z—Cartesian coordinate system with origin on the pipe centerline. x Axial coordinate, y, z transverse coordinates

$u_{max} = \kappa a^2/4\mu$. At any position between the wall and the tube axis, the velocity varies smoothly with r, with no step change at any point.

From physical analysis, it can be shown that the parabolic velocity profile results from a balance of the forces on the fluid in the pipe. The pressure gradient along the pipe accelerates fluid in the forward direction through the pipe. At the same time, viscous shear stress retards the fluid motion. A parabolic profile is created by the balance of these effects.

Although the velocity profile is important and informative, in practice one is more apt to be more concerned with measurement of the *discharge rate*, or total rate of flow in the pipe, Q, which can far more easily be accessed. The volume flow rate is given by area-integration of the velocity across the tube cross-section:

$$Q = \int_A \mathbf{u} \cdot d\mathbf{A} = \frac{\pi \kappa a^4}{8\mu} \quad (46.9)$$

which is Poiseuille's Law.

For convenience, the relationship between pressure and flow rate is often reexpressed in an Ohm's Law form, driving force = flow × resistance, or:

$$\kappa = -\frac{\partial P}{\partial x} = Q \times \frac{8\mu}{\pi a^4} \quad (46.10)$$

from which the resistance to flow, $8\mu/\pi a^4$, is seen to be inversely proportional to the fourth power of the tube radius.

A further point about Poiseuille flow concerns the area-average velocity, U. Clearly, $U = Q$/cross-sectional area $= (\pi \kappa a^4/8\mu)/\pi a^2/8\mu$. But, as was pointed out, the maximum velocity in the tube is $u_{max} = \kappa a^2/4\mu$. Hence $U = u_{max}/2 = (1/2)u|_{r=0} = (1/2)u_{CL}$.

Finally, the shear stress exerted by the flow on the wall can be a critical parameter, particularly when it is desired to control the wall's exposure to shear. From the solution for $u(r)$, it can be shown that wall shear stress, τ_w, is given by:

$$\tau_w = -\mu \frac{du}{dr}\bigg|_{r=a} = \frac{\partial P}{\partial x}\frac{a}{2} = \frac{4\mu Q}{\pi a^3} \quad (46.11)$$

To summarize, Poiseuille's Law, equation (Eq. 46.9), provides a relationship between the pressure drop and net laminar flow in any tube, while equation (Eq. 46.11) provides a relationship between the flow rate and wall shear stress. Thus, physical forces on the wall may be calculated from knowledge of the flow fields.

46.3.2.2
Entrance Flow

It can be shown that a Poiseuille velocity profile is the velocity distribution that minimizes energy dissipation in steady laminar flow through a rigid tube. Consequently, it is not surprising that if the flow in a tube encounters a perturbation that alters its profile, such as a branch vessel or a region of stenosis, immediately downstream of the perturbation the velocity profile will be disturbed away from a parabolic form, perhaps highly so. However, if the Reynolds number is low enough for the flow to remain stable as it convects downstream from the site of the original distribution, a parabolic form is gradually recovered. Consequently, at a sufficient distance downstream, a fully developed parabolic velocity profile again emerges. This process is depicted in Fig. 46.2.

Both the blood vessels and bronchial tubes of the lung possess an enormous number of branches, each of which produces its own flow disturbance. As a result, many physiologic flows may not be fully de-

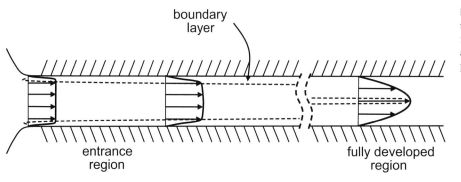

Fig. 46.2 Progression of the flow field from uniform to fully developed at the entrance of a round pipe

veloped over a significant fraction of their length. It therefore becomes important to ask, what length of tube is required for a perturbed velocity profile to recover its parabolic form, (i.e., how long is the entrance length in a given tube?)? This question can be formally posed as, if x is the coordinate along the tube axis, for what value of x does $u|_{r=0} = 2U$? Through dimensional analysis it can be shown that $x/d = const. \times (\rho dU/\mu) = const. \times Re$, where d is the tube diameter [13]. Thus the length of tube over which the flow develops is $const \times Re \times d$. The constant must be determined by experiment, and is found to be in the range 0.03–0.04.

Since the entrance length, in units of tube diameters, is proportional to the Reynolds number and the mean Reynolds number for flow in large tubes such as the aorta and trachea is of the order of 500–1000, the entrance length in these vessels can be as much as 20–30 diameters. In fact, there are few segments of these vessels even close to that length without a branch or curve that perturbs their flow. Consequently, flow in them can be expected to almost never be fully developed. In contrast, flow in the smallest bronchioles, arterioles, and capillaries may take place with $Re < 1$. As a result, their entrance length is $<< 1$ diameter, and flow in them will virtually always be nearly or fully developed.

46.3.2.3
Mechanical Energy Equation

Flow fields in tubes with more complex shapes than simple straight pipes, such as those possessing bends, curves, orifices, and other intricacies, are often analyzed with an *energy balance* approach, since they are not well described by Poiseuille's Law. Understanding such flow fields is significant for in vitro studies and perfusion devices, to establish dynamic similitude parameters, as well as for in vivo studies of curved and/or branched vessel flows. For any system of total energy E, the first law of thermodynamics states that any change in the energy of the system ΔE must appear as either heat transferred to the system in unit time, Q, or as work done by the system, W, so that $\Delta E = Q - W$. Here, a sign convention is taken such that Q, when positive, represents heat transferred *to* the system and W, when positive, is the work done *by* the system on its surroundings. The general form of the energy equation for a fluid system is:

$$\dot{Q} - \dot{W}_S = \frac{d}{dt} \int_V \left(\frac{U^2}{2} + gz + e \right) \rho dV + \int_S \left(\frac{p}{\rho} + \frac{U^2}{2} + gz + e \right) \rho \mathbf{u} \cdot d\mathbf{S} \quad (46.12)$$

where W_s, the "shaft work," represents work done on the fluid contained within a volume V bounded by a surface S (Fig. 46.3) by pumps, turbines, or other external devices through which power is often transmitted by means of a shaft, $U^2/2$ is the kinetic energy per unit mass of the fluid within V, gz is its potential energy per unit mass, with z the vertical coordinate and g gravitational acceleration, e is its internal energy per unit mass, and the density ρ is assumed to be constant.

The general equation can be simplified greatly when the flow is steady, since the total energy contained within any prescribed volume is then constant, and $d/dt = 0$. Applying Eq. 12 to steady flow through a control volume whose end faces are denoted 1 and 2, respectively, then gives:

$$\frac{p_1}{\gamma} + \beta_1 \frac{U_1^2}{2g} + z_1 + h_P = \frac{p_2}{\gamma} + \beta_2 \frac{U_2^2}{2g} + z_2 + h_L, \quad (46.13)$$

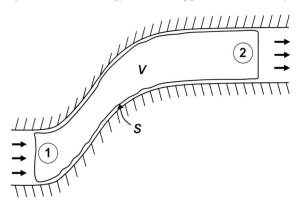

Fig. 46.3 Control volume V for steady flow through a rigid wall pipe of arbitrary shape. S Boundary of V

where p_1 and p_2 are the pressures at faces 1 and 2, z_1 and z_2 are the vertical positions of those faces, $\gamma = \rho g$, h_p represents head supplied by a pump, and the coefficients β_1 and β_2 are kinetic energy correction factors introduced to simplify notation. Calculations show that $\beta = 1$ when the velocity is uniform across the section, and $\beta = 2$ for laminar Poiseuille flow. Mechanical energy lost from the system is lumped together as a single term called *head loss*, h_L. For flow in a rigid pipe of length L and diameter d, h_L is well represented by $h_L = f(L/d)(U^2/2g)$, where f is called the friction factor of the pipe, and depends on both the pipe roughness and the flow Reynolds number. It can be shown that for laminar flow, $f = 64/Re$. Then $h_L = (32\mu LU/\gamma d^2)$. Forms that h_L can take on in turbulent flows are given in a variety of texts [20, 21].

It is worth repeating that Eq. 46.13 is only correct when the fluid density is constant, as is normally the case in tissue-engineering applications and even for air flow in the lung. Compressibility effects require separate energy considerations.

46.4
Pulsatile Flow

Flow in a straight, round tube driven by an axial pressure gradient that varies in time is the basis for blood transport in the arterial tree as well as respiratory gas transport in the trachea and bronchi. When the flow is confined within a tube of rigid, undeformable walls, its direction will always be parallel to the tube axis, so that there will only be an axial component of velocity $\mathbf{u} = (u(r,t),0,0)$ (Fig. 46.2). In that case, all of the fluid elements in the tube, regardless of axial position, will respond to any change in the pressure magnitude instantaneously and in unison. Consequently, the velocity profile will be the same at all positions along the tube. It is as if all of the fluid in the tube moves as a single rigid body.

As a result of the flow field acceleration and decelerations in pulsatile flows, a special type of boundary layer known as the Stokes layer develops. When the pressure gradient varies sinusoidally with time, as the pressure increases to its maximum, the flow increases, and as the pressure decreases, the flow does also. If the oscillations are of very low frequency, the pressure varies only gradually, the velocity field will essentially be in phase with the pressure gradient, and the boundary layers will have adequate time to grow into the tube core region. In the limit of very low frequency, the velocity field must therefore approach that of a steady Poiseuille flow. As frequency increases, however, the pressure gradient changes more rapidly and the flow begins to lag behind it due to the inertia of the fluid. The Stokes layers then become confined to a region near the wall, lacking the time required for further growth. In addition, the flow amplitude decreases with increasing oscillation frequency, as pressure gradient reversals occur more and more rapidly. In the limit of very high frequency, fluid in the tube center hardly moves at all and the Stokes layers are confined to a very thin region along the wall.

Because of the inertia of the fluid, the Stokes layer thickness, δ, is inversely related to the flow frequency, with $\delta \propto (\nu/\omega)^{1/2}$, where ω is the flow angular frequency (in rads/s).

46.4.1
Hemodynamics in Rigid Tubes: Womersley's Theory

The rhythmic contractions of the heart produce a pressure distribution in the arterial tree that includes both a steady component, P_s, and a purely oscillatory component, P_{osc}, as does the velocity field. In contrast, flow in the trachea and bronchi has no steady component, and thus is purely oscillatory. It is common practice to refer to these components of pressure and flow as steady and oscillatory, respectively, and to use the term pulsatile to refer to the superposition of the two. A very useful feature of these flows, when they occur in rigid tubes, is that the governing equation (Eq. 46.4) is linear, since the flow field is unidirectional and independent of axial position. The steady and oscillatory components can therefore be decoupled from each other and analyzed separately. This gives:

$$P(x,t) = P_s(x) + P_{osc}(x,t)$$

$$u(r,t) = u_s(r) + u_{osc}(r,t). \qquad (46.14)$$

The oscillatory component of this flow may be analyzed assuming the flow to be fully developed, so that entrance effects may be neglected, and to be driven by a purely oscillatory pressure gradient, $-(1/\rho)(\partial P/\partial x) = K\cos(\omega t) = \text{Re}(Ke^{i\omega t})$, where $i=\sqrt{-1}$ and here "Re" indicates the Real part of $Ke^{i\omega t}$. It is also convenient to introduce a new dimensionless parameter, the Womersley number, α [22], defined as $\alpha = a(\omega/\nu)^{1/2}$. Thus defined, α represents the ratio of the tube radius to the Stokes layer thickness.

The velocity field is then governed by:

$$\frac{\partial u}{\partial t} = -\frac{1}{\rho}\frac{\partial P}{\partial x} + \upsilon\left(\frac{\partial^2 u}{\partial r^2} + \frac{1}{r}\frac{\partial u}{\partial r}\right) \quad (46.15)$$

subject to the no-slip boundary condition at the tube wall, which for a round tube takes the form $\mathbf{u} = 0$ for $r = a$.

The particular solution to Eq. 46.15 under this condition is most easily expressed in terms of complex ber and bei functions, which themselves are defined through [23] $ber(r) + i\cdot bei(r) = J_0(r\cdot i\sqrt{i})$, where J_0 represents the complex Bessel function of the first kind. Then:

$$u(r,t) = \frac{K}{\omega}(B\cos\omega t + (1-A)\sin\omega t) \quad (46.16)$$

where

$$A = \frac{ber\alpha \times ber\alpha\frac{r}{a} + bei\alpha \times bei\alpha\frac{r}{a}}{ber^2\alpha + bei^2\alpha} \quad (46.17)$$

and

$$B = \frac{bei\alpha \times ber\alpha\frac{r}{a} - ber\alpha \times bei\alpha\frac{r}{a}}{ber^2\alpha + bei^2\alpha} \quad (46.18)$$

Representative velocity profiles derived from these expressions are shown in Fig. 46.4 for two values of α, at four phases of the flow cycle. In these figures the radial position, r, has been normalized by the tube radius, a. At $\alpha = 3$ (Fig. 46.4a), a value that under resting conditions can occur in the smallest arteries and larger arterioles as well as the middle airways, Stokes layers can occupy a significant fraction of the tube radius. The velocity at the wall is zero, as required by the no-slip condition, and as in steady flow the velocity varies smoothly with r, with no step change at any point. However, even at this low α, the velocity profile resembles a parabola only during peak flow rates. At other flow phases, a more uniform profile forms across the tube core.

In contrast, at $\alpha = 13$ (Fig. 46.4b), which characterizes rest state flow in the aorta and trachea, the velocity profile of the pipe core is nearly uniform at all flow phases. Flow in the boundary layer is out of phase with that in the core, and flow reversals are

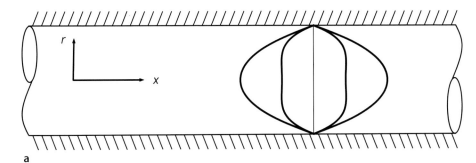

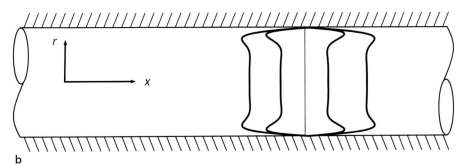

Fig. 46.4 Representative velocity profiles of laminar, oscillatory flow in a straight, rigid tube, at four phases of the flow cycle. **a** $\alpha = 3$; **b** $\alpha = 13$

possible in the Stokes layer. These changes in the velocity fields result from the inertia of the fluid, since as the flow frequency increases, less time is available in each flow cycle to accelerate the fluid.

To these flow fields of course must be added a steady component if the flow field is pulsatile rather than purely oscillatory.

As with steady flows, it is important to be able to use these expressions for the velocity field to determine the instantaneous total volume flow rate, Q_{inst}, or equivalently the instantaneous mean velocity, U_{inst}, since $Q_{inst} = U_{inst} \times$ pipe area. This may seem odd, in that a purely oscillatory flow field simply moves back and forth with no net fluid translation. Certainly, over a full flow cycle the *net* volume flow of Q_{osc} will be zero. However, oscillatory motion substantially alters the instantaneous shear stress to which the wall is exposed, compared to steady flow at an equivalent Reynolds number. It therefore has the potential to significantly alter the wall response. Following [24], it can be shown that the mean velocity is:

$$U(t) = \frac{K}{\omega}\left(\frac{2D}{\alpha}\cos\omega t + \left(\frac{1-2C}{\alpha}\right)\sin\omega t\right)$$
$$= \frac{K}{\omega}\sigma\cos(\omega t - \delta), \qquad (46.19)$$

where

$$C = \frac{ber\alpha \cdot bei'\alpha - bei\alpha \cdot ber'\alpha}{ber^2\alpha + bei^2\alpha} \qquad (46.20)$$

$$D = \frac{ber\alpha \cdot ber'\alpha + bei\alpha \cdot bei'\alpha}{ber^2\alpha + bei^2\alpha} \qquad (46.21)$$

$$\sigma^2 = \left(\frac{1-2C}{\alpha}\right)^2 + \left(\frac{2D}{\alpha}\right)^2 \qquad (46.22)$$

$$\tan\delta = \frac{(1-2C/\alpha)}{(2D/\alpha)} \qquad (46.23)$$

The oscillatory shear stress at the wall, $\tau_{w,osc}$, is given by $\tau_{w,osc} = -\mu(\partial u_{osc}/\partial r)|_{r=a}$.
This results in:

$$\tau_{w,osc} = \text{Re}\left(\frac{\rho K a \sqrt{i}}{\alpha} \frac{J_1\left(a\sqrt{\frac{i\omega}{\upsilon}}\right)}{J_0\left(a\sqrt{\frac{i\omega}{\upsilon}}\right)} \cdot e^{i\omega t}\right) \qquad (46.24)$$

As with the oscillatory flow rate, the oscillatory wall shear stress lags the pressure gradient, reaching a maximum during peak flow.

46.4.2
Hemodynamics in Elastic Tubes

Because of the mathematical complexity of analysis of pulsatile flows in elastic tubes, and the variety of physical phenomena associated with them, space does not permit a full description of this topic. More complete treatments are given in several references [13, 25–27]. Here we briefly summarize the most important features of these flows, to give the reader a sense of the richness of the physics underlying them.

In brief, in a tube with a nonrigid wall, any pressure change within the tube will lead to localized bulging of the tube wall in the high-pressure region (Fig. 46.5). Fluid can then flow in the radial direction into the bulge. Hence, not only is the radial velocity v no longer zero, but both u and v can no longer be independent of x even far from the tube ends. Thus the flow field is governed by the continuity condition along with the full Navier-Stokes equations. Assuming axial symmetry of the tube, these become:

$$\frac{\partial u}{\partial x} + \frac{\partial v}{\partial r} + \frac{v}{r} = 0, \qquad (46.25)$$

$$\frac{\partial u}{\partial t} + u\frac{\partial u}{\partial x} + v\frac{\partial u}{\partial r} = -\frac{1}{\rho}\frac{\partial P}{\partial x} +$$

$$\nu\left(\frac{\partial^2 u}{\partial x^2} + \frac{\partial^2 u}{\partial r^2} + \frac{1}{r}\frac{\partial u}{\partial r}\right)$$

$$\frac{\partial v}{\partial t} + u\frac{\partial v}{\partial x} + v\frac{\partial v}{\partial r} = -\frac{1}{\rho}\frac{\partial P}{\partial r} +$$

$$\nu\left(\frac{\partial^2 v}{\partial x^2} + \frac{\partial^2 v}{\partial r^2} + \frac{1}{r}\frac{\partial v}{\partial r} - \frac{v}{r^2}\right) \qquad (46.26)$$

The most important consequence of this is that even if the inlet pressure gradient depends only on t, within the tube the pressure gradient depends on x as

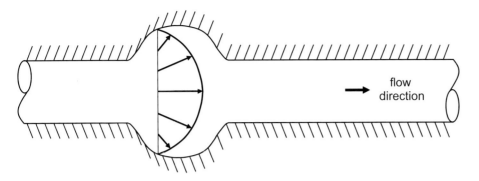

Fig. 46.5 Local bulging of the tube wall at regions of high pressure in pulsatile flow in an elastic tube

well as *t*. An oscillatory pressure gradient applied at the tube entrance therefore propagates down the tube in a wave motion. Both the pressure and the velocity fields therefore take on wave characteristics.

The speed with which these waves travel down the tube can be expected to depend on the fluid inertia (i.e., on its density, and on the wall stiffness). If the wall thickness is small compared to the tube radius and the effect of viscosity is neglected, the wave speed c_0 is given by the Moen-Korteweg formula $c_0 = (Eh/\rho d)^{1/2}$, where E is the stiffness, or Young's modulus, of the tube wall and h is its thickness. As can be expected on physical grounds, the wave speed increases as the wall stiffness rises. When the wall is rigid, E becomes infinite. Thus, oscillatory motion in a rigid tube, in which all the fluid moves together in bulk, may be thought of as resulting from a wave traveling with infinite speed, so that any change in the pressure gradient is felt throughout the whole tube instantaneously. In an elastic tube, by contrast, pressure changes are felt locally at first and then propagate downstream at finite speed.

Because of the action of the pressure and shear stress on the wall position and displacement, oscillatory flow in an elastic tube is inherently a *coupled* problem, in the sense that it is not possible in general to determine the fluid motion without also determining the resulting wall motion; the two are intrinsically linked. That is, the components of the wall motion are determined by the wall pressure and shear stress, which are themselves properties of the flow. Furthermore, the layer of fluid in contact with the wall must have the same velocity as the wall. Hence the no-slip boundary condition imposes the constraint that at the fluid-wall interface, $u(x,a,t)$ must equal the axial velocity component of the wall, and $v(x,a,t)$ must equal the radial velocity component of the wall.

With these governing equations and boundary conditions in place, and if the input pressure distribution that drives the flow field is known, it is possible to develop a formal solution for the axial velocity. For an oscillatory flow, the input pressure would normally be expected to be of a sinusoidal form $P(x,r,t) = const \times e^{i\omega t}$. Following [25], the method of characteristics shows the pressure distribution throughout the tube to be $P(x,r,t) = A(x,r)e^{i\omega(t-x/c)}$, where c is the wave speed in the fluid and A is the pressure amplitude. Since the fluid must be taken to be viscous, c is not equal to c_0, the inviscid fluid wave speed. Instead, $c = c_0(2/(1-\sigma^2)z)^{1/2}$, with z a parameter of the problem that depends on a, ω, ν, σ, ρ, ρ_w, and h. It can also be shown that the pressure amplitude A depends on x, but not on r, and therefore the pressure is uniform across any axial position in the tube [25]. Under these conditions, the solution for u, the principal velocity component of interest, can be stated as:

$$u(x,r,t) = \mathrm{Re}\left(\frac{A}{\rho c}\left\{1 - G\frac{J_0\left(r\sqrt{-\frac{i\omega}{\upsilon}}\right)}{J_0\left(a\sqrt{-\frac{i\omega}{\upsilon}}\right)}\right\} \cdot e^{i\omega(t-x/c)}\right)$$

(46.27)

with G a factor that modifies the velocity profile shape compared to that in a rigid tube due to the wall elasticity. G is given by:

$$G = \frac{2 + z(2\nu - 1)}{z(2\nu - g)} \qquad (46.28)$$

with

$$g = \frac{2J_1\left(a\sqrt{-\frac{i\omega}{\nu}}\right)}{\left(a\sqrt{-\frac{i\omega}{\nu}}\right)J_0\left(a\sqrt{-\frac{i\omega}{\nu}}\right)}. \qquad (46.29)$$

It is apparent from inspection of Eq. 46.28 that the difference between the velocity field in a rigid tube and that in an elastic tube is contained in the factor G. However, since G is complex, and both its real and imaginary parts depend on the flow frequency ω, the difference is by no means readily evident. The reader is referred to [25] for a detailed depiction of representative velocity profiles.

A final word about oscillatory flow in an elastic tube concerns the possibility of wave reflections. In a rigid tube, there is no wave motion as such, and flow arriving at an obstruction or branch is disturbed in some way, but otherwise progresses through the obstruction. In contrast, the wave nature of flow in an elastic tube leads to entirely different behavior at an obstacle. At an obstruction such as a bifurcation or a branch, some of the energy associated with pressure and flow is transmitted through the obstruction, while the remainder is reflected. This leads to a highly complex pattern of superposing primary and reflected pressure and flow waves, particularly in the arterial tree, since blood vessels are elastic and vessel branchings are ubiquitous throughout the vascular system. Such wave reflections may be analyzed in terms of transmission line theory [14, 25].

46.4.3
Turbulence in Pulsatile Flow

Transition to turbulence in oscillatory pipe flows occurs through fundamentally different mechanisms than transition in steady flows, for two reasons. The first is that the oscillatory nature of the flow leads to a unique base state, the most important feature of which is the formation of an oscillatory Stokes layer on the tube wall. This layer has its own stability characteristics, which are not comparable to the stability characteristics of the boundary layer of steady flow. The second reason is that temporal deceleration destabilizes the whole flow field, so that perturbations of the Stokes layer can cause the flow to break down into unstable, random fluctuations. In steady flow, perturbations either grow or decay. In contrast, even when unstable, oscillatory flow can present fully laminar motion during its acceleration and peak flow phases. Instability often occurs during the deceleration phase of the flow cycle, and is immediately followed by relaminarization as the net flow decays to zero prior to reversal. Because of these characteristics, during deceleration phases of the flow cycle instabilities are observable in the Stokes layer even at much lower Reynolds numbers than those for which they would be found in steady flow [28].

Since the Stokes layer thickness δ itself depends on the flow frequency, transition to turbulence depends on the Womersley number as well as the Reynolds number. Experimental measurements of the velocity made in rigid tubes by noninvasive optical techniques [22] have shown that over a range of values of $\alpha \geq 8$, the flow was found to be fully laminar for $Re_\delta \leq 500$, where Re_δ is the Reynolds number based on the Stokes layer thickness rather than tube diameter. That is, $Re_\delta = U\delta/\nu$. For $500 < Re_\delta < 1300$, the core flow remained laminar while the Stokes layer became unstable during the deceleration phase of fluid motion. This turbulence was most intense in an annular region near the tube wall. These results are in accord with theoretical predictions of instabilities in Stokes layers [29, 30]. For higher values of Re_δ, instability can be expected to spread across the tube core.

A capstone illustration of the principles of oscillatory flow and turbulence development using the hemodynamics of aortic aneurysms as a descriptive example is given in [13].

46.5
Models and Computational Techniques

46.5.1
Approximations to the Navier–Stokes Equations

The Navier–Stokes equations, Eqs 46.4–46.7, together with the continuity condition, Eq. 46.3, provide a complete set of governing equations for the motion of an incompressible Newtonian fluid. If appropriate boundary and initial conditions can be specified for the motion of such a fluid in a given flow system, in principle a full set of governing equations and conditions for the system will be known. It may then be expected that the fluid motion can be deduced simply by solution of the resulting boundary value problem. Unfortunately, however, the mathematical difficulties resulting from the nonlinear character of the acceleration terms $D\mathbf{u}/Dt$ in the Navier–Stokes equations are so great that only a very limited number of exact solutions have ever been found. The simplest of these pertain to cases in which the velocity has the same direction at every point in the flow field, as in the steady and pulsatile pipe flows discussed above.

Accordingly, there is a strong incentive to seek conditions under which one or more of the terms in Eqs. 46.4–46.7 are negligible or nearly so, and therefore an approximate and much simpler governing equation can be generated by neglecting them altogether. For example, the Reynolds number represents the ratio of inertial to viscous forces in the flow field. Accordingly, in flows for which $Re \gg 1$, it can be shown that the viscous term $\nu \nabla^2 \mathbf{u}$ is very much smaller than the acceleration $D\mathbf{u}/Dt$. Consequently, it can be omitted from the governing equation, which leads to solutions that are approximately valid at least outside the boundary layer. Conversely, when $Re \ll 1$, the viscous term $\nu \nabla^2 \mathbf{u}$ is much larger than the acceleration $D\mathbf{u}/Dt$.

In summary, these approximations show that viscosity is important in three situations:
1. When the overall Reynolds number is *low*, since then viscous effects act over the full flow field.
2. When the overall Reynolds number is *high*, viscosity is important in thin boundary layers.
3. When the flow is *enclosed*, as in a pipe flow, since then the available diffusion time is very large, and viscous effects can become important in the whole flow after some initial region or time.

An alternative approach to seeking simplifications to the Navier–Stokes equations is to accept the full set of equations, but approximate each term in the equation with a simpler form that permits solutions to be developed. Although the resulting equations are only *approximately* correct, the advent of modern digital computers has allowed them to be written with great fineness, so that highly accurate solutions are achieved. These techniques are called CFD.

46.5.2
Computational Fluid Dynamics

In the last few decades, CFD has become a very powerful and versatile tool for the analysis of complex problems of interest in the engineering biosciences. By providing a cost-effective means to simulate real flows in detail, CFD permits studying complex problems combining thermodynamics, chemical reaction kinetics, and transport phenomena with fluid flow aspects. In addition, such problems often arise in highly complex geometries. Consequently, they may be far too difficult to study accurately without computational model approaches.

Furthermore, CFD offers a means for testing flow conditions that are unachievable or prohibitively expensive to test experimentally. For example, most flow loops and wind tunnels are limited to a fixed range of flow rates and governing parameter values. Such limits generally do not apply to CFD analyses. Moreover, flow under a wide range of parameter values may be tested with far less cost than performing repeated experiments.

A representative example of widespread interest to biomedical engineers is the analysis of hemodynamics in blood vessel models. When analyzing biologic responses to flow, or before employing newly developed surgical procedures, characterization studies need to be conducted to substantiate applicability. Cellular metabolic rates in encapsulated and free states, as well as pertinent transport phenomena, can be evaluated in anatomically realistic vessel configurations. These data, coupled with CFD modeling, provide the basis for redesign/reconfigurations as apropos. CFD is a very powerful and versatile tool for an analysis of this type.

At present, CFD methods are finding many new and diverse applications in bioengineering and biomimetics. For example, CFD techniques can be used to predict:

1. Velocity and stress distribution maps in complex reactor performance studies as well as in vascular and bronchial models.
2. Strength of adhesion and dynamics of detachment for mammalian cells.
3. Transport properties for nonhomogeneous materials and nonideal interfaces.
4. Multicomponent diffusion rates using the Maxwell-Stefan transport model, as opposed to the limited traditional Fickian approach, incorporating interactive molecular immobilizing sites.
5. Materials processing capabilities useful in encapsulation technology and designing functional surfaces.

Although a full description of CFD techniques is beyond the scope of this chapter, thorough descriptions of the methods and procedures may be found in many texts, for example [31–33].

References

1. Bird RB, Stewart WE, Lightfoot EN (2002) Transport Phenomena, 2nd edn. Wiley, New York
2. Lightfoot EN (1974) Transport Phenomena and Living Systems. Wiley-Interscience, New York
3. Cooney DO (1976) Biomedical Engineering Principles. Dekker, New York
4. Lightfoot EN, Duca KA (2000) The Roles of Mass Transfer in Tissue Function. In: Bronzino JD (ed) The Biomedical Engineering Handbook, 2nd edn. CRC, Boca Raton, Ch. 115
5. Fisher RJ, Peattie RA (2006) Controlling tissue microenvironments: biomimetics, transport phenomena and reacting systems. In: Kaplan DL, Lee K (eds) Advances in Biochemical Engineering/Biotechnology, Special Volume: Tissue Engineering
6. Ratner BD, Bryant SJ (2004) Ann Rev Biomed Eng 6:41
7. Lauffenburger DA, Linderman JJ (1993) Receptors: Models for Binding, Trafficking, and Signaling. Oxford University Press, New York
8. Tarbell JM, Qui Y (2006) Arterial wall mass transport: the possible role of blood phase resistance in the localization of arterial disease. In: Bronzino JD (ed) The Biomedical Engineering Handbook, 3rd edn. CRC, Boca Raton, Ch. 9
9. Galletti PM, Colton CK, Jaffrin M, Reach G (2006) Artificial pancreas. In: Bronzino JD (ed) The Biomedical Engineering Handbook, 3rd edn. Tissue Eng and Art Organs. CRC, Boca Raton, Ch. 71
10. Lewis AS, Colton CK (2006) Tissue engineering for insulin replacement in diabetes. In: Ma PX, Elisseeff J (eds) Scaffolding in Tissue Engineering. CRC, Boca Raton, Ch. 37
11. Freshney RI (2000) Culture of Animal Cells: A Manual of Basic Technique, 4th edn. Wiley-Liss, New York
12. Palsson B (2005) Tissue engineering. In: Enderle J, Blanchard S, Bronzino JD (eds) Introduction to Biomedical Engineering, 2nd edn. Academic Press, Orlando, Ch. 12
13. Peattie RA, Fisher RJ (2006) Perfusion effects and hydrodynamics. In: Kaplan DL, Lee K (eds) Advances in Biochemical Engineering/Biotechnology, Special Volume: Tissue Engineering
14. Fung YC (1997) Biomechanics: Circulation. Springer-Verlag, New York
15. Lamb H (1945) Hydrodynamics. Dover, New York
16. Schlichting H (1979) Boundary Layer Theory, 7th edn. McGraw-Hill, New York
17. Hinze JO (1986) Turbulence, (Reissued) McGraw-Hill, New York
18. Tennekes H, Lumley JL (1972) A First Course in Turbulence. MIT Press, Cambridge
19. Poiseuille JLM (1840) Comptes Rendus 11:961
20. Fox RW, McDonald AT, Pritchard, PJ (2008) Introduction to Fluid Mechanics, 7th edn. John Wiley and Sons, New York
21. Crowe CT, Elger DF, Roberson JA, (2006) Engineering Fluid Mechanics, 8th edn. John Wiley and Sons, New York
22. Womersley JR (1955) J Physiol 127:553
23. Dwight HB (1961) Tables of Integrals and Other Mathematical Data. McMillan, New York
24. Gerrard JH (1971) J. Fluid Mech 46:43
25. Zamir M (2000) The Physics of Pulsatile Flow. AIP Press, Springer-Verlag, New York
26. Womersley JR (1955) Phil Mag 46:199
27. Atabek SC, Lew HS (1966) Biophys J 6:481
28. Eckmann DM, Grotberg JB (1991) J Fluid Mech 222:329
29. Davis SH, von Kerczek C (1973) Arch Rat Mech Anal 188:112
30. von Kerczek C, Davis SH (1974) J Fluid Mech 62:753
31. Fletcher CA (1991) Computational Techniques for Fluid Dynamics, Volume I, 2nd edn. Springer-Verlag, Berlin
32. Fletcher CA (1991) Computational Techniques for Fluid Dynamics, Volume II, 2nd edn. Springer-Verlag, Berlin
33. Chung TJ (2002) Computational Fluid Dynamics. Cambridge University Press, Cambridge

Ex Vivo Formation of Blood Vessels

R. Y. Kannan, A. M. Seifalian

Contents

47.1 Introduction 685
47.2 Ideal Vascular Graft 686
47.3 Tissue-Engineered Vascular Grafts ... 686
47.3.1 Scaffold 687
47.3.2 Matrix 688
47.3.3 Endothelialisation 689
47.4 Current Problems 689
47.5 Conclusion 690
References 691

47.1 Introduction

Vascular bypass grafts represent a well-grounded technological advance in the new millennium, this being related to the high prevalence of atherosclerosis and the ensuing treatment of myocardial infarction [1]. However, more recently a paradigm shift has occurred casting the subject of vascular tissue engineering in new light. The development of cardiovascular devices both for high- and low-flow scenarios in the form of functional vascular prostheses has made this possible. While the patency rates of high-flow vessels [2], such as polyethylene terephthalate (Dacron) and expanded polytetrafluoroethylene (ePTFE) grafts following prolonged clinical use are acceptable, these figures are less satisfactory at lower flow rates [3]. Given the limitations with synthetic materials alone, a biological or biohybrid vascular prosthesis could provide us with the ideal blood vessel substitute.

The current interest in developing artificial organs such as the tissue-engineered bladders [4] and composite tissue has sparked renewed interest in this field. In the latter case, the limitations of immunogenicity with allotransplantation [5] have put the development of these blood vessels, and in particular the development of artificial capillary systems [6], at the forefront. Such integrated vascular networks would serve to nourish the various tissue planes within a tissue-engineered construct [7]. This has formed the basis for the emphasis of developing small-calibre (<6 mm) cardiovascular bypass conduits. In addition, the poor long-term patency of ePTFE grafts used in coronary and distal infrainguinal bypasses [8] mean the added morbidity of harvesting vein or arterial grafts [9].

In this chapter, a broad overview of the formation of both macro- and microvascular systems will be discussed. These would include biological, biohybrid and synthetic vessel formation and construction. This in particular would be useful in the field of paediatric vascular surgery, eliminating the need for sequential replacement of synthetic grafts in paediatric patients [10].

47.2
Ideal Vascular Graft

The ideal characteristics of vascular grafts are shown in Table 47.1. These indicate the various perspectives that challenge the tissue engineer attempting to mimic nature. While Dacron grafts, for instance, have been successful in high-flow states such as aortic bypass grafts, as implied above, blood at lower flow rates is more unforgiving, and therefore vascular grafts would need to have most, if not all of these ideal characteristics.

Seeding vascular grafts with endothelial cells has been extensively researched [11]. There are three possible sources for endothelialisation, namely (1) transanastomotic pannus ingrowth from the native artery, (2) transmural tissue ingrowth and (3) "fallout" endothelialisation by circulating progenitor cell precursors in blood [12]. In spite of successful seeding techniques, the absence of a metabolically active smooth muscle cell layer affects the sustenance of the overlying endothelial cells. In addition, the relative non-elasticity of the vascular prosthesis compared to native vessels affects compliance and encourages intimal hyperplasia [13]. This only strengthens the argument to develop tissue-engineered vascular grafts.

One of the first attempts to create small-diameter tissue-engineered vascular grafts was with the Sparks' mandril vascular graft. This is a fibrocollagenous tube formed by implanting a silicone tube in vivo and allowing the foreign-body reaction to create a neovessel [14]. However, high rates of thrombosis and aneurismal change posed unattractive features. The same held true for other alternatives such as cryopreserved vessels [15] and arterial allografts [16]. These methods also require prolonged in vivo implantation that is not clinically compatible.

47.3
Tissue-Engineered Vascular Grafts

A tissue-engineered artery would need to satisfy all the requirements for cardioavascular haemodynamics by means of an inherent composite structure, wherein each layer contributes a particular characteristic to the superstructure. This is depicted graphically in Fig. 47.1, which shows the multi-layered arterial wall structure, with the inner layer conferring haemocompatibility and the outer layer accounting for mechanical strength. The aim of tissue engineering is to replicate the structural and cellular organisation of an artery so as to confer the diverse range of properties necessary for its function.

Table 47.1 The ideal vascular bypass graft specification

Parameters	Essential	Desirable
Mechanical properties	Mechanical strength to withstand physiological haemodynamic pressure.	Compliance matching with native artery. Modulation of mechanical properties depending on changes in haemodynamics in short and long term.
Blood compatibility	Low thrombogenicity.	Non-thrombogenic. Localised coagulation at site of injury to vessel wall.
Biocompatibility	Long-term biostability. Non-immunogenic.	Good graft healing and endothelialisation without excessive fibrovascular infiltration compromising vessel calibre and compliance.
Functional properties		Incorporation of homeostatic/regulatory mechanisms—e.g. nitric oxide, prostacyclin release. Infection-resistant.
Availability	Available quickly in urgent cases.	Available without further invasive procedures for the patient.

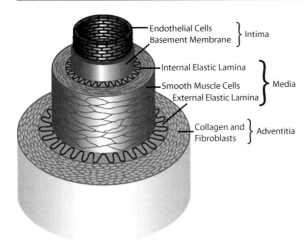

Fig. 47.1 The multiple laminar organisation of the arterial wall allows it to maintain a wide range of properties

47.3.1
Scaffold

The basic components of a tissue-engineered blood vessel are the scaffold, matrix and the endothelial cells. A scaffold provides the necessary shape and strength to allow both endothelial and smooth muscle cells to adhere and proliferate on it. In addition, it has to allow for the vascular permeation of the construct.

Many approaches have been used to achieve the common goal of viable confluent sheets of cells anchored onto a tubular scaffold. The role of the scaffold is to provide shape as well as initial strength. Some investigators have done away with the need for a scaffold altogether, using a mandrel on which cellular attachment, adherence and confluence are encouraged using cell culture techniques in a bioreactor (see Fig. 47.2) [17]. However, the use of a matrix or scaffold into which cells can embed is generally more successful [18]. The scaffolds used are either synthetic or biologically sourced.

Naturally derived vascular substitutes are, for example, acellular materials like intestinal submucosa [19], tanned carotid arteries [20] and cryopreserved vein grafts [21]. Bell and colleagues in 1979 developed a completely biological vascular graft based on fibroblast-mediated contraction of hydrated protein lattices. Contraction of these lattices could also be achieved by smooth muscle cells (SMCs) in order to form a tubular mandrel capable of producing prostacyclin [22]. However, these natural scaffolds are vulnerable to unintended biodegradation, leading to graft-wall weakness, aneurysm and failure. This is especially problematic with elastin in decellularised xenograft scaffolds, which degrades despite glutaraldehyde fixation, due to its low opportunities for cross-linking [23].

Ratcliffe transposed this concept onto an ether-based polyurethane scaffold seeded with SMCs and then subjected it to preconditioning [24]. The resulting construct was capable of synthesising collagen, elastin and proteoglycans. Following endothelialisation, these vessels remained patent in vivo for up to 4 weeks. As ether-based polyurethanes are susceptible to in vivo degradation with the potential release of toluene diamine, they have been superceded by carbonate-based polyurethanes have been introduced

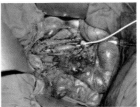

Fig. 47.2 A living cellular matrix, comprising type I collagen and porcine smooth muscle cells, which is contracted onto a central mandrel within a bioreactor, and after endothelialisation being evaluated in vivo using a porcine infrarenal aortic model

instead. These vascular grafts have the additional advantage of similar radial compliance to native vessels [25] and would be optimal as a vascular scaffold [26].

Biodegradable scaffolds on the other hand can provide a suitable environment for adherence, proliferation and organisation of SMCs and endothelial cells. By titrating the rate of degradation against cell proliferation, primary mechanical stability can be maintained while the cellular structures mature. As demonstrated in Fig. 47.3, the scaffold initially withstands the stresses of pulsatile flow. As the scaffold begins to degrade and formation of matrix occurs, the newly regenerating cells are gradually loaded with physiological stress, further simulating tissue regeneration. Eventually, the scaffold completely degrades and the regenerated tissue bears the stress with cells such as fibroblasts, SMCs and vascular endothelial cells (Fig. 47.3). The attachment of progenitor stem cells is an exciting possibility [27], aimed at improving cellular retention and adherence from endogenous blood flow.

Polyglycolic acid (PGA) scaffolds seeded with SMCs under pulsatile pressure showed similar morphology and function to native arteries, with 100% patency at 4 weeks [28]. These grafts could also respond to serotonin and endothelin-1, in addition to withstanding high burst pressures. However, the vast majority of these grafts were implanted into high-flow systems such as the aorta [29] and it remains to be seen whether the same results would hold for lower-flow vessels. Hybrid SMC-gel scaffolds have been developed, but their mechanical strength is in question [30].

47.3.2
Matrix

An artery has a degree of mechanical anisotropy wherein it exhibits high elasticity at low pressures and greater stiffness along with low distensibility at high pressures [8]. This is a result of the multi-laminar arrangement of collagen and elastin. Collagen provides the high tensile strength necessary to withstand high pressures, while elastin accounts for high distensibility. The synergism of both strength and elasticity has been attempted with materials such as segmented polyurethanes. This is to minimise the compliance mismatch between the vascular graft and the native vessel, which should reduce intimal hyperplasia, which is a long-term cause of vessel occlusion. A problem with polyurethane vessels is that while compliance is far greater compared to conventional vascular grafts, its compliance values are isomorphic across the pressure ranges. This problem was overcome by the addition of polyhedral oligomeric silsesquioxane (POSS) nanofillers into a carbonate-based polyurethane. The resulting microvessel (800 μm internal diameter) exhibited dynamic compliance characteristics secondary to the intrinsic flexibility of the POSS nanocage [6].

Using biodegradable hydrophilic polymers, which would be eventually replaced by incoming extracellular matrix (ECM) via the porous system of the scaffold, it may be possible to aim for a biohybrid matrix. These systems would be based on PGA, polydioxanone or polylactide. PGA forms a porous structure but is resorbed within 8 weeks. In order to slow degradation, it is often combined with an adjunct such as polyhydroxyalkanoate, poly-4-hydroxybutyrate or polyethylene glycol [31]. Cellular infiltration is dependent on a highly porous open structure.

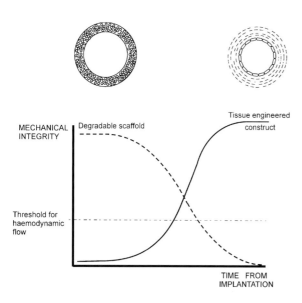

Fig. 47.3 The timeline for the development of a tissue-engineered bypass graft using a synthetic biodegradable scaffold, which is gradually replaced by arterial matrix with an endothelial lining [59]

Collagen by itself could also serve as the foundation for biological scaffolds, primarily due to its strength. Pioneering work on collagen-based scaffolds, however, showed that they were very weak, often requiring a Dacron mesh to bolster its strength [32]. Berglund [33] has shown that elastin makes collagen tubes stronger, by means of redistributing the stress across the tissue-engineered vessel wall. Elastin also improves the viscous nature, which again is generally thought to be due to collagen. Without elastin's gradual stretching influence, the abrupt pressure wave on the collagen construct may not allow sufficient time for the viscous component to manifest itself. Conventionally, although elastin can be extracted from animal sources, further application is hindered by its low solubility. Leach's group [34] have overcome this by using water-soluble α-elastin in conjunction with a "molecular glue" to form "elastin-like" structures with similar mechanical and cell-adherent properties.

The mechanical strength of a vessel may also be boosted by the circumferentially arranged SMC layers, which require prolonged culture to achieve, as well as a porous collagen network. This orientation can be further optimised by pulsatile flow in bioreactors with circumferential stress acting as a mechanical stimulus [35].

Other derivatives of biological structures are hyaluronan, which acts as a temporary scaffold. Lepidi's work on hyaluronan [36] showed rapid accumulation of organised arterial cellular and ECM components in an animal model, with biodegradation of the scaffold over 4 months. This cellular reorganisation is augmented by the scaffold's inherent ability to maintain a loosely packed ECM. However, this success is yet to be replicated in the larger 3- to 6-mm grafts where the strength of the construct is strongly dependent on the ECM formed [37].

47.3.3
Endothelialisation

It is difficult to render a synthetic material non-thrombogenic. The current best hope is for the development of an endothelial lining. This is not due to a physical smooth surface presented by the endothelial layer, but rather due to the haemostatic mechanism incorporated into each endothelial cell as well as an efficient intracellular signalling system triggered by the prevailing haemodynamic conditions. Furthermore, it also acts as a physical barrier separating blood from the moderately thrombogenic basement membrane and the highly thrombogenic collagen of the ECM. Unfortunately, although some animal models readily endothelialise [38], a synthetic graft with long confluent endothelialised sections in clinical practice is yet to be achieved. Recognising this, some groups have accepted the need for blood interaction with collagen, turning to surface modifications of collagen such as heparin-bonding, which can be partially successful in reducing thrombogenicity [39].

Tissue engineering aims to provide low graft thrombogenicity via active seeding of the blood interface with endothelial cells. This may be performed either as a single- [40] or a two-stage procedure. Efforts to improve cell adherence then led to specific peptide coating [41]. Recently, the possibility of recruiting endothelial progenitor cells (EPCε) circulating in the blood, which subsequently differentiate into endothelial cells, is being realised in vitro, as is the seeding of pluripotent stem cells from bone marrow and adipose tissue [42]. An alternative theory evolving is the existence of "vasculogenic zones" within the vessel walls of large- and medium-sized vessels, as shown in Fig. 47.4. In either case, EPCs have the unique ability of providing purified cells capable of very high cell-doubling capacity for long periods of time, which in time would promote vasculogenesis [43].

47.4
Current Problems

A major limitation of tissue-engineered grafts is the length of time required before an individualised graft can be made available. Attempts to speed the fabrication process have already been put into place. For instance, ready-to-use ECM components have been made available [34, 33]. Collagen hydrogels are also convenient for forming confluent scaffolds, in spite of the danger of shrinkage. This effect can be minimised by including whole collagen fibres into the hydrogel matrix [44].

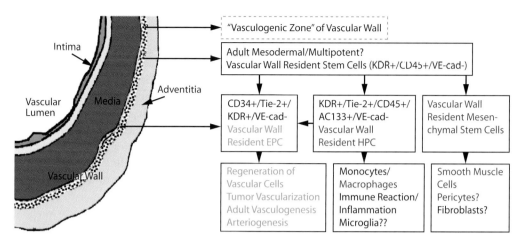

Fig. 47.4 The vasculogenic zones in the perivessel region, which is abundant in multipotent vascular stem cells. These cells have the capacity to convert into inflammatory, vascular, smooth muscle cells as well as pericytes. *EPC* Endothelial progenitor cells, *HPC* haematopoietic progenitor cells. Reprinted with permission from reference [43]

Another limiting factor is the long time necessary for endothelialisation, particularly in the two-stage setting. Niklason's group [45] have demonstrated a direct correlation between the time for maturation and the strength of the resultant graft. Single-stage seeding on the other hand involves autoseeding cells rather than relying on physical extraction of cells by enzymatic or physical means where cell yields are low and adhesive efficiency and then proliferation is limited. In addition, this also yields different levels of differentiation of SMC lines depending on the method of extraction [46]. Alternatively, seeding certain biodegradable synthetic polymers with poor cell adherence properties following collagen binding could improve overall endothelialisation [47, 48], leading to the possibility of autoseeding.

An alternative for autoseeding is with EPCs. These stem cells, circulating in peripheral blood, contribute to rapid endothelialisation of these scaffolds [49]. However, the number of circulating EPCs in whole blood is low and the time taken for confluent endothelialisation by autoseeding without cell culture would be high. Nevertheless, certain groups have shown effective endothelialisation with human EPCs on collagen-coated polyurethane conduits [50]. Moreover, the problem of low EPC numbers in circulating blood can be circumvented by rapid seeding of stem-cell-rich bone marrow, as discussed previously. A further alternative may be to persuade the bone marrow to mobilise larger numbers of EPCs, using granulocyte colony stimulating factor and granulocyte-macrophage colony stimulating factor [51]. Other techniques to improve endothelialisation include surface modifications with RGD-containing peptides [52] and growth factors such as basic fibroblast growth factor [53]. The RGD moiety is also independently antithrombogenic [54] and increases the feasibility of a single-stage seeding procedure [55]. Combining this with nanocomposite technology, dendritic POSS nanocomposites covalently bound to RGD-peptides have now been developed and would have great potential as vascular bypass grafts [56].

47.5
Conclusion

Until recently, there have only been a few cases wherein tissue-engineered vascular grafts have been used. Shin'oka and colleagues [57] seeded a biodegrading scaffold before implanting the biohybrid device into a child as a pulmonary artery graft. Other groups have also reported the use of tissue-engineered tissue (both using SMC and bone-marrow-derived stem cells) for pulmonary arterial and venous congenital anomaly surgery [10, 58, 59]. Thus, it can be clearly seen that the immediate benefit of this technology would be in paediatric vascular surgery. The next step would be to introduce a tissue-engineered vascular graft into a critical small-vessel arterial system such as the coronary circulation. Although extensive in vivo testing is in store for this technology, the future is bright for vascular tissue engineering.

References

1. British Heart Foundation (2003) Coronary heart disease statistics
2. Kannan, R. Y, Salacinski, H. J, Butler, P. E, Hamilton, G, and Seifalian, A. M (2005) J Biomed Mater Res B Appl Biomater 74, 570–581
3. Kashyap, V. S, Ahn, S. S, Quinones-Baldrich, W. J, Choi, B. U, Dorey, F, Reil, T. D, Freischlag, J. A, and Moore, W. S (2002) Vasc Endovasc Surg 36, 255–262
4. Atala, A, Bauer, S. B, Soker, S, Yoo, J. J, and Retik, A. B (2006) Lancet 367, 1241–1246
5. Gautam, M, Cheruvattath, R, and Balan, V (2006) Liver Transpl 12, 1813–1824
6. Kannan, R. Y, Salacinski, H. J, Edirisinghe, M. J, Hamilton, G, and Seifalian, A. M (2006) Biomaterials 27, 4618–4626
7. Kannan, R. Y, Salacinski, H. J, Sales, K, Butler, P, and Seifalian, A. M (2005) Biomaterials 26, 1857–1875
8. Sarkar, S, Salacinski, H. J, Hamilton, G, and Seifalian, A. M (2006) Eur J Vasc Endovasc Surg 31, 627–636
9. Veith, F. J, Moss, C. M, and Sprayregen, S (1979) Surgery 85, 253–256
10. Matsumura, G, Hibino, N, Ikada, Y, Kurosawa, H, and Shin'oka, T (2003) Biomaterials 24, 2303–2308
11. Seifalian, A. M, Tiwari, A, Hamilton, G, and Salacinski, H. J (2002) Artif Organs 26, 307–320
12. Sales, K. M, Salacinski, H. J, Alobaid, N, Mikhail, M, Balakrishnan, V, and Seifalian, A. M (2005) Trends Biotechnol 23, 461–467
13. Salacinski, H. J, Goldner, S, Giudiceandrea, A, Hamilton, G, Seifalian, A. M, Edwards, A, and Carson, R. J (2001) J Biomater Appl 15, 241–278
14. Sparks, C. H (1972) Am J Surg 124, 244–249
15. Harris, L, O'Brien-Irr, M, and Ricotta, J. J (2001) J Vasc Surg 33, 528–532
16. Castier, Y, Leseche, G, Palombi, T, Petit, M. D, and Cerceau, O (1999) Am J Surg 177, 197–202
17. Baguneid, M. S, Siefalian, A. M, Hamilton, G, and Walker, M. G (2000) Br J Surg 87, 87 (abstract)
18. Abilez, O, Benharash, P, Mehrotra, M, Miyamoto, E, Gale, A, Picquet, J, Xu, C, and Zarins, C (2006) J Surg Res 132, 170–178
19. Lantz, G. C, Badylak, S. F, Hiles, M. C, Coffey, A. C, Geddes, L. A, Kokini, K, Sandusky, G. E, and Morff, R. J (1993) J Invest Surg 6, 297–310
20. Rosenberg, N, Martinez, A, Sawyer, P. N, Wesolowski, S. A, Postlethwait, R. W, and Dillon, M. L, Jr (1966) Ann. Surg 164, 247–256
21. Neufang, A, Espinola-Klein, C, Dorweiler, B, Reinstadler, J, Pitton, M, Savvidis, S, Fischer, R, Vahl, C, and Schmiedt, W (2005) Eur J Vasc Endovasc Surg 30, 176–183
22. Weinberg, C. B. and Bell, E (1986) Science 231, 397–400
23. Isenburg, J. C, Simionescu, D. T, and Vyavahare, N. R (2004) Biomaterials 25, 3293–3302
24. Ratcliffe, A (2000) Matrix Biol 19, 353–357
25. Tai, N. R, Salacinski, H. J, Edwards, A, Hamilton, G, and Seifalian, A. M (2000) Br J Surg 87, 1516–1524
26. Salacinski, H. J, Punshon, G, Krijgsman, B, Hamilton, G, and Seifalian, A. M (2001) Artif Organs 25, 974–982
27. Riha, G. M, Lin, P. H, Lumsden, A. B, Yao, Q, and Chen, C (2005) Tissue Eng 11, 1535–1552
28. Niklason, L. E, Gao, J, Abbott, W. M, Hirschi, K. K, Houser, S, Marini, R, and Langer, R (1999) Science 284, 489–493
29. Shum-Tim, D, Stock, U, Hrkach, J, Shin'oka, T, Lien, J, Moses, M. A, Stamp, A, Taylor, G, Moran, A. M, Landis, W, Langer, R, Vacanti, J. P, and Mayer, J. E, Jr (1999) Ann Thorac Surg 68, 2298–2304
30. Hirai, J, Kanda, K, Oka, T, and Matsuda, T (1994) ASAIO J 40, M383–M388
31. Kakisis, J. D, Liapis, C. D, Breuer, C, and Sumpio, B. E (2005) J Vasc Surg 41, 349–354
32. Weinberg, C. B. and Bell, E (1986) Science 231, 397–400
33. Berglund, J. D, Nerem, R. M, and Sambanis, A (2004) Tissue Eng 10, 1526–1535
34. Leach, J. B, Wolinsky, J. B, Stone, P. J, and Wong, J. Y (2005) Acta Biomater 1, 155–164
35. L'Heureux, N, Germain, L, Labbe, R, and Auger, F. A (1993) J Vasc Surg 17, 499–509
36. Lepidi, S, Abatangelo, G, Vindigni, V, Deriu, G. P, Zavan, B, Tonello, C, and Cortivo, R (2006) FASEB J 20, 103–105
37. Remuzzi, A, Mantero, S, Columbo, M, Morigi, M, Binda, E, Camozzi, D, and Imberti, B (2004) Tissue Eng 10, 699–710
38. Rashid, S. T, Salacinski, H. J, Hamilton, G, and Seifalian, A. M (2004) Biomaterials 25, 1627–1637
39. Keuren, J. F, Wielders, S. J, Driessen, A, Verhoeven, M, Hendriks, M, and Lindhout, T (2004) Arterioscler Thromb Vasc Biol 24, 613–617
40. Herring, M, Gardner, A, and Glover, J (1978) Surgery 84, 498–504
41. Wissink, M. J, Beernink, R, Scharenborg, N. M, Poot, A. A, Engbers, G. H, Beugeling, T, van Aken, W. G, and Feijen, J (2000) J Control Release 67, 141–155
42. Wang, H, Riha, G. M, Yan, S, Li, M, Chai, H, Yang, H, Yao, Q, and Chen, C (2005) Arterioscler Thromb Vasc Biol 25, 1817–1823
43. Zengin, E, Chalajour, F, Gehling, U.M, Ito, W.D, Treede, H, Lauke, H, Reichenspurner, H, Kilic, N, Ergun, S (2006) Development 133, 1543–1541
44. Lewus, K. E and Nauman, E. A (2005) Tissue Eng 11, 1015–1022
45. Niklason, L. E, Abbott, W. M, Gao, J, Klagges, B, Hirschi, K. K, Ulubayram, K, Conroy, N, Jones, R, Vasanawala, A, Sanzgiri, S, and Langer, R (2001) J Vasc Surg 33, 628–638

46. Opitz, F, Schenke-Layland, K, Richter, W, Martin, D. P, Degenkolbe, I, Wahlers, T, and Stock, U. A (2004) Ann Biomed Eng 32, 212–222
47. Ma, Z, He, W, Yong, T, and Ramakrishna, S (2005) Tissue Eng 11, 1149–1158
48. Iwai, S, Sawa, Y, Ichikawa, H, Taketani, S, Uchimura, E, Chen, G, Hara, M, Miyake, J, and Matsuda, H (2004) J Thorac Cardiovasc Surg 128, 472–479
49. He, H, Shirota, T, Yasui, H, and Matsuda, T (2003) J Thorac Cardiovasc Surg 126, 455–464
50. Shirota, T, He, H, Yasui, H, and Matsuda, H (2003) Tissue Eng 9, 127–136
51. Cho, S. W, Lim, J. E, Chu, H. S, Hyun, H. J, Choi, C. Y, Hwang, K. C, Yoo, K. J, Kim, D. I, and Kim, B. S (2006) J Biomed Mater Res A 76, 252–263
52. Hsu, S. H, Sun, S. H, and Chen, D. C (2003) Artif Organs 27, 1068–1078
53. Conklin, B. S, Wu, H, Lin, P. H, Lumsden, A. B, and Chen, C (2004) Artif Organs 28, 668–675
54. Eriksson, A. C. and Whiss, P. A (2005) J Pharmacol Toxicol Methods 52, 356–365
55. Tiwari, A, Kidane, A, Salacinski, H. J, Punshon, G, Hamilton, G, and Siefalian, A. M (2003) Eur J Vasc Endovasc Surg 25, 325–329
56. Alobaid, N, Salacinski, HJ, Sales, KM, Ramesh, B, Kannan, RY, Hamilton, G, Seifalian, AM, Eur J Vasc Endovasc Surg (2006) Eur J Vasc Endovasc Surg 32, 76–83
57. Shin'oka, T, Imai, Y, and Ikada, Y (2001) N Engl J Med 344, 532–533
58. Shin'oka, T, Matsumura, G, Hibino, N, Naito, Y, Watanabe, M, Konuma, T, Sakamoto, T, Nagatsu, M, and Kurosawa, H (2005) J Thorac Cardiovasc Surg 129, 1330–1338
59. Berglund, J. D, Mohseni, M. M, Nerem, R. M, and Sambanis, A (2003) Biomaterials 24, 1241–1254

Biomechanical Function in Regenerative Medicine

B. David, J. Pierre, C. Oddou

Contents

48.1	Introduction	693
48.2	Critical Structural and Mechanical Properties of a Scaffold for Regenerative Medicine	694
48.2.1	Pore Radius, Porosity, Specific Pore Area and Tortuosity	695
48.2.2	Topography and Surface Chemistry	695
48.2.3	Intrinsic MechanicalProperties	696
48.2.4	Scaffold as an Ad Hoc Microenvironment for Cells	697
48.3	Critical Cell Culture Methods for Bone Tissue Engineering	697
48.3.1	Matrix Seeding Quality	698
48.3.2	Solving Problems of Mass Transfer	698
48.3.3	Mastering the Applied Mechanical Forces	701
	References	702

48.1 Introduction

The essential aim of regenerative medicine is to obtain a sufficient mass of whatever specific type of cell, or more organised entity such as tissue or organ, needed to restore the normal physiology of a part of the body damaged by physical, chemical or ischaemic insult, or as a consequence of infectious or genetic disease. This is necessitated by the fact that the intrinsic regenerative capacity of most tissues and organs is very limited in mammals compared to many lower vertebrates. Even where damage does induce significant cellular proliferation, the newly formed cells often fail to differentiate appropriately to replace those that have been lost. The generation of autologous grafts then requires the development of procedures to quickly expand human mesenchymal stem cells (MSCs) in 3D systems and to promote their differentiation in a controlled way in order to maintain their tissue genesis potential.

As illustrated on the Fig. 48.1, this can be achieved if these cells are immersed in a porous scaffold, whose specificity with regard to the nature of the materials and geometry of the microscopic architecture will be reviewed here, with particular attention focussed on the bone tissue-engineering domain. Moreover, efficient nutrient delivery and mechanical stimulation can be provided if such scaffolds are adequately perfused, as will be discussed from the engineering sciences and biomechanics viewpoints. These standpoints, which are becoming increasingly important in tissue engineering, are expected to promote a better fundamental and quantitative understanding of how cells and tissues respond to nutriment feeding and mechanical stresses. Indeed, quantification by means of computer modelling avers to be essential in interpreting experimental results and identifying the most relevant mechanisms. Such models are likely to offer the possibility of predicting and optimising culture conditions in order to design new experimental protocols [1–5].

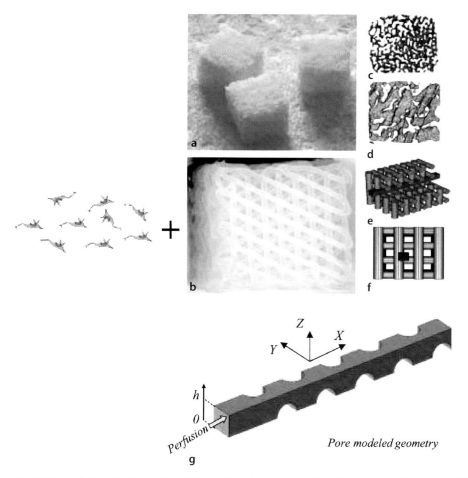

Fig. 48.1a–g Schematic illustration of the methodology and technique used in bone tissue engineering. Prior to implantation mesenchymal stem cells are seeded onto a 3D and porous scaffold of natural coral (**a**) or artificial polymeric composite (**b**). Microscopic view of some scaffolds: plane sections of two natural coralline scaffolds corresponding to Porite (**c**) and Acropora (**d**) coral showing large interconnected pores; schematic drawings of a sample of Poly-Capro-Lactone manufactured by the rapid prototyping technique and displaying simpler ultrastructure geometry (**e** and **f**). Then, cells are cultured under dynamic conditions in a perfusion bioreactor until implantation. **g** Simplified schematic diagram of a pore as used in models

48.2 Critical Structural and Mechanical Properties of a Scaffold for Regenerative Medicine

The use of porous materials has been a long-established approach for providing integration with the surrounding tissue in ocular [6] and orthopaedic applications [7], and most recently in tissue engineering [8]. In tissue engineering, a porous scaffold is required to accommodate mammalian cells and guide their growth and tissue regeneration in a 3D environment. Design variables for producing optimum scaffold architecture include the provision of adequate space for growth and the development of sufficient transport pathways within the porous material; these properties can dictate the development of tissue and the provision of appropriate nutritional conditions.

Despite the wide use of porous scaffolds in manufacturing implants, an ad hoc optimisation of the materials has to be carefully treated for successful integration. Criteria that must be considered in the design of materials include provision of adequate mechanical strength, maximising the surface area available for cell attachment and growth, inclusion of large pore volumes to accommodate and deliver a cell mass sufficient for tissue repair, strong pore interconnectivity for transport of nutrients and waste products to and from the implant, timely resorption of the scaffold [9], and adequate development of vascularisation [10]. These criteria are heavily influenced by the pore microarchitecture within the scaffolds. For this reason, a key parameter in producing new implants has been the control of this pore structure (pore size, pore shape, pore volume and pore interconnectivity) of the scaffold.

To design the ideal scaffold it is necessary to both accurately characterise this structure in a 3D environment and to understand exactly how the scaffold microscopic (length scale of a pore surface including the close environment of the cells) structure influences important mechanical and transport macroscopic (length scale of the implant including many pores) properties. These major properties are reviewed hereafter.

48.2.1
Pore Radius, Porosity, Specific Pore Area and Tortuosity

Pore and fibre characteristic sizes (denoted a), which are fairly large as compared with the sizes of the cells, have worked most satisfactorily in these applications. Indeed, a large pore open space enables not only osteoprogenitor cells, for example, but also endothelial cells to migrate into the overall structure, and contributes to the long-term development of the vascular beds necessary to nourish the newly formed tissue. 3D scaffolds with sufficiently high porosity (designated φ[1]) promoting tissue formation are required to have specific internal microarchitectural features.

These features that characterise highly porous interconnected structures include significant tortuosity T[2] of the pores associated with large surface-to-volume ratios (named specific pore area A_v[3]) favouring cell ingrowth and cell distribution throughout the matrix [11].

48.2.2
Topography and Surface Chemistry

Particle size, shape, and surface roughness all affect the adhesion, proliferation, and phenotype of cells. More specifically, cells are sensitive and responsive to the chemistry, topography and surface energy of the material substrates with which they come into contact. The type, amount and conformation of the specific proteins that adsorb onto material surfaces have therefore functional effects on the cells they encounter [12]. Hydrophobic biomaterials promote the adsorption and retention of the proteins (such as fibrinogen and fibronectin, initially derived from blood serum), which are involved in cell-adhesion processes [13]. Moreover, the state of the mechanical environment to the cell as well as its response to a variety of stimuli is strongly dependent upon the mechanical coupling exerted by the extracellular matrix (ECM). The importance of dimensionality in cell–ECM associations for controlling such a coupling has recently been reported [4]: organisation and composition of the adhesion molecular assemblies and transmission in mechanical signalling forces are generally different when the cells are on a 2D or within a 3D environment.

---

[1] Viewed as a porous biphasic medium with total volume VT and pore volume VP, the substrate has the following porosity: $\phi = \dfrac{V_P}{V_T}$

[2] Due to the definition of the tortuosity from electric conductivity concepts, it can be shown that, for conveying fluid within the porous substrate, channels crossing a thickness L of the medium in the wise stream perfusion direction display a pore length L_P such that [1]: $L_P^2 = T \times L^2 (T \geq 1)$.

[3] In case of tortuous cylindrical channels of radius a, the specific area is given by the following approached expression [11]: $A_v \cong \dfrac{\phi}{a}$.

48.2.3
Intrinsic Mechanical Properties

The ECM provides cells not only with structural support, but also with vital information for cell adhesion, spreading, migration, apoptosis, survival, proliferation and differentiation. Recent studies have suggested that these vital cues are conveyed not only via biochemical entities, but also via the intrinsic mechanical characteristics of the ECM itself [14]. It was suggested, for example, in a recent study that MSC morphology, RNA profiles and affinity for a specific lineage are all influenced by matrix elasticity [15].

48.2.3.1
Composite Porous and Fibrous Nature of the Substrates

The structure of porous and fibrous scaffolds and their mechanical properties, as well as those of the biological tissue they are expected to replace [16, 17], have to be similar in many respects to that of the engineering foams and granular materials [18]. They are made of inhomogeneous, multiphase, fibrous composite materials. Their fibrous polymeric component associated with mineral ceramics in the case of hard tissue confers to them hyperelastic, nonlinear and often anisotropic rheological behaviour. Thus, whatever be the nature or sizes of the samples, these "mechanically complex structures" display characteristic properties depending upon the porous and composite nature of the substrates.

The mechanical properties of open cellular foams, models of the substrate, can be described by the relationship:

$$\frac{E^*}{E_s} = C_1 \times (1-\varphi)^k,$$

where E^* and E_s are respectively the macroscopic (apparent) and microscopic (solid phase) elastic moduli, C_1 and k two constants depending upon the microarchitecture of the medium and the way the sample is mechanically loaded. C_1 is typically of the order of 1, whereas k should lie between 1 and 3, depending on the architecture of the adopted model. For example, for a prismatic sheet-like structure, $k \cong 1$ in the "longitudinal" direction (along the prism axis) and $k \cong 3$ in the transverse direction. Moreover, for isotropic and cubic beam-like structures, the dependence in porosity is such that $k \cong 2$.

These simple and idealised models of a cellular solid provide a useful insight into the mechanical behaviour of the porous and fibrous substrates. They give a clear interpretation of the dependence of the mechanical properties on both their porosity and microarchitecture, and are useful for predicting the general effects of their changes in magnitude during culture processes. These models also provide insight into the mechanisms by which the substrates deform and ultimately fail. Thus, predictions for the dependency of compressive strength on the porosity of a cubic beam-like model show that the failure occurs primarily by elastic buckling of the constituting struts. Once these elements buckle and collapse, the structure is permanently deformed. The compressive strength at which buckling occurs is called the yield strength, which is the stress at which the tested structure no longer behaves as a linear elastic material sample. Generally, the yield strength is less than the ultimate strength, which is the maximum stress that can be supported before catastrophic failure. Although the porous structure usually continues to support loads as it collapses plastically and densifies, there is a consensual interest in the yield strength because it represents the limit of elastic or non-permanent deformation. For the beam-like model, it can be shown that the compressive yield strength is a quadratic function of the porosity:

$$\frac{\sigma^*}{\sigma_{ys}} = C_2 \times (1-\varphi)^2,$$

where σ^* is the apparent compressive yield strength of the sample and σ_{ys} is the compressive yield strength of the solid matrix itself. Analysis of different structural models suggests that the exponent of the porosity term in the compressive strength-density relationship should lie between 1 (linear dependence) and 2 (square dependence), depending on the structure geometry.

48.2.3.2
Multiphasic Nature of the Cultured Organoids

The recurrent question in the biomechanical aspects of tissue engineering is whether it is justified to perfectly match the initial unsteady rheological properties of the substitutes with those of normal living tissue, knowing that they will necessarily undergo during the culture processing a slow evolution of their structure towards either a more rigid calcified state or degradation by enzymatic chemical reactions. Generally speaking, biological tissues often respond in a time-dependent manner characterised by hysteresis, creep or relaxation effects, according to the applied unsteady loading mode. Associated with this question is the detailed knowledge of these phenomena, often named under the generic term viscoelasticity, and its interpretation in terms of several intimate physical mechanisms. One of these influent processes is the transient movement of liquid through the medium due to the non-uniformity of interstitial pressure following sudden and localised application of stresses [19]. Such a process, known as the poroelasticity of the porous and deformable medium, depends not only on the elastic properties of the solid matrix, but also to its hydraulic properties such as the permeability $K^{(d)}$ defined by Darcy's law, which relates the perfusion velocity, U, of the liquid (dynamic viscosity η) to the pressure gradient ($\Delta p/L$) through the following relationship:

$$U = \frac{K}{\eta} \times \frac{\Delta p}{L}.$$

Thus, the response to unsteady mechanical loading of such biomaterials (cubic samples of volume $\approx L^3$ tested in unconfined uniaxial compression) is very similar to the behaviour of a viscoelastic system with the following characteristic time [5]:

$$\tau \approx \frac{\eta \times L^2}{E \times K},$$

which reveals behaviour in time not only as a function of the elasticity and viscosity of the medium, as in classical viscoelasticity, but also depending upon the size of the tested samples as well as the specific boundary conditions imposed on the fluid motion.

48.2.4
Scaffold as an Ad Hoc Microenvironment for Cells

Transplantation is a critical step that involves transferring cells from a well-oxygenated environment enriched in growth factors with physiological pH and osmolarity levels, to an avascular one. To master this critical stage, which continues up to vascular bed establishment, it would be of great interest to develop scaffold materials that can provide transplanted cells with a physiological environment in terms of, for example, the pH, oxygen tension, growth factor concentration and waste removal processes. However, in the case of degradable scaffold, material should not generate toxic products or trigger osmotic injury.

48.3
Critical Cell Culture Methods for Bone Tissue Engineering

Although experimental studies combining MSCs with porous scaffolds implanted into ectopic or orthotopic animal models have given encouraging results in bone tissue engineering [20], the present methods of expanding MSCs (i.e. those involving the use of cell culture flasks and static culture conditions) need to be simplified because they are far too time-consuming and labour-intensive. Last but not least, there is a non-negligible risk of infection because of the many manipulations involved in these procedures. In addition, osteocompetent cells expanded directly onto porous scaffold have been found to favour in vivo osteogenesis. However, the cell colonisation patterns and the uniformity and homogeneity of these bone constructs do not yet meet the standards required to be able to reasonably consider using them for routine therapeutic purposes. In the next sections we will analyse how to overcome the limitations of 2D static cell culture methods by perfusion of a cell population immersed within a porous substrate exhibiting large specific area.

48.3.1
Matrix Seeding Quality

The initial distribution of the cells inside the matrix is critical for the ultimate homogeneity of the bone construct [21, 22]. Several techniques have therefore been developed for obtaining an even initial pattern of cell distribution throughout the scaffold upon cell loading. The procedure most commonly used so far has consisted of seeding cells into the scaffold under static conditions, before they are introduced into the bioreactor. However, this method does not give very satisfactory results [23, 24] as it yields an uneven pattern of cell distribution inside the matrices [25–27]. By contrast, a high level of efficiency and uniformity has been obtained when seeding is performed under dynamic conditions (i.e. using a stirred flask [28]). In comparison with conventional static seeding methods, dynamic methods have yielded not only highest initial levels of adherent cells, but also a more uniform pattern of cell distribution [25]. Direct perfusion of a cell suspension through the pores of a 3D microporous matrix has also been attempted and has led to a more even distribution of cells throughout the scaffold [29].

48.3.2
Solving Problems of Mass Transfer

Mass-transfer processes are known to be a limiting and decisive factor in the context of tissue engineering. These mass-transfer processes depend on the ability of a solution: (1) to move from the aqueous phase (the culture medium) into the scaffold, as occurs in the case of nutrients, and (2) conversely, to move from the scaffold into the culture medium (as in the case of the waste material produced by metabolic processes). How to improve these transfers is often said to be one of the greatest challenges to be met with regard to producing spare human body parts of a therapeutically relevant size. Generally speaking, the transfer of a nutrient from one compartment to another may depend on two factors, namely (1) the diffusion (i.e. the movement of the nutrient relying only on concentration gradients) and (2) the convection, in the case of dynamic systems (i.e. the movement of the nutrient relying on the fluid convection). Under static growth conditions, the balance between the nutrient supply and the requirements of the cells in the scaffold relies only on diffusion processes, whereas under dynamic culture conditions, it relies on different mechanisms. Under the latter conditions, the problems are associated with more complex limiting laws of coupled phenomena, including convection, diffusion and reaction.

First of all, the scaffold itself is critical for establishing an appropriate oxygen and nutrient distribution, as the overall size of the bone construct, the mean diameter and the length of the micropores, and their interconnections (and the number of interconnections) all affect oxygen and nutrient distribution. Nevertheless, under static culture conditions, the cell distribution is often highly inhomogeneous in bone constructs, which show a hypoxic central region and an extremely dense peripheral layer of viable cells. These obstacles to mass transfer encountered under static conditions can be overcome by growing cells in dynamic systems. The aim of using systems of this kind is to promote optimum nutrient distribution and to generally increase the molecular mass transfers. In order to improve these culture processes, it is then of importance to know what culture conditions are experienced by the cells in their close environment. In this framework of osteoarticular tissue engineering, taking into account the multiscale and multiphysics interplaying processes (cf Fig. 48.2), modelling is useful to help identify the dominant mechanisms occurring during tissue formation and evaluate a large range of independent variables.

For instance, modelling parametric studies are necessary to determine the effects of diffusion and convection processes on the cells' penetration length within the porous substrate. Such a length is determined through a somewhat arbitrarily fixed critical concentration of oxygen molecules above which cells can survive. The proposed mathematical model has to include research for solutions of coupled differential equations characterising both fluid mechanics within an "active" porous medium and transport of solute molecules. The presence of cells within the medium is taken into account by nonlinear processes on pore boundary conditions relating flux of oxygen

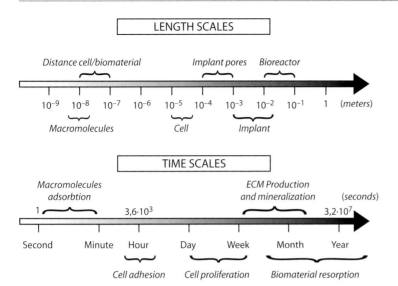

Fig. 48.2 Typical length and time scales in biology

to individual cell consumption rate and their surface density, on the one part, and to the local volume concentration of the gas molecules, on the other.

Under static conditions, the underlying diffusion processes determine the penetration distance of oxygen and nutrients in the aqueous solvent phase for the different chemical species. For biological tissues viewed as porous biphasic media, it is easy to show, by considerations based upon mass conservation, that the penetration length of a given species (for instance oxygen with diffusion coefficient D, entrance molar volume concentration C_0, oxygen molar flux uptake by individual cells V, surface concentration of cells σ within pores) can be evaluated by the following relationship:

$$\overline{L}_D = \sqrt{\frac{D \times C_0}{\sigma \times V \times A_V}} \approx 1\,mm$$

for oxygen with $D \cong 3 \times 10^{-9}\,m^2 s^{-1}$, $C_0 = 0.2\,mol \times m^{-3}$ and for biological tissues with cell density $\sigma \approx 2 \times 10^9\,cells \times m^{-2}$, oxygen consumption rate $V = 4 \times 10^{-17}\,mol \times cell^{-1} \times s^{-1}$ and porous structure characterised by pore radius scale of the order $a \cong 100\,\mu m$.

Interesting enough to mention here is the fact that such an order of magnitude is always found for the distance between microvessels within different types of physiological microcirculatory systems or sizes of cancer tumour without internal necrosis or angiogenesis.

For small nutrient molecules that pass directly across the cell membrane, the kinetics of their uptake generally follows a Michaelis-Menten law, which stipulates that at low concentration ($C \leq K_M \cong 6 \times 10^{-3}\,mol \times m^{-3}$ Michaelis-Menten constant) the chemical reaction rate is of first order type in concentration, whereas at increasing concentrations the rate asymptotically approaches a constant value. Then, the rate R of oxygen consumption by unit area of cell layer takes the following expression:

$$R = -\sigma \times V \times \frac{C}{C + K_M}.$$

In convective perfusion, an evaluation of this length could be obtained from simple considerations about the homogeneity in physical dimensions of the mathematical equation describing the processes of convection-diffusion-reaction (cell consumption being considered here as a chemical reaction). Such an equation, written for a 1D model, gives the following balance for the variation rate of the oxygen molar local concentration $C(X,T)$:

$$\frac{\partial C}{\partial T} = -U_0 \frac{\partial C}{\partial X} + D \frac{\partial^2 C}{\partial X^2} + \frac{R}{a}, \tag{48.1}$$

where U_0 is the perfusion velocity inside the medium. Such a relationship can be written in terms of dimensionless variable such as the reduced concentration $c = C/C_0$ (where C_0 is the upstream entry oxygen concentration) as a function of the normalised coordinates in space $x = X/a$ and time $t = T \times U_0/a$, as:

$$\frac{\partial c}{\partial t} = -\frac{\partial c}{\partial x} + \frac{1}{Pe}\frac{\partial^2 c}{\partial x^2} - \frac{Da}{Pe} f(c),$$

where the Peclet number Pe represents the ratio of convective and diffusive effects for problems in mass transport (it is the equivalent of Reynolds number, Re, for the fluid dynamics phenomena of momentum transfer), the Damkohler number, Da, represents the ratio of reactive (for instance, oxygen consumption of the cell layer) and diffusive effects and $f(c)$ the normalised Michaelis-Menten relationship (approximately, $f(c) = c$ for $c \leq 1$ and $f(c) = 1c$ for $c \geq 1$). These dimensionless numbers are defined in Table 48.1 with their order of magnitude and usual domain of variation in case of the bone tissue engineering.

Under steady conditions, neglecting diffusion and considering only oxygen convection ($Pe >> 1$) a brief examination of this equation leads to:

$$\left|\frac{\partial c}{\partial x}\right| \approx \frac{1}{\overline{L_C}/a} \approx \frac{Da}{Pe} \Rightarrow \frac{\overline{L_C}}{a} \approx \frac{Pe}{Da} \Rightarrow \overline{L_C} = a \times \frac{U_0 \times C_0}{\sigma \times V}.$$

This expression gives a fair approximation of the penetration length for convective perfusion (reference values of the order of 1 cm) and shows that the depth of cell viability due to oxygen molecular penetration is directly proportional to both perfusion velocity and upstream entry concentration of oxygen. On the contrary, and as expected, this length is inversely proportional to the cell oxygen consumption. Figure 48.3 illustrates these results, thus predicting the gross features of the cell culture under different conditions of bioreactor functioning. Moreover, such a model taking into account coupled phenomenon of the growth of cell population in presence of oxygen is able to predict the evolution in time and space (in the streamwise direction) of both the mean volumic density of cells and nutrients [30].

To further characterise the hydrodynamic microenvironment and local transport phenomena, a 3D model of the "idealised" implant geometry (for instance, the one corresponding to fibrous polymeric substrate [31] as shown on Fig. 48.1) can be considered. This analysis, made at the scale of the implant pore length (mesoscopic description), allows evidence to be included and thus quantification of the local variations in the conditions of oxygenation and mechanical stimulation to which cells are subjected during the first days of culture.

The oxygen repartition within the representative domain is given by an oxygen diffusion convection similar to Eq. 48.1, taking into account the 3D nature of the flow and the fact that the oxygen consumption takes place at the limit of the domain and play then the role of boundary conditions. To solve this equation, it is necessary to know the velocity field that is obtained from the solution of the classical Navier-Stokes equations describing the dynamics of the solvent. Calculations have to be made by numerical techniques, for instance using finite-element methods such as usually found in commercial software of type FIDAP [32] or COMSOL [33].

As expected for such a flow dominated by viscous effects (Reynolds number around 10^{-2} as seen in Table 84.1), the spiralling structure of the velocity field reproduces the waviness, periodicities and symmetries of the substrate microarchitecture. Particularly, an unexpected secondary flow due to the cross-cren-

Table 48.1 Definitions, typical range and reference values of the dimensionless numbers characterising molecular transport to cells within the framework of the engineered osteoarticular tissue

Dimensionless numbers	Reynolds	Peclet*	Damkohler
Definitions*	$Re = \dfrac{a \times U_0}{v}$	$Pe = \dfrac{a \times U_0}{D}$	$Da = \dfrac{\sigma \times V \times a}{C_0 \times D}$
Range	$10^{-3} \leftrightarrow 10^{-1}$	$1 \leftrightarrow 10^2$	$10^{-3} \leftrightarrow$
Reference values	10^{-2}	10	10^{-2}

*Here, a is the average pore width, $U_0 \approx 100\,\mu m/s$ the mean perfusion velocity within the substrate, v the kinematic viscosity of the fluid, D the diffusion coefficient of nutrient molecules carried by the solvent fluid, V the molar flux uptake by individual cells, σ the surface concentration of cells on the pore's wall and K_M the Michaelis-Menten constant (see text)

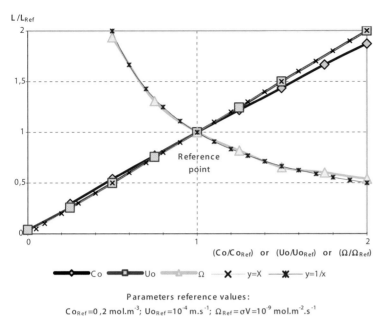

Fig. 48.3 Plot of the oxygen entry length ($L_{Ref} \approx 1$ cm) due to convective effects in a perfused substrate as a function of the perfusion velocity U_0 (*red line*), oxygen concentration C_0 at the upstream site (*dark line*) and consumption flux rate $\Omega = \sigma \times V$ (*green line*): comparison between numerical results of transport phenomena within a 2D model of a crenulated pore [43] and analytical approach (refer to the text)

ulation of the channel is generated and significantly contributes to the increase in the nutriment transport processes from the centre of the channel towards its periphery where the oxygen-consuming cells are lying. This phenomenon enhances diffusive effects in the vicinity of the fibre surface, leading to a relative homogeneous repartition of the oxygen concentration in the transverse direction of the pore channel (less than 10% of variation around mean values calculated for the reference case), as shown on Fig. 48.4

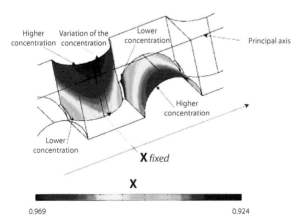

Fig. 48.4 Calculated distribution of the dimensionless oxygen concentration in the vicinity of two adjacent fibres, in the case of the 3D substrate model considered here

48.3.3
Mastering the Applied Mechanical Forces

The importance of using cells responding to the local forces within the medium and ensuring rapid tissue growth has already been underlined in order to define new approaches for bone tissue engineering in which organoids capable of functional load bearing are created [34]. The effects of the mechanical loads applied to osteocompetent cells pose particularly delicate problems, which need to be described and mastered when cells are exposed to dynamic conditions in a bioreactor. Indeed, much debate exists regarding the critical components in the mechanical loading stimulation of the cells and whether different components, such as fluid shear, tension or compression can have an influence in differing ways [35]. Since the fluid moving inside the bioreactors comes into contact with the scaffold surfaces, the tangential friction forces exerted on these surfaces of the matrices tend to slow down the flow rates. Forces of this kind are referred to as shear stress τ multiplied by the surface area and depend on the local orientation of the surface element under consideration.

Osteocompetent cell proliferation and differentiation processes depend on mechanical loads such as the pressure and shear loads. The studies available so far have focussed mainly on the latter type of mechanical loading, which is now known to have decisive effects on the pattern of tissue development. Several studies [36–40] carried out on cell cultures in a 2D channel have focussed on the responses to shear loads covering an extremely large domain of variations in magnitude (from 1 µPa to 1 Pa) and time scale (from 1 s to 10^6 s or more). The authors of these studies reported that both the cell proliferation and mineralisation rates increased under these experimental conditions. Little information is available so far, however, regarding the optimum shear values that can be used in bioreactors. All in all, these findings indicate that mechanical shear loads as small as 1–10 µPa in magnitude can stimulate the growth of osteocompetent cells under continuous dynamic conditions [36]. At these very low shear values, membrane receptors of the integrin type may be stimulated and may thus contribute to regulating the expression of target genes [36, 38].

Even more subtle is the situation when a 3D microflow across the porous medium is considered. As shown on Fig. 48.5, the perfusion flow inside the porous substrate generates in the vicinity of the fibre surface a non-unidirectional and non-homogeneous repartition of viscous stresses, the magnitude of which being of the order of 10^{-3}–10^{-2} Pa on the major part of the fibres. These results are significantly different from those of a 2D model, where lower amplitudes and more uniform distribution of shear stresses were found. As was pointed out when dealing with cell engineering and mechanobiology [4, 41, 42], environmental mechanical actions are generally differently exerted when the cells are on 2D or within 3D matrices, and the dimensionality of the extracellular environment is undoubtedly a key factor for further study.

References

1. Grodzinsky AJ, Kam RD, Lauffenburger DA (2000) Quantitative aspects of tissue engineering: basic issues in kinetics, transport, and mechanics. In: Lanza RP, Langer R, Vacanti J (eds) Principles of Tissue Engineering. Academic, San Diego, pp 195–205
2. Sengers BG, Taylor M, Please CP, et al (2007) Computational modeling of cell spreading and tissue regeneration in porous scaffolds. Biomaterials 28:1926–1940
3. Boschetti F, Raimondi MT, Migliacacca F, et al (2006) Prediction of the micro fluid dynamic environment imposed to three-dimensional engineered cell systems in bioreactors. J Biomech 39:418–425
4. Pedersen JA, Swartz MA (2005) Mechanobiology in the third dimension. Ann Biomed Eng 33:1469–1490
5. Oddou C, Pierre J (2005) Biomechanical aspects in tissue engineering. Clin Hemorheol Microcirc 33:189–195
6. McNab A (1995) Hydroxyapatite orbital implants. Experience with 100 cases. Aust N Z J Ophthalmol 23:117–123

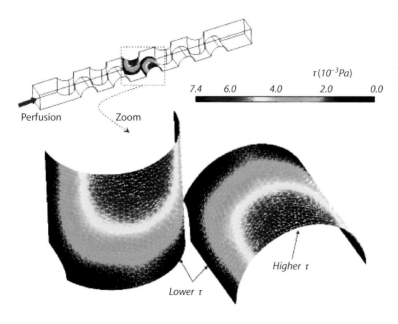

Fig. 48.5 Calculated distribution of the magnitude of the viscous shear stresses on the surface of two adjacent fibres, in the case of the 3D substrate model considered here

7. Shors EC, Holmes RE (1993) Porous hydroxyapatite. In: Hench LL, Wilson J (1993) Introduction to Bioceramics. World Scientific, Singapore, pp 181–198
8. Langer R, Vacanti JP (1993) Tissue engineering. Science 260:920–926
9. Fricain JC, Roudier M, Rouais F, et al (1996) Influence of the structure of three corals on their resorption kinetics. J Periodontal Res 31:463–469
10. Demers C, Hamdy CR, Corsi K, et al (2002) Natural coral exoskeleton as a bone graft substitute: a review. Biomed Mater Eng 12:15–35
11. Guyon E, Hulin JP, Petit L (2001) Hydrodynamique Physique. EDP Sciences, CNRS Editions Paris
12. Boyan BD, Hummert TW, Dean DD, et al (1996) Role of material surfaces in regulating bone and cartilage cell response. Biomaterials 17:137–146
13. Fisher JP, Lalani Z, Bossano CM, et al (2004) Effect of biomaterial properties on bone healing in a rabbit tooth extraction socket model. J Biomed Mater Res A 68:428–438
14. Discher DE, Janmey P, Wang YL (2005) Tissue cells feel and respond to the stiffness of their substrate. Science 310–5751:1139–1143
15. Engler AJ, Sen S, Sweeney HL, et al (2006) Matrix elasticity directs stem cell lineage specification. Cell 126:677–689
16. Fung YC (1993) Biomechanics Mechanical Properties of Living Tissues, 2nd edition Springer-Verlag, New York
17. Ethier CR, Simmons CA (2005) Introductory Biomechanics: From Cells to Organisms, Cambridge University Press, Cambridge
18. Gibson LJ, Ashby MF (1997) Cellular Solids: Structure and Properties, 2nd edition Cambridge University Press, Cambridge
19. Swartz MA, Fleury ME (2007) Interstitial flow and its effects in soft tissues. Annu Rev Biomed Eng 9:229–256
20. Petite H, Viateau V, Bensaïd W, et al (2000) Tissue-engineered bone regeneration. Nat Biotechnol 18:959–963
21. Kim BS, Putnam AJ, Kulik TJ, et al (1998) Optimizing seeding and culture methods to engineer smooth muscle tissue on biodegradable polymer matrices. Biotechnol Bioeng 57:46–54
22. Holy CE, Shoichet MS, Davies JE (2000) Engineering three-dimensional bone tissue in vitro using biodegradable scaffolds: investigating initial cell-seeding density and culture period. J Biomed Mater Res 51–3:376–382
23. Kim BS, Putnam AJ, Kulik TJ, et al (1998) Optimizing seeding and culture methods to engineer smooth muscle tissue on biodegradable polymer matrices. Biotechnol Bioeng 57–1:46–54
24. Bruinink A, Siragusano D, Ettel G, et al (2001) The stiffness of bone marrow cell-knit composites is increased during mechanical load. Biomaterials 22–23:3169–3178
25. Li Y, Ma T, Kniss DA, et al (2001) Effects of filtration seeding on cell density, spatial distribution, and proliferation in nonwoven fibrous matrices. Biotechnol Prog 17–5:935–944
26. Burg KJ, Delnomdedieu M, Beiler RJ (2002) Application of magnetic resonance microscopy to tissue engineering: a polylactide model. J Biomed Mater Res 61–3:380–390
27. Xie Y, Yang ST, Kniss DA (2001) Three-dimensional cell-scaffold constructs promote efficient gene transfection: implications for cell-based gene therapy. Tissue Eng 7–5:585–598
28. Vunjak-Novakovic G, Freed LE, Biron RJ, et al (1996) Effects of mixing on the composition and morphology of tissue-engineered cartilage. Bioeng Food Nat Prod 42:850–860
29. Wendt D, Marsano A, Jakob M, et al (2003) Oscillating perfusion of cell suspensions through three-dimensional scaffolds enhances cell seeding efficiency and uniformity. Biotechnol Bioeng 84–2:205–214
30. Pierre J, Oudina K, Petite H, et al (2005) Modeling of transport phenomena in porous media: bone tissue engineering application. In: Bennacer R (ed) Progress in Computationl Heat and Mass Transfer. Lavoisier, Paris, pp 1074–1079
31. Zein I, Hutmacher D, Cheng Tan K, et al (2002) Fused deposition modeling of novel scaffold architectures for tissue engineering applications. Biomaterials 23:1169–1185
32. Raimondi MT, Boschetti F, Falcone l, et al (2002) Mechanobiology of engineered cartilage cultured under a quantified fluid-dynamic environment. Biomechan Model Mechanobiol 1:69–82
33. Pierre J (2007) Analyse Théorique de Bioréacteurs et d'Implants Utilisés en Génie Tissulaire Osseux et Cartilagineux. PhD Thesis, University Paris 12
34. Klein-Nulend J, Bacabac RG, Mullender MG (2005) Mechanobiology of bone tissue. Pathol Biol (Paris) 53:576–580
35. Mullender M, El Haj AJ, Yang Y, et al (2004) Mechanotransduction of bone cells in vitro: mechanobiology of bone tissue. Med Biol Eng Comp 42:14–21
36. Glowacki J, Mizuno S (2001) Histogenesis in three-dimensionnal scaffolds in vitro. Orthopaetic J Harvard Med School 3:58–60
37. Liegibel UM, Sommer U, Bundschuh B, et al (2004) Fluid shear of low magnitude increases growth and expression of TGFbeta1 and adhesion molecules in human bone cells in vitro. Exp Clin Endocrinol Diabetes 112–7:356–363
38. Hillsley MV, Frangos JA (1997) Alkaline phosphatase in osteoblasts is down-regulated by pulsatile fluid flow. Calcif Tissue Int 60–1:48–53
39. Scaglione S, Wendt D, Miggino S, et al (2008) Effects of fluid flow and calcium phosphate coating on human bone marrow stromal cells cultured in a defined 2D model system. J Biomed Mater Res A 86:411–419
40. Nauman EA, Satcher RL, Keaveny TM, et al (2001) Osteoblasts respond to pulsatile fluid flow with short-term in PGE(2) but no change in mineralization. J Appl Physiol 90:1849–1854
41. Leclerc E, David B, Griscom L, et al (2006) Study of osteoblastic cells in microfluidic environment. Biomaterials 27:586–595
42. Peng CA, Palsson BO (1996) Cell growth and differentiation on feeder layers is predicted to be influenced by bioreactor geometry. Biotechnol Bioeng 50:479–492
43. Pierre J, Oddou C (2007) Engineered bone culture in a perfusion bioreactor: a 2D computational study of stationary mass and momentum transport. Comput Methods Biomech Biomed Eng 10–6:429–438

Influence of Biomechanical Loads

U. Meyer, J. Handschel

Contents

49.1 Introduction 705
49.2 Bone Biomechanics 708
49.3 Cartilage Biomechanics 712
     References 714

49.1 Introduction

Biomechanics is a critical issue in bone and cartilage tissue engineering. It is important for the in vitro development of tissues as well as for the in vivo fate of implanted cellular scaffold constructs. The biomechanical impact of skeletal tissue can be considered as a highly complex and dynamic process [16]. Physical and biological parameters affecting the success of regeneration and remodelling include the macro- and microscopical anatomy, the direction and amount of the applied forces, and the regenerative capacity of the tissues involved. Force transduction via adjacent structures (joints, ligaments, muscles, and soft tissue) influences the regeneration of the tissue by modulating the stress produced within the tissue environment. Skeletal biomechanics combines physical signals with biological reactions, thereby having a complex and dynamic interaction. To better understand the complex field of biomechanics, reaction to external forces can be conceptualised as the reaction of loads to non-living compound materials (matrix/minerals) in a simplified way, and the reaction of cells towards the biophysical signals that are present in the microenvironment of the tissue.

The reaction of non-living skeletal parts (e.g. an explanted femur, acellular scaffold materials) to loads can be analysed at least two levels. Firstly, we can assess the material properties of bone and cartilage (or the replacement constructs) independent of its anatomical structure and geometry by performing standardised mechanical tests on uniform specimens [103]. Second, the mechanical behaviour of bone as an anatomical entity and the contributions made by its structural properties can be assessed. Mechanically, these properties determine how bone (or bone substitute materials) responds to forces in a clinical setting.

Load is a commonly used and poorly defined term. It is not defined in terms of a physical determinable condition (Table 49.1). It is mostly used when a complex mechanically related state is meant. In contrast to load, various other terms are precisely defined. The internal resistance to an applied force is known as *stress*. Stress, defined as equal in magnitude but opposite in direction to the applied *force*, is distributed over the cross-sectional area of bone and cartilage. It is expressed in units of force per unit area: stress = δ = force/area. In loaded skeletal sites, mechanical forces are transduced through the soft-tissue envelope. Due to the different material properties of bone, cartilage and other tissues, the force/stress relationship is in most anatomical locations complex and therefore difficult to assess.

Table 49.1 Nomenclature of physical parameters and their SI units

Term	Definition	Units
Load	Not precisely defined	–
Force	Mass × acceleration	Newtons
Stress	Force/area	Newtons/m²
Strain	Change in length/original length	%
Young's modulus	Stress/strain	Newtons/m²
Resilience	Stored energy	Newtons × m

Most stress patterns are combinations of three stress types, namely tension, compression and shear [27]. Bending, for example, produces a combination of tensile forces on the convex side of a material and compression on the concave side. Torsion or twisting produces shear stresses along the entire length of a material, while tensile stress elongates it and compressive stress shortens it. In order to simplify the complex loading situation in distinct skeletal sites, tissue deformation can be regarded as an axial elongation of the cell/matrix complex, leading to the development of tensile stresses. The measurement of deformation (displacement of a defined point, measured in millimetres) normalised to the original length of the specimen is called *strain* (Fig 49.1): strain = E = change in length/original length = (deformed length–original length)/(original length). Strain is dimensionless and expressed as a percentage change.

At low levels of stress there is a linear relationship between the stress applied and the resultant deformations in terms of strain [23]. This proportionality is called the modulus of elasticity or Young's modulus. It is a measure of the slope of the linear portion of the curve and is calculated by dividing the stress by the strain at any point along this portion of the curve. If the stress/strain area were to be generated by testing a whole bone as opposed to a uniform specimen, this measurement of the linear slope would give the *stiffness* or *rigidity* of the bony or cartilaginous tissue. With respect to tissue engineering, it is important to recognise that the Young's modulus is nearly stable in mature skeletal sites, but varies in regenerated tissue during its ex vivo formation as well as after implantation, due to the degradation and remodelling process. Thus, the modulus of elasticity of the ex-vivo-generated tissue construct can be expected not only to alter during the in vitro period but to a more profound extent in the post-implantation period.

Energy put into deformation of an elastic material just prior to reaching the yield point can be recovered by removing the stress [20]. The energy recovered is known as *resilience* and is a measure of the ability of the material to store energy. Although this energy is not always recoverable in a useful form, it will not be lost as long as the material does not undergo permanent deformation.

The understanding of the relationship between load application and resultant tissue deformation is crucial for the outcome of tissue-engineering approaches. It is obvious that the biomechanical behaviour of whole bones is more complex than that of its individual components (cortical and cancellous bone, cartilage). A variety of powerful approaches

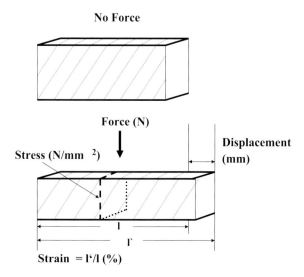

Fig. 49.1 Scheme of strain and stress definition

have generated a comprehensive picture of how a tissue responses to load, but no in vitro measures are able to accurately predict the in vivo situation. In order to assess the load-related tissue deformation in the in vivo setting, various approaches have been introduced (Table 49.2). Indirect measurements of deformations are measured by means of topographical sensors [2, 87] or strain gauges [4, 5]. In addition, holographic interferometry has been performed to assess the impact of mechanical forces on tissue deformation [28]. One major disadvantage of all experimental studies using either strain gauges or holographic interferometry is the inability to determine strains at defined positions within the specimen. Biomechanical research by these methods is also limited to surface deformations, and neither stresses nor dislocations within the tissue can be measured directly. Computer-aided simulation based on finite element analysis (FEA) has addressed the adequacy of mathematical models to relate mechanical factors such as load transfer to the biomechanical behaviour of bone and cartilage specimens.

Various authors have reported that given the high correlation between a calculated strain and stress state and the measured experimental data, various biomechanical parameters can be determined within a deformed bone tissue [82, 104]. Because of the complex challenges involved in measuring the in situ loads and deformations of cartilage under realistic in vivo conditions, several investigators have used theoretical models of joint contact to predict load-related deformations in cartilage tissue [8, 13, 26]. Theoretical and experimental studies of cell–matrix interactions in cartilage, from the macroscopic to the microscopic levels [44, 89], constitute the basis for the calculation of biomechanical parameters that often cannot be measured. In combination with experimental measurements of cell–matrix interactions within cartilage tissue [43, 45], theoretical models (e.g. FEAs) may allow researchers to investigate hypotheses on the influence of various mechanical factors on the biology or biomechanics of cartilage. Whereas the accuracy of FEA in describing the biomechanical behaviour of non-living specimens has been demonstrated by various authors [49, 61, 107], it is unknown how accurately the numerical methods mirror the real in vivo situation. In addition to the use of FEA in the prediction of load-related deformations, FEA methods have recently been used to model load situations in bioreactors [85] in an effort to enhance tissue formation by creating an optimal strain/stress environment.

In contrast to non-living specimens (explanted bone, cartilage, scaffolds), living tissue has a very different structure due to the high dynamics of tissue remodelling. Whereas a number of in vitro studies have investigated the relationship between the applied forces and the resulting tissue deformation, much less is known about the in vivo features of loaded bones and joints. Bone as well as cartilage undergoes deformation in a changing mechanical environment, and reacts biologically when a force is applied [6]. If the skeletal sites are fixed so that they cannot move, or if equal and opposed forces are applied to them, deformations will occur, resulting in the generation of an internal resistance to the applied force [14]. Because of the complex structure of both tissues (non-homogenous, non-isotropic tissue properties) and the technical problems involved, it is difficult to determine directly the force/tissue deformation relationship.

Table 49.2 Common measures to determine physical parameters

Method	Method characteristics
Holographic inferometry	Indirect, not quantifiable
Topographic sensors	Direct, difficult measurements
Pressure sensors	Direct, difficult to miniaturize
Strain gauges	Direct, only usable at surfaces
Finite-element analyses	Indirect, complex possibilities
Theorethical modelling	Indirect, difficult to perform

49.2
Bone Biomechanics

The purpose of bone is to provide structural strength corresponding to its mechanical usage (Fig. 49.2) [14]. This means that bones provide enough strength to keep voluntary physical loads from causing pain or damage. Bone deformation through mechanical loads is a complex feature. It is dependent not only on the bone structure, but also on the action of various tissues (joints, muscles, ligaments, bones; Fig. 49.3). Recent research has suggested that bones display an extraordinary adaptive behaviour towards a changing mechanical environment, which is often regarded as phenotype plasticity [1, 36, 37, 39, 67]. Specific strain-dependent signals are thought to control this adaptive mode of bony tissue modelling [79]. The adaptive mechanisms include basic multicellular units (BMUs) of bone remodelling. Effector cells within BMUs have been shown to function in an interdependent manner. While hormones may bring about as much as 10% of the post-natal changes in bone strength and mass, 40% are determined by mechanical effects. This has been shown by the loss of extremity bone mass in patients with paraplegia (more than 40%).

Modelling occurs by separate formation and resorption drifts to reshape, thicken, and strengthen a bone or trabecula by moving its surfaces around in tissue space [30–33, 101]. Remodelling also involves both the resorption and formation of bone. BMUs turn bone over in small packets through a process in which an activating event causes some bone resorption followed by bone formation [31, 57, 73]. This BMU-based remodelling operates in two modes: "conservation mode" and "disuse mode" (Fig. 49.4) [38]. A specific strain threshold range controls which of these two modes is active at any given time [73].

As load-related skeletal modelling and remodelling is a highly dynamic cellular process, tissue deformations in the sense of tensile strains are the leading stimuli for bone regeneration [34]. It is generally suggested that forces leading to cellular deformation are signalled to the cellular genome through mechanotransduction mechanisms. Mechanotransduction, or the conversion of a biophysical force into a cellular response, is an essential mechanism in bone biology [80]. It allows bone cells to respond to a changing

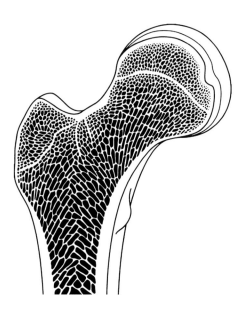

Fig. 49.2 Schematic drawing of a femur head. The femur is a bone that can be considered to be shaped by external forces

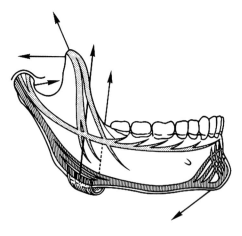

Fig. 49.3 Schematic drawing of the interaction between the various components influencing mandibular form (teeth, joint, ligaments, muscles)

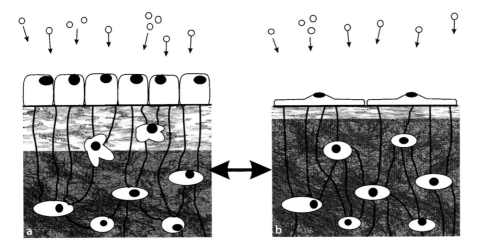

Fig. 49.4 Bone histology in the resting (b) and active (a) state

mechanical environment. Mechanotransduction can be categorised in an idealised manner into:
1. Mechanocoupling, which means the transduction of mechanical force applied to the tissue into a local mechanical signal perceived by a bone cell.
2. Biochemical coupling, the transduction of a local mechanical signal into biochemical signal cascades, altering gene expression or protein activation.
3. Transmission of signals from the sensor cells to effector cells, which actually form or remove bone.
4. The effector cell response: when loads are applied to bone, the tissue begins to deform causing local strains (typically reported in units of microstrain; 10,000 microstrain = 1% change in length).

It is well known that osteoblasts and osteocytes act as the sensors of local bone strains and that they are appropriately located in the bone for this function. Various investigators have revealed that reactive loads give rise to relatively high strains at fundamental frequencies that extend from 1 to 10 Hz. It was found that peak strain magnitudes measured in a wide variety of species are remarkably similar, ranging from 2,000 to 3,500 microstrain. Lanyon et al. [64] showed that, within a single period of loading, the remodelling process seemed to be saturated after only a few (<50) loading cycles. Further strain repetitions then produced no extra effect.

There have been many suggestions for biophysical transduction mechanisms on a cellular level [3, 12, 58, 102]. Most studies have suggested that mechanical strain transduction is directly related to the mechanical deformation of ultrastructural organelles or proteins, thereby converting the mechanical information into a biochemical signal. Although hydrostatic compression has been proposed to be analogous to physiological strain and to recreate physiological strain effects, no significant distortion or compression of the fluid-filled cell can occur until very high pressures are attained [9]. Cell elongation seems to be the driving force for signal transduction. Studies of cell elongation in vitro have demonstrated that physiological loading of osteoblast-like cells induces the differentiation of osteoblasts, whereas hyperphysiological load tends to dedifferentiate cells towards a fibroblastic phenotype [84]. It is assumed that the strain sensor is linked to the cytoskeleton, although other hypotheses have been suggested. If the strain sensor is located in the cytoskeleton, then deformations of this structure would tend not to be homogeneous because different compartments would have different mechanical properties, and the weakest link would deform the most. In this model, elongation of bone cells appears to influence subsequent transcriptional events.

The functional biological environment of any bone tissue is thus derived from a dynamic interaction between various active basic multicellular units exposed to a biophysical microenvironment undergoing continuous load-related changes. Bone formation occurs through separate formation and resorption drifts to reshape, thicken, and strengthen a bone or trabecula by moving its surfaces around in tissue space [35]. While the load-related bone-modelling theory was long assumed to be only applicable to intact bones [36], recent research has shown that this adaptive behaviour is also present in bone-healing processes [39]. The oestrogen receptor was shown to be involved in the adaptive response of osteoblasts to mechanical strain [24], and so biophysical force transduction seems also to be closely connected with the hormonal regulation of bone turnover, indicative of the complex situation of bone-cell behaviour towards molecular-level signals. The evidence supporting the concept that biophysical signals are able to positively regulate bone-cell behaviour is accepted, but the relationship between defined biophysical forces and resultant effects on bone-cell behaviour is to some extent contradictory. Whereas some investigations offered evidence that biophysical forces, including substrate deformation, hydrostatic pressure, fluid flow or hypergravity, increase the expression and/or synthesis of markers of bone-cell differentiation [50, 59, 74, 83, 91, 111], others failed to do so, or even reported negative effects of biophysical stimuli on cell proliferation and differentiation [53, 78, 92]. It can be assumed that this contradictory evidence is associated with the fact that various biophysical stimuli (physiological and non-physiological) were exerted on different bone-cell lines (primary cells, altered cells) in the different experimental investigations.

When biomechanical effects are considered at the tissue level (Figs. 49.5 and 49.6), it is important to recognise that mechanical and electrical effects are exerted simultaneously in bone (Table 49.3) [100]. Mechanical and electrical interactions, being quite complex during the modelling and remodelling processes, integrate the action of several signals to form the final response. Tissue deformation as a key stimulus in bone physiology leads to a complex,

Table 49.3 Biophysical effectors of bone

Mechanical effects
Electromechanical effects
Thermic effects
Radiation effects
Pressure
Pressure gradients
Distortion
Enhancement of molecular transport
Energy dissipation

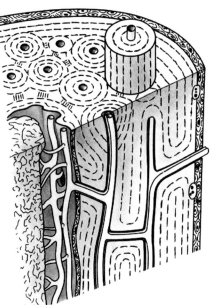
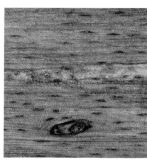
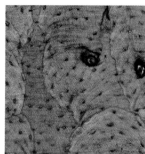

Fig. 49.5 Histology and scheme of cortical bone

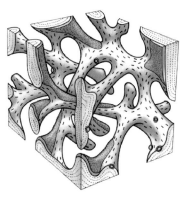 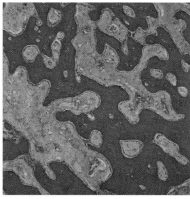

Fig. 49.6 Histology and scheme of cancellous (spongy) bone

non-uniform biophysical environment within bone tissue, consisting of fluid flow, direct mechanical strain, and electrokinetic effects on bone cells. At the tissue level, some researchers, assuming the differentiating tissue to be a continuous material, evaluated biophysical signals by characterising the stimulus in terms of engineering quantities. Based on the tissue material properties and approximations of tissue loading in the different experimental or clinical situation, these quantities were calculated throughout the tissue and were related to various patterns of tissue differentiation. The biophysical mechanisms underlying the tissue response were directly related to mechanical effects [81], electromechanical effects [42, 94] or enhancement of molecular transport mechanisms [93]. Pressure, distortion, pressure gradients and energy dissipation are additional engineering level quantities that were quantified and related to tissue responses by other authors (e.g. [27]).

It has been suggested that load-induced interstitial fluid flow provides a basic convergence feature between electrical and mechanical signals resulting from affecting the cellular level, thus triggering the subsequent hard-tissue formation in adaptation [21]. Bone loading leads to a deformation of cells in the microenvironment of the tissue, and simultaneously to a fluid-flow-related generation of electrical potentials. Pressure gradients from mechanical loading of bone tissue elongate cells and the mineralised matrix, and move extracellular fluid radially outward. The fluid flow stream generates electrical potentials that exert effects on adjacent cells [94]. Electric fields of a magnitude and frequency that would occur endogenously as a result of physiological bone load via piezoelectric effects, streaming potentials or a combination of these [40, 51], have also been demonstrated to modulate bone formation in the craniofacial skeleton [70]. The sensitivity of the biophysically driven bone response was found to be much greater when the cells are part of a tissue than when they are evaluated as single entities. The bone tissue system integrates and amplifies a complex physical signal such as mechanical strain or functionally induced fluid flow, by transmitting the signal from the signal-detecting cells to the change-effecting cells. Bone cells are functionally coupled both in vivo [56] and in vitro [98] by gap junctions. Gap junctions, which are well suited to the integration and amplification of the response of bone-cell networks to biophysical signals, are therefore important mediators of the effector response. The role of gap junctions in biophysical signal transduction at the tissue level was exemplified in the bone-cell response to electromagnetic fields (EMFs). In contrast to single cells, the intercellular network was found to contribute to the ability of EMFs to stimulate alkaline phosphatase activity, a marker for osteoblastic differentiation, suggesting that gap junctions are involved in the enhancement of EMF-stimulated osteoblastic differentiation. Gap junctions were also demonstrated to contribute to bone-cell responsiveness to other biophysical signals such as cell deformation or fluid flow [110], indicating their special role in transmitting single cell signals into the tissue.

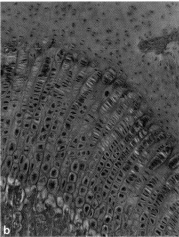

Fig. 49.7 Histology of growing (**b**) and mature (**a**) cartilage

49.3 Cartilage Biomechanics

Cartilage biomechanics varies within the different subsets of cartilage. The development and maintenance of cartilage structure and mechanical characteristics are closely correlated to the effect of mechanical loading. The impact of load on cartilage structure and function is of utmost importance, especially in hyaline articular cartilage (Figs. 49.7 and 49.8). Physiological joint loading maintains cartilage structure and function. In the context of tissue engineering, it is important to recognise that stresses in a cartilage defect or the surrounding tissue may be altered significantly from their normal mechanical environment, and therefore impair tissue integrity before and after cell/scaffold implantation. The histological structure of articular cartilage is influenced by the local mechanical loading of chondrocytes in the different zones (for review see [108]). In contrast to bone, where considerable quantitative information is available on the in vivo stresses and strains that are encountered under different conditions [11, 22, 48, 62, 69, 86, 109], for articular cartilage there are unique challenges with respect to the magnitude and rate of loading within the joint that make determinations of load parameters difficult. Cartilage can be considered as a compound material. It can be conceptualised as a mixture of fluid and solid constituents (despite the close interrelations of both components).

Chondrocytes are directly attached to the extracellular matrix (ECM) and can be considered a part of the fluid continuum of cartilage tissue. The zonal structure of articular cartilage is essential to its ability to support physiological joint forces. Patterns of stress, strain and fluid flow created in the joint result in spatial and temporal alterations in chondrocyte function. The external forces are transmitted through the joint and the adjacent tissues, resulting in the generation of pressure over the "contact area" of the two opposing cartilage surfaces (Fig. 49.9).

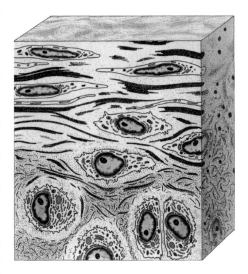

Fig. 49.8 Scheme of cartilage histology

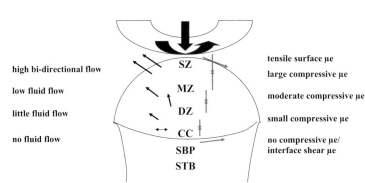

Fig. 49.9 Schematic drawing of the biomechanical environment of a joint. *SZ* Superficial zone, *MZ* medial zone, *DZ* deep zone, *CC* calcified cartilage, *SBP* subchondral bone plate, *STB* subchondral trabecular bone

There is only limited information on the normal in vivo mechanical environment. Stresses in a normal joint are difficult to determine in vivo, but evidence from experimental studies indicate that normal stresses may range from 5 to 10 MPa in human and other animal joints [15, 97, 106]. Pressure magnitudes in articular cartilage have been shown to exceed 18 MPa under distinct experimental conditions. In areas of peak pressure, most of the force is transmitted through a thin fluid film layer that is created between the solid matrices of the superficial cartilage layers on opposite areas of the joint (for review see [108]). Loads tend to deform the cartilage tissue, but these compressive loads and the subsequent tissue deformations are resisted by:
1. Stresses in the solid phase.
2. The generation of fluid pressure.
3. The restriction of tissue deformation by the nearly impermeable subchondral bone and the surrounding adjacent cartilage surfaces.

Intermittent loads (created by normal joint movements) transfer cyclic hydrostatic pressure in the interstitial fluid. In mature joints, cyclic loads produce cyclic hydrostatic fluid pressure through the entire cartilage thickness that is comparable in magnitude to the applied joint pressure. The pressure in the interstitial fluid component of cartilage transfers the loads to the surrounding tissues by distributing the forces to the underlying bone tissues in an efficient manner [8, 52, 68, 72, 76, 77, 88, 90]. The superficial zone of cartilage exudes interstitial fluid and consolidates, forming an effective seal at the joint surface. In the radial zone of articular cartilage, fluid flow is lower than in the superficial zone, being restricted laterally by adjacent tissue, below by the subchondral bone layer and above by the consolidated surface zone.

An outside-directed fluid flow is therefore seen mainly in areas of the superficial layer just outside the contact area [88]. Under these conditions compressive strains are substantial. The matrix of the superficial layer consolidates the cells in this region, leading to substantial compressive deformations (strains). The collagen network resists tensile and shear stresses and strains, while the fluid component resists the high hydrostatic compressive stresses that are generated. Shear stresses created in cartilage are always associated with tensile strains in some direction. In addition to the load-related compressive strains, the rolling movement of opposite joint surfaces impose cyclic tensile strains that are tangential to the surface of cartilage in the superficial area [7]. In this thin zone, the collagen fibrils are oriented parallel to the articular surface. Stresses are inhomogeneously distributed in the cartilage tissue. Peak compressive strains are present in superficial layers, whereas the strain environment decreases near the calcified cartilage layer. It is important to note that there is a contrasting effect of shear and pressure on chondrocyte function, indicative of the complex mechanobiology of cartilage.

It is important to recognise that chondrocytes (like bone cells) are not compressible and they do not provide any special structural resistance to loading.

Because of the structural and compositional characteristics of articular cartilage and the intrinsic coupling between the mechanical and physiochemical properties of the tissue [63], it has been, in contrast to bone tissue, more difficult to achieve a complete understanding of the mechanical signal-transduction pathways used by chondrocytes. Fundamental to this issue is an understanding of the mechanical environment of the chondrocytes within the articular cartilage ECM. For example, compressive loading of the cartilage in the ECM exposes the chondrocytes to spatially and time-varying stress, strain, fluid flow and pressure, osmotic pressure and electric fields [29, 63, 71]. The relative contribution of each of these factors to the regulation of chondrocyte activity is an important consideration that is being studied by several investigators [46]. It was found that under loading conditions, the shapes, pressures and chemical environments of chondrocytes are altered by deformations created in the ECM as well as by local fluid pressure and fluid flow. Chondrocytes in articular cartilage undergo large changes in shape and intercellular spacing as the ECM is deformed. The cells were found to recover their morphology upon the removal of compression and to have a stiffness that was less than or equal to that of the surrounding tissues. Furthermore, collagen fibre orientation was shown to change with compression and recover upon removal of compression.

New microscope techniques have allowed three-dimensional imaging of fluorescently labelled chondrocytes [45] or subcellular structures [43] within a tissue explant. Volumetric images of chondrocytes revealed decreases in cell height of 26%, 19% and 20% and decreases in cell volume of 22%, 16% and 17% of control values in different zones of an osteochondral explant, thereby differing from mechanically induced deformations of bone cells. In responding to mechanical signals the chondrocyte responds to the several different types of information using various signal-transduction mechanisms (for review see [47]). It was demonstrated by in vivo and in vitro studies that chondrocytes are able (in a similar manner to bone cells) to detect an array of physical parameters (tissue pressure, deformation, fluid flow). The sum of cellular reactions in response to loading is a gene activation state of the cell that determines the histomorphology of articular cartilage remodelling and repair [10, 17–19, 25, 41, 54, 60, 65, 66, 75, 95, 96, 99, 105]. The factors, pathways and receptors of cell sensing and the response of cells concerning the effect on its external environment have not yet been identified, but as indicated for bone cells, chondrocytes respond to several external signals.

The chondrocyte perceives its mechanical environment through complex biological and biophysical interactions with the cartilage ECM. This matrix consists of several distinct regions, termed the pericellular, territorial and interterritorial matrices [55]. The bulk of the tissue is made up of the interterritorial matrix, which consists primarily of water containing dissolved small electrolytes (e.g. Na^+, Cl^-, Ca^{2+}). The transfer of information proceeds to some extent in a similar manner as described for bone cells, from the cell membrane to the cell nucleus, also interacting with other signalling pathways. Current experimental studies in tissue engineering try not only to evaluate the complex mechanobiology that underlies load-related cartilage histogenesis, but also use biophysical stimuli to promote cartilage-like tissue growth in specially designed bioreactors [85].

References

1. Aegerter E, Kirkpatrick JA (1973) Orthopedic Diseases; Physiology, Pathology, Radiology. Saunders, Philadelphia
2. Ahmad R, Bates JF, Lewis TT (1982) Measurement of strain rate behaviour in complete mandibular dentures. Biomaterials 3:87–92
3. Ando J, Komatsuda T, Kamiya A (1988) Cytoplasmic calcium response to fluid shear stress in cultured vascular endothelial cells. In Vitro Cell Dev Biol 24:871–7
4. Arendts FJ , Sigolotto C (1989) Standard measurements, elasticity values and tensile strength behavior of the human mandible, a contribution to the biomechanics of the mandible-I. Biomed Tech (Berl) 34:248–55
5. Arendts FJ, Sigolotto C (1990) Mechanical characteristics of the human mandible and study of in vivo behavior of compact bone tissue, a contribution to the description of biomechanics of the mandible-II. Biomed Tech (Berl) 35:123–30
6. Aronson J (1992) Mechanical Factors Generated During Distraction Osteogenesis. The International Society for Fracture Repair, Ottrot, France, April, 1992
7. Askew MJ, Mow VC (1978) The biomechanical function of the collagen fibril ultrastructure of articular cartilage. J Biomech Eng 100:105–15
8. Ateshian GA, Wang H (1995) A theoretical solution for the frictionless rolling contact of cylindrical biphasic articular cartilage layers. J Biomech 28:1341–55

9. Bagi C, Burger EH (1989) Mechanical stimulation by intermittent compression stimulates sulfate incorporation and matrix mineralization in fetal mouse long-bone rudiments under serum-free conditions. Calcif Tissue Int 45:342–7
10. Beaupre GS, Stevens SS, Carter DR (2000) Mechanobiology in the development, maintenance, and degeneration of articular cartilage. J Rehabil Res Dev 37:145–51
11. Biewener AA (1993) Safety factors in bone strength. Calcif Tissue Int 53:S68–74
12. Binderman I, Zor U, Kaye AM, Shimshoni Z, Harell A, Somjen D (1988) The transduction of mechanical force into biochemical events in bone cells may involve activation of phospholipase A2. Calcif Tissue Int 42:261–6
13. Blankevoort L, Kuiper JH, Huiskes R, Grootenboer HJ (1991) Articular contact in a three-dimensional model of the knee. J Biomech 24:1019–31
14. Brown TD, Ferguson AB Jr (1978) The development of a computational stress analysis of the femoral head. Mapping tensile, compressive, and shear stress for the varus and valgus positions. J Bone Joint Surg Am 60:619–29
15. Brown TD, Shaw DT (1983) In vitro contact stress distributions in the natural human hip. J Biomech 16:373–84
16. Burger EH, Veldhuijzen JP (1993) Influence of mechanical factors on bone formation, resorption, and growth in vitro. In: Hall BK (ed) Bone, vol 7. CRC, Boca Raton, pp 37–56
17. Buschmann MD, Gluzband YA, Grodzinsky AJ, Hunziker EB (1995) Mechanical compression modulates matrix biosynthesis in chondrocyte/agarose culture. J Cell Sci 108:1497–508
18. Carter DR, Wong M (1990) Mechanical stresses in joint morphogenesis and maintenance. In: Mow VC, Ratcliffe A, Woo SLY (eds) Biomechanics of Diarthrodial Joints. Springer-Verlag, New York, pp 155–74
19. Carter DR, Orr TE, Fyhrie DP, Schurman DJ (1987) Influences of mechanical stress on prenatal and postnatal skeletal development. Clin Orthop Relat Res (219):237–50
20. Chamay A (1970) Mechanical and morphological aspects of experimental overload and fatigue in bone. J Biomech 3:263–70
21. Chao EY, Inoue N (2003) Biophysical stimulation of bone fracture repair, regeneration and remodelling. Eur Cell Mater 6:72–84; discussion 84–5
22. Chuong CJ, Fung YC (1983) Three-dimensional stress distribution in arteries. J Biomech Eng 105:268–74
23. Curey JD (1962) Stress concentration in bone. Q J Micros Sci 103:111–33
24. Damien E, Price JS, Lanyon LE (1998) The estrogen receptor's involvement in osteoblasts' adaptive response to mechanical strain. J Bone Miner Res 13:1275–82
25. Davisson T, Kunig S, Chen A, Sah R, Ratcliffe A (2002) Static and dynamic compression modulate matrix metabolism in tissue engineered cartilage. J Orthop Res 20:842–8
26. Donzelli PS, Spilker RL, Ateshian GA, Mow VC (1999) Contact analysis of biphasic transversely isotropic cartilage layers and correlations with tissue failure. J Biomech 32:1037–47
27. Einhorn TA (1996) Biomechanics of bone. In: Bilezikian JP, Raisz LG, Rodan GA (eds) Principles of Bone Biology. Academic Press, New York, pp 25–37
28. Ferre JC, Legoux R, Helary JL, Albugues F, Le Floc'h C, Bouteyre J, Lumineau JP, Chevalier C, Le Cloarec AY, Orio E et al (1985) Study of the mandible under static constraints by holographic interferometry. New biomechanical deductions. Anat Clin 7:193–201
29. Frank EH, Grodzinsky AJ (1987) Cartilage electromechanics – II. A continuum model of cartilage electrokinetics and correlation with experiments. J Biomech 20:629–39
30. Frost HM (1964) Laws of Bone Structure. Charles C. Thomas, Springfield,
31. Frost HM (1964) Mathematical Elements of Lamellar Bone Remodeling. Charles C. Thomas, Springfield
32. Frost HM (1983) The regional acceleratory phenomenon: a review. Henry Ford Hosp Med J 31:3–9
33. Frost HM (1986) Intermediary Organization of the Skeleton, vols I and II. CRC, Boca Raton
34. Frost HM (1992) Perspectives: bone's mechanical usage windows. Bone Miner 19:257–71
35. Frost HM (1996) Perspectives: a proposed general model of the mechanostat (suggestions from a new skeletal-biologic paradigm) Anat Rec 244:139–47
36. Frost HM (2000) The Utah paradigm of skeletal physiology: an overview of its insights for bone, cartilage and collagenous tissue organs. J Bone Miner Metab 18:305–16
37. Frost HM (2000) Why the ISMNI and the Utah paradigm? Their role in skeletal and extraskeletal disorders. J Musculoskelet Neuronal Interact 1:5–9
38. Frost HM, Schonau E (2000) The muscle-bone unit in children and adolescents: a 2000 overview. J Pediatr Endocrinol Metab 13:571–90
39. Frost HM, Meyer U, Joos U, Jensen OT (2002) Dental alveolar distraction and the Utah Paradigm. In: Jensen OT (ed) Alveolar Distraction Osteogenesis. Quintessence, Chicago, pp 1–16
40. Fukada E, Yasuda I (1957) On the piezoelectric effect of bone. J Phys Soc 12:1158–62
41. Giannoni P, Siegrist M, Hunziker EB, Wong M (2003) The mechanosensitivity of cartilage oligomeric matrix protein (COMP). Biorheology 40:101–9
42. Gross D, Williams WS (1982) Streaming potential and the electromechanical response of physiologically-moist bone. J Biomech 15:277–95
43. Guilak F (1995) Compression-induced changes in the shape and volume of the chondrocyte nucleus. J Biomech 28:1529–41
44. Guilak F, Mow VC (2000) The mechanical environment of the chondrocyte: a biphasic finite element model of cell-matrix interactions in articular cartilage. J Biomech 33:1663–73
45. Guilak F, Ratcliffe A, Mow VC (1995) Chondrocyte deformation and local tissue strain in articular cartilage: a confocal microscopy study. J Orthop Res 13:410–21
46. Guilak, F, Sah RL, Setton LA (1997) Physical regulation of cartilage metabolism In: Mow VC, Hayes WC (eds) Basic Orthopaedic Biomechanics. Lippincott-Raven, Philadelphia, pp 179–207

47. Guilak F, Butler DL, Goldstein SA (2001) Functional tissue engineering: the role of biomechanics in articular cartilage repair. Clin Orthop Relat Res (391):S295–305
48. Han HC, Fung YC (1995) Longitudinal strain of canine and porcine aortas. J Biomech 28:637–41
49. Hart RT, Hennebel VV, Thongpreda N, Van Buskirk WC, Anderson RC (1992) Modeling the biomechanics of the mandible: a three-dimensional finite element study. J Biomech 25:261–86
50. Harter LV, Hruska KA, Duncan RL (1995) Human osteoblast-like cells respond to mechanical strain with increased bone matrix protein production independent of hormonal regulation. Endocrinology 136:528–35
51. Hastings GW, Mahmud FA (1988) Electrical effects in bone. J Biomed Eng 10:515–21
52. Higginson GR, Snaith J (1979) The mechanical stiffness of articular cartilage in confined oscillating compression. Eng Med 8:11–4
53. Hillsley MV, Frangos JA (1994) Bone tissue engineering: the role of interstitial fluid flow. Biotechnol Bioeng 43:573–81
54. Holmvall K, Camper L, Johansson S, Kimura JH, Lundgren-Akerlund E (1995) Chondrocyte and chondrosarcoma cell integrins with affinity for collagen type II and their response to mechanical stress. Exp Cell Res 221:496–503
55. Hunziker EB (1992) Articular cartilage structure in humans and experimental animals. In: Kuettner KE, Schleyerbach R, Peyron JG, Hascall VC (eds), Articular Cartilage and Osteoarthritis. Raven Press, New York, pp 183–99
56. Jeansonne BG, Feagin FF, McMinn RW, Shoemaker RL, Rehm WS (1979) Cell-to-cell communication of osteoblasts. J Dent Res 58:1415–23
57. Jee WSS (1999) The interactions of muscles and skeletal tissue: In: Lyritis GP (ed) Musculoskeletal Interactions, vol II. Hylonome, Athens, pp 35–46
58. Jones DB, Nolte H, Scholubbers JG, Turner E, Veltel D (1991) Biochemical signal transduction of mechanical strain in osteoblast-like cells. Biomaterials 12:101–10
59. Kawashima K, Shibata R, Negishi Y, Endo H (1998) Stimulative effect of high-level hypergravity on differentiated functions of osteoblast-like cells. Cell Struct Funct 23:221–9
60. Knudson W, Aguiar DJ, Hua Q, Knudson CB (1996) CD44-anchored hyaluronan-rich pericellular matrices: an ultrastructural and biochemical analysis. Exp Cell Res 228:216–28
61. Korioth TW, Versluis A (1997) Modeling the mechanical behavior of the jaws and their related structures by finite element (FE) analysis. Crit Rev Oral Biol Med 8:90–104
62. Korvick DL, Cummings JF, Grood ES, Holden JP, Feder SM, Butler DL (1996) The use of an implantable force transducer to measure patellar tendon forces in goats. J Biomech 29:557–61
63. Lai WM, Hou JS, Mow VC (1991) A triphasic theory for the swelling and deformation behaviors of articular cartilage. J Biomech Eng 113:245–58
64. Lanyon LE, Hampson WG, Goodship AE, Shah JS (1975) Bone deformation recorded in vivo from strain gauges attached to the human tibial shaft. Acta Orthop Scand 46:256–68
65. Lee DA, Noguchi T, Frean SP, Lees P, Bader DL (2000) The influence of mechanical loading on isolated chondrocytes seeded in agarose constructs. Biorheology 37:149–61
66. Loeser RF (2000) Chondrocyte integrin expression and function. Biorheology 37:109–16
67. Losos JB (2001) Evolution: a lizard's tale. Sci Am 284:64–9
68. Macirowski T, Tepic S, Mann RW (1994) Cartilage stresses in the human hip joint. J Biomech Eng 116:10–8
69. Malaviya P, Butler DL, Korvick DL, Proch FS (1998) In vivo tendon forces correlate with activity level and remain bounded: evidence in a rabbit flexor tendon model. J Biomech 31:1043–9
70. Marino AA, Gross BD, Specian RD (1986) Electrical stimulation of mandibular osteotomies in rabbits. Oral Surg Oral Med Oral Pathol 62:20–4
71. Maroudas A (1979) Physicochemical properties of articular cartilage. In: Freeman M (ed) Adult Articular Cartilage. Pitman Medical, Tunbridge Wells, pp 215–90
72. Maroudas A, Bullough P (1968) Permeability of articular cartilage. Nature 219:1260–1
73. Martin RB, Burr DB, Sharkey NA (1998) Skeletal Tissue Mechanics. Springer, New York
74. Matsuda N, Morita N, Matsuda K, Watanabe M (1998) Proliferation and differentiation of human osteoblastic cells associated with differential activation of MAP kinases in response to epidermal growth factor, hypoxia, and mechanical stress in vitro. Biochem Biophys Res Commun 249:350–4
75. Mauck RL, Soltz MA, Wang CC, Wong DD, Chao PH, Valhmu WB, Hung CT, Ateshian GA (2000) Functional tissue engineering of articular cartilage through dynamic loading of chondrocyte-seeded agarose gels. J Biomech Eng 122:252–60
76. McCutchen CW (1959) Mechanisms of animal joints: sponge-hydrostatic and weeping bearings. Nature 184:1284–5
77. McCutchen CW (1962) The frictional properties of animal joints. Wear 1:1–7
78. Meazzini MC, Toma CD, Schaffer JL, Gray ML, Gerstenfeld LC (1998) Osteoblast cytoskeletal modulation in response to mechanical strain in vitro. J Orthop Res 16:170–80
79. Meyer U, Kleinheinz J, Szulczewski DH, Jones DB, Joos U (1997) Strain application for evaluation of mechanotransduction in osteoblasts. In: Diner PA, Vasquez MP (eds) International Congress on Craniofacial and Facial Bone Distraction Processes, Paris, France. Monduzzi, Bologna, Italy
80. Meyer U, Meyer T, Jones DB (1997) No mechanical role for vinculin in strain transduction in primary bovine osteoblasts. Biochem Cell Biol 75:81–7
81. Meyer U, Meyer T, Vosshans J, Joos U (1999) Decreased expression of osteocalcin and osteonectin in relation to high strains and decreased mineralization in mandibular distraction osteogenesis. J Craniomaxillofac Surg 27:222–7

82. Meyer U, Vollmer D, Homann C, Schuon R, Benthaus S, Vegh A, Felszegi E, Joos U, Piffko J (2000) Experimental and finite-element models for the assessment of mandibular deformation under mechanical loading. Mund Kiefer Gesichtschir 4:14–20
83. Meyer U, Joos U, Szuwart T, Wiesmann HP (2001) Mineralized 3D bone tissue engineered by osteoblasts cultured in a collagen gel. Tissue Eng 7:671
84. Meyer U, Terodde M, Joos U, Wiesmann HP (2001) Mechanical stimulation of osteoblasts in cell culture. Mund Kiefer Gesichtschir 5:166–72
85. Meyer U, Nazer N, Wiesmann HP (2006) Design and performance of a bioreactor system for mechanically promoted three-dimensional tissue engineering. Br J Oral Maxillofac Surg 44:134–140
86. Milgrom C, Finestone A, Simkin A, Ekenman I, Mendelson S, Millgram M, Nyska M, Larsson E, Burr D (2000) In-vivo strain measurements to evaluate the strengthening potential of exercises on the tibial bone. J Bone Joint Surg Br 82:591–4
87. Mosley JR, March BM, Lynch J, Lanyon LE (1997) Strain magnitude related changes in whole bone architecture in growing rats. Bone 20:191–8
88. Mow VC, Ateshian GA (1997) Lubrication and wear of diarthrodial joints In: Mow VC, Hayes WC (eds) Basic Orthopaedic Biomechanics. Lippincott-Raven, Philadelphia, pp 275–315
89. Mow VC, Ratcliffe A, Poole AR (1992) Cartilage and diarthrodial joints as paradigms for hierarchical materials and structures. Biomaterials 13:67–97
90. Oloyede A, Broom N (1993) Stress-sharing between the fluid and solid components of articular cartilage under varying rates of compression. Connect Tissue Res 30:127–41
91. Owan I, Burr DB, Turner CH, Qiu J, Tu Y, Onyia JE, Duncan RL (1997) Mechanotransduction in bone: osteoblasts are more responsive to fluid forces than mechanical strain. Am J Physiol 273:810–5
92. Ozawa H, Imamura K, Abe E, Takahashi N, Hiraide T, Shibasaki Y, Fukuhara T, Suda T (1990) Effect of a continuously applied compressive pressure on mouse osteoblast-like cells (MC3T3-E1) in vitro. J Cell Physiol 142:177–85
93. Piekarski K, Munro M (1977) Transport mechanism operating between blood supply and osteocytes in long bones. Nature 269:80–2
94. Pienkowski D, Pollack SR (1983) The origin of stress-generated potentials in fluid-saturated bone. J Orthop Res 1:30–41
95. Ragan PM, Chin VI, Hung HH, Masuda K, Thonar EJ, Arner EC, Grodzinsky AJ, Sandy JD (2000) Chondrocyte extracellular matrix synthesis and turnover are influenced by static compression in a new alginate disk culture system. Arch Biochem Biophys 383:256–64
96. Reid DL, Aydelotte MB, Mollenhauer J (2000) Cell attachment, collagen binding, and receptor analysis on bovine articular chondrocytes. J Orthop Res 18:364–73
97. Ronsky JL, Herzog W, Brown TD, Pedersen DR, Grood ES, Butler DL (1995) In vivo quantification of the cat patellofemoral joint contact stresses and areas. J Biomech 28:977–83
98. Schirrmacher K, Schmitz I, Winterhager E, Traub O, Brummer F, Jones D, Bingmann D (1992) Characterization of gap junctions between osteoblast-like cells in culture. Calcif Tissue Int 51:285–90
99. Smith RL, Thomas KD, Schurman DJ, Carter DR, Wong M, van der Meulen MC (1992) Rabbit knee immobilization: bone remodeling precedes cartilage degradation. J Orthop Res 10:88–95
100. Spadaro JA (1997) Mechanical and electrical interactions in bone remodelling. Bioelectromagnetics 18:3–202
101. Takahashi HE (1995) Spinal Disorders in Growth and Aging. Springer, Tokyo102.
102. Takei T, Mills I, Arai K, Sumpio BE (1998) Molecular basis for tissue expansion: clinical implications for the surgeon. Plast Reconstr Surg 102:247–58
103. Turner CH, Chandran A, Pidaparti RM (1995) The anisotropy of osteonal bone and its ultrastructural implications. Bone 17:85–9
104. Vollmer D, Meyer U, Joos U, Vegh A, Piffko J (2000) Experimental and finite element study of a human mandible. J Craniomaxillofac Surg 28:91–6
105. von der Mark K, Mollenhauer J (1997) Annexin V interactions with collagen. Cell Mol Life Sci 53:539–45
106. von Eisenhart R, Adam C, Steinlechner M, Muller-Gerbl M, Eckstein F (1999) Quantitative determination of joint incongruity and pressure distribution during simulated gait and cartilage thickness in the human hip joint. J Orthop Res 17:532–9
107. Voo K, Kumaresan S, Pintar FA, Yoganandan N, Sances A Jr (1996) Finite-element models of the human head. Med Biol Eng Comput 34:375–81
108. Wong M, Carter DR (2003) Articular cartilage functional histomorphology and mechanobiology: a research perspective. Bone 33:1–13
109. Woo SL, Debski RE, Wong EK, Yagi M, Tarinelli D (1999) Use of robotic technology for diathrodial joint research. J Sci Med Sport 2:283–97
110. Yellowley CE, Li Z, Zhou Z, Jacobs CR, Donahue HJ (2000) Functional gap junctions between osteocytic and osteoblastic cells. J Bone Miner Res 15:209–17
111. Yoshikawa T, Peel SA, Gladstone JR, Davies JE (1997) Biochemical analysis of the response in rat bone marrow cell cultures to mechanical stimulation. Biomed Mater Eng 7:369–77

XI Immune System Issues

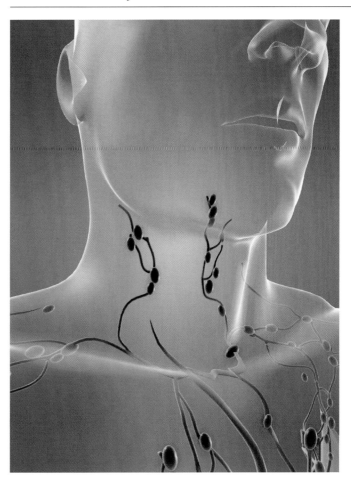

Innate and Adaptive Immune Responses in Tissue Engineering

L. W. Norton, J. E. Babensee

Contents

50.1 Introduction 721
50.2 Innate Immunity 722
50.2.1 PRRs and Pathogen Detection by Leukocytes 722
50.2.2 Monitoring of Tissue Injury by PRRs 725
50.2.3 Protein Adsorption to Implanted Materials and Biomaterial Recognition Using PRRs 726
50.2.4 Wound Healing 727
50.3 Adaptive Immunity 730
50.3.1 Antigen Presentation 730
50.3.2 Antigens in Tissue Engineering 732
50.3.3 Adjuvants in Tissue Engineering 734
50.3.4 Strategies to Prevent Immune Rejection of Tissue-Engineered Devices 736
References 737

50.1 Introduction

Tissue engineering holds great promise for the regeneration of completely natural living tissues and organs, thus addressing the most challenging problem of donor tissue shortage in modern medicine [1, 2]. Traditionally, the tissue engineering concept has been that a relatively small number of cells are obtained from patients, expanded in culture, seeded onto an appropriate biomaterial scaffold, and then transplanted. In addition to autologous differentiated cell sources, current research is investigating the utility of autologous adult stem cells and cells of allogeneic and xenogeneic origin, when autologous differentiated cells are not available [2]. Critical to the success of the engraftment of these cells is the control of the local microenvironment. By engineering the cellular niche, tissue-engineered devices can provide an environment suitable to modify host responses and direct cell differentiation, growth, migration, and survival [3]. Tissue engineering scaffolds have been designed to direct cell function by controlling scaffold material properties and incorporating delivery of cell-instructive biomolecules by immobilization or controlled release [2]. Several tissue-engineered products that incorporate these approaches are already on the market for treating extensive burns [4, 5]. Realizing their potential, new therapeutic strategies resulting from scientific progress in tissue engineering will not only significantly improve the quality of life for the affected individuals, but also save the billions of dollars spent annually to treat patients with lost, damaged, or failing organs.

A major impediment to the development of tissue-engineered products is a destructive host response. Therefore, a current focus of research effort is to improve the host acceptance of tissue-engineered devices. Furthermore, through an understanding of the cellular and molecular aspects of the host response, there is an aim to control healing. Once implanted in vivo, tissue-engineered devices elicit an innate immune reaction toward the biomaterial component, known as the foreign body response. In addition to stimulating an innate immune response, combination devices also can elicit an adaptive immune reaction if it includes an immunogenic biological component. This chapter explores the innate and adaptive immune responses to implanted devices. After implantation of devices, a non-specific response to the device may affect the quality and intensity of an adaptive immune response when immunogens are present. This chapter includes a general description of the innate immune response, highlighting the relevance of patternrecognition receptors (PRRs) in the detection of pathogens and tissue injury and in potentiating adaptive immune responses. In the second half of the chapter, the adaptive immune response is discussed, highlighting the role of antigens and adjuvants in tissue-engineered devices. The interconnections between an innate and adaptive immune response in the context of combination products is also highlighted wherein the biomaterial acts as an adjuvant. The experimental basis for this biomaterial adjuvant effect is also presented, examining the effect of biomaterials on the phenotype of key antigen presenting cells (APCs), dendritic cells (DCs).

50.2
Innate Immunity

Implantation of a tissue-engineered device requires surgical disruption of the surrounding tissues, initiating an innate immune response. The innate immune response is a non-specific response by the host in defense from foreign pathogens or materials, repair of tissue damage (e.g., removal of necrotic cells and cellular debris), and removal of apoptotic cells [6]. Innate immune responses are highly conserved events, exhibiting similar characteristics between insects and mammals [7]. Innate immunity provides a rapid response to infection, which precedes the slower, but specific, adaptive immune response. Key to the innate immune response is the participation of leukocytes, which initially engulf and eliminate infectious pathogens and clear cellular/tissue debris.

Tissue injury from device implantation initiates a foreign-body response, which is an abnormal wound-healing sequence that typically results in encapsulation of the device in fibrous tissue [8]. The general progression of events following implantation, as summarized by Anderson, includes: "injury, blood-material interactions, provisional matrix formation, acute inflammation, chronic inflammation, granulation tissue, foreign body reaction, and fibrosis" [8].

50.2.1
PRRs and Pathogen Detection by Leukocytes

Leukocytes distinguish self from non-self in order to mount an innate response to foreign pathogens. Foreign pathogens are recognized by leukocytes by several motifs: conserved structures not found in eukaryotic cells, carbohydrate structures characteristic of pathogens, or opsonizing proteins. These recognition sequences are known as pathogen-associated molecular patterns (PAMPs). Leukocytes have specific receptors for PAMPs on their cell surface, known as PRRs [9]. These receptors include C-type lectins [10, 11], caspase-recruiting domain (CARD) helicases [12, 13], CD14 [14], complement receptors [15], Toll-like receptors (TLRs) [14], nucleotide-binding oligomerization domain (NOD)-like receptors (NLRs) [12], receptor of advanced glycation end-products (RAGE) [16], and scavenger receptors (SRs) [17] In addition to recognizing foreign pathogens, PRRs can detect tissue injury and adsorbed protein layers on implanted biomaterials, as discussed in section 50.2.3.

50.2.1.1
C-Type Lectins

The C-type lectin receptors recognize pathogen carbohydrate structures through calcium-dependent binding to their carbohydrate recognition domains (CRDs) [10, 18] to mediate cellular processes, including: migration [19]; participate in processes for adaptive immunity, such as interaction with lymphocytes [20]; and internalization of pathogens for antigen processing and presentation [21]. These receptors have been categorized on the basis of the number of CRDs that they possess (i.e., type I and II C-type lectins with multiple and single CRDs, respectively

[10]) or by their affinity to specific carbohydrates, particularly lectins binding mannose-type or galactose-type carbohydrates [22, 23]. The C-type lectin receptors are expressed on fibroblasts, T-lymphocytes (T-cells), polymorphonuclear leukocytes, endothelial cells (ECs), and APCs, which include B-lymphocytes (B-cells), macrophages, and DCs [22, 24, 25].

Endogenous and exogenous carbohydrate ligands are recognized by C-type lectin receptors [26]; however, C-type lectins that function as PRRs are capable of distinguishing self from non-self ligands [21, 22]. Carbohydrate specificity by C-type lectins is conferred by several mechanisms including: their recognition of pathogenic carbohydrate moieties not frequently found on host cells (e.g., terminal mannose, fucose, and N-acetylglucosamine [22, 23]) and formation of multimeric complexes facilitating secondary binding (e.g., tetramers of DC-specific intercellular adhesion molecule (ICAM)-3-grabbing nonintegrin (DC-SIGN) [27, 28]), which increases their affinity for repetitive, dense carbohydrate arrays on pathogens due to multivalent interactions [22, 23]. Ligand binding to some C-type lectin receptors (e.g., mannose receptor and dectin-1) induces phagocytosis [29]. The C-type lectin receptors can also work in concert with other PRRs. For example, dectin-1 (β-glucan receptor), collaborates with TLRs, particularly TLR2 [30] and TLR6 [29], to mount an inflammatory response to pathogens [21, 31]. Signaling of dectin-1 and TLR2 induces production of reactive oxygen species and cytokines, such as tumor necrosis factor (TNF) and interleukin (IL)-12 [32]. The C-type lectins receptors have also been implicated in immune tolerance processes including differential expression of macrophage galactose-like lectin on tolerogenic DCs [33] and participation of DC-SIGN ligation in the production of regulatory T-cells by DCs [34].

50.2.1.2
CARD Helicases

The CARD helicases are intracellular PRRs and include retinoic acid-inducible protein I (RIG-I) and melanoma differentiation-associated gene 5 (MDA5). Intracellular RIG-I and MDA5 detect viral double stranded RNA (dsRNA) and polyinosinic:polycytidylic acid (poly I:C), a synthetic dsRNA [13, 31, 35]. After activation by ligands, RIG-1 and MDA5 activate the transcription factor nuclear factor-κB (NF-κB) and induce the production of the inflammatory cytokine, type I interferon (IFN) [13, 36]. NF-κB is a transcription factor that regulates many genes [37], including: those encoding proinflammatory and immune regulatory cytokines; inducible enzymes (e.g., inducible nitric oxide synthase); cell adhesion molecules (e.g., ICAM); apoptosis regulators (e.g., Fas ligand); expression of major histocompatibility complex (MHC) molecules [38]; and members of the B7 costimulatory family of molecules [39]—indicating a link between innate and adaptive immunity.

50.2.1.3
CD14

The leucine-rich repeat (LRR) protein, CD14, is a PRR that aids in the detection of bacteria. Lipopolysaccharide (LPS) is found on the cell surface of Gram-negative bacteria. As an opsonic receptor, CD14 binds to LPS binding protein, as well as anionic phospholipids and peptidoglycan [14, 40]. CD14 can be found in soluble form in serum or expressed on the cell membranes of monocytes, macrophages, and neutrophils [41]. To aid in TLR responsiveness, CD14 transfers LPS to MD-2 for MyD88-independent signaling by TLR4 and enhances signaling for TLR1, TLR2, TLR3, and TLR6 [40].

50.2.1.4
Complement Receptors

Complement components are plasma proteins that aid in the detection and removal of foreign pathogens in a non-specific manner [42]. Complement component C3 can be activated indirectly by pathogens through antibody–antigen interactions (classical pathway), pathogen surfaces directly (alternate pathway), and by mannose-binding lectin [13]. Complement activation pathways, classical, alternative, and lectin, lead to the production of C3 convertase, which in turn produces C3b and C3a from C3. In the classical pathway, the complement component C1q binds to pathogens primarily through attachment to the Fc portion of bound antibodies. The C1q component can also directly bind to pathogens by other mechanisms such as binding to C-reactive protein and pathogen polysaccharides [43]. In the alternate pathway, spontaneous cleavage of C3 results in opsonization of pathogens, by covalent C3b attachment to hydroxyl groups on surfaces [44]. Spontaneously generated C3b can also bind to host cells, but membrane-bound complement regulatory proteins (e.g., CD46, CD55, and CD59) prevent the destruction of host cells [45, 46]. Activation of the classical or lectin pathways results in the cleavage of C4 into C4a and C4b. Complement component C4b is capable of opsonizing pathogens by covalently attaching to cells through binding of amide groups, hydroxyl groups, or carbohydrates [47]. All complement activation pathways induce the cleavage of C5 by C5-convertases into C5a and C5b. Complement component C5b participates in the formation of the membrane attack complex, causing pore formation in target cells and cell lysis. The anaphylatoxin C5a induces: leukocyte chemotaxis; oxidative bursts; enhanced phagocytosis; release of granule enzymes; increased leukocyte proinflammatory cytokine production of TNF-α, IL-1β, IL-6, and IL-8; and increased expression of Fc receptors [47].

Several complement receptors are involved in the innate recognition of pathogens. Complement receptor 1 (CR1, CD35) binds to opsonins, including C1q, C4b, and C3b, and is expressed on erythrocytes, B-cells, 10–15% of peripheral T-cells, monocytes, neutrophils, eosinophils, and DCs [48, 49]. Complement receptor 2 (CR2, CD21) binds to opsonizing C3 fragments, iC3b, C3d, and C3dg. These C3 fragments are produced by factor I-mediated cleavage of C3b bound to factor H, CR1, or membrane cofactor protein [50]. CR1 and CR2 play a role in adaptive immunity, by increasing B-cell function and enhancing opsonized antigen retention on follicular DCs [51, 52]. Complement receptors 3 and 4 (CR3 and CR4, respectively) are β2-integrins, specifically $\alpha_M\beta_2$-integrin (macrophage receptor-1, Mac-1, CR3, CD11b/CD18) and $\alpha_X\beta_2$-integrin (CR4, CD11c/CD18), respectively. The β2-integrins are found on monocytes, macrophages, neutrophils, granulocytes, DCs, and natural killer (NK) cells [48]. The $\alpha_M\beta_2$-integrin is also found on 2–10% of peripheral T-cells [49]. The β2-integrins on phagocytes recognize pathogens opsonized by iC3b [13, 53]. Another integrin, $\alpha_2\beta_1$ integrin, has recently been found to be a receptor of C1q receptor on peritoneal mast cells [54]. The C3a anaphylatoxin receptor is found on all peripheral blood leukocytes and expressed on activated lymphocytes and ECs during inflammation [47]. The receptor for C5a (C5aR) is found on neutrophils, eosinophils, basophils, monocytes, and mast cell [47]. Complement receptor of the immunoglobulin (Ig) superfamily is required for removal of circulating pathogens opsonized by C3b and iC3b by Kupffer cells [55].

50.2.1.5
Toll-Like Receptors

Each TLR recognizes specific pathogen components, such as LPS, teichoic acids, lipoprotein, dsRNA, bacterial DNA, flagellin, and mannuronic acid polymers, which are complex lipids and carbohydrates [9, 14, 31]. The TLRs are expressed on various cell types including: macrophages, DCs, B-cells, T-cells, fibroblasts, and epithelial cells [31]. Expression of TLRs 1, 2, 4, 5, and 6 is on the cell surface, while TLRs 3, 7, 8, and 9 are expressed on intracellular compartments, requiring the internalization of pathogen components [31]. The most significant consequence of ligand binding to TLR is an intracellular signaling cascade leading to NF-κB, activator protein-1, and mitogen-activated protein kinase activation via a MyD88-dependent pathway [31, 56]. By signaling via a MyD88-independent pathway, TLR3 and TLR4 can activate the transcription factor IFN-regulating factor 3, which induces production of IFN-β and IFN-inducible gene products [57]. NF-κB also participates in gene regulation in the MyD88-independent pathway of TLR signaling.

50.2.1.6
NOD-Like Receptors

The NLRs are cytoplasmic PRRs that contain three domains: (1) a C-terminus LRR; (2) nucleotide-binding domain (e.g. NACHT, which is an acronym for four NLRs: neuronal apoptosis inhibitor protein (NAIP), MHC class II transactivator (CIITA), plant het gene product involved in vegetative incompatibility (*HET-E*), and telomerase-associated protein 1 (TP-1)); and (3) an N-terminus effector domain (e.g. pyrin domain) [58]. The NLR family includes: NOD1, NOD2, and the NACHT, LRR and pyrin domain containing (NALP) subfamily. The receptors, NOD1 and NOD2, respond to components of bacterial peptidoglycans, meso-diaminopimelic acid and muramyl dipeptide, respectively [30, 57]. The NALP subfamily form multiprotein complexes called inflammasomes, which induce the activation of IL-1β, and IL-18 from precursors in response to bacterial components [13, 31]. Though few of the NALP subfamily have been well characterized, NALP3 has been shown to respond to the bacterial peptidoglycan muramyl dipeptide [59]. Expression of NLRs has not been elucidated in great detail; however, it has been shown that NALP is expressed mainly in immune cells [59].

50.2.1.7
Receptor of Advanced Glycation End-Products

RAGE is expressed in tissue and vasculature by several cell types including ECs, smooth muscle cells, monocytes, macrophages, and neurons [60]. This receptor is upregulated in certain pathological conditions (e.g., diabetes, atherosclerosis, and Alzheimer's disease) and upon binding of its ligands [16, 61]. Ligands for RAGE include advanced glycation end-products, amyloid-β peptide, amyloid β-sheet fibrils, proinflammatory cytokine-like mediators of the S100/calgranulins, high-mobility group box-1 (HMGB1), and the $\alpha_M\beta_2$-integrin [16, 60, 62]. Advanced glycation end-products are proteins and lipids that have been post-translationally modified by glycation and oxidation due to conditions such as oxidative stress, hyperglycemia, and inflammation [60]. Another group of RAGE ligands produced during inflammation are the S100/calgranulins that are released by neutrophils, monocytes, macrophages, lymphocytes, and DCs [60]. Signaling by RAGE occurs via the NF-κB pathway.

50.2.1.8
Scavenger Receptors

SRs are a family of receptors that bind polyanionic ligands, such as oxidized or acetylated low-density lipoproteins (LDL) [63] and bacterial products, LPS and lipoteichoic acid [17]. Macrophages, DCs, platelets, smooth muscle cells, and ECs express SRs [17, 63, 64]. The SRs include: SR-AI, SR-AII, macrophage receptor with a collagenous structure (MARCO), SR with C-type lectin I/collectin placenta 1, lectin-like oxidized LDL receptor 1 (LOX-1), SR expressed by ECs (SREC), stabilin-1 (also known as FEEL-1 and CLEVER-1), and SR class B type I (found in mice)/CD36 (found in humans) [17, 63]. Ligand binding to SRs increases cell-surface expression of MARCO, cytokine secretion (e.g., TNF-α and IL-1β [65]), and production of reactive oxygen species [64].

50.2.2
Monitoring of Tissue Injury by PRRs

By recognizing the virally infected and necrotic cell products, PRRs are also internal monitors of tissue well-being. Tissue injury from device implantation can liberate endogenous ligands for TLRs, NLRs, and RAGE. These endogenous ligands are known as "danger signals" or "alarmins" [13, 66, 67]. Damage-associated molecular patterns encompass PAMPs and alarmins, which activate immune cells [66]. In addition to the liberation of alarmins, tissue injury can prime the innate immune response through enhanced responsiveness of TLR2 and TLR4, in particular, without increased expression [68]. Due to receptor promiscuity, TLR2 and TLR4 are capable of responding to a variety of ligands, possibly due to the exposure of hydrophobic residues [6].

Among the endogenous TLR ligands are the heat-shock proteins (HSPs), which are molecular chaperones that are released by necrotic cells in response to stress [69]. Several studies have shown that HSP60 and HSP70 are ligands for TLRs that generates a proinflammatory response [70–73]. In particular, HSP90 has been shown to be involved in the intracellular signaling of TLR4 and TLR9, possibly controlling the inflammatory response [74]. However, more recent studies have indicated the necessity of ensuring that HSP preparations are not contaminated with bacterial TLR ligands [75–78]. These particular TLRs, TLR2 and TLR4, also recognize the alarmin HMGB1 [60, 79, 80], which is released by necrotic cells, cytolytic cells, and cells stimulated by proinflammatory stimuli [66]. However, signaling initiated by HMGB1 has been shown to require $\alpha_M\beta_2$-integrin for neutrophil recruitment [62]. Other endogenous ligands for TLRs that act as a "danger signals" to activated DCs include: small hyaluronan fragments, released upon degradation of extracellular matrix (ECM) at sites of inflammation [81–83]; fibrinogen [84]; fibronectin [85]; and heparan sulfate proteoglycan [86]. Necrotic (but not apoptotic) cells themselves induce expression of genes involved in inflammatory and tissue-repair responses in an NF-κB activation-dependent manner through TLR2 [87].

Other PRRs have been shown to respond to endogenous ligands. The S100/calgranulins, secreted by phagocytes at sites of inflammation, and HMGB1 are endogenous ligands for RAGE [60, 66, 79, 80]. Complement receptor, $\alpha_x\beta_2$ integrin, has been shown to bind to exposed acidic residues on denatured or proteolyzed fibrinogen, which can act as a "danger signal" for tissue damage [88]. The NLR, NALP3, has been shown to respond to extracellular ATP and uric acid crystals from dying cells [89, 90]. The NLR inflammasomes have also been shown to be activated by hypotonic stress in vitro [59]. In addition, the SR, LOX-1, is a HSP-binding structure on DCs, binding to HSP60 and HSP90 [69], and is involved in HSP70-induced antigen cross-presentation [63]. The SRs, SREC-1 and FEEL-1/CLEVER-1, have also been shown to bind HSP70 [69].

50.2.3
Protein Adsorption to Implanted Materials and Biomaterial Recognition Using PRRs

Initial injury due to the implantation of materials causes a disruption of the local connective tissue and vasculature [8, 91]. Proteins adsorb to implant surfaces immediately after device implantation, with the implant acquiring a layer of adsorbed proteins from local bleeding [8, 92]. The adsorbed protein layer has been shown to change in protein composition over time, known as the Vroman effect, with adsorbed fibrinogen replaced with factor XII and high-molecular-weight kininogen [93]. The adsorbed protein layer on implanted materials is cell adhesive and opsonizing, including complement activation fragments, immunoglobulin, fibronectin, vitronectin, and fibrinogen [92, 94–97]. The protein layer is recognized by the cognate receptors, including PRRs, on host inflammatory/immune cells. Complement activation by biomaterials via the alternative pathway results in the coating of the implant with complement activation fragments [98, 99], and/or release of anaphylatoxins (C3a, C4a and C5a), which are chemoattractants for leukocyte infiltration and cause leukocyte activation [100]. The classical pathway of complement activation with adsorbed IgG has also been demonstrated [101] and implicates C1q generation in platelet adhesion and activation [102, 103]. Platelet adhesion and activation, through the adsorbed IgG and fibrinogen, mediates neutrophil reactive oxygen generation [104] in a P-selectin-dependent manner [105], and monocyte tissue factor expression [106]. Fibrinogen adsorption has been shown to be the responsible for neutrophil and monocyte adhesion to implanted biomaterials [107, 108]. Adsorption of IgG facilitates neutrophil adhesion to surfaces under flow conditions via Fc receptors [109–111], which aids in immune-mediated inflammation [112]; however, IgG adsorption has not been found to facilitate macrophage adhesion [113]. Endogenous ligands for TLRs are associated with biomaterial implantation including adsorption of known TLR4 activators, fibrinogen [84], and fibronectin [85], and modulation in the level of mRNA expression of HSPs by biomaterial surface chemistry [114]. In addition, SRs have been shown to facilitate uptake of nanoparticles and carbon nanotubes with adsorbed proteins, particularly albumin, in the liver [115, 116].

50.2.4
Wound Healing

50.2.4.1
Tissue Injury, Leukocyte Recruitment, and Inflammation

After surgical disruption of tissue, blood loss is minimized by release of vasoconstrictors (e.g., thromboxane A2 and prostaglandin 2-α) and activation of the coagulation cascade by platelets exposed to collagen [117, 118]. Blood coagulation, after tissue injury, deposits a fibrin clot, which serves as a provisional extracellular matrix (ECM) [8, 91, 119]. The resulting clot consists of collagen, platelets, thrombin, and fibronectin [117, 119]. The ECM also provides a scaffold for neutrophils, monocytes, fibroblasts and ECs, as well as being a reservoir for mediators facilitating controlled regulation of the inflammatory response [117, 119, 120].

Inflammation is characterized by fluid and plasma protein accumulation, as well as the emigration of leukocytes, primarily neutrophils and monocytes [8, 9]. Acute inflammation is initiated by tissue injury and is mediated by factors released from platelets, mast cells, ECs, and damaged cells. In addition, endogenous "danger signals", liberated by tissue injury, activate host inflammatory/immune cells toward mounting an immune response [121]. Secreted inflammatory mediators include: prostaglandins, which induce vasodilation; leukotrienes, which induce vascular permeability; and cytokines and chemokines, which induce chemotaxis of leukocytes such as IL-1, platelet-derived growth factor (PDGF), transforming growth factor (TGF)-β, platelet factor 4, TNF-α, and monocyte chemoattractant protein-1 (MCP-1 or CCL2) [120, 122, 123]. Mast cells release histamine in response to implants, encouraging the recruitment of inflammatory cells to the wound site [124]. Inflammatory stimuli at the wound site (e.g., IL-1β, TNF-α, endotoxins, hemodynamic factors, viruses, and thrombin [125]) activate ECs lining the nearby vasculature [126]. Activated ECs perform several functions key to the inflammatory response, including: cytokine secretion, such as IL-1 and TNF-α [120]; sequestering cytokines on their apical surface for presentation to leukocytes in the blood; and providing ligands on their cell surface to aid the emigration of leukocytes to the wound site [127].

Ley et al. describe the events for migration of leukocytes to sites of inflammation as: "leukocyte capture, rolling, slow rolling, arrest, adhesion strengthening, intraluminal crawling, paracellular and transcellular migration, and migration through the basement membrane" [128]. Leukocyte recruitment and adhesion is enhanced by chemokine secretion and binding to proteoglycans (e.g., heparan sulfate) on activated ECs [127]. Structural and inflammatory cells (e.g., leukocytes, epithelial cells, and fibroblasts) can be induced by ligand binding to PRRs to secrete chemokines [129]. Platelets also deposit inflammatory chemokines on the endothelium [120, 128].

Leukocyte rolling is primarily dependent on selectins expressed on leukocytes (L-selectin) and activated ECs (P-selectin and E-selectin), which interact with glycosylated selectin ligands (e.g., P-selectin glycoprotein ligand 1, PSGL-1) [123, 128]. Leukocyte PSGL-1 expression facilitates secondary tethering of other leukocytes, improving leukocyte recruitment [130–132]. Since P-selectin and PSGL-1 is also present on the surface of activated platelets, platelets may also play a role in leukocyte tethering [123]. In addition to selectin-dependent rolling of leukocytes, one mechanism of activated T-cell rolling is selectin-independent and occurs by binding of CD44 on T-cells to hyaluronate on activated ECs [127, 133]. Slow leukocyte rolling involves selectins, α4-integrins ($\alpha_4\beta_7$-integrin and $\alpha_4\beta_1$-integrin), and β2-integrins ($\alpha_L\beta_2$-integrin and $\alpha_M\beta_2$-integrin) [127, 128]. For example, E-selectin induces conformational change in $\alpha_L\beta_2$-integrin from low to intermediate affinity for ligand (e.g., ICAM) binding [128].

Two mechanisms are involved in the firm adhesion of leukocytes to the endothelium: chemokine activation of β1- and β2-integrins, and selectin-dependent activation of β2-integrins [125]. Integrins interact with cell-adhesion molecules, including: $\alpha_4\beta_1$-integrin (very late antigen 4) binding to vascular cell-adhesion molecule 1 (VCAM-1) [134] and $\alpha_L\beta_2$-integrin (lymphocyte function associated antigen-1, CD11a/CD18) binding to ICAM-1 [135]. Upon encountering chemoattractant signal presented on ECs, integrin affinity on most leukocytes is modulated from a low-affinity to a high-affinity state [136], which triggers leukocyte arrest [128]. Though many chemokines (e.g., IL-8) have been shown to arrest rolling leukocytes in vitro, only a few chemokines have shown this activity under physiologic conditions. These chemokines include, but are not limited to: secondary lymphoid tissue chemokines (CCL21); keratinocyte-derived chemokines (mouse GRO-α, CXCL1); MCP-1; and regulated on activation, normal T-cell exposed and secreted (CCL5) [137].

Although much is known about leukocyte capture, rolling, and arrest, the process of diapedesis, the mechanisms of migration of leukocytes across the EC barrier, are less well-known [126]. Chemokine signaling induces leukocyte polarization, which establishes cell orientation for transmigration [136]. Leukocyte transmigration may occur via the paracellular route (between ECs) or transcellular route (through the body of ECs) [128, 138, 139]. Activated ECs express adhesion proteins involved in leukocyte transmigration either at tight junctions (e.g., junctional adhesion molecules, poliovirus receptor, and endothelial selective adhesion molecule) or at endothelial contacts (e.g., platelet-EC adhesion molecule-1, ICAM-1, ICAM-2, and CD99) [138]. In both paracellular and transcellular routes, leukocyte diapedesis is associated with a "ring-like" or "cup-like" structure containing ICAM-1, VCAM-1, $\alpha_L\beta_2$-integrin, moesin, and ezrin [140–142]. Leukocyte migration through endothelial junctions may take place at tricellular contacts, where unique cell contact proteins are found at the junction of three ECs [138]. ECs can create an adhesive haptotactic gradient of junctional molecules during paracellular transmigration [128]. The transcellular route may occur by organization of vesiculo-vacuolar organelles, which provide a passage through the body of ECs [128]. After migrating across the endothlelium, leukocytes cross the basement membrane, possibly through gaps between pericytes and less dense ECM [128]. At the site of inflammation, leukocytes interact with the ECM through surface integrin molecules, such as the β1 integrin family for monocytes (i.e. $\alpha_1\beta_1$, $\alpha_2\beta_1$, $\alpha_4\beta_1$, $\alpha_5\beta_1$, and $\alpha_6\beta_1$-integrins) and other myeloid cell integrins ($\alpha_v\beta_3$, $\alpha_x\beta_2$, and $\alpha_4\beta_7$) [143]. Integrin molecule engagement on leukocytes promotes leukocyte survival, activation, and differentiation [143].

Within hours of the onset of inflammation, neutrophils are the most abundant leukocyte in the tissue [120]. In inflammation, the role of neutrophils is primarily phagocytosis of foreign pathogens after detection by PRRs and clear cellular debris [117]. Afterwards, macrophages are the more numerous cell type at the site of inflammation, owing to the short-lived nature of neutrophils and the emigration of monocytes into the tissue [91, 121, 144]. Emigrating monocytes differentiate into tissue macrophages or DCs [144]. Inflammatory mediators stimulate leukocytes to perform several functions, including: proinflammatory and angiogenic cytokine release [105, 120, 145], proteolytic enzyme release [146], upregulation of cell adhesion molecules, expression of a procoagulant phenotype [147], arachidonic acid metabolites [8, 148], and release of reactive oxygen intermediates [149] as effector functions aimed at the elimination of associated bacteria [150]. Other leukocytes and lymphocytes may participate in the innate inflammatory response, including: eosinophils, which combat parasitic infections and mediate allergic inflammation; mast cells, which enhance inflammatory cell recruitment; and NK cells, which can be activated directly by pathogens or indirectly by DCs to eliminate pathogens [144, 151].

50.2.4.2
Tissue Regeneration and Repair

In cutaneous wound healing, inflammation proceeds into the proliferative and the remodeling phases. Mesenchymal cells proliferate during the healing of wounds, directed by cytokines secreted by nearby cells (e.g., activated platelets and macrophages) and by ECM components such as collagen peptides and fibronectin [8, 117, 120]. During angiogenesis, ECs proliferate and migrate to form new capillaries, vascularizing the wound bed and forming granulation tissue from ECM components [8, 152]. Migration and proliferation of epithelial cells facilitates epithelialization of the wound, stimulated by epidermal growth factor (EGF) and TGF-α [117]. Keratinocytes migrate and proliferate in the wound, stimulated by keratinocyte growth factors and IL-6 [117]. After stimulation by PDGF and EGF, fibroblasts participate in the healing of wounds by: proliferating; migrating; producing ECM components, such as collagen type

III, glycoasminoglycans, and fibronectin; and assembling ECM components into granulation tissue [117, 152]. Leukocytes direct the remodeling of the ECM through the release of matrix metalloproteinases (MMPs) for the closure of wounds [143]. Fibroblasts continue remodeling of the tissue through secretion of MMPs and collagen deposition [120]. T-cells, attracted by secretion of INF-γ, regulate wound repair by cytokine secretion [153, 154]. Resolution of inflammation is mediated by glucocorticoids, prostaglandins, nitric oxide, soluble receptors, and anti-inflammatory cytokines [121]. However, unresolved acute inflammation, yielding chronic inflammation, can lead to scar formation or tissue destruction via TLR signaling [155].

A heterogeneous population of macrophages is involved in the inflammatory processes, particularly classically and alternatively activated macrophages. Classically activated macrophages are stimulated by proinflammatory mediators, including: proinflammatory cytokines (e.g., INF-γ), PAMPs, endogenous "danger signals", hypoxia, and abnormal matrix, such as pathological collagen deposition [156, 157]. Classically activated macrophages are proinflammatory, secrete reactive oxygen species, and mediate the removal of intracellular pathogens [157, 158]. Classically activated macrophages also participate in delayed-type hypersensitivity by secreting inflammatory cytokines and chemokines [157]. Classically activated macrophages support T-helper (T$_H$)-1 responses by producing low levels of IL-10 and high levels of IL-12 [157, 159]. Alternatively activated macrophages are stimulated by T$_H$2 cytokines (IL-4 and IL-13), anti-inflammatory cytokines (e.g., IL-10 and TGF-β), glucocorticoids, vitamin D3, and apoptotic cells [157, 159]. Alternatively activated macrophages are associated with fusion and granuloma formation [158]. Alternatively activated macrophages promote T$_H$2 responses, produce anti-inflammatory cytokines and chemokines, promote angiogenesis, and increase the fibrogenic activity of fibroblasts, and are considered reparative [157].

Wound healing is not limited to repair of dermal lesions; placement of tissue-engineered devices may induce wounds in tissues other than the dermis [160]. A thorough description of the complex wound healing events in organs and tissues potentially replaced by tissue-engineered devices is beyond the scope of this chapter.

50.2.4.3
Wound Healing After Device Implantation

Wound healing from implanted devices progresses by an abnormal sequence of events. In vitro studies have shown that materials with adsorbed proteins influence monocyte cytokine secretion profiles [145] and macrophage adhesion [161]. Phagocytosis of large, non-degradable implanted materials usually does not occur due to the size disparity between the implant and phagocytes [8, 91]. Instead, frustrated phagocytosis may occur, a process whereby leukocytes secrete reactive oxygen intermediates that can degrade the devices [105, 150]. However, particulates delivered (e.g., nanoparticle or microparticle (MP) drug delivery [162–165]) or generated upon implant degradation (e.g., wear debris [166]) are phagocytosable if of a size of <10 µm [167–169].

After device implantation, chronic inflammation can be induced by a persistent inflammatory stimuli; however, the exact nature of these stimuli has not been determined [170]. Proposed chronic inflammatory stimuli from implants have included: motion of the implant and material properties of implants that promote macrophage survival and proliferation [8, 91, 170]. In chronic inflammation, the tissue surrounding implants is characterized by the presence of blood vessels, connective tissue, and immune cells, including mononuclear cells lymphocytes and plasma cells [91]. As with normal wound healing, granulation tissue forms in the implantation wound site, characterized by neovascularization and deposition of ECM by fibroblasts [8, 91]. Typically, over the following weeks, the tissue reaction around the implant continues to remodel into the classic foreign body capsule: an avascular fibrous encapsulating tissue that surrounds a persistent layer of foreign-body giant cells (FBGCs) [8]. Alternatively activated macrophage fusion into FBGCs, stimulated by IL-4 [171] and IL-13 [172], is commonly associated with biomaterial chronic inflammation [172–174] in an integrin-dependent manner [175]. The FBGCs have been shown to degrade implanted materials by stress cracking and pitting [176]. IL-4 prevents apoptosis of biomaterial-adherent macrophages by inducing shedding of TNF-α receptor I, preventing this TNF-α-mediated process [174]. In addition, the macrophage phenotype has been shown to play a role in biomaterial scaffold remodeling [177].

Multifaceted approaches to influencing the progression of a foreign-body reaction by developing "materials that heal," including addressing vascularity and fibrosis, are receiving much attention from the biomaterials research communities [178–181]. In order to improve the integration of tissue-engineered devices, Lumelsky suggests examining healing systems that support tissue regeneration [155], which include: regeneration of multiple tissues in Murthy Roths Large mice [182, 183]; limb regeneration in amphibians and scarless healing of mammalian fetal tissue [184]; and the role of inflammatory mediators in murine organ regeneration (e.g., IL-6, TNF-α, and TGF-β). In addition, strategies employed to change the nature of the foreign-body capsule, from a thin, fibrous matrix to a highly vascularized matrix, have included surface texture, material microarchitecture, and surface chemistry [185]. Pillared and porous surfaces have been shown to be encapsulated by vascularized tissue [186, 187]. Vascularity surrounding porous implants were influenced by pore size and the grafting of angiogenic molecules [187–189]. However, compared to smooth implants, textured implants have had higher presence of macrophages and FBGCs at the wound site [8]. The dimensions of the implanted materials have also been shown to alter the foreign-body capsule [190]. Increased capsule vascularity has been stimulated through the local sustained release of growth factors from biomaterial drug-delivery systems. Angiogenic growth factors (e.g., vascular endothelial growth factor (VEGF) [191–194] and fibroblast growth factors [195–199]) have been delivered from multiple materials to increase the vascularity of the encapsulation tissue. Release of glucocorticoid and VEGF have been shown to mediate inflammation and increase capsule vascularity, respectively [200, 201]. However, since newly formed vessels may regress after the reservoir for the delivered growth factor becomes depleted [193], additional strategies (e.g., formation of mature vessels or genetic modification of tissue) may need to be employed to maintain capsule vascularity. Gene delivery of antisense thrombospondin 2 has also been shown to improve capsule vascularity near implanted membranes [202]; and dual delivery of PDGF and VEGF from implanted scaffolds induced the formation of mature vessels near the implant [203].

50.3
Adaptive Immunity

While the innate immune response is a non-specific immune response aimed at the removal of pathogens and other foreign entities such as biomaterials, the adaptive immune response, or acquired immunity, is an antigen-specific immune response aimed at the elimination of pathogens and cells displaying non-self antigens. The adaptive immune response may also play a role in regulating innate immune responses. Recently, the adaptive immune cells, specifically T-cells, have been shown to temper the TLR-stimulated innate immune response to viral infections that can lead to host death due to unregulated cytokine activity [204].

Tissue-engineered devices are typically composed of biomaterial and biologic components. While implantation of a biomaterial will cause tissue injury leading to an innate immune response (or a foreign body response as described above), the presence of an immunogenic biological component can lead to an adaptive immune response. Immunogens, antigens capable of initiating a specific immune response, may be shed from or delivered as part of the biologic component of the tissue-engineered device (e.g., shed antigens from allogeneic cells—cells from one person into a genetically different person—in a tissue-engineering construct). For this reason, the potential for an adaptive immune response must be considered in the design of tissue-engineered devices.

50.3.1
Antigen Presentation

MHC molecules are expressed on most cell surfaces. There are three classes of MHC: MHC class I, MHC class II, and MHC class III. The MHC class I and MHC class II molecules play important roles in adaptive immunity. While MHC class I molecules are displayed on most nucleated cells, MHC class II are displayed on APCs, such as macrophages, DCs, and B-cells. Typically, MHC class I molecules display self-peptides or foreign antigens from pathogen-infected cells such as virally infected cells. Peptides displayed on MHC class I molecules are typically derived from proteins produced within a cell that are degraded by the ubquitin-proteosome pathway in the cytoplasm or nucleus [205]. Some of the peptides generated during degradation are trafficked to the endoplasmic reticulum (ER) by transporter associated with antigen processing (TAP). In the ER, these peptides are bound to MHC class I molecules, which are then transported to the plasma membrane. The MHC class II molecules on APCs display exogenous foreign antigens, such as peptides derived from bacteria. The MHC class II molecules are formed in the ER, where the invariant chain (Ii) occupies the peptide-binding groove of MHC class II molecules. After exiting the ER and the Golgi apparatus, Ii is degraded in the endosomal compartment by cathepsins, leaving class II-associated peptides (CLIP) in the peptide-binding groove [206]. After endocytosis of exogenous foreign pathogens by APCs, CLIP is removed from the peptide-binding groove by H-2M in mice or human leukocyte antigen (HLA)-DM for substitution with foreign antigens. The MHC class II molecules with bound antigens are then transported to the APC plasma membrane. In cross-presentation, foreign antigens are presented on MHC class I molecules, which is performed primarily by DCs. Antigen processing for cross-presentation occurs via two pathways, although the exact mechanisms of these pathways have yet to be determined [205]. The first pathway is proteosome- and TAP-dependent. In this pathway, endocytosed antigens are transported to the cytosol and hydrodrolyzed by proteosomes [205]. The peptides generated in the cytosol may be reinserted into phagosomes for binding to MHC class I molecules, or may be imported to the ER for binding to MHC class I molecules [205]. The peptides are then transferred to MHC class I molecules. The second pathway is TAP- and proteosome-independent. In this pathway, peptides are produced from foreign antigens degraded by endosomal proteases (e.g., cathepsin S) and may bind to MHC class I molecules in the endosome [205].

The T-cell receptors (TCRs) recognize peptide fragments of antigens bound to MHC molecules, followed by secondary binding of co-receptors, CD4 or CD8 molecules, to MHC molecules. Naïve T_H cells require a second signal for development into protective effector cells after TCR stimulation; this is the ligation of costimulatory molecules with their counterreceptor on DCs (i.e., B7 molecules on APCs to the CD28 ligand on T-cells). Low levels of costimulatory molecules, B7.2 (CD86), are present on resting B-cells, DCs, and macrophages; upon APC activation, B7.2 and B7.1 (CD80) expressions are upregulated and induced, respectively [207]. A third signal, polarizing cytokine secretion, directs the production of T-cell subsets (e.g., polarization towards $T_H 1$ and $T_H 2$ phenotypes by secretion of IL-12 and IL-4, respectively). Alternatively, a lack of costimulation leads to T-cell anergy.

Activation of effector T-cells requires only the MHC-peptide complex. Cytotoxic (CD8+) and helper (CD4+) T-cells recognize antigens bound to MHC class I and MHC class II molecules, respectively. Cytotoxic T-cells kill pathogen-infected cells, which is a cell-mediated response. T_H cells perform several functions by activating other cells: $T_H 1$ cells activate macrophages to kill antigen-bearing pathogens (cell-mediated response); $T_H 2$ cells activate B-cells to produce antibodies against target antigens (humoral response). In addition to the production of effector T-cells and B-cells, long-lived memory T-cells and B-cells are produced from activated cells that enhance immune responses, which are greater in magnitude and longer in duration after secondary exposure to antigens. T-cells participate in the resolution of immunity through expression of cytotoxic T-lymphocyte antigen 4 (CTLA4), a molecule homologous to CD28, which binds to costimulatory B7 molecules [208].

APCs initiate the production of antibodies against non-self antigens or a cell-mediated immune response by the activation of T-lymphocytes. Antigen uptake by B-cells occurs after antigen binding to membrane-bound immunoglobulin. Activated B-cells serve as APCs by processing antigens from their receptors and presenting the antigens on MHC class II molecules. Macrophages and DCs, resident in peripheral tissues, use PRRs to recognize foreign pathogens and tissue injury through endogenous "danger signals." The "danger" model of immunity proposes that key to triggering adaptive immunity is an association between an antigen and a "danger" signal [209]. "Danger signals" can be considered adjuvants, substances that enhance the immune response to antigens by activating APCs that have upregulated expression of MHC and costimulatory molecules. Natural adjuvants (necrotic fibroblasts, smashed vessels, and

INF-α released by virally infected cells) enhanced the adaptive immune response initiated by DCs to a weak immunogen, ovalbumin (OVA) [210]. Endogenous "danger signals" released from injured tissues can promote immune responses (e.g., transplantation, autoimmunity, tumor immunity, and infection) in the presence or absence of exogenous adjuvants [211]. In this way, macrophages and DCs provide an important link between adaptive and innate immunity. Macrophages, activated by the phagocytosis of pathogens, present peptide bound to MHC class II with costimulatory molecules to T-cells. Resting macrophages do not express costimulatory molecules on their cell surface. Of the APCs, DCs are professional APCs with an extraordinary capacity to initiate primary immune responses due to in their ability to stimulate naïve T-cells.

Precursors of DCs differentiate from hematopoietic stem cells in the bone marrow, and migrate through the blood to non-lymphoid tissues, guided by chemokines [212, 213]. Immature DCs reside in peripheral tissues functioning as sentinels, sampling the local microenvironment by capturing and processing antigens. Immature DCs are efficient at capturing antigens but inefficient at stimulating T-cells due to the rapid internalization and recycling of MHC molecules [214, 215]. Upon encountering signals that drive their maturation (e.g., PRR-induced phagocytosis of pathogens, disruption in the balance of inflammatory and anti-inflammatory cytokines or chemokines, or T-cell-derived signals), they migrate to the lymph nodes where they present antigens for T-cell priming. Mature DCs express high levels of surface MHC, costimulatory molecules (CD40, CD80, CD86), CD83, which is involved in DC-mediated immune responses, and ICAM-1 [214, 216, 217]. Compared to the high expression on immature DCs, mature DCs express lower levels of macrophage marker CD14 and of Fcγ receptors (CD32 and CD64), which facilitate endocytosis of antibody-coated material [218]. Maturation of DCs is normally accompanied by DC production of selective sets of cytokines, the T-cell polarizing third signal, to drive the T-cells to become T_H1, T_H2, or regulatory T-cells. Cytokine secretion by mature DCs is dependent on the activation of particular PRRs by PAMPs and other inflammatory tissue factors [219].

50.3.2
Antigens in Tissue Engineering

The biologic components of tissue-engineered devices are typically the functional elements in reparative medicine. Cells used in tissue-engineered devices can be autologous, isogenic, allogeneic, or xenogeneic in origin. Autologous cells have been expanded in culture for use in tissue-engineered devices (e.g., keratinocytes, fibroblasts, chondrocytes, myoblasts, and mesenchymal stem cells, MSCs) [220] and provide the potential for "off-the-shelf" construct availability. Autologous cell sources pose particular challenges in tissue engineering, including: time delay in expansion of autologous cells for sufficient cell number, difficulty in expanding cells from older patients, inability to correct some genetic defects, potential for a shift in cell phenotype with culture, donor site morbidity, and the high burden of good manufacturing practices during commercial scale-up [221]. Novel sources of MSCs include adipose tissue [222], amniotic fluid [223], dental pulp [224], placenta [223], and bone marrow [225]. Alternatives to autologous tissue-specific cell sources for tissue engineering include: genetic engineering of another autologous cell type, use of cell lines, and allogeneic or xenogeneic cells [220]. However, cell lines and allogeneic or xenogeneic cells introduce a significant concern regarding the immunogenicity of the biologic component. Extensive culturing, genetic modification, or cell death can render autologous cells immunogenic, indicating that lack of immunogenicity of autologous sources should not be taken for granted [220]. However, use of allogeneic or xenogeneic sources may eliminate the issue of lack of sufficient cell number and certain bioprocessing issues; but the host immune response to the foreign material could impede the function of a tissue engineered device.

Immune rejection of allogeneic or xenogeneic material is well known from cell/organ transplantation biology [208, 220, 226, 227]. Graft rejection can be classified in three categories: hyperacute rejection, which occurs within 1 day for vascularized organs; acute rejection, which occurs within a few weeks of transplantation; and chronic rejection, which occurs within months to years of transplantation [226].

Hyperacute rejection is usually mediated by complement activation and preformed antibodies [228]. Rejection is due to histoincompatibility between donor and host antigens and is T-cell dependent [208]. For the most part, histocompatibility is determined by alloantigens (i.e., MHC molecules); however, minor histocompatibility antigens presented in the MHC groove can play a role in rejection, requiring host immunosuppression for MHC-matched grafts [208, 226]. Immune rejection of transplants occurs in the following manner: activation of host T-cells, T-cell proliferation in response to alloantigens, and activation of immune effector cells resulting destruction of the graft. Pathways to allorecognition may be direct or indirect. In the direct pathway, host T-cells are activated by donor APCs (passenger leukocytes) by recognizing donor MHC molecules, with or without presentation of donor antigens. Tissue injury and stress to the allograft can activate passenger APCs, particularly through TLR4 signaling [229]. Passenger APCs migrate to host lymphoid tissues and activate of naïve CD4+ T-cells (T_H0). In addition, passenger DCs are effective at activating naïve CD8+ T-cells. However, passenger leukocyte activation of T-cells is a less likely scenario in tissue engineering, since passenger APCs, can be removed upon extensive culturing, as for islets of Langerhans [230]. In the indirect pathway, host APCs present antigens from the allogeneic or xenogeneic material to host T-cells; this is considered the more dominant pathway in immune rejection [208]. The activated T-cells produce cytokines, particularly IL-2, and increase expression of the IL-2 receptor (CD25). Autocrine and paracrine signaling by IL-2 causes T-cell clonal expansion. Activated T-cells also secrete IL-3 and granulocyte-macrophage colony stimulating factor, which stimulate the production of granulocytes and macrophages. Cytokine polarization signals from APCs induce activated CD4+ T-cells to differentiate into effector cells: T_H1 cells, which secrete IFN-γ, TNF-α, and TNF-β; and T_H2 cells, which secrete IL-4, IL-5, IL-6, IL-10, IL-13, and TGF-β. Cell-mediated responses are stimulated by T_H1, which activates cytotoxic CD8+ T-cells and macrophages, leading to tissue destruction. The T_H2-driven response can activate clonal expansion and differentiation of B-cells, isotype switching, eosinophilia, and induce tolerance [208]. Donor vascular endothelium also plays a role in immune rejection, since blood-borne host immune cells and soluble mediators encounter and cross donor endothelium upon reperfusion [208]. In hyperacute rejection, ECs are activated before complement-mediated death, inducing thrombosis and edema [228]. In more delayed rejection, the endothelium is activated by inflammatory cells (e.g., monocytes, macrophages, and NK cells) or antibody binding, inducing antibody-dependent cell-mediated cytotoxicity, which involves NK cells [228]. EC activation in delayed rejection leads to macrophage and NK accumulation, as well as thrombosis and fibrin deposition [228].

The immunogenicity of xenografts and allografts can be altered by removing the cellular constituents of the graft. Decellularized allogeneic and xenogeneic ECM provides resorbable scaffolds for tissue repair (e.g., clinical use of decellularized porcine small intestinal submucosa for tendon repair, and de-cellularized human dermis and small intestinal mucosa as wound dressings) [231]. Immune responses to native decellularized allogeneic and xenogeneic ECM (i.e., decellularized grafts without crosslinking) is markedly reduced compared to immune-mediated rejection of transplanted tissues and whole organs [231]. Decellularized allogeneic ECM has been shown to induce similar host responses as decellularized syngeneic ECM in rats, such as T-cell infiltrate density and antibody production levels [232]. Decellularized native xenograft ECM has been shown to elicit a T_H2-mediated immune response [233–235], inducing production of IgG1 antibodies [233] and expression of IL-4 mRNA at the implantation site [233]. Post-processing of allogeneic or xenogeneic grafts after decellularization can impact the host response. For example, cross-linking improves tensile properties but can also prevent remodeling, causing a foreign-body response [231], such as prolonged presence of multinucleated giant cells around the implant [236].

Although ECM is rendered less immunogenic by decellularization, implanted ECM by may contain cellular components due to insufficient cell removal or by design. Since complete removal of cellular constituents from porcine allografts is difficult [231], cells and/or cellular components may remain in the graft after processing, impacting the host response [237]. ECM and its components have been used as scaffolds for cell seeding; but introducing cellular components may render tissue-engineered devices immunogenic [231].

Interestingly, embedding allogeneic or xenogeneic ECs in collagen-based matrices in three dimensions has been shown to induce an EC phenotype with reduced immunogenicity, compared to two-dimensionally cultured xenogeneic and allogeneic ECs [238–241].

50.3.3
Adjuvants in Tissue Engineering

50.3.3.1
Biomaterial Adjuvants

Since DCs are a critical link between the innate and adaptive immune responses, characterizing DC maturation upon biomaterial contact and the role of biomaterials as adjuvants proposes a significant novel biocompatibility consideration in biomaterial selection and design for tissue engineering. Among the natural biomaterials used in tissue engineering, chitosan has adjuvant activity: stimulates T- and B-cells [242], modulates of nitric oxide release by macrophages [243, 244], and is chemotactic for leukocytes [244]. Soluble fragments of hyaluronic acid [245] and collagen type I [246] support maturation of DCs. Alginate with high mannuronic acid composition was shown to be more immunogenic than alginate with high guluronic acid [247], presumably through polymannuronic acid binding to CD14 [248]. The downstream signaling and cell activation by the polymannuronic acid component of alginate seems to be mediated by TLR2 and TLR4 [249]. Agarose appears to elicit minimal humoral and cellular responses in vivo [250, 251]. Poly(lactide-co-glycolide) (PLGA) has also been used as an adjuvant for the MP delivery of vaccines [252, 253].

Contact with biomaterials has been shown to induce maturation of DCs in vitro. As mature DCs are effective stimulators of T-cell proliferation, human peripheral blood monocyte-derived DCs (hDCs) matured by exposure to PLGA MPs had higher allostimulatory capacity than immature hDCs, as measured by an allogeneic mixed lymphocyte reaction, but had a lower allostimulatory capacity than hDCs matured with LPS treatment [254]. The maturation of hDCs by PLGA MPs required direct contact with immature hDCs. In that study, however, hDCs in contact with PLGA films did not induce significant T-cell proliferation over that induced by control immature hDCs. Treatment of murine bone-marrow-derived DCs (mDCs) with PLGA films and MPs induced significantly increased secretion of IL-6 and TNF-α, but only a slight increase in co-stimulatory molecule (CD80 and CD86) expression, indicating that mDCs also responded to biomaterial with a maturation effect [255]. This can lend support to the biomaterial-induced maturation of DCs as a mechanism for the biomaterial adjuvant effect. This effect was observed when a model antigen, OVA, was delivered with PLGA MPs or scaffolds, which supported an enhanced humoral immune response as compared to antigen delivered with phosphate-buffered saline (as described in more detail below). Further illustrating the biomaterial-dependent effect on DC maturation, phagocytosis of PLGA or polystyrene MPs (both 3 μm in diameter) was found to have a differential effect on maturation of hDCs. The PLGA MPs, but not polystyrene MPs, were shown to induce DC maturation, as measured by increased TNF-α secretion by hDCs [256]. Differential DC maturation has been demonstrated after hDC contact with agarose, alginate, chitosan, hyaluronic acid, or PLGA [257]. Treatment with chitosan or PLGA films was shown to support DC maturation by increased DC expression levels of MHC class II (HLA-DQ) and costimulatory molecules (CD80), and CD40 compared to immature DCs. The increased expression levels of cell-surface DC maturation markers were found to be similar to expression levels on DCs matured with LPS. Agarose film-treated DCs had higher expression of DC maturation markers than immature DCs, but to a lesser extent than DCs treated with chitosan or PLGA films. In contrast, DCs treated with alginate or hyaluronic acid films had decreased expression of DC maturation markers. The effect of phagocytosis or simply biomaterial contact on the extent of DC maturation was assessed by treating DCs with PLGA or agarose MPs of phagocytosable (3 μm) versus nonphagocytosable (20 μm) sizes, while keeping the total biomaterial surface area constant. The DCs treated with PLGA or agarose MPs of the different sizes did not change their expression level of mDC markers, implying that exposed PLGA biomaterial surface area, not phagocytosis, was the main contributor to the MP-induced DC maturation [258].

In order to elucidate the biomaterial adjuvant effect in vivo, biomaterials injected or implanted in mice with model antigen, OVA, were found to enhance OVA-specific antibody production, OVA-specific T-cell proliferation, and delayed-type hypersensitivity (DTH, a T_H1-cell-mediated immune response), as compared to OVA delivered with phosphate-buffered saline as the negative control. In a simplified model system of shed antigens from tissue-engineered devices, OVA was preadsorbed to or encapsulated in biomaterial carrier vehicles used in tissue engineering [259, 260]. The biomaterials tested were nonbiodegradable phagocytosable polystyrene MPs (6 µm in diameter), biodegradable phagocytosable 50:50 or 75:25 PLGA MPs (6 µm in diameter) [259] or PLGA scaffolds (0.7 cm diameter, 0.2 cm thick) [260]. Biomaterial carrier vehicles supported a moderate OVA-specific humoral immune response after injection or implantation in C57BL/6 mice that was maintained for the 18 week duration of the studies. This humoral immune response was primarily T_H2 helper T-cell-dependent as indicated by the predominant IgG1 isotype. Furthermore, this humoral immune response was not material chemistry-dependent. Co-delivery of OVA with a biomaterial stimulated similar in vivo proliferation of fluorescently-labeled OVA-specific CD4+ T-cells from transgenic OT-II mice to that observed for antigen delivered with the strong adjuvant, complete Freund's adjuvant (CFA) [261]. Another measure of the adjuvant activity of a polymer is its ability to support a DTH reaction against the model antigen, OVA [254]. OVA delivered with PLGA MPs induced a similar level of DTH as OVA delivered with the strong adjuvant, CFA, and a lower response than PLGA MPs without OVA. In addition, DCs have been shown to infiltrate implanted scaffolds; twice as many DCs were observed in PLGA scaffolds with OVA than without OVA (A. Paranjpe and J.E. Babensee, unpublished observations). DCs isolated from within polyvinyl sponges after subcutaneous implantation for 6 days and 2 weeks have been shown to change from an inflammatory to a tolerogenic phenotype, with reduced allostimulatory capacity in vitro [262].

Biomaterial-Associated "Danger Signals"

Considering the effect of "danger signals" on the adaptive immune response to shed antigens from biomaterials, the immune response toward an associated antigen may depend on the form of the carrier vehicle, which is proportional to the extent of tissue damage associated with its insertion. To examine the effect of "danger signals" enhancing the biomaterial adjuvant effect, OVA was incorporated into polymeric biomaterial carriers made of PLGA in the form of MP or scaffolds [260]. To allow for the assessment of the differential level of enhancement of the immune response depending on the form of carrier vehicle (MP vs. scaffold), the total amounts of polymer and OVA delivered were kept constant, as was the release rate of OVA for both carrier vehicles. The materials, MPs and scaffolds, were injected and implanted into C57BL/6 mice, respectively. The level of the humoral immune response was higher and sustained for OVA released from PLGA scaffolds, which were implanted with associated tissue damage; a lower and transient immune response was observed when the same amount of polymer and OVA was delivered from PLGA MPs, which were delivered by injection, a minimally invasive procedure. Although the immune response was primarily T_H2-, T-cell-dependent, for the strong adjuvant, CFA, and PLGA scaffold carriers, there was both a T_H1 and T_H2 response contribution. These results implicate "danger signals" associated with the implantation of the scaffolds due to tissue injury, which primed the system for an enhanced immune response. These results indicate that it is important to deliver/implant a tissue-engineered construct as non-invasively as possible to minimize any potential immune responses. Furthermore, there is the potential to "hide" an immunological tissue-engineered construct if it is delivered in such a manner as to avoid "danger signals".

50.3.3.2
Endotoxin Contamination

In the development of tissue-engineered devices, one should consider potential endotoxin contamination of commercial reagents and biomaterials and aim to minimize it. Endotoxin acts as an adjuvant in immune responses by activating immune cells through binding to PRRs. Endotoxin induces the release of mediators such as ILs, prostaglandins, free radicals, and cytokines from activated immune cells [263]. Scaffolds for tissue-engineered devices can incorporate recombinant growth factor delivery to improve parenchymal cell development or engraftment of the device [264]. Recombinant protein preparations may include agonists for TLR, such as endotoxin, inducing a proinflammatory response [76]. Commercial preparations of non-recombinant and recombinant proteins have been found to be contaminated by bacterial products [76, 77, 265]. Recombinant proteins produced in Gram-negative bacteria can become contaminated with endotoxins, thus necessitating endotoxin removal [263]. For in vivo assessment of tissue-engineered devices, steps should be taken to minimize and test for the endotoxin content of materials used in tissue-engineered devices to prevent endotoxin-mediated contributions to the immune/inflammatory response to the implant.

50.3.4
Strategies to Prevent Immune Rejection of Tissue-Engineered Devices

Strategies from transplantation biology to prevent graft rejection, such as host immunosuppression or tolerance induction, can be employed to prevent the immune rejection of tissue-engineered devices. Tolerance-induction strategies have included costimulatory blockade, APC and lymphocyte depletion at the time of implantation, and donor-marrow or stem-cell transplantation [227]. In addition, DCs have been genetically modified to induce tolerance by expressing: chimeric fusion protein CTLA4-Ig [266–268]; the immunomodulatory cytokines TGF-β [269] and IL-10 [270]; and T-cell apoptosis-inducing Fas ligand [271] and TNF receptor apoptosis-inducing ligand [272]. Strategies targeting PRRs may reduce immune rejection. Blockade of endogenous danger signaling from RAGE delayed allograft rejection [229]. The presence of TLR agonists in allotransplantation has prevented the induction of tolerance [229], whereas functional polymorphisms in TLR4 inducing hyporesponsiveness to endotoxin were shown to be associated with reduced acute allograft rejection [273, 274]. Modifications to the tissue-engineered device may also prevent immune rejection (e.g., immunoisolation devices [220] and donor-specific use of stem cells [275]). Although immunoisolation devices prevent direct interaction between host APCs and the biologic component, shed antigens capable of diffusing across the immunoisolation membrane can induce a specific immune response that is detrimental to device function [276, 277].

Stem cells have tremendous therapeutic potential for tissue engineering. Embryonic stem cells are most versatile for producing differentiated cells; however, allogeneic embryonic stem cells may become immunogenic after differentiation and culture with non-human feeder cells [275]. Allogeneic MSCs have been shown to engraft and promote tolerance to allogeneic tissue [278]. In addition, third-party allogeneic MSCs have been shown to induce immunosuppression in grade IV graft-versus-host disease [279]. The MSCs have been shown to have immunoregulatory effects on T-cells [280] by several mechanisms, including: inhibition of naïve and memory T-cells that have encountered an antigen [281], which may be TGF-β1 mediated [282], and altered production of T_H1 and T_H2 cytokines [283, 284]. However, IL-2 administration reversed the immunosuppressive effects of MSCs [285]. In other studies, administration of allogeneic MSCs required immunosuppressive therapy for bone tissue engineering [286]. Differentiation may alter or remove the immunomodulatory effects of MSCs. Xenogeneic bone marrow-derived MSCs have been shown to have a differential effect on DC maturation after differentiation. Chondrogenic-differentiated rat bone marrow-derived MSCs induced maturation of human DCs; undifferentiated, osteogenic-, and adipogenic-differentiated MSCs had an inhibitory effect on human DC maturation [287]. While MSCs differentiate into a limited number cell types, pluripotent stem cells have the potential to differentiate into cells of any of the three germ layers. Novel strategies have been developed to produce pluripotent stem cells from somatic cells, introducing the possibility of new cell sources for tissue-engineered devices. Induced pluripotent stem cells were produced from adult fibroblasts by the introduction of transcription factors [288–291]. Recent studies have shown that primate pluripotent stem cells could be produced by somatic cell nuclear transfer, a result achieved previously only in murine cells [292].

In addition to therapies developed in transplantation biology, immune rejection in tissue engineering may be mitigated by addressing potential adjuvants incorporated into the device. Endotoxin contamination should be prevented or removed of from all device components, since it is a strong adjuvant through multiple PRR signaling pathways. In addition, biomaterial selection could also contribute to enhancing a potential adaptive immune response. Some biomaterials, such as hyaluronic acid, may actually minimize an immune response to associated antigens. By understanding the molecular aspects of the mechanisms by which DCs recognize and respond to biomaterials, the biomaterial design, biomaterial selection criteria, and device implantation route can be tailored to improve the tolerance of the biologic components of tissue-engineered devices.

Abbreviations

APC	Antigen presenting cell
CARD	Caspase-recruiting domain
CFA	Complete Freund's adjuvant
CLIP	Class II-associated peptides
CR1	Complement receptor 1
CR2	Complement receptor 2
CR3	Complement receptor 3
CR4	Complement receptor 4
CRD	Carbohydrate recognition domain
CTLA4	Cytotoxic T-lymphocyte antigen 4
DC	Dendritic cell
DC-SIGN	ICAM-3-grabbing nonintegrin
dsRNA	Double stranded RNA
DTH	Delayed-type hypersensitivity
EC	Endothelial cell
ECM	Extracellular matrix
EGF	Epidermal growth factor
ER	Endoplasmic reticulum
FBGC	Foreign body giant cell
hDC	Human peripheral blood monocyte-derived DCs
HLA	Human leukocyte antigen
HMGB1	High-mobility group box-1
HSP	Heat shock protein
ICAM	Intercellular adhesion molecule
IFN	Interferon
Ii	Invariant chain
IL	Interleukin
LDL	Low-density lipoproteins
LOX-1	Lectin-like oxidized LDL receptor 1
LPS	Lipopolysaccharide
LRR	Leucine-rich repeat
MARCO	Macrophage receptor with a collagenous structure
MCP-1	Monocyte chemoattractant protein-1
MDA5	Melanoma differentiation-associated gene 5
mDC	Murine bone marrow-derived DC
MHC	Major histocompatibility complex
MMP	Matrix metalloproteinase
MP	Microparticle
MSC	Mesenchymal stem cell
NACHT	Acronym for: neuronal apoptosis inhibitor protein (NAIP), MHC class II transactivator (CIITA), plant het gene product involved in vegetative incompatibility (HET-E), and telomerase-associated protein 1 (TP-1)
NALP	NACHT, LRR and pyrin domain
NF-κB	Nuclear factor-κB
NK	Natural killer
NLR	NOD-like receptor
NOD	Nucleotide-binding oligomerization domain
OVA	Ovalbumin
PDGF	Platelet-derived growth factor
PAMP	Pathogen-associated molecular pattern
PLGA	Poly(lactide co glycolide)
Poly I:C	Polyinosinic:polycytidylic acid
PRR	Pattern recognition receptor
PSGL-1	P-selectin glycoprotein ligand 1
RAGE	Receptor of advanced glycation end-products
RIG-I	Retinoic acid-inducible protein I
SR	Scavenger receptor
SREC	SR expressed by ECs
TAP	Transporter associated with antigen processing
TCR	T-cell receptor
TGF	Transforming growth factor
TH	T-helper
TLR	Toll-like receptor
TNF	Tumor necrosis factor
VCAM-1	Vascular cell-adhesion molecule 1
VEGF	Vascular endothelial growth factor

References

1. Langer, R, Vacanti, J.P. Tissue Engineering. Science 260, 920–926 (1993)
2. Lavik, E, Langer, R. Tissue engineering: current state and perspectives. Applied Microbiology and Biotechnology 65, 1–8 (2004)
3. Kong, H.J, Mooney, D.J. Microenvironmental regulation of biomacromolecular therapies. Nature Reviews Drug Discovery 6, 455–463 (2007)
4. Lysaght, M.J, Reyes, J. The growth of tissue engineering. Tissue Engineering 7, 485–493 (2001)
5. Mansbridge, J. Commercial considerations in tissue engineering. Journal of Anatomy 209, 527–532 (2006)

6. Seong, S.Y, Matzinger, P. Hydrophobicity: an ancient damage-associated molecular pattern that initiates innate immune responses. Nature Reviews Immunology 4, 469–478 (2004)
7. Hoffmann, J.A, Kafatos, F.C, Janeway, C.A, Ezekowitz, R.A.B. Phylogenetic perspectives in innate immunity. Science 284, 1313–1318 (1999)
8. Anderson, J.M. Biological responses to materials. Annual Review of Materials Research 31, 81–110 (2001)
9. Janeway, C.A, Medzhitov, R. Innate immune recognition. Annual Review of Immunology 20, 197–216 (2002)
10. Figdor, C.G, van Kooyk, Y, Adema, G.J. C-type lectin receptors on dendritic cells and Langerhans cells. Nature Reviews Immunology 2, 77–84 (2002)
11. van Kooyk, Y, Geijtenbeek, T.B.H. DC-sign: Escape mechanism for pathogens. Nature Reviews Immunology 3, 697–709 (2003)
12. Meylan, E, Tschopp, J, Karin, M. Intracellular pattern recognition receptors in the host response. Nature 442, 39–44 (2006)
13. Lee, M.S, Kim, Y.J. Pattern-recognition receptor signaling initiated from extracellular, membrane, and cytoplasmic space. Molecules and Cells 23, 1–10 (2007)
14. Aderem, A, Ulevitch, R.J. Toll-like receptors in the induction of the innate immune response. Nature 406, 782–787 (2000)
15. Medzhitov, R, Janeway, C.A. Innate immune recognition and control of adaptive immune responses. Seminars in Immunology 10, 351–353 (1998)
16. Chavakis, T, Bierhaus, A, Al Fakhri, N, Schneider, D, Witte, S, Linn, T, Nagashima, M, Morser, J, Arnold, B, Preissner, K.T, Nawroth, P.P. The pattern recognition receptor (RAGE) is a counterreceptor for leukocyte integrins: a novel pathway for inflammatory cell recruitment. Journal of Experimental Medicine 198, 1507–1515 (2003)
17. Peiser, L, Mukhopadhyay, S, Gordon, S. Scavenger receptors in innate immunity. Current Opinion in Immunology 14, 123–128 (2002)
18. McGreal, E.P, Miller, J.L, Gordon, S. Ligand recognition by antigen-presenting cell C-type lectin receptors. Current Opinion in Immunology 17, 18–24 (2005)
19. Geijtenbeek, T.B.H, Krooshoop, D.J.E.B, Bleijs, D.A, van Vliet, S.J, van Duijnhoven, G.C.F, Grabovsky, V, Alon, R, Figdor, C.G, van Kooyk, Y. DC-SIGN-ICAM-2 interaction mediates dendritic cell trafficking. Nature Immunology 1, 353–357 (2000)
20. Geijtenbeek, T.B.H, Torensma, R, van Vliet, S.J, van Duijnhoven, G.C.F, Adema, G.J, van Kooyk, Y, Figdor, C.G. Identification of DC-SIGN, a novel dendritic cell-specific ICAM-3 receptor that supports primary immune responses. Cell 100, 575–585 (2000)
21. Engering, A, Geijtenbeek, T.B.H, van Kooyk, Y. Immune escape through C-type lectins on dendritic cells. Trends in Immunology 23, 480–485 (2002)
22. McGreal, E.P, Martinez-Pomares, L, Gordon, S. Divergent roles for C-type lectins expressed by cells of the innate immune system. Molecular Immunology 41, 1109–1121 (2004)
23. Weis, W.I, Taylor, M.E, Drickamer, K. The C-type lectin superfamily in the immune system. Immunological Reviews 163, 19–34 (1998)
24. Gordon, J. B-Cell Signaling Via the C-type Lectins Cd23 and Cd72. Immunology Today 15, 411–417 (1994)
25. He, B, Qiao, X.G, Klasse, P.J, Chiu, A, Chadburn, A, Knowles, D.M, Moore, J.P, Cerutti, A. HIV-1 envelope triggers polyclonal Ig class switch recombination through a CD40-independent mechanism involving BAFF and C-type lectin receptors. Journal of Immunology 176, 3931–3941 (2006)
26. Geijtenbeek, T.B.H, van Vliet, S.J, Engering, A, 't Hart, B.A, van Kooyk, Y. Self- and nonself-recognition by C-type lectins on dendritic cells. Annual Review of Immunology 22, 33–54 (2004)
27. Mitchell, D.A, Fadden, A.J, Drickamer, K. A novel mechanism of carbohydrate recognition by the C-type lectins DC-SIGN and DC-SIGNR—subunit organization and binding to multivalent ligands. Journal of Biological Chemistry 276, 28939–28945 (2001)
28. Guo, Y, Feinberg, H, Conroy, E, Mitchell, D.A, Alvarez, R, Blixt, O, Taylor, M.E, Weis, W.I, Drickamer, K. Structural basis for distinct ligand-binding and targeting properties of the receptors DC-sign and DC-signR. Nature Structural Molecular Biology 11, 591–598 (2004)
29. Brown, G.D. dectin-1: a signalling non-TLR pattern-recognition receptor. Nature Reviews Immunology 6, 33–43 (2006)
30. Gantner, B.N, Simmons, R.M, Canavera, S.J, Akira, S, Underhill, D.M. Collaborative induction of inflammatory responses by Dectin-1 and toll-like receptor 2. Journal of Experimental Medicine 197, 1107–1117 (2003)
31. Akira, S, Uematsu, S, Takeuchi, O. Pathogen recognition and innate immunity. Cell 124, 783–801 (2006)
32. Brown, G.D, Herre, J, Williams, D.L, Willment, J.A, Marshall, A.S.J, Gordon, S. Dectin-1 mediates the biological effects of beta-glucans. Journal of Experimental Medicine 197, 1119–1124 (2003)
33. van Vliet, S.J, van Liempt, E, Geijtenbeek, T.B.H, van Kooyk, Y. Differential, regulation of C-type lectin expression on tolerogenic dendritic cell subsets. Immunobiology 211, 577–585 (2006)
34. Smits, H.H, Engering, A, van der Kleij, D, de Jong, E.C, Schipper, K, van Capel, T.M.M, Zaat, B.A.J, Yazdanbakhsh, M, Wierenga, E.A, van Kooyk, Y, Kapsenberg, M.L. Selective probiotic bacteria induce IL-10-producing regulatory T cells in vitro by modulating dendritic cell function through dendritic cell-specific intercellular adhesion molecule 3-grabbing nonintegrin. Journal of Allergy and Clinical Immunology 115, 1260–1267 (2005)
35. Melchjorsen, J, Jensen, S.B, Malmgaard, L, Rasmussen, S.B, Weber, F, Bowie, A.G, Matikainen, S, Paludan, S.R. Activation of innate defense against a paramyxovirus is mediated by RIG-I and TLR7 and TLR8 in a cell-type-specific manner. Journal of Virology 79, 12944–12951 (2005)
36. Yoneyama, M, Kikuchi, M, Natsukawa, T, Shinobu, N, Imaizumi, T, Miyagishi, M, Taira, K, Akira, S, Fujita, T. The RNA helicase RIG–I has an essential function in double-stranded RNA-induced innate antiviral responses. Nature Immunology 5, 730–737 (2004)
37. May, M.J, Ghosh, S. Signal transduction through NF-kappa B. Immunology Today 19, 80–88 (1998)
38. Ghosh, S, May, M.J, Kopp, E.B. NF-kappa B and rel proteins: evolutionarily conserved mediators of immune

responses. Annual Review of Immunology 16, 225–260 (1998)
39. Medzhitov, R, Janeway, C.A. A human homologue of the Drosophila Toll protein signals activation of adaptive immunity. Nature 388, 394–397 (1997)
40. Jerala, R. Structural biology of the LPS recognition. International Journal of Medical Microbiology 297, 353–363 (2007)
41. Martinez, F.D. CD14, endotoxin, and asthma risk. Proceedings of the American Thoracic Society 4, 221–225 (2007)
42. Sacks, S.H, Zhou, W.D. Allograft rejection: effect of local synthesis of complement. Springer Seminars in Immunopathology 27, 332–344 (2005)
43. Gasque, P. Complement: a unique innate immune sensor for danger signals. Molecular Immunology 41, 1089–1098 (2004)
44. Ajees, A.A, Gunasekaran, K, Volanakis, J.E, Narayana, S.V.L, Kotwal, G.J, Murthy, H.M.K. The structure of complement C3b provides insights into complement activation and regulation. Nature 444, 221–225 (2006)
45. Russell, S. CD46: a complement regulator and pathogen receptor that mediates links between innate and acquired immune function. Tissue Antigens 64, 111–118 (2004)
46. Gelderman, K.A, Blok, V.T, Fleuren, G.J, Gorter, A. The inhibitory effect of CD46, CD55, and CD59 on complement activation after immunotherapeutic treatment of cervical carcinoma cells with monoclonal antibodies or bispecific monoclonal antibodies. Laboratory Investigation 82, 483–493 (2002)
47. Haas, P.J, van Strijp, J. Anaphylatoxins—Their role in bacterial infection and inflammation. Immunologic Research 37, 161–175 (2007)
48. Underhill, D.M, Ozinsky, A. Phagocytosis of microbes: complexity in action. Annual Review of Immunology 20, 825–852 (2002)
49. Wagner, C, Hansch, G.M. Receptors for complement C3 on T-lymphocytes: relics of evolution or functional molecules? Molecular Immunology 43, 22–30 (2006)
50. Janssen, B.J.C, Gros, P. Structural insights into the central complement component C3. Molecular Immunology 44, 3–10 (2007)
51. Chen, Z.B, Koralov, S.B, Kelsoe, G. Regulation of humoral immune responses by CD21/CD35. Immunological Reviews 176, 194–204 (2000)
52. Roozendaal, R, Carroll, M.C. Complement receptors CD21 and CD35 in humoral immunity. Immunological Reviews 219, 157–166 (2007)
53. Schymeinsky, J, Mocsai, A, Walzog, B. Neutrophil activation via beta(2) integrins (CD II/CD 18): Molecular mechanisms and clinical implications. Thrombosis and Haemostasis 98, 262–273 (2007)
54. Edelson, B.T, Stricker, T.P, Li, Z.Z, Dickeson, S.K, Shepherd, V.L, Santoro, S.A, Zutter, M.M. Novel collectin/C1q receptor mediates mast cell activation and innate immunity. Blood 107, 143–150 (2006)
55. Helmy, K.Y, Katschke, K.J, Gorgani, N.N, Elliott, J.M, Diehl, L, Scales, S.J, Ghilardi, N, Campagne, M.V. CRIg: A macrophage complement receptor required for phagocytosis of circulating pathogens. Cell 124, 915–927 (2006)
56. Zhang, G.L, Ghosh, S. Toll-like receptor-mediated NF-kappa B activation: a phylogenetically conserved paradigm in innate immunity. Journal of Clinical Investigation 107, 13–19 (2001)
57. Akira, S, Takeda, K. Toll-like receptor signalling. Nature Reviews Immunology 4, 499–511 (2004)
58. Fritz, J.H, Girardin, S.E. How Toll-like receptors and Nod-like receptors contribute to innate immunity in mammals. Journal of Endotoxin Research 11, 390–394 (2005)
59. Martinon, F, Tschopp, J. NLRs join TLRs as innate sensors of pathogens. Trends in Immunology 26, 447–454 (2005)
60. Herold, K, Moser, B, Chen, Y.L, Zeng, S, Yan, S.F, Ramasamy, R, Emond, J, Clynes, R, Schmidt, A.M. Receptor for advanced glycation end products (RAGE) in a dash to the rescue: inflammatory signals gone awry in the primal response to stress. Journal of Leukocyte Biology 82, 204–212 (2007)
61. Schmidt, A.M, Yan, S.D, Yan, S.F, Stern, D.M. The biology of the receptor for advanced glycation end products and its ligands. Biochimica et Biophysica Acta—Molecular Cell Research 1498, 99–111 (2000)
62. Orlova, V.V, Choi, E.Y, Xie, C.P, Chavakis, E, Bierhaus, A, Ihanus, E, Ballantyne, C.M, Gahmberg, C.G, Bianchi, M.E, Nawroth, P.P, Chavakis, T. A novel pathway of HMGB1-mediated inflammatory cell recruitment that requires Mac-1-integrin. EMBO Journal 26, 1129–1139 (2007)
63. Adachi, H. Tsujimoto, M. Endothelial scavenger receptors. Progress in Lipid Research 45, 379–404 (2006)
64. Murphy, J.E, Tedbury, P.R, Homer-Vanniasinkam, S, Walker, J.H, Ponnambalam, S. Biochemistry and cell biology of mammalian scavenger receptors. Atherosclerosis 182, 1–15 (2005)
65. Hsu, H.Y, Chiu, S.L, Wen, M.H, Chen, K.Y, Hua, K.F. Ligands of macrophage scavenger receptor induce cytokine expression via differential modulation of protein kinase signaling pathways. Journal of Biological Chemistry 276, 28719–28730 (2001)
66. Bianchi, M.E. DAMPs, PAMPs and alarmins: all we need to know about danger. Journal of Leukocyte Biology 81, 1–5 (2007)
67. Gallucci, S, Lolkema, M, Matzinger, P. Natural adjuvants: endogenous activators of dendritic cells. Nature Medicine 5, 1249–1255 (1999)
68. Bethke, K, Staib, F, Distler, M, Schmitt, U, Jonuleit, H, Enk, A.H, Galle, P.R, Heike, M. Different efficiency of heat shock proteins (HSP) to activate human monocytes and dendritic cells: superiority of HSP60. Journal of Immunology 169, 6141–6148 (2002)
69. Calderwood, S.K, Mambula, S.S, Gray, P.J, Therlault, J.R. Extracellular heat shock proteins in cell signaling. FEBS Letters 581, 3689–3694 (2007)
70. Kol, A, Bourcier, T, Lichtman, A.H, Libby, P. Chlamydial and human heat shock protein 60s activate human vascular endothelium, smooth muscle cells, and macrophages. Journal of Clinical Investigation 103, 571–577 (1999)
71. Asea, A, Kraeft, S.K, Kurt-Jones, E.A, Stevenson, M.A, Chen, L.B, Finberg, R.W, Koo, G.C, Calderwood, S.K. HSP70 stimulates cytokine production through a CD14-dependant pathway, demonstrating its dual role as a

chaperone and cytokine. Nature Medicine 6, 435–442 (2000)
72. Chen, W, Syldath, U, Bellmann, K, Burkart, V, Kolb, W. Human 60-kDa heat-shock protein: A danger signal to the innate immune system. Journal of Immunology 162, 3212–3219 (1999)
73. Basu, S, Binder, R.J, Suto, R, Anderson, K.M, Srivastava, P.K. Necrotic but not apoptotic cell death releases heat shock proteins, which deliver a partial maturation signal to dendritic cells and activate the NF-kappa B pathway. International Immunology 12, 1539–1546 (2000)
74. De Nardo, D, Masendycz, P, Ho, S, Cross, M, Fleetwood, A.J, Reynolds, E.C, Hamilton, J.A, Scholz, G.M. A central role for the Hsp90(.)Cdc37 molecular chaperone module in interleukin-1 receptor-associated-kinase-dependent signaling by toll-like receptors. Journal of Biological Chemistry 280, 9813–9822 (2005)
75. Bausinger, H, Lipsker, D, Ziylan, U, Manie, S, Briand, J.P, Cazenave, J.P, Muller, S, Haeuw, J.F, Ravanat, C, de la Salle, H, Hanau, D. Endotoxin-free heat-shock protein 70 fails to induce APC activation. European Journal of Immunology 32, 3708–3713 (2002)
76. Gao, B.C, Tsan, M.F. Endotoxin contamination in recombinant human heat shock protein 70 (Hsp70) preparation is responsible for the induction of tumor necrosis factor a release by murine macrophages. Journal of Biological Chemistry 278, 174–179 (2003)
77. Gao, B.C, Tsan, M.F. Recombinant human heat shock protein 60 does not induce the release of tumor necrosis factor alpha from murine macrophages. Journal of Biological Chemistry 278, 22523–22529 (2003)
78. Tsan, M.F, Gao, B.C. Endogenous ligands of toll-like receptors. Journal of Leukocyte Biology 76, 514–519 (2004)
79. Park, J.S, Svetkauskaite, D, He, Q.B, Kim, J.Y, Strassheim, D, Ishizaka, A, Abraham, E. Involvement of toll-like receptors 2 and 4 in cellular activation by high mobility group box 1 protein. Journal of Biological Chemistry 279, 7370–7377 (2004)
80. Park, J.S, Gamboni-Robertson, F, He, Q.B, Svetkauskaite, D, Kim, J.Y, Strassheim, D, Sohn, J.W, Yamada, S, Maruyama, I, Banerjee, A, Ishizaka, A, Abraham, E. High mobility group box 1 protein interacts with multiple Toll-like receptors. American Journal of Physiology—Cell Physiology 290, C917–C924 (2006)
81. Termeer, C.C, Hennies, J, Voith, U, Ahrens, T, Weiss, J.M, Prehm, P, Simon, J.C. Oligosaccharides of hyaluronan are potent activators of dendritic cells. Journal of Immunology 165, 1863–1870 (2000)
82. Termeer, C, Benedix, F, Sleeman, J, Fieber, C, Voith, U, Ahrens, T, Miyake, K, Freudenberg, M, Galanos, C, Simon, J.C. Oligosaccharides of hyaluronan activate dendritic cells via toll-like receptor 4. Journal of Experimental Medicine 195, 99–111 (2002)
83. Scheibner, K.A, Lutz, M.A, Boodoo, S, Fenton, M.J, Powell, J.D, Horton, M.R. Hyaluronan fragments act as an endogenous danger signal by engaging TLR2. Journal of Immunology 177, 1272–1281 (2006)
84. Smiley, S.T, King, J.A, Hancock, W.W. Fibrinogen stimulates macrophage chemokine secretion through toll-like receptor 4. Journal of Immunology 167, 2887–2894 (2001)
85. Okamura, Y, Watari, M, Jerud, E.S, Young, D.W, Ishizaka, S.T, Rose, J, Chow, J.C, Strauss, J.F. The extra domain A of fibronectin activates toll-like receptor 4. Journal of Biological Chemistry 276, 10229–10233 (2001)
86. Johnson, G.B, Brunn, G.J, Kodaira, Y, Platt, J.L. Receptor-mediated monitoring of tissue well–being via detection of soluble heparan sulfate by toll-like receptor 4. Journal of Immunology 168, 5233–5239 (2002)
87. Li, M, Carpio, D.F, Zheng, Y, Bruzzo, P, Singh, V, Ouaaz, F, Medzhitov, R.M, Beg, A.A. An essential role of the NF-kappa B/Toll-like receptor pathway in induction of inflammatory and tissue-repair gene expression by necrotic cells. Journal of Immunology 166, 7128–7135 (2001)
88. Vorup-Jensen, T, Carman, C.V, Shimaoka, M, Schuck, P, Svitel, J, Springer, T.A. Exposure of acidic residues as a danger signal for recognition of fibrinogen and other macromolecules by integrin alpha x beta(2). Proceedings of the National Academy of Sciences of the United States of America 102, 1614–1619 (2005)
89. Martinon, F, Petrilli, V, Mayor, A, Tardivel, A, Tschopp, J. Gout-associated uric acid crystals activate the NALP3 inflammasome. Nature 440, 237–241 (2006)
90. Mariathasan, S, Weiss, D.S, Newton, K, McBride, J, O'Rourke, K, Roose-Girma, M, Lee, W.P, Weinrauch, Y, Monack, D.M, Dixit, V.M. Cryopyrin activates the inflammasome in response to toxins and ATP. Nature 440, 228–232 (2006)
91. Llull, R. Immune considerations in tissue engineering. Clinics in Plastic Surgery 26, 549–568 (1999)
92. Horbett, T.A. Principles underlying the role of adsorbed plasma-proteins in blood interactions with foreign materials. Cardiovascular Pathology 2, S137–S148 (1993)
93. Vroman, L, Adams, A.L, Fischer, G.C, Munoz, P.C. Interaction of high molecular-weight kininogen, factor-XII, and fibrinogen in plasma at interfaces. Blood 55, 156–159 (1980)
94. Wilson, C.J, Clegg, R.E, Leavesley, D.I, Pearcy, M.J. Mediation of biomaterial-cell interactions by adsorbed proteins: a review. Tissue Engineering 11, 1–18 (2005)
95. Zdolsek, J, Eaton, J.W, Tang, L. Histamine release and fibrinogen adsorption mediate acute inflammatory responses to biomaterial implants in humans. Journal of Translational Medicine 5, 31 (2007)
96. Tang, L.P, Eaton, J.W. Natural responses to unnatural materials: A molecular mechanism for foreign body reactions. Molecular Medicine 5, 351–358 (1999)
97. Gorbet, M.B, Sefton, M.V. Biomaterial-associated thrombosis: roles of coagulation factors, complement, platelets and leukocytes. Biomaterials 25, 5681–5703 (2004)
98. McNally, A.K, Anderson, J.M. Complement C3 participation in monocyte adhesion to different surfaces. Proceedings of the National Academy of Sciences of the United States of America 91, 10119–10123 (1994)
99. Cheung, A.K, Hohnholt, M, Gilson, J. Adherence of neutrophils to hemodialysis membranes—role of complement receptors. Kidney International 40, 1123–1133 (1991)

100. Cheung, A.K, Parker, C.J, Janatova, J. Analysis of the complement C-3 fragments associated with hemodialysis membranes. Kidney International 35, 576–588 (1989)
101. Tengvall, P, Askendal, A, Lundstrom, I. Complement activation by IgG immobilized on methylated silicon. Journal of Biomedical Materials Research 31, 305–312 (1996)
102. Wettero, J, Askendal, A, Tengvall, P, Bengtsson, T. Interactions between surface-bound actin and complement, platelets, and neutrophils. Journal of Biomedical Materials Research Part A 66A, 162–175 (2003)
103. Gemmell, C.H. Platelet adhesion onto artificial surfaces: Inhibition by benzamidine, pentamidine, and pyridoxal-5-phosphate as demonstrated by flow cytometric quantification of platelet adhesion to microspheres. Journal of Laboratory and Clinical Medicine 131, 84–92 (1998)
104. Nimeri, G, Ohman, L, Elwing, H, Wettero, J, Bengtsson, T. The influence of plasma proteins and platelets on oxygen radical production and F-actin distribution in neutrophils adhering to polymer surfaces. Biomaterials 23, 1785–1795 (2002)
105. Wettero, J, Tengvall, P, Bengtsson, T. Platelets stimulated by IgG-coated surfaces bind and activate neutrophils through a selectin-dependent pathway. Biomaterials 24, 1559–1573 (2003)
106. Gorbet, M.B, Sefton, M.V. Material-induced tissue factor expression but not CD11b upregulation depends on the presence of platelets. Journal of Biomedical Materials Research Part A 67A, 792–800 (2003)
107. Tang, L.P, Eaton, J.W. Fibrin(ogen) mediates acute inflammatory responses to biomaterials. Journal of Experimental Medicine 178, 2147–2156 (1993)
108. Tang, L.P, Ugarova, T.P, Plow, E.F, Eaton, J.W. Molecular determinants of acute inflammatory responses to biomaterials. Journal of Clinical Investigation 97, 1329–1334 (1996)
109. Darrigo, C, Candalcouto, J.J, Greer, M, Veale, D.J, Woof, J.M. Human neutrophil Fc receptor-mediated adhesion under flow—a hollow-fiber model of intravascular arrest. Clinical and Experimental Immunology 100, 173–179 (1995)
110. Katz, D.A, Haimovich, B, Greco, R.S. Fc-Gamma-RII, Fc-Gamma-RIII, and CD18 receptors mediate in part neutrophil activation on a plasma coated expanded polytetrafluoroethylene surface. Surgery 118, 154–161 (1995)
111. Skilbeck, C.A, Lu, X.M, Sheikh, S, Savage, C.O.S, Nash, G.B. Capture of flowing human neutrophils by immobilised immunoglobulin: roles of Fc-receptors CD16 and CD32. Cellular Immunology 241, 26–31 (2006)
112. Coxon, A, Cullere, X, Knight, S, Sethi, S, Wakelin, M.W, Stavrakis, G, Luscinskas, F.W, Mayadas, T.N. Fc gamma RIII mediates neutrophil recruitment to immune complexes: a mechanism for neutrophil accumulation in immune-mediated inflammation. Immunity 14, 693–704 (2001)
113. Jenney, C.R, Anderson, J.M. Adsorbed IgG: A potent adhesive substrate for human macrophages. Journal of Biomedical Materials Research 50, 281–290 (2000)
114. Kato, S, Akagi, T, Sugimura, K, Kishida, A, Akashi, M. Evaluation of biological responses to polymeric biomaterials by RT-PCR analysis III: Study of HSP 70; 90 and 47 mRNA expression. Biomaterials 19, 821–827 (1998)
115. Dutta, D, Sundaram, S.K, Teeguarden, J.G, Riley, B.J, Fifield, L.S, Jacobs, J.M, Addleman, S.R, Kaysen, G.A, Moudgil, B.M, Weber, T.J. Adsorbed proteins influence the biological activity and molecular targeting of nanomaterials. Toxicological Sciences 100, 303–315 (2007)
116. Furumoto, K, Nagayama, S, Ogawara, K, Takakura, Y, Hashida, M, Higaki, K, Kimura, T. Hepatic uptake of negatively charged particles in rats: possible involvement of serum proteins in recognition by scavenger receptor. Journal of Controlled Release 97, 133–141 (2004)
117. Broughton, G, Janis, J.E, Attinger, C.E. Wound healing: an overview. Plastic and Reconstructive Surgery 117, 294S (2006)
118. Witte, M.B, Barbul, A. General principles of wound healing. Surgical Clinics of North America 77, 509–528 (1997)
119. Martin, P. Wound healing—aiming for perfect skin regeneration. Science 276, 75–81 (1997)
120. Henry, G, Garner, W.L. Inflammatory mediators in wound healing. Surgical Clinics of North America 83, 483–507 (2003)
121. Ayala, A, Chung, C.S, Grutkoski, P.S, Song, G.Y. Mechanisms of immune resolution. Critical Care Medicine 31, S558–S571 (2003)
122. Broughton, G, Janis, J.E, Attinger, C.E. The basic science of wound healing. Plastic and Reconstructive Surgery 117, 12S–34S (2006)
123. von Hundelshausen, P. Weber, C. Platelets as immune cells—Bridging inflammation and cardiovascular disease. Circulation Research 100, 27–40 (2007)
124. Tang, L.P, Jennings, T.A, Eaton, J.W. Mast cells mediate acute inflammatory responses to implanted biomaterials. Proceedings of the National Academy of Sciences of the United States of America 95, 8841–8846 (1998)
125. Rao, R.M, Yang, L, Garcia-Cardena, G, Luscinskas, F.W. Endothelial-dependent mechanisms of leukocyte recruitment to the vascular wall. Circulation Research 101, 234–247 (2007)
126. Muller, W.A. Leukocyte-endothelial-cell interactions in leukocyte transmigration and the inflammatory response. Trends in Immunology 24, 327–334 (2003)
127. Garrood, T, Lee, L, Pitzalis, C. Molecular mechanisms of cell recruitment to inflammatory sites: general and tissue-specific pathways. Rheumatology 45, 250–260 (2006)
128. Ley, K, Laudanna, C, Cybulsky, M.I, Nourshargh, S. Getting to the site of inflammation: the leukocyte adhesion cascade updated. Nature Reviews Immunology 7, 678–689 (2007)
129. Coelho, A.L, Hogaboam, C.M, Kunkel, S.L. Chemokines provide the sustained inflammatory bridge between innate and acquired immunity. Cytokine Growth Factor Reviews 16, 553–560 (2005)
130. Tu, L.L, Chen, A.J, Delahunty, M.D, Moore, K.L, Watson, S.R, McEver, R.P, Tedder, T.F. L-selectin binds to P-selectin glycoprotein ligand-1 on leukocytes—interactions between the lectin, epidermal growth factor, and consensus repeat domains of the selectins determine ligand binding specificity. Journal of Immunology 157, 3995–4004 (1996)

131. Eriksson, E.E, Xie, X, Werr, J, Thoren, P, Lindbom, L. Importance of primary capture and L-selectin-dependent secondary capture in leukocyte accumulation in inflammation and atherosclerosis in vivo. Journal of Experimental Medicine 194, 205–217 (2001)
132. Sperandio, M, Smith, M.L, Forlow, S.B, Olson, T.S, Xia, L.J, McEver, R.P, Ley, K. P-selectin glycoprotein ligand-1 mediates L-selectin-dependent leukocyte rolling in venules. Journal of Experimental Medicine 197, 1355–1363 (2003)
133. DeGrendele, H.C, Estess, P, Siegelman, M.H. Requirement for CD44 in activated T cell extravasation into an inflammatory site. Science 278, 672–675 (1997)
134. Berlin, C, Bargatze, R.F, Campbell, J.J, Vonandrian, U.H, Szabo, M.C, Hasslen, S.R, Nelson, R.D, Berg, E.L, Erlandsen, S.L, Butcher, E.C. Alpha-4 integrins mediate lymphocyte attachment and rolling under physiological flow. Cell 80, 413–422 (1995)
135. Chesnutt, B.C, Smith, D.F, Raffler, N.A, Smith, M.L, White, E.J, Ley, K. Induction of LFA-1-dependent neutrophil rolling on ICAM-1 by engagement of E-selectin. Microcirculation 13, 99–109 (2006)
136. Rose, D.M, Alon, R, Ginsberg, M.H. Integrin modulation and signaling in leukocyte adhesion and migration. Immunological Reviews 218, 126–134 (2007)
137. Ley, K. Arrest chemokines. Microcirculation 10, 289–295 (2003)
138. Vestweber, D. Adhesion and signaling molecules controlling the transmigration of leukocytes through endothelium. Immunological Reviews 218, 178–196 (2007)
139. Petri, B, Bixel, M.G. Molecular events during leukocyte diapedesis. Febs Journal 273, 4399–4407 (2006)
140. Barreiro, O, Yanez-Mo, M, Serrador, J.M, Montoya, M.C, Vicente-Manzanares, M, Tejedor, R, Furthmayr, H, Sanchez-Madrid, F. Dynamic interaction of VCAM-1 and ICAM-1 with moesin and ezrin in a novel endothelial docking structure for adherent leukocytes. Journal of Cell Biology 157, 1233–1245 (2002)
141. Shaw, S.K, Ma, S, Kim, M.B, Rao, R.M, Hartman, C.U, Froio, R.M, Yang, L, Jones, T, Liu, Y, Nusrat, A, Parkos, C.A, Luscinskas, F.W. Coordinated redistribution of leukocyte LFA-1 and endothelial cell ICAM-1 accompany neutrophil transmigration. Journal of Experimental Medicine 200, 1571–1580 (2004)
142. Carman, C.V, Springer, T.A. A transmigratory cup in leukocyte diapedesis both through individual vascular endothelial cells and between them. Journal of Cell Biology 167, 377–388 (2004)
143. Molteni, R, Fabbri, M, Bender, J.R, Pardi, R. Pathophysiology of leukocyte-tissue interactions. Current Opinion in Cell Biology 18, 491–498 (2006)
144. Luster, A.D, Alon, R, von Andrian, U.H. Immune cell migration in inflammation: present and future therapeutic targets. Nature Immunology 6, 1182–1190 (2005)
145. Bonfield, T.L, Colton, E, Marchant, R.E, Anderson, J.M. Cytokine and growth factor production by monocytes macrophages on protein preadsorbed polymers. Journal of Biomedical Materials Research 26, 837–850 (1992)
146. Erfle, D.J, Santerre, J.P, Labow, R.S. Lysosomal enzyme release from human neutrophils adherent to foreign material surfaces: enhanced release of elastase activity. Cardiovascular Pathology 6, 333–340 (1997)
147. Gorbet, M.B, Sefton, M.V. Leukocyte activation and leukocyte procoagulant activities after blood contact with polystyrene and polyethylene glycol-immobilized polystyrene beads. Journal of Laboratory and Clinical Medicine 137, 345–355 (2001)
148. Aderem, A, Underhill, D.M. Mechanisms of phagocytosis in macrophages. Annual Review of Immunology 17, 593–623 (1999)
149. Falck, P. Characterization of human neutrophils adherent to organic polymers. Biomaterials 16, 61–66 (1995)
150. Kaplan, S.S, Basford, R.E, Jeong, M.H, Simmons, R.L. Biomaterial-neutrophil interactions: Dysregulation of oxidative functions of fresh neutrophils induced by prior neutrophil-biomaterial interaction. Journal of Biomedical Materials Research 30, 67–75 (1996)
151. Lodoen, M.B, Lanier, L.L. Natural killer cells as an initial defense against pathogens. Current Opinion in Immunology 18, 391–398 (2006)
152. Midwood, K.S, Mao, Y, Hsia, H.C, Valenick, L.V, Schwarzbauer, J.E. Modulation of cell-fibronectin matrix interactions during tissue repair. Journal of Investigative Dermatology Symposium Proceedings 11, 73–78 (2006)
153. Eming, S.A, Krieg, T, Davidson, J.M. Inflammation in wound repair: molecular and cellular mechanisms. Journal of Investigative Dermatology 127, 514–525 (2007)
154. Schaffer, M, Barbul, A. Lymphocyte function in wound healing and following injury. British Journal of Surgery 85, 444–460 (1998)
155. Lumelsky, N.L. Commentary: engineering of tissue healing and regeneration. Tissue Engineering 13, 1393–1398 (2007)
156. Duffield, J.S. The inflammatory macrophage: a story of Jekyll and Hyde. Clinical Science 104, 27–38 (2003)
157. Van Ginderachter, J.A, Movahedi, K, Ghassabeh, G.H, Meerschaut, S, Beschin, A, Raes, G, De Baetselier, P. Classical and alternative activation of mononuclear phagocytes: picking the best of both worlds for tumor promotion. Immunobiology 211, 487–501 (2006)
158. Ma, J, Chen, T, Mandelin, J, Ceponis, A, Miller, N.E, Hukkanen, M, Ma, G.F, Konttinen, Y.T. Regulation of macrophage activation. Cellular and Molecular Life Sciences 60, 2334–2346 (2003)
159. Goerdt, S, Politz, O, Schledzewski, K, Birk, R, Gratchev, A, Guillot, P, Hakiy, N, Klemke, C.D, Dippel, E, Kodelja, V, Orfanos, C.E. Alternative versus classical activation of macrophages. Pathobiology 67, 222–226 (1999)
160. Lindblad, W.J. Wound healing, regenerative medicine and tissue engineering: a continuum. Wound Repair and Regeneration 10, 345 (2002)
161. Jenney, C.R, Anderson, J.M. Adsorbed serum proteins responsible for surface dependent human macrophage behavior. Journal of Biomedical Materials Research 49, 435–447 (1999)
162. Ahsan, F.L, Rivas, I.P, Khan, M.A, Suarez, A.I.T. Targeting to macrophages: role of physicochemical properties of particulate carriers-liposomes and microspheres-on the phagocytosis by macrophages. Journal of Controlled Release 79, 29–40 (2002)

163. Prior, S, Gander, B, Blarer, N, Merkle, H.P, Subira, M.L, Irache, J.M, Gamazo, C. In vitro phagocytosis and monocyte-macrophage activation with poly(lactide) and poly(lactide-co-glycolide) microspheres. European Journal of Pharmaceutical Sciences 15, 197–207 (2002)
164. Owens, D.E, Peppas, N.A. Opsonization, biodistribution, and pharmacokinetics of polymeric nanoparticles. International Journal of Pharmaceutics 307, 93–102 (2006)
165. Hirota, K, Hasegawa, T, Hinata, H, Ito, F, Inagawa, H, Kochi, C, Soma, G.I, Makino, K, Terada, H. Optimum conditions for efficient phagocytosis of rifampicin-loaded PLGA microspheres by alveolar macrophages. Journal of Controlled Release 119, 69–76 (2007)
166. Amstutz, H.C, Campbell, P, Kossovsky, N, Clarke, I.C. Mechanism and clinical significance of wear debris-induced osteolysis. Clinical Orthopaedics and Related Research 7–18 (1992)
167. Tabata, Y, Ikada, Y. Effect of the size and surface-charge of polymer microspheres on their phagocytosis by macrophage. Biomaterials 9, 356–362 (1988)
168. Rudt, S, Muller, R.H. In vitro phagocytosis assay of nanoparticles and microparticles by chemiluminescence. 1. Effect of analytical parameters, particle-size and particle concentration. Journal of Controlled Release 22, 263–271 (1992)
169. Gonzalez, O, Smith, R.L, Goodman, S.B. Effect of size, concentration, surface area, and volume of polymethylmethacrylate particles on human macrophages in vitro. Journal of Biomedical Materials Research 30, 463–473 (1996)
170. Hamilton, J.A. Nondisposable materials, chronic inflammation, and adjuvant action. Journal of Leukocyte Biology 73, 702–712 (2003)
171. McNally, A.K, Anderson, J.M. Interleukin-4 induces foreign body giant cells from human monocytes macrophages—differential lymphokine regulation of macrophage fusion leads to morphological variants of multinucleated giant cells. American Journal of Pathology 147, 1487–1499 (1995)
172. Defife, K.M, Jenney, C.R, McNally, A.K, Colton, E, Anderson, J.M. Interleukin-13 induces human monocyte/macrophage fusion and macrophage mannose receptor expression. Journal of Immunology 158, 3385–3390 (1997)
173. McNally, A.K, Defife, K.M, Anderson, J.M. Interleukin-4-induced macrophage fusion is prevented by inhibitors of mannose receptor activity. American Journal of Pathology 149, 975–985 (1996)
174. Brodbeck, W.G, Shive, M.S, Colton, E, Ziats, N.P, Anderson, J.M. Interleukin-4 inhibits tumor necrosis factor-alpha-induced and spontaneous apoptosis of biomaterial-adherent macrophages. Journal of Laboratory and Clinical Medicine 139, 90–100 (2002)
175. McNally, A.K, Anderson, J.M. beta 1 and beta 2 integrins mediate adhesion during macrophage fusion and multinucleated foreign body giant cell formation. American Journal of Pathology 160, 621–630 (2002)
176. Zhao, Q, Topham, N, Anderson, J.M, Hiltner, A, Lodoen, G, Payet, C.R. Foreign body giant cells and polyurethane biostability: in vivo correlation of cell adhesion and surface cracking. Journal of Biomedical Materials Research 25, 177–183 (1991)
177. Badylak, S. F, Gilbert, T. W. Immune response to biologic scaffold materials. Seminars in Immunology 20, 109–116 (2008)
178. Sieminski, A.L, Gooch, K.J. Biomaterial-microvasculature interactions. Biomaterials 21, 2233–2241 (2000)
179. Castner, D.G, Ratner, B.D. Biomedical surface science: foundations to frontiers. Surface Science 500, 28–60 (2002)
180. Ratner, B.D. Reducing capsular thickness and enhancing angiogenesis around implant drug release systems. Journal of Controlled Release 78, 211–218 (2002)
181. Marshall, A.J, Irvin, C.A, Barker, T, Sage, E.H, Hauch, K.D, Ratner, B.D. Biomaterials with tightly controlled pore size that promote vascular in-growth. Polymer Preprints 45, 100–101 (2004)
182. Clark, L.D, Clark, R.K, Heber-Katz, E. A new murine model for mammalian wound repair and regeneration. Clinical Immunology and Immunopathology 88, 35–45 (1998)
183. Leferovich, J.M, Bedelbaeva, K, Samulewicz, S, Zhang, X.M, Zwas, D, Lankford, E.B, Heber-katz, E. Heart regeneration in adult MRL mice. Proceedings of the National Academy of Sciences of the United States of America 98, 9830–9835 (2001)
184. Harty, M, Neff, A W, King, M.W, Mescher, A.L. Regeneration or scarring: an immunologic perspective. Developmental Dynamics 226, 268–279 (2003)
185. Ward, W.K, Slobodzian, E.P, Tiekotter, K.L, Wood, M.D. The effect of microgeometry, implant thickness and polyurethane chemistry on the foreign body response to subcutaneous implants. Biomaterials 23, 4185–4192 (2002)
186. Picha, G.J, Drake, R.F. Pillared-surface microstructure and soft-tissue implants: effect of implant site and fixation. Journal of Biomedical Materials Research 30, 305–312 (1996)
187. Brauker, J.H, Carr-Brendel, V.E, Martinson, L.A, Crudele, J, Johnston, W.D, Johnson, R.C. Neovascularization of synthetic membranes directed by membrane microarchitecture. Journal of Biomedical Materials Research 29, 1517–1524 (1995)
188. Sharkawy, A.A, Klitzman, B, Truskey, G.A, Reichert, W.M. Engineering the tissue which encapsulates subcutaneous implants. 1. Diffusion properties. Journal of Biomedical Materials Research 37, 401–412 (1997)
189. Sanders, J.E, Baker, A.B, Golledge, S.L. Control of in vivo microvessel ingrowth by modulation of biomaterial local architecture and chemistry. Journal of Biomedical Materials Research 60, 36–43 (2002)
190. Sanders, J.E, Stiles, C.E, Hayes, C.L. Tissue response to single-polymer fibers of varying diameters: Evaluation of fibrous encapsulation and macrophage density. Journal of Biomedical Materials Research 52, 231–237 (2000)
191. Elcin, Y.M, Dixit, V, Gitnick, T. Extensive in vivo angiogenesis following controlled release of human vascular endothelial cell growth factor: implications for tissue engineering and wound healing. Artificial Organs 25, 558–565 (2001)

192. Tabata, Y, Miyao, M, Ozeki, M, Ikada, Y. Controlled release of vascular endothelial growth factor by use of collagen hydrogels. Journal of Biomaterials Science—Polymer Edition 11, 915–930 (2000)
193. Ravin, A.G, Olbrich, K.C, Levin, L.S, Usala, A.L, Klitzman, B. Long- and short-term effects of biological hydrogels on capsule microvascular density around implants in rats. Journal of Biomedical Materials Research 58, 313–318 (2001)
194. Zisch, A.H, Lutolf, M.P, Ehrbar, M, Raeber, G.P, Rizzi, S.C, Davies, N, Schmokel, H, Bezuidenhout, D, Djonov, V, Zilla, P, Hubbell, J.A. Cell-demanded release of VEGF from synthetic, biointeractive cell-ingrowth matrices for vascularized tissue growth. FASEB Journal 17, 2260–2262 (2003)
195. Prokop, A, Kozlov, E, Nun, N.S, Dikov, M.M, Sephel, G.C, Whitsitt, J.S, Davidson, J.M. Towards retrievable vascularized bioartificial pancreas: induction and long-lasting stability of polymeric mesh implant vascularized with the help of acidic and basic fibroblast growth factors and hydrogel coating. Diabetes Technology Therapeutics. 3, 245–261 (2001)
196. Lee, K.Y, Peters, M.C, Anderson, K.W, Mooney, D.J. Controlled growth factor release from synthetic extracellular matrices. Nature 408, 998–1000 (2000)
197. Fujita, M, Ishihara, M, Simizu, M, Obara, K, Ishizuka, T, Saito, Y, Yura, H, Morimoto, Y, Takase, B, Matsui, T, Kikuchi, M, Maehara, T. Vascularization in vivo caused by the controlled release of fibroblast growth factor-2 from an injectable chitosan/non-anticoagulant heparin hydrogel. Biomaterials 25, 699–706 (2004)
198. Pieper, J.S, Hafmans, T, van Wachem, P.B, van Luyn, M.J.A, Brouwer, L.A, Veerkamp, J.H, van Kuppevelt, T.H. Loading of collagen-heparan sulfate matrices with bFGF promotes angiogenesis and tissue generation in rats. Journal of Biomedical Materials Research 62, 185–194 (2002)
199. Perets, A, Baruch, Y, Weisbuch, F, Shoshany, G, Neufeld, G, Cohen, S. Enhancing the vascularization of three-dimensional porous alginate scaffolds by incorporating controlled release basic fibroblast growth factor microspheres. Journal of Biomedical Materials Research Part A 65A, 489–497 (2003)
200. Norton, L.W, Koschwanez, H.E, Wisniewski, N.A, Klitzman, B, Reichert, W.M. Vascular endothelial growth factor and dexamethasone release from nonfouling sensor coatings affect the foreign body response. Journal of Biomedical Materials Research Part A 81A, 858–869 (2007)
201. Patil, S.D, Papadmitrakopoulos, F, Burgess, D.J. Concurrent delivery of dexamethasone and VEGF for localized inflammation control and angiogenesis. Journal of Controlled Release 117, 68–79 (2007)
202. Kyriakides, T.R, Hartzel, T, Huynh, G, Bornstein, P. Regulation of angiogenesis and matrix remodeling by localized, matrix-mediated antisense gene delivery. Molecular Therapy 3, 842–849 (2001)
203. Richardson, T.P, Peters, M.C, Ennett, A.B, Mooney, D.J. Polymeric system for dual growth factor delivery. Nature Biotechnology 19, 1029–1034 (2001)
204. Kim, K.D, Zhao, J, Auh, S, Yang, X.M, Du, P.S, Tang, H, Fu, Y.X. Adaptive immune cells temper initial innate responses. Nature Medicine 13, 1248–1252 (2007)
205. Rock, K.L, Shen, L. Cross-presentation: underlying mechanisms and role in immune surveillance. Immunological Reviews 207, 166–183 (2005)
206. Busch, R, Rinderknecht, C.H, Roh, S, Lee, A.W, Harding, J.J, Burster, T, Hornell, T.M.C, Mellins, E.D. Achieving stability through editing and chaperoning: regulation of MHC class II peptide binding and expression. Immunological Reviews 207, 242–260 (2005)
207. Collins, M, Ling, V, Carreno, B.M. The B7 family of immune-regulatory ligands. Genome Biology 6, 223 (2005)
208. Trivedi, H.L. Immunobiology of rejection and adaptation. Transplantation Proceedings 39, 647–652 (2007)
209. Matzinger, P. Tolerance, danger, and the extended family. Annual Review of Immunology 12, 991–1045 (1994)
210. Gallucci, S, Lolkema, M, Matzinger, P. Natural adjuvants: Endogenous activators of dendritic cells. Nature Medicine 5, 1249–1255 (1999)
211. Rock, K.L, Hearn, A, Chen, C.J, Shi, Y. Natural endogenous adjuvants. Springer Seminars in Immunopathology 26, 231–246 (2005)
212. Banchereau, J, Steinman, R.M. Dendritic cells and the control of immunity. Nature 392, 245–252 (1998)
213. Gunn, M.D. Chemokine mediated control of dendritic cell migration and function. Seminars in Immunology 15, 271–276 (2003)
214. Cella, M, Engering, A, Pinet, V, Pieters, J, Lanzavecchia, A. Inflammatory stimuli induce accumulation of MHC class II complexes on dendritic cells. Nature 388, 782–787 (1997)
215. Watts, C. Immunology—Inside the gearbox of the dendritic cell. Nature 388, 724–725 (1997)
216. Lechmann, M, Berchtold, S, Hauber, J, Steinkasserer, A. CD83 on dendritic cells: more than just a marker for maturation. Trends in Immunology 23, 273–275 (2002)
217. Tsuji, S, Matsumoto, M, Takeuchi, O, Akira, S, Azuma, I, Hayashi, A, Toyoshima, K, Seya, T. Maturation of human dendritic cells by cell wall skeleton of Mycobacterium bovis bacillus Calmette-Guerin: Involvement of Toll-like receptors. Infection and Immunity 68, 6883–6890 (2000)
218. Sallusto, F, Lanzavecchia, A. Efficient presentation of soluble-antigen by cultured human dendritic cells is maintained by granulocyte-macrophage colony-stimulating factor plus interleukin-4 and down-regulated by tumor-necrosis-factor-alpha. Journal of Experimental Medicine 179, 1109–1118 (1994)
219. Kapsenberg, M.L. Dendritic-cell control of pathogen-driven T-cell polarization. Nature Reviews Immunology 3, 984–993 (2003)
220. Babensee, J.E, Anderson, J.M, McIntire, L.V, Mikos, A.G. Host response to tissue engineered devices. Advanced Drug Delivery Reviews 33, 111–139 (1998)
221. Mason, C. Hoare, M. Regenerative medicine bioprocessing: building a conceptual framework based on early studies. Tissue Engineering 13, 301–311 (2007)
222. Zuk, P.A, Zhu, M, Mizuno, H, Huang, J, Futrell, J.W, Katz, A.J, Benhaim, P, Lorenz, H.P, Hedrick, M.H. Mul-

tilineage cells from human adipose tissue: Implications for cell-based therapies. Tissue Engineering 7, 211–228 (2001)
223. Guillot, P.V, O'Donoghue, K, Kurata, H, Fisk, N.M. Fetal stem cells: Betwixt and between. Seminars in Reproductive Medicine 24, 340–347 (2006)
224. Mao, J.J, Giannobile, W.V, Helms, J.A, Hollister, S.J, Krebsbach, P.H, Longaker, M.T, Shi, S. Craniofacial tissue engineering by stem cells. Journal of Dental Research 85, 966–979 (2006)
225. Pittenger, M.F, Mackay, A.M, Beck, S.C, Jaiswal, R.K, Douglas, R, Mosca, J.D, Moorman, M.A, Simonetti, D.W, Craig, S, Marshak, D.R. Multilineage potential of adult human mesenchymal stem cells. Science 284, 143–147 (1999)
226. Kuby, J. Immunology (W.H. Freeman and Company, New York, NY; 1997)
227. Hale, D.A. Basic transplantation immunology. Surgical Clinics of North America 86, 1103–1125 (2006)
228. Auchincloss, H, Sachs, D.H. Xenogeneic transplantation. Annual Review of Immunology 16, 433–470 (1998)
229. Larosa, D.F, Rahman, A.H, Turka, L.A. The innate immune system in allograft rejection and tolerance. Journal of Immunology 178, 7503–7509 (2007)
230. Kuttler, B, Hartmann, A, Wanka, H. Long-term culture of islets abrogates cytokine-induced or lymphocyte-induced increase of antigen expression on beta cells. Transplantation 74, 440–445 (2002)
231. Badylak, S.F. Xenogeneic extracellular matrix as a scaffold for tissue reconstruction. Transplant Immunology 12, 367–377 (2004)
232. Meyer, S.R, Nagendran, J, Desai, L.S, Rayat, G.R, Churchill, T.A, Anderson, C.C, Rajotte, R.V, Lakey, J.R.T, Ross, D.B. Decellularization reduces the immune response to aortic valve allografts in the rat. Journal of Thoracic and Cardiovascular Surgery 130, 469–476 (2005)
233. Allman, A.J, McPherson, T.B, Badylak, S.F, Merrill, L.C, Kallakury, B, Sheehan, C, Raeder, R.H, Metzger, D.W. Xenogeneic extracellular matrix grafts elicit a Th2-restricted immune response. Transplantation 71, 1631–1640 (2001)
234. Allman, A.J, McPherson, T.B, Merrill, L.C, Badylak, S.F, Metzger, D.W. The Th2–restricted immune response to xenogeneic small intestinal submucosa does not influence systemic protective immunity to viral and bacterial pathogens. Tissue Engineering 8, 53–62 (2002)
235. Meyer, T, Meyer, B, Schwarz, K, Hocht, B. Immune response to xenogeneic matrix grafts used in pediatric surgery. European Journal of Pediatric Surgery 17, 420–425 (2007)
236. Courtman, D.W, Errett, B.F, Wilson, G.J. The role of crosslinking in modification of the immune response elicited against xenogenic vascular acellular matrices. Journal of Biomedical Materials Research 55, 576–586 (2001)
237. Zheng, M.H, Chen, J, Kirilak, Y, Willers, C, Xu, J, Wood, D. Porcine small intestine submucosa (SIS) is not an acellular collagenous matrix and contains porcine DNA: Possible implications in human implantation. Journal of Biomedical Materials Research Part B—Applied Biomaterials 73B, 61–67 (2005)
238. Methe, H, Nugent, H.M, Groothuis, A, Seifert, P, Sayegh, M.H, Edelman, E.R. Matrix embedding alters the immune response against endothelial cells in vitro and in vivo. Circulation 112, I89–I95 (2005)
239. Methe, H, Edelman, E.R. Tissue engineering of endothelial cells and the immune response. Transplantation Proceedings 38, 3293–3299 (2006)
240. Methe, H, Hess, S, Edelman, E.R. Endothelial cell-matrix interactions determine maturation of dendritic cells. European Journal of Immunology 37, 1773–1784 (2007)
241. Methe, H, Groothuis, A, Sayegh, M.H, Edelman, E.R. Matrix adherence of endothelial cells attenuates immune reactivity: induction of hyporesponsiveness in allo- and xenogeneic models. FASEB Journal 21, 1515–1526 (2007)
242. Seferian, P.G, Martinez, M.L. Immune stimulating activity of two new chitosan containing adjuvant formulations. Vaccine 19, 661–668 (2000)
243. Hwang, S.M, Chen, C.Y, Chen, S.S, Chen, J.C. Chitinous materials inhibit nitric oxide production by activated RAW 264.7 macrophages. Biochemical and Biophysical Research Communications 271, 229–233 (2000)
244. Peluso, G, Petillo, O, Ranieri, M, Santin, M, Ambrosio, L, Calabro, D, Avallone, B, Balsamo, G. Chitosan-mediated Stimulation of Macrophage Function. Biomaterials 15, 1215–1220 (1994)
245. Yang, R, Yan, Z, Chen, F, Hansson, G.K, Kiessling, R. Hyaluronic acid and chondroitin sulphate A rapidly promote differentiation of immature DC with upregulation of costimulatory and antigen-presenting molecules, and enhancement of NF-kappa B and protein kinase activity. Scandinavian Journal of Immunology 55, 2–13 (2002)
246. Brand, U, Bellinghausen, I, Enk, A.H, Jonuleit, H, Becker, D, Knop, J, Saloga, J. Influence of extracellular matrix proteins on the development of cultured human dendritic cells. European Journal of Immunology 28, 1673–1680 (1998)
247. Kulseng, B, Skjak-Braek, G, Ryan, L, Andersson, A, King, A, Faxvaag, A, Espevik, T. Transplantation of alginate microcapsules—Generation of antibodies against alginates and encapsulated porcine islet-like cell clusters. Transplantation 67, 978–984 (1999)
248. Espevik, T, Otterlei, M, Skjakbraek, G, Ryan, L, Wright, S.D, Sundan, A. the Involvement of CD14 in stimulation of cytokine production by uronic-acid polymers. European Journal of Immunology 23, 255–261 (1993)
249. Flo, T.H, Ryan, L, Latz, E, Takeuchi, O, Monks, B.G, Lien, E, Halaas, O, Akira, S, Skjak-Braek, G, Golenbock, D.T, Espevik, T. Involvement of Toll-like receptor (TLR) 2 and TLR4 in cell activation by mannuronic acid polymers. Journal of Biological Chemistry 277, 35489–35495 (2002)
250. Starke, J.R, Edwards, M.S, Langston, C, Baker, C.J. A mouse model of chronic pulmonary infection with Pseudomonas aeruginosa and Pseudomonas cepacia. Pediatric Research 22, 698–702 (1987)
251. Rahfoth, B, Weisser, J, Sternkopf, F, Aigner, T, der Mark, K, Brauer, R. Transplantation of allograft chondrocytes

embedded in agarose gel into cartilage defects of rabbits. Osteoarthritis and Cartilage 6, 50–65 (1998)
252. Ertl, H.C.J, Varga, I, Xiang, Z.Q, Kaiser, K, Stephens, L.D, Otvos, L. Poly(DL-lactide-co-glycolide) microspheres as carriers for peptide vaccines. Vaccine 14, 879–885 (1996)
253. O'Hagan, D.T, Jeffery, H, Davis, S.S. Long–Term Antibody-responses in mice following subcutaneous immunization with ovalbumin entrapped in biodegradable microparticles. Vaccine 11, 965–969 (1993)
254. Yoshida, M, Babensee, J.E. Poly(lactic-co-glycolic acid) enhances maturation of human monocyte-derived dendritic cells. Journal of Biomedical Materials Research Part A 71A, 45–54 (2004)
255. Yoshida, M, Mata, J, Babensee, J.E. Effect of poly(lactic-co-glycolic acid) contact on maturation of murine bone marrow-derived dendritic cells. Journal of Biomedical Materials Research Part A 80A, 7–12 (2007)
256. Yoshida, M, Babensee, J.E. Molecular aspects of microparticle phagocytosis by dendritic cells. Journal of Biomaterials Science—Polymer Edition 17, 893–907 (2006)
257. Babensee, J.E, Paranjpe, A. Differential levels of dendritic cell maturation on different biomaterials used in combination products. Journal of Biomedical Materials Research Part A 74A, 503–510 (2005)
258. Yoshida, M, Babensee, J.E. Differential effects of agarose and poly(lactic-co-glycolic acid) on dendritic cell maturation. Journal of Biomedical Materials Research Part A 79A, 393–408 (2006)
259. Matzelle, M.M, Babensee, J.E. Humoral immune responses to model antigen co–delivered with biomaterials used in tissue engineering. Biomaterials 25, 295–304 (2004)
260. Bennewitz, N.L, Babensee, J.E. The effect of the physical form of poly(lactic-co-glycolic acid) carriers on the humoral immune response to co-delivered antigen. Biomaterials 26, 2991–2999 (2005)
261. Babensee, J.E, Stein, M.M, Moore, L.K. Interconnections between inflammatory and immune responses in tissue engineering. Annals of the New York Academy of Sciences 961, 360–363 (2002)
262. Vasilijic, S, Savic, D, Vasilev, S, Vucevic, D, Gasic, S, Majstorovic, I, Jankovic, S, Colic, M. Dendritic cells acquire tolerogenic properties at the site of sterile granulomatous inflammation. Cellular Immunology 233, 148–157 (2005)
263. Magalhaes, P.O, Lopes, A.M, Mazzola, P.G, Rangel-Yagui, C, Penna, T.C.V, Pessoa, A. Methods of endotoxin removal from biological preparations: a review. Journal of Pharmacy and Pharmaceutical Sciences 10, 388–404 (2007)
264. Babensee, J.E, McIntire, L.V, Mikos, A.G. Growth factor delivery for tissue engineering. Pharmaceutical Research 17, 497–504 (2000)
265. Weinstein, J.R, Swarts, S, Bishop, C, Hanisch, U.K, Moller, T. Lipopolysaccharide is a frequent and significant contaminant in microglia-activating factors. Glia 56, 16–26 (2007)
266. Lu, L, Gambotto, A, Lee, W.C, Qian, S, Bonham, C.A, Robbins, P.D, Thomson, A.W. Adenoviral delivery of CTLA4Ig into myeloid dendritic cells promotes their in vitro tolerogenicity and survival in allogeneic recipients. Gene Therapy 6, 554–563 (1999)
267. Bonham, C.A, Peng, L.S, Liang, X.Y, Chen, Z.Y, Wang, L.F, Ma, L.L, Hackstein, H, Robbins, P.D, Thomson, A.W, Fung, J.J, Qian, S.G, Lu, L. Marked prolongation of cardiac allograft survival by dendritic cells genetically engineered with NF-kappa B oligodeoxyribonucleotide decoys and adenoviral vectors encoding CTLA4-Ig. Journal of Immunology 169, 3382–3391 (2002)
268. O'Rourke, R.W, Kang, S.M, Lower, J.A, Feng, S, Ascher, N.L, Baekkeskov, S, Stock, P.G. A dendritic cell line genetically modified to express CTLA4-IG as a means to prolong islet allograft survival. Transplantation 69, 1440–1446 (2000)
269. Lee, W.C, Zhong, C, Qian, S, Wan, Y, Gauldie, J, Mi, Z, Robbins, P.D, Thomson, A.W, Lu, L. Phenotype, function, and in vivo migration and survival of allogeneic dendritic cell progenitors genetically engineered to express TGF-beta. Transplantation 66, 1810–1817 (1998)
270. Takayama, T, Nishioka, Y, Lu, L, Lotze, M.T, Tahara, H, Thomson, A.W. Retroviral delivery of viral interleukin-10 into myeloid dendritic cells markedly inhibits their allostimulatory activity and promotes the induction of T-cell hyporesponsiveness. Transplantation 66, 1567–1574 (1998)
271. Min, W.P, Gorczynski, R, Huang, X.Y, Kushida, M, Kim, P, Obataki, M, Lei, J, Suri, R.M, Cattral, M.S. Dendritic cells genetically engineered to express Fas ligand induce donor-specific hyporesponsiveness and prolong allograft survival. Journal of Immunology 164, 161–167 (2000)
272. Liu, Z.G, Xu, X, Hsu, H.C, Tousson, A, Yang, P.A, Wu, Q, Liu, C.R, Yu, S.H, Zhang, H.G, Mountz, J.D. CII-DC-AdTRAIL cell gene therapy inhibits infiltration of CII-reactive T cells and CII-induced. Journal of Clinical Investigation 112, 1332–1341 (2003)
273. Palmer, S.M, Burch, L.H, Davis, R.D, Herczyk, W.F, Howell, D.N, Reinsmoen, N.L, Schwartz, D.A. The role of innate immunity in acute allograft rejection after lung transplantation. American Journal of Respiratory and Critical Care Medicine 168, 628–632 (2003)
274. Palmer, S.M, Burch, L.H, Mir, S, Smith, S.R, Kuo, P.C, Herczyk, W.F, Reinsmoen, N.L, Schwartz, D.A. Donor polymorphisms in Toll-like receptor-4 influence the development of rejection after renal transplantation. Clinical Transplantation 20, 30–36 (2006)
275. Batten, P, Rosenthal, N.A, Yacoub, M.H. Immune response to stem cells and strategies to induce tolerance. Philosophical Transactions of the Royal Society B—Biological Sciences 362, 1343–1356 (2007)
276. Loudovaris, T, Mandel, T.E, Charlton, B. CD4+ T cell mediated destruction of xenografts within cell-impermeable membranes in the absence of CD8(+) T cells and B cells. Transplantation 61, 1678–1684 (1996)
277. Babensee, J.E, Sefton, M.V. Viability of HEMA-MMA microencapsulated model hepatoma cells in rats and the host response. Tissue Engineering 6, 165–182 (2000)
278. Deng, W.M, Han, Q, Liao, L.M, Li, C.H, Ge, W, Zhao, Z.G, You, S.G, Deng, H.Y, Zhao, R.C.H. Allogeneic bone marrow-derived flk-1(+)Sca-(1–) mesenchymal stem cells leads to stable mixed chimerism and donor-

specific tolerance. Experimental Hematology 32, 861–867 (2004)
279. Le Blanc, K, Rasmusson, I, Sundberg, B, Gotherstrom, C, Hassan, M, Uzunel, M, Ringden, O. Treatment of severe acute graft-versus-host disease with third party haploidentical mesenchymal stem cells. Lancet 363, 1439–1441 (2004)
280. Maitra, B, Szekely, E, Gjini, K, Laughlin, M.J, Dennis, J, Haynesworth, S.E, Koc, O. Human mesenchymal stem cells support unrelated donor hematopoietic stem cells and suppress T-cell activation. Bone Marrow Transplantation 33, 597–604 (2004)
281. Krampera, M, Glennie, S, Dyson, J, Scott, D, Laylor, R, Simpson, E, Dazzi, F. Bone marrow mesenchymal stem cells inhibit the response of naive and memory antigen-specific T cells to their cognate peptide. Blood 101, 3722–3729 (2003)
282. Di Nicola, M, Carlco-Stella, C, Magni, M, Milanesi, M, Longoni, P.D, Matteucci, P, Grisanti, S, Gianni, A.M. Human bone marrow stromal cells suppress T-lymphocyte proliferation induced by cellular or nonspecific mitogenic stimuli. Blood 99, 3838–3843 (2002)
283. Aggarwal, S, Pittenger, M.F. Human mesenchymal stem cells modulate allogeneic immune cell responses. Blood 105, 1815–1822 (2005)
284. Klyushnenkova, E, Mosca, J.D, Zernetkina, V, Majumdar, M.K, Beggs, K.J, Simonetti, D.W, Deans, R.J, McIntosh, K.R. T cell responses to allogeneic human mesenchymal stem cells: immunogenicity, tolerance, and suppression. Journal of Biomedical Science 12, 47–57 (2005)
285. Bartholomew, A, Sturgeon, C, Siatskas, M, Ferrer, K, McIntosh, K, Patil, S, Hardy, W, Devine, S, Ucker, D, Deans, R, Moseley, A, Hoffman, R. Mesenchymal stem cells suppress lymphocyte proliferation in vitro and prolong skin graft survival in vivo. Experimental Hematology 30, 42–48 (2002)
286. Dai, F, Shi, D.W, He, W.F, Wu, J, Luo, G.X, Yi, S.X, Xu, J.Z, Chen, X.W. hCTLA4-gene modified human bone marrow-derived mesenchymal stem cells as allogeneic seed cells in bone tissue engineering. Tissue Engineering 12, 2583–2590 (2006)
287. Chen, X, McClurg, A, Zhou, G.Q, McCaigue, M, Armstrong, M.A, Li, G. Chondrogenic differentiation alters the immunosuppressive property of bone marrow-derived mesenchymal stem cells, and the effect is partially due to the upregulated expression of B7 molecules. Stem Cells 25, 364–370 (2007)
288. Takahashi, K, Tamanaka, S. Induction of pluripotent stem cells from mouse embryonic and adult fibroblast culture by defined factors. Cell 126, 663–676 (2006)
289. Takahashi, K, Tanabe, K, Ohnuki, M, Narita, M, Ischisaka, T, Tomoda, K, Tamanaka, S. Induction of pluripotent stem cells from adult human fibroblasts by defined factors. Cell 131, 861–872 (2007)
290. Wernig, M, Meissner, A, Foreman, R, Brambrink, T, Ku, M.C, Hochedlinger, K, Bernstein, B.E, Jaenisch, R. In: vitro reprogramming of fibroblasts into a pluripotent ES-cell-like state. Nature 448, 318–324 (2007)
291. Okita, K, Ichisaka, T, Yamanaka, S. Generation of germline-competent induced pluripotent stem cells. Nature 448, 313–317 (2007)
292. Byrne, J.A, Pedersen, D.A, Clepper, L.L, Nelson, M, Sanger, W.G, Gokhale, S, Wolf, D.P, Mitalipov, S.M. Producing primate embryonic stem cells by somatic cell nuclear transfer. Nature 450, 497–502 (2007)

Toll-Like Receptors: Potential Targets for Therapeutic Interventions

S. Pankuweit, V. Ruppert, B. Maisch, T. Meyer

Contents

51.1 Introduction 749
51.2 Inflammatory Heart Disease 749
51.3 Induction of Autoreactivity 750
51.4 The Role of the Innate Immune System in Controlling Infection 751
51.5 Summary 754
      References 754

51.1 Introduction

Members of the Toll-like receptor (TLR) family play critical roles as regulators of innate immune responses. TLRs have been described to bind to pathogen-associated molecular patterns (PAMPs) expressed on a broad variety of microbes including bacteria and viruses. Recognition of pathogen-derived ligands by TLRs triggers the activation of nuclear factor κB (NF-κB) and type I interferon pathways that lead to the production of proinflammatory cytokines, an increase in natural killer cell cytotoxicity, and activation of T cells. In addition to the innate immune system, TLRs are also involved in shaping adaptive immune reactions. TLRs function as integral membrane glycoproteins that are characterized by extracellular domains containing variable numbers of leucine-rich repeat (LRR) motifs and a cytoplasmic Toll/interleukin (IL)-1 receptor (TIR) homology domain. In humans and mice, at least 12 TLRs have been identified which are engaged in different gene expression programs. Since TLRs act as key players in the pathogen-induced stimulation of dendritic cells, including antigen uptake and processing, as well as differentiation of CD4+ T cells, they may become interesting targets for future pharmacological interventions. In the following, we will briefly discuss the role of TLRs in the pathogenesis of inflammatory heart disease and particularly focus on their potential as targets for therapeutical interventions.

51.2 Inflammatory Heart Disease

Both myocarditis and inflammatory cardiomyopathy reflect ongoing inflammatory processes in cardiac tissue and were introduced in 1837 by Sobernheim [1]. Myocarditis is the major cause of sudden unexpected death in patients younger than 40 years of age and may account for up to 20% of mortality from cardiovascular causes [2, 3]. Chronic myocarditis on the other hand is usually asymptomatic, and only a limited number of reports are available concerned with the incidence and prevalence of this disease [4]. However, large unselected autopsy series indicate that 1% to 5% of the general population suffer from undiagnosed myocarditis, i.e., the disease might be more common than generally assumed [5]. In up to

30% of cases, myocarditis can give rise to dilated cardiomyopathy with progression to heart failure and a poor prognosis with a 10-year survival rate of less than 40% [6]. Introduced in the early 1980s, endomyocardial biopsies provided a clinically applicable means for assessing inflammatory reactions in human heart tissue [7]. However, early biopsy studies elicited highly variable results, with incidence of myocarditis ranging from 0 to 80% [8, 9]. To overcome the underlying problems in diagnosing the disease, the Dallas criteria for histological diagnosis were established in the year 1987 [10].

The term "inflammatory cardiomyopathy" was introduced in 1995 by the World Health Organization/International Society and Federation of Cardiology Task Force (ISFC, presently the World Heart Federation [WHF]) on the Definition and Classification of Cardiomyopathies [11, 12]. Inflammatory cardiomyopathy was defined as "myocarditis in association with cardiac dysfunction" and includes conditions of heart failure due to idiopathic, autoimmune, or infectious etiologies. Diagnostic criteria were refined by the introduction of immunohistochemical methods. Particularly, chronic inflammation of the heart was characterized by defining a cutoff for the number of infiltrating leukocytes necessary for the diagnosis (≥ 14 infiltrating lymphocytes/mm², up to 4 macrophages may be included) and by histopathologically subtyping a given inflammatory infiltrate.

So far, the treatment of myocarditis is mainly symptomatic and based on the management of heart failure including diuretics, angiotensin-converting enzyme inhibitors, beta-blockers, digitalis, and in severe cases, inotropic support, mechanical circulatory assistance, and heart transplantation. However, new therapeutic options aimed at interfering with the pathophysiology of the inflammatory reaction are currently being investigated in several centers.

Although the etiology of myocarditis in a given patient often remains unknown, a large variety of infectious agents, systemic diseases, drugs, and toxins have been associated with the development of the disease [13, 14]. In general, infectious agents, including viruses such as enteroviruses, cytomegalovirus, and adenoviruses, bacteria such as *Borrelia burgdorferi* or *Chlamydia pneumoniae*, protozoa, and even fungi can cause myocarditis. It is generally accepted that viral infections are the most important causes of myocarditis in North America and Europe.

Initially, rising antibody titers against specific viruses in the patient's serum obtained in the florid phase of myocarditis or during reconvalescence were used to demonstrate etiologic relevance [15]. More recently, genomes of enteroviruses, adenoviruses, cytomegalovirus, influenza virus, hepatitis C virus, and parvovirus B19 have been identified in endomyocardial biopsies from patients with myocarditis and dilated cardiomyopathy, using polymerase chain reactions [16–22]. Whether these viruses play a pathogenetic role or act merely as an innocent bystander is still a matter of debate.

51.3
Induction of Autoreactivity

Murine studies on coxsackievirus B3-induced myocarditis have provided insight into the pathophysiology of inflammatory heart disease [23, 24]. Infiltration of the myocardium with immunocompetent cells is the histopathological hallmark of the disease (Fig. 51.1). Normally, self-reactive lymphocytes generated in the bone marrow and thymus are deleted or inactivated before moving into the periphery. Activation of mature T lymphocytes and B lymphocytes in the periphery depends on how antigen is presented to them. T cells become activated when the T-cell receptor recognizes an antigenic peptide located inside an MHC molecule expressed on antigen-presenting cells (dendritic cells or macrophages). The antigen-presenting cell itself must be activated to express costimulatory signals which are essential for efficient T-cell activation. In addition, the concentration of antigenic peptides is critical for T-cell activation, as the antigens must be presented in quantities large enough to activate T cells.

Similarly, a B cell requires antigens and costimulation delivered by an activated T cell to become activated. After antigen presentation, an immune response is initiated which specifically reacts against microbial peptides, leading to the elimination of the invader and finally recovery from the disease. Three hypotheses on the trigger mechanism of heart-specific pathogenic autoreactivity have been considered: (i) *Activation by a foreign antigen*: Processing of a foreign antigen might result in the presentation

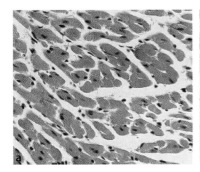
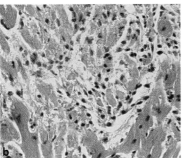

Fig. 51.1a,b Normal murine myocardial tissue (**a**) and myocardium from a mouse with experimentally induced acute myocarditis due to infection with coxsackievirus B3 (**b**). Sections were stained with hematoxylin and eosin. Note the infiltrating immunocompetent cells in (**b**) as a typical sign of acute heart inflammation

of an epitope highly homologous to cardiac myosin or other cardiac-specific proteins. As a consequence cross-reactive T cells are activated. (ii) *Increased presentation of self-peptides*: Viral infection and necrosis of myocytes may lead to the release of buried intracellular antigens whose extracellular concentrations are normally very low. Release of these proteins in larger quantities decreases the threshold necessary for an immune response, resulting in activation of self-reactive T cells. (iii) *Presentation of cryptic self-peptides*: In addition, altered processing of intracellular proteins might occur, resulting in activation of self-reactive T cells. This mechanism may then initiate unregulated immune reactions against cardiac antigens in the absence of further infectious triggers.

Inflammatory dilated cardiomyopathy is characterized by different phases of the disease: phase 1 is dominated by viral infection itself, phase 2 by the onset of (probably) multiple autoimmune reactions, and phase 3 by the progression toward heart failure in the absence of an infectious agent or severe cardiac inflammation. In selected groups of patients, pharmacological treatment should be based on immunohistochemical and molecular biological investigation of endomyocardial biopsies. Modern diagnostic techniques aim to identify the infective agent and focus on host interactions to reveal the etiology of cardiac manifestation. This may change the management of patients with myocarditis or inflammatory heart disease in the near future. One of the hopes is to discern the underlying dominant pathomechanisms in a given patient to make an early decision for the most promising therapy.

51.4
The Role of the Innate Immune System in Controlling Infection

Unlike adaptive immunity, the innate immune response appears in almost all organisms in evolution. The induction time of innate responses is short (a few hours to 3 days), before the adaptive immune response is activated [25]. The innate immune response is more static than the adaptive (acquired) immune response due to its limitation in recognizing only so-called PAMPs (pathogen-associated molecular patterns). The principal effector cells of the innate immune response are macrophages, natural killer cells, and mast cells. Receptors for the recognition of PAMPs include the Toll-like receptors, nuclear oligomerization receptors, scavenger receptors, CD11b, and CD14 [26]. Most of these are proteins with a leucine-rich repeat domain, which is important for ligand recognition, and an intracellular Toll/IL-1-receptor (TIR) domain responsible for intracellular signaling. TLRs transmit transmembrane signals that activate NF-κB and MAP kinases, finally resulting in cytokine production, leukocyte recruitment, and phagocytosis.

Ligands of these pattern recognition receptors, the PAMPs, consist of highly conserved pathogen motifs, for example bacterial wall components (lipopolysaccharides, lipoproteins, peptidoglycans, and bacterial DNA), viral constituents and viral DNA and RNA, fungal cell wall components, putative endogenous ligands such as heat shock protein 70, and autoantigens [27]. Viruses, bacteria, and fungi produce a broad range of PAMPs that can be detected by the host using specific receptor molecules—for example, Toll-like-receptors. Ligand binding of these receptors initiates host signaling pathways that activate innate and adaptive immunity and eliminate the infectious agent. In addition to intracellular pattern recognition receptors such as retinoic acid-inducible protein 1 (RIG-1) or the nucleotide-binding oligomerization domain (NOD)-like receptors (NLRs), Toll-like receptors take part in a highly effective protection system responsible for the rapid recognition and neutralization of invading microbial organisms that bridge and modulate innate and adaptive immune reactions. RIG-1 is an interferon-inducible protein containing caspase-activating and recruitment domains (CARDs) and an RNA helicase domain that acts as a cytoplasmic receptor for double-stranded RNA. The NOD proteins represent a large family of cytoplasmic receptors harboring an LRR domain for ligand binding and have been implicated in the recognition of bacterial components.

As mentioned before, TLRs are engaged in signal transduction from cytoplasmic and endosomal membranes to downstream nuclear effector molecules that function as transcription factors. Natural and synthetic ligands for these receptors have been identified: TLR2 recognizes peptidoglycan found in gram-positive bacteria, TLR3 recognizes double-stranded RNAs produced during viral infection, TLR4 binds to lipopolysaccharide in gram-negative bacteria, TLR7 and TLR8 recognize single-stranded viral RNAs, and TLR9 recognizes an unmethylated cytosine-phosphate-guanine (CpG) DNA motif of prokaryotic genomes and DNA viruses.

The most important intracellular molecule for TLR signaling protein is myeloid differentiation factor 88 (MyD88), used by most of the TLR, although MyD88-independent activation pathways also exist for some TLRs (Fig. 51.2). MyD88 has a carboxy-terminal TIR-containing portion that is associated with the TIR domain of TLRs. Other adapter molecules involved in the downstream signal transduction include TIR domain-containing adaptor protein/ MyD88-adapter-like (TIRAP/Mal), TIR domain-containing adapter inducing interferon-β (TRIF), and TRIF-related adapter molecule (TRAM). For example, binding of double-stranded viral RNA to TLR3 leads to the recruitment of TRIF, resulting in the activation of the transcription factor interferon-regulatory factor 3 (IRF-3) that increases secretion of type I interferon. Recruitment of MyD88 to the receptor triggers a cascade of signaling events that for some TLRs result in the phosphorylation of IL-1R-associated kinase (IRAK) and complex formation with the E3 ubiquitin ligase TRAF6 (TNF receptor-associated factor 6). Activated TRAF6 then induces the activation of the TGF-β-activated kinase 1 (TAK1)-TAB1/2/3 complex which in turn phosphorylates and activates the IκB kinase (IKK) complex culminating in the phosphorylation and proteasomic degradation of IκBα. This step induces the release and transloca-

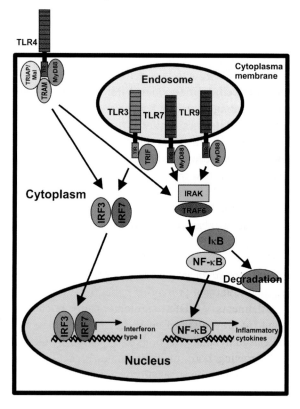

Fig. 51.2 Toll-like receptor (TLR) signaling by pathogen-associated molecular patterns (PAMPs). Binding of lipopolysaccharide to TLR4 localized in the plasma membrane activates MyD88-dependent and MyD88-independent pathways. The latter results in the activation of interferon-regulatory factors that function as transcription factors for type I interferon. Recognition of double-stranded RNA by TLR3, single-stranded RNA by TLR7, and bacterial DNA containing CpG motifs by TLR9 leads via IRAK and TRAF6 to the phosphorylation and proteasomic degradation of IκB. NF-κB is released and translocated into the nucleus where it activates genes coding for inflammatory cytokines. For details, see text

tion of transcriptionally active NF-κB to the nucleus where it binds to the promoter region of inflammatory cytokine genes. Activation of both pathways results in the secretion of different cytokines, expression of costimulatory molecules important for T-cell activation and an enhanced MHC class II expression. As a consequence of the TLR stimulation, the adaptive immune response is activated. Overactivation of these receptors may cause severe disease, such as has been reported in infectious diseases, autoimmunity, and chronic inflammation [25].

The functional importance of the signaling protein MyD88 was investigated in a well-established mouse model [28]. Immunization of mice with myosin heavy chain leads to severe myocarditis which mimics that observed in humans. In contrast to control littermates, MyD88$^{-/-}$ mice were protected from myocarditis after immunization with myosin heavy chain-derived peptide. MyD88$^{-/-}$ but not MyD88$^{+/+}$ primary antigen-presenting dendritic cells were defective in their capacity to activate CD4 T cells. This defect mainly resulted from the inability of MyD88$^{-/-}$ dendritic cells to release tumour necrosis factor. The critical role of MyD88 signaling in dendritic cells in the peripheral lymphatic compartments was confirmed by repeated injection of activated, myosin heavy chain-loaded MyD88$^{+/+}$ dendritic cells that fully restored T-cell expansion and myocarditis in MyD88$^{-/-}$ mice. It was concluded that induction of autoimmune myocarditis depends on MyD88 signaling in self-antigen presenting cells in the peripheral compartments. MyD88 might, therefore, become a target for prevention of heart-specific autoimmunity and cardiomyopathy.

A growing body of evidence suggests there are fundamental cellular functions of TLRs in the cardiovascular system. Expression of the receptors has been found in most human cells of cardiovascular origin, such as myocytes, endothelial cells, fibroblasts, and macrophages. Recently, it has been shown that human cardiomyocytes express most known TLRs (Fig. 51.3). Of these, TLR2, TLR4, and TLR5 signal via NF-κB, resulting in decreased contractility and a concerted inflammatory response including the expression of the proinflammatory cytokine IL-6, chemokines, and the cell surface adhesion molecule ICAM-1 in cardiomyocytes [29].

In patients with coxsackievirus B3-induced myocarditis, it has been shown that cardiac inflammatory responses triggered by viral infection are mainly TLR8-dependent [30]. In infected individuals, TLR7 and TLR8 are continuously activated by the presence of viral single-stranded RNA resulting from enteroviral replication. Upon cell lysis during acute infection, TLR4 is able to trigger an immune response against the released newly synthesized virions. The synergic inflammatory response of TLR4, TLR7, and TLR8 seems to produce a chronic inflammatory reaction against the virus which could eventually lead to irreversible myocardial injury [31, 32].

Recently, an association was found between higher TLR4 levels and enteroviral RNA in patients with dilated cardiomyopathy. VP1/TLR4 double staining showed extensive colocalization of VP1 and TLR4 proteins in the cytoplasm of cardiac myocytes. In patients with dilated cardiomyopathy, high TLR4 expression levels showed lower left ventricular ejection fractions (LVEFs) and larger left ventricular end-diastolic diameters (LVEDDs), as assessed by echocardiography. Thus, myocardial expression of TLR4 appears to be associated with enteroviral replication, suggesting a functional role of TLR4 in

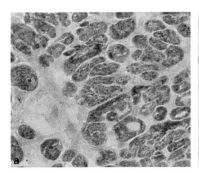
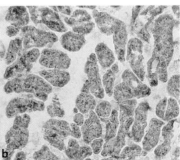

Fig. 51.3a,b Immunohistochemical staining of Toll-like receptor 7 (TLR7, **a**) and TLR9 (**b**) in a patient diagnosed with acute myocarditis. Tissue samples were taken from endomyocardial biopsies and stained with antibodies against TLR7 and TLR9, respectively. Nuclei were counterstained with hematoxylin

the pathogenesis of enterovirus-induced dilated cardiomyopathy [31, 32]. What are the clinical implications of these findings? Viruses and bacteria can trigger innate immunity, which strengthens the host to fight against the infection—for example, through activation of STAT proteins and gp130 signaling for survival, and IRF triggering interferon production for systemic clearance of the virus [33]. Unregulated activation of innate immunity may have adverse consequences of production of cytokines and costimulation of T cells, leading to clonal expansion and inflammatory infiltrate. Excessive inflammation may result in destruction of heart tissue and paradoxical impairment of myocardial function, resulting in a more unfavorable outcome as compared with patients with less intense inflammatory reactions [34].

51.5 Summary

TLRs constitute a family of pattern recognition receptors that have emerged as key mediators of innate immunity, which sense invading microbes and initiate prompt immune responses. MyD88 has been shown to play a crucial role as an adaptor protein for some of the TLR-mediated signal transduction pathways. Recruitment of MyD88 results in activation and nuclear translocation of NF-κB, resulting in the subsequent release of different cytokines, expression of costimulatory molecules required for T-cell activation and an enhanced MHC class II expression. TLR-mediated signaling is an important pathogenic link between innate and adaptive immune responses and contributes to the inflammation seen in inflammatory heart disease. A delicate balance exists between the initial fight against viral propagation and the suppression of autodestructive immune reactions, which has to be taken into account when developing new diagnostic and therapeutic concepts to prevent progression of myocarditis to cardiomyopathy with severe heart failure. Interventions into TLR signaling promise to slow down the progression of inflammatory heart disease in dilated cardiomyopathy. In the near future the potential of such novel treatment options for the management of patients with inflammatory heart diseases remains to be explored.

References

1. Mattingly TW. Changing concepts of myocardial diseases. JAMA 1965; 191: 33–37
2. Huber SA, Gauntt CJ, Sakkinen P. Enteroviruses and myocarditis: viral pathogenesis through replication, cytokine induction, and immunopathogenicity. Adv Virus Res 1989; 51: 35–80
3. Drory Y, Turetz Y, Hiss Y, Lev B, Fisman EZ, Pines A, Kramer MR. Sudden unexpected death in persons less than 40 years of age. Am J Cardiol 1991; 68: 1388–1392
4. No authors listed. Coxsackie B5 virus infections during 1965: a report to the Director of the Public Health Laboratory Service from various laboratories in the United Kingdom. BMJ 1967; 4: 575–577
5. Gore I, Saphir O. Myocarditis: a classification of 1402 cases. Am Heart J 1947; 34: 827–830
6. Feldman AM, McNamara D. Myocarditis. N Engl J Med 2000; 343: 1388–1398
7. Mason JW. Techniques for right and left ventricular endomyocardial biopsy. Am J Cardiol 1978; 41: 887–892
8. Parrillo JE, Aretz HT, Palacios I, Fallon JT, Block PC. The results of transvenous endomyocardial biopsy can frequently be used to diagnose myocardial diseases in patients with idiopathic heart failure. Endomyocardial biopsies in 100 consecutive patients revealed a substantial incidence of myocarditis. Circulation 1984; 69: 93–101
9. Shanes JG, Ghali J, Billingham ME, Ferrans VJ, Fenoglio JJ, Edwards WD, Tsai CC, Saffitz JE, Isner J, Furner S. Interobserver variability in the pathologic interpretation of endomyocardial biopsy results. Circulation 1987; 75: 401–405
10. Aretz HT, Billingham ME, Edwards WD, Factor SM, Fallon JT, Fenoglio JJ, Olsen EG, Schoen FJ. Myocarditis: a histopathologic definition and classification. Am J Cardiovasc Path 1987; 1: 3–4
11. Richardson P, McKenna W, Bristow M, Maisch B, Mautner B, O'Connell J, Olsen E, Thiene G, Goodwin J, Gyarfas I, Martin I, Nordet P. Report of the 1995 World Health Organization/International Society and Federation of Cardiology Task Force on the Definition and Classification of cardiomyopathies. Circulation 1996, 93: 841–842
12. Ristić AD, Maisch B, Hufnagel G, Seferovic PM, Pankuweit S, Ostojic M, Moll R, Olsen E. Arrhythmias in acute pericarditis: An endomyocardial biopsy study. Herz 2000; 25: 729–733
13. Liu P, Martino T, Opavsky MA, Penninger J. Viral myocarditis: balance between viral infection and immune response. Can J Cardiol 1996; 12: 935–943
14. Anandasabapathy S, Frishman WH. Innovative drug treatments for viral and autoimmune myocarditis. J Clin Pharmacol 1998; 38: 295–308
15. Cambridge G, MacArthur CG, Waterson AP, Goodwin JF, Oakley CM. Antibodies to Coxsackie B viruses in congestive cardiomyopathy. Br Heart J 1979; 41: 692–696
16. Baboonian C, Treasure T. Meta-analysis of the association of enteroviruses with human heart disease. Heart 1997; 78: 539–543

17. Grumbach IM, Heim A, Pring-Akerblom P, Vonhof S, Hein WJ, Müller G, Figulla HR. Adenoviruses and enteroviruses as pathogens in myocarditis and dilated cardiomyopathy. Acta Cardiol 1999; 54: 83–88
18. Maisch B, Schönian U, Crombach M, Wendl I, Bethge C, Herzum M, Klein HH. Cytomegalovirus associated inflammatory heart muscle disease. Scan J Infect Dis 1993; Suppl 88: 135–148
19. Martin AB, Webber S, Fricker FJ, Jaffe R, Demmler G, Kearney D, Zhang YH, Bodurtha J, Gelb B, Ni J. Acute myocarditis. Rapid diagnosis by PCR in children. Circulation 1994; 90: 330–339
20. Matsumori A, Matoba Y, Sasayama S. Dilated cardiomyopathy associated with hepatitis C virus infection. Circulation 1995; 92: 2519–2525
21. Pankuweit S, Moll R, Baandrup U, Portig I, Hufnagel G, Maisch B. Prevalence of the parvovirus B19 genome in endomyocardial biopsy specimens. Hum Pathol 2003; 34: 497–503
22. Kühl U, Pauschinger M, Noutsias M, Seeberg B, Bock T, Lassner D, Poller W, Kandolf R, Schultheiss HP. High prevalence of viral genomes and multiple viral infections in the myocardium of adults with "idiopathic" left ventricular dysfunction. Circulation. 2005; 111: 887–893
23. Woodruff JF. Viral myocarditis. A review. Am J Pathol 1980; 101: 425–484
24. Malkiel S, Kuan AP, Diamond B. Autoimmunity in heart disease: mechanisms and genetic susceptibility. Mol Med Today 1996; 2: 336–342
25. de Kleijn D, Pasterkamp G. Toll-like receptors in cardiovascular diseases. Cardiovasc Res 2003; 60: 58–67
26. Janeway CA, Medzhitov R. Innate immune recognition. Annu Rev Immunol 2002; 20: 197–216
27. Akira S, Takeda K. Toll-like receptor signalling. Nat Rev Immunol 2004; 4: 499–511
28. Marty RR, Dirnhofer S, Mauermann N, Schweikert S, Akira S, Hunziker L, Penninger JM, Eriksson U. MyD88 signaling controls autoimmune myocarditis induction. Circulation 2006; 113: 258–265
29. Boyd JH, Mathur S, Wang Y, Bateman RM, Walley KR. Toll-like receptor stimulation in cardiomyocytes decreases contractility and initiates an NF-κB dependent inflammatory response. Cardiovasc Res 2006; 72: 384–393.
30. Triantafilou K, Orthopoulos G, Vakakis E, Ahmed MA, Golenbock DT, Lepper PM, Triantafilou M. Human cardiac inflammatory responses triggered by Coxsackie B viruses are mainly Toll-like receptor (TLR) 8-dependent. Cell Microbiol 2005; 7: 1117–1126
31. Satoh M, Nakamura M, Akatsu T, Shimoda Y, Segawa I, Hiramori K. Toll-like receptor 4 is expressed with enteroviral replication in myocardium from patients with dilated cardiomyopathy. Lab Invest 2004; 84: 173–181
32. Satoh M, Shimoda Y, Maesawa C, Akatsu T, Ishikawa Y, Minami Y, Hiramori K, Nakamura M. Activated toll-like receptor 4 in monocytes is associated with heart failure after acute myocardial infarction. Int J Cardiol 2006; 109: 226–234
33. Maekawa Y, Ouzounian, Opavsky MA, Liu PP. Connecting the missing link between dilated cardiomyopathy and viral myocarditis: Virus, cytoskeleton, and innate immunity. Circulation 2007; 115: 5–8
34. Liu PP, Mason JW. Advances in the understanding of myocarditis. Circulation 2001; 104: 1076–1082

XII Study Design Principles

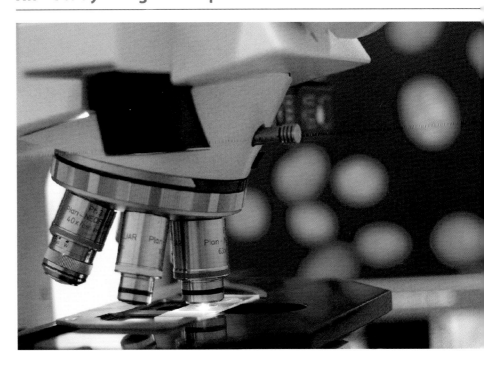

Tissue Engineered Models for In Vitro Studies

C. R. McLaughlin, R. Osborne, A. Hyatt, M. A. Watsky,
E. V. Dare, B. B. Jarrold, L. A. Mullins, M. Griffith

Contents

52.1	Tissue Engineered In Vitro Models	759
52.2	Scaffold Materials and Cells	760
52.2.1	Tissue Engineered Scaffolds	760
52.2.2	Cell Line Derivation and Characterization	760
52.3	In Vitro Models of Corneas	761
52.3.1	Tissue Engineered Three-Dimensional Models	761
52.3.2	Innervated Corneal Model	762
52.4	In Vitro Models of Angiogenesis and Vasculogenesis	763
52.5	In Vitro Models of Articular Cartilage	764
52.6	In Vitro Models of Skin	765
52.6.1	Tissue Engineered Three-Dimensional Models	765
52.6.2	Characterization of Skin Barrier	767
52.7	Summary	767
	References	768

52.1 Tissue Engineered In Vitro Models

The field of tissue engineering is multidisciplinary, seeking to fabricate artificial organs or substitutes in order to replace failing or damaged organs. These engineered tissues and organs are mostly targeted as substitutes for human donor tissues and as such are developed to avoid the complications associated with donor matching and immune rejection [1]. Examples include substitutes of organs including liver, heart, kidney, skin, teeth, and cornea [2].

However, engineered tissues are more recently being developed as in vitro models for toxicology testing, as animal alternatives. This is important in the context of the ban on animal testing for consumer products that is currently in place in Europe (European Union Directive 76/768/EEC), which is expected to expand worldwide. However, unlike developing tissue substitutes for transplantation, to create in vitro models for testing purposes, care must be taken to assure that the alternative not only reproduces the function of the target organ, but also mimics key biochemical, morphological, physiological, and even genetic components.

The aim of this review is to provide a synopsis of techniques and materials commonly used to engineer in vitro organotypic models for testing purposes rather than transplantation, and to provide examples of commonly used models.

52.2
Scaffold Materials and Cells

52.2.1
Tissue Engineered Scaffolds

The majority of in vitro organ tissue engineering approaches require a three-dimensional scaffold upon which cells can attach, proliferate, and differentiate into physiologically functional tissues that are structurally similar to the native tissue. Many tissue engineering strategies rely on the use of extracellular matrix (ECM) proteins to provide a suitable scaffold. The ECM is a product secreted by cells into the surrounding organ that helps to direct and maintain the cell phenotype [3]. Typically, ECM comprises both structural and functional proteins, glycosaminoglycans, glycoproteins, and other small molecules that are arranged into a specific three-dimensional structure [4].

The cells used in in vitro engineered tissues can either be stem cells, primary cells, or cells with extended life-spans. Cells with extended life-spans or established cell lines are used most often, as these can be expanded in culture to provide a homogenous source of cells for constructing in vitro models, especially for use in safety and efficacy testing for drugs, chemicals, and/or compounds. Often, cofactors for enhancing cellular differentiation and maintenance of the differentiated phenotype and viability, such as cell adhesion proteins or growth factors, are added to the gels to promote and maintain proper cellular differentiation.

52.2.2
Cell Line Derivation and Characterization

Cell lines, i.e., cells with extended life-spans, can be developed by transfection of the genes that code for telomerase (TERT) [5, 6]. Telomerase is an enzyme that synthesizes DNA at the ends of chromosomes (telomeres), thereby conferring replicative immortality to cells. In addition, cell can also be "immortalized" by transfection with a number of virally derived genes, including those from DNA tumor viruses such as the gene for the large T antigen that lies within the early regions of simian virus 40 (SV40), or the *E6E7* oncogenes from human papillomavirus 16 (HPV16E6E7). SV40 large T inactivates *p53* and *Rb* genes, whose products are believed to be involved in normal senescence of cells [7, 8], thereby extending the life-spans of cells. HPV *E6E7* genes also degrade the p53 protein, and are also able to reactivate the telomerase enzyme, which in turn extends the telomeric complement of cells, thereby extending their life-spans [9].

Screening of cell lines with extended life-spans is required to ensure that they retain key characteristics of the primary cells from which they were derived. Cells are screened for a range of criteria including morphology, cell culture characteristics, expression of various cell type specific markers, and cytogenetic analyses. For in vitro cornea development, a novel functional test was developed utilizing patch clamp technology to compare the electrophysiological features of genetically altered cells to those of freshly dissociated or low passage corneal cells obtained from post mortem human corneas [10]. This method employed the amphotericin perforated-patch whole cell patch clamp technique [11] to definitively identify cell lines with "normal" phenotypes, as compared with transformed cells, which showed unique currents. Using this technique, corneal epithelial, stromal, and endothelial cell lines were all effectively screened, with an excellent correlation between the electrophysiology data and data from the other screening assays mentioned above. This was corroborated by experiments in which corneal constructs showed the most in vivo–like morphological appearance and physiological function when cells with the best electrophysiological screening scores were used. However, when cells expressing the electrophysiological characteristics of transformed cells were used in the corneal constructs, the constructs were unsuccessful because cells expressing the transformed cell-type current invaded the matrix and did not differentiate like the "normal" cells. It is also important to note that cells deemed "transformed" by electrophysiology often expressed what was thought to be the appropriate cell-specific protein markers for the desired cell type, demonstrating that cell-specific markers are not always sufficient as a screening tool for cell line identification.

52.3
In Vitro Models of Corneas

The cornea is the transparent, anterior-most surface of the eye and the major refractive element of the eye, being responsible for refracting 70% of the light onto the surface of the retina to allow for focused vision [12]. Although the human cornea is only 500-μm thick centrally [13], it is extremely tough due to its stromal extracellular matrix (ECM), consisting primarily of collagen. The mainly type I collagen has a unique infrastructure that provides both optical clarity and mechanical strength [14]. The cornea is an immune-privileged, avascular tissue that has an innervation density 300–600 times that of skin [15, 16]. The cornea consists of three main cellular layers: an outermost nonkeratinized epithelium comprising 5–7 layers of cells, a network of keratocytes interspersed within the middle stromal layer, and an innermost single-cell-thick endothelial layer. A primary function of the cornea is to act as a barrier to the external environment, and the epithelium provides this barrier. The basal layer of the epithelium proliferates to replace the superficial cells lost at the anterior surface [17], and in turn, these cells are replaced by corneal stem cells that live within the corneal limbus [18]. The epithelium produces an insoluble mucous layer, which contains several antiinflammatory and antimicrobial factors [19]. This insoluble mucous also assists in stabilizing the tear layer. The stroma consists mainly of type I collagen (70%) and keratocytes that form a network via gap junctions [20–22]. The remaining stromal ECM components include proteoglycans such as keratan sulphates (lumican, mimecan, and keratocan) [23]. Lastly, the function of the endothelium is to regulate the hydration level of the corneal stroma. This is accomplished via a Na^+,K^+-ATPase-driven pump leak mechanism that circulates aqueous humour between the anterior chamber and stroma [12].

Innervation of the cornea occurs primarily from the ophthalmic branch of the trigeminal nerve, and consists mainly of sensory nerves [24]. Corneal nerves enter the cornea through the sclera as unmyelinated bundles, parallel to the corneal surface. They are often found in proximity to keratocytes, with occasional contacts through cytoplasmic extensions. As they pass into the basal epithelium, the nerves form a network of beaded and nonbeaded subepithelial nerves, and upon penetration into the more superficial layers, individual beaded fibers are observed invaginating individual epithelial cells [25].

The cornea's superficial location in the eye renders it highly susceptible to mechanical and chemical injury. Traditionally, rabbit eye test methods such as the Draize test and low volume eye test (LVET) have been the benchmark standards for assessment. These tests are performed by instilling a test material into one eye of albino rabbits, and monitoring the ocular response of the cornea (e.g., opacification) [26]. Due to the push to develop alternative methods to animal testing, a range of tests ranging from the use of excised animal corneas to detect chemical-induced opacifications and chick chorioallantoic membrane vascularization assays [27, 28], to monolayer and, more recently, monotypic cellular cultures have been developed [29]. Monotypic, stratified three-dimensional cornea epithelial cultures from cell lines that are now being used include SV40 large T immortalized cultures derived from the HCE-T line [29] and more recently, TERT and HPV immortalized cells from Clonetics™. The EpiOcular™ model (MatTek Corporation), on the other hand is a "corneal model" based on substitute cells derived from human foreskin epithelial cells [30]. While the models are gaining acceptance, a disadvantage is that they are not able to predict responses to chemicals that may affect the other cell layers, or that are dependent upon the interactions of the epithelial cells with other cell types within the cornea.

52.3.1
Tissue Engineered Three-Dimensional Models

One of the earliest three-dimensional models was developed by Zieske et al., who utilized a stroma comprising neutralized type I collagen blended with rabbit stromal fibroblasts, and seeded with rabbit corneal epithelial cells. On the reverse side, immortalized mouse endothelial cells were seeded, making a true three-dimensional corneal construct [31]. In 1999, our group reported the fabrication of a full thickness human corneal equivalent that reproduced key morphological, biochemical, and physiological functions of human corneas [32]. These corneas were based on immortalized human cornea epithelial, stromal, and endothelial cell lines (prescreened as described in Section 52.2.2, for retention of primary cell phenotypes), seeded within and on either side of a tissue engineered scaffold. The scaffold comprised glutaraldehyde cross-linked type I collagen with chondroitin sulphate stroma, and provided an adequately robust, physiologically relevant structure. These constructs were successfully tested for opacity changes in response to irritating chemicals, as well as for changes in the expression of wound-healing genes such as c-*fos*, and *IL-1 alpha* [32]. However, this model was limited in that it lacked functional innervation, and could not address the important role of "pain or irritation" in chemical injury, which can be obtained from direct animal testing.

52.3.2
Innervated Corneal Model

While pain or irritation requires signaling to the brain to be perceived, it is nevertheless possible to make a determination based on examination of the release of neurotransmitters from nerves that are known pain receptors.

We demonstrated in Suuronen et al. [33] that our previous model (described in Section 52.3.1 above) could be functionally innervated with a sensory nerve source. These nerves traversed the sclera, formed a subbasal network, and then terminated within the epithelium [33]. Along with the type I rat-tail collagen and chondroitin sulphate, laminin and nerve growth factor (NGF) were introduced into the scaffolds. Both laminin and NGF had both previously been shown to promote the differentiation and guidance of nerves [34, 35]. For the innervated construct, chicken dorsal root ganglia explants (DRGs) were implanted within the gel prior to gelling. In addition, the medium was supplemented with retinoic acid, which has been shown to induce neurite outgrowth from embryonic mouse DRGs [36]. Within the innervated constructs, DRGs were shown to extend neurites consisting of smooth fibers traversing the corneal stroma, with smaller beaded entering the epithelium, similar to those found in the human cornea [25, 37]. Transmission electron microscopy demonstrated direct innervation of the individual epithelial cells [33].

Action potentials recorded from surface terminals following ganglia stimulation demonstrated that nerves in the tissue engineered constructs were physiologically functional. These action potentials were lidocaine sensitive, and similar to those found in guinea-pig corneal polymodal nociceptors [38].

Substance P (SP) is one of the prevalent neurotransmitters found in the cornea and is released during wounding, where it is implicated in nerve-stimulated enhancement of epithelial healing in conjunction with other growth factors, such as IGF [12, 39]. In tissue engineered corneas [40], the nerves were shown to differentially release SP in response to chemicals, such as capsaicin, veratridine, and ouabain. We were able to utilize this innervated in vitro corneal construct to develop a more physiologically relevant toxicological assay [41], measuring a differential response in the intracellular sodium concentration following exposure to veratridine and ouabain, using a sodium-sensitive dye and two-photon microscopy. Previous work had shown that the variability of the Na^+ current is correlated with the amplitude of the nerve response [42]. The innervated corneal constructs also had a neuroprotective effect on the epithelium in response to surfactants, as compared with noninnervated corneas, as determined using a live/dead analysis of the epithelium. This confirmed in vivo observations of the role of nerves in the homeostasis of epithelial cells in human corneas [43]. The innervated corneal model can therefore be utilized as an alternative to animal ocular irritancy testing, through assessment of the degree of cellular death, levels of neurotransmitter release, and the change in internal neuron sodium concentration following exposure to different sodium transport or channel blockers.

More recently, we showed that complex models of the entire anterior ocular surface are possible, with the development of corneal-scleral models [44]. In these models, an avascular cornea is surrounded by a pseudosclera containing vessel-like structures; both cornea and sclera were innervated. Nerves within the sclera formed a ring around the cornea with branches penetrating into the cornea (Fig. 52.1), similar to what was previously described in Bee [45].

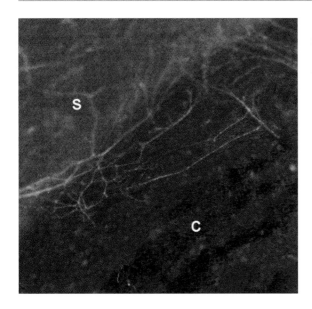

Fig. 52.1 Tissue engineered corneal-scleral model, showing neurofilament-20 stained nerves (*green*) from the pseudosclera penetrating (*s*) into reconstructed cornea (*c*). Nerves from chick embryo dorsal root ganglia embedded within the sclera, circle the cornea prior to entering the stromal matrix. Bar, 50 μm

52.4 In Vitro Models of Angiogenesis and Vasculogenesis

Angiogenesis refers to the formation of new blood vessels from preexisting vessels, a process that occurs during growth, development, and wound healing, as well as pathological conditions such as diabetes, macular degeneration, rheumatoid arthritis, and tumor growth [46, 47]. Vasculogenesis is the process by which vascular networks form from endothelial progenitors; it is therefore crucial during development. In vitro models in which endothelial cells are embedded within three-dimensional matrices such Matrigel™, collagen, and fibrin have been widely used to study different aspects of both angiogenesis and vasculogenesis. Figure 52.2 shows the development of capillary-like structures within a collagen matrix.

In vitro tissue engineered models of angiogenesis or vasculogenesis have been used to study the roles played by integrins, adhesion molecules, supporting cells, growth factors, angiogenic inhibitors, and matrix metalloproteinases [48]. To examine the role of integrins during vasculogenesis, Davis and Camarillo [49] suspended endothelial cells within type I collagen matrices. They showed that vacuole and lumen formation within endothelial cells, two key steps during the development of capillary-like structures, are dependent on the collagen-binding integrin α2β1. Korff and Augustin [50] also used collagen matrices to show that integrin-dependent tensional forces within matrices influence the direction of capillary sprouting and the formation of interconnecting networks of cells. Furthermore, Yang et al. [51] demonstrated that CD31 and vascular endothelial cadherin (VE-cadherin), two endothelial cell adhesion molecules, were required for cell–cell interactions leading to network formation within three-dimensional collagen constructs.

Interactions between endothelial cells and other cell types during capillary morphogenesis have also been investigated using three-dimensional matrices. It has been shown that inclusion of supporting cell types, such as fibroblasts, can improve endothelial cell survival and increase cell migration [52, 53]. Growth factors, including fibroblast growth factor-2 (FGF-2), vascular endothelial growth factor (VEGF), and transforming growth factor β (TGF-β), have also been evaluated through the use of matrices. FGF-2 was found to facilitate tube formation, enhance cell migration, and promote cell survival [54, 55]. Simi-

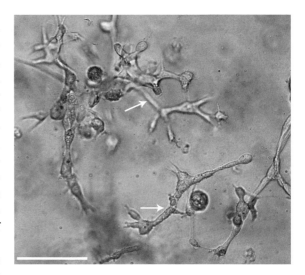

Fig. 52.2 Human umbilical vein endothelial cells begin to form capillary-like structures (*arrows*) within a collagen I matrix at 8 days in culture. Bar, 100 μm

larly, VEGF and TGF-β have been shown to induce the formation of capillary-like tubules [56, 57]. VEGF was also shown to work synergistically with FGF-2 to enhance the angiogenic response [57].

Angiogenesis is known to be important in the progression of a number of different pathologies, and angiogenesis inhibitors have emerged as promising drugs for dealing with a number of angiogenesis-dependent diseases. In vitro angiogenesis models can serve as invaluable tools for evaluating potential inhibitors. Park et al. [58] used three-dimensional collagen matrices to examine the ability of statins (3-hydroxy-3-methylglutaryl coenzyme A reductase inhibitors) to interfere with angiogenesis. They found that simvastatin and mevastatin inhibited capillary-like tube formation in these matrices, and therefore could be useful drugs for the treatment of certain cancers and other angiogenic diseases. Another group of researchers used a collagen-based in vitro model to evaluate the angiogenic inhibitor SU5416, a VEGF receptor-2 selective inhibitor [59]. This work could lead to a new treatment for diseases such as macular degeneration, which utilize this signaling pathway.

Another area of endothelial cell research that has taken advantage of in vitro models is matrix metalloproteinase (MMP) research. MMPs play a critical role in vascular remodeling during angiogenesis by degrading various ECM components [60, 61]. Membrane type 1 matrix metalloproteinase (MT1-MMP) expression has been shown to increase when endothelial cells are exposed to angiogenic growth factors; furthermore, development of capillary-like structures within collagen gels is delayed when endothelial cells are exposed to anti-MT1-MMP antibodies [62]. Similarly, Lafleur et al. [61] showed that MT-MMPs play a critical role during tubulogenesis within fibrin matrices. This type of in vitro data has lead to MMP inhibitors being tested to treat various angiogenic diseases [63, 64].

52.5
In Vitro Models of Articular Cartilage

Hyaline articular cartilage of the synovial joints is composed of an ECM encapsulating chondrocytes [65]. Type II collagen is the major component of the ECM, and forms a highly cross-linked network that entraps large, charged, aggregating proteoglycans, such as aggrecan. The interaction between type II collagen and proteoglycans contributes to the unique biomechanical properties of articular cartilage [66]. The current consensus is that full- or partial-thickness defects in mature articular cartilage do not heal. There are several surgical procedures that have been developed to treat cartilage lesions, including osteochondral grafting; however, these techniques have all failed to restore function [67, 68]. This largely stems from the inferior biomechanical properties of the fibrocartilage repair tissue that often results from these procedures [69].

To study the healing and regeneration of articular cartilage and its integration with bone, it is also possible to culture articular chondrocytes within a biomaterial matrix. Figure 52.3 shows that human articular chondrocytes cultured within a fibrin hydrogel that is underlaid with a bone substitute, will form an articular cartilage-like structure (Dare et al. unpublished results). Such three-dimensional biomaterial coculture systems have also been developed to model complex osteochondral diseases. For example, to study cartilage destruction due to rheumatoid arthritis, Schultz et al. [70] developed a three-dimensional model where human articular chondrocytes were suspended in a fibrin scaffold and cocultured with a three-dimensional gel seeded with synovial cells harvested from normal patients or patients afflicted with rheumatoid arthritis. It was found that the diseased synovial cells were highly invasive and invaded the chondrocyte-seeded scaffolds where they induced disease marker MMP-1 and VCAM-1 expression by the chondrocytes.

Biomaterial scaffolds have also been utilized to model disease onset and progression. Cortial et al. [71] seeded bovine chondrocytes into type I/III collagen sponges and then induced the osteoarthritic phenotype by culturing with interleukin-1β. The authors then proposed testing the effect of potential antiosteoarthritic drugs such as glucocorticosteroids [72] on the disease model.

In vitro culture systems have also been developed as tools to study normal embryonic bone formation. A model recapitulating the entire mammalian chondrocyte differentiation pathway was created by treating epiphyseal chondrocytes with 5-azacytidine [73]. Such models may provide a better understanding of the underlying mechanisms of some skeletal malformations and pathological disorders.

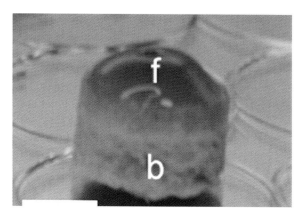

Fig. 52.3 Human fibrin gel (f) seeded with human articular chondrocytes form an articular cartilage-like structure, when cultured over a TruFit® BGS plug bone substitute (b) (Osteo-Biologics Inc.). Bar, 5.5 mm

52.6
In Vitro Models of Skin

Skin acts as a protective barrier to external assaults such as those from toxins, solar radiation, and infectious agents, and prevents dehydration by limiting water loss from the surface of the skin. Central to its role is its capacity to reepithelialize to restore the skin barrier after injury, through a wound healing process [74–76]. The stratified, cornified squamous epithelium, the epidermis, comprises layers of keratinocytes that continually divide in the basal layer and undergo programmed cell differentiation in the suprabasal layers. The outermost layer of the epidermis, the stratum corneum, contains terminally differentiated corneocytes, which in this layer are enucleated cells with intracellular cross-linked keratins and cell envelopes and extracellular lamellar lipids [77, 78]. Underlying the basal cells of the epidermis is a basement membrane that interfaces with the connective tissue of the dermis [79]. The dermis contains fibroblasts that form the ECM of the dermis, with type I and III collagens forming over 90% of the ECM [80–82]. In addition to keratinocytes and fibroblasts, skin also contains blood vessels, nerves, dendritic cells such as melanocytes and Langerhans cells, mast cells, and adnexa such as hair follicles with associated sebaceous glands, and sweat glands.

The need for skin for transplantation in patients with skin burns or chronic ulcers has driven research on tissue engineered three-dimensional skin constructs [83–86]. Constructs containing autologous or allogeneic skin cells have been used successfully as biological dressings to treat various types of wounds including burns; several of these are available commercially [40, 83–94].

In addition, in vitro skin models have been used to study diseases of the skin and as nonanimal models for toxicology [95–98]. To date, the major cell types within the skin from both human and animal sources have been successfully cultured [99–105].

Epidermal constructs have been grown using epidermal keratinocytes grown on cell culture inserts coated with collagen and basement membrane proteins to enhance the attachment, spreading, and proliferation of the keratinocytes. When raised to an air-liquid interface, the keratinocytes differentiate to form a stratified squamous epidermis. The stratum corneum surface of these cultures can be dosed with test chemicals and products to simulate topical exposures in vivo. This approach has been the basis for skin corrosion, irritation, and genotoxicity-type tests [95–98, 106–112], some of which have been accepted as international regulatory standards as alternatives to rabbit skin corrosion and irritation tests [113].

52.6.1
Tissue Engineered Three-Dimensional Models

Three-dimensional human skin coculture models have been developed, including epidermal-dermal and epidermal-melanocyte constructs. Epidermal-dermal constructs have been grown with human fibroblasts embedded in matrices containing animal-derived collagen such as bovine, or semisynthetic scaffolds [87, 89–91, 114–118]. Other models have incorporated fibroblasts in self-formed dermal matrices [119, 120]. Upon completion, the completed dermal matrices have been seeded on their surface with epidermal keratinocytes, which are allowed to grow and differentiate to form a stratified squamous epithelium at an air-liquid interface. Epidermal-dermal constructs can express major structural collagens in the dermal matrix, such as type I and III collagens and procollagen I, and basement membrane zone-specific markers such as type IV collagen, in addition to epidermal hyaluronic acid (Fig. 52.4). Interestingly, collagen

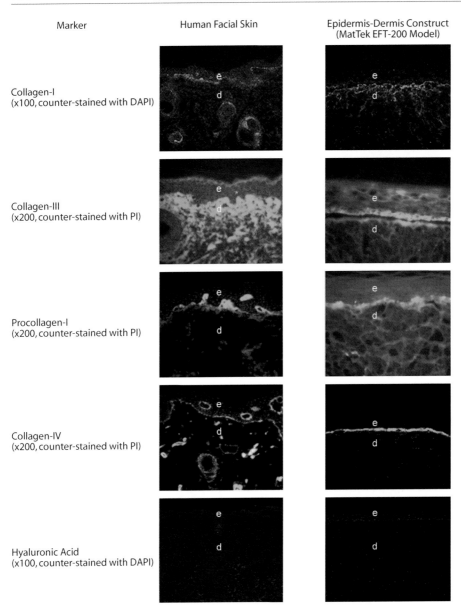

Fig. 52.4 Expression of skin markers by immunostaining of frozen sections of human facial skin (left panels) or human epidermal (*e*)—dermal (*d*) skin constructs (*right panels*). Eight micron sections were stained with monoclonal antibodies to specific human collagens, or biotinylated hyaluronic acid–binding protein. Secondary antibodies were labeled with fluorescein (collagens) or Texas red (hyaluronic acid). The sections were counterstained with 4',6-diamidino-2-phenylindole (DAPI) or propidium iodide (PI). Major dermal collagens expressed in facial skin (procollagen-I, collagens I and III) are also expressed in the skin constructs. A basement membrane zone (BMZ) collagen (IV) is localized to the BMZ in the skin constructs. Hyaluronic acid is primarily localized to the epidermis of both natural skin and the skin constructs

staining is observed to a lesser extent in the epidermis, for example type III collagen III in Figure 52.4. Some reports of epidermal expression of collagens have been made, such as type XVIII collagen [121].

52.6.2
Characterization of Skin Barrier

In all skin models, the aim is to reproduce key physiological functions. The best characterized is the skin barrier or stratum corneum barrier. The stratum corneum consists of flattened, terminally differentiated corneocytes surrounded by specialized lamellar lipids. The lipids are predominantly nonpolar and include cholesterol and ceramides [122, 123]. The lipids act to retard loss of water from the skin's surface, which can be measured functionally by use of an evaporimeter instrument [124, 125]. This instrument contains two water sensors mounted vertically in a sampling probe; when the probe is applied to skin, the difference in water vapor values between the two sensors is an indication of the flow rate of water from the surface of the skin. The end point measured has been termed transepidermal water loss (TEWL) and is expressed in units of grams of water per meter square of skin per hour. Evaporimeter instruments are commercially available and are used routinely in clinical studies to assess the extent of recovery of the stratum corneum barrier in wound healing studies [126, 127].

TEWL measurements can also be applied to in vitro skin systems. Epidermal cultures from a commercial supplier (Epiderm 201 cultures; MatTek Corp, Ashland, MA, USA) were grown in culture for 4 days. During this period there was an increase in the number of differentiated cell layers and stratum corneum thickness (insets in Fig. 52.5). In parallel, evaporimeter measurements indicated a decrease in TEWL values (Fig. 52.5), to levels comparable to those measured in intact human skin in vivo [128]. This finding indicates that in vitro epidermal constructs form a functional water barrier. One approach to understanding the molecular regulation of skin barrier formation is through analysis of mRNA expression with microarrays or gene chips, such as those available from Affymetrix (Santa Clara, CA, USA). Analysis of epidermal constructs (MatTek Epiderm model) reveals expression of pathways of genes involved in cholesterol synthesis that are also expressed in vivo [129] (Table 52.1). This approach confirms the value of these in vitro skin constructs to study the molecular biology of stratum corneum formation [130–132].

52.7
Summary

A number of tissue engineered in vitro models have been developed for use in research as well as in vitro alternatives to animal testing. We have provided four examples and their utility to date. In the development of all these models, the in vitro engineered construct of a desired tissue or organ is very much dependent upon the use of the appropriate biomaterial(s) as scaffolds, coupled with the use of the appropriate cell source(s). In most cases, replication of the natural microenvironment of the target tissue or organ appears to be the key to development of a mechanistically accurate in vitro model.

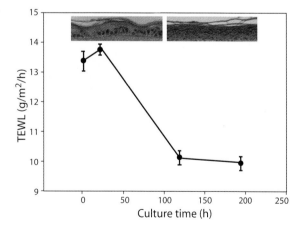

Fig. 52.5 Transepidermal water loss (TEWL) measurements of epidermal constructs. Epidermal cells grown at an air-liquid interface (*inset, top left*) developed stratified cell layers and a thickened stratum corneum (*inset, top right*) in parallel with a decrease in TEWL (*graph*), an indication of barrier development in vitro

Table 52.1. Human epidermal cholesterol metabolism genes expressed in human epidermal constructs, as determined by gene chip analysis

HMGCS1	3-Hydroxy-3-methylglutaryl-coenzyme A synthase 1
MMGCR	3-Hydroxy-3-methylglutaryl-coenzyme A reductase
MVK	Mevalonate kinase
PMVK	Phosphomevalonate kinase
MVD	Mevalonate (diphospho) decarboxylase
IDI1	Isopentenyl-diphosphate delta isomerase 1
FDPS	Farnesyl diphosphate synthase
FDFT1	Farnesyl-diphosphate farnesyltransferase 1
SQLE	Squalene epoxidase
LSS	Lanosterol synthase
CYP51A1	Cytochrome P450, family 51, subfamily A, polypeptide 1
SC4MOL	Sterol-C4-methyl oxidase-like
SC5DL	Sterol-C5-desaturase
NSDHL	NAD(P) dependent steroid dehydrogenase-like
KHCR7	7-Dehydrocholesterol reductase

References

1. Vacanti JP, Langer R (1999) Tissue engineering: the design and fabrication of living replacement devices for surgical reconstruction and transplantation. Lancet 354 Suppl 1:SI32–4
2. Atala A, Lanza RP, editors (2002) Methods of Tissue Engineering. San Diego: Academic Press;
3. Boudreau N, Myers C, Bissell MJ (1995) From laminin to lamin: regulation of tissue-specific gene expression by the ECM. Trends Cell Biol 5:1–4
4. Baldwin HS (1996) Early embryonic vascular development. Cardiovasc Res 31 Spec No:E34–45
5. Bodnar AG, Ouellette M, Frolkis M, Holt SE, Chiu CP, Morin GB, et al. (1998) Extension of life-span by introduction of telomerase into normal human cells. Science 279:349–52
6. Jiang XR, Jimenez G, Chang E, Frolkis M, Kusler B, Sage M, et al. (1999) Telomerase expression in human somatic cells does not induce changes associated with a transformed phenotype. Nat Genet 21:111–4
7. Huschtscha LI, Reddel RR (1999) p16(INK4a) and the control of cellular proliferative life span. Carcinogenesis 20:921–6
8. Wynford-Thomas D (1996) p53: guardian of cellular senescence. J Pathol 180:118–21
9. Rhim JS, Tsai WP, Chen ZQ, Chen Z, Van Waes C, Burger AM, et al. (1998) A human vascular endothelial cell model to study angiogenesis and tumorigenesis. Carcinogenesis 19:673–81
10. Watsky MA, Griffith M (2000) Whose Naughty or Nice: Electrophysiological Screening of Cells for Use in Tissue-Engineered Corneas. e-biomed: The Journal of Regenerative Medicine 1:115–20
11. Rae J, Cooper K, Gates P, Watsky M (1991) Low access resistance perforated patch recordings using amphotericin B. J Neurosci Methods 37:15–26
12. Nishida T. Cornea (1997) Fundamentals of cornea and external disease. In: Krachmer JJ, Mannis MJ, Holland EJ, editors. Cornea. St. Louis, Missouri: Mosby-Year Book Inc. p. 3–27
13. Jonas JB, Holbach L (2005) Central corneal thickness and thickness of the lamina cribrosa in human eyes. Invest Ophthalmol Vis Sci 46:1275–9
14. Freegard TJ (1997) The physical basis of transparency of the normal cornea. Eye 11 (Pt 4):465–71
15. Niederkorn JY (1990) Immune privilege and immune regulation in the eye. Adv Immunol 48:191–226
16. Rozsa AJ, Beuerman RW (1982) Density and organization of free nerve endings in the corneal epithelium of the rabbit. Pain 14:105–20
17. Ren H, Wilson G (1996) Apoptosis in the corneal epithelium. Invest Ophthalmol Vis Sci 37:1017–25
18. Dua HS, Azuara-Blanco A (2000) Limbal stem cells of the corneal epithelium. Surv Ophthalmol 44:415–25
19. Sack RA, Nunes I, Beaton A, Morris C (2001) Host-defense mechanism of the ocular surfaces. Biosci Rep 21:463–80
20. Komai Y, Ushiki T (1991) The three-dimensional organization of collagen fibrils in the human cornea and sclera. Invest Ophthalmol Vis Sci 32:2244–58
21. Ueda A, Nishida T, Otori T, Fujita H (1987) Electron-microscopic studies on the presence of gap junctions between corneal fibroblasts in rabbits. Cell Tissue Res 249:473–5
22. Watsky MA (1995) Keratocyte gap junctional communication in normal and wounded rabbit corneas and human corneas. Invest Ophthalmol Vis Sci 36:2568–76
23. Michelacci YM (2003) Collagens and proteoglycans of the corneal extracellular matrix. Braz J Med Biol Res 36:1037–46
24. ten Tusscher MP, Klooster J, van der Want JJ, Lamers WP, Vrensen GF (1989) The allocation of nerve fibres to the anterior eye segment and peripheral ganglia of rats. I. The sensory innervation. Brain Res 494:95–104
25. Muller LJ, Pels L, Vrensen GF (1996) Ultrastructural organization of human corneal nerves. Invest Ophthalmol Vis Sci 37:476–88
26. Draize JH, Woodard G, Calvery HO (1944) Method for the study of irritation and toxicity of substances applied topically to the skin and mucous membranes. J Pharmacol Exp Ther 82:377–89
27. Bagley DM, Waters D, Kong BM (1994) Development of a 10-day chorioallantoic membrane vascular assay as an alternative to the Draize rabbit eye irritation test. Food Chem Toxicol 32:1155–60
28. Gautheron P, Dukic M, Alix D, Sina JF (1992) Bovine corneal opacity and permeability test: an in vitro assay of ocular irritancy. Fundam Appl Toxicol 18:442–9

29. Ward SL, Walker TL, Dimitrijevich SD (1997) Evaluation of chemically-induced toxicity using an in vitro model of human corneal epithelium. Toxicol In Vitro 11
30. Klausner M, Hayden PJ, Breyfogle BA, Bellavance KL, Osborn MM, Cerven DR, et al (2003) The EpiOcular Prediction Model: A Reproducible In Vitro Means of Assessing Ocular Irritancy. In: Salem H, Katz SA, editors. Alternative toxicological methods Boca Raton, Fla.: CRC Press. p. 591
31. Zieske JD, Mason VS, Wasson ME, Meunier SF, Nolte CJ, Fukai N, et al. (1994) Basement membrane assembly and differentiation of cultured corneal cells: importance of culture environment and endothelial cell interaction. Exp Cell Res 214:621–33
32. Griffith M, Osborne R, Munger R, Xiong X, Doillon CJ, Laycock NL, et al. (1999) Functional human corneal equivalents constructed from cell lines. Science 286:2169–72
33. Suuronen EJ, Nakamura M, Watsky MA, Stys PK, Muller LJ, Munger R, et al. (2004) Innervated human corneal equivalents as in vitro models for nerve-target cell interactions. Faseb J 18:170–2
34. Chan KY, Haschke RH (1982) Isolation and culture of corneal cells and their interactions with dissociated trigeminal neurons. Exp Eye Res 35:137–56
35. Riggott MJ, Moody SA (1987) Distribution of laminin and fibronectin along peripheral trigeminal axon pathways in the developing chick. J Comp Neurol 258:580–96
36. Corcoran J, Shroot B, Pizzey J, Maden M (2000) The role of retinoic acid receptors in neurite outgrowth from different populations of embryonic mouse dorsal root ganglia. J Cell Sci 113 (Pt 14):2567–74
37. Muller LJ, Vrensen GF, Pels L, Cardozo BN, Willekens B (1997) Architecture of human corneal nerves. Invest Ophthalmol Vis Sci 38:985–94
38. Brock JA, McLachlan EM, Belmonte C (1998) Tetrodotoxin-resistant impulses in single nociceptor nerve terminals in guinea-pig cornea. J Physiol 512 (Pt 1):211–7
39. Nakamura M, Nishida T, Ofuji K, Reid TW, Mannis MJ, Murphy CJ (1997) Synergistic effect of substance P with epidermal growth factor on epithelial migration in rabbit cornea. Exp Eye Res 65:321–9
40. Trent JF, Kirsner RS (1998) Tissue engineered skin: Apligraf, a bi-layered living skin equivalent. Int J Clin Pract 52:408–13
41. Suuronen EJ, McLaughlin CR, Stys PK, Nakamura M, Munger R, Griffith M (2004) Functional innervation in tissue engineered models for in vitro study and testing purposes. Toxicol Sci 82:525–33
42. Johansson S (1994) Graded action potentials generated by differentiated human neuroblastoma cells. Acta Physiol Scand 151:331–41
43. Araki-Sasaki K, Aizawa S, Hiramoto M, Nakamura M, Iwase O, Nakata K, et al. (2000) Substance P-induced cadherin expression and its signal transduction in a cloned human corneal epithelial cell line. J Cell Physiol 182:189–95
44. Suuronen EJ, Muzakare L, Doillon CJ, Kapila V, Li F, Ruel M, et al. (2006) Promotion of angiogenesis in tissue engineering: developing multicellular matrices with multiple capacities. Int J Artif Organs 29:1148–57
45. Bee JA (1982) The development and pattern of innervation of the avian cornea. Dev Biol 92:5–15
46. Adamis AP, Aiello LP, D'Amato RA (1999) Angiogenesis and ophthalmic disease. Angiogenesis 3:9–14
47. Deroanne CF, Lapiere CM, Nusgens BV (2001) In vitro tubulogenesis of endothelial cells by relaxation of the coupling extracellular matrix-cytoskeleton. Cardiovasc Res 49:647–58
48. Sieminski AL, Hebbel RP, Gooch KJ (2005) Improved microvascular network in vitro by human blood outgrowth endothelial cells relative to vessel-derived endothelial cells. Tissue Eng 11:1332–45
49. Davis GE, Camarillo CW (1996) An alpha 2 beta 1 integrin-dependent pinocytic mechanism involving intracellular vacuole formation and coalescence regulates capillary lumen and tube formation in three-dimensional collagen matrix. Exp Cell Res 224:39–51
50. Korff T, Augustin HG (1999) Tensional forces in fibrillar extracellular matrices control directional capillary sprouting. J Cell Sci 112 (Pt 19):3249–58
51. Yang S, Graham J, Kahn JW, Schwartz EA, Gerritsen ME (1999) Functional roles for PECAM-1 (CD31) and VE-cadherin (CD144) in tube assembly and lumen formation in three-dimensional collagen gels. Am J Pathol 155:887–95
52. Dietrich F, Lelkes PI (2006) Fine-tuning of a three-dimensional microcarrier-based angiogenesis assay for the analysis of endothelial-mesenchymal cell co-cultures in fibrin and collagen gels. Angiogenesis 9:111–25
53. Sieminski AL, Padera RF, Blunk T, Gooch KJ (2002) Systemic delivery of human growth hormone using genetically modified tissue-engineered microvascular networks: prolonged delivery and endothelial survival with inclusion of nonendothelial cells. Tissue Eng 8:1057–69
54. Montesano R, Vassalli JD, Baird A, Guillemin R, Orci L (1986) Basic fibroblast growth factor induces angiogenesis in vitro. Proc Natl Acad Sci U S A 83:7297–301
55. Satake S, Kuzuya M, Ramos MA, Kanda S, Iguchi A (1998) Angiogenic stimuli are essential for survival of vascular endothelial cells in three-dimensional collagen lattice. Biochem Biophys Res Commun 244:642–6
56. Madri JA, Pratt BM, Tucker AM (1988) Phenotypic modulation of endothelial cells by transforming growth factor-beta depends upon the composition and organization of the extracellular matrix. J Cell Biol 106:1375–84
57. Pepper MS, Ferrara N, Orci L, Montesano R (1992) Potent synergism between vascular endothelial growth factor and basic fibroblast growth factor in the induction of angiogenesis in vitro. Biochem Biophys Res Commun 189:824–31
58. Park HJ, Kong D, Iruela-Arispe L, Begley U, Tang D, Galper JB (2002) 3-hydroxy-3-methylglutaryl coenzyme A reductase inhibitors interfere with angiogenesis by inhibiting the geranylgeranylation of RhoA. Circ Res 91:143–50
59. Takeda A, Hata Y, Shiose S, Sassa Y, Honda M, Fujisawa K, et al. (2003) Suppression of experimental choroidal neovascularization utilizing KDR selective receptor tyrosine kinase inhibitor. Graefes Arch Clin Exp Ophthalmol 241:765–72

60. Coussens LM, Werb Z (1996) Matrix metalloproteinases and the development of cancer. Chem Biol 3:895–904
61. Lafleur MA, Handsley MM, Knauper V, Murphy G, Edwards DR (2002) Endothelial tubulogenesis within fibrin gels specifically requires the activity of membrane-type-matrix metalloproteinases (MT-MMPs). J Cell Sci 115:3427–38
62. Chan VT, Zhang DN, Nagaravapu U, Hultquist K, Romero LI, Herron GS (1998) Membrane-type matrix metalloproteinases in human dermal microvascular endothelial cells: expression and morphogenetic correlation. J Invest Dermatol 111:1153–9
63. Overall CM, Kleifeld O (2006) Towards third generation matrix metalloproteinase inhibitors for cancer therapy. Br J Cancer 94:941–6
64. Thabet MM, Huizinga TW (2006) Drug evaluation: apratastat, a novel TACE/MMP inhibitor for rheumatoid arthritis. Curr Opin Investig Drugs 7:1014–9
65. Buckwalter JA, Mankin HJ (1998) Articular cartilage: tissue design and chondrocyte–matrix interactions. Instr Course Lect 47:477–86
66. Freemont AJ, Hoyland J (2006) Lineage plasticity and cell biology of fibrocartilage and hyaline cartilage: its significance in cartilage repair and replacement. Eur J Radiol 57:32–6
67. Hunziker EB (2002) Articular cartilage repair: basic science and clinical progress. A review of the current status and prospects. Osteoarthritis Cartilage 10:432–63
68. Marlovits S, Singer P, Zeller P, Mandl I, Haller J, Trattnig S (2006) Magnetic resonance observation of cartilage repair tissue (MOCART) for the evaluation of autologous chondrocyte transplantation: determination of interobserver variability and correlation to clinical outcome after 2 years. Eur J Radiol 57:16–23
69. Buckwalter JA, Mankin HJ (1998) Articular cartilage repair and transplantation. Arthritis Rheum 41:1331–42
70. Schultz O, Keyszer G, Zacher J, Sittinger M, Burmester GR (1997) Development of in vitro model systems for destructive joint diseases: novel strategies for establishing inflammatory pannus. Arthritis Rheum 40:1420–8
71. Cortial D, Gouttenoire J, Rousseau CF, Ronziere MC, Piccardi N, Msika P, et al. (2006) Activation by IL-1 of bovine articular chondrocytes in culture within a 3D collagen-based scaffold. An in vitro model to address the effect of compounds with therapeutic potential in osteoarthritis. Osteoarthritis Cartilage 14:631–40
72. Augustine AJ, Oleksyszyn J (1997) Glucocorticosteroids inhibit degradation in bovine cartilage explants stimulated with concomitant plasminogen and interleukin-1 alpha. Inflamm Res 46:60–4
73. Cheung JO, Grant ME, Jones CJ, Hoyland JA, Freemont AJ, Hillarby MC (2003) Apoptosis of terminal hypertrophic chondrocytes in an in vitro model of endochondral ossification. J Pathol 201:496–503
74. Slavin J (1996) The role of cytokines in wound healing. J Pathol 178:5–10
75. Smola H, Stark HJ, Thiekotter G, Mirancea N, Krieg T, Fusenig NE (1998) Dynamics of basement membrane formation by keratinocyte-fibroblast interactions in organotypic skin culture. Exp Cell Res 239:399–410
76. Werner S, Krieg T, Smola H (2007) Keratinocyte-fibroblast interactions in wound healing. J Invest Dermatol 127:998–1008
77. Elias PM (2005) Stratum corneum defensive functions: an integrated view. J Invest Dermatol 125:183–200
78. Fuchs E (2007) Scratching the surface of skin development. Nature 445:834–42
79. McMillan JR, Akiyama M, Shimizu H (2003) Epidermal basement membrane zone components: ultrastructural distribution and molecular interactions. J Dermatol Sci 31:169–77
80. Kielty CM, Shuttleworth CA (1997) Microfibrillar elements of the dermal matrix. Microsc Res Tech 38:413–27
81. Olsen BR (1991) Collagen Biosynthesis. In: Hay ED, editor. Cell Biology of the Extracellular Matrix. New York: Plenum Press. p. 177–220
82. Ushiki T (2002) Collagen fibers, reticular fibers and elastic fibers. A comprehensive understanding from a morphological viewpoint. Arch Histol Cytol 65:109–26
83. Bello YM, Falabella AF, Eaglstein WH (2001) Tissue-engineered skin. Current status in wound healing. Am J Clin Dermatol 2:305–13
84. Horch RE, Kopp J, Kneser U, Beier J, Bach AD (2005) Tissue engineering of cultured skin substitutes. J Cell Mol Med 9:592–608
85. Shakespeare PG (2005) The role of skin substitutes in the treatment of burn injuries. Clin Dermatol 23:413–8
86. Supp DM, Boyce ST (2005) Engineered skin substitutes: practices and potentials. Clin Dermatol 23:403–12
87. Boyce ST (1998) Skin substitutes from cultured cells and collagen-GAG polymers. Med Biol Eng Comput 36:791–800
88. Compton CC (1992) Current concepts in pediatric burn care: the biology of cultured epithelial autografts: an eight-year study in pediatric burn patients. Eur J Pediatr Surg 2:216–22
89. Eaglstein WH, Falanga V (1997) Tissue engineering and the development of Apligraf, a human skin equivalent. Clin Ther 19:894–905
90. Eaglstein WH, Falanga V (1998) Tissue engineering and the development of Apligraf, a human skin equivalent. Cutis 62:1–8
91. Eaglstein WH, Falanga V (1998) Tissue engineering and the development of Apligraf a human skin equivalent. Adv Wound Care 11:1–8
92. Hansbrough JF, Dore C, Hansbrough WB (1992) Clinical trials of a living dermal tissue replacement placed beneath meshed, split-thickness skin grafts on excised burn wounds. J Burn Care Rehabil 13:519–29
93. Higham MC, Dawson R, Szabo M, Short R, Haddow DB, MacNeil S (2003) Development of a stable chemically defined surface for the culture of human keratinocytes under serum-free conditions for clinical use. Tissue Eng 9:919–30
94. Kirsner RS (1998) The use of Apligraf in acute wounds. J Dermatol 25:805–11
95. Fentem JH, Botham PA (2002) ECVAM's activities in validating alternative tests for skin corrosion and irritation. Altern Lab Anim 30 Suppl 2:61–7

96. Osborne R, Perkins MA (1994) An approach for development of alternative test methods based on mechanisms of skin irritation. Food Chem Toxicol 32:133–42
97. Ponec M (2002) Skin constructs for replacement of skin tissues for in vitro testing. Adv Drug Deliv Rev 54 Suppl 1:S19–30
98. Welss T, Basketter DA, Schroder KR (2004) In vitro skin irritation: facts and future. State of the art review of mechanisms and models. Toxicol In Vitro 18:231–43
99. Bell E, Ivarsson B, Merrill C (1979) Production of a tissue-like structure by contraction of collagen lattices by human fibroblasts of different proliferative potential in vitro. Proc Natl Acad Sci U S A 76:1274–8
100. Boyce ST, Ham RG (1983) Calcium-regulated differentiation of normal human epidermal keratinocytes in chemically defined clonal culture and serum-free serial culture. J Invest Dermatol 81:33s–40s
101. Coulomb B, Lebreton C, Dubertret L (1989) Influence of human dermal fibroblasts on epidermalization. J Invest Dermatol 92:122–5
102. Hennings H, Michael D, Cheng C, Steinert P, Holbrook K, Yuspa SH (1980) Calcium regulation of growth and differentiation of mouse epidermal cells in culture. Cell 19:245–54
103. Rheinwald JG, Green H (1975) Serial cultivation of strains of human epidermal keratinocytes: the formation of keratinizing colonies from single cells. Cell 6:331–43
104. Wille JJ, Jr., Pittelkow MR, Shipley GD, Scott RE (1984) Integrated control of growth and differentiation of normal human prokeratinocytes cultured in serum-free medium: clonal analyses, growth kinetics, and cell cycle studies. J Cell Physiol 121:31–44
105. Yuspa SH, Kilkenny AE, Steinert PM, Roop DR (1989) Expression of murine epidermal differentiation markers is tightly regulated by restricted extracellular calcium concentrations in vitro. J Cell Biol 109:1207–17
106. Curren RD, Mun GC, Gibson DP, Aardema MJ (2006) Development of a method for assessing micronucleus induction in a 3D human skin model (EpiDerm). Mutat Res 607:192–204
107. Fentem JH (1999) Validation of in vitro Tests for Skin Corrosivity. Altex 16:150–3
108. Fentem JH, Briggs D, Chesne C, Elliott GR, Harbell JW, Heylings JR, et al. (2001) A prevalidation study on in vitro tests for acute skin irritation. results and evaluation by the Management Team. Toxicol In Vitro 15:57–93
109. Hoffmann S, Hartung T (2006) Designing validation studies more efficiently according to the modular approach: retrospective analysis of the EPISKIN test for skin corrosion. Altern Lab Anim 34:177–91
110. Kandarova H, Liebsch M, Spielmann H, Genschow E, Schmidt E, Traue D, et al. (2006) Assessment of the human epidermis model SkinEthic RHE for in vitro skin corrosion testing of chemicals according to new OECD TG 431. Toxicol In Vitro 20:547–59
111. Netzlaff F, Lehr CM, Wertz PW, Schaefer UF (2005) The human epidermis models EpiSkin, SkinEthic and EpiDerm: an evaluation of morphology and their suitability for testing phototoxicity, irritancy, corrosivity, and substance transport. Eur J Pharm Biopharm 60:167–78
112. Perkins MA, Osborne R, Johnson GR (1996) Development of an in vitro method for skin corrosion testing. Fundam Appl Toxicol 31:9–18
113. OECD (2004) OECD guideline for testing of chemicals. Paris, France: Organisation for Economic Co-operation and development;
114. Bell E, Ehrlich HP, Buttle DJ, Nakatsuji T (1981) Living tissue formed in vitro and accepted as skin-equivalent tissue of full thickness. Science 211:1052–4
115. El Ghalbzouri A, Jonkman MF, Dijkman R, Ponec M (2005) Basement membrane reconstruction in human skin equivalents is regulated by fibroblasts and/or exogenously activated keratinocytes. J Invest Dermatol 124:79–86
116. Hansbrough JF, Morgan JL, Greenleaf GE, Bartel R (1993) Composite grafts of human keratinocytes grown on a polyglactin mesh-cultured fibroblast dermal substitute function as a bilayer skin replacement in full-thickness wounds on athymic mice. J Burn Care Rehabil 14:485–94
117. Hayden PJ, Ayehunie S, Jackson GR, Kupfer-Lamore S, Last TJ, Klausner M, et al (2003) In vitro skin equivalent models for toxicity testing. In: Salem H, Katz SA, editors. Alternative Toxicological Methods. Boca Raton, FL, USA: CRC Press. p. 229–47
118. Stark HJ, Szabowski A, Fusenig NE, Maas-Szabowski N (2004) Organotypic cocultures as skin equivalents: A complex and sophisticated in vitro system. Biol Proced Online 6:55–60
119. Funk WD, Wang CK, Shelton DN, Harley CB, Pagon GD, Hoeffler WK (2000) Telomerase expression restores dermal integrity to in vitro-aged fibroblasts in a reconstituted skin model. Exp Cell Res 258:270–8
120. Wang CK, Nelson CF, Brinkman AM, Miller AC, Hoeffler WK (2000) Spontaneous cell sorting of fibroblasts and keratinocytes creates an organotypic human skin equivalent. J Invest Dermatol 114:674–80
121. Saarela J, Rehn M, Oikarinen A, Autio-Harmainen H, Pihlajaniemi T (1998) The short and long forms of type XVIII collagen show clear tissue specificities in their expression and location in basement membrane zones in humans. Am J Pathol 153:611–26
122. Uchida Y, Hamanaka S (2006) Stratum corneum ceramides: Function, origins and therepeutic applications. In: Elias PM, Feingold KR, editors. Skin Barrier. New York, NY, USA: Taylor & Francis. p. 43–64
123. Wertz PW. Biochemistry of human stratum corneum lipids. In: Elias PM, Feingold KR, editors. 2006 Skin Barrier. New York, NY, USA: Taylor & Francis. p. 33–42.
124. Oestmann E, Lavrijsen AP, Hermans J, Ponec M (1993) Skin barrier function in healthy volunteers as assessed by transepidermal water loss and vascular response to hexyl nicotinate: intra- and inter-individual variability. Br J Dermatol 128:130–6
125. Pinnagoda J, Tupker RA, Agner T, Serup J (1990) Guidelines for transepidermal water loss (TEWL) measurement. A report from the Standardization Group of the European Society of Contact Dermatitis. Contact Dermatitis 22:164–78
126. Levy JJ, von Rosen J, Gassmuller J, Kleine Kuhlmann R, Lange L (1995) Validation of an in vivo wound healing

model for the quantification of pharmacological effects on epidermal regeneration. Dermatology 190:136–41
127. Visscher M, Hoath SB, Conroy E, Wickett RR (2001) Effect of semipermeable membranes on skin barrier repair following tape stripping. Arch Dermatol Res 293:491–9
128. Breternitz M, Flach M, Prassler J, Elsner P, Fluhr JW (2007) Acute barrier disruption by adhesive tapes is influenced by pressure, time and anatomical location: integrity and cohesion assessed by sequential tape stripping. A randomized, controlled study. Br J Dermatol 156:231–40
129. Wertz PW (2000) Lipids and barrier function of the skin. Acta Derm Venereol Suppl (Stockh) 208:7–11
130. Koria P, Brazeau D, Kirkwood K, Hayden P, Klausner M, Andreadis ST (2003) Gene expression profile of tissue engineered skin subjected to acute barrier disruption. J Invest Dermatol 121:368–82
131. Ponec M, Weerheim A, Kempenaar J, Mulder A, Gooris GS, Bouwstra J, et al. (1997) The formation of competent barrier lipids in reconstructed human epidermis requires the presence of vitamin C. J Invest Dermatol 109:348–55
132. Ponec M, Boelsma E, Weerheim A, Mulder A, Bouwstra J, Mommaas M (2000) Lipid and ultrastructural characterization of reconstructed skin models. Int J Pharm 203:211–25

53 In Vivo Animal Models in Tissue Engineering

J. Haier, F. Schmidt

Contents

53.1	In Vivo Animal Models in Tissue Engineering: General Considerations	773
53.2	Experimental Animals	774
53.2.1	Pigs	774
53.2.2	Dogs	774
53.2.3	Rodents	775
53.2.4	Rabbits	775
53.2.5	Primates	775
53.3	Animal Models in Selected Areas of Tissue Engineering	775
53.3.1	Cardiovascular Tissue Engineering: Creation and Testing of Vascular Grafts	775
53.3.2	Intestinal Tissue Engineering	777
53.3.3	Biological Patches and Grafts	777
53.4	Prosthetics and Biomaterials	777
	References	778

53.1 In Vivo Animal Models in Tissue Engineering: General Considerations

Medical, legal, and ethical considerations require intense preclinical investigations of new biomedical products before introduction into clinical applications. For pharmacological developments these steps are well defined and regulated, but tissue engineering also necessitates comparable testing and developmental strategies. After animal cell lines have been used to establish new tissue engineering strategies in vitro, living animals are used to test the viability of engineered tissues in living organisms. Factors which cannot be assessed on a smaller scale include, but are not limited to, the role of angiogenesis in newly created tissues, complex immune reactions to the graft eventually leading to rejection and inflammation, as well as functional considerations such as rheological properties of vascular grafts, innervation, and kinetic properties of engineered intestinal tissues or the effects of surgical interventions. Each of these factors can significantly influence the experimental results, but all of the available models have limitations in one or more aspects.

To determine the model that is best suited for a specific task, anatomical, physiological, and pathophysiological properties of different species must be considered and compared with advantages and disadvantages of each model. Careful review of the existing literature is necessary to discover what has been done by others in the field asking a similar question. What has worked, and which techniques have not? What were the shortcomings of a model animal in a given context?

The following chapter attempts to provide the reader with a starting point by giving examples of in vivo animal models used in a variety of tissue engineering studies. A comprehensive review of this growing field is beyond the scope of this book. However, there are numerous excellent reviews concerned with the use of in vivo animal models in the different branches of tissue engineering and links can be found throughout this text.

53.2 Experimental Animals

53.2.1 Pigs

Pig anatomy, physiology, and immunological make-up resemble those of humans to a remarkable degree, and the costs for housing of these animals are relatively modest. However, pigs have the unfortunate tendency to grow quickly and reach considerable final sizes. This not only leads to increased costs of care and feeding, but also to experimental difficulties, such as matching engineered grafts and tissues to quickly changing animal size.

Strains bred specifically to reduce final body size and rate of growth, such as the Danish Landrace pig and the aptly named Göttingen minipig [1], are options to circumvent some of these problems. However, these races are significantly more expensive and harder to acquire than standard animals (for an approximation of the costs involved in care and feeding of different experimental animals, refer to Table 53.1).

The surgical and anesthesiological skills required to perform operations on pigs are more challenging, with complication rates exceeding those from equivalent procedures in humans.

53.2.2 Dogs

Dogs are of convenient size (10–30 kg), and the ease of care and modest growth rates (see Table 53.2) circumvent many of the problems involved in other experimental animals. While a variety of tissue engineering studies have been performed on dogs, their use has been sharply reduced by ethical considerations raised particularly in Western countries.

While mongrel dogs can and have been used in research, their varying sizes, growth rates, as well as other differences between individuals can lead to considerable experimental difficulties. Dogs bred specifically for experimental research, such as beagles, foxhounds, or Labradors, are more uniform, and are also often more cooperative and therefore easier to handle.

Table 53.1 Purchasing cost, cost of care and feeding, and cost of surgical procedures for several model animals used in tissue engineering studies

		Initial cost/animal	Monthly upkeep	Cost of anesthesia
Rat	Outbreed	10–20€	5–10€	15–20€
	Inbreed	20–30€	5–10€	
	Immunodeficient	80–100€	10–15€	
Mouse	Outbreed	4–6€	2–5€	15–20€
	Inbreed	10–17€	4–7€	
	Immunodeficient	45–60€	6–8€	
Guinea pig		40–50€	15–25€	15–20€
Hamster		20–30€	15–25€	15–20€
Rabbit		70–100€	20–30€	20–30€
Pig		150–200€	50–70€	50–100€
	Minipig	1000–1500€		
Dog	Variable (!)	~200–300€	150–200€	50–100€

Table 53.2 Advantages and disadvantages of in-vivo animal models

	Costs	Surgical skills	Susceptibility during anesthesia	Genetic variants (Transgenic/ Knock-outs)	Immunodeficient available	Anatomy comparable to humans	Physiology comparable to humans
Mouse	↓↓↓	↑↑	↑	↑↑↑	↑↑	↓↓	↓↓
Rat	↓↓	↓	↓	↑	↑	↓↓	↓↓
Rabbit	↓	↓↓	↓	-	-	↓	↓
Pig	↑	↑↑	↑↑	-	-	↑↑	↑↑
Dog	↑↑	↑↑	↓	-	-	↑↑	↑

53.2.3 Rodents

Rats and mice are widely used in all fields of tissue engineering research. Their ubiquity and ease of care make them ideal animal models where a close resemblance to the human organism is not required. This includes the study of biological features of cells and engineered tissues, preliminary proof-of-concept experiments and serial trials which would be too expensive or otherwise impossible using larger animals.

Commercially available rodent strains are not only highly standardized, leading to smaller biological variability, but are also frequently adapted to specific purposes by inbreeding and genetic modification.

Due to the small size of the animals, however, surgical procedures are often difficult to perform and limited in scope, and, therefore, results cannot easily be transferred to humans. Thus, while many new concepts can be tested successfully in rodent models, encouraging results must usually lead to further studies involving larger animals.

53.2.4 Rabbits

Rabbits as models for tissue engineering have the advantage of being readily available, relatively cheap, and reasonably easy to maintain. Surgical procedures, however, are more difficult as compared with larger animals, but the size of rabbits enables reasonable success rates even without specific surgical training. Comparable to other rodents, the transfer of research results to humans may not be easily accomplished.

53.2.5 Primates

Apes and monkeys can be considered the animal model most closely resembling human anatomy and physiology, and should therefore be most suited for many questions studied in the field of tissue engineering. In addition, a slow and predictable rate of growth makes them ideally suited for experiments determining the long-term behavior of tissue engineered grafts. Ethical and legal restrictions, however, combined with excessive cost involved in care and feeding of animals over long periods of time (see Table 53.1), have prevented a widespread use in the field.

53.3 Animal Models in Selected Areas of Tissue Engineering

53.3.1 Cardiovascular Tissue Engineering: Creation and Testing of Vascular Grafts

Before tissue engineered replacements for diseased blood vessels, heart valves, or cardiac muscle can be considered for use in humans suffering from one of the many prominent diseases of the cardiovascular system, extensive testing must be performed in experimental animals.

While tissue growth, histological, and immunological properties of engineered materials in many cases can be successfully studied using small animal models such as rodents, rabbits, or guinea pigs, surgical implantation of full-size grafts and long-term surveillance of adverse effects may require the use of bigger animals more closely resembling human anatomy and (patho-)physiology. This allows detailed monitoring and surveillance of adverse effects such as the rejection of prostheses, compromised blood supply, or thrombosis formation due to changes in blood flow.

Cardiovascular tissue engineering often involves the creation and testing of vascular grafts; either taken directly from organs such as the small intestine or made from synthetic materials enhanced with cell layers to prevent adverse effects due to thrombus formation or immunological reactions.

To protect synthetic endografts from rejection and to provide a more physiological inner lining small intestinal mucosa from pigs has been used to grow venous grafts in vivo. For example, when interposed into the superior vena cava, jejunal segments integrated into the surrounding vessel and developed an endothelial lining indistinguishable from the native venous endothelium [2]. Artificial mesh stents coated with porcine smooth muscle cells have been successfully implanted back into the animal, showing very little inflammatory reaction during a surveillance period of 1 month [3]. In another study, viability could be increased even more by pre-stressing coated grafts in bioreactors for 14 days prior to implantation [4].

In vivo dog models have been used to determine the viability of a wide variety of vascular grafts, such as artificial vascular prostheses coated with autologous cells derived from adipose tissue, epithelium, smooth muscle, and small intestinal submucosa. A combination of canine smooth muscle and epithelial cells, supported by a synthetic scaffold to provide necessary stability, showed extraordinary patency 26 weeks after implantation in the carotid artery position of dogs [5]. Dogs have also been transplanted with grafts constructed from decellularized vessels of bovine and porcine origin, showing patency and freedom from complications over considerable periods of time if used as a scaffold for autologous cells [6, 7]. Intestinal submucosa served as an autograft, allograft, and xenograft in a variety of experiments, demonstrating significant advantages over artificial materials in respect to patency, immunological reactions, and thrombosis formation [8]. Clowes et al. [9] used baboons to show proliferation of autologous epithelial and smooth muscle cells seeded on artificial iliacal grafts, with the newly formed cell layer covering up to 60% of the graft surface area after 12 months. Previous work by Zilla et al. [10], seeding polytetrafluoroethylene (PTFE) grafts with epithelial cells and RDG and leading to persistent confluency on the inner graft surface, has led to large clinical trials on human patients needing bypass surgery for peripheral artery disease and thrombosis.

In addition to these and other experiments in the field of cardiac tissue engineering, numerous studies have used rodents to determine mechanical, immunological, and rheological properties of engineered vascular grafts. Liu et al. [11] successfully reduced the complication rates of PTFE aortic prostheses in rats by incorporating a jugular vein to form the inner surface of the graft, thereby combining strength and biological compatibility. In another study, patency of carotid artery grafts was greatly increased by coating the synthetic material with epithelial cells [12].

In a series of studies conducted by Campbell et al. [13], vascular grafts have been created by implanting synthetic tubing into rabbit peritoneum. The resulting inflammatory reaction led to a complete covering of the tubes by a layer of mesothelial cells, myofibroblasts, and collagen after 2 weeks. After removal from the peritoneum, these layers could be stripped from the tubing, everted, and used as interponates in various vascular positions. Patency was high, while immunological reactions were virtually nonexistent.

In a study by Eschenhagen et al. [14], artificially engineered cardiac muscle originating from neonatal rat myocytes became well vascularized and survived for up to 14 days after being implanted into the peritoneum of model rats. Rat myocytes, made into three-dimensional constructs by a process of stacking cell-coated polymer sheets, were successfully reimplanted. Besides showing sufficient vascularization in vivo and viability for up to 12 weeks, these constructs exhibited electric potentials detectable in surface ECGs, and rhythmical contractions as well [15].

53.3.2
Intestinal Tissue Engineering

While there have been attempts at engineering almost every intestinal organ, such as the stomach, esophagus, or even liver and pancreas, living animal models are mainly used for the testing of engineered small intestinal grafts or patches.

53.3.3
Biological Patches and Grafts

Efforts to generate new intestinal mucosa by patching defects of the intestine with serosa from other parts of the organism, as well as from different species, were among the first attempts at intestinal tissue engineering. This patching with small intestine submucosa (SIS) taken from the same or different species has been studied in a variety of animal models and graft-host combinations, including pig, dog, and rabbit [16, 17]. For example, Chen and Badylak [18] used dogs as a model to evaluate the repair of defects in the animal's small bowel with porcine small intestinal submucosa, either as patches to repair smaller defects, or as tubular interponates. Regeneration of a mucosal epithelial layer, however, took place in a rather unorganized fashion, and was not feasible for larger patched areas. Animals grafted with tubular SIS usually died during the observation period. Submucosa from porcine small intestine was also used by Demirbilek et al. [19] to repair jejunal defects in rabbits with much better results. Six weeks after surgery, grafts showed a well-organized mucosal and submucosal layer. Problems reported included a certain amount of contraction in the grafts, while the anastomoses were not tested at all. In a study by Wang et al. [20], 2-cm segments of rat SIS were implanted into animals of the same species as isolated ileal loops, showing not only smooth muscle and serosa after 24 weeks, but biocompatibility and neovascularization. Reinnervation, however, did not take place. The New Zealand white rabbits were used in a series of experiments involving the repair of intestinal openings by patching them with SIS, abdominal wall, and peritoneum. While results were encouraging, showing complete coverage in case of peritoneal grafts, outcome was less favorable when neovascularization, nutritional transport, and were examined.

Patching of a defect in the terminal ileum with vascularized rectus abdominis muscle by Lillemoe et al. [17] led to a complete neomucosal cover of the patched area after 4 weeks, and the neomucosa exhibited electrophysiological parameters, glucose, and bile salt absorptions closely resembling the adjacent resident mucosa.

Collagen sponges derived from porcine skin were used to patch the wall of the stomach in a series of experiments conducted by Hori et al. [21, 22]. To protect the patch from gastric acid, a layer of silicone was applied on the luminal side. After 16 weeks, the graft was completely covered by functionally competent gastric mucosa. This led to further studies, in which silicone tubing covered with collagenous material was interposed in canine jejunum. Following the removal of the silicone stent after 1 month, intestinal tissue completely covered the luminal side of the collagen sheath by month 4.

53.4
Prosthetics and Biomaterials

These commonly consist of a scaffold made from synthetic materials (PGA, PTFE, etc.), coated with a variety of submucosal, epithelial, or other cells. Vacanti and colleagues were the first to use a rat animal model to grow neomucosa from disaggregated intestinal tissue (organoids) transferred to synthetic, absorbable biomaterials [23]. The process involves seeding polyglycolic acid (PGA)–polylactic acid (PLLA) scaffolds with organoids and implanting them into a rat's omentum, where a cystic, vascularized structure evolves. This neomucosa resembled the original tissue, showing all cell types present in normal intestinal mucosa, as well as a crypt-villi structure and functional microenvironment. Reimplantation into the animal's intestine did not lead to alimentary problems, and anastomoses displayed a great level of competence.

These promising first results have since led to further studies, investigating many aspects of biological properties of the organoid-scaffold model. Most of these, as the original research, have been carried out in rats. PGA scaffolds, combined with PLLA and coated with a variety of intestinal mucosal and epithelial cells, have undergone extensive testing. Aspects of biological graft behavior studied in vivo include angiogenesis [24], nutritional transport mechanisms [25], as well as lymphangiogenesis [26] and the patency of operative anastomoses [27, 28].

References

1. Hansen SB, Nielsen SL, Christensen TD, Gravergaard AE, Baandrup U, Bille S, Hasenkam JM. Latissimus dorsi cardiomyoplasty: A chronic experimental porcine model. feasibility study of cardiomyoplasty in Danish Landrace pigs and Göttingen minipigs. Laboratory Animal Science 1998;48(5):483–9.
2. Robotin-Johnson MC, Swanson PE, Johnson DC, Schuessler RB, Cox JL, Kennedy JH, Chachques JC, Crawford FAJ, Verrier ED. An experimental model of small intestinal submucosa as a growing vascular graft. Journal of Thoracic and Cardiovascular Surgery 1998;116(5):805–11.
3. Panetta CJ, Miyauchi K, Berry D, Simari RD, Holmes DR, Schwartz RS, Caplice NM. A tissue-engineered stent for cell-based vascular gene transfer. Human Gene Therapy 2002;13(3):433–41.
4. Tiwari A, Kidane A, Punshon G, Hamilton G, Seifalian AM. Extraction of cells for single-stage seeding of vascular-bypass grafts. Biotechnology and Applied Biochemistry 2003;38(1):35–41.
5. Matsuda T, Miwa H. A hybrid vascular model biomimicking the hierarchic structure of arterial wall: Neointimal stability and neoarterial regeneration process under arterial circulation. Journal of Thoracic and Cardiovascular Surgery 1995;110(4 I):988–97.
6. Conklin BS, Richter ER, Kreutziger KL, Zhong D-, Chen C. Development and evaluation of a novel decellularized vascular xenograft. Medical Engineering and Physics 2002;24(3):173–83.
7. Clarke DR, Lust RM, Sun YS, Black KS, Ollerenshaw JD. Transformation of nonvascular acellular tissue matrices into durable vascular conduits. Annals of Thoracic Surgery 2001;71(5 SUPPL.).
8. Lantz GC, Badylak SF, Hiles MC, Coffey AC, Geddes LA, Kokini K, Sandusky GE, Morff RJ. Small intestinal submucosa as a vascular graft: A review. Journal of Investigative Surgery 1993;6(3):297–310.
9. Clowes AW, Kirkman TR, Clowes MM. Mechanisms of arterial graft failure. II. chronic endothelial and smooth muscle cell proliferation in healing polytetrafluoroethylene prostheses. Journal of Vascular Surgery 1986;3(6):877–84.
10. Zilla P, Preiss P, Groscurth P, Rosemeier F, Deutsch M, Odell J, Heidinger C, Fasol R, Von Oppell U. In vitro-lined endothelium: Initial integrity and ultrastructural events. Surgery 1994;116(3):524–34.
11. Liu SQ, Moore MM, Yap C. Prevention of mechanical stretch-induced endothelial and smooth muscle cell injury in experimental vein grafts. Journal of Biomechanical Engineering 2000;122(1):31–8.
12. Kobayashi H, Kabuto M, Ide H, Hosotani K, Kubota T. An artificial blood vessel with an endothelial-cell monolayer. Journal of Neurosurgery 1992;77(3):397–402.
13. Campbell JH, Efendy JL, Campbell GR. Novel vascular graft grown within recipient's own peritoneal cavity. Circulation Research 1999;85(12):1173–8.
14. Eschenhagen T, Didié M, Heubach J, Ravens U, Zimmermann W-. Cardiac tissue engineering. Transplant Immunology 2001;9(2–4):315–21.
15. Shimizu T, Yamato M, Isoi Y, Akutsu T, Setomaru T, Abe K, Kikuchi A, Umezu M, Okano T. Fabrication of pulsatile cardiac tissue grafts using a novel 3-dimensional cell sheet manipulation technique and temperature-responsive cell culture surfaces. Circulation Research 2002;90(3).
16. Binnington HB, Sumner H, Lesker P. Functional characteristics of surgically induced jejunal neomucosa. Surgery 1974;75(6):805–10.
17. Lillemoe KD, Berry WR, Harmon JW. Use of vascularized abdominal wall pedicle flaps to grow small bowel neomucosa. Surgery 1982;91(3):293–300.
18. Chen MK, Badylak SF. Small bowel tissue engineering using small intestinal submucosa as a scaffold. Journal of Surgical Research 2001;99(2):352–8.
19. Demirbilek S, Kanmaz T, Özardali I, Edali MN, Yücesan S. Using porcine small intestinal submucosa in intestinal regeneration. Pediatric Surgery International 2003;19(8):588–92.
20. Wang ZQ, Watanabe Y, Toki A. Experimental assessment of small intestinal submucosa as a small bowel graft in a rat model. Journal of Pediatric Surgery 2003;38(11):1596–601.
21. Hori Y, Nakamura T, Kimura D, Kaino K, Kurokawa Y, Satomi S, Shimizu Y. Functional analysis of the tissue-engineered stomach wall. Artificial Organs 2002;26(10):868–72.
22. Hori Y, Nakamura T, Matsumoto K, Kurokawa Y, Satomi S, Shimizu Y. Tissue engineering of the small intestine by acellular collagen sponge scaffold grafting. International Journal of Artificial Organs 2001;24(1):50–4.
23. Vacanti JP, Morse MA, Saltzman WM, Domb AJ, Perez-Atayde A, Langer R. Selective cell transplantation using bioabsorbable artificial polymers as matrices. Journal of Pediatric Surgery 1988;23(1):3–9.
24. Gardner-Thorpe J, Grikscheit TC, Ito H, Perez A, Ashley SW, Vacanti JP, Whang EE. Angiogenesis in tissue-engineered small intestine. Tissue Engineering 2003;9(6):1255–61.
25. Tavakkolizadeh A, Berger UV, Stephen AE, Kim BS, Mooney D, Hediger MA, Ashley SW, Vacanti JP, Whang EE. Tissue-engineered neomucosa: Morphology,

enterocyte dynamics, and SGLT1 expression topography. Transplantation 2003;75(2):181–5.
26. Duxbury MS, Grikscheit TC, Gardner-Thorpe J, Rocha FG, Ito H, Perez A, Ashley SW, Vacanti JP, Whang EE. Lymphangiogenesis in tissue-engineered small intestine. Transplantation 2004;77(8):1162–6.
27. Kim SS, Kaihara S, Benvenuto MS, Choi RS, Kim B-, Mooney DJ, Vacanti JP. Effects of anastomosis of tissue-engineered neointestine to native small bowel. Journal of Surgical Research 1999;87(1):6–13.
28. Kaihara S, Kim S, Benvenuto M, Kim B-, Mooney DJ, Tanaka K, Vacanti JP. End-to-end anastomosis between tissue-engineered intestine and native small bowel. Tissue Engineering 1999;5(4):339–46.

Assessment of Tissue Responses to Tissue-Engineered Devices

K. Burugapalli, J. C. Y. Chan, A. Pandit

Contents

54.1	Introduction	781
54.2	Tissue Responses to Tissue-Engineered Devices	781
54.2.1	Innate and Adaptive Defence Systems	782
54.2.2	Pathologies Associated with Tissue Responses to Tissue-Engineered Devices	785
54.2.3	Adverse Affects of Pathologies on Device Function	788
54.2.4	Tissue-Engineered Device Design	788
54.3	Assessment Considerations	788
54.4	Tests for Assessment of Tissue Responses to Tissue-Engineered Devices	789
54.4.1	Organ and Body Weights	789
54.4.2	Haematology	790
54.4.3	Clinical Biochemistry	790
54.4.4	Histopathology	790
54.4.5	Immune Function Tests	791
54.4.6	Other Tests	792
54.5	Evaluation Tools	792
54.6	Future Directions	793
	References	796

54.1 Introduction

Tissue-engineered devices are functional substitutes for damaged, dysfunctional or lost tissues. The basic structural framework for these devices is provided by biodegradable biomaterials. In some instances, the biomaterials are used alone, especially when the materials are capable of recruiting host cells that will, in time, proliferate and differentiate to form desired functional tissues. However, for most tissue-engineering applications, biomaterials are commonly combined with biological components like cells, proteins, polysaccharides or DNA to facilitate the desired tissue regeneration. Such devices that combine biomaterial and biological components are referred to as combination devices. In an in-vivo environment, the various components of the combination devices could cause complex tissue responses that require rigorous pre-clinical testing. Nevertheless, before commencing any tests, it is essential to understand and define the necessary parameters required to detect the possible tissue responses. Hence, the objective of this chapter is to provide a fundamental overview of tissue responses to tissue-engineered devices, outline the tests needed to assess these responses and, finally, give recommendations for testing.

54.2 Tissue Responses to Tissue-Engineered Devices

A host responds to the implantation of a tissue-engineered device by activating its innate and adaptive defence systems. These defences are capable of recognising some or all components of the implanted

54.2.1
Innate and Adaptive Defence Systems

54.2.1.1
Innate Defence System

The innate defence system, referred to as natural, native or innate immunity, is a relatively non-specific response. This system has developed as a defence against common infectious microbial agents. The principal cellular components of this system, their properties and functions are listed in Table 54.1. The innate immunity constitutes the first line of defence triggered by tissue injury and subsequent binding of host cells and proteins to the foreign implant. These host proteins act as opsonins. Opsonised device components of approximately 10 µm in size or smaller are phagocytosed by phagocytic cells such as neutrophils and macrophages. After phagocytosis, these components are degraded by lysozymal enzymes and oxidative agents within these cells.

Non-phagocytosable components may be attacked by the extracellular release of hydrolytic enzymes and oxidative agents. Macrophages come in contact with each other and form multinucleated giant cells on the surface of the device. These giant cells also participate in extracellular degradation similar to macrophages and neutrophils. In addition, soluble components of the innate immunity, such as cytokines and the cells that secrete them, play a critical role in mobilising subsequent effectors, including components of adaptive immunity, in the effort to degrade and eliminate the tissue-engineered device components.

54.2.1.2
Adaptive Defence System

The adaptive defence system, referred to as specific, acquired or adaptive immunity, is an evolutionarily advanced system developed to adapt and respond to diverse foreign materials that come in contact with the tissues. Exquisite antigen/antibody specificity and memory for distinct macromolecules distinguish adaptive immunity from the primitive innate immunity. The cellular components of the adaptive immunity, their properties and functions are summarised in Table 54.2. A host has a pool of circulating cells and antibodies with specificity for most antigens (individual macromolecules on foreign materials) that the host tissues encounter during the host's lifetime. Any component (antigen) of the tissue-engineered device that is recognised by the existing pool of antibodies and/or memory cells is attacked first and, if possible, degraded and eliminated. However, the recognition of new antigens and subsequent development of antibodies against them requires a specialised mechanism called antigen presentation. Antigen-presenting cells (APCs) function to phagocytose the foreign material, process and present the processed antigens to T-helper cells, which activate B cells to develop antibodies and memory cells specific to the new antigens.

The innate and adaptive immune systems exist primarily to protect the body against infectious agents. However, when exposed to a tissue-engineered device, the components (antigens) of the device may not have been previously encountered by the host's defence system. The resultant tissue responses are, therefore, quite complex and difficult to predict. The immune activation may lead to the development of responses that adversely affect the host tissues, the device, and/or the device function. Hence, the assessments of tissue responses should encompass all possible pathologies associated with introduction of tissue-engineered devices in the body.

Table 54.1 Cellular components of the innate immune system, their properties and functions

Cell type	Properties	Chemotactic/activation factor	Secreted growth factors and cytokines	Functions
Neutrophils	12–15 µm, 4 h to 2 days lifespan; 70% of WBCs in blood; Multilobed nucleus joined by fine strands of nuclear material and colourless or pale eosinophilic stained cytoplasm	PF-4, IL-3, IL-8, IFN-γ, C5a, CCL3, CCL4	VEGF, FGF, CSF	Phagocytosis; degranulation—extracellular release of proteolytic and oxidative enzymes, free radicals, microbicidal agents that cleanse the wounds
Basophils	10–12 µm, 0.01–0.3% of WBCs in blood, contain large basophilic cytoplasmic granules which obscure the cell nucleus under the microscope; when unstained, the nucleus is visible and it usually has two lobes	IgE, antigen, GM-CSF, IL-3, CCL3, CCL4, CCL5	Histamine, serotonin, PG-D2, LT-C4, LT-E4, LT-D4, PAF, TNF-α, IL-4, IFN-α, ECF	Immediate hypersensitivity reaction (minutes of exposure to allergen): vasodilation, causing severe inflammation by vasoactive amines and lipid mediators. Late-phase reaction (2–4 h of exposure to allergen) of initiation: airway contraction, vasoconstriction, anaphylaxis, increased vascular permeability, recruitment of neutrophils, monocytes, and eosinophils, in response to cytokines
Mast cells	5–20 µm, present in connective tissues, contain large basophilic cytoplasmic granules			
Eosinophils	12–17 µm, 8 h to 12 days lifespan; 1–6% of WBCs in blood; Bilobed nucleus with strongly eosinophilic granules in the cytoplasm	CCL11, CCL24, CCL5 (RANTES), LT-B4, IL-5, GM-CSF, CCL-3, CCL4, ECF and IL-3	TGF-β, VEGF, PDGF; IL-1, IL-2, IL-4, IL-5, IL-6, IL-8, IL-13, TNF-α; LTC4, LTD4, LTE4, PG-E2	Degranulation releases histamine, free radicals, proteins such as eosinophil peroxidase, RNase, DNases, lipase, plasminogen, and major basic protein toxic to parasites and host tissues; phagocytosis and antigen presentation; involved in allergic responses, asthma pathogenesis, allograft rejection and neoplasia
Monocytes	16–20 µm, 1–3 days in blood, then migrate to tissues to differentiate into tissue macrophages, 3–8% of WBCs in blood, large bilobate eccentric nucleus, pale cytoplasm	PF-4, MCP-1, CCL-3, CCL-4, CCL-5, CSF	IL-8	Migrate to tissues and differentiate into macrophages in response to inflammatory stimuli
Macrophages	Varied morphologic appearance, normally round with large number of lysosomes. Cytoplasm can display slight granularity and pseudopodia. Small darkly stained nuclei normally eccentrically positioned, cell surface markers—CD68, CD163	IL-1, IL-3, GA-CSF, M-CSF, GM-CSF, IFN-α, IFN-β	IL-1, IL-6, IL-8, TNF-α, IFN-γ, CCL-3, CCL4, VEGF, TGF-α, PG, LT, CSF, PGDF, C, CF, HSP, FR, Enz	Wide range of functions—chemotaxis, phagocytosis, antigen presentation, immune regulation, fuse to form giant cells if they encounter nonphagocytosable particles/materials/cells, critical role in wound healing and resorption of tissue debris and engineered device components
Giant cells	Large cell with large number of nuclei and pale cytoplasm	IL-4 and IL-13		Phagocytosis and extracellular digestion of foreign materials
Natural killer (NK) cells	Large granular lymphocytes. Express cell surface receptors—CD16 and CD56	CCL-5, IL-1, IFN-α, IFN-β	IFN-γ, TNF-β	Defend the host from both tumours and virally infected cells without previous sensitization; kill cells which are missing MHC class I

Table 54.1 *(continued)* Cellular components of the innate immune system, their properties and functions

Cell type	Properties	Chemotactic/activation factor	Secreted growth factors and cytokines	Functions
Fibroblasts	Large oval nucleus with one or two nucleolus, tapering spindle-shaped morphology, basophilic cytoplasm, abundant rough ER when active. Inactive fibroblasts, called fibrocytes, are smaller and spindle shaped	CCL-5, PF-4, FGF	IL-1, IL-6, IL-8, IFN-β, TNF-α	Fibroblasts make collagens, glycosaminoglycans, reticular and elastic fibres, and glycoproteins found in the extracellular matrix. Tissue damage stimulates fibrocytes and induces the mitosis of fibroblasts
Endothelial cells	Cells lining the lumen of blood vessels and capillaries, cell surface markers—CD31 and von-Willibrand factor	CCL-5, PF-4, FGF, VEGF, hypoxia	ICAM-1, ELAM-1, VCAM-1	Involved in angiogenesis, and tissue regeneration, aid blood cell migration into tissues upon appropriate stimulus

C complement, *CCL* chemokines, *CD* cluster differentiation, *CF* clotting factors, *CSF* colony stimulating factors (prefix: *GA* granulocyte, *M* macrophage, *GM* granulocyte-macrophage), *Enz* enzymes, *ECF* eosinophilic chemotactic factor, *ELAM* endothelial-leukocyte adhesion molecule, *ER* endoplasmic reticulum, *FGF* fibroblast growth factor, *FR* free radicals, *HSP* heat shock proteins, *ICAM* intercellular adhesion molecule, *Ig* immunoglobulin, *IL* interleukins, *IFN* interferon, *LT* leukotriene, *PAF* platelet activating factor, *PDGF* platelet derived growth factor, *PF* platelet factor, *PG* prostaglandin, *TGF* transforming growth factor, *TNF* tumour necrosis factors, *VCAM* vascular cell adhesion molecule, *VEGF* vascular endothelial growth factor, *WBCs* white blood cells

Table 54.2 Cellular components of the adaptive immune system, their properties and functions

Cell type	Properties	Chemotactic/activation factor	Secrete growth factors and cytokines	Cell surface markers	Functions
TH1 helper cells	~7 μm diameter; 20–40% of WBCs in blood, mostly small lymphocytes with large, dark-staining nucleus with little to no basophilic cytoplasm, with the exception of plasma cells that are large lymphocytes (8–15 μm) with basophilic cytoplasm and an eccentric nucleus with heterochromatin in a characteristic cartwheel arrangement. Their cytoplasm also contains a pale zone next to nucleus and abundant rough endoplasmic reticulum	MCP-1, CCL-3, CCL-4, CCL-5; IL-1, IL-2, IL-4, IL-5, IL-6, IL12, IL15, IFN-β, TGF-b	IL-2, IFN-γ, IL-3, TNF-α, TNF-β GA-CSF, MCF	TCRαβ, CD3, CD4	Interact with APCs and are activated through cell membrane interactions, induces IgG2a (assists humoral immunity)
TH2 helper cells			IL-4, IL-5, IL-6, IL-10, IL-3, GM-CSF, TNF-β, IL-13	TCRαβ, CD3, CD4	Interact with APCs and are activated through cell membrane interactions, Induce IgG1 (assists humoral immunity)
Cytotoxic T cells			IFN-γ, TNF-β, TNF-α	TCRαβ, CD3, CD8	Lysis (apoptosis) of virally infected cells, tumour cells and allografts; participate in transplant rejection and Type IV hypersensitivity
Suppressor T cells			IL-10, TGF-β	TCRγ,δ, CD3	Immunoregulation and cytotoxicity
B cells		IL-1, IL-2, IL-4, IL-6	IL-4, TNF-α, IgG, IgA, IgM, IgE, IgD	MHC class II, CD19, CD21	Differentiate into plasma, resting or memory cells that secrete immunoglobulins (e.g. IgG, IgM, IgE)

APCs antigen-presenting cells, *MCP* monocyte chemotactic factor, *MHC* major histocompatibility complex, *TCR* T-cell receptor

54.2.2
Pathologies Associated with Tissue Responses to Tissue-Engineered Devices

A wide range of pathologies associated with innate and adaptive immune systems are known and well characterised [11]. These pathologies usually occur as secondary injury as a result of the innate/adaptive immune responses. The resulting disease states are commonly observed with microbial infections and toxins, such as tuberculosis, viral infections and bacterial lipopolysaccharides. It is crucial to assess if tissue-engineered devices would cause any pathology to the host. The common pathologies include tissue injury, necrosis, hypersensitivity, granuloma formation, foreign body reaction, fibrosis, scarring, calcification, and systemic pathology.

54.2.2.1
Tissue Injury and Necrosis

Injury to host tissue can be either direct (primary) or indirect (secondary). Irrespective of the cause, injury leads to necrosis (cell/tissue death). Histologically, necrosis may not be visible unless it is severe. If observed macroscopically or histologically, it is an important parameter in the assessment of tissue responses.

Primary/Direct Injury

The primary injury is the breach of integrity of the host tissue. From the perspective of assessment of tissue responses, the primary injury is attributed to the surgical implantation of the tissue-engineered device in the host. The physical presence of the device can further cause direct mechanical injury to the surrounding host tissue. In a healthy host, the normal wound healing mechanisms take place (involving haemostasis, inflammation, proliferation and/or fibroplasia) to return the injured tissue to tissue homeostasis. This process normally leads to the integration of the tissue-engineered device with the surrounding host tissue.

Secondary/Indirect Injury

Secondary injuries are caused by the host's innate and adaptive immune mechanisms against the tissue-engineered device. The innate immune mechanisms involve the extracellular release of degradative enzymes, oxidative agents and other chemicals by neutrophils, macrophages and foreign body giant cells. These agents can cause injury to the surrounding healthy host tissues. Such injury can also be triggered by leachable cytotoxic agents (including residual sterilising agents) from the device. If the device causes sustained tissue injury, neutrophils can be recruited in large quantities to cleanse the implantation site of deleterious agents (cytotoxic and infectious agents). Since neutrophils are short-lived and undergo apoptosis (programmed cell death), they become cell debris that requires clearance by macrophages. This debris is necrotic tissue that is usually visible histologically and macroscopically.

Tissue injury occurring secondary to interaction of antigen with humoral antibody or cell-mediated immune mechanisms is denoted as hypersensitivity. Traditionally hypersensitivity reactions are categorised into four types (Type I, Type II, Type III and Type IV). Type I, the most common hypersensitivity reaction (also called allergy) is mediated by IgE. The first exposure to an antigen produces memory B cells that, once activated, are capable of secreting IgG. Subsequent or prolonged exposure to the antigen usually mounts a rapid IgG production. Sometimes, the B cells can undergo spontaneous mutation to switch class and produce IgE (to any one of the millions of antigens that the host encounters in its lifetime). The antigen binds to IgE bound on mast cells or basophils to trigger an immediate hypersensitivity (allergy), potentially culminating in anaphylaxis (often lethal).

Type II hypersensitivity (cytotoxic) is mediated by antibodies directed toward antigens adsorbed on the host cell surface or other host tissue components (e.g. extracellular matrix). Three different antibody-dependant mechanisms may be involved in this type of reaction: (1) complement-dependent reactions, (2) antibody-dependent cell-mediated cytotoxicity, or (3) antibody-mediated cellular dysfunction. The former two induce phagocytosis by macrophages and neutrophils. Firstly, complement activation causes direct cytolysis of host cells that are adsorbed with antibody-antigen complexes, by the formation of membrane attack complex. Secondly, phagocytes bind to antibody-antigen complexes adsorbed on host cells and induce phagocytosis. Large non-phagocytosable cells and particles promote frustrated phagocytosis (extracellular release of degradative agents) by neutrophils, macrophages or giant cells causing indirect tissue injury. In addition, bound antibodies can also attract non-sensitised natural killer (NK) cells. Finally, antibodies bound on antigens adsorbed on receptors of host cells (e.g. receptors on neuronal synaptic junctions) cause tissue dysfunction without injury. Examples of medical conditions involving Type II hypersensitivity include Graves' disease (causing hyperthyroidism) and myasthenia gravis (causing muscular weakness and fatigue).

Type III hypersensitivity (immune complex mediated) reaction is induced by circulating (insoluble aggregates of) antigen-antibody immune complexes (ICs). ICs are usually cleared by macrophages in the spleen and the liver. Occasionally, ICs can deposit in vascular beds, on antigens in the extracellular matrix or on tissue-engineered devices. These IC deposits activate the complement system to recruit neutrophils and macrophages, leading to secondary tissue injury.

Type IV hypersensitivity (cell-mediated) reactions are initiated by specifically sensitised T-lymphocytes. This reaction includes the direct cell cytotoxicity mediated by $CD8^+$ T cells and the classic delayed-type hypersensitivity (DTH) reaction initiated by CD4+ T cells. The former is highly specific, without significant "innocent bystander" injury. However, the latter is relatively non-specific, mediated by non-specific cytokines released by $CD4^+$ T-helper cells that are stimulated through the class II major histocompatibility complex (MHC) on APCs. The soluble cytokines induce recruitment and proliferation of T-lymphocytes. They also attract and activate antigen non-specific macrophages that phagocytose and clear cell/tissue/device components present at the site. Evolutionarily, this system was developed to clear intracellular infections not accessible to antibodies (e.g. tuberculosis, leishmania, histoplasmosis), as well as a variety of large infectious agents not well controlled by antibodies alone (e.g. fungi, protozoans, parasites). The cytokines and activated macrophages are responsible for the tissue injury seen in delayed-type hypersensitivity.

Occasionally (but fortunately, rarely), the immune system loses tolerance for endogenous (self) antigens. Such conditions, called autoimmune diseases, result in secondary injury, involving specific cross-reactions to host tissue (self antigens). One example is the anti-cardiac antibodies that developed following certain streptococcal infections, causing rheumatic heart disease.

54.2.2.2
Granulation Tissue Formation

Injured and necrotic tissue is cleared by macrophages and the tissue deficit is replaced by granulation tissue. Macrophages initiate this healing response. As inflammation subsides, the fibroblasts and vascular endothelial cells at the injury site proliferate and begin to form granulation tissue. By budding/sprouting from blood vessels in the surrounding tissues, endothelial cells proliferate, mature and organise to form new capillaries and extend into the wound site or pores of the implanted device. This process is known as neovascularisation/angiogenesis. In addition, fibroblasts actively synthesise collagen and proteoglycans, and progressively fill the wound with new extracellular matrix. Macrophages are normally present in this newly forming tissue, while other cells may be present if chemotactic stimuli are generated by any residual/surrounding inflammatory cells.

From the perspective of a tissue-engineered device, different stages of granulation tissue formation may simultaneously be present in different regions of the implantation site [1]. Within minutes to hours of the surgical insertion of the tissue-engineered device, haemostasis and coagulation are triggered, resulting in the deposition of a provisional fibrin matrix between the tissue and the device. The provisional matrix provides a framework for migration of inflammatory cells on to the surface of the device. Eventually, macrophages degrade this matrix and simultaneously trigger granulation tissue formation. This occurs within 3–5 days following implantation. In some instances, depending on the surface and bulk properties of the biomaterial utilised, the provisional matrix may not be replaced by granulation tissue [14]. Granulation tissue formation within the actual device takes much longer and depends on the resorption rate of the biomaterial component of the device and resolution of inflammation. The resorbing biomaterial is gradually replaced by granulation tissue. Granulation tissue is only a provisional extracellular matrix formed to fill tissue deficits. Depending on the location of injury, the type of surrounding tissue and/or the bioactive agents present in the tissue-engineered device, the granulation tissue eventually remodels to form fibrous scar tissue or, sometimes, regenerates to form native tissue.

54.2.2.3
Foreign Body Reaction

During granulation tissue formation, macrophages at the surface of the non-phagocytosable device components fuse to form foreign body giant cells (FBGCs). The FBGCs together with granulation tissue components (macrophages, fibroblasts and capillaries) are referred to as a foreign body reaction. The foreign body reaction is considered deleterious to the device function, and may persist at the tissue/device interface for the lifetime of the implanted device.

54.2.2.4
Fibrous Encapsulation and Calcification

At the tissue/implant interface, if there is excessive or prolonged activation of the phagocytes, either by the sheer presence of the device and associated immune reactions or due to toxic agents leaching from the device, there would be continual host-tissue damage (chronic inflammatory reaction), leading to deposition of granulation tissue that eventually matures into fibrosis and scarring. The more severe the irritation and injury, the thicker and tightly packed would be the fibrous tissue (fibrous capsule). The capsule becomes avascular over a period of time, and the fibroblasts become scant and they differentiate into fibrocytes. Device encapsulation is a common problem encountered with biosensors.

Depending on the (surface) properties of the device and the presence of dead or degenerated tissue on the surface and/or within the device, deposition of calcium can be induced, leading to calcification. An example of pathological calcification is observed with glutaraldehyde crosslinked xenogenic/allogenic heart valves.

54.2.2.5
Systemic Pathology

Tissue-engineered devices are designed to contain both biomaterial and biological components, with the objective that the device provides initial functional support, gradually assisting host-tissue regeneration and, finally, allowing the regenerated tissue to assume the original tissue function. During this process, the components of the device that have served their purpose and no longer needed are removed/resorbed from the site. Essentially, their degradation and removal is aided by macrophages. The cocktail of enzymes, oxidative and other agents produced by macrophages breaks the device/tissue components (intracellularly or extracellularly) into oligomeric/monomeric units. The hydrolytic/oxidative degradation of the device components can potentially expose the host to new antigenic sites for immune recognition.

The breakdown products are cleared through the lymphatic system or they enter the blood stream. Breakdown products entering the lymphatic system reach the lymph node by either the intracellular or extracellular route. Intracellular route entry is facilitated by the APCs that transport the phagocytosed degraded products with the objective of sensitising T cells through the presented antigen. The lymphatic fluid that solubilises and transports the residual degraded products extracellularly is drained through lymphatic ducts. The lymphatic fluid is eventually drained into a subclavian vein and enters the blood circulatory system. All degraded products entering the blood stream are then transported to the liver and the spleen. Any antigenic determinants present in the degraded products are sensitised in the spleen. In the liver, these products are further metabolised by macrophages. The final metabolites are excreted through the lungs, kidneys or alimentary tract. Potentially any of the degradation products can cause systemic pathology. These potential problems occur either by excessive stimulation of inflammatory or immune cells, through circulating immune complexes at inappropriate sites causing injury, by activation of the complement cascade, or by facilitating binding of neutrophils and macrophages. Known examples include excessive stimulation by bacterial toxins and post-streptococcal glomerulonephritis [11].

54.2.3
Adverse Affects of Pathologies on Device Function

Pathologies (Sect. 54.2.2) can principally affect the device function through four mechanisms [2]. Firstly, macrophages and foreign body giant cells, of the foreign body response, can degrade the biomaterial component of the device. This principally affects the fundamental mechanical strength essential for the device to function. Secondly, the formation of a fibrous capsule prevents or significantly delays the resorption of the device components and subsequent restoration of host-tissue phenotype, architecture and function. Thirdly, a delay in angiogenesis and reduction in diffusion of nutrients can cause necrosis of exogenous cells within the device. Finally, immune responses can cause rejection of the biological components of the device. Hence, the device design parameters should also incorporate additional factors to circumvent the adverse affects. Examples include the use of growth factors, drugs, immunosuppressants or gene delivery agents to ameliorate the adverse effects. The use of such agents further complicates the resultant tissue responses.

54.2.4
Tissue-Engineered Device Design

The regenerative capacity of most adult human tissues is limited and if the structural framework of the tissue in question is damaged, the common result is fibrosis (usually leading to partial or complete loss of tissue function). The most challenging objective in tissue-engineering device design is the restoration of tissue architecture and function of damaged or lost tissue. In addition, the design parameters must also consider impediments like the affects of adverse host pathologies discussed in Sect. 54.2.3.

The basic structural framework for the device is provided by the biomaterial. In order to improve the regenerative potential of the device, three general approaches are followed in the design. The first approach is the incorporation of bioactive agents like growth factors, peptides and genes within the scaffold that are capable of stimulating the recruitment of appropriate cell types (stable/expandable cells and/or labile or stem cells) to restore the tissue phenotype. Secondly, cells with appropriate phenotype are expanded in vitro and incorporated in the device prior to implantation. Finally, the appropriate tissue is grown in vitro and the fully grown tissue is implanted. The tissue responses to such multi-component devices are complex and require a battery of tests to be able to predict the potential tissue responses in humans.

54.3
Assessment Considerations

The principal goal in tissue engineering is to restore or replace damaged or diseased tissue. In this direction, the challenge also lies in the effort to manipulate and obtain a favourable tissue response towards the tissue-engineered construct. Hence, it is essential not only to assess the safety related to the biological and chemical interactions of all device components with the host tissues (biocompatibility), but also to assess the structural and functional efficacy of the regenerated tissues (outcome variables).

Tissue engineered devices are combination devices containing biomaterial and biological components. The biomaterial is usually bioresorbable and requires assessments for both local and systemic effects. In contrast, the biological components are potential immunogens and require immunotoxicity assessments. Hence, tissue response assessments to tissue-engineered devices must include local, systemic and immune responses. The relevant experimental studies should be planned and implemented according to international standard (e.g. ISO 10993) guidelines.

Typically, assessment of safety and efficacy of a tissue-engineered device involves an implantation test in a clinically relevant animal model. The choice of the animal model usually depends on the tests to be performed. For example, systemic toxicity assessments are generally carried out in rodents, sensitisation assessments in rabbits or guinea pigs, and bone implants usually in large animal models like dogs, pigs or sheep.

In addition, the implantation sites for general safety assessments are performed in subcutaneous, muscle or bone tissues. Furthermore, since the tissue response assessments of a tissue-engineered device involve functional efficacy, the device must be evaluated using a clinically relevant functional model to validate its use and purpose.

Local, systemic and immunotoxicity assessments can be combined in one animal study, where each test device is placed in separate animals. The number of animals and treatments should be determined such that results for all parameters being assessed are statistically relevant. General guidelines in ISO 10993 can be used in determining the animals and groupings.

The test periods of tissue response assessments are usually long (e.g. months) for tissue-engineered devices because, they contain degrading materials and several device components are in continuous interaction with the surrounding host tissue. In general, the tissue response does not reach a steady state until after the degradation of the biomaterial components and local tissue heals/regenerates and returns to normal.

The inclusion of controls is essential for comparison of device efficacy and statistical comparison of results. For systemic and immune responses, tests on healthy animals are essential to reach any meaningful conclusion about the safety of the device. Sham treatments can also be used. The multi-component tissue-engineered devices are usually unique and may not have a suitable standard (or commercial) control device for comparison. In addition to sham treatments, (non-device) standard therapy for the targeted condition should be included. When available, suitable control device and/or commercial control should also be included as controls in the same study to show substantial equivalence or advantage.

Consideration should be given for the use of assays that could also be used in clinical trials. Non-destructive functional assays (e.g. blood tests) that are established in animal models can be useful in human studies if relevant. However, destructive means of assessing tissue responses and biomechanical testing of regenerating tissue are not feasible in patients. In such cases, non-invasive and non-destructive surrogate methods should be developed and validated in animal models and then be utilised in clinical trials.

Finally, adequate sample size is required for statistical testing of chronological changes in tissue responses for local, systemic and immunotoxicity assessments.

54.4
Tests for Assessment of Tissue Responses to Tissue-Engineered Devices

A wide range of tests are established for local, systemic and immune response assessments [6]. However, routine screening tests are used first. Then, further tests are selected when needed. The primary objective of this section is to summarise the different tests used for assessment of local, systemic and immune responses. Not all tests are essential for the application of a regulatory approval for the device. It is advisable to consult regulatory experts for specific testing requirements based on the device and the application in question.

54.4.1
Organ and Body Weights

Changes in organ and body weights are used as general (non-specific) indicators of potential systemic (adrenals, brain, heart, kidneys, liver, ovaries/testes, body) and immune (spleen, thymus, lymph nodes) toxicity.

54.4.2
Haematology

Differential peripheral blood cell counts are used as screening tests for systemic and immunological pathologies. Decrease in red blood cells and haemoglobin indicate anaemia, while increase indicates polycythemia, shock or trauma (often due to surgery). Decrease or increase in white blood cells and platelets indicate immune suppression or stimulation, respectively. Increase in neutrophils can indicate infections or tissue necrosis. Increase in lymphocytes can indicate viral infections or leukaemia, while the presence of excess eosinophils indicates allergy or parasitism.

Significant changes in any haematological parameters can be followed up by more sophisticated methods, such as flow cytometric analyses or immunostaining techniques useful in sorting of cells, based on unique cell surface receptors. Decreases or increases in percentages of any of the cell population relative to controls or in the ratios of B cells to T cells or $CD4^+$ to $CD8^+$ cells may indicate immunotoxicity. If immunotoxicity is suggested, phenotyping of cells of lymphoid organs like the bone marrow, spleen, thymus and draining lymph nodes can provide further insights into immune pathways (e.g. Th1, Th2) and/or pathologies (e.g. hypersensitivity reactions). Access to these organs will involve tissue biopsies or organ retrieval from the test animals.

54.4.3
Clinical Biochemistry

Routine clinical biochemistry markers, like serum proteins, enzymes and triglycerides in the blood, and urinalysis markers, like ions, proteins and pH of urine, are indicators of organ function and immune changes. Some examples of organ function abnormalities include: (1) very significant decrease in albumin, indicating extensive liver damage; (2) elevation of alkaline phosphatase and bilirubin, indicating cholestasis (may results in liver damage); (3) increase of lactic acid dehydrogenase, indicating skeletal muscle, cardiac muscle or liver damage; (4) absolute alterations in total protein, indicating decreased production in liver or increased loss through the kidneys. These systemic effects are the results of organ system damage and may be transient in nature.

Changes in albumin to globulin ratio in the blood can indicate immune dysfunction. The specifics of changes can be assayed using immunoelectrophoresis, radioimmunoassay (RIA) or enzyme-linked immunosorbent assays (ELISA). The relative proportions of IgG, IgM, IgA and IgE are indicators of B-cell activity. Serum levels of autoantibodies to host DNA, mitochondria and cells can be used to assess autoimmunity. Activation of complement in the presence of autoantibodies is also an indicator for autoimmunity. Furthermore, serum cytokines such as interleukins and interferons can be assayed to evaluate macrophage and lymphocyte activities. Prostaglandin E2 is used as a marker for macrophage function. Assays for components of the complement system (e.g. C3, C4) in the blood are used as measures of activity of the complement system.

54.4.4
Histopathology

Histopathology is an invasive and destructive method that cannot be used in assessment of patients, except where tissue biopsies are possible. However, this method provides the most significant data needed to understand the local, systemic and immune responses. At pre-determined time intervals, the test animals are euthanized, and the remaining device and/or implant sites and organs are collected for histopathology examinations.

For local tissue responses, histopathology is the key and direct method for evaluation. The parameters for histological assessment of local tissue responses, as a function of time, include:
1. The presence, thickness and quality of fibrous tissue surrounding the device (commonly referred as fibrous encapsulation)
2. The presence and extent of calcification
3. The number and distribution of inflammatory cells (including neutrophils, monocytes, macrophages, fibroblasts, lymphocytes, plasma cells, eosinophils, mast cells and giant cells), within and surrounding the device

4. The presence, extent and types of necrosis
5. The degree of vascularisation (key parameter where device contains exogenous cells)
6. The degradation of biomaterial component: rate, fragmentation and/or debris presence, form and location of remnants of degrading material
7. The quality of granulation tissue, granuloma formation (foreign body reaction), fatty infiltration, bone formation, and terminal restoration of local tissue configuration or fibrosis.

Routine haematoxylin and eosin and Masson's trichrome stained sections are sufficient for preliminary screening of local responses. If significant recruitment of lymphocytes is observed, specific immunohistochemical (IHC) staining for phenotyping of subsets of immune cells (e.g. $CD4^+$, $CD8^+$, $CD45^+$, $CD68^+$, $CD163^+$, $CD31^+$) can be utilised to elucidate the immune mechanisms. In addition, in-situ hybridization methods can be used to detect cytokines. Cytokines are produced in small quantities (picomolar), short-lived and are difficult to detect by simple IHC staining. In-situ hybridization on histology slides is used to amplify cytokine content (for subsequent IHC detection) from corresponding mRNA expressed within cells.

For systemic and immune responses, organs are histologically examined for clinical pathologies and abnormalities which include necrosis, atrophy, cysts, granulomas, focal aggregations of mononuclear cells in non-lymphoid organs and infections. In lymphoid organs, changes in relative volumes of germinal centres (B cells) and periarteriolar lymphocyte sheath (T cells) are used as indicators for immunotoxicity. Changes in cellularity are tested with immunostaining of T cells and their subsets in the lymphoid organs.

54.4.5
Immune Function Tests

Immune insults resulting from host innate and adaptive immune mechanisms may not always produce apparent morphological changes, unless the host is subjected to undue stress, repeated or prolonged exposure to the device. In addition, several mechanisms can produce the same outcome. Hence, routine toxicological and pathological tests are sometimes not conclusive to predict all possible immune pathologies. A wide range of immune function tests have been developed and recommended to diagnose immune dysfunctions and the parameters evaluated include immune suppression, immune stimulation, hypersensitivity and autoimmunity [6, 12].

Immune suppression is the down-modulation of the immune system caused by cell depletion, dysfunction or dysregulation that can result in decreased resistance to infections and neoplasia (tumours). Immune stimulation, in contrast, is an exacerbated immune reaction that may be apparent in the form of tissue damaging hypersensitivity reactions or pathologic autoimmunity. Differential cell counting and histopathology examination of lymphoid organs can be indicative of immune suppression or stimulation. Functional tests for immune suppression involving humoral immunity include plaque forming and natural killer (NK) cell assays, while that of adaptive immunity include the mixed lymphocyte response (MLR) and cytotoxic T-lymphocyte (CTL)-mediated assays [6]. Dose-related reduction in antibody producing B cells and NK cell activity is correlated to increased tumorigenesis and infectivity. MLR and CTL assays are predictive of host responses to transplantation and general immune competence. Lymphoproliferative assays for B and T cells are used to diagnose immune stimulation [6]. In addition, flow-cytometric analysis of lymphocyte subpopulations in the blood and lymphoid organs can also be used to diagnose immune suppression or stimulation.

Hypersensitivity reactions are exacerbated immune responses to an antigen leading to tissue damage and require at least two exposures to the antigen. Various methods are established to evaluate hypersensitivity reactions [3, 6]. Type I hypersensitivity reactions are studied using passive or active cutaneous or systemic anaphylaxis (PCA/ACA/ASA) and the shock reaction or skin reaction is monitored. There are no reliable pre-clinical models available to predict the effects for Type II and III hypersensitivity reactions. IgE and immune complexes in the sera of exposed animals can be assayed using ELISA or RIA techniques, if specific antibodies against the device components are available. Known pathologies like glomerulonephritis in the kidneys can be detected by histopathology. Type IV or delayed hypersensitivity reactions are tested using contact sensitisation tests, including guinea pig maximization test (GPMT), mouse ear swelling test (MEST) or local lymph node assay (LLNA). The latter two tests are quantitative.

Autoimmunity occurs when lymphocytes produce antibodies against the host's own body constituents. The mechanisms for autoimmunity are not fully elucidated. Traditional methods for testing autoimmunity involve the quantification of auto-antibodies in serum. Popliteal lymph node assay (PLNA) has been proposed as a test for autoimmunity [12].

54.4.6
Other Tests

Tests for evaluation of macrophage and neutrophil functions should also be included in the study, because they are the main effector cells responsible for the resorption of device components. The assays include differential counts of resident peritoneal cells, antigen presentation, cytokine production, phagocytosis, intracellular production of oxygen-free radicals and direct tumour killing potential.

In order to evaluate the quality of regenerated tissue, suitable biomarkers (e.g. antibodies to cell surface receptors/antigen) of the relevant tissue can be used. Biomechanical analysis of the regenerated tissue is also essential.

54.5
Evaluation Tools

The primary objective for tissue response assessments is to rule out any of the pathologies described in Sect. 54.2.2. A selection of tests (summarised in Sect. 54.4) may be chosen for tissue-response assessments as needed. The results of the majority of the tests (e.g. functional assays and clinical biochemistry) are quantitative and can be used for direct statistical power analysis. However, an assessment method such as histopathology requires tedious and time-consuming tools to acquire quantitative data for comparison. Hence, most researchers restrict their reporting of histopathology data to qualitative descriptions.

International standards (e.g. ISO 10993) recommend the use of scoring systems as tools for the evaluation of histopathological data. The scores are used for objective qualitative descriptions. However, scorings can become subjective and ambiguous in assessing parameters like necrosis, collagen and fatty cell contents. Therefore, such scores may not provide statistically relevant data.

Objects like cells of interest and blood vessels can be counted per microscopic field (e.g. $\times 400$ high-power field) on two-dimensional (2D) histology sections to obtain absolute numbers for statistical analysis. Often, such counts are also used for quantitative scoring systems. In addition, the availability and ease of use of image analysis software has made it possible to obtain quantitative data for assessing parameters like fibrosis, necrosis, collagen, and fatty cell infiltrations (e.g area fractions of collagen to total implant area) from 2D histology sections.

A more objective way of obtaining quantitative data from histology sections is to use stereology [5, 7, 9, 10, 13]. In stereology, quantitation is considered the objective method to describe a structure, compare two structures, study structural change and relate structure to function. Stereology allows access to three-dimensional (3D) information about geometrical structures based on observations made on 2D sections. Estimations of parameters of geometrical structures are made using sampled information, allowing inference of geometrical parameters such as volume, surface area, number, thickness and spacing. In biology, stereology provides a spatial framework upon which to lay physiological and molecular information. The nature of the structure under study in itself does not matter. It could include any macro- or microstructure in biology or tissue engineering, even holes can also be considered as structures.

The wide applicability of stereological approach is due to its reliance on basic geometric and statistical principles. Fields such as neurobiology, reproductive biology and cancer cell biology have applied stereological approaches to the interpretation of a wide variety of problems. To date, the use of stereology on wound healing and tissue engineering has been limited and mainly confined to the area of bone tissue engineering.

Stereological methods are simple, powerful, fast, accurate, objective, reproducible and verifiable. Their application for the quantitative evaluation of tissue responses to tissue-engineered devices has been recently reviewed by Garcia et al. [7]. Comprehensive evaluation of tissue responses and wound healing using stereological approaches have been reported by Burugapalli et al. [4] and Garcia et al. [8], respectively.

54.6
Future Directions

Tissue-engineered devices are combination devices that induce complex tissue responses in the host. The regulatory guidelines for evaluation of such devices are still under development and the device developers should consult regulatory experts for custom testing. Generally, functional assays are recommended to take precedence over tests for soluble mediators, followed by phenotyping, because they act as diagnostic assays for different pathologies associated with tissue responses to a tissue-engineered device. Tests for soluble mediators and phenotyping aid in elucidating the underlying immune mechanisms. The different pathologies associated with tissue-response assessments to tissue-engineering devices and testing recommendations are illustrated in Table 54.3.

The stereological approach is an efficient and objective method for obtaining quantitative information from histological sections. Accurate information can be obtained from a minimum amount of slices and counts. Furthermore, the increasing availability and affordability of powerful computing have made data collection, storage and analysis effortless. Hence, the use of stereology for evaluation of tissue responses to tissue-engineered devices is strongly recommended.

The immune mechanisms involve genetic components for the development of the immune pathologies. The relevance of the findings from pre-clinical animal models is questionable when they do not correlate with clinical pathologies. Alternative testing of tissue-engineered devices using microarray technology can be used to elucidate the underlying genetic pathways. However, extensive research is still required to develop and validate the microarray techniques customised for assessment of tissue responses to tissue-engineered devices.

Table 54.3 Pathologies associated with tissue responses to tissue engineered devices and tests for their assessments

Pathologies	Screening tests	Function tests	Soluble mediators	Phenotyping
Primary/direct injury	Local histopathology	—	—	—
Secondary injury				
Humoral				
Immune suppression	Histopathology of lymphoid organs and haematology	Plaque forming assay for antibody producing B cells, NK cell activity	Immunoassays for immunoglobulins,	Lymphocyte subsets in blood and lymphoid organs, using FACS, IHC
Immune stimulation	Histopathology of lymphoid organs and haematology	B cell lymphoproliferative assay	Immunoassays for immunoglobulins,	Lymphocyte subsets in blood and lymphoid organs, using FACS, IHC
Adaptive immunity				
Immune suppression	Histopathology of lymphoid organs and haematology	Mixed lymphocyte response and T-lymphocyte mediated assays	Cytokines in serum and tissue	Lymphocyte subsets in blood and lymphoid organs, using FACS, IHC
Immune stimulation	Histopathology of lymphoid organs and haematology	T cell lymphoproliferative assay	Cytokines in serum and tissue	Lymphocyte subsets in blood and lymphoid organs, using FACS, IHC
Type I hypersensitivity	—	Active or passive cutaneous or systemic anaphylaxis	Serum IgE	—
Type II hypersensitivity	—	No reliable testing methods	—	—
Type III hypersensitivity	—	No reliable testing methods	—	—
Type IV hypersensitivity	—	Guinea pig maximization test, mouse ear swelling test, local lymph node assay	—	—
Auto-immunity	—	—	Immunoassays for serum auto-antibodies	Routine diagnostic assays for auto immune diseases using IHC
Granulation tissue formation	Local histopathology on H&E and MT stained sections	—	—	—
Foreign body reaction	Local histopathology on H&E stained sections	—	—	—
Fibrous encapsulation and calcification	Local histopathology on H&E, MT, von Kossa/ Alizarin red stained histology sections	—	—	—

Table 54.3 (continued) Pathologies associated with tissue responses to tissue engineered devices and tests for their assessments

Pathologies	Screening tests	Function tests	Soluble mediators	Phenotyping
Systemic pathology	Body and organ weights, Clinical symptoms, haematology, clinical chemistry, urinalysis, Histopathology	–	Immune assays for immunoglobulins, complement proteins, autoantibodies and immune complexes	IHC for diagnosis of clinical pathologies
Acute and chronic inflammation	Local histopathology on H&E and MT stained sections	–	–	–
Angiogenesis	Local histopathology on IHC stained sections	–	–	–
Other tests				
Macrophage function tests	Local histopathology on H&E and MT stained sections, differential counts of resident peritoneal cells	Phagocytosis, intracellular production of oxygen-free radicals, direct tumour killing potential, antigen presentation, chemotaxis	Cytokines in serum and tissue	IHC for local tissues, local lymph nodes, for evaluation of phenotypic differentiation of macrophages
Neutrophil function tests	Local histopathology on H&E stained sections	Phagocytosis, intracellular production of oxygen-free radicals, chemotaxis	–	–

FACS fluorescence-activated cell sorting, *H&E* haematoxylin and eosin, *IHC* immunohistochemistry, *MT* Masson's trichrome

References

1. Anderson JM (2004) Inflammation, wound healing, and the foreign-body response. In: Ratner BD, Hoffman AS, Schoen FJ, Lemons JE (eds) Biomaterials science: an introduction to materials in medicine. Elsevier, London, pp. 296–304
2. Anderson JM (2006) Inflammatory and immune responses to tissue engineered devices. In: Bronzino JD (ed) Tissue engineering and artificial organs. Taylor & Francis, Boca Raton
3. Burleson GH, Dean JH, Munson AE (1995) Methods in immunotoxicity. Wiley, New York
4. Burugapalli K, Pandit A (2007) Characterization of tissue response and in vivo degradation of cholecyst-derived extracellular matrix. Biomacromolecules 8:3439–3451
5. Cruz-Orive LM, Weibel ER (1990) Recent stereological methods for cell biology: a brief survey. Am J Physiol 258:L148–L156
6. Gad SC (2002) Safety evaluation of medical devices. Marcel Dekker, New York
7. Garcia Y, Breen A, Burugapalli K, Dockery P, Pandit A (2007) Stereological methods to assess tissue response for tissue-engineered scaffolds. Biomaterials 28:175–186
8. Garcia Y, Wilkins B, Collighan RJ, Griffin M, Pandit A (2008) Towards development of a dermal rudiment for enhanced wound healing response. Biomaterials 29:857–868
9. Mandarim-de-Lacerda CA (2003) Stereological tools in biomedical research. An Acad Bras Cienc 75:469–486
10. Mayhew TM (1991) The new stereological methods for interpreting functional morphology from slices of cells and organs. Exp Physiol 76:639–665
11. Mitchell RN (2004) Innate and adaptive immunity: the immune response to foreign materials. In: Ratner BD, Hoffman AS, Schoen FJ, Lemons JE (eds) Biomaterials science: an introduction to materials in medicine. Elsevier, London, pp 304–318
12. Putman E, van der Laan JW, van Loveren H (2003) Assessing immunotoxicity: guidelines. Fundam Clin Pharmacol 17:615–626
13. Russ JC, Dehoff RT (1999) Practical stereology. Plenum Press, New York
14. Sanders JE, Lamont SE, Mitchell SB, Malcolm SG (2005) Small fiber diameter fibro-porous meshes: Tissue response sensitivity to fiber spacing. J Biomed Mater Res 72A:335–342

Part F
Clinical Use

XIII Clinical Application

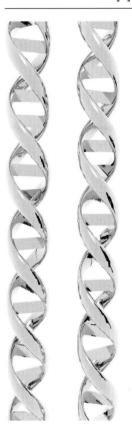
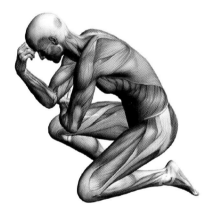

Evidence-based Application in Tissue Engineering and Regenerative Medicine

U. Meyer, J. Handschel

Contents

55.1 Introduction 801
55.2 Study Design Parameters 801
55.3 Criteria for Evidence-Based
      Clinical Studies 804
55.4 Clinical Use 805
      References 813

55.1 Introduction

Tissue engineering and regenerative medicine therapies are now being used in all medical specialities. Currently, efforts are being made to introduce new treatment options and also to establish clinical standards for new therapies. The further development of this new biomedical field necessitates that clinicians participate actively in the design, development, clinical introduction and informed use of this treatment option. The active participation requires that clinicians have a solid understanding of the biological and material-based principles of ex-vivo or in-vivo tissue generation. Several issues are important for assessing the clinical use of such therapies:

- Which parameters should be considered when such treatments are introduced?
- Which kind of clinical study in the light of evidence-based medicine should be performed?
- What is the recent state and evidence-based knowledge of clinical use in different medical fields?

55.2 Study Design Parameters

For assessing the efficacy and safety of tissue engineering, clinical studies must include a number of defined outcome measures. In contrast to animal studies, in which the euthanized animals are subjected to a variety of different investigations ranging from macroscopic to molecular in scale, clinical studies in humans use restricted outcome parameters (Tables 55.1, 55.2). Animal studies should, therefore, include histological and biomechanical evaluations in addition to outcome measures similar to those that are anticipated in clinical situations. When incorporating a cellular substitute in regenerative medicine, radiographs and computed tomography should be routinely performed in order to link high-resolution data with clinical findings. The repair response to a cellular construct varies substantially according to the complex biological and immunological situation of the defect site. In-vivo remodelling either reinforces, maintains, or degrades the tissue formed after transplantation. Both the initial and long-term biological responses to tissue-engineered constructs are strongly influenced by a variety of inherent patient factors. The integration of an extracorporally generated hybrid material into the surrounding tissue, for example, depends clinically to a large extent on a sufficient fixation of the transplanted construct and biologically to a great extent on the subsequent remodelling phase. Fixation of tissue constructs can be performed by plating, suturing or by a fit-in-fit system, or coverage of the cell-containing hybrid material with membranes or soft tissues (e.g. periosteal flaps). The loosening or loss of transplanted constructs should be considered as a recently unsolved aspect of tissue engineering.

Additionally, some specialised tissues, like bone or cartilage, remodel and form new tissue in the dependence of the mechanical environment. Implant integration of a variety of substitutes into the host site depends on the strain and stress environment acting on the scaffold construct as well as on the surrounding implantation bed. An incorrect fixation may result in the loss or destruction of scaffold integrity and also may induce cellular necrosis and scar formation [9]. Repeated disruption of vascular invasion at the construct interface resulting in impaired tissue formation is a clinically relevant problem for nearly all clinical applications. In the complete absence of adequate mechanical stimuli, however, implant integration may also fail in various tissue transplantation approaches. Thus, environmental factors like fixation precision and others can elicit both positive and negative effects on the biological response to implanted cell/scaffold constructs. Environmental host effects should, therefore, be anticipated in clinical studies when substitutions with extracorporal grown devices are planned [5]. An understanding of how host signals affect the integration of the tissue construct may provide clinically relevant information for improved transplantation strategies. Models that take these environmental parameters into account may be more predictive for human clinical trials.

Despite the remarkable progress towards the generation of tissue-engineered substitutes for therapeutic use, immunological problems associated with the recipient's response to the bioengineered tissue will have to be overcome. Although little has been published on the immune reaction towards bioengineered bone or cartilage tissue, it is known that the cell/scaffold complex will most likely induce a host reaction [6]. Both inflammatory and immune reactions can be anticipated, even with autologous tissue (Tables 55.3,

Table 55.1 Measures to evaluate tissue engineering and regenerative medicine therapies in preclinical and clinical studies

Animal experimantal and clinical evaluation methods	
Method (in vivo)	Determination
– Clinic	– Form, Function
– X-ray	– Hard tissue structure/overview
– Microradiography	– Hard tissue structure/details
– CT	– Hard and soft tissue structure/overview
– μCT	– Hard and soft tissue structure/details
– NMR	– Hard and soft tissue structure/details
– Densiometry	– Hard and soft tissue structure
– Serum analysis	– Mineral content
– Urine analysis	– Bone formation and resorption marker, hormones, cytokines
Method (in vitro)	
– Histology	– Tissue structure

Table 55.2 Outcome parameters and common investigations

Evaluation of in-vivo models
Clinically
– Form
– Function
Radiographically
– X-rays
– Ultrasound
– Computed tomography
– Magnetic reconance tomography
– Positron emission tomography
– Szintigraphy
Structurally
– Histology
– Immunohistochemistry
– Ultrastructure
Mecanically
– E-modulus
– Compression resistance

55.4). Clinical trials with allografts and xenografts have demonstrated that innate and adaptive immunity represent serious barriers. Inflammation likewise influences cell/scaffold engraftment and may also be pivotal for proper tissue remodelling. Therefore, it should not be assumed that an extracorporally bioengineered tissue will be inert. Rather, the host immune system will have a profound impact on the implanted bioengineered construct, and vice versa.

The two main components of an extracorporal tissue-engineered hybrid construct are, in general, the transplanted cells and the scaffold material. The implantation of a polymer/cell construct combines concepts of both biomaterials and cell transplantation. Immune rejection is a common host response towards the cellular component of tissue-engineered devices containing allogeneic, xenogeneic or ex-vivo-manipulated immunogenic cells [10].

The implantation of pre-formed scaffold material without cells can initiate a sequence of events related to a foreign body reaction. Clinically, it may present as an acute inflammatory response and in other situations as a chronic inflammatory rejection [2]. The duration and intensity of the clinical response is dependent on patient-associated parameters (extent of tissue injury, age, immunological status) and scaffold parameters (chemical composition, porosity, roughness, implant size and shape) [1]. It has been shown that the intensity of the response is modulated by the biodegradation process itself [7, 11], determining to a great extent the clinical fate of the hybrid material [3, 12, 18]. The immune response to common scaffold materials implanted into a defect region may result in a fibrous capsule containing macrophages and foreign body giant cells with needle-like, crystalline particles [14]. Some in-vivo studies have suggested that particles released from a degrading implant may affect tissue regeneration or repair by direct interaction with cells [7, 18]. Indirect interactions through inflammatory cytokines released by macrophages upon particle phagocytosis will have to be examined in more detail.

Immunological reactions against the transplanted cells and the host's response towards the hybrid material are critical after transplantation of the extracorporally fabricated construct [10]. Implant rejection can be avoided or at least diminished in clinical practice by using autologous or isogeneic cells as the principal cell source. Obviously, cells derived from an autologous source can be rendered immunogenic if they are genetically modified [17] or extensively cultured in vitro [15] prior to their transplantation. If allogeneic or xenogeneic cells are incorporated into the device, the prevention of immune rejection by immune suppressive treatment, the induction of tolerance in the host or immunomodulation of the graft become important clinical issues [8]. Immune rejection is the most important host response to the cellular component of tissue-engineered constructs containing non-autologous cells. Xenografts are rejected hyperacutely by pre-formed natural antibodies and complement activation [13] or more slowly by cellular immunity [4], when preventive measures had not been taken. It should be emphasised that the immune response to a cellular construct is critical for

Table 55.3 Aspects and components of immune reactions towards tissue engineered devices

Inflammatory/immune responses	
Aspect	**Components**
– Non-specific (innate)	– Cellular (Granulocytes Macrophages, Lymphocytes) – Non-cellular (Complement system, Lysozyme)

– Specific (acquired)
  – Naturally acquired
  – Artificially acquired
	– Humoral (B-cells, Immunoglobulins)
	– Cell-mediated (T cells)

Table 55.4 Factors contributing to immunity

Non-specific	Specific
Mechanical barriers (Ext. Surface)	T-cells
Chemical barriers (pH)	B-cells
Physiological factors (Temperature)	Immunoglobulins
Phagocytosis	
Antibody	
Complement	
Chemotactic factors	
Interferons	
Beta lysin	

the clinical success. Thus, the immunogenic nature of the biomaterial and the cellular components must be considered in the selection of both the scaffold material and the cell source. Clinical trials should (parallel to preclinical animal studies) evaluate thoroughly the immunological host response in order to improve clinical outcomes.

The use of allogeneic or xenogeneic cells has the clinical advantage of avoiding cell harvesting from the patient. Additionally, a time delay between the initial cell explantation and the final implantation is not given when allogeneic or xenogeneic cells are employed. Methods for the use of these cells in extracorporal tissue engineering have been developed. Microencapsulation is a clinically applied method to prevent immune reaction against allogeneic or xenogeneic cells [16]. It places cells within a core, such that the cells are sequestered by a material from direct contact with host cells and huge macromolecules. The wall thus formed, typically consisting of a polymer, is thought to provide immunoisolation by preventing contact between the encapsulated cells and the host immune system to allow for the transplantation of allogeneic or xenogeneic cells without extensive immunosuppressive therapy. Cartilage repair by xenogeneic chondrocytes placed in a polymer core, for example, is a recently established method in joint reconstruction. Similar approaches may also be successful in other cellular tissue engineering strategies.

55.3
Criteria for Evidence-Based Clinical Studies

Evidence-based medicine categorizes different types of clinical evidence and ranks them according to the strength of their freedom from the various biases that beset medical research. For example, the strongest evidence for therapeutic interventions is provided by systematic review of randomized, double-blind, placebo-controlled trials involving a homogeneous patient population and medical condition. In contrast, patient testimonials, case reports, and even expert opinion have little value as proof because of the placebo effect, the biases inherent in observation and reporting of cases, difficulties in ascertaining who is an expert, and so on. Systems to stratify evidence by quality have been developed, such as the following by the U.S. Preventive Services Task Force for ranking evidence about the effectiveness of treatments or screening:

- Level I: Evidence obtained from at least one properly designed randomized controlled trial.
- Level II-1: Evidence obtained from well-designed controlled trials without randomization.
- Level II-2: Evidence obtained from well-designed cohort or case-control analytic studies, preferably from more than one centre or research group.
- Level II-3: Evidence obtained from multiple time series with or without the intervention. Dramatic results in uncontrolled trials might also be regarded as this type of evidence.
- Level III: Opinions of respected authorities, based on clinical experience, descriptive studies, or reports of expert committees.

When possible, clinical studies of tissue engineering or regenerative medicine should be performed incorporating good clinical practice, ideally with Level I or II study designs. As these new therapies may also impose some risks, another classification system may be used to assess the treatment outcomes. Recommendation for a clinical service can also be classified by the balance of risk versus benefit of the service *and* the level of evidence on which this information is based. The uses are:

Level A: Good scientific evidence suggests that the benefits of the clinical service substantially outweigh the potential risks. Clinicians should discuss the service with eligible patients.

Level B: At least fair scientific evidence suggests that the benefits of the clinical service outweigh the potential risks. Clinicians should discuss the service with eligible patients.

Level C: At least fair scientific evidence suggests that there are benefits provided by the clinical service, but the balance between benefits and risks are too close for making general recommendations. Clinicians need not offer it unless there are individual considerations.

Level D: At least fair scientific evidence suggests that the risks of the clinical service outweigh potential benefits. Clinicians should not routinely offer the service to asymptomatic patients.

Level I: Scientific evidence is lacking, of poor quality, or conflicting, such that the risk versus benefit balance cannot be assessed. Clinicians should help patients understand the uncertainty surrounding the clinical service.

The inclusion of a second level is a distinct and conscious improvement on other recommendation systems and may be applied when such biomedical treatment options are compared with conventional measures.

55.4 Clinical Use

An unknown number of tissue engineering and regenerative medicine therapies were performed in humans (as single case applications or even in controlled clinical trials) during the last decade and a high number of them were published in biomedical journals. Current publications include an overview of:

- The dynamics of this biomedical field in clinical medicine over time.
- The research and clinical efforts concerning the various tissues and organs.
- The use of cell sources (stem cells versus differentiated cells).
- The distribution of clinical studies between the various medical specialities.
- The quality of the presented clinical study data.

Figures 55.1 to 55.9 provide an overview of the current state of publications concerning tissue engineering and regenerative medicine. The evaluation of the publication behaviour in respect of different criteria was done according to standardized searches in major libraries (PubMed, Medline, Cochrane). Analysis of the publication efforts in this new biomedical field mirrors the high dynamics of basic and clinical research. It is obvious that basic research in the field of tissue engineering evolved following invention of the term by Vacanti and Langer in 1998. As expected, basic research (in-vitro investigations and animal research) was conducted and published in the first phase of this new field. Significant numbers of clinical-based publications were first seen as late as in the year 2002. Table 55.5 gives a selection of publications meeting criteria of evidence-based medicine. Study groups now undertake clinical studies with advanced and challenging study designs (EBM Level 1). The results of these studies are at the forefront of the recent knowledge in tissue engineering and regenerative medicine. They can be used to discuss and determine how this new biomedical field is situated in clinical medicine.

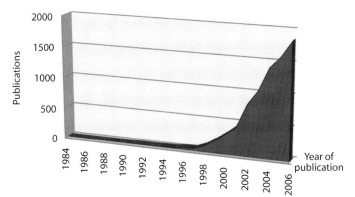

Fig. 55.1 Publications since 1984

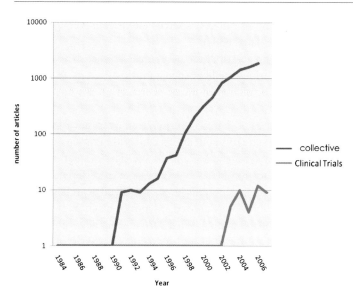

Fig. 55.2 Fraction of clinical trials among whole publications since 1984

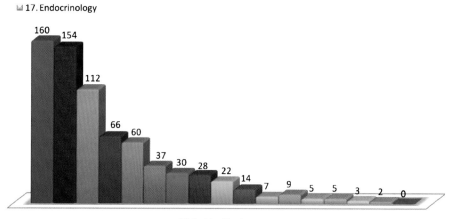

Fig. 55.3 Classification of publications according to clinical application

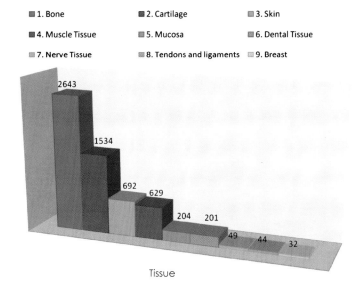

Fig. 55.4 Classification of publications according to tissue

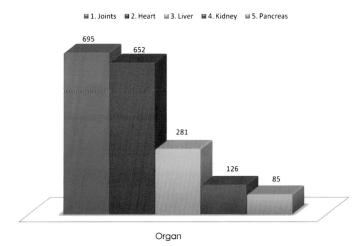

Fig. 55.5 Classification of publications according to organs

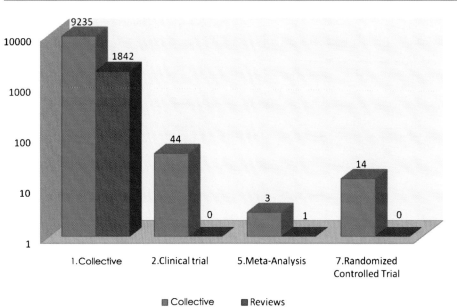

Fig. 55.6 Abstract of articles about tissue engineering

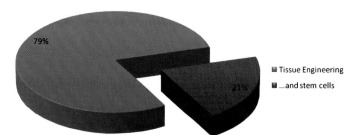

Fig. 55.7 Collective of publications about tissue engineering

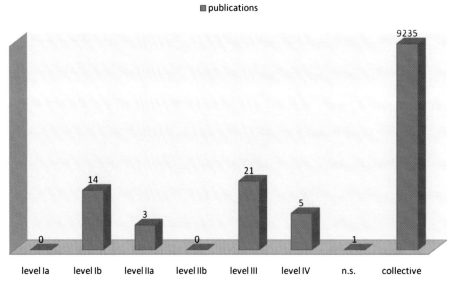

Fig. 55.8 Tissue engineering according to level of evidence

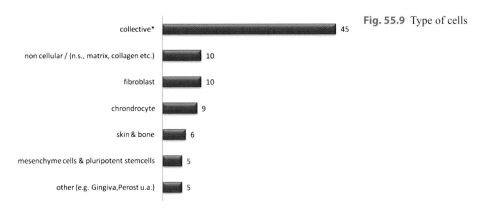

Fig. 55.9 Type of cells

Table 55.5 Selection of publications with specified evidence level (especially those with high evidence grades), in order to give the reader an insight into the current clinical evidence knowledge

Evidence	No.	Author(s)	Title	Pat. no.	Institution
Ib	1.	McGuire et al.	A randomized, double-blind, placebo-controlled study to determine the safety and efficacy of cultured and expanded autologous fibroblast injections for the treatment of interdental papillary insufficiency associated with the papilla priming procedure	20	Dept. of Periodontics, Dental Branch at Houston, University of Texas, Houston, USA
	2.	Weiss et al.	Autologous cultured fibroblast injection for facial contour deformities: a prospective, placebo-controlled, Phase III clinical trial	215	Dept. of Dermatology, Johns Hopkins University School of Medicine, Baltimore, Maryland, USA
	3.	Price et al.	A comparison of tissue-engineered hyaluronic acid dermal matrices in a human wound model	0	Center of Cutaneous Research, University of London, London, UK
	4.	Brigido et al.	The use of an acellular dermal regenerative tissue matrix in the treatment of lower extremity wounds: a prospective 16-week pilot study	28	Foot and Ankle Center at Coordinated Health, East Stroudsburg, PA, USA
	5.	Cooke et al.	Effect of rhPDGF-BB delivery on mediators of periodontal wound repair	16	Dept. of Periodontics and Oral Medicine, School of Dentistry, University of Michigan, Michigan, USA
	6.	Sarment et al.	Effect of rhPDGF-BB on bone turnover during periodontal repair	47	Dept. of Periodontics and Oral Medicine, School of Dentistry, University of Michigan, Michigan, USA
	7.	Richmon et al.	Effect of growth factors on cell proliferation, matrix deposition, and morphology of human nasal septal chondrocytes cultured in monolayer	6	Div. of Head an Neck Surgery, University of California-San Diego, San Diego, USA

Table 55.5 *(continued)* Selection of publications with specified evidence level (especially those with high evidence grades), in order to give the reader an insight into the current clinical evidence knowledge

Evidence	No.	Author(s)	Title	Pat. no.	Institution
	8.	McGuire et al.	Evaluation of the safety and efficacy of periodontal applications of a living tissue-engineered human fibroblast-derived dermal substitute. I. Comparison to the gingival autograft: a randomized controlled pilot study	25	Dept. of Periodontics, University of Texas Dental Branch, Houston, USA
	9.	Omar et al.	Treatment of venous leg ulcers with dermagraft	18	Dept. of General Surgery, Faculty of Medicine, Shebinel-Kom, Egypt
	10.	Yoshikawa et al.	Bone regeneration by grafting of cultured human bone	27	Dept. of Pathology, Nara Medical University, Kashihara City, Nara, Japan
	11.	Brigido et al.	Effective management of major lower extremity wounds using an acellular regenerative tissue matrix: a pilot study	n.a.	St. Agnes Medical Center, Philadelphia, Pa, USA
	12.	Tausche et al.	An autologous epidermal equivalent tissue-engineered from follicular outer root sheath keratinocytes is as split-thickness skin autograft in recalcitrant vascular leg ulcers	n.a.	Dept. of Dermatolgy, University Hospital of TU Dresden, Germany
	13.	Falanga	Tissue engineering in wound repair	240	Dept. of Dermatology, Boston University, MA, USA
	14.	Naughton et al.	A metabolically active human dermal replacement for the treatment of diabetic foot ulcers	n.a.	Advanced Tissue Sciences, La Jolla, California, USA
IIa	15.	Pradel et al.	Bone regeneration after enucleation of mandibular cysts: comparing autogenous grafts from tissue-engineered bone and iliac bone	20	Dept. of Oral and Maxillofacial Surgery, University Hospital of TU Dresden, Germany
	16.	Bachmann et al.	MRI in the follow up of matrix- supported autologous chondrocyte transplantation (MACI) and microfracture	27	Dept. of Radiology, Kerckhoff-Klinik Bad Nauheim, Germany
	17.	van Zuijlen et al.	Dermal substitution in acute burns and reconstructive surgery: a subjective and objective long-term follow-up	42	Dept. of Surgery, Red Cross Hospital and the Dutch Burns Foundation, Beverwijk, Netherlands
III	18.	Scala et al.	Clinical applications of autologous cryoplatelet gel for the reconstruction of the maxillary sinus. A new approach for the treatment of chronic oro-sinusal fistula	13	S.C. Oncologia Chirurgica, NCRI, Genoa, Italy
	19.	Zheng et al.	Matrix-induced autologous chondrocyte implantation (MACI): biological and historical assessment	56	Dept. of Orthopaedics, University of Western Australia, Nedlands, Perth, Australia

Table 55.5 *(continued)* Selection of publications with specified evidence level (especially those with high evidence grades), in order to give the reader an insight into the current clinical evidence knowledge

Evidence	No.	Author(s)	Title	Pat. no.	Institution
	20.	Filho Cerruti et al.	Allogenous bone grafts improved by bone marrow stem cells and platelet growth factors: clinical case reports	32	Clinica CERA LTDA, University of Sao Paulo Medical School, Brazil
	21.	Ochs et al.	Treatment of osteochondritis dissecans of the knee: one-step procedure with bone grafting and matrix-supported autologous chondrocyte transplantation	22	BG Unfallklinik Tübingen, Germany
	22.	Andereya et al.	Treatment of patellofemoral cartilage defects utilizing a 3D collagen gel: two-year clinical results	14	Orthopädische Universitätsklinik der RWTH Aachen, Germany
	23.	Yonezawa et al.	Clinical study with allogenic cultured dermal substitutes for chronic leg ulcers	13	Dept. of Dermatology, Kyoto University, Japan
	24.	Mavilio et al.	Correction of junctional epidermolysis bullosa by transplantation of genetically modified epidermal stem cells	1	Dept. of Biomedical Sciences, University of Modena and Reggio Emilia, Modena, Italy
	25.	Cebotari et al.	Clinical application of tissue engineered human heart valves using autologous progenitor cells	2	Dept. of Thoracic and Cardiovascular Surgery, Hannover Medical School, Germany
	26.	Hollander et al.	Maturation of tissue engineered cartilage implanted in injured and osteoarthritic human knees	23	University of Bristol Academic Rheumatology, Bristol, UK
	27.	Liu et al.	Ex-vivo expansion and in-vivo infusion of bone marrow-derived Flk-1⁺CD31⁻CD34⁻ mesenchymal stem cells: feasibility and safety from monkey to human	n.a.	Dept. of Hematology, Affiliated Hospital of Academy of Military Medicine Science, Beijing, China
	28.	Andereya et al.	First clinical experiences with a novel 3D-collagen gel (CaReS) for treatment of focal cartilage defects in the knee	22	Orthopädische Universitätsklinik der RWTH Aachen, Germany
	29.	Perovic et al.	New perspectives of penile enhancement surgery: tissue engineering with biodegradable scaffolds	204	Dept. of Urology, University Children's Hospital, Belgrad, Serbia & Montenegro
	30.	Beele et al.	A prospective multicenter study of the efficacy and tolerability of cryopreserved allogenic human keratinocytes to treat venous leg ulcers	27	Dept. of Dermatology, University Hospital, Ghent, Belgium
	31.	Marcacci et al.	Articular cartilage engineering with Hyalograft C: 3-year clinical results	141	Instituti Ortopedici Rizzoli, Laboratorio di Biomeccanica, Bologna, Italy
	32.	Lewalle et al.	Growth factors and DLI in adult haploidentical transplant: a three-step pilot study towards patient and disease status adjusted management	33	Dept. of Hematology, University of Brussels, Belgium

Table 55.5 *(continued)* Selection of publications with specified evidence level (especially those with high evidence grades), in order to give the reader an insight into the current clinical evidence knowledge

Evidence	No.	Author(s)	Title	Pat. no.	Institution
	33.	Strasser et al.	Stem cell therapy for urinary incontinence	42	Klinik für Urologie, Med. Universität Innsbruck, Austria
	34.	Nishida et al.	Corneal reconstruction with tissue-engineered cell sheets composed of autologous oral mucosal epithelium	4	Dept. of Ophtalmology, Osaka University Med. School, Osaka, Japan
	35.	Cherubino et al.	Autologous chondrocyte implantation using a bilayer collagen membrane: a preliminary report	13	Institute of Orthopeadics and Traumatology, University of Insubria, Italy
	36.	Camelo et al.	Periodontal regeneration in human class II furcations using purified recombinant human platelet-derived growth factor-BB (rhPDGF-BB) with bone allograft	4	Harvard School of Dental Medicine, Boston, USA
	37.	Humes et al.	Renal cell therapy is associated with dynamic and individualized responses in patients with acute renal failure	9	Dept. of Medicine, University of Michigan, Ann Arbor, USA
	38.	Rodkey et al.	A clinical study of collagen meniscus implants to restore the injured meniscus	8	REGEN Biologics, Vail, Colo., USA
IV	39.	Marcacci et al.	Stem cell associated with macroporous bioceramics for long bone repair: 6- to 7- year outcome of a pilot clinical study	4	Instituti Ortopedici Rizzoli, Laboratorio di Biomeccanica, Bologna, Italy
	40.	Gobbi et al.	Patellofemoral full-thickness chondral defects treated with Hyalograft-C: a clinical, anthroscopic, and histologic review	32	Orthopeadic Arthroscopic Surgery International, Milan, Italy
	41.	Schimming et al.	Tissue-engineered bone for maxillary sinus augmentation	27	Dept. of Maxillofacial Surgery, University Hospital Freiburg, Germany
	42.	Kashiwa et al.	Treatment of full-thickness skin defect with concomitant grafting of 6-fold extended mesh auto-skin and allogeneic cultured dermal substitute	5	Dept. of Plastic and Reconstructive Surgery, Kagawa Prefectural Central Hospital, Takamatsu, Kagawa, Japan
	43.	El-Kassaby et al.	Urethral stricture repair with an off-the-shelf collagen matrix	28	Center of Genitourinary Tissue Reconstruction, Dept. of Urology, Harvard Medical School; Boston, USA
	44.	Hasegawa et al.	Clinical trial of allogeneic cultured dermal substitute for the treatment of intractable skin ulcers in 3 Patients with recessive dystrophic epidermolysis bullosa	3	n.a.
n.a.					

References

1. Anderson JM (1988) Inflammatory response to implants. ASAIO Trans 34(2):101–107
2. Anderson JM (1993) Mechanisms of inflammation and infection with implanted devices. Cardiovasc Pathol 2:33–41
3. den Dunnen WF, Robinson PH, van Wessel R, Pennings AJ, van Leeuwen MB, Schakenraad JM (1997) Long-term evaluation of degradation and foreign-body reaction of subcutaneously implanted poly(DL-lactide-epsilon-caprolactone). J Biomed Mater Res 36(3):337–346
4. Dorling A, Riesbeck K, Lechler RI (1996) The T-cell response to xenografts: molecular interactions and graft-specific immunosuppression. Xeno 4:68–76
5. Guldberg RE (2002) Consideration of mechanical factors. Ann N Y Acad Sci 961:312–314
6. Harlan DM, Karp CL, Matzinger P, Munn DH, Ransohoff RM, Metzger DW (2002) Immunological concerns with bioengineering approaches. Ann N Y Acad Sci 961:323–330
7. Lam KH, Schakenraad JM, Groen H, Esselbrugge H, Dijkstra PJ, Feijen J, Nieuwenhuis P (1995) The influence of surface morphology and wettability on the inflammatory response against poly(L-lactic acid):a semiquantitative study with monoclonal antibodies. J Biomed Mater Res 29(8):929–942
8. Meiser BM, Reichart B (1994) New trends in clinical immunosuppression. Transplant Proc 26(6):3181–3183
9. Meyer U, Joos U, Wiesmann HP (2004) Biological and biophysical principles in extracorporal bone tissue engineering, Part I. Int J Oral Maxillofac Surg 33(4):325–332
10. Mikos AG, McIntire LV, Anderson JM, Babensee JE (1998) Host response to tissue engineered devices. Adv Drug Deliv Rev 33(1–2):111–139
11. Nakaoka R, Tabata Y, Ikada Y (1996) Production of interleukin 1 from macrophages incubated with poly (DL-lactic acid) granules containing ovalbumin. Biomaterials 17(23):2253–2258
12. Peter SJ, Miller ST, Zhu G, Yasko AW, Mikos AG (1998) In vivo degradation of a poly(propylene fumarate)/beta-tricalcium phosphate injectable composite scaffold. J Biomed Mater Res 41(1):1–7
13. Platt JL (1992) Mechanisms of tissue injury in hyperacute xenograft rejection. Asaio J 38(1):8–16
14. Rozema FR, Bergsma JE, Bos RRM, Boering G, Nijenhuis AJ, Pennings AJ, De Bruijn WC (1994) Late tissue response to bone-plates and screws of poly(L-lactide) used for fixation of zygomatic fractures. J Mater Sci Mater Med 5:575–581
15. Sato GH (1997) Animal cell culture. In: Lanza R, Langer R, Chick W (eds) Principles of tissue engineering. RG Landes, Austin, pp 101–109
16. Sefton MV, May MH, Lahooti S, Babensee JE (2000) Making microencapsulation work: conformal coating, immobilization gels and in vivo performance. J Control Release 65(1–2):173–186
17. Verna IM, Somia N (1997) Gene therapy—promises, problems and prospects. Nature 389:239–242
18. Winet H, Bao JY (1997) Comparative bone healing near eroding polylactide-polyglycolide implants of differing crystallinity in rabbit tibial bone chambers. J Biomater Sci Polym Ed 8(7):517–532

56. Tissue Engineering Applications in Neurology

E. L. K. Goh, H. Song, G.-Li Ming

Contents

56.1	Introduction	815
56.2	Cell-Based Tissue Engineering in CNS Repair	816
56.3	Applications of Tissue Engineering in Therapeutic Approaches to Nervous System Injury and Neurodegenerative Diseases	818
56.3.1	Spinal Cord Injury	818
56.3.2	Neurodegenerative Diseases	819
56.4	Conclusion	821
	References	822

56.1 Introduction

Tissue engineering has tremendous potential to revolutionize regenerative medicine. Research activities related to tissue engineering have expanded from the repair, replacement and regeneration of bone [1], blood vessels [2, 3] and organs such as the kidneys and heart [2–4] to the delicate nervous system. Initial research efforts mainly focused on providing three-dimensional (3D) scaffolds to facilitate the growth of nerve cells in vitro. The ultimate goal is to generate desirable cell types and create niches to support the growth and regeneration of endogenous cells and/or transplanted cells by recapitulating an in-vivo-like environment. Advances in basic neuroscience and biomaterials engineering have resulted in significant progress in this direction over the past few years. This chapter will discuss the potential and applications of tissue engineering approaches in regenerative neurological medicine.

The nervous system comprises of two interconnected but fundamentally different parts: the central nervous system (CNS) and the peripheral nervous system (PNS). The CNS consists of the brain and the spinal cord, whereas the PNS consists of somatic and autonomic nervous systems, which connect peripheral organs and tissues to the CNS. Neurons in the adult PNS retain some ability to regenerate. Thus, tissue engineering for neurological applications in vivo has been dominated by research on developing polymeric nerve guidance conduits to facilitate nerve regeneration and prevention of fibrous tissue in-growth that impedes the regenerating nerve. In contrast, adult CNS neurons are in general refractory to regeneration upon physical or chemical damage [5–7]. Several factors contribute to this lack of regenerative capacity. These include reduced intrinsic growth ability in adult neurons [8], physical barriers formed by the scar tissue, and the presence of inhibitory molecules that may exhibit further elevated expression and/or release upon injury [9]. Injuries and degenerative neurological diseases of the CNS eventually lead to the loss of neurons and supporting tissues. While glial cells are continuously generated in the adult nervous system, neurons in most regions of the brain are not. Therefore, cell replacement strategies, including the transplantation of stem cells or differentiated neural cells as well as promoting regeneration of endogenous neurons, have been the active lines of research [10–12]. The development of a combinatory approach of biodegradable polymeric channels used for PNS and the incorporation of growth factors and Schwann cells has provided excellent foundations for advances in tissue engineering for CNS regeneration [13].

56.2 Cell-Based Tissue Engineering in CNS Repair

The main purpose of tissue engineering is to improve or replace the biological functions of damaged or missing tissues or organs. In order to enhance the efficiency of tissue engineering, scaffolds with biocompatible and biodegradable materials are normally used in combination with biological active molecules, cells or tissues (Fig. 56.1).

Cell-based therapy has been attempted in injuries of the CNS and in some neurological disorders that are associated with neuronal cell death [14, 15]. Biomedical engineering approaches were initially applied in vitro, largely to optimize cell culture processes in order to obtain the desired cell types suitable for cell therapy. One particularly promising area involves the use of stem cells. Different types of scaffolds that can influence the survival, proliferation, differentiation, and fate determination of stem cells for various applications have been engineered [16–18]. Different types of stem cells, including embryonic stem cells, neural stem cells, hematopoietic stem cells and mesenchymal stem cells, have all been considered as transplantable cell sources for cell replacement therapy owing to their capability to differentiate into different cell types [10, 19–23]. The differentiation and regenerative potential of these stem cells have been extensively reviewed elsewhere [22, 23] and will not be elaborated upon here. Embryonic stem cells have been considered to be an ideal cell type for regeneration therapy due to their pluripotency, i.e., the potential to differentiate into almost all cell types found in adult tissues [24–30]. However, widespread clinical use of embryonic stem cells requires overcoming various practical hurdles and scientific and ethical issues associated with the acquisition of these cells. Therefore, efforts have also focused on finding sources of expandable and transplantable adult stem cells. Recent studies have indeed shown the existence of stem cells in many parts of the body, such as the brain, blood, skin, adipose tissue, bone, testis and amniotic fluid [10]. Other than

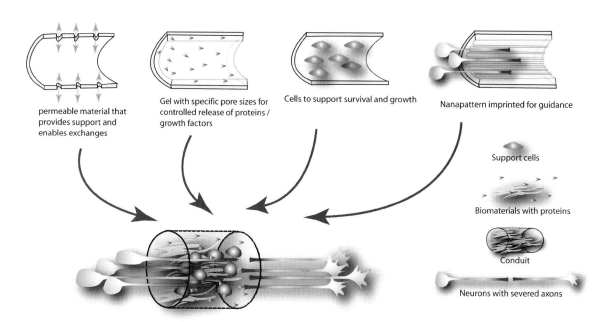

Fig. 56.1 Nerve conduits for regeneration of the PNS and CNS

these stem cell sources, the induced pluripotent stem cells (iPSC) derived from reprogramming of somatic cells has tremendous clinical potential for cell transplantations [31]. The potentially unlimited supply of somatic cells and also the potential of autologous transplantations are the two main advantages among many. However, issues such low reprogramming efficiency and the concern of genetic alterations will need to be solved before iPSC can be used for regenerative medicine.

Studies over the past decade have firmly established that new neurons are continuously generated in restricted regions of the adult mammalian brain, including the subventricular zone of the lateral ventricles and dentate gyrus of the hippocampus [32–37]. Through a process termed adult neurogenesis, new neurons from adult neural stem cells go through a sequential process of neuronal development and eventually become incorporated into the existing neuronal circuitry [37] and are believed to contribute to specific brain functions [38]. Other than these two regions, the generation of new neurons in the adult CNS appears to be very limited or non-existing [11, 39, 40]. On the other hand, a variety of adult neural stem cells have now been identified and isolated not only from regions with active neurogenesis but also from other brain regions, including the cortex and spinal cord [41–53]. With the novel tools and strategies developed, we are now able to manipulate adult neural stem cells from experimental animals and from humans in vitro [10, 37, 50, 54], and to study how these cells develop in vivo in animal transplantation models [55]. While there is a potential of using adult neurogenesis for regenerative applications, how these endogenous progenitor or stem cells could be mobilized or utilized in clinical settings remains to be explored.

One major challenge in developing stem cell-based transplantation therapy relates to the number and types of transplantable cells that can be realistically derived from adult stem cells expanded in culture [56, 57]. Tissue availability, suitable (preferably autologous) stem cell types and their expandability in culture are major issues that are currently being explored. There are also many obstacles associated with stem cell transplantation. Immune acceptance of allogenic grafts, effective means of transplantation, and the survival, differentiation and integration of transplanted cells, are all critical issues that need to be addressed. Advances in bioengineering could greatly facilitate efforts in developing stem cell-based therapy by providing permissive graft environments [17]. For example, neural stem cells supported by suitably engineered scaffolds were able to interact with host tissue when implanted into the infarct cavities of mouse brains from induced hypoxia-ischemia [58] and promote functional recovery after traumatic spinal cord injury [59].

In another aspect of cell-based therapy, the effective delivery of biological factors is also a main focus. Many biological factors, including neurotransmitters and neurotrophic factors, are known to influence the survival, neurite outgrowth, guidance and cellular function of implanted cells or endogenous cells in the nervous system. During development, axons and dendrites can be guided to their targets by a wide array of guidance cues, possibly present in the developing CNS as concentration gradients [60–62]. Multiple membrane-associated and soluble factors work in synergy to guide axons to their precise target. This guidance environment that exists during development is, however, lost in the adult CNS. Neurotrophic factors have been applied in animal models of CNS injury or degenerative neurological diseases to promote cell survival, axon regeneration and guidance [63–65]. In addition, micro- and nano-patterning have been explored in the bioengineering of scaffolds for neurite guidance in vitro (Fig. 56.1). For example, axons of peripheral neurons could be guided by nanopatterns on polymethylmethacrylate (PMMA) or protein-micropatterned surfaces in vitro [66]. While some positive effects on neuronal survival and neurite outgrowth have been observed upon delivery of these factors, either via osmotic pump or through implantation of silicon reservoirs, there are several caveats. These include potential failure of devices, infections, inflammations and poor in-vivo stability of the protein factors. In most cases, these factors can only be effectively delivered through transplanted cells engineered to release them in order to circumvent the blood-brain barrier (BBB) [67].

The immune response of the host is also a major issue in implantation/transplantation therapy. Cell or protein encapsulation can potentially overcome this problem [68]. The membrane of the capsules provides a barrier to prevent the immune system from neutralizing the proteins or killing the implanted cells. In addition, this approach also allows a slow and sustained release of molecules from the capsules [69].

In summary, cell-based engineering approaches serve to provide physical support, suitable microenvironment for survival, proliferation and appropriate differentiation as well as efficient cell and growth factor delivery in vivo. At the same time, these approaches also work on minimizing problems associated with the invasiveness of transplantation, including the host immune response.

56.3
Applications of Tissue Engineering in Therapeutic Approaches to Nervous System Injury and Neurodegenerative Diseases

The development of suitable biomaterials is expected to play an important role in the feasibility of therapeutic success, and is therefore a main engineering activity that would aid regenerative medicine. Some of these biomaterials developed for various applications in the nervous system are briefly discussed below.

Nerve guidance channels (NGCs) have been shown to facilitate axonal regeneration after transection injury to the peripheral nerve [13]. The guidance channels, as the name implies, serve as a physical guide for nerve growth and also confine the migration of cells within the channels. NGCs with growth factors incorporated onto the walls or within the matrix of their inner lumen were engineered to enhance regeneration [70, 71]. Neurotrophic factors (such as nerve growth factor) that promote neuronal survival and axonal outgrowth have been shown to improve regeneration in the PNS when used in combination with the channels [72]. Although NGCs have been shown to be able to repair short gaps, axonal regeneration over longer distances remains incomplete [4, 73]. One of the few reports showing significant regeneration over the distance of 30 mm in the PNS was published in 2000 [73]. In that study, 80-mm conduits consisting of a polyglycolic acid (PGA) collagen tube filled with laminin coated collagen fibers were implanted in the peroneal nerve in canine subjects. Numerous myelinated nerves were observed across the long gap and significantly enhanced functional recovery was recorded over controls.

56.3.1
Spinal Cord Injury

Despite the tremendous development and success of NGC in promoting the regeneration of the PNS, functional regeneration of the CNS remains very limited due to its physically and molecularly unfavorable environment [5–7]. The application of blockers of growth-inhibitor signaling, antibodies to inhibitory molecules, digestive enzymes of the glial scar or anti-inflammatory agents that prevent the formation of glial scar have shown some promises in improving the regeneration of CNS neurons [9, 74]. Tissue engineering using a variety of polymers has been evaluated for their biocompatibility as potential implants and cell carriers for the repair of spinal cord injury (SCI). These include natural polymers (such as alginate hydrogel and collagen), synthetic biodegradable materials [such as matrigel, fibronectin, fibrin glue, polyethylene glycol and poly (α-hydroxy acids)], synthetic non-biodegradable polymers [such as NeuroGel and poly 2-hydroxyethyl methacrylate (PHEMA) and poly 2-hydroxyethyl methacrylate-co-methyl methacrylate (PHEMM)] [75, 76]. Encouraging results have been observed from procedures implanting composite biosynthetic conduits combining the more rigid scaffolds with hydrogel or extracellular matrix molecules [77]. The rigid scaffolds mainly serve as a physical support, while the hydrogel can be impregnated with growth and guidance factors and also serve as extracellular matrix for implanted cells. For example, PHEMA soaked with brain-derived neurotrophic factor (BDNF) [78] or N-(2-hydroxypropyl) methacrylamide (HPMA) incorporated with the cell-adhesive region of fibronectin Arg-Gly-Asp (RGD) tripeptide sequence [77, 78] and implanted into lesion optic tract or cerebral cortex [78] of the adult rats promoted angiogenesis and axonal penetration into the gel and within the microstructure of the tissue network. Similar observations were made with poly-[N-(2-hydroxypropyl)-methacrylamide] (PHPMA)-RGD [79] implanted into injured spinal cord [79]. In addition, the hydrogel reduced nerve tissue necrosis and cavitation in the adjacent white and gray matter [79]. RGD coupled to HPMA hydrogels also encouraged growth of glial tissue when implanted into rat brains [80].

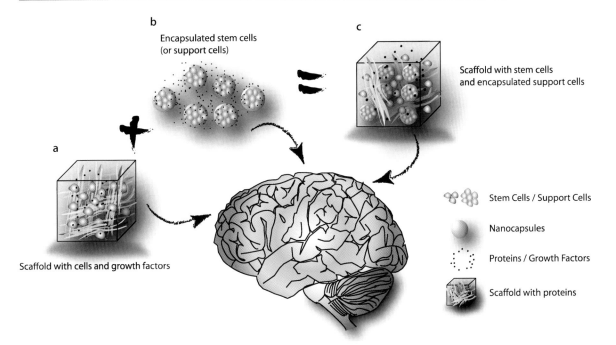

Fig. 56.2a–c The various systems for the regeneration of the CNS. **a** Growth factors or proteins of interest incorporated in biomaterials made up the scaffold that support growth and survival of the cells. **b** Encapsulation of cells genetically engineered to release growth factors or proteins of interest can also avoid an immune response in vivo. **c** The combination of the biomaterial scaffold and cell encapsulation approach allows more sophisticated manipulation of microenvironment that increases the efficiency of implantation

It is now widely accepted that a combinatory approach is necessary to facilitate functional recovery from SCI. This means a combination of scaffold architecture, transplanted cells and locally delivered molecular agents (Fig. 56.2). Different cell types have been used as cell replacements, to release growth-promoting factor and to facilitate myelination. A multi-channel, biodegradable scaffold [polylactic-co-glycolic acid (PLGA) with copolymers] containing Schwann cells implanted after SCI was shown to contain regenerating axons 1 month post-operation [59]. In a sophisticated study, neural stem cells were seeded into a multicomponent polymer scaffold modeled after the intact spinal cord. Implantation of this scaffold-neural stem cells unit into an adult rat hemisection model of SCI promoted long-term functional improvement (persistent for 1 year in some animals) [81]. The regrowth of axons within injured CNS is also promoted by implanted hydrogel matrices containing BDNF and ciliary neurotrophic factor (CNTF) producing fibroblasts [82]. As mentioned earlier, encapsulation has been suggested to alleviate the host immune response. Indeed, a study using alginate encapsulated BDNF-producing fibroblast grafts showed it to permit recovery of function without immune suppression [83, 84].

56.3.2
Neurodegenerative Diseases

56.3.2.1
Parkinson's Disease (PD)

PD is generally characterized by the loss of dopaminergic neurons at the substantia nigra pars compacta (SN), ultimately resulting in non-motor and motor-related symptoms such as tremor, rigidity and bradykinesia [85]. The current treatments for PD include conventional drug replacement therapy, surgical treatment, and cell transplantation [85]. The use of cell transplantation in patients with PD has proven to be the most successful treatment cases of all neurodegenerative diseases to date. The loss of dopaminergic neurons in PD patients can be overcome by cell replacement therapy using cells engineered to release dopamine. The major drawbacks associated with this treatment are poor survival of these transplanted cells and the lack of effective control in the amount of dopamine released by these cells [85].

Tissue engineering approaches have been used to improve the means of drug and cell delivery to the CNS. Cultured astrocytes grown on polyethylene terephthalate (PET) matrices have been evaluated as a therapeutic treatment for PD because PET supports high-density growth of astrocytes with stable glial cell line-derived neurotrophic factor (GDNF) production over a period of 18 days in vitro [86]. GDNF has been shown to promote survival and fiber outgrowth of the degenerating dopaminergic neurons [87]. Intrastriatal injections of GDNF in the striatum of PD-rats after transplantation of fetal dopaminergic cells results in significant behavioral improvement [88–90]. In fact, GDNF infused unilaterally into the putamen for 6–12 months has also been shown to significantly improve motor function and quality of life measures in patients with PD in a phase-I trial [91]. GDNF-loaded PLGA microspheres, when implanted in the brains of PD rats, were well tolerated and induced sprouting of the preserved dopaminergic fibers with synaptogenesis, accompanied by functional improvement [92]. GNDF-loaded pharmacologically active microcarriers (PAM) carrying fetal ventral mesencephalic cells also improve the survival of grafted cells and fiber outgrowth [93]. Another study demonstrated that intraventricular chronic infusion of low-dose GDNF from encapsulated genetically engineered mouse myoblast cells (C2C12) demonstrated efficacious benefits on the "Parkinsonian" baboons [93]. These baboons showed transient recovery of motor deficits (hypokinesia), significant protection of intrinsic striatal dopaminergic function in the immediate vicinity of the site of implantation of the capsule in the caudate nucleus, and significant long-lasting neurotrophic properties at the nigral level with an increased volume of the cell bodies. All these studies involve the transplantation of cells (mostly fetal SN cells) into the stratium where dopamine is released and not into the SN where dopaminergic neurons normally reside. Therefore, the next desirable advance will be to reconstruct a functional dopaminergic nigrostriatal pathway using a PD animal model, and in the future, in human patients with PD.

56.3.2.2
Alzheimer's Disease (AD)

AD is characterized by dementia with progressive cognitive deterioration mainly due to neuronal loss or atrophy, principally involving cells in the hippocampus and temporoparietal cortex, as well as the frontal cortex in advance stages. This atrophy is often accompanied by inflammatory response with the deposition of amyloid plaques and neurofibrillary tangles [94]. Earlier strategies for treatment of AD include delivery of drugs, neuroprotective and anti-amyloidogenic agents [94]. Many of the existing FDA-approved drugs for the treatment of AD have a short half-life and adverse side effects at higher doses [95–100]. In order to prolong the half-life and also to enable sustained release of drugs, encapsulation of drugs by biodegradable polymers was desirable. The pharmacokinetics of several commonly used drugs encapsulated in microparticles had been evaluated both in vitro and in vivo. For example, Tacrine encapsulated by poly-(D,L-lactide-co-glycolide) (PLG), huperzine A by PLG or PLGA and denepezil by PLGA were tested in animal models of AD and showed sustained release over a few weeks [95–100]. The tissue engineering pioneered by developments in PD treatment discussed above may help to develop therapies for other neurodegenerative diseases, including AD. While preliminary and animal-based results are encouraging, the ultimate goal for the treatment of AD is the replacement of lost neurons and the restoration of cognitive function. Cell therapy involving the use of neural stem cells can potentially replace these lost cells in regions of the brain [12, 40]. One can also attempt to modulate neurogenesis in AD patients to replace the lost neurons. Modulating neurogenesis, in combination with strategies targeting the underlying causes of AD, including the formation and accumulation of amyloid plaques, may be most effective in the treatment of AD.

56.3.2.3
Amytrophic Lateral Sclerosis (ALS)

ALS results from a gradual loss of motor neurons throughout the brain and spinal cord, resulting in muscle weakness, atrophy and eventual death. Currently, there are limited choices for treating ALS patients. While studies have suggested that CNTF and GDNF can promote motor neuron survival [101–103], the short half-life and lack of effective ways for their delivery into the CNS, had limited the clinical use of these factors. The Aebischer laboratory has reported the use of polymer-encapsulated cells engineered to produce CNTF or GDNF constitutively [104]. The authors demonstrated that GDNF significantly reduced the loss of facial motoneurons [105]. In addition, using models of progressive motoneuronopathy (pmn) mice displaying motoneuron degeneration, it was shown that CNTF but not GDNF increases the life span of these animals [106].

Treatment with neurotrophic factors, while able to delay the progression of the disease, is not useful for late-stage ALS. Recently, motor neurons were successfully derived from embryonic stem cells [107], and paralyzed adult rats transplanted with these stem cell-derived motor neurons showed partial recovery from paralysis [108]. This restoration of function was greatly enhanced in the presence of various factors, such as phosphodiesterase type 4 inhibitor and dibutyryl cyclic adenosine monophosphate (cAMP), which could overcome myelin-mediated repulsion. In addition, GDNF has been shown to attract transplanted embryonic stem cell-derived axons toward skeletal muscle targets [109].

The main challenges in stem cell-based treatment of ALS patients are the availability of the cells, the ability to maintain cell survival and recognizing the appropriate factors that encourage nerve growth and guidance at the strategic location of transplant. Improvement of tissue engineering approaches will hopefully be able to refine and bring current research and treatment for ALS to the next level.

56.3.2.4
Huntington's Disease (HD)

HD is an autosomal dominant inherited neurological disorder [110]. The genetic mutation affects neurons in the basal ganglia and results in devastating clinical effects on cognitive, psychological, and motor functions. These clinical symptoms primarily relate to the progressive loss of medium-spiny GABAergic neurons of the striatum. There is currently no treatment available, but likely therapeutic candidates include the use of several neurotrophic factors that are able to prolong the survival of striatal neurons. The use of CNTF as gene therapy for HD has been examined in different animal models, such as monkeys [111, 112] and mice [64]. The encouraging results in animal models have led to a phase-I clinical study. Baby hamster kidney cells (BHK) cells engineered to synthesize human CNTF were encapsulated by a semipermeable membrane and implanted into the lateral ventricle. Improvements in electrophysiological results were observed in subjects with higher amounts of CNTF detected, and with no obvious toxicity over a period of 2 years, indicating the safety and feasibility of the approach. However, the heterogeneity in terms of cell survival and levels of CNTF secreted remains to be refined.

56.4
Conclusion

The human nervous system is exceedingly complex, and is therefore perhaps the most difficult to repair after injury and degeneration. Damage to the nervous system by injury or diseases is highly resistant to treatments and there is a lack of effective therapies. Advances in tissue engineering and basic neuroscience have brought new hopes and strategies for repair and regeneration. The field of CNS regeneration has still a long way to go compared with the regeneration of other tissues or organs such as the liver. An ideal situation for the regeneration and treatment of injuries of the nervous system or neurological disorders is to activate the endogenous regenerative capability and allow self-regulated regrowth and repair. However, we do not yet have a sufficient understanding of adult neurogenesis, and future studies focusing on uncovering the mechanisms involved in the activation of endogenous repair are highly desired. In addition, one can create artificial microniches where transplanted cells can grow, develop and differentiate in vivo. Bioengineering approaches play an important part in both the design and the construction of such "pseudo in-vivo" systems, which may serve a variety of different purposes. One would hope to eventually build a scaffold that could provide physical support for different types of neural cells to grow on, and which would encapsulate proteins factors that would ensure the cells survive, form connections and integrate into the host tissue.

Acknowledgments

The research in the authors' laboratory was supported by grants from the Culpeper Scholarship in Medical Science, March of the Dimes, Klingenstein Fellowship Award in the Neuroscience, Adelson Medical Research Foundation and National Institute of Health and Maryland Stem Cell Research Fund.

References

1. Dickson G et al (2007) Orthopaedic tissue engineering and bone regeneration. Technol Health Care 15(1):57–67
2. Baguneid MS et al (2006) Tissue engineering of blood vessels. Br J Surg 93(3):282–290
3. Nerem RM, Ensley AE (2004) The tissue engineering of blood vessels and the heart. Am J Transplant 4(Suppl 6):36–42
4. Schmidt CE, Leach JB (2003) Neural tissue engineering: strategies for repair and regeneration. Annu Rev Biomed Eng 5:293–347
5. Fournier AE, Strittmatter SM (2001) Repulsive factors and axon regeneration in the CNS. Curr Opin Neurobiol 11(1):89–94
6. Schwab ME, Kapfhammer JP, Bandtlow CE (1993) Inhibitors of neurite growth. Annu Rev Neurosci 16:565–595
7. Filbin MT (2003) Myelin-associated inhibitors of axonal regeneration in the adult mammalian CNS. Nat Rev Neurosci 4(9):703–713
8. Goldberg JL (2004) Intrinsic neuronal regulation of axon and dendrite growth. Curr Opin Neurobiol 14(5):551–557
9. Yiu G, He Z (2006) Glial inhibition of CNS axon regeneration. Nat Rev Neurosci 7(8):617–627
10. Goh EL et al (2003) Adult neural stem cells and repair of the adult central nervous system. J Hematother Stem Cell Res 12(6):671–679
11. Lie DC et al (2004) Neurogenesis in the adult brain: new strategies for central nervous system diseases. Annu Rev Pharmacol Toxicol 44:399–421
12. Sailor KA, Ming GL, Song H (2006) Neurogenesis as a potential therapeutic strategy for neurodegenerative diseases. Expert Opin Biol Ther 6(9):879–890
13. Xu XM et al (1999) Regrowth of axons into the distal spinal cord through a Schwann-cell-seeded mini-channel implanted into hemisected adult rat spinal cord. Eur J Neurosci 11(5):1723–1740
14. Martino G, Pluchino S (2006) The therapeutic potential of neural stem cells. Nat Rev Neurosci 7(5):395–406
15. Rossi F, Cattaneo E (2002) Opinion: neural stem cell therapy for neurological diseases: dreams and reality. Nat Rev Neurosci 3(5):401–409
16. Levenberg S et al (2003) Differentiation of human embryonic stem cells on three-dimensional polymer scaffolds. Proc Natl Acad Sci U S A 100(22):12741–12746
17. Teixeira AI, Duckworth JK, Hermanson O (2007) Getting the right stuff: controlling neural stem cell state and fate in vivo and in vitro with biomaterials. Cell Res 17(1):56–61
18. Liu CY, Apuzzo ML, Tirrell DA (2003) Engineering of the extracellular matrix: working toward neural stem cell programming and neurorestoration—concept and progress report. Neurosurgery 52(5):1154–1165; discussion 1165–1167
19. Weissman IL (2000) Stem cells: units of development, units of regeneration, and units in evolution. Cell 100(1):157–168
20. Weissman IL, Anderson DJ, Gage F (2001) Stem and progenitor cells: origins, phenotypes, lineage commitments, and transdifferentiations. Annu Rev Cell Dev Biol 17:387–403
21. Smith AG (2001) Embryo-derived stem cells: of mice and men. Annu Rev Cell Dev Biol 17:435–462
22. Eiges R, Benvenisty N (2002) A molecular view on pluripotent stem cells. FEBS Lett 529(1):135–141
23. Prelle K, Zink N, Wolf E (2002) Pluripotent stem cells—model of embryonic development, tool for gene targeting, and basis of cell therapy. Anat Histol Embryol 31(3):169–186
24. Prusa AR et al (2003) Oct-4-expressing cells in human amniotic fluid: a new source for stem cell research? Hum Reprod 18(7):1489–1493
25. Kern S et al (2006) Comparative analysis of mesenchymal stem cells from bone marrow, umbilical cord blood, or adipose tissue. Stem Cells 24(5):1294–301
26. Toma JG et al (2001) Isolation of multipotent adult stem cells from the dermis of mammalian skin. Nat Cell Biol 3(9):778–784

27. Yamamoto N et al (2007) Isolation of multipotent stem cells from mouse adipose tissue. J Dermatol Sci 48(1):43–52
28. Kanatsu-Shinohara M et al (2004) Generation of pluripotent stem cells from neonatal mouse testis. Cell 119(7):1001–1012
29. Guan K et al (2006) Pluripotency of spermatogonial stem cells from adult mouse testis. Nature 440(7088):1199–1203
30. Williams N, Jackson H, Meyers P (1979) Isolation of pluripotent hemopoietic stem cells and clonable precursor cells of erythrocytes, granulocytes, macrophages and megakaryocytes from mouse bone marrow. Exp Hematol 7(10):524–534
31. Yu J, Thomson JA (2008) Pluripotent stem celllines. Genes Dev 22, 1987–1997
32. Altman J (1962) Are new neurons formed in the brains of adult mammals? Science 135:1127–1128
33. Gage FH (2000) Mammalian neural stem cells. Science 287(5457): 1433–1438
34. Alvarez-Buylla A, Temple S (1998) Stem cells in the developing and adult nervous system. J Neurobiol 36(2):105–110
35. Anderson DJ (2001) Stem cells and pattern formation in the nervous system: the possible versus the actual. Neuron 30(1):19–35
36. Temple S (2001) The development of neural stem cells. Nature 414(6859):112–117
37. Ming GL, Song H (2005) Adult neurogenesis in the mammalian central nervous system. Annu Rev Neurosci 28:223–250
38. Kitabatake Y et al (2007) Adult neurogenesis and hippocampal memory function: new cells, more plasticity, new memories? Neurosurg Clin N Am 18(1):105–113
39. Rakic P (2002) Neurogenesis in adult primate neocortex: an evaluation of the evidence. Nat Rev Neurosci 3(1):65–71
40. Emsley JG et al (2005) Adult neurogenesis and repair of the adult CNS with neural progenitors, precursors, and stem cells. Prog Neurobiol 75(5):321–341
41. Reynolds BA, Weiss S (1992) Generation of neurons and astrocytes from isolated cells of the adult mammalian central nervous system. Science 255(5052):1707–1710
42. Xu Y et al (2003) Isolation of neural stem cells from the forebrain of deceased early postnatal and adult rats with protracted post-mortem intervals. J Neurosci Res 74(4):533–540
43. Davis SF et al (2006) Isolation of adult rhesus neural stem and progenitor cells and differentiation into immature oligodendrocytes. Stem Cells Dev 15(2):191–199
44. Gritti A et al (2002) Multipotent neural stem cells reside into the rostral extension and olfactory bulb of adult rodents. J Neurosci 22(2):437–445
45. Seri B et al (2006) Composition and organization of the SCZ: a large germinal layer containing neural stem cells in the adult mammalian brain. Cereb Cortex 16 (Suppl 1):i103–i111
46. Pagano SF et al (2000) Isolation and characterization of neural stem cells from the adult human olfactory bulb. Stem Cells 18(4):295–300
47. Nunes MC et al (2003) Identification and isolation of multipotential neural progenitor cells from the subcortical white matter of the adult human brain. Nat Med 9(4):439–447
48. Roy NS et al (2000) Promoter-targeted selection and isolation of neural progenitor cells from the adult human ventricular zone. J Neurosci Res 59(3):321–331
49. Roy NS et al (2000) In vitro neurogenesis by progenitor cells isolated from the adult human hippocampus. Nat Med 6(3):271–277
50. Gage FH et al (1995) Survival and differentiation of adult neuronal progenitor cells transplanted to the adult brain. Proc Natl Acad Sci U S A 92(25):11879–11883
51. Arsenijevic Y et al (2001) Isolation of multipotent neural precursors residing in the cortex of the adult human brain. Exp Neurol 170(1):48–62
52. Palmer TD et al (1999) Fibroblast growth factor-2 activates a latent neurogenic program in neural stem cells from diverse regions of the adult CNS. J Neurosci 19(19):8487–8497
53. Gottlieb DI (2002) Large-scale sources of neural stem cells. Annu Rev Neurosci 25:381–407
54. Song HJ, Stevens CF, Gage FH (2002) Neural stem cells from adult hippocampus develop essential properties of functional CNS neurons. Nat Neurosci 5(5):438–445
55. Ge S et al (2006) GABA regulates synaptic integration of newly generated neurons in the adult brain. Nature 439(7076):589–593
56. Zlokovic BV, Apuzzo ML (1997) Cellular and molecular neurosurgery: pathways from concept to reality—part I: target disorders and concept approaches to gene therapy of the central nervous system. Neurosurgery 40(4):789–803; discussion 803–804
57. Corti S et al (2003) Neuronal generation from somatic stem cells: current knowledge and perspectives on the treatment of acquired and degenerative central nervous system disorders. Curr Gene Ther 3(3):247–272
58. Park KI, Teng YD, Snyder EY (2002) The injured brain interacts reciprocally with neural stem cells supported by scaffolds to reconstitute lost tissue. Nat Biotechnol 20(11):1111–1117
59. Teng YD et al (2002) Functional recovery following traumatic spinal cord injury mediated by a unique polymer scaffold seeded with neural stem cells. Proc Natl Acad Sci U S A 99(5):3024–3029
60. Dontchev VD Letourneau PC (2003) Growth cones integrate signaling from multiple guidance cues. J Histochem Cytochem 51(4):435–444
61. Tessier-Lavigne M, Goodman CS (1996) The molecular biology of axon guidance. Science 274(5290):1123–1133
62. Song H, Poo M (2001) The cell biology of neuronal navigation. Nat Cell Biol 3(3):E81–E88
63. Paves H, Saarma M (1997) Neurotrophins as in vitro growth cone guidance molecules for embryonic sensory neurons. Cell Tissue Res 290(2):285–297
64. Bloch J et al (2004) Neuroprotective gene therapy for Huntington's disease, using polymer-encapsulated cells engineered to secrete human ciliary neurotrophic factor: results of a phase I study. Hum Gene Ther 15(10):968–975
65. Moore K, MacSween M, Shoichet M (2006) Immobilized concentration gradients of neurotrophic factors guide neurite outgrowth of primary neurons in macroporous scaffolds. Tissue Eng 12(2):267–278

66. Gustavsson P et al (2007) Neurite guidance on protein micropatterns generated by a piezoelectric microdispenser. Biomaterials 28(6):1141–1151
67. Emerich DF, Winn SR (2001) Immunoisolation cell therapy for CNS diseases. Crit Rev Ther Drug Carrier Syst 18(3):265–298
68. Aebischer P, Winn SR, Galletti PM (1988) Transplantation of neural tissue in polymer capsules. Brain Res 448(2):364–368
69. Goraltchouk A et al (2006) Incorporation of protein-eluting microspheres into biodegradable nerve guidance channels for controlled release. J Control Release 110(2):400–407
70. Hadlock T et al (1999) A novel, biodegradable polymer conduit delivers neurotrophins and promotes nerve regeneration. Laryngoscope 109(9):1412–1416
71. Aebischer P, Guenard V, Brace S (1989) Peripheral nerve regeneration through blind-ended semipermeable guidance channels: effect of the molecular weight cutoff. J Neurosci 9(10):3590–3595
72. Piotrowicz A, Shoichet MS (2006) Nerve guidance channels as drug delivery vehicles. Biomaterials 27(9):2018–2027
73. Matsumoto K et al (2000) Peripheral nerve regeneration across an 80-mm gap bridged by a polyglycolic acid (PGA)-collagen tube filled with laminin-coated collagen fibers: a histological and electrophysiological evaluation of regenerated nerves. Brain Res 868(2):315–328
74. Harel NY, Strittmatter SM (2006) Can regenerating axons recapitulate developmental guidance during recovery from spinal cord injury? Nat Rev Neurosci 7(8):603–616
75. Novikov LN et al (2002) A novel biodegradable implant for neuronal rescue and regeneration after spinal cord injury. Biomaterials 23(16):3369–3376
76. Novikova LN, Novikov LN, Kellerth JO (2003) Biopolymers and biodegradable smart implants for tissue regeneration after spinal cord injury. Curr Opin Neurol 16(6):711–715
77. Bakshi A et al (2004) Mechanically engineered hydrogel scaffolds for axonal growth and angiogenesis after transplantation in spinal cord injury. J Neurosurg Spine 1(3):322–329
78. Woerly S et al (2001) Spinal cord repair with PHPMA hydrogel containing RGD peptides (NeuroGel). Biomaterials 22(10):1095–1111
79. Plant GW, Woerly S, Harvey AR (1997) Hydrogels containing peptide or aminosugar sequences implanted into the rat brain: influence on cellular migration and axonal growth. Exp Neurol 143(2):287–299
80. Moore MJ et al (2006) Multiple-channel scaffolds to promote spinal cord axon regeneration. Biomaterials 27(3):419–429
81. Loh NK et al (2001) The regrowth of axons within tissue defects in the CNS is promoted by implanted hydrogel matrices that contain BDNF and CNTF producing fibroblasts. Exp Neurol 170(1):72–84
82. Tobias CA et al (2005) Alginate encapsulated BDNF-producing fibroblast grafts permit recovery of function after spinal cord injury in the absence of immune suppression. J Neurotrauma 22(1):138–156
83. Tomac A et al (1995) Protection and repair of the nigrostriatal dopaminergic system by GDNF in vivo. Nature 373(6512):335–339
84. Akerud P et al (2001) Neuroprotection through delivery of glial cell line-derived neurotrophic factor by neural stem cells in a mouse model of Parkinson's disease. J Neurosci 21(20):8108–8118
85. Betchen SA, Kaplitt M (2003) Future and current surgical therapies in Parkinson's disease. Curr Opin Neurol 16(4):487–493
86. Jollivet C et al (2004) Striatal implantation of GDNF releasing biodegradable microspheres promotes recovery of motor function in a partial model of Parkinson's disease. Biomaterials 25(5):933–942
87. Rosenblad C, Martinez-Serrano A, Bjorklund A (1998) Intrastriatal glial cell line-derived neurotrophic factor promotes sprouting of spared nigrostriatal dopaminergic afferents and induces recovery of function in a rat model of Parkinson's disease. Neuroscience 82(1):129–137
88. Slevin JT et al (2007) Unilateral intraputamenal glial cell line-derived neurotrophic factor in patients with Parkinson disease: response to 1 year of treatment and 1 year of withdrawal. J Neurosurg 106(4):614–620
89. Patel NK et al (2005) Intraputamenal infusion of glial cell line-derived neurotrophic factor in PD: a two-year outcome study. Ann Neurol 57(2):298–302
90. Gill SS et al (2003) Direct brain infusion of glial cell line-derived neurotrophic factor in Parkinson disease. Nat Med 9(5):589–595
91. Basu S, Yang ST (2005) Astrocyte growth and glial cell line-derived neurotrophic factor secretion in three-dimensional polyethylene terephthalate fibrous matrices. Tissue Eng 11(5–6):940–952
92. Tatard VM et al (2007) Pharmacologically active microcarriers releasing glial cell line-derived neurotrophic factor: survival and differentiation of embryonic dopaminergic neurons after grafting in hemiparkinsonian rats. Biomaterials 28(11):1978–1988
93. Kishima H et al (2004) Encapsulated GDNF-producing C2C12 cells for Parkinson's disease: a pre-clinical study in chronic MPTP-treated baboons. Neurobiol Dis 16(2):428–439
94. Spencer B et al (2007) Novel strategies for Alzheimer's disease treatment. Expert Opin Biol Ther 7(12):1853–1867
95. Yang Q et al (2001) Controlled release tacrine delivery system for the treatment of Alzheimer's disease. Drug Deliv 8(2):93–98
96. Fu XD et al (2005) Preparation and in vivo evaluation of huperzine A-loaded PLGA microspheres. Arch Pharm Res 28(9):1092–1096
97. Gao P et al (2007) Controlled release of huperzine A from biodegradable microspheres: In vitro and in vivo studies. Int J Pharm 330(1–2):1–5
98. Zhang P et al (2007) In vitro and in vivo evaluation of donepezil-sustained release microparticles for the treatment of Alzheimer's disease. Biomaterials 28(10):1882–1888
99. Chu DF et al (2006) Pharmacokinetics and in vitro and in vivo correlation of huperzine A loaded poly(lactic-co-glycolic acid) microspheres in dogs. Int J Pharm 325(1–2):116–123

100. Liu WH et al (2005) Preparation and in vitro and in vivo release studies of Huperzine A loaded microspheres for the treatment of Alzheimer's disease. J Control Release 107(3):417–427
101. Henderson CE et al (1994) GDNF: a potent survival factor for motoneurons present in peripheral nerve and muscle. Science 266(5187):1062–1064
102. Zurn AD et al (1996) Combined effects of GDNF, BDNF, and CNTF on motoneuron differentiation in vitro. J Neurosci Res 44(2):133–141
103. Turgeon VL, Houenou LJ (1999) Prevention of thrombin-induced motoneuron degeneration with different neurotrophic factors in highly enriched cultures. J Neurobiol 38(4):571–580
104. Sagot Y et al (1996) GDNF slows loss of motoneurons but not axonal degeneration or premature death of pmn/pmn mice. J Neurosci 16(7):2335–2341
105. Tan SA et al (1996) Rescue of motoneurons from axotomy-induced cell death by polymer encapsulated cells genetically engineered to release CNTF. Cell Transplant 5(5):577–587
106. Harper JM et al (2004) Axonal growth of embryonic stem cell-derived motoneurons in vitro and in motoneuron-injured adult rats. Proc Natl Acad Sci U S A 101(18):7123–7128
107. Wichterle H et al (2002) Directed differentiation of embryonic stem cells into motor neurons. Cell 110(3):385–397
108. Deshpande DM et al (2006) Recovery from paralysis in adult rats using embryonic stem cells. Ann Neurol 60(1):32–44
109. Emerich DF et al (1997) Protective effect of encapsulated cells producing neurotrophic factor CNTF in a monkey model of Huntington's disease. Nature 386(6623):395–399
110. Shao J, Diamond MI (2007) Polyglutamine diseases: emerging concepts in pathogenesis and therapy. Hum Mol Genet 16(Spec No. 2):R115–R123
111. Emerich DF et al (1998) Cellular delivery of CNTF but not NT-4/5 prevents degeneration of striatal neurons in a rodent model of Huntington's disease. Cell Transplant 7(2):213–225
112. Emerich DF, Winn SR (2004) Neuroprotective effects of encapsulated CNTF-producing cells in a rodent model of Huntington's disease are dependent on the proximity of the implant to the lesioned striatum. Cell Transplant 13(3):253–259

57 Tissue Engineering in Maxillofacial Surgery

H. Schliephake

Contents

57.1	Introduction	827
57.2	Tissue Engineering of Bone Tissue	828
57.2.1	Growth Factors in Maxillofacial Skeletal Reconstruction	828
57.2.2	Cell-Based Devices in Maxillofacial Skeletal Reconstruction	830
57.3	Tissue Engineering of Cartilage	831
57.4	Tissue Engineering of Mucosa	832
57.5	Tissue Engineering of Salivary Glands	833
57.6	Concluding Remarks	833
	References	833

57.1 Introduction

Orofacial rehabilitation after ablative surgery or trauma is one of the main objectives of oral and maxillofacial surgery. Many treatment strategies in this field thus fall into the realm of reconstructive surgery. A standard measure in the field of oral and maxillofacial reconstruction is the use of autogenous grafts for the replacement of tissue parts, which is associated with a significant increase in surgical effort and morbidity. The growing understanding of tissue healing and the achievements of biotechnology have raised hopes that defects of the jaw bones, articular cartilage, or oral mucosa can be repaired without the transfer of autogenous tissue and that custom made body parts can be cultivated from the patients own cells in the lab as a substitute for autogenous grafts.

During the past decade, the principles of tissue engineering have also been applied to oral and maxillofacial surgery. In particular, bone tissue is in major focus, but oral mucosa, cartilage, and salivary glands have also received some attention. In oral reconstruction, there are specific conditions that need to be considered such as possible exposure to the oral or paranasal cavities with subsequent bacterial contamination as well as specific problems of mechanical loading during mastication. This makes approaches for tissue engineering in oral and maxillofacial surgery different from other areas.

Additionally, oral tissues are extremely well vascularized. This accounts for a remarkably high ability of the oral tissues to heal by secondary intention and leads to a rather low frequency of chronic infections in case of wound breakdown. For instance, jaw bone tissue has a regenerative capacity that allows for the successful use of purely osteoconductive scaffold materials without cells or growth factors for the repair of smaller rim defects in conjunction with dental implants. It is therefore not easy to identify useful "tissue engineered" applications for the repair of oral bone or mucosa defects that justify the effort and costs associated with the use of in vitro technology, since smaller defects may heal spontaneously or with the help of inexpensive conventional biomaterials.

57.2
Tissue Engineering of Bone Tissue

A key procedure in bone regeneration is the specialization of mesenchymal cells that undergo differentiation into bone-forming cells. These cells start to form bone matrix that eventually becomes mineralized. Osteogenic differentiation and bone matrix production is induced by a complex system of signaling molecules as well as local mechanical factors such as mechanical loading and deformation. The growth and differentiation factors required for osteogenic induction are supplied by local cells on the one hand and originate from the adjacent bone matrix on the other. The bone matrix itself functions not only as a storage place for these growth factors that are released during bone remodeling and repair, but also acts as a scaffold that contributes to the three-dimensional arrangement of bone-forming cells. In this way, cells, signaling molecules, and bone matrix are intimately linked together in the process of bone tissue regeneration and repair. Bone replacement through a tissue engineering approach that mimics the natural process of bone regeneration would thus use cultivated autogenous bone cells, osteogenic growth factors, and synthetic or natural scaffolds as a bone matrix substitute.

Many tissue engineering approaches for the reconstruction of the facial skeleton yet have employed only two of the three components. On the one hand they have used cells and scaffolds relying on the host tissue to supply the necessary cytokines that provide appropriate signals for survival, proliferation, and differentiation of the implanted/transplanted cells within the scaffolds. On the other hand, scaffold materials have been implanted together with growth factors that were supposed to recruit undifferentiated mesenchymal cells required for bone tissue repair. Combinations of scaffolds, cells, and growth factors have not yet been extensively explored. Most approaches have employed recombinant cells which were designed to produce the desired growth factor after implantation in vivo [34]; very few have used scaffolds loaded with growth factors and cells [111].

57.2.1
Growth Factors in Maxillofacial Skeletal Reconstruction

A multitude of growth and differentiation factors have been employed with an even larger variety of scaffold materials for maxillofacial bone repair on the preclinical level. There are at least six growth factors involved in bone regeneration that have been used in maxillofacial reconstruction in a large number of animal models:

1. Platelet-derived growth factor (PDGF) [27, 64, 73]
2. Basic Fibroblast Growth Factor (bFGF) [53, 79, 80]
3. Insulin-like growth factors (IGF) [50, 62, 87, 92]
4. Transforming Growth Factor beta (TGFβ) [13, 24, 90]
5. Vascular Endothelial Growth Factor (VEGF) [43, 49]
6. Bone Morphogenetic Proteins (BMPs) [69, 70, 82, 102]

Apart from human recombinant growth factor preparations, autogenous growth factors derived from platelet rich plasma (PRP) have been extensively tested in preclinical models. From the above-mentioned list of growth factors, PRP contains PDGF, TGFβ, and VEGF but as well has additional amounts of epithelial growth factor (EGF). PRP has been used in conjunction with both autogenous bone grafts and synthetic or natural carriers such as tricalcium phosphate (TCP), bovine bone mineral, collagen and demineralized freeze dried bone [4, 26, 44, 58, 95] in peri-implant bone defects, sinus floor augmentations, onlay grafts, and segmental mandibular defects [17, 25, 56, 78, 88]. The results of all these studies were rather ambiguous, as those reports that indicated a positive effect were opposed by an even larger number of studies which failed to show a significant enhancement of bone regeneration after the use of PRP. The confounding data were discussed as to result from the unavailability of appropriate technology to prepare PRP from animal blood with adequate quality [52]. However, also for human PRP preparations, only poor reliability with respect to the degree of platelet concentration as well as the growth factor content has been found [106]. Despite a large number of clinical case reports or case series, clinical data has not yet been able to show a clear advantage for the use of PRP, as only few controlled trials have been conducted to prove the effect of this growth factor formulation [67, 93, 94, 107]. Only one of these studies have shown a significant positive effect on bone regeneration [107] while the remaining reports could not find a significant contribution to bone formation in the augmented areas.

Isolated applications of bone growth factors as human recombinant molecules in preclinical settings as well have not shown truly osteoinductive properties for the most of them. Platelet-derived growth factor (PDGF) proved to enhance peri-implant bone regeneration in conjunction with other growth factors such as IGF or TGFβ in preclinical models [11, 50, 64]. Positive effects on periodontal regeneration were also registered in combination with barrier membranes and in conjunction with IGF in preclinical and clinical settings [27, 33]. The same applied vice versa to Insulin-like growth factor (IGF) and basic fibroblast growth factor (bFGF). While isolated local administration did not result in significantly increased bone formation [80], combined use in conjunction with membranes [50, 62, 87] and other growth factors were associated with enhanced bone regeneration. Vascular endothelial growth factor (VEGF) has only recently exhibited osteopromotive properties in calvarial defects, where it also enhanced bone regeneration additionally to increasing vascularization [49]. Moreover, VEGF has been considered to act as an important co-factor for bone formation in conjunction with bone morphogenic proteins (BMP) [43]. In this respect, it is comparable to transforming growth factor beta (TGFβ) that in itself also has shown variable enhancement of bone formation in experimental OMF bone reconstruction depending on the carrier used [55, 109, 112], but when combined with PDGF or BMP has resulted in significant bone formation [82].

The only growth factors that have consistently promoted bone formation in almost all preclinical settings are the bone morphogenic proteins (BMPs), a family of more than 30 different molecules that have widespread homology among each other and to TGFβ molecules with whom they share a superfamily. There are three isoforms of bone morphogenic proteins, BMP-2, BMP-4, and BMP-7 [also called osteogenic protein 1 (OP-1)], that have been explored in the vast majority of studies. Experimental repair of alveolar bone defects, sinus floor augmentations, and mandibular reconstructions have been successfully performed using these molecules [14, 20, 91, 96]. In particular, critical size defect models of the mandible and segmental defects have been evaluated in a number of experimental models with a large variety of carriers. Collagen, collagen/HA/TCP, anorganic bovine bone, hyaluronic acid, PLA/PGA coated gelatine sponges, and PGLA beads have been employed with various doses of either rhBMP-2, rhBMP4, or rhBMP7 in rodents, minipigs, and nonhuman primates [9, 10, 51, 59, 71, 72, 113].

The excellent preclinical results have fostered the conviction that clinically successful bone regeneration by the application of BMPs is just around the corner. However, dosage studies had shown that moving from the preclinical environment into the clinical arena was associated with a huge increase in the amount of BMPs necessary to induce a reliable tissue response [102]. In a recent study about sinus floor augmentations more than 24 mg/ml of BMP2 per sinus have been reported to be necessary to produce bone of sufficient quality and quantity to accommodate implants in a second stage operation [15]. Compared to the BMP content in native bone of approximately 7 ng/g demineralized bone matrix [12], the dosage of BMPs required for clinical use is of a magnitude of 10^7–10^8 higher than what is provided during natural bone regeneration. The need for such excessive doses in clinical therapy has raised concerns about safety and costs on the one hand and has prompted the search for slow release carriers that could provide a more physiological release profile of the growth factor on the other. In order to induce bone formation in a reliable manner with predictable volume, the carrier material should allow for a controlled retarded release of growth factors while at the same time provide adequate strength to withstand pressure from overlying or surrounding soft tissue. Unfortunately, the carrier material collagen that has received FDA approval as carrier for BMP2 in clinical use does not fulfill any of these criteria. Bovine bone mineral has also been used in combination with BMP-2 in a guided bone regeneration approach to lateral peri-implant augmentation, but has not shown significantly enhanced bone volume [41]. In order to provide a degradable scaffold with slow release characteristics, resorbable polymers of synthetic (poly-DL-Lactid) or natural origin (Chitosan) have been explored. The volumes tested in preclinical models indicate that these carriers may hold some promise for the use as anatomically preformed osteoinductive implants in bone reconstructive procedures [83]. Nevertheless, the promising preclinical results are still awaiting clinical testing for use in daily routine.

Another growth factor from the BMP family that has been introduced into clinical testing is the growth and differentiation factor 5 (GDF-5). GDF-5 in conjunction with particulate TCP has shown to increase bone formation in sinus floor augmentations in minipigs where it has also proven to be superior to a mixture of the carrier material and autogenous minced bone [29]. Clinical testing is currently underway and there is evidence that bone regeneration in TCP sinus floor augmentations combined with GDF-5 is equivalent to the carrier in combination with minced autogenous bone.

57.2.2
Cell-Based Devices in Maxillofacial Skeletal Reconstruction

The use of bone-forming cells seeded into three-dimensional carriers has been also used in maxillofacial reconstruction to achieve bone formation in areas that would otherwise not be able to produce bone. This approach is closer to the original idea of tissue engineering in that a tissue is "built" using engineering principles that combine specifically designed scaffolds together with specialized cells to form a lab-based artificial biohybrid construct. This approach is by far more complex than the use of growth factors as the bioactive part is made up of living cells that are difficult to monitor both in terms of quality and quantity once they have been seeded onto the scaffold [54]. Moreover, the functionality of the constructs, which is determined by cell survival and continuous bone formation after implantation, is much more dependent on acute contributions from the recipient bed by supply of cytokines and oxygen (vascularization) than the use of growth factors and scaffolds.

Despite considerable preclinical research activity for more than a decade, there are a number of basic questions that are still unanswered. These questions relate to degree of specialization that the seeded cells are supposed to have in order to proliferate and to produce bone in vivo on the one hand, and the mode of cell seeding as well as the technique and duration of cell culturing in the lab on the other. Successful preclinical testing in cranial defects and in both rim and segmental defects of the mandible has been performed using autogenous cells in a number of animal models such as rodents, dogs, sheep, and minipigs. Also human cells have been evaluated in xenogenic models in immunocompromised rodents. While the autogenous models have almost consistently shown an enhancing effect on bone formation when compared with nonseeded scaffolds [1, 22, 28, 68, 81, 85], xenogenic models using human cells have produced ambiguous results in that some reports have been unable to show that bone formation in scaffolds seeded with native bone marrow stroma cells has been superior to control scaffolds undergoing osteoconductive bone ingrowth [34, 65, 84, 99].

A major impediment to valid and sufficient testing of human tissue-engineered constructs is the fact that appropriately sized models in immunocompromised animals are not available. Thus, the use of cell-based devices is currently less advanced in clinical tissue-engineered reconstruction of bone in maxillofacial surgery than the use of osteogenic growth factors. Most of the studies and reports that exist are case series or case reports without controls, which impairs the assessment of the efficacy of this approach. As the source of cells that have been used is different in these studies a comprehensive analysis is even more difficult. Polyglycolic fleeces, collagen matrices, and bovine bone mineral have been used in conjunction with periosteal cells from the mandibular ramus and bone-derived cells from the maxillary tuberosity to produce bone in sinus lift procedures and lateral rim augmentations in preimplant surgery [77, 86]. Only in 18 of 27 sinus lifts with seeded polymer fleeces bone formation was found to be sufficient to allow for secondary implant placement [77]. Resorption of tissue-engineered bone after sinus lift using polymer fleeces was found to be 90% compared to 29% resorption after using autogenous bone [78]. Bone formation in the sinuses grafted with seeded collagen matrix equaled that found in sinuses augmented with seeded bovine bone mineral [86].

Suspension of human bone marrow stroma cells in a fibrin matrix containing PRP as "injectable bone" without defined three-dimensional scaffold has been successful in the support of bone formation in distraction gaps of alveolar bone distractions [31] as well as in sinus floor augmentations [101]. The fact that all patients in this study had been successfully treated using PRP and cells compared to only 18/27 patients treated with polyglycolic scaffolds and cells alone suggests that the presence of growth factors may enhance the functionality of cell-based devices, in that proliferation and vascularization of the seeded cells is enhanced after implantation in the in vivo environment.

Cell-based devices in combination with growth factors—in particular BMPs—have been used for flap prefabrication in major reconstructive procedures. Bovine bone mineral has been used together with 7 mg of BMPs and bone marrow aspirates in a preformed titanium mesh to produce bone tissue by implantation into the latissimus dorsi muscle and subsequent revascularized transfer to a segmental defect of the mandible [104]. A similar approach has been used by Heliotis et al. [30], who used coralline hydroxylapatite implanted into the pectoralis major muscle with pedicled transfer to the mandible [30]. From these two approaches, only revascularized transfer appeared to be finally successful. These labor-intensive but spectacular procedures may not be cost effective yet when compared to the established methods of revascularized tissue transfer for mandibular reconstruction, but they show a proof of principle that adds to the growing knowledge of how tissue engineering of bone can be accommodated in the current strategies of maxillofacial reconstruction.

57.3
Tissue Engineering of Cartilage

Cartilage tissue has been one of the first tissues to be addressed by tissue engineering approaches. The ear-shaped cartilage construct that has been produced in the back of nude rats by transplantation of cartilage cells in a polymer scaffold has been one of the incentives of tissue engineering. Cartilage tissue as a graft has the advantage of not being vascularized and therefore being less demanding with respect to biologic requirements for integration after transfer. Its high content of intercellular matrix makes it well suitable to be used as a tissue-engineered graft as revascularization of the seeded scaffold, which is a critical point in the functionality of other tissue-engineered constructs after transfer, is not as crucial for the success of tissue-engineered cartilage. The entire challenge to successfully incorporate tissue-engineered cartilage is, therefore, made up by the mechanical requirements that these biohybrids have to fulfill.

In oral and maxillofacial surgery, one area of development has been the engineering of nasal cartilage [42, 66] using cartilage cells in polymer scaffolds, which were implanted into nude mice. Recent preclinical studies work with injectable preparations of chondrocyte macroaggregates and fibrin sealant to form nasal cartilage structures [18, 108]. There have been very few clinical applications yet, such as augmentation of the nasal dorsum using injection of cultured autogenous auricular cartilage cells as a gelatinous mass [110]. Most of the recent work in tissue engineering of cartilage in OMF surgery is directed towards replacement of the TMJ disc and condylar cartilage. When designing tissue engineering methodology for TMJ cartilage construction, it is important to realize that TMJ condylar cartilage appears to be an intermediate between fibrocartilage and hyaline cartilage [103] and that the native TMJ disc contains half as much glycosaminoglycans than usual articular cartilage but has a six times higher tensile modulus [38].

Other than usual hyaline cartilage, the TMJ disc cartilage also has only minor amounts of Collagen II but almost exclusively Collagen I fibers that run longitudinally in an anterior-posterior direction [21]. The mechanical properties of these types of cartilage thus indicate that it is not only pressure that tissue-engineered constructs will have to take but also a high amount of tension [38]. The characteristics of common hyaline cartilage that many tissue engineering approaches aim at to replace articular cartilage will therefore not be appropriate to replace TMJ condylar cartilage or disc cartilage [8].

The specific structural requirements of the TMJ cartilage to withstand the in vivo forces also indicate that mechanical properties of tissue-engineered TMJ cartilage as well as fixation and connection to remaining ligaments is of particular significance. Hence, research for tissue engineering of condylar cartilage or disc replacement is still in its infancy. So far, no clinical studies are available that would have been able to test a tissue-engineered cartilage construct and there are not even preclinical studies that have evaluated tissue-engineered TMJ cartilage in an appropriate animal model. In vitro engineering of a polymer scaffold in condylar shape using porcine mesenchymal stem cells had resulted in formation of mineralized tissue inside the scaffold following the shape of the condyle. Moreover, a few successful attempts have been made to engineer an osteochondral condylar implant in ectopic sites in rodent models using biphasic scaffolds, transduced fibroblasts (BMP-7), and chondrocytes on the one hand [76], and hydrogel and rat bone marrow stroma cells in immunodeficient mice on the other [6, 7]. Future research has to focus on developing appropriate preclinical models for testing of cultured or engineered osteochondral tissue components before tissue-engineered replacement of cartilage in oral maxillofacial surgery will become a clinical option.

57.4
Tissue Engineering of Mucosa

Cultured mucosal cells had been used for the repair of epithelial defects in the oral cavity already in the 1980s [46, 100]. These epithelial sheets that were applied as cell monolayers without supporting scaffolds were very vulnerable and tended to shrink considerably during healing. The approach for tissue-engineered repair of oral mucosa is different from these early attempts in that replacement of all components of the oral mucosa should be accomplished. A stratified epithelium, a continuous basement membrane, and a fibrous connective tissue layer that supports the epithelium and gives three-dimensional stability during handling and healing is what the in vitro process is supposed to bring about. These engineered full-thickness oral mucosa constructs are commonly referred to as ex-vivo produced oral mucosa equivalent (EVPOME) [36].

The "backbone" for these engineered full-thickness mucosa constructs is made up by synthetic or naturally derived scaffolds that are seeded from both sides for generation of a multilayered construct.

Numerous custom made or commercially available products have been used for this purpose, many of them combining both synthetic and natural components [57]. The cell sources for both fibroblasts and epithelial cells required for the tissue-engineered mucosa are commonly the palatal or buccal mucosa; in some instances also the dermal layer of the skin has been used. There is some evidence that the proliferation rate and growth of mucosal epithelial cells is reduced in older donors [48].

In vitro technology using cell culturing on permeable membranes at an air/liquid interface has been able to produce a multilayered construct that quite closely resembles that of native mucosa in terms of morphological, histochemical, and genetic patterns [5, 74]. Clinically EVPOME preparations have been employed for some extraoral applications such as urethroplasty and conjunctival reconstruction. The majority of tissue-engineered oral mucosa equivalents has been used for intraoral applications such as vestibuloplasty, repair of superficial postablative mucosal defects, and for prelamination of free radial forearm flaps with subsequent transfer to the oral cavity [32, 47, 75].

Clinical and histopathological monitoring of these grafts has shown a good-to-excellent take rate with vascular ingrowth from the recipient bed and persistence of grafted cultured keratinocytes on the EVPOME surface after 6 days [35]. The grafted epithelium started to unite with the adjacent local epithelium after 10 days [32]. Total wound healing time was found to be shorter than in the control defects covered with unseeded scaffolds only [32] and histologically the EVPOME grafts had enhanced maturation of the underlying submucosal layer when compared to defects covered with the control scaffolds [37].

The present data thus show a sound proof of principle for clinical applications of tissue-engineered mucosa. However, EVPOME has not yet found widespread use in clinical therapy probably due to the fact that limited mucosal defects tend to heal spontaneously by secondary intention within 3–4 weeks. Moreover, if the rapid removal of relevant pathology is indicated, substantial delay of the ablative procedure by the process of EVPOME fabrication may be difficult to justify. As mucosal defects following laser resection of small- and even medium-sized oral tumors have shown to heal quite smoothly and to produce excellent functional results by secondary healing, the need for coverage of even larger superficial defect areas by EVPOME may not be readily apparent, if costs and time are taken into consideration. Tissue-engineered mucosal grafts, therefore, will have to establish themselves in clinical routine by proving their added value in selected indications.

57.5
Tissue Engineering of Salivary Glands

Salivary glands have not yet been much in focus of tissue engineering endeavors in oral and maxillofacial surgery. However, the need to repair or replace salivary gland tissue is obvious in patients suffering from severe hyposalivation, as hyposecretion of saliva is not only associated with functional impairment and dysphagia but also with diminished antibacterial and antifungal defense. The most common cause for these conditions is pharmaceutically induced hyposecretion which is not a domain for tissue-engineered repair. Also, the second most frequent reason being autoimmune exocrinopathy such as Sjögren's syndrome may not be amenable to engineering of autogenous tissue. The large cohort of patients, however, having undergone radiotherapy for head and neck cancer could benefit from advances in this field.

Gene transfer using intraductal delivery of a recombinant adenovirus has been applied in animal models to increase secretion of saliva as part of an in situ repair strategy [23, 63]. More recently, in vitro research has shown that nonhuman primate salivary gland cells can be isolated and expanded while preserving essential secretory cellular features [98] and that acini and ducts can be formed from single human salivary cells in three-dimensional culture systems [40]. Currently, numerous approaches are explored on the in vitro level to construct a functional salivary gland equivalent using different cell sources, scaffolds, and modifications of in vitro technology [2, 3, 16, 19, 39, 89, 97, 105].

What makes tissue engineering of salivary gland tissue a more complex endeavor than tissue engineering of many other tissues is the need to construct a fully differentiated and structured organ in vitro. Other than tissue-engineered replacement of mesenchymal tissues such as bone or cartilage, for instance, tissue engineering of salivary glands cannot rely on the support from the recipient bed that provides morphogenetic guidance for final tissue maturation through cells or signals. The road to clinical application of engineered functional salivary gland tissue will therefore be longer than for many other tissues in the oral and maxillofacial area.

57.6
Concluding Remarks

One and a half decades have passed since the breakthrough into a new era of reconstructive surgery was launched by papers of Nerem and others [60]. From the clinical perspective not many of the numerous promises that have been anticipated in early years have become reality in our daily routine yet [61]. Currently, less complex approaches like the application of growth factors to enhance in situ regeneration of tissues are clinically more advanced than the cell-based strategies. However, the growing body of knowledge creates a much deeper understanding of tissue physiology and organ function. This increased comprehension as well as the steady energy of research groups around the globe in conjunction with a continued enthusiasm for the issue are likely to overcome yet insurmountable problems in a slow growing evolutionary process of success rather than a revolutionary breakthrough.

References

1. Abukawa H, Shin M, Williams WB, Vacanti JP, Kaban LB, Troulis MJ. Reconstruction of mandibular defects with autologous tissue-engineered bone. J Oral Maxillofac Surg 2004; 62: 601–605
2. Aframian DJ, David R, Ben-Bassat H, Shai E, Deutsch D, Baum BJ, Palmon A. Characterization of murine autologous salivary gland cells: a model for use with an artificial salivary gland. Tissue Eng 2004; 10: 914–920
3. Aframian DJ, Amit D, David R, Shai E, Deutsch D, Honigman A, Panet A, Palmon A. Reengineering salivary gland cells to enhance protein secretion for use in developing artificial salivary gland device. Tissue Eng 2007; 13: 995–1001

4. Aghaloo TL, Moy PK, Freymiller EG. Evaluation of platelet-rich plasma in combination with freeze-dried bone in the rabbit cranium. A pilot study. Clin Oral Implants Res 2005; 16: 250–257
5. Alaminos M, Garzon I, Sanchez-Quevedo MC, Moreu G, Gonzalez Andrades M, Fernandez Montoya A, Campos A. Time-course study of histological and genetic patterns of differentiation in human engineered oral mucosa. J Tissue Eng Reg Med 2007; 1: 350–359
6. Alhadlaq A, Mao JJ. Tissue engineered neogenesis of human-shaped mandibular condyle from rat mesenchymal stem cells. J Dent Res 2003; 82: 951–956
7. Alhadlaq A, Mao JJ. Tissue-engineered osteochondral constructs in the shape of an articular condyle. J Bone Joint Surg Am 2005; 87: 936–944
8. Almarza AJ, Athanasiou KA. Design characteristics for the tissue engineering of cartilaginous tissue. An Biomed Eng 2004; 32: 2–17
9. Arosarena O, Collins W. Comparison of BMP-2 and -4 for rat mandibular bone regeneration at various doses. Othod Craniofac Res 2005; 8: 267–276
10. Arosarena O, Collins W. Bone regeneration in the rat mandible with bone morphogenetic protein-2: a comparison of two carriers. Otolaryngol Head Neck Surg 2005; 132: 592–597
11. Becker W, Lynch SE, Lekholm U, Becker BE, Caffesse R, Donath K, Sanchez R. A comparison of ePTFE membranes alone or in combination with platelet-derived growth factors and insulin-like growth factor-I or demineralized freeze-dried bone in promoting bone formation around immediate socket implants. J Periodontol 1992: 63: 929–940
12. Blum B, Moseley J, Müller L, Richelsoph K, Haggard W. Measurement of bone morphogenetic proteins and other growth factors in demineralised bone matrix. Orthopedics 2004; 27: s161–s165
13. Bosch C, Melsen B, Gibbons R, Vargervik K. Human recombinant transforming growth favctor-beta 1 in healing of calvarial bone defects. J Craniofac Surg 1996; 7: 300–310
14. Boyne PJ. Animal studies of application of rhBMP-2 in maxillofacial reconstruction. Bone 1996; 19: 83S–92S
15. Boyne PJ, Lilly LC, Marx RE, Moy PK, Nevins M, Spagnoli DB, Triplett RG.: De novo bone induction by recombinant human bone morphogenetic protein-2(rhBMP-2) in maxillary sinus floor augmentation: J Oral Maxillofac Surg 2005; 63: 1693–1707
16. Bücheler M, Wirz C, Schütz A, Bootz F. Tissue engineering of human salivary gland organoids. Acta Otolaryngol 2002; 122: 541–555
17. Butterfield KJ, Bennett J, Gronowicz, Adams D. Effect of platelet-rich plasma with autogenous bone graft for maxillary sinus augmentation in a rabbit model. J Oral Maxillofacial Surg 2005; 63: 370–375
18. Chang J, Rasamny JJ, Park SS. Injectable tissue-engineered cartilage using a fibrin sealant. Arch Facial Plast Surg 2007; 9: 161–166
19. Chen MH, Chen RS, Hsu YH, Chen YJ, Young TH.:Proliferation and phenotypic preservation of rat parotid acinar cells. Tissue Eng 2005; 11: 526–534
20. Chu TM, Warden SJ, Turner CH, Stewart RL. Segmental bone regeneration using a load-bearing biodegradable carrier of bone morphogenetic protein-2. Biomaterials 2007; 28: 459–467
21. Detamore MS, Athanasiou KA. Motivation, characterization and strategy for tissue engineering the temporomandibular joint disc. Tissue Eng 2003; 9: 1065–1087
22. De Kok IJ, Drapeau SJ, Young R, Cooper LF. Evaluation of mesenchymal stem cells following implantation in alveolar sockets: A canine safety study. Int J Oral Maxillofacial Impl 2005; 20: 511–518
23. Delporte C, O'Connell BC, He X, Lancaster HE, O'Connell AC, Agre P, Baum BJ. Increased fluid secretion following adenoviral mediated transfer of the auquporin-1 cDNA to irradiated rat salivary glands. Proc Natl Acad Sci USA 1997; 94: 3286–3273
24. Duneas N, Crooks J, Ripamonti U. Transforming growth factor-beta 1: Induction of bone morphogenetic protein genes expression during endochondral bone formation in the baboon and synergistic interaction with osteogenic protein-1. Growth Factors 1998; 15: 259–277
25. Fennis JP, Stoelinga PJ, Jansen JA. Reconstruction of the mandible with an autogenous irradiated cortical scaffold, autogenous corticocancellous bone graft and autogenous platelet-rich plasma: an animal experiment. Int J Oral Maxillofac Surg 2005; 34: 158–166
26. Gerard D, Carlson ER, Gotcher JE, Jacobs M. Effects of platelet-rich plasma on the healing of autologous bone grafted Mandibular defects in dogs. J Oral Maxillofac Surg 2006; 64: 443–451
27. Giannobile WV, Hernandez RA, Finkelman RD, Ryan S, Kiritsy CP, D'Andrea M, Lynch SE.: Comparative effect of platelet-derived growth factor-BB and insulin-like growth factor -I, individually and in combination, on periodontal regeneration in Macaca fascicularis. J Periodontal Res 1996: 31: 301–312
28. Groger A, Klaring S, Merten HA, Holste J, Kaps C, Sittinger M. Tissue engineering of bone for mandibular augmentation in immunocompetent minipigs: preliminary study. Scand J Plast Reconstr Surg Hand Surg 2003; 37: 129–133
29. Gruber, R., Ludwig, A., Achilles, M., Poehling, S., Merten, H.A., Schliephake, H. Sinus floor augmentation with recombinant human growth and differentiation factor-5 (rhGDF-5): a histological and histomorphometric study in the Goettingen miniature pig. Clin Oral Impl Res 2008, in print
30. Heliotis M, Lavery KM, Ripamonti U, Tsiridis E, de Silvio L. Transformation of a prefabricated hydroxylapatite/osteogenic protein-1 implant into a vascularized pedicled bone flap in the human chest.: Int J Oral Maxillofac Surg 2006; 35: 265–269
31. Hibi H, Yamada Y, Kagami H, Ueda M. Distraction osteogenesis assisted by tissue engineering in an irradiated mandible : a case report. Int J Oral Maxillofac Implants 2006; 21: 141–147
32. Hotta T, Yokoo S, Terashi H, Komori T. Clinical and histopathological analysis of healing process of intraoral reconstruction with ex vivo produced oral mucosa equivalent. Kobe J Med Sci 2007; 53: 1–14

33. Howell TH, Fiorellini JP, Paquette DW, Ofenbacher S, Giannobile WV, Lynch SE. A phase I/II trial to evaluate a combination of recombinant human platelet-derived growth factor-BB and recombinant human insulin-like growth factor-I in patients with periodontal disease. J Periodontal Res 1997; 68: 1186–1193
34. Huang YC, Kaihler D, Rice KG, Krebsbach PH, Mooney DJ. Combined angiogenic and osteogenic factor delivery enhances bone marrow stroma cell-driven bone regeneration. J Bone Miner Res 2005; 20: 848–857
35. Izumi K, Feinberg SE, Ida A, Yoshizawa M. Intraoral grafting of an ex vivo produced oral mucosa equivalent: a preliminary report. Int J Oral Maxillofac Surg 2003; 32: 188–197
36. Izumi K, Song J, Feinberg SE. Development of a tissue engineered human oral mucosa: from the bench to the bed side. Cells Tissues Organs 2004; 176: 134–152
37. Izumi K, Tobita T, Feinberg SE. Isolation of human oral keratinocyte progenitor/stem cells. J Dent Res 2007; 86: 341–347
38. Johns DE, Athansiou KA. Design characteristics for temporomandibular joint disc tissue engineering: learning from tendon and articular cartilage. Proc inst Mech Eng 2007; 221: 509–526
39. Joraku A, Sullivan CA, Yoo JJ, Atala A. Tissue engineering of functional salivary gland tissue. Laryngoscope 2005; 115: 244–248
40. Joraku A, Sullivan CA, Yoo J, Atala A. In-vitro reconstitution of three-dimensional human salivary gland tissue structures. Differentiation. 2007 Apr;75(4):318–324
41. Jung RE, Glauser R, Schärer P, Hämmerle CH, Sailer HF, Weber FE. Effect of rhBMP-2 on guided bone regeneration in humans. Clin Oral Implants Res 2003; 14 : 556–568
42. Kamil SH, Kojima K, Vacanti MP, Bonassar LJ, Vacanti CA, Eavey RD. In itro engineering to generate a humans sized auricle and nasal tip. Laryngoscope 2003; 113: 90–94
43. Kleinzheinz J, Stratmann U, Joos U, Wiesmann HP. VEGF-activated angiogenesis during bone regeneration. J Oral Maxillofac Surg 2005; 63: 1310–1316
44. Klongnoi B, Rupprecht S, Kessler P, Thorwarth M, Wiltfang J, Schlegel KA.: Influence of platelet-rich plasma on a bioglass and autogenous bone in sinus augmentation. An explorative study. Clin Oral Impl Res 2006; 17: 312–320
45. Kaigler D, Wang Z, Horger K, Mooney DJ, Krebsbach PH. VEGF Scaffolds enhance angiogenesis and bone regeneration in irradiated osseous defects. J Bone Miner Res 2006; 21: 735–744
46. Lauer G.: Autografting of feeder-cell free cultured gingival epithelium. Method and clinical application. J Craniomaxillofac Surg 1994; 22:18–22
47. Lauer G, Schimming R, Gellrich NC, Schmelzeisen R.: Prelaminating the fascial radial forearm flap by using tissue-engineered mucosa: improvement of donor and recipient site. Plast Reconstr Surg 2001; 108: 1564–1572
48. Lauer G, Siegmund C, Hübner U. Influence of donor age and culture conditions on tissue engineering of mucosa autografts. Int J Oral Maxillofac Surg 2003; 32: 305–312
49. Leach JK, Kaigler D, Wang Z, Krebsbach PH, Mooney DJ. Coating of VEGF-releasing scaffolds with bioactive glass for angiogenesis and bone regeneration. Biomaterials 2006; 27: 3249–3255
50. Lynch SE, Buser D, Hernandez RA, Weber HP, Stich H, Fox CH, Williams RC. Effects of the platelet-derived growth factor/insulin-like growth factor-I combination on bone regeneration around dental implants. Results of a pilot study in beagle dogs. J Periodontol 1991: 62: 710–716
51. Marukawa E, Asahina I, Oda M, Seto I, Alam MI, Enomoto S. Bone regeneration using recombinant human bone morphogenetic protein-2 (rhBMP-2) in alveolar defects of primate mandibles. Br J Oral Maxillofac Surg 2001; 39: 452–459
52. Marx RE. Commentary, Int J Oral Maxillofac Implants 2006; 21: 190–191
53. Merten HA, Wiltfang J. Ectopic bone formation with the help of growth factor bFGF. J Craniomaxillofac Surg 1996; 24: 300–304
54. Materna T, Rolf, HJ, Napp J, Schulz J, Gelinsky M, Schliephake H. In-vitro characterization of three-dimensional scaffolds seeded with human bone marrow stromal cells for tissue engineered growth of bone—mission impossible ? A methodological approach. Clin Oral Impl Res 2008, in print
55. Meraw SJ, Reeve CM, Lohse CM, Sioussat TM. Treatment of peri-implant defects with combination growth factor cement. J Periodontol 2000: 71: 81–13
56. Miranda SR, Nary Fllho H, Padovan LE, Ribelro DA, Nicolielo D, Matsumoto MA. Use of platelet-rich plasma under autogenous onlay bone grafts. Clin Oral Impl Res 2006; 17: 694–697
57. Moharamzadeh K, Brook IM, van Noort R, Scutt AM, Thornhill MH. Tissue-engineered oral mucosa: a review of the scientific literature. J Dent Res 2007; 86: 115–124
58. Mooren RE, Merkx MA, Bronkhorst EM, Jansen JA, Stoelinga PJ. The effect of platelet-rich plasma on early and late bone healing: An experimental study in goats. Int J Oral Maxillofac Surg 2007; 36: 626–631
59. Nagao H, Tachikawa N, Miki T, Oda M, Mori M, Takahashi K, Enomoto S. Effect of recombinant human bone morphogenetic protein-2 on bone formation in alveolar ridge defects in dogs. Int J Oral Maxillofac Surg 2002; 31: 66–72
60. Nerem RM. Tissue engineering in the USA. Med Biol Eng Comput 1992; 30: CE8-CE12
61. Nerem RM.: Tissue engineering: the hope, the hype, and the future. Tissue Eng 2006; 12: 1143–1150
62. Nociti FH Jr, Stefani CM, Machado MA, Sallum EA, Toledo S, Sallum AW.: Histometric evaluation of bone regeneration around immediate implants partially in contact with bone: a pilot study in dogs. Implant Dent 2000: 9: 321–328
66. O'Connell AC, Baggalini L, Fox PC, O'connell BC, Kenshalo D, Oweisy H, Hock AT, Sun D, Herscher LL, Braddon VR, Delporte C, Baum BJ..: Safety and efficacy of adenovirus-mediated transfer of he human aquaporin-1 cDNA to irradiated parotid glands of non-human primates. Cancer Gene Ther 1999; 6: 505–513

64. Park JB, Matsuura M, Han KY, Norderyd O, Lin WL, Genco RJ, Cho MI. Periodontal regeneration in class III furcation defects of beagle dogs using guided tissue regenerative therapy with platelet-derived growth factor. J Periodontol 1995: 66: 462–477
65. Peterson B, Iglesias R, Zhang J, Wang JC, Lieberman JR. Genetically modified human derived bone marrow cells for posterolateral lumbar spine fusion in athymic rats. Beyond conventional autologous bone grafting. Spine 2005; 30: 283–290
66. Puelacher WC, Mooney DJ, Langer R, Upton J, Vacanti JP. Design of nasoseptal cartilage replacements synthesized from biodegradable polymers and chondrocytes. Biomaterials 1993; 15: 774–778
67. Raghoebar GM, Schortinghuis J, Liem RS, Ruben JL, van der Wal JE, Vissink A. Does platelet-rich plasma promote remodelling of autologous bone grafts used for augmentation of the maxillary sinus floor? Clin Oral Impl Res 2005; 16: 349–356
68. Ren T, Ren J, Jia X, Pan K. The bone formation in vitro and mandibular defect repair using PLGA porous scaffolds. J Biomed Mater Res A 2005; 74: 562–569
69. Riley EH, Lane JM, Urist MR, Lyons KM, Lieberman JR.. Bone Morphogenetic Proetin-2 Biology and Applications. Clin Orthop Rel Res 1996; 324: 39–46
70. Ripamonti U, Ferrettii C, Heliotis M. Soluble and insoluble signals and the induction of bone formation: molecular therapeutics recapitulating development. J Anat 2006; 209: 447–468
71. Roldan JC, Jepsen S, Miller J, Freitag S, Rueger DC, Acil Y, Terheyden H. Bone formation in the presence of platelet-rich plasma vs. bone morphogenetic protein-7. Bone 2004; 34: 80–90
72. Roldan JC, Jepsen S, Schmidt C, Knüppel H, Rueger DC, Acil Y, Terheyden H. Sinus floor augmentation with simultaneous placement of dental implants in the presence of platelet-rich plasma or recombinant human bone morphogenetic protein-7. Clin Oral Implants Res 2004; 15: 716–723
73. Rutherford RB, Ryan ME, Kennedy JE, Tucker MM, Charette MF. Platelet-derived growth factor and dexamethasone combined with a collagen matrix induce regeneration of the periodontium in monkeys. J Clin Periodontol 1993; 20: 537–544
74. Sanchez-Quevedo MC, Alaminos M, Capitan LM, Moreu G, Garzon I, Crespo PV, Campos A. Histological and histochemical evaluation of human oral mucosa constructs developed by tissue engineering. Histol Histopathol 2007; 22: 631–640
75. Sauerbier S, Gutwald R, Wiedemann-Al-Ahmad M, Lauer G, Schmelzeisen R. Clinical application of tissue-engineered transplants. Part I: Mucosa. Clin Oral Implants Res 2006; 17: 625–632
76. Schek RM, Taboas JM, Hollister SJ, Krebsbach PH.Tissue engineering osteochondral implants for temporomandibular joint repair. Orthod Craniofac Res. 2005; 8: 313–319
77. Schimming R, Schmelzeisen R. Tissue-engineered bone for maxillary sinus augmentation. J Oral Maxillofac Surg 2004; 62: 724–729
78. Schlegel KA, Zimmerman R, Thorwarth M, Neukam, FW, Klngnoi B, Nkenke E, Felszeghy E. Sinus floor elevation using autologous bone or bone substitute combined with platelet-rich plasma. Oral Surg Oral Med Oral Pathol Oral Radiol Endod 2007; 104:15–25
79. Schliephake H, Neukam FW, Löhr A, Hutmacher D. The use of basic fibroblast growth factor (bFGF) for enhancement of bone ingrowth into pyrolized bovine bone. Int J Oral Maxillofac Surg 1995; 24:181–185
80. Schliephake, H., M. U. Jamil, J. W. Knebel. Reconstruction of the mandible using bioresorbable membranes and basic fibroblast growth factor in alloplastic scaffolds. J Oral Maxillofac Surg 1998 56: 6160–626
81. Schliephake H, Knebel J, Aufderheide M, Tauscher M. Use of cultivated osteoprogenitor cells to increase bone formation in segmental mandibular defects—an experimental pilot study in mini pigs. Int J Oral Maxillofac Surg 2001 30:531–537
82. Schliephake H. Bone Growth Factors in Maxillofacial Skeletal Reconstruction. Int J Oral Maxillofac Surg 2002 31: 469–484
83. Schliephake, H., Weich, H., Dullin, C, Gruber R, Frahse, S. Mandibular reconstruction by implantation of rhBMP in a slow release system of polylactic acid. Biomaterials 2008;2 9:103–110
84. Schliephake H, Zghoul N, Jäger V, Gelinsky M, Szubtarsky N.: Bone formation in bone marrow stroma cell seeded scaffolds used for reconstruction of the rat mandible. Int J Oral Maxillofac Surg 2008, submitted
85. Shang Q, Wang Z, Liu W, Shi Y, Cui L, Cao Y. Tissue-engineered bone repair of sheep cranial defects with autologous bone marrow stromal cells. J Craniofac Surg 2001; 12: 586–593
86. Springer IN, Nocini PF, Schlegel KA, de Santis D, Park J, Warncke PH, Terheyden H, Zimmermann R, Chiarini L, Gardner K, Ferrari F, Wiltfang J. Two techniques for the preparation of cell-scaffold constructs suitable for sinus augmentation: steps into clinical application. Tissue Eng 2006; 12: 2649–2656
87. Stefani CM, Machado MA, Sallum EA, Toledo S, Nocti H Jr. Platelet-derived growth factor/insulin-like growth factor-1 combination and bone regeneration around implants placed into extraction sockets: a histometric study in dogs. Implant Dent 2000: 9:126–131
88. Suba Z, TAkacs D, Gyulai-Gaal S, Kovascs K. Facilitation of beta-tricalcium phosphate-induced alveolar bone regeneration by platelet-rich plasma in beagle dogs: A histologic and histomorphometric study. Int J Oral Maxillofac Surg 2004; 19:832–838
89. Sun T, Zhu J, Yang X, Wang S. Growth of miniature pig parotid cells on biomaterials on vitro. Arch Oral Biol 2006; 51: 351–358
90. Tatakis DN, Wikesjo UM, Razi SS, Sigurdsson TJ, Lee MB. Periodontal repair in dogs: effect of transforming growth factor-beta 1 on alveolar bone and cementum regeneration. J Clin Periodontol 2000: 27: 698–704
91. Terheyden H, Jepsen S, Rueger DR. Mandibular reconstruction in miniature pigs with prefabricated vascularized bone grafts using recombinant human osteogenic protein-1: a preliminary study. Int J Oral Maxillofac Surg 1999; 28: 461–463

92. Thaller SR, Salzhauer MA, Rubinstein AJ, Thion A, Tesluk H. Effect of insulin-like growth factor type I on critical size calvarial bone defects in irradiated rats. J Craniofac Surg 1998: 9: 138–141
93. Thor A, Franke-Stenport V, Johansson CB, Rasmusson L. Early bone formation in human bone grafts with platelet-rich plasma: preliminary histometric results. Int J Oral Maxillofac Surg 2007; 36: 1164–1167
94. Thor A, Wanfors K, Sennerby L, Rasmusson L. Reconstruction of the severely resorbed maxilla with autogenous bone, platelet-rich plasma, and implants: 1-year results of a controlled prospective 5-year study. Clin Implant Dent Relat Res. 2005; 7(4): 209–20
95. Thorwarth M, Wehrhan F, Schultze-Mosgau S, Wiltfang J, Schlegel KA. PRP modulates expression of bone matrix proteins in vivo without long-term effects on bone formation. Bone 2006; 38: 30–40
96. Toriumi DM, Kotler HS, Luxenberg DP, Holtrop ME, Wang EA. Mandibular reconstruction with a recombinant bone-inducing factor. Functional, histologic, and biomechanical evaluation. Arch Otolaryngol Head Neck Surg 1991; 117: 1101–1112
97. Tran SD, Wang J, Bandyopadhyay BC, Redman RS, Dutra A, Pak E, Swaim WD, Gertsenhaber JA, Bryant JM, Zheng C, Goldsmith CM, Kok MR, Welölner RB, Baum BJ. Primary culture of polarized human salivary epithelial cells for use in developing an artificial salivary gland. Tissue Eng 2005; 11: 172–181
98. Tran SD, Sugito T, Disquale G, Cotrim AP, Bandyopadhyay BC, Riddle K, Mooney D, Kok MR, Chiorini JA, Baum BJ. Re-engineering primary epithelial cells from rhesus monkey parotid glands for use in developing an artificial salivary gland. Tissue Eng 2006; 12: 2939–2948
99. Tsuda H, Wada T, Yamashita T, Hamada H. Enhanced osteoinduction by mesenchymal stem cells transfected with a fiber-mutant adenoviral BMP2 gene. J Gene Med 2005; 7: 1322–1334
100. Ueda M, Ebata K, Kaneda T.: IN vitro fabrication of bioartificial mucosa for reconstruction of oral mucosa: basic research and clinical application. Ann Plast Surg 1991; 27: 540–549
101. Ueda M, Yamada Y, Ozawa R, Okazaki Y. Clinical case reports of injectable tissue-engineered bone for alveolar augmentation with simultaneous implant placement. Int J Periodontics Retsorative Dent 2005 ; 25 : 129–137
102. Valentin-Opran A, WOzney J, Csimma C, Lilly L, Riedel GE. Clinical Evaluation of recombinant human bone morphogenetic Protein-2.: Clin Orthop Rel Res 2002; 395: 110–120
103. Wang L, Detamore M. Tissue engineering of the mandibular condyle. Tissue Eng 2007; 13:1955–1971
104. Warncke PH, Springer IN, Wiltfang J, Acil Y, Eufinger H, Wehmöller M, Russo PA, Bolte H., Sherry E, Behrens E, Terheyden H. Growth and transplantation of a custom vascularized bone graft in a man. Lancet 364; 766–770, 2004
105. Wei C, Larsen M, Hoffman MP, Yamada KM. Self-organization and branching morphogenesis of primary salivary epithelial cells. Tissue Eng 2007; 13: 721–735
106. Weibrich G, Kleis WK: Curasan PRP kit vs. PCCS PRP system. Collection efficiency and platelet counts of two different methods for the preparation of platelet-rich plasma. Clin Oral Impl Res 2002; 14: 357–362
107. Wiltfang J, Schlegel KA, Schultze-Mosgau S, Nkenke E, Zimmermann R, Kessler P: Sinus floor augmentation with beta-tricalciumphosphate (beta-TCP): does platelet-rich plasma promote ist osseous integration and degradation? Clin Oral Implants Res 2003; 14: 213–218
108. Wu W, Chen F, Feng X, Liu Y, Mao T. Engineering cartilage tissue with the shape of human nasal alar by using chondrocyte macroaggregates—Experimental study in rabbit model. J Biotechnol 2007; 130: 75–84
109. Yamamoto M, Tabata Y, Hing L, Miyamoto S, Hashimoto N, Ikada Y. Bone regeneration by transforming growth factor-beta 1 released from a biodegradable hydrogel. J Controlled Release 2000 64: 133–142
110. Yanaga H, Koga M, Imai K, YanagaK.: Clinical Application of biotechnologically cultured autologous chondrocytes as novel graft material for nasal augmentation. Arch Plast Surg 2004; 28: 212–221
111. Yang X, Whitaker MJ, Sebald W, Clarke N, Howdle SM, Shakesheff KM, Oreffo ROC. Human osteoprogenitor bone formation using encapsulated bone morphogenetic protein 2 in porous polymer scaffolds. Tissue Eng 2004; 10: 1037–1045
112. Zellin G, Beck S, Hardwick R, Linde A. Opposite effects of recombinant human transforming growth factor-beta-1 on bone regeneration in vivo: effects of exclusion of periosteal cells by microporous membrane. Bone 1998 22: 613–620
113. Zellinn G, Linde A. Importance of delivery systems for growth-stimulatory factors in combination with osteopromotive membranes. An experimental study using rh-BMP-2 in rat mandibular defects. J Biomed Mater Res 1997; 35: 181–190

Tissue Engineering Strategies in Dental Implantology

U. Joos

Contents

58.1	Introduction	839
58.2	Influence of Bone Properties on Osseointegration	841
58.3	Influence of Bone Biology on Osseointegration	843
58.4	Analytical Methods	844
58.5	Features of In Vivo Mineralization at Implants	845
58.6	Features of In Vitro Mineralization at Implants	846
58.7	Engineering Approaches	847
	References	850

58.1 Introduction

Tissue engineering and regenerative medicine strategies have gained increased attention in dental implantology, since dental implantology faces new challenges in clinical dentistry. Whereas conventional implant placement and implant loading protocols under ideal anatomical and biological conditions seem to be clinically solved, engineering and regenerative strategies are aimed at allowing implant placement under compromised conditions (lack of bone or mucosa tissue, impaired regenerative capacity, need for early loading protocols). To understand the new approaches in engineering and regenerative medicine therapies, it is important to recapitulate the history and basics of implant dentistry in order assess the clinical impact of new approaches.

The establishment of dental implants in clinical routine has profoundly changed the possibilities of oral rehabilitation [1]. The long-term success of bone-interfacing implants requires rigid fixation of the implant within the host bone site. This condition, known as functional osseointegration, is achieved in various implant systems by an interlock between the surface features of the implant (threaded, porous, or textured surfaces) and the bone tissue. Clinical and experimental studies demonstrate that osseointegration can be achieved when implants are placed under distinct circumstances in bone of different quantity and quality [63]. It was shown that implants cannot only become stable in bone of compromised size and structure but also have the ability to remain stable when implants are loaded [9]. Recent research indicates that an undisturbed osseointegration can be achieved even when healing under load is present [69]. In contrast, evidence from a multitude of clinical and experimental studies reveal that implant failures do also occur [46]. The failure of osseointegrated implants in the treatment of completely and partially edentulous patients with a sufficient amount and quality of bone has been well documented in the literature [1]. Various studies indicated a higher rate of long-term implant loss in the maxilla in comparison to the mandible. It has been suggested that the amount and quality of maxillary bone is responsible for the higher rate of implant failure in this area [2].

A failure of osseointegration or a disintegration of a formerly stable anchored implant can be conceptualized as a failure of the mineralized extracellular matrix directly attached to the artificial surface since a mechanically competent implant/bone bond is dependant on an intact mineralized interface structure. A main research area, to date, is therefore the application of tissue engineering and regenerative medicine approaches to improve the overall implant success, even under compromised anatomical or biological circumstances. In order to understand recent engineering strategies to improve dental implantology therapies, we give a detailed description of the principles of the maintenance and emergence of mineralized bone tissue and of the implant/bone interface reactions [92].

The structural and functional tissue properties adjacent to the implant surface can be related to the interaction between an artificial material (e.g., titanium, calcium phosphates) and the microenvironment at the host site. The dynamic interaction between artificial materials and bone are interrelated as one object affects the other. The interaction is different in the subareas of the bony host site, since the cellular as well as the biophysical microenvironment between, for example, the cortical and spongiosal layer and even within such a layer is different.

It is in this respect important to note that osseointegration of implants, a term that was initially defined by Branemark [13] as a direct bone-to-implant contact and later on defined on a more functional basis as a direct bone-to-implant contact under load, is in its details not definitively determined. Especially the dynamic cellular and acellular processes at the interface at a micro- and nanoscale level are not fully elucidated. Additionally, early aspects of the bone/biomaterial interaction in terms of seconds to minutes are not well known in the in vivo environment. The recent knowledge in both aspects of implant osseointegration is even more limited when the process of biomineral formation is under consideration. To gain insight into the state of mineralization at implant interfaces the various levels of bone structure and physiology are considered and evaluated in light of the recent tissue engineering strategies to improve implant osseointegration.

The success of dental implants depends on their placement in bone of sufficient density and volume in order to achieve primary stability. Optimal esthetics of implants requires their placement in a position approximating that of the natural teeth they replace. However, there is generally at least some degree of atrophy in the sites where implants are placed and often the viability of oral hard and soft tissues may be compromised by pathologic conditions. All of these conditions may potentially compromise the survival rate of implants or at least the final esthetics and function of implant-supported restorations. During the initial years of the development of the osseointegration protocol, the implants themselves were as described before modified in material and structure. Therefore, various engineering attempts have been made to improve bone formation around implants by influencing implant-specific or bone-specific aspects. Most notably, implant surface characteristics [24, 48, 70] (material, surface topography, surface chemistry) and implant geometries were altered to improve osseointegration [41, 69]. Although application of these improved implants resulted in better osteoconduction, bone formation, and bone remodeling, they are still not ideal. The challenges for designing successful dental implants are even greater than those for designing orthopedic implants because of the proximity of the implant site to other anatomical structures (i.e., the maxillary sinus and the mandibular inferior alveolar canal and associated nerve), poor bone quality around the implant (especially trabecular bone in the posterior maxilla), and prevention of acute inflammation around the implant. Implant materials, design, and surface topographies have been described in detail previously. As osseointegration was successfully achieved, and esthetic outcomes were convincing with new implant systems, a paradigm shift occurred in implant dentistry from merely placing implants in healthy implant sites to enabling implant therapy also in bone sites of insufficient bone volume or in bone of impaired biological quality. These objectives have been materialized by advancements in surgical techniques, as well as availability of biomaterials to enable regeneration of oral hard tissues. As surgical methods to augment lost bone is accompanied by a harvesting morbidity, and as biomaterials do not possess biological activity, more biological-oriented strategies were elaborated to overcome the limitations of conventional reconstruction therapies. Different strategies can be considered to be the main options to improve peri-implant bone healing: first, activation of implant surfaces, second, the use of cell-driven regeneration protocols, and third, the use of new materials (or as a newly elaborated option: the renouncement of artificial materials by new tissue technologies). As it was demonstrated that alterations in the biophysical properties of the outermost implant structure (e.g., an increased hydrophilicity) of Ti implants might accelerate the initial healing reaction at the biomaterial/biosystem interface in vivo, and as it was currently shown that high energetic Ti implant surfaces (the modSLA-type surface) cause stimulatory effects on osteoblasts in vitro, activation and fictionalization of implants seems to be a promising strategy to increase bone formation. A promising measure to activate surface structures is the adhesion of defined molecules at the implant surface. Hydrogels are a group of biomolecules that have unique properties concerning the adjustment of biophysical parameters. These materials have therefore gained special interest in the field of tissue engineering. They are applied as space-filling agents, as delivery vehicles for bioactive molecules, and as three-dimensional structures that organize cells and present stimuli to direct the formation of a desired tissue. Much of the success of scaffolds in these roles hinges on finding an appropriate material to address the critical physical, mass transport, and biological design variables inherent to each application. Advancements in hydrogel chemistry (elaborated in the microarray technology) have enabled users to define precisely the biophysical properties of these materials. In addition, hydrogels are an appealing coating material because they are structurally similar to the extracellular matrix of many tissues, can often be processed under relatively mild conditions, and may be delivered in a minimally invasive manner. Additionally, new hydrogels allow a highly defined adhesion and release of proteins, and they can easily be attached to metal surfaces. Consequently, cell-based transplantation protocols may offer new advances in peri-implant bone loss situations.

With the advancement of a cell-driven concept of "regenerative medicine" and in particular the field of tissue engineering, a completely new strategy of bone generation can be expected with the potential to change treatment regimes profoundly. Cell-driven bone tissue engineering aims at regenerating defected or lost bone and highlights a transition from the historically biomaterial-based approaches in which mechanically stable, biocompatible materials were used to augment lost bone, to a focus on cell-based devices. Therefore, the use of ex vivo grown cell containing tissue specimens offer new therapeutic options in bone regeneration. The use of such transplanted cells in peri-implant bone augmentation is highlighted by the fact that cells are unique in having the driving force to create new tissue even in sites of impaired host site conditions. Whereas the single use of hydrogels or cells is assumed to improve peri-implant bone healing, complex approaches (using a cell containing tissue at functionalized implants) have not been performed until now.

58.2 Influence of Bone Properties on Osseointegration

The properties of bone are directly related to the features of the mineralized extracellular matrix adjacent to implants in two ways. First, the macroscopic and microscopic implant geometry and the insertion approach (as characterized by the preparation of the implant bed) determine the principal bone/implant relation. Second, the properties of bone have a major impact on the load-related characteristics of the microenvironment adjacent to implants.

Bone is defined as a bone-specific mineralized hard tissue [3]. Mineralization is therefore not only the defining feature of bone presence or formation, it is also the fundamental aspect that enables implants to remain stable in place even when forcibly loaded. Bone can be considered on a basic level as a compound material (a soft tissue network that is reinforced by minerals), possessing rigid as well as elastic properties [54]. It is composed of a variety of cell types and an organic matrix that is strengthened by matrix-associated calcium minerals (primarily calcium and phosphate in the form of hydroxyapatite). Cells, matrix, and minerals are connected in a special way to give bone its unique biophysical and biological properties [55]. Morphologically there are two forms of bone, which impose different structural and functional features: cortical and cancellous bone. Both aspects of bone tissue interact differently towards implants [38]. The structure of the cortical layer and the trabecular system is optimized to transfer the loads through the bone by a dynamic feedback between load perception of cells and their subsequent cellular reaction. The differences in their histologic and ultrastructural appearance of the two tissue types are related to some extent to their functions: the cortical part of bone provides the mechanical and protective functions, whereas cancellous bone is also involved in metabolic functions (e.g., calcium homeostasis). Both aspects (structural and metabolic) are closely related to the features of the mineralized extracellular matrix at implant surfaces.

One guiding principle in implant bone interaction is that the fixture design should be coincident with primary stability. A second emerging principle is that the implant must allow the transmission of forces without threading the biomechanical competence of the bone's material properties, leading to microfractures of the mineralized matrix. Third, implants should have an intimate contact with the bone directly after insertion. All three prerequisites, interacting mainly with the bone considered as a compound material and leaving behind the biological reactivity of bone, are closely related with the shape of implants.

With dental implants, where axial symmetry is possible, symmetrical implant forms have proved effective for achieving secure implant fixation within bone [1]. During the development of implant dentistry, root-form implants of screw, coated cylinder, and to a minor extent, hollow-basket geometries were introduced in clinical treatment protocols [29, 93]. In the past decade, a convergence to threaded screw designs has been observed [81]. More recently, parabolic "root" shaped implants have been demonstrated to possess advantages in respect to the biomechanical features of load transfer from implants to bone [69, 73].

The threads of implants are representative of macroscopic surface features that allow mechanical interlocking of implant within bone [18]. Thread-containing implants can be inserted in bone by a self-cutting procedure or by preparation of the implant bed through a thread cutter. Histological analysis of probes indicate that self-cutting screws are associated with a generally higher bone-to-implant contact pronounced at the crestal part, when compared to preparation of the bony implantation bed [17]. The results of various experimental studies suggest that the quality of the primary implant stability is dependent to a large extent on the geometric relation between implant shape and the surgically created host side. The reason that an intimate contact between implants and bone directly after insertion can be achieved is based on the fact that cortical bone has an elasticity of up to 5% (with cancellous bone having even a higher elasticity). If the implant insertion is not accompanied by an extension of the cortical layer over this threshold, a direct contact between the implant and the present mineralized matrix can be assured. The core diameter of the implant bed should therefore be adjusted to the core implant diameter. By a slight expansion of fully mineralized bone, a direct contact can be achieved over large areas. Therefore, bone remodeling more than new bone formation (modeling) would take place. Experimental studies reveal that implant systems having a conceptual geometric approach of insertion may therefore affect the tissue response in a positive way. A histological evaluation of screw-shaped parabolic implant systems revealed a high congruency between the implant and the surrounding bone tissue [68]. A direct contact between implants and bone was achieved over large surface areas directly after insertion when parabolic implant systems were used. Cylindrical implant systems in contrast possess the disadvantage of a crestally pronounced incongruence [17]. Excellent adaptation of the host bone to titanium surfaces was observed on an ultrastructural level in a comparable manner after insertion of self-tapping screws in calvaria bone by Sowden and Schmitz [96]. Several studies demonstrated that when self-tapping parabolic-shaped screws were placed in loading or nonloading positions the long-term histology showed that the bone tissue around the implants was maintained in both situations [31].

In contrast, if implants are inserted in a mechanically preconditioned implant bed, threading the elasticity of cortical bone (e.g., by using an osteotome technique, or by having a larger discrepancy between the diameter of the final burr and the implant), the primary stability will decrease. This was shown by recent investigations, demonstrating the occurrence of microfractures in condensed mineralized peri-implant bone [16]. The observation of a higher bone-to-implant contact under such circumstances [91] is not accompanied by an improved stability [16], indicative for the fact that the bone-to-implant contact itself must be carefully analyzed as a predictive factor of a good osseointegration.

58.3
Influence of Bone Biology on Osseointegration

The high clinical success rates of implant osseointegration are based on the fact that bone biology has some unique features. Bone has the ability to recapitulate specific aspects of its initial developmental processes and thereby undergo regeneration to a stage of a repair ad integrum. Osseointegration at implants within the bone repair and regeneration process occurs between the implant surface and the tissues covering the implantation bed. Repair and regeneration, despite commonly used interchangeable terms, are different processes with respect to mineral formation since repair (modeling) is associated with a matrix vesicle initiated mineralization of the newly synthesized extracellular matrix [54]. Regeneration (remodeling), in contrast, is not dependant on matrix vesicle formation. It works through a balance of dissolution and formation of collagen-associated biominerals. Various sources of cells (periosteal cells, cortical cells, cells derived from the surrounding soft tissues, and marrow cells) and signals that set up these fields are responsible for the features of biomineral formation. Osteoblasts and osteocytes adjacent to the implant produce various morphogens starting from the onset of tissue disintegration, suggesting that these signal molecules act locally to recruit and induce skeletogenic cells to proliferate and differentiate. Whereas the nature and origins of the initiating morphogenetic signals for the skeletogenic cells in the repair tissue are not completely determined, a multitude of signals have been identified.

The impact of bone biology on mineral formation can be considered as a highly complex and dynamic cell-driven process. Biological and biophysical parameters affect the success of modeling or remodeling at implant surfaces by modulating bone cell reactions [104]. The condition of the bone microenvironment (temperature, O_2-tension, vascularity, load) affects the cell response at implant surfaces [71]. Two main factors are of special originating relevance. Tissue maintenance and emergence is dependent on the extent of surgical trauma directly at insertion as well as through load-related deformations under implant load, especially when immediate or early loading protocols are applied. The initial mode of osseointegration critically depends therefore to a greater extent on the state of cells and matrix at the surface of the artificially created implant site [2, 77]. It is controversially discussed in the literature whether the surgical trauma itself leads to a disturbance of the cellular or noncellular components at the surgical site [77]. Considering the recent data, it is until now not convincingly shown on an ultrastructural level that the surgical procedure, if properly performed, is accompanied by a disturbance of cell activity at the surface of the implant bed nor with a disintegration of bone minerals. The extent of bone deformation under load (in sense of resulting stress and strain distributions) is perhaps the more important regulating factor, dependent on the physical properties of the bone tissue (e-modulus, elasticity, strength), the direction and amount of the applied forces, and also to a main extent by the geometry of the implant used [8, 21, 51]. Because of the complex structure of cortical and cancellous tissues (nonhomogenous, nonisotropic tissue properties) and the technical problems involved, it is difficult to directly determine the force/tissue deformation relationship. Despite the difficulties in directly determining accurately the deformations in the vicinity of implants, the extent of cell deformations in the microenvironment of implants determine to a great extent the fate of the subsequent mineralization process.

Recent research has suggested that bones display an extraordinary adaptive behavior towards a changing mechanical environment, which is often regarded as phenotype plasticity [34, 90]. Specific strain-dependent signals are thought to control this adaptive mode of bony tissue modeling [30, 35]. The adaptive mechanisms include basic multicellular units (BMUs) of bone remodeling. Effector cells within BMUs have been shown to function in an interdependent manner. The importance of load-related bone reactions over hormonal- or cytokine-related factors is mirrored by the fact that while hormones or cytokines may bring about as much as 10% of the postnatal changes in bone strength and mass, 40% are determined by mechanical effects. This has been shown by the loss of extremity bone mass in patients with paraplegia (more than 40%). That peri-implant tissue formation and mineralization by osteoblasts are strongly dependent on the local mechanical environment in the interface zone is therefore generally accepted [80]. Carter and Giori [21] suggested that proliferation and differentiation of the osteoblasts responsible for peri-implant tissue formation are regulated by the local mechanical environment according to the tissue differentiation hypothesis proposed by Frost and his colleagues for callus formation [35]. According to Frost's theory [34], bone cells within the bone tissue that experience a loading history of physiological strain are likely to be osteogenic, assuming an adequate blood supply [15]. However, if the healing tissue is exposed to excessive strains, fibrogenesis will result [50]. The relationship of defined cell deformation and bone remodeling has been documented in various in vivo studies. Loading of intact bone after osteotomy, during growth, in fracture healing, and during distraction osteogenesis resulted in strain-related tissue responses [22, 37, 65, 74]. Whereas physiological bone loading (500–3,000 microstrains) leads to mature bone formation, higher peak strains (>5,000 microstrains), known to be hyperphysiologic, result in immature bone mineral formation and fibroblastic cell pattern.

With respect to micromotion-related mineralization phenomena at implant interfaces, a second aspect is of importance. Micromotion does not only influence osteoblast cell behavior but has also a profound effect on the physicochemical -driven mineralization process. It is known that collagen mineralization in an acellular environment is regulated by the strains exposed to collagen [33]. Over a distinct threshold collagen mineralization will fail through a disturbance of the spatial orientation of collagen fibers. The experimental findings of a defect mineral formation [66] are in agreement with theoretical calculations of strain-related mineral formation at organic molecules.

That loads regulate the bone healing process around implants has been confirmed by various experimental in vivo studies. Bone healing under nonloading and loading conditions was demonstrated by various authors [7, 20, 25, 42, 46, 78, 79, 89] and Szmukler-Moncler and colleagues have given a literature review concerning the timing of loading and the effect of micromotion on the bone-implant interface formation [98].

In order to gain a critical approach to study reports on mineralization at implant surfaces some issues must be stressed. Most studies on bone healing have shown the bone reaction on a light microscopical level and a lot of these studies were undertaken on decalcified samples, thereby limiting the possibility to evaluate the features of mineralization during bone healing. Special, limited data are available on the microstructure of the healing tissue at the implant interface at an ultrastructural and crystallographic level [102]. In addition, the complex interactions of cells and extracellular matrix components with mineral deposits directly after insertion and under immediate mechanical loads are poorly understood [44, 45]. Various analytical methods allow nowadays a more precise insight into bone mineralization.

58.4
Analytical Methods

Mineralization of the extracellular matrix combines acellular mineralogical aspects with biological phenomena. Investigations on mineralization at implant surfaces in the (artificial) in vitro environment or in the in vivo situation must therefore not only be evaluated by histological, histochemical, and biochemical techniques but in addition by mineralogical investigations. Histological and biochemical criteria include all aspects from molecular features of the cell to the composition of the extracellular matrix in the microenvironment of the implant [3, 87, 88, 101]. Commonly, histological and biochemical assays used traditional methods of histology and protein chemistry (e.g., staining, immunostaining, electrophoresis, and blot techniques) [43, 49]. More recently, the availability of genetic probes for many of the bone's proteins made it easy to assess gene expression of multiple phenotypic markers by cells gene chips. Mineralogical aspects range from ultrastructural analysis of cells to the time and spatial relationships of early (matrix vesicle formation) to late stages of mature biomineral formation [105].

Various mineralogical studies have led to an improved understanding of mineralization at implant surfaces [82]. Advancements in imaging and analytical techniques enables morphological and structural insight into the formation of a mineralized extracellular matrix [26]. Morphological investigations span the whole range from μCT evaluations [75] over gross microscopical investigations of undecalcified probes to high resolution pictures of mineralized probes (transmission electron microscopy, scanning electron microscopy, atomic force microscopy). In addition, analytical investigations enable the insight into the anorganic aspect of minerals. X-ray or electron diffraction analyses [9, 33] as well as advanced element mapping measures [72] allow to discriminate between a bone-related, mature, or otherwise comprised mineral formation. Advancements in imaging [57] and analytical techniques [19, 59] allow a more detailed insight into the features of mineral formation at implants. Some of the new techniques additionally enable the combined biochemical and crystallographic analysis of implant probes (TOF-SIMS), important with respect to fully understanding the process of osseointegration. Whereas most of these techniques were introduced and evaluated concerning their methodical applicability, limited information is gained through the use of these techniques in respect to preclinical or clinical implant investigations. Reports on the in vivo or in vitro mineralization should therefore be critically viewed in light of the methodological approach.

58.5
Features of In Vivo Mineralization at Implants

In vivo mineralization at implant surfaces is closely related to the features of the extracellular matrix with special respect to the mineralization of collagens [12, 54]. Ultrastructural investigations on the collagen mineralization in the different calcifying tissues (bone, dentin, enamel) have revealed principal similarities in general but also some differences from the stage of crystal nucleation to tissue maturation [44]. In order to critically evaluate the maintenance or emergence of mineral at implant surfaces, the physiological processes of in vivo mineral formation must be considered. It has been found that new bone formation is connected with matrix vesicles [5], which spreads over the border of the vesicle membrane and includes the extra- and intracollagenous mineralization. Bone modeling at implants surfaces implies therefore the synthesis of matrix vesicles [5]. Anderson [5] demonstrated that matrix vesicles serve as the initial site of calcification in all skeletal tissues. Matrix vesicles as membrane-invested particles of 100nm diameter that are selectively located within the extracellular matrix [105], were found to be present at implant surfaces at various stages of implant osseointegration [92]. They are regarded as nucleation core complexes that are responsible for mineral induction by matrix vesicles. This complex process of bone mineralization is then associated with the occurrence and distribution of various types of collagen. Healing of bone is accompanied by the expression of collagen types I, II, and III in bone healing, collagen type I being the most important protein [105]. Concerning the mineralizing process in bone it is known that not only the type of collagen but the spatial relationship with noncollagenous proteins gives rise to a structure configuration which is necessary for the process of mineralization. The complex interactions between collagen and the noncollagenous proteins seem to facilitate crystal formation and subsequent crystal growth [86].

In light of what is known about the physiological bone mineral formation it is controversially discussed what kind of hard tissue formation is present at implant surfaces of in vivo. Two different hypotheses were stated in the literature. Davies was the first to propose a theory for a de novo hard tissue formation in the vicinity of implants [26]. He reported on a granular afibrillar zone directly at the implant surfaces. It was hypothesized that a zone, which is less than 1 μm thick, derived through the deposition of calcified afibrillar accretions by osteoblasts. The features of mineralization should include the initial deposition of calcium, phosphorus, sulfur, and noncollagenous matrix proteins. As these layers have morphological similarities to the cement lines, found at discontinuities in natural bone tissue, the term "cement line" was used for the interfacial zone at implants by some authors [27]. The presence of such cement lines was mainly observed at ceramic implant surfaces.

In contrast, Albrektseson and colleagues [9] suggested that mineralization started through the initial deposition of a collagen-containing matrix. This matrix is then mineralized to a state of mature bone mineral. A thin 20–50 nm electro-dense deposit was found to separate the implant from the mineralized bone matrix [97]. The electron dense zone, suggested to have a "glue" like function, and not consistently confirmed by other authors [68], requires further investigation. Various investigations have shown that osteoblasts and osteocytes were present in the vicinity of implant surfaces or even contacted the implant surface by cell protrusions [76]. Ultrastructural investigation of high resolution confirmed the presence of adhered cells, surrounded by a fully mineralized bone matrix, at smooth implant surfaces [36, 68, 97] indicative for a naturally derived mineralization process at smooth artificial surfaces. It was found in other studies that the ultrastructure of mineralized tissue was the same whether the apposing implant was ceramic or titanium [28, 83]. Concerning their review on the primary mineralization at the surface of implanted materials, Sela and coworkers [92] confirmed the hypothesis that mineral spherites are likely to be the initiation sites of mineral formation when a de novo bone formation at implants is present. It was also observed that matrix vesicle formation in vivo depends on the properties of the material surface [14]. It was concluded that bone bonding materials increase matrix vesicle enzyme activity to a higher extend than nonbonding materials; however, details of the relation between the substrate surface and mineral formation are not solved yet. A striking finding of implants containing rough surfaces (e.g., SLA surfaces), inserted in trabecular bone, is the formation of a mineralized layer covering the complete surface of the implant [11]. As this mineralized tissue is present in nonphysiologic microenvironments (trabecular bone does not completely fill up the marrow space) it remains controversial whether this tissue is laid down due to the effects of the surface or the load transfer towards the peri-implant space. More important, it is not solved whether this mineralized tissue is mineralized in a mature bone-like manner, or if this tissue can be considered as a nonphysiologic mineralized matrix (e.g., comparable to an artificial "cement" line).

A common finding in dental implantology is the loss of mineralized matrix in crestal parts of the implant. Loss of crestal bone mineral can be seen in clinical practice, a phenomenon which has commonly been observed in different stages of implant loading as a decrease of radiological marginal bone height around oral implants. Animal experimental studies proved that increased marginal bone loss may be associated with occlusal overloading [45, 47]. Clinical reports also suggested that implant design, parafunction, and high bending moments are among others associated with marginal bone loss [8, 46, 47, 84]. Whereas no correlation was detected between implant surface structure and marginal bone loss, various experimental studies indicated a strong correlation between micromotion at the tissue level and the bone remodeling answer. Strains exceeding 4,000 microstrain, demonstrated to be present in the crestal part of bone [67], was assumed to account for the loss of mineralized matrix. Repetitive loading of hyperphysiological dimensions was therefore discussed to be a factor of impaired implant survivals in various anatomical locations [80, 98]. The findings of hyperphysiological strain environments (>4,000 microstrain) with subsequent fibrous tissue development [30] especially at implants inserted in bone of impaired quality may be the underlying cause for many of the long-term implant failures. Implant failure caused by overload seems to be mainly dependant on the underlying bone quantity and quality. The amount of crestal height plays another but slightly minor role on the effects of micromotion under mechanical loading. Numerical evaluation of load distribution at implant surfaces reveal that implant insertion in bone of impaired quantity or quality leads to elevated strains at the bone/implant interface even when a bicortical fixation of implants is present. Bicortical fixation is generally considered to eliminate apical peak strains and reduce trabecular bone stresses and strains [100], thereby preserving present bone minerals or by allowing the bone's matrix to mineralize. A failure of matrix mineralization can become substantial when the size of the implant bed is decreased or the Young's modulus of bone is low. Optimized implant geometries are able to reduce crestal peak strains [51] and therefore may improve the clinical implant success by homogenization of local strains [69, 94, 95].

58.6
Features of In Vitro Mineralization at Implants

Osteoblast culture models as well as noncellular models are commonly used to evaluate the effect of implant surface characteristics on mineralizing matrix formation [23, 39, 58, 62]. In conventional osteoblast cultures, mineralizing osteoid is only observed in multilayered structures that form nodules after an extended period of time [103]. With respect to assessing bone-like mineral formation at biomaterial surfaces in vitro [4, 39, 64], a detailed analytical evaluation of the crystal structure has to be performed. One disadvantage of most studies dealing with in vitro mineralization at implant surfaces is the lack of analytical data concerning the surface structure as well as the mineral structure. The difference between a precipitation of calcium phosphate or carbonate at a surface and a bone-like apatite formation is obvious [53, 99]. It is important to note that some general controversies exist about the ability of biomineral formation in cell culture systems. The controversy is not only related to mineral formation at implant surfaces but is mentioned in a much broader sense. Only when maintained under suitable culture conditions, some cells form bone-like nodules in cell culture [103]. There is growing evidence that some cell populations appear capable of finally differentiating in vitro. Due to the finding that in cell culture matrix vesicle mediated-mineralization is not followed by collagen mineralization, bone-like extracellular mineral formation on artificial surfaces is not demonstrated up to now [105]. This should be considered as a general problem of bone-like structure formation on implant surfaces in vitro. It was found that microtophographic characteristics of surfaces regulate not only cell proliferation and migration [58] but also influence mineral deposition in culture [10]. Coating of titanium implants by calcium-containing layers (hydroxyapatite, tri-calcium phosphate, calcium-carbonate) are common techniques aimed to improve mineral formation at surfaces [29, 56]. Different authors, for example, reported on mineral formation at the surfaces of calcium polyphosphate. It was reported that the formation of "cement" lines by cells in culture was quite different to the "cement" lines found in vivo. It is therefore not solved whether these mineralization processes are culture artifacts or not. The presence of initial calcified structures on material surfaces, either precipitated from the extracellular fluid or induced by cellular activity [32, 85], is supposed to improve the subsequent osteoblast activity on these materials [52, 60]. New approaches, performed to precoat implant surfaces with calcified collagen or other biomimetic matrices bound to implant surfaces [6, 40], are currently under intensive investigation. Whereas ultrastructural analysis have not convincingly shown a bone-like collagen mineralization, precoating of implant surfaces impose some promises.

58.7
Engineering Approaches

Biomaterial and biomolecular engineering strategies represent approaches to increase biofunctional activity and provide promising strategies for the engineering of robust biofunctional materials that control bone cell adhesion and signaling and promote osteoblast proliferation, differentiation, and final matrix mineralization. Several means are currently being developed for biomaterials improvement like pattering material surfaces or surface modification of materials by selectively adsorbing adhesion peptides, proteins, or growth factors (Fig. 58.1).

The in vivo effect of surface topography and chemistry on the nature of the adherent cell population, its diversity, and its activity has not been fully elucidated. Cooper reviewed the influence of surface structure on implant healing at titanium implants (one of the most commonly used and to some extent best-investigated artificial material). He concluded that it is difficult to state precisely how surface topography and chemistry affects the bone-implant interface, but the sum of the in vitro and in vivo findings are indicative that an increased titanium implant surface topography improves the bone-to-implant contact and the mechanical properties of the interface. Growing clinical evidence for increased bone-to-implant contact at altered titanium implants confirms the temporally limited observations made in preclinical studies. More recent experimental results indicate that it may be possible to impart titanium surfaces with osteoconductive behavior [49]. Recent studies demonstrating high bone-to-implant contact formation in humans (approaching more than 70% bone-implant contact at initial healing periods) indicate that the modified topography and chemistry of titanium implants may impart favorable osteoconductive behavior [50–52]. In the absence of controlled comparative clinical trials, the aggregate experimental evidence supports therefore the use of titanium implants with

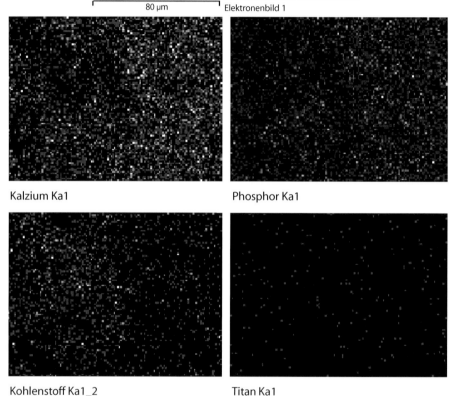

Fig. 58.1 Principle of defect regeneration by means of tissue-engineered implant-hydrogel-cell complex implantation

increased surface topography. The question remains what kind of a special surface structure is able to enhance osteoconduction at implants.

A new approach is to coat implants with hydrogels of defined physic-chemical characteristics (Figs. 58.2, 58.3). These hydrogels enable the precoating of implants with cells (e.g., the adhesion of bone micromasses with a hydrogel integrin layer) and the subsequent strong adhesion of cells after implant placement.

Fabrication of patterned surface topographies seems interesting to be applied on bone-anchored implants. Precise evaluation of the features of cells in culture and in vivo via precise engineering of the material surface properties can offer new insights into surface-structure-related tissue formation in vitro and in vivo. Substrate patterning by control of the surface physicochemical and topographic features enables selective localization and phenotypic and genotypic control of living cells. Patterned arrays of single or multiple cell types in culture can serve as model systems for exploration of cell–cell and cell–matrix interactions. This research has, until now, been restricted to microtopographies due to the high cost of producing controlled nanotopography. New research is, however, starting to make controlled nanotopography available in the quantities required for in vitro and in vivo research. A limited number of studies have begun to assess the response of various cellular phenotypes to nanotopography surfaces. A review of these studies indicates that different cell phenotypes have different levels of sensitivity [22–24]. For example the macrophage cell line P338D1 can react to features (grooves and ridges) within the 30–282 nm dimensions, a width that is comparable to a single collagen fiber [23]. Novel advances in material processing techniques are also allowing higher quality and improved definition nanotopography surfaces to be produced [25, 26]. The incorporation of such techniques allows the formation of chemically homogenous controlled patterned surfaces on the nanotopography scale, accepting changes in the wet chemistry due to the topography itself, therefore providing the basis of studies that can be used to more fully assess the effects of nanotopography on cellular response. No statistically significant difference between viable osteoblast adherences to the range of nanotopographies (the resulting titanium dioxide surfaces were covered with 111+5 nm high and 159+9 nm wide hemispherical protrusions) was observed in a recent study. New fabrication techniques allow nowadays the fabrication of ordered micro- and nanotopographies over large surface areas. The new manufacturing techniques allow the coating of surfaces with different biological elements such as molecules, proteins, and cytokines (Fig. 58.4). This offers new perspectives in evaluating cellular and supracellular reactions towards topographical surface structures that are perceived by bone cells in vivo.

Based on the knowledge that protein adsorption on surfaces influences the cell adhesion behavior, selective coating of surfaces with adhesion proteins offers improvements of osteoblast behavior on artificial surfaces. Therefore, approaches have been performed to develop materials containing surface bound proteins in order to improve the subsequent cell attachment process. As it is known that surface chemistry and topography influences the adherence

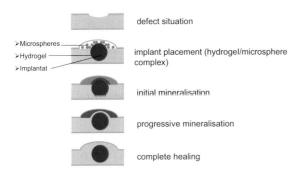

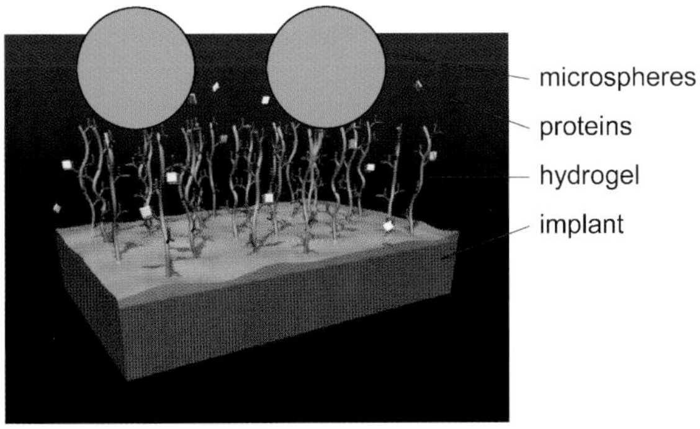

Fig. 58.2 Adhesion of cell complexes (bone cell microspheres) to implants through a coating of implants with a hydrogel-protein layer

Fig. 58.3 Progressive interaction and on-growth of cells to the hydrogel-coated implant surface

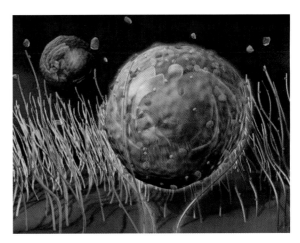

Fig. 58.4 Fabrication process of surface coating

of proteins at the material surface, much work was carried out to selectively bind protein on patterned surfaces. In recent years, considerable attention has been given to the behavior of protein-containing fluids spreading on surfaces at the nanofeature level. It is assumed that the adhesion characteristics of the proteins/fluid environment are also very relevant to the reactions of biological cells to surfaces bearing nanotopography. Some authors have shown nonlinear behavior of interfacial forces on phospholipid strips deposited at a glass-water interface. The different wetting behavior of the hydrophilic channels and the hydrophobic stripes was used to deposit proteins along the channels, for example, by either an anisotropic wetting/dewetting process (determined by the surface structure, or by using capillary and electrostatic forces). The technique of molecular imprinting creates specific protein recognition sites in polymers by using template molecules. Molecular recognition is attributed to binding sites that complement molecules in size, shape, and chemical functionality. Whereas attempts to imprint defined proteins have met with only limited success, a recent study reported on a new method for imprinting surfaces with specific protein-recognition sites. These emerging surface engineering approaches now focus on creating protein-containing biomimetic substrates that target integrins to activate signaling pathways directing cell differentiation programs.

Considerable research efforts have concentrated also on creating biomimetic substrates that target integrins to activate signaling pathways directing the osteoblast differentiation program. The identification of recognition sequences, such as RGD for FN, that mediate integrin-mediated adhesion has stimulated the development of adhesive surfaces. As previously described, RGD-peptides have shown their efficacy in promoting osteoblast adhesion. This biomolecular strategy involves incorporation of short adhesive peptides onto material surfaces in order to produce biofunctional surfaces [33, 34]. Several groups have demonstrated that incorporation of RGD peptides into synthetic polymers, alginate hydrogels, silk films, and silicon-based substrates promotes integrin-mediated cell adhesion and migration [35–47]. Whereas promotion of osteoblast adhesion by RGD-peptides was successful, the cell adhesion was not associated with a subsequent enhancement of cellular functions [135]. Model FN-mimetic surfaces were engineered through immobilizing FNIII7–10, a recombinant fragment of FN encompassing the PHSRN and RGD binding domains, onto BSA supports. Osteoblasts adhered to these functionalized surfaces over a range of ligand densities via integrin. Furthermore, adherent cells spread and assembled focal adhesions containing vinculin and talin.

A challenging approach in biomolecular engineering is the coating of surfaces with growth and/or differentiation factors. Among the growth factors, some are at the present time being tested for their ability to promote bone regeneration. Notably, members of the Transforming Growth Factor- family are being widely studied: TGF-$_1$, BMP-2, BMP-7 [or osteogenic protein-1 (OP-1)]. BMPs were identified following purification of bovine bone proteins after it was discovered that demineralized bone segments or extracts of demineralized bone-induced bone formation in ectopic sites [128, 129]. BMPs associated with various carriers have experimentally shown their efficacy [130–133] and are currently under clinical investigation [134].

References

1. Adell R, Eriksson B, Lekholm M, Branemark PI, Jemt T. A long-term follow-up study of osseointegrated implants in the treatment of totally edentulous jaws. Int J Oral Maxillofac Impl 1990:5:347–359
2. Adell R, Lekholm U, Grfndahl K, Branemark PI, Lindström J, Jacobsson M. Reconstruction of severely resorbed edentulous maxillae using osseointegrated fixtures in immediate autogenous bone grafts. Int J Oral Maxillofac Impl 1990:5:233–246
3. Aerrsens J, Dequeker J, Mbuyi-Muamba JM. Bone tissue composition: biochemical anatomy of bone. Clin Rheumatol 1994:13:54–62
4. Ahmad M, McCarthy MB, Gronowic G. An in vitro model for mineralization of human osteoblast-like cells on implant materials. Biomat 1999:20:211–220
5. Anderson HC. Molecular biology of matrix vesicles. Clin Orthop 1995:314:266–280
6. Andre-Frei V, Chevallay B, Orly I, Boudeulle M, Huc A, Herbage D. Acellular mineral deposition in collagen-based biomaterials incubated in cell culture media. Laborat Invest 2000:66:204–211
7. Akagawa Y, Ichikawa Y, Nikai H, Tsuru H. Interface histology of early loaded partially stabilized zirconia endosseous implant in initial bone healing. J Prosth Dent 1993:69:599–604
8. Akin-Nergiz N, Nergiz I, Sschulz A, Arpak N, Niedermeier W. Reactions of peri-implant tissues to continu-

ous loading of osseointegrated implants. Am J Orthodont Dent Orthop 1998:114:292–298
9. Albrektsson T, Branemark PI, Hansson HA, Lindström J. Osseointegrated titanium implants: Requirements for ensuring a long-lasting direct bone anchorage in man. Acta Orthop Scan 1981:52:155–170
10. Boyan BD, Bonewald LF, Paschalis EP, Lohmann CH, Rosser J, Cochran DL, Dean DD, Schwartz Z, Boskey AL. Osteoblast-mediated mineral deposition in culture is dependent on surface microtopography. Calcif Tissue Int 2002:71:519–529
11. Boyan BD, Hummert TW, Dean DD, Schwartz Z. Role of material surfaces in regulating bone and cartilage cell response. Biomat 1996:17:137–146
12. Boyan BD, Weesner TC, Lohmann CH, Andreacchio D, Carnes DL, Dean DD, Cochran DL, Schwartz Z. Porcine fetal enamel matrix derivative enhances bone formation induced by demineralized freeze dried bone allograft in vivo. J Periodontol 2000:71:1278–1286
13. Branemark PI, Adell R, Albrektsson T, Lekholm U, Lundkvist S, Rockler B. Osseointegrated titanium fixtures in the treatment of edentulousness. Biomat 1983:4:25–28
14. Braun G, Kohavi D, Amir D, Luna M, Caloss R, Sela J, Dean DD, Boyan BD, Schwartz Z. Markers of primary mineralization are correlated with bone-bonding ability of titanium or stainless steel in vivo. Clin Oral Impl Res 1995:6 :1–7
15. Brunski JB. In vivo bone response to biomechanical loading at the bone/dental- implant interface. Adv Dent Res 1999:13:99 119
16. Büchter A, Kleinheinz J, Wiesmann HP, Kersken J, Nienkemper M, Weyhrother H, Joos U, Meyer U. Biological and biomechanical evaluation of bone remodeling and implant stability after using an osteotome technique. Clin Oral Impl Res 2005:16:1–8
17. Büchter A, Kleinheinz J, Wiesmann HP, Seper L, Joos U, Meyer U. Peri-implant bone formation around cylindrical and conical implant systems. Mund Kiefer Gesichtschir 2004:8:282–288
18. Buser D, Nydegger T, Hirt HP, Cochran DL, Nolte LP. Removal torque values of titanium implants in the maxilla of miniature pigs. Int J Oral Maxillofac Implants 1998 :13:611–619
19. Carden A, Morris MD. Application of vibrational spectroscopy to the study of mineralized tissues. J Biomed Opt 2000:5:259–268
20. Carr AB, Gereard DA, Larsen PE. The response of bone in primates around unloaded dental implants supporting prostheses with different levels of fit. J Prost Dent 1996:76.500–509
21. Carter DR, Giori NJ. Effect of mechanical stress on tissue differentiation in the bony implant bed. In: Davies JE, editor. The bone-biomaterial interface. Toronto: University of Toronto Press 1991;p. 367
22. Cheal EJ, Mansmann KA, Digioia III AM, Hayes WC, Pereen SM. Role of interfragmentary strain in fracture healing: Ovine model of a healing osteotomy. J Orthop Res 1991:9:131–142
23. Cooper LF, Masuda T, Yliheikkila PK, Felton DA Generalizations regarding the process and phenomenon of osseointegration. Part II. In vitro studies. Int J Oral Maxillofac Implants 1998:13:163–174
24. Cooper LF. A role for surface topography in creating and maintaining bone at titanium endosseous implants. J Prosthet Dent 2000:84:22–34
25. Corso M, Sirota C, Fiorelline J, Rasool F, SZMUKLER-Moncler S, Weber HP. Evaluation of the osseointegration of early loaded free-standing dental implants with various coatings in the dog model: Periostest and radiographic results. J Prosth Dent 1999:82:428–435
26. Davies JE, Baldan N. Scanning electron microscopy of the bone-bioactive implant interface. J Biomed Mater Res 1997:36:429–440
27. Davies JE. Understanding peri-implant endosseous healing. J Dent Educ. 2003:67:932–49
28. Deporter DA, Watson PA, Pilliar RM, Melcher AH, Winslow J, Howley TP, Haisel P, Maniatopoulos C, Rogriguez A. A histological assessment of the initial healing response adjacent to porous surfaced Ti alloy dental implants in dogs. J Dent Res 1986:65:1064–1070
29. Ducheyne P. Titanium and calcium phosphate ceramic dental implants, surfaces, coatings and interfaces. J Oral Implantol 1988:14:325–340
30. Duncan RL, Turner CH. Mechanotransduction and the functional response of bone to mechanical strain. Calcif Tissue Int 1995:57:344–358
31. Duyck J, Van Oosterwyck H, Vander Sloten J, De Cooman M, Puers R, Naert I. Preload on oral implants after screw tightening fixed full prostheses: an in vivo study. J Oral Rehabil 2001.28.226–233
32. Fan Y, Duan K, Wang R. A composite coating by electrolysis-induced collagen self assembly and calcium phosphate mineralization. Biomat 2005:26:1623–1632
33. Fratzl P, Schreiber S, Klaushofer K. Bone mineralization as studied by small-angle x-ray scattering. Connect Tissue Res 1996:34:247–254
34. Frost HM. The Utah paradigm of skeletal physiology: An overview of its insights for bone, cartilage and collagenous tissue organs. J Bone Miner Metab 2000:18:305–316
35. Frost H, Joos U, Meyer U, Jensen OT. Distraction osteogenesis based on the Utah paradigm. Alveolar Distraction Osteogenesis. Edited by Jensen OT, Quintessence Publisher, 2002, Illinois, USA
36. Futami T, Fujii N, Ohnishi H, Taguchi N, Kusakari H, Ohshima H, Maeda T. Tissue response to titanium implants in the rat maxilla: ultrastructural and histochemical observations of the bone-titanium interface. J Periodontol 200:71:287–298
37. Garces GL, Garcia-Castellano JM, Nogales J. Longitudinal overgrowth of bone after osteotomy in young rats: Influence of bone stability. Calcif Tissue Int 1997:60:391–399
38. Gorski JP. Is all bone the same? Distinctive distributions and properties of non-collagenous matrix proteins in lamellar vs. woven bone imply the existence of different underlying osteogenic mechanisms. Crit Rev Oral Biol Med 1998:9:201–223
39. Groessner-Schreiber B, Tuan. Enhanced extracellular matrix production and mineralization by osteoblasts cultured on titanium surfaces in vitro. J Cell Sci 1992:101:209–217

40. Hartgerink JD, Beniash E, Stupp SI. Self-assembly and mineralization of peptide-amphiphile nanofibers. Science 2001:294:1684–1688
41. Hayakawa T, Yoshinari M, Kiba H, Yamamoto H, Nemoto K, Jansen JA. Trabecular bone response to surface roughened and calcium phosphate (Ca-P) coated titanium implants. Biomat 2002:23:1025–1031
42. Hashimoto M, Akagawa Y, Hashimoto N, Nikai H, Tsuru H. Single crystal sapphire endosseous implant loaded with functional stress: Clinical and histological evaluation of peri-implant tissues. J Oral Rehab 1988:15:65–76
43. Hemmerle J, Voegel JC. Ultrastructural aspects of the intact titanium implant-bone interface from undecalcified ultrathin sections. Biomat 1996:17:1913–1920
44. Höhling HJ, Barckhaus RH, Krefting ER, Althoff J, Quint P. Collagen mineralization: Aspects of the structural relationships between collagen and the apatitic crystallites. In: Ultrastructure of skeletal tissues. Academic Press, 1990, pp. 41–62
45. Hoshaw SJ, Brunski JB, Cochran GVB. Mechanical loading of Branemark implants affects interfacial bone modeling and remodeling. J Oral Maxillofac Surg 1994:9:345–360
46. Isidor F. Loss of osseointegration caused by occlusal load of oral implants. A clinical and radiographic study in monkeys. Clin Oral Impl Res 1996:7:143–152
47. Isidor F. Histological evaluation of peri-implant bone at implants subjected to occlusal overload or plaque accumulation. Clin Oral Impl Res 1997:8:1–9
48. Jayaraman M, Meyer U, Bühner M, Joos U, Wiesmann HP Influence of titanium surfaces on attachment of osteoblast-like cells in vitro. Biomat 2003:25:625–631
49. Johansson CB, Roser K, Bolind P, Donath K, Albrektsson T. Bone-tissue formation and integration of titanium implants: an evaluation with newly developed enzyme and immunohistochemical techniques. Clin Implant Dent Relat Res 1999:1:33–40
50. Jones D, Leivseth G, Tenbosh J. Mechano-reception in osteoblast-like cells. J Biochem Cell Biol 1995:73.:525–532
51. Joos U, VollmeR D, Kleinheinz J. Effect of implant geometry on strain distribution in peri-implant bone. Mund Kiefer Gesichtschir 2000:4:143–147
52. Knabe C, Driessens FCM, PlanelL JA, Gildenhaar R, Berger G, Reif D, Fitzner R, Radlanski RJ, Gross U. Evaluation of calcium phosphates and experimental calcium phosphate bone cements using osteogenic cultures. J Biomed Mater Res 2000:52:498–508
53. Lavos-Valereto IC, Wolynec S, Deboni MC, Konig B Jr. In vitro and in vivo biocompatibility testing of Ti-6AI-7Nb alloy with and without plasma-sprayed hydroxyapatite coating. J Biomed Mater Res 2001:58:727–733
54. Landis WJ. Mineral characterization in calcifying tissues: Atomic, molecular and macromolecular perspectives. Connect Tissue Res 1996:34:239–246
55. Landis WJ. An overview of vertebrate mineralization with emphasis on collagen-mineral interaction. Gravit Space Biol Bull 1999:12:15–26
56. Le Geros RZ. Properties of osteoconductive biomaterials: calcium phosphates. Clin Orthop Relat Res 2002:395:81–98
57. Legrand AP, Bresson B, Guidoin R, Famery R. Mineralization followup with the use of NMR spectroscopy and others. J Biomed Mater Res 2002:63:390–359
58. Lenhert S, Meier MB, Meyer U, Chi L, Wiesmann HP Osteoblast alignment, elongation and migration on grooved polystyrene patterned by Langmuir-Blodgett lithography. Biomat 2004:
59. Leung Y, Walters MA, Blumenthal NC, Ricci JL, Spivak JM. Determination of the mineral phases and structure of the bone-implant interface using Raman spectroscopy. J Biomed Mater Res 1995:29:591–594
60. Loty C, Sautier JM, Boulekbache H, Kokubo T, Kim HM, Forest N. In vitro bone formation on a bone-like apatite layer prepared by a biomimetic process on a bioactive glass-ceramic. J Biomed Mater Res 2000:15:423–434
61. Lum LB, Beirne OR, Curtis DA. Histological evaluation of Ha-coated vs. uncoated titanium blade implants in delayed and immediately loaded applications. Int J Oral Maxillofac Impl 1991:6:456–462
62 Martin JY, Sschwartz Z, Hummert TW, Schraub DM, Simpson J, Lankford J Jr, Dean DD, Cochran DL, Boyan BD. Effect of titanium surface roughness on proliferation, differentiation, and protein synthesis of human osteoblast–like cells (MG63). Biomed Mater Res 1995:29:389–397
63. Masuda T, Yliheikkila PK, Felton DA, Cooper LF. Generalizations regarding the process and phenomenon of osseointegration. Int J Oral Maxillofac Impl 1998:13:17–29
64. Matzusuka K, Walboomers F, De Ruijter A, Jansen JA. Effect of microgrooved poly-l-lactic (PLA) surfaces on proliferation, cytoskeletal organization, and mineralized matrix formation of rat bone marrow cells. Clin Oral Impl Res 2000:11:325–333
65. Meyer U, Wiesmann HP, Kruse-Lösler B, Handschel J, Stratmann U, Joos U. Strain related bone remodeling in distraction osteogenesis of the mandible. Plast Reconstr Surg 1999:103:800–807
66. Meyer U, Wiesmann HP, Meyer T, Schulze-Osthoff D, Jäsche J, Kruse-Lösler B, Joos U. Microstructural investigations of strain-related collagen mineralization. Br J Oral Maxillofac Surg 2001:39:381–389
67. Meyer U, Vollmer D, Runte C, Bourauel C, Joos U. Bone loading pattern around implants in average and atrophic edentulous maxillae: a finite-element analysis. J Craniomaxillofac Surg 2001:29:100–105
68. Meyer U, Joos U, Jayaraman, Stamm T, Hohoff A, Fillies T, Stratmann U, Wiesmann HP. Ultrastructural characterization of the implant/bone interface of immediately loaded dental implants. Biomat 2003:25:1959–1967
69. Meyer U, Wiesmann HP, Fillies T, Joos U. Early tissue reaction at the interface of immediately loaded dental implants. Int J Oral Maxillofac Impl 2003:18:489–499
70. Meyer U, Büchter U, Wiesmann HP, Joos U, Jones DB. Basic reactions of osteoblasts on structured material surfaces. Europ Cells Mater 2005:9:39–49

71. Meyer U, Joos U, Wiesmann HP, Biological and biophysical principles in extracorporal bone tissue engineering. Part III. Int J Oral Maxillofac Surg 2004:33:635–641
72. Meyer U, Büchter A, Bühner M, Wiesmann HP. A new fast element mapping method for titanium wear around oral implants. Clin Oral Impl Res (2005: in press)
73. Misch CE, Bidez MW, Sharawy M. A bioengineered implant for a predetermined bone cellular response to loading forces. A literature review and case report. J Periodontol 2001:72:1276–1286
74. Mosley JR, March BM, Lynch J, Lanyon LE. Strain magnitude related changes in whole bone architecture in growing rats. Bone 1997:20:191–198
75. Muller R, Ruegsegger P. Micro-tomographic imaging for the nondestructive evaluation of trabecular bone architecture. Stud Health Technol Inform 1997:40:61–79
76. Murai K, Takeshita F, Ayukawa Y, Kiyoshima T, Suetsugu T, Tanaka T. Light and electron microscopic studies of bone-titanium interface in the tibiae of young and mature rats. J Biomed Mater Res 1996:30:523–533
77. Oh TJ, Yoon J, Misch CE, Wang HL. The causes of early implant bone loss: myth or science? J Periodontol 2002:73:322–333
78. Piatelli A, Ruggierie A, Franchi M, Romasco N, Trisi P. A histologic and histomorphometric study of bone reactions to unloaded and loaded non-submerged single implants in monkeys: A pilot study. J Oral Implant 1993 :19:314–320
79. Piatelli A, Paoloantonio M, Corigliano M, Scarano A. Immediate loading of titanium plasma-sprayed screw-shaped implants in man: A clinical and histological report of 2.cases. J Periodontol 1997:68:591–597
80. Pillar RM. Quantitative evaluation of the effect of movement at a porous coated implant-bone interface. In: Davies JE, editor. The bone-biomaterial interface. Toronto: University of Toronto Press 1991:pp. 380
81 Pilliar RM, Deporter DA, Watson PA, Valquette N. Dental implant design: effect on bone remodeling. J Biomed Mat Res 1991:25:467–483
82. Plate U, Arnold S, Stratmann U, Wiesmann HP, Höhling HJ. General principle of ordered apatitic crystal formation in enamel and collagen rich hard tissues. Connect Tissue Res 1998:38:149–157
83. Porter AE, Hobbs LW, Rosen VB, Spector M. The ultrastructure of the plasma-sprayed hydroxyapatite-bone interface predisposing to bone bonding. Biomat 2002:23:725–733
84. Quirynen M, Naert I, Van Steenberghe D. Fixture design and overload influence marginal bone loss and fixture success in the Branemark system. Clin Oral Impl Res 1992:3:104–111
85. Rezania A, Healy KE. The effect of peptide surface density on mineralization of a matrix deposited by osteogenic cells. J Biomed Mater Res 2000:52:595–600
86. Roach HI. Why does bone matrix contain non-collagenous proteins? The possible roles of osteocalcin, osteonectin, osteopontin and bone sialoprotein in bone mineralization and resorption. Cell Biol Int 1994:18:617–628
87. Röser K, Johansson CB, Donath K, Albrektsson T. A new approach to demonstrate cellular activity in bone formation adjacent to implants. J Biomed Mater Res 2000:51:280–291
88. Rosengren A, Johansson BR, Danielsen N, Thomsen P, Ericson LE. Immunohistochemical studies on the distribution of albumin, fibrinogen, fibronectin, IgG and collagen around PTFE and titanium implants. Biomat 1996:17:1779–1786
89. Sagara M, Akagawa Y, Nikai H, Tsuru H. The effects of early occlusal loading on one-stage titanium implants in beagle dogs: A pilot study. J Prosth Dent 1993:69:281–288
90. Sasaguri K, Jiang H, Chen J. The affect of altered functional forces on the expression of bone-matrix proteins in developing mouse mandibular condyle. Arch Oral Biol 1998:43:83–92
91. Schlegel KA, Kloss FR, Kessler P, Schultze-Mosgau S, Nkenke E, Wiltfang J. Bone conditioning to enhance implant osseointegration: an experimental study in pigs. Int J Oral Maxillofac Impl 2003:18:505–511
92. Sela J, Gross UM, Kohavi D, Shani J, Dean DD, Boyan BD, Schwartz Z. Primary mineralization at the surfaces of implants. Crit Rev Oral Biol Med 2000:11:423–436
93. Simmons CA, Valiquette N, Pilliar RM. Osseointegration of sintered porous-surfaced and plasma-spray coated implants: an animal model study of early post-implantation healing response and mechanical stability. J Biomed Mater Res 1999:47:127–138
94. Simmons CA, Meguid SA, Pilliar RM. Mechanical regulation of localized and appositional bone formation around bone-interfacing implants. J Biomed Mater Res 2001:55:63–71
95. Simmons CA, Meguid SA, Pilliar RM. Differences in osseointegration rate due to implant surface geometry can be explained by local tissue strains. J Orthop Res 2001:19:187–194
96. Sowden D, Schmitz JP. AO self-drilling and self-tapping screws in rat calvarial bone: an ultrastructural study of the implant interface. J Oral Maxillofac Surg 2002:60:294–299
97. Steflik DE, Corpe RS, Lake FT, Young TR, Sisk AL, Parr GR, Hanes PJ, Berkery DJ. Ultrastructural analyses of the attachment (bonding) zone between bone and implanted biomaterials. J Biomed Mater Res 1998 :39:611–620
98. Szmukler-Moncler S, Salama S, Reingewirtz Y, Dubruille JH. Timing of loading and effect of micro-motion on bone-implant interface: A review of experimental literature. J Biomed Mat Res 1998:43:193–203
99 Thoma RJ, Phillips RE. The role of material surface chemistry in implant device calcification: a hypothesis. J Heart Valve Dis 1995:4:214–221
100. Van Oosterwyck H, Duyck J, Vander Sloten J, Van Der Perre G, De Cooman M, Lievens S, Puers R, Naert I. The influence of bone mechanical properties and implant fixation upon bone loading around oral implants. Clin Oral Impl Res 1998:9:407–418
101. Walters MA, Blumenthal NC, Leung Y, Wang Y, Ricci JL, Spivak JM. Molecular structure at the bone-implant

interface: a vibrational spectroscopic characterization. Calcif Tissue Int 1991:48:368–369
102. Wen HB, Cui FZ, Feng QL, Li HD, Zhu XD. Microstructural investigation of the early external callus after diaphyseal fractures of human long bone. J Struct Biol 1995:114:115–122
103. Wiesmann HP, Joos U, Klatt K, Meyer U. Mineralized 3-D bone tissue engineered by osteoblasts cultured in a collagen gel. J Oral Maxillofac Surg 2003:61:1455–1462
104. Wiesmann HP, Joos U, Meyer U. Biological and biophysical principles in extracorporal bone tissue engineering. Part II. Int J Oral Maxillofac Surg 2004:33:523–530
105. Wiesmann HP, Meyer U, Plate U, Höhling HJ. Aspects of collagen mineralization in hard tissue formation. Int Rev Cytol 2005:242:121–156

Tissue Engineering Application in General Surgery

Y. Nahmias, M. L. Yarmush

Contents

59.1	Introduction	855
59.2	Historical Overview	855
59.3	Cardiovascular System	856
59.4	Musculoskeletal System	859
59.5	Artificial Skin	860
59.6	Digestive System	861
59.7	Current Challenges and Opportunities	863
	References	864

59.1 Introduction

The field of tissue engineering encompasses efforts to construct functional biological elements using cells, biomaterials, or their combination [1]. While most definitions of tissue engineering cover a broad range of applications, in practice the term is closely associated with efforts to repair or replace tissue function that was lost due to trauma, disease, or genetic disorder. The term has also been applied to efforts to construct extracorporeal support systems such as kidney dialysis or bioartificial liver [2, 3]. The term regenerative medicine is sometimes used synonymously with tissue engineering, although the former term emphasizes the use of stem and progenitor cells to produce tissues. Finally, cellular therapies are closely related approaches which rely on the ability of individual cells to engraft and adapt to the host microenvironment without the support of biomaterials or a well-defined architecture.

This chapter aims to review current tissue engineering applications which found their way into clinical practice. We will begin with a short historical review of the field and continue by describing tissue engineering applications in clinical practice. This work will examine applications in the cardiovascular system, musculoskeletal system, skin, and finally the digestive system. We will conclude this work by summarizing the current challenges and expanding opportunities of the field of tissue engineering.

59.2 Historical Overview

Although some would suggest that the field of tissue engineering was born at some point at middle of the 20th century [1], it is well worth noting that the roots of tissue engineering and some of its basic principles lie in ancient history. For example, artificial limbs were crafted in the 15th and 16th centuries out of iron, to replace leg function following amputation [4]. Over the following centuries, wood replaced metal, due to its lighter weight, allowing greater maneuverability. During the 16th century, Ambrose Paré is considered the first surgeon to describe the use of artificial eyes to fit an eye socket [5]. These ocular prosthetics were first made from gold with colored enamel but later produced from cryolite glass [5]. Perhaps most intriguing is the ancient Egyptian description of wound care. The ancient Papyrus of Ebers (1500 B.C.) details the use of lint, grease, and honey for the treatment for skin wounds [6, 7]. It is thought that the lint provided a fibrous base promoting wound regeneration, the grease provided a semiepidermal barrier to environmental pathogens, while the honey served as an antibiotic agent [6, 7]. The striking resemblance between these components and contemporary dermal substitutes such as Integra, described later in the chapter, is nothing less than astounding.

Like many other fields, the actual origin of the modern tissue engineering field is not clear. Some point to a 1938 publication by Nobel laureate Alexis Carrel and Charles Lindbergh titled *The Culture of Organs* [8, 9], while others point to an NSF panel in 1987 which defined tissue engineering as an emerging field [10]. Regardless of its origin, the field has captured the imagination of scientists in academic and industrial institutes throughout the world and has brought about a change of focus from the purely mechanical, device-based approach to one that emphasizes the biological, cell-based part of the equation. This transition has brought about a complimentary change in the review process of tissue-engineered products by the Food and Drug Administration. This federal agency began reviewing tissue-engineered products under the Center for Biological Evaluation and Research (CBER) rather than the Center for Devices and Radiological Health (CDRH).

The promise of a cell-based approach to tissue engineering lays in the opportunity to create and transplant tissues indistinguishable from the host, tissues that grow with the patient and restore the myriad functions of the original organ [1, 11]. However, the significant time and cost associated with acquiring and propagating donor cells in *real time*, the increased chance of infection due to the inability to sterilize living cells, and the complexity of handling living tissues has limited clinical application of novel tissue-engineered products [12]. It is a testament to the challenges still facing tissue engineers that the majority of tissue-engineered products, currently in use, don't contain cells.

59.3 Cardiovascular System

Cardiovascular disease is the leading cause of death and disability in the USA and Western Europe [13]. The underlying causes of cardiovascular disease, atherosclerosis and high blood pressure, have become widespread over the last decade. A recent study suggests that almost 1 billion people worldwide have high blood pressure [14]. Diseases of the heart cause 30% of all related deaths, necessitating the development of complex cardiac surgery procedures [15]. It is thought that relatively simple cardiac surgeries were preformed at various times throughout the 19th century [16]. However, the repair of complex intracardiac pathologies requires that the heart be stopped and drained of blood for a significant period of time. Therefore, there was a need to develop a device that will provide the function of both the heart and the lungs during such an operation.

The first cardiopulmonary bypass device was developed by Dr. John Heysham Gibbon in 1953, and popularized by Dr. John W. Kirklin. This heart-lung machine was composed of a pump and a bubble oxygenator, simulating a function of the heart and lungs [17, 18]. Although the device caused severe hemolysis, it allowed for a whole new range of operations to be carried out, including the repair of congenital heart defects, valve replacement, as well as coronary bypass surgery [17, 18]. The original device has been much improved over the years, as a new generation of centrifugal pumps replaced peristaltic pump reducing hemolysis, caused by mechanical forces. In addition, membrane oxygenators replaced bubble oxygenators, significantly reducing cell death, thought to occur due to high shear rates at the air-liquid interface [19]. In the recent decade, heparin-coated membrane oxygenators have been introduced by Medtronic (Minneapolis, MN) including Carmeda Bioactive Surface, and Trillium Biopassive Surface while Baxter (Deerfield, IL) introduced Duraflo II mimicking the endothelial lining of the blood vessels [20, 21]. The Carmeda Bioactive Surface for example is generated by cross linking polyethyleneimine (PEI) and dextran sulfate with glutaraldehyde to the surface, and treating it with heparin and HNO_2 [22, 23]. The process aims to generate end-point immobilization of heparin, similar to the orientation of heparin sulfate molecules on vascular endothelium [22, 23]. Heparin-coated oxygenators are thought to reduce the need of systemic heparinization, reduce thrombin and complement activation, reduce blood cell activation, and maintain platelet count [20–22].

Recent years have seen the development of a synthetic implantable artificial heart currently used to support patients awaiting heart transplantation. The AbioCor Implantable Replacement Heart (AbioMed, Danvers, MA) was the first successful implant of a self-contained device implanted in the University of Louisville on 2001, the first patient survived for 17 months (Fig. 59.1) [24]. Another device developed by CardioWest (currently SynCardia Systems, Tucson, AZ) called temporary Total Artificial Heart (TAH-t) was based on the older Jarvik design [25]. The device was reportedly successfully implanted 79% of the time, with patient's 1-year survival rate being 86%. Both devices have been approved for use by the FDA [26]. The main challenge of both devices is to function flawlessly for over a year while pumping over 5 liters per minute, leading to significant wear and tear [27]. Anticoagulation therapy is required for regular use [24]. Finally, the enormous energy requirement of the pump limits the life of the implanted internal battery to less than 1.5 hours, necessitating the implantation of a transcutaneous energy transfer coil, as well as an internal controller unit in the abdominal cavity [25, 27]. Patients therefore have to rely on cumbersome batteries and chargers weighing several kilograms.

One of the essential components of an artificial heart is the heart valve. Artificial heart valves have been developed separately since the early 1950s to replace function lost due to congenital defects or diseases, such as rheumatic fever [28]. There are two main types of tissue-engineered valves in clinical use: (1) a mechanical valve constructed from synthetic material, or (2) a bioprosthetic valve (tissue valve) made from a combination of sterilized tissue and synthetic materials [29]. Mechanical valves, made from pyrolytic carbon and titanium, can last for the lifespan of the patient but require long-term anticlotting medication. Designs for mechanical valves include a caged-ball made by Edwards Lifesciences, a tilting-disk design popularized by Medtronic, and a bileaflet valve designed by St. Jude Medical [29]. The bileaflet valve consists of two semicircular leaflets that rotate about struts attached to the valve housing [30]. It offers the greatest effective opening area and is the least thrombogenic of the artificial valves and therefore requires the least amount of anticoagulant drugs [30, 31]. However, the design is vulnerable to backflow and therefore cannot be considered ideal [29, 31].

On the other hand, bioprosthetic valves (tissue valve) utilize sterilized biological tissue to make leaflets, which are sewn into a metal frame [32]. The tissue can be harvested from human cadavers, porcine aortic valve, or from bovine or equine pericardium. The tissue is sterilized during processing, removing many of the biological markers which stimulate the host's immune response. The leaflets are flexible and are well tolerated, removing the need for long-term anticlotting medication [32, 33]. However, it is thought that the valves have a limited lifespan (about 15 years), necessitating a second surgery [32–34].

Fig. 59.1 AbioCor I is the first fully artificial heart developed by AbioMed (Danvers, MA). This 2-pound titanium and polyurethane artificial heart consists of 2 artificial ventricles, 4 mechanical trileaflet valves, and a motor-driven hydraulic pump. The hydraulic pump oscillates between driving blood to the lung, to driving it to the rest of the body, therefore providing complete left- and right-sided pulsatile circulatory support. (Image published with permission from AbioMed.)

Both the mechanical and bioprosthetic valves are not composed of living cells, and therefore are inadequate for use in children with congenital heart defects as the valves fail to grow with the child [35]. A myriad of approaches are currently investigated in an attempt to generate a functional tissue-engineered heart valve. Scaffolds for these tissue-engineered heart valves include synthetic, biodegradable, or decellularized matrix. Cells are either harvested from the host [36], grown from stem or progenitor cells, or allowed to migrate from the surrounding tissue [35]. In one clinical procedure a decellularized pulmonary allograft was seeded with autologous vascular endothelial cells, conditioned in a bioreactor, and used to reconstruct the right ventricular outflow [37]. Mechanical strength was maintained based on a 1-year follow up, with no evidence of calcification or thrombogenesis [37].

A similar effort has been devoted to the development of a tissue-engineered vascular graft for use in coronary or peripheral vascular bypass procedures [38, 39]. The current standard of care in peripheral vascular surgery is the autologous grafting of the long saphenous vein. However, in patients that require a second or third surgery, vessels of suitable quality are unavailable [38, 39]. The first successful use of tubes made of synthetic fabric as arterial prostheses was demonstrated in the early 1950s. The technique quickly advanced to vessels of woven polytetrafluoroethylene (PTFE) or Dacron which are still in use today [38, 40]. However, prosthetic grafts are liable to fail due to graft occlusion caused by surface thrombogenicity and lack of elasticity [41]. Seeding the graft lumen with endothelial cells has been shown to improve the healing process and reduce graft failure [41, 42]. However, developing a completely biological vascular graft has remained a significant challenge. Completely biological grafts developed to date differ from natural vessels by their lack of essential extracellular matrix molecules such as elastin, and insufficient circumferential alignment of cells and matrix leading to low burst strengths and permanent creep [43].

One approach, reported by Shin'oka et al. currently undergoing clinical trials is the combination of a biodegradable polymer scaffolds and autologous bone marrow stem cells [44, 45]. The technique was reported to form endothelialized vessels within 2 weeks of implantation in low pressure vessels such as the pulmonary artery [44, 45]. A similar technique was recently reported to generate vascular grafts which have been reported to grow with the host following implantation in lambs [46]. In that case, polyglycolic acid (PGA) mesh was seeded with vascular endothelial cells and myofibroblasts and conditioned for 14 days in pulse duplicator bioreactor to induce cellular differentiation [46]. The vessels were reported to grow over the 2-year length of the study by 30% in diameter and 45% in length [46]. It is hoped that such tissue-engineered vessels could be used to repair congenital defects in children.

Blood is an important component of the cardiovascular system, delivering oxygen, sustaining pH balance, and maintaining vascular volume [47, 48]. Blood transfusion from allogeneic sources is a common practice during or following surgery. The use of a tissue-engineered blood substitute during surgery could potentially eliminate disease transmission and minimize immunomodulatory reactions to donor's blood [47, 48]. Two approaches have been developed based on either hemoglobin or fluorocarbon blood substitutes. Hemoglobin-based oxygen carriers can potentially reproduce the oxygen carrying capacity of blood [9, 19, 20]. However, outside the erythrocyte hemoglobin is unstable, auto-oxidizing in hours and releasing free radicals and reactive oxygen species in the process [9, 21]. In addition, human hemoglobin undergoes a marked increase in oxygen binding affinity outside the erythrocyte, necessitating the binding of a 2,3-DPG analog to reduce its affinity to that of normal blood. HemoLink (Hemosol, Canada), PolyHeme (Northfield Laboratories, Evanston, IL), and Hemospan (Sangart, San Diego, CA) are three examples of human hemoglobin-based blood substitute currently undergoing clinical trials [47, 49]. A new arrival to the scene is HemoZyme (SynZyme Technologies, Irvine, CA) a hemoglobin-based blood substitute which includes an antioxidant enzyme mimic in the form of a caged nitric oxide compound [50]. Hemopure (Biopure, Cambridge, MA) is a hemoglobin-based blood substitute from bovine origin. In contras with human hemoglobin, the bovine protein's oxygen affinity depends on chloride ions and therefore does not require the DGP analog [51]. Hemopure received regulatory approval in South Africa in 2001 for clinical use in acute anemia due to the prevalence of HIV in the population [47].

Fluorocarbon oils are highly stable molecules, which can carry over twenty-fold more oxygen than water [52, 53]. Fluorocarbons are not subject to oxidation or free radical reactions, and cannot be metabolized [47, 53]. Although fluorocarbons are not miscible with water, they can be emulsified by a number of surfactants such as egg yolk lecithin [54, 55]. The stability of fluorocarbons and their ability to carry oxygen and carbon dioxide led to their development as blood substitutes in clinical settings [47, 56]. The first commercial fluorocarbon emulsion to be used as a blood substitute in a clinical setting was Fluosol, Alpha Therapeutic (Los Angeles, CA), which received regulatory approval in 1989 in the USA and Europe for oxygenating the heart during percutaneous translumenal coronary angioplasty [56]. Other first-generation fluorocarbon emulsions have been tested in humans and are still in use in Russia and China [47, 53, 55]. One fluorocarbon emulsion named Oxygent, (Alliance Pharmaceutical, San Diego, CA) is currently undergoing clinical trials (Table 59.1).

59.4 Musculoskeletal System

Arthritis and other rheumatic conditions are a growing medical problem affecting over 46 million people across the USA [57]. The Center for Disease Control and Prevention reports that the total annual cost of care for arthritis patients is in excess of $81 billion [57]. There are two major types of arthritis. Rheumatoid arthritis is a chronic, inflammatory, autoimmune disorder which affects the joints. Osteoarthritis is a degenerative joint disease with a potential hereditary cause. In both cases, chronic pain and disability necessitate joint replacement surgery [58].

Total hip replacement surgery was pioneered by Sir John Charnley in 1962 [59]. The procedure replaces joint surfaces with a stainless steel head working against an ultra high-density polyethylene acetabular cup cemented to the underlying bone with methyl methacrylate [59]. Stainless steel was soon replaced by cobalt-chromium-molybdenum alloy (Vitallium) to mitigate the excessive friction of the polyethylene and loosening [60, 61]. Fixation screws subsequently replaced the cement [60, 61]. These advances in materials increased success rates in joint replacement surgeries to >95% at 10 years post surgery. Over the years, alumina-zirconia ceramics were shown to cause less damage over time to the polyethylene parts [61, 62]. The last decade has seen the development of metal-on-metal technologies (without polymer) demonstrating lower wear rate than standard techniques [61].

The advent of stem and progenitor cell culture techniques has brought renewed interest to the field as many laboratories attempt to engineer bone-like structures in vitro [63]. In a recent clinical study, bone marrow samples from 27 orthopedic patients were expanded and cultured in a porous ceramic which was subsequently implanted in vivo [64]. Bone was shown to grow independently of patient's median age (56.1 years) [64]. Novel scaffold and bioreactor designs offer new opportunities to differentiate these bone marrow cells under conditions which mimic bone regeneration in vivo [63]. Marolt et al. recently demonstrated the differentiation of human bone marrow stromal cells seeded on a silk scaffold and cultured in a rotating bioreactor [65]. This design is reported to generate bone constructs resembling

Table 59.1 Composition and characteristics of oxygen. Oxygen is a second-generation blood substitute developed by Alliance Pharmaceutical (San Diego, CA). Perflubron ($C_8F_{17}Br$) is the main oxygen carrier, a highly lipophilic fluorocarbon which is excreted from the body exceptionally quickly. Perfluorodecyl bromide ($C_{10}F_{21}Br$) stabilizes the emulsion against Oswald ripening due to its low water solubility. The final emulsion has a viscosity which is 30% greater than that of normal blood

Characteristic	Value
Perflubron (%, w/v)	58
Perfluorodecyl bromide (%)	2
Egg yolk phospholipids (%)	3.6
Mean droplet diameter (nm)	170
pH	7.0
Viscosity (cP)	4.0
Osmolarity (mOsm/kg)	300
Oxygen solubility (vol %)	19.3
Carbon dioxide solubility (vol %)	157
Sterilization	Heat sterilized
Stability	1 year at 8°C

trabecular bone with a mineralized volume fraction of 12% [65].

Concomitantly with the development of tissue-engineered bone, far-reaching advances have been made in the engineering of cartilage. Hyaline cartilage is an avascular tissue lining the joints composed primarily of collagen type II, hyaluronan, and bound aggrecan [66]. One tissue engineering approach shown to have significant promise is Hyalograft C (Fidia Advanced Biopolymers, Italy), a tissue-engineered product composed of autologous chondrocytes grown on a scaffold entirely made of HYAFF 11, an esterified derivative of hyaluronic acid [66, 67]. Hyalograft C received regulatory approval in Europe in 2000 and has been used extensively to restore joint function [66]. Results from extensive clinical studies demonstrate the deposition of collagen type II and glycosaminoglycans as well as the formation of hyaline-like cartilage [67, 68]. Furthermore, close to 96% of the patients had recovered use of their knee at the end of a 3-year clinical trial [68]. A similar technique termed Matrix-induced Autologous Chondrocyte Implantation (MACI) has been initiated by Verigen, a subsidiary of Genzyme GmbH, (Cambridge, MA) utilizing cells seeded on a collagen membrane [69]. It is important to note that a recent partnership between Fidia Advanced Biopolymers and TiGenix, which recently developed specific markers for hyaline-specific chondrocytes, offers a unique promise of an integrated therapy [70].

The challenge of engineering muscle is significantly greater than the development of cartilage as functional muscle tissue is both highly vascularized and innervated. Although we are unaware of clinical studies at this time, two approaches show promise. Levenberg et al. recently demonstrated the engineering and implantation of a vascularized skeletal muscle by seeding myoblasts, embryonic fibroblasts, and endothelial cells on highly porous, biodegradable polymer scaffolds [71]. A similar technique was recently reported to generate vascularized cardiac muscle from embryonic stem cells (Fig. 59.2) [72]. An alternative approach, propagated by Dr. Bar-Cohen is the use of advanced electroactive polymers as artificial muscles [73].

59.5 Artificial Skin

A staggering 1 million burn injuries occur in the USA every year. Severe extensive burns covering over 60% of the total body surface area account for 4% of hospital admissions. Since the introduction of skin grafts by Reverdin in 1871, the transplantation of the patient's own skin from a healthy donor site, either as a full thickness or a split thickness skin graft, has become the surgical gold standard to cover skin wounds [74]. However, in massive burns the mere extent of wound surfaces necessitated the invention of various temporary or permanent skin substitutes, simply because there are not enough autologous skin resources that would allow for recovery [74]. To achieve long-term recovery, a skin substitute needs to replace both the epidermal barrier function, and the dermal matrix allowing for the ingrowth of granulation tissue and vascularization [74, 75].

One popular dermal substitute that has seen widespread use in burn injuries is Integra (Integra Life Sciences, Plainsboro, NJ). Integra is a dermal template made of bovine collagen and chondroitin-6-sulphate (GAG), covered by a silicone membrane functioning as an epidermal cover [75]. When placed on full thickness wound, the take of the collagen-GAG is re-

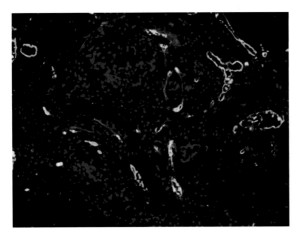

Fig. 59.2 Stem-cell-derived vascularized muscle. C2C12 mouse myoblast cells implanted with human umbilical vein endothelial cells and mouse embryonic fibroblast. The vascular structures which form in the muscle are stained for PECAM-1 (*red*) and Von Willebrand factor (*green*). [Image courtesy of Dr. Shulamit Levenberg (Technion, Israel).]

ported to be greater than 80% [76]. Fourteen days after grafting, the dermal template is characterized by fibrovascular ingrowth resembling the dermis [76]. Once the collagen layer has been integrated into the wound, the silicone is replaced with an ultra-thin, split-thickness skin graft. This treatment is the current state of the art for massive burns [77]. Integra's completely acellular approach stands in contrast to alternative tissue-engineered skin products using autologous or allogeneic cells. Examples include: TransCyte, formerly Dermagraft-TC, (Advanced Tissue Sciences, Coronado, CA) a cryopreserved human fibroblast-derived dermal substitute on which neonatal fibroblasts are seeded. A thin silicon membrane provides barrier function [78]. Epicel (Genzyme, Cambridge, MA) is an autologous skin graft [79], while AlloDerm (LifeCell, Branchburg, NJ) is a fibroblast-seeded allogeneic skin replacement composed of decellularized human dermis [80].

Regardless of their approach, all grafts fail to regenerate hair follicles, sweat and sebaceous glands in the newly restored skin. One approach to regenerate these important structures is to isolate autologous skin stem cells [81] and engraft them into a dermal substitute prior to epidermal engraftment for the treatment of full thickness wounds. Research in this field has been greatly stimulated by the discovery and extensive characterization of integrin $\alpha 6^+/CD34^+$ skin stem cells, which reside in the underlying basal lamina of the skin in a region of the hair follicle known as the bulge. When implanted in nude mice the cells were shown to regenerated hair follicles [81, 82] offering the promise of complete regeneration.

59.6
Digestive System

End-stage kidney disease occurs when kidney function falls below 10% of normal, usually as a result of advanced diabetes [83]. It is estimated that over 400,000 people are currently on long-term kidney dialysis in the USA [84].

The first kidney dialysis extracorporeal device was developed by Dr. Willem Kolff in 1943 [85]. The device was quickly improved by the application of a negative pressure against the blood allowing for the removal of fluids in 1946 [86]. The last part of the design has been the development of Teflon shunts connected to silicone tubing allowing for easy access to patient's blood without significant damage to the vasculature [87]. The development of new dialyzer membranes enables high-flux dialysis, with a cutoff pore between 10,000 to 60,000 daltons, allowing for the removal of medium-sized proteins such as beta-2-microglobulin (MW 11,600 daltons) without depleting the albumin content of blood (MW ~66,400 daltons) [88]. Manufacturers of hemodialysis machines include Fresenius, Gambro, Baxter, and Bellco.

In contrast to the relatively simple function of the kidney, the liver is ascribed with over 500 interrelated functions including albumin secretion, urea synthesis, bile excretion, glycogen storage, and lipid metabolism among others [89]. Loss of liver function causes 25,000 deaths per year and is the 10th most frequent cause of death in the USA [90]. A simulation for the US population for the years 2010–2019 predicts nearly 200,000 incidents of liver failure associated with the prevalence of Hepatitis C Virus infection, and direct medical expenditures in excess of $10 billion. It is for these reasons that a significant effort has been devoted to the development of an extracorporeal liver assist device over the past 50 years [3, 91].

The first device to be approved by the FDA for liver support was the Liver Dialysis Unit (HemoTherapies, San Diego, CA) which was based on a kidney dialysis machine in which toxins were deactivated by passing through activated charcoal [92]. HemoTherapies went bankrupt in 2001. An alternative approach, developed by Teraklin (Gambro, Lakewood, CO), was the Molecular Adsorbent Recycling System (MARS) in which albumin in the counter flowing dialysis solution attracts bound toxins from the blood, facilitating their removal [93]. Clinical studies are currently ongoing but suggest an improvement of patient survival [93, 94].

However, the real promise in the field is the development of a Bioartificial Liver (BAL), a reactor containing functional hepatocytes which can potentially restore the myriad functions of the liver [3, 91]. HepatAssist (Arbios, Pasadena, CA) was the first BAL tested in FDA-approved Phase II/III clinical trials which included 171 patients [95] (Fig. 59.3). The device is composed of a hollow fiber bioreactor in which primary porcine hepatocytes are attached to the outer layer of the fiber and carbon-filtered plasma flows inside [96]. The study showed only a marginal increase in survival but little evidence of protein synthesis [95], leading to the bankruptcy of Circe Biomedical and its purchase by Arbios. Another BAL system utilizing porcine hepatocytes in a hollow fiber bioreactor is the BAL Support System, or BLSS (Excorp Medical, Minneapolis, MN) which is currently undergoing Phase I/II clinical trials [97]. An alternative approach to porcine hepatocytes is the extracorporeal liver assist device termed ELAD (Vital Therapies, San Diego, CA), which uses an immortalized human liver C3A cell line [98]. In two separate Phase I/II clinical trials ELAD was shown to be safe but failed to demonstrate improved survival [98, 99], leading to the bankruptcy of VitaGen and its purchase by Vital Therapies.

A promising new BAL device is the modular extracorporeal liver support device (MELS), developed by Gerlach et al. [100]. The core unit is the CellModule, a bioreactor containing primary human hepatocytes in which the cells form tissue-like aggregates while exposed to plasma. Secondary modules include a DetoxModule with albumin dialysis for the removal of albumin-bound toxins, and a Dialysis Module for continuous hemofiltration for parallel kidney support [100, 101]. The device is based on hollow fiber membranes and is used for cocultures of parenchymal and nonparenchymal cells, giving rise to three-dimensional, liver-like tissue which maintains its function for over 3 weeks [101]. The device is showing promise in Phase I clinical studies using either primary porcine or primary human hepatocytes from discarded tissue [102, 103].

According to the World Health Organization, over 170 million people worldwide suffer from diabetes. Diabetes incidence is increasing rapidly, possibly linked to obesity, and it is estimated that the number will double in the next 2 decades [104]. Lifelong care is provided in the form of daily insulin injections [105]. Alternatively, pancreas or islet transplantation offer relief at the cost of systemic immune suppression which is associated with significant side effects [106]. One potential therapy that has been under intense investigation for the past few decades is the transplantation of encapsulated islets which offer cellular function while being protected from the host immune response [107].

Two approaches show promise for future transplantation [108]. The islet sheet (Cerco Medical, San Francisco, CA) is several centimeters in diameter, with a 250-μm thick shell, composed of highly purified alginate, which contains $2-3 \times 10^6$ cells (Fig. 59.4) [108, 109]. Four to six sheets implanted

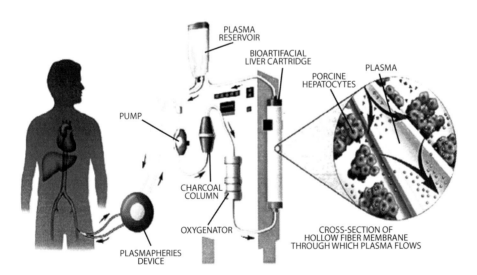

Fig. 59.3 HepatAssist is bioartificial liver developed by Arbios (Los Angeles, CA). This extracorporeal device uses carbon filtration to deactivate toxins from plasma as well as a hollow fiber bioreactor loaded with primary porcine hepatocytes. Hepatocytes are attached to the outer layer of the fiber and carbon-filtered plasma flows inside. (Image published with permission from Arbios.)

in the peritoneal cavity of the patient are sufficient to restore pancreatic function [109]. The main advantage of this design is the small number of objects which need to be implanted and the ability to retrieve the artificial pancreas in case of emergency or in a follow-up surgery [109]. The second approach is microencapsulation of individual islets in alginate by Living Cell Technologies (Australia) and AmCyte (Santa Monica, CA) [110, 111], alginate with polyamino acid shell (MicroIslet, San Diego, CA), or with polyethylene glycol (Novocell, San Diego, CA) [108, 112]. Sources of cells include allogeneic cells from cadavers, xenogeneic porcine islets, insulin-secreting cell lines (BetaGene, Dallas, TX), or human embryonic stem cells [111–113]. Living Cell Technologies (formerly Diatranz) has extensively tested their DiabeCell product in primates and is currently proceeding with Phase I/IIA clinical trials in Russia and New Zealand using neonatal porcine cells [111, 114]. A similar study is currently being initiated by AmCyte in Canada, while Novocell is currently completing a Phase I/II study in which allogeneic islets are transplanted subcutaneously [113].

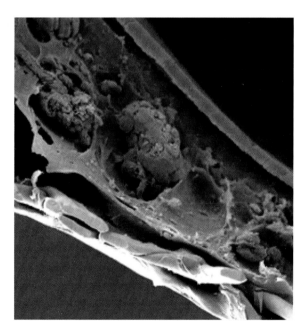

Fig. 59.4 A pseudo-colored electron micrograph of an islet sheet developed by Cerco Medical (San Francisco, CA). The sheet is 250 μm thick, composed of highly purified alginate (*red*), and contains cellular aggregates (*brown*). Supporting mesh is shown in *blue*. [Image courtesy of Dr. Scott King (Cerco Medical).]

59.7 Current Challenges and Opportunities

The field of tissue and bioengineering is rapidly expanding as critical new technologies are developed and emerge into the market [11]. It is important to note that to date, the most clinically useful products have been the most simple. Dacron grafts, collagen sponge, hyaluronan matrix, kidney dialysis, and a rotating carbon disc are some of the examples mentioned in this chapter. In this context, the main challenge of new tissue-engineered products will be to demonstrate that they are significantly better, while being just as safe, as the aforementioned established artificial tissues.

The main advantage of cell-based engineered tissues is their ability to grow and develop with the patient. This ability is critically important for the repair of congenital defects in children, who have yet to grow into adulthood. In addition, cell-based tissues offer the opportunity to completely restore tissue function. Cardiac muscle will not require external batteries, while encapsulated islet will not require a new dose of insulin. However, the disadvantages of using living cells are many and in some cases outweigh their usefulness. Xenogeneic and cells can be rejected and require immune suppressive therapy, while autologous cells need to time grow and proliferate in culture. In both cases the product is simply not available off-the-shelf, limiting its usefulness. In addition, cell-based tissues cannot be properly sterilized, increasing chances of infection.

However, developing research can potentially alleviate some of these concerns. Embryonic Stem (ES) cells, derived from the inner cell mass of the embryo, can be induced to differentiate into all cell types in the body [115, 116]. ES cells offer the hope of creating a library of histocompatible cell lines which will not elucidate an immune response but could differentiate into the full gamete of cells for potential therapeutics [117]. Currently, ES cells offer a limitless allogeneic cell source for tissue engineering and regenerative medicine. However, undifferentiated ES cells are known to form tumor-like anomalies called teratomas, and therefore still pose a risk for clinical applications [116]. Finally, cryopreservation, vitrification, and future drying of cells and tissues offer to solve the problem of off-the-shelf availability of tissue-engineered products [118, 119].

In summary, tissue engineering and regenerative medicine offer significant promise for the extension of human life expectancy. Tissue-engineered products are becoming increasingly available for clinical applications, slowly changing the practice of medicine. It is hoped that cell-based approaches will take an increasing central stage in the field, supplementing artificial devices in clinical applications.

References

1. Langer, R. & Vacanti, J.P. Tissue engineering. Science 260, 920–926 (1993)
2. Yarmush, M.L. et al. Hepatic tissue engineering. Development of critical technologies. Ann N Y Acad Sci 665, 238–252 (1992)
3. Strain, A.J. & Neuberger, J.M. A Bioartificial Liver—State of the Art. Science 295, 1005–1009 (2002)
4. Orr, J.F., James, W.V. & Bahrani, A.S. The history and development of artificial limbs. Eng Med. 11, 155–161 (1982)
5. Danz, W.S. Ancient and contemporary history of artificial eyes. Adv Ophthalmic Plast Reconstr Surg. 8, 1–10 (1990)
6. Majno, G. The Healing Hand: Man and Wound in the Ancient World, 600 (Harvard University Press, Cambridge, Ma, 1991)
7. Lindblad, W.J. The 1996 lifetime achievement award. Wound Repair and Regeneration 6, 1–7 (1998)
8. Bing, R.J. Lindbergh and the biological sciences (a personal reminiscence). Tex Heart Inst J. 14, 230–237 (1987)
9. Carrel, A. & Lindbergh, C.A. The Culture of Organs, 221 (P. B. Hoeber, New York, 1938)
10. Vacanti, J.P. Tissue Engineering: A 20-Year Personal Perspective. Tissue Engineering 13, 231–232 (2007)
11. Griffith, L.G. & Naughton, G. Tissue Engineering—Current Challenges and Expanding Opportunities. Science 295, 1009–1016 (2002)
12. Nerem, R.M. Tissue engineering: the hope, the hype, and the future. Tissue Engineering 12, 1143–1150 (2006)
13. Mokdad, A.H., Marks, J.S., Stroup, D.F. & Gerberding, J.L. Actual causes of death in the United States, 2000. JAMA 291, 1238–1245 (2004)
14. Chockalingam, A. Impact of world hypertension day. Can J Cardiol 23, 517–519 (2007)
15. CDC. Chronic Disease Overview. (National Center for Chronic Disease Prevention and Health Promotion, 2005)
16. Shumacker, H.B. When did cardiac surgery begin? J Cardiovasc Surg (Torino) 30, 246–249 (1989)
17. Hill, J.D. John H. Gibbon, Jr. Part I. The development of the first successful heart-lung machine. Ann Thorac Surg 34, 337–341 (1982)
18. Kirklin, J.W. Open-heart surgery at the Mayo Clinic. The 25th anniversary. Mayo Clin Proc. 55, 339–341 (1980)
19. Chalmers, J.J. & Bavarian, F. Microscopic visualization of insect cell-bubble interactions. II: The bubble film and bubble rupture. Biotechnol Prog. 7, 151–158 (1991)
20. Gravlee, G.P. Heparin-coated cardiopulmonary bypass circuits. J Cardiothorac Vasc Anesth. 8, 213–222 (1994)
21. Zimmermann, A.K., Weber, N., Aebert, H., Ziemer, G. & Wendel, H.P. Effect of biopassive and bioactive surface-coatings on the hemocompatibility of membrane oxygenators. J Biomed Mater Res B Appl Biomater. 80, 433–439 (2007)
22. Baksaas, S.T. et al. In vitro evaluation of new surface coatings for extracorporeal circulation. Perfusion 14, 11–19 (1999)
23. Larm, O., Larsson, R. & Olsson, P. A new non-thrombogenic surface prepared by selective covalent binding of heparin via a modified reducing terminal residue. Biomater Med Devices Artif Organs 11, 161–173 (1983)
24. Dowling, R.D. et al. Initial experience with the AbioCor implantable replacement heart at the University of Louisville. ASAIO J. 46, 579–581 (2000)
25. Copeland, J.G. et al. Total artificial hearts: bridge to transplantation. Cardiol Clin. 21, 101–113 (2003)
26. Nosé, Y. FDA approval of totally implantable permanent total artificial heart for humanitarian use. Artif Organs 31, 1–3 (2007)
27. Gray, N.A. & Selzman, C.H. Current status of the total artificial heart. Am Heart J. 152, 4–10 (2006)
28. Frater, R.W. Artificial heart valves. Lancet 2, 1171 (1962)
29. Yoganathan, A.P., He, Z. & Jones, S.C. Fluid mechanics of heart valves. Annu Rev Biomed Eng. 6, 331–362 (2004)
30. Emery, R.W. & Nicoloff, D.M. St. Jude Medical cardiac valve prosthesis: in vitro studies. J Thorac Cardiovasc Surg. 78, 269–276 (1979)
31. Masters, R.G., Helou, J., Pipe, A.L. & Keon, W.J. Comparative clinical outcomes with St. Jude Medical, Medtronic Hall and CarboMedics mechanical heart valves. J Heart Valve Dis. 10, 403–409 (2001)
32. Lund, O. & Bland, M. Risk-corrected impact of mechanical versus bioprosthetic valves on long-term mortality after aortic valve replacement. J Thorac Cardiovasc Surg. 132, 20–26 (2006)
33. Colli, A., Verhoye, J.P., Leguerrier, A. & Gherli, T. Anticoagulation or antiplatelet therapy of bioprosthetic heart valves recipients: an unresolved issue. Eur J Cardiothorac Surg. 31, 573–577 (2007)
34. Doenst, T., Borger, M.A. & David, T.E. Long-term results of bioprosthetic mitral valve replacement: the pericardial perspective. J Cardiovasc Surg (Torino) 45, 449–454 (2004)
35. Mendelson, K. & Schoen, F.J. Heart valve tissue engineering: concepts, approaches, progress, and challenges. Ann Biomed Eng. 34, 1799–1819 (2006)
36. Schmidt, D. et al. Living autologous heart valves engineered from human prenatally harvested progenitors. Circulation 114, I125–I131 (2006)
37. Dohmen, P.M., Lembcke, A., Hotz, H., Kivelitz, D. & Konertz, W.F. Ross operation with a tissue-engineered heart valve. Ann Thorac Surg. 74, 1438–1442 (2002)

38. Daly, C.D., Campbell, G.R., Walker, P.J. & Campbell, J.H. Vascular engineering for bypass surgery. Expert Rev Cardiovasc Ther. 3, 659–665 (2005)
39. Vara, D.S. et al. Cardiovascular tissue engineering: state of the art. Pathol Biol (Paris) 53, 599–612 (2005)
40. Campbell, C.D., Goldfarb, D. & Roe, R. A small arterial substitute: expanded microporous polytetrafluoroethylene: patency versus porosity. Ann Surg. 182, 138–143 (1975)
41. Schmidt, S.P. et al. Small-diameter vascular prostheses: two designs of PTFE and endothelial cell-seeded and nonseeded Dacron. J Vasc Surg. 2, 292–297 (1985)
42. Herring, M., Gardner, A. & Glover, J. Seeding endothelium onto canine arterial prostheses. The effects of graft design. Arch Surg. 114, 679–682 (1979)
43. Isenberg, B.C., Williams, C. & Tranquillo, R.T. Small-diameter artificial arteries engineered in vitro. Circ Res. 98, 25–35 (2006)
44. Shin'oka, T., Imai, Y. & Ikada, Y. Transplantation of a tissue-engineered pulmonary artery. N Engl J Med 344, 532–533 (2001)
45. Matsumura, G., Hibino, N., Ikada, Y., Kurosawa, H. & Shin'oka, T. Successful application of tissue engineered vascular autografts: clinical experience. Biomaterials 24, 2303–2308 (2003)
46. Hoerstrup, S.P. et al. Functional growth in tissue-engineered living, vascular grafts: follow-up at 100 weeks in a large animal model. Circulation 114, I159–I166 (2006)
47. Lowe, K.C. Engineering Blood: Synthetic Substitutes from Fluorinated Compounds. Tissue Engineering 9, 389–399 (2003)
48. Riess, J.G. Oxygen carriers ("blood substitutes")—raison d'etre, chemistry, and some physiology. Chem Rev. 101, 2797–2920 (2001)
49. Niiler, E. Setbacks for blood substitute companies. Nature Biotechnology 20, 962–963 (2002)
50. Buehler, P.W., Haney, C.R., Gulati, A., Ma, L. & Hsia, C.J. Polynitroxyl hemoglobin: a pharmacokinetic study of covalently bound nitroxides to hemoglobin platforms. Free Radic Biol Med 37, 124–135 (2004)
51. Sprung, J. et al. The use of bovine hemoglobin glutamer-250 (Hemopure) in surgical patients: results of a multicenter, randomized, single-blinded trial. Anesth Analg. 94, 799–808 (2002)
52. King, A.T., Mulligan, B.J. & Lowe, K.C. Perfluorochemicals and Cell Culture. Nature Biotechnology 7, 1037–1042 (1989)
53. Riess, J.G. & Krafft, M.P. Fluorinated materials for in vivo oxygen transport (blood substitutes), diagnosis and drug delivery. Biomaterials 19, 1529–1539 (1998)
54. Magdassi, S., Royz, M. & Shoshan, S. Interactions between collagen and perfluorocarbon emulsions. International Journal of Pharmaceutics 88, 171–176 (1992)
55. Riess, J.G. Understanding the fundamentals of perfluorocarbons and perfluorocarbon emulsions relevant to in vivo oxygen delivery. Artif Cells Blood Substit Immobil Biotechnol 33, 47–63 (2005)
56. Cowley, M.J. et al. Perfluorochemical perfusion during coronary angioplasty in unstable and high-risk patients. Circulation 81, IV27-IV34 (1990)
57. Yelin, E. et al. Medical care expenditures and earnings losses among persons with arthritis and other rheumatic conditions in 2003, and comparisons with 1997. Arthritis & Rheumatism 56, 1397–1407 (2007)
58. Tutuncu, Z. & Kavanaugh, A. Rheumatic disease in the elderly: rheumatoid arthritis. Rheum Dis Clin North Am. 33, 57–70 (2007)
59. Charnley, J. The long-term results of low-friction arthroplasty of the hip performed as a primary intervention. 1970. Clin Orthop Relat Res. 430, 3–11 (2005)
60. McKee, G.K. & Watson-Farrar, J. Replacement of arthritic hips by the McKee-Farrar prosthesis. J Bone J Surg 48-B, 245–259 (1996)
61. Long, M. & Rack, H.J. Titanium alloys in total joint replacement—a materials science perspective. Biomaterials 19, 1621–1639 (1998)
62. Saikko, V. Wear of polyethylene acetabular cups against zirconia femoral heads studied with a hip joint simulator. Wear 176, 207–212 (1994)
63. Kretlow, J.D. & Mikos, A.G. Review: Mineralization of Synthetic Polymer Scaffolds for Bone Tissue Engineering. Tissue Engineering 13, 927–938 (2007)
64. Yoshikawa, T., Ohgushi, H., Ichijima, K. & Takakura, Y. Bone Regeneration by Grafting of Cultured Human Bone. Tissue Eng. 10, 688–698 (2004)
65. Marolt, D. et al. Bone and cartilage tissue constructs grown using human bone marrow stromal cells, silk scaffolds and rotating bioreactors. Biomaterials 27, 6138–6149 (2006)
66. Tognana, E., Borrione, A., Luca, C.D. & Pavesio, A. Hyalograft(R) C: Hyaluronan-Based Scaffolds in Tissue-Engineered Cartilage. Cells Tissues Organs [Epub ahead of print](2007)
67. Hollander, A.P. et al. Maturation of tissue engineered cartilage implanted in injured and osteoarthritic human knees. Tissue Eng. 12, 1787–1798 (2006)
68. Marcacci, M. et al. Articular cartilage engineering with Hyalograft C: 3-year clinical results. Clin Orthop Relat Res 435, 96–105 (2005)
69. Zheng, M.H. et al. Matrix-induced autologous chondrocyte implantation (MACI): biological and histological assessment. Tissue Eng. 13, 737–746 (2007)
70. Callegaro, L., O'Regan, M., Beyen, G. & Motmans, K. TiGenix and Fidia Advanced Biopolymers enter into strategic partnership. (TiGenix, 2007)
71. Levenberg, S. et al. Engineering vascularized skeletal muscle tissue. Nat Biotechnol. 23, 879–884 (2005)
72. Caspi, O. et al. Tissue engineering of vascularized cardiac muscle from human embryonic stem cells. Circ Res. 100, 263–272 (2007)
73. Bar-Cohen, Y. Current and future developments in artificial muscles using electroactive polymers. Expert Rev Med Devices 2, 731–740 (2005)
74. Horch, R.E., Kopp, J., Kneser, U., Beier, J. & Bach, A.D. Tissue engineering of cultured skin substitutes. J. Cell. Mol. Med. 9, 592–608 (2005)
75. Yannas, I.V., Burke, J.F., Orgill, D.P. & Skrabut, E.M. Wound tissue can utilize a polymeric template to synthesize a functional extension of skin. Science 215, 174–176 (1982)

76. Jeng, J.C. et al. Seven years' experience with Integra as a reconstructive tool. J Burn Care Res. 28, 120–126 (2007)
77. Winfrey, M.E., Cochran, M. & Hegarty, M.T. A new technology in burn therapy: INTEGRA artificial skin. Dimens Crit Care Nurs. 18, 14–20 (1999)
78. Noordenbos, J., Doré, C. & Hansbrough, J.F. Safety and efficacy of TransCyte for the treatment of partial-thickness burns. J Burn Care Rehabil. 20, 275–281 (1999)
79. Carsin, H. et al. Cultured epithelial autografts in extensive burn coverage of severely traumatized patients: a five year single-center experience with 30 patients. Burns 26, 379–387 (2000)
80. Lattari, V. et al. The use of a permanent dermal allograft in full-thickness burns of the hand and foot: a report of three cases. J Burn Care Rehabil. 18, 147–155 (1997)
81. Blanpain, C., Lowry, W.E., Geoghegan, A., Polak, L. & Fuchs, E. Self-Renewal, Multipotency, and the Existence of Two Cell Populations within an Epithelial Stem Cell Niche. Cell 118, 635–648 (2004)
82. Rhee, H., Polak, L. & Fuchs, E. Lhx2 Maintains Stem Cell Character in Hair Follicles. Science 312, 1946–1949 (2006)
83. Kramer, H. & Luke, A. Obesity and kidney disease: a big dilemma. Curr Opin Nephrol Hypertens 16, 237–241 (2007)
84. Robin, J. et al. Renal dialysis as a risk factor for appropriate therapies and mortality in implantable cardioverter-defibrillator recipients. Heart Rhythm 3, 1196–1201 (2006)
85. Kolff, W.J. Lasker Clinical Medical Research Award. The artificial kidney and its effect on the development of other artificial organs. Nat Med. 8, 1063–1065 (2002)
86. Kurkus, J., Nykvist, M., Lindergard, B. & Segelmark, M. Thirty-five years of hemodialysis: two case reports as a tribute to Nils Alwall. Am J Kidney Dis. 49, 471–476 (2007)
87. Blagg, C.R. The early history of dialysis for chronic renal failure in the United States: a view from Seattle. Am J Kidney Dis. 49, 482–496 (2007)
88. Cheung, A.K. et al. Effects of high-flux hemodialysis on clinical outcomes: results of the HEMO study. J Am Soc Nephrol. 14, 3251–3263 (2003)
89. Nahmias, Y., Berthiaume, F. & Yarmush, M.L. Integration of technologies for hepatic tissue engineering. in Advances in Biochemical Engineering/Biotechnology, Vol. 103 (eds. Lee, K. & Kaplan, D.) 309–329 (Springer Berlin, Heidelberg, 2006)
90. Popovic, J.R. & Kozak, L.J. National Hospital Discharge Survey: Annual summary, 1998. National Center for Health Statistics. Vital Health Stat 13 (2000)
91. Tsiaoussis, J., Newsome, P.N., Nelson, L.J., Hayes, P.C. & Plevris, J.N. Which Hepatocyte Will it be ? Hepatocyte choice for bioartificial liver support systems. Liver Transplantation 7, 2–10 (2001)
92. Ash, S.R. Extracorporeal blood detoxification by sorbents in treatment of hepatic encephalopathy. Adv Ren Replace Ther. 9, 3–18 (2002)
93. Mitzner, S.R. et al. Improvement of hepatorenal syndrome with extracorporeal albumin dialysis MARS: results of a prospective, randomized, controlled clinical trial. Liver Transpl. 6, 277–286 (2000)
94. Heemann, U. et al. Albumin dialysis in cirrhosis with superimposed acute liver injury: a prospective, controlled study. Hepatology 36, 949–958 (2002)
95. Demetriou, A.A. et al. Prospective, randomized, multicenter, controlled trial of a bioartificial liver in treating acute liver failure. Ann Surg. 239, 660–667 (2004)
96. Custer, L. & Mullon, C.J. Oxygen delivery to and use by primary porcine hepatocytes in the HepatAssist 2000 system for extracorporeal treatment of patients in end-stage liver failure. Adv Exp Med Biol. 454 (1998)
97. Mazariegos, G.V. et al. First clinical use of a novel bioartificial liver support system (BLSS). Am J Transplant. 2, 260–266 (2002)
98. Ellis, A.J. et al. Pilot-controlled trial of the extracorporeal liver assist device in acute liver failure. Hepatology 24, 1446–1451 (1996)
99. Millis, J.M. et al. Safety of continuous human liver support. Transplant Proc 33, 1954 (2001)
100. Gerlach, J.C. Bioreactors for extracorporeal liver support. Cell Transplantation 15, S91–103 (2006)
101. Gerlach, J.C. et al. Use of primary human liver cells originating from discarded grafts in a bioreactor for liver support therapy and the prospects of culturing adult liver stem cells in bioreactors: a morphologic study. Transplantation 76, 781–786 (2003)
102. Sauer, I.M. et al. Clinical extracorporeal hybrid liver support—phase I study with primary porcine liver cells. Xenotransplantation 10, 460–469 (2003)
103. Sauer, I.M. et al. Primary human liver cells as source for modular extracorporeal liver support—a preliminary report. Int J Artif Organs 25, 1001–1005 (2002)
104. Engelgau, M.M. et al. The evolving diabetes burden in the United States. Ann Intern Med. 140, 945–950 (2004)
105. Rother, K.I. Diabetes treatment—bridging the divide. N Engl J Med. 356, 1517–1526 (2007)
106. Shapiro, J. et al. Islet transplantation in seven patients with type 1 diabetes mellitus using a glucocorticoid-free immunosuppressive regimen. N Engl J Med 343, 230–238 (2000)
107. Beck, J. et al. Islet Encapsulation: Strategies to Enhance Islet Cell Functions. Tissue Engineering 13, 589–599 (2007)
108. Orive, G., Hernández, R.M., Gascón, A.R., Igartua, M. & Pedraz, J.L. Encapsulated cell technology: from research to market. Trends Biotechnol. 20, 382–387 (2002)
109. Storrs, R., Dorian, R., King, S.R., Lakey, J. & Rilo, H. Preclinical Development of the Islet Sheet. Annals of the New York Academy of Sciences 944, 252–266 (2001)
110. Tsang, W.G. et al. Generation of functional islet-like clusters after monolayer culture and intracapsular aggregation of adult human pancreatic islet tissue. Transplantation 83, 685–693 (2007)
111. Elliott, R.B. et al. Live encapsulated porcine islets from a type 1 diabetic patient 9.5 yr after xenotransplantation. Xenotransplantation 14, 157–161 (2007)
112. D'Amour, K.A. et al. Production of pancreatic hormone-expressing endocrine cells from human embryonic stem cells. Nat Biotechnol 24, 1392–1401 (2006)

113. Scharp, D.W. Encapsulated Human Islet Allografts: Providing Safety with Efficacy. in Cellular Transplantation: From Laboratory to Clinic (eds. Halberstadt, C. & Emerich, D.F.) 135–154 (Academic Press, Burlington, 2006)
114. Elliott, R. et al. Intraperitoneal alginate-encapsulated neonatal porcine islets in a placebo-controlled study with 16 diabetic cynomolgus primates. Transplantation Proceedings 37, 3505–3508 (2005)
115. Evans, M.J. & Kaufman, M.H. Establishment in culture of pluripotential cells from mouse embryos. Nature 292, 154–156 (1981)
116. Thomson, J.A. et al. Embryonic stem cell lines derived from human blastocysts. Science 282, 1145–1147 (1998)
117. Kim, K. et al. Histocompatible embryonic stem cells by parthenogenesis. Science 315, 482–486 (2007)
118. Fowler, A. & Toner, M. Cryo-injury and biopreservation. Ann N Y Acad Sci. 1066, 119–135 (2005)
119. Fahy, G.M. et al. Cryopreservation of organs by vitrification: perspectives and recent advances. Cryobiology 48, 157–178 (2004)

Regeneration of Renal Tissues

T. Aboushwareb, J. J. Yoo, A. Atala

Contents

60.1	Introduction	869
60.2	Basic Components	869
60.3	Approaches for Renal Tissue Regeneration	870
60.3.1	Developmental Approaches	870
60.3.2	Tissue Engineering Approaches	871
60.4	Regeneration of Functional Tissue In Vivo	872
60.5	Summary	874
	References	874

60.1 Introduction

The kidney is a vital and complex organ that performs many critical functions [1–3]. It is responsible for filtering the body's wastes, such as urea, from the blood and excreting them as urine. In addition to the excretory function, the kidney maintains the body's homeostasis by regulating acid-base balance, blood pressure, and plasma volume. Moreover, it synthesizes 1, 25 vitamin D3, erythropoietin, glutathione, and free radical scavenging enzymes. It is also known that the kidney participates in the catabolism of low molecular weight proteins and in the production and regulation of cytokines [4–6].

There are many conditions where the kidney functions are diminished and these lead to renal failure. End stage renal failure is a devastating condition which involves multiple organs in affected individuals. Although dialysis can prolong survival via filtration, other kidney functions are not replaced, leading to long-term consequences such as anemia and malnutrition [2, 3, 7]. Currently, renal transplantation is the only definitive treatment that can restore full kidney function. However, transplantation has several limitations, such as critical donor shortage, complications due to chronic immunosuppressive therapy and graft failure [2, 3, 7].

The limitations of current therapies for renal failure have led investigators to explore the development of alternative therapeutic modalities that could improve, restore, or replace renal function. The emergence of cell-based therapies using tissue engineering and regenerative medicine strategies has presented alternative possibilities for the management of pathologic renal conditions [1, 7–13]. The concept of kidney cell expansion followed by cell transplantation using tissue engineering and regenerative medicine techniques has been proposed as a method to augment either isolated or total renal function. Despite the fact that the kidney is considered to be one of the more challenging organs to regenerate and/or reconstruct, investigative advances made to date have been promising [8, 10–12].

60.2 Basic Components

The unique structural and cellular heterogeneity present within the kidney creates many challenges for tissue regeneration. The system of nephrons and collecting ducts within the kidney is composed of multiple functionally and morphologically distinct segments. For this reason, appropriate conditions need to be provided for the long-term survival, differentiation, and growth of many types of cells. Recent efforts in the area of kidney tissue regeneration have focused on the development of a reliable cell source [14–19]. In addition, optimal growth conditions have been extensively investigated to provide an adequate enrichment to achieve stable renal cell expansion systems [20–24].

Isolation of particular cell types that produce specific factors, such as erythropoietin, may be a good approach for selective cell therapies. However, total renal function would not be achieved using this approach. To create kidney tissue that would deliver full renal function, a culture containing all of the cell types comprising the functional nephron units should be used. Optimal culture conditions to nurture renal cells have been extensively studied and the cells grown under these conditions have been reported to maintain their cellular characteristics [25]. Furthermore, renal cells placed in a three-dimensional culture environment are able to reconstitute renal structures.

Recent investigative efforts in the search for a reliable cell source have been expanded to stem and progenitor cells. The use of these cells for tissue regeneration is attractive due to their ability to differentiate and mature into specific cell types required for regeneration. This is particularly useful in instances where primary renal cells are unavailable due to extensive tissue damage. Bone marrow-derived human mesenchymal stem cells have been shown to be a potential source because they can differentiate into several cell lineages [14, 15, 18]. These cells have been shown to participate in kidney development when they are placed in a rat embryonic niche that allows for continued exposure to a repertoire of nephrogenic signals [19]. These cells, however, were found to contribute mainly to regeneration of damaged glomerular endothelial cells after injury. In addition, the major cell source of kidney regeneration was found to originate from intrarenal cells in an ischemic renal injury model [14, 17]. Another potential cell source for kidney regeneration is circulating stem cells, which have been shown to transform into tubular and glomerular epithelial cells, podocytes, mesangial cells, and interstitial cells after renal injury [15, 16, 26–30]. These observations suggest that controlling stem and progenitor cell differentiation may lead to successful regeneration of kidney tissues.

Although isolated renal cells are able to retain their phenotypic and functional characteristics in culture, transplantation of these cells in vivo may not result in structural remodeling that is appropriate for kidney tissue. In addition, cell or tissue components cannot be implanted in large volumes as diffusion of oxygen and nutrients is limited [31]. Thus, a cell-support matrix, preferably one that encourages angiogenesis, is necessary to allow diffusion across the entire implant. A variety of synthetic and naturally derived materials have been examined in order to determine the ideal support structures for regeneration [9, 32–35]. Biodegradable synthetic materials, such as poly-lactic and poly-glycolic acid polymers, have been used to provide structural support for cells. Synthetic materials can be easily fabricated and configured in a controlled manner, which make them attractive options for tissue engineering. However, naturally derived materials, such as collagen, laminin, and fibronectin, are much more biocompatible and provide a similar extracellular matrix environment to normal tissue. For this reason, collagen-based scaffolds have been used in many applications [36–39].

60.3
Approaches for Renal Tissue Regeneration

60.3.1
Developmental Approaches

Transplantation of a kidney precursor, such as the metanephros, into a diseased kidney has been proposed as a possible method for achieving functional restoration. In an animal study, human embryonic metanephroi, transplanted into the kidneys of an immune deficient mouse model, developed into mature kidneys [40]. The transplanted metanephroi produced urine-like fluid but failed to develop ureters. This study suggests that development of an in vitro system in which metanephroi could be grown may lead to transplant techniques that could produce a small replacement kidney within the host. In another study, the metanephros was divided into mesenchymal tissue and ureteral buds, and each of the tissue segments was cultured in vitro [41]. After 8 days in culture, each portion of the mesenchymal tissues had grown to the original size. A similar method was used for ureteral buds, which also propagated. These results indicate that if the mesenchyme and ureteral buds were placed together and cultured in vitro, a metanephros-like structure would develop. This suggests that the metanephros could be propagated under optimal conditions.

In another study, transplantation of metanephroi into a nonimmunosuppressed rat omentum showed that the implanted metanephroi are able to undergo differentiation and growth that is not confined by a tight organ capsule [42]. When the metanephroi with an intact ureteric bud were implanted, the metanephroi were able to enlarge and become kidney-shaped tissue within 3 weeks. The metanephroi transplanted into the omentum were able to develop into kidney tissue structures with a well-defined cortex and medulla. Mature nephrons and collecting system structures are shown to be indistinguishable from those of normal kidneys by light or electron microscopy [43]. Moreover, these structures become vascularized via arteries that originate at the superior mesenteric artery of the host [43]. It has been demonstrated that the metanephroi transplanted into the omentum survive for up to 32 weeks post implantation [44]. These studies show that the developmental approach may be a viable option for regenerating renal tissue for functional restoration.

60.3.2
Tissue Engineering Approaches

The ability to grow and expand cells is one of the essential requirements in engineering tissues. The feasibility of achieving renal cell growth, expansion, and in vivo reconstitution using tissue engineering techniques was investigated [9]. Donor rabbit kidneys were removed and perfused with a nonoxide solution which promoted iron particle entrapment in the glomeruli. Homogenization of the renal cortex and fractionation in 83 and 210 micron sieves with subsequent magnetic extraction yielded three separate purified suspensions of distal tubules, glomeruli, and proximal tubules. The cells were plated separately in vitro and after expansion, were seeded onto biodegradable polyglycolic acid scaffolds and implanted subcutaneously into host athymic mice. This included implants of proximal tubular cells, glomeruli, distal tubular cells, and a mixture of all three cell types. Animals were sacrificed at 1 week, 2 weeks, and 1 month after implantation and the retrieved implants were analyzed. An acute inflammatory phase and a chronic foreign body reaction were seen, accompanied by vascular ingrowth by 7 days after implantation. Histologic examination demonstrated progressive formation and organization of the nephron segments within the polymer fibers with time. Renal cell proliferation in the cell-polymer scaffolds was detected by in vivo labeling of replicating cells with the thymidine analog bromodeoxyuridine [17]. BrdU incorporation into renal cell DNA was confirmed using monoclonal anti-BrdU antibodies. These results demonstrated that renal specific cells can be successfully harvested and cultured, and can subsequently attach to artificial biodegradable polymers. The renal cell-polymer scaffolds can be implanted into host animals where the cells replicate and organize into nephron segments, as the polymer, which serves as a cell delivery vehicle, undergoes biodegradation.

Initial experiments showed that implanted cell-polymer scaffolds gave rise to renal tubular structures. However, it was unclear whether the tubular structures reconstituted de novo from dispersed renal elements, or if they merely represented fragments of donor tubules which survived the original dissociation and culture processes intact. Further investigation was conducted in order to examine the process [45]. Mouse renal cells were harvested and expanded in culture. Subsequently, single isolated cells were seeded on biodegradable polymers and implanted into immune competent syngeneic hosts. Renal epithelial cells were observed to reconstitute into tubular structures in vivo. Sequential analyses of the retrieved implants over time demonstrated that renal epithelial cells first organized into a cord-like structure with a solid center. Subsequent canalization into a hollow tube could be seen by 2 weeks. Histologic examination with nephron-segment-specific lactins showed successful reconstitution of proximal tubules, distal tubules, loop of Henle, collecting tubules, and collecting ducts. These results showed that single suspended cells are capable of reconstituting into tubular structures, with homogeneous cell types within each tubule.

60.4 Regeneration of Functional Tissue In Vivo

The kidneys are critical to body homeostasis because of their excretory, regulatory, and endocrine functions. The excretory function is initiated by filtration of blood at the glomerulus, and the regulatory function is provided by the tubular segments. Although our prior studies demonstrated that renal cells seeded on biodegradable polymer scaffolds are able to form some renal structures in vivo, complete renal function could not be achieved in these studies. In a subsequent study we sought to create a functional artificial renal unit which could produce urine [46]. Mouse renal cells were harvested, expanded in culture, and seeded onto a tubular device constructed from polycarbonate [39]. The tubular device was connected at one end to a silastic catheter which terminated into a reservoir. The device was implanted subcutaneously in athymic mice. The implanted devices were retrieved and examined histologically and immunocytochemically at 1, 2, 3, 4 and 8 weeks after implantation. Fluid was collected from inside the implant, and uric acid and creatinine levels were determined.

Histological examination of the implanted device demonstrated extensive vascularization as well as formation of glomeruli and highly organized tubule-like structures. Immunocytochemical staining with antiosteopontin antibody, which is secreted by proximal and distal tubular cells and the cells of the thin ascending loop of Henle, stained the tubular sections. Immunohistochemical staining for alkaline phosphatase stained proximal tubule-like structures. Uniform staining for fibronectin in the extracellular matrix of newly formed tubes was observed. The fluid collected from the reservoir was yellow and contained 66 mg/dl uric acid (as compared to 2 mg/dl in plasma) suggesting that these tubules are capable of unidirectional secretion and concentration of uric acid. The creatinine assay performed on the collected fluid showed an 8.2-fold increase in concentration, as compared to serum. These results demonstrated that single cells from multicellular structures can become organized into functional renal units that are able to excrete high levels of solutes through a urine-like fluid [46].

To determine whether renal tissue could be formed using an alternative cell source, nuclear transplantation (therapeutic cloning) was performed to generate histocompatible tissues, and the feasibility of engineering syngeneic renal tissues in vivo using these cloned cells was investigated [25]. Nuclear material from bovine dermal fibroblasts was transferred into unfertilized enucleated donor bovine eggs. Renal cells from the cloned embryos were harvested, expanded in vitro, and seeded onto three-dimensional renal devices (Fig. 60.1a). The devices were implanted into the back of the same steer from which

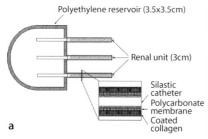

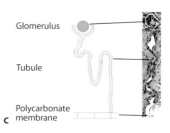

Fig. 60.1a–c Formation of functional renal tissue in vivo. **a** Renal device. **b** Tissue-engineered renal unit shows the accumulation of urine-like fluid. **c** There was a clear unidirectional continuity between the mature glomeruli, their tubules, and the polycarbonate membrane

the cells were cloned, and were retrieved 12 weeks later. This process produced functioning renal units. Urine production and viability were demonstrated after transplantation back into the nuclear donor animal (Fig. 60.1b). Chemical analysis suggested unidirectional secretion and concentration of urea nitrogen and creatinine. Microscopic analysis revealed formation of organized glomeruli and tubular structures (Fig. 60.1c). Immunohistochemical and RT-PCR analysis confirmed the expression of renal mRNA and proteins, whereas delayed-type hypersensitivity testing and in vitro proliferative assays showed that there was no rejection response to the cloned cells. This study indicates that the cloned renal cells are able to form and organize into functional tissue structures which are genetically the same as the host. Generating immune-compatible cells using therapeutic cloning techniques is feasible and may be useful for the engineering of renal tissues for autologous applications.

In our previous study, we showed that renal cells seeded on synthetic renal devices attached to a collecting system are able to form functional renal units that produce urine-like fluid. However, a naturally derived tissue matrix with existing three-dimensional kidney architecture would be preferable, because it would allow for transplantation of a larger number of cells, resulting in greater renal tissue volumes. Thus, we developed an acellular collagen-based kidney matrix, which is identical to the native renal architecture. In a subsequent study we investigated whether these collagen-based matrices could accommodate large volumes of renal cells and form kidney structures in vivo [47].

Acellular collagen matrices, derived from porcine kidneys, were obtained through a multistep decellularization process. During this process, serial evaluation of the matrix for cellular remnants was performed using histochemistry, scanning electron microscopy (SEM) and RT-PCR. Mouse renal cells were harvested, grown, and seeded on 80 of the decellularized collagen matrices at a concentration of 30×10^6 cells/ml. Forty cell-matrix constructs grown in vitro were analyzed 3 days, 1, 2, 4, and 6 weeks after seeding. The remaining 40 cell-containing matrices were implanted in the subcutaneous space of 20 athymic mice. The animals were sacrificed 3 days, 1, 2, 4, 8, and 24 weeks after implantation for analyses.

Gross, SEM, histochemical, immunocytochemical, and biochemical analyses were performed.

Scanning electron microscopy and histologic examination confirmed the acellularity of the processed matrix. RT-PCR performed on the kidney matrices demonstrated the absence of any RNA residues. Renal cells seeded on the matrix adhered to the inner surface and proliferated to confluency 7 days after seeding, as demonstrated by SEM. Histochemical and immunocytochemical analyses performed using H & E, periodic acid Schiff, alkaline phosphatase, antiosteopontin and anti-CD-31 identified stromal, endothelial, and tubular epithelial cell phenotypes within the matrix. Renal tubular and glomerulus-like structures were observed 8 weeks after implantation. MTT proliferation and titrated thymidine incorporation assays performed 6 weeks after cell seeding demonstrated a population increase of 116% and 92%, respectively, as compared to the 2-week time points. This study demonstrates that renal cells are able to adhere to and proliferate on the collagen-based kidney matrices. The renal cells reconstitute renal tubular and glomeruli-like structures in the kidney shaped matrix. The collagen-based kidney matrix system seeded with renal cells may be useful in the future for augmenting renal function.

However, creation of renal structures without the use of an artificial device system would be preferable, as implantation procedures are invasive and may result in unnecessary complications. We also investigated the feasibility of creating three-dimensional renal structures for in situ implantation within the native kidney tissue. Primary renal cells from 4-week-old mice were grown and expanded in culture. These renal cells were labeled with fluorescent markers and injected into mouse kidneys in a collagen gel for in vivo formation of renal tissues. Collagen injection without cells and sham-operated animals served as controls. In vitro reconstituted renal structures and in vivo implanted cells were retrieved and analyzed. The implanted renal cells formed tubular and glomerular structures within the kidney tissue, as confirmed by the fluorescent markers. There was no evidence of renal tissue formation in the control and the sham-operated groups. These results demonstrate that single renal cells are able to reconstitute kidney structures when placed in a collagen-based scaffolding system. The implanted renal cells are able to

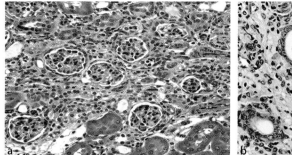
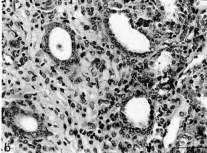

Fig. 60.2a,b Reconstitution of kidney structures. The implanted renal cells self assembled into (**a**) glomerular and (**b**) tubular structures within the kidney tissue

self assemble into tubular and glomerular structures within the kidney tissue (Fig. 60.2). These findings suggest that this system may be the preferred approach to engineer functional kidney tissues for the treatment of end stage renal disease.

60.5
Summary

The increasing demand for renal transplantation, combined with the critical shortage of donor organs, has ignited intense interest in seeking alternative treatment modalities. Cell-based approaches have been proposed as a viable method for kidney tissue regeneration. Various tissue engineering and regenerative medicine approaches that aim to achieve functional intra- and extracorporeal kidney support were presented in this chapter. Although the progress in this area has been somewhat successful toward regeneration of functional kidney tissue, clinical application of this technology is still distant.

The concept of renal cell transplantation has been demonstrated on multiple occasions. It has been shown that renal cells are able to reconstitute into functional kidney tissues in vivo, but numerous challenges must be addressed and solved in order for these techniques to be applied clinically. Some of these challenges include the generation of a large tissue mass that could augment systemic renal function, the formation of adequate vascularization in the engineered renal tissue, and the development of a reliable renal failure model system for testing cell-based technologies.

References

1. Amiel, G. E., Yoo, J. J., Atala, A.: Renal therapy using tissue-engineered constructs and gene delivery. World J Urol, 18: 71, 2000
2. Chazan, J. A., Libbey, N. P., London, M. R. et al.: The clinical spectrum of renal osteodystrophy in 57 chronic hemodialysis patients: a correlation between biochemical parameters and bone pathology findings. Clin Nephrol, 35: 78, 1991
3. Cohen, J., Hopkin, J., Kurtz, J.: Infectious complications after renal transplantation Philadelphia, WB Saunders, 1994
4. Frank, J., Engler-Blum, G., Rodemann, H. P. et al.: Human renal tubular cells as a cytokine source: PDGF-B, GM-CSF and IL-6 mRNA expression in vitro. Exp Nephrol, 1: 26, 1993
5. Maack, T.: Renal handling of low molecular weight proteins. Am J Med, 58: 57, 1975
6. Stadnyk, A. W.: Cytokine production by epithelial cells. Faseb J, 8: 1041, 1994
7. Amiel, G. E., Yoo, J. J. and Atala, A.: Renal tissue engineering using a collagen-based kidney matrix. Tissue Engineering suppl, 2000
8. Amiel, G. E., Atala, A.: Current and future modalities for functional renal replacement. Urol Clin North Am, 26: 235, 1999
9. Atala, A., Schlussel, R. N., Retik, A. B: Renal cell growth in vivo after attachment to biodegradable polymer scaffolds. J Urol, 153, 1995

10. Atala, A.: Tissue engineering in the genitourinary system. Vol chap 8, 1997
11. Atala, A.: Future perspectives in reconstructive surgery using tissue engineering. Urol Clin North Am, 26: 157, 1999
12. Humes, H. D., Buffington, D. A., MacKay, S. M. et al.: Replacement of renal function in uremic animals with a tissue-engineered kidney. Nat Biotechnol, 17: 451, 1999
13. Humes, H. D., Fissell, W. H., Weitzel, W. F. et al.: Metabolic replacement of kidney function in uremic animals with a bioartificial kidney containing human cells. Am J Kidney Dis, 39: 1078, 2002
14. Ikarashi, K., Li, B., Suwa, M. et al.: Bone marrow cells contribute to regeneration of damaged glomerular endothelial cells. Kidney Int, 67: 1925, 2005
15. Kale, S., Karihaloo, A., Clark, P. R. et al.: Bone marrow stem cells contribute to repair of the ischemically injured renal tubule. J Clin Invest, 112: 42, 2003
16. Lin, F., Cordes, K., Li, L. et al.: Hematopoietic stem cells contribute to the regeneration of renal tubules after renal ischemia-reperfusion injury in mice. J Am Soc Nephrol, 14: 1188, 2003
17. Lin, F., Moran, A., Igarashi, P.: Intrarenal cells, not bone marrow-derived cells, are the major source for regeneration in postischemic kidney. J Clin Invest, 115: 1756, 2005
18. Prockop, D. J.: Marrow stromal cells as stem cells for nonhematopoietic tissues. Science, 276: 71, 1997
19. Yokoo, T., Ohashi, T., Shen, J. S. et al.: Human mesenchymal stem cells in rodent whole-embryo culture are reprogrammed to contribute to kidney tissues. Proc Natl Acad Sci U S A, 102: 3296, 2005
20. Carley, W. W., Milici, A. J., Madri, J. A.: Extracellular matrix specificity for the differentiation of capillary endothelial cells. Exp Cell Res, 178: 426, 1988
21. Horikoshi, S., Koide, H., Shirai, T.: Monoclonal antibodies against laminin A chain and B chain in the human and mouse kidneys. Lab Invest, 58: 532, 1988
22. Humes, H. D., Cieslinski, D. A.: Interaction between growth factors and retinoic acid in the induction of kidney tubulogenesis in tissue culture. Exp Cell Res, 201: 8, 1992
23. Milici, A. J., Furie, M. B., Carley, W. W.: The formation of fenestrations and channels by capillary endothelium in vitro. Proc Natl Acad Sci U S A, 82: 6181, 1985
24. Schena, F. P.: Role of growth factors in acute renal failure. Kidney Int Suppl, 66: S11, 1998
25. Lanza, R. P., Chung, H. Y., Yoo, J. J. et al.: Generation of histocompatible tissues using nuclear transplantation. Nat Biotechnol, 20: 689, 2002
26. Gupta, S., Verfaillie, C., Chmielewski, D. et al.: A role for extrarenal cells in the regeneration following acute renal failure. Kidney Int, 62: 1285, 2002
27. Ito, T., Suzuki, A., Imai, E. et al.: Bone marrow is a reservoir of repopulating mesangial cells during glomerular remodeling. J Am Soc Nephrol, 12: 2625, 2001
28. Iwano, M., Plieth, D., Danoff, T. M. et al.: Evidence that fibroblasts derive from epithelium during tissue fibrosis. J Clin Invest, 110: 341, 2002
29. Poulsom, R., Forbes, S. J., Hodivala-Dilke, K. et al.: Bone marrow contributes to renal parenchymal turnover and regeneration. J Pathol, 195: 229, 2001
30. Rookmaaker, M. B., Smits, A. M., Tolboom, H. et al.: Bone-marrow-derived cells contribute to glomerular endothelial repair in experimental glomerulonephritis. Am J Pathol, 163: 553, 2003
31. Folkman, J., Hochberg, M.: Self-regulation of growth in three dimensions. J Exp Med, 138: 745, 1973
32. Atala, A., Bauer, S. B., Soker, S. et al.: Tissue-engineered autologous bladders for patients needing cystoplasty. Lancet, 367: 1241, 2006
33. El-Kassaby, A. W., Retik, A. B., Yoo, J. J. et al.: Urethral stricture repair with an off-the-shelf collagen matrix. J Urol, 169: 170, 2003
34. Oberpenning, F., Meng, J., Yoo, J. J. et al.: De novo reconstitution of a functional mammalian urinary bladder by tissue engineering. Nat Biotechnol, 17: 149, 1999
35. Tachibana, M., Nagamatsu, G. R., Addonizio, J. C.: Ureteral replacement using collagen sponge tube grafts. J Urol, 133: 866, 1985
36. Freed, L. E., Vunjak-Novakovic, G., Biron, R. J. et al.: Biodegradable polymer scaffolds for tissue engineering. Biotechnology (N Y), 12: 689, 1994
37. Hubbell, J. A., Massia, S. P., Desai, N. P. et al.: Endothelial cell-selective materials for tissue engineering in the vascular graft via a new receptor. Biotechnology (N Y), 9: 568, 1991
38. Mooney, D. J., Mazzoni, C. L., Breuer, C. et al.: Stabilized polyglycolic acid fibre-based tubes for tissue engineering. Biomaterials, 17: 115, 1996
39. Wald, H. L., Sarakinos, G., Lyman, M. D. et al.: Cell seeding in porous transplantation devices. Biomaterials, 14: 270, 1993
40. Dekel, B., Burakova, T., Arditti, F. D. et al.: Human and porcine early kidney precursors as a new source for transplantation. Nat Med, 9: 53, 2003
41. Steer, D. L., Bush, K. T., Meyer, T. N. et al.: A strategy for in vitro propagation of rat nephrons. Kidney Int, 62: 1958, 2002
42. Rogers, S. A., Lowell, J. A., Hammerman, N. A. et al.: Transplantation of developing metanephroi into adult rats. Kidney Int, 54: 27, 1998
43. Hammerman, M. R.: Transplantation of embryonic kidneys. Clin Sci (Lond), 103: 599, 2002
44. Rogers, S. A., Powell-Braxton, L., Hammerman, M. R.: Insulin-like growth factor I regulates renal development in rodents. Dev Genet, 24: 293, 1999
45. Fung, L. C. T., Elenius, K., Freeman, M., Donovan, M. J., Atala, A.: Reconstitution of poor EGFr-poor renal epithelial cells into tubular structures on biodegradable polymer scaffold. Pediatrics, 98 (Suppl): S631, 1996
46. Yoo, J. J., Ashkar, S., Atala, A.: Creation of functional kidney structures with excretion of kidney-like fluid in vivo. Pediatrics, 98S: 605, 1996
47. Yoo, J. J., Atala, A.: A novel gene delivery system using urothelial tissue engineered neo-organs. J Urol, 158: 1066, 1997

Tissue Engineering Applications in Plastic Surgery

M. D. Kwan, B. J. Slater, E. I. Chang, M. T. Longaker, G. C. Gurtner

Contents

61.1	Introduction	877
61.2	Skeletal Tissue Engineering	878
61.3	Cartilage Tissue Engineering	879
61.4	Skin Tissue Engineering	881
61.5	Adipose Tissue Engineering	881
61.6	Vascular Tissue Engineering	882
61.7	Conclusion	884
	References	884

61.1 Introduction

Plastic surgery, a derivative of the Greek word plastikos, is in essence a specialty of "forming or molding" to repair defects that may arise from traumatic mechanisms, extirpation of malignancies, or congenital malformations [20]. As one of the broadest surgical specialties, plastic surgeons are confronted with defects in all regions of the body including the integument, craniomaxillofacial region, extremities, breast, and trunk. When addressing large defects which are unlikely to heal on their own, surgeons have a wide array of tools for reconstruction, including the use of autogenous grafts, allogeneic substitutes, and synthetic materials. Each of these reconstructive options has a corresponding clinical situation that it addresses best. However, these resources are also accompanied by their inherent disadvantages. When wounds are assessed to be unlikely to heal in a timely fashion because of size or inadequate blood supply, plastic surgeons have over the years devised a plethora of local flaps, and with the advent of microsurgical technology, a host of microvascular free tissue transfers. Autogenous grafts are considered ideal because of their lack of immunogenicity. However, the use of autogenous grafts is tempered by consideration for donor site morbidity and relatively limited quantities. Allogeneic substitutes offer an alternative, but the risks of disease transmission, immunologic rejection, and graft-versus-host disease are not negligible. Finally, synthetic materials are frequently used but are associated with concerns for infection, tissue incorporation, and structural integrity.

The recent intersection of advances in stem cell biology, molecular biology, materials sciences, and bioengineering has brought regenerative medicine into the forefront. Regenerative medicine seeks to recapitulate developmental processes by providing progenitor cells, seeded on a compatible scaffold, with the appropriate molecular and environmental cues. Given the focus of plastic surgery and regenerative medicine on addressing absent or injured tissues, this shared objective offers plastic surgery the unique opportunity for a paradigm shift away from repair and reconstruction to that of tissue regeneration.

The specifics for tissue engineering efforts differ depending on the application, but the fundamental elements remain constant. Tissue engineering involves the delivery of progenitor cells while providing the proper environmental signals to regenerate injured or missing tissue. In broad strokes, progenitor cells can be obtained from embryonic and postnatal sources. Embryonic stem cells are considered the gold standard because of their totipotentiality and hence, the ability to differentiate into tissues from all three germ layers. Due to the political and ethical debate surrounding embryonic stem cells, significant attention has been directed towards identifying postnatal progenitor cells. Tissue-specific progenitor cells, such as osteoblasts for bone or chondrocytes for cartilage, seem to be the most obvious choice because of their genetic commitment along their respective lineage. However, the use of these lineage-specific progenitor cells is in many cases severely curbed by limited autogenous quantities, the need to harvest from the very tissue that requires reconstruction, and often times their restricted ability to proliferate. As a result, efforts have been directed towards identifying populations of postnatal, multipotent cells, including those present in bone marrow, umbilical cord blood, and adipose tissue, which may be more abundant sources of progenitor cells. Bone marrow mesenchymal stem cells (MSCs) have been the most studied [48]. Recent identification of adipose-derived stem cells (ASCs) offer yet another promising source of postnatal progenitor cells [71, 72]. This chapter will focus on recent efforts to combine cells, cytokines, and scaffolds to engineer bone, cartilage, skin, adipose, and blood vessels as they pertain to plastic surgery.

61.2
Skeletal Tissue Engineering

Plastic surgeons are often confronted with skeletal defects, resulting from trauma, nonunion of fractures, tumor resection, or congenital anomalies. Considering craniofacial skeletal injuries alone, the Healthcare Cost and Utilization Project reported over 50,000 skull and facial fractures in 2004, resulting in an aggregate cost exceeding 500 million dollars [23]. Fractures of the axial skeleton accounted for an additional one million hospitalizations in 2004 [23].

The current options for surgical repair of skeletal defects consist of autogenous bone grafts, allogeneic materials, and prosthetic substitutes. Autogenous bone grafts are typically harvested from the iliac crest, ribs, and calvaria. Bone autografts have traditionally been the gold standard because of their nonimmunogenicity and the robust bone formation that results from their implantation. However, use of autogenous grafts is limited by donor-site morbidity, lack of sufficient bone for harvesting, resorption of the transplanted graft, and growth deformity [51].

Allogeneic bone grafts, derived from cadaveric donors, provide another option for reconstruction of large osseous defects. Although often used, allogeneic materials are accompanied by the risk for infection, disease transmission, and immunologic rejection [55, 68]. Alternatively, a large number of prosthetic materials, such as metals alloys, polymers, and ceramics, have been developed by material scientists to serve as bone substitutes [16, 45, 51]. Concern for structural integrity, lack of incorporation into surrounding bone, contouring abnormalities, and risk of infection are among some of the disadvantages of these synthetic materials. Given the many different techniques for repair of bone defects, as well as their attendant disadvantages, it is clear that an enormous clinical need exists for improved bone regeneration techniques.

Tissue engineering has emerged as a promising field for the creation of novel and targeted treatments of skeletal defects while avoiding the problems discussed above. The ideal strategy for bone formation would be to combine a biomimetic scaffold with osteoprogenitor cells and molecular cues to allow for finely controlled bone formation. Many types of scaffolds have been developed to serve as a three-dimensional matrices for tissue reconstruction by facilitating cellular delivery and creating an environmental niche [15, 54, 62]. The main categories of scaffolds include natural polymers, inorganic materials, synthetic polymers, and composites [66]. Ideally, these scaffolds should have the capacity for osteoinduction or recruitment of osteoprogenitor cells, biocompatibility, and controlled biodegradation [15, 55, 68]. Material scientists are finding that combinations of these different materials allow for the individual components to provide what the other is lacking. For example, blends of polylactic acid with hydroxyapatite provide the constructs with sufficient structural integrity from the polymer and osteoinductive qualities from the inorganic phosphate.

Substantial work has been directed towards identifying postnatal sources of osteoprogenitor cells for bone tissue engineering applications. Notably, bone marrow mesenchymal stem cells (MSC) have been the most studied for regenerating bone [47, 55]. Bone marrow MSCs are easily isolated based on their adherence to culture dishes and have been shown to differentiate along multiple lineages, including the osteogenic pathway [22, 47]. Furthermore, the implantation of bone marrow MSCs into osseous defects has been shown to accelerate healing in numerous animal models [44, 57].

The recent identification of ASCs as a source of osteoprogenitor cells has provided another avenue for bone tissue engineering. ASCs are an ideal source because of the simplicity of adipose tissue harvest and the relatively large number of progenitor cells obtained from harvest [12]. Our laboratory has demonstrated the ability of ASCs seeded onto apatite-coated, polylactic-co-glycolic acid scaffolds to heal critical-sized (4 mm) calvarial defects in mice [12]. Using histology and micro-CT, Cowan et al. showed bone formation in defects implanted with ASCs after 2 weeks and substantial bony bridging after 12 weeks. Of importance, implanted donor ASCs contributed to 98% of the new bone formation in the healed defects of adult mice, as evidenced by chromosomal detection.

In order to create the perfect bone substitute, a number of growth factors have been combined with scaffolds in order to produce increased host regeneration. These growth factors include bone morphogenetic proteins (BMP), platelet-derived growth factors (PDGF), transforming growth factors (TGF), and fibroblast growth factors (FGF) [51, 66]. Several BMP products, containing BMP-2 or -7, are now commercially available and have been demonstrated in clinical trials to contribute to successful healing in delayed and nonunion fractures [66].

Our laboratory has examined in depth the molecular biology of BMP and FGF signaling involved in osteodifferentiation of ASCs. Wan et al. demonstrated the requirement of BMP receptor IB (BMPR-IB) for osteogenic differentiation [69]. During osteogenic differentiation of mouse-derived ASCs in vitro, a seven-fold increase in BMPR-IB expression was observed. Conversely, Wan et al. demonstrated that silencing of BMPR-IB expression via RNA interference resulted in significant inhibition of the bone forming capacity of ASCs. In terms of FGF signaling, Quarto et al. demonstrated the ability of FGF-2 to serve as a potent mitogen for mouse-derived ASCs, while preserving their osteogenic potential [49]. After culturing ASCs for ten passages with or without FGF-2, ASCs cultured in the presence of FGF-2 maintained their ability to proliferate and differentiate into osteoblasts. High-passage ASCs cultured in the absence of FGF-2 proliferated more slowly and displayed weak potential for bone formation. As the biology of osteodifferentiation of progenitor cells becomes elucidated, it will be possible to incorporate key inductors of bone formation in cell-based, bone tissue engineering constructs.

61.3
Cartilage Tissue Engineering

In 2004, there were 30,000 hospitalizations related to head and neck cancers and approximately 100,000 hospitalizations related to head and neck trauma [23]. In many of these patients, plastic surgeons are faced with cartilage defects. Cartilage defects afflicting the craniofacial region most commonly involve the nasal septum, auricle, and temporomandibular joint. The auricle and temporomandibular joint, especially, are very complex, three-dimensional structures that are challenging to reconstruct. Cartilage possesses minimal potential for regeneration on its own. This has been attributed to the relatively low number of chondrocytes in the extracellular matrix and the dependence of cartilage on diffusion for transport of nutrients and waste due to its avascular nature [21]. As a result, plastic surgeons frequently use autogenous grafts from the ribs or a host of local flaps to address these defects. Use of autogenous grafts is accompanied by concerns for donor site morbidity and inconsistent outcomes.

In reviewing current efforts at recapitulating cartilage formation it is important to keep in mind the three types of cartilage: hyaline, elastic, and fibrocartilage. Common ingredients span across these three types of cartilage, including the presence of chondrocytes, glycosaminoglycans, fibronectin, and water.

However, variations in the content of the extracellular matrix distinguishes these cartilage types from each other and, hence, imparts their unique biomechanical characteristics [50]. Hyaline cartilage, consisting of large amounts of collagen II, is found mostly on articular surfaces. It is able to withstand repetitive mechanical load-bearing while providing a low-friction interface. Elastic cartilage, containing high concentrations of elastic fibers, possesses more flexible material properties and maintains the shape of the ear, nasal septum, pharynx, trachea, and ribs. Finally, fibrocartilage is highly specialized with thick fibers of collagen I. Its structure provides tensile strength in regions such as the temporomandibular joint, intervertebral discs, and tendons. Ultimately, tissue engineering efforts hope to tailor regeneration of cartilage to the mechanical demands of the defect environment.

Much of the current literature on cartilage tissue engineering involves the use of cultured chondrocytes. The use of autogenous cultured chondrocytes is already in clinical use under the patented name of Carticel (Genzyme Corporation; Cambridge, MA). Chondral plugs are harvested in an initial surgery, from which chondrocytes are isolated, expanded in vitro, and subsequently reimplanted into the articular defect. Nevertheless, the use of chondrocytes is arguably suboptimal because of their low replicative ability and the limited donor sites available for their harvest. Cartilage tissue engineering efforts are encouraged by the potential of both bone marrow MSCs and ASCs in high-density alginate, agar, or micromass pellet cultures to produce extracellular matrix molecules consistent with cartilage formation [58, 71].

Studies in cartilage tissue engineering have underscored the importance of spatial orientation in prompting chondrogenic differentiation. When chondrocytes are cultured in monolayer, they take on fibroblast-like qualities, with a spindle-like morphology, cessation of type II collagen synthesis, and increased production of fibroblast-like collagen, types I and III [9, 67]. However, when chondrocytes are cultured in three-dimensional conditions, such as an agarose gel, they retain their rounded morphology and produce collagen type II and proteoglycans [7].

Many growth factors are also known to stimulate chondrocytes. Perhaps the most well-known of these cytokines is transforming growth factor β (TGF-β). When chondrocytes are in three-dimensional culture, exposure to TGF-β1 is known to promote cartilaginous differentiation, with increased production of type II collagen and sulfated glycosaminoglycans [33]. TGF-β3 has similarly been demonstrated to promote chondrogenic differentiation. Na et al. implanted rabbit chondrocytes in a hydrogel either with or without TGF-β3 into subcutaneous pockets of nude mice [41]. Constructs with TGF-β3 demonstrated greater cartilage formation than those without. Fibroblast growth factor 2 (FGF-2) is also known to drive the differentiation of mesenchymal cells into chondrocytes, which produce collagen II and proteoglycans [4].

The image of a human ear construct on the back of a nude mouse has generated much public attention. However, this work by Vacanti and colleagues highlights the ability of current scaffold materials to be molded into the shape of the target tissue [27, 59]. Early work on cartilage tissue engineering utilized natural scaffolds, such as agarose, alginate, hyaluronic acid, gelatin, fibrin glue, and collagen derivatives. However, natural materials have proven to be suboptimal because of their inability to sustain a mechanical load as well as concerns for antigenicity [50]. Scaffolds composed of synthetic materials are preferred because of the ability to fine tune their properties based on polymer composition. Hydrogels have shown promise as scaffolds conducive to chondrogenic applications. Hydrogels offer an attractive option because they are injectable and can be molded to the shape of the defect. Furthermore, the intimate, three-dimensional relationship between the hydrogels and the implanted cells promotes the cell to take on a more rounded morphology, which is essential for chondrogenic differentiation. Finally, hydrogels, allow for easy incorporation of growth factors into the scaffold. Recently, Mikos and colleagues have described encapsulating TGF-β1 within hydrogel scaffolds [24, 46]. They found that delivery of TGF-β1 within the hydrogel increased the number of glycosaminoglycan-producing cells within the construct.

Most critical to cartilage tissue engineering are efforts to elucidate the molecular regulators of the chondrogenic differentiation process of progenitor cells. This is especially pertinent given the studies which suggest that cartilage formed from MSC-derived chondrocytes is inferior in quantity and mechanical properties when compared to cartilage formed from articular chondrocytes [37]. Further work examining the process of chondrogenic differentiation will be essential in determining whether these mesenchymal progenitor cells are inherently incapable of forming functional cartilage or whether they are lacking the necessary cues.

61.4
Skin Tissue Engineering

A significant portion of plastic surgery is devoted to providing timely skin coverage in patients who suffer from burn injuries, chronic ulcers, or resection of large soft tissue and skin neoplasms. Annually, there are approximately 13,000 patients who have severe burns, approximately 600,000 patients afflicted with diabetic ulcers, one million patients with venous ulcers, and two million with decubitus ulcers [3]. Open wounds subject patients to a multitude of infectious complications. As a result, surgeons strive to provide coverage for these wounds in a timely manner, often using split-thickness skin grafts. In 2004, skin grafts alone accounted for approximately 42,000 hospitalizations, costing over 840 million dollars in aggregate costs [23]. Furthermore, scarring from large skin defects, which heal by secondary intention, can lead to aesthetic problems, functional losses, contractures, and adverse physiological effects [38]. Thus, the demand for enhanced skin products in plastic surgery is great.

Because of the highly specialized structures within skin, split thickness skin grafts, which contain the epidermis and a portion of the dermis, remain the gold standard for coverage of large deficits. However, the limited amount of skin available for harvest is a major disadvantage of this approach. With key advances in skin tissue engineering, cutaneous substitutes have evolved from single epidermal sheets, cultured from keratinocytes or fibroblasts, to complex, bilayered constructs [6]. The optimal characteristics of a skin substitute include a bilayered structure with a dermal layer for rapid repair of the wound and an epidermal layer to function as a barrier [6]. Furthermore, it must adhere to the wound bed, allow for angiogenesis, be nonimmunogenic, and possess the ability for self repair throughout the patient's life [35]. The two major types of skin substitutes currently being investigated in the field of skin tissue engineering include biomaterials and engineered tissues containing skin cells [3].

The FDA has approved several acellular skin substitutes for clinical use, including AlloDerm (Life Cell; Branchburg, NJ), a dermal matrix derived from cadaveric skin; Xenoderm (Biometica; St. Gallen, Switzerland), another acellular matrix derived from porcine skin; and Integra (Integra Life Sciences; Plainsboro, NJ), an artificial dermal and epidermal layer composed of bovine collagen and chondroitin-6-sulfate on a silastic membrane [61]. These skin substitutes essentially constitute scaffolds that promote cellular in-growth, attachment, and differentiation [3]. The FDA has also approved several skin substitutes containing cells. These include TransCyte (Advanced BioHealing; La Jolla, CA), which contains newborn human fibroblasts cultured on nylon mesh and is primarily used for treatment of burns; and Dermagraft (Advanced BioHealing; La Jolla, CA), which consists of human fibroblasts cultured on a biodegradable matrix for diabetic ulcers [14]. Apligraf (Organogenesis Inc.; Canton, MA) represents one of the more advanced tissue-engineered skin constructs that is approved for treatment of venous ulcers. It is a bilayered skin substitute, with the lower dermal layer composed of bovine type 1 collagen and human fibroblasts and the upper epidermal layer containing human keratinocytes. However, Apligraf lacks melanocytes, Langerhans' cells, hair follicles, or sweat glands.

As evidenced by the many skin products commercially available, much progress has been achieved in the field of skin tissue engineering. However, significant challenges remain. Most of the current skin substitutes incorporate allogeneic- or xenogenic-derived materials, which are accompanied by infection and immunogenic risks. Work directed at creating novel matrices for cells and growth factors to accelerate integration and repair of the wound bed will help to circumvent these issues.

61.5
Adipose Tissue Engineering

Over the last 20 years, plastic surgeons have increasingly used autologous fat harvested from the abdomen, thigh, and flank for various body-contouring procedures. Autologous lipoinjections have most commonly been used for facial contouring, gluteal-enhancement procedures, and breast resurfacing [8, 39]. The use of autologous lipoinjection, however, has been tempered by concerns for the lack of long-term durability of these grafts [19]. Some reports have estimated 40–60% graft resorption over time. The fragility of adipocytes and the lack of adequate vascularization are common reasons cited for these long-term outcomes. Due to these concerns, most fat grafting procedures are restricted to small volumes of lipoinjection and are directed at surface irregularities.

Efforts in adipose tissue engineering could provide the latitude needed for addressing larger soft tissue deficits in a more lasting manner. Adipose tissue engineering has been spurred forward by in vitro studies documenting the potential for postnatal mesenchymal cells to differentiate along an adipogenic lineage under the appropriate condition [47, 71, 72]. When cultured in vitro, ASCs express elevated levels of the adipogenic marker genes, adipsin and PPARγ, and form lipid-containing vesicles when exposed to a cocktail containing insulin, dexamethasone, indomethacin, and methylxanthine [71]. Much of the current literature on adipose tissue engineering involves differentiating ASCs or MSCs cells into adipocytes, seeding them onto a variety of scaffolds, and implanting this construct into animal models [19].

A major hurdle that remains to be surmounted for future adipose tissue engineering efforts, especially for large adipose constructs, is providing this tissue with an adequate vascular supply. One proposed idea that may induce neovascularization of adipose tissue constructs is to coculture preadipocytes with endothelial progenitors. Frerich and colleagues observed formation of capillary-like structures when culturing ASCs with umbilical vein endothelial cells together on a fibrin scaffold [17]. They further observed increase of these capillary-like structures with application of vascular endothelial growth factor.

61.6
Vascular Tissue Engineering

In the process of reconstructing damaged tissue, plastic surgeons must often restore adequate blood flow. Plastic surgeons are most frequently involved in revascularization procedure involving the upper extremity, often using saphenous vein grafts as a bypass conduit. However, given the high incidence of vascular disease in the USA, the usual choices for vascular conduits are all too often not available. The American Heart Association (AHA) reports that in 2004 an estimated 16 million Americans were afflicted with coronary artery disease, and 427,000 coronary artery bypass operations were performed [1]. In this subset of patients, the saphenous vein often times has already been harvested for coronary bypass. An estimated 8–12 million Americans also suffer from peripheral artery disease, many of whom will require revascularization procedures utilizing small caliber vessels [2].

While synthetic grafts are readily available for large vessels such as the aorta, iliac arteries, and femoral arteries, grafts for vessels <6 mm often fail due to thrombosis, atherosclerosis, or intimal hyperplasia [11]. Autologous vessels such as the saphenous vein continue to be the gold standard for revascularization operations to treat peripheral arterial disease, claudication, or chronic ulcers. When microvascular free flaps are needed for coverage of such ulcers, often the bypass operation must be performed first in order to restore blood flow to the extremity. However, frequently there is a shortage of suitable vein as a result of other comorbidities, peripheral vascular disease, smoking, amputation, or prior harvests. Consequently, a number of approaches have emerged to engineer suitable vascular conduits.

The ideal replacement conduit should satisfy three specific criteria: (1) mechanical plasticity including adequate tensile strength and elastic compliance to allow the vessel to withstand the physiological hemodynamics and pressure of the circulation without rupture or aneurysm formation; (2) physiological function that allows appropriate permeability to cells, nutrients, and water, responsiveness to intrinsic and exogenous vasoactive agents, and resistance to thrombosis and infection; and (3) practical properties such as ease of production, availability, cost, ease of handling and suturability, and nonimmunogenicity [26, 40, 42, 43]. A variety of approaches have been developed to address these criteria.

Rosenberg et al. first pioneered the use of acellular bioscaffolds using bovine carotid artery which was subsequently expanded by Lantz et al. using small intestine submucosa for fashioning vascular conduits [32, 53]. These tissues, now commercially available, are completely devoid of cells and are composed entirely of extracellular matrix. Once the extracellular matrix is implanted, it provides the necessary tensile strength needed to withstand the pressure of arterial circulation as well as the scaffold structure for migration of host endothelial cells. Since these tissues are acellular, the risk of immunogenicity is theoretically eliminated [13, 25]. However, the decellularization process involves treating the tissue with detergents, buffers, and solvents that can negatively impact the extracellular matrix making them more prone to aneurysm, thrombosis, or infection [34, 63]. While canine models employing grafts fashioned from acellular tissues have demonstrated comparable patency rates to saphenous vein grafts, issues regarding long-term viability and thrombosis remain to be elucidated [25, 56].

A number of biodegradable scaffolds have been developed using various polymers, most commonly polyglycolic acid (PGA). The concept is based on the premise that endothelial cells and/or fibroblasts can be seeded onto a scaffold that supports cellular proliferation, migration, and remodeling. As the polymer is resorbed, cells populating the scaffold produce sufficient extracellular matrix to sustain the mechanical properties of the conduit [29, 60, 73]. Niklason et al. cultured bovine smooth muscle cells onto a PGA scaffold and then incubated the graft under pulsatile pressure, prior to seeding the scaffold with endothelial cells. The conduits were then reanastomosed in continuity with porcine saphenous arteries with 100% patency and burst pressures exceeding 2000 mm Hg. The conduits also behaved similarly to native arteries with comparable contractility in response to endothelin I and serotonin [43]. Watanabe et al. engineered a similar scaffold using a hybrid polymer of PGA and polycaprolactone-co-polylactic acid which was seeded with cells derived from recipient vessels. Seven months after implanting the bioengineered graft into the 4-year-old child, the graft remained patent with no evidence of stenosis or aneurysm formation [70]. Follow-up studies on 22 patients undergoing cardiovascular surgery using these grafts also demonstrated 100% patency without evidence of thrombosis or stenosis [36].

Attractive aspects of this technique include the ability to carefully titrate the culture environment and to manipulate the scaffold and cells. Endothelial cell dynamics can be stimulated using exogenous growth factors, like VEGF or FGF, or pulsed electromagnetic fields [18, 64]. The freedom of cell selection adds another appealing facet to this technique where researchers have been able to embed endothelial cells, smooth muscle cells, and fibroblasts. With the discovery and characterization of bone-marrow-derived endothelial progenitor cells that home to areas of ischemia, bioengineered grafts can now be seeded with stem cells harvested from patients requiring revascularization procedures [5, 10, 18, 65].

Auger et al. pioneered the purest technique of vascular tissue engineering by the fabrication of vessels de novo using cellular sheets. Human-derived smooth muscle cells were cultured in the presence of supraphysiologic levels of ascorbic acid to stimulate collagen synthesis which ultimately produced a "sheet" of smooth muscle cells. This sheet was then wrapped around a porous tubular construct that allowed exchange of nutrients and formed the media layer of an artery. Fibroblasts cultured under similar conditions also produced a sheet that was wrapped around the sheet of smooth muscle cells to create the adventitia. Finally, the inner porous tube was removed, and the lumen was seeded with endothelial cells producing a multilayered vascular conduit that was responsive to vasoactive substances (i.e., bradykinin, histamine, ATP, and UTP), expressed von Willebrand factor, and resisted platelet aggregation. Subsequent implantation into a canine model has produced promising results [30, 31].

While tremendous advances have occurred in the field of vascular bioengineering, a number of obstacles continue to limit the widespread clinical application of these techniques. Aside from the three ideal criteria described above, the grafts need to tolerate the hemodynamic forces of physiologic circulation without failure. These conduits will be subjected to the shear stress of blood flow, luminal perfusion pressure, cyclical mechanical stretch, and longitudinal tension. Unfortunately, in order to generate grafts with sufficient burst strength, compliance is often sacrificed and vice versa. Likewise, interactions between the cellular components themselves and between the cells and the extracellular matrix are largely unknown. Alternatively, research in the field of vasculogenesis and the mobilization of bone-marrow-derived endothelial progenitor cells offers another means of augmenting blood flow to areas of ischemia. Combining the technology of the bioscaffolds with stem cells such as endothelial progenitor cells has been attempted with promising results [28, 52]. With the developing technology of available scaffolds and continuing advances with stem cell biology, the goal of engineering a small-diameter vascular conduit seems promising.

61.7
Conclusion

With a burgeoning list of new scaffold materials and readily available populations of progenitor cells, the ability to engineer bone, cartilage, skin, adipose, and blood vessels appears to be within reach. This is especially exciting for plastic surgeons as they stand on the cusp of being able to regenerate absent or deficient tissues, rather than reconstruct or replace them. Cell-based tissue engineering, using progenitor cells, arguably offers the most durable approach. Fundamental to further developments in this area, however, is research aimed at elucidating the molecular and cellular mechanisms regulating the differentiation of progenitor cells along a particular lineage. A fuller understanding of the molecular processes underlying lineage specific differentiation will enable greater control of tissue engineering efforts. Furthermore, it is recognized that the mesenchymal cells currently being used for tissue engineering applications represent a heterogenous mixture. While cell surface markers of embryonic stem cells and hematopoietic stem cells have been well defined, such characterization of mesenchymal cells is lacking. It is foreseeable that the identification of such markers will also serve to optimize cell-based tissue engineering applications. Ultimately, the ability to regenerate tissue, without posing additional risks to the patient, will prove to be highly relevant to plastic surgeons.

References

1. AHA (2007) Heart Disease and Stroke Statistics, http://www.americanheart.org.
2. AHA (2007) Peripheral Arterial Disease Statistics; http://www.amercanheart.org.
3. Andreadis ST (2007) Gene-modified tissue-engineered skin: the next generation of skin substitutes. Adv Biochem Eng Biotechnol 103:241–274
4. Arevalo-Silva CA, Cao Y, Weng Y, Vacanti M, Rodriguez A, Vacanti CA, Eavey RD (2001) The effect of fibroblast growth factor and transforming growth factor-beta on porcine chondrocytes and tissue-engineered autologous elastic cartilage. Tissue Eng 7:81–88
5. Asahara T, Murohara T, Sullivan A, Silver M, van der Zee R, Li T, Witzenbichler B, Schatteman G, Isner JM (1997) Isolation of putative progenitor endothelial cells for angiogenesis. Science 275:964–967
6. Auger FA, Berthod F, Moulin V, Pouliot R, Germain L (2004) Tissue-engineered skin substitutes: from in vitro constructs to in vivo applications. Biotechnology and applied biochemistry 39:263–275
7. Aulthouse AL, Beck M, Griffey E, Sanford J, Arden K, Machado MA, Horton WA (1989) Expression of the human chondrocyte phenotype in vitro. In Vitro Cell Dev Biol 25:659–668
8. Cardenas-Camarena L, Lacouture AM, Tobar-Losada A (1999) Combined gluteoplasty: liposuction and lipoinjection. Plast Reconstr Surg 104:1524–1531; discussion 1532–1523
9. Castagnola P, Dozin B, Moro G, Cancedda R (1988) Changes in the expression of collagen genes show two stages in chondrocyte differentiation in vitro. J Cell Biol 106:461–467
10. Ceradini DJ, Kulkarni AR, Callaghan MJ, Tepper OM, Bastidas N, Kleinman ME, Capla JM, Galiano RD, Levine JP, Gurtner GC (2004) Progenitor cell trafficking is regulated by hypoxic gradients through HIF-1 induction of SDF-1. Nat Med 10:858–864
11. Conte MS (1998) The ideal small arterial substitute: a search for the Holy Grail? Faseb J 12:43–45
12. Cowan CM, Shi Y-Y, Aalami OO, Chou Y-F, Mari C, Thomas R, Quarto N, Contag CH, Wu B, Longaker MT (2004) Adipose-derived adult stromal cells heal critical-size mouse calvarial defects. Nature biotechnology 22:560–567
13. Dahl SL, Koh J, Prabhakar V, Niklason LE (2003) Decellularized native and engineered arterial scaffolds for transplantation. Cell Transplant 12:659–666
14. Ehrenreich M, Ruszczak Z (2006) Update on tissue-engineered biological dressings. Tissue engineering 12:2407–2424
15. El-Ghannam A, Ning CQ (2006) Effect of bioactive ceramic dissolution on the mechanism of bone mineralization and guided tissue growth in vitro. J Biomed Mater Res A 76:386–397
16. Eppley BL, Pietrzak WS, Blanton MW (2005) Allograft and alloplastic bone substitutes: a review of science and technology for the craniomaxillofacial surgeon. The Journal of craniofacial surgery 16:981–989
17. Frerich B, Lindemann N, Kurtz-Hoffmann J, Oertel K (2001) In vitro model of a vascular stroma for the engineering of vascularized tissues. Int J Oral Maxillofac Surg 30:414–420
18. Galiano RD, Tepper OM, Pelo CR, Bhatt KA, Callaghan M, Bastidas N, Bunting S, Steinmetz HG, Gurtner GC (2004) Topical vascular endothelial growth factor accelerates diabetic wound healing through increased angiogenesis and by mobilizing and recruiting bone marrow-derived cells. Am J Pathol 164:1935–1947

19. Gomillion CT, Burg KJ (2006) Stem cells and adipose tissue engineering. Biomaterials 27:6052–6063
20. Grabb WC, Smith JW, Aston SJ (1991) Plastic surgery. Little, Brown, Boston
21. Grande DA, Breitbart AS, Mason J, Paulino C, Laser J, Schwartz RE (1999) Cartilage tissue engineering: current limitations and solutions. Clin Orthop Relat Res:S176–185
22. Haynesworth SE, Goshima J, Goldberg VM, Caplan AI (1992) Characterization of cells with osteogenic potential from human marrow. Bone 13:81–88
23. HCUP (2007) Healthcare Cost and Utilization Project. Agency for Healthcare Research and Quality, Rockville, MD, http://www.ahrq.gov/data/hcup/
24. Holland TA, Tabata Y, Mikos AG (2005) Dual growth factor delivery from degradable oligo(poly(ethylene glycol) fumarate) hydrogel scaffolds for cartilage tissue engineering. J Control Release 101:111–125
25. Huynh T, Abraham G, Murray J, Brockbank K, Hagen PO, Sullivan S (1999) Remodeling of an acellular collagen graft into a physiologically responsive neovessel. Nat Biotechnol 17:1083–1086
26. Isenberg BC, Williams C, Tranquillo RT (2006) Small-diameter artificial arteries engineered in vitro. Circ Res 98:25–35
27. Kamil SH, Vacanti MP, Aminuddin BS, Jackson MJ, Vacanti CA, Eavey RD (2004) Tissue engineering of a human sized and shaped auricle using a mold. Laryngoscope 114:867–870
28. Kaushal S, Amiel GE, Guleserian KJ, Shapira OM, Perry T, Sutherland FW, Rabkin E, Moran AM, Schoen FJ, Atala A, Soker S, Bischoff J, Mayer JE, Jr. (2001) Functional small-diameter neovessels created using endothelial progenitor cells expanded ex vivo. Nat Med 7:1035–1040
29. Kim BS, Putnam AJ, Kulik TJ, Mooney DJ (1998) Optimizing seeding and culture methods to engineer smooth muscle tissue on biodegradable polymer matrices. Biotechnol Bioeng 57:46–54
30. L'Heureux N, Paquet S, Labbe R, Germain L, Auger FA (1998) A completely biological tissue-engineered human blood vessel. Faseb J 12:47–56
31. L'Heureux N, Stoclet JC, Auger FA, Lagaud GJ, Germain L, Andriantsitohaina R (2001) A human tissue-engineered vascular media: a new model for pharmacological studies of contractile responses. Faseb J 15:515–524
32. Lantz GC, Badylak SF, Hiles MC, Coffey AC, Geddes LA, Kokini K, Sandusky GE, Morff RJ (1993) Small intestinal submucosa as a vascular graft: a review. J Invest Surg 6:297–310
33. Lee JE, Kim SE, Kwon IC, Ahn HJ, Cho H, Lee SH, Kim HJ, Seong SC, Lee MC (2004) Effects of a chitosan scaffold containing TGF-beta1 encapsulated chitosan microspheres on in vitro chondrocyte culture. Artif Organs 28:829–839
34. Livesy S, del Campo A, Nag A, Nichols K, Coleman C (1994) Method for Processing and Preserving Collagen-Based Tissues for Transplantation. US Patent 5336616
35. MacNeil S (2007) Progress and opportunities for tissue-engineered skin. Nature 445:874–880
36. Matsumura G, Hibino N, Ikada Y, Kurosawa H, Shin'oka T (2003) Successful application of tissue engineered vascular autografts: clinical experience. Biomaterials 24:2303–2308
37. Mauck RL, Yuan X, Tuan RS (2006) Chondrogenic differentiation and functional maturation of bovine mesenchymal stem cells in long-term agarose culture. Osteoarthritis Cartilage 14:179–189
38. Metcalfe AD, Ferguson MWJ (2005) Harnessing wound healing and regeneration for tissue engineering. Biochemical society transactions 33:413–417
39. Missana MC, Laurent I, Barreau L, Balleyguier C (2007) Autologous fat transfer in reconstructive breast surgery: Indications, technique and results. Eur J Surg Oncol
40. Mitchell SL, Niklason LE (2003) Requirements for growing tissue-engineered vascular grafts. Cardiovasc Pathol 12:59–64
41. Na K, Kim S, Woo DG, Sun BK, Yang HN, Chung HM, Park KH (2007) Synergistic effect of TGFbeta-3 on chondrogenic differentiation of rabbit chondrocytes in thermo-reversible hydrogel constructs blended with hyaluronic acid by in vivo test. J Biotechnol 128:412–422
42. Niklason LE (1999) Techview: medical technology. Replacement arteries made to order. Science 286:1493–1494
43. Niklason LE, Gao J, Abbott WM, Hirschi KK, Houser S, Marini R, Langer R (1999) Functional arteries grown in vitro. Science 284:489–493
44. Ohgushi H, Goldberg VM, Caplan AI (1989) Repair of bone defects with marrow cells and porous ceramic. Experiments in rats. Acta Orthopaedica Scandinavica 60:334–339
45. Parikh SN (2002) Bone graft substitutes: past, present, future. Journal of postgraduate medicine 48:142–148
46. Park H, Temenoff JS, Holland TA, Tabata Y, Mikos AG (2005) Delivery of TGF-beta1 and chondrocytes via injectable, biodegradable hydrogels for cartilage tissue engineering applications. Biomaterials 26:7095–7103
47. Pittenger MF, Mackay AM, Beck SC, Jaiswal RK, Douglas R, Mosca JD, Moorman MA, Simonetti DW, Craig S, Marshak DR (1999) Multilineage potential of adult human mesenchymal stem cells. Science 284:143–147
48. Pittenger MF, Mosca JD, McIntosh KR (2000) Human mesenchymal stem cells: progenitor cells for cartilage, bone, fat and stroma. Curr Top Microbiol Immunol 251:3–11
49. Quarto N, Longaker MT (2006) FGF-2 inhibits osteogenesis in mouse adipose tissue-derived stromal cells and sustains their proliferative and osteogenic potential state. Tissue Eng 12:1405–1418
50. Raghunath J, Rollo J, Sales KM, Butler PE, Seifalian AM (2007) Biomaterials and scaffold design: key to tissue-engineering cartilage. Biotechnol Appl Biochem 46:73-84
51. Rah DK (2000) Art of replacing craniofacial bone defects. Yonsei medical journal 41:756–765
52. Riha GM, Lin PH, Lumsden AB, Yao Q, Chen C (2005) Review: application of stem cells for vascular tissue engineering. Tissue Eng 11:1535–1552
53. Rosenberg N, Martinez A, Sawyer PN, Wesolowski SA, Postlethwait RW, Dillon ML, Jr. (1966) Tanned colla-

gen arterial prosthesis of bovine carotid origin in man. Preliminary studies of enzyme-treated heterografts. Ann Surg 164:247–256
54. Saadeh PB, Khosla RK, Mehrara BJ, Steinbrech DS, McCormick SA, DeVore DP, Longaker MT (2001) Repair of a critical size defect in the rat mandible using allogenic type I collagen. The Journal of craniofacial surgery 12:573–579
55. Salgado AJ, Coutinho OP, Reis RL (2004) Bone tissue engineering: state of the art and future trends. Macromol Biosci 4:743–765
56. Sandusky GE, Jr., Badylak SF, Morff RJ, Johnson WD, Lantz G (1992) Histologic findings after in vivo placement of small intestine submucosal vascular grafts and saphenous vein grafts in the carotid artery in dogs. Am J Pathol 140:317–324
57. Schantz J-T, Hutmacher DW, Lam CXF, Brinkmann M, Wong KM, Lim TC, Chou N, Guldberg RE, Teoh SH (2003) Repair of calvarial defects with customised tissue-engineered bone grafts II. Evaluation of cellular efficiency and efficacy in vivo. Tissue engineering 9 Suppl 1:S127–S139
58. Shakibaei M, De Souza P (1997) Differentiation of mesenchymal limb bud cells to chondrocytes in alginate beads. Cell Biol Int 21:75–86
59. Shieh SJ, Terada S, Vacanti JP (2004) Tissue engineering auricular reconstruction: in vitro and in vivo studies. Biomaterials 25:1545–1557
60. Shum-Tim D, Stock U, Hrkach J, Shinoka T, Lien J, Moses MA, Stamp A, Taylor G, Moran AM, Landis W, Langer R, Vacanti JP, Mayer JE, Jr. (1999) Tissue engineering of autologous aorta using a new biodegradable polymer. Ann Thorac Surg 68:2298–2304; discussion 2305
61. Stern R, McPherson M, Longaker MT (1990) Histologic study of artificial skin used in the treatment of full-thickness thermal injury. Journal of burn care & rehabilitation 11:7–13
62. Stile RA, Healy KE (2001) Thermo-responsive peptide-modified hydrogels for tissue regeneration. Biomacromolecules 2:185–194
63. Teebken OE, Haverich A (2002) Tissue engineering of small diameter vascular grafts. Eur J Vasc Endovasc Surg 23:475–485
64. Tepper OM, Callaghan MJ, Chang EI, Galiano RD, Bhatt KA, Baharestani S, Gan J, Simon B, Hopper RA, Levine JP, Gurtner GC (2004) Electromagnetic fields increase in vitro and in vivo angiogenesis through endothelial release of FGF-2. Faseb J 18:1231–1233
65. Tepper OM, Capla JM, Galiano RD, Ceradini DJ, Callaghan MJ, Kleinman ME, Gurtner GC (2005) Adult vasculogenesis occurs through in situ recruitment, proliferation, and tubulization of circulating bone marrow-derived cells. Blood 105:1068–1077
66. Vaibhav, Nilesh, Vikram, Anshul (2007) Bone morphogenic protein and its application in trauma cases: A current concept update. Injury
67. von der Mark K, Gauss V, von der Mark H, Muller P (1977) Relationship between cell shape and type of collagen synthesised as chondrocytes lose their cartilage phenotype in culture. Nature 267:531–532
68. Wan DC, Nacamuli RP, Longaker MT (2006) Craniofacial Bone Tissue Engineering. Dental Clinics of North America 50:175–190
69. Wan DC, Shi YY, Nacamuli RP, Quarto N, Lyons KM, Longaker MT (2006) Osteogenic differentiation of mouse adipose-derived adult stromal cells requires retinoic acid and bone morphogenetic protein receptor type IB signaling. Proc Natl Acad Sci U S A 103:12335–12340
70. Watanabe M, Shin'oka T, Tohyama S, Hibino N, Konuma T, Matsumura G, Kosaka Y, Ishida T, Imai Y, Yamakawa M, Ikada Y, Morita S (2001) Tissue-engineered vascular autograft: inferior vena cava replacement in a dog model. Tissue Eng 7:429–439
71. Zuk PA, Zhu M, Ashjian P, De Ugarte DA, Huang JI, Mizuno H, Alfonso ZC, Fraser JK, Benhaim P, Hedrick MH (2002) Human adipose tissue is a source of multipotent stem cells. Mol Biol Cell 13:4279–4295
72. Zuk PA, Zhu M, Mizuno H, Huang J, Futrell JW, Katz AJ, Benhaim P, Lorenz HP, Hedrick MH (2001) Multilineage cells from human adipose tissue: implications for cell-based therapies. Tissue Eng 7:211–228
73. Zund G, Hoerstrup SP, Schoeberlein A, Lachat M, Uhlschmid G, Vogt PR, Turina M (1998) Tissue engineering: a new approach in cardiovascular surgery: Seeding of human fibroblasts followed by human endothelial cells on resorbable mesh. Eur J Cardiothorac Surg 13:160–164

Tissue Engineering Applications for Cardiovascular Substitutes

62

M. Cimini, G. Tang, S. Fazel, R. Weisel, R.-K. Li

Contents

62.1	Introductory Remarks	887
62.2	Principles of Tissue Engineering	888
62.3	Components of Tissue-Engineered Constructs	888
62.3.1	Cell Types	888
62.3.2	Cardiomyocytes and Cardiac-Derived Progenitors	889
62.3.3	Noncardiac Adult Muscle Cells	890
62.3.4	Bone Marrow-Derived or Circulating Adult Stem and Progenitor Cells	890
62.3.5	Embryonic Stem Cells	891
62.3.6	Scaffolds	892
62.4	Tissue Engineering for the Replacement or Repair of Cardiac Muscle	893
62.4.1	Cardiomyocytes for the Replacement or Repair of Cardiac Muscle	895
62.4.2	Smooth Muscle Cells for the Replacement or Repair of Cardiac Muscle	895
62.4.3	Skeletal Myoblasts for the Replacement or Repair of Cardiac Muscle	896
62.4.4	Mesenchymal Stem Cells for the Replacement or Repair of Cardiac Muscle	896
62.4.5	Embryonic Stem Cells for the Replacement or Repair of Cardiac Muscle	896
62.4.6	Limitations and Future Perspectives on Tissue-Engineered Cardiac Muscle	897
62.5	Tissue-Engineered Cardiovascular Structures	898
62.5.1	Vascular Grafts	901
62.5.2	Heart Valves	902
62.6	Concluding Remarks	903
	References	904

62.1 Introductory Remarks

Tissue engineering is a rapidly growing field at the junction between biological and physical sciences. Tissue creation in vitro and in vivo holds the potential to revolutionize modern medicine by providing therapies for a vast array of diseases. Diseases that affect the myocardium, coronary vasculature, and heart valves, or congenital cardiac defects, are targets for tissue engineering strategies. Of these, the creation of replacement myocardium poses the greatest challenge because currently there are few treatment options for a failing heart [1]. This chapter will focus on the principles of tissue engineering, candidate cells that can be used in construct creation, and biomaterial options for three dimensional scaffolding as they specifically relate to engineered myocardial substitutes. A brief discussion on engineered heart valve and vascular substitutes will follow.

62.2 Principles of Tissue Engineering

Tissue engineering aims to restore and/or replace abnormal or diseased tissues. Cell-based therapies have been an intense arena for investigation with many successes and limitations [2–4]. Cell suspensions injected into damaged myocardial tissue hold great promise, but limited cell survival may reduce treatment efficacy [5]. Cell transplantation may also be difficult to apply to children born with congenital heart defects. Tissue engineering offers advantages and may provide better cardiac repair [1, 6]. Therefore, for the purpose of this text, the distinction between cellular cardiomyoplasty without the use of scaffolding (i.e., injectable liquids or rigid matrices) will be referred to as cell transplantation, while those strategies employing such additions will be referred to as tissue engineering.

Every tissue consists of a matrix for structural support, stromal cells that maintain the matrix, and specialized cells that perform the function of the tissue in question. A variety of tissue engineering models have been developed, but most approaches can be explained by the traditional paradigm (Fig. 62.1) [7], which involves in vitro cell seeding into a natural or synthetic scaffold. Culture conditions may be optimized for tissue maturation which could involve incubation in a bioreactor. The cell preseeded tissue construct is then implanted into the host to allow remodeling in vivo. An alternative paradigm involves the implantation of an acellular scaffold into the host. This approach depends solely on endogenous cells for tissue maturation in vivo, but may include modification of the scaffold with chemical mediators such as cytokines, chemokines, and growth factors embedded within the biomaterial or administered systemically in order to promote endogenous cell infiltration [8].

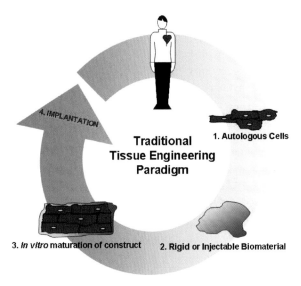

Fig. 62.1 The traditional tissue engineering paradigm. The outlined model involves a four-step process: *1* Cells are explanted, isolated, and expanded in vitro (if necessary differentiation protocols may be implemented). *2* After sufficient cell numbers are acquired, cells are seeded into a biomaterial that serves as a scaffold which can be either rigid or liquid in nature. *3* The cell-seeded construct can be matured in vitro by either treatment with soluble factors or incubation in a bioreactor. *4* Implantation procedure via surgery or via a catheter (if injectable). This traditional paradigm serves as a basis for all approaches with additions or deletions or modifications of steps

62.3 Components of Tissue-Engineered Constructs

62.3.1 Cell Types

Tissue-engineered cardiac grafts (TECGs) provide specific problems. In theory, cardiomyocytes are a natural choice because they possess the electrophysiological, structural, and contractile properties necessary for cardiac function [1]. Since cardiomyocytes do not readily proliferate [9], use of these cells is limited by cell number. Therefore, much attention has been given to potentiate differentiation of progenitor cells into cardiomyocytes in vitro [10, 11] and in vivo [12–14]. To date, no ideal adult autologous cell type has been identified for complete cardiac regeneration. This has directed the development of diverse strategies to aid in the repair of injured myocardium, including: (i) increasing angiogenesis; (ii) limiting cardiomyocyte loss in stunned, hibernating, or scarred myocardium [13, 15–20]; and (iii) modulating the extracellular matrix to limit left ventricular (LV) dilatation and heart failure [6, 21–24].

For TECGs, the ideal cell type [25, 26] should be nonimmunogenic, easy to harvest and highly proliferative, with the ability to differentiate into cell types well-suited to survive under ischemic conditions (Table 62.1). In this respect, autologous cell appear to be the ideal source to seed biomaterials, thereby limiting immunorejection and the need for immunosuppressive therapy. However, autologous cells have their own challenges, especially when they are harvested from sick and/or aging patients [27] who have limited numbers of stem cells and a diminished proliferative capacity [9]. The following sections serve to identify candidate cell types and discuss the characteristics that make each an attractive option for the development of tissue-engineered constructs (summarized in Table 62.2).

62.3.2 Cardiomyocytes and Cardiac-Derived Progenitors

62.3.2.1 Cardiomyocytes

The natural source for myocardial TECG is the cardiomyocyte. Cardiomyocytes in the young and adult myocardium are differentiated, and essentially lose their proliferative capacity. Since the total number of cardiomyocytes is relatively stable after the early postnatal stage, cardiomyocyte hypertrophy is the only means of increasing cardiac mass in the face of cardiomyocyte loss after injury. Fetal cardiomyocytes do proliferate in vitro, mainly due to their ability to disassemble and reassemble sarcomeres [9]. Neonatal cardiomyocyte populations may have decreased contractility after expansion in vitro, but they do not lose their ability to proliferate [28]. Therefore, fetal and neonatal cardiomyocytes may be sources of cells for tissue engineering therapies to repair congenital heart defects, but the use of these cells in adults may not permit repair of infarcted or dilated myocardium.

Table 62.1 Desirable characteristics of donor cells for tissue engineering efforts

Desirable characteristics of cells
1. Contractile potential with propensity for cardiomyogenic differentiation
2. Autologous
3. Easily isolated and highly expandable
4. Increased survival in ischemic microenvironment
5. No adverse effects—calcification, neoplastic transformation, etc.

Table 62.2 Candidate cell characteristics for tissue engineering efforts (Adapted from [1])

Cell phenotype		Autologous	Cardiomyogenic potential	Ease of expansion	Immunogenic	Safety
Cardiomyocytes	Fetal	−	+	−	+	+
	Neonatal	−	+	−	+	+
Adult—cardiac progenitors		+	+	+	+	+
Skeletal myoblast and satellite cells		+	+	+	+	− Arrhythmia
Smooth muscle cells		+	−	+	+	+
Hematopoietic stem cells		+	+/−	+	+	−
Mesenchymal stem cells		+	+	+/−	−	+ Calcification
Endothelial progenitor cells		+	+/−	+/−	+	−
Embryonic stem cells		−	+	+	+/−	+ Teratoma

62.3.2.2
Cardiac-Derived Progenitor Cells

Adult cardiomyocytes were traditionally believed to be terminally-differentiated cells with very low proliferative potential and little capacity for self-renewal, but this notion has been recently challenged. Putative adult cardiac stem cells [29–32] have been recently described as small, round cells that express the stem cell factor receptor, c-kit. After isolation and cloning, these cells were found to be multipotent, giving rise to smooth muscle cells, endothelial cells, and cardiomyocytes in vitro [29]. Like neural stem cells, they form embryoid-body-like clusters termed "cardiospheres" in culture, composed of cells with multilineage fates [33, 34]. In addition, a Sca-1-expressing population termed, cardiac side-population (SP) cells, has also been identified with putative regenerative potential in vivo [35, 36]. Unlike the c-kit+ population of cardiac resident stem cells, SP cells lack c-kit expression. A third cardiac progenitor cell type was also identified, characterized by the expression of Islet1 without concomitant expression of c-kit or Sca-1 [37]. Unlike c-kit and Sca-1, Islet1 is a transcription factor whose abnormal expression and absence has been shown to lead to abnormal development of the right ventricle and its outflow tract [38]. Islet1-expressing cells isolated from neonatal mice demonstrated maturation into beating cardiomyocytes [37]. Although these cells have not been employed in tissue-engineered constructs because they were so recently characterized, they may be an ideal cell type for the creation of TECG.

62.3.3
Noncardiac Adult Muscle Cells

62.3.3.1
Skeletal Myoblasts

Skeletal myoblasts may be well-suited for cell transplantation and tissue engineering [2, 21, 39–43]. These cells can be isolated and expanded to yield a large number of cells, and in vitro and in vivo studies have demonstrated their ability to fuse and form myotubes [43–45]. Therefore, preclinical studies justified clinical trials which are well underway [46–50]. Skeletal myoblasts were able to survive in hostile ischemic environments and engrafted in human hearts. Yet, the primary concern after transplantation has been their isolation from the host myocardium and their induction of arrhythmias, which has been suggested in the initial clinical trials [46, 49].

62.3.3.2
Smooth Muscle Cells

Smooth muscle cells (SMCs) are good candidates for myocardial tissue engineering because of their unique ability to modulate the ECM and their potential to contribute to the developing vasculature [51]. Our laboratory has completed several studies using SMCs in cell transplantation [52, 53] and in TECGs [24, 54–57]. The most significant limitation of SMC-seeded constructs is that they are noncontractile. Nevertheless, SMCs possess advantages because they are easy to isolate, culture, and expand from vessels currently used in coronary artery bypass grafts [51]. The rationale for using such cells in tissue-engineered grafts for the repair of myocardial injury is that smooth muscle cells will enhance myocardial elastic properties that are reduced after cardiac injury.

62.3.4
Bone Marrow-Derived or Circulating Adult Stem and Progenitor Cells

62.3.4.1
Hematopoietic Stem Cells

Hematopoietic stem cells (HSCs) produce blood cells and are responsible for the regeneration of blood-forming tissue [58]. HSCs are characterized by the expression of c-kit, thy-1, sca-1 (mouse), and CD34 (human), and are lineage negative (lin-) [59, 60]. It is possible to isolate HSCs from bone marrow aspirates or peripheral blood, but they cannot be expanded in vitro to provide a sufficient quantity to repair injured myocardium. HSC transdifferentiation into cardiomyocytes has been demonstrated [14, 29, 61, 62], but the ability of these cells to commit to cardiac lineages remains unresolved [63–67].

62.3.4.2
Mesenchymal Stem Cells

Mesenchymal stem or stromal cells (MSCs) are multipotent stem cells in the bone marrow. MSCs are attractive candidates for tissue engineering because they lack cell surface expression of MHC-II [68], the receptor responsible for initiating immune rejection. In addition, these cells are easily isolated and expanded in culture, and they can transdifferentiate into cardiomyocytes [69–72]. The transdifferentiation of MSCs to cardiomyocytes was suggested by Makino and colleagues [10], who showed that these cells could develop into beating cardiomyocytes in vitro when grown in the presence of 5-azacytidine, a chromosomal demethylating agent that removes epigenetic restrictions on cell differentiation pathways. Some of the cells began to spontaneously beat in culture. The beating cells resemble adult cardiomyocytes in their transcription factors, protein expression, electron microscopic structure, and electromechanical activity [10]. Positive results in the initial preclinical studies sparked several clinical trials using MSCs, which have suggested that these cells can improve cardiac function in humans [73–76]. In addition, MSCs have been proposed to be immunoprivileged in that they can induce tolerance [70–73]. Therefore, allogeneic MSCs may be employed for tissue engineering constructs in humans.

62.3.4.3
Endothelial Progenitor Cells

Angiogenesis has been proposed to be a major contributing factor to functional improvement after cell therapy for myocardial ischemia or infarction [8, 77, 78]. Protein and gene angiogenic therapies have been shown to improve recovery after myocardial ischemia [79–81]. Proposed protein-based therapies are currently limited by the short half-life of proteins [82, 83], while gene therapies are limited by inefficient expression of the gene product and brief physiological affects [84]. By extension, the induction of angiogenesis with endothelial progenitor cells (EPCs) in TECGs may be a valid approach for cardiac repair. It is well established that EPCs are capable of regulating angiogenic events after recruitment to sites of injury [85–87]. In injured tissue, they differentiate into endothelial cells and SMCs [88], and induce the formation of the vasculature [26]. Among the cell surface markers expressed by EPCs are c-kit/CD133/CD34/VEGF-R2 [87, 89–91]. The best markers to delineate the true progenitor phenotype may be CD34 and VEGF-R2, both of which are expressed by mature endothelial cells [90]. In addition, EPCs have been suggested to transdifferentiate into cardiomyocytes in cocultures with cardiomyocytes [11], although these results have been questioned [92]. Collectively, these characteristics make EPCs an attractive cell choice for tissue-engineered grafts. They may be able to acquire cardiomyocyte contractile function, and they can also vascularize tissue-engineered grafts. These two features may help achieve construct thickness with enhanced diffusion of metabolites.

62.3.5
Embryonic Stem Cells

Despite the associated ethical controversy, embryonic stem cells (ESCs) have consistently been shown to possess strong potential for cardiogenic differentiation [93–96], and provide a compelling therapeutic option for TECG creation. ESCs are self-renewing, can be directed toward differentiation, and have been shown to be nonimmunogenic, perhaps even when employed in xenogenic transplantation. Most importantly, ESCs possess a high tolerance to hypoxic conditions [97]. Well-established protocols have been developed for the isolation, propagation, and cardiogenic differentiation of ESCs. These protocols may also permit coupling to the host myocardium by gap and adherent junctions, which is essential for TECG integration [93, 98, 99]. The use of ES cells in cardiac regenerative therapies is not without concerns. While Mènard et al. [97] found no significant immune response in a model of xenogenic ES transplantation, others have reported [100, 101] a significant T-lymphocyte and dendritic cell population around transplanted allogeneic ESCs. Another major concern with the use of ESCs is their propensity for teratoma formation [102, 103]. Menasche's group demonstrated that ES cells precommitted towards a cardiomyocyte lineage may circumvent this possibility, while others remain wary of their teratogenic potential [102, 104].

62.3.6
Scaffolds

The scaffold plays an important role in tissue engineering strategies because it controls the immediate microenvironment of transplanted cells. The scaffold may augment the microenvironment of the injured myocardium when applied without cells. The scaffold provides structural support for cells seeded into it, but more importantly, a well-suited scaffold may provide the biological cues necessary for tissue formation [105]. Thus, an ideal biomaterial should be nonimmunogenic, biocompatible, and preferably biodegradable if synthetic (Table 62.3) [106]. In addition, mechanical stability at physiological loading conditions is needed to maintain cardiac integrity. Rigid biomaterials provide a 3D scaffold for cellular seeding and some success has been reported, yet forced structural adaptation, propensity for poor vascularization, stimulation of host immune responses, and toxic degradation are all substantial challenges for these biomaterials [5].

Both synthetic and naturally-derived scaffolds have been developed for tissue engineering therapy. Synthetic scaffolds are generally made of polymers including polyglycolic acid (PGA), polylactic acid (PLA), and copolymers such as poly-lactide-co-glycolic acid (PLGA) [24, 54, 55, 107–113]. Naturally derived scaffolds used in experimental studies often consist of ECM components including collagen, gelatin, fibrin, and alginate [17, 40, 74, 109, 114–119].

Recent studies suggest that the ECM, once thought to be a static entity, is rather dynamic. It is essential in modulating cellular proliferation, apoptosis, cell-specific responses, and differentiation [8]. An ideal biomaterial promotes all these aspects and "mimics" the native ECM for optimal tissue maturation. Such native ECM-resembling biomaterials, termed "biomimetic materials" [1], promote native cell-ECM interactions and are commonly modified with bioactive molecules such as the fibronectin signaling domain RGD (Arginine-glycine-aspartate) [8]. RGD incorporation into biomaterials has been beneficial for cell retention and, therefore, for cell survival [120, 121]. Yet, simply immobilizing RGD does not always stimulate cell adhesion and direct cellular function (e.g., endothelial cell–cell junctions to promote angiogenesis). Spacing and orientation have been found to be critical [122].

In addition, liquid injectable scaffolds have been used in a variety of therapeutic approaches. Collagen type I, Matrigel, and fibrin gels have been used to deliver neonatal cardiomyocytes [119], bone-marrow-derived MSCs [74], bone marrow mononuclear cells [123], skeletal myoblasts [118], and endothelial cells [124]. Interestingly, fibrin glue injection, with or without coinjection of skeletal myoblasts, preserved LV geometry and cardiac function in an acute myocardial infarction model [40], demonstrating the significant impact biomaterials may have on cardiac geometry.

As discussed above, biomaterials can control cellular differentiation and function by augmenting the structural characteristics that influence cellular adhesion, degradation, and porosity. These properties will influence not only coinjected/transplanted cells,

Table 62.3 Desirable characteristics of biomaterials for tissue engineering efforts

Desirable characteristics of biomaterials to be used as scaffolds
1. Mechanical stability
2. Nonimmunogenic
3. Biodegradable
4. Nontoxic
5. Promotes cell adherence
6. Maintains cellular phenotype
7. Conducive to vascularization

but also endogenous cells which home to the biomaterial. Porosity (i.e., pore size and connectivity) is a key factor. Pore size influences the trafficking of cells important to both cell seeding and construct vascularization [125, 126]. Critical thresholds of pore size are required because the size may influence mechanical strength and stability of the biomaterial [125, 127, 128]. Immediately after implantation, when the biomaterial is devoid of vasculature, diffusion of metabolites is largely influenced by pore size [129]. Increased diffusion increases seeded cell viability and the thickness of the graft [6].

In addition to delivering cells to mediate improvements in cardiac geometry and function, scaffolds may also be embedded with proteins, drugs, genes, and other factors that can influence angiogenesis, cardiomyogenesis, and attenuation of cardiac fibrosis. In this area, the Lee group has made significant advances with the use of self-assembling peptide nanofibers conjugated with growth factors for controlled delivery of the factors to the infarcted myocardium [17, 130–132]. Self-assembling peptides, when exposed to physiological osmolarity and pH, will rapidly assemble into small nanofibers (10 nm) which can form 3D cellular microenvironments after they are injected into the myocardium. These peptide nanofibers and their microenvironments then recruit a variety of cells [131]. To date, this group has used nanofibers for the controlled delivery of platelet-derived growth factor-BB (PDGF-BB) [132] and insulin-like growth factor-1 (IGF-1) [130], both with promising results. In the PDGF-BB study, the nanofibers were injected into the infarct border zones after coronary ligation in rat hearts. PDGF-BB nanofibers imparted beneficial effects by reducing cardiomyocyte apoptosis via the PI3/Akt pathway, with concomitant beneficial effects on cellular proliferation and angiogenesis/arteriogenesis. The increased survival of cardiomyocytes resulted in decreased infarct size and improved fractional shortening. In a similar study, IGF-1 biotin conjugated to nanofibers with and without neonatal cardiomyocytes was coinjected into healthy and infarcted rat hearts [130]. Coinjection increased growth and survival of the transplanted cells, decreased apoptosis, improved cardiac geometry and fractional shortening, and reduced cardiac dilatation. These studies demonstrated that "smart" biomaterials that can control the delivery of conjugated mediators warrant further study for their potential as powerful tools in cardiac tissue engineering.

62.3.6.1
Soluble Factors for Use in Cell Preparation or in Biomaterial Scaffolds

Soluble factors can be used to induce the differentiation of various cells types into a cardiomyogenic phenotype. In addition, they may promote maturation, engraftment, and integration by stimulating host cell homing to the graft, upregulating matrix proteins, and inducing neovascularization. ESC differentiation towards a cardiomyogenic phenotype can be initiated by members of the transforming growth factor (TGF) superfamily including TGF-b1 [133] and bone morphogenetic proteins (BMPs) [95]. PDGF-BB [134] and vascular endothelial growth factor (VEGF) [95, 135] have been found to have similar effects. PDGF-BB was also found to stimulate ESC differentiation to SMCs, and VEGF can direct ESC differentiation to endothelial cells.

In adult stem cells, c-kit+ HSCs were found to upregulate cardiomyogenic genes in response to TGF-b1 [136]. HSC stimulation with FGF, IGF, or both resulted in an increased propensity of these cells toward an endothelial cell lineage [137]. MSCs treated with TGF-beta induced SMC differentiation [138, 139], and 5-azacytidine induced MSC differentiation toward a cardiac phenotype [10, 140]. Collectively, these studies demonstrate that soluble factors may be used to induce the cellular maturation of engineered tissues either in vitro or in situ.

62.4
Tissue Engineering for the Replacement or Repair of Cardiac Muscle

The following sections contain a review of the literature highlighting successes in the field of cardiac tissue engineering and discuss issues that remain (Table 62.4).

Table 62.4 Representative studies of tissue engineered cardiac grafts

Cell type	Author/publication year	Reference	Animal model	Biomaterial	Implantation site	Bioreactor type	Cell source
Cardiomyocytes	Sakai, 2001	143	Rat	Gelatin	RVOT defect	NA	Fetal/adult syngeneic
	Birla, 2005	144	Rat	NA	Subcutaneous	NA	Neonatal syngeneic
	Shimizu, 2002	145	Rat	NA	Subcutaneous	NA	Neonatal syngeneic
	Radisic, 2004	147	In vitro	Collagen sponge + matrigel	NA	Biomimetic	Neonatal (rat)
	Zimmermann, 2006	118	Rat	Collagen + matrigel	LV infarct	Dynamic	Neonatal syngeneic
Skeletal myoblasts	Memon, 2005	21	Rat	NA	LV infarct	NA	Neonatal syngeneic
	Fuchs, 2006	149	Lamb	Collagen	LV defect onlay	NA	Fetal autologous
	Hata, 2006	43	Canine	NA	LV wall in pacing-induced heart failure	NA	Adult autologous
Smooth muscle cells	Ozawa, 2002	24	Rat	PCLA, gelatin, PGA	RVOT defect	NA	Syngeneic
	Matsubayashi, 2003	55	Rat	PCLA	LV infarct, repaired with EVCPP	NA	Syngeneic
Mesenchymal stem cells	Krupnick, 2002	115	Rat	PTFE, + PLLA + collagen I and IV	LV wall via heterotopic heart transplantation	NA	Syngeneic heart
	Miyahara, 2006	148	Rat	NA	LV infarct	NA	Syngeneic
Embryonic stem cells	Guo, 2006	152	In vitro/mouse	Collagen I + matrigel	Subcutaneous	Dynamic	Mouse ES-D3
	Kofidis, 2005	121	Rat	Collagen I	LV infarct, pouch	NA	Mouse undifferentiation
	Caspi, 2007	154	In vitro	PLGA + PLLA	NA	NA	huESCs + HUVEC or huESCs + EmF

PCLA, ε-caprolactone + L-lactide; *PGA*, polyglycolic acid; *PLLA*, poly (L-lactic) acid; *PTFE*, polytetrafluoethylene

62.4.1
Cardiomyocytes for the Replacement or Repair of Cardiac Muscle

In our laboratory, we seeded adult and fetal cardiomyocytes onto a gelatin mesh. This TECG was then used to repair a defect in the right ventricular outflow tract (RVOT) in a rat model, with the unseeded biodegradable biomaterial serving as a control. Endocardial seeding quickly covered both patches, but the cell-seeded grafts were thicker [141]. Birla et al. also created 3D cardiac tissue, which they called "cardioids" [142]. In their study, they focused on in vivo neovascularization of the tissue after subcutaneous implantation. After 3 weeks, they observed increased thickness and blood vessel density, and contractile proteins were present in the explanted grafts. Interestingly, the authors found a substantial increase in force generation following in vivo incubation, and were able to increase the pulse rate of the tissue in response to epinephrine treatment.

In a similar study, Shimizu et al. utilized a unique cell sheet technology to create a 3D bioengineered graft with neonatal rat cardiomyocytes [143]. This technology eliminates the use of a scaffold by plating cells to form monolayers on temperature-responsive dishes. Engineered cell sheets from this technique revealed cardiomyocyte-like phenotypes, extracellular matrix, and integrative adhesive agents. Cardiomyocyte cell sheets were engineered in vitro and implanted subcutaneously for up to 1 year. Grafts matured with well-defined characteristics of cardiac tissue, including elongated cardiomyocytes with well differentiated sarcomeres and gap junctions resulting in increased size, conduction velocity, and contractile force generation [144]. Collectively, these studies demonstrate that ectopic implantation of grafts, with the animals acting as an in vivo bioreactor, shows promise as a method to induce maturation before implantation into a diseased organ. In addition, these data suggest that graft maturation time may play a key role, with longer in vivo incubation resulting in better-engineered heart tissue properties.

To circumvent the maturation time, Radisic et al. developed a rapid method of in vitro maturation that resulted in TECGs with functional physiological and structural properties in merely 8 days from cell isolation to final product [145]. Maturation was achieved by exposing the graft to electrical signals designed to mimic those of the native heart. Over the 8-day period, the construct showed a seven-fold increase in contraction amplitude compared to nonstimulated constructs. In addition, stimulated grafts demonstrated better cellular organization and increased cardiomyocyte specific proteins (Cx43, aMHC, bMHC, cardiac a-actin, and troponin I). Ultrastructurally, stimulated grafts matured more rapidly with well-organized sarcomeres, adherens junctions, and a high concentration of mitochondria. The maximal thickness of these stimulated constructs was only 100 μm, probably because of the limited oxygen diffusion. Limited diffusion may be the limiting factor for this approach, and the implementation of neovascularization strategies may permit constructs with greater thickness.

62.4.2
Smooth Muscle Cells for the Replacement or Repair of Cardiac Muscle

In our laboratory, we have completed studies to evaluate the efficacy of TECGs composed of vascular SMCs in combination with various scaffolds to repair either a defect in the RVOT or a myocardial infarct (MI). We hypothesized that SMCs may be the most appropriate cell type for cardiac repair because they possess contractile properties and induce rapid ECM production, which stabilizes cardiac geometry and prevents cardiac dilatation and congestive heart failure. In 2002, we used syngeneic rat aortic SMCs engrafted into three different biodegradable scaffolds (PCLA, GEL, PGA) to repair a RVOT defect [24]. PCLA patches performed better than the other biomaterials. PCLA-TECGs were associated with increased cellular penetration, increased elastin production, and decreased patch thinning, resulting in diminished cardiac dilatation, without inducing a significant inflammatory response. In a subsequent long-term study, this construct induced greater vessel density with no evidence of thinning in the RVOT [55]. Next, we evaluated the capacity of this TECG to repair infarcted segments of the LV after coronary artery ligation [54]. Compared to unseeded controls, SMC-seeded TECGs were thicker and showed evidence of prominent elastic tissue formation. The cell-seeded grafts stabilized LV geometry and prevented the increase in LV volumes seen in the non cell-seeded controls, resulting in improved LV function. These studies demonstrate that SMC-derived TECGs created with biodegradable scaffolds could be used to prevent cardiac dilatation and repair both RVOT defects and infarcted LV segments. Future studies to design TECGs of greater thickness and improved vascularization are warranted.

62.4.3
Skeletal Myoblasts for the Replacement or Repair of Cardiac Muscle

The major issue with skeletal myoblast cell transplantation has been limited cell engraftment and survival. Tissue engineering approaches aim to resolve these issues because scaffolds can improve these outcomes. In a study by Memon et al. [21], a direct comparison was performed between cell transplantation and cell-seeded TECGs, which demonstrated improved cell engraftment and survival in TECGs. The authors implanted two-layers of autologous myoblast sheets onto the surface of the infarcted rat heart, and compared this treatment with myoblast injections into the infarcted area. They found tissue-engineered grafts reduced fibrosis, increased infarct cellularity, and improved cardiac function compared to the cell injections. They also demonstrated striking differences in HSC recruitment to the infarcted area after TECG application.

Skeletal myoblasts have also been used in other studies to repair cardiac defects. Fuchs et al. [146] isolated skeletal myoblasts from fetal lambs and seeded them into collagen hydrogels, which were then implanted after birth over the left ventricle. After 30 weeks, troponin I-positive cells were found within the construct, suggesting transdifferentiation of the cells into a cardiomyocyte-like phenotype. This approach clearly demonstrates the efficacy of using skeletal myoblasts as an autologous cell source for repair of cardiac defects. But translation of this strategy into the clinical setting may be problematic.

62.4.4
Mesenchymal Stem Cells for the Replacement or Repair of Cardiac Muscle

The potential of MSCs to differentiate to a cardiac phenotype has been demonstrated [10, 12]. Krupnick et al. combined bone-marrow-derived MSCs with collagen and Matrigel, seeded the mixture onto a porous poly(L-lactide acid) and poly(tetrafluoroethylene) composite, and implanted the TECG into the infarcted region after a ventriculotomy [113]. The authors showed reduced aneurysm formation and improved cardiac geometry. But bone marrow is not the only source for MSCs; adipose tissue has also been consistently demonstrated to harbor these cells. Miyahra et al. used adipose-derived MSCs to repair infarcted segments in rat hearts [147]. Monolayered MSC-TECG were created and implanted onto the epicardial surface of the scarred myocardium. The grafts showed evidence of in situ growth with increased thickness, vascularization, and some cardiomyocyte differentiation. More importantly, at 4 weeks after implantation, the grafts improved LV geometry, function and survival [147].

62.4.5
Embryonic Stem Cells for the Replacement or Repair of Cardiac Muscle

Embryonic stem cells have consistently been demonstrated to differentiate into cardiomyocytes. However, differences in cardiomyocyte yield have been reported depending on species and clones [99, 148, 149], and future studies to determine suitable methods of creating TECGs using ESCs are warranted. Also, the use of ESCs in clinical practice may be problematic. Precommitment of these cells has been found to limit teratoma formation [97], but inclusion of just one undifferentiated cell with the ESC graft could, at least in theory, result in teratoma growth. The major concern with the use of ESCs for cardiac repair is the continuing ethical debate about the harvesting of human embryos to create these constructs.

Current literature indicates that ESCs can be successfully used to create tissue-engineered constructs. In one study by Guo et al. [150], embryoid body directed differentiation from mouse ES-D3 clones with 1% ascorbic acid was achieved. The cells were enriched with a Percoll gradient, mixed with liquid type I collagen, and the mixture was cast into a ring-like structure. The graft was then matured in vitro, exposed to unidirectional stretch cycles for 7 days. The authors found regular contraction intervals in organ baths, and the ECTs (engineered cardiac tissue) responded to pharmaceutical interventions. ECTs demonstrated the ultrastructural hallmarks of immature neonatal cardiomyocytes: sarcomeric organization, a high density of mitochondria, adherens junctions, desmosomes, Z bands, and gap junctions. Immunohistochemical analysis confirmed cardiomyocyte survival and contractility with troponin-I+ after subcutaneous implantation. No teratomas were reported. Interestingly, ECTs were composed of fibroblasts, endothelial cells, and nestin-positive neural cells, demonstrating that the ES cell enrichment still allowed for organoid multilineage diversification. This is an important characteristic of this strategy, which would reduce concerns of vascularization and extracellular matrix maturation and maintenance as described by other studies [151, 152].

Kofidis et al. employed both a rat heterotopic transplant model [153] and a mouse infarct model [119] to demonstrate that in situ TECG created with the coinjection of ESCs and Matrigel was more effective at improving cardiac geometry and function than ESCs or Matrigel alone. Studies by the Robbins group at Stanford University demonstrate that an injectable matrix can reduce cell loss and improve ESC engraftment.

Caspi of Levenberg's group created a vascularized, 3D tissue-engineered human cardiac muscle in vitro [151]. The construct was derived from hESCs H9.2 clone-derived cardiomyocytes, HUVECs, hESC-derived ECs, and embryonic fibroblasts seeded into porous biodegradable sponges consisting of 50% poly-L-lactic acid and 50% polylactic-glycolic acid. In this landmark study, the authors described the cardiac-specific molecular (RT-PCR of immature and mature cardiomyocytes), ultrastructural (Z-bodies, presence of mitochondria, t-tubules, sarcoplasmic reticulum, desmosomes, and Cx43), and functional (impulse propagation with calcium imaging) properties. Unique to this study was the demonstration that coculturing with cardiomyocytes, endothelial cells, and embryonic fibroblasts resulted in improved vascularization. Embryonic fibroblasts appeared to promote vessel maturation by contributing to the mural cell population. The embryonic fibroblasts also stabilized the TECG, increased EC proliferation, reduced EC apoptosis, upregulated cardiac-specific genes, and increased cardiomyocyte proliferation. This model may circumvent the limitations of metabolite diffusion because of its high vascularization. The study emphasizes the potential of embryonic cells to induce all of the structures necessary to create a cardiac construct designed to repair or replace damaged or impaired myocardium.

62.4.6
Limitations and Future Perspectives on Tissue-Engineered Cardiac Muscle

Neonatal and fetal cardiomyocytes provide proof-of-concept for tissue engineering applications, but there are concerns associated with these cell types. First, both are better suited as autologous sources for tissue to treat children born with heart defects. Further, cells taken from a biopsy from the fetus or the neonate must be expanded and scaled for human applications. Finally, the amount of tissue that can be obtained from in utero biopsies is currently limited [146]. Therefore, studies with cardiomyocytes derived from fetal or neonatal biopsies should address these concerns.

Tissue engineering is associated with two common underlying concerns for clinical application: contractility and thickness [1]. Both the thickness and the contractility of the derived cardiac tissue are dependent on the vascularity of the construct. Several recent studies have suggested new methods to circumvent these problems. Studies from the Zimmermann and Eschenhagen groups have made significant contributions to the field of cardiac tissue engineering with distinct methods to create engineered heart tissue with different tailor-made constructs [5, 6, 25, 107, 116, 154–156]. These constructs demonstrated contractility in vitro and after implantation in a rat model of myocardial infarction, along with improved cardiac geometry and function. The improvements were associated with consistent graft integration, no apparent development of arrhythmias, maintenance and maturation of the cardiomyocyte phenotype, and vascularization of the construct. Therefore, engineered heart tissue can provide a measurable functional benefit. Yet, as with every technology, there are drawbacks. Engineered heart tissue was initially constructed with serum and Matrigel, which may not be applicable for the clinical situation [116]. Recently, the group created their engineered heart tissue in serum- and Matrigel-free conditions with promising results [152]. In the same study, the authors also demonstrated improvements in ESC culture with the addition of nonmyocyte cardiac cells. Their results are supported by the recent studies from Caspi et al. [151]. In addition, the authors are evaluating different sources of cardiogenic cell types because of the inherent limitations of neonatal cardiomyocytes to repair infarcts [157]. The next generation of engineered heart tissue may offer great promise.

Other groups have also made significant steps towards creating unique tissue-engineered grafts. The Okano laboratory pioneered a unique layering strategy to create mono- and multilayer grafts by employing temperature-responsive culture dishes [110, 143, 158]. Neonatal cardiomyocytes [143, 158] and adipose-derived MSCs [147] were used to create the grafts. As discussed above, adipose-derived MSC grafts became vascularized and improved cardiac geometry and function compared to untreated controls. Graft thickness was a concern due to the increasing appearance of necrotic cardiomyocytes in the multilayered sheets that demonstrated the need for vascularization to maintain tissue integrity and contractility. A novel polysurgery technique was developed that placed grafts subcutaneously at daily intervals [159]. In this manner, the grafts sustained thicknesses of 1 mm in comparison to the maximum thickness of single subcutaneous implantations (about 80 µm). The potential for this technology to improve cardiac repair and incite regeneration has been demonstrated using skeletal myoblasts [21, 42], cardiomyocytes [110, 143, 144, 158–161], and MSCs [147] in small and large animal models. Therefore, the efficacy of this unique system may allow for a multidimensional approach whereby grafted cell types could be customized for a specific clinical situation.

Although results have been encouraging, the issues involved with strategies discussed in this section demonstrate the many challenges in creating a tissue-engineered cardiac tissue. To date, no single technique has been shown to generate tissue with all the desirable characteristics of TECGs: consistent and meaningful contractility, stable electrophysiological properties, vascularization, and an autologous cell source. Hybrid approaches using appropriate biomaterials, ESC- or MSC-derived cardiomyocytes and noncardiomyocyte supporting cells, possibly from the host, may represent meaningful combinations. In addition, grafts used to treat myocardial infarcts and congenital heart defects may be required to have different properties due to the divergent nature of these two conditions. Since much of the success thus far has been achieved with injectable liquid biomaterials [5], overcoming the requirement for a rigid tissue graft to repair a cardiac defect should be addressed.

62.5 Tissue-Engineered Cardiovascular Structures

The previous sections focused on tissue-engineered myocardial grafts, but other cardiovascular structures have also been studied, such as vascular grafts and heart valves. The following sections will briefly examine the many strategies employed over the past decade to create these tissues (Tables 62.5, 62.6).

Table 62.5 Representative studies of tissue engineered vascular grafts

Author	Year published	Ref.	Model	Cell type	Mode of engineering	Scaffold type	Bioreactor type	Implantation site
Shinoka et al.	1998	166	Lamb	Vascular cells	Autologous	Synthetic	Static	Pulmonary artery
Shum-Tim et al.	1999	170	Lamb	Vascular cells	Autologous	Synthetic	Static	Abdominal aorta
Teebken et al.	2000	171	In vitro	Human vascular cells	N/A	Decellularized porcine aorta	Biomimetic	None
Clarke et al.	2001	172	Dog	None	Allogenic	Decellularized bovine ureter	Static	Abdominal aorta
Watanabe et al.	2001	173	Dog	Vascular cells	Autologous	Synthetic	Static	Inferior vena cava
Hoerstrup et al.	2002	167	In vitro	Human umbilical cord cells	N/A	Synthetic	Biomimetic	None
Matsumura et al.	2003	174	Dog	Bone marrow cells	Autologous	Synthetic	Static	Inferior vena cava
Cho et al.	2005	168	Dog	Bone marrow cells	Autologous	Decellularized canine vena cava	Static	Inferior vena cava
Hibino et al.	2005	169	Dog	Bone marrow cells	Autologous	Synthetic	Static	Inferior vena cava
Iwai et al.	2005	175	Pig, dog	None	None	Synthetic	Static	Abdominal aorta, PA, RVOT
Hoerstrup et al.	2006	176	Lamb	Vascular cells	Autologous	Synthetic	Biomimetic	Pulmonary artery
Cho et al.	2006	177	Dog	Bone marrow cells	Autologous	Decellularized canine aorta	Static	Abdominal aorta
Sekine et al.	2006	181	Rat	Neonatal cardiomyocytes	Syngeneic	NA	In vivo	Thoracic aorta into abdominal aorta

Table 62.6 Representative studies of tissue engineered cardiac valves

Author	Year published	Ref.	Model	Cell type	Mode of engineering	Scaffold type	Bioreactor type	Implantation site
Shin'oka et al.	1995	189	Lamb	Vascular cells	Autologous/allogenic	Synthetic	Static	Pulmonic valve leaflet
Shin'oka et al.	1996	190	Lamb	Vascular cells	Autologous	Synthetic	Static	Pulmonic valve leaflet
Bader et al.	1998	191	In vitro	Human vascular cells	N/A	Decellularized xenogenic valve	Static	None
Hoerstrup et al.	2000	187	Lamb	Vascular cells	Autologous	Synthetic	Biomimetic	Pulmonic valve
Matheny et al.	2000	192	Pig	Host cells	Autologous	Decellularized SIS	None	Pulmonic valve leaflet
Sodian et al.	2000	193	Lamb	Vascular cells	Autologous	Synthetic	Biomimetic	Pulmonic valve
Sodian et al.	2000	194	In vitro	Vascular cells	N/A	Synthetic	Biomimetic	None
Steinhoff et al.	2000	195	Sheep	Vascular cells	Autologous	Decellularized allogenic valve	Static	Pulmonic valve
Stock et al.	2000	196	Lamb	Vascular cells	Autologous	Synthetic	Static	Pulmonic valve
Cebotari et al.	2002	197	In vitro	Human vascular cells	N/A	Decellularized allogenic valve	Biomimetic	None
Hoerstrup et al.	2002	188	In vitro	Human BMCs	N/A	Synthetic	Biomimetic	None
Dohmen et al.	2003	198	Sheep	Vascular cells	Autologous	Decellularized xenogenic valve	Static	Pulmonic valve
Leyh et al.	2003	199	Sheep	Vascular cells	Autologous	Decellularized xenogenic valve	Static	Pulmonic valve
Leyh et al.	2003	200	Sheep	Host cells	Allogenic/xenogenic	Decellularized allogenic/xenogenic valve	Static	Pulmonic valve
Perry et al.	2003	201	In vitro	Sheep BMCs	N/A	Synthetic	Biomimetic	None
Schenke-Layland et al.	2003	202	In vitro	Vascular cells	N/A	Decellularized xenogenic valve	Biomimetic	None
Knight et al.	2005	203	In vitro	Human mesenchymal stem cells	N/A	Decellularized xenogenic valve	Static	None
Dohmen et al.	2006	204	Sheep	Vascular cells vs. None	Autologous	Decellularized xenogenic valve	Static	Pulmonic valve

Table 62.6 *(continued)* Representative studies of tissue engineered cardiac valves

Author	Year published	Ref.	Model	Cell type	Mode of engineering	Scaffold type	Bioreactor type	Implantation site
Lichtenberg et al.	2006	205	Lamb	Vascular cells vs. None	Autologous	Decellularized xenogenic valve	Static	Pulmonic valve
Schmidt et al.	2006	206	In vitro	Human fetal progenitor cells	N/A	Synthetic	Biomimetic	None
Bin et al.	2006	207	In vitro	Human mesenchymal stem cells	N/A	Decellularized allogenic valve	Static	None
Takagi et al.	2006	208	Dog	None	N/A	Decellularized xenogenic (rabbit) valve	Static	Abdominal aorta (aortic valve)

62.5.1
Vascular Grafts

Larger-caliber conduits with growth potential are particularly applicable for the repair of congenital heart defects. Synthetic vascular grafts, such as Dacron or expanded polytetrafluoroethylene (ePTFE) are currently used for surgical repair. Although they function reasonably well in high-flow, low-resistance conditions, they are limited by the risk of thromboembolism, infection and lack of growth as the recipient child becomes an adult. Initial efforts to seed synthetic grafts with autologous endothelial cells were limited by inadequate cell retention after in vivo placement. Clinical studies have also revealed conflicting midterm graft patency rates using large conduits [162].

Allografts and homografts, though biological, have limited or no viability and can calcify, inducing stenosis of the graft. Clinical demands have stimulated efforts to create a vascular graft created from autologous cells and a biodegradable scaffold. Construction of a bioengineered vascular conduit was first reported by Weinberg and Bell in 1986, using collagen and bovine vascular cells [163]. Subsequently, a number of studies have been performed to create large-diameter vascular grafts [164–175] (Table 62.5). For example, autologous vascular cells and bone marrow cells were seeded into synthetic scaffolds in a static bioreactor system [164, 167, 168, 171, 172, 174, 175]. Short- to midterm results showed that bioengineered grafts could survive in vivo, with histological evidence of remodeling. Using decellularized porcine aorta as a natural scaffold, Teebken et al. seeded human vascular cells in vitro and subjected the vascular construct to pulsatile flow [169]. The porcine matrix was seeded with a monolayer of human endothelial cells, and the vascular graft sustained biomechanical stability under in vitro physiologic conditions. Hoerstrup et al. similarly seeded human umbilical cord cells in a bioabsorbable polymeric tubular scaffold in vitro in a pulsatile bioreactor [165]. The morphological features and mechanical stability of the tubular construct were found to be similar to those of a human pulmonary artery, but ECM components and tissue elasticity were reduced. The same group recently reported a 100-week implantation of a vascular-seeded graft in the lamb pulmonary artery [174]. The graft demonstrated significant growth in length and diameter, with histological and mechanical properties similar to those of the native vessel. They found no evidence of thrombosis, calcification or neointimal formation, and the construct appears promising for future clinical application.

Shin'oka et al. recently reported on the first clinical series of tissue-engineered vascular autografts for the repair of congenital heart defects [176–178]. In one trail, 42 patients received an autologous BMC-derived biodegradable patch or conduit. The biomaterial was composed of a co-polymer of L-lactide and e-caprolactone reinforced with a woven PGA fabric. Midterm follow-up (mean of 490 days with a maximum of 31 months) revealed a single mortality unrelated to graft function, and no morbidity or prosthesis-related complications. Follow-up angiography and computed tomography revealed two patch stenose, but no dilation or rupture of the implanted autografts. All conduits were patent without signs of calcification and demonstrated an increase in graft diameter. However, histological evidence of BMC survival, differentiation, or proliferation within the autograft could not be evaluated. While the results were encouraging, longer follow-up will be necessary to assess the feasibility of this procedure.

Finally, one strategy deserves attention because it documents a unique approach for circulatory support. Sekine et al. constructed myocardial tubes from neonatal rat cardiomyocytes that were grafted over a resected portion of the thoracic aorta, and then implanted then into the abdominal aortas of athymic rats. The pulsatile grafts demonstrated differentiated cardiomyocytes with contractions independent of the host heart. This method demonstrates a novel approach for cardiac assistance using engineered cardiac tissue [179].

62.5.2
Heart Valves

Despite the emergence of valve repair to preserve native anatomy and physiology, valve replacement remains the most common treatment for advanced valvular disease [180]. Both mechanical and biological prostheses have been developed for clinical use. Mechanical valves offer the advantage of durability, but suffer from potential complications of mechanical failure, thromboembolism, and the side effects of life-long anticoagulation [181]. Bioprosthese offer the advantage of a more physiologic hemodynamic profile and avoid anticoagulation, but may suffer from a limited lifespan secondary to calcification and structural deterioration [182]. This limitation is particularly relevant in younger patients given their higher susceptibility to calcification. Cryopreserved cadaveric homografts offer better biocompatibility than mechanical or traditional bioprosthese, but have the disadvantage of chronic rejection, limited donor supply, and early failure in the pediatric population [183, 184]. Most importantly, none of the currently-available valve prostheses have growth potential. Pediatric patients often require multiple reoperations for prosthesis-related valvular stenosis, causing additional mortality and morbidity. To overcome these limitations, the development of an autologous, biological valve has been an intense focus of cardiovascular tissue engineering investigations [185–206] (Table 62.6).

As with the construction of bioengineered vascular conduits, two main approaches have been pursued in the creation of tissue-engineered heart valves: (1) in vitro cell seeding into a biological or biodegradable synthetic scaffold, and (2) decellularization of an allogenic/xenogenic heart valve with in vitro or in vivo cell engraftment. Using the first approach, Shin'oka et al. reported the first feasibility study of that attempted to create a tissue-engineered heart valve by respectively seeding autologous and allogenic fibroblasts and endothelial cells into a biodegradable polymer, and then implanting the engineered leaflets in the pulmonic position in lambs [187]. Subsequently, a number of cell types have been seeded onto synthetic biomaterials, including arterial and venous myofibroblasts, vascular cells, umbilical cord cells, and bone marrow cells [204, 207, 208]. Although cell survival has been demonstrated in vitro and in vivo, accurate immunohistochemical characterization of engrafted valve tissue remains challenging. A number of biodegradable polymers have been evaluated as appropriate building blocks for tissue-engineered heart valves. PGA, PLA, and copolymer PLGA were found to be viable scaffolds for cell seeding and cell survival, but in vivo performance of these biomaterials was limited by inadequate flexibility of the valve leaflets [188]. Other more flexible materials such as polyhydroxyalkanoate (PHA) and poly-4-hydroxybutyrate (P4HB)-coated PGA polymers had been investigated, but as with other biodegradable synthetic scaffolds, the engineered valve leaflet surface lacked complete endothelialization following autologous pulmonic valve replacement, serving as a nidus for inflammation or thrombosis [185, 191].

The decellularization approach may reduce the host immune response to the implanted tissue-engineered heart valves. An in vivo study performed by Wilson et al. demonstrated that decellularized valve constructs could recruit some host cells [209]. Cell seeding of an acellularized valve matrix prior to implantation might create a more complete endothelial coverage on the leaflet surface and reduce the risk of thrombosis [196], but early in vitro and in vivo results indicated evidence of inflammation and calcification on the valvular constructs [189, 193]. Recent efforts have focused more on careful in vitro reseeding of decellularized valves with bone marrow or circulating progenitor cells in different bioreactor conditions, in order to improve cell alignment and biomechanical properties [199, 201, 205, 210, 211]. However, few recent investigations have evaluated short- or mid-term in vivo remodeling of these recellularized valves. Lichtenberg et al. compared the in vivo effects of decellularized ovine valves with or without vascular cell reseeding for 3 months, and found a lack of host endothelial cell coverage in the nonseeded valves, even though both valves were populated with host interstitial cells [203]. In contrast, Dohmen et al. found that both nonseeded and vascular cell-seeded valves were coated with an endothelial layer at 3 months after implantation in a sheep model [202]. Their results suggested that preseeding a decellularized valve might not be necessary. One possible explanation for this discrepancy might be differences in the decellularization techniques and animal models employed in these studies. Nonetheless, our previous reports also confirmed that endothelialization occurred in acellular scaffolds when implanted in the RVOT or the left ventricle [54, 141].

Decellularized porcine heart valves treated with SynerGraft technology were first implanted in humans in 2001 in the pulmonic position, but unfortunately the prosthesis was associated with significant mortality and morbidity, including a strong host inflammatory response and evidence of early calcification [212]. No endogenous cell repopulation was identified upon explantation. This result was in contrast with that of the SynerGraft pulmonary valve homograft, which had an explantation rate of less than 1% in two and a half years in the Ross registry series [212]. These positive results with the SynerGraft homografts were recently confirmed by Zehr et al., who reported transvalvular gradients similar to those in standard cryopreserved homografts among 22 SynerGraft patients during an average 30-month follow-up [213]. Most recently, Cebotari et al. reported the first clinical implantation of a decellularized homograft reseeded with autologous EPCs [214]. Two patients that received pulmonary valve replacements had growth of their prosthese with satisfactory function at 3.5 years. Further comparative studies, such as those designed by Lichtenberg and Dohmen, will be necessary to determine if cell reseeding is beneficial in maintaining the clinical longevity of decellularized valves. For now, it appears that reseeding may not be necessary to increase homograft durability, but may be necessary for decellularized xenografts.

The discrepancy in clinical outcomes obtained with SynerGraft xenografts and homografts might be attributable to the difference in substrate materials (porcine vs. human tissue). Rieder et al. recently reported a significant difference in monocyte response with decellularized porcine versus human pulmonary heart valves [215]. Significant protein remnants in the porcine valves may have induced the reaction. Furthermore, the same group evaluated three different decellularization techniques and found that only one could completely remove the toxic cellular fragments that inhibited human cell reseeding of the decellularized porcine matrix [216]. Both remnants in the decellularized matrix and specific decellularization protocols may play critical roles in the success of in vivo cell engraftment. Ongoing investigations in large animal models will be necessary to guide future clinical trial design.

62.6
Concluding Remarks

Tissue-engineered grafts to repair have shown great potential to repair the damaged myocardium, cardiac vasculature and heart valves. Many advances have been reported. Yet these successes have raised further questions about the clinical applicability, function, and potential clinical benefits of these constructs. Scalability of current tissue engineering strategies may limit their clinical application, and new cell sources may be necessary to achieve rapid expandability in vitro. Continuing studies will eventually provide tissue-engineered constructs to repair and replace most cardiac defects.

Acknowledgements

Most of our own research described in this chapter was supported by grants from the Heart and Stroke Foundation of Ontario and the Canadian Institutes for Health Research to R-KL. MC was supported by the Heart and Stroke Foundation of Canada (HSFC)/Transamerica Life Canada Research Fellowship Award; R-KL is a HSFC Career Investigator, and holds a Canada Research Chair in cardiac regeneration.

References

1. Leor J, Amsalem Y, Cohen S. (2005) Cells, scaffolds, and molecules for myocardial tissue engineering. Pharmacol Ther 105:151–163
2. Laflamme MA, Murry CE. (2005) Regenerating the heart. Nat Biotechnol 23:845–856
3. Murry CE, Field LJ, Menasche P. (2005) Cell-based cardiac repair: reflections at the 10-year point. Circulation 112:3174–3183
4. Dimmeler S, Zeiher AM, Schneider MD. (2005) Unchain my heart: the scientific foundations of cardiac repair. J Clin Invest 115:572–583
5. Zimmermann WH, Didie M, Doker S, Melnychenko I, Naito H, Rogge C, Tiburcy M, Eschenhagen T. (2006) Heart muscle engineering: an update on cardiac muscle replacement therapy. Cardiovasc Res 71:419–429
6. Zimmermann WH, Melnychenko I, Eschenhagen T. (2004) Engineered heart tissue for regeneration of diseased hearts. Biomaterials 25:1639–1647
7. Rabkin E, Schoen FJ. (2002) Cardiovascular tissue engineering. Cardiovasc Pathol 11:305–317
8. Davis ME, Hsieh PC, Grodzinsky AJ, Lee RT. (2005) Custom design of the cardiac microenvironment with biomaterials. Circ Res 97:8–15
9. Li RK. (2001) Cardiomyocytes. In: Human Cell Culture Vol. V, edited by Koller MR, Palsson BO and Masters JRW. Kluwer Academic Publishers, p. 103–124
10. Makino S, Fukuda K, Miyoshi S, Konishi F, Kodama H, Pan J, Sano M, Takahashi T, Hori S, Abe H, Hata J, Umezawa A, Ogawa S. (1999) Cardiomyocytes can be generated from marrow stromal cells in vitro. J Clin Invest 103:697–705
11. Badorff C, Brandes RP, Popp R, Rupp S, Urbich C, Aicher A, Fleming I, Busse R, Zeiher AM, Dimmeler S. (2003) Transdifferentiation of blood-derived human adult endothelial progenitor cells into functionally active cardiomyocytes. Circulation 107:1024–1032
12. Tomita S, Li RK, Weisel RD, Mickle DA, Kim EJ, Sakai T, Jia ZQ. (1999) Autologous transplantation of bone marrow cells improves damaged heart function. Circulation 100:II247–II256
13. Tomita S, Mickle DA, Weisel RD, Jia ZQ, Tumiati LC, Allidina Y, Liu P, Li RK. (2002) Improved heart function with myogenesis and angiogenesis after autologous porcine bone marrow stromal cell transplantation. J Thorac Cardiovasc Surg 123:1132–1140
14. Orlic D, Kajstura J, Chimenti S, Jakoniuk I, Anderson SM, Li B, Pickel J, McKay R, Nadal-Ginard B, Bodine DM, Leri A, Anversa P. (2001) Bone marrow cells regenerate infarcted myocardium. Nature 410:701–705
15. Fedak PW, Verma S, Weisel RD, Mickle DA, Li RK. (2001) Angiogenesis: protein, gene, or cell therapy? Heart Surg Forum 4:301–304
16. Huang ML, Tian H, Wu J, Matsubayashi K, Weisel RD, Li RK. (2006) Myometrial cells induce angiogenesis and salvage damaged myocardium. Am J Physiol Heart Circ Physiol 291:H2057–H2066
17. Huang NF, Yu J, Sievers R, Li S, Lee RJ. (2005) Injectable biopolymers enhance angiogenesis after myocardial infarction. Tissue Eng 11:1860–1866
18. Kim C, Li RK, Li G, Zhang Y, Weisel RD, Yau TM. (2007) Effects of cell-based angiogenic gene therapy at 6 months: persistent angiogenesis and absence of oncogenicity. Ann Thorac Surg 83:640–646
19. Yau TM, Fung K, Weisel RD, Fujii T, Mickle DA, Li RK. (2001) Enhanced myocardial angiogenesis by gene transfer with transplanted cells. Circulation 104:I218–I222
20. Yau TM, Kim C, Li G, Zhang Y, Fazel S, Spiegelstein D, Weisel RD, Li RK. (2007) Enhanced angiogenesis with multimodal cell-based gene therapy. Ann Thorac Surg 83:1110–1119
21. Memon IA, Sawa Y, Fukushima N, Matsumiya G, Miyagawa S, Taketani S, Sakakida SK, Kondoh H, Aleshin AN, Shimizu T, Okano T, Matsuda H. (2005) Repair of impaired myocardium by means of implantation of engineered autologous myoblast sheets. J Thorac Cardiovasc Surg 130:1333–1341
22. Mizuno T, Mickle DA, Kiani CG, Li RK. (2005) Overexpression of elastin fragments in infarcted myocardium attenuates scar expansion and heart dysfunction. Am J Physiol Heart Circ Physiol 288:H2819–H2827
23. Mizuno T, Yau TM, Weisel RD, Kiani CG, Li RK. (2005) Elastin stabilizes an infarct and preserves ventricular function. Circulation 112:I81–I88
24. Ozawa T, Mickle DA, Weisel RD, Koyama N, Ozawa S, Li RK. (2002) Optimal biomaterial for creation of autologous cardiac grafts. Circulation 106:I176–I182
25. Zimmermann WH, Eschenhagen T. (2003) Cardiac tissue engineering for replacement therapy. Heart Fail Rev 8:259–269
26. Huang NF, Lee RJ, Li S. (2007) Chemical and Physical Regulation of Stem Cells and Progenitor Cells: Potential for Cardiovascular Tissue Engineering. Tissue Eng
27. Zhang H, Fazel S, Tian H, Mickle DA, Weisel RD, Fujii T, Li RK. (2005) Increasing donor age adversely impacts beneficial effects of bone marrow but not smooth muscle

28. Atherton BT, Meyer DM, Simpson DG. (1986) Assembly and remodelling of myofibrils and intercalated discs in cultured neonatal rat heart cells. J Cell Sci 86:233–248
29. Beltrami AP, Barlucchi L, Torella D, Baker M, Limana F, Chimenti S, Kasahara H, Rota M, Musso E, Urbanek K, Leri A, Kajstura J, Nadal-Ginard B, Anversa P. (2003) Adult cardiac stem cells are multipotent and support myocardial regeneration. Cell 114:763–776
30. Dawn B, Stein AB, Urbanek K, Rota M, Whang B, Rastaldo R, Torella D, Tang XL, Rezazadeh A, Kajstura J, Leri A, Hunt G, Varma J, Prabhu SD, Anversa P, Bolli R. (2005) Cardiac stem cells delivered intravascularly traverse the vessel barrier, regenerate infarcted myocardium, and improve cardiac function. Proc Natl Acad Sci U S A 102:3766–3771
31. Urbanek K, Torella D, Sheikh F, De Angelis A, Nurzynska D, Silvestri F, Beltrami CA, Bussani R, Beltrami AP, Quaini F, Bolli R, Leri A, Kajstura J, Anversa P. (2005) Myocardial regeneration by activation of multipotent cardiac stem cells in ischemic heart failure. Proc Natl Acad Sci U S A 102:8692–8697
32. Urbanek K, Rota M, Cascapera S, Bearzi C, Nascimbene A, De Angelis A, Hosoda T, Chimenti S, Baker M, Limana F, Nurzynska D, Torella D, Rotatori F, Rastaldo R, Musso E, Quaini F, Leri A, Kajstura J, Anversa P. (2005) Cardiac stem cells possess growth factor-receptor systems that after activation regenerate the infarcted myocardium, improving ventricular function and long-term survival. Circ Res 97:663–673
33. Messina E, De Angelis L, Frati G, Morrone S, Chimenti S, Fiordaliso F, Salio M, Battaglia M, Latronico MV, Coletta M, Vivarelli E, Frati L, Cossu G, Giacomello A. (2004) Isolation and expansion of adult cardiac stem cells from human and murine heart. Circ Res 95:911–921
34. Smith RR, Barile L, Cho HC, Leppo MK, Hare JM, Messina E, Giacomello A, Abraham MR, Marban E. (2007) Regenerative potential of cardiosphere-derived cells expanded from percutaneous endomyocardial biopsy specimens. Circulation 115:896–908
35. Oh H, Chi X, Bradfute SB, Mishina Y, Pocius J, Michael LH, Behringer RR, Schwartz RJ, Entman ML, Schneider MD. (2004) Cardiac muscle plasticity in adult and embryo by heart-derived progenitor cells. Ann N Y Acad Sci 1015:182–189
36. Oh H, Bradfute SB, Gallardo TD, Nakamura T, Gaussin V, Mishina Y, Pocius J, Michael LH, Behringer RR, Garry DJ, Entman ML, Schneider MD. (2003) Cardiac progenitor cells from adult myocardium: homing, differentiation, and fusion after infarction. Proc Natl Acad Sci U S A 100:12313–12318
37. Laugwitz KL, Moretti A, Lam J, Gruber P, Chen Y, Woodard S, Lin LZ, Cai CL, Lu MM, Reth M, Platoshyn O, Yuan JX, Evans S, Chien KR. (2005) Postnatal isl1+ cardioblasts enter fully differentiated cardiomyocyte lineages. Nature 433:647–653
38. Cai CL, Liang X, Shi Y, Chu PH, Pfaff SL, Chen J, Evans S. (2003) Isl1 identifies a cardiac progenitor population that proliferates prior to differentiation and contributes a majority of cells to the heart. Dev Cell 5:877–889
39. Al Radi OO, Rao V, Li RK, Yau T, Weisel RD. (2003) Cardiac cell transplantation: closer to bedside. Ann Thorac Surg 75:S674–S677
40. Christman KL, Fok HH, Sievers RE, Fang Q, Lee RJ. (2004) Fibrin glue alone and skeletal myoblasts in a fibrin scaffold preserve cardiac function after myocardial infarction. Tissue Eng 10:403–409
41. Fazel S, Angoulvant D, Desai N, Yang S, Weisel RD, Li RK. (2005) Cardiac restoration: frontier or fantasy? Can J Cardiol 21:355–359
42. Hata H, Matsumiya G, Miyagawa S, Kondoh H, Kawaguchi N, Matsuura N, Shimizu T, Okano T, Matsuda H, Sawa Y. (2006) Grafted skeletal myoblast sheets attenuate myocardial remodeling in pacing-induced canine heart failure model. J Thorac Cardiovasc Surg 132:918–924
43. Murry CE, Wiseman RW, Schwartz SM, Hauschka SD. (1996) Skeletal myoblast transplantation for repair of myocardial necrosis. J Clin Invest 98:2512–2523
44. Rubart M, Soonpaa MH, Nakajima H, Field LJ. (2004) Spontaneous and evoked intracellular calcium transients in donor-derived myocytes following intracardiac myoblast transplantation. J Clin Invest 114:775–783
45. Reinecke H, Minami E, Poppa V, Murry CE. (2004) Evidence for fusion between cardiac and skeletal muscle cells. Circ Res 94:e56–e60
46. Smits PC, van Geuns RJ, Poldermans D, Bountioukos M, Onderwater EE, Lee CH, Maat AP, Serruys PW. (2003) Catheter-based intramyocardial injection of autologous skeletal myoblasts as a primary treatment of ischemic heart failure: clinical experience with six-month follow-up. J Am Coll Cardiol 42:2063–2069
47. Hagege AA, Carrion C, Menasche P, Vilquin JT, Duboc D, Marolleau JP, Desnos M, Bruneval P. (2003) Viability and differentiation of autologous skeletal myoblast grafts in ischaemic cardiomyopathy. Lancet 361:491–492
48. Menasche P, Hagege AA, Vilquin JT, Desnos M, Abergel E, Pouzet B, Bel A, Sarateanu S, Scorsin M, Schwartz K, Bruneval P, Benbunan M, Marolleau JP, Duboc D. (2003) Autologous skeletal myoblast transplantation for severe postinfarction left ventricular dysfunction. J Am Coll Cardiol 41:1078–1083
49. Pagani FD, DerSimonian H, Zawadzka A, Wetzel K, Edge AS, Jacoby DB, Dinsmore JH, Wright S, Aretz TH, Eisen HJ, Aaronson KD. (2003) Autologous skeletal myoblasts transplanted to ischemia-damaged myocardium in humans. Histological analysis of cell survival and differentiation. J Am Coll Cardiol 41:879–888
50. Herreros J, Prosper F, Perez A, Gavira JJ, Garcia-Velloso MJ, Barba J, Sanchez PL, Canizo C, Rabago G, Marti-Climent JM, Hernandez M, Lopez-Holgado N, Gonzalez-Santos JM, Martin-Luengo C, Alegria E. (2003) Autologous intramyocardial injection of cultured skeletal muscle-derived stem cells in patients with non-acute myocardial infarction. Eur Heart J 24:2012–2020
51. Li RK, Jia ZQ, Weisel RD, Merante F, Mickle DA. (1999) Smooth muscle cell transplantation into myocardial scar tissue improves heart function. J Mol Cell Cardiol 31:513–522
52. Li RK, Yau TM, Weisel RD, Mickle DA, Sakai T, Choi A, Jia ZQ. (2000) Construction of a bioengineered cardiac graft. J Thorac Cardiovasc Surg 119:368–375

53. Liu M, Montazeri S, Jedlovsky T, Van Wert R, Zhang J, Li RK, Yan J. (1999) Bio-stretch, a computerized cell strain apparatus for three-dimensional organotypic cultures. In Vitro Cell Dev Biol Anim 35:87–93
54. Matsubayashi K, Fedak PW, Mickle DA, Weisel RD, Ozawa T, Li RK. (2003) Improved left ventricular aneurysm repair with bioengineered vascular smooth muscle grafts. Circulation 108 Suppl 1:II219–II225
55. Ozawa T, Mickle DA, Weisel RD, Matsubayashi K, Fujii T, Fedak PW, Koyama N, Ikada Y, Li RK. (2004) Tissue-engineered grafts matured in the right ventricular outflow tract. Cell Transplant 13:169–177
56. Yoo KJ, Li RK, Weisel RD, Mickle DA, Li G, Yau TM. (2000) Autologous smooth muscle cell transplantation improved heart function in dilated cardiomyopathy. Ann Thorac Surg 70:859–865
57. Yoo KJ, Li RK, Weisel RD, Mickle DA, Tomita S, Ohno N, Fujii T. (2002) Smooth muscle cells transplantation is better than heart cells transplantation for improvement of heart function in dilated cardiomyopathy. Yonsei Med J 43:296–303
58. Weissman IL. (2000) Translating stem and progenitor cell biology to the clinic: barriers and opportunities. Science 287:1442–1446
59. Bradfute SB, Graubert TA, Goodell MA. (2005) Roles of Sca-1 in hematopoietic stem/progenitor cell function. Exp Hematol 33:836–843
60. Murray L, DiGiusto D, Chen B, Chen S, Combs J, Conti A, Galy A, Negrin R, Tricot G, Tsukamoto A. (1994) Analysis of human hematopoietic stem cell populations. Blood Cells 20:364–369
61. Kajstura J, Rota M, Whang B, Cascapera S, Hosoda T, Bearzi C, Nurzynska D, Kasahara H, Zias E, Bonafe M, Nadal-Ginard B, Torella D, Nascimbene A, Quaini F, Urbanek K, Leri A, Anversa P. (2005) Bone marrow cells differentiate in cardiac cell lineages after infarction independently of cell fusion. Circ Res 96:127–137
62. Quaini F, Urbanek K, Beltrami AP, Finato N, Beltrami CA, Nadal-Ginard B, Kajstura J, Leri A, Anversa P. (2002) Chimerism of the transplanted heart. N Engl J Med 346:5–15
63. Laflamme MA, Myerson D, Saffitz JE, Murry CE. (2002) Evidence for cardiomyocyte repopulation by extracardiac progenitors in transplanted human hearts. Circ Res 90:634–640
64. Murry CE, Soonpaa MH, Reinecke H, Nakajima H, Nakajima HO, Rubart M, Pasumarthi KB, Virag JI, Bartelmez SH, Poppa V, Bradford G, Dowell JD, Williams DA, Field LJ. (2004) Haematopoietic stem cells do not transdifferentiate into cardiac myocytes in myocardial infarcts. Nature 428:664–668
65. Balsam LB, Wagers AJ, Christensen JL, Kofidis T, Weissman IL, Robbins RC. (2004) Haematopoietic stem cells adopt mature haematopoietic fates in ischaemic myocardium. Nature 428:668–673
66. Jackson KA, Majka SM, Wang H, Pocius J, Hartley CJ, Majesky MW, Entman ML, Michael LH, Hirschi KK, Goodell MA. (2001) Regeneration of ischemic cardiac muscle and vascular endothelium by adult stem cells. J Clin Invest 107:1395–1402
67. Deten A, Volz HC, Clamors S, Leiblein S, Briest W, Marx G, Zimmer HG. (2005) Hematopoietic stem cells do not repair the infarcted mouse heart. Cardiovasc Res 65:52–63
68. Le Blanc K, Tammik C, Rosendahl K, Zetterberg E, Ringden O. (2003) HLA expression and immunologic properties of differentiated and undifferentiated mesenchymal stem cells. Exp Hematol 31:890–896
69. Pittenger MF, Mackay AM, Beck SC, Jaiswal RK, Douglas R, Mosca JD, Moorman MA, Simonetti DW, Craig S, Marshak DR. (1999) Multilineage potential of adult human mesenchymal stem cells. Science 284:143–147
70. Pittenger MF, Martin BJ. (2004) Mesenchymal stem cells and their potential as cardiac therapeutics. Circ Res 95:9–20
71. Shake JG, Gruber PJ, Baumgartner WA, Senechal G, Meyers J, Redmond JM, Pittenger MF, Martin BJ. (2002) Mesenchymal stem cell implantation in a swine myocardial infarct model: engraftment and functional effects. Ann Thorac Surg 73:1919–1925
72. Toma C, Pittenger MF, Cahill KS, Byrne BJ, Kessler PD. (2002) Human mesenchymal stem cells differentiate to a cardiomyocyte phenotype in the adult murine heart. Circulation 105:93–98
73. Chen SL, Fang WW, Ye F, Liu YH, Qian J, Shan SJ, Zhang JJ, Chunhua RZ, Liao LM, Lin S, Sun JP. (2004) Effect on left ventricular function of intracoronary transplantation of autologous bone marrow mesenchymal stem cell in patients with acute myocardial infarction. Am J Cardiol 94:92–95
74. Thompson CA, Nasseri BA, Makower J, Houser S, McGarry M, Lamson T, Pomerantseva I, Chang JY, Gold HK, Vacanti JP, Oesterle SN. (2003) Percutaneous transvenous cellular cardiomyoplasty. A novel nonsurgical approach for myocardial cell transplantation. J Am Coll Cardiol 41:1964–1971
75. Galinanes M, Loubani M, Davies J, Chin D, Pasi J, Bell PR. (2004) Autotransplantation of unmanipulated bone marrow into scarred myocardium is safe and enhances cardiac function in humans. Cell Transplant 13:7–13
76. Assmus B, Schachinger V, Teupe C, Britten M, Lehmann R, Dobert N, Grunwald F, Aicher A, Urbich C, Martin H, Hoelzer D, Dimmeler S, Zeiher AM. (2002) Transplantation of Progenitor Cells and Regeneration Enhancement in Acute Myocardial Infarction (TOPCARE-AMI). Circulation 106:3009–3017
77. Agbulut O, Vandervelde S, Al Attar N, Larghero J, Ghostine S, Leobon B, Robidel E, Borsani P, Le Lorc'h M, Bissery A, Chomienne C, Bruneval P, Marolleau JP, Vilquin JT, Hagege A, Samuel JL, Menasche P. (2004) Comparison of human skeletal myoblasts and bone marrow-derived CD133+ progenitors for the repair of infarcted myocardium. J Am Coll Cardiol 44:458–463
78. Dimmeler S, Zeiher AM. (2004) Wanted! The best cell for cardiac regeneration. J Am Coll Cardiol 44:464–466
79. Haider HK, Ye L, Jiang S, Ge R, Law PK, Chua T, Wong P, Sim EK. (2004) Angiomyogenesis for cardiac repair using human myoblasts as carriers of human vascular endothelial growth factor. J Mol Med 82:539–549
80. Retuerto MA, Beckmann JT, Carbray J, Patejunas G, Sarateanu S, Kane BJ, Smulevitz B, McPherson DD,

Rosengart TK. (2007) Angiogenic pretreatment to enhance myocardial function after cellular cardiomyoplasty with skeletal myoblasts. J Thorac Cardiovasc Surg 133:478–484
81. Retuerto MA, Schalch P, Patejunas G, Carbray J, Liu N, Esser K, Crystal RG, Rosengart TK. (2004) Angiogenic pretreatment improves the efficacy of cellular cardiomyoplasty performed with fetal cardiomyocyte implantation. J Thorac Cardiovasc Surg 127:1041–1049
82. Ruel M, Laham RJ, Parker JA, Post MJ, Ware JA, Simons M, Sellke FW. (2002) Long-term effects of surgical angiogenic therapy with fibroblast growth factor 2 protein. J Thorac Cardiovasc Surg 124:28–34
83. Ruel M, Sellke FW. (2003) Angiogenic protein therapy. Semin Thorac Cardiovasc Surg 15:222–235
84. Ruel M, Song J, Sellke FW. (2004) Protein-, gene-, and cell-based therapeutic angiogenesis for the treatment of myocardial ischemia. Mol Cell Biochem 264:119–131
85. Hristov M, Erl W, Weber PC. (2003) Endothelial progenitor cells: mobilization, differentiation, and homing. Arterioscler Thromb Vasc Biol 23:1185–1189
86. Rehman J, Li J, Orschell CM, March KL. (2003) Peripheral blood "endothelial progenitor cells" are derived from monocyte/macrophages and secrete angiogenic growth factors. Circulation 107:1164–1169
87. Fazel S, Cimini M, Chen L, Li S, Angoulvant D, Fedak P, Verma S, Weisel RD, Keating A, Li RK. (2006) Cardioprotective c-kit+ cells are from the bone marrow and regulate the myocardial balance of angiogenic cytokines. J Clin Invest 116:1865–1877
88. Wang CH, Anderson N, Li SH, Szmitko PE, Cherng WJ, Fedak PW, Fazel S, Li RK, Yau TM, Weisel RD, Stanford WL, Verma S. (2006) Stem cell factor deficiency is vasculoprotective: unraveling a new therapeutic potential of imatinib mesylate. Circ Res 99:617–625
89. Asahara T, Kawamoto A. (2004) Endothelial progenitor cells for postnatal vasculogenesis. Am J Physiol Cell Physiol 287:C572–C579
90. Peichev M, Naiyer AJ, Pereira D, Zhu Z, Lane WJ, Williams M, Oz MC, Hicklin DJ, Witte L, Moore MA, Rafii S. (2000) Expression of VEGFR-2 and AC133 by circulating human CD34(+) cells identifies a population of functional endothelial precursors. Blood 95:952–958
91. Gehling UM, Ergun S, Schumacher U, Wagener C, Pantel K, Otte M, Schuch G, Schafhausen P, Mende T, Kilic N, Kluge K, Schafer B, Hossfeld DK, Fiedler W. (2000) In vitro differentiation of endothelial cells from AC133-positive progenitor cells. Blood 95:3106–3112
92. Gruh I, Beilner J, Blomer U, Schmiedl A, Schmidt-Richter I, Kruse ML, Haverich A, Martin U. (2006) No evidence of transdifferentiation of human endothelial progenitor cells into cardiomyocytes after coculture with neonatal rat cardiomyocytes. Circulation 113:1326–1334
93. Kehat I, Kenyagin-Karsenti D, Snir M, Segev H, Amit M, Gepstein A, Livne E, Binah O, Itskovitz-Eldor J, Gepstein L. (2001) Human embryonic stem cells can differentiate into myocytes with structural and functional properties of cardiomyocytes. J Clin Invest 108:407–414
94. Laflamme MA, Gold J, Xu C, Hassanipour M, Rosler E, Police S, Muskheli V, Murry CE. (2005) Formation of human myocardium in the rat heart from human embryonic stem cells. Am J Pathol 167:663–671
95. Behfar A, Zingman LV, Hodgson DM, Rauzier JM, Kane GC, Terzic A, Puceat M. (2002) Stem cell differentiation requires a paracrine pathway in the heart. FASEB J 16:1558–1566
96. Kattman SJ, Huber TL, Keller GM. (2006) Multipotent flk-1+ cardiovascular progenitor cells give rise to the cardiomyocyte, endothelial, and vascular smooth muscle lineages. Dev Cell 11:723–732
97. Mènard C, Hagege AA, Agbulut O, Barro M, Morichetti MC, Brasselet C, Bel A, Messas E, Bissery A, Bruneval P, Desnos M, Puceat M, Menasche P. (2005) Transplantation of cardiac-committed mouse embryonic stem cells to infarcted sheep myocardium: a preclinical study. Lancet 366:1005–1012
98. Kehat I, Khimovich L, Caspi O, Gepstein A, Shofti R, Arbel G, Huber I, Satin J, Itskovitz-Eldor J, Gepstein L. (2004) Electromechanical integration of cardiomyocytes derived from human embryonic stem cells. Nat Biotechnol 22:1282–1289
99. Mummery C, Ward D, van den Brink CE, Bird SD, Doevendans PA, Opthof T, Brutel dlR, Tertoolen L, van der HM, Pera M. (2002) Cardiomyocyte differentiation of mouse and human embryonic stem cells. J Anat 200:233–242
100. Kofidis T, deBruin JL, Tanaka M, Zwierzchoniewska M, Weissman I, Fedoseyeva E, Haverich A, Robbins RC. (2005) They are not stealthy in the heart: embryonic stem cells trigger cell infiltration, humoral and T-lymphocyte-based host immune response. Eur J Cardiothorac Surg 28:461–466
101. Kolossov E, Bostani T, Roell W, Breitbach M, Pillekamp F, Nygren JM, Sasse P, Rubenchik O, Fries JW, Wenzel D, Geisen C, Xia Y, Lu Z, Duan Y, Kettenhofen R, Jovinge S, Bloch W, Bohlen H, Welz A, Hescheler J, Jacobsen SE, Fleischmann BK. (2006) Engraftment of engineered ES cell-derived cardiomyocytes but not BM cells restores contractile function to the infarcted myocardium. J Exp Med 203:2315–2327
102. Nussbaum J, Minami E, Laflamme MA, Virag JA, Ware CB, Masino A, Muskheli V, Pabon L, Reinecke H, Murry CE. (2007) Transplantation of undifferentiated murine embryonic stem cells in the heart: teratoma formation and immune response. FASEB J 21:1345–1357
103. Swijnenburg RJ, Tanaka M, Vogel H, Baker J, Kofidis T, Gunawan F, Lebl DR, Caffarelli AD, de Bruin JL, Fedoseyeva EV, Robbins RC. (2005) Embryonic stem cell immunogenicity increases upon differentiation after transplantation into ischemic myocardium. Circulation 112:I166–I172
104. Xie CQ, Zhang J, Xiao Y, Zhang L, Mou Y, Liu X, Akinbami M, Cui T, Chen YE. (2007) Transplantation of human undifferentiated embryonic stem cells into a myocardial infarction rat model. Stem Cells Dev 16:25–29
105. Langer R, Tirrell DA. (2004) Designing materials for biology and medicine. Nature 428:487–492
106. Christman KL, Lee RJ. (2006) Biomaterials for the treatment of myocardial infarction. J Am Coll Cardiol 48:907–913

107. Zimmermann WH, Didie M, Wasmeier GH, Nixdorff U, Hess A, Melnychenko I, Boy O, Neuhuber WL, Weyand M, Eschenhagen T. (2002) Cardiac grafting of engineered heart tissue in syngenic rats. Circulation 106:I151–I157

108. Ozawa T, Mickle DA, Weisel RD, Koyama N, Wong H, Ozawa S, Li RK. (2002) Histologic changes of nonbiodegradable and biodegradable biomaterials used to repair right ventricular heart defects in rats. J Thorac Cardiovasc Surg 124:1157–1164

109. Leor J, Aboulafia-Etzion S, Dar A, Shapiro L, Barbash IM, Battler A, Granot Y, Cohen S. (2000) Bioengineered cardiac grafts: A new approach to repair the infarcted myocardium? Circulation 102:III56–III61

110. Shimizu T, Yamato M, Kikuchi A, Okano T. (2003) Cell sheet engineering for myocardial tissue reconstruction. Biomaterials 24:2309–2316

111. McDevitt TC, Woodhouse KA, Hauschka SD, Murry CE, Stayton PS. (2003) Spatially organized layers of cardiomyocytes on biodegradable polyurethane films for myocardial repair. J Biomed Mater Res A 66:586–595

112. Pego AP, Siebum B, Van Luyn MJ, Van Seijen XJ, Poot AA, Grijpma DW, Feijen J. (2003) Preparation of degradable porous structures based on 1,3-trimethylene carbonate and D,L-lactide (co)polymers for heart tissue engineering. Tissue Eng 9:981–994

113. Krupnick AS, Kreisel D, Engels FH, Szeto WY, Plappert T, Popma SH, Flake AW, Rosengard BR. (2002) A novel small animal model of left ventricular tissue engineering. J Heart Lung Transplant 21:233–243

114. Akhyari P, Fedak PW, Weisel RD, Lee TY, Verma S, Mickle DA, Li RK. (2002) Mechanical stretch regimen enhances the formation of bioengineered autologous cardiac muscle grafts. Circulation 106:I137–I142

115. Dar A, Shachar M, Leor J, Cohen S. (2002) Optimization of cardiac cell seeding and distribution in 3D porous alginate scaffolds. Biotechnol Bioeng 80:305–312

116. Zimmermann WH, Melnychenko I, Wasmeier G, Didie M, Naito H, Nixdorff U, Hess A, Budinsky L, Brune K, Michaelis B, Dhein S, Schwoerer A, Ehmke H, Eschenhagen T. (2006) Engineered heart tissue grafts improve systolic and diastolic function in infarcted rat hearts. Nat Med 12:452–458

117. Boublik J, Park H, Radisic M, Tognana E, Chen F, Pei M, Vunjak-Novakovic G, Freed LE. (2005) Mechanical properties and remodeling of hybrid cardiac constructs made from heart cells, fibrin, and biodegradable, elastomeric knitted fabric. Tissue Eng 11:1122–1132

118. Christman KL, Vardanian AJ, Fang Q, Sievers RE, Fok HH, Lee RJ. (2004) Injectable fibrin scaffold improves cell transplant survival, reduces infarct expansion, and induces neovasculature formation in ischemic myocardium. J Am Coll Cardiol 44:654–660

119. Kofidis T, de Bruin JL, Hoyt G, Ho Y, Tanaka M, Yamane T, Lebl DR, Swijnenburg RJ, Chang CP, Quertermous T, Robbins RC. (2005) Myocardial restoration with embryonic stem cell bioartificial tissue transplantation. J Heart Lung Transplant 24:737–744

120. Cannizzaro SM, Padera RF, Langer R, Rogers RA, Black FE, Davies MC, Tendler SJ, Shakesheff KM. (1998) A novel biotinylated degradable polymer for cell-interactive applications. Biotechnol Bioeng 58:529–535

121. Wang DA, Ji J, Sun YH, Shen JC, Feng LX, Elisseeff JH. (2002) In situ immobilization of proteins and RGD peptide on polyurethane surfaces via poly(ethylene oxide) coupling polymers for human endothelial cell growth. Biomacromolecules 3:1286–1295

122. Lee KY, Alsberg E, Hsiong S, Comisar W, Linderman J, Ziff R, Mooney DJ. (2004) Nanoscale adhesion ligand organization regulates osteoblast proliferation and differentiation. Nano Letters 4:1501–1506

123. Ryu JH, Kim IK, Cho SW, Cho MC, Hwang KK, Piao H, Piao S, Lim SH, Hong YS, Choi CY, Yoo KJ, Kim BS. (2005) Implantation of bone marrow mononuclear cells using injectable fibrin matrix enhances neovascularization in infarcted myocardium. Biomaterials 26:319–326

124. Chekanov V, Akhtar M, Tchekanov G, Dangas G, Shehzad MZ, Tio F, Adamian M, Colombo A, Roubin G, Leon MB, Moses JW, Kipshidze NN. (2003) Transplantation of autologous endothelial cells induces angiogenesis. Pacing Clin Electrophysiol 26:496–499

125. Karande TS, Ong JL, Agrawal CM. (2004) Diffusion in musculoskeletal tissue engineering scaffolds: design issues related to porosity, permeability, architecture, and nutrient mixing. Ann Biomed Eng 32:1728–1743

126. Salem AK, Stevens R, Pearson RG, Davies MC, Tendler SJ, Roberts CJ, Williams PM, Shakesheff KM. (2002) Interactions of 3T3 fibroblasts and endothelial cells with defined pore features. J Biomed Mater Res 61:212–217

127. Guan L, Davies JE. (2004) Preparation and characterization of a highly macroporous biodegradable composite tissue engineering scaffold. J Biomed Mater Res A 71:480–487

128. Wei HJ, Liang HC, Lee MH, Huang YC, Chang Y, Sung HW. (2005) Construction of varying porous structures in acellular bovine pericardia as a tissue-engineering extracellular matrix. Biomaterials 26:1905–1913

129. Botchwey EA, Dupree MA, Pollack SR, Levine EM, Laurencin CT. (2003) Tissue engineered bone: measurement of nutrient transport in three-dimensional matrices. J Biomed Mater Res A 67:357–367

130. Davis ME, Hsieh PC, Takahashi T, Song Q, Zhang S, Kamm RD, Grodzinsky AJ, Anversa P, Lee RT. (2006) Local myocardial insulin-like growth factor 1 (IGF-1) delivery with biotinylated peptide nanofibers improves cell therapy for myocardial infarction. Proc Natl Acad Sci U S A 103:8155–8160

131. Davis ME, Motion JP, Narmoneva DA, Takahashi T, Hakuno D, Kamm RD, Zhang S, Lee RT. (2005) Injectable self-assembling peptide nanofibers create intramyocardial microenvironments for endothelial cells. Circulation 111:442–450

132. Hsieh PC, Davis ME, Gannon J, MacGillivray C, Lee RT. (2006) Controlled delivery of PDGF-BB for myocardial protection using injectable self-assembling peptide nanofibers. J Clin Invest 116:237–248

133. Kofidis T, de Bruin JL, Yamane T, Tanaka M, Lebl DR, Swijnenburg RJ, Weissman IL, Robbins RC. (2005) Stimulation of paracrine pathways with growth factors enhances embryonic stem cell engraftment and host-specific differentiation in the heart after ischemic myocardial injury. Circulation 111:2486–2493

134. Sachinidis A, Gissel C, Nierhoff D, Hippler-Altenburg R, Sauer H, Wartenberg M, Hescheler J. (2003) Identification of plateled-derived growth factor-BB as cardiogenesis-inducing factor in mouse embryonic stem cells under serum-free conditions. Cell Physiol Biochem 13:423–429
135. Yamashita J, Itoh H, Hirashima M, Ogawa M, Nishikawa S, Yurugi T, Naito M, Nakao K, Nishikawa S. (2000) Flk1-positive cells derived from embryonic stem cells serve as vascular progenitors. Nature 408:92–96
136. Li TS, Hayashi M, Ito H, Furutani A, Murata T, Matsuzaki M, Hamano K. (2005) Regeneration of infarcted myocardium by intramyocardial implantation of ex vivo transforming growth factor-beta-preprogrammed bone marrow stem cells. Circulation 111:2438–2445
137. Shi Q, Rafii S, Wu MH, Wijelath ES, Yu C, Ishida A, Fujita Y, Kothari S, Mohle R, Sauvage LR, Moore MA, Storb RF, Hammond WP. (1998) Evidence for circulating bone marrow-derived endothelial cells. Blood 92:362–367
138. Wang D, Park JS, Chu JS, Krakowski A, Luo K, Chen DJ, Li S. (2004) Proteomic profiling of bone marrow mesenchymal stem cells upon transforming growth factor beta1 stimulation. J Biol Chem 279:43725–43734
139. Kinner B, Zaleskas JM, Spector M. (2002) Regulation of smooth muscle actin expression and contraction in adult human mesenchymal stem cells. Exp Cell Res 278:72–83
140. Xu W, Zhang X, Qian H, Zhu W, Sun X, Hu J, Zhou H, Chen Y. (2004) Mesenchymal stem cells from adult human bone marrow differentiate into a cardiomyocyte phenotype in vitro. Exp Biol Med (Maywood) 229:623–631
141. Sakai T, Li RK, Weisel RD, Mickle DA, Kim ET, Jia ZQ, Yau TM. (2001) The fate of a tissue-engineered cardiac graft in the right ventricular outflow tract of the rat. J Thorac Cardiovasc Surg 121:932–942
142. Birla RK, Borschel GH, Dennis RG. (2005) In vivo conditioning of tissue-engineered heart muscle improves contractile performance. Artif Organs 29:866–875
143. Shimizu T, Yamato M, Isoi Y, Akutsu T, Setomaru T, Abe K, Kikuchi A, Umezu M, Okano T. (2002) Fabrication of pulsatile cardiac tissue grafts using a novel 3-dimensional cell sheet manipulation technique and temperature-responsive cell culture surfaces. Circ Res 90:e40
144. Shimizu T, Sekine H, Isoi Y, Yamato M, Kikuchi A, Okano T. (2006) Long-term survival and growth of pulsatile myocardial tissue grafts engineered by the layering of cardiomyocyte sheets. Tissue Eng 12:499–507
145. Radisic M, Park H, Shing H, Consi T, Schoen FJ, Langer R, Freed LE, Vunjak-Novakovic G. (2004) Functional assembly of engineered myocardium by electrical stimulation of cardiac myocytes cultured on scaffolds. Proc Natl Acad Sci U S A 101:18129–18134
146. Fuchs JR, Nasseri BA, Vacanti JP, Fauza DO. (2006) Postnatal myocardial augmentation with skeletal myoblast-based fetal tissue engineering. Surgery 140:100–107
147. Miyahara Y, Nagaya N, Kataoka M, Yanagawa B, Tanaka K, Hao H, Ishino K, Ishida H, Shimizu T, Kangawa K, Sano S, Okano T, Kitamura S, Mori H. (2006) Monolayered mesenchymal stem cells repair scarred myocardium after myocardial infarction. Nat Med 12:459–465
148. Kehat I, Gepstein L. (2003) Human embryonic stem cells for myocardial regeneration. Heart Fail Rev 8:229–236
149. Xu C, Police S, Rao N, Carpenter MK. (2002) Characterization and enrichment of cardiomyocytes derived from human embryonic stem cells. Circ Res 91:501–508
150. Guo XM, Zhao YS, Chang HX, Wang CY, LL E, Zhang XA, Duan CM, Dong LZ, Jiang H, Li J, Song Y, Yang XJ. (2006) Creation of engineered cardiac tissue in vitro from mouse embryonic stem cells. Circulation 113:2229–2237
151. Caspi O, Lesman A, Basevitch Y, Gepstein A, Arbel G, Habib IH, Gepstein L, Levenberg S. (2007) Tissue engineering of vascularized cardiac muscle from human embryonic stem cells. Circ Res 100:263–272
152. Naito H, Melnychenko I, Didie M, Schneiderbanger K, Schubert P, Rosenkranz S, Eschenhagen T, Zimmermann WH. (2006) Optimizing engineered heart tissue for therapeutic applications as surrogate heart muscle. Circulation 114:I72–I78
153. Kofidis T, de Bruin JL, Hoyt G, Lebl DR, Tanaka M, Yamane T, Chang CP, Robbins RC. (2004) Injectable bioartificial myocardial tissue for large-scale intramural cell transfer and functional recovery of injured heart muscle. J Thorac Cardiovasc Surg 128:571–578
154. Zimmermann WH, Eschenhagen T. (2005) Questioning the relevance of circulating cardiac progenitor cells in cardiac regeneration. Cardiovasc Res 68:344–346
155. Zimmermann WH, Fink C, Kralisch D, Remmers U, Weil J, Eschenhagen T. (2000) Three-dimensional engineered heart tissue from neonatal rat cardiac myocytes. Biotechnol Bioeng 68:106–114
156. Zimmermann WH, Eschenhagen T. (2003) Tissue engineering of aortic heart valves. Cardiovasc Res 60:460–462
157. Zimmermann WH, Eschenhagen T. (2007) Embryonic stem cells for cardiac muscle engineering. Trends Cardiovasc Med 17:134–140
158. Shimizu T, Yamato M, Akutsu T, Shibata T, Isoi Y, Kikuchi A, Umezu M, Okano T. (2002) Electrically communicating three-dimensional cardiac tissue mimic fabricated by layered cultured cardiomyocyte sheets. J Biomed Mater Res 60:110–117
159. Shimizu T, Sekine H, Yang J, Isoi Y, Yamato M, Kikuchi A, Kobayashi E, Okano T. (2006) Polysurgery of cell sheet grafts overcomes diffusion limits to produce thick, vascularized myocardial tissues. FASEB J 20:708–710
160. Furuta A, Miyoshi S, Itabashi Y, Shimizu T, Kira S, Hayakawa K, Nishiyama N, Tanimoto K, Hagiwara Y, Satoh T, Fukuda K, Okano T, Ogawa S. (2006) Pulsatile cardiac tissue grafts using a novel three-dimensional cell sheet manipulation technique functionally integrates with the host heart, in vivo. Circ Res 98:705–712
161. Miyagawa S, Sawa Y, Sakakida S, Taketani S, Kondoh H, Memon IA, Imanishi Y, Shimizu T, Okano T, Matsuda H. (2005) Tissue cardiomyoplasty using bioengineered contractile cardiomyocyte sheets to repair damaged myocardium: their integration with recipient myocardium. Transplantation 80:1586–1595
162. Meinhart JG, Deutsch M, Fischlein T, Howanietz N, Froschl A, Zilla P. (2001) Clinical autologous in vitro endothelialization of 153 infrainguinal ePTFE grafts. Ann Thorac Surg 71:S327–S331

163. Weinberg CB, Bell E. (1986) A blood vessel model constructed from collagen and cultured vascular cells. Science 231:397–400
164. Shin'oka T, Shum-Tim D, Ma PX, Tanel RE, Isogai N, Langer R, Vacanti JP, Mayer JE, Jr. (1998) Creation of viable pulmonary artery autografts through tissue engineering. J Thorac Cardiovasc Surg 115:536–545
165. Hoerstrup SP, Kadner A, Breymann C, Maurus CF, Guenter CI, Sodian R, Visjager JF, Zund G, Turina MI. (2002) Living, autologous pulmonary artery conduits tissue engineered from human umbilical cord cells. Ann Thorac Surg 74:46–52
166. Cho SW, Park HJ, Ryu JH, Kim SH, Kim YH, Choi CY, Lee MJ, Kim JS, Jang IS, Kim DI, Kim BS. (2005) Vascular patches tissue-engineered with autologous bone marrow-derived cells and decellularized tissue matrices. Biomaterials 26:1915–1924
167. Hibino N, Shin'oka T, Matsumura G, Ikada Y, Kurosawa H. (2005) The tissue-engineered vascular graft using bone marrow without culture. J Thorac Cardiovasc Surg 129:1064–1070
168. Shum-Tim D, Stock U, Hrkach J, Shinoka T, Lien J, Moses MA, Stamp A, Taylor G, Moran AM, Landis W, Langer R, Vacanti JP, Mayer JE, Jr. (1999) Tissue engineering of autologous aorta using a new biodegradable polymer. Ann Thorac Surg 68:2298–2304
169. Teebken OE, Bader A, Steinhoff G, Haverich A. (2000) Tissue engineering of vascular grafts: human cell seeding of decellularised porcine matrix. Eur J Vasc Endovasc Surg 19:381–386
170. Clarke DR, Lust RM, Sun YS, Black KS, Ollerenshaw JD. (2001) Transformation of nonvascular acellular tissue matrices into durable vascular conduits. Ann Thorac Surg 71:S433–S436
171. Watanabe M, Shin'oka T, Tohyama S, Hibino N, Konuma T, Matsumura G, Kosaka Y, Ishida T, Imai Y, Yamakawa M, Ikada Y, Morita S. (2001) Tissue-engineered vascular autograft: inferior vena cava replacement in a dog model. Tissue Eng 7:429–439
172. Matsumura G, Miyagawa-Tomita S, Shin'oka T, Ikada Y, Kurosawa H. (2003) First evidence that bone marrow cells contribute to the construction of tissue-engineered vascular autografts in vivo. Circulation 108:1729–1734
173. Iwai S, Sawa Y, Taketani S, Torikai K, Hirakawa K, Matsuda H. (2005) Novel tissue-engineered biodegradable material for reconstruction of vascular wall. Ann Thorac Surg 80:1821–1827
174. Hoerstrup SP, Cummings Mrcs I, Lachat M, Schoen FJ, Jenni R, Leschka S, Neuenschwander S, Schmidt D, Mol A, Gunter C, Gossi M, Genoni M, Zund G. (2006) Functional growth in tissue-engineered living, vascular grafts: follow-up at 100 weeks in a large animal model. Circulation 114:I159–I166
175. Cho SW, Lim JE, Chu HS, Hyun HJ, Choi CY, Hwang KC, Yoo KJ, Kim DI, Kim BS. (2006) Enhancement of in vivo endothelialization of tissue-engineered vascular grafts by granulocyte colony-stimulating factor. J Biomed Mater Res A 76:252–263
176. Matsumura G, Hibino N, Ikada Y, Kurosawa H, Shin'oka T. (2003) Successful application of tissue engineered vascular autografts: clinical experience. Biomaterials 24:2303–2308
177. Shin'oka T, Imai Y, Ikada Y. (2001) Transplantation of a tissue-engineered pulmonary artery. N Engl J Med 344:532–533
178. Shin'oka T, Matsumura G, Hibino N, Naito Y, Watanabe M, Konuma T, Sakamoto T, Nagatsu M, Kurosawa H. (2005) Midterm clinical result of tissue-engineered vascular autografts seeded with autologous bone marrow cells. J Thorac Cardiovasc Surg 129:1330–1338
179. Sekine H, Shimizu T, Yang J, Kobayashi E, Okano T. (2006) Pulsatile myocardial tubes fabricated with cell sheet engineering. Circulation 114:I87–I93
180. Schoen FJ, Levy RJ. (1999) Tissue heart valves: current challenges and future research perspectives. J Biomed Mater Res 47:439–465
181. Vongpatanasin W, Hillis LD, Lange RA. (1996) Prosthetic heart valves. N Engl J Med 335:407–416
182. Hammermeister K, Sethi GK, Henderson WG, Grover FL, Oprian C, Rahimtoola SH. (2000) Outcomes 15 years after valve replacement with a mechanical versus a bioprosthetic valve: final report of the Veterans Affairs randomized trial. J Am Coll Cardiol 36:1152–1158
183. Rajani B, Mee RB, Ratliff NB. (1998) Evidence for rejection of homograft cardiac valves in infants. J Thorac Cardiovasc Surg 115:111–117
184. Arrington CB, Shaddy RE. (2006) Immune response to allograft implantation in children with congenital heart defects. Expert Rev Cardiovasc Ther 4:695–701
185. Hoerstrup SP, Sodian R, Daebritz S, Wang J, Bacha EA, Martin DP, Moran AM, Guleserian KJ, Sperling JS, Kaushal S, Vacanti JP, Schoen FJ, Mayer JE, Jr. (2000) Functional living trileaflet heart valves grown in vitro. Circulation 102:III44–III49
186. Hoerstrup SP, Kadner A, Melnitchouk S, Trojan A, Eid K, Tracy J, Sodian R, Visjager JF, Kolb SA, Grunenfelder J, Zund G, Turina MI. (2002) Tissue engineering of functional trileaflet heart valves from human marrow stromal cells. Circulation 106:I143–I150
187. Shin'oka T, Breuer CK, Tanel RE, Zund G, Miura T, Ma PX, Langer R, Vacanti JP, Mayer JE, Jr. (1995) Tissue engineering heart valves: valve leaflet replacement study in a lamb model. Ann Thorac Surg 60:S513–S516
188. Shin'oka T, Ma PX, Shum-Tim D, Breuer CK, Cusick RA, Zund G, Langer R, Vacanti JP, Mayer JE, Jr. (1996) Tissue-engineered heart valves. Autologous valve leaflet replacement study in a lamb model. Circulation 94:II164–II168
189. Bader A, Schilling T, Teebken OE, Brandes G, Herden T, Steinhoff G, Haverich A. (1998) Tissue engineering of heart valves—human endothelial cell seeding of detergent acellularized porcine valves. Eur J Cardiothorac Surg 14:279–284
190. Matheny RG, Hutchison ML, Dryden PE, Hiles MD, Shaar CJ. (2000) Porcine small intestine submucosa as a pulmonary valve leaflet substitute. J Heart Valve Dis 9:769–774
191. Sodian R, Hoerstrup SP, Sperling JS, Daebritz S, Martin DP, Moran AM, Kim BS, Schoen FJ, Vacanti JP, Mayer JE, Jr. (2000) Early in vivo experience with tissue-engineered trileaflet heart valves. Circulation 102:III22–III29
192. Sodian R, Hoerstrup SP, Sperling JS, Daebritz SH, Martin DP, Schoen FJ, Vacanti JP, Mayer JE, Jr. (2000) Tis-

sue engineering of heart valves: in vitro experiences. Ann Thorac Surg 70:140–144
193. Steinhoff G, Stock U, Karim N, Mertsching H, Timke A, Meliss RR, Pethig K, Haverich A, Bader A. (2000) Tissue engineering of pulmonary heart valves on allogenic acellular matrix conduits: in vivo restoration of valve tissue. Circulation 102:III50–III55
194. Stock UA, Nagashima M, Khalil PN, Nollert GD, Herden T, Sperling JS, Moran A, Lien J, Martin DP, Schoen FJ, Vacanti JP, Mayer JE, Jr. (2000) Tissue-engineered valved conduits in the pulmonary circulation. J Thorac Cardiovasc Surg 119:732–740
195. Cebotari S, Mertsching H, Kallenbach K, Kostin S, Repin O, Batrinac A, Kleczka C, Ciubotaru A, Haverich A. (2002) Construction of autologous human heart valves based on an acellular allograft matrix. Circulation 106:I63–I68
196. Dohmen PM, Ozaki S, Nitsch R, Yperman J, Flameng W, Konertz W. (2003) A tissue engineered heart valve implanted in a juvenile sheep model. Med Sci Monit 9:BR97–BR104
197. Leyh RG, Wilhelmi M, Rebe P, Fischer S, Kofidis T, Haverich A, Mertsching H. (2003) In vivo repopulation of xenogeneic and allogeneic acellular valve matrix conduits in the pulmonary circulation. Ann Thorac Surg 75:1457–1463
198. Leyh RG, Wilhelmi M, Walles T, Kallenbach K, Rebe P, Oberbeck A, Herden T, Haverich A, Mertsching H. (2003) Acellularized porcine heart valve scaffolds for heart valve tissue engineering and the risk of cross-species transmission of porcine endogenous retrovirus. J Thorac Cardiovasc Surg 126:1000–1004
199. Perry TE, Kaushal S, Sutherland FW, Guleserian KJ, Bischoff J, Sacks M, Mayer JE. (2003) Bone marrow as a cell source for tissue engineering heart valves. Ann Thorac Surg 75:761–767
200. Schenke-Layland K, Opitz F, Gross M, Doring C, Halbhuber KJ, Schirrmeister F, Wahlers T, Stock UA. (2003) Complete dynamic repopulation of decellularized heart valves by application of defined physical signals—an in vitro study. Cardiovasc Res 60:497–509
201. Knight RL, Booth C, Wilcox HE, Fisher J, Ingham E. (2005) Tissue engineering of cardiac valves: re-seeding of acellular porcine aortic valve matrices with human mesenchymal progenitor cells. J Heart Valve Dis 14:806–813
202. Dohmen PM, da Costa F, Yoshi S, Lopes SV, da Souza FP, Vilani R, Wouk AF, da Costa M, Konertz W. (2006) Histological evaluation of tissue-engineered heart valves implanted in the juvenile sheep model: is there a need for in-vitro seeding? J Heart Valve Dis 15:823–829
203. Lichtenberg A, Tudorache I, Cebotari S, Suprunov M, Tudorache G, Goerler H, Park JK, Hilfiker-Kleiner D, Ringes-Lichtenberg S, Karck M, Brandes G, Hilfiker A, Haverich A. (2006) Preclinical testing of tissue-engineered heart valves re-endothelialized under simulated physiological conditions. Circulation 114:I559–I565
204. Schmidt D, Mol A, Breymann C, Achermann J, Odermatt B, Gossi M, Neuenschwander S, Pretre R, Genoni M, Zund G, Hoerstrup SP. (2006) Living autologous heart valves engineered from human prenatally harvested progenitors. Circulation 114:I125–I131
205. Bin F, Yinglong L, Nin X, Kai F, Laifeng S, Xiaodong Z. (2006) Construction of tissue-engineered homograft bioprosthetic heart valves in vitro. Asaio J 52:303–309
206. Takagi K, Fukunaga S, Nishi A, Shojima T, Yoshikawa K, Hori H, Akashi H, Aoyagi S. (2006) In vivo recellularization of plain decellularized xenografts with specific cell characterization in the systemic circulation: histological and immunohistochemical study. Artif Organs 30:233–241
207. Kadner A, Hoerstrup SP, Tracy J, Breymann C, Maurus CF, Melnitchouk S, Kadner G, Zund G, Turina M. (2002) Human umbilical cord cells: a new cell source for cardiovascular tissue engineering. Ann Thorac Surg 74:S1422–S1428
208. Kadner A, Hoerstrup SP, Zund G, Eid K, Maurus C, Melnitchouk S, Grunenfelder J, Turina MI. (2002) A new source for cardiovascular tissue engineering: human bone marrow stromal cells. Eur J Cardiothorac Surg 21:1055–1060
209. Wilson GJ, Courtman DW, Klement P, Lee JM, Yeger H. (1995) Acellular matrix: a biomaterials approach for coronary artery bypass and heart valve replacement. Ann Thorac Surg 60:S353–S358
210. Lichtenberg A, Cebotari S, Tudorache I, Sturz G, Winterhalter M, Hilfiker A, Haverich A. (2006) Flow-dependent re-endothelialization of tissue-engineered heart valves. J Heart Valve Dis 15:287–293
211. Lichtenberg A, Tudorache I, Cebotari S, Ringes-Lichtenberg S, Sturz G, Hoeffler K, Hurscheler C, Brandes G, Hilfiker A, Haverich A. (2006) In vitro re-endothelialization of detergent decellularized heart valves under simulated physiological dynamic conditions. Biomaterials 27:4221–4229
212. Simon P, Kasimir MT, Seebacher G, Weigel G, Ullrich R, Salzer-Muhar U, Rieder E, Wolner E. (2003) Early failure of the tissue engineered porcine heart valve SYNERGRAFT in pediatric patients. Eur J Cardiothorac Surg 23:1002–1006
213. Zehr KJ, Yagubyan M, Connolly HM, Nelson SM, Schaff HV. (2005) Aortic root replacement with a novel decellularized cryopreserved aortic homograft: postoperative immunoreactivity and early results. J Thorac Cardiovasc Surg 130:1010–1015
214. Cebotari S, Lichtenberg A, Tudorache I, Hilfiker A, Mertsching H, Leyh R, Breymann T, Kallenbach K, Maniuc L, Batrinac A, Repin O, Maliga O, Ciubotaru A, Haverich A. (2006) Clinical application of tissue engineered human heart valves using autologous progenitor cells. Circulation 114:I132–I137
215. Rieder E, Seebacher G, Kasimir MT, Eichmair E, Winter B, Dekan B, Wolner E, Simon P, Weigel G. (2005) Tissue engineering of heart valves: decellularized porcine and human valve scaffolds differ importantly in residual potential to attract monocytic cells. Circulation 111:2792–2797
216. Rieder E, Kasimir MT, Silberhumer G, Seebacher G, Wolner E, Simon P, Weigel G. (2004) Decellularization protocols of porcine heart valves differ importantly in efficiency of cell removal and susceptibility of the matrix to recellularization with human vascular cells. J Thorac Cardiovasc Surg 127:399–405

Tissue Engineering Applications in Orthopedic Surgery

A. C. Bean, J. Huard

Contents

63.1 Introduction 913
63.2 Stem Cells and Orthopedic Surgery 913
63.3 Muscle-Derived Stem Cells 914
63.4 Tissue Engineering in Orthopedic Surgery 916
63.4.1 Bone 916
63.4.2 Cartilage 916
63.4.3 Muscle 917
63.5 Summary 917
References 918

63.1 Introduction

Orthopedic surgeons have traditionally utilized either mechanical devices or allograft materials to replace damaged tissues. Limitations of these methods include wear of mechanical devices over time, limited graft materials, infection, and graft rejection, all of which can result in significant morbidity and mortality. In recent years, however, the focus of treating orthopedic-related injuries or diseases has shifted towards the regeneration of functional tissues using autologous cells, which are cells obtained from the patients themselves. This method of tissue regeneration could eliminate the previous problems of mechanical device failure and graft rejection. The potential applications of tissue engineering in orthopedic surgery are vast, and include treatment for many different types of tissues. In this chapter, an overview of the different types of cells being investigated for tissue engineering in orthopedic applications will be presented, followed by an in-depth discussion of muscle-derived stem cells (MDSCs), which have been found to have potential for treatment of a variety of orthopedic-related problems.

63.2 Stem Cells and Orthopedic Surgery

Many current tissue engineering research strategies in orthopedics are focusing on the use of stem cells for regeneration of tissues. Stem cells have an increased capability for expansion in comparison to fully differentiated cells, as well as the ability to differentiate towards multiple tissue lineages [1, 2]. Two general types of stem cells that are currently being investigated are embryonic stem cells, which are derived from embryos, and postnatal stem cells. Each type of stem cell has its benefits and limitations. Embryonic stem cells have the capability of replicating indefinitely as well as the ability to differentiate into any cell type [1], while postnatal stem cells appear to be more limited in their replication and differentiation potential [3, 4]. Previous studies, however, have demonstrated that embryonic stem cells can induce an immune response as well as become tumorigenic when utilized in vivo [1], which may limit their future clinical application. Additionally, ethical concerns about obtaining stem cells from embryos have resulted in restrictions on federal funding for embryonic stem cells to research on cell lines in existence prior to August 9, 2001 [1]. While postnatal stem cells may have limited replication and differentiation potential, there are several advantages of these cells in comparison to embryonic stem cells: (1) they are easily obtained from human tissue at any age, (2) they do not create an immune response when implanted autologously into the donor, (3) they are not as tumorigenic, and (4) they do not raise substantial ethical concerns [3, 4].

Postnatal stem cells have been isolated from many different tissues, including blood [5], bone marrow [6], periosteum [4], synovial membrane [7], adipose tissue [3], and skeletal muscle [8, 9]. There is currently no consensus on whether these cells are derived from separate origins or from a single type of stem cell, but there does appear to be evidence that these stem cells may originate from blood vessels, and subsequently that the availability of stem cells correlates with the amount of angiogenesis in the regenerating tissue [10]. In orthopedic applications, stem cells isolated from bone marrow have been most extensively investigated. These cells are retrieved through bone marrow aspiration and then expanded through in vitro culture prior to injecting them back into the donor. Bone-marrow-derived stem cells (BMSCs) have shown to be capable of differentiating into osteoblasts, chondrocytes, myoblasts, and adipocytes in vitro [6], and in vivo experiments have shown that these cells have the ability to repair long bone defects [11] as well as create spinal fusions in athymic rats [12].

While BMSCs have the ability to differentiate toward various lineages, obtaining them is a somewhat invasive process, and only a limited number of cells can be acquired with each bone marrow biopsy. Due to these factors, recent attention has been given to investigating the use of stem cells obtained from fat, also called adipose tissue. Adipose-derived stem cells are easily obtained and can differentiate towards osteogenic, chondrogenic, and myogenic lineages in addition to an adipocytic cell type [3]. However, while these cells are easily obtained, they appear to have a narrower differentiation potential in comparison to BMSCs, which may restrict their future use in tissue engineering applications [13].

Our laboratory has isolated and characterized another postnatal stem cell type known as muscle-derived stem cells (MDSCs). These cells overcome the limitations of other postnatal stem cells in that they are easily obtained through a small punch biopsy, can be manipulated in vitro or in vivo, and can differentiate toward a variety of different lineages. The remainder of this chapter will focus on the use of these cells in applications for orthopedic surgery in the future.

63.3
Muscle-Derived Stem Cells

Skeletal muscle is an ideal tissue for obtaining stem cells since it is very abundant in the adult human body and cells can be easily harvested through a minimally invasive punch biopsy. Additionally, skeletal muscle has an intrinsic ability to regenerate, as it is constantly undergoing cycles of microinjury followed by repair with normal activities such as running [14].

There are several types of progenitor cells that reside in skeletal muscle. One of these cell types are satellite cells, which are quiescent under normal circumstances, lying beneath the basal lamina of the muscle. Following injury, satellite cells are activated, and then divide and fuse with other progenitor cells to form new myofibers [15]. Satellite cells have been shown to have the ability to differentiate towards osteoblastic, chondrogenic, and adipogenic lineages when stimulated with growth factors in vitro [16]. However, satellite cells express the surface markers MyoD and Pax7, indicating commitment towards a myogenic lineage, suggesting that they have limited differentiation potential. Other types of cells found within the skeletal muscle include side population (SP) cells [17] and cells that are positive for the marker $CD133^+$ [18], both of which have been shown to have the ability to engraft into skeletal muscle when injected into the circulation of mice. A detailed overview of these cell types can be found in the recent review by Peault et al. [19].

MDSCs are another type of progenitor cell that are distinct from and display several advantages over satellite cells. While satellite cells express surface markers of a myogenic lineage, MDSC surface markers include CD34 and Sca-1, which are commonly described markers of stem cells [15]. Additionally, while satellite cells appear to have limited differentiation potential, MDSCs have shown the ability to differentiate into cell types originating from all three primary germ layers, including myogenic, hematopoietic, osteogenic, chondrogenic, adipogenic, neuronal, adipogenic, and endothelial cell types [2, 20] (Fig. 63.1). MDSCs have also displayed the ability to undergo long-term proliferation and self-renewal, making them an ideal cell type for regeneration of tissues. Previous studies have also demonstrated that MDSCs can be easily manipulated through gene therapy techniques to further improve their use as tools for regeneration [17, 21].

As mentioned previously, MDSCs are obtained through a muscle biopsy. The cells obtained in the tissue are enzymatically digested and then subjected to serial plating onto collagen-coated flasks, a method known as the preplate technique (Fig. 63.2). The basis for this method is that fully differentiated cells such as fibroblasts tend to adhere to the flasks quickly, while progenitor cells, including MDSCs, adhere more slowly. The isolated cells can then be cultured and manipulated in vitro followed by in vivo implantation.

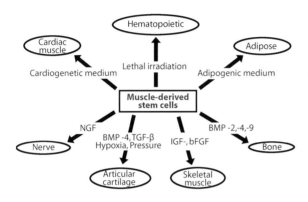

Fig. 63.1 Multipotent differentiation capacity of MDSCs

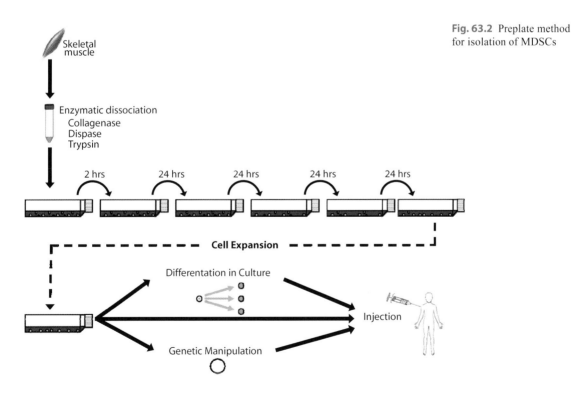

Fig. 63.2 Preplate method for isolation of MDSCs

63.4 Tissue Engineering in Orthopedic Surgery

63.4.1 Bone

While bone generally heals well, it is estimated that the rate for failure of bone repair (nonunion) and delayed bone formation may be as high as 10% [22], potentially causing clinically significant problems. Current bone substitutes include demineralized bone matrix as well as nonbiological materials such as titanium, neither of which has the same biological or mechanical properties as native, living bone. In order to regenerate bone using stem-cell-based therapies, it is useful to genetically alter the cells to be transplanted to express proteins that increase their ability to differentiate towards the osteogenic lineage. Previous research strategies have utilized genetically modified mesenchymal stem cells (MSCs) combined with ceramic scaffolds to generate bone tissue [23–26]. These experiments have demonstrated the repair of both segmental long bone defects as well as generation of spinal fusion, though the mechanical properties did not match those of uninjured bone.

MDSCs are also easily manipulated through viral transduction to express specific genes, making them an ideal cell type for bone regeneration. Bone morphogenetic proteins (BMPs), including BMP-2, BMP-4, and BMP-7, have been shown to induce bone formation when injected into ectopic areas [15]. Research in our laboratory has shown that MDSCs transduced with BMP-4 can heal critical-sized defects in rat femurs and there is no significant difference in torsional stiffness or energy-to-failure between nondamaged and healed bones [27].

Another key factor in successful bone healing is a good blood supply. Vascular endothelial growth factor (VEGF) is the most important factor in angiogenesis, and combining MDSCs that express BMP-2 or BMP-4 with MDSCs that express VEGF resulted in both greater angiogenesis as well as greater bone healing than MDSCs expressing either type of BMP alone [28, 29]. However, a high ratio of VEGF to BMP-2/4 results in decreased healing, suggesting that the ratio of the factors is critically important to maximize the synergistic effects of VEGF and BMPs [28, 29].

In addition to nonunions or delayed unions, growth of bone in undesirable locations such as in muscle, tendons or ligaments may cause clinically relevant problems. This abnormal growth of bone is called heterotopic ossification and most often occurs following trauma and surgical procedures. Studies have demonstrated that the occurrence of heterotopic ossification following total hip replacement may range from 15 to 90% in those patients experiencing clinically relevant pain, with loss of motion occurring in 1–27% of cases [30]. MDSCs transduced with the growth factor Noggin, which prevents BMP binding to its receptor, blocked expected bone formation in several environments: (1) following trauma to the Achilles tendon, (2) following insertion of a demineralized bone matrix, and (3) following implantation of MDSCs expressing BMP-4 [31]. This suggests that the use of genetically modified MDSCs may prove to be successful for the treatment of heterotopic ossification.

63.4.2 Cartilage

Treatment of damaged articular cartilage has been a particularly challenging problem in orthopedic surgery. Current treatments including arthroplasty, autologous chondrocyte implantation, microfracture and meniscal/osteochondral allografts have all proven to be relatively ineffective in long-term prevention of degenerative joint disease [32, 33]. Much of the difficulty in the repair of damaged articular cartilage is because cartilage is unable to regenerate on its own, likely due to lack of a substantial blood supply. While autologous articular chondrocytes would be ideal for repairing cartilage, their effectiveness is severely limited due to difficulty in obtaining a large number of cells as well as loss of the chondrocytic phenotype during long-term in vitro culture [33, 34]. MSCs have been thoroughly investigated for use in cartilage tissue engineering with some positive results. However, it has been shown that autologous MSCs isolated from patients with osteoarthritis have significantly less regeneration potential in comparison to those from healthy controls [35], limiting their effectiveness in patients who have already developed degenerative joint disease. Another study examined the ability of muscle-derived cells (MDCs) to heal full-thickness articular cartilage defects [36]. MDCs were shown to have similar potential to repair the articular cartilage while having increased proliferative potential in comparison to chondrocytes [36]. However, the ability of MDCs to acquire a chondrocytic phenotype was not evaluated.

Follow-up studies have shown that MDSCs transduced with BMP-4 resulted in an increased number of cells expressing type II collagen in vitro. Additionally, BMP-4-transduced MDSCs mixed with fibrin glue and injected into full thickness defects within the trochlear groove of the femur of 12-week-old rats showed histologically superior regeneration of cartilage in comparison to controls over a 24-week period [37]. The transplanted MDSCs were found to colocalize with type II collagen producing cells at 12 weeks, demonstrating that the MDSCs survive for long periods of time and acquire a chondrocytic phenotype. Ideally, application of these methods in human patients will utilize cells obtained from the patients themselves in order to eliminate the problem of host rejection. However, the clinical relevance of this study may be limited due to the fact that the MDSCs were obtained from very young animals (3-week-old mice), which may have an improved ability to regenerate tissues in comparison to older animals. Also, there is a possibility that these cells may contribute to ossification at times after 24 weeks. However, this study still demonstrates that MDSCs may potentially provide an improved treatment method for damaged and degenerating articular cartilage.

63.4.3
Muscle

Skeletal muscle injuries are a common occurrence, especially for persons participating in sports. Once activated following muscle damage, satellite cells divide and differentiate into myotubes and myofibers to replace the damaged muscle tissue. However, this regeneration is often limited by the proliferation of fibroblasts and the deposition of extracellular matrix, an undesired process known as fibrosis [38].

Recent evidence has suggested that the use of anti-inflammatory drugs including nonsteroidal anti-inflammatory drugs and cyclooxygenase-2 inhibitors may increase fibrosis following injury and therefore, these drugs should be used with caution. The hypothesized mechanism of increased fibrosis is due to the drugs' upregulation of transforming growth factor-β1 (TGF-β1), which promotes fibrosis and inhibits proliferation and differentiation of precursor cells [39]. Efforts to enhance regeneration and reduce fibrosis by blocking the effects of TGF-β1 have shown positive results. Experiments using insulin-like growth factor-1 [40], γ-interferon [41], decorin [42, 43], and suramin [44, 45] to block TGF-β1 have demonstrated a decreased amount of fibrosis and increased muscle fiber regeneration following injury in animal models.

Previous experiments have demonstrated that MDSCs are capable of directly participating in muscle healing and exhibit long-term survival in vivo. Additionally, MDSCs may be manipulated through viral transduction to overexpress the proregenerative and antifibrotic agents discussed above, thus enhancing their ability to regenerate damaged muscle. It has been shown that following laceration, TGF-β1 promotes the differentiation of MDSCs toward myofibroblasts, resulting in increased fibrosis [38]. Thus, inhibition of profibrotic factors may result in decreased fibrosis and improved regeneration. MDSCs also have demonstrated the ability to differentiate into neuronal and endothelial cells, thus enhancing nerve and blood vessel regeneration within the muscle, respectively [21]. These characteristics make MDSCs an attractive cell type for research and development of an effective treatment for muscle injuries.

63.5
Summary

Tissue engineering has become a popular field of research for the treatment of many medical conditions. Strategies in tissue engineering for orthopedic disorders including those of bone, cartilage and muscle have been intensively studied. Stem cells are particularly attractive for tissue engineering applications due to their ability to self-renew, high proliferation capacity and ability to differentiate into many different cell types. In particular, MDSCs may be especially well suited for use in tissue engineering for orthopedic applications. MDSCs are easily obtained and have demonstrated the capability to differentiate into a variety of types of tissue, including those important for orthopedics: bone, cartilage, and skeletal muscle. Additionally, they are easily manipulated to express various factors through viral transduction, which is likely important in order to generate the optimal in vivo environment for tissue regeneration. Current evidence suggests that research efforts in tissue engineering, and in particular the results that have been obtained utilizing MDSCs, may revolutionize the treatment of orthopedic injuries in the near future.

References

1. Tuan, R.S., Boland, G. & Tuli, R. Adult mesenchymal stem cells and cell-based tissue engineering. Arthritis Research and Therapy 5, 32–45 (2003)
2. Deasy, B.M., Jankowski, R.J. & Huard, J. Muscle-derived stem cells: Characterization and potential for cell-mediated therapy. Blood Cells, Molecules, and Diseases 27, 924–933 (2001)
3. Zuk, P.A. et al. Multilineage cells from human adipose tissue: Implications for cell-based therapies. Tissue Engineering 7, 211–228 (2001)
4. Nakahara, H., Goldberg, V.M. & Caplan, A.I. Culture-expanded periosteal-derived cells exhibit osteochondrogenic potential in porous calcium phosphate ceramics in vivo. Clinical Orthopedics and Related Research, 291–298 (1992)
5. Zvaifler, N.J. et al. Mesenchymal precursor cells in the blood of normal individuals. Arthritis Research 2, 477–488 (2000)
6. Pittenger, M.F. et al. Multilineage potential of adult human mesenchymal stem cells. Science 284, 143–147 (1999)
7. De Bari, C., Dell'Accio, F., Tylzanowski, P. & Luyten, F.P. Multipotent mesenchymal stem cells from adult human synovial membrane. Arthritis and Rheumatism 44, 1928–1942 (2001)
8. Qu-Petersen, Z. et al. Identification of a novel population of muscle stem cells in mice: Potential for muscle regeneration. Journal of Cell Biology 157, 851–864 (2002)
9. Gussoni, E. et al. Dystrophin expression in the mdx mouse restored by stem cell transplantation. Nature 401, 390–394 (1999)
10. Thompson, R.B. et al. Comparison of intracardiac cell transplantation: autologous skeletal myoblasts versus bone marrow cells. Circulation. 108, II264–271. (2003)
11. Bruder, S.P. et al. Bone regeneration by implantation of purified, culture-expanded human mesenchymal stem cells. Journal of Orthopedic Research 16, 155–162 (1998)
12. Peterson, B., Iglesias, R., Zhang, J., Wang, J.C. & Lieberman, J.R. Genetically modified human derived bone marrow cells for posterolateral lumbar spine fusion in athymic rats: Beyond conventional autologous bone grafting. Spine 30, 283–289 (2005)
13. Huang, J.I. et al. Chondrogenic potential of progenitor cells derived from human bone marrow and adipose tissue: A patient-matched comparison. Journal of Orthopedic Research 23, 1383–1389 (2005)
14. Irintchev, A. & Wernig, A. Muscle damage and repair in voluntarily running mice: strain and muscle differences. Cell and Tissue Research 249, 509–521 (1987)
15. Peng, H. & Huard, J. Muscle-derived stem cells for musculoskeletal tissue regeneration and repair. Transplant Immunology 12, 311–319 (2004)
16. Asakura, A., Komaki, M. & Rudnicki, M. Muscle satellite cells are multipotential stem cells that exhibit myogenic, osteogenic, and adipogenic differentiation. Differentiation; research in biological diversity 68, 245–253 (2001)
17. Gussoni, E. et al. Dystrophin expression in the mdx mouse restored by stem cell transplantation. Nature 401, 390–394 (1999)
18. Torrente, Y. et al. Human circulating AC133(+) stem cells restore dystrophin expression and ameliorate function in dystrophic skeletal muscle. The Journal of clinical investigation 114, 182–195 (2004)
19. Peault, B. et al. Stem and progenitor cells in skeletal muscle development, maintenance, and therapy. Mol Ther 15, 867–877 (2007)
20. Deasy, B.M. & Huard, J. Gene therapy and tissue engineering based on muscle-derived stem cells. Current Opinion in Molecular Therapeutics 4, 382–389 (2002)
21. Qu-Petersen, Z. et al. Identification of a novel population of muscle stem cells in mice: potential for muscle regeneration. J Cell Biol. 157, 851–864. Epub 2002 May 2020. (2002)
22. Cattermole, H.R. The footballer's fracture. British Journal of Sports Medicine 30, 171–175 (1996)
23. Arinzeh, T.L. et al. Allogeneic mesenchymal stem cells regenerate bone in a critical-sized canine segmental defect. The Journal of bone and joint surgery 85-A, 1927–1935 (2003)
24. Bruder, S.P., Kraus, K.H., Goldberg, V.M. & Kadiyala, S. The effect of implants loaded with autologous mesenchymal stem cells on the healing of canine segmental bone defects. The Journal of bone and joint surgery 80, 985–996 (1998)
25. Bruder, S.P. et al. Bone regeneration by implantation of purified, culture-expanded human mesenchymal stem cells. J Orthop Res 16, 155–162 (1998)
26. Cinotti, G. et al. Experimental posterolateral spinal fusion with porous ceramics and mesenchymal stem cells. J Bone Joint Surg Br 86, 135–142 (2004)

27. Shen, H.C. et al. Structural and functional healing of critical-size segmental bone defects by transduced muscle-derived cells expressing BMP4. Journal of Gene Medicine 6, 984–991 (2004)
28. Peng, H. et al. VEGF improves, whereas sFlt1 inhibits, BMP2-induced bone formation and bone healing through modulation of angiogenesis. J Bone Miner Res 20, 2017–2027 (2005)
29. Peng, H. et al. Synergistic enhancement of bone formation and healing by stem cell-expressed VEGF and bone morphogenetic protein-4. The Journal of clinical investigation 110, 751–759 (2002)
30. Ahrengart, L. Periarticular heterotopic ossification after total hip arthroplasty: Risk factors and consequences. Clinical Orthopedics and Related Research, 49–58 (1991)
31. Hannallah, D. et al. Retroviral Delivery of Noggin Inhibits the Formation of Heterotopic Ossification Induced by BMP-4, Demineralized Bone Matrix, and Trauma in an Animal Model. Journal of Bone and Joint Surgery—Series A 86, 80–91 (2004)
32. Minas, T. & Nehrer, S. Current concepts in the treatment of articular cartilage defects. Orthopedics 20, 525–538 (1997)
33. Brittberg, M. et al. Treatment of deep cartilage defects in the knee with autologous chondrocyte transplantation. New England Journal of Medicine 331, 889–895 (1994)
34. Magne, D., Vinatier, C., Julien, M., Weiss, P. & Guicheux, J. Mesenchymal stem cell therapy to rebuild cartilage. Trends in Molecular Medicine 11, 519–526 (2005)
35. Sethe, S., Scutt, A. & Stolzing, A. Aging of mesenchymal stem cells. Ageing research reviews 5, 91–116 (2006)
36. Adachi, N. et al. Muscle derived, cell based ex vivo gene therapy for treatment of full thickness articular cartilage defects. The Journal of rheumatology 29, 1920–1930 (2002)
37. Kuroda, R. et al. Cartilage repair using bone morphogenetic protein 4 and muscle-derived stem cells. Arthritis and Rheumatism 54, 433–442 (2006)
38. Li, Y. & Huard, J. Differentiation of muscle-derived cells into myofibroblasts in injured skeletal muscle. The American journal of pathology 161, 895–907 (2002)
39. Shen, W., Li, Y., Tang, Y., Cummins, J. & Huard, J. NS-398, a cyclooxygenase-2-specific inhibitor, delays skeletal muscle healing by decreasing regeneration and promoting fibrosis. The American journal of pathology 167, 1105–1117 (2005)
40. Sato, K. et al. Improvement of muscle healing through enhancement of muscle regeneration and prevention of fibrosis. Muscle & nerve 28, 365–372 (2003)
41. Foster, W., Li, Y., Usas, A., Somogyi, G. & Huard, J. Gamma interferon as an antifibrosis agent in skeletal muscle. J Orthop Res 21, 798–804 (2003)
42. Li, Y. et al. Transforming growth factor-beta1 induces the differentiation of myogenic cells into fibrotic cells in injured skeletal muscle: a key event in muscle fibrogenesis. The American journal of pathology 164, 1007–1019 (2004)
43. Fukushima, K. et al. The use of an antifibrosis agent to improve muscle recovery after laceration. The American journal of sports medicine 29, 394–402 (2001)
44. Chan, Y.S., Li, Y., Foster, W., Fu, F.H. & Huard, J. The use of suramin, an antifibrotic agent, to improve muscle recovery after strain injury. The American journal of sports medicine 33, 43–51 (2005)
45. Chan, Y.S. et al. Antifibrotic effects of suramin in injured skeletal muscle after laceration. J Appl Physiol 95, 771–780 (2003)

64 Tissue Engineering and Its Applications in Dentistry

M. A. Ommerborn, K. Schneider, W. H.-M. Raab

Contents

64.1	Introduction	921
64.2	Natural Tissue Regeneration of the Dentine–Pulp Complex	922
64.3	Dental Pulp Stem Cells	923
64.4	Scaffolds	925
64.5	Growth Factors	927
64.6	Gene Therapy	929
64.6.1	Viral and Nonviral Vectors	929
64.6.2	In Vivo and Ex Vivo Gene Therapy	931
64.7	Periodontal Regeneration and Approaches for Periodontal Tissue Engineering	931
64.8	Conclusions	933
	References	934

64.1 Introduction

In dentistry, caries, pulpitis, apical periodontitis, and periodontitis represent a substantial burden for the health care system worldwide. Traditionally, clinical dental medicine is based on the removal of impaired or damaged tissues and the restoration of reduced tooth function using restorative materials. However, little of them reveal chemical, biological, or physical characteristics similar to natural enamel and/or dentine. This may result in the loss of cavity restorations [80] or may contribute to microleakage at the tooth restoration interface that will allow the infiltration of bacteria into the pulp. The barrier properties of reparative dentine were found to protect the pulp tissue better than any artificial material [131].

The successful integration of preventive-oriented strategies and the increased application of endodontic treatment have prevented premature tooth loss in millions of cases. For example, a large epidemiological study in the USA analyzed 1.4 million cases and found 97% of the endodontically treated teeth to be functional over an 8-year follow-up period [108]. Other treatment approaches intervene at an earlier point of damage and intend to preserve the vitality of the dentine-pulp complex. For this purpose the pulp capping with calcium hydroxide has been used widely to induce dentine regeneration through formation of dentine bridges at sites of pulp exposure after dental tissue injury [40, 61]. It has been known for many years that the use of calcium hydroxide stimulates reparative dentinogenesis to bridge pulpal exposures in dentine. Consequently, dentistry has long been a pioneer in regenerative medicine. The results of a re-examination of patients, a minimum of 4 years after they had received direct pulp capping treatment from students, showed a success rate of 80%. In skilled hands, this rate increased up to 90% [7]. However, depending on the progression of caries and the resulting inflammation of the pulp, incompletely formed roots, fractures, apical root resorption, in many cases the vitality of a tooth or the tooth itself is not preservable. Beside conventional treatment approaches, the application of tissue engineering in dentistry presents a prospective alternative.

Since it does not only focus on the elimination of diseased tissue but also on the replacement and regeneration of the dentine-pulp complex [81, 89, 126], this evolving concept would transcend currently existing approaches.

In an early definition by Langer and Vacanti, tissue engineering was described as "an interdisciplinary field that applies the principles of engineering and life sciences toward the development of biological substitutes that restore, maintain, or improve tissue function" [65]. Depending on the scientific development over the years, definitions were modified and other aspects were emphasized. MacArthur and Oreffo defined tissue engineering as "understanding the principles of tissue growth, and applying this to produce functional replacement tissue for clinical use" [71]. Recently Murray and colleagues stated that "tissue engineering is the employment of biological therapeutic strategies aimed at the replacement, repair, maintenance, and/or enhancement of tissue function" [81]. Although each definition holds its specific accentuation, all these statements underline tissue engineering strategies to be the essential foundation for regenerative medicine that utilizes cells, engineering materials, and suitable biochemical factors to repair or replace structures damaged due to disease, trauma, or cancer. These principles can be transferred to dentistry in the field of endodontic tissue engineering as well as in the field of periodontal tissue engineering [112] and are based on three essential components of biologic tissues: responding cells such as adult stem cells, growth factors, and an extracellular matrix scaffold to maintain and preserve the microenvironment.

64.2
Natural Tissue Regeneration of the Dentine–Pulp Complex

Natural tissue regeneration requires an adequate cellular basis for regeneration. However, due to the nonvital nature of mature dental enamel and the loss of the ameloblasts that once secreted enamel during tooth development, the regeneration of enamel includes many problems with the exception of physicochemical remineralization processes. The principle of physicochemical remineralization of enamel has been extensively investigated and, nowadays, represents a well-established principle of preventive dentistry that has been routinely used in dental practice for many years. In contrast to enamel, the vitality of the dentine-pulp complex provides substantial opportunities for regeneration. However, to successfully initiate regenerative processes in the dentine–pulp complex, the presence of a conducive tissue environment is mandatory. In particular, the presence of pulpal inflammation might have a significant influence on the response of the dentine–pulp complex. Previous investigations have stressed the context between bacteria and inflammatory events in the dental pulp, and the presence of a reversible pulpitis was found to effectively prevent the initiation of regeneration [104]. Therefore, to promote natural tissue regeneration, the control of the bacterial infection and the successive inflammation in the tooth is of essential importance. Classically, the former has often been accomplished by means of extensive removal of infected dental hard tissues during restorative procedures. Modern approaches aimed toward minimal invasive therapy, however, potentially include the risk of uncertain bacterial control due to incomplete bacterial removal.

The first report of tertiary dentinogenesis following dental injury was presented by Hunter in the 18th century and since then the concept of tissue regeneration in the dentine–pulp complex has been kept in mind. Despite the apparent quiescence of odontoblast activity after primary dentinogenesis and, thus, the completion of tooth formation, regeneration or wound healing in the form of tertiary dentinogenesis can result in response to injury [134]. This provides the basis for dentine bridge formation at sites of pulp exposure. After odontoblast death at these sites, dentine bridge formation results from the differentiation of a new generation of odontoblast-like cells from pulp stem or progenitor cells. This process involves a sequence of cellular events including stem cell recruitment, cytodifferentiation, and subsequent activation or upregulation of the secretory activity of the cells and has been termed reparative dentinogenesis [67, 129]. As aforementioned, the intensity of the bacterial irritation and the consequent inflammatory response of the exposed pulp due to carious events may unfavorably influence reparative dentinogenesis and dentine bridge formation [104, 105] and, thus, may compromise pulp vitality and opportunities for repair and regeneration to take place. If the injury is only slight such as early caries, the odontoblasts underlying the injured surface survive and are upregulated to secrete a tertiary dentine matrix at the pulp–dentine interface called reactionary dentine [67, 129]. From a physiological point of view, this mechanism acts as an expedient protective function, contributing to increasing the dental hard tissue barrier between the cells of the pulp and the invading injury. In contrast to reparative dentinogenesis, the important feature of reactionary dentinogenesis is that there is no cell renewal. Thus, the cellular events during reactionary dentinogenesis are much less complex than during reparative dentinogenesis since dentine-secreting cells are already differentiated and their secretory activity simply has to be upregulated.

64.3
Dental Pulp Stem Cells

In general, stem cells are defined as clonogenic cells capable of both self-renewal and multilineage differentiation. Postnatal stem cells have been isolated from various tissues, including bone marrow, neural tissue, skin, retina, and dental epithelium [11, 12, 19, 31, 32, 48, 49, 94]. These cells offer enormous therapeutic chances for future regenerative medicine and, in particular, for regenerative endodontics. However, currently available research has left many questions unanswered that need to be addressed before these specialized cells can be routinely applied to patients. Due to the potential danger of teratoma formation the guarantee of safety and efficacy of this approach is strongly required. To receive the approval of the Food and Drug Administration (FDA), further extensive investigations are needed applying animal models, in-vitro studies, and clinical trials.

The postnatal dental pulp tissue contains a variety of cells such as pulp cells, endothelial cells, undifferentiated mesenchymal cells, fibroblasts, pericytes, odontoblast differentiation hierarchy as well as immune cells and neurocytes. During tooth formation, epithelial/mesenchymal interactions within the early tooth germ initiate the differentiation of a population of ectomesenchymal cells in the dental papilla into odontoblasts. Odontoblast cells are responsible for the secretion of primary dentine and are postmitotic terminally differentiated cells that are not able to proliferate to replace subjacent irreversibly injured odontoblasts [77]. Following primary dentinogenesis, these odontoblasts, however, maintain their function to further secrete physiological secondary dentine, even though in a reduced quantity. Moreover, they retain their ability to react to mild environmental stimuli with an increased secretory activity during reactionary dentinogenesis leading to dentinal regeneration [129]. However, after a more intense stimulus such as severe pulp damage or mechanical or caries exposure, the odontoblast population is often irreversibly injured beneath the local wound site [78, 79]. In this situation, dentine regeneration is performed by the differentiation of a new generation of odontoblast-like cells from a precursor population in a process referred to as reparative dentinogenesis [127]. The term odontoblast-like cells is due to the ability of these stem/progenitor cells to synthesize and secrete dentine matrix similar to the odontoblast cells they replace [63]. The origin of the odontoblast-like cells during reparative dentinogenesis remains vague. It was assumed that the cells within the subodontoblast cell-rich layer or zone of Höhl adjacent to the odontoblasts [22, 53], perivascular cells, undifferentiated mesenchymal cells, and fibroblasts [101] differentiate into odontoblast-like cells. In the dental pulp tissue of human adult teeth, Gronthos and colleagues have identified a putative population of postnatal stem cells, dental pulp stem cells (DPSCs) [43, 44]. The most important feature of DPSCs is their capability to differentiate into odontoblasts and adipocytes after in vivo transplantation. Moreover, their ability to regenerate a dentine–pulp-like complex that is composed of mineralized matrix with tubules lined with odontoblasts and fibrous tissue containing blood vessels in an arrangement similar to the dentine-pulp complex found in normal human teeth, has been demonstrated [43]. Collectively, these studies suggest a hierarchy of progenitors in adult dental pulp, including a minor population of self-renewing, highly proliferative, multipotent stem cells, among a larger compartment of perhaps more committed progenitors [44]. The same group has also identified a population of potential mesenchymal stem cells originating from human exfoliated deciduous teeth (SHED), capable of extensive proliferation and multipotential differentiation [74].

A further important part in the allocation of DPSCs for tissue engineering approaches is seen in the identification and isolation of these cells. However, to identify DPSCs, no specific markers are available so far [60]. In fact, it has been suggested that stem cell niche identified in many different adults tissues are often highly vascularized sites [136]. In order to identify a stem cell niche in the dental pulp, different markers of the microvasculature networks have been analyzed. Human DPSCs have been isolated by immunomagnetic-activated cell sorting using antibodies reactive to either STRO-1 or CD146 [45, 115, 116]. STRO-1 is an antibody which recognizes an antigen on perivascular cells in human bone marrow and dental pulp tissue [42] and is regarded as a marker for stromal stem cells. CD146 is a perivascular cell marker and is expressed by endothelial and smooth muscle cells. It has also been reported that DPSCs linked with pulp vasculature express CD146 and a proportion of these cells was also found to be positive for a smooth muscle actin and a pericyte-associated antigen (3G5) [45, 75].

Deciduous teeth, however, differ significantly from permanent teeth regarding their developmental processes, tissue structure, and function. Consequently, it is not surprising that SHED were found distinct from DPSCs with regard to their higher proliferation rate, increased cell-population doublings, sphere-like cell-cluster formation, osteoinductive capacity in vivo, and failure to reconstitute a dentine-pulp-like complex [74]. With respect to the expression of dentine sialophosphoprotein (DSPP), DPSCs and SHEDs show similar properties. It has been demonstrated that in xenogenic transplants containing hydroxyapatite/tricalcium phosphate engrafted into immunocompromised mice for 8 weeks, both DPSCs and SHEDs express DSPP [116]. In contrast to the former, it has been shown that in bone formed by bone marrow stromal cells in similar transplants, this expression was lacking, leading to the assumption that the clonogenic dental pulp–derived cells represent an undifferentiated preodontogenic phenotype in vitro. For many years, some proteins such as dentine phosphoprotein (DPP) and dentine sialoprotein (DSP) have long been considered as unique phenotypic markers of dentine and secretory odontoblasts [72]. In 2001, the first report appeared describing the expression of the DSPP gene outside dental tissues, in the inner ear and jaw [156]. A subsequent investigation provided strong evidence for the expression of DSPP in bone. Analyses have provided evidence that DSP in bone and dentine is very similar in nature but different in quantity. The amount of DSP and probably DPP in bone is estimated to be about 1/400 of that in dentine [95]. Dentine matrix protein-1 (DMP-1) was also expressed by both bone and dentine [34]. A group of four molecules have been grouped as a family of small integrin-binding ligand N-linked glycoproteins (SIBLINGs) [29]. These combine DSPP, and accordingly DPP and DSP, DMP-1, bone sialoprotein (BSP), and osteopontin (OPN). The highest levels of BSP and OPN are found in bone but both proteins are also present in odontoblasts and dentine. The matrix extracellular phosphorylated protein (MEPE) is a more distantly related member of the SIBLING family. Although it has been known for a while that bone marrow cells have indicated their potential to give rise to dental mesenchymal-like cells [92, 135], it has been recently found for the first time that bone marrow cells have the ability to differentiate into odontoblast-like cells and ameloblast-like cells as well. These findings offer novel opportunities for tooth-tissue engineering and the study of the simultaneous differentiation of one bone marrow cell subpopulation into cells of two different embryonic lineages [58].

An alternative approach for the isolation of a distinct population of adult pulp stem cells has also been presented [37]. Hematopoietic stem cells were first isolated from bone marrow by their ability to pump out Hoechst 33342, a DNA binding fluorescence dye. A distinct population definable by flow cytometry is called the "side population" (SP) [90]. SP cells have also been isolated from various adult tissues such as skeletal muscle, brain, liver, pancreas, lung, heart, and kidney, indicating the common features of the SP phenotype. Moreover, SP cells have been detected in adult pulp tissue from several species such as human, bovine, canine, and porcine species. As three key characteristics of the SP cells self-renewal capacity, expression of stem cells surface markers, and multilineage differentiation have been determined. It has been shown that SP cells isolated from porcine dental pulp tissue have the plasticity to differentiate into adipogenic, chondrogenic, and neurogenic lineages. Bone morphogenetic protein (BMP)-2 stimulated the expression of DSPP and enamelysin in three-dimensional pellet cultures, indicating that

BMP-2 enhanced differentiation of pulp SP cells into odontoblasts. First experiences in an animal model (dogs) with autogenous transplantation of the BMP-2-supplemented pulp SP cells on the amputated pulp stimulated the reparative dentine formation. In view of its potential clinical utility for regenerative endodontics, future research on pulp–derived SP cells in vitro and in vivo might be promising [60].

64.4 Scaffolds

The scaffold serves as a physicochemical and biological three-dimensional microenvironment for cell growth and differentiation, promoting cell adhesion, and migration (Fig. 64.1). Scaffold design is a fundamental step for any tissue engineering procedure. In order to adequately manage cells and tissues for research purposes, appropriate culture media and specific surface structures specific for every single cell type, are needed. The scaffold should serve as a carrier for growth factors such as BMPs to aid stem cell proliferation and differentiation, resulting in improved and faster tissue development [93]. Furthermore, it should allow for effective nutrition, oxygen supply, and waste removal. To prevent any bacterial in-growth in the root canal systems, the scaffold may possibly contain antibiotics. The "ideal" scaffold should be nontoxic, biocompatible, biodegradable, and have an individually structured intra- and extrageometry [50]. Biodegradability is essential, since scaffolds need to be absorbed by the surrounding tissues without the necessity of surgical removal [109]. If the tissue matrix is regenerated, the scaffold will be replaced while retaining the morphological feature of the final tissue architecture and organization. A high porosity and an adequate pore size are required to allow for easy cell seeding and diffusion of nutrients throughout the whole structure of cells [106].

The extracellular matrix of dentine represents a scaffold for adhesion, proliferation, and differentiation of odontoblasts. Pulp cells adhere to osteodentine before differentiation into odontoblasts to form tubular dentine in calcium hydroxide pulp capping [111]. Calcium hydroxide is one of the most commonly used pulp-capping agents, and although it is not acidic, it does have some solubilizing effect on dentine matrix [130]. During this process components of the dentine matrix are solubilized including transforming growth factor 1 (TGF-1). Its release may contribute to the initiation of dentine bridge formation as a result of pulp capping with this agent. Consequently, there may be a more rational biological explanation for the effect of calcium hydroxide rather then the more traditional view that was emphasized for many years that calcium hydroxide would "irritate" the pulp cells and stimulate odontoblast-like cell differentiation for dentine bridge formation [111]. Tubular matrix lined with odontoblasts is also formed when demineralized or native dentine matrix is implanted into the pulp tissue [2, 82, 83, 146]. Recombinant human BMP2 with 4M-guanidine chloride extracted inactivated demineralized dentine matrix as a scaffold induced tubular dentine formation in the cavity of the exposed pulp. On the other hand, only osteodentine formed with type I collagen as the scaffold [84, 85]. Therefore, the presence of an insoluble dentine matrix seems to be an important tool for pulp cells to attach and differentiate into odontoblasts during reparative dentinogenesis. However, the exact requirements on the physicochemical surface have, to date, not yet been elucidated. Moreover, the scaffold may exert essential mechanical and biological functions needed by replacement tissue [13]. In pulp-exposed teeth, dentine chips have been found

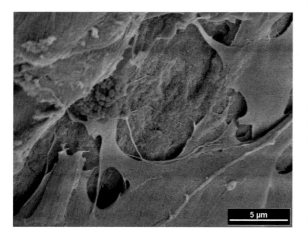

Fig. 64.1 Scanning electron microscope image of an insoluble collagenous bone matrix with murine embryonic stem cells. Image clearly demonstrates that the stem cells were located in direct contact to the scaffold surface

to stimulate reparative dentine bridge formation [63]. Dentine chips may serve as a matrix for pulp stem cell attachment [76] and also represent a reservoir of growth factors [117]. Growth factors will be described in detail in the following section.

Tooth structure can be regenerated by seeding dissociated tooth cells onto a suitable scaffold consisting of natural polymers or synthetic polymers, biodegradable or permanent. Most of the scaffold materials used in tissue engineering have been known for a long time due to their use in medicine as bioresorbable sutures and meshes applied in wound dressings [3]. To date, various biodegradable synthetic polymers such as polylactic acid (PLA), polyglycolic acid (PGA) [30, 62], polycaprolactone (PCL) [54–56], and poly(lactide-co-glycolide) acid (PLGA), which are all common polyester materials, have been successfully used as scaffolds for tissue engineering purposes since they are degradable fibrous structures with the ability to support the growth of various different stem cell types, allowing for precise control over the physicochemical properties such as degradation rate [114]. The major disadvantages of these materials are seen in the difficulty of preserving a high porosity and regular pore size. Scaffolds were also made from natural materials; in particular, several derivatives of the extracellular matrix have been investigated regarding their capability to support cell growth [150]. Proteinic materials such as collagen and fibrin [144] combine the advantages of good biocompatibility and biodegradability. In this context, chitosan should also be mentioned, since it is a biocopolymer comprising glucosamine and N-acetylglucosamine, and gelatin (Gel) and is a partially denatured derivative of collagen, which has low immunogenicity, low cost, and can be degraded entirely in vivo [47]. Collagen sponge has a number of advantages as a scaffold including its similarity in composition to the extracellular matrix, its low immunogenicity and cytotoxicity, and the efficiency with which it can form different shapes [20, 118]. The pore structure of collagen sponge was found to be beneficial for the colonization of seeded cells. Furthermore, collagen sponge was reported to enhance bone formation by promoting the differentiation of osteoblasts [113, 141]. To assess the performance of collagen sponge scaffold with respect to cell adhesion and differentiation in vitro, a recent study has compared the performance of collagen sponge scaffold with PGA fiber mesh scaffold. For this purpose, porcine third molar teeth at the early stage of crown formation were enzymatically dissociated into single cells and the heterogeneous cells were seeded either onto collagen sponge or PGA fiber mesh scaffolds. Scaffolds were then cultured to evaluate cell adhesion and alkaline phosphatase (ALP) activity, indicating the level of cell differentiation, in vitro. Moreover, to explore the potential of collagen sponge as a three-dimensional scaffold for tooth-tissue engineering, an in vivo analysis was performed by implanting the harvested constructs into the omentum of immunocompromised rats up to 25 weeks. As a result, the initial cell attachment and the ALP activity on collagen sponge scaffold were found to be higher than those of PGA scaffold after 7 days of culture. Furthermore, the comparison of the two scaffold types to regenerate tooth-like structures revealed that the success rate of tooth-tissue production was higher for the collagen sponge scaffold than for the PGA fiber mesh scaffold. Interestingly, complete tooth morphology with root-like structures was found only in the implants obtained from the collagen sponge scaffolds [139]. Previous studies using PGA fiber mesh as scaffolds have only demonstrated the formation of the tooth crown, which has been confirmed in the aforementioned investigation [57, 158]. To generate bone and tooth-like tissues, another approach was designed evaluating the utility of a hybrid tooth-bone tissue-engineering model [159]. Histological and immunohistochemical analyses of the hybrid tooth-bone constructs demonstrated collagen type III-positive connective tissue resembling periodontal ligament (PDL) and early tooth root formation and, consequently, supports the usefulness of this approach for eventual clinical treatment of tooth loss accompanied by alveolar bone resorption.

Fibronectin might mediate the binding of signaling molecules and is involved in the interactions between extracellular matrix and cells to reorganize the cytoskeleton of polarizing preodontoblasts in the pulpal wound healing process [149], whereas the aspartic acid site in fibronectin is critical for cell attachment. Synthetic extracellular matrix may also be a suitable scaffold for reparative dentinogenesis [14]. In regenerative endodontics, in particular in root canal systems, a tissue-engineered pulp is not required to provide structural support for the tooth. For this reason, hydrogels that serve as an injectable scaffold have been constructed to be applied by syringe [25, 143]. Alginate hydrogel facilitates pulpal wound healing with hydration properties and tethering growth factors [27]. Mineral trioxide aggregate (MTA) powder composed of fine hydrophilic particles of tricalcium silicate, tricalcium aluminate, tricalcium oxide, and silicate oxide has been recently investigated as a potential alternative restorative material. MTA is biocompatible, sets in the presence of moisture, prevents microleakage, and it supports reparative dentine formation [142]. Considering the role of the three-dimensional environment on cell behavior, future progress in the development of scaffolds for regenerative endodontics should be focused on the improvement of the design to regulate and facilitate dentine regeneration. Although the involvement of growth factors as components of the scaffold is likely to be critical in providing the necessary regulation of cell behavior, such tissue engineering approaches may provide very promising clinical options.

64.5
Growth Factors

Growth factors represent a group of proteins that bind to receptors on the cell and induce cellular proliferation and/or differentiation [134, 153]. These extracellularly secreted molecules play a pivotal role in signaling numerous processes of tissue engineering. In particular, events being associated with initial injury of a tissue and succeeding defense mechanisms are significantly influenced on their regulatory abilities. They may exhibit a degree of specificity regarding the type of cells they act upon, whereas many of them are quite versatile, stimulating cellular division in various cell types. Many names of the individual growth factors were attributed to the source they were originally identified from and have little to do with their main functions. Moreover, since a more widespread occurrence has often been found, several names may appear a little confusing (Table 64.1).

Since the influence of growth factors in cell signaling during tooth development and repair has been recognized, growth factors are found responsible for signaling many of the pivotal mechanisms in tooth morphogenesis and differentiation, and the consideration of these processes offers exciting opportunities for the development of new treatment approaches and, in particular, for completely novel strategies in endodontic treatment. During tooth development, tooth initiation resulting in differentiation of the odontoblasts and ameloblasts secreting dentine and enamel respectively, is due to epithelial-mesenchymal interactions. These interactions essentially affect the early morphogenetic processes in tooth development

Table 64.1 The origin and function of some commonly recognized growth factors

Name	Abbreviation	Primary origin	Function
Bone morphogenetic proteins	BMP	Bone matrix	BMP induces differentiation of osteoblasts and mineralization of bone
Epidermal growth factor	EGF	Submaxillary glands	EGF promotes proliferation of mesenchymal, glial and epithelial cells
Fibroblast growth factor	FGF	A wide range of cells	FGF promotes proliferation of many cells
Insulin-like growth factor-I or II	IGF	I- liver II-variety of cells	IGF promotes proliferation of many cell types
Interleukins IL-1 to IL-13	IL	Leukocytes	IL are cytokines which stimulate the humoral and cellular immune responses
Nerve growth factor	NGF	A protein secreted by a neuron's target tissue	NGF is critical for the survival and maintenance of sympathetic and sensory neurons
Platelet-derived growth factor	PDGF	Platelets, endothelial cells, placenta	PDGF promotes proliferation of connective tissue, glial and smooth muscle cells
Transforming growth factor-α	TGF-α	Macrophages, brain cells, and keratinocytes	TGF-α may be important for normal wound healing
Transforming growth factor-β	TGF-β	Dentine matrix, activated TH_1 cells (T-helper) and natural killer (NK) cells	TGF-β is anti-inflammatory, promotes wound healing, inhibits macrophage and lymphocyte proliferation

[21] and also the subsequent processes leading to odontoblast [100] and ameloblast [161] differentiation. Growth factors are diffusible bioactive molecules that move between the epithelial and mesenchymal compartments of the tooth germ and lead to these interactions. In particular, growth factors of the TGF-β family seem to be of major importance for signaling of odontoblast differentiation [8, 9, 100]. During the late bell stage of tooth development, the inner dental epithelium secrets these factors and, thus, signals the peripheral cells of the dental papilla to differentiate into odontoblasts. Considering these mechanisms of odontoblast differentiation during tooth development, various scientific investigators supposed similar processes during reparative dentinogenesis. However, in the mature tooth, due to the lack of dental epithelium, other signaling mechanisms have been assumed for the delivery of bioactive molecules.

Although it has been assumed for a long time that dentine is a comparatively inert matrix, many studies have been performed evaluating the potential contributions of growth factors to dental tissue repair. These investigations largely resulted from the observations that despite the absence of dental epithelium to signal odontoblast-like cell differentiation dentine matrix is autoinductive. It has been shown that demineralized dentine matrix [2, 82, 98, 145] or soluble dentine matrix extracts [127] implanted at sites of pulpal exposure as well as dentine chips arising from operative debris [137] are capable of stimulating reparative dentinogenesis. The implantation of isolated dentine matrix components in nonexposed and exposed cavities can stimulate both reactionary [128] and reparative [127, 147] dentinogenic responses, respectively. Experimental application of recombinant TGF-β1 or TGF-β3 to cultured tooth slices revealed reactionary and reparative dentinogenic responses [123]. The expression of receptors to these growth factors has been shown in odontoblasts and pulpal cells [122, 125]. This indicates that the dentine matrix contains bioactive components and is not as inert as assumed in former times. Beside members of the TGF-β family, insulin-like growth factor (IGF) I and II [18, 28] as well as various angiogenic growth factors [97] have been identified to be present in dentine matrix. Since dentine matrix has been found to contain growth factors, and allowing for their putative role in signaling odontoblast differentiation during embryogenesis, further examination of the effects of TGF-βs, BMPs, fibroblast growth factors (FGFs), and IGFs in reparative dentinogenesis has been performed [84–86, 102, 103, 123, 148]. In all of these numerous studies reparative responses have been observed ranging from initial secretion of relatively nonspecific atubular fibrodentine to apparently more specific tubular dentinogenic responses.

Another important family of growth factors in tooth development and dentinogenic reparation is presented by the BMPs. BMPs were originally isolated from demineralized bone matrix. BMP-7 (OP-1) was one of the first growth factors that have been examined regarding their influence on pulp repair. In the monkey, BMP-7 (OP-1) induced widespread osteodentine formation throughout the pulp, including the total occlusion of the roots of treated teeth [102, 103]. An irregular reparative osteodentine was deposited in the coronal pulp of rat molars 30 days after BMP-7 treatment, and extensive reparative dentine was observed in the roots of these teeth, beneath a calcio-traumatic line, in many cases with complete occlusion of the mesial root canal [120]. These observations point to an important aspect of reparative dentinogenesis, namely, that the coronal and root regions of the pulp may represent different environments for repair. Although the reasons for these differences are not yet clarified, the above studies provide promising opportunities for the development of novel endodontic treatment strategies, including the application of growth factors. Moreover, these findings also indicate the need for a detailed understanding of the underlying mechanisms of action of molecules such as BMP, before they can be effectively applied to clinical use. Particularly to assess the dose-dependency of reparative processes, further investigations are required. Recombinant human BMP-2 stimulates differentiation of the bovine and human adult pulp cells into odontoblasts in monolayer cultures [86, 107] and in three-dimensional pellet cultures [59]. Comparable effects of TGF β1-3 and BMP-7 have been demonstrated in cultured tooth slices [123, 124]. Some BMPs such as recombinant human BMP-2, BMP-4, and BMP-7 have demonstrated to induce reparative dentine formation in vivo [84, 85, 119]. The application of recombinant human IGF-1 together with collagen has been found to induce complete dentine bridging and tubular dentine formation [70]. This indicates the potential of growth factors to be included into endodontic treatment strategies. Thus, for instance growth factors might be added before pulp capping or they might be incorporated into restorative and endodontic materials to stimulate dentine and pulp regeneration.

Growth factors may be released from the dentine matrix as a result of both injury events such as a carious demineralization process and clinical restorative procedures such as the application of cavity etchants. For example, in human dentine, TGF-β1 is the predominant isoform and is distributed relatively equally between the soluble and insoluble matrices [18]. Accordingly, the process of a carious demineralization of dentine will solubilize the soluble pool of TGF-β1 and expose the insoluble pool. Solubilized TGF-βs may be able to diffuse to the odontoblasts and pulpal cells and signal tertiary dentinogenesis. While the TGF-βs immobilized on the insoluble matrix may have the ability to induce odontoblast-like cell differentiation during reparative dentinogenesis, they, thus, act in a similar manner to basement membrane-immobilized TGF-βs during tooth development [132]. This once more emphasizes the close similarities in the molecular signaling of events during tooth development and repair [133]. Other growth factors such as the group of angiogenic growth factors or any cytokines sequestered in the dentine matrix [97] will also be released due to caries demineralization and might contribute to the overall repair process. *Smith* concluded from his wealth of findings that dentine matrix may be regarded as a powerful cocktail of bioactive molecules which, if released following tissue injury, has the potential to dramatically influence cellular events in the dentine–pulp complex [133, 134].

64.6
Gene Therapy

The application of gene therapy in regenerative endodontics is a relatively new field. Although only few investigations are described in the literature so far, this new therapeutic approach should also be addressed within this chapter.

The use of gene therapy to treat exposed pulp by delivering DNA, RNA, or antisense sequences leads to an altered gene expression within a target cell population in pulp tissue. Gene therapy manipulates cellular processes and responses. Thus, the transfected genes stimulate immune response, modify cellular information or developmental program, or produce a therapeutic protein with specific functions [99]. Gene delivery should accomplish a stable expression of the transgene into target cell populations in an appropriate form without side effects such as interaction with the host genome, toxicity, carcinogenic transformation, and insertional mutagenesis [90]. Therefore, the development of safe and controllable vectors is considered as one of the biggest challenges. Studies of vector uptake, intracellular trafficking, and gene regulation have contributed to a better understanding of the involved mechanisms. New techniques involving viral or nonviral vectors have been employed for gene delivery.

64.6.1
Viral and Nonviral Vectors

Viral vectors, genetically modified to deliver genes, are derived from viruses with either RNA or DNA genomes such as retrovirus, lentivirus, adenovirus, adeno-associated virus, and herpes simplex virus [33, 151, 160]. Their modification aimed to prohibit the possibility of causing disease, but to still retain the capacity for infection [81]. Integrating vectors (retroviruses and lentiviruses) have the potential to express the gene product over long periods of time. On the other hand, nonintegrating vectors (adenovirus and herpes simplex virus) that are maintained as episome can be useful for efficient gene transduction in nondividing cells. The potential serious health risks of using viruses include immunogenicity and cytotoxicity and have been drastically emphasized in recent clinical trials [36]. Insertional mutagenesis might be another potential health hazard that compellingly needs to be considered, in which the ectopic chromosomal integration of viral DNA disrupts the expression of a tumor-suppressor gene or activates an oncogene [110, 154], leading to the malignant transformation of cells [6, 155].

Nonviral gene delivery methods have important safety advantages over viral approaches, including their reduced pathogenicity and capacity for insertional mutagenesis, as well as their low cost and ease of production. The application of nonviral vectors to humans has, however, been hampered by the poor efficiency of their delivery to cells and the transient expression of their transgenes [36]. Conventional nonviral methods of delivering transgenes comprise cationic polymer methods and physical methods. Cationic polymer methods, which include diverse liposomal formulations, basic proteins, and polymers such as polyethylenimine, interact with DNA to assist cell entry by binding or enveloping DNA through a charge interaction. Although these vectors work reasonably well in vitro, they have poor in vivo transfection efficiency and only confer transient gene expression. The plasmid DNA that is delivered by cationic liposomes is transient as the DNA is lost during changes at mitosis in the nuclear membrane. Lipofection involves the use of cationic lipids that are mixed with DNA in an aqueous solution. The lipids form the liposomes that encapsulate the DNA. The liposomes with the gene are taken up by the cells by pinocytosis and phagocytosis. However, a high toxicity and the fact that the liposomes are rapidly cleared from the plasma have resulted in a limited use of liposomes in vivo. Cationic peptides consist either of consecutive basic amino acids (for example polylysine), which compact DNA into spherical complexes, or of chromatin components such as histones or protamine, which compact DNA in a structured manner. The cellular entry of these cationic peptide-DNA complexes takes place by endocytosis. Linear or branched polymers such as polyethylenimine contain in addition to a high cationic charge, allowing an efficient plasmid DNA condensation and cell entry, many protonable amino nitrogen atoms, which make the polymer an effective "proton sponge." This property leads to a rupture of the endosomes by osmotic swelling, and the release of these polyethylenimine-DNA complexes into the cytoplasm [36]. Receptor-mediated delivery involves the use of the aforementioned polycations conjugated to a cell-specific ligand to provide receptor-mediated endocytosis of the polycation-DNA complex by target cells.

Physical methods for gene delivery include hydrodynamic pressure, particle bombardments, electroporation, and sonoporation. The hydrodynamic pressure technique uses the intravascular injection of high volumes of plasmid DNA to drive DNA molecules out of the blood circulation and into the tissue. Although the principle of this method might be applicable to humans, it would be limited to regions of the body which tolerate a temporarily increased pressure, for example by using a blood-pressure cuff applied to a limb [162]. Ballistic delivery, also called particle bombardment or gene gun, uses DNA-coated metal microparticles that have been accelerated to a high velocity to penetrate cell membranes [23]. Currently, in vivo applications are limited to cutaneous targets in animals. The application of regulated electric pulses to deliver genes to cells is referred to as electroporation [52, 152]. Electroporation uses electric fields to create transient pores to facilitate entry of plasmid DNA. Although it used to be restricted to in vitro applications, the development of new electrodes designed for in vivo applications has allowed electroporation-based transfection in contained areas of the body such as tumors, muscles, and liver tissue in animal models [1]. Sustained long-term expression of the plasmid DNA has been shown especially in skeletal and cardiac muscle [24]. In several studies, electroporation increased the gene expression over 100-fold compared to infection of the plasmid DNA. Therefore, electroporation seems to be an efficient technique. However, tissue damage has to be considered as one of its limitations [90]. Recognizing the disadvantage of electroporation being an invasive method, other methods have been explored such as ultrasound to deliver DNA. The application of ultrasound produces acoustic cavitation and, consequently, leads to an increased cell membrane permeability promoting the delivery of plasmid DNA. It was found that cavitation can be improved by application of ultrasound contrast agents [10, 26, 140]. The use of a combination of ultrasound and electroporation has also been investigated and it was concluded that the combined application was superior to each of these two methods alone [157]. The choice of gene delivery system depends on the accessibility and physiological characteristics of the target cell population. With respect to the application for human gene therapy, recent advances achieved by the variety of physical methods as described above indicate that efficient, long-term gene expression can be achieved by applying nonviral methods. Although each of these approaches certainly demonstrates a promising development, it still remains to be seen if any of *these approaches* will find its way into clinical gene therapy trials [152].

64.6.2
In Vivo and Ex Vivo Gene Therapy

Regarding the application of gene therapy in dentistry, in particular in regenerative endodontics, one use of gene delivery could be seen in the delivery of mineralizing genes into pulp tissue to promote tissue mineralization. However, to date, a review of the currently available literature reveals little research in this field using either in vivo or ex vivo gene transfer. Rutherford [105] tested these strategies for tissue regeneration in an adult ferret model of vital pulp therapy. In this model a reversible pulpitis was induced first. Three days later, the pulps were directly infected with a recombinant adenovirus containing a full-length cDNA encoding mouse BMP-7 gene or implanted with ex vivo transduced autologous dermal fibroblasts. The BMP-7 ex vivo transduced dermal fibroblasts induced reparative dentinogenesis with apparent regeneration of the dentine-pulp complex, whereas in vivo infection with the recombinant adenovirus containing BMP-7 gene failed to produce reparative dentine. Moreover, in a previous investigation it has been demonstrated that a single application of a therapeutic dressing comprising BMP-7 and a bovine bone-derived collagen carrier induced reparative dentinogenesis in ferret dental pulps, but failed to induce reparative dentine formation in ferret teeth with reversible pulpitis [104]. Taking into consideration the advantages of nonviral techniques, one study transduced growth/differentiation factor 11 (Gdf-11) cDNA plasmid by electroporation into pulp cells in vitro and induced the expression of DSPP, a differentiation marker for odontoblasts. Gdf-11 gene transfer in the amputated dental pulp in vivo, however, failed as there was a lake of erythrocytes in a plasma clot due to thermal damage and tissue invasiveness of electrode and the induced incomplete and non-homogeneous reparative dentine covered the exposed pulp [87]. However, transfer of Gdf-11 plasmid by ultrasound-mediated gene delivery with microbubbles induced differentiation of pulp stem cells into odontoblast-like cells in vitro and homogeneous and complete reparative dentine formed in vivo [88]. Concluding from these very few available data, there remains much to be accomplished to use gene therapy as part of endodontic treatment [81]. In particular, with respect to the apparent high risk of health hazards, research must be focused on the accurate control of gene therapy allowing its safe clinical application. Thus, since gene therapy is at an early stage of development, its use as an established method for regenerative endodontic treatment seems very unlikely in the near future.

64.7
Periodontal Regeneration and Approaches for Periodontal Tissue Engineering

The periodontium is a topographically complex organ consisting of epithelial tissue and soft and mineralized connective tissue. The structures comprising the periodontium include the gingiva, periodontal ligament (PDL), cementum, and alveolar bone (Fig. 64.2). Several diseases affect the composition and integrity of periodontal structures leading to the destruction of the connective tissue matrix and cells, the loss of the fibrous attachment, and the resorption of alveolar bone. Periodontitis, an inflammatory periodontal disease, is the most frequently occurring periodontal disease worldwide and is associated with bacterial deposits (dental plaque) on the tooth root surface. If left untreated, plaque accumulation may cause the aforementioned processes, and will ultimately result in tooth loss. Although during the early phases of periodontal disease the periodontium is capable of performing some minor regeneration. Once damaged, the periodontium has only limited capacity for regeneration. By the formation of new cementum, remodeling of the PDL and new bone formation is possible within the process of orthodontic tooth movement, but this represents a physiological response to a force affecting a tooth and its surrounding tissues rather than a reparative mechanism due to pathologically induced tissue damage [5].

The ultimate goal of periodontal treatment is to prevent further attachment loss and regenerate supporting tissues. To make a successful periodontal regeneration possible, several prerequisites are required such as the formation of a functional epithelial seal, the insertion of new connective fibers into the root surface, and the restoration of alveolar bone height [4]. However, following conventional treat-

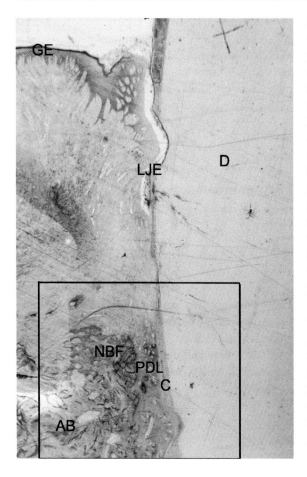

Fig. 64.2 Histological nondecalcified section (toluidine-blue) from a guinea pig tooth illustrating dental hard tissue and periodontal tissues: Gingiva (gingival epithelium: *GE*), dentine (*D*), cementum (*C*), periodontal ligament (*PDL*), and alveolar bone (*AB*). Moreover, image shows long junctional epithelium (*LJE*) and new bone formation (NBF) at the alveolar crest following periodontal regeneration

ment (scaling, root planning, and curettage), achieving these conditions is, to date, quite challenging. This is due to the fact that the proliferation of epithelial tissues into the defect is much faster than that of mesenchymal tissues [69]. Consequently, this results in the formation of a long junctional epithelium that prevents the selective migration, proliferation, and differentiation of cells derived from the PDL [91]. Since numerous investigations have demonstrated that the exclusion of epithelial cells improve periodontal regeneration and, in this context, assuming that only PDL cells might have the potential to create a new connective tissue attachment, the clinical application of the concept of guided tissue regeneration (GTR) has been advocated. Moreover, in recent years, procedures including root surface application of various growth factors as well as bone and enamel matrix proteins have been suggested [15, 35, 38, 68, 96, 121]. The use of these factors is thought to stimulate repopulation of the defect with appropriate endogenous progenitor cells in order to induce regeneration. These procedures have shown encouraging but limited potential in promoting regeneration of periodontal tissue due to inconsistent and unpredictable outcomes [17, 51, 66]. Nevertheless, the scientific development over the past years has contributed to an increased understanding of the requirements for successful periodontal regeneration, namely the need to control and coordinate both cellular and molecular processes.

Progress in tissue engineering has made it possible to regenerate damaged tissues and organs based on the generation of the necessary prerequisites supporting natural healing potential of tissues. Considering the former, to overcome the limitations of conventional periodontal regenerative treatment approaches and to develop a more reliable and predictable periodontal regeneration, the use of tissue engineering was suggested to offer promising opportunities for developing novel strategies for periodontal regeneration.

As mentioned before, mesenchymal stem cells were first identified in aspirates of adult bone marrow by Friedenstein and colleagues [19, 31, 94] by their capacity to form clonogenic clusters of adherent fibroblast-like cells or fibroblastic colony-forming units with the potential to undergo extensive proliferation in vitro and to differentiate into different stromal lineages. Using the above criteria, in an impressive work, Seo and colleagues have recently identified cells that could be classified as mesenchymal stem cells, for the first time derived from adult PDL. These PDL stem cells (PDLSCs) exhibited the capacity to generate clonogenic adherent cell colonies when plated under the same growth conditions as described for bone marrow stromal stem cells (BMSSCs) [112]. To attain information about the origin and the characterization of PDLSCs, the expression of putative stem cell markers has been performed and was compared with other postnatal stem cells. Previous investigations have shown that human bone marrow and dental pulp tissues contain postnatal stem cells that are capable of differentiating into osteoblasts/odontoblasts, adipocytes, and neuronal-like cells. These stem cells were characterized as STRO-1/CD146-positive progenitors derived from a perivascular niche within the bone marrow and dental pulp microenvironments [43, 44, 115]. In fact, PDLSCs were found to be similar to other mesenchymal stem cells with respect to their expression of STRO-1/CD146, suggesting that PDLSCs might also be derived from a population of perivascular cells [39, 73]. In contrast to DPSCs and BMSSCs, which were capable of generating a dentine/pulp-like structure and a lamellar bone/marrow-like structure, respectively [43, 45, 64], PDLSCs represent a novel population of multipotent stem cells, as shown by their capacity to develop into cementoblast-like cells, adipocytes in vitro, and cementum/PDL-like tissue in vivo, and by their high expression of scleraxis, a specific transcription factor associated with tendon cells [16]. PDLSCs also demonstrated the capacity to form collagen fibers, similar to Sharpey's fibers, connecting to the cementum-like tissue, suggesting the potential to regenerate PDL attachment. Consequently, these findings support the notion that PDLSCs are a unique population of postnatal stem cells [112]. In consideration of this knowledge, the use of PDLSCs to produce functional tissues represents an attractive therapeutic option for regenerating periodontal tissue given the recent advances in the field of tissue engineering [41, 138].

To test whether periodontal PDLSCs reveal a tissue regenerative capacity similar to that of BMSSCs, Gronthos and colleagues have initiated a number of studies to investigate the use of these cells for periodontal regeneration [5]. To date, cultured human PDLSCs have been implanted into surgically created periodontal defects in nude rats [112]. Results indicated that the PDLSCs attached to both the alveolar bone and cementum surfaces and there was evidence of formation of a PDL-like structure. Moreover, the same group has identified the ovine counterparts of human BMSSCs and PDLSCs, which demonstrate similar functional properties when transplanted into immunocompromised mice with the hydroxyapatite/tricalcium phosphate carrier particles [46]. Table 64.2 shows a list of biomaterials for tissue repair in periodontics. Taking these recent findings into account, an ovine preclinical model has been established to determine the clinical utility of these cells in future investigations before progressing to human trials [5]. Accordingly, further studies are now required to determine the efficacy of ex vivo expanded stem cells to repair periodontal defects. Growth and differentiation conditions that induce lineage-specific commitment will need to be established. In addition, suitable carriers and inductive factors capable of helping implants to integrate into the surrounding environment for the reconstruction of functional complex organ systems will need to be developed [5].

64.8
Conclusions

Tissue engineering strategies offer exciting possibilities for the development of novel, biologically-based clinical treatment approaches that are aimed at regenerating dental tissues. In dentistry, including both disciplines, regenerative endodontics and regenerative periodontics, developmental strategies have the potential for regenerating several tissues such as pulp,

Table 64.2 Scaffolds used to aim periodontal regeneration

Type	Biomaterial
Allograft	Calcified freeze-dried bone, decalcified freeze-dried bone
Alloplasts	Hydroxyapatite, tricalcium phosphate, hard-tissue polymers, bioactive glass
Polymers and collagens	Collagen, poly(lactide-co-glycolide), methylcellulose, hyaluronic acid, chitosan
Xenograft	Bovine mineral matrix, bovine-derived hydroxyapatite

dentine, and periodontal structures. Consequently, regenerative dentistry may offer alternative methods to save teeth that may have a compromised structural integrity. Due to the enormous tissue regenerative capacity of the dentine-pulp complex, regenerative endodontics can be expected to occupy an exceptional position in regenerative medicine in the future. While the engineering of a whole tooth represents the ultimate goal of regenerative dental purposes, this aim certainly represents a tremendous challenge for future scientific efforts and will hardly become reality to an extent of clinical significance in the near future. Tissue engineering of the pulp, however, may be achieved over a shorter period and might represent an essential improvement of currently available clinical treatment approaches in endodontics [134]. Different regenerative strategies have been described and each of them comprises advantages and disadvantages. Some of them are largely theoretical. Although regenerative dentistry is at an noteworthy point of development harnessing the biological potential of the dental tissues to facilitate wound healing and tissue regeneration, there is much to be learned from nature. Future progress is necessary to understand how to control and regulate the various processes and components that are involved in the respective regenerative mechanisms. With respect to the risk of health hazards, these are essential prerequisites that definitely need to be clarified and integrated to produce the outcome of a regenerated dentine-pulp complex before these promising principles might be applied to clinical trials. To implement these future prospects into regenerative endodontics, interdisciplinary collaborations involving different areas of expertise are required. The further development of tissue engineering strategies for the regeneration of dental tissues has now the potential to substantially modify clinical treatment strategies, and, thus, to lead to a tremendous benefit for millions of patients.

References

1. Andre F, Mir LM (2004) DNA electrotransfer: its principles and an updated review of its therapeutic applications. Gene Ther 11:33–42
2. Anneroth G, Bang G (1972) The effect of allogeneic demineralized dentin as a pulp capping agent in Java monkeys. Odontol Revy 23:315–328
3. Athanasiou KA, Niederauer GG, Agrawal CM (1996) Sterilization, toxicity, biocompatibility and clinical applications of polylactic acid/polyglycolic acid copolymers. Biomaterials 17:93–102
4. Bartold PM, McCulloch CA, Narayanan AS et al. (2000) Tissue engineering: a new paradigm for periodontal regeneration based on molecular and cell biology. Periodontol 2000 24:253–269
5. Bartold PM, Shi S, Gronthos S (2006) Stem cells and periodontal regeneration. Periodontol 2000 40:164–172
6. Baum C, Kustikova O, Modlich U et al. (2006) Mutagenesis and oncogenesis by chromosomal insertion of gene transfer vectors. Hum Gene Ther 17:253–263
7. Baume LJ, Holz J (1981) Long term clinical assessment of direct pulp capping. Int Dent J 31:251–260
8. Begue-Kirn C, Smith AJ, Ruch JV et al. (1992) Effects of dentin proteins, transforming growth factor beta 1 (TGF beta 1) and bone morphogenetic protein 2 (BMP2) on the differentiation of odontoblast in vitro. Int J Dev Biol 36:491–503
9. Begue-Kirn C, Smith AJ, Loriot M et al. (1994) Comparative analysis of TGF beta s, BMPs, IGF1, msxs, fibronectin, osteonectin and bone sialoprotein gene expression during normal and in vitro-induced odontoblast differentiation. Int J Dev Biol 38:405–420
10. Bekeredjian R, Grayburn PA, Shohet RV (2005) Use of ultrasound contrast agents for gene or drug delivery in cardiovascular medicine. J Am Coll Cardiol 45:329–335
11. Bianco P, Riminucci M, Gronthos S et al. (2001) Bone marrow stromal stem cells: nature, biology, and potential applications. Stem Cells 19:180–192
12. Blau HM, Brazelton TR, Weimann JM (2001) The evolving concept of a stem cell: entity or function? Cell 105:829–841
13. Boccaccini AR, Blaker JJ (2005) Bioactive composite materials for tissue engineering scaffolds. Expert Rev Med Devices 2:303–317
14. Bohl KS, Shon J, Rutherford B et al. (1998) Role of synthetic extracellular matrix in development of engineered dental pulp. J Biomater Sci Polym Ed 9:749–764
15. Bosshardt DD, Zalzal S, McKee MD et al. (1998) Developmental appearance and distribution of bone sialoprotein and osteopontin in human and rat cementum. Anat Rec 250:13–33
16. Brent AE, Schweitzer R, Tabin CJ (2003) A somitic compartment of tendon progenitors. Cell 113:235–248
17. Carlson NE, Roach RB, Jr. (2002) Platelet-rich plasma: clinical applications in dentistry. J Am Dent Assoc 133:1383–1386
18. Cassidy N, Fahey M, Prime SS et al. (1997) Comparative analysis of transforming growth factor-beta isoforms 1–3 in human and rabbit dentine matrices. Arch Oral Biol 42:219–223
19. Castro-Malaspina H, Gay RE, Resnick G et al. (1980) Characterization of human bone marrow fibroblast colony-forming cells (CFU-F) and their progeny. Blood 56:289–301
20. Chevallay B, Abdul-Malak N, Herbage D (2000) Mouse fibroblasts in long-term culture within collagen three-dimensional scaffolds: influence of crosslinking with diphenylphosphorylazide on matrix reorganization,

growth, and biosynthetic and proteolytic activities. J Biomed Mater Res 49:448–459
21. Cobourne MT, Sharpe PT (2003) Tooth and jaw: molecular mechanisms of patterning in the first branchial arch. Arch Oral Biol 48:1–14
22. Cotton WR (1968) Pulp response to cavity preparation as studied by the method of thymidine 3H autoradiography. In: Finn SB (ed) Biology of the dental pulp organ. Univ Alabama Press, Tuscaloosa, pp 69–101
23. Davidson JM, Krieg T, Eming SA (2000) Particle-mediated gene therapy of wounds. Wound Repair Regen 8:452–459
24. Dean DA (2005) Nonviral gene transfer to skeletal, smooth, and cardiac muscle in living animals. Am J Physiol Cell Physiol 289:C233–245
25. Dhariwala B, Hunt E, Boland T (2004) Rapid prototyping of tissue-engineering constructs, using photopolymerizable hydrogels and stereolithography. Tissue Eng 10:1316–1322
26. Dijkmans PA, Juffermans LJ, Musters RJ et al. (2004) Microbubbles and ultrasound: from diagnosis to therapy. Eur J Echocardiogr 5:245–256
27. Dobie K, Smith G, Sloan AJ et al. (2002) Effects of alginate hydrogels and TGF-beta 1 on human dental pulp repair in vitro. Connect Tissue Res 43:387–390
28. Finkelman RD, Mohan S, Jennings JC et al. (1990) Quantitation of growth factors IGF-I, SGF/IGF-II, and TGF-beta in human dentin. J Bone Miner Res 5:717–723
29. Fisher LW, Fedarko NS (2003) Six genes expressed in bones and teeth encode the current members of the SIBLING family of proteins. Connect Tissue Res 44 Suppl 1:33–40
30. Freed LE, Marquis JC, Nohria A et al. (1993) Neocartilage formation in vitro and in vivo using cells cultured on synthetic biodegradable polymers. J Biomed Mater Res 27:11–23
31. Friedenstein AJ, Ivanov-Smolenski AA, Chajlakjan RK et al. (1978) Origin of bone marrow stromal mechanocytes in radiochimeras and heterotopic transplants. Exp Hematol 6:440–444
32. Fuchs E, Segre JA (2000) Stem cells: a new lease on life. Cell 100:143–155
33. Gardlik R, Palffy R, Hodosy J et al. (2005) Vectors and delivery systems in gene therapy. Med Sci Monit 11:RA110–121
34. George A, Sabsay B, Simonian PA et al. (1993) Characterization of a novel dentin matrix acidic phosphoprotein. Implications for induction of biomineralization. J Biol Chem 268:12624–12630
35. Giannobile WV (1996) Periodontal tissue engineering by growth factors. Bone 19:23S–37S
36. Glover DJ, Lipps HJ, Jans DA (2005) Towards safe, nonviral therapeutic gene expression in humans. Nat Rev Genet 6:299–310
37. Goodell MA, Brose K, Paradis G et al. (1996) Isolation and functional properties of murine hematopoietic stem cells that are replicating in vivo. J Exp Med 183:1797–1806
38. Gottlow J, Nyman S, Karring T et al. (1984) New attachment formation as the result of controlled tissue regeneration. J Clin Periodontol 11:494–503
39. Gould TR, Melcher AH, Brunette DM (1977) Location of progenitor cells in periodontal ligament of mouse molar stimulated by wounding. Anat Rec 188:133–141
40. Graham L, Cooper PR, Cassidy N et al. (2006) The effect of calcium hydroxide on solubilisation of bio-active dentine matrix components. Biomaterials 27:2865–2873
41. Griffith LG, Naughton G (2002) Tissue engineering—current challenges and expanding opportunities. Science 295:1009–1014
42. Gronthos S, Graves SE, Ohta S et al. (1994) The STRO-1+ fraction of adult human bone marrow contains the osteogenic precursors. Blood 84:4164–4173
43. Gronthos S, Mankani M, Brahim J et al. (2000) Postnatal human dental pulp stem cells (DPSCs) in vitro and in vivo. Proc Natl Acad Sci U S A 97:13625–13630
44. Gronthos S, Brahim J, Li W et al. (2002) Stem cell properties of human dental pulp stem cells. J Dent Res 81:531–535
45. Gronthos S, Zannettino AC, Hay SJ et al. (2003) Molecular and cellular characterisation of highly purified stromal stem cells derived from human bone marrow. J Cell Sci 116:1827–1835
46. Gronthos S, Mrozik K, Shi S et al. (2006) Ovine periodontal ligament stem cells: isolation, characterization, and differentiation potential. Calcif Tissue Int 79:310–317
47. Guo T, Zhao J, Chang J et al. (2006) Porous chitosan-gelatin scaffold containing plasmid DNA encoding transforming growth factor-beta1 for chondrocytes proliferation. Biomaterials 27:1095–1103
48. Handschel J, Wiesmann HP, Depprich R et al. (2006) Cell-based bone reconstruction therapies—cell sources. Int J Oral Maxillofac Implants 21:890–898
49. Harada H, Kettunen P, Jung HS et al. (1999) Localization of putative stem cells in dental epithelium and their association with Notch and FGF signaling. J Cell Biol 147:105–120
50. Harlan DM, Karp CL, Matzinger P et al. (2002) Immunological concerns with bioengineering approaches. Ann N Y Acad Sci 961:323–330
51. Heijl L, Heden G, Svardstrom G et al. (1997) Enamel matrix derivative (EMDOGAIN) in the treatment of intrabony periodontal defects. J Clin Periodontol 24:705–714
52. Heller LC, Ugen K, Heller R (2005) Electroporation for targeted gene transfer. Expert Opin Drug Deliv 2:255–268
53. Höhl E (1896) Beitrag zur Histologie der Pulpa und des Dentins. Arch Anat Physiol 32:31–54
54. Honda M, Yada T, Ueda M et al. (2000) Cartilage formation by cultured chondrocytes in a new scaffold made of poly(L-lactide-epsilon-caprolactone) sponge. J Oral Maxillofac Surg 58:767–775
55. Honda M, Morikawa N, Hata K et al. (2003) Rat costochondral cell characteristics on poly (L-lactide-co-epsilon-caprolactone) scaffolds. Biomaterials 24:3511–3519
56. Honda MJ, Yada T, Ueda M et al. (2004) Cartilage formation by serial passaged cultured chondrocytes in a new scaffold: hybrid 75:25 poly(L-lactide-epsilon-caprolactone) sponge. J Oral Maxillofac Surg 62:1510–1516

57. Honda MJ, Sumita Y, Kagami H et al. (2005) Histological and immunohistochemical studies of tissue engineered odontogenesis. Arch Histol Cytol 68:89–101
58. Hu B, Unda F, Bopp-Kuchler S et al. (2006) Bone marrow cells can give rise to ameloblast-like cells. J Dent Res 85:416–421
59. Iohara K, Nakashima M, Ito M et al. (2004) Dentin regeneration by dental pulp stem cell therapy with recombinant human bone morphogenetic protein 2. J Dent Res 83:590–595
60. Iohara K, Zheng L, Ito M et al. (2006) Side population cells isolated from porcine dental pulp tissue with self-renewal and multipotency for dentinogenesis, chondrogenesis, adipogenesis, and neurogenesis. Stem Cells 24:2493–2503
61. Kardos TB, Hunter AR, Hanlin SM et al. (1998) Odontoblast differentiation: a response to environmental calcium? Endod Dent Traumatol 14:105–111
62. Kim WS, Vacanti JP, Cima L et al. (1994) Cartilage engineered in predetermined shapes employing cell transplantation on synthetic biodegradable polymers. Plast Reconstr Surg 94:233–237; discussion 238–240
63. Kitasako Y, Shibata S, Pereira PN et al. (2000) Short-term dentin bridging of mechanically-exposed pulps capped with adhesive resin systems. Oper Dent 25:155–162
64. Kuznetsov SA, Krebsbach PH, Satomura K et al. (1997) Single-colony derived strains of human marrow stromal fibroblasts form bone after transplantation in vivo. J Bone Miner Res 12:1335–1347
65. Langer R, Vacanti JP (1993) Tissue engineering. Science 260:920–926
66. Laurell L, Gottlow J (1998) Guided tissue regeneration update. Int Dent J 48:386–398
67. Lesot H, C. B-K, Kübler MD et al. (1993) Experimental induction of odontoblast differentiation and stimulation during reparative processes. Cell Mater 3:201–217
68. Lindskog S (1982) Formation of intermediate cementum. I: early mineralization of aprismatic enamel and intermediate cementum in monkey. J Craniofac Genet Dev Biol 2:147–160
69. Listgarten MA, Rosenberg MM (1979) Histological study of repair following new attachment procedures in human periodontal lesions. J Periodontol 50:333–344
70. Lovschall H, Fejerskov O, Flyvbjerg A (2001) Pulp-capping with recombinant human insulin-like growth factor I (rhIGF-I) in rat molars. Adv Dent Res 15:108–112
71. MacArthur BD, Oreffo RO (2005) Bridging the gap. Nature 433:19
72. MacDougall M, Simmons D, Luan X et al. (1997) Dentin phosphoprotein and dentin sialoprotein are cleavage products expressed from a single transcript coded by a gene on human chromosome 4. Dentin phosphoprotein DNA sequence determination. J Biol Chem 272:835–842
73. McCulloch CA (1985) Progenitor cell populations in the periodontal ligament of mice. Anat Rec 211:258–262
74. Miura M, Gronthos S, Zhao M et al. (2003) SHED: stem cells from human exfoliated deciduous teeth. Proc Natl Acad Sci U S A 100:5807–5812
75. Miura M, Chen XD, Allen MR et al. (2004) A crucial role of caspase-3 in osteogenic differentiation of bone marrow stromal stem cells. J Clin Invest 114:1704–1713
76. Mjor IA, Dahl E, Cox CF (1991) Healing of pulp exposures: an ultrastructural study. J Oral Pathol Med 20:496–501
77. Murray PE, Lumley PJ, Ross HF et al. (2000) Tooth slice organ culture for cytotoxicity assessment of dental materials. Biomaterials 21:1711–1721
78. Murray PE, About I, Lumley PJ et al. (2002) Cavity remaining dentin thickness and pulpal activity. Am J Dent 15:41–46
79. Murray PE, Hafez AA, Smith AJ et al. (2003) Histomorphometric analysis of odontoblast-like cell numbers and dentine bridge secretory activity following pulp exposure. Int Endod J 36:106–116
80. Murray PE, Garcia-Godoy F (2004) Stem cell responses in tooth regeneration. Stem Cells Dev 13:255–262
81. Murray PE, Garcia-Godoy F, Hargreaves KM (2007) Regenerative endodontics: a review of current status and a call for action. J Endod 33:377–390
82. Nakashima M (1989) Dentin induction by implants of autolyzed antigen-extracted allogeneic dentin on amputated pulps of dogs. Endod Dent Traumatol 5:279–286
83. Nakashima M (1990) An ultrastructural study of the differentiation of mesenchymal cells in implants of allogeneic dentine matrix on the amputated dental pulp of the dog. Arch Oral Biol 35:277–281
84. Nakashima M (1994) Induction of dentin formation on canine amputated pulp by recombinant human bone morphogenetic proteins (BMP)-2 and -4. J Dent Res 73:1515–1522
85. Nakashima M (1994) Induction of dentine in amputated pulp of dogs by recombinant human bone morphogenetic proteins-2 and -4 with collagen matrix. Arch Oral Biol 39:1085–1089
86. Nakashima M, Nagasawa H, Yamada Y et al. (1994) Regulatory role of transforming growth factor-beta, bone morphogenetic protein-2, and protein-4 on gene expression of extracellular matrix proteins and differentiation of dental pulp cells. Dev Biol 162:18–28
87. Nakashima M, Mizunuma K, Murakami T et al. (2002) Induction of dental pulp stem cell differentiation into odontoblasts by electroporation-mediated gene delivery of growth/differentiation factor 11 (Gdf11). Gene Ther 9:814–818
88. Nakashima M, Tachibana K, Iohara K et al. (2003) Induction of reparative dentin formation by ultrasound-mediated gene delivery of growth/differentiation factor 11. Hum Gene Ther 14:591–597
89. Nakashima M, Akamine A (2005) The application of tissue engineering to regeneration of pulp and dentin in endodontics. J Endod 31:711–718
90. Nakashima M, Iohara K, Zheng L (2006) Gene therapy for dentin regeneration with bone morphogenetic proteins. Curr Gene Ther 6:551–560
91. Nyman S, Lindhe J, Karring T et al. (1982) New attachment following surgical treatment of human periodontal disease. J Clin Periodontol 9:290–296
92. Ohazama A, Modino SA, Miletich I et al. (2004) Stem-cell-based tissue engineering of murine teeth. J Dent Res 83:518–522

93. Oringer RJ (2002) Biological mediators for periodontal and bone regeneration. Compend Contin Educ Dent 23:501–504, 506–510, 512 passim; quiz 518
94. Owen M, Friedenstein AJ (1988) Stromal stem cells: marrow-derived osteogenic precursors. Ciba Found Symp 136:42–60
95. Qin C, Brunn JC, Cadena E et al. (2002) The expression of dentin sialophosphoprotein gene in bone. J Dent Res 81:392–394
96. Quinones CR, Caffesse RG (1995) Current status of guided periodontal tissue regeneration. Periodontol 2000 9:55–68
97. Roberts-Clark DJ, Smith AJ (2000) Angiogenic growth factors in human dentine matrix. Arch Oral Biol 45:1013–1016
98. Robson WC, Katz RW (1992) Preliminary studies on pulp capping with demineralized dentin. Proc Finn Dent Soc 88 Suppl 1:279–283
99. Rubanyi GM (2001) The future of human gene therapy. Mol Aspects Med 22:113–142
100. Ruch JV, Lesot H, Begue-Kirn C (1995) Odontoblast differentiation. Int J Dev Biol 39:51–68
101. Ruch JV (1998) Odontoblast commitment and differentiation. Biochem Cell Biol 76:923–938
102. Rutherford RB, Wahle J, Tucker M et al. (1993) Induction of reparative dentine formation in monkeys by recombinant human osteogenic protein-1. Arch Oral Biol 38:571–576
103. Rutherford RB, Spangberg L, Tucker M et al. (1994) The time-course of the induction of reparative dentine formation in monkeys by recombinant human osteogenic protein-1. Arch Oral Biol 39:833–838
104. Rutherford RB, Gu K (2000) Treatment of inflamed ferret dental pulps with recombinant bone morphogenetic protein-7. Eur J Oral Sci 108:202–206
105. Rutherford RB (2001) BMP-7 gene transfer to inflamed ferret dental pulps. Eur J Oral Sci 109:422–424
106. Sachlos E, Czernuszka JT (2003) Making tissue engineering scaffolds work. Review: the application of solid freeform fabrication technology to the production of tissue engineering scaffolds. Eur Cell Mater 5:29–39; discussion 39–40
107. Saito T, Ogawa M, Hata Y et al. (2004) Acceleration effect of human recombinant bone morphogenetic protein-2 on differentiation of human pulp cells into odontoblasts. J Endod 30:205–208
108. Salehrabi R, Rotstein I (2004) Endodontic treatment outcomes in a large patient population in the USA: an epidemiological study. J Endod 30:846–850
109. Schopper C, Ziya-Ghazvini F, Goriwoda W et al. (2005) HA/TCP compounding of a porous CaP biomaterial improves bone formation and scaffold degradation—a long-term histological study. J Biomed Mater Res B Appl Biomater 74:458–467
110. Schroder AR, Shinn P, Chen H et al. (2002) HIV-1 integration in the human genome favors active genes and local hotspots. Cell 110:521–529
111. Schröder U (1985) Effects of calcium hydroxide-containing pulp-capping agents on pulp cell migration, proliferation, and differentiation. J Dent Res 64 Spec No:541–548
112. Seo BM, Miura M, Gronthos S et al. (2004) Investigation of multipotent postnatal stem cells from human periodontal ligament. Lancet 364:149–155
113. Seol YJ, Lee JY, Park YJ et al. (2004) Chitosan sponges as tissue engineering scaffolds for bone formation. Biotechnol Lett 26:1037–1041
114. Sharma B, Elisseeff JH (2004) Engineering structurally organized cartilage and bone tissues. Ann Biomed Eng 32:148–159
115. Shi S, Gronthos S (2003) Perivascular niche of postnatal mesenchymal stem cells in human bone marrow and dental pulp. J Bone Miner Res 18:696–704
116. Shi S, Bartold PM, Miura M et al. (2005) The efficacy of mesenchymal stem cells to regenerate and repair dental structures. Orthod Craniofac Res 8:191–199
117. Silva TA, Rosa AL, Lara VS (2004) Dentin matrix proteins and soluble factors: intrinsic regulatory signals for healing and resorption of dental and periodontal tissues? Oral Dis 10:63–74
118. Silver FH, Pins G (1992) Cell growth on collagen: a review of tissue engineering using scaffolds containing extracellular matrix. J Long Term Eff Med Implants 2:67–80
119. Six N, Decup F, Lasfargues JJ et al. (2002) Osteogenic proteins (bone sialoprotein and bone morphogenetic protein-7) and dental pulp mineralization. J Mater Sci Mater Med 13:225–232
120. Six N, Lasfargues JJ, Goldberg M (2002) Differential repair responses in the coronal and radicular areas of the exposed rat molar pulp induced by recombinant human bone morphogenetic protein 7 (osteogenic protein 1). Arch Oral Biol 47:177–187
121. Slavkin HC, Bringas P, Jr., Bessem C et al. (1989) Hertwig's epithelial root sheath differentiation and initial cementum and bone formation during long-term organ culture of mouse mandibular first molars using serumless, chemically-defined medium. J Periodontal Res 24:28-40
122. Sloan AJ, Matthews JB, Smith AJ (1999) TGF-beta receptor expression in human odontoblasts and pulpal cells. Histochem J 31:565–569
123. Sloan AJ, Smith AJ (1999) Stimulation of the dentine-pulp complex of rat incisor teeth by transforming growth factor-beta isoforms 1-3 in vitro. Arch Oral Biol 44:149–156
124. Sloan AJ, Rutherford RB, Smith AJ (2000) Stimulation of the rat dentine-pulp complex by bone morphogenetic protein-7 in vitro. Arch Oral Biol 45:173–177
125. Sloan AJ, Couble ML, Bleicher F et al. (2001) Expression of TGF-beta receptors I and II in the human dental pulp by in situ hybridization. Adv Dent Res 15:63–67
126. Sloan AJ, Smith AJ (2007) Stem cells and the dental pulp: potential roles in dentine regeneration and repair. Oral Dis 13:151–157
127. Smith AJ, Tobias RS, Plant CG et al. (1990) In vivo morphogenetic activity of dentine matrix proteins. J Biol Buccale 18:123–129
128. Smith AJ, Tobias RS, Cassidy N et al. (1994) Odontoblast stimulation in ferrets by dentine matrix components. Arch Oral Biol 39:13–22

129. Smith AJ, Cassidy N, Perry H et al. (1995) Reactionary dentinogenesis. Int J Dev Biol 39:273–280
130. Smith AJ, Garde C, Cassidy N et al. (1995) Solubilisation of dentine extracellular matrix by calcium hydroxide. J Dent Res 74 (abstr 59):829
131. Smith AJ, Sloan AJ, Matthews JB et al. (2000) Reparative processes in dentine and pulp. In: Addy M, Embery G, Edgar WM et al. (eds) Toothwear and sensitivity. Martin-Dunitz London, pp 53–66
132. Smith AJ, Lesot H (2001) Induction and regulation of crown dentinogenesis: embryonic events as a template for dental tissue repair? Crit Rev Oral Biol Med 12:425–437
133. Smith AJ (2002) Pulpal responses to caries and dental repair. Caries Res 36:223–232
134. Smith AJ (2003) Vitality of the dentin-pulp complex in health and disease: growth factors as key mediators. J Dent Educ 67:678–689
135. Smith AJ (2004) Tooth tissue engineering and regeneration—a translational vision! J Dent Res 83:517
136. Spradling A, Drummond-Barbosa D, Kai T (2001) Stem cells find their niche. Nature 414:98–104
137. Stanley HR (1989) Pulp capping: conserving the dental pulp—can it be done? Is it worth it? Oral Surg Oral Med Oral Pathol 68:628–639
138. Stock UA, Vacanti JP (2001) Tissue engineering: current state and prospects. Annu Rev Med 52:443–451
139. Sumita Y, Honda MJ, Ohara T et al. (2006) Performance of collagen sponge as a 3-D scaffold for tooth-tissue engineering. Biomaterials 27:3238–3248
140. Tachibana K (2004) Emerging technologies in therapeutic ultrasound: thermal ablation to gene delivery. Hum Cell 17:7–15
141. Takahashi Y, Yamamoto M, Tabata Y (2005) Enhanced osteoinduction by controlled release of bone morphogenetic protein-2 from biodegradable sponge composed of gelatin and beta-tricalcium phosphate. Biomaterials 26:4856–4865
142. Torabinejad M, Chivian N (1999) Clinical applications of mineral trioxide aggregate. J Endod 25:197–205
143. Trojani C, Weiss P, Michiels JF et al. (2005) Three-dimensional culture and differentiation of human osteogenic cells in an injectable hydroxypropylmethylcellulose hydrogel. Biomaterials 26:5509–5517
144. Trubiani O, Scarano A, Orsini G et al. (2007) The performance of human periodontal ligament mesenchymal stem cells on xenogenic biomaterials. Int J Immunopathol Pharmacol 20:87–91
145. Tziafas D, Kolokuris I, Alvanou A et al. (1992) Short-term dentinogenic response of dog dental pulp tissue after its induction by demineralized or native dentine, or predentine. Arch Oral Biol 37:119–128
146. Tziafas D, Lambrianidis T, Beltes P (1993) Inductive effect of native dentin on the dentinogenic potential of adult dog teeth. J Endod 19:116–122
147. Tziafas D, Alvanou A, Panagiotakopoulos N et al. (1995) Induction of odontoblast-like cell differentiation in dog dental pulps after in vivo implantation of dentine matrix components. Arch Oral Biol 40:883–893
148. Tziafas D, Alvanou A, Papadimitriou S et al. (1998) Effects of recombinant basic fibroblast growth factor, insulin-like growth factor-II and transforming growth factor-beta 1 on dog dental pulp cells in vivo. Arch Oral Biol 43:431–444
149. Tziafas D, Smith AJ, Lesot H (2000) Designing new treatment strategies in vital pulp therapy. J Dent 28:77–92
150. van Amerongen MJ, Harmsen MC, Petersen AH et al. (2006) The enzymatic degradation of scaffolds and their replacement by vascularized extracellular matrix in the murine myocardium. Biomaterials 27:2247–2257
151. Verma IM, Weitzman MD (2005) Gene therapy: twenty-first century medicine. Annu Rev Biochem 74:711-738
152. Wells DJ (2004) Gene therapy progress and prospects: electroporation and other physical methods. Gene Ther 11:1363–1369
153. Wingard JR, Demetri GD (1999) Clinical applications of cytokines and growth factors. Springer, New York
154. Woods NB, Muessig A, Schmidt M et al. (2003) Lentiviral vector transduction of NOD/SCID repopulating cells results in multiple vector integrations per transduced cell: risk of insertional mutagenesis. Blood 101:1284–1289
155. Woods NB, Bottero V, Schmidt M et al. (2006) Gene therapy: therapeutic gene causing lymphoma. Nature 440:1123
156. Xiao S, Yu C, Chou X et al. (2001) Dentinogenesis imperfecta 1 with or without progressive hearing loss is associated with distinct mutations in DSPP. Nat Genet 27:201–204
157. Yamashita Y, Shimada M, Tachibana K et al. (2002) In vivo gene transfer into muscle via electro-sonoporation. Hum Gene Ther 13:2079–2084
158. Young CS, Terada S, Vacanti JP et al. (2002) Tissue engineering of complex tooth structures on biodegradable polymer scaffolds. J Dent Res 81:695–700
159. Young CS, Abukawa H, Asrican R et al. (2005) Tissue-engineered hybrid tooth and bone. Tissue Eng 11:1599–1610
160. Young LS, Searle PF, Onion D et al. (2006) Viral gene therapy strategies: from basic science to clinical application. J Pathol 208:299–318
161. Zeichner-David M, Diekwisch T, Fincham A et al. (1995) Control of ameloblast differentiation. Int J Dev Biol 39:69–92
162. Zhang G, Budker V, Williams P et al. (2001) Efficient expression of naked dna delivered intraarterially to limb muscles of nonhuman primates. Hum Gene Ther 12:427–438

Tissue Engineering Applications in Endocrinology

M. R. Hammerman

Contents

65.1	Introduction	939
65.2	Type 1 and Type 2 Diabetes Mellitus	940
65.2.1	Treatment Modalities for Diabetes—Limitations of Pancreas or Islet Transplantation	940
65.2.2	Xenotransplantation Therapy for Diabetes Mellitus	941
65.3	Pancreas Development	941
65.4	Transplanting Pancreatic Primordia to Treat Streptozotocin (STZ) Type 1 Diabetes	941
65.4.1	Rat-to-Rat Isotransplantation and Allotransplantation	941
65.4.2	Rat-to-Mouse Xenotransplantation of Pancreatic Primordia	943
65.4.3	Pig-to-Rat Xenotransplantation of Pancreatic Primordia	943
65.4.4	Pig-to-Mouse Xenotransplantation of Pancreatic Primordia	945
65.5	Pancreatic Primordia for ZDF (Type 2) Diabetes	945
65.6	Summary and Conclusions	949
	References	949

65.1 Introduction

Successful engineering of a new endocrine pancreas to treat diabetes mellitus represents a major challenge in endocrinology. One strategy is to isolate or generate insulin-secreting cells in vitro and transplant them in sufficient numbers to normalize glucose tolerance in a diabetic host [31]. Transplantation of embryonic organ primordia to replace the function of diseased organs followed by growth and differentiation in vivo (organogenesis) offers theoretical advantages relative to transplantation of either pluripotent embryonic stem (ES) cells, or of fully differentiated (adult) organs (whole pancreas or islets) [15]:

1. Unlike ES cells, organ primordia differentiate along defined organ-committed lines. There is no requirement to steer differentiation and no risk of teratoma formation [11]. *In the case of embryonic pancreas*, the glucose sensing and insulin releasing functions of beta cells that differentiate from primordia are functionally linked. Glucose tolerance in formerly diabetic hosts is rendered normal and hypoglycemia does not occur [3–5, 12, 17, 25–28].

2. The growth potential of cells within embryonic organs is enhanced relative to those in terminally-differentiated organs. *In the case of embryonic pancreas*, beta cell mass expands in situ [3, 5, 17]. A diabetic rat host can be rendered euglycemic post isotransplantation of a single primordium [5].

3. The cellular immune response to transplanted primordia obtained early during embryogenesis is attenuated relative to that directed against adult organs [15]. *In the case of embryonic pancreas*, a lesser host immune response is generated post transplantation relative to adult pancreas [12]. If obtained sufficiently early during development, xenotransplantation of pancreatic primordia (chick-to-rat [10] or pig-to-rat [26–28]) is possible without the need for host immunosuppression.

4. Early organ primordia are avascular. The ability of cellular primordia to attract a host vasculature renders them less susceptible to humoral rejection than are adult organs with donor blood vessels transplanted across a discordant xenogeneic barrier. *In the case of embryonic pancreas* (as for islets), this permits transplantation from pig-to-primate without humoral rejection [6, 14, 18].
5. Organ primordia differentiate selectively. *In the case of embryonic pancreas*, exocrine pancreatic tissue does not differentiate following transplantation, obviating complications that can result from exocrine components such as the enzymatic autodigestion of host tissues [3, 5].

65.2
Type 1 and Type 2 Diabetes Mellitus

Type 1 diabetes mellitus (juvenile diabetes), is a syndrome resulting from beta cell destruction that usually results in insulin deficiency presenting during childhood or adolescence. Consistent with an autoimmune etiology, antibodies directed against islet cell antigens are present in the circulation of most type 1 diabetics [2]. Type 2 diabetes (adult onset diabetes) that accounts for the majority of adult patients is thought to result from a progressive insulin secretory defect in the setting of insulin resistance [2]. The number of individuals with diabetes mellitus is rising dramatically in modern societies such as the USA. In 2000 there were approximately 17 million diabetics in the USA, all but approximately 1 million type 2. The total will increase to more than 30 million by 2030 [32].

65.2.1
Treatment Modalities for Diabetes—Limitations of Pancreas or Islet Transplantation

Hyperglycemia represents the major health problem for the diabetic patient. When inadequately controlled, chronic hyperglycemia can lead to microvascular complications such as retinopathy and blindness, nephropathy and renal failure, and macrovascular complications that include atherosclerotic cardiovascular, peripheral vascular and cerebrovascular disease [22, 31]. Use of oral hypoglycemic agents and exogenous insulin are cornerstones of treatment [2, 22, 31]. Administration of drugs or insulin is regulated by external monitoring of glucose levels. Unfortunately, treatment based on external monitoring cannot recapitulate the glycemic control attainable on the basis of glucose sensing and insulin release within the pancreatic beta cell. As a result, control of circulating glucose levels sufficient to prevent complications cannot be attained by most patients. Attempts to maintain euglycemia through intensive insulin therapy lead to an increased incidence of hypoglycemia [22].

Glucose intolerance can be rendered normal after whole pancreas or islet allotransplantation in humans. Given existing technology, a major limitation to the use of either modality is the insufficient supply of human organs. There are fewer than 1,000 whole pancreas transplants performed per year in the USA [9]. Compounding this limitation for islet transplantation is the need to transplant large quantities of islets from multiple donors to achieve even short-term insulin sufficiency [2]. It is likely that problems inherent in the isolation of islets from cadaveric donors or individuals maintained on life support result in diminished viability. Coupled with the limited ability of beta cells within mature islets to replicate, the diminished viability results in a declining pool of functioning engrafted islets over time [2]. An additional problem with pancreas or islet transplantation is the need to immunosuppress recipients. In effect one disease (diabetes) is traded for another (immunosuppression) [2].

Islet transplantation is generally regarded as a therapy for type 1 diabetes only [2, 13]. Selection criteria at most US centers for pancreas transplantation dictate a conservative approach that excludes most type 2 diabetics, traditionally older and poorer surgical risks than type 1 patients. In 1999, type 2 diabetics comprised less than 2% of pancreas recipients [13].

65.2.2
Xenotransplantation Therapy for Diabetes Mellitus

In that they are plentiful and because porcine insulin works well in humans, the pig has been suggested as a pancreas organ donor for human diabetics [19]. Unfortunately, hyperacute and acute vascular rejection (humoral rejection) occurs following transplantation of vascularized pig organs into old world primates and humans. The genesis of humoral rejection reflects largely the reaction of preformed circulating antibodies with antigens present on porcine vascular endothelium. One major target is Galα1-3Galβ1-4GlcNAc;Gal (Gal) [6, 15, 18]. The severity of humoral rejection effectively precludes the use of pigs as whole pancreas donors. However, because they are vascularized by the host post transplantation, islets like other cell transplants are not subject to humoral rejection [6, 15, 18]. Recent experience with pig-to-primate islet [18] or neonatal islet [6] transplantation shows that sustained insulin independence can be achieved, but only through the use of immunosuppressive agents that are not approved for human use or result in an unacceptable morbidity in diabetic primates.

65.3
Pancreas Development

The pancreas is derived from two separate primordia (the dorsal and ventral pancreas) [1, 30]. In the rat that has a 21-day gestation, the primordia arise from the duodenum during embryonic day 11 (E11) of development (dorsal pancreas) and from the endoderm of the hepatic diverticulum E12.1 (ventral pancreas), and later fuse (E13). The head of the pancreas develops from the ventral anlage and the tail from the dorsal anlage. The glandular tissue of the pancreas is formed by the budding and rebudding of cords of cells derived from this primordia mass. Islets are identifiable on approximately E15 [30]. The terminal parts of the cords gradually take on the characteristics of pancreatic acini. Individual endocrine cells, at first indistinguishable from ductal cells, first migrate away from primitive ducts prior to coalescing into islets. A number of cell and interstitial adhesion molecules are thought to guide the migration and coalescence [30]. In pigs (approximately 120 day gestation), the dorsal and ventral primordia are formed on E20–29. The pig pancreas develops much more slowly than the rat pancreas. Cells moderately immunoreactive for insulin can be detected at approximately 4 weeks of gestation. At 10 weeks of gestation, insulin-staining cells appear in the portion of the pancreas derived from the ventral bud. Beginning at 13 weeks of gestation, cells that are intensely immunoreactive for insulin are distributed throughout the parenchyma. However, it is not until approximately 10–13 days after birth that the cells cluster in islets [1].

65.4
Transplanting Pancreatic Primordia to Treat Streptozotocin (STZ) Type 1 Diabetes

65.4.1
Rat-to-Rat Isotransplantation and Allotransplantation

Experimental type-1 diabetes induced using STZ in rodent hosts has been treated successfully using embryonic rodent pancreas transplants. If rat pancreatic primordia are obtained sufficient early during embryogenesis (prior to E17), only the endocrine component differentiates post transplantation [3–5, 17, 25]. Selective endocrine differentiation obviates the problem of host tissue digestion by exocrine tissue that can occur post transplantation of rodent primordia obtained at later times [3, 5]. Brown et al. showed that a partial reversal of STZ diabetes in rats into which fetal pancreases were isotransplanted beneath the renal capsule, was rendered complete following shunting of the venous drainage from the transplants to the liver. Levels of circulating insulin in transplanted rats fell following imposition of the shunt as a result of increased extraction of insulin passing into the liver as well as diminished secretion by the transplants [4]. Hegre et al. proposed that intraperitoneal transplantation is advantageous relative to subrenal-capsular transplantation for diabetes control in that the former: (1) involves more limited surgery, (2) provides a large surface area for implantation, and (3) recapitulates an orthotopic site physiologically, as secreted insulin enters the portal system (via the superior mesenteric vein) rather than the systemic venous system (via the renal vein) [17].

In studies that compared directly the fates of E17–18 rat pancreatic primordia isografts and allografts, allotransplanted nonimmunosuppressed recipients showed only a transient recovery from the diabetes (3–13 days) followed by a return to the diabetic state with graft rejection [17]. However, Eloy et al. showed that xenotransplantation of pancreas from chick embryos can normalize levels of glucose in diabetic rats without the need for *any host immunosuppression* if the embryonic tissue is obtained sufficiently early during chick development [10].

We transplanted whole pancreatic primordia into the mesentery of STZ-diabetic rats [25–27]. STZ produces a rapid beta cell necrosis. If given in sufficiently large amounts to a rat, a single dose will destroy most of the beta cells and induce permanent severe diabetes. A lower dose will kill fewer cells and provoke less severe diabetes [7]. The diabetogenic action of the same dose of STZ varies from animal to animal. However, neither severe STZ-induced diabetes nor the less severe form is reversible in the rat [7, 21], and there is no significant recovery of beta cell numbers after the use of STZ in rats we used in experiments [16]. On E12.5 the rat pancreas is relatively undifferentiated. Dorsal and ventral components remain separate [25]. Shown in Fig. 65.1a is a photograph of a section of duodenum (duo) freshly removed from an E12.5 Lewis rat embryo.

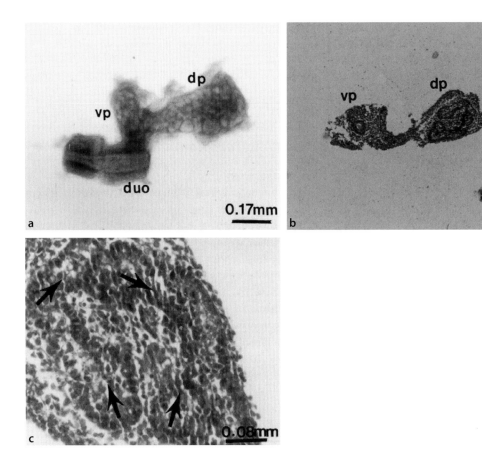

Fig. 65.1a–c Photograph **a** and photomicrographs **b, c** of pancreatic primordia freshly dissected from E12.5 rat embryos. **a** A section of duodenum (*duo*) from an E12.5 Lewis rat embryo. The dorsal pancreas primordium (*dp*) and ventral pancreas primordium (*vp*) are labeled; **b** An H&E-stained section of the dp and vp with the duodenum removed; **c** high-power view of the dorsal pancreas shown in **a** and **b**. *Arrows* delineate a condensing cord of tubulo-acinar cells. Magnifications are shown. (Reproduced with permission [25])

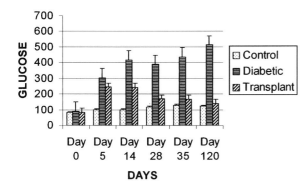

Fig. 65.2 Levels of glucose (mg/dl) measured over time in control rats (Control), rats rendered diabetic using streptozotocin (Diabetic) and streptozotocin-diabetic rats into which 10 E12.5 rat pancreatic primordia were transplanted after glucose measurements were made on day 5 (Transplant). Data are expressed as mean ±SEM. (Reproduced with permission [25].)

The dorsal pancreas (dp) and ventral pancreas (vp) are labeled. Shown in Fig. 65.1b is an H&E-stained section of the dorsal and ventral pancreatic primordia with the duodenum removed. Figure 65.1c provides a higher-power view of the dorsal pancreas shown in Fig. 65.1a, b. A condensing cord of tubulo-acinar cells is delineated (*arrows*).

By 4 weeks post transplantation of whole pancreatic primordia from E 12.5 Lewis rats into the mesentery of a STZ-diabetic Lewis rat, the tissue has undergone differentiation and insulin-positive islets of Langerhans can be delineated amidst stroma. There is no differentiation of exocrine tissue [25]. Abnormal glucose tolerance in STZ diabetic rat hosts is normalized within 2–4 weeks post isotransplantation of pancreatic primordia (Fig. 65.2) [25], as is the pattern of abnormal weight gain over time characteristic of diabetic animals [26]. No host immunosuppression is required for isotransplantation. The normalization of glucose tolerance is life-long [26].

65.4.2
Rat-to-Mouse Xenotransplantation of Pancreatic Primordia

Xenotransplantation of E12.5 Lewis rat pancreatic primordia was carried out in C57Bl/J6 mice. Growth and differentiation post transplantation occurs exactly as for isotransplantation into rats if mice are immunosuppressed. Rat embryonic pancreata do not differentiate if mouse hosts do not receive immunosuppression [25].

65.4.3
Pig-to-Rat Xenotransplantation of Pancreatic Primordia

To establish the feasibility of pig-to-rodent xenotransplantation of pancreatic primordia, we implanted pancreatic primordia from E28–29 pig embryos into STZ-diabetic adult Lewis rats. On E28–29 the pig pancreas is relatively undifferentiated. Dorsal and ventral components remain separate [26]. Glucose tolerance in STZ-diabetic Lewis rats is normalized permanently by the physiological secretion of porcine insulin from transplanted pig pancreatic primordia [26, 27]. As is the case following transplantation of E12.5 rat pancreatic primordia into STZ-diabetic adult Lewis rats or C57Bl/J6 mice, exocrine tissue does not differentiate after transplantation of pig pancreatic primordia into STZ-diabetic adult Lewis rats. However, rather than islets surrounded by stroma that are observed following transplantation of rat pancreatic primordia into rat or mouse mesentery [25], individual alpha and beta cells engraft within the mesentery by 6 weeks after pig-to-rat pancreatic primordia transplantation as demonstrated by light and electron microscopy [26, 27]. While we do not know for certain why this is the case, it may reflect the inability of individual pig islet cells—once they have migrated away from the primitive ducts—to coalesce into islets in the setting of a xenogeneic (rat) extracellular matrix [15].

Figure 65.3a is a photomicrograph of a Gomori-stained pancreas from a normal Lewis rat. Islets stain purple (*arrow*) and exocrine tissue stains pink (*arrowhead*). Figure 65.3b shows a Lewis rat mesentery 6 weeks post transplantation of E12.5 rat pancreatic primordia. A Gomori-stained islet is delineated (*arrow*), but in contrast to what is shown in Fig. 65.3a, no exocrine tissue is present. Figure 65.3c depicts a Gomori-positive endocrine cell (*arrow*) in the mesentery of a Lewis rat 6 weeks post transplantation of E28 pig pancreatic primordia. Since: (1) mRNA for porcine insulin and porcine insulin itself is present only in the mesentery of previously-diabetic Lewis rats following pig pancreatic primordia transplantation [26], (2) porcine insulin but not rat insulin can be detected in circulation [27], and (3) destruction of beta cells and no evidence for regeneration of native beta cells is found in the native rat pancreas following administration of STZ [16], we conclude that glucose tolerance in STZ-diabetic Lewis rats is normalized via secretion of porcine insulin from the pig fetal pancreatic implants [26, 27].

Remarkably and consistent with the findings of Eloy et al. [10], if obtained from E28 [27] or E29 [26] pig embryos, within a developmental "window" prior to E35 [27], pancreatic primordia engraft sufficiently well such that glucose tolerance is normalized in *nonimmunosuppressed immunocompetent* diabetic rat hosts. In contrast, levels of glucose are not reduced post transplantation of E35 pig pancreatic primordia into diabetic rats and transplanted tissue is rejected. In contrast to the case for E28 pancreatic primordia, *renal* primordia obtained from E28 pig embryos do not engraft unless hosts are immunosuppressed [27], and the successful transplantation of rat pancreatic primordia obtained at a comparable developmental stage (E12.5) into mice requires that hosts be immunosuppressed [25].

To determine whether Lewis rats are rendered anergic following transplantation of pig pancreatic primordia, we transplanted E28 pig *renal* primordia each into the mesentery of two nonimmunosuppressed Lewis rats that had been rendered STZ diabetic and into which E28 pig p*ancreatic* primordia had been implanted previously—immediately after glucose levels were measured to confirm diabetes. The first rat was normoglycemic at 2 months post transplantation of E28 pig pancreatic primordia, at

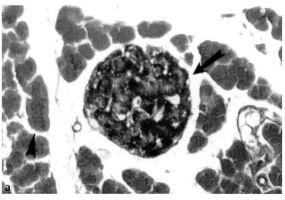
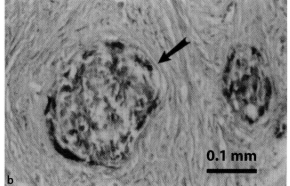
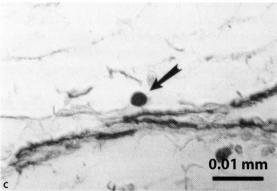

Fig. 65.3a–c Photomicrographs of a Gomori-stained tissue **a** pancreas from a normal Lewis rat. Islets stain purple (*arrow*) and exocrine tissue stains pink (*arrowhead*). **b** Lewis rat mesentery 6 weeks posttransplantation of E12.5 rat pancreatic primordia. A Gomori-stained islet is delineated (*arrow*); **c** A Gomori-positive endocrine cell (*arrow*) in the mesentery of a Lewis rat 6 weeks posttransplantation of E28 pig pancreatic primordia (Reproduced with permission [16, 25, 26])

which time the E28 kidneys were implanted. The rat was sacrificed at 8 months at which time the mesentery was examined and no trace of the implanted E28 renal primordia was observed [27]. Histology of mesentery revealed isolated alpha and beta endocrine cells. Normoglycemia was maintained post-E28 kidney transplantation until the time of death. The second rat was normoglycemic at 12 months post transplantation of pig pancreatic primordia, at which time the E28 pig renal primordia were implanted (Fig. 65.4). Four weeks later a laparotomy was performed and no trace of the transplanted renal primordia could be detected. The laparotomy was closed, and the rat remained normoglycemic [27]. Thus rats transplanted with E28 pig *pancreatic* primordia are not anergic as evidenced by rejection of E28 pig *renal* primordia transplanted *subsequently*. However, engrafted pancreatic primordia remain viable after kidneys are rejected and rats remain normoglycemic throughout [27].

We do not know why the host response to E28 pig pancreatic primordia transplanted into rats differs in this way from that to E28 pig renal primordia transplanted into rats or E12.5 rat renal primordia transplanted into mice. It is possible that the pattern of the growth and differentiation of E28 pig pancreatic primordia as a function of time following implantation in rats (no acinar tissue, no islets, no stroma) (Fig. 65.3c), so radically different from what occurs during normal porcine pancreas development (Fig. 65.3a) [1], and different from what happens after transplantation of rat pancreatic renal primordia into rats (Fig. 65.3b) or mice [25], results in a pattern of antigen expression that is not recognized as foreign by the host. Antigens expressed as a function of E28 pig *renal* primordia differentiation post transplantation might be more widely representative of pig antigens and as such, better stimulators of the host immune response than those expressed after E28 pig pancreatic primordia implantation. Similarly, antigens expressed by *E35* pig pancreatic primordia (that are rejected), perhaps some induced by relative ischemia post transplantation, may render the process of transplanting E35 pig pancreatic primordia more immunogenic [27].

65.4.4
Pig-to-Mouse Xenotransplantation of Pancreatic Primordia

Eventov-Friedman and coworkers implanted embryonic pig pancreatic tissues of different gestational ages beneath the kidney capsule of immunodeficient (NOD-SCID) diabetic mice and immunocompetent diabetic mice that were immunosuppressed. Using NOD-SCID animals, they showed that pancreatic tissue obtained from E42 embryos exhibits reduced immunogenicity relative to that obtained from E56 embryos as determined by a lesser reduction in levels of circulating porcine insulin following immune reconstitution by infusion of human peripheral blood mononuclear cells. In both models, it was possible to normalize the level of glucose following pig pancreatic primordia transplantation [12].

65.5
Pancreatic Primordia for ZDF (Type 2) Diabetes

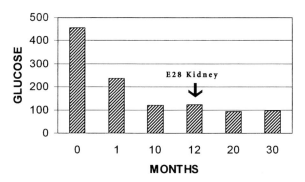

Fig. 65.4 Levels of glucose (mg/dl) in a STZ-diabetic rat measured as a function of time post transplantation of 5 E28 pig pancreatic primordia on day 0. E28 pig renal primordia were transplanted where delineated (E28 Kidney) (Reproduced with permission [27].)

The ZDF rat is an inbred strain derived from a colony of Zucker fatty rats. ZDF and Zucker fatty animals have an autosomal recessive mutation in the gene (fa) that encodes the leptin receptor [24].

Homozygous Zucker fatty rats (fa/fa) manifest hyperphagia, obesity, and severe insulin resistance, but remain normoglycemic. ZDF homozygous males (fa/fa) become hyperglycemic starting after 6 weeks of age and thereafter spontaneously develop overt diabetes [24]. Homozygous females become overtly diabetic beginning at age 6–8 weeks if maintained on a diabetogenic high fat diet. In ZDF males and females, hyperglycemia occurs concomitant with markedly elevated levels of circulating insulin and failure of insulin secretion in response to a glucose challenge, mimicking the pathophysiology of human type 2 diabetes mellitus, Diet restriction permits the use of ZDF rats as breeders, but does not reverse glucose intolerance or insulin resistance [8, 23]. Homozygous dominant (+/+) and heterozygous (fa/+) ZDF rats are lean and do not become diabetic [24].

Once developed, diabetes in ZDF males and females is irreversible [8, 23, 24]. Aging is initially associated with enlargement of islets. From 18 weeks onward, no further enlargement is noted, but islet boundaries become increasingly irregular leading to the appearance of projections of endocrine cells into surrounding exocrine tissue. At the islet boundaries as well as within islets progressive fibrosis is observed. Staining intensity for insulin increases slightly until 10 weeks of age, and thereafter decreases rapidly. Histopathologic changes become increasingly severe in both males and females eventually leading to islet disintegration [20].

To define the utility for transplantation of pig pancreatic primordia in an animal model of human type 2 diabetes, embryonic pancreas from E28 pig embryos was implanted into the mesentery of diabetic ZDF rats. In combination with a standard diet, transplantation of E28 pig pancreatic primordia normalizes glucose tolerance in diabetic ZDF males and females and ameliorates (ZDF diabetic females) or eliminates (ZDF diabetic males) insulin resistance in formerly diabetic rats [28]. Porcine insulin is detectable in plasma of formerly diabetic ZDF rats that receive pig pancreatic primordia transplants. Levels peak at 15 min after an oral glucose load [28]. To localize porcine insulin producing cells following implantation of pig pancreatic primordia into rats in situ hybridization was performed using a porcine proinsulin-specific antisense probe. Cells expressing porcine proinsulin mRNA are present in liver, mesenteric lymph nodes, and pancreas at 40 weeks post transplantation [28].

Shown in Fig. 65.5 are cells within liver (a, b) germinal centers (c, d, *arrows*), and medullary sinuses (c, d, *arrowheads*, e, f) of mesenteric lymph nodes that stain with the antisense probe (a, c, e, f, *red*), but not with the sense probe (b, d). The cells have rounded nuclei with prominent nucleoli and abundant cytoplasm (f, *arrowhead*), consistent with the morphology of engrafted pig beta cells in our previous studies [26, 27] and of pig beta cells in vivo [1]. No staining is observed in pancreas if an excess of unlabeled antisense probe is added to labeled antisense probe (Fig. 65.5g). Scattered cells within pancreas stain positive with the antisense probe (Fig. 65.5h, *arrowheads*), but not with the sense probe (Fig. 65.5i). To confirm that transcripts identified in transplanted ZDF rat tissues by in-situ hybridization are for porcine proinsulin, RT-PCR was performed using primers designed to amplify a porcine proinsulin RNA sequence different from the one recognized by the antisense probe. A band is amplified from RNA originating from transplanted ZDF rat liver that corresponds to a transcript present in adult pig pancreas [28]. Sequencing confirms that transcripts are for porcine proinsulin [28]. A first phase insulin release characteristic of beta cells results within 1 min of glucose addition to media in which mesenteric lymph nodes from transplanted ZDF rats are incubated [28]. Electron microscopy (Fig. 65.6) confirms the endocrine identity of cells within the mesenteric lymph nodes of ZDF rats into which E28 pig pancreatic primordia had been transplanted. Shown are cells that contain granules, some of which have a crystalline core surrounded by a clear space as shown before in Lewis rats [25, 27].

Figure 65.7 shows the characteristic islet disintegration [20] (Fig. 65.7b, c relative to a) in a formerly diabetic ZDF fa/fa rat transplanted with pig pancreatic primordia. The appearance of native islets in transplanted rats with normal glucose tolerance did not differ from that of nontransplanted diabetic ZDF rats. Thus, there is no evidence for regeneration of native islets.

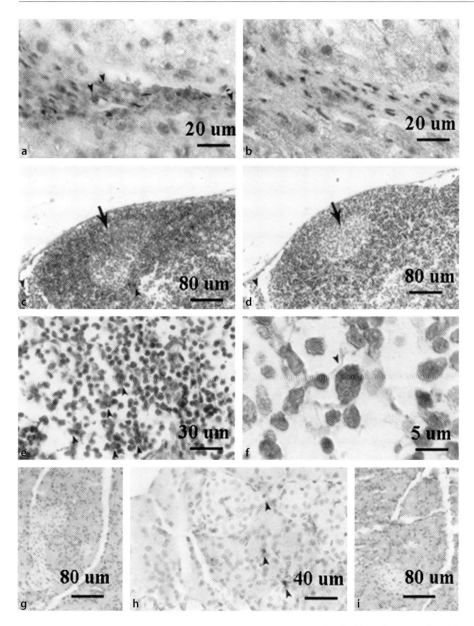

Fig. 65.5a–i In-situ hybridization was performed using pig proinsulin antisense or sense probes on tissue originating from diabetic ZDF rats into which pig pancreatic primordia had been transplanted 40 weeks previously: **a** Liver stained using antisense probe; **b** Liver stained using sense probe; **c** Mesenteric lymph node stained using antisense probe; **d** Mesenteric lymph node stained using sense probe; **e, f** Mesenteric lymph node stained using antisense probe; **g** Pancreas stained with antisense probe and excess of unlabeled antisense probe; **h** Pancreas stained with antisense probe; **i** Pancreas stained with sense probe. *Arrow* delineates germinal centers in **c** and **d**; *Arrowheads* delineate cells that stain positive for porcine proinsulin RNA in **a, c, e, f,** and **h**, and negative staining cells in **d**. Magnifications are shown. (Reproduced with permission [28])

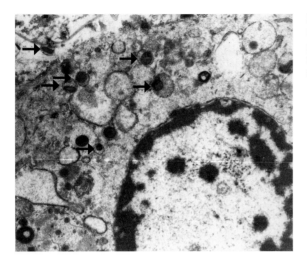

Fig. 65.6 Electron micrograph of a mesenteric lymph node from a ZDF rat transplanted 40 weeks previously with E28 pig pancreatic primordia. *Arrows* delineate granules with crystalline core surrounded by a clear space. Magnification: ×15,000

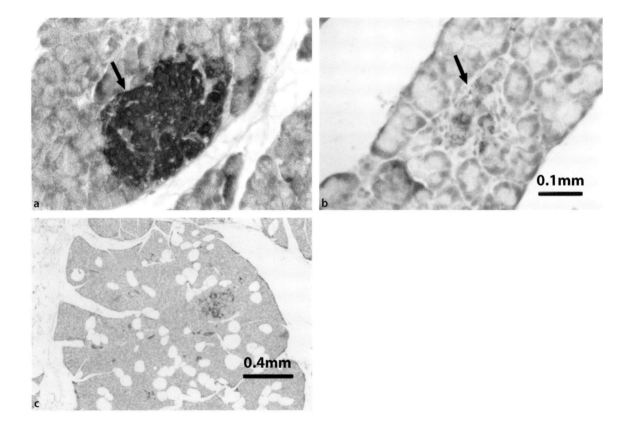

Fig. 65.7a–c Sections of rat pancreas stained using anti-insulin antibodies. **a** A section from a nondiabetic rat. **b, c** Sections from a 30-week-old ZDF fa/fa female with normal glucose tolerance posttransplantation with E28 pig pancreatic primordia. Magnifications are shown for **a** and **b** (in **b**) and for **c**. There is reduced staining intensity for insulin in **b** and **c** relative to **a** and islet structure is severely disrupted in **b** and **c** relative to **a**

65.6 Summary and Conclusions

Immunosuppression of humans following whole pancreas or islet transplantation introduces a set of comorbidities resulting from the use of immunosuppressive agents per se [2]. Our studies [26–28] and those of others [10, 12] indicate that under selected circumstances pancreatic primordia elicit a muted immune response relative to more differentiated tissue such that host immunosuppression is not required for engraftment. These findings are important because they open a window for further exploration of the utility of pig pancreatic primordia transplantation (organogenesis of the endocrine pancreas) as a means to increase functioning beta cell mass and normalize glucose tolerance in human diabetics, possibly in the absence of immunosuppression. In fact, our findings in rats are reproducible in an embryonic pig-to-nonhuman primate xenotransplantation model currently being tested [29]. If endocrine pancreas organogenesis proves to be safe for humans, its implementation will provide an essentially unlimited source of material for transplantation available whenever and wherever needed. This will result in a paradigm shift in how transplantation therapy is carried out for diabetes: (1) transplantation can be done electively at a convenient time; (2) transplantation can be offered to high-risk individuals and repeated as required; (3) transplantation can be offered to individuals who are presently not candidates, such as type 2 diabetics; and (4) transplantation can be carried out in nonimmunosuppressed hosts.

Acknowledgments

MRH is supported by grant 1-2005-110 from the Juvenile Diabetes Research Foundation and by the NIH Washington University George M. O'Brien Center P30 DK079333.

References

1. Alumets J, Hakanson R, Sundler F (1983) Ontogeny of endocrine cells in porcine gut and pancreas. Gastroenterology; 85:1359–1372
2. Bottino R, Trucco M (2005). Multifaceted therapeutic approaches for a multigenic disease. Diabetes 54 (Supplement 2) S79–S86
3. Brown J, Clark WR, Molnar IG et al. (1976) Fetal pancreas transplantation for reversal of STZ-induced diabetes in rats. Diabetes 25: 56–64
4. Brown J, Mullen Y, Clark W et al. (1979) Importance of hepatic portal circulation for insulin action in STZ-diabetic rats transplanted with fetal pancreases. J. Clin. Invest. 64:1688–1694
5. Brown J, Heninger D, Kuret J, Mullen YS (1981) Islet cells grow after transplantation of fetal pancreas and control of diabetes. Diabetes 30: 9–13
6. Cardona K, Korbutt GS, Milas Z et al. (2006) Long-term survival of neonatal porcine islets in rhesus macaques by targeting costimulation pathways. Nature Medicine 12:304–306
7. Chang AY, Dinai AR (1985) Chemically and hormonally induced diabetes mellitus. In: Volk BW, Arquilla ER editors. The Diabetic Pancreas. New York & London. Plenum; 415–438
8. Clark JB, Palmer CI, Shaw WN (1983) The diabetic Zucker fatty rat. Proc Soc. Experimental Biology and Medicine 173:68–75
9. Danovitch GM, Cohen DJ, Weir MR et al. (2005) Current status of kidney and pancreas transplantation in the United States 1994-2003. Am. J. Transplantation 5(Part 2):904–915
10. Eloy R, Haffen K, Kedinger M, Grenier JF (1979) Chick embryo pancreatic transplants reverse experimental diabetes of rats. J. Clin. Invest. 64:361–373
11. Eventov-Friedman S, Katchman H, Shezen E et al. (2005) Embryonic pig liver, pancreas and lung as a source for transplantation: Optimal organogenesis without teratoma depends on distinct time windows. Proc. Natl. Acad. Sci. U.S.A. 102: 2928–2933
12. Eventov-Friedman S, Tchorsh D, Katchman H et al. (2006) Embryonic pig pancreatic tissue transplantation for the treatment of diabetes. PLoS Medicine 7:1165–1177
13. Friedman AL, Friedman EA (2002) Pancreas transplantation for type 2 diabetes at US transplant centers. Diabetes Care 25:1896
14. Groth CG, Korsgren O, Tibell A (1994) Transplantation of porcine fetal pancreas to diabetic patients. The Lancet 344:1402–1404
15. Hammerman MR (2005) Windows of opportunity for organogenesis 15:1–8
16. Hammerman MR (2006) Growing new endocrine pancreas in situ Clin. Exp. Nephrol. 10:1–7
17. Hegre OD, Leonard RJ, Erlandsen SL et al. (1976) Transplantation of islet tissue in the rat. Acta Endocrinol. Suppl 205:257–278

18. Hering B, Wijkstrom M, Graham M et al. (2006) Prolonged diabetes reversal after intraportal xenotransplantation of wild-type porcine islets in immunosuppressed nonhuman primates. Nature Medicine 12:301–303
19. Ibrahim Z, Busch J, Awwad M et al. (2006) Selected physiologic compatibilities and incompatibilities between human and porcine organ systems. Xenotransplantation 13: 488–499
20. Janssen SWJ, Hermus ARMM, Lange WPH et al. (2001) Progressive histopathological changes in pancreatic islets of Zucker Diabetic fatty rats. Exp. Clin. Endocrinol Diabetes 109:273–282
21. Johansson KA, Grappin-Botton, A (2002) Development and diseases of the pancreas. Clinical Genetics 62: 14–23
22. Peck, AB, Chaudhari M, Cornelius JG et al. (2001) Pancreatic stem cells: building blocks for a better surrogate islet to treat type 1 diabetes. Annals of Medicine:33:186–192
23. Peterson RG, Shaw WN, Neel MA (1990) Zucker diabetic fatty rat as a model for non-insulin-dependent diabetes ILAR News 32:16–19
24. Phillips MS, Hammond HA, Dugan V (1996) Leptin receptor missense mutation in the fatty Zucker rat Nat. Genetics 13: 18–19
25. Rogers SA, Liapis H, Hammerman MR (2003) Intraperitoneal transplantation of pancreatic primordia ASAIO Journal 49: 527–532
26. Rogers SA, Chen F, Talcott M et al. (2004) Islet cell engraftment and control of diabetes in rats following transplantation of pig pancreatic primordia. Am. J. Physiol. 286:E502–509
27. Rogers SA, Liapis H, Hammerman MR (2005) Normalization of glucose post-transplantation of pig pancreatic primordia into non-immunosuppressed diabetic rats depends on obtaining primordia prior to embryonic day 35. Transplant Immunology 14: 67–75
28. Rogers SA, Chen F, Talcott M, Liapis H et al. (2006) Glucose tolerance normalization following transplantation of pig pancreatic primordia into non-immunosuppressed diabetic ZDF rats. Transplant Immunology 16:176–184
29. Rogers SA, Chen F, Talcott MR, Faulkner C et al. (2007) Long term engraftment following transplantation of pig pancreatic primordia into non-immunosuppressed diabetic rhesus macaques. Xenotransplantation 14:591–602
30. Slack JMW (1995) Developmental biology of the pancreas. Development 121:1569–1580
31. Thivolet C (2001) New therapeutic approaches to type 1 diabetes: from prevention to cellular or gene therapies. Clinical Endocrinology. 55:565–574
32. Wild S, Roglic G, Green A et al. (2004) Global prevalence of diabetes. Estimates for the year 2000 and projections for 2030. Diabetes Care 27:1047–1053

Regenerative Medicine Applications in Hematology

A. Wiesmann

Contents

66.1 Introduction 951
66.2 Assays for HSCs 952
66.3 Donors 954
66.4 Sources of HSCs 955
66.5 Complications of HSC Transplantations ... 956
66.6 Manipulation of the Graft 956
66.7 Conclusions 960
     References 961

66.1 Introduction

Hematopoietic stem cell (HSC) transplantation provides a potential cure for a variety of hematological and non-hematological diseases. In the 1950s studies demonstrated that total-body irradiation caused lethal damages but that animals could be rescued by the transfusion of bone marrow from another animal. The first attempt to transfer these results to patients with leukemia was performed in 1957 [92]. After the identification and typing of the human leucocyte antigens (HLAs) in 1958, the major histocompatibility complex and the development of immunosuppressive drugs, allogeneic HSC transplantation became possible. Due to matched HLA antigens, the first long-term survivors after allogeneic bone marrow transplantation were reported in 1968 [5]. During the 1980s autologous bone marrow transplantations were successfully performed in patients with lymphomas after conventional chemotherapy had failed. With a better understanding of stem cell biology and transplant immunology, as well as better supportive care, the number of transplants has increased. Today, more than 20,000 allogeneic and more than 30,000 autologous HSC transplantations are performed yearly worldwide. The underlying diseases for autologous HSC transplantation are most often multiple myeloma or Non-Hodgkin lymphoma. About 50% of allogeneic transplantations are done in patients with acute leukemias. Table 66.1 shows additional indications for HSC transplantation, including non-malignant diseases such as severe aplastic anemia and hemoglobinopathies.

HSC transplantations result in higher remission rates and cures for many patients compared to conventional treatments. Unfortunately, morbidity and transplant-related mortality rate is also higher compared to alternative therapies, ranging from below 2% for some autologous transplant settings to up to 40% for patients with advanced malignant diseases who undergo allogeneic stem cell transplantation. The protocol for allogeneic transplantation is essential for success and is more and more individualized with the aim to eliminate malignant cells, suppress host immunity, prevent graft rejection, to minimize graft-versus-host disease (GVHD) but to allow graft-versus-tumor (GVT) reaction, and to reduce organ toxicity. High-dose myeloablative conditioning regimens, reduced intensity conditioning (RIC), different combinations of cytotoxic drugs, whole body

Table 66.1 Indications for HSC transplantations

	Autologous HSCT	**Allogeneic HSCT**
Malignant diseases	Non-Hodgkin's Lymphoma (NHL) Multiple myeloma (MM) Hodgkin's Disease (HD) Acute myeloid leukemia (AML) Germ-cell tumor	Acute myeloid leukemia (AML) Acute lymphoblastic leukemia (ALL) Chronic myeloid leukemia (CML) Non-Hodgkin's Lymphoma (NHL) Myelodysplastic syndromes (MDS) Myeloproliferative diseases (MPS) Hodgkin's Disease (HD) Chronic lymphatic leukemia (CLL) Multiple myeloma (MM)
Non-malignant diseases	Autoimmune diseases	Aplastic anemia (AA) Paroxysmal nocturnal hemoglobinuria (PNH) Thalassemia major Sickle cell anemia Severe combined immunodeficiency (SCID)

irradiation, total lymphoid irradiation, immunosuppressive agents (purine analoga), and T-cell depleting antibodies can be part of the transplant protocols. Allogeneic transplantation became possible with unrelated mismatched or related haploidentical grafts even for heavily pretreated high risk patients and patients older than 50 years of age.

HSCs are a rare population in the bone marrow and characterized by self-renewal and multilineage differentiation potential. They can proliferate and differentiate into all hematopoietic lineages. Furthermore, they can produce daughter cells, which retain stem-cell features. Both abilities provide a lifelong production of all blood cells (Fig. 66.1).

Murine HSCs were demonstrated to express the phenotype Sca1+ (stem cell antigen) c-kit+ and are negative for mature lineage markers such as B220, T-cell receptors, CD8, GR-1, and Mac-1 (Lin-). The first studies showed that as few as 1,000 purified HSCs can rescue irradiated animals from lethal aplasia [83]. More recently, it was proven that a single stem cell (Sca-1+, c-kit+, Lin-, CD34-) can reconstitute long-term lymphohematopoiesis [64]. In contrast to murine HSCs, human HSCs are most often characterized and isolated in clinics as well as in research by the expression of CD34 [14]. Despite all efforts to purify a homogenous stem cell population by using morphological markers, cell-cycle status, monoclonal antibodies to cellsurface molecules, the isolated cell population remains heterogenous [101].

66.2 Assays for HSCs

After the isolation of HSCs, stem cell function can be tested by several assays [17]. There are different in vivo and in vitro tests to detect distinct cell populations of the hematopoietic cell hierarchy (Table 66.2).

In the culture colony-forming unit or CFU-C assay, isolated cells are tested for their ability to produce colonies in a cytokine supported semisolid medium. Colony formation is not uniquely associated with HSCs but also with hematopoietic progenitor cells, and the results are fairly quantitative. Detected colonies can be differentiated into erythroid (CFU-E), granulocyte-macrophage (CFU-GM), granulocyte-erythroid-macrophage-megakaryocyte (CFU-GEMM), etc. Another in vitro assay—usually termed the long-term culture initiating cell (LTCIC) assay—analyzes HSCs upon bone marrow derived stroma cells (feeder layers). During this first phase, HSCs are expanded without differentiation into committed progenitor cells [23]. After harvesting the expanded cells, CFU-C assays read out the products of the initial expansion [68, 87]. As the co-culture is initiated under limiting dilution conditions, the frequency of HSCs can be estimated. More colonies are interpreted with a higher number of initiating HSCs. Co-culture of HSC and stroma cells allows mobile stem

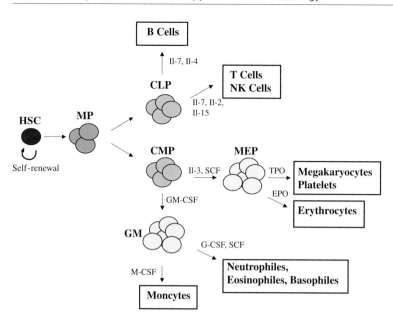

Fig. 66.1 Model for the hematopoietic stem cell hierarchy. HSC have the potential to divide and self-renew or differentiate into common progenitor cells. By subsequent further divisions, lymphoid cells, NK cells, megacaryocytes, platelets, erythrocytes, and myeloid cells are produced. Indicated cytokines play a critical role during differentiation into all blood cell lineages. *HSC* hematopoietic stem cell, *MP* multipotent progenitor cell, *CLP* common lymphoid progenitor cell, *CMP* common myeloid progenitor cell, *MEP* megakaryocyte/erythroid progenitor cell, *GMP* granulocyte/macrophage progenitor cell, *NK* natural killer, *Il* interleukin, *SCF* stem cell factor, *TPO* thrombopoietin, *EPO* erythropoietin, *G-CSF* granulocyte colony-stimulating factor, *GM-CSF* granulocyte-macrophage colony-stimulating factor, *M-CSF* macrophage colony-stimulating factor

Table 66.2 Hematopoietic cell assays

Assay	HSC	Primitive progenitor	Committed progenitor	Lineage committed precursor
CFU-C	+	+	+	+
CAFC (day 7)	–	(+)	+	(+)
CFU-S (day 8)	–	(+)	+	(+)
CFU-S (day 12)	(+)	+	(+)	–
CAFC (day 28)	+	(+)	–	–
CRU	+	–	–	–
SRC	+	–	–	–

CFU-C culture colony-forming unit; *CAFC* cobblestone-area-forming cell; *CFU-S* colony-forming unit spleen; *CRU* competitive repopulating units, *SRC*: NOD/SCID repopulating cells

cells to interdigitate between and beneath the stroma cell layer. These "covered" HSC can proliferate and form clusters of packed cells with a non-refractile appearance by phase contrast microscopy. These cell clusters are called cobblestone areas, and the assay to quantitate these clusters is called the cobblestone-area-forming cell (CAFC) assay [60, 68]. The number of cobblestone areas can be correlated to HSC activity in vivo and the kinetics of cobblestone area formation reflects the maturation of the initiating cells. More primitive hematopoietic cells require a longer period of time to establish a cobblestone area.

These in vitro liquid culture assays do not measure long-term engraftment, and therefore may overestimate progenitor cell expansion. Furthermore, the complexity and the length of the culture procedure hinder the reproducibility.

Although in vitro assays are easy to perform, and some correlation can be made between in vitro assays and the long-term reconstitution of lethally irradiated animals, a better way to determine HSC function is an in vivo transplantation. In vitro assays are limited to support only a few hematopoietic lineages and mammalian cells are known to adapt to tissue culture systems. The so-called colony-forming units spleen (CFU-S) assay was first described in 1961 [93]. Irradiated animals were examined for clonally derived colonies of proliferating hematopoietic cells in the spleens on defined days after transplantation. In the second phase of transplant experiments self-renewal and multilineage potential was demonstrated [79, 80] but a broad heterogeneity in the ability to initiate new splenic colonies was observed. During the following decades, it was shown that early CFU-S (day 8) disappear a few days later and that most of the late CFU-S (day 11–14) arise from a different group of hematopoietic cells [51]. Certain HSC preparations fail to form spleen colonies. Possible explanations include either inhibitory cytokines or splenic seeding considerations. Furthermore, it became obvious that the formation of spleen colonies is not a unique characteristic for primitive HSCs. An in vivo clonal assay for long-term repopulation is the limiting of dilution competitive repopulation to derive a measure of competitive repopulating units (CRUs). Small numbers of genetically marked cells are transplanted into irradiated mice. Normal bone marrow cells for radioprotection are added and transplanted mice are analyzed for donor-derived blood cells over long periods of times [10, 88]. The isolation of mouse HSCs from bone marrow resulted in a repopulation of recipient animals after the transplantation of a single injected cell [81, 84]. Although seeding efficiencies are unknown, this syngeneic model provides a valuable system for HSC function and one can compare engraftment at 12–15 weeks post-transplant. For evaluating human HSC function in vivo, enriched stem cells can be transplanted into immunosuppressed mice, e.g., irradiated NOD/SCID (non-obese diabetic/severe combined immunodeficient) mice. Peripheral blood cells of transplanted animals are then analyzed for the expression of human CD45 antigen over time [44]. Although in vivo assays examine true stem cell function, direct conclusions for human HSC transplantations should be considered with caution, as the first assay is an animal model with mouse HSCs, and the latter is a xenotransplantation model. The marrow microenvironment of NOD/SCID mice has an altered murine immune system, which does not represent the human stem cell niche. In addition, stem cell homing to the murine bone marrow may not reproduce the homing of human HSC to human bone marrow.

66.3
Donors

High-dose chemotherapy supported by autologous HSC transplantation has become a widely applied therapy in many hematological malignancies such as Non-Hodgkin's lymphoma (NHL) and multiple myeloma. After conventional chemotherapy, followed by the application of a granulocyte colony-stimulating factor (G-CSF), HSC are mobilized into the peripheral blood, and can be harvested by leukapheresis. In this autologous transplant setting, donor and patient are the same person. Hematopoietic recovery occurs fast and lethal complications are rare. Transplant related mortality has been in the order of or below 2%. Therefore, patients up to 65 years of age have been included in recent clinical trials. Unfortunately, the contamination of the graft with malignant cells remains the main cause of relapse.

For allogeneic HSC transplantation, a sibling with the same two HLA haplotypes as the recipient is the best donor (MRD = matched related donor). Transplantations between twins (syngeneic transplantation) are complicated by an increased risk of relapse due to a lack of GVT reaction [38]. With no sibling, there is a chance of about 70% to find a matched unrelated donor (MUD). Large registers of unrelated donors are established and more than 10 million donors are available today. Recently, genetic methods have replaced the phenotypic method for tissue typing. As a consequence, the number of identified alleles has increased substantially. Many transplant centers are aiming for a 10/10 match regarding HLA-A, B, C, DR, and DQ. Mismatches have been associated with graft rejection as well as GVHD [66]. If no matched family or suitable unrelated donor is available, a transplantation from a parent, sibling, or child of the patient with only one identical HLA haplotype can be considered (a haploidenitical donor). Initially associated with extremely high rates of GVHD and graft failure, pharmacoprophylaxis of GVHD, supportive care, patient selection, and technical advantages have improved the transplant-related mortality of this approach. Promising results have been reported after vigorously T-cell depletion combined with high-dose HSC transplantations [4]. Recent data demonstrates the involvement of natural killer (NK) cells (CD56+CD3−) into alloreactive mechanisms of a haploidentical HSC transplantation [56, 57]. NK cells express combinations of activating and inhibitory killer-cell immunoglobin-like receptors (KIRs), which can interact with class I HLA epitopes. Cytolytic activity of NK cells is determined by the balance of signals. NK cells are the first lymphoid cells that reconstitute after HSC transplantation 2–3 weeks after transplantation and are activated by the absence of self-MHC class I molecules on the surface of target cells [62]. The expression of appropriate self-MHC molecules on target cells delivers an inhibitory signal to NK cells. Donor-versus-recipient NK cell alloreactivity derives from a mismatch between inhibitory receptors for self-MHC class I molecules on donor NK cells and MHC class I ligands on recipient cells. This alloreactivity results in better engraftment, less GVHD, and reduced relapse rates [74]. Furthermore, missing a KIR ligand was shown to be associated with a higher risk of severe acute GVHD [55]. Among transplantations with unrelated donors, about 50% are mismatched for 1 or 2 HLA alleles. Therefore, the alloreactivity of NK cells can be considered when choosing the best donor.

66.4 Sources of HSCs

Bone marrow was the first source for HSCs. It is obtained by the repeated aspiration of the posterior iliac crests while the donor is under general anesthesia. A discomfort from the harvest disappears within one or two weeks and severe complications are rare.

Besides bone marrow, HSCs appear in the peripheral blood. HSCs detach from their stem cell niche in the bone marrow continuously, and enter the circulation before returning to the bone marrow. The number of circulating HSCs can be increased by applying either the cytokines G-CSF or the granulocyte-macrophage colony-stimulating factor (GM-CSF) prior to stem cell collection. For autologous transplantation, the cytokines are given after chemotherapy to increase the yield of the stem cell harvest. AMD3100 is a reversible inhibitor to the CXC chemokine receptor 4 (CXCR4), a receptor on CD34+ cells, which mediates the adhesion to the bone marrow. Recent data suggests that the combination of G-CSF and AMD3100 is superior in HSC mobilization than G-CSF alone [28]. Leukapheresis is performed through antecubital veins. Up to 25 liters of blood are processed within 4 hours, which usually result in enough CD34+ peripheral blood stem cells for a rapid engraftment after transplantation (Fig. 66.2).

Peripheral blood HSC can be frozen and stored at temperatures below –120°C and therefore be viable for years. Compared to bone marrow, mobilized peripheral blood HSC result in faster neutrophil and platelet recovery. Another source of HSC is cord blood, which was first used in a child with Fanconi anemia [32]. Cord blood is collected immediately after birth, rich in HSCs but limited in volume. After being frozen, it is usually transplanted to unrelated patients, when no matched donor is available. Today more than 6,000 cord blood transplantations

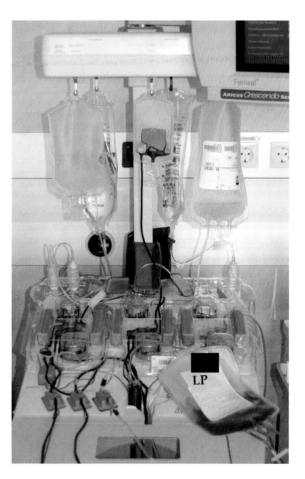

Fig. 66.2 Peripheral blood stem cell collection by leukapheresis (*LP* leukapheresis product)

are performed worldwide and cord blood banks in more than 20 countries store about 170,000 units. Due to low stem cell numbers, hematopoietic recovery is slowly and infection often occurs soon after transplantation. On the other hand, mismatched HSC from cord blood are less likely to cause GVHD, without loosing the graft-versus-leukemia effect. Today, about 20% of all pediatric transplantations are performed with umbilical cord blood HSCs. However, the prolonged immune reconstitution after cord blood HSC transplantation still remains a significant barrier.

66.5
Complications of HSC Transplantations

HSC transplantation is a curative therapy for many neoplastic and non-neoplastic diseases. However, relapse, GVHD, graft rejection, and opportunistic infections still remain the major complications. The rate of infectious death after allogeneic unrelated transplantations is about 20% and after cord blood transplantation it is as high as 40% [45, 71]. The conditioning regimen damages the thymic microenvironment resulting in a decreased Il-7 production, which is necessary for thymocyte development. Therefore, T-cell reconstitution from naïve T-cells in the thymus is reduced. Due to prolonged immunodeficiency viral (CMV, EBV) and fungal infections (Candida, Aspergillus) are now the major contributors for transplant mortality, usually occurring after day 50, post-transplant.

GVHD occurs when immunocompetent cells of the graft-recognizing histocompatilility antigens start an immune attack against the cells of the transplant recipient. GVHD can be classified as either acute (within the first 100 days) or chronic GVHD, which are different entities. Chronic GVHD is less well understood, but there are good animal models for acute GVHD. Acute GVHD involves skin, intestine, liver, lung, thymus, and secondary lymphoid tissues. The pathophysiology of this complex disease can be divided into several stages, starting with the conditioning regimen. The tissue damage is followed by a release of proinflammatory cytokines (cytokine storm), which leads to an expansion of donor T-cells mediated by antigen presenting cells. Alloreactive T-cells then differentiate into either Th1/Tc1 or Th2/Tc2 cells and subsequently migrate to the target tissues. The recruitment of other effector leucocytes results in further tissue injury and cytokine release. Post-transplant pharmacological agents used to prevent GVHD include cyclosporine A (a calcineurin inhibitor), tacrolimus (blocks Il-2 production and T-cell expansion), rapamycin (inhibitor of cytokine responsiveness), and lympholytic agents such as corticosteroids. In vivo, T-cell depletion can be performed by antithymocyte globulin (ATG) or anti-CD52. Due to the minor histocompatibility antigens the risk of GVHD is higher among matched unrelated donors compared to matched related donors. Recent results also suggest the importance of human KIRs, which are expressed on NK and T-cells, for the incidence of GVHD. The gene complex is located on chromosome 19, and therefore independent of HLA on chromosome 6. The direction of KIR/KIR ligand incompatibility may also influence graft rejection and GVT effect, but clinical data is conflicting [74].

Graft rejection (host-versus-graft reaction, or HVD) occurs when the conditioning regimen was not sufficient and remaining immunocompetent cells in the recipient attack the graft. Mediated by NK cells, graft rejection occurs most often after the transplantation of an HLA-mismatched graft and/or after T-cell depletion.

66.6
Manipulation of the Graft

Autologous HSC transplantation is often performed in elderly and heavily pretreated patients. HSC frequency decreases with age and chemotherapy reduces the stem cell pool further. Therefore, a sufficient stem cell number, which is important for rapid hematopoietic reconstitution to avoid serious infectious complications, can be a limiting factor. Cytokine-induced ex vivo expansion of HSC was shown to result in a higher number of phenotypically defined stem cells [35]. An expansion of HSC and progenitor cells of 50-fold or more was achieved. Murine transplant models demonstrated that the stem cell factor (SCF), the Flt-3 ligand (FL), and thrombopoietin (TPO) are crucial cytokines [9, 16, 67]. The SCF improves the homing capacity in animal models, whereas FL supports short-term expansion and alters the expression of adhesion molecules such as the very late antigens (VLA-) 4 and VLA-5 [19, 20, 39, 65, 82]. The combination of TPO and FL seems to reduce apoptosis and support self-renewal by preventing telomere degradation [29, 59]. However, even in short-term cultures with only a moderate expansion of cells, a loss of stem cell function is observed [100]. In vitro cultured cells enter the cell cycle [30, 31], and induce apoptosis by the increased expression of the Fas ligand (CD95), downregulation of antiapoptotic proteins (Bcl-2), and increased activation of caspases [25, 49]. Furthermore, the homing of HSC to the bone marrow can be disrupted by the altered expression of integrins [63]. So far, it does not seem to be possible to expand peripheral blood HSCs without losing stem cell function. However, long-term hematopoietic reconstitution was achieved in breast cancer patients treated with high-dose chemotherapy followed by a transplantation of cytokine induced in vitro-expanded bone marrow [86]. Another problem among autologous HSC transplantation is the contamination of the graft with malignant cells. Depletion of the tumor cells from the graft is a form of negative selection. Peripheral blood HSCs can be incubated in vitro with mafosfamide, washed and further cultured in the presence of cytokines to allow cell proliferation before transplantation [53]. Using this two-step culture technique, residual leukemic cells can be eradicated during the first culture period, which also results in a 90% death of progenitor cells while SCID repopulating cells can be maintained. During the second culture period, cell proliferation can be achieved. A triple purging strategy combined positive CD34+ selection with short-term imatinib and mafosfamide incubation followed by cytokine supported in vitro culture. CML patient aphereses were eliminated from BCR-ABL+ malignant cells while preserving normal progenitors [102]. Another attempt to decrease acute myeloid leukemia blast cells from the graft is the application of hyperthermia at 42°C before kryoconservation [27]. More often, harvested peripheral HSCs are positively selected for CD34 expression. Although the tumor burden can be reduced in cell numbers by several logs, no difference of disease-free and overall survival was observed in patients with multiple myeloma after CD34+ selection compared to unmanipulated grafts [85]. One problem for using CD34 expression as a marker for positive selection remains the expression by tumor cells, e.g., Ewing's sarcoma, and acute myeloid leukemia [27, 103]. In addition, the expression of cell surface molecules can change depending on the activation status of the cell. Therefore, a direct correlation between phenotype and stem cell function is not allowed. CD34 is also expressed on early progenitor cells. Some animal models suggest that CD34+ cells are not the cells responsible for long-term hematopoietic reconstitution [18, 37]. Furthermore, differences between CD34+ CD38– cell numbers and SCID-repopulating capacity have been described after the in vitro manipulation of HSC [26], and it is likely that HSCs posses a CD34+ and CD34– phenotype [105]. Another—probably more immature—marker is CD133, which even can be combined with CD34 to differentiate between leukemic cells and HSCs [27, 104]. Nevertheless, positive selection techniques are routinely performed for patients with many malignant diseases undergoing autologous stem cell transplantation. Despite all efforts, purging techniques do not improve survival after autologous transplantation and the failure to eradicate cancer still remains the main cause of relapse.

Allogeneic umbilical cord blood transplantation has become an established therapy for patients without a matched donor. The main problem is a sufficient HSC number in the cord blood unit to allow engraftment for an average-sized adult. More than 2×10^7 nucleated cells/kg is recommended and several efforts have tried to achieve cell dose guidelines. Improving the harvesting of cord blood units and the post-thaw procedures can increase the number of total nucleated cells [12, 73]. Another way to overcome the problem of the stem cell number is the pooling of two cord blood units, which has already been performed with promising results [6, 22]. Many preclinical results demonstrated the feasibility of cord blood stem cell expansion in vitro. Non-contact and stroma-free liquid cultures were shown to maintain long-term engrafting cells, which engraft secondary or tertiary murine hosts [48]. Furthermore, cytokine-supported cultures of cord blood cells resulted in increased total cell numbers, CFCs, and CD34+ cells [54]. However, an increase in stem cell function measured by long-term culture potential or in vivo engrafting potential in SCID mice was not observed [24]. Therefore, it was not astonishing that initial efforts to expand umbilical cord blood stem cells and to transplant the expanded cells with or without unmanipulated cord blood were largely unsuccessful. More mature and differentiated cells rather than true stem cells were expanded [70]. In vitro expanded cells were shown to result in faster engraftment at the expense of long-term engraftment in animal models [52]. Patients, who were transplanted with ex vivo manipulated cord blood cells failed to show improvement in their hematopoietic recovery after high-dose chemotherapy [78]. In contrast to bone marrow cultures, cord blood cell cultures do not develop a stroma cell layer, which may be critical for the maintenance of stem cell function. Although Schofield hypothesized a stem cell niche in the bone marrow as early as 1978 [76], it was only in the last decade that we began to understand molecular pathways influencing the stem cell niche. Mesenchymal stem cells (MSCs) are pluripotent cells that can differentiate into osteoblasts, chondroblasts, adipocytes, and myelosupportive stroma cells, thereby providing cytokines, adhesion molecules, and extracellular matrix molecules and influencing proliferation, apoptosis, homing, and the differentiation of hematopoietic cells. MSCs are known to support HSCs in co-culture systems and they facilitate engraftment after the transplantation of multidonor cord blood HSC transplantation into NOD/SCID mice [42, 61]. The first clinical studies demonstrated, that MSCs are able to facilitate HSC engraftment even after graft failure/rejection [47]. Furthermore, MSCs have been shown to prevent or arrest the progression of treatment-resistant acute GVHD [46, 94]. Other key players to regulate hematopoiesis are osteoblasts. Osteoblasts express many molecules (G-CSF, GM-CSF, M-CSF, Il-1, TGF-ß, SDF-1) on their surface. In a co-culture of HSC on an osteoblast monolayer, HSC were increased measured by LTC-IC [90]. Osteoblastic cells not only regulate the HSC niche (Fig. 66.3), but also support B-lymphocyte commitment and differentiation from HSCs [13, 41, 89, 98, 108].

Several molecular pathways have been shown to influence the HSC niche, such as bone morphogenetic proteins (BMPs), Wnt/ß-catenin, Notch, and Tie2—a receptor tyrosine kinase [2]. BMPs are members of the TGF-ß family, which influence self-renewal of HSCs. BMP receptor activation of spindle shaped N-cadherin+CD45- osteoblastic cells control the size of stem cell niche by N-cadherin and ß-catenin binding [107]. Wnt-mediated signaling involves the activation of target genes, such as cyclin D1 and c-Myc. When CD34+ bone marrow cells were co-cultured on Wnt transduced stroma feeder layers, the proliferation of HSCs was induced [97]. Direct evidence for the involvement of Wnt was observed by maintaining the naïve phenotype of HSC in long-term cultures and increased long-term hematopoietic engraftment ability [58, 69]. Agents such as the glycogen synthase kinase-3 inhibitor 6-bromoindurubin-3-oxime, which activates the Wnt pathway, may enhance HSC repopulation [75, 95]. Notch signaling influences differentiation, apoptosis, proliferation, and the adhesion of cells. In mammals, five Notch ligands have been identified: Jaged-1/2 and Delta-1/3/4. Depending on Notch ligand concentration during the ex vivo

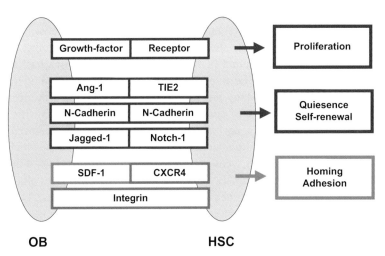

Fig. 66.3 Interactions between osteoblasts and HSCs in the niche. Osteoblasts influence HSCs in the bone marrow by various mechanisms. Different growth factors expressed on OB bind to their corresponding receptors on HSCs, which regulates HSC proliferation. Quiescence and self-renewal of HSC are influenced via Ang-1/Tie2, N-cadherin, and Jagged-1/Notch-1 mediated signaling pathways. Several integrins and SDF-1/CXCR4 interaction facilitate adhesion and homing from HSC to the niche. (*OB* osteoblast, *HSC* hematopoietic stem cell, *Ang-1* angiopoietin-1, *TIE2* endothelial tyrosine kinase receptor, *SDF-1* stromal cell derived factor-1, *CXCR4* chemokine receptor 4)

culture of HSCs, the hematopietic recovery of NOD/SCID mice was enhanced (low ligand concentrations) or decreased (higher concentrations) due to apoptosis [21]. The receptor tyrosine kinase Tie2 on HSC and its ligand angiopoietin-1 (Ang-1) expressed by osteoblasts in the marrow may induce the adhesion of HSCs to the niche and keep the stem cells in a quiescence state. The overexpression of Ang-1 was shown to protect HSCs from myelosuppressive stress [1]. Recent attempts to expand HSCs in vitro include the inhibition of transcription. The methyltransferase inhibitor 5-aza-2'-deoxycytidine (5azaD) and the histone deacetylase inhibitor trichostatin A (TSA) sustain the self-renewal of gene transcription and may expand HSCs in vitro [3].

Immune reconstitution plays an important role after autologous and allogeneic transplantation, and infectious complication still remains a major clinical complication. The delayed immune reconstitution, especially in adults, after the transplantation of cord blood HSCs may be due to immature cells in the cord blood and a decreased lymphopoiesis with age. However, ex vivo expansion does not seem to alter the capacity of umbilical cord blood CD34+ cells to generate functional T lymphocytes and dendritic cells [43, 72]. On the other hand, it seems to be attractive to reduce the immunological process of GVHD without affecting the GVT reaction by selective expansion strategies. T regulatory (Treg) cells have been identified to reduce GVHD and prevent graft rejection without affecting the relapse of the underlying disease [15, 33, 36, 91]. Treg cells express CD4, CD25, and the intracellular protein Foxp3. With the more recently used combination of the markers CD4, CD25, and CD127 (Il-7Rα) highly suppressive Tregs can be isolated without impairing the cell function due to the permeabilization of the cells [50, 77]. Treg cell function and expansion can be influenced by immunosuppressive drugs, which are given to prevent or treat GVHD. Cyclosporine A, a calcineurin inhibitor, was shown to reduce Treg cell proliferation and function by decreased Il-2 production, whereas rapamycin increased Treg cell numbers [40, 96, 106].

In an allogeneic transplantat setting, GVHD remains a major complication. Donor T-cell depletion from the graft reduced the incidence of GVHD, but graft rejection, viral infection and the relapse of the malignant underlying disease increased [99]. With no available matched related or unrelated donor, virtually every patient has a suitable haploidentical related donor (parents or children). This transplant modality has a high risk of GVHD and graft rejection. To overcome the HLA barrier, a megadose of HSC ($>10 \times 10^6$ CD34+ cells/kg) combined with T-cell depletion by agglutination and E-rosetting resulted in a low rate of GVHD, but the relapse rate was high among patients transplanted who were not in complete remission [4, 34]. A promising approach of graft manipulation seems to be the CD3/CD19 depletion with anti-CD3 and anti-CD19-coated microbeads on a CliniMACS device (Fig. 66.4).

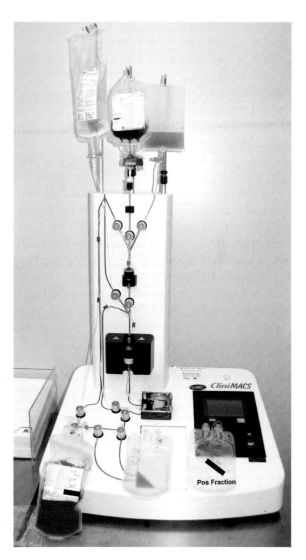

Fig. 66.4 CD3 and CD19 depletion on a CliniMACS device

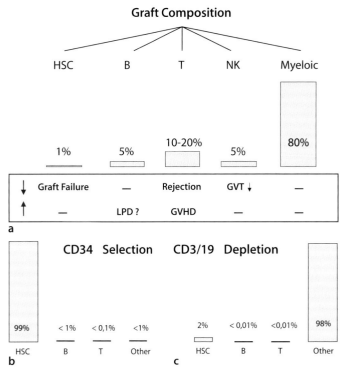

Fig. 66.5a–c Influence of manipulation techniques on the cell composition of the graft. **a** Unmanipulated HSC products from the peripheral blood contain the indicated percentages of HSC, myeloic, T, B, and NK cells. Several complications can be induced by different cell types post-transplant. Low numbers of HSC may result in graft failure; high B cell numbers in the transplant are discussed as a reason for lymphoproliferative disease (LPD). In contrast, reduced numbers of NK cells can result in an insufficient GVT effect followed by a relapse of the underlying disease. Most critical are T cells, which are key players for GVHD and rejection of the graft. **b** Graft composition including HSC, B and T cells after CD34 positive selection. **c** Graft composition including HSC, B and T cells after CD3 and CD19 negative selection

This technique does not remove CD34- HSC with either a repopulating ability or other defined graft-facilitating cells, such as CD8+ T-cells, NK cells, monocytes, and antigen-presenting cells while reducing the risk of GVHD and Epstein-Barr Virus-related lymphoproliferative disease (Fig. 66.5) [8].

In Fig. 66.6 different positive (CD34 and CD133) and negative (CD3, CD19) selection techniques are compared by fluorescence-activated cell sorting (FACS).

Other selective depletion techniques have used anti-CD25 immunotoxine, magnetic bead depletion of CD25+ and CD69+ T-cells or photodynamic purging to selectively eliminate activated T-cells, and therefore reduce or prevent GVHD while permitting immune reconstitution [7]. Another strategy to modify GVHD is the transfection of transplanted T-cells with a HSV-TK suicide gene. Patients developing severe GVHD use ganciclovir as the suicide agent to eliminate the alloreacting cells. The first clinical trials have shown the feasibility in patients with haploidentical HSC transplantations for advanced malignancies [11].

Other interesting manipulation techniques need to be further validated in clinical trials: to reduce post-transplant infectious complications, cord-blood T-cells may be expanded by using anti-CD3/anti-CD28 coated beads in vitro and reinfused post-transplant; naïve T-cells, differentiated into CD19-specific cytolytic effector cells and expanded in vitro may eliminate residual CD19+ tumor cells; thymic function may be improved by the overexpression of FOXN1 in mesenchymal cord blood cells; T-cell reconstitution may be improved by the co-culture of cord blood HSCs on stroma cell lines overexpressing Notchligand Delta-like 1. With the identification of leukemia-specific antigens such as WT-1, proteinase 3, and HA-1 and HA-2, the adoptive transfer of antigen-specific T-cells may enhance the GVT reaction and subsequently reduce relapse rates.

66.7
Conclusions

The enhanced understanding of stem cell biology and hematopoiesis has changed HSC transplantation in the past decades. Growth factors are now available

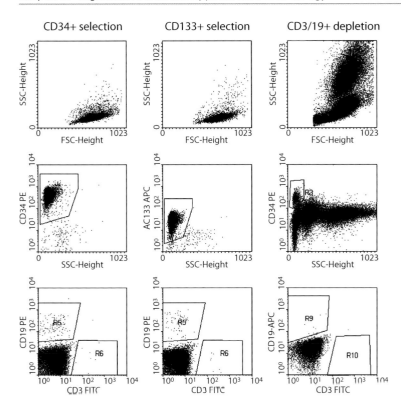

Fig. 66.6 FACS (fluorescence activated cell sorting) analysis after CD34+ or CD133+ selection compared to CD3/19 depletion. Cells were stained with fluorochromes conjugated antibodies (FITC, PE, APC). (*FSC* forward scatter, *SSC* side scatter, *FITC/PE* fluorochromes)

to support the production of erythrocytes, leucocytes, and platelets. With cytokine-supported peripheral HSC harvest, autologous stem cell transplantation is now available even for older or heavily pretreated patients.

Oral agents supporting hematopoiesis may be available in the near future. Tumor cell contamination is still a concern in autologous transplant settings. Different purging techniques show promising results, but remaining tumor cells are still the major reason for relapse. Allogeneic HSC transplantation has changed from an experimental treatment of end-stage leukemia patients to a routine procedure for many malignant and non-malignant diseases. Reduced intensity conditioning (RIC) offers this treatment for patients older than 50 years with comorbidities, but the best ablative regimens for specific conditions still have to be determined. Cord blood transplantation is often restricted to pediatric patients due to the limiting cell numbers. The in vitro expansion of isolated HSC would provide sufficient cells numbers from cord blood units for allogeneic HSC transplantation in adults. Although the in vitro expansion of hematopoietic stem cells without losing stem cell potential is theoretically appealing, the clinical breakthrough is still being awaited. The pooling of cord blood units seems to be possible and safe and may solve this problem. Graft manipulation by positive and/or negative selection techniques opened the way for haploidentical HSC transplantation. Infused T regulatory cells, cytokine antagonists, or suicide genes may reduce the life-threatening acute GVHD. HSC transplantation has now evolved from bone marrow transplantation into an adoptive cellular immunotherapy. Future work will have to balance enhanced immune reconstitution with GVHD, graft rejection, and relapse.

References

1. Arai F, Hirao A, Ohmura M et al. (2004) Tie2/angiopoietin-1 signaling regulates hematopoietic stem cell quiescence in the bone marrow niche. Cell 118:149–61
2. Arai F, Hirao A, Suda T (2005) Regulation of hematopoietic stem cells by the niche. Trends Cardiovasc Med 15:75–9
3. Araki H, Yoshinaga K, Boccuni P et al. (2007) Chromatin-modifying agents permit human hematopoietic stem cells to undergo multiple cell divisions while retaining their repopulating potential. Blood 109:3570–8

4. Aversa F, Tabilio A, Terenzi A et al. (1994) Successful engraftment of T-cell-depleted haploidentical "three-loci" incompatible transplants in leukemia patients by addition of recombinant human granulocyte colony-stimulating factor-mobilized peripheral blood progenitor cells to bone marrow inoculum. Blood 84:3948–55
5. Bach FH, Albertini RJ, Joo P et al. (1968) Bone-marrow transplantation in a patient with the Wiskott-Aldrich syndrome. Lancet 2:1364–6
6. Ballen KK (2005) New trends in umbilical cord blood transplantation. Blood 105:3786–92
7. Barrett J (2006) Improving outcome of allogeneic stem cell transplantation by immunomodulation of the early post-transplant environment. Curr Opin Immunol 18:592–8
8. Bethge WA, Haegele M, Faul C et al. (2006) Haploidentical allogeneic hematopoietic cell transplantation in adults with reduced-intensity conditioning and CD3/CD19 depletion: fast engraftment and low toxicity. Exp Hematol 34:1746–52
9. Bhatia M, Wang JC, Kapp U et al. (1997) Purification of primitive human hematopoietic cells capable of repopulating immune-deficient mice. Proc Natl Acad Sci USA 94: 5320–5
10. Boggs DR, Boggs SS, Saxe DF et al. (1982) Hematopoietic stem cells with high proliferative potential. Assay of their concentration in marrow by the frequency and duration of cure of W/Wv mice. J Clin Invest 70:242–53
11. Bondanza A, Valtolina V, Magnani Z et al. (2006) Suicide gene therapy of graft-versus-host disease induced by central memory human T lymphocytes. Blood 107:1828–36
12. Bornstein R, Flores AI, Montalban MA et al. (2005) A modified cord blood collection method achieves sufficient cell levels for transplantation in most adult patients. Stem Cells 23:324–34
13. Calvi LM, Adams GB, Weibrecht KW et al. (2003) Osteoblastic cells regulate the haematopoietic stem cell niche. Nature 425:841–6
14. Civin CI, Strauss LC, Brovall C et al. (1984) Antigenic analysis of hematopoiesis. III. A hematopoietic progenitor cell surface antigen defined by a monoclonal antibody raised against KG-1a cells. J Immunol 133:157–65
15. Cohen JL, Trenado A, Vasey D et al. (2002) CD4(+)CD25(+) immunoregulatory T Cells: new therapeutics for graft-versus-host disease. J Exp Med 196:401–6
16. Conneally E, Cashman J, Petzer A et al. (1997) Expansion in vitro of transplantable human cord blood stem cells demonstrated using a quantitative assay of their lympho-myeloid repopulating activity in nonobese diabetic-scid/scid mice. Proc Natl Acad Sci USA 94:9836–41
17. Coulombel L (2004) Identification of hematopoietic stem/progenitor cells: strength and drawbacks of functional assays. Oncogene 23:7210–22
18. Danet GH, Lee HW, Luongo JL et al. (2001) Dissociation between stem cell phenotype and NOD/SCID repopulating activity in human peripheral blood CD34(+) cells after ex vivo expansion. Exp Hematol 29:1465–73
19. De Felice L, Di Pucchio T, Breccia M et al. (1998) Flt3L enhances the early stem cell compartment after ex vivo amplification of umbilical cord blood CD34+ cells. Bone Marrow Transplant 22 Suppl 1:66–7
20. De Felice L, Di Pucchio T, Mascolo MG et al. (1999) Flt3LP3 induces the ex-vivo amplification of umbilical cord blood committed progenitors and early stem cells in short-term cultures. Br J Haematol 106:133–41
21. Delaney C, Varnum-Finney B, Aoyama K et al. (2005) Dose-dependent effects of the Notch ligand Delta1 on ex vivo differentiation and in vivo marrow repopulating ability of cord blood cells. Blood 106:2693–9
22. de Lima M, Shpall E (2006) Strategies for widening the use of cord blood in hematopoietic stem cell transplantation. Haematologica 91:584–7
23. Dexter TM, Lajtha LG (1974) Proliferation of haemopoietic stem cells in vitro. Br J Haematol 28:525–30
24. DiGiusto DL, Lee R, Moon J et al. (1996) Hematopoietic potential of cryopreserved and ex vivo manipulated umbilical cord blood progenitor cells evaluated in vitro and in vivo. Blood 87:1261–71
25. Domen J, Cheshier SH, Weissman IL (2000) The role of apoptosis in the regulation of hematopoietic stem cells: Overexpression of Bcl-2 increases both their number and repopulation potential. J Exp Med 191:253–64
26. Dorrell C, Gan OI, Pereira DS et al. (2000) Expansion of human cord blood CD34(+)CD38(-) cells in ex vivo culture during retroviral transduction without a corresponding increase in SCID repopulating cell (SRC) frequency: dissociation of SRC phenotype and function. Blood 95:102–10
27. Feller N, van der Pol MA, Waaijman T et al. (2005) Immunologic purging of autologous peripheral blood stem cell products based on CD34 and CD133 expression can be effectively and safely applied in half of the acute myeloid leukemia patients. Clin Cancer Res 11:4793–801
28. Flomenberg N, Devine SM, Dipersio JF et al. (2005) The use of AMD3100 plus G-CSF for autologous hematopoietic progenitor cell mobilization is superior to G-CSF alone. Blood 106:1867–74
29. Gammaitoni L, Weisel KC, Gunetti M et al. (2004) Elevated telomerase activity and minimal telomere loss in cord blood long-term cultures with extensive stem cell replication. Blood 103:4440–8
30. Giet O, Huygen S, Beguin Y et al. (2001) Cell cycle activation of hematopoietic progenitor cells increases very late antigen-5-mediated adhesion to fibronectin. Exp Hematol 29:515–24
31. Glimm H, Oh IH, Eaves CJ (2000) Human hematopoietic stem cells stimulated to proliferate in vitro lose engraftment potential during their S/G(2)/M transit and do not reenter G(0). Blood 96:4185–93
32. Gluckman E, Broxmeyer HA, Auerbach A et al. (1989) Hematopoietic reconstitution in a patient with Fanconi's anemia by means of umbilical-cord blood from an HLA-identical sibling. N Engl J Med 321:1174–8
33. Hanash AM, Levy RB (2005) Donor CD4+CD25+ T cells promote engraftment and tolerance following MHC-mismatched hematopoietic cell transplantation. Blood 105:1828–36
34. Handgretinger R, Schumm M, Lang P et al. (1999) Transplantation of megadoses of purified haploidentical stem cells. Ann NY Acad Sci 872:351–61
35. Haylock DN, To LB, Dowse TL et al. (1992) Ex vivo expansion and maturation of peripheral blood CD34+ cells into the myeloid lineage. Blood 80:1405–12

36. Hoffmann P, Ermann J, Edinger M et al. (2002) Donor-type CD4(+)CD25(+) regulatory T cells suppress lethal acute graft-versus-host disease after allogeneic bone marrow transplantation. J Exp Med 196:389–99
37. Horn PA, Thomasson BM, Wood BL et al. (2003) Distinct hematopoietic stem/progenitor cell populations are responsible for repopulating NOD/SCID mice compared with nonhuman primates. Blood 102:4329–35
38. Horowitz MM, Bortin MM (1990) Current status of allogeneic bone marrow transplantation. Clin Transpl 41–52
39. Jiang Y, Prosper F, Verfaillie CM (2000) Opposing effects of engagement of integrins and stimulation of cytokine receptors on cell cycle progression of normal human hematopoietic progenitors. Blood 95:846–54
40. June CH, Blazar BR (2006) Clinical application of expanded CD4+25+ cells. Semin Immunol 18:78–88
41. Jung Y, Wang J, Song J et al. (2007) Annexin II expressed by osteoblasts and endothelial cells regulates stem cell adhesion, homing, and engraftment following transplantation. Blood 110:82–90
42. Kim DW, Chung YJ, Kim TG et al. (2004) Cotransplantation of third-party mesenchymal stromal cells can alleviate single-donor predominance and increase engraftment from double cord transplantation. Blood 103:1941–8
43. Kobari L, Giarratana MC, Gluckman JC et al. (2006) Ex vivo expansion does not alter the capacity of umbilical cord blood CD34+ cells to generate functional T lymphocytes and dendritic cells. Stem Cells 24.2130–7
44. Larochelle A, Vormoor J, Hanenberg H et al. (1996) Identification of primitive human hematopoietic cells capable of repopulating NOD/SCID mouse bone marrow: implications for gene therapy. Nat Med. 2:1329–37
45. Laughlin MJ, Eapen M, Rubinstein P et al. (2004) Outcomes after transplantation of cord blood or bone marrow from unrelated donors in adults with leukemia. N Engl J Med 351:2265–75
46. Le Blanc K, Pittenger M (2005) Mesenchymal stem cells: progress toward promise. Cytotherapy 7:36–45
47. Le Blanc K, Samuelsson H, Gustafsson B et al. (2007) Transplantation of mesenchymal stem cells to enhance engraftment of hematopoietic stem cells. Leukemia. May 31; [Epub ahead of print]
48. Lewis ID, Almeida-Porada G, Du J et al. (2001) Umbilical cord blood cells capable of engrafting in primary, secondary, and tertiary xenogeneic hosts are preserved after ex vivo culture in a noncontact system. Blood 97:3441–9
49. Liu B, Buckley SM, Lewis ID et al. (2003) Homing defect of cultured human hematopoietic cells in the NOD/SCID mouse is mediated by Fas/CD95. Exp Hematol 31:824–32
50. Liu W, Putnam AL, Xu-Yu Z et al. (2006) CD127 expression inversely correlates with FoxP3 and suppressive function of human CD4+ T reg cells. J Exp Med 203:1701–11
51. Magli MC, Iscove NN, Odartchenko N (1982) Transient nature of early haematopoietic spleen colonies. Nature 295:527–9
52. McNiece IK, Almeida-Porada G, Shpall EJ et al. (2002) Ex vivo expanded cord blood cells provide rapid engraftment in fetal sheep but lack long-term engrafting potential. Exp Hematol 30:612–6
53. McNiece I, Civin C, Harrington J et al. (2006) Ex vivo expansion of mafosfamide-purged PBPC products. Cytotherapy 8:459–64
54. McNiece I, Kubegov D, Kerzic P et al. (2000) Increased expansion and differentiation of cord blood products using a two-step expansion culture. Exp Hematol 28:1181–6
55. Miller JS, Cooley S, Parham P et al. (2007) Missing KIR ligands are associated with less relapse and increased graft-versus-host disease (GVHD) following unrelated donor allogeneic HCT. Blood 109:5058–61.
56. Moretta L, Bottino C, Pende D et al. (2002) Human natural killer cells: their origin, receptors and function. Eur J Immunol 32:1205–11R
57. Moretta L, Moretta A. (2004) Killer immunoglobulin-like receptors. Curr Opin Immunol 16:626–33
58. Murdoch B, Chadwick K, Martin M et al. (2003) Wnt-5A augments repopulating capacity and primitive hematopoietic development of human blood stem cells in vivo. Proc Natl Acad Sci USA 100:3422–7
59. Murray LJ, Young JC, Osborne LJ et al. (1999) Thrombopoietin, flt3, and kit ligands together suppress apoptosis of human mobilized CD34+ cells and recruit primitive CD34+ Thy-1+ cells into rapid division. Exp Hematol 27:1019–28
60. Neben S, Anklesaria P, Greenberger J et al. (1993) Quantitation of murine hematopoietic stem cells in vitro by limiting dilution analysis of cobblestone area formation on a clonal stromal cell line. Exp Hematol 21:438–43
61. Noort WA, Kruisselbrink AB, Anker PS et al. (2002) Mesenchymal stem cells promote engraftment of human umbilical cord blood-derived CD34(+) cells in NOD/SCID mice. Exp Hematol 30:870–8
62. Oberg L, Johansson S, Michaelsson J et al. (2004) Loss or mismatch of MHC class I is sufficient to trigger NK cell-mediated rejection of resting lymphocytes in vivo: role of KARAP/DAP12-dependent and -independent pathways. Eur J Immunol 34:1646–53
63. Orschell-Traycoff CM, Hiatt K, Dagher RN et al. (2000) Homing and engraftment potential of Sca-1(+)lin(−) cells fractionated on the basis of adhesion molecule expression and position in cell cycle. Blood 96:1380–7
64. Osawa M, Hanada K, Hamada H et al. (1996) Long-term lymphohematopoietic reconstitution by a single CD34-low/negative hematopoietic stem cell. Science 273:242–5
65. Papayannopoulou T, Priestley GV, Nakamoto B (1998) Anti-VLA4/VCAM-1-induced mobilization requires cooperative signaling through the kit/mkit ligand pathway. Blood 91:2231–9
66. Petersdorf EW, Longton GM, Anasetti C et al. (1995) The significance of HLA-DRB1 matching on clinical outcome after HLA-A, B, DR identical unrelated donor marrow transplantation. Blood 86:1606–13
67. Piacibello W, Sanavio F, Severino A et al. (1999) Engraftment in nonobese diabetic severe combined immunodeficient mice of human CD34(+) cord blood cells after ex vivo expansion: evidence for the amplification and self-renewal of repopulating stem cells. Blood 93:3736–49
68. Ploemacher RE, van der Sluijs JP, Voerman JS et al. (1989) An in vitro limiting-dilution assay of long-term

repopulating hematopoietic stem cells in the mouse. Blood 74:2755–63
69. Reya T, Duncan AW, Ailles L et al. (2003) A role for Wnt signalling in self-renewal of haematopoietic stem cells. Nature 423:409–14
70. Rice A, Flemming C, Case J et al. (1999) Comparative study of the in vitro behavior of cord blood subpopulations after short-term cytokine exposure. Bone Marr Trans 23:211–20
71. Rocha V, Labopin M, Sanz G et al. (2004) Transplants of umbilical-cord blood or bone marrow from unrelated donors in adults with acute leukemia. N Engl J Med 351:2276–85
72. Rosenzwajg M, Canque B, Gluckman JC (1996) Human dendritic cell differentiation pathway from CD34+ hematopoietic precursor cells. Blood 87:535–44
73. Rubinstein P, Dobrila L, Rosenfield RE et al. (1995) Processing and cryopreservation of placental/umbilical cord blood for unrelated bone marrow reconstitution. Proc Natl Acad Sci USA 92:10119–22
74. Ruggeri L, Aversa F, Martelli MF et al. (2006) Allogeneic hematopoietic transplantation and natural killer cell recognition of missing self. Immunol Rev 214:202–18
75. Sato N, Meijer L, Skaltsounis L et al. (2004) Maintenance of pluripotency in human and mouse embryonic stem cells through activation of Wnt signaling by a pharmacological GSK-3-specific inhibitor. Nat Med 10:55–63
76. Schofield R (1978) The relationship between the spleen colony-forming cell and the haemopoietic stem cell. Blood Cells 4:7–25
77. Seddiki N, Santner-Nanan B, Martinson J et al. (2006) Expression of interleukin (IL)-2 and IL-7 receptors discriminates between human regulatory and activated T cells. J Exp Med 203:1693–700
78. Shpall EJ, Quinones R, Giller R et al. (2002) Transplantation of ex vivo expanded cord blood. Biol Blood Marr Trans 8:368–76
79. Siminovitch L, McCulloch A, Till JE (1963) The distribution of colony-forming cells among spleen colonies. J Cell Physiol 62:327–36
80. Siminovitch L, Till JE, McCulloch EA (1964) Decline in colony-forming ability of marrow cells subjected to serial transplantation into irradiated mice. J Cell Physiol 64:23–31
81. Smith LG, Weissman IL, Heimfeld S (1991) Clonal analysis of hematopoietic stem-cell differentiation in vivo. Proc Natl Acad Sci USA 88:2788–92
82. Solanilla A, Grosset C, Duchez P et al. (2003) Flt3-ligand induces adhesion of haematopoietic progenitor cells via a very late antigen (VLA)-4- and VLA-5-dependent mechanism. Br J Haematol 120:782–6
83. Spangrude GJ, Heimfeld S, Weissman IL (1988) Purification and characterization of mouse hematopoietic stem cells. Science 241:58–62
84. Spangrude GJ, Brooks DM, Tumas DB (1995) Long-term repopulation of irradiated mice with limiting numbers of purified hematopoietic stem cells: in vivo expansion of stem cell phenotype but not function. Blood 85:1006–16
85. Stewart AK, Vescio R, Schiller G et al. (2001) Purging of autologous peripheral-blood stem cells using CD34 selection does not improve overall or progression-free survival after high-dose chemotherapy for multiple myeloma: results of a multicenter randomized controlled trial. J Clin Oncol 19:3771–9
86. Stiff P, Chen B, Franklin W et al. (2000) Autologous transplantation of ex vivo expanded bone marrow cells grown from small aliquots after high-dose chemotherapy for breast cancer. Blood 95:2169–74
87. Sutherland HJ, Lansdorp PM, Henkelman DH et al. (1990) Functional characterization of individual human hematopoietic stem cells cultured at limiting dilution on supportive marrow stromal layers. Proc Natl Acad Sci U S A 87:35848
88. Szilvassy SJ, Humphries RK, Lansdorp PM et al. (1990) Quantitative assay for totipotent reconstituting hematopoietic stem cells by a competitive repopulation strategy. Proc Natl Acad Sci USA 87:8736–40
89. Taichman RS (2005) Blood and bone: two tissues whose fates are intertwined to create the hematopoietic stem-cell niche. Blood 105:2631–9
90. Taichman RS, Reilly MJ, Emerson SG (1996) Human osteoblasts support human hematopoietic progenitor cells in vitro bone marrow cultures. Blood 87:518–24
91. Taylor PA, Lees CJ, Blazar BR (2002) The infusion of ex vivo activated and expanded CD4(+)CD25(+) immune regulatory cells inhibits graft-versus-host disease lethality. Blood 99:3493–9
92. Thomas ED, Lochte HL, Lu WC, Ferrebee JW (1957) Intravenous infusion of bone marrow in patients receiving radiation and chemotherapy. N Engl J Med 257:491–6
93. Till JE, McCulloch EA (1961) A direct measurement of the radiation sensitivity of normal mouse bone marrow cells. Radiat Res 14:213–22
94. Tisato V, Naresh K, Girldstone J et al. (2007) Mesenchymal stem cells of cord blood origin are effective at preventing but not treating graft-versus-host disease. Leukemia. Jul 12; Epub ahead of print]
95. Trowbridge JJ, Xenocostas A, Moon RT et al. (2006) Glycogen synthase kinase-3 is an in vivo regulator of hematopoietic stem cell repopulation. Nat Med 12:89–98
96. Valmori D, Tosello V, Souleimanian NE et al. (2006) Rapamycin-mediated enrichment of T cells with regulatory activity in stimulated CD4+ T cell cultures is not due to the selective expansion of naturally occurring regulatory T cells but to the induction of regulatory functions in conventional CD4+ T cells. J Immunol 177:944–9
97. Van Den Berg DJ, Sharma AK, Bruno E et al. (1998) Role of members of the Wnt gene family in human hematopoiesis. Blood 92:3189–202
98. Visnjic D, Kalajzic Z, Rowe DW et al. (2004) Hematopoiesis is severely altered in mice with an induced osteoblast deficiency. Blood 103:3258–64
99. Wagner JE, Thompson JS, Carter SL et al. (2005) Effect of graft-versus-host disease prophylaxis on 3-year disease-free survival in recipients of unrelated donor bone marrow (T-cell depletion trial): a multi-centre, randomised phase II-III trial. Lancet 366:733–41
100. Wiesmann A, Kim M, Georgelas A et al. (2000) Modulation of hematopoietic stem/progenitor cell engraftment by transforming growth factor beta. Exp Hematol 28:128–39
101. Wiesmann A, Phillips RL, Mojica M et al. (2000) Expression of CD27 on murine hematopoietic stem and progenitor cells. Immunity 12:193–9

102. Yang H, Eaves C, de Lima M et al. (2006) A novel triple purge strategy for eliminating chronic myelogenous leukemia (CML) cells from autografts. Bone Marr Trans 37:575–82
103. Yaniv I, Stein J, Luria D et al. (2007) Ewing Sarcoma tumor cells express CD34: implications for autologous stem cell transplantation. Bone Marr Trans 39:589–94
104. Yin AH, Miraglia S, Zanjani ED et al. (1997) AC133, a novel marker for human hematopoietic stem and progenitor cells. Blood 90:5002–12
105. Zanjani ED, Almeida-Porada G, Livingston AG et al. (1998) Human bone marrow CD34– cells engraft in vivo and undergo multilineage expression that includes giving rise to CD34+ cells. Exp Hematol 26:353–60
106. Zeiser R, Nguyen VH, Beilhack A et al. (2006) Inhibition of CD4+CD25+ regulatory T-cell function by calcineurin-dependent interleukin-2 production. Blood 108:390–9
107. Zhang J, Niu C, Ye L et al. (2003) Identification of the haematopoietic stem cell niche and control of the niche size. Nature 425:836–41.
108. Zhu J, Garrett R, Jung Y et al. (2007) Osteoblasts support B-lymphocyte commitment and differentiation from hematopoietic stem cells. Blood 109:3706–12

The Reconstructed Human Epidermis Models in Fundamental Research

A. Coquette, Y. Poumay

Contents

67.1	Introduction	967
67.2	Key Developments in In Vitro Human Epidermis Reconstruction	967
67.3	The Culture and Morphology of Reconstructed Human Epidermis (RHE) Models	968
67.4	Reconstructed Human Epidermis Models and Skin Physiopathology	969
67.5	Uses of RHE as Alternatives to Animal/Human Testing	971
67.5.1	In Vitro Skin Irritation	971
67.5.2	In Vitro Skin Sensitization	971
67.5.3	In Vitro Skin Permeation and Metabolization	972
	References	973

67.1 Introduction

Large skin trauma allows pathogens to invade, and causes water, electrolyte and protein loss, inducing adverse events often incompatible with sustained life if the wound is not rapidly covered and the lost tissue replaced. This last point is often critical in wound management because of the poor availability of skin samples. The recent advances in the in vitro techniques of keratinocyte culture have led to the development of several commercial skin substitutes. Although they represent significant advances in wound treatment, they are of relative efficacy and can induce adverse events such as infectious diseases and allergic reactions in patients with known sensitivity to materials of bovine origin [46]. Moreover, a risk of prior diseases is also present even if serum and/or collagen are sourced from countries theoretically free from bovine spongiform encephalitis [44]. Therefore, a more efficient and safe means of cultivation are needed in order to ensure the safety of the patient, such as the use of serum-free culture media. Moreover, for significant progresses in clinical efficacy, the understanding of the complex process of wound healing and development of the novel growth factor and gene therapies would benefit from research models that mimic closely the human skin epidermis/dermis structure, and in which studies of the cell-cell and cell-matrix interactions could be possible.

67.2 Key Developments in In Vitro Human Epidermis Reconstruction

The first stratified colonies of human keratinocytes from single cells were obtained by Rheinwald and Green [72], and these cultures were even used for grafting in patients with extensive burns or for treatment of giant naevi [29, 30]. Although this model remains as a reference for human therapy because of the selection of cells of high clonogenic potential, it is limited for fundamental and applied research. In fact, differentiating keratinocytes exhibit a strong involucrin expression in the first suprabasal cell layer, a situation that differs from in vivo differentiation where involucrin is mainly induced in upper spinous and granular layers [2]. Moreover, the simultaneous presence of proliferative and differentiating cells makes difficult the discrimination between events occurring in proliferative and differentiating keratinocytes, or simply normal phenomenon in fibroblasts. Finally, the presence of serum, and particularly vitamin A, as well as the addition in the culture medium of epidermal growth factor (EGF) with insulin, make studies of epidermal cytokines and growth factors quite difficult because of the interference with molecules already present in the culture medium [65].

This problem can be avoided by autocrine culture conditions allowing stimulation of the keratinocyte EGF receptor [55]. Indeed, the culture medium conditioned by growing keratinocytes contains amphiregulin, a growth factor regulated by heparin sulfate, which is able to bind and activate the keratinocyte EGF receptor [9, 10]. When keratinocytes cultured in autocrine conditions reach confluence, they demonstrate a very strong concomitant induction in the expression and accumulation of suprabasal keratins 1 and 10 [64, 65]. Simultaneously, the expression of involucrin increases mimicking the sequence of events described during the stratification and differentiation of keratinocytes in vivo. Nevertheless, in the models produced by these techniques, the stratification of the basal, spinous, granular, and cornified layers is not the same as in normal tissue and is described as basal, intermediate, and upper zones [37].

The first well-differentiated model was developed by Prunieras and co-workers. They have demonstrated that it was possible to get full differentiation simply by raising cells up to the air-liquid interface [68]. Apparently, the interface with air stimulates the synthesis of profillagrin in keratinocytes and thus the appearance of the granular phenotype when keratohyalin granules develop. These cultures exhibit a well-stratified epithelium and cornified epidermis with significantly improved barrier function [35, 73]. Differentiation markers such as suprabasal keratins, $\alpha6\beta4$ integrin, fibronectin, involucrin, filaggrin, trichohyalin, type I, III, IV, V and VII collagen, laminin, heparan sulfate, and membrane-bound transglutaminase are expressed similarly to their respective expression within the normal epidermis [24, 59, 74]. Moreover, keratin synthesis and the production of cornified envelopes parallel the same event observed in vivo. Spinous cells display abundant glycogen deposits and keratohyalin granules are more abundant in the granular layer. Both the size and the number of hemidesmosomes increase during maturation in vitro and anchoring fibrils are observed [24, 53, 73, 86, 80, 85]. Based on this development, the European Center for the Validation of Alternative Methods, set up by the European Union in Ispra (Italy), has considered skin reconstruction among the most advanced projects for the development of in vitro pharmacotoxicology and research programs (76/768/EEC, Feb. 2003).

67.3 The Culture and Morphology of Reconstructed Human Epidermis (RHE) Models

Despite the similarity in tissue architecture, some deviations from normal epidermis have been reported in the first models. In fact, the stratum corneum lipids appear to be organized in multilamellar structures with a periodicity of 12 nm in contrast to native epidermis, in which two lamellar phases with periodicities of 6.4 nm and 13.4 nm are typically detected [6, 56]. Furthermore, the reduction of ceramides 4–7 and 6–7, integrin overexpression, the premature expression of specific differentiation markers, and abnormal lipid composition have been observed under normal in vitro culture conditions [49, 58, 88].

Improvements of the culture conditions, such as maintenance of the cultures in delipidized serum, reduction of the relative humidity, [69] and the use of a chemically defined medium [74] have led to the further optimization of initial models. Epidermal tissues generated at 33°C in absence of EGF [27, 58] and retinoic acid, [74] but in the presence of vitamin C, exhibit an improved stratum corneum architecture and lipid profile [59]. In fact, EGF supplementation has a deleterious effect on epidermal morphogenesis and differentiation. Moreover, the synthesis of K1 and K10 is suppressed at both protein and mRNA levels

[59]. In the absence of serum, the relative amounts of ceramides, free fatty acids, and cholesterol are comparable to native epidermis, while vitamin C increases the content of glucosylceramides and of ceramides 6 and 7 [27]. However, irrespective of the culture conditions, involucrin is aberrantly expressed [59]. Among the most important existing artificial epidermal models (Episkin®, SkinEthic® and Epiderm®), ceramide 7 is missing, ceramides 5–6 are usually low, and ceramide 2 is found in much larger amount than in the human epidermis [61]. Very recently, the genes involved in the process of stratification were identified in RHE. The data demonstrates that, in fact, the exposition of cultures to air increases the expression of genes that are known to protect cells against oxidative damages, whereas submerged tissues enhance the expression of genes that are involved in hypoxy. Most of the genes that are expressed differentially between air-liquid and submerged tissues are related to carbohydrates metabolism, lipid biosynthesis and oxidative phosphorylation suggesting a possible link between these metabolic pathways and stratification [42]. Therefore, it seems that high rate proliferation could be obtained by hyperbaric oxygen treatment in relationship with an appropriate gradient of nutrients during the culture.

The development of keratinocyte culture systems using inert filters [73], collagen matrix [57], or de-epidermized dermis [27], has led to the production of commercially available cultured human skin models, including: (1) EpiDerm (MatTek Corporation, USA), (2) Episkin (L'Oréal, France), (3) the human reconstructed epidermis from SkinEthic (SkinEthic, France), (4) Living Skin Equivalent (Organogenesis, USA), (5) Skin2TM (Advanced Tissue Sciences, USA), (6) StratiCell reconstructed human epidermis (StratiCell, Belgium) and Phenion Full Thickness Model (Phenion, Germany) [16, 25, 76, 91]. For in vitro experiments, they represent reproducible models in a controlled environment [74, 60]. Consequently, factors such as the variability of skin source, i.e., different donors or different anatomical body sites, which could influence the final results, can be ruled out. Nevertheless, because of their commercial origin, the discontinuation of production for financial reasons, and the difficulties in controlling the culture parameters can be a problem for many laboratories [67]. Partial information on methods allowing the preparation of RHE was available in the literature [reviewed in 12] until a procedure using easily available materials and reagents was published [66]. In this procedure, primary cultures of keratinocytes are initiated in the commercially available KGM-2medium (Biowhittaker), supplemented with $CaCl_2$, human EGF, human insulin, transferrin, epinephrine, hydrocortisone, and bovine pituitary extract. The EpiLife medium (Cascade Biological) is then used for subculture and tissue reconstruction. The RHE exhibit histological features similar to those observed in vivo. The response of the model to estradiol, irritants, and sensitizers reveals that the model behavior is similar to the SkinEthic® model in term of barrier function and cytokines release [11, 66]. From a clinical point of view, this procedure opened to everyone discards completely the risk of viral transmission of viral diseases due to the use of bovine sera. Moreover, KGM-2 is more efficient in expanding keratinocytes just isolated from the epidermis which could reduce the time needed to obtain human autologous keratinocyte sheet grafts. This culture method could be a valuable tool in the treatment of patients having a limited amount of healthy skin.

67.4
Reconstructed Human Epidermis Models and Skin Physiopathology

One of the advantages of in vitro 3D cultures over monolayers include either the possibility of incorporating various additional cell types in combination with normal keratinocytes, or to recreate pathological skin reconstructs by using abnormal cells. For instance, an in vitro 3D model of psoriasis was developed recently in a serum-free environment by using psoriatic keratinocytes cultured on the surface of a gel comprising fibroblasts from the same origin [3]. In this model, the distribution of differentiation markers such as involucrin, fillagrin, keratin 16 and 10 is similar in the psoriatic model and in the culture derived from normal cells. But, in the psoriatic model, the chemokine receptor CXCR2 is overexpressed in the granular layer as seen in vivo. In parallel, pro-inflammatory genes, particularly IL-8, TNFα and IFNγ are expressed at a high level in the model of psoriasis compared to the normal model. Thus, the release of these chemokines and the overexpression of the CXCR2 receptor may be the regulatory factors influencing the growth of keratinocytes in psoriasis [3]. This model could provide a starting point to investigate the mechanisms of abnormal keratinocyte growth, gene expression, and novel therapies by using blocking antibodies or gene transfer.

On the other side, the introduction of melanocytes into RHE [84] has allowed the understanding of hypopigmentation often observed in patients grafted with skin substitutes [82]. Melanocytes can indeed be successfully isolated and cultured from the human epidermis [54] and diluted into a suspension of keratinocytes when preparing a RHE in 3D culture [5]. As in the epidermis in vivo, melanocytes appear as dendritic cells polarized to the basal layers, and maintaining close contacts with both the basal lamina and neighboring keratinocytes. Such advances in tissue culture have demonstrated that keratinocyte factors are needed in order to maintain the growth and differentiation of melanocytes, and that a mature epidermis regulates the epidermal melanization unit [14, 33, 78]. For instance, the presence of an organized basement membrane is critical; otherwise the melanocytes fail to develop a secure attachment and are thus rapidly lost [15]. On the contrary, the presence of fibroblasts in the basement membrane results in amelanotic skin [36]. Following UV irradiation, the number of dopa-positive melanocytes increases in the basal layer, resulting in increased pigmentation within the irradiated skin equivalent. It provides an attractive in vitro system to study the regulation of melanogenesis and melanocyte-keratinocyte interactions, and to investigate, in a better defined model, how these processes are affected by UV irradiation. In addition, this epidermal model can be used to test the phototoxic or photoprotective potential of various compounds such as sunscreens [63]. Régnier and coworkers were the first to develop a 3D epidermal model in which the pigmentary response to diverse agents can be studied [71]. This possibility affords a distinct advantage over other animal models [19]. Hence, the future of RHE models that include melanocytes looks promising. Firstly, with better standardization, assays for pharmacologic and cosmetic agents suspected to interfere with pigmentation may be developed in order to screen interesting molecules, with a better relevance than on monolayers of cultured melanocytes. Secondly, pathophysiological and therapeutic studies in human pigmentary disorders could be approached directly ex vivo or after grafting the reconstructed epidermis onto nude mice. Finally, RHE models have provided insight into the interaction between tumor cells and healthy skin cells. A number of research groups are examining the cell interactions involved in early-stage melanoma invasion [47]. Such models can thus be used to look at the extent of invasion by different melanoma cell lines, to examine the deliberate upregulation or downregulation of adhesion molecules on these melanomas, and to study matrix metalloproteinases and inflammation. Invasion has been shown to be dependent on the interaction of the tumor cells with the adjacent skin cells [21].

More recent advances in culture techniques have allowed the development of RHE containing not only keratinocytes but also melanocytes and Langerhans cells as well. Langerhans cells can be produced in vitro from CD34+ cells [7]. For this purpose, cord blood-derived CD34+ hematopoietic progenitor cells induced to differentiate by GM-CSF and TNFα are seeded onto a RHE composed of keratinocytes and melanocytes. This culture system gives rise to RHE models displaying a pigmented epidermis with melanocytes in the basal layer and resident epidermal Langherhans cells located suprabasally and expressing major histocompatibility complex class II, CD1 antigen, and Birbeck granules [22, 71]. More recently, an endothelialized skin equivalent coculture model was used to study the behavior of hematopoietic progenitor cells in epidermal and dermal environments. CD34(+) cells were cultured and seeded in the endothelialized skin equivalent at different time points to favor either dermal or epidermal integration. This integration gave rise to both cutaneous and dermal dendritic cells while endothelial cells are sufficiently activated to acquire HLA-DR expression. For the first time, this model provides a more complex environment integrating vascular components to study the differentiation of interstitial dendritic cells in a dermis equivalent. Such sophisticated skin equivalent may clarify some intriguing aspects of the numerous regulatory mechanisms controlling skin homeostasis, and opens the possibility of preparation of immunocompetent reconstructed skin [17].

Finally, unexpected results have been obtained in relation to the contraction of human skin grafts.

The data from works using reconstructed skin from sterilized mature cross-linked dermis show that keratinocytes, rather than fibroblasts, contract this dermis [8]. Both keratinocytes and fibroblasts can contract dilute collagen gels, but the finding that keratinocytes have a marked ability to contract human dermis sheds new light on the problem of skin-graft contracture. This model is being used to identify new pharmacological targets to block contraction. Recent evidence shows that inhibitors of the collagen cross-linking enzyme lysyl oxidase can block contraction to a large degree without having adverse effects on skin morphology [31, 78].

67.5 Uses of RHE as Alternatives to Animal/Human Testing

67.5.1 In Vitro Skin Irritation

In vitro RHE allows investigations of products over a wide range of concentrations and in formulations identical to those used in vivo. For these reasons, RHE can be widely exploited for various research purposes, including studies of cutaneous biogenesis, skin wound healing, investigation of the regulation of keratinocyte differentiation, pharmaceutical agent metabolism studies and absorption properties, the assessment of cutaneous immunotoxicological responses, and responses to irritants and/or to sensitizers [48]. Furthermore, the lower part of the tissue is bathed in the culture medium that can be withdrawn for analysis of released mediators and/or metabolites providing quantifiable and objective endpoint measurements compared to those in vivo studies where subjective parameters, such as erythema and oedema are often evaluated by different investigators.

Based on these developments and the ethical problems which arise from the use of animals, the European Center for the Validation of Alternative Methods (ECVAM), set up by the European Union, has considered skin reconstruction among the most advanced projects for the development of in vitro pharmacotoxicology and research programs (76/768/EEC, Feb. 2003).

Moreover, changes in the European chemical policy (REACH) and the 7th Amendment to the EC Cosmetic Directive prohibiting the use of animals for gathering toxicological data for cosmetics have made validated skin tests mandatory. Therefore, very rapidly, some of these models have undergone various validation programs in order to evaluate their suitability in cutaneous toxicology (corrosivity, irritation, and sensitization) and pharmacology, considering that EU regulation will prohibit the use of animals for gathering toxicological data for cosmetics [12, 48, 76, 90].

The most frequently measured endpoints include the histological analysis of tissue damage, cell membrane damage estimated by the released lactate dehydrogenase activity, cell viability determination either by MTT conversion assay or Neutral Red assay, the modulation of the stratum barrier function, and the measurement of the release of proinflammatory mediators, such as IL-1α, IL-β, IL-6, IL-8, TNFα, prostaglandins, hydroxyeicosanotetraeno and leukotriene B4, plasminogen activator, cytokine mRNA expression, antileukoproteinase synthesis, intercellular adhesion molecule-1 expression, integrin receptor modulation, measurement of intracellular ATP, and corneosurfametry [12].

Based on a MTT conversion assay, the use of RHE has been approved by the OECD for skin corrosivity testing [51]. For skin irritation, the performances of the Episkin®, SkinEthic® and Epiderm® models were judged as sufficient regarding sensitivity, specificity, and accuracy [23, 40, 62, 85].

67.5.2 In Vitro Skin Sensitization

The development of an in vitro test able to replace the in vivo assays intended for the detection of sensitizers remains a challenge. We have demonstrated that the prediction of skin sensitization and/or irritation potential of different topically applied chemicals is mainly based on the differential expression and release of IL-1α and IL-8 [11, 66]. Particularly, the ratio of IL8/IL1α has been found indicative of the sensitizing vs. irritant potential of the chemicals [13]. Other studies have examined the effect of allergens on the IL-1β mRNA expression in RHE containing Langerhans cells and the expression of CD86 cell surface antigen expression in these cells. The results have demonstrated that all tested allergens induce an IL-1β mRNA and CD86 mRNA overexpression while the irritants have no effects [22].

However, RHE does not fully represent the in vivo situation. In vivo, the different cell types interact with each other through signaling molecules and cell-cell contact. The initiation and amplification of the immune response after exposure to a sensitizing chemical also requires the participation of other epidermal cells, in particular mediators of the innate immunity, such as macrophages and natural killer lymphocytes [1, 52]. Recent data based on transcriptome analysis of CD34-dendritic cells exposed to four sensitizing chemicals (DNCB, Oxazolone, Eugenol, and Nickel Sulphate) and two non-sensitizing chemicals (SDS and BC), and further real-time RT-PCR and statistical analysis, demonstrated that 13 genes have a highly significant capability to distinguish between sensitizers and non-sensitizers [76]. However, the exact role of these marker genes in the sensitization process and how they are influenced by other cell types in the skin after exposure to sensitizers is unknown [26]. Therefore, an investigation of an in vitro coculture model consisting of dendritic cells, keratinocytes and peripheral blood mononuclear cells should provide many of the currently lacking insights.

67.5.3
In Vitro Skin Permeation and Metabolization

The non-invasive character of in vivo transdermal drug delivery systems represents a patient-friendly option for the administration of pharmaceutical products that is able to improve compliance with therapy [80]. Obviously, the investigation of the percutaneous uptake is of the highest importance in the development of this kind of therapeutic approach [39]. For this purpose, in vitro protocols using excised human or animal skin samples have been developed [34] and adopted in OECD guideline 428 [51]. However, for legal and ethical reasons, the use of such skin samples is often limited and closely regulated. Moreover, the relevance of conclusions drawn from animal data to human skin has always been limited and questionable. Therefore, OECD has stated that artificial human skin models also could be used in those studies, given that the equivalency is proven [50].

Comparative studies performed on the most important existing artificial epidermal models (Episkin®, SkinEthic® and Epiderm®) have confirmed their suitability for transport experiments of drugs and xenobiotics across skin [48]. Percutaneous penetration studies performed with human skin recombinant models have revealed that the stratum corneum forms a substantial barrier to 3H-water [69, 91], pindolol, calcitonin, toluene, carbazole, benzopyrene [91], testosterone [18, 20, 70], estradiol [20], hydrocortisone [20, 28, 70], benzoic acid [18], cyclosporin [88], salicylic acid, provitamin B5, theophylline, and scopolamine [18]. The results obtained are more consistent and reproducible than cadaver skin and correlate well with those recorded for hairless guinea pig skin [88]. Nevertheless, the relative permeability of normal human skin compared to reconstructed skin is different and is likely to vary considerably from one compound to another. A good correlation for one class of chemicals is not necessarily indicative of a similar relationship with other chemicals [18, 20, 69, 91].

However, for topical drug development, one other important issue is the metabolizing capacity of the tissue where the drug is delivered. In fact, the human skin and particularly its epidermis is a metabolically active organ containing enzymes able to catalyze the metabolism not only of endogenous chemicals but also xenobiotics [38]. Therefore, regarding pharmaceuticals which undergo biotransformation within the epidermis into metabolites that are able to penetrate to a greater extent than the parent compound, their metabolism is a critical determinant affecting their benefit/risk ratio [83, 41]. Despite their still controversial barrier function, several studies have demonstrated that RHE has exhibited a metabolic activity towards prednicarbate, betamethasone, testosterone, and estradiol [4, 32, 43, 45]. To date, RHE represent powerful tools in the development of molecules that can be topically applied. The relatively weak barrier function of commercially available RHE models is considered as their biggest limitation for comparison with human epidermis. Thus, their improvement regarding this function is considered as an important challenge for the future of in vitro reconstructed skin. Nevertheless, if the permeation/penetration of reconstructed epidermis by substances of different physicochemical characteristics is higher than the uptake by human skin by a defined ratio, a higher permeability could be considered in the interpretation of the pharmacokinetic results, especially since the variations in transport rates are low compared to those seen with human skin. Therefore, for bioequivalence or bioavailability investigations, RHE can contribute significantly to saving time and costs in the pre-clinical development of topically applied pharmaceuticals delivery systems and/or formulations.

In summary, the RHE models have attained a level of characterization that allows their production either in specialized commercial companies or by research laboratories. Their current availability and cost make their utilization and analysis more frequent every day, bringing to knowledge data that stimulates new refinements in the analytical approaches of the engineered tissue.

References

1. Albanesi C, Scarponi C, Giustizieri ML, Girolomoni G (2005) Keratinocytes in inflammatory skin diseases. Curr Drug Targets Inflamm Allergy 4(3):329–34
2. Banks-Schlegel S, Green H (1981) Involucrin synthesis and tissue assembly by keratinocytes in natural and culture human epithelia. J Cell Biol 90: 732–737
3. Barker CL, McHale MT, Gillies AK, Waller J, Pearce DM, Osborne J, Hutchinson PE, Smith GM, Pringle JH (2004) The development and characterization of an in vitro model of psoriasis. J Invest Dermatol 123(5):892–901
4. Bernard F-X, Barrault C, Deguercy A, De Wever B, Rosdy M (2000) Development of a highly sensitive in vitro phototoxicity assay using the SkinEthic reconstructed human epidermis. Cell Biology and Toxicology. 16(6): 391–400
5. Bessou S, Surlève-Bazeille JE, Sorbier E, Taïeb A (1995) Ex-vivo reconstruction of epidermis with melanocytes and UVB influence. Pigm Cell Res 8(5):241–249
6. Bouwstra JA, Gooris GS, van der Spek JA, Bras W (1991) Structural investigations of human stratum corneum by small-angle X-ray scattering. J Invest Dermatol 97(6):1005–12
7. Caux C, Dezutter-Dambuyant C, Schmitt D, Banchereau J (1992) GM-CSF and TNF-alpha cooperate in the generation of dendritic Langerhans cells. Nature 360:258–261
8. Chakrabarty KH, Heaton M, Dalley AJ, Dawson RA, Freedlander E, Khaw PT, Mac Neil S (2001) Keratinocyte-driven contraction of reconstructed human skin. Wound Repair Regen 9(2):95–106
9. Cook PW, Mattox PA, Keeble WW, Pittelkow MR, Plowman GD, Shoyab M, Adelman JP, Shipley GD (1991) A heparin sulfate-regulated human keratinocyte autocrine factor is similar or identical to amphiregulin. Mol Cell Biol 11: 2547–2557
10. Cook PW, Pittelkow MR, Shipley GD (1991) Growth factor-independent proliferation of normal human neonatal keratinocytes: production of autocrine- and paracrine-acting mitogenic factors. J Cell Physiol 146: 277–289
11. Coquette A, Berna N, Rosdy M, Vandenbosch A, Poumay Y (1999) Differential expression and release of cytokines by an in vitro reconstructed human epidermis following exposure to skin irritant and sensitising chemicals. Toxicol In Vitro 13:867–877
12. Coquette A, Berna N, Poumay Y, Pittelkow MR (2000) The keratinocyte in cutaneous irritation and sensitization. In: Kydonieus AF, Wille JJ (eds) Biochemical Modulation of Skin Reaction. CRC Press, New York, pp 125–143
13. Coquette A, Berna N, Vandenbosch A, Rosdy M, De Wever B, Poumay Y (2003) Analysis of the induction and release of interleukin-1alpha (IL-1α) and interleukin-8 (IL-8) in in vitro reconstructed human epidermis in order to predict in vivo skin irritation and/or sensitisation. Toxicol In Vitro 17:311–321
14. De Luca M, D'Anna F, Bodanza S, Franzi AT, Cancedda R. (1988) Human epithelial cells induce human melanocyte growth in vitro but only skin keratinocyte regulates its proper differentiation in the absence of dermis. J Cell Biol 107:1919–1926
15. De Luca M, Franzi AT, D'Anna F, Zicca A, Albanese E, Bondanza S, Cancedda R (1988) Coculture of human keratinocytes and melanocytes: differentiated melanocytes are physiologically organised in the basal layer of the cultured epithelium. Eur J Cell Biol 46:176–180
16. De Wever B, Rheins LA (1994) Skin 2TM: An in vitro skin analog. Alternative Methods in Toxicology 10: 121–131
17. Dezutter-Dambuyant C, Black A, Bechetoille N, Bouez C, Marechal S, Auxenfans C, Cenizo V, Pascal P, Perrier E, Damour O (2006) Evolutive skin reconstructions: From the dermal collagen-glycosaminoglycan-chitosane substrate to an immunocompetent reconstructed skin. Biomed Mater Eng 16:S85–S94
18. Doucet O., Garcia N, Zastrow L (1998) Skin culture model:a possible alternative to the use of excised human skin for assessing in vitro percutaneous absorption. Toxicol In Vitro 12:423–430
19. Duval C, Schmidt R, Regnier M, Facy V, Asselineau D, Bernerd F (2003) The use of reconstructed human skin to evaluate UV-induced modifications and sunscreen efficacy. Exp Dermatol 12 Suppl 2:64–70
20. Ernesti AM, Swiderek M, Gay R (1992) Absorption and metabolism of topically applied testosterone in organotypic skin culture. Skin Pharmacol 5:146–153

21. Eves P, Katerinaki E, Simpson C, Layton C, Dawson R, Evans G, MacNeil S (2003) Melanoma invasion in reconstructed human skin is influenced by skin cells—investigation of the role of proteolytic enzymes. Clin Exp Metast 20(8):685–700
22. Facy V, Flouret V, Régnier M, Schmidt R (2005) Reactivity of Langerhans cells in human reconstructed epidermis to known allergens and UV radiation. Toxicol In Vitro 19: 787–795
23. Faller C, Bracher M, Dami N, Roguet R (2002) Predictive ability of reconstructed human epidermis equivalents for the assessment of skin irritation of cosmetics. Toxicol In Vitro 16: 557–572
24. Fleischmajer R, MacDonald ED, Contard P, Perlish JS (1993) Immunochemistry of a keratinocyte-fibroblast co-culture model for reconstruction of human skin. J Histochem Cytochem 41(9):1359–1366
25. Gay R, Swiderek M, Nelson D, Ernesti A (1992) The living skin equivalent as a model in vitro for ranking the toxic potential of dermal irritants. Toxicol In Vitro 6(4):303–315
26. Gazel A, Ramphal P, Rosdy M, De Wever B, Tornier C, Hosein N, Lee B, Tomic-Canic M (2003) Transcriptional profiling of epidermal keratinocytes: comparison of genes expressed in skin, cultured keratinocytes, and reconstructed epidermis, using large DNA microarrays. J Invest Dermatol 121:1459–1468
27. Gibbs S, Vicanova J, Bouwstra J, Valstar D, Kempenaar J, Ponec M (1997) Culture of reconstructed epidermis in a defined medium at 33 degrees C shows a delayed epidermal maturation, prolonged lifespan and improved stratum corneum. Arch Dermatol Res 289(10): 585–595
28. Godwin DA, Michniak BB, Creek KE (1997) Evaluation of transdermal penetration enhancers using a novel skin alternative. J Pharm Sci 86(9):1001–5
29. Green H, Kehinde O, Thomas J (1979) Growth of cultured epidermal cells into multiple epithelia suitable for grafting. Proc Natl Acad Sci USA 76(11):5665–68
30. Green H (1991) Cultured cells for the treatment of disease. Sci Am 265:96–102
31. Grinnell F (1994) Fibroblasts, myofibroblasts, and wound contraction. J Cell Biol 124(4):401–4
32. Gysler A, Kleuser B, Sippl W, Lange K, Korting HC, Höltje H-D, Schäfer-Korting M (1999) Skin penetration and metabolism of topical glucocorticoids in reconstructed epidermis and excised human skin. Pharm Res 16(9):1386–1391
33. Haake AR, Scott GA (1991) Physiologic distribution and differentiation of melanocytes in human fetal and neonatal skin equivalents. J Invest Dermatol 96:71–77
34. Hadgraft J (2001) Modulation of the barrier function of the skin. Skin Pharma Appl Skin Physiol 17:72–81
35. Harriger MD, Hull BE (1992) Cornification and basement membrane formation in a bilayered human skin equivalent maintained at an air-liquid interface. J Burn Care Rehabil 13:187–93
36. Hedley SJ, Layton C, Heaton M, Chakrabarty KH, Dawson RA, Gawkrodger DJ, MacNeil S (2002) Fibroblasts play a regulatory role in the control of pigmentation in reconstructed human skin from skin types I and II. Pigment Cell Res 15(1):49–56
37. Holbrook KA, Hennings H (1983) Phenotypic Expression of Epidermal Cells in Vitro: A Review. J Investig Dermatolog 81:11s–24s
38. Hotchkiss S (1998) Dermal metabolism..In: Roberts MS, Walter KA (eds.) Dermal absorption and toxicity assessment. Drug and the pharmaceutical sciences. New York, Marcel Dekker Inc. Vol 91:43–101
39. Howes D, Guy R, Hadgraft J, Heylings J, Hoeck U, Kemper F, Maibach H, Marty JP, Merck H, Parra J, Rekkas D, Rondelli I, Schaefer H, Tauber U, Verbiese N (1996) Methods for assessing percutaneous absorption. Altern Lab Anim 24:81–106
40. Kandárová H, Liebsch M, Genschow E, Gerner I, Traue D, Slawik B, Spielmann H (2004) Optimisation of the EpiDerm test protocol for the upcoming ECVAM validation study on in vitro skin irritation tests. ALTEX 21(3):107–14
41. Kanikkannan N, Kandimalla K, Lamba SS and Singh M (2000) Structure-activity relationship of chemical penetration enhancers in transdermal drug delivery. Curr Med Chem 7(6):593–608
42. Koria P, Andreadis ST (2006) Epidermal morphogenesis: the transcriptional program of human keratinocytes during stratification. J Invest Dermatol 126:1834–1841
43. Lombardi Borgia S, Schlupp P, Mehnert W and Schäfer-Korting M (2008). In vitro skin absorption and drug release—A comparison of six commercial prednicarbamate preparations for topical use. Eur J Pharm Biopharm 68(2):380–389
44. MacNeil S (2007) Progress and opportunities for tissue-engineered skin. Nature 445:874–880
45. Mahmoud A, Haberland A, Dürrfeld M, Heydeck D, Wagner S, Schäfer-Korting M (2005) Cutaneous estradiol permeation, penetration and metabolism in pig and man. Skin Pharm Phys 18:27–35
46. Marston WA, Hanft J, Norwood P, Pollak R (2003) The efficacy and safety of Dermagraft in improving the healing of chronic diabetic foot ulcers: Results of a prospective randomized trial. Diabetes Care 26:1701–1705
47. Meier F, Nesbit M, Hsu MY, Martin B, Van Belle P, Elder DE, Schaumburg-Lever G, Garbe C, Walz TM, Donatien P, Crombleholme TM, Herlyn M (2000) Human melanoma progression in skin reconstructs: biological significance of βFGF. Am J Pathol 156(1):193–200
48. Netzlaff F, Lehr CM, Wertz PW, Schaefer UF (2005) The human epidermis models Episkin, SkinEthic and Epi-Derm: An evaluation of morphology and their suitability for testing phototoxicity, irritancy, corrosivity, and substance transport. Eur J Pharm Biopharm 60:167–178
49. Nickoloff BJ and Naidu Y (1994) Perturbation of epidermal barrier function correlates with initiation of cytokine cascade in human skin. J Am Acad Dermatol 30(4):535–46.
50. OECD (2003) Guidance document no. 28 for the conduct of skin absorption studies. Adopted at 35th Joint Meeting
51. OECD (2004) Test guideline 428: Skin absorption: In vitro method. Adopted on 13th April 2004, Paris, France
52. O'Leary JG, Goodarzi M, Drayton DL, von Andrian UH (2006) T cell- and B cell-independent adaptive immunity mediated by natural killer cells. Nat Immunol 7(5):507–16

53. Parnigotto PP, Bernuzzo S, Bruno P, Conconi MT, Montesi F (1998) Chracterization and applications of human epidermis reconstructed in vitro on de-epidermized derma.
54. Pittelkow MR, Shipley GD (1989) Serum-free culture of human melanocytes: growth kinetics and growth factor requirements. J Cell Physiol 140:565–576
55. Pittelkow MR, Cook PW, Shipley GD, Derynck R, Coffey RJ Jr (1993) Autonomous growth of human keratinocytes requires epidermal growth factor receptor occupancy. Cell Growth Diff 4:513–521
56. Ponec M (1993) Glucocorticoid receptors. Curr Probl Dermatol 21:20–8.
57. Ponec M, Kempenaar J (1995) Use of human skin recombinants as an in vitro model for testing the irritation potential of cutaneous irritants. Skin Pharmacology 8: 49–59
58. Ponec M, Gibbs S, Weerheim A, Kempenaar J, Mulder A, Mommaas AM (1997) Epidermal growth factor and temperature regulate keratinocyte differentiation. Arch Dermatol Res. 289(6): 317–326
59. Ponec M, Weerheim A, Kempenaar J, Mulder A, Gooris GS, Bouwstra J, Mommaas AM (1997) The formation of competent barrier lipids in reconstructed human epidermis requires the presence of vitamin C. J Invest Dermatol 109: 348–355
60. Ponec M, Boelsma E, Weerheim A, Mulder A, Bouwstra J, Mommaas M (2000) Lipid and ultrastructural characterization of reconstructed skin models. Int J Pharm 203: 211–225
61. Ponec M, Boelsma E, Gibbs S, Mommaas M (2002) Characterization of reconstructed skin models. Skin Pharmacol Appl Skin Physiol 15 Suppl 1:4–17
62. Portes P, Grandidier MH, Cohen C, Roguet R (2002) Refinement of the Episkin protocol for the assessment of acute skin irritation of chemicals: follow-up to the ECVAM prevalidation study. Toxicol In Vitro 16(6):765–70
63. Portes P, Pygmalion MJ, Popovic E, Cottin M, Mariani M (2002) Use of human reconstituted epidermis Episkin for assessment of weak phototoxic potential of chemical compounds. Photodermatol Photoimmunol Photomed 18(2):96–102
64. Poumay Y, Pittelkow MR (1995) Cell density and culture factors regulate keratinocyte commitment to differentiation and expression of suprabasal K1/K10 keratins. J Invest Dermatol 104: 271–276
65. Poumay Y, Herphelin F, Smits P, De Potter IY, Pittelkow MR (1999) High-cell-density, phorbol ester and retinoic acid upregulate involucrin and downregulate suprabasal keratin 10 in autocrine cultures of human epidermal keratinocytes. Mol Cell Biol Res Commun 2:138–144
66. Poumay Y, Dupont F, Marcoux S, Leclercq-Smekens M, Hérin M, Coquette A (2004) A simple reconstructed human epidermis: preparation of the culture model and utilization in in vitro studies. Arch Dermatol Res 296:203–211
67. Poumay Y, Coquette A (2006) Modeling the human epidermis in vitro: Tools for basic and applied research. Arch. Dermatol. Res 298(8):361–369
68. Pruniéras M, Régnier M, Woodley D (1983) Methods for cultivation of keratinocytes with an air-liquid interface. J Invest Dermatol 81:28s–33s
69. Régnier M, Caron D, Reichert U, Schaeffer H (1992) Reconstructed human epidermis: A model to study in vitro the barrier function of the skin. Skin Pharmacol 5:42–56
70. Régnier M, Caron D, Reichert U, Schaeffer H (1993) Barrier function of human skin and human reconstructed epidermis. Journal of Pharmaceutical sciences 82(4), 404–407.
71. Régnier M, Staquet MJ, Schmitt D, Schmidt R (1997) Integration of Langerhans cells into a pigmented reconstructed human epidermis. J Invest Dermatol 109:510–512
72. Rheinwald JG, Green H (1975) Serial cultivation of strains of human epidermal keratinocytes: the formation of keratinising colonies from single cells. Cell 6(3): 331–343
73. Rosdy M, Claus LC (1990) Terminal epidermal differentiation of human keratinocytes grown in chemically defined medium on inert filter substrates at the air-liquid interface. J Invest Dermatol 95:409–414
74. Rosdy M., Pisani A, Ortonne JP (1993) Production of basement membrane components by a reconstructed epidermis cultured in the absence of serum and dermal factors. Brit J Dermatol 129:227–234
75. Rosdy M., Fartasch M, Ponec M (1996) Structurally and biochemically normal permeability barrier of human epidermis reconstituted in chemically defined medium. J Invest Dermatol 107:664
76. Schoeters E, Verheyen GR, Nelissen I, Van Rompay AR, Hooyberghs J, Van Den Heuvel, RL (2007) Microarray analyses in dendritic cells reveal potential biomarkers for chemical-induced skin sensitization. Mol Immunol 44(12):3222–3233
77. Schreiber S, Mahmoud A, Vuia A, Rübbelke MK, Schmidt E, Schaller M, Kandarova H, Haberland A, Schäfer UF, Bock U, Korting HC, Liebsch M, Schäfer-Korting M (2005) Reconstructed epidermis versus human and animal skin in skin absorption studies. Toxicol In Vitro 19:813–822
78. Scott GA, Haake AR (1991) Keratinocytes regulate melanocytes number in human fetal and neonatal skin equivalents. J Invest Dermatol 97:776–781
79. Souren JM, Ponec M, van Wijk R (1989) Contraction of collagen by human fibroblasts and keratinocytes In Vitro Cell Dev Biol 25(11):1039–45
80. Stiel D (2000) Exploring the link between gastrointestinal complications and over-the-counter analgesics: current issues and considerations. Am J Therapeut 7:91–98
81. Stoppie P, Borghraef P, De Wever B, Geysen J, Borgers M (1993) The epidermal architecture of an in vitro reconstructed human skin equivalent (Advanced Tissue Sciences Skin2 Models ZK 1300/2000). Eur J Morphol 31(1–2): 26–29
82. Swope VB, Supp AP, Schwemberger S, Babcock G, Boyce S (2006) Increased expression of integrins and decreased apoptosis correlate with increased melanocyte retention in cultured skin substitutes. Pigment Cell Res 19(5):424–33
83. Tauber U (1989) Drug metabolism in the skin. In: Hadgraft J, Guy RH (eds.) Transdermal drug delivery. Marcel Dekker, New York, pp. 99–112
84. Todd C, Hewitt SD, Kempenaar J, Noz K, Thody AJ, Ponec M (1993) Coculture of human melanocytes and

keratinocytes in a skin equivalent model: effect of ultraviolet radiation. Arch Dermatol Res 285:455
85. Tornier C, Rosdy M, Maibach HI.(2006) In vitro skin irritation testing on reconstituted human epidermis: reproducibility for 50 chemicals tested with two protocols. Toxicol In Vitro 20(4):401–16
86. van de Sandt JJ, Bos TA, Rutten AA (1995) Epidermal cell proliferation and terminal differentiation in skin organ culture after topical exposure to sodium dodecyl sulphate. In Vitro Cell Dev Biol Anim 31(10):761–766
87. Vicanova J, Mommaas AM, Mulder AA, Koerten HK, Ponec M (1996) Impaired desquamation in the in vitro reconstructed human epidermis. Cell and Tiss Res 286(1):115–122
88. Vickers AEM, Biggi WA, Dannecker R, Fisher V (1995) Uptake and metabolism of cyclosporin A and SDZ IMM 125 in the human in vitro Skin 2TM dermal and barrier function models. Life Sci 57(3):215–224
89. von den Driesch P, Fartasch M., Huner A, Ponec M (1995) Expression of integrin receptors and ICAM-1 on keratinocytes in vivo and in an in vitro reconstructed epidermis: effect of sodium dodecyl sulphate. Arch Dermatol Res 287(3–4):249–253
90. Welss T, Basketter DA, Schroder KR (2004) In vitro skin irritation: facts and future. State of the art review of mechanisms and models. Toxicol In Vitro 18:231–243
91. Yang JJ, Krueger A (1992) Evaluation of two commercial human skin cultures for in vitro percutaneous absorption. In Vitro Toxicol 5(4):211–217

XIV Clinical Handling and Regulatory Issues

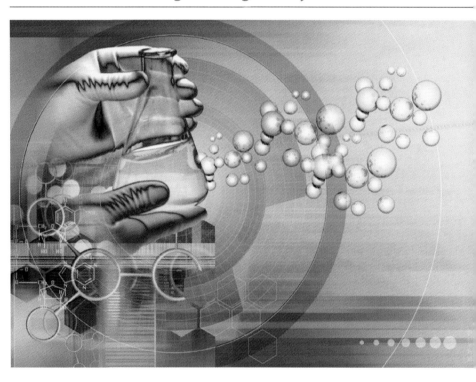

Regulatory Issues

B. Lüttenberg

Contents

Clinical Handling and Control 980

Tissue engineering is a new field of research which generates products and applications in a very wide range, and is developing fast. Knowledge of the legal situation is important for all scientists and physicians working in all aspects of this new life science. Tissue engineering products are on the one hand not true medical devices, and on the other hand, not medicinal products. Additionally, autologous tissue engineering or tissue-derived products by allogeneic tissues/cells face very different issues. The outcome of this fact is that the new products of tissue engineering are not sufficiently covered by the established regulatory frameworks; especially the existing uncertainty regarding the classification of tissue engineering products for human use has complicated the commercialization process and therefore probably even impeded its initial development during the last years. To overcome this situation the existing legal arrangements have had to be adjusted to the new circumstances. Therefore, the purpose of the needed regulations is to clarify the requirements for the processes and the products themselves, as well as to guarantee all public health issues.

The USA, Canada and Australia have already developed regulatory frameworks concerning human-derived tissue engineering products for human use before 2002. In contrast to new regulatory frameworks, some countries, like Japan, decided to adopt existing legislation to manage future products.

To gain a common European regulatory framework, the European Commission and the European Council have published new directives since the late 1990s, namely Directive 2001/83/EC, 2004/23/EC and the latest ones, 2006/17/EC and 2006/86/EC. The aim of these was to get a common framework for the so-called advanced therapies, including gene therapy, cell therapy and tissue engineering. The present member states of the European Union now are on their way to implement the provisions of these directives into countries' laws.

In Germany, for example, the already existing laws (TPG, AMG, TFG), regulating transplantations, drugs, and transfusions, have been modified and adjusted in many ways into the so called "Tissue Law" ("Gewebe-Gesetz"). The lower house of the German parliament (the Bundestag) has passed this "Tissue Law" in May, 2007 and it became operative on August 1st, 2007. It contains the main provisions of the European tissue Directive 2004/23/EC. In contrast to former regulations, the permission for the collection of tissues and cells has changed and is now regulated by section 20 of the German Drug Law (AMG) with two exceptions. This does not apply for collected autologous tissue from one patient that is re-inserted within a cycle of treatment (even if this treatment covers a long time). The second exception is the field of basic research far away from any product for human use that could be brought to market. However, all other tissues and cells that are more or less designated for tissue engineering, including industrial processing, apply to section 20 of the AMG (or section 13 of the AMG) and therefore need permissions before collection. This permission is accompanied with standardization and validation of the processes and an intensive documentation of each single step. The procurement of tissue for application in humans, e.g., for the purpose of generating a medicinal product, should be performed according to the "good professional practice" (GPP). For this reason, special practical guidelines must be generated by different joint advisory boards (for clinical trials, the principles and guidelines for "good clinical practice:" see directive 2005/28/EC). In order to provide safe products of reliable quality, the principles and guidelines of the "good manufacturing practice" (GMP; see also the 2003/94/EC directive of the European commission) in respect of medicinal products for human use and investigational medicinal products for human use should be followed. The various guidelines cover the selection of donors, the retrieval of cells, the testing of cells (e.g., after subcultures), the processing of cell/scaffold constructs, the storage and delivery of finished tissues as well as quality management, personnel, premises, and equipment and documentation. As tissue engineering is more or less just beginning, the test of both the regulations and the new products is still to come.

Clinical Handling and Control

In order to provide safe cell/scaffold products of reliable quality, the principles of GMP must be followed in the selection of donors, retrieval of cells, testing of cells (e.g. after subculture), processing of cell/scaffold constructs, storage and delivery of finished tissues. To date, the regulations are established individually by country. If autologous cells are used in Germany, for example, manufacturing is regulated by the AMG and by the German Operation Ordinance for Pharmaceutical Entrepreneurs.

Additionally, some institutions have developed regulatory frameworks to guide the use of cell-containing products, initially implemented for tissue bank products. Codes of practice (COP) for tissue banks specify the requirements for the activities of tissue banks that store and/or process human tissue for therapeutic use. The scope of these codes includes in most instances all human tissue used for therapeutic purposes, including those used in clinical trials. The various tissues of the body, including haemopoietic progenitor cells (e.g. bone marrow, peripheral blood, cord/placental blood), are all examples that are included in this scope. Professional institutions and specialist societies have, in many instances, prepared standards and guidelines for specific procedures in order to introduce a framework for developing and supporting pre-existing standards and guidelines. Currently, no specific guidance exists in the area of tissue engineering, but when therapeutic products (including cells of human origin, as it is the case for tissue engineering applications) are processed by commercial or non-commercial organisations, this code applies. In addition, the following principles apply:

- If autologous cells are cultured in vitro prior to implantation, the procedures must be validated or monitored, e.g. to demonstrate vitality of cells and sterility as well as identification of samples. In addition, lack of malignant transformation and maintenance of relevant biological properties are desired properties of autologous cells, which should ideally be documented by proper test systems.
- If allogeneic cells are used in tissue engineering applications (a strategy that is not yet introduced with clinical bone or cartilage use), donors should be thoroughly evaluated to reduce the risk of transmitting diseases. The use of allogenic stem cells is commonly regulated by laws and regulations set up for tissue banks.

The intention to establish a code is based on quality management systems (in Germany, for example, International Organization for Standardization [ISO] 9000 series: ISO 9001–2000). This system addresses cell-based therapeuticals. The principles of GMP concerning tissue banks differ in detail among countries (for example, in *Rules and Guidance for Pharmaceutical Manufacturers and Distributors*, UK Department of Health, 1997). The complexity of the ex vivo generation of artificial tissues and organs emphasises the need for compliance with an integrated quality assurance program. As for tissue banks, commercial or non-commercial organisations working in the cell-based tissue engineering field must adhere to the principles set forth in other documents (e.g. *Guidance on the Microbiological Safety of Human Organs, Tissues and Cells used in Transplantation*,

UK Department of Health, 2000). All stages of the selection and testing of donors, retrieval (harvesting) of cells or tissues, culturing, coaxing with materials, processing, storage and delivery of cell/scaffold constructs should meet the established criteria. It is important to note that all institutions working in this field have a general duty and responsibility to comply with the statutory legislation of the countries that are involved in the performance and processing of tissue-engineered products. Physicians and involved personnel must ensure that proper consent has been obtained for the retrieval of cells or tissue, as addressed by the Human Tissue Act 1961, the Human Fertilisation and Embryology Act 1990, the Human Fertilisation and Embryology (disclosure of information) Act 1992 and other professional guidelines.

The processes that are relevant for the clinical use of extracorporeally fabricated tissue constructs should be monitored by a quality system. Today, autologous cells are the sole cell source used in clinical bone and cartilage tissue engineering strategies. Quality management should include all aspects of tissue engineering: formal issues of tissue banks, managerial responsibilities, donor selection and testing, written agreements or contracts with third parties, specified testing of tissues, standards for the processing and storage of tissue, documentation and record keeping, and transport of tissues. Formal issues of cell-based tissue engineering facilities include buildings and premises, environmental controls, processing areas, personnel, ancillary materials, contamination from tissues of donors and equipment. A variety of donor criteria should be documented if allogenic or autologous cells are used, for example, medical and behavioural history, pathology, microbiological testing, exclusion factors and consent. The above-mentioned parameters are also relevant for donors donating tissue for autologous and/or directed allogeneic use. Ideally, the following parameters should be included in the entire process of clinical tissue engineering, with respect to transplantation of an ex vivo–grown cell/scaffold complex:

- Manufacture
  - Product consistency
  - Product stability
- Preclinical safety
  - Material sourcing
  - Adventitious agents (testing-process validation)
  - Toxicity testing
  - Sterility
  - Sterilisation by-products
- Preclinical activity
  - Material characterisation: structural/functional
  - Biomaterial compatibility testing
  - In vitro animal models (efficacy measures)
- Clinical
  - Indications
  - Efficacy endpoints
  - Safety monitoring
  - Population exposure
  - Post-market report

It is important to note that it seems difficult to fulfil each of these criteria, since many of them are not defined in detail. Future developments should stress the need to monitor and eventually to refine such quality systems in a changing scientific environment.

Subject Index

^{13}C NMR   509
^{1}H NMR   509
2D culture   650
2-tetrahydropyridine (MPTP)   281
3D cell culture   650
3D collagen matrix   290
3D culture   969
3-dioxygenase   186
3D network   473
3D scaffold   613, 657
3-tetrahydropyridine (MPTP)   281
5-azacytidine   764
5′ cap structure   99
6-hydroxydopamine   281
6-tetrahydropyridine (MPTP)   281
$\alpha_1\beta_1$ integrin   393
α2-macroglobulin promoter   77
α-actin   143, 430
α-cell   411
α-elastin   689
α-fetoprotein   172, 398
α-hydroxy polyester   183
α-smooth muscle actin   568
β1-integrin   141, 144, 728
β2-integrin   724
β2-microglobulin   65
β-actin promoter   355
β-casein   77
β-catenin   958
β-cell   411, 412, 417, 418, 419, 423
β-glycerolphosphate   162, 171, 221, 553
β-glycerophosphate   490, 573
β-TCP. *see* beta-tricalcium phosphate
β-tricalcium phosphate   353, 354
γ-actin   430
γ-interferon   917

δ-cell   411
μCT   802, 844
μstrain   635, 640

A

AA. *see* aplastic anemia
AAV. *see* adeno-associated vector
AAV transduction   351
AAV-VegF   351
abdomen   881
abdominal cavity   857
abdominal wall   777
abdominoplasty specimen   292
ablation   476
absorbing film-assisted LIFT   618
ACA. *see* active cutaneous anaphylaxis
Academic Medical Center   400, 403
acellular collagen matrice   873
acetabular cup   859
acetaldehyde   96, 102, 103
acetaldehyde-responsive promoter   103
acetylcholine   281
acetylcholine receptor   250
acetylcholine receptor-inducing activity
   (ARIA)   123
Achilles tendon   261, 320, 916
ACI. *see* autologous chondrocyte implantation
acid-base balance   869
acid etching   479, 480
acidic FGF   124
ACL. *see* anterior cruciate ligament
ACL fibroblast   320
acoustic cavitation   930
acquired immunity   782
acrylamide gel   652
acrylic splint   377

acryloyl chloride 505
actin 87
actin cytoskeleton 85, 477, 652
actin filament 243, 477, 478, 655
activated charcoal 861
activated leukocyte cell adhesion molecule 177
activator-based expression control 95
active cutaneous anaphylaxis 791
active systemic anaphylaxis 791
activin 123
activin A 65, 297
activin receptor-like kinase 6 352, 354
acute graft versus host disease 167, 956
acute leukemia 951
acute lymphoblastic leukemia 952
acute myeloid leukemia 952, 957
acute myeloid leukemia blast cell 957
acute periodontitis model 360
acute traumatic herniation 313
AD. see Alzheimer's disease
adaptive defence system 781, 782
adaptive immune response 753, 754
adaptive immunity 730, 751, 782, 791, 803
ADCT. see autologous disc chondrocyte transplantation
adeno-associated vector 351
adeno-associated virus 104, 149, 929
adeno-BMP2 vector 353
adenocarcinoma 432
adenoma 205
adenoviral construct 184
adenoviral gene transfer 415
adenoviral system 103
adenoviral vector 353, 359, 415
adenovirus 93, 104, 110, 111, 149, 163, 183, 265, 300, 360, 750, 833, 929
adenovirus-derived promoter 97
adhesion 182, 478, 648, 650, 652, 684, 695
adhesion molecule 185, 570, 763, 958, 970
adhesion molecule-1 971
adhesion peptide 847
adhesion site 498, 500
adhesiveness 507
adhesive surface 850
adipocyte 101, 105, 126, 168, 172, 179, 180, 204, 205, 263, 290, 292, 294, 299, 300, 360, 391, 392, 393, 882, 914, 933, 958
adipocyte differentiation 179
adipogenesis 130, 141, 180, 295, 300, 391, 392, 393, 394, 652, 656

adipogenic cell type 915
adipogenic differentiation 125, 172, 180, 294, 300
adipogenic lineage 105, 294, 882
adipogenic marker gene 882
adipogenic precursor cell 290
adipose 354, 878
adipose-derived stem cell 182, 290, 878, 914
adipose tissue 84, 185, 290, 292, 293, 294, 300, 301, 302, 382, 389, 390, 391, 392, 393, 394, 419, 732, 776, 816, 878, 881, 914
adipose tissue engineering 290, 298, 389, 390, 392, 393, 394, 395
adipose tissue function 290
adipsin 882
adjacent cell 470
adjuvant 734, 735, 736
adnexa 204, 765
adoptive cellular immunotherapy 961
ADSC 292, 293, 294, 296, 297, 298, 299, 300, 301, 302
adult onset diabetes 940
adult stem cell 51, 54, 139, 140, 147, 160, 172, 197, 292, 418, 555, 560, 562, 659, 721
adventitia 883
AF. see amniotic fluid; see annulus fibrosus
affinity 472
aFGF 124
agar 880
agarose 181, 182, 183, 184, 185, 565, 651, 657, 734, 735, 880
agarose gel 659, 880
agarose hydrogel 182
agglutination 959
aggrecan 180, 181, 182, 184, 239, 240, 257, 309, 311, 764, 860
aggregation 504
AGO. see argonaute
AGO2 110
air-liquid interface 856
albumin 172, 173, 297, 393, 398, 478, 726, 790, 861
albumin dialysis 862
albumin secretion 172
albumin to globulin ratio 790
ALCAM 177
alchemy 6, 10
Alcian blue staining 172
alcohol dehydrogenase 102
AlcR 97
algae 499

Subject Index

alginate   129, 182, 183, 184, 247, 248, 277, 299, 392, 500, 505, 541, 566, 657, 734, 863, 880, 926
alginate amalgam   651
alginate gel   182
alginate hydrogel   181, 651, 818, 850
alginate-poly-l-lysine-sodium   402
alginate/poly-l-ornithine   419
alginate-poly-l-ornithine-alginate   420
alimentary tract   787
Alizarin red   171
alkaline phosphatase   172, 180, 294, 485, 490, 564, 568, 642, 652, 711, 790, 872, 873, 926
alkaline phosphatase activity   105
alkaline phosphatase staining   101
alkanethiol   476
alkyl chain   476
alkylsiloxane   247
ALL. *see* acute lymphoblastic leukemia
allele   954
allergy   785, 790
alloantibody   413, 414
alloantigen recognition   185
allocation   13, 14, 20, 45
allogeneic material   878
allogeneic stem cell transplantation   951
allogeneic substitute   877
allogeneic transplantation   959
allogeneic transplantat setting   959
allogeneic umbilical cord blood   957
allogeneic versus autologous   149
allogenic cell   560
allogenic graft   817
allograft   261, 262, 265, 346, 733, 776, 803, 933, 942
allograft rejection   78
alloplast   933
alloreactive mechanism   954
alloreactive T-cell   956
allosensitization   413, 414
allospecific tolerance   415
allotransplantation   941
allow graft-versus-tumor   951
ALP. *see* alkaline phosphatase
alpha actin   297
alpha cell   943
alpha endocrine cell   945
ALS. *see* amyotrophic lateral sclerosis/Lou Gehrig's disease
alumina   475
alumina-zirconia ceramic   859

aluminium oxide   476
alveolar bone   931
alveolar bone defect   829
alveolar bone distraction   831
Alzheimer's disease   50, 273, 820
amalgam   183, 360
AMC. *see* Academic Medical Center
AMC-BAL   400, 403
AMD3100   955
amelanotic skin   970
ameloblast   922, 928
amelogenesis   351
amelogenin gene splice product   360
aminoglycoside antibiotic   98
aminopeptidase N   145
AML. *see* acute myeloid leukemia
ammonia   398, 400, 404
amnion   144, 276
amnion membrane   141
amniotic-derived cell   382
amniotic epithelial cell   147, 153
amniotic fluid   140, 141, 143, 144, 149, 153, 732, 816
amniotic stem cell   153
amorphous calcium phosphate   486
amphibian limb regeneration   199
amphiregulin (AR)   123, 968
amplification cascade   103
amyloid plaque   820, 821
amyotrophic lateral sclerosis/Lou Gehrig's disease   273, 279, 821
anaemia   790
anaphylaxis   785
anastomosis   777
anemia   413, 869
anephric rat   65, 66
anergic   944
aneurysm   882, 883
Ang-1. *see* angiopoietin
angiogenesis   103, 104, 123, 124, 222, 260, 264, 301, 359, 393, 394, 431, 436, 437, 575, 576, 699, 763, 764, 773, 778, 786, 788, 818, 881, 916
angiogenic control   11
angiogenic disease   764
angiogenic growth factor   928, 929
angiogenic inhibitor   763
angiopoietin   93, 301, 958, 959
angiotensin-converting enzyme inhibitor   750
animal cloning   161
anion   505

anisotropic dewetting process   850
anisotropic wetting process   850
ankylosis   359
annulus fibrosus   308, 311, 313
annulus ring   314
anorganic bovine bone   829
antecubital vein   955
anterior chamber   761
anterior cruciate ligament   256, 258, 261, 317, 318, 319, 320, 321, 322, 323, 324, 325
anti-amyloidogenic agent   820
antiapoptotic protein (Bcl-2)   957
antiarrhythmic   26
antibiotic   96, 97, 101
antibiotic-responsive promoter   103
antibody   586, 669, 733, 782, 803
antibody-dependent cell-mediated cytotoxicity   785
antibody library   586
anti-cardiac antibody   786
anti-CD3/anti-CD28   960
anti-CD25   960
anti-CD-31   873
anti-CD45RO/RB   415
anti-CD52   956
anti-CD154   415
anticlotting medication   857
anticoagulation therapy   857
antigen   782, 783, 785
antigen-antibody immune complex   786
antigen expression   945
antigenicity   162, 569
antigen presentation   782, 792
antigen-presenting cell   66, 185, 186, 187, 414, 722, 782, 784, 786, 960
antigen-specific T-suppressor-cell   414
anti-HLA antibody   413
anti-inflammatory drug   917
anti-LFA   415
anti-MT1-MMP antibody   764
antiosteopontin   872, 873
antioxidant enzyme   858
antisense probe   946, 947
antithymocyte globulin   956
antiviral defense   72
antiviral response   72
anxiety   389
aorta   670, 679, 882
aortic cell culture   124

aortic prosthesis   776
aorto-gonad-mesonephros (AGM) region   140
apatite-coated, polylactic-co-glycolic acid scaffold   879
apatite crystal   636
APC. *see* antigen-presenting cell
apheresis   957
apical periodontitis   921
apical root resorption   921
aplasia   952
aplastic anemia   167, 951, 952
apoptosis   71, 72, 78, 99, 104, 186, 203, 415, 723, 785, 957, 958
apoptotic cell   722, 729
apoptotic cell death   72
apoptotic event   413
aquaporin 2   65
aqueous humour   761
aqueous phase   698
AR. *see* amphiregulin
arginate-poly-l-lysine-encapsulated   421
arginine-glycine-aspartate (RGD) motif   476
arginine-glycine-aspartic acid   357, 500
argonaute (AGO)   110
ARIA. *see* acetylcholine receptor-inducing activity
aromatic ring   476
arterial circulation   883
arterial prosthesis   858
arteriovenous (AV) loop   250
artery   883
arthritis   859
arthrofibrosis   321
arthroplasty   916
arthropod exoskeleton   501
arthrosis   32
articular cartilage   126, 185, 233, 361, 489, 501, 575, 713, 714, 764, 827, 916
articular cartilage regeneration   361
articular cartilage repair   356, 361
articular surface   880
artificial bone ECM   492
artificial device   159
artificial eye   856
artificial heart   857
artificial heart valve   857
artificial limb   855
artificial muscle   860
artificial pancreas   554, 863
artificial renal unit   872

artificial skin   860
artificial tissue substitute   159
ASA. *see* active systemic anaphylaxis
ASC. *see* adipose-derived stem cell
ascorbate   649
ascorbate-2-phosphate   264
ascorbic acid   162, 171, 221, 553, 568, 573, 883
ascorbic acid 2-phosphate   490
aspartic acid   926
Aspergillus nidulans-derived acetaldehyde-responsive transactivator   97
astrocyte   122, 125, 143, 153, 179, 271, 272, 275, 276, 277, 278, 279, 280, 284, 820
asymmetric unit membrane   430, 434, 436
ATG. *see* antithymocyte globulin
atherosclerosis   123, 856, 882, 940
athymic mouse   125, 871, 872
athymic rat   914
atomic force microscopy   577, 845
ATP   726, 883, 971
atrophy   791, 820, 821, 840
attachment   471, 649
auditory function   282
AUM. *see* asymmetric unit membrane
auricle   879
autoantibody   790
autoaugmentation   432
autocatalytic mRNA destruction   98
autogenous bone graft   878
autogenous graft   827, 877, 879
autograft   159, 211, 261, 262, 265, 776
autografting   334
autograft tissue transplantation   159
autoimmune disease   147, 786, 952
autoimmune exocrinopathy   833
autoimmune inflammatory disorder   188
autoimmune myocarditis   111
autoimmune reaction   751
autoimmunity   732, 790, 791
autologous bone graft   491
autologous cell   560, 803, 863
autologous cell-based graft   608
autologous cell-based product   606
autologous chondrocyte   860
autologous chondrocyte implantation   16, 18, 19, 20, 21, 916
autologous chondrocyte spheroid technology   234
autologous disc chondrocyte transplantation   307, 313, 314
autologous graft   606
autologous matrix-induced chondrogenesis   234
autologous spongiosa   491
autologous stem cell transplantation   957
autologous transplant   562
autologous transplantation   959
automation   544, 606, 607, 608
autonomic nervous system   815
auto-phosphorylation   74, 121
autosomal recessive mutation   945
availability   13, 412
AviTag   94
axial force   616
axon   271, 272, 274, 275, 277, 278, 279, 280, 284, 285, 817, 819
axonal outgrowth   273
axonal penetration   818
axonal regeneration   817, 818
axonal sprouting   273, 274
axotomy   271

B

B7   414
B7-1   144, 185, 187
B7-2   144, 185, 187
B7-pathway   414
B220   952
baboon   776, 820
baby hamster kidney cell   821
back pain   307
bacteria   100, 623, 750, 921
bacterial DNA   751
bacterial infection   922
bacterial lipopolysaccharide   785
bacterial toxin   788
BAEC. *see* bovine aortic endothelial cell
BAL. *see* bioartificial liver
BALB/cByJ mouse   84
ballistic delivery   930
BAMG. *see* bladder acellular matrix graft
bands of Bungner   274, 275
banking   167
barrier membrane   829
basal cell   373
basal ganglia   821
basal layer   761, 765, 968, 970
basal stratum   374
basal transcription complex   77
basement membrane   654, 765, 832, 970

basic FGF 124
basic fibroblast growth factor 124, 160, 172, 264, 297, 300, 828, 829
basic multicellular unit 708, 843
basic protein 930
basic research 559
basophil/basophile 783, 785, 953
BBB. *see* blood-brain barrier
B-cell 73, 724, 730, 734, 782, 784, 790, 803, 953
B-cell activity 790, 791
bcl-x$_L$ 73, 99
BCR-ABL 957
BDNF. *see* brain-derived neurotrophic factor
benefit-cost analysis 20
benzoic acid 972
benzopyrene 972
beta-2-microglobulin 861
beta-blocker 750
beta cell 105, 939, 942, 943, 944, 946
beta cell mass 949
betacellulin (BTC) 123
beta endocrine cell 945
beta-glycerophosphate (beta-GP) 127
beta-GP. *see* beta-glycerophosphate
beta lysin 803
betamethasone 972
beta-tricalcium phosphate 125
bFGF. *see* basic fibroblast growth factor
BFU-E 175
BHAP. *see* biohybrid artificial pancreas
BHK. *see* baby hamster kidney cell
biglycan 309
bile 861
bilirubin 188, 398, 400, 790
bioabsorbable glass fibre 248
bioabsorbable material 124
bioactive agent 787
bioactive chamber 16
bioactive glass 265, 346, 933
bioactive molecule 841, 928
bioactive peptide 650
bioartificial liver 397, 398, 400, 402, 403, 404, 405, 406, 407, 855, 861
Bioartificial Liver Support System 398
bioavailability 129
biocompatibility 247, 255, 259, 299, 321, 423, 475, 498, 503, 513, 551, 650, 788, 841, 878
biodegradability 262, 274, 551, 925
biodegradable polymer 124, 820
biodegradable polymeric channel 816

biodegradable scaffold 181, 321, 424
biodegradation 383, 503, 803, 878
bio-electrospray 623
biohybrid 831
biohybrid artificial pancreas 419
bioinert spacer 476
biological laser printing 618
biological matrix 469
biological patch 777
biological polymer 247
BioLP. *see* biological laser printing
biomaterial 475, 480, 633, 729, 734, 777, 788, 933
biomaterial component 791
biomaterial property 598
biomaterial scaffold 181, 649, 721
biomaterial surface 847
biomechanic 121
biomimetic 485
biomimetic matrice 847
biomimetic scaffold 878
biomimetic substrate 850
biomineral formation 840, 843, 844, 847
biomineralisation 486
biomolecular strategy 850
biomolecule 614
Bio-Oss 362
biopaper 615, 617, 618, 620, 621, 627
biophysical effect 213
biophysical force 214
biophysical property 841
biophysical signal 634, 635, 636
biophysical stimulus 214, 636, 644, 710
biopolitics 37, 38
biopolymer 423
bioprosthetic valve 857
biopsy 604
bioreactivity 475
bioreactor 10, 64, 218, 320, 539, 552, 559, 595, 596, 598, 599, 600, 601, 602, 603, 605, 606, 607, 608, 633, 637, 638, 639, 640, 641, 644, 671, 701, 707, 714, 776, 859, 861, 862
bioreactor design 859
biosensor 787
biosynthetic activity 320, 321
biotechnology 827
biotin 94, 96, 104
biotin ligase 96
BirA 97
Birbeck granule 970
bisphosphonate 345

Subject Index

bladder 21, 65, 66, 429, 430, 431, 432, 433, 434, 435, 436, 437
bladder acellular matrix graft 434, 435
bladder augmentation 433
bladder carcinoma 17, 110
bladder engineering 431
bladder organ culture 431
bladder reconstruction 65, 432
bladder smooth muscle 431
blastema 10, 198, 199
blastocyst 28, 139, 160, 164, 204, 560
blastocyst implantation 123
blastocyst stage 416
bleomycin 179
blood 141, 204, 219, 669, 670, 689, 727, 790, 791, 816, 858, 861, 872, 914
blood-brain barrier 817
blood capillary 486
blood capillary ingrowth 489
blood cell 290, 952
blood cell activation 857
blood cell lineage 953
blood circulation 83, 930
blood coagulation 500, 727
blood-compatibility 476
blood flow 670, 776, 883
blood platelet 476
blood pressure 869
blood serum 695
blood stream 787
blood substitute 858, 859
blood supply 844, 877, 916
blood test 789
blood transfusion 858
blood vessel 500, 588, 595, 598, 599, 675, 676, 682, 685, 687, 729, 763, 775, 786, 878, 884, 914, 917, 940
blood vessel formation 577
blot technique 844
B-lymphocyte 72, 750, 958
BMMSC. *see* bone marrow mesenchymal cell
BMP. *see* bone morphogenetic protein
BMP-1 128
BMP-2 101, 103, 105, 124, 128, 129, 180, 182, 184, 187, 294, 300, 351, 352, 353, 354, 357, 359, 360, 829, 850, 879, 916, 924, 925, 928
BMP-3 128
BMP-4 65, 128, 180, 184, 296, 829, 916, 917, 928
BMP-5 128, 184
BMP-6 128, 162, 180, 184, 296

BMP-7 65, 128, 129, 162, 180, 184, 296, 353, 357, 359, 360, 362, 829, 850, 879, 916, 928
BMP-7 gene 931
BMP-9 180
BMP-11 360
BMP receptor 180, 296, 879
BMP signaling 123
BMSC. *see* bone marrow-derived stem cell
BMSSC. *see* bone marrow stromal stem cell
BMU. *see* basic multicellular unit
body mass 84
body mass index 313
body temperature 504
body weight 789
bone 85, 105, 122, 141, 151, 161, 163, 182, 187, 211, 213, 214, 218, 219, 220, 221, 222, 259, 345, 541, 543, 545, 553, 561, 595, 599, 603, 633, 634, 635, 641, 644, 705, 707, 708, 710, 764, 802, 807, 809, 816, 845, 859, 878, 916, 924
bone-anterior cruciate ligament (ACL)-bone allograft 123
bone biology on osseointegration 843
bone/biomaterial interaction 840
bone cell 85, 635, 636, 637, 638, 640, 641, 708, 711, 714
bone cell cultivation 485
bone chip 573
bone-chip culturing 573
bone construct 859
bone defect 345, 485
bone differentiation 128
bone formation 84, 85, 103, 104, 127, 129, 130, 159, 180, 182, 294, 634, 658, 840
bone-forming cell 636
bone fracture 152
bone growth 126
bone healing 636, 845
bone implant 789
bone-implant contact 354
bone-ligament 323
bone-like apatite 847
bone-like nodule formation 564
bone marrow 65, 105, 129, 140, 141, 142, 144, 151, 167, 168, 174, 177, 178, 179, 180, 185, 188, 202, 263, 292, 294, 320, 382, 486, 561, 562, 567, 570, 576, 732, 750, 790, 859, 870, 878, 879, 914, 923, 924, 932, 933, 951, 954, 955, 957
bone marrow aspirate 607, 649
bone marrow aspiration 914
bone marrow cell 143, 147, 218, 859, 954

bone marrow culture   958
bone marrow-derived mesenchymal cell   125
bone marrow-derived stem cell   65, 567, 570, 914
bone marrow mesenchymal cell   172, 174
bone marrow mesenchymal stem cell   294, 736, 878, 879
bone marrow mononuclear cell   171
bone marrow niche   177
bone marrow stem cell   65, 607
bone marrow stroma   293
bone marrow stromal cell   127, 831, 932
bone marrow transplantation   65, 187, 204, 951, 961
bone mass   843
bone matrix   122, 125, 127, 828
bone matrix protein   932
bone mechanosensing   83
bone mechanosensor   86
bone micromass   848
bone mineral   480, 843
bone modelling   635
bone morphogenetic protein   62, 93, 122, 127, 128, 129, 162, 180, 184, 214, 215, 216, 220, 264, 349, 350, 351, 357, 359, 360, 361, 362, 563, 649, 828, 829, 879, 916, 927, 958
bone morphogenetic protein 2/4   239
bone-muscle flap   129
bone neovascularization   576
bone physiology   635
bone property   841
bone quality   840, 846
bone quantity   846
bone reabsorption   123
bone reconstruction   159
bone regeneration   103, 105, 126, 127, 128, 346, 347, 349, 351, 352, 353, 354, 356, 357, 358, 359, 574, 708, 828
bone regenerative capacity   355
bone remodeling   126, 129, 635, 828, 840
bone repair   574, 828
bone replacement material   489
bone resorption   84, 127, 634
bone sialoprotein   172, 296, 360, 924
bone substitute   878
bone tissue   129, 184, 789, 828
bone tissue engineering   570
bone tissue regeneration   129
bone-to-implant contact   847
Boolean logic operation   103

bovine   857, 924
bovine aortic endothelial cell   620, 621, 628
bovine bone mineral   828, 829, 830
bovine collagen   321
bovine-derived hydroxyapatite   933
bovine dermal fibroblast   872
bovine mineral matrix   933
Bp50   144
bradykinin   883
brain   50, 111, 143, 147, 148, 149, 153, 161, 179, 198, 204, 271, 789, 815, 816, 821, 924
brain death criterion   31
brain-derived neurotrophic factor   279, 282, 359, 818
brain function   817
breast   389, 390, 393, 394, 807, 877
breast cancer   389, 390, 394, 957
breast reconstruction   389
breast resurfacing   882
breast tissue engineering   389
BRG1   77
bromodeoxyuridine   871
bronchial epithelium   205
bronchial tube   676
bronchoalveolar lavage   203
bronchus   678
brown algae   501
brown fat   290
Brownian motion   87
BSP. see bone sialoprotein
BTC. see betacellulin
buccal mucosa   17, 374, 832
buccal mucosa graft   373
budding   786
bulk corticospongiosal graft   216
burn   334, 335, 338, 765, 860, 881, 967
burn surgery   331
burst strength   883
butyrolactone   96, 97
butyrolactone inducer   97
bypass   685
bypass operation   882

C

C. see complement
C2C12 myoblast   105
C3   790
C3A   398, 404
C3H/HeJ mouse   84

C4   790
C5a   783
C57BL/6J mouse   84
Ca. see calcium
Ca^{2+} 86, 505
Ca^{2+} binding domain   178
$Ca_{10}(PO_4)_6(OH)_2$ 346
caALK6. see activin receptor-like kinase 6
CAAT enhancer binding protein α (C/EBPα)   101, 105
$CaCl_2$ 969
CAD/CAM. see computer-aided design/computer-aided manufacturing
cadherin-11   178
cadherin V   170
Caenorhabditis elegans   109, 647
CAFC. see cobblestone-area-forming cell
CAFC-assay   953
calcification   171, 785, 787, 790, 845, 858
calcified cartilage   713
calcified collagen   847
calcified freeze-dried bone   933
calcified matrix deposition   598
calcineurin inhibitor   956
calcitonin   972
calcium   84, 86, 87, 180, 486, 501, 568
calcium-carbonate   847
calcium channel   599
calcium-containing layer   847
calcium deposition   171
calcium homeostasis   842
calcium hydroxide   921, 925
calcium influx   85
calcium ion   86, 505
calcium mineral   841
calcium phosphate   357, 486, 840, 847
calcium phosphate cement   129
caldesmon   297
callus formation   844
calponin   297
calvaria   878
calvaria bone   842
calvarial critical size defect   124
calvarial defect   127, 829
calvarian bone   216
CAM. see chorioallantoic membrane
cAMP. see cyclic adenosine monophosphate
cAMP-responsive element-binding protein   77
canaliculi   486

cancellous bone   570, 706, 841
cancellous bone chip   216
cancer   148, 160
cancer ablation   289
cancer cell biology   792
canine   924
capillary   786, 787
capillary force   850
capillary-like structure   763
capillary morphogenesis   763
capillary sprouting   763
capsaicin   762
CAR. see coxsackie B- and adenovirus receptor
carbazole   972
carbohydrate   724
carbohydrate epitope   162
carbohydrates metabolism   969
carbon dioxide   600, 602, 669, 859
carbon fiber   261
carbon filament   280
carcinogenic transformation   929
carcinoma   173
CARD. see caspase-recruiting domain; see caspase-activating and recruitment domain
cardiac   295, 296
cardiac cycle   670
cardiac defect   381
cardiac muscle   775, 776, 790, 863, 930
cardiac myosin   111
cardiac surgery   856
cardiac tissue   141, 749
cardiac tissue engineering   776
cardiology   806
cardiomyoblast   141
cardiomyocyte   111, 141, 161, 163, 168, 205, 296
cardiomyogenic differentiation   296
cardiomyopathy   750, 753, 754
cardiopulmonary bypass device   856
cardiotrophin-1   73
cardiovascular disease   412
cardiovascular structure   381, 382
cardiovascular substitute   381
cardiovascular surgery   883
cardiovascular system   776, 855, 856
cardiovascular tissue engineering   383, 776
caries   921, 923
carotid artery   776, 882
carrier vehicle   735

cartilage   16, 105, 124, 126, 128, 139, 141, 144, 151, 161, 163, 181, 182, 183, 184, 187, 218, 219, 233, 234, 235, 236, 237, 238, 239, 240, 241, 361, 541, 543, 545, 553, 554, 561, 568, 575, 595, 599, 615, 641, 644, 648, 649, 658, 705, 706, 707, 712, 802, 807, 831, 860, 878, 879, 916
cartilage biopsy   607
cartilage damage   181
cartilage defect   16, 181, 361, 879
cartilage-derived morphogenetic protein (CDMP)   264
cartilage destruction   764
cartilage differentiation   127
cartilage engineering   568, 575
cartilage graft   604
cartilage implant   575
cartilage lesion   184
cartilage proteoglycan core protein   180
cartilage regeneration   126, 361, 567
cartilage remodelling   714
cartilage repair   16, 17, 18, 21, 181, 183, 188, 320, 714
cartilage tissue   126, 127, 185, 541
cartilage tissue engineering   500, 563, 575
cartilaginous matrix accumulation   182
cartilaginous tissue   182
Cascade Platelet-Rich Fibrin Matrix   324
case-control analytic study   804
case report   804
$CaSO_4$ scaffold   125
caspase   957
caspase-activating and recruitment domain   752
caspase-recruiting domain   722, 723
catecholamine   283
cathepsin K   492
cationic polymer method   930
caudate nucleus   820
caveolae   85
CB   168. *see* cord blood
CB bank   167, 174
CBER. *see* Center for Biological Evaluation and Research
CB sample   171
CB transplantation   167
CCAAT/enhancer binding protein   180
CCD video camera   603
CCL. *see* chemokine
CCL-3   783, 784
CCL-4   783, 784
CCL-5   783, 784
CCL-11   783
CCL-24   783
CD. *see* cluster differentiation
CD1 antigen   970
CD3   187, 784, 960
CD3/CD19 depletion   959
CD4   170, 187, 414, 415, 784, 790, 791, 959
CD4 regulatory T-cell   186, 187, 414
$CD4^+$ T-cell   414, 749, 786
CD4 T-helper (Th)   414
CD4 T-lymphocyte   414
CD8   170, 187, 414, 415, 784, 790, 791, 952
$CD8^+$ cytotoxic T-cell   414
$CD8^+$ T-cell   786, 960
CD10   170
CD11   751
CD11a   145, 170
CD11b   170
CD11c   170
CD13   145, 147, 168, 170
CD14   141, 144, 147, 170, 177, 178, 561, 723, 751
CD15   170
CD16   170
CD18   170
CD19   784, 960
CD21   784
CD23   73
CD25   145, 170, 959
$CD25^+$ 187, 960
CD27   170
CD28   186, 187, 414
CD29   141, 143, 144, 168, 170, 178, 561
CD31   141, 144, 170, 297, 763, 791
CD33   170
CD34   141, 143, 144, 170, 174, 175, 177, 185, 915, 952, 957
$CD34^-$ 561
$CD34^+$ 105
$CD34^+$ cell   140, 955, 970
$CD34^-$ dendritic cell   972
$CD34^-$ HSC   960
$CD34^+$ selection   957
CD38   140, 144
CD40   144, 170, 185, 187, 414, 415
CD40 ligand   185, 414, 415
CD44   141, 143, 144, 170, 177, 562
CD45   78, 141, 144, 168, 170, 178, 185, 203, 791
$CD45^-$ 561

CD45 antigen 954
CD45RO/RB mAb 415
CD49a 177, 178
CD49b 170
CD49d 145, 170
CD49e 145, 147, 170, 178
CD50 144, 170
CD51 144
CD54 145
CD56 170, 954
CD58 144, 170
CD62E 170
CD62L 170
CD62P 170
CD68 144, 791
CD69 187, 960
CD71 144, 170
CD73 141, 144, 145, 168, 170, 177, 178
CD80 144, 170, 187, 414
CD86 144, 170, 187, 414, 972
CD90 141, 143, 146, 170, 177, 178, 185, 562, 567
CD105 141, 144, 147, 168, 170, 177, 178, 185, 294
CD106 141, 144, 170, 185
CD113 143
CD117 145, 170, 294
CD120a,b 145
CD123 145, 170
CD127 145
CD127 (Il-7Rα) 959
CD133 143, 145, 914, 957, 960
CD140b 178
CD146 146, 170, 924
CD147 145
CD154 414, 415
CD163 791
CD166 141, 144, 170, 177, 294
CD235a 146
CD271 178
CD340 178
CD349 178
cDNA 931
CDRH. see Center for Devices and Radiological Health
CEA. see cost-effectiveness analysis; see cultured epithelial autograft
CEA sheet 333
C/EBP 180
C/EBPα. see CAAT enhancer binding protein α

cell 803, 863, 878
cell adhesion 469, 539, 568, 569, 577, 654, 696, 849
cell adhesion molecule 178, 723, 727
cell adhesion protein 760
cell-adhesive region of fibronectin 818
cell alignment 478, 480
cell amplification 552
cell apoptosis 696
cell array 618
cell attachment 478, 695
cell bank 171
cell behaviour 470
cell-biomaterial-signal 317
cell carrier 498, 508
cell-cell adhesion 178, 179
cell-cell interaction 182
cell-conditioned media 567
cell confluence 605
cell construct 803
cell cultivation 637
cell culture 9, 470, 559, 573
cell culture characteristics 760
cell culture condition 488, 564
cell culture incubator 598
cell culture medium 488
cell cycle 187
cell-cycle progression 72
cell-cycle regulation 72
cell-cycle regulator 93
cell-cycle-related protein 570
cell damage 617, 628
cell debris 785
cell deformation 844
cell density 569, 617, 624, 639, 648
cell depletion 791
cell differentiation 552, 567, 568, 580, 620, 628, 639, 659, 696, 710, 721
cell division 125, 570
cell encapsulation 817
cell engraftment 803
cell environment 469
cell filopodia 478
cell fusion 10
cell growth 94, 123, 507
cell harvest 320, 504, 804
cell homing 174
cell-impermeable barrier 358
cell ingrowth 488

cell junction   565
cell line   560, 563, 564, 575, 604, 732, 760
cell lysis   724
cell marker   586
cell-matrix adhesion   178
cell-matrix interaction   183, 319
cell membrane   77, 588, 670, 714, 930
cell migration   123, 470, 539, 628, 650, 696, 721, 728, 763
cell morphology   470, 652
cell motility   178
cell movement   178
cell nucleus   714
cell nucleus transfer   54
cell number   600, 639
cell nutrition   576
cell passage   605
cell printer   614, 617, 624, 625, 626, 627
cell printing   613, 614, 615, 620, 621, 623, 625, 628
cell proliferation   567, 598, 654, 696, 710, 728
cell receptor   471
cell receptor ligand   476
cell regulation   650
cell scaffold   606, 623, 629
cell seed   488
cell sheet   606
cell sorting   561
cell source   803, 805
cell spreading   477
cell surface   509
cell surface adhesion molecule ICAM-1   753
cell surface molecule   957
cell surface receptor   178, 498, 790
cell survival   471, 763
cell suspension   64, 66, 539
cell/tissue death   785
cell/tissue encapsulation   281
cell/tissue transplantation   281
cell-to-cell communication   102, 103
cell-to-cell contact   110, 160, 182
cell transplantation   139
cellular alignment   655
cellular cross-talk   492
cellular debris   722
cellular immunity   803
cellular ingrowth   125, 507
cellular necrosis   802
cellular pathway   470

cellular recognition   471
cellular response   482
cellular sphere   552
cellular transformation   72
cellulose   182, 398, 476
cell viability   116, 117, 182, 539, 600, 615, 626, 638, 651, 670, 971
cement   859
cement line   845
cementoblast-like cell   933
cementum   358, 359, 360, 931
cementum/PDL-like tissue   933
Center for Biological Evaluation and Research   856
Center for Devices and Radiological Health   856
central nervous system   50, 143, 161, 179, 271, 275, 419, 815
CEP68   309
ceramic   159, 211, 475, 476, 859, 878
ceramic implant surface   845
ceramic material   181
ceramic scaffold   607, 916
ceramide   767, 969
ceramide 2   969
ceramide 7   969
ceramides 4–7   968
ceramides 5–6   969
ceramides 6–7   968
cerebral cortex   818
cerebrovascular disease   940
cervical cancer   148
CF. *see* clotting factor
CFC   957
CFD. *see* computational fluid dynamic
c-fos   762
CFU-C assay   952
CFU-F   178, 185
CFU-GM   175
CFU-O assay   221
CFU-S. *see* colony-forming unit spleen
C-GAG. *see* collagen-glycosaminoglycan
cGMP   84
channel blocker   762
chaperone   725
charge   473, 478, 507
cheek   369
chemical barrier (pH)   803
chemical cross-link   497, 500
chemical-induced opacification   761

chemiluminescence 113
chemokine 727, 728, 753, 784
chemokine receptor CXCR2 969
chemotactic factor 803
chemotactic stimuli 786
chemotaxis 122, 125, 724
chemotherapy 954, 955, 956, 958
chick chorioallantoic membrane vascularization assay 761
chick embryo 942
chicken chorioallantoic membrane 104, 761
chimerism 151
chitosan 500, 501, 541, 734, 830, 926, 933
chitosan-alginate 321
chitosan/coral composite 125
chitosan scaffold 127
chloride channel 7 492
cholestasis 790
cholesterol 767, 969
choline acetyltransferase 172
chondral defect 181, 182
chondroblast 126, 160, 168, 218, 555, 958
chondrocyte 16, 122, 124, 126, 127, 168, 179, 180, 181, 182, 183, 184, 188, 204, 218, 222, 233, 234, 235, 236, 237, 239, 241, 256, 258, 263, 307, 308, 311, 313, 314, 321, 361, 552, 553, 554, 563, 565, 568, 574, 575, 604, 638, 641, 654, 712, 713, 714, 732, 764, 878, 879, 880, 914
chondrocyte cell line 575
chondrocyte culture 9, 124
chondrocyte differentiation 179
chondrocyte hypertrophy 179
chondrocyte implantation 361
chondrocyte-like cell 258, 563
chondrocyte macroaggregate 831
chondrocytic marker 184
chondrogenesis 126, 128, 130, 141, 179, 180, 181, 183, 184, 185, 295, 296, 553, 651, 656, 657, 658, 659
chondrogenic cell type 915
chondrogenic differentiation 179, 182, 292, 296, 648, 657
chondrogenic factor 124
chondrogenic growth factor 240
chondrogenic lineage 172
chondrogenic or osteogenic lineage 181
chondrogenic precursor cell 567
chondrogenic stem cell 570
chondroitin-6-sulfate 860, 881

chondroitinase ABC 280
chondroitin sulfate 182, 762
chondroitin sulfate proteoglycan 275, 296
chondroprogenitor mesenchymal cell 179
chondrospheres 234, 236, 237, 238, 239, 240, 241, 554
chorioallantoic membrane 104, 124
chorioallantois membrane assay 491
chorionic villus sampling 141
chorion villi 143
chromatin 202, 930
chromatin remodeling 77
chromatin structure 161
chromosomal integration 929
chromosomal region maintenance 1 75
chromosome 760, 956
chromosome separation 85
chrondrocyte 205, 809
chronic GVHD 956
chronic inflammatory reaction 787
chronic inflammatory rejection 803
chronic lymphatic leukemia 952
chronic myeloid leukemia 952
chronic myocarditis 749
chronic ulcer 881, 882
CIITA. see MHC class II transactivating protein
cilia 85, 636
ciliary neurotrophic factor 73, 105, 819
circulating antibody 941
circulating stem cell 65
circulation 168
circulatory system 675
cistron 100
CITR. see Collaborative Islets Transplant Registry
c-Jun 77, 275
CK. see cytokeratin
CK 1 373, 375, 376, 377
CK 2 373, 375, 376, 377
CK 4 377
CK 5 373, 374, 375
CK 6 373, 374, 375
CK 10 373, 375, 376, 377
CK 11 373, 375, 376, 377
CK 13 374, 376, 377
CK 14 376
CK 17 373
CK18 146
CK 19 377
c-kit 141, 952

class I and II major histocompatibility antigen 143
class I HLA epitope 955
claudication 882
CLC 7. *see* chloride channel 7
clinic 802
clinical biochemistry 790, 792
clinical dentistry 839
clinical pathology 791
clinical practice 855
clinical study 801, 804, 805
clinical trial 14, 569, 604, 802, 804, 808, 980
clinical ultrasound scanning 603
clinical use 489, 577, 639, 801, 805
CLL. *see* chronic lymphatic leukemia
clonal anergy 187
clonal expansion 187
clonal sarcoma cell line 564
cloned frog 161
cloning 10, 11, 27, 31, 66, 140, 161, 201
clonogenic cluster 932
clotting factor 783, 784
CLSM. *see* confocal laser scanning microscope
cluster differentiation 784
CML. *see* chronic myeloid leukemia
c-Myc 73, 201, 417, 958
CNS 67, 271, 272, 273, 276, 277, 278, 279, 280, 282, 284. *see* central nervous system
CNS neuron 271, 815
CNS regeneration 816
CNS repair 816
CNTF. *see* ciliary neurotrophic factor
CO_2 637
coactivator 77
coagulation 786
coating 248, 847
coating layer 474
coating material 841
cobalt-chromium-molybdenum alloy 859
cobblestone area 953
cobblestone-area-forming cell 953
cochlear implant 282, 283
coculture 375, 376, 566, 567, 573, 764, 862
cohort-control study 804
coiled–coil domain 73
Col II 181, 182, 184
Collaborative Islets Transplant Registry 413
collagen 62, 125, 126, 129, 151, 182, 184, 247, 248, 250, 256, 257, 258, 262, 308, 318, 320, 321, 335, 337, 346, 347, 352, 353, 357, 371, 372, 374, 375, 377, 378, 383, 392, 393, 434, 469, 541, 554, 621, 636, 644, 688, 689, 713, 727, 729, 734, 761, 763, 776, 786, 818, 828, 829, 860, 870, 880, 881, 926, 928, 933, 967
collagen/alginate 354
collagenase 160, 393, 565
collagenase digestion 308
collagen-associated biomineral 843
collagen-based scaffold 182, 870
collagen-binding integrin α2β1 763
collagen carrier 931
collagen/chitosan composite microgranule 127
collagen crosslinking 971
collagen fiber/collagen fibre 259, 689, 714
collagen fibril 257, 258, 259, 713
collagen gel 182, 233, 393, 657, 873, 971
collagen-glycosaminoglycan 335, 336
collagen hydrogel 393, 689
collagen I 244
collagen IV 244
collagen layer 861
collagen matrice 830
collagen matrix 128, 969
collagen membrane 233, 488, 860
collagen microarchitecture 636
collagen mineralization 844
collagen-PGA 436
collagen scaffold 124
collagen sponge 65, 184, 488
collagen synthesis 124
collagen tube 276
collagen type I 179, 237, 257, 258, 260, 262, 263, 308, 309, 311, 357, 362, 371, 377, 402, 485, 486, 564, 568, 625, 657, 734, 761, 763, 765, 766, 845, 880, 925, 968
collagen type I gel 621
collagen type II 172, 179, 237, 239, 240, 257, 258, 296, 308, 309, 311, 554, 764, 845, 860, 880, 917
collagen type III 257, 260, 263, 309, 311, 568, 765, 766, 845, 880, 926, 968
collagen type II, VI, IX, X, XI, and XII 257
collagen type IV 568, 765, 766, 968
collagen type IX 296
collagen type V 257, 568, 968
collagen type VI 296, 568
collagen type VII 968
collagen type X 258
collagen type XVIII 767
collecting duct 872

collecting tubule   871
collective responsibility   39
colloidal lithography   479, 482
colon   188
colony-forming precursors cell   575
colony-forming unit   185, 221, 952
colony forming units assay   570
colony-forming unit spleen   954
colony stimulating factor   72, 783, 784
commercial funding   44
common lymphoid progenitor cell   953
common myeloid progenitor cell   953
compact bone   486
competitive repopulating unit   954
complement   723, 783, 784, 790, 803, 857
complement activation   733, 785, 803
complement-dependent reaction   785
complement system   790, 803
composite   129, 360, 878
composite biosynthetic conduit   818
composite construct   377
composite cystoplasty   434
composite multi-layered dermal-epidermal analogue   332
compound material   841
compression resistance   802
compressive modulus   512
computational fluid dynamic   671, 683, 684
computed tomography   546, 801, 802
computer-aided design/computer-aided manufacturing   616, 617
concanavalin   186
concentration gradient   603
conditioned medium   567
conductivity   475
condylar cartilage   361, 831
condyle   832
condylectomy   361
confluence   564, 606
confocal fluorescence micrograph   622
confocal laser scanning microscope   491, 511
conformation   471, 472, 478
conformational change   471, 473
congenital anomaly   878
congenital defect   863
congenital heart defect   856, 858
congenital malformation   877
conjunctival reconstruction   832
connective fiber   931

connective tissue   219, 565, 729, 931
connective tissue layer   832
connective tissue stem cell   561
connexin 43   178
constitutive promoter   94
contact guidance   480
contact lense   498
contact sensitisation test   791
contractile protein isoform   244
contraction rate   243
controlled trial without randomization   804
convection   539, 637, 698
cooling rate   587
cooperative binding   73
coordinated expression of several gene   99
copolymer   183
coralline hydroxylapatite   831
coral scaffold   125
cord blood   140, 167, 185, 955, 959
cord blood bank   956
cord blood cell   562, 960
cord blood-derived CD34$^+$ hematopoietic progenitor cell   970
cord blood transplantation   955, 961
cord blood unit   957
cord blood vessel   143
cornea   760, 761
cornea epithelial culture   760, 761, 762
corneal cell   760
corneal limbus   338, 761
corneal-scleral model   762
corneal stem cell   761
corneocyte   767
corneosurfametry   971
cornified layer   968
coronary artery   111
coronary artery bypass operation   856, 858, 882
coronary artery disease   882
coronary disease   50
corrosion   475
corrosivity   971
cortex   817
cortical bone   573, 706, 841
cortical bone chip   216
cortical cell   843
cortical layer   840, 841
corticospongiosal graft   216
corticosteroid   956
Co-Smad   123

cost   15, 20
cost-effectiveness   15, 17
cost-effectiveness analysis (CEA)   15
cost estimation   15
co-stimulatory molecule   185, 187
co-stimulatory signal   187
costochondral graft   361
cost-outcome relationship   15
cost-utility   17
cost-utility analysis   15, 17
co-transplantation   174
coumermycin   96
covalent binding   475
COX. *see* cyclooxygenase
coxsackie B- and adenovirus receptor   109, 110, 111, 112, 114, 115, 116, 117, 118
coxsackievirus   110, 111, 112, 116, 118, 750, 753
coxsackievirus B3   750, 753
CP-690550   78
C-peptide   413, 417, 418
CpG. *see* cytosine-phosphate-guanine
CpG island   200
cranial defect   830
craniofacial bone   635
craniofacial muscle   245
craniofacial region   243, 879
craniofacial skeletal injury   878
craniofacial skeleton   635
craniomaxillofacial region   877
creatinine   66, 872, 873
creatinine assay   872
CREB   77
crestal bone   846
crestal height   846
critical cell mass   552
critically sized calvarial defect   105, 879
critical size defect   223, 916
critical size defect model   829
CRM1   74. *see* chromosomal region maintenance 1
cross hemodialysis   397
cross-link   321, 496, 504
cross-linked dermis   971
cross-linked keratin   765
cross-linked type I collagen   762
CRU. *see* competitive repopulating unit
cryobiology   588
cryopreservation   377, 378, 382, 412, 587, 606, 863
cryoprotectant   587, 588

crypt-villi structure   777
crystal nucleation   845
crystal structure   847
CSF. *see* colony stimulating factor
CSPG   280
CT. *see* computed tomography
CTL. *see* cytotoxic T-cell; *see* cytotoxic T-lymphocyte
CTL-mediated assay   791
CTNF   282
C-type lectin receptor   722, 723
cultivation   504
culture artifact   847
culture chamber   601
culture condition   648
cultured epithelial autograft   333, 334, 335
culture vessel   565
cumate   96, 97
cumate-responsive repressor   97
CuO   97
curettage   932
custom testing   793
CVS. *see* chorionic villus sampling
CXC chemokine receptor 4   955
CXCR4   958. *see* CXC chemokine receptor 4
cyclic adenosine monophosphate   821
cyclical mechanical stretch   883
cyclic RGD peptide   491
cyclin D1   73, 958
cyclin D2   162
cyclin-dependent kinase inhibitor   93
cyclooxygenase-2 inhibitor   917
cyclooxygenase (COX)   84
cyclosporine   412, 414, 972
cyclosporine A   956, 959
cylindrical organoid   402
CymR   96, 97
CymR-specific operator element   97
cynomolgus   78
Cyp2B6   172
Cyp3A4   172
cyst   791
cystectomy   17, 436
cystoplasty   17, 434
cytochrome 2E1   297
cytochrome P-450 activity   398
cytodifferentiation   922
cytogenetic analysis   760
cytokeratin   373, 375, 376, 377

cytokeratin 8   170, 172
cytokeratin 18   170, 172
cytokeratin 19   172
cytokine   71, 72, 75, 76, 78, 84, 103, 121, 124,
    125, 174, 186, 187, 569, 570, 669, 723, 727, 729,
    736, 749, 782, 786, 791, 828, 830, 843, 849, 869,
    878, 880, 929, 952, 953, 957, 958
cytokine antagonist   961
cytokine mRNA expression   971
cytokine production   792
cytokine receptor   73, 74
cytokine-related factor   843
cytokine release   956, 969
cytolysis   785, 955
cytomegalovirus   355, 750
cytomegalovirus immediate early promoter   102
cytomegalovirus promoter   96, 97
cytopathic effect   117
cytoplasm   97, 376, 930
cytoplasmic extension   761
cytoplasmic membrane   752
cytoplasmic protein   471
cytosine-phosphate-guanine   752
cytoskeletal movement   84
cytoskeletal protein   245
cytoskeletal reorganisation   482
cytoskeleton   471, 637, 655, 709, 926
cytotoxic   785
cytotoxic agent   785
cytotoxic drug   951
cytotoxicity   622, 626, 733, 926, 929
cytotoxic T-cell   414, 784
cytotoxic T-lymphocyte   791

D

daclizumab   413, 414
Dacron   685, 858
DALY. *see* disability-adjusted life years
Damkohler number   700
DA neuron   281, 282
Darcy's law   697
DBM   299. *see* demineralized bone matrix
DC. *see* dendritic cell
deacetylation   200
decalcified bone channel   276
decalcified freeze-dried allogeneic bone   359
decalcified freeze-dried bone   933
deciduous tooth   924
decision analysis   15, 16, 17, 20, 21

decision-making strategy   13
decision tree   16, 17, 19
Declaration of Helsinki   24
decorin   257, 309, 917
decoy oligonucleotide   78
decubitus ulcer   881
de-epidermized dermis   969
defect site   489, 801
deformation   843
degenerative joint disease   916
degenerative neurological disease   815
degradability   503
degradation   259, 261, 492, 543, 639, 787, 791
degradation of complementary mRNA   100
degradation rate   182, 183, 321
degradative agent   786
degrading material   789
dehydration   500
deiodination   63
delamination   475
delayed-type hypersensitivity   786, 791
deletion   200
dementia   820
demineralized bone matrix   299, 357, 362, 916
demineralized dentine matrix   928
demineralized freeze-dried bone   357, 828
denaturation   491
dendrite   271, 817
dendritic cell   187, 414, 415, 575, 722, 749, 750,
    753, 765, 959, 970, 972
denepezil   820
densiometry   802
density gradient centrifugation   567
density gradient separation   569
dental caries   360
dental epithelium   923, 928
dental hard tissue   346, 348, 349, 350, 355, 360
dental implant   354, 356, 357, 372, 827
dental implantology   8, 806, 839, 840, 846
dental impression material   8
dental papilla   923
dental plaque   931
dental pulp   351, 353, 360, 732, 933
dental pulp stem cell   923
dental surgical   350
dental tissue   807
dentate gyrus   278, 817
dentin   353, 360, 845
dentine   921, 924, 925, 928, 934

dentine chip   928
dentine matrix   925
dentine matrix extracts   928
dentine matrix protein-1   924
dentine phosphoprotein   924
dentine-pulp complex   922, 931, 934
dentine/pulp-like structure   933
dentine regeneration   921
dentine sialophosphoprotein   924, 931
dentine sialoprotein   924
dentin formation   360
dentin matrix extract   360
dentinoblast differentiation   360
dentinogenesis   351, 360, 921, 922, 928, 931
dentin-pulp   361
dentin-pulp complex   360, 361
dentin-pulp regeneration   360
dentin-pulp stem cell   360
dentistry   806, 921
dentoalveolar bone   358
denucleation   307
dephosphorylation   75
dermal cell   66
dermal fibroblast   9, 263, 377
dermal matrix   860
dermal substitute   861
dermal template   861
dermatology   806
dermis   198, 765, 861, 881
dermis and epidermis biopsy   148
dermis structure   967
dermoepidermal junction   333
descriptive study   804
design   801
designer baby   27
design variable   841
desmin   143, 245, 249, 430
detachment   504
determined cell   552, 570, 575
detoxification   400, 402, 406
detrusor instability   17
development   43, 65, 72, 103, 127, 181, 185, 199, 200
device design   788
device efficacy   789
DEX. *see* dexamethasone
dexamethasone   127, 162, 169, 170, 171, 172, 180, 184, 221, 490, 553, 568, 573, 649, 882
dextran   511

dextran sulfate   856
DFDB. *see* decalcified freeze-dried allogeneic bone
diabetes   129, 411, 412, 413, 763, 862, 946, 949
diabetes mellitus   159, 411, 419, 562, 940
diabetic mouse   105
diabetic ulcer   881
diabetogenic effect   413
diagnostic assay   793
dialysis   861, 869
dialysis machine   64
dialysis solution   861
dialyzer membrane   861
dibutyryl cAMP   172
dibutyryl cyclic adenosine monophosphate   821
Dicer   109
dicistronic   102
dicistronic retroviral vector   101
Dieffenbach, Johann Friedrich   8
diethylenetriamine [DETA]   247
differential count   792
differentially regulated gene expression   101
differentiated cell   568
differentiated neural cell   815
differentiation   53, 62, 71, 83, 93, 94, 101, 102, 103, 104, 105, 121, 122, 123, 124, 125, 129, 140, 141, 143, 147, 153, 161, 162, 163, 170, 171, 173, 174, 178, 179, 180, 182, 184, 187, 198, 200, 203, 204, 469, 478, 539, 554, 560, 562, 564, 567, 574, 608, 636, 638, 649, 650, 656, 701, 709, 711, 736
differentiation capacity   220, 292
differentiation factor   354, 569, 828, 850
diffraction analysis   636
diffusion   539, 698, 699, 788
diffusion chamber   573
diffusion rate   588
diffusivity   652
digestive enzyme   818
digestive system   855, 861
digitalis   750
dignity   43, 44
dilated cardiomyopathy   111, 750, 751
dilution   504
dimerization   74, 78, 94, 111
dimerization-based expression control   95
dimer-specific nuclear import signal   73
dimer-specific nuclear localization signal (dsNLS)   75
dimethyl sulfoxide   587, 588
dipole–dipole interaction   496

direct allorecognition   414
disability-adjusted life years   15
disc chondrocyte   309, 313
disc-derived chondrocyte   308, 309, 311, 313
discectomy   307
disc herniation   307, 313
diseased bladder   429
disease transmission   359, 877, 878
dispersant   496
dissociation   95
dissociation rate   75
dissolution   488
dissolved oxygen   601
distal tubule   871
distensability   429
distentum stratum   374
distortion   709
distraction osteogenesis   213, 214, 844
diuretic   750
divalent cation   501
divalent ion   496
Dlx5   180
DMD. see Duchenne muscular dystrophy
DMEM. see Dulbecco's modified essential medium
DMP-1. see dentine matrix protein-1
DMSO. see dimethyl sulfoxide
DNA-binding   73
DNA-binding domain   73, 75, 95
DNA-binding protein   71, 72, 94, 95
DNA-binding protein-transactivator fusion   94
DNA-coated metal microparticle   930
DNA condensation   930
DNA element   71
DNA microarray analysis   431
DNA synthesis   187
DNA tumor virus   760
DNA virus   752
DOCT. see Doppler optical coherence tomography
documentation   980
dog   774, 777, 789, 830
dog model   776
Dolly   10, 11, 161, 200
donor matching   759
donor shortage   159
donor site morbidity   159, 181, 182, 211, 216, 217, 877, 878, 879
donor tissue   559
donor tubule   871
donor-versus-recipient NK cell alloreactivity   955

dopamine   172, 281, 820
dopaminergic neuron   281, 819, 820
dopaminergic pathway   172
doping   32
Doppler optical coherence tomography   602
dorsal ectoderm   245
dorsal pancreas   941, 943
dorsal root ganglia explant   762
double-blind-controlled trial   804
double bond   505
double immunostaining   172
double-stranded RNA   109, 112, 117
doxycycline   96, 97
doxycycline-inducible expression   105
doxycycline-regulated expression   105
DPP. see dentine phosphoprotein
DRG. see dorsal root ganglia explant
Drosha-DGCR8 (DiGeorge syndrome critical region gene 8)   110
drug   476, 498, 979
drug delivery   498, 508, 729
drug-delivery application   498
drug delivery system   730, 972
drug discovery   93
drug metabolism   406
drug-release   508
drug-releasing matrix material   495
drugs   760, 788
dsNLS. see dimer-specific nuclear localization signal
DSP. see dentine sialoprotein
DSPP. see dentine sialophosphoprotein
DSPP gene   924
DTH. see delayed-type hypersensitivity
dual-regulated gene expression   101
Duchenne muscular dystrophy (DMD)   151
duct   833
ductal cell   941
ductal epithelial cell   418
Dulbecco's modified essential medium   370, 490, 618
duodenum   941, 942
durability   474
dynamic seeding system   597
dysphagia   833
dystrophin   151, 296
DYZ1   148

E

E3 ubiquitin ligase TRAF6 (TNF receptor-associated factor 6)   752
E6E7 oncogene   760
ear   880
EB. *see* embryoid body
EBL. *see* electron-beam lithography
ECF. *see* eosinophilic chemotactic factor
ECG   776
echocardiography   753
ECM   181, 182, 183, 184, 244, 257, 258, 260, 262, 263, 264, 275, 277, 280. *see* extracellular matrix
ECM component   764
ECM production   182
E.coli biotin ligase   97
economic evaluation   13, 20
economic model   21
economic modeling   15
ectactin   244
ectoderm   647
ectodermal   416
ectodermal cell   168
ectodermal lineage   170
ectodermal stem cell   161
ectomesenchymal cell   923
EDC (N-(3-dimethylaminopropyl)-N'-ethylcarbodiimide)   488
EDTA. *see* ethylenediaminetetraacetic acid
effectiveness   13, 14, 15, 16, 17, 21
efficacy   13, 17, 789, 801
efficacy testing   760
efficiency   13
EGE. *see* European Group on Ethics
EGF   264, 968, 969. *see* epidermal growth factor
eGFP. *see* enhanced green fluorescent protein
EGF receptor   75, 968
egg donation   42
egg extraction   42
egg/sperm donation   42
EHDJ   616
EHS   51
ejection fraction   153
ELAD. *see* Extracorporeal Liver Assist Device
ELAM. *see* endothelial-leukocyte adhesion molecule
ELAM-1   784
elastic cartilage   879, 880
elastic compliance   882
elastic fiber   258, 880
elasticity   258, 338, 383, 507, 706, 842, 843
elastic modulus   602
elastic property   841
elastin   258, 259, 383, 469, 858
electrical field   634, 644
electrical potential   635
electrical stimulation   214, 273, 599
electrical stimulus   634
electric field   635, 714, 930
electric pulse   930
electroactive polymer   860
electrode   931
electromagnetic field   636, 711
electromagnetic signal   213
electromechanical device   67
electron-beam lithography   479, 481
electron diffraction analysis   845
electron microscopy   636
electrophoresis   844
electrophysiological characteristics   760
electroporation   149, 163, 930
electrospinning   623, 654
electrospun polystyrene   436
electrostatic force   473, 850
electrostatic interaction   496
element mapping measure   845
ELISA. *see* enzyme-linked immunosorbent assay
ELISA technique   791
embossing   479
embryo   37, 38, 39, 41, 42, 43, 44, 45, 49, 52, 53, 73, 161, 647, 655, 863, 872, 914
embryo donor   40
embryo-fetoscope   150
embryogenesis   73, 103, 128, 599, 928, 939, 941
embryoid body   64, 163, 416, 417, 561
embryoid-body-derived ES cell   652
embryonic bone formation   764
embryonic cell   37, 38, 39
embryonic development   39, 124, 179
embryonic fibroblast   160, 860
embryonic formation   10
embryonic germ layer   141
embryonic human stem cell   49, 53
embryonic kidney   64, 66
embryonic lethal phenotype   72
embryonic metanephro   65, 66
embryonic metanephroi   870
embryonic organ   63, 66
embryonic pancreas   939, 943, 946
embryonic stem cell   11, 27, 29, 42, 51, 54, 64, 65, 105, 139, 140, 160, 161, 162, 163, 172, 197, 218,

219, 278, 292, 337, 351, 382, 404, 416, 417, 419, 553, 555, 560, 604, 649, 657, 658, 816, 860, 863, 878, 884, 913, 925
embryonic tissue   39, 61
emdogain   359
EMF. *see* electromagnetic field
E-modulus   802, 843
enamel   351, 845, 921, 922
enamel-forming cell   360
enamel matrix protein   932
enamel regeneration   360
enamelysin   924
encapsulated cell   496, 501
encapsulated stem cell   819
encapsulation   820
encephalopathy   397
endochondral bone   216, 222
endochondral bone formation   182, 184, 574
endochondral ossification   179, 222, 575
endocrine cell   941, 946
endocrine pancreas   949
endocrine pancreas organogenesis   949
endocrine tissue   123
endocrinology   806, 939
endocytosis   492, 930
endoderm   160, 416, 417, 647, 941
endodermal differentiation   172
endodermal hepatic cell   168
endodermal lineage   170
endodermal marker   172
endodermal stem cell   161
endodontic treatment   360, 921
endogenous gene network   102
endogenous neuron   815
endogenous promoter   98
endoglin   177, 178, 185
endograft   776
endometrial cell   142
endomyocardial biopsy   750
endoneurial tube   271, 272
endoplasmic reticulum   274, 731, 784
endosomal membrane   752
endosome   930
endostatin   103, 105
endothelial cell   63, 86, 122, 124, 125, 275, 290, 382, 476, 546, 553, 566, 576, 577, 640, 686, 687, 695, 753, 763, 784, 786, 858, 860, 873, 883, 923, 924
endothelial cell line   760, 762
endothelial cell marker   148

endothelial cell type   915
endothelialisation   690
endothelialized vessel   858
endothelial layer   761
endothelial-leukocyte adhesion molecule   784
endothelial lineage   141
endothelial lining   776
endothelial progenitor   763
endothelial progenitor cel   127
endothelial progenitor cell   576, 689, 690, 882, 883
endothelin   688, 883
endothelium   205
endotoxin   735, 737
energy   473
energy transfer coil   857
engineered bone   508
engineered graft   774
engraftment   140, 141, 143, 151, 153, 174, 179, 185, 202, 203, 949, 954
enhanced green fluorescent protein   163
enhancer   95
enhancer binding protcin beta   297
enhancer element   94
enhancer/promoter   355
ensheath axon   277
ENT   806
enterocystoplasty   432, 433, 437
enterovesical anastomosis   432
enterovirus   750
entubulation   274, 275, 280
entubulation tube   280
environmental host effect   802
enzymatic activity   507
enzymatic degradation   182
enzymatic dissociation   160
enzymatic inactivation   509
enzyme   476, 785, 790
enzyme deficiency syndrome   150
enzyme digestion   570
enzyme-linked antibody   586
enzyme-linked immunosorbent assay   113, 173, 790
E-OFF   96
E-ON   96
eosinophilic chemotactic factor   783, 784
EPC. *see* endothelial progenitor cell
epidermal cell   122
epidermal cover   860
epidermal cytokine   968
epidermal-dermal construct   765

epidermal engraftment 861
epidermal growth factor 73, 75, 105, 122, 123, 124, 347, 369, 370, 927, 968
epidermal-melanocyte construct 765
epidermal morphogenesis 968
epidermal sheet 881
epidermal stem cell 130, 334
epidermis 161, 198, 334, 765, 967, 968, 972
epigen 123
epigenetic bistability 102
epigenetic modification 200
epilepsy 273
epinephrine 969
epiphyseal chondrocyte 764
epiregulin (EPR) 123
episome 929
epithelia 62
epithelial cell 63, 122, 179, 334, 336, 369, 605, 724, 727, 728, 832, 932
epithelial defect 832
epithelial development 62
epithelial growth factor 828
epithelial layer 371
epithelial mucosa graft 372
epithelial regeneration 330
epithelial tissue 931
epithelial tube 61, 63
epithelial tubule 64
epithelium 66, 332, 333, 334, 335, 336, 351, 358, 369, 373, 376, 761, 776
epitope 586, 751
EPO 953
EPR. *see* epiregulin
Epstein-Barr Virus-related lymphoproliferative disease 960
ePTFE. *see* expanded polytetrafluoroethylene
equine pericardium 857
equity 14, 20
ER. *see* endoplasmic reticulum
E.REX 104
ERK. *see* extracellular signal-regulated kinase
ERK1/2 180
ER-LBD. *see* estrogen receptor ligand-binding domain
E-rosetting 959
erythema 971
erythroblast 171
erythrocyte 171, 931, 953, 961
erythrocyte hemoglobin 858

erythroid progenitor cell 953
erythromycin 100, 102
erythropoiesis 72, 73
erythropoietin 63, 73, 869, 870, 953
ESC 279. *see* embryonic stem cell
esophagus 777
established cell line 760
esthetic of implants 840
estradiol 969, 972
estrogen 97
estrogen receptor ligand-binding domain 97
ethanol 102, 488
ethical assessment 47
ethical controversy 37
ethical discourse 38
ethical issue 38, 562
ethics committee 38
ethylenediaminetetraacetic acid (EDTA) 160
ethylene glycol 588
euglycemia 412, 419, 423
EuroDISC 313
European Convention on Human Rights and Biomedicine 55
European Group on Ethics (EGE) 30, 32
evaporimeter instrument 767
evidence-based clinical study 804
evidence-based medicine 801
evidence level 809
E-VP16 100
EVPOME. *see* ex-vivo produced oral mucosa equivalent
Ewing's sarcoma 957
excitation 586
exendin-4 297
exocrine tissue 941, 943, 946
expanded polytetrafluoroethylene 685
expansion 170, 171, 495
experimental animal 774
expert opinion 804
explant outgrowth 570
exportin-5 110
export receptor 75
external bioreactor 433
extracellular matrix 62, 122, 127, 140, 141, 159, 174, 178, 244, 256, 271, 321, 382, 384, 434, 435, 469, 551, 552, 553, 554, 564, 565, 577, 598, 600, 603, 613, 634, 636, 642, 644, 649, 650, 651, 654, 655, 657, 658, 670, 671, 689, 695, 696, 712, 714, 726, 727, 728, 729, 733, 786, 841

extracellular matrix molecule   818, 958
extracellular matrix protein   580
extracellular matrix scaffold   922
extracellular matrix structure   129
extracellular protein   599
extracellular signal-regulated kinase   84
extracorporally generated tissue construct   159
extracorporal strategy   218
extracorporal tissue engineering   218
extracorporeal dialysis   65
extracorporeal liver assist device   861
Extracorporeal Liver Assist Device   398, 403
extraction socket   353
extrahepatic stem cell   404
extremity bone mass   843
ex-vivo expansion   174
ex-vivo produced oral mucosa equivalent   832
eye   243, 761
eye surgery   806

F
facial contouring   881
facial fracture   878
facial motoneuron   821
facial skeleton   828
facial wound reconstruction   8
FACS. *see* fluorescence-activated cell sorting machine; *see* fluorescence-activated cell sorting
FACS analysis   175
factor XII   726
FAH-null mouse   203
FAK. *see* focal adhesion kinase
Fanconi anemia   955
fascicle   256
fasciculation   277
Fas ligand (CD95)   185, 957
fast Fourier transform   86
fat   139, 141, 182, 219, 561
fat cell   298, 565
fat graft   290
fat grafting procedure   882
fat tissue   126, 289
fatty acid   969
FBGC. *see* foreign body giant cell
FDA   125. *see* U.S. Food and Drug Administration
FDA approval   829
FDM. *see* fused deposition modelling
FEA. *see* finite element analysis
feedback control   85

feeder cell   369
feeder layer   370, 560, 952
femoral artery   882
femur   140, 152, 184, 491, 705, 917
fermentation   595
fertilization   39, 52, 53
fetal blood   140
fetal bone marrow   149, 219
fetal bradycardia   150
fetal calf serum   369, 370, 371, 375, 490
fetal cardiac cell   153
fetal circulation   140
fetal development   61, 71
fetal growth   126
fetal heart   648
fetal hepatocyte   404
fetal liver   149, 168, 185
fetal stem cell   139, 143, 144, 147, 149, 153
fetal tolerance   186
fetal ventral mesencephalic cell   820
fetoscopy   150
fetus   37, 66, 198
FFT. *see* fast Fourier transform
FGF   184, 185. *see* fibroblast growth factor
FGF-1   124
FGF-2   124, 160, 181, 182, 264, 359, 763, 764, 880
FGF-8   162
FGF receptor   62, 121, 124
FHF. *see* fulminant hepatic failure
fiber diameter   321
fiber/fibre   248, 650
fibril   256
fibrin   129, 181, 182, 258, 321, 334, 336, 500, 541, 651, 727, 763, 926
fibrin gel   383, 622, 765
fibrin glue   126, 184, 299, 300, 353, 392, 880, 917
fibrin glue matrix   182
fibrin hydrogel   764
fibrin matrix   184, 764, 786, 831
fibrinogen   250, 500, 622, 695, 726
fibrinogen adsorption   478
fibrinolysis   500
fibrin scaffold   882
fibrin sealant   831
fibroblast   122, 123, 124, 125, 126, 160, 179, 185, 221, 244, 246, 256, 257, 258, 260, 263, 264, 265, 275, 290, 320, 321, 322, 333, 336, 347, 348, 354, 359, 360, 369, 370, 375, 376, 377, 378, 382, 435,

476, 625, 647, 724, 727, 729, 731, 732, 736, 753, 763, 765, 784, 786, 787, 790, 809, 832, 861, 881, 917, 923, 931, 968, 970, 971
fibroblast colony-forming unit (CFU-F)   177, 932
fibroblast growth factor   62, 65, 121, 122, 124, 184, 216, 321, 763, 783, 784, 879, 883, 927, 928
fibroblast growth factor-2   181, 214, 239, 296
fibroblastic cell   141
fibroblast-like cell   85, 179, 932
fibroblast proliferation   124, 125
fibrocartilage   18, 20, 257, 258, 259, 265, 831, 879, 880
fibrocartilage formation   20
fibrocartilage repair tissue   764
fibrocartilage tissue   181
fibrocartilaginous   258, 325
fibrocyte   435
fibrogenesis   844
fibronectin   62, 141, 145, 178, 244, 247, 248, 258, 260, 275, 280, 299, 370, 383, 392, 393, 477, 562, 568, 651, 695, 726, 727, 729, 870, 872, 879, 926, 968
fibronectin-coated   393
fibroplasia   785
fibrosis   347, 348, 722, 785, 917
fibrotic disorder   125
fibrous biomaterial meshe   183
fibrous capsule   787, 788, 803
fibrous encapsulation   787
fibrous meshe   182
fibrous tissue   790
fibrous tissue formation   124
fibula   216, 358
filaggrin   968
filament   243, 280
filament bundle   280
fillagrin   969
finite element analysis   639, 641, 707
FISH. *see* fluorescence-in-situ-hybridization
FITC. *see* fluorescein isothiocyanate
fit-in-fit system   802
fixateur externe   489
fixation   802, 831
fixation precision   802
fixation screw   859
fixture design   842
FKBP   94, 96
flagellin   724
flank   881

flap   9
flap prefabrication   217, 218, 831
flat-membrane bioreactor   402
flexibility   488
flow chamber   642
flow cytometric analysis   790, 791
flow cytometry   178, 924
flow rate   598
Flt-3 ligand   175, 956
fluid-driven mechanical stimulation   599
fluid flow   541, 634, 635, 636, 637, 669, 672, 710, 711, 712, 713, 714
fluid shear   85, 701
fluid shear flow   86
fluorescein isothiocyanate   511
fluorescence-activated cell sorting   178, 960
fluorescence-activated cell sorting machine   163
fluorescence-in-situ-hybridization   147, 148, 152
fluorescence microscope   586
fluorescence microscopy   202, 477, 626, 639
fluorescence recovery after photobleaching   508, 511
fluorescent dye   586
fluorescent marker   873
fluorescent molecule   511
fluorescent tagging   202
fluorimetry   602
fluorocarbon blood substitute   858
fluorocarbon emulsion   859
fluorocarbon oil   859
FMB. *see* flat-membrane bioreactor
FMB-BAL   402
fMRI. *see* functional magnetic resonance imaging
focal adhesion   577, 850
focal adhesion kinase   655
focal adhesion point   471
focal aggregations of mononuclear cell   791
Food and Drug Administration   856
foot ulcer   412
foreign body giant cell   785, 787, 803
foreign body reaction   785, 787, 871
foreign body response   721, 788
Fos   77
four-helix bundle   73
Fourier-transform infrared spectroscopy   487
FOXN1   960
foxP3+   187
FR. *see* free radical
fracture   151, 878, 879

fracture activation   214
fracture healing   126, 844
FRAP. see fluorescence recovery after photobleaching
FRB   94
free flap   358
free radial forearm flap   832
free radical   783, 784, 858
free radical reaction   859
freeze drying   544, 546
frequency   711
frontal cortex   820
FT-IR. see Fourier-transform infrared spectroscopy
Fu   124
fulminant hepatic failure   397, 403
fumaric acid   505
function   802
functional assay   792, 793
functional efficacy   788
functional genomic   93
functionality   255
functional model   789
functional osseointegration   839
functional renal unit   873
functional substitute   781
fungi   750, 786
fused deposition modelling   546
FZD-9   141

G

GA. see granulocyte
GABAalpha(6)r   162
GA-CSF   783, 784
GAG   258, 260
Gal4   94
Gal4 DNA-binding domain   97
Gal4 operator   97
gamete   43
gamma-activated site   72, 74
gamma irradiation   488
ganciclovir   960
ganglia   762
GAP-43   275
GAPDH   77, 112, 114
gap junction   85, 86, 257, 635, 636, 711, 761
GAS   75. see gamma activated site
gas bubble   496
gastrointestinal tract   204
gastrulation   73

GATA4   172
GBP-1. see guanylate-binding protein-1
GBR   212, 213
G-CSF   174, 175. see granulocyte colony-stimulating factor
GDF   184, 215. see growth and differentiation factor
GDF-5   127, 265, 830
GDF-6   265
GDF-7   162, 265
Gdf-11. see growth and differentiation factor
Gdf-11-cDNA plasmid   931
GDNF   62, 282. see glial cell line-derived neurotrophic factor
gel   614, 650, 760, 762
gelatin   125, 127, 182, 183, 500, 651, 880
gelatin carrier   357
gelatin (Gel)   926
gelatin scaffold   353
gelatin sponge   124, 392
gelation   501
gel-fiber amalgam   181
Gelfoam   392
gel stability   504
GEM. see growth-factor enhanced matrix
gene activation   714
gene activity   641
gene amplification   587
gene chip   577, 580, 767, 844
gene circuitry   103
gene complex   956
gene delivery   150, 356, 929
gene delivery agent   788
gene delivery system   930
gene delivery vehicle   126
gene-enhanced tissue engineering   347, 359, 360
gene expression   77, 83, 94, 98, 102, 103, 110, 117, 123, 140, 163, 180, 184, 200, 202, 356, 470, 471, 479, 564, 580, 657, 709, 844, 929, 930
gene gun   149, 930
gene network   103
gene product   99, 929
general immune competence   791
general surgery   806, 855
gene regulation   929
gene repression   200
gene silencing   109, 112, 113, 116
gene therapy   25, 93, 149, 150, 153, 347, 563, 821, 915, 929, 931, 967, 979

genetically engineered mouse myoblast cell   820
genetic circuitry   102
genetic diagnosis   37, 44
genetic modification   775
genetic probe   844
genetic variant   775
gene transcription   959
gene transduction   929
gene transfer   28, 103, 201, 350, 415, 833, 931
gene vector   153
genitourinary system   21
genome   929
genomic alteration   160
genotoxicity-type test   765
genotype   628
geometrical structure   792
GER   500, 508
germ cell   29, 44, 416
germ cell donor   38
germ-cell tumor   952
germinal center   791, 946, 947
germ layer   143, 561, 736
germ line   160
germ-line therapy   28
gestation   140, 143, 147, 150
GETE. *see* gene-enhanced tissue engineering
GFP   206. *see* green fluorescent protein
GGF. *see* glial growth factor
ghrelin   418
giant cell   783, 786, 790
giant naevi   967
gingiva   809, 931
gingival epithelial layer   371
gingival fibroblast   125, 359
gingival fibroblast DNA synthesis   126
gingival keratinocyte   371, 372, 375, 376, 378
gingival mucosa   372
glandular tissue   126, 941
glass fibre   248
glass slide   623
glass transition temperature   545
glial cell   122, 123, 125, 204, 205, 272, 273, 278, 279, 280, 815
glial cell line-derived neurotrophic factor   820
glial fibrillary acidic protein   277
glial growth factor   123
glia limitan   284
glial scar   818
glioblastoma multiforme   110

glomerular epithelial cell   870
glomeruli   66
glomerulonephritis   791
glomerulus   871, 872, 873
glucagon   297, 418
glucagon-secreting   411
glucocorticoid receptor   77
glucocorticosteroid   764
glucokinase   105
gluconeogenesis   398
glucosamine   926
glucosaminoglycan   641
glucose   600, 601, 637, 669, 946
glucose 6-phosphate dehydrogenase   85
glucose concentration   602
glucose intolerance   940, 946
glucose sensing   939
glucose tolerance   939, 946, 948, 949
glucosylceramide   969
glutamine synthetase   404
glutaraldehyde   762, 856
glutathione   869
glycerol   587, 588
glycine-glutamic acid-arginine   500
glycoasminoglycan   729
glycogen   861, 968
glycogen storage   172
glycogen synthase kinase-3 inhibitor   958
glycohemoglobin   421
glycophorin A   146, 170
glycoprotein   258, 383, 760
glycoprotein laminin   244
glycosaminoglycan   126, 181, 257, 319, 335, 469, 501, 505, 760, 831, 860, 879
GM. *see* granulocyte-macrophage
GM-CSF   73, 174, 783, 784, 970. *see* granulocyte-macrophage colony-stimulating factor
GMP   170, 174, 234. *see* good manufacturing practice
GNDF-loaded pharmacologically active microcarrier   820
gold   476
gold surface   477
Golgi apparatus   430, 731
good clinical practice   44, 804
good manufacturing practice   313, 596, 607, 980
good professional practice   980
good scientific evidence   804
G_0 phase   187

gp 105-120   144
gp130   73, 754
GPMT. *see* guinea pig maximization test
GPP. *see* good professional practice
GR   77. *see* glucocorticoid receptor
GR-1   952
gracilis muscle   433
graft   773, 777, 803
graft failure   858, 869, 954
graft lumen   858
graft manipulation   961
graft rejection   732, 736, 913, 942, 951, 954, 956, 959
graft resorption   882
graft shrinkage   434
graft-versus-host disease   162, 187, 877, 951, 954, 959, 960
graft-versus-tumor   951, 959
Gram-negative bacteria   736
granular layer   968
granulation tissue   786, 791, 860
granulation tissue formation   786
granulocyte   171, 174, 297, 724, 784, 803, 953
granulocyte colony-stimulating factor   953, 954, 958
granulocyte-erythroid-macrophage-megakaryocyte (CFU-GEMM)   952
granulocyte-macrophage   784
granulocyte-macrophage (CFU-GM)   952
granulocyte-macrophage colony-stimulating factor   953, 955, 958
granulocyte/monocyte   297
granuloma   791
granuloma formation   785
granuloma formation (foreign body reaction)   791
Graves' disease   786
gray matter   275, 818
Greek mythology   5
green fluorescent protein   76, 99, 203, 352
grinding   479, 480
grit blasting   480
groove   478
groove depth   480
growth   101, 103, 844
growth and differentiation factor   264, 830, 850, 931
growth arrest   99
growth deformity   878
growth differentiation factor   184

growth factor   63, 65, 72, 121, 124, 128, 163, 181, 184, 185, 188, 321, 469, 498, 507, 546, 547, 553, 562, 568, 569, 828, 637, 671, 697, 730, 760, 763, 788, 816, 819, 828, 847, 850, 879, 880, 883, 914, 922, 925, 926, 927, 929, 932, 958, 960, 967, 968
growth-factor enhanced matrix   125
growth factor receptor   570
growth hormone   73, 126
growth-inhibitor signaling   818
growth rate   774
growth regulation   127
GRP   278
GSC   172
GS domain   122
GTP. *see* guanosine triphosphate
GTPase   75, 655
GTR. *see* guided tissue regeneration
guanine nucleotide exchange factor RCC1   74, 75
guanosine triphosphate   75
guanylate-binding protein-1   72
guidance   817
guidance cue   817
guided bone regeneration   212
guided tissue regeneration   358
guiding axon   280
guinea pig   774, 776, 789
guinea pig maximization test   791
guluronic acid   734
guluronic acid (G-block)   501
gut   161, 188
GVHD   187, 188. *see* graft-versus-host disease
GVT. *see* graft-versus-tumor
GVT reaction   954
gyrase B   96
GyrB   96, 97
Gys2   172

H

HA   657
HA-1   960
HA-2   960
haemangioblast   140
haematological parameter   790
haematoxylin   791
haemoglobin   790, 858
haemoglobinopathy   150, 167, 951
haemopoiesis   140, 141
haemopoietic cell   140
haemopoietic lineage   140

haemopoietic stem cell 139, 149
haemopoietic system 139, 140
haemostasis 785, 786
hair follicle 765, 861, 881
half-life 356, 359
hamartomatous tooth-like tissue 362
hammerhead ribozyme 97
hamster 774
hamstring 261, 318
HAP. see hydroxyapatite
haploidentical graft 952
haploidentical HSC transplantation 960
HAP nanocrystal 485
hard tissue 633
hard-tissue formation 711
hard-tissue polymer 933
harvesting 292
harvesting morbidity 840
harvest technique 292
HbA 413, 414
HB-EGF. see heparin-binding EGF-like growth factor
hBMSC. see human mesenchymal stromal cell
HCAM 177
HCAM-1 141, 144, 562
HCE. see human corneal epithelial
HCE-T line 761
HD. see Huntington's disease; see Hodgkin's disease
HDAC. see histone deacetylasis
HDAC4 97
head and neck cancer 833, 879
head and neck trauma 879
healing capacity 265
healing tissue 844
health care 13, 17, 20, 40, 559
healthcare costs 15, 16
healthcare outcome 15
healthcare provision 40, 44
health technology assessment 17
hearing 83, 282, 283
heart 50, 65, 111, 126, 161, 198, 669, 675, 678, 789, 807, 856, 924
heart defect 382
heart failure 750, 751
heart-lung machine 856
heart or respiration failure 31
heart-specific autoimmunity 753
heart surgery 806

heart tissue 750
heart transplantation 9, 750, 857
heart valve 25, 378, 381, 382, 384, 554, 597, 599, 775, 787, 857
heat shock protein 627, 725, 726, 783, 784
heat shock protein 60 725, 726
heat shock protein 70 725, 726, 751
heat shock protein 90 95, 96, 97, 725, 726
Heisenberg's uncertainty principle 201
hematological malignancy 954
hematology 806, 951
hematopoiesis 174, 292, 295, 297, 960
hematopoietic banking 169
hematopoietic cell 168, 177, 203, 953, 954, 958
hematopoietic cell hierarchy 952
hematopoietic cell type 915
hematopoietic differentiation 297
hematopoietic lineage 65
hematopoietic niche 178
hematopoietic progenitor 174, 178
hematopoietic progenitor cell 178, 952
hematopoietic progenitor cell antigen 177
hematopoietic reconstitution 170, 956
hematopoietic recovery 956, 958
hematopoietic regeneration 174
hematopoietic stem cell 129, 167, 168, 204, 278, 293, 404, 575, 816, 884, 924, 951, 953
hematopoietic stem-cell transplantation 187
hematopoietic surface marker 185
hematopoietic system 167, 219
heme oxygenase-1 104
hemidesmosome 968
hemimandibulectomy 129
hemiparkinsonian rat 153
hemodialysis 65, 397, 407
hemodynamic 882
hemodynamic force 883
hemofiltration 64
hemoglobin-based blood substitute 858
hemolysis 856
hemoperfusion 397, 402
heparan sulfate 968
heparan sulfate proteoglycan 244, 726
heparin 124, 129, 258, 856
heparin-binding EGF-like growth factor (HB-EGF) 123
heparin binding growth factor 124
heparin-coated membrane oxygenator 856
heparin sulfate 280, 968

Subject Index

HepatAssist   398, 400, 403
hepatectomy   202
hepatic failure   403, 407
hepatic stem cell   123, 405
hepatitis C virus   750, 861
hepatoblast   404
hepatocyte   64, 123, 141, 204, 205, 397, 398, 400, 402, 403, 404, 405, 406, 407, 624, 861
hepatocyte-colony-forming unit   404
hepatocyte growth factor   186, 203, 297
hepatocyte growth factor (HGF)   63, 123
hepatocyte-like cell   173
hepatoma cell line HepG2   398
hepato-renal injury   147
hepatotoxin   202
hepatotoxin retrorsine   202
HepG2   173, 404
hereditary tyrosinemia   203
heregulin (HRG)   123
herpes simplex virus   94, 97, 929
hESC. see human embryonic stem cell
heterochromatin   200, 202
heterodimer   74, 75, 125
heterokaryon   566, 567
heterotopic ossification   916
HFM   277, 278, 280, 281, 282, 283
HFM entubulation   280
HGF   172, 173, 174, 186, 301. see hepatocyte growth factor
high blood pressure   856
high density cell pellet   183
high-mobility group protein   162
high performance liquid chromatography   172
high-resolution TEM   487
hippocampus   143, 817, 820
hip replacement surgery   859
histamine   727, 783, 883
histochemistry   873
histocompatible cell line   863
histocompatilility antigen   956
histone   161, 930
histone acetyltransferase activity   77
histone deacetylase inhibitor trichostatin A   959
histone deacetylasis (HDAC)   77
histone protein   200
histopathology   790, 792
histoplasmosis   786
history   5
HIV   301

HLA   147, 167, 185. see human leukocyte antigen
HLA-A   954
HLA antigen   951
HLA-B   954
HLA barrier   959
HLA-C   954
HLA-class I   170
HLA-class II   170, 186
HLA-DQ   954
HLA-DR   140, 145, 187, 294, 954
HLA-DR expression   970
HLA haplotype   954
HLA I   140
HLA II   140
HLA-mismatched graft   956
HNF1   172
HNF3b   172
HNF4α   172
Hodgkin's disease   952
Hohenheim, Theophrastus von   6
hollow fiber   398, 400, 402
hollow fiber membrane   277, 398
hollow tube   871
homeobox protein   198
homeodomain   162
homeostasis   598, 785
homing   140, 958
homing-associated cell adhesion molecule   177
homing capacity   957
homodimer   74, 75, 125
homograft   381
Homunculus   6, 7
hormonal (over)stimulation   42
hormonal regulation   710
hormonal-related factor   843
hormone   72, 84, 103, 570, 635, 675, 708, 843
Hoshin process   11
host cell   782
host DNA   790
host immune response   819
host immune system   803
host immunity   951
host innate immune mechanism   791
host pathology   788
host respond   781
host response   791, 803
host side   842
host tissue   785
host-tissue phenotype   788

host-versus-graft reaction 956
HPCA 177
HPLC. *see* high performance liquid chromatography
HPRT 97
HPV 105
HPV-16-E6/E7 105
HPV16E6E7 760
HPV E6E7 gene 760
HPV immortalized cell 761
HRG. *see* heregulin
HRTEM. *see* high-resolution TEM
HSC 278. *see* haemopoietic stem cell
HSC engraftment 958
HSC harvest 961
HSC niche 958
HSC transplantation 951, 960, 961
HSP. *see* heat shock protein
HSP60 725, 726
HSP60/70 628, 751
HSP70 725, 726
HSP90. *see* heat shock protein 90
HSP90 95, 96, 97, 725, 726
HSPG 244
HSV-TK suicide gene 960
hTGF-β2 183
human co-culture system 492
human corneal epithelial (HCE) 105
human deacetylase 4 97
human dignity 49, 52, 53, 54
human embryo 164
human embryonic stem cell 416, 417, 560, 863
human endothelial cell 476
human glial cell 623
human hepatocyte cell line C3A 398, 862
humanity 7
humanized "animal" liver 404
human leukocyte antigen 185, 412, 951
human liver C3A cell line 398, 862
human mesenchymal stem cell 693
human mesenchymal stromal cell 489
human osteoblast 478
human osteosarcoma cell 618
human papilloma virus 105, 760
human papilloma virus 16 760
human papilloma virus 16-E6-E7 105
human preproinsulin gene 418
human proinsulin gene 418
human recombinant molecule 829
human rights 49

human stem cell 292
human tissue engineering 563
human umbilical cord blood cell 553, 554
humeral head 318
humerus 152
humoral immune response 735
humoral rejection 940, 941
huntingtin 105
Huntington's disease 105, 273, 821
huperzine A 820
HYAFF 11 392
hyaline articular cartilage 712, 764
hyaline cartilage 18, 20, 21, 183, 184, 233, 234, 237, 300, 831, 860, 879
hyaline-like cartilage 860
hyaline repair tissue 21
hyaline-specific proteoglycan 239, 240
hyaluronan 181, 183, 392, 469, 500, 541, 568, 651, 689, 860
hyaluronan sponge 392
hyaluronic acid 129, 182, 183, 233, 299, 321, 500, 734, 737, 765, 766, 829, 880, 933
hyaluronic acid-based scaffold 321
hyaluronic acid modified carrier 392
hyaluronic acid receptor 177
hybrid cell 416
hybrid material 802
hybrid scaffold 130
hydrochloric acid 486
hydrocortisone 370, 969, 972
hydrodynamic pressure 930
hydrogel 127, 182, 183, 184, 277, 392, 393, 394, 437, 495, 541, 547, 614, 615, 618, 620, 621, 832, 841, 848, 880, 926
hydrogel chemistry 841
hydrogel integrin layer 848
hydrogel network 508
hydrogel property 504
hydrogel scaffold 126, 880
hydrogen bond 473
hydrolysis 248
hydrolytic degradation 787
hydrolytic enzyme 782
hydrophilic channel 850
hydrophilicity 841
hydrophilic polyester matrix 400
hydrophilic polymer network 495
hydrophilic surface 472
hydrophobic interaction 478

hydrophobicity   183
hydrophobic stripes   850
hydrophobic surface   472
hydrostatic pressure   637, 710, 713
hydrostatic stress   658
hydroureteronephrosis   432
hydroxyapatite   129, 130, 346, 357, 485, 541, 841, 847, 878, 933
hydroxyapatite/tricalcium phosphate   924
hydroxyapatite/tricalcium phosphate carrier   933
hydroxyeicosanotetraeno   971
hydroxyl group   478, 724
hyperbaric oxygen treatment   969
hyperglycemia   105, 180, 420, 421, 940, 946
hyperphagia   946
hyperplasia   686
hypersensitivity   785, 791
hypersensitivity reaction   785, 786, 790, 791
hypersensitivity test   873
hyperthermia   957
hyperthyroidism   786
hypertrophic cartilage   180
hypoglycemia   411, 413, 414, 419, 939
hypoglycemic agent   940
hypopigmentation   970
hyposalivation   833
hyposecretion of saliva   833
hypoxanthine phosphoribosyl transferase   97
hypoxia   104, 413, 640, 784
hypoxia-inducible expression vector   104
hypoxic condition   640
hysteretic switch   102, 103

I

IC. *see* immune complex
ICAM. *see* intercellular adhesion molecule
ICAM-1   145, 732, 784
ICAM-3   144, 723
ICAM promoter   77
ice crystal   587
ICER. *see* incremental cost-effectiveness ratio
IDO   186. *see* indoleamine 2,3-dioxygenase
Id protein   180
IFN. *see* interferon
IFN-α   783
IFN-β   783, 784
IFN-γ   186, 783, 784, 969
Ig. *see* immunoglobulin
IgA   784, 790
Igai   124
IgD   784
IgE   73, 783, 784, 785, 790, 791
IGF   184, 762. *see* insulin-like growth factor
IGF-I   185, 250, 300, 359, 360
IGF-II   264, 359
IgG   784, 785, 790
IgM   784, 790
IKK. *see* IκB kinase
IKVAV   394
IL. *see* interleukin
IL-1   783, 784, 958
IL-1 alpha   762
IL-1 beta   174
IL-1R-associated kinase   752
IL-1α   971
IL-1β mRNA   972
IL-2   73, 187, 783, 784, 953
IL-2 production   959
IL-2R   145
IL-3   73, 296, 783, 784, 953
IL-3R   145
IL-4   73, 75, 783, 784, 953
IL-5   73, 783, 784
IL-6   72, 73, 174, 296, 414, 783, 784, 971
IL-7   73, 953
IL-7 production   956
IL-7R   145
IL-8   174, 783, 784, 969, 971
IL-9   73
IL-10   186, 187, 784
IL-10 signaling   186
IL-11   72, 174
IL-12   73, 174, 414, 784
IL-13   73, 783, 784
IL-15   73, 174, 784, 953
IL-23   73
IL-27   73
ileal loop   777
ileocystoplasty   17, 20, 21
ileum   777
iliacal graft   776
iliac artery   882
iliac crest   216, 346, 348, 353, 354, 358, 362, 567, 878, 955
ILT3   414, 415
ILT3-Fc protein   415
ILT4   414
IL-β   971

image analysis software   792
imaging technique   844
imatinib   957
immobilization   498
immortalized cell   160, 247, 560, 564
immortalized cell line   404
immortalized human osteoblast cell line   564
immune acceptance   560
immune complex   786, 791
immune deficient mouse model   870
immune dysfunction   790
immune escape   185
immune function test   791
immune incompatibility   9
immune insult   791
immune pathology   791
immune pathway   790
immune privilege   186
immune reaction   773
immune reconstitution   945, 956, 959, 960
immune regulation   71
immune rejection   355, 732, 736, 737, 759, 803
immune response   71, 359, 560, 722, 730, 736, 788, 803, 863, 913, 939, 945, 949
immune stimulation   790, 791
immune suppression   185, 790, 791, 862
immune suppressive therapy   863
immune suppressive treatment   803
immune surveillance   72, 188
immune system   126, 147, 669, 675
immunization   111
immunoblotting   113, 114
immunocompatibility   419
immunocompromised mouse   924
immunocompromised rodent   830
immunocytochemical staining   622
immunodeficiency   72, 167, 774, 956
immunodeficiency disorder   149
immunodeficient mouse   153, 832
immunodeficient (NOD-SCID) diabetic mouse   945
immunoelectrophoresis   790
immunofluoerescence   142
immunogen   788
immunogenicity   153, 219, 499, 877, 926, 929
immunogenic reaction   499
immunoglobulin   110, 784, 803
immunoglobulin fold   73
immunoglobulin-like domain   122

immunoglobulin switching   73
immunohistochemistry   586, 791, 802
immunohistology   586
immunoisolation   804
immunological barrier   149
immunological reaction   776
immunological rejection   54
immunological status   803
immunologic rejection   877, 878
immunomodulation   9, 415, 803
immunophenotype   147, 168
immunostaining   580, 586, 844
immunostaining technique   790
immunosuppressant   788
immunosuppression   162, 188, 331, 412, 421, 562, 940, 942, 949
immunosuppressive agent   949, 952
immunosuppressive drug   187, 951, 959
immunosuppressive therapy   804, 869
immunosuppressive treatment   31
immunosupression   9
immunotoxicity   790, 791
immunotoxicity assessment   788, 789
implant   370, 839
implantable artificial heart   857
implantation   124, 153, 161, 489
implantation site   789
implantation test   789
implant bed   802, 841, 843
implant/bone bond   840
implant/bone interface reaction   840
implant failure   839
implant geometry   840, 841, 846
implant healing   847
implant integration   802
implant loading   839, 846
implant loss   839
implant material   840
implant placement   839
implant shape   803, 842
implant size   803
implant structure   841
implant-supported restoration   840
implant surface   840, 844, 845
implant surface characteristic   840
import factor   75
importin   75
importin α   74
importin α5   75

importin β   74, 75
inbreed   774, 775
incremental cost-effectiveness ratio (ICER)   16
indirect allorecognition   414
indoleamine 2   186
indoleamine 2,3-dioxygenase   414, 415
indomethacin   172, 882
induced pluripotent stem cell   417, 649
inducible expression control   96
inducible knockdown of target genes   100
inducible nitric oxide synthase (iNOS)   84
inducible one-promoter-based short hairpin RNA expression   101
inducible promoter   99, 100, 103
industrial processing   979
infarct cavity   817
infarct zone   111
infection   117, 732, 790, 791, 863, 877, 878, 882, 883
infectious agent   765, 782
infectious disease   967
infectious disease transmission   355
infiltration   182
inflammation   72, 347, 725, 727, 728, 773, 785, 803, 922, 970
inflammatory cardiomyopathy   749
inflammatory cell   790
inflammatory chemical   260
inflammatory cytokine   803
inflammatory heart disease   749, 754
inflammatory response   726, 803
influenza virus   750
informed consent   26, 41, 42, 143
INFUSE® Bone Graft   357
inhibitor of DNA binding/differentiation helix-loop-helix protein   180
inhibitory molecule   815, 818
injection   735
injured cornea   179
injury   199, 203, 204, 787, 815
injury of the CNS   816
injury site   786
innate defence system   781, 782
innate immune   751
innate immune response   749, 754
innate immune system   749, 751
innate immunity   754, 782, 803, 972
inner cell mass   160, 162
inner ear   924

innervation   773
inorganic crystal   469
inorganic material   878
inorganic phosphate   878
inorganic polymer   247
iNOS. *see* inducible nitric oxide synthase
insemination   41, 42, 66
insertional oncogenesis   150
in situ hybridization   203, 946, 947
in-situ hybridization method   791
insomnia   389
institutional ethical approval   143
insulin   126, 172, 264, 297, 411, 412, 413, 417, 418, 419, 421, 423, 862, 863, 882, 939, 940, 941, 943, 945, 948, 968, 969
insulin-dependent   419
insulin-like growth factor   122, 126, 184, 215, 216, 251, 321, 828, 829
insulin-like growth factor-1   214, 264, 917
insulin-like growth factor I   927, 928
insulin-like growth factor II   927, 928
insulinoma   417
insulin pump   159
insulin release   946
insulin resistance   940, 946
insulin-secreting cell   105
insulin-secreting cell line   863
insulin secretion   946
insulin therapy   940
integrin   62, 140, 178, 258, 471, 508, 656, 727, 763, 850, 957, 958, 968
integrin binding   599
integrin-mediated adhesion   850
integrin-mediated cell adhesion   850
integrin receptor   178, 655, 971
integrin α4/β1   178
integrin α6+/CD34+   861
integrin-β1   178
integument   877
intention-to-treat   14
intercalated disc   111
intercellular adhesion molecule   784
intercellular matrix   831
interconnecting pore   491
interconnectivity   490, 541
interface structure   840
interfacial zone   845
interference   498
interferon   71, 72, 414, 754, 784, 790, 803

interferon-regulated factor-9 (IRF-9)   72
interferon-regulatory factor   77, 752, 754
interferon-regulatory factor-1 (IRF-1)   72
interferon-stimulated gene factor 3 (ISGF3)   72
interferon α/β   72
interferon γ   72, 76, 77
interferon-γ   186
interleukin   72, 723, 784, 790, 927, 953
interleukin 1 beta   239
interleukin 1β   764
interleukin 6   297
interleukin 7   297
interleukin 8   297
interleukin 10   72
interleukin 11   297
internal bioreactor   433
internal ribosome entry site   99, 100
International Islet Transplant Registry   413
International Society for Stem Cell Research (ISSCR)   41
international standard guideline   789
interstitial cell   870
interstitial cystitis   432
interstitial fluid   635, 713
intervertebral disc   307, 308, 311, 313, 314, 880
intervertebral disc regeneration   307
intestinal epithelium   10
intestinal mucosa   369, 776
intestinal stem cell   129
intestinal submucosa   776
intestinal tissue engineering   777
intestine   956
intimal hyperplasia   882
intracerebral haemorrhage   153
intra/extra-synovial   325
intrahepatic stem cell   404
intramuscular implantation   105
intraoral mucosa   373
intra-synovial   319, 320, 323, 324
intra-uterine transplantation   140, 149, 185
intrinsic growth ability   815
intrinsic tyrosine kinase   75
intron   97
intron-encoded shRNA   100
in vitro assay   563
in vitro culture   650
in vitro fertilization (IVF)   37, 38, 39, 40, 41, 42, 43
in vitro investigation   805

in vitro model   761, 767
in vitro organogenesis   600
in vitro study   470
in vivo animal model   773
in vivo bioreactor   608
in vivo remodelling   801
in vivo transplantation   953
involucrin   968, 969
ion beam   475
ion channel   84, 85, 637
ionic milieu   505
ionic strength   486, 510
IP3   84
iPSC. see induced pluripotent stem cell
IRAK. see IL-1R-associated kinase
IRES. see internal ribosome entry site
IRF. see interferon-regulatory factor
IRF-1. see interferon-regulatory factor-1
IRF-3   752
IRF-9. see interferon-regulated factor-9
irradiated NOD/SCID   954
ischemia   817, 883
ischemia/perfusion injury   65
ischemic renal injury model   870
ISGF3. see interferon-stimulated gene factor 3
ISL-1   172
islet   862, 941, 946
islet allotransplantation   940
islet cell antigen   940
islet-containing-arginate-poly-l-lysine/poly-l-ornithine   421
islet disintegration   946
islet of Langerhans   411, 419, 733, 943
islet sheet   863
islet structure   948
islet transplantation   412, 413, 414, 862, 940, 949
I-Smad   123
isogeneic cell   803
isograft   942
isolation technique   292
isotransplantation   939, 941, 943
ISSCR. see International Society for Stem Cell Research
ITR. see International Islet Transplant Registry
IVF. see in vitro fertilization
IκB kinase   752
IκBα   752

J

JAG1   162
Jagged-1   958
Jagged-1/2   958
JAK   75, 78. *see* Janus kinase
JAK1   72
JAK2   72
JAK3   72
JAK3 inhibitor   78
JAK homology (JH) domain   72
JAK kinase   74
JAK-STAT pathway   72, 78
JAK-STAT signaling   71
Janus kinase (JAK)   71, 72
jaw   362, 924
jaw bone   827
jejunal defect   777
jejunal segment   776
joint   181, 634, 641, 705, 708, 713, 807, 859, 860
joint cartilage   124
joint disease   859
joint function   260, 321, 860
joint kinematic   256
joint replacement   16
joint replacement surgery   859
joint surface   859
jugular vein   776
junctional epithelium   359, 932
Jurkat cell   623
juvenile diabetes   940
juxtacrine signal   470

K

kainic acid/ibotenic acid   282
karyopherin β   75
karyotype   417
karyotypic abnormality   160
karyotypic stability   160
K-cell   418
keratan sulphate   761
keratin   968
keratin 1   968
keratin 10   968, 969
keratin 16   969
keratinization   372, 374
keratinocyte   9, 123, 332, 333, 334, 335, 336, 337, 338, 358, 369, 370, 371, 374, 375, 376, 377, 378, 728, 732, 765, 832, 881, 967, 968, 969, 971, 972
keratinocyte EGF receptor   968

keratinocyte graft   371
keratinocyte growth factor   123, 430, 431
Keratinocyte-SFM   375
keratocan   761
keratocyte   761
keratohyalin   968
KFS   335
KGF. *see* keratinocyte growth factor
kidney   50, 61, 63, 65, 66, 85, 126, 143, 144, 161, 412, 430, 432, 615, 787, 789, 790, 791, 807, 861, 869, 870, 871, 873, 924, 945
kidney cell   65
kidney cell therapy   65
kidney development   66
kidney dialysis   855
kidney disease   861
kidney repair   65
kidney support   862
kidney transplant   413
kidney transplantation   9, 945
killer-cell immunoglobin-like receptor   955, 956
kinase activity   76
kinetic property   773
KIR. *see* killer-cell immunoglobin-like receptor
KIR/KIR ligand incompatibility   956
klf4   201, 417
knee   361, 860
knee joint   181
knee replacement   17
knockout animal   124
knock-out mouse   72, 73, 111
KRAB. *see* Krüppel-associated box protein
Krause   198
Krüppel-associated box protein   94, 96, 97
Krüppel-associated box protein-derived transcriptional repressor   97
kryoconservation   957
kynurenine   186
kyphoscoliosis   151

L

lachrymal gland   62
lactate   600, 603, 637
lactate dehydrogenase activity   971
lactic acid dehydrogenase   790
lactogenesis   73
lacunae   486
lamellar bone/marrow-like structure   933
lamin   112, 113, 114, 115, 117

laminar flow   638, 670, 674, 675
laminin   62, 141, 145, 244, 248, 275, 280, 299, 370, 393, 430, 478, 562, 568, 762, 818, 870, 968
laminin-1   393
Langerhans cell   765, 881, 970, 972
laparotomy   945
large-scale production   606
laser beam   511, 616
laser-guided direct write   616
laser-induced forward transfer   616, 617, 626, 627
laser pulse   618
laser scanning microscopy   76
lateral mesoderm   245
lateral ventricle   817, 821
latissimus dorsi   216
latissimus dorsi muscle   129, 831
latissimus dorsi musculocutaneous flap   390
lawgiver   39
LCA   144, 177
lecithin   859
lectin   723
left ventricle   111
left ventricular ejection fraction   753
left ventricular end-diastolic diameter   753
LEFTYA   162
LEFTYB   162
legal regulation   38
legal restriction   775
leishmania   786
lentiviral system   103
lentiviral vector   140
lentivirus   104, 105, 149, 163, 929
leptin   294
leptin receptor   945
leucine-rich repeat   749
leucine-rich repeat domain   751
leucocyte   956, 961
leucopenia   413
leukaemia   790
leukaemia inhibitory factor (LIF)   160
leukapheresis   954, 955
leukemia   65, 150, 167, 188, 951
leukemia inhibitory factor (LIF)   72
leukemia-specific antigen   960
leukemic cell   957
leukocyte   669, 722, 727, 728
leukocyte common antigen   177
leukotriene   783, 784
leukotriene B4   971

level of evidence   804, 808
lewy body   281
LFA. see lymphocyte function-associated antigen
LFA-1   145, 415
LFA-3   144
LGDW. see laser-guided direct write
LGPA   393
library   805
LIF   174. see leukemia inhibitory factor; see leukaemia inhibitory factor
life expectancy   159
life style   32
LIFT. see laser-induced forward transfer
ligament   255, 256, 257, 258, 259, 260, 261, 262, 263, 264, 265, 266, 317, 318, 319, 321, 322, 323, 325, 634, 657, 705, 708, 831, 916
ligament ECM   264
ligament fibroblast   263, 264
ligand binding   74
ligand-gated channel   172
light absorption   618
LIN 28   417
lineage control   101
lineage differentiation   147
lineage marker   140
lineage-specific progenitor cell   878
lining cell   220, 221, 563, 570
linker   500
link protein   180
lipid   392, 765, 767, 968
lipid accumulation   180
lipid biosynthesis   969
lipid-containing vesicle   882
lipid metabolism   861
lipid profile   968
lipoaspirate   292, 301, 394
lipofection   930
lipoinjection   881
lipophilicity   507
lipopolysaccharide   125, 751
lipoprotein   751
lipoprotein lipase   294
liposomal formulation   930
liposome   149, 184, 351, 930
liposuction aspirate   182
liquid culture assay   953
lithogenesis   434
lithography   481

liver   50, 64, 65, 73, 122, 126, 140, 141, 143, 153, 161, 172, 188, 197, 198, 202, 204, 219, 397, 398, 413, 615, 726, 777, 786, 787, 789, 790, 807, 861, 924, 941, 946, 956
liver cell   64, 400, 402
liver damage   790
liver failure   153, 398, 400, 405, 407, 861
liver function   861
liver injury   147, 153
liver lobule-like structure module   402
liver reconstruction   554
liver-slice perfusion   397
Liver Support System   398
liver tissue   588, 930
liver transplantation   397
L-lactide   129
LLNA. *see* local lymph node assay
LLS. *see* liver lobule-like structure module
LLS-BAL   402
LNGFR   178
load   83, 843
loading protocol   839
load perception   842
load transfer   842
local assessment   789
local flap   877, 879
local lymph node assay   792
long bone   635
long bone defect   914
longitudinal tension   883
long saphenous vein   858
long-term culture initiating cell   952, 958
long-term stability   500, 504
loop of Henle   871
loss of motion   916
lower forearm flap   374
low volume eye test   761
LPS-R   144
LRR. *see* leucine-rich repeat
LRR domain   752
LRR-motif   749
LSS. *see* Liver Support System
LT. *see* leukotriene
LT-B4   783
LT-C4   783
LTC-IC. *see* long-term culture initiating cell
LTC-IC-assay   952
LT-D4   783
LTD_4 783

LTE_4 783
luciferase   77
lumbar arthrodesis   125
lumbar disc herniation   313
lumican   761
luminal perfusion pressure   883
lumpectomy   389
lung   61, 62, 83, 85, 126, 140, 143, 144, 161, 204, 787, 856, 924, 956
lung injury   179, 203
lung squamous cell cancer   205
LVEDD. *see* left ventricular end-diastolic diameter
LVEF. *see* left ventricular ejection fraction
LVET. *see* low volume eye test
lymphangiogenesis   778
lymphatic duct   787
lymphatic fluid   787
lymphatic system   787
lymph node   787, 789
lymphocyte   185, 186, 187, 728, 750, 790, 791, 803
lymphocyte activity   790
lymphocyte function-associated antigen   415
lymphocyte subpopulation   791
lymphohematopoiesis   952
lymphoid cell   953
lymphoid development   72
lymphoid lineage   140
lymphoid organ   790, 791
lymphoid tissue   147, 956
lympholytic agent   956
lymphoma   167, 951
lymphopoiesis   126, 959
lymphoproliferative assay   791
lysosomal storage disease   150
lysozymal enzyme   782
lysozyme   803
lysyl oxidase   971

M
M. *see* macrophage
mAb. *see* monoclonal antibody
Mac-1 (Lin-)   952
macroarchitecture   650
macrolide   96, 97, 101
macromolecule   782
macrophage   122, 125, 492, 669, 724, 732, 733, 750, 751, 753, 782, 783, 784, 785, 786, 787, 788, 790, 803, 972

macrophage activity   790
macrophage cell line P338D1   849
macrophage colony-stimulating factor   297, 491, 953
macrophage function   790
macrophage infiltration   124
macrophage progenitor cell   953
macroporosity   489
macrosialin   144
macrovascular complication   940
macular degeneration   763, 764
mafosfamide   957
magnesium   501
magnetic bead depletion   960
magnetic reconance tomography   802
magnetic resonance imaging   509
major burn   330, 333, 334, 337
major histocompatibility complex   185, 414, 723, 730, 731, 732, 733, 734, 750, 784, 951
major histocompatibility complex class II   970
major histocompatibility complex (MHC)   72
major histocompatibility complex (MHC) class II protein   72
major histocompatibility complex (MHC)-mismatched mouse   162
major transplantation antigen   174
malignancy   125, 877, 960
malignant cell   954, 957
malnutrition   869
mammalian cell   613, 616, 623, 684, 694, 953
mammary   485, 501, 647, 693, 722
mammary gland   64
mammary gland development   73
mammary gland factor   73
mammary stem cell   64
mammopoiesis   73
mandible   830, 831, 839
mandibular condyle   361
mandibular defect   129, 828
mandibular discontinuity defect   129
mandibular inferior alveolar canal   840
mandibular ramus   830
mandibular reconstruction   129, 829
manganese superoxide dismutase   415
mannuronic acid block (M-block)   501
MAP kinase   751
MAPLE DW. *see* matrix-assisted pulsed laser evaporation direct write
market   21, 863

market analysis   14
Markov model   16
marrow   221
marrow aspiration   567, 570
marrow cell   843
marrow-derived hematopoietic stem cell   204
marrow-derived mesenchymal stem cell   204
marrow-derived osteoblast   598
marrow-derived stem cell   570
marrow stroma   561
marrow stromal cell   562, 567
MASH1   279
masseter muscle   245
Masson's trichrome   791
mass transport   841
mast cell   260, 727, 751, 765, 783, 785, 790
mastectomy   289, 389, 390
master control gene   356
mastication   827
masturbation   42
matched related donor   954
matched unrelated donor   954
material   840
materials science   877
material surface   469
material surface property   470
maternal blood   147
MATH1   162
mathematical modeling   103
Matrigel   123, 248, 618
matrillin   239
matrix   485, 614
matrix-assisted pulsed laser direct write   618
matrix-assisted pulsed laser evaporation direct write   618
matrix-associated chondrocyte implantation   233
matrix deposition   62
matrix destruction   62
matrix extracellular phosphorylated protein   924
matrix metalloproteinase   203, 729, 763, 764
matrix metalloproteinase-2   437
matrix metalloproteinasis   970
matrix mineralization   105, 179, 846
matrix protein   570, 580
matrix protein expression   181, 564
matrix receptor   63
matrix stiffness   648
matrix synthesis   127, 492
matrix vesicle   843, 845

matrix vesicle formation 844
maturation 648, 690
mature cell 555, 570
mature hematopoietic lineage marker 205
mature macrophage 575
mature osteoblast 220, 563
maxilla 839
maxillary sinus 840
maxillary tuberosity 830
maxillofacial reconstruction 831
maxillofacial surgery 8, 806, 827
M-cadherin 245
MCF 784
MCM-5 77
MCP. see monocyte chemotactic factor
MCP-1 783, 784
M-CSF 174, 783, 953, 958
MCSF. see macrophage colony-stimulating factor
MDS. see myelodysplastic syndrome
MDSC. see muscle-derived stem cell
MDSC surface marker 915
mdx mouse 151
mechanical approach 249
mechanical barrier (ext. surface) 803
mechanical circulatory assistance 750
mechanical deformation 828
mechanical dissociation 160
mechanical environment 802
mechanical force 707
mechanical injury 785
mechanical interlocking 842
mechanical load-bearing 880
mechanical loading 85, 86, 213, 827, 828, 846
mechanical property 470, 507, 831
mechanical signal 264
mechanical stimulation 214, 262, 641, 693
mechanical stimulus 598, 802
mechanical strength 469, 500
mechanical stress 470
mechanical valve 857
mechanosensor 83, 84, 85
mechanotransduction 85, 470, 708, 709
mechanotransductive process 479
medical device 979
medical field 801
medical research 804
medical speciality 805
medicinal product 979, 980
medium exchange 605

medium flow rate 597
medium-spiny GABAergic neuron 821
medullary sinus 946
megakaryocyte 953
megalin 65
megaureter 432
melanocyte 765, 881, 970
melanogenesis 970
melanoma 970
melanoma cell 477
MELS. see Modular Extracorporeal Liver Support
membrane 567, 606, 802
membrane attack complex 786
membrane oxygenator 856
membrane-spanning domain 110
membrane technique 212
membrane type 1 matrix metalloproteinase 764
memory cell 782, 785
meningoencephalitis 111
meniscal/osteochondral allograft 916
meniscus 648, 649
MEPE. see matrix extracellular phosphorylated protein
mesangial cell 870
mesenchymal 62, 125, 292, 416
mesenchymal cell 63, 66, 122, 636, 828, 882
mesenchymal cell lineage 562
mesenchymal construct 431
mesenchymal-like cell 924
mesenchymal lineage 171, 179
mesenchymal progenitor 160
mesenchymal progenitor cell 320, 881
mesenchymal stem cell 105, 125, 130, 139, 140, 142, 143, 148, 160, 161, 168, 169, 172, 177, 204, 219, 233, 263, 278, 292, 404, 560, 561, 567, 649, 651, 652, 655, 656, 657, 658, 659, 696, 697, 728, 732, 809, 816, 870, 923, 932, 958
mesenchymal tissue 63, 871
mesenchyme 64
mesenteric lymph node 946, 947, 948
mesentery 943, 946
mesh size 503, 507
mesh stent 776
mesoderm 124, 160
mesodermal 416
mesodermal cell 126
mesodermal/ectodermal differentiation 168
mesodermal osteoblast 168
mesothelial cell 776

mesothelium   276
MEST. *see* mouse ear swelling test
meta-analysis   808
metabolic storage disease   167
metabolism   602
metabolite   669
metachromatic feature   564
metal   159, 211, 360, 475
metal alloy   878
metal ion   500
metalloproteinase   198
metal-on-metal   859
metal surface   841
metanephric kidney   66
metanephro   66
metanephros   871
metaplasia   374
methylation   200
methylcellulose   933
methyl methacrylate   859
methyltransferase inhibitor 5-aza-2'-deoxycytidine (5azaD)   959
methylxanthine   882
mevastatin   764
Mg^{2+}. *see* magnesium
MGF. *see* mammary gland factor
MHC   185, 297. *see* major histocompatibility complex
MHC class I   414
MHC class II   73, 185, 414, 753, 784, 786
MHC class II expression   754
MHC class II transactivating protein (CIITA)   72
MHC class I ligand   955
MHC gene expression   162
Michaelis-Menten constant   699
Michaelis-Menten law   699
Michaelis-Menten relationship   700
microarchitecture   650, 700
microarray   767
microarray technology   793, 841
microbead   959
microbial agent   782
microbial infection   785
microbubble   931
microcapsule   402, 419, 420, 421, 423
microchimerism   147, 153
micro-computed tomography   603
microdialysis   603
microeconomic   13

micro-encapsulated aggregate   64
microencapsulated islet   421
microencapsulation   412, 419, 420, 421, 804, 863
microenvironment   178
micro-fibril   256
microfluidic channel   183
microfracture   17, 18, 21, 181, 842, 916
microglia   271
microglial cell   275
microgranule   127
microgroove   657
microinjury   914
micromass   182
micromass body   163
micromass culture   163, 181, 182, 553, 555, 565, 651
micromass pellet culture   880
micromass technique   552
micromass technology   552
micromass tissue-culture technique   361
micromotion   844
micro-patterning   817
microporosity   492
microporous hollow fiber   64
microradiography   802
microRNA (miRNA)   110
microscopic field   792
microscopic protein deposit   281
microskin grafting   334
microsphere   126, 127, 284, 423, 849
microstrain   709
microsurgery   8, 9
microsurgical approach   250
microsurgical technology   877
microtopography   849
microtubular-associated-protein-2   622
microvascular anastomosis   375
microvascular free flap   882
microvascular free tissue transfer   877
microvascular surgery   9
microvessel   670, 699
microwick   283
Middle Ages   6
mifepristone (RU486)   95, 97, 104
migration   143, 147, 149, 168, 182, 469, 649
mimecan   761
mineral   636, 841
mineral deposition   847
mineral formation   843, 845, 846

Subject Index

mineralisation 636
mineralization 129, 180, 346, 843, 845
mineralized bone tissue 840
mineralized collagen 487, 489
mineralized extracellular matrix 840
mineralized hard tissue 841
mineralized matrix 846, 923
mineralized tissue 485
mineralizing gene 931
mineralogical investigation 844
mineralogical study 844
mineral spherite 846
mineral trioxid aggregate 360
mineral trioxide aggregate 927
minimal promoter 94, 95, 98, 99, 100, 101
miniosmotic pump 283
minipig 774, 829, 830
minor histocompatibility antigen 956
miRNA. see microRNA
mismatch 954
mitochondria 274, 790
mitogen 121, 122, 123, 125, 126, 162, 186
mitogen-activated protein kinase pathway 414
mitosis 930
mixed lymphocyte reaction 186
mixed lymphocyte response 791
MLR 186. see mixed lymphocyte response
MM. see multiple myeloma
MMP. see matrix metalloproteinase
MMP-1 764
MNC. see bone marrow mononuclear cell
MnSOD. see manganese superoxide dismutase
model animal 773
modeling 843
modified mesenchymal stem cell 916
modular building block 102
Modular Extracorporeal Liver Support 398, 400, 403
molecular agent 819
molecular biology 877
molecular cue 878
molecular imprinting 850
monitoring technique 596
monkey 775, 821
monoclonal antibody 415, 587
monocrotaline 202
monocyte 187, 491, 575, 724, 733, 783, 790, 953, 960
monocyte chemotactic factor 784

monocyte differentiation antigen 177
monolayer 565
monolayer culture 127, 604, 637
monomer 73
monomeric unit 787
mononuclear cell 198, 575, 945
mononuclear cell separation 171
Monte Carlo simulation 16
moral dignity 37
moral right 37
morphogen 121, 354, 355, 843
morphogenesis 62, 121, 128
morphogenetic mechanism 67
morphogenetic signal 843
morphogen gene 347, 350, 354, 355
morphology 375, 760
mortality 15, 111
motor deficit 820
motor function 820
motor neuron 821
mouse 774, 775, 821, 954
mouse calvaria 573
mouse ear swelling test 792
mouse melanoma cell 477
MPC 245, 248, 250, 320, 321
MPS. see myeloproliferative disease
MRD. see matched related donor
MRI. see magnetic resonance imaging
MRL mouse 198
mRNA 97, 98, 99, 100, 110
mRNA destruction 97
mRNA expression 114, 767
mRNA stability 94
mRNA transcript 99, 110
MSC 181, 182, 183, 185, 187, 188, 204, 219, 262, 263, 264, 278, 279, 293, 296, 297, 299, 301, 736. see mesenchymal stem cell
Msx-1 105, 198
Msx-2 180
MT1-MMP. see membrane type 1 matrix metalloproteinase
MTA. see mineral trioxid aggregate; see mineral trioxide aggregate
MTJ 250, 259, 260
MTT conversion assay 971
MTT proliferation 873
mucopolysaccharidose 150
mucosa 369, 372, 373, 374, 376, 377, 378, 807, 832

mucosa graft   372, 373, 374, 375
mucosa keratinocyte   374
mucosal defect   832
mucosal epithelium   373, 374
mucosal graft   833
MUD. *see* matched unrelated donor
multicapillary PUF-BAL   402
multicistronic expression vector   99, 100
multicistronic mRNA   99
multicistronic vector   99
multigene-multiregulated expression control   99, 100, 101, 102
multilayered cultured epithelial sheet   333
multilayered epithelial transplant   332
multi-lineage   185
multi-lineage differentiation   150, 179
multilineage potential   954
multimerization   105
multinucleated giant cell   782
multinucleated myotube   248, 250
multinucleated syncytia   248
multiple myeloma   73, 951, 952, 954
multiple sclerosis   50, 273
multiple time serie   804
multiplication   200
multipotency   140, 170, 172, 188
multipotential stem cell   552
multipotent progenitor cell   953
multipotent stem cell   933
multiwell plate   599
murine embryonic stem cell   553
murine transplant model   956
muscle   83, 129, 140, 144, 161, 204, 219, 243, 561, 634, 705, 708, 860, 916, 917, 930
muscle biopsy   915
muscle cell   126
muscle contraction   648, 657
muscle-derived stem cell   913, 914
muscle fibre   243, 249
muscle tissue   789, 807
muscle weakness   821
muscular dystrophy   150
musculoskeletal disease   318
musculoskeletal system   855, 859
mutagenesis   929
mutation   151, 200
myasthenia gravis   786
MyD88. *see* myeloid differentiation factor 88
myelin-associated glycoprotein (MAG)   275

myelinated nerve   818
myelinating capability   277
myelination   277, 279
myeloablation   149
myeloablative conditioning   174, 951
myelodysplasia   167
myelodysplastic syndrome   952
myeloid cell   953
myeloid differentiation factor 88   752, 753, 754
myeloid lineage   140
myelomeningocele   432
myelopoiesis   72
myeloproliferative disease   952
myelosupportive stroma cell   958
myelosuppressive stress   959
myf5   296
myf6   296
myoblast   101, 122, 125, 246, 249, 250, 860, 914
myocardial infarction   104, 111, 153
myocardial injury   753
myocarditis   111, 749, 754
myocyte   111, 122, 753
MyoD   77, 105, 914
MyoD1   296
myofiber/myofibre   249, 250, 914, 917
myofibril   111
myofibroblast   143, 179, 259, 435, 776, 858
myofibroblast-like   382
myogenesis   105, 126, 141, 244, 245, 247, 295, 296
myogenic cell type   915
myogenic differentiation   151, 247, 248
myogenic lineage   914
myogenic precursor cell   244
myogenin   296
myoid   249
myopathy   151
myosin   87, 143, 243, 751
myosin heavy chain   296, 297
myotendinous junction   243, 250, 259
myotube   249, 917
myotube formation   248

N
Na   184
N-acetylglucosamine   723, 926
naïve T-cell   960
Na$^+$,K$^+$-ATPase   761
nanocage   688

nanocapsule 819
nanocomposite 485
nanofibrous meshe 181
nanofibrous scaffold 183
nanog 141, 142, 146, 162, 201, 417
nanog-3 141
nanogroove 479
nanoparticle 284, 729
nanopattern 817
nano-patterning 817
nanoprinting 479
nanoscale 470
nanosized sphere 478
nanostructure 478, 510
nanotopography 478, 849
nasal cartilage 831
nasal cartilage structure 831
nasal septum 879, 880
native hyaline cartilage 182
native tissue 470, 787
natural antibody 803
natural killer 953
natural killer cell 185, 186, 724, 751, 783, 786
natural killer cell cytotoxicity 749
natural killer lymphocyte 972
natural organogenesis 61
natural polymer 129, 321, 818, 878, 926
natural tissue regeneration 922
Navier–Stokes equation 672, 673, 674, 675, 680, 683
N-cadherin 178, 179, 651, 958
NDF. *see* Neu differentiation factor
necrectomy 331
necrosis 671, 699, 785, 788, 791
necrotic cell 722, 726
need 14, 20
need assessment 14
needle-shaped crystal 486
negative pressure ventilator 63
neobladder 432, 436
neomucosa 777
neomycin 97, 98, 104, 565
neonatal islet transplantation 941
neoplasia 198, 199, 791
neoplastic disease 956
neovascularization 351, 361, 576, 777, 786, 882
neovessel 686
nephron 869, 870
nephronectin 62

nephron segment 871
nephropathy 412, 940
nephrotoxicity 413
nerve 271, 762, 763, 765
nerve cell 85, 86, 815
nerve growth factor 172, 762, 818, 927
nerve growth factor receptor 178
nerve guidance channel 818
nerve regeneration 815
nerve repair 284
nerve tissue 807
nervous system 271, 272, 273, 815
nervous system injury 818
nestin 141, 146, 297, 417
net charge 475
Neu differentiation factor 123
NeuN 297
neural cell 168, 172, 173
neural differentiation 153, 172
neural injury 148
neural-like cell 360
neural lineage 168, 170, 172
neural marker 143
neural-restricted precursor 278
neural stem cell 143, 197, 278, 282, 351, 654, 816, 819, 820
neural tissue 923
neural tube 162, 245
neuregulin 123
neurite 275, 276, 280
neurite outgrowth 762, 817
neurobiology 792
neurocyte 923
NeuroD 172
neurodegenerative disease 273, 279, 285, 818, 819, 820
neurodegenerative disorder 282
neuroectoderm 124, 162, 204
neuroendocrine-cell-like 418
neurofibrillary tangle 820
neurofilament 172, 173, 274
neurofilament-20 763
neurogenesis 128, 817, 821
neurogenic capacity 278
neurogenic lineage 141
neurogenin 3 105
neuroglial cell 271
neurological application 815
neurological disorder 816

neurological medicine   815
neurology   806, 815
neuromuscular-like junction   250
neuron   61, 67, 123, 141, 143, 162, 163, 172, 179, 204, 205, 272, 273, 274, 275, 278, 279, 280, 281, 282, 622, 815, 817
neuronal cell   478
neuronal cell death   816
neuronal cell type   915
neuronal circuitry   817
neuronal development   817
neuronal differentiation   278
neuronal-like cell   933
neuronal lineage   297
neuronal loss   820
neuronal stem cell   141, 297
neuronal synaptic junction   786
neuronal tissue   437
neuron-specific enolase   143
neuron-specific marker   173
neuropathic bladder   432, 436
neuropathy   412
neuroprotection   105
neuroscience   815, 821
neurosurgery   806
neurothelin   145
neurotransmitter   85, 762, 817
neurotrophic factor   283, 817, 818, 821
neurotrophin   283
neurotropin/neurotrophic factor   280
Neutral Red assay   971
neutropenia   174
neutrophic factor   72
neutrophil   724, 782, 783, 785, 786, 788, 790, 953
neutrophil recovery   955
new blood vessel growth   576
Newton's second law (F = ma)   672
NF-κB   95, 97. see nuclear factor-κB; see nuclear factor κB
NGC. see nerve guidance channel
NGF. see nerve growth factor
NHL. see Non-Hodgkin's lymphoma
niche   178
nicotinamide   297
nidation   44
niobium   478
nipple   390
nitric oxide   858
NK cell   953, 954, 960

NK-cell   186
NK-cell activity   791
Nkx2.2   418
Nkx2.230   417
Nkx6.1   172, 418
NLR. see NOD-like receptor
NMR   508, 802. see nuclear magnetic resonance
N-Myc interactor (Nmi)   77
NO   86
nociceptor   762
NOD. see nucleotide-binding oligomerization domain
NODAL   162
NOD-like receptor   722, 752
NOD/SCID mouse   174, 958, 959
nodule   564, 847
noggin   916
non-collagenous protein   485
non-destructive functional assay   789
non-fusion technique   314
non-hematopoietic stem cell   168
Non-Hodgkin's lymphoma   951, 952, 954
non-human feeder cell   736
non-human primate   829
non-mineralized fibrocartilage   323
non-neoplastic disease   956
non-polar interface   473
non-specific (innate)   803
nonsteroidal anti-inflammatory drug   917
non-transformed clonal cell line   160, 564
nonunion   878, 916
nonviral vector   929
normative validity   47
normoglycemia   412, 423, 945
normoglycemic   946
Notch   129, 958
Notch 1   172, 958
Notch 4   172
Notch ligand   958
Notch ligand Delta-like 1   960
Notch signaling   958
notochord   245
NP. see nucleus pulposus
NPC. see nuclear pore complex
NPI-1/hSrp1   75
NRP   278
NRP-GRP   278
NSC. see neural stem cell
NT-3   282

nuclear accumulation 75, 76
nuclear cloning 416, 417
nuclear envelope 75
nuclear factor-kappa β ligand 324
nuclear factor κB 749, 751, 753, 754
nuclear factor-κB 77
nuclear import 73, 74, 75
nuclear magnetic resonance 509
nuclear oligomerization receptor 751
nuclear pore complex 74, 75
nuclear transfer 10, 161, 416, 424
nuclear transfer technique 423
nuclear transplantation 416, 872
nucleated cell 570
nucleation site 508
nucleocytoplasmic shuttling 74, 77, 78
nucleoplasm 202
nucleoporin 75
nucleotide base 476
nucleotide-binding oligomerization domain 752
nucleotide hydrolysis 75
nucleus 71, 95, 97, 180, 309
nucleus pulposus 308, 311
nude mouse 65, 880
nutrient 507, 669, 788, 883, 925
nutrient delivery 693
nutrient supply 698
nutrition 925
nutritional transport 777
nutritional transport mechanism 778
nylon membrane 623
nylon mesh 881

O

O_2-tension 843
obesity 946
occlusal overloading 846
Oct-4 141, 142, 143, 146, 162, 201, 417
Oct-4 positive cell 141
ocular aqueous 670
ocular irritancy testing 762
ocular prosthetic 856
ODF. see osteoclast differentiation factor
ODI. see Oswestry Disability Index
odontoblast 922, 923, 924, 925, 928, 931, 933
odontoblastic cell 360
odontoblast-like cell 360
OEC 277, 278, 284
oedema 971

oestrogen receptor 637, 710
olfactory bulb 277
olfactory bulb ensheathing cell 277
oligodendrocyte 140, 141, 143, 271, 275, 278, 279, 284
oligomeric unit 787
oligopeptide sequence 478
omental flap 433
omentocystoplasty 433
omentum 777
oncogene 929
oncogenicity 139, 419
oncology 50
onco-retrovirus 104
oncostatin M (OSM) 73
onlay graft 828
ontogenesis 53
oocyte 27, 38, 42, 43, 128
OP-1. see osteogenic protein-1
opacification 761
open reading frame 98
operative anastomosis 778
operator 94, 95, 96, 99, 101, 102
OPG 324
OPGL. see osteoprotegerin ligand
ophthalmic branch 761
opioid peptide 283
OPN. see osteopontin
opportunistic infection 956
opsonin 782
optical coherence tomography 603
optic tract 818
oral cavity 369, 832
oral hard tissue 840
oral implant 846
oral mucosa 332, 369, 373, 377, 488, 827
oral mucosal keratinocyte 376
oral rehabilitation 839
oral soft tissue 840
oral surgery 827
oral tumor 832
organelle 637, 709
organ function 790
organic matrix 841
organic polymeric fibre 248
organogenesis 61, 62, 63, 64, 125, 249, 423, 424, 470, 671, 939, 949
organoid 403, 777
organoid-scaffold model 778

organotypic model 759
organ primordia 939
organ printing 613
organ regeneration 161
organ retrieval 790
organ supply 30
organ system damage 790
organ tissue engineering 760
organ toxicity 951
organ transplantation 9
organ transplant rejection 188
organ weight 789
orientation 471
oropharyngeal mucosa 123
orthopaedic surgery 806
orthopedic implant 840
orthopedic surgery 913
oscillatory frequency 513
OSM 173. see oncostatin M
osmotic imbalance 473
osmotic potential 588
osmotic pressure 84, 714
osmotic pump 817
osmotic swelling 930
osseointegration 357, 480, 845
osseous defect 125, 879
ossification 103
osteoarthritis 16, 233, 318, 859, 916
osteoblast 101, 105, 122, 124, 125, 126, 127, 129, 130, 151, 160, 162, 170, 171, 179, 204, 220, 221, 258, 263, 265, 296, 354, 553, 555, 563, 564, 566, 570, 573, 636, 637, 638, 641, 644, 652, 709, 843, 846, 878, 914, 933, 958, 959
osteoblast adhesion 480, 850
osteoblast culture 847
osteoblast differentiation 105, 129, 171, 180, 294, 552, 640, 711
osteoblastic cell 127, 642
osteoblastic marker 130
osteoblastic phenotype 130
osteoblastic progenitor 346
osteoblast-like cell 564, 570, 573, 709
osteoblast-like cell line 489
osteoblast lineage 180
osteoblast phenotype 570
osteoblast precursor cell 168
osteocalcin 172, 296, 481, 485, 642
osteochondral condylar implant 832
osteochondral defect 181, 489

osteochondral defect repair 183
osteochondral differentiation 183
osteochondral disease 764
osteochondral grafting 764
osteochondrogenic progenitor 180
osteoclast 218, 258, 324, 566, 575
osteoclast differentiation factor 576
osteoclast inhibitor 324
osteoclast-like cell 491
osteoconduction 350, 840
osteoconductive behavior 847
osteoconductive component 349
osteoconductive environment 348
osteoconductive scaffold material 827
osteocyte 85, 168, 205, 218, 220, 221, 258, 486, 563, 570, 636, 709, 843, 846
osteocyte-like cell 563
osteodentine 925, 928
osteodifferentiation 879
osteogenesis 105, 130, 141, 179, 180, 221, 295, 355, 359, 652, 656, 697
osteogenesis imperfecta 150, 151, 152
osteogenic cell 162, 219, 348, 349, 350, 561, 570
osteogenic cell type 915
osteogenic condition 547
osteogenic differentiation 130, 162, 169, 172, 180, 296, 299, 300, 652, 828, 879
osteogenic lineage 180, 294, 296, 490
osteogenic pathway 879
osteogenic potential 124, 179, 292
osteogenic protein 359
osteogenic protein-1 129, 829, 850, 928
osteogenic protein factor 357
osteogenic supplement 480
osteogenic tissue 573
osteogenin 128
osteoid 847
osteoinduction 878
osteoinductive activity 129
osteoinductive agent 349, 350, 351, 353
osteoinductive component 350
osteoinductive environment 348
osteoinductive factor 352
osteoinductive graft 607
osteoinductive protein factor 357
osteonectin 172, 214, 296
osteopenia 151
osteopetrosis 151
osteopontin 172, 294, 481, 485, 642, 924

osteoporosis 126, 129, 345
osteoprogenitor 220, 570
osteoprogenitor cell 122, 130, 162, 212, 221, 349, 570, 695, 878
osteopromotive property 829
osteoprotegerin 324
osteoprotegerin ligand 576
osteosarcoma cell line 564
osteotendinous junction 260
osteotomy 844
Oswestry Disability Index 314
ouabain 762
outbreed 774
outcome measure 801
outcome parameter 801
outcome probability 17
outcome variable 788
ovary 789
oviduct 85
ovocyte 66
ovum 52, 200
oxidation 507
oxidative agent 782, 785
oxidative degradation 787
oxidative phosphorylation 969
oxidative stress 725
oxygen 264, 400, 598, 602, 633, 637, 638, 639, 640, 669, 698, 699, 700, 701, 729, 830, 858
oxygenating hollow-fiber bioreactor 400
oxygenation 400, 402, 406
oxygenator 857
oxygen binding affinity 858
oxygen carrier 858
oxygen-free radical 792
oxygen supply 925
oxygen tension 633, 639, 697
OXY-HFB. *see* oxygenating hollow-fiber bioreactor
OXY-HFB-BAL 400, 402

P

P4HB. *see* poly(4-hydroxybutyrate)
P-15 352
p21 99
p27^{Kip1} 99
p48. *see* interferon-regulated factor-9 (IRF-9)
p53 77, 760
p65 95, 96, 97
p300/CREB-binding protein (CBP) 77
pacemaker 159

pacemaker activity 296
PACT. *see* protein activator of protein kinase PKR
PAF. *see* platelet activating factor
PAGE. *see* polyacrylamide gel electrophoresis
pain reliever 283, 284
palatal mucosa 832
palate 369
palmitylation 110
PAM. *see* pharmacologically active microcarrier
PAMP. *see* pathogen-associated molecular pattern
pancreas 62, 111, 140, 143, 144, 411, 412, 413, 414, 418, 423, 777, 807, 862, 924, 940, 943, 946, 947, 948, 949
pancreas development 941
pancreas transplant 940
pancreas transplantation 412
pancreatic acini 941
pancreatic cell 416, 417
pancreatic duct 419
pancreatic embryonic cell 123
pancreatic function 863
pancreatic islet 411, 412, 413, 415, 420
pancreatic polypeptide 418
pancreatic polypeptide-secreting 411
pancreatic primordia 941, 942, 943, 944, 945, 947, 948
pancreatic β-cell 417
pancreatitis 111
pancreatogenesis 417
PAN/PVC 277
Papyrus of Ebers 856
Paracelsus 6, 93
paracrine factor 566
paraplegia 635, 708, 843
parasite 786
parasitism 790
parathyroid hormone 180, 357
Parkinson's disease 50, 105, 272, 273, 281, 282, 819
paroxysmal nocturnal hemoglobinuria 952
parthenogenesis 42
partial pressure of oxygen 600
particle 496
particle bombardment 930
parvovirus B19 750
passive cutaneous anaphylaxis 791
PAS-staining 172, 173
patch-clamp 622
patch-clamp analysis 172

patch clamp technology 760
patellar 261
patellar ligament 318
patellar tendon 261, 320
patellar tendon fibroblast 320
patency 776, 778, 883
pathogen-associated molecular pattern 722, 725, 749, 751
pathogenicity 501
pathological test 791
patient testimonial 804
patocyte aggregate 403
pattering material surface 847
patterned array 848
patterned surface 849
patternrecognition receptor 722
PAX2/6 162
Pax4 172, 418
Pax4,29 417
Pax6 418
Pax7 914
PBMC. *see* peripheral blood mononuclear cell
PCA. *see* passive cutaneous anaphylaxis
PCL. *see* polycaprolactone
PCR 147, 170
PD. *see* population doubling; *see* Parkinson's disease
PD 168393 279
PDGF. *see* platelet-derived growth factor
PDGF-A 125
PDGF-B 125, 360
PDGF receptor 75
PDL. *see* periodontal ligament
PDLLA. *see* poly(D,L-lactide)
PDLSC. *see* periodontal ligament stem cell
PDL stem cell 932
PDS 322
pDuoRex vector 101
Pdx-1 105, 172, 418
PDZ-domain 110
PECAM 144
Peclet number Pe 700
pectoralis major muscle 831
pectoralis muscle 245
pediatrics 806
pediatric transplantation 956
PEG. *see* poly(ethylene glycol)
PEG diacrylate hydrogel 394
PEI. *see* polyethyleneimine

pellet high cell density culture 181
penetration 94, 183
penicillin G 370, 371
pentagastrin 297
PEP. *see* polymorphic eruption of pregnancy
peptide 476, 500, 788
peptidoglycan 751
peptidomimetic 78
perfusion 597, 670, 671
perfusion bioreactor 598, 599, 602
perfusion bioreactor system 607
perfusion culture 375, 376
perfusion seeding 596
perfusion system 565
periarteriolar lymphocyte sheath (T-cell) 791
perichondrium 184, 648
pericyte 728, 923
peri-implant bone 842
peri-implant bone defect 828
peri-implant bone healing 840
peri-implant bone regeneration 829
perinatal lethality 72
perineurial cell 275
periodontal defect 125
periodontal disease 125, 345, 358
periodontal ligament 353, 358, 359, 926, 931
periodontal ligament cell 125
periodontal ligament fibroblast 359
periodontal ligament stem cell 359
periodontal precursor cell 359
periodontal regeneration 125, 356, 358, 359, 829, 931
periodontal structure 934
periodontal surgery 369
periodontal tissue 358
periodontal tissue engineering 125
periodontitis 921, 933
periost 809
periosteal cell 574, 830, 843
periosteal-derived mesenchymal precursor cell 221
periosteal flap 16, 216, 802
periosteum 129, 221, 570, 576, 649, 914
peripheral artery disease 776, 882
peripheral blood 955
peripheral blood cell 954
peripheral blood cell count 790
peripheral blood mononuclear cell 187, 415, 972
peripheral blood stem cell 174

peripheral nerve   272, 818
peripheral nervous system   271, 815
peripheral organ   815
peripheral vascular   940
peripheral vascular bypass   858
periphereous NS   161
peritoneal cavity   863
peritoneum   776
perivascular cell   923, 924
perivascular niche   933
permeability   280, 281, 507, 600, 602, 652
permeable membrane   832
peroneal nerve   818
peroxisome proliferator-activated receptor gamma (PPAR-γ)   294
perspiration   670
PERV. *see* porcine endogenous retrovirus
PET. *see* polyethylene terephthalate
Petri dish   595, 599
PF. *see* platelet factor
PF-4   783, 784
PG   257, 258, 259, 263, 783. *see* prostaglandin
PGA. *see* polyglycolic acid
PGA-collagen tube   818
PGA degradation   323
PGA scaffold   124
PG-D2   783
PGDF   783
PG-E$_2$ 783
PGL   322
PGLA bead   829
pH   598, 601, 602, 633, 790
phage λ-derived DNA-binding domain   96
phagocyte   786
phagocytic cell   782
phagocytosis   491, 492, 728, 729, 734, 782, 785, 792, 803, 930
pharmacokinetic   820
pharmacological agent   956
pharmacological development   773
pharmacologically active microcarrier   820
pharmacology   971
pharmacoprophylaxis   954
pharynx   880
phase separation   482
pH balance   858
PHEMA. *see* poly(hydroxy ethyl methacrylate)
PHEMM. *see* poly 2-hydroxyethyl methacrylate-co-methyl methacrylate

phenotype   628
phenotype plasticity   843
phenotyping   793
phenotyping of cells   790
phosphate   847
phosphate-buffered saline   734
phosphate ion   486
phosphodiesterase type 4 inhibitor   821
phospholipase C   84
phospholipid strip   850
phosphorylation   71, 72, 78, 110
photodynamic purging   960
photolithographic technique   478
photolithography   479, 481
photo-polymerizing gel   181
photoreceptor   283
physical barrier   815
phytohemagglutinin   186
PI3K/Akt pathway   180
PIAS. *see* protein inhibitor of activated STAT
Picornaviridae   111
piercing   67
piezoelectric   623
piezoelectric effect   711
piezo-tip   615
pig   774, 777, 789
pigmentary disorder   970
pigmentation   970
pigmented epidermis   970
pig pancreatic primordia   946
pindolol   972
pinocytosis   930
pioneer axon   280
pioneer neuron   280
PIP   97
PIP-KRAB   102
PIP-OFF   96, 105
PIP-ON   96
PIP-system   101
pituitary   62, 969
PIWI-Argonaute-Zwille (PAZ) domain   110
PIWI domain   110
pixel   615
PKC   86
PLA. *see* polylactide acid; *see* polylactic acid
PLA2   84
placebo-controlled trial   804
placebo effect   804
placenta   140, 141, 144, 147, 149, 162, 732

placental cell   141
placental derived stem cell   153
placental growth factor   301
placental secreted alkaline phosphatase   102
PLA/PGA coated gelatine sponge   829
plaque assay   112
plaque forming and natural killer (NK) cell assay   791
plasma   122, 872
plasma cell   790
plasma exchange   397
plasma membrane   71, 76, 587, 731
plasma oxidation   247
plasmapheresis   397
plasma volume   869
plasmid   77, 102
plasmid BMP-2   351
plasmid DNA   350, 351, 930
plasmid transfection   351
plasminogen activator   404, 971
plastic   436
plastic and reconstructive surgery   8
plastic facial surgery   8
plasticity   139, 351
plastic surgery   9, 50, 806, 877
platelet   122, 125, 669, 727, 790, 883, 953, 961
platelet activating factor   783, 784
platelet count   857
platelet-derived growth factor   75, 122, 125, 215, 216, 321, 347, 359, 727, 730, 783, 784, 828, 879, 927
platelet-derived growth factor-BB (PDGF-BB)   264
platelet-derived growth factors AA/AB   347
platelet factor   784
platelet recovery   955
platelet-rich fibrin matrix   324
platelet-rich plasma   322, 354, 359
plating   802
PLC   84, 86
pleiotropic effect   71
pleiotropy   72
PLG   353
PLGA   183, 185, 265, 299, 300. *see* polylactic-co-glycolic acid; *see* poly(lactide-co-glycolide)
PLGA-BG   265
PLLA   248, 263, 322. *see* poly-L-lactic acid; *see* polyglycolic acid (PGA)–polylactic acid
PLLA [poly(L-lactic acid)]   129

PLNA. *see* popliteal lymph node assay
pluripotency   37, 64, 160, 162, 816
pluripotency marker   140
pluripotency stem cell marker   141
pluripotent   560
pluripotent cell   555
pluripotent cell population   337
pluripotent embryonal carcinoma cell   618
pluripotent embryonic stem cell   939
pluripotential cell   164
pluripotent stem cell   51, 65, 294, 417, 736, 809
PMMA. *see* polymethylmethacrylate
pneumocyte   179, 203
pneumonia   188
PNH. *see* paroxysmal nocturnal hemoglobinuria
PNS   271, 272, 273, 274, 279, 284. *see* peripheral nervous system
PNS axon   273, 275, 284
PNS-CNS transition zone (TZ)   272
PNS-CNS TZ   272, 273
podocyte   870
Poiseuille's Law   675, 676, 677
Poland syndrome   289
poliovirus   728
poly 2-hydroxyethyl methacrylate-co-methyl methacrylate   818
poly(4-hydroxybutyrate)   383
polyacrylamide gel electrophoresis   113
polyadenylation site   97
polyaminoacid   419
poly-amino acid shell   863
polycaprolactone   247, 248, 393, 480, 926
polycaprolactone-co-polylactic acid   883
polycarbonate chamber   250
polycarbonate membrane   66, 376
polycation   930
polycation-DNA complex   930
polycythemia   790
poly(desaminotyrosyl-tyrosine ethyl carbonate) [poly (DTE carbonate)]   263
polydimethylsiloxane   247
polydioxanone   263
poly(DL-lactic-co-glycolic acid) scaffold   182
poly-DL-lactid   830
poly(D,L-lactide)   129
polyester   247, 926
polyester amalgam   651
polyester foam   181, 651
polyesterurethane   248

Subject Index

polyethylene 476
poly(ethylene glycol) 299, 352, 393, 394, 476, 501, 503, 651, 657, 863
poly(ethylene glycol/oxide) 182
polyethyleneimine 856
polyethylene terephthalate 261, 393, 685, 820
polyethylene terephthalate-polypropylene 261
polyethylenimine 930
polyglycolic acid 124, 183, 247, 248, 299, 321, 383, 393, 436, 641, 688, 777, 858, 883, 926
polyglycolic acid mesh 858
polyglycolic acid (PGA)–polylactic acid 777
poly-glycolic acid polymer 870
polyglycolic acid scaffold 871
polyglycolic fleece 830
polyglycolic scaffold 831
polyhedral oligomeric silsesquioxane 688
polyhydroxy acid 541
poly(hydroxy ethyl methacrylate) 503
polylactic acid 124, 321, 501, 878, 926
poly-lactic acid polymer 870
polylactic-co-glycolic acid 263, 299, 353, 354, 393, 436
polylactide 371
polylactide acid 183, 277, 299, 383
poly(lactide-co-glycolide) 183, 545, 734, 735, 933
poly(lactide-co-glycolide) acid 926
polylactide/glycolide 352
polylactide scaffold 124
poly(L-aspartic acid) 487
poly-L-lactic acid 182, 247, 263, 393, 545, 654
poly-L-lactic acid:polyglycolic acid composite 353
poly-L-lysine 419
poly-l-ornithine 420
polylysine 930
polymer 159, 211, 321, 346, 383, 475, 804, 850, 878, 930, 933
polymerase chain reaction 580, 750
polymer chain 501, 504
polymer composition 880
polymer construct 803
polymer demixing 482
polymeric film 247
polymeric nerve guidance conduit 815
polymerization 508
polymer network 496
polymer ratio 482
polymer scaffold 860

polymethylmethacrylate 480, 817
polymorphic eruption of pregnancy (PEP) 148
polymorphism 85
poly(NiPAAn) 182
poly(N-isopropylacrylamide) 606
polyornithine 419
polypropylene 247, 248, 261
poly(propylene fumarate) 182, 503
polysaccharide 184, 500
polysaccharide group 476
polysiloxane polymer (silicone) 335
polystyrene 476, 490
polytetrafluoroethylene 261, 393, 776, 777, 858
polytetrafluoroethylene mesh 393
polytetrafluoroethylene (Teflon) 436
polyurethane 183, 247, 283, 336, 476, 688
polyurethane foam module 402
poly(vinyl alcohol) 503
polyvinyl sponge 436
poly(α-hydroxy ester) 182, 495
popliteal lymph node assay 792
population doubling 142, 169
population doubling time 169
porcelain crown 360
porcine aortic valve 857
porcine endogenous retrovirus 420
porcine hepatocyte 861
porcine insulin 941
porcine mesenchymal stem cell 832
porcine smooth muscle cell 776
porcine specie 924
porcine wound model 105
pore 930
pore diameter 488
pore size 321, 488, 507, 543, 925, 926
pore-size distribution 654
pore system 491
porosity 183, 274, 280, 299, 357, 543, 803, 925, 926
porous material 476
porous scaffold 694
porphyrin 476
positive charge 501
positron emission tomography 802
POSS. see polyhedral oligomeric silsesquioxane
posterior cruciate ligament 318
post-ischemic kidney 179
postnatal multipotent cell 878
postnatal stem cell 913, 914, 923

post-streptococcal glomerulonephritis   788
posttranscriptional gene silencing   109, 110
POU domain   162
PPAR-γ   180, 882
PPAR-γ-2   294
PP-cell   411
PPF. *see* poly(propylene fumarate)
preadipocyte   180, 290, 292, 391, 392, 393, 394, 882
precipitation   486
preclinical animal model   793
preclinical animal study   804
preclinical environment   829
preclinical investigation   564
preclinical model   791, 832
preclinical testing   781
pre-confluent cell graft   332
precursor cell   123, 220, 563, 570, 575
precursor miRNA   110
prednicarbate   972
prefabrication   218
pregnancy   39, 40, 140, 143, 147, 148, 153
preimplant surgery   830
prelamination   374, 832
pre-miRNA. *see* precursor miRNA
preodontoblast   926
preosteoblast   218
preplate technique   915
preproinsulin   411
pressure   635
preterm infant   140
primary cell   563, 564, 637, 760
primary human monocyte   491
primary injury   785
primary microRNA   110
primary myogenesis   245
primary osteoblast   489
primate   129, 775, 863
pri-miRNA. *see* primary microRNA
primordia mass   941
printed drop   623
pristinamycin   101, 102
PR-LBD. *see* progesterone receptor ligand-binding domain
procollagen type I   765, 766
procurement of tissue   980
professionalization   48
profillagrin   968
progenitor cell   129, 141, 199, 290, 559, 858, 877, 878, 914, 922, 957

progesterone receptor ligand-binding domain (PR-LBD)   97
progressive motoneuronopathy   821
proinflammatory cytokine   956
proinflammatory cytokine IL-6   753
proinflammatory mediator   971
proinsulin   126, 946, 947
proinsulin RNA   946, 947
prokaryotic genome   752
prolactin   73
proliferation   64, 72, 73, 78, 105, 110, 121, 122, 124, 125, 127, 162, 178, 180, 182, 184, 186, 187, 469, 554, 564, 574, 600, 602, 636, 641, 649, 650, 652, 695, 701, 785
proliferation capacity   292, 382
proliferation rate   320, 490
prolin   172
Prometheus   5
promoter   71, 72, 74, 77, 78, 93, 94, 95, 102, 123
promoter repression   103
proof-of-concept   775
prostaglandin   784, 971
prostaglandin E2   790
prostaglandin synthesis   84
prostate   61, 62
prostate carcinoma   110
prosthesis   381, 384, 776
prosthetic device   8, 9
prosthetic graft   858
prosthetic material   878
prosthetic substitute   878
protamine   930
protease   62
protease inhibitor   62
protein A   186
protein activator of protein kinase PKR (PACT)   109
protein adsorption   471, 478, 849
proteinase 3   960
protein binding   471
protein biosynthesis   94
protein chemistry   844
protein encapsulation   817
protein factory   355
protein folding   500
protein inhibitor of activated STAT (PIAS)   78
protein kinase   724
protein-micropatterned surface   817
protein–protein interaction   73
protein recognition   850

protein scaffold 9
proteins/fluid environment 850
protein synthesis 186, 862
protein translation 97
protein-tyrosine-phosphatase-1B (PTP1B) 78
proteoglycan 184, 237, 244, 257, 308, 383, 554, 568, 727, 761, 764, 786, 880
proteolytic enzyme 504
proton sponge 930
proto-oncogene 150
protozoan 750, 786
provitamin B5 972
proximal tubule 871
PRP. see platelet-rich plasma
PRR. see pattern-recognition receptor
Pseudomonas putida 97
Pseudomonas putida-derived cumate-responsive repressor 97
pseudosynovial sheath 276
psoriasis 969
PTFE. see polytetrafluoroethylene
PTGS. see posttranscriptional gene silencing
PTH receptor 296
PTP1B. see protein-tyrosine-phosphatase-1B
pTRIDENT vector 99
public 38
public health 13, 20
PUF. see polyurethane foam module
PUF-BAL 402
pulmonary allograft 858
pulmonary artery 858
pulmonary disease 50
pulp 361, 921, 925, 933
pulpal cell 928
pulpal inflammation 922
pulpal necrosis 360
pulp capping 360
pulp capping/dentin 356
pulp cell 923, 925
pulp damage 923
pulp-derived SP cell 925
pulpitis 360, 921, 922, 931
pulp progenitor cell 360
pulp stem cell 922, 931
pulp tissue 360
pulsatile pressure 883
pulsed electromagnetic field 883
pulse duplicator bioreactor 858
purine analoga 952
Purkinje cell 204

putamen 820
PVA. see poly(vinyl alcohol)
pyrolytic carbon 857
pyrrolizidine alkaloid 202

Q
QALY. see quality-adjusted life years
QBPD. see Quebec Back Pain Disability Scale
Q-mate 104
Q-ON 96
quadricep 265
quadricep tendon 261, 318
quality-adjusted life years 15, 17
quality control 170
quality management 980
quality of life 15, 17, 559, 721
quartz surface 475
Quebec Back Pain Disability Scale 314
quiescence 178
quinolinic acid 105

R
RA 162
rabbit 774, 775, 776, 777, 789
radial flow bioreactor 398
radial force 616
radial forearm flap 374, 375
radiation injury 203
radiation pneumonitis 203
radical mastectomy 389
radical starter 505
radiograph 801
radiographic analysis 636
radioimmunoassay 790
radioprotection 954
radiotherapy 833
radius 358
RAGE. see receptor of advanced glycation end-products
Ran 75, 110
randomized controlled trial 804, 808
RanGAP. see RanGTPase-activating protein
RanGDP 74
RanGTPase-activating protein 74, 75
RANKL. see RANK ligand; see receptor activator of nuclear factor kappa B ligand
RANK ligand 324, 491
Ran nucleotide exchange factor 75
rapalog 105
rapamycin 94, 96, 104, 105, 415, 956

rapid prototyping 546
rat 774, 775
rat bone marrow stroma cell 832
rat femur 916
rat model 490
rat myocyte 776
rat omentum 871
Rawls' model 31
Rb gene 760
reaction mixture 487
reactive chemical group 475
ready-to-use BAL 403
real-time PCR 113
real-time RT-PCR 114, 972
receptor 71, 74, 958
receptor-2 selective inhibitor 764
receptor activator of nuclear factor kappa B ligand 576
receptor binding 637
receptor-ligand binding 474
receptor of advanced glycation end-products 722, 725, 726, 736
receptor recruitment 78
receptor tyrosine kinase Tie2 959
recognition 782
recognition sequence 850
recombinant adenovirus 931
recombinant cell 828
reconstitution 174
reconstructed human epidermis 968
reconstructed human epidermis model 968
reconstructive surgery 8, 9, 159, 827
red blood cell 790
reduced intensity conditioning 951, 961
reflex bladder 432
Reformation 7
Regenafil 357
regeneration 128, 181, 348
regeneration capability 275
regeneration of bone 570
regenerative capacity 223, 815
regenerative medicine 93, 608, 613, 629, 648, 659, 693
regenerative medicine application 474
regenerative therapy 485
regulation 37, 39
regulatory approval 789
regulatory framework 979
regulatory guideline 793

regulatory T-cell 186, 187, 959, 961
reinnervation 777
rejection 31, 66, 147, 162, 413, 773, 776, 788, 958
relapse rate 955
relaxation 86, 87
release of growth factor 829
remineralization 922
remodeling 347, 489, 491, 843
remodelling process 492
Renaissance 6
renal barrier 509
renal capsule 66, 941
renal cell 64, 872, 873
renal cell growth 871
renal cell-polymer scaffold 871
renal cortex 871
renal development 62
renal device 872
renal dialysis 63
renal disease 874
renal epithelial cell 871
renal failure 869
renal function 869, 872, 874
renal injury 870
renal parenchyma 66
renal primordia 944, 945
renal tissue 869, 872
renal transplantation 869, 874
renal tubular structure 871
renal unit 872
renal vein 942
renin 63
repair and regeneration of tissues 159
repair capacity 323
replacement construct 246
replication 110, 187
replicative immortality 760
reporter gene 102
reporter gene assay 77
repressor 96, 102, 103
repressor-based expression control 95
repressor protein 94
reproductive biology 792
reproductive cloning 161
reproductive medicine 40, 44
reprogramming 50, 51, 54, 198, 417
resident peritoneal cell 792
resilience 706
resolution 615

resorption rate   787
respiratory distress   161
respiratory gas   675
respiratory system   669
responsive promoter   102
reticular cell   221, 570
retina   105, 761, 923
retinal endothelium   205
retinal ganglion neuron   283
retinal neovascularization   105
retinal pigment epithelium cell   283
retinoic acid   65, 172, 762, 968
retinoic acid-inducible protein   752
retinopathy   412, 940
retrorsine   202
retroviral delivery   130
retroviral gene transfer   206
retroviral transduction   187
retroviral vector   65, 105
retrovirus   93, 149, 163, 929
reversine   201
Rex-1   141, 142
Reynolds number   674, 676, 677, 678, 680, 682, 683
RFB. *see* radial flow bioreactor
RFB-BAL   400
RFF. *see* radial forearm flap
RGD   357, 394, 500, 508, 690
RGD-peptide   850
rhBMP   216
rhBMP-2   350, 352, 357, 359, 361, 491, 829
rhBMP-4   829
rhBMP-7   357, 359, 829
RHE. *see* reconstructed human epidermis
rheological property   773
rheumatic condition   859
rheumatic fever   857
rheumatic heart disease   786
rheumatoid arthritis   763, 764
RIA. *see* radioimmunoassay
RIA technique   791
rib   878, 879, 880
ribbon   478
ribosome   98
ribosome assembly   100
ribozyme   97, 104
ribozyme inhibitor   98
RIC. *see* reduced intensity conditioning
ridge   478

RIG. *see* retinoic acid-inducible protein
RIG-1   752
rigidity   706
RISC   110
risk assessment   11, 24, 29, 32, 33
risk-benefit relation   25
risk versus benefit balance   805
RNA   696, 929
RNA helicase domain   752
RNAi. *see* RNA interference
RNA interference   109, 110, 112, 117
RNA polymerase   77, 94
RNA polymerase II   110
RNase-III endonuclease   109
robotic system   604
robustness   17, 21
rodent   775, 776, 789, 829, 830
rodent model   832
rodent strain   775
Roller tube culture system   552
Romberg's disease   289, 301
root canal   925
root ganglia   763
root/periodontal complex   353
root planning   932
root surface   931
rotating-wall vessel   596
rotating-wall vessel bioreactor   603, 644
rotating-wall vessel reactor   642
rotator cuff   318, 319, 322, 323, 324
rotator cuff tear   318
rotator cuff tendon   324, 325
roughness   480, 652, 695, 803
RP. *see* rapid prototyping
R-Smad   123
RT-PCR   112, 116, 171, 172, 873
RT-PCR analyse   492
runt-related transcription factor 1   172, 296
runt-related transcription factor 2   105, 130, 180, 294, 352, 354
Runx-1. *see* runt-related transcription factor 1
Runx-2. *see* runt-related transcription factor 2

S
S100   239, 309, 311, 313
S100 antigen   240
SA. *see* streptavidin
saccharose   545
safety   760, 788, 789, 801

safety assessment   789
SafraninO   311, 313
safranin O-staining   237
salicylic acid   972
saliva   122, 833
salivary gland   61, 62, 63, 419, 833
salivary gland cell   833
SAM. *see* self-assembled monolayer
sample size   789
sandblasting   479
sandwich construct   376, 377
sandwich culture   402
sandwich graft   377
saphenous vein   882
saphenous vein graft   882
sarcolemma   111
sarcoma cell   160, 560
sarcomere   243
satellite cell   245, 246, 914, 915, 917
Sca-1   915
Sca1+ (stem cell antigen)   952
scaffold   64, 65, 66, 121, 127, 153, 163, 182, 183, 185, 487, 539, 540, 541, 543, 544, 547, 564, 576, 633, 637, 638, 639, 640, 650, 654, 659, 707, 712, 727, 735, 788, 828, 830, 841, 859, 860, 877, 878, 925
scaffold construct   802
scaffold engraftment   803
scaffold-free tissue   552
scaffold integrity   802
scaffold material   262, 614, 650, 760, 803, 828
scaffold structure   539, 551
scaling   932
scanning electron microscopy   479, 577, 845, 873
scapula   216, 358
scar   347, 348, 802
scarring   17, 785
scar tissue   787
scavenger receptor   722, 751
sCD40   416
sCD40-Ig   416
SCF   174, 175, 953. *see* stem cell factor
SCFR, c-kit   145
Schiff's base   505
Schwann cell   271, 274, 275, 276, 277, 278, 280, 284, 816, 819
SCI. *see* spinal cord injury
SCID. *see* severe combined immune deficiency; *see* severe combined immunodeficiency

SCID-repopulating capacity   957
sclera   762, 763
scleraxis   933
sclerostin   85
SCNT. *see* somatic cell nuclear transfer
scopolamine   972
scoring system   792
screening test   790
SDF   203
SDF-1   958
SDF-1α   174
SDS-polyacrylamide gel electrophoresis (PAGE)   113
SEAP. *see* secreted alkaline phosphatase
sebaceous gland   765
second messenger   71, 72, 84, 87
secreted alkaline phosphatase   102
seeding method   596
segmental defect   830
selection marker   141
selectively permeable membrane   566
self antigen   786
self-assembled monolayer   247, 475, 477
self-association   473
self-cleaving hammerhead ribozyme   97, 98
self-determination   48, 53
self-esteem   50
self-MHC class I molecule   955
self-MHC molecule   955
self-organization   63
self-organized tendon (SOT)   250
self-renewal   129, 140, 150, 162, 178, 198, 560, 561
self-renewal capacity   178
self-renewal potential   218
self renewing property   153
SEM. *see* scanning electron microscopy
seminal vesicle   62
sense probe   946, 947
sensing nerve cell   83
sensitisation assessment   789
sensitivity   338
sensitivity analysis   17, 19, 20, 21
sensitizer   971
sensor   601, 602
sensory nerve   761
Sepax procedure   171
sequestration-based expression control   95
sequestrectomy   308, 313, 314

sera 791
serine threonine 122
serine/threonine kinase 122
serine/threonine kinase domain 122
serosa 777
serotonin 172, 688, 783, 883
Sertoli cell 419
serum 967, 968
serum analysis 802
serum-containing media 568
serum cytokine 790
serum-free medium 569
serum protein 508, 790
serum-response factor 431
severe combined immune deficiency (SCID) 28
severe combined immunedeficient bg mice 105
severe combined immunodeficiency 952
sexual dimorphism 73
sexual reproduction 5
SFF. see solid-free form fabrication technique
SH2 143, 144, 177, 185, 562
SH2-containing phosphatae-1 (SHP-1) 78
SH2 domain 71, 73, 74, 78
SH3 143, 144, 177, 185
SH4 145, 177
sham treatment 789
Sharpey's fiber 258, 933
shear flow sensor 85
shear force 618, 625
shear rate 670
shear stress 599, 627, 628, 638, 642, 658, 670, 676, 681, 702, 706, 713, 883
sheep 789, 830
sheer 787
sheet graft 333, 334
SHH. see sonic hedgehog
shock 790
shock reaction 791
short hairpin RNA (shRNA) 97, 98, 100
shoulder 318
SHP-1. see SH2-containing phosphatae-1
SHP2 78
shRNA. see short hairpin RNA
SIBLING. see small integrin-binding ligand N-linked glycoprotein
sickle cell anemia 952
side population 924
side population cell 914
signaling 321

signaling molecule 828, 926
signaling pathway 84, 655, 714
signal molecule 669, 843
signal transducer and activator of transcription (STAT) 71, 72
signal transduction 71, 72, 102, 103, 567, 636, 714
signal transduction pathway 564, 566
silane 474
silane molecule 475
silanisation 475
silastic membrane 336, 881
silicate oxide 927
silicon-based substrate 850
silicone 475, 861
silicone gel 498
silicone membrane 860
silicone stent 777
silicone tubing 861
silicone wafer 481
silicon membrane 861
silicon reservoir 817
silk 263, 321
silk film 850
silk hydrogel 500
silk protein 500
silver-containing polymer film 476
simian virus 40 760
simian virus 40 promoter 96, 99, 102
simian virus 40 T antigen 105
simvastatin 764
single-celled organism 675
single-cell resolution 615
sinuatrial valve 111
sinus augmentation 354, 357
sinus floor augmentation 828, 829, 831
sinus lift 357
sinus lift procedure 830
siRNA. see small interfering RNA
sirolimus 413, 414
SIS. see small intestinal submucosa
Sjögren's syndrome 833
skeletal defect 878
skeletal development 180
skeletal muscle 111, 141, 243, 244, 245, 249, 250, 251, 790, 821, 914, 924, 930
skeletal muscle cell 296
skeletal myoblast 105, 130
skeletal myocyte 205

skeletal reconstruction 828
skeletal tissue 878
skeletogenic cell 843
skeleton 179
skin 50, 147, 204, 206, 329, 330, 331, 332, 333, 334, 335, 336, 338, 369, 374, 595, 613, 765, 807, 809, 816, 832, 855, 878, 881, 923, 956, 967
skin barrier 767
skin cell 50
skin defect 124, 331
skin epidermis 967
skin epithelium 369
skin fibroblast 320
skin flap 8
skin graft 8, 187, 330, 331, 334, 335, 337, 860, 881
skin-graft contracture 971
skin inflammation 148
skin mucosa 488
skin neoplasm 881
skin permeation 972
skin reaction 791
skin reconstruct 969
skin sensitization 971
skin stem cell 861
skin substitute 860, 881, 967
skin substitution 329, 330, 331, 332, 334, 335, 336, 338
skin transplantation 8
skin ulcer 332
skin wound 856, 860
skin wound healing 971
skull fracture 878
SM22 297
Smad 1 123, 180
Smad 2 123, 180
Smad 3 123, 180
Smad 4 123, 180
Smad 5 123, 180
Smad 6 123
Smad 7 123
Smad 8 123, 180
Smad transcription factor 180
small bowel 777
small hepatocyte 404
small integrin-binding ligand N-linked glycoprotein 924
small interfering RNA (siRNA) 109, 113, 114, 115, 116, 117, 118, 163

small intestinal submucosa 434, 435, 437, 776
small intestine 776
small intestine submucosa 777
small-molecule-responsive ribozyme 98
smoking 129
smooth and skeletal muscle 204
smoothelin 297
smooth muscle 61, 205, 250, 290, 430, 431, 434, 435, 436, 437, 776
smooth muscle actin 924
smooth muscle actin-alpha 319
smooth muscle cell 122, 125, 162, 295, 297, 382, 476, 687, 883, 924
smooth muscle differentiation 431
smooth muscle myosin 430
SN. see substantia nigra pars compacta
social acceptability 47
social environment 50
social institution 49
society 27, 40, 43, 47
SOCS. see suppressor of cytokine signaling
sodium-channel protein 172
sodium pyruvat 172
soft lithography 657
soft tissue 513, 705
solar radiation 765
solid-free form fabrication technique 544, 546, 547
solubility 501
somatic cell nuclear transfer 417, 562, 737
somatic cell nuclear transfer methodology 220
somatic cell nuclear transfer (SCNT) 29
somatic nervous system 815
somatic stem cell 292
somatostatin 105, 297, 418
sonic hedgehog 62, 162, 354
sonoporation 930
Sox-2 162, 201, 417
Sox-9 182, 239
Sox-17 172
SP. see substance P; see side population
Sp1 77
spacing 792
spatial distribution 476
spatial topographical order 480
spectrophotometry 602
Spemann, Hans 10
spermatozoon 27
sperm cell 42, 52

sperm donation   42, 43
sphere   650
spheroid   233, 234, 239, 240, 309, 311, 402
spheroid body   553
spinal cord   271, 272, 273, 275, 276, 277, 278, 279, 280, 284, 815, 817, 818, 821
spinal cord injury   818
spinal fusion   129, 492, 914, 916
spinal fusion surgery   129, 307
spine   313
spinner flask   596, 637, 644
spinous layer   968
spinous stratum   374
spleen   140, 143, 144, 168, 174, 419, 786, 787, 789, 790, 954
splenic colony   954
splice variant   73
splicing   97, 100, 122
split skin graft   369
split-thickness allograft   335
split-thickness skin graft   329, 336, 861, 881
spondylolisthesis   313
sponge   182, 183, 184
spongialization   181
spongiosal layer   840
spongy bone graft   573
sprouting   786
squamous cell carcinoma   73
SR. *see* scavenger receptor
src   75
src homology 2 domain   73, 177
src homology 3 and 4 domains   177
SRF. *see* serum-response factor
SRF-regulated angiotensin receptor II   431
SRY   148
SSEA-1   162
SSEA-3   141, 142, 162
SSEA-4   141, 142, 143, 146, 162
SSG. *see* split-thickness skin graft
staining   844
stainless steel   859
start codon   98
start-up company   604
STAT. *see* signal transducer and activator of transcription
STAT1   72, 75, 76, 77
STAT2   72
STAT3   72, 73, 75, 77
STAT4   72, 73

STAT5   73, 77
STAT5a   72
STAT5b   72
STAT6   72, 73, 75
statistical power analysis   792
statistical testing   789
staurosporine   76
steady state   789
steer   872, 939
stem cell   10, 11, 37, 49, 63, 64, 66, 124, 125, 126, 129, 164, 167, 181, 185, 188, 219, 416, 913, 560, 561, 562, 563, 569, 570, 613, 648, 656, 658, 669, 688, 690, 760, 815, 819, 858, 913, 922, 932
stem cell biology   951
stem cell culture   570
stem cell differentiation   470
stem cell factor   141, 203, 953, 956
stem cell homing   178, 954
stem cell line   39, 41
stem cell marker   932
stem cell medicine   39
stem cell niche   178, 924, 954, 955, 958
stem cell plasticity   10, 197, 198, 206
stem cell pool   956
stem cell recruitment   922
stem cell research   10, 37, 38, 39, 41, 42, 44, 45
stem cell science   39
stem cell technique   7
stem cell technology   43
stem cell therapy   65
stem cell transplantation   174, 204, 817
stenosis   883
stereological approach   792, 793
stereology   792
sterilising agent   785
sterility   600
sternocleidomastoid muscle   216
sternum   567
steroid hormone   95
steroid hormone receptor   95
steroid/thyroid hormone receptor   77
stiffness   507, 512, 600, 652, 706, 714
Stokes layer   678, 680, 682
stomach   777
stop codon   98
strain   635, 774, 802, 843, 844, 846
strain-dependent signal   843
strain transduction   709
stratified epithelium   832

stratium 820
stratum corneum 373, 765, 767, 968
strength 600, 602, 843
streptavidin 94, 96, 97
streptococcal infection 786
streptogramin 96, 97, 101
streptomyces coelicolor-derived butyrolactone-responsive repressor 97
streptomyces pristinaespiralis 97
streptomycin 370
streptozotocin (STZ) type 1 diabetes 941
stress 635, 639, 705, 707, 712, 843
stress environment 802
stress fibre 477
striatal neuron 821
striatum 105, 281, 282, 820, 821
stricture 17
Stro-1 146, 178, 294, 924
STRO-1/CD146-positive progenitor 933
stroma 61, 187, 761
stroma cell 174
stroma cell line 960
stroma feeder layer 958
stromal cell 873
stromal cell line 760, 762
stromal derived factor 203
stromal extracellular matrix 761
stromal fibroblastic cell 219, 561
stromal patch 434
stromal stem cell 219, 561
structural efficacy 788
study design 804, 805
study design parameter 801
SU5416 764
subchondral drilling 181
subcutaneous 302
subcutaneous adipose tissue 292
subcutaneous ADSC 296
subcutaneous fat tissue 292
subcutis 206
sub-fibril 256
submucosa 777
substance P 762
substantia nigra 281, 282
substantia nigra pars compacta 819
substrate patterning 848
substrate stiffness 85
substratum 477
subventricular zone 143, 278, 817

suicide 389
suicide gene 961
sulfated glycosaminoglycan 182
sulfated proteoglycan-rich matrix 296
superficial burn 330
superior mesenteric artery 871
superior mesenteric vein 942
supply of nutrients 495
support cell 819
suppressive T cell 186
suppressor of cytokine signaling 77
suppressor T-cell 784
suprabasal cell layer 968
suramin 917
surface 472, 477
surface antigen expression 178
surface area 792
surface chemistry 471, 840, 849
surface coating 475
surface migration 620
surface modification 847
surface property 470
surface roughness 480
surface structure 847
surface-structuring technique 481
surface topography 840, 847
surfactant 475, 762, 859
surgery 67, 633
surgical implantation 785
surgical trauma 843
surrogate cell 418
surrogate islet cell 418
surrogate method 789
surveillance 776
survival rate 174
survival rate of implants 840
suspension 670
suturing 802
SV40. see simian virus 40
SV40 large T immortalized culture 761
SV40 T 105
SV40-Tag 565
SVZ 282
sweat gland 765, 881
SYBR-green 112, 113, 114
synaptogenesis 820
synaptophysin 172
synchronization 103
syncytium 247, 249

syngeneic host   871
syngeneic transplantation   954
synovial adipose tissue   292
synovial cell   256, 764
synovial fluid   670
synovial joint   764
synovial membrane   185, 914
synovitis   261
synovium-derived stem cell   182
synthesis   406
synthetic biodegradable material   818
synthetic biology   66, 93, 102
synthetic gene network   99, 100, 101, 102
synthetic graft   265, 882
synthetic material   292, 870, 877, 880
synthetic morphology   66, 67
synthetic non-biodegradable polymer   818
synthetic polymer   129, 321, 501, 850, 878, 926
synthetic polymer scaffold   183
synthetic promoter   103
systemic assessment   789
systemic pathology   785, 787
systemic toxicity assessment   789
szintigraphy   802

T

T9   144
T10   144
TA. *see* transient amplifying
tacrine   820
tacrolimus   413, 414, 956
talin   850
tamoxifen   95, 97, 104, 105
T-antigen   760
target gene   958
TAR-RNA binding protein (TRBP)   109
tartrate-resistant acid phosphatase   491
TATA box   94
tattooing   67
TBP (TATA-binding protein)   77
TBSA. *see* total body surface area
Tc45   74, 78
Tc45 phosphatase   75
T-cell   73, 147, 185, 186, 187, 414, 724, 730, 731, 732, 733, 734, 736, 787, 790, 803, 953, 956, 960
T-cell activation   753
T-cell depleting antibody   952
T-cell depletion   954, 959
T-cell expansion   956

T-cell proliferation   186, 187
T-cell receptor   731, 784, 952
T-cell reconstitution   956
TCP. *see* tricalcium phosphate
TCR. *see* T-cell receptor
TCRαβ   784
TCRγδ   784
tear   670
tear layer   761
Teflon shunt   861
telomerase   565
telomerase activity   140, 142, 417
telomerase enzyme   760
telomerase (TERT)   760
telomere   140, 167, 760
telomere degradation   957
TEM. *see* transmission electron microscopy
template   486
template molecule   850
temporalis muscle   216
temporomandibular joint   361, 879, 880
temporomandibular joint dysfunction   361
temporoparietal cortex   820
tenascin-C   258
tendon   219, 243, 255, 256, 257, 258, 259, 260, 261, 262, 263, 264, 265, 266, 317, 318, 319, 320, 322, 323, 561, 595, 603, 648, 657, 880, 916
tendon and ligament   807
tendon cell   320, 933
tendon ECM   264
tendon fibroblast   263, 264
tendon insertion   323
tenocyte   250
tensile modulus   831
tensile strength   880, 882, 883
tension stress   636
teratocarcinoma   162, 562
teratoma   65, 139, 140, 162, 562, 649, 863, 923, 939
TERT   761
test animal   790
testis   816
testosterone   251, 972
test period   789
TET   104
TET-OFF   96, 105
TET-ON   96
TetR. *see* tetracycline repressor
tetracycline   96, 97, 99, 101, 105

tetracycline repressor 96
tetracycline-responsive repressor 97
tetracycline-responsive transactivator 99
tetramer formation 73
tetramerization 73
tetraploid nucleus 199
TetR-VP16 99
TET-system 101
TEWL. see transepidermal water loss
TFIIB 77
TGF. see transforming growth factor
TGF-1 850, 925
TGF-1β 174
TGF-b 784
TGF-α. see transforming growth factor-α
TGF-β 301, 727, 736, 763, 764, 828, 880. see transforming growth factor-β
TGF-β1 172, 182, 185, 359, 360, 649, 880, 917, 928
TGF-β2 214, 431
TGF-β2/3 431
TGF-β3 214, 880, 928
TGF-β-activated kinase 1 (TAK1)-TAB1/2/3 complex 752
TGF-β receptor 127, 178, 180
TGF-β type II receptor 180
TGF-β type I receptor 180
Th 282
Th1 790
Th1 helper cell 784
Th1/Tc1 cell 956
Th2 790
Th2 helper cell 784
Th2/Tc2 cell 956
thalassemia major 952
T-helper cell 782
theophylline 972
therapeutic cloning 66, 416, 417, 872, 873
therapy-research interface 37
thermoplasticity 383
thickness 792
thigh 881
thiol 474, 507
three-dimensional matrice 763
three-dimensional scaffold 596, 598
thrombin 500, 622, 727, 857
thromboembolism 381
thrombogenesis 858
thrombopoietin 73, 175, 953, 956

thrombopoietin receptor 105
thrombosis 476, 576, 686, 776, 882, 883
thrombus 776
Thy-1 141, 143, 146, 177, 185, 562
thymic function 960
thymic microenvironment 956
thymidine 871
thymidine incorporation assay 873
thymidine kinase promoter 97
thymocyte 186
thymocyte development 956
thymus 168, 750, 789, 790, 956
thymus tumor 148
thyroid hormone 63
thyroid tumor 148
tibia 152, 261, 358
tibialis anterior muscle 105
tibia shaft fracture 129
Tie2 958
tight junction 110, 430
Ti implant surface 841
time-of-flight secondary ion mass spectroscopy 577
TIR. see toll/interleukin (IL)-1 receptor
TIRAP/Mal. see TIR domain-containing adaptor protein/MyD88-adapter-like
TIR domain-containing adapter inducing interferon-β 752
TIR domain-containing adaptor protein/MyD88-adapter-like 752
tissue 469, 807
tissue architecture 788
tissue availability 817
tissue biopsy 790
tissue culture 479
tissue-culture flask 604
tissue damage 147
tissue design 559
tissue differentiation 648
tissue donation 38
tissue-engineered device 781
tissue-engineering firm 608
tissue-engineering strategy 606
tissue flap 433
tissue growth 470
tissue injury 147, 782, 785, 786, 803
tissue-like aggregate 862
tissue maturation 656
tissue necrosis 790

tissue organisation   469
tissue phenotype   788
tissue plasminogen activator   319
tissue regeneration   174, 181, 188, 478
tissue remodeling   111, 803
tissue repair   128, 147, 148, 151, 153, 174, 183
tissue scaffold   470
tissue typing   954
tissue valve   857
titanium   357, 362, 840, 857, 916
titanium cage   491
titanium dioxide surface   849
titanium implant   847
titanium mesh cage   129
titanium orthopaedic implant   475
titanium surface   842
TLR. *see* toll-like receptor
TLR2   752, 753
TLR3   752
TLR4   752, 753
TLR5   753
TLR7   752
TLR8   752
TLR8-dependent   753
TLR9   752
TLR-family   749
TLRs trigger   749
T-lymphocyte   72, 186, 750, 786, 959
TMJ. *see* temporomandibular joint
TMJ disc   831
TNF. *see* tumor necrosis factor
TNF-R1,2   145
TNF-α   239, 729, 783, 784, 969, 970, 971
TNF-β   783, 784
TOF-SIMS   845
tolerance   162, 803
tolerogenic   414
toll/interleukin (IL)-1 receptor   749, 751
toll-like receptor   722, 749, 751, 753, 754
toluene   972
tongue   243
tooth   351, 352, 353, 356, 358, 359, 360, 361, 362, 634, 840
tooth crown   926
tooth development   922, 928, 929
tooth eruption   360
tooth formation   923
tooth germ   923
tooth loss   921

tooth restoration   921
tooth root   356, 359
topical drug development   972
topographical scale   478
topography   470, 471, 695
torsion   706
total body surface area   333, 334
total hip replacement   916
total lymphoid irradiation   952
totipotency   160, 294, 560, 649
totipotentiality   878
toxic agent   787
toxicity   929, 930
toxicological assay   762
toxicological test   759, 791
toxicology   765
toxin   765, 785, 861
toxin adsorption   64
toyocamycin   97
TPO   953. *see* thrombopoietin
Tra-1-60   162
Tra-1-61   141, 142
Tra-1-81   141, 142, 162
trabecular bone   545, 573, 640, 840, 846, 860
trabecular bone formation   180
trabecular structure   489
trabecular system   841
trabecule   486
trachea   678, 679, 880
tracheal cartilage   124
TRAM. *see* TRIF-related adapter molecule
TRAM flap   390
transactivation domain   73, 95, 96
transactivator   95, 96, 100, 102
transcription   101, 959
transcriptional activation domain   94
transcriptional activity   75, 77, 78
transcriptional modulator   100, 102
transcriptional silence   94
transcription factor   71, 73, 74, 77, 78, 130, 162, 180, 200, 202, 570, 736, 752, 933
transcriptome analysis   972
transdifferentiation   10, 185, 351, 354, 418, 419
transduction   350
transepidermal water loss   767
transfected gene   929
transfection   65, 114, 117, 118, 163, 350, 353, 760, 960
transfection agent   353

transfection efficiency 163, 930
transfection marker 99
transferrin 969
transformation 71, 201
transformed cell 760
transforming growth factor 62, 264, 430, 784, 879
transforming growth factor-α 123, 430, 431, 783, 927
transforming growth factor-β 122, 124, 126, 128, 180, 181, 184, 186, 214, 215, 216, 220, 239, 264, 321, 347, 359, 430, 563, 783, 784, 927, 958
transfusion 979
transgene 93, 101, 103, 150
transgene expression 94
transgenic mouse 105, 203
transglutaminase 968
transient amplifying 338
transitional epithelium 429
translation 94, 100
translation termination 98
translocation 95, 200
transmission electron microscopy 487, 577, 845
transplantation 66, 139, 151, 153, 159, 162, 188, 569, 697, 732, 737, 759, 803, 979
transplantation medicine 8, 39
transplantation of teeth 8
transplantation strategy 802
transplant failure 671
transplant immunology 951
transrepressor 100
transthyretin 297
transwell insert 492
TRAP. *see* tartrate-resistant acid phosphatase
trapezius 216
trauma 181, 504, 790, 878, 916
traumatic spinal cord injury 817
TRBP. *see* TAR-RNA binding protein
Treg. *see* regulatory T-cell; *see* CD4 regulatory T-cell
T-REX 96
tricalcium 847
tricalcium aluminate 927
tricalcium oxide 927
tricalcium phosphate 129, 346, 541, 828, 829, 830, 933
tricalcium silicate 927
trichohyalin 968
TRIF. *see* TIR domain-containing adapter inducing interferon-β

TRIF-related adapter molecule 752
trigeminal nerve 761
trigger-inducible gene activation 103
triglyceride 790
triglyceride droplet staining 101
trochlear groove 917
trophoblast 139, 147
trophoblast differentiation 162
tropocollagen 256
trunk 877
trypsin 160, 565, 606
tryptophan 186
Ts 414
TSA. *see* histone deacetylase inhibitor trichostatin A
T-suppressor 187
TTF-1 203
tube formation 763
tuberculosis 785, 786
tubular device 872
tubular epithelial cell 870, 873
tubular interponate 777
tubular kidney 66
tubular segment 872
tubular structure 873
tubule 764
tubule-like structure 872
tubulin 85, 275
tubulo-acinar cell 943
tubulogenesis 764
tumor 187, 791, 930
tumor-derived cell line 404
tumor formation 355
tumor growth 123, 763
tumorigenesis 417, 791
tumorigenicity 110, 419
tumor-like anomaly 863
tumor necrosis factor 279, 297, 723, 753, 784
tumor resection 878
tumor suppression 72
tumor suppressor 110
tumor-suppressor gene 929
two-dimensional (2D) histology section 792
two-vector-based inducible knockdown 101
Tyk2 72
type I collagen 249, 309, 319, 320, 322
type II collagen 308
type III collagen 309, 319
type-III β-tubulin 297
type I interferon 749

type I rat-tail collagen 762
type I-type III collagen ratio 321
tyrosine 71
tyrosine dephosphorylation 73, 76
tyrosine hydroxylase 105, 172
tyrosine kinase 72, 121, 122, 958
tyrosine kinase receptor 141
tyrosine phosphatase 78
tyrosine phosphorylation 73, 74, 76
TZ 284

U

UCB. see umbilical cord blood
UCLA. see University of California, Los Angeles
UCLA-BAL 402
ulna 152
ultrasound 86, 802, 930
ultrasound-guided cardiocentesis 143
ultrasound-guided cord puncture 143
ultrasound-guided fetal blood sampling 150
ultrasound stimulation 214
ultrastructure 802
ultraviolet exposure 618
umbilical CB 167, 168
umbilical cord 149, 186, 382
umbilical cord blood 140, 141, 278, 382, 419, 562, 878
umbilical cord blood CD34+ cell 297, 959
umbilical cord blood cell 153, 164, 555
umbilical cord blood HSC 956
umbilical cord cell 382
umbilical cord matrix cell 153
umbilical cord stem cell 160
umbilical cord tissue 140
umbilical vein 150, 882
umbilical vein endothelial cell 763
umbrella cell 430
uncontrolled trial 804
undifferentiated cell 560
undifferentiated mesenchymal cell 923
unfolding change 473
University of California, Los Angeles 402
unrestricted somatic stem cell 168, 169, 170, 171, 173, 174
urea 66, 601, 637, 869
urea nitrogen 873
ureogenesis 398
ureter 870
ureteral bud 66, 871

ureterocystoplasty 432
urethra 372, 373
urethroplasty 17, 20, 21, 832
uric acid 872
urinalysis marker 790
urinary barrier 434
urinary bladder 429
urinary continence 429
urine 66, 122, 429, 430, 670, 790, 869, 872, 873
urine analysis 802
urine collecting duct system 66
urogenital tissue 17
urokinase 404
urology 806
uroplakin 430, 431
urothelial cell 430, 431, 433, 435, 436
urothelial cell carcinoma 432
urothelial construct 431
urothelial development 431
urothelial muscle 431, 434, 435
urothelial stem cell 436
urothelial strip 434
urothelium 429, 430, 431, 434, 435, 436, 437
US Food and Drug Administration 11, 129
US National Bioethics Advisory Committee 39
USSC. see unrestricted somatic stem cell
utah array 283
uterocystoplasty 434
uterus 37, 40, 161, 186
utility 15, 16, 17, 19, 20, 21
UTP 883
UV irradiation 124, 970
UV-light 503

V

value generation 14
valve 381
valve replacement 856
valvular heart disease 159, 381
valvular interstitial cell 127
vanadium 478
Van der Waals force 472, 473, 496
VAS. see Visual Analog Pain Scale
vascular anastomosis 9
vascular bed 786
vascular cell 218
vascular cell adhesion molecule 178, 185, 784
vascular conduit 882, 883
vascular disease 882

vascular endothelia cell   61
vascular endothelial cadherin   763
vascular endothelial cell   786, 858
vascular endothelial growth factor   264, 301, 351, 359, 430, 431, 437, 730, 763, 764, 783, 784, 828, 829, 882, 883, 916
vascular endothelial growth factor (VEGF)   93, 103, 104, 105
vascular endothelium   857, 941
vascular graft   773, 775, 776, 858
vascular grafting   206
vascular ingrowth   832
vascular invasion   802
vascularisation   250, 302, 695, 791
vascularity   843
vascularization   65, 103, 124, 393, 394, 419, 829, 830, 860
vascularized bone   216
vascularized bone graft   129
vascularized capsule   392
vascularized cardiac muscle   860
vascularized rectus abdominis   777
vascularized skeletal muscle   860
vascular prosthesis   776
vascular rejection   941
vascular remodeling   764
vascular supply   882
vascular tissue   882
vascular tissue engineering   883
vascular volume   858
vasculature   861, 940
vasculogenesis   577, 689, 763, 883
vasoactive agent   882
VCAM. *see* vascular cell adhesion molecule
VCAM-1   141, 144, 178, 185, 764, 784
VE-cadherin. *see* vascular endothelial cadherin
vector   356, 929, 930
vector system   356
VEGF   63, 174. *see* vascular endothelial growth factor (VEGF); *see* vascular endothelial growth factor
velocity   673, 674, 675, 676, 678, 679, 680, 681, 684, 930
velocity field   700
vena cava   776
venous endothelium   776
venous graft   776
venous ulcer   881
ventral pancreas   941, 943

veratridine   762
very late antigen   177, 957
vesicle membrane   845
vesiculo stomatitis virus (VSV)   77
vessel   129, 161, 670, 685, 730, 731, 858, 882
vessel-like structure   620
vessel wall   689
vestibuloplasty   372, 373, 377, 832
viability   598, 602
viable cell   625
vibration   83, 86
Vicryl   436
Vicryl net   377
vimentin   141, 143, 145, 148, 170, 562
vinculin   430, 655, 850
vinyl-chloride   147
viral attachment   110
viral DNA   929
viral gene delivery   104
viral gene transfer vector   103
viral infection   72, 116, 163, 785, 790, 959
viral myocarditis   111
viral plaque assay   77
viral protein 16-derived transactivation domain   97
viral transduction   93, 103, 163, 916, 917
viral transmission   969
viral vector   93, 94, 103, 129, 149, 163, 350, 351, 353, 929
Virchow, Rudolf   9
virus   112, 167, 500, 727, 929
virus capsid   110
virus propagation   117
visceral   302
visceral adipose tissue   292
visceral ADSC   296
viscoelastic gel system   497
viscoelasticity   697
viscoelastic material   259
vision   283
Visual Analog Pain Scale   314
vitamin A   968
vitamin C   968
vitamin D3   729, 869
vitrification   863
vitro model   759
vitronectin   726
VLA. *see* very late antigen
VLA-4   145, 178, 957
VLA-5   145, 957

VLA-α1 177
VOF. *see* virtual operating field
void space 496, 508
voltage gated sodium channel 172
voltage-sensitive 172
volume 792
voluntariness 26
von Willebrand factor 170, 883
VP1/TLR4 753
VP16 94, 95, 96, 97
VP16 transactivation domain 97
Vroman effect 726
VSV. *see* vesiculo stomatitis virus
vWF 141, 145

W

Wallerian degeneration 274
waste product removal 598
waste removal 633, 925
Watson-Crick base pairing 110
wavy-walled reactor 596
WBC. *see* white blood cell
western blot 76, 114, 116, 171
western blot immunophenotype 170
wet chemistry 849
wetting behavior 850
Wharton's jelly 140, 142, 144
Wharton's jelly cell 143
white blood cell 784, 790
white fat 290
white matter 275, 818
whole body irradiation 951
whole-liver perfusion 397
willingness-to-pay (WTP) 20
Wilmut, Ian 10
wind tunnel 683
Wnt 62, 65, 66, 129, 162, 184
Wnt signaling 85
Wnt/β-catenin 958
wound 500, 765
wound healing 123, 124, 500, 671, 722, 727, 728, 729, 763, 792
wound repair 124
woven bone 485
WT-1 960
WTP. *see* willingness-to-pay

X

X chromosome 200
xenobiotic 972
xenogeneic 863
xenogeneic barrier 940
xenogeneic embryo 423
xenogeneic extracellular matrix 943
xenogeneic porcine islet 863
xenogeneic recipient 140
xenogenic cell 560
xenogenic model 830
xenogenous serum 553
xenograft 261, 334, 381, 420, 421, 687, 733, 776, 803, 933
xenotransplantation 30, 404, 406, 420, 423, 424, 942, 943, 945
xenotransplantation model 949, 954
xenotransplantation therapy 941
X inactivation 200
X-linked immunodeficiency 65
X-ray 802
X-ray analysis 845
X-ray diffraction 487, 577
XRD. *see* X-ray diffraction

Y

Y-chromosome 147, 148, 203
yield strength 696
YIGSR 393, 394
Y-microchimerism 147
yolk sac 140, 168
Young's modulus 259, 512, 681, 706, 846

Z

ZDF (type 2) diabetes 945
Zetos system 86
ZIC1 162
ZIC2 162
zinc finger domain 96
zinc finger protein 94
zone of Höhl 923
zygote 11, 41, 53, 161

Printing and Binding: Stürtz GmbH, Würzburg